EARTH SCIENCE

The Earth, the Atmosphere, and Space

THIRD EDITION

EARTH SCIENCE

The Earth, the Atmosphere, and Space

THIRD EDITION

Stephen Marshak
UNIVERSITY OF ILLINOIS URBANA-CHAMPAIGN

Robert Rauber
UNIVERSITY OF ILLINOIS URBANA-CHAMPAIGN

W. W. NORTON & COMPANY
Independent Publishers Since 1923

W. W. Norton & Company has been independent since its founding in 1923, when William Warder Norton and Mary D. Herter Norton first published lectures delivered at the People's Institute, the adult education division of New York City's Cooper Union. The firm soon expanded its program beyond the Institute, publishing books by celebrated academics from America and abroad. By midcentury, the two major pillars of Norton's publishing program—trade books and college texts—were firmly established. In the 1950s, the Norton family transferred control of the company to its employees, and today—with a staff of five hundred and hundreds of trade, college, and professional titles published each year—W. W. Norton & Company stands as the largest and oldest publishing house owned wholly by its employees.

Editor: Andy Blitzer
Senior Project Editor: Jennifer Barnhardt
Associate Director of Production, College: Benjamin Reynolds
Associate Editor: Elaina Sassine
Copy Editor: Norma Sims Roche
Associate Copy Editor: Susan McColl
Consulting Editor for Author Team: Kathryn Gollin Marshak
Managing Editors, College: Carla Talmadge and Kim Yi
Associate Managing Editor: Marcus Van Harpen
Digital Media Editor: Ariel Eaton
Assistant Editor, Digital Media: Rachel S. Bass
Marketing Manager, Geology: Marlee Lisker

Design Director: Rubina Yeh
Designer (Interior and Cover): Anne DeMarinis
Director of College Permissions: Megan Schindel
Permissions Associate: Patricia Wong
Photo Editor: Stephanie Romeo
Photo Researcher: Fay Torresyap
Composition: MPS North America LLC
MPS Project Manager: Jackie Strohl
Illustrations for the First Edition: Stan Maddock and Joanne Brummett
Illustrations for the Second and Third Editions: Troutt Visual Services
Manufacturing: Transcontinental Printing

Permission to use copyrighted material is included in the backmatter of this book.

Library of Congress Control Number: 2023947212

ISBN: 978-0-393-89354-0 (cl)

W. W. Norton & Company, Inc., 500 Fifth Avenue, New York, NY 10110
wwnorton.com

W. W. Norton & Company Ltd., 15 Carlisle Street, London W1D 3BS
1 2 3 4 5 6 7 8 9 0

To Kathy, Emma, David, Michelle, Sam, and Robert
(My bedrock, through uplift and subsidence.)
—STEVE MARSHAK

To Ruta, Carolyn, Josh, Molly, Max, Stacy, Fabian, Henry, and Robert
(My sunshine, no matter how big the storms.)
—BOB RAUBER

BRIEF CONTENTS

CONTENTS

Introduction
WELCOME TO EARTH SCIENCE!   2

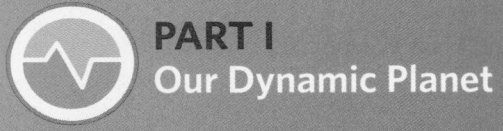

PART I
Our Dynamic Planet

We begin our introduction to geology, the study of the Earth, with a discussion of how the Earth, and the elements that constitute it, formed. We then introduce the theory of plate tectonics—the grand, unifying idea that provides a foundation for understanding how and why geologic processes operate. We next examine the materials that make up the solid Earth—minerals (the building blocks of rock), then igneous, sedimentary, and metamorphic rocks. This background allows us to discuss mountain building and earthquakes, the Earth's history, and its energy and mineral resources.

Chapter 1
FROM THE BIG BANG TO THE BLUE MARBLE   22

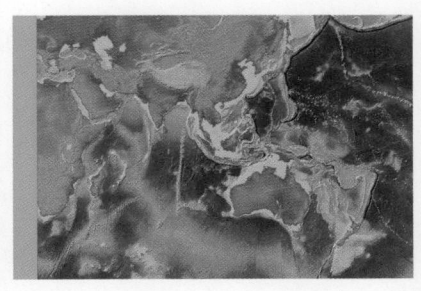

Chapter 5
THE REST OF THE ROCK CYCLE
Sedimentary and Metamorphic Rocks 164

Chapter 6
CRACKS, CRUMPLES, AND CRAGS
Geologic Structures and Mountain Building 210

Chapter 7

A VIOLENT PULSE

Earthquakes　242

Chapter 8

DEEP TIME

How Old Is Old?　288

PART II
Ever-Changing Landscapes

Picture the jagged peaks of a mountain range, the grassy plains of a continental interior, the ice of a glacier, the sand and stone of a desert, or the shrubs of wetlands. Clearly, our planet hosts an amazing variety of landscapes. In Part II, we explore the characteristics of these landscapes and the processes that produce them, such as erosion and deposition, the hydrologic cycle, downslope movement, and the impact of humanity. We then conclude with a focus on the Earth's extreme environments, desert and glacial landscapes.

PART III
Restless Seas

Most of our planet's surface lies hidden beneath a vast global ocean of saltwater that, in some places, is deeper than mountains are high. Part III focuses on this watery realm. We begin by introducing oceanography, the study of seawater (its composition, and its currents, tides, waves, and life). Then we turn our attention to marine geology (the study of ocean basins and the seafloor). We conclude with a focus on coasts, the boundaries between land and sea, and show how coastal landforms take shape and evolve.

PART IV
A Blanket of Gas: The Earth's Atmosphere

A thin blanket of gas—an atmosphere—surrounds our planet. Without it, there could be no life on the Earth. Part IV begins by exploring the properties and behavior of the atmosphere. We describe its composition and character, and show how it changes with altitude and over time. Next, we turn our attention to clouds, precipitation, wind, and solar radiation, the key factors that define weather conditions. Our discussion addresses both the monsoonal weather of the tropics and the nature of air circulation in mid-latitudes. We also review the formation and evolution of storms, such as thunderstorms, tornadoes, hurricanes, and mid-latitude cyclones. We conclude this part by characterizing the Earth's climate. We explain why temperature and precipitation vary across the globe and through the Earth's history, and why accelerated change during recent centuries poses a hazard to society.

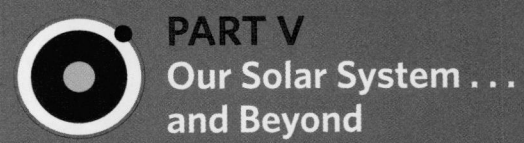

PART V
Our Solar System . . . and Beyond

In Part V, we delve into deep space. First, we review the development of astronomy and look at some key discoveries made before and after modern instruments were available. This discussion covers planetary motions, phases of the Moon, eclipses, distance scales, and telescopes. Next, we explore our Solar System, visiting each of the planets, their moons, and other objects that orbit our Sun. Then we direct our attention to the stars, and we follow their evolution over time. This discussion starts by examining our own star, the Sun, to see what it's made of and how it generates energy. We then discuss other stars, both large and small. Finally, we probe deep space, inspecting galaxies, supermassive black holes, and quasars. We conclude with a look at models that address the origin, overall structure, and fate of the Universe.

Chapter 22
THE SUN, THE STARS, AND DEEP SPACE 854

PREFACE

Narrative Themes

Why do earthquakes, volcanoes, floods, and landslides happen? What causes mountains to rise? Does climate change over time? Where can we find the record of evolution? When did the Earth form, and by what process? How can we locate valuable metals, and where do we drill to find oil? What drives violent storms? Is the size of the Universe constant? Could life exist on other planets or moons? Why are there oceans? The study of Earth Science addresses these important questions and many, many more.

Earth Science, Third Edition, introduces the reader to the study of our planet and its context in space. It addresses the essence of four different disciplines: geology, oceanography, atmospheric science (including meteorology), and astronomy. As such, this book will help you understand what's beneath your feet, over the horizon, up in the sky, and far beyond the highest clouds.

Because of the diversity of topics covered in *Earth Science*, it's important to keep a set of narrative themes in mind as you read this book. These themes, listed below, serve as the building blocks from which you can construct your personal understanding of our amazing planet and of the Universe around it.

1. Our planet's land, water, atmosphere, and living inhabitants are dynamically interconnected. Materials cycle constantly among living and nonliving reservoirs on, above, and within the planet. Researchers refer to this complex network of interconnecting features and phenomena as the *Earth System*.

2. The Earth is a planet. Like many other planets, it was formed from dust and gas that once floated in space. But, in contrast to other planets, the Earth is a dynamic place where new geologic features continue to form and old ones continue to be destroyed. A combination of special conditions, such as its distance from the Sun, its chemical composition, the presence of liquid oceans, and the presence of life, makes our planet unique in our Solar System. Quite literally, there's no place like home! While exoplanets (planets orbiting other stars) that resemble the Earth may exist, we'll probably never interact with them.

3. The Universe, the Solar System, and the Earth are all very old. Researchers have concluded that the Universe began with the Big Bang about 13.8 billion years ago, and that our Solar System originated around 4.57 billion years ago. The Earth's birth occurred between 4.56 and 4.54 billion years ago. Since our planet's origin, its surface, subsurface, and atmosphere have changed, and life has evolved.

4. Unlike all other planets or moons in the Solar System, the Earth's outer shell, the lithosphere, consists of a number of plates. The *theory of plate tectonics* states that these plates slowly move relative to one another so that the map of our planet continuously changes. Plate interactions cause earthquakes and volcanoes, build mountains, shift oceans, provide gases that make up the atmosphere, and affect the distribution of life on the Earth.

5. Internal processes (driven by the Earth's internal heat—and manifested by plate motions) and external processes (driven by heat from the Sun) interact at the Earth's surface to produce complex landscapes.

6. Knowledge of the Earth, its oceans, and its atmosphere can help society to understand dangerous natural hazards such as earthquakes, tsunamis, volcanoes, landslides, storms, and floods. In some cases, this knowledge can help to reduce the casualties and devastation that these hazards can cause.

7. Energy and mineral resources come from the Earth and are formed by geologic phenomena. Geologic studies can help locate these resources and mitigate the consequences of their use.

8. Physical features of the Earth are interwoven with life processes, and vice versa. Therefore, the history of life links intimately to the history of the physical Earth. The careful study of rocks can provide a rich record of this history.

9. The constant movement of water in the oceans moderates the climate by transporting heat around the world. And where the ocean interacts with the land, fascinating coastal landscapes develop.

10. Our planet's atmosphere has changed dramatically over the course of the Earth's history. If you went back a billion years in time, the atmosphere would be unbreathable, and you would suffocate. In modern times, this precious blanket of air has been significantly affected by the activities of society.

11. The atmosphere moves constantly, sometimes smoothly, sometimes violently. A variety of dangerous storms can develop in association with atmospheric movements, and these events cannot be predicted more than a few days in advance.

12. Climates on the Earth vary with latitude, elevation, and distance from the sea, and the positions of continents with respect to climate belts have changed over the course of geologic time. Consequently, both climate and sea level can change, sometimes very significantly, so that regions that were once dry can become shallow seas, and places that were once warm can become covered in ice.

13. Human society, during the past few centuries, has become a major agent of *climate change* on our planet. By burning fossil fuels, we have sent significant quantities of carbon into the atmosphere. This carbon, previously stored in rocks and reservoirs underground, and therefore isolated from the Earth's atmosphere, accumulated over millions of years but is being released within a human time frame. The resulting atmospheric greenhouse effect has led to an increase in global temperature. Since this rapid change influences many geologic and atmospheric phenomena, it impacts our society.

14. We have much to learn by looking upward and examining the Universe beyond the Earth. Modern instruments and spacecraft have answered many questions, but with each answer, more questions arise about our Solar System, stars, galaxies, and the array of bizarre and beautiful objects that lie at inconceivable distances from us.

15. Science comes from observation, and people make scientific discoveries through those observations. Earth Science incorporates ideas from physics, chemistry, and biology, so the study of Earth Science provides an excellent means to improve science literacy.

These narrative themes serve as this book's take-home message, a message that we hope students will remember long after they finish an introductory Earth Science course. In effect, the themes provide a mental framework on which students can organize and connect ideas to develop a modern, coherent image of our planet in its context.

Pedagogical Approach

Many students learn best from textbooks when they can actively engage with a combination of narrative text and narrative art. Some students respond more to words, which help them to organize information, provide answers to questions, fill in the essential steps that link ideas together, and develop a personal context for understanding information. Other students learn most effectively through images that illustrate processes. Still others benefit from question-and-answer-based active learning, an approach by which students practice their knowledge in real time. *Earth Science*, Third Edition, provides all these learning tools. The text of this book has been crafted to be engaging and to carry students forward along a narrative arc. The book's abundant line art and photographs have been configured to tell a story (see "Narrative Art and *What an Earth Scientist Sees*"). Chapters are laid out to help students internalize key principles (see "Organization"). Chapter Reviews and online activities have been designed to engage students, to provide active feedback (see "Guided Learning Explorations" and "Smartwork Online Activities"), and to guide students toward a richer understanding of important concepts (see "Pedagogical Features Designed to Help Students Comprehend and Retain Concepts").

Of note, scientists throughout the world use the metric system for describing distances, weights, temperatures, and other physical features. But many of the students using this book grew up using the English system of units. To help all students to visualize and appreciate these units, and to see the relationship between metric and English units, we provide measurements in both systems. The conversions we provide are sometimes exact and sometimes approximate, in order to reflect the precision of the measurement provided. For example, if we describe a distance as "1,000 kilometers," with the implication that this is an approximate figure, we'll provide the conversion as 600 miles (instead of 621.37 miles), because it would be misleading to present one unit as an approximation and the other as a precise one.

Organization

The topics covered in this book have been arranged so that students can build their knowledge of Earth Science on a foundation of overarching principles. This book contains five parts:

Part I begins by considering how the Earth formed, and how it's structured, overall, from surface to center. With this basic background, students can delve into plate tectonics, the grand unifying theory of Earth Science. Plate tectonics appears early in the book, so that students can use the theory of plate tectonics as a foundation for interpreting and linking the ideas presented in the subsequent geology chapters. Knowledge of plate tectonics, for example, helps students understand the suite of chapters on minerals, rocks, and the rock cycle. Knowledge of plate tectonics and rocks together, in turn, provides a basis for understanding earthquakes and mountains. And then, with this background, students can see how the Earth's map, and its inhabitants, have changed throughout the vast expanse of geologic time. We dedicate one chapter to energy and mineral resources, an especially relevant topic given present-day global interest in sustainability.

Part II characterizes the landscapes of the Earth's surface, the visible part of our planet. It explains how landscapes result from a never-ending battle between uplift (ultimately driven by plate tectonics) and erosion (ultimately driven by energy from the Sun). We begin by discussing surface movements in general, manifested most dramatically by landslides. We then consider our planet's freshwater, the landscapes associated with water movement, and the challenges presented by floods and the loss of water supplies. We conclude this part by looking at the dramatic landscapes developed in extreme climates—the realms of glaciers and deserts.

Part III takes us offshore, into the ocean. We first examine the water of this realm, focusing on its characteristics, its movements, and the life within it. Then, we turn our attention to the geology of the seafloor. This group of chapters concludes by considering the interface between water and the solid Earth, both beneath the ocean and along its shores.

Part IV of the book introduces students to the Earth's atmosphere. We begin by describing the character and evolution of the atmosphere and the overall composition of air. We then develop an understanding of how and why air moves, and how to characterize the weather. Next, the book focuses on the devastating dramas of thunderstorms, tornadoes, hurricanes, and mid-latitude cyclones, all of which impact human society. This part culminates with an examination of the Earth's climate, and of scientific and social issues related to climate change.

Part V goes beyond the confines of the Earth and heads into outer space. We'll see how new discoveries provide answers to age-old questions about where our planet came from and what its distant future holds. Students will learn how humanity's knowledge of space has advanced over the centuries, what the other objects of our Solar System look like, and how the Sun and the stars produce energy. This part ends by taking a voyage to the farthest reaches of the Universe. We learn about these distant areas by analyzing light that began its journey to the Earth billions of years before our planet had even formed.

Although we have structured the chapters in *Earth Science* to provide an overall direction to our narrative, we have ensured that this book is flexible, and that instructors can assign chapters in a different sequence. Therefore, each chapter is self-contained, and we reiterate relevant material where necessary. This flexibility allows instructors to choose their own strategies for teaching Earth Science.

Special Features of This Text
Narrative Art and *What an Earth Scientist Sees*

To help students visualize topics, this book is lavishly illustrated. Each figure provides a realistic context for interpreting features and phenomena without overwhelming students with extraneous detail. The authors worked directly with the artists to ensure that renderings convey the intended content. Many drawings and photographs have been integrated into ***narrative art*** that is laid out, labeled, and annotated to tell a story—the figures are drawn to teach! Subcaptions are positioned adjacent to the relevant parts of a figure, labels point out key features, and balloons provide important detail. Subparts are arranged to convey time progression, where relevant. The color palette in illustrations stays consistent through the book, so that students have an easier time keeping track of key features. Also, colors in drawings are tied to those of related photos, so that students can easily visualize the relationships between drawings and photos. In many places, photographs are accompanied by annotated sketches labeled **What an Earth Scientist Sees**, which help students to be certain that they actually see the specific features that the photo was intended to show. The illustrations also serve as the foundation for a robust suite of videos, animations, and simulations that give students a fuller and more dynamic means for visualizing complicated processes that happen over long periods (see "Narrative Art Videos, Animations, and Simulations").

Earth Science at a Glance

In addition to individual figures, each chapter contains at least one dramatic **Earth Science at a Glance** illustration that spreads over two full pages. Some of these spreads expand on a particular topic, while others provide a synopsis of many topics in a beautiful, artistic rendering. These illustrations provide a way for students to visualize key concepts . . . at a glance.

Pedagogical Features Designed to Help Students Comprehend and Retain Concepts

Evidence-based pedagogical research emphasizes the importance of setting objectives and working out answers to questions. Each chapter begins with a series of *learning objectives* that frame the major concepts of the chapter for students and carry through to the Chapter Review. Active-learning engagement is not limited to the Chapter Reviews, however. Rather, at relevant points in the text, **Did you ever wonder . . .** questions prompt students to connect new information to their existing knowledge base by asking simple Earth-Science-related questions that they have probably already thought about in everyday life. Each numbered section of a chapter ends with a **Take-home message**, like the one below, a brief summary that helps students identify and remember the highlights of the section before moving on to the next section. Each of these summaries also offers

Take-home message...

The theory of plate tectonics explains why igneous activity occurs where it does. Decompression melting occurs at hot spots, mid-ocean ridges, and rifts. Flux melting produces melts at convergent boundaries. Heat-transfer melting also takes place at rifts and volcanic arcs. Particularly large quantities of lava erupt at large igneous provinces.

Quick Questions

- Do the same kinds of volcanoes form at all plate boundaries?
- At what type of plate boundary do batholiths form?
- What is a flood basalt?

three or four basic-level (low on Bloom's taxonomy) **Quick Questions** to encourage students to think through key messages from the section before they move forward. **See for Yourself** panels, provided throughout the chapters, guide students on virtual mini-field trips, via Google Earth, to locations around the globe or in the sky where they can apply their newly acquired knowledge to interpret real-world geologic features. **How can I explain . . .** features stimulate student interaction with the book's narrative by providing simple hands-on exercises that illustrate important concepts. **Putting Earth Science to Use** boxes help students connect the concepts they are learning to real-life applications. **A Deeper Look** boxes provide additional content to explain key concepts more thoroughly. **Science Toolboxes** give students brief, accessible summaries of basic science terms and concepts that can help them better grasp the chapter material, especially if they need a review. The **Chapter Review** at the end of each chapter provides easy reference and serves as a study guide.

Up-to-Date Coverage of Current Topics

Earth Science, Third Edition, reflects the latest research in the discipline to help students understand events and discoveries that have been featured in news headlines. For example, the coverage of natural hazards has

---

BOX 18.3 **Putting Earth Science to Use**

Interpreting weather maps

Weather impacts us daily, whether it's a hot day outdoors, a brief rain shower requiring an umbrella, heavy snow that clogs a driveway, or severe weather that threatens a neighborhood. Today, most people get weather information from television or the internet, presented in the form of maps (Fig. Bx18.3a, b). Armed with the information you've learned in this and previous chapters, you should now be able to interpret these maps and understand the weather impacting your life. The maps of Figure Bx18.3, which depict different aspects of the weather at the same time, illustrate a mid-latitude cyclone, whose low-pressure center lies over Lake Michigan. Note that the pattern of isobars (showing air pressure at sea level) forms the basis for identifying the low-pressure center.

Let's examine the maps and see what they tell us about weather at particular locations. Note that a cold front extends southward from the low-pressure center. Across the front, wind shifts direction

(indicated by a change in the symbols used for indicating wind direction; see Fig. 16.12), but temperature changes only slightly. As a result, clouds and some precipitation happen right along the frontal boundary, but no severe thunderstorms develop. So, if you lived in Ohio, Indiana, or Tennessee, you would experience a few light showers, followed by a slight temperature drop and a wind shift from southerly to westerly as the front passed by. If you lived farther north, in southern Minnesota or Wisconsin, your weather experience would be very different. There, temperatures are below freezing, snow is falling, and a much stronger pressure gradient exists. This gradient drives strong winds, which have generated blizzard conditions. Farther east, north of the warm front, the pink colors in Figure Bx18.3b indicate mixed precipitation, either rain mixed with snow, or possibly freezing rain. Given your knowledge at this point in the chapter, you should now be able to look at these maps and give a description of the weather at any given location.

FIGURE Bx18.3 Interpreting weather maps.

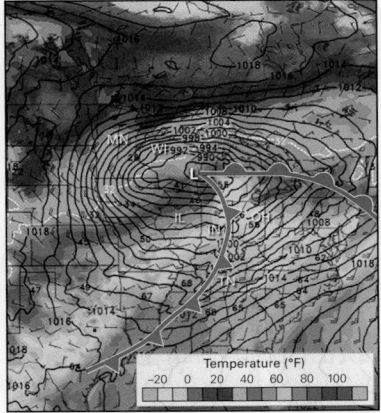

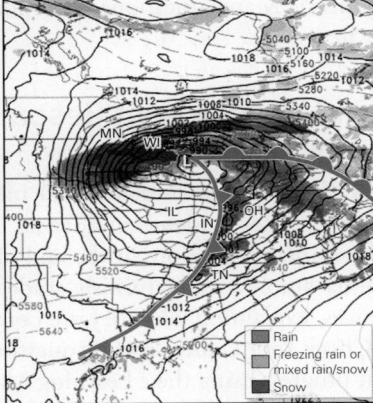

(a) A weather map of the eastern United States and Canada showing isobars of sea-level pressure (black lines), temperatures (colors), and fronts.

(b) The same area, but instead of temperature and wind, it shows precipitation type.

been substantially updated by drawing on the authors' research related to their other textbook, *Natural Disasters: Hazards of a Dynamic Planet*. This edition of *Earth Science* highlights increased awareness of the links between hazards and climate change, and it presents many new discoveries across the discipline of Earth Science. For example, the book provides the most recent changes to the geologic time scale and supercontinent reconstructions, and it discusses recent disastrous weather events. In the astronomy chapters, images from the James Webb Space Telescope bring new features of the Universe into focus.

Chapter Reconfigurations

In this Third Edition, we have reconfigured the chapters on Earth materials. This updated organization merges the discussion of the rock cycle with the presentation of minerals and the introduction to rock groups, and it combines treatment of sedimentary rocks and metamorphic rocks into one chapter. These revisions better emphasize that the processes of sedimentary rock formation and metamorphic rock formation are both manifestations of change in the Earth System over time (though, of course, in very different environments). The more efficient treatment of this subject allows more time for other topics. In addition to the organizational updates, the text and illustrations have been meticulously revised to clarify discussions and descriptions and to correct any errors. We are very grateful to readers—both instructors and students—who helped draw our attention to places in the book that would benefit from revision.

Remake of the Chapter Reviews

In this Third Edition, the material that concludes each chapter has been expanded and reorganized. Each topic in a Chapter Review corresponds to one learning objective from the start of that chapter, and to one section within that chapter, so students can quickly find the relevant section of the Chapter Review if they wish to practice their knowledge before proceeding. The review unit for each section starts with a reiteration of the learning objective, then covers *Key Concepts*, lists *Earth-Science Vocabulary* terms from that section, and concludes with *Review Questions*. Review Questions are designed to test basic knowledge, to stimulate critical thinking, and to reinforce key visual images of topics. The questions range from those seeking straightforward reiteration of concepts (at the base of the Bloom's taxonomy pyramid) to those requiring students to analyze and evaluate situations (at the top of the pyramid). Many questions incorporate modified art from the chapter, or new art.

Media and Assessment Resources

New Norton Illumine Ebook

The *Norton Illumine Ebook* delivers an immersive, interactive reading experience. To bolster student confidence, it provides formative *Check Your Understanding* questions at the end of each main chapter section. The questions automatically furnish students with feedback on their answer choices, explaining why they are correct or incorrect. Instructors can assign Check Your Understanding questions for completion and monitor student progress using their Learning Management System (LMS).

The Norton Illumine Ebook also engages students with a wealth of embedded multimedia resources, including new interactive 3D models and Earth Science in Action GIFs. These are supplemented by a wealth of animations and real-world videos embedded in the ebook.

The Check Your Understanding questions were authored by Brian Zimmer (Appalachian State University) and J. Cory Pettijohn (University of Illinois Urbana-Champaign), and they have been reviewed by Liana Boop (San Jacinto College). The Earth Science in Action GIFs were carefully selected by Paul Cutlip (St. Petersburg College). The placements for our videos and animations were recommended by Paul Cutlip (St. Petersburg College) and J. Cory Pettijohn (the University of Illinois Urbana-Champaign).

New Explore Earth Science in 3D Models

Explore Earth Science in 3D models allow students to examine virtual specimens and structures as if they were in the field or the lab. Included in both the Norton Illumine Ebook and Smartwork, these models are also available at digital.wwnorton.com/earthscience3.

The 3D models were created and curated by Scott Wilkerson (DePauw University). Scott also wrote the associated questions in Smartwork.

Narrative Art Videos, Animations, and Simulations

Earth Science, Third Edition, is supported by a rich collection of media resources. The animations depict natural processes in vivid, three-dimensional detail.

Explore Earth Science in 3D
Pillow Basalt

Additionally, interactive simulations allow students to change the variables in a model—such as the rate of ice flow within a glacier, or the temperature of magma—and see the result. Students are encouraged to ask "what if" questions, fostering curiosity and deeper understanding. These multimedia resources are supplemented by **Narrative Art videos**, in which author Stephen Marshak explains important concepts and figures from each chapter. The media program for *Earth Science* is easy to access. The videos, animations, and simulations require no special hardware and are available in a variety of settings to offer maximum flexibility for students and instructors. They are on the book's digital landing page, they are integrated into Smartwork activities and Guided Learning Explorations, and they can be imported into any major LMS.

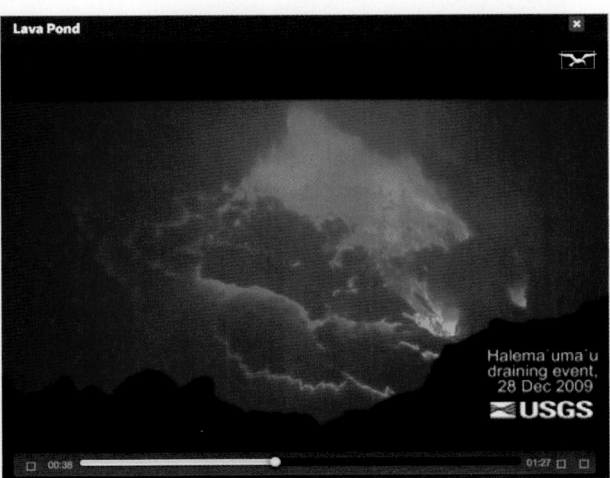

Earth Science in Action Videos

In addition to the media resources described above, every new copy of *Earth Science*, Third Edition, unlocks a library of **Earth Science in Action videos**. Curated by Tobin Hindle (Florida Atlantic University) and Christine Clark (Eastern Michigan University), the purpose of these videos is to show Earth Science and its practitioners at work. Taken from many sources—including NASA, the U.S. Geological Survey, and independent photojournalists—these videos demonstrate the many ways in which earth science is relevant to our daily lives. Teaching notes and classroom discussion questions to accompany the videos are also available on the Norton Teaching Tools instructors' page. The videos themselves are available at digital.wwnorton.com/earthscience3.

Guided Learning Explorations

Guided Learning Explorations (GLEs) are thematic online activities structured around three thoughtfully scaffolded stages: (1) foundational concept review; (2) application questions enriched by scientific data, video clips, and interactive simulations; and (3) in-depth inquiry that challenges students to apply what they have learned to real-world sites. The GLEs deliver an outstanding learning experience. Students don't simply complete the exercises; they receive feedback on every answer, right or wrong.

The GLEs address 20 fundamental topics in the Earth Science curriculum, including plate tectonics, earthquakes, coastal processes, extreme weather, and stellar evolution. These activities are available for free with all new copies of the text, in any format. They can also be purchased separately at digital.wwnorton.com/earthscience3.

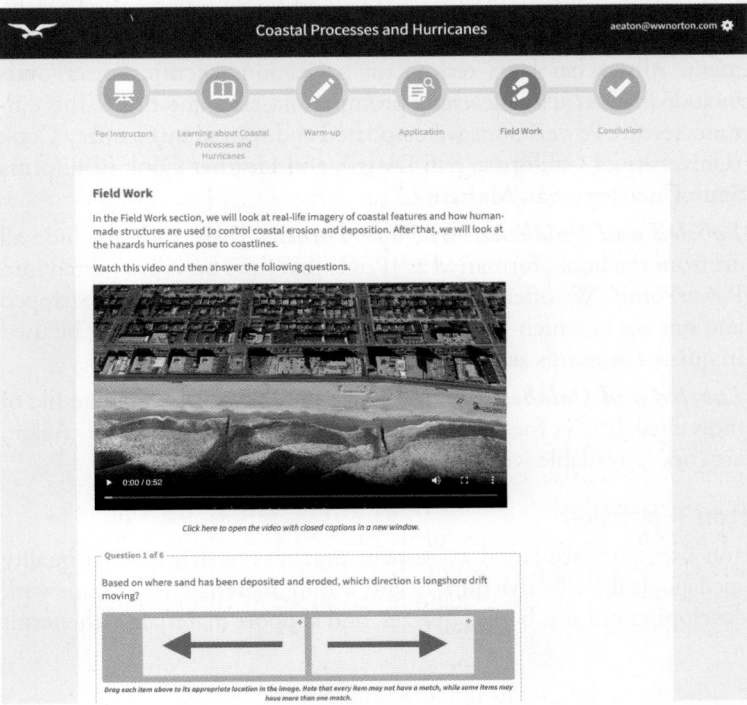

Smartwork Online Activities

The **Smartwork** online assessment system features visual assignments that provide students with answer-specific feedback and links to the relevant ebook section for every question. Students get the coaching they need to work through assignments, while instructors get real-time insights into student progress via automatic grading and detailed item analysis. Smartwork can also be integrated directly into instructors' campus LMS, so that students benefit from single sign-on, and instructors benefit from automatic reporting of assignment results to the LMS gradebook.

Smartwork questions are written by and for Earth-Science educators to support the textbook's learning objectives. Among the many question types available in Smartwork are ranking, sorting, and labeling tasks that build directly on the text's art. Questions based on the Narrative Art videos, animations, and simulations and on the Earth Science in Action videos help students engage with Earth-Science processes in action. Questions based on the Explore Earth Science in 3D models allow students to examine and analyze specimens and sites as if they were in the field or lab. Short-answer questions based on the *Putting Earth Science to Use* and *How Can I Explain . . .* text boxes encourage students to explain Earth-Science phenomena and applications in their own words. And finally, Smartwork also provides basic reading quizzes that help students come to class better prepared, as well as guided inquiry activities from the *Geotours Workbook* that use Google Earth.

Smartwork is free with all new copies of *Earth Science*, Third Edition, in any format. It can also be purchased standalone at digital.wwnorton.com/earthscience3. Randye Rutberg (City University of New York) has revised and updated the Smartwork course for the Third Edition.

Additional Instructor Materials

Lecture PowerPoints and Art Files

- **Lecture PowerPoints**—Designed for instant classroom use, these highly visual slides use photographs and line art from the book in a form that has been optimized for use in the PowerPoint environment. All art has been resized for projection. Lecture PowerPoints include in-class active-learning prompts and class questions. This edition's lecture PowerPoints were updated and revised by Geoffrey Cook (University of California, San Diego), and Heather Cook (California State University, San Marcos).

- **Labeled and Unlabeled Art PowerPoints**—These slides include all art from the book, formatted as JPEGs that have been pre-pasted into PowerPoints. We offer one set in which all labeling has been stripped and one set in which labeling remains, so that the slides may be used in quizzes or exams as well as in presentations.

- **Labeled and Unlabeled Art JPEGs**—We provide a complete file of individual JPEGs for art and photographs used in the book. Again, artwork is available with and without labels.

Norton Testmaker

Norton uses evidence-based assessment practices to deliver high-quality and pedagogically effective quizzes and testing materials. The framework for developing our test banks, quizzes, and support materials is the result of a collaboration with leading academic researchers and advisers. Our test bank includes over 60 questions per chapter, including a variety of questions classified by section and difficulty, making it easy to construct tests and quizzes that are meaningful and diagnostic.

Norton Testmaker brings Norton's high-quality testing materials online. Create assessments for your course from anywhere with an Internet connection, without downloading files or installing specialized software. Search and filter test-bank questions by chapter, type, difficulty, learning objectives, and other criteria. You can also customize test-bank questions to fit your course. Easily export your tests to Microsoft Word or Common Cartridge files for your LMS.

The Third Edition's test questions have been revised and updated by Liana Boop at San Jacinto College. The previous editions' test bank was originally authored by Maureen Lemke (Texas State University).

Norton Teaching Tools

The **Norton Teaching Tools instructors' page**, prepared for the Third Edition by Liana Boop (San Jacinto College), is designed to help instructors prepare lectures, homework, and exams. It contains the following materials for each chapter:

- Learning objectives
- Chapter summaries
- Complete answers to end-of-chapter review questions
- Descriptions, teaching notes, and discussion questions for the Narrative Art videos, animations, and simulations and the Earth Science in 3D videos
- Activity ideas to implement active learning in class

This resource also includes a guide to transitioning between Marshak and Rauber's *Earth Science*, Third Edition, and Tarbuck and Lutgens's *Earth Science*, Fifteenth Edition, for those looking to make the switch.

Geotours Workbook, Second Edition

Created by Scott Wilkerson and Beth Wilkerson (DePauw University), and Stephen Marshak, the **Geotours Workbook** contains active-learning exercises, arranged by topic, that take students on virtual field trips in Google Earth to see outstanding examples of geologic features at locations around the world. Each Geotour is accompanied by a worksheet that includes instructions and multiple-choice questions. The workbook accompanies a custom-made Google Earth kmz file created by Scott and Beth Wilkerson, available for free download by all instructors and students using *Earth Science*, Third Edition, at digital.wwnorton.com/earthscience3. The *Geotours Workbook* can be packaged for free with the text and includes complete user instructions and advanced instruction. Request a sample copy from your local Norton representative to preview each worksheet.

See for Yourself Google Earth Sample Sites

Earth Science users who simply want access to the *See for Yourself* field sites for classroom presentations or distribution to students can download the sites from digital.wwnorton.com/earthscience3.

See for yourself
Using *Google Earth*

Visiting Field Sites Identified in the Text

There's no better way to appreciate geology than to see it firsthand in the field. The challenge is that the great variety of geologic features that we discuss in this book can't be visited from any one locality. So, even if your class can take geology field trips during the semester, at most you'll see just a few geologic settings. Fortunately, Google Earth makes it possible for you to fly to spectacular geologic field sites anywhere in the world in a matter of seconds—you can take a virtual field trip electronically. Using related options, you can also look at stars in the sky.

In each chapter in this book, at least one *See for Yourself* provides sites that you can explore on your own computer (Mac or PC) using Google Earth software, or on your Apple/Android smartphone or tablet with the appropriate Google Earth app.

To get started, follow these three simple steps:

1 Check to see whether Google Earth is installed on your personal computer, smartphone, or tablet. If not, download the software from **earth.google.com** or the app from the Apple or Android app store.

2 Each *See for Yourself* site provides a thumbnail photo and brief description of the site (highlighting what you will see), as well as the latitude and longitude of the site.

3 Open Google Earth, and enter the coordinates of the site in the search window. As an example, let's find Mt. Fuji, a beautiful volcano in Japan. We specify the coordinates in the book as follows:

Latitude   35°21′41.78′N   Longitude   138°43′50.74′E

Type these coordinates into the search window as:

35 21 41.78N, 138 43 50.74E

Note that the degree (°), minute (′), and second (″) symbols are optional and can be left as blank spaces.

When you click Enter or Return, your device will bring you to the viewpoint right above Mt. Fuji illustrated by the thumbnail on the left. Note that you can use the tools built into Google Earth to vary the elevation, tilt, orientation, and position of your viewpoint. The thumbnail on the right shows the view you'll see of the same location if you tilt your viewing direction and look north.

View looking down.

View looking north.

Need More Help?

Please visit https://digital.wwnorton.com/earthscience3 to find a video showing you how to download and install Google Earth, more detailed instructions on how to find the *See for Yourself* sites, additional sites not listed in this book, links to Google Earth videos describing basic functions, and links to any hardware and software requirements. Also, notes addressing important Google Earth updates will be available at this site.

We also offer a separate book—the *Geotours Workbook*, Second Edition (ISBN 978-1-324-00096-9)—that identifies additional interesting geologic sites to visit, provides active-learning exercises linked to the sites, and explains how you can create your own virtual field trips.

ACKNOWLEDGMENTS

Many people helped the authors bring this book from the concept stage to the shelf, and we greatly appreciate their roles in continuing the forward momentum of this sometimes overwhelming project.

First and foremost, we wish to thank Kathy Marshak, who served as the book's in-home production and editorial coordinator. She worked with the authors to review and revise text, monitored the flow of manuscript and proofs between authors to meet schedule expectations, and helped the authors respond to queries about content and design. Her efforts helped make this book possible.

As always, the people at W. W. Norton & Company have been wonderful to work with, and we thank them for their insightful input and enduring patience over this book's three editions. It remains a privilege to work with an employee-owned company whose editorial staff remains willing to work directly with authors. A strategy to write this book originated in discussion with the late Jack Repcheck, a mentor whose impact will always be remembered. Eric Svendsen's experience and skill guided the book through the starting gate. Eric injected many ideas that contributed to its style and organization, connecting the project to new trends in science pedagogy and book design. Jake Schindel took over the reins during the First Edition's journey and guided us through the Second Edition. His abilities to respond quickly and thoughtfully to every issue that came up, to solve problems skillfully, and to keep the project moving forward were more than impressive, and we greatly appreciate his enthusiasm and hard work. Andy Blitzer joined Norton in time to guide the Third Edition's journey. Andy has brought new ideas to the table, and he continues to seek pedagogically sound approaches to integrating traditional print-book approaches with the digital world.

Thom Foley, our project editor for over two decades at Norton, put intense and much-appreciated effort into every book that he managed. For this edition of *Earth Science*, we have worked with a new production team at Norton and have been delighted with the effort and guidance that they brought to this complex project. Jen Barnhart skillfully took over as project editor. She has done a superb job of overseeing the complicated process of managing the manuscript, proofs, and schedule, all while remaining incredibly calm. We appreciate the contributions of Elaina Sassine and Mia Davis for their support in this effort. Norma Sims Roche has, as always, done a spectacular job of copy editing, patiently and sensitively sorting through and resolving many layers of edits. Her eye for detail and consistency is amazing. We also greatly appreciate copy-editing support from Susan McColl. We are very grateful to Ben Reynolds. Not only did he coordinate the back-and-forth between the publisher and various suppliers, he also served as the institutional memory for the book and, with Jen, kept a close watch on the schedule. Stephanie Romeo's expert and thoughtful management of the photo program and tireless efforts to resolve permissions issues was invaluable. We also thank Fay Torresyap for her very insightful photo research. Many of the book's photos were taken by Steve Marshak, and he thanks his family and students for being willing to pose as scales for geologic features.

The development of digital pedagogy and ancillaries has become an integral part of textbook development. We are very pleased to have been able to work with Ariel Eaton and her team, who oversee the development of digital media. Many thanks to Debra Morton Hoyt and Anne DeMarinis for overseeing the book's cover and design, to Anne DeMarinis for the interior design, and to Megan Schindel for overseeing permissions. We are grateful to the book's marketing director, Marlee Lisker, and to Norton's staff of campus sales representatives, who have done an outstanding job of bringing this book to the attention of a broad audience.

We consider the art in this book to be as important as the text in conveying ideas. For this edition, Troutt Visual Services (TVS) in Champaign, Illinois, managed the art program. We were delighted to work again with Jan Troutt and Joanne Brummett on new figures and on revisions for this edition. As always, art development is a close partnership between authors and artists, and Jan and Joanne make this process work very smoothly.

We also wish to thank the many artists with whom we worked in the past, whose contributions continue in this book. Our enduring thanks go to Stan Maddock, who was responsible for conceptualizing the book's art style and for producing many of the figures, at all levels of complexity, in earlier editions. It was great fun to interact with Gary Hincks, who painted many of the two-page spreads, in part using his own designs and geologic insights. Some of Gary's paintings originally appeared in *Earth Story* (produced by BBC Worldwide, 1998) and were conceived by Simon Lamb and Felicity Maxwell. Others were produced for *Earth: Portrait of a Planet*. New two-page spreads for *Earth Science* were developed by Stan Maddock, Joanne Brummett, Andrew Recher, and Craig Durant. Their amazing artistic talents will continue to help students see the big picture.

This book and its related texts, *Earth: Portrait of a Planet*, *Essentials of Geology*, and *Natural Disasters: Hazards of a Dynamic Planet*, have benefited greatly from input by expert reviewers of specific chapters, by general reviewers of the entire book, and by other reviewers who have provided helpful feedback for this and previous editions:

Mary Abercrombie, *Florida Gulf Coast University*
Kevin Barrett, *Tarrant County College, Northwest Campus*
Erica Barrow, *Ivy Tech Community College*
Karin Block, *City College of New York*
Brett Burkett, *Collin College*
Karen Busen, *Tallahassee Community College*
Marianne Caldwell, *Hillsborough Community College*
Cinzia Cervato, *Iowa State University*
Jennifer J. Charles-Tollerup, *Rowan College of South Jersey*
Laura Chartier, *University of Alaska, Anchorage*

Robert L. Dennison, *Heartland Community College*

Melissa Driskell, *University of North Alabama*

Todd Feeley, *Montana State University*

Kerry Workman Ford, *California State University, Fresno*

Nicholas Frankovits, *University of Akron*

Kenneth Galli, *Boston College*

Bryan Gibbs, *Dallas College, Richland Campus*

Alessandro Grippo, *Santa Monica College*

Amy Gross, *University of North Carolina at Pembroke*

Mary Hall-Brown, *University of North Carolina at Greensboro*

Tobin Hindle, *Florida Atlantic University*

Mark Horrell, *Northwest Florida State College*

Ashanti Johnson, *University of Texas, Arlington*

Felix Jose, *Florida Gulf Coast University*

Leslie Kanat, *Northern Vermont University, Johnson campus*

Zoran Kilibarda, *Indiana University Northwest*

Michael Kozuch, *California State University, East Bay*

Michael Kruge, *Montclair State University*

Kody Kuehnl, *Franklin University*

Kristine Larsen, *Central Connecticut State University*

Ana Larson, *University of Washington*

Brett C. Latta, *Franklin University*

Rita Leafgren, *University of Northern Colorado*

Maureen Lemke, *Texas State University*

Judy McIlrath, *University of South Florida*

Jamie Mitchem, *University of North Georgia, Gainesville*

Daniel Murphy, *Dallas College, Eastfield campus*

Jacob Napieralski, *University of Michigan, Dearborn*

Barbara O'Grady, *Montana State University*

Stacy Palen, *Weber State University*

Mark Peebles, *St. Petersburg College*

Melanie Roberti, *Santa Fe College*

Steven H. Schimmrich, *SUNY Ulster County Community College*

Marcia Schulmeister, *Emporia State University*

Jennifer K. Sheppard, *Moraine Valley Community College*

Andrew Smith, *Vincennes University*

Steven Stemle, *Palm Beach State College*

Keith Sverdrup, *University of Wisconsin-Milwaukee*

Tawny Tibbits, *Broward College*

Anthony J. Vega, *Clarion University*

Adil M. Wadia, *The University of Akron*

Shizuko Watanabe, *Dallas College, Eastfield campus*

Terry R. West, *Purdue University*

We also wish to thank the instructors who joined us for focus groups regarding this text and the courses it serves, whose input helped shape the form and content of this book:

Mary I. Abercrombie, *Florida Gulf Coast University*

Randall Adsit, *East Los Angeles College*

Karyn Alme, *Kennesaw State University*

Kathleen Bardsley, *Hillsborough Community College*

Kevin Barrett, *Tarrant County College, Northeast Campus*

Erica Barrow, *Ivy Tech Community College*

Gregory Barrows, *Macomb Community College*

Bulent Bas, *San Diego Mesa College*

Nicholas Bianco, *Ivy Tech Community College*

Liana M. Boop, *San Jacinto College District*

Mark Boryta, *Mt. San Antonio College*

Wayne Brew, *Montgomery County Community College*

Brandon Brooks, *Chaffey College*

Timmy Butts, *Hillsborough Community College*

Haluk Cetin, *Murray State University*

Johnsely S. Cyrus, *North Carolina Agricultural and Technical State University*

Frank DeCourten, *Sierra College*

Paul DeLaLuz, *Lee University*

Jeremy Dillon, *University of Nebraska at Kearney*

David Douglass, *Pasadena City College*

Todd Feeley, *Montana State University*

Alex Geddes, *Lane Community College*

Carresse Gerald, *North Carolina Central University*

Kayla Gibbs, *Texas A&M University-Commerce*

Amy Gross, *University of North Carolina at Pembroke*

Stanley Hatfield, *Southwestern Illinois College*

Emily Hayes, *Ivy Tech Community College*

Garry Hayes, *Modesto Junior College*

Sven Holbik, *Harford Community College*

Mark Holland, *West Texas A&M University*

Tara Holmberg, *Northwestern Connecticut Community College*

Richard W. Hurst, *California Lutheran University*

Esosa Iriowen, *Wilmington University*

Asaad Istephan, *Madonna University*

Ashley Johnson, *Jacksonville University*

Jeri Jones, *Messiah University*

Alyssa Kaess, *California State University, Bakersfield*

Brian Kirchner, *Henry Ford College*

Maureen Lemke, *Texas State University*

Melissa Lobegeier, *Middle Tennessee State University*

John Marino, *Bradley University*

Sylvain Masclin, *University of California, Merced*

John McDaris, *Carleton College*

Jonathan McKenzie, *Florida SouthWestern State College*

Rosemary Millham, *SUNY New Paltz*

Ann Moulding, *Collin County Community College District*

Roann Mulvihill-Cobb, *Wilmington University*

Kent Murray, *University of Michigan, Dearborn*

Jeffrey Myers, *Western Oregon University*

Claudia Owen, *Lane Community College*

Tamara Pavelec, *Ivy Tech Community College*

Caroline Pew, *University of Washington*

Jason Pittman, *Folsom Lake College*

Nancy Price, *University of Washington*

Eric Jonathan Pyle, *James Madison University*

M. Hassan Rezaie Boroon, *Los Angeles Mission College/LACCD*

Dr. Jeffery G. Richardson, *Columbus State Community College*

Melanie Roberti, *Santa Fe College*

Cyrus Screwvala, *Ivy Tech Community College*

Upul Senaratne, *Wor-Wic Community College*

Jennifer Sheppard, *Moraine Valley Community College*

Rene Shroat-Lewis, *University of Arkansas, Little Rock*

Richard Sleezer, *Emporia State University*

Alexander Smirnov, *Lone Star College System*

Carol Anne Stein, *University of Illinois Chicago*

Richard Sumner, *Ivy Tech Community College*

Huma Syed, *Collin County Community College District*

Rebecca Teed, *Wright State University*

LeAnne Sue Teruya, *San Jose State University*

Suzanne Traub-Metlay, *Front Range Community College*

Jagruti Vedamati, *Evergreen Valley College*

David Voorhees, *Waubonsee Community College*

Jamieson Webb, *Gulf Coast State College*

Matthew Wiesner, *Benedictine University*

ABOUT THE AUTHORS

Stephen Marshak is a Professor Emeritus of Geology at the University of Illinois Urbana-Champaign, where he was on the faculty for 35 years. He served as Department Head of Geology and as Director of the School of Earth, Society, and Environment. Steve holds an A.B. from Cornell University, an M.S. from the University of Arizona, and a Ph.D. from Columbia University. His research interests lie in structural geology and tectonics, and he has participated in field projects on a number of continents. Steve loves teaching and has won his college's and university's highest teaching awards, as well as the Neil Miner Award from the National Association of Geoscience Teachers (NAGT) for "exceptional contributions to the stimulation of interest in the Earth sciences." Steve is a Fellow of the Geological Society of America. In addition to research papers and *Earth Science*, he has authored *Earth: Portrait of a Planet* and *Essentials of Geology*, and has coauthored *Laboratory Manual for Introductory Geology*, *Earth Structure: An Introduction to Structural Geology and Tectonics*, and *Basic Methods of Structural Geology*. Steve coauthored *Natural Disasters: Hazards of a Dynamic Planet* with Bob Rauber, and he coauthored *Laboratory Manual for Earth Sciences* with Jessica Olney, Allan Ludman, and Bob Rauber.

Robert Rauber is a Professor Emeritus of Atmospheric Sciences at the University of Illinois Urbana-Champaign, where he was a long-term faculty member. He served as Department Head of Atmospheric Sciences and as Director of the School of Earth, Society, and Environment. He holds a B.S. in Physics and a B.A. in English from the Pennsylvania State University, as well as M.S. and Ph.D. degrees in Atmospheric Science from Colorado State University. In addition to teaching and writing, for years Bob oversaw a research program that focuses on the development and behavior of storms. To carry out this work, Bob would fly into hazardous weather in specially equipped airplanes—during these flights, he's so intent on recording data that he usually doesn't notice the bouncing or lightning. Bob has won several campus teaching awards, is a Fellow of the American Meteorological Society (AMS), and served as Publication Commissioner for the AMS. In addition to research papers and *Earth Science*, his textbook writing credits include lead authorship of *Severe and Hazardous Weather: An Introduction to High Impact Meteorology* and *Radar Meteorology: A First Course*. Bob coauthored *Natural Disasters: Hazards of a Dynamic Planet* with Steve Marshak, and he coauthored *Laboratory Manual for Earth Sciences* with Jessica Olney, Allan Ludman, and Steve Marshak.

EARTH SCIENCE

The Earth, the Atmosphere, and Space

THIRD EDITION

INTRODUCTION
Welcome to Earth Science!

After studying this introduction, you should be able to . . .

1. list the subject areas covered in an Earth-Science course, and the types of scientists who contribute to these areas.

2. describe key themes that serve as foundations for understanding the Earth and the Universe.

3. identify issues and challenges that the study of Earth Science can help you to understand and address.

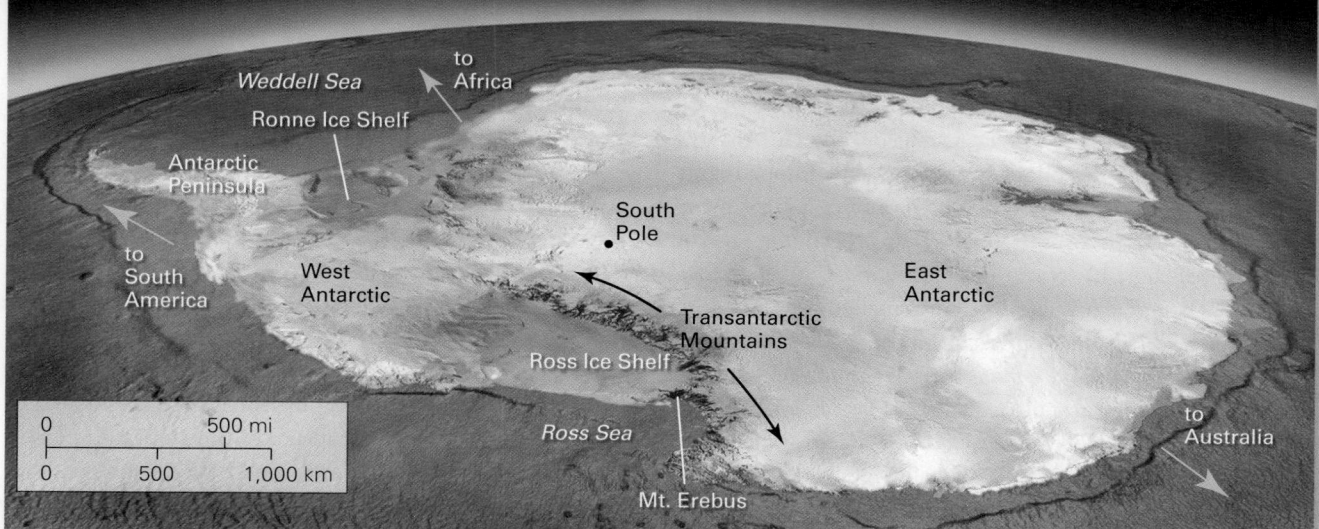

FIGURE I.1 Fieldwork in Antarctica unlocks the mysteries of an icebound continent.

(a) The Transantarctic Mountains separate East Antarctica from West Antarctica.

(b) The smoking summit of Mt. Erebus seen from the cockpit of a plane.

(c) A plane dropping geologists on a snowfield.

The C-130 transport plane rose from the frozen surface of the Ross Sea, near the coast of Antarctica, and turned south. We were heading to a field site about 250 km (155 mi) away where, with luck, we'd be able to spend the next month studying cliff exposures of some very unusual rocks **(Fig. I.1)**. The plane climbed past the smoking summit of Mt. Erebus, the Earth's southernmost volcano, and for the next hour it flew along the Transantarctic Mountains, a long range of rugged peaks that divides the continent into East Antarctica and West Antarctica. Over millions of years, snow falling in Antarctica's frigid climate accumulated to build *glaciers,* sheets and rivers of solid ice that slowly flow across the land. The glacial ice

(d) Sledding to a field site with crates of supplies.

<< Along the Colorado River at the bottom of the Grand Canyon, we can see interactions among many components of the Earth System—sunlight, air, water, rock, soil, and life—and we can see a few pages of the Earth's long history as recorded by characteristics of the rock layers in the canyon's walls.

sheet covering East Antarctica attains a thickness of over 3 km (2 mi), so its top surface is a high-elevation plain called the Polar Plateau.

As we flew south, we marveled at the stark panorama of ice and rock, so different from the forests, grasslands, farms, and cities of our planet's more populated regions. Then we

3

heard the engines slow and felt the C-130 descend. As the plane approached the glacier just below the cliff that we hoped to study, the pilot lowered the ski-equipped landing gear. Shouting above the engine noise, a crew member reminded us of the emergency procedure: "If you hear three short blasts of the siren, hold on for dear life!" Seconds later, the skis slammed into small frozen snowdrifts that formed wave-like ripples on the surface of the glacier. *Wham, wham, wham, wham!* It felt as though a fairy-tale giant was shaking the plane. Then, as fast as it began, the shaking stopped, for we were airborne again, looking for a softer landing surface. We finally touched down at a location where snow blankets the Polar Plateau. Bundled against the frigid gale generated by the plane's still-roaring propellers, we jumped out and helped unload food, fuel, stoves, tents, sledges, and snowmobiles. The instant the cargo was out, the engine noise grew, and trailing clouds of snow, the C-130 accelerated and rose skyward. When the glint of the plane's metal skin passed beyond the horizon, the silence of Antarctica hit us. No dogs barked, no leaves rustled, and no traffic rumbled in this land of black rock, white snow, and blue ice. It would take us almost two days, even with the aid of snowmobiles, to haul sledges of food

and equipment to our field site, where we would spend the next month collecting rock samples.

Why go to so much effort and expense to study a rock exposure? The Scottish writer Walter Scott (1771–1832) gave a colorful answer: "Some rin uphill and down dale, knapping the chucky stanes to pieces wi' hammers, like sae mony road-makers run daft—they say it is to see how the warld was made!" Indeed, to see how the world was made. Field expeditions like the one we've just described, along with analyses carried out using instruments in laboratories, measurements collected from ships and planes, surveys made by drones and satellites, calculations run on computers, and scans by telescopes and space probes, have led to an explosion of discoveries about the land, sea, and atmosphere of the Earth, and about objects that populate the rest of the *Universe*—all of space, and everything within it. Such discoveries provide the foundations of Earth Science, and they lead to an understanding not only of our home planet but also of its context in space (Fig. I.2). To help you explore these discoveries, this Introduction outlines the topics covered by a course in Earth Science, introduces themes that tie these topics together, and suggests ways that you can use what you'll learn.

FIGURE I.2 Astronauts aboard the International Space Station have this view of space beyond the Earth's horizon as they orbit the planet.

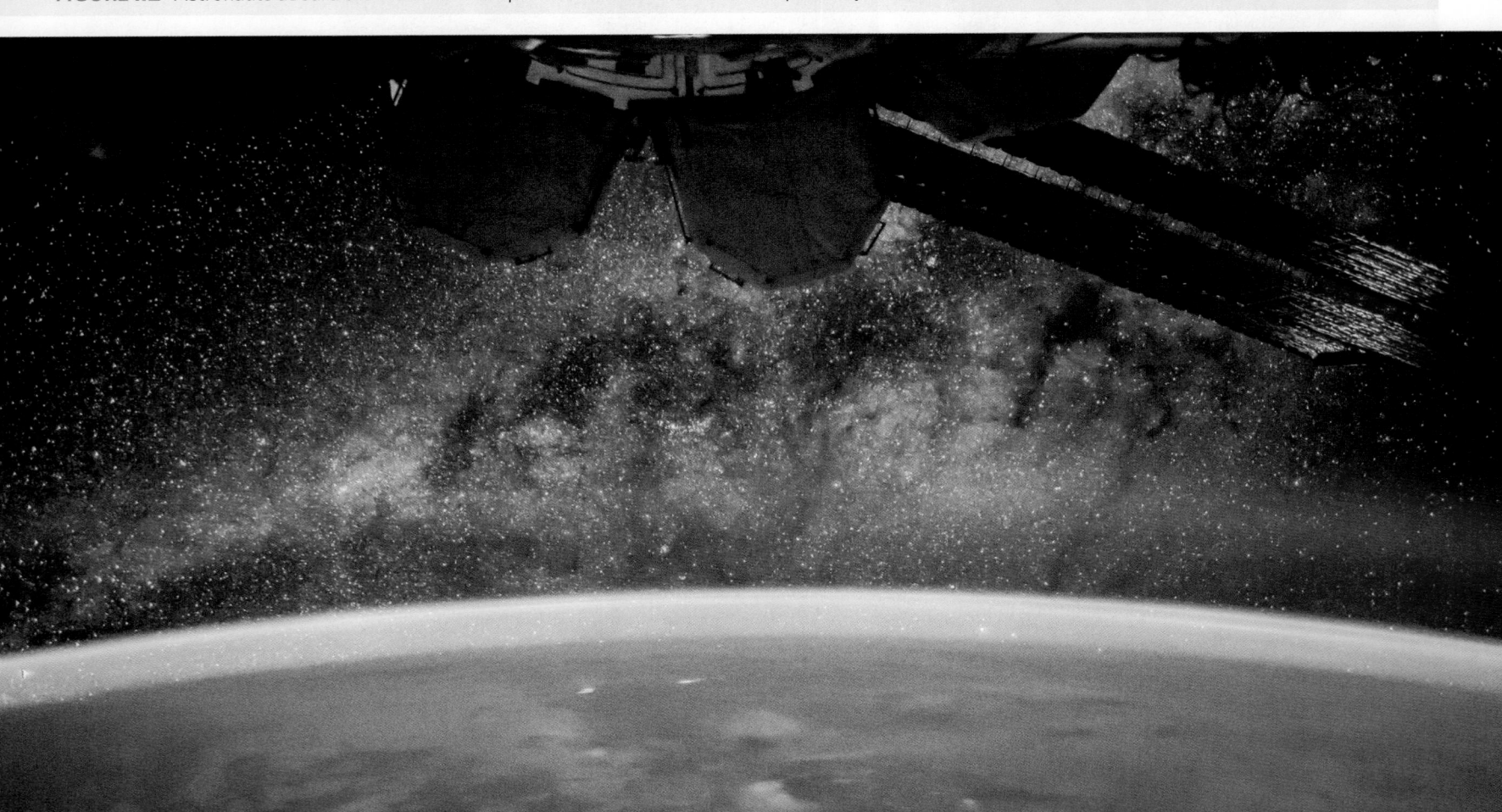

I.1 What's in an Earth-Science Course?

Defining Earth Science and Its Components

For most of humanity's existence, speculation about the Earth and space lay in the realms of philosophy or religion, and people attributed puzzling natural features in their surroundings to supernatural phenomena. In the past few centuries, however, study of our planet and its context has become a focus of **science**, the systematic analysis of natural phenomena and processes based on observation, experiment, and calculation. As the volume of scientific knowledge grew, scientists began to distinguish among distinct subjects, most of which are familiar to you. For example, if someone asks, "What's chemistry?" you might respond that it's the study of chemicals and the reactions among them. Similarly, if someone asks, "What's physics?"

you might respond that it's the study of matter and energy **(Box I.1)**, and if someone asks, "What's biology?" you might respond that it's the study of life. But if someone asks, "What's Earth Science?" you might be stumped.

Earth Science defies easy description because the subject encompasses many disciplines. Specifically, a typical Earth-Science course covers the basics of geology, oceanography, atmospheric science, and astronomy **(Fig. I.3)**. What does each of these disciplines address? **Geology** focuses on our planet's structure, composition, landscapes, motions (as manifested by earthquakes and volcanoes), and history. It also considers the many processes and phenomena that cause the planet, as well as the living organisms that call it home, to change. **Oceanography** addresses the characteristics of water in the oceans, as well as the way in which this water moves, interacts with land and air, and serves as the home for countless life forms. **Atmospheric science** examines the composition and behavior of the air layer that surrounds the

---

FIGURE I.3 The various disciplines covered in an Earth-Science course.

(a) A geologist studies a rock face.

(b) An oceanographer samples the sea.

(c) An atmospheric scientist monitors and analyzes the atmosphere.

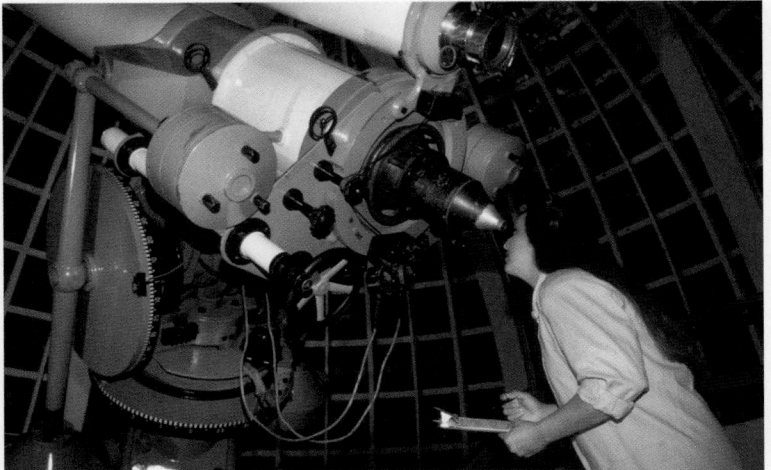

(d) An astronomer explores the cosmos.

BOX I.1 **Science Toolbox**

Matter and energy

The Universe consists of matter and energy. What do matter and energy consist of? **Matter** makes up the material substance of the Universe—it's the stuff that occupies space. We refer to the amount of matter in an object as its *mass,* so an object with greater mass contains more matter. As we'll discuss further in Chapter 1, matter consists of tiny particles called *atoms,* which can attach to each other to form *molecules.*

Scientists recognize four different *states of matter*, meaning the forms in which matter exists. A **solid** can retain its shape, regardless of changes in the size of its container, because its atoms and molecules remain locked in position **(Fig. BxI.1a)**. A **liquid** can flow and conform to the shape of its container because clusters of atoms and molecules within it can move relative to one another **(Fig. BxI.1b)**. A **gas** not only can flow, but also expands to fill its container, because its atoms and molecules are not attached to one another and can move about freely **(Fig. BxI.1c)**. We'll discuss a fourth state of matter, *plasma,* later in the book—for now, we'll simply say that it resembles a very hot gas.

From your everyday experience, you know that it's more difficult to lift a block of rock than it is to lift a block of Styrofoam of the same size. That's because the two blocks have different densities. **Density** refers to the amount of matter within a given volume, or, formally stated, the mass per unit volume. A volume of space that contains hardly any matter, meaning that its density approaches zero, is a **vacuum** (from the Latin word *vacuus,* meaning vacant).

Energy, in the jargon of physics, is the ability to cause an object to move or to cause a physical system to change. Scientists distinguish among several forms of energy, including *kinetic energy*, the energy of motion; *potential energy*, the energy stored in a material that can be released later; *radiant energy*, the energy carried from one location to another in the form of *electromagnetic radiation* (such as light); and *thermal energy*, or *heat*, the energy that can cause a material to warm up.

FIGURE BxI.1 Three states of matter.

Solid

(a) A solid retains its shape and density.

Liquid

(b) A liquid conforms to its container, but retains its density.

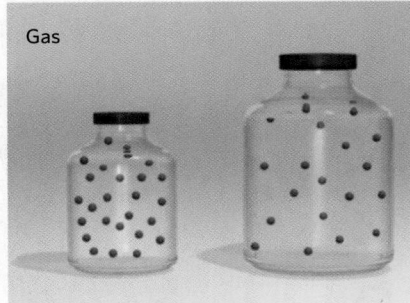

Gas

(c) A gas expands to fill its container, so its density can change.

Earth. It includes both *meteorology* (the study of *weather*, which focuses on the condition of the atmosphere at a given location and time, as well as on the movement of air) and *climate science* (which focuses on *climate*, the range of typical weather conditions and weather extremes in a region over the course of a year, as averaged over many years). **Astronomy**, or *space science*, considers the nature and behavior of the **Universe**, meaning space and all the material and energy within it. Studying astronomy provides a framework of knowledge on which to build an understanding of the formation and fate of our own planet. All subjects covered

in an Earth-Science course incorporate knowledge gleaned from physics, chemistry, and biology. In sum, an Earth-Science course can cover the nature, origin, and evolution of our natural surroundings. It's a broad subject, indeed!

The Scientists Who "Do" Earth Science

The information presented in this book comes from the work of scientists. Popular media often characterize scientists as awkward loners with poor taste in clothing. Who are they, really? Simply put, **scientists** are people who search for ideas that explain how natural systems, phenomena, and processes operate. At any given time, at locations all around the world, *field scientists* scale cliffs (**Fig.I.4a**), fly into storms (**Fig. I.4b**), or plow through stormy seas to examine natural features directly. And in a broader sense, field scientists also include people who use various instruments—telescopes, lasers, cameras, radar, and various types of sensors (located in observatories or carried by satellites, drones, or airplanes) to examine natural features from a distance. Meanwhile, *laboratory scientists* peer down microscopes, monitor high-tech analytical equipment, and observe reactions in test tubes (**Fig. I.4c**), and *computational scientists* program computers and solve equations. When relevant, laboratory and computational scientists develop **models** (simulations) to visualize processes that take place too slowly or too quickly to see in real time, or to characterize objects that are too small or too large to examine directly (**Fig. I.5**).

FIGURE I.4 Scientific research takes place in many environments.

(a) Cliff exposures in the desert of Utah provide a record of the Earth's past.

(b) To learn about severe weather, researchers fly into storms and take measurements.

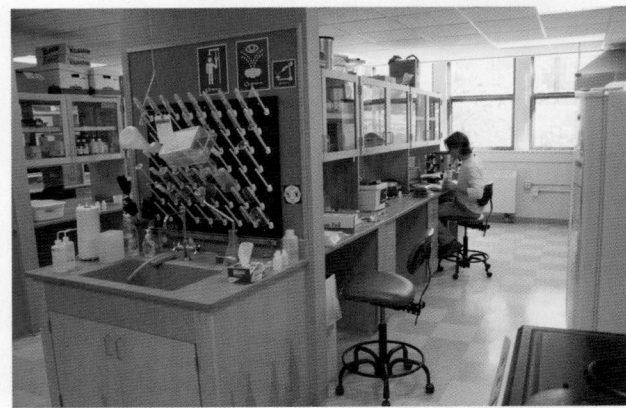

(c) Laboratories provide scientists with the opportunity to carry out controlled experiments.

FIGURE I.5 Use of models in Earth Science.

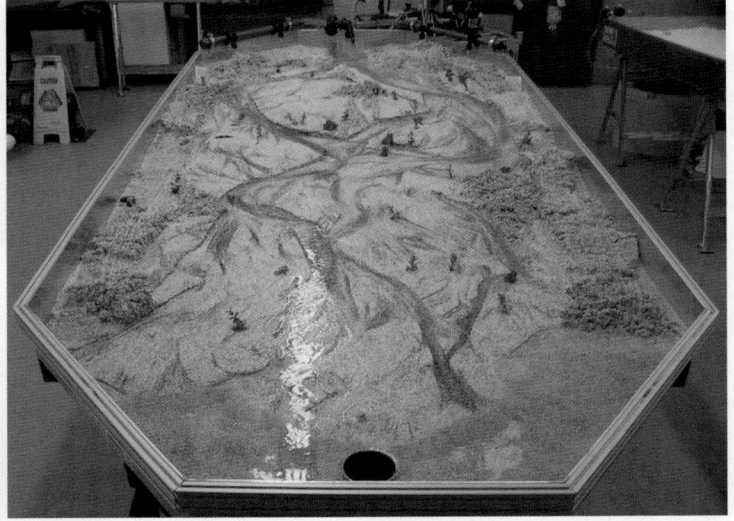

(a) A stream of water flowing over sand can simulate the evolution of a river.

(b) The flow of air through a severe thunderstorm can be simulated by a supercomputer.

BOX I.2 Science Toolbox

consider the idealized components of the scientific method and see how researchers applied them to propose and verify the meteorite-impact story.

The scientific method

Sometime in the distant past, a large block of rock or metal zoomed in from space and slammed into the Earth's surface at a site in what is now a flat cornfield in the midwestern United States. The impact of this *meteorite* blasted debris skyward and carved a deep, bowl-shaped depression called a *crater*. This event also shattered rock beneath the crater and caused layers of rock from deep underground to spring upward and tilt on end. Over time, *erosion* (the grinding away and removal of material at the Earth's surface by flowing water, moving ice, or blowing wind) wore down the crater until the depression disappeared entirely. But erosion did not carve deeply enough to remove the fractures and tilted rock layers deeper down. Some 15,000 years ago, *glacial till*—deposits of clay, sand, and gravel left by melting glaciers—buried the area and hid all evidence of the impact from view **(Fig. BxI.2)**.

Where did the scenario we've just described come from? It's the outcome of scientific research guided by the **scientific method**. Let's

- *Recognizing the problem:* Any scientific project, like any detective story, generally begins by identifying a problem. Our story began when workers drilling a well in the cornfield discovered that limestone, a rock commonly made of shell fragments, lay just below the layer of glacial deposits. In surrounding regions, sandstone, a rock made of sand grains that have been cemented together, lay directly beneath the glacial deposits, and limestone lay beneath the sandstone. Limestone can be used to make cement, roadbeds, and lime, so at a site near the well, workers dug a quarry to excavate the limestone. They were surprised to find that the limestone layers exposed in the quarry were tilted steeply and contained many large fractures. In contrast, the rock layers beneath the glacial till of surrounding regions were horizontal, like the layers in a birthday cake, and contained relatively few cracks. What phenomenon moved the limestone upward, tilted its layering, and produced large fractures? Curious scientists journeyed to the quarry to find out.

FIGURE BxI.2 An ancient meteorite impact excavated a crater and permanently changed rock beneath the Earth's surface.

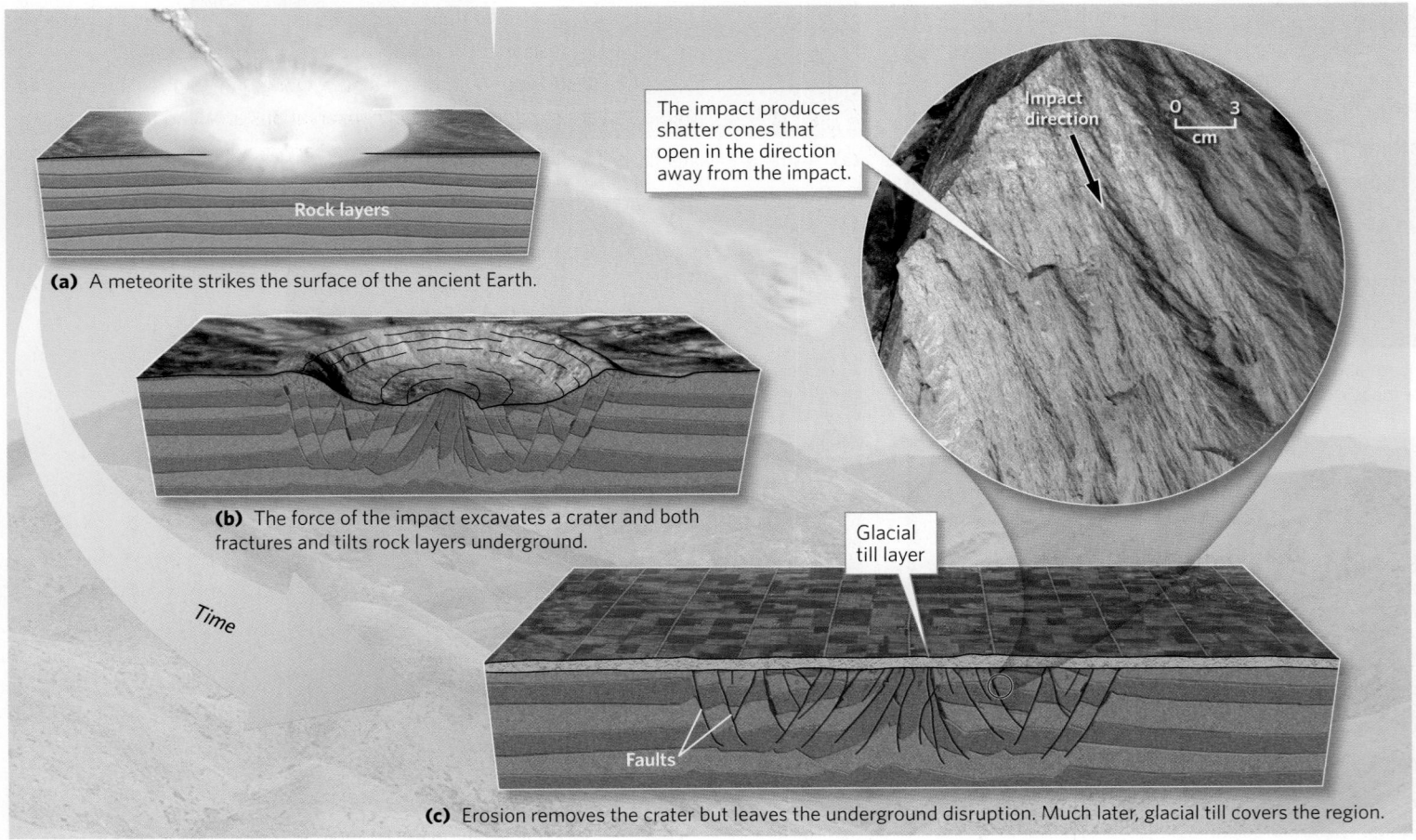

Rock layers

(a) A meteorite strikes the surface of the ancient Earth.

The impact produces shatter cones that open in the direction away from the impact.

Impact direction

0   3
cm

Time

(b) The force of the impact excavates a crater and both fractures and tilts rock layers underground.

Glacial till layer

Faults

(c) Erosion removes the crater but leaves the underground disruption. Much later, glacial till covers the region.

- *Collecting data:* To better characterize and, hopefully, solve a problem, scientists collect *data*—sets of observations, measurements, or calculations. Scientists working at the quarry measured the orientation of the rock layers and *documented* (made a written or photographic record of) the fractures that broke up the rocks. In the study of this quarry, data came from natural features formed long ago. In other situations, data might come from processes currently taking place, or from experiments and models. To be trustworthy, observations, calculations, and experiments should be *repeatable*, in the sense that another researcher going to the same location or following the same procedure should obtain the same result.
- *Proposing hypotheses:* A scientific **hypothesis** is merely a possible explanation for a set of observations or calculations. Scientists may propose multiple hypotheses before, during, or after their initial data collection. In this example, the scientists came up with two alternative hypotheses to explain the features in the quarry: (1) the features formed during an ancient volcanic explosion; and (2) the features were the consequence of an ancient meteorite impact.
- *Testing hypotheses:* Because a hypothesis is just a proposal that can be either right or wrong, scientists must test a hypothesis to determine its validity. Often, such tests involve making *predictions* about what additional observations, experiments, or calculations will show if the hypothesis is right. If a hypothesis fails these tests, scientists discard it. If a hypothesis passes these tests, it might be right.

The scientists studying the quarry compared their field observations with published observations made at known sites of volcanic explosions and meteorite impacts, and with the results of models. They realized that the quarry did not contain the type of rock formed at volcanoes. Because no such rocks were found there, the volcanic-explosion hypothesis failed the test. The scientists also learned that if the features in the quarry were the result of an impact, rocks there should contain *shatter cones*, arrays of tiny cone-shaped cracks. Shatter cones can be overlooked, so the scientists returned to the quarry specifically to search for them—and found them in abundance. The impact hypothesis passed this test and remained a possibility!

Scientists don't always see the entire path to solving a problem when they start working on it. In fact, at first they might not recognize the problem to be solved, so they don't always follow the components of the scientific method in a specific sequence. Furthermore, serendipity often plays a role in research, in that scientists may stumble onto a new problem or solution without planning on it. And while, ideally, scientists try to confirm their results by repeating observations, calculations, or experiments, in Earth Science such work isn't always possible because the phenomena under study may have happened in the very distant past, or in locations that are far away. In fact, some questions just can't be answered fully using the available resources and methods, so scientists must accept that some interpretations remain uncertain.

In common English, the word *theory* often substitutes for the word *hypothesis*. For example, you may see sentences like, "The detective's proposal is only a theory," or "The reporter proposed many theories," with the implication that a theory is just an idea that's as likely to be wrong as it is to be right. In scientific discussion, however, a **theory** is an idea that is supported by strong evidence and has passed many tests. Consequently, scientists have much more confidence in the validity of a scientific theory than they do in the validity of a hypothesis. Continued study in the quarry eventually yielded so much evidence favoring the impact hypothesis that the proposal came to be viewed as a theory. Scientists continue to test theories over a long time. Successful theories, those that are supported by many observations and lead to many successful predictions, eventually become part of a discipline's foundation.

In some cases, scientists have been able to devise concise statements that completely describe a specific relationship or phenomenon observed in nature. Such a statement, called a *scientific law*, applies without exception over a defined range of conditions. But unlike a theory, a law doesn't provide an explanation of a phenomenon. For example, Newton's *law of gravity* states that the gravitational attraction between two objects depends on the mass of the objects and the distance between them, but it does not explain why gravity exists.

As you study Earth Science, you'll have an opportunity to see how scientists conduct *scientific research*, the process of seeking to understand natural phenomena. You'll see that such research does not rely on subjective guesses or preconceived notions, but rather anchors itself in the development of consistent, testable explanations obtained by following the *scientific method* (Box I.2). Some scientists seek to define new principles that can profoundly change our understanding of nature—think of Isaac Newton, Charles Darwin, Marie Curie, or Albert Einstein—whereas others focus on practical matters, using the results of their research to improve our lives, find our resources, and protect our environment. When scientists finish a project, they present their results to others. Presentations may be in the form of a talk at a professional meeting, or a *peer-reviewed publication* in a professional

EXPLORE EARTH SCIENCE IN 3D

Meteorite Impact: Shatter Cones

journal (meaning an article that has been evaluated by other scientists to ensure that it makes sense and incorporates trustworthy data). Scientists who work in industry may present their results to company executives, for their research influences decisions about where to invest company funds. Increasingly, scientists are distributing their results to the public as lectures, podcasts, blogs, newsletters, or magazine articles, in order help everyone make informed choices.

Who are the scientists that make the discoveries we'll be focusing on in this book? When a headline begins with "Scientists say . . ." and then continues with "an earthquake shook Japan today," or "the supply of oil may be running out," the scientists under discussion are *geologists*. If the headline continues instead with "severe storms are approaching," or "the length of the growing season will change," the scientists referred to are *atmospheric scientists*. If the headline describes how "a shift in the pattern of ocean currents was detected," or "seawater has become more acidic in recent years," the scientists involved are *oceanographers*. And if the headline announces, "a new planet has been found in orbit around a distant star," or "a solar storm might disrupt communications," the scientists making news are *astronomers* or *space scientists*. (In this book, for simplicity, we refer to geologists, atmospheric scientists, and oceanographers together as *Earth scientists*.)

Take-home message . . .

An Earth-Science course encompasses geology, oceanography, atmospheric science, and astronomy. Field, laboratory, and computational scientists try to understand our surroundings using the scientific method.

Quick Questions ───────────────

- Provide a basic definition of the Universe.
- Why is it harder to define Earth Science than it is to define, say, chemistry?
- Name the four states of matter.
- How does geology differ from oceanography?
- Explain how a hypothesis differs from a theory, in a scientific context.

I.2 Narrative Themes of This Book

To enjoy a novel, you need to learn the names of the characters, and you need to pay attention to how the characters interact and relate to one another. Similarly, to enjoy a textbook, you need to learn some new vocabulary to discuss ideas efficiently, and you need to pay attention to how ideas connect to one another. In this section,

we introduce some of these connected ideas, or *narrative themes*, that tell the story of the Earth and its context in space. These themes represent this book's overall take-home message.

Matter and Energy Behave in Predictable Ways, but Uncertainty Exists

Natural interactions between matter and energy happen in predictable ways (Box I.3). For example, when the Sun shines on a dark-colored car, the car's surface becomes very hot, and when you place liquid water in a freezer, it turns into solid ice. Because such phenomena happen in the same way under the same circumstances every time, scientists can predict, to some extent, how natural processes change the Universe over time.

We say "to some extent" here because nature often displays *unpredictability*. For example, while we expect a boulder pushed over a cliff to tumble downslope, we can't pinpoint exactly where it will land, for it may ricochet off the ground or off other boulders on its way down. Similarly, when we make a measurement repeatedly, we may find that the values vary slightly, either because of the randomness inherent in some natural phenomena, or because our measurement technique isn't perfect. Scientists describe such variability by characterizing the *degree of uncertainty* associated with a measurement or prediction. Specifically, scientists distinguish between **accuracy**, representing how close a measured value is to the true value, and **precision**, representing how close a succession of measurements of the same feature are to one another.

The Earth System Contains Many Interacting Realms

As you study the Earth, you'll discover that our home planet contains a variety of **realms**, or *domains*, meaning regions or volumes with a distinct character, composition, and behavior. These realms include the following:

- *Geosphere:* Earth scientists refer to the part of our planet that extends from its solid surface down to its center—6,371 km (3,959 mi) below—as the **geosphere**, and they divide it into concentric layers nested one within the other like the layers of an onion. We live on the surface of the *crust*, the outermost layer. The crust overlies the *mantle*, which in turn surrounds the outer and inner *core* (Fig. I.6).

- *Hydrosphere:* The **hydrosphere** includes the liquid, solid (frozen), and gaseous water of the Earth. Liquid water includes *surface water* (which occurs in oceans, lakes, streams, rivers, and swamps) and *groundwater* (which fills open holes and cracks underground in the upper crust).

- *Cryosphere:* Sometimes, Earth scientists refer to the frozen component (snow and ice) of the hydrosphere separately as the **cryosphere** (from *cryo,* the Greek word for icy cold). The cryosphere includes regions where glaciers cover the land surface, where the ground remains frozen all year, and where sea ice covers the ocean.

- *Atmosphere:* The **atmosphere** is a layer of gas that extends from the Earth's surface upward. We refer to the mixture of gases in the atmosphere as *air.* The density of air decreases with increasing altitude, so 99% of the mass of the atmosphere lies below an altitude of about 30 km (18 mi).

- *Biosphere:* All living organisms on the Earth, together with the portions of the Earth in which organisms exist, constitute the **biosphere**. The biosphere encompasses the land surface, the hydrosphere, the uppermost several kilometers of the geosphere, and the lower several kilometers of the atmosphere. The bottom of the biosphere lies at depths where groundwater becomes too hot for life to survive.

Earth scientists refer to the realms that we've just described, along with the ways in which realms interact with one another over time, as the **Earth System** (**Earth Science at a Glance**, pp. 12–13). Some materials move from realm to realm in the Earth System over time. In this context, a realm can be thought of as a *reservoir,* a place that can contain a quantity of material for a period of time. In some cases, a succession of transfers from realm to realm, so that the material returns to the realm

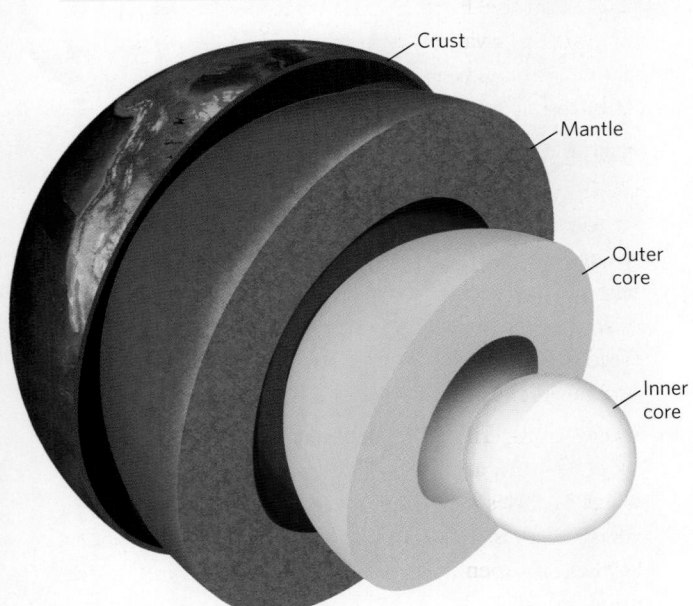

FIGURE I.6 Simplistically, the geosphere can be pictured as a set of concentric shells or layers.

Crust

Mantle

Outer core

Inner core

BOX I.3 ▶ How can I explain . . .

The difference between potential and kinetic energy

What are we learning?
That one kind of energy can transform into another.

What you need:
- Two golf balls
- Two full cups of water
- Paper towels

Instructions:
- Drop one golf ball into the first cup from a couple of centimeters (about an inch) above the water's surface.
- Drop the second golf ball into the second cup of water from a meter (about one yard) above the water's surface.
- Remove both balls and compare the amounts of water that remain in the two cups.

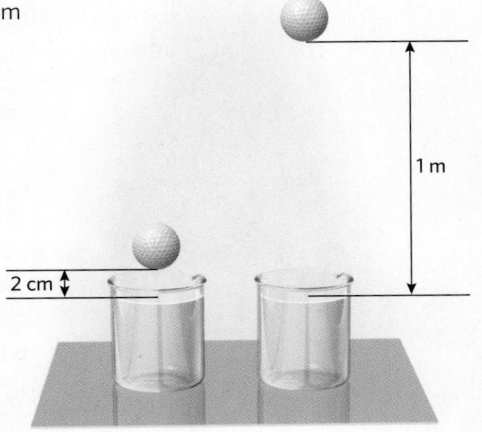
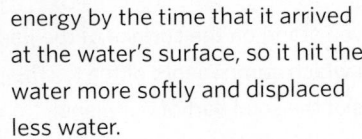
1 m
2 cm

What did we see?
- The ball falling from just above the water's surface displaced only a small amount of water, whereas the ball falling from a meter up hit the water much harder and therefore displaced a lot more water.
- The heights of the balls represent their potential energy, and the amount of water displaced represents an amount of kinetic energy. The golf ball that fell from the higher position had more energy to start with because by raising the ball a meter, you provided it with more potential energy. When it dropped, that potential energy was converted into kinetic energy. The ball that fell from the lower height had less potential energy to convert to kinetic

energy by the time that it arrived at the water's surface, so it hit the water more softly and displaced less water.
- The potential energy of the balls exists because of gravity. An object being pulled on by gravity stores energy until it can move. The higher it is, the more energy it stores. Why? Gravity causes objects to *accelerate* (speed up) as they fall. So the higher ball hits the water at a higher *velocity* (speed) than does the lower ball, because the higher ball falls farther. Because both balls have the same mass, the difference in kinetic energy is due to their different velocities.

The Earth System

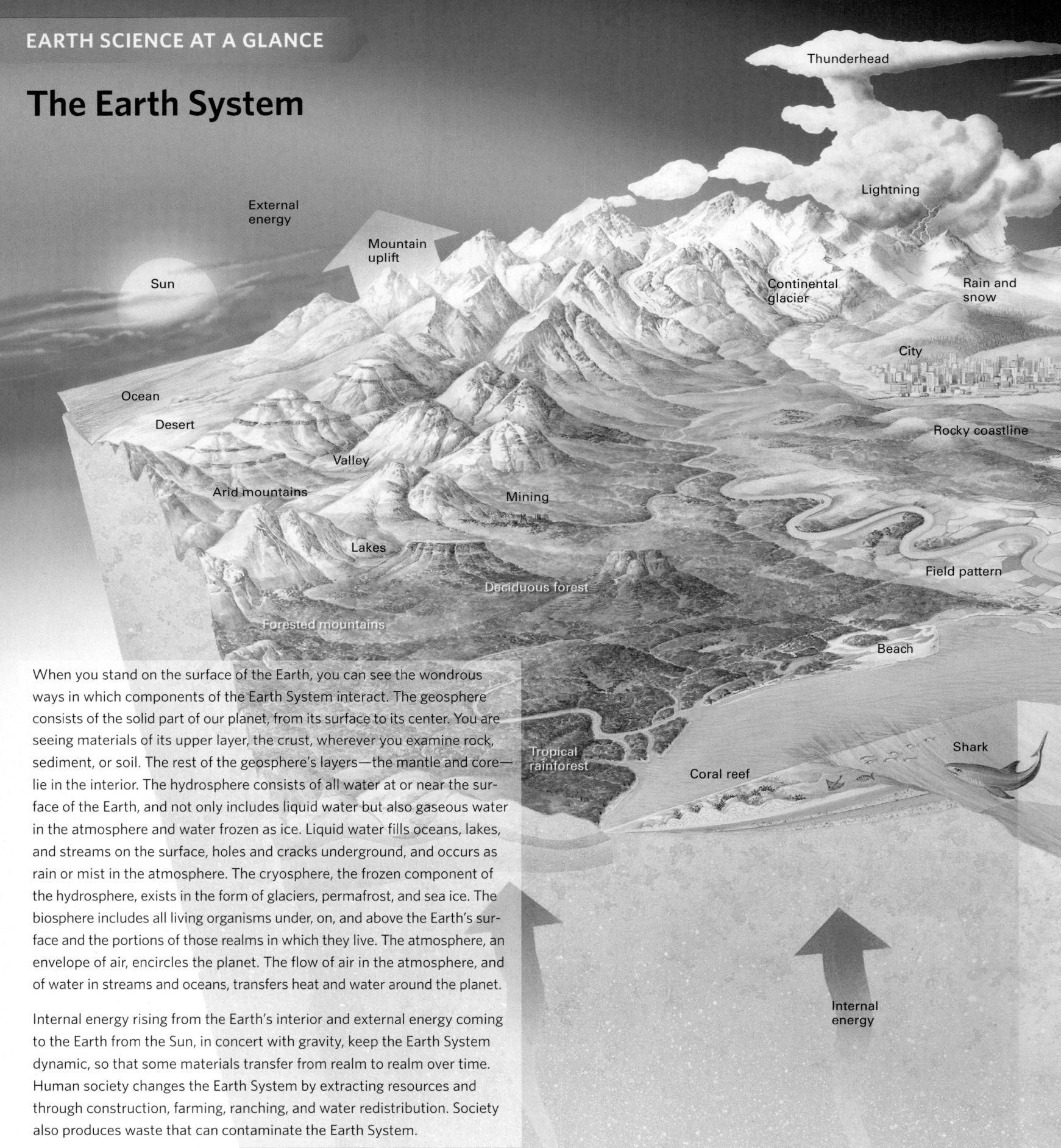

Thunderhead

Lightning

External energy

Mountain uplift

Continental glacier

Rain and snow

Sun

City

Ocean

Desert

Rocky coastline

Valley

Arid mountains

Mining

Lakes

Field pattern

Deciduous forest

Forested mountains

Beach

When you stand on the surface of the Earth, you can see the wondrous ways in which components of the Earth System interact. The geosphere consists of the solid part of our planet, from its surface to its center. You are seeing materials of its upper layer, the crust, wherever you examine rock, sediment, or soil. The rest of the geosphere's layers—the mantle and core— lie in the interior. The hydrosphere consists of all water at or near the surface of the Earth, and not only includes liquid water but also gaseous water in the atmosphere and water frozen as ice. Liquid water fills oceans, lakes, and streams on the surface, holes and cracks underground, and occurs as rain or mist in the atmosphere. The cryosphere, the frozen component of the hydrosphere, exists in the form of glaciers, permafrost, and sea ice. The biosphere includes all living organisms under, on, and above the Earth's surface and the portions of those realms in which they live. The atmosphere, an envelope of air, encircles the planet. The flow of air in the atmosphere, and of water in streams and oceans, transfers heat and water around the planet.

Tropical rainforest

Coral reef

Shark

Internal energy

Internal energy rising from the Earth's interior and external energy coming to the Earth from the Sun, in concert with gravity, keep the Earth System dynamic, so that some materials transfer from realm to realm over time. Human society changes the Earth System by extracting resources and through construction, farming, ranching, and water redistribution. Society also produces waste that can contaminate the Earth System.

Moon

Jet stream

Cirrus clouds

Aurora

Wind system

Ice and snow

Coniferous
forest

Evaporation

Volcanic islands

Industrial pollution

Cold surface current

Surface waters

Delta

Swamps

Warm surface current

Twilight zone

Abyssal zone

Whale

Seafloor

Bacteria and
plankton

Giant squid

Deep-sea
current

Black
smokers

13

FIGURE I.7 Time scales in Earth Science. Cosmologic time (left) goes back to the beginning of the Universe. Geologic time (right) goes back to the birth of the Earth. It is subdivided into eons and eras.

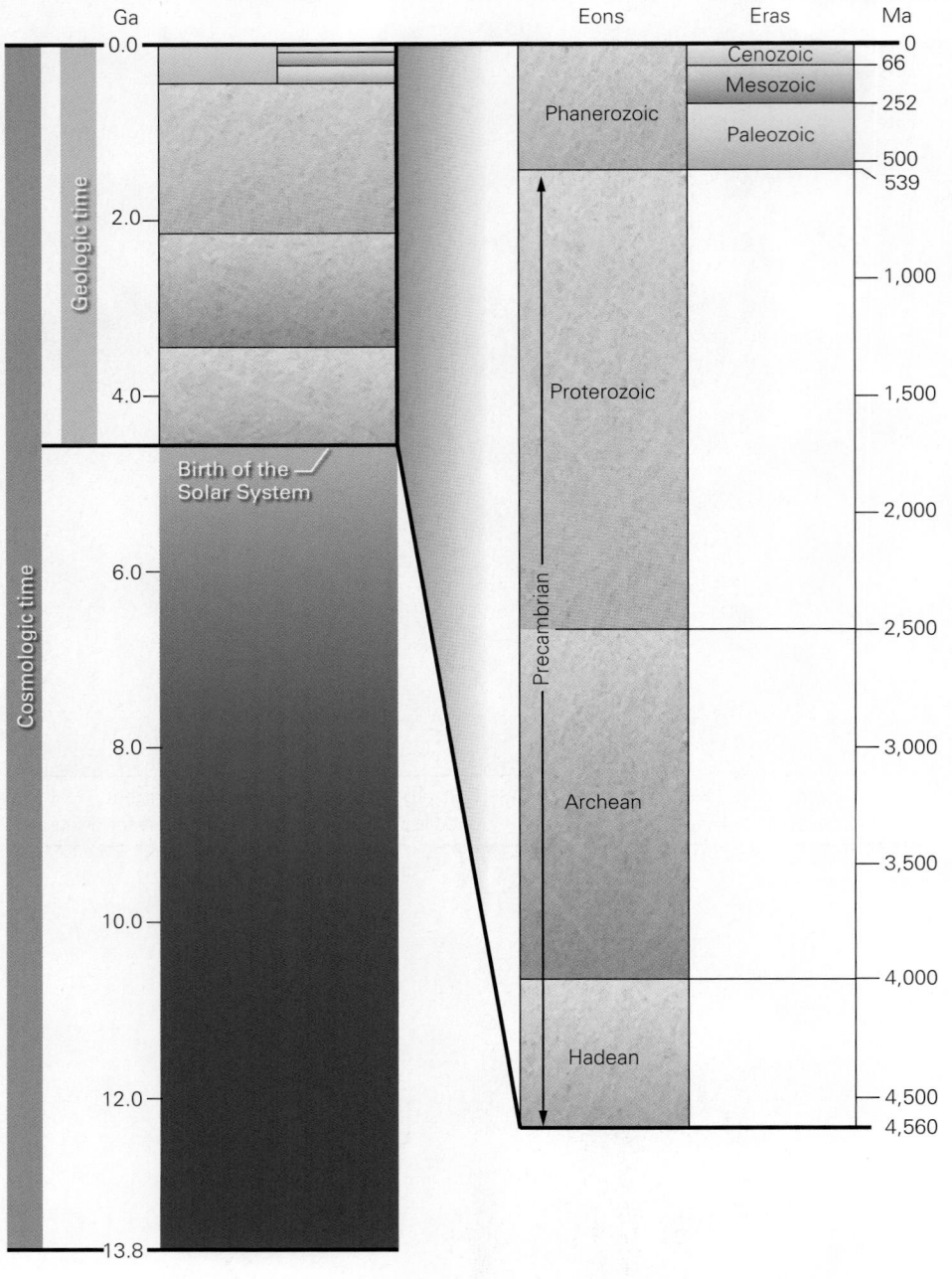

formed 4.56 to 4.54 billion years ago, so **geologic time** starts when the Universe was already over 9 billion years old (Fig. I.7). Such large numbers emphasize that our planet existed long before human history began. The immense span of cosmologic time means that more than enough time has passed for stars to form, mature, and die, and the immense span of geologic time means that more than enough time has passed for mountain ranges to rise and later be eroded away. Because the numbers used in discussing cosmologic or geologic time can be so huge, researchers use the following abbreviations: Ka for thousands of years (Ka stands for *kilo-annum*), Ma for millions of years (Ma stands for *mega-annum*), and Ga for billions of years (Ga stands for *giga-annum*). For example, we can say, "The dinosaurs went extinct at 66 Ma," meaning the dinosaurs went extinct 66 million years ago, or we can say "The cliff exposes 145 Ma granite," meaning that the rock in the cliff face has existed for 145 million years. As we'll see in Chapter 8, the **geologic time scale** divides the Earth's history into *eons, eras,* and other subdivisions of time.

The Theory of Plate Tectonics Explains Major Geologic Phenomena

According to the **theory of plate tectonics** (or, simply, *plate tectonics;* see Chapter 2), the outer, relatively rigid shell of the Earth consists of roughly 20 pieces, called *plates,* that move relative to one another (Fig. I.8). Plates interact with one another along *plate boundaries*—the breaks between adjacent plates—and those interactions produce earthquakes, volcanoes, and mountain ranges. Because of plate motions, **continents** (large land areas) move relative to one another as the **ocean basins** (places where the Earth's surface lies submerged beneath deep seawater) between them open and, later, close over the course of geologic time. As a result, the map of the Earth's surface showing the positions of continents constantly changes. For example, at 200 Ma, the Atlantic Ocean didn't exist, and a dinosaur could have walked from New York to Paris without getting its feet wet. Plate motions take place so slowly that they're almost imperceptible during the course of a human lifetime. But over geologic time, displacements become large. For example, North America has moved over 5,800 km (3,600 mi) away from Africa during the past 180 million years.

The Earth Is a Planet, and Each Planet Is Unique

Eight planets orbiting the Sun, together with moons and many smaller objects, make up the *Solar System.* The inner four planets, the ones that lie closest to the Sun, are known as the *terrestrial planets* because they have solid,

it started from, defines a *cycle* in the Earth System. Heat also transfers among realms, so warming of one part of the system can lead to warming of another.

The Universe and the Earth Are Very Old

Scientific research indicates that the Universe as we know it originated about 13.8 billion years ago, so **cosmologic time** (from the word *cosmos,* a synonym for the Universe) starts then. Our home planet, the Earth,

FIGURE I.8 Major plate boundaries and plates on the Earth. A plate is an area surrounded by plate boundaries. There are three types of plate boundaries (see Chapter 2). At transform boundaries, plates slide horizontally in a direction parallel to the boundary between them. At divergent boundaries, plates move away from each other. And at convergent boundaries, plates move toward each other. A place where landmasses converge with each other is a collision zone, and a place where a continent is splitting apart is a rift. Orange arrows indicate the velocity of plate movement: the longer the arrow, the faster the motion.

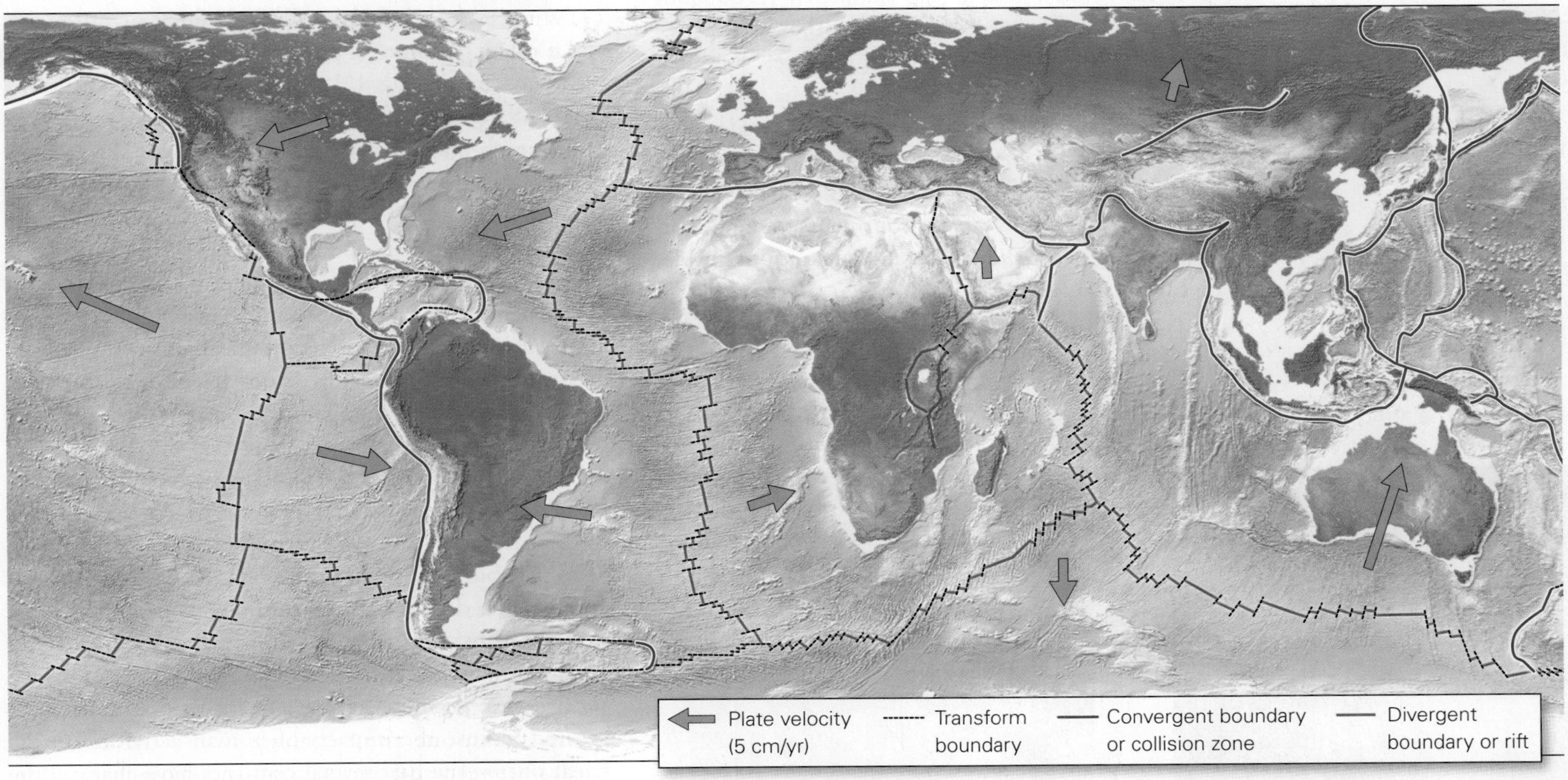

← Plate velocity (5 cm/yr)	----- Transform boundary	—— Convergent boundary or collision zone	—— Divergent boundary or rift

rocky surfaces like that of the Earth, the third planet out from the Sun. The outer four planets are known as the *Jovian planets* because, like Jupiter, they contain vast amounts of gas and ice and do not have rocky surfaces. No two planets look the same. Of the terrestrial planets, only the Earth has liquid water at its surface, an oxygen-rich atmosphere, and moving plates, and only the Earth, so far as we know, harbors life.

Various Sources of Energy Affect the Earth

The Earth is a dynamic planet because interactions among moving plates continue to generate earthquakes, volcanoes, and mountains, and the flow of water, ice, and air causes erosion of our planet's surface and redistributes materials. Three sources of energy drive this work **(Fig. I.9)**. **Internal energy** comes from heat within the Earth. Plate motion, with all its consequences, can happen because this heat keeps much of the Earth's interior soft enough to flow slowly. **External energy** travels to the Earth from the Sun in the form of light. This energy causes the ground, the atmosphere, and the oceans to

FIGURE I.9 The difference between internal and external energy in the Earth System.

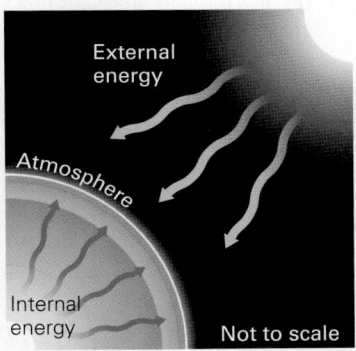

warm. **Gravity**, the pull that one mass exerts on another, holds the Earth together and keeps it circling the Sun. It also causes tides to rise and fall, rivers to flow, and rocks to tumble downslope. Together with gravity, internal energy causes flow of rock in the mantle, and external energy drives wind and ocean currents.

Change Happens

In space, stars and planets form, objects collide, stars explode, and the Universe expands. On the Earth, mountains rise, continents move, landscapes erode, the climate warms and cools, sea level rises and falls, and life evolves. Clearly, change has happened throughout the Earth System—and elsewhere in the Universe—throughout geologic and cosmologic time! Rates of change vary immensely. For example, some stars have remained virtually unchanged since the Universe began, whereas others burned furiously for just a few million years and then violently exploded. On the Earth, it takes hundreds of millions of years for continents to merge and drift apart, but a landslide may remove a mountainside, or an earthquake may collapse a town, in a matter of

seconds. Air circulates around the Earth over the course of days (Fig I.10a), but a tornado can produce a trail of destruction in minutes (Fig. I.10b). And if you study the ocean, you'll see that currents carry water across ocean basins over the course of weeks to months, but 600 waves may arrive at a beach every hour (Fig. I.10c).

Witness the aftermath of a large flood or hurricane, and it's clear that natural phenomena can change human-made structures (Fig. I.11a). Watch a bulldozer clear a swath of forest, hear dynamite blast apart a cliff, walk in a human-made canyon between rows of skyscrapers, or measure the rapid increase of carbon dioxide in the atmosphere, and it's clear that people similarly play a key role in changing our planet (Fig. I.11b). When discussing change on the Earth, therefore, we distinguish between **natural change**, meaning change due to geologic, oceanographic, meteorological, or astronomical events that would happen whether or not people existed, and **anthropogenic change**, meaning change that results from human activity.

For all but the past 0.000001% of geologic time, all change that took place in the Earth System was natural. In the past few centuries, however, our species has developed the ability to modify the Earth System in ways, and at rates, that do not happen naturally. During that time, we have transformed prairies and forests into fields and cities, we have moved mountains and redirected rivers, and we have synthesized new substances that didn't exist before. Because the impacts of human activities on the Earth during the past several centuries have changed the Earth profoundly, and will leave a record that will remain identifiable for millions of years into the future, some scientists refer to all or part of this interval of Earth history as the *Anthropocene*.

FIGURE I.10 The atmosphere and oceans of the Earth System are in constant motion.

(a) A satellite view shows swirling regions of clouds.

(b) In a tornado, air spirals upward in the form of a funnel at immense speed.

(c) Ocean waves crash on the rocky shore of a Hawaiian island.

Take-home message . . .

Many narrative themes tie together concepts in Earth Science. For example, matter and energy behave in predictable ways; the Earth is a complex system with interacting realms; the Universe and the Earth are so old that there has been plenty of time for significant change to take place; plate tectonics causes geologic phenomena on the Earth; and three key sources of energy do work in the Earth System.

Quick Questions —————————

- Distinguish between cosmologic and geologic time.
- Name three differences between the Earth and the other terrestrial planets.
- What sources of energy cause movements in the Earth System?
- What examples emphasize that humans now change the Earth at significant rates?

FIGURE I.11 Natural change and anthropogenic change.

(a) The natural power of a hurricane can level a beachside community.

(b) The human-generated power of a chain saw and a bulldozer can level a forest.

I.3 Why Study Earth Science?

Hundreds of thousands of people pursue careers as professionals in various aspects of geology, oceanography, atmospheric science, and astronomy. These professionals work in corporations, universities, government agencies, observatories, consulting firms, and nongovernmental organizations. Nevertheless, most students reading this book don't have such careers as a goal. Why should people study Earth Science if they do not plan to be Earth scientists? This question has many answers, all of which emphasize that an Earth Science course may be among the most useful courses anyone can take.

Earth Science Guides Decision Making

Knowledge of Earth Science can help you and your community make a variety of decisions, both big and small. Let's consider a few examples:

- *How to respond to natural hazards:* Is the location of an apartment that you'd like to buy or rent safe from **natural hazards**, meaning conditions that could lead to events such as flooding, landslides, or earthquakes that can cause casualties and damage **(Fig. I.12)**? Study of Earth Science can help you to assess the risk of encountering these hazards.

FIGURE I.12 Earth Science provides insight into natural hazards.

Colombia mudslide, 2017

(a) Knowledge of Earth Science might have prevented the building of homes on a slope subject to landslides.

Indonesia earthquake, 2018

(b) Knowledge of Earth Science could guide the establishment of building codes that allow buildings to withstand earthquake shaking.

FIGURE I.13 A thunderstorm approaches, bringing with it wind, rain, and lightning.

- *Whether a water supply will be safe:* Does your community pump drinking water from a well? Study of Earth Science can help you understand why a proposed new landfill project might contaminate the water supply.
- *Whether a parade should be canceled:* Are you worried that thunderstorms may disrupt an upcoming parade **(Fig. I.13)**? Study of Earth Science can help you interpret a radar display showing a storm's approach.
- *How much to pay:* The price of agricultural commodities depends on the supply. Therefore, predictions about rainfall and temperature over the course of a season, factors that control the supply, can help farmers and investors to determine whether to buy or sell.

- *Whether to take a news story seriously:* Have you seen a news article about climate change, or about the need for upgrading building codes, or about a local celebrity who predicts that an earthquake will happen next Tuesday at 4:05 P.M.? Study of Earth Science provides the basis for interpreting such articles and for determining whether you need to act in response to them.

If we had room in this book, we could continue adding to this list for pages. We hope the examples we've mentioned emphasize that by the end of your Earth-Science course, you'll have insights into many issues that could affect your livelihood, your family, your community, and your planet. All citizens of the 21st century have a stake in the Earth's future, and the study of Earth Science can help you understand how personal choices affect this future.

Earth Science Helps You Understand Our Context in the Universe

Reading this book can help give you a scientific perspective on how your home planet, you, and all your surroundings came to be. You'll be able to learn not only about the Earth, but also about the dazzling variety of other objects out in the immense Universe. The limits of our home planet—a spinning ball only 12,742 km (7,918 mi) in diameter—came fully into human consciousness when astronauts of the Apollo program sent back the first photo of the Earth rising above the horizon of the Moon **(Fig. I.14)**. All materials that humans have ever touched, seen, interacted with, or changed—not counting Moon rocks brought back by astronauts or meteorites that collide with the Earth—lie on, in, or just above that glowing island in space.

FIGURE I.14 Earth Science may help you understand the context of your home planet in space. This view of the Earth, taken by an astronaut orbiting the Moon, emphasizes that we live on an island in space.

(a) Coal from this mine can be burned to produce energy.

(b) Gases emitted from this power plant can affect air quality.

FIGURE I.15 The activities of human society can impact environmental quality and the climate.

Earth Science Addresses Resources, Sustainability, and the Environment

We take it for granted that we can construct buildings, roads, and dams out of concrete and glass; that we can drive to a store and buy metal cooking pots, paper clips, or plastic wrap; and that we can light a city all day. Where do we obtain the cement in concrete, the clay in bricks, the copper in wires, or the coal burned to generate electricity? All such resources come from the Earth **(Fig. I.15a)**. Because Earth Science addresses the origins and supplies of Earth materials from which we extract the resources that society uses, as well as the consequences of using these materials, the study of Earth Science provides a basis for understanding **sustainability** (the ability to maintain or improve our standard of living without running out of

resources) and *environmental quality* (the breathability of air, the drinkability of water, and the ability of a region to support life) **(Fig. I.15b)**.

Earth Science Helps You Appreciate Your Surroundings

We hope this book will help you see your world differently. When you drive down a highway or hike in the mountains, the rocks you see will no longer be merely ruddy gray walls. They will become books narrating the story of the Earth's long history **(Fig. I.16)**. When you lie on a beach, you won't just enjoy the warmth of the sand, but you may wonder about where the sand comes from, and marvel at the ability of waves to move it. When you gaze at clouds drifting by on a windy day, you won't just

FIGURE I.16 Geologic exploration provides not only beautiful views, but a look into the Earth's past. The features of this cliff tell us that the location of the cliff in the past was a low area in which enough sand collected for the material be transformed into sandstone beds (layers), and that later, the beds underwent bending due to plate motion, which caused them to trace out curves called folds.

A cliff in California

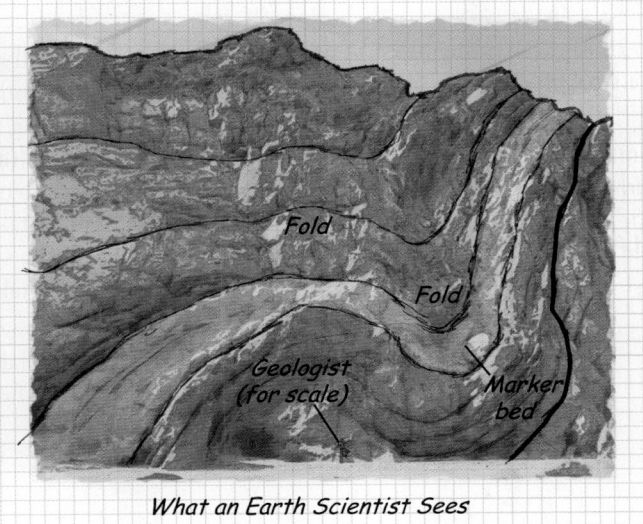

Fold

Fold

Geologist (for scale)

Marker bed

What an Earth Scientist Sees

see puffs of white, but will think about their importance as a water source. And when you peer at the night sky, you'll know that the stars aren't just points of light, but that each is a sun lighting its own planetary system.

Now, it's onward to Earth Science. We'll begin the book at the beginning, with the formation of the Universe and the elements in it. With this background, we can then describe our planet's internal architecture, and the grand unifying theory of geology, plate tectonics. The remainder of Part I covers topics related to the geosphere, the materials beneath it, the features formed on its surface, and the portion of the hydrosphere that flows over it and lies under it. Subsequent parts address the oceans, then the atmosphere, and finally, space. Read on, and you will develop your own personal sense of our island in space, and its context.

Take-home message . . .

The study of Earth Science can help you understand many practical issues, such as whether your home is subject to natural hazards. It can also help you to appreciate your surroundings and our place in the Universe, and to understand sustainability and environmental issues.

Quick Questions

- Provide an example of a natural hazard that could influence the safety of a specific home site.
- Why does an Earth-Science book mention materials such as concrete and plastic wrap?
- Define *sustainability*.

INTRODUCTION REVIEW

Objective I.1

List the subject areas covered in an Earth-Science course, and the types of scientists who contribute to these areas.

KEY CONCEPTS

- Earth-Science courses provide an overview of geology, oceanography, atmospheric science, and typically, astronomy. Therefore, such courses provide a foundation for understanding our planet and its context in the Universe.
- Earth Science incorporates ideas from other sciences (physics, chemistry, and biology).
- Scientists who contribute to an understanding of Earth Science work in a variety of settings. Some work in the field, some in a laboratory, and some at a computer.
- Scientists interpret the natural world using observations, measurements, and calculations, following the scientific method.

EARTH-SCIENCE VOCABULARY

astronomy (p. 6)
atmospheric science (p. 5)
density (p. 6)
energy (p. 6)
gas (p. 6)
geology (p. 5)

hypothesis (p. 9)
liquid (p. 6)
matter (p. 6)
model (p. 7)
oceanography (p. 5)
science (p. 5)

scientific method (p. 8)
scientist (p. 6)
solid (p. 6)

theory (p. 9)
Universe (p. 6)
vacuum (p. 6)

REVIEW QUESTIONS

1. **(a)** What general types of issues does science address? **(b)** Describe the focus of each discipline that an Earth-Science course covers. **(c)** Which type of scientist does **Figure A** show?

2. Distinguish among matter, mass, and density. Define energy, and describe different types of energy.

3. Not all scientists wear a white coat and spend their days looking through a microscope. **(a)** What other types of activities do scientists undertake? **(b)** Which activity are the scientists in **Figure B** undertaking?

4. **(a)** In a scientific context, can there be more than one hypothesis to explain a phenomenon? Explain your answer. **(b)** How does a scientific theory differ from a scientific law?

Objective I.2

> Describe key themes that serve as foundations for understanding the Earth and the Universe.

KEY CONCEPTS

- Matter and energy in the Universe behave predictably, to a large extent. There is, however, a degree of uncertainty in all measurements and predictions.
- The several realms (the geosphere, hydrosphere, cryosphere, atmosphere, and biosphere) of the Earth System interact in a variety of ways.
- The interior of the Earth consists of the crust, mantle, and outer and inner core.
- Scientific research indicates that the birth of the Universe happened 13.8 billion years ago, while the Earth formed 4.56–4.54 billion years ago, so cosmologic time is much longer than geologic time.
- Geologic time has been divided into named increments, such as eons and eras.
- The crust of the Earth is divided into plates that move relative to one another. This theory of plate movement, called plate tectonics, explains earthquakes, volcanoes, and mountain building.
- On the Earth, internal energy, together with gravity, drives plate tectonics; external energy, together with gravity, drives the flow of oceans and the air; and gravity alone drives river flow and downslope movements.
- The Earth is a dynamic planet where change is the norm. Over geologic time, the planet has had time to change in major ways. Human society now causes significant anthropogenic change in the Earth System.

EARTH-SCIENCE VOCABULARY

accuracy (p. 10)	**geologic time scale** (p. 14)
anthropogenic change (p. 16)	**geosphere** (p. 10)
	gravity (p. 15)
atmosphere (p. 11)	**hydrosphere** (p. 10)
biosphere (p. 11)	**internal energy** (p. 15)
continents (p. 14)	**natural change** (p. 16)
cosmologic time (p. 14)	**ocean basins** (p. 14)
cryosphere (p. 11)	**precision** (p. 10)
Earth System (p. 11)	**realms** (p. 10)
external energy (p. 15)	**theory of plate**
geologic time (p. 14)	**tectonics** (p. 14)

REVIEW QUESTIONS

5. How does accuracy differ from precision of a measurement?

6. **(a)** Distinguish between the cryosphere and the hydrosphere. **(b)** What features define the top and bottom of the geosphere? **(c)** How thick is the geosphere?

7. Provide a brief definition of the theory of plate tectonics.

8. **(a)** Why will a map of the Earth's surface showing the positions of continents change over time? **(b)** What do the red and black lines on **Figure C** represent, and what do the arrows represent?

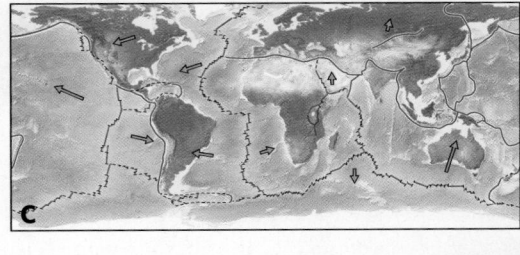

9. **(a)** How old was the Universe at the time that the Earth formed? **(b)** What is the difference between 600 Ma and 0.8 Ga? (Express your answer in millions of years.) **(c)** On **Figure D**, a column representing the geologic time scale, label the present day and the birth of the Earth. **(d)** Do ages get younger moving upward on the time scale, or downward?

10. Distinguish between internal energy and external energy in the Earth System.

11. Provide two examples of change in the Earth System that happen at different rates.

Objective I.3

> Identify issues and challenges that the study of Earth Science can help you to understand and address.

KEY CONCEPTS

- Knowledge of Earth Science can help people to understand and prepare for natural phenomena, ranging from minor (a thunderstorm) to major (a disastrous flood), and to interpret issues in the news.
- The study of Earth Science gives us a sense of the Earth's place in the Universe.
- By understanding Earth Science, you will understand the challenges inherent in finding and extracting resources that society depends on.
- Knowledge of Earth Science can make your natural surroundings more interesting to you, because you will be able to interpret what you see.

EARTH-SCIENCE VOCABULARY

natural hazard (p. 17)	**sustainability** (p. 19)

REVIEW QUESTIONS

12. Describe some practical problems facing your community that involve aspects of the Earth System.

13. How can knowledge of Earth Science decrease vulnerability to natural hazards?

14. What types of resources that society depends upon are discussed in an Earth-Science course?

1 FROM THE BIG BANG TO THE BLUE MARBLE

After studying this chapter, you should be able to . . .

1. identify celestial objects visible to the naked eye, list the principal kinds of objects within our Solar System, and explain what the study of cosmology encompasses.

2. summarize the expanding Universe theory and the Big Bang theory, describe how the first stars came into existence, and explain how and where different elements formed.

3. describe how solar systems like ours formed, and why terrestrial planets develop internal layering and a nearly spherical shape.

4. characterize the Earth's magnetic field, distinguish the realms of the Earth System, and list the key materials found within each realm.

5. name and characterize the principal internal layers of the Earth.

Sometime in the distant past, perhaps more than 50,000 generations ago, our ancestors developed a capacity for complex thought, and began to ponder broad questions about their natural surroundings. What process separated land from sea? When did mountains rise? Why does wind blow and rain fall? How does the Sun produce heat and light? Where do the bright objects speckling the night sky come from, and how do they move? In early cultures, such musings spawned legends in which supernatural heroes sculpted the landscape or pushed stars across the heavens. In recent centuries, *science*—the systematic study of natural phenomena based on observation, experiment, and calculation—has yielded fascinating new answers to these questions.

With such questions in mind, this chapter sets the stage for thinking about the Earth and its context. First, we briefly introduce the objects of the night sky and describe scientific ideas about how they came into existence. Next, we provide an initial picture of the Earth's immediate surroundings and surface by taking an imaginary journey aboard a spacecraft that zooms toward our planet and then goes into orbit around it. Finally, we peek deep inside the Earth and characterize its internal layering. The concepts introduced here provide a foundation for the remaining chapters of this book.

FIGURE 1.1 The Milky Way, as seen in the dark night sky of Washington State.

1.1 A Basic Image of the Universe

During the day, the Sun shines brightly. And on a cloudless night, the Moon, along with many twinkling points of light, decorates the sky. If you head into the dark countryside on a moonless night, you'll see not only countless points of light, but also a hazy, glowing band arcing from horizon to horizon (Fig. 1.1). Our distant ancestors were mystified by this vista. What are these **celestial objects**, these objects beyond the clouds, and where do they lie in relationship to our home? This question is one of many addressed by modern **cosmology**, the scientific analysis of the overall structure, formation, and evolution of the **Universe**, which consists of the Earth, Moon, Sun, and all celestial objects, as well as all the space that contains them and the energy that governs interactions among them (see Box I.1 in the Introduction).

<< About 4 billion years ago, the Earth was pummeled by meteorite impacts. The thermal energy produced by these impacts was probably enough to melt much of the planet's surface and evaporate any liquid water that might have existed. The Moon, seen at the horizon, endured intense bombardment as well.

Introducing Stars, Galaxies, and Nebulae

Stare at the night sky over the course of a few hours, and you'll notice that nearly all of the twinkling points of light that you can see move at the same speed and in the same direction. In fact, the position of each point remains fixed, relative to the others, year after year. Early observers referred to these points as *stars*. The Sun itself is a **star**, an incredibly immense and amazingly hot sphere of gas and plasma that emits vast amounts of light. Here on the Earth, it looks like a disk instead of a point of light because it lies much closer to us than do other stars (Fig. 1.2a). Observations eventually revealed that stars accumulate in vast, spinning clusters called **galaxies**, held together by the force of gravity (Box 1.1). Some galaxies, such as Andromeda (Fig. 1.2b), look like a flattened spiral. By some estimates, over a trillion galaxies populate the Universe. Our Sun, along with over 300 billion other stars, belongs to the **Milky Way Galaxy**, which, if seen from a great distance, resembles Andromeda. To observers on the Earth, the Milky Way looks like a hazy, glowing band because its stars are so far away that our eyes cannot resolve them to see each as a separate point. Most of the individual bright points of light that we do see with our naked eyes are individual stars that lie within relatively nearby parts of the Milky Way.

BOX 1.1   Science Toolbox

What is a force?

Simplistically, we can think of a **force** as a push or pull acting on an object. Physicists define *force* more formally with an equation, $F = ma$. In words, this equation states that force equals mass times acceleration. *Acceleration* means a change in the velocity of an object over a given time interval. Note that the *velocity* of an object depends on both the speed (defined as: speed = distance ÷ time) at which the object moves and the direction in which the object travels. So an acceleration happens when an object's speed, direction, or both undergo a change. Application of force can cause an object to accelerate.

Scientists distinguish between contact forces and field forces. A *contact force* acts when one object pushes or pulls another after the objects touch. To picture a contact force, imagine kicking a ball or pulling a wagon. A *field force* (also known as an *action-at-a-distance force*) emanates invisibly from an object and can cause another object to move even if the objects aren't touching. Examples of field forces include **electromagnetism**, produced by magnets or by electric currents, and **gravity**, produced by the mass of an object. Electromagnetism can cause either *attraction* (causing objects to move toward each other) or *repulsion* (causing objects to move away from each other), whereas gravity can cause only attraction. In practice, the "field" of a field force refers to the region in which the force can cause a measurable consequence. The *magnetic field* surrounding a handheld magnet can move a nearby metal paper clip, and the *gravitational field* of the Earth keeps you from floating out of your chair.

The material substance, or **matter**, making up objects in the Universe consists of atoms and molecules **(Box 1.2)**. Matter, as we noted in this book's Introduction, can exist in four different states: solid, liquid, gas, and plasma. Most, but not all, matter exists as gas or plasma within stars. The remainder occurs in smaller solid objects such as planets and moons (to be introduced shortly), or in **nebulae**, huge, wispy clouds floating in space **(Fig. 1.3)**. A nebula can contain three kinds of matter: (1) gases, mostly made up of hydrogen molecules (H_2) and helium atoms (He); (2) **ices**, solids composed of frozen *volatile materials*, such as water (H_2O), ammonia (NH_3), carbon dioxide (CO_2), and methane (CH_4); and (3) **dust**, tiny particles of *refractory materials*, such as silica (SiO_2), iron oxide (Fe_2O_3), and magnesium oxide (MgO). (Note that we provide the chemical formulas of materials in parentheses; see Box 1.2.) **Volatile materials** can turn into liquid or gas under the relatively low temperatures found at

FIGURE 1.2 Stars and galaxies.

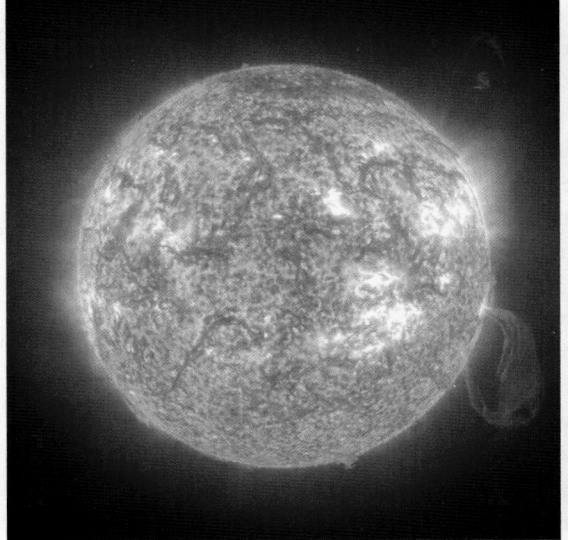

(a) The Sun, like all stars, radiates immense amounts of energy.

(b) The Andromeda Galaxy has a spiral shape. The Milky Way has a similar shape.

FIGURE 1.3 The Orion Nebula. The Hubble Space Telescope captured this image of the Orion Nebula, a cloud of gas and dust 24 light-years across, in which new stars are forming.

FIGURE 1.4 The relative order of the planets in the Solar System. Note that their orbits all lie approximately in the same plane, which has been highlighted. (Neither the distances between the planets nor their sizes relative to the Sun are drawn to scale. Their sizes relative to one another are roughly to scale.)

the Earth's surface, whereas **refractory materials** undergo such changes only at high temperatures. Note that in the context of cosmology, the term *ice* has a broader meaning than it does in everyday English, as it refers not only to water but also to other frozen volatile materials; in contrast, the term *dust* has a more restricted meaning.

Ancient observers noticed that a few celestial objects move relative to the backdrop of stars and relative to one another. Because of that movement, these objects came to be known as *planets*, from the Greek word *planan*, meaning wanderer. Modern researchers define a **planet** as a large

spherical body that not only follows an **orbit** (a closed circular to elliptical path) around a star, but also has incorporated all the matter that lies in its orbit. Each planet occupies its own orbit **(Fig. 1.4)**. The orbit of a planet defines an imaginary surface called an *orbital plane*; notably, the orbital planes of all planets in our Solar System are nearly the same. By this definition, eight planets, including the Earth, orbit our Sun. Mercury, Venus, the Earth, and Mars are known as the *terrestrial planets* because they resemble the Earth by being relatively small (the Earth is the largest), and by having rocky surfaces. These planets are also

Science Toolbox

The components of matter

Developing a scientific understanding of phenomena that happened during the formation of the Universe (and the stars and planets within it) requires familiarity with the tiny particles that make up matter and how these particles attach to one another. Here's a quick review.

Elements, Atoms, and Nuclear Bonds

An **element** is a piece of matter that cannot be subdivided into smaller components with different properties. Chemists have assigned a one- or two-letter *chemical symbol* to each element—for example, H = hydrogen; He = helium; C = carbon; O = oxygen; and Fe = iron. Copper (Cu) fits the definition of an element because it can't be subdivided into other materials. Common table salt, in contrast, doesn't fit the definition because chemists can separate salt into other elements (sodium and chlorine), neither of which looks or behaves like salt. If you keep subdividing a piece of an element into smaller and smaller pieces, you end up with an *atom*, the smallest piece of an element that still has the properties of the element. Modern studies have shown that atoms are so small that a single gram (0.035 oz) of hydrogen contains about 602,000,000,000,000,000,000,000 atoms. In fact, 5–10 trillion atoms fit within the area of the period at the end of this sentence.

In the early 20th century, scientists realized that an atom can be further subdivided into still smaller *subatomic particles*—namely, *protons*, which have a positive charge; *neutrons*, which have a neutral charge; and *electrons*, which have a negative charge (Fig. Bx1.2a). (Simplistically, *charge* refers to the electrical behavior of a particle. Think of a positive charge as the "+" end of a battery and a negative charge as the "−" end of a battery. Particles with opposite charges *attract*, meaning they pull toward each other, whereas those with like charges *repel*, meaning they push away from each other.) A neutron has slightly more mass than a proton, and each has about 1,800 times the mass of an electron.

Protons and neutrons bind together to form a *nucleus* at the center of an atom (see Fig. Bx1.2a). Electrons swirl around the nucleus in an *electron cloud*, so the outer edge of the electron cloud defines the surface of an atom. Considering that a nucleus occupies only a very tiny volume within the electron cloud, atoms actually consist mostly of empty space!

Scientists refer to the attractive force that keeps protons and neutrons together in a nucleus, despite electrical repulsion between protons, as a **nuclear bond**. Nuclear bonds act only over subatomic distances. A **nuclear reaction** happens when nuclear bonds form or

FIGURE Bx1.2 Atoms and nuclear reactions.

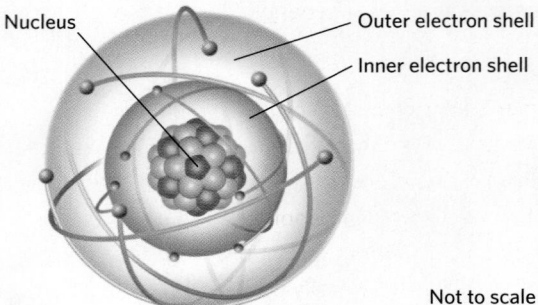

Not to scale

(a) A schematic image of an atom with a nucleus orbited by electrons.

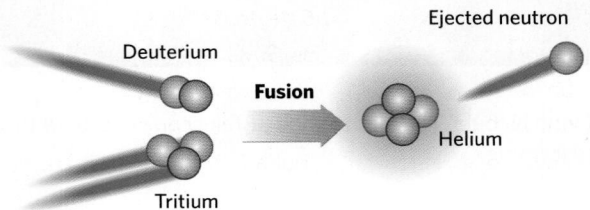

(b) Fusion happens when colliding particles combine to form a new, larger nucleus. (Deuterium and tritium are isotopes of hydrogen.)

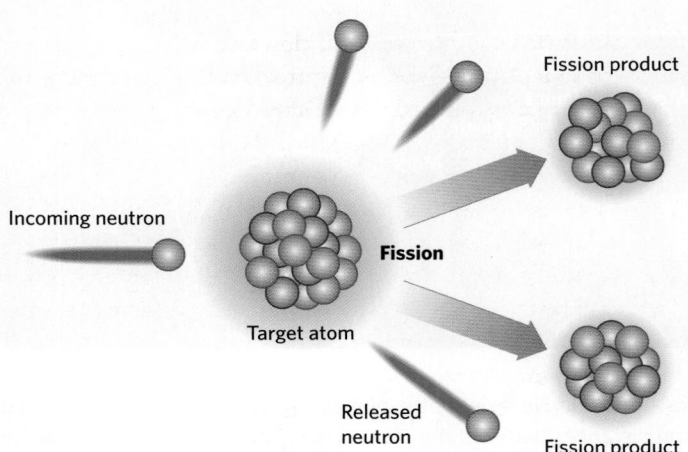

(c) Fission can take place when an incoming particle strikes an atom and splits it, producing smaller fragments that become the nuclei of lighter elements.

break. Specifically, during *nuclear fusion*, two smaller nuclei that have been stripped of their electron clouds, and then collide at very high velocity, stick together to form a single larger nucleus (Fig. Bx1.2b). Only an extremely high-velocity collision brings particles close enough to one another for nuclear bonds to operate. During *nuclear fission*, a large nucleus splits into two smaller ones (Fig. Bx1.2c). We can think of nuclear fission as the opposite of fusion. Both fusion and fission reactions release immense amounts of energy.

Atomic Number and Atomic Mass

An atom of one element differs from an atom of another element because it contains a different number of protons. Scientists refer to the number of protons in an element's nucleus as its **atomic number**. For example, hydrogen has an atomic number of 1, helium has an atomic number of 2, carbon has an atomic number of 6, and iron has an atomic number of 26. The largest naturally occurring atom, uranium, has an atomic number of 92. The **atomic mass** of an atom, in contrast, approximately represents the sum of the number of protons plus the number of neutrons. For example, carbon has an atomic mass of 12, because its nuclei contain 6 protons and 6 neutrons. All atoms, except for most hydrogen atoms, contain neutrons. Small atoms generally have the same number of neutrons as protons, whereas large atoms contain more neutrons than protons. **Isotopes** of an element are versions that have the same atomic number, but a different atomic mass.

For example, all hydrogen atoms contain only one proton, and most hydrogen atoms contain no neutrons. But some isotopes of hydrogen do contain neutrons. Specifically, deuterium contains one proton and one neutron, and tritium contains one proton and two neutrons.

Molecules, Compounds, and Chemical Bonds

Chemical bonds attach atoms to one another to form **molecules**. Such bonds, unlike nuclear bonds, involve interactions between the electron clouds of atoms, such as the sharing or exchanging of electrons—they do not affect atomic nuclei. Some molecules contain atoms of only one element, whereas others, known as *compounds*, contain atoms of two or more different elements. Chemists specify the composition of a molecule by providing its **chemical formula**, a recipe that uses abbreviations to represent the elements in a compound as well as their relative proportions. For example, a hydrogen molecule (H_2) consists of two atoms of hydrogen, while a water molecule (H_2O) consists of two hydrogen atoms bonded to a single oxygen atom. Note that a molecule is the smallest piece of a compound that has the characteristics of that compound.

---

called the *inner planets* because their orbits lie closest to the Sun. The remaining four planets—Jupiter, Saturn, Uranus, and Neptune—are the *Jovian planets*, named after Jupiter, the largest of them. These planets are also called the *outer planets* because their orbits lie farther away from the Sun, or the *giant planets* because they are much larger than the terrestrial planets. Unlike the terrestrial planets, the giant planets consist mostly of gases and ices. Each planet rotates on an *axis* (an imaginary line passing through the planet's center and its two poles) like a spinning top.

The stars appear to move across the sky over the course of a night because of the Earth's rotation on its axis. The planets of our Solar System, though they are far from the Earth, lie much closer to us than do stars. They move relative to one another and relative to the backdrop of stars because they orbit the Sun at different distances and at different speeds. We can't see planets around other stars with our naked eyes because they are too small and dim. But observations made using powerful telescopes confirm their existence.

Each of our Solar System's planets, except Mercury and Venus, has one or more moons. Astronomers define a **moon** as a sizable solid object that orbits a planet. The Earth has a large one—we call it *the Moon*—whereas Jupiter has 95 known moons that range in size from larger than Mercury to less than 10 km (6 mi) across. The Sun,

planets, and moons—as well as various other objects—orbiting the Sun constitute the **Solar System**. Prior to the emergence of modern scientific thinking in the 16th century, people in Western cultures assumed that the stars were anchored to the surface of a distant, invisible sphere, and that this sphere, along with all other celestial objects—including the Sun—slowly circled around the Earth. Scientific observations over succeeding centuries proved this *geocentric model* to be wrong.

What lies between celestial objects? Most of this empty space is a **vacuum**, meaning a volume that contains virtually no matter. The vacuum of *interstellar space*, the region between stars, contains only about 1–3 atoms or molecules per cubic meter (per 35 cubic feet). By comparison, air that you breathe at sea level contains 300,000,000,000,000,000,000,000 atoms or molecules per cubic meter. A nebula typically contains thousands of atoms per cubic meter, but even so, a nebula's density remains vastly less than that of the best vacuum ever produced in a laboratory on the Earth.

Measuring Distances in Space

When we see the Sun, or any star, we are seeing light that has traveled through space. (Visible light is only one of many types of *electromagnetic radiation*; other types include radio waves, X-rays, and ultraviolet light.)

To avoid having to use colossal numbers, astronomers specify distances between stars, or between galaxies, in terms of light-years. One **light-year** represents the distance that light travels in one year. Light travels at a constant speed of about 300,000 km/s (186,000 mi/s) in the vacuum of space, so one light-year equals a huge distance, roughly 9.5 trillion km (5.9 trillion mi). The nearest star to our Sun lies 4.2 light-years away, and the Andromeda Galaxy lies over 2.5 million light-years away. This means that when we look at Andromeda, we are seeing light that began its journey toward the Earth 2.5 million years ago—we are looking back in time!

Take-home message...

The Universe includes stars (such as the Sun), as well as planets, moons, and nebulae. The Earth is one of eight planets that orbit the Sun. Our Solar System lies in the Milky Way Galaxy, one of over a trillion galaxies in the Universe.

Quick Questions ──────────────

- What is the difference between a volatile and a refractory material?
- Which planet has to travel farther to complete an orbit around the Sun: the Earth or Jupiter?
- Why do astronomers use light-years to measure the distances among stars?
- Why, when you look at the night sky, are you seeing into the past?

1.2 Formation of the Universe and Its Elements

The Expanding Universe

In the 1920s, astronomers such as Edwin Hubble (1889–1953) braved many a frosty night gazing through telescopes at mountaintop observatories to chart distant galaxies that were too faint to be seen with the naked eye.

At first, the astronomers just wanted to record the locations and shapes of the galaxies. But eventually they learned how to determine whether the galaxies are moving toward or away from the Earth, and to calculate how fast they are moving (Chapter 21 describes these methods). While pondering such measurements, Hubble noticed that all distant galaxies—regardless of their direction from the Earth—are moving away from us, and that the farther away they are, the faster they appear to be moving. Hubble quickly realized that this result meant that the Universe must be expanding. Up until Hubble's eureka moment, astronomers had assumed that the size of the Universe remained unchanged. Hubble's hypothesis eventually passed so many tests that it gained the status of a theory (see Box I.2), now known as the **expanding Universe theory**.

To picture an expanding Universe, imagine raisins in a ball of fresh dough being baked to make raisin bread (Fig. 1.5). As the dough bakes, the volume of the ball expands, so all the raisins move away from one another, regardless of direction. Significantly, raisins that were farther apart prior to baking move away from each other at a higher velocity than do raisins that were closer to each other prior to baking. To see why, keep the definition of speed (speed = distance ÷ time) in mind. If the diameter of the loaf doubles during an hour of baking, then raisins initially positioned 1 cm apart move away from each other at a speed of 1 cm/h, while raisins initially positioned 3.5 cm apart move away from each other at 3.5 cm/h.

The Big Bang

Hubble's expanding Universe theory immediately triggered a question: When did the expansion begin? Further research eventually allowed astronomers to conclude that the Universe's expansion began with a specific event, now known as the **Big Bang**, about 13.8 billion years ago (Fig. 1.6). According to this idea, which has been supported by so much evidence that it's now referred to as the *Big Bang theory*, everything that constitutes our Universe initially existed within an extremely small, unfathomably dense and hot volume. At the moment of the Big Bang, which marks the origin of the Universe, this volume began to expand at high velocity. Of course, no human saw the Big Bang directly. But by combining clever calculations with careful observations, researchers have developed a model of how matter came into existence shortly after the expansion started.

A moment after the Big Bang, the Universe remained so dense and hot that matter as we know it couldn't exist (Box 1.3). Scientists hypothesize that after a few seconds of expansion, the Universe had cooled sufficiently so that the smallest atoms, those of hydrogen, began to form. Only a few

FIGURE 1.5 A raisin-bread analogy for the expanding Universe. As the dough expands, each raisin moves away from all the others. The speed of travel between pairs of raisins depends on the initial distance between the raisins.

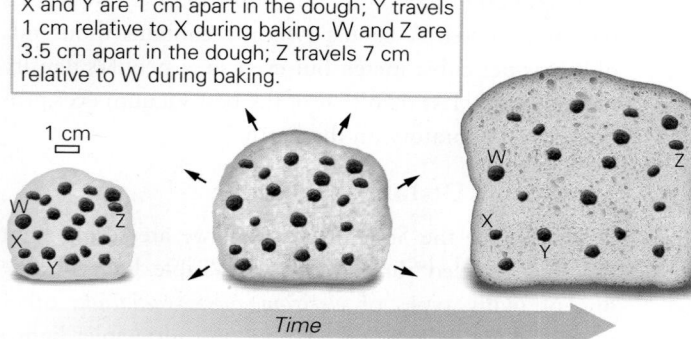

X and Y are 1 cm apart in the dough; Y travels 1 cm relative to X during baking. W and Z are 3.5 cm apart in the dough; Z travels 7 cm relative to W during baking.

1 cm

Time

Heat and temperature

The atoms and molecules that make up a solid object do not stay rigidly fixed in place. Rather, they vibrate without breaking the chemical bonds that hold them to their neighbors. Atoms or molecules in a liquid or gas not only vibrate, but can also move about with respect to one another. Their vibrations and other movements represent **thermal energy**, defined as the total kinetic energy (energy of motion; see Box I.1) of all atoms and molecules in a given volume of matter. Addition of thermal energy to a volume causes the atoms or molecules to move faster, while removal of thermal energy causes the atoms or molecules to slow down. Scientists use the term **heat** for the thermal energy transferred from one material to another.

When we say that one material is hotter or colder than another, we are describing its **temperature**, a measure of warmth relative to a known reference value. Temperature represents the average kinetic energy of atoms in the material. Heat always flows from a location of higher temperature to a location of lower temperature. To specify temperature—that is, to indicate relative hotness or coldness—most of the world uses the *Celsius scale* (also called the *centigrade scale*), which is divided into *degrees* (°C). Scientists have calibrated the Celsius scale so that at sea level, the boiling point of water is 100°C and its

freezing point is 0°C. An increment of 1°C, therefore, equals 1/100 of the temperature difference between the boiling and freezing points of water. People in the United States generally use the *Fahrenheit scale* for temperature, except in scientific literature. On this scale, water boils at 212°F and freezes at 32°F at sea level. Note that a degree on the Fahrenheit scale represents a smaller change than does a degree on the Celsius scale. Adding heat to a material increases its temperature, while letting heat flow out of a material decreases its temperature. Similarly, compressing a material increases its temperature, while allowing a material to expand decreases its temperature.

Scientists sometimes use the *Kelvin scale*, named for Lord Kelvin (1824–1907, a British physicist), to specify temperature. On the Kelvin scale, the coldest possible temperature, called *absolute zero*, is designated as 0 K. (Note that numbers on the Kelvin scale do not use the degree sign.) A material can't get colder than absolute zero, meaning that you cannot extract any thermal energy from a substance at 0 K because all the motion of its atoms and molecules has ceased. An increment of temperature on the Kelvin scale has the same magnitude as one on the Celsius scale. 0 K equals −273.15°C, so on the Kelvin scale, the freezing point of water is 273.15 K, and the boiling point of water is 373.15 K.

---

minutes later, when the Universe's temperature had fallen below 1 billion °C (1.8 billion °F), most of the hydrogen atoms that exist today had formed. For the next several minutes, some of the hydrogen atoms collided and underwent nuclear fusion to form helium atoms as well as small quantities of other elements with atomic numbers of less than 5. By the time the Universe was a mere 20 minutes old, this process of atom formation, known as **Big Bang nucleosynthesis**, had concluded. It yielded mostly hydrogen and helium at a ratio of about 3 to 1.

Birth of the First Stars from Nebulae

By the time the Universe reached its 200 millionth birthday, large volumes of hydrogen and helium gas had collected into immense, slowly swirling nebulae. These first nebulae were relatively dense because the young Universe was still fairly small. Significantly, early nebulae differed from the nebulae of the Universe today in that they consisted only of gas, for elements with larger atomic

FIGURE 1.6 An artist's rendering of Universe expansion from the Big Bang through the present and on into the future.

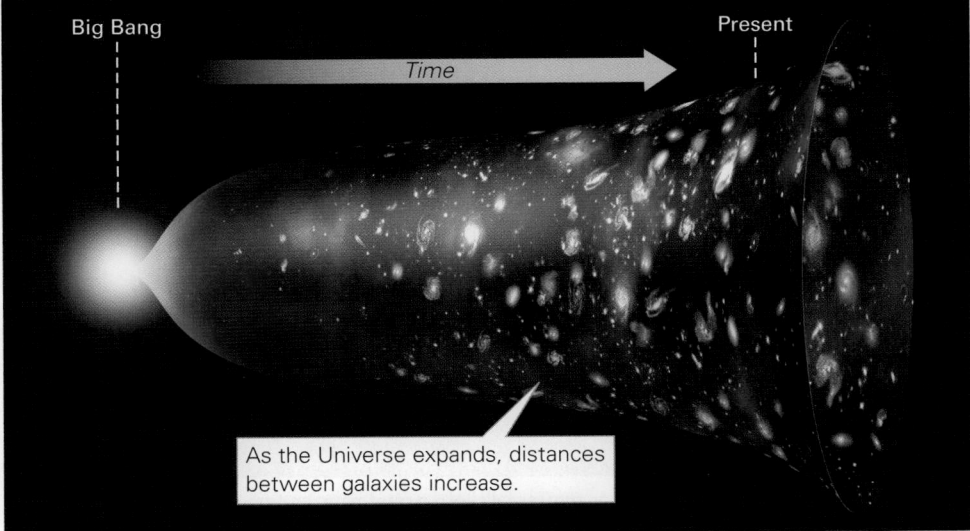

Big Bang

Present

Time

As the Universe expands, distances between galaxies increase.

numbers, which make up particles of ice and dust, had not yet formed. Eventually, at many locations, patches of nebulae became relatively dense. These denser regions then began to collapse inward due to the pull of gravity. During this collapse, atoms and molecules of gas moved closer together, so the collapsing region became denser still. Compression of a material into a smaller volume causes its temperature to rise (see Box 1.3), so as a collapsing patch within a nebula got denser and denser, it also became hotter and hotter. In addition, its initial swirling motion evolved into rotational motion around an axis. Eventually, the collapsing patch of nebula took the shape of a rotating flattened disk with a bulbous center (Fig. 1.7a). Astronomers refer to such a disk as an **accretion disk** because matter accreted (attached) to the disk over time. Matter that was moving more slowly ended up in the bulbous center, while matter that was moving faster ended up in the flattened outer part of the disk, orbiting around the center.

When the central bulge of an accretion disk became sufficiently massive and dense, it became hot enough to glow as a **protostar**. For a given protostar, what happened next depended on the amount of matter the protostar contained. That's because an increase in the temperature of a gas causes the gas to expand or spread out, producing an outward push called *thermal pressure*. (You can see an example of thermal pressure in action by placing an inflated balloon outside on a hot day: expansion of the warming gas in the balloon stretches the balloon.) If a protostar contained a relatively small amount of matter, the outward push of thermal pressure would balance the inward pull of gravity before the protostar could collapse any further. Therefore, though these protostars glowed, they could not become true stars, for in a true star, nuclear fusion reactions produce energy (see Box 1.2). But in the early Universe, some nebulae contained so much mass that protostars formed from them were extremely massive. The inward pull of gravity in an extremely massive protostar could overcome thermal pressure, allowing the star to continue collapsing inward until its interior became so dense, and so extremely hot (about 10 million °C), that hydrogen nuclei were moving fast enough to overcome electrical repulsion, slam together, and form nuclear bonds. These collisions resulted in the fusion of hydrogen nuclei to form helium nuclei and generated prodigious amounts of energy (see Box 1.2). When fusion began in massive protostars, those giant nuclear furnaces became true stars, and light from these first stars lit the Universe (Fig. 1.7b). Such *first-generation stars* ignited at countless locations throughout the many dense nebulae of the young Universe, and by about 800 million years after the Big Bang, vast numbers of first-generation stars were radiating intense light. Gravitational pull among these stars caused them to collect into the first galaxies. The concept that the earliest stars formed within collapsing patches of nebulae formed exclusively of atoms produced by Big Bang nucleosynthesis has come to be known as the **nebular theory** of star formation.

This truth within thy mind rehearse, / That in a boundless universe / Is boundless better, boundless worse.

—ALFRED TENNYSON
(BRITISH POET, 1809–1892)

▶ **Narrative Art Video**
Formation of the Solar System

FIGURE 1.7 The nebular theory.

(a) Artist's rendition of a star forming as matter falls into a rotating accretion disk. As the mass collapses inward, the orbital velocity increases, as occurs when a spinning skater pulls in her arms.

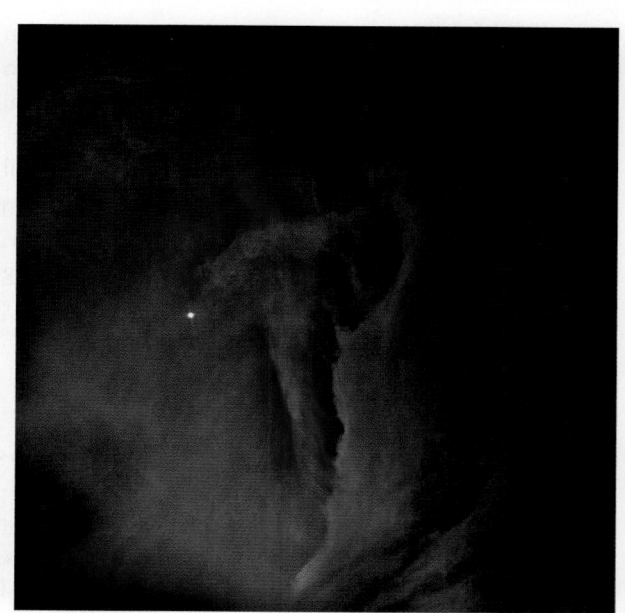

(b) An early nebula might have looked like this when the first star formed. It's mostly swirling gas.

We Are All Made of Stardust!

The Big Bang produced all the matter in the Universe, but it did not produce all the 92 naturally occurring elements that exist today. As we've seen, Big Bang nucleosynthesis yielded only hydrogen, helium, and trace amounts of other elements with atomic numbers of less than 5. Where did the elements with larger atomic numbers come from? Moderately heavy elements form inside stars by **stellar nucleosynthesis**, a succession of nuclear reactions during which smaller nuclei fuse together to form larger ones. In other words, stars are element factories! Initially, hydrogen atoms fuse to produce helium atoms. As hydrogen begins to run out, fusion reactions in small to medium-sized stars yield elements up through carbon (atomic number 6). In very large stars, temperatures become hot enough for fusion reactions to generate elements up to iron (atomic number 26).

Once formed, some of the atoms of a star escape into space during the star's lifetime by moving fast enough to overcome the star's gravitational pull. These atoms make up *stellar wind*, a stream of charged particles that flows from a star. Astronomers refer to stellar wind from our own Sun as the **solar wind** (Fig. 1.8a). Most atoms, however, do not escape until the star has used up most of its nuclear fuel and starts to undergo its death throes. As we'll see in Chapter 22, the way a star dies depends on its size. A small dying star ejects matter after it has evolved into a *red giant*, whereas a large star ejects matter during a violent explosion. Astronomers refer to a giant exploding star as a **supernova** (Fig. 1.8b), from the Latin *nova*, meaning new, because to observers on the Earth, the explosion appears as a new, bright, but short-lived point of light in the night sky. Most elements with atomic numbers larger than 26 form by **supernova nucleosynthesis** during these explosions, but some are products of even more energetic events.

First-generation stars, initially containing only elements produced during Big Bang nucleosynthesis, were huge—some contained hundreds or thousands of times the mass of our Sun. Astronomers calculate that temperatures in such massive stars would have become so extreme that fusion reactions could consume nuclear fuel very quickly. As a result, many first-generation stars may have survived only a few million years before becoming supernovae. Supernova explosions, therefore, peppered the young Universe, and these explosions injected elements that had formed by stellar nucleosynthesis, as well as elements produced by supernova nucleosynthesis, into surrounding nebulae. Consequently, as first-generation stars died, atoms of heavier elements mixed with hydrogen and helium left over from the Big Bang, and the chemical composition of nebulae in the Universe changed. *Second-generation stars* formed from these new, compositionally diverse nebulae. Those second-generation stars lived and died and contributed elements to nebulae from which third-generation stars formed, and so on.

Because progressively more products of stellar nucleosynthesis and supernova nucleosynthesis mix into the nebulae from which stars form, the accretion disks of successive generations of stars contain an increasing proportion of heavier elements. At any given time, the Universe contains many generations of stars. Our own Sun may be a third-, fourth-, or fifth-generation star, so molecules in our bodies contain atoms produced in the hearts of stars or during cataclysmic stellar explosions (as well as hydrogen produced during the Big Bang). Think of it—we are all made of stardust!

Did you ever wonder...

where the atoms that make up your body first formed?

(a) When the Moon blocks the Sun during an eclipse, we can see the gases of the solar wind streaming into space.

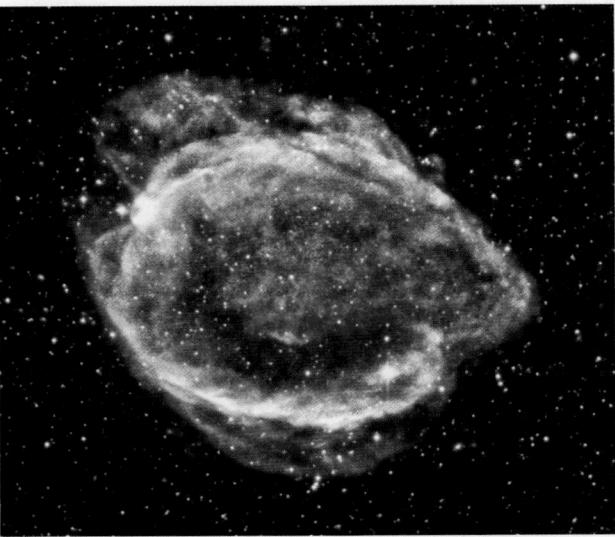

(b) This expanding ring-like nebula represents the remnant of a supernova explosion that happened 4,500 years ago.

FIGURE 1.8 Element factories in space.

1.3 Birth of the Earth

Condensation and the Birth of Planets

We've seen that first-generation stars had to be very massive, or they wouldn't have collapsed enough for fusion to begin. How, then, do smaller stars, like our Sun, form? Development of smaller stars could take place only in nebulae that contained tiny particles of dust and various ices, in addition to gases. Such particles formed from the condensation of matter into solids, and such condensation could take place only in nebulae that contained larger atoms (such as oxygen, silicon, and iron), elements produced by stellar nucleosynthesis and supernova nucleosynthesis. Ice and dust particles disperse heat efficiently, so their presence decreased the thermal pressure in protostars and, therefore, permitted smaller protostars to collapse and become dense and hot enough for fusion reactions to commence. As such smaller stars grew, gas, ice, and dust remaining farther out in the accretion disk provided the material from which planets formed. Researchers refer to the model describing how an ice- and dust-containing nebula can evolve into a star surrounded by planets as the **condensation theory** because of the important role played by condensation of matter into solids during the process.

With the condensation theory in mind, let's examine the scientific model for the birth of our own Solar System from a nebula located in an arm of the Milky Way Galaxy (Fig. 1.9a). The nebula contained not only hydrogen and helium, but also larger atoms produced by earlier generations of stars and their demise. It collapsed into an accretion disk with a central bulge. That bulge continued collapsing until, around 4.57 billion years ago, it became the Sun, and began radiating energy produced by nuclear-fusion reactions. As this happened, the part of the accretion disk that did not get drawn into the Sun by gravity continued swirling around the Sun fast enough to remain in orbit around it. Scientists refer to this remaining disk as a **protoplanetary disk** because it provided the matter that formed planets. As soon as the Sun became a nuclear furnace, the solar wind began to blow. In the inner part of the protoplanetary disk, solar radiation warmed frozen volatile materials sufficiently to transform them into gases, and the solar wind blew the volatile gases into the outer part of the disk, where some refroze into ices. A boundary called the *frost line* developed that separated the part of the protoplanetary disk closer to the Sun (where volatile material became gaseous) from the part farther out (where volatile materials could exist as ices).

How did planetary growth take place? Under the force of gravity, specks of dust and ice acted as "seeds" to which other atoms and particles could attach, gradually growing into soot-sized particles, then dusty balls, and eventually boulder-sized blocks (Fig. 1.9b, c). Eventually, some of these bodies became big enough for their gravity to pull in other blocks and pack them together, producing a sizable, coherent mass (Fig. 1.9d). Such a solid mass that grows to a diameter greater than about 1 km is called a **planetesimal**. A few planetesimals won the competition to attract other objects into their orbits and grew into **protoplanets**, bodies approaching the size of today's planets. Astronomers estimate that the early Solar System may have had up to 80 such protoplanets. But their gravitational pulls tugged on one another, causing some to fall into the Sun, some to escape the Solar System, some to break apart, and some to collide with one another. Eventually, only eight protoplanets succeeded in collecting all the material in their orbits. These bodies became the planets that exist today (Earth Science at a Glance, pp. 34–35).

The condensation theory explains many characteristics of the Solar System. For example, all eight planets lie in roughly the same orbital plane because they all formed from the same protoplanetary disk. Those planets occupying inner orbits, where the protoplanetary disk contained mostly refractory dust, became the terrestrial planets, consisting mostly of rock and metal. In contrast, those planets that grew in the outer part of the Solar System, beyond the frost line, where the protoplanetary disk contained large amounts of volatile materials, grew into the giant planets, consisting mostly of gases and ices.

Not all solid materials of the protoplanetary disk ended up within planets, however. Large numbers of

FIGURE 1.9 Formation of solid bodies in the Solar System.

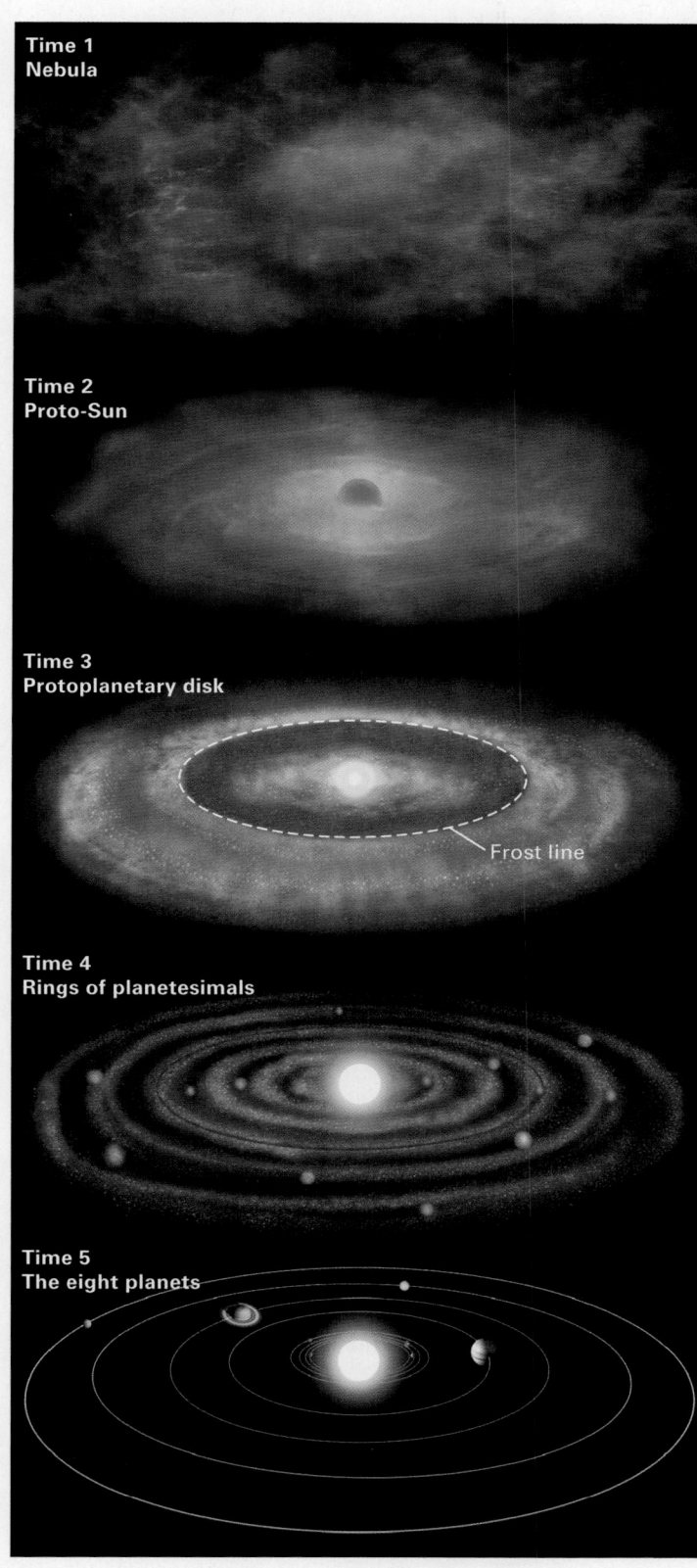

Time 1
Nebula

Time 2
Proto-Sun

Time 3
Protoplanetary disk

Frost line

Time 4
Rings of planetesimals

Time 5
The eight planets

(a) Stages during the evolution of the Solar System.

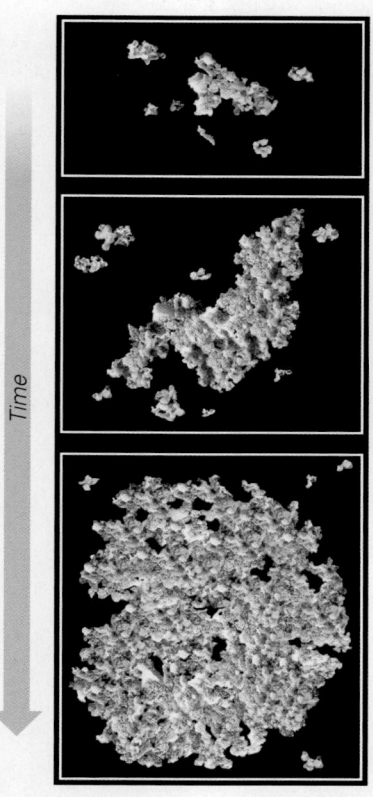

Time

(b) A close-up image, showing dust and ice particles collecting into progressively larger bodies.

(c) An artist's impression of planetesimals orbiting the newborn Sun.

(d) This photo of the asteroid Bennu illustrates the grainy texture of a growing planetesimal. The Empire State Building indicates scale.

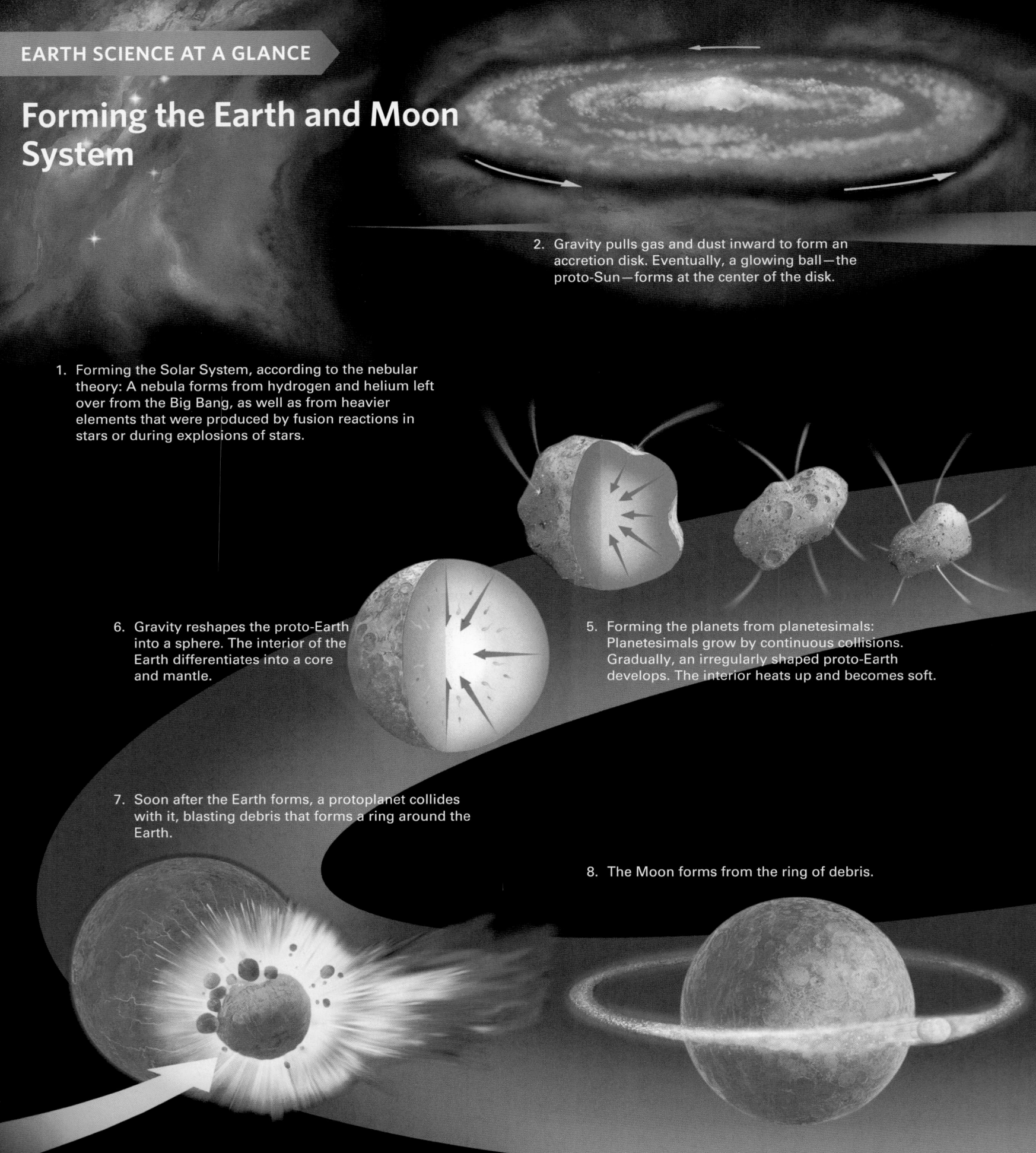

Forming the Earth and Moon System

2. Gravity pulls gas and dust inward to form an accretion disk. Eventually, a glowing ball—the proto-Sun—forms at the center of the disk.

1. Forming the Solar System, according to the nebular theory: A nebula forms from hydrogen and helium left over from the Big Bang, as well as from heavier elements that were produced by fusion reactions in stars or during explosions of stars.

6. Gravity reshapes the proto-Earth into a sphere. The interior of the Earth differentiates into a core and mantle.

5. Forming the planets from planetesimals: Planetesimals grow by continuous collisions. Gradually, an irregularly shaped proto-Earth develops. The interior heats up and becomes soft.

7. Soon after the Earth forms, a protoplanet collides with it, blasting debris that forms a ring around the Earth.

8. The Moon forms from the ring of debris.

3. "Dust" (particles of refractory materials) concentrates in the inner part of the disk, while "ices" (particles of volatile materials) concentrate in the outer part. Eventually, the dense ball of gas at the center of the disk becomes hot enough for fusion reactions to begin. When it ignites, it becomes the Sun.

4. Dust and ice particles collide and stick together, forming planetesimals.

9. Eventually, the atmosphere develops from volcanic gases. When the Earth becomes cool enough, moisture condenses and precipitates to produce the oceans. Some gases may be added by passing comets.

BOX 1.4   How can I explain . . .

Differentiation of the Earth's interior

What are we learning?

Why the Earth, and other planets, separated into distinct internal layers.

What you need:

- An empty 16 fl oz (0.5 L) glass bottle with a screw cap
- Vegetable oil (about 1½ C, or 0.350 L), to represent the mantle
- Red vinegar (about ¼ C, or 0.60 L), to represent the core

Instructions:

- Pour the oil and the vinegar into the bottle. The amounts represent the approximate proportion, by volume, of mantle material to core material in the Earth.
- Place the bottle in a refrigerator until cold. Remove the bottle and shake it hard to homogenize the contents. While the bottle remains cold, the oil and vinegar remain homogeneous (thoroughly mixed).
- Heat the bottle to a warm temperature (about 30°C, or 86°F). If it's a warm, sunny day, just put it outside; otherwise, place it in a pan of hot water.
- Watch, and you will see the oil and vinegar separate as the denser material (vinegar) sinks to the bottom and the less dense material (oil) rises to the top. The contents of the bottle will no longer be homogeneous.

What did we see?

This exercise is a model of the Earth's differentiation. Warming a material increases its ability to flow, so that materials of different densities can separate into layers. The exercise also demonstrates the difference between homogeneous and inhomogeneous materials.

Homogeneous oil and vinegar

Oil has separated from the vinegar

planetesimals and planetesimal fragments lie in the *asteroid belt* between the orbits of Mars and Jupiter. These objects, called *asteroids* (see Fig. 1.9d), which consist of rock and metal, may have originated either as planetesimals, as solid fragments that never became incorporated

in a planetesimal, or as protoplanets or planetesimals that broke apart. A few of these, now known as *dwarf planets*, are quite large, with diameters exceeding about 400 km (250 mi). Beyond the orbit of Neptune, millions of icy fragments orbit the Sun. Several of these are also large enough to be considered dwarf planets; one of the largest, Pluto, was once classified as a planet. *Comets*, glowing objects with long tails, come from these distant reaches of the Solar System. Notably, the moons that orbit planets didn't all form in the same way. Some are captured asteroids or comets, some formed from rings of ice and dust that remained in orbit around a planet after the planet formed, and at least one—the Earth's Moon—formed from debris ejected into space after the collision of our planet with a protoplanet about 4.3 billion years ago.

Internal Differentiation of the Earth

When a terrestrial planetesimal starts to grow, it's fairly homogeneous. As it grows, it becomes progressively warmer inside, for three reasons. First, as the mass of a planetesimal increases, gravity squeezes its interior together more tightly, and compression of matter causes it to warm up. Second, during the early history of the Solar System, planetesimals frequently collided, and at each impact, kinetic energy (see Box I.1) transformed into thermal energy. (To see an example of this phenomenon, hit a nail repeatedly with a hammer, and then feel the nail—it has become hot.) And third, planetesimals contained *radioactive atoms*, meaning atoms that can spontaneously split into smaller atoms or emit subatomic particles. These nuclear reactions not only change atoms of the original element into atoms of a different element, but also release energy, which heats up their surroundings.

Heating in the interiors of large planetesimals produced temperatures high enough for some materials to begin melting. Such melting produced droplets of molten *iron alloy* (a blend of iron and other elements). These droplets were denser than the remaining material around them, so they sank downward due to the inward pull of gravity and accumulated into a very hot sphere—a *core*—at the center of the planetesimal (Box 1.4). The remaining material became a rocky shell, called the *mantle*, surrounding the core. By this process of **differentiation**, large planetesimals, including the one that eventually became the Earth, developed internal compositional layering early in their history (Fig. 1.10). The growth and differentiation of the Earth occurred between 4.56 and 4.54 billion years ago, so this age range represents the time of the Earth's birth. Note that when the Earth formed, the Universe was already over 9 billion years old. Notably, Jovian planets also underwent differentiation at the same time, as we will discuss in Chapter 21.

FIGURE 1.10 Differentiation of the Earth's interior. Early on, the Earth was fairly homogeneous inside (left). When the temperature got hot enough (center), iron began to melt. The iron accumulated at the center of the planet to form a metallic core (right).

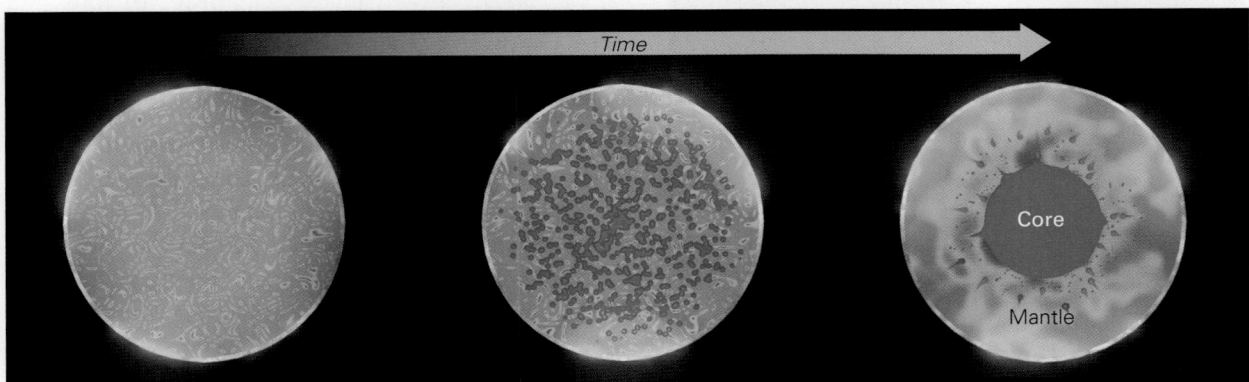

(a) Early on, the Earth was fairly homogeneous inside.

(b) When the temperature got hot enough, iron began to melt.

(c) The iron accumulated at the center of the planet to form a metallic core.

Making the Earth Round

Small planetesimals had irregular shapes, as do asteroids and small moons today. Planets, dwarf planets, and large moons, on the other hand, are nearly spherical. Why? A small planetesimal is cool and rigid, so the pull of gravity within it cannot overcome the strength of its material and cause it to flow. But once a planetesimal reaches a diameter of 400–800 km (250–500 mi), its interior becomes warm and soft enough, and the force of gravity becomes large enough, for material in the object's interior to flow very slowly. When this happens, surface bulges sink and dimples rise until the planetesimal evens out into a nearly spherical shape whose mass lies evenly distributed around its center, so that the force of gravity is nearly the same at all points on its surface (see Earth Science at a Glance, pp. 34–35).

Take-home message . . .

Small stars, like the Sun, grew from nebulae that contained ice and dust particles, as explained by the condensation theory. The planets of the Solar System formed from material in the protoplanetary disk that surrounded the newborn Sun. Gravity brought materials together into planetesimals, then protoplanets, and finally planets. In the outer part of the Solar System, large amounts of gases and ices collected to form the Jovian planets. Once they were large enough, planetesimals differentiated internally and became spherical.

Quick Questions

- How does the condensation theory differ from the nebular theory?
- Why did the Jovian planets form only in the outer part of the protoplanetary disk?
- What material sinks to the center of a terrestrial planet during differentiation, and why?

1.4 The Blue Marble: Introducing the Earth System

Imagine that you're an interplanetary explorer on board a spacecraft flying toward the Earth, and you've reached the final phase of your voyage. As you approach the Earth, you'll first detect the influence of the planet on the space around it. Then, when you go into orbit around it, you'll marvel at the beauty and complexity of the **Earth System**, the web of interacting **realms** of the planet: the atmosphere, the hydrosphere, the cryosphere, the geosphere, and the biosphere (see the Introduction). Once in orbit, you can analyze the atmosphere and aspects of the planet's surface. Finally, with probes sent down to the surface, you'll even be able to define the basic internal structure of the planet. Let's examine your discoveries.

Welcome to the Neighborhood!

As you pass the orbit of Mars, about 81,000,000 km (50,000,000 mi) away from the Earth at its closest, the Earth looks like a large star with a bluish tint. From 30,000,000 km (20,000,000 mi) away, the Earth begins to resemble a bluish glass marble, and from 1,000,000 km (600,000 mi), you can start to see the complexities of the planet's surface and atmosphere (Fig. 1.11). As you traverse this distance, your instruments determine that the Earth produces a *magnetic field* that can influence the paths of charged particles and magnetic objects (see Box 1.1). After you pass the orbit of the Moon, you cross into the **magnetosphere**, the region in which the strength of the magnetic field exceeds the strength of other magnetic fields in space. Like a toy bar magnet, the Earth's magnetic field is a *dipole*, meaning that it

See for yourself

The Whole Earth

Latitude: 49°18′51.35″ N
Longitude: 36°48′46.13″ W

Zoom to an elevation of 20,000 km (12,400 mi) and look straight down. You can see a view of the whole Earth, showing land, sea, and ice. In this view, clouds have been removed. Green areas are vegetated, tan areas are not. Spin the globe on your screen and zoom in to explore specific localities.

FIGURE 1.11 The Earth, as seen from space. During the day, the Earth and clouds reflect sunlight. Oceans, land, and ice are all visible.

Did you ever wonder...

why compasses work?

has a north pole and a south pole (Fig. 1.12a, b). Invisible *magnetic-field lines*, which represent the directions along which magnetic materials such as compass needles align when placed in the field, curve around the Earth and intersect the planet at its north and south magnetic poles. The solar wind, as it approaches the Earth, warps the Earth's magnetosphere, making it fairly narrow on the side of the Earth facing the Sun and extremely wide on the side opposite the Sun, so overall, it has a teardrop shape (Fig. 1.12c). Notably, the magnetosphere acts like a shield and deflects most, though not all, of the solar wind. Because the solar wind's more energetic particles can be dangerous to organisms, the magnetosphere helps make the Earth hospitable to life.

The magnetic field isn't strong enough to affect your spacecraft, so you continue speeding toward the Earth. At a distance of between 1,000 and 60,000 km (620 to 37,000 mi) from its surface, you encounter the *Van Allen radiation belts*, named for the physicist who first recognized them, which trap energetic solar-wind particles. If solar winds are particularly strong, you may admire spectacular curtains of glowing red and green light known as *aurorae* at high latitudes (Fig. 1.12d, e). These curtains form when electrons flow along magnetic-field lines toward the magnetic poles and interact with gases in the upper atmosphere.

Introducing the Atmosphere

As you descend below an altitude of about 1,000 km (620 mi), the Earth fills your field of view. Your instruments detect that the concentration of gas outside the

spacecraft has started to increase slightly. This change means that you've reached the uppermost limit of the **atmosphere**, the blanket of gas that surrounds the Earth and represents the outermost realm of the Earth System (Fig. 1.13). From the outer edge of the atmosphere on down, the density of gas progressively increases, because the weight of the gas higher in the atmosphere pushes down on the gas below and squeezes molecules closer together. The atmosphere remains barely detectable down to an altitude of about 100 km (62 mi). In fact, 99.9% of the molecules in the atmosphere lie below an altitude of only 50 km (30 mi), less than 1% of the Earth's radius—it's a thin blanket, indeed! To avoid having your spacecraft slow down and heat up due to interaction with air molecules, you go into orbit at an altitude of 400 km (250 mi).

Now that you've arrived in orbit, you first analyze the composition of **air**, the distinct mixture of gases making up the Earth's atmosphere, and find that 78% consists of nitrogen molecules (N_2) and 21% consists of oxygen molecules (O_2). The remaining 1% consists of *trace gases*, including argon (Ar), carbon dioxide (CO_2), and methane (CH_4). Air also contains varying amounts of **water vapor** (gaseous H_2O). Next, you measure **atmospheric pressure**, the push that air exerts on its surroundings. *Pressure* in any material (air, water, or rock) can be described in units of *force per unit area*. One such unit, a *bar*, represents the approximate average pressure of the atmosphere at sea level. Because atmospheric pressure decreases with increasing elevation, it has a value of only 0.3 bars at the peak of the Earth's highest mountain, Mt. Everest (Fig. 1.14). At this elevation, 8.85 km (5.5 mi) above sea level, people can't survive long without breathing bottled oxygen.

Finally, you analyze the distribution of water vapor and measure air movement. You learn that in the atmosphere below an altitude of about 12 km (7.5 mi), the concentration of water vapor in the air varies between nearly 0% and slightly over 4%. (Averaged globally, water vapor accounts for about 0.25% of air.) In places, molecules of water stick together to form tiny liquid droplets or tiny ice crystals, which accumulate to form *clouds*. As you watch the atmosphere, you see that the shapes and distribution of clouds change constantly, providing a visual clue that air moves (Fig. 1.15). In fact, you determine that the speed of *wind*, the component of air movement parallel to the Earth's surface, generally ranges between 1 and 50 km/h (0.6–30 mph) near the ground, but that locally, especially in swirling, turbulent cloud masses called *storms*, wind velocity can increase to over 257 km/h (160 mph).

The Hydrosphere and Cryosphere

Having made a quick analysis of the atmosphere, you turn your attention to the Earth's surface. Immediately, you realize that 70% of its surface is covered by *oceans*

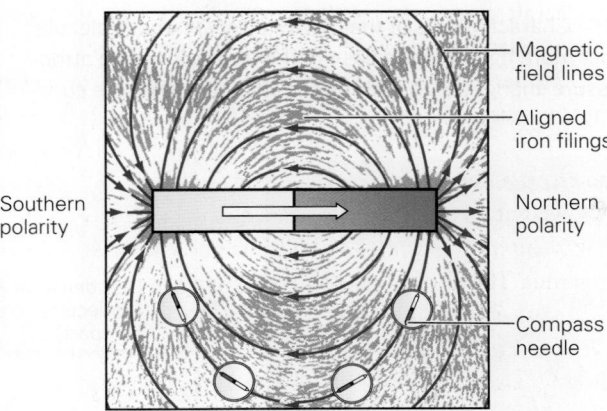

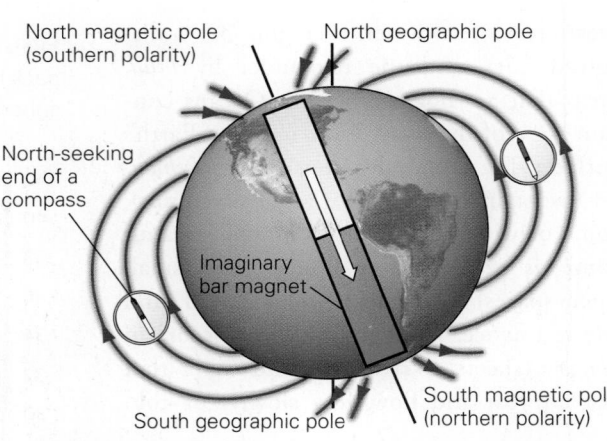

FIGURE 1.12 The Earth's magnetic field.

(a) Magnetic-field lines produced by a bar magnet point out of the magnet's north pole (the end with "northern polarity") and into the magnet's south pole (the end with "southern polarity"). On a curving magnetic-field line, arrows point from the north to the south pole. Compass needles align with the field lines. The south pole of one magnet attracts the north pole of another.

(b) We can represent the Earth's magnetic field as an imaginary bar magnet inside the planet. As the arrows associated with the field lines show, the northern polarity of the magnetic field currently lies near the south geographic pole, and the southern polarity of the field currently lies near the north geographic pole. (Admittedly, this can be confusing. But because of this convention, the red end of a compass needle points to the north geographic pole.)

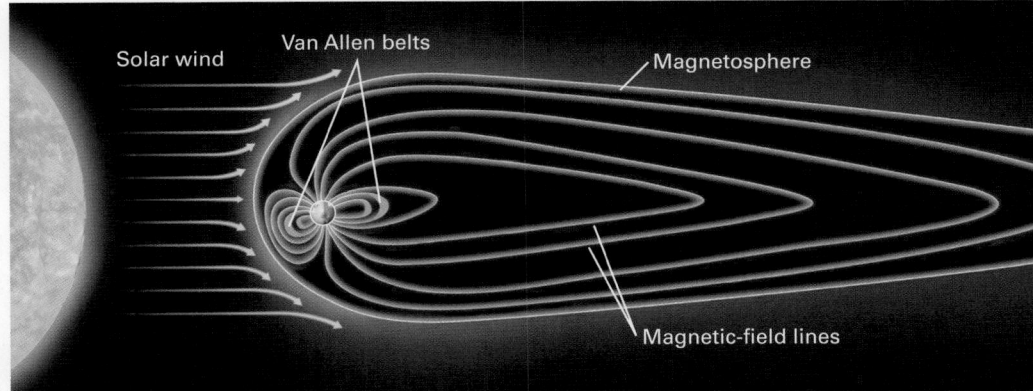

(c) The Earth behaves like a magnetic dipole, but its magnetic-field lines are distorted by the solar wind. The Van Allen radiation belts trap charged particles.

(d) A view of aurorae as seen from space.

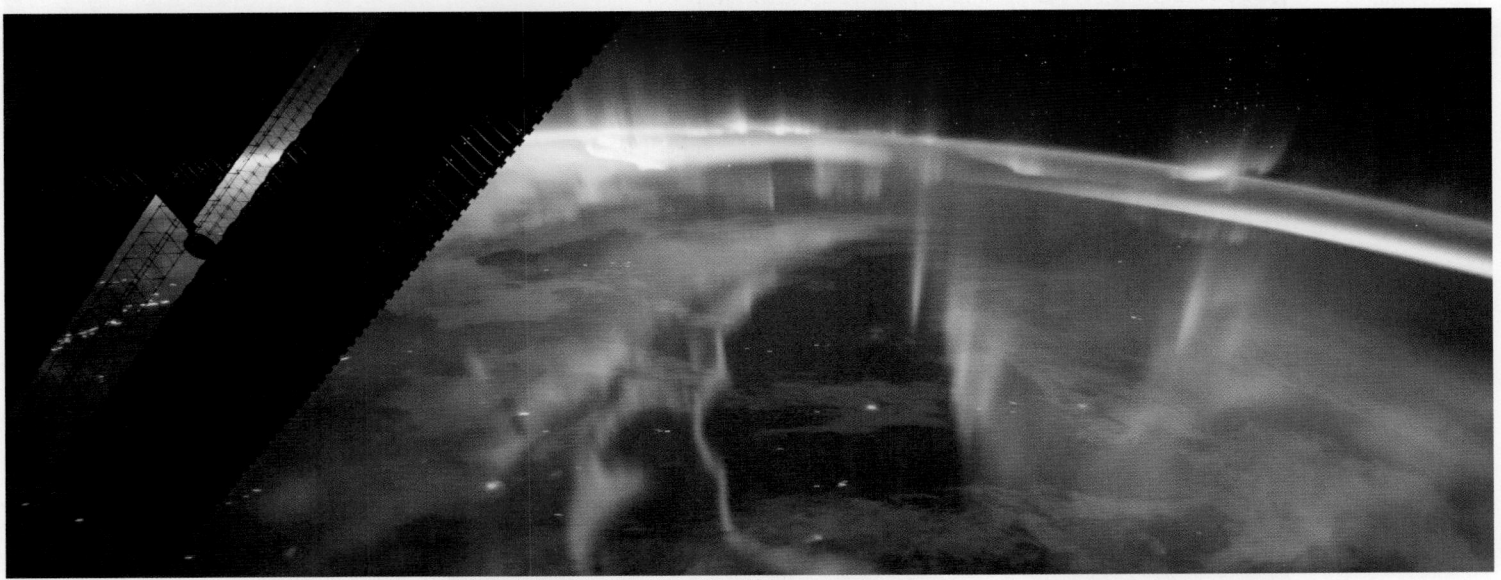

(e) Aurorae, as seen from the International Space Station orbiting the Earth.

(vast areas of saltwater, averaging 3.5% dissolved salt), and the remainder by *land*, areas that are not submerged. Oceans contain 97% of the water at or near the Earth's surface. The remainder consists of *freshwater* (averaging less than 0.1% dissolved salt), some of which occurs in lakes and streams, some as **groundwater** filling *pores* (small open spaces) and cracks underground, and the rest as ice. Most ice, in turn, occurs in *glaciers* (sheets or rivers of solid water that last all year and flow very slowly) of cold

FIGURE 1.14 Characteristics of the Earth's atmosphere. Molecules pack together more tightly at the base of the atmosphere, so atmospheric pressure changes with altitude (as shown by the blue curve).

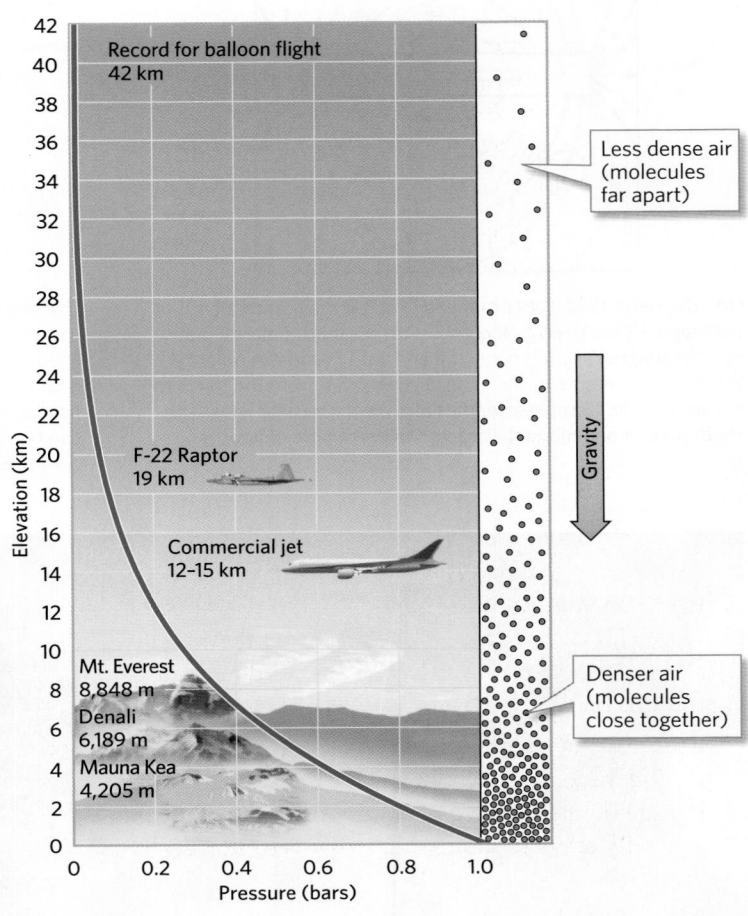

FIGURE 1.13 An orbiting astronaut's photograph shows the haze of the atmosphere fading up into the blackness of space. The International Space Station (ISS) is enlarged to be visible. The inset represents the composition of the atmosphere.

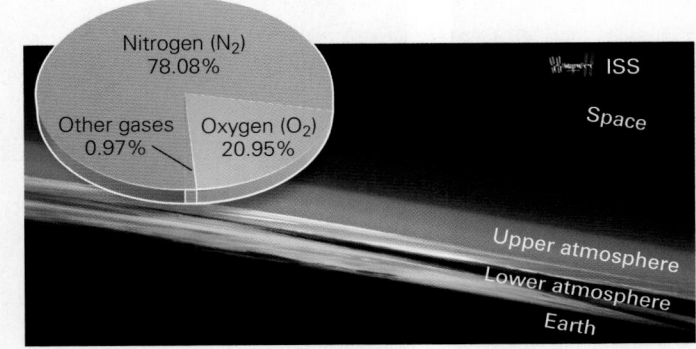

FIGURE 1.15 Clouds take many forms in the Earth's atmosphere.

polar or mountainous regions (Fig. 1.16). Glaciers cover about 10% of the land area. In cold regions, near-surface groundwater also freezes, producing *permafrost* (permanently frozen ground), and as much as 15% of the sea surface freezes to form a thin layer of *sea ice*. The realm containing water on or near our planet's surface constitutes the **hydrosphere** of the Earth System. The Earth's frozen-water realm makes up the **cryosphere**, a part of the hydrosphere.

Materials of the Geosphere

If you sent enough probes down to the Earth's surface to sample and analyze the composition of the solid realm of the Earth System, the **geosphere**, you would eventually detect all 92 naturally occurring elements. You would find, however, that only four of these—iron, oxygen, silicon, and magnesium—make up 90% of the Earth's solid mass (Fig. 1.17). The geosphere is not homogeneous, but rather it contains a great variety of different kinds of materials, which differ from one another in terms of *composition* (the chemicals present) and *texture* (the way in which solid pieces are oriented relative to each other, and connect to each other). Let's take a quick look at some basic categories of these materials (Fig. 1.18)—all of which will be discussed more fully later in this book.

- *Melts:* A **melt**, or *molten material*, forms when a solid becomes hot enough to transform into a liquid. In the context of Earth Science, the word generally refers to molten rock that is so hot that it can flow.

- *Minerals:* A **mineral** is a naturally occurring solid in which atoms are arranged in an orderly pattern. A piece of a mineral that has naturally grown into its present shape is a *crystal*.

- *Glasses:* A solid in which atoms are not arranged in an orderly pattern is a **glass**.

- *Grains:* Geologists refer to a small fragment of a solid as a **grain**. A grain can consist of a crystal, a part of a crystal, or a piece of glass or rock.

- *Sediment:* **Sediment** is loose material made up of grains, shells or shell fragments, and/or crystals that have precipitated from a water solution.

- *Soil:* Sediment that has been altered by reactions with air, water, and life at the surface of the Earth and has mixed with the remains of organisms becomes a **soil**.

- *Rocks:* A **rock** is a coherent aggregate (collection) composed of mineral crystals or grains or of a mass of natural glass. Geologists recognize three principal groups of rocks: (1) *Igneous rock* forms by the solidification of molten rock, either underground or at the Earth's

FIGURE 1.16 Water of the hydrosphere occurs in both liquid and solid form. This satellite image of the Arctic Ocean and Greenland displays both. The frozen part of the hydrosphere can also be called the cryosphere. In this image, the cryosphere consists of both a huge ice sheet covering Greenland and a vast area of sea ice covering the Arctic Ocean.

FIGURE 1.17 The proportions of elements making up the solid mass of the Earth. Note that iron and oxygen account for most of the Earth's mass.

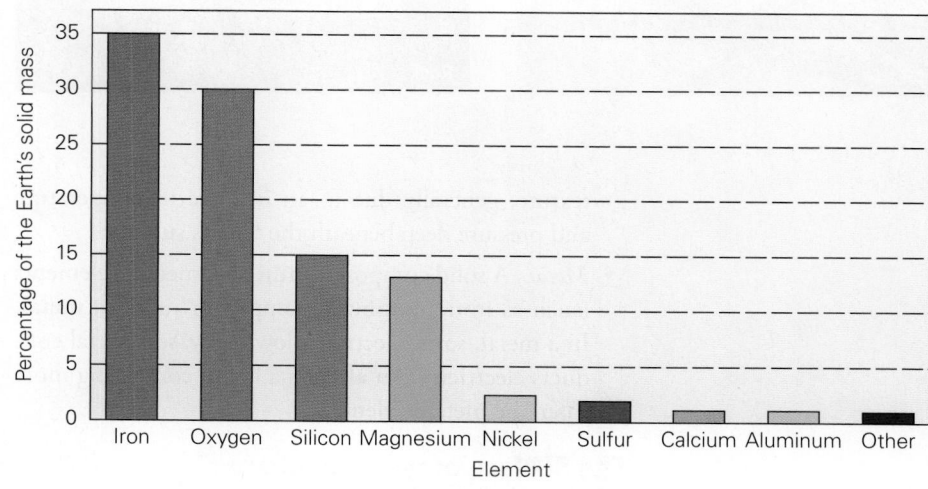

surface. (2) *Sedimentary rock* forms at or near the Earth's surface either by the cementing together of sediment, or by the precipitation of minerals from water solutions. (3) *Metamorphic rock* forms when pre-existing rock undergoes changes in its mineral content and

FIGURE 1.18 Materials of the geosphere.

Melt

Mineral

Glass

Sediment

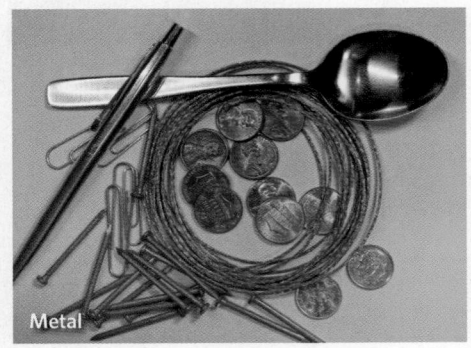
Metal

Rock

texture, generally due to an increase in temperature and pressure deep beneath the Earth's surface.

- *Metals:* A solid composed entirely of metallic elements (such as iron, aluminum, copper, or tin) is a **metal**. In a metal, some electrons flow freely, so a metal conducts electricity. An **alloy** is a blend containing more than one metallic element.

The Surface of the Geosphere

Overall, the Earth has the shape of a very slightly flattened sphere. But in detail, its surface has many ups and downs, which you can detect from orbit and represent on a map that indicates vertical distances relative to *sea level*, the average elevation of the ocean surface (Fig. 1.19).

An analysis of **bathymetry**, the variation in the depth of a water body (the vertical distance from the water surface to the bottom), reveals that the deepest ocean floor lies at 6–11 km (4–7 mi), but on average, the seafloor has a depth of about 4 km (2.5 mi). Oceans overlie broad *ocean basins*, which can be divided into distinct provinces that differ in their bathymetric characteristics. These provinces include abyssal plains (wide, fairly flat areas of deep seafloor), trenches (deep, elongate troughs), mid-ocean ridges (long submarine mountain ranges), seamounts (isolated submarine peaks), and continental shelves (broad areas of shallow ocean fringing the continents).

An analysis of **topography**, the variation in the elevation (height) of the land surface above sea level, reveals that the land has a maximum elevation of about 8.8 km (5.5 mi) and an average elevation of about 0.8 km (0.5 mi). Though there are over 900,000 islands on the Earth, most of our planet's land lies within **continents**, large landmasses ranging from 4,000 to 8,000 km (2,500–5,000 mi) across. (By comparison, the circumference of the Earth is about 40,000 km, or 25,000 mi.) Mapping of the land surface reveals a diverse assemblage of topographic features such as plains (relatively smooth areas), mountain ranges (long chains of mountains), valleys, canyons (steep-sided valleys), and plateaus (high-elevation plains). While the various bathymetric and topographic features that you can detect from orbit look impressive, the vertical distance between the highest mountain peak and the deepest seafloor is only about 19.9 km (12.4 mi), about 0.16% of the Earth's 12,742 km (7,915 mi) diameter.

Together, the bathymetry and topography of the Earth make it look very different from the other terrestrial planets (Fig. 1.20). Most notably, the surfaces of the other planets remain pockmarked by *impact craters*, bowl-shaped indentations caused by the impact of meteorites (see Fig. BxI.2), but very few craters are visible on the Earth's surface. This difference indicates that processes must be taking place that can change the Earth's surface over time. The changes involve **uplift** (movement of the land surface upward to higher elevations), **erosion** (the grinding or beveling away of the land), and **deposition** (the burying of features by sediment). In other words, unlike the other terrestrial planets, the Earth remains a dynamic place, where natural phenomena continue to modify its surface.

The Biosphere

The Earth System hosts an amazingly complex **biosphere**, defined to include not only the great variety of living organisms on the planet, but also the portion of the Earth in which life can survive. This realm extends from a few kilometers below the Earth's surface to a few kilometers above it, so it overlaps with the lower atmosphere, all of the hydrosphere, and the top of the geosphere. From your orbiting spacecraft, you can see plants and animals on the land surface, you can detect chemicals produced by living organisms, and you can see evidence of *microbes* (microscopic organisms). You can also see that human society has transformed forest, prairie, and desert lands into farm fields, cities, highways, canals, dams, and mines. And you can identify the abundance of *anthropogenic* (human-produced) materials, such as concrete, asphalt, brick, lumber, tile, and sheet metal placed on the Earth's surface. A view of the Earth at night, showing the lights of cities, emphasizes the reach of present-day civilization (Fig. 1.21).

Take-home message . . .

A spacecraft approaching the Earth would learn that the Earth has a magnetic field and an atmosphere composed mainly of nitrogen and oxygen. The elevation of its surface varies, both on land and on the seafloor. Overall, the Earth System consists of several realms: the atmosphere, hydrosphere, cryosphere, geosphere, and biosphere. These realms contain a wide range of Earth materials.

Quick Questions

- How does the thickness of the atmosphere compare with the radius of the Earth?
- Distinguish between freshwater and saltwater. Which occurs in greater quantities worldwide?
- What are the various materials that make up the geosphere?

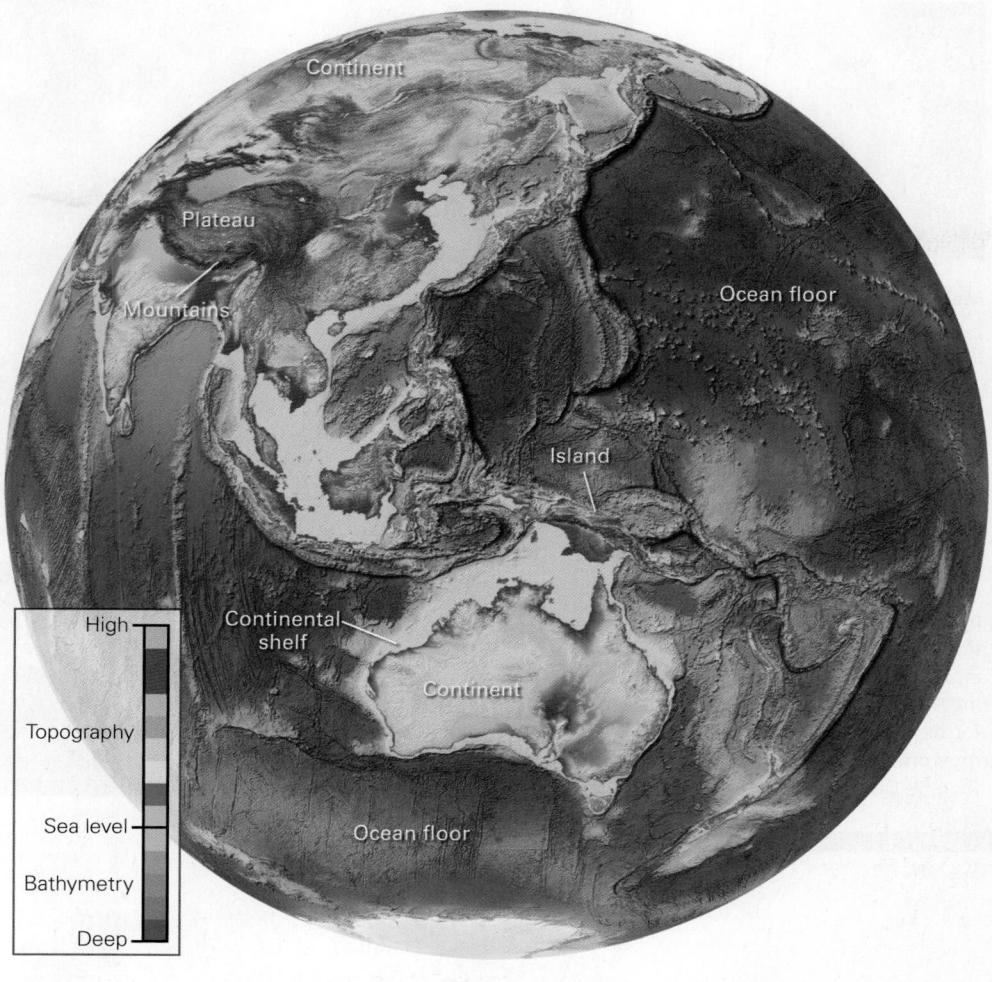

FIGURE 1.19 Bathymetry and topography. This map shows the Earth's bathymetry (variation in seafloor depth) and topography (variation in land elevation).

FIGURE 1.20 Comparing the topography of the Earth, Mars, and Venus shows that each planet is unique.

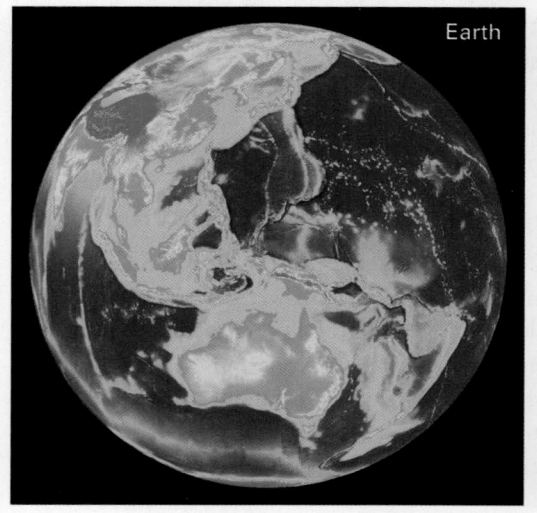
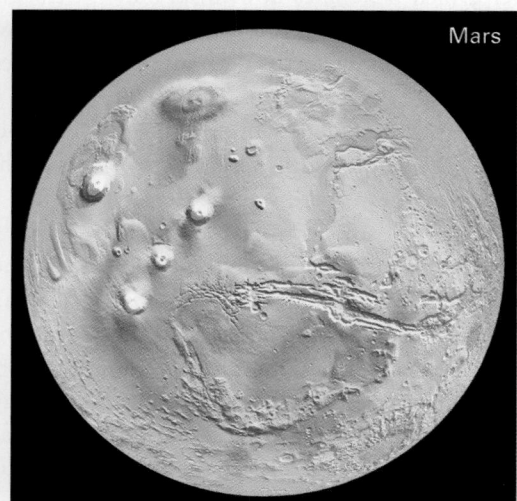
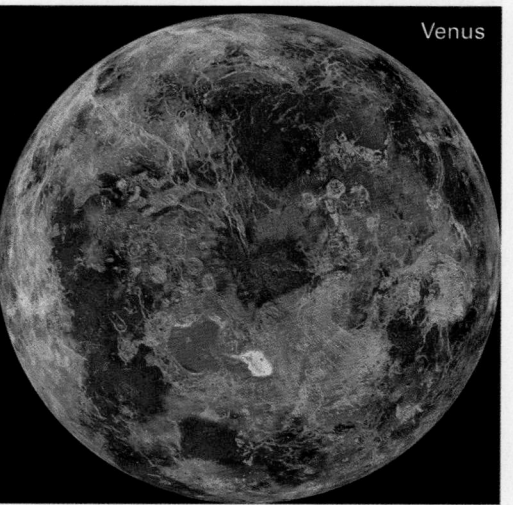

Low — Elevation (not to scale) — High

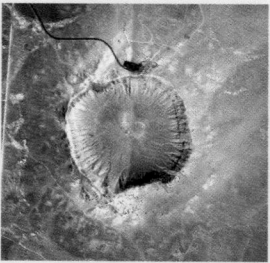

Meteor Crater

Latitude: 35°1′37.18″ N
Longitude: 111°1′20.17″ W

A space probe mapping Mars from orbit reveals thousands of craters, depressions that mark the locations of meteorite impacts. On the Earth, most meteorite craters have eroded away, but a few remain. Zoom to an elevation of 5.7 km (19,000 ft) and look down. You are seeing Meteor Crater, Arizona, formed by the impact of a meteorite 50 m wide less than 50,000 years ago.

FIGURE 1.21 City lights at night, as seen from space.

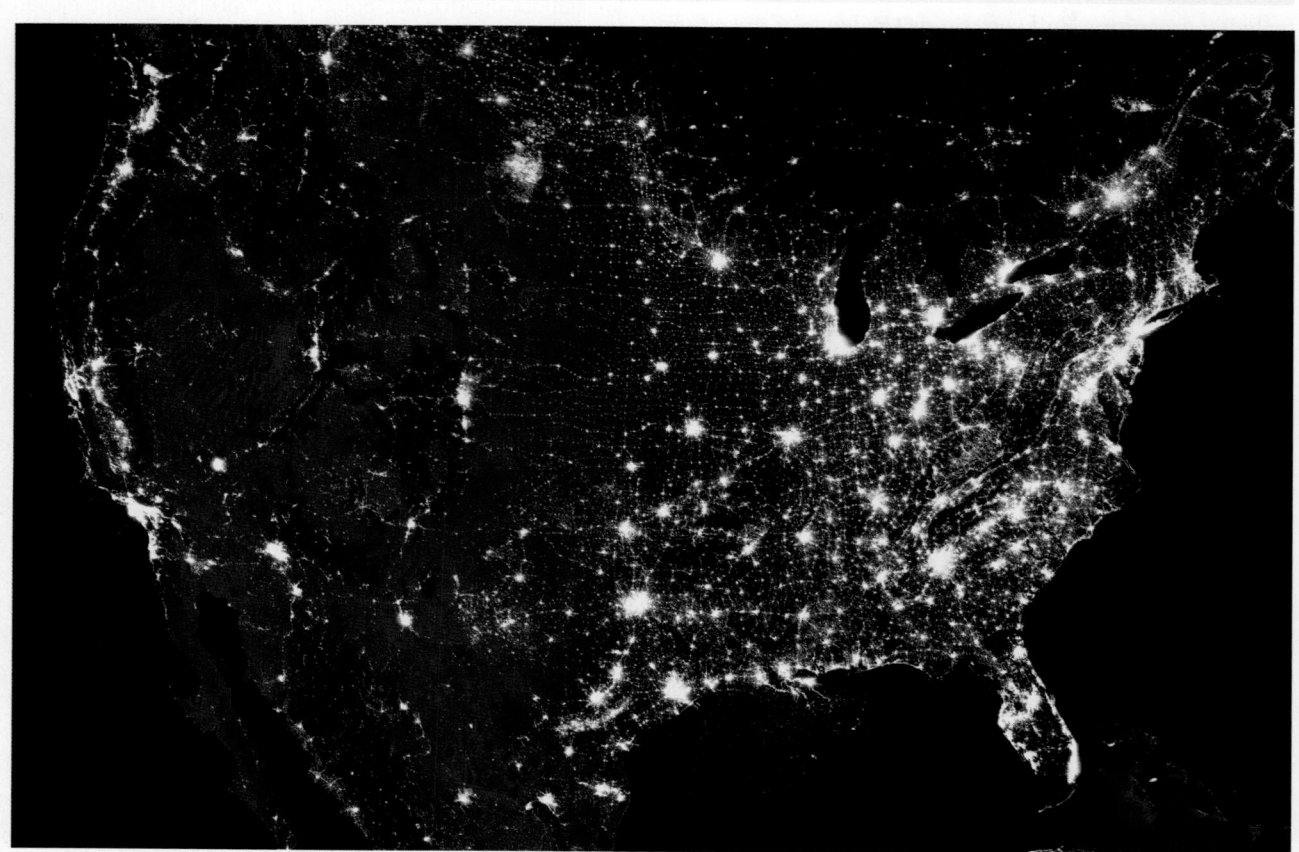

1.5 A First Glance at the Earth's Interior

Hints about What's Inside the Earth

The deepest hole ever drilled penetrates downward only about 12.3 km (7.6 mi)—a mere pinprick on the skin of our planet—so we can't examine the Earth's deep interior directly. Fortunately, there are other ways to gain insight into this portion of the geosphere. First, by measuring the mass and volume of the whole Earth, researchers in the 19th century calculated the planet's average density (5.5 g/cm³), and by analyzing samples from rock exposures worldwide, they directly measured the average density of surface rocks (2.6 g/cm³). A comparison of these two measurements shows that the average density of the whole Earth is about twice that of rocks found at its surface. Consequently, the material inside the planet must be much denser than the material at its surface. Second, the Earth is a very slightly flattened sphere, even though it spins rapidly on its axis. Calculations demonstrate that if the Earth were mostly molten beneath its surface, or if its mass were uniformly distributed throughout, its rotation would cause it to flatten much more, so it would become a disk. The Earth's nearly spherical shape means not only that its interior must be mostly solid, but also that the densest material must be concentrated near its center. Taken together, these results led 19th-century researchers to conclude that the Earth resembles a hard-boiled egg, with three principal layers (Fig. 1.22): (1) a thin *crust* (the eggshell) composed mostly of relatively low-density rock; (2) a solid *mantle* (the white) composed of relatively high-density rock; and (3) a *core* (the yolk) consisting of very high-density metal alloy.

Refining the Picture of the Earth's Interior

As the 20th century dawned, key questions about the Earth's interior remained unanswered: How thick are the internal layers? What exactly do the layers consist of? Are the boundaries between layers sharp or gradational? Can the layers be subdivided in a meaningful way? Researchers had to wait for the invention of new techniques and instruments in order to have the data needed to answer these questions and to calculate how temperature and pressure vary inside the planet (Box 1.5). The study of *seismic waves*, vibrations produced by earthquakes, proved

BOX 1.5 ▶ **A Deeper Look**

Temperatures and pressures inside the Earth

At times during the first few hundred million years of its existence, the Earth was probably so hot that its surface and much of its interior were molten (see the image at the opening of this chapter). As heat escaped into space, the planet cooled, and much of it solidified. However, it remains hot enough in the interior that much of the mantle can still flow. Much of this heat remains from the Earth's formation and early days, but some comes from the decay of radioactive atoms. Consequently, though the Earth's average ground-surface temperature hovers around 14°C (57°F), the temperature at a depth of 40 km (25 mi) generally ranges between 450°C and 900°C (840°F–1,650°F), and the temperature at the center may reach 6,000°C (11,000°F). Geologists refer to the rate at which temperature increases with depth as the **geothermal gradient**, and to the line on a graph representing the change in temperature with depth as the *geotherm* (Fig. Bx1.5). The slope (or steepness) of the geotherm indicates the rate of temperature change with increasing depth. Note that this rate isn't the same at all depths.

Pressure also increases with depth due to the weight of overlying material. Because the materials of the solid Earth (the geosphere) are denser than the gases of the atmosphere, pressure increases much more rapidly with depth in the geosphere than in the atmosphere. Geologists estimate that at a depth of 40 km, the pressure reaches 10,000 bars, and at the center of the Earth, it reaches about 3,500,000 bars.

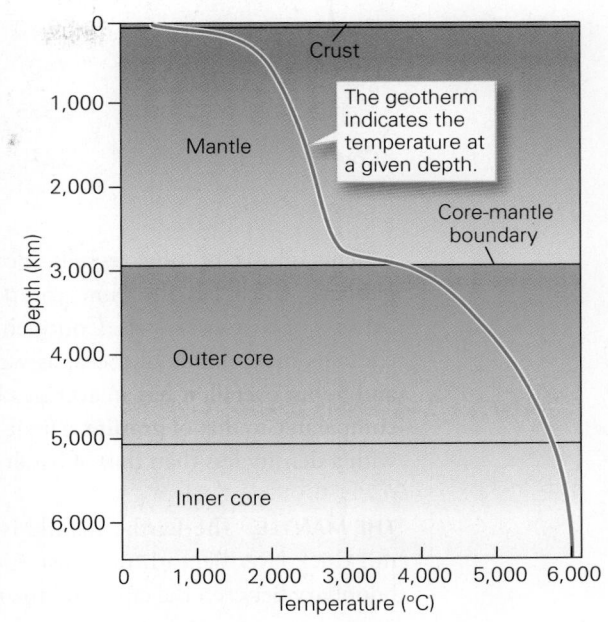

FIGURE Bx1.5 The geothermal gradient is the rate at which the temperature of the Earth's interior changes with depth.

The geotherm indicates the temperature at a given depth.

to be a particularly helpful source of data. Measuring the path and speed of these waves as they travel through the Earth has provided a much clearer image of its interior, which we'll explore further in Chapter 7. Below, we introduce the key characteristics of the Earth's interior layers revealed by these studies, so that we can use this information in our discussion of plate tectonics in Chapter 2. **Table 1.1** summarizes key characteristics of the layers.

THE CRUST. When you stand on the Earth's solid surface, you're on top of its outermost layer, the **crust** (Fig. 1.23a). Oceanic crust and continental crust differ markedly in average thickness and composition (see Table 1.1). Compared with the radius of the Earth, continental crust is so thin that if the Earth were the size of an inflated balloon, the crust would be the thickness of the balloon's skin. What does the crust consist of? A blanket of sediment, generally less than 1 km (0.6 mi) thick, forms the top surface of the oceanic crust. Beneath this blanket, oceanic crust contains of two kinds of dark gray, relatively dense igneous rock: **basalt**,

FIGURE 1.22 An early image of the Earth's internal layers.

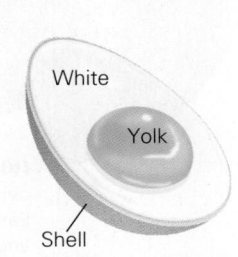

(a) The hard-boiled egg analogy for the Earth's interior.

White

Yolk

Shell

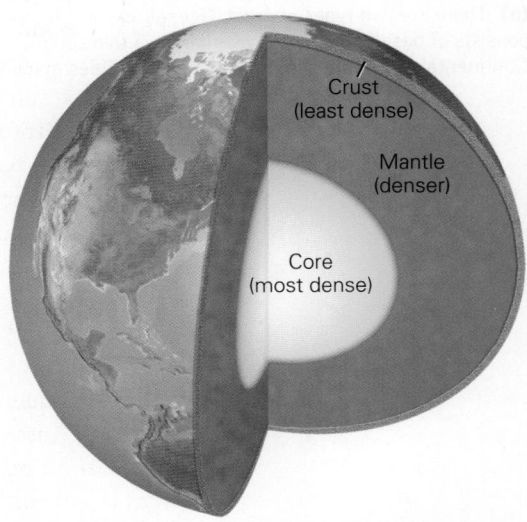

(b) In the 19th century, researchers recognized that the Earth has three principal layers.

Crust (least dense)

Mantle (denser)

Core (most dense)

TABLE 1.1 The Principal Layers Inside the Earth

Layer name	Thickness (km)	Thickness (mi)	Material
Oceanic crust	7–10	4–6	Basalt and gabbro (density = 3.0 g/cm³)
Continental crust (average)	25–70 (40)	15–43 (25)	Average composition comparable to that of granite (density = 2.7 g/cm³)
Mantle	2,885	1,790	Peridotite (density = 4.5 g/cm³)
Core	3,470	2,155	Iron alloy (density = 10–13 g/cm³)

which consists of tiny crystals, occurs as a layer over **gabbro**, which has the same composition as basalt but consists of coarser crystals. Continental crust, in contrast, contains many kinds of rocks, as we'll see in Chapters 4 and 5, but overall, it has an average chemical composition comparable to that of **granite**, a light-colored igneous rock with a density less than that of basalt or gabbro.

THE MANTLE. The Earth's mantle is a 2,885 km (1,792 mi) thick layer beneath the crust. Geologists refer to the boundary between the crust and the mantle as the **Moho**.

The mantle makes up the largest part of the geosphere, in terms of volume (**Fig. 1.23b**), and it consists of various versions of a very dense igneous rock called **peridotite**. That means that peridotite, though rare at the Earth's surface, is the most abundant rock on our planet! Geologists divide the mantle into two sublayers: the *upper mantle,* from the Moho down to a depth of 660 km (410 mi), and the *lower mantle,* below that depth. Nearly the entire mantle is solid, but most remains hot enough to be able to flow very slowly, at a rate of less than 15 cm/yr (6 in/yr). (This flow resembles the flow of soft—but not liquid—plastic,

FIGURE 1.23 A modern view of the Earth's interior layers.

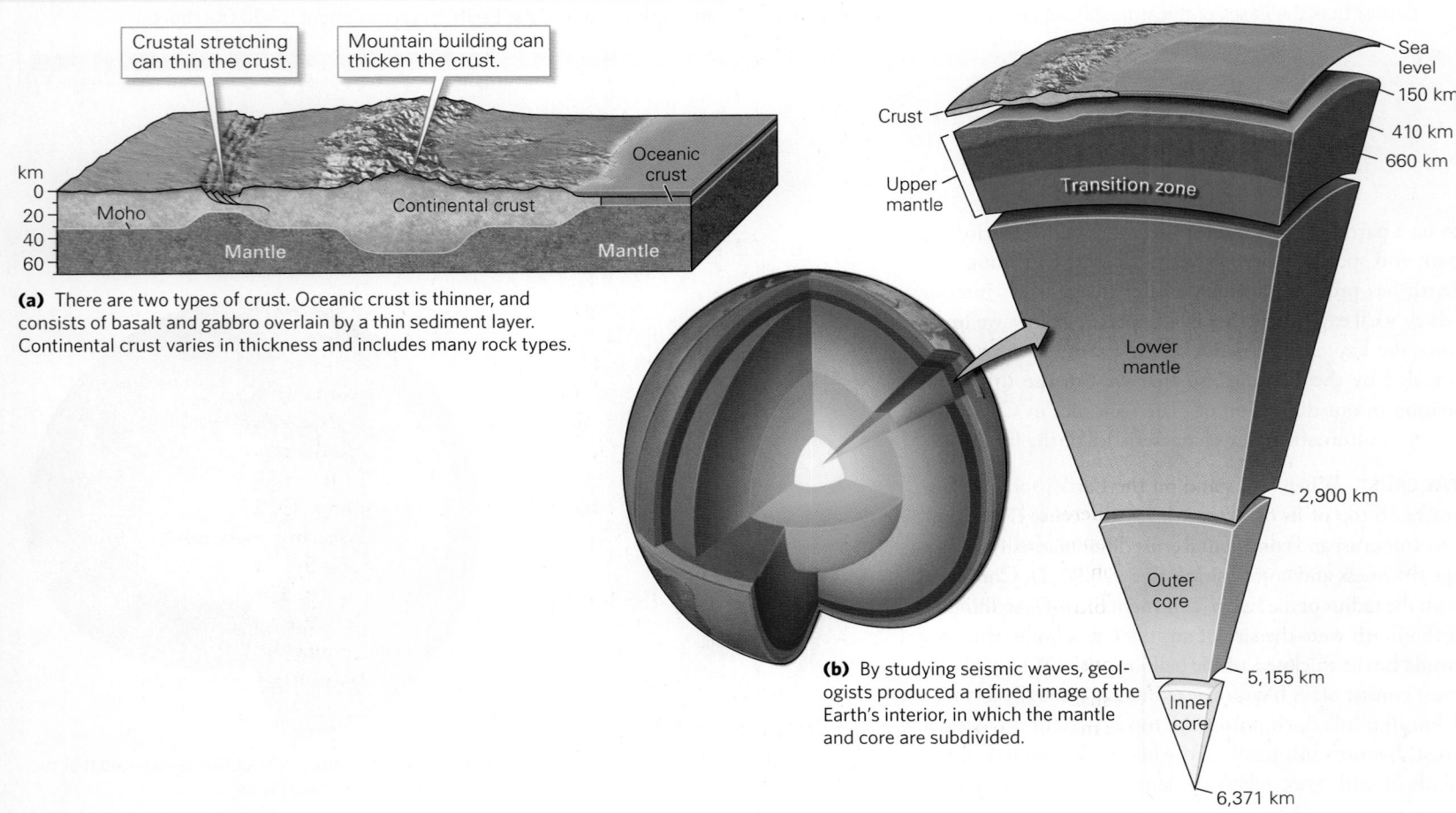

(a) There are two types of crust. Oceanic crust is thinner, and consists of basalt and gabbro overlain by a thin sediment layer. Continental crust varies in thickness and includes many rock types.

(b) By studying seismic waves, geologists produced a refined image of the Earth's interior, in which the mantle and core are subdivided.

so scientists refer to it as *plastic flow*.) Notably, in specific localities, generally above a depth of about 150 km, mantle rock does melt slightly, as we'll see in Chapter 4.

THE CORE. Early calculations suggested that the core, the central ball deep inside the Earth, had the same density as gold, so for many years people held the fanciful hope that vast riches lay at the heart of our planet. Alas, geologists eventually concluded that the core consists of a far less glamorous material: iron alloy, a mix of iron with 4% nickel and up to 10% oxygen, silicon, or sulfur. The *outer core* exists in liquid form and flows rapidly, whereas the *inner core* is solid. Notably, flow in the outer core generates the Earth's magnetic field.

LITHOSPHERE AND ASTHENOSPHERE. As emphasized by the descriptions above, the traditional division of the Earth's interior into layers reflects variation in the composition of the Earth with depth. In the early 20th century, geologists realized that the outer portion of the Earth could also be divided into layers based on how the layers behave when subjected to force. Using this alternative basis, geologists distinguish between an outermost layer, called the **lithosphere**, that behaves somewhat like a rigid shell, in that it can bend and fracture, but overall does not flow; and an underlying layer, called the **asthenosphere**, that can undergo plastic flow very slowly. The lithosphere consists of all the crust plus the uppermost part of the upper mantle. The asthenosphere extends from the base of the lithosphere at least down to the base of the upper mantle; its lower boundary is not well defined. We'll refine this distinction in the next chapter and see how an understanding of lithosphere and asthenosphere behavior contributed to the development of plate tectonics.

Take-home message . . .

The Earth's interior can be divided into three layers with differing compositions: the crust, mantle, and core. The crust beneath the oceans differs from the crust beneath the continents. The mantle consists of very dense, almost entirely solid rock, most of which is hot enough to flow very slowly. The core consists of iron alloy. Flow in the liquid outer core generates the Earth's magnetic field. The Earth's inner core is solid. The outer part of the Earth can also be divided into lithosphere and asthenosphere.

Quick Questions —————————

- What are the three principal layers inside the Earth?
- What is the temperature at the center of the Earth?
- Which of these rocks is the densest: granite, peridotite, or basalt?

1 CHAPTER REVIEW

Objective 1.1

Identify celestial objects visible to the naked eye, list the principal kinds of objects within our Solar System, and explain what the study of cosmology encompasses.

KEY CONCEPTS

- The Universe includes numerous different types of celestial objects, including stars, galaxies, planets, moons, and nebulae. Cosmology addresses the structure and evolution of the Universe.
- The Earth is one of eight planets orbiting our Sun in the Solar System. The four terrestrial planets are relatively small and have rocky surfaces. The large Jovian planets consist mostly of gases and ices.
- Our Solar System lies in the spiral-shaped Milky Way Galaxy, which is one of over a trillion galaxies. We measure large distances between stars in light-years.

EARTH-SCIENCE VOCABULARY

atomic mass (p. 27)
atomic number (p. 27)
celestial object (p. 23)
chemical bond (p. 27)
chemical formula (p. 27)
cosmology (p. 23)
dust (p. 24)
electromagnetism (p. 24)
element (p. 26)
force (p. 24)
galaxy (p. 23)
gravity (p. 24)
ice (p. 24)
isotope (p. 27)
light-year (p. 28)
matter (p. 24)

Milky Way Galaxy (p. 23)
molecule (p. 27)
moon (p. 27)
nebula (p. 24)
nuclear bond (p. 26)
nuclear reaction (p. 26)
orbit (p. 25)
planet (p. 25)
refractory material (p. 25)
Solar System (p. 27)
star (p. 23)
Universe (p. 23)
vacuum (p. 27)
volatile material (p. 24)

REVIEW QUESTIONS

1. **(a)** How did ancient astronomers distinguish between a star and a planet? **(b)** What does the Milky Way consist of, and why does it look like a hazy band to observers on the Earth?

2. **(a)** Which has a higher melting temperature: a volatile material or a refractory material? **(b)** What is the difference between a nebula and a vacuum? **(c)** Label the images in **Figure A** to indicate which shows a nebula and which shows a galaxy.

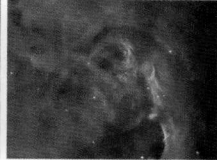

A

3. Compare a contact force with a field force, and provide an example of each.

4. **(a)** Distinguish between an element and a compound, and name the smallest component of each. **(b)** What is the difference between the atomic number and atomic mass of an element? **(c)** What is an isotope? **(d)** Label the nuclear reactions shown in **Figure B**.

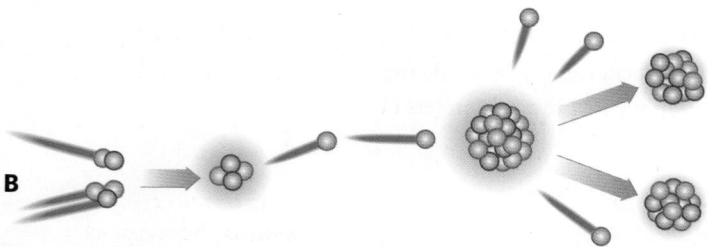

B

5. **(a)** Provide a modern definition of a planet. **(b)** How do terrestrial planets differ from Jovian planets? **(c)** How do both kinds of planets differ from moons?

6. **(a)** How long does it take light emitted from a star that lies 10 light-years away to reach the Earth? **(b)** How long would it take you to fly to that star in a spacecraft traveling at 17,000 mph (the speed of the International Space Station)?

Objective 1.2

Summarize the expanding Universe theory and the Big Bang theory, describe how the first stars came into existence, and explain how and where different elements formed.

KEY CONCEPTS

• The Universe has been expanding since the Big Bang 13.8 billion years ago, so distant galaxies are moving away from one another. Big Bang nucleosynthesis produced small atoms, which collected into early nebulae.

• Massive early nebulae could collapse inward, due to gravity, to produce massive protostars, which could continue to collapse until nuclear fusion began. When this happened, the protostars became true stars. These first-generation stars exploded as supernovae.

• Stellar nucleosynthesis produced atoms as large as iron, and supernova nucleosynthesis produced even larger atoms—so stars, from birth through death, are element factories. Heavier atoms, once formed, were ejected into space by stellar winds or during the death of stars, and mixed into nebulae.

EARTH-SCIENCE VOCABULARY

accretion disk (p. 30)
Big Bang (p. 28)
Big Bang nucleosynthesis (p. 29)
expanding Universe theory (p. 28)
heat (p. 29)
nebular theory (p. 30)
protostar (p. 30)
solar wind (p. 31)
stellar nucleosynthesis (p. 31)
supernova (p. 31)
supernova nucleosynthesis (p. 31)
temperature (p. 29)
thermal energy (p. 29)

REVIEW QUESTIONS

7. **(a)** According to the expanding Universe theory, why does the speed at which galaxies move away from each other depend on the distance between the galaxies, as illustrated by a raisin-bread analogy (**Figure C**)? **(b)** What does the name "Big Bang" refer to? **(c)** According to the Big Bang theory, when did the Universe begin to expand?

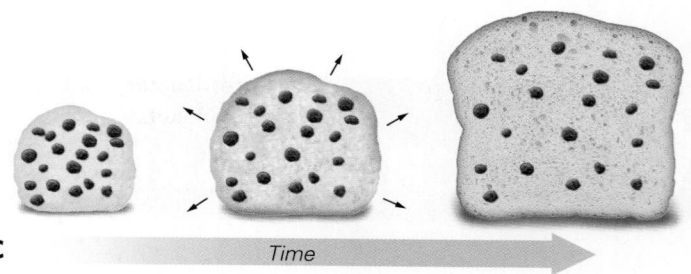

C   Time

8. **(a)** Name a common element that could have formed during stellar nucleosynthesis. **(b)** Could the Earth have formed from atoms in the earliest nebulae? **(c)** Which of these processes produced elements with the biggest atomic number: Big Bang nucleosynthesis, stellar nucleosynthesis, or supernova nucleosynthesis?

9. **(a)** How did the first stars form according to the nebular theory? **(b)** Why were the first stars so much larger than our Sun? **(c)** Why can't our Sun be considered a first-generation star?

10. **(a)** How do some atoms produced by stellar nucleosynthesis end up traveling through space far from the star in which they formed? **(b)** Which process does **Figure D** show? Draw arrows to indicate the general direction of movement of the glowing material.

D

Objective 1.3

Describe how solar systems like ours formed, and why terrestrial planets develop internal layering and a nearly spherical shape.

KEY CONCEPTS

- According to the condensation theory, smaller stars such as our Sun can form only from nebulae that contain particles of ice and dust, for these particles disperse heat and allow small protostars to collapse to become true stars.
- The planets of our Solar System began to grow in the Sun's accretion disk when ice and dust collected into clumps, which then merged to form planetesimals.
- The completion of planetary growth happened when planetesimals collided with one another, or collected all the other matter in their orbits, and grew into protoplanets. Eight protoplanets cleared their orbits to become today's eight planets.
- Terrestrial planets formed mostly from dust nearer the Sun, while farther from the Sun, the Jovian planets incorporated vast amounts of gases and ices.
- When a planetesimal became large enough for its interior to become very hot, it underwent differentiation into a metallic core and a rocky mantle. Heating and softening of the interior allowed these bodies to become spherical.

EARTH-SCIENCE VOCABULARY

condensation theory (p. 32) **protoplanetary disk** (p. 32)
differentiation (p. 36) **protoplanets** (p. 32)
planetesimals (p. 32)

REVIEW QUESTIONS

11. **(a)** Why must large atoms such as silicon and iron be present for stars like our Sun to form? **(b)** How does the energy source of a protostar differ from that of a true star? **(c)** Why can't a small protostar evolve into a true star?

12. **(a)** Distinguish among a planetesimal, a protoplanet, and a planet. **(b)** What other solid objects, besides planets and moons, formed from the protoplanetary disk?

13. Why don't terrestrial planets contain large volumes of volatile elements, as do Jovian planets?

14. **(a)** Why isn't the Earth homogeneous? **(b)** Explain what **Figure E** shows.

15. **(a)** What process caused the Earth to become a sphere? **(b)** Why aren't small asteroids sphere-shaped?

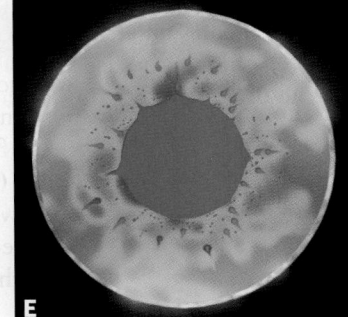

E

Objective 1.4

Characterize the Earth's magnetic field, distinguish the realms of the Earth System, and list the key materials found within each realm.

KEY CONCEPTS

- A magnetic field extends into space around the Earth. The magnetic field shields the Earth from solar wind.
- The atmosphere, an envelope of air, consists mostly of nitrogen and oxygen gas. Nearly all the air—99.9%—lies within only 50 km (30 mi) of the Earth's surface.
- The surface of the Earth consists of land and water. Most of the water (97%) is salty and fills the oceans and seas of the world. The remaining liquid freshwater is water that lies either on the land in lakes, streams, and wetlands, or underground in pores and cracks.
- A great variety of different materials—melts, minerals, glasses, grains, sediment, soil, rocks, and metals—make up the geosphere.
- The elevation of the geosphere varies across its surface. Variation in land-surface elevation is topography, and variation in seafloor depth is bathymetry.
- The biosphere, living organisms and the realm in which living organisms can survive, extends from a few kilometers below the Earth's surface to a few kilometers above it.

EARTH-SCIENCE VOCABULARY

air (p. 38)	**groundwater** (p. 40)
alloy (p. 42)	**hydrosphere** (p. 41)
atmosphere (p. 38)	**magnetosphere** (p. 37)
atmospheric pressure (p. 38)	**melt** (p. 41)
bathymetry (p. 42)	**metal** (p. 42)
biosphere (p. 42)	**mineral** (p. 41)
continent (p. 42)	**realm** (p. 37)
cryosphere (p. 41)	**rock** (p. 41)
deposition (p. 42)	**sediment** (p. 41)
Earth System (p. 37)	**soil** (p. 41)
erosion (p. 42)	**topography** (p. 42)
geosphere (p. 41)	**uplift** (p. 42)
glass (p. 41)	**water vapor** (p. 38)
grain (p. 41)	

REVIEW QUESTIONS

16. **(a)** What causes the Earth's magnetic field? **(b)** How does the magnetic field interact with the solar wind? **(c)** On **Figure F**, label the magnetosphere and the solar wind.

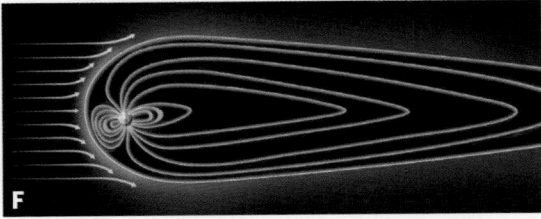

F

17. **(a)** What are the two principal gases in the Earth's atmosphere? **(b)** Describe how the density of air changes with increasing altitude, and why this change happens.

18. **(a)** List the realms of the Earth System. **(b)** Which realm(s) does **Figure G** show?

19. **(a)** What is the proportion of land area to ocean area on the Earth? **(b)** What is the average depth of the seafloor, and what is the average elevation of land?

20. **(a)** On a bathymetric map of the seafloor, is the seafloor completely flat? **(b)** On a topographic map of the continents, are mountains randomly distributed, or do they occur in distinct belts?

21. **(a)** How does rock differ from sediment? **(b)** How does glass differ from a mineral? **(c)** How does metal differ from other solid materials?

Objective 1.5

Name and characterize the principal internal layers of the Earth.

KEY CONCEPTS

- The interior of the Earth consists of a thin crust, a thick mantle, and a central core. A variety of data sources have led to this model.
- Pressure and temperature both increase with depth in the Earth. The rate of temperature increase is the geothermal gradient.
- The crust beneath the oceans differs in thickness and composition from the crust beneath the continents. Rocks of the crust are less dense than those of the mantle.
- The mantle can be subdivided into the upper and lower mantle. The core, which consists of iron alloy, can be subdivided into a liquid

outer core and a solid inner core. Flow in the outer core produces the Earth's magnetic field.

EARTH-SCIENCE VOCABULARY

asthenosphere (p. 47)   **granite** (p. 46)
basalt (p. 45)   **lithosphere** (p. 47)
crust (p. 45)   **Moho** (p. 46)
gabbro (p. 46)   **peridotite** (p. 46)
geothermal gradient (p. 45)

REVIEW QUESTIONS

22. Researchers estimate that the geothermal gradient in South Africa is fairly low, averaging about 10°C/km (18°F/km). The deepest mine in South Africa descends 4 km. **(a)** What is the temperature of rock at the base of this mine, assuming that the temperature at the ground averages 20°C? (By comparison, the hottest temperature ever recorded on land is 56.7°C, or 134°F.) **(b)** At this geothermal gradient, how deep could the mine go before the temperature reached the boiling point of water?

23. **(a)** Label the principal layers of the Earth on **Figure H**. **(b)** What sources provide geologists with information about the character of the Earth's interior?

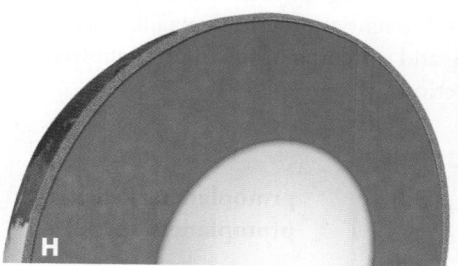

24. **(a)** What is the Moho? **(b)** Describe basic differences between continental crust and oceanic crust. **(c)** At highway speeds (100 km/h), how long would it take to drive a distance equal to the thickness of the continental crust?

25. **(a)** What is the mantle composed of? **(b)** Is the entire mantle rigid and unmoving? **(c)** In what way do lithosphere and asthenosphere differ from each other? **(d)** Which occurs at a shallower depth, the Moho or the lithosphere-asthenosphere boundary?

26. **(a)** What does the core consist of, and how do the inner and outer cores differ? **(b)** Which part produces the magnetic field? **(c)** At highway speeds (100 km/h), how long would it take you to drive along a radius of the Earth to the planet's center?

ANOTHER VIEW A view looking from the coast of Nova Scotia across the Bay of Fundy highlights many aspects of the Earth System. We can see the atmosphere, biosphere, hydrosphere, and geosphere, and the interactions among them.

2 THE WAY THE EARTH WORKS
Plate Tectonics

After studying this chapter, you should be able to...

1. discuss the geologic evidence that Alfred Wegener used to justify the proposal that continents moved following the breakup of the supercontinent Pangaea.

2. describe key observations about the nature of the seafloor that paved the way for development of the seafloor-spreading hypothesis.

3. contrast lithosphere with asthenosphere, characterize the relationship between earthquakes and plate boundaries, and summarize the basic tenets of plate tectonics.

4. name the three types of plate boundaries, describe the geologic features that form at each type, and recognize examples of plate boundaries.

5. explain how new plate boundaries can form and how existing ones can eventually cease activity.

6. point out examples of triple junctions and hot spots, and explain how hot spots differ from plate boundaries.

7. characterize the forces that drive plate motion and the rates at which plates move.

8. provide examples of data, from studies of paleomagnetism and other sources, that serve as proofs of plate tectonics.

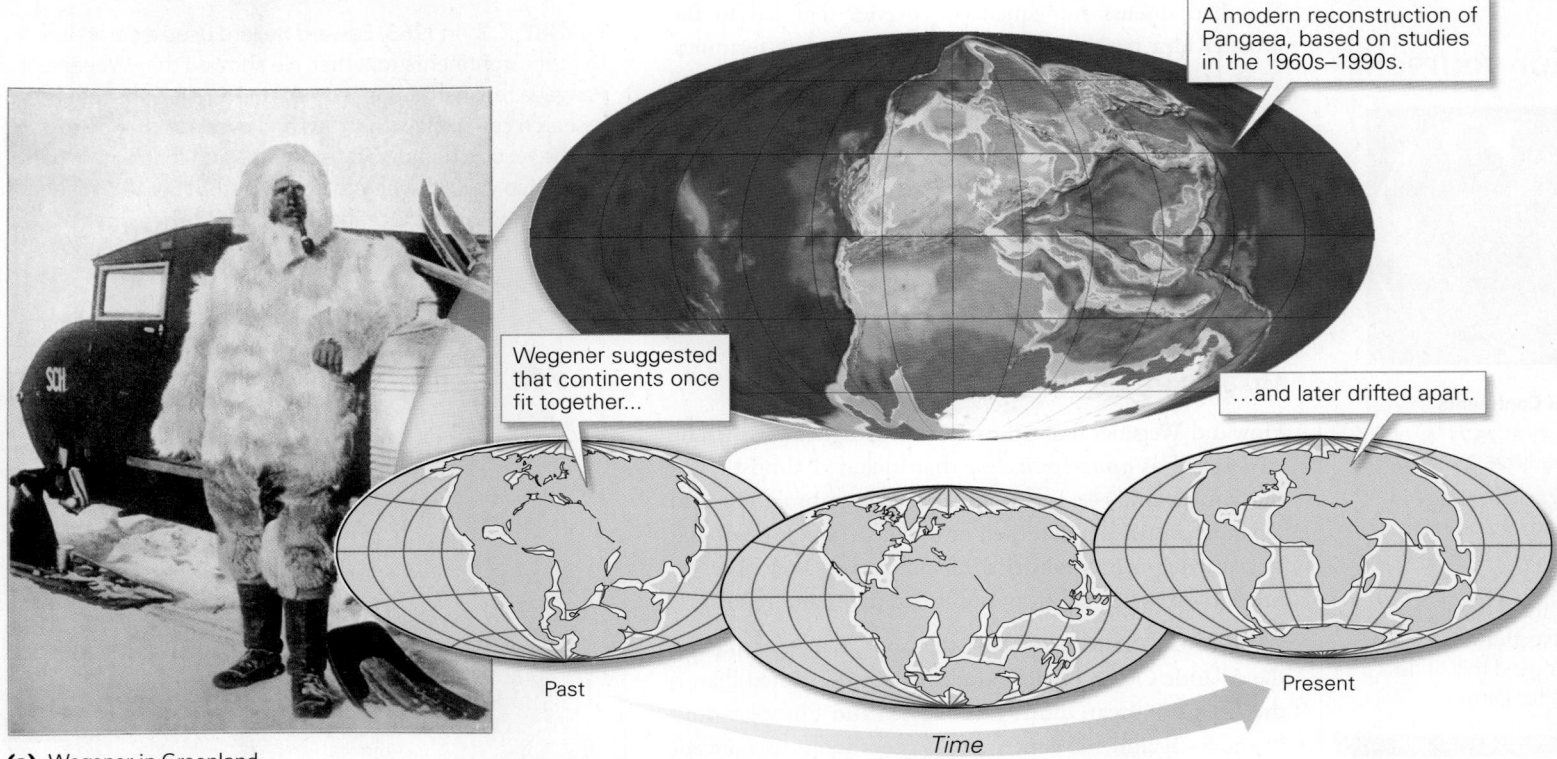

A modern reconstruction of Pangaea, based on studies in the 1960s–1990s.

Wegener suggested that continents once fit together...

...and later drifted apart.

Past

Present

Time

(a) Wegener in Greenland.

(b) Wegener's image of Pangaea and its breakup. The larger globe shows a modern reconstruction of Pangaea.

FIGURE 2.1 Alfred Wegener and his model of continental drift.

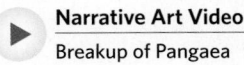 **Narrative Art Video**
Breakup of Pangaea

In September 1930, a German meteorologist named Alfred Wegener, together with several companions, sledged from Greenland's coast across its vast ice sheet to resupply weather observers stranded at a remote camp (Fig. 2.1a). The outward trip went well, and Wegener's group found the observers and dropped off several crates of much-needed food. But because of pressing obligations, Wegener and one companion decided to trek back to the coast immediately, and set off into a storm. They were never seen alive again. At the time of his death, Wegener was well known, not only to meteorologists studying arctic weather, but also to geologists. His geologic notoriety arose because some 15 years earlier he had published a book in which he presented evidence challenging the long-held assumption that continents remained fixed in position through all geologic time. Wegener argued, instead, that continents had once been joined together, like pieces of a giant jigsaw puzzle, to form a vast "supercontinent," which he named *Pangaea* (pronounced Pan-JEE-ah, Greek for

<< This map shows the topography and bathymetry of the African, Asian, and Australian continents, along with many islands. Land colors represent elevation (light tan is highest) and ocean colors represent water depth (dark blue is deepest). Plate tectonics provides a basis to interpret the shape of the Earth's surface.

all land), and that Pangaea later fragmented into separate continents that moved apart to reach their present positions. He named this process *continental drift* (Fig. 2.1b).

In 1930, hardly any geologists agreed with Wegener. Sadly, the scientist died without knowing that three decades later, his ideas would become the foundation of a *scientific revolution* in geology, a complete change in thinking about how geologic phenomena operate and how the Earth's surface has evolved. Today, geologists take it for granted that the map of the Earth constantly changes. This scientific revolution began in 1960, when an American geologist, Harry Hess (1906–1969), suggested that new seafloor forms as continents separate from each other, and that old seafloor sinks into the Earth's interior as continents move toward each other. Geologists worldwide immediately began to explore the implications of this proposal, and by 1968, they had developed a model describing these movements. In this model, the Earth's *lithosphere*—its relatively rigid outer shell—consists of several pieces, or *plates*, that slowly move relative to one another. Because geologists have confirmed this model in many ways, it has gained the status of a theory, now called the *theory of plate tectonics* or, simply, *plate tectonics*. The name comes from the Greek word *tekton*, which means builder, because plate interactions "build" regional geologic features.

We begin this chapter by introducing the observations that led Wegener to propose that continents move.

See for yourself

The Fit of Continents

Latitude: 33°31′25.77″ N
Longitude: 39°26′23.32″ W

Zoom to an elevation of 14,800 km (9,200 mi) and look straight down. Note that northwestern Africa could fit snugly along eastern North America. Wegener used this fit as evidence for Pangaea.

We then discuss subsequent discoveries that led to the broader idea now known as the theory of plate tectonics. Next, we describe the nature of lithosphere plates and the boundaries between them, and the way geologists describe plate motions. This chapter concludes by summarizing the evidence that proves the theory of plate tectonics. In succeeding chapters, we'll see how plate tectonics provides a basis for understanding many aspects of our dynamic planet.

2.1 Continental Drift

Wegener's Evidence

How did Wegener make the case that **Pangaea**—which he envisioned as a *supercontinent* that included almost all of the Earth's land—existed, and that it later broke up into separate continents that then moved apart—a process that he named **continental drift**? He used several key observations, mostly gleaned from published maps and reports.

FIT OF THE CONTINENTS. Almost as soon as maps depicting the Atlantic Ocean became available, scholars noted that, if the Atlantic Ocean didn't exist, Africa and Europe would fit snugly against the Americas. Wegener took this concept further by showing that all the continents could fit together. To Wegener, this arrangement appeared to be too good to be coincidental. Later studies demonstrated that remarkably few overlaps or gaps occurred between modern continents when they were fitted together in Pangaea (Fig. 2.2).

DISTRIBUTION OF CLIMATE BELTS IN THE PAST. Different climate belts occur at different latitudes (see Chapter 19). For example, polar climates lie at high latitudes, hot and dry climates at subtropical latitudes, and steamy climates at tropical latitudes. Wegener speculated that if a continent moved from one latitude to another, the climate at a particular location on the continent would change over time, and that environmental indicators preserved in a succession of sedimentary-rock beds (layers) at the location would record this change (see Chapter 5).

To confirm his speculation, Wegener first looked for reports describing the distribution of rocks formed from sediment left when vast glaciers that formed several times between 330 and 295 million years ago (during the late Paleozoic Era) finally melted away. His search paid off. Wegener found such rocks in South America, Africa, India, and Australia, land areas that now lie at nonpolar latitudes. He also found *glacial striations* (scratches carved by glaciers in a direction parallel to ice movement) on rock surfaces buried by the glacially deposited sedimentary beds (Fig. 2.3a). Significantly, the striations seem to indicate that late Paleozoic glaciers moved from what is now ocean onto land. Wegener knew that glaciers couldn't have moved in such directions, for glaciers form on land and flow downward toward the sea. Clearly, these late Paleozoic glacial features could not have

FIGURE 2.2 In 1965, Edward Bullard used a computer to fit the continents together. He showed that Wegener's Pangaea reconstruction worked quite well. Subsequent research can explain the gaps and overlaps that do exist on the reconstruction. For example, land that is now part of Mexico was attached to northwestern South America before the Atlantic Ocean and Gulf of Mexico opened.

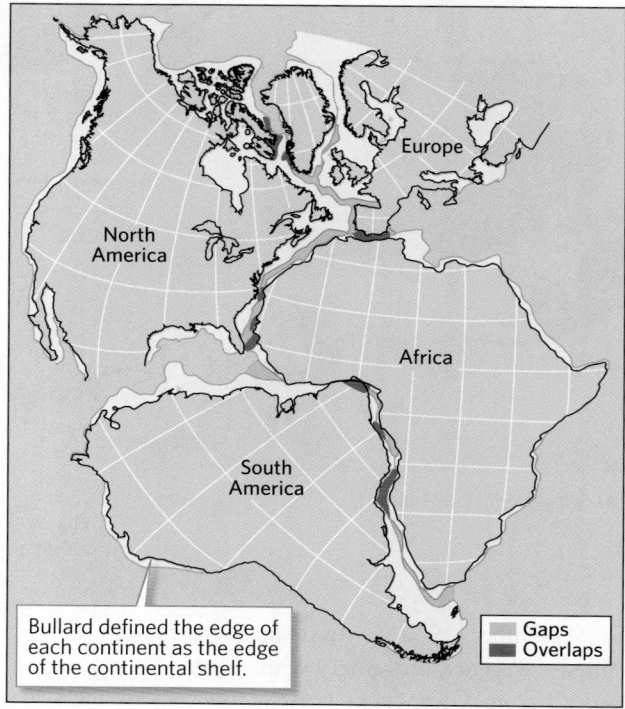

Bullard defined the edge of each continent as the edge of the continental shelf.

Gaps
Overlaps

developed at the latitudes where they now occur, or with the continents separated as they are today (Fig. 2.3b). Wegener showed that the distribution of rocks formed from glacially deposited sediment, as well as the orientation of striations, make sense if the continents were once united in Pangaea, and if Pangaea were positioned so that the southern part of the supercontinent lay at south-polar latitudes (Fig. 2.3c).

If the southern part of Pangaea straddled the south-polar region during the late Paleozoic, then at that time, land that now lies in southern North America, southern Europe, and northwestern Africa would have straddled the equator and would have hosted tropical climates, in which densely vegetated swamps developed on land and shell-producing organisms built reefs offshore. Such tropical climate belts would have been bordered on either side by subtropical desert climate belts, in which lakes or shallow seas contained very salty water and the land surface was locally covered by sand dunes. In the belt of Pangaea that Wegener predicted was equatorial, late Paleozoic sedimentary beds do include coal (rock formed from plant remains) as well as limestone (a type of sedimentary rock commonly made of shells). And in the belt of Pangaea that Wegener predicted was subtropical, late Paleozoic sedimentary beds include evaporites

FIGURE 2.3 Evidence of glaciation on Pangaea.

(a) Glacial striations of late Paleozoic age along the southern coast of Australia. The striations were carved into the surface of bedrock that had been polished by sand embedded in the ice of a moving glacier.

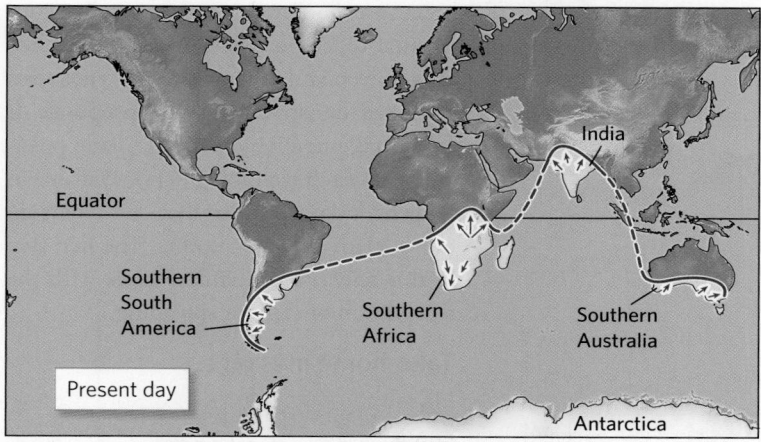

(b) A map showing the distribution of late Paleozoic glacial deposits. The arrows show the direction of glacial flow indicated by striations.

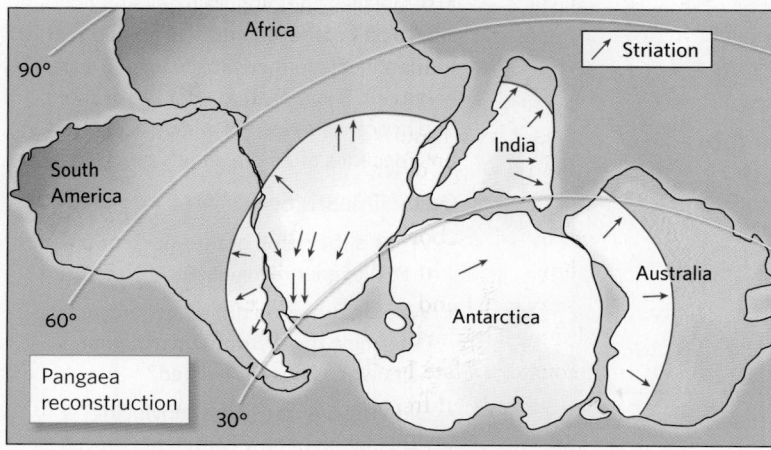

(c) In Wegener's reconstruction of Pangaea, the glaciated areas connect, thereby outlining a late Paleozoic south-polar ice cap.

FIGURE 2.4 The distribution of climate belts in the late Paleozoic, as indicated by distinct sedimentary rock types, makes sense on a map of Pangaea.

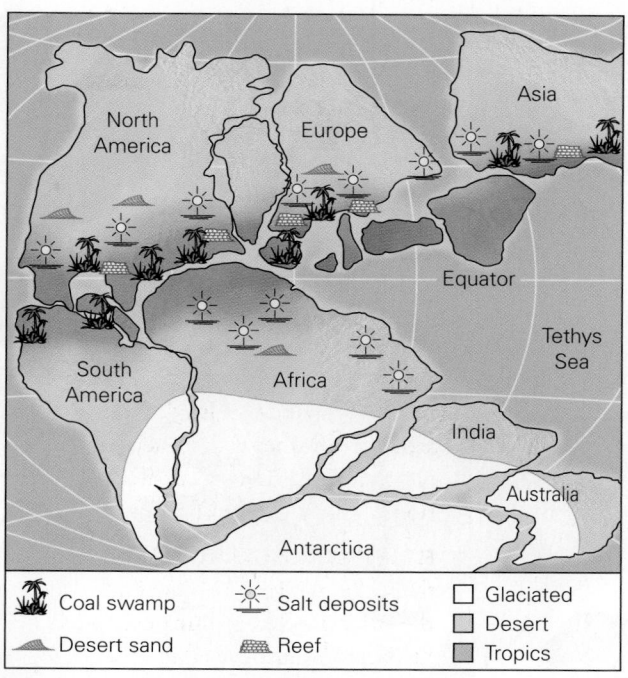

Coal swamp   Salt deposits   ☐ Glaciated
Desert sand   Reef   ☐ Desert
                                     ☐ Tropics

(rock composed of minerals that grow in very salty water) and sandstone (rock composed of cemented-together sand) **(Fig. 2.4)**.

DISTRIBUTION OF FOSSILS. Today, different continents provide homes for different species. Kangaroos, for example, live in Australia but nowhere else, because they cannot swim across oceans to other continents. Wegener inferred that while Pangaea existed, land animals and plants could have colonized regions now separated by wide oceans. To test this idea, he plotted locations of *fossils* (relicts of past life preserved in rock) to show where land-dwelling species lived when Pangaea existed. As predicted, fossils of the same species occur in sedimentary beds on different continents that, though now separated by oceans, were adjacent in Pangaea **(Fig. 2.5)**.

MATCHING ROCK ASSEMBLAGES AND MOUNTAIN BELTS. Wegener speculated that distinct assemblages of very ancient rocks should be traceable from the coast of one continent to the coast of the adjacent continent on Pangaea. So he searched for such matches, and he found them. For example, exposures of Precambrian rocks in eastern South America look just like those that occur in western Africa **(Fig. 2.6a)**. Wegener also proposed

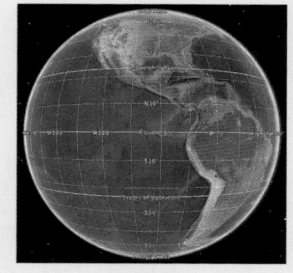

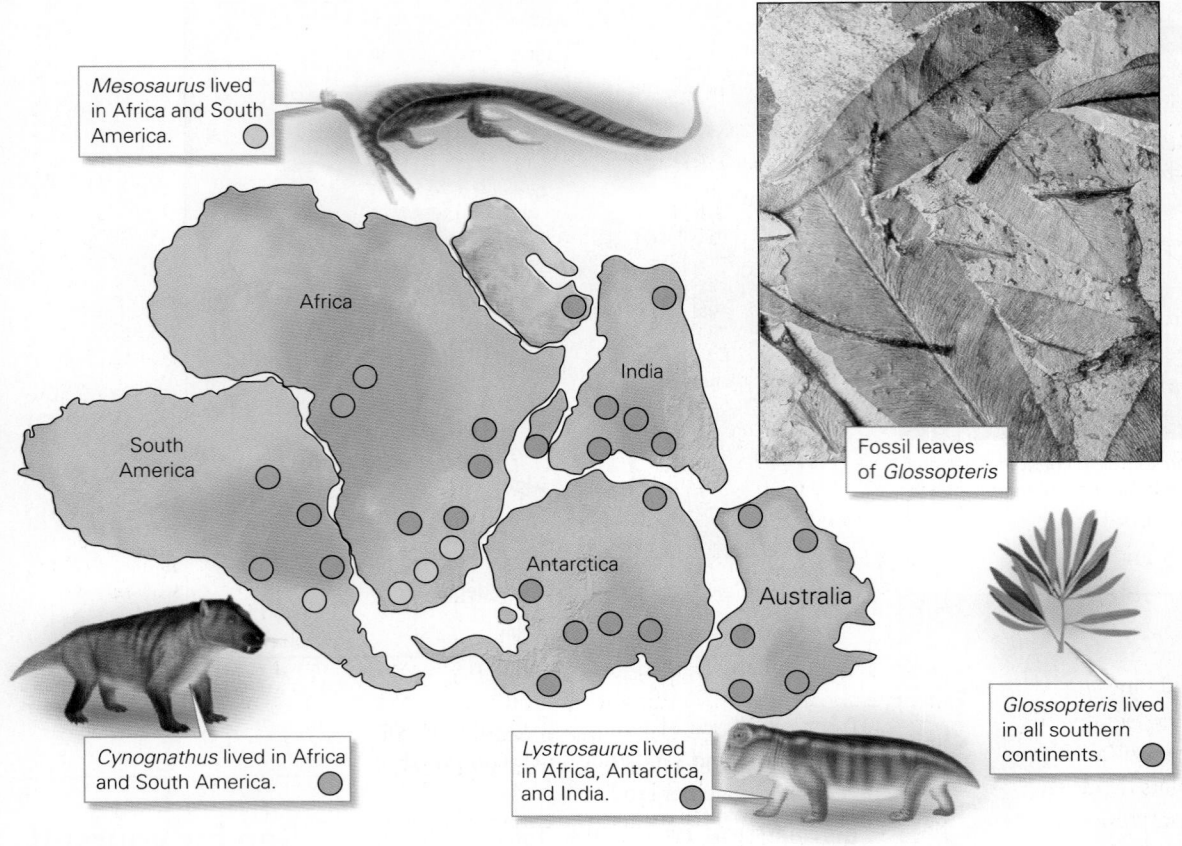

that Paleozoic mountain belts along the coastline on one side of the Atlantic Ocean should be similar in age to mountain belts bordering the opposite side. He found that the Appalachian Mountains of the United States and Canada do indeed align with similar-aged mountain belts of Great Britain, Scandinavia, and northwestern Africa on his map of Pangaea (Fig. 2.6b).

The Opposing View: Drift Denial

Though Wegener's evidence for continental drift seemed compelling, leading geologists of his day argued that it could not possibly occur because no known forces could accomplish the work of moving a continent. Wegener speculated that continents somehow moved by plowing through the rock of the ocean floor like a ship plowing through water. His opponents insisted that such a process simply could not take place, because seafloor rocks are stronger than continental rocks. So, when Wegener perished in a Greenland blizzard, most geologists remained unconvinced of his proposals. It would take new observations, made possible by research technologies not available in Wegener's day, to prove that continents do move relative to one another. The first step in this scientific revolution came with the recognition of seafloor spreading.

Take-home message...

Alfred Wegener justified his hypothesis of continental drift by comparing coastline shapes, by examining the record of past climates and fossils preserved in sedimentary rocks, and by matching rock assemblages and mountain belts on opposite coasts of the Atlantic. He argued that the continents formerly made up one supercontinent, Pangaea, that later broke apart. The hypothesis was not widely accepted until decades after Wegener's death.

Quick Questions

- What is the name of the supercontinent that Wegener recognized?
- What information did Wegener use to determine the distribution of climate belts when Pangaea existed?
- Did Wegener provide a convincing explanation of how continents moved relative to the seafloor?

FIGURE 2.6 Matching up rocks on different sides of the ocean.

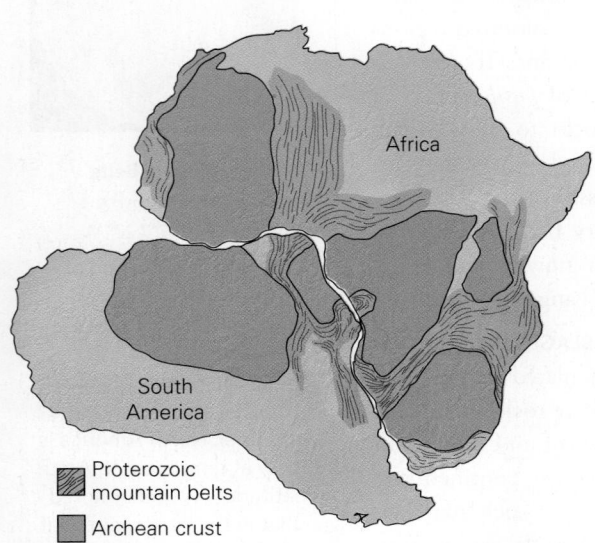

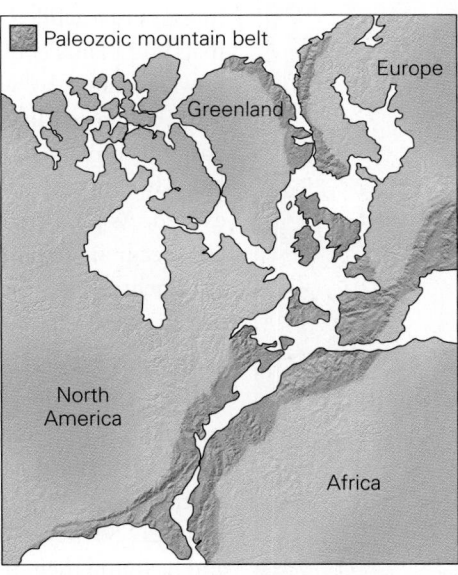

(a) On Pangaea, distinctive belts of Precambrian rock in South America align with similar ones in Africa. Note that the Precambrian includes the Archean Eon (4.0–2.5 Ga) and the Proterozoic Eon (2.5–0.5 Ga).

(b) If the Atlantic Ocean didn't exist, Paleozoic mountain belts on both of its coasts would be adjacent. The Paleozoic mountain belts shown on the map formed in stages between 460 Ma and 250 Ma (see Chapter 9).

2.2 The Discovery of Seafloor Spreading

New Geologic Observations of the Mid-20th Century

The invention of new technologies made the mid-20th century an age of discovery in Earth Science, allowing researchers to examine realms of the Earth System that previously had been barely studied. Here, we review some of the astounding discoveries that resulted from such efforts.

BATHYMETRIC FEATURES OF THE SEAFLOOR. Prior to the 20th century, the study of **bathymetry**—variation in the depth of a water body—was an extremely tedious undertaking because it required sailors to make many *soundings*, each of which involved lowering a heavy weight, attached to a cable, to the water body's floor. In deeper parts of the ocean, such a measurement—which gave the depth at only one location—could take hours. The development of modern *sonar* (echo-sounding) instruments during World War II greatly sped up the process of making depth measurements. Sonar measures the depth of the seafloor by timing how long it takes for a pulse of sound sent into the water from a ship to bounce off the seafloor and return to the ship. Using sonar, scientists can obtain a continuous record of the seafloor depth while a ship cruises, and from this record, they can quickly produce a *bathymetric profile*, a diagram that plots depth on the vertical axis against location on the horizontal axis.

In 1977, Marie Tharp, an American geologist and cartographer who worked with a marine geologist, Bruce Heezen, compiled many bathymetric profiles. From these data, she drew the first bathymetric maps that depicted the seafloor in a visually meaningful way. These maps highlight several bathymetric features:

- **Continental shelves**, bands of shallow water 50–250 km (30–150 mi) wide, underlain by submerged margins of continents, that fringe many, but not all, continental coasts

- **Ocean basins**, broad areas of the Earth's surface that underlie the deeper-water areas of oceans beyond the edges of continental shelves

- **Abyssal plains**, extensive flat regions of ocean basins that lie at depths of 4.5–5.5 km (2.8–3.4 mi)

- **Mid-ocean ridges**, long submarine mountain ranges that lie halfway between the two margins of the ocean basin in some, but not all, oceans, whose centerline, or *ridge axis*, lies at a depth of 2.0–2.5 km (1.2–1.5 mi)

- **Fracture zones**, narrow linear bands of broken-up oceanic crust that cross mid-ocean ridges and trend roughly at right angles to the ridge axis

- **Deep-sea trenches**, long troughs in which the ocean floor reaches depths of 7–11 km (4–7 mi), some, but not all, of which border continents

- **Seamounts** and **oceanic islands**, peaks of igneous rock that rise above the surrounding seafloor; if the top of a peak pokes above sea level, it forms an oceanic island, whereas if its top lies below sea level, it's a seamount.

Beginning in the 1980s, satellites carrying instruments capable of measuring ocean depth have surveyed the seafloor from space and have yielded a new generation of precise bathymetric maps **(Fig. 2.7)**. These new

See for yourself

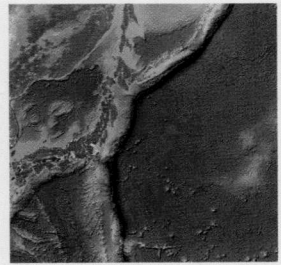

Deep-Sea Trenches
Latitude: 35°37′59.86″ N
Longitude: 145°36′12.78″ E

Fly to an elevation of 5,500 km (3,500 mi) and look straight down. Zoom in. On a laptop, you can use the elevation tool to measure trench depth. The dark bands that you see are the trenches of the western Pacific Ocean, near Japan.

FIGURE 2.7 Bathymetry of the world's oceans.

Trench · Active continental margin · Mountain range · Glacial ice sheet · Plain · Seamount chain · Plain · Continental shelf · Mid-ocean ridge · Mountain range · Trench · Abyssal plain · Active continental margin · Fracture zone · Rift · Mid-ocean ridge · Oceanic island · Mid-ocean ridge · Trench · Passive continental margin · Passive continental margin · Mid-ocean ridge · Passive continental margin · Fracture zone

Deep — Sea level — High

Bathymetry — Topography

FIGURE 2.8 Faulting and earthquake distribution.

(a) When rock inside the Earth suddenly breaks and slips, forming a fracture called a fault, vibrations radiate outward.

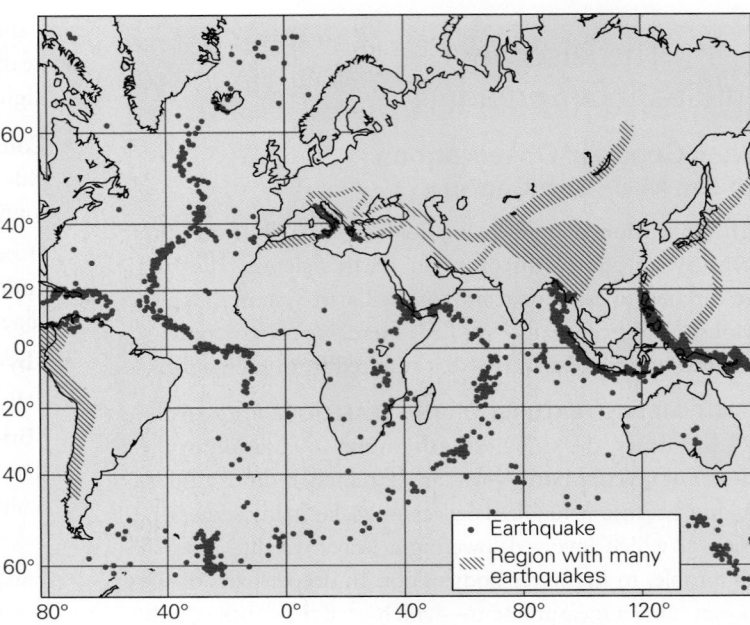

(b) A 1953 map showing the distribution of earthquake locations in the ocean basins, as recorded over a period of several years. Each red dot represents the point on the Earth's surface directly above the location where, deep underground, an episode of fault slip took place. Note that most of the red dots occur in belts.

maps confirm the existence of seafloor features that Marie Tharp first depicted, and they reveal many new ones.

CHARACTER OF OCEANIC CRUST. In addition to obtaining information about the shape of the seafloor surface, scientists dredged the seafloor, drilled into the seafloor, and used various instruments to characterize subsurface layers of the oceanic crust. This work led to the realization that sediment covers most of the ocean floor. This sediment settles out of the overlying seawater and accumulates over time to form a layer that thickens progressively away from a mid-ocean ridge. In fact, hardly any sediment occurs along the ridge axis, but a layer up to 1 km thick underlies portions of the abyssal plains. Beneath the sediment, oceanic crust consists of *basalt* and *gabbro*. Both of these names apply to dense, dark-gray igneous rocks; they differ in that basalt consists of very tiny grains, whereas gabbro consists of moderately large grains.

Scientists also measured the temperature at various depths beneath the seafloor. From these measurements, they calculated **heat flow**, the rate at which heat rises from the Earth's interior. By comparing results from many locations, scientists discovered that more heat rises beneath mid-ocean ridges than beneath abyssal plains.

DISTRIBUTION OF EARTHQUAKES. A **fault** is a fracture surface on which *slip* (sliding) takes place. **Figure 2.8a** shows a *cross section* of a fault, meaning a diagram that represents a vertical slice across the fault. The heavy line on the diagram represents the *fault trace,* meaning the intersection between the fault plane and the cross-sectional plane, and the half arrows indicate the *sense of slip* on the fault, meaning the direction in which rock on one side of the fault moved relative to rock on the other. Sudden slip on a fault generates an **earthquake**,

which produces vibrations that travel through the Earth and along the Earth's surface. These vibrations, known as *seismic waves*—from the Greek word *seismos*, which means earthquake (see Chapter 7)—cause the ground to shake. When scientists learned how to find the location of an earthquake, it became possible to produce maps that show the global distribution of earthquakes. Even the early versions of such maps, produced in the 1950s, demonstrated that earthquakes do not occur randomly, but rather cluster in distinct bands called **seismic belts** (Fig. 2.8b). When earthquake-distribution maps were compared with bathymetric maps, it immediately became clear that most seismic belts overlie distinct bathymetric features: trenches, mid-ocean ridge axes, and parts of fracture zones.

The "Essay in Geopoetry"

Harry Hess realized that the new geologic discoveries of the mid-20th century were clues that could ultimately help answer the question of how ocean basins evolved over time. He speculated that if all ocean basins were as old as the Earth itself, they should be covered with very thick sediment layers. The relative thinness of the sediment on the ocean floor implied that present-day ocean basins could not be as old as the Earth. Furthermore, the progressive increase in sediment thickness outward from a mid-ocean ridge axis implied that ocean floor gets progressively older away from a ridge, because older ocean floor has more time to accumulate sediment. These observations led Hess to conclude that new ocean floor

FIGURE 2.9 Harry Hess's 1962 concept of seafloor spreading. We will see that this sketch is an oversimplification and contains errors. For example, the sketch implied, incorrectly, that only the crust moved. But it did introduce, for the first time, the concept that ocean basins grow by seafloor spreading and that seafloor sinks into the mantle at a trench.

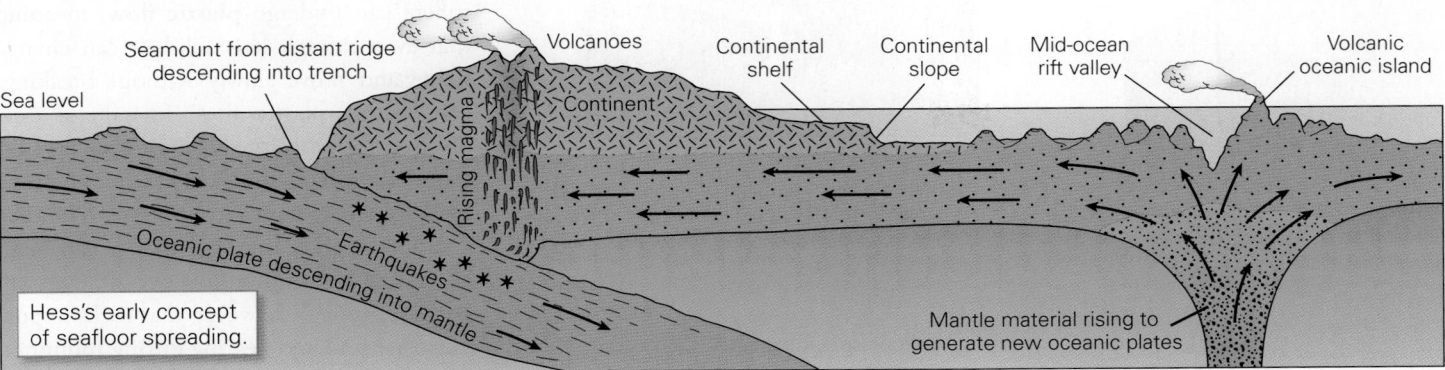

forms at a mid-ocean ridge and then moves away from the ridge, so that an ocean basin grows wider with time. How could this happen? Hess realized that the association of earthquakes with mid-ocean ridges provided an answer: the seafloor must be actively splitting apart by faulting along mid-ocean ridges. In addition, the high heat flow measured along mid-ocean ridge axes implied that hot molten rock rises beneath the zone where this splitting takes place. When solidified, this molten rock becomes new gabbro and basalt on the seafloor.

Slightly after Hess had written a manuscript describing this process of seafloor formation at mid-ocean ridges, another geologist, Robert Dietz, came to a similar conclusion and named the process **seafloor spreading**. In sum, the *seafloor-spreading hypothesis*, as proposed in the early 1960s, states that oceanic crust stretches apart and splits along the axis of a mid-ocean ridge, and that new oceanic crust forms from molten rock that rises and solidifies along the split. As it forms, new seafloor moves away from the ridge, so the ocean basin grows wider over time.

Researchers realized that if new ocean floor forms at mid-ocean ridges, then old seafloor must be consumed elsewhere—otherwise, the Earth's circumference, and therefore its volume, would have to be increasing. Hess's concept of seafloor spreading proposed that old seafloor sinks back into the mantle at deep-sea trenches, and that this movement generates the seismic belts observed adjacent to trenches (Fig. 2.9). This sinking process later came to be known as *subduction*. The Earth does not expand because the amount of new seafloor produced by seafloor spreading at mid-ocean ridges equals the amount consumed by subduction at trenches.

Hess, along with other geologists, realized that seafloor spreading provided the long-sought explanation of how continental "drift" occurs, the explanation that had eluded Wegener. Simply put, continents move apart due

to seafloor spreading at mid-ocean ridges, and continents move toward each other as the seafloor between them undergoes subduction at trenches. This overall concept seemed so elegant to Hess that he dubbed his description of it "an essay in geopoetry."

Take-home message...

Observations of seafloor bathymetry, sediment thickness, heat flow, and seismic belts led to the proposal that new seafloor forms at mid-ocean ridges, causing seafloor spreading. Old seafloor sinks back into the mantle at deep-sea trenches by the process of subduction. Continents move due to these processes.

Quick Questions —————

- What is the relationship between seismic belts and bathymetric features of ocean basins?
- What is the relationship between heat flow and a mid-ocean ridge axis?
- Which feature underlies the deepest water: a mid-ocean ridge, an abyssal plain, or a trench?

2.3 The Modern Theory of Plate Tectonics

Over the course of the 1960s, while the Beatles were dominating the pop-music charts, geologists realized that their long-held interpretations of geologic phenomena, based on the premise that the positions of continents do not change and that the ocean basins visible today date back to the birth of the Earth, were wrong! If seafloor spreading and subduction take place, and if the continents move, then the outer layer of the Earth, the shell that we live on, must be quite mobile indeed.

FIGURE 2.10 The nature and behavior of the lithosphere.

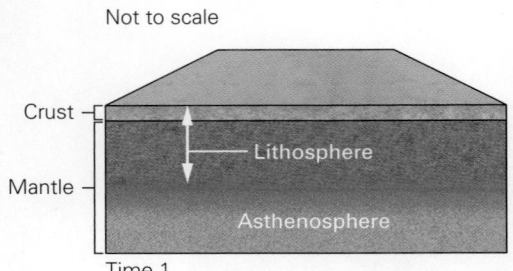

(a) The lithosphere consists of the crust plus the uppermost mantle.

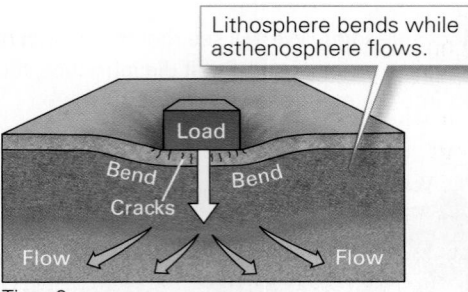

Lithosphere bends while asthenosphere flows.

(b) When a heavy load builds on the surface of the lithosphere, the lithosphere bends downward. Underlying asthenosphere can flow plastically out of the way.

Geologists distinguish between lithosphere and asthenosphere to emphasize that the two layers behave differently when acted on by a force. Specifically, asthenosphere can undergo **plastic flow**, meaning that even though it is solid, it can change shape and move slowly without breaking. (To picture plastic flow, imagine a candle transforming into a mound of wax on a hot day, even though it has not melted.) Lithosphere, in contrast, behaves overall like a **rigid** material, meaning that it can bend and break, but cannot flow overall. (We will see in Chapter 5, however, that rocks in portions of the crust that undergo heating during mountain building can undergo local plastic flow.) To picture this difference, imagine placing a large, heavy load (such as a volcano or glacier) on the surface of the lithosphere—the lithosphere would bend downward and might undergo some cracking, while the underlying asthenosphere would slowly flow out of the way (Fig. 2.10b).

The lithospheric mantle consists of the same kind of rock (a dark and very dense rock known as *peridotite*) as the underlying asthenosphere. Why, then, does the lithospheric mantle behave like a rigid material while the asthenosphere can flow plastically? At the lithosphere's base, peridotite reaches a temperature of about 1,280°C (2,340°F). At temperatures hotter than 1,280°C, peridotite can flow plastically, whereas at cooler temperatures, it cannot. So, although the lithospheric mantle and the asthenosphere consist of the same type of rock, the layers behave differently (one rigidly, the other plastically) because of their respective temperatures.

Confusingly, the depth of the base of the asthenosphere has not been well defined. Some geologists equate it with the upper mantle down to a depth of just a few hundred kilometers, the depth to which flow of the asthenosphere accommodates displacements of the lithosphere-asthenosphere boundary (see Fig. 2.10b). Others extend its depth down to around 660 km, below which the key characteristics of the mantle change. Still others apply the term asthenosphere to the entire mantle (including the lower mantle) below the lithosphere, to emphasize that all of the sublithospheric mantle flows plastically. For simplicity, in this book, we will use the definition that extends the asthenosphere down to a depth of about 660 km.

PLATES OF THE EARTH. The Earth's lithosphere contains several major breaks that separate it into *lithosphere plates,* or simply **plates** (Fig. 2.11a). Geologists refer to the dividing line between one plate and its neighbor as a **plate boundary**, and to the portion of a plate that does not lie near a plate boundary as a *plate interior*. Presently, the Earth's lithosphere can be divided into seven major plates,

By 1968, ideas concerning seafloor spreading and continental drift coalesced into a new model of how the Earth works. In this model, our planet's outer shell consists of separate pieces, or *plates*, that move relative to one another. After the model had passed many tests, it came to be known as the *theory of plate tectonics*, or simply **plate tectonics**. Eventually, this theory not only explained all of Wegener's observations, but also explained many major geologic phenomena, such as earthquakes, volcanoes, and mountain building. In effect, plate tectonics became the "grand unifying theory" of geology. In this section, we describe several key aspects of the theory. We begin by clarifying what we mean by a plate.

What Is a Lithosphere Plate?

DISTINGUISHING LITHOSPHERE FROM ASTHENOSPHERE. Geologists back in the 19th century realized that the interior of the Earth consists of distinct layers (crust, mantle, and core) that differ in composition, as we discussed in Chapter 1. Not surprisingly, when Harry Hess began thinking about seafloor spreading, subduction, and moving continents, he assumed that movement involved only the crust, the outermost layer (as depicted in Fig. 2.9). Soon, though, geologists realized that this image is incorrect. Rather, the movement involves the **lithosphere**, which consists of the entire crust together with the underlying **lithospheric mantle**, which is the topmost part of the upper mantle. Together, the crust and the uppermost mantle behave as a relatively rigid layer that moves (Fig. 2.10a). The Earth's crust, as described in Chapter 1, has a thickness of 7–10 km (4–6 mi) beneath oceans and 25–70 km (15–43 mi) beneath continents. The lithosphere, in contrast, attains a thickness of up to 100 km (60 mi) beneath the oceans and typically 150 km (90 mi) beneath the continents. The lithosphere sits on another layer, called the **asthenosphere**, which is entirely within the mantle. (Notably, geologists now realize that, beneath continents, at least the upper 100 km (62 mi) of the asthenosphere may move with the plates. To simply our discussion, in this book we will depict plates as consisting only of lithosphere.)

each several thousand kilometers across, and eight minor plates, which are hundreds of kilometers to 2,000 km across. In addition, some geologically complex regions of our planet include several *microplates,* most of which are less than a few hundred kilometers across, but some of which approach the size of minor plates. Some plates, such as the North American Plate and the African Plate, have familiar-sounding names, but others, such as the Cocos Plate and the Juan de Fuca Plate, don't. Significantly, as Figure 2.11a shows, some plates consist entirely of oceanic lithosphere, whereas others consist of both oceanic and continental lithosphere.

Geologists refer to the boundary region between a continent and the adjacent ocean basin as a **continental margin**. Some continental margins align with plate boundaries, but others do not. Consequently, geologists distinguish between **active margins** of continents, which delineate plate boundaries, and **passive margins** of continents, which do not. A passive margin forms when continental lithosphere is stretched, becomes thinner, and then breaks during the initial formation of an ocean basin. (We'll discuss this process, called *rifting,* in Section 2.5.) The stretched continental lithosphere slowly sinks to form a depression called a **passive-margin basin** (Fig. 2.11b). Deposits of sediment accumulate in the passive-margin basin, as its floor sinks, to form a thick wedge of sediment. These deposits include sand and mud carried to the sea by rivers, as well as shells of dead marine organisms. The surface of a passive-margin basin sediment wedge tends to be a broad continental shelf (Fig. 2.11c).

Identifying Plate Boundaries

How do we recognize the location of a plate boundary? All plate boundaries coincide with seismic belts. Earthquakes happen at plate boundaries because slip along a plate boundary accommodates the movement of a plate relative to its neighbors (Fig. 2.12). Plate interiors are relatively earthquake-free.

Geologists classify plate boundaries into three types based on the motion of the plate on one side of the boundary relative to the plate on the other (Fig. 2.13). A boundary at which two plates move apart from each other and new oceanic lithosphere forms is a *divergent boundary,* a boundary at which two plates move toward each other so that an oceanic plate sinks beneath another plate is a *convergent boundary,* and a boundary at which one plate slips horizontally along the edge of another plate is a *transform boundary.*

Plate Tectonics Today: A Synopsis

It took years of work by many researchers around the world to refine the theory of plate tectonics (Earth Science at a Glance, pp. 62–63). Considering our introduction

FIGURE 2.11 Plate boundaries, plate interiors, and continental margins.

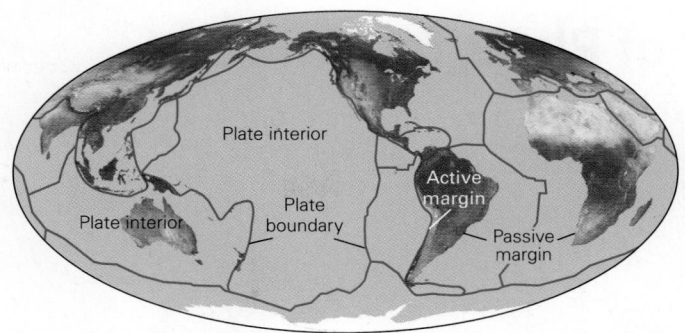

(a) An image of the globe on which some of the major plate boundaries are highlighted. Active continental margins lie along plate boundaries; passive margins do not.

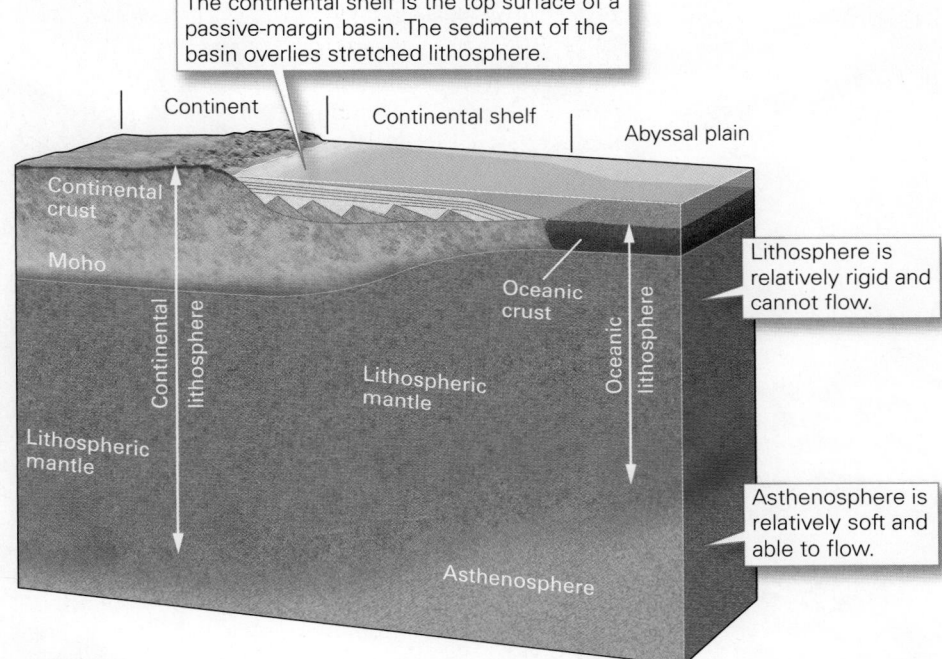

The continental shelf is the top surface of a passive-margin basin. The sediment of the basin overlies stretched lithosphere.

Lithosphere is relatively rigid and cannot flow.

Asthenosphere is relatively soft and able to flow.

(b) A passive-margin basin forms where continental lithosphere is stretched, thinned, and buried by sediment.

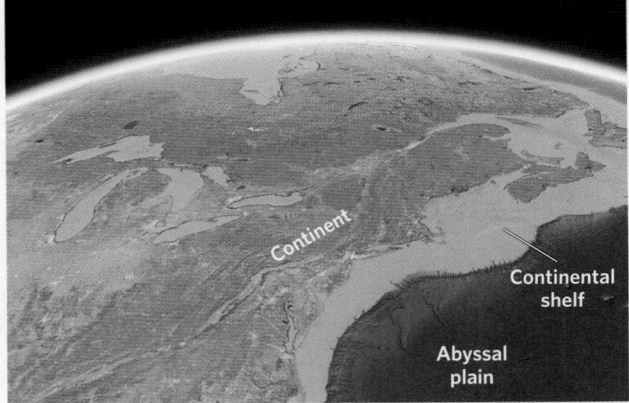

(c) A bathymetric image of the continental shelf along the east coast of North America. This bathymetric feature overlies a broad passive-margin basin.

The Theory of Plate Tectonics

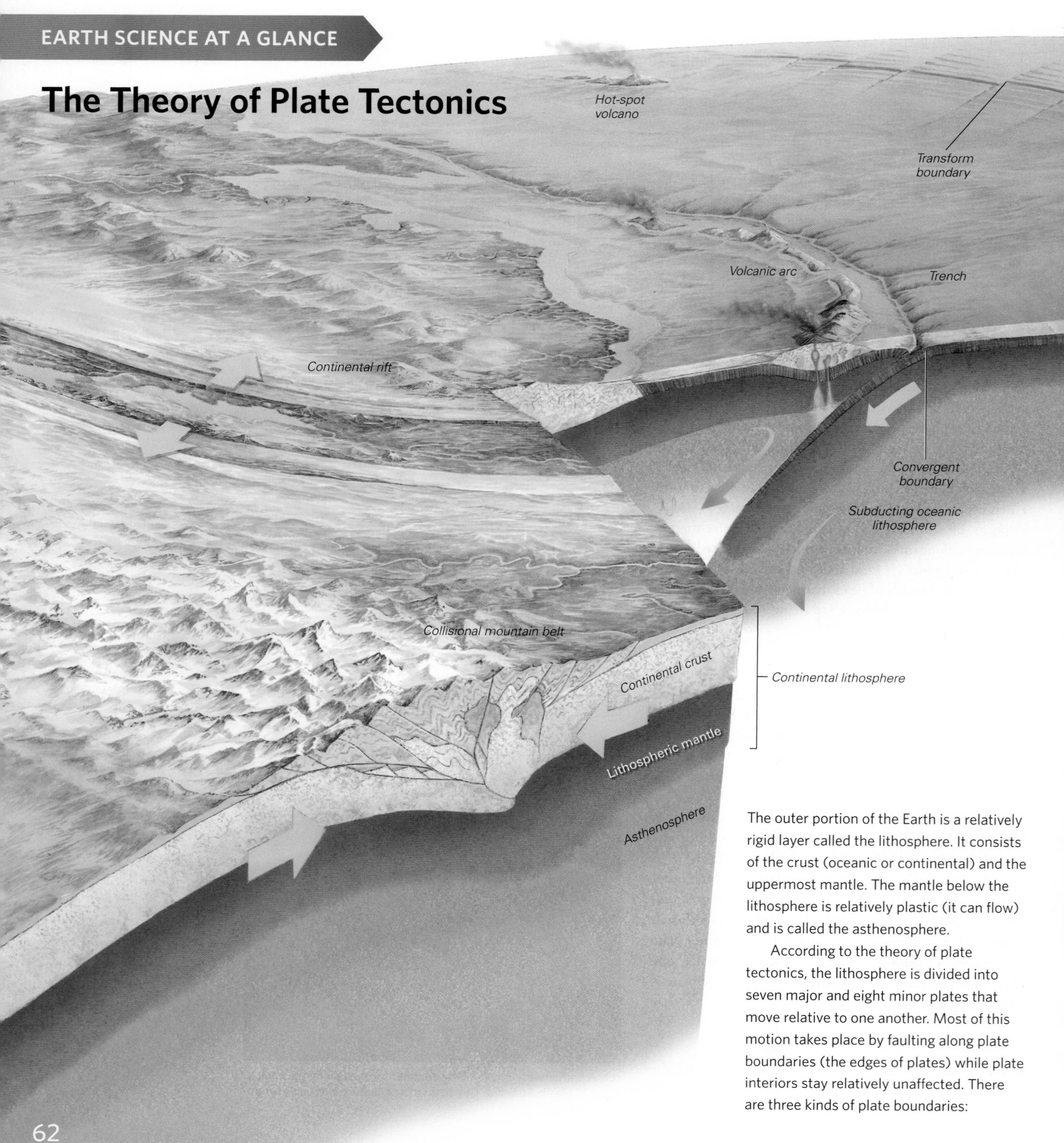

Hot-spot volcano

Transform boundary

Volcanic arc

Trench

Continental rift

Convergent boundary

Subducting oceanic lithosphere

Collisional mountain belt

Continental crust

Continental lithosphere

Lithospheric mantle

Asthenosphere

The outer portion of the Earth is a relatively rigid layer called the lithosphere. It consists of the crust (oceanic or continental) and the uppermost mantle. The mantle below the lithosphere is relatively plastic (it can flow) and is called the asthenosphere.

According to the theory of plate tectonics, the lithosphere is divided into seven major and eight minor plates that move relative to one another. Most of this motion takes place by faulting along plate boundaries (the edges of plates) while plate interiors stay relatively unaffected. There are three kinds of plate boundaries:

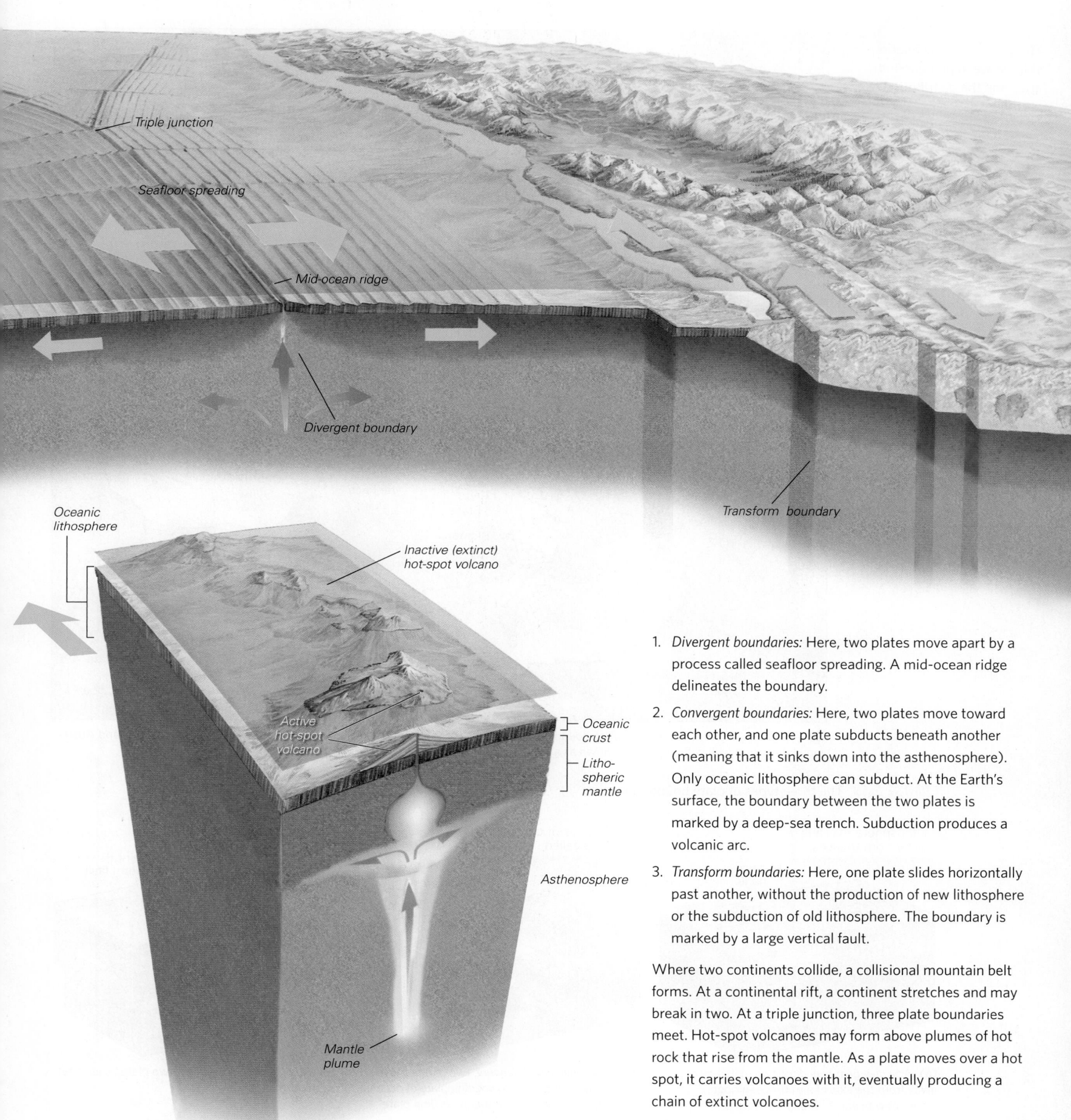

Triple junction

Seafloor spreading

Mid-ocean ridge

Divergent boundary

Transform boundary

Oceanic lithosphere

Inactive (extinct) hot-spot volcano

Active hot-spot volcano

Oceanic crust

Lithospheric mantle

Asthenosphere

Mantle plume

1. *Divergent boundaries:* Here, two plates move apart by a process called seafloor spreading. A mid-ocean ridge delineates the boundary.

2. *Convergent boundaries:* Here, two plates move toward each other, and one plate subducts beneath another (meaning that it sinks down into the asthenosphere). Only oceanic lithosphere can subduct. At the Earth's surface, the boundary between the two plates is marked by a deep-sea trench. Subduction produces a volcanic arc.

3. *Transform boundaries:* Here, one plate slides horizontally past another, without the production of new lithosphere or the subduction of old lithosphere. The boundary is marked by a large vertical fault.

Where two continents collide, a collisional mountain belt forms. At a continental rift, a continent stretches and may break in two. At a triple junction, three plate boundaries meet. Hot-spot volcanoes may form above plumes of hot rock that rise from the mantle. As a plate moves over a hot spot, it carries volcanoes with it, eventually producing a chain of extinct volcanoes.

FIGURE 2.12 The relation-
ship between plate bound-
aries and the distribution of
earthquakes.

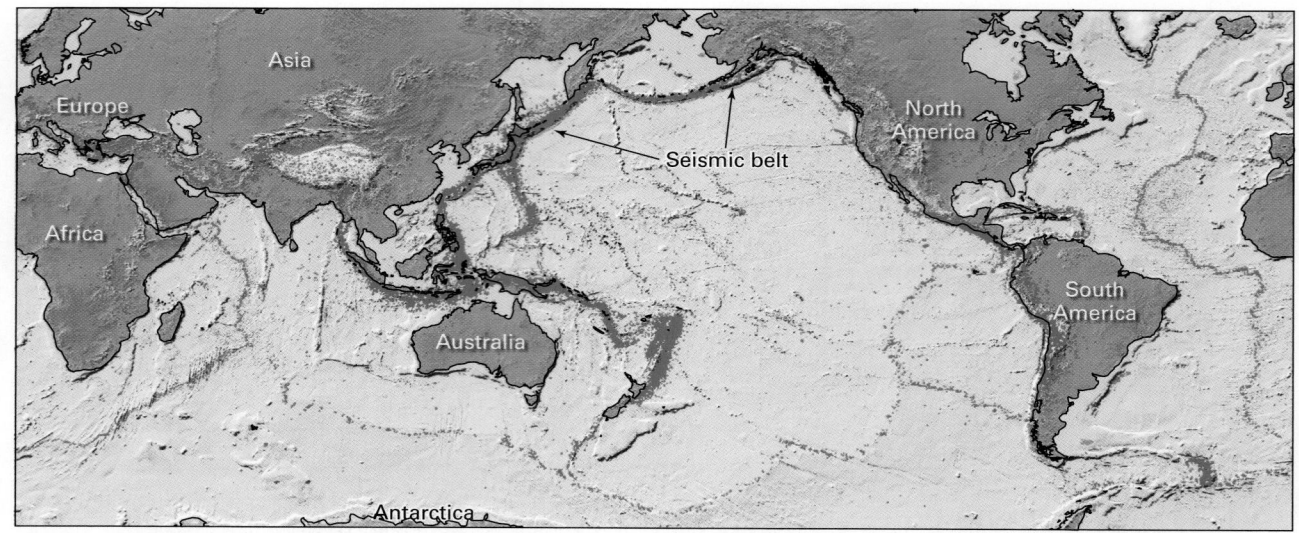

(a) The locations of most earthquakes (red dots) fall into distinct seismic belts that correspond to plate boundaries. Relatively few earthquakes occur in the stabler plate interiors.

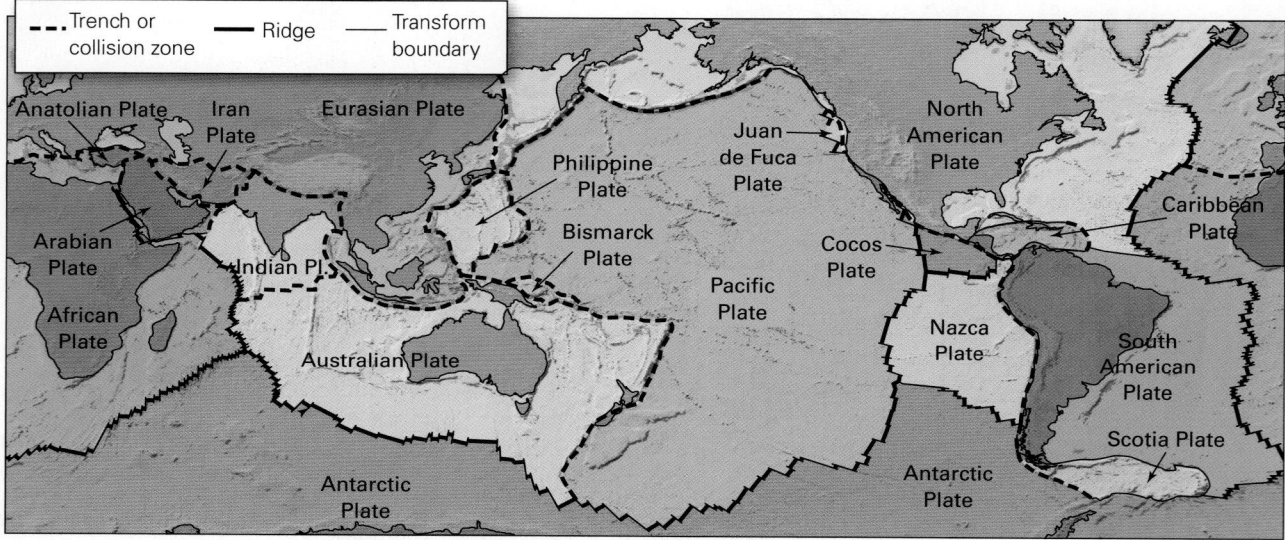

(b) A map of the plates shows that plate boundaries match the locations of seismic belts. The boundary between the Indian and Austra-lian plates is not well defined, so it is shown as a dashed line.

FIGURE 2.13 The three types of plate boundaries.

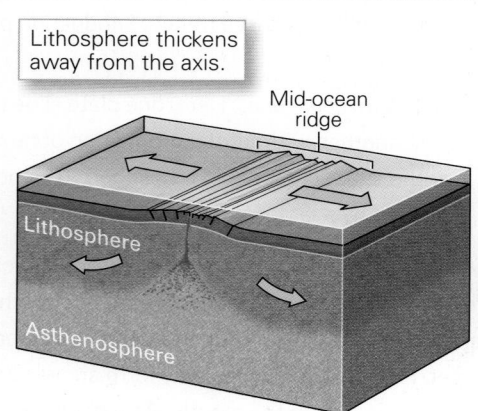

(a) At a *divergent boundary*, two plates move away from the axis of a mid-ocean ridge, where new oceanic lithosphere forms.

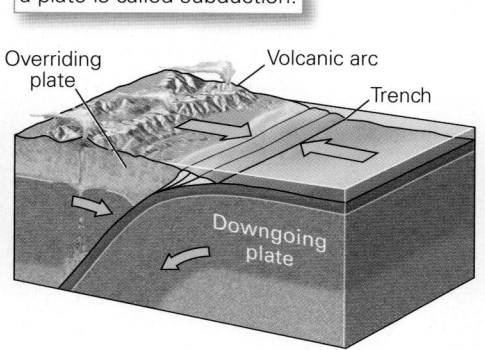

(b) At a *convergent boundary*, two plates move toward each other; the downgoing plate of oceanic lithosphere sinks beneath the overriding plate.

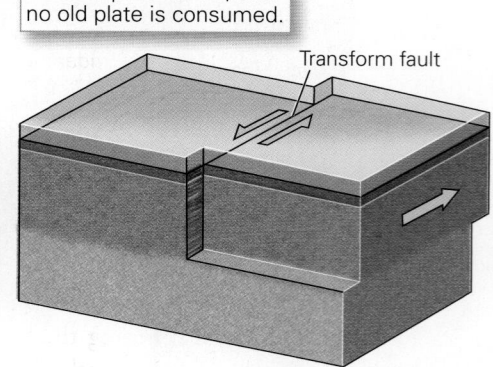

(c) At a *transform boundary*, two plates slide past each other on a vertical fault.

to lithosphere, asthenosphere, plates, and plate boundaries, we can summarize the key components of the theory as follows:

- The Earth's lithosphere, its rigid outer shell, consists of the crust and underlying lithospheric mantle. The lithosphere is divided into plates that move relative to one another.

- Plates are on the order of 100–150 km (60–90 mi) thick and range from hundreds to thousands of kilometers across. Some consist of only oceanic lithosphere, whereas others consist of both oceanic and continental lithosphere.

- Plates sit on the underlying asthenosphere, which is warm enough to flow plastically.

- As plates move, interactions at plate boundaries cause earthquakes and other geologic phenomena, but plate interiors remain mostly intact. The distribution of seismic belts, therefore, defines the positions of plate boundaries.

- Geologists distinguish among three types of plate boundaries—convergent, divergent, and transform—based on relative plate motions across the boundary.

- The movement of plates results in the movement of continents relative to one another (what Wegener called "continental drift"). Because of plate-tectonic processes, the map of the Earth's surface constantly changes.

Take-home message...

According to the theory of plate tectonics, the Earth's lithosphere—its rigid outer shell—is divided into plates that move relative to one another. As a plate moves, its interior remains mostly unchanged, but slip occurs along the plate boundary, causing earthquakes. There are three types of plate boundaries, distinguished from each other by relative plate movement at the boundary.

Quick Questions

- What does the lithosphere consist of? How does it differ from the asthenosphere?
- Name the seven major plates.
- Name the three types of plate boundaries.

2.4 Geologic Features of Plate Boundaries

We've seen that geologists distinguish among three types of plate boundaries by characterizing the motion of the plates on either side relative to each other. Geologic processes taking place at plate boundaries generate many distinct features, which we now explore.

Divergent Boundaries

In 1872, HMS *Challenger* set sail from England to begin the world's first oceanographic research cruise. Every now and then the ship's crew took a sounding. The results hinted that a submerged mountain range ran down the center of the Atlantic Ocean basin. This feature, now known as the *Mid-Atlantic Ridge*, rises about 2 km (1.2 mi) above abyssal plains, so its crest typically lies at depths of 2.5–3.0 km (1.6–1.9 mi). It extends from the latitude of far northern Greenland to the latitude of far southern South America (Fig. 2.14a, b). Significantly, the axis of this ridge is not a continuous line. Rather, along its length, it consists of distinct *ridge segments* that terminate at fracture zones. Furthermore, the ridge segments do not line up, so one segment's end may be a few kilometers to several hundred kilometers east or west of the next segment's end (Fig. 2.14c).

Bathymetric studies over the next several decades found that all the major oceans contain such ridges, which came to be known in general as *mid-ocean ridges*, since they are all within ocean basins even though not all follow the center line of the basins. The ridge of the eastern Pacific Ocean was named the *East Pacific Rise* because, in contrast to other mid-ocean ridges, it has a relatively broad, smooth crest. All mid-ocean ridges are segmented, with segments ending at fracture zones.

Until the theory of plate tectonics came along, no one knew why mid-ocean ridges existed. Plate tectonics, however, clearly explains their origin. A mid-ocean ridge delineates a **divergent boundary**, also known as a *spreading boundary*, because at this type of plate boundary, two oceanic plates move apart by seafloor spreading. (Note that we use the term "divergent boundary" only for mid-ocean ridges. Places where two parts of a continent are moving apart, but no ocean lithosphere has formed, are called continental rifts, as we discuss later.) Note that during spreading, open space never develops between diverging plates. Rather, as the plates move apart, new oceanic lithosphere forms between them (Fig. 2.15a). Generally, a narrow valley, bordered by faults, delineates the centerline of the ridge axis (Fig. 2.15b), and numerous faults occur on both sides of the ridge axis. Overall, the seafloor on each side of the ridge axis slopes gently away from it, reaching abyssal-plain depth at 700–1,200 km (430–750 mi) from the ridge axis. Mid-ocean ridges are roughly symmetrical, in that the bathymetry on one side mirrors the bathymetry on the other side.

How does oceanic crust form at a mid-ocean ridge? As seafloor spreading takes place, hot asthenosphere rises beneath the ridge. As it rises, it begins to melt, for reasons we'll explain in Chapter 4; geologists refer to such molten rock, while it resides underground, as **magma**. The magma produced in the asthenosphere rises still higher and accumulates to form a *magma chamber*, below the

See for yourself

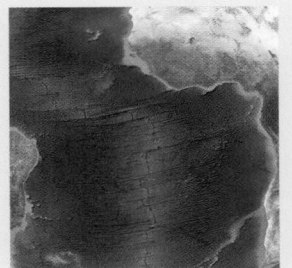

The Mid-Atlantic Ridge

Latitude: 2°34′26.28″ S
Longitude: 15°37′59.86″ W

Fly to an elevation of 11,300 km (6,800 mi) and look straight down to see a view of the Mid-Atlantic Ridge in the equatorial Atlantic Ocean. Zoom in to see a fracture zone.

FIGURE 2.14 Bathymetry of a mid-ocean ridge.

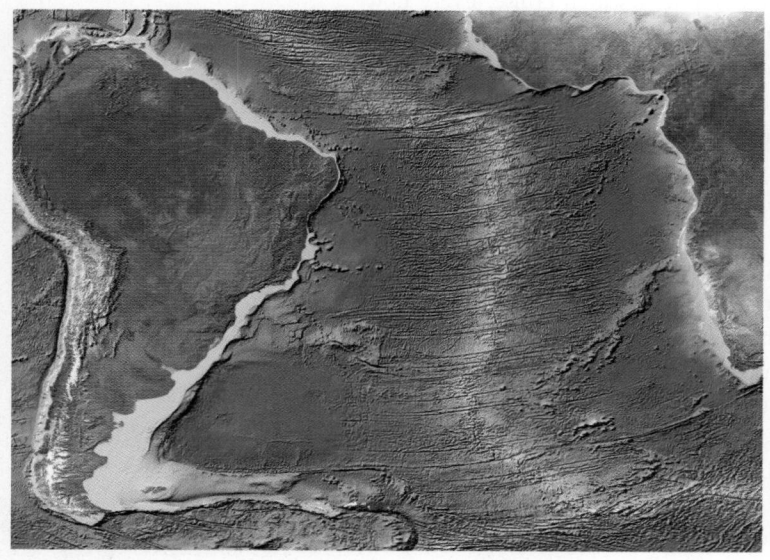
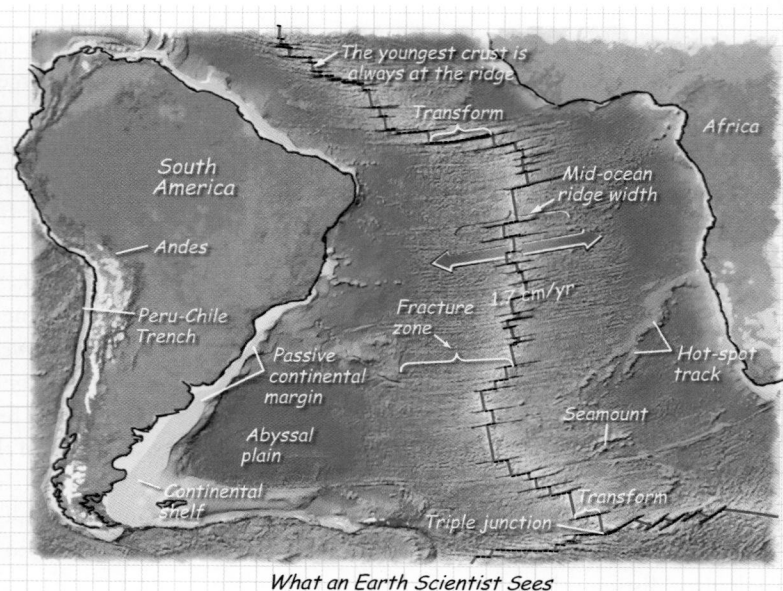

What an Earth Scientist Sees

(a) A bathymetric map of the Mid-Atlantic Ridge in the South Atlantic Ocean. The lighter shades of blue are shallower water depths. Red lines indicate the axis of the ridge. The sketch on the right labels key features.

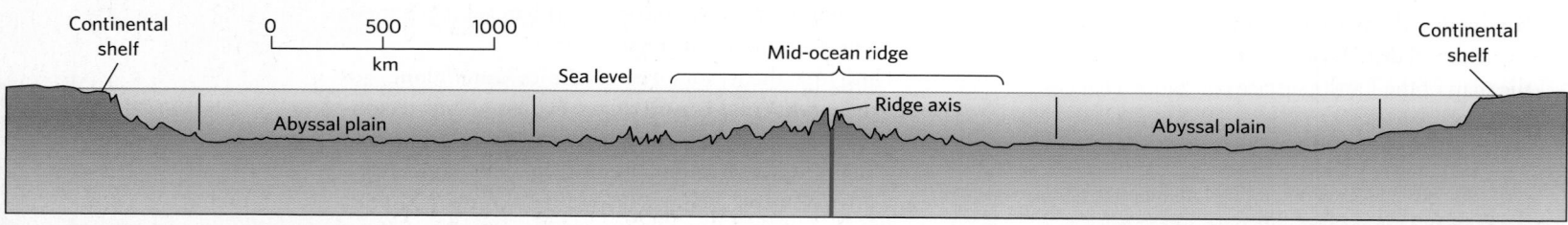

(b) A bathymetric profile (vertical slice) across the Mid-Atlantic Ridge, from continent to continent. Note that the ridge is higher than the abyssal plain. The red line indicates the location of rising magma.

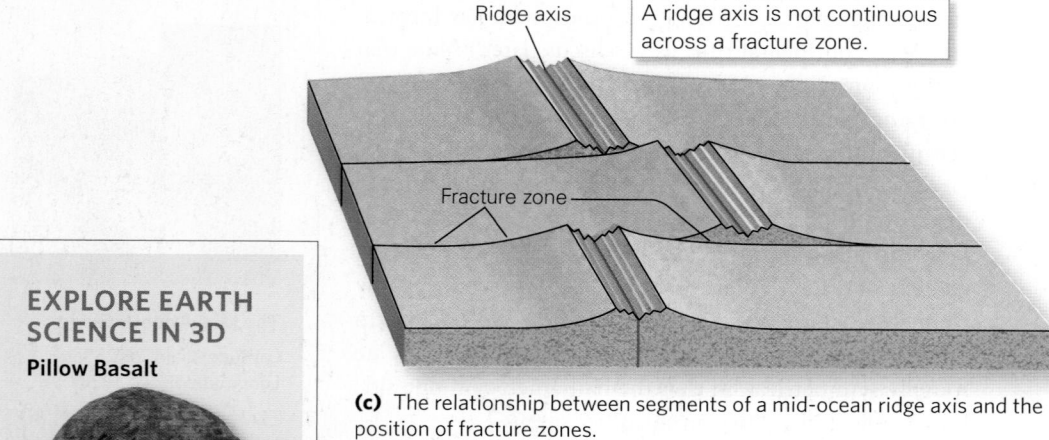

(c) The relationship between segments of a mid-ocean ridge axis and the position of fracture zones.

EXPLORE EARTH SCIENCE IN 3D

Pillow Basalt

ridge axis (see Fig. 2.15b). Some of this magma solidifies into gabbro along the sides of the chamber, but some is injected into vertical cracks that trend parallel to the ridge axis, where it solidifies into wall-like sheets, called *dikes*, made of basalt. (As discussed in Chapter 4, basalt and gabbro have the same composition, but the former has smaller grains because it cools faster.) Some magma rises all the way up to the seafloor and seeps out on the surface; geologists refer to molten rock that has come out onto the surface as **lava**. The lava, which emerges from small submarine volcanoes along the ridge axis, cools quickly when it comes in contact with cold seawater and forms a layer of rounded basalt blobs known as *pillows*, which accumulate into mounds of *pillow basalt*. The entire layer of igneous rock formed at the ridge axis, from the base of the gabbro, up through the basalt dikes, to the top of the pillow basalt, represents new oceanic crust.

As it forms, oceanic crust moves away from the ridge axis, and as this happens, more asthenosphere rises, more melt forms and rises, and still more crust forms. In other words, as if driven by a vast conveyor belt, oceanic crust forms at the ridge and then moves away from the ridge.

FIGURE 2.15 The process of seafloor spreading at a mid-ocean ridge.

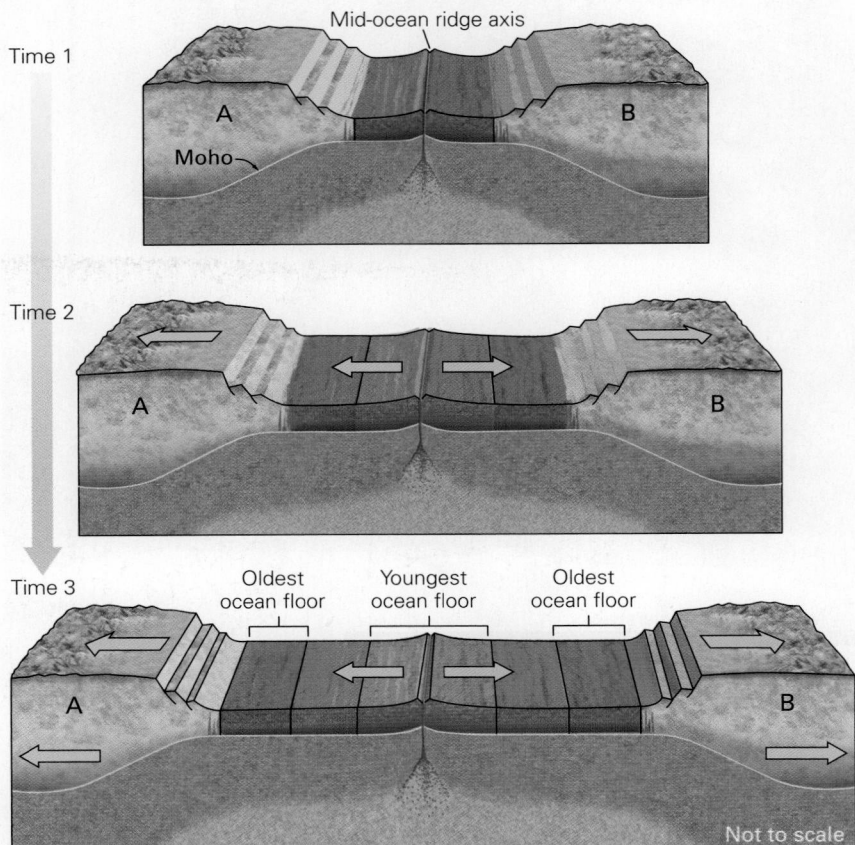

(a) During seafloor spreading, the ocean floor gets wider and continents on either side move apart. New oceanic crust forms at the ridge axis.

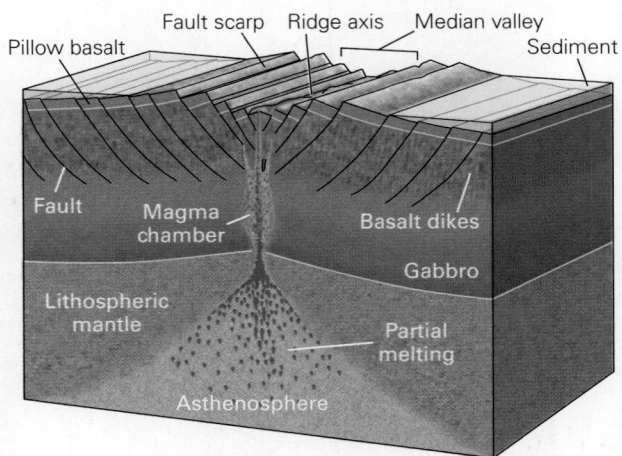

(b) Beneath a mid-ocean ridge is a magma chamber. Gabbro forms on the side of the magma chamber. Basalt dikes protrude upward. Pillow basalt forms on the seafloor surface.

FIGURE 2.16 Changes accompanying the aging of lithosphere.

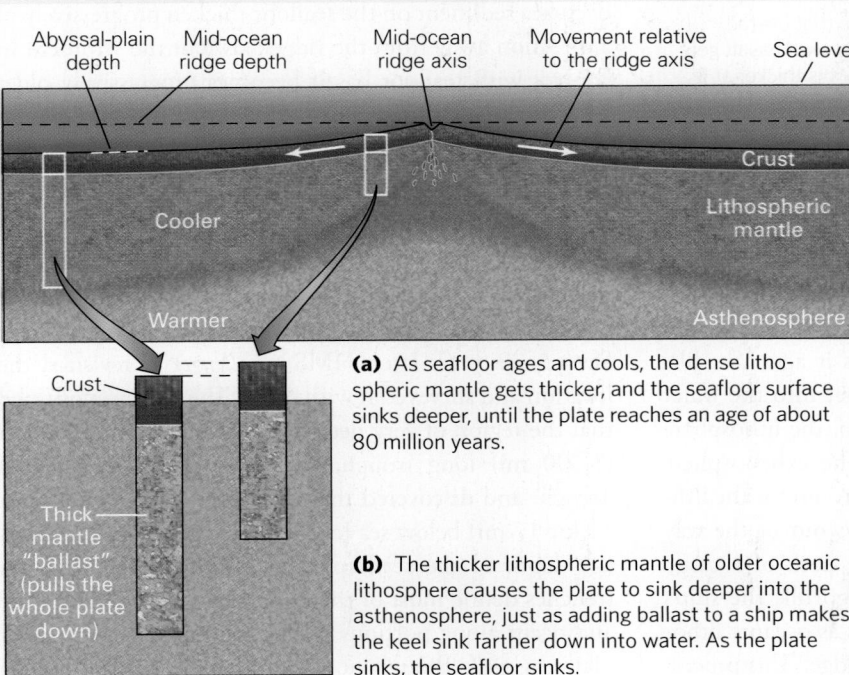

(a) As seafloor ages and cools, the dense lithospheric mantle gets thicker, and the seafloor surface sinks deeper, until the plate reaches an age of about 80 million years.

(b) The thicker lithospheric mantle of older oceanic lithosphere causes the plate to sink deeper into the asthenosphere, just as adding ballast to a ship makes the keel sink farther down into water. As the plate sinks, the seafloor sinks.

The stretching force applied to newly formed solid crust as spreading takes place breaks up this new crust, resulting in the formation of faults bordering and running parallel to the ridge axis. Slip on these faults accounts for many divergent-boundary earthquakes.

Right at the ridge axis, a lithosphere plate consists only of new crust formed by solidification of molten rock that rose from the asthenosphere. Once this crust moves away from the ridge axis, no additional igneous rock is added to it, so the crust's thickness stays the same. But the *peridotite* (dark, dense mantle rock) directly beneath the crust progressively loses heat as it moves with the crust away from the ridge axis. Eventually, as this heat loss causes the peridotite to cool below about 1,280°C, it begins to behave rigidly and becomes lithospheric mantle. The lithospheric mantle thickens progressively as the plate continues to move away from the ridge axis and continues to cool. Therefore, oceanic lithosphere, overall, thickens as it ages (Fig. 2.16). The process of thickening stops once a plate is about 80 million years old

FIGURE 2.17
Age of the
seafloor and
its relation
to sediment
thickness.

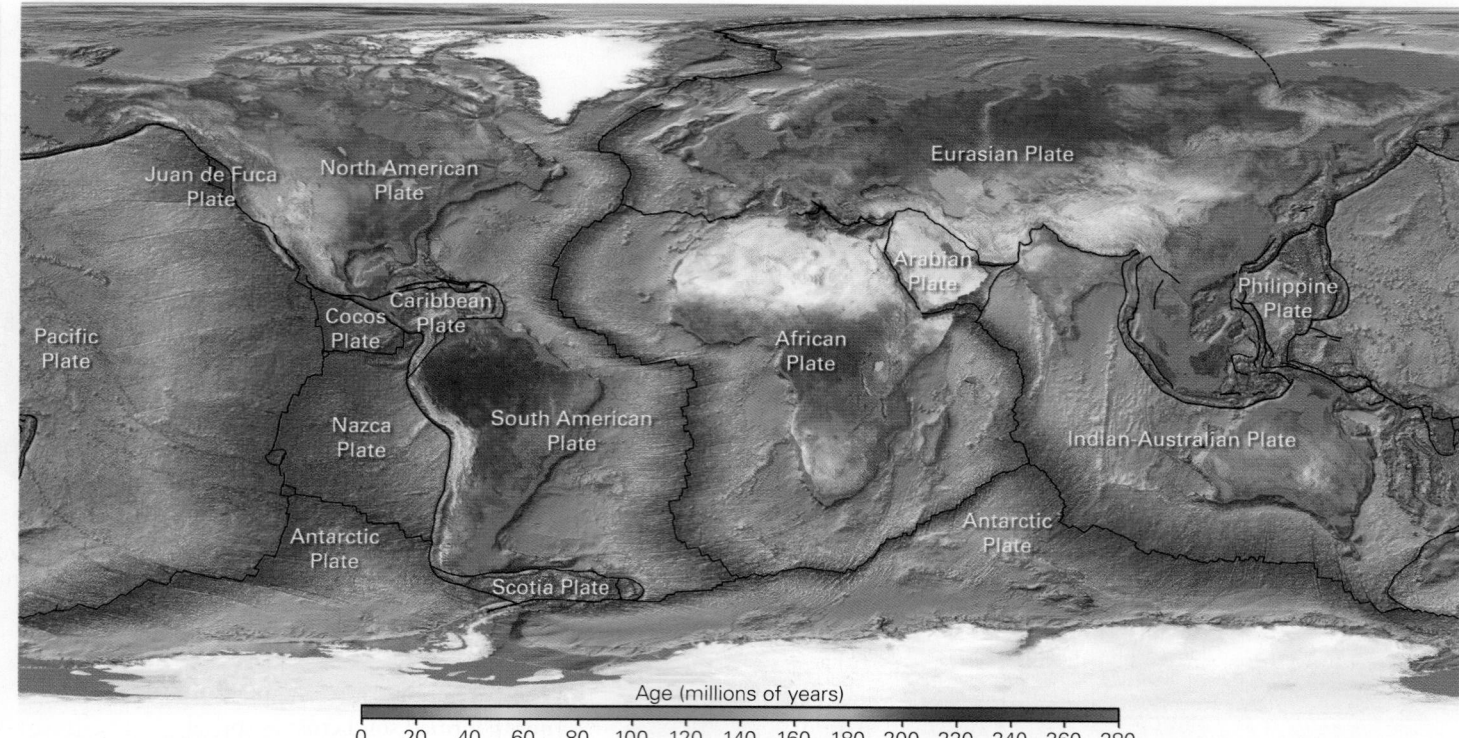

Age (millions of years)

0 20 40 60 80 100 120 140 160 180 200 220 240 260 280

(a) Note that the seafloor grows older with distance from a mid-ocean ridge axis.

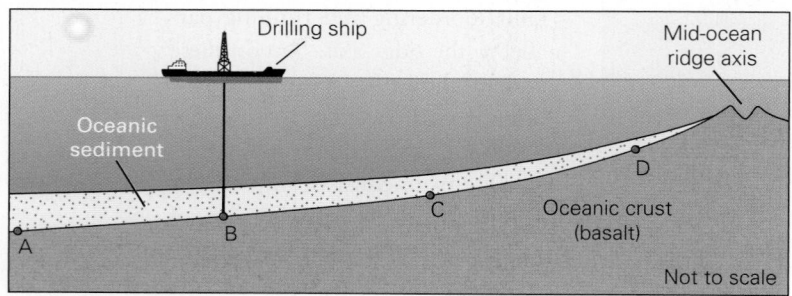

(b) Drilling into the sediment layer of the ocean floor confirms that overall, seafloor sediment gets thicker, and the basal sediment in contact with basalt gets older, the farther away a location lies from the ridge. So, sediment is thicker at B than at C or D, and the oldest sediment at A is older than the oldest sediment at the other points.

because at that time, the heat being added to the base of the plate by the underlying asthenosphere equals the heat being lost by the surface of the plate to the ocean. Significantly, because rock becomes denser as it cools, the weight of a column of lithosphere increases as it ages. So, like the keel of a cargo ship that sinks deeper into the water as workers fill the hold with heavy cargo, the lithosphere settles deeper into the asthenosphere. (The asthenosphere flows plastically out of the way to make room for the lithosphere, somewhat like water that flows out of the way of a cargo ship.) Therefore, just as the deck of a cargo ship moves downward when heavy cargo fills the ship's hold, the depth of the seafloor increases as oceanic lithosphere moves away from a mid-ocean ridge. This process

explains why abyssal plains, which are underlain by old, cold lithosphere, are deeper than mid-ocean ridges, which are underlain by young, warm lithosphere.

Because all seafloor forms at mid-ocean ridges, the youngest seafloor in an ocean basin borders the ridge axis, and the oldest seafloor lies farthest from the ridge (Fig. 2.17a). For this reason, not only does the layer of deep-sea sediment on the seafloor thicken progressively in a direction away from the ridge axis, but the sediment in contact with seafloor basalt becomes progressively older with increasing distance from the ridge. This observation serves as a proof of seafloor spreading (Fig. 2.17b).

Convergent Boundaries

During its epic voyage of discovery, HMS *Challenger* made many soundings in the western Pacific, and it found a location where ocean depth exceeded 8 km (5 mi). Seventy-six years later, HMS *Challenger II* revisited the location and surveyed it with sonar. The survey confirmed that the region of very deep water delineated a 2,500 km (1,500 mi) long trough, now known as the Mariana Trench, and discovered that its deepest point lies almost 11 km (7 mi) below sea level—more than twice the average depth of the seafloor (Fig. 2.18a)! Similar deep-sea trenches define most of the boundary of the Pacific Plate and occur along portions of the margins of several other plates as well, though none are as deep as the Mariana

FIGURE 2.18 The bathymetry of a convergent boundary.

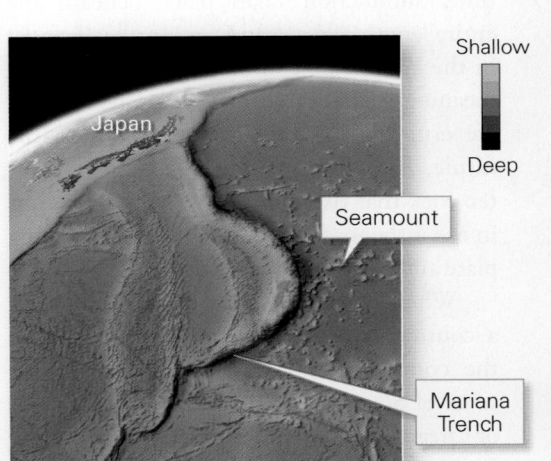

(a) A bathymetric map of the western Pacific, showing the trace of the Mariana Trench. Dark blue represents very deep water.

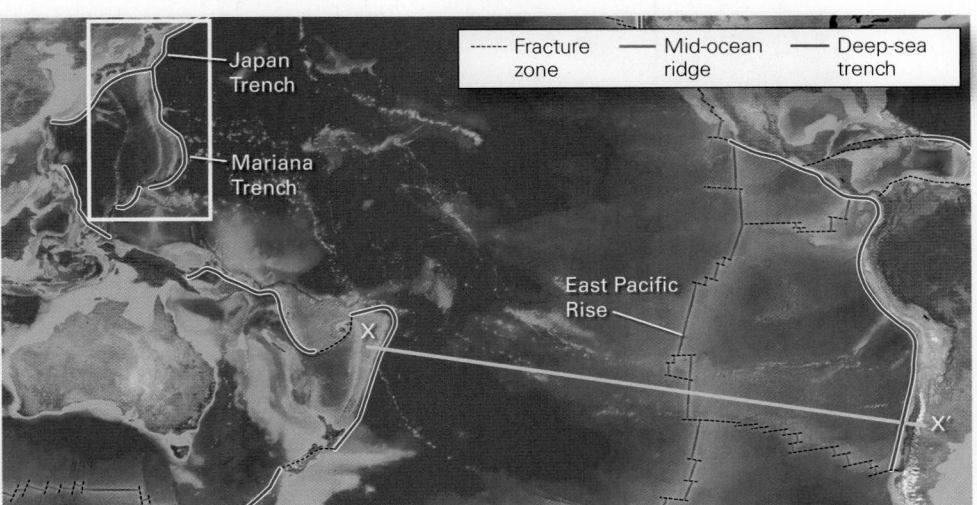

(b) Trenches occur on both sides of the Pacific Ocean. The yellow line shows the location of the bathymetric profile in part (c). The white rectangle shows the area of (a).

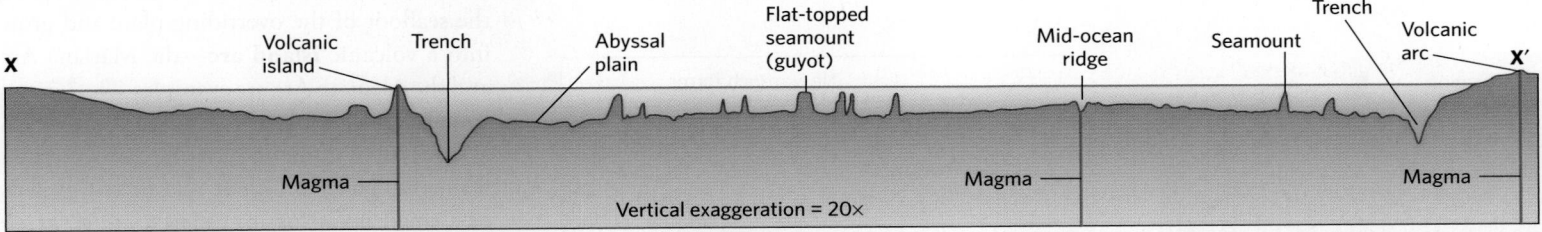

(c) A bathymetric profile showing the contrast between the depth of a trench and the depth of an abyssal plain. This profile extends from the Tonga Trench, east of Australia, across the Pacific to the coast of South America. Red lines indicate rising magma.

Trench (Fig. 2.18b,c). A **volcanic arc**, a chain of volcanoes that, in many locations, traces out a curve in map view, borders one side of a trench. Massive earthquakes happen with unnerving frequency in the vicinity of trenches and volcanic arcs.

As was the case with mid-ocean ridges, the origin and geologic meaning of deep-sea trenches remained a mystery before the advent of plate tectonics. But once the theory had been proposed, geologists realized that trenches mark locations where the lithosphere of an oceanic plate bends downward, slides under the edge of the adjacent plate, and sinks into the asthenosphere. As a consequence, plates on either side of a trench move toward each other. Due to this relative motion, geologists refer to the plate boundary associated with a trench as a **convergent boundary** (Fig. 2.19a). Geologists refer to the bending downward and sinking of the *downgoing plate* beneath the *overriding plate* as **subduction**, so a convergent boundary can also be called a **subduction zone**. (As we will see later in this chapter, when relatively buoyant, non-oceanic lithosphere enters a subduction zone,

subduction ceases and the convergent boundary evolves into a collisional boundary.)

Subduction starts when the edge of one plate pushes over the edge of another plate along a fault. Once the edge of the downgoing oceanic plate has entered the asthenosphere, the plate begins to sink like an anchor, because the rock of the plate is denser than the rock in the surrounding asthenosphere (Fig. 2.19b, c). To make room for the sinking lithosphere, asthenosphere flows plastically out of its way, just as water flows out of the way of a sinking anchor. Asthenosphere can flow only very slowly, however, so subduction cannot proceed any faster than about 15 cm/yr (6 in/yr). Note that if the asthenosphere were not soft enough to flow, subduction couldn't happen. All ocean floor eventually undergoes subduction, unless it gets trapped between colliding continents (as discussed in Section 2.5), so the oldest ocean floor on the Earth is only about 200 million years old.

Let's look more closely at the geologic activity of a convergent boundary. As subduction takes place, seafloor

FIGURE 2.19 During subduction, oceanic lithosphere sinks into the asthenosphere.

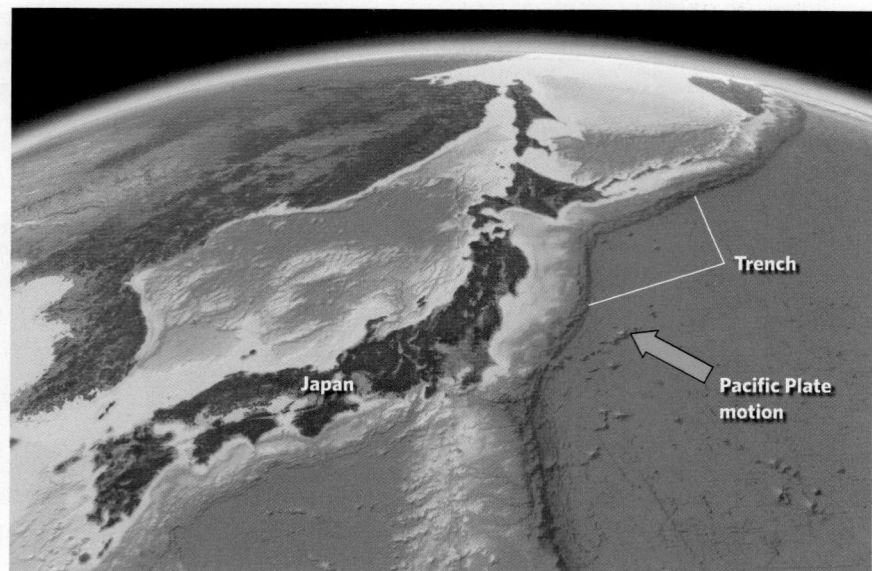

(a) A trench delineates the place where the Pacific Plate is being subducted beneath Japan.

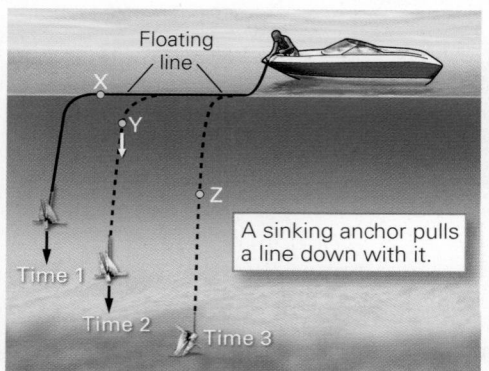

(b) Sinking of the downgoing slab resembles the sinking of an anchor.

A sinking anchor pulls a line down with it.

(c) Once subduction begins, the downgoing plate sinks very slowly, like an anchor.

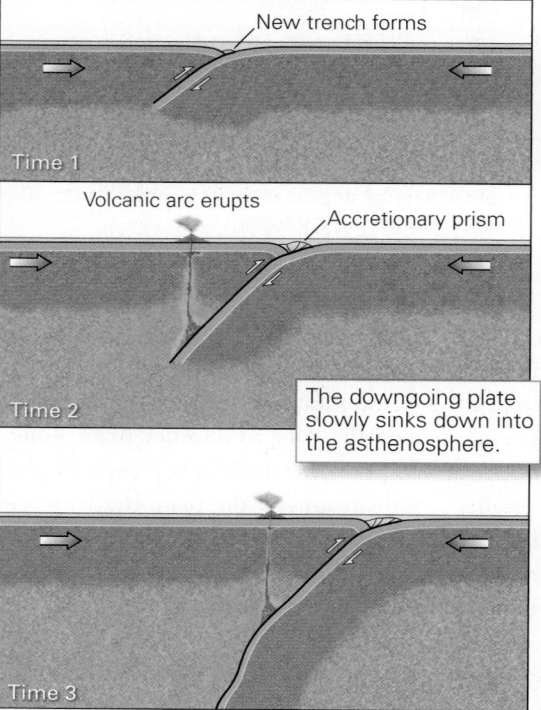

The downgoing plate slowly sinks down into the asthenosphere.

On the far side of the accretionary prism, relative to the trench, a volcanic arc develops on the overriding plate. Note that, at any given time, subduction takes place beneath the entire length of a volcanic arc, so all volcanoes in the arc can be considered *active volcanoes*, meaning that they all have recently erupted, are erupting, or are likely to erupt in the future. As we'll see in Chapter 4, most molten rock that rises under a volcanic arc forms in the asthenosphere between the downgoing plate and the overriding plate.

Where an oceanic plate subducts beneath a continent, a **continental arc** develops on the continent—the Andean Arc along the coast of South America and the Cascade Arc of Oregon and Washington serve as examples. In some cases, faults develop on the continental side of the arc (see Fig. 2.20a). At places where an oceanic plate subducts beneath another oceanic plate, volcanoes build up on the seafloor of the overriding plate and grow into a volcanic **island arc**—the Mariana Arc and the Aleutian Arc are examples **(Fig. 2.20c)**. In a few locations, a volcanic arc grows on a sliver of continental crust that separated from the main continent due to the growth of a small ocean basin, called a *marginal sea* **(Fig. 2.20d)**—the Japan Arc represents such a location.

The boundary zone between the downgoing plate and the base of the overriding plate at a convergent boundary is a zone of faulting; slip on faults in this zone generates earthquakes at fairly shallow depths. Earthquakes also happen at greater depths beneath a convergent boundary. These earthquakes define a sloping band, called a *Wadati-Benioff zone*, that occurs in the downgoing plate; geologists can detect such earthquakes down to a depth of 660 km (410 mi) **(Fig. 2.21)**. Downgoing plates eventually sink below 660 km, but they do so without generating earthquakes because, simplistically, the downgoing plate becomes too warm and soft at that depth to be able produce and transmit seismic waves. As the downgoing plate continues to sink, it may separate into pieces, or it may become contorted into curving shapes. Because they consist of relatively dense rock, these pieces may eventually sink to the base of the mantle, a region that may effectively serve as a "plate graveyard."

sediment, as well as sand and mud that washed into the trench from nearby land, gets scraped off the downgoing plate and piles up to form a wedge-shaped mass, known as an **accretionary prism**, along the margin of the overriding plate **(Fig. 2.20a, b)**. In effect, an accretionary prism resembles a pile of sand that builds in front of a bulldozer.

FIGURE 2.20 The nature of a convergent boundary varies with location.

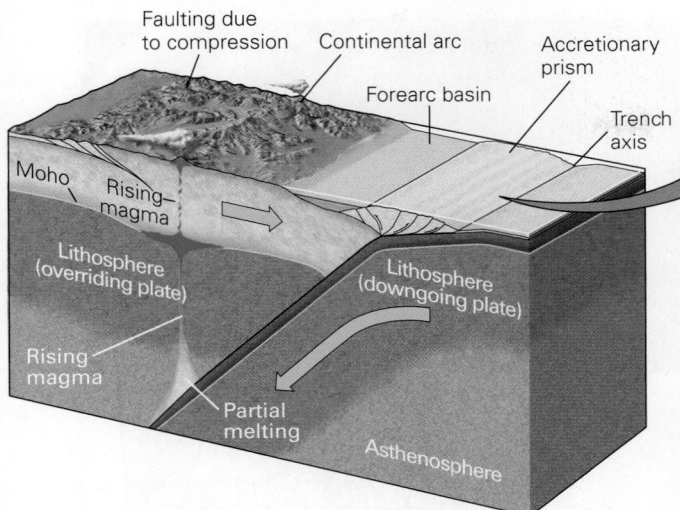

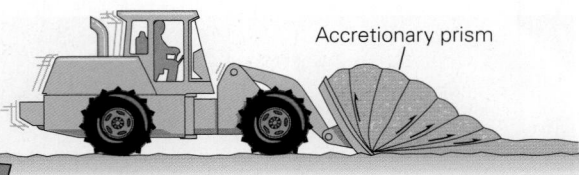

(b) The overriding plate acts like a bulldozer, scraping sediment off the downgoing plate to build an accretionary prism between the trench and volcanic arc.

(a) A subduction zone along the edge of a continent. In this example, faults have developed behind the continental arc.

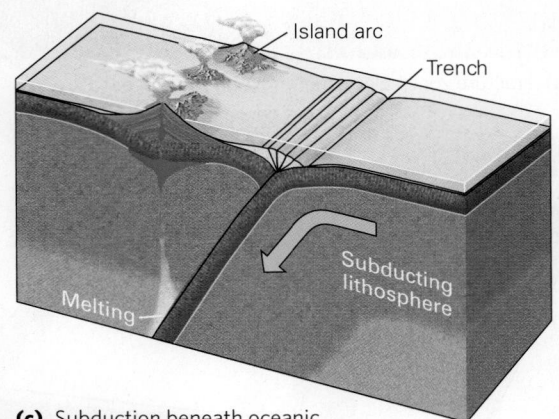

(c) Subduction beneath oceanic lithosphere produces an island arc.

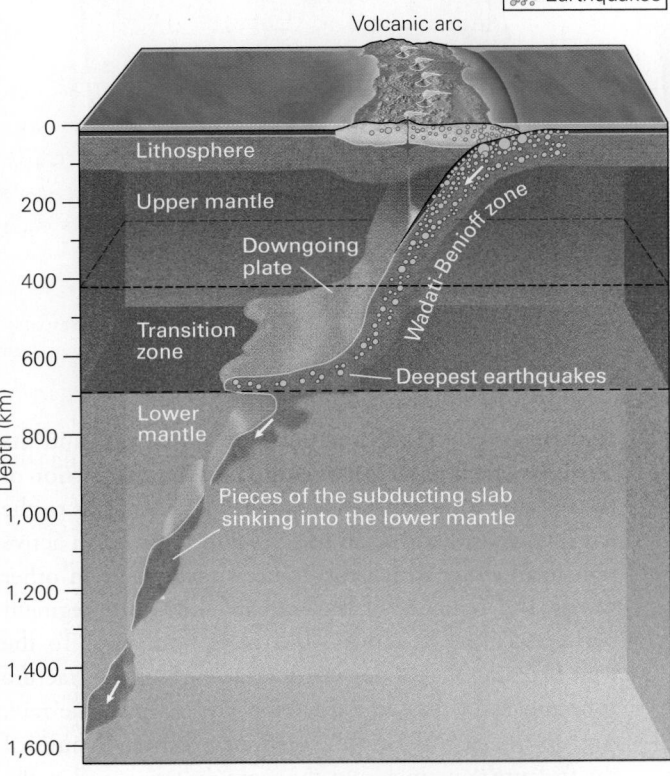

FIGURE 2.21 Subducted lithosphere sinks into the mantle like a denser fluid sinking into a less dense fluid. The downgoing plate may eventually separate into pieces as it passes through the layers of the mantle (see Chapter 1). Earthquakes occur in the downgoing plate down to a depth of about 660 km (410 mi), a region known as the Wadati-Benioff zone.

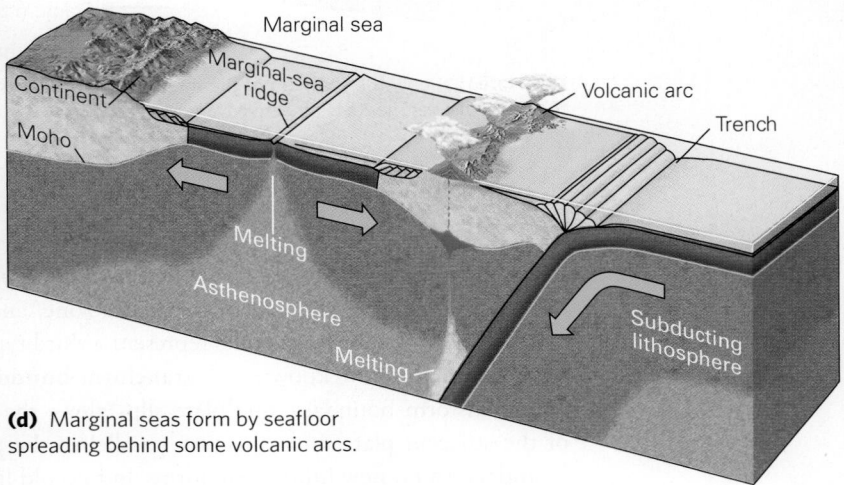

(d) Marginal seas form by seafloor spreading behind some volcanic arcs.

Transform Boundaries

We noted earlier that mid-ocean ridges are segmented and that the ridge segments intersect fracture zones (Fig. 2.22a). In the early 1960s, geologists began to think about these fracture zones in the context of seafloor spreading. A Canadian researcher, J. Tuzo Wilson (1908–1993), proposed that the portion of a fracture zone between the ends of two ridge segments is a plate boundary, whereas the rest of the fracture zone is not. Wilson introduced the

FIGURE 2.22 Oceanic transform faulting.

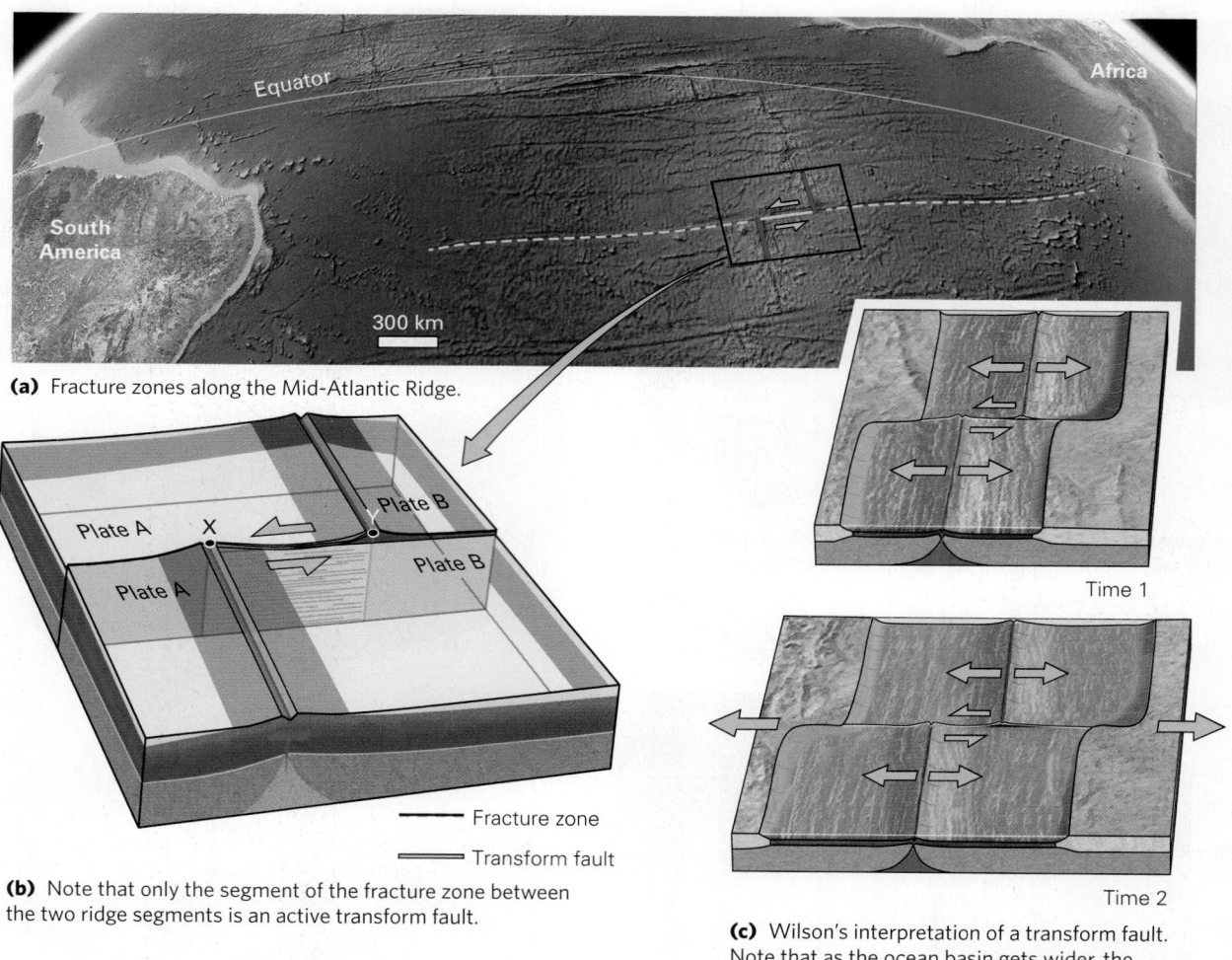

(a) Fracture zones along the Mid-Atlantic Ridge.

(b) Note that only the segment of the fracture zone between the two ridge segments is an active transform fault.

Fracture zone
Transform fault

Time 1

Time 2

(c) Wilson's interpretation of a transform fault. Note that as the ocean basin gets wider, the length of the transform fault remains the same.

term **transform fault** (often simplified to *transform*) for the actively slipping segment of a fracture zone, and he emphasized that transform faults represent a third type of plate boundary, now known as a **transform boundary**. At a transform boundary, one plate slips along the side of the adjacent plate, in a direction parallel to the plate boundary, so no new lithosphere forms and no old lithosphere undergoes subduction. Transform boundaries are, therefore, large vertical faults on which the slip direction is horizontal (**Fig. 2.22b**).

To understand the nature of motion on an oceanic transform fault, consider a map showing two north-south-trending mid-ocean ridge segments linked by an east-west-trending fracture zone. (**Figure 2.22c** shows the *trace* of the fracture zone on the map, meaning the intersection between the plane of the vertical fracture zone and the seafloor surface.) Plate A moves west and Plate B moves east, relative to the ridge axis. The southern ridge

segment intersects the fracture zone at Point X, and the northern one intersects it at Point Y. Along the portion of the fracture zone between X and Y, Plate A moves to the left relative to Plate B, so this portion must be an active transform along which earthquakes take place. In other words, this segment of the fracture zone—the segment between Points X and Y—is a plate boundary. To the west of Point X, seafloor north and south of the fracture zone moves in the same direction and at the same rate. And to the east of Point Y, seafloor north and south of the fracture zone moves in the same direction and at the same rate. These portions of the fracture zone, therefore, are not active faults. When you cross these segments, you're staying on the same plate, although the lithosphere on one side of the boundary is older than that on the other side.

Note the orientation of the half arrows in Figure 2.22 that indicate the sense of slip along a transform fault

FIGURE 2.23 The San Andreas Fault is a continental transform boundary.

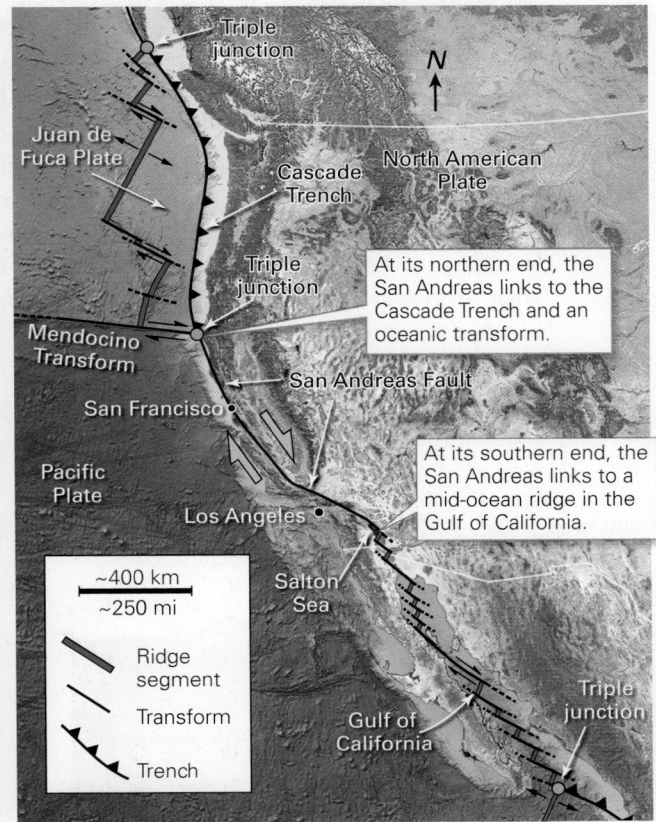

Callout: At its northern end, the San Andreas links to the Cascade Trench and an oceanic transform.

Callout: At its southern end, the San Andreas links to a mid-ocean ridge in the Gulf of California.

Legend:
~400 km / ~250 mi
Ridge segment
Transform
Trench

(a) In southern California, the San Andreas Fault cuts across a dry landscape. The fault trace, meaning the intersection of the fault with the ground surface, lies in a narrow valley.

(b) The San Andreas Fault is a continental transform boundary between part of the North American Plate and part of the Pacific Plate. The Pacific Plate is moving northwest relative to North America.

linking segments of a mid-ocean ridge. If, as geologists once incorrectly thought, slip on the transform fault displaced one part of a once-continuous ridge segment relative to the other, these arrows would be backward. But since the arrows actually indicate the direction in which the plates are moving relative to each other, and the plates must move away from the ridge axis, they are correct. Studies of the slip direction associated with earthquakes along oceanic transform faults confirmed the orientation of the arrows shown in Figure 2.22; this discovery served as another proof of seafloor spreading.

Not all transforms link mid-ocean ridge segments, and not all transforms are submarine. For example, the Alpine Fault is a transform boundary that cuts across New Zealand and links two trenches. The San Andreas Fault of California serves as a transform boundary between parts of the North American Plate and the Pacific Plate and links a ridge segment at its southern end to a trench at its northern end. The portion of California that lies to the west of the fault moves with the Pacific Plate, whereas the portion to the east of the fault moves with the North American Plate (Fig. 2.23).

Take-home message...

Geologists distinguish among three types of plate boundaries based on relative plate movement: divergent, convergent, and transform boundaries. Seafloor spreading occurs at divergent boundaries. At convergent boundaries, an oceanic plate sinks beneath the edge of another plate. At transform boundaries, one plate slips sideways along the edge of another.

Quick Questions ────────

- Does oceanic lithosphere sink into the asthenosphere at a divergent or a convergent boundary?

- At which type of boundary does seafloor spreading take place?

- Describe the relative plate motion that takes place along a transform boundary.

Animation
Plate Boundaries

2.5 The Birth and Death of Plate Boundaries

The configuration of plates and plate boundaries visible on our planet today has not existed for all geologic history, and it will not stay the same in the future. Because of plate motion, new oceanic plates form, only to be consumed by subduction later, and continents merge, then later separate. In this section, we first examine how a convergent boundary ceases to exist when two continents collide. Then we look at how continents split apart, or *rift*, a process that may produce a new divergent boundary.

When Continents Converge: The Process of Collision

India was once a small, separate continent that lay far to the south of Asia. Over time, however, subduction consumed ocean floor between India and Asia, and India moved northward. By 40–50 million years ago, all the ocean floor between the two continents had been subducted, and the Indian continent pushed into the Asian continent. This type of event—the merging of two landmasses after subduction of the intervening ocean floor—is called **collision**, and the boundary between the once-separate landmasses is called a **suture**. If the event involves two continents, geologists refer to it as *continental collision*.

Collision occurs when two pieces of relatively buoyant crust, such as continental crust or an island arc, are carried into a convergent boundary, because buoyant crust cannot be subducted (Fig. 2.24a, b). It may seem strange to use the word *buoyant* for a landmass, for in everyday English, the adjective usually applies to very low-density materials like cork or Styrofoam. But, in fact, a **buoyancy force** develops wherever a less dense material is placed in a denser material, and drives the less dense material upward relative to the denser material. Put another way, a less dense material is *buoyant* relative to a denser material. If the denser material is weak and can flow out of the way of the less dense material, the latter can rise until it floats on top. When collision happens, the convergent boundary that once lay between the buoyant landmasses gradually ceases to exist, and the subducted oceanic plate eventually breaks off and sinks. But the colliding buoyant crustal pieces

FIGURE 2.24 Continental collision.

(a) Subduction consumes oceanic lithosphere between two continents until the two continents collide.

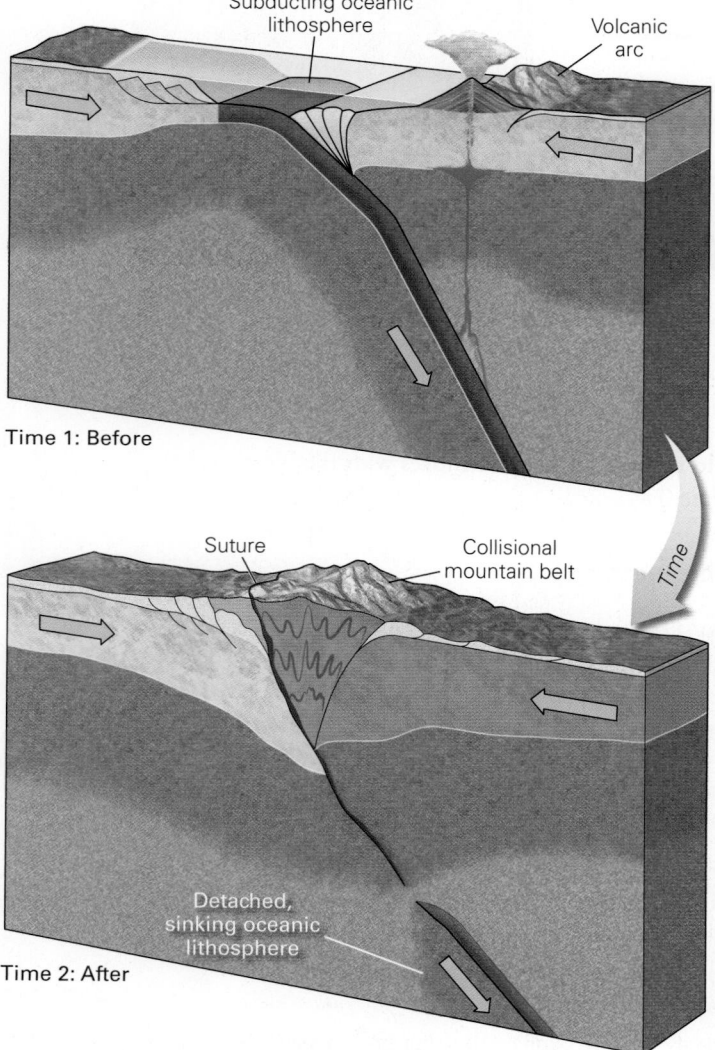

(b) After collision, the downgoing oceanic lithosphere detaches and sinks deep into the mantle. Rock caught in the collision zone gets broken, bent, and squashed, and uplifts to form a mountain range.

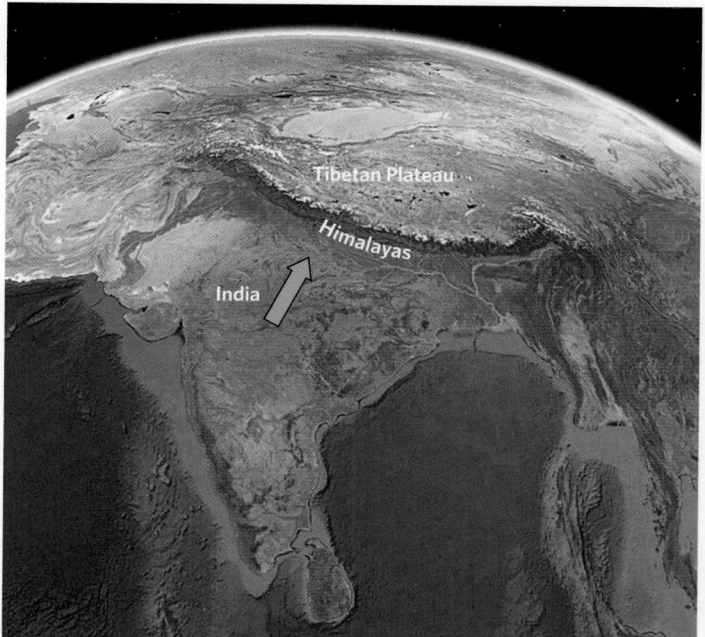

(c) An oblique satellite view showing the collision of India with Asia to produce the Himalayas.

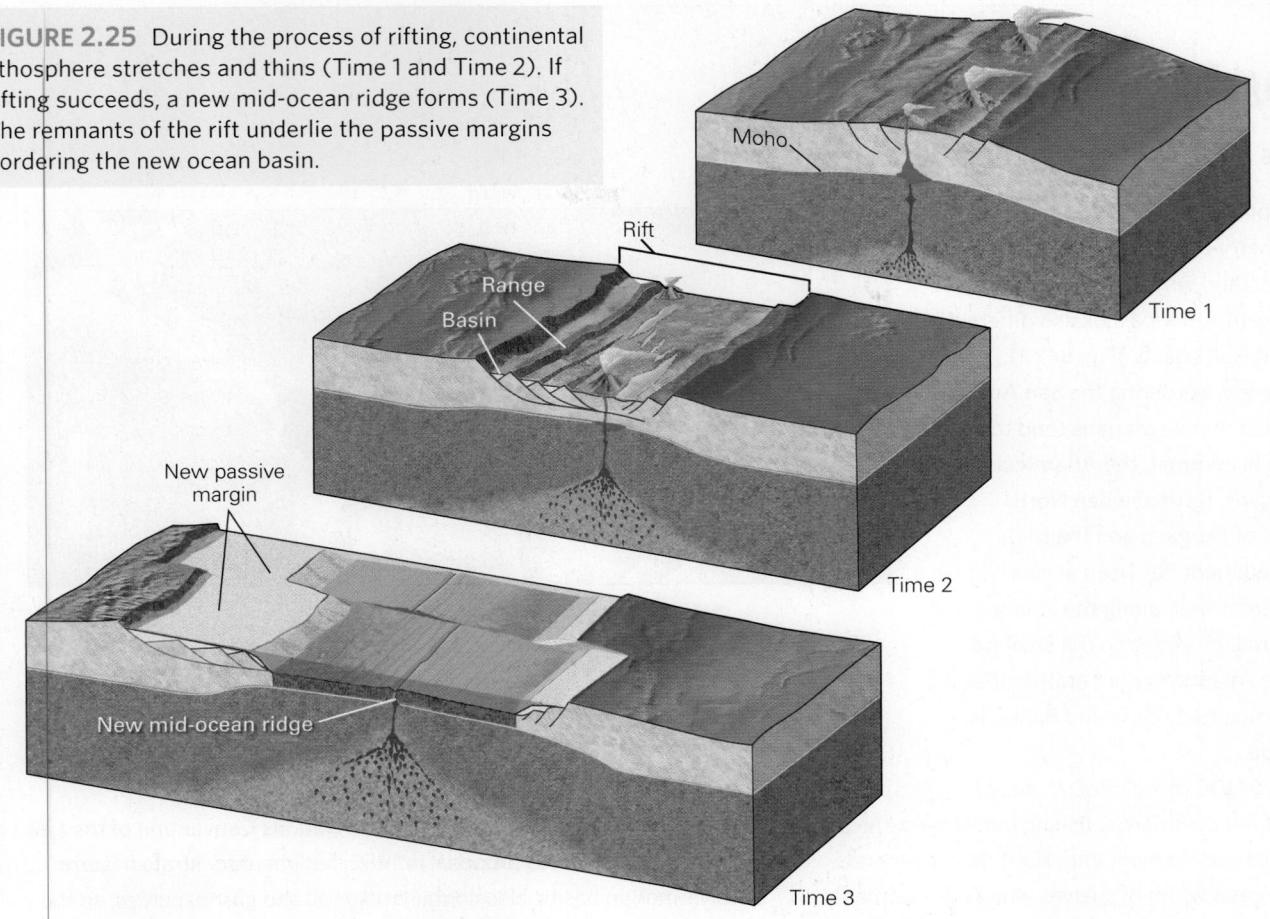

FIGURE 2.25 During the process of rifting, continental lithosphere stretches and thins (Time 1 and Time 2). If rifting succeeds, a new mid-ocean ridge forms (Time 3). The remnants of the rift underlie the passive margins bordering the new ocean basin.

Moho

Rift

Range

Basin

Time 1

New passive margin

Time 2

New mid-ocean ridge

Time 3

stay near the surface of the Earth. They move toward one another and squeeze together, but subduction no longer takes place, as it does at a plate boundary. Collision typically results in the development of mountain ranges because, as landmasses push into each other, they shorten horizontally and thicken vertically. The Himalayas represent a present-day *collisional mountain belt*—the range rises to an elevation of 8.85 km (5.5 mi) **(Fig. 2.24c)**.

Stretching a Continent: The Process of Rifting

Continents not only can collide, but can also break apart, as Wegener realized when he proposed that Pangaea broke into several smaller continents. The process that stretches a continent and may eventually break it apart, yielding a divergent plate boundary, is called **rifting** **(Fig. 2.25)**. This process tends to be confined to a linear belt called a **rift**. The manner in which rock responds to rifting varies with depth. Near the surface of the continent, stretching breaks the crust, leading to the development of many faults. Blocks of crust tilt and slip downward on these faults so that low areas, called *rift basins*, develop. The basins fill with eroded sediment from the rift's margin as well as from the tilted blocks. Deeper down, in places where crustal rock has become

warmer and softer, stretching can take place by plastic flow (see Fig. 2.25). As continental lithosphere thins during rifting, hot asthenosphere flows plastically and rises beneath the rift, so no open space develops. Some of this rising asthenosphere melts, producing molten rock that erupts from volcanoes in the rift (see Chapter 4).

A rift that continues to stretch until the continent breaks apart and a new mid-ocean ridge forms can be called a *successful rift*. Once seafloor spreading begins, the remnants of a successful rift, now abandoned along the margins of the now-separate continents that border the young ocean, become inactive and evolve into passive-margin basins **(Box 2.1)**. Along an *unsuccessful rift*, rifting ceases before it splits a continent. The trace of an unsuccessful rift remains as a scar in the crust, delineated by faults and by rift-related sedimentary and igneous rocks.

Rifting is actively taking place today at several localities on the Earth. For example, along the *East African Rift*, easternmost Africa is moving away from central Africa. A fault-bounded valley, containing elongate lakes, delineates the axis of this rift **(Fig. 2.26a)**. The *Basin and Range Province* of the western United States formed when rifting produced many long, linear mountain ranges (the edges of tilted crustal blocks) separated from each other by narrow,

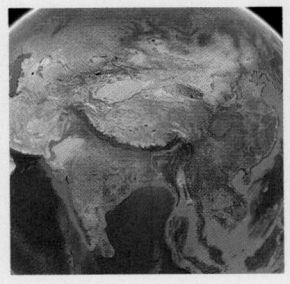

BOX 2.1

Putting Earth Science to Use

Continental shelves and exclusive economic zones

With a basic knowledge of plate tectonics, you can look at a map depicting the bathymetry of coastal areas bordering the United States and understand why the continental shelf along the Pacific coast of North America looks so different from that of the Atlantic and Gulf coasts (Fig. Bx2.1). The Pacific coast is an active margin, bordering the San Andreas Fault and the Cascadia Trench. Active margins tend to have narrow continental shelves. In contrast, the Atlantic and Gulf coasts are passive margins, formed when North America rifted from the rest of Pangaea and the Mid-Atlantic Ridge developed. Sediment has been accumulating and building out the continental shelf along the Atlantic and Gulf coasts for almost 200 million years, so the shelf has grown very wide. In fact, the Atlantic continental shelf east of Newfoundland, a region known as the Grand Banks, is about 400 km (250 mi) wide.

Notably, shallow seas (<300 m, or 1,000 ft, deep) provide most of the fish that the commercial fishing industry catches. Therefore, continental shelves are the most important fishing grounds in the world. Because of the great width of shelves along passive margins, some governments proposed that *exclusive economic zones* (EEZs), the offshore areas whose resources the bordering country owns, be widened from 12 nautical miles (22 km) to 200 nautical miles (370 km)

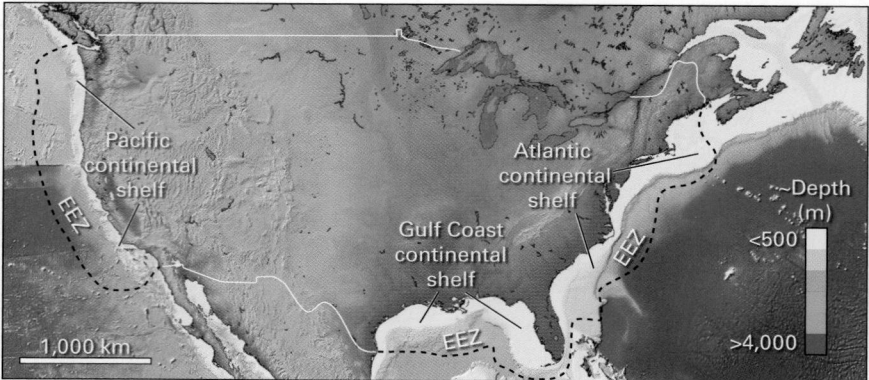

FIGURE Bx2.1 The continental shelves and exclusive economic zones bordering the continental United States. Note that the Atlantic and Gulf coast EEZs roughly correspond to the wide continental shelves underlain by passive-margin basins, but the Pacific coast EEZ is wider than the continental shelf of the active margin.

as measured from the shore. The United Nations Convention of the Law of the Sea adopted this proposal in 1982. Sedimentary strata in some passive-margin basins also contain major oil and gas resources, so the definition of EEZs has implications for offshore drilling as well. Because of their economic significance, boundaries of EEZs have, unfortunately, become points of conflict between nations.

---

sediment-filled basins (Fig. 2.26b). Earthquakes in the region indicate that rifting is still happening there. The Rio Grande Rift, which produced the low area in which the Rio Grande flows today, has been volcanically active within the past 10,000 years. Since no oceanic lithosphere has yet formed, these rifts are not yet plate boundaries. Geologists have no way of knowing whether or not any of these rifts will succeed and evolve into a new mid-ocean ridge. Figure 2.26a shows two rifts that have succeeded during the past several million years. Specifically, both the Red Sea and the Gulf of Aden are very narrow oceans that began as rifts. Another example of a successful rift, the Gulf of California, lies between Baja California and the mainland of Mexico (see Fig. 2.23b). Geologic studies have also identified several unsuccessful rifts. For example, rifting occurred in what is now the middle of the United States, forming a linear scar in the crust that can be traced from

Lake Superior southwest through Kansas (Fig. 2.26c). This feature, known as the Midcontinent Rift, has mostly been buried by Paleozoic sedimentary rock.

Take-home message...

Where two buoyant pieces of continental crust collide, a collisional mountain belt forms and the convergent boundary that once lay between the landmasses ceases to exist. Rifting can split a continent in two and can lead to the formation of a new divergent boundary.

Quick Questions

- What is the difference between a successful and an unsuccessful rift?
- Why can't a continent be subducted at a convergent boundary?

FIGURE 2.26 Examples of present-day rifting.

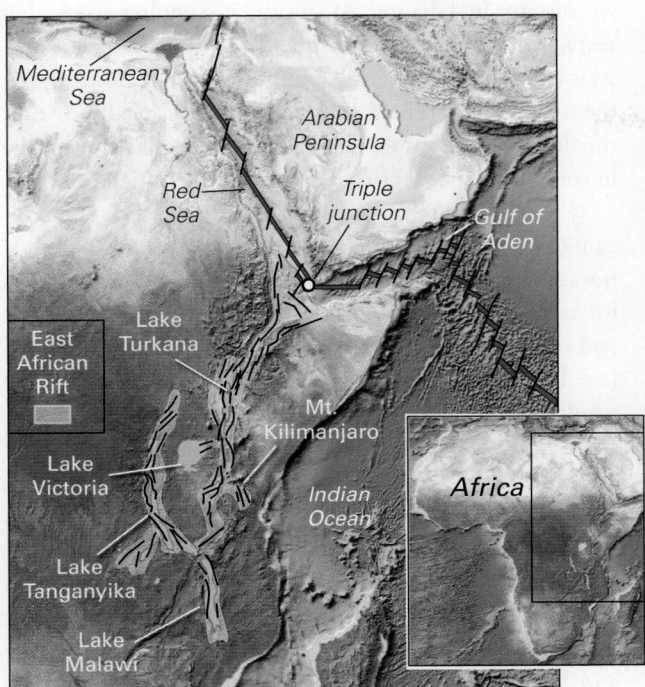

(a) The East African Rift is growing today. The inset shows the map's location. The Red Sea and the Gulf of Aden formed where rifting succeeded.

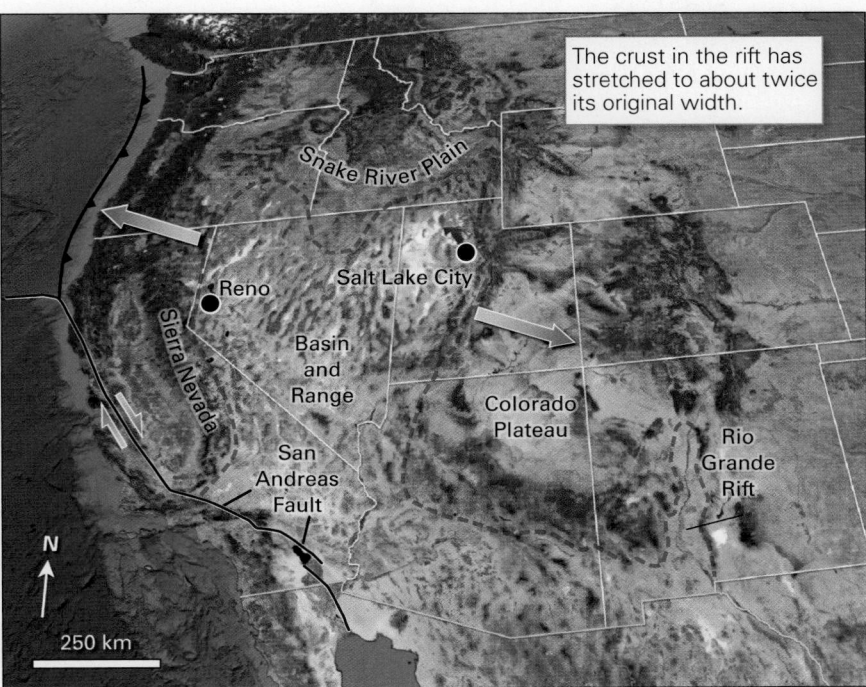

(b) The Basin and Range Province is a rift. The narrow north-south-trending mountain ranges, separated by basins, formed because of faulting. The arrows indicate the direction of stretching.

2.6 Special Features in the Plate Mosaic

Triple Junctions

At several localities on the Earth, three plate boundaries intersect. Each of these intersections, a **triple junction**, stands out on a global map of plate boundaries. For example, a triple junction occurs at the point where three ridges intersect in the western Indian Ocean **(Fig. 2.27a)**. Another occurs along the coast of California, where the San Andreas Fault intersects the Cascadia subduction zone and an oceanic transform fault **(Fig. 2.27b)**.

Hot Spots

On a global basis, most volcanic activity occurs along plate boundaries. But geologists have found about a hundred locations where volcanoes do not form as a direct consequence of plate interactions. These locations are called **hot spots**. Most hot spots occur in plate interiors, away from plate boundaries. Examples include the Hawaiian hot spot **(Fig. 2.28a)**, located in the interior of the Pacific Plate, and the Yellowstone hot spot, located in the interior of the North American Plate. A few hot spots underlie mid-ocean ridges. Such hot spots can be distinguished from normal segments of a mid-ocean ridge because they produce much

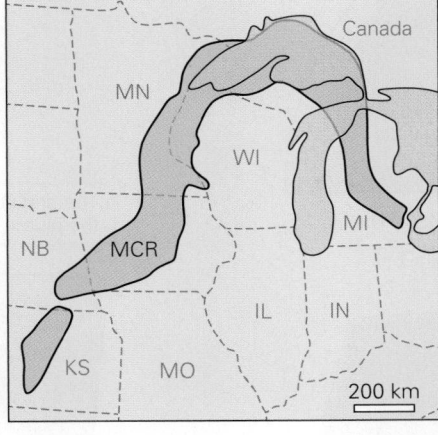

(c) The Midcontinent Rift (MCR) formed at 1.1 Ga. Rifting ceased before a new ocean basin formed, so this rift was unsuccessful. The shading shows the area underlain by the sedimentary and igneous rocks that filled the rift. These rocks, except along the southwestern tip of Lake Superior, are completely buried beneath younger sedimentary rock.

more igneous rock than do normal segments. Such hot spots, therefore, build an *oceanic plateau*, an area of particularly shallow seafloor whose crest may protrude from the water as an island. Iceland sits on an oceanic plateau that rises 2 km (1.2 mi) above the rest of the Mid-Atlantic Ridge, and it was formed by such a hot spot.

Why do hot spots exist? Researchers still debate this question, but most favor a model in which most hot spots develop above **mantle plumes**, narrow columns of particularly hot rock that is buoyant relative to the surrounding, cooler asthenosphere, so it flows upward through the asthenosphere to the base of the lithosphere **(Fig. 2.28b)**.

FIGURE 2.27 Triple junctions.

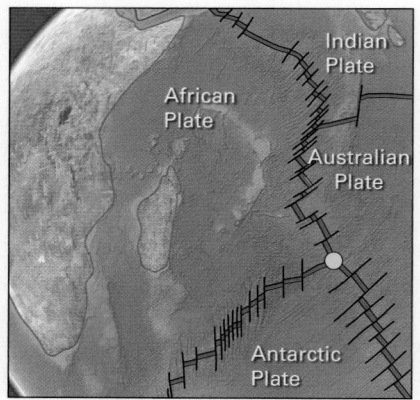

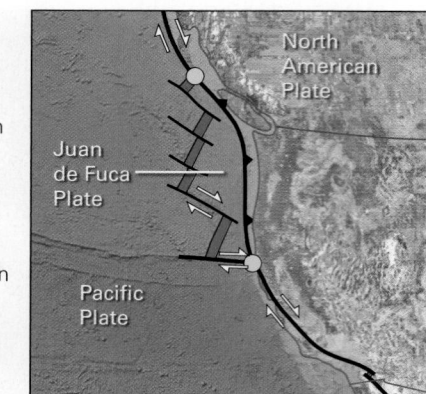

(a) A ridge-ridge-ridge triple junction occurs in the western Indian Ocean.

Ridge

Transform

Triple junction

Subduction zone

Fracture zone

(b) A trench-transform-transform triple junction occurs at the northern end of the San Andreas Fault. A ridge-trench-transform triple junction occurs near the west coast of Canada. (Figure 2.23b shows another ridge-trench-transform triple junction that lies near the southern end of the Gulf of California.)

FIGURE 2.28 Hot spots.

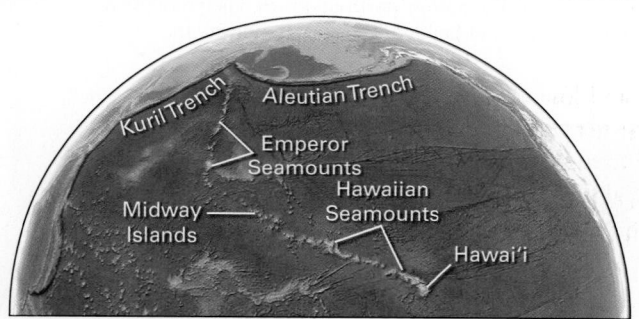

(a) A bathymetric map showing the hot-spot tracks of the North Pacific Ocean.

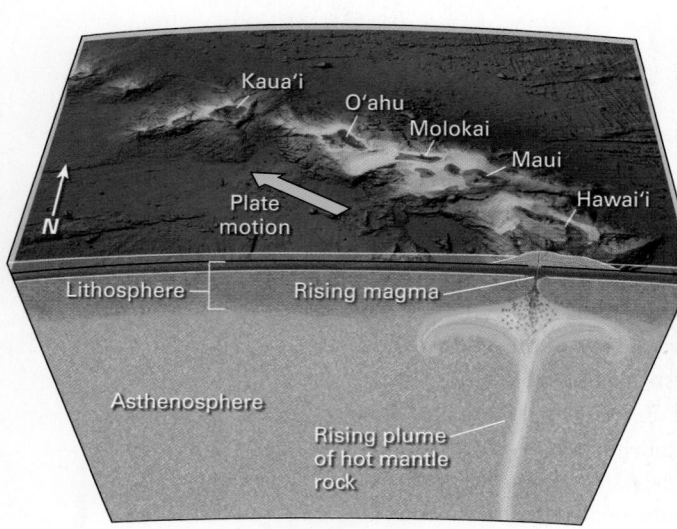

(b) According to the mantle-plume model, Hawai'i lies above a rising column of hot mantle rock.

When the hot peridotite in a mantle plume reaches the base of the lithosphere, it begins to melt (for reasons that we discuss in Chapter 4), producing molten rock that seeps up through the lithosphere and spills out of a hot-spot volcano. Some mantle plumes may start in the upper mantle, perhaps where subducting plates locally disrupt the flow of asthenosphere, whereas others may start in the lower mantle, perhaps even at the core-mantle boundary.

J. Tuzo Wilson noticed that most active hot-spot volcanoes lie at the end of a chain of *extinct volcanoes* (volcanoes that will never erupt again). In the Hawaiian Islands, for example, active volcanoes erupt at the southeastern end of a chain of extinct volcanic islands and seamounts (see Fig. 2.28a). This configuration differs from that of a volcanic arc bordering a subduction zone, for in a volcanic arc, several volcanoes can be concurrently active. To explain the configuration of the Hawaiian-Emperor seamount chain, Wilson proposed that the position of a hot spot remains fixed, relative to the Earth, while a plate moves over the top of it. As the plate moves, a volcano that forms above the hot spot eventually moves off the hot spot and, therefore, away from its supply of molten rock, and becomes extinct **(Fig. 2.28c)**. A new volcano then starts to grow over the hot spot and this volcano remains active for a while, until it, too, gets carried off the hot spot

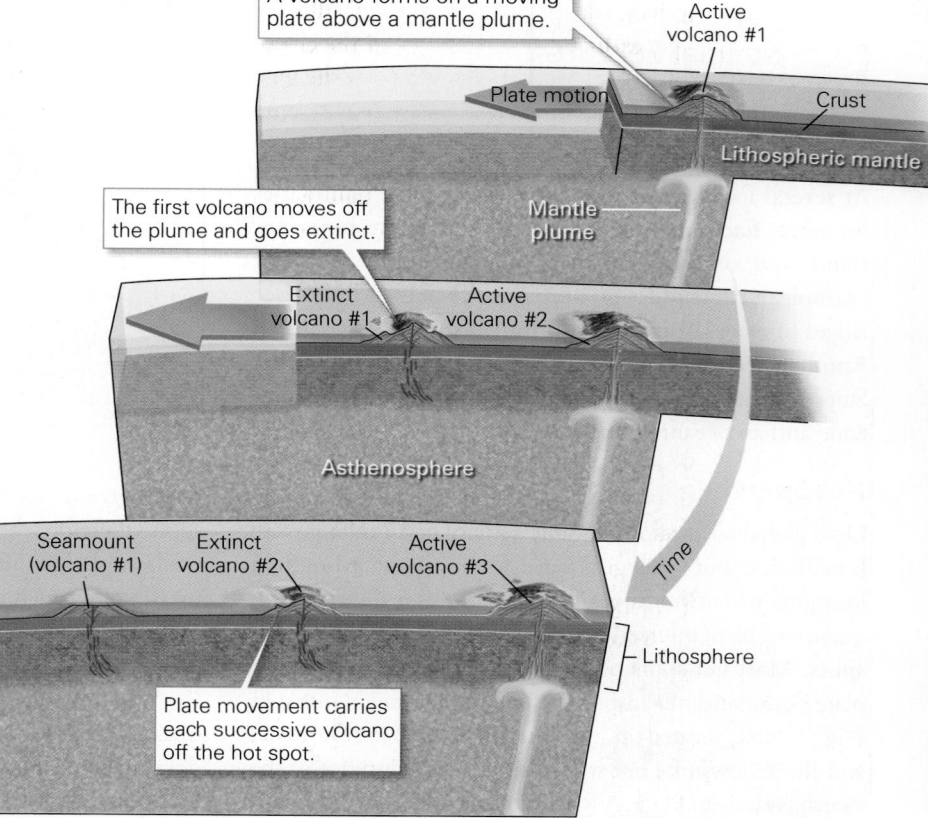

(c) Progressive stages in the development of a hot-spot track, according to the mantle-plume model. Note that the Pacific Plate is moving progressively from right to left in these drawings.

How can I explain . . .

The evolution of hot-spot tracks

What are we learning?

How a hot-spot track forms in the interior of a plate.

What you need:

• A disposable aluminum-foil baking pan
• Cornstarch and water
• A short candle and some matches
• Two wood blocks that are slightly taller than the candle

Instructions:

• Mix the water and cornstarch to make a paste.
• Cover the base of the pan with a 2 cm (1 in) thick layer of paste.
• Configure the blocks so that they border the candle and can support the pan.
• Light the candle.

• Slide the pan slowly over the candle. Bubbles should form in the paste above the candle and remain as mounds in the paste. Number these "volcanoes" from oldest to youngest.

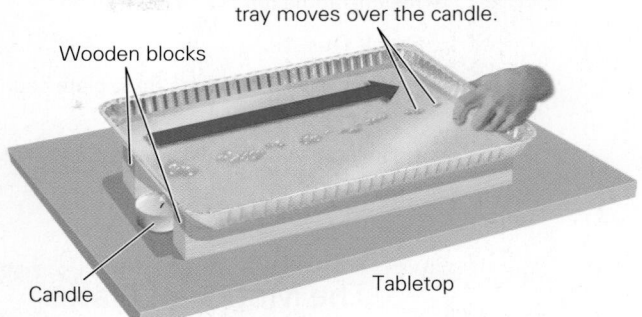

Bubbles form in cornstarch as the tray moves over the candle.

Wooden blocks

Candle

Tabletop

What did we see?

As the paste layer (representing a plate) moves over the fixed heat source (representing a mantle plume), bubbles (representing volcanoes) form. But no "volcano" erupts for long, because as soon as it gets carried off the heat source, it becomes inactive (extinct).

and goes extinct (Box 2.2). The process repeats for as long at the hot spot survives, which can be for tens of millions of years. As they age, extinct volcanic islands of the chain slowly become submerged, because the surface of the seafloor that they formed on becomes deeper as it ages, and because the islands erode and lose chunks to submarine landslides. Therefore, an extinct volcanic island eventually becomes a seamount. The overall chain of inactive volcanic islands and seamounts, along with the currently active volcano at its end, represents a **hot-spot track** (Fig. 2.29).

FIGURE 2.29 Locations of hot-spot volcanoes and hot-spot tracks. The most recent volcano (red dot) is at one end of each track (red line). Hot-spot tracks on mid-ocean ridges leave tracks in both directions.

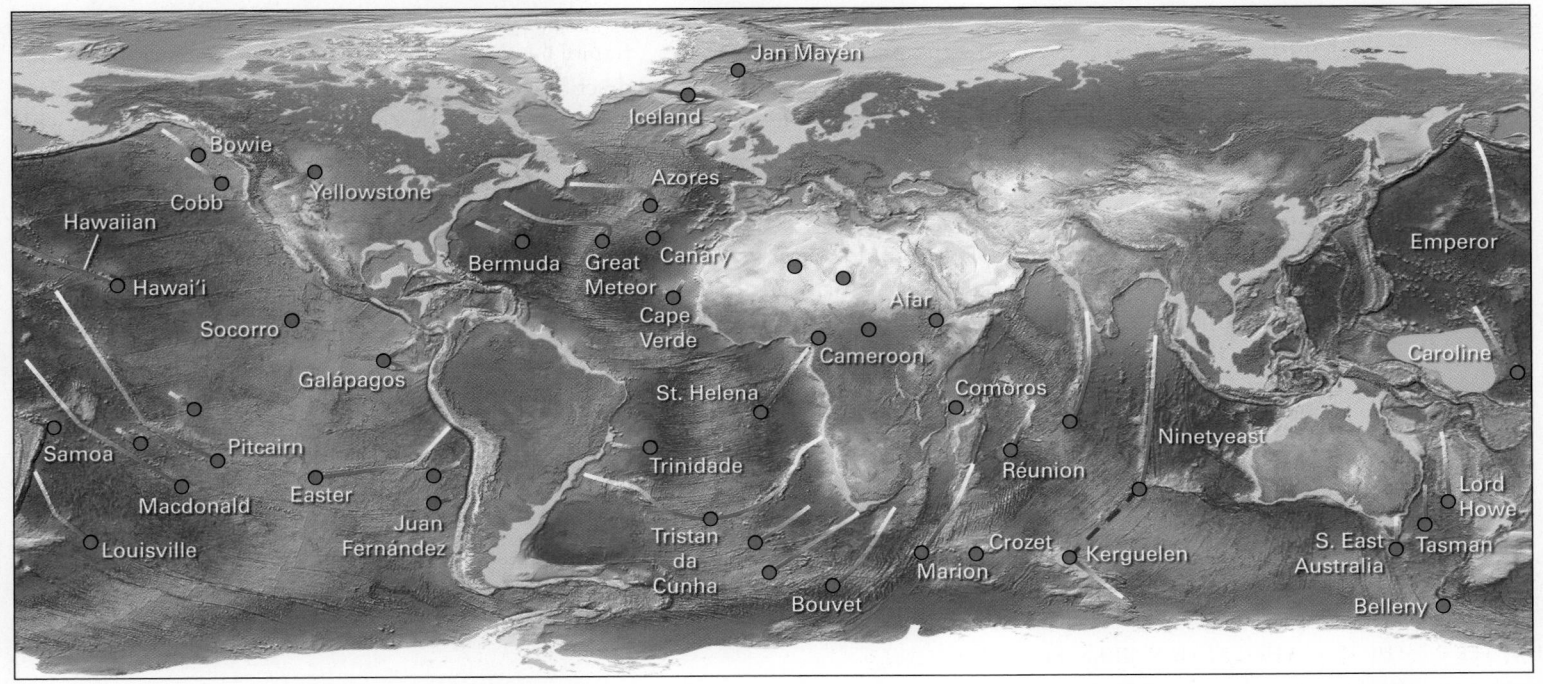

2.7 The Motion of Plates

What Drives the Plates?

Because Alfred Wegener had not been able to answer the question of what makes continents move, most geologists of his era didn't accept his hypothesis of continental drift. But the minority who did agree with Wegener continued to speculate about processes that could cause mobility of the Earth's outer layer. In 1928, a British geologist, Arthur Holmes, suggested that continental drift may be related to convection in the mantle. **Convection**, in general, refers to the process of circulation and heat transfer that happens when a deeper layer of material warms up and, therefore, expands and becomes less dense than the overlying layer of cooler material. As a result, the deeper material becomes buoyant. If both the denser material and the buoyant material are weak enough to flow, the buoyant material flows upward, and the cooler material flows downward to take the place of the rising warm material. The overall pattern of convective flow is called a *convective cell*. You can see convection in action by heating a pot of water on a stove. Convection in the mantle happens for the same reason—the presence of a cooler, denser layer over a warmer, buoyant layer—but because the mantle consists mostly of solid rock, mantle convection takes place very slowly, by plastic flow. Holmes suggested that continents are carried along like rafts on the top of convective cells.

Right after the theory of plate tectonics was proposed, geologists built on Holmes's ideas and initially attributed the motion of lithosphere plates entirely to convection in the mantle. As a result, early drawings depicting plate motion implied that warm mantle rock rises at mid-ocean ridges and cool rock sinks at subduction zones, and that the geometry of the resulting *convection cells* (pathways of convective flow) was quite simple. (Fig. 2.30a). But subsequent studies concluded that it's not possible to draw a global system of convection cells that can explain the complexity of observed plate-boundary configurations.

Studies over the past two decades reveal, instead, that *mantle upwelling* (the part of a convection cell where warm mantle rises) and *mantle downwelling* (the part of a convection cell where cooler mantle sinks) take place over very broad, irregularly shaped areas that aren't all spatially associated with specific mid-ocean ridges and deep-sea trenches (Fig. 2.30b). Therefore, plates aren't simply rafts being moved only by flow underneath them. Geologists now realize that the details of plate motion are affected not only by convective flow, but also by other forces applied along plate boundaries. Here, we describe two of these forces: ridge-push force and slab-pull force.

Ridge-push force develops because the young, warm lithosphere beneath mid-ocean ridges sits higher than the old, cold lithosphere beneath adjacent abyssal plains. The elevated lithosphere tries to spread sideways due to gravity, much as a mound of honey spreads sideways. This motion drives plates in a direction pointing away from the ridge axis (Fig. 2.30c). **Slab-pull force** develops because oceanic lithospheric mantle is denser than the asthenosphere below. So, once a plate starts to subduct, it continues to sink, and the subducted section pulls the rest of the plate along behind it (Fig. 2.30d). Geologists still debate the relative roles of these different forces acting on plates. Most likely, the motions that we see reflect a complex combination of forces.

Relative versus Absolute Plate Velocities

How fast do plates move? The answer depends on your reference frame. To illustrate this concept, imagine two cars speeding in the same direction down the highway. From the viewpoint of a tree along the side of the road, Car A zips by at 100 km/h, while Car B moves at 80 km/h. But relative to Car B, Car A moves at only 20 km/h. How we describe the velocity of a car depends on whether another moving car or a fixed tree serves as the reference frame. Geologists use two different reference frames for describing plate velocity. We are specifying **relative plate velocity** when we describe the movement of Plate A with respect to Plate B, whereas we are specifying **absolute plate velocity** when we describe the movement of both plates with respect to a fixed reference point that doesn't lie on either of the plates (Fig. 2.31).

Recall from Chapter 1 that to specify velocity, we need to indicate both the rate of movement and the direction of movement. Physicists define rate of motion (speed) by the equation speed = distance ÷ time. We can determine the rate of relative motion across a mid-ocean ridge by measuring the distance that seafloor of a known age has moved away from the ridge axis at which it formed. Assuming that the spreading rate remained constant during that time, the rate of motion of a point on the plate relative to the ridge axis can be calculated from the

FIGURE 2.30 Forces driving plate motions.

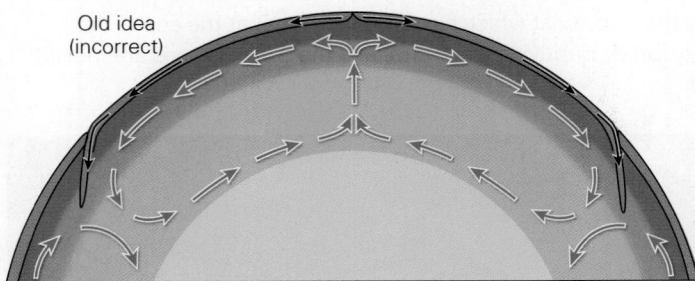

Old idea
(incorrect)

(a) The old, incorrect image of simple convective cells in the astheno-sphere. In this model, the plates were viewed as passive rafts carried along by convective flow in the mantle.

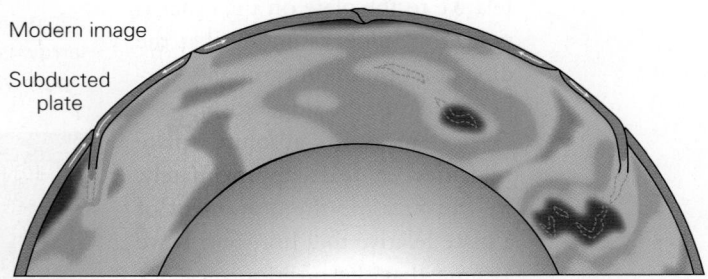

Modern image

Subducted plate

(b) Modern studies, which use seismic-wave velocities to help character-ize materials in the mantle, show that convective flow in the mantle is not easily correlated with the positions of specific plate boundaries. Here, red indicates warm (rising) areas, and blue indicates cool (sinking) areas.

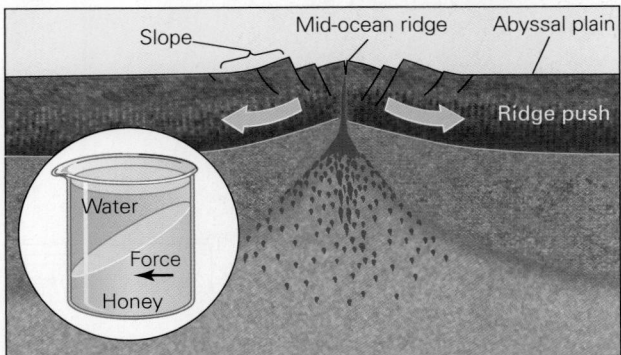

Slope   Mid-ocean ridge   Abyssal plain

Ridge push

Water

Force

Honey

(c) Ridge push develops because the region of a ridge is elevated. Like a mass of honey with a sloping surface, the mass of the ridge pushes sideways.

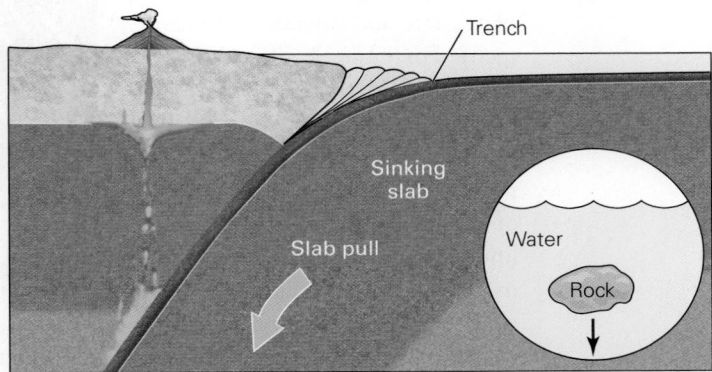

Trench

Sinking slab

Slab pull

Water

Rock

(d) Slab pull develops because lithosphere is denser than the underlying asthenosphere and sinks like a rock in water (though much more slowly).

FIGURE 2.31 Plate velocities worldwide. The black arrows show relative plate velocities and the red arrows show absolute plate velocities. The specific length of the red arrow in the legend corresponds to a speed of 5 cm/yr. Therefore, red arrows on the map that are longer than the arrow in the legend represent a speed faster than 5 cm/yr, and arrows that are shorter represent a speed slower than 5 cm/yr. Relative plate velocities are indicated by the numbers (in cm/yr) next to the black arrows on the map.

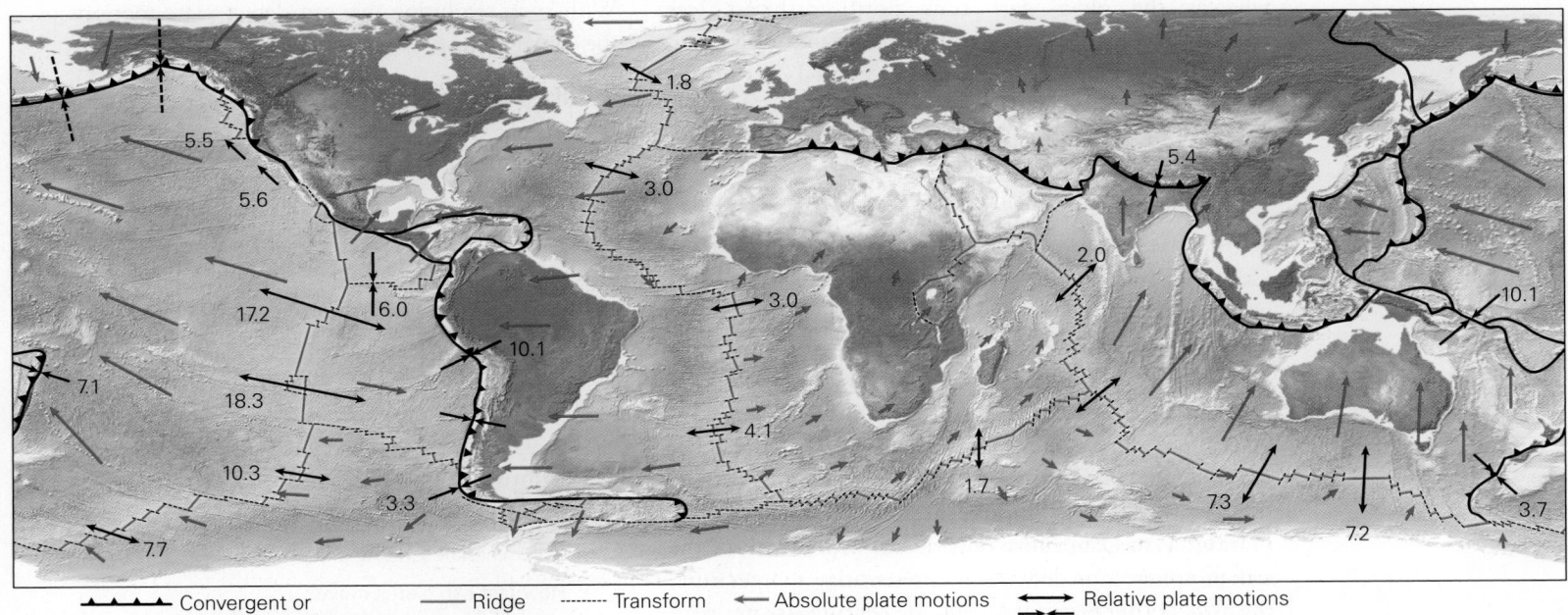

▲▲▲ Convergent or collision zone boundary   ── Ridge or rift   ------- Transform   ← Absolute plate motions (5 cm/yr)   ⇄ Relative plate motions

equation above. The rate of motion of the plate on one side of the ridge relative to the plate on the other is twice this value, assuming that the direction of motion is roughly perpendicular to the ridge axis.

To measure absolute plate velocities, at least approximately, we can measure the movement of a plate relative to a hot spot. If we assume that the position of a hot spot does not change much for a long time, then a hot-spot track on a plate moving over a mantle plume provides a record of the plate's rate and direction of movement. For example, if we assume that the Hawaiian hot spot's position remains fixed, then the orientation of the Hawaiian-Emperor hot-spot track (consisting of the Hawaiian Islands and the line of seamounts linked to them) indicates the absolute direction of Pacific Plate motion. Further, by dating the rocks of an inactive volcano in the chain (which gives us the time the volcano was over the hot spot) and by measuring the distance between that volcano and the presently active volcano on the Big Island of Hawai'i (which indicates the present location of the hot spot), we can calculate the speed of plate movement using the equation in the previous paragraph. Note that the Hawaiian-Emperor hot-spot track contains a distinct bend: the younger part of the track runs northwest, whereas the older part runs north-northwest (see Fig. 2.28a). This bend represents a change in the direction of the Pacific Plate's absolute motion. Based on the ages of volcanic rocks collected from seamounts on either side of the bend, it appears that this change happened about 43 million years ago.

GPS: Observing Plate Motions in Real Time

Using the methods described above, geologists determined that relative plate motions on the Earth today occur at a speed of 1–15 cm/yr (0.4–6 in/yr); by comparison, your fingernails grow at an average rate of about 3 cm/yr. These speeds, though low, can yield large displacements during the immensity of geologic time. For example, in 1 million years, a plate moving at 10 cm/yr travels 1,000 km (620 mi). What limits the speed of plate motion? Plates can move only as fast as the rock of the asthenosphere can flow. At its present-day temperature, the asthenosphere can flow only by centimeters per year.

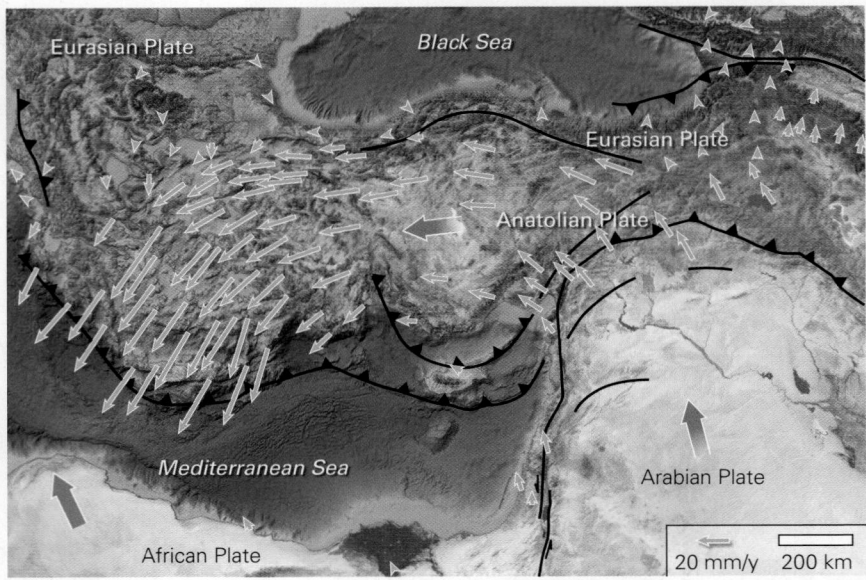

FIGURE 2.32 GPS measurements can now track modern-day plate motions accurately. The length of each arrow on this map of the eastern Mediterranean region indicates the velocity at which a point on the crust at the end of the arrow is moving. Turkey is moving westward, and the Greek islands are moving southwestward.

Eurasian Plate
Black Sea
Eurasian Plate
Anatolian Plate
Mediterranean Sea
Arabian Plate
African Plate
20 mm/y   200 km

(If it were hotter and softer, it could flow faster, so plate movement could be faster.)

Can we detect these very slow plate motions by direct observation? Until the 1990s, the answer was no, for traditional surveying systems were not accurate enough. But now, by using the modern **global positioning system** (**GPS**)—the same technology that automobile drivers use to find destinations—geologists can detect relative displacements of a few millimeters per year (Fig. 2.32). With such resolution, we can detect plate motions over the course of just a year!

Take-home message...

Plate motion takes place because plates are acted on by ridge push, slab pull, and convective flow. This motion happens at rates of 1–15 cm/yr. Relative plate velocity refers to the rate at which a plate moves relative to its neighbor, whereas absolute plate velocity refers to the rate at which a plate moves relative to a fixed reference point. GPS measurements can now detect plate motions directly.

Quick Questions ——————————

- Explain why convection takes place.
- What is the source of slab-pull force?
- Distinguish between absolute and relative velocity, as used when describing plate motion.
- How fast can plates move?

2.8 Proving Plate Tectonics Using Studies of Paleomagnetism

We've seen that plate tectonics not only can explain Wegener's observations, but also can explain why seafloor sediment thickens progressively away from the ridge axis in an ocean basin, why oceanic transform faults slip in the direction they do, why the depth of the ocean varies, why earthquakes and volcanoes happen where they do, why GPS measurements indicate that continents are moving, and even why the rocks of the Hawaiian Islands get progressively older with increasing distance from the presently active volcanoes. During the 1960s, several eye-opening tests of plate tectonics came from studies of *paleomagnetism*. In this section, we introduce paleomagnetism, then show how key paleomagnetic observations successfully tested plate tectonics and helped convince geologists that the theory is correct.

What Is Paleomagnetism?

More than 2,000 years ago, sailors in China recognized that a lodestone bar, when allowed to rotate, points toward the poles of the Earth. In Europe, starting about 1200 C.E., sailors began to rely on such compasses for navigation. But no one understood why compasses worked until 1600 C.E., when a British doctor, William Gilbert, proposed that the Earth generates a magnetic field. In the 20th century, researchers attributed this field to the flow of liquid iron alloy in the Earth's outer core.

Like a common toy magnet, the Earth's magnetic field is *dipolar*, meaning that it has two **magnetic poles**, a north magnetic pole and a south magnetic pole. Such a pair of opposite magnetic poles is called a **dipole**; physicists represent it symbolically by an arrow that points from the south pole of the magnet to its north pole. The Earth's dipole can be represented by an imaginary arrow that passes through the center of the Earth. Invisible *magnetic-field lines* curve through space from one end of the Earth's dipole to the other, and at the magnetic poles, magnetic-field lines point straight down or straight up **(Fig. 2.33a)**. By convention, geologists refer to the point where the tail of the dipole arrow intersects the surface of the Earth as the *north magnetic pole* and to the point where the head of the arrow intersects the surface of the Earth as the *south magnetic pole*. The line that passes through the Earth from magnetic pole to magnetic pole is the *magnetic axis*. Note that the Earth's dipole points from the Earth's north magnetic pole to its south magnetic pole. This labeling convention can seem a bit confusing, for it means that the south end of the dipole, as defined by physicists, lies, at present, at the north magnetic pole of the Earth. This labeling convention developed because it means that the north-seeking end of a compass needle

FIGURE 2.33 Features of the Earth's magnetic field.

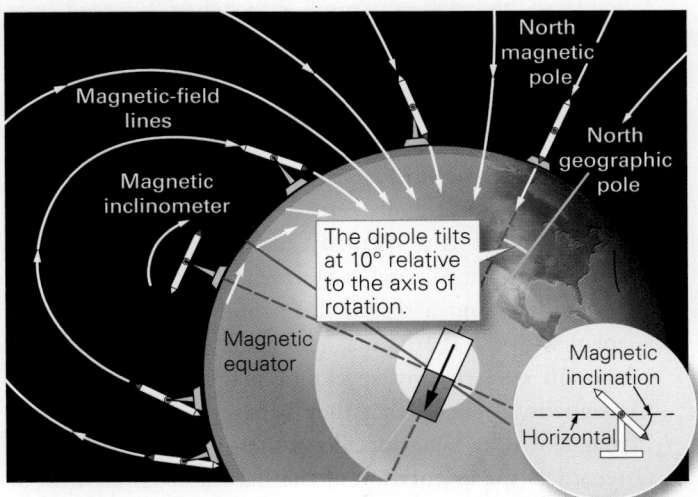

(a) The Earth's magnetic axis is not parallel to its axis of rotation. Magnetic inclination (inset) changes with latitude because of the shape of magnetic-field lines.

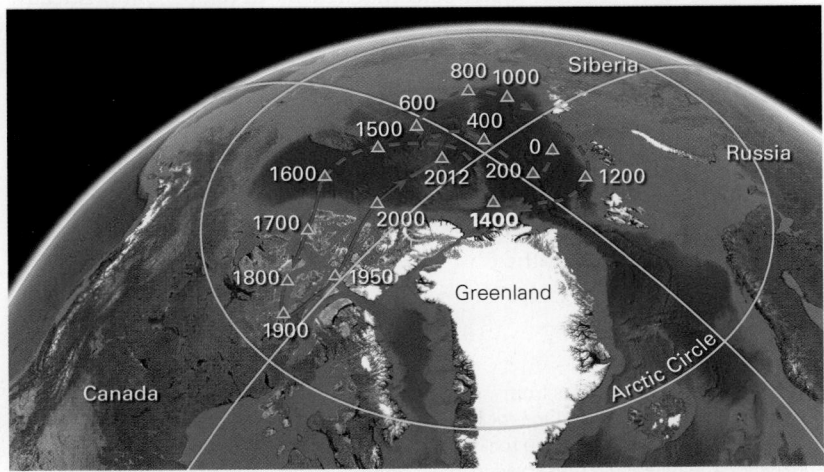

(b) A simplified map showing the changing position of the north magnetic pole over the past 2,000 years. Before about 1600, its position was not as well known, so the path is dashed.

points toward geographic north when placed in the Earth's magnetic field. Note, too, that because of the curvature of magnetic-field lines in space, a magnetic needle that can pivot on a horizontal axis will be horizontal (parallel to the Earth's surface) at the equator, tilted at a steep angle at high latitudes, and vertical (perpendicular to the Earth's surface) at the magnetic pole. The angle of tilt is called *magnetic inclination* (see the inset of Fig. 2.33a).

At any given time, the magnetic poles of the Earth do not exactly overlie our planet's **geographic poles**, the points at which the rotational axis of the Earth intersects its surface. In fact, during recorded history, the north magnetic pole has followed a circuitous route to its present location in the Arctic Ocean off the northern coast of Canada, and it continues to move today at about 55 km/yr **(Fig. 2.33b)**. Generally, magnetic poles don't stray far from the geographic poles, however, because the spin of the Earth plays a role in controlling the magnetic field. So, researchers assume that, when averaged over several thousand years, the position of a magnetic pole roughly corresponds with that of the geographic pole. The angle between the direction to *true north* (geographic north) and the direction that a compass points at a given time is called *magnetic declination*. This angle exists whenever true north and magnetic north do not coincide.

Lodestone is a naturally magnetic rock because it consists of the mineral *magnetite* (iron oxide), so when suspended from a thread, lodestone swivels into alignment with the Earth's magnetic field. Certain rock types contain enough tiny specks of magnetite, or of other magnetic minerals, that they behave, overall, like weak magnets. This behavior allows these rocks to preserve a record of the orientation of the Earth's magnetic field at the time the rock formed. This record of past magnetism is called **paleomagnetism**.

How does paleomagnetism develop? Simplistically, during a rock's formation, the dipoles of magnetic mineral specks in the rock align with the Earth's magnetic field. Once the rock-forming process has finished, the rock's overall dipole represents the sum of the dipoles of all the mineral specks. This overall dipole will not change, even if the orientation of the Earth's magnetic field relative to the rock, or the orientation of the rock relative to the Earth's magnetic field, changes. For example, as basalt forms by the solidification of lava, tiny magnetite crystals grow along with other minerals. The dipoles of the magnetite crystals align with the Earth's magnetic field while the solidifying basalt remains hot. But once the new rock cools below about 600°C, the orientation of the tiny dipole in each crystal becomes fixed. The *paleomagnetic dipole* for the rock, which we represent with an arrow, is the sum of all the tiny locked-in dipoles, each associated with one crystal, scattered through the rock.

Apparent Polar Wander

In the early 20th century, researchers developed instruments capable of detecting paleomagnetism in basalt as well as in certain types of sedimentary rock. When they determined the paleomagnetic dipoles in rocks, they made a surprising discovery: in rocks that formed long ago, the paleomagnetic dipole does not point to the Earth's present-day magnetic poles. At first, these researchers interpreted this discovery to mean that at the time the rock formed, the Earth's magnetic poles were in different locations than they are today. With that concept in mind, they called the location to which a paleomagnetic dipole pointed a **paleopole**. Additional work showed that paleopole positions, as recorded in rocks of different ages from the same region, appeared to change over time. In fact, when plotted on a map, paleopole positions for successively younger rocks from a region trace out a curving line on the surface of the Earth. This line is called an **apparent polar-wander path** (Fig. 2.34a).

To interpret apparent polar-wander paths determined by analyzing paleomagnetism in a sequence of rocks, researchers first assumed that the position of a continent from which a sequence of rocks had been collected had stayed the same through geologic time. If that assumption was correct, then the apparent polar-wander path would represent how the position of the Earth's magnetic pole had migrated over time. But the researchers were in for a surprise! When they obtained apparent polar-wander paths from different continents, they found that each was different from the others (Fig. 2.34b). The hypothesis that continents are fixed cannot explain this observation, for if the magnetic pole moved while all the continents stayed fixed, then paleomagnetic records from all continents should produce the same apparent polar-wander paths.

The researchers realized that they were looking at apparent polar-wander paths in the wrong way. It's not the magnetic pole that moves relative to fixed continents—rather, it's the continents that move relative to a more or less fixed magnetic pole! Moreover, since each continent has a unique apparent polar-wander path,

FIGURE 2.34 Apparent polar-wander paths and their interpretation.

(a) Imagine that a geologist collects six rocks from near Point X on the continental block shown. One rock formed at 600 Ma, one at 500 Ma, and so on, up to the present. By determining the paleopole preserved in each rock (using methods discussed in more advanced books), the geologist can draw an apparent polar-wander path (the red line) that ends at the present-day pole. At first, researchers thought that the red line actually represented the track of the Earth's magnetic pole over time. This interpretation turned out to be incorrect.

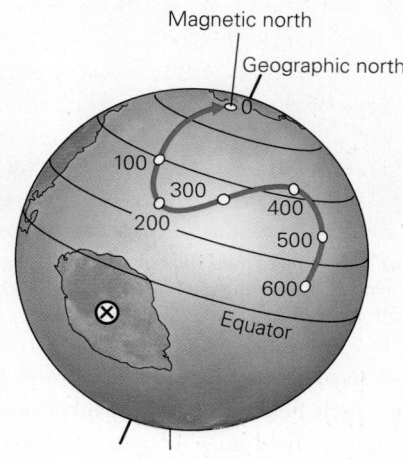

(b) Geologists later discovered that the apparent polar-wander paths for different continents are not the same. For example, on this map of the Earth (looking down on the North Pole), we see apparent polar-wander paths for three continents (North America, Europe, and Africa). They do not lie on top of each other, as would be expected if the positions of continents were fixed over time. Conclusion: continents must move relative to one another at times when they are not joined in a supercontinent.

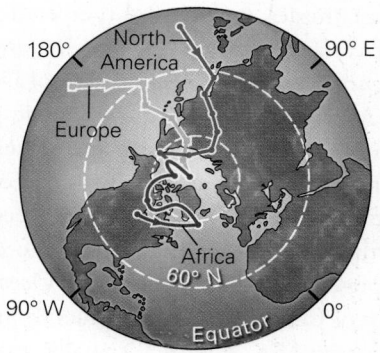

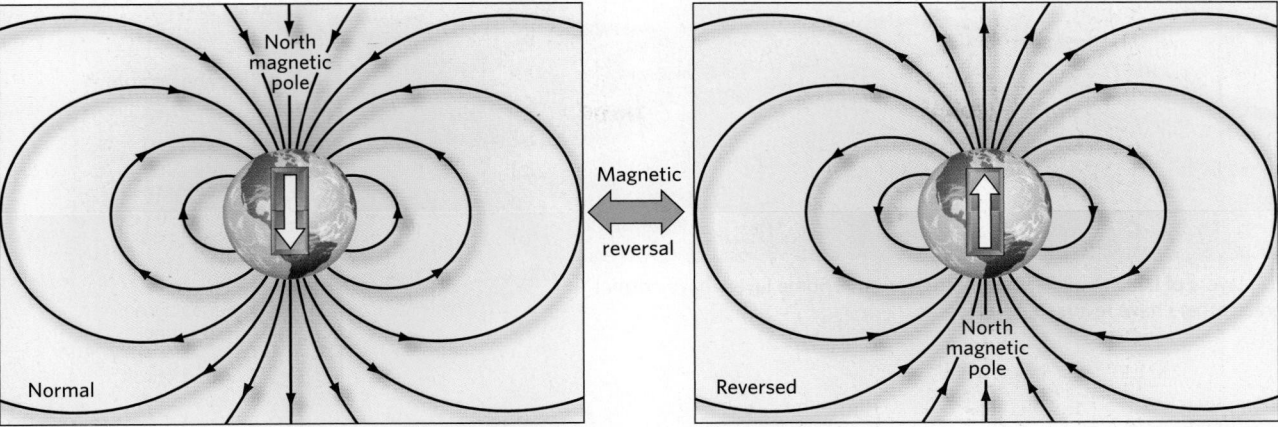

(a) During normal polarity, the dipole points to the south.

(b) During reversed polarity, the dipole points to the north.

continents must move not only relative to the magnetic pole, but also relative to one another. This discovery proved that Wegener was right all along—continents do move! Therefore, the discovery served as a positive test of plate tectonics. (Subsequent studies show that, at times, the lithospheric shell as a whole has moved relative to the magnetic poles, due to redistribution of continental mass. Researchers account for such *true polar wander* when calculating apparent polar-wander paths.)

Magnetic Reversals and Marine Magnetic Anomalies

Has the north-seeking end of a compass always pointed north? No. Study of paleomagnetism has revealed that at various times in the past, the Earth's magnetic field has suddenly flipped. Sometimes the Earth has *normal polarity*, as it does today, with the north magnetic pole near the north geographic pole, and sometimes it has *reversed polarity*, with the north magnetic pole near the south geographic pole (Fig. 2.35). During times of reversed polarity, the dipole pointed in the direction opposite the direction it points in today. Changes in the Earth's magnetic field from normal to reversed polarity, or from reversed back to normal polarity, are called **magnetic reversals**. They may be due to changes in the circulation pattern of liquid iron alloy in the outer core. Note that during a magnetic reversal, only the magnetic-field polarity changes—the Earth itself does not turn upside down.

Not long before the discovery of magnetic reversals, geologists had developed methods for obtaining numerical ages of rocks (see Chapter 8). By recording the paleomagnetic polarity in successions of progressively older layers of basalt whose numerical ages had been determined, researchers in the 1960s established a *magnetic-reversal chronology*, a chart showing when reversals had happened

and, therefore, the durations of polarity intervals between reversals (Fig. 2.36a). A major polarity interval is called a *polarity chron*. Some chrons include a few short-duration intervals, known as *polarity subchrons* (Fig. 2.36b). Because of limitations on the precision of dating, the first magnetic-reversal chronology charts covered only the most recent 4.5 million years of Earth history.

At the same time that some geologists were studying the timing of magnetic reversals, other geologists were measuring regional variations in the strength of the Earth's magnetic field by using an instrument called a *magnetometer*. At any given location on the surface of the Earth, the magnetic field includes two components: one produced by the main magnetic field of the Earth (represented by the dipole), and the other produced by the local magnetism of near-surface rocks. Geologists found that the strength of the magnetism that magnetometers detect varies slightly with location, and they coined the term **magnetic anomaly** for the difference between the expected strength of the Earth's main magnetic field at a location and the actual observed strength of the magnetic field at that location. Field strengths stronger than expected are *positive anomalies*, whereas field strengths weaker than expected are *negative anomalies*. From data collected by towing magnetometers back and forth across the ocean, geologists mapped **marine magnetic anomalies**, variations in magnetic-field strength above the seafloor (Fig. 2.37a). These data defined a stripe-like pattern in which alternating positive and negative anomalies on the seafloor were aligned parallel to the mid-ocean ridge axis at which the seafloor formed (Fig. 2.37b, c).

What does the pattern of marine magnetic anomalies mean? In 1963, geologists proposed that it develops as magnetic reversals take place during seafloor spreading. Simply put, a positive anomaly occurs where the

FIGURE 2.36 Magnetic reversals and their chronology.

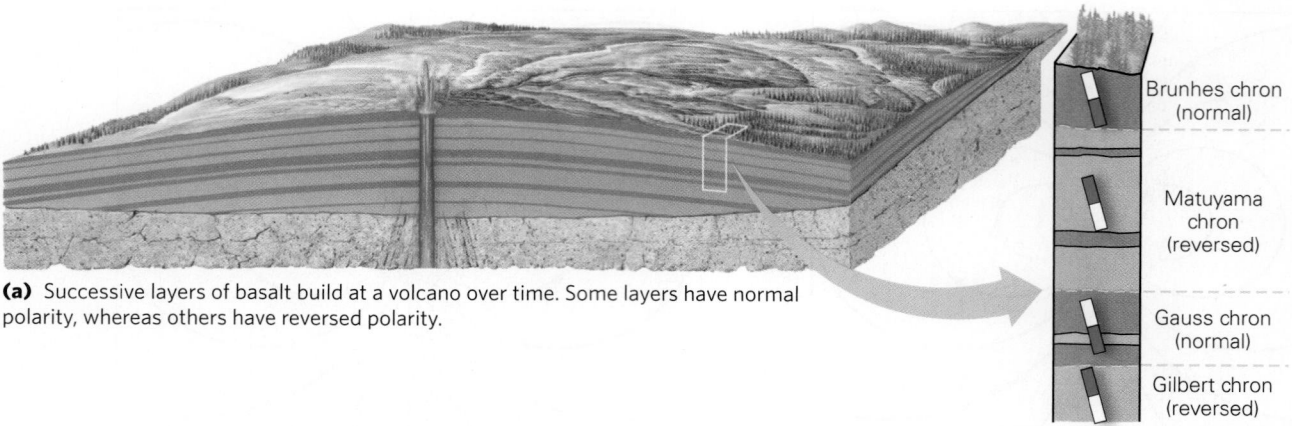

(a) Successive layers of basalt build at a volcano over time. Some layers have normal polarity, whereas others have reversed polarity.

Brunhes chron (normal)

Matuyama chron (reversed)

Gauss chron (normal)

Gilbert chron (reversed)

(b) Observations of such layers led to the production of a magnetic-reversal chronology with named polarity intervals, or chrons.

FIGURE 2.37 Marine magnetic anomalies.

(a) A ship towing a magnetometer detects changes in the strength of the magnetic field on the seafloor.

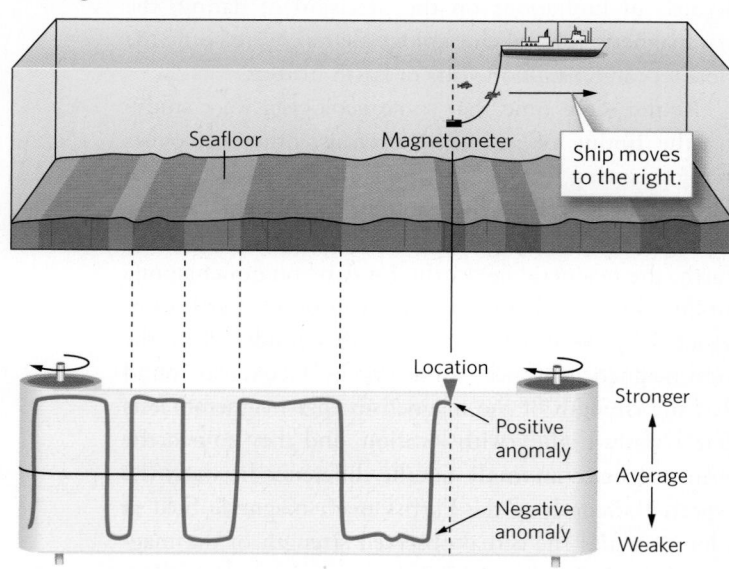

Seafloor · Magnetometer · Ship moves to the right.

Location · Positive anomaly · Stronger · Average · Negative anomaly · Weaker

(b) On a paper record, intervals of stronger magnetism (positive anomalies) alternate with intervals of weaker magnetism (negative anomalies).

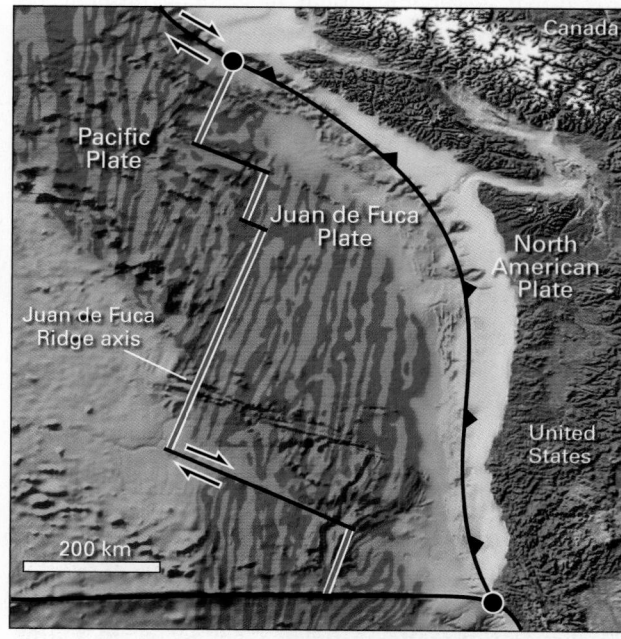

Canada · Pacific Plate · Juan de Fuca Plate · Juan de Fuca Ridge axis · North American Plate · United States · 200 km

(c) An early map showing areas of positive anomalies (dark) and negative anomalies (light) off the west coast of North America. At the time the map was made, the anomalies in the southwest corner of the map had not been measured.

magnetism of the basalt that makes up the seafloor has normal polarity, for in these areas, the magnetic force produced by the basalt adds to the force produced by the Earth's main magnetic field, so a magnetometer measures a stronger signal than expected. A negative anomaly occurs over regions of the seafloor where the basalt has reversed polarity. In these areas, the magnetic force produced by the basalt subtracts from the force produced by the Earth's main magnetic field, so

the magnetometer measures a weaker signal (Fig. 2.38a). Therefore, seafloor yielding positive anomalies forms at times when the Earth has normal polarity, whereas seafloor yielding negative anomalies forms when the Earth has reversed polarity. As seafloor spreading takes place, bands of seafloor with different polarities form and then move away from the ridge axis (Fig. 2.38b).

Researchers realized that for this interpretation of marine magnetic anomalies to be correct, variations in

FIGURE 2.38 The progressive development of marine magnetic anomalies.

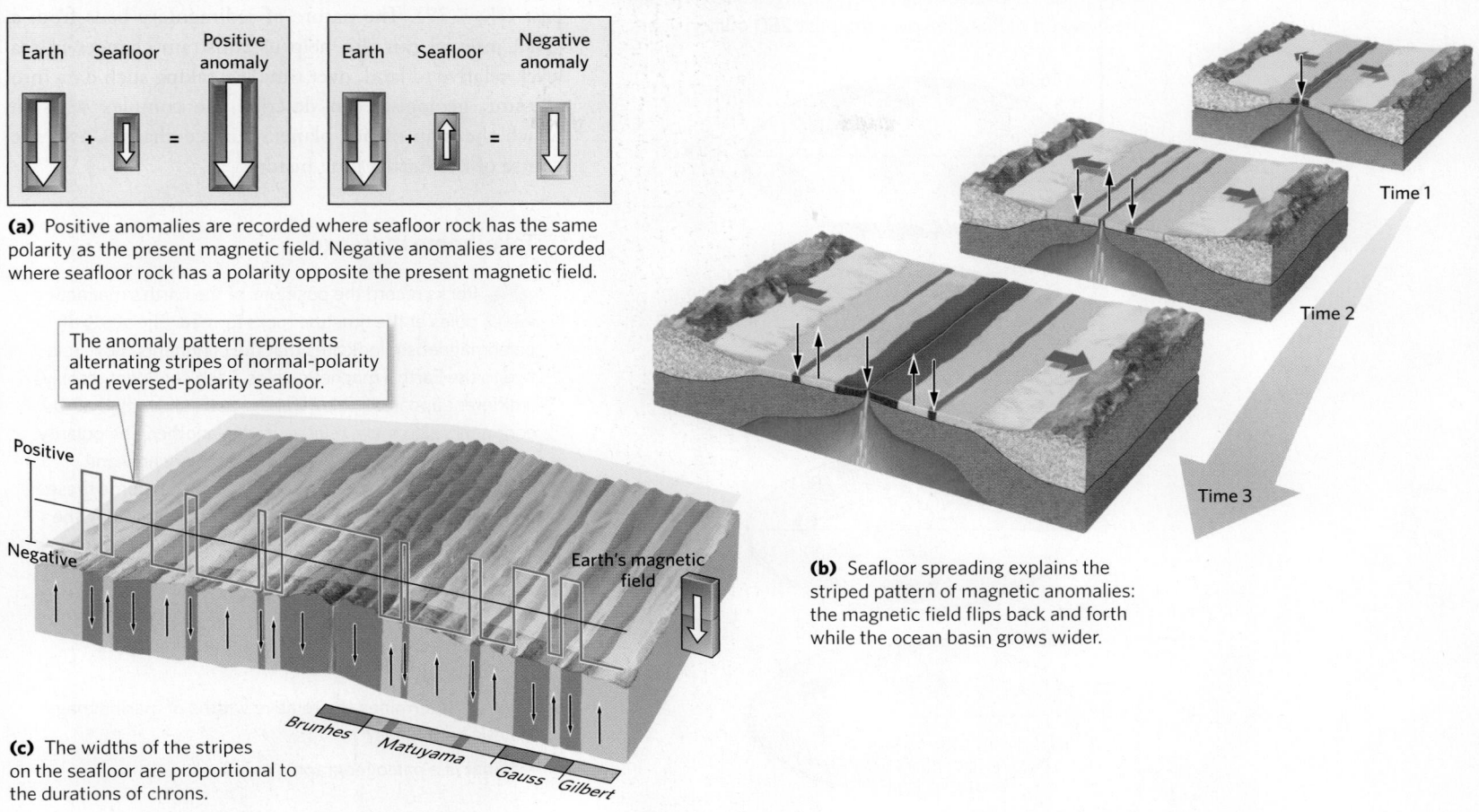

(a) Positive anomalies are recorded where seafloor rock has the same polarity as the present magnetic field. Negative anomalies are recorded where seafloor rock has a polarity opposite the present magnetic field.

The anomaly pattern represents alternating stripes of normal-polarity and reversed-polarity seafloor.

(c) The widths of the stripes on the seafloor are proportional to the durations of chrons.

(b) Seafloor spreading explains the striped pattern of magnetic anomalies: the magnetic field flips back and forth while the ocean basin grows wider.

the widths of stripes on one side of a mid-ocean ridge should be a mirror image of those on the other side and, further, that the relative widths of the anomalies closest to the ridge axis should correspond to the relative durations of polarity chrons for the past 4.5 million years. To test this prediction, they planned an oceanographic research cruise across a mid-ocean ridge to obtain magnetic-anomaly data (Fig. 2.38c). Their results matched the predictions exactly, and therefore provided another key proof of plate tectonics. Notably, by assuming constant rates of plate motion, geologists can estimate the age of seafloor and, therefore, the times of magnetic reversals recorded by seafloor older than 4.5 million years. These estimates indicate that the oldest seafloor on the Earth today is only about 200 million years old, less than 5% of the age of the Earth.

Paleogeography

An understanding of plate tectonics provided a whole new way of thinking about our planet's history. For example, ocean basins don't last forever; they open and later close. Mountain belts don't just rise arbitrarily, but instead form due to continental collisions or due to the development of a convergent boundary along the margin of a continent. The climate at a given location doesn't stay the same over geologic time, but instead changes as the location moves across latitudes. And the map of the Earth evolved as supercontinents formed by collisions and later broke apart by rifting (a process that has happened several times over the Earth's long history). Geologists represent these changes by drawing **paleogeographic maps**—maps depicting the Earth's surface long before human history. Such maps represent, in effect, hypotheses about the configuration of land and sea on ancient Earth.

Where do the data used for the construction of paleogeographic maps come from? By studying marine magnetic anomalies, geologists can reconstruct the opening of present-day oceans, and the positions of the continents bordering those oceans, confidently back to about 100 million years, and in a few locations back to about 200 million years. Studies of apparent polar-wander paths, as well as of the distributions of fossils and particular rock types, provide partial constraints on the relative positions of continents earlier in Earth history, before the seafloor we can see today formed. These features can tell us when continents were merged into supercontinents

The least movement is of importance to all nature.

—*BLAISE PASCAL (FRENCH MATHEMATICIAN, 1623–1662)*

FIGURE 2.39 Due to plate-tectonic processes, the map of the Earth's surface slowly changes. Here we see the breakup of Pangaea over the past 250 million years.

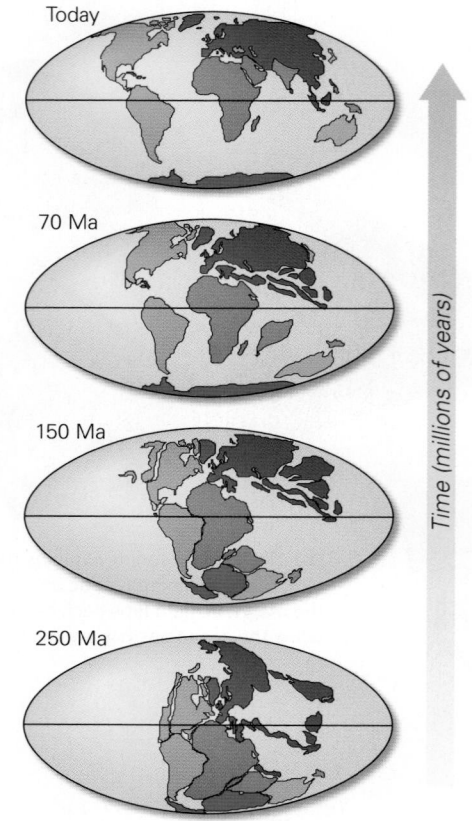

Today

70 Ma

150 Ma

250 Ma

Time (millions of years)

and when they were separate, and they can tell us the latitude at which a landmass resided at a given time in the past (Fig. 2.39). The nature of sedimentary beds from a given interval can also help us constrain changes in sea level, relative to land, over time. By taking such data into account, geologists can describe the complex ways in which the map of our planet's surface changes over the course of the Earth's long history.

Take-home message...

Rocks record the positions of the Earth's magnetic poles at the time the rocks formed. The study of paleomagnetism indicates that the continents move relative to the Earth's magnetic poles. Each continent displays a different apparent polar-wander path, showing that the continents also move relative to one another. The polarity of the Earth's magnetic field reverses every now and then. Seafloor formed at mid-ocean ridges records these reversals as marine magnetic anomalies, confirming the existence of seafloor spreading. Paleomagnetic data allow geologists to draw paleogeographic maps.

Quick Questions

• How do geologists now interpret apparent polar-wander paths?

• What determines the relative widths of marine magnetic anomalies?

• What is a paleogeographic map?

2 CHAPTER REVIEW

Objective 2.1

Discuss the geologic evidence that Alfred Wegener used to justify the proposal that continents moved following the breakup of the supercontinent Pangaea.

KEY CONCEPTS

• Wegener's continental-drift hypothesis states that continents were once joined together to form a single supercontinent, Pangaea, that subsequently broke into smaller continents that moved apart.

• Observations from geology, geography, and paleontology supported this hypothesis; nevertheless, very few geologists accepted the hypothesis in Wegener's day.

EARTH-SCIENCE VOCABULARY

continental drift (p. 54)   **Pangaea** (p. 54)

REVIEW QUESTIONS

1. On what basis did Alfred Wegener fit smaller continents together to form the supercontinent shown on **Figure A**?

2. **(a)** Describe several of the key observations that Wegener used as evidence for continental drift. **(b)** Label the climate belts (tropical; subtropical; polar) and the equator on **Figure A**.

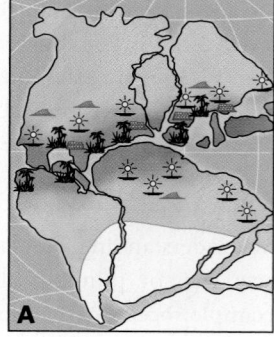

A

3. Why did most geologists in Wegener's day refuse to accept the continental-drift hypothesis?

Objective 2.2

Describe key observations about the nature of the seafloor that paved the way for development of the seafloor-spreading hypothesis.

KEY CONCEPTS

- The invention of sonar allowed researchers to map seafloor bathymetry efficiently.
- Bathymetric maps of the seafloor allow Earth scientists to visualize mid-ocean ridges, deep-sea trenches, fracture zones, and abyssal plains.
- Researchers discovered that more heat rises from mid-ocean ridges than from abyssal plains, that earthquakes occur in distinct belts, and that the sediment layer on oceanic crust varies in thickness.
- These new data led to the seafloor-spreading hypothesis, which states that new seafloor forms at a mid-ocean ridge axis, moves away from the ridge axis, and undergoes subduction at a deep-sea trench.

EARTH-SCIENCE VOCABULARY

abyssal plain (p. 57)	**heat flow** (p. 58)
bathymetry (p. 57)	**mid-ocean ridge** (p. 57)
continental shelf (p. 57)	**ocean basin** (p. 57)
deep-sea trench (p. 57)	**oceanic island** (p. 57)
earthquake (p. 58)	**seafloor spreading** (p. 59)
fault (p. 58)	**seamount** (p. 57)
fracture zone (p. 57)	**seismic belt** (p. 58)

REVIEW QUESTIONS

4. (a) How high does a mid-ocean ridge rise above surrounding abyssal plains? (b) How deep is the deepest trench? (c) What is the average depth of an abyssal plain? (d) What is the orientation of a fracture zone, relative to a mid-ocean ridge axis?

5. (a) Is sediment thicker at a mid-ocean ridge or on an abyssal plain? Explain your answer. (b) Why is heat flow higher along a mid-ocean ridge axis than at locations on an abyssal plain? (c) What bathymetric feature is subduction associated with?

6. (a) Outline the basic concepts in the seafloor-spreading hypothesis, as proposed by Hess. (b) On **Figure B** (a map that extends from South America to Africa), label the locations where seafloor spreading takes place, and where subduction takes place.

7. Why did Hess and others conclude that if seafloor spreading takes place, subduction must also take place?

Objective 2.3

Contrast lithosphere with asthenosphere, characterize the relationship between earthquakes and plate boundaries, and summarize the basic tenets of plate tectonics.

KEY CONCEPTS

- Lithosphere, the Earth's relatively rigid outer shell, consists of the crust and the uppermost mantle. It overlies asthenosphere, warmer mantle that behaves plastically and can flow.
- The lithosphere is divided into seven major and eight minor plates, which move relative to one another. There are three types of plate boundaries, distinguished from one another by the relative motion of the plates.
- Because movement takes place at plate boundaries, they are delineated by earthquakes. Plate interiors remain fairly intact.

EARTH-SCIENCE VOCABULARY

active margin (p. 61)	**passive-margin basin** (p. 61)
asthenosphere (p. 60)	**plastic flow** (p. 60)
continental margin (p. 61)	**plate** (p. 60)
lithosphere (p. 60)	**plate boundary** (p. 60)
lithospheric mantle (p. 60)	**plate tectonics** (p. 60)
passive margin (p. 61)	**rigid** (p. 60)

REVIEW QUESTIONS

8. (a) Explain the difference between rigid behavior and plastic flow. (b) Which layer of the Earth is relatively rigid, and which is capable of plastic flow? (c) Label the layers in **Figure C**.

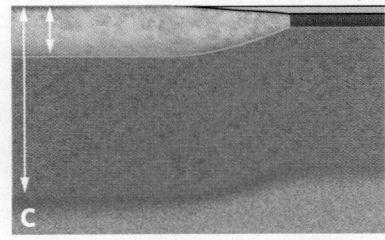

9. (a) Do lithosphere and asthenosphere have the same or different composition? (b) What factor determines whether rock of the mantle flows plastically, or not? (c) What happens to the position of the lithosphere-asthenosphere boundary when a heavy load develops on the Earth's surface?

10. (a) Distinguish between an active continental margin and a passive continental margin. (b) Label these features on **Figure B**.

11. What is the relationship of a passive-margin basin to a continental shelf?

Objective 2.4

Name the three types of plate boundaries, describe the geologic features that form at each type, and recognize examples of plate boundaries.

- At divergent boundaries, seafloor spreading takes place, so plates on either side of the boundary move away from each other. A mid-ocean ridge delineates a divergent boundary.

- Beneath the ridge axis, rising asthenosphere melts to produce molten rock. Some of this molten rock solidifies at depth under the ridge axis, but some spills out onto the seafloor.

- At convergent boundaries, plates on either side of the boundary move toward each other. Oceanic lithosphere on one side undergoes subduction and sinks back into the mantle under the overriding plate. A trench delineates the boundary.

- As subduction takes place, sediment gets scraped off the downgoing plate and a volcanic arc forms.

- At transform boundaries, one plate slides horizontally along the boundary, so no new lithosphere forms and no existing lithosphere gets subducted.

EARTH-SCIENCE VOCABULARY

accretionary prism (p. 70) **magma** (p. 65)
continental arc (p. 70) **subduction** (p. 69)
convergent boundary (p. 69) **subduction zone** (p. 69)
divergent boundary (p. 65) **transform boundary** (p. 72)
island arc (p. 70) **transform fault** (p. 72)
lava (p. 66) **volcanic arc** (p. 69)

REVIEW QUESTIONS

12. **(a)** On **Figure D**, trace out the plate boundaries that are visible. Explain the basis for your placement of lines. **(b)** Label a divergent boundary, a convergent boundary, and a transform boundary. **(c)** Name the plates that are visible, and indicate which ones consist only of oceanic lithosphere.

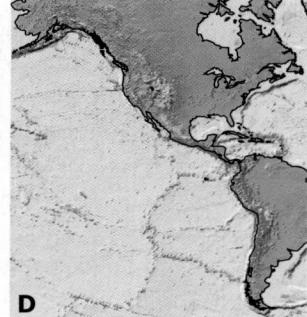

D

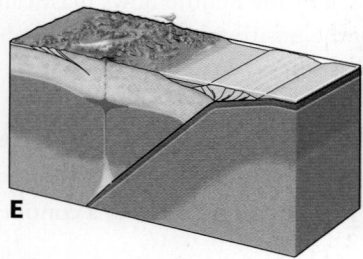

E

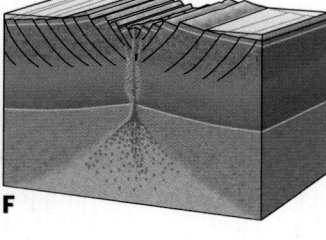

F

13. **(a)** Which type of plate boundary does **Figure E** show? **(b)** Add arrows to the figure to indicate the relative plate motion.

14. **(a)** Explain how oceanic crust forms along a mid-ocean ridge. **(b)** On **Figure F**, label the ridge axis and the magma chamber.

15. **(a)** How does lithospheric mantle form? **(b)** Does lithospheric mantle become thicker or thinner away from a ridge axis? **(c)** Why does ocean-floor sediment get progressively thicker away from the ridge axis?

16. **(a)** Where does an accretionary prism form, and why does it form? **(b)** Distinguish between a continental arc and an island arc. Which type does **Figure E** show? **(c)** What is a Wadati-Benioff zone?

17. **(a)** Which part of an oceanic fracture zone is considered to be a transform fault? **(b)** Why don't tall mountain ranges form along a transform boundary? **(c)** On **Figure G**, indicate the sense of slip on the transform fault with arrows. **(d)** How many different plates are visible in **Figure G**? Explain the basis for your answer.

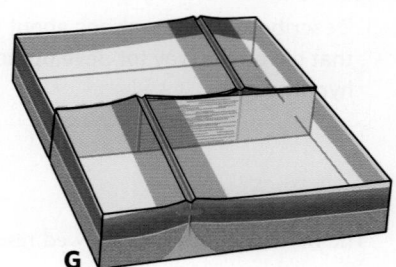

G

Objective 2.5

Explain how new plate boundaries can form and how existing ones can eventually cease activity.

KEY CONCEPTS

- Subduction ceases when two buoyant pieces of crust converge at the subduction zone and collide. The collision can produce a mountain belt.

- A continent can split by the process of rifting. During rifting, continental lithosphere stretches and thins.

EARTH-SCIENCE VOCABULARY

buoyancy force (p. 74) **rifting** (p. 75)
collision (p. 74) **suture** (p. 74)
rift (p. 75)

REVIEW QUESTIONS

18. **(a)** During continental collision, does the crust become thicker or thinner? **(b)** On **Figure H**, label the suture, and indicate the relative motion of the two continents on either side of the suture.

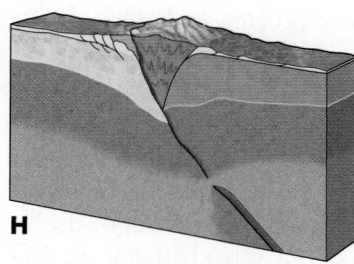

H

19. **(a)** During rifting, does the lithosphere become thicker or thinner? **(b)** How might you recognize a rift at the surface of the Earth? **(c)** Name the rift that appears in **Figure I**. **(d)** Label the Red Sea in **Figure I**. Is the Red Sea a successful rift or an unsuccessful rift? How can you tell? **(e)** What is the relationship between rifting and the development of a passive-margin basin?

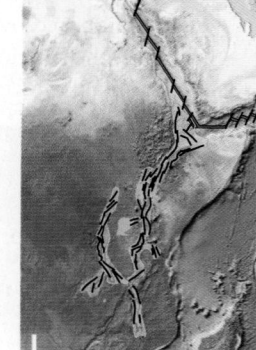

I

Objective 2.6

> Point out examples of triple junctions and hot spots, and explain how hot spots differ from plate boundaries.

KEY CONCEPTS

- At a triple junction, three plate boundaries intersect.

- At a hot spot, asthenosphere rises, probably in a narrow column called a mantle plume. Melting that takes place at the top of the mantle plume provides molten rock that rises to cause volcanic activity at the Earth's surface. A plate can move over a hot spot.

EARTH-SCIENCE VOCABULARY

hot spot (p. 77) **mantle plume** (p. 77)
hot-spot track (p. 79) **triple junction** (p. 77)

REVIEW QUESTIONS

20. **(a)** A triple junction occurs along the west coast of California. Identify the location of this triple junction. **(b)** List the types of plate boundaries that meet at this triple junction.

21. **(a)** On **Figure J**, which shows the Hawaiian Islands, indicate which of the islands is the oldest, and add an arrow to show the direction in which

    the Pacific Plate is moving relative to the hot spot. **(b)** Why do the islands get progressively smaller going from southeast to northwest?

Objective 2.7

> Characterize the forces that drive plate motion and the rates at which plates move.

KEY CONCEPTS

- Plate motion happens due to several forces, including ridge push, slab pull, and convective flow in the asthenosphere. Plates move at rates of 1–15 cm/yr (0.4–6 in/yr).

- Modern GPS methods can measure plate motions in real time—we can effectively observe plates moving!

EARTH-SCIENCE VOCABULARY

absolute plate velocity (p. 80) **relative plate velocity** (p. 80)
convection (p. 80) **ridge-push force** (p. 80)
global positioning system (GPS) **slab-pull force** (p. 80)
(p. 82)

REVIEW QUESTIONS

22. Describe the major forces that move lithosphere plates.

23. **(a)** Explain the difference between relative plate velocity and absolute plate velocity. **(b)** **Figure K** is a map of southern California. The black lines represent the traces of faults; the

San Andreas Fault (SAF) is labeled. What do the red arrows represent, and why are the arrows east of the San Andreas Fault shorter than those to the west?

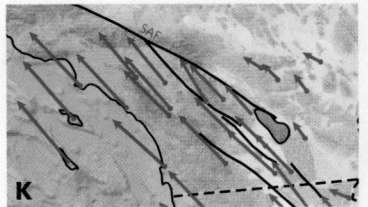

Objective 2.8

> Provide examples of data, from studies of paleomagnetism and other sources, that serve as proofs of plate tectonics.

KEY CONCEPTS

- Observations about sediment thickness, the slip on transform faults, and the velocity of plate motions, among many others, serve as successful tests of the theory of plate tectonics.

- Paleomagnetic studies, which analyzed the record of the Earth's ancient magnetic field as preserved in rock, were particularly important in testing the theory.

- Apparent polar-wander paths are different for different continents because continents move relative to one another, while the Earth's magnetic poles remain roughly fixed.

- Marine magnetic anomalies develop because magnetic reversals happen while seafloor spreading is taking place. The pattern of magnetic anomalies reflects seafloor spreading.

- Because of plate motion, continents move relative to one another, so the map of the Earth's surface constantly changes over geologic time.

EARTH-SCIENCE VOCABULARY

apparent polar-wander path **magnetic reversal** (p. 85)
(p. 84) **marine magnetic anomaly**
dipole (p. 83) (p. 85)
geographic pole (p. 83) **paleogeographic map** (p. 87)
magnetic anomaly (p. 85) **paleomagnetism** (p. 84)
magnetic pole (p. 83) **paleopole** (p. 84)

REVIEW QUESTIONS

24. **(a)** What is paleomagnetism? **(b)** How were apparent polar-wander paths originally interpreted? **(c)** How did study of apparent polar-wander paths from different continents serve as a proof that continents move?

25. **(a)** Does the Earth turn upside down during times of reversed polarity? **(b)** How did geologists figure out the magnetic-reversal chronology of the last few million years?

26. **(a)** What is a marine magnetic anomaly? **(b)** How do geologists now interpret the pattern of these anomalies relative to a mid-ocean ridge axis?

27. Based on the paleogeographic map of **Figure L**, which ocean opened first: the North Atlantic or the South Atlantic? **L**

3 WHAT IS THE EARTH MADE OF?
Introducing Minerals and the Nature of Rock

After studying this chapter, you should be able to . . .

1. define minerals and crystals in the context of geology, and describe how minerals form.

2. recognize key physical characteristics of minerals, and explain how minerals can be identified and classified.

3. discuss why some minerals are gemstones, and explain how jewelers modify gemstones to make jewelry.

4. recognize key characteristics of rock, and distinguish among the three rock groups.

5. sketch a diagram of the rock cycle, and explain why this cycle takes place on the Earth.

A dusty caravan of four-wheel-drive vehicles slowly climbed a desert hill, following a track that was rapidly getting narrower and bumpier. The jouncing students inside were hoping to visit an abandoned mine rumored to contain beautiful mineral crystals. But when the group finally reached the mine and got out, they suddenly realized . . . it wasn't abandoned. A one-armed watchman glared out from behind an old refrigerator door on which he'd painted the words, "Is there life after death? Pass this sign and find out!" After a few moments of awkward silence, one of the students spoke up. "We're just poking around looking for minerals," she said. The watchman replied, "I know, I've been watching." Undeterred, the student quietly asked, "Any chance we can take a look in the mine?" There was another long silence. Then, the man whispered, "You forgot the magic word!" In unison, all the students looked up and said, "Please?" With a grin, the watchman said, "What the heck, I could use some company. You can look but don't collect." He led them into the mine, a tunnel that had been carved into a rock cliff. With the headlamps on their hard hats shining and their eyes open wide, the students explored. On some walls, the rocks were streaked by bright blue and green minerals (Fig. 3.1). The trip had not been wasted!

We briefly introduced minerals and rocks in Chapter 1, in the context of discussing the geosphere, the solid part of the Earth System. Here in Chapter 3, we refine our definition of a mineral and describe features used to distinguish among and classify minerals. After reviewing minerals, we turn our attention to rocks and present the basis for classifying rocks into three groups. We conclude this chapter by characterizing the rock cycle. Chapter 3 sets the stage for the following two chapters.

3.1 What Is a Mineral?

Defining Minerals

In everyday English, people commonly use the word *mineral* for a great variety of different materials. **Mineralogists**, researchers who study minerals, apply the word in a more restricted sense: a **mineral** is a naturally occurring, homogeneous, crystalline solid that has a definable chemical composition and, in most cases, is inorganic. Let's pull apart this mouthful of a definition.

- *Naturally occurring:* Real minerals form in nature, not in factories. We emphasize this point because in recent decades, chemists have synthesized materials with

<< A museum specimen of a relatively rare mineral, wulfenite, a lead-containing oxide mineral. Crystals of wulfenite grow in beautiful clusters of brilliant yellow plates, making this mineral popular with collectors.

FIGURE 3.1 Colorful minerals on the wall of a mine. This example contains copper.

characteristics identical to those of real minerals. Mineralogists refer to such materials as *synthetic minerals*.

- *Homogeneous:* A piece of a mineral has roughly the same composition and structure throughout.
- *Solid:* A mineral cannot be a liquid (such as oil or water) or a gas (such as air).
- *Crystalline:* Minerals are **crystalline**, meaning that the atoms within them are arranged in a specific, orderly pattern.
- *Definable chemical composition:* A given mineral contains specific elements in specific proportions, so we can write a *chemical formula* for a mineral (Box 3.1). For example, a sample of calcite has the formula $CaCO_3$, and a sample of quartz has the formula SiO_2.
- *Inorganic:* To understand this term, we must first distinguish between organic and inorganic chemicals. **Organic chemicals** are compounds containing carbon and hydrogen (and generally other elements, too, such as oxygen and nitrogen), in which carbon atoms typically connect in chains or rings. Carbon-containing molecules in living organisms are organic chemicals, but not all organic chemicals occur in organisms. Plastic and oil, for example, consist of organic chemicals. Almost all minerals consist of **inorganic chemicals**—chemicals that do not fit the definition of being organic. The exceptions include rare crystalline compounds that grow in ancient bird droppings.

With the geologic definition of a mineral in mind, let's distinguish between a mineral and a glass. Atoms, ions, or molecules in a mineral lock into a regular arrangement, like soldiers standing in formation (Fig. 3.2a), whereas those in a glass connect in a semi-chaotic way, like people standing around at a party (Fig. 3.2b). In other words, a **glass** is an inorganic solid that does not have a crystalline structure. Most glass you encounter in everyday life comes from a factory. But some glass, as we'll see in Chapter 4, forms by natural geologic processes at volcanoes.

▶ **Animation**
Minerals

BOX 3.1

Science Toolbox

Basic chemistry terminology

To describe minerals, we need to use many terms defined by chemists. We provide a list of these terms here for ready reference. Terms appear in an order that allows us to use previous terms to define later ones.

- *Element:* An *element* is a substance that cannot be separated into other elements (see Box 1.2). Chemists represent each element with a symbol.

- *Atom:* An **atom**, the smallest piece of an element that retains the characteristics of that element, consists of a nucleus surrounded by an electron cloud. Except for the common form of hydrogen, whose nucleus contains one proton and no neutrons, an atomic nucleus contains both protons and neutrons. Electrons have a negative charge, protons have a positive charge, and neutrons have a neutral charge. The number of protons in an atom's nucleus is its *atomic number*, and the number of protons plus neutrons is approximately its *atomic mass*.

- *Ion:* An atom with the same number of electrons as protons has a neutral charge. If an atom gains or loses electrons, it becomes an **ion**. An *anion* has a negative charge because it has more electrons than protons, while a *cation* has a positive charge because it has fewer electrons than protons. A superscript sign in the chemical formula indicates the charge. For example, Cl^- (the chloride anion) has gained one extra electron, whereas Fe^{2+} (the ferrous cation) has lost two electrons.

- *Chemical bond:* An attractive force holding two atoms together is a *chemical bond. Covalent bonds* form when atoms share electrons. *Ionic bonds* form when a cation and an anion get close to and bond to each other because their opposite charges attract. In materials with *metallic bonds*, some electrons can move freely.

- *Molecule:* Two or more atoms bonded together form a **molecule**. The atoms may be of the same element or of different elements.

- *Compound:* A substance whose molecules contain two or more elements.

- *Chemical:* A general name for a pure substance made from atoms or molecules of a single element, or from molecules of a single compound.

- *Chemical reaction:* A process that involves the breaking or forming of chemical bonds. Chemical reactions can break existing molecules apart or produce new molecules out of atoms or other molecules.

- *Mixture:* A combination of two or more elements or compounds that can be separated without requiring chemical reactions.

- *Solution:* A type of material in which one chemical (the *solute*) dissolves in another (the *solvent*). A solute may separate into ions during the process of dissolution. For example, when the mineral halite (NaCl), common table salt or rock salt, dissolves in water, it separates into sodium (Na^+) and chloride (Cl^-) ions.

- *Precipitate* (verb): To form a solid from ions in a solution. For example, when saltwater evaporates, salt crystals precipitate from the water.

- *Precipitate* (noun): The solid formed when ions in a solution bond together and separate from the solution is a **precipitate**.

If you ever need to figure out whether a substance is a mineral or not, just check it against the criteria of the definition. Is gasoline a mineral? No—it's a liquid consisting of organic chemicals. Is table salt a mineral? Yes—it's a natural solid crystalline compound, with the formula NaCl. Is sugar in a sugar bowl a mineral? No—it's a processed organic chemical.

Beauty in Patterns: Crystals and Their Structure

As is the case for the word mineral, mineralogists use a more limited definition of the word crystal than we use in daily life. Specifically, a **crystal** is a single, continuous (uninterrupted) solid inside which atoms, molecules, or ions have an orderly arrangement. Geologists use another word, **grain**, in a more general sense, for any small, natural, solid particle. Some grains are crystals that grew into their present shape, but others are fragments of larger crystals, fragments of rocks containing many crystals, or shards of glass.

Some crystals, known as *euhedral crystals*, have smooth, flat surfaces, known as **crystal faces**, that intersect at sharp edges. In a euhedral crystal of a given mineral, each face makes a specific angle relative to the adjacent face; that *interfacial angle* is a characteristic of the mineral. For example, each face of a quartz crystal makes an angle of 120° relative to the adjacent face. Euhedral crystals come in many distinctive geometric shapes—cubes, trapezoids, pyramids, octahedrons, hexagonal columns, prisms, blades, needles, columns, and obelisks—that look like they belong on the pages of a geometry book (Fig. 3.3). *Anhedral crystals*, in contrast, have irregularly shaped surfaces. A given mineral can occur either in euhedral or anhedral forms; the difference in shape depends on the availability of space when it grew, as we will discuss below.

The orderly arrangement of atoms inside a crystal—its **crystal structure**—serves as one of nature's most spectacular examples of *pattern*, the repetition of a shape in a geometric arrangement (Fig. 3.4a, b). Mineralogists refer to the imaginary geometric grid defining the relative positions of atoms in the crystal as a *crystal lattice*, and represent it visually in a variety of ways (Fig. 3.4c). Notably, in

Did you ever wonder...

what makes the shiny faces on a quartz crystal?

FIGURE 3.2 Distinguishing crystalline from noncrystalline solids.

(a) Crystalline solids, such as quartz, contain an orderly arrangement of atoms. The geometry of this arrangement defines a mineral's crystal structure. That structure resembles scaffolding, but at an atomic scale.

A crystal face.

A crystal structure resembles scaffolding.

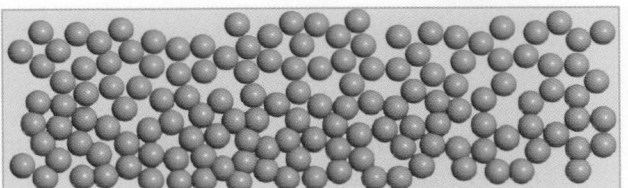

(b) A "cut crystal" bowl is not actually made from a crystal, but rather consists of glass. The atoms within it do not have an orderly arrangement. The shiny surfaces are made by grinding away the glass.

FIGURE 3.3 Crystals come in a variety of shapes. Each example shown here illustrates an ideal crystal of the specified mineral.

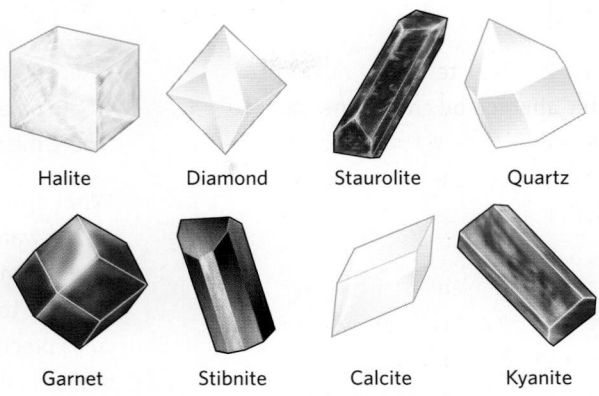

Halite   Diamond   Staurolite   Quartz

Garnet   Stibnite   Calcite   Kyanite

FIGURE 3.4 The internal structure of minerals.

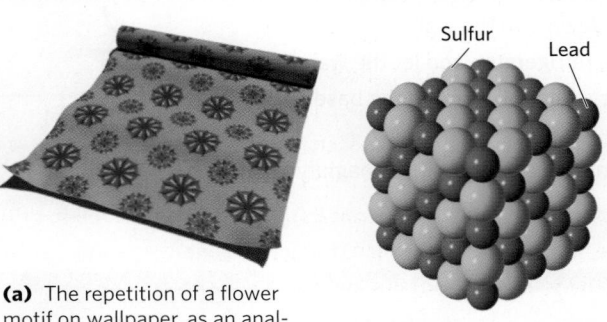

(a) The repetition of a flower motif on wallpaper, as an analogy for the pattern of atoms in a mineral.

(b) The repetition of alternating sulfur and lead atoms in the mineral galena (PbS).

Sulfur   Lead

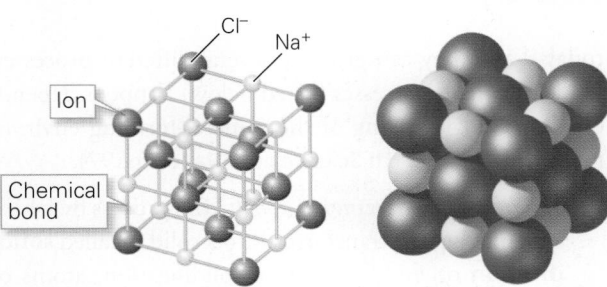

Cl^- Na^+

Ion

Chemical bond

(c) Models of halite (NaCl). The ball-and-stick model on the left portrays ions as spheres and chemical bonds as straight bars. The ball model on the right gives a sense of how ions fit together.

a few cases, two or more minerals with different crystal structures share the same chemical formula. The two or more versions are called *polymorphs* of the chemical. For example, calcite and aragonite are polymorphs of $CaCO_3$.

How Do Minerals Form?

Minerals in the Earth System have formed at various times—some have been around for billions of years, some formed thousands of years ago, and some grew

How can I explain…

Crystal growth

What are we learning?

Crystals grow over time at rates affected by environmental conditions.

What you need:

- Water and table salt
- A 0.5 L (16 fl oz) glass jar
- A spoon and a string

Instructions:

- Heat enough water to fill the jar. Pour the hot water into the jar and dissolve as much salt in the water as it can hold.
- Tie the string to the spoon and place the spoon across the top of the jar, so that the string hangs down into the water.
- Place the jar, uncovered, in a refrigerator, and let the water evaporate over a period of days. Salt crystals will form on the base of the jar and on the string.
- Remove the string to examine the crystals with a magnifying glass.

- Repeat the experiment, only this time, place the jar in a warm-water bath on a hot plate. Let the water in the jar evaporate quickly. Compare the sizes of the crystals produced by the two experiments.

What did we see?

As the water evaporates, the solution becomes supersaturated, so that it has no more capacity to hold dissolved salt. The Na^+ and Cl^- ions combine to form salt molecules, and salt crystals grow as those molecules precipitate. Under magnification, the characteristic cubic shape of the crystals may be visible. Evaporation rates affect crystal size.

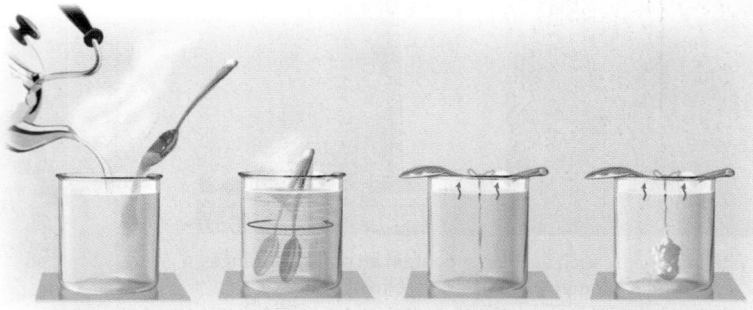

today—and they can grow by several different processes. Which of these processes (listed below) happens depends on the geologic setting of the mineral-forming environment **(Box 3.2)** (**Earth Science at a Glance**, p. 97).

- *Solidification (freezing) of a melt:* The process by which a melt (a liquid) transforms into a solid is called **solidification** or *freezing*. During solidification, atoms or ions moving about in the melt progressively bond together to form solid crystals, which grow until all the melt has transformed into a solid. Formation of ice from water in a freezer represents this process.

- *Precipitation from a liquid solution:* During precipitation, ions of a solute bond together to form solid crystals that separate from the solvent. In other words, the minerals are precipitates. Salt crusts left when seawater evaporates form by precipitation. Some precipitated crystals are attached to a pre-existing solid surface as they grow, while others form while still suspended in a solution and then settle out due to gravity. Precipitation happens when a solution becomes

supersaturated, meaning that it contains more ions than can remain in solution.

- *Biomineralization:* The process of *biomineralization* takes place when metabolism in living cells, or the release of chemicals by cells, causes dissolved atoms in the surrounding solution to precipitate. Extraction of calcium (Ca^{2+}) and carbonate (CO_3^{2-}) ions by marine organisms to make shells of calcite or aragonite (polymorphs of $CaCO_3$) serves as an example.

- *Precipitation from a gas:* Minerals may precipitate at vents (openings) where volcanic gases enter the atmosphere and cool quickly. Deposits of bright yellow sulfur crystals—known in literature as brimstone—grow by this mechanism.

- *Solid-state diffusion:* **Diffusion** involves the migration of atoms or molecules through a material. Atoms can diffuse through water or air to reach the surfaces of a growing crystal during precipitation. Diffusion can also take place in solids, but much more slowly. During such *solid-state diffusion*, atoms migrate from

The Formation of Minerals

A melt starts to cool.

At first, a few crystals form. They remain surrounded by melt.

Eventually, all the melt solidifies.

Temperature decreases over time

Precipitation of minerals happens when water of a salty desert lake undergoes evaporation, so that the water becomes supersaturated with salt.

Solidification from a melt happens when lava, erupted by a volcano, cools.

A close-up of salt crystals

Precipitation of minerals from volcanic gas can also occur. Yellow sulfur crystals form this way.

Fumarole

Sulfur crystal

Microstructure of crystals in a shell

Biomineralization refers to the production of minerals by organisms.

Diffusion can happen in a solid rock, though very slowly. During the process, atoms migrate through the crystal. New minerals, such as garnet, can grow in the rock.

$CaCO_3$ shell

Ions

Reef organisms extract ions from water to make shells.

Garnet schist

FIGURE 3.5 The growth of crystals.

(a) A crystal grows as atoms in the surrounding material attach themselves to its faces. As the crystal grows, its faces move outward into the available space. This example shows precipitation from a water solution.

Atoms attach to the crystal face.

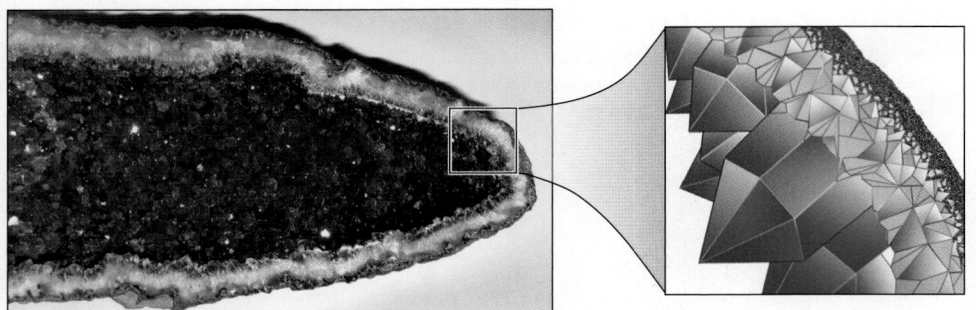

(b) Crystals that grow into an open space, like the purple quartz crystals (amethyst) in this geode from Brazil, develop a euhedral form (inset).

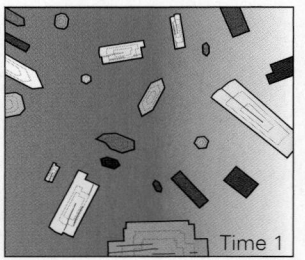

(c) Growing crystals maintain their shapes until they interfere with one another.

(d) A crystal growing in a confined space will be anhedral.

point to point through a crystal lattice. The process, which requires chemical bonds to break and re-form, can allow new crystals to grow in solid rock from atoms that diffused away from pre-existing different minerals. Garnet, for example—a dark purple mineral that grows into a 12-sided crystal—grows in solid rock by incorporating atoms released by chemical reactions involving other minerals.

How does a mineral crystal attain the size and shape that you see in a rock? The process starts with the chance appearance of a microscopic crystal called a *seed*. Other atoms or ions in the surrounding material then attach themselves to the faces of the seed, and the seed grows large enough to be called a crystal. During growth, the youngest part of the crystal lies on the crystal's surface **(Fig. 3.5a)**.

The crystal faces of euhedral crystals maintain the same angle relative to one another as they grow. A crystal's shape (needle-like, sheet-like, blade-like) depends both on the geometry of the crystal lattice and on whether the crystal grows faster in one direction than in others.

If a mineral grows without being inhibited by its surroundings, it remains a euhedral crystal, so the flat faces of euhedral crystals grew that way—they are not a consequence of breaking. For example, crystals protruding from the walls of a **geode**, a mineral-lined open cavity in rock, commonly have a euhedral form because they grow outward into water **(Fig. 3.5b)**. A crystal that grows within an enclosed space, such as the space between other crystals that have already formed, commonly becomes anhedral because the walls of the space inhibit its growth, and the crystal fills in the irregularly shaped space between the other, already formed, crystals **(Fig. 3.5c, d)**.

Take-home message . . .

Minerals are naturally occurring, homogeneous solids with a crystal structure and a definable chemical composition; most are inorganic. They grow in many ways, including solidification from a melt and precipitation from a solution. Crystal faces of a given mineral have a characteristic orientation relative to one another.

Quick Questions

- Do all crystals have the same shape?
- How does glass differ from a mineral?
- What is the difference between solidification and precipitation?
- What is a euhedral crystal?

3.2 How Can We Tell One Mineral from Another?

Identifying Minerals by Their Physical Properties

Mineralogists have identified about 4,000 different minerals so far, and they discover new ones every year. Minerals differ from one another because they have different chemical compositions, different crystal structures, or both. Amateurs and professionals alike get a kick out of identifying a *mineral specimen*, meaning an intact piece consisting of one or more crystals or grains of an individual mineral. How do they do it? The skill lies in learning to recognize *physical properties*—visual and material characteristics—that distinguish one mineral from another. Some of these properties, such as shape and color, can be determined from a distance. Others can be determined only by handling the specimen or by performing an identification test on it. *Identification tests* include scratching the mineral against another object,

FIGURE 3.6 Physical properties of minerals.

(a) Color is diagnostic of some minerals, but not all. Quartz, for example, can come in many colors.

(b) To obtain the streak of a mineral, rub it against an unglazed ceramic plate. The streak consists of mineral powder.

(c) Pyrite has a metallic luster because it gleams like metal.

(d) Feldspar has a nonmetallic luster. Two common versions of feldspar, plagioclase and orthoclase (potassium feldspar), differ in terms of the cations they contain.

(e) Calcite fizzes as it reacts with hydrochloric acid to produce carbon dioxide gas.

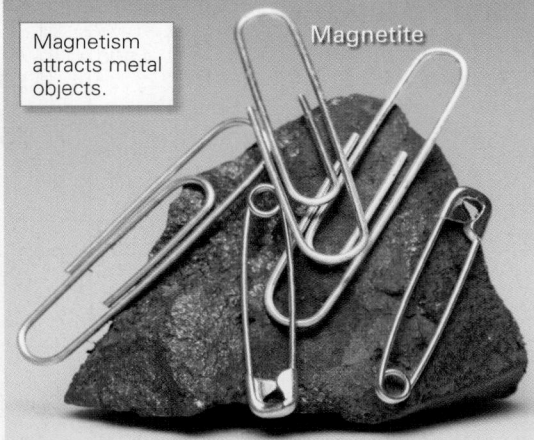

(f) Magnetite is magnetic.

placing it near a magnet, lifting it, tasting it, or placing a drop of acid on it. Let's consider some of the common physical properties used in mineral identification.

INTERACTION WITH LIGHT. The wavelengths of visible light that a mineral does not absorb, and your eye detects, determine the **color** of a mineral. Some minerals always have the same color, but others come in a range of colors (Fig. 3.6a). Typically, color variations depend on impurities in the crystal—the presence of iron atoms in quartz, for example, makes the quartz red or purple. The **streak** of a mineral refers to the color of a powder produced by scraping the mineral against an unglazed ceramic plate (Fig. 3.6b). A streak for a given mineral tends to be less variable than the color of a whole specimen. **Luster** refers to the way a mineral's surface reflects light. You can describe luster by comparing the appearance of the mineral with that of a familiar substance. For example, a mineral that looks like a metal has a *metallic luster*, and one that does not has a *nonmetallic luster* (Fig. 3.6c, d). Various adjectives used for different types of nonmetallic luster include silky, glassy, satiny, resinous, pearly, and earthy.

HARDNESS AND DENSITY. **Hardness** indicates the relative ability of a mineral to resist scratching—simplistically, it depends on the strength and abundance of bonds in a mineral. Recognizing that harder minerals can scratch softer ones, but not vice versa, a mineralogist named Friedrich Mohs made a ranked list of several minerals in sequence of relative hardness and assigned numbers (1 to 10) to these minerals. In the resulting **Mohs hardness scale**, a mineral with a hardness of 5, for example, can scratch all minerals with a hardness of less than 5. We can add familiar objects (fingernails, copper coins, window glass, and steel knives) to the scale to provide context for judging the hardness of unfamiliar minerals (Table 3.1). For example, if window glass has a hardness of about 5.5, then any mineral that can scratch glass has a hardness greater than 5.5, and any mineral that cannot scratch glass has a hardness of less than 5.5. The *density* (mass per

EXPLORE EARTH SCIENCE IN 3D
Orthoclase

TABLE 3.1 Mohs hardness scale. The table gives Mohs's hardness numbers. Note that these numbers are relative, and that they do not have units. The graph indicates hardness as measured by indenting mineral specimens with a tool of a specific shape. The scale is in units of force per unit area. "Kgf" is the force of gravity acting on a mass of 1 kg at the Earth's surface. Note that diamond is 3.5 times harder than corundum.

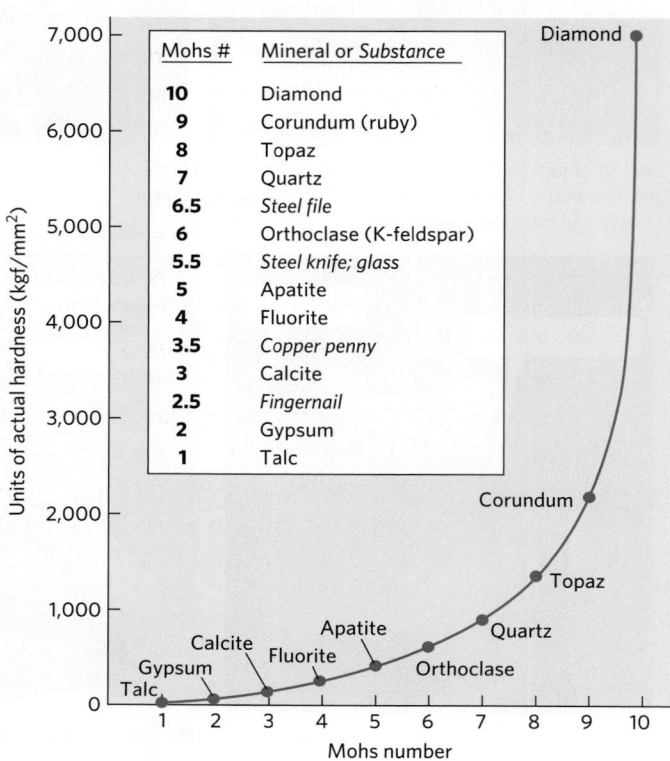

Mohs #	Mineral or *Substance*
10	Diamond
9	Corundum (ruby)
8	Topaz
7	Quartz
6.5	*Steel file*
6	Orthoclase (K-feldspar)
5.5	*Steel knife; glass*
5	Apatite
4	Fluorite
3.5	*Copper penny*
3	Calcite
2.5	*Fingernail*
2	Gypsum
1	Talc

unit volume) of a mineral depends on the atomic mass of atoms in the mineral and on how closely packed those atoms are. We can represent the density of a mineral by stating its **specific gravity**, which is the ratio of the weight of a volume of the mineral and the weight of an equal volume of water. For example, 1 cm³ of quartz has a weight of 2.7 g, whereas 1 cm³ of water has a weight of 1.0 g, so the specific gravity of quartz is 2.7. In practice, you can develop a sense for specific gravity by hefting mineral specimens in your hand—minerals with higher specific gravity feel heavier.

DISTINCTIVE PROPERTIES. Some minerals have properties that readily distinguish them from other minerals. For example, calcite ($CaCO_3$) and dolomite [$CaMg(CO_3)_2$] react with dilute hydrochloric acid to produce carbon dioxide (CO_2) gas **(Fig. 3.6e)**, graphite makes a gray mark on paper, magnetite attracts iron objects **(Fig. 3.6f)**, halite tastes salty, and plagioclase has striations (thin parallel corrugations) on crystal faces. Some minerals can be recognized by the distinctive appearance of specimens, as

determined by the mineral's **crystal habit**, the shape of a single euhedral crystal or the shape of an aggregate of many well-formed crystals that grew together as a group. A description of crystal habit uses familiar adjectives. For example, crystals that are roughly the same length in all directions are *blocky*, those that are much longer in one dimension than in others are *columnar* or, in the extreme, *needle-like*. Similarly, specimens shaped like sheets of paper are *platy*, and those shaped like knives are *bladed*. The shape of their crystals makes some minerals, such as asbestos, hazardous **(Box 3.3)**.

HOW A MINERAL BREAKS. Different minerals break in different ways, depending on their crystal structure. If a mineral breaks to form *planar* (flat) surfaces that have a specific orientation relative to the crystal structure, then we say that the mineral has **cleavage** and refer to each surface as a *cleavage plane*. These minerals break along planes where the bonds holding atoms together in the crystal are weakest **(Fig. 3.7a–f)**. Some minerals have one direction of cleavage. For example, mica has very weak bonds in one direction but strong bonds in other directions, so it splits easily into parallel sheets, each bounded by a cleavage plane. Other minerals have two or three directions of cleavage that intersect at a specific angle. In halite, for example, three directions of cleavage intersect at right angles, so halite crystals break into little cubes. Cleavage planes are not crystal faces, though sometimes the two types of surfaces can be mistaken for one another. By examining a mineral sample carefully, you can generally distinguish between cleavage planes and crystal faces: cleavage planes of a given orientation are repeated throughout the specimen, whereas a crystal face does not repeat **(Fig. 3.7g)**. Minerals that have no cleavage at all break by forming either irregular fractures or smoothly curving, clamshell-shaped *conchoidal fractures* **(Fig. 3.7h)**.

Classifying Minerals

Organizing minerals into *classes*, groups that share key features, has proved to be a challenge. Physical properties, such as color, can't serve as a basis for classification, for many otherwise unrelated minerals can have the same color. Mineralogists eventually concluded that the chemical makeup of minerals provides the most useful basis for mineral classification. They distinguish among several mineral classes based on the anion (negative ion) or anionic group (negatively charged molecule) that the minerals contain. Different minerals within a class differ from one another in terms of the cations (positively charged ions) they contain, their crystal structures,

BOX 3.3 **Putting Earth Science to Use**

Hazardous minerals

Most minerals are harmless, and some we ingest routinely. For example, we use halite as table salt and calcite as an antacid. But certain minerals pose hazards to human health due to their composition, crystal habit, or grain size.

Some minerals contain poisonous chemicals. For example, a version of pyrite (FeS_2) called *arsenopyrite* (FeAsS) contains arsenic, a deadly poison. Arsenopyrite can reside unchanged in rock buried deep in the Earth for millions of years. But when exposed to oxygen-bearing groundwater or to the air, the mineral dissolves in water. Drinking arsenic-bearing groundwater from wells can cause illness or death.

Some minerals are carcinogens. For example, concerns about the cancer-causing consequences of inhaling asbestos have been featured in the news for decades. That's because asbestos was once used to manufacture insulation, brakes, roof shingles, and floor tiles, so many people have encountered it. *Asbestos* refers to a group of several minerals that grow in clumps of very fine, flexible needles or fibers (Fig. Bx3.3a, b). Minerals in this group have a very high melting temperature and are very strong, which is why they were used

for so many different products. Unfortunately, some (not all) kinds of asbestos can become embedded in human lungs, where they can cause cancer. For this reason, asbestos has been banned for most applications, and most remodeling or demolition projects involving pre-1980s buildings require contractors to follow strict rules for *asbestos abatement*. They must enclose the areas where asbestos is being removed and must train workers to wear protective clothing and minimize production of asbestos dust (Fig. Bx3.3c).

Many common minerals, such as quartz and feldspar, are safe under most circumstances, but can pose a danger when pulverized. Specifically, when inhaled, the dust of these minerals can become embedded in lungs. Because these minerals do not dissolve or decay, and cannot be coughed out, they remain there permanently. Irritation caused by this dust triggers the growth of fibrous masses in lungs. These masses eventually block access to air and make breathing difficult, producing an illness known as *silicosis*. Everyone breathes a little silicate dust now and then with no ill effects, but prolonged exposure can be dangerous.

FIGURE Bx3.3 Asbestos has characteristics that can make it both useful and hazardous.

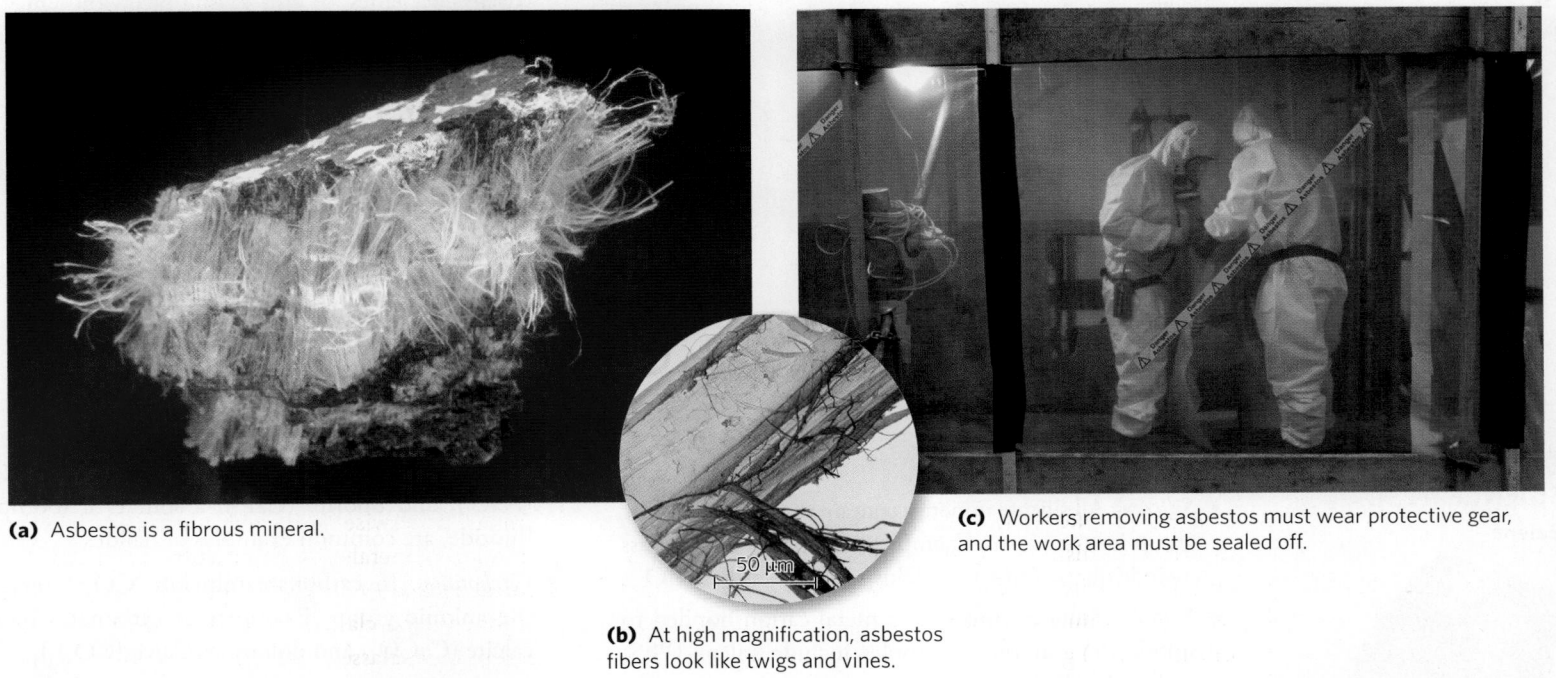

(a) Asbestos is a fibrous mineral.

(b) At high magnification, asbestos fibers look like twigs and vines.

50 μm

(c) Workers removing asbestos must wear protective gear, and the work area must be sealed off.

FIGURE 3.7 The nature of mineral cleavage and fracture.

(a) Mica has one plane of cleavage and splits into sheets.

(b) Pyroxene has two planes of cleavage that intersect at 90°.

(c) Amphibole has two planes of cleavage that intersect at 60°.

Halite breaks into cubes.

(d) Halite has three mutually perpendicular planes of cleavage.

Calcite breaks into rhombs.

(e) Calcite has three planes of cleavage, one of which is inclined.

(f) Diamond has four planes of cleavage, each inclined relative to the others.

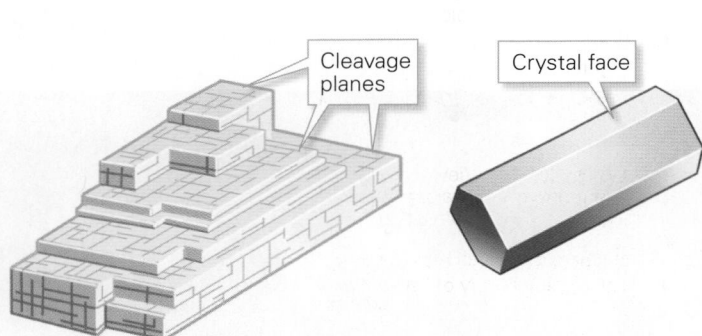

Cleavage planes

Crystal face

(g) How do you distinguish between cleavage planes and crystal faces? Cleavage planes may be repeated, whereas a crystal face is a single surface.

Irregular fracture

Crystal face

Conchoidal fracture

Garnet

Quartz

(h) Minerals without cleavage form irregular or conchoidal fractures.

or both. The most common mineral classes include the following:

- *Silicates:* All silicate minerals contain the SiO_4^{4-} anionic group in some form. Examples of minerals in this class include quartz (SiO_2) and feldspars (such as $KAlSi_3O_8$).
- *Sulfides:* Sulfides consist of a metal cation bonded to sulfide (S^{2-}) anions. Examples include galena (PbS) and pyrite (FeS_2).
- *Oxides:* Oxides consist of metal cations bonded to oxygen anions. Typical oxide minerals include hematite (Fe_2O_3) and magnetite (Fe_3O_4).

- *Halides:* The anion in a halide is a halogen, such as chloride (Cl^-) or fluoride (F^-). Halite, or rock salt (NaCl), and fluorite (CaF_2), a source of toothpaste fluoride, are common examples of halides.
- *Carbonates:* In carbonate minerals, CO_3^{2-} serves as the anionic group. Examples of carbonates include calcite ($CaCO_3$) and dolomite [$CaMg(CO_3)_2$].
- *Sulfates:* Sulfates consist of metal cations bonded to SO_4^{2-} anionic groups. Many sulfates, such as gypsum ($CaSO_4 \cdot 2H_2O$), form by precipitation out of salty water at or near the Earth's surface.

Most of our planet consists of **silicate minerals**. The SiO_4^{4-} anionic group, which serves as the fundamental building block of this mineral class, is known as the **silicon-oxygen tetrahedron** or, informally, the *silica tetrahedron*. Each tetrahedron consists of a silicon atom surrounded by four oxygen atoms, so it has a pyramid-like shape with four triangular faces (**Fig. 3.8a**).

Mineralogists distinguish among different silicate minerals (such as olivine, pyroxene, amphibole, mica, quartz, feldspar) by the way the silica tetrahedra within a mineral link together. The nature of linkages determines how many oxygen atoms are shared, which in turn determines the ratio of silicon to oxygen atoms in the mineral (**Fig. 3.8b**). In olivine, for example, the tetrahedra do not share any oxygen atoms and are held together only by the attraction of cations between tetrahedra; in pyroxene,

tetrahedra link to form a single chain; in amphibole, they link to form double chains; in mica, they link to form two-dimensional sheets; and in quartz or feldspar, they link to form a three-dimensional jungle-gym-like network.

Notably, the mineral names we've just used, except for quartz, represent groups or subcategories of silicate minerals. Minerals in each group share the same general structure, but contain different cations. For example, forsterite and fayalite are both olivines, but the former contains Mg and the latter contains Fe. Similarly, plagioclase and orthoclase are both feldspars, but the former contains either Na or Ca, while the latter contains K. And mica consists of two forms: dark brown to black biotite, which contains Fe or Mg, and light tan to silvery-grey muscovite, which does not.

FIGURE 3.8 The structure of silicate minerals.

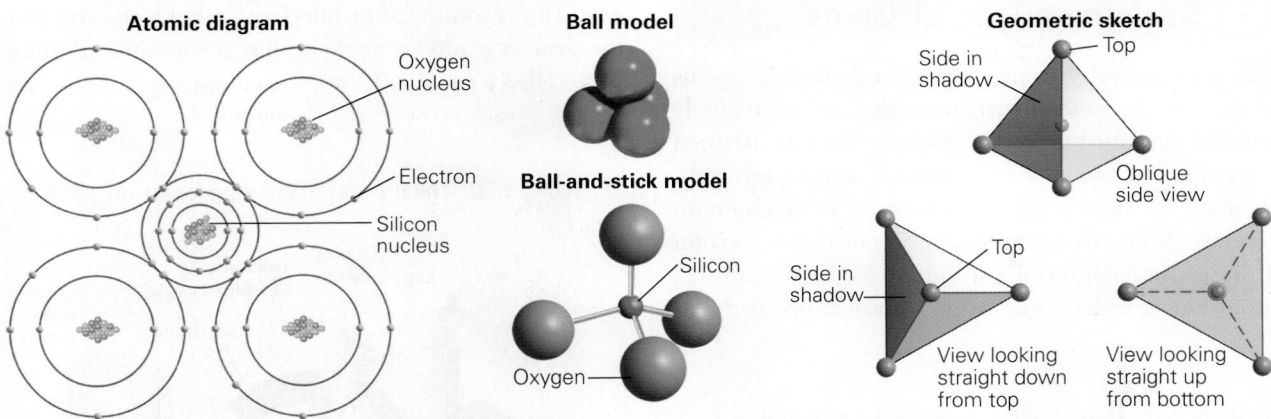

(a) The fundamental building block of a silicate mineral is the silicon-oxygen tetrahedron. Oxygen atoms occupy the corners of the tetrahedron, and a silicon atom lies at the center. Geologists portray the tetrahedron in a several different ways. In many of these models, the silicon atom isn't visible. Note that SiO_4^{4-} is an anionic group with a charge of −4.

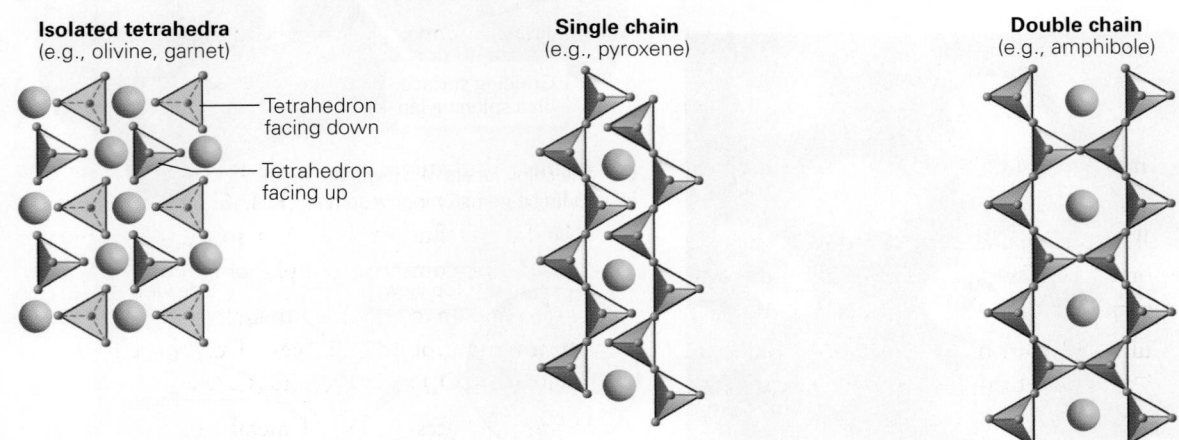

(b) Groups of silicate minerals differ from one another in the way in which the silicon-oxygen tetrahedra are linked. Where the tetrahedra link, they share an oxygen atom. Oxygen atoms are shown in blue. Positive ions (depicted by yellow balls) occupy spaces between tetrahedra.

Take-home message . . .

We can distinguish among different minerals by physical properties such as luster, color, cleavage, hardness, and specific gravity. Minerals can be grouped into classes based on their chemical composition. Silicate minerals are the most abundant minerals in the Earth.

Quick Questions

- How can you test the hardness of a mineral specimen?
- What is the difference between streak and color?
- What is the fundamental building block of a silicate mineral?
- Do all minerals have distinctive cleavage?
- Which criterion do mineralogists use to group minerals into classes?

Did you ever wonder . . .

how gems in jewelry get all those shiny faces?

See for yourself

Kimberley Diamond Mine

Latitude: 28°44'17.06" S
Longitude: 24°46'30.77" E

Zoom to an altitude of 13 km (~8 mi) and look straight down. The field of view shows the town of Kimberley, South Africa, and its inactive diamond mine. The mine looks like a circular pit. You can also see the tailings pile of excavated rock debris.

3.3 Something Special: Gems

Walk past a jewelry store window, and you'll see **gems** (or *gemstones*), mineral specimens that are particularly beautiful or valuable, set in gold or silver to be used as jewelry (**Fig. 3.9**). Some gems are unique minerals, but many are merely special versions of more common minerals. Ruby, for example, is a transparent version of the common mineral corundum (Al_2O_3). (Crushed corundum is widely used as an abrasive, because it is

FIGURE 3.9 Gem-quality versions of corundum (Al_2O_3) can be fashioned into gems of many shapes and colors. Rubies consist of red corundum. Typical sapphires are blue corundum.

very hard.) The quality of a gem depends on its color as well as on the concentration of *inclusions* (tiny flecks of other minerals or bubbles filled with air or water) that it contains. Typically, a gem containing many inclusions has less value.

The gems with glittery surfaces that you see in a store window did not form that way—they are not *raw* or *rough gems*, in their natural form as originally found in rock or sediment. Rather, they have been cut and polished to make them interact with light in a visually pleasing way. Such modified gems are generally known as *jewels*. Some jewels act like tiny prisms, splitting the light that enters them into a spectrum of colors—flashes of these colors give such jewels their "fire."

A "cut" gemstone is one whose surface contains many **facets**, polished planes that make sharp angles with their neighbors and make the jewel sparkle when rotated in a light beam. The facets of a jewel are not natural cleavage planes or crystal faces. Rather, gem cutters produce facets by attaching a rough gem (or a piece that has been carefully broken off a larger rough gem) to the end of a *doping arm*. This is connected to a device that holds the arm and the gem at a precise angle against a spinning grinding plate called a *lap* (**Fig. 3.10a**). The abrasive grit on the lap

FIGURE 3.10 The process of creating facets on gems.

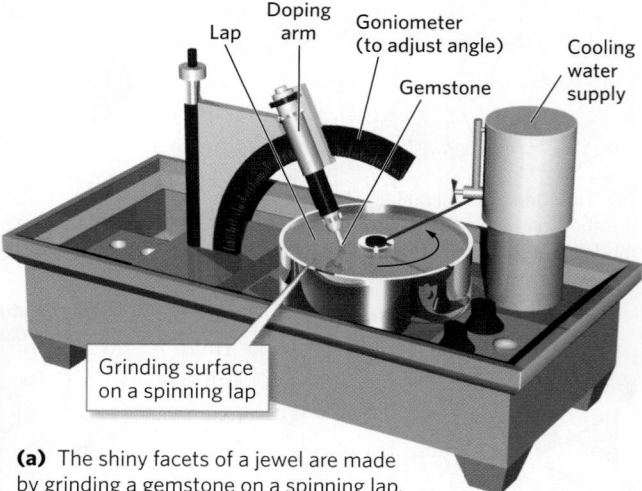

(a) The shiny facets of a jewel are made by grinding a gemstone on a spinning lap.

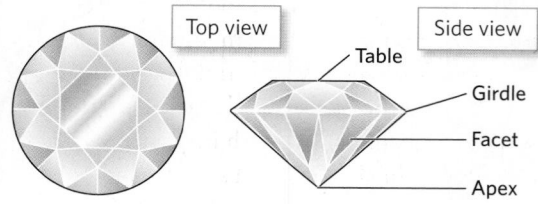

(b) There are many different "cuts" for a jewel. Here we see the top and side views of a brilliant-cut diamond with 57 facets. If the apex were ground to form a tiny flat surface, it would have 58.

FIGURE 3.11 Where diamonds come from.

(a) Surface diamond mines in northern Canada.

(b) A diamond embedded in solid kimberlite.

(c) Panning for diamonds in a streambed.

acts like sandpaper and produces a planar surface on the gem. Once a surface has been finished, the gem cutter rotates the gem by a specific angle and then places it back against the lap to produce another facet. Repeating this process many times facets the entire surface of the gem. The configuration and shape of the facets defines the *cut* of the resulting jewel—a "brilliant-cut" jewel has 57 or 58 facets (Fig. 3.10b).

Where do rough gems come from? They can form in many ways, like many other minerals. Some solidify from a melt, some form by diffusion, and some result from the chemical interaction of rock with hot-water solutions. Several types of gems occur in an unusual igneous rock called *pegmatite*, which forms by rapid mineral growth in water-rich melts. Rocks containing gems are relatively rare, so mining companies or individual prospectors may have to search for many years before finding a promising source.

Diamond—the hardest mineral known, and the one most commonly used for wedding rings—grows from pure carbon that was carried to a depth of over 100 km in the Earth by subduction. The great pressure at this depth causes carbon atoms to arrange into a compact and very strong crystal lattice. (At shallower depths, carbon crystallizes into soft graphite, the "lead" used in pencils; diamond and graphite are polymorphs of carbon.) Geologists speculate that gem diamonds return to the Earth's surface when rifting thins the crust and causes some of the underlying mantle to melt—the rising melt carries the diamonds up with it. Solidification of diamond-containing melt produces a special kind of rock called *kimberlite* (named for Kimberley, South Africa, where it was first found). Diamonds can be separated from kimberlite by crushing the rock once it has been mined (Fig. 3.11a, b). The natural breakdown of kimberlite into fragments at the Earth's surface can also release the diamonds. Some diamonds remain as solid grains in river sand, where prospectors can collect them by swirling the sand, along with water, in a pan. Diamonds are denser than the other minerals (such as quartz and feldspar) in sand, so such *panning* washes away most sand and leaves behind the diamonds (Fig. 3.11c).

Take-home message . . .

Gems are particularly rare and beautiful versions of minerals. Converting a gem into a jewel typically involves faceting the gem by grinding it with a lap; the facets are not natural crystal faces or cleavage planes.

Quick Questions

- Did the shiny facets on a jewel grow to their present shapes?
- What causes the fire of a jewel?
- At what depths in the Earth do pressures become great enough to form a diamond?

3.4 Introducing Rocks

What Is a Rock?

In the mid-19th century, as cities along the west coast of the United States expanded, the need for a transcontinental railroad became overwhelming. Thousands of workers faced death all too frequently as they blasted and chiseled their way through the Sierra Nevada range of California to make a path for the tracks. If you had

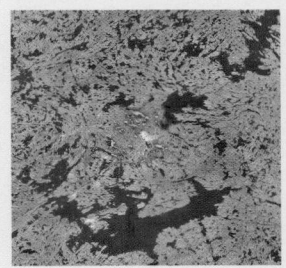

FIGURE 3.12 The contrast between clastic and crystalline rocks.

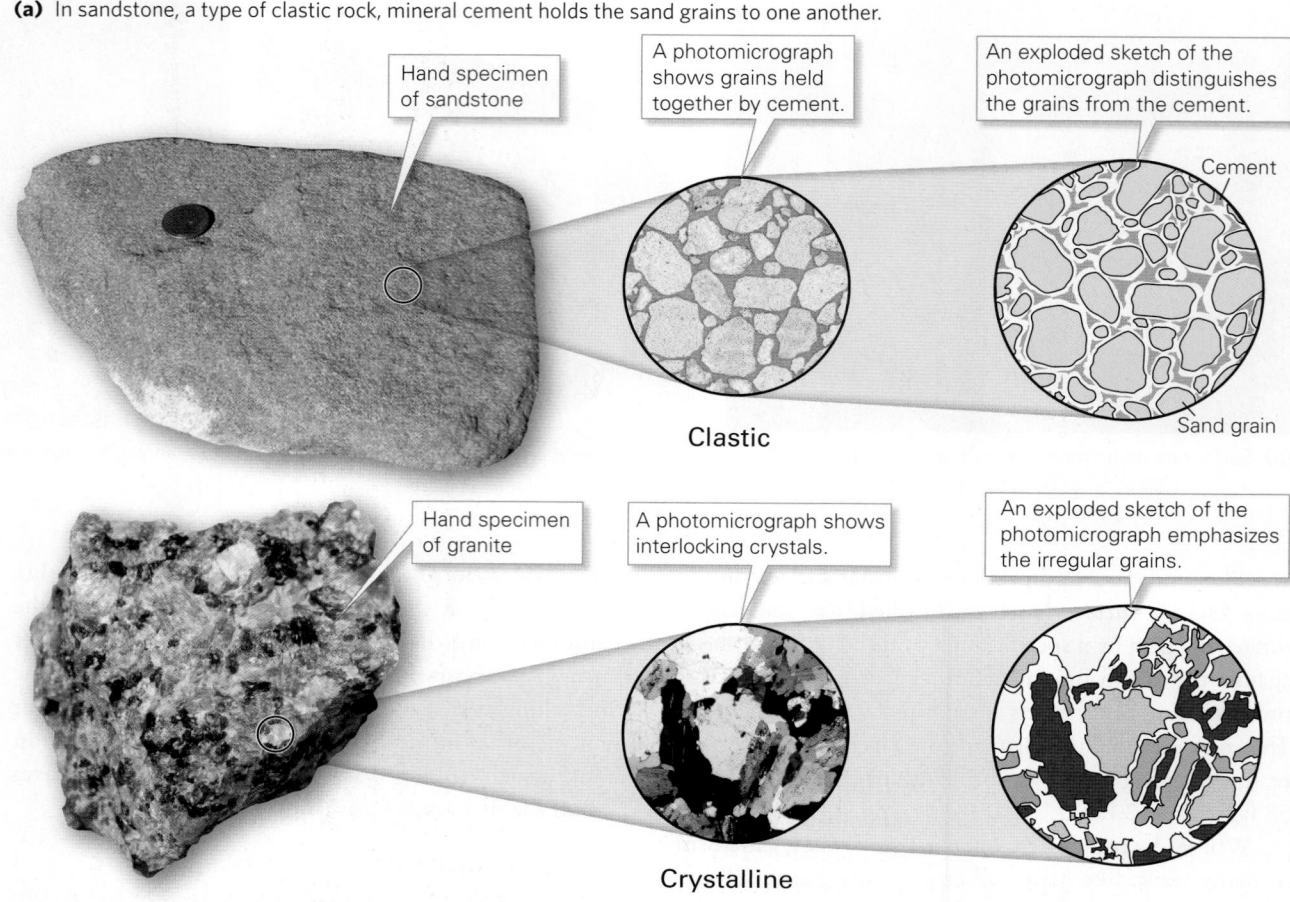

(a) In sandstone, a type of clastic rock, mineral cement holds the sand grains to one another.

Hand specimen of sandstone

A photomicrograph shows grains held together by cement.

An exploded sketch of the photomicrograph distinguishes the grains from the cement.

Cement

Clastic

Sand grain

Hand specimen of granite

A photomicrograph shows interlocking crystals.

An exploded sketch of the photomicrograph emphasizes the irregular grains.

Crystalline

(b) In granite, a crystalline rock, the grains interlock like pieces of a jigsaw puzzle.

 **Animation**
Rock Groups

asked those workers to define a rock, they would probably have responded with the obvious: a rock is a hard, heavy, solid mass that's really difficult to excavate. Geologists use a more technical definition: a **rock** is a coherent, naturally occurring solid consisting of an aggregate of minerals or, much less commonly, of glass. To clarify this definition, let's look at its components.

- *Coherent:* A rock holds together as a solid mass. It's not a pile of loose grains.

- *Naturally occurring:* Rocks form by geologic processes. Manufactured materials, such as concrete and brick, are not rocks, even though they resemble rocks.

- *Aggregate of minerals or glass:* Most rocks consist of an *aggregate* (a collection) of many crystals, grains, or larger fragments. Some rocks contain only one kind of mineral, whereas others contain several different kinds. A few rock types, formed at volcanoes, consist of glass (see Chapter 4).

What holds a rock together? **Clastic rocks** consist of **clasts** (grains or larger fragments) held together by **cement**, a binder composed of mineral crystals that precipitated from water in the space between clasts **(Fig. 3.12a)**. **Crystalline rocks** consist of crystals that interlocked with one another, like pieces in a jigsaw puzzle, as they grew **(Fig. 3.12b)**. **Glassy rocks** consist mostly of a continuous mass of glass; a relatively small percentage of tiny crystals may be dispersed through the glass.

The word *bedrock*, in everyday English, implies something that seems solid and stable. This usage makes sense because, geologically speaking, **bedrock** is rock that remains attached to the Earth's crust below **(Fig. 3.13)**. Loose clasts are not bedrock, for they have broken free from the crust and have moved away from their point of origin. Geologists refer to an exposure of bedrock as an **outcrop**. An outcrop may be a knob out in a field or in the woods, a cliff or ridge on a mountainside, or the wall or bed of a stream **(Fig. 3.14a)**. In the past few centuries, people have created outcrops by carving road cuts, railroad cuts, and

FIGURE 3.13 What is bedrock, and what isn't? The loose boulders on the ground at the base of this cliff in Utah are not bedrock, but the layer at the top of the cliff behind them is.

Outcrop
(bedrock)

Boulders
(not bedrock)

FIGURE 3.14 Common types of outcrops.

Stream cut

(a) A stream cut near Catskill, New York, exposes rock that would otherwise be hidden by trees.

Road cut

(b) To keep the slope (grade) of this highway gentle, engineers cut an artificial canyon through bedrock west of Denver, Colorado.

Natural cliff outcrop

(c) Natural cliff exposures on a mountain face near Salt Lake City, Utah. The cliff slopes are too steep for trees to take root.

foundations **(Fig. 3.14b)**. To people who live in cities or forests or farmland, outcrops may be unfamiliar, for they lie hidden beneath vegetation, sand, gravel, mud, soil, water, asphalt, concrete, or buildings. But in mountainous areas or deserts, outcrops may be widespread **(Fig. 3.14c)**.

Basics of Rock Classification

For centuries, geologists struggled to develop a meaningful approach to rock classification. Eventually, it became clear that a genetic approach—one that focuses on the origin (genesis) of rocks—works best. The modern *genetic classification* of rocks recognizes three rock groups:

- **Igneous rocks** form by the solidification (freezing) of molten rock, or *melt* **(Fig. 3.15a)**. Solidification can take place underground or at the Earth's surface.
- **Sedimentary rocks** form at or near the Earth's surface from minerals that precipitate out of a water solution, from the shells of organisms, or from loose clasts **(Fig. 3.15b)**. Clasts are produced by **weathering** (the breakup of pre-existing rock due to cracking and/or interaction with air and water) followed by **erosion** (the grinding away and removal of materials at the Earth's surface by moving water, air, or ice). Once formed, clasts are transported to a new location where they settle out, undergo burial, and eventually become cemented together.

FIGURE 3.15 The three rock groups.

Igneous

(a) Lava freezes at the Earth's surface to form igneous rock. Here, the molten tip of a brand-new lava flow still glows red.

Sedimentary

(b) Sand, formed from grains eroded from the cliffs above, has collected on this beach. If buried and cemented together, it would become a layer of sandstone, like those making up the cliff.

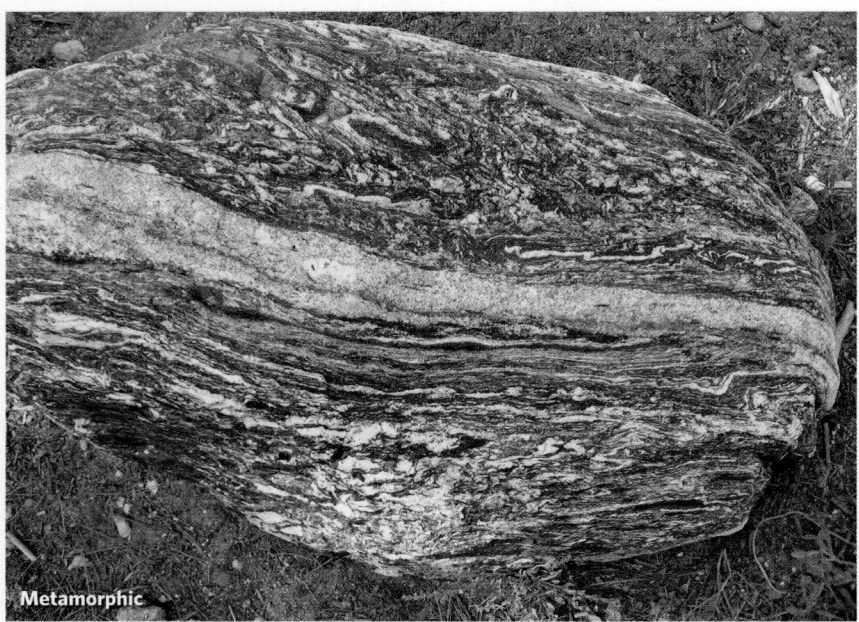

Metamorphic

(c) Metamorphic rock forms when pre-existing rocks undergo changes in specific environmental conditions.

- **Metamorphic rocks** form when pre-existing rocks undergo changes at depth in the crust, without first melting. The changes take place in response to a rise in pressure and/or temperature and/or interaction with very hot solutions. *Metamorphism*, the process of producing a metamorphic rock, may change a rock's appearance radically (Fig. 3.15c).

Each of the three rock groups contains many different individual rock types, each of which has a name. Some of these names may be familiar to you (such as granite and sandstone), but others may not (such as peridotite and arkose). Rock names come from various sources: some reflect the dominant component making up the rock, some indicate a region where the rock was first discovered, some derive from a word of Latin origin, and some come from local traditional names. All told, there are about 200 different rock names.

Fundamental Characteristics of Rocks

When studying rocks, geologists focus on describing the following characteristics:

- *Composition:* The minerals in a rock determine the rock's chemical makeup, a key characteristic used to identify a rock.

- *Grains or clasts:* In some rocks, the grains or clasts are so small that they can't be seen without a microscope, whereas in others, they may be centimeters or even tens of centimeters across (Fig. 3.16a). Some rocks contain grains or clasts of only one size, whereas others contain many different sizes.

- *Grain or clast shape:* In some rocks, the grains or clasts are **equant**, meaning that they have similar dimensions in all directions, whereas in others they are **inequant**, meaning that their dimensions are not the same in all directions (Fig. 3.16b).

- *Texture:* The term **texture** refers to the way the material in a rock is connected to provide coherence. As we've seen, rocks can have clastic, crystalline, or glassy textures.

- *Fabric:* Geologists use the term **fabric** when describing the degree to which grains or clasts are equant or inequant, and the degree to which inequant grains align parallel to one another.

- *Layering:* Some rock bodies appear to contain distinct layering. Layering in sedimentary rocks is called *bedding*. Beds can differ from one another in terms of composition, texture, or color (Fig. 3.17a). Layering in metamorphic rock is called *foliation*. It can be defined by the alignment of inequant grains and/or by alternating bands of different composition (Fig. 3.17b) (see also Fig. 3.16b).

Studying Rocks

Ideally, the study of a rock begins at an outcrop. If the outcrop is big enough, you'll be able to see large-scale layering, and you can study relationships between the rock you're interested in and other rocks around it (Fig. 3.18a). To gain insight into the details of the composition, texture, and fabric of the rock, you may need to break off a **hand specimen**, a fist-sized piece that you can look at more closely with a hand lens (Fig. 3.18b).

To examine rocks in even more detail, geologists can bring rock samples back to the laboratory, where they can prepare a **thin section**, a very thin slice (about 0.03 mm, the thickness of a human hair) glued to a glass slide (Fig. 3.19a). By placing the thin section beneath the lens of a *petrographic microscope*, a special microscope used to study rocks (Fig. 3.19b), geologists can see grains at high magnification. A petrographic microscope differs from an ordinary microscope in that it illuminates the thin section with *transmitted light*. This means that the light passes up through the thin section from beneath. Why use transmitted light? Many minerals, when sliced to the thinness of a thin section, become transparent or semitransparent, meaning that light can pass through them as if they were a window. Commonly, geologists use both plane light (normal light) and *polarized light*—a beam in which light waves all align in the same direction—when viewing thin sections. Polarized light interacts with minerals to display a unique suite of colors or gray tones (Fig. 3.19c). The specific color or gray tone that the observer sees for a grain

FIGURE 3.16 Describing the components of a rock.

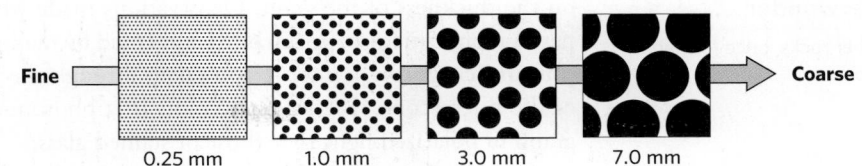

Fine   Coarse

0.25 mm   1.0 mm   3.0 mm   7.0 mm

(a) Geologists define grain size by using a comparison chart like this one.

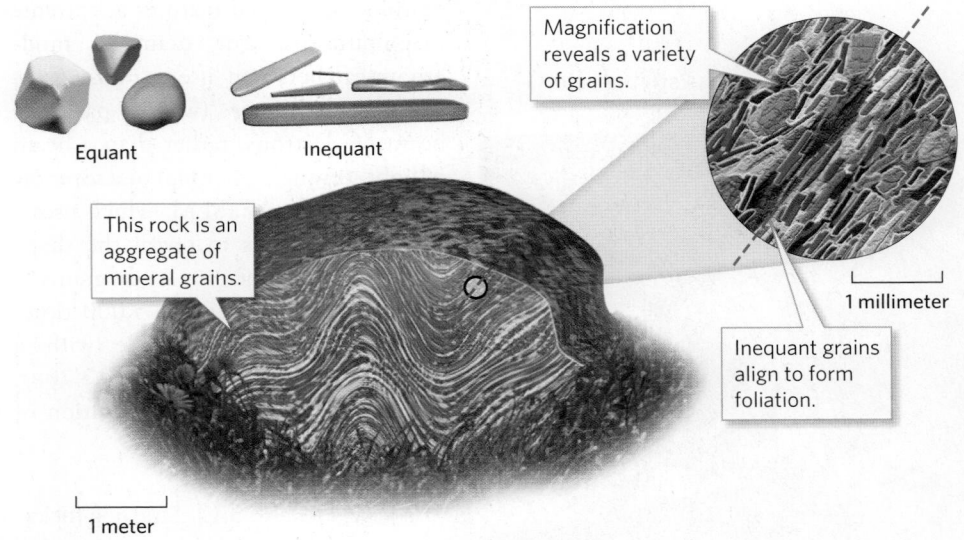

Equant   Inequant

Magnification reveals a variety of grains.

This rock is an aggregate of mineral grains.

Inequant grains align to form foliation.

1 millimeter

1 meter

(b) Grains in rock come in a variety of shapes. Some are equant, whereas others are inequant. In this example of metamorphic rock, inequant grains align to define metamorphic foliation.

FIGURE 3.17 Layering in rock.

Bedding planes

(a) Bedding in sedimentary rock, here defined by alternating layers of coarser and finer grains, as exposed on a cliff in northwestern China. Here, bedding is horizontal. Parts of a few bedding planes are highlighted.

Foliation planes

(b) Alternation of light and dark layers defines the foliation in this outcrop of metamorphic rock near Mecca, California. The color of each layer depends on the minerals it contains. Here, foliation has a steep dip (tilt). Parts of a few foliation planes are highlighted.

depends on the identity of the mineral, on the orientation of the grain's crystal lattice relative to the light beam, and on the thickness of the grain. Observations made with a petrographic microscope can be documented by making a **photomicrograph**, a photograph taken through the lens of the microscope. The brilliant colors in a photomicrograph in polarized light rival those of stained glass.

Some grains and their internal features are too small to see even with a petrographic microscope. Fortunately, high-tech instruments now enable geologists to examine grains and textures at extreme magnifications. For example, modern researchers can use a *transmission electron microscope* (which passes a beam of electrons, rather than a beam of light, through a sample) or a *scanning electron microscope* (SEM, which uses a beam of electrons to reveal the shape of a sample's surface) to study grains at magnifications more than 5,000 times greater than those possible with a petrographic microscope (Fig. 3.20a). To study the chemical composition of grains, geologists employ *electron microprobes*, which analyze X-rays produced when a beam of electrons strikes a grain (Fig. 3.20b). And to study the crystal structure of grains, they use *X-ray diffractometers*, which send X-rays through a crystal.

Take-home message...

Geologists classify rocks into three groups—igneous, sedimentary, and metamorphic—based on the way they form. Each group includes many different rock types, distinguished from one another by physical characteristics and composition. We can study rocks at outcrops in the field and by taking samples back to the lab for more detailed analysis.

Quick Questions ——————

- Which of these criteria is used to classify rocks into groups: color, grain size, process of origin, or composition?
- Why isn't sand a rock? Why isn't concrete?
- Explain what a thin section is, and why geologists make them.

FIGURE 3.18 Studying rocks in the field.

(a) A field geologist examining a hand specimen (on a cold day).

(b) A hand specimen.

FIGURE 3.19 Studying rocks in thin section.

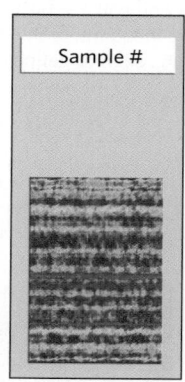

Sample #

(a) A geologist glues a chip of rock to a glass slide and grinds it down until it is so thin that light can pass through it.

(b) With a petrographic microscope, it's possible to view thin sections with light that shines through the sample from below.

500 μm

(c) In polarized light, different grains display different colors when viewed through the microscope. The color depends on the type of mineral and on the orientation of the grain.

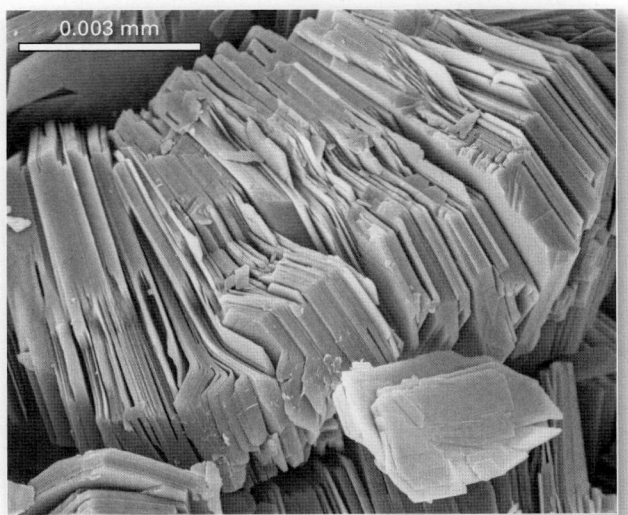

(a) An SEM image of a type of clay called kaolinite. (10 μm = 0.01 mm.)

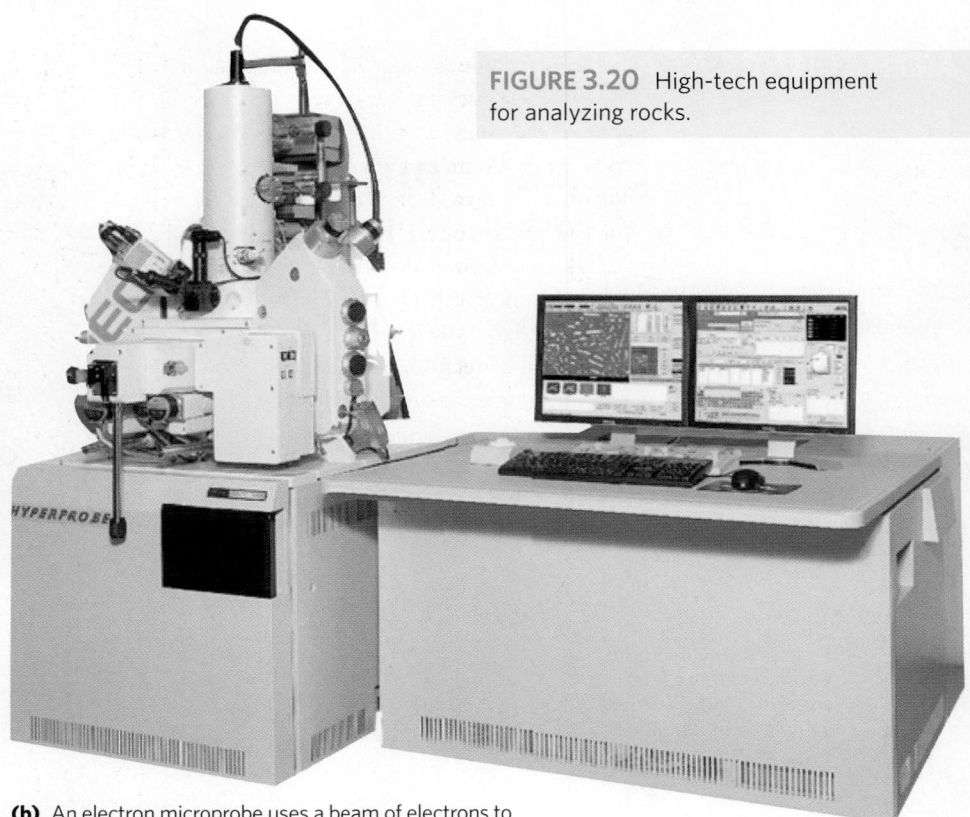

FIGURE 3.20 High-tech equipment for analyzing rocks.

(b) An electron microprobe uses a beam of electrons to analyze the chemical composition of minerals.

3.5 The Rock Cycle

Even though rock seems to be the embodiment of strength and durability, most rocks don't last forever in the context of Earth history. Due to many geologic processes, the atoms making up minerals of one rock type may eventually be released, relocated, and rebonded to become incorporated in other minerals that are parts of different rocks, at the same locality or perhaps far away. Put another way, material that resides in a given rock type at one point in geologic time may later end up in another rock type of the same group, or a rock type of another group. Geologists refer to the transformations that result in the passage of atoms through different rock types over geologic time as the **rock cycle** (Fig. 3.21). The rock cycle takes place because the Earth remains a dynamic planet where many rock-forming environments exist. External energy, internal energy, and gravitational energy all play a role in driving the rock cycle by keeping the mantle, crust, atmosphere, streams, and oceans in motion (see Chapter 1).

To picture a path around the rock cycle, imagine that magma rises from the mantle, comes out of a volcano as lava, cools, and becomes a new igneous rock (Fig. 3.22a). If that igneous rock later undergoes weathering at the Earth's surface, it may break down into sediment that, after being eroded and transported, undergoes deposition, burial, and lithification to become part of a new sedimentary rock (Fig. 3.22b). Sediment may also come from the shells produced by organisms, so life can play a key role in this stage of the rock cycle. If a sedimentary rock later becomes buried deeply and endures high pressures and temperatures, it may be transformed into metamorphic rock (Fig. 3.22c). At extreme temperatures, metamorphic rock may melt and produce new magma, which

may solidify to form new igneous rock, completing a loop around the rock cycle. The path we've just described, however, isn't the only path through the rock cycle—there can be shortcuts (see Fig. 3.21). For example, metamorphic rock may be uplifted to the Earth's surface, where it breaks down to form new sediment that later becomes incorporated in new sedimentary rock. Likewise, an igneous rock may be metamorphosed directly, without first being incorporated in a sedimentary rock.

Whether materials in a rock go through stages of the rock cycle quickly or slowly depends on the proximity of the materials to plate boundaries, collision zones, or rifts—places where dynamic processes take place. For example, materials located in parts of continents that have rifted and later collided multiple times have probably passed through the rock cycle more than once, whereas materials located in stable parts of continents, away from plate boundaries or collision zones, have not. Because not all materials pass through the rock cycle at the same rate, we can find rocks of many different ages at the surface of the Earth. Some rocks remain in one form for less than a few million years, while others have stayed unchanged for billions of years.

As you read the next two chapters of this book, keep in mind that every rock-forming and rock-changing environment on the Earth contributes to a stage of the rock cycle (Earth Science at a Glance, pp. 112–113). Understanding the processes involved at each stage of the rock cycle will help you to interpret the Earth's amazing history.

Narrative Art Video
Rock Cycle

Rock-Forming Environments and the Rock Cycle

Rocks form in many different environments. Igneous rocks develop where melt rises from depth and cools either underground or at the surface. Weathering and erosion break up existing rock and produce sediment. When this sediment eventually gets deposited and lithified, new sedimentary rocks form. Pre-existing rocks can undergo change in the solid state to produce metamorphic rocks.

Drainage networks collect surface water that can transport sediment to the ocean.

In a desert environment, rock weathers and fragments. Debris moves downslope in landslides.

Sand dunes form from grains carried by the wind.

Flash floods carry sediment out of canyons to form an alluvial fan.

Volcanic eruptions produce new igneous rocks.

Sedimentary rocks may form a cover over basement.

In humid climates, thick soils develop.

Deep levels of continents consist of ancient metamorphic and igneous rocks. These rocks form the basement of a continent.

Heat from magma causes contact metamorphism.

Magma rises from the asthenosphere.

Partial melting in the asthenosphere produces new magma.

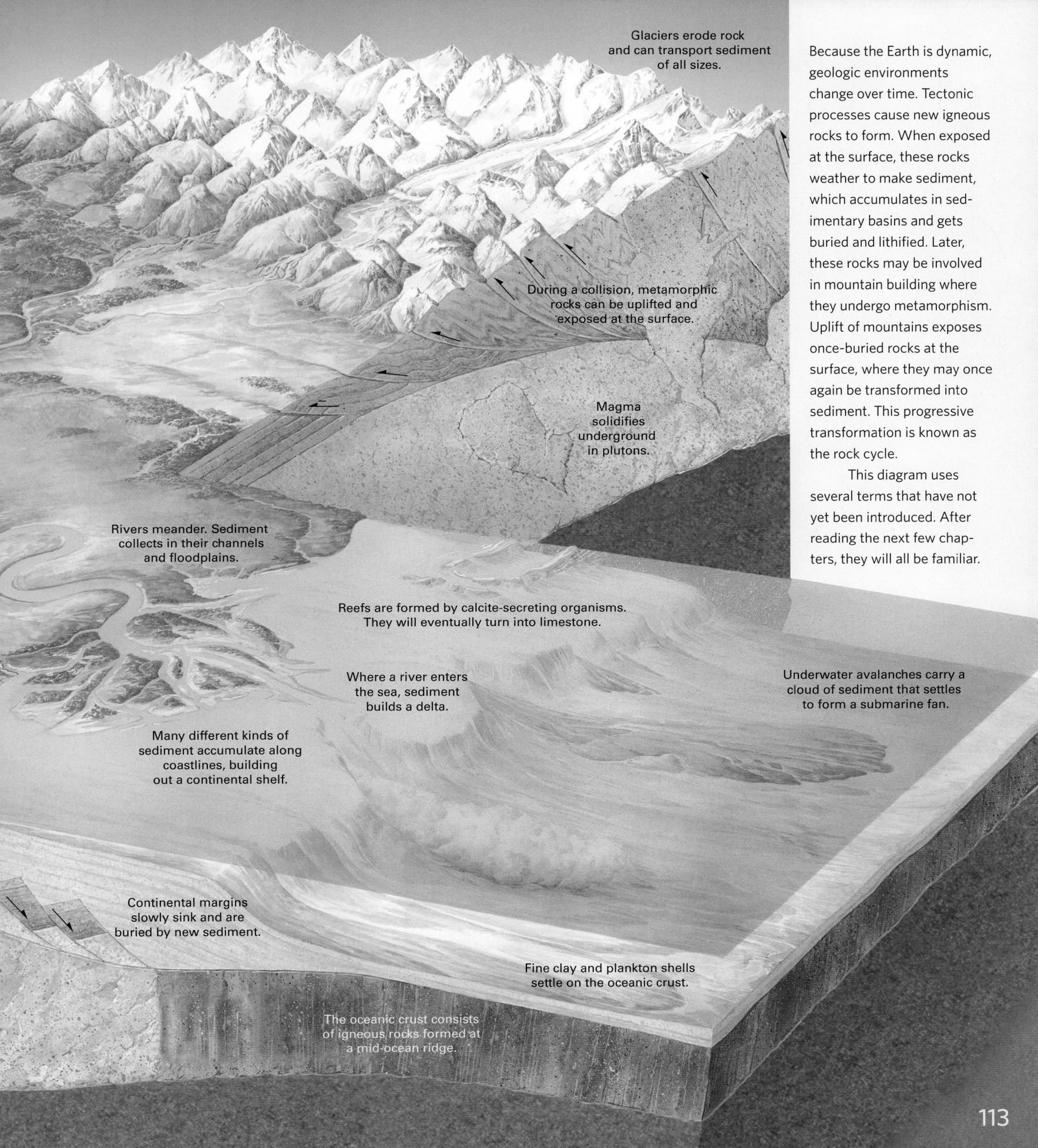

Glaciers erode rock and can transport sediment of all sizes.

During a collision, metamorphic rocks can be uplifted and exposed at the surface.

Magma solidifies underground in plutons.

Rivers meander. Sediment collects in their channels and floodplains.

Reefs are formed by calcite-secreting organisms. They will eventually turn into limestone.

Where a river enters the sea, sediment builds a delta.

Underwater avalanches carry a cloud of sediment that settles to form a submarine fan.

Many different kinds of sediment accumulate along coastlines, building out a continental shelf.

Continental margins slowly sink and are buried by new sediment.

Fine clay and plankton shells settle on the oceanic crust.

The oceanic crust consists of igneous rocks formed at a mid-ocean ridge.

Because the Earth is dynamic, geologic environments change over time. Tectonic processes cause new igneous rocks to form. When exposed at the surface, these rocks weather to make sediment, which accumulates in sedimentary basins and gets buried and lithified. Later, these rocks may be involved in mountain building where they undergo metamorphism. Uplift of mountains exposes once-buried rocks at the surface, where they may once again be transformed into sediment. This progressive transformation is known as the rock cycle.

This diagram uses several terms that have not yet been introduced. After reading the next few chapters, they will all be familiar.

113

FIGURE 3.21 The rock cycle.

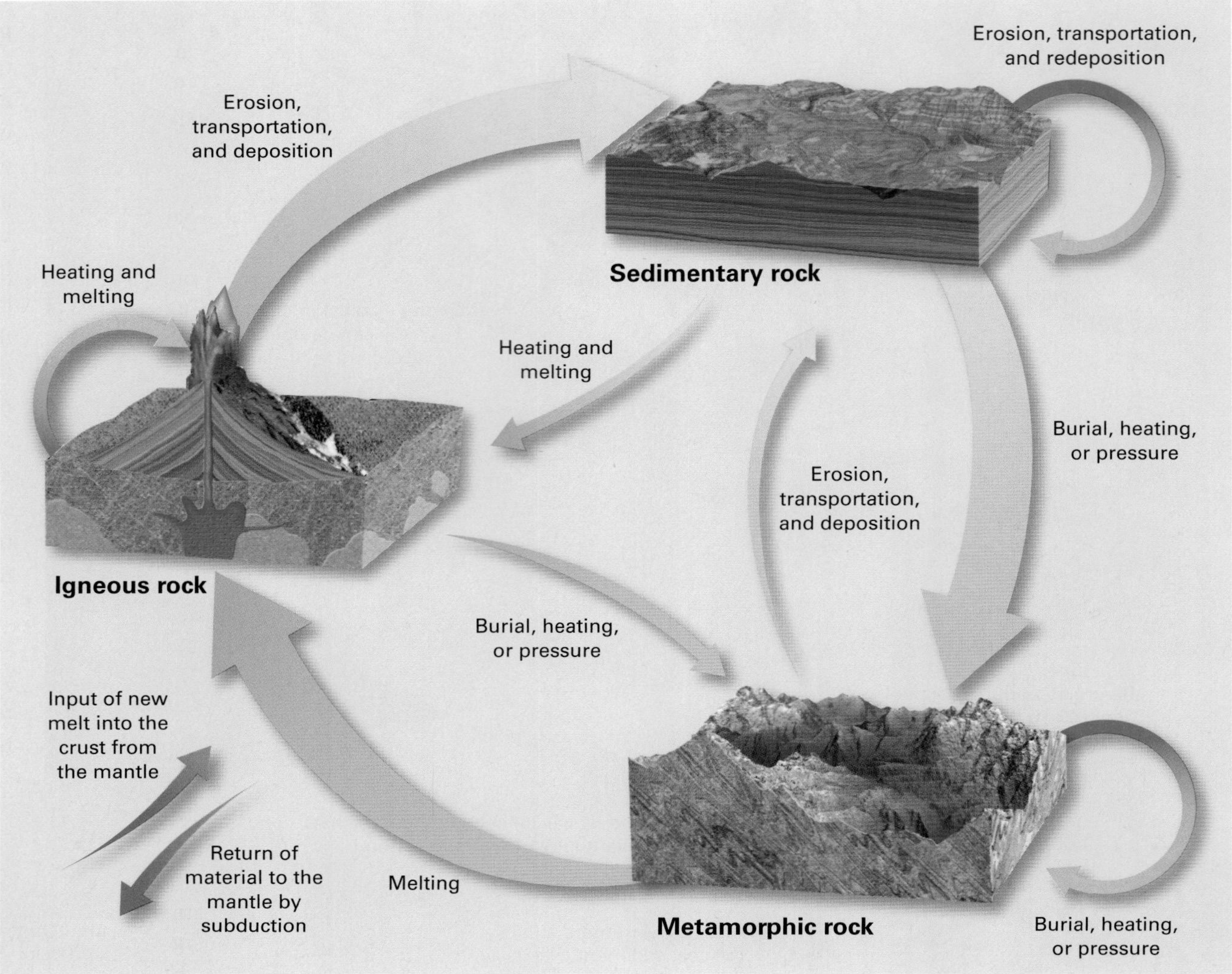

Erosion, transportation, and deposition

Erosion, transportation, and redeposition

Sedimentary rock

Heating and melting

Heating and melting

Burial, heating, or pressure

Erosion, transportation, and deposition

Igneous rock

Burial, heating, or pressure

Input of new melt into the crust from the mantle

Return of material to the mantle by subduction

Melting

Metamorphic rock

Burial, heating, or pressure

FIGURE 3.22 Examples of the three rock groups.

(a) Igneous rock forming, Hawaii.

(b) Sedimentary strata, Utah.

(c) Metamorphic rock, Brazil.

Take-home message . . .

Rocks on the Earth don't last forever. Melting, weathering, burial, and metamorphism can change rock so that the atoms in one rock group end up in another and, later, in another still. This progression, called the rock cycle, is driven by both internal and external energy as well as by gravity.

Quick Questions

• Do all rock-forming materials follow the same path through the rock cycle?

• Does a rock cycle currently take place on the Moon? To answer this question, keep in mind that many craters on the moon are over 2 billion years old.

Objective 3.1

Define minerals and crystals in the context of geology, and describe how minerals form.

KEY CONCEPTS

- Minerals are naturally occurring, homogeneous crystalline solids formed by geologic processes. They have a definable chemical composition and an orderly crystal structure. Most minerals are inorganic.
- The crystal structure of a mineral is defined by the orderly arrangement of atoms, ions, or molecules in a specific pattern.
- Minerals can form by solidification of a melt, precipitation from a water solution, diffusion through a solid, metabolism of organisms, or precipitation from a gas.

EARTH-SCIENCE VOCABULARY

atom (p. 94)	**inorganic chemical** (p. 93)
crystal (p. 94)	**ion** (p. 94)
crystal face (p. 94)	**mineral** (p. 93)
crystalline (p. 93)	**mineralogist** (p. 93)
crystal structure (p. 94)	**molecule** (p. 94)
diffusion (p. 96)	**organic chemical** (p. 93)
geode (p. 98)	**precipitate** (p. 94)
glass (p. 93)	**solidification** (p. 96)
grain (p. 94)	**supersaturated** (p. 96)

REVIEW QUESTIONS

1. How does the geologic definition of a mineral differ from the usage of the word in common English?

2. **(a)** Why aren't sugar or plastic considered to be minerals, while table salt is? **(b)** How does the material making up a "cut crystal" pitcher differ from the material in a crystal of quartz?

3. **(a)** Describe the process by which a mineral crystal grows, in a general sense. **(b)** Explain the difference between precipitation and biomineralization. **(c)** What process can allow a new mineral crystal to grow in an already solid rock? **(d)** What process leads to the formation of minerals in lava?

4. In **Figure A**, representing the crystal structure of halite, what do the different-colored balls represent? What do the sticks represent?

A

5. **(a)** How do you distinguish a euhedral crystal from an anhedral one? **(b)** What conditions determine whether a crystal has an anhedral or a euhedral form?

Objective 3.2

Recognize key physical characteristics of minerals, and explain how minerals can be identified and classified.

KEY CONCEPTS

- Mineralogists have discovered about 4,000 different minerals, of which only a few dozen occur commonly. Many can be identified by their distinct physical properties (such as hardness, streak, and specific gravity).
- Minerals can be grouped into classes (such as silicates, oxides, sulfides, and carbonates) based on their chemical composition.
- The silicon-oxygen tetrahedron serves as the building block of all silicate minerals, the most common minerals on the Earth. Groups of silicate minerals can be distinguished from one another by the way in which the tetrahedra link together.

EARTH-SCIENCE VOCABULARY

cleavage (for a mineral) (p. 100)	**Mohs hardness scale** (p. 99)
color (p. 99)	**silicate mineral** (p. 103)
crystal habit (p. 100)	**silicon-oxygen tetrahedron** (p. 103)
hardness (p. 99)	**specific gravity** (p. 100)
luster (p. 99)	**streak** (p. 99)

REVIEW QUESTIONS

6. List and define the key physical properties that can be used to identify a mineral specimen.

7. **(a)** If mineral sample A can scratch mineral sample B, which sample is harder? **(b)** What do the numbers on the Mohs hardness scale represent? **(c)** Would it be correct to assume that a mineral with a Mohs hardness of 8 is twice as hard as one with a Mohs hardness of 4?

8. **(a)** How do you distinguish cleavage planes from crystal faces on a mineral? **(b)** Label the types of surfaces shown in **Figure B**, and explain how each formed. **(c)** What is a conchoidal fracture?

B

9. Imagine that someone gives you a mineral specimen to identify. The specimen is somewhat transparent, has three planes of cleavage, each oriented at 90° to the others, is generally colorless to white, and is softer than a copper penny. Which of these minerals—calcite, halite, quartz, or gypsum—might it be? Indicate which characteristic led you to your answer, and what other tests you could do to confirm your answer.

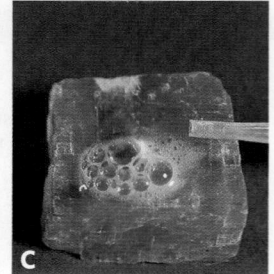

10. (a) Which mineral does **Figure C** show? (b) What gas do the bubbles consist of?

11. (a) Which class of minerals composes most of the Earth? (b) What mineral class does calcite belong in? (c) The mineral bornite (Cu_5FeS_4) is a source of copper. Based on its chemical formula, what is the anion in the mineral, and what mineral class does it belong in?

12. (a) What is the fundamental building block of silicate minerals? (b) How do these building blocks link in mica? (c) Which group of silicate minerals has the structure shown in **Figure D**?

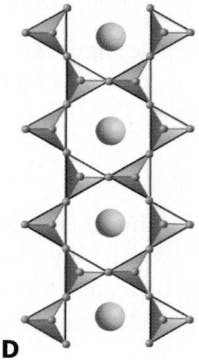

D

Objective 3.3

Discuss why some minerals are gemstones, and explain how jewelers modify gemstones to make jewelry.

KEY CONCEPTS

• Gemstones are versions of minerals known for their beauty and rarity. They form in a variety of geologic settings.

• The facets on jewels are made by grinding and polishing gemstones.

EARTH-SCIENCE VOCABULARY

facet (p. 104)   **gem** (p. 104)

REVIEW QUESTIONS

13. (a) Why are some versions of the mineral beryl considered to be gems, whereas others are not? (b) Corundum powder is used for grinding other materials. What gem has the same chemical formula?

14. (a) Describe the process by which facets on a diamond on a wedding ring are produced. (b) Does **Figure E** show a raw diamond or a cut diamond? (c) How are diamonds formed, and what process probably brought them to the Earth's surface?

E

Objective 3.4

Recognize key characteristics of rock, and distinguish among the three rock groups.

KEY CONCEPTS

• Rocks are coherent, naturally occurring aggregates of minerals or masses of glass. A given rock can be classified into one of three rock groups—igneous, sedimentary, or metamorphic—based on the way it formed.

• Each rock group contains many different types of rocks, each with a name. Rocks can be identified by physical characteristics such as grain size, texture, and composition.

• Rocks are exposed in outcrops, from which they can be collected for detailed studies that use petrographic microscopes and a variety of high-tech analytical equipment.

EARTH-SCIENCE VOCABULARY

bedrock (p. 106)   **igneous rock** (p. 107)
cement (p. 106)   **inequant** (p. 108)
clast (p. 106)   **metamorphic rock** (p. 108)
clastic rock (p. 106)   **outcrop** (p. 106)
crystalline rock (p. 106) **photomicrograph** (p. 110)
equant (p. 108)   **rock** (p. 106)
erosion (p. 107)   **sedimentary rock** (p. 107)
fabric (p. 108)   **texture** (in general) (p. 108)
glassy rock (p. 106)   **thin section** (p. 109)
hand specimen (p. 109)   **weathering** (p. 107)

REVIEW QUESTIONS

15. (a) How does a clastic rock differ from a crystalline rock, and how do both differ from a glassy rock? (b) Identify the two textures shown in **Figure F**.

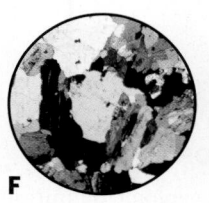 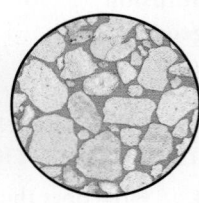

F

16. (a) Distinguish between equant and inequant grains. (b) Which type of grain occurs in **Figure G**? What does the red dashed line represent?

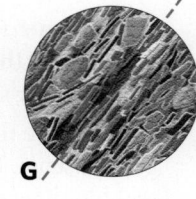

G

17. (a) On what basis do geologists classify rocks into three groups? (b) Name the three groups, and indicate which can form at or near the Earth's surface and which can form deep underground.

18. (a) What is an outcrop? (b) Provide examples of several different types of outcrops. (c) Which type of outcrop does **Figure H** show?

H

19. **(a)** What tools can geologists use to study rocks in the field or lab? **(b)** What does the image in **Figure I** show?

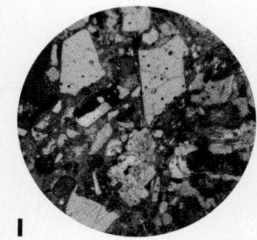

I

Objective 3.5

Sketch a diagram of the rock cycle, and explain why this cycle takes place on the Earth.

KEY CONCEPTS

- Over geologic time, atoms in one rock type may end up in different rock types. This progressive transfer of atoms from one rock type to another, and then to yet another, is called the rock cycle.

- Not all material follows the same path through the rock cycle, or passes through the cycle at the same rate.

- The rock cycle operates because the Earth System remains dynamic. External, internal, and gravitational energy, as well as life, play roles in driving the rock cycle.

EARTH-SCIENCE VOCABULARY

rock cycle (p. 111)

REVIEW QUESTIONS

20. Once formed, does a rock necessarily last for all of Earth history? Explain your answer.

21. **(a)** Label the three rock groups shown in **Figure J**. Also, label the processes indicated by the large arrows. **(b)** Can an igneous rock become metamorphosed without first weathering and eroding?

J

22. Have all continental rocks passed through the rock cycle the same number of times? Explain your answer.

23. **(a)** Describe a pathway through the rock cycle. **(b)** Why does the rock cycle happen on the Earth?

ANOTHER VIEW Looking out the window of an airplane, you can see the rock cycle at work. Cliffs of bedrock of all three rock groups stand out. The white rock in the top right is igneous, the dark peaks are metamorphic, and the red layers are sedimentary. All of these rocks formed at different times and under different conditions. Present-day weathering and erosion are turning all three rock types into sediment.

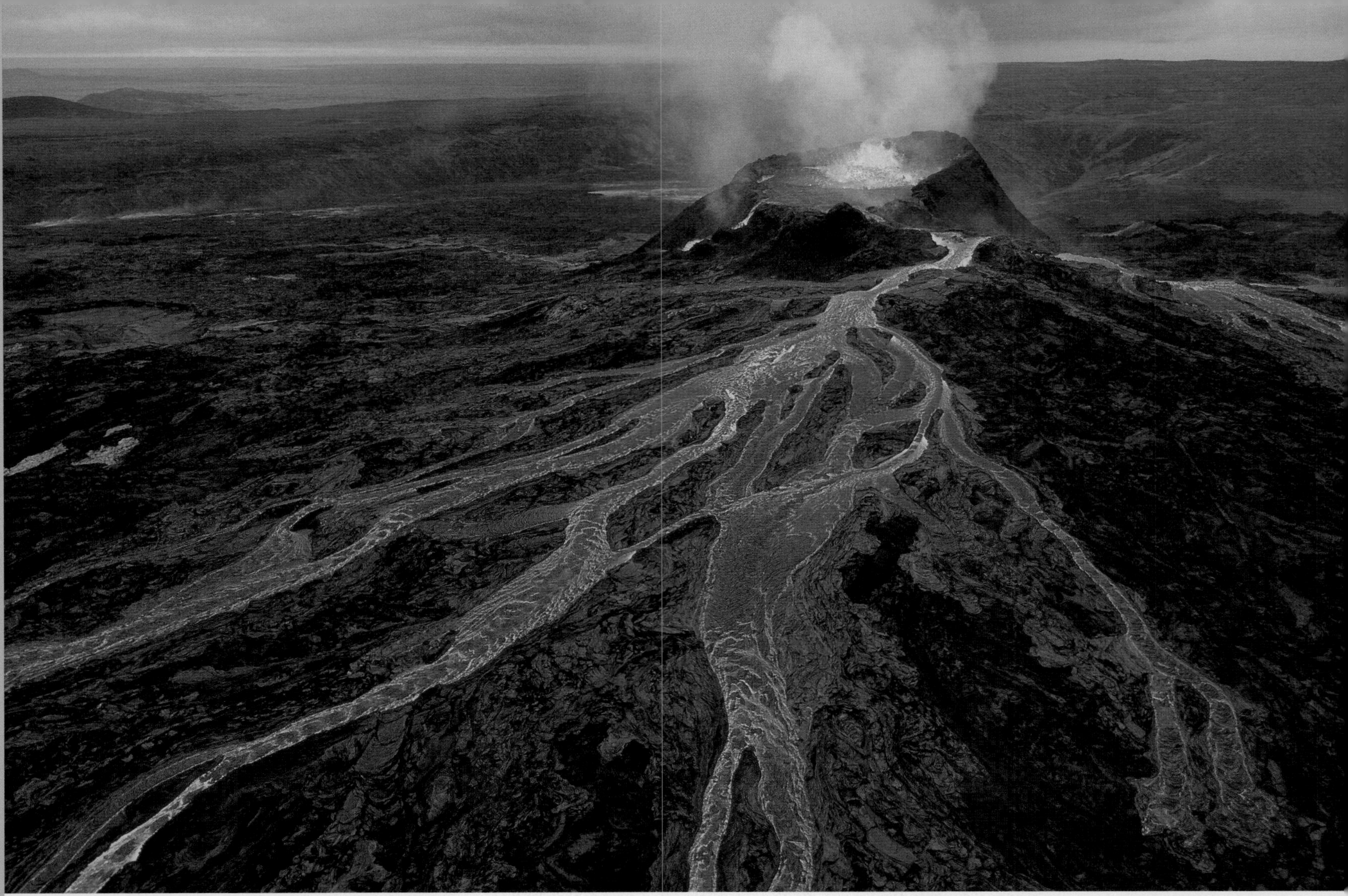

4 UP FROM THE INFERNO
Igneous Rocks and Volcanoes

Nearly two thousand years ago, the Roman town of Pompeii prospered at the foot of Mt. Vesuvius, in what is now southern Italy. Pompeiians thought that the imposing peak, which at the time towered 3 km (9,800 ft) above the nearby sea, was just another scenic mountain. Sadly, they were wrong—very wrong. Mt. Vesuvius is a *volcano*, capable of emitting molten rock (an incandescent liquid) as well as solid rock fragments, dust-like ash, and acrid gas. During the spring of 79 C.E., what looked like smoky steam began to drift upward from an opening at the peak, earthquakes started to jolt Pompeii with unnerving frequency, and the mountain began to grumble like distant thunder. Then, at 1:00 P.M. on August 24, the mountain suddenly roared. Molten rock, gas, ash, and rock fragments burst from its summit and spewed into the sky in a dark, turbulent plume. As lightning sparked in its crown, the plume billowed upward like a furious storm cloud, then spread like an umbrella over Pompeii, turning day into night. In the town, choking fumes filled the air, ash sifted down like snow, and rock fragments fell like hail. Terrified inhabitants rushed toward the shore, hoping to sail out of danger. But for about 2,000 people, it was too late. Mt. Vesuvius suddenly exploded, blasting out vastly more molten rock, debris, and gas at high velocity. Much of this material swirled skyward, but some swept downslope in a scalding avalanche that, in mere minutes, had buried Pompeii entirely (Fig. 4.1a).

The blanket of volcanic debris that spread over Pompeii protected the ruins of the town so well that 18 centuries later, when archaeologists began digging, they found an amazingly well-preserved record of daily life in the Roman Empire (Fig. 4.1b). Occasionally, they also found open spaces in the debris that, when filled with plaster, provided casts of Pompeii's inhabitants, twisted in agony or huddled in despair, at the moment of death (Fig. 4.1c). The heat of the ash had instantly vaporized their bodies. On that fateful day in 79 C.E., Pompeii had experienced a *volcanic eruption*, an event during which molten rock and/or solid debris (ash and other fragments) rises from the Earth's interior and emerges from a vent. About 1,350 volcanoes with the potential to erupt exist on the Earth today. In a given year, about 50 do erupt.

We begin this chapter by explaining why and where solid rock inside the Earth melts, and why the melt seeps upward to cooler environments where it solidifies to form new rock known as igneous rock. We next describe various types of *igneous rocks*, and the features of these rocks

<< The drama of an eruption at a volcano in Iceland. Red-hot molten rock burbles out of volcanic vents and flows in streams downslope. When this melt eventually cools, it solidifies to become new igneous rock.

FIGURE 4.1 The eruption of Mt. Vesuvius buried Pompeii in 79 C.E.

(a) A painting depicting the massive eruption.

Ruts carved into the street by wagon wheels

(b) Streets and buildings of Pompeii, excavated from the ash blanket, are well preserved.

(c) Plaster casts of unfortunate victims who were buried beneath volcanic ash.

that provide clues both to the origin of the melt from which they formed and to the environment in which that melt solidified. The chapter then shifts gears to focus on volcanoes and their eruptions. We characterize different types of eruptions, explain their relationship to plate tectonics, and discuss how volcanoes impact human society.

Glowing waves rise and flow, burning all life on their way, and freeze into black, crusty rock which... builds the land, thereby adding another day to the geologic past... I became a geologist forever, by seeing with my own eyes—the Earth is alive!

—*HANS CLOOS (GEOLOGIST, 1886–1951), ON SEEING AN ERUPTION OF MT. VESUVIUS*

Did you ever wonder...

where the molten rock erupting from a volcano comes from?

4.1 Forming Molten Rock and Volcanic Gas inside the Earth

During a **volcanic eruption**, as the unfortunate Pompeiians discovered, molten rock (or *melt*), ash, rock fragments, and acrid *volcanic gas* come from underground. At the time, people thought that eruptions happened when Vulcan, the god of fire, fueled his subterranean forges to manufacture weapons for other gods. No one believes that myth anymore, but the god's name became the root of the English word **volcano**, now used in reference both to an opening or vent from which material from inside the Earth emerges and to a landscape feature built from this material.

To distinguish between molten rock underground and molten rock that has emerged at the surface, geologists refer to the former as **magma** and the latter as **lava**. *Igneous rock* forms when magma solidifies underground, or when the products of an eruption (including lava, ash, and coarser fragments) solidify. The process of melting, the rise of melt and volcanic gas, *volcanism* (the activity of volcanoes), and the production of igneous rocks, all taken together, constitute **igneous activity** (Fig. 4.2). Considering the fiery heat of the molten rock involved in igneous activity, the name, derived from the Latin *ignis* (meaning fire), certainly seems apt. Igneous activity doesn't take place everywhere—otherwise, our planet's surface would be speckled everywhere with volcanoes. Let's examine the locations inside the Earth where the mantle or crust melts sufficiently to trigger igneous activity.

Causes of Melting to Produce Magma

Popular media stories about volcanoes may leave an impression that the Earth's mantle consists entirely of molten rock, and that volcanoes erupt wherever a conduit through the crust connects this underground molten sea to the Earth's surface. That image is fantasy! The crust and mantle both consist almost completely of solid rock. (The only molten layer of the Earth's interior is the outer core, and it could not possibly be the source of magma because the outer core consists of iron alloy and igneous rocks do not.) What special conditions lead to magma production from the solid rocks of the mantle and crust? Laboratory studies of the pressure and temperature conditions under which mantle rock melts provide an answer.

If you empty a packet of sugar in a pan and turn up the heat, the sugar melts, for heat causes atoms of the sugar to vibrate faster and break free of the sugar crystals. The same concept applies to rocks—raising the temperature sufficiently causes them to melt. But the temperatures needed to melt rocks are much higher than that needed to melt sugar. At the Earth's surface, *peridotite* (the dark, dense rock making up the mantle) starts to melt at temperatures in excess of 1,200°C (2,200°F). Deeper in the Earth, the temperature at which melting begins is even higher, because pressure inhibits melting. Why? Simplistically, increasing pressure prevents vibrating atoms from breaking free of crystals. Because pressure increases with depth due to the weight of overlying rock, melting temperature also increases with depth.

We can represent the pressure and temperature conditions at which melting begins with a curve, called the *solidus*, on a graph (Fig. 4.3a). While melting begins at the solidus, complete melting won't take place at a given pressure unless the temperature gets still hotter. The pressure and temperature conditions at which peridotite completely melts, called the *liquidus*, can be represented by another line on the graph. Under pressure and temperature conditions between the liquidus and the solidus, peridotite undergoes **partial melting**, meaning that only certain components of the peridotite have become liquid. The melt that does form occurs in between the remaining, still-solid mineral grains. Note that Figure 4.3a includes a third curving line. This line, called the **geotherm**, represents the actual increase in temperature with increasing depth in the Earth. With this graph in mind, we see that three conditions can trigger melting and produce magma.

- *Melting due to a pressure decrease:* As pressure decreases, atoms in crystals are squeezed together less tightly, so bonds between atoms break more easily. Therefore, if the pressure squeezing very hot rock decreases, melting may begin. Such **decompression melting** happens when hotter rock from deeper in the mantle rises to shallower depths without cooling significantly (Fig. 4.3b). Rock can stay hot as it rises because surrounding rock provides good insulation.

- *Melting due to volatile addition:* Compounds that can exist as gases near the Earth's surface are *volatile*

FIGURE 4.2 Molten rock spewing from Kilauea, a Hawaiian volcano, in 2018 flowed over a neighborhood. The dark band on either side of the orange-red molten rock consists of newly formed igneous rock.

FIGURE 4.3 Decompression melting.

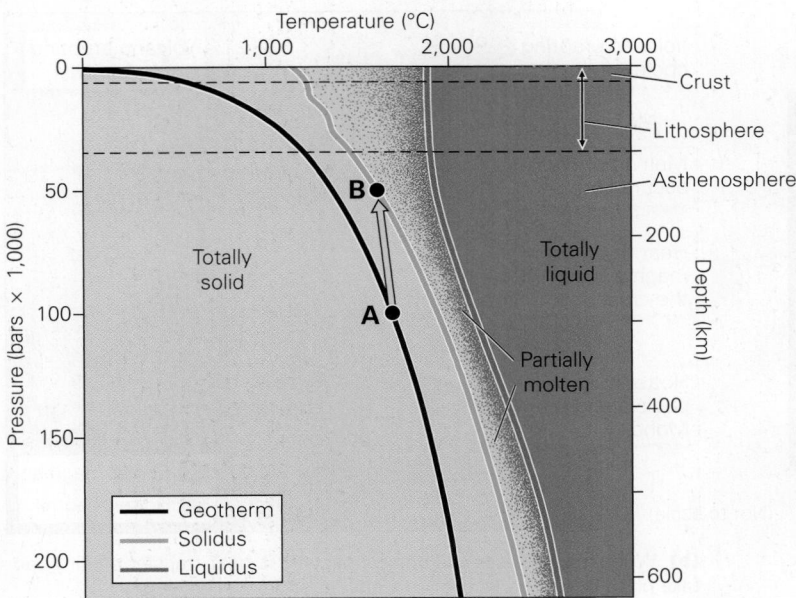

(a) This graph of pressure and temperature conditions inside the Earth shows that when rock rises from deep in the Earth (Point A) to a shallower location (Point B), the pressure acting on it decreases significantly, but the rock cools only a little. Therefore, the rock reaches conditions on the solidus and starts to melt.

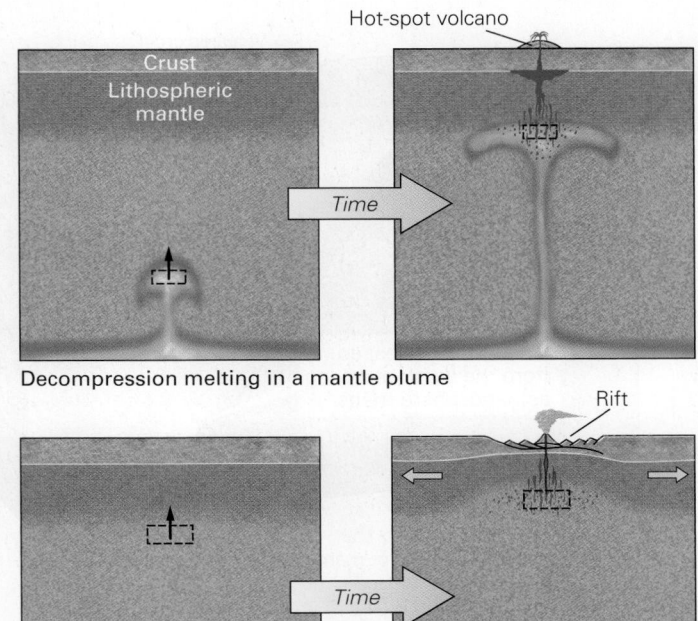

Decompression melting in a mantle plume

Decompression melting beneath a rift

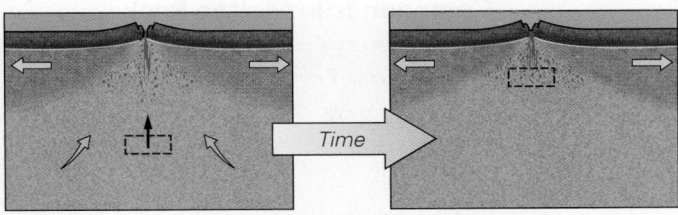

Decompression melting beneath a mid-ocean ridge

(b) Conditions leading to decompression melting occur in several different geologic settings. In each case, when a volume of hot asthenosphere (outlined by dashed rectangle) rises to a shallower depth and, therefore, is subjected to lower pressure, melt (red blobs) starts to form. In a mantle plume, hot rock rises because it's buoyant compared with cooler rock surrounding it. During rifting, the lithosphere thins, and asthenosphere rises to keep space filled. Formation of new lithosphere by seafloor spreading occurs as the asthenosphere rises, keeping the space filled as plates move apart.

materials. When volatiles, such as H_2O and CO_2, seep into hot rock, the rock can begin to melt, for volatiles promote the separation of atoms from crystals. In other words, the addition of volatiles lowers the melting temperature of rock, somewhat like sprinkling salt on ice makes the ice melt. Geologists refer to this process as **flux melting** (Fig. 4.4a); in this context, a *flux* is a chemical that, when added to a material, lowers the material's melting temperature. Subducting oceanic crust releases H_2O and CO_2 when it sinks into the hot asthenosphere, causing molten rock to form in the overlying asthenosphere.

• *Melting due to heat transfer:* Not all rock types start to melt at the same temperature. Magma rising from the mantle can be 500°C (900°F) hotter than the melting temperature of continental crustal rocks. As heat flows from the hot magma into the continental crust, it can make crustal rocks warm enough to melt and produce additional magma (Fig. 4.4b). (To picture the process, imagine what would happen if you placed ice cream on top of hot fudge.) We call this process **heat-transfer melting**.

Compositional Varieties of Molten Rock

Molten rock ranges from yellowish to reddish-orange, and is very hot and quite gooey. Unlike water, which contains a single compound, H_2O, it contains several different *oxides*, compounds consisting of oxygen anions (see Box 3.1) bonded to metal cations (Table 4.1). Different occurrences of molten rock contain different proportions of these oxides. Geologists distinguish among four compositional varieties of molten rock by the proportion of silica (silicon dioxide, SiO_2), relative to the sum of iron and magnesium oxides, that the melt contains (Table 4.2). *Felsic* (or *silicic*) melts contain the highest proportion of silica, as well as the other oxides (namely, Al_2O_3, K_2O, or Na_2O) that are common in the minerals feldspar and quartz. *Mafic* melts contain more magnesium and iron

FIGURE 4.4 Flux melting and heat-transfer melting.

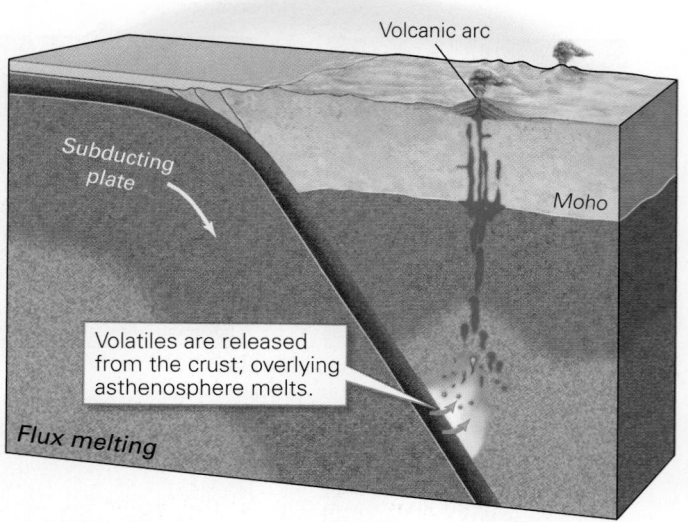

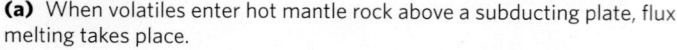

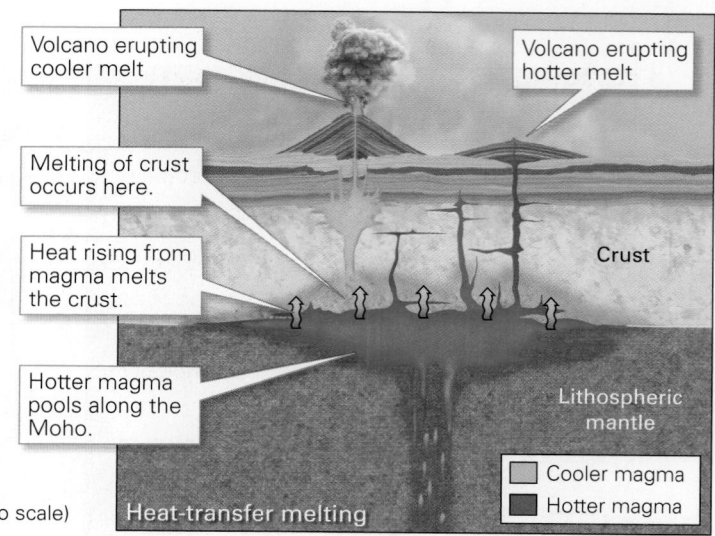

(a) When volatiles enter hot mantle rock above a subducting plate, flux melting takes place.

(b) When magma rises and brings heat with it, heat-transfer melting can take place in the overlying or surrounding rock.

TABLE 4.1 Examples of Common Oxide Compounds in Molten Rock

Silica	SiO_2	Iron oxide	FeO or Fe_2O_3
Magnesium oxide	MgO	Aluminum oxide	Al_2O_3
Sodium oxide	Na_2O	Calcium oxide	CaO
Potassium oxide	K_2O		

▶ **Animation**
Lava Composition
(Interactive)

▶ **Narrative Art Video**
Partial Melting

oxides (namely, MgO, FeO, or Fe_2O_3), which are common in the minerals olivine, amphibole, and pyroxene. Felsic melts are less dense than mafic melts because the oxides in felsic melts contain atoms with lower atomic masses than do those in mafic melts. *Intermediate* magmas, as their name suggests, have a composition between

TABLE 4.2 The Four Compositional Varieties of Molten Rock

Variety	Ratio of silica to iron and magnesium oxide	Silica (%)	Temperature (°C)	Viscosity
Felsic	High	66–76	700	High
Intermediate		52–66	900	
Mafic		45–52	1,100	
Ultramafic	Low	38–45	1,300	Low

that of felsic and mafic magmas, and *ultramafic* magmas have a composition richer in magnesium and iron oxides than do mafic magmas. The specific composition of the molten rock at a given location depends on many factors, as described in **Box 4.1**.

Notably, the composition of molten rock affects the temperature at which a melt solidifies when cooled. For example, a felsic magma solidifies at a temperature hundreds of degrees lower than that at which an ultramafic magma solidifies. The proportion of silica in a melt also affects its **viscosity**, meaning the ease with which it can flow, because silica tends to form chain-like molecules that tangle with one another and slow the flow (see Table 4.2). Therefore, felsic melts have relatively high viscosity and flow less easily, whereas mafic melts have relatively low viscosity and flow more easily.

Origin of Volcanic Gas

The volatiles released from minerals during melting rise with magma and, in some cases, bubble up through magma. This **volcanic gas** can emerge from the principal vent of a volcano or from *fumaroles* (cracks that emit only gas). Sampling volcanic gas isn't easy. While most consists of H_2O and CO_2, it may also include sulfur dioxide (SO_2), hydrogen sulfide (H_2S), and other stinky and poisonous fumes. To capture samples in specially designed containers, or to measure the gas with detecting devices, researchers must wear respirators

BOX 4.1 ▶ **A Deeper Look**

Why does magma composition vary?

Not all magma (or lava) has the same chemical makeup. Geologists suggest that the following conditions and phenomena affect magma composition:

- *Chemical makeup of the magma source:* The chemicals available to go into a magma depend on the composition of the original rock (the *magma source*) from which the magma was extracted. Because magma sources differ from place to place, not all melting provides the same proportions of oxides.

- *Partial melting:* Under the temperature and pressure conditions inside our planet, only a small percentage of the rock at a given location can melt before the resulting magma rises and migrates away. In other words, magma is the product of *partial melting*. The percentage of the magma source that melts to form magma affects the proportions of chemicals in the magma because different minerals start to melt at different temperatures. Significantly, partial melting always yields magma that is more felsic than the magma source, because felsic minerals start releasing oxides into the magma before mafic minerals do. Partial melting of ultramafic rock, for example, produces mafic magma.

- *Chemical interaction with surroundings:* Moving magma may incorporate chemicals dissolved from the solid rock through which it passes. This process is called *assimilation*.

- *Fractional crystallization:* Because magma contains many different oxides, many different minerals can form when it solidifies (freezes). In the 1920s, a Canadian geologist named Norman Bowen discovered that these minerals do not all solidify at the same time. Specifically, when a mafic magma starts to freeze, crystals of olivine and pyroxene, minerals with a high proportion of iron and magnesium oxide, grow first (Fig. Bx4.1a). Their formation removes some iron and magnesium from the melt. Therefore, as cooling continues, the next minerals to form contain relatively less iron and magnesium oxide. Bowen observed that amphiboles start to form next, whereas micas, orthoclase (a feldspar containing potassium), and quartz start to form last. Meanwhile, crystals of plagioclase (a feldspar containing calcium or sodium) also grow. As the melt progressively cools, the composition of the plagioclase crystals forming also changes. Specifically, early-formed plagioclase contains more CaO, whereas later-formed plagioclase contains more Na_2O. The overall sequence of crystallization as molten rock cools is known as *Bowen's reaction series* (Fig. Bx4.1b), and the process of removing chemicals from the melt as a succession of minerals crystallize is called *fractional crystallization*. The succession that involves different minerals is the *discontinuous series*, and the succession that involves the evolving composition of plagioclase is the *continuous series*. Fractional crystallization progressively extracts iron and magnesium from magma and locks it into crystals, so the composition of the remaining magma reflects the degree of fractional crystallization.

FIGURE Bx4.1 Bowen's reaction series represents the sequence of mineral crystallization in cooling magma.

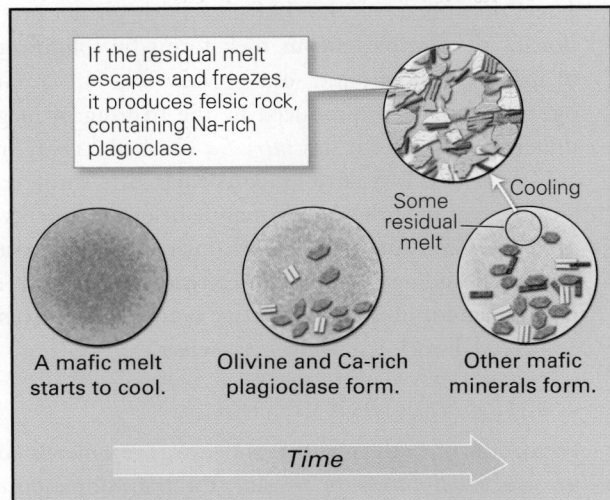

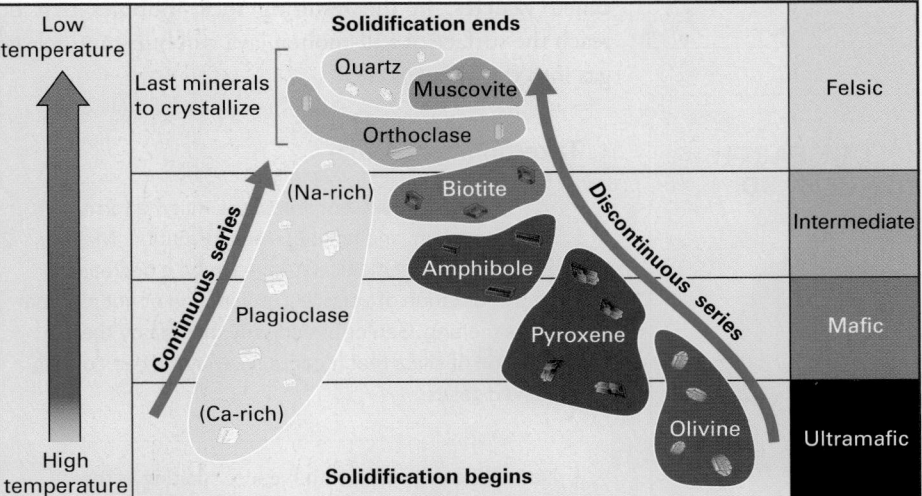

(a) As fractional crystallization takes place, the composition of the remaining magma becomes increasingly more felsic.

(b) This chart displays the discontinuous and continuous sides of Bowen's reaction series. Rocks formed from minerals at the beginning of the series are mafic, whereas rocks formed at the end of the series (at cooler temperatures) are felsic.

FIGURE 4.5 A geologist wears protective gear while sampling gases from a fumarole on a volcano.

4.2 Production of Igneous Rock

The basalt of the seafloor, the granite of a mountain cliff, and the solidified ash that covers the land around Mt. Vesuvius are all examples of igneous rocks. In fact, any rock whose production involves the formation and later solidification of a melt is an **igneous rock**, the final product of igneous processes. The process of igneous rock formation begins when magma moves away from its source and ends when the last drop of melt turns into a solid.

What Causes Magma to Move?

Magma, once formed, doesn't stay put, but tends to seep upward, migrating away from the magma source and into cooler locations, where it solidifies. Why does such **magma migration** take place? First, when a rock transitions from solid to liquid, its molecules end up farther apart, making it less dense, so buoyancy force pushes magma upward. Second, the weight of overlying rock produces pressure at depth that squeezes the magma upward toward regions of lower pressure. Initially, rising magma percolates along the surfaces of grains, but as more magma forms and rises, it surrounds crystals, producing *crystal mush*, a mixture of crystals suspended in melt.

Extrusive versus Intrusive Realms

Most people have seen photos of volcanic eruptions, dramatic images highlighting glowing floods of lava or blasts of fragmented rock debris and fine volcanic ash. Geologists refer to new igneous rock formed from these materials as **extrusive igneous rock** (Fig. 4.6a, b), and to the Earth's surface as the *extrusive realm* of igneous rock formation. Not all igneous rock, however, originates after extrusion from a volcano. In fact, a vastly greater proportion of the Earth's igneous rock forms by solidification of magma underground, out of view (Fig. 4.6c). Geologists refer to rock produced by the freezing of magma underground, after it has *intruded* (pushed its way into) cooler rock, as **intrusive igneous rock**. The entire crust, and some of the lithospheric mantle below, represents the *intrusive realm* for igneous rock formation. Pre-existing rock surrounding an intrusion constitutes *wall rock*, and the surface forming the boundary between the intrusion and the wall rock is an *intrusive contact*.

Transforming Melt into Rock

Formation of both intrusive and extrusive igneous rocks involves *solidification*, or *freezing*, the transformation of a liquid to the solid state. In our everyday experience, freezing generally refers to the transformation of liquid water to solid ice at a temperature of 0°C (32°F) or below. Solidification of magma or lava into igneous rock indeed represents the same phenomenon, but at a much higher temperature. For

(Fig. 4.5), or use drones that can fly through a volcanic gas cloud.

Where does volcanic gas come from? Volatiles exist in rock as ions bonded to minerals. When rock starts to melt, volatiles separate from solid minerals and become independent molecules. But under the high pressure found deep underground, these molecules remain dissolved in the magma, just as carbon dioxide remains dissolved in an unopened can of cola. When magma migrates up to within about 5 km (3 mi) of the Earth's surface, the pressure decreases enough for volatiles to come out of solution and form bubbles, just as bubbles form in an open can of cola. Bubbles that don't escape from magma or lava while it solidifies remain as gas-filled openings, called **vesicles**, in the resulting rock. Bubbles that do reach the surface of still-molten lava will burst and release gas into the atmosphere.

EXPLORE EARTH SCIENCE IN 3D

Basalt

EXPLORE EARTH SCIENCE IN 3D

Granite

Take-home message...

Igneous activity encompasses magma formation, volcanism, and igneous-rock formation. Magma forms in specific geologic settings where a decrease in pressure, addition of volatiles, or injection of hot magma triggers melting. Geologists classify magma by the proportion of silica that it contains. Magma can contain dissolved gases.

Quick Questions ————————

- Name three conditions that cause melting in the Earth's interior.
- What are the four compositional varieties of magma?
- Name some compounds that commonly occur in volcanic gas.

FIGURE 4.6 Formation of extrusive and intrusive igneous rocks.

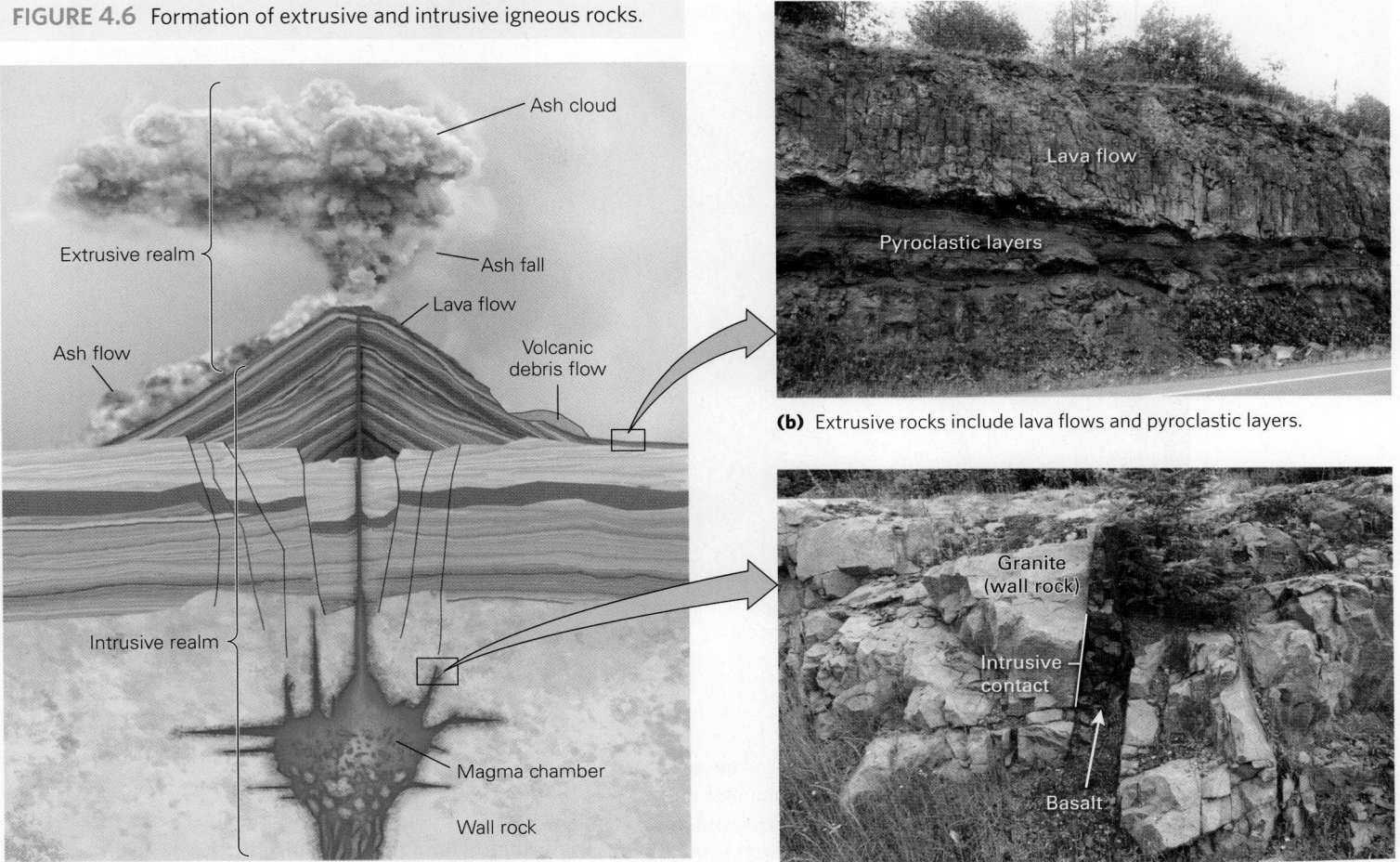

Ash cloud

Extrusive realm

Ash fall

Lava flow

Ash flow

Volcanic debris flow

Intrusive realm

Magma chamber

Wall rock

(a) The intrusive realm lies underground, and the extrusive realm lies on the Earth's surface and in the atmosphere. Lava flows, as well as various types of solid volcanic debris, all produce extrusive rocks.

Lava flow

Pyroclastic layers

(b) Extrusive rocks include lava flows and pyroclastic layers.

Granite (wall rock)

Intrusive contact

Basalt

(c) An intrusion of basalt (dark rock) cuts across a mass of granite (light rock).

example, Hawaiian lava starts to solidify at about 1,200°C (2,190°F) and becomes completely solid at about 900°C (1,650°F). To put these temperatures in perspective, keep in mind that your home oven reaches a top temperature of 260°C (500°F). Why does igneous rock solidify over a range of temperatures, instead of at one specific temperature? This behavior happens because a magma or lava contains several different chemical compounds, and the minerals formed from them start to crystallize at different temperatures. Water, in contrast, freezes at one temperature, 0°C, because it consists of only one compound, H_2O, that forms only one type of mineral (ice). As indicated by Figure 4.3a, solidification begins at conditions along the liquidus curve and ends at conditions along the solidus curve, so the specific temperature of complete solidification for a melt of a given composition depends on depth in the Earth.

How long it takes a volume of molten rock to solidify depends on how fast heat transfers from the melt into its surroundings. Consequently, an intrusion that cools deep in the crust, surrounded by warm wall rock, cools slowly, like coffee in a thermos bottle, because wall rock, like a

thermos bottle, serves as a good insulator. Lava flowing onto the surface of the Earth cools quickly because it is less well insulated. In fact, if lava flows into water, it cools especially fast, because water removes heat very efficiently. The cooling rate of a body of melt also depends on its shape and size, because a body loses heat to its surroundings only at its surface. As a result, the ratio of surface area to volume affects the rate of heat loss. This relationship means that a pancake-shaped intrusion cools faster than a spherical intrusion of the same volume because the pancake has more surface area. Similarly, a shoebox-sized intrusion cools more quickly than a building-sized intrusion, and small lava blobs sprayed into the air cool faster than a thick lava flow.

A newly formed igneous rock takes time to cool down to the temperature of its surroundings. During the final stages of cooling, igneous rock shrinks, for as temperature decreases, volume decreases. Shrinkage can cause the rock to form cracks, called *joints*. In some basalt intrusions, or in the thick interior of a lava layer, joints outline roughly hexagonal columns, producing a visually striking pattern called *columnar jointing* (Fig. 4.7).

FIGURE 4.7 Columnar jointing.

Hexagonal column top

Column faces
(joint surfaces)

What an Earth Scientist Sees

(a) Columnar jointing exposed in Northern Ireland. Vertical joints bound each column. The top of each column is roughly hexagonal.

(b) This three-dimensional sketch shows the shape of idealized columns.

Take-home message...

Magma rises because it is buoyant and because pressure squeezes it upward. When molten rock enters a cooler environment, it solidifies into rock. Intrusive igneous rock forms underground, whereas extrusive igneous rock forms on the Earth's surface. The rate of cooling depends on the environment in which the rock cools and on the size and shape of the melt body.

Quick Questions

- What is magma migration?
- Distinguish between intrusive and extrusive igneous rock.
- Does all igneous rock solidify at the same temperature?

4.3 The Products of Igneous Activity

Igneous Intrusions

Igneous intrusions form from bodies of magma that solidify underground. They aren't visible at the Earth's surface while they form, but over time, erosion of overlying rock can expose them so that they can be studied. Geologists distinguish among several different types of intrusions.

A *sheet intrusion* (or *tabular intrusion*) has relatively planar, parallel contacts on both sides and a somewhat uniform thickness. Such intrusions range in thickness from centimeters to hundreds of meters. Geologists distinguish between two geometries of sheet intrusions based on the orientation of the intrusion relative to layering in the wall rock bordering it. A **dike** cuts across pre-existing layering (bedding or foliation) of the adjacent wall rock, whereas a **sill** intrudes parallel to pre-existing layering (Fig. 4.8). By convention, in places where wall rock does not have layering, geologists refer to steeply sloping or vertical sheet intrusions as dikes, and to nearly horizontal sheet intrusions as sills.

A **pluton** is a nontabular body of igneous rock, meaning that it doesn't have parallel sides. Plutons come in a variety of shapes (blob-shaped, lens-shaped, irregularly shaped with many branches, or light-bulb-shaped) and range from tens of meters to several kilometers across. The process of pluton formation remains controversial. Plutons may form by solidification of a **magma chamber**, a large volume consisting mostly of crystal mush, meaning crystals mixed with melt (Fig. 4.9a). Or, they may grow by successive intrusions of sills, with each new sill positioned beneath the previous one (Fig. 4.9b). Plutons tend to form rounded hills or mountains when exposed at the Earth's surface (Fig. 4.9c). Intrusion of numerous plutons in a region produces a huge composite igneous body, known as a **batholith** (Fig. 4.9d). The spectacular cliffs of Yosemite National Park expose rock of the Sierra Nevada batholith in California, which is about 600 km long and up to 100 km wide. The many plutons in the batholith intruded between 105 and 85 million years ago (Fig. 4.9e).

FIGURE 4.8 Igneous sills and dikes are examples of tabular intrusions.

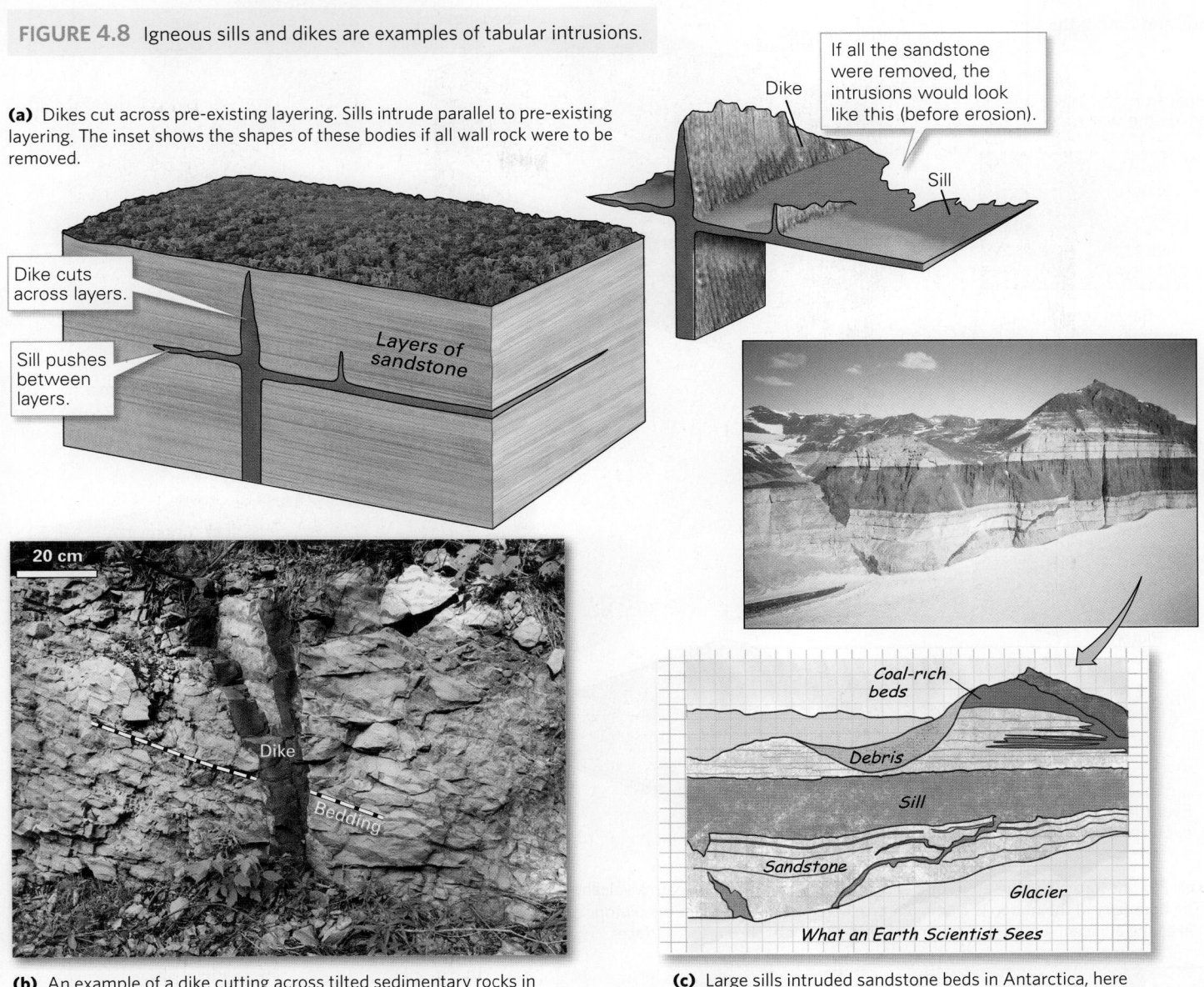

(a) Dikes cut across pre-existing layering. Sills intrude parallel to pre-existing layering. The inset shows the shapes of these bodies if all wall rock were to be removed.

Dike cuts across layers.

Sill pushes between layers.

Layers of sandstone

Dike

If all the sandstone were removed, the intrusions would look like this (before erosion).

Sill

20 cm

Dike

Bedding

Coal-rich beds

Debris

Sill

Sandstone

Glacier

What an Earth Scientist Sees

(b) An example of a dike cutting across tilted sedimentary rocks in Montreal, Canada. The dike split into two branches.

(c) Large sills intruded sandstone beds in Antarctica, here exposed at Finger Mountain.

Where does the room for an intrusion come from? When sills intrude, they push the Earth's surface upward to make room. Igneous dikes intrude into cracks that form where the crust undergoes horizontal stretching. Horizontal stretching of the crust may also provide space for some plutons. Alternatively, pluton-forming magma may produce its own space by pushing aside soft wall rock as it rises, by pushing up the Earth's surface, by assimilating wall rock, or by *stoping*, the process of incorporating blocks of wall rock that fall into the magma chamber.

Lava Flows

When Kilauea, a volcano on the Big Island of Hawai'i, erupted in 2018, streams of lava spilled out of vents and headed downslope, covering fields, burning through homes and forests, and encasing highways. Eventually, the lava turned into layers of brand-new rock. Both a layer or stream of molten lava moving over the Earth's surface and the layer of solid igneous rock formed from that layer or stream can be called a **lava flow**. The characteristics of a lava flow reflect both its viscosity and its volume **(Box 4.2)**. For example, a relatively cool lava of felsic composition has such high viscosity that it cannot flow easily, and instead builds into a bulbous mound, called a *lava dome*, over the vent **(Fig. 4.10a)**. Lava of intermediate composition extrudes in short, blocky flows; still-glowing semisolid chunks may break free from the flow and tumble down

FIGURE 4.9 Plutons and batholiths.

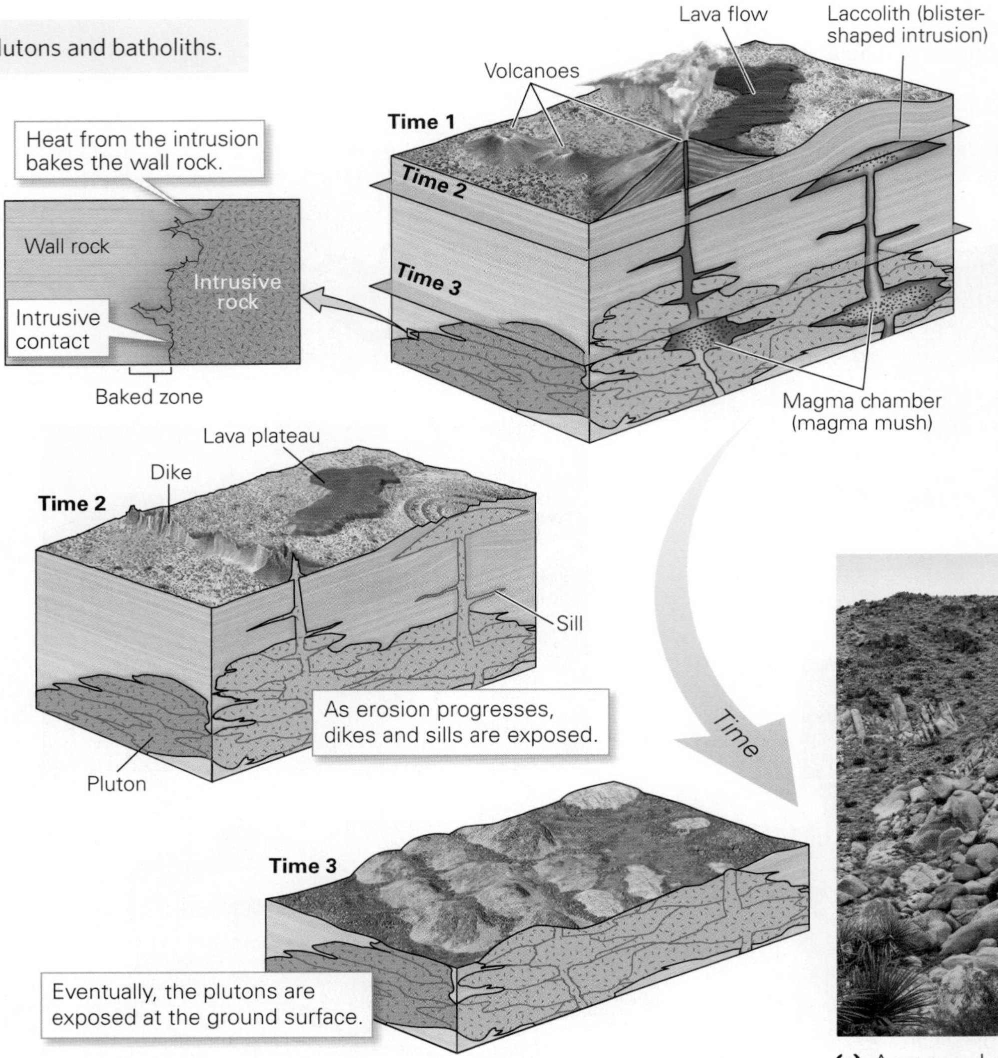

Heat from the intrusion bakes the wall rock.

Wall rock

Intrusive rock

Intrusive contact

Baked zone

Time 1

Volcanoes

Lava flow

Laccolith (blister-shaped intrusion)

Time 2

Time 3

Magma chamber (magma mush)

Older sill

Younger sill

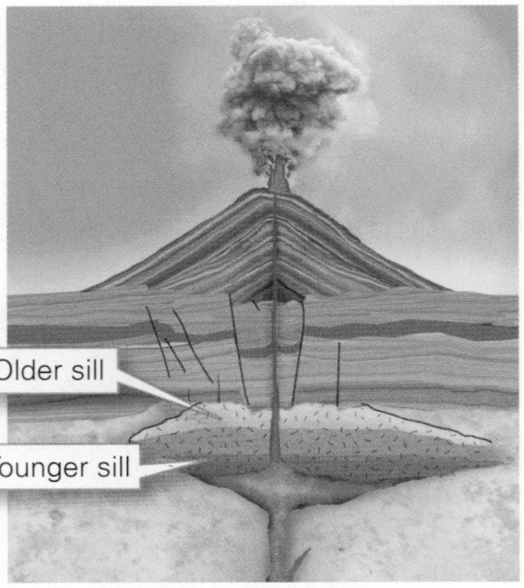

(b) Some plutons may form from a succession of sills.

Time 2

Lava plateau

Dike

Sill

As erosion progresses, dikes and sills are exposed.

Pluton

Time 3

Eventually, the plutons are exposed at the ground surface.

Time

Intrusive contact

Wall rock

Granite

(c) An exposed pluton from the Mojave Desert shows the contact with overlying wall rock.

(a) Plutons form when magma cools slowly at depth, perhaps beneath a volcano. The boundary of the pluton is a type of intrusive contact. Heat from the pluton bakes the adjacent wall rock. Erosion exposes plutons at the Earth's surface.

(d) During the Mesozoic and early Cenozoic, subduction produced a long volcanic arc in what is now western North America. Several batholiths intruded the crust beneath the arc. Erosion and uplift subsequently exposed them.

Coast Ranges batholith

Canada

United States

Idaho batholith

Sierra Nevada batholith

Peninsular batholith

Exposed batholith

(e) The Sierra Nevada batholith formed under the portion of the volcanic arc that erupted in what is now California. The huge cliffs of Yosemite National Park expose part of this batholith.

How can I explain…

The meaning of viscosity

What are we learning?

The viscosity of a magma or lava affects how it flows and how easily gas can escape from it.

What you need:

- A squeeze bottle of ketchup
- A jar of smooth peanut butter
- A large plastic cutting board, a stopwatch, a scoop or spoon, a cup, and plastic straws
- A microwave oven

Instructions:

- Place the cutting board on a counter with one end propped up so that the surface of the board tilts at an angle of 45° into a sink or tray.
- Place a scoop of ketchup near the upslope end of the cutting board. See how far it flows down the cutting board in a given amount of time, and estimate its ending thickness. Repeat with the peanut butter. Which is more viscous?
- Based on your reading of this chapter, which substance symbolizes felsic magma?
- Insert one straw into a cup of ketchup and another into a jar of peanut butter, then blow hard into each one. In which substance do bubbles form and rise more easily? Based on this observation, how does viscosity affect the escape of gas?
- Heat each food briefly in a microwave oven. Repeat the experiment on the tilted cutting board, using the warmed foods. How does rising temperature affect viscosity?

What did we see?

This exercise illustrates the concept of using physical models to represent the behavior of real systems. In detail, it illustrates the contrast between high viscosity and low viscosity and the effect of temperature on viscosity.

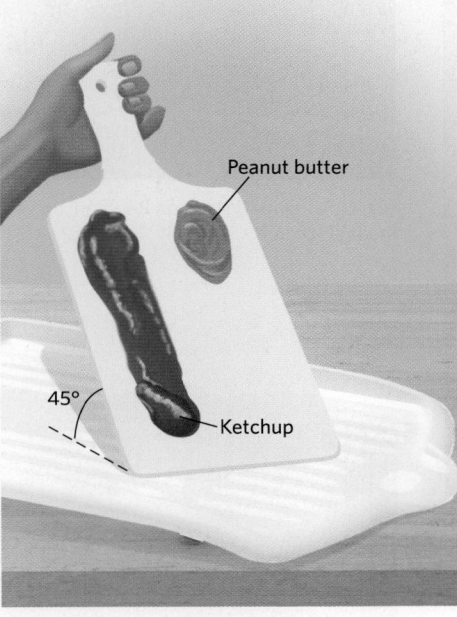

Peanut butter

45°

Ketchup

Bubbles in ketchup

Bubbles don't form easily in peanut butter

the slope of a volcano (Fig. 4.10b). Hot, mafic lava has relatively low viscosity, so it forms relatively long, thin flows (Fig. 4.10c–f). Cooling eventually causes the surface of a mafic flow to crust over while the internal part remains molten and continues to move. As the flow continues to solidify, the lava eventually can move only through *lava tubes*, tunnel-like conduits in the interior of a larger flow (Fig. 4.10g). Lava tubes that drain become empty tunnels (Fig. 4.10h).

The surface texture of a mafic lava flow depends on the volume and rate of the flow. Slower-moving, low-volume flows wrinkle as the flow moves. Such flows, known by a Polynesian name, *pāhoehoe* (pronounced pa-hoy-hoy), solidify to have surfaces covered with rope-like ridges (Fig. 4.11a, b). Flows containing a larger volume of faster-moving lava develop a solid crust that breaks up into a jumble of jagged fragments before the flow stops moving. The resulting rubbly-surfaced flow is known as *ʻaʻā* (pronounced ah-ah) (Fig. 4.11c).

Mafic flows that erupt underwater look quite different from those that erupt on land because lava cools particularly quickly when in contact with water. Moments after extrusion, submarine mafic lava forms a glass-encrusted blob, or *pillow* (Fig. 4.11d). The rind of the

EXPLORE EARTH SCIENCE IN 3D
Pāhoehoe

FIGURE 4.10 Lava flows.

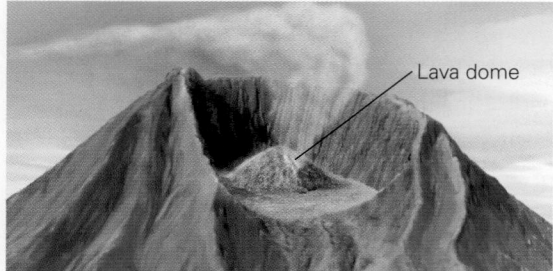

(a) Felsic lava has high viscosity and may build a dome over a vent.

(b) Intermediate lava may produce glowing blocks that tumble downslope.

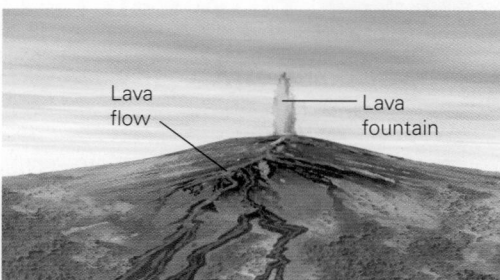

(c) Mafic lava flows can travel long distances.

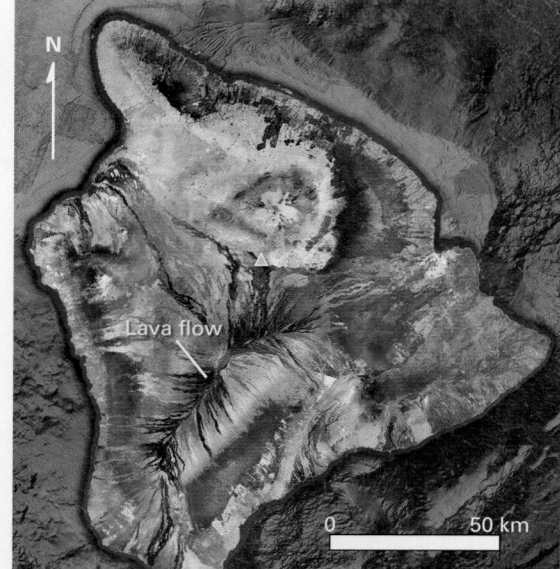

(d) This satellite image of Hawai'i shows the numerous lava flows (dark streaks) that flowed down from the crest of Kilauea.

(e) Flowing mafic lava on Réunion, an island in the Indian Ocean.

(g) Molten lava once flowed through this tunnel under a crust of solid rock. When the lava drained, an open tunnel remained. Part of the roof collapsed, providing an entrance to the tunnel.

(f) A highway cuts through a mafic lava flow on the Big Island of Hawai'i. (Cinder cones are visible in the background.) The location is indicated by the yellow triangle in part (d).

(h) A drained lava tube at Lava Beds National Monument, California.

pillow stops the flow's advance until the pressure in the lava squeezing into the pillow breaks the rind. A new blob of lava squirts out, and then freezes into another pillow. Geologists refer to layers or mounds composed of many blobs as **pillow basalt**.

Pyroclastic Debris

In 1943, as Dionisio Pulido prepared to sow a field west of Mexico City, the ground started to shake, and the surface of the field bulged upward and cracked. Suddenly, fragments of rock and ash spewed out of the crack and into the air, and Dionisio fled. By the following morning,

the field lay buried beneath a mound of gray-black debris 40 m (130 ft) high. Dionisio had witnessed the birth of a new volcano, Paricutín. It continued erupting until 1952, eventually yielding a cone of debris rising 420 m (1,380 ft) high.

Geologists refer to all kinds of fragmental material ejected by an eruption, such as the material that built Paricutín, as **pyroclastic debris** (from the Greek word for fire). Pyroclastic debris can form from the breakup or pulverization of new lava that partially solidified in the volcano's vent, or of pre-existing igneous rocks surrounding the vent. It can also form from

FIGURE 4.11 Surface textures of mafic lava flows.

"Ropes" of pāhoehoe

0.5 m

(a) Pāhoehoe developing on a new lava flow in Hawai'i.

"Ropes" of pāhoehoe

0.5 m

(b) The surface of a pāhoehoe flow after it has cooled and solidified.

(c) The rubbly surface of an 'a'ā flow at Sunset Crater, Arizona.

(d) Pillow basalt develops when lava erupts underwater. Later uplift can raise pillows above sea level, as in this Oregon outcrop.

drops or blobs of lava of a **lava fountain** (lava erupted forcefully enough to spew upward for tens to hundreds of meters above a vent) (Fig. 4.12a), or from a mist of lava sprayed into the air. Some pyroclastic debris blasts skyward to high altitudes and then falls to the ground like rain or hail (Fig. 4.12b), but some debris flows like an avalanche down the side of the volcano in a *pyroclastic flow.*

Specific terms used for fragments of pyroclastic debris indicate their size and shape. **Ash** consists of tiny flakes or slivers, commonly composed of volcanic glass (Fig. 4.12c). **Lapilli** consist of marble- to golf-ball-sized fragments of erupted rock, or of ash balls that form when ash mixes with water in the air (Fig. 4.12d), or of small blobs of lava that freeze in the air (Fig. 4.12e). The term *cinders* can be used specifically for lapilli or small blocks consisting of *scoria* (a glassy, bubble-filled type of basalt or andesite, which we'll discuss later). *Blocks* are relatively large, angular chunks (Fig. 4.12f), and *volcanic bombs* are blocks that

were still soft during eruption so that they became stream-lined as they erupted and flattened into a cow-pie shape when they landed (Fig. 4.12g). The term *tephra* refers exclusively to pyroclastic debris that falls through the air, not to the debris of pyroclastic flows.

Take-home message...

Igneous intrusions form underground and occur in a variety of shapes. Extrusion during a volcanic eruption produces lava flows, whose character depends on composition, and pyroclastic debris, which comes in many different sizes.

Quick Questions ———————

- Distinguish among a dike, a sill, and a pluton.
- How does a mafic lava flow differ from a felsic one?
- Define pyroclastic debris.

EXPLORE EARTH SCIENCE IN 3D

Volcanic Bomb

FIGURE 4.12 Pyroclastic debris.

(a) Clots of lava being ejected in a lava fountain. In this time exposure, we can see the arc of each clot's path through the sky.

(b) Pyroclastic debris billowing from an eruption of Chaitén, in Chile.

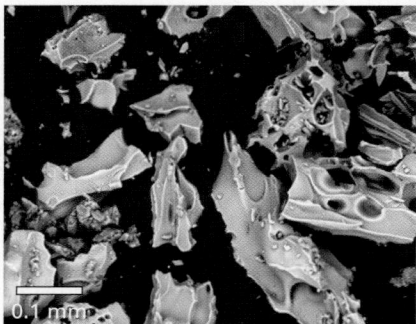

(c) Electron photomicrograph of ash.

(d) These lapilli formed from ash that clotted together in moist air.

(e) Tephra at Sunset Crater, Arizona; note that this sample includes lapilli and small blocks.

(f) Large blocks blasted out of a Hawaiian volcano.

(g) Blobs of lava that become streamlined during eruption are volcanic bombs.

4.4 Classifying Igneous Rocks

Because molten rock comes in a variety of compositions, and because it can solidify in many different environments, igneous activity produces a wide range of igneous rocks. Two characteristics serve, in general, as the basis for classifying these rocks:

1. *Composition:* A description of composition distinguishes among ultramafic, mafic, intermediate, and felsic igneous rocks. Composition provides clues that help characterize the magma source as well as the way in which the magma evolved before completely solidifying.

2. *Texture:* A description of texture distinguishes among *crystalline* (consisting of interlocking crystals), *fragmental* (consisting of pieces that are either welded or cemented to one another), or *glassy* (consisting mostly of glass) igneous rocks (Fig. 4.13a–c). In the case of crystalline rocks, a description of texture also indicates crystal size. Texture tells us about the environment in which the rock solidified.

Crystalline Igneous Rocks

A **crystalline igneous rock** consists of mineral crystals that grow in a melt and fit together like pieces in a jigsaw puzzle. As the rock solidifies, later-formed crystals fill spaces between, or grow around, earlier-formed crystals. Geologists distinguish between coarse-grained (technically known as *phaneritic*) igneous rocks, in which the minerals making up individual grains can be distinguished with the naked eye, and fine-grained (*aphanitic*) igneous rocks, in which the crystals are too small to be identified without a microscope (Fig. 4.13d). Some igneous rocks have a uniform texture, in that they contain mostly grains of about the same size. Others, known as *porphyritic* rocks, contain large grains (*phenocrysts*) surrounded by a matrix of tiny grains (*groundmass*). Simplistically, the grain size of a crystalline igneous rock reflects the cooling rate: finer-grained rocks cooled quickly, whereas coarser-grained rocks cooled slowly. Porphyritic rocks started cooling slowly (to produce the large grains) and then cooled quickly (to produce the fine grains). Because of the relationship between grain size and cooling rate, plutons that cooled slowly deep in the crust have large grains, while lava flows

FIGURE 4.13 Igneous rock textures as viewed through a microscope. The field of view for the inset photomicrographs is about 3 mm.

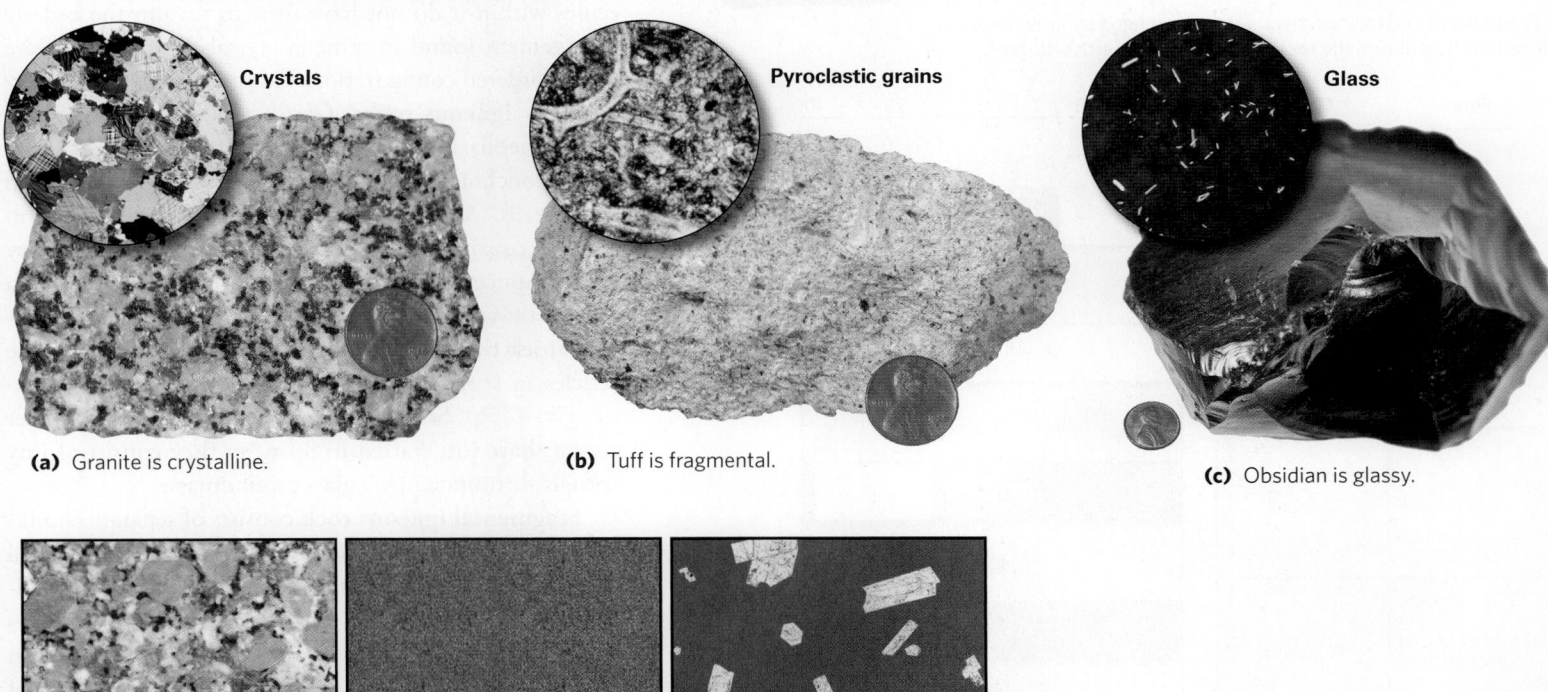

(a) Granite is crystalline.

(b) Tuff is fragmental.

(c) Obsidian is glassy.

Coarse-grained (phaneritic)

Fine-grained (aphanitic)

Phenocrysts in groundmass (porphyritic)

(d) The three textures of crystalline igneous rock have different grain sizes.

FIGURE 4.14 Classification of crystalline igneous rocks.

(a) Examples of crystalline igneous rocks, arranged by grain size and composition. The ultramafic rocks are rare at the Earth's surface.

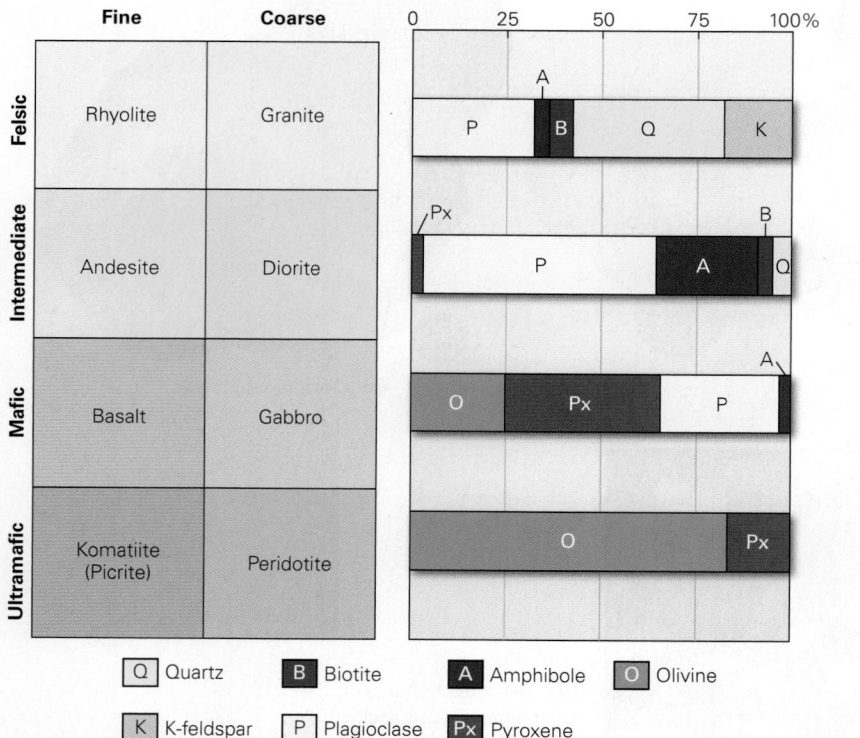

(b) The average mineral composition of different types of crystalline igneous rocks. The chart at the right indicates the percentages of minerals in the various rock types.

that cooled at the surface have small grains. An exception to the above relationship happens when crystals solidify from very water-rich magma, for the presence of water allows very large crystals to grow rapidly; such cooling produces *pegmatite*, a very coarse-grained rock usually found in dikes or sills.

There are many different types of crystalline igneous rocks. The name applied to a given rock sample depends on both the composition and the grain size of the sample (Fig. 4.14). For example, a fine-grained felsic rock is a *rhyolite* while a coarse-grained one is a *granite*; a fine-grained intermediate rock is an *andesite* while a coarse-grained one is a *diorite*; a fine-grained mafic rock is a *basalt* while a coarse-grained one is a *gabbro*; and a fine-grained ultramafic rock is a *komatiite* (an uncommon rock) while a coarse-grained one is a *peridotite*. Note that the two members of each pair have the same chemical composition, but they cooled at different rates. The density, and sometimes the color, of an igneous rock provide clues to its composition. For example, samples of felsic rock have relatively low density and tend to be light tan or pinkish maroon, whereas samples of mafic rock tend to have relatively high density and tend to be dark gray.

Glassy and Fragmental Igneous Rocks

In some cases, lava cools so quickly that atoms or molecules within it do not have time to fit into the orderly arrangement found in mineral crystals, but rather freeze in a disordered configuration. This process yields a variety of **glassy igneous rocks**. *Obsidian* is a black or brown, homogeneous mass of felsic glass (see Fig. 4.13c) that breaks conchoidally, so that it can be fashioned into sharp arrowheads. A frothy felsic melt may solidify to form *pumice*, consisting of tiny vesicles separated by glassy walls. Pumice contains so much gas that it's light enough to float on water (Fig. 4.15a). *Scoria* is a glassy mafic volcanic rock that also contains abundant vesicles, but the vesicles in scoria tend to be bigger than those in pumice (Fig. 4.15b). Some glassy rocks solidify into glass after crystals have just started to form, so they consist of tiny crystals surrounded by a glass groundmass.

Fragmental igneous rock consists of separate chunks or grains that either become welded together while still extremely hot, or become cemented together after cooling by minerals precipitated from groundwater. Examples of such rocks include *volcanic breccia*, which forms in an 'a'ā flow, and *tuff*, a very fine-grained rock composed of volcanic ash (formed, in part, from the glassy walls of pulverized pumice) mixed with pumice fragments (Fig. 4.16). Notably, pyroclastic debris may move downslope after initial deposition, but before cementation, if it gets transported by streams or landslides. Geologists, therefore, make the subtle distinction between *pyroclastic rocks*,

FIGURE 4.15 Glassy igneous rocks containing vesicles.

(a) Pumice is so light that it can float. This sample was erupted by a submarine volcano. The inset shows a close-up of the vesicles inside a similar sample. The vesicles are 0.01–0.05 mm in diameter.

(b) Scoria, like this sample from Sunset Crater, Arizona, looks like a dark sponge, though it is very hard.

formed directly from pyroclastic debris, and *volcaniclastic rocks*, formed from transported pyroclastic debris.

Take-home message…

Igneous rocks form from material that solidified from a melt. They can be crystalline, glassy, or fragmental. Grain size in a crystalline igneous rock can reflect the rock's cooling rate. The name used for a given crystalline rock depends on its grain size and composition. Fragmental igneous rocks form from pyroclastic debris.

Quick Questions ─────────────

- What rock type forms from felsic magma cooled in a large pluton at depth?
- Which has smaller grains—a basalt or a gabbro?
- How does obsidian differ from tuff?

4.5 The Nature of Volcanoes

Volcanic eruptions are the most dramatic, visible manifestations of igneous activity. Now that we have introduced igneous materials and their origins, let's focus our attention on volcanoes and their behavior. We begin by examining the different ways in which volcanoes erupt.

Will It Flow, or Will It Blow?

The character of a volcanic eruption—whether it emits lava flows or billows of pyroclastic debris—defines a volcano's *eruptive style*. Geologists recognize many different eruptive styles, distinguished from one another by the proportion of lava to pyroclastic debris, the presence or absence of steam, and the total volume of material erupted. Here, for simplicity, we focus on two main classes of eruptive styles: effusive

A golf-ball-sized fragment of pumice

FIGURE 4.16 Fragmental igneous rocks.

(a) This cliff consists of thick tuff layers. Each layer formed from a huge volcanic eruption.

(b) A close-up emphasizes that tuff consists mostly of ash, but can also include pumice lapilli.

FIGURE 4.17 Examples of effusive eruptions.

(a) An effusive eruption on Hawai'i. Lava spills out of a small crater and collects in a fast-moving flow.

(b) A lava flow from the 2018 eruption of Kilauea, on the Big Island of Hawai'i.

(c) A lava fountain from the 2018 eruption.

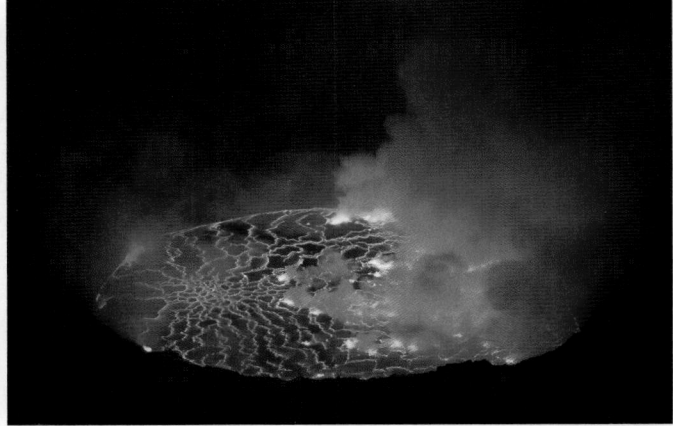

(d) A lava lake at the summit of a volcano. A thin crust of new rock forms over still-molten lava.

and explosive. The difference between these styles largely depends on the silica content of lava because, as we've seen, silica content controls viscosity.

Effusive eruptions, or lava-dominated eruptions, primarily produce mafic lava that spills or fountains out of a vent or, in some cases, fills a *lava lake* (a pool of molten rock) over the vent **(Fig. 4.17)**. Mafic lava has low viscosity, so it not only flows relatively easily, but allows bubbles of volcanic gas to rise and escape, so that explosive pressures don't build up inside the volcano. During effusive eruptions, pressure in the lava can drive a lava fountain, but it won't blow the volcano apart. Lava fountains eject clots of lava (which solidify to produce cinders), bombs, and blocks.

As the name implies, an **explosive eruption** involves an energetic blast that forcefully ejects cloud-like plumes and/or avalanches of pyroclastic debris **(Box 4.3)**.

Explosive eruptions occur in a range of sizes. They can be assigned a number on the **volcanic explosivity index (VEI)**, a logarithmic scale based on the volume of material erupted **(Table 4.3)**. The largest of these eruptions, ranked at VEI 8, have major global consequences. Volcanoes that produce VEI 8 eruptions have come to be known as *supervolcanoes*. Explosive eruptions take place for two reasons.

EXPLOSIVE ERUPTIONS DUE TO TRAPPED GAS. Explosive eruptions happen when countless bubbles of volcanic gas become trapped in intermediate or felsic magma because the high viscosity of these magmas prevents the bubbles from reaching the vent. As magma containing trapped bubbles rises inside a volcano, the pressure due to the weight of overlying magma decreases, so the gas bubbles try to expand. But due to the high viscosity of the lava, or because the lava has already partially solidified into glass,

TABLE 4.3 Volcanic Explosivity Index

VEI	Erupted volume	Height of eruptive cloud
0		
	0.0001 km^3	<0.1 km
1		
	0.001 km^3	0.1–1 km
2		
	0.01 km^3	1–5 km
3		
	0.1 km^3	3–15 km
4		
	1 km^3	10–25 km
5		
	10 km^3	>25 km
6		
	100 km^3	"
7		
	1,000 km^3	"
8		

the bubbles can't grow. As a result, pressure within the bubbles increases. If a lava dome at the summit of the volcano, or the rock forming the flank of the volcano, suddenly breaks or slumps away, the pressure pushing down on the lava suddenly decreases, so gas in the bubbles tries to expand even more. Eventually, the outward pressure in the bubbles breaks the glassy walls surrounding the bubbles, the trapped gas expands explosively, blasting ash (derived from the shattered glass of bubble walls) and fragmented rock out of the vent as an **eruptive jet** that can carry debris and lava skyward for several hundred meters (Fig. 4.18a). In some cases, explosive eruptions may have enough force to blow the top or side off the volcano, producing particularly immense volumes of debris, including pulverized volcanic rocks that had been produced during earlier eruptions.

Shortly after an explosive eruption begins, the scalding ash and gas in the jet heat the surrounding air, producing a cloud-like plume of hot ash and air. The plume thus becomes buoyant relative to the surrounding cold air and produces an *updraft* (an upward flow of air). The resulting **convective plume** can reach stratospheric heights, where it no longer remains buoyant, but instead spreads out like an umbrella or mushroom cap (Fig. 4.18b). If the updraft in the plume is strong enough, some ash gets pushed up into the stratosphere itself. Intense lightning commonly develops in the convective plume, as it does in thunderstorms. Soon, high-altitude winds spread the umbrella downwind. Pyroclastic debris falls like snow and hail from the convective plume and umbrella, burying the surrounding landscape in gray ash and tephra.

Coarser or slightly cooler pyroclastic debris doesn't rise in the convective plume, but rather collapses from the base of the plume as well as from the sides of the eruptive jet to become a **pyroclastic flow**, an avalanche of hot ash and gas that rushes down the slope of the volcano at 80 km/h (50 mph) (see Fig. 4.18a). The ash carried by a pyroclastic flow hides the still-glowing, incredibly hot material within the flow (Fig. 4.18c). In fact, flow remains so hot that it can vaporize or ignite everything in its path. And because ash is heavier than snow, the flow applies enough force to knock down buildings. Once the flow stops and cools, it covers the ground with a gray blanket (Fig. 4.18d).

EXPLOSIVE ERUPTIONS INVOLVING STEAM. Some explosive eruptions take place when water (seawater or groundwater) encounters hot magma within a volcano, or within a magma chamber just below a volcano. Heat from the magma instantly converts the water into high-pressure steam, and the expansion of this steam blasts pyroclastic debris out of the vent. During the 2022 eruption of a volcano in Tonga (a nation of volcanic islands in the western Pacific), an influx of seawater caused the volcano to explode cataclysmically, producing a convective plume that rose 58 km (36 mi) and a shock wave that circled the globe (Fig. 4.18e). In 1883, a similar explosion destroyed Krakatau, a volcano in Indonesia (see Box 4.3). During the 2018 eruption of Kilauea, a lava lake that had partially filled the volcano's summit caldera drained, so the floor of the caldera collapsed and clogged the vent with debris. Steam pressure generated by the interaction of groundwater with magma built up until, together with pressure from volcanic gas, it explosively pushed the debris out of the vent. Ash turned the lush green of the nearby rainforest to dull gray, and a mix of blocks and lapilli pelted the nearby landscape. When seawater gains access to magma in a submarine vent, dramatic eruptions of steam, ash, and pumice boil out of the ocean, even where the summit of the volcano remains hidden by water (Fig. 4.18f).

The Architecture of Volcanoes

All volcanoes share the same basic components. Beneath the volcano, a mush of crystals and melt accumulates to form a magma chamber. Magma and gas rise from the chamber through a conduit to the Earth's surface and erupt from a vent. Over time, pyroclastic debris and lava flows build up around the vent to produce a hill or mountain, known as the **volcanic edifice**, with a peak, or *summit*, and sides, or *flanks*. In some volcanoes, the conduit through which lava and gas reach the surface has a chimney-like shape and is topped by a circular depression, or **crater**, that resembles a

Did you ever wonder . . .

whether all volcanoes look the same?

BOX 4.3 ▶ A Deeper Look

Explosive eruptions to remember

Mt. St. Helens, a snow-crested volcano in the Cascade Range of the northwestern United States, had not erupted since 1857. However, geologic evidence suggested that the mountain had endured a violent past, punctuated by explosive eruptions. On March 20, 1980, earthquakes announced that the volcano was awakening once again, and a week later, the summit began emitting gas and pyroclastic debris. Geologists who set up monitoring stations to observe the volcano noted that its north side was beginning to bulge, suggesting that rising magma was inflating the volcano like a balloon. Concern that an eruption was imminent led local authorities to evacuate people from the area. All but about 60 left.

The main eruption came suddenly. At 8:32 A.M. on May 18, David Johnston, a US Geological Survey geologist monitoring the volcano from a station 10 km away, shouted over his two-way radio, "Vancouver, Vancouver, this is it!" An earthquake had caused 3 km³ (0.7 mi³) of the volcano's weakened north side to slide away. The massive landslide released pressure on the magma, causing a sudden and violent expansion of gases that blasted through the side of the volcano (Fig. Bx4.3a). Rock, steam, and ash screamed north at the speed of sound and flattened a forest and everything in it over an area of 600 km² (230 mi²) (Fig. Bx4.3b). Tragically, Johnston, along with those who hadn't evacuated, vanished forever beneath the debris.

Seconds after the sideways blast, a convective plume carried over 500 million tons of ash (about 1 km³) up to the stratosphere, where strong high-altitude winds eventually transported it around the globe. In towns near the volcano, a blizzard of ash buried fields, and water-saturated ash flooded river valleys, carrying away everything in its path. When the eruption was finally over, Mt. St. Helens was 440 m (1,450 ft) lower, and it had a gaping gouge in one side. The snow-covered landscape of the mountain had turned to gray.

An even greater explosive eruption happened in 1883. Krakatau, a volcano in the sea between Java and Sumatra, where the Indian Ocean floor subducts beneath Southeast Asia, had grown to become an island 9 km (6 mi) long, rising 800 m (2,600 ft) above the sea. On May 20, a series of large explosions shook the mountain, yielding ash that settled as far as 500 km (300 mi) away. Smaller explosions continued through June and July, and steam and ash rose from the island, forming a huge black cloud that rained ash into the surrounding straits. Krakatau's demise came on August 27, perhaps when the volcano cracked, flooding the magma chamber with seawater that flashed to steam. The resulting blast released five thousand times more energy than the Hiroshima atomic-bomb explosion, and could be heard as far as 4,800 km (3,000 mi) away. Large *tsunamis* (broad water waves caused by the sudden push of the blast and

the pyroclastic flows against water) raced outward from the explosion, killing over 36,000 people when they slammed into nearby coastal towns. When the air finally cleared, Krakatau was gone, replaced by a depression some 300 m (1,000 ft) deep. All told, the eruption sent 12 km³ (about 3 mi³) of rock skyward. Ash that circulated in the lower stratosphere caused spectacular sunsets for the next several years.

The 79 C.E. eruption of Mt. Vesuvius, the 1980 eruption of Mt. St. Helens, and the 2022 eruption in Tonga each had a volcanic explosivity index of 5 (see Table 4.3). Krakatau's 1883 eruption had a VEI of 6, for it ejected 12–21 km³ (3–5 mi³) of debris. The largest directly observed eruption—that of Tambora, in Indonesia, in 1815—also had a VEI of 6, but was much larger, for it erupted up to 50 km³ (12 mi³) of debris. (It was once thought to have erupted as much as 150 km³ of debris, but recent research has lowered the estimate.) All of these explosions pale in comparison to supervolcanic eruptions that happened before human history began. About 74,000 years ago, Toba, in Indonesia, ejected 2,800 km³ (670 mi³) of debris in a VEI 8 eruption. Another VEI 8 explosion, which took place over 630,000 years ago in what is now Yellowstone National Park, Wyoming (Fig. Bx4.3c), ejected more than 1,000 km³ (240 mi³), and an eruption there 2.1 million years ago produced almost 2,500 km³ (600 mi³). At least 10 supervolcanic eruptions have occurred in the Yellowstone region, all of which covered much of the United States in ash. The most recent volcanic event that approached the size of a supervolcanic eruption occurred at New Zealand's Taupo Volcano about 26,500 years ago, when it blasted out 530 km³ (130 mi³) of debris.

FIGURE Bx4.3 Memorable explosive eruptions.

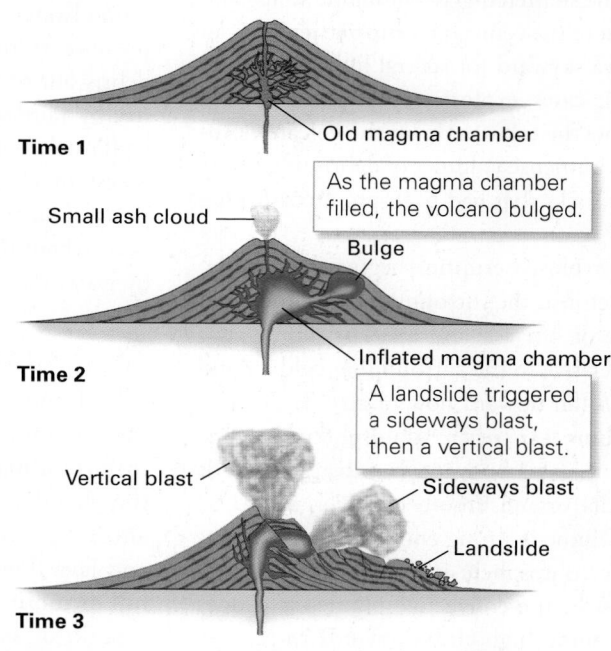

Time 1 — Old magma chamber

Small ash cloud

As the magma chamber filled, the volcano bulged.

Bulge

Time 2 — Inflated magma chamber

A landslide triggered a sideways blast, then a vertical blast.

Vertical blast

Sideways blast

Landslide

Time 3

(a) Stages during the eruption of Mt. St. Helens, 1980.

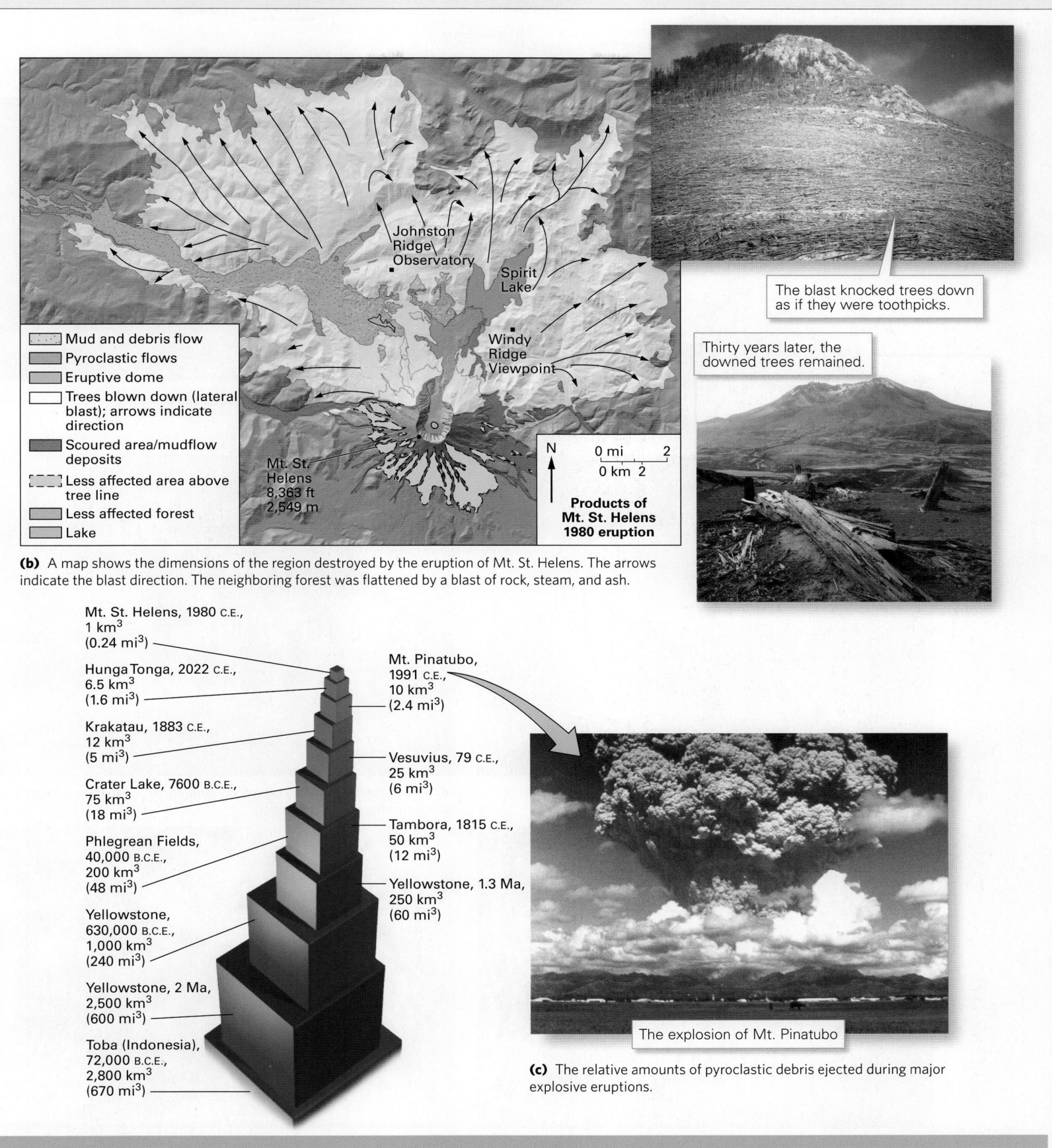

The blast knocked trees down as if they were toothpicks.

Thirty years later, the downed trees remained.

(b) A map shows the dimensions of the region destroyed by the eruption of Mt. St. Helens. The arrows indicate the blast direction. The neighboring forest was flattened by a blast of rock, steam, and ash.

Mud and debris flow
Pyroclastic flows
Eruptive dome
Trees blown down (lateral blast); arrows indicate direction
Scoured area/mudflow deposits
Less affected area above tree line
Less affected forest
Lake

Johnston Ridge Observatory

Spirit Lake

Windy Ridge Viewpoint

Mt. St. Helens
8,363 ft
2,549 m

N

0 mi 2
0 km 2

Products of Mt. St. Helens 1980 eruption

Mt. St. Helens, 1980 C.E.,
1 km³
(0.24 mi³)

Hunga Tonga, 2022 C.E.,
6.5 km³
(1.6 mi³)

Krakatau, 1883 C.E.,
12 km³
(5 mi³)

Crater Lake, 7600 B.C.E.,
75 km³
(18 mi³)

Phlegrean Fields,
40,000 B.C.E.,
200 km³
(48 mi³)

Yellowstone,
630,000 B.C.E.,
1,000 km³
(240 mi³)

Yellowstone, 2 Ma,
2,500 km³
(600 mi³)

Toba (Indonesia),
72,000 B.C.E.,
2,800 km³
(670 mi³)

Mt. Pinatubo,
1991 C.E.,
10 km³
(2.4 mi³)

Vesuvius, 79 C.E.,
25 km³
(6 mi³)

Tambora, 1815 C.E.,
50 km³
(12 mi³)

Yellowstone, 1.3 Ma,
250 km³
(60 mi³)

The explosion of Mt. Pinatubo

(c) The relative amounts of pyroclastic debris ejected during major explosive eruptions.

FIGURE 4.18
Explosive eruptions.

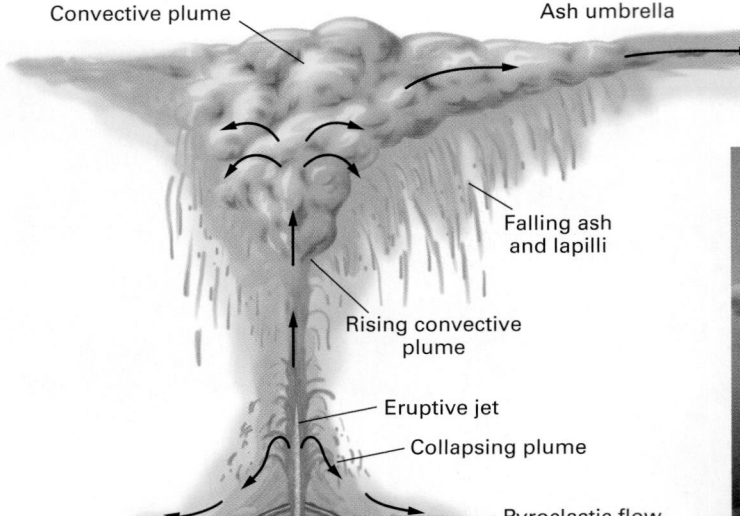

(a) The large convective plume emitted by an explosive eruption contains several components.

Labels in (a): Convective plume · Ash umbrella · Wind · Stratospheric haze · Falling ash and lapilli · Rising convective plume · Eruptive jet · Collapsing plume · Pyroclastic flow

(b) The 1989 eruption of Redoubt, a volcano in Alaska, produced a broad umbrella.

(c) This eruption produced both a convective plume and a pyroclastic flow. The active flow does not appear to be hot, but it is.

Labels in (c): Convective plume · Ash cloud · Pyroclastic flow

(d) An artist's sketch of an explosive eruption. Note the now-cold pyroclastic flow from a previous eruption.

Labels in (d): Convective plume · Pyroclastic debris · Lava flow · Solidified lava flow · Pyroclastic-flow deposits · Moving pyroclastic flow

(e) The giant eruption in Tonga in 2022, as viewed looking straight down from space. We are seeing the top of the convective plume, spreading out like an umbrella at the base of the stratosphere.

(f) A submarine eruption blasts out steam and pyroclastic debris.

Label in (f): Sea surface

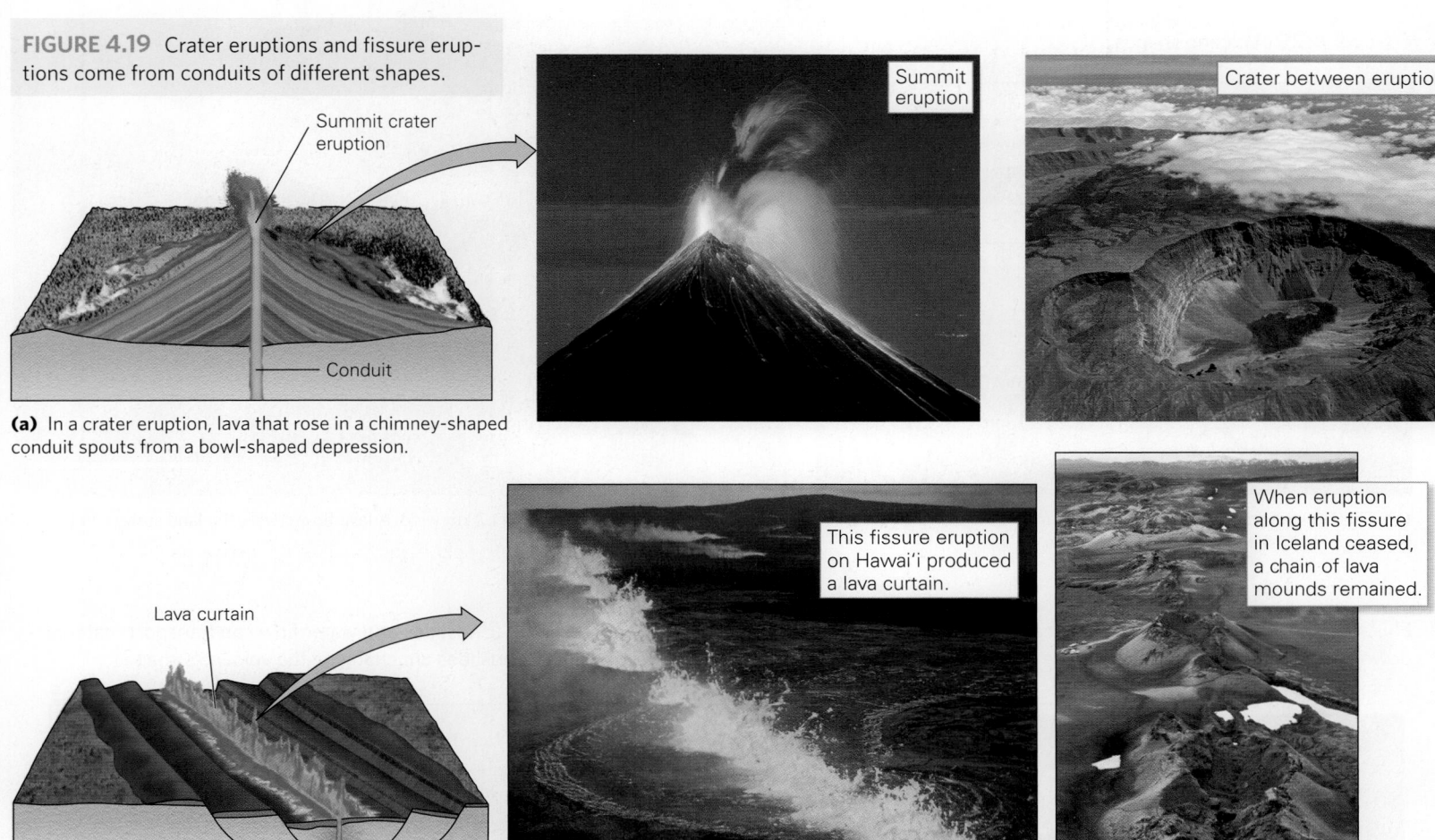

FIGURE 4.19 Crater eruptions and fissure eruptions come from conduits of different shapes.

Summit crater eruption

Conduit

Summit eruption

Crater between eruptions

(a) In a crater eruption, lava that rose in a chimney-shaped conduit spouts from a bowl-shaped depression.

Lava curtain

Fissure

This fissure eruption on Hawai'i produced a lava curtain.

When eruption along this fissure in Iceland ceased, a chain of lava mounds remained.

(b) During a fissure eruption, lava fountains out like a curtain along the length of a crack. After the eruption has ceased, a chain of small edifices remains.

bowl (Fig. 4.19a). Craters, which may be up to 500 m (1,600 ft) across and 200 m (650 ft) deep, may form during an eruption when pyroclastic debris builds up around the vent, or after an eruption when part of the vent area collapses into the drained conduit. Notably, not all volcanoes erupt from a single vent at the summit. Some eruptions, known as *flank eruptions*, happen on the slopes of a volcano. During a crater eruption on a volcano's flank, the vent produces a crater. During a **fissure eruption**, curtains of lava spew from an elongate crack, or *fissure* (Fig. 4.19b).

The overall shapes of volcanic edifices vary significantly, but geologists distinguish among three principal shapes. **Shield volcanoes**, so named because they resemble a soldier's shield lying on the ground, are broad, gentle domes—the largest may be tens of kilometers across. They form from layer upon layer of low-viscosity mafic lava produced by effusive eruptions (Fig. 4.20a). Thin layers of pyroclastic debris, formed when lava in the conduit interacts with groundwater and triggers a steam explosion, occur between some lava flows. **Cinder cones** consist of symmetrical, cone-shaped piles of volcanic cinders

(scoria lapilli and small blocks) ejected by a lava fountain, and *spatter cones* consist of blobs of still-molten lava that rise only a short distance before landing near the vent (Fig. 4.20b). In some cases, cinder cones or spatter cones occur in a cluster, but they may also form in a line, following the trace of a fissure (see Fig. 4.19b). **Stratovolcanoes**, also known as *composite volcanoes*, tend to be large (up to a few kilometers high and 25 km, or 15 mi, across) somewhat cone-shaped mountains made from layers of lava and pyroclastic debris erupted during alternating effusive and explosive eruptions of mostly intermediate lava (Fig. 4.21). The prefix *strato–* emphasizes the stratification defined by these alternating layers. In many stratovolcanoes, the upper, steeper part of the cone consists of layers of lava and tephra that accumulated in place, whereas the lower, less steep part consists of deposits of pyroclastic flows or volcaniclastic debris (pyroclastic debris transported by water or landslides after its initial deposition). Particularly large explosive eruptions may blast off the summit or side of a stratovolcano. When this happens, a new volcanic edifice grows over the remnants of its predecessor.

FIGURE 4.20 Volcano shapes.

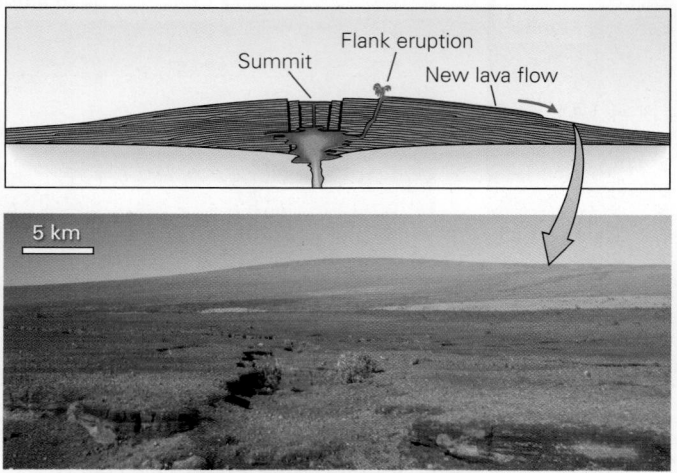

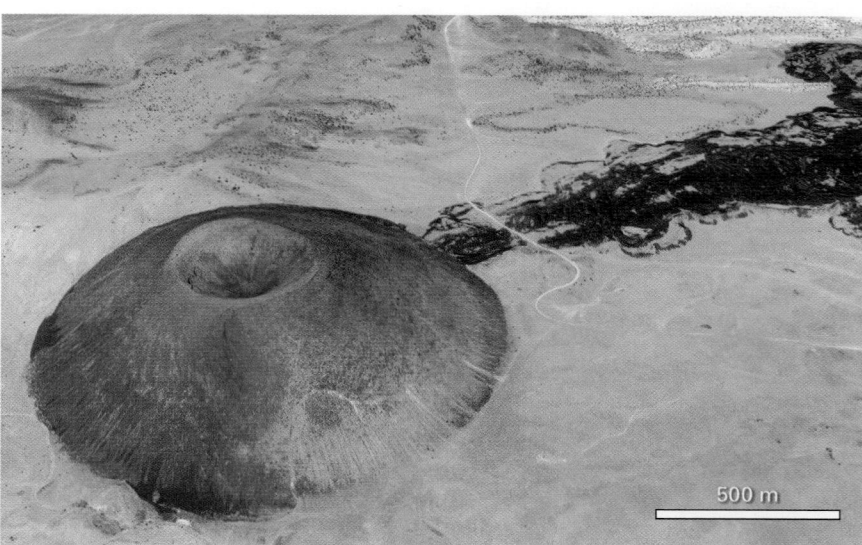

(a) This 50 km wide shield volcano, made from successive layers of low-viscosity basaltic lava, has very gentle slopes.

(b) This cinder cone in Arizona is 1.2 km wide. A lava flow covers the land surface in the distance. Note the road for scale.

FIGURE 4.21 Anatomy of a stratovolcano. A stratovolcano consists of layers of tephra and lava. Landslides and water flow can transport material downslope as volcaniclastic debris. The inset photos show Mt. Fuji, Japan, a famous example. Landslides are changing the volcano's shape.

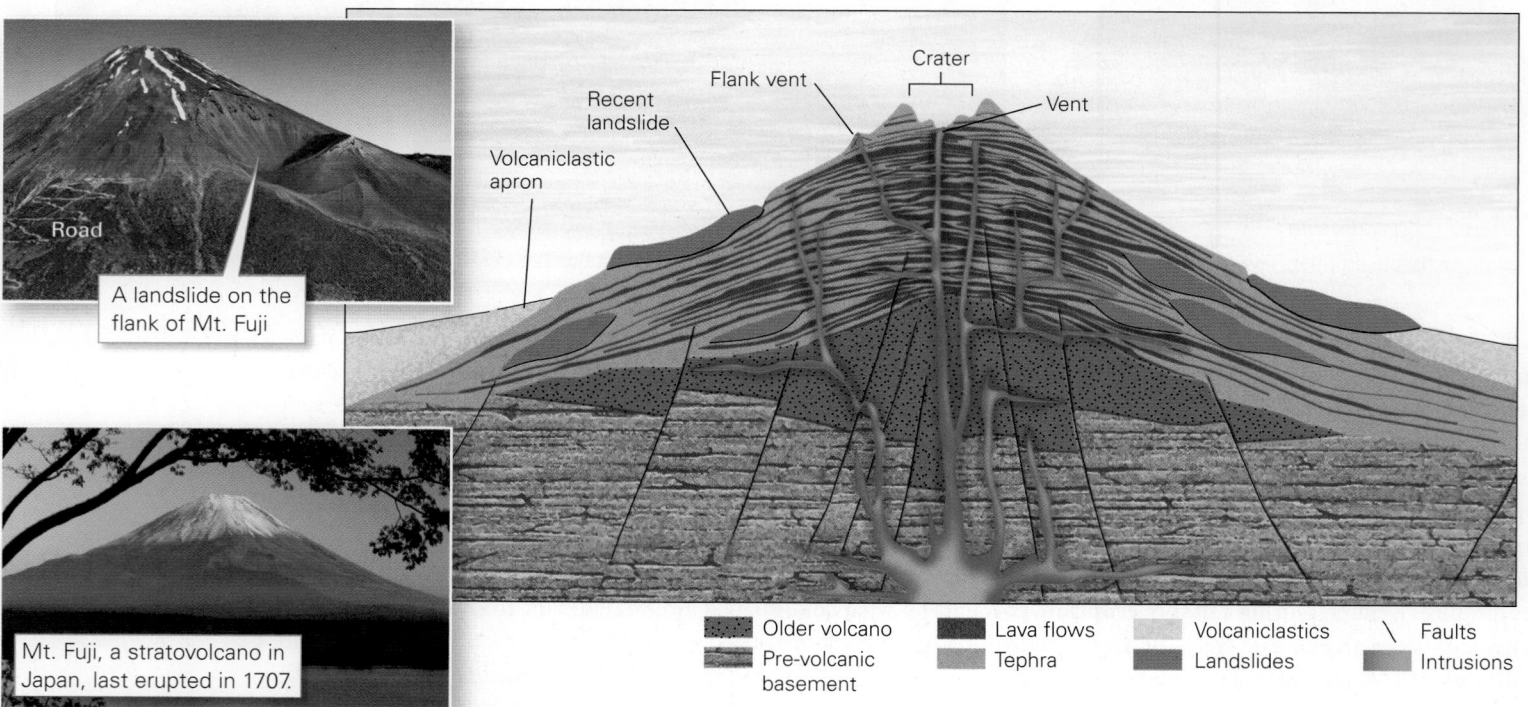

A landslide on the flank of Mt. Fuji

Mt. Fuji, a stratovolcano in Japan, last erupted in 1707.

Older volcano   Lava flows   Volcaniclastics   Faults
Pre-volcanic basement   Tephra   Landslides   Intrusions

After large eruptions, part or all of the remaining edifice of a volcano may collapse into the drained magma chamber below to produce a **caldera**, a circular to elliptical depression several kilometers to tens of kilometers across and hundreds of meters deep. Calderas generally have a fairly flat floor covered by lava or pyroclastic debris (Fig. 4.22a–c). Supervolcanic eruptions produce immense calderas (see Box 4.3). The largest of these on land, formed during the prehistoric explosion of Toba, a volcano in Indonesia, has a diameter of 100 km (62 mi), and the next largest, the Yellowstone caldera, has a diameter of 72 km (45 mi). Some calderas have filled with water to become deep lakes. Crater Lake, Oregon, for example, fills the caldera left by the collapse of Mt. Mazama following an eruption 7,700 years ago (Fig. 4.22d).

FIGURE 4.22 Formation of a volcanic caldera.

Time →

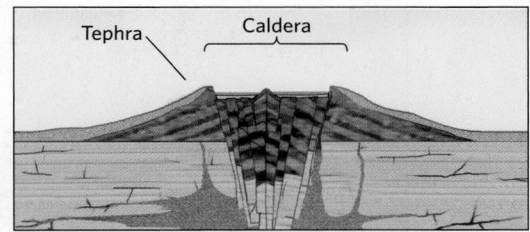

(a) As an eruption begins, the magma chamber inflates with magma. There may be a central vent and one or more flank vents.

(b) During the eruption, the magma chamber drains, and the central portion of the volcano collapses downward.

(c) The collapsed area becomes a caldera. Later, a new volcano may begin to grow within the caldera.

(d) This caldera in Oregon formed about 7,700 years ago. Afterward, it filled with water to become Crater Lake. Wizard Island, protruding from the lake, is a small volcano that grew on top of the caldera floor.

Take-home message...

At a volcano, magma rises from a magma chamber and erupts from a chimney-like conduit or a crack-like fissure. During effusive eruptions, a volcano erupts mostly lava, whereas during explosive eruptions, large volumes of pyroclastic debris blast skyward. Volcanic edifices can be classified as shields, cinder cones, or stratovolcanoes. Collapse of a volcano following a major explosive eruption can produce a caldera.

Quick Questions

- What is the difference between an effusive eruption and an explosive eruption?
- What volume of pyroclastic debris does a volcano have to erupt to be called a supervolcano?
- Which type of volcano has steeper flanks—a shield volcano or a cinder cone?

4.6 Where Does Igneous Activity Occur?

Why does igneous activity happen where it does? Before the theory of plate tectonics was proposed, geologists had no good answer to this question. Now, in the context of the theory, it has become clear that most igneous activity occurs along divergent and convergent boundaries, but some occurs at hot spots or in rifts (Fig. 4.23). In this section, we relate the character of igneous activity to different geologic settings and explain why melting takes place in these settings.

Fissures along Mid-Ocean Ridges

During seafloor spreading at a mid-ocean ridge, the underlying hot asthenosphere moves upward from deeper in the mantle (see Fig. 4.3b). When this hot peridotite

FIGURE 4.23 The geologic settings of volcanoes.

(a) The distribution of volcanoes around the world.

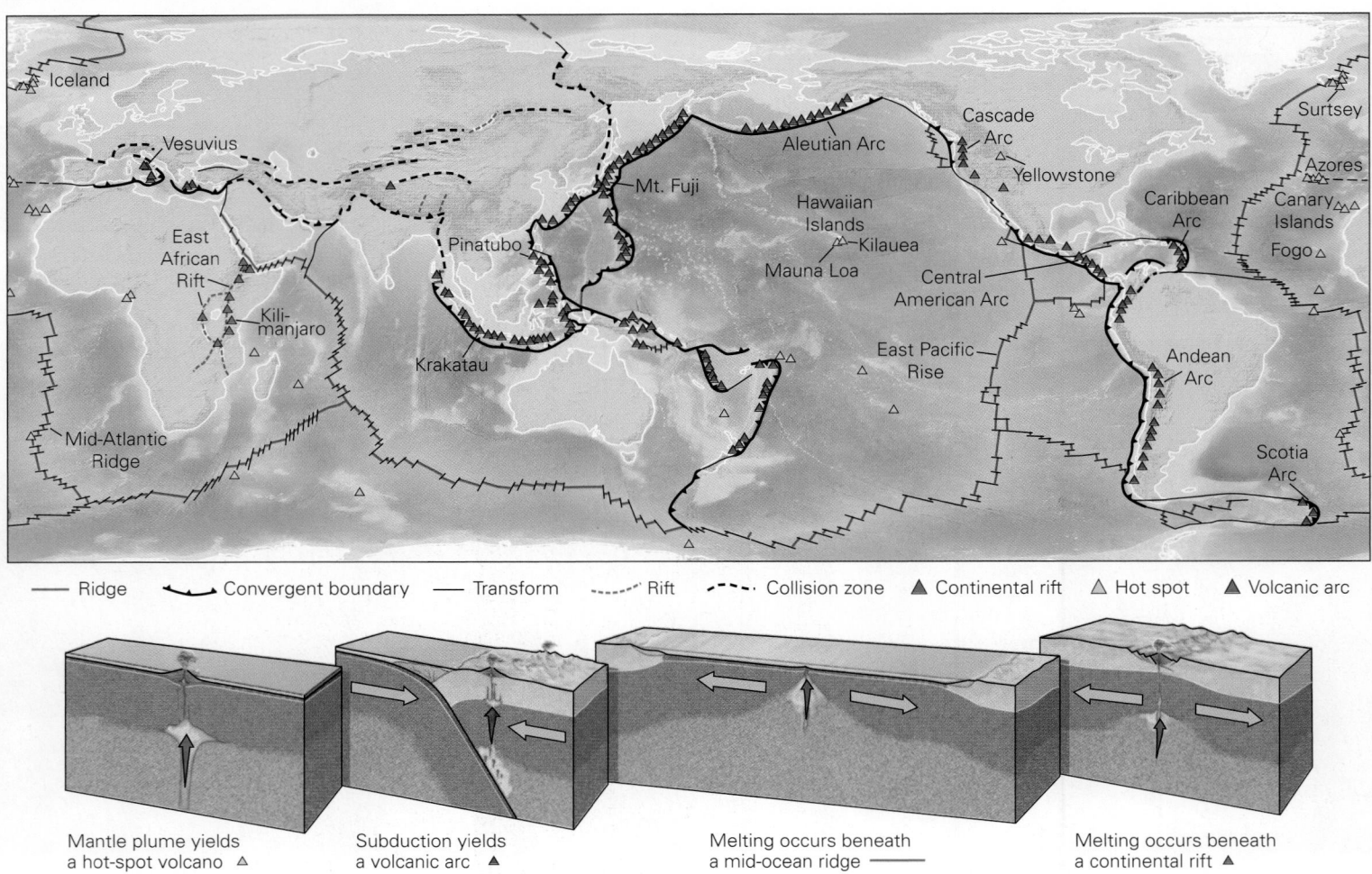

| — Ridge | Convergent boundary | — Transform | Rift | Collision zone | ▲ Continental rift | △ Hot spot | ▲ Volcanic arc |

Mantle plume yields a hot-spot volcano △

Subduction yields a volcanic arc ▲

Melting occurs beneath a mid-ocean ridge —

Melting occurs beneath a continental rift ▲

(b) The four basic types of geologic settings in which volcanoes form, in the context of plate tectonics.

rises above a depth of about 100 km (60 mi), it undergoes decompression, which triggers partial melting. The resulting, relatively buoyant, mafic magma rises still farther. Some of the magma accumulates in a magma chamber about 5 km (3 mi) beneath the ridge axis. Solidification of crystal mush along the sides of the magma chamber yields gabbro. Some magma continues farther upward and solidifies in vertical cracks to form basalt dikes parallel to the plate boundary. Lava that reaches the seafloor erupts to produce pillow basalt.

We don't generally see the volcanic activity of mid-ocean ridges because most of it occurs beneath 2 km (1.2 mi) of seawater. Geologists in submersibles, however, have found that eruptions take place along fissures parallel to the ridge axis (Fig. 4.24a, b). Notably, cracks in the newly formed oceanic crust admit seawater, which, as it

enters the still-hot rock, heats up and dissolves minerals. Convection causes the resulting hot solution to rise and spurt out at *hydrothermal vents* (cracks emitting hot water) on the seafloor. When this hot water encounters seawater, it cools quickly. Consequently, the dissolved minerals precipitate to form a dark cloud of tiny suspended grains. Because of these clouds, hydrothermal vents at mid-ocean ridges are known as **black smokers** (Fig. 4.24c).

Volcanic Arcs at Convergent Boundaries

Most *subaerial volcanoes* (those that erupt above sea level) occur along convergent boundaries. The magma that feeds these volcanoes forms because volatile compounds seep from the crust of the downgoing plate into the overlying asthenosphere and trigger flux melting in the asthenosphere (see Fig. 4.4a). (The crust contains

FIGURE 4.24 Igneous activity along a mid-ocean ridge.

(a) Mounds of pillow basalt erupt along fissures.

(b) Pillow basalt on the seafloor along the Juan de Fuca Ridge.

(c) An example of an active black smoker along the Mid-Atlantic Ridge.

volatiles that became incorporated into its minerals by hydrothermal activity at mid-ocean ridges.) The escape of volatiles happens at a depth of about 150 km (90 mi), for at this depth, the downgoing plate becomes hot enough that volatile compounds can no longer remain attached to minerals.

Once magma forms in the asthenosphere above the downgoing plate, it rises and eventually seeps through the lithosphere of the overriding plate. Some magma freezes underground in dikes, sills, or plutons, but some rises to the Earth's surface and erupts from a chain of stratovolcanoes called a **volcanic arc**. If the overriding plate consists of oceanic lithosphere, the chain can be called an *island arc* (Fig. 4.25a), whereas if the overriding plate consists of continental lithosphere, the chain can be called a *continental arc* (Fig. 4.25b). Continental and

FIGURE 4.25 Volcanic arcs at convergent boundaries.

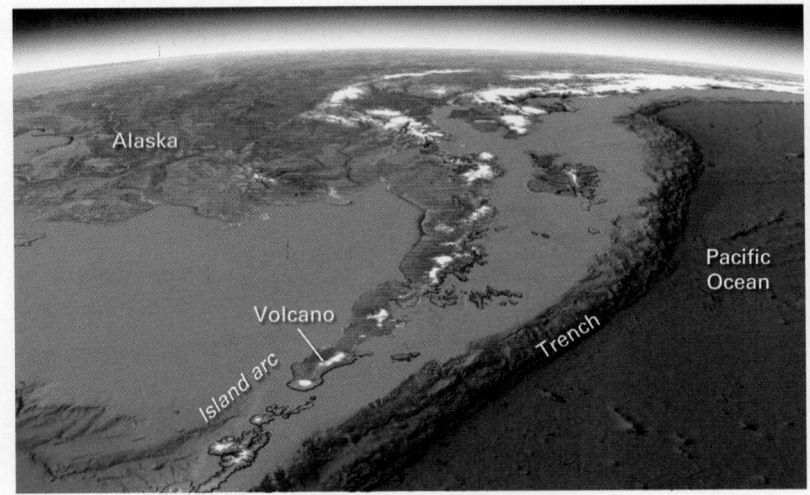

(a) The Aleutian Islands of Alaska, an island arc, as viewed looking northeast. The white peaks are volcanoes.

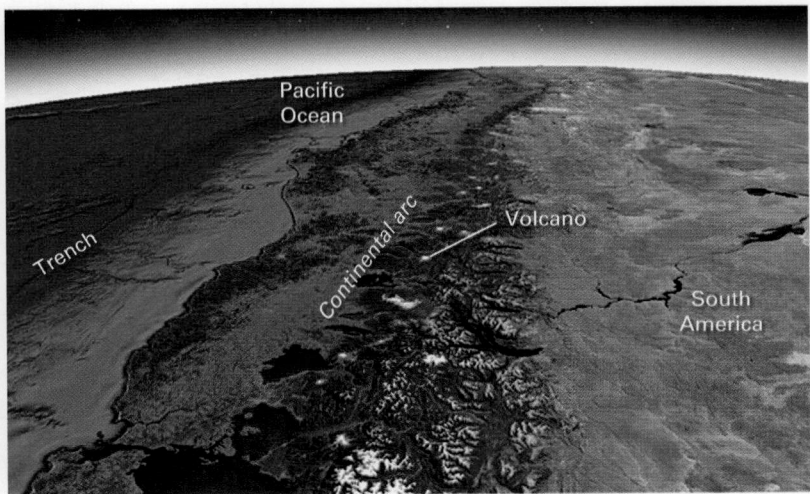

(b) The Andean Arc, a continental arc, formed near the western coast of South America. Forces due to convergence at this plate boundary have also produced the broader Andes mountain chain.

island arcs lie along much of the Pacific Ocean's border, so that this border has come to be known, colloquially, as the *Ring of Fire*.

Many kinds of igneous rock can form at volcanic arcs. If magma rises directly from the mantle, it remains mafic in composition and produces basalt. Rising magma, however, may undergo fractional crystallization, and may also assimilate compounds from wall rock as it moves upward; both processes make the rising magma more felsic (see Box 4.1). In addition, rising magma brings so much heat with it that it may trigger heat-transfer melting of the crust itself. As a result of these processes, volcanic arcs—especially continental arcs—also erupt andesite and rhyolite. Since both effusive and explosive eruptions take place in volcanic arcs, such igneous activity yields stratovolcanoes. As we noted earlier, not all magma reaches the Earth's surface—large volumes of intermediate and felsic intrusive rock solidify beneath continental arcs to produce plutons or even batholiths.

Two Kinds of Eruptions in Continental Rifts

During rifting, continental lithosphere stretches horizontally and thins vertically (see Fig. 4.3b). The thinning of the lithosphere causes decompression melting of the underlying asthenosphere and the production of mafic magma. Some of this magma rises straight to the surface and erupts effusively, but some undergoes fractional crystallization, and/or causes heat-transfer melting within the crust, to yield felsic magma that erupts explosively (Fig. 4.26a). Therefore, rifts typically contain basalt lava flows and layers of rhyolitic tuff. In a few locations, rifts also host stratovolcanoes, such as Mt. Kilimanjaro. Continued stretching in a rift leads to the development of faults that cut across the volcanoes (Fig. 4.26b).

Melting at Hot Spots

According to the mantle-plume model (see Chapter 2), hot-spot igneous activity occurs above a column of hot asthenosphere that rises to the base of the lithosphere. The depth at which such mantle plumes originate remains a subject of debate; some may rise from the base of the mantle, whereas others may originate in the upper mantle or may be due to local circulation patterns around subducted plates. Hot rock at the top of the plume undergoes decompression melting to yield mafic magma. Initially, mafic magma beneath oceanic crust erupts underwater, yielding a mound of pillow basalt and fragmental basalt. Over time, a volcano grows up above the sea surface and becomes an island. When the vent emerges from the sea, successive effusive eruptions produce thousands of basalt flows that build a shield volcano (Fig. 4.27a). Eventually, plate motion carries the volcano off the hot spot, and its eruptions cease. A new volcanic island then forms over the hot spot. As we've seen, the process eventually yields a hot-spot track (Fig. 4.27b).

Not all oceanic hot-spot volcanoes occur in the middle of a plate. Iceland, for example, formed over a hot spot under the axis of the Mid-Atlantic Ridge. Because of this hot spot, far more magma erupts in Iceland than at other places along the ridge, and these eruptions have built a broad oceanic plateau (Fig. 4.28a). Iceland

Did you ever wonder . . .

whether we can see all the volcanic eruptions taking place on the Earth today?

See for yourself

Mauna Loa, Hawai'i

Latitude: 19°27′0.50″ N
Longitude: 155°36′2.94″ W

Look straight down from 25 km (15.5 mi).

You can see the NNE-trending crest of Mauna Loa, a shield volcano associated with the Hawaiian hot spot. A large, elliptical caldera dominates the view. Basaltic lava flows have spilled down its flank.

FIGURE 4.26 Igneous activity in continental rifts.

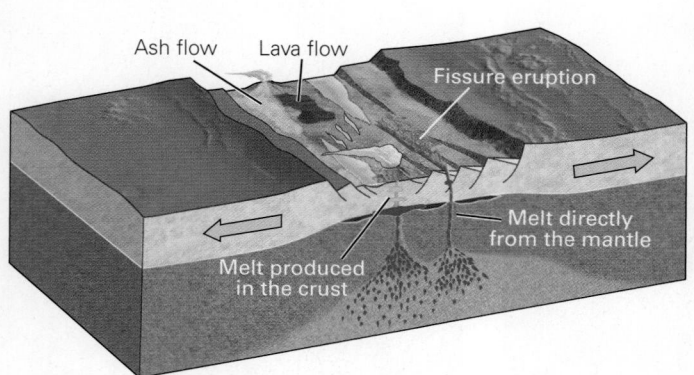

(a) As the lithosphere thins during rifting, asthenosphere melts to produce mafic magma. Some of this magma rises directly to the surface and erupts to produce basalt along fissures. Some pauses in magma chambers at depth and causes melting of the crust. The resulting felsic magma erupts explosively.

(b) These recent cinder cones have been cut by faults associated with rifting. Note that the line of cinder cones parallels the faults.

FIGURE 4.27 An oceanic hot-spot volcano.

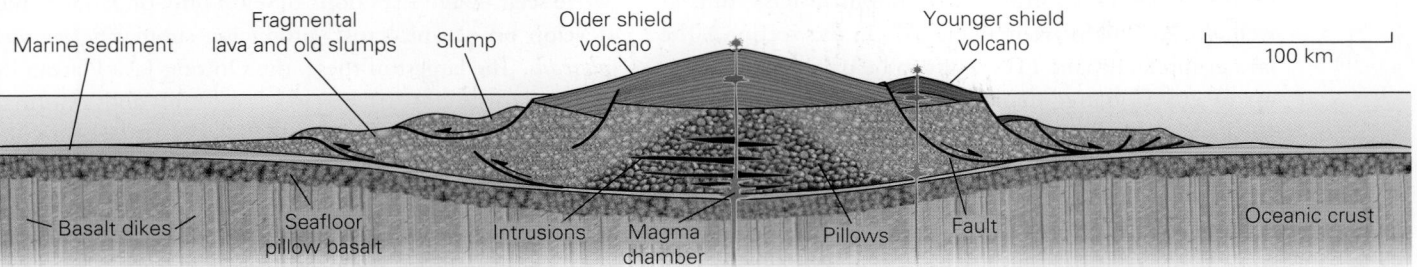

Marine sediment — Fragmental lava and old slumps — Slump — Older shield volcano — Younger shield volcano — 100 km

Basalt dikes — Seafloor pillow basalt — Intrusions — Magma chamber — Pillows — Fault — Oceanic crust

(a) The structure of an oceanic hot-spot volcano is complicated. Its formation begins with submarine eruptions, which produce pillow basalt. When it becomes subaerial, it produces a shield volcano. Gravity causes parts of the volcano to slump into the sea. This cross section approximates features along section line XY in part (b).

200 km

Older igneous rock

Youngest igneous rock

N

Pacific Plate motion

X

Y

(b) The Hawaiian hot-spot track.

straddles a divergent boundary, so plate motion actively stretches the island apart. Indeed, the central part of the island is a rift marking the trace of the Mid-Atlantic Ridge (Fig. 4.28b).

Hot-spot igneous activity also takes place in continental lithosphere. The best-known example produced the supervolcanoes of Yellowstone National Park. Unlike those at the Hawaiian hot spot, eruptions at the Yellowstone hot spot have produced both basaltic and rhyolitic lava as well as pyroclastic debris, for basaltic magma rising from the mantle causes heat-transfer melting in the continental crust. Eruptions of felsic supervolcanoes produce thick layers of tuff (Fig. 4.29a). Movement of the North American Plate over the hot spot that now lies beneath Yellowstone has produced a string of calderas over the past 16 million years (Fig. 4.29b, c).

In several locations around the world, particularly large volumes of magma have intruded or erupted at hot

Animation

Hot-Spot Formation

FIGURE 4.28 Iceland lies over a hot spot on the Mid-Atlantic Ridge.

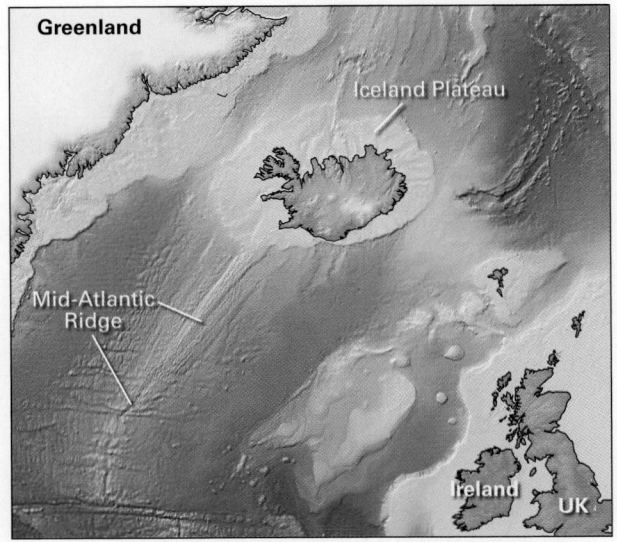

Greenland

Iceland Plateau

Mid-Atlantic Ridge

Ireland

UK

(a) A bathymetric map shows that Iceland sits atop a huge oceanic plateau straddling the Mid-Atlantic Ridge. Light blue is shallower water; dark blue is deeper.

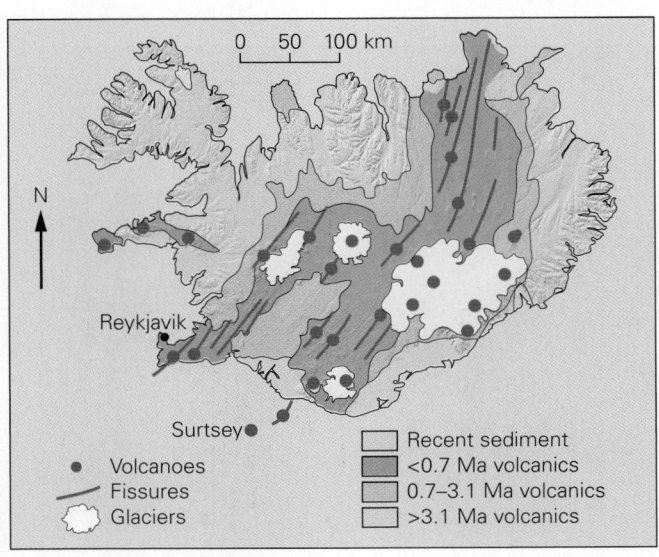

0 50 100 km

N

Reykjavik

Surtsey

• Volcanoes
— Fissures
Glaciers

Recent sediment
<0.7 Ma volcanics
0.7–3.1 Ma volcanics
>3.1 Ma volcanics

(b) A geologic map of Iceland shows that the youngest volcanoes occur in the central rift, which is effectively the on-land portion of the Mid-Atlantic Ridge.

spots, producing regions now known as **large igneous provinces**, or **LIPs** (Fig. 4.30a). (Geologists now use this term in reference to immense eruptions of felsic tuffs as well, but we will focus on mafic LIPs in this section.) The lava erupted at mafic LIPs tends to be particularly hot—up to 200°C hotter than normal mafic lava—so it has particularly low viscosity and spreads out in vast flows, some of which extend 500 km (300 mi) from the vent. Geologists refer to the rock formed from these flows as **flood basalts** (Fig. 4.30b). Successive flood-basalt eruptions on land have built broad *basalt plateaus*, such as the Columbia River Plateau in Oregon and Washington (see Fig. 4.29b). The largest continental basalt plateau contains 175,000 km³ (42,000 mi³) of basalt. But, as we've seen, mafic LIPs don't develop only on land. Some develop on oceanic crust, producing submarine *oceanic plateaus*. The largest of these, the Ontong-Java Plateau in the western Pacific, rises over 3 km (2 mi) above normal abyssal-plain depths.

What causes eruptions of flood basalts? According to one hypothesis, flood basalts form when a large mantle plume rises beneath a region undergoing rifting (Fig. 4.30c). The plume head can be quite broad—much wider than the underlying column of the plume—so it produces a particularly large volume of magma when it undergoes decompression melting. The resulting magma feeds immense fissure eruptions at the surface. Because the magma source of such eruptions (the asthenosphere in the plume) is particularly hot, the lava spewing from these fissures has very low viscosity.

FIGURE 4.29 Volcanic activity at the Yellowstone hot spot.

(a) Felsic tuffs form the walls of Yellowstone Canyon. The color of the tuff gave the park its name.

Take-home message...

The theory of plate tectonics explains why igneous activity occurs where it does. Decompression melting occurs at hot spots, mid-ocean ridges, and rifts. Flux melting produces melts at convergent boundaries. Heat-transfer melting also takes place at rifts and volcanic arcs. Particularly large quantities of lava erupt at large igneous provinces.

Quick Questions

- Do the same kinds of volcanoes form at all plate boundaries?
- At what type of plate boundary do batholiths form?
- What is a flood basalt?

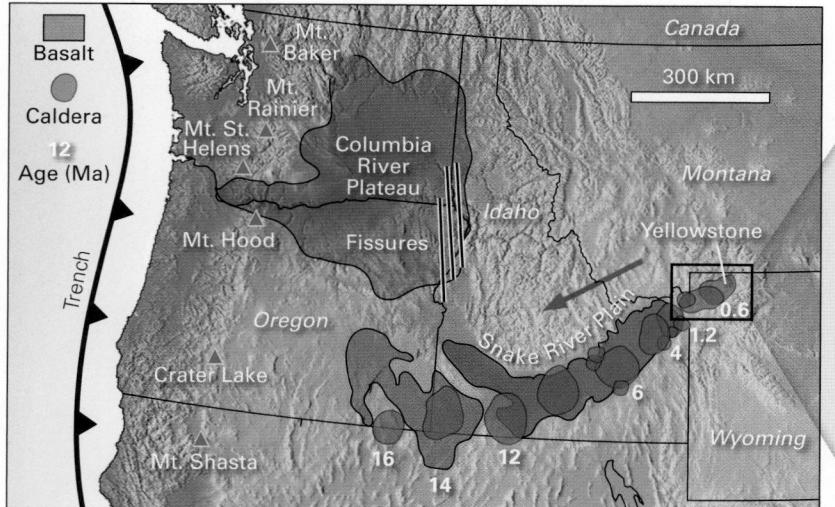

(b) Yellowstone National Park lies at the end of a continental hot-spot track. Progressively older calderas follow the Snake River Plain to the west. The blue arrow indicates plate motion.

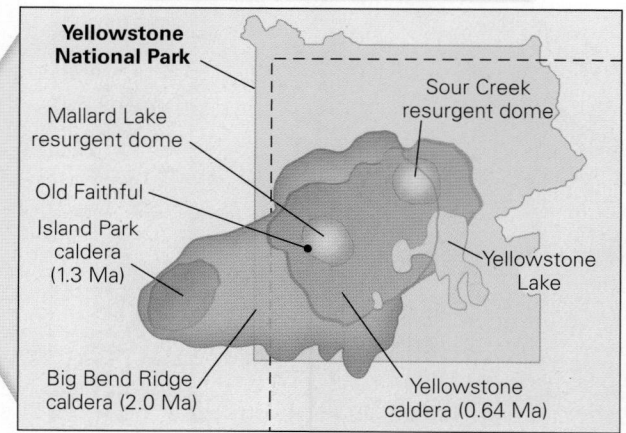

(c) The park itself overlies a huge caldera.

Airplane encountering
ash; St. Elmo's fire

Volcanoes pose several kinds of hazards to life and property. In this composite illustration, the explosive eruption at the summit of a volcano produces an eruptive jet. Some of the erupted material collapses and flows down the flank of the volcano in pyroclastic flows, which devastate towns. The rest rises in a convective plume. Intense lightning flashes in this plume, and ash from the plume blankets the countryside. An airliner encountering the plume glows with St. Elmo's fire, caused by static electricity; its engines will shut down if they ingest too much ash. Fumaroles along a fissure emit sulfurous gas, which accumulates to produce a smog-like mist called vog. Where ash mixes with water, it forms a lahar, a viscous slurry that can bury areas downslope. Lava flows may spill down a volcano's flanks and encroach on towns. If lava reaches the sea, it can turn seawater into laze, steam containing hydrochloric acid. Explosions, such as the lateral blast happening near the summit, can destroy the volcanic edifice. A new edifice may later grow over the remnants of the old one. Volcanoes are indeed a danger to be reckoned with!

Vog

Old cinder cone

Fumaroles
along a fissure

Lava
fountain

Active cinder
cone

Old lava flow

Active lava
flow

Burning
buildings

Lava spilling into the
sea to produce laze

Delta of
pyroclastic debris

FIGURE 4.31 Volcanic hazards: lava and pyroclastic debris.

Lava Flows

(a) A lava flow reaches a house in Hawai'i and sets it on fire.

(b) Lava from Mt. Etna threatens a town and olive grove in Sicily.

(c) Residents rescue household goods after a lava flow filled the streets of Goma, along the East African Rift.

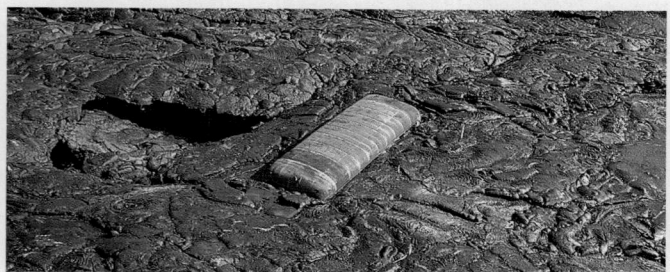

(d) This empty school bus was engulfed by lava in Hawai'i.

Pyroclastic Debris

(e) A pyroclastic flow rushes down the slope of the Soufrière Hills volcano, on the island of Montserrat.

(f) A blizzard of ash fell from the plume erupted by Kelud in Solo, Central Java.

(g) Lapilli fall from an eruption in Iceland.

(h) A lahar submerges farmland in Colombia.

buildings and forests. These eruptions also send pyroclastic debris skyward in an eruptive jet. Closer to the volcano, lapilli and blocks in the jet tumble back down, smashing through roofs or even crushing buildings. The mixture of hot debris and scalding air that collapses from the jet yields pyroclastic flows that race down the flank of the volcano at speeds of up to a few hundred kilometers per hour, sometimes reaching and engulfing land several kilometers out from the base of the volcano (Fig. 4.31e). A pyroclastic flow can be so hot and poisonous that it causes instant death to anyone caught in its path. For example, the 1902 explosion of Mt. Pelée, on the island of Martinique, produced a pyroclastic flow that swept through the city of St. Pierre, killing all but two of the city's 28,000 inhabitants. Explosive blasts, together with pyroclastic flows that rush into the sea, generate tsunamis. While the ash fall tends to be thickest near the volcano, high-altitude winds can carry ash that has reached the stratosphere over a broad region (Fig. 4.31f, g). An ash fall can kill crops and insidiously infiltrate homes and machinery.

Volcanic ash and gas that reach stratospheric heights can also be a hazard to air traffic. Ash particles produced by volcanoes are very sharp, and *aerosols* (extremely tiny solid particles or solution droplets that can remain suspended in air) produced by volcanoes are quite acidic, so when a plane flies through a high-altitude convective plume, its windows quickly become opaque, and its fuselage is damaged. Ash sucked into a jet engine melts to produce a glassy coating that restricts air flow, making the engine shut down. This happened to a British Airways 747 that flew through the ash cloud over a volcano on Java in 1982. All four engines failed, and for 13 minutes, the plane glided downward from its cruising altitude of 11.5 km (37,000 ft), as the pilots frantically tried to restart them. Finally, at 3.7 km (12,000 ft) altitude, enough of the glassy coating had broken off, and air had become dense enough, that the engines could roar back to life. The plane headed to Jakarta, where, without functioning instruments or a view through the windshield, the pilot brought the 263 passengers and crew in for a safe landing. A similar event affected a KLM 747 flying across the volcanic arc of southern Alaska in 1989. Because of this risk, officials now shut down airspace near an eruption. For example, all airspace in Europe was closed for six days following the 2010 ash-rich eruption of Iceland's Eyjafjallajökull volcano. The closure stranded millions of passengers and temporarily disrupted the global economy.

When ash from an explosive eruption mixes with water from heavy rains or melting snow and ice, it produces a **lahar**, a viscous slurry of ash and water that flows down the volcano's flanks and can travel tens of kilometers along stream channels beyond the base of the volcano (Fig. 4.31h). Because lahars are much denser than water,

they can knock down or carry away heavy objects, such as bridges, boulders, and trees, in their path. Perhaps the most destructive lahar of recent times accompanied the eruption of a snow-crested volcano in Colombia in 1985. The lahar surged down the mountainside into a valley, burying the sleeping town of Armero, 60 km (40 mi) away, and entombing its 25,000 citizens in a layer of muddy debris 5 m (16 ft) thick.

Hazards due to Volcanic Gas

Volcanoes emit significant amounts of gas, both during eruptions and between eruptions. Volcanic gas plays an important role in the Earth System because it includes H_2O, essential to life, and CO_2, which serves an important role in regulating the atmosphere's temperature. On a few occasions, however, local accumulations of CO_2 have become deadly. Such an event happened in Cameroon, when CO_2 seeping from a hot-spot volcano's vent dissolved in a layer of cold water at the bottom of a lake that filled the crater. One night, a landslide disturbed the cold water, which released the dissolved CO_2 suddenly. Large volumes of CO_2 rose to the lake's surface and into the air, forming an invisible cloud (Fig. 4.32a). Since the gas's density exceeds that of air, the CO_2 cloud silently drifted down the flank of the volcano and suffocated the sleeping inhabitants of villages downslope.

Sulfur-bearing volcanic gases (SO_2 and H_2S) and dust can also be problematic. They can form choking clouds that make breathing difficult or impossible near vents or fumaroles (see Fig. 4.5), and they undergo reactions and dissolve in atmospheric moisture to produce acidic droplets. If the weather conditions are appropriate, sulfurous aerosols accumulate with volcanic dust in the atmosphere to produce hazy *vog*, so named because it resembles the smog over cities caused by pollution from cars, power plants, and factories (Fig. 4.32b). Dangerous acidic fumes can also form where lava flows into seawater—the steam rising from these locations contains chlorine, which reacts with water to form *laze*, a mist of dilute hydrochloric acid aerosols (Fig. 4.32c).

Volcanoes and Climate

In the summer of 1783, Benjamin Franklin, then serving as the American ambassador to France, noticed that the weather was unusually cool there, and that the air was particularly hazy. Franklin, an accomplished scientist as well as a statesman, sought an explanation for this phenomenon. He learned that a volcanic eruption had taken place in Iceland in June of 1783 and wondered if the "smoke" from the eruption prevented sunlight from reaching the Earth. Franklin reported this idea at a scientific meeting, and by doing so, may have become the first scientist to suggest a link between volcanic eruptions and climate.

FIGURE 4.32 Volcanic hazards: gases.

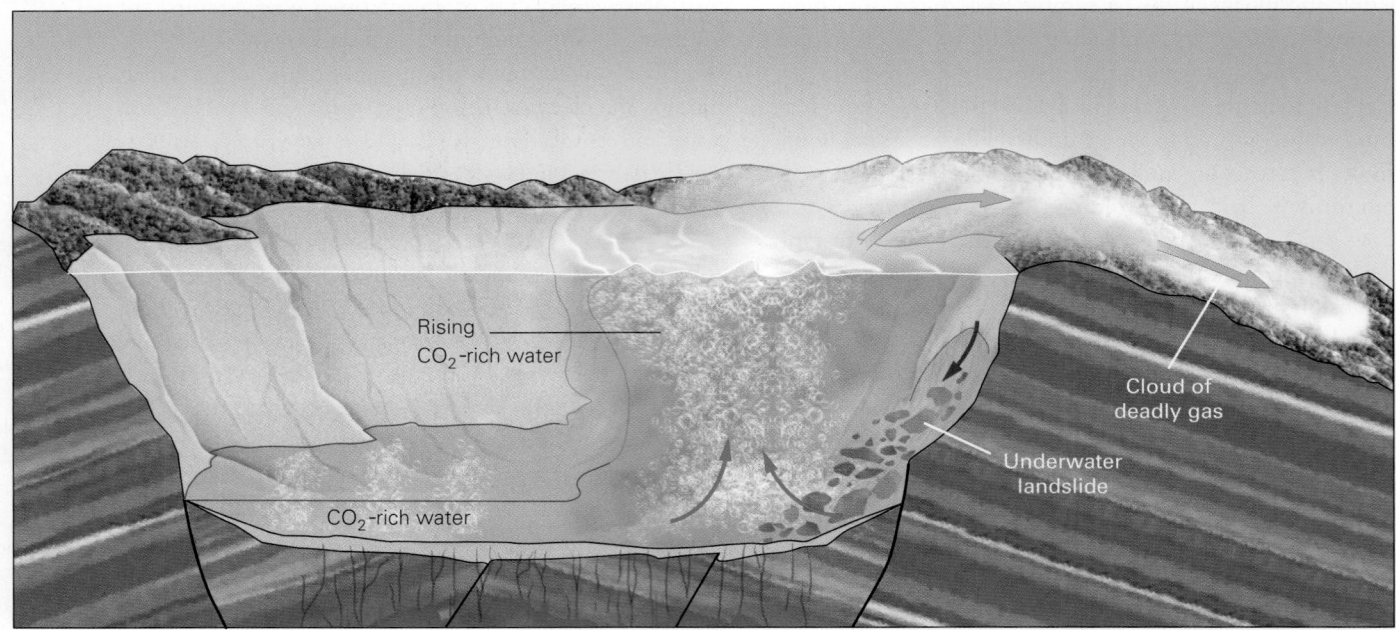

(a) In Cameroon, CO_2 emitted by a volcanic vent dissolved in the cold water at the bottom of a lake in the crater. A landslide caused the CO_2 to bubble out all at once and flow downslope.

(b) Vog surrounds the base of Mauna Loa during the 2018 eruptions at Kilaueaʻi, Hawaiʻi.

(c) Where lava enters the sea, it produces laze, steam that contains hydrochloric acid.

Franklin's idea was confirmed in 1815, when Mt. Tambora, in Indonesia, exploded, making the sky so hazy that temperatures worldwide dipped by several degrees. It remained so cold in the northern hemisphere that 1816 became known as "the year without a summer." The dreary weather had far-reaching catastrophic effects, for it caused crops to fail and people in many locations to starve. Gloomy weather also confined Mary Shelley and her literary friends indoors, where they passed the time first by reading ghost stories and then by writing them. Mary Shelley took the task to heart and penned *Frankenstein*.

Modern studies of atmospheric conditions following the 1991 eruption of Mt. Pinatubo in the Philippines provide clear documentation of the short-term effects that an eruption can have on global atmospheric temperature. How does volcanic activity cool the climate? When a large explosive eruption takes place, ash and sulfurous droplets enter the stratosphere and, within weeks, encircle the planet. This material may remain suspended in the atmosphere for as long as a few years, producing a high-altitude haze that reflects sunlight back to space during the day, so that the Earth's surface doesn't warm as much as it would if the eruption hadn't occurred.

As we will see in Chapter 19, CO_2 sent skyward during the eruption of a LIP, in contrast to high-altitude dust and sulfurous gases, may trigger long-term global warming. CO_2 is a greenhouse gas that retains heat in the atmosphere, so adding large quantities to the atmosphere during production of flood basalts—a process that may continue for hundreds of thousands of years or more—could change the climate sufficiently to cause widespread species extinction.

Take-home message...

Volcanoes can be dangerous. Lava flows, pyroclastic flows, ash falls, explosions, lahars, landslides, and tsunamis produced during eruptions can destroy cities and farmland. Ash in the air can be a hazard for air travel. Ash and sulfurous aerosols thrown high into the atmosphere reflect sunlight and can cause a temporary drop in global temperature.

Quick Questions ——————————————

- Why can a mafic lava flow cause nearby buildings to ignite?
- Is a pyroclastic flow or an ash fall more hazardous to people at the base of a volcano?
- Why is volcanic gas a hazard?

4.8 Protecting Ourselves from Vulcan's Wrath

After thriving for centuries on islands in the eastern Mediterranean during the Bronze Age, the Minoan culture vanished. Its demise may be linked to the explosive eruption of Santorini around 1600 B.C.E., which left behind a large submarine caldera. (Part of the caldera's rim remains above sea level today as the Greek island of Thera.) Archaeologists speculate that ash clouds and tsunamis generated by the eruption ruined crops, buried towns, and damaged ports and fishing fleets, disrupted trade, and destroyed wealth, causing populations to disperse. The eruption may also have served as the basis for disaster legends in ancient cultures of the Mediterranean region. Clearly, volcanic eruptions can be extremely hazardous to society. Let's consider the evidence that geologists use to determine whether a volcano has the potential to erupt, and the steps people can take to mitigate its destructive effects.

Active or Extinct?

How do we know if a volcano has the potential to erupt? The first clue comes from determining whether a volcano is active or extinct. Geologists refer to volcanoes that have erupted during the last several thousand years and might erupt in the future as **active volcanoes**, to distinguish them from **extinct volcanoes**, meaning ones that

have shut off entirely and will never erupt again because the geologic conditions leading to eruption no longer exist. Active volcanoes include both currently erupting volcanoes and **dormant volcanoes**, ones not currently erupting, but capable of erupting in the future. Extinct volcanoes are not a worry—active volcanoes are!

To decide whether a volcano is active or extinct, geologists start by determining when past eruptions took place. If a volcano has erupted during, roughly, the past few millennia, it could be active. In regions where the historical record doesn't extend back more than a few centuries, geologists date volcanic rocks, or wood buried by pyroclastic debris, using radioisotopic methods (see Chapter 8). Landscape characteristics of a volcano also provide clues to its activity (Fig. 4.33a, b). For example, if the lava flows or pyroclastic debris that cover the volcano's surface have not been vegetated or incised by streams, then the volcano erupted recently enough to be considered active. In contrast, a volcano whose flanks have become vegetated and eroded could be dormant. A volcano that has completely eroded away, so that underlying intrusions crop out, has likely become extinct (Fig. 4.33c, d). Studies of volcanic landscapes and igneous rocks provide a detailed image of volcanism in the United States over time (Box 4.4).

Forecasting Eruptions

Eruptions at a given volcano do not occur periodically. But by knowing the timing of the past several eruptions at an active volcano, geologists can calculate the average time between eruptions, a number known as the **recurrence interval** (or *return period*) of eruptions. Because the recurrence interval merely indicates the average time between eruptions, it does not provide a basis for forecasting the exact timing of a future eruption. However, short-term (weeks to months) forecasts of impending volcanic activity are sometimes feasible because in advance of an eruption, a volcano become "restless" and sends out distinct warning signals:

- *Earthquake activity:* When magma flows into a magma chamber, rocks surrounding the magma chamber break and slip, and small gas explosions may take place. This movement causes earthquakes, so in the days or weeks preceding an eruption, the frequency of earthquakes beneath the volcano increases.

- *Changes in heat flow:* The intrusion of magma into a volcano increases local *heat flow*, the amount of heat passing up through rock. In some cases, the increase in heat flow melts snow or ice on the volcano.

- *Increases in gas and steam emission:* Even though magma remains below the surface, gases bubbling out

Time 1

Time 2

Erosion carves gullies, vegetation grows, and landslides happen.

Time 3

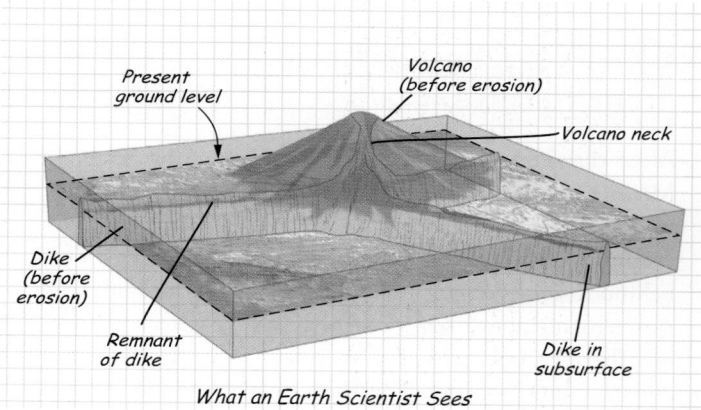

(b) The degree of vegetation and stream incision distinguishes the edifice that eruptions that Mount Vesuvius built in the 20th century from the exploded remnant of an older edifice. (V.E. = vertical exaggeration)

Eroded volcano   Growing volcano

Pompeii

1944 lava flow

Naples

~ 3 km
VE = 1.5X

(a) Time 1 shows a recently active volcano. Time 2 shows a long-dormant volcano. Time 3 shows the insides of a likely extinct volcano.

Aerial photo from 11 km altitude.

Volcanic neck   Dike

Dike   Dike

Viewpoint

Present ground level   Volcano (before erosion)

Volcano neck

Dike (before erosion)

Remnant of dike

Dike in subsurface

What an Earth Scientist Sees

(c) Shiprock, in New Mexico, seen here from the air, is the eroded remnant of an extinct volcano. What we see is rock formed by solidification in a shallow magma chamber in or just under the now-eroded volcano. Dikes radiating outward from the center of the volcano have also been exposed.

(d) A ground view of a dike at Shiprock, from the viewpoint indicated on the aerial photo.

A Deeper Look

Volcanism in the United States

The portion of the *North American Cordillera* (the name used for all the mountainous area between the eastern edge of the Rocky Mountains and the coast) in the United States, together with Alaska and Hawai'i, hosts an impressive record of volcanism. When dinosaurs stomped about the landscape, the region of the Sierra Nevada range, as well as a large portion of Idaho, was a continental volcanic arc, much like the Andes is today. The immense granite batholiths exposed in these areas today formed beneath that arc. Between 80 and 40 million years ago, subduction-related volcanic activity swept eastward almost as far as present-day Denver. During this time interval, supervolcanoes erupted in southwestern Colorado. As subduction ceased, widespread felsic volcanism covered much of what are now the western states.

After subduction ceased, the San Andreas transform boundary became established and crustal stretching initiated. The stretching led to formation of the Basin and Range Rift and the Rio Grande Rift. Both of these regions hosted both felsic and mafic eruptions. Though rift-related volcanoes haven't erupted in these regions since written historical records began, geologists still consider some of their volcanic areas to be active, especially along the boundary between the Sierra Nevada and the Basin and Range.

Volcanism has also happened in the northern US portion of the North American Cordillera during the past 17 million years. Flood-basalt eruptions yielded the Columbia River Plateau, and explosive eruptions left a track of calderas, defining the Yellowstone hot-spot track, along the Snake River Plain in Idaho (see Fig. 4.29b). While the most recent eruption at this hot spot happened about 70,000 years ago, the occurrence of seismic activity, hot springs, and other features of volcanism suggests that the Yellowstone region remains active.

Three areas of the United States host clearly active volcanoes. First, the Cascade Range, a chain of about 20 volcanoes that runs north from northern California across Oregon and Washington to the Canadian border, has hosted three major eruptions since 1900. This continental arc remains active because the small Juan de Fuca Plate continues to subduct beneath the margin of the North American Plate. Second, the Aleutian Arc, which lies along the southern border of the Alaska mainland and out along the Aleutian Islands, accommodates subduction of the Pacific Plate beneath North America. Finally, effusive volcanic eruptions occur relatively frequently on Mauna Loa and Kilauea, on the Big Island of Hawai'i, which lie over the Hawaiian hot spot.

Did volcanism ever happen in the eastern United States? Yes. While Mesozoic volcanic arcs erupted in the west, rifting took place along the Atlantic margin as North America separated from the rest of Pangaea. Volcanoes associated with this rifting have long since eroded away, but the basalt produced during the rifting remains. For example, the towering cliffs on the New Jersey side of the Hudson River, facing New York City, form the edge of a huge basalt sill that intruded between layers of sediment that contain dinosaur footprints.

---

of the magma, or steam formed as the magma heats groundwater, may be emitted at fumaroles or at the main vent. An increase in the volume of gas emission, or the formation of new hot springs, can indicate that magma has entered the ground below.

- *Changes in shape:* As magma accumulates inside the volcano, it pushes outward and can cause the surface of the volcano to bulge (Fig. 4.34). This change can be detected by accurate surveying or, more recently, by satellite measurements.

Mitigating Volcanic Hazards

There's no way to stop an eruption, so what can we do to prevent the loss of life and property near an active volcano? As an important first step, geologists compile a volcanic-hazard map (Box 4.5), which delineates areas that lie in the path of potential lava flows, lahars, or pyroclastic flows. If an eruption seems imminent, people within these danger zones should evacuate.

In a few cases, threatened communities have tried to divert or stop a lava flow. For example, during a 1669 eruption of Mt. Etna, an active volcano on the island of Sicily, a basaltic lava flow approached the town of Catania. Fifty townspeople boldly hacked through the solidified side of the flow to create an opening through which lava could exit. They hoped to cut off the supply of lava feeding the end of the flow that was approaching their homes. Their strategy worked, but unfortunately, the diverted flow began to move toward the neighboring town of Paterno. Five hundred Paternoans chased away the Catanians. The hole closed, and the flow headed

BOX 4.5 **Putting Earth Science to Use**

Volcanic hazard maps

By studying the consequences of past eruptions, geologists can develop volcanic-hazard maps to show the potential impact of a volcanic eruption on the surrounding region. **Figure Bx4.5** provides an example of such a map for the area surrounding Mt. Rainier in Washington State. Both lava flows and pyroclastic flows present a significant threat close to the volcano. Lahars from a Mt. Rainier eruption could travel much farther—they could even reach Tacoma, nearly 100 km (60 mi) away. The United States Geological Survey (USGS) publishes long-term hazard-assessment reports that describe the types and likelihoods of hazards at specific sites of volcanism and show where the affected areas in those regions would be. Such maps can help you determine whether you reside, work, or attend school in a region that could be affected by a volcanic eruption, and if so, what hazards your region may be vulnerable to. Municipalities that lie within a volcanic-hazard area should post hazard maps, emergency plans, and evacuation routes in public places and on websites.

FIGURE Bx4.5 A volcanic-hazard map for Mt. Rainier, Washington.

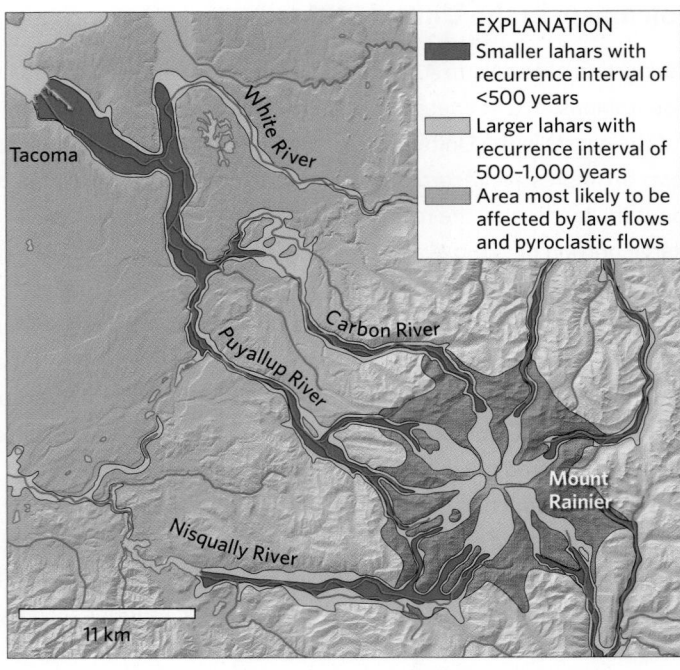

EXPLANATION
- Smaller lahars with recurrence interval of <500 years
- Larger lahars with recurrence interval of 500–1,000 years
- Area most likely to be affected by lava flows and pyroclastic flows

11 km

FIGURE 4.34 Ground movements may reflect the intrusion of magma below a volcano. Measurements of Kilauea, made by satellite-based radar called InSAR, indicate that the surface of the volcano is rising. Each color band represents 3 cm (1 in) of uplift. The inset map of Hawai'i shows the location of the measurements. Red areas are relatively recent lava flows. (K. = Kama'ehuakanaloa, an active submarine volcano on the Lō'ihi seamount.)

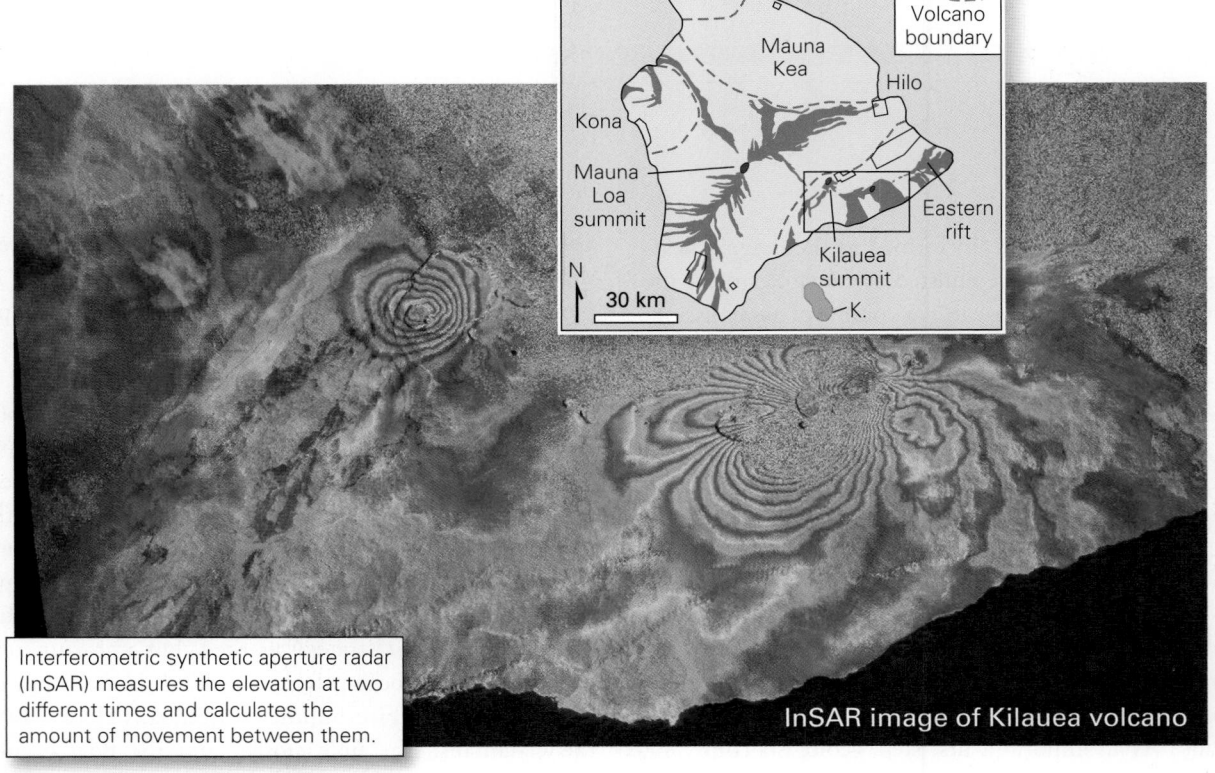

Interferometric synthetic aperture radar (InSAR) measures the elevation at two different times and calculates the amount of movement between them.

InSAR image of Kilauea volcano

back toward Catania, eventually overrunning part of the town. More recently, emergency workers have used high explosives to blast breaches in the flanks of lava flows, have employed bulldozers to build dams and channels to divert lava, and have even sprayed lava flows with cold seawater to solidify them. Efforts to divert flows from eruptions of Mt. Etna in 1983 and in 1992 were successful (Fig. 4.35).

Take-home message...

 Geologists can distinguish between active and extinct volcanoes, and they can provide near-term predictions of eruptions, allowing people to take precautions. Volcanic-hazard maps help determine what areas should be evacuated. Rarely, it's possible to divert lava flows.

Quick Questions

- If a volcano's surface has a covering of fresh tephra, with no vegetation, is it active or extinct?
- What can precise surveying tell us about the potential for a volcanic eruption?
- What features does a volcanic-hazard map show?

FIGURE 4.35 Workers use bulldozers to build an embankment to divert a lava flow on the flanks of Mt. Etna.

4 CHAPTER REVIEW

Objective 4.1

Define igneous activity, explain why melting happens inside the Earth, and distinguish among melts based on their composition.

KEY CONCEPTS

- Igneous activity refers to the formation, movement, and evolution of magma underground, to volcanism, and to the formation of igneous rock underground or at the Earth's surface.
- Magma formation can be triggered by decompression, addition of volatiles, or heat transfer. Generally, only a small proportion of a magma source melts to become magma. Temperatures needed to cause melting depend on pressure.
- The composition of molten rock reflects both the composition of the magma source and the way the magma evolves.

- Geologists recognize four compositional varieties of molten rock (felsic, intermediate, mafic, and ultramafic) by the proportion of silica to iron and magnesium oxide in the melt.
- Various types of volatiles exist in magma. At depth, they remain dissolved in magma, but near the Earth's surface they come out of solution and form bubbles of volcanic gas.

EARTH-SCIENCE VOCABULARY

decompression melting (p. 120)
flux melting (p. 121)
geotherm (p. 120)
heat-transfer melting (p. 121)
igneous activity (p. 120)
lava (p. 120)

magma (p. 120)
partial melting (p. 120)
vesicle (p. 124)
viscosity (p. 122)
volcanic eruption (p. 120)
volcanic gas (p. 122)
volcano (p. 120)

1. **(a)** What phenomena does the term igneous activity encompass? **(b)** Identify geologic settings in which melting takes place,

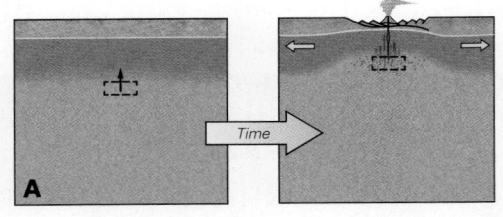

and indicate the type of melting that takes place in each setting. **(c)** Which type of melting does **Figure A** represent?

2. **(a)** Explain the difference between magma and lava. **(b)** What characteristic distinguishes the compositional varieties of magma from one another? **(c)** Which compositional variety has the greatest viscosity? **(d)** Where do the compounds in volcanic gas come from?

3. **(a)** In the context of Bowen's reaction series, explain how magma composition can evolve as magma cools. **(b)** What does fractional crystallization refer to?

Objective 4.2

Describe how igneous rocks form underground and at the Earth's surface.

KEY CONCEPTS

- An igneous rock is any rock whose production involves the formation and later solidification of melt.
- Once formed, magma rises because of its buoyancy and because of pressure caused by the weight of overlying rock.
- Intrusive igneous rocks form when magma intrudes into pre-existing rock underground. Extrusive igneous rocks form when lava or debris emerges from a volcano at the Earth's surface.
- Magma or lava becomes rock by solidification, which takes place over a range of temperatures. The rate at which magma cools depends on the depth, size, and shape of a body of melt.

EARTH-SCIENCE VOCABULARY

extrusive igneous rock (p. 124)

igneous rock (p. 124)

intrusive igneous rock (p. 124)

magma migration (p. 124)

REVIEW QUESTIONS

4. **(a)** Can you produce magma from solid rock by baking it in your home oven? Explain your answer. **(b)** Why doesn't magma stay put once melting begins at a magma source? **(c)** Label the extrusive and intrusive realms in **Figure B**.

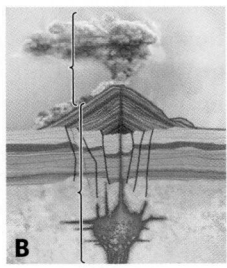

5. **(a)** Which cools faster, a large blob of magma intruded at depth or a thin flow of lava extruded at the Earth's surface? Why? **(b)** Imagine two cubes of magma. Each edge of the first cube is 1 m long, and each edge of the second, larger, cube is 2 m long. Calculate the ratio of surface area to volume for each. Which cools faster? Explain why, using your calculation.

Objective 4.3

Recognize the various types of igneous intrusions, lava flows, and pyroclastic debris.

KEY CONCEPTS

- Intrusions are distinguished from one another by shape. Tabular intrusions include dikes and sills, nontabular intrusions are called plutons, and huge bodies composed of many plutons are batholiths.
- How plutons form remains uncertain. They may represent solidified magma chambers or composites of sills.
- The character of a lava flow depends on its viscosity, which in turn depends on its composition. Viscous felsic lava flows tend to build a dome over a volcano's vent, whereas less viscous mafic lava can flow great distances.
- Volcanic eruptions can produce large quantities of pyroclastic debris, including very fine ash, marble- to golf-ball-sized lapilli, angular blocks, and smooth bombs.

EARTH-SCIENCE VOCABULARY

ash (p. 131)

batholith (p. 126)

dike (p. 126)

igneous intrusion (p. 126)

lapilli (p. 131)

lava flow (p. 127)

lava fountain (p. 131)

magma chamber (p. 126)

pillow basalt (p. 130)

pluton (p. 126)

pyroclastic debris (p. 130)

sill (p. 126)

REVIEW QUESTIONS

6. **(a)** What are igneous intrusions, and how do they form? **(b)** How does a tabular intrusion differ from a nontabular intrusion? Which type appears in **Figure C**? **(c)** How does a pluton differ from a batholith?

7. **(a)** What key factor controls the viscosity of a melt, and how does viscosity affect the behavior of lava? **(b)** Does **Figure D** show a mafic flow or a felsic flow? **(c)** Describe three different surface textures of mafic lava flows. Which forms underwater? **(d)** How can a mafic lava flow keep growing at its end even when its surface has frozen over?

8. **(a)** Describe different ways in which lapilli can form. **(b)** What does volcanic ash consist of? **(c)** Which type of tephra has the finest grain size, and which has the coarsest?

Objective 4.4

Identify different kinds of igneous rocks, and interpret characteristics of igneous rocks.

KEY CONCEPTS

- Igneous rocks are classified according to texture and composition.
- Crystalline igneous rocks contain interlocking crystals, glassy igneous rocks consist mostly of glass, and fragmental igneous rocks consist of fragments welded or cemented together.
- Depending primarily on their cooling history, crystalline igneous rocks can be fine-grained, coarse-grained, or porphyritic (with large crystals surrounded by fine-grained groundmass).
- Fragmental igneous rocks form from pyroclastic debris.
- Gas bubbles trapped in molten rock can form vesicles in solid igneous rock.

EARTH-SCIENCE VOCABULARY

crystalline igneous rock (p. 133)

fragmental igneous rock (p. 134)

glassy igneous rock (p. 134)

REVIEW QUESTIONS

9. **(a)** How does a crystalline igneous rock differ from a glassy rock or fragmental rock? **(b)** Which texture does the photomicrograph of **Figure E** show? **(c)** Which texture does the photo in **Figure F** show?

E

F

10. **(a)** What is the basic relationship between grain size and cooling rate in igneous rocks? **(b)** Name two types of intermediate-composition igneous rock and two types of felsic igneous rock, and indicate which of these are coarse-grained and which are fine-grained.

11. **(a)** What holds a fragmental igneous rock together in a coherent mass? **(b)** How does obsidian differ from scoria, and why can obsidian be used to make arrowheads? **(c)** Explain why pumice can float.

Objective 4.5

Distinguish among various kinds of volcanic eruptions, and among the various shapes of volcanoes.

KEY CONCEPTS

- Effusive volcanic eruptions are dominated by lava flows or fountains. They typically produce mafic lava.
- Explosive volcanic eruptions blast large quantities of pyroclastic debris skyward. Explosions can be triggered by sudden decompression or by sudden interactions of water with magma.
- The size of an explosive eruption can be represented by a VEI number. The largest of these eruptions—rated VEI 8 and known as supervolcanic eruptions—blast out over 1,000 km³ of pyroclastic debris.
- A volcano's shape depends on its eruptive style. Shield volcanoes are broad, gently sloped domes formed by effusive eruptions. Cinder cones are symmetrical hills of lapilli. Stratovolcanoes consist of alternating layers of pyroclastic debris and lava flows.
- Eruptions may occur in a crater at a volcano's summit, from vents on its flanks, or from elongate fissures. Collapse of a volcano produces a huge, bowl-shaped depression called a caldera.

EARTH-SCIENCE VOCABULARY

caldera (p. 142)

cinder cone (p. 141)

convective plume (p. 137)

crater (p. 137)

effusive eruption (p. 136)

eruptive jet (p. 137)

explosive eruption (p. 136)

fissure eruption (p. 141)

pyroclastic flow (p. 137)

shield volcano (p. 141)

stratovolcano (p. 141)

volcanic edifice (p. 137)

volcanic explosivity index (VEI) (p. 136)

REVIEW QUESTIONS

12. **(a)** Describe characteristic features that can develop during an effusive volcanic eruption. **(b)** Will an effusive eruption produce broad flows of rhyolite? Explain your answer.

13. **(a)** What causes extreme pressures, sufficient to cause an explosive eruption, to develop in a volcano? **(b)** Label the components of the large explosive eruption depicted in **Figure G**.

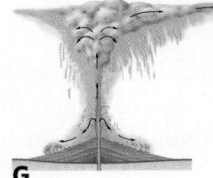

G

14. **(a)** How many VEI 3 explosive eruptions would have to take place to produce the same volume of erupted material as a single VEI 4 explosive eruption? **(b)** Has a VEI 8 (supervolcanic) eruption ever happened in historical time? **(c)** What evidence can geologists use to identify VEI 8 eruptions? **(d)** What event triggered the explosive eruption of Mt. St. Helens?

15. **(a)** Define the following terms as they pertain to volcanoes: edifice, crater, fissure, summit, flank. **(b)** Explain how a caldera forms.

16. **(a)** What are the three principal shapes of volcanic edifices, and how does each form? **(b)** Do all volcanic edifices have the same compositions? **(c)** Which shape does **Figure H** depict?

H

Objective 4.6

> Relate occurrences of igneous activity to plate boundaries and hot spots.

KEY CONCEPTS

- The theory of plate tectonics explains where volcanoes occur. Most volcanoes result from processes along plate boundaries, but some occur at hot spots.
- At mid-ocean ridges, decompression melting produces mafic lava that erupts from submarine fissures. Hydrothermal activity along ridges produces black smokers.
- Most of the world's subaerial stratovolcanoes form due to flux melting in the asthenosphere above subducting oceanic lithosphere; the resulting magma builds volcanic arcs.
- Decompression melting and heat-transfer melting produce both mafic and felsic magmas at continental rifts.
- Hot spots form where hot asthenosphere rises to the base of the lithosphere in a mantle plume. The depth at which plumes originate remains uncertain.
- Most hot spots occur in plate interiors, but some lie along mid-ocean ridges. Particularly large hot spots have produced large igneous provinces (LIPs), some of which are manifested by thick flood basalts.

EARTH-SCIENCE VOCABULARY

black smoker (p. 144)
flood basalt (p. 148)

large igneous province (LIP) (p. 148)
volcanic arc (p. 145)

REVIEW QUESTIONS

17. **(a)** Why does melting take place beneath the axis of a mid-ocean ridge? **(b)** What type of magma forms at mid-ocean ridges, and what rock types solidify from it? **(c)** Explain why black smokers form.

18. **(a)** Why do magmas form in association with subduction? **(b)** Does a volcanic arc form on the overriding plate or on the downgoing plate? **(c)** How does a continental arc differ from an island arc?

19. **(a)** What type of plate boundary do the volcanoes that are considered to be part of the Ring of Fire form along? **(b)** Why are there no subaerial volcanoes along the Pacific-Antarctic Plate boundary?

20. **(a)** What process in the mantle may be responsible for causing hot-spot volcanoes to form? **(b)** Name a hot-spot volcano that lies along a mid-ocean ridge, and name one that lies in the interior of an oceanic plate.

21. **(a)** Describe how magmas are produced at continental rifts. **(b)** Why can you find both basalt and rhyolite in such settings?

22. **(a)** What is a large igneous province (LIP), and how might it form? **(b)** What is the large sheet of black on the ground surface in **Figure I**?

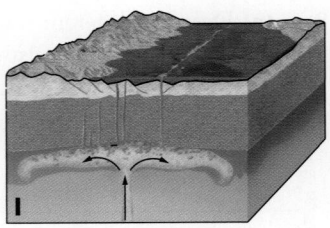

Objective 4.7

> Characterize the hazards presented by volcanoes.

KEY CONCEPTS

- Several types of natural hazards—geologic phenomena that could cause casualties or damage property—exist in association with volcanoes.
- Effusive eruptions of mafic lava can produce flows that cover homes, roads, fields, and forests.
- Explosive eruptions, if large, can produce a blast that flattens forests and can produce tsunamis.
- Pyroclastic flows can incinerate towns, and falls of ash and lapilli can ruin crops, crush buildings, block roads, and foul machinery. Fine ash reaching high altitudes can cause airliner engines to fail. Lahars, which form when ash mixes with water and flows downslope and along river valleys, cause extensive destruction.
- Volcanic gases, such as SO_2 and H_2S, can be poisonous. Carbon dioxide produced by volcanoes and stored in lakes can cause suffocation if released suddenly. Dust and sulfurous aerosols from eruptions can cause short-term global cooling.

EARTH-SCIENCE VOCABULARY

lahar (p. 153)

natural hazard (p. 149)

REVIEW QUESTIONS

23. **(a)** Describe the hazards associated with effusive eruptions. **(b)** Describe the hazards associated with explosive eruptions. **(c)** What volcanic hazard does **Figure J** show? **(d)** Which volcanic hazard—ash fall or pyroclastic flow—causes damage over a larger area? Explain why.

24. In what way can a volcanic eruption affect climate over the short term?

25. Do people living near the volcanoes of Hawai'i face the same kinds of volcanic hazards as do people living near a volcano in the Cascade Range?

Objective 4.8

> Evaluate efforts to forecast volcanic eruptions and mitigate volcanic hazards.

KEY CONCEPTS

- We can distinguish between active and extinct volcanoes based on their eruption history, and in some cases, by the degree to which a volcano has been eroded. Active volcanoes are natural hazards.

- Short-term forecasting of eruptions can sometimes be successful, for the rise of magma into a volcano's magma chamber triggers detectable increases in earthquake activity, heat flow, and emissions of gas and steam, and can change the shape of the volcano.

- By looking at the consequences of past eruptions, geologists can develop hazard maps for the region surrounding a volcano. Such maps highlight those areas most likely to be affected during an eruption.

- In a few cases, people have successfully diverted lava flows to keep them from encroaching on populated areas.

REVIEW QUESTIONS

26. **(a)** Explain how you can distinguish between the edifice of an active volcano that has erupted very recently, and one that has been dormant for quite a while. **(b)** Which does **Figure K** show? **(c)** What is an extinct volcano?

27. **(a)** If eruptions at a given volcano have a recurrence interval of 200 years, and the volcano last erupted 197 years ago, will it necessarily erupt in the next few years? Explain your answer. **(b)** List several indicators that might cause geologists to forecast that an eruption is imminent at a particular volcano. For each, indicate the change that would raise concern.

28. Describe how people can decrease the negative impact that the eruption of a volcano can have on their community.

ANOTHER VIEW   The people of Heimaey, Iceland, know what it's like to live on a volcano. This small island, perched on the Mid-Atlantic Ridge, consists entirely of volcanic rock produced during the past 2 million years. During a 1973 eruption, lava and pyroclastic debris (seen in this view) caused extensive damage and reshaped the island's coast. Heimaey's residents sprayed advancing lava with icy seawater to prevent the lava from blocking the harbor's entrance.

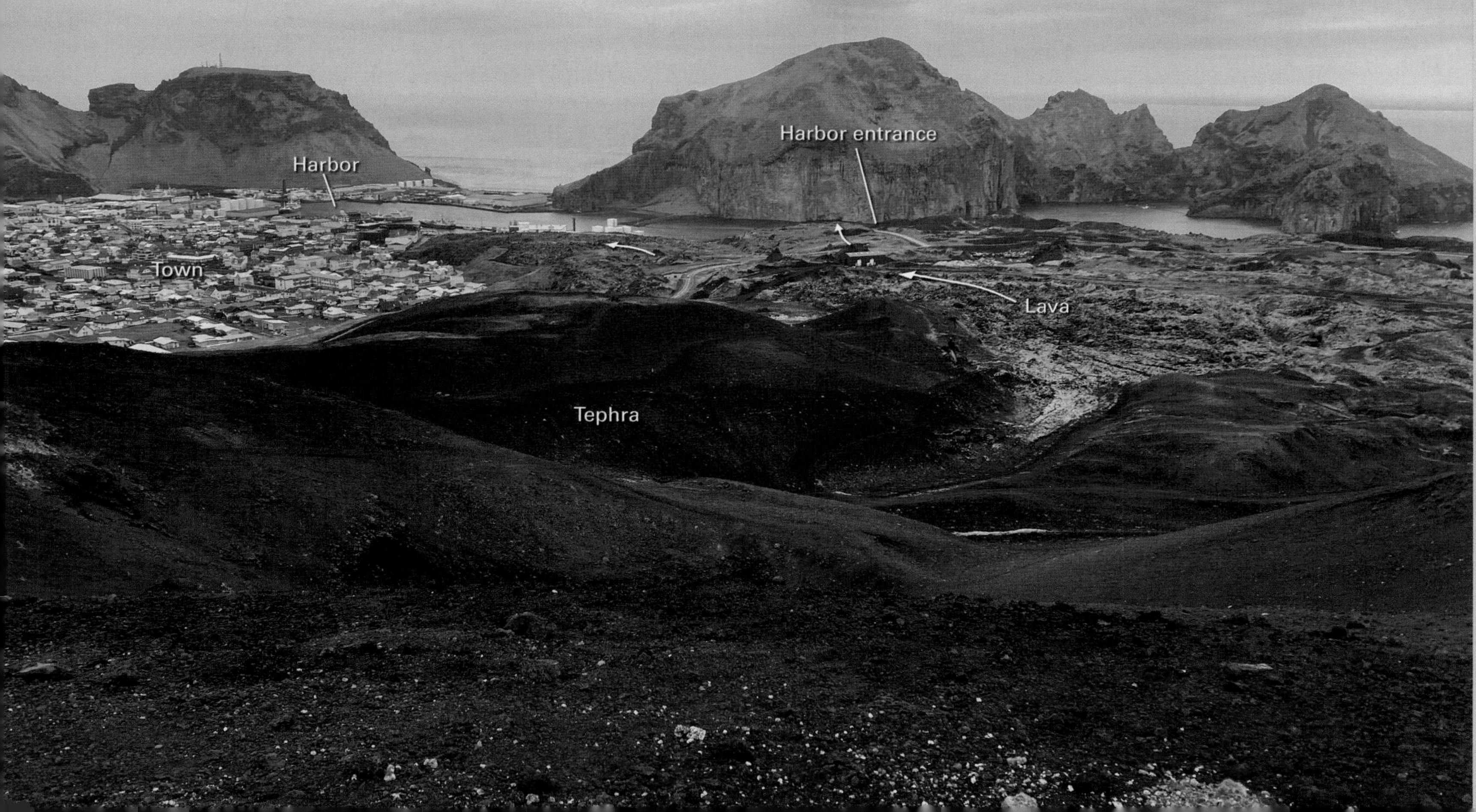

Harbor

Harbor entrance

Town

Lava

Tephra

5 THE REST OF THE ROCK CYCLE
Sedimentary and Metamorphic Rocks

After studying this chapter, you should be able to...

1. describe the process of weathering and sediment production, and explain how soil forms.

2. name examples of clastic sedimentary rocks, and outline the processes that produce them.

3. provide examples of biochemical, organic, and chemical sedimentary rocks and discuss how each can form.

4. characterize sedimentary structures, and discuss how they develop.

5. associate characteristics of sedimentary strata and sedimentary structures with specific depositional settings, and discuss phenomena that determine where sediment accumulates.

6. define metamorphism, characterize key features of metamorphic rocks, and describe processes that happen during metamorphism.

7. classify and name metamorphic rocks, and distinguish among different metamorphic grades.

8. characterize geologic phenomena that cause metamorphism, and identify subsurface environments where metamorphism takes place.

Countless outcrops of barren rock decorate the blustery hills of Scotland. During the mid-18th century, James Hutton, who would one day be celebrated as a founder of science-based geology, delighted in studying these outcrops, hoping to learn how rock formed. He saw that some rocks resembled the products of lava solidification, and that some looked like layers of *sediment*—mud, silt, sand, gravel, and whole or broken seashells—that settle from water **(Fig. 5.1a, b)**. Such observations led to the genetic classification of rocks that we use today (see Chapter 3), which considers rocks formed by cooling of a melt to be *igneous rocks* (see Chapter 4) and rocks formed from accumulations of sediment to be *sedimentary rocks*. Hutton also came upon puzzling rocks that didn't look either igneous or sedimentary. He described these puzzling rocks as "a mass of matter which had evidently formed originally in the ordinary manner . . . but is now extremely distorted in its structure . . . and variously changed in its composition." He speculated that such rocks formed when pre-existing rock underwent modification in response to heat and pressure deep underground **(Fig. 5.1c)**. Decades later, this third category came to be known as *metamorphic rocks* (from the Greek words *meta*, meaning change, and *morphe*, meaning form), to emphasize that they started out as one kind of rock and were later changed into another kind.

Eventually, Hutton came to a bold conclusion: over time, rocks can be altered or transformed in two general ways. At or near the Earth's surface, rocks can undergo *weathering* (modification due to interaction with water and the atmosphere) to produce sediment, which later accumulates and hardens to form new *beds* (layers) of sedimentary rock. Deep below the Earth's surface, rocks can undergo *metamorphism*, chemical reactions and shape changes that take place at elevated temperatures and pressures, producing new rocks without involving melting or weathering. Hutton realized that transitions from rock type to rock type take vast amounts of time, and that they had happened several times in the past and would continue in the future. He suggested that these transitions represent a "great geological cycle," which we now call the rock cycle (see Chapter 3).

Chapter 3 discussed minerals, and Chapter 4 introduced igneous rocks and processes. In this chapter, we introduce sedimentary and metamorphic rocks and the processes that form them, thereby completing our introduction to Earth materials. The first half of this chapter discusses sediment production by weathering. Then we turn our attention to how the various classes of sedimentary rocks form, and how the study of sedimentary rocks allows us to characterize ancient environmental conditions.

FIGURE 5.1 Hutton learned that rocks could form in three different ways. Each photo shows an example from Scotland.

(a) These igneous rocks near Edinburgh, Scotland, solidified from melt.

(b) These sedimentary rocks, on the coast of the North Sea, formed from layers of sand.

(c) These metamorphic rocks in the Scottish highlands underwent changes deep underground.

<< A staircase of horizontal beds forms the walls of the Grand Canyon from the rim down to the edge of the inner gorge. The inner gorge looks different—it consists of vertical layers whose texture and composition differ from those of the overlying beds. The horizontal beds consist of sedimentary rock, while the inner gorge consists of metamorphic rock. Both rock types formed from pre-existing materials, but in very different ways.

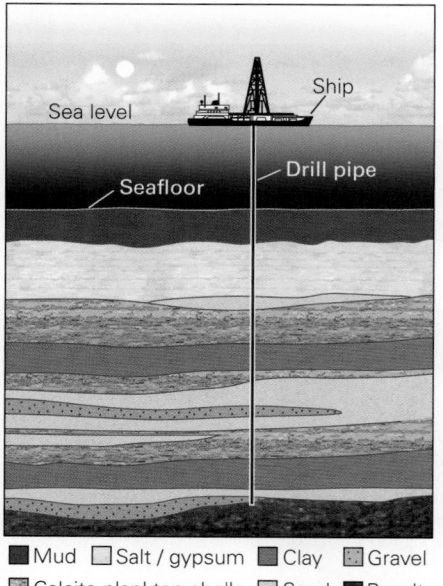

■ Mud ☐ Salt / gypsum ■ Clay ▨ Gravel
▨ Calcite plankton shells ☐ Sand ■ Basalt

The second half of this chapter delves into metamorphism. We'll see how this process can modify a rock so profoundly that it ends up looking nothing like the rock from which it was derived. Consideration of sedimentary and metamorphic rocks brings us full circle around the rock cycle on our dynamic home planet.

5.1 Weathering and the Production of Sediment

In the 1970s, researchers drilled a deep hole into the floor of the Mediterranean Sea. Before reaching the basalt layer of oceanic crust, the hole penetrated many layers, or *beds*, of **sediment**, material that consists of varying amounts of *unconsolidated* (loose) mineral or rock fragments of various sizes (mud, silt, sand, and gravel), whole and broken shells of marine organisms, and crystals of various salts **(Fig. 5.2)**. Where did the sediment found in the drillhole come from?

Sediment ultimately forms during the weathering of rock. **Weathering** encompasses processes that alter and weaken *fresh rock* (rock that has not yet weathered) at or near the Earth's surface, and ultimately yields **clasts** (a general name for fragments of rocks, mineral crystals, or shells) and ions **(Fig. 5.3)**. Geologists refer to the cracking and breaking up of rock as **physical weathering** (or *mechanical weathering*), and to chemical reactions during which rock interacts with air and water as **chemical weathering**.

Prying Rock Apart: Physical Weathering

The process of physical weathering generally begins with the formation of **joints**, natural cracks in bedrock. Joints develop for many reasons. For example, when rock from deep below the Earth's surface undergoes **exhumation**—meaning that erosion strips off *overburden* (overlying material) as rock undergoes uplift, so that the once-deep rock ends up at or near the ground surface—the pressure and temperature acting on the rock decrease. Consequently, the rock expands and cools, and therefore changes shape slightly. The shape change produces enough force to crack the rock and produce joints. Joints also form when

FIGURE 5.3 Evidence of weathering.

(a) A statue that shows the effects of weathering.

(b) The contrast between weathered and fresh granite. Mineral grains in the weathered rock (above the white dashed line) look dull and are starting to separate from one another.

(c) Close-up photos illustrating the textures of fresh and weathered granite.

mountain-building forces bend or stretch rock slightly.

The character of joints varies with rock type (Fig. 5.4a). Sedimentary rocks typically break not only along *bedding planes* (the surfaces between beds), but also along sets of roughly planar, parallel joints oriented perpendicular to bedding planes (Fig. 5.4b). Homogeneous crystalline rock, such as granite, tends to split into onion-like sheets along *exfoliation joints* oriented roughly parallel to the local ground surface (Fig. 5.4c), as well as along joints oriented perpendicular to exfoliation planes. Local cracking also develops in rock, producing irregular joints that break the rock into jagged pieces.

Once initiated, joints can *propagate* (grow longer) over time. In some cases, joints may be wedged open and forced to propagate. For example, when water trapped in a joint later expands during freezing, ice pushes the walls of the joint apart, causing *frost wedging* (Fig. 5.5a). Similarly, when a tree root expands as it grows in a joint, it causes *root wedging* (Fig. 5.5b). Eventually, propagating joints intersect each other or other planes of weakness in a rock, separating intact bedrock into detached blocks. These blocks may undergo further cracking in place, or they may tumble downslope (Fig. 5.5c). When tumbling blocks strike one another or bedrock, they smash apart into smaller fragments.

Altering Minerals in Rock: Chemical Weathering

A variety of different chemical-weathering reactions affect rock. For example, during *dissolution*, ions break free from minerals and dissolve in groundwater or surface water (Fig. 5.6a). Rocks made of salt or carbonate minerals dissolve relatively rapidly (Fig. 5.6b), but even quartz-rich

(a) Different patterns of joints develop in different rock types.

FIGURE 5.4 Joints can break bedrock into loose blocks or sheets.

(b) Vertical joints in beds of sedimentary strata (Grand Canyon).

(c) Exfoliation joints on a sloping granite mountain face (Sierra Nevada).

rocks dissolve slightly. During *hydrolysis*, minerals chemically react with water molecules to produce new minerals. For example, hydrolysis transforms feldspar and many other silicate minerals—but not quartz—into clay. **Clay** refers to extremely tiny grains, so small that you can't see them even with an ordinary microscope; most clay consists of *clay minerals*, a class of silicate minerals whose crystals look like extremely tiny booklets of hexagonal flakes, resembling flakes of mica. (Clay minerals differ from one another based on chemical composition and the details of their crystal structure.) During *oxidation*, iron-bearing minerals effectively "rust" and turn into a reddish-brown mixture of weak iron oxide and iron hydroxide minerals.

FIGURE 5.5 Examples of physical weathering processes.

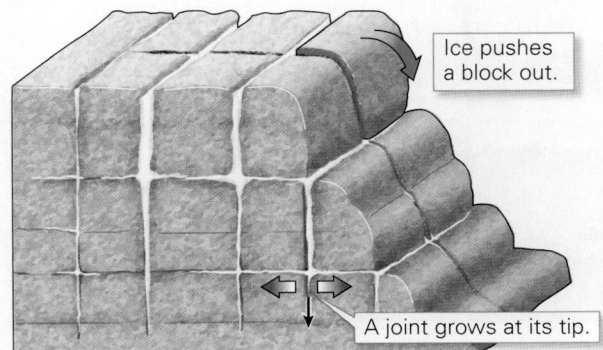

(a) Frost wedging occurs when water fills joints, freezes, expands, and pushes a joint open.

(b) Root wedging pushes open a joint, separating a block from the cliff.

(c) Blocks eventually tumble down the slope and break up further.

FIGURE 5.6 Dissolution is a form of chemical weathering.

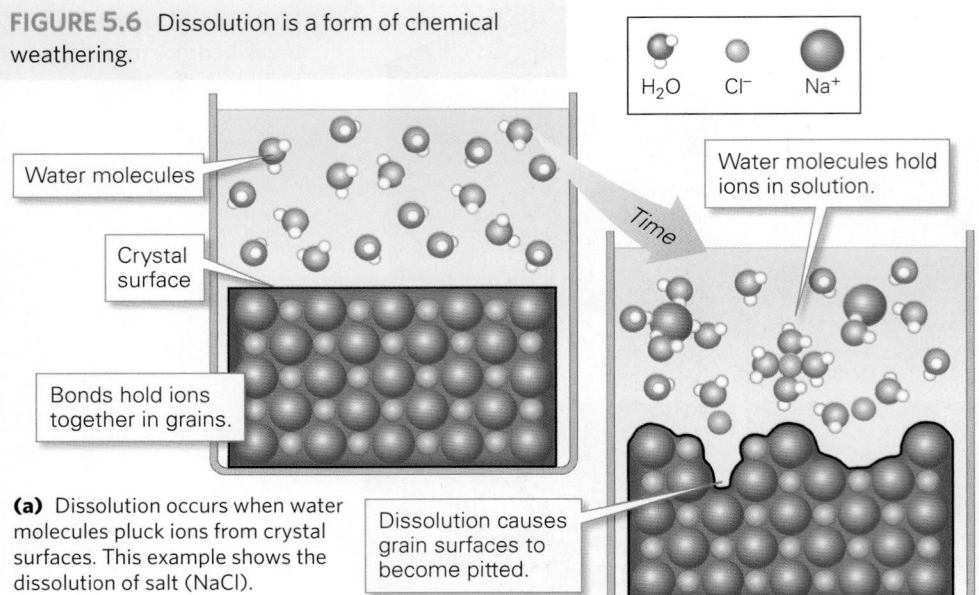

H_2O Cl^- Na^+

Water molecules

Crystal surface

Bonds hold ions together in grains.

Water molecules hold ions in solution.

Time

(a) Dissolution occurs when water molecules pluck ions from crystal surfaces. This example shows the dissolution of salt (NaCl).

Dissolution causes grain surfaces to become pitted.

(b) Dissolution along joints in this limestone bedrock in Ireland produced troughs and rounded corners.

Organisms can contribute to the process of chemical weathering. For example, plant roots, fungi, and lichens secrete acids that help dissolve minerals. And some bacteria and archaea, tiny single-celled organisms, literally eat minerals for lunch, in that they pluck molecules from minerals and use the energy from the molecules' chemical bonds to sustain their metabolism.

Physical and Chemical Weathering Working Together

In the natural world, physical and chemical weathering can operate simultaneously and accelerate each other. Physical weathering speeds up chemical weathering because, by breaking up rock, it provides more surface area that can encounter air or water (Fig. 5.7a). In turn, chemical weathering speeds up physical weathering, either by dissolving cements that hold grains in rock together or by transforming strong minerals such as feldspar into weak clay. Both processes allow rock to disintegrate more easily (Fig. 5.7b).

Not all rocks weather at the same rate. *Resistant rocks* weather more slowly, whereas *nonresistant rocks* weather more rapidly. Due to such *differential weathering*, cliffs composed of alternating layers of resistant sandstone and nonresistant shale develop a stair-step or sawtooth profile (Fig. 5.8a). Similarly, cemetery headstones made of resistant granite retain sharp inscriptions for centuries, whereas soluble marble headstones become unreadable

FIGURE 5.7 Physical and chemical weathering processes work together.

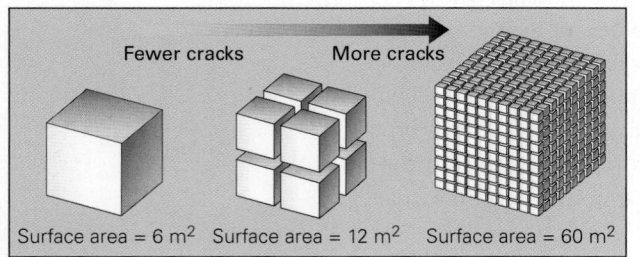

Fewer cracks   More cracks

Surface area = 6 m²   Surface area = 12 m²   Surface area = 60 m²

(a) As rock breaks apart due to physical weathering, the surface area increases relative to the volume; therefore, more rock surface is exposed to chemical weathering. In this example, a cube has a surface area of 6 m². If it breaks into four cubes, they have a combined surface area of 12 m².

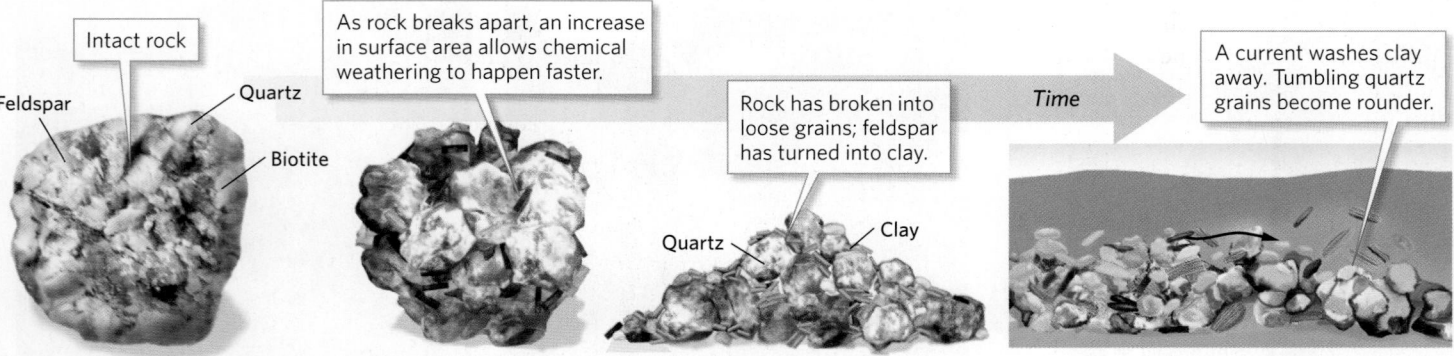

Intact rock

Feldspar   Quartz

Biotite

As rock breaks apart, an increase in surface area allows chemical weathering to happen faster.

Rock has broken into loose grains; feldspar has turned into clay.

Quartz   Clay

Time

A current washes clay away. Tumbling quartz grains become rounder.

(b) Chemical weathering weakens rock, such as this piece of granite, so it breaks apart. As this happens, the surface area increases, so chemical weathering happens even faster. Eventually, the rock completely disintegrates to form sediment. Weathering of granite produces quartz sand (which resists chemical weathering) and clay (produced by the weathering of feldspar, mica, and other silicates).

FIGURE 5.8 Differential weathering.

(a) A sawtooth weathering profile has developed due to the alternating resistant and nonresistant layers on this outcrop in New Mexico. The nonresistant layers have become indented.

(b) Inscriptions on a granite headstone (left) last for centuries, but those on a marble headstone (right) may weather away in decades.

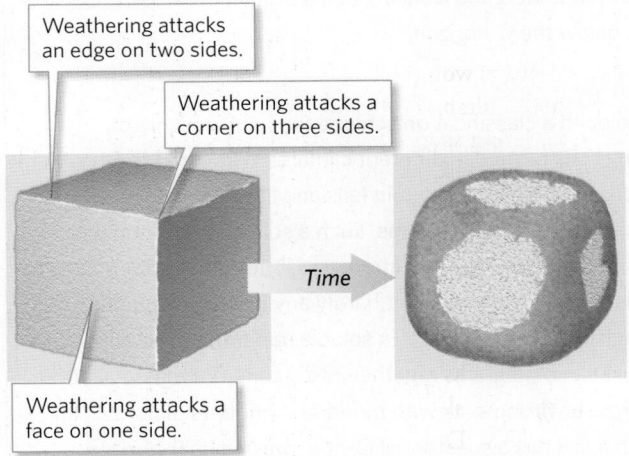

Weathering attacks an edge on two sides.

Weathering attacks a corner on three sides.

Time

Weathering attacks a face on one side.

(c) Weathering attacks a block of rock from one side on the flat face, two sides at the edges, and three sides at the corners, so homogeneous rocks tend to weather into rounded blocks.

(d) Weathering of these granite blocks in California made them rounded.

within decades (Fig. 5.8b). Notably, weathering tends to blunt edges and round corners over time because weathering attacks a flat rock face from only one direction, an edge from two directions, and a corner from three directions (Fig. 5.8c, d).

Sediment: The End Product of Weathering

The various processes of sediment production that we've just described can yield a great variety of sediment types. For simplicity, we can distinguish among four general classes of sediment: (1) clastic sediment, which consists of fragments broken off pre-existing rocks—notably, we can also use the term *grain* for a relatively small clast; (2) biochemical sediment, which consists of shells produced by organisms, and/or fragments of shells; (3) organic sediment, which consists of organic material that once made up the cells of organisms; and (4) chemical sediment, which consists of minerals precipitated directly from water solutions. When relevant, geologists also distinguish different sediment types, and the sedimentary rocks formed from them, by their mineral composition. For example, *siliciclastic* sediment contains mostly silicate-mineral grains, *argillaceous* sediment contains mostly clay, and *carbonate* sediment contains mostly calcite or dolomite (minerals that contain the carbonate ion). The products of weathering not only serve as a source of sediment, but they also become the raw materials from which soil can form (Box 5.1).

Narrative Art Video
Forming Soil

BOX 5.1 ▶ Putting Earth Science to Use

Soil, the foundation of terrestrial life

If you've ever had the chance to dig in a garden, field, or forest floor, you've seen firsthand that the soil in which flowers, crops, and trees grow looks and feels different from bedrock, beach sand, or potter's clay. **Soil** consists of sediment that has been modified over time by interaction with rainwater, air, organisms, and decaying organic matter. Because of soil's significance to society, let's examine how it forms, why its character varies with location, and why conserving it should be a priority for society.

Producing Soil

Soil formation can begin on weathered bedrock or on a layer of sediment. Three overlapping processes contribute to soil formation. First, chemical and physical weathering break up and alter pre-existing material at the Earth's surface to produce loose debris. Second, rainwater sinks into and percolates downward through the debris (like water in a drip coffee maker). As this water moves, it incorporates ions by dissolving them, and it picks up microscopic clay flakes. The interval from which downward-percolating water removes ions and clay is called the *zone of leaching*. Farther underground, in the *zone of accumulation*, water percolation slows, so new mineral crystals precipitate from the water and clay stops moving (Fig. Bx5.1a). Third, the altered debris interacts with organisms and their by-products. Specifically, at the ground surface, organic material (decayed plants) mixes in with mineral grains, while underground, organisms (rodents, earthworms, insects, fungi, plant roots, and microorganisms) break up solid masses, absorb and release ions, and leave behind waste.

Controls on Soil Development

The soil in one locality can differ greatly from the soil in another in terms of composition, thickness, and texture. Such diversity exists because the makeup of a soil depends on several factors:

- *Source composition:* Soils formed from different parent materials have different chemical compositions.
- *Climate:* Rainfall and temperature affect the rate of chemical weathering, the amount of leaching, and the abundance of organisms in soil.
- *Slope steepness:* Soil depends on slope angle. If a slope is too steep, soil washes or slides away before it can accumulate. Therefore, thicker soils tend to develop on level ground.
- *Time:* Soil takes time to form, so a young soil tends to be thinner and less developed than an old soil. A thick soil may take millennia to form.
- *Organism type:* Different kinds of organisms extract or add different nutrients and quantities of organic matter, and different plants have different kinds of root systems.

Soil Horizons

Because soil-forming processes change with depth, soils develop distinct zones, known as **soil horizons**, arranged in a vertical sequence known as a *soil profile*. As an example, let's consider a soil profile formed in a temperate forest (Fig. Bx5.1b, c). The highest horizon, the *O-horizon*, consists mostly of *humus* (decayed plant debris). Below the O-horizon lies the *A-horizon*, in which decayed humus has mixed with mineral grains. In some locations, the A-horizon grades down into an *E-horizon*, in which leaching has taken place but no organic matter has mixed in. The O- and A-horizons together lie within the zone of leaching and form the *topsoil*. The underlying E-horizon does not contain organic matter, although it has undergone leaching. It is not part of the topsoil, but instead is considered a transition zone. Beneath the A-horizon (or E-horizon) lies the B-horizon, or *subsoil*, representing the zone of accumulation. At the base of a soil profile lies the C-horizon, consisting of weathered rock or sediment that has not undergone leaching or accumulation. Fresh rock or sediment occurs below the C-horizon.

Soil Types

Researchers developed a classification scheme for soils to highlight differences among soils formed in different climates (Fig. Bx5.1d). For example, in regions where hardly any rain falls and little or no vegetation grows, a desert soil, or *aridisol*, forms. Such a soil has no O-horizon because deserts host almost no vegetation, and it has almost no A-horizon because so little rain falls that hardly any leaching happens. Because of the lack of leaching, calcite (a soluble mineral) remains in the B-horizon, where it can bind grains together into a coherent mass called *caliche*. In temperate environments with moderate amounts of rainfall, an *alfisol* forms. Such a soil has a substantial O-horizon, and its B-horizon can accumulate only insoluble minerals (minerals that do not dissolve), because enough water percolates down through the B-horizon to carry calcite and other soluble minerals still deeper. In the tropics, where heavy rains fall during much of the year, a distinct brick-red tropical soil called an *oxisol* (traditionally known as laterite) develops. So much water constantly percolates downward through such soil that the A-horizon ends up consisting largely of insoluble iron and aluminum oxides.

Soil Conservation

Without soil, society could not grow food, raise animals, or obtain wood. Unfortunately, humans have been damaging this essential resource at an astounding rate. For example, by replacing deep-rooted native grasses and trees with annual crops, by leaving farmland bare after harvest, and by plowing deeply, we have exposed soil to wind and rain and destroyed root networks. As a result, **soil erosion**, the process

FIGURE Bx5.1 Formation and character of soil.

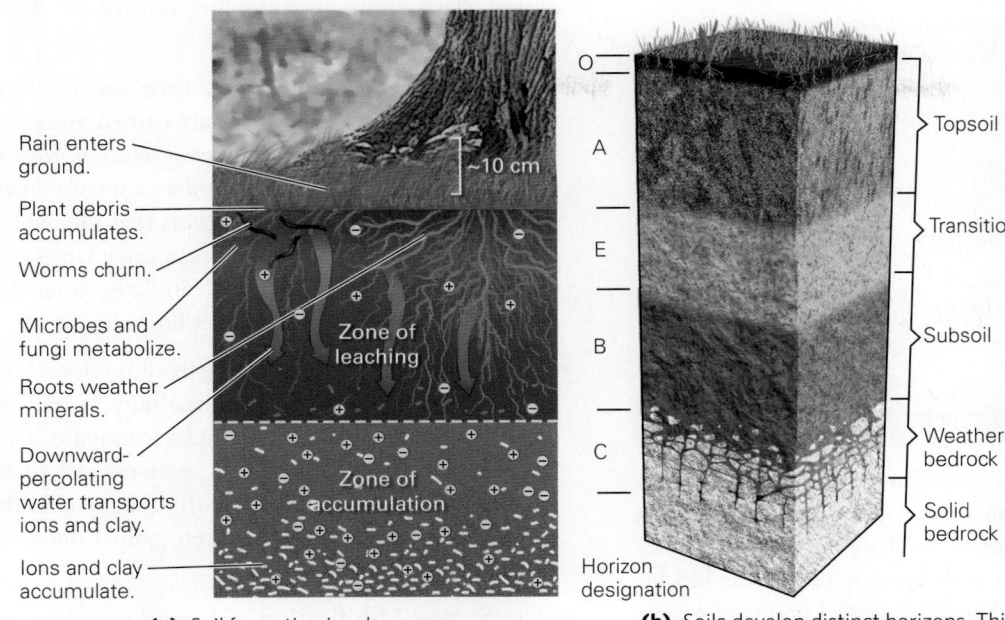

Rain enters ground.

Plant debris accumulates.

Worms churn.

Microbes and fungi metabolize.

Roots weather minerals.

Downward-percolating water transports ions and clay.

Ions and clay accumulate.

~10 cm

Zone of leaching

Zone of accumulation

(a) Soil formation involves many processes.

O
A
E
B
C

Horizon designation

Topsoil

Transition

Subsoil

Weathered bedrock

Solid bedrock

(b) Soils develop distinct horizons. This example is from a temperate forest.

This soil hosts a temperate forest. It displays a thick O-horizon.

(c) Photo of a soil exposed by an excavation in a temperate forest. Note the color changes.

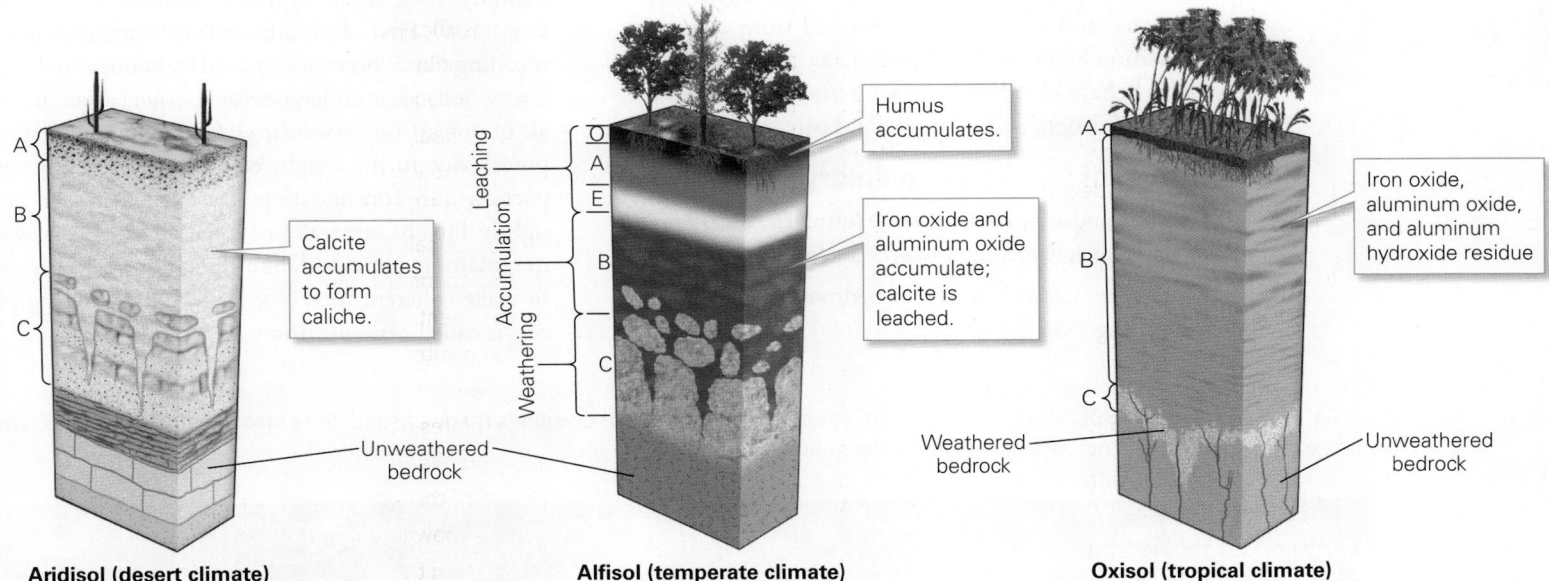

A
B
C

Calcite accumulates to form caliche.

Unweathered bedrock

Aridisol (desert climate)

Leaching
Accumulation
Weathering

O
A
E
B
C

Humus accumulates.

Iron oxide and aluminum oxide accumulate; calcite is leached.

Alfisol (temperate climate)

A
B
C

Iron oxide, aluminum oxide, and aluminum hydroxide residue

Weathered bedrock

Unweathered bedrock

Oxisol (tropical climate)

(d) Examples of soil classification. Soils formed in different climates have different characteristics.

of soil removal by wind and water, has accelerated. By some estimates, about half of the Earth's topsoil has vanished during the past century and a half. Further, the nutrient content of the soil that remains has decreased substantially, so farmers must add fertilizer to grow crops. And the proportion of carbon in soil (derived from organic matter) has diminished because agricultural practices and grazing take away organic matter that, in natural forests and prairies, would have mixed back into soil. Carbon lost from soil ultimately ends up incorporated in atmospheric CO_2 or CH_4, greenhouse gases that contribute to global warming. Clearly, to sustain food supplies and to keep more carbon from entering the atmosphere, *soil conservation*—saving and restoring soil—must be a priority for society.

Take-home message...

During physical weathering, rock breaks into pieces, and during chemical weathering, air and water react with minerals in rock. Working together, chemical and physical weathering cause rock to disintegrate over time, producing materials (solid fragments and dissolved ions) that become sediment.

Quick Questions ─────────────

- What can sediment consist of?
- How does physical weathering differ from chemical weathering?
- How does soil differ from sediment?

5.2 Clastic Sedimentary Rocks

Nine hundred years ago, a community of Native Americans living on the Mesa Verde Plateau built stone-block buildings beneath ledges of overhanging rock (Fig. 5.9). If you were to rub your thumb along one of the blocks, it would feel gritty, and small, rounded grains of quartz would break free. The blocks consist of *sandstone*, made from sand-sized clasts cemented together. Geologists refer to sandstone, and all other rocks formed from clasts, as **clastic sedimentary rocks**. Though, as we'll see, shell fragments behave like clastic sediment, the term *clastic sedimentary rock* generally implies a siliciclastic composition.

Formation of Clastic Sedimentary Rock

How did the sandstone of Mesa Verde form? The overall process of clastic rock formation involves the following steps:

- *Weathering:* Clasts form from bedrock broken up by weathering (Fig. 5.10a).

- *Erosion:* **Erosion** happens when moving air, water, or ice breaks off and removes grains or larger fragments of bedrock, or picks up already-loose clasts. In some cases, clasts move by tumbling downslope due to gravity.

- *Transportation:* After formation, clasts undergo **transportation**, meaning that they are carried away from their site of origin by a *transporting medium* (moving air, water, or ice). The viscosity and velocity of a flowing medium determine the size of clasts that the medium can carry. Denser fluids (water) can carry larger clasts than less dense fluids (air) can. Similarly, faster flows can carry larger clasts than slower flows can.

- *Deposition:* When a transporting medium slows (or, in the case of ice, melts), it loses its ability to transport clasts, so the clasts settle out and accumulate, a process called **deposition**. Clastic sediment, when first deposited, is loosely packed and contains abundant pores (small open spaces between grains) filled with air and/or water.

- *Lithification:* **Lithification**, the transformation of an unconsolidated sediment layer into a coherent sedimentary rock layer, typically involves three stages (Fig.5.10b). First, the sediment layer undergoes burial, meaning that it becomes covered by younger sediment layers. Second, it undergoes *compaction* (squeezing out air in pores) and *consolidation* (squeezing out water in pores), due to the weight of overburden. After compaction and consolidation, clasts fit together more tightly. Finally, minerals precipitate from groundwater in remaining pores and, like glue, bind clasts together to make coherent rock (Fig. 5.10c). This binding process is called **cementation**.

FIGURE 5.9 Between 600 and 1300 C.E., Pueblo people inhabited these dwellings made of sandstone blocks beneath a protective ledge of sandstone bedrock. The inset shows the grainy character of the rock.

Sandstone layer

FIGURE 5.10 Steps in clastic sedimentary rock formation.

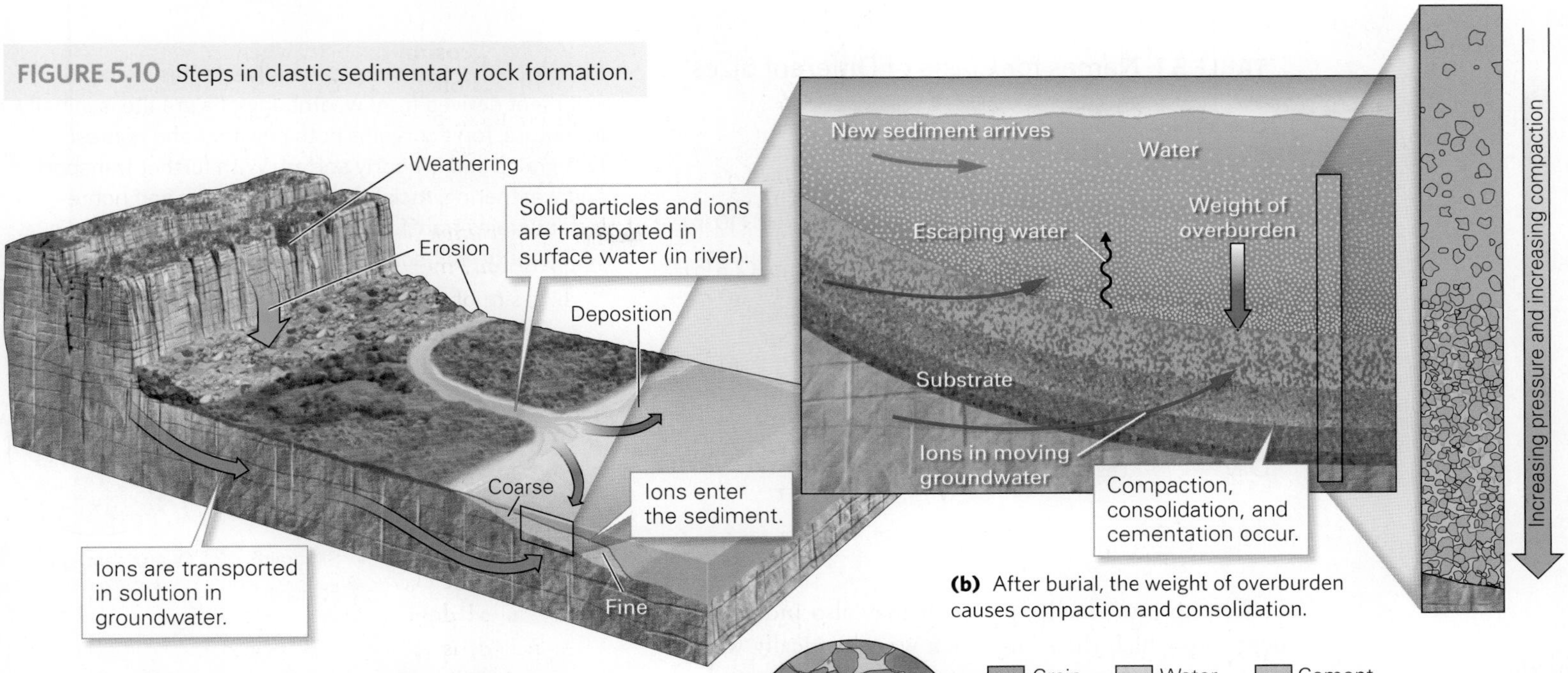

Weathering

Erosion

Solid particles and ions are transported in surface water (in river).

Deposition

Coarse

Ions enter the sediment.

Ions are transported in solution in groundwater.

Fine

(a) Sediment eroded from a cliff gets transported to a site where it is deposited.

New sediment arrives

Water

Escaping water

Weight of overburden

Substrate

Ions in moving groundwater

Compaction, consolidation, and cementation occur.

Increasing pressure and increasing compaction

(b) After burial, the weight of overburden causes compaction and consolidation.

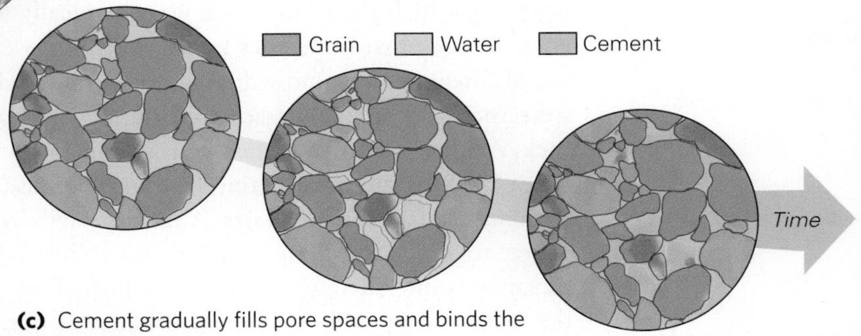

Grain   Water   Cement

Time

(c) Cement gradually fills pore spaces and binds the clasts together.

Describing the Sediment in Clastic Sedimentary Rocks

When describing clasts in clastic sedimentary rocks, geologists characterize several features: (1) *clast size*, by specifying clast diameter (Fig. 5.11a) or by using appropriate names (Table 5.1)—note that in this context, familiar names (clay, silt, sand, pebble, cobble, and boulder) correspond to specific size ranges; (2) *clast angularity*, by specifying the degree to which clasts have sharp edges or rounded edges (Fig. 5.11b); (3) *clast sorting*, by specifying the degree to which the rock contains grains of only one size or grains of many different sizes (Fig. 5.11c), and (4) *clast composition*, by specifying whether grains are parts of single mineral crystals or are rock fragments containing many minerals, and by identifying the minerals or rock types present.

What can a description of the above characteristics tell you? Clast size gives a sense of transport distance, for sediment that has been carried farther will probably have

FIGURE 5.11 Clast characteristics.

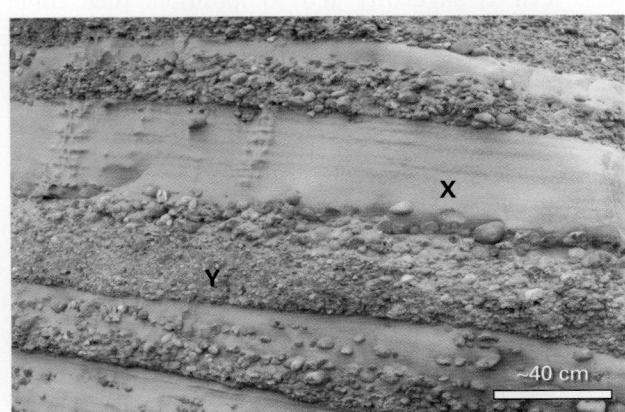

(a) A close-up view of a cliff face exposing contrasting layers of clastic sediment. Layer X consists of quartz sand grains, deposited by a slower current, and Layer Y consists of pebbles, deposited by a faster current.

~40 cm

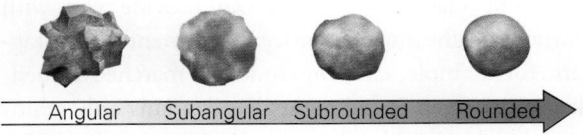

Angular   Subangular   Subrounded   Rounded

(b) Different degrees of clast angularity.

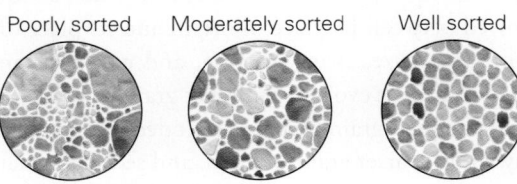

Poorly sorted   Moderately sorted   Well sorted

(c) Different degrees of clast sorting.

TABLE 5.1 Names for Clasts of Different Sizes

Clast Name	Diameter
Boulder	More than 256 mm (10 in)
Cobble	Between 64 mm (2.5 in) and 256 mm (10 in)
Pebble	Between 2 mm (0.08 in) and 64 mm (2.5 in)
Sand	Between $\frac{1}{16}$ mm (0.0025 in) and 2 mm (0.08 in)
Silt	Between $\frac{1}{256}$ mm (0.00015 in) and $\frac{1}{16}$ mm (0.0025 in)
Clay	Less than $\frac{1}{256}$ mm (0.00015 in)

FIGURE 5.12 Sediment can evolve. In this example, sediment derived from weathering of a granite is initially immature, for it contains both resistant and nonresistant grains and is poorly sorted. With further transport and weathering, rock fragments break up and nonresistant grains transform into clay, producing a mature sediment of well-sorted, well-rounded, resistant grains.

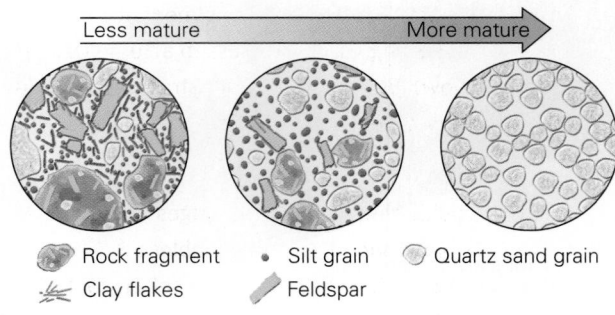

Less mature   More mature

Rock fragment   • Silt grain   Quartz sand grain
Clay flakes   Feldspar

Did you ever wonder...

where beach sand comes from?

See for yourself

Murray River Floodplain, Australia

Latitude: 34°7′4.68″ S
Longitude: 140°46′9.75″ E

Look down from an altitude of 15 km (9.3 mi).

The Murray River meanders across the flat-lying landscape of South Australia. In this view, the water is restricted to the channel, exposing the sand and silt deposits that have covered its floodplain.

broken up into smaller pieces. It may also indicate the degree to which the source rock was chemically weathered, for rock that undergoes intense chemical weathering disintegrates into individual mineral grains, whereas fresh rock will probably crack to form larger chunks of rock. And, as we noted earlier, clast size tells you about the velocity of the transporting medium, because faster currents can carry larger clasts. Angularity tells you the degree to which the clasts tumbled and collided during transport, and therefore provides information about the transporting medium and the distance of transport; specifically, sediment that became lithified before being transported very far tends to be angular, whereas sediment that tumbled in a fast-moving stream tends to be rounded. **Sorting**, the degree to which a rock contains clasts of a single grain size or a range of grain sizes, also tells you about the transporting medium and the distance of transport, because the ability to carry sediment depends on the velocity and viscosity of the medium (Box 5.2); for example, glacial ice can carry sediment of all sizes, so it leaves poorly sorted sediment, whereas waves on a beach wash away clay and silt, leaving behind well-sorted sand.

Studying clast composition can provide you with important insight into the source of sediment. Rock fragments, for example, can sometimes be matched to bedrock sources. Even in sediment that contains only grains of individual minerals, the numerical age of the grains can be matched to that of a bedrock source. Clast composition also gives insight into the extent to which a sediment evolved between its place of origin and its place of deposition. Grain size, shape, sorting, and composition all change as sediment evolves, for larger grains break into smaller ones, angular grains become rounded, finer grains wash away while coarser grains remain, and sediment that

once contained a variety of minerals may end up consisting only of resistant minerals. The transformation of large granite fragments into quartz beach sand illustrates such evolution (see Fig. 5.7b). Initially, the fragments may break up into smaller ones, and weathered ones may disintegrate into grains of quartz, feldspar, and mica. Sediment that contains a mixture of clast sizes and includes nonresistant minerals is known as *immature sediment* (Fig. 5.12). Over time, hydrolysis transforms feldspar and mica into clay, and remaining large clasts disintegrate. Flowing water then carries the clay away. Eventually, a *mature sediment*, consisting only of resistant quartz grains of similar size, remains.

Types of Clastic Sedimentary Rocks

When assigning a name to a clastic sedimentary rock, geologists focus first on the rock's clast size. Other factors, such as angularity and composition, then come into play to distinguish among different rocks of a given grain size. Table 5.2 lists the principal types of clastic sedimentary rocks, from the coarsest clast size down to the finest. Note that a rock formed from angular blocks yields a *breccia* (Fig. 5.13a), and one formed from rounded blocks yields a **conglomerate** (Fig. 5.13b). A rock formed by lithification of sand yields *arkose* if it contains quartz and feldspar grains (Fig. 5.13c) or **sandstone** if composed primarily of quartz sand grains (Fig. 5.13d). *Siltstone* forms by lithification of silt (mostly tiny quartz grains). Lithification of *mud* (a mixture of clay minerals, extremely tiny quartz grains, and water) produces **shale** if the rock splits into thin sheets (Fig. 5.13e), and *mudstone* if it does not.

BOX 5.2 ▶

How can I explain…

Sediment sorting

What are we learning?

The size of sediment that water or wind can transport depends on fluid velocity, so fluid flow sorts sediment. It's difficult to demonstrate water sorting without access to a flume (a long tank in which water can flow). But you can demonstrate wind sorting using a common blow-dryer.

What you need:

- A large cookie sheet with raised edges
- A cup of mixed sand and small pebbles
- A blow-dryer, with the temperature set on cool
- Eye protection

Instructions:

- Spread the sand-pebble mixture across the cookie sheet near one end. Give the sheet a tap to smooth the layer.
- Put the dryer on low speed, and bring it just close enough to the end of the cookie sheet with the sediment that the sediment starts moving.
- Gently blow back and forth and watch how the sand moves to the other end of the cookie sheet, while the pebbles remain close to where they started.
- Increase the fan speed to high and watch the pebbles start to move.

Mixed sand and pebbles

Before

Sand   Pebbles

After

What did we see?

- At low speed, the moving air sorts the sand from the pebbles. This process resembles the way in which an air or water current carries away sand and leaves behind gravel.
- When the fan speed increases, even the pebbles move. This observation emphasizes that faster air or water currents move larger clasts.

Take-home message…

🏠 Clastic sedimentary rocks consist of fragments weathered and eroded from pre-existing rocks. Erosion followed by transportation carries grains to a site of deposition. Lithification transforms sediment into rock. Describing clasts in sedimentary rock provides useful information about the history of the sediment from which it formed. Different types of clastic sedimentary rocks are distinguished primarily by grain size, but also by composition.

Quick Questions

- What does a clastic sedimentary rock consist of?
- What processes take place during lithification?
- Distinguish among shale, sandstone, conglomerate, and breccia.

TABLE 5.2 Clastic Sedimentary Rocks

Rock Name	Clast size	Sediment	Clast character
Breccia	Very coarse	Gravel	Angular
Conglomerate	Very coarse	Gravel	Rounded
Sandstone	Medium	Sand	Mostly quartz
Arkose	Medium	Sand	Quartz and feldspar
Siltstone	Fine	Silt	Mostly quartz
Shale	Very fine	Mud	Mostly clay
Mudstone	Very fine	Mud	Mostly clay

FIGURE 5.13 Different kinds of sediments lithify into different kinds of sedimentary rocks.

Sediment ——— Lithification ——→ Sedimentary rock

(a) Angular clasts, when cemented together, produce breccia.

(b) Rounded gravel (pebbles and cobbles) lithifies into conglomerate.

(c) Lithification of immature sediment yields arkose.

(d) Layers of sand lithify into sandstone.

Sandstone

Shale

(e) Layers of mud lithify to form shale.

5.3 Biochemical, Organic, and Chemical Sedimentary Rocks

So far, we've discussed processes that form new sedimentary rocks from clasts derived from pre-existing rocks. Recall that weathering not only yields such grains, but also produces ions in solution. These ions can precipitate as cement in the pores of clastic sedimentary rocks, can be extracted from solution by organisms to form shells, or can precipitate directly to form mineral grains. In some settings, sediments also contain a substantial proportion of organic material. Below, we introduce the variety of sedimentary rocks that consist of sediments other than siliciclastic grains **(Table 5.3)**.

Biochemical Sedimentary Rocks

A vast array of invertebrate organisms—ranging from microscopic plankton to giant clams—extract dissolved ions from the water around them to make shells. When the organisms die, these shells remain. Accumulations of shells, or fragments of shells, when lithified, become **biochemical sedimentary rocks**. Geologists use composition and texture to distinguish among different types of such rocks.

A snorkeler gliding above a coral reef sees an incredibly diverse community of corals and algae, around which other shelled creatures such as clams, oysters, and snails live, and above which plankton float **(Fig. 5.14a)**. These organisms all share an important characteristic: they make shells of calcite and/or aragonite (a polymorph of calcite) **(Fig. 5.14b)**, which therefore consist of $CaCO_3$ (calcium carbonate). Any rock formed predominantly from $CaCO_3$ is called **limestone**. Limestone consisting of shells or shell fragments is known, technically, as *biochemical limestone*. Some examples of biochemical limestone formed directly from reefs of shells anchored to the seafloor, while others consist of loose shells and shell fragments that accumulated

TABLE 5.3 Common Biochemical, Chemical, and Organic Sedimentary Rocks

Rock Name	Minerals contained	Category
Limestone	Calcite	Biochemical or chemical
Dolostone	Dolomite	Biochemical or chemical
Evaporite	Halite; gypsum	Chemical
Chert	Cryptocrystalline quartz	Biochemical or chemical
Coal	Carbon molecules	Organic
Organic shale	Clay and hydrocarbons	Organic

FIGURE 5.14 The formation of biochemical limestone.

(a) Corals and other organisms living on this reef make their shells out of CaCO₃.

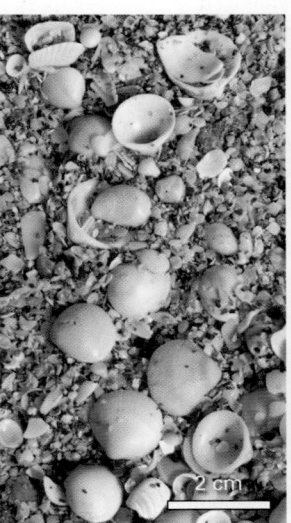

(b) Shells on a beach. If lithified, this material becomes a type of limestone called coquina.

(c) Broken shells were cemented together to form this 415-million-year-old fossiliferous limestone.

and later become cemented together. We can refer to limestone containing abundant identifiable fossil shells as *fossiliferous limestone* (Fig. 5.14c). Limestone that consists of particles that are too fine to be seen with the naked eye is called *micrite*; when this rock consists of white plankton shells it is also known as *chalk*. In an outcrop, most limestone typically looks like a massive light gray to dark bluish-gray rock (Fig. 5.14d), and breaks into chunky blocks that internally appear to have a crystalline texture. That's because interaction with groundwater, over time, causes the original calcite or aragonite of limestone to partially dissolve and new calcite crystals and cement to precipitate.

Not all biochemical sedimentary rocks consist of carbonate minerals because not all organisms make carbonate shells. For example, a type of plankton called radiolaria produces extremely tiny silica shells. When radiolaria die, their shells collect on the seafloor, where they eventually dissolve to produce a layer of gel-like ooze that, when lithified,

(d) Ancient limestone tends to occur in gray, blocky layers.

becomes a type of chert known as *biochemical chert*. (**Chert**, in general, is a rock composed of *cryptocrystalline quartz*, meaning that although its grains are too small to be seen with a petrographic microscope, it is crystalline, not glass.) In an outcrop, biochemical chert typically occurs in beds 3–8 cm thick with porcelain-like surfaces (Fig. 5.15a).

(a) This chert developed from deep-sea sediment made up of the shells of silica-secreting plankton.

(b) Coal is deposited in layers (beds), just like other kinds of sedimentary rocks.

FIGURE 5.15 Examples of biochemical and organic sedimentary rocks.

FIGURE 5.16 The formation of evaporite deposits.

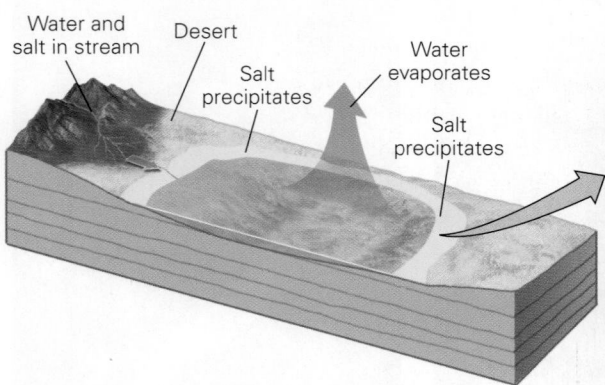

(a) When saltwater evaporates, various mineral salts, including halite and gypsum, precipitate.

(b) Under the desert sun, this salt lake on the floor of Death Valley has dried up, leaving a white crust of new salt crystals.

(c) A close-up of the white crust shows that it's made of tiny halite crystals.

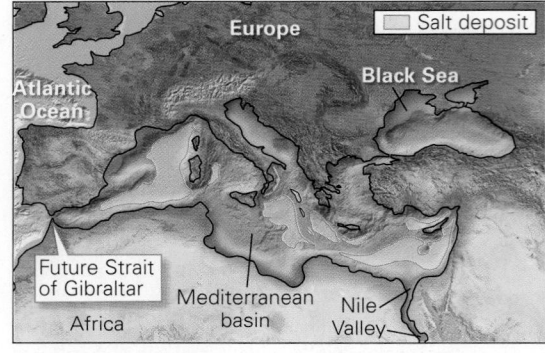

(d) The dried-up Mediterranean Sea. All the dissolved salt crystallized to form thick layers on the seafloor.

Organic Sedimentary Rocks

We've seen that the shells of organisms can become biochemical sedimentary rock. What becomes of the cellulose, fat, carbohydrate, or protein of those organisms? Generally, this material either gets eaten or decomposes at the Earth's surface. But in oxygen-poor environments, organic debris accumulates along with clay and undergoes burial before completely decomposing. When lithified, such sediment becomes **organic sedimentary rock**. The most common example, **coal**, forms from woody plant remains (Fig. 5.15b). *Organic shale*, the source of oil and gas, forms when the remains of plankton mix with clay to make mud, which then becomes shale.

Chemical Sedimentary Rocks

In 1965, a vehicle driven on land set a speed record of 600 mph (966 km/h). Such high-speed trials must take place on extremely large, flat surfaces, such as the Bonneville Salt Flats of Utah, which formed when an ancient salt lake

evaporated. During evaporation, water molecules escape from a water body and drift into the atmosphere, but salts in the water remain behind. As evaporation progresses, saltwater becomes *supersaturated*, meaning that it exceeds its capacity to contain dissolved ions. In a supersaturated solution, ions bond together to form solid grains that either settle out of the water or grow on the floor of the water body (Fig. 5.16a). This process, which takes place in desert salt lakes and along oceanic coasts in hot climates, produces layers of crystalline rock salt (halite) and gypsum (Fig. 5.16b, c). Because of the way these rocks form, geologists refer to them as **evaporites**. Evaporites serve as an example of a **chemical sedimentary rock**, a rock formed by precipitation from water without the activity of organisms.

Looking back at Figure 5.2, we find thick layers of salt beneath the seafloor of the Mediterranean Sea. These salt layers are evaporites, so their existence led researchers to conclude that at times in the past, the Mediterranean Sea must have dried up (Fig. 5.16d). How could that have happened? Studies documenting that over the course of geologic time, global sea level has gone up and down by over 100 m, relative to the land surface, provide an answer. When global sea level dropped sufficiently, the Mediterranean became a hot desert basin whose floor lay 3.5 km (2.2 mi) below sea level. During such times, the Strait of Gibraltar served as a dam, keeping the water of the Atlantic Ocean from entering the Mediterranean basin. Therefore, all water entering the basin at that time came from rain or rivers, and these sources did not supply enough water to keep the sea filled, so it dried up. When sea level later rose, the Strait of Gibraltar became an immense waterfall, and the Mediterranean filled again. The process repeated multiple times.

Did you ever wonder...

how the chert used for arrowheads first formed?

Not all chemical sedimentary rocks are evaporites, however. Another type, known as *travertine*, consists of calcite precipitated from groundwater that seeps out of the ground. Travertine accumulates at hot springs, because hot groundwater degasses (loses some dissolved CO_2) when it comes out of the ground and cools, and it also evaporates; both of these processes make the water supersaturated. Hot-spring travertine may build visually stunning terraces and mounds (Fig. 5.17a). Travertine also forms where water containing dissolved calcite seeps

out of the walls or ceilings of limestone caves and precipitates to produce icicle-like stalactites and other elaborate cave deposits, called *speleothems* (Fig. 5.17b) (see Chapter 12).

Another type of chemical sedimentary rock, *replacement chert*, forms when silica precipitates from groundwater and replaces other minerals within limestone. Replacement chert occurs in lumpy blobs called *nodules* (Fig. 5.18), which may concentrate in layers parallel to the limestone's bedding. It differs from biochemical chert

FIGURE 5.17 Examples of travertine deposits.

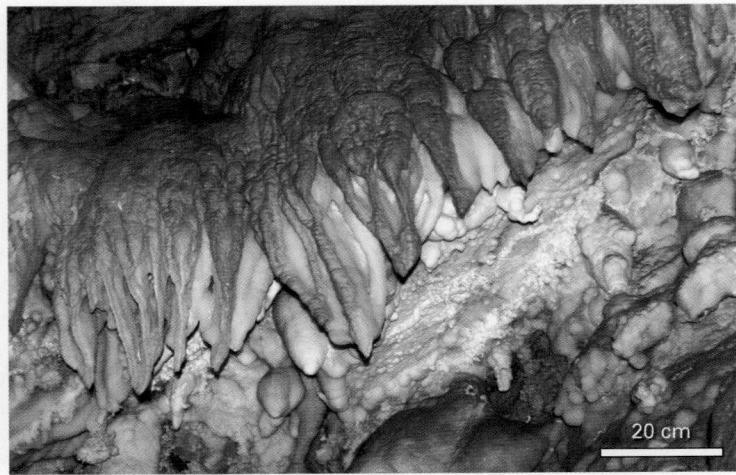

(a) Travertine accumulates in terraces at Mammoth Hot Springs in Yellowstone National Park, Wyoming.

(b) Travertine speleothems that grew as water dripped from the ceiling of Timpanogos Cave in Utah.

FIGURE 5.18 Replacement chert exposed on a cliff along the coast of southern England.

(a) The chert occurs as a layer of black nodules, parallel to the bedding plane, within a chalk bed. The inset photo shows a close-up of the nodules.

(b) As the chalk cliffs erode, the resistant chert collects to form pebbles on the beach, while the nonresistant chalk breaks up and washes away, or dissolves.

See for yourself

Grand Canyon, Arizona

Latitude: 36°8′8.94″ N
Longitude: 112°15′48.56″ W

Look straight down from an altitude of 15 km (9.3 mi).

You can see the spectacular color banding caused by the succession of stratigraphic formations exposed on the walls of the Grand Canyon.

EXPLORE EARTH SCIENCE IN 3D

Mudcracks A

EXPLORE EARTH SCIENCE IN 3D

Mudcracks B

only in how it forms (recall that the former precipitates as a sedimentary layer derived from silica-containing shells of plankton). All chert, whether biochemical or chemical, served as an important resource for cultures that did not have metalworking because the rock fractures conchoidally, so a skilled worker could use it to produce sharp-edged tools or arrowheads.

Some carbonate rocks contain a large proportion of *dolomite*, a mineral with the chemical formula $CaMg(CO_3)_2$. These rocks, known as **dolostone** or *dolomite*, form when limestone chemically reacts with magnesium-bearing groundwater, causing magnesium ions to exchange places with calcite ions, so that calcite crystals are replaced or overgrown by dolomite crystals. This chemical process of *dolomitization* can take place soon after the limestone has formed or long after the limestone has been buried.

Take-home message…

Biochemical sedimentary rocks incorporate minerals used by organisms to produce shells. Sediment derived from calcite shells produces various types of limestone, and sediment from silica-shelled plankton produces biochemical chert. Organic debris produces coal and organic shale. Precipitation from solutions produces a variety of chemical sedimentary rocks, such as evaporites.

Quick Questions

- Why are limestone and dolostone referred to as carbonate rocks?
- What does chert consist of?
- How does an evaporite form?

5.4 Sedimentary Structures

Sedimentary rock isn't just a homogeneous, featureless mass. Rather, it contains fascinating layering, textures, and shapes that form during or immediately after sediment deposition. Geologists refer to such features as **sedimentary structures** and distinguish among several types.

Bedding and Bed-Surface Markings

If you look at a typical outcrop of sedimentary rock, you'll see that it consists of a stack of layers. Geologists refer to the layering as *bedding*, each layer as a **bed**, and the boundary between two beds as a *bedding plane*. The term **strata** refers to a succession of several beds (Fig. 5.19a).

What features cause one bed to look different from adjacent ones, or causes sedimentary strata to split along bedding planes?

The development of bedding commonly reflects changes in the composition of sediment, which can happen when the velocity of flow in the transporting medium changes so that different grain sizes settle out, or when the source of sediment changes so that sediment composition changes as well. To visualize the role of velocity change in causing grain size to change, picture sediment layers being deposited by a stream (Fig. 5.19b). On a normal day, the stream flows slowly and transports silt, which collects on the streambed. After a rainstorm, the stream flows faster and can carry gravel, which accumulates over the silt layer. Later, when the stream slows again, another layer of silt covers the gravel. When later buried and lithified, this succession of sediment will contain beds of siltstone and conglomerate.

Not all bedding represents a change in the character of sediment, however. For example, in some cases, a bedding plane represents a time during which deposition ceases, so that the sediment layer that was already deposited undergoes weathering and soil formation before the overlying layer gets deposited. The resulting changes in sediment texture, composition, and color due to soil formation can be visibly preserved. Finally, compaction and consolidation of mud composed mostly of clay minerals will cause the platy (pancake-shaped) clay flakes to lie flat and parallel to the bedding so that, after lithification, the resulting rock will split into thin sheets.

Notably, the surface of a mud layer exposed to air right after deposition may still be soft, so the impact of raindrops can leave small impact craters called *raindrop impressions* that can survive burial and lithification. If a creature crawls across the soft bed, it may leave prints or tracks. And if the bed dries out and shrinks, it commonly breaks into roughly hexagonal plates. The gaps between the plates are known as **mudcracks** (Fig. 5.20). All such features, because they occur on a bedding plane, are called *bed-surface markings*.

Current-Related Structures

In quiet water or calm air, sediment settles like snow, forming a layer that has a smooth surface and a homogeneous texture. A sediment layer deposited beneath flowing air or water, however, typically displays wave-like ridges and troughs, which can be preserved (Fig. 5.21). Such **ripple marks** tend to be relatively small and are aligned perpendicular to the current flow. Much larger ridges formed by currents of wind or flowing water are called **dunes** (Fig. 5.22a, b). If you slice into a ripple or

FIGURE 5.19 Bedding in sedimentary rock.

(a) Beds of sedimentary rock exposed along a road in Utah.

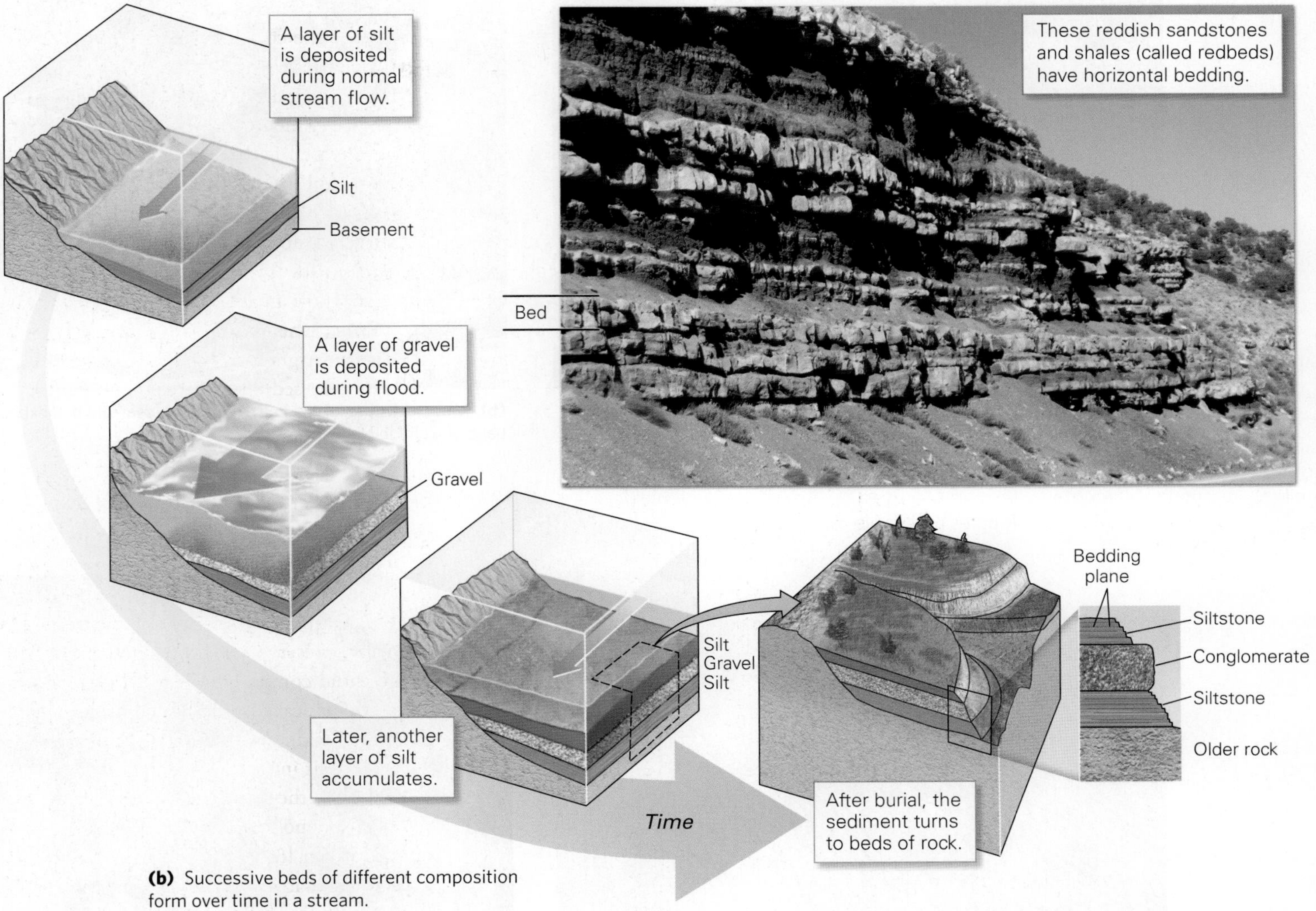

These reddish sandstones and shales (called redbeds) have horizontal bedding.

Bed

A layer of silt is deposited during normal stream flow.

Silt

Basement

A layer of gravel is deposited during flood.

Gravel

Later, another layer of silt accumulates.

Silt
Gravel
Silt

Time

After burial, the sediment turns to beds of rock.

Bedding plane

Siltstone

Conglomerate

Siltstone

Older rock

(b) Successive beds of different composition form over time in a stream.

Distinctive beds can sometimes be traced for long distances.

~250 m

FIGURE 5.20 Formation of mudcracks.

(a) Mudcracks in red mud at Bryce Canyon, Utah. As the mud dries, it contracts, and the cracks form.

(b) Mudcracks preserved in a 410-million-year-old bed exposed at the base of a cliff in New York.

FIGURE 5.21 Ripples and ripple marks.

(a) Modern ripples, exposed at low tide, on a sandy beach on Cape Cod, Massachusetts.

(b) Ripples on a tilted, 145-million-year-old sandstone bed.

EXPLORE EARTH SCIENCE IN 3D

Ripple Marks

dune, you'll find thin layers or laminations inclined at an angle to the main bedding planes. These laminations, known as **cross beds**, develop when a current of water or wind picks up sand from the upstream, or windward, face of the ripple or dune and deposits it on the downstream, or leeward, face (**Fig. 5.22c**). Sand builds up until the slope of the face becomes too steep to remain stable and gravity causes the sand to slide down the leeward face, which geologists therefore call the *slip face*. As more sand buries each successive slip face, it preserves the slip face as a cross bed; during the process, the active slip face migrates downstream. When new beds build over existing ones, a succession of cross-bedded layers, separated from each other by main bedding planes, accumulates. Notably, the orientation of cross beds provides a clue to the current direction at the time of deposition (**Fig. 5.22d**).

You may have seen videos of avalanches, showing a turbulent cloud of snow mixed with air rushing down a slope. The same phenomenon can happen on submarine slopes where unconsolidated sediment has accumulated. A storm or earthquake may cause the sediment to start sliding. It can then mix with water and become a turbulent cloud of sediment suspended in water. The

FIGURE 5.22 The formation of dunes and cross beds.

(a) A small dune developing in Death Valley. Ripples have formed on the top of the dune.

(b) Large sand dunes in a windstorm.

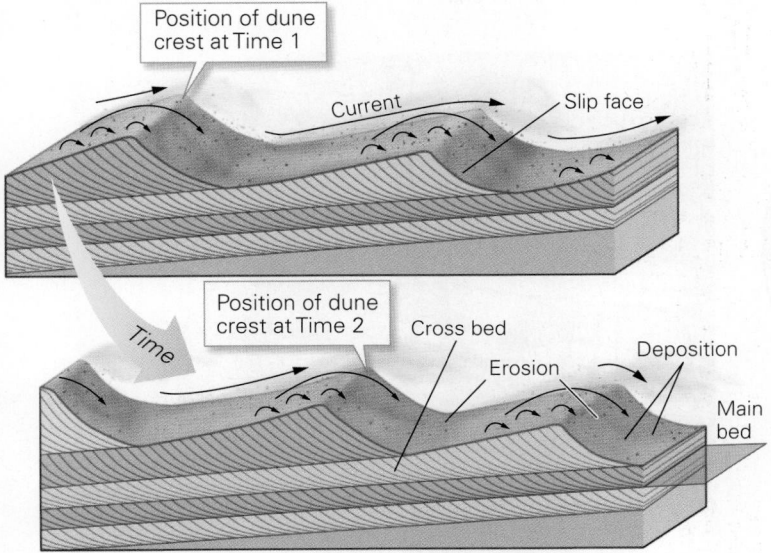

(c) Cross beds form when a wind current carries sand up the windward side of a ripple or dune, where it builds up until it falls down the slip face.

What an Earth Scientist Sees

(d) This sandstone exposed on a cliff face in Zion Canyon, Utah, contains large cross beds formed 200 million years ago, when the area was a sandy desert.

density of this cloud exceeds that of clear water, so it flows downslope along the seafloor as a **turbidity current** (Fig. 5.23). At the base of the slope, the turbidity current slows, so the sediment that it carries settles out. Larger grains sink through water faster than smaller grains, so the coarsest sediment accumulates first, progressively finer-grained sediment accumulates above it, and the finest-grained sediment (clay) settles out last. As a result, the new sediment layer, known as a *turbidite*, contains **graded bedding**, meaning that its grain size varies from large at the bottom to small at the top. A succession of turbidites can accumulate to build a thick wedge of sediment called a *submarine fan*.

Take-home message...

Sedimentary structures include bedding, ripple marks, dunes, cross bedding, mudcracks, and turbidites. All of these structures form during or soon after deposition of sediment. Some of them form on the surfaces of beds, some within beds.

Quick Questions

- How can a change in current velocity be represented by a succession of beds?
- Name two sedimentary structures formed by the flow of water.
- Why do turbidity currents stay fairly close to the seafloor as they move?

In every grain of sand, there is the story of the Earth.

—*RACHEL CARSON (AMERICAN NATURALIST, 1907–1964)*

FIGURE 5.23 The development of turbidites.

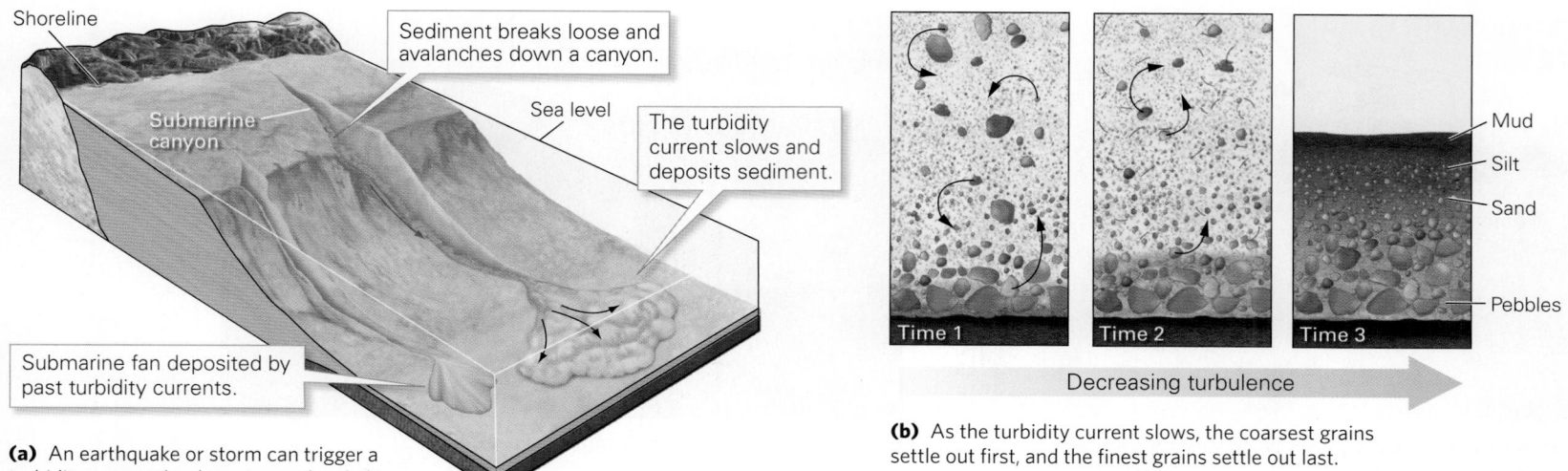

Shoreline

Sediment breaks loose and avalanches down a canyon.

Submarine canyon

Sea level

The turbidity current slows and deposits sediment.

Submarine fan deposited by past turbidity currents.

(a) An earthquake or storm can trigger a turbidity current (underwater avalanche). These currents commonly flow down submarine canyons.

Mud
Silt
Sand
Pebbles

Time 1   Time 2   Time 3

Decreasing turbulence

(b) As the turbidity current slows, the coarsest grains settle out first, and the finest grains settle out last.

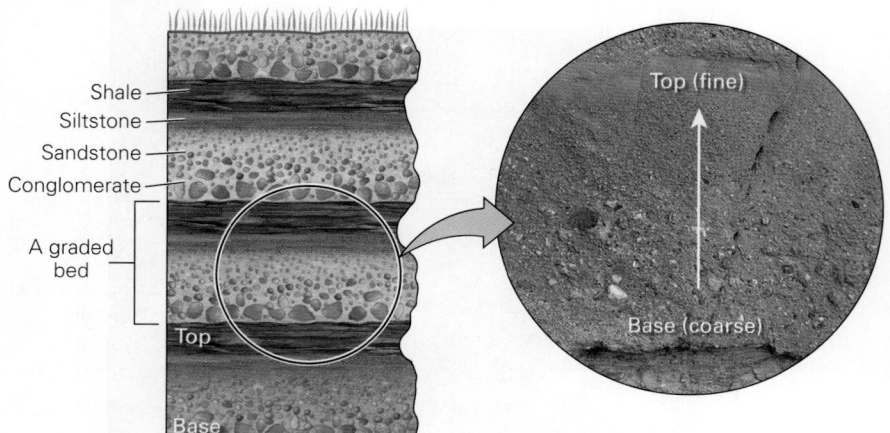

Shale
Siltstone
Sandstone
Conglomerate

A graded bed

Top

Base

Top (fine)

Base (coarse)

(c) Repetition of the process produces a succession of turbidites; each bed is coarser at the bottom and finer at the top.

(d) An outcrop of turbidites in western Italy.

5.5 Depositional Settings and Sedimentary Basins

Can we determine the **depositional setting** (environmental conditions at the time and place the sediments accumulated) by studying sedimentary rock? Fortunately, the answer is yes. To do this, geologists, like crime-scene investigators, look for clues in the rock. Detectives check fingerprints and DNA to identify a culprit, while geologists examine grain size, composition, sorting, sedimentary structures, and *fossils* (relics of organisms buried with the sediment) to identify a depositional setting. We group depositional settings into two main categories: **nonmarine deposition** occurs on land or in lakes and streams, whereas **marine deposition**

occurs in oceans or along the coasts of oceans (**Earth Science at a Glance**, pp. 188–189).

Nonmarine Deposition

The solid ice of glaciers can carry clasts of any size, so when glacial ice melts away, it leaves unsorted sediment known as *glacial till*. Therefore, if you see a sedimentary rock bed composed of large clasts surrounded by finer-grained sediment, you may be seeing lithified glacial till (**Fig. 5.24a**). Flowing water, in contrast, does sort sediment. In a flowing stream, coarser deposits (gravel and sand) typically collect on the streambed while finer sediment (silt and mud) remains suspended in the water. During a flood, however, the water level in the stream may rise high enough to submerge the stream's *floodplain*, the flat land

FIGURE 5.24 Examples of terrestrial depositional settings.

(a) Glacial till deposited by a glacier in France.

(b) Sediment in a Maryland stream.

(c) An alluvial fan in Death Valley, California.

on either side of the channel. Because of friction with the surface of the floodplain, this floodwater slows as it flows over the floodplain, so silt and mud settle out onto the submerged ground. Therefore, if you see strata in which thin lenses of conglomerate lie between ripple-marked siltstone beds (Fig. 5.24b), you may be looking at lithified deposits of a stream channel and a floodplain. In a desert, wind sorts and rounds sand grains, and it may build dunes containing cross beds. Therefore, if you see thick sandstone beds containing mature sediment and large cross beds, you're probably seeing sediments deposited in dunes (see Fig. 5.22d). Finally, at locations where temporary streams flow out of mountain canyons and onto desert plains after a heavy downpour, sand and gravel accumulate in a wedge called an *alluvial fan* (Fig. 5.24c). This sand has not traveled very far from its source, so it remains immature (see Fig. 5.12) and contains a variety of silicate minerals, instead of just quartz. So, if you find poorly sorted beds of arkose and conglomerate, you may be looking at lithified deposits of an alluvial fan. Notably, nonmarine clastic deposits of all grain sizes may develop a reddish tint during lithification because, in the presence of oxygen, reddish iron oxide minerals precipitate with the cement. Such reddish strata are known, in a general sense, as *redbeds* (see Fig. 5.19).

Marine Deposition

Where large rivers empty into the sea, a *marine delta* develops. Such deltas host various depositional settings, including swamps, channels, floodplains, and submarine slopes (Fig. 5.25a, b). Ocean currents and waves pick up and transport some of the clastic sediment in the delta and distribute it along the coast. The intensity of wave action controls the type of sediment deposited at a given location. Where sand washes back and forth in the surf, it becomes well sorted and well rounded, and sand beds may host ripple marks. In broad tidal flats, in lagoons,

and in deeper water offshore—regions less well sorted by waves—silt and mud also accumulate.

In warm shallow-marine environments lacking a supply of clastic sediment, clear water can nurture organisms that produce carbonate shells. Locally, *coral reefs*, mounds or ridges composed largely of coral shells, can grow. Storm waves break up some of the coral and distribute fragments around the reefs. Inland of the reefs, lagoons accumulate carbonate mud, and beaches collect sand made of small shell fragments, (Fig. 5.25c, d). All these deposits, if buried and lithified, transform into various types of limestone.

On slopes that drop from continental shelves to abyssal plains or trenches, turbidity currents carve deep submarine canyons. Where turbidites drain from these canyons, submarine fans composed of turbidites accumulate. Far from land, only fine clay and plankton are available to settle out of the ocean onto abyssal plains (Fig. 5.25e, f), so deposits of deep-marine sediment, when buried, yield shale, biochemical chert, or micrite. Locally, metal-oxide nodules, commonly called manganese nodules, precipitate on the seafloor.

Why Do Depositional Settings Change over Time?

Look once more at the photo of the Grand Canyon cliff face at the opening of this chapter. Some layers consist of marine strata and some of nonmarine strata (redbeds). This rock record tells us that the depositional setting at this location didn't remain the same over geologic time! Three key phenomena explain such changes, in a general sense. First, the climate at a location changes, both because of global climate change over time and because continents drift across climate belts. Second, sediment sources change over time because different areas of land are uplifted and start eroding, because river systems change their courses, or because wind patterns change direction.

See for yourself

Alluvial Fans and Evaporites, Death Valley, California

Latitude: 36°7'35.28" N
Longitude: 116°45'24.44" W

Look obliquely east from an altitude of 1.5 km (1 mi).

Death Valley is a narrow rift whose floor lies below sea level. You can see semi-circular alluvial fans and white evaporites at many localities along the length of the valley.

FIGURE 5.25 Examples of shallow-marine depositional settings.

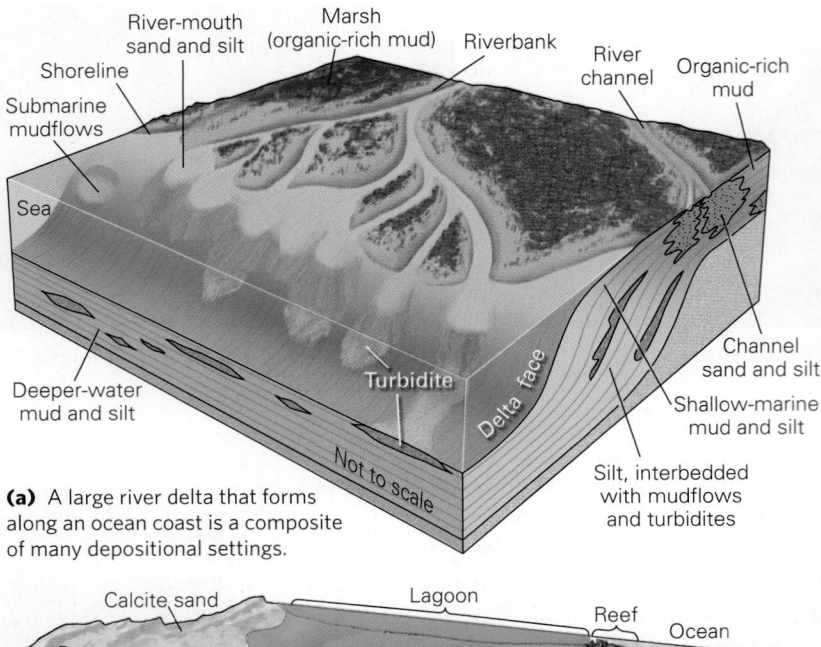

(a) A large river delta that forms along an ocean coast is a composite of many depositional settings.

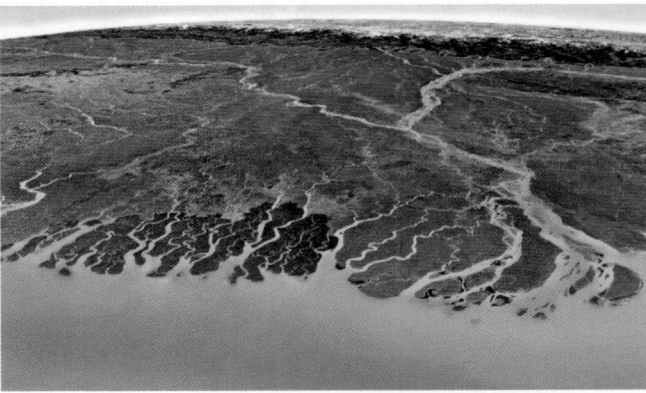

(b) A satellite photo of the Ganges Delta, on the southern coast of Bangladesh, illustrates a variety of depositional settings.

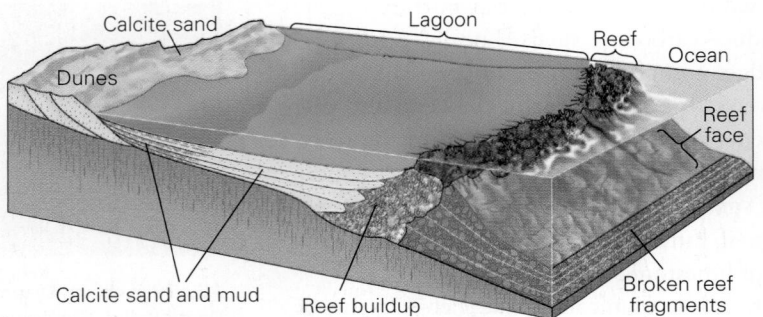

(c) Many depositional settings occur in association with coral reefs in clear, warm water.

(d) A coral reef and lagoon bordering a tropical island.

(e) The deep seafloor hosts deposits of mud and plankton shells, covered locally by metal-oxide nodules.

(f) Carbonate shells of plankton, as seen with a scanning electron microscope. Such sediment accumulates on the deep seafloor.

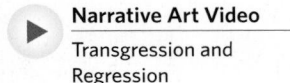

Narrative Art Video

Transgression and Regression

Third, **relative sea level**—the sea-surface elevation specified with respect to the land surface—goes up and down over time for various reasons (see Chapter 9), and land undergoes uplift and later sinks. When relative sea level rises, ocean shorelines migrate inland—a process called **transgression**—and when relative sea level falls, ocean shorelines migrate seaward—a process called **regression** (Fig. 5.26). During a transgression, marine deposits bury terrestrial deposits. The resulting sediment layers can blanket a broad region. Notably, at times during Earth history, continental interiors became submerged, and marine strata accumulated in places that now lie far from the sea.

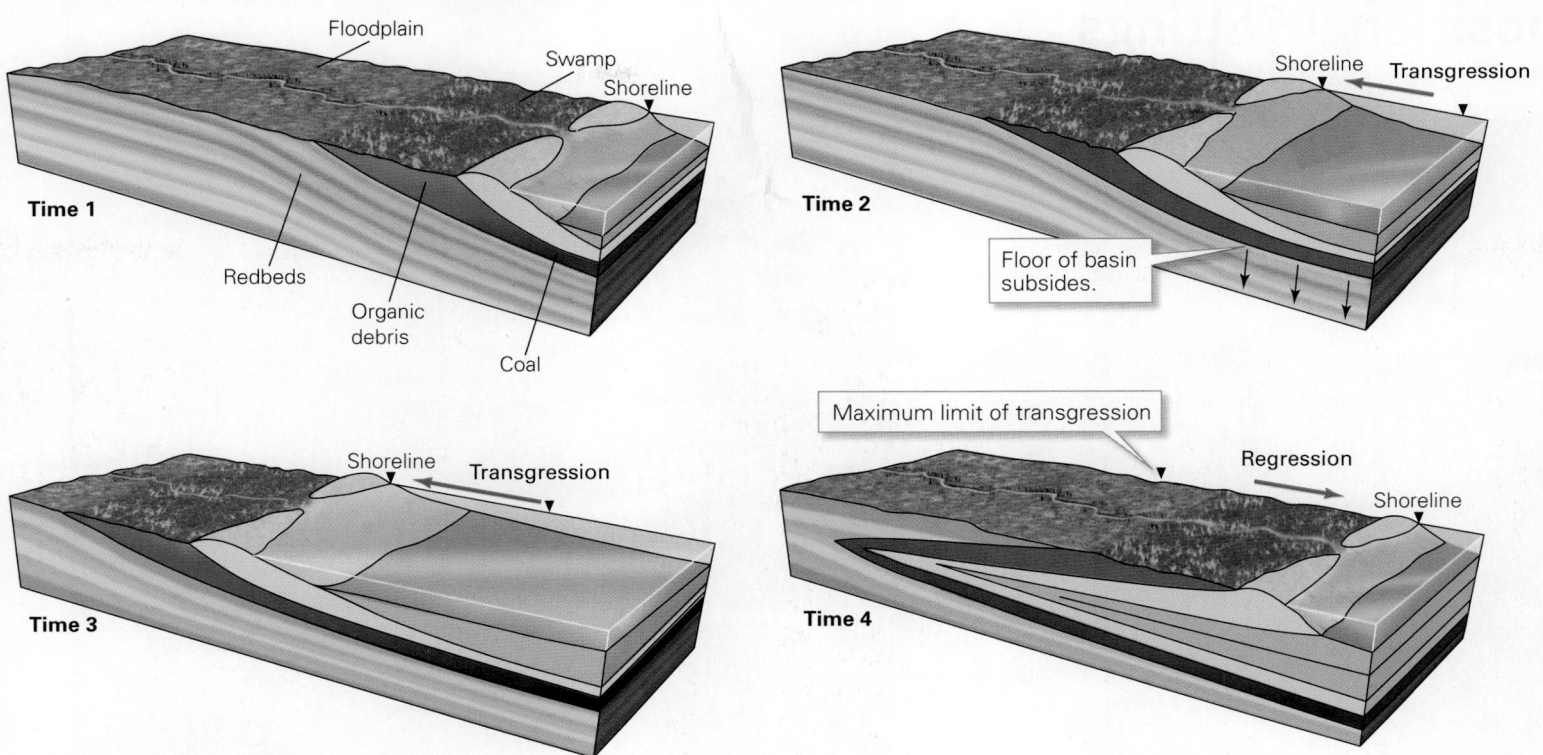

Sedimentary Basins

Globally, sedimentary strata form a *cover* that overlies igneous or metamorphic *basement* (Fig. 5.27). Generally, cover strata vary in thickness from 0 km (where basement lies exposed at the Earth's surface) to about 2 km (1.2 mi), a small fraction of typical continental crustal thickness (about 43 km, or 27 mi). In some regions, however, strata thickness ranges from 2 km to as much as 20 km (12 mi). Geologists refer to such regions, where the lithosphere has undergone **subsidence** (sinking) to produce a depression in which a particularly thick succession of sedimentary strata has collected, as a **sedimentary basin**.

FIGURE 5.27 Sedimentary strata form a cover overlying basement.

(a) A photo taken from the edge of the inner gorge in the Grand Canyon.

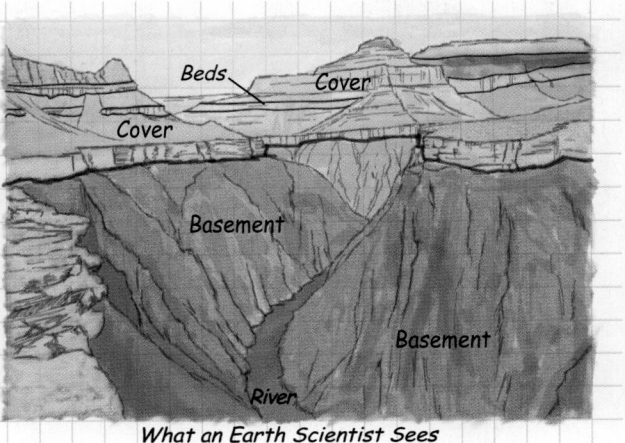

What an Earth Scientist Sees

(b) A sketch of the photo, emphasizing the distinction between basement and cover.

See for yourself

Shallow-Marine Environments, Red Sea, Egypt

Latitude: 22°38′17.70″ N
Longitude: 36°13′21.27″ E

Look straight down from an altitude of 4 km (2.5 mi).

The desert sands of the Sahara border the blue waters of the Red Sea. What types of sedimentary rocks would form if the sediments visible in this view were to be buried and preserved?

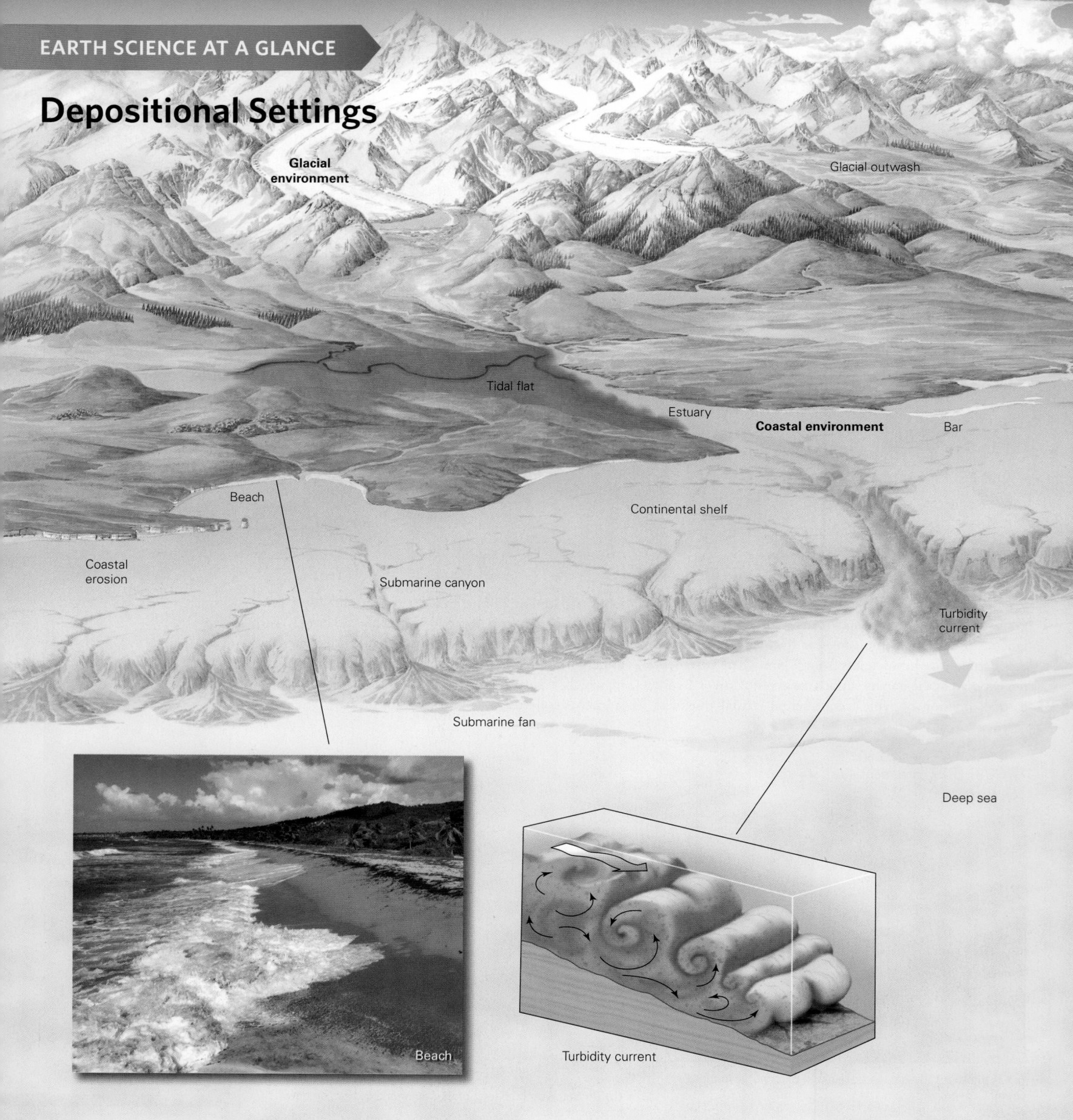

Depositional Settings

Glacial environment

Glacial outwash

Tidal flat

Estuary

Coastal environment

Bar

Beach

Continental shelf

Coastal erosion

Submarine canyon

Turbidity current

Submarine fan

Deep sea

Beach

Turbidity current

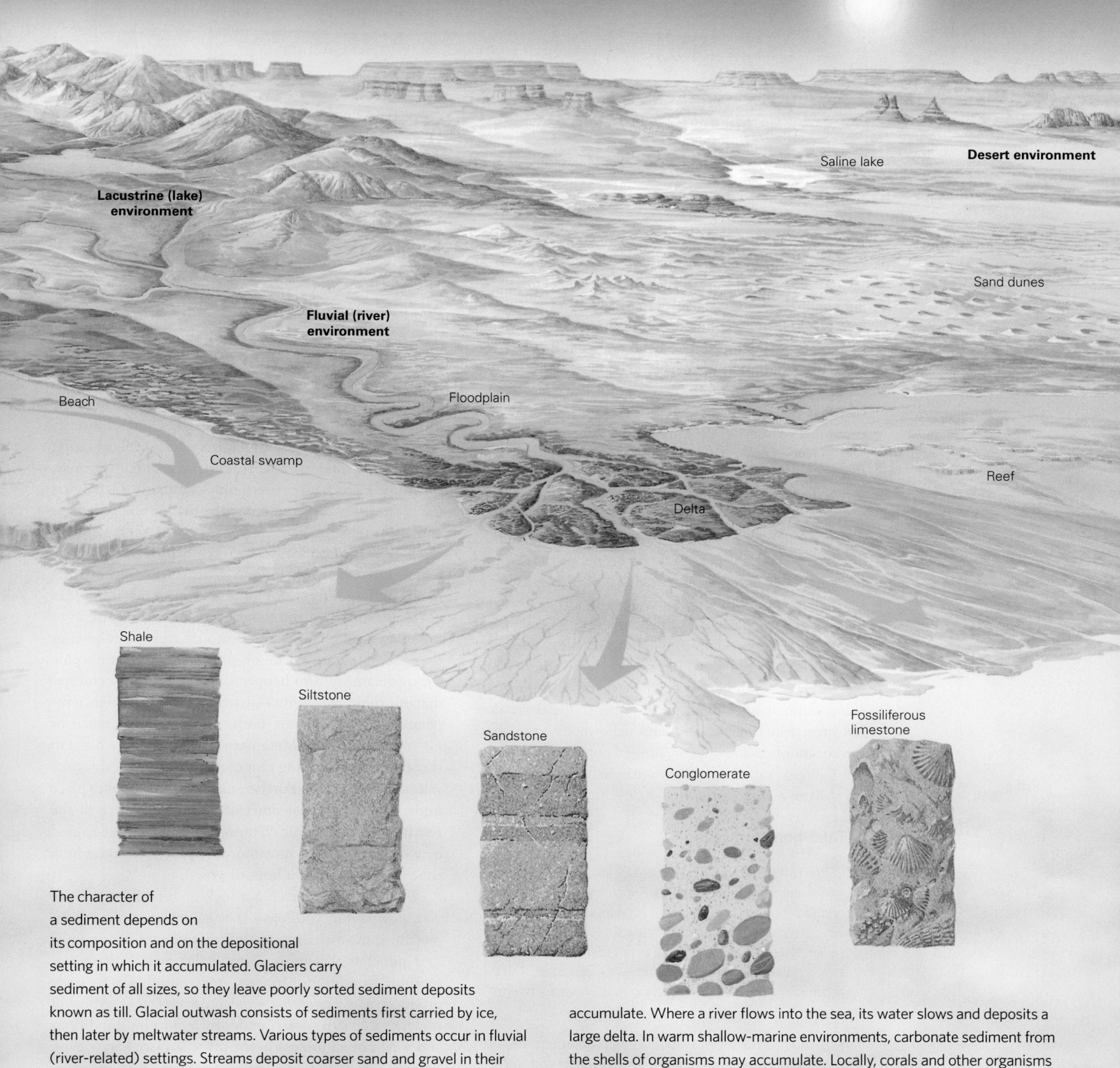

Lacustrine (lake) environment

Saline lake

Desert environment

Sand dunes

Fluvial (river) environment

Beach

Floodplain

Coastal swamp

Reef

Delta

Shale

Siltstone

Sandstone

Conglomerate

Fossiliferous limestone

The character of a sediment depends on its composition and on the depositional setting in which it accumulated. Glaciers carry sediment of all sizes, so they leave poorly sorted sediment deposits known as till. Glacial outwash consists of sediments first carried by ice, then later by meltwater streams. Various types of sediments occur in fluvial (river-related) settings. Streams deposit coarser sand and gravel in their channels and finer sediment on floodplains. In lacustrine (lake-related) settings, fine-grained mud accumulates in quiet water. In desert environments, sand builds into dunes, evaporites precipitate in saline lakes, and gravel and sand gather in alluvial fans. Wave action along coastal beaches leaves behind well-sorted sand. In swampy areas, large volumes of plant matter accumulate. Where a river flows into the sea, its water slows and deposits a large delta. In warm shallow-marine environments, carbonate sediment from the shells of organisms may accumulate. Locally, corals and other organisms build carbonate reefs. Offshore, submarine canyons direct turbidity currents out to the deep seafloor. Far from shore, deep-marine sediment, commonly consisting of plankton shells, slowly settles out of ocean water. Burial and lithification of all these various types of sediment turn them into different kinds of sedimentary rocks: clastic, biochemical, organic, and chemical.

189

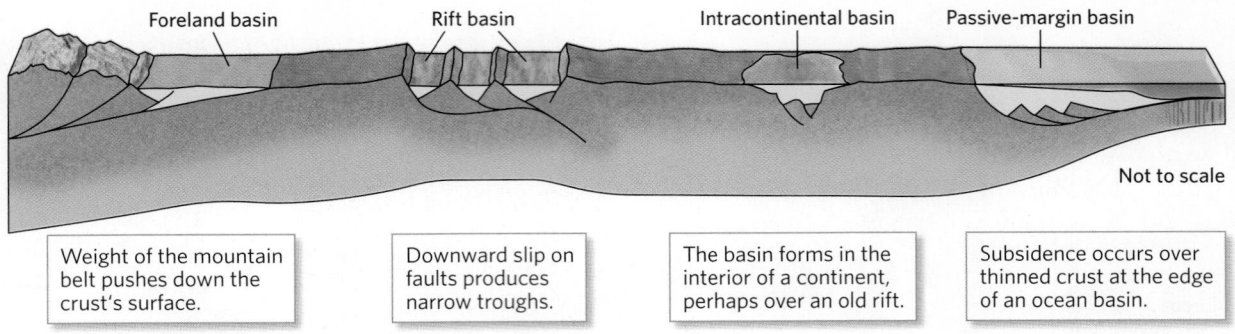

FIGURE 5.28 The geologic settings of sedimentary basins.

Foreland basin   Rift basin   Intracontinental basin   Passive-margin basin

Not to scale

| Weight of the mountain belt pushes down the crust's surface. | Downward slip on faults produces narrow troughs. | The basin forms in the interior of a continent, perhaps over an old rift. | Subsidence occurs over thinned crust at the edge of an ocean basin. |

Not all sedimentary basins form for the same reason (Fig. 5.28). *Foreland basins* form where the weight of a growing mountain belt presses down on the lithosphere, causing it to bend downward (just as the weight of a person causes the surface of a trampoline to bend downward). *Rift basins* form where lithosphere stretches and thins during rifting. *Intracontinental basins* form over very ancient, unsuccessful rifts that initiated in the interior of a continent and then continued to subside for a long time after rifting ceased. *Passive-margin basins*, the largest sedimentary basins, develop along the edges of continents after rifting has succeeded in breaking apart a supercontinent. Subsidence happens at passive-margin basins because the lithosphere heats up during rifting, so when rifting ceases, the remaining thinned lithosphere cools, thickens, and gets denser. As the resulting thicker, denser lithosphere subsides, underlying asthenosphere flows out of its way to make room. (To picture why, imagine two wooden boats of the same size and design, but built of different wood, floating in the sea. The one made of denser wood floats with its keel deeper in the water.) Notably, sediment continuously accumulates within sedimentary basins during basin subsidence, so water in the basins does not necessarily become deep, and sedimentary-basin strata can consist of shallow-marine, or even nonmarine, deposits. The sediment surface of a passive-margin basin, for example, is a shallow-marine continental shelf.

Take-home message...

Different types of sedimentary rocks form in different depositional settings. For example, strata deposited along a river differ from strata deposited by ocean waves or in the deep sea. Geologists study features such as sedimentary structures to figure out the depositional setting in which the sediments under study were deposited. Thick accumulations of sediment develop in sedimentary basins.

Quick Questions

- A layer consisting of evaporites and sand-dune deposits represents what type of depositional setting?
- Distinguish between transgression and regression.
- What is a sedimentary basin?

5.6 Causes and Consequences of Metamorphism

What Is a Metamorphic Rock?

After a caterpillar grows to full size, it hangs from a branch and becomes encased in a chrysalis. Inside, the protoplasm of the caterpillar reorganizes and a butterfly takes shape (Fig. 5.29a). Biologists refer to this amazing change as *metamorphosis*. In certain geologic settings, rocks undergo changes every bit as dramatic. Geologists use a similar word, **metamorphism**, for such changes and use the term *metamorphic rock* for their product (Fig. 5.29b). Specifically, a **metamorphic rock** is one that forms when minerals in a **protolith** (pre-existing rock) undergo reactions and disappear as new ones grow, and/or when the shape, size, and arrangement of mineral grains change. The protolith can be an igneous rock, sedimentary rock, or pre-existing metamorphic rock. Metamorphism happens when a protolith endures metamorphic conditions—namely, relatively high temperatures, pressures, or both, and/or the presence of hot-water solutions. Metamorphic changes take place in the solid state, meaning that metamorphic rock does not solidify from a melt. Further, metamorphism takes place under conditions that don't occur at the surface of the Earth, so metamorphic changes differ from those caused by weathering. Most metamorphic processes take a long time (thousands to millions of years).

If someone were to put a rock on a table in front of you, how would you know that it's metamorphic? First, metamorphic rocks may possess **metamorphic minerals**, new minerals that grow within solid rock under metamorphic conditions. In fact, metamorphism typically produces a *metamorphic mineral assemblage*, a group of minerals that can coexist under metamorphic conditions and were not all present in the protolith (see Fig. 5.29b). Second, metamorphic rocks display **metamorphic texture**, a relationship among grains that did not occur in the protolith. Typically, metamorphic textures are crystalline (composed of interlocking grains), regardless of whether the protolith was crystalline (Fig. 5.29c). Third, some (not all) metamorphic rocks display a **metamorphic fabric**, a directionality that develops due to the alignment of inequant grains (see Chapter 3) that did not occur in the protolith (Fig. 5.29d). Because of metamorphic minerals, textures, and fabrics, a metamorphic rock can look as different from its protolith as a butterfly does from a caterpillar.

How Does Metamorphism Take Place?

To understand metamorphism, we need to picture the interior of a mineral crystal. Under the conditions found at the Earth's surface, temperatures are low enough that atoms in the crystal vibrate only very slowly and remain bonded to one another. Under metamorphic conditions, however, various atomic-scale mechanisms (discussed in more advanced books) allow atoms to shift positions. For example, during **solid-state diffusion**, the bonds within a solid that hold an atom to its neighbors break, the atom moves slightly, and then it reattaches to new neighbors. When this process is repeated countless times at countless locations in a rock, significant overall changes can take place, such as the following:

- *Recrystallization:* Recrystallization changes the shape and size of mineral grains in a rock, as well as the way in which the grains are held together, without changing the identity of the minerals involved (Fig. 5.30a). The process, during which some crystals disappear while others grow, can transform a fine-grained clastic rock into a coarse-grained crystalline rock.

- *Metamorphic reactions:* Under metamorphic conditions, solid-state diffusion allows crystals of new minerals to grow by incorporating atoms released from pre-existing minerals of the protolith. Such metamorphic reactions can, for example, transform a rock composed of clay and quartz into a rock composed of mica, quartz, and garnet (Fig. 5.30b).

- *Phase change:* When one mineral transforms into another mineral that has the same chemical composition but a different crystal structure, a phase change has taken place. The transformation of graphite into its polymorph, diamond, is a phase change—both minerals consist entirely of carbon, but the arrangement of carbon atoms in diamond differs from that in graphite (see Chapter 3).

- *Pressure solution:* If a film of water that is bonded to the grain surfaces lies between the grains of a protolith, the grains will dissolve preferentially at the contact points where they are being squeezed against neighboring grains. Ions released at the contact points then migrate along the water film, and either precipitate on a part of the grain that

FIGURE 5.29 Dramatic transformations.

(a) A caterpillar undergoes metamorphosis to become a butterfly.

Before · After

(b) Shale undergoes metamorphism to become garnet schist. The metamorphic rock contains minerals that were not in the protolith

Fossiliferous limestone · Marble

(c) A sedimentary-rock protolith, fossiliferous limestone (left), consists of very small calcite fossils, tiny grains of calcite, and calcite cement. Marble (right), formed by metamorphism of limestone, has a crystalline texture.

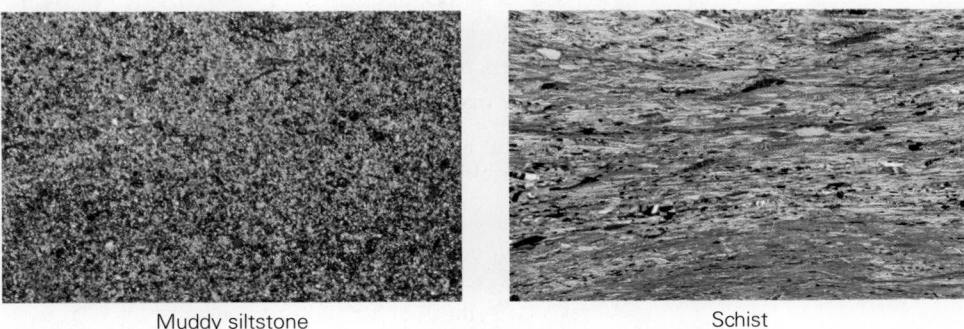

Muddy siltstone · Schist

(d) This schist (right), formed by metamorphism of muddy siltstone (left), contains large aligned crystals of mica.

is not being squeezed or dissolve in groundwater. This process, called *pressure solution*, modifies the shapes of grains but doesn't change their composition (Fig. 5.30c).

- *Plastic deformation:* Solid-state diffusion and other atomic-scale mechanisms can allow mineral grains

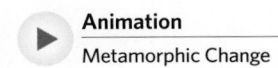

Animation
Metamorphic Change

FIGURE 5.30 Metamorphic processes as seen through a microscope.

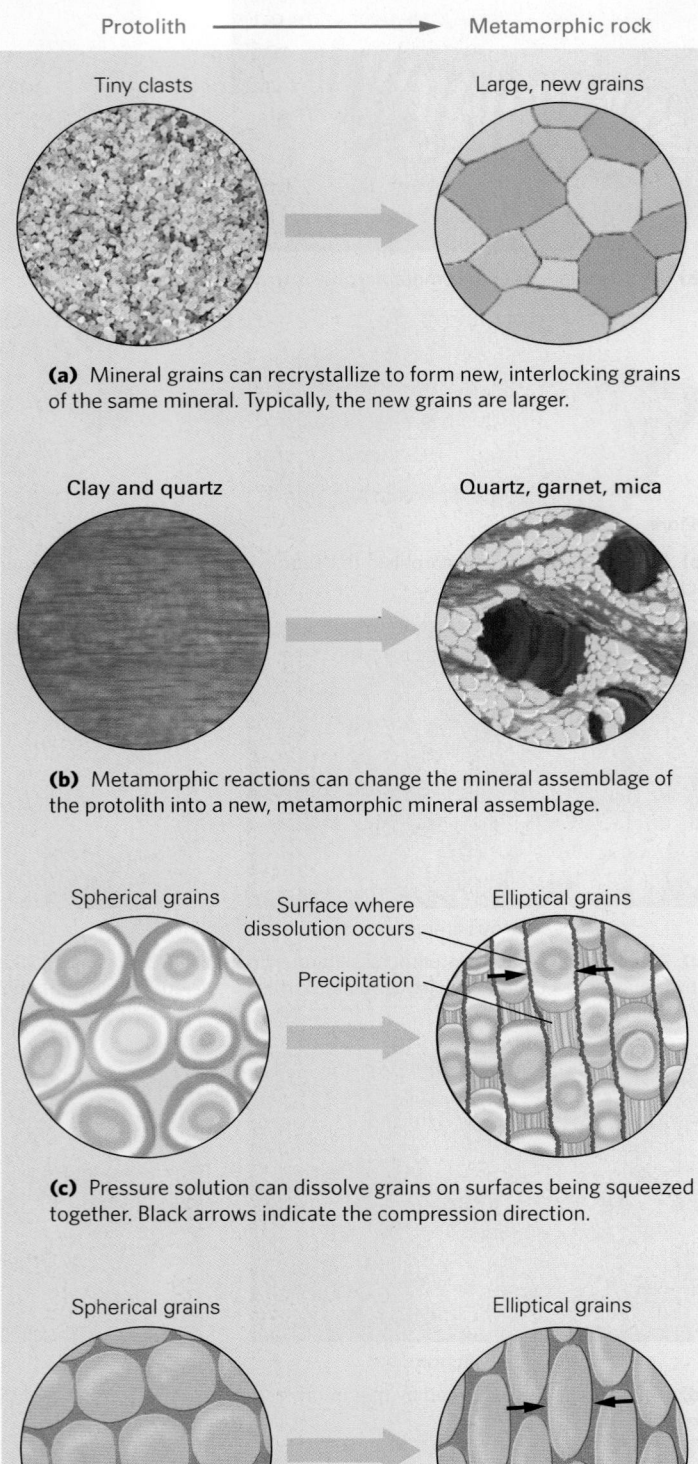

Protolith ⟶ Metamorphic rock

Tiny clasts   Large, new grains

(a) Mineral grains can recrystallize to form new, interlocking grains of the same mineral. Typically, the new grains are larger.

Clay and quartz   Quartz, garnet, mica

(b) Metamorphic reactions can change the mineral assemblage of the protolith into a new, metamorphic mineral assemblage.

Spherical grains   Surface where   Elliptical grains
                    dissolution occurs
                    Precipitation

(c) Pressure solution can dissolve grains on surfaces being squeezed together. Black arrows indicate the compression direction.

Spherical grains   Elliptical grains

(d) Plastic deformation can change the shapes of grains without breaking them. Black arrows indicate the compression direction.

to change shape very slowly, without changing composition, breaking, or dissolving. This process of *plastic deformation* can allow spherical grains to flatten into pancake shapes or stretch into cigar shapes (Fig. 5.30d).

Agents of Metamorphism

Metamorphic change cannot happen at the Earth's surface. Rather, it must be driven by one or more of the following *agents of metamorphism* that occur deep underground.

INCREASE IN TEMPERATURE. As a mineral grain warms, heat energy makes its atoms vibrate faster, so the chemical bonds locking atoms to their neighbors stretch and bend. If a bond stretches or bends too much, it breaks, allowing the atom that it once held to move. Increases in temperature, therefore, accelerate solid-state diffusion and other atomic-scale processes, and thus play an important role in driving metamorphism.

INCREASE IN PRESSURE. When you swim underwater, water squeezes against you equally from all sides, so your body feels *pressure*. The deeper you go, the greater the pressure you experience, because the weight of the overlying water increases. Pressure also increases with depth in the solid rock of the Earth's crust. The squeeze caused by increasing pressure drives atoms to fit together more tightly than they did in the protolith minerals, and this process results in the formation of a new crystal structure. The pressures at which metamorphism takes place are so great that geologists specify them in *kilobars* (1 kbar = 1,000 bars) or similar units.

INCREASE IN BOTH TEMPERATURE AND PRESSURE. Inside the Earth, at any location, pressure and temperature both increase with increasing depth (Fig. 5.31). Experiments show that the specific metamorphic reactions taking place in a rock depend on both temperature and pressure. In fact, certain minerals are stable, meaning that they can survive unchanged, only within a given range of pressure and temperature. If conditions change, these minerals become unstable, and their atoms are rearranged or released so that new, stable minerals can grow.

INTERACTION WITH HYDROTHERMAL FLUIDS. The presence of hot-water solutions, known as **hydrothermal fluids**, speeds up metamorphic reactions because atoms can diffuse faster through fluids than they can through a solid mineral grain. In addition, hydrothermal fluids passing through a rock may pick up some dissolved ions and drop others off, just as a bus picks up and drops off passengers. So, interaction with hydrothermal fluids can change the

FIGURE 5.31 The average change in pressure and temperature with depth in the Earth's crust. This chart assumes that temperature increases by 20°C/km. In reality, the rate of change, the geotherm (see Chapter 4), varies between 10°C/km (in old, cold crust) and 30°C/km; where igneous activity exists, it can be even greater.

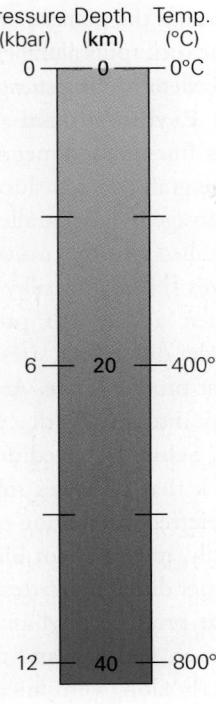

Pressure (kbar)	Depth (km)	Temp. (°C)
0	0	0°C
6	20	400°
12	40	800°

overall chemical composition of a rock during metamorphism. Geologists refer to such changes as **metasomatism**.

APPLICATION OF DIFFERENTIAL STRESS. We noted above that *pressure* refers to a condition in which the push acting on an object is the same in all directions (Fig. 5.32a). You feel pressure if you are standing in still air or floating under nonflowing water. However, in a solid or a moving fluid, a push in one direction can be greater than a push in another direction. You can picture why if you put a ball of dough on the floor and step on it—the ball flattens into a pancake, oriented parallel to the floor, because the downward push you apply with your foot exceeds the push provided by air in other directions (Fig. 5.32b). And if you press the ball of dough between a wall and a vertical book, the dough flattens into a pancake shape parallel to the wall, emphasizing that the orientation of flattening depends on the direction of the greatest push. If you grab each end of the dough ball and pull, it will become longer and flatter. And, if you place the dough on a table, set your hand on top of it, and move your hand parallel to the table, the ball will become longer and flatter. Scientists refer to the pushing or squeezing of an object as **compression**, and to the pulling or stretching of an object as **tension**. By smearing the dough ball across a table, you're applying **shear**. Compression, tension, and shear are all examples of *stress*, the force applied per unit area at a location. Simplistically, therefore: stress = force ÷ area (see Chapter 7). Geologists use the term **differential stress** to distinguish

conditions in which the push or pull in one direction differs from that in another, or in which shearing is taking place.

Differential stress can affect the orientation and shape of grains under metamorphic conditions and, therefore, can produce metamorphic fabrics. Specifically, it can transform equant grains, which have roughly the same dimensions in all directions, into inequant grains, which have different dimensions in different directions. In addition, it can cause inequant grains to lie parallel to one another. For example, during plastic deformation, the application of differential stress can change spherical quartz grains into *platy* (pancake-shaped) (Fig. 5.33a) or *elongate* (cigar-shaped) grains, and can cause these inequant grains to align with one another (Fig. 5.33b). Similarly, it can cause new inequant crystals of metamorphic minerals to grow in a particular direction and align with one another. Specifically, platy crystals tend to grow faster in the direction perpendicular to the direction of greatest compression, and elongate grains tend to lie parallel to the direction of shear. When platy and/or elongate grains in a rock become aligned, we say that they display a **preferred orientation** (Fig. 5.33c).

Take-home message...

A metamorphic rock contains mineral assemblages and textures that differ from those found in its protolith. Many processes, such as recrystallization, metamorphic reaction, phase change, pressure solution, and plastic deformation, can take place during metamorphism. These processes happen in response to changes in temperature and pressure, interaction with hydrothermal fluids, or the application of differential stress.

Quick Questions

- What is the difference between pressure and differential stress?
- Contrast metamorphic reactions with recrystallization.
- What process can change the overall composition of rock during metamorphism?

FIGURE 5.32 The types of stress affecting metamorphic rock can be demonstrated with a ball of dough.

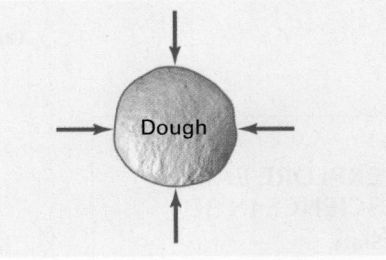

(a) An object subjected to pressure experiences the same stress from all directions.

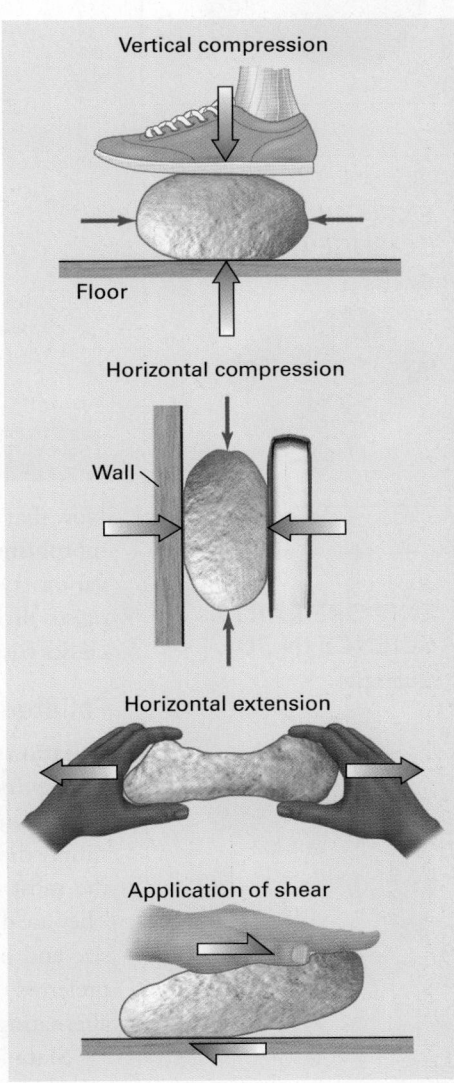

Vertical compression

Floor

Horizontal compression

Wall

Horizontal extension

Application of shear

(b) An object subjected to differential stress experiences different stresses in different directions.

FIGURE 5.33 Effects of differential stress on grain shapes in metamorphic rocks.

Parallel platy grains

(a) Platy minerals oriented parallel to one another.

Parallel elongated grains

(b) Elongate minerals oriented parallel to one another.

EXPLORE EARTH SCIENCE IN 3D

Slate

Compression

Direction of preferred orientation

(c) A photomicrograph showing preferred orientation of quartz and mica crystals.

EXPLORE EARTH SCIENCE IN 3D

Gneiss

5.7 Types of Metamorphic Rocks

Now that we've introduced the processes involved in metamorphism and the agents that cause it, let's examine the various types of rocks formed by metamorphism. Geologists divide metamorphic rocks into two fundamental classes based on whether or not the rocks contain foliation.

Foliated Metamorphic Rocks

Foliation (from the Latin *folium*, meaning leaf) refers to any type of planar fabric (layering) that forms during metamorphism. **Foliated metamorphic rocks**, therefore, are simply those that contain a planar fabric. Here, we describe the most common types of foliated metamorphic rocks. They are distinguished from one another by the rock's grain size and by whether the planes of foliation represent the preferred orientation of inequant grains or the existence of alternating layers of different metamorphic minerals.

Slate is a very fine-grained rock consisting of aligned microscopic clay flakes. It forms from a shale or mudstone that undergoes metamorphism at relatively low pressures and temperatures in the presence of differential stress. During such metamorphism, tiny clay flakes rotate or grow to become parallel to one another in a direction perpendicular to that of greatest compression. Geologists generally refer to the fabric in slate as *slaty cleavage*

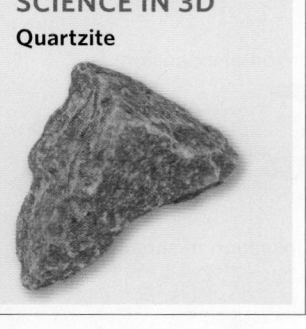

EXPLORE EARTH SCIENCE IN 3D

Quartzite

Did you ever wonder...

why slate makes such nice roofing shingles?

because the planes parallel to clay flakes tend to be weak. The rock splits along cleavage planes into sheets that make excellent roofing shingles (Fig. 5.34).

Phyllite (from the Greek word *phyllon*, meaning leaf) is a fine-grained metamorphic rock consisting of aligned fine-grained, translucent mica. This rock splits along cleavage planes parallel to the mica flakes, so its foliation is called *phyllitic cleavage*. The translucent mica of phyllite gives the rock a silky luster (Fig. 5.35a). Phyllite forms when a clay-rich protolith undergoes metamorphism under differential stress at temperatures higher than those that produce slate. At these temperatures, clay recrystallizes into mica with a preferred orientation.

Schist is a medium- to coarse-grained metamorphic rock that possesses *schistosity*, a foliation defined by the preferred orientation of relatively large mica crystals (typically, muscovite or biotite) (Fig. 5.35b). This rock forms under differential stress at temperatures higher than those that produce phyllite. Other minerals—such as quartz, garnet, feldspar, and amphibole—also typically grow in schist, along with mica.

Gneiss is a metamorphic rock in which alternating layers of dark-colored and light-colored minerals define foliation (Fig. 5.36a, b). Such *compositional layering*, or *gneissic banding*, can form in several ways. In some cases, it develops when a protolith containing different rock types undergoes extreme shearing at high temperatures and pressures. Such banding may be inherited from layering in the protolith. But it may also form when plastic deformation smears out bodies of different rock types into aligned sheets (Fig. 5.36c), or when metamorphic differentiation takes place. During *metamorphic differentiation*, chemical reactions cause felsic minerals and mafic minerals to concentrate in different layers (Fig. 5.36d).

Nonfoliated Metamorphic Rocks

In **nonfoliated metamorphic rocks**, metamorphic minerals are either equant or, if inequant, have a random orientation. Geologists distinguish among nonfoliated metamorphic rocks by their composition. **Hornfels**, for example, is a fine- to medium-grained nonfoliated rock containing a variety of different minerals—the minerals present depend on the protolith's composition. Inequant grains in hornfels do not have a preferred orientation because the rock grows without being subjected to differential stress (Fig. 5.37a). **Quartzite**, in contrast, is a fine- to coarse-grained metamorphic rock composed almost entirely of one mineral—quartz. It forms from pure quartz sandstone, under a wide range of temperature and pressure conditions, and can be nonfoliated if it develops in the absence of differential stress. Metamorphism destroys the distinction between cement and grains, so quartzite has a crystalline texture and is held together by interlocking

FIGURE 5.34 Slate is a foliated metamorphic rock that forms from shale or mudstone at relatively low temperatures and pressures.

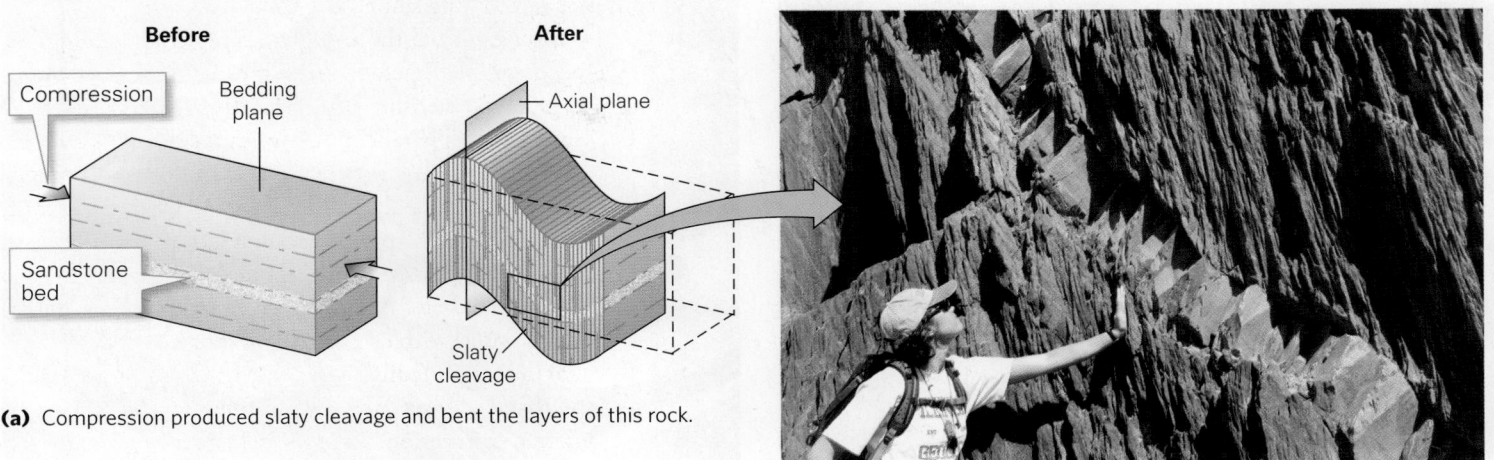

(a) Compression produced slaty cleavage and bent the layers of this rock.

(b) An outcrop of slate (black rock) containing a tan sandstone layer.

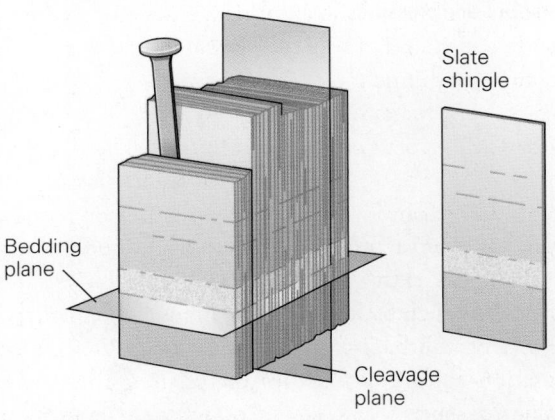

(c) A block of slate splits easily along cleavage planes.

(d) A worker splits slate to produce shingles for the roof of a home.

FIGURE 5.35 Examples of foliated metamorphic rocks.

(a) During the formation of this phyllite, clay recrystallized to form tiny mica flakes that reflect light, giving the rock a silky sheen. Where it formed, in Brazil, compression wrinkled the foliation.

(b) This schist contains coarse mica flakes along with other metamorphic minerals. Compare the grain size with that in phyllite.

FIGURE 5.36 The nature of gneiss.

(a) This block of gneiss, from the Appalachian Mountains, displays compositional layering.

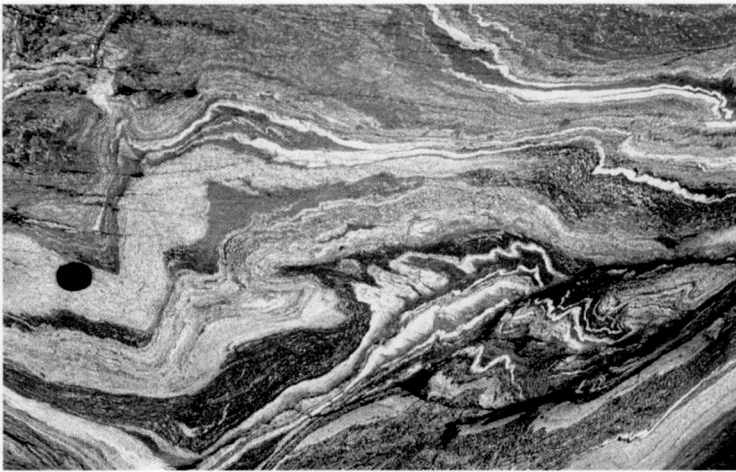

(b) An outcrop of 2.7-billion-year-old gneiss in Ontario, Canada, displays contorted banding due to plastic deformation of the rock when it was at metamorphic temperatures and pressures.

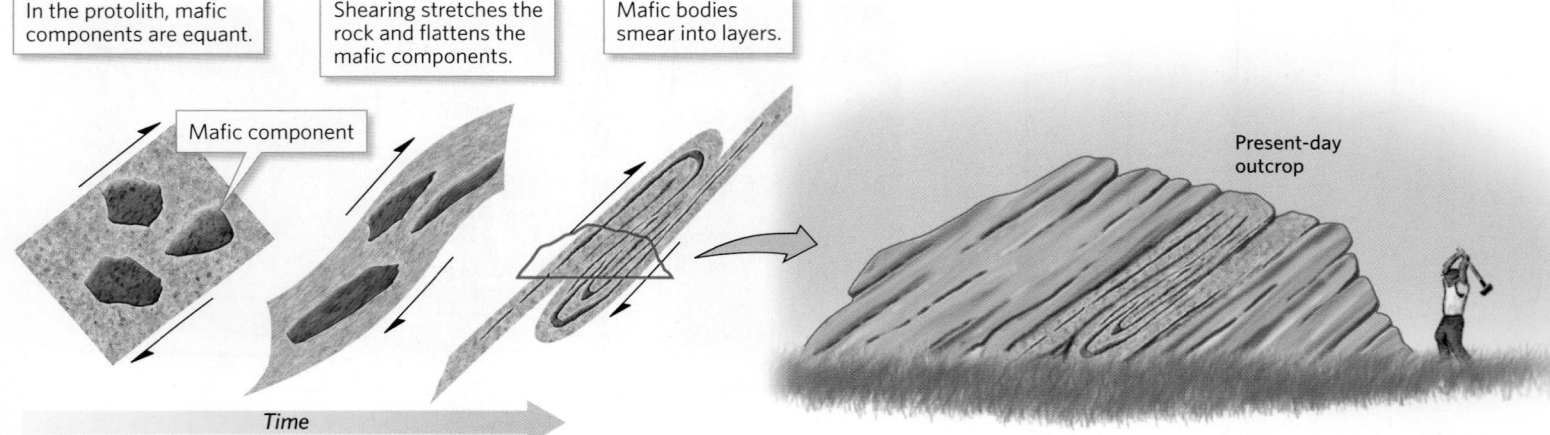

In the protolith, mafic components are equant.

Shearing stretches the rock and flattens the mafic components.

Mafic bodies smear into layers.

Mafic component

Present-day outcrop

Time

(c) Formation of gneiss, in some cases, involves extreme shearing. In this example, bodies of mafic rock are embedded in a felsic rock. Due to shearing, the original contrasting rock types are smeared into parallel layers.

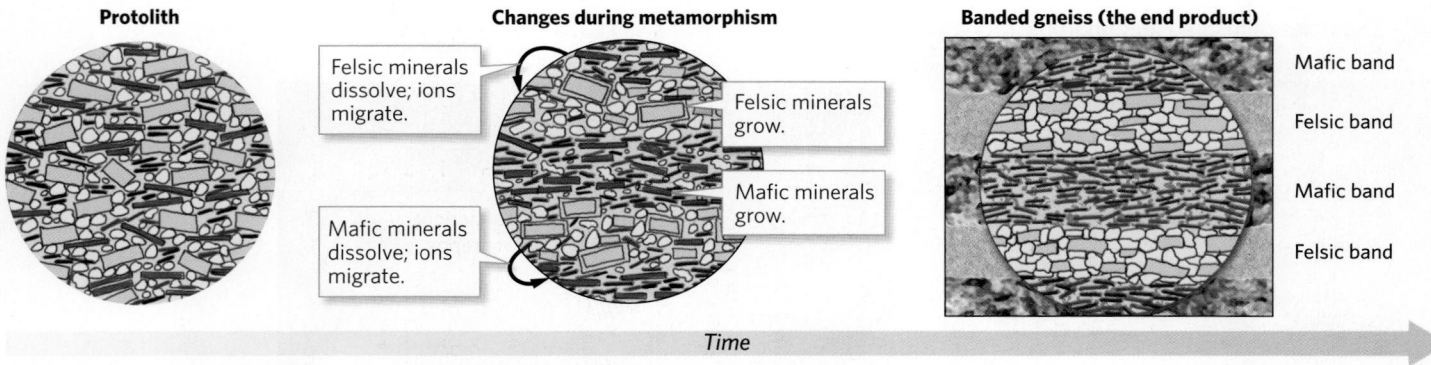

Protolith

Changes during metamorphism

Banded gneiss (the end product)

Felsic minerals dissolve; ions migrate.

Felsic minerals grow.

Mafic minerals grow.

Mafic minerals dissolve; ions migrate.

Mafic band

Felsic band

Mafic band

Felsic band

Time

(d) Gneiss may also form by metamorphic differentiation.

quartz crystals (Fig. 5.37b). **Marble**, like quartzite, is a metamorphic rock composed almost entirely of one mineral. But in contrast to quartzite, marble consists of interlocking calcite crystals (Box 5.3). It forms by metamorphic recrystallization of limestone under a range of temperature and pressure conditions. As in quartzite, the distinction between grains and cement disappears, and a crystalline texture develops (Fig. 5.37c). Marble that forms in the absence of differential stress is nonfoliated. In marble that develops in the presence of differential stress, impurities in

FIGURE 5.37 Examples of nonfoliated metamorphic rocks.

Elongate metamorphic crystal

Cross section of a crystal

Fine-grained matrix

What an Earth Scientist Sees

EXPLORE EARTH SCIENCE IN 3D
Marble

(a) An example of hornfels. The sketch emphasizes that its inequant grains have random orientations.

Quartz sandstone—the protolith of quartzite

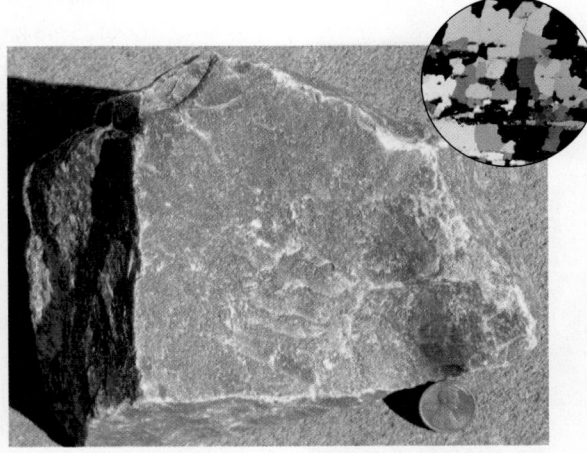

(b) In this sandstone (left), the sand grains stand out. In contrast, this nonfoliated maroon quartzite (right) looks glassy. Note the contrast in texture revealed by the photomicrographs.

(c) In this limestone from New York (left), you can see bedding as well as fossils and shell fragments. Such distinctions are not visible in this white marble (right) exposed in an Italian quarry. Note the texture contrast in the photomicrographs.

(d) The marble tiles in this floor display color banding.

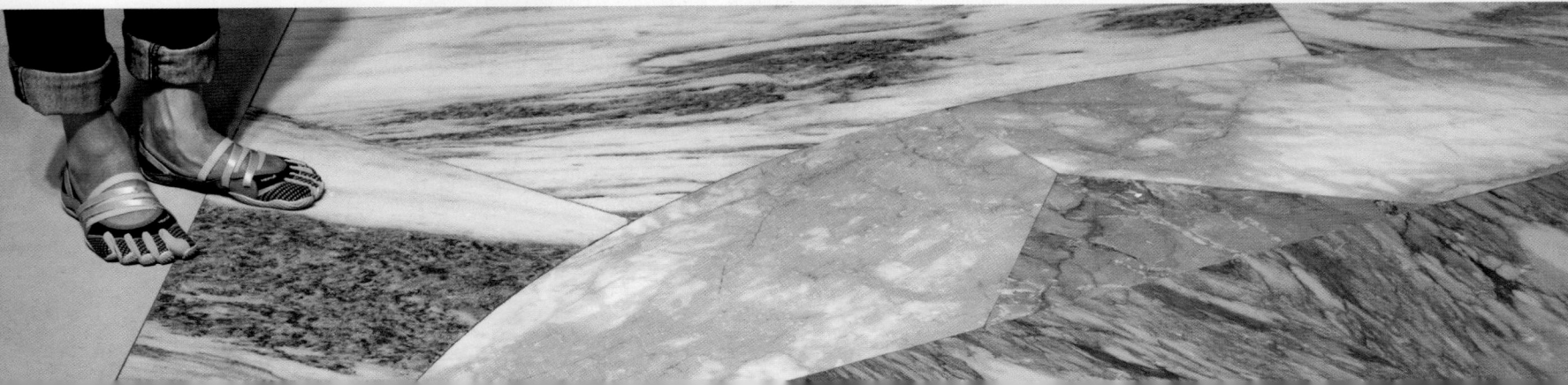

BOX 5.3 ▶ **Putting Earth Science to Use**

The sculptures of Michelangelo

Michelangelo (1475–1564), one of the most famous Italian artists of the Renaissance, once said, "Every block of stone has a statue inside it, and it is the task of the sculptor to discover it." Michelangelo's fame as a sculptor came from the drama and realism of his creations in marble **(Fig. Bx5.3a)**. He did not use just any blocks of marble for his works.

Rather, he worked exclusively with marble from the Carrara quarries, which perch at the crest of the Apuan Alps, a 1.6 km (1 mi) high mountain range in northwestern Italy. The quarries, some of which date back to the Roman Empire, expose cliffs of nearly pure white stone **(Fig. Bx5.3b)**.

How did Carrara marble form? Its story began over 200 million years ago, when the region that is now northern Italy hosted shallow seas in which marine organisms with calcite shells lived and died. The shells were buried and lithified to form limestone. During mountain building, about 25 million years ago, the limestone ended up at depths of 15–18 km (9–11 mi) below the Earth's surface, where it underwent metamorphism at a temperature of around 400°C (750°F), and recrystallized. The lack of minerals other than calcite in the protolith meant that the resulting marble was very pure. Continued mountain building eventually resulted in exhumation of the marble. Exhumation began about 20 million years ago and continues today.

For over two millennia, people have sweated and struggled to break chunks of this marble out of bedrock **(Fig. Bx5.3c)**. In Michelangelo's day, workers wedged blocks of rock free from cliffs by pounding a row of chisels into the rock to produce a crack. Beginning in the 19th century, wire saws became popular tools. These saws consist of an abrasive wire that revolves around two pulleys. Movement of the wire slowly grinds a slot into the rocks.

Why did Michelangelo choose Carrara marble to sculpt? The rock, because of its purity, has a uniform color. In addition, its texture of interlocking crystals makes it resistant to shattering during carving, and the softness of calcite (3 on the Mohs hardness scale) allows marble to be carved and polished relatively easily. Light can penetrate slightly into Carrara marble, so the surfaces of sculptures made from it are slightly translucent, a quality that gives the sculptures a waxy glow like that of human skin.

FIGURE Bx5.3 Michelangelo's marble.

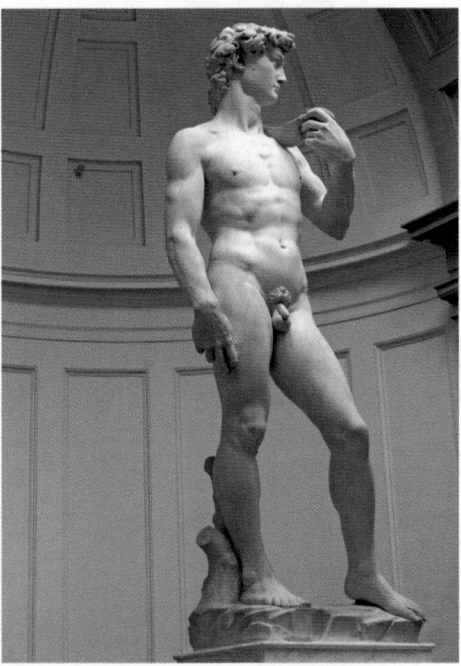

(a) Michelangelo's famous sculpture *David* was carved from Carrara marble.

(b) The marble for *David* came from quarries in the Apuan Alps of northwestern Italy.

(c) An example of a marble quarry sliced into a mountainside.

the marble smear out into beautiful color banding due to shear (Fig. 5.37d).

Metamorphic Grade

Not all metamorphism takes place under the same conditions of temperature and pressure. For example, rock at a depth of 40 km (24 mi) beneath a mountain range undergoes metamorphism at a higher temperature and pressure than does rock at a depth of 20 km (12 mi). Geologists use the concept of **metamorphic grade** in an informal way to characterize the intensity of metamorphism. Grades are distinguished from one another primarily by the temperature of metamorphism: low-grade metamorphic rocks form at 200°C–400°C (392°F–750°F), intermediate-grade metamorphic rocks at 400°C–600°C (750°F–1,100°F), and high-grade metamorphic rocks at 600°C–850°C (1,100°F–1,560°F) (Fig. 5.38a). Slate and phyllite are low-grade metamorphic rocks, most schists and

FIGURE 5.38 Intensity of metamorphism is indicated by metamorphic grade.

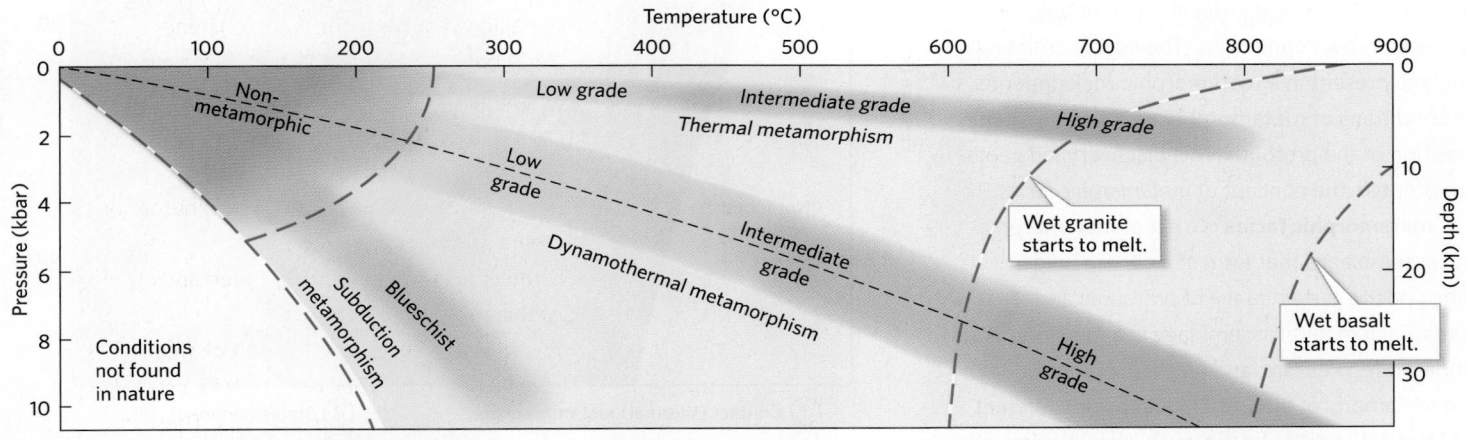

(a) This graph depicts the approximate temperatures and pressures associated with metamorphic grades. Different metamorphic conditions occur in different geologic settings. *Wet* means that the protolith contains water and therefore melts at a lower temperature than does dry (water-free) rock.

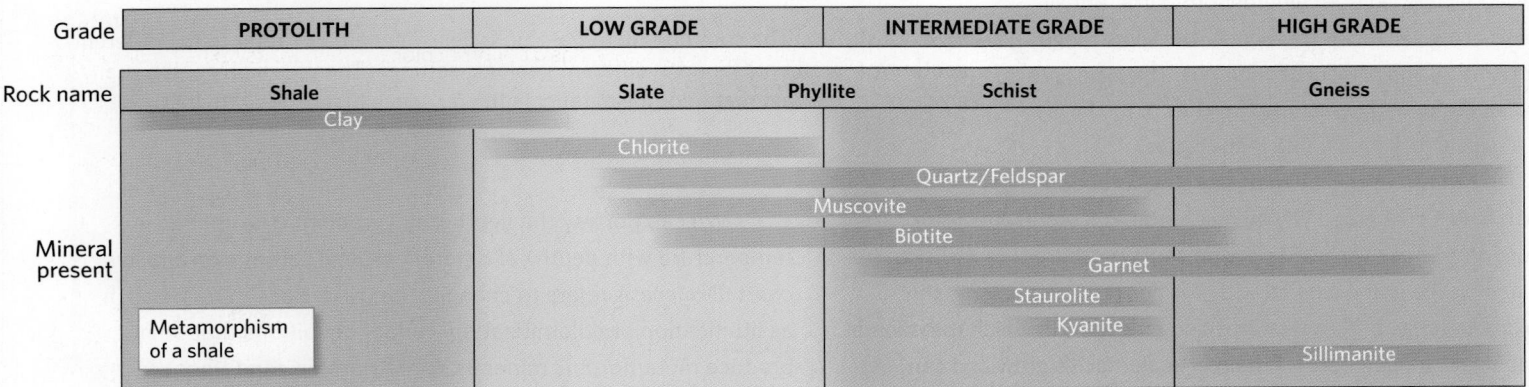

(b) The metamorphic minerals that form in a rock depend on metamorphic grade and protolith composition. This chart lists the metamorphic minerals that form from a shale protolith at different metamorphic grades.

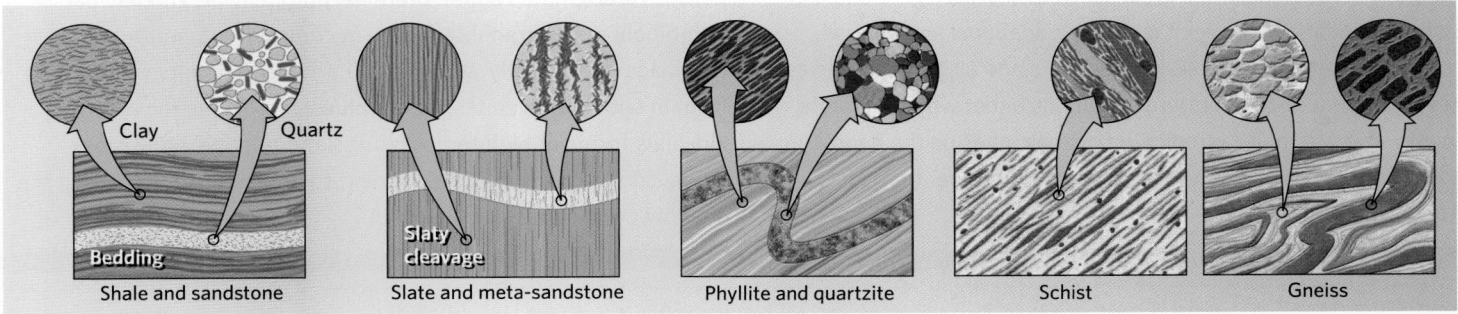

(c) Here we see the consequences of the progressive metamorphism of shale and sandstone during mountain building. The background colors indicate the range of grades at which key metamorphic minerals form. Note how rock textures change with increasing grade.

BOX 5.4 | A Deeper Look

Metamorphic facies

In the early 20th century, geologists realized that the mineral assemblage in a metamorphic rock approximately represents a condition of *chemical equilibrium* that existed in the rock at the time of metamorphism. Simplistically, this means that metamorphic reactions reorganize the chemicals making up the rock into a set of minerals that can coexist and can remain stable under metamorphic conditions. The specific mineral assemblage present in a metamorphic rock depends on the conditions of metamorphism as well as on the composition of the protolith. This discovery led geologists to propose the concept of *metamorphic facies*.

A **metamorphic facies** is a set of metamorphic mineral assemblages that form at a certain grade, meaning under a certain range of pressures and temperatures. Each mineral assemblage in a facies depends on temperature, pressure, and protolith composition. A given metamorphic facies includes several different kinds of rocks that differ from one another in terms of chemical composition, and therefore mineral content, because they formed from different protoliths. But all the rocks of a given facies form under the same range of temperature and pressure conditions. The facies are named either for a mineral found in rocks of the facies or for a typical rock type found in the facies.

We can represent the metamorphic conditions for each facies on a graph whose horizontal axis represents temperature and whose vertical axis represents pressure (Fig. Bx5.4). Each area on the graph labeled with a facies name represents the approximate range of temperatures and pressures in which the mineral assemblages characteristic of that particular facies grow and can remain stable. For example, rock subjected to the pressure and temperature at Point A (4.5 kbar and 400°C, or 750°F) develops a mineral assemblage characteristic of the greenschist facies. As Figure Bx5.4 implies, the boundaries between facies are gradational and approximate. Note that some amphibolite-facies rocks and all granulite-facies rocks form at pressure-temperature conditions under which a granite containing water will melt (as indicated by the curves on the figure). Therefore, metamorphic rocks of these very high-temperature facies can form only from dry rocks.

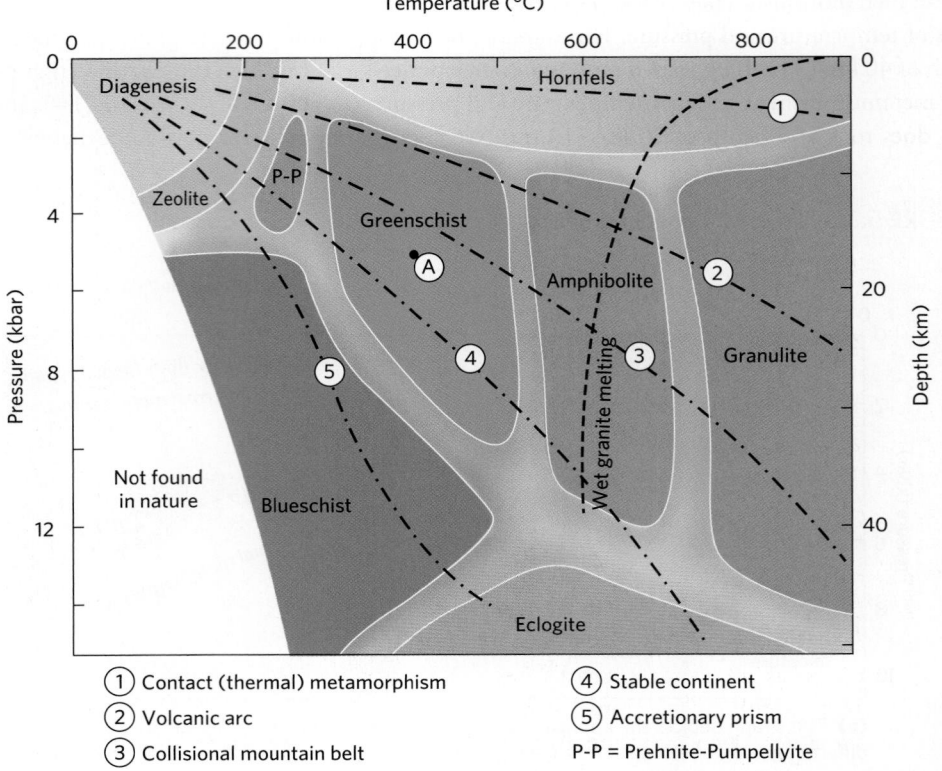

FIGURE Bx5.4 The common metamorphic facies. The numbers refer to the different geothermal gradients found in different geologic settings. The very high geothermal gradient for contact metamorphism reflects the presence of igneous activity.

We can portray the geothermal gradient (the change in temperature with depth) of different crustal settings on a facies chart. *Diagenesis* refers to changes that take place in rock—such as lithification or dolomitization—under conditions that do not produce metamorphic minerals. Metamorphic conditions start at temperatures of 200°C–250°C (400°F–482°F). Beneath collisional mountain belts and volcanic arcs, the geothermal gradient passes through the zeolite, prehnite-pumpellyite, greenschist, amphibolite, and granulite facies. In contrast, temperature increases more slowly with depth in the accretionary prisms that form in subduction zones, where blueschist-facies rocks form. Hornfels forms at high temperatures and relatively low pressures, conditions that can exist next to an igneous intrusion in the shallow crust.

some gneisses are intermediate-grade rocks, and some schists and most gneisses are high-grade rocks. Because different minerals are stable under different conditions, most metamorphic rock types formed at different grades contain different minerals (Fig. 5.38b, c). The exceptions, quartzite and marble, contain only quartz and calcite, respectively. However, not all rocks of a given grade contain the same assemblage of minerals, because the minerals that can form during metamorphism depend on protolith composition. The set of mineral assemblages that forms at a given grade is called a *metamorphic facies* (Box 5.4).

A protolith that starts at the Earth's surface and eventually ends up 40 km beneath a mountain belt doesn't transform directly from the protolith to a high-grade metamorphic rock. Rather, it passes progressively through successively higher grades: first it becomes a low-grade rock, then an intermediate-grade rock, and finally a high-grade rock (see Fig. 5.38c). Geologists refer to this overall transition as *prograde metamorphism*. Prograde metamorphic reactions generally release volatiles, such as water, from rock. Can a high-grade rock become progressively lower grade as it cools? Sometimes. In order for minerals characteristic of low-grade metamorphism to form in a high-grade rock carried upward to lower-grade conditions—a process called *retrograde metamorphism*—water must be added back by circulating hydrothermal fluids. Added water not only replaces water lost during prograde metamorphism, but it also accelerates metamorphic reactions, which under dry conditions would occur too slowly to change the rock at low temperatures. Significantly, high-grade rocks can remain unchanged near the Earth's surface for billions of years if they do not react with hydrothermal fluids.

By considering the temperature range at which metamorphic minerals grow, geologists estimate that most metamorphic conditions exist at depths greater than 10–15 km (6–9 mi) in the continental crust, depending on geothermal gradient. The greatest depth at which metamorphism takes place is the depth at which rocks start to melt. This depth depends on the rock's mineral composition and water content. For a dry granite, melting occurs at depths of 40–65 km (25–40 mi). Notably, a felsic melt can start to form in a rock subjected to high-grade metamorphism while the mafic components of the rock remain solid. If the new melt solidifies before migrating

FIGURE 5.39 This road cut in Ontario contains a nearly vertical layer of migmatite. The light bands consist of felsic igneous rock, and the dark bands are mafic metamorphic rock.

Light (felsic) bands formed from melt.

Dark (mafic) bands remained solid.

Nothing in the world lasts, save eternal change.

—*HONORAT DE BUEIL* (FRENCH POET, *1589–1650*)

away, a rock known as a *migmatite*—a mixture of igneous and metamorphic rock—forms (Fig. 5.39).

Take-home message...

Geologists divide metamorphic rocks into classes based on whether the rock contains foliation. Foliated rocks include slate, schist, and gneiss. Nonfoliated rocks include marble and quartzite. The type of metamorphic rock that forms depends on the conditions of metamorphism.

Quick Questions ——————

• What is the difference between schist and gneiss?

• What is the difference between quartzite and marble?

• How do geologists describe the intensity of metamorphism at a location?

5.8 Where Does Metamorphism Occur?

Thermal or Contact Metamorphism

Imagine a place where hot magma intrudes into cooler wall rock. Heat flows from the magma into the wall rock because heat always flows from hotter to colder

FIGURE 5.40 Geologic settings of metamorphism.

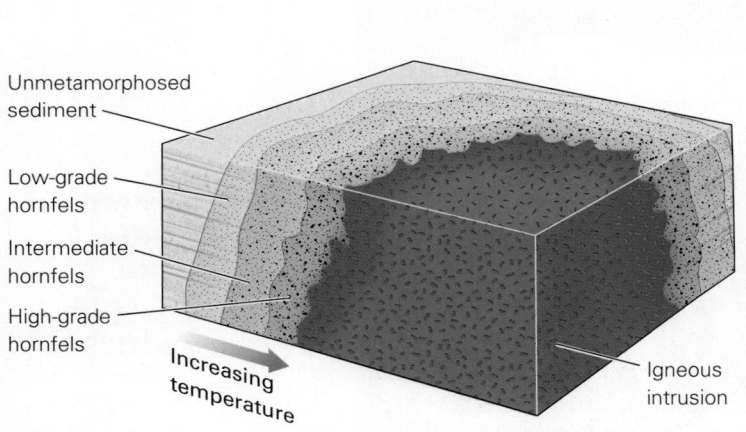

Unmetamorphosed sediment

Low-grade hornfels

Intermediate hornfels

High-grade hornfels

Increasing temperature

Igneous intrusion

(a) Thermal metamorphism occurs when heat from a large intrusion of magma produces a metamorphic aureole, in which hornfels develops.

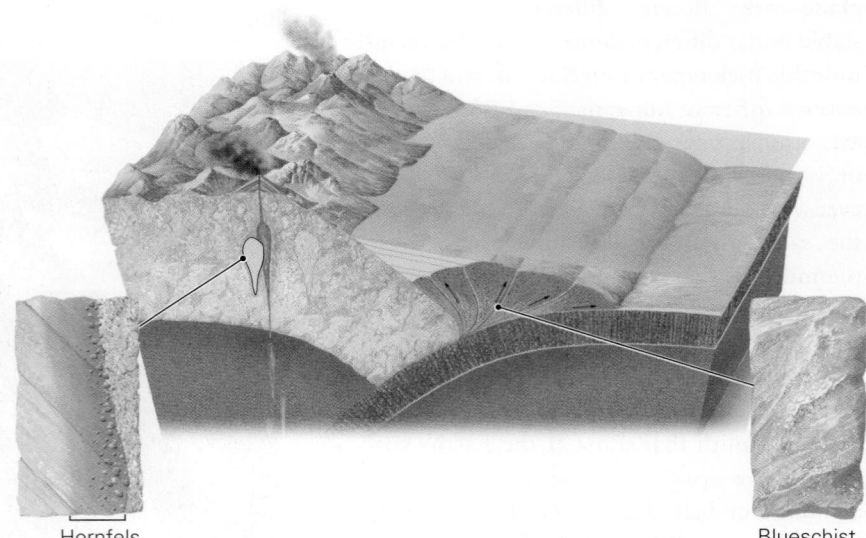

Hornfels

Blueschist

(b) Blueschist forms at the base of an accretionary prism at a subduction zone.

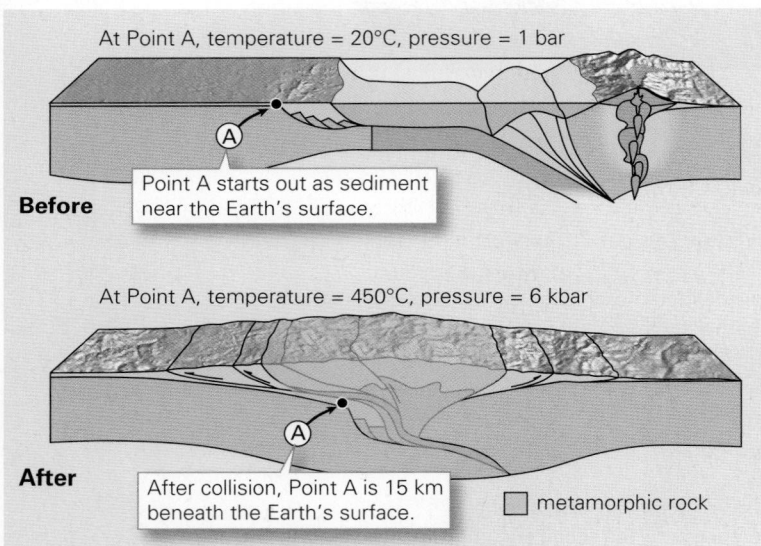

At Point A, temperature = 20°C, pressure = 1 bar

A

Point A starts out as sediment near the Earth's surface.

Before

At Point A, temperature = 450°C, pressure = 6 kbar

A

After

After collision, Point A is 15 km beneath the Earth's surface.

☐ metamorphic rock

(c) Dynamothermal metamorphism occurs during mountain building. The process is also called regional metamorphism.

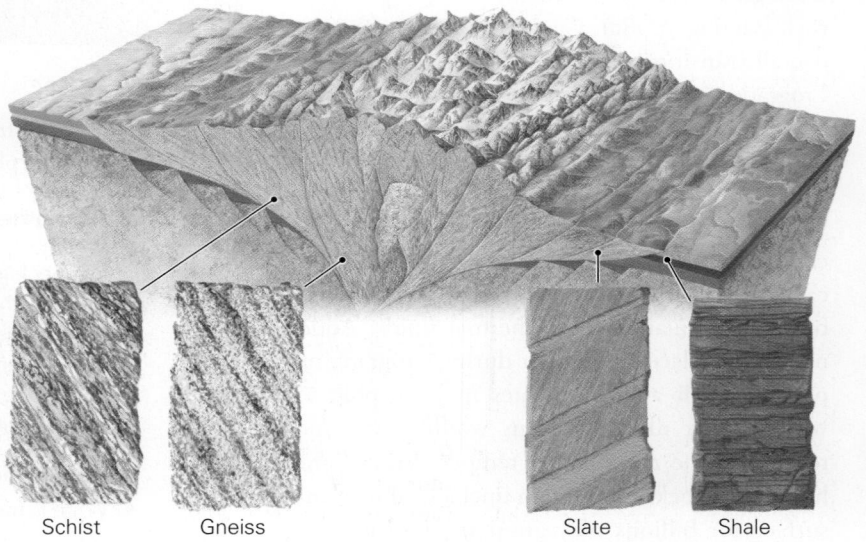

Schist

Gneiss

Slate

Shale

(d) A variety of foliated metamorphic rocks form in a collisional mountain belt.

materials. Consequently, the magma cools and solidifies while the wall rock heats up. As this happens, hydrothermal fluids may be circulating through both the intrusion and the wall rock. Heat and hydrothermal circulation cause the wall rock to undergo metamorphism. The highest-grade metamorphic rock forms immediately adjacent to the intrusion, where the temperature is highest, and progressively lower-grade rock forms farther away. The band of metamorphic rock that forms around an igneous intrusion is called a **metamorphic aureole** (from the Latin *aureola*, meaning crown or halo) **(Fig. 5.40a, b)**. Aureole width depends on the temperature, size, and shape of the intrusion, as well as on the amount of hydrothermal circulation that takes place.

We can refer to the local metamorphism caused by heat from an igneous intrusion either as **thermal metamorphism**, to emphasize that it develops in response to heat without a change in pressure and without differential stress, or as *contact metamorphism*, to emphasize that it develops adjacent to the contact between an intrusion and its wall rock. Because this type of metamorphism takes place without differential stress, no preferred orientation develops, and the metamorphic aureole contains hornfels. Such metamorphism develops adjacent to plutons such as those intruded beneath continental volcanic arcs. It can form at depths of less than 10 km.

Dynamothermal (Regional) Metamorphism

During continental collision, large slices of continental crust slip along faults and move up and over other portions of the crust. Consequently, a protolith that was once near the Earth's surface along the margin of a continent can end up at great depth beneath a mountain range. In this environment, the protolith heats up (because temperature increases with depth and because igneous activity happens nearby), it endures greater pressure (because the weight of the overburden increases with depth), and it is subjected to differential stress (because of crustal movements during the collision) (Fig. 5.40c, d). The protolith, therefore, transforms into foliated metamorphic rock. Which type of metamorphic rock it becomes depends on the depth at which metamorphism takes place—specifically, slate forms at shallower depths, while schist and gneiss form at greater depths. Because the metamorphism we've just described involves changes due to elevated temperature and differential stress, geologists refer to it as **dynamothermal metamorphism** (from the Greek *dynamis*, meaning power). Typically, such metamorphism affects a large region, such as the entire length and breadth of a mountain belt, so it's also known as *regional metamorphism*.

A special type of dynamothermal metamorphism occurs in accretionary prisms that form along subduction zones. Because the downgoing plate is relatively cold, rocks at the base of the prism endure high pressures and shearing, but at relatively low temperatures. These conditions produce *blueschist*, a foliated rock whose color comes from the presence of a bluish amphibole that forms only at high pressures and relatively low temperatures (see Fig. 5.40b).

Where Do We Find Metamorphic Rocks Today?

When you stand on an outcrop of metamorphic rock, you are standing on rock that once lay many kilometers beneath the surface of the Earth. The rocks you see have ended up back at the Earth's surface because of exhumation. Where can you see exhumed metamorphic rocks today? You can start your quest by hiking into a mountain range, because the processes involved in mountain building can lead to significant amounts of uplift and erosion (Fig. 5.41a, b). You can find even more metamorphic rocks by walking across a **craton**, which consists of Precambrian crust that has not been affected by mountain building for at least a billion years. The Precambrian rocks of cratons were metamorphosed during the succession of collisions involving island arcs, oceanic plateaus, and continental fragments that resulted in formation of continental crust. Cratons include **shields**, where Precambrian rocks are exposed over a broad region, and *cratonic platforms*, where the Precambrian basement has been covered by Paleozoic and/or younger strata (Fig. 5.41c). A shield occupies much of the area of northern Canada, so metamorphic rocks are exposed extensively there (Fig. 5.41d). In the United States, similar Precambrian rocks form the basement on which a cover of younger strata was deposited. To see these Precambrian rocks, you must look into canyons that have cut down deeply through the cover (Fig. 5.41e), or find places where faulting has brought the rock up to the surface.

Take-home message...

Thermal metamorphism develops around an igneous intrusion due to heat from the intrusion. Dynamothermal metamorphism develops beneath mountain belts, where rock undergoes compression and shear at high temperatures and pressures. Most exposures of metamorphic rocks occur in mountain ranges and shields.

Quick Questions
- Why are we able to see metamorphic rocks at the Earth's surface today?
- What is a shield?

See for yourself

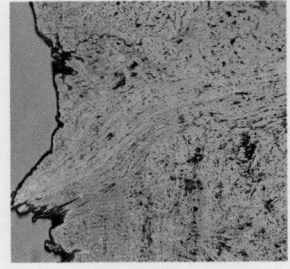

Canadian Shield

Latitude: 61°20′29.27″ N
Longitude: 75°47′23.60″ W

Look down from an altitude of 400 km (250 mi).

This region of Canada has sparse vegetation. The foliation of the metamorphic rocks stands out even in this satellite photo.

See for yourself

Wind River Mountains, Wyoming

Latitude: 43°6′7.22″ N
Longitude: 109°21′36.45″ W

Look down obliquely from an altitude of 20 km (12.5 mi).

This rock, the basement of western Wyoming, underwent high-grade metamorphism over 2 billion years ago. Faulting uplifted it during mountain building 40–80 million years ago, and overlying cover strata were eroded away. Remaining cover strata are still visible on the east side of the range.

FIGURE 5.41 Exposures of metamorphic rock.

(a) Coastal outcrops in Maine expose metamorphic rocks produced during Paleozoic collisional mountain building.

(b) The cliffs of the Teton Range in Wyoming expose gneiss formed in the Precambrian and brought back to the Earth's surface by faulting during the Cenozoic.

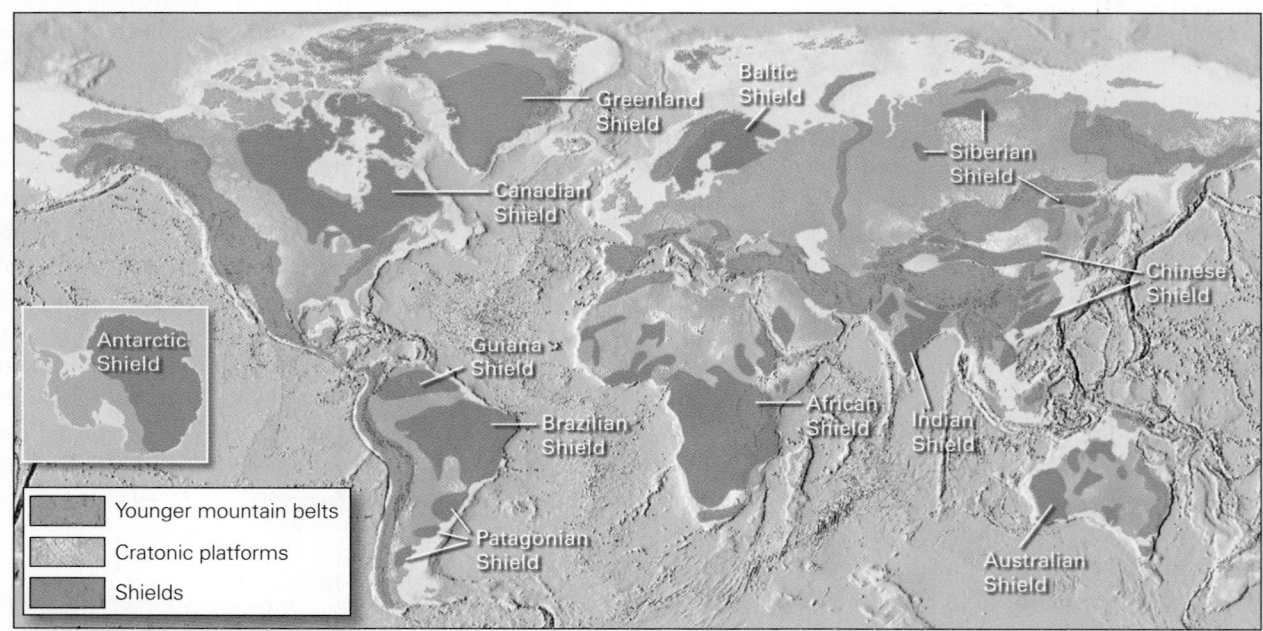

(c) A map showing the global distribution of shields, areas where broad expanses of Precambrian crust, including Precambrian metamorphic rocks, crop out.

(d) An aerial photograph of the flat landscape of the eastern Canadian Shield. All the exposed basement visible in this view consists of metamorphic rock.

(e) The Gunnison River has carved a deep canyon into the Precambrian basement in Colorado, exposing metamorphic rock.

Objective 5.1

> Describe the process of weathering and sediment production, and explain how soil forms.

KEY CONCEPTS

- Rocks at or near the Earth's surface undergo weathering and, as a result, they eventually disintegrate into loose clasts and ions in solution. These are the raw materials from which sedimentary rock forms.

- Physical weathering breaks rock into fragments. The process begins with joint formation, followed by frost and root wedging, which pushes joints open and separates blocks from bedrock. Blocks may tumble downslope and break up.

- Chemical weathering involves reactions of minerals with air and water. These reactions can dissolve grains, transform minerals into new ones, or "rust" iron-bearing minerals.

- The interaction of rain and organic material with sediment, over time, can produce soil.

- Clastic sediment consists of grains or fragments, biochemical sediment consists of shells made by organisms, chemical sediment comes from precipitation of minerals, and organic sediment comes from buried organic matter.

EARTH-SCIENCE VOCABULARY

chemical weathering (p. 166)	**sediment** (p. 166)
clast (p. 166)	**soil** (p. 170)
clay (p. 167)	**soil erosion** (p. 170)
exhumation (p. 166)	**soil horizon** (p. 170)
joint (p. 166)	**weathering** (p. 166)
physical weathering (p. 166)	

REVIEW QUESTIONS

1. **(a)** If you see a large block of rock break off an outcrop, are you seeing chemical or physical weathering in action? **(b)** Why does weathering of a cliff face sometimes produce a sawtooth profile?

2. **(a)** Provide examples of chemical reactions that can be involved in weathering. **(b)** What happens to feldspar and quartz during chemical weathering? **(c)** The joints

in **Figure A** have become narrow troughs with rounded edges. Why?

3. Hieroglyphics on a rock obelisk remained sharp for 2,000 years while the obelisk stood in the desert. In a matter of decades after the obelisk was moved to a city, the hieroglyphics were too blurred to read. Why?

4. **(a)** Describe the process of soil formation and explain what soil horizons are. **(b)** Why does the type of soil that develops at an area depend on the climate? **(c)** How do human activities affect soil?

5. **(a)** How does clastic sediment differ from biochemical sediment? **(b)** How does chemical sediment differ from organic sediment? **(c)** How does siliciclastic sediment differ from carbonate sediment?

Objective 5.2

> Name examples of clastic sedimentary rocks, and outline the processes that produce them.

KEY CONCEPTS

- Clastic sedimentary rocks form when solid fragments or grains, produced by weathering and erosion, undergo transportation, deposition, burial, and finally, lithification (compaction, consolidation, and cementation). The cement minerals precipitate from groundwater.

- Descriptions of sedimentary rocks typically include characterization of clast composition, size, sorting, and angularity. Such description provides information about sediment source and maturity.

- Clastic rocks are classified initially by grain size, but composition and grain shape also are taken into consideration. Common examples include shale, siltstone, sandstone, arkose, conglomerate, and breccia.

EARTH-SCIENCE VOCABULARY

cementation (p. 172)	**lithification** (p. 172)
clastic sedimentary rock (p. 172)	**sandstone** (p. 174)
conglomerate (p. 174)	**shale** (p. 174)
deposition (p. 172)	**sorting** (p. 174)
erosion (p. 172)	**transportation** (p. 172)

6. **(a)** Describe each step leading to the production of a clastic sedimentary rock from fresh bedrock. **(b)** Name three media that transport clasts. **(c)** What processes contribute to lithification? **(d)** On **Figure B**, label the clasts and the cement.

B

7. **(a)** How do grain size, sorting, and angularity change during the transportation and weathering of sediment? **(b)** In **Figure B**, are the grains well sorted or poorly sorted?

8. Why does deposition of clasts take place?

9. **(a)** Shale forms from the lithification of what type of sediment? **(b)** How does arkose differ from sandstone? **(c)** Which type of rock does **Figure C** show?

C

Objective 5.3

Provide examples of biochemical, organic, and chemical sedimentary rocks and discuss how each can form.

KEY CONCEPTS

- Biochemical limestone forms from calcium carbonate shells. Biochemical chert forms from silica shells of plankton.
- Organic sedimentary rocks consist of organic remains of organisms, such as plant debris.
- Chemical sedimentary rocks precipitate directly from water solutions. This group includes evaporites, made of salts.

EARTH-SCIENCE VOCABULARY

biochemical sedimentary rock (p. 176)
chemical sedimentary rock (p. 178)
chert (p. 177)
coal (p. 178)

dolostone (p. 180)
evaporite (p. 178)
limestone (p. 176)
organic sedimentary rock (p. 178)

REVIEW QUESTIONS

10. **(a)** Explain the processes involved in forming biochemical sedimentary rocks. **(b)** How do these processes differ from those that form chemical sedimentary rocks? **(c)** What does coal form from?

11. **(a)** Provide examples of different types of limestone, and indicate which are biochemical and which are chemical. **(b)** What is dolostone, and how does it form? **(c)** Name the rock shown in **Figure D**.

D

12. **(a)** What is chert, and in what two ways can it form? **(b)** Which type of sediment does **Figure E** show? **(c)** Where and why do evaporite deposits form?

E

Objective 5.4

Characterize sedimentary structures, and discuss how they develop.

KEY CONCEPTS

- Sedimentary structures form during or shortly after sediment deposition.
- Bedding can develop for a variety of reasons, including a change in sediment source, a change in the depositional environment, an interruption in deposition, or compaction and consolidation to cause alignment of clay flakes.
- Some sedimentary structures, such as mudcracks, develop on bedding planes just after the bedding plane forms.
- Ripple marks, dunes, and cross bedding develop in sediment deposited by a current. Graded bedding forms from sediment settling out of a turbidity current.

EARTH-SCIENCE VOCABULARY

bed (p. 180)
cross bed (p. 182)
dune (p. 180)
graded bedding (p. 183)
mudcrack (p. 180)

ripple mark (p. 180)
sedimentary structure (p. 180)
strata (p. 180)
turbidity current (p. 183)

REVIEW QUESTIONS

13. **(a)** Explain how bedding develops in sedimentary strata. **(b)** What visual features can you use to recognize bedding?

14. **(a)** What does the presence of ripple marks indicate? **(b)** What sedimentary structure does **Figure F** show, and how did it form?

F

15. **(a)** How do cross beds form? **(b)** In **Figure G**, show the direction of the current that produced the cross beds.

G

16. Explain how sediment containing graded bedding develops, and give an example of a setting where such sediment might accumulate.

Objective 5.5

Associate characteristics of sedimentary strata and sedimentary structures with specific depositional settings, and discuss phenomena that determine where sediment accumulates.

KEY CONCEPTS

- The character of sedimentary strata reflects the environment at the site of deposition. Geologists distinguish between two general categories of depositional settings: marine and nonmarine.

- Glaciers, streams, alluvial fans, deserts, rivers, lakes, deltas, coastal areas, and the deep seafloor represent different depositional settings. Each setting accumulates distinctive types of sedimentary strata.

- Depositional settings change over time due to movement of continents, climate change, and relative sea-level change. Transgressions occur when relative sea level rises, and regressions occur when relative sea level falls.

- Particularly thick deposits of sedimentary strata accumulate in sedimentary basins.

EARTH-SCIENCE VOCABULARY

depositional setting (p. 184)
marine deposition (p. 184)
nonmarine deposition (p. 184)
regression (p. 186)
relative sea level (p. 186)
sedimentary basin (p. 187)
subsidence (p. 187)
transgression (p. 186)

REVIEW QUESTIONS

17. **(a)** Distinguish between nonmarine and marine depositional settings, and provide examples of each. **(b)** Which does **Figure H** show?

18. **(a)** What clues can you examine to identify the depositional setting that a bed of sediment accumulated in? **(b)** How might you distinguish between a sandstone formed from dune deposits and a turbidite?

19. Distinguish between transgression and regression, and explain why these phenomena happen.

20. **Figure I** shows the top three layers in the Grand Canyon. Layer X consists of well-sorted sandstone containing large cross beds; Layer Y consists of gypsum-bearing shale interlayered with thin beds of marine sandstone; and Layer Z consists of gray limestone containing marine fossils. Does this succession of strata record a transgression or a regression? Explain your reasoning. Keep in mind that Layer X is the youngest of the three layers.

21. **(a)** What process must happen in order for a very thick succession of sedimentary strata to accumulate in a sedimentary basin? **(b)** Name geologic settings in which sedimentary basins can form.

Objective 5.6

Define metamorphism, characterize key features of metamorphic rocks, and describe processes that happen during metamorphism.

KEY CONCEPTS

- Metamorphism refers to change that takes place in rock deep in the crust to produce new minerals or textures and fabrics in the rock without melting or weathering. The product of this change is a metamorphic rock.

- The protolith from which a metamorphic rock forms can be igneous, sedimentary, or metamorphic.

- Metamorphism involves recrystallization, phase changes, metamorphic reactions, pressure solution, and/or plastic deformation. Some of these processes involve solid-state diffusion.

- Metamorphism is driven by various agents, including increases in temperature and pressure, reactions with hydrothermal fluids, and application of differential stress.

- If hydrothermal fluids deposit or remove ions so that the chemical composition of a rock changes during metamorphism, metasomatism has occurred.

- A preferred orientation of grains develops in metamorphic rock where differential stress causes inequant grains to become aligned.

EARTH-SCIENCE VOCABULARY

compression (p. 193)
differential stress (p. 193)
hydrothermal fluid (p. 192)
metamorphic fabric (p. 191)
metamorphic mineral (p. 190)
metamorphic rock (p. 190)
metamorphic texture (p. 190)
metamorphism (p. 190)
metasomatism (p. 193)
preferred orientation (p. 193)
protolith (p. 190)
shear (p. 193)
solid-state diffusion (p. 191)
tension (p. 193)

22. **(a)** How does metamorphism differ from weathering? **(b)** How do metamorphic rocks differ from igneous and sedimentary rocks?

23. **(a)** What is a protolith? **(b)** Do metamorphic rocks necessarily contain the same minerals that are present in a protolith? Explain your answer. **(c)** How does recrystallization differ from a metamorphic reaction? **(d)** Why is solid-state diffusion important during metamorphism?

24. **(a)** Describe the various agents of metamorphism. **(b)** Why is differential stress important during metamorphism? **(c)** What is the relationship between differential stress and preferred orientation? Draw the preferred orientation on the thin section in **Figure J**.

J

Objective 5.7

Classify and name metamorphic rocks, and distinguish among different metamorphic grades.

KEY CONCEPTS

- Geologists separate metamorphic rocks into two classes. Foliated rocks, which have a planar fabric, include slate, phyllite, schist, and gneiss. Nonfoliated rocks include hornfels, quartzite, and marble.

- Metamorphic conditions vary with location in the crust. Metamorphic rocks formed under relatively low temperatures are low-grade rocks, whereas those formed under high temperatures are high-grade rocks. Intermediate-grade rocks develop between these two extremes. Different minerals form at different metamorphic grades.

- A metamorphic facies is a set of metamorphic mineral assemblages that develop under a specified range of temperature and pressure conditions.

EARTH-SCIENCE VOCABULARY

foliated metamorphic rock (p. 194)
foliation (p. 194)
gneiss (p. 194)
hornfels (p. 194)
marble (p. 196)
metamorphic facies (p. 200)

metamorphic grade (p. 199)
nonfoliated metamorphic rock (p. 194)
phyllite (p. 194)
quartzite (p. 194)
schist (p. 194)
slate (p. 194)

25. **(a)** How does slate differ from phyllite? **(b)** How does phyllite differ from schist? **(c)** Which of these rock types does **Figure K** show? Highlight the orientation of the foliation by drawing a line parallel to it. **(d)** How does the compositional layering in gneiss form?

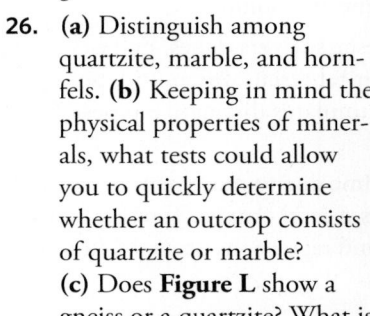

K

26. **(a)** Distinguish among quartzite, marble, and hornfels. **(b)** Keeping in mind the physical properties of minerals, what tests could allow you to quickly determine whether an outcrop consists of quartzite or marble? **(c)** Does **Figure L** show a gneiss or a quartzite? What is the basis for your answer?

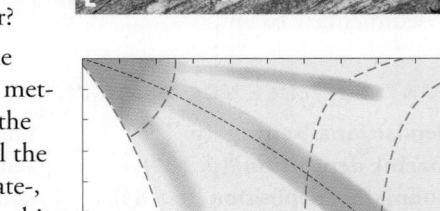

L

27. **(a)** What is meant by the metamorphic grade of a metamorphic rock? **(b)** On the graph in **Figure M**, label the areas of low-, intermediate-, and high-grade metamorphic conditions. **(c)** Would metamorphism of a basalt and a quartz sandstone produce the same mineral assemblage under the pressure and temperature conditions in which greenschist-facies rocks develop?

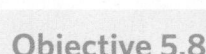

M

Objective 5.8

Characterize geologic phenomena that cause metamorphism, and identify subsurface environments where metamorphism takes place.

KEY CONCEPTS

- Thermal (contact) metamorphism occurs in a metamorphic aureole surrounding an igneous intrusion. Dynamothermal (regional) metamorphism results when rocks undergo heating, compression, and shearing during mountain building.

- We find belts of metamorphic rocks exposed in mountain ranges, shields (which expose broad areas of Precambrian basement), and deep canyons that cut through cover strata.

EARTH-SCIENCE VOCABULARY

craton (p. 203)
dynamothermal metamorphism (p. 203)
metamorphic aureole (p. 202)
shield (p. 203)
thermal metamorphism (p. 203)

REVIEW QUESTIONS

28. Describe the geologic settings where thermal and dynamothermal metamorphism take place.

29. On **Figure N**, label the metamorphic aureole, and label the highest-grade metamorphic rock in the aureole.

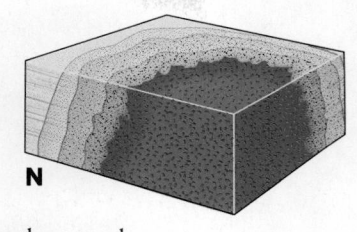

N

30. **(a)** Where would you go if you wanted to find exposed metamorphic rocks? **(b)** How did such rocks get to the surface of the Earth after undergoing metamorphism at depth in the crust? **(c)** Do you think that you would be likely to find a broad region (hundreds of kilometers across) in which outcrops consist of high-grade hornfels? Explain your answer.

31. **(a)** Could you find a layer of metamorphic rock between the layers of sedimentary rock in a sedimentary basin? Explain your answer. **(b)** The geothermal gradient of Mars is 8°C/km (23°F/mi), and the crust of Mars is 30 km (19 mi) thick. Can high-grade metamorphic rocks form in this crust?

ANOTHER VIEW Visitors to toadstool hoodoos, in the Grand Staircase-Escalante National Monument, Utah, will see some rather unusual rock formations. What phenomenon produced the tall, somewhat mushroom-shaped column in the photo? After reading this chapter, you can interpret the rocks and landscape in this photo.

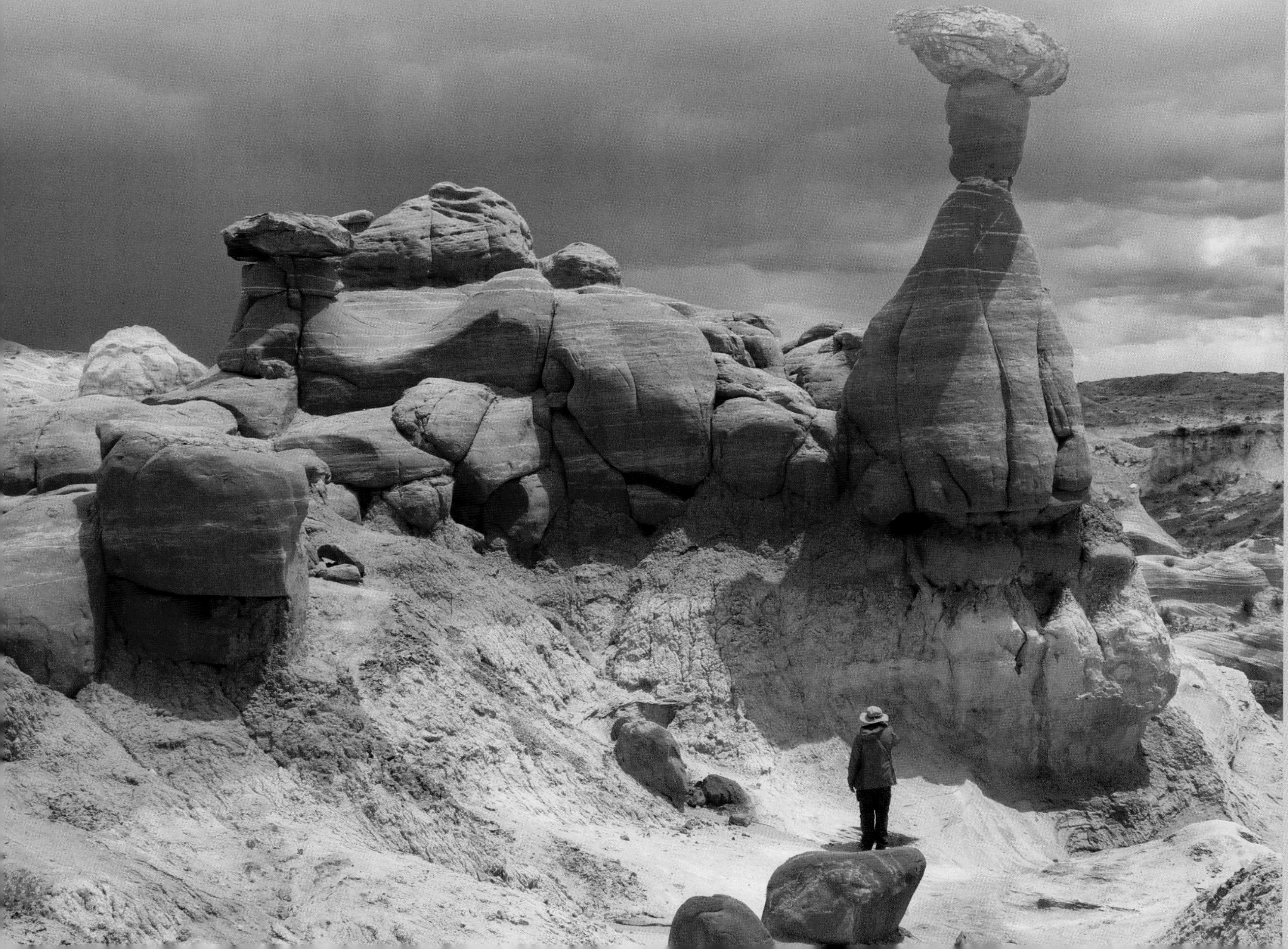

6 CRACKS, CRUMPLES, AND CRAGS
Geologic Structures and Mountain Building

After studying this chapter, you should be able to...

1. distinguish between force and stress, explain the relationship between geologic structures and deformation, and characterize different types of stress and the strain that they cause.

2. identify and describe geologic structures formed by brittle deformation, such as joints and faults, and distinguish among different types of each.

3. characterize geologic structures (folds and foliation) formed by ductile deformation, distinguish among different types, and discuss their development.

4. explain why mountain building takes place, in the context of plate tectonics.

5. relate the formation of certain rock types, high elevations, and rugged topography to mountain building.

6. distinguish between a craton and a mountain belt, and recognize basins and domes in cratons.

Mt. Everest's summit reaches an elevation 8.85 km (29,029 ft) above sea level, almost the cruising height of jet planes. In 1953, Edmund Hillary and Tenzing Norgay became the first people to reach that summit. Since then, thousands of other people have also succeeded . . . but over 310 have died trying. Mountains challenge climbers, but they lure nonclimbers as well, for the stark cliffs, clear air, flower-dotted meadows, roaring streams, and glistening glaciers of their rugged landscapes provide an escape from the mundane.

Apart from volcanic edifices formed over hot spots, mountains do not occur in isolation. Rather, they align in elongate ranges of elevated land called *mountain belts*. For a mountain belt to form, mountain-building processes must uplift thousands of cubic kilometers of rock skyward, against the pull of gravity! Mountain building, which takes millions to tens of millions of years, not only produces high topography, but also causes rock to bend, break, or flow. Such *deformation* yields distinct features—joints, faults, folds, and foliation—known as *geologic structures*. Presently, the Earth hosts about a dozen major mountain belts and numerous smaller ones (Fig. 6.1), each containing fascinating geologic structures. In the geologic past, many other mountain belts developed on our planet. But mountain belts don't last forever—as soon as land rises, erosion attacks, and once a mountain-building episode ceases, erosion eventually bevels away the peaks. Erosion exposes deformed rocks that once lay at depth beneath the range, for once formed, geologic structures survive unless they are eroded. Geologic structures influence the distribution and character of rock assemblages, which in turn control later topography as well as the distribution of valuable resources. Furthermore, geologic structures provide clues that characterize plate interactions happening today as well as in the past, so their characteristics and ages define important events in Earth history.

In this chapter, we first learn about the processes that deform rocks, and the geologic structures produced by deformation. Next, we address the question of why mountain belts develop, in the context of plate tectonics, and relate different kinds of geologic structures to different kinds of mountain belts. We also see how mountain building can produce distinctive rock types and generate topography. Finally, we distinguish between the structure of mountain belts and that of broad, relatively stable regions of continents known as *cratons*.

<< This cliff along the southwestern coast of England displays once-horizontal layers of sedimentary rock that were bent into dramatic folds during a mountain-building episode millions of years ago.

6.1 Rock Deformation

What Is a Tectonic Geologic Structure?

If you were to drive past a road cut through bedrock in the plains of the midwestern United States, you would see nearly horizontal beds of sandstone, shale, and limestone, cut here and there by natural cracks called *joints* (Fig. 6.2a). These beds have the same orientation that they had when first deposited. Microscopic study would reveal that grains in the sandstone retain the nearly spherical shape that they had when deposited, and that clay flakes in the shale lie roughly parallel to the bedding. As we will see, such joints form when removal of overlying rock by erosion causes the remaining rock to expand slightly.

If instead, you were to stand at the base of a cliff in the Alps of Switzerland, the rocks you would see would look quite different. The cliff illustrated in Figure 6.2b exposes layers of quartzite, slate, and marble (the metamorphic equivalents of sandstone, shale, and limestone) that have been contorted into wave-like shapes called *folds*, and the slate contains slaty cleavage. Note that the quartzite and slate layers terminate abruptly at a *fault*, a fracture on which slip has occurred, below which bedrock consists of folded marble. Folds, cleavage, faults, and joints develop when rock undergoes *deformation*. The resulting features are **tectonic geologic structures** if they develop in geologic materials, subsequent to the formation of those materials, during *tectonic deformation*, meaning in association with mountain building or related processes. The adjective *tectonic* distinguishes the structures discussed in this chapter from those we discussed in Chapters 4 and 5—such as bedding in sedimentary rock or the breakup of a lava flow—which result from the process of rock formation, or from structures formed by expansion following erosion. Clearly, the character of geologic structures varies with location, as illustrated by the contrast between the two locations depicted in Figure 6.2: contemporary mountain belts and remnants of ancient mountain belts contain a greater variety of geologic structures than do rocks that were never incorporated in a mountain belt.

Stress, Deformation, and Strain

If you visit a national park in a mountainous area, you may find an exhibit stating something like, "These mountains, and the geologic structures within them, were produced by forces deep within the Earth." What does this sentence mean? Isaac Newton defined the word *force* by the equation force = mass × acceleration, which means that applying a force to an object can cause the velocity of the object to change. The actual consequences of applying a force, however, depend

FIGURE 6.1 This digital map of world topography shows the locations of major mountain belts.

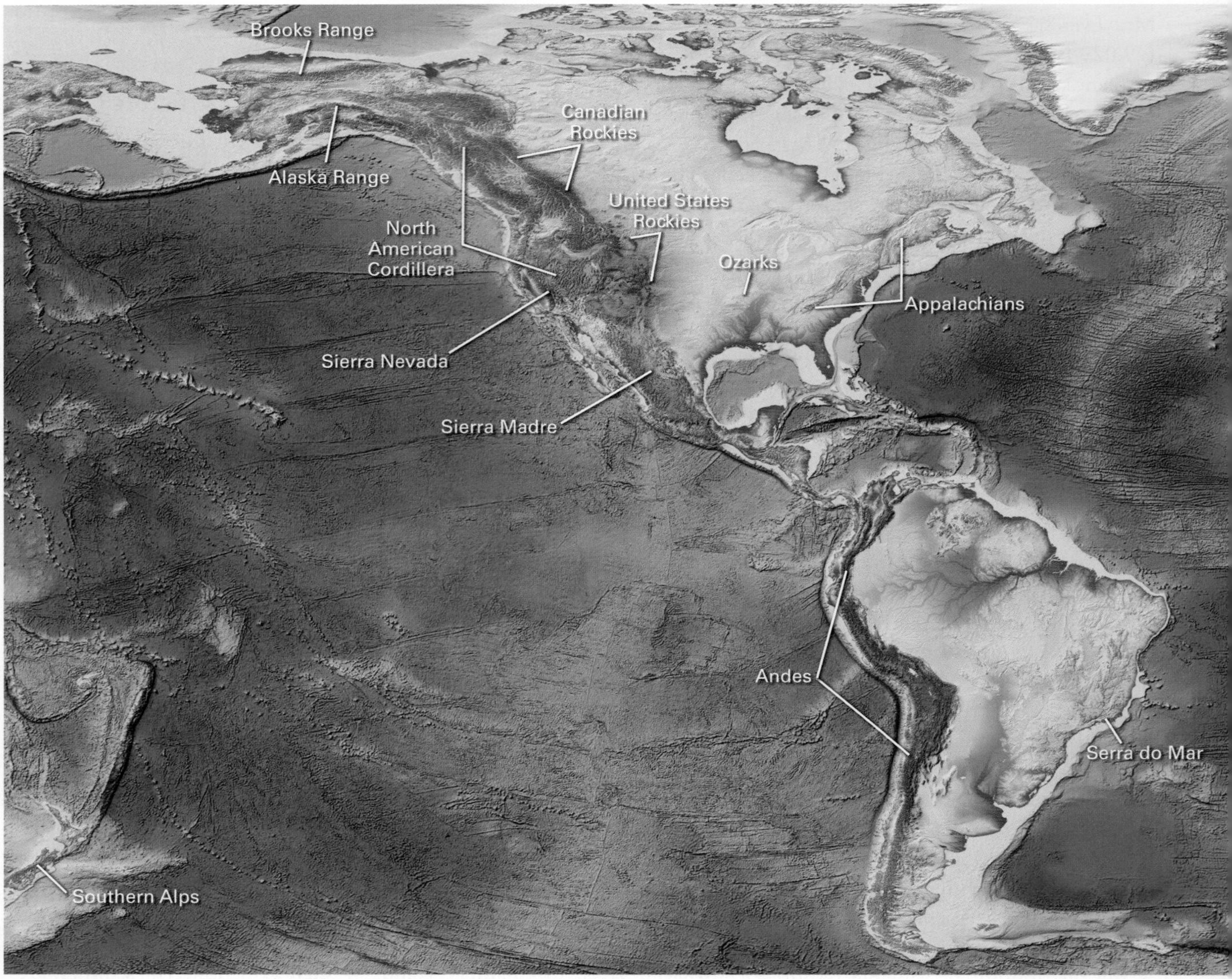

Brooks Range

Canadian Rockies

Alaska Range

North American Cordillera

United States Rockies

Ozarks

Appalachians

Sierra Nevada

Sierra Madre

Andes

Serra do Mar

Southern Alps

FIGURE 6.2 Deformation changes the character and configuration of rocks.

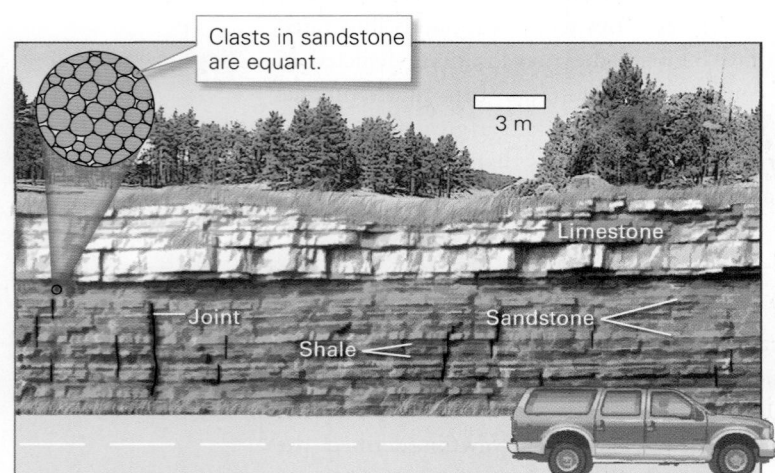

Clasts in sandstone are equant.

3 m

Limestone

Joint

Sandstone

Shale

(a) In these flat-lying beds, joints, natural cracks in rock, are the only geologic structures visible. These joints formed when the rock layers expanded slightly, due to removal of overlying rocks, and are not a consequence of mountain building.

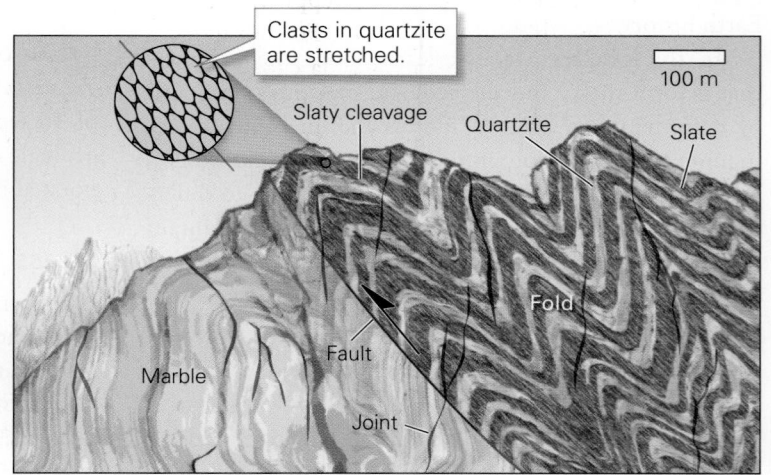

Clasts in quartzite are stretched.

100 m

Slaty cleavage

Quartzite

Slate

Fold

Marble

Fault

Joint

(b) In this Alpine cliff, deformation has caused rock layers to undergo folding and faulting, and to develop foliation (in this case, slaty cleavage). On a microscopic scale, sand grains have been flattened into ellipsoids. Joints, which formed later, cut across all the other structures.

Caledonides

Urals

Verkhoyansk

Pyrenees

Carpathians

Caucasus

Tien Shan

Alps

Atlas

Tibetan Plateau

Zagros

Makran

Himalayas

East African Rift

Central Range

Flinders & Mt. Lofty Ranges

Great Dividing Range

not only on the total amount of force applied, but also on the area over which the force acts. For example, if you stand on top of a single empty soda can, the force caused by the weight of your body will crush the can. But if you place a board on top of 100 empty cans and then stand on the board, the cans won't crush because the force caused by your weight spreads over a broad area **(Fig. 6.3a)**. Geologists, therefore, generally use the word *stress* instead of *force* when talking about the formation of geologic structures. We define **stress** as force per unit area, a relationship that can be written as stress = force ÷ area. Whether or not geologic structures form at a given location depends on the stress in rock at that location, not on the overall force driving one lithosphere plate to interact with another.

We can distinguish among different types of stress. Squeezing a block of rock by pushing one face toward the opposite one generates **compression** in the direction

of the push **(Fig. 6.3b)**. Stretching the block by pulling one face away from the other yields **tension (Fig. 6.3c)**. And moving one face of the block in a direction parallel to the opposite face produces **shear stress (Fig. 6.3d)**. We can represent these types of stress simplistically with pairs of arrows: arrows pointing toward each other signify compression, arrows pointing away from each other signify tension, and half arrows pointing in opposite directions signify shear stress. The stress types that we've just described are all examples of **differential stress**, meaning, simplistically, that the stress applied in one direction dominates over the stresses applied in other directions. (In Fig. 6.3b–d, the pairs of arrows show only the stress responsible for the structure shown; the stresses in other directions exist, but they are smaller and are not shown.)

What's the difference between *differential stress* and *pressure*? An object experiences **pressure** when the same amount of compression applies in all orientations

FIGURE 6.3 Types of stress.

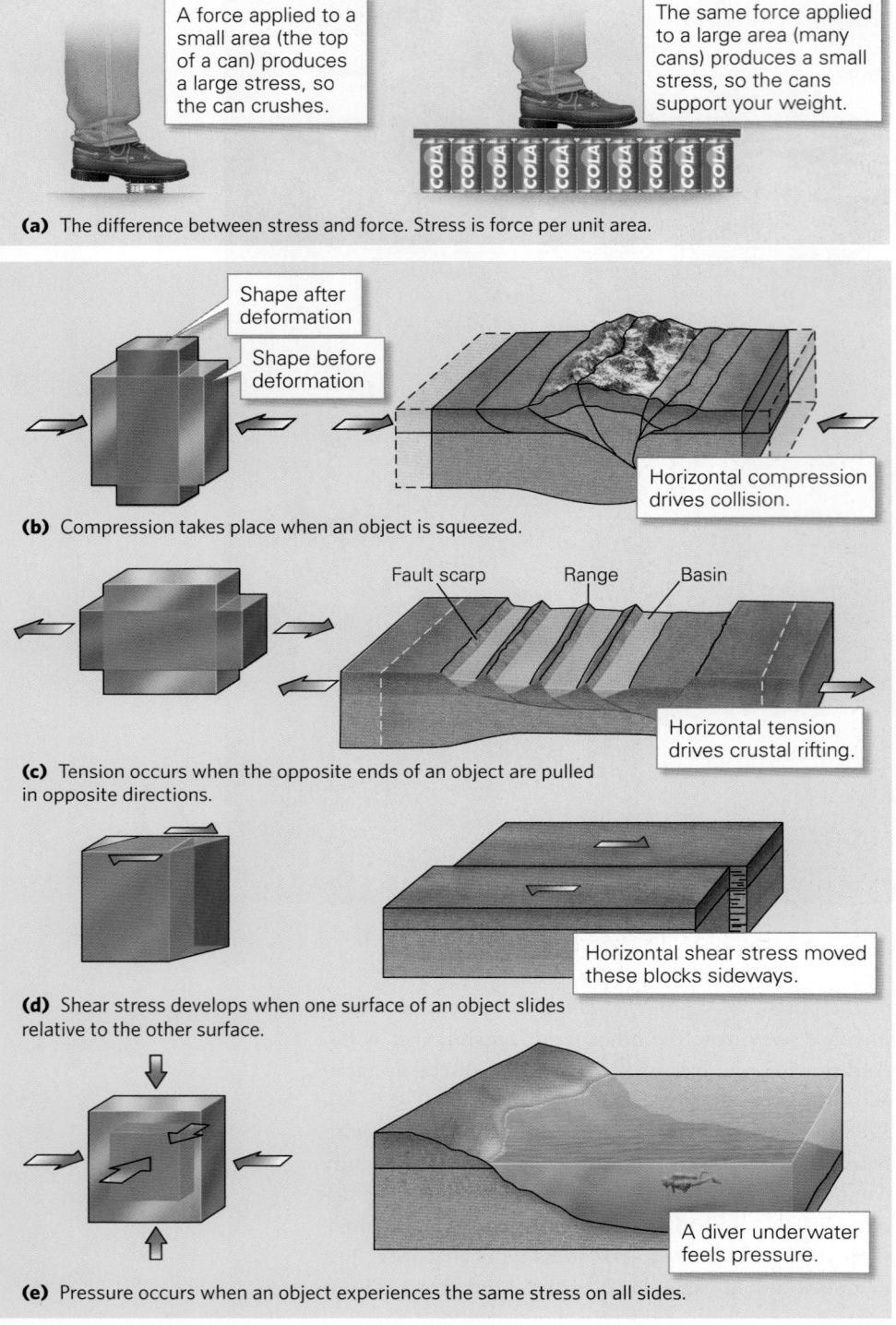

(a) The difference between stress and force. Stress is force per unit area.

Shape after deformation

Shape before deformation

Horizontal compression drives collision.

(b) Compression takes place when an object is squeezed.

Fault scarp   Range   Basin

Horizontal tension drives crustal rifting.

(c) Tension occurs when the opposite ends of an object are pulled in opposite directions.

Horizontal shear stress moved these blocks sideways.

(d) Shear stress develops when one surface of an object slides relative to the other surface.

A diver underwater feels pressure.

(e) Pressure occurs when an object experiences the same stress on all sides.

(Fig. 6.3e). Pressure can change the volume of an object, but it will not cause it to move, rotate, change shape, or develop a fabric. Therefore, pressure alone does not produce geologic structures.

When geologic structures do form, we can say that differential stress has caused **deformation**. In a general sense, deformation produces one or more of the following changes (Fig. 6.4): a *displacement* (change in location); a *rotation* (change in orientation); a *distortion* (change in shape); a *loss of coherence* (development of breaks); and/or a *change in fabric* (reorientation or reshaping of grains).

A change in shape due to deformation is called a *strain*. In everyday English, the words stress and strain seem interchangeable. A TV ad, for example, may encourage you to take an aspirin to get relief from a headache brought on by the "stress and strain of your busy day." In geologic discussion, however, the two words have very different meanings. Formally defined, **strain** is a measure of distortion, the shape change that develops in a body of rock or in a region of crust during deformation (Fig. 6.5). If a body becomes longer in response to differential stress, it has undergone *extension* (or *stretching*); if the body becomes shorter, it has undergone *shortening* (or *contraction*); and if the body changes shape because of shearing, it has developed *shear strain*. Simply put, we can think of differential stress as the cause and strain as a consequence. So, we can say that tension causes stretching, compression causes shortening, and shear stress causes shear strain.

Elastic vs. Brittle vs. Ductile Deformation

Step on a rubber ball and it changes shape. Lift your foot, and the ball returns to its original shape. You've just observed **elastic deformation**, during which a material develops a strain because chemical bonds within the material stretch or bend without breaking. When you remove the stress, the bonds return to their original shape and orientation. Now, drop a porcelain plate on the floor, so that it breaks apart. You've just observed **brittle deformation**, the process by which a material permanently separates into pieces along fracture planes. During brittle deformation, all the chemical bonds along one or more fracture planes break (Fig. 6.6a). Finally, grab a thin metal pipe at each end and bend it. You've just observed **ductile deformation**, the process during which a material changes shape permanently without forming visible cracks or breaking apart. As we'll discuss shortly, some tectonic geologic structures form by brittle deformation, and others by ductile deformation; elastic deformation does not produce permanent structures. (In Chapter 7, we will see that elastic deformation plays a role in earthquakes.)

Notably, in most rocks, ductile deformation takes place only under metamorphic conditions, meaning conditions of elevated temperature and/or pressure. Ductile deformation under these conditions generally involves movement of atoms within grains or along the surfaces of grains. This movement happens in such a way that new chemical bonds form as soon as pre-existing bonds break

FIGURE 6.4 The changes produced by deformation include displacement, rotation, and distortion.

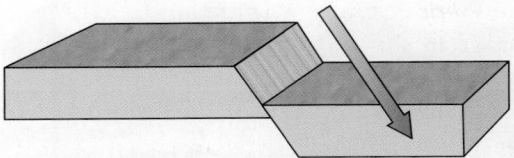

(a) Displacement occurs when a body of rock moves from one location to another.

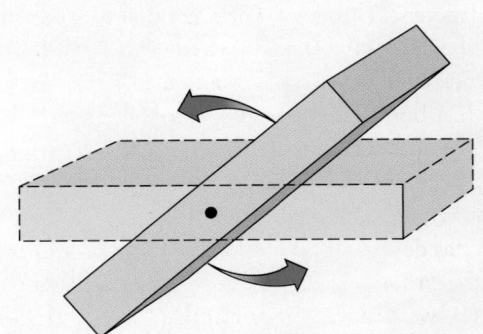

(b) Rotation occurs when a body of rock undergoes tilting.

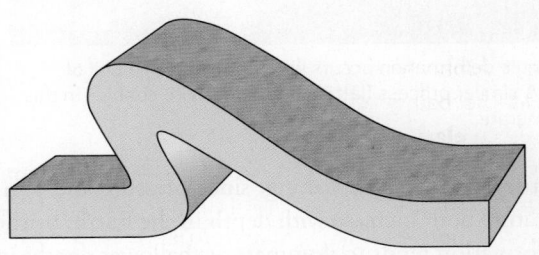

(c) Distortion occurs when a body of rock changes shape.

Slip on a fault transported this rock down from a higher position.

>2.7-Ga greenschist

~2.2-Ga quartzite

These beds were horizontal when deposited. They are now tilted.

These beds were once horizontal and of constant thickness. They are now distorted.

FIGURE 6.5 Strain is a measure of the change in shape that takes place during deformation.

Unstrained

(a) An unstrained cube and an unstrained fossil shell of a brachiopod. The white circle is a reference shape on the face of the cube.

Shortening

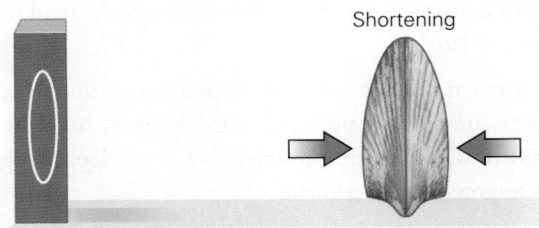

(b) Horizontal shortening changes the cube into a vertical brick and makes the shell narrower. Here, the circle becomes a vertical ellipse.

Extension

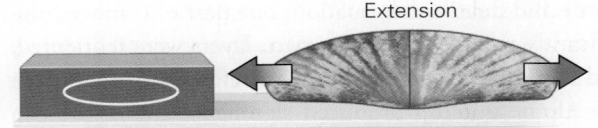

(c) Horizontal stretching changes the cube into a horizontal brick and elongates the shell. Here, the circle becomes a horizontal ellipse.

Shear strain

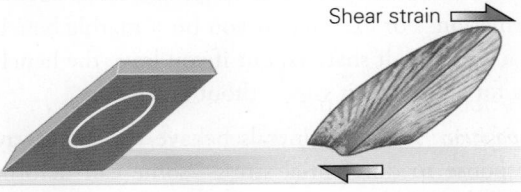

(d) Shear strain transforms the cube into a parallelogram and changes angular relationships in the shell. The circle becomes a tilted ellipse.

Before

After

Cracks separate the plate into pieces.

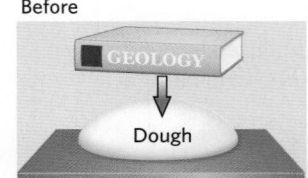

Before

After

The dough remains a single, coherent piece during deformation.

FIGURE 6.6 Brittle versus ductile deformation.

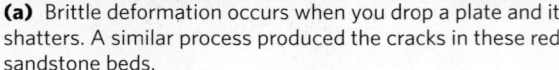

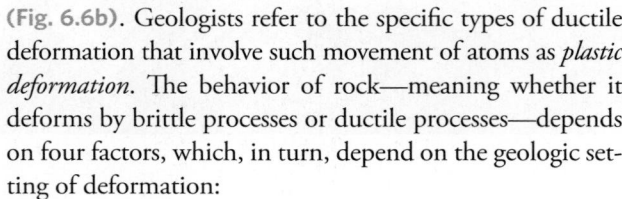

(a) Brittle deformation occurs when you drop a plate and it shatters. A similar process produced the cracks in these red sandstone beds.

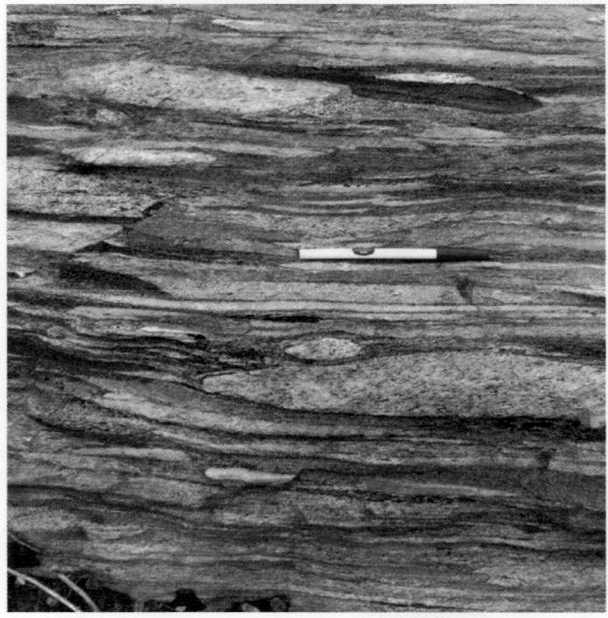

(b) Ductile deformation occurs when you squash a ball of dough. A similar process flattened the quartzite cobbles in this conglomerate.

(Fig. 6.6b). Geologists refer to the specific types of ductile deformation that involve such movement of atoms as *plastic deformation*. The behavior of rock—meaning whether it deforms by brittle processes or ductile processes—depends on four factors, which, in turn, depend on the geologic setting of deformation:

- *Temperature:* Heat makes rock softer, so warmer rocks tend to deform ductilely, whereas colder rocks tend to deform brittlely.

- *Pressure:* Under high pressure, rock deforms ductilely, whereas under low pressure, rock deforms brittlely; simplistically, pressure prevents rock from separating into fragments as it deforms.

- *Deformation rate:* At a given pressure and temperature, a sudden change in shape can cause brittle deformation, whereas a slow change in shape can cause ductile deformation. For example, if you hit a marble bench with a hammer, it shatters, but if you leave the bench alone for a century, it sags without breaking.

- *Composition:* Not all minerals behave the same way in response to differential stress—some can deform ductilely under conditions in which others deform

brittlely. Generally speaking, since pressure and temperature both increase with depth in the Earth, brittle deformation tends to dominate at shallower depths in the crust, and ductile deformation tends to dominate at greater depths.

The overall amount, or *intensity*, of deformation and the type of deformation that takes place depend on the magnitude, orientation, and duration of differential stress, as well as on the conditions of deformation and the composition of the deforming rock. That's why the midwestern outcrop and the Alpine outcrop in Figure 6.2 look so different. The midwestern outcrop was subjected to relatively low differential stress under conditions that favored brittle deformation, so only joints formed. In fact, because stress did not move or reorient the exposed rocks, no strain developed, and no fabric developed, we can say that the rock in that outcrop is nearly *undeformed*. In contrast, the Alpine outcrop has geologic structures formed by both brittle and ductile deformation: one part of it moved significantly relative to another part, layers were reoriented, and fabrics developed. Overall, therefore, deformation in the Alpine outcrop produced significant strain. We can say that the Alpine outcrop is *intensely deformed*.

▶ **Animation**
Rock Deformation

Take-home message...

Because of differential stress that develops in the Earth, the position, orientation, shape, and/or fabric of rock can change. All of these changes are manifestations of deformation. Stress, defined as force per unit area, drives deformation. Strain is a measure of change in shape. During brittle deformation, rocks break, whereas during ductile deformation, rocks bend and distort without breaking.

Quick Questions

- When you sit on a chair, does the weight of your body apply vertical compression or vertical tension to the seat of the chair?
- What aspect of deformation does strain measure?
- Which of the following illustrates brittle deformation: bending a stick or breaking a stick?

6.2 Geologic Structures Formed by Brittle Deformation

Joints: Natural Cracks in Rocks

The term *joint* comes from a Latin word meaning to join together. Ironically, in a geologic context, the word has nearly the opposite meaning—a geologic **joint** is a natural crack in rock across which once-intact rock has permanently separated into pieces, but no sliding or shear has taken place (Fig. 6.7a, b). Joints develop by application of tension to brittle rock. This tension may develop for a variety of geologic reasons. Some joints develop in response to stresses that develop when rock changes shape slightly due to the erosion of overlying rock (see Chapter 5), or when rock cools and shrinks (see Chapter 4). Others form when layers of rock bend or stretch during mountain building. Numerous joints may occur together to

FIGURE 6.7 Examples of joints.

(a) Prominent vertical joints cut these red sandstone beds in Arches National Park, Utah.

(b) Vertical joints in shale on a cliff face near Ithaca, New York.

(c) Two sets of joints breaking up a bed of limestone.

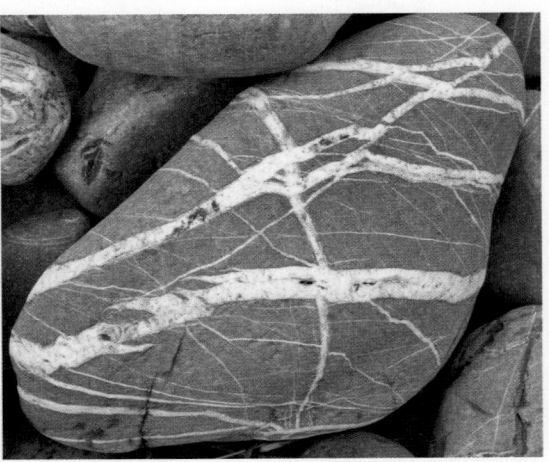

(d) Quartz veins in a boulder.

form an array called a *joint set*. The joints in a joint set have similar dimensions and orientations, and somewhat uniform spacing. A rock body may contain multiple joint sets; their presence allows the body to break into blocks. Sedimentary rocks commonly contain two joint sets intersecting at right angles, which cause beds to break into rectangular blocks (Fig. 6.7c).

While still at depth underground, some joints open up and provide passageways for hot-water solutions to flow through. Minerals may precipitate from the solutions and fill the passageways. The resulting mineral-filled crack is called a **vein**. Veins commonly contain milky white quartz or calcite, so they stand out in contrast to the darker host rock on an outcrop (Fig. 6.7d). Formation of veins accommodates extension in a rock body.

Faults: Surfaces of Slip

WHAT IS A FAULT? In the 19th century, coal miners sometimes came upon fractures that abruptly terminated the coal layer that they were excavating. The miners dubbed such fractures *faults*, for they appeared to be "flaws" in the regular succession of sedimentary beds. To find the coal layer on the far side of the fault, miners would have to dig a vertical shaft (Fig. 6.8a). Geologists adopted the miners' term, and now formally define a **fault** as a fracture surface on which sliding, or *slip*, takes place. The amount of slip along a fault is called the **displacement** on the fault. Faults have been studied intensely, not only because their presence affects the distribution of *rock units* (distinctive layers, assemblages, or bodies of rock that can be traced and correlated) and, therefore, the distribution of energy and mineral resources, but also because they play a key role in earthquakes and in mountain building (see Chapter 2 and Chapter 7).

THE MAJOR FAULT TYPES. Motion on a fault takes place when differential stress becomes sufficient either to rupture intact rock or to overcome friction on the surface of a pre-existing fault. To classify faults, geologists focus on two characteristics: (1) the *dip*, or slope, of the fault plane (Box 6.1) and (2) the *shear sense* (or *sense of slip*)

FIGURE 6.8 The concept of a fault.

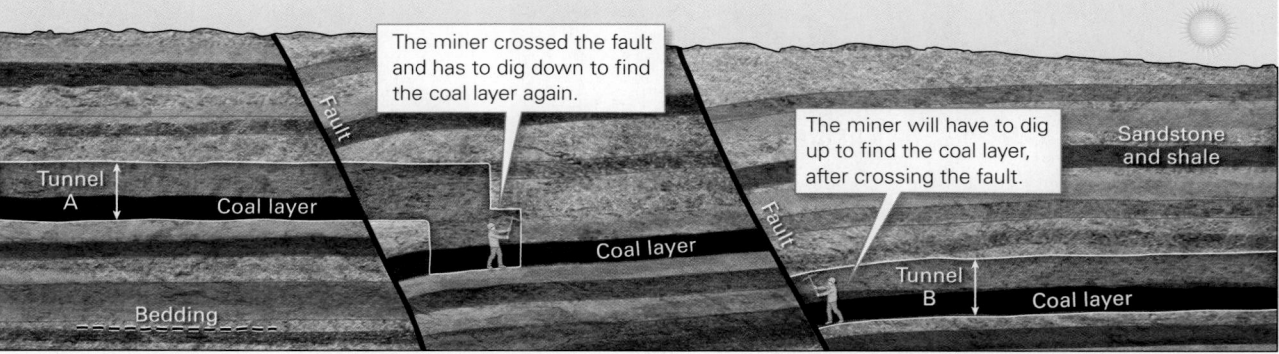

(a) The miner in Tunnel A was able to follow the coal layer until the location where the fault cuts it. The miner then had to dig a vertical shaft downward to find the displaced layer. The miner in Tunnel B will have to dig a shaft upward, after crossing the fault, to reach the coal layer.

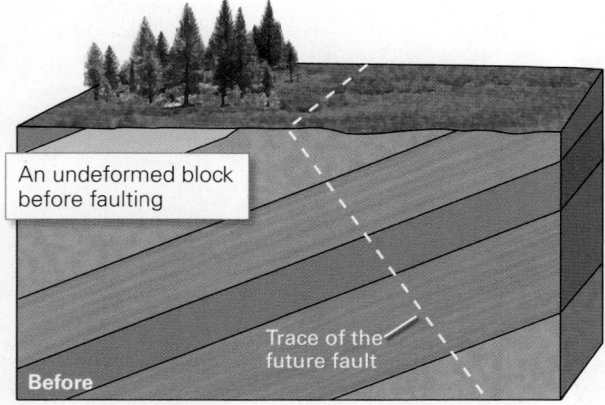

(b) Before a fault forms and slips, the layers are uninterrupted across the width of this block of rock.

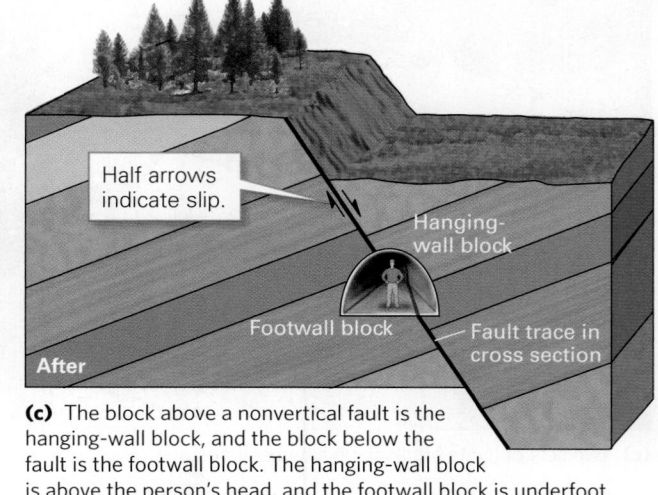

(c) The block above a nonvertical fault is the hanging-wall block, and the block below the fault is the footwall block. The hanging-wall block is above the person's head, and the footwall block is underfoot.

BOX 6.1 ▸ **Science Toolbox**

Describing the orientation of geologic structures

When discussing geologic structures, it's important to be able to communicate information about their orientation. For example, does a fault exposed in an outcrop at the edge of town continue beneath the nuclear power plant 3 km to the north, or does it go beneath the hospital 2 km to the east? If we know the fault's orientation, we may be able to answer such questions. To describe the orientation of a geologic structure, geologists picture the structure as a simple geometric shape, then they specify the angles that the shape makes with respect to a horizontal plane, a vertical plane, and the north direction.

Let's start by considering planar structures such as faults, beds, joints, and foliation. We call these *planar structures* because they are two-dimensional surfaces, so they can be represented as a geometric plane. A planar structure's orientation can be specified by its *strike* and *dip*. The **strike** is the angle between an imaginary horizontal line on the planar structure, the *strike line*, and an imaginary horizontal line pointing to true north **(Fig. Bx6.1a)**.

The **dip** is the angle indicating the plane's slope. In technical terms, dip is the angle between a horizontal plane and the *dip line* (an imaginary line parallel to the steepest slope on the planar structure), as measured in a vertical plane perpendicular to the strike. A horizontal planar structure has a dip of 0°, whereas a vertical one has a dip of 90°. Geologists measure strike with a special type of compass **(Fig. Bx6.1b)** and measure dip with an inclinometer, a type of protractor that uses a bubble to indicate the horizontal. A *strike-and-dip symbol* represents the orientation of a planar structure at a location.

We can picture a *linear structure* as a geometric line, rather than a plane. Examples of linear structures include grooves scratched on a rock surface and aligned elongate mineral crystals. Geologists specify the orientation of a linear structure by stating its *plunge* and *bearing* **(Fig. Bx6.1c)**. The **plunge** is the angle between a linear structure and an imaginary horizontal line in the vertical plane that contains the structure. A horizontal linear structure has a plunge of 0°, and a vertical one has a plunge of 90°. The **bearing** is the angle between true north and the projection of the linear structure on a horizontal plane. (A *projection* is the shadow of a line on a horizontal plane if lit directly from above or below.)

FIGURE Bx6.1 Specifying the orientation of planar and linear structures.

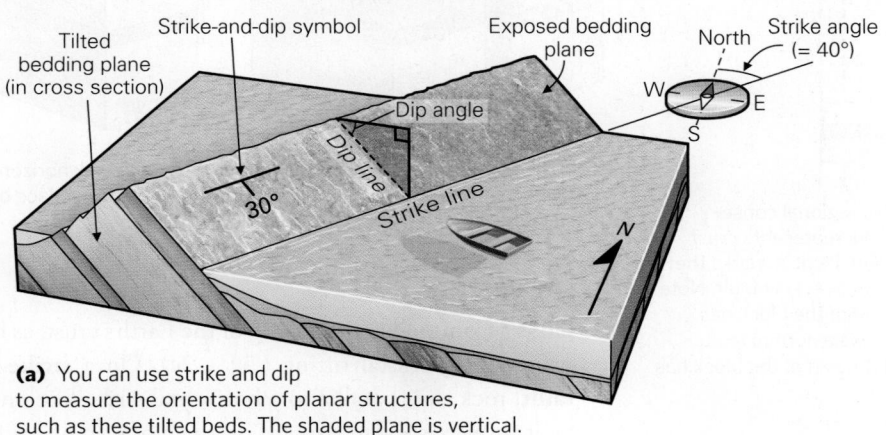

(a) You can use strike and dip to measure the orientation of planar structures, such as these tilted beds. The shaded plane is vertical.

The edge of the compass is parallel to the strike.

The intersection of the water and the rock is horizontal.

(b) Geologists use a special compass to measure strike. The compass includes an inclinometer.

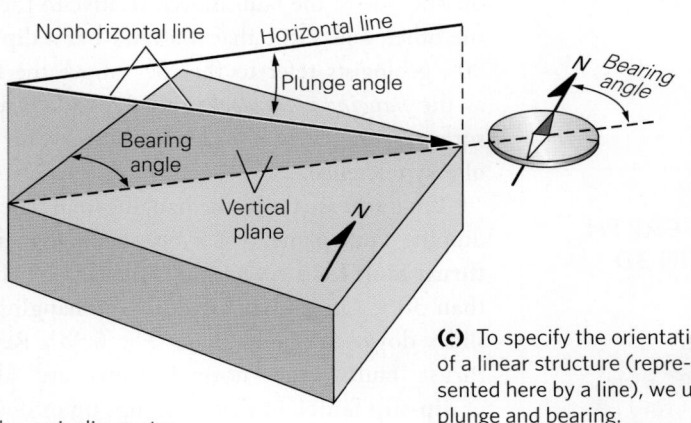

(c) To specify the orientation of a linear structure (represented here by a line), we use plunge and bearing.

FIGURE 6.9 The principal types of faults.

Dip-slip faults

Hanging wall up

Reverse (steep dip)   Thrust (gentle dip)

Hanging wall down

Normal

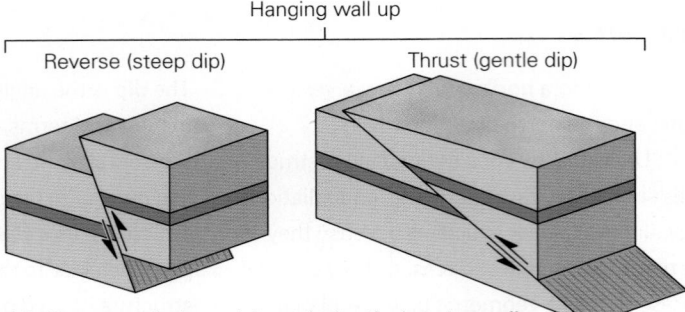
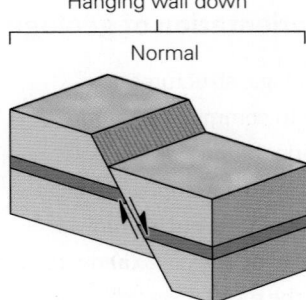

(a) On reverse faults and thrust faults, the hanging-wall block moves up. Both types of faults accommodate crustal shortening, but their dips are different.

(b) On normal faults, the hanging-wall block slips down. Normal faults accommodate crustal stretching.

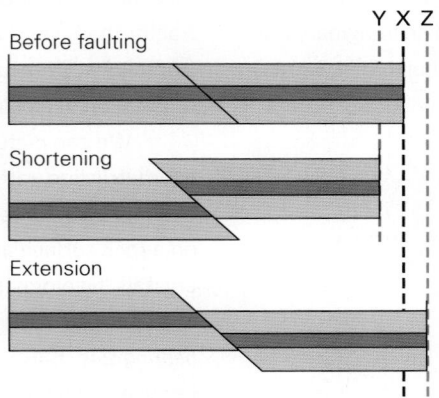

Y X Z

Before faulting

Shortening

Extension

(c) These cross sections emphasize the regional consequences of dip-slip faulting. The top block represents crust containing a fault that has not yet slipped. Point X marks the end of the block. The middle block shows a reverse fault. Note that the crust has shortened, and the end of the block has moved to Point Y. The bottom block shows a normal fault. Note that the crust has stretched, and the end of the block has moved to Point Z.

Strike-slip faults

Left-lateral   Right-lateral

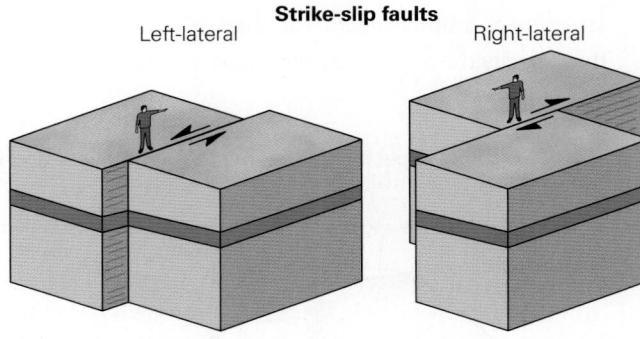

(d) Displacement on a strike-slip fault moves one block horizontally with respect to the other. No up-and-down motion takes place on such faults.

▶ **Animation**
Faults

EXPLORE EARTH SCIENCE IN 3D

Fault

across the fault, meaning the direction in which material on one side of the fault moved relative to the material on the other side. Note that if a fault has a dip of less than 90°, geologists refer to the rock above the fault surface as the *hanging-wall block*, and the rock below the fault surface as the *footwall block* (Fig. 6.8b, c), for the purpose of easy reference.

On a **reverse fault**, the hanging-wall block slides up a dipping fault plane (Fig. 6.9a); geologists use the term **thrust fault** for a reverse fault that has a gentle dip (less than 30°). On a **normal fault**, the hanging-wall block slides down the fault plane (Fig. 6.9b). Reverse faults, thrust faults, and normal faults are all examples of **dip-slip faults**, in that rock slips up or down the slope of the fault surface. Significantly, both reverse faults and thrust faults accommodate shortening of the crust, as happens during continental collision, whereas normal

faults accommodate stretching of the Earth's crust, as happens during crustal rifting (Fig. 6.9c). On a **strike-slip fault**, rock slips parallel to a horizontal strike line on the fault surface (see Fig. Bx6.1a), so the block on one side of the fault moves horizontally along its contact with the other block. Strike-slip faults tend to be roughly vertical, so the terms *hanging wall* and *footwall* generally aren't used when describing their motion. Instead, to describe the shear sense across a strike-slip fault, imagine that you're standing on one side of the fault and looking across it. On a *left-lateral strike-slip fault*, the block on the far side slips to your left, whereas on a *right-lateral strike-slip fault*, the block on the far side slips to your right (Fig. 6.9d).

RECOGNIZING FAULTS. How do you recognize a fault when you see one? A strike-slip fault that intersects the ground surface, and has slipped relatively recently, breaks up the soil and displaces roads, fences, and stream valleys (Fig. 6.10a). Because faulting breaks up and weakens rock, the trace of an ancient strike-slip fault may erode over time to become a linear valley. A dip-slip fault that intersects the ground surface and has slipped relatively recently produces a **fault scarp**, a step in the land surface whose face represents the exposed surface of the fault (Fig. 6.10b). Not all faults intersect the ground surface when they slip, however. You can see subsurface faults only when erosion later exposes them, or if they are detected using techniques described in Chapter 10. A fault trace exposed on a cliff face or road cut looks like a break along which rock on one side has shifted relative to rock on the other side (Fig. 6.10c, d). You can recognize this displacement by looking for distinctive offset markers, such as beds, veins, or tabular intrusions.

The presence of a fault may also be indicated by the occurrence of a band of broken or crushed rock. Geologists refer to this broken or crushed rock as *fault gouge* if it consists of fine powder, and as *fault breccia* if it consists of visible angular fragments (Fig. 6.11a, b). Breccia and gouge often occur together. Some fault surfaces are polished and grooved by movement on the fault. Polished fault surfaces are called **slickensides**, and linear grooves on fault surfaces are called **slip lineations** (Fig. 6.11c). The features we've mentioned in this paragraph so far develop during brittle deformation. Some faults, however, develop deeper in the crust under conditions that favor ductile deformation. Instead of breccia and gouge, these faults—generally known as *shear zones*—contain strongly foliated and fine-grained rock (see Chapter 5); the foliation in shear zones nearly parallels the shear-zone boundaries (Fig. 6.11d).

FIGURE 6.10 Recognizing fault displacement in the field.

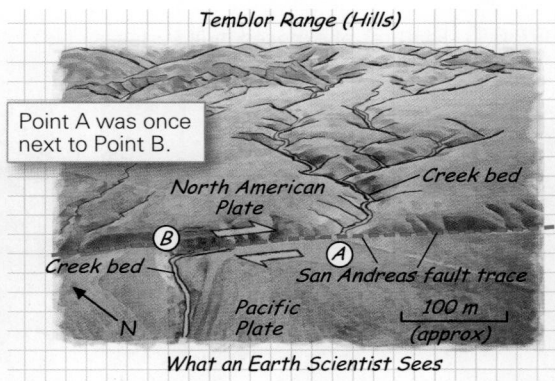

(a) An aerial photograph of a portion of the San Andreas Fault where it offsets a stream channel. This offset indicates that the San Andreas is a right-lateral strike-slip fault.

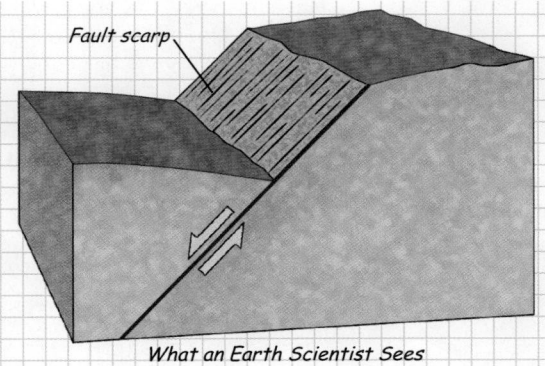

(b) This normal fault displaced the ground surface during an earthquake in 1954. The sketch emphasizes that the fault scarp is an exposed portion of a fault surface.

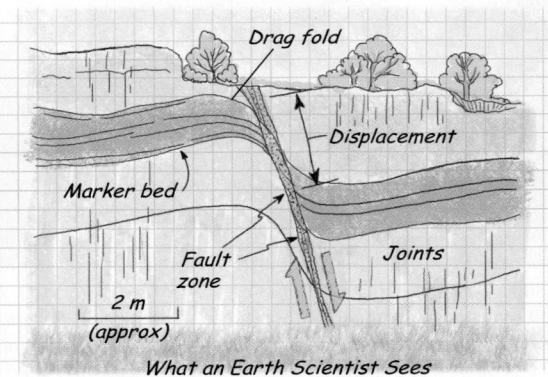

(c) A steep normal fault has displaced a distinctive marker bed (in this case, a redbed).

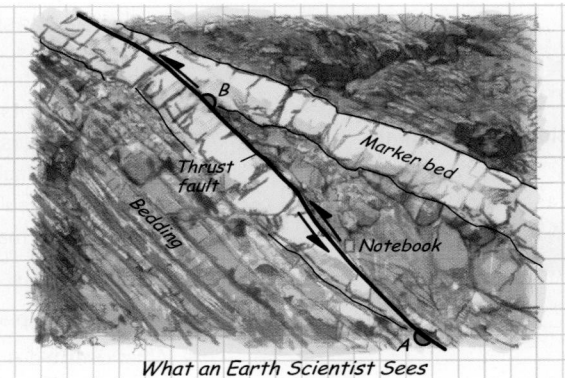

(d) Slip on a thrust fault has caused one part of this rock body to be shoved over another part, as emphasized by the displacement of the marker bed. The distance between the two red dots in the sketch is the displacement. The notebook in the photo, for scale, is 19 cm (7.5 in) long.

FIGURE 6.11 Features of exposed fault surfaces.

(a) Fault gouge consists of powder-sized grains formed when fault movement crushes brittle rock. Note the larger fragments surrounded by gouge.

(b) Fault breccia consists of irregular angular fragments.

(c) Slip lineations (fault striations) on the surface of a fault may look like grooves or scratches. They are linear geologic structures.

Striation orientation

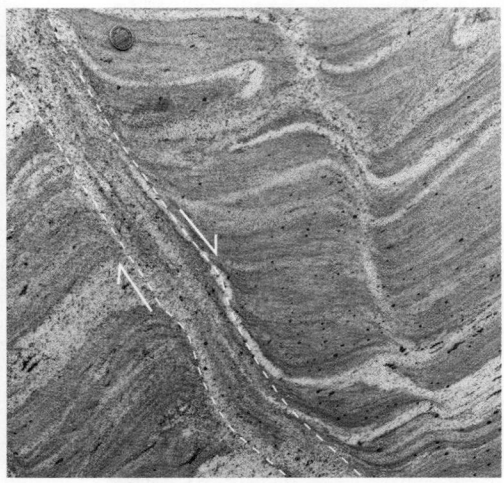

(d) A shear zone in gneiss, outlined by dashed lines. The arrows give the shear sense. Note that the foliation in the shear zone has a different orientation than gneissic banding outside the shear zone.

EXPLORE EARTH SCIENCE IN 3D

Fold Hinge and Limbs

Take-home message...

Geologic structures formed by brittle deformation include joints (natural cracks) and faults (fractures on which slip occurs). Geologists distinguish among different kinds of faults based on the relative displacement across the fault. Faulting can break up rock, and fault surfaces themselves may be polished and grooved due to slip.

Quick Questions ——————————

- Does slip take place along a joint when it forms?
- What does a description of the shear sense across a fault tell you?
- What does fault breccia consist of?

6.3 Folds and Foliation

Fold Shapes

We've seen that newly deposited sediments occur in nearly horizontal layers and, when they are lithified, become horizontal beds. During mountain building, differential stress can cause sedimentary-rock beds to bend. The result—a curve, bend, or wrinkle defined by the shape of a rock layer or a succession of rock layers—is called a **fold** (see the chapter opening photo and Figs. 6.2b and 6.4c). Folding doesn't affect only sedimentary rocks, however—it can also reorient foliation or veins in metamorphic rocks, as well as the shapes of tabular igneous intrusions. Regardless of the rock type being folded, all folds are defined by the presence of curving surfaces.

Not all folds look the same. To describe fold shapes, we must first label the components of a fold. The **limbs** are the relatively straight outer parts of a folded layer, while the *hinge zone* refers to the portion of the fold between the two limbs, where the fold has tighter curvature (Fig. 6.12a). The **hinge** itself is a geometric line along which a folded surface has its tightest curvature, and the **axial plane** is an imaginary surface that contains the hinges of successive layers affected by the fold. The axial plane effectively separates the two limbs of the fold.

Using the above terminology, geologists distinguish among the following types of folds: (1) **Anticlines** have an arch-like shape in which the limbs dip away from the hinge (Fig. 6.12b). (2) **Synclines** have a trough-like shape in which the limbs dip toward the hinge (Fig. 6.12c). (3) **Domes** slope downward in all directions away from a central point, so that they resemble an overturned bowl (Fig. 6.12d). (4) **Basins** slope inward in all directions toward a central point, so that they resemble an upright bowl (Fig. 6.12e). (5) **Monoclines** have the shape of a carpet draped over a stair step (Fig. 6.12f). Note that the hinges of the folds that we've just illustrated are represented by horizontal lines. Not all natural folds have horizontal hinges, however. Geologists refer to folds with plunging (tilted) hinges as *plunging folds* (Fig. 6.12g). Note that a fold hinge is a linear structure, so we can describe its orientation by giving its plunge and bearing. All this vocabulary may seem a bit burdensome, but it really streamlines descriptions of folds. See if you can use the terms to describe the types and components of the folds illustrated in Figure 6.13.

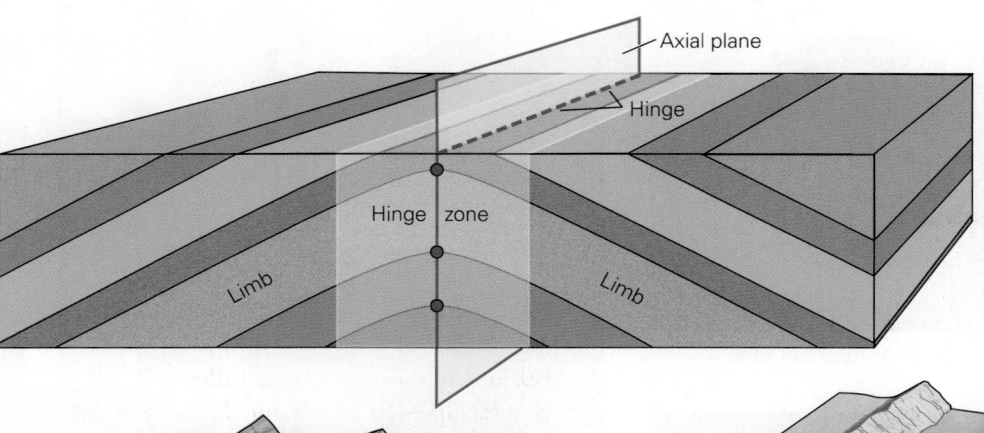

Axial plane

Hinge

Hinge zone

Limb

Limb

(a) This block diagram shows folded layers of sedimentary rock. On the front face of the block, we see the limbs and the hinge zone, and on the top of the block, we can see the hinge. The intersections of hinges with the cross-section face are represented by dots. The axial plane contains the hinges of successive layers.

FIGURE 6.12 Geometric characteristics of folds.

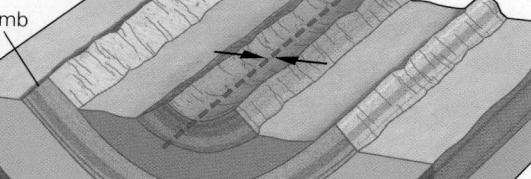

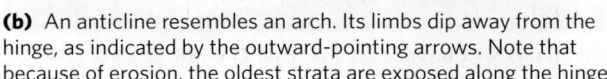

Limb

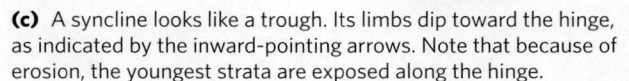

Limb

▶ **Animation**
Folds

(b) An anticline resembles an arch. Its limbs dip away from the hinge, as indicated by the outward-pointing arrows. Note that because of erosion, the oldest strata are exposed along the hinge.

(c) A syncline looks like a trough. Its limbs dip toward the hinge, as indicated by the inward-pointing arrows. Note that because of erosion, the youngest strata are exposed along the hinge.

EXPLORE EARTH SCIENCE IN 3D
Fold

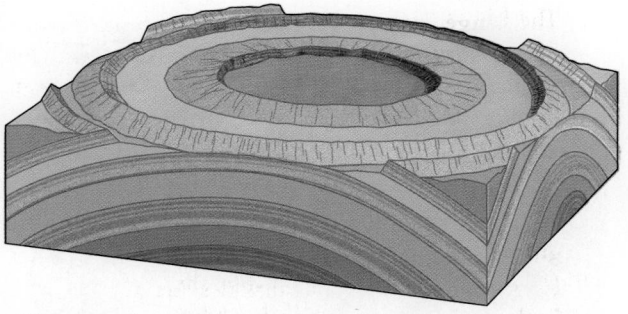

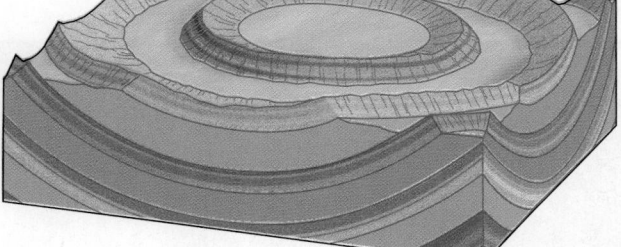

(d) A dome has the shape of an overturned bowl.

(e) A basin has the shape of an upright bowl.

See for yourself

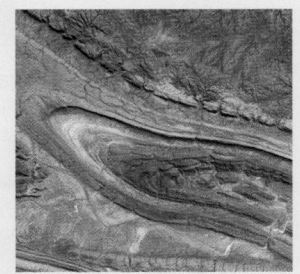

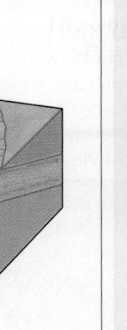

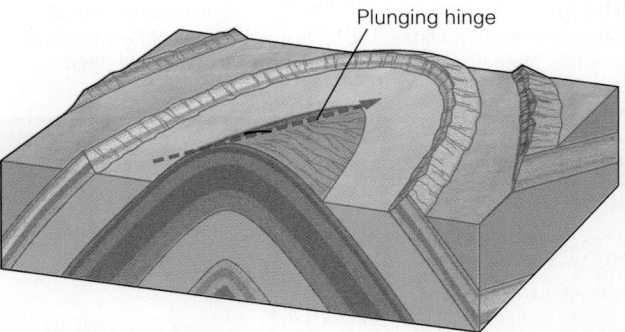

Plunging hinge

(f) A monocline looks like a stair step. Monoclines are commonly draped over the edge of an uplifted fault block.

Basement

(g) A plunging anticline has a tilted hinge.

Folds, Central Australia

Latitude: 24°18′44.08″ S
Longitude: 132°10′34.02″ E

Zoom to an altitude of 30 km (18.5 mi) and look down.

Alternating beds of resistant and nonresistant sedimentary strata have been warped into plunging folds.

Formation of Folds

Folds can develop under many circumstances, including (1) when layers wrinkle up, or *buckle*, in response to end-on compression, much like a rug when you push on its end (Fig. 6.14a); (2) when end-on compression shoves beds in sedimentary rock up and over a step-like segment of a thrust fault; the step, called a *ramp*, cuts across bedding and connects bedding-parallel segments of the fault, each of which is called a *flat* (Fig. 6.14b); (3) when shear of one part of a rock body relative to another part

FIGURE 6.13
Examples of folds on outcrops and in the landscape.

(a) This anticline, exposed by a road cut near Kingston, New York, involves beds of Paleozoic limestone.

(b) This syncline, exposed by a road cut in Maryland, involves beds of Paleozoic sandstone and shale.

(d) This train of several folds, exposed on sea cliffs in eastern Ireland, includes anticlines and synclines. The folds involve beds of Paleozoic sandstone and shale. Note that in this location, the axial planes are tilted.

Fold hinge

Fold limb

(c) This fold, exposed along the coast of Brazil, involves compositional bands in Precambrian gneiss.

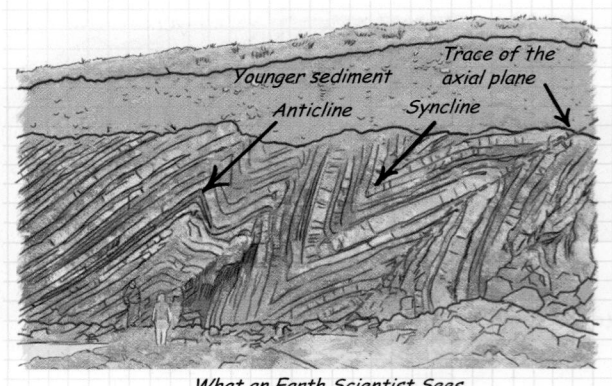

Younger sediment

Trace of the axial plane

Anticline

Syncline

What an Earth Scientist Sees

Aerial view

(e) The plunging anticline of Sheep Mountain, Wyoming, is easy to see because of the lack of vegetation. Resistant rock layers (sandstone) stand out as ridges, whereas weaker rock layers (shale) have eroded away. A block diagram shows how the surface exposures relate to underground structures.

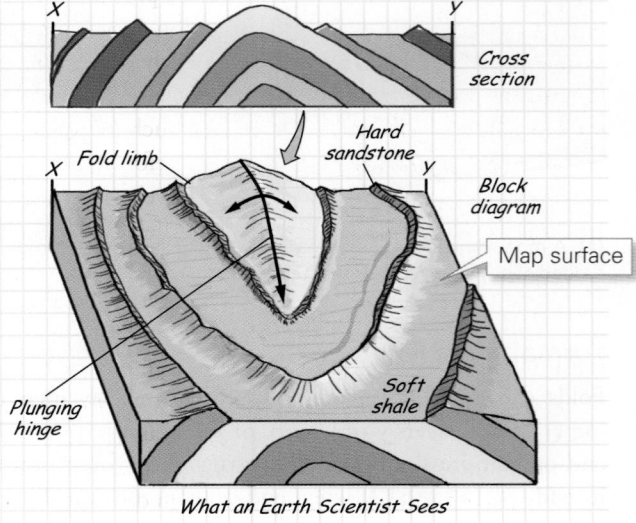

X

Y

Cross section

Fold limb

Hard sandstone

X

Y

Block diagram

Map surface

Plunging hinge

Soft shale

What an Earth Scientist Sees

Before   **After**

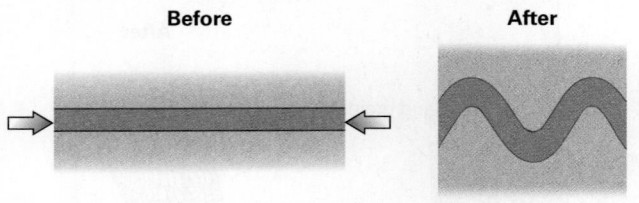

(a) Folding due to buckling, which happens when a layer of stronger rock embedded in weaker rock undergoes end-on compression.

Before   **After**

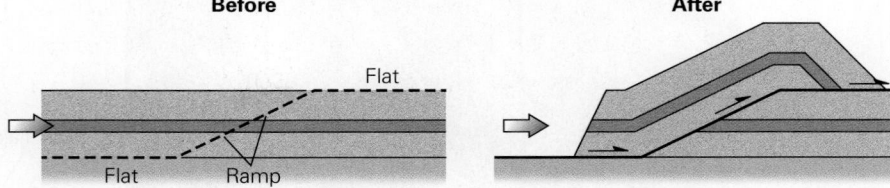

(b) In many localities, thrust faults consist of different segments, some of which lie along a bedding plane and some of which cut across bedding. A segment that cuts across bedding is called a *ramp*. When compression shoves rock layers up and over a ramp, the layers undergo folding.

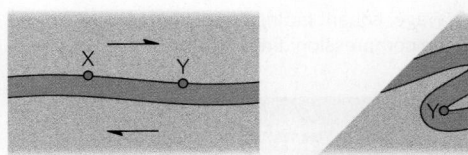

(c) Shear stress applied to a layer may cause it to fold by bringing one part of the layer over another part.

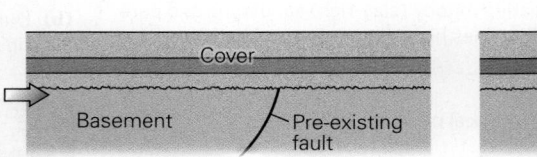

(d) In places where sedimentary strata overlie basement, renewed slip on a fault in the basement may cause overlying strata to fold into a monocline.

causes one part of a layer to bend and move over another (Fig. 6.14c); or (4) when slip on a fault causes a block of crystalline basement to push up into layers of overlying sedimentary-rock cover (Fig. 6.14d).

Rock Foliation Produced by Deformation

As we discussed in Chapter 5, when metamorphism takes place in the presence of differential stress, the resulting metamorphic rock develops layering, or *foliation*. In low-grade rocks, foliation can be due to pressure solution (see Chapter 5), which concentrates insoluble clay along distinct planes, or due to reorientation of clay flakes and growth of new ones so that the flakes are aligned parallel to one another. Notably, in low-grade rocks, foliation planes containing clay are weak, so when exposed by erosion, the rock tends to split along these planes. Because of this characteristic, low-grade foliation is also known as *cleavage*. In a deformed bed of muddy sandstone or limestone, discrete cleavage planes tend to be separated by rock in which no cleavage planes occur; such cleavage can be called *spaced cleavage* (Fig. 6.15a). In deformed shale at low metamorphic grade, *slaty cleavage* develops, in which all the clay flakes become aligned parallel to the foliation plane so that all of the rock is foliated (Fig. 6.15b). The resulting slate can split easily into very thin sheets. In phyllite, clay metamorphoses to become fine-grained mica. Cleavage in slate and phyllite commonly develops in association with folds, in which case it attains an orientation close to that of the fold's axial plane (Fig. 6.15c); this orientation implies that the cleavage planes develop roughly perpendicular to the compression direction.

At higher metamorphic grades, schistosity and gneissic banding develop, generally due to intense shear accompanied by metamorphic recrystallization and various metamorphic reactions (see Chapter 5). Eventually, shearing causes the foliation to become parallel or nearly parallel to compositional bands, and if the rock contains remnants of folds, the fold limbs are also parallel to foliation (Fig. 6.15d).

Foliation development of all types represents a type of ductile deformation because it occurs without the development of brittle cracks or fractures. As we've noted, platy minerals (such as clay in low-grade rocks, or mica in higher-grade rocks) become aligned parallel to one another, either because they rotate during deformation or because they grow by metamorphic recrystallization in the plane of foliation. Crystals of non-platy minerals may undergo plastic deformation, causing them to flatten into pancake shapes parallel to foliation.

EXPLORE EARTH SCIENCE IN 3D
Folded Limestone

Take-home message...

Folds, such as anticlines and synclines, are bends or curves in rock layers that form in response to differential stress. Stress can also cause the shape and orientation of grains to change, yielding foliation.

Quick Questions —

- Do beds have the same orientation at all points on the surface of a folded bed?
- What processes cause rock layers to undergo folding?
- Why does foliation develop during deformation?

FIGURE 6.15 Examples of foliation.

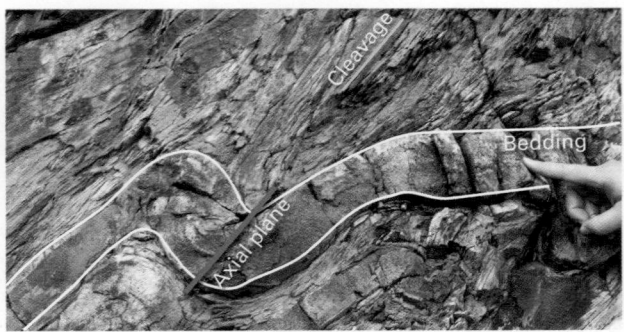

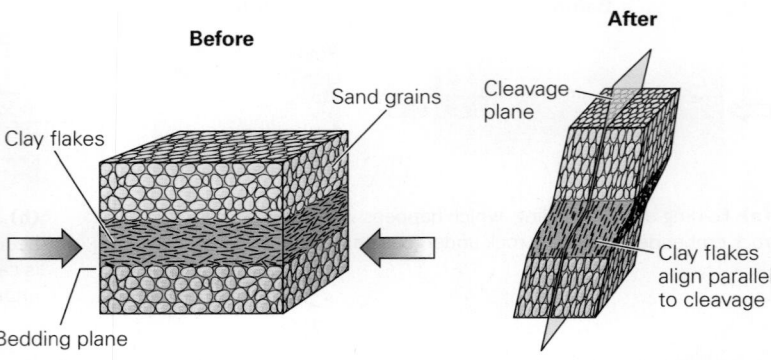

(a) Spaced cleavage has formed in muddy sandstone beds in this outcrop of alternating muddy (clay-rich) sandstone and pure sandstone beds. (White lines highlight the top and bottom of a pure sandstone bed.)

(b) During development of slaty cleavage, equant grains flatten and platy grains align in a direction perpendicular to the compression direction.

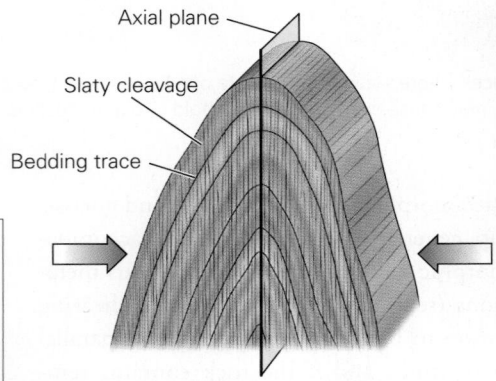

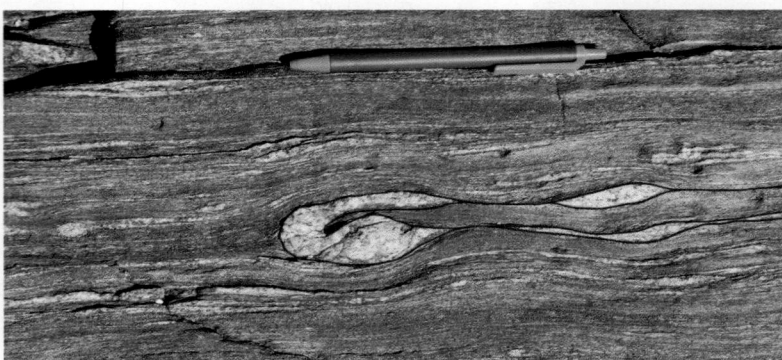

(c) Slaty cleavage that forms during folding tends to be parallel to the axial plane of the fold.

(d) In higher-grade rocks, aligned minerals and compositional layers tend to be parallel. In this example, pieces of a folded vein (outlined by black lines) are visible. Parts of the limbs became so thin that the limbs separated into lens-shaped pieces.

See for yourself

Appalachian Fold Belt

Latitude: 40°53′24.28″ N
Longitude: 77°4′28.11″ W

Zoom to an altitude of 60 km (37 mi). Tilt the view and look west.

This landscape, the Valley and Ridge Province of Pennsylvania, is the relict of the fold-thrust belt formed when Africa collided with North America, producing the Appalachian orogen about 280 million years ago. Erosion removed the high mountains. Resistant strata underlie the ridges, and nonresistant strata underlie the valleys.

6.4 Causes of Mountain Building

Why do mountain belts, and the structures within them, exist? Before the scientific era, myths attributed mountains to the actions of supernatural beings. Early geologists attributed them either to shrinkage and consequent wrinkling of the Earth, or to the upward push of igneous intrusions that formed due to melting at the bases of deep sedimentary basins. Such ideas, however, could not explain many key features of mountain belts, and eventually they had to be discarded. In fact, it's fair to say that before the theory of plate tectonics became established in the 1960s, geologists were just plain confused about why mountain belts, which they refer to as **orogens**, formed. In light of plate tectonics (see Chapter 2), the processes driving mountain building, or **orogeny**, suddenly became clear: mountain building happens in response to continental collision, subduction-related deformation, or rifting. Orogeny happens very slowly—at rates of less than a couple of millimeters to a few centimeters per year— because plate movements happen very slowly.

Collisional Orogens

Once the oceanic lithosphere between two blocks of relatively buoyant continental crust subducts completely, the blocks, which were once separated by an ocean, collide with each other and produce a **collisional orogen** (Fig. 6.16a, b). A **suture**, the boundary between what had once been separate crustal blocks, may be delineated by slivers of oceanic crust that were caught up in the collision. Intense deformation and metamorphism take place in the *internal zone*, or central part of the mountain belt. Here, rocks that undergo shearing and compression at depth, under metamorphic conditions, eventually end up exposed in the mountain belt because as they get squashed horizontally, they get pushed up vertically toward the Earth's surface, where erosion removes overlying rock. The process of collision also tends to thicken the crust substantially. In fact, the crust beneath a collision zone may end up becoming twice as thick as normal continental crust.

On either side of the internal zone lies the *foreland* portion of the orogen. A **fold-thrust belt** typically

FIGURE 6.16 Collisional orogeny.

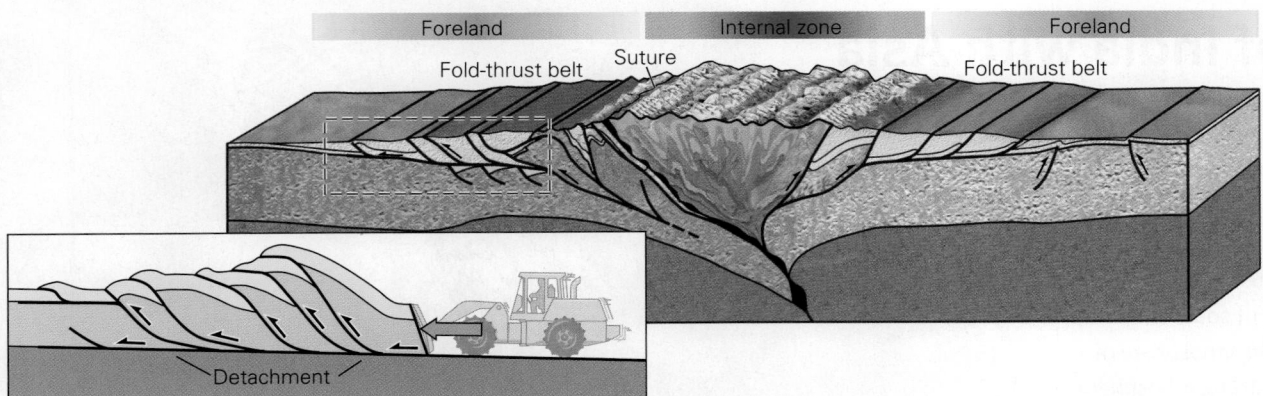

(a) During continental collision, continents squeeze together and deform. Faulting brings metamorphic rock up from deep beneath the resulting mountain belt to shallower crustal levels. In the foreland on either side, a fold-thrust belt develops in which faults displace rock toward the foreland. The inset shows a conceptual model of a fold-thrust belt: a bulldozer that pushes on layers of rock so they overlap.

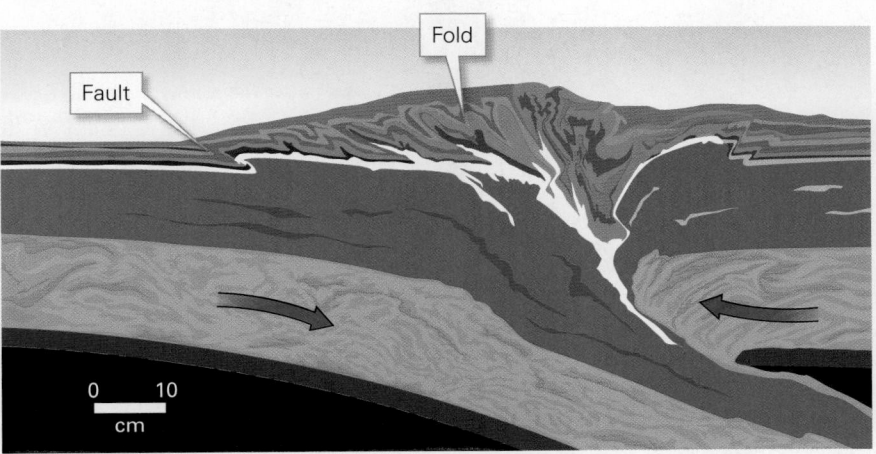

(b) Geologists can simulate collisional orogeny in the laboratory using layers of colored sand. Dragging the left side of such a model under the right side produces geologic structures and uplift, as shown in this sketch based on a museum display.

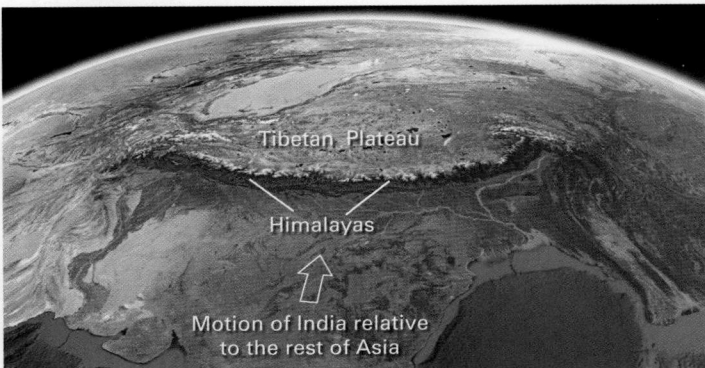

(c) This view of the Himalayas from space emphasizes that these mountains are being uplifted as India collides with and pushes into Asia. This collision also triggered the uplift of the Tibetan Plateau.

develops within the foreland. This belt consists of an array of numerous thrust faults, most of which dip toward the internal zone. The folds that develop in this belt typically form where layers of strata are pushed up and over ramps that occur along thrust faults (see Fig. 6.14b); the body of rock carried by a thrust fault, in such a context, is called a *thrust sheet*. Growth of a fold-thrust belt substantially shortens strata in the upper part of the crust, much like layers of sediment being pushed from the rear by a giant bulldozer (see Fig. 6.16a, inset). Typically, the thrust faults of a fold-thrust belt in the foreland merge with a broad subhorizontal fault known as a **detachment**; rock below the detachment is not involved in the thrusting and associated folding.

The most intense collisions happen when two large continents come together. For example, the collision of India with the rest of Asia yielded the Himalayas during the Cenozoic Era (Fig. 6.16c) (**Earth Science at a Glance**, pp. 228–229). But collisional orogens also develop when small continental blocks, island arcs, or oceanic plateaus collide with continents or with each other. In fact, along some convergent boundaries, numerous collisions over geologic time have sutured several crustal blocks to the edge of the overriding continent. Geologists refer to the suturing of smaller crustal blocks to a continental margin as **accretion** (Fig. 6.17a), and to the blocks themselves as *accreted terranes*. The process of accretion can add new crust to the edge of a continent (Fig. 6.17b).

Subduction-Related Orogens

The highest peak in the western hemisphere, Aconcagua, which lies just east of the Chile-Argentina border, reaches an elevation of nearly 7 km (22,838 ft). This mountain lies in the majestic Andes, a mountain belt that fringes the entire west coast of South America. Why does this mountain belt exist? It began to grow about 100 million years ago, in association with the subduction of the Pacific Ocean floor beneath South America when the west coast of South America became a convergent boundary. Recall from Chapter 4 that subduction triggers melting in the

Did you ever wonder...

why mountains occur in distinct belts?

▶ **Narrative Art Video**
Continental Collision

The Collision of India with Asia

The Himalayas and other highlands of southern Asia are a consequence of the collision of India—a block of old and, therefore, cold continental lithosphere—with Asia between 55 and 40 million years ago. At the time of the collision, the southern margin of Asia consisted of several smaller crustal blocks that had accreted to Asia, so the lithosphere of southern Asia was relatively warmer and weaker than the lithosphere of India. Since the initial collision, the strong lithosphere of India has continued to push slowly into Asia.

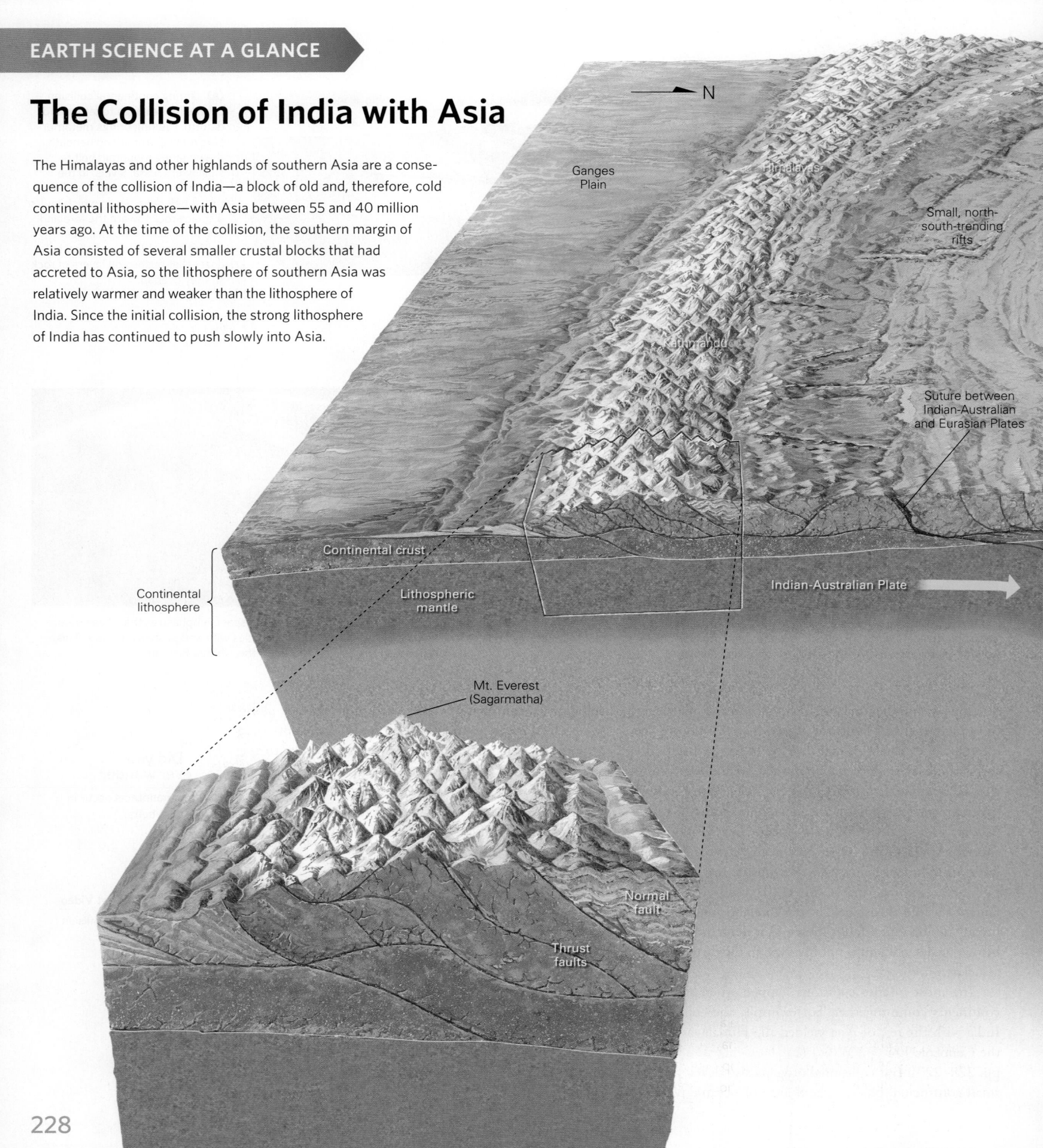

N

Ganges Plain

Himalayas

Small, north-south-trending rifts

Kathmandu

Suture between Indian-Australian and Eurasian Plates

Continental crust

Continental lithosphere

Lithospheric mantle

Indian-Australian Plate

Mt. Everest (Sagarmatha)

Normal fault

Thrust faults

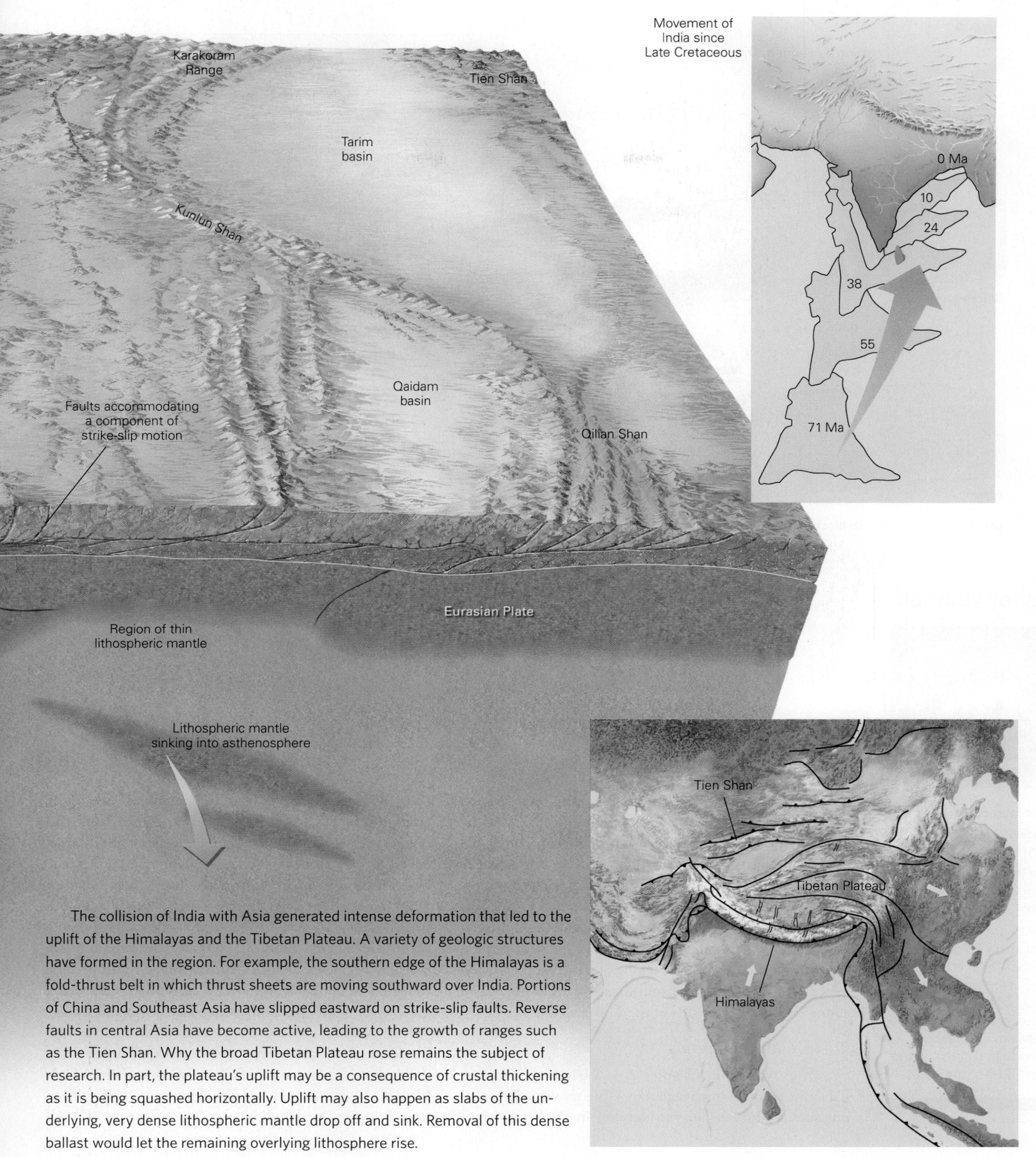

Karakoram Range

Tien Shan

Tarim basin

Kunlun Shan

Faults accommodating a component of strike-slip motion

Qaidam basin

Qilian Shan

Eurasian Plate

Region of thin lithospheric mantle

Lithospheric mantle sinking into asthenosphere

Movement of India since Late Cretaceous

0 Ma

10

24

38

55

71 Ma

Tien Shan

Tibetan Plateau

Himalayas

The collision of India with Asia generated intense deformation that led to the uplift of the Himalayas and the Tibetan Plateau. A variety of geologic structures have formed in the region. For example, the southern edge of the Himalayas is a fold-thrust belt in which thrust sheets are moving southward over India. Portions of China and Southeast Asia have slipped eastward on strike-slip faults. Reverse faults in central Asia have become active, leading to the growth of ranges such as the Tien Shan. Why the broad Tibetan Plateau rose remains the subject of research. In part, the plateau's uplift may be a consequence of crustal thickening as it is being squashed horizontally. Uplift may also happen as slabs of the underlying, very dense lithospheric mantle drop off and sink. Removal of this dense ballast would let the remaining overlying lithosphere rise.

FIGURE 6.17 Accretion can widen a mountain belt.

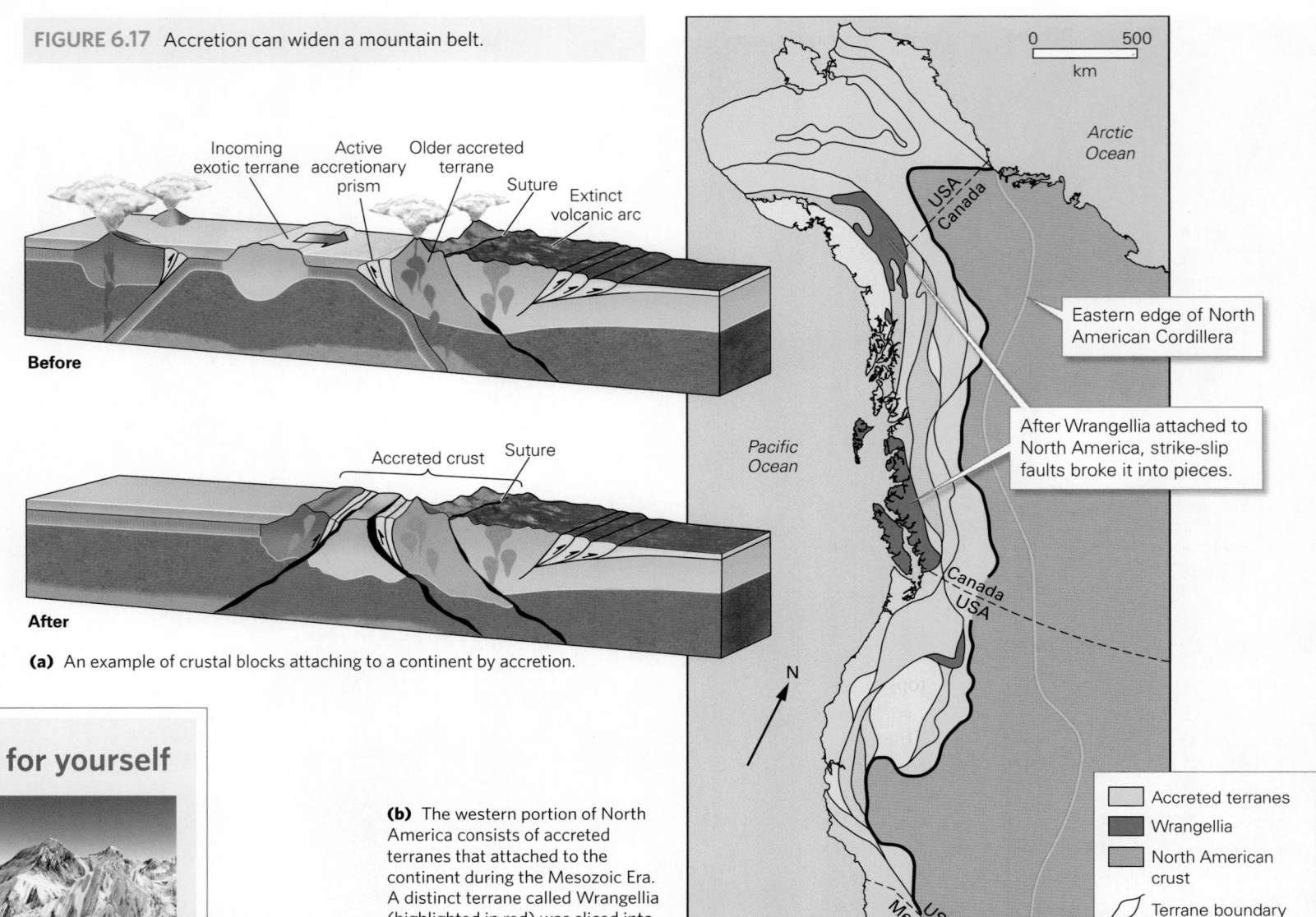

(a) An example of crustal blocks attaching to a continent by accretion.

Before

After

Incoming exotic terrane
Active accretionary prism
Older accreted terrane
Suture
Extinct volcanic arc

Accreted crust
Suture

Eastern edge of North American Cordillera

After Wrangellia attached to North America, strike-slip faults broke it into pieces.

Arctic Ocean

Pacific Ocean

USA / Canada

Canada / USA

Mexico / USA

N

0   500
km

Accreted terranes
Wrangellia
North American crust
Terrane boundary

(b) The western portion of North America consists of accreted terranes that attached to the continent during the Mesozoic Era. A distinct terrane called Wrangellia (highlighted in red) was sliced into pieces by strike-slip faults.

asthenosphere above the downgoing plate, and magma from this melting rises to form a volcanic arc along the edge of the overriding plate. Volcanoes erupt along the length of the Andes. Interaction between the downgoing plate and the overriding plate also generates horizontal compressive stress in the continental crust. This compression has produced a **subduction-related orogen** (or *convergent-boundary orogen*) in which horizontal shortening and vertical thickening of the crust have taken place (Fig. 6.18). A broad fold-thrust belt has developed along the continental side of the Andes.

Rift-Related Orogens

A *rift*, as we saw in Chapter 2, is a band of continental crust undergoing stretching. This movement causes numerous normal faults to develop in the upper crust

(Fig. 6.19a). Some of these faults, which break the upper crust into blocks, curve to a shallower dip at depth and merge with a detachment that underlies these blocks. All faults in a rift generally have roughly the same strike, but not all dip in the same direction. Where adjacent faults have alternating dip directions, horsts and grabens develop: a *horst* is a block that stays high because the normal faults bounding it dip away from each other, whereas a *graben* is a block that drops downward because the faults bounding it dip toward each other. Over time, grabens fill with sediment eroded from adjacent horsts. Where adjacent faults have parallel dips, hanging-wall blocks tend to rotate and become tilted as they slip down dip; in other words, faulting produces **tilted fault blocks**. Wedge-shaped troughs, known as *half grabens*, form between tilted fault blocks. One boundary of a half

FIGURE 6.18 Subduction-related orogeny.

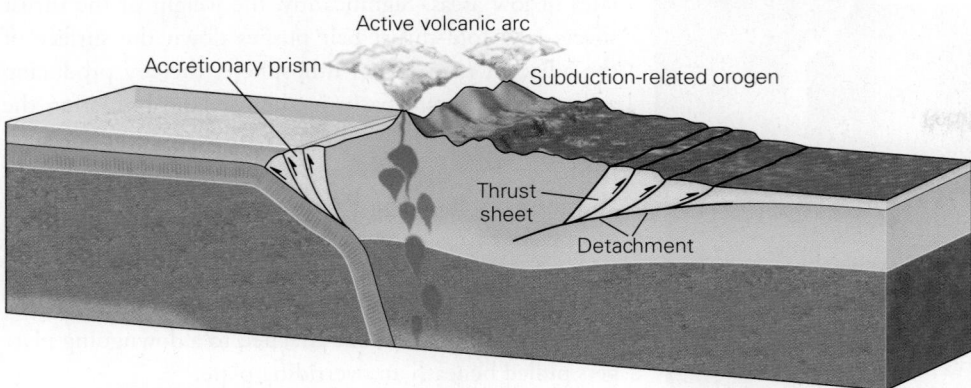

(a) At some convergent boundaries, compression uplifts a mountain belt. The continental side of the orogen includes a fold-thrust belt.

(b) These mountains in Chile are part of the Andes, which formed along a convergent boundary.

graben is a normal fault, while the other is the top surface of a tilted fault block. Sediment from erosion of tilted fault blocks slowly fills half grabens.

The presence of horsts and grabens, and of tilted fault blocks and half grabens, yields a landscape containing narrow, elongate mountain ranges (the tops of horsts or of tilted fault blocks) separated by deep, sediment-filled basins (the grabens or half grabens). The Basin and Range Province in Utah, Nevada, and Arizona, which formed by rifting during the past 25 million years, serves as an example of such a landscape (Fig. 6.19b). Because the ranges associated with rifts form due to slip on faults, they are traditionally known as *fault-block mountain ranges*.

Measuring Mountain Building in Progress

Mountain building happens too slowly for you to be able to see it happening just by staring at a mountain range. But satellite technology can detect where it's happening, and how fast movements take place. Specifically, research-quality GPS devices can locate points on the ground surface to an accuracy of ±1 mm (0.04 in), so a change in the distance between two points of greater than 1 mm can be detected. In Chapter 2, we mentioned that GPS can be used to measure plate motions. Here, we note that geologists can also compare the precise positions of different locations on two sides of a mountain belt over a few years to detect the strain developing in the mountain belt. For example, GPS measurements demonstrate that the central Andes currently shortens horizontally at a rate of a couple of centimeters per year relative to a location in the interior of South America (Fig. 6.20).

FIGURE 6.19 Rift-related orogeny.

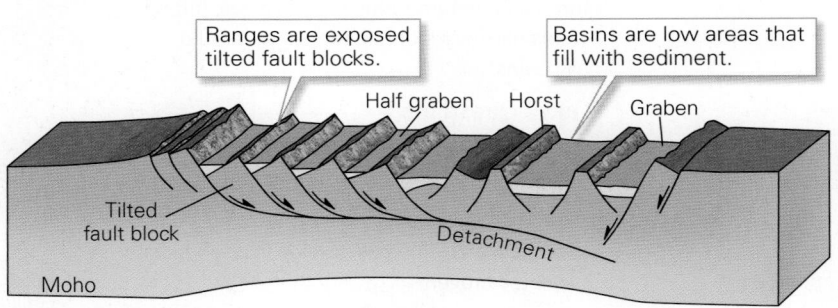

(a) Rifting leads to the development of numerous narrow mountain ranges separated by rift basins.

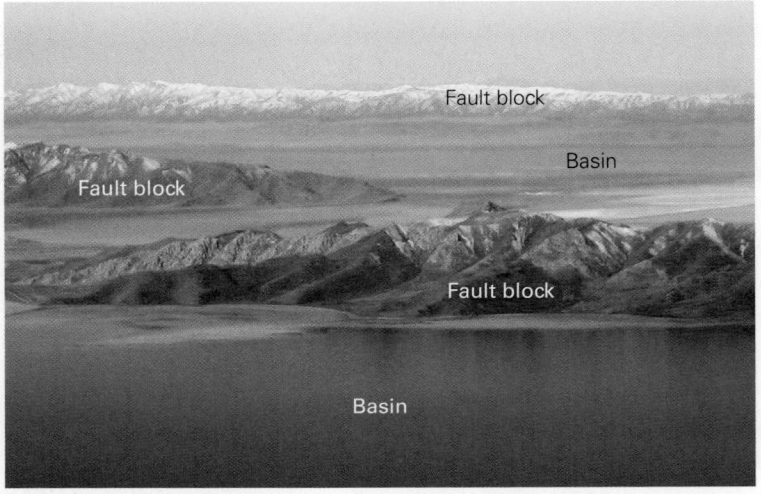

(b) The Basin and Range Province of the western United States formed during Cenozoic rifting.

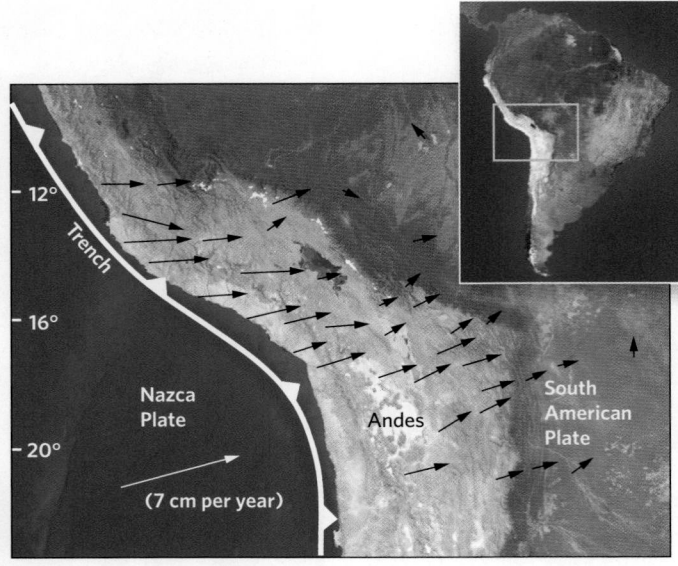

tuff. Weathering and erosion in mountain belts generate vast quantities of sediment, which eventually accumulates in low areas. Significantly, the weight of the thrust sheets in a fold-thrust belt pushes down the surface of the adjacent continental lithosphere, thereby producing a deep sedimentary basin in the foreland bordering the mountains (Fig. 6.21b). Because of their location, such basins are known as *foreland basins*. Metamorphic rocks also form in abundance during orogeny: contact metamorphism forms aureoles around plutons, and dynamothermal metamorphism affects vast volumes of rock in places where thrust faults emplace stacks of thrust sheets, or where continental crust attached to a downgoing plate gets pulled beneath an overriding plate.

Processes That Cause Uplift

Leonardo da Vinci, the great Renaissance artist and scientist, enjoyed walking in the mountains, where he sketched ledges and examined the rocks he found there. To his surprise, he discovered fossil marine shells in limestone beds a kilometer above sea level. He puzzled over this observation and finally concluded that the limestone must have been uplifted from below sea level to its present elevation. What processes can cause the surface of the Earth to rise, or undergo **uplift**? To understand why uplift takes place, we must first introduce the concept of *isostasy*.

Take-home message...

Mountain belts form in association with continental collisions, subduction, and rifting. During collisional and subduction-related orogeny, crust thickens, thrust faults and folds form, and metamorphism takes place. Rifting yields narrow mountain ranges separated by basins.

Quick Questions —————

- Name three geologic settings in which mountain belts form.
- Which types of faults develop in a rift-related orogen? Which types develop in a subduction-related orogen?
- Can the present-day motions of mountain building be detected?

FIGURE 6.21 Rocks of all three basic groups may be formed by orogeny.

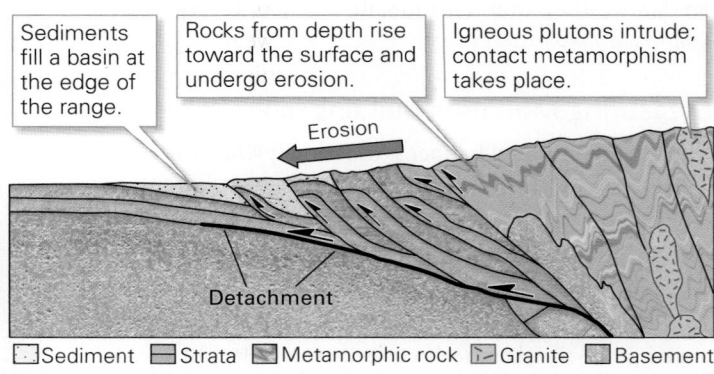

Sediments fill a basin at the edge of the range.

Rocks from depth rise toward the surface and undergo erosion.

Igneous plutons intrude; contact metamorphism takes place.

Erosion

Detachment

☐ Sediment ☐ Strata ☒ Metamorphic rock ☒ Granite ☐ Basement

(a) In the internal zone of a mountain range, metamorphic rocks and igneous rocks form. At the edge of the range, sedimentary rocks form.

6.5 Other Consequences of Mountain Building

Formation of Rocks in and near Mountains

Rocks of all three basic groups can form in mountain belts (Fig. 6.21a). Specifically, igneous rocks intrude subduction-related mountain belts because melting takes place at depth in regions where mountains are forming. In collisional orogens, some intrusion takes place after collision. And, as we discussed in in Chapter 4, abundant igneous activity takes place during rifting. Therefore, rifted crust contains dikes and sills of basalt as well as thick layers of basaltic lava and felsic

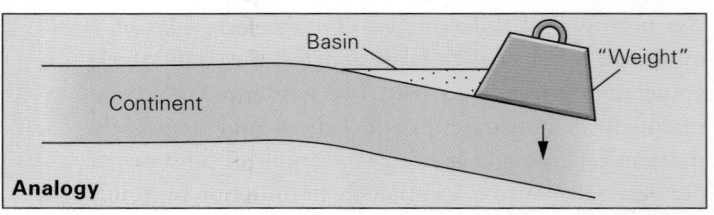

Basin

"Weight"

Continent

Analogy

(b) A sedimentary basin develops where the orogen acts as a weight that pushes down the surface of the lithosphere.

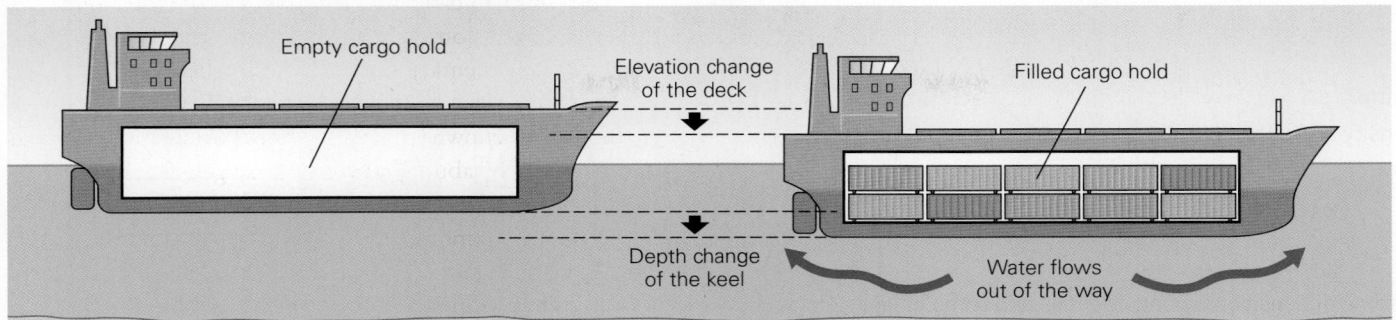

FIGURE 6.22 A ship analogy for the concept of isostasy. The elevation of a ship's deck above the sea surface changes when a load is added to the ship. Water flows out of the way as the ship moves downward.

Imagine a ship anchored in a port. Its deck lies at a specific elevation above the sea surface. If we make the ship heavier by adding cargo, the elevation of its deck becomes lower. In contrast, if we make the ship lighter by removing cargo, the elevation of its deck becomes higher. Some vertical movements of the Earth's surface are similar, because the lithosphere, which consists of relatively rigid crust and lithospheric mantle, rests on the softer asthenosphere below. Consequently, the elevation of the land's surface, like the elevation of a ship's deck, depends on a balance between forces pulling the lithosphere down and forces pushing it up (Fig. 6.22). This balance is known as **isostasy**. Put another way, isostasy exists where the elevation of the Earth's surface reflects the level at which the lithosphere naturally sits. Anything that changes the lithosphere's density, or the thickness of its layers, affects the elevation of the land surface (Box 6.2). So, to answer the question of why mountain belts can rise, we must identify geologic processes that can change the thickness or density of layers in the lithosphere.

- *Crustal thickening:* When the crust in a collision zone shortens horizontally, the mass must go somewhere, so the crust thickens vertically. As a result, the surface of the crust goes up, and the base of the crust goes down (Fig. 6.23). We can think of the downward protrusion of crust as a **crustal root**. The presence of a thickened crust effectively buoys up the Earth's surface.

- *Removal of lithospheric mantle:* The weight of the very dense lithospheric mantle effectively pulls the land surface down, just as heavy cargo makes a ship settle deeper into the water. Therefore, removing lithospheric mantle from the base of a plate causes the land surface to rise, even if the crust's thickness remains unchanged. Such removal—a process known as **delamination**—resembles removal of cargo from a ship's hold. When delamination happens, the detached lithospheric mantle sinks into the lower mantle, and hot asthenosphere fills the space left behind. This may

trigger igneous activity in the mantle. (Such activity could produce plutons in collisional orogens.)

- *Thinning and heating of the lithosphere:* In rifts, the lithosphere undergoes stretching and, therefore, thinning. Consequently, hot asthenosphere—which is less dense than cooler lithospheric mantle—rises beneath the rift. This rising hot rock heats the overlying thinned lithosphere and makes it less dense. The rise of asthenosphere and the heating of remaining lithosphere act together to cause the overall region of the rift and its margins to rise. Later, when rifting ceases, and the lithosphere cools and thickens, it sinks.

Did you ever wonder . . .

whether mountains last forever?

FIGURE 6.23 The concept of isostasy as applied to collisional mountain building.

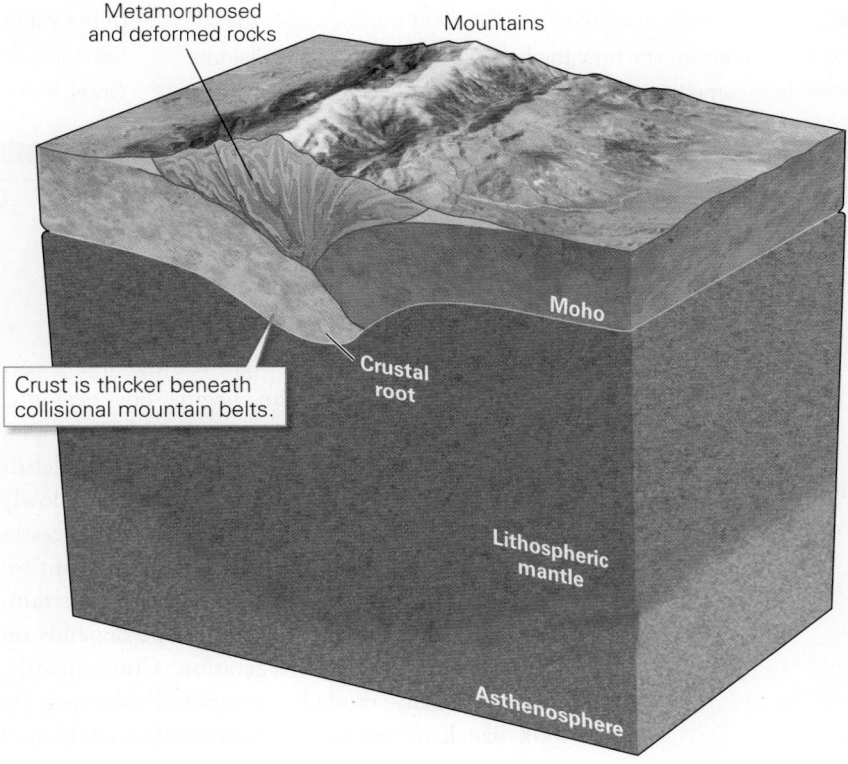

Metamorphosed and deformed rocks

Mountains

Moho

Crustal root

Crust is thicker beneath collisional mountain belts.

Lithospheric mantle

Asthenosphere

Isostasy

What are we learning?

How changes in density and changes in thickness are accommodated to maintain isostasy.

What you need:

- A tub of water
- Wooden blocks, some of dense hardwood, like oak, and some of less dense wood, like pine (begin with two pine blocks and one oak block, all of the same thickness)
- A ruler

Instructions:

- Place a pine block in the water. When it stops bobbing up and down, isostasy has been achieved. Use the ruler to measure the distance between the water surface and the top of the block. Record your result.
- Add a second pine block on top of the first. (You may have to prop the blocks against the corner of the tub to keep them from tipping.) With the second block in place, measure the distance between the water surface and the top of the top block. You will see that this distance has increased.
- Now, place a block of oak in the water. Using the ruler, measure the distance from the top of this block to the water surface, and determine whether it is greater or less than the distance you measured for the single pine block.
- By lining up a row of floating blocks of different thickness but the same density, you can see how the highest elevations of a collisional mountain belt come to overlie the thickest crust.

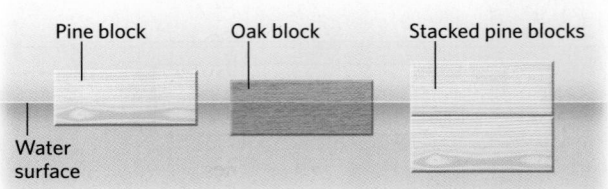

Pine block   Oak block   Stacked pine blocks

Water surface

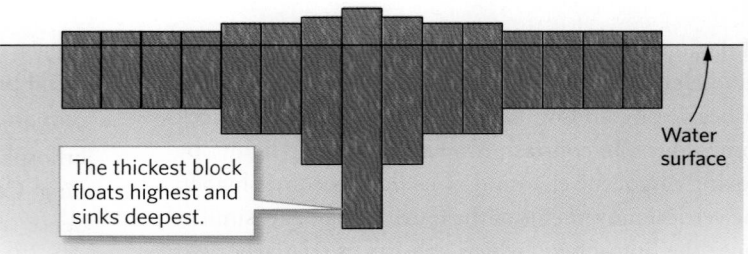

The thickest block floats highest and sinks deepest.

Water surface

What did we see?

The distance between a block's top surface and the water surface depends on both the thickness and the density of the block. Increase the block's thickness (by adding a second block), and the top is higher. Increase the block's density (by replacing one block with another of denser wood), and the top is lower. If you were able to measure thicknesses, volumes, and densities, you would see that a wooden block sinks until the mass of the block below the water is equal to the mass of the water displaced. This relationship, explaining buoyancy, bears the name *Archimedes's principle*, recognizing its discoverer, an ancient Greek scientist.

▶ **Animation**

Isostasy

What Goes Up Must Come Down

When the land surface rises significantly and slopes form, weathering and erosion begin. For example, landslides cause rock and debris to tumble down slopes; water collects in streams whose flow carries away debris and sculpts valleys and canyons; and glaciers slowly carve peaks and deepen valleys. Together, these processes grind away elevated areas and produce the beautiful landscapes that we associate with mountainous terrain. The character of a mountainous landscape depends on the agent of erosion and on vegetation. Consequently, very high, glacially eroded mountain landscapes do not look like lower-elevation, river-eroded landscapes

(Fig. 6.24). For every kilometer of rock removed by erosion, isostasy causes the remaining crust to rise by about 0.3 km, so **exhumation**, the combination of erosion and uplift happening together, eventually exposes rocks that were once deep in the crust. Note that uplift and erosion happen simultaneously in growing mountain belts, so uplift must happen faster than erosion for the height of the range to increase. If the uplift rate becomes less than the erosion rate, the elevation of the range decreases. As we mentioned earlier, after tens of millions of years, the high peaks may be gone, but remnants of the orogen remain visible as a belt of deformed and metamorphosed rock.

FIGURE 6.24 As soon as land rises, water and ice begin eroding it.

(a) Glaciers carved these rugged peaks in Switzerland.

(b) Streams cut these valleys into weathered bedrock in Brazil.

Take-home message...

Mountain building is typically accompanied by igneous activity and metamorphism, as well as by deposition of sediment. Beneath some mountain belts, the continental crust has thickened, so the surface of the crust rises higher. Once uplift has occurred, erosion sculpts rugged topography.

Quick Questions

- Why does sediment accumulate along the margins of a mountain belt?
- What does isostasy refer to?
- What process leads to the exposure of rocks from deep in the crust at the surface?

6.6 Basins and Domes in Cratons

Continents can be divided into two general kinds of crust (**Box 6.3**). Phanerozoic (< 531 Ma) mountain belts consist of crust that has been affected by mountain building recently enough that the crust remains relatively weak. **Cratons** (from the Greek word *kratos*, meaning strength), in contrast, consist of crust that has not been affected by mountain building for so long that it has become relatively cold and, therefore, strong. (Recall that colder rock cannot undergo ductile deformation.) How young is the youngest continental crust that can be called a craton? Not all geologists offer the same answer. Some restrict the definition of *craton* to regions containing crust older than 2.5 billion years. In North America, the word *craton* is generally applied to the region between the Rocky Mountains on the west and the Appalachian Mountains on

the east. This region has not been affected by widespread metamorphism or igneous activity for at least the last 1 billion years **(Fig. 6.25)**. Because of the lack of younger mountain belts, cratons generally have relatively low relief (that is, no dramatic mountain peaks).

FIGURE 6.25 North America's craton consists of a shield, where Precambrian rock is exposed, and a cratonic platform, where Phanerozoic sedimentary rock covers the Precambrian rock.

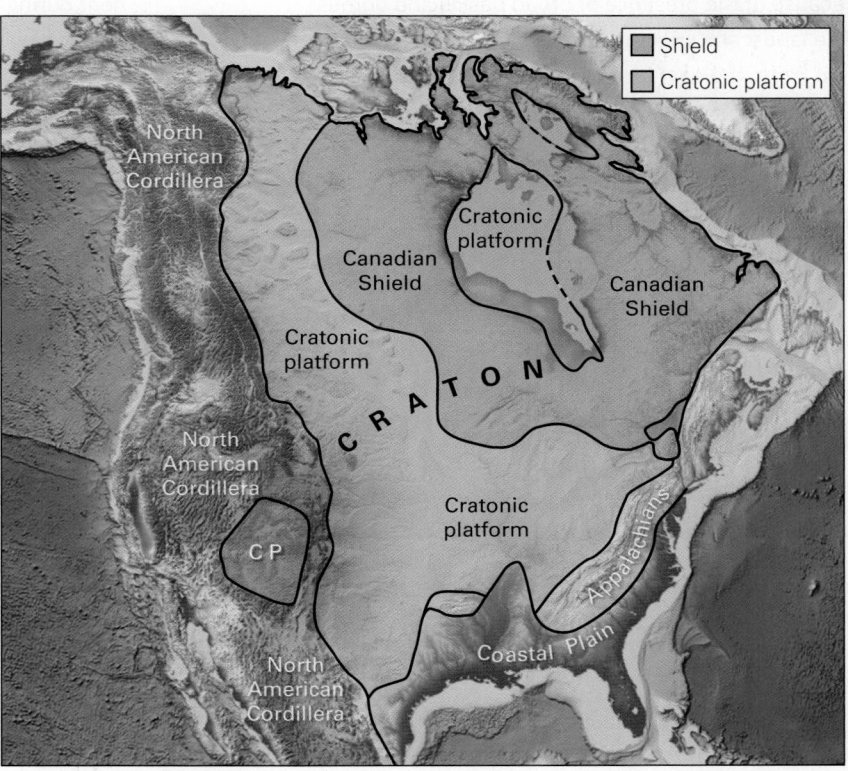

BOX 6.3 **Putting Earth Science to Use**

Seeing the mountain belts of North America

If you ever have an opportunity to drive coast-to-coast across the USA, take it . . . and don't ignore the road cuts and landscapes that you pass! If you begin on the east coast and head west, you'll first cross the coastal plain, where sediment buried the margin of the continent when sea level was high during the Cretaceous and early Cenozoic (Fig. Bx6.3). Then, you'll cross the Piedmont, a deeply eroded, low-lying area underlain by remnants of the internal zone of the Appalachian Mountains. Structures in these mountains formed during three distinct orogenies. During the last of these events, North America collided with Africa and South America. Compression during this collision produced the Appalachian Valley and Ridge Province, an eroded remnant of a fold-thrust belt (see Fig. 6.16a). Driving through the Valley and Ridge, you'll see that the bedding dip changes from one road cut to the next as strata curve around folds. The valleys formed where bedrock is weak, and the ridges remain where bedrock is strong. The distribution of strong and weak units reflects the geometry of large folds.

You'll know you've reached the western edge of the Valley and Ridge, and have entered the Appalachian Plateau, when road cuts expose horizontal strata. As you cross the interior plains, part of North America's relatively stable craton, the bedrock you'll see in most outcrops has remained nearly horizontal. Joints may be the only structures that you see in these rocks. But you'll cross through units of different ages, because of the presence of broad basins and domes.

The landscape changes radically when you reach the Rocky Mountain Front, the east face of the Rocky Mountains in the United States. The name "Rocky Mountains" refers only to the mountains along the eastern side of the North American Cordillera. The Rockies

began to rise in the late Mesozoic and continued to grow through the early Cenozoic. They formed when faults uplifted basement and caused large monoclines to form in overlying strata. As you drive into the foothills on the east side of the Rockies, you'll observe that strata tilt to a near-vertical dip. These strata are part of a monocline formed by slip on a large reverse fault. Traveling at highway speeds, you'll cross these steeply dipping strata quickly, and will then enter the Precambrian metamorphic rocks of the fault's hanging wall.

Continuing westward, you'll arrive on the relatively flat surface of the Colorado Plateau, where horizontal beds have eroded into flat-topped mesas bounded by joints. As you cross the plateau, which is an uplifted block of cratonic crust, you'll cross several monoclines. The Wasatch Mountains define the boundary between the Colorado Plateau and the Basin and Range Rift. Due to crustal stretching, the floor of the Basin and Range has dropped to lower elevations, so as you drive down the west flank of the Wasatch Mountains, the climate changes from alpine to desert. In the next day of driving, you'll cross desert plains, underlain by sediment-filled grabens or half grabens, alternating with narrow, rugged mountains (the unburied tips of tilted fault blocks; see Fig. 6.19).

The landscape and rocks change abruptly once again when you reach the front of the Sierra Nevada, at the west edge of the Basin and Range. You now enter a province where road cuts expose Mesozoic granite that once lay beneath a continental volcanic arc. From here to the west coast, outcrops contain crust that was accreted to the continent during the Mesozoic and early Cenozoic. Just before reaching the Pacific, you'll cross a present-day plate boundary—the San Andreas Fault—and pass from the North American Plate to the Pacific Plate. Your road trip will have traversed over 4,000 km (2,500 mi) and introduced you to a remarkable record of the Earth's long history!

FIGURE Bx6.3 A traverse across the United States, from Washington, DC, to San Francisco, crosses many geologic provinces. Geologic features visible at various locations speak to the long geologic history of the North American continent.

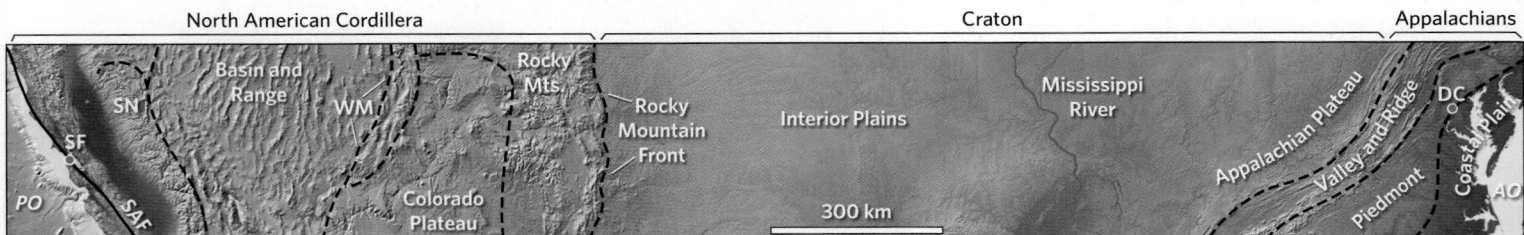

Note: SF = San Francisco; SAF = San Andreas Fault; PO = Pacific Ocean; SN = Sierra Nevada; WM = Wasatch Mts.; DC = Washington, DC; AO = Atlantic Ocean

Colorado Plateau strata

Rocky Mountain Front

Valley and Ridge dipping strata

FIGURE 6.26 Domes and basins of the North American cratonic platform.

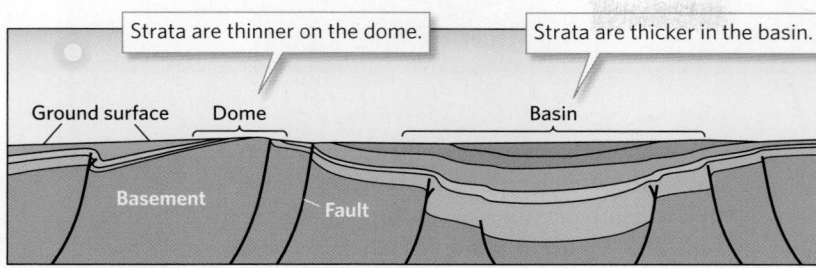

Strata are thinner on the dome.

Strata are thicker in the basin.

Ground surface Dome Basin

Basement Fault

(a) A cross section showing how strata thin toward the crest of a dome and thicken toward the center of a basin. (The cross section is vertically exaggerated.)

Wisconsin arch Michigan basin

Appalachian basin

Illinois basin Cincinnati arch

Ozark dome

Nashville dome

Precambrian basement is deeper beneath basins and closer to the surface in arches and domes.

Coastal plain

Atlantic Ocean

Gulf of Mexico

| 0 | mi | 400 |
| 0 | km | 600 |

N

(b) A geologic map of the mid-continent platform region, showing the locations of basins and domes. The terms in the legend refer to geologic time periods (see Chapter 8 for more information).

Cretaceous-Paleogene Mississippian Precambrian
Permian Devonian Appalachian Mountains
Pennsylvanian Silurian Mountain front
  Ordovician
  Cambrian

Geologists divide cratons into **shields**, broad regions of subdued topography in which Precambrian metamorphic and igneous rocks (basement) are exposed at the ground surface, and **cratonic platforms**, where a relatively thin layer of Phanerozoic strata (cover) buries the basement (see Fig. 6.25). Because the basement of cratons was affected by orogeny during the Precambrian, it consists of plutons and metamorphic rock, and much of it displays the consequences of intense deformation. But the orogenies that caused this deformation, metamorphism, and igneous activity happened so long ago that the crust of the craton has undergone significant erosion, cooling, and strengthening since then.

The cover strata of North America's cratonic platform were deposited during transgressions, when shallow seas flooded the interior of the continent (see Chapter 5). These strata are generally flat-lying and nearly undeformed (see Figure 6.2a), but they are not perfectly horizontal. In fact, over broad areas (a few hundred kilometers across), bedding dips define regional-scale basins and domes that formed in pulses during the last half billion years (Fig. 6.26a). *Cratonic basins* are regions that gradually subsided (sank) over geologic time. Strata thicken toward a basin's center, indicating that the floor of the basin was subsiding at the same time sediment was accumulating; this subsidence provided space that the sediment filled. *Cratonic domes* are regions that did not subside, or even rose slightly, over geologic time. The thickness of strata decreases toward the top of a dome because less sediment could accumulate on higher crust. Note that the pattern of contacts between rock layers in both basins and domes, as viewed on the plane of a map, displays a bull's-eye shape (Fig. 6.26b); the youngest exposed rocks lie at the center of the bull's-eye in a basin, whereas the oldest exposed rocks lie at the center of the bull's-eye in a dome (see Fig. 6.12). At some locations, faults cut across strata of cratonic platforms, and movement on these faults causes monoclines to develop in overlying strata. These faults initially formed during episodes of Precambrian rifting, and have remained weak enough to slip during the Phanerozoic at times when collisions along the continental margins applied stress to the craton.

Take-home message...

Cratons are portions of continents that consist of old (Precambrian) and relatively stable crust. Shields are broad regions of exposed Precambrian basement. Cratonic platforms have been buried by Phanerozoic cover strata. A basin occurs where the strata are warped downward into a bowl shape, and a dome occurs where the strata are warped upward. Sediment accumulates in basins as they undergo subsidence.

Quick Questions —————————

- Are there any cratons composed of Mesozoic basement?
- What is the difference between a cratonic platform and a shield?
- Why are strata thicker beneath the center of a basin than beneath the center of a dome?

Objective 6.1

Distinguish between force and stress, explain the relationship between geologic structures and deformation, and characterize different types of stress and the strain that they cause.

KEY CONCEPTS

- Tectonic geologic structures—such as joints, faults, folds, and fabrics—develop after a geologic material has formed. Different types of structures form in different locations. More complex structures form in mountain belts.

- Stress is the force per unit area applied to a body. Compression, tension, and shear are all examples of stress. Deformation happens in response to differential stress, which exists when the stress in one direction is different from the stress in other directions.

- Deformation results in one or more of the following: a displacement, a rotation, a distortion, a loss of coherence, and/or a change in fabric.

- Strain is a measure of the distortion that happens during deformation. Shortening, extension, and shear strain are all types of strain.

- Elastic deformation, accommodated by bending and stretching of bonds, disappears from rock when stress is removed. Brittle deformation involves the breaking of all bonds across fracture planes, so that the rock permanently breaks into pieces. Ductile deformation involves change in shape without cracking or breaking; plastic deformation, a type of ductile deformation, involves movement of atoms within grains or along the surfaces of grains.

- The type of deformation that takes place depends on temperature, pressure, and mineral composition.

EARTH-SCIENCE VOCABULARY

brittle deformation (p. 214)	**shear stress** (p. 213)
compression (p. 213)	**strain** (p. 214)
deformation (p. 214)	**stress** (p. 213)
differential stress (p. 213)	**tectonic geologic structure**
ductile deformation (p. 214)	(p. 211)
elastic deformation (p. 214)	**tension** (p. 213)
pressure (p. 213)	

REVIEW QUESTIONS

1. **(a)** Provide a definition of geologic structures and give a few examples. **(b)** Is bedding in sedimentary rock considered to be a tectonic geologic structure? Explain your answer. **(c)** Do more geologic structures develop within a mountain belt, or in crust outside of a mountain belt?

2. **(a)** What is the relationship between force and stress? **(b)** Which type of stress causes shortening, and which type causes extension? **(c)** Which stress type does **Figure A** show? **(d)** Characterize the types of changes in rock that can develop due to deformation.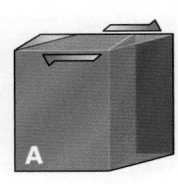

3. **(a)** Imagine a body of crust subjected to east-west compression. If the stress is great enough to cause deformation, will the crust become longer or shorter in an east-west direction? **(b)** Name the type of strain that forms in the situation just described. **(c)** What type of crustal strain accompanies the development of a rift?

4. **(a)** How does elastic deformation differ from brittle deformation? **(b)** What factors determine whether a rock will deform brittlely or ductilely?

Objective 6.2

Identify and describe geologic structures formed by brittle deformation, such as joints and faults, and distinguish among different types of each.

KEY CONCEPTS

- A joint is a natural crack in rock that forms in response to tension. A mineral-filled crack in rock is a vein. Joints can form outside of a mountain belt when overlying rock erodes away so that a once deeply buried rock body can expand; they can also form in response to cooling. Joints form during mountain building when layers of rock are bent or stretched.

- Joint sets consist of many joints with the same orientation and roughly uniform spacing. Intersecting joint sets can break rock bodies into blocks.

- Faults are fractures on which slip takes place. Types of faults (normal, reverse, thrust, strike-slip) can be distinguished from one another based on their dip and sense of shear.

- On dip-slip faults, the hanging wall moves up or down the slope of the fault. On strike-slip faults, slip is parallel to a horizontal strike line on the fault.

- Strike-slip faults that intersect the ground surface may offset features on the ground surface. Dip-slip faults may produce a fault scarp. In outcrops, faults may appear as fracture planes across which displacement can be recognized.

- Exposed fault planes may host slickensides and slip lineations, and may appear as a band of fault breccia or gouge.

EARTH-SCIENCE VOCABULARY

bearing (p. 219)
dip (p. 219)
dip-slip fault (p. 220)
displacement (p. 218)
fault (p. 218)
fault scarp (p. 221)
joint (p. 217)
normal fault (p. 220)

plunge (p. 219)
reverse fault (p. 220)
slickenside (p. 221)
slip lineation (p. 221)
strike (p. 219)
strike-slip fault (p. 220)
thrust fault (p. 220)
vein (p. 218)

REVIEW QUESTIONS

5. **(a)** How does a fault differ from a joint? **(b)** How does a vein differ from a joint? **(c)** How does a normal fault differ from a reverse fault? **(d)** How does a thrust fault differ from other types of reverse faults? **(e)** Does **Figure B** show a right-lateral or a left-lateral strike-slip fault?

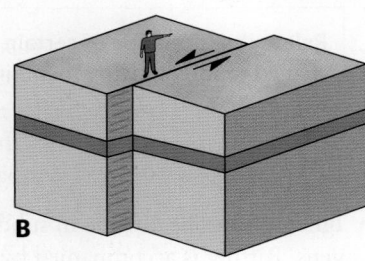

6. **(a)** How do fault gouge and fault breccia form? **(b)** What is the difference between them? **(c)** Which does **Figure C** show? **(d)** Why do valleys sometimes form along the trace of a strike-slip fault?

Objective 6.3

Characterize geologic structures (folds and foliation) formed by ductile deformation, distinguish among different types, and discuss their development.

KEY CONCEPTS

- A fold is a curve or bend in a rock layer. Each fold has two limbs and a hinge zone; the hinge is the line where the fold has the tightest curvature, and the axial plane is an imaginary surface that contains the hinges of successive beds.

- Folds come in a variety of shapes: anticlines are arch shaped, synclines are trough shaped, monoclines look like a stair step, domes look like an overturned bowl, and basins look like an upright bowl.

- Folds form in response to buckling, slip over a bend in a fault, shear, or the uplift of a basement block.

- Foliation, such as slaty cleavage, can develop due to the alignment and flattening of mineral grains during deformation. Commonly, foliation is parallel to the axial plane of associated folds.

EARTH-SCIENCE VOCABULARY

anticline (p. 222)
axial plane (p. 222)
basin (p. 222)
dome (p. 222)
fold (p. 222)

hinge (p. 222)
limb (p. 222)
monocline (p. 222)
syncline (p. 222)

REVIEW QUESTIONS

7. **(a)** How can you recognize the presence of a fold in an outcrop? **(b)** How does an anticline differ from a syncline? **(c)** Imagine a location where the horizontal ground surface is intersected by nonplunging folds. Are the oldest beds exposed in the hinge zone of an anticline or of a syncline? **(d)** How does a basin differ from a dome? **(e)** Which type of fold does **Figure D** show? **(f)** Distinguish between a plunging and a nonplunging fold.

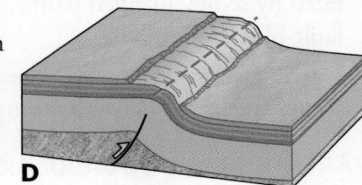

8. **(a)** Draw a sketch to show how buckling produces folds. **(b)** Draw a sketch to show how shearing produces folds.

9. Imagine that a geologist sees two outcrops containing the same resistant sandstone, as depicted in **Figure E**. The region between the outcrops is covered by soil. The curving lines in the bed indicate the shape of the cross beds in the sandstone. Keeping in mind how cross beds form (see Chapter 5), sketch the connection of the cross-bedded layer from one outcrop to the other, as it was before erosion. What geologic structure have you drawn?

10. **(a)** How does the foliation in low-grade metamorphic rocks form, and why is it also known as cleavage? **(b)** How does the foliation in higher-grade rocks form? **(c)** What is the geometric relationship between folds and slaty cleavage?

Objective 6.4

Explain why mountain building takes place, in the context of plate tectonics.

KEY CONCEPTS

- Geologists refer to a mountain-building event as an orogeny. Orogenies can occur because of collision between relatively buoyant crustal blocks, subduction along a continental margin, or rifting of continents.

- When the oceanic crust between two regions of continental crust has been completely subducted, the crustal blocks collide. Intense metamorphism and deformation develop in the internal zone of the resulting orogen, and fold-thrust belts develop in the foreland. A suture marks the boundary between the once-separate continents.

- At some convergent boundaries along the margin of a continent, subduction produces compression in the overriding plate and builds a mountain belt. During protracted subduction, pieces of relatively buoyant crust may collide with and accrete to the margin of a continent, enlarging the continent.

- Continental rifting forms a region of narrow mountain ranges separated by sediment-filled basins. The ranges are traditionally known as fault-block mountains.

- By using very accurate GPS measurements, geologists can watch strain developing across a mountain belt over the course of a year.

EARTH-SCIENCE VOCABULARY

accretion (p. 227)
collisional orogen (p. 226)
detachment (p. 227)
fold-thrust belt (p. 226)
orogen (p. 226)

orogeny (p. 226)
subduction-related orogen (p. 230)
suture (p. 226)
tilted fault block (p. 230)

REVIEW QUESTIONS

11. **(a)** Why do collisional orogens develop when two continents collide, as opposed to one of the continents simply being subducted? **(b)** On **Figure F**, label the internal zone and the foreland of the orogen, and circle the fold-thrust belts.

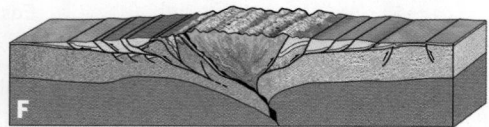

12. **(a)** Describe the strain that develops in a collisional mountain belt. **(b)** What is a suture, and how can it be recognized? **(c)** What kinds of faults are most common in a collisional mountain belt?

13. **(a)** In order for a large orogen to develop during subduction, must the overriding plate be continental, or oceanic? **(b)** Must the overriding plate be under compression or under tension? **(c)** Name a present-day example of such an orogen. **(d)** How can present-day horizontal deformation in such an orogen be measured directly?

14. The Pyrenees, an east-west-trending mountain range along the border between Spain and France, uplifted during the Cenozoic. On both sides of the range, fold-thrust belts have developed. In the interior of the range, metamorphic rocks are exposed. **(a)** In what direction did rocks in the thrust sheets on the south side of the range move due to slip on thrust faults? **(b)** In what direction did Spain move, relative to France, to cause the compression that built the range?

15. **(a)** Is the crust in a rift being subjected to horizontal tension or horizontal compression? **(b)** What is the overall strain that

develops in crust that is undergoing rifting? **(c)** Distinguish among a horst, a graben, a tilted crustal block, and a half graben. **(d)** Label the basins and ranges, and the detachment, in **Figure G**.

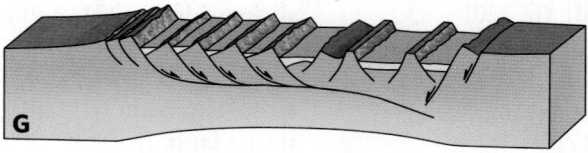

Objective 6.5

Relate the formation of certain rock types, high elevations, and rugged topography to mountain building.

KEY CONCEPTS

- Igneous plutons intrude in subduction-related and collisional orogens. Rifting is accompanied by intrusions of basalt and eruptions of both basalt and tuff.

- Erosion of mountain belts produces large volumes of sediment that fill foreland basins along the margins of the mountains, which form because the weight of the fold-thrust belt pushes the surface of the Earth down.

- Metamorphic rocks develop around plutons, and also in broad regions where crustal thickening and thrust faulting result in placing kilometers of rock over rock that was once at shallower depths.

- Isostasy exists when the lithosphere sits on the asthenosphere at an appropriate level for its thickness and density, so that upward buoyancy force and downward gravitational force balance. Changes in the thickness and/or density of the components of the lithosphere can contribute to the uplift of mountain belts.

- Due to exhumation, the combination of uplift and erosion, rocks that were once deep below a mountain belt can eventually be exposed at the surface of the Earth.

EARTH-SCIENCE VOCABULARY

crustal root (p. 233)
delamination (p. 233)
exhumation (p. 234)

isostasy (p. 233)
uplift (p. 232)

REVIEW QUESTIONS

16. **(a)** Why does igneous activity occur in a subduction-related orogen? **(b)** What are the products of igneous activity in a rift? **(c)** Why does a foreland basin develop along the margin of a collisional orogen? **(d)** Why do metamorphic rocks form in collisional mountain belts? **(e)** Where are these rocks exposed in an orogen?

17. **(a)** What processes can lead to thickening of the crust and thinning of the lithospheric mantle? **(b)** What effects do each of these processes have on land elevation? **(c)** Explain the relationship of the movements you just described to isostasy.

18. (a) Why does rugged topography develop in a mountain belt? (b) What two processes contribute to causing exhumation of deep crust beneath mountain belts? (c) How do the relative rates of these two processes determine whether the height of a mountain belt increases or decreases over time?

Objective 6.6

> Distinguish between a craton and a mountain belt, and recognize basins and domes in cratons.

KEY CONCEPTS

- Cratons are broad areas of continents underlain by Precambrian rocks that are relatively strong, stable, and cool, and have not been involved in mountain building for a very long time.

- Cratons include shields, low-relief regions where bedrock consists exclusively of Precambrian basement, and cratonic platforms, where Phanerozoic strata cover the Precambrian basement.

- The strata of cratonic platforms vary in thickness: over cratonic domes, the cover is very thin, whereas in cratonic basins, it can be a few kilometers thick. Basins and domes look like bull's-eyes on maps depicting the contacts between strata.

EARTH-SCIENCE VOCABULARY

craton (p. 235)   shield (p. 237)
cratonic platform (p. 237)

REVIEW QUESTIONS

19. (a) In what way(s) is a craton different from a mountain belt? (b) Distinguish between the basement and the cover of a craton. (c) Explain how a shield differs from a cratonic platform. (d) Has the Phanerozoic cover of cratons undergone metamorphism?

20. (a) Which feature contains thicker strata, a cratonic basin or a cratonic dome? Explain your answer. (b) Identify the basins and domes on the geologic map of **Figure H**. Note that the letter O overlies the oldest rocks in the map area, and the letter Y overlies the youngest rocks (not counting the yellow or white areas). (c) Are the youngest strata exposed at the ground surface in the center of a basin, or in the center of a dome?

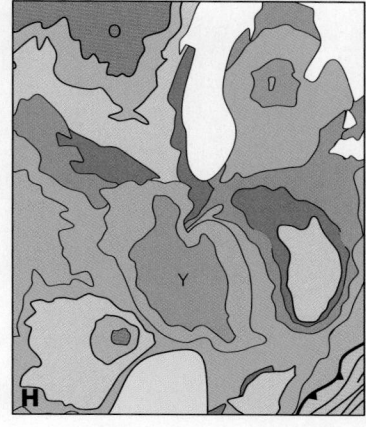

ANOTHER VIEW Mountain belts display amazing textures, as viewed from an airplane on a clear day. The texture seen here, in China, reflects both the structure of the rock and the consequences of erosion.

7 A VIOLENT PULSE
Earthquakes

After studying this chapter, you should be able to...

1. explain how faulting can release earthquake energy, and distinguish between the focus and the epicenter of an earthquake.

2. describe the different types of seismic waves and the methods used to measure and locate earthquakes.

3. interpret a news story that mentions an earthquake's magnitude and intensity.

4. relate earthquakes to specific geologic settings, in the context of plate tectonics.

5. distinguish among the different ways in which earthquakes cause casualties and damage, and differentiate between a tsunami and a storm wave.

6. critique a prediction of an earthquake, and interpret a seismic-hazard map.

7. list the steps that communities and individuals can take to reduce casualties and destruction due to earthquakes.

8. explain how seismic waves behave as they pass through the Earth's interior, and discuss what they can tell us about layering inside the Earth.

It was midafternoon on March 11, 2011, and in harbor towns of Tōhoku, a province along the Pacific coast of northern Japan, fishing fleets unloaded their catch, shoppers browsed the stores, and office workers tapped at computers. No one realized that their surroundings were about to change forever. Tōhoku lies near a convergent boundary at which the Pacific Plate slides beneath Japan. Over millions of years, subduction along this boundary built a wide accretionary prism (see Chapter 2). Averaged over time, the Pacific Plate and Japan move toward each other at about 8 cm/yr (3 in/yr), but the movement doesn't happen smoothly. Rather, for a while, stressed rock adjacent to the boundary slowly bends and warps. Then, suddenly, like a wooden stick that snaps when you bend it too far, the rock breaks, and one part slips relative to another part along a fault (Fig. 7.1). On March 11, at 2:46 P.M., the motion on the fault started about 130 km (80 mi) east of the coast and about 24 km (15 mi) below the Earth's surface. When slip finished, Japan had lurched eastward by a few meters relative to the Pacific Plate and, in places, the surface of the accretionary prism rose vertically by several centimeters.

The instant that fault slip took place, vibrations known as *seismic waves* (from the Greek word *seismos*, meaning shock or earthquake) began to pass through the surrounding rock, just like the vibrations that pass along a stick as it snaps. Seismic waves traveling at an average speed of 11,000 km/h (7,000 mph)—10 times the speed of sound in air—carried energy away from the slipped fault through the Earth and to the surface, where that energy caused the ground to shake. Geologists refer to both the energy-generating event and the resulting ground motion as an *earthquake*.

When seismic waves reached and shook the harbor towns and other communities of Tōhoku, people lost their balance as buildings twisted and swayed and, in some cases, collapsed (Fig. 7.2a). Inside, plates flew off shelves, ceilings fell, bookshelves tipped, and furniture seemed to dance. Outside, dust clouded the air, and landslides tumbled down hillslopes. Here and there, gas from broken pipelines ignited, sending up billows of flame and smoke. Unfortunately, when the ground motion had mostly ceased, the disaster wasn't over. The sudden displacement of the accretionary prism's surface had not only generated seismic waves, but had also pushed against the ocean's water. This movement ultimately produced several

FIGURE 7.1 What happens during an earthquake?

Stick

(a) Before deformation, the rock layers in this example are not bent.

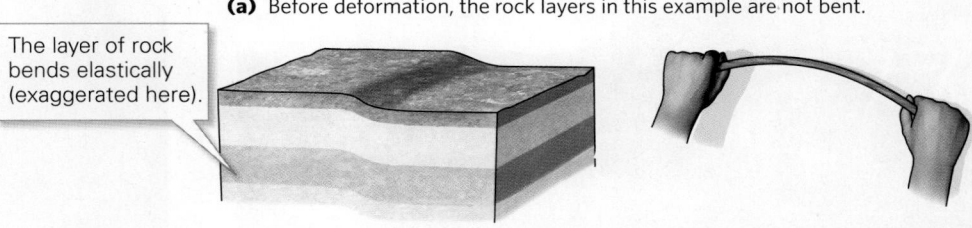

The layer of rock bends elastically (exaggerated here).

(b) As deformation occurs, rock bends, like a stick that you arch between your hands.

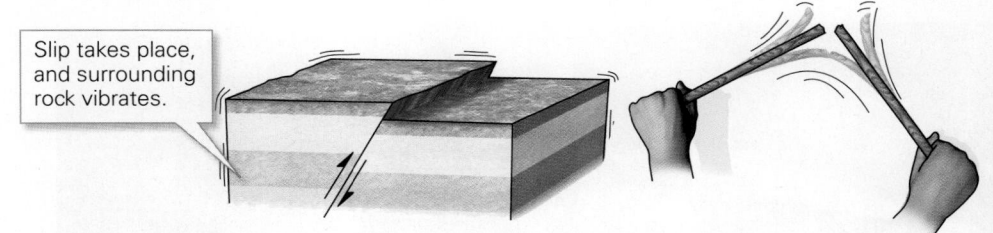

Slip takes place, and surrounding rock vibrates.

(c) Eventually, rock breaks, just like a stick that snaps if it bends too far. Immediately, slip takes place on a fault and vibrations pass through the surrounding rock.

large *tsunamis*, each a very broad, rapidly traveling water wave. When each tsunami approached the shore, it grew in height, so its water overtopped seawalls and surged inland, stripping the land of buildings, cars, and people (Fig. 7.2b, c).

Earthquakes are a fact of life on our dynamic planet—an estimated 50,000 to 100,000 earthquakes large enough to be felt happen every year. Fortunately, most cause no damage or casualties, either because they release only a small amount of energy and produce only tiny vibrations, or because they affect only unpopulated areas. But every year, a few hundred earthquakes rattle the ground sufficiently to crack or topple buildings and injure people, and about a hundred of these cause enough damage to disrupt affected communities significantly. Every 5–20 years, on average, an earthquake results in a horrific calamity. What geologic phenomena trigger earthquakes? Why do earthquakes take place where they do? How do they cause damage? Can we predict when earthquakes will happen, or even prevent them from happening? What can earthquakes tell us about the interior of the Earth? Read on, and you'll be able to answer these questions.

<< After a major earthquake shook the ground in southwestern China, kindergarten children were evacuated to the outside playground. They covered their heads with pillows as a precaution against falling debris.

FIGURE 7.2 The Tōhoku earthquake and tsunami, Japan, 2011.

(a) Some buildings collapsed due to shaking.

(b) The rising water of the tsunamis filled harbors. Here it spills over a seawall.

Coast

(c) An aerial view shows a tsunami advancing across the shore and then inland.

7.1 What Causes Earthquakes?

As we've seen, an **earthquake** is an event during which sudden movement underground produces vibrations in rock, known as **seismic waves**, which can transport energy long distances. The term also applies to the **ground motion** (shaking of the land) that happens when seismic waves reach the Earth's surface. Ancient cultures offered a variety of myths to explain earthquake activity, or **seismicity**. Most involved the restless gyrations of giant subterranean beasts. Studies in the late 19th and early 20th centuries led scientists to conclude that, in reality, most earthquakes happen when a body of rock suddenly moves past neighboring rock on a **fault**, a fracture surface on which sliding, or *slip*, takes place **(Fig. 7.3)**. The amount of slip on the

FIGURE 7.3 This fault surface in Arizona became exposed because the rock above the fault surface eroded away. Slip formed on this fault surface during movement.

Slip lineations

Fault surface

Fault trace

What an Earth Scientist Sees

fault is its **displacement** (Fig. 7.4). (Volcanic eruptions, meteorite impacts, and nuclear explosions can also produce earthquakes, but because fault-related earthquakes are so dominant, we address only these in this chapter.)

How Does Slip on a Fault Generate an Earthquake?

Seismic waves can be produced either when previously intact rock suddenly breaks to form a new fault, or when a pre-existing fault suddenly slips again. Let's look at both of these processes in turn.

To picture fault development in intact rock, imagine gripping each side of a brick-shaped block of rock with a clamp. Now, you apply a slight upward push (a stress) on one clamp and a slight downward push on the other. This action causes the rock to bend, but not rupture. Recall from Chapter 6 that such **elastic deformation** happens when application of stress causes chemical bonds in a material to stretch and bend, but not break. If you remove the stress when only elastic deformation has taken place, the rock "relaxes" and returns to its original shape, like a rubber band that you release. Now imagine that you repeat the experiment, but this time apply more stress to push one side of the block up and the other side down by a greater amount. This time, small cracks start to develop in the brittle rock (Fig. 7.5a). With continued pushing, the cracks begin to connect to one another until, suddenly, a fracture cuts through the entire block of rock (Fig. 7.5b). The instant that happens, the block breaks in two, and rock on one side of the fracture slides past rock on the other side (Fig. 7.5c). Because sliding has taken place, the fracture has become a fault.

FIGURE 7.4 A wooden fence that straddled the San Andreas Fault was offset during the 1906 San Francisco earthquake. The black line represents the fault trace, meaning the intersection between the fault and the ground, and the half arrows indicate the relative motion across the fault.

What an Earth Scientist Sees

If you look closely at Figure 7.5b, you'll notice that the band of rock in which bending takes place and elastic deformation develops is wider than the fault itself. When the new fault suddenly forms and slips, the bent rock on either side straightens out, or **rebounds**, so that the elastic deformation relaxes and stress decreases. When relaxation takes place, however, the rock doesn't move smoothly back to its original shape; rather, it swings back

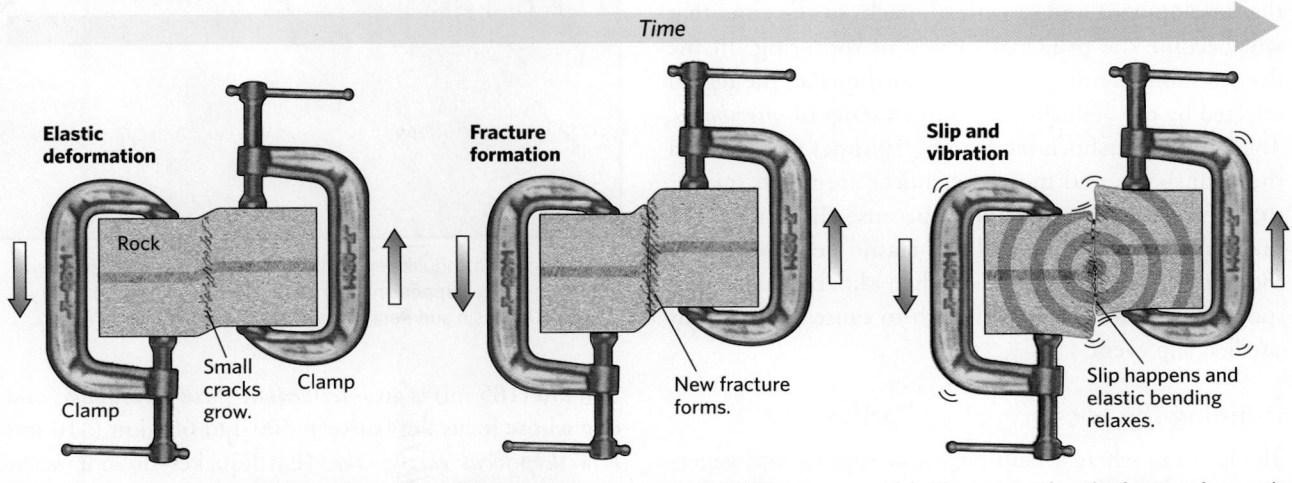

Time

(a) Imagine a block of rock gripped by two clamps. Move one clamp upward, and the rock starts to bend. Small cracks develop along the bend. (Displacements are greatly exaggerated in this sketch.)

(b) Eventually, the cracks link. When this happens, a fracture cuts completely through the rock.

(c) The instant that the fracture forms, the rock breaks into two pieces that slide past each other.

FIGURE 7.5 A model representing the development of a new fault.

and forth, like the ends of a snapped stick (see Fig. 7.1). This back-and-forth movement of rock on either side of the fault, together with the breaking of rock, generates seismic waves, which travel through the Earth or along the Earth's surface and can cause the ground to shake. **Seismologists**, researchers who specialize in the study of earthquakes, refer to this concept as the **elastic-rebound theory** of earthquake generation. The amount of energy released by the earthquake depends on the amount of elastic strain released. (To picture why, imagine launching marbles with a slingshot—the farther you stretched the elastic band, the farther the marbles went.)

Once a fault forms, it doesn't continue to slip for long because **friction**—the resistance to sliding caused by bumps and irregularities on a surface that act as tiny anchors and resist shear—eventually slows and stops the movement. But once a fault has formed, it remains weaker than the surrounding, intact rock. So, when stress builds up, slip will generally take place on a pre-existing fault before the stress becomes large enough to generate a new fault. Since a pre-existing fault won't slip until the stress overcomes friction, elastic deformation can build up in rock adjacent to the fault, so that when the fault finally does slip, elastic rebound takes place. Seismologists refer to the cycle of stress buildup, then sliding and stress release, on a pre-existing fault as **stick-slip behavior**. You can find pre-existing faults at many locations in the Earth's crust. But don't panic! Fortunately, only some of these are *active faults*, those that have moved relatively recently or might move in the future. Most are *inactive faults*, those that slipped in the distant geologic past, but probably won't slip again.

The main rupturing event along a fault produces the *mainshock*. Commonly, smaller earthquakes, called *foreshocks*, precede the mainshock. They may be due to the development and growth of cracks in the zone that will become the principal surface of rupturing. In the days to months following a large earthquake, the region affected by the mainshock endures a series of *aftershocks*. The largest aftershock tends to be 10 times smaller than the mainshock, and most aftershocks are much smaller than that. Aftershocks happen because slip during the mainshock doesn't relax all the elastic deformation in rock adjacent to the fault. Rather, slip produces new spots where stress is large enough to cause another, but smaller, slip event.

Defining the Location of an Earthquake

The location where a fault begins to rupture and generate seismic waves is called the **focus** (or *hypocenter*) of an earthquake (Fig. 7.6a). An earthquake whose focus lies at a depth of less than 70 km (43 mi) is a *shallow-focus earthquake*, one that occurs at a depth between 70 km and

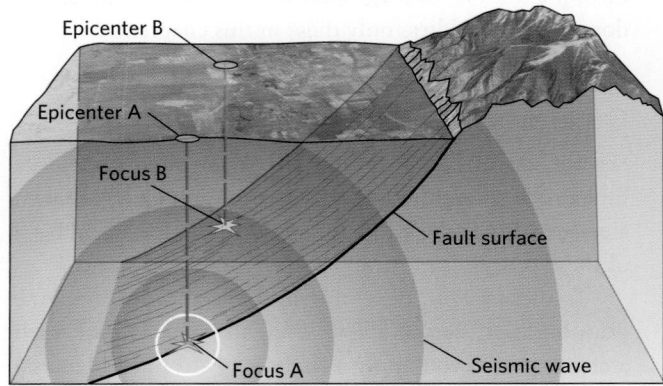

FIGURE 7.6 Earthquake foci and epicenters.

(a) The focus is the point on a fault where slip begins and from which seismic waves begin radiating. The epicenter is the point on the Earth's surface directly above the focus.

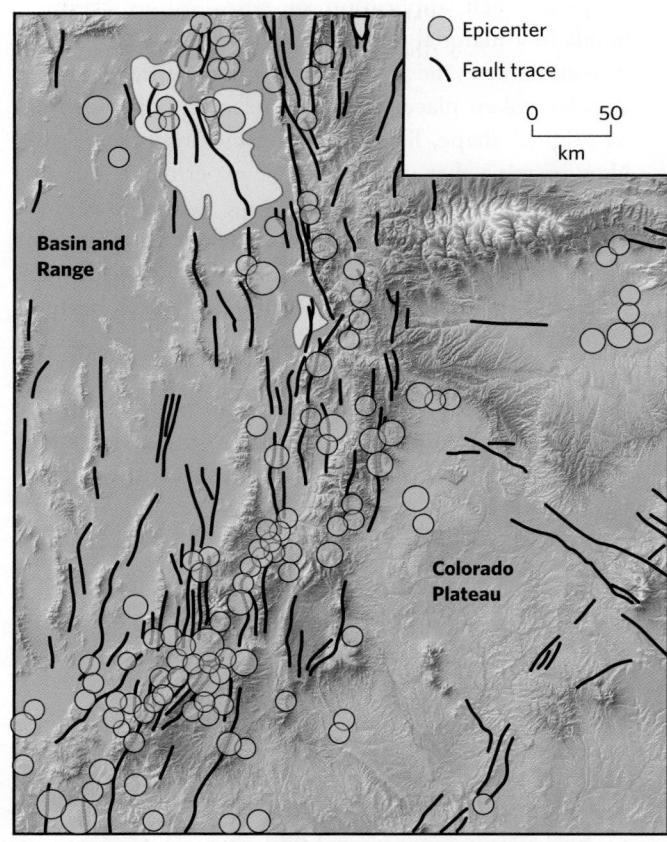

(b) A map of earthquake epicenters in Utah, recorded over several years. Seismicity happens mostly in a belt following the boundary between the Basin and Range Province and the Colorado Plateau.

300 km (185 mi) is an *intermediate-focus earthquake*, and one whose focus lies between 300 and 660 km (410 mi) is a *deep-focus earthquake*. (Earthquakes do not occur below a depth of 660 km because at the temperatures and pressures that exist at such depths, rock doesn't rupture seismically.) Because earthquake foci do not lie on the Earth's surface, we can't plot their positions directly on

a map. A dot on a map representing the location of an earthquake is the **epicenter** of the earthquake, the point on the surface of the Earth that lies vertically above the focus (Fig. 7.6b).

Take-home message...

Most earthquakes happen when stress causes either sudden formation of a new fault or slip on a pre-existing fault. When the slip takes place, elastically deformed rock on either side of the fault rebounds and generates seismic waves. The focus of an earthquake is the point within the Earth where slip begins, and the epicenter is the point on the Earth's surface that is vertically above the focus.

Quick Questions

- What is elastic deformation?
- Do all earthquakes happen when intact rock breaks?
- What is an earthquake epicenter?

7.2 Seismic Waves and Their Measurement

The Different Types of Seismic Waves

The energy released by slip on a fault moves through the Earth in the form of seismic waves, as we noted earlier. You've felt such vibrations if you've dropped a heavy book on the floor and felt a thud in your feet, for the impact causes vibrations to pass through the floor. Seismic waves are similar to the thud because they also involve the passage of vibrations through a material. Notably, seismic waves are an elastic phenomenon, in that when they pass through the Earth's interior, bonds stretch and bend, but do not break. (Passage of large seismic waves along the ground surface can crack the ground.) Seismologists distinguish between two basic categories of seismic waves depending on where they move, and within each of these categories, different types of waves can be distinguished from each other based on how they move.

- **Body waves** are seismic waves that pass through the interior of the Earth. They come in two different types. During movement of a *compressional wave*, rock vibrates back and forth in a direction parallel to the direction in which the wave itself moves (Fig. 7.7a). As a compressional wave passes, rock first contracts (squeezes together), then dilates (expands). To see this kind of motion, push on the end of a spring—you'll see that a pulse of contraction moves along the length of the spring. Compressional body waves are called **P-waves** (*P* for primary). During movement of a

shear wave, in contrast, the back-and-forth movement takes place in a direction perpendicular to the direction in which the wave travels (Fig. 7.7b). To see shear-wave motion, jerk the end of a rope up and down and watch how the up-and-down motion travels along the rope. Shear body waves are called **S-waves** (*S* for secondary).

- **Surface waves** are seismic waves that travel along the Earth's surface. They also come in two different types. *R-waves* (*R* for Rayleigh) are surface waves that cause the ground to move up and down in rolling undulations. In contrast, L-waves (*L* for Love) cause the ground surface to shimmy back and forth sideways, like a snake (Fig. 7.7c). (Rayleigh waves and Love waves were named for their discoverers.)

The different types of seismic waves travel at different velocities. P-waves move fastest, so they are the first seismic waves to arrive at a location, which is why they are called "primary." S-waves travel at about 60% of the speed of P-waves, so they arrive after the P-waves, which is why they are called "secondary." Both R-waves and L-waves are slower than body waves.

Using Seismographs to Detect and Record Earthquakes

In 1889, a German scientist noticed that a pendulum (a heavy weight suspended from a wire) in his lab appeared to sway slightly about an hour after an earthquake shook Tokyo, over 9,000 km (5,600 mi) away. What happened? *Inertia* (the tendency of an object at rest to remain at rest, or an object in motion to remain in motion, unless acted on by an outside force) kept the pendulum in place, while the lab moved due to very slight ground motion. This observation demonstrated that seismic waves can pass through the entire Earth and be detected, even when they produce motion so slight that people don't feel it. It led to the development, first, of a **seismometer**, an instrument that detects ground motion caused by seismic waves, and then a **seismograph**, a device that combines a seismometer with a recording device that documents the movements. Seismographs can be configured in two ways: a *vertical-motion seismograph* detects and records up-and-down ground motion (Fig. 7.8a), whereas a *horizontal-motion seismograph* detects and records back-and-forth ground motion (Fig. 7.8b).

The heart of a traditional mechanical seismograph consists of a heavy weight. In a vertical-motion seismograph, the weight hangs from a spring that is attached to a sturdy frame bolted to the ground. A pen extends from the weight and touches a vertical revolving paper-covered cylinder that is also attached to the frame (see Fig. 7.8a). In a

FIGURE 7.7 Different types of seismic waves.

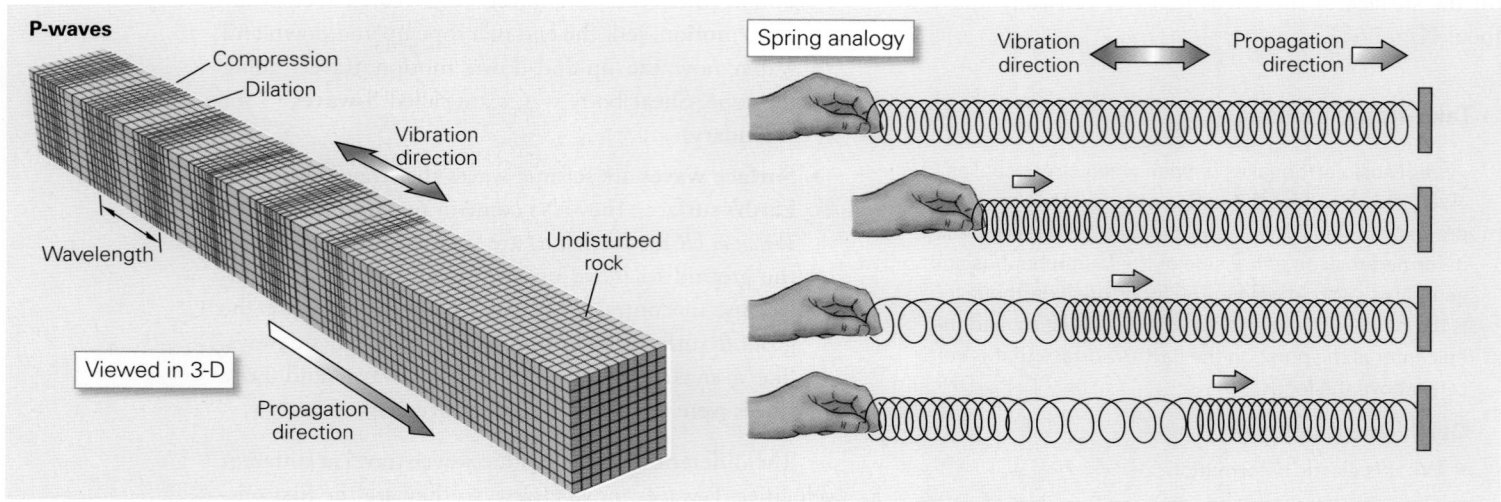

(a) P-waves are compressional waves, so the vibration direction is parallel to the direction of wave propagation. Such waves can be pictured by pushing and pulling on the end of a spring.

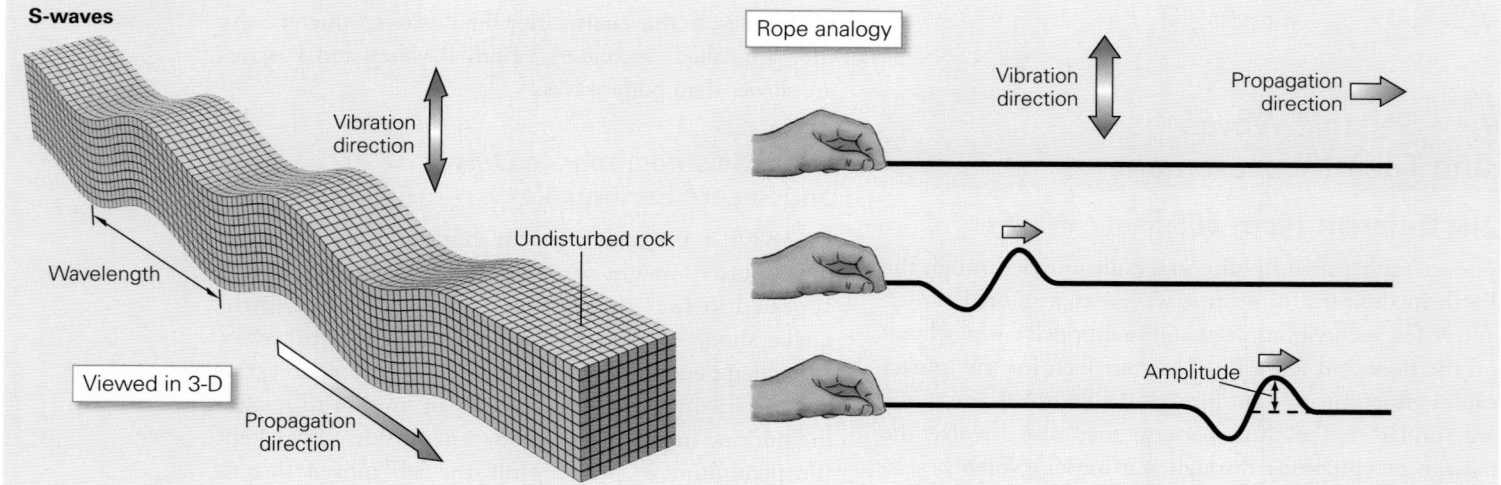

(b) S-waves are shear waves, so the vibration direction is perpendicular to the direction of wave propagation. Such waves can be pictured by moving the end of a rope up and down.

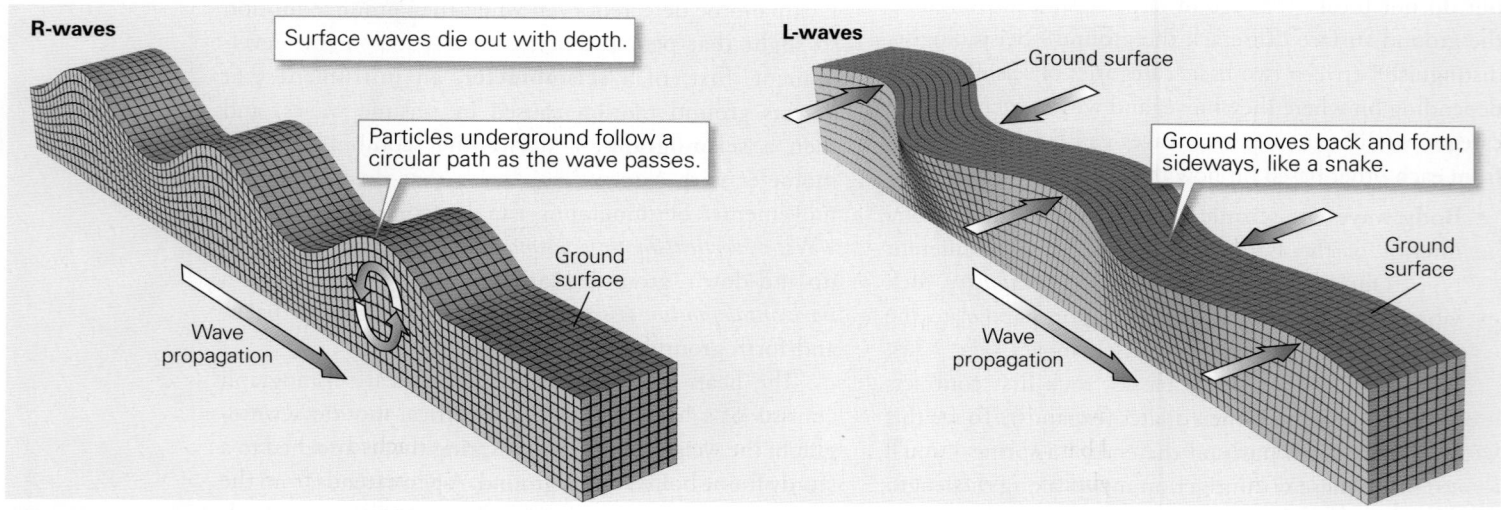

(c) There are two types of surface waves. R-waves make the ground surface go up and down in a rolling motion. When an L-wave passes, the ground surface moves back and forth like a slithering snake.

FIGURE 7.8 The basic operation of a mechanical seismograph.

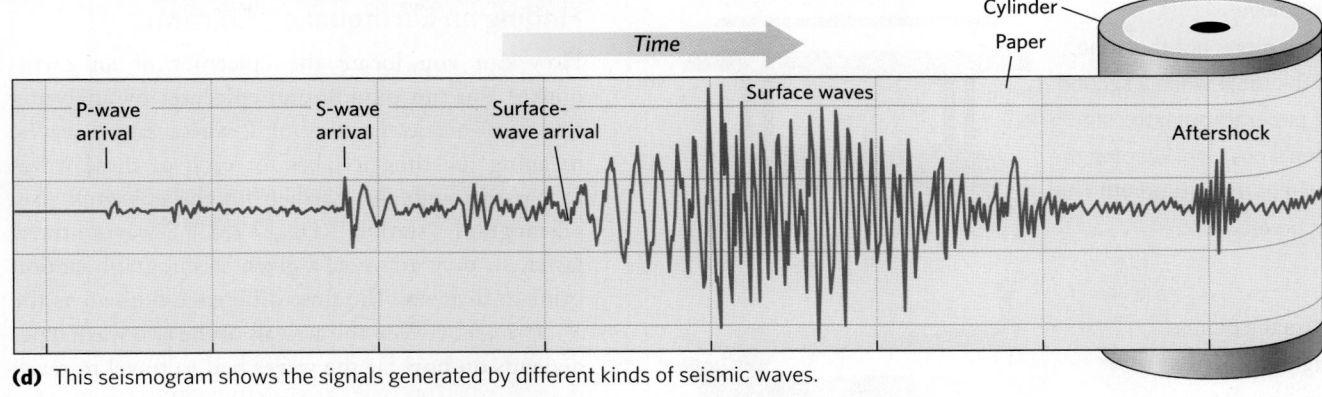

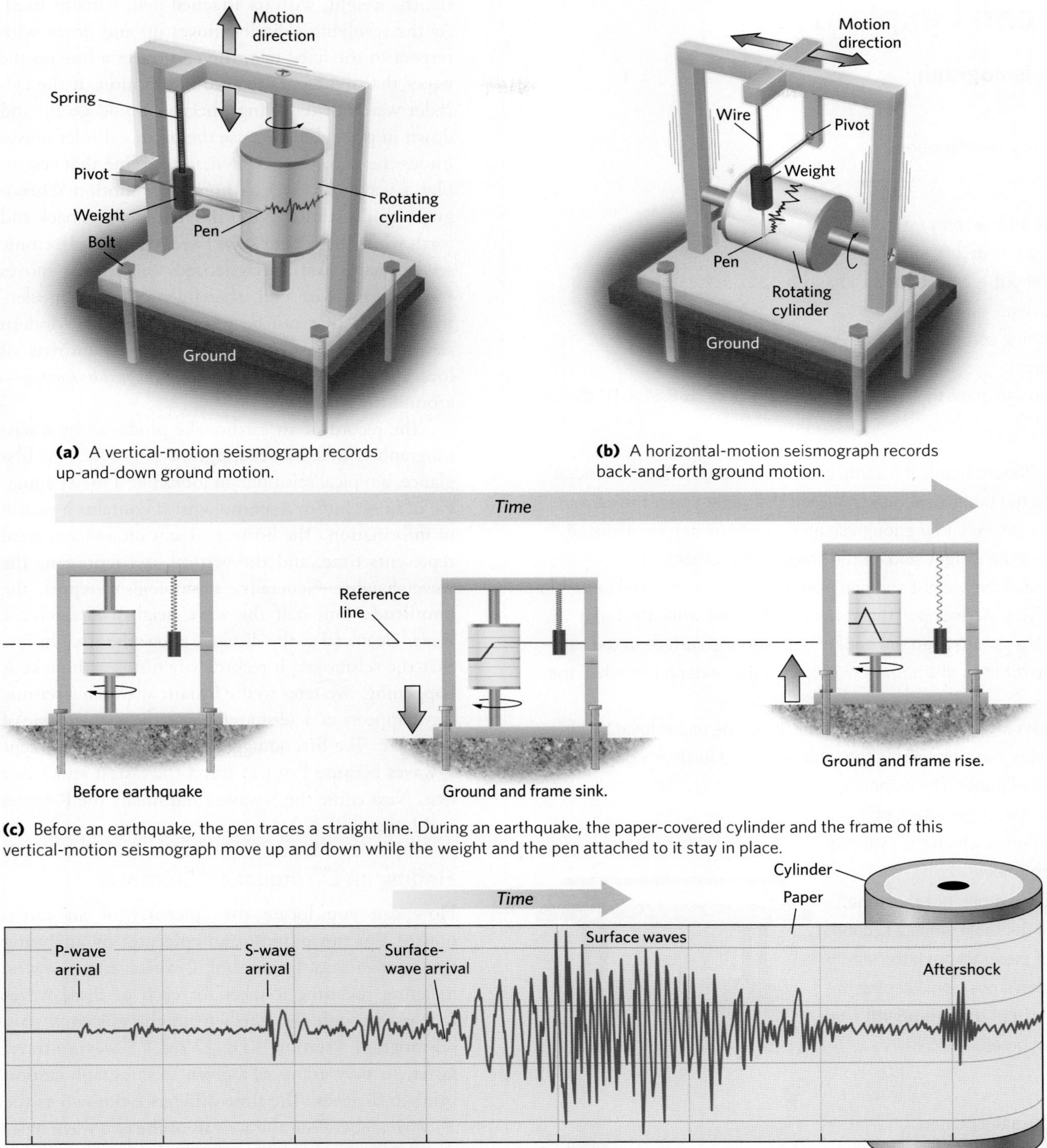

Motion direction

Spring
Pivot
Weight
Bolt
Pen
Rotating cylinder
Ground

(a) A vertical-motion seismograph records up-and-down ground motion.

Motion direction

Wire
Pivot
Weight
Pen
Rotating cylinder
Ground

(b) A horizontal-motion seismograph records back-and-forth ground motion.

Time

Reference line

Before earthquake

Ground and frame sink.

Ground and frame rise.

(c) Before an earthquake, the pen traces a straight line. During an earthquake, the paper-covered cylinder and the frame of this vertical-motion seismograph move up and down while the weight and the pen attached to it stay in place.

Cylinder
Paper

Time

P-wave arrival
S-wave arrival
Surface-wave arrival
Surface waves
Aftershock

(d) This seismogram shows the signals generated by different kinds of seismic waves.

horizontal-motion seismograph, the weight hangs from a wire, and is connected to a horizontal bar anchored to the frame at a pivot, and the revolving cylinder is horizontal.

To better understand how a seismograph works, let's follow a vertical-motion seismograph as it responds to up-and-down ground motion. Before an earthquake, when the ground is steady, the pen attached to the suspended weight traces out a straight reference line on the paper as the cylinder turns. When a seismic wave arrives and causes the ground surface to move, the seismometer

BOX 7.1

How can I explain . . .

The workings of a seismograph

What are we learning?

How a seismograph detects ground motion.

What you need:

- A small table (preferably a bit wobbly)
- A strong string (about 60 cm, or 2 ft, long)
- A 500–1,000 g weight (about 1–2 lb), such as a hand-held exercise weight
- A medium-point felt-tip pen
- Adhesive tape, or something comparable
- A sheet of sturdy cardboard
- A long sheet of paper; you can make this by taping together five or six 8.5″ × 11″ sheets

Instructions:

- Cut a slit through the cardboard that is the same width as the paper. Tape the sides of the cardboard sheet to the table, near one end. Feed the paper under the cardboard and up through the slit. Pull it far enough so about eight inches are exposed.
- Tape the pen to one end of the weight, and tie the string to the other end.
- Hold the string, and suspend the weight and pen from the string so that the pen touches the paper. Have your partner pull the paper past the pen while the table remains stationary, making sure to keep the paper flush with the surface of the cardboard. This simulates the rotation of a paper-covered cylinder under a pen when the ground is stable.
- Now return the paper to its original position. This time, keep the paper fixed by taping it to the table. Have your partner wobble the table back and forth in a direction perpendicular to the strip of paper. The wobbling simulates an earthquake. The paper's lack of motion simulates a situation in which the cylinder of a seismograph does not move.
- Repeat the experiment once more, but this time, have one partner wobble the table while a second partner pulls the sheet of paper through the slit and slowly beneath the pen's position, keeping the paper flush with the surface of the cardboard. The pen traces out wave-like curves.

What did we see?

A seismograph works because the pen, which is connected to a weight, stays fixed due to inertia, while the cylinder and frame of the seismograph move with the shaking Earth. The cylinder must rotate so that the pen's trace doesn't keep overprinting itself.

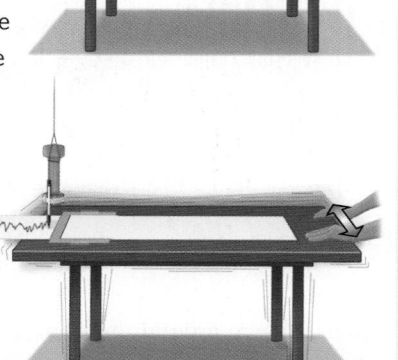

frame—along with the paper-covered cylinder—moves with it (Fig. 7.8c). However, because of inertia, the weight, with its attached pen, remains fixed. As the revolving cylinder moves up and down with respect to the fixed pen, the pen traces a line on the paper that represents the ground motion. If the cylinder were not revolving, the pen would go up and down in place, but because the paper cylinder moves under the pen, the pen traces out a line that resembles a wave (Box 7.1). A horizontal-motion seismograph works similarly, but the frame moves back and forth while the weight stays fixed. Modern electronic seismometers use a magnetized weight that moves relative to a wire coil, thereby producing an electrical signal that can be recorded digitally. Modern seismographs have been installed at hundreds of locations—each known as a *seismograph station*—around the world.

The record of an earthquake produced by a seismograph is called a **seismogram** (Fig. 7.8d). At first glance, a typical seismogram looks like a messy squiggle of lines, but to a seismologist it contains a wealth of information. The horizontal axis on a seismogram represents time, and the vertical axis represents the wave height. Generally, seismologists report the **amplitude** (one-half the wave height) of a seismic wave, measured as the distance above the straight line that the seismograph records when no earthquake is happening. We refer to the instant at which a seismic wave appears at a seismograph as the *arrival time* of the wave. The first squiggles on the record represent P-waves because P-waves travel the fastest and arrive first. Next come the S-waves, and finally the R-waves and L-waves.

Finding an Earthquake's Epicenter

How can you locate the epicenter of an earthquake? You can pinpoint an epicenter by analyzing the relative **travel times** of P-waves and S-waves, meaning the time it takes for each of these waves to pass through the Earth from the epicenter to a seismograph station (Fig. 7.9a). P-waves travel faster, so they arrive at a given seismograph station prior to S-waves. The time difference (known as the *S – P time*) between the arrivals of the two wave types depends on how far the waves had to travel to reach the seismograph station (Fig. 7.9b). (To picture why, imagine two cars traveling in the same direction, one at 60 km/h and the other at 50 km/h. At 60 km from the starting line, the faster car will be 10 km ahead, but at 120 km from the starting line, the faster car will be 20 km ahead.) Therefore, because the speed of both S-waves and P-waves is known, the S – P time

FIGURE 7.9 Locating an earthquake's epicenter.

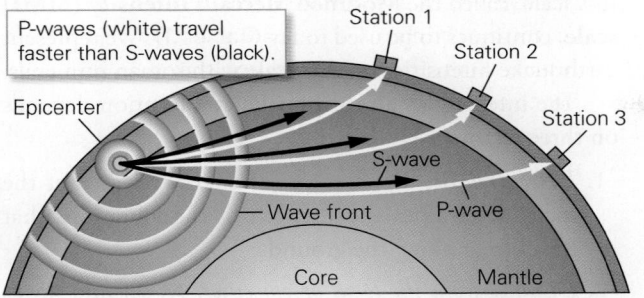

P-waves (white) travel faster than S-waves (black).

Station 1
Station 2
Station 3
Epicenter
S-wave
Wave front
P-wave
Core
Mantle

(a) The greater the distance between the epicenter and a seismograph station, the longer the path of a body wave traveling through the Earth.

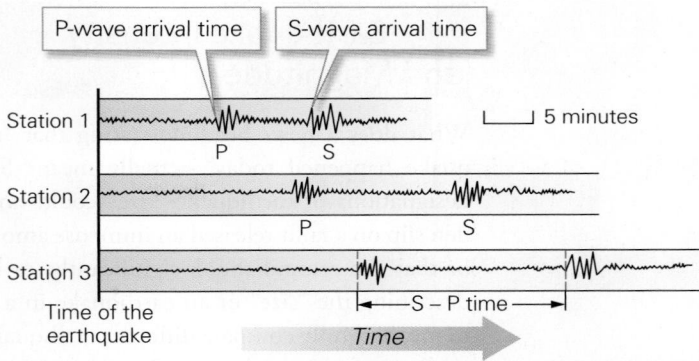

P-wave arrival time   S-wave arrival time

Station 1 |— 5 minutes
P   S
Station 2
P   S
Station 3
Time of the earthquake
←— S – P time —→
Time

(b) S-waves are slower than P-waves, so they arrive later. The difference between their arrival times, called the S – P time, increases with distance from the epicenter.

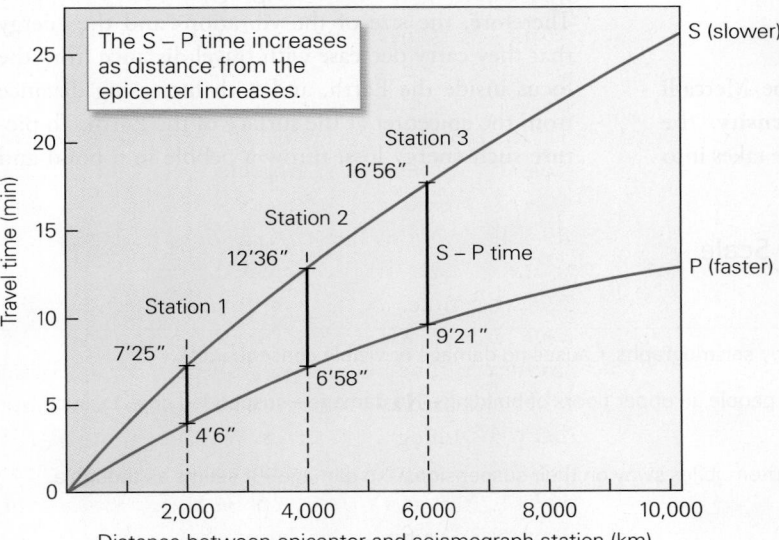

The S – P time increases as distance from the epicenter increases.

S (slower)

Station 3
16'56"

Station 2
12'36"
S – P time

P (faster)

Station 1
7'25"

9'21"

6'58"

4'6"

Travel time (min)

Distance between epicenter and seismograph station (km)

(c) Arrival times of P-waves and S-waves can be plotted on a graph as travel-time curves. The S – P time at a given distance from the epicenter is represented by the vertical distance between the S-wave and P-wave curves.

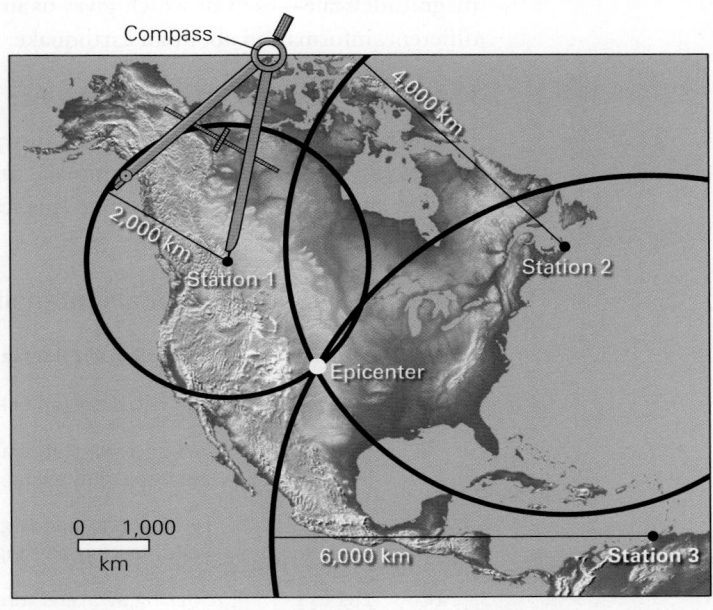

Compass
4,000 km
2,000 km
Station 1
Station 2
Epicenter
0   1,000
km
6,000 km
Station 3

(d) If an earthquake epicenter lies 2,000 km from Station 1, we draw a circle with a radius of 2,000 km around that station at the scale of the map. We repeat for the other two stations. The intersection of the three circles is the epicenter.

recorded at one station tells you the distance between the epicenter and the station.

Note, however, that the S – P time does not give the direction from the station to the epicenter. To find the epicenter location, you must calculate the distance from the epicenter to three different stations. A graph plotting the relationship between distance traveled and time since initiation, for both S-waves and P-waves, provides a quick way to obtain this information from S – P times **(Fig. 7.9c)**. (Lines on the graph curve because waves that travel farther also go deeper into the Earth, where they move faster.) You can then draw a circle around each station on a map such that the radius of the circle represents the distance between that station and the epicenter at the scale of the map. The epicenter lies at the intersection of the three circles, for this is the only point that has the appropriate measured distance from all three stations **(Fig. 7.9d)**.

Take-home message...

  Earthquake energy travels as seismic waves. Body waves travel through the interior of the Earth, whereas surface waves travel along its surface. Seismologists distinguish between two kinds of body waves (P-waves and S-waves) and between two kinds of surface waves (R-waves and L-waves). Seismographs measure and record ground shaking. Because different wave types travel at different speeds, the difference in arrival time between S-waves and P-waves indicates the distance between a seismograph and an earthquake epicenter.

Quick Questions

- Name the four kinds of seismic waves. Which two are body waves?

- Which part of a seismograph moves with the ground, and which stays fixed in position, during an earthquake?

- Can you locate an epicenter from data obtained at one seismograph station?

▶ **Animation**
Seismic Waves and Epicenter Location (Interactive)

7.3 Earthquake Intensity and Magnitude

What does a news headline stating that "a large earthquake happened today" actually mean? Such informal designations of earthquake "size" could imply that sudden slip on a fault released an immense amount of energy, or that the ground shook significantly, or both! Indeed, describing the "size" of an earthquake in a uniform way to meaningfully compare different earthquakes has been a challenge for researchers. Ultimately, seismologists developed two different scales—the intensity scale and the magnitude scale—each of which gives us important, but different, information about an earthquake.

Modified Mercalli Intensity Scale

In 1902, an Italian scientist named Giuseppe Mercalli devised a scale for defining earthquake **intensity**, the amount and vigor of ground motion. This scale takes into account both the damage that the earthquake caused and people's perception of the ground motion. A version of this scale, called the **Modified Mercalli Intensity (MMI) scale**, continues to be used today (Table 7.1). We represent earthquake intensities on this scale with roman numerals.

The intensity of an earthquake at a location depends on three factors:

1. *Energy release:* The amount of energy released at the focus determines the strength of the vibrations that travel to or along the ground.

2. *Distance from the focus or epicenter:* As seismic waves move outward from the focus, they spread out over a progressively larger volume and, because rock isn't perfectly elastic, some energy gets lost to friction. Therefore, the size of the vibrations and the energy that they carry decrease with travel distance from the focus inside the Earth, and with increasing distance from the epicenter at the surface of the Earth. To picture such energy loss, throw a pebble in a pond and

TABLE 7.1 Modified Mercalli Intensity Scale

MMI	Shaking	Perception and Damage
I	None	Not felt; detected only by seismographs. Causes no damage or visible consequences.
II	Weak	Felt by a few stationary people, in upper floors of buildings. No damage—suspended objects, such as lamps, may swing.
III	Weak	Felt indoors; standing automobiles sway on their suspensions. No damage—it seems as though a heavy truck is passing.
IV	Light	Shaking awakens some sleepers. No damage—dishes and windows rattle.
V	Moderate	Most people awaken. Very light damage—some dishes and windows break; unstable objects tip over; trees and poles sway.
VI	Strong	Shaking frightens some people. Light damage—plaster walls crack; heavy furniture moves slightly; a few chimneys crack.
VII	Very strong	Most people are frightened. Moderate damage—plaster walls crack; windows break; some chimneys topple; unstable furniture overturns; poorly built buildings sustain considerable damage.
VIII	Severe	Everyone is frightened. Moderate to heavy damage—many chimneys and factory smokestacks topple; heavy furniture overturns; substantial buildings sustain damage; poorly built buildings suffer severe damage.
IX	Violent	Terrifying. Heavy damage—frame buildings separate from their foundations; most buildings sustain damage; some buildings collapse; the ground cracks; underground pipes break; rails bend; some landslides occur.
X	Extreme	Terrifying. Very heavy damage—most masonry structures are destroyed; the ground cracks in places; landslides occur; some bridges collapse; facades on buildings collapse; railways and roads are severely damaged.
XI	Extreme	Terrifying. Nearly complete destruction; few masonry buildings remain; many bridges collapse; broad fissures form in the ground; most pipelines break; severe liquefaction of sediment occurs; some dams collapse.
XII	Extreme	Terrifying. Seismic waves cause visible undulations of the ground surface; objects fly up off the ground; buildings and bridges of all types are completely destroyed.

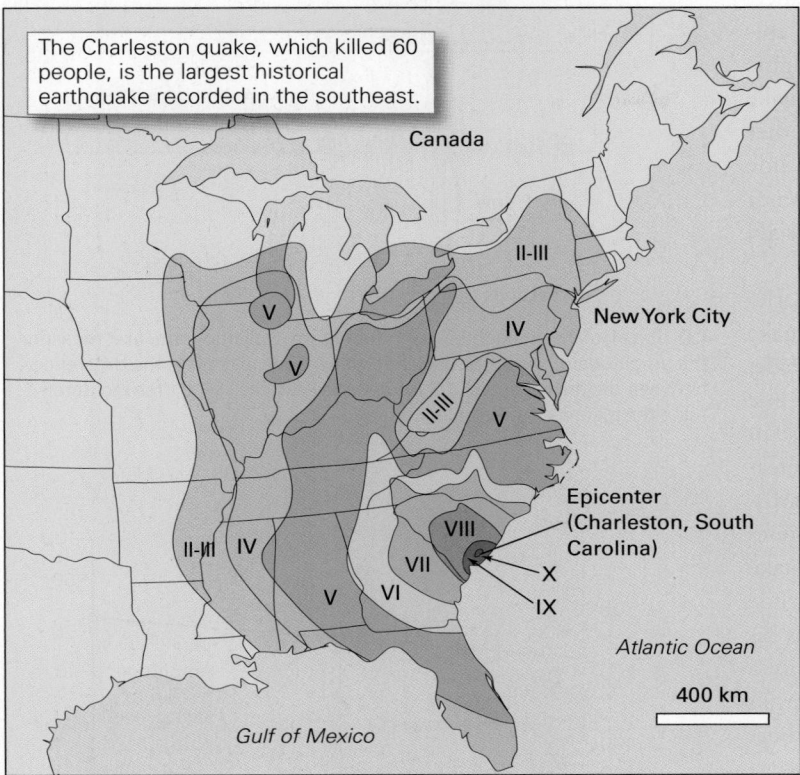

The Charleston quake, which killed 60 people, is the largest historical earthquake recorded in the southeast.

(a) A contoured map of intensity for the 1886 Charleston, South Carolina, earthquake. Ground shaking reached an MMI of X at the epicenter, but was II-III in New York City.

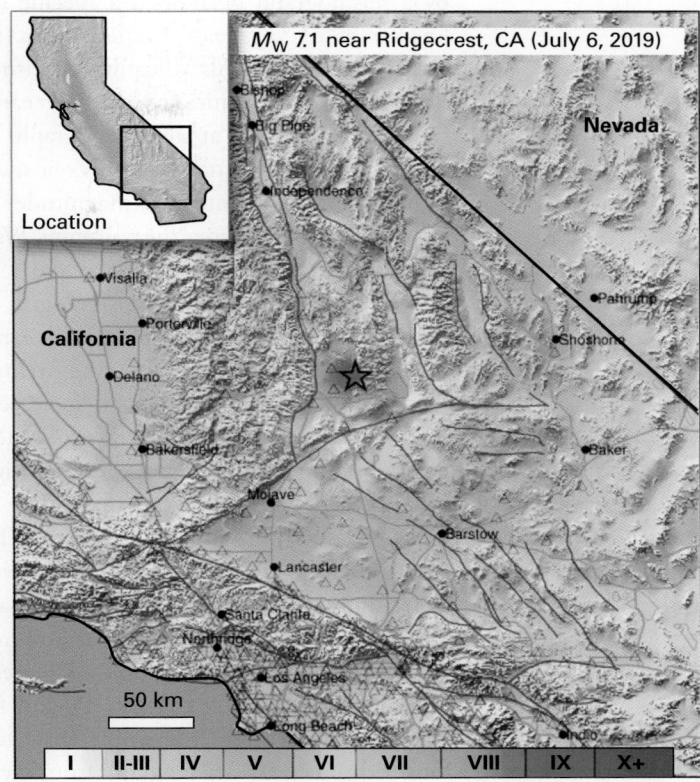

M_W 7.1 near Ridgecrest, CA (July 6, 2019)

| I | II-III | IV | V | VI | VII | VIII | IX | X+ |

(b) A USGS ShakeMap for an earthquake in California. The inset map shows the location.

watch how the heights of the resulting waves diminish as the waves move away from the pebble's impact location.

3. *Composition of Earth materials:* The character of the material through which the seismic waves travel also affects earthquake intensity. On a regional scale, stronger crust transmits seismic waves more efficiently than does weaker crust. On a local scale, passage of waves through soft sediment results in *amplification* of the waves. (Amplification happens because the waves slow as they pass into the sediment, so their amplitude must increase.)

Because of these three factors, we can make the following generalizations about earthquake intensity: the intensity of a shallow-focus earthquake exceeds that of a deep-focus earthquake for a given amount of energy released; intensity tends to be greatest near the epicenter and to decrease progressively away from the epicenter; ground motion for an earthquake of a given intensity affects a much larger area in regions of strong crust than in regions of weak crust; and intensity tends to be greater where the substrate consists of weak sediment than where it consists of hard bedrock.

Seismologists portray the variation of an earthquake's intensity over a region on a **seismic-intensity map** (Fig. 7.10a). Each color on the map represents a different intensity. The US Geological Survey now rapidly produces seismic-intensity maps, branded as *ShakeMaps*, for most significant earthquakes, using a computer that estimates intensity from seismograph measurements (Fig. 7.10b), and *"Did You Feel It?"* maps, which plot crowd-sourced data obtained when people log into a website to register that they felt an earthquake.

Earthquake Magnitude Scales

News stories about earthquakes generally do not provide an MMI number, but rather include a statement like, "An earthquake with a magnitude of 7.2 struck the city yesterday." What does this number mean? An earthquake's **magnitude** represents the amount of energy released by the earthquake, rather than the consequences of that energy release. Very simply, to calculate a magnitude, seismologists first measure the height of the largest spike on a seismogram, for it represents the maximum amplitude of the ground motion. Then, after they have determined the distance between the epicenter and the seismograph, they adjust the measurement to be equivalent to the maximum amplitude that would be recorded by a seismograph positioned a specific distance, known as the *reference distance*, from the epicenter. Because of this adjustment, seismologists obtain the same magnitude for a given earthquake from a measurement at any seismograph. In other words, a given earthquake has only one magnitude number, so, unlike intensity, magnitude does not depend on distance from the epicenter. We specify magnitude (*M*) using arabic numerals.

In 1935, an American seismologist, Charles Richter, developed a scale for defining earthquake magnitude. This scale, known as the **Richter scale**, is logarithmic, meaning that an increase of one unit of magnitude represents a tenfold increase in the maximum amplitude of ground motion measured by a seismograph. So, a magnitude 8 earthquake results in ground motion that is 10 times greater than that of a magnitude 7 earthquake and 1,000 times greater than that of a magnitude 5 earthquake. Richter used 100 km (62 mi) as his reference distance. Since there's not necessarily a seismograph station at exactly 100 km from an epicenter, he developed a simple chart to adjust for the distance of a seismograph from the epicenter (Fig. 7.11).

Later research showed that the Richter scale does not accurately represent the size of very large earthquakes. Therefore, seismologists now use a different scale, called the **moment magnitude scale**, to better represent an earthquake's size. To calculate an earthquake's moment magnitude (abbreviated M_w), seismologists measure the amplitudes of several different seismic waves, and take into account the dimensions of the slipped area on the fault and the amount of displacement that occurred. The moment magnitude scale, which, like the Richter scale, is logarithmic, has become the number associated with an earthquake in official reports. News reports sometimes describe an earthquake as being "a 6 on the Richter scale," because the name is familiar, but in fact, the number they provide likely comes from the moment magnitude scale. Similarly, news reports sometimes incorrectly state that the magnitude scale "goes from 1 to 10." In fact, microearthquakes (M_w –1 or M_w –2) can be detected by seismographs positioned close to the epicenter, and there is no defined upper limit to the magnitude scale. That said, seismologists estimate that M_w 9.5 is about as big as a fault-generated earthquake can get on the Earth, given the known dimensions of faults. Only one M_w 9.5 earthquake has happened in recorded history—it struck coastal Chile in 1960.

FIGURE 7.11 Using the Richter magnitude scale.

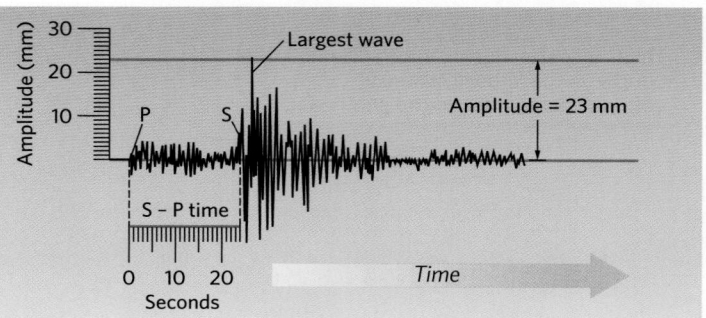

(a) To calculate the Richter magnitude from a seismogram, first measure the amplitude of the largest seismic wave. Then, calculate the difference between the arrival times of S-waves and P-waves (S – P time) to determine the distance to the epicenter (see Fig. 7.9).

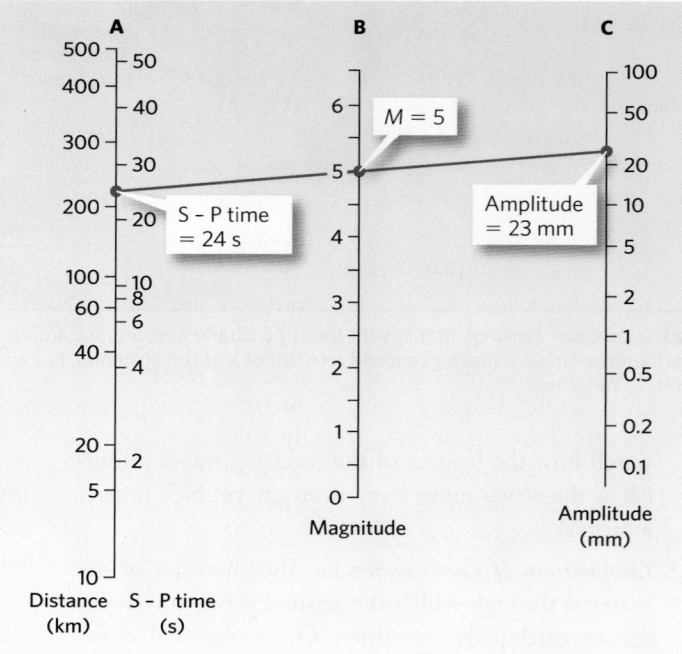

(b) Next, draw a line from the point on Column A representing the S – P time (or distance to the epicenter) to the point on Column C representing the wave amplitude. Then read the Richter magnitude off Column B.

To make discussion of earthquakes easier, seismologists use familiar adjectives to describe an earthquake's magnitude and its effects (Table 7.2). Most people standing on the ground near the epicenter of a light shallow-focus earthquake can feel slight shaking. Buildings near the epicenter will be significantly damaged by moderate earthquakes. Great earthquakes are catastrophic near the epicenter and can cause damage up to a few hundred kilometers from the epicenter. For shallow-focus earthquakes, the intensity at the epicenter roughly correlates with magnitude. Notably, the greater the magnitude of a shallow-focus earthquake, the broader the area of high intensity. Deep-focus earthquakes also cause damage, but generally less than an equivalent shallow-focus earthquake, because seismic waves weaken as they travel upward.

Energy Release by Earthquakes

How much energy does an earthquake release? To give a sense of the amount of energy released by an earthquake, seismologists compare earthquakes to other energy-releasing events. For example, according to some estimates, an M_W 5.3 earthquake releases about as much energy as the Hiroshima atomic bomb. The largest hydrogen bomb ever detonated (the Tsar Bomba, in 1961) released about as much energy as an M_W 8.0 earthquake, so an M_W 9.0 earthquake releases vastly more energy than that bomb. Notably, although an increase of one unit of magnitude represents a tenfold increase in the maximum amplitude of ground motion, it represents a 32-fold increase in energy release. Therefore, an M_W 8 earthquake releases about 1 million times more energy than an M_W 4 earthquake (Fig. 7.12). Fortunately, great earthquakes occur much less frequently than minor earthquakes!

The amount of energy released by an earthquake depends on both the area and the displacement of the fault surface that slips. Generally, the larger the earthquake, the larger the slipped area and the greater the displacement. For example, during the M_W 9.0–9.1 Tōhoku earthquake, an area 300 km (180 mi) long by 100 km (60 mi) wide slipped, and up to 30 m (100 ft) of displacement occurred. In contrast, during a light earthquake, an area of less than a square kilometer may slip, and maximum displacement may be only a few centimeters. The amount of displacement varies with location along a fault; it tends to be greatest near the focus and to die out progressively toward the edge of the slipped area, beyond which the displacement is zero.

Take-home message...

We can specify earthquake size by intensity (a measure based on perception of damage caused and shaking felt) or by magnitude (a representation of energy released by the earthquake). An increase of one magnitude unit represents a 10-fold increase in the amplitude of shaking and a 32-fold increase in energy. Some great earthquakes release more energy than the largest hydrogen bomb.

Quick Questions

- Can you specify the size of an earthquake by giving just one intensity number on the MMI scale?
- How many magnitude numbers can be assigned to a single earthquake?
- Do modern seismologists use the Richter scale to define the magnitude of an earthquake?

TABLE 7.2 Adjectives for Describing Earthquakes

Adjective	Magnitude	Intensity at Epicenter	Effects
Great	>8.0	X–XII	Total destruction
Major	7.0–7.9	IX–X	Extreme damage
Strong	6.0–6.9	VII–VIII	Moderate to serious damage
Moderate	5.0–5.9	VI–VII	Slight to moderate damage
Light	4.0–4.9	IV–V	Felt by most; slight damage
Minor	<3.9	III or smaller	Felt by some; hardly any damage

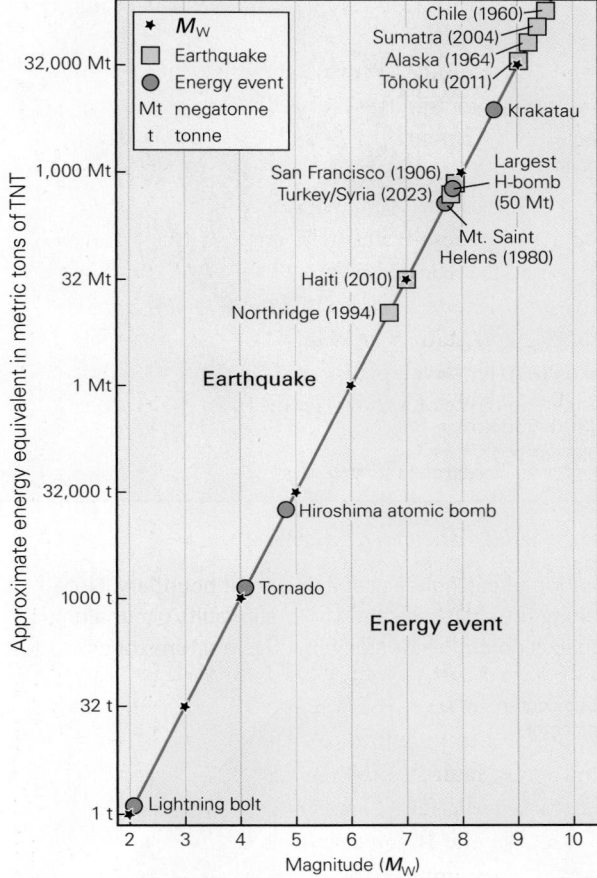

FIGURE 7.12 The energy released by earthquakes increases dramatically with magnitude. Some great earthquakes (>M_W 8.0) release more energy than the largest hydrogen bombs.

7.4 Where and Why Do Earthquakes Occur?

Earthquakes do not take place everywhere on the globe. By plotting the distribution of earthquake epicenters on a map, seismologists have found that most, but not all, earthquakes occur in **seismic belts**, or *seismic zones* (Fig. 7.13). Most seismic belts, as we've seen, correspond to plate boundaries, and earthquakes within these belts are known as *plate-boundary earthquakes*. Seismic belts also follow active *rifts* (where continents are breaking apart)

Did you ever wonder . . .

whether an earthquake will happen near where you live?

FIGURE 7.13 A map of epicenters over a period of several decades emphasizes that most earthquakes occur in distinct seismic belts along plate boundaries. The color of an epicenter dot represents the focus depth of the earthquake.

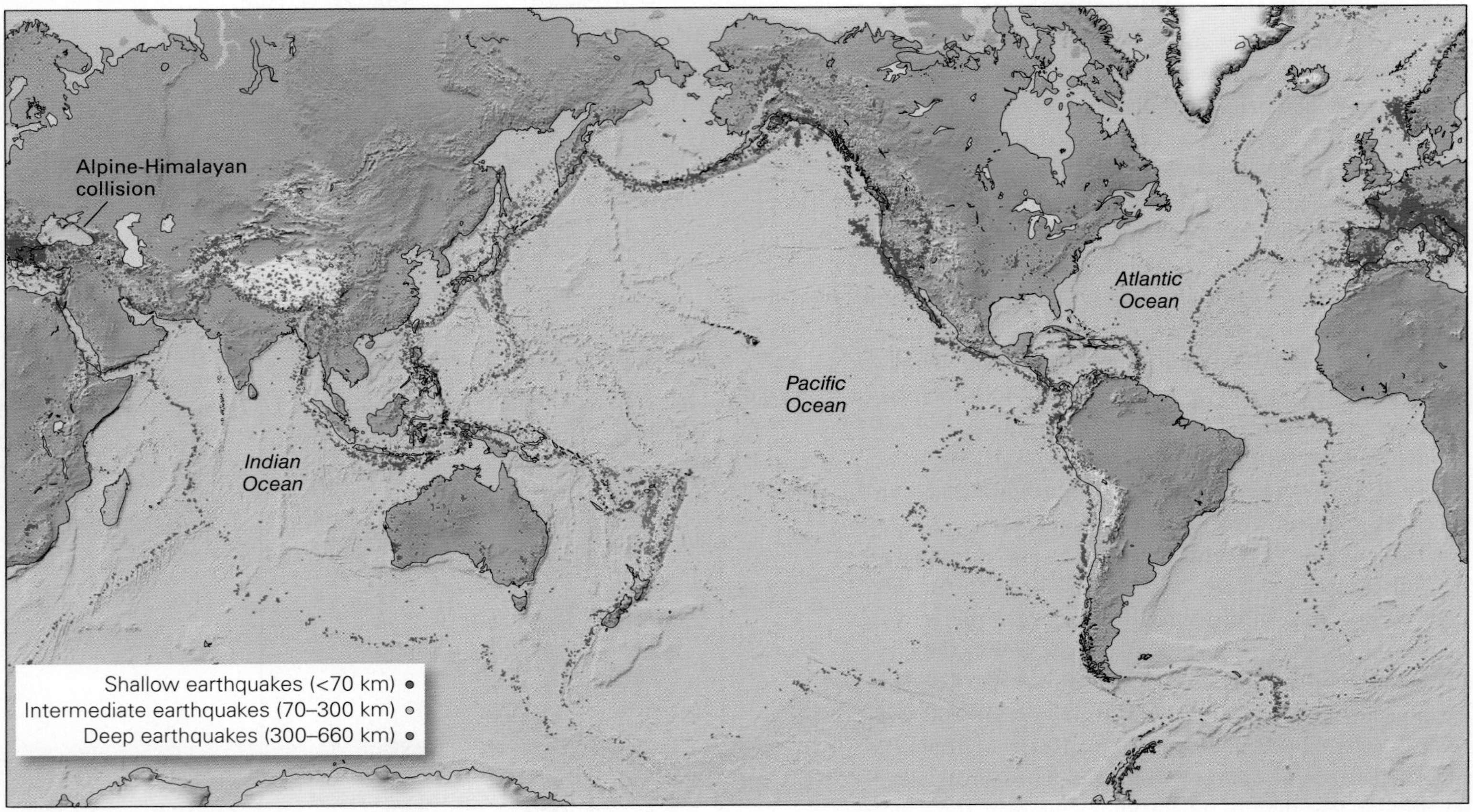

Alpine-Himalayan collision

Atlantic Ocean

Pacific Ocean

Indian Ocean

Shallow earthquakes (<70 km) ●
Intermediate earthquakes (70–300 km) ◦
Deep earthquakes (300–660 km) ●

FIGURE 7.14 The distribution of earthquakes at a divergent boundary. Note that normal faults occur along the ridge axis and strike-slip faults occur along active transforms. Earthquakes do not occur along inactive fracture zones.

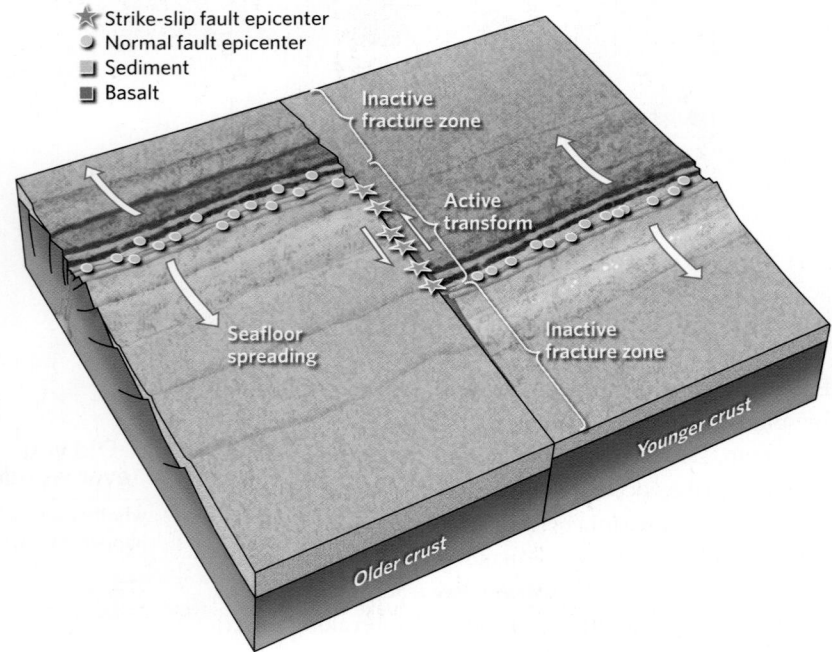

★ Strike-slip fault epicenter
● Normal fault epicenter
▢ Sediment
▣ Basalt

Inactive fracture zone

Active transform

Seafloor spreading

Inactive fracture zone

Older crust

Younger crust

and *continental collision zones* (where two continents once separated by oceanic lithosphere have pushed into each other). Those earthquakes that occur away from plate boundaries, rifts, or collision zones are called *intraplate earthquakes*. Different types of faults—normal, reverse, thrust, and strike-slip (see Chapter 6)—occur in these different geologic settings.

Plate-Boundary Earthquakes

Plates move relative to their neighbors at rates of 1–15 cm/yr (0.4–6 in/yr). So, over time, large stresses build along plate boundaries, and these stresses trigger sudden slip on faults.

DIVERGENT-BOUNDARY SEISMICITY. At a divergent boundary (a mid-ocean ridge), two plates form and move away from the ridge axis. Divergent boundaries are broken into ridge segments linked by transform faults. Along ridge segments, stretching generates normal faults, whereas along the transforms, strike-slip displacement occurs (Fig. 7.14). Earthquakes on all these faults have shallow foci. But most do not cause casualties or damage because they take place beneath the ocean, far from population centers.

TRANSFORM-BOUNDARY SEISMICITY ON CONTINENTS.
At transform boundaries, strike-slip displacement takes place. Most transform faults on the Earth link segments of mid-ocean ridges, which, as we've noted, are far from population centers. But a few transform faults, such as the San Andreas Fault of California, the Alpine Fault of New Zealand, and the North and South Anatolian Faults in Turkey, cut through continental crust. Large earthquakes on continental transform faults can be very destructive because they have shallow foci, so the seismic waves they produce may still be carrying a large amount of energy when they reach the ground surface. Furthermore, because they're on land, they may lie near population centers. Let's consider a couple of examples.

In 2023, a catastrophic earthquake along the South Anatolian Fault, one of two major faults that accommodate the westward movement of Turkey (see Fig. 2.32), slipped and generated a shallow M_W 7.8 earthquake in southern Turkey and northern Syria, in a heavily populated region. Due to design flaws and poor construction in many concrete buildings, not only did old buildings fail, but many new high-rise apartment buildings collapsed like pancakes. The resulting catastrophe killed about 60,000 people and injured over 120,000. In the recent past, strike-slip faulting along the North Anatolian Fault has produced similar destruction and casualties. For example, an earthquake near Istanbul in 1999 led to the deaths of over 18,000 people.

In the United States, faulting associated with a continental transform boundary takes places along the San Andreas Fault. The San Andreas is the longest continuous fault in a system of strike-slip faults in western California, which together accommodate the displacement of the Pacific Plate, relative to the North American Plate, at an average rate of 6 cm/yr (2.4 in/yr). Stick-slip behavior causes numerous earthquakes to happen along this fault and on subsidiary faults near it (Fig. 7.15a). Some of these events have been large and destructive, or would be very destructive were they to happen today. In 1857, for example, the Fort Tejon earthquake (M_W 7.9) ruptured

FIGURE 7.15 Seismicity along the San Andreas Fault, a continental transform fault.

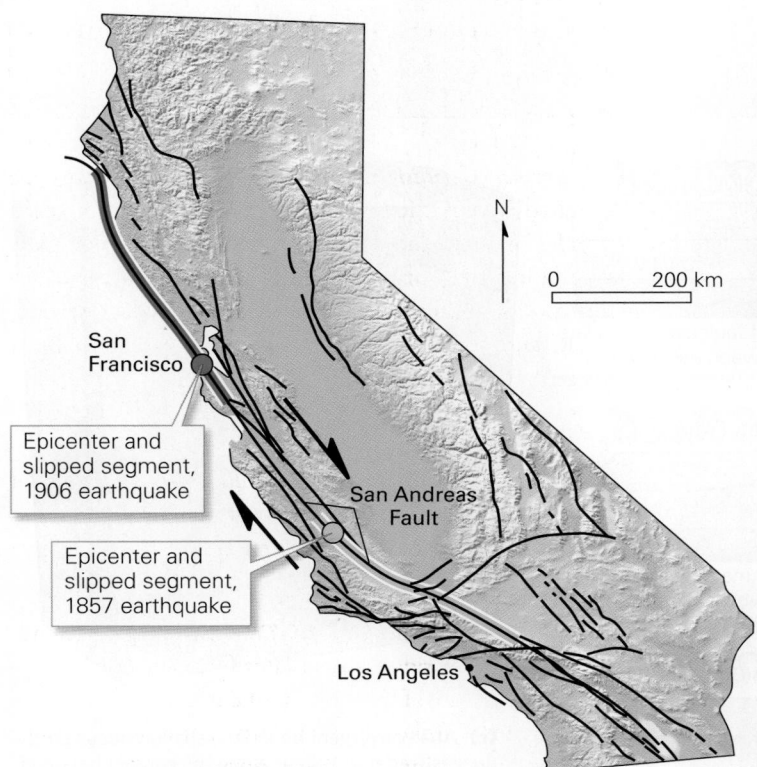

(a) The San Andreas Fault system in California. Note that the system includes many faults in a band 100 km wide. Colored lines indicate the portions of the fault that ruptured in 1857 and in 1906.

(b) A street in San Francisco after the 1906 earthquake and the related fire.

(c) The 1989 Loma Prieta earthquake caused a two-level freeway in San Francisco to collapse.

a 350 km (220 mi) segment of the southern San Andreas Fault. The MMI for the event reached IX (extreme damage) just north of Los Angeles, but given that fewer than 5,000 people lived in Los Angeles then, it did not cause a catastrophe. A repeat of such an event could be devastating for the region today. In 1906, a 477 km (300 mi) segment of the fault suddenly slipped by up to 7 m (23 ft) and produced an M_W 7.9 earthquake (see Figure 7.4). When seismic waves struck nearby San Francisco, streets in the city undulated, buildings swayed and collapsed, and toppled stoves and lamps sparked a fire that consumed the ruins (Fig. 7.15b). In 1989, an M_W 6.9 event struck Loma Prieta, 100 km to the south of San Francisco. This event caused the tragic collapse of a double-decked freeway and disrupted a World Series baseball game (Fig. 7.15c).

CONVERGENT-BOUNDARY SEISMICITY. Convergent boundaries are complicated regions that produce earthquakes at a variety of depths (Fig. 7.16a). Most slip takes place along thrust faults. Notably, more than 80% of the most devastating earthquakes to take place on the Earth happen at locations on the convergent boundaries that surround much of the Pacific Plate—the same boundaries that produce the belt of volcanic arcs constituting the Ring of Fire.

The most dangerous convergent-boundary earthquakes occur along the large, relatively shallow thrust fault that delineates the boundary between the base of the overriding plate (and its accretionary prism) and the top of the downgoing plate. Because the foci of these earthquakes are shallow, much of the seismic energy produced by elastic rebound reaches the ground surface. Notable examples of devastating convergent-boundary earthquakes include the largest recorded earthquakes: the 1960 M_W 9.5 earthquake in Chile, the 1964 M_W 9.2 Good Friday earthquake in Alaska, the 2004 M_W 9.3 earthquake in Sumatra, and the 2011 M_W 9.0 Tōhoku earthquake in Japan. In 1700, a similar earthquake happened in the *Cascadia subduction zone* bordering Washington and Oregon, where the Juan de Fuca Plate subducts beneath North America. Each of these earthquakes released huge amounts of energy because each involved a large displacement on a very broad area of the thrust fault between the overriding and downgoing plates. Seismologists now refer to such earthquakes as **megathrust earthquakes**.

Prior to a megathrust earthquake, the motion of the downgoing plate drags the base of the accretionary prism landward, and elastic deformation builds, as if a giant spring were being shortened (Fig. 7.16b). The accretionary prism shortens in the direction perpendicular to

FIGURE 7.16 Convergent-boundary seismicity.

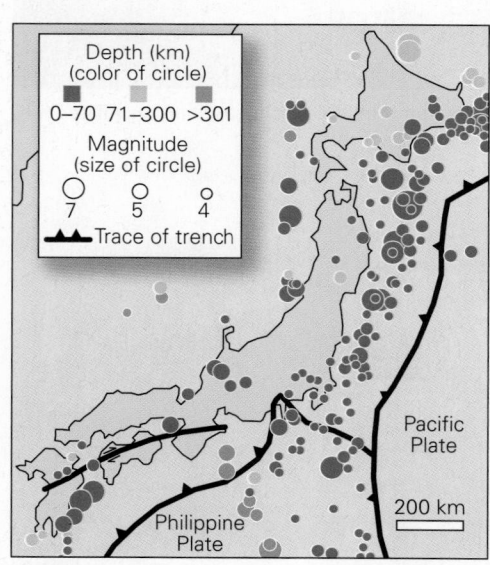

(a) A map of earthquake epicenters near Japan shows how complex convergent boundaries can be. The color of each circle indicates focus depth, and its diameter indicates magnitude.

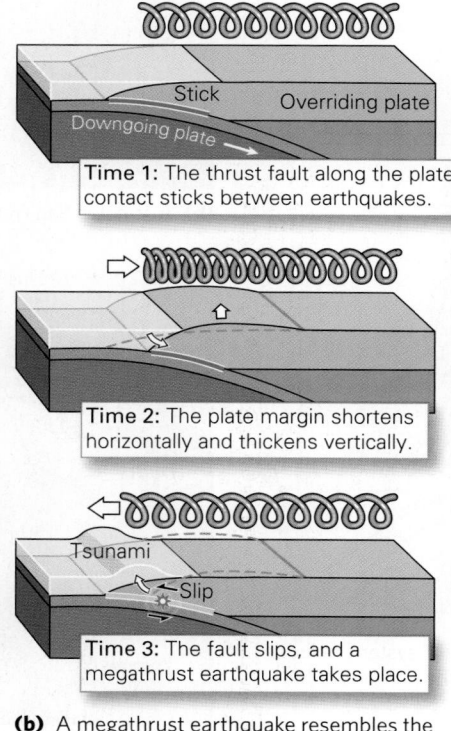

(b) A megathrust earthquake resembles the shortening and then the rebound of a giant horizontal spring.

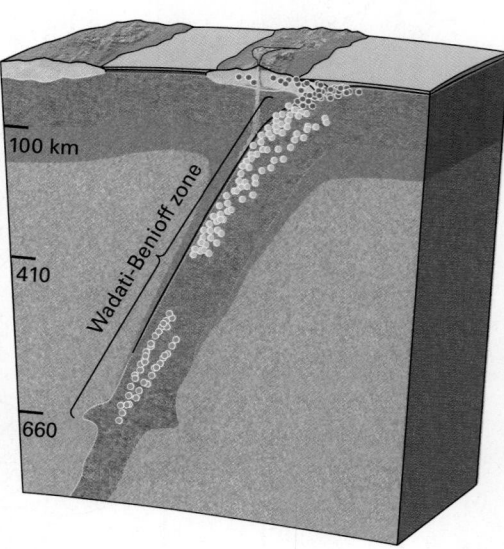

(c) At a convergent boundary, shallow-focus earthquakes (red dots) occur along the contact between the downgoing plate and the overriding plate, as well as within the two plates. Intermediate-focus (yellow dots) and deep-focus (blue dots) earthquakes define the Wadati-Benioff zone.

FIGURE 7.17 A composite diagram portraying the geologic settings in which earthquakes occur in continental lithosphere. (Subduction-related earthquakes in continental crust are not shown.)

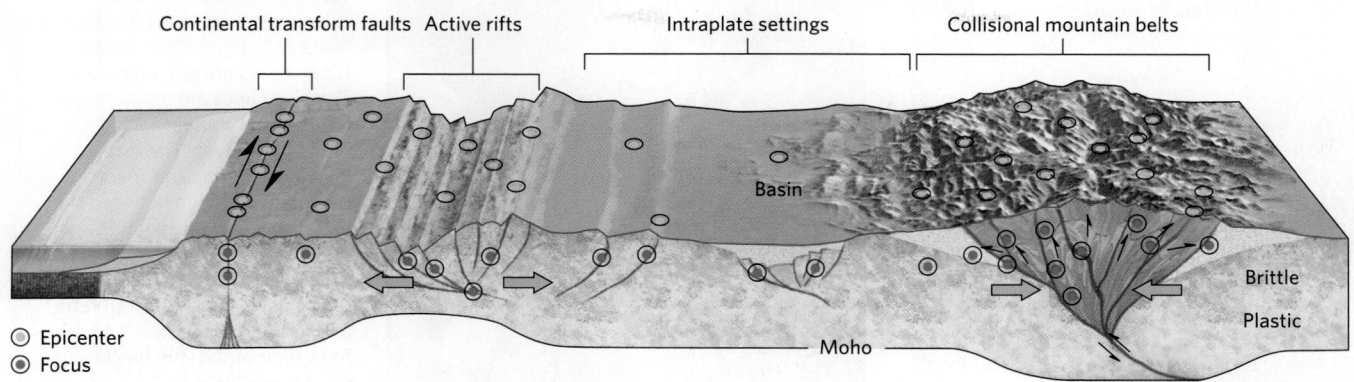

the trench, and thickens vertically. (These motions, while too small to see just by watching, can be detected by precise GPS measurements.) When slip abruptly takes place along much of the thrust fault, the prism twangs forward, as if the spring were released. The face of the prism lurches up and seaward, and the prism thins vertically.

Unlike divergent or transform boundaries, convergent boundaries also host many intermediate- and deep-focus earthquakes (Fig. 7.16c). A plot of the foci of such earthquakes on a cross section through the Earth defines a sloping band of seismicity called a *Wadati-Benioff zone* (named for the seismologists who first recognized it), which extends to a depth of 660 km (410 mi). Earthquakes in this zone occur within subducted lithosphere as it sinks down through the asthenosphere. The physical cause of these earthquakes remains a topic of research. Note in Figure 7.16c that earthquakes also happen in the overriding plate, and in the downgoing plate on the ocean side of the trench.

Earthquakes within Continents

Not all earthquakes on continents are associated with plate boundaries. Here, we consider a few other geologic settings within continents in which earthquakes occur (Fig. 7.17).

CONTINENTAL RIFTS. The stretching of crust at a continental rift generates normal faults, and slip on these faults produces earthquakes. Active rifts today include the East African Rift, the Basin and Range Province of Utah (see Fig. 7.6b), Nevada, and Arizona, and the Rio Grande Rift of New Mexico. In all these places, shallow-focus earthquakes occur, some of which cause major damage. Fault slip during some major earthquakes has produced fault scarps more than 2 m (6 ft) high (see Fig. 6.10b).

CONTINENTAL COLLISION ZONES. In 2015, a large earthquake shook the ground in the mountainous landscape of

Nepal. Monuments that had stood for centuries collapsed in Kathmandu (Fig. 7.18), and throughout the remote countryside, landslides buried communities and roads. Earthquakes are frequent in Nepal, as well as elsewhere along the Himalayas, because the range is actively growing due to the collision of India with Asia, and compression caused by this collision drives slip on thrust faults.

INTRAPLATE SEISMICITY. Some earthquakes, known as **intraplate earthquakes**, occur in the interiors of plates, away from plate boundaries, collision zones, or active rifts. The foci of almost all intraplate earthquakes lie at depths of less than 25 km (15 mi). Many of these earthquakes take place on weak pre-existing faults in the upper crust. Some of these faults initially formed during rifting events in the Precambrian and have remained as weak zones ever since. Slip on these faults today appears to be a response to relatively small compressional stresses in the continent, so the faults now slip either as reverse faults or as a combination of reverse and strike-slip faults.

Intraplate earthquakes are not uniformly distributed. In North America, most take place in southeastern Missouri, eastern Tennessee, eastern South Carolina, and southern Quebec. Even though the total energy release due to intraplate earthquakes is small—less than 5% of the total energy released by earthquakes globally in a given year—a few have been historic. We've already mentioned the 1886 Charleston earthquake (see Fig. 7.10a), which rang bells as far away as New York City. Three similar-sized earthquakes (M_W 7.0–7.4) struck in the New Madrid seismic zone, a seismically active region in southernmost Missouri, during the winter of 1811–1812. The earthquakes, due to slip on faults beneath the Mississippi Valley, displaced the ground surface, temporarily reversed the flow of the Mississippi River, and toppled cabins. The region, which now includes large cities, remains seismically active today (Fig. 7.19).

FIGURE 7.18 Seismicity in a collision zone.

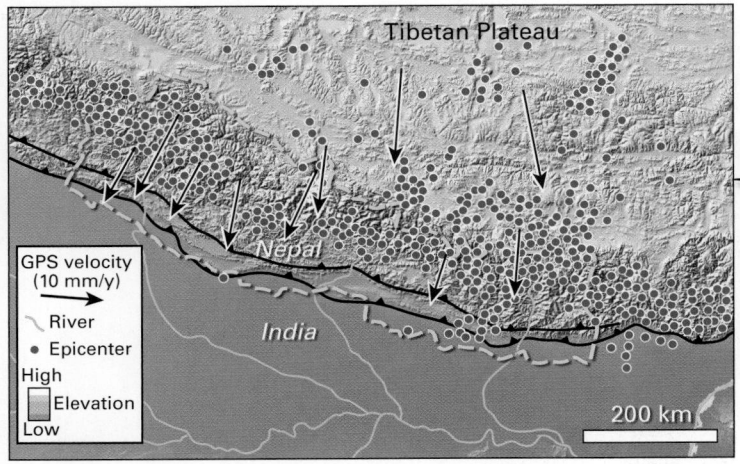

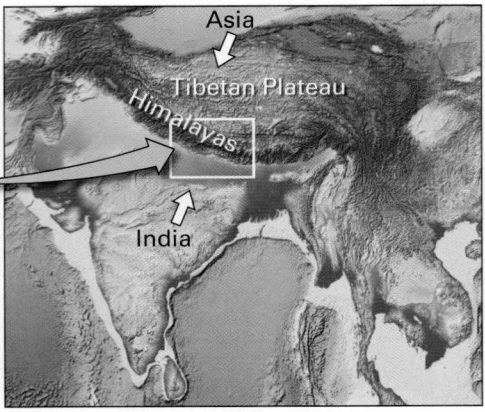

Location map

(a) A simplified map of earthquake epicenters over a period of years in Nepal (yellow border). Magnitudes are not indicated. The black lines are major thrust faults at the southern edge of the Himalayas. The arrows indicate GPS measurements of the relative motion between the region north of these faults and the interior of India. The location map emphasizes that Nepal is in the collision zone between Asia and India. The collision led to the formation of the Himalayas and the Tibetan Plateau.

(b) The 2015 Nepal earthquake destroyed many towns.

(c) Due to the destruction of their homes, about 2 million people were displaced, and many had to move to tent camps.

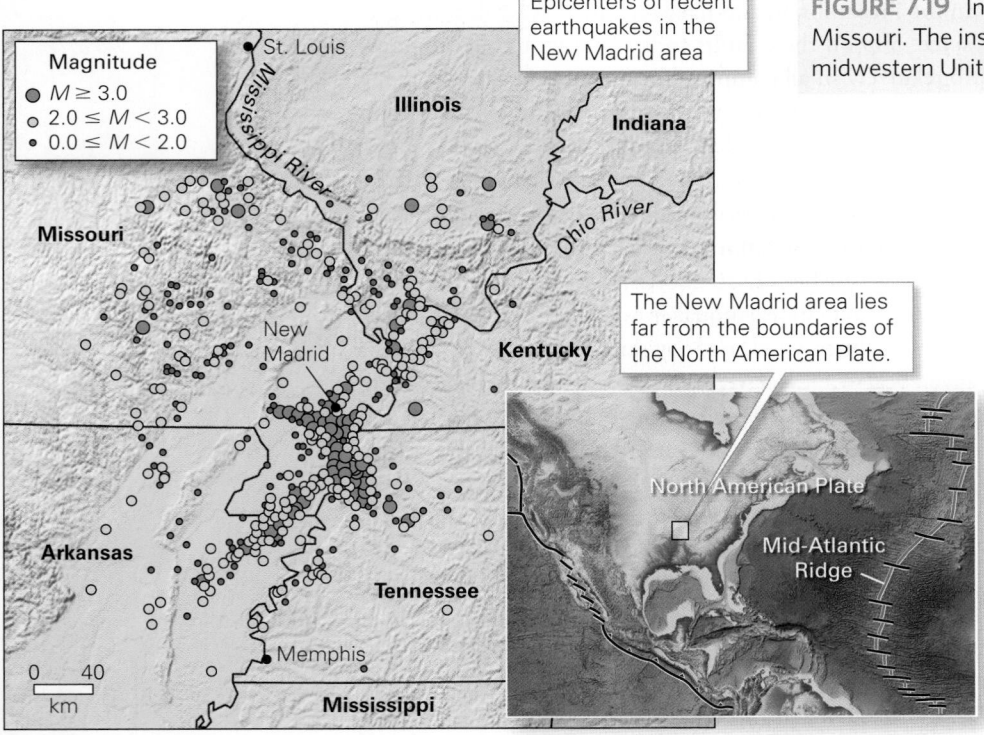

Epicenters of recent earthquakes in the New Madrid area

The New Madrid area lies far from the boundaries of the North American Plate.

FIGURE 7.19 Intraplate seismic activity in the New Madrid seismic zone, Missouri. The inset map shows the location of this seismic zone in the midwestern United States, far from any plate boundary.

Take-home message...

Most, but not all, earthquakes happen along plate boundaries. Earthquakes on faults along convergent boundaries or continental transforms can cause catastrophic damage because they often have shallow foci and occur near population centers. Large earthquakes also occur in rifts and collision zones, and a few take place along weak pre-existing faults within plate interiors.

Quick Questions

- Why are continental transform faults so dangerous?
- At which type of plate boundary does a Wadati-Benioff zone occur?
- What is a megathrust earthquake?

7.5 How Do Earthquakes Cause Damage?

If earthquakes affected only isolated regions of the Earth, such as oceanic crust in the middle of an ocean, they would not make news. But unfortunately, earthquakes often occur near or even directly beneath population centers, and when that happens, they can cause many kinds of damage that affect communities. Major and great earthquakes can be historically significant natural disasters. Let's examine some of the ways earthquakes trigger destruction (**Earth Science at a Glance**, pp. 262–263).

Ground Displacement and Shaking

Many faults occur deep underground, so while seismic waves generated by slip on those faults can rattle the ground, the fault surface itself does not break through the ground surface. In places where faults are close to the ground surface, they can cut through it and displace it. The line marking the intersection of a fault with the ground surface is called a *fault trace*. Strike-slip faults shift one part of the ground sideways relative to another (**Fig. 7.20a**; see also Fig. 7.4), while dip-slip faults (normal or reverse) change the elevation of the land surface on one side of the fault relative to that on the other side (**Fig. 7.20b**). Ground displacement can not only offset natural features, such as stream channels, and create fault scarps (see Chapter 6), but can also displace or break buildings and infrastructure. The sudden formation of a scarp cutting a road, needless to say, becomes a deadly traffic hazard. Expensive damage happens when ground displacement cuts buildings in half, buckles rail lines, and snaps power lines. If ground displacement causes the failure of a dam, populations downstream face the hazard of a flood.

During an earthquake, the ground shakes, both when body waves passing through the solid Earth reach the surface and when surface waves moving along the ground pass by. Ground motion starts suddenly and may last from a few seconds to a few minutes. The duration of ground motion at a given locality depends not only on how much time passes between the beginning and ending of sliding along the fault, but also on the distance between the locality and the earthquake's focus. The second factor reflects the fact that not all seismic waves travel at the same velocity, so they don't all reach the locality at the same time; thus, the time delay between the fastest and slowest waves increases with distance from the epicenter (see Fig. 7.9).

The nature of ground motion during an earthquake can be quite variable and can change as time passes, as different kinds of seismic waves cause different kinds

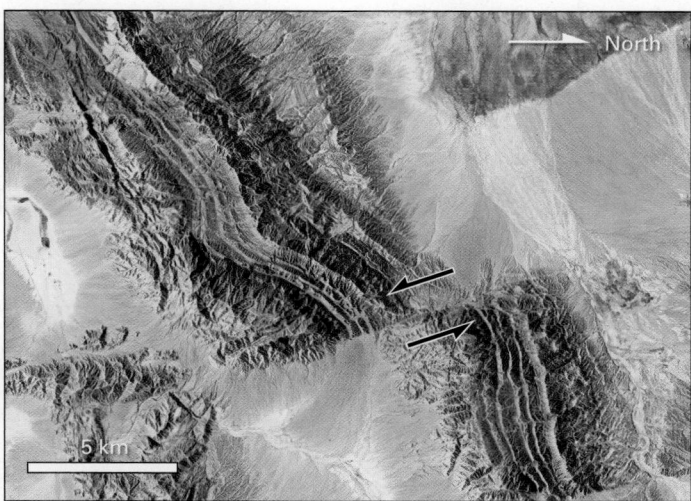

FIGURE 7.20
Ground displacement during an earthquake.

(a) This satellite image shows a strike-slip fault offseting a ridge of sedimentary beds in the desert of northwestern China.

(b) The trace of a thrust fault in Taiwan that uplifted part of a running track to produce a fault scarp. The hanging wall lies to the left of the fault trace. The red arrow indicates relative movement of the hanging wall.

of ground motion (**Fig. 7.21**). As we noted earlier, the intensity of shaking at a given location depends on several factors, including earthquake magnitude, distance from the focus, and the strength of the substrate. Because of substrate-strength variations, the damage caused by ground motion can vary within a region. Specifically, ground motion tends to be greater over a sediment-filled basin than over nearby areas underlain by bedrock.

As a seismic wave passes, the ground surface, and everything on it, goes from being stationary to moving rapidly up and down and side to side. In other words, the ground surface undergoes *acceleration*, meaning simply a change in speed and/or direction. You'll feel an earthquake if the ground suddenly moves by as little as 0.1 mm (0.004 in), and sudden motions of as little as several millimeters to a few centimeters cause damage. During a major quake, the ground may move by tens of centimeters to a few meters. Note that for ground motion

Did you ever wonder . . .

how long an earthquake lasts?

Damage Due to Earthquakes

Tsunami

Dam failure

Landslide

Landslide

Ruptured pipeline

Fire

New fault scarp

Offset roads and buildings

Epicenter

Home damage

Building collapse

Inactive fault

Active fault

Most earthquakes happen due to the sudden slip of a fault. In some cases, a fault offsets the ground surface, displacing roads and buildings. Ground displacement may also break pipelines, water mains, and fuel tanks. Ground shaking damages buildings, roads, and bridges. Milder shaking shatters windows and causes facades to crumble, but intense shaking can result in complete building collapse. In some cases, damage due to shaking leads to the ignition of horrific fires. The failure of bridges and highways disrupts transportation, and the failure of dams causes floods. Shaking can lead to landslides on steep slopes, and it can trigger liquefaction of sediment that produces fissures and sand volcanoes. Buildings sitting on liquefied sediment may topple or collapse. If landslides fall into water, or occur underwater, they can produce local tsunamis in lakes, reservoirs, bays, or coastal areas. When fault slip causes displacement of large areas of the seafloor (not shown here), large tsunamis develop. These waves devastate nearby coasts, and can also cross an ocean and damage coasts thousands of kilometers away.

Earthquakes can happen in a variety of tectonic settings. This illustration depicts an earthquake along a normal fault in a rift.

Surface of tilted fault block

Less damage farther from the epicenter

Flooding

Shutdown of airport

Collapsed bridge

Oil spill

Toppled structures

WATER

Sand volcanoes and fissures

Tilted buildings due to liquefaction

Ruptured tank

Rubble-covered street

Worse damage due to amplification

Fissures and lateral spreading

Wet sand

Sediment-filled basin

Older sedimentary basin fill

Old crystalline bedrock

FIGURE 7.21 Different types of seismic waves move the ground in different directions. During an earthquake, the ground can shake in many ways, causing surface structures to move. The net movement can be quite chaotic.

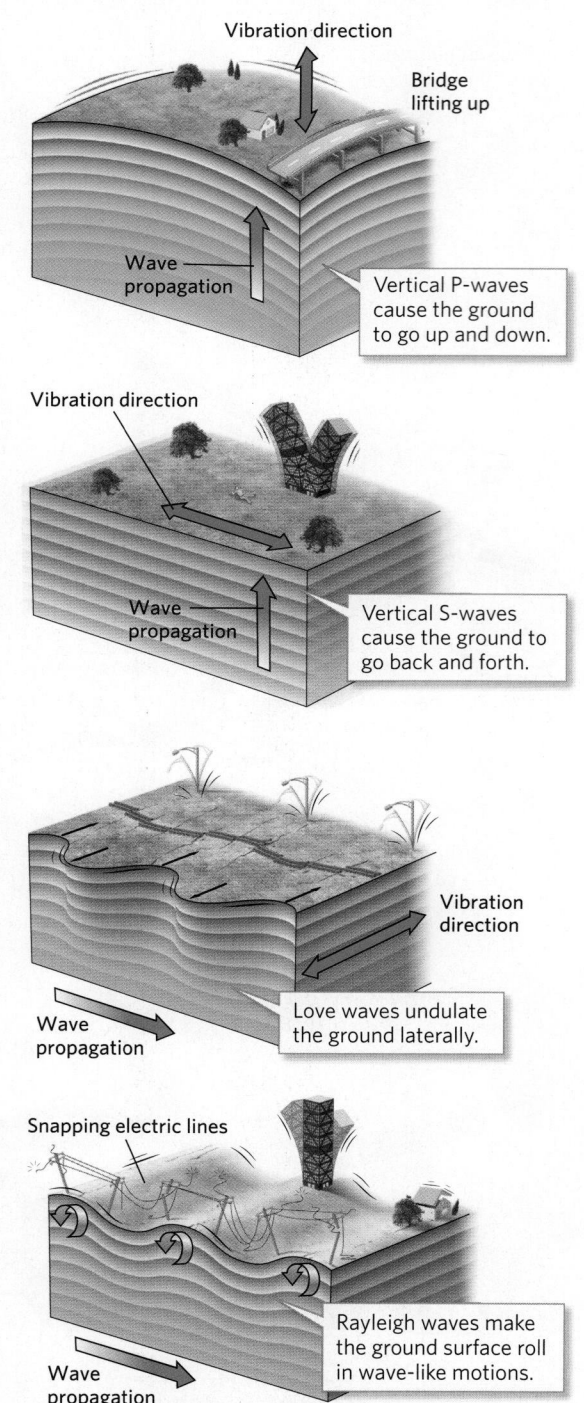

Vibration direction

Bridge lifting up

Wave propagation

Vertical P-waves cause the ground to go up and down.

Vibration direction

Wave propagation

Vertical S-waves cause the ground to go back and forth.

Vibration direction

Wave propagation

Love waves undulate the ground laterally.

Snapping electric lines

Rayleigh waves make the ground surface roll in wave-like motions.

Wave propagation

Did you ever wonder...

whether California will fall into the sea during an earthquake?

to cause damage, not only must the movements be large, but they must happen suddenly. If the ground were to accelerate slowly during shifts in direction, the movement wouldn't constitute an earthquake. Put another way, the ground acceleration must be extremely rapid. (To picture

this contrast, think about how different it would feel to suddenly step on a treadmill that is set for "slow walk" as opposed to one set for a "sprint"; you could step on the former one without losing your balance, but if you stepped on the latter one, your feet would go out from under you, and you'd fall.) Seismologists compare the **peak ground acceleration (PGA)** during an earthquake with the acceleration produced by gravity ($g = 9.8$ m/s^2). During a moderate earthquake, acceleration near the epicenter approaches 0.2 g, enough to knock you off your feet. During a great earthquake, acceleration near the epicenter can range between 1.7 g and 3.0 g, comparable to the acceleration you feel during a terrifying roller-coaster ride.

Even during a major earthquake, ground motion alone generally won't hurt you if you're out in an open field. You may bounce around a bit, but your body is too flexible to break. Human-made structures aren't so lucky, however (Fig. 7.22). When seismic waves pass, roads, rail lines, and pipelines buckle or rupture. Buildings sway, twist back and forth, or lurch up and down, so windows shatter, roofs fail, and building facades crash to the ground. Acceleration can cause the floors of a multistory building, or the deck of a bridge, to apply enough downward stress to crush and buckle support columns, and when this happens, the structure collapses. Falling debris and structural failures of buildings and bridges generally cause most earthquake-related casualties.

Landslides and Sediment Liquefaction

Seismic shaking can cause ground on steep slopes, or ground underlain by weak sediment, to give way. This movement results in a **landslide**, the tumbling or flow of soil and rock downslope (see Chapter 11). Landslides are often a major cause of earthquake-related destruction. During major earthquakes in mountainous areas, hundreds of landslides may tumble down slopes. Sadly, some of these landslides bury communities in the valleys below (Fig. 7.23a). When landslides happen along the cliff-rimmed coast of California, they block roads, bury parts of seaside communities, and turn expensive homes built to enjoy sunset views into rubble piles (Fig. 7.23b). Images of such coastal landslides lead to the misperception that "California west of the San Andreas Fault will fall into the sea during an earthquake." Fortunately, although relatively small portions of the coastal cliffs do collapse to the beach, the state remains firmly attached to the continent, despite what Hollywood scriptwriters say.

In 1964, an M_W 7.5 earthquake struck Niigata, Japan, a city that had been partly built on land underlain by wet sand. During the ground shaking, the foundations of over 15,000 buildings sank into the ground, causing walls and roofs to crack. Several apartment buildings

FIGURE 7.22 Examples of earthquake damage due to ground shaking.

tilted dramatically, and some tipped over (Fig. 7.23c). A 2011 earthquake in Christchurch, New Zealand, caused sand beneath the ground surface to squirt through cracks onto the surface. The loss of underground sand caused overlying land (including roads) to collapse into pit-like sinkholes, some of which were large enough to swallow cars (Fig. 7.23d). During a 2018 earthquake in Indonesia, the ground under a large neighborhood turned to liquid and began to flow, carrying away the entire neighborhood and tearing apart and burying the scores of homes that had been built on it (Fig. 7.23e).

The above examples illustrate the process of **sediment liquefaction**. Liquefaction in wet sand happens when shaking causes the sand grains to try to settle together more tightly. This movement increases pressure in the groundwater between the grains, and that force, in turn, pushes the grains apart. As a result, what had been stable, load-bearing sand turns into a sand-water slurry incapable of supporting weight. Where cracks open between the liquefied sand layer and the ground surface, pressure caused by the weight of the overlying sediment squeezes the wet sand upward and out onto the ground surface. In some cases, the sand accumulates to form cone-shaped mounds called *sand volcanoes* (Fig. 7.23f). The settling of sediment layers into the liquefied layer disrupts bedding and leads to formation of open fissures and pits on the land surface.

A similar phenomenon happens in a special type of clay called **quick clay**. Ground motion transforms what had been a gel-like solid mass into liquid, slippery mud by destroying cohesion among the grains. This phenomenon happened during the 1964 Good Friday earthquake in southern Alaska, beneath the coastal Turnagain Heights neighborhood of Anchorage. When ground shaking began, a layer of quick clay beneath the neighborhood lost cohesion, and the overlying layers of sediment, along with the houses built on top of them, slumped seaward and transformed into a chaotic jumble (Fig. 7.24).

Fire

The shaking during an earthquake can make lamps, stoves, or candles with open flames tip over, and can break wires or topple power lines and produce sparks. Flames and sparks ignite both rubble and undamaged buildings. Ruptured gas pipelines and

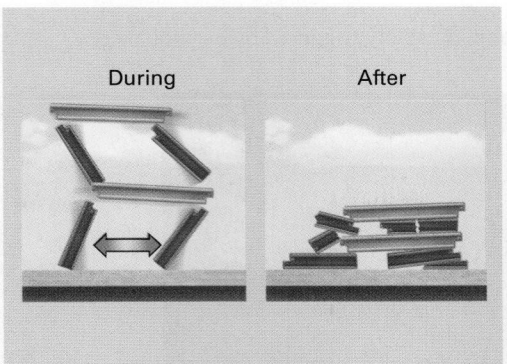

(a) During a 2023 earthquake in Turkey, concrete buildings collapsed when supports gave way and floors piled on one another like pancakes.

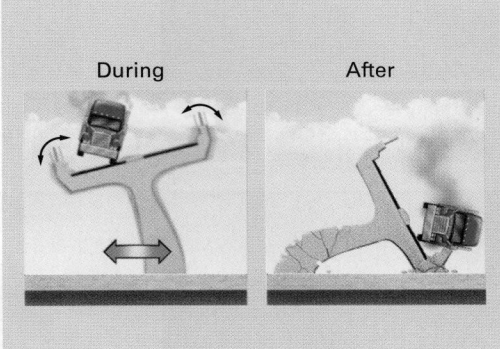

(b) Shaking causes swaying. An elevated bridge tipped over during the 1995 earthquake in Kobe, Japan.

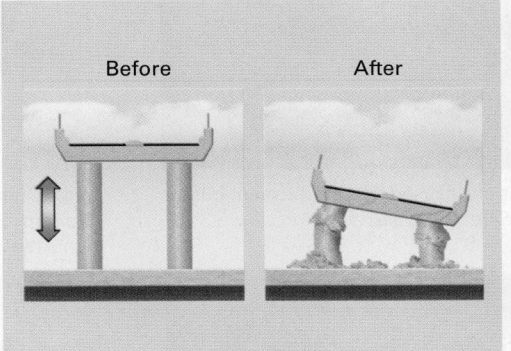

(c) Concrete bridge supports were crushed during the 1994 earthquake in Northridge, California, when the overlying bridge bounced up and slammed back down.

(d) Masonry cracks and crumbles in response to ground motion. A neighborhood of masonry buildings in Armenia collapsed during a 1999 earthquake.

FIGURE 7.23 Examples of ground instability triggered by earthquakes.

(a) During a 2008 earthquake in China, many landslides tumbled down steep slopes and overran the towns at their base.

(b) Collapse of this steep slope along the California coast carried part of a home with it.

(c) Liquefaction under their foundations caused these apartment buildings in Japan to tip over during a 1964 earthquake.

(d) During the 2011 New Zealand earthquake, the pavement collapsed to form a sinkhole and partly swallowed a car when sand liquefied and spurted out of the ground.

(e) An entire neighborhood on an Indonesian island collapsed when the sediment beneath it liquefied.

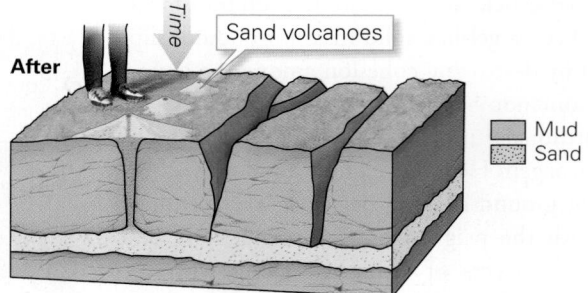

(f) Sand liquefaction causes fissures and sand volcanoes to develop.

FIGURE 7.24 The Turnagain Heights disaster. During the 1964 earthquake, the land on which the neighborhood was built broke up and slid seaward on a weak layer of quick clay.

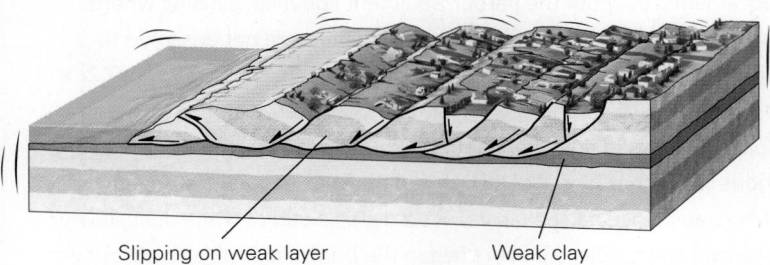

Slipping on weak layer   Weak clay

fuel tanks feed the flames, sending columns of fire erupting skyward (Fig. 7.25a). Once a fire starts to spread, it may become an unstoppable inferno, especially since debris-filled streets block firefighting equipment and broken pipes make fire hydrants inoperable. If the fire significantly heats overlying air, the hot air rises, allowing cooler air from outside the burning area to rush in, generating wind gusts of over 160 km/h (100 mph). The winds stoke the blaze and produce a *firestorm*. As we've seen, fire destroyed much of San Francisco after the 1906 earthquake. An even larger fire swept through Tokyo after an earthquake in 1923, when coals from cooking stoves set buildings made of wood and paper alight. It evolved into a firestorm that consumed the city and took the lives of tens of thousands of residents who had survived the ground shaking (Fig. 7.25b).

Disease

After an earthquake, disease may threaten lives in the damaged region. Rupturing of pipes and sewers contaminates clean-water supplies and exposes the public to pathogens (Box 7.2). Debris blocks roads and railroads and makes harbors and airports unusable, thereby preventing food and medicine from reaching stricken areas. And people who have lost their homes have no choice but to crowd into unsanitary camps. The severity of such problems may exceed the ability of emergency services to cope, so it may take months to years for daily life in a damaged area to return to normal.

Tsunamis

The azure waters and palm-fringed islands of the Indian Ocean's eastern coast border the Sunda Trench, one of the most seismically active plate boundaries on the Earth. Just before 8:00 A.M. on December 26, 2004, an area 1,300 km (800 mi) long by 100 km (62 mi) wide on the principal thrust fault defining this convergent boundary broke. The break caused a megathrust earthquake during which the overriding plate (including the accretionary prism along the plate's edge) lurched westward by as much as 15 m (50 ft) relative to the Indian Ocean floor (see Fig. 7.16b for an analogy). This movement pushed up the seafloor surface under the portion of the accretionary prism adjacent to the trench by as much as a few tens of centimeters, and led to sinking of the top of the accretionary prism closer to the volcanic arc. Elastic rebound, along with the rupturing of rock, released an immense amount of energy, so the great M_W 9.3 Sumatra earthquake shattered buildings and triggered many landslides. The movement of such a broad area of seafloor also displaced an immense amount of water in the overlying ocean and produced large tsunamis, just as happened during the 2011 Tōhoku earthquake described at the

FIGURE 7.25 Fire sometimes follows an earthquake.

(a) Broken gas tanks erupt in fountains of flame after the 2011 Tōhoku, Japan, earthquake.

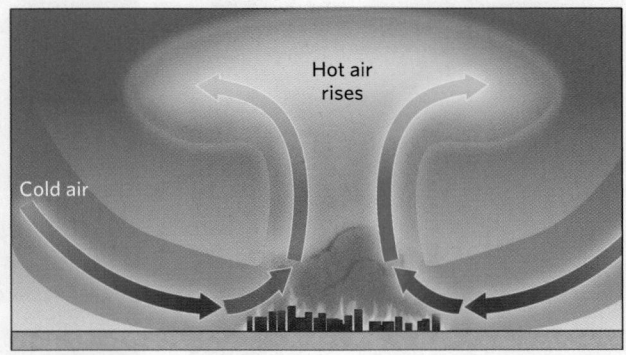

Hot air rises

Cold air

(b) A firestorm develops when cool air rushes in to replace rising hot air above a huge fire. The cool air stokes the blaze.

BOX 7.2 ▶ A Deeper Look

The 2010 Haiti catastrophe

Haiti sits astride a transform boundary along which the North American Plate moves westward at about 2 cm/yr (0.8 in/yr) relative to the Caribbean Plate (**Fig. Bx7.2a**). Therefore, earthquakes in Haiti are inevitable. But the last great earthquake on this plate boundary had happened in 1770, so by 2010, elastic deformation and associated stress had been accumulating for 240 years. On the sunny afternoon of January 12, a 70 km (45 mi) segment of a large strike-slip fault suddenly slipped by as much as 4 m (12 ft). The motion began 25 km (15 mi) west-southwest of Port-au-Prince, the capital of Haiti, and 13 km (8 mi) beneath the ground surface (**Fig. Bx7.2b**). The seismic waves of the resulting M_W 7 earthquake caused severe ground motion for about 35 seconds.

The impact of an earthquake on society depends not only on the earthquake's size, but also on the nature of the ground (sediment or hard rock), on the steepness of the slopes, on construction practices, and on the quality of emergency services in the affected area. Much of Port-au-Prince sits on a basin of weak sediment, which amplified ground movements. Beneath the harbor, sediment liquefied, causing wharfs to sink into the sea. Sadly, the city's buildings were not designed to withstand ground motion, so many collapsed into rubble (**Fig. Bx7.2c**). In addition, some of the city's neighborhoods were perched on steep slopes, which slid downhill during the quake, carrying the neighborhoods with them (**Fig. Bx7.2d**). When the shaking finally stopped, most of Port-au-Prince had collapsed. As a dense cloud of white dust slowly rose over the rubble, survivors began the frantic scramble to dig out victims, a task made more hazardous by aftershocks, of which over 50 had magnitudes between 4.5 and 6.1. The aftershocks caused still-standing but weakened structures to collapse on rescuers. Some estimates place the death toll at 230,000. In the days that followed, local emergency services were overwhelmed. An air caravan of aid arrived to help out, but even so, in the months that followed, cholera spread. Years after the event, recovery from the earthquake remains far from complete, and unfortunately, political and social unrest, as well as hurricanes and floods, have slowed or stopped recovery efforts.

FIGURE Bx7.2 The January 2010 earthquake in Haiti and its geologic setting.

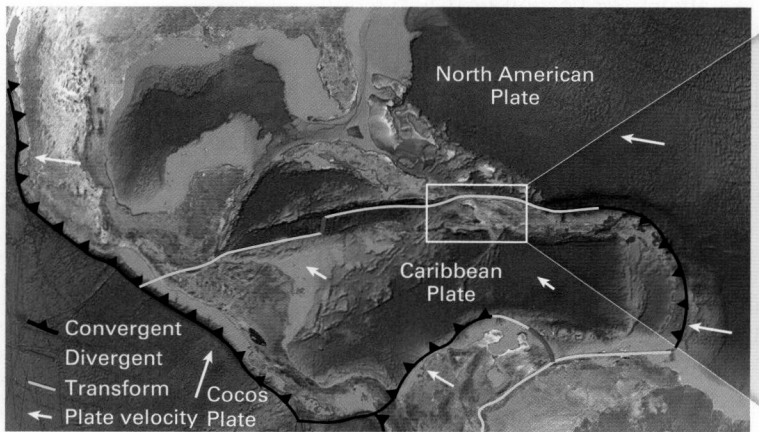

(a) The Caribbean Plate has complex boundaries delineated by bathymetric features. The white rectangle shows the location of Haiti.

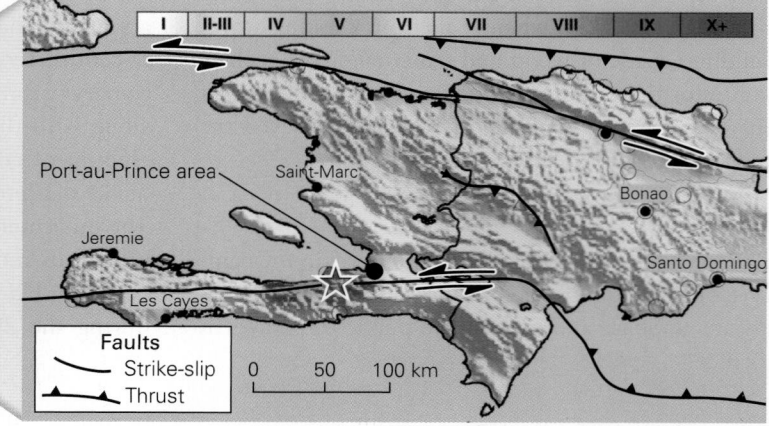

(b) An earthquake-intensity map for the 2010 earthquake. The star marks the epicenter of the quake.

(c) Survivors salvage what they can in Port-au-Prince, the capital of Haiti, after the earthquake.

(d) Ground shaking caused most of the houses in this residential neighborhood to collapse.

FIGURE 7.26 Formation of a tsunami.

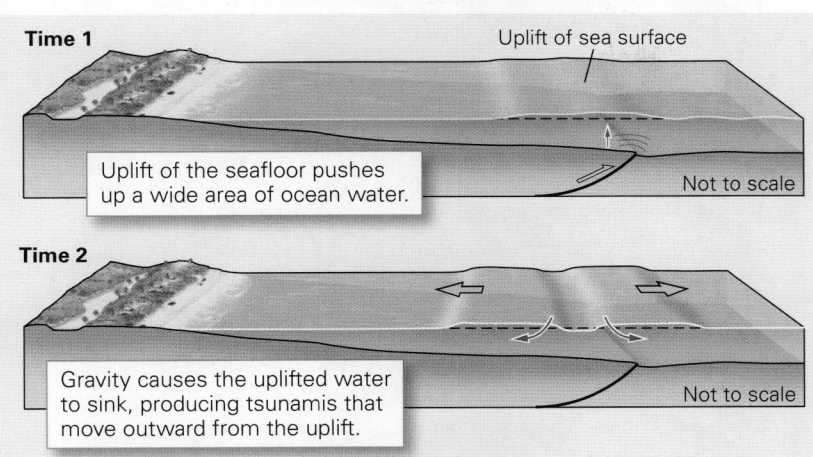

(a) After initial uplift of a mound of water, gravity causes the mound to collapse and a wave to spread outward.

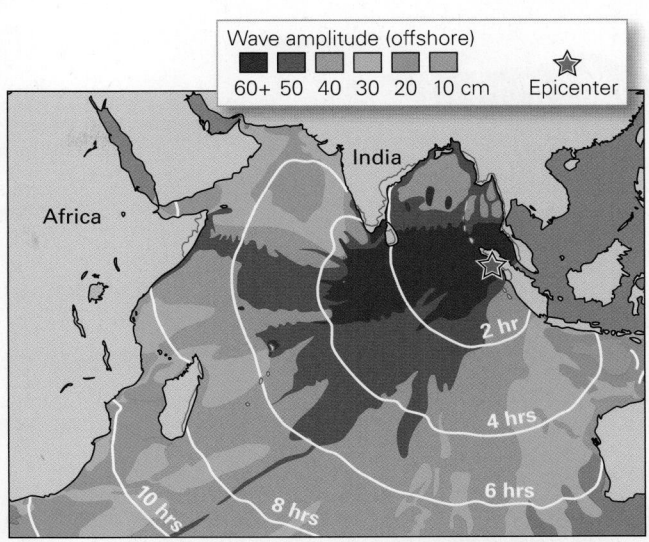

(b) A map showing the variation in the height of the first 2004 tsunami as it crossed the Indian Ocean. The white lines indicate the position of the wave front at the time specified.

opening of this chapter. Let's now examine why such waves are produced and why they are so devastating.

As soon as the vast mound of water off Sumatra rose, gravity caused the mound to sink downward (Fig. 7.26a). The ocean surface undulated up and down several times before returning to equilibrium. This movement produced several very broad waves that traveled outward at a velocity of up to 800 km/h (500 mph)—almost the speed of a jet plane (Fig. 7.26b). Geologists use the term **tsunami** for a wave produced by the push of a large solid mass against water. The displacement can be due to an earthquake, as we've just seen, a volcanic explosion (see Chapter 4), or a landslide that occurs underwater or along the coast (see Chapter 11). Notably, tsunamis do not occur alone, but rather as a group of successive waves, as the sea surface rises and falls several times at the site of tsunami generation before it comes back to equilibrium. The set of several waves produced by the motion constitutes a *tsunami train*.

Tsunamis differ significantly from more familiar wind-driven waves (Fig. 7.27a, b). Wind-driven waves form when moving air applies shear to the surface of the sea and pushes water into narrow ridges. Large wind-driven waves formed during storms can reach heights of 10–30 m (32–100 ft) in the open ocean. But even such monsters are only tens of meters wide, as measured perpendicular to wave motion, so they contain a relatively small volume of water. In contrast, although the sea surface rises by at most only a few tens of centimeters at the site where a large tsunami forms, the resulting wave may be tens to hundreds of kilometers wide (as measured in

the direction parallel to the wave movement), so it contains an immense volume of water. In fact, a large tsunami can be 100 to 1,000 times wider than a wind-driven wave. And while the wavelength of a storm-wave train is about 100 m, the wavelength of a tsunami train is about 200 km. So, while a large storm wave advances over the shore for 5–30 seconds, and several storm waves arrive over the course of a few minutes, the front of a large tsunami can advance over the shore for five minutes to a half hour, and the arrival of successive tsunamis in a tsunami train can take a few hours (Fig. 7.27c). So, if you picture a storm wave as a narrow ridge of high water, you can picture a tsunami as a broad plateau of high water. Large storm waves and tsunamis, therefore, have very different consequences.

Note that we use the adjective "large" above to emphasize that our discussion focuses on tsunamis that are big enough to cause damage. Not all tsunamis are large—the small ones just don't make headlines. Assigning the name *tsunami* to a wave implies its cause, its width, its velocity, and its wavelength. The name alone does not require that the wave be high or that it result in major destruction.

When any wave (tsunami or storm wave) approaches the shore, friction between the base of the wave and the seafloor slows the bottom of the wave, so the rear of the wave catches up to the front. The added water volume builds the wave higher (see Fig. 7.27b). Near the beach, the top edge of an arriving wave falls over the front face of the wave, forming a turbulent breaker. In the case of a wind-driven wave, the breaker may be tall when it washes onto the beach, but because the wave doesn't contain

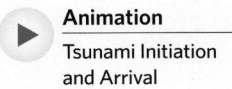

Animation

Tsunami Initiation and Arrival

FIGURE 7.27 The difference between a storm wave and a tsunami.

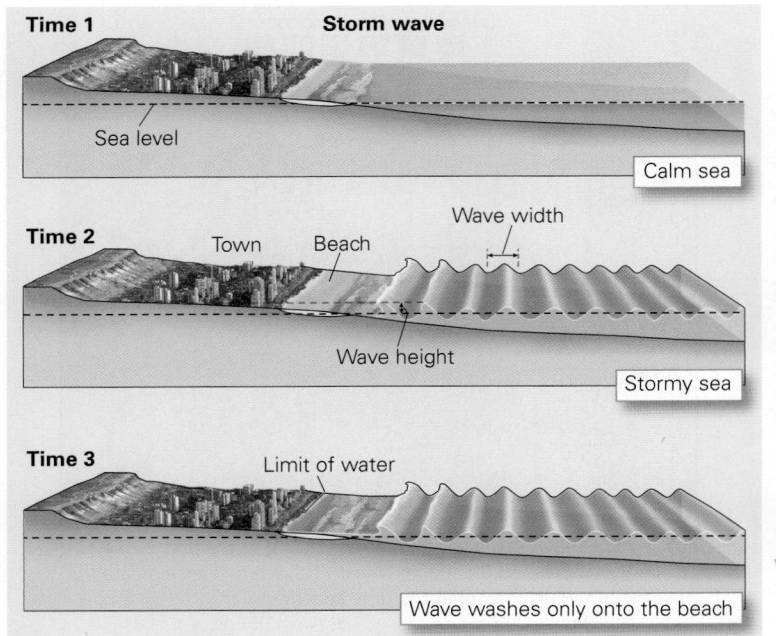

(a) Storm waves can be high, but because they are not very wide, they contain relatively little water.

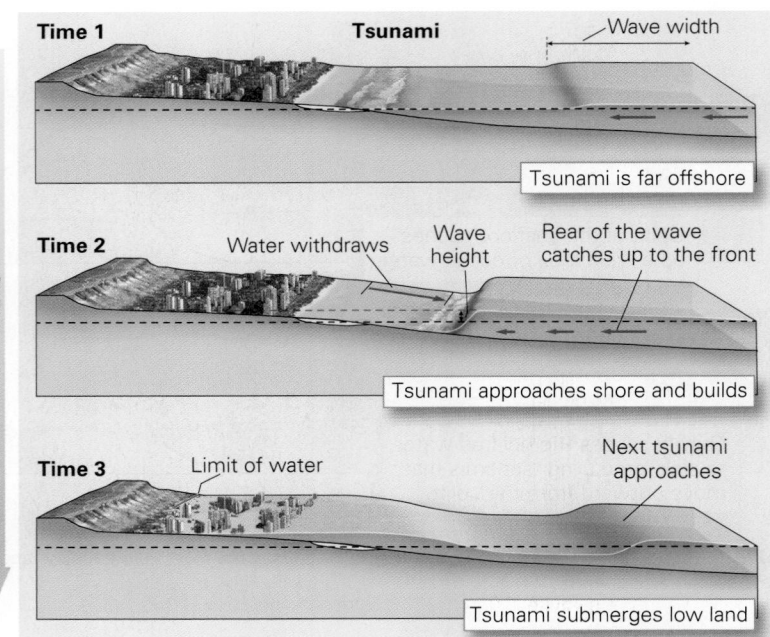

(b) A tsunami is not very high out in the open ocean, but as it approaches the land, the rear catches up to the front, and the wave grows higher. Each wave is very wide.

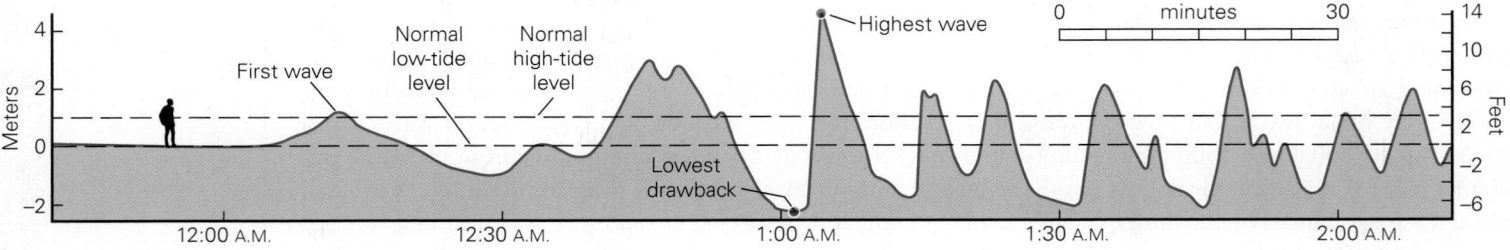

(c) A graph showing changes in the sea-surface level over time reveals the arrival of a train of tsunamis in Hilo, Hawai'i, in 1960, after the M_W 9.5 earthquake in Chile. Note that the highest wave was not the first. The sea surface drops far below the low-tide level in the times between successive wave crests.

much water, the flow of water stops at the landward edge of the beach. (In fact, a beach exists because the wind-driven waves frequently wash over it so that vegetation cannot take root.) In the case of a large tsunami, however, when friction slows the wave and increases its height near the shore, the wave doesn't run out of water after it has crossed the beach. So, where coastal land rises only a few meters above sea level, the water of a tsunami can move far inland—the largest tsunamis grow to heights of about 30 m (100 ft) and can submerge land several kilometers inland from the shore. Significantly, the water in a tsunami advances faster than that in a storm wave, and instead of rolling in like the water inside a storm wave, the water in a tsunami moves more uniformly forward. Therefore, tsunamis apply more stress to homes and cars in their path when they arrive, and can cause more destruction.

Videos of the 2004 Indian Ocean tsunamis alerted the world to the kind of damage large tsunamis can cause. When the first tsunami approached Banda Aceh, a city at the northern end of the island of Sumatra, the sea receded much farther than anyone had ever seen, exposing large areas of reefs that normally remained submerged even at low tide. People walked out onto the exposed reefs in wonder to look at the coral and sea stars (starfish). Then, a wall of frothing water—the turbulent breaker at the front of a tsunami—appeared offshore (Fig. 7.28a). With a rumble that grew to a roar, the tsunami approached. Friction with the seafloor slowed it to less than 30 km/h (20 mph), but it still moved faster than people could run. In places, the wave front reached heights of 15–30 m (45–100 ft) as it overwhelmed the beach. The impact of the water at Banda Aceh ripped boats from their moorings, snapped trees (Fig. 7.28b), flattened or floated away

FIGURE 7.28 The great Indian Ocean tsunamis of 2004.

(a) This snapshot shows a tsunami rushing toward the coast of Sumatra. Recession of water in advance of the wave exposed a reef.

(b) The wave blasts through a grove of palm trees as it strikes Thailand.

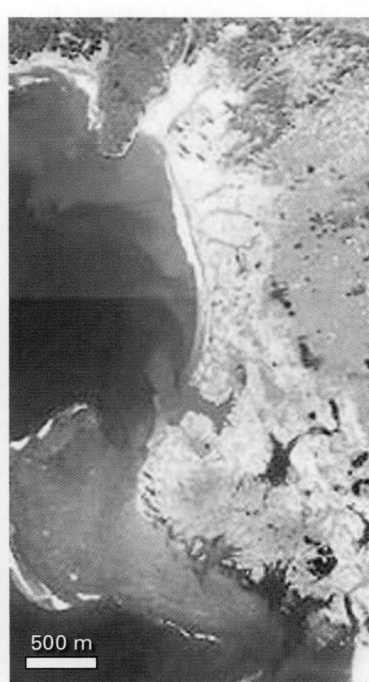

500 m

(c) Satellite photos of the Indonesian province of Aceh before and after the tsunamis struck. Note that the city has been washed away and the beach has vanished.

buildings, and tossed cars and trucks like toys. Because the tsunami was so wide, water just kept coming, eventually flooding land as far as 7 km (4 mi) inland. When gravity finally pulled the water back seaward, it carried with it a massive jumble of flotsam as well as the bodies of its unfortunate victims. Other tsunamis in the train followed—and, as Figure 7.27c shows, the first wave in a tsunami train isn't necessarily the highest. When the last wave had subsided, the land had been stripped of buildings, farms, and forest (Fig. 7.28c). Sadly, the horror of Banda Aceh was just a preamble to the devastation that would soon visit other stretches of Indian Ocean coast. Tsunamis crossed the Indian Ocean, striking southern India and Sri Lanka within two hours after the earthquake, and the eastern coast of Africa within about seven hours. In the end, nearly 230,000 people died that day.

The tsunamis that struck Japan soon after the 2011 Tōhoku earthquake generated additional international awareness of tsunami hazards. Existing seawalls 10 m (30 ft) high that had been built to protect Japan's coastal towns were no match for the advance of the waves, which locally rose to heights up to 33 m (100 ft) when they reached shore. Outside the seawalls, water rose, like an incredibly high and rapid tide, until it spilled over them. The water then proceeded inland (see Fig. 7.2), picking up dirt and debris as it moved, so that it evolved into a muddy slurry. Notably, many Japanese harbor towns were built on a flat estuary at the head of a coastal recess whose sides were bordered by cliffs. The energy of arriving tsunamis funneled into the estuaries, so the highest waves arrived there. The viscous, rapidly moving water completely flattened the towns (Fig. 7.29). In fact, the word *tsunami* comes from a Japanese word meaning harbor wave, because these waves have caused particularly severe devastation in harbor towns.

Unfortunately, even when the Tōhoku tsunamis receded, the catastrophe was not over. Rising water inundated the Fukushima Daiichi nuclear power plant, where it not only destroyed power lines, cutting the plant off from the electrical grid, but also drowned the backup diesel generators. As a result, the water pumps driving the plant's cooling system stopped functioning, so water surrounding the plant's hot reactor cores boiled away. Some

See for yourself

Banda Aceh before the Tsunamis

Latitude: 5°33′31.82″ N
Longitude: 95°17′13.01″ E

Fly to this location using Google Earth. Zoom to an altitude of 1.5 km (1 mi) and look down.

With the historical imagery tool (represented by a clock symbol), you can look at the area today, and at various times in the past. Set the clock back to June 22, 2004, and you'll see dense housing. Then set the clock to 2006. Compare the views and see how the land has changed.

FIGURE 7.29 Damage to a harbor town along the coast of Japan caused by the 2011 tsunamis. Only a few reinforced buildings remained standing.

of the water separated into H_2 and O_2 gas, which, when ignited by a spark, exploded. Explosions blew the tops off three of the plant's four containment buildings, contaminating the surroundings with radioactivity.

> We learn geology the morning after the earthquake.
>
> —*RALPH WALDO EMERSON*
> *(AMERICAN POET, 1803–1882)*

Take-home message...

Earthquakes cause devastation in many ways. Ground motion, landslides, and sediment liquefaction disrupt landscapes and cause buildings to crumble. Though ground motion itself cannot kill people, falling debris and collapsing buildings can. Fire and disease may follow earthquakes, causing even more loss of life. In coastal areas, tsunamis may wash over broad areas of low land.

Quick Questions ———————————————

- Define sediment liquefaction.
- What conditions cause a firestorm to develop in a city?
- Are all tsunamis over 5 m high?

7.6 Can We Predict the "Big One?"

Can seismologists predict earthquakes? The answer depends on the time frame of the prediction. With our present understanding of the distribution of seismic belts and the frequency at which earthquakes occur, we can make *long-term predictions* (on the time scale of decades to centuries). For example, we can say with some (but not complete) confidence that an earthquake will rattle Istanbul, but not north-central Canada, sometime during the next century, because Istanbul lies in a seismic belt while north-central Canada does not. Seismologists cannot, however, make accurate *short-term predictions* (on the time scale of hours to years). It's not possible to say, for example, that an earthquake will strike San Francisco 40 days from now, or that one will happen on July 22, 2095. New technologies, however, do allow seismologists to give people in some cities an early warning after an earthquake has been generated, but only seconds to minutes before seismic waves arrive. In this section, we look at the scientific basis of long-term predictions and of earthquake early-warning systems.

Long-Term Predictions

An assessment of **seismic risk** gives a long-term prediction of the probability, or likelihood, that an earthquake of a given size (defined as producing a specified PGA, or a specified intensity) will happen during a stated time period in a given region. For example, a seismologist may say, "The probability that an earthquake capable of producing a PGA of 0.8 *g* will occur in this state sometime during the next 50 years is 20%." This sentence implies that there's a 1-in-5 chance that a great earthquake will happen in the state before 50 years have passed. The basic premise of seismic-risk assessment can be stated as follows: A region where many earthquakes have occurred in the past is likely to experience earthquakes in the future. Seismic belts are, therefore, regions of higher seismic risk (see Figure 7.13). This doesn't mean that disastrous earthquakes can't happen far from a seismic belt—they can and do—but the probability that an earthquake will happen in such places in any given time window is less. Urban planners use such assessments to write building codes for a region. Regulations requiring earthquake-resistant buildings make sense for regions with high seismic risk, for those buildings are more likely to be shaken significantly during their lifetimes. Such a code might not be proposed for a region of low seismic risk, however, given that the funding needed to construct earthquake-resistant buildings might be better applied elsewhere.

Seismologists sometimes state the recurrence interval of an earthquake of a given size, on a fault or in a region, to give a sense of seismic risk. The **recurrence interval (RI)** is the average time between successive events of a given size. Because earthquakes do not happen periodically (meaning at predictably spaced time intervals), a recurrence interval does not provide the exact time between successive events. People unfamiliar with the term, however, may ignore the word "average," and make the incorrect deduction that if the RI is 100 years, and an earthquake happened 45 years ago, then the next one will happen in 55 years. To avoid such confusion, many seismologists prefer to specify the **annual probability** of an earthquake, which gives the likelihood that an earthquake will happen in any given year. Annual probability relates to RI by the following equation: Annual probability = 1 ÷ recurrence interval. For example, if the recurrence interval for an earthquake with an MMI of IX (a violent earthquake) in a region is 100 years, then the annual probability of such an earthquake is $\frac{1}{100}$, or 1%. This means that there is a 1-in-100 chance of MMI IX ground motion happening in a given year. Note, however, that because stress builds up on a fault over time, elastic-rebound theory implies that the annual probability of an earthquake may progressively increase as time passes.

To determine the recurrence interval for major earthquakes within a given seismic belt, seismologists must determine when previous earthquakes happened within that belt. For places where the historical record does not extend far enough back in time to reveal multiple major events, researchers look for evidence of large earthquakes preserved in the geologic record. For example, in places where sediment has accumulated over a fault, researchers may dig a trench and look for buried layers of sand volcanoes or disrupted bedding in the stratigraphic record. Each such layer, whose age can be determined by using radiocarbon dating of plant fragments (see Chapter 8), records the time of an earthquake (Fig. 7.30). Such **paleoseismology**

studies allow researchers to calculate the number of years between successive events and then calculate the average to obtain the recurrence interval.

Paleoseismology studies provided the basis for stating that the Cascadia subduction zone off Washington and Oregon hosted an M_W 9 earthquake a few hundred years ago. The evidence came from the dating of a "ghost forest," an area of trees that died when the land was submerged by a tsunami. There is no written historical record of this event in English, but the event was recorded in the oral histories of Native Americans and was documented by Japanese historians who recorded the arrival times of tsunamis generated by the event that crossed the Pacific. The tsunamis are now attributed to a megathrust earthquake produced when a large area of the fault separating the downgoing Juan de Fuca Plate from the overriding North American Plate suddenly slipped on January 26, 1700. Further paleoseismology studies indicate that megathrust earthquakes in the Cascadia subduction zone have a recurrence interval of 200–500 years, so the region faces significant seismic risk.

Maps that convey information about seismic risk help the public visualize the hazard that they may be facing in their community. These maps are based on studies of recurrence intervals together with other data. Seismologists have developed a few different kinds of such *seismic-hazard maps*. Some outline regions that have a 2% chance of experiencing a given PGA within 50 years. (This estimate effectively states that such a PGA will happen in the region, on average, about once in a 2,500-year period.) Other maps depict the estimated number of damaging earthquakes that might occur in a given region within a 10,000-year period. Color coding on seismic-hazard maps reveals, at a glance, which locations are more seismically active than others—generally, red indicates areas of relatively high risk (Fig. 7.31). The maps show that seismic hazards are greatest along plate boundaries and within rifts and collision zones. They also highlight regions of intraplate seismicity. In the continental United States, the greatest risk lies along the San Andreas Fault and the Cascadia subduction zone, and to a lesser extent, in a region called the *Intermountain seismic belt* in the western states. The New Madrid seismic zone is the main area of intraplate seismic risk in the eastern United States, but seismologists disagree about whether this area, in which earthquakes have

FIGURE 7.30 Evidence of past earthquakes in the geologic record is used to determine recurrence intervals. The block diagram shows subsurface sediment layers near an active fault. Buried sand volcanoes and disrupted beds (indicated by numbers) reveal when earthquakes happened in the past. Dating wood fragments in the layers gives the ages of the events. Tree rings can indicate when trees tilted.

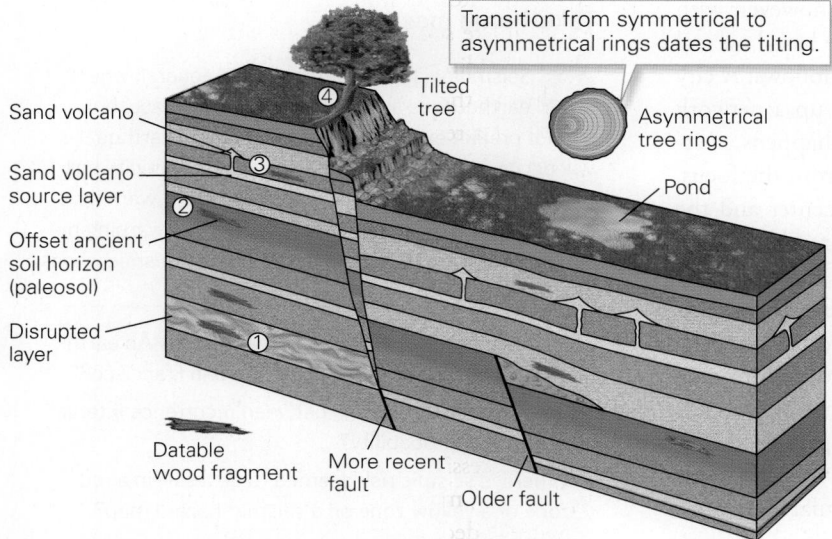

Sand volcano

Sand volcano source layer

Offset ancient soil horizon (paleosol)

Disrupted layer

Datable wood fragment

More recent fault

Older fault

Tilted tree

Transition from symmetrical to asymmetrical rings dates the tilting.

Asymmetrical tree rings

Pond

FIGURE 7.31 Examples of seismic-hazard maps.

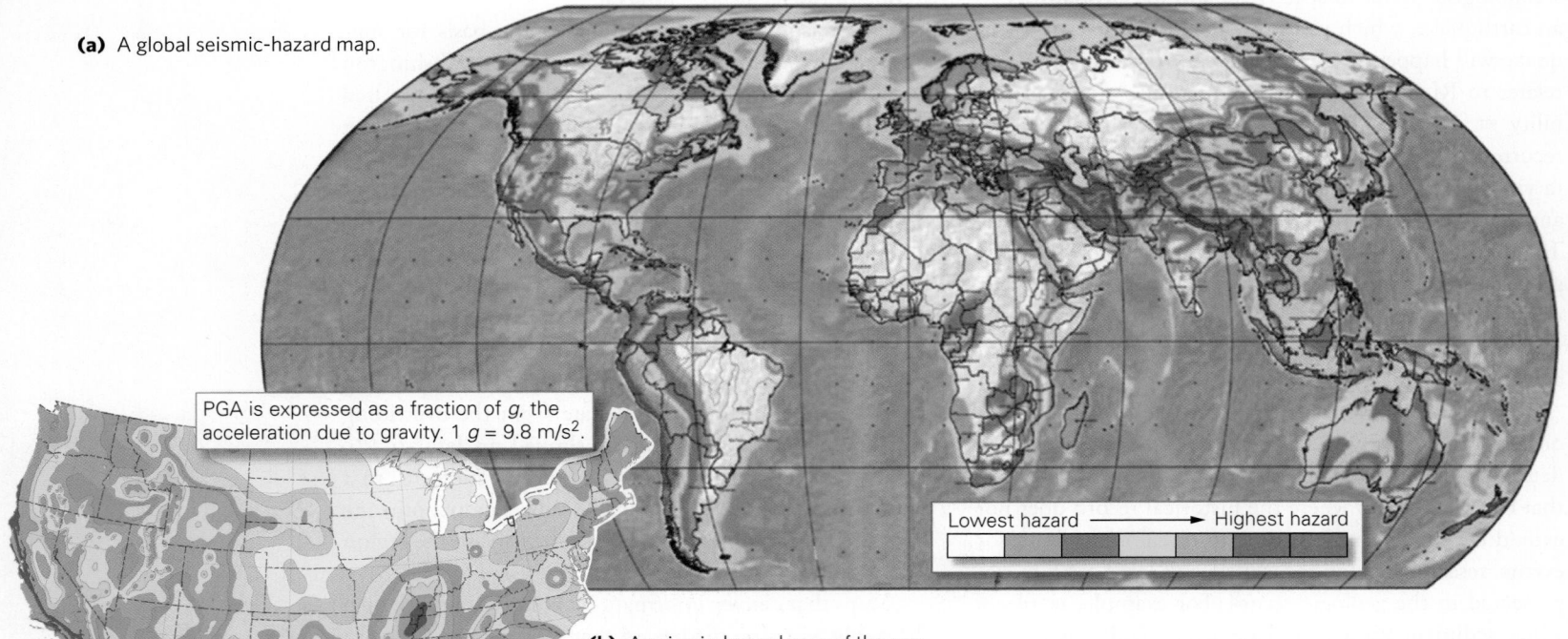

(a) A global seismic-hazard map.

PGA is expressed as a fraction of g, the acceleration due to gravity. 1 g = 9.8 m/s^2.

Lowest hazard ⟶ Highest hazard

Fraction of g
0.02 0.06 0.14 0.30 0.80

(b) A seismic-hazard map of the continental United States. This version of the map indicates areas where there is a 2% probability that PGA will exceed the indicated value during a 50-year period.

relatively low recurrence intervals, should be assigned a risk as high as what the map shows.

Earthquake Early-Warning Systems

We've pointed out that short-term predictions specifying that an earthquake will happen on a given date, or within a time window of days to years, are bogus. Such short-term predictions should not be confused, however, with *earthquake early-warning systems*, which are based on real signals. An early-warning system works as follows: A city in a region of significant seismic risk sets up a network of seismometers. When an earthquake happens, seismic waves start traveling toward the city from the focus. Seismometers positioned between the epicenter and the city detect the seismic waves before they reach the city. The instant that a seismometer detects the earthquake, it transmits a radio signal to a control center, which automatically broadcasts an emergency signal to the city. The emergency signal travels at the speed of light, so it reaches the city seconds to minutes before the seismic waves, and can trigger automatic shutdowns of gas pipelines, trains, nuclear reactors, and power lines. It also sets off sirens and triggers alerts on radio, TV, and cellular networks, warning people to take precautions.

Because tsunamis are so dangerous, predicting their arrival can save thousands of lives in coastal communities. The US National Oceanic and Atmospheric Administration (NOAA) has installed tsunami-detection buoys in oceans worldwide, and it maintains tsunami early-warning centers in Hawai'i and Alaska. At these centers, observers keep track of earthquakes and use data relayed from tide gauges and seafloor pressure gauges to determine whether a particular earthquake has generated a tsunami (Fig. 7.32). If the observers detect a tsunami, they flash warnings to authorities who can send out alerts urging people to evacuate to higher ground.

Take-home message...

Seismologists can determine, in general, where earthquakes are most likely to take place, but cannot predict exactly when and where an earthquake will occur. Seismic risk is greater where seismicity has happened more frequently in the past. Early-warning systems can provide seconds to minutes of warning by sending out signals that travel faster than seismic waves.

Quick Questions ——————

- Should you believe a warning that states, "An earthquake will happen next Tuesday in San Francisco?"
- What is the relationship between recurrence interval and annual probability?
- Where is seismic risk deemed to be less—in a red zone or a yellow zone on a seismic-hazard map?

FIGURE 7.32 Sending out a warning.

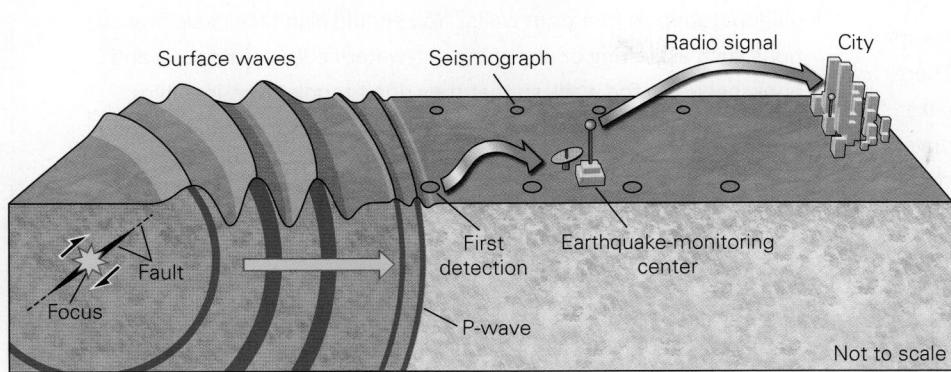

(a) The configuration of an earthquake early-warning system.

(b) Buoys can detect a tsunami in the open ocean so that people on land can be warned to evacuate low-lying coastal areas.

7.7 Preparing for Earthquakes

About 60% of the world's human population lives in locations subject to significant seismic risk. But the consequences that people and communities suffer when an earthquake of a given size happens vary dramatically from place to place. They depend on many factors, including the proximity of the epicenter to a population center; the depth of the focus; the style of building construction in the epicentral region; the steepness of slopes; whether faulting displaces the seafloor; the proximity of the affected region to the sea; whether building foundations are on solid bedrock or on weak sediment; whether the earthquake happens when people are outside or inside; and whether the government is able to provide emergency services promptly. Fortunately, individuals and communities can mitigate the dire consequences of earthquakes by taking sensible precautions, such as those we describe in this section, which together improve **earthquake preparedness**.

Earthquake Engineering

Designing *earthquake-resistant buildings* (ones that can withstand ground motion), a practice called **earthquake engineering**, can help save lives and property. In regions prone to large earthquakes, buildings and other structures of any size should be constructed to be sturdy but somewhat flexible (so that ground shaking cannot crack them easily). Facades, decorative features, and interior structures should be anchored properly so that they don't detach. Vertical supports must be strong enough to stand up to loads that exceed the normal weight of overlying floors, because accelerations during ground motion add to the acceleration due to gravity and make the weight of the floors effectively much greater.

How can homes, buildings, and bridges of any size be made more earthquake resistant (Box 7.3)? In regions with significant seismic risk, certain kinds of construction should be avoided. For example, concrete-block, unreinforced-concrete, and unreinforced-brick buildings crack and crumble under conditions in which wood-frame, steel-girder, or reinforced-concrete buildings remain standing. Traditional heavy, brittle tile roofs that can shatter and bury the inhabitants should also be avoided. During building construction, support columns should be encased in metal or cable sheaths so that they don't split and buckle; floors or spans should be bolted to foundations or support columns to prevents them from bouncing up or sideways; and diagonal braces should be added to frames to prevent buildings from twisting and shearing (Fig. 7.33a, b). In the case of large buildings, new technologies now allow engineers to install shock absorbers in the foundation of a building to counteract vibrations, or suspended weights near the top of a building to counteract swaying (Fig. 7.33c, d). Unfortunately, because less wealthy communities generally can afford only relatively weak construction materials, casualties due to collapse of structures in a vulnerable developing nation can be orders of magnitude greater than those caused by a similar-sized earthquake in a wealthy industrialized country.

If funds are available, structures can be made safer by **seismic retrofitting**, the process of strengthening existing structures that were not originally designed to be earthquake resistant. Seismic-retrofitting methods may include

Did you ever wonder . . .

if buildings can be designed to survive earthquakes?

BOX 7.3 ▸ **Putting Earth Science to Use**

Staying safe in earthquake country

Because so many communities are built on or near seismic belts, many people will endure an earthquake at some point in their lives. Therefore, everyone needs to take personal responsibility for *earthquake preparedness*. As a starting point, pay attention to maps that convey information about land stability in your area. For example, you may wish to avoid living in a house or apartment that straddles a fault zone or sits on land that may be susceptible to landslides, liquefaction, or inundation due to dam failure or tsunamis. Check with local officials and/or visit your town's website to ensure that satisfactory emergency plans exist, that evacuation routes are clear, and that your region has access to supplies for rescue and recovery.

In your dwelling, you can proactively determine whether simple earthquake retrofitting efforts have been completed by asking these questions: Is the building properly anchored to its foundation? Are floors properly anchored to support columns (Fig. Bx7.3a)? Are there diagonal support beams in walls? You should also take basic precautions such as bolting or strapping hot-water heaters, furnaces, and bookshelves to the walls so that they don't topple over; installing latches on cabinets so that they don't pop open during shaking; and knowing how to shut off gas, water, and electricity. Hold an earthquake preparedness drill for your family, and ask your local schools, offices, and factories to do the same. As part of your drill, know where to go if shaking begins. In general, if you're outside, move away from buildings that could shed debris. If you're inside, it's best to drop to the floor on your hands and knees and seek cover beneath or next to a very sturdy piece of furniture that won't get crushed if the ceiling falls, and isn't adjacent to glass windows (Fig. Bx7.3b). As long as lithosphere plates continue to move, earthquakes will continue to shake us. But we can reduce the chances of damage or injury by being prepared.

FIGURE Bx7.3 Personal earthquake preparedness.

Frame and foundation strengthening   **Interior stabilization**

Straps or anchors holding refrigerator and stove to the wall

Bookshelves bolted to the wall; straps restrain books

Heavy wall items (mirrors; paintings) anchored to the wall

Chandeliers and fans anchored to the ceiling

TV bolted to table; table bolted to floor

Lightweight, non-masonry, well-anchored chimney

Shear wall

Lightweight flexible roofing material

Cabinet latches

Framing anchor

Diagonal braces

Hold-down attaching the studs to the base plate.

Furnace and water heater strapped to the wall

Automatic gas shutoff valve

Bolts to attach the base plate of the frame securely to the foundation.

Deep footings to secure the foundation

(a) Freestanding homes can apply earthquake-engineering practices to strengthen the structure and common-sense measures to stabilize the interior. Interior-stabilization methods can also make apartments safer.

(b) If an earthquake strikes and you're inside, take cover under a sturdy piece of furniture near a strong wall.

FIGURE 7.33 Examples of earthquake engineering.

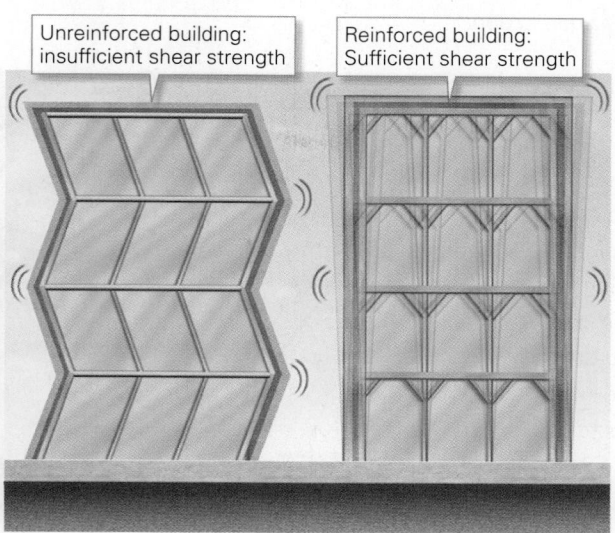

(a) When ground motion takes place, poorly braced buildings may sway and collapse.

(b) Adding diagonal braces to support columns can prevent buildings from breaking apart.

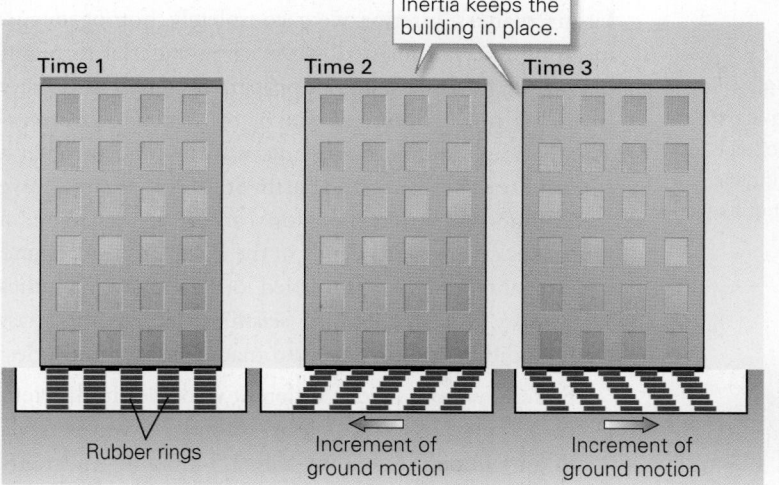

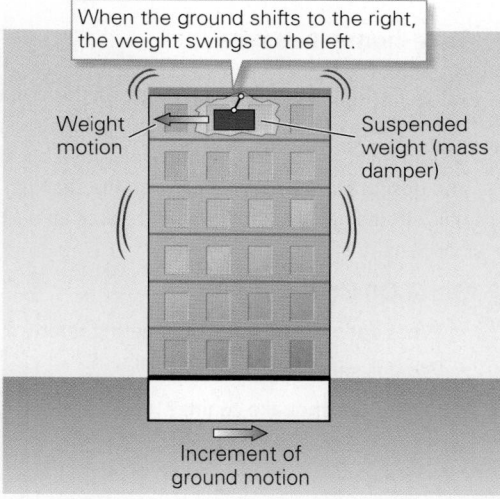

(c) To prevent swaying and shearing, a building can be set on rubber rings, or other types of shock absorbers, which allow the building to stay in the same place while the ground underneath it shifts.

(d) A large weight hanging from a pivot at the top of a building can counteract ground motion.

adding internal braces to a building's structure, removing poorly anchored facades, adding anchors to hold floors or spans to support columns, and encasing existing support columns to make them stronger.

Earthquake Zoning and Building Codes

Urban planners in seismically active areas can decrease seismic risk by instituting thoughtful **earthquake zoning**, meaning the establishment of regulations limiting construction in dangerous areas. Zoning relies on an assessment of where land is stable and where it is not. For example, zoning can discourage building on land underlain by weak mud or wet sand, or along or below steep slopes. Similarly, zoning should discourage construction of buildings (especially critical buildings such as schools, hospitals, fire stations, communications centers, and power plants) on the trace of an active fault, where fault slip could displace one part of a building relative to another. And in coastal areas, zoning should prevent construction in lowlands prone to flooding by tsunamis. Maps of seismic risk, and of associated risks such as liquefaction **(Fig. 7.34)**, can help officials make

FIGURE 7.34 This simplified map of the San Francisco area shows where the land might be susceptible to liquefaction.

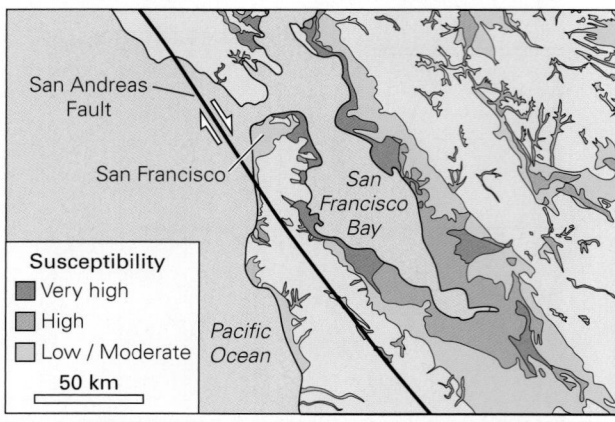

FIGURE 7.35 A simplified image of the Earth's interior, as first proposed in the 19th century. The numbers indicate the range of densities within each layer.

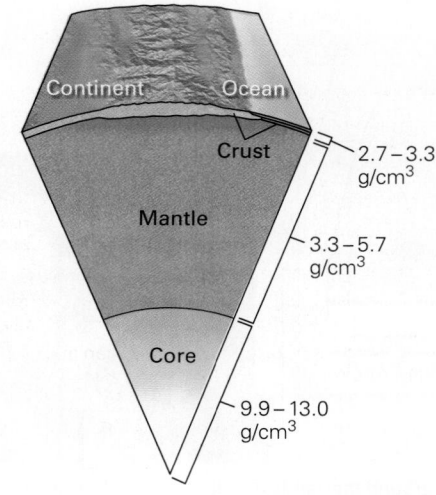

zoning decisions. In seismically active areas where construction is allowed, establishment of *seismic building codes* can require contractors to design sturdier and safer buildings.

Take-home message...

🏠 Earthquakes are a fact of life on this dynamic planet. People in regions facing high seismic risk should build on stable ground, avoid unstable slopes, and design structures that can survive shaking. Individuals should have a plan for what to do if an earthquake happens.

Quick Questions ─────────

• What does earthquake-engineering refer to?

• What is seismic retrofitting?

• What is earthquake zoning?

7.8 Seismic Study of the Earth's Interior

So far in this chapter, we've focused on understanding the causes, consequences, and locations of earthquakes. Seismologists continue to study these issues, but they also use earthquakes as a tool to "see" inside the Earth. Decades before the invention of the seismograph, researchers had concluded, based on measurements of the Earth's mass and shape, that the Earth consists of three concentric layers that differ from one another in terms of their composition and density: the crust; the mantle; and the core (Fig. 7.35; see Chapter 1). By measuring how fast seismic waves travel through the Earth, and how they reflect or bend as they travel, seismologists have subsequently developed a much more refined image of the Earth's interior.

Learning to Interpret Paths of Seismic Waves Inside the Earth

If our planet's interior were completely homogeneous, meaning that it consisted of the same material throughout, with the same material properties—such as density (mass per unit volume) as well as resistance to compression, bending, or shear—seismic waves would travel at a constant speed through the Earth. Studies of seismic-wave travel times show that they don't, meaning that material properties change with depth in the Earth. To understand these changes, scientists carried out laboratory studies to measure what happens to seismic waves as they pass through different materials, and made several discoveries:

• Seismic waves move at different velocities in different rock types (Fig. 7.36a). For example, P-waves travel 8 km/s in peridotite, but only 3.5 km/s in sandstone. Therefore, waves speed up or slow down as they pass from one rock type into another.

• P-waves travel more slowly in a liquid than in a solid of the same composition. Therefore, P-waves travel more slowly in magma than in solid rock, and more slowly in molten iron alloy than in solid iron alloy (Fig. 7.36b).

• Both P-waves and S-waves can travel through a solid, but only P-waves can travel through a liquid (Fig. 7.36c).

With this understanding of how material properties control seismic-wave velocity, seismologists could characterize the paths that seismic waves take through the Earth by taking into account a key principle from physics—namely, that if a wave travels at different velocities in two different materials, it can both bounce, or *reflect*, off the boundary between those materials, and bend, or *refract*,

FIGURE 7.36 The propagation of seismic waves.

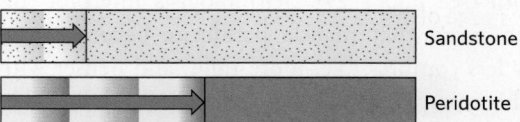

(a) Seismic waves travel at different velocities in different rock types. After a given time, a wave will have traveled farther in peridotite than in sandstone.

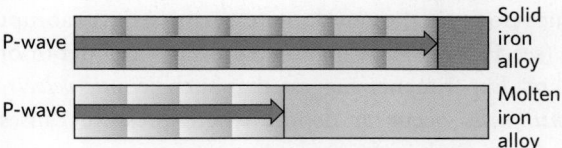

(b) P-waves travel faster in solid iron alloy than in molten iron alloy.

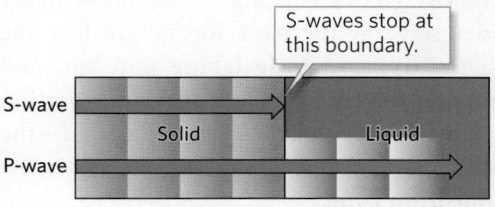

(c) Both P-waves and S-waves can travel through a solid, but only P-waves can travel through a liquid.

at that boundary. The angle between a line drawn perpendicular to the boundary plane and an incident (incoming) wave is the same as the angle between the line and a reflected wave (**Fig. 7.37a**). But the angle by which a refracted wave bends at a boundary depends both on the difference between the wave's velocities in the materials before and after the boundary and on the angle at which the wave strikes the boundary. As a rule, as waves pass from a material through which they travel rapidly into one through which they travel more slowly, they bend away from the boundary (see Fig. 7.37a). Alternatively, if waves pass into a material through which they travel more rapidly, they bend toward the boundary (**Fig. 7.37b**). If the boundary between the two materials is abrupt, the waves bend abruptly, but if the boundary is gradual, the waves bend gradually. With this knowledge in hand, seismologists could begin to use seismic waves as a tool to see the Earth's interior, somewhat like a doctor who uses X-rays or ultrasound to see inside your body.

Identifying the Earth's Layers with Seismic Waves

DISCOVERING THE CRUST-MANTLE BOUNDARY.
In 1909, Andrija Mohorovičić, a Croatian scientist, noted that P-waves arriving at seismograph stations less than 200 km (125 mi) from the epicenter of an earthquake traveled at an average speed of 6 km/s (4 mi/s), whereas P-waves

FIGURE 7.37 Refraction and reflection of seismic waves.

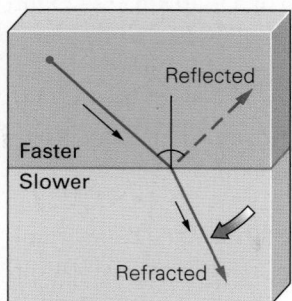

(a) Part of the energy of an incoming wave reflects off the boundary between two different materials, and part crosses the boundary. A wave that enters a material through which it travels more slowly bends away from the boundary.

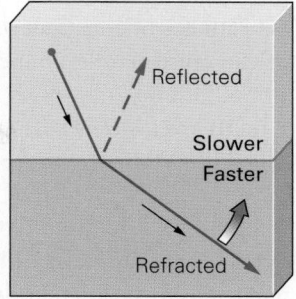

(b) A wave that enters a material through which it travels more rapidly bends toward the boundary.

arriving at seismographs more than 200 km from the epicenter traveled at an average speed of 8 km/s (5 mi/s). To explain this observation, he suggested that P-waves reaching nearby seismographs followed a shallow path that kept them entirely within the crust in which they traveled relatively slowly, whereas P-waves reaching distant seismographs traveled through the mantle for part of their route, in which their velocity was faster. Therefore, the waves refracted as they crossed the crust-mantle boundary (Fig. 7.38). Mohorovičić was able to calculate the depth at

FIGURE 7.38 Discovery of the Moho.

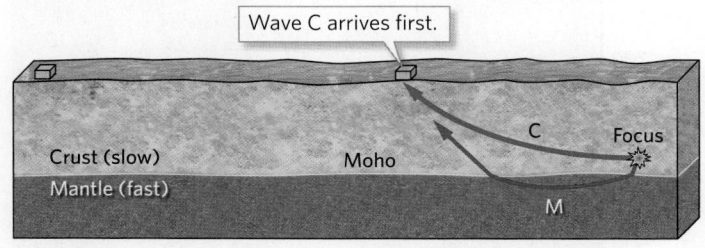

(a) P-waves traveling mostly through the crust reach a nearby seismograph first.

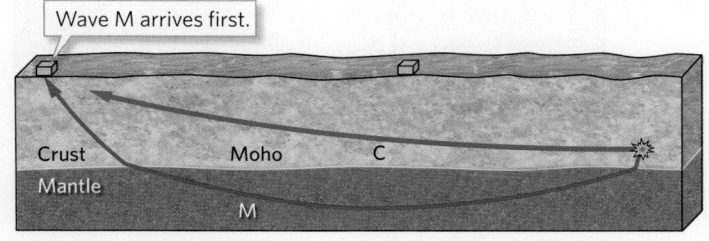

(b) P-waves traveling for most of their path through the mantle reach a distant seismograph first.

which refraction took place and, therefore, the depth of the crust-mantle boundary, from this observation. He proposed that it occurred at a depth of 35–40 km (22–25 mi) beneath continents. Later studies showed that the depth of the crust-mantle boundary beneath continents varies from 25 to 70 km (15–45 mi), and beneath oceans from 7 to 10 km (4–6 mi). The crust-mantle boundary is now called the **Moho**, in honor of Mohorovičić.

DISCOVERING THE STRUCTURE OF THE MANTLE. Seismologists eventually learned that seismic waves travel at different speeds at different depths in the mantle (Fig. 7.39). Specifically, at depths of 100–200 km (60–125 mi) beneath the seafloor, seismic velocities are lower than in the overlying portion of the mantle. This 100–200 km deep layer is now known as the **low-velocity zone (LVZ)**. Researchers suggest that the LVZ corresponds to a layer in which mantle rock has undergone slight partial melting. Because P-waves travel more slowly through liquids than through solids, the presence of even a tiny bit of melt slows them down. Simplistically, the top of the LVZ delineates the base of the lithosphere and the top of the asthenosphere beneath oceanic plates. Seismologists do not find a well-developed LVZ beneath continents.

FIGURE 7.39 The velocity of P-waves changes with depth in the mantle because the physical properties of the mantle change with depth.

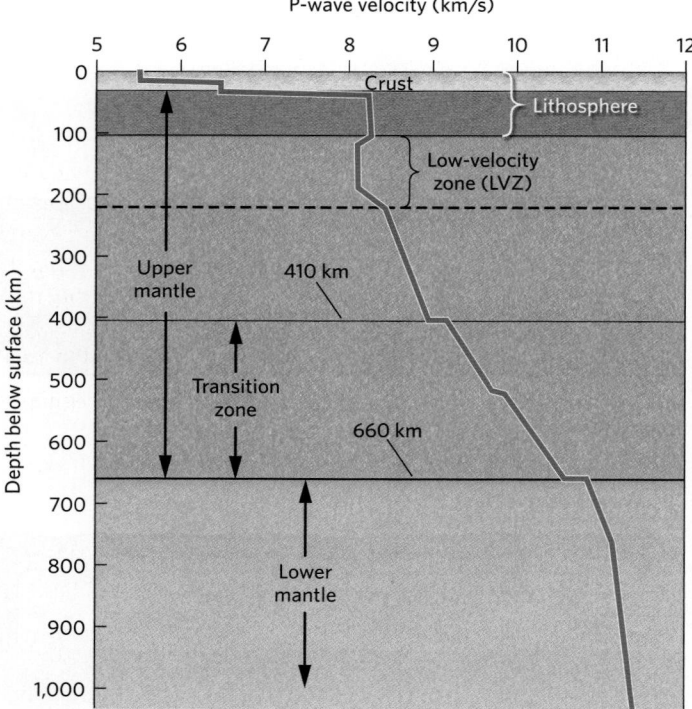

Below about 200 km, seismic-wave velocities in the mantle, beneath both continents and oceans, increase with depth (see Fig. 7.39). Seismologists interpret this change to mean that mantle peridotite becomes progressively less compressible, more rigid, and denser with depth. This interpretation makes sense, considering that the weight of overlying rock increases with depth, and that as pressure increases, the atoms making up minerals squeeze together more tightly and are less free to move.

At depths between 410 and 660 km (255–410 mi), seismic velocity in the mantle increases in a series of abrupt steps (see Fig. 7.39). A major step occurs at a depth of 660 km. Experiments suggest that such *seismic-velocity discontinuities* occur at depths where pressure causes atoms in minerals to rearrange into more compact crystal structures of the same composition, a phenomenon called a *phase change* (see Chapter 5). Seismic-velocity discontinuities serve as the basis for subdividing the mantle into the **upper mantle** (above 660 km) and the **lower mantle** (below 660 km). The lower portion of the upper mantle, between 410 and 660 km—the region in which the seismic discontinuities occur—is called the **transition zone**.

DISCOVERING THE STRUCTURE OF THE CORE. In the early 20th century, researchers installed seismographs at many stations around the world. In 1914, measurements by these instruments revealed that P-waves from a given earthquake did not arrive at seismographs within a band between 103° and 143°, as measured along the curve of the Earth's surface, from the earthquake epicenter. This band is now called the *P-wave shadow zone* (Fig. 7.40a). The presence of the P-wave shadow zone means that deep in the Earth, a major boundary exists where seismic waves abruptly refract downward because their velocity suddenly decreases. The dimensions of the shadow zone allowed seismologists to calculate that this boundary lies at a depth of 2,900 km (1,800 mi), and they consider it to be the **core-mantle boundary**.

Seismologists also found that S-waves did not arrive at seismograph stations located between 103° and 180° from the epicenter (a band called the *S-wave shadow zone*)—which means that S-waves cannot pass through the core at all (Fig. 7.40b). If they could, an S-wave headed straight down through the Earth should reach the ground surface on the other side of the planet. Recall that S-waves cannot pass through liquid, so the fact that S-waves do not pass through this zone led seismologists, at first, to conclude that the entire core might be liquid iron alloy. But in 1936, a Danish seismologist, Inge Lehmann, discovered that P-waves passing through the core reflect off a boundary within the core (Fig. 7.40c). She proposed that the

FIGURE 7.40 Shadow zones and the discovery of the Earth's core. (Note that the circumference of a circle is 360°, so it is 180° from a given point to a point on the other side of the planet.) The black arrows indicate the various paths that seismic waves propagating from the focus can take.

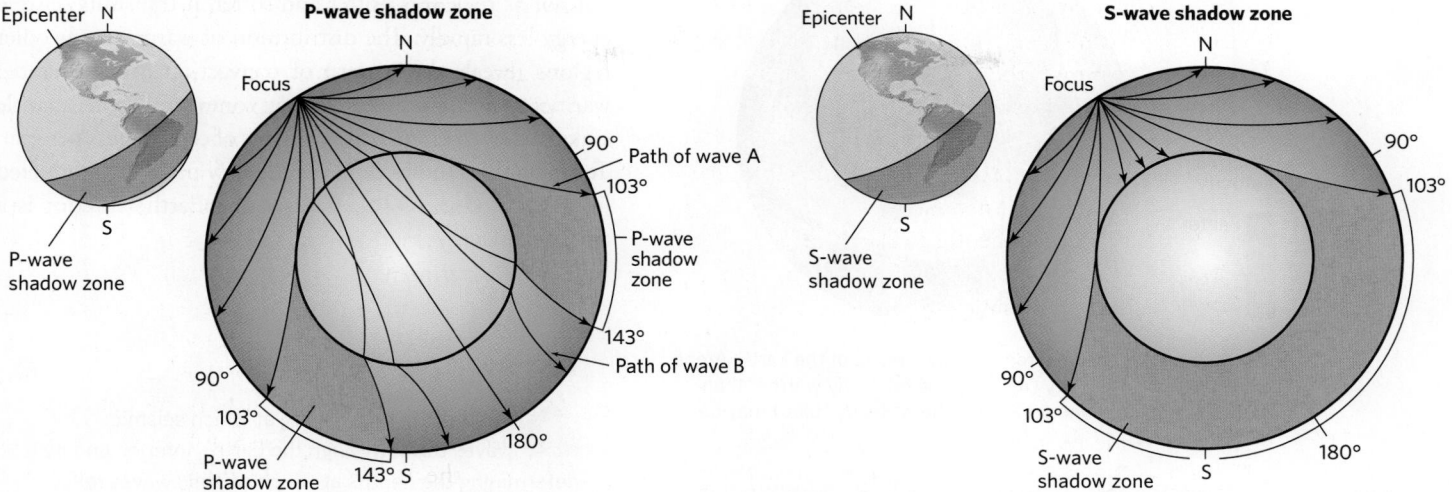

(a) P-waves do not arrive in the P-wave shadow zone because they are refracted at the core-mantle boundary.

(b) S-waves do not arrive in the S-wave shadow zone because they cannot pass through the liquid outer core.

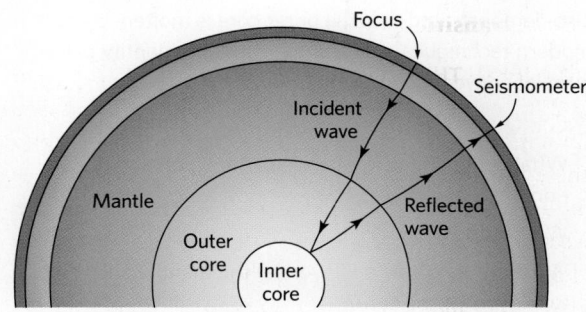

(c) A P-wave reflects off the inner core–outer core boundary.

called **seismic tomography**, seismologists can produce three-dimensional images of seismic-velocity variation in the Earth's interior, somewhat like doctors who produce three-dimensional CT (computerized tomography) scans of the human body **(Fig. 7.42a)**. Tomographic studies allow seismologists to identify regions in the mantle where seismic waves travel faster or slower than expected, and these studies have led to the realization that

core includes two parts: an **outer core** consisting of liquid iron alloy, and an **inner core** consisting of solid iron alloy. (Recall that circulation in the liquid outer core generates the Earth's magnetic field.) The boundary between the inner and outer core was eventually located at a depth of about 5,155 km (3,203 mi).

A MODERN IMAGE OF THE EARTH'S LAYERS. Through painstaking effort, seismologists have compiled data on seismic-wave travel times to develop a graph, known as a *velocity-versus-depth curve*, that shows the average depths at which seismic-wave velocity suddenly changes and the average amounts of those changes. Depths at which major changes take place define the principal layers and sublayers of the Earth down to its center **(Fig. 7.41)**.

More detailed studies in recent years have shown that the onion-like layered model of the Earth we've described so far is an oversimplification. Using a technique

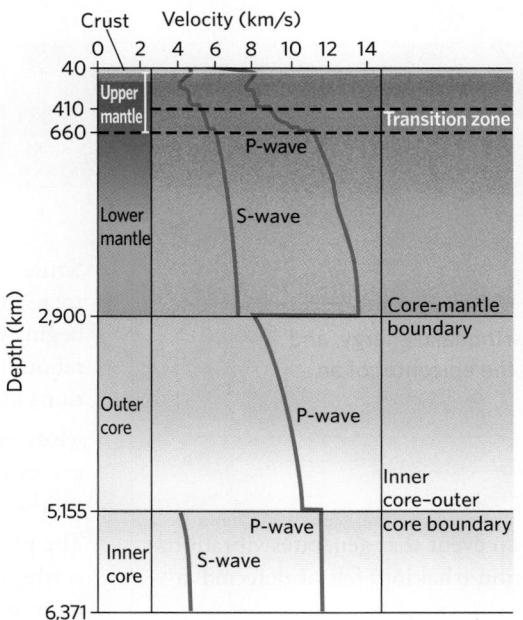

FIGURE 7.41 The velocity-versus-depth curve is a graph showing how velocities of P-waves and S-waves vary with increasing depth in the Earth. Note that the graph does not show a velocity for S-waves in the outer core because S-waves cannot travel through molten iron alloy (a liquid).

FIGURE 7.42 Modern images of the Earth's interior.

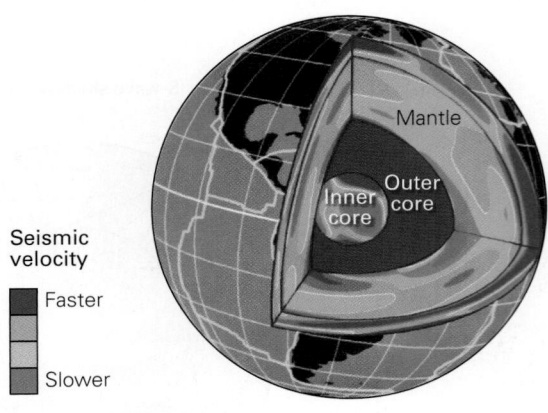

Seismic velocity

Faster

Slower

(a) A three-dimensional tomographic image of the Earth. Areas of lower seismic velocity (red) may be relatively warmer than their surroundings, and areas of higher velocity (blue) may be relatively cooler.

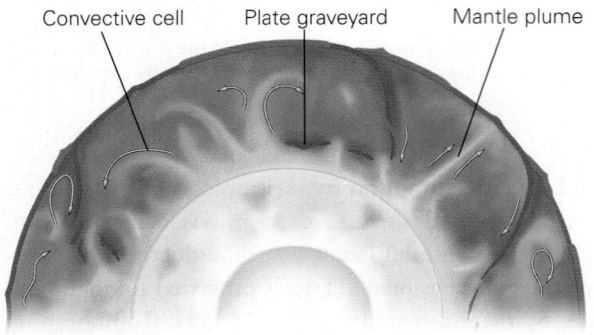

Convective cell   Plate graveyard   Mantle plume

(b) A conceptual image of what the Earth's interior looks like. Hotter mantle material undergoes upwelling, and cooler mantle material sinks. Plates form at the surface and sink back down into the mantle.

the velocities of seismic waves vary significantly with location at a given depth. The regions of lower velocity probably represent warmer mantle material, and the regions of higher velocity probably represent cooler mantle material, for as rock gets hotter and softer, it transmits seismic energy less rapidly. The distribution of warmer and cooler regions reveals the pattern of convection in the mantle: warmer mantle rises in upwelling zones, and cooler mantle sinks in downwelling zones, at rates of centimeters per year. Tomographic studies can even identify pieces of subducted lithosphere (Fig. 7.42b). Clearly, the Earth's interior is a dynamic place!

Take-home message…

By studying the velocity at which seismic waves travel through the Earth's interior, and by determining the depths at which seismic waves reflect or refract, seismologists have been able to locate the boundaries between the Earth's layers, such as the Moho between the crust and mantle. Such work has also demonstrated that the outer core is molten. Using modern techniques, researchers can even identify patterns of convection in the mantle.

Quick Questions

• What is the difference between reflection and refraction of seismic waves?

• What is the Moho?

• What are the three sublayers of the mantle, and the two layers of the core?

---

7 CHAPTER REVIEW

Objective 7.1

Explain how faulting can release earthquake energy, and distinguish between the focus and the epicenter of an earthquake.

KEY CONCEPTS

• The term *earthquake* refers both to an event that generates vibrations in the Earth and to the ground motion (shaking) felt or detected at the Earth's surface.

• Some earthquakes occur due to the formation of a fault in intact rock. Prior to slip, rock bordering the fault deforms elastically. Slip begins when rock along the fault breaks. The bordering rock then rebounds to its original shape; this elastic rebound produces vibrations known as seismic waves.

• Most earthquakes happen when stress overcomes friction on a pre-existing fault so that the fault slips again until friction stops the motion. Once formed, faults can exhibit stick-slip behavior.

• The place where an earthquake begins is called the focus of the earthquake, and the point on the Earth's surface directly above the focus is the earthquake's epicenter.

EARTH-SCIENCE VOCABULARY

displacement (p. 245)
earthquake (p. 244)
elastic deformation (p. 245)
elastic-rebound theory
  (p. 246)
epicenter (p. 247)
fault (p. 244)
focus (p. 246)

friction (p. 246)
ground motion (p. 244)
rebound (p. 245)
seismicity (p. 244)
seismic wave (p. 244)
seismologist (p. 246)
stick-slip behavior (p. 246)

REVIEW QUESTIONS

1. **(a)** What is an earthquake? **(b)** Do all earthquakes happen on faults? **(c)** In what form does earthquake energy travel through the Earth? **(d)** Explain how elastic rebound plays a role in the generation of earthquakes.

2. **(a)** What phenomenon or force prevents a fault from continuing to slip forever once slip begins? **(b)** Explain how stick-slip behavior operates. **(c)** Distinguish among a mainshock, a fore-shock, and an aftershock. **(d)** Why do aftershocks take place?

3. **(a)** Label the focus and epicenter of the earthquakes shown in **Figure A**. **(b)** When you look at a map showing the "locations of earthquakes," are you looking at a map of epicenters or a map of foci? **(c)** Distinguish between shallow-focus and deep-focus earthquakes.

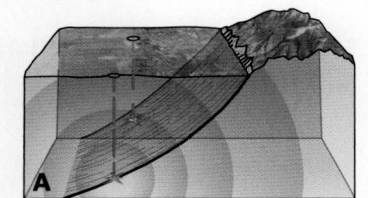

Objective 7.2

Describe the different types of seismic waves and the methods used to measure and locate earthquakes.

KEY CONCEPTS

- Seismic waves include body waves (P- and S-waves), which pass through the interior of the Earth, and surface waves (L- and R-waves), which move along the surface of the Earth.

- A seismograph can detect and record seismic waves. The resulting seismograms demonstrate that different types of seismic waves travel at different velocities.

- Using the difference between P-wave and S-wave arrival times at three locations, seismologists can pinpoint the epicenter of an earthquake.

EARTH-SCIENCE VOCABULARY

amplitude (p. 250)
body wave (p. 247)
P-wave (p. 247)
seismogram (p. 250)
seismograph (p. 247)

seismometer (p. 247)
surface wave (p. 247)
S-wave (p. 247)
travel time (p. 250)

REVIEW QUESTIONS

4. **(a)** What is the difference between a body wave and a surface wave? **(b)** How do P-waves and S-waves differ from each other? **(c)** How do R-waves and L-waves differ from each other? **(d)** Which type of wave does **Figure B** show? Draw an arrow to indicate the direction in which the waves are moving.

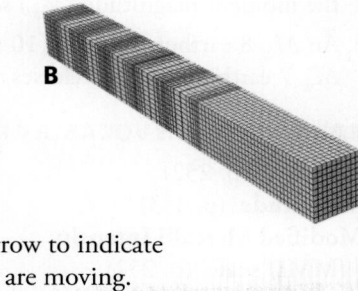

5. **(a)** Explain how the ground move-ments produced by an earthquake are detected by a mechanical seis-mograph. **(b)** Does **Figure C** show a horizontal- or vertical-motion seis-mograph? Label the parts that move with the Earth's surface, and the parts that don't.

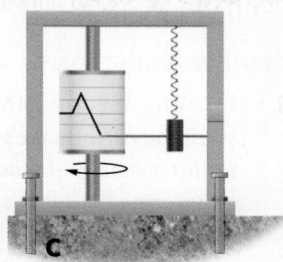

6. **(a)** Label P-, S-, and surface waves on the seismogram in **Figure D**, and label the axes of the graph. **(b)** Which type of wave travels the fastest? **(c)** Some surface waves travel faster than others. Given this information, how does the duration of vibrations due to surface waves at a seismograph station depend on the distance between an earthquake epicenter and that station?

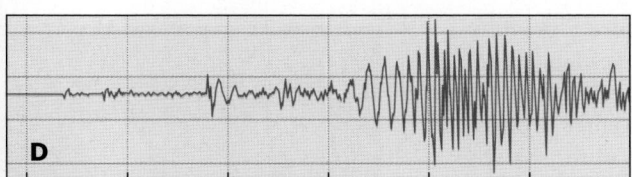

7. **(a)** Explain how seismologists determine the location of an earth-quake's epicenter. **(b)** Assuming that the spacing between the ver-tical lines represents five minutes, determine the distance between the seismograph station where the seismogram was recorded and the earthquake epicenter, using the appropriate graph in the text.

Objective 7.3

Interpret a news story that mentions an earthquake's magnitude and intensity.

KEY CONCEPTS

- The Modified Mercalli Intensity (MMI) scale characterizes the vigor of ground motion at a locality by taking into account the damage caused by an earthquake and people's perception of the motion.

- Magnitude scales characterize the amount of energy released at the focus of an earthquake, as indicated by the amplitude of ground motion at a reference distance from the epicenter. The Richter scale is an early version of a magnitude scale. These days, seismologists use the moment magnitude (M_w) scale instead.

- An M_w 8 earthquake yields 10 times as much ground motion as an M_w 7 earthquake, and releases about 32 times as much energy.

EARTH-SCIENCE VOCABULARY

intensity (p. 252)
magnitude (p. 253)
Modified Mercalli Intensity (MMI) scale (p. 252)

moment magnitude scale (p. 254)
Richter scale (p. 254)
seismic-intensity map (p. 253)

REVIEW QUESTIONS

8. **(a)** Why does the MMI of an earthquake depend on distance from the earthquake's epicenter? **(b)** Will the MMI at the epicenter for a shallow-focus earthquake be greater or less than that of an intermediate-focus earthquake that releases the same amount of energy? Why?

9. **(a)** Explain what the Richter scale measures and reports as a magnitude. **(b)** How does the modern moment magnitude scale differ from the traditional Richter scale? **(c)** Will the M_w of a shallow-focus and a deep-focus earthquake be the same or different if the two earthquakes release the same amount of energy? **(d)** The largest hydrogen bomb ever detonated represented a release of energy equal to about what earthquake magnitude?

10. **Figure E** provides a ShakeMap of an M_w 7.1 earthquake whose epicenter (white dot) was in the sparsely populated Mojave Desert. The black line is the segment of the fault that slipped.
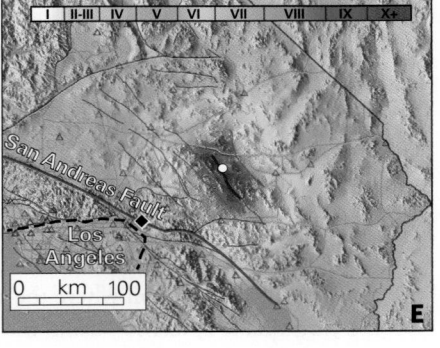
(a) Circle the area that felt an MMI of IX or greater and describe the effects that people in this area felt. Note the relationship between the location and orientation of area you circled and the trace of the fault. **(b)** At the scale of the map, about how far from the epicenter would a standing person feel an intensity of less than V?

11. **(a)** Look again at **Figure E**. Would the maximum intensity be similar if the earthquake's focus had been at the same depth, but its epicenter were on the San Andreas Fault at the black diamond? **(b)** Would the consequences to society be different if the epicenter were at the black diamond? Explain why. **(c)** The earthquake mapped in **Figure E** had a focus depth of 20 km. How would the ShakeMap look different if the focus had been at 200 km?

12. **(a)** How many earthquakes of M_w 4 would have to happen to release the same total energy as one earthquake of M_w 8? **(b)** If both were shallow-focus earthquakes, what adjective would be used to describe each?

Objective 7.4

Relate earthquakes to specific geologic settings, in the context of plate tectonics.

KEY CONCEPTS

- Earthquake activity takes place mainly in seismic belts, the majority of which lie along plate boundaries. Abundant earthquakes also happen in rifts and collision zones. Intraplate earthquakes happen in the interior of a plate, away from plate boundaries, rifts, or collision zones.

- Different kinds of faults are active at different kinds of plate boundaries. Shallow-focus earthquakes happen along mid-ocean ridges and along transform faults. Shallow-focus earthquakes also happen along convergent boundaries, but so do intermediate- and deep-focus earthquakes, which occur in the Wadati-Benioff zone.

- At convergent boundaries, slip on a broad area of the fault separating the base of the accretionary prism from the downgoing plate can produce huge megathrust earthquakes.

EARTH-SCIENCE VOCABULARY

intraplate earthquake (p. 259)

megathrust earthquake (p. 258)
seismic belt (p. 255)

REVIEW QUESTIONS

13. **(a)** On the map of North America in **Figure F**, label a geologic setting where seismicity is due to slip on normal faults. Do the same for thrust faults and strike-slip faults. **(b)** At which type of plate boundary do megathrust earthquakes take place? **(c)** Why are megathrust earthquakes so dangerous?

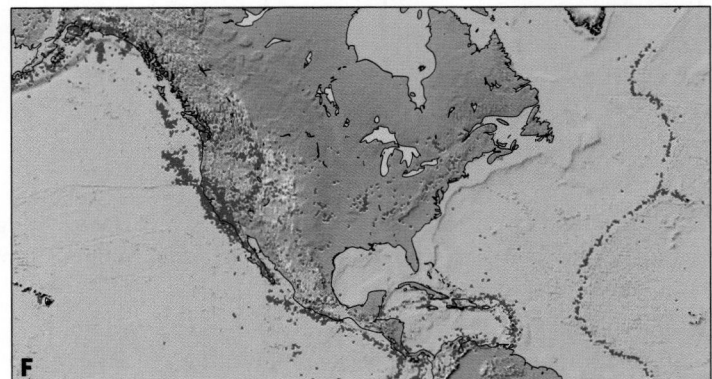

14. **(a)** Label the Mid-Atlantic Ridge on **Figure F**. Would a person standing in North America feel most of the earthquakes that take place along the Mid-Atlantic Ridge? Why or why not? **(b)** Which type of fault (normal, reverse, thrust, or strike-slip) is most common along a transform boundary? **(c)** Name three major continental transform faults.

15. **(a)** Label the Wadati-Benioff zone on **Figure G**. **(b)** Circle the dots that represent shallow-focus earthquakes. **(c)** What is the

relationship between a Wadati-Benioff zone and a subducting plate?

16. **(a)** What is an intraplate earthquake? **(b)** Name two intraplate earthquakes that have affected the United States.

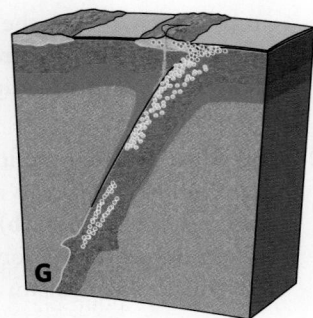

17. Label the geologic settings of seismicity on **Figure H**.

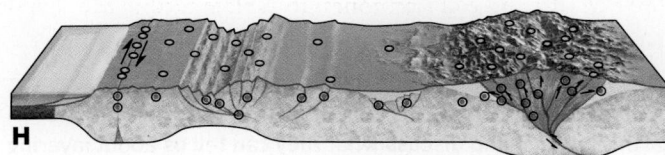

Objective 7.5

Distinguish among the different ways in which earthquakes cause casualties and damage, and differentiate between a tsunami and a storm wave.

KEY CONCEPTS

• Faults that intersect the ground surface can displace the ground surface and anything on it. The ground motion of an earthquake won't hurt you if you're out in an open field; most casualties are caused by falling debris and collapsing buildings, as well as by landslides and sediment liquefaction. Earthquakes can also cause fires and lead to outbreaks of disease.

• Coastal areas are subject to tsunami trains, which form when the seafloor suddenly pushes against the water of the ocean. Tsunamis differ from wind-driven waves in a number of ways that make them especially devastating when they are large.

EARTH-SCIENCE VOCABULARY

landslide (p. 264)
peak ground acceleration (PGA) (p. 264)
quick clay (p. 265)

sediment liquefaction (p. 265)
tsunami (p. 269)

REVIEW QUESTIONS

18. **(a)** Do all types of seismic waves cause the same type of ground motion? **(b)** Which type of wave caused the ground motion shown in **Figure I**? **(c)** Which substrate, soft sediment or solid granite, will host greater ground motion at a locality during a given earthquake?

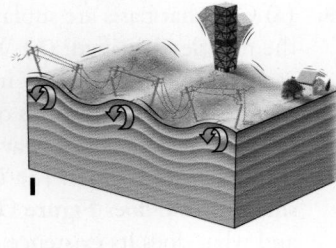

19. **(a)** Define peak ground acceleration (PGA), and compare the acceleration you would feel when standing on the ground during a great earthquake with that you might feel on a roller coaster. **(b)** Can ground motion alone injure you severely? **(c)** What can ground motion do to buildings?

20. **(a)** Why are landslides a major cause of destruction during an earthquake? **(b)** What is sediment liquefaction, and how can it affect buildings? **(c)** Why did the neighborhood shown in **Figure J** break up and slide seaward?

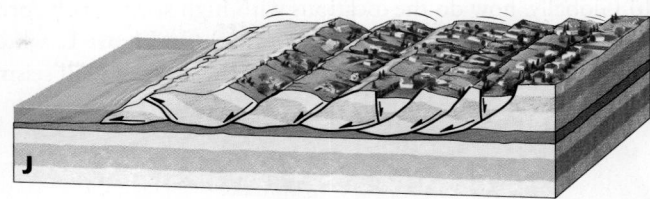

21. **(a)** Why are the fires ignited during an earthquake difficult to fight, and what phenomenon can lead to the growth of such a fire into a firestorm? **(b)** Why might disease spread in the wake of an earthquake?

22. **(a)** What causes a tsunami train to form? **(b)** How does a train of large tsunamis differ from a train of large storm waves? Label each type of wave in **Figure K**. **(c)** How can a large tsunami cause more damage over a larger area than does a large storm wave? **(d)** Is the wave height of all tsunamis large?

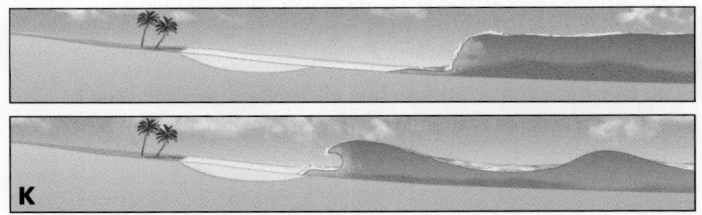

Objective 7.6

Critique a prediction of an earthquake, and interpret a seismic-hazard map.

KEY CONCEPTS

• Seismologists can determine the recurrence interval for earthquakes in a particular region. From such information, it is possible to provide long-term estimates stating the probability that an earthquake will happen during a specified time interval. Earthquakes are more likely to take place in seismic belts than elsewhere.

• Seismologists cannot provide short-term estimates giving the exact time and place at which an earthquake will happen, but earthquake early-warning systems can provide seconds to minutes of advance notice that an earthquake is coming.

EARTH-SCIENCE VOCABULARY

annual probability (p. 273)
paleoseismology (p. 273)

recurrence interval (RI) (p. 272)
seismic risk (p. 272)

23. **(a)** What does it mean to say that the recurrence interval of great earthquakes on a fault is 200 years? **(b)** If the recurrence interval for a great earthquake is 200 years, what is the annual probability that such an earthquake will happen? **(c)** What information can paleoseismology studies provide?

24. **(a)** What information can be shown on a seismic-hazard map? **(b)** Globally, how do the locations with high seismic risk correlate with the locations of plate boundaries? **(c)** On **Figure L**, which color represents the greatest risk? Label the areas of high seismic risk that are on plate boundaries.

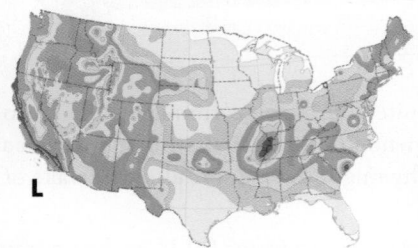

L

25. **(a)** Is there any trusted basis for providing a warning that an earthquake will happen on a certain day, two months in the future? **(b)** How does an earthquake early-warning system operate?

Objective 7.7

List the steps that communities and individuals can take to reduce casualties and destruction due to earthquakes.

KEY CONCEPTS

- Earthquake preparedness can help mitigate the consequences of earthquakes. Preparedness measures include incorporating earthquake engineering into building codes, requiring seismic retrofitting, and mandating earthquake zoning.

- Earthquake engineering guides construction of earthquake-resistant buildings that can withstand ground motion. Features that make existing structures sturdier can be added during seismic retrofitting.

- Earthquake zoning indicates where ground motion and land instability are likely to be most hazardous during an earthquake.

EARTH-SCIENCE VOCABULARY

earthquake engineering (p. 275)

earthquake preparedness (p. 275)

earthquake zoning (p. 277)

seismic retrofitting (p. 275)

REVIEW QUESTIONS

26. **(a)** What factors influence the degree of devastation during an earthquake? **(b)** Describe the features that can be incorporated in the design of a tall building to make the building earthquake resistant. **(c)** How can a bridge that sits on vertical reinforced-concrete columns be constructed to be more resistant

to collapse? **(d)** Note the pan-caked buildings in **Figure M**. Why might such structural failure have happened?

M

27. **(a)** What seismic-retrofitting steps can people take to make their residences safer? **(b)** What factors influence decisions about earthquake and tsunami zoning?

Objective 7.8

Explain how seismic waves behave as they pass through the Earth's interior, and discuss what they can tell us about layering inside the Earth.

KEY CONCEPTS

- Seismic waves travel at different velocities in different rock types. S-waves cannot travel through liquids.

- Seismic waves reflect or refract at boundaries between layers of different materials inside the Earth. By studying seismic waves, seismologists can identify the depths at which boundaries between layers occur, and they can identify where material is solid or molten.

- Seismic studies have yielded a refined view of the Earth's layering. More recently, seismic tomography studies show that at a given depth, some regions are warmer than others.

EARTH-SCIENCE VOCABULARY

core-mantle boundary (p. 280)

inner core (p. 281)

lower mantle (p. 280)

low-velocity zone (LVZ) (p. 280)

Moho (p. 280)

outer core (p. 281)

seismic tomography (p. 281)

transition zone (p. 280)

upper mantle (p. 280)

REVIEW QUESTIONS

28. **(a)** Why do seismic waves refract at specific depths within the Earth? **(b)** In **Figure N**, in which layer do seismic waves travel faster (X or Y)? **(c)** What clue led to the detection of the Moho?

29. **(a)** On what basis are sublayers in the mantle identified? **(b)** What causes changes in velocity in the mantle? **(c)** What features of the Earth's interior cause P-wave and S-wave shadow zones? **(d)** Which shadow zone does **Figure O** show, and what does its existence require?

30. **(a)** On what basis was the inner core identified? **(b)** What does a tomographic image of the Earth's interior reveal?

ANOTHER VIEW A swarm of earthquakes struck the southwestern coast of Puerto Rico in late 2019 and early 2020. The largest had a magnitude of M_w 6.4, and caused damage to many buildings. Reinforced columns holding the upper story of this two-story home broke apart and the steel rebar within buckled, so that the columns now curve like snakes. The ceiling now tilts at an angle of about 30°. Puerto Rico lies in the seismic belt along the plate boundary between the North American and Caribbean Plates.

8 DEEP TIME
How Old Is Old?

After studying this chapter, you should be able to...

1. define geologic time, explain the difference between relative and numerical ages, and discuss key geologic principles used to decipher the geologic history visible in an outcrop.

2. identify basic types of fossils, explain how fossils form and are preserved, and relate fossil succession to the theory of evolution.

3. interpret unconformities, describe the basis for correlating stratigraphic formations, and explain how correlation led to development of the geologic column.

4. discuss how to determine numerical ages by using radioisotopic dating, and interpret what the age of a rock means.

5. explain the basis for establishing the geologic time scale and for determining the age of the Earth.

In 1869, a one-armed Civil War veteran named John Wesley Powell, along with nine companions, set out to explore the Grand Canyon, the greatest gorge on the Earth. For three months, they drifted down the thundering rapids of the Colorado River, which flows along the floor of the canyon (Fig. 8.1). During their voyage, seemingly insurmountable walls of rock both imprisoned and amazed the explorers and led them to pose important questions about the Earth and its history, the same questions that tourists visiting the canyon ponder today: How long did it take to carve the canyon? When did the rocks making up the walls of the canyon form? Was there a time before these rocks accumulated? Thinking about such questions opens a door to thinking about *geologic time*, the span of time since the Earth's formation.

Our modern understanding of geologic time stems from 19th-century research that established geologic principles for determining the relative ages of rock units and other geologic features. A *relative age* specifies whether a given rock or feature is older than or younger than another. Scientists also carried out systematic studies of fossils and learned how to use the fossil record to track the evolution of life on the Earth and to determine the relative ages of rocks worldwide. Such work set the stage for developing the *geologic column*, a chart that divides geologic time into intervals. When geologists refined methods for dating rocks in the mid-20th century, it became possible to define the *numerical age*—the age in years—of rocks. These dates ultimately allowed geologists to assign numerical ages to intervals on the geologic column, which they could then transform into the *geologic time scale*. Such dates also allowed geologists to calculate a numerical age for the Earth, and showed that our planet's history extends billions of years into the past. This chapter introduces these amazing ideas and their implications. Clearly, the geologic discovery of what science popularizers call *deep time*—time measured in millions to billions of years—has changed humanity's perception of the Universe as profoundly as the astronomical discovery that the limit of deep space extends billions of light-years beyond the edge of our Solar System.

<< At the Mammoth Site in South Dakota, over 100 fossil mammoth skeletons have been found. Visitors standing at the edge of the excavation are looking back 26,000 years, to a time when, over the course of 700 years, mud and silt buried generations of the beasts after they had fallen into a sinkhole.

FIGURE 8.1 This aerial view of the Grand Canyon shows layer upon layer of sedimentary strata that were deposited over a basement of metamorphic rock. Differential erosion produced the stair steps of the canyon walls. Rocks in this view provide insight into the Earth's history.

8.1 Geologic Principles, Relative Ages, and Geologic History

Sir Isaac Newton's proposal of physical laws, such as the law of gravity and the laws of motion, helped initiate the Age of Enlightenment in Europe in the 18th century. During this time scientists began to seek natural, rather than supernatural, explanations for the features and phenomena in the world around them. Geology, as a discipline, came into existence at this time, and in this free-thinking environment, early geologists realized that the Earth's history began long before human history. James Hutton (1726–1797), a Scottish farmer and physician, played a key role in this discovery. While wandering in the rugged highlands of Scotland, he noticed that many features in outcrops of sedimentary rocks resembled features that form in modern depositional environments. For example, the surfaces of some sandstone beds displayed ripple marks identical to those visible on a modern beach. These observations led Hutton to speculate that ancient rocks and landscapes were a product of the same natural processes that were operating in modern times.

Hutton's idea came to be known as the principle of **uniformitarianism**. This principle emphasizes that physical processes operating in the modern world also operated in the past, at roughly the same rates—or, more concisely, that "the present is the key to the past" (Fig. 8.2). Hutton

FIGURE 8.2 The principle of uniformitarianism.

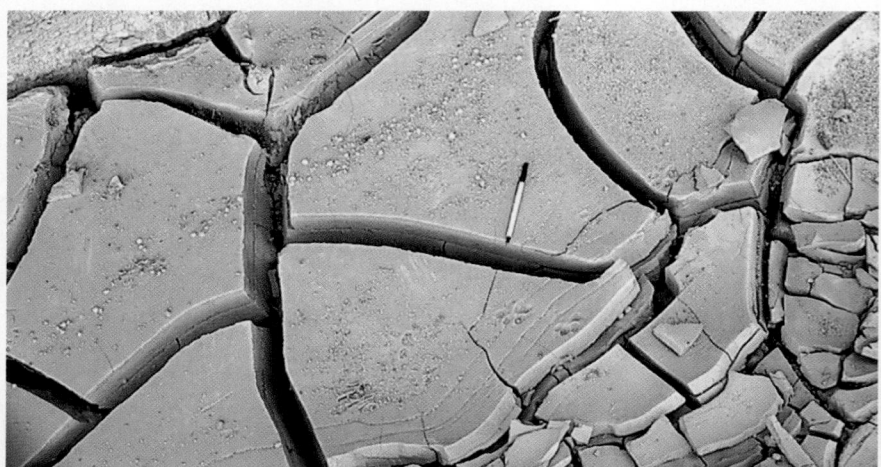

(a) Mudcracks form when a modern-day mud puddle dries up. Note the pen for scale.

(b) A 300-million-year-old rock containing preserved mudcracks. According to the principle of uniformitarianism, they formed in the same manner as modern mudcracks.

also deduced that not all the rocks he observed, or the structures that affected them, formed at the same time. Because no one had observed the entire process of rock formation and mountain building, Hutton concluded that these processes happen very, very slowly. Uniformitarianism, therefore, led to the idea that **geologic time**—the time since the formation of the Earth—must be very long indeed.

A determination of the temporal order in which events take place defines the **relative ages** of the events. For example, when historians say that World War II happened after World War I, without stating the years during which the wars happened, they are specifying the relative ages of the two wars. When geologists say that one rock formed before another, they are specifying the relative ages of the two rocks. How can the relative ages of two rock layers be determined? Hutton's discoveries, along with the work of other early geologists, established several more principles, in addition to uniformitarianism, that made it possible to determine relative ages of rocks, structures, and geologic events:

- **Original horizontality**: This principle states that layers of sediment, when first deposited, are roughly horizontal. Why? Sediments accumulate on relatively flat surfaces, such as floodplains or the seafloor. If they collect on a steep slope, they slide downslope before they can be buried and lithified (Fig. 8.3a). With this principle in mind, geologists conclude that folds and tilted beds represent the products of deformation events that happened after deposition.

- **Superposition**: This principle states that each layer of sedimentary rock must be younger than the one

below it, for a layer of sediment cannot accumulate unless there is already a surface on which it can collect. Therefore, in a sedimentary sequence, the oldest layer lies at the bottom and the youngest is at the top (Fig. 8.3b).

- **Cross-cutting relations**: This principle states that if one geologic feature cuts across another, the feature that has been cut is older. For example, if an igneous dike cuts across a sequence of sedimentary beds, the beds must be older than the dike (Fig. 8.3c). If a layer of sediment buries the dike, that sediment must be younger than the dike. And if a fault cuts across and displaces layers of sedimentary rock, then the fault must be younger than the layers.

- **Baked contacts**: This principle states that if an igneous intrusion thermally metamorphoses ("bakes") the wall rock that it intruded, the rock that was metamorphosed must be older than the intrusion (Fig. 8.3d). Geologists informally refer to an intrusive contact along which the intrusion metamorphosed wall rock as a *baked contact*.

- **Inclusions**: This principle states that a rock containing an *inclusion* (fragment of another rock) must be younger than the inclusion. So, a conglomerate containing basalt pebbles is younger than the basalt, whereas a lava flow containing sandstone fragments must be younger than the sandstone (Fig. 8.3e).

Geologists apply these geologic principles to determine the relative ages of geologic events, such as deposition, erosion, intrusion or extrusion of igneous rocks, and deformation (folding or faulting). The sequence of

FIGURE 8.3 Geologic principles used for determining relative ages.

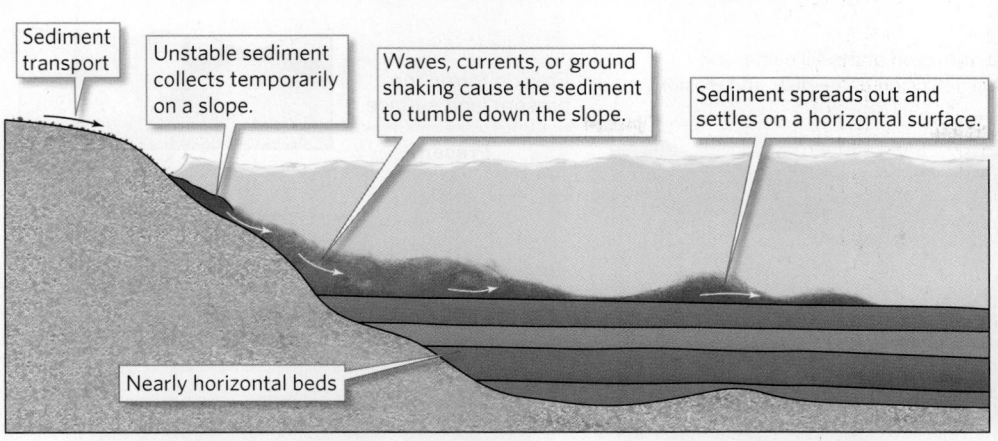

Horizontal sandstone beds in Wisconsin

Bedding plane

Cross beds

What an Earth Scientist Sees

Sediment transport

Unstable sediment collects temporarily on a slope.

Waves, currents, or ground shaking cause the sediment to tumble down the slope.

Sediment spreads out and settles on a horizontal surface.

Nearly horizontal beds

(a) Original horizontality: Gravity causes sediment to accumulate in roughly horizontal sheets (left). So, when we see these horizontal beds of rock in Wisconsin (right), we assume that the beds are in their original orientation.

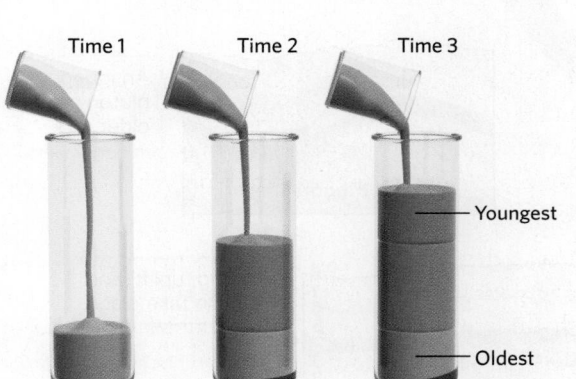

Time 1 Time 2 Time 3

Youngest

Oldest

Younger

Older

(b) Superposition: Pouring sand into a glass cylinder (left) illustrates that in a sedimentary sequence, the oldest bed is on the bottom and the youngest on the top. In the photo (right), beds get younger going up.

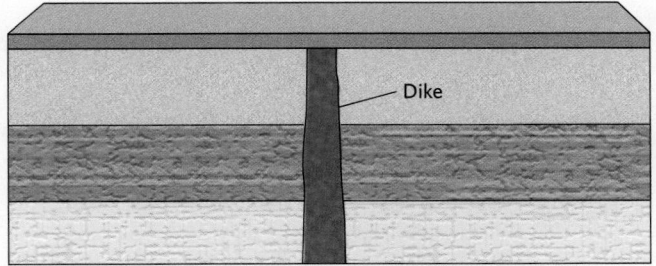

Dike

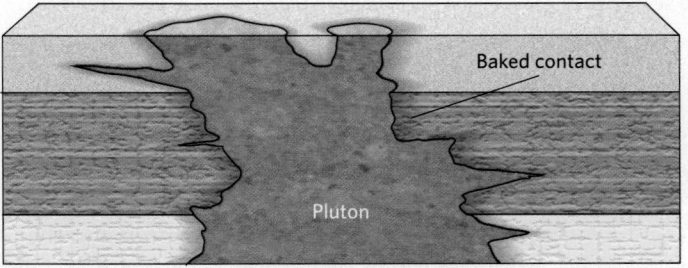

Baked contact

Pluton

(c) Cross-cutting relations: The dike is younger than the beds it cuts across. The sediment layer that buries the dike is younger than the dike.

(d) Baked contacts: A metamorphic aureole (green area) surrounds a pluton. The pluton is younger than the rock it metamorphosed.

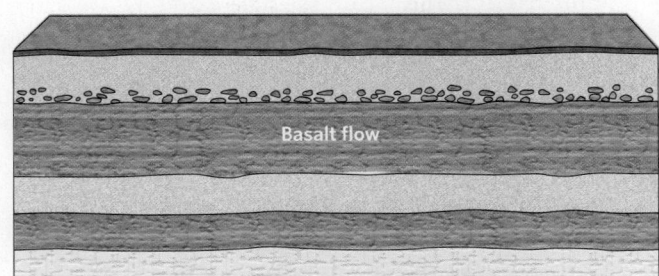

Basalt flow

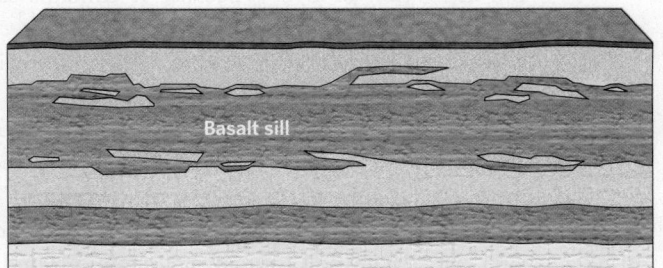

Basalt sill

(e) Inclusions: Pebbles derived from the underlying basalt flow must be older than the conglomerate that contains them (left). The chunks of sandstone incorporated in a basalt sill must be older than the sill that contains them (right).

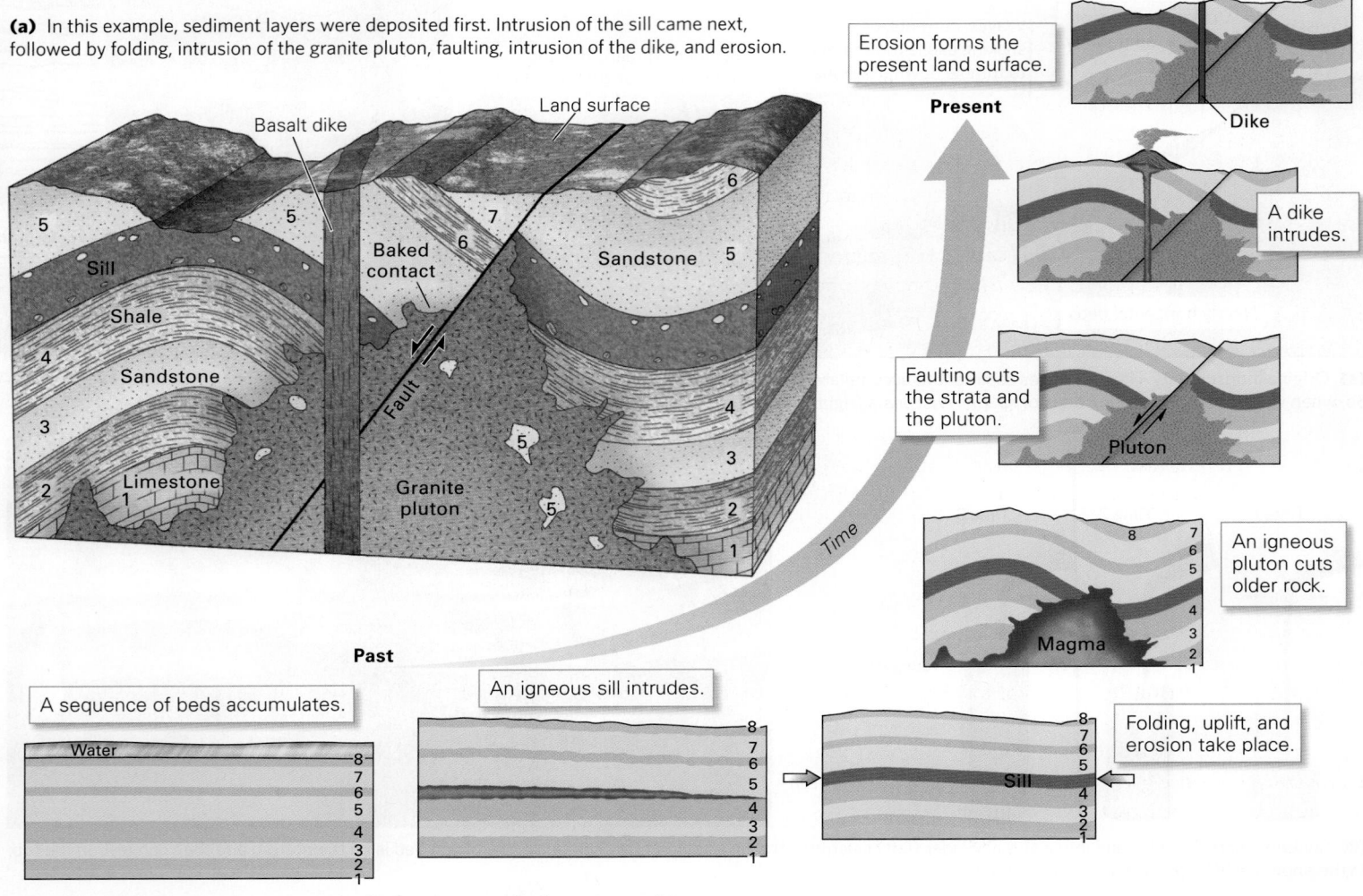

(a) In this example, sediment layers were deposited first. Intrusion of the sill came next, followed by folding, intrusion of the granite pluton, faulting, intrusion of the dike, and erosion.

Erosion forms the present land surface.

Present

Dike

A dike intrudes.

Faulting cuts the strata and the pluton.

Pluton

An igneous pluton cuts older rock.

Magma

Time

Past

A sequence of beds accumulates.

Water

An igneous sill intrudes.

Folding, uplift, and erosion take place.

Sill

Basalt dike

Land surface

Sill

Shale

Sandstone

Limestone

Baked contact

Fault

Granite pluton

Sandstone

(b) The sequence of geologic events leading to the example shown in part (a).

events, ordered by their relative ages, defines the **geologic history** of the region.

To visualize how we can unravel the geologic history of a region, let's decipher the relative ages of the geologic events depicted in Figure 8.4. According to the principle of superposition, Bed 1 was deposited first, followed by Beds 2 through 8. We know that the sill intruded after the deposition of Bed 5 because it contains sandstone inclusions from that bed. Then all the beds, together with the sill, underwent folding. The granite intruded after the folding, because we see that the pluton cuts the fold and that the folded rocks have been metamorphosed along the baked contact. The fault slipped after the pluton intruded, because it cuts and offsets the pluton. The dike intruded after the fault slipped, because it cuts across the fault. Finally, erosion formed the present ground surface, which cuts across all other features. Note that Bed 8 has been completely eroded away.

▶ Animation
Relative Age Determination

Take-home message...

🏠 The principle of uniformitarianism—that the present is the key to the past—provides a basis for interpreting geologic features and implies that the Earth must be very old. This principle, together with other geologic principles derived from it, allows geologists to determine the relative ages of features and construct the geologic history of a region.

Quick Questions

• Explain the difference between a relative age and a numerical age.

• What simple phrase conveys the essence of the principle of uniformitarianism?

• If a basalt dike cuts through granite, is the basalt or the granite older?

8.2 Memories of Past Life: Fossils and Evolution

When looking at sedimentary strata or lithified volcanic ash layers, you may come across features that resemble shells, bones, leaves, or footprints (Fig. 8.5a). Such **fossils** (from the Latin word *fossilis*, which means dug up) are remnants or traces of ancient living organisms that have been preserved in geologic materials. The 19th century saw **paleontology**, the study of fossils, ripen into a science as museum drawers filled with cataloged specimens (Fig. 8.5b). As we'll see, *paleontologists*—researchers who specialize in studying the fossil record (Fig. 8.5c)—eventually realized that the assemblage of fossils in one sequence of strata differs from that in another, so fossils can be used as a basis for determining the relative ages of sedimentary strata that do not crop out in the same location. Consequently, fossils have become an indispensable tool for studying geologic history and the evolution of life.

Formation and Preservation of Fossils

Fossils form when organisms die and become buried by sediment or ash, or when organisms travel over or through sediment and leave imprints or feces. Paleontologists refer to the process of fossil formation as **fossilization**. To see how a typical fossil develops in sedimentary rock, let's follow the fate of an old dinosaur as it searches for food along a muddy riverbank (Fig. 8.6). On a scorching summer day, the dinosaur succumbs to the heat and collapses dead in the mud. Soon after, scavengers strip the skeleton of meat and may scatter the bones. But before the bones have had time to weather away, the river floods and buries the bones, along with the dinosaur's footprints, under a layer of silt. More sediment buries the bones and prints still deeper, until eventually, the sediment containing the bones and footprints undergoes lithification to become siltstone and shale. The footprints remain outlined by the *contact* (the boundary surface) between the siltstone and the shale, and the bones become encased in the siltstone. Over time, minerals precipitated from groundwater gradually replace some of the chemicals in the bones, until the bones themselves become rock-like. The buried bones and footprints are now fossils. A hundred million years later, uplift and erosion have exposed the dinosaur's grave and a lucky paleontologist finds them and excavates them. The dinosaur rises again, but this time in a museum.

Not all living organisms become fossils when they die. For example, a dead squirrel on a road doesn't get fossilized, because scavengers come along and eat the carcass, insects and microbes infest the carcass and gradually digest it, or oxidation (chemical reaction with oxygen)

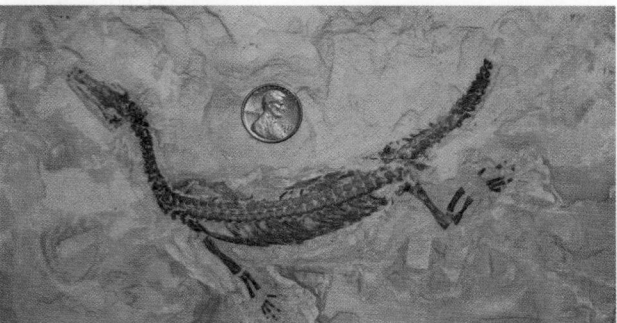

(a) A fossil skeleton in 200-million-year-old sandstone.

(b) A drawer of labeled fossil specimens in a collection.

(c) A paleontologist collecting specimens.

FIGURE 8.5 Examples of fossils and fossil collections.

breaks it down into gases that mix with the atmosphere. In fact, only an exceedingly small number of organisms do undergo fossilization, for it takes special circumstances to produce a fossil and allow it to survive:

- *Death in an anoxic environment:* An organism has a better chance of being preserved if its body ends up in an *anoxic* (oxygen-poor) environment, because in such an environment most scavenging organisms can't survive, microbial metabolism takes place only very slowly, and oxidation reactions don't happen.

- *Rapid burial:* If an organism dies in a low-energy depositional environment (one without strong waves or currents) and if sediment accumulates rapidly, it has a better chance of being buried and protected before disintegrating. In a low-energy environment, the

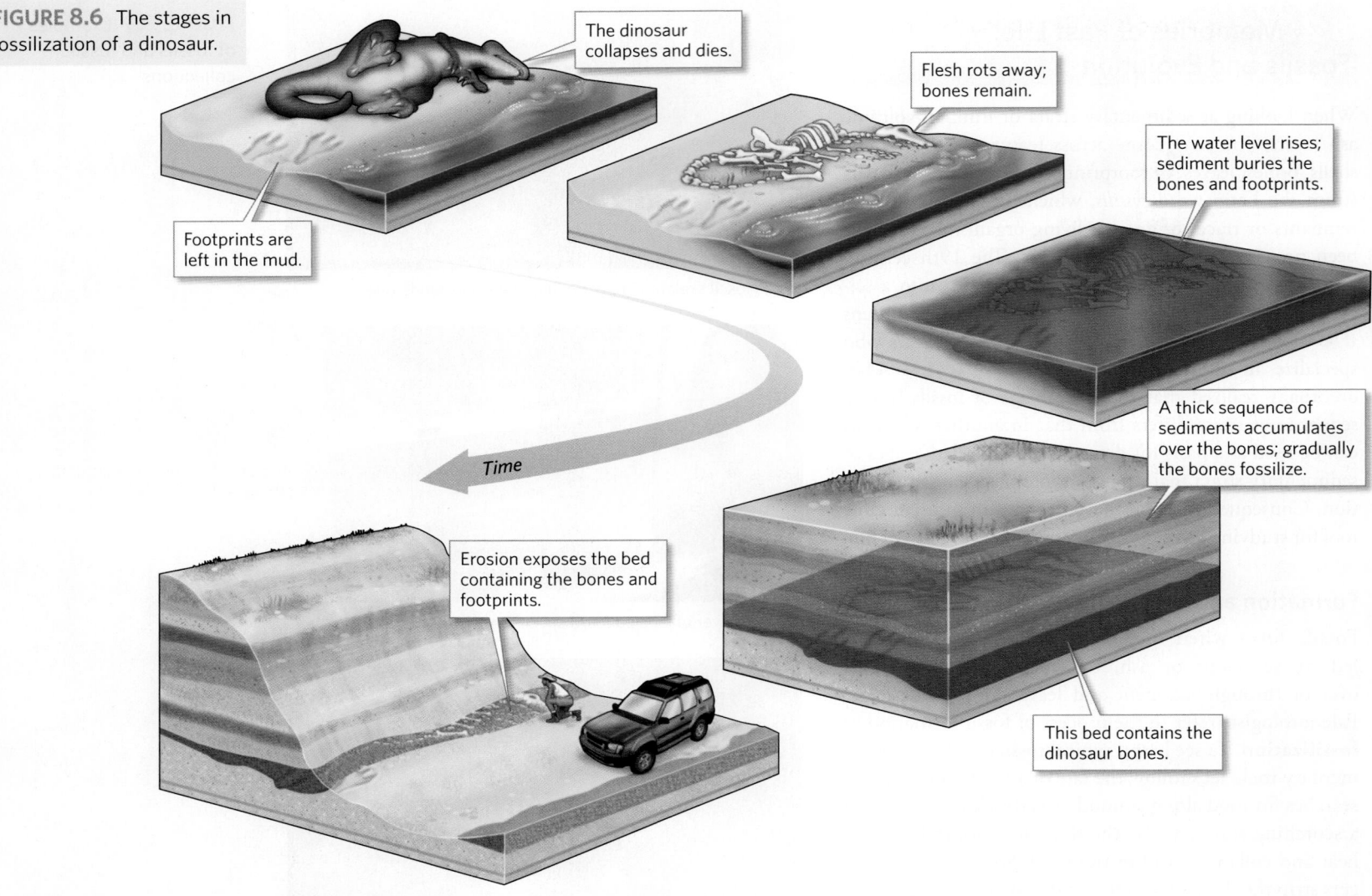

FIGURE 8.6 The stages in fossilization of a dinosaur.

The dinosaur collapses and dies.

Footprints are left in the mud.

Flesh rots away; bones remain.

The water level rises; sediment buries the bones and footprints.

Time

A thick sequence of sediments accumulates over the bones; gradually the bones fossilize.

Erosion exposes the bed containing the bones and footprints.

This bed contains the dinosaur bones.

▶ **Animation**
Fossils and Fossilization

body of an organism won't break up, and if sediment accumulates over the body quickly, it will be buried before decomposing or being eaten by scavengers.

- *The presence of hard parts:* Organisms without *hard parts*—durable shells or skeletons—are less likely to be fossilized, because soft flesh decays long before hard parts under most depositional conditions. For this reason, paleontologists have found many more fossil clams than, say, fossil jellyfish.

By considering factors that influence fossil preservation, and by studying modern organisms and depositional environments, paleontologists can estimate the **preservation potential** of organisms, meaning the likelihood that an organism will be buried and transformed into a fossil. Only a small fraction of organisms have a high preservation potential. Of these organisms, only a few die in a depositional setting where they actually become fossilized. So, fossilization is the exception rather than the rule.

The Many Different Kinds of Fossils

Perhaps when you picture a fossil, you imagine a bone embedded in rock. In fact, preserved bones are just one of several types of fossils, distinguished from one another by the process of fossilization they undergo:

- *Frozen or desiccated body fossils:* In a few environments, whole bodies of organisms may be preserved. Most of these fossils are very young by geologic standards—their ages can be measured in thousands of years. Frozen bodies of woolly mammoths have been found in permafrost (permanently frozen ground) **(Fig. 8.7a)**, and desiccated (dried or "mummified") fossils of giant ground sloths have been found in caves.

- *Fossils preserved in amber:* Insects landing on trees may become trapped in the sticky sap that trees produce. This sap envelops the insects and, over time, hardens into amber **(Fig. 8.7b)**. The oldest insect-containing amber has an estimated age of 230 Ma.

FIGURE 8.7 Examples of different kinds of fossils.

(a) This 1 m long baby mammoth, found in permafrost in Siberia, died 37,000 years ago.

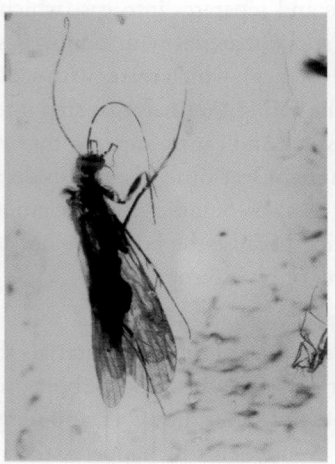

(b) This insect became embedded in amber at about 200 Ma.

(c) A fossil skeleton of a 2 m high giant ground sloth from the La Brea Tar Pits in California.

(d) Casts of brachiopod shells.

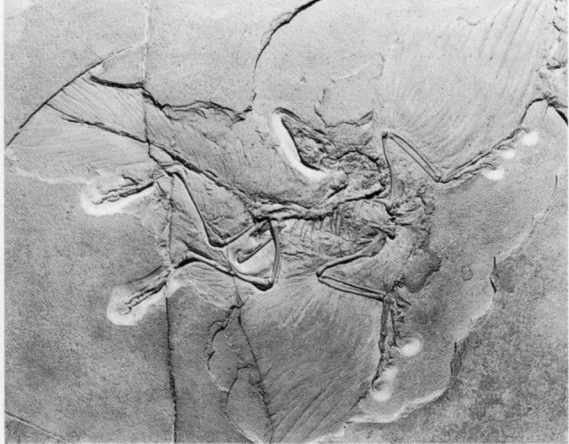

(e) A 150 Ma mold of *Archaeopteryx* from Germany. This is an extraordinary fossil, for the imprints of feathers are clearly visible.

(f) Carbonized impressions of fern fronds in shale.

(g) Petrified wood from Arizona. Chert replaced the original wood.

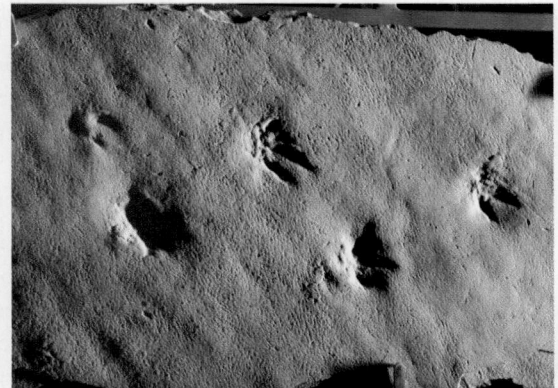

(h) These dinosaur footprints from Connecticut are one form of trace fossil.

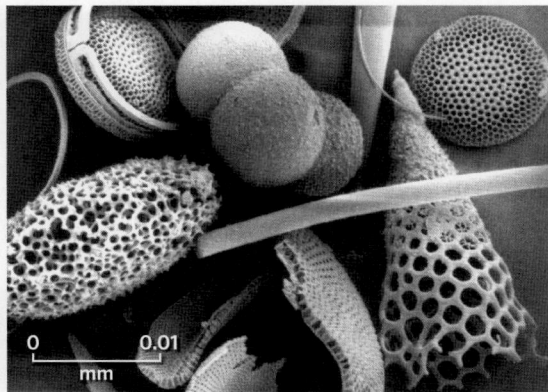

(i) Very tiny fossils, formed from plankton shells like these, are called microfossils.

- *Fossils preserved in tar:* An *oil seep* is a spring from which both water and underground oil bubble up. Over time, the oil degrades and separates into gas (which evaporates) and *tar* (sticky material with a molasses-like to nearly solid consistency), which mixes with sediment at the bottom of the spring. Animals that step into such a *tar pit* become stuck and can't escape, so they soon die and become buried. Over time, their flesh decays, but their skeletons remain. Significantly, skeletons in tar pits remain intact, because the bones are not scattered by running water or by scavengers.

- *Preserved or replaced bones, teeth, and shells:* Bones, teeth, and shells consist of durable materials that may survive in tar or rock (Fig. 8.7c). Some bone, tooth, or shell minerals are not stable, however, and recrystallize over time. But even when this happens, the shape of the original hard part may be preserved.

- *Molds and casts:* As sediment compacts around an organism, it conforms to the shape of the organism and produces an indentation called a *mold* in the shape of the organism. If the organism leaves an imprint, but then washes away or decomposes so that sediment fills

the indentation, the infill will also preserve the shape of the organism as a *cast* (Fig. 8.7d). Commonly, shells remain in molds through the burial process, but later recrystallize. Usually only hard parts turn into molds or casts. Rarely, soft parts leave detailed imprints of skin, feathers, or hair; these particularly informative fossils are known as *extraordinary fossils* (Fig. 8.7e).

- *Carbonized impressions of soft bodies:* Impressions are flattened molds created when soft or semisoft organisms or their parts (leaves, insects, invertebrates, sponges, feathers, jellyfish) get pressed between layers of sediment. Chemical reactions eventually remove most organic material, leaving only a thin film of carbon on the surface of the impression (Fig. 8.7f).

- *Permineralized organisms: Permineralization* refers to the process by which minerals precipitate from groundwater that has seeped into the pores of porous material, such as wood or bone. Petrified wood, for example, forms when wood permineralizes to become chert (Fig. 8.7g).

- *Trace fossils:* All the fossil types that we described above are *body fossils,* in the sense that they represent the shape of an organism's body. *Trace fossils,* in contrast, are marks or debris that organisms leave behind in sediment. Examples include footprints, feeding traces, burrows, and feces (Fig. 8.7h). Fossil feces are known as *coprolites.*

- *Chemical fossils:* Living things consist of complex organic chemicals. Over geologic time, most of these chemicals break down. Some become different, but distinctive, chemicals. A distinctive chemical derived from an organism and preserved in rock is called a *chemical fossil* or *biomarker.*

Paleontologists also find it useful to distinguish among fossils on the basis of their size. *Macrofossils* (like those in Fig. 8.7a–h) are fossils large enough to be seen with the naked eye. But some rocks and sediments also contain abundant *microfossils,* which can be seen only with a microscope (Fig. 8.7i). Microfossils include remnants of plankton, bacteria, and pollen.

Classifying Fossils

Paleontologists identify and classify fossils using the same principles that biologists use to classify modern organisms (Box 8.1). Indeed, there's nothing magical about identifying fossils. You can recognize common ones by examining their *morphology,* meaning their form or shape (Fig. 8.8). If the fossil has complete and distinctive features, the process can be straightforward, but if it has broken into fragments, or if parts are missing, identification can be a challenge. Many fossil organisms resemble modern ones, so it is relatively easy to figure out how to classify

FIGURE 8.8 Common types of invertebrate fossils.

Trilobite Gastropod Bivalve

Brachiopod Bryozoan

Crinoid

Graptolite Ammonite Coral

BOX 8.1 ► A Deeper Look

Taxonomy: The organization of life

The study of how to identify and name organisms is called **taxonomy**. Traditionally, taxonomic analysis could be done only by comparing physical characteristics. Now, relationships among some organisms can be determined by comparing DNA. The same principles apply to fossil organisms, though DNA studies can be used only for organisms well preserved in permafrost for less than about 2 million years.

On the basis of DNA studies, biologists divide life into three *domains* (Archaea, Bacteria, and Eukarya). *Archaea* and *Bacteria* are microscopic single-celled organisms whose cells are *prokaryotic* (meaning that the cells do not have a central nucleus containing DNA). *Eukarya*, whose cells are *eukaryotic* (meaning that the cells have a nucleus), have traditionally been sorted into kingdoms (Protista, Fungi, Plantae, and Animalia), each of which consists of one or more phyla. Each phylum is divided into one or more classes, each class consists of one or more orders, each order of one or more families, each family into one or more genera, and each genus of one or more species. The taxonomic relationships among organisms can be portrayed on a *phylogenetic tree*, informally known as the *tree of life*. In a *rooted phylogenetic tree* (**Fig. Bx8.1a**), groups of organisms radiate from a common ancestor. An *unrooted phylogenic tree* does not imply a common ancestor (**Fig. Bx8.1b**).

How are modern humans classified according to the phylogenetic tree? All present-day humans are assigned to the genus

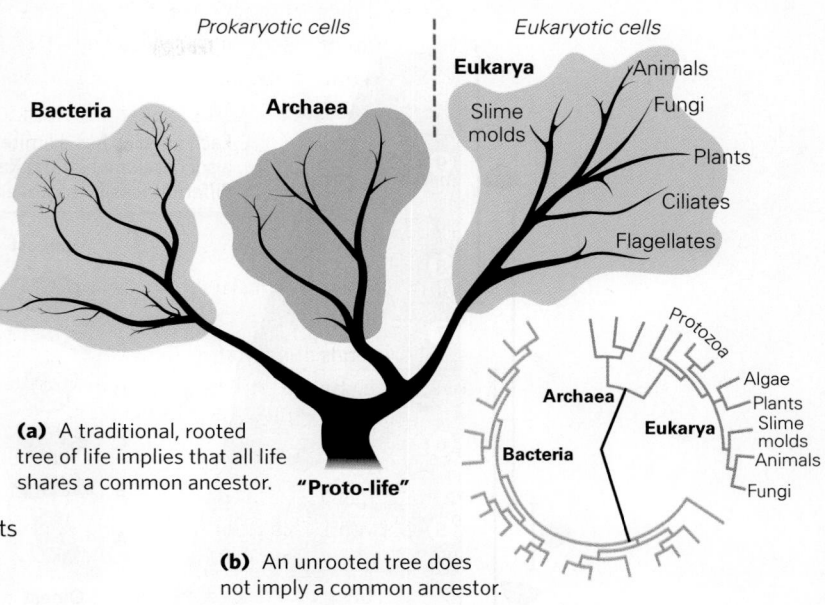

FIGURE Bx8.1 Simplified phylogenetic trees depicting taxonomic relationships among all organisms.

(a) A traditional, rooted tree of life implies that all life shares a common ancestor.

(b) An unrooted tree does not imply a common ancestor.

Homo and the species *sapiens*. Humans, together with apes and monkeys, make up the order Primates, which, together with other warm-blooded organisms with hair, constitute the class Mammalia. Mammals, along with all animals with backbones, are in the phylum Chordata, and all animals of any type are in the kingdom Animalia.

---

them. For example, a fossil clam looks like a clam and not like, say, a snail. To classify a fossil more specifically, down to genus or species level, identification may require documenting details such as the number of ridges on the surface of its shell. Fossils that do not resemble known living organisms can be hard to identify, and figuring out their taxonomic relationships (see Box 8.1) remains even more of a challenge. Natural-history museums and books often include models or paintings depicting fossil organisms as they would have appeared when alive (**Fig. 8.9**). When possible, these depictions take into account studies of extraordinary fossils and comparisons with modern organisms . . . but they often require a lot of informed imagination!

The Concept of Extinction

In the 18th century, paleontologists recognized that not all fossils represented the remains of observed living species. But they tacitly assumed that, since they had not

FIGURE 8.9 An artist's portrayal of what some fossil organisms looked like when alive. The painting is based on fossils found in the Burgess Shale of Canada.

FIGURE 8.10
The principle of fossil succession. The block diagram shows a succession of 10 beds and the ranges of different fossil species within those beds. The inset shows the relative ages of the fossil species.

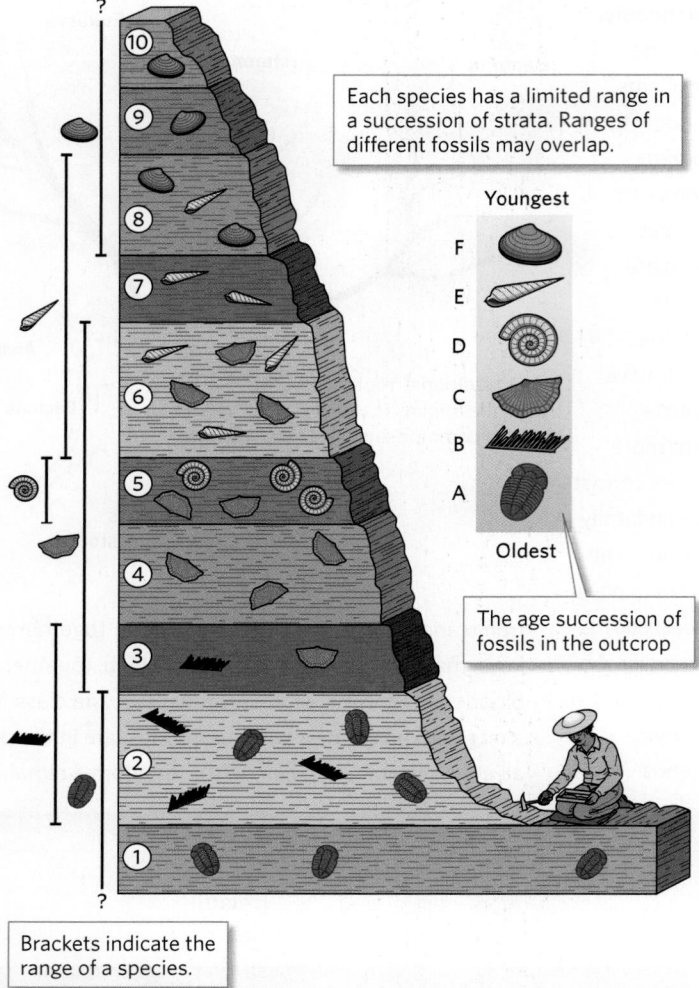

Each species has a limited range in a succession of strata. Ranges of different fossils may overlap.

Youngest

F
E
D
C
B
A

Oldest

The age succession of fossils in the outcrop

Brackets indicate the range of a species.

Using Fossils to Determine Relative Ages: Fossil Succession

As Britain entered the industrial revolution in the late 18th century, factories demanded coal to fire their steam engines, and they needed an inexpensive means to transport raw materials and manufactured goods. Investors decided to construct a network of canals, and they hired an engineer named William Smith (1769–1839) to survey the excavations. Canal digging provided fresh exposures of bedrock that had previously been covered by vegetation. Smith learned to recognize distinct layers of sedimentary rock and to identify the **fossil assemblage** (the group of fossil species) that each layer contained. He also realized that a particular assemblage could be found in only a limited sequence of beds and not in beds above or below. In other words, a fossil species appears at a specific level in a sequence of strata and did not exist before then, and once a fossil species disappears at a specific, higher, level in a sequence of strata, it never reappears in still younger strata, because extinction is forever.

Smith's observation, which has been repeated at millions of locations around the world, has become known as the principle of **fossil succession**. To see how this principle works, examine Figure 8.10, which depicts a sequence of strata. Bed 1 at the base contains Species A, Bed 2 contains Species A and B, Bed 3 contains B and C, Bed 4 contains C, and so on. From these data, we can define the *range* for each species, meaning the interval in the sequence in which fossils of that species occur—each species appears at the beginning of its range and goes extinct at the end of its range. In Figure 8.10, the succession of fossils, from oldest to youngest, is A, B, C, D, E, F. Note that the range of one species may overlap with that of another.

Once the relative ages of fossil species have been determined, those species can be used to determine the relative ages of the beds containing them. For example, if a bed contains Fossil A (from Fig. 8.10), geologists can say that the bed is older than a bed containing Fossil F, even if the two beds do not crop out in the same area.

Tracing the History of Life's Evolution

Knowledge of the *fossil record*—the documented array of identified species—allows paleontologists to reconstruct the history of life over time. Paleontologists realized that, to explain the observed fossil succession, some species would have to go extinct and new species would have to appear. In other words, the fossil record documents life's *evolution* on the Earth. The fossil record, however, does not explain how evolution happens.

Charles Darwin (1809–1882) and Alfred Wallace (1823–1913) addressed the question of how evolution happens in an 1858 paper that introduced the *theory of*

Animation
Biostratigraphy
(Interactive)

yet explored the world completely, all fossils represented species still living somewhere on the planet. By the 19th century, however, it became clear that this interpretation could not be true, for scientific explorations had by then covered the globe without finding giant fossil animals, such as mastodons or dinosaurs, anywhere. Given their size, such creatures would have been hard to miss! Based on this realization, the French paleontologist Georges Cuvier (1769–1832) argued that most fossil species had undergone **extinction** (or had gone *extinct*), meaning that all individuals of those species had died. Most people take the phenomenon of extinction for granted today, since numerous animals have become extinct in historical time, but Cuvier's proposal was revolutionary in his day.

evolution by natural selection. (Darwin expanded greatly on this idea in a book called *On the Origin of Species*, which was published the following year.) In essence, the **theory of evolution** states that the traits that make an individual organism successful in competing for survival and in attracting mates are traits that succeeding generations can inherit, because successful individuals are more likely to have offspring. Individuals whose traits make them less capable of surviving or of attracting a mate are less likely to produce offspring, so their traits do not appear as frequently in succeeding generations and eventually no longer exist. This phenomenon, known as *survival of the fittest,* eventually leads to the appearance of new species over the course of many, many generations. During the last several decades, studies of DNA have emphasized that new traits come from natural *mutations* (changes in the structure of the DNA molecule), as well as from the transfer of bits of DNA from viruses or from one species to another.

Though Darwin's theory of evolution, and subsequent genetic studies, provide a basis for explaining fossil succession, many mysteries still surround the path of life's evolution on the Earth, for known fossils cannot account for every intermediate step in the evolution of every species. Why is the fossil record so incomplete? First, despite all the fossil-collecting efforts of the past 350 years, paleontologists will never be able sample every cubic centimeter of sedimentary rock exposed on the Earth. Just as biologists have not yet identified every living species, paleontologists have not come close to identifying every fossil species. (Of the up to 50 billion fossil species that might exist, only about 250,000 have been identified and catalogued.) Second, not all species are represented in the fossil record because not all species have a high preservation potential—only a small fraction of species have left a fossil record. Finally, as we'll see below, the sequences of sedimentary strata that exist on the Earth today do not account for every minute of geologic time.

Take-home message...

Fossils are the remnants or traces of ancient organisms. Fossils form in many ways, but the fossil record is incomplete. Fossils provide both a basis for geologists to determine the relative ages of rock layers and a record of life's evolution.

Quick Questions ──────────────

- On what basis can you identify a fossil that you come across on an outcrop?
- When you pick up a fossil bone or shell, does it have the same composition as a living bone or shell?
- Define the principle of fossil succession.

8.3 Establishing the Geologic Column

Stratigraphic Formations and Contacts

Geologists divide the strata of a region into distinct units called stratigraphic formations. Formally defined, a **stratigraphic formation** (or simply a *formation*) is a sequence of beds—and, in some cases, extrusive igneous deposits and/or flows—that can be traced over a broad region. For example, a view from the rim of the Grand Canyon shows successions of strata traceable across the Grand Canyon region. Each succession, which looks like a stripe on the wall of the canyon, represents a stratigraphic formation (Fig. 8.11a). Resistant formations stand out as cliffs, whereas nonresistant formations form gentle slopes that are covered with debris falling from higher cliffs.

A formation can consist of a single rock type, or it can include interlayered beds of two or more rock types. In the latter case, individual beds might not be traceable over the entire region in which the formation occurs, but as individual beds pinch out, other similar beds appear, so the overall sequence can still be recognizable. Typically, a formation represents the products of deposition (or eruption) during a definable interval of time. Therefore, in a succession of formations, different formations contain different fossil assemblages and, as we'll see, a given formation can be assigned a specific age range. Not all formations have the same thickness, and the thickness of an individual formation can vary with location. The characteristics of rocks in a formation at a given location provide clues about the environment at the location during that age range.

Commonly, geologists name a formation after a locality where it was first identified or first studied. If a formation consists of only one rock type, that rock type may appear as part of the name (for example, the Kaibab Limestone), but if a formation contains more than one rock type, the word *formation* becomes part of the name (for example, the Toroweap Formation). Note that all words in the name are capitalized (Fig. 8.11b). Several formations in a succession may be lumped together as a *stratigraphic group*, and several adjacent groups, in turn, may be lumped together as a *supergroup*.

The boundary surface between any two rock bodies—regardless of whether one or both are sedimentary, igneous, or metamorphic—is called a **contact**. Geologists distinguish among three major types of contacts (Fig. 8.11c): Where a bed of sedimentary rock has been deposited over pre-existing rock, the boundary surface between two formations, or between beds within a formation, is a *depositional contact*. A *fault contact* occurs where a fault cuts through and displaces pre-existing rock such that

See for yourself

Grand Canyon, Arizona

Latitude: 36°5′53.99″ N
Longitude: 112°10′58.58″ W

Zoom to an altitude of 1.5 km (1 mi) and look obliquely downstream.

The strata you see represent an immense amount of geologic time. The dark rocks of the inner gorge form the Vishnu Schist, the oldest horizontal layer is the Tapeats Sandstone, and the youngest unit (the layer at the top) is the Kaibab Limestone.

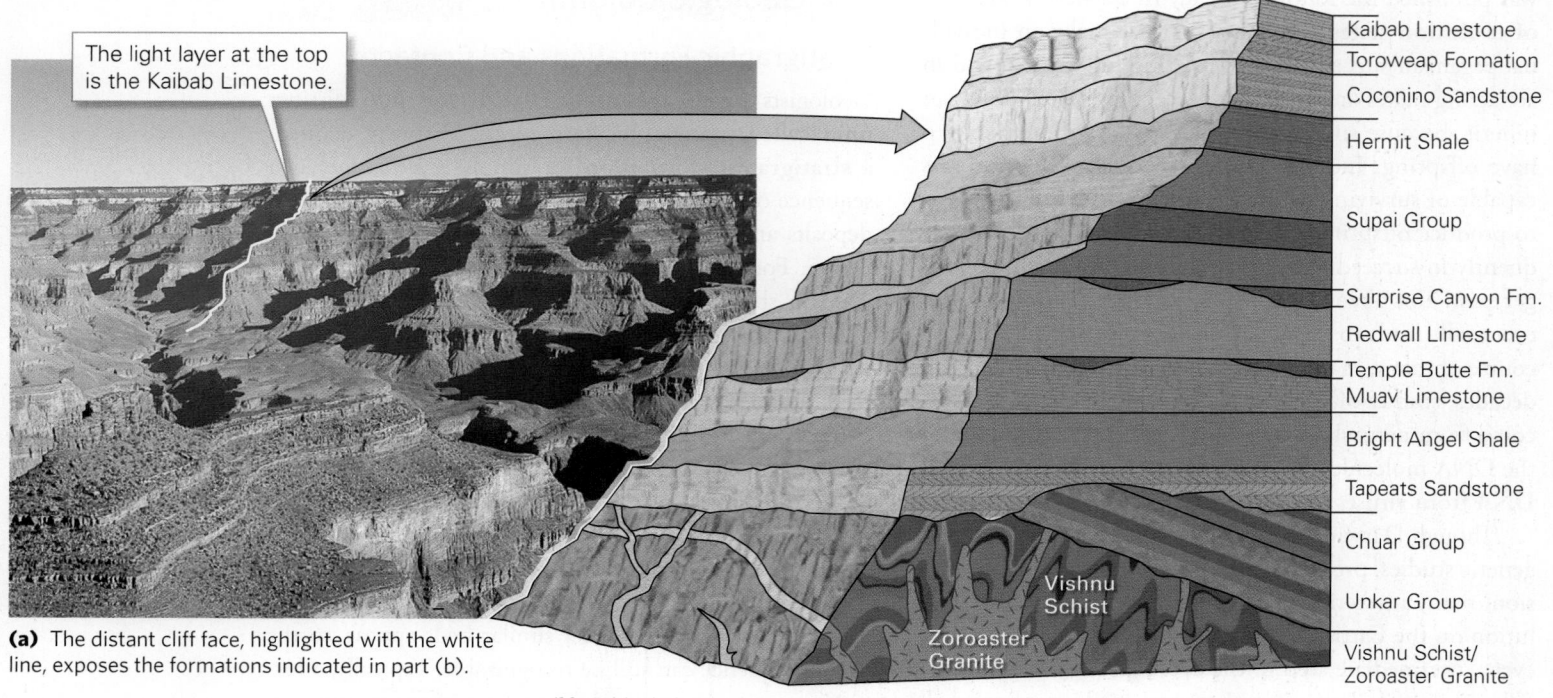

The light layer at the top is the Kaibab Limestone.

Kaibab Limestone
Toroweap Formation
Coconino Sandstone
Hermit Shale
Supai Group
Surprise Canyon Fm.
Redwall Limestone
Temple Butte Fm.
Muav Limestone
Bright Angel Shale
Tapeats Sandstone
Chuar Group
Vishnu Schist
Unkar Group
Zoroaster Granite
Vishnu Schist/ Zoroaster Granite

(a) The distant cliff face, highlighted with the white line, exposes the formations indicated in part (b).

(b) A block diagram representing the stratigraphic formations of the Grand Canyon.

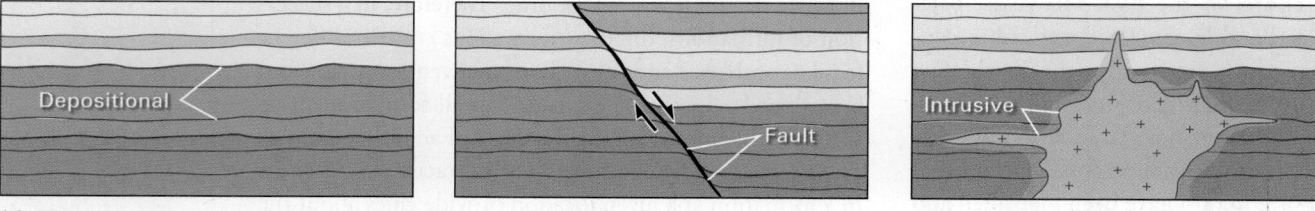

Depositional

Fault

Intrusive

(c) The major types of geologic contacts: depositional, fault, and intrusive.

rock on one side of the fault is in contact with rock on the other side of the fault that it was not in contact with prior to faulting. The contact between an igneous intrusion and the wall rock that it cuts across is an *intrusive contact*.

Unconformities: Gaps in the Record

To find good exposures of rock, James Hutton sometimes boated along the eastern coast of Scotland, where waves of the stormy North Sea strip away soil and shrubbery. He was particularly puzzled by an outcrop at Siccar Point, which exposes a contact between two different stratigraphic formations (Fig. 8.12). In the lower portion of the outcrop, beds of gray sandstone and shale are nearly vertical, whereas in the upper portion, beds of red sandstone display a dip of less than 15°. The gently dipping layers seem to have been deposited across the truncated ends of the vertical layers, like a handkerchief lying across the top of a row of books. We can imagine that, as Hutton was examining this odd relationship, the waves deposited a new layer of sand on top of a nearby stretch of rocky

shore. With the principle of uniformitarianism in mind, he suddenly realized the significance of what he saw.

Hutton deduced that the sediments that now make up the gray sandstone shale sequence had been deposited, lithified, tilted, and truncated by erosion before the sediments that now form the red sandstone were deposited. Therefore, the depositional contact between the gray and red stratigraphic formations represents a time interval during which no new strata accumulated at Siccar Point, and older strata were eroded away. Geologists now refer to such a depositional contact, representing a time period of nondeposition and possibly erosion, as an **unconformity**. The gap in the geologic record represented by an unconformity, meaning the period of time not represented by strata, is known as a *hiatus*. Geologists distinguish among three common types of unconformities:

- *Angular unconformity:* Beds below an **angular unconformity** were tilted or folded before the unconformity developed (Fig. 8.13a), so an angular unconformity cuts across the underlying beds, and the orientation

FIGURE 8.12 James Hutton deduced that the red sandstone beds above the unconformity at Siccar Point were deposited after the gray siltstone and shale beds below had been lithified, tilted, and eroded.

(a) Siccar Point juts out into the North Sea.

(b) The beds of reddish sandstone above the unconformity dip (are tilted) gently. The beds of gray sandstone and shale below are vertical.

Devonian "Old Red sandstone"

Unconformity

Silurian sandstone and shale

What an Earth Scientist Sees

FIGURE 8.13 The three kinds of unconformities and their formation.

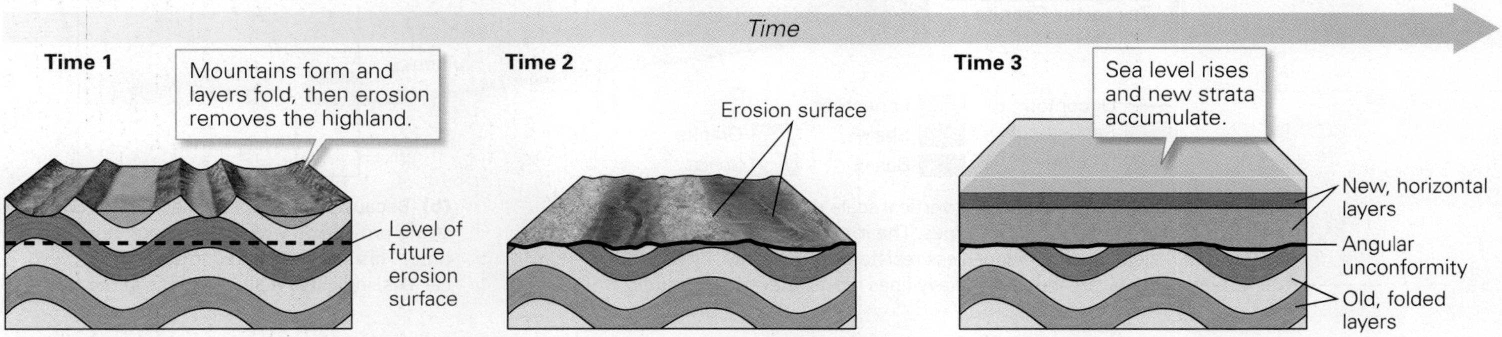

Time

Time 1 Mountains form and layers fold, then erosion removes the highland.

Level of future erosion surface

Time 2 Erosion surface

Time 3 Sea level rises and new strata accumulate.

New, horizontal layers

Angular unconformity

Old, folded layers

(a) Angular unconformity: (1) layers undergo folding; (2) erosion produces a flat surface; (3) sea level rises, and new layers of sediment accumulate.

Time 1 Granite

Future erosion surface

Time 2 Erosion removes cover, so basement lies exposed at the Earth's surface.

Erosion surface

Time 3 Sea level rises and new strata accumulate.

Nonconformity

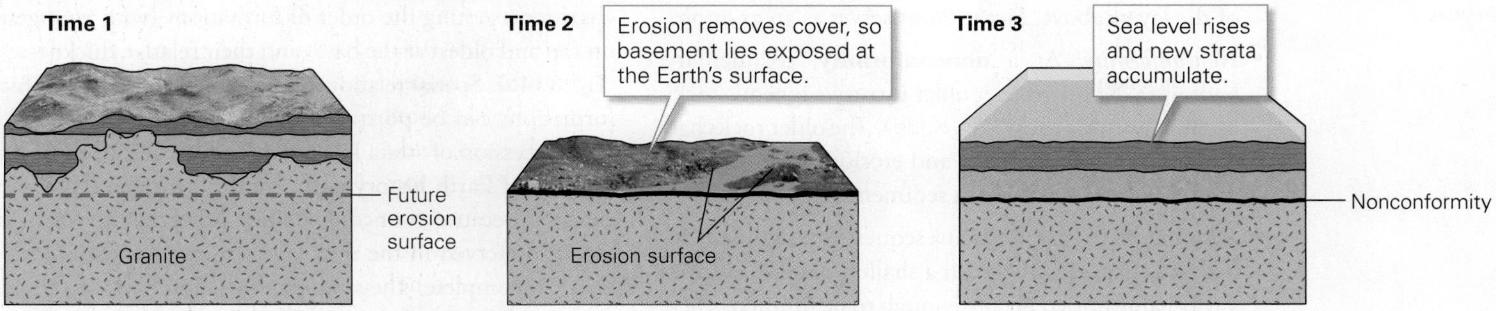

(b) Nonconformity: (1) a pluton intrudes; (2) erosion cuts down into the crystalline rock; (3) new sediment layers accumulate above the erosion surface.

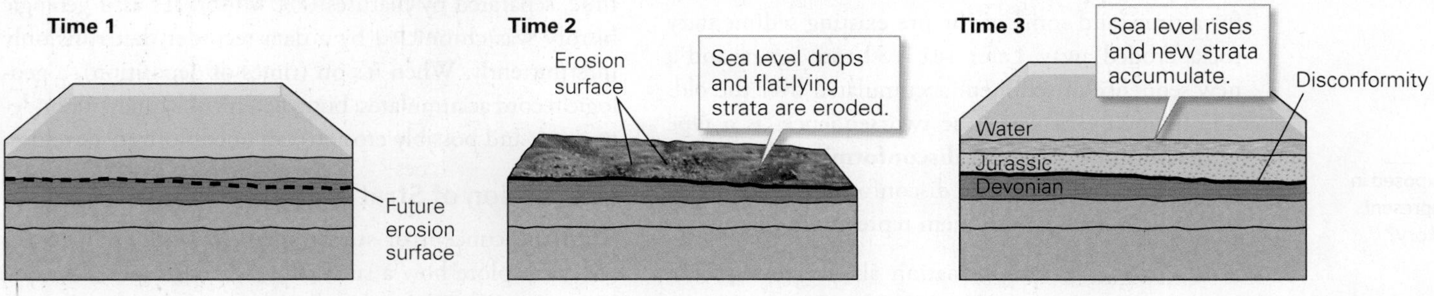

Time 1 Future erosion surface

Time 2 Erosion surface

Sea level drops and flat-lying strata are eroded.

Time 3 Sea level rises and new strata accumulate.

Disconformity

Water
Jurassic
Devonian

(c) Disconformity: (1) layers of sediment accumulate; (2) sea level drops and an erosion surface forms; (3) sea level rises and new sediment layers accumulate.

FIGURE 8.14 Interpreting the stratigraphic column of the Grand Canyon.

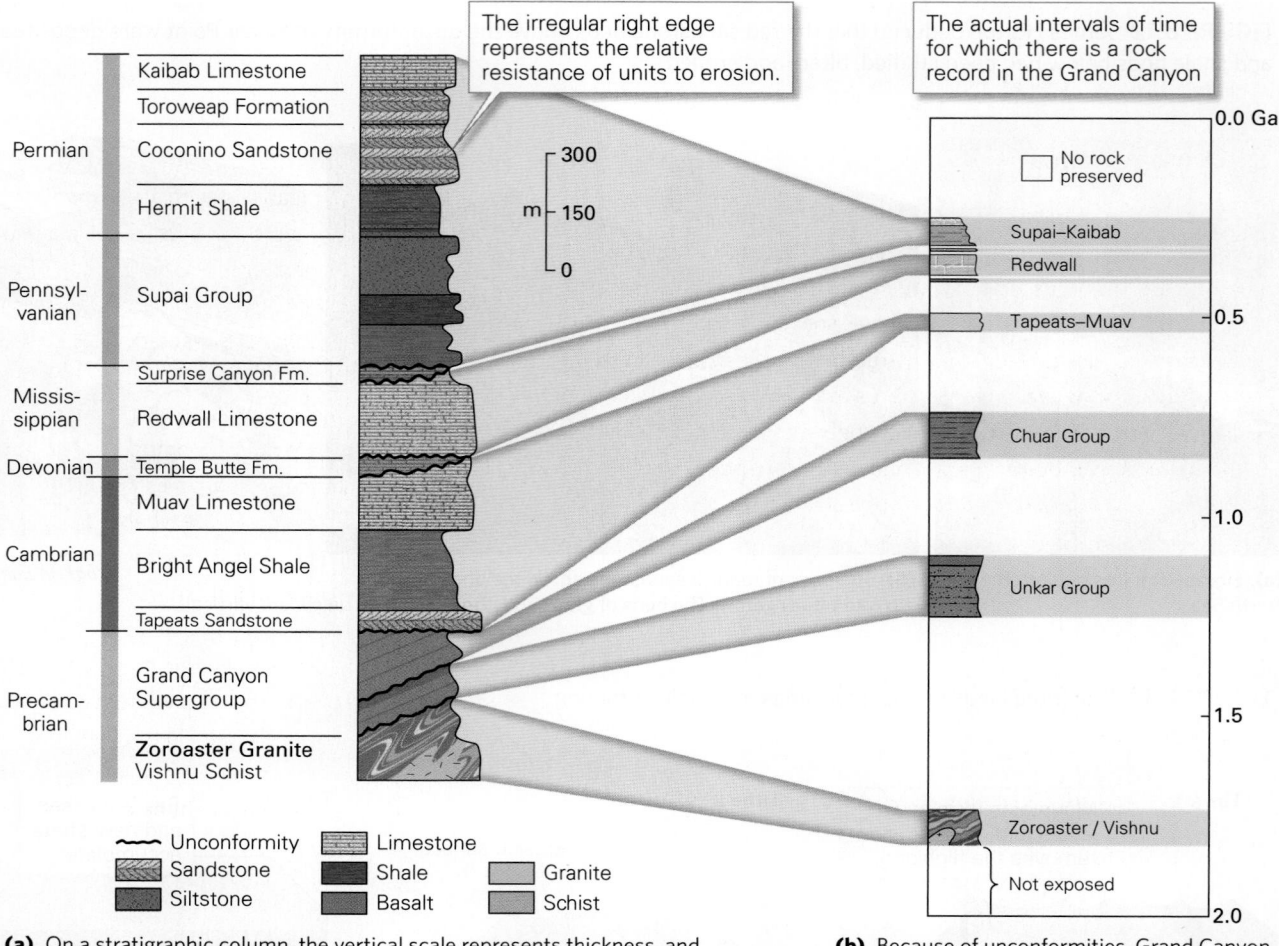

The irregular right edge represents the relative resistance of units to erosion.

The actual intervals of time for which there is a rock record in the Grand Canyon

(a) On a stratigraphic column, the vertical scale represents thickness, and patterns represent different rock types. The irregular right edge of the profile represents resistance to erosion—less resistant formations, such as those consisting of shale, are indented. Wavy lines represent unconformities.

(b) Because of unconformities, Grand Canyon strata provide only a partial record of geologic time. This chart, with a time scale as the vertical axis, indicates hiatuses (gaps in the record).

Animation

Unconformities

Did you ever wonder...

whether the strata exposed in the Grand Canyon represent all of the Earth's history?

of the beds below the unconformity differs from that of the layers above. Siccar Point serves as an example.

- *Nonconformity:* At a **nonconformity**, sedimentary beds were deposited over older intrusive igneous rocks or metamorphic rocks **(Fig. 8.13b)**. The older rocks had undergone cooling, uplift, and erosion before becoming the substrate on which sediment accumulated.

- *Disconformity:* Imagine that a sequence of sedimentary beds was deposited beneath a shallow sea and eventually became buried deeply enough to be lithified. Then sea level dropped, so no new sediment accumulated for a time, and some of the pre-existing sedimentary rocks eroded away. Later, sea level rose again, and a new sequence of sediment accumulated over the old. The boundary between the two sequences is a type of unconformity called a **disconformity** **(Fig. 8.13c)**. Beds above and below the disconformity are parallel, but the contact between them represents a hiatus.

We can summarize information about the sequence of formations, and any unconformities that exist, at a particular location by drawing a **stratigraphic column**, a chart representing the order of formations (with youngest on top and oldest at the base) and their relative thicknesses **(Fig. 8.14a)**. Spatial relationships among such stratigraphic formations can be portrayed by geologic maps **(Box 8.2)**. The succession of strata in a stratigraphic column provides a record of Earth history at the locality represented by the column. Because of unconformities, however, the geologic record preserved in the rock layers at any particular location is incomplete. The stratigraphic column of the Grand Canyon, for example, represents only portions of geologic time, separated by hiatuses **(Fig. 8.14b)**. It's as if geologic history was chronicled by a data recorder that runs only intermittently. When it's on (times of deposition), a geologic record accumulates, but when it's off (times of nondeposition and possibly erosion), an unconformity develops.

Correlation of Stratigraphic Formations

With the concept of stratigraphic formations in mind, we can explore how a succession of beds exposed in one location relates, in terms of relative age, to a succession

BOX 8.2 ▶ **Putting Earth Science to Use**

Geologic maps

A **geologic map** portrays the distribution of bedrock stratigraphic units, intrusions, and/or metamorphic units that would be exposed at the Earth's surface if all soil, vegetation, and human-built structures were removed (Fig. Bx8.2a, b). (Other maps, such as soil maps, show the distribution of soil on the land surface.) To construct a geologic map, geologists study outcrops and, if available, drill cores that provide samples of bedrock.

On a geologic map, contacts between stratigraphic formations, or between igneous intrusions and wall rock, appear as lines. In other words, each line is the *map trace* of a contact. With practice, you can learn to read a geologic map and recognize the patterns of contacts that portray features such as flat-lying strata, unconformities, igneous intrusions, folds, and faults. Specific symbols on a map indicate the orientation of layers and the location of geologic structures.

Geologic maps are important tools for conveying information about the geologic character and history of a location. As such, they are not only used in academic studies, but have important practical applications. Information about the distribution and orientation of rock units may help energy companies find reserves of oil, gas, or coal, and may help mining companies find reserves of ores. Geologic maps also help planners determine where geologic hazards exist. For example, if the map shows that much of the bedrock in a region consists of very weak rock, of if the map shows that the region contains active faults, planners may decide to look elsewhere for building sites. You may come across geologic maps on a display at a visitors' center in a national park. The map allows you to figure out which stratigraphic formations are exposed in the park (Fig. Bx8.2c). And if you're searching for fossils, a geologic map can show you where outcrops of fossil-bearing formations might be exposed.

FIGURE Bx8.2 A geologic map depicts the distribution of rock units.

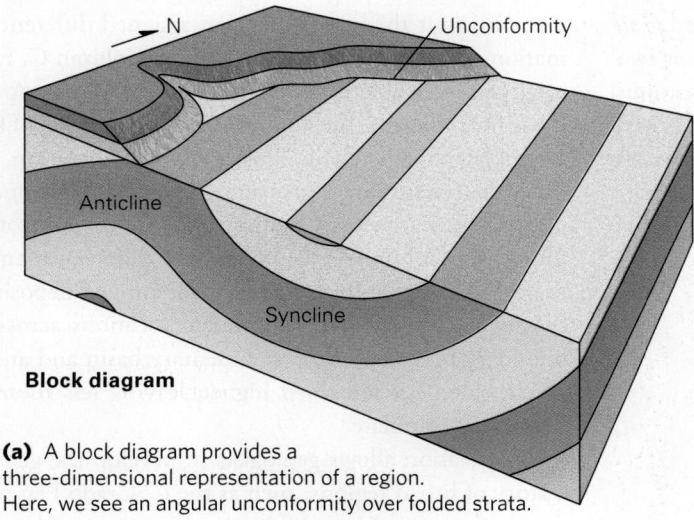

Block diagram

(a) A block diagram provides a three-dimensional representation of a region. Here, we see an angular unconformity over folded strata.

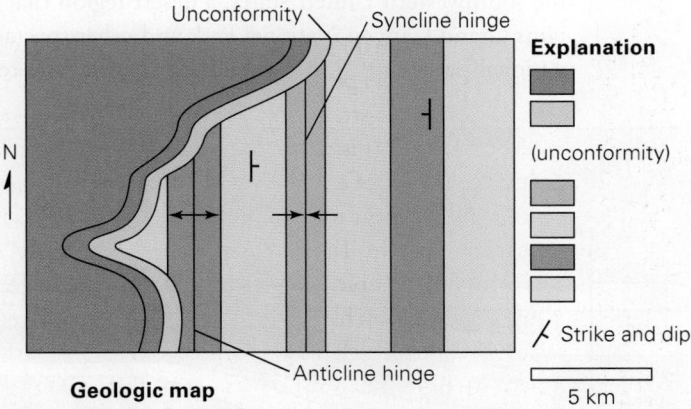

Geologic map

(b) A geologic map portrays the top surface of the block diagram, as viewed looking straight down from above. The black lines represent the traces of contacts between units. Strike and dip symbols indicate bed orientation, and fold symbols (paired arrows) indicate the traces of fold hinges.

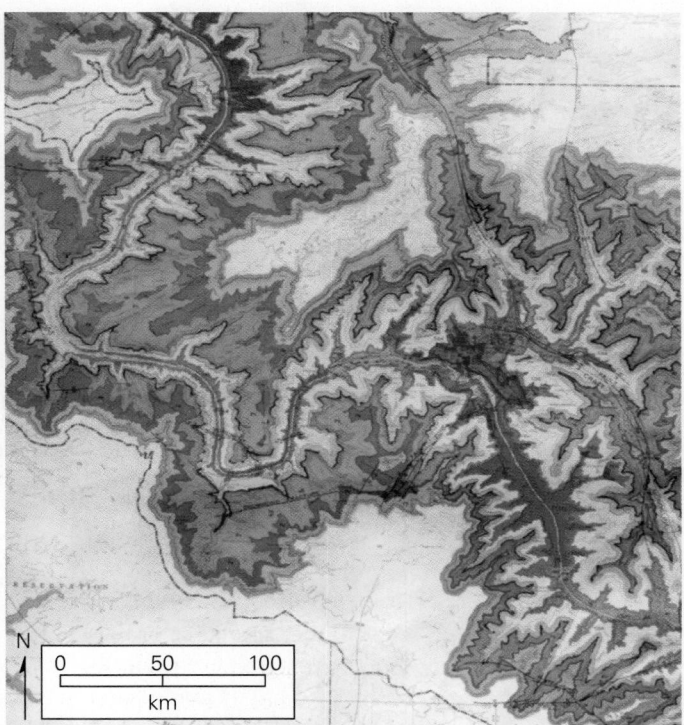

(c) This geologic map portrays the distribution of stratigraphic formations in a portion of the Grand Canyon. Each color represents a different formation. The oldest units are at the floor of the canyon.

exposed in another location. This process of determining the age relationships between successions of strata at different locations is called **correlation**.

How does correlation work? Typically, geologists correlate formations between nearby localities on the basis of similarities in rock type, a method known as *lithologic correlation*. For example, the sequence of strata on the southern rim of the Grand Canyon correlates with the sequence on the northern rim because they contain the same rock types in the same order. You can be more confident of a proposed correlation if you can trace a distinctive layer, called a *marker bed*, between localities. To correlate strata over broader regions, however, we may not be able to rely on lithologic correlation. Depositional settings at a given time may be different at different locations in a sedimentary basin, so widely separated localities may contain different successions of strata that probably do not share formation names. To overcome this problem, geologists use fossils to define the relative ages of sedimentary strata at different locations, a method called *fossil correlation*. If fossils of the same age range appear at two locations, we can say that the strata at the two locations correlate. Fossil species that are widespread but survived for a relatively short interval of geologic time serve as

index fossils, meaning that their presence in a bed characterizes the bed's relative age fairly precisely.

To illustrate correlation, imagine that you set out to draw stratigraphic columns for three locations, using data from your field studies (Fig. 8.15). You first visited Location A and divided the sequence of strata visible there into several distinct formations, to which you assigned names, and you drew a stratigraphic column (Column A) representing those formations. You then traveled east and found another exposure of strata at Location B. The formations you drew in Column B are thinner and coarser than those in Column A, but otherwise are fairly similar, and the succession of units in both columns is the same. So lithologic correlation allowed you to assign formation names from Column A to units of strata in Column B. When you traveled even farther east to Location C, however, you found that, while there are some similarities between strata in Location C and those in Location A, the differences are significant enough (in terms of rock type and unit thickness) that you assigned different formation names to units when you drew Column C. However, you were able to correlate units in Column A with those in Column C by looking at the relative ages of fossil assemblages. When you finished your correlation, you discovered that some units from Column A have pinched out, producing unconformities (specifically, nonconformities) in Column C. Why do the differences among the columns exist? In this case, at the time of deposition, the sequences occurred at different locations across the boundary between a deep sedimentary basin and an area where basement remained higher, leaving less room for strata to accumulate.

Correlation allows geologists to develop the geologic history of broad regions, such as the Colorado Plateau of the southwestern United States, a desert region that contains Grand Canyon National Park and other spectacular national parks (Fig. 8.16). The oldest rocks of the region

FIGURE 8.15 The principles of correlation.

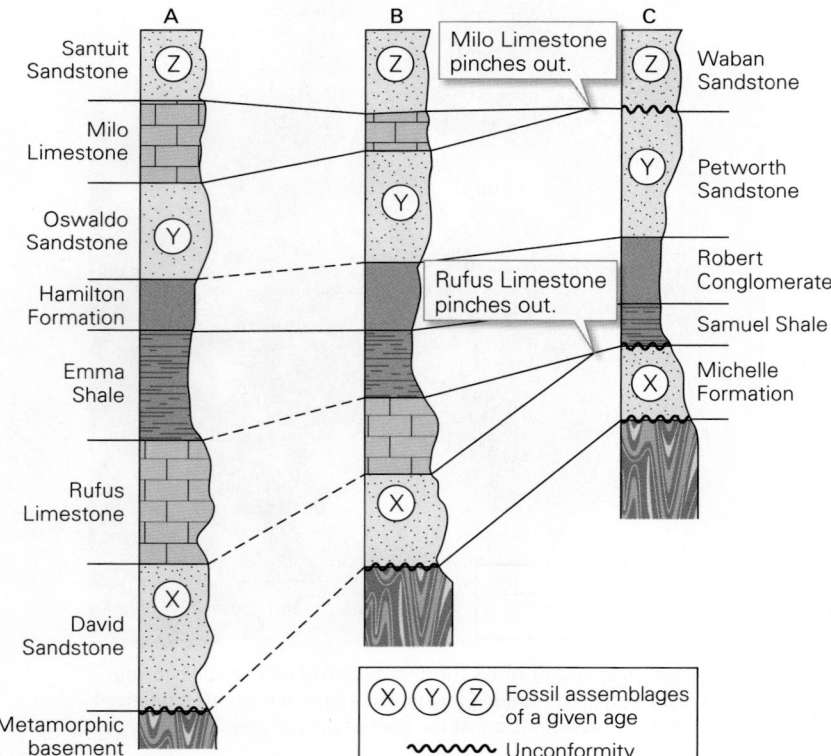

(a) Stratigraphic columns can be correlated by matching rock types (lithologic correlation) or by matching fossils of the same age (fossil correlation).

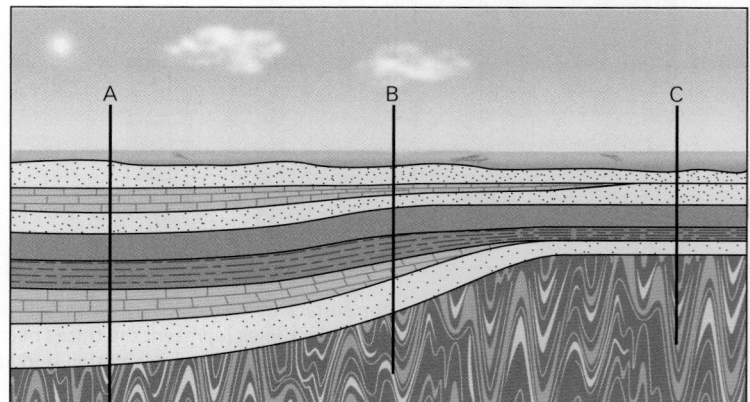

(b) At the time of deposition, Locations A, B, and C [which correlate with the columns in part (a)] were in different parts of a sedimentary basin. The basin floor was subsiding fastest at A.

FIGURE 8.16 Correlation of strata among the national parks of Arizona and Utah.

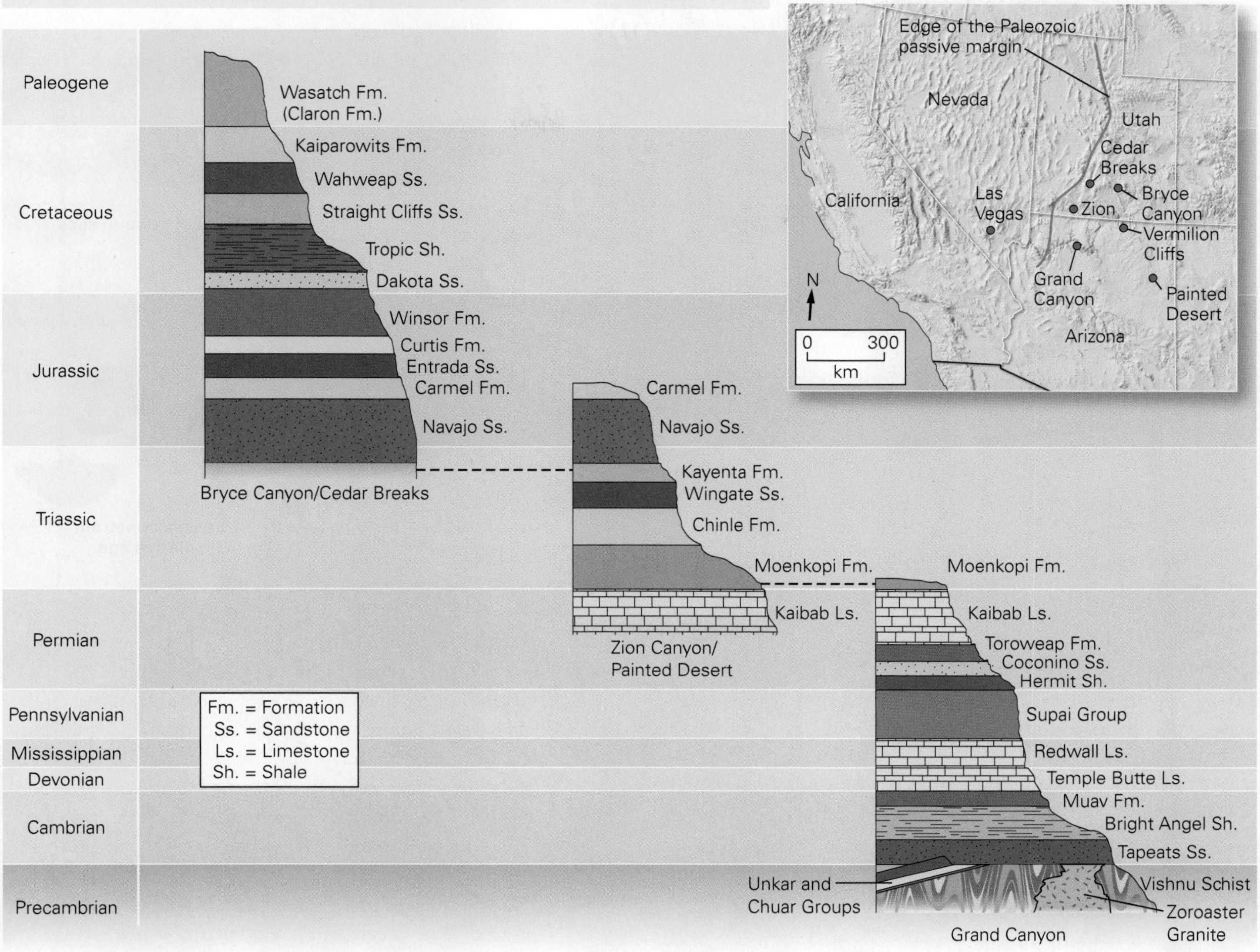

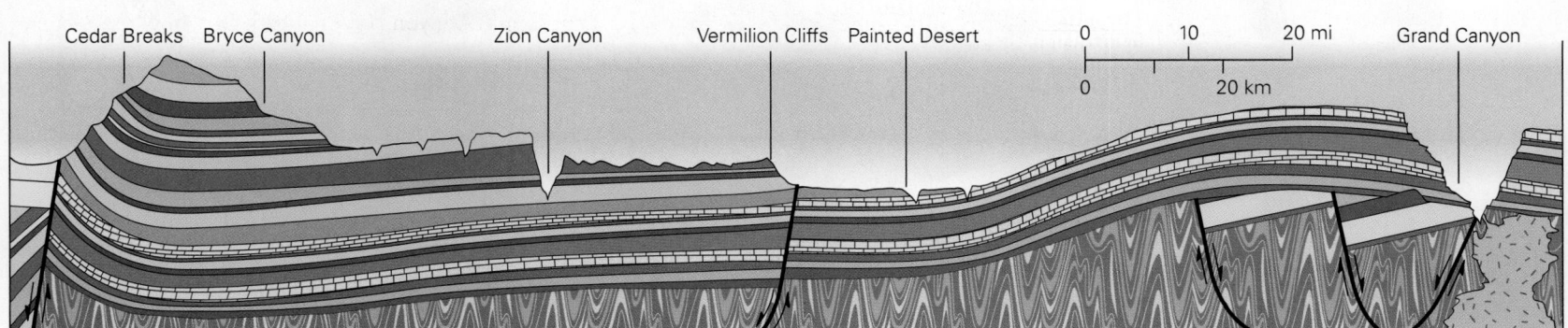

Paleogene	Wasatch Fm. (Claron Fm.)
Cretaceous	Kaiparowits Fm. / Wahweap Ss. / Straight Cliffs Ss. / Tropic Sh. / Dakota Ss.
Jurassic	Winsor Fm. / Curtis Fm. / Entrada Ss. / Carmel Fm. / Navajo Ss.

Bryce Canyon/Cedar Breaks

(a) Different intervals of geologic time are represented by the stratigraphic columns for the different parks. The inset map shows the locations of the parks. All of the parks lie on uplifted cratonic platform (see Chapter 6).

Edge of the Paleozoic passive margin

Nevada — Utah — Cedar Breaks — Bryce Canyon
California — Las Vegas — Zion — Vermilion Cliffs
Grand Canyon — Painted Desert — Arizona

N
0   300
km

Carmel Fm. / Navajo Ss. / Kayenta Fm. / Wingate Ss. / Chinle Fm. / Moenkopi Fm. / Kaibab Ls.

Zion Canyon/ Painted Desert

Moenkopi Fm. / Kaibab Ls. / Toroweap Fm. / Coconino Ss. / Hermit Sh. / Supai Group / Redwall Ls. / Temple Butte Ls. / Muav Fm. / Bright Angel Sh. / Tapeats Ss.

Unkar and Chuar Groups — Vishnu Schist — Zoroaster Granite

Grand Canyon

Fm. = Formation
Ss. = Sandstone
Ls. = Limestone
Sh. = Shale

Paleogene / Cretaceous / Jurassic / Triassic / Permian / Pennsylvanian / Mississippian / Devonian / Cambrian / Precambrian

(b) This cross-sectional sketch (vertically exaggerated) shows how Phanerozoic strata cover the basement of the Colorado Plateau region. Faults cut and warp these strata, and canyons cut into them.

Cedar Breaks   Bryce Canyon   Zion Canyon   Vermilion Cliffs   Painted Desert   0   10   20 mi   Grand Canyon
                                                                                                0   20 km

(Precambrian gneiss, schist, and granite) crop out near the base of the Grand Canyon, while the youngest rocks (red mudstone and siltstone) form the cliffs of Cedar Breaks and Bryce Canyon. Each rock layer gives an indication of the climate and topography of the region at a time in the past, so taking a tour of the area effectively summarizes the geologic history in the region (**Earth Science at a Glance**, pp. 306–307). For example, when the Precambrian metamorphic and igneous rocks exposed in the Grand Canyon formed, the region was a high mountain range, perhaps as

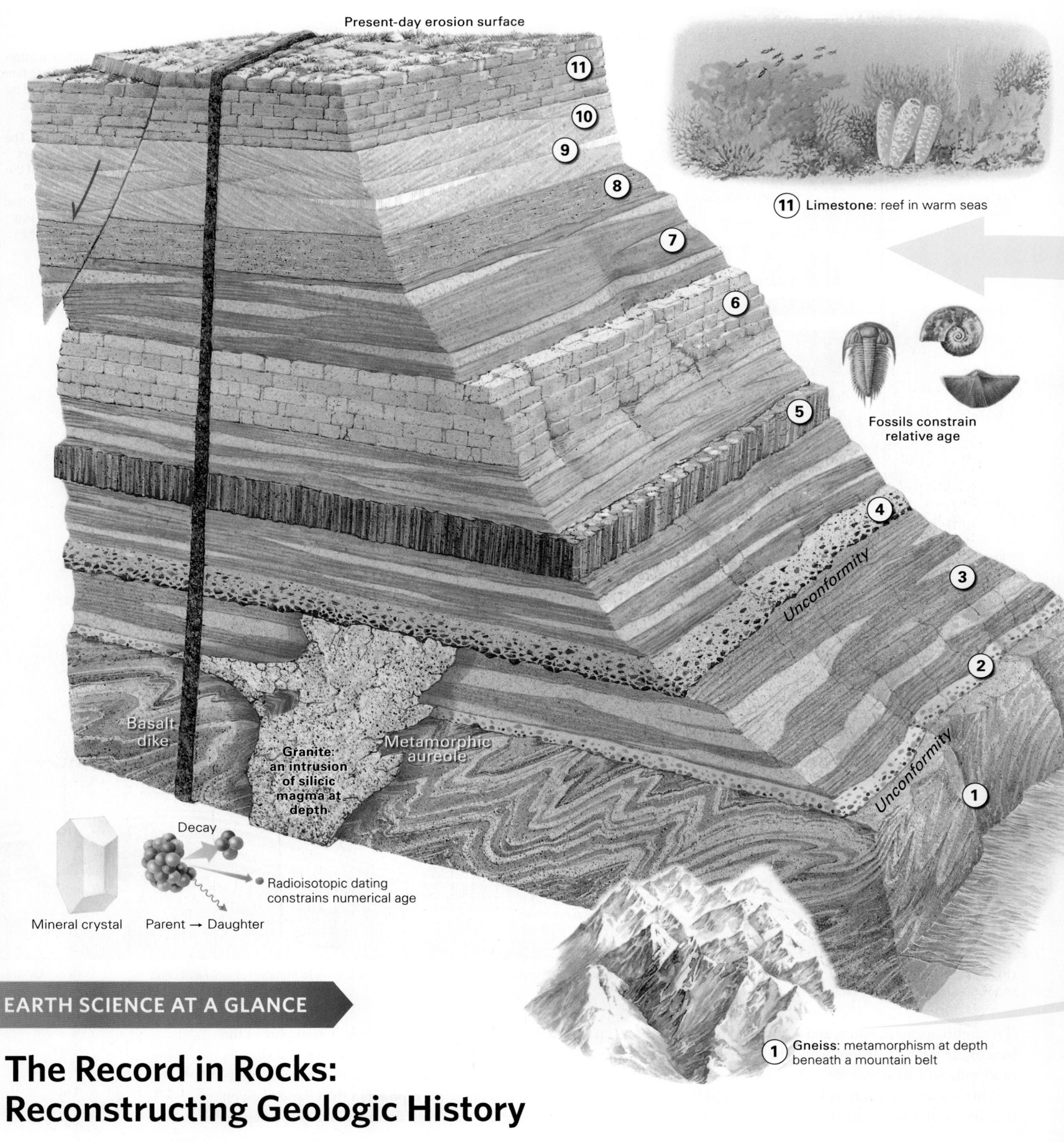

Present-day erosion surface

11

10

9

8

7

6

5

4

3

2

1

Unconformity

Unconformity

Basalt
dike

Granite:
an intrusion
of silicic
magma at
depth

Metamorphic
aureole

Decay

Radioisotopic dating
constrains numerical age

Mineral crystal   Parent → Daughter

11 Limestone: reef in warm seas

Fossils constrain
relative age

1 Gneiss: metamorphism at depth
beneath a mountain belt

The Record in Rocks:
Reconstructing Geologic History

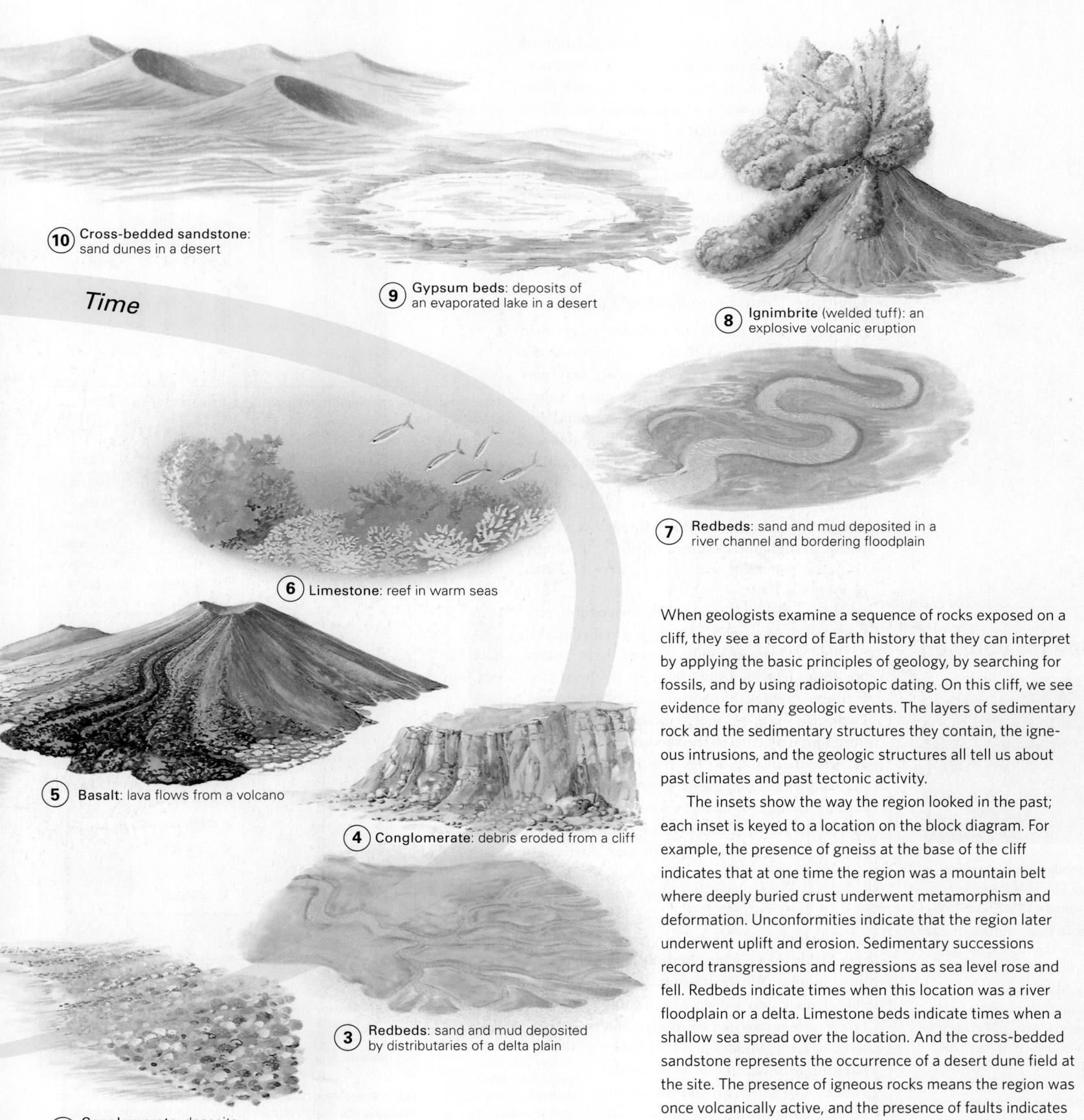

10 Cross-bedded sandstone: sand dunes in a desert

Time

9 Gypsum beds: deposits of an evaporated lake in a desert

8 Ignimbrite (welded tuff): an explosive volcanic eruption

7 Redbeds: sand and mud deposited in a river channel and bordering floodplain

6 Limestone: reef in warm seas

5 Basalt: lava flows from a volcano

4 Conglomerate: debris eroded from a cliff

3 Redbeds: sand and mud deposited by distributaries of a delta plain

2 Conglomerate: deposits of a pebble beach

When geologists examine a sequence of rocks exposed on a cliff, they see a record of Earth history that they can interpret by applying the basic principles of geology, by searching for fossils, and by using radioisotopic dating. On this cliff, we see evidence for many geologic events. The layers of sedimentary rock and the sedimentary structures they contain, the igneous intrusions, and the geologic structures all tell us about past climates and past tectonic activity.

The insets show the way the region looked in the past; each inset is keyed to a location on the block diagram. For example, the presence of gneiss at the base of the cliff indicates that at one time the region was a mountain belt where deeply buried crust underwent metamorphism and deformation. Unconformities indicate that the region later underwent uplift and erosion. Sedimentary successions record transgressions and regressions as sea level rose and fell. Redbeds indicate times when this location was a river floodplain or a delta. Limestone beds indicate times when a shallow sea spread over the location. And the cross-bedded sandstone represents the occurrence of a desert dune field at the site. The presence of igneous rocks means the region was once volcanically active, and the presence of faults indicates that it was seismically active. Fossils constrain the ages of the sedimentary rocks and, therefore, the times of different depositional environments, and radioisotopic studies provide the times of the igneous activity and of metamorphism.

dramatic as today's Himalayas. Accumulation of the sediment that makes up the Kaibab Limestone, now visible at the rim of the canyon, took place when the region was a warm, shallow sea. When the sand in the rocks that constitute the towering red cliffs of sandstone in Zion Canyon was deposited, the region was a Sahara-like desert, blanketed with huge sand dunes. And when the sediment that became the mudstone and siltstone of Bryce Canyon settled out, the region hosted lakes fed by freshwater streams.

Dividing Time: From Eons to Epochs

No outcrop on the Earth provides a complete record of our planet's history, for two reasons: first, because a given outcrop exposes only part of the stratigraphic succession in a given region, for older strata remain hidden underground and younger strata have been eroded away; and second, because stratigraphic successions contain unconformities. But by correlating rocks from locality to locality at millions of places around the world, geologists have pieced together a composite stratigraphic column, called the **geologic column**, that symbolizes the entirety of Earth history (Fig. 8.17).

The geologic column can be divided into segments, each of which represents a specific interval of time. The largest subdivisions of Earth history are **eons**, named, from oldest to youngest, the Hadean, Archean, Proterozoic, and Phanerozoic. Geologists refer to the Hadean, Archean, and Proterozoic together as the **Precambrian**. The suffix –*zoic* means life, so *Phanerozoic* means visible life, and *Proterozoic* means first life. (This 19th-century terminology can be somewhat confusing because in recent decades, geologists have shown that the earliest life—Bacteria and Archaea—appeared in the Archean.) The Phanerozoic Eon is subdivided into **eras**, named, from oldest to youngest, the Paleozoic (ancient life), Mesozoic (middle life), and Cenozoic (recent life). We further divide each era into **periods** and each period into **epochs**. Where relevant, geologists subdivide periods into Upper, Middle, and Lower (when referring to stratigraphic level) or Early, Middle, and Late (when referring to time). If you look back at Figures 8.14 and 8.16, you will see the periods along the left side of the stratigraphic columns shown.

Where do the names of the periods come from? They refer either to a

FIGURE 8.17 Global correlation of rock units led to the development of the geologic column.

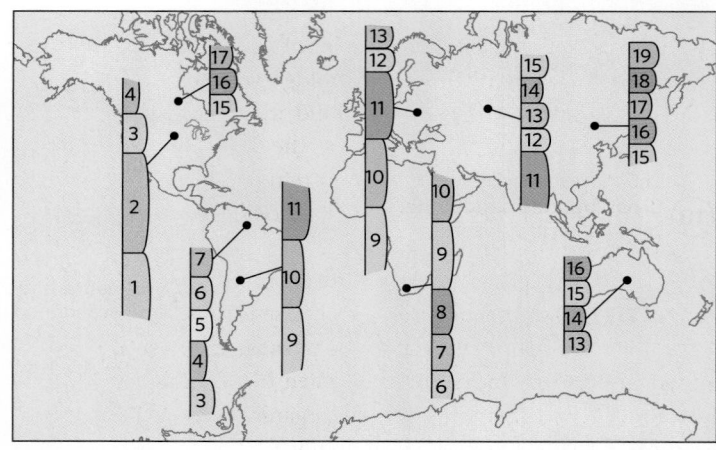

(a) Each of the small columns on this map symbolizes the stratigraphy at a given location.

(b) By correlating these columns, geologists determined the relative ages of rock units and filled in the gaps in the record to make a complete column with no gaps.

Eon	Era	Period		Epoch
Phanerozoic	Cenozoic	Quaternary	"Tertiary"	Holocene
				Pleistocene
		Neogene		Pliocene
				Miocene
		Paleogene		Oligocene
				Eocene
				Paleocene
	Mesozoic	Cretaceous		
		Jurassic		
		Triassic		
	Paleozoic	Permian		
		Carboniferous	Pennsylvanian	
			Mississippian	
		Devonian		
		Silurian		
		Ordovician		
		Cambrian		
Proterozoic				
Archean				
Hadean				

Precambrian

(c) The geologic column. Geologists divide the column into intervals and assign names to them, but since the column was built without knowledge of numerical ages, it does not show the actual durations of these intervals. (Note that the term "Tertiary" is no longer used, but it still appears in older literature.)

locality where a fairly complete succession of strata representing the interval was first identified (for example, rocks representing the Devonian Period crop out near Devon, England), or to a characteristic of the time (rocks from the Carboniferous Period contain a lot of coal, a material that consists of carbon). The terminology was not organized in a planned fashion that would make it easy to learn. Instead, it grew haphazardly in the years between 1760 and 1845 as geologists began to refine their understanding of geologic history and fossil succession.

Early geologists defined boundaries between eras, and between some periods, of the Phanerozoic based on where particularly major changes in fossil assemblages took place. These changes mark the times of **mass-extinction events**, intervals of short duration when **biodiversity**—the number of different types of organisms living at the same time—decreased substantially (Fig. 8.18). The causes of these mass-extinction events remain a subject of research. Some are the consequence of asteroid impacts. Others may be due to intense volcanic activity, severe climate change, or a combination of these factors. Life diversified following mass-extinction events, as new species appeared and evolved to occupy vacated environments. Therefore, fossil assemblages in strata deposited before a mass-extinction event differ significantly from those in strata deposited after the event.

In light of the principle of fossil succession, paleontologists can use fossil assemblages to determine where a particular stratigraphic formation fits into the geologic column. For example, if you know what fossils of the Cambrian Period look like, and you find those fossils in a stratigraphic formation, you can say that the formation

is Cambrian. Note that identification of a rock type alone generally does not indicate where a unit fits into the column. That's because many of the same kinds of rocks (such as sandstone, shale, and basalt) have been forming for most of the past few billion years. An Eocene sandstone can look identical to a Jurassic sandstone, but they will contain very different fossil assemblages.

Take-home message...

A stratigraphic formation is a sequence of beds that can be mapped over a broad region. Unconformities represent times during which deposition didn't occur or bedrock underwent erosion. They represent gaps in the stratigraphic record. Geologists correlate formations on the basis of rock type and fossil content, and they can portray the relationships among formations on a geologic map. Correlation led to the production of the geologic column, which represents the entirety of Earth history. It is subdivided into eons, eras, periods, and epochs. Mass-extinction events delineate the boundaries of some subdivisions.

Quick Questions

- What is a stratigraphic formation?
- How does an angular unconformity differ from a disconformity?
- Which occurred first: the Precambrian or the Phanerozoic?

8.4 Determining Numerical Ages

Geologists since the time of Hutton could determine the relative ages of rocks, but they had no way to specify a rock's **numerical age** (called *absolute age* in older literature), meaning its age in years. Therefore, they could not determine the durations of time intervals on the geologic column. The technology that ultimately changed this situation began with the discovery in 1896 of **radioactive elements**, elements whose atoms spontaneously decay to form atoms of different elements. Scientists soon discovered that a population of radioactive atoms decays at a constant rate, and by 1905 they suggested that they could date rocks by measuring the proportion of atoms that have decayed. The invention of the mass spectrometer by the middle of the 20th century allowed geologists to put this idea into practice, and modern **radioisotopic dating** (also known as *radiometric dating*) became available. This technique uses the ratio between the number of radioactive atoms and the number of atoms of a stable decay product as a basis for calculating numerical ages of rocks. The overall study of numerical ages of rocks developed

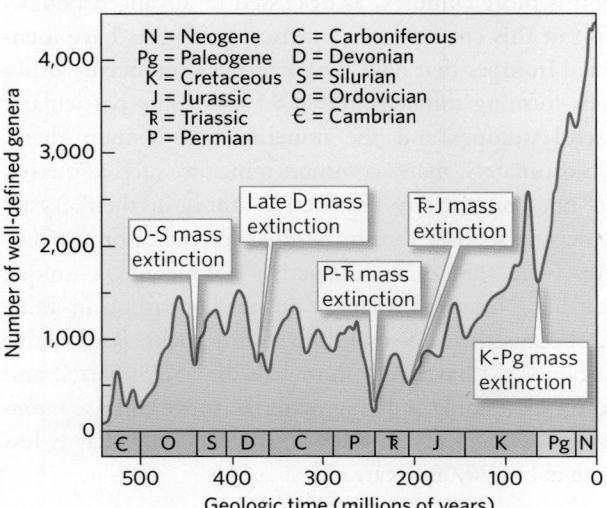

FIGURE 8.18 Changes in the diversity of life over time. Sudden drops in the number of genera indicate mass-extinction events.

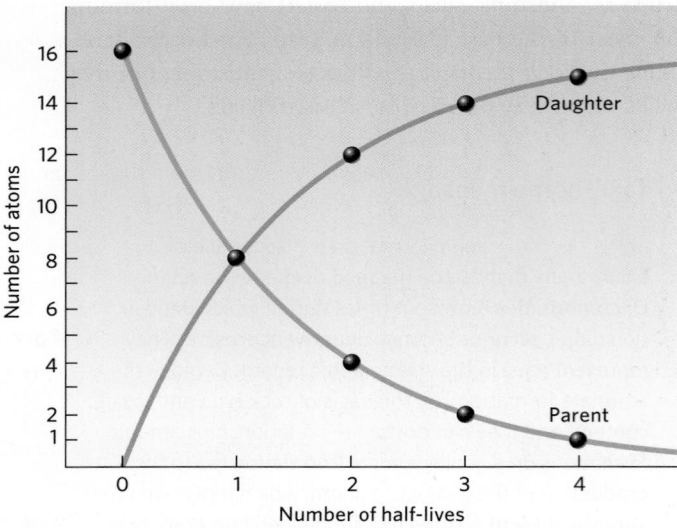

(a) This graph shows how the number of parent atoms of a radioactive isotope decreases and the number of daughter atoms of a stable isotope increases as time passes. (Note that in a real mineral crystal, the number of atoms would be immensely larger.) The rate of change decreases with time.

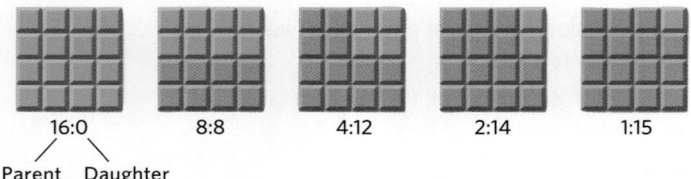

16:0   8:8   4:12   2:14   1:15

Parent Daughter

(b) The ratio of parent atoms to daughter atoms changes with the passage of each successive half-life.

(c) In a cluster of parent atoms of a radioactive isotope undergoing decay, there is no way to predict which specific atom will decay next.

element have the same number of neutrons in their nucleus, which means that not all atoms of an element have the same *atomic mass* (approximately, the number of protons plus neutrons). The different versions of an element, with the same atomic number but different atomic masses, are called **isotopes** of the element. For example, all uranium atoms have 92 protons, but the uranium-238 isotope (abbreviated ^{238}U) has an atomic mass of 238, whereas the ^{235}U isotope has an atomic mass of 235.

Some isotopes are stable, meaning that they last essentially forever. *Radioactive isotopes*, however, are unstable, in that they eventually undergo **radioactive decay**, a process that changes the atomic number of the atom. Consequently, radioactive decay changes an atom of one element into an atom of a different element. Scientists refer to the radioactive atom that undergoes decay as the **parent atom**, and to the decay product as the **daughter atom**. For example, the radioactive isotope ^{238}U is a parent atom that decays, through a series of steps, to become a stable isotope, lead-206 (^{206}Pb), which is the daughter atom.

Scientists cannot specify how long an individual atom of a radioactive isotope will survive before it decays, but they can measure the **half-life**, the time it takes for half of a group of atoms of a given radioactive isotope to decay. The half-life for a given decay reaction is constant, so the ratio of parent to daughter atoms can serve as a natural clock (Fig. 8.19). Box 8.3 helps you visualize the concept of a half-life.

Radioisotopic Dating Techniques

Because radioactive decay proceeds at a known rate, it provides a basis for telling time. In other words, because an element's half-life is a constant, we can calculate the age of a mineral by measuring the ratio of parent atoms to daughter atoms in the mineral. (The actual calculation is more complex, as discussed in advanced books.) To put this concept into practice, geologists have identified isotopes that have long half-lives and occur within rock-forming minerals. Table 8.1 lists some particularly useful isotopes and the minerals that contain them. Unfortunately, many common minerals, such as quartz, do not contain any radioactive atoms in their crystal structure, so they cannot be used for radioisotopic dating. Note that each radioactive isotope has a unique half-life. Note, too, that we do not list carbon in Table 8.1 because it is not useful for dating rocks. Radioactive carbon (^{14}C) has a very short half-life (5,730 years) and accumulates only in living material, so we can use *radiocarbon dating* only to date organic material that is less than about 70,000 years old.

Did you ever wonder...

how geologists specify the ages of rocks in years?

into the discipline of **geochronology**. Radioisotopic dating techniques have improved greatly over the past half century, and geologists can now make very precise measurements from very small samples. But the underlying concept of these modern techniques remains the same, and to explain it, we first review the process of radioactive decay. In our discussion of numerical ages, recall that geologists use abbreviations—Ga for billion years, and Ma for million years—when stating ages.

Radioactive Decay

All atoms of a given element have the same *atomic number*, meaning that they have the same number of protons in their nucleus. However, not all atoms of a given

BOX 8.3 ► How can I explain . . .

The concept of a half-life

What are we learning?

How the ratio of radioactive parent to stable daughter atoms changes over time, and how we can represent this ratio as a half-life.

What you need:

- 32 small, disk-shaped objects (such as checkers), all of the same color
- Colored sticky-note paper, cut into squares
- A cake pan (representing a mineral crystal)
- Graph paper and a pencil

Instructions:

- Spread the disks around the cake pan. These disks represent parent atoms of a radioactive isotope.
- Create a graph that plots number of disks on the vertical y-axis (from 0 to 32) and number of half-lives (from 0 to x) on the horizontal x-axis.
- Plot a point on the graph representing the number of unmarked disks at time 0. Since there are 32 disks, $y = 32$ and $x = 0$. These unmarked disks represent parent atoms.
- Place sticky notes randomly on half the disks. The sticky note indicates an atom of the stable daughter product. Count the number of unmarked disks that remain. The answer, 16, represents the number of parents remaining after the first half-life has passed. Plot the numbers of parents and daughters at 1 half-life on the graph.
- To represent the number of parent atoms remaining after the second half-life, randomly mark half the remaining unmarked disks (8 disks), then record the numbers of marked and unmarked disks. Repeat

the procedure for the remaining half-lives. Continue until all parents have been marked.

- At each stage, record the numbers of parents and daughters on the graph.

What did we see?

- The number of parent atoms decreases and the number of daughter atoms increases as time passes, so the ratio of parents to daughters decreases progressively. If you connect the points on your graph, you see that they lie along a curve, not a straight line.
- Note that after enough time passes, the method no longer works, for no more parents remain to decay.
- To check your understanding, imagine that the half-life for this experiment is 10 million years. How old is a mineral after 2.5 half-lives? What is the ratio of parent to daughter atoms?

Randomly distributed radioactive parent atoms in a mineral

After one half-life, half of the parents have decayed to become daughters.

The procedure for dating minerals uses the following steps. (We have greatly simplified the description of these steps for presentation in this book—in reality, the procedure is challenging.)

- *Collecting appropriate samples:* For radioisotopic dating methods to work, you generally need to obtain unweathered rock samples, because chemical weathering reactions may cause the loss of parent or daughter atoms.

- *Separating datable minerals:* Once samples have been collected, crush them so that you can pick out and isolate the datable minerals.

TABLE 8.1 Isotopes Used in the Radioisotopic Dating of Rock

Parent → daughter	Half-life (years)	Minerals containing the isotopes
$^{147}Sm \rightarrow {}^{143}Nd$	106.0 billion	Garnets, micas
$^{87}Rb \rightarrow {}^{87}Sr$	48.8 billion	Potassium-bearing minerals (mica, feldspar, hornblende)
$^{238}U \rightarrow {}^{206}Pb$	4.5 billion	Uranium-bearing minerals (zircon, uraninite)
$^{40}K \rightarrow {}^{40}Ar$	1.3 billion	Potassium-bearing minerals (mica, feldspar, hornblende)
$^{235}U \rightarrow {}^{207}Pb$	713.0 million	Uranium-bearing minerals (zircon, uraninite)

FIGURE 8.20 In a radioisotopic dating laboratory, samples are analyzed using a mass spectrometer.

• *Analyzing the parent-daughter ratio:* To determine the ratio of parent to daughter atoms, you must first dissolve the minerals in acid and then evaporate the liquid, or vaporize a spot on the mineral with a laser. Then, send the resulting gas, containing parent and daughter atoms along with other atoms, through a *mass spectrometer.* The magnet in this instrument separates atoms from one another based on atomic mass, allowing the number of atoms of a specific element to be counted (Fig. 8.20).

At the end of the laboratory process, you know the ratio of parent to daughter atoms in a mineral. By plugging these numbers into a set of equations that allow for not knowing the total number of parent atoms before decay began, you can calculate the age of the mineral.

What Does a Radioisotopic Date Mean?

At high temperatures, atoms in a crystal vibrate so rapidly that chemical bonds break and re-form relatively easily. Consequently, atoms can escape from or move into crystals. Because radioisotopic dating depends on knowing the parent-daughter ratio, the isotopic clock for a crystal starts only when it becomes cool enough for both parent and daughter atoms to be locked into it. The temperature below which atoms can no longer freely move into and out of a given mineral is called the **closure temperature** of the mineral. When we give a radioisotopic date for a mineral, we are defining the time at which the mineral cooled below its closure temperature.

The concept of closure temperature provides a basis for interpreting the meaning of radioisotopic dates. In the case of igneous rocks, radioisotopic dating tells us when a magma or lava cooled to form a solid igneous rock. In the case of metamorphic rocks, radioisotopic dating tells us when a rock cooled from a metamorphic temperature above the closure temperature to a temperature below it. Because different minerals have different closure temperatures, different minerals yield different dates for the same rock, if the rock cooled slowly.

Can we use radioisotopic dating to obtain the numerical age of a clastic sedimentary rock directly? No. Dating a mineral crystal in a sedimentary rock gives the time when the crystal grew as part of an igneous or metamorphic rock, but not when the crystal, or a fragment of it, was deposited as sediment. For example, if zircon crystals that we extract from a sandstone yield a radioisotopic date of 1.1 Ga, then we know that the sand was deposited sometime during the past 1.1 billion years, and that the source of the sediment in the sand included rocks as old as 1.1 Ga. While that information is useful, it doesn't tell us the time when sand that eventually became the sandstone was deposited, nor does it tell us when the sand was lithified to become rock. The sand in the sandstone may have been deposited, for example, at 50 Ma, and lithification may have happened only at 40 Ma. To obtain dates for sedimentary rocks, geologists must look for cross-cutting relations between sedimentary rocks and datable igneous or metamorphic rocks. For example, if a sequence of sedimentary strata was deposited unconformably on datable granite, the strata must be younger than the granite (Fig. 8.21). Similarly, if a datable basalt dike cuts the strata, the strata must be older than the dike, and if datable volcanic ash buried the strata, then the strata must be older than the ash.

Take-home message...

Radioisotopic dating of rocks specifies numerical ages in years. To obtain a radioisotopic date, we measure the ratio of radioactive parent atoms to stable daughter atoms in a mineral. A radioisotopic date gives the time at which a mineral cooled below its closure temperature. We can directly date igneous and metamorphic rocks. Numerical ages for sedimentary rocks can be deduced from radioisotopic dating of cross-cutting datable rocks.

Quick Questions ———————

- What is the difference between parent and daughter atoms?

- If the half-life of a radioactive parent is 10 million years, what percentage of parent atoms remain after 20 million years?

- Does the numerical age of a mineral grain in a sedimentary rock give us the time of deposition?

FIGURE 8.21 Using cross-cutting relations to date sedimentary rocks.

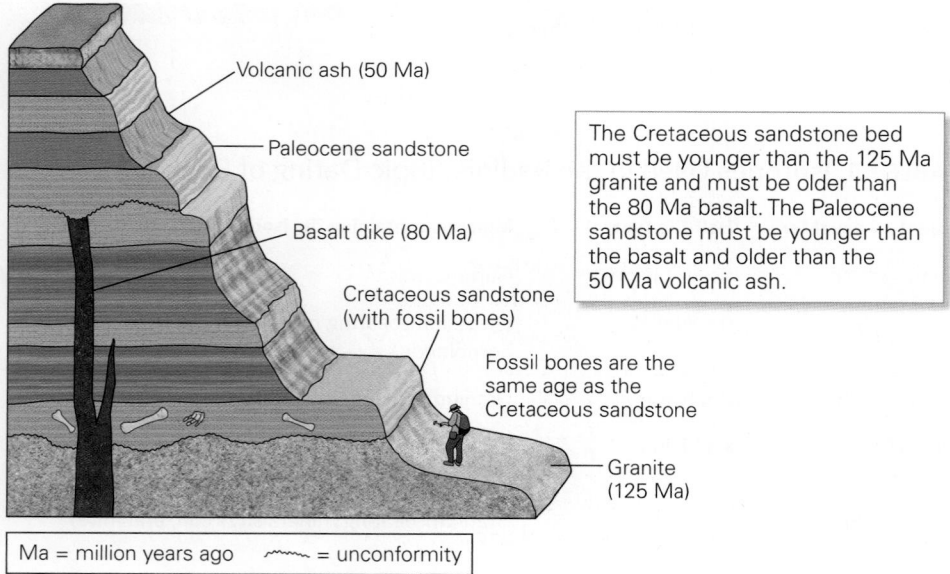

Volcanic ash (50 Ma)

Paleocene sandstone

Basalt dike (80 Ma)

The Cretaceous sandstone bed must be younger than the 125 Ma granite and must be older than the 80 Ma basalt. The Paleocene sandstone must be younger than the basalt and older than the 50 Ma volcanic ash.

Cretaceous sandstone (with fossil bones)

Fossil bones are the same age as the Cretaceous sandstone

Granite (125 Ma)

Ma = million years ago   ⌇⌇⌇ = unconformity

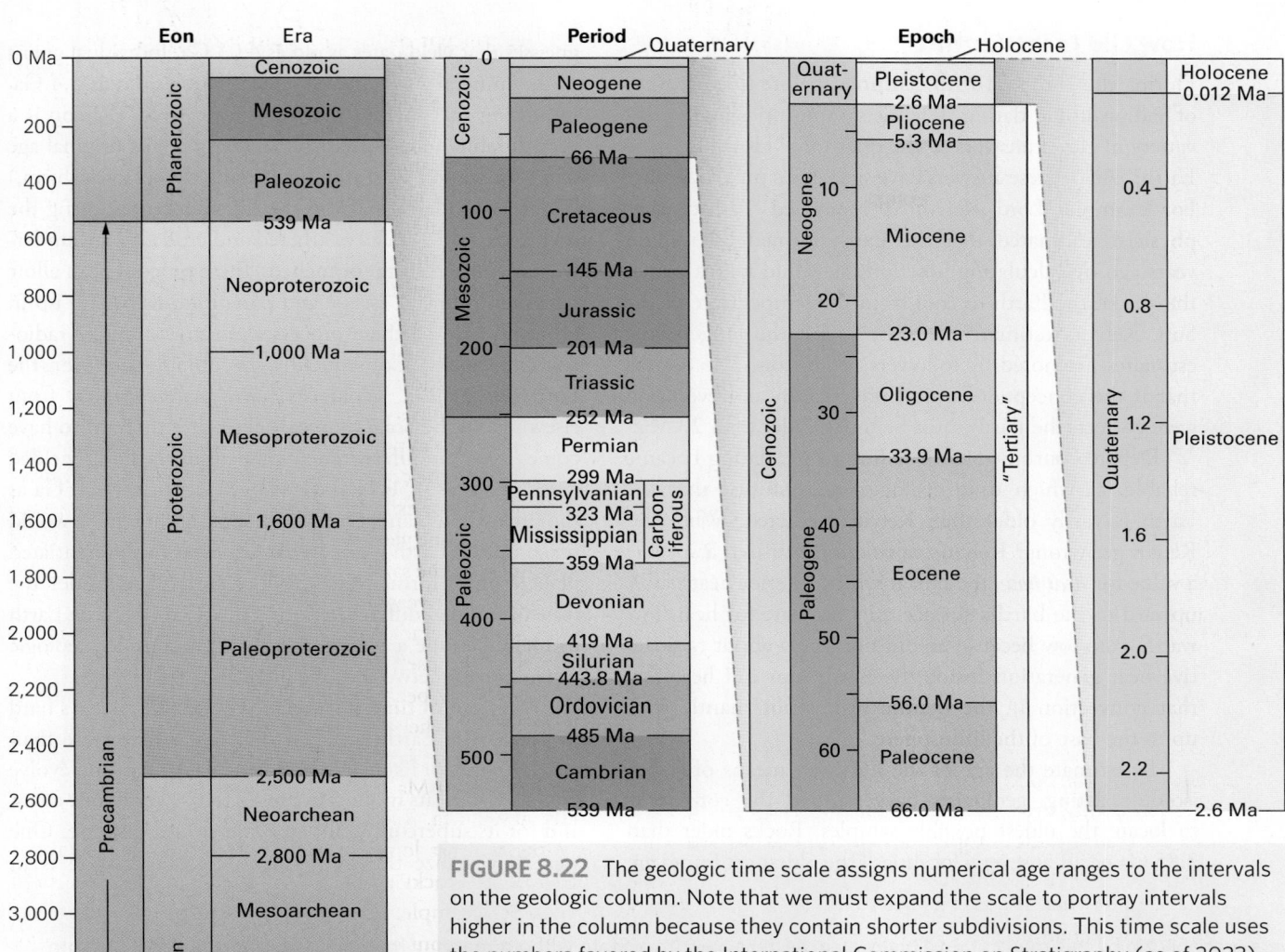

FIGURE 8.22 The geologic time scale assigns numerical age ranges to the intervals on the geologic column. Note that we must expand the scale to portray intervals higher in the column because they contain shorter subdivisions. This time scale uses the numbers favored by the International Commission on Stratigraphy (as of 2023).

could not be assigned to the boundaries between intervals on the geologic column. Once good dating methods became available, however, the possibility of specifying—in years—when each interval on the column began and ended became reality. Geologists searched the world for localities where they could recognize cross-cutting relations between sedimentary rocks and datable igneous rocks. The study of such datable cross-cutting relations revealed, for example, that the Cretaceous Period began at 145 Ma and ended at 66 Ma.

As new data become available and dating methods become more accurate, the specific dates defining the boundaries of geologic intervals may shift. For example, around 1995, new studies on ash layers above and below the Precambrian-Cambrian boundary placed this boundary at 542 Ma, in contrast to previous, less definitive studies that had placed it at 570 Ma. Subsequent studies moved the date to 541 Ma, and in the last decade, to 539 Ma. The chart showing the currently favored ages of boundaries between intervals in the geologic column is called the **geologic time scale** (Fig. 8.22).

> If the Eiffel Tower were now representing the world's age, the skin of paint on the pinnacle-knob at its summit would represent man's share of that age; and anybody would perceive that that skin was what the tower was built for... I reckon they would, I dunno.
>
> —*MARK TWAIN (AMERICAN WRITER, 1835–1910)*

8.5 The Geologic Time Scale

Adding Numerical Ages to the Geologic Column

In Section 8.4, we pointed out that the geologic column was established a century before researchers had developed radioisotopic dating methods. Therefore, numerical ages

How Old Is the Earth?

Did you ever wonder...

how old the Earth's oldest rock is?

During the 18th and 19th centuries, before the discovery of radioisotopic dating, scientists came up with a great variety of clever answers to the question, "How old is the Earth?" All of these answers have since been proved wrong. For example, Lord Kelvin, a renowned 19th-century physicist, estimated that the Earth formed 20 million years ago by calculating how long it would take a planet the size of the Earth to cool from the temperature of the Sun. Kelvin's estimate contrasted with the longer time estimates promoted by followers of Hutton, who argued that if the concepts of uniformitarianism and evolution were correct, the Earth must be much older than 20 Ma.

Debate continued until radioisotopic dating became reliable, at which time geologists determined that the Earth is vastly older than Kelvin predicted. What did Kelvin get wrong? Kelvin's calculations utilized a specific a value for *heat flow*, the rate at which interior heat passes upward to the Earth's surface. His estimate for heat flow was far too low because he did not know about radioactive heat generation inside the Earth, nor did he know that convection in the mantle brings hot mantle rock up to the base of the lithosphere.

To estimate the age of the Earth by means of radioisotopic dating, geologists have scoured the continents to locate the oldest possible samples. Rocks older than 3.85 Ga occur at several localities. The oldest yet found are gneisses that yield dates as old as 4.03 Ga. Individual grains of the mineral zircon have yielded dates as old as 4.4 Ga, indicating that rock of this age once existed. (Zircon is a very durable mineral that can "remember" its original age of crystallization even after weathering or metamorphism.) Older rock doesn't exist on the Earth because during the rock cycle, older rocks weathered and eroded away, underwent metamorphism, or melted. These processes can allow daughter atoms to escape and parent atoms to end up in other crystals, so these processes effectively "reset" radioisotopic clocks. Consequently, to obtain dates for the Earth's formation, geologists have analyzed rocks from elsewhere in the Solar System. Meteorites thought to have come from undifferentiated planetesimals have yielded dates as old as 4.56 Ga, so geologists identify 4.56 Ga as the time when planetesimals—including those that would make up the Earth—grew. Meteorites from differentiated planetesimals formed at 4.54 Ga, so if we consider the time of internal differentiation as the time when the Earth formally became a planet, then we can consider geologic time to begin between 4.56 and 4.54 Ga.

This span of time is so staggeringly large that it's hard to grasp. The Earth is so old that it has had more than enough time for its rocks and life forms to form and evolve (Fig. 8.23), for its mountain ranges to rise and erode away, and for its supercontinents to coalesce and disperse. One way to visualize the immensity of geologic time is to

FIGURE 8.23 Most drivers probably zip by this road cut on a highway in eastern New York State without realizing that it provides amazing insights into events during the Earth's history. Thanks to radioisotopic dating techniques, we can specify the times of these events in years. The history we see here includes (1) deposition of Middle Ordovician strata; (2) folding and tilting; (3) uplift and erosion (forming an unconformity); (4) deposition of Upper Silurian and Lower Devonian strata; (5) another episode of folding and tilting; and (6) uplift and erosion. Put simply, we are seeing the record of two mountain-building events, separated by a time when the mountains were eroded away and the area was submerged by a shallow sea. This history spans about 45 million years. That is a long time... but it is only about 0.001% of geologic time.

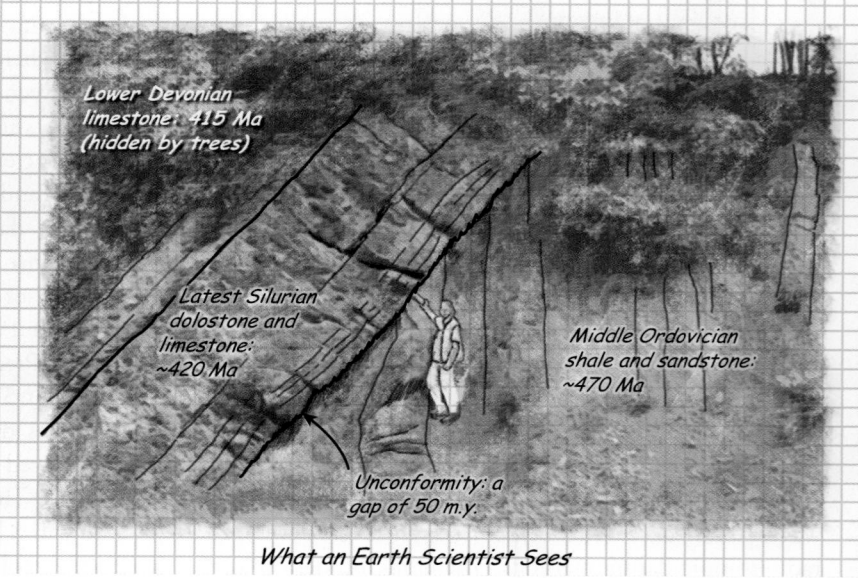

Lower Devonian limestone: 415 Ma (hidden by trees)

Latest Silurian dolostone and limestone: ~420 Ma

Middle Ordovician shale and sandstone: ~470 Ma

Unconformity: a gap of 50 m.y.

What an Earth Scientist Sees

equate the Earth's history to a single calendar year. On this scale, the oldest rocks preserved on the Earth appear in early February, the first bacteria evolve on February 21, the first invertebrates to grow shells appear on October 25, the first amphibians crawl out of the sea and onto land on November 20, and the continents coalesce into Pangaea on December 7. Dinosaurs appear on December 15 and go extinct on December 25. The last week of December represents the Cenozoic Era, the last 66 million years of Earth history. The first human-like ancestor appears on December 31 at 3 P.M. Our species, *Homo sapiens*, shows up an hour before midnight, and all of recorded human history takes place in the last 30 seconds. To put it another way, human history occupies only the last 0.000001% of Earth history.

Take-home message...

Radioisotopic dating techniques were used to develop the geologic time scale. The oldest rock of the Earth's crust is about 4 billion years old. Dating of meteorites indicates that the Earth was forming at 4.56 Ga and had differentiated by 4.54 Ga.

Quick Questions

- Is it possible for a Jurassic sandstone to contain an Ordovician zircon grain? Explain your answer.
- Which are older, differentiated or undifferentiated planetesimals?
- Why do the dates assigned to the boundaries between intervals on the geologic time scale change?

8 CHAPTER REVIEW

Objective 8.1

Define geologic time, explain the difference between relative and numerical ages, and discuss key geologic principles used to decipher the geologic history visible in an outcrop.

KEY CONCEPTS

- Geologic time refers to the time span since the Earth's formation.
- Relative age specifies whether one geologic feature is older or younger than another, whereas numerical age gives the age of a geologic feature in years.
- The principle of uniformitarianism, along with other geologic principles derived from it, serves as the foundation for interpreting relative ages and for deciphering the geologic history of a region.

EARTH-SCIENCE VOCABULARY

baked contact (p. 290)
cross-cutting relations (p. 290)
geologic history (p. 292)
geologic time (p. 290)
inclusion (p. 290)

original horizontality (p. 290)
relative age (p. 290)
superposition (p. 290)
uniformitarianism (p. 289)

REVIEW QUESTIONS

1. **(a)** What is the principle of uniformitarianism? **(b)** How does this principle lead to the premise that the Earth is very old? **(c)** Define geologic time.

2. **(a)** If you are given a list of historical events in the 20th century, in order from youngest to oldest, can you determine from the list alone what percentage of that century the list represents? Explain your answer. **(b)** From that list, can you determine the relative ages or the numerical ages of the events?

3. **(a)** How does the principle of cross-cutting relations allow you to determine the relative ages of two rocks visible in the same outcrop? **(b)** If a conglomerate contains clasts of granite, is the granite older than the conglomerate, or vice versa? Which geologic principle justifies this interpretation? **(c)** A cliff exposes a sequence of horizontally bedded strata. Where will you find the oldest strata? **(d)** How can you tell if an igneous rock is older or younger than another rock that it is in contact with?

4. In **Figure A,** number the beds to the right of the fault in sequence from oldest (1) to youngest. (The black rock is basalt.) Then, list the succession of events that constitute the geologic history represented by the rocks depicted in the figure in order from youngest (at the top) to oldest (at the bottom).

Objective 8.2

Identify basic types of fossils, explain how fossils form and are preserved, and relate fossil succession to the theory of evolution.

KEY CONCEPTS

- Fossils form when organisms or traces of organisms are buried and preserved in rock. There are many types of fossils. Typically, hard parts of organisms are most likely to be preserved. Fossils can be classified based on the same taxonomic principles used in biology. Many fossils can be identified by their morphology.

- The assemblage of fossils changes with level in a stratigraphic succession, so different fossils occur in younger strata than in older strata. Fossils, therefore, provide a way of determining the ages of rock layers relative to one another, as well as a record of life's evolution.

- Darwin's theory of evolution by natural selection states that the fittest organisms survive and pass on their traits to their offspring.

- The fossil record is incomplete because paleontologists have found only a tiny fraction of fossil species, because not all organisms are preserved, and because preserved strata do not record all of geologic time.

EARTH-SCIENCE VOCABULARY

extinction (p. 298)
fossil (p. 293)
fossil assemblage (p. 298)
fossilization (p. 293)
fossil succession (p. 298)
paleontology (p. 293)
preservation potential (p. 294)
taxonomy (p. 297)
theory of evolution (p. 299)

REVIEW QUESTIONS

5. **(a)** Provide a definition of a fossil. **(b)** Describe the steps that lead to the formation of a fossil. **(c)** Describe the different kinds of fossils.

6. **(a)** Describe the conditions or characteristics that give an organism a high preservation potential. **(b)** Which organism is more likely to be preserved as a fossil: a spider or a clam? Why?

7. **(a)** On what basis are fossils classified? **(b)** What does a phylogenetic tree depict? **(c)** Which type of fossil organism does **Figure B** show?

8. **(a)** Which of the fossil assemblages in **Figure C** is the youngest? **(b)** Based on the principle of fossil succession, would you expect the same fossil assemblage to occur in strata at the top of the Grand Canyon and in strata at the base? **(c)** Why is the fossil record incomplete?

9. **(a)** Define extinction. **(b)** How does the theory of evolution explain the observations that led to the principle of fossil succession? **(c)** Why is it that paleontologists haven't yet identified all fossil species?

B

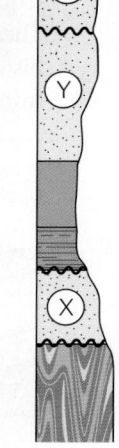

C

Objective 8.3

Interpret unconformities, describe the basis for correlating stratigraphic formations, and explain how correlation led to development of the geologic column.

KEY CONCEPTS

- A given sequence of beds that can be traced over a fairly broad region is a stratigraphic formation.

- Strata are not necessarily deposited continuously at a location. An interval of nondeposition or erosion is called an unconformity. Geologists distinguish among angular unconformities, nonconformities, and disconformities.

- A stratigraphic column shows the succession of strata in a region and provides a record of Earth history in that region. A geologic map shows the distribution of formations and geologic structures.

- Correlation defines the age relationships between formations at one location and formations at another.

- The geologic column represents the entirety of geologic time. Its largest subdivisions are eons. Eons are subdivided into eras, eras into periods, and periods into epochs.

- At several times during the Earth's history, a high percentage of species went extinct. Such mass-extinction events may be a consequence of catastrophic events.

EARTH-SCIENCE VOCABULARY

angular unconformity (p. 300)
biodiversity (p. 309)
contact (p. 299)
correlation (p. 304)
disconformity (p. 302)
eon (p. 308)
epoch (p. 308)
era (p. 308)
geologic column (p. 308)
geologic map (p. 303)
mass-extinction event (p. 309)
nonconformity (p. 302)
period (p. 308)
Precambrian (p. 308)
stratigraphic column (p. 302)
stratigraphic formation (p. 299)
unconformity (p. 300)

REVIEW QUESTIONS

10. **(a)** How many distinct Paleozoic stratigraphic formations occur in the Grand Canyon? **(b)** Would you expect any of the same formations to be visible in outcrops found near Zion National Park? On what basis could you prove your answer? **(c)** What type of contact occurs between sedimentary beds in a formation? **(d)** How are stratigraphic formations and contacts portrayed on geologic maps?

11. **(a)** How does an unconformity develop? **(b)** What is the difference between a disconformity and a nonconformity? **(c)** Which type of unconformity does **Figure D** show?

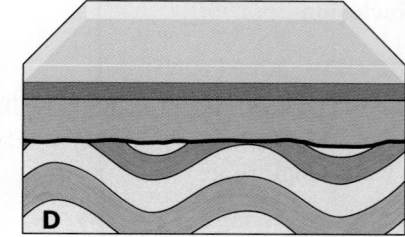

D

12. (a) Describe two different methods of correlating rock units. (b) How was correlation used to develop the geologic column? (c) Each stratigraphic column in **Figure E** portrays the

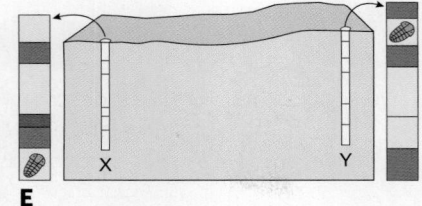
E

stratigraphic column obtained by studying a drill core. The beds are all horizontal. The colors represent different rock types. Note that each drill core contains a unit in which a specific index fossil has been found. Draw lines to correlate column X with column Y. Explain your interpretation?

13. (a) What is a mass-extinction event, and why might such events happen? (b) What is the relationship between mass-extinction events and major boundaries between time intervals on the geologic column? (c) What feature of a stratigraphic formation allows you to determine the period on the geologic column in which it was deposited?

Objective 8.4

Discuss how to determine numerical ages by using radioisotopic dating, and interpret what the age of a rock means.

KEY CONCEPTS

- The numerical ages of rocks can be determined by radioisotopic dating. This method is based on the observation that radioactive atoms decay at a rate characterized by a constant half-life.

- The radioisotopic date of a mineral specifies the time at which the mineral cooled below a closure temperature. So, geologists use radioisotopic dating to determine when an igneous rock solidified and when a metamorphic rock cooled.

- To date sedimentary strata, geologists examine cross-cutting relations between the strata and datable igneous or metamorphic rock.

EARTH-SCIENCE VOCABULARY

closure temperature (p. 312)
daughter atom (p. 310)
geochronology (p. 310)
half-life (p. 310)
isotope (p. 310)

numerical age (p. 309)
parent atom (p. 310)
radioactive decay (p. 310)
radioactive element (p. 309)
radioisotopic dating (p. 309)

REVIEW QUESTIONS

14. (a) What distinguishes one isotope from another of the same element? (b) How does a radioactive isotope differ from a stable isotope? (c) What does the half-life of an element indicate? (d) In the model shown in **Figure F**, each unmarked checker represents a parent atom and each checker

F

with yellow paper on it represents a daughter atom. How many half-lives would produce the number of daughter atoms in the image?

15. (a) Describe the steps used to obtain a radioisotopic date. (b) What do the numerical ages of igneous and metamorphic rocks mean? (c) Can we use radiocarbon dating to determine the ages of rocks?

16. (a) Why can't geologists date sedimentary rocks directly? (b) Given the ages (in Ma) of the rocks shown in **Figure G**, during what time interval did the shale form? (c) How were the numerical age ranges of intervals on the geologic column obtained?

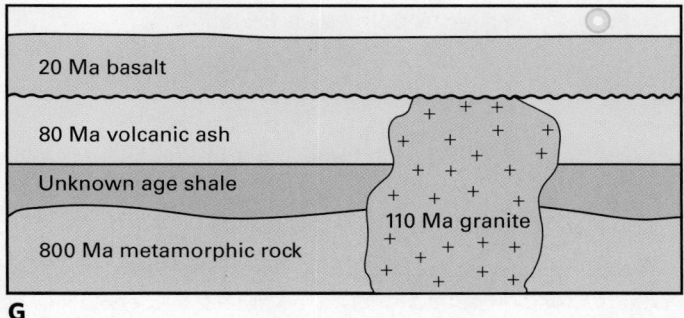

20 Ma basalt

80 Ma volcanic ash

Unknown age shale

110 Ma granite

800 Ma metamorphic rock

G

Objective 8.5

Explain the basis for establishing the geologic time scale and for determining the age of the Earth.

KEY CONCEPTS

- Because the oldest rock that exists on the Earth formed after the planet itself had formed, scientists have studied the ages of meteorites to obtain an age for planet formation.

- Radioisotopic dating of meteorites indicates that the Earth started to form at about 4.56 Ga and differentiated at about 4.54 Ga.

- Human history occupies only the last 0.000001% of the Earth's history.

EARTH-SCIENCE VOCABULARY

geologic time scale (p. 313)

REVIEW QUESTIONS

17. (a) Could the age of the Earth be determined before radioisotopic dating was developed? (b) What is the age of the oldest rock yet found on the Earth? Why don't even older rocks exist? (c) What is the age of the oldest meteorites? Do they come from differentiated or undifferentiated planetesimals?

18. (a) How does the geologic time scale differ from the geologic column? (b) On what basis have dates been assigned to the boundaries between intervals on the geologic time scale? (c) A 33 Ma dike cuts across a sandstone bed, and the sandstone bed overlies a 42 Ma basalt flow. During what period was the sandstone deposited?

9 HISTORY BEFORE HISTORY
A Biography of the Earth

After studying this chapter, you should be able to...

1. characterize how the Earth formed and changed during the Hadean Eon, the time from our planet's birth until the rock record begins.

2. provide a geologic interpretation of how early continents, oceans, and life forms came to be and evolved during the Archean Eon.

3. explain how life, the atmosphere, and climate changed during the Proterozoic Eon, and how the first supercontinents came into existence.

4. identify major Paleozoic continents and the supercontinent that formed at the end of the era, relate Paleozoic orogenies to plate interactions, and discuss sea-level change and life evolution during this era.

5. describe how the map of the world changed during the Mesozoic Era, characterize the dominant life forms of the time, and outline a widely accepted hypothesis to explain the sudden end of the era.

6. discuss the origin of present-day mountain belts during the Cenozoic Era, and place the Pleistocene Ice Age and the appearance of *Homo sapiens* in a geologic time context.

7. define global change, provide examples of various types of change that have happened during Earth history, and consider how humanity has become an agent of global change.

In 1868, Thomas Henry Huxley, a British biologist, presented a public lecture to an audience in Norwich, England. Seeking a way to convey his fascination with the Earth's history, he focused the audience's attention on the piece of chalk he'd been drawing with. And what a tale the chalk had to tell! Chalk consists of microscopic marine plankton shells. The specific chalk that Huxley held came from 90 Ma beds now exposed along the dramatic white cliffs of England's southeastern coast. Geologists in Huxley's day knew that similar chalk beds cropped out throughout much of Europe and that some of these beds contained fossils of bizarre swimming reptiles that were very different from those of modern times. Clues in his humble piece of chalk allowed Huxley to demonstrate that the geography and inhabitants of the Earth in the past differed markedly from those of today. In other words . . . the Earth has a history!

This chapter introduces the Earth's history by offering a brief geological and paleontological biography of our home planet, from its birth to the present. You will see that over time, continents move, mountain belts appear and erode, climates warm and cool, sea level rises and falls, and life evolves, for ours is a world of constant change. This chapter's coverage proceeds chronologically, from older to younger. As you read, watch for several unifying themes: continents didn't begin to form until well after the Earth itself had; at several times during the past few billion years continents merged to form a supercontinent, which then eventually broke apart; the fossil record reveals that the biosphere hosted only single-celled organisms during most of geologic time, for complex organisms with distinct body parts did not appear until around 600 Ma; as time passes, life diversifies and becomes increasingly complex; the composition of the atmosphere has changed over time, reflecting exchanges with oceans as well as inputs from volcanoes and life; mountain building takes place during distinct events (*orogenies*) due to plate interactions; and the succession of sedimentary strata reflects sea-level rises and falls, proximity to sources of sediment, and climate change. This chapter concludes by summarizing types of change and by discussing ways in which humanity will leave its imprint on what, in the distant future, will be the geologic record of today.

<< Rafters in the Grand Canyon drift past layers of rock that, like pages of a book, record stories in the long history of the Earth. The rocks next to the rafts were metamorphosed during mountain building 1.8 billion years ago. Horizontal beds in the distance tell us that, at successive times in the past, this location was a mudflat, a desert, and a shallow sea.

9.1 The Hadean Eon (4.56–4.00 Ga): Before the Rock Record Began

When James Hutton, the 18th-century Scottish geologist, pondered the implications of his principle of uniformitarianism (see Chapter 8), he realized that, since most geologic processes happen very slowly, our planet must be very old indeed. Without access to numerical ages, however, he had no way to measure the actual duration of its history, and thought that maybe there was no "vestige of a beginning." The advent of radioisotopic (radiometric) dating overcame this limitation. Radioisotopic dating of meteorites indicates that the earliest solids of the Solar System formed just after 4.57 Ga. Fairly quickly, solid flecks clumped together, the clumps in turn merged, and soon, larger pieces called planetesimals speckled the protoplanetary disk. After many more collisions and combinations, the eight planets that now exist took shape. Accretion of planetesimals had produced the proto-Earth by 4.56 Ga (Fig. 9.1a). This body became so hot inside that its interior separated into an iron core and rocky mantle (Fig. 9.1b). By 4.54 Ga, the Earth was a differentiated body, so we use the age range of 4.56–4.54 Ga as the time of the Earth's birth.

At its birth, so much heat flowed from the Earth's interior to the planet's surface that much of the planet, including the surface, was molten. This heat came from the transformation of the kinetic energy of impacting objects into thermal energy as they struck the growing planet, as well as from the compression of its material into a denser mass, friction between molecules colliding as they came in contact, the release of energy by decay of radioactive atoms, and the release of energy due to differentiation. Eventually, however, radiation of heat from the Earth's surface into space allowed the surface to start solidifying. Initially, rafts of mantle rock developed, but they were dense enough to sink and remelt. Finally, a thin, volcano-speckled crust of frozen mantle—like the skin of ice that forms on a pond in winter—probably covered the surface. This crust lasted until about 4.53 Ga, when the collision of a protoplanet, called *Theia*, with the Earth caused catastrophic change. The impact blasted away and vaporized any crust that existed, as well as a significant fraction of the Earth's mantle and much of Theia. The debris collected in an orbit around the infant Earth and soon coalesced to form our Moon. At first, the Moon orbited the Earth at a distance of only about 30,000 km (18,000 mi), less than 10% of its present orbital distance, so its gravitational pull caused immense tides, which also contributed to heating the interior (Fig. 9.1c; see Chapter 14).

The man who should know the true history of the bit of chalk which every carpenter carries about in his breeches pocket, though ignorant of all other history, is likely . . . to have a truer and therefore a better conception of this wonderful universe and of man's relation to it than the most learned student who [has] deep-read the records of humanity [but is] ignorant of those of nature.

—*THOMAS HENRY HUXLEY (1825–1895)*

FIGURE 9.1 Scenes from the birth of the Earth and the Hadean Eon.

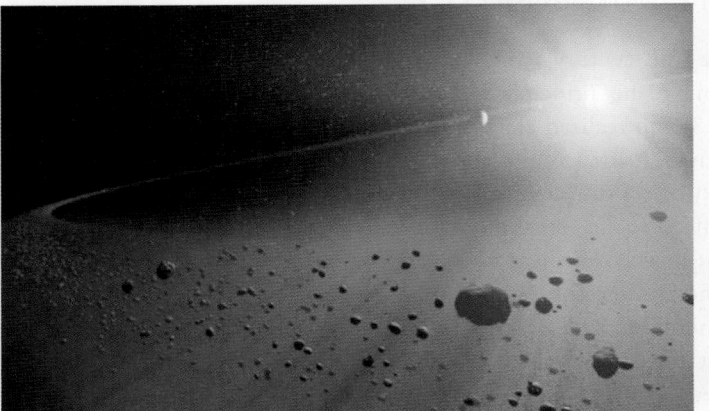

(a) The Earth grew from the accumulation of planetesimals.

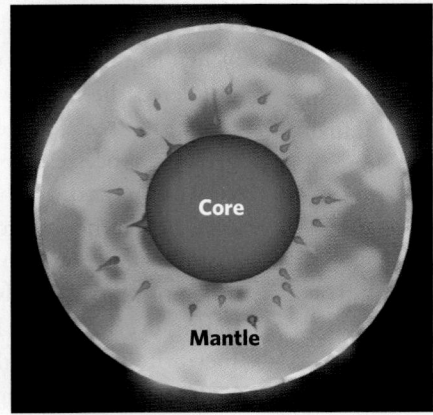

(b) During differentiation, iron sank to the center of the Earth to form its core, leaving a rocky mantle behind.

(c) The Moon formed from material blasted from the Earth by a collision with a protoplanet.

(d) A solid crust and an early water ocean may have developed during the Hadean.

(e) At times, the surface of the Earth was covered with molten rock, as it and the nearby Moon were battered by the late heavy bombardment.

The thermal energy generated during the Theia impact left the Earth with a molten surface. Within a few million years, however, a new skin-like crust of solidified mantle rock developed; this rock was ultramafic, with the same composition as the peridotite of the mantle. Because the planet's interior remained very hot, volcanism continued to be widespread. Volatile elements and compounds that had been attached to mantle minerals began to emerge from volcanic vents. Gravity prevented the products of this *volcanic outgassing* from escaping into space, so they accumulated to form the Earth's *first atmosphere*, a murky, unbreathable mixture consisting mostly of carbon dioxide (CO_2), water vapor (H_2O), nitrogen (N_2), methane (CH_4), ammonia (NH_3), hydrogen (H_2), hydrogen sulfide (H_2S), and sulfur dioxide (SO_2). Some researchers speculate that comets

colliding with the Earth may have contributed additional volatile materials, and perhaps even organic compounds, to this first atmosphere. When the planet's solid surface and atmosphere had cooled sufficiently, water rained from the first atmosphere, filling low areas of the solid surface with seas of liquid water (Fig. 9.1d). These early seas, which appeared only a few hundred million years after Moon formation, were temporary, because between 4.1 and 3.8 Ga, in an interval called the **late heavy bombardment** (Fig. 9.1e), meteorites again began to pummel the Earth. The kinetic energy of these impacts produced enough thermal energy to vaporize any water that had formed, and probably to remelt the planet's surface once again. Craters formed by this event still pockmark the Moon's surface, because these features have not been eroded or buried on the Moon.

No minerals or rocks formed on the Earth during the first 180 million years of our planet's history remain today because the Earth remained so hot that radioisotopic clocks couldn't operate. The oldest mineral grain yet found on the Earth dates to about 4.36 Ga, and the oldest whole rock yet found yields an age of about 4.03 Ga, meaning that it formed a half billion years after the Earth formed. Geologists refer to this mysterious half-billion-year interval before the rock record began as the **Hadean Eon**, named for Hades, the Greek god of the underworld, to convey the rather hellacious conditions that existed for part or much of that time. By convention, this end of this eon has been assigned a date of 4.0 Ga.

Take-home message...

The Hadean Eon began when the Earth differentiated. Initially, the planet had a molten surface, but eventually, a skin-like crust of solidified ultramafic mantle rock formed. Collision of a protoplanet with the Earth sent debris into orbit and remelted the surface. The debris coalesced to form the Moon. Abundant meteorite impacts a few hundred million years later remelted the surface again. Consequently, no rock record exists for the first half billion years of the Earth's history.

Quick Questions

- What is Theia?
- Where did the gases of the Earth's first atmosphere come from?
- What event marks the beginning of the Hadean?

9.2 The Archean Eon (4.0–2.5 Ga): Birth of Continents and Life

Geologists place the start of the **Archean Eon** (from the Greek word *archē*, meaning beginning) at 4.0 Ga, about the age of the oldest known rock. As we noted above, the late heavy bombardment probably pulverized and melted any crust that existed before that time. In addition, because early crustal rocks were ultramafic, they were so dense that they could eventually sink back into the hot, plastic, and partly molten mantle beneath. Finally, any rocks that did survive the late heavy bombardment, and didn't sink into the mantle, have probably passed through the rock cycle (see Chapter 3) and have become components of younger rocks.

By 3.85 Ga, the near-surface realm of the Earth was cool and stable enough for the rock record to become more complete. The cooling Earth System allowed water in the atmosphere to condense, so rains filled permanent oceans, and the earliest marine sediments were deposited. Most atmospheric CO_2 dissolved into the new oceans. Removal of H_2O and CO_2 from the atmosphere left N_2 gas behind, for nitrogen does not react chemically with or dissolve in other substances. So, at the dawn of the Archean, the Earth had a semitransparent *second atmosphere*, composed of 80% N_2 and 20% CO_2 (see Chapter 16).

Continental Crust Appears

As the Earth began to cool, it was no longer hot enough in the mantle for complete melting of ultramafic rock to produce ultramafic magma. Instead, *partial melting* (during which only some components of a rock enter a melt, which then migrates elsewhere, leaving unmelted components behind) took place. As we discussed in Chapter 4, partial melting produces melt that is more felsic than the source rock. So partial melting of ultramafic mantle peridotite produced mafic melt, which, when solidified, yielded mafic igneous rock. Therefore, at hot spots, layers of basalt piled up into thick plateaus, and when plate-tectonic processes began to operate, mafic igneous activity also began to take place at convergent boundaries to produce island arcs of mafic rocks **(Fig. 9.2a)**. The earliest land to rise above primordial oceans may have formed at hot spots and volcanic arcs.

Mafic oceanic plateaus and island arcs could not be subducted because they were less dense than the ultramafic rock of the underlying mantle. So, when these relatively buoyant blocks collided with one another, they sutured together to form larger crustal blocks that remained at the Earth's surface **(Fig. 9.2b)**. Such collisions also caused crustal thickening, so that rocks formed near the surface ended up deep underground, where they underwent metamorphism. Consequently, the first metamorphic rocks came into existence during the Archean. During subsequent rifting of these early crustal blocks, flood basalts extruded at various locations. Eventually, melting at depth in the newly formed crust produced intermediate and felsic igneous rocks, and erosion of crust that had risen above sea level yielded sediments. As a result of all these processes, newly formed composite crustal blocks, called **protocontinents**, eventually included a variety of rock types.

As collisions continued, protocontinents sutured together to form even larger continents. Eventually, some continental blocks became so wide that their interiors were isolated from the heat of plate-boundary volcanism and collisions. Such continental interiors cooled, so the crust they contained became resistant to deformation by plastic flow and, therefore, became durable. By 3.2 Ga, several durable small continents probably existed, and by 2.7 Ga, some of them had been sutured together to form the first large continents. Continental crustal production continued through the Archean, and by the end of the eon, perhaps

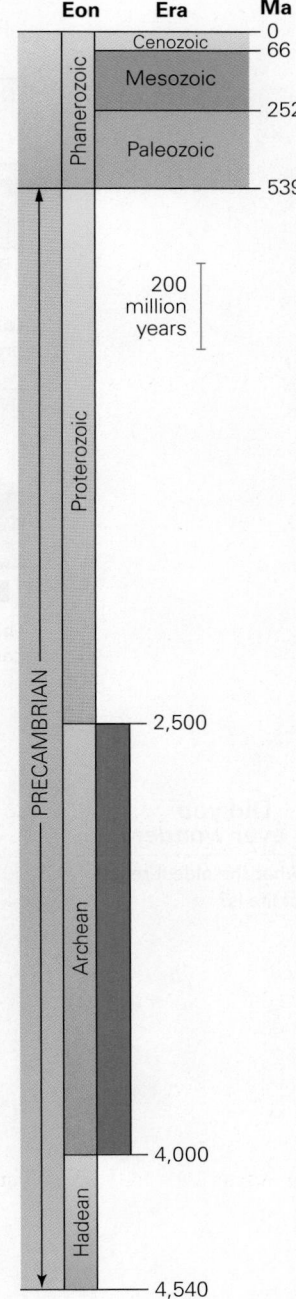

FIGURE 9.2 A model for crust formation during the Archean Eon.

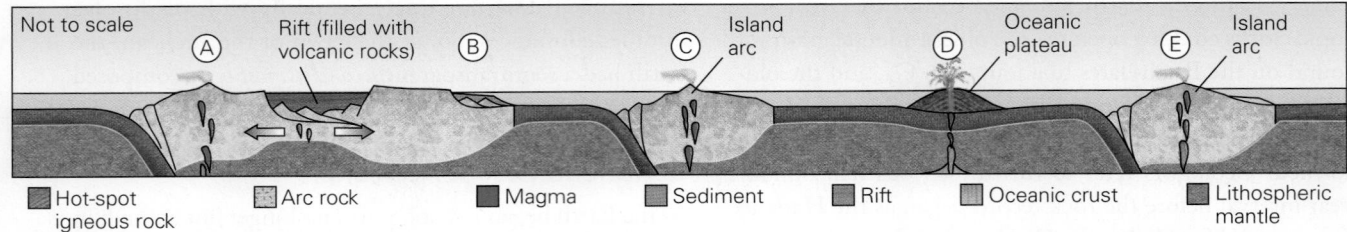

(a) In the Archean, convergent boundaries and hot spots built small blocks of relatively buoyant crust. Rifting of these blocks may have produced flood basalts, and erosion of the blocks produced sediment.

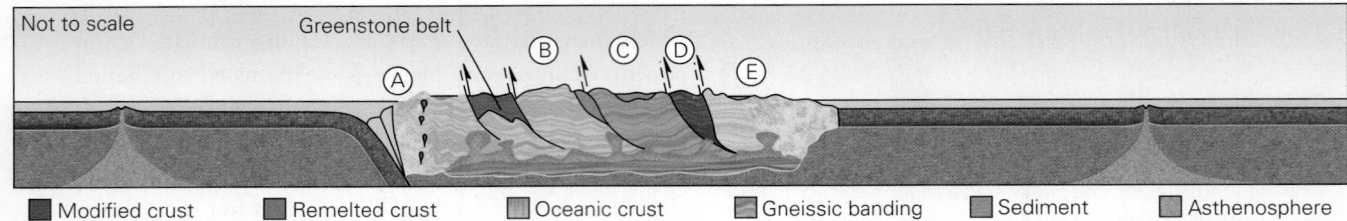

(b) Buoyant blocks collided and were sutured together, forming protocontinents, which were intruded by granite. Eventually, regions of crust cooled, stabilized, and became continents.

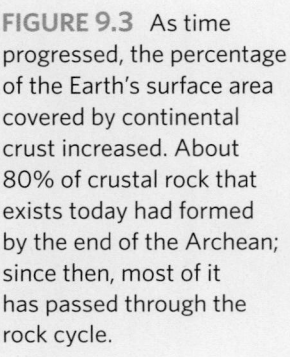

Did you ever wonder...

what the oldest relict of life is?

80% of the rock-forming chemicals that have remained part of the Earth's continental crust ever since had separated from the mantle (Fig. 9.3). The process of continent formation effectively distilled the chemicals that make up lower-density minerals out of the mantle, and these chemicals remained in continental crust even after passing through the rock cycle one or more times.

Early Life Appears

The search for the earliest life on the Earth makes headlines in the popular media. While some evidence hints that life appeared before the late heavy bombardment, convincing chemical fossils representing life appear only in strata younger than about 3.8 Ga. Shapes resembling cells appear in rocks dated as early as 3.5 Ga, but the oldest undisputed fossil cells of bacteria and archaea, the most primitive living organisms, occur in strata dating from 3.2 Ga (Fig. 9.4a). Some of these fossils occur in **stromatolites**, distinct layered mounds interpreted to be made up of fossilized mats of *cyanobacteria* (also known as

blue-green algae, an organism capable of photosynthesis). The mounds form because sediment settling out of ocean water sticks to a mucus-like substance secreted by the mats of cyanobacteria. As the mat becomes buried, new cyanobacteria colonize the top of the sediment, building the mound upward (Fig. 9.4b–d).

What specific environment of the Archean Earth served as the cradle of life? Laboratory experiments of the 1950s led researchers to speculate that life began in warm pools of surface water, shocked by bolts of lightning, beneath a methane- and ammonia-rich atmosphere. More recent studies, however, suggest that hydrothermal vents, perhaps along mid-ocean ridges, may have hosted the first organisms. These vents emit clouds of ion-charged solutions from which sulfide minerals precipitate. The earliest Archean life may well have been heat-loving bacteria or archaea that dined on sulfides emitted from hydrothermal vents, perhaps in the dark depths of the ocean. Only later in the Archean, when organisms evolved the ability to carry out photosynthesis (between 3.5 and 3.2 Ga), could life survive in shallow, brightly lit ocean water. The oxygen produced by these organisms began to change the composition of the atmosphere.

As the Archean Eon came to a close, the first continents had formed, an early form of plate tectonics was underway, collisional mountain belts were forming, and sediment production was taking place. In addition, life had colonized both the depths of the sea and the shallow-marine realm, and the atmosphere had transformed from an H_2O- and CO_2-rich one into an N_2-rich one containing traces of O_2. By about 2.5 Ga, the stage was set for another major change in the Earth System.

FIGURE 9.3 As time progressed, the percentage of the Earth's surface area covered by continental crust increased. About 80% of crustal rock that exists today had formed by the end of the Archean; since then, most of it has passed through the rock cycle.

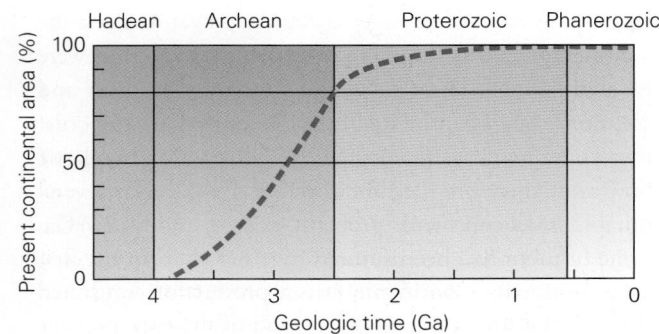

FIGURE 9.4 Examples of Archean life.

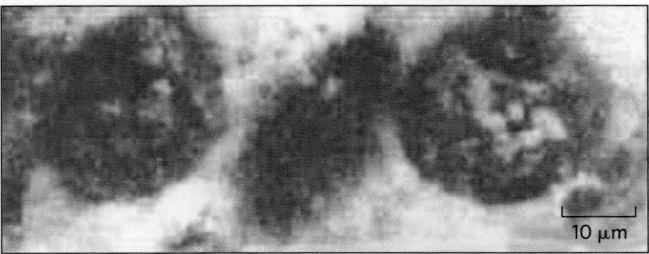

(a) These shapes, found in chert dated at 3.2 Ga, are thought to be fossil bacteria or archaea.

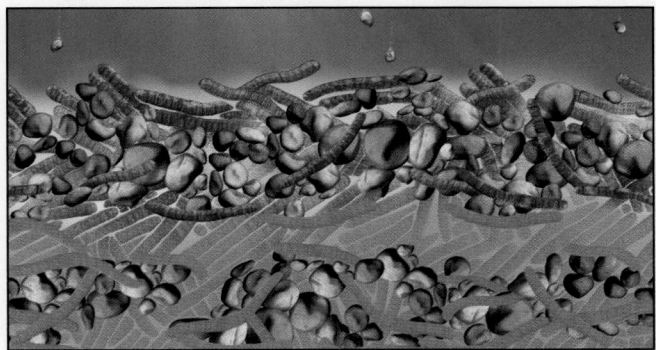

(b) Stromatolites form when sediment sticks to layers of cyanobacteria. As one layer dies and compacts, a new one grows above.

(c) This weathered outcrop of dolostone near Marquette, Michigan, reveals the layering in stromatolites. The ridges represent individual mats. Similar stromatolites occur in exposures of Archean rocks.

(d) Modern stromatolites in Shark Bay, Western Australia.

Take-home message...

During the Archean (4.0–2.5 Ga), the first continental crust formed from colliding island arcs and oceanic plateaus. As oceans filled with water, the atmosphere lost its H_2O and CO_2 and became nitrogen rich. Early life may have appeared at hydrothermal vents. Some early forms built mounds called stromatolites.

Quick Questions

- How does the first Archean continental crust differ from Hadean crust?
- What processes removed H_2O and CO_2 from the atmosphere during the Archean?
- What is a stromatolite?

9.3 The Proterozoic Eon (2.50–0.539 Ga): The Earth in Transition

The **Proterozoic Eon**, the last time interval of the Precambrian, spans roughly 2 billion years—almost half of the Earth's history—from 2.5 Ga to the beginning of the Cambrian Period at 539 Ma. During the Proterozoic, the Earth's surface environment changed from a strange world of small continents and an oxygen-poor atmosphere into a world of large continents, with an atmosphere containing a significant proportion of oxygen, that would seem much more familiar to us.

Large Continents Form

New continental crust continued to form during the Proterozoic, but at progressively slower rates. By the middle of the eon, over 90% of the Earth's present continental crustal material had formed (see Fig. 9.3). But a glance at a simplified crustal-province map of the Earth today shows that Precambrian rocks currently underlie much less than 90% of the continental crust (Fig. 9.5a). That's because during the more recent Phanerozoic Eon, much of the rock initially formed during the Precambrian has gone through various stages of the rock cycle, so the atoms in those very ancient rocks have since been incorporated in new Phanerozoic-aged rocks.

Recall from Chapter 6 that a block of crust that has not been pervasively affected by metamorphism, igneous activity, or plastic deformation for at least a billion years is called a **craton**. Cratons are relatively cool, strong, and stable. Modern continents consist of one or more cratons, bordered by either Phanerozoic orogens or Phanerozoic rifted crust. The youngest cratonic crust in existence today formed at about 1 Ga. Notably, each present-day craton

FIGURE 9.5 Different parts of the continental crust are of different ages.

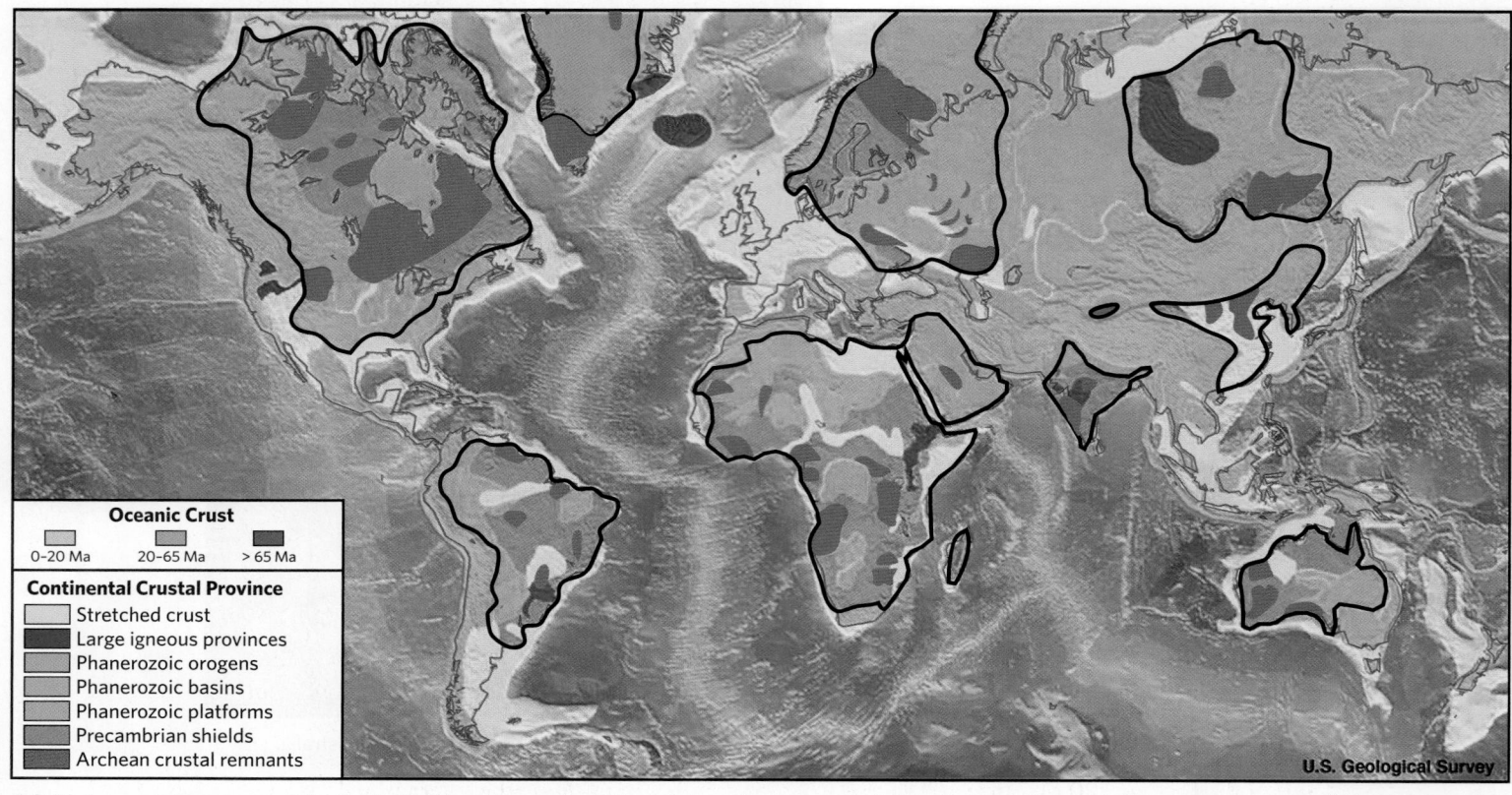

Oceanic Crust
0–20 Ma   20–65 Ma   > 65 Ma

Continental Crustal Province
Stretched crust
Large igneous provinces
Phanerozoic orogens
Phanerozoic basins
Phanerozoic platforms
Precambrian shields
Archean crustal remnants

U.S. Geological Survey

(a) Major crustal provinces of the Earth. The black lines indicate the borders of regions underlain by Precambrian crust. Phanerozoic mountain belts surround the Precambrian regions.

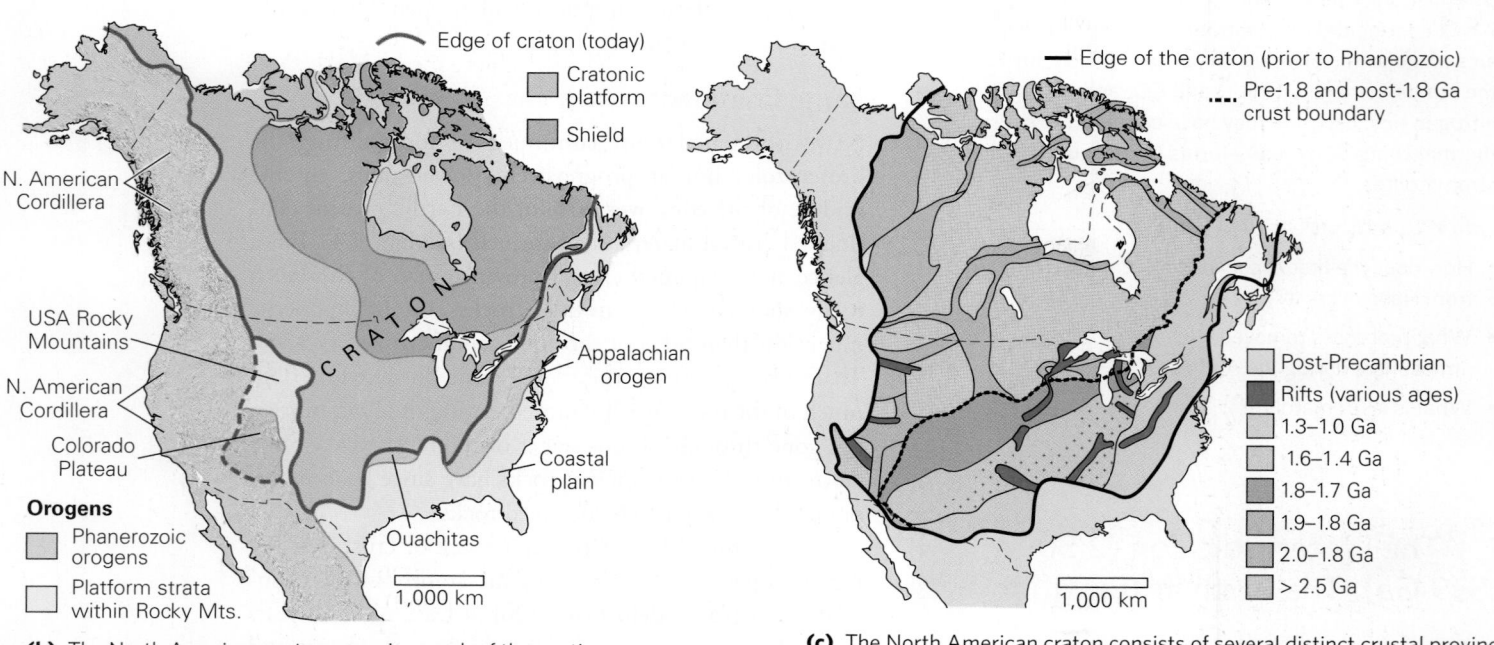

Edge of craton (today)
Cratonic platform
Shield

N. American Cordillera
USA Rocky Mountains
N. American Cordillera
Colorado Plateau

CRATON

Appalachian orogen
Coastal plain
Ouachitas

Orogens
Phanerozoic orogens
Platform strata within Rocky Mts.

1,000 km

(b) The North American craton occupies much of the continent's interior. In the shield, Precambrian rock is exposed at the surface. In the cratonic platform, Phanerozoic strata cover Precambrian rock.

Edge of the craton (prior to Phanerozoic)
Pre-1.8 and post-1.8 Ga crust boundary

Post-Precambrian
Rifts (various ages)
1.3–1.0 Ga
1.6–1.4 Ga
1.8–1.7 Ga
1.9–1.8 Ga
2.0–1.8 Ga
> 2.5 Ga

1,000 km

(c) The North American craton consists of several distinct crustal provinces, distinguished from one another by their radioisotopic ages. Most of what is now Canada and the northwestern United States had assembled into a fairly large continent by 1.8 Ga. Belts of younger rocks accreted to this continent between 1.8 and 1.1 Ga.

may consist of many once-separate Precambrian blocks, which served as smaller cratons before being incorporated into larger ones (Fig. 9.5b, c).

As we discussed in Chapter 2, the map of the Earth constantly changes due to plate interactions and the associated movement of continents. Using a variety of data sources, including correlation of rock units and studies of paleomagnetism, geologists have been able to provide constraints on these movements and plot them on **paleogeographic maps**, maps of the Earth's surface representing times in the past. Such maps emphasize that during some time intervals, the Earth's surface hosted several continents and smaller crustal blocks, but during others, collisions sutured together most continental crust to produce a **supercontinent**. Arguably the earliest large supercontinent, called **Nuna** (also known as *Columbia*), had formed by about 1.9 Ga and lasted until about 1.3 Ga, when rifting broke it apart. Continents that separated from Nuna dispersed until about 1.2 Ga, when they began to suture together into a new configuration.

By about 900 Ma, assembly of a new supercontinent, named **Rodinia**, was complete. On a paleogeographic map of Rodinia (Fig. 9.6), we can begin to identify a few crustal blocks that remain linked together

FIGURE 9.6 Rodinia had formed by 900 Ma, and it broke up between 800 and 600 Ma. Laurentia, which consists of the landmasses that are now North America and Greenland, is thought to have been near the center of Rodinia. Note the location of the Grenville orogen along the eastern margin of Laurentia. (RdP = Rio de la Plata; GL = Greenland.)

■ Grenville orogen
⌇ Reference coastline

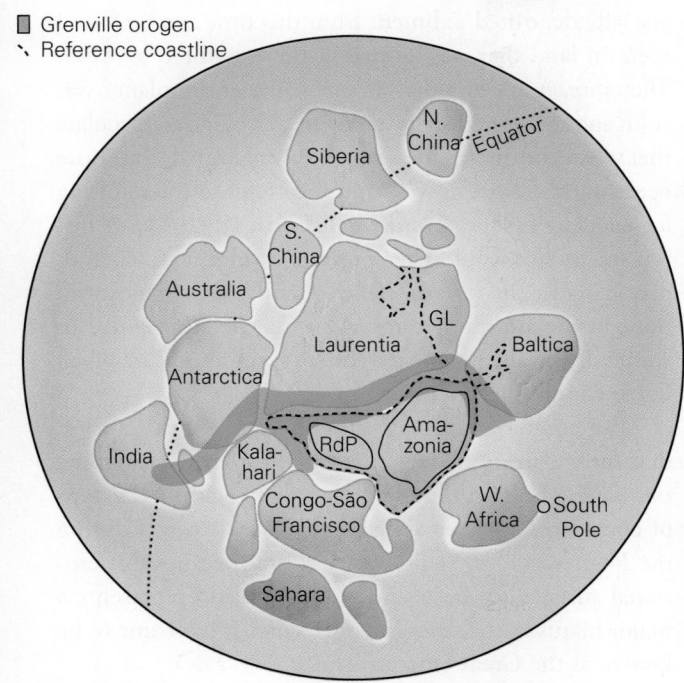

in the familiar continents of today. Not all geologists agree on the arrangement of continental crustal blocks in Rodinia, but in a favored interpretation, **Laurentia** (which consists of Precambrian North America, together with Greenland) lay near the center of the assembly. Antarctica and Australia were attached to what is now the western margin of North America. The last major collision during the formation of Rodinia, an event called the **Grenville orogeny**, produced the rocks of the Grenville orogen that now underlie parts of eastern and southern North America. Rodinia began to break apart about 200 million years after it formed. Antarctica and Australia, for example, broke away from Laurentia and eventually collided with India and what is now southern Africa. This assembly may have produced a short-lived supercontinent, known as *Pannotia*. But many geologists question this interpretation, for new dating suggests that the Grenville side of Laurentia rifted away from its neighbors before the collisions thought to have constructed Pannotia were complete.

Life and Climate in the Proterozoic World

Fossil evidence emphasizes that the Proterozoic saw important steps in the evolution of life. When this eon began, most life was *prokaryotic*, meaning that it consisted of single-celled organisms—archaea and bacteria—that do not contain a cell nucleus (the internal, membrane-surrounded region containing DNA). Though studies of chemical fossils hint that *eukaryotic* life (consisting of cells that do have nuclei) originated as early as 2.7 Ga, the first possible body fossil of a eukaryotic organism occurs in rocks dated at 2.1 Ga, and abundant body fossils of eukaryotic organisms can be found only in rocks younger than about 1.2 Ga. Therefore, establishment of eukaryotic life, the foundation from which complex organisms (including plants and animals) eventually evolved, took place during the Proterozoic.

Multicellular organisms, in the form of organized clusters of bacteria, had appeared by 2 Ga, as represented by spiral-shaped fossils called *Grypania spiralis*. The last half billion years of the Proterozoic Eon saw the remarkable transition from such simple organisms to *complex organisms*, multicellular organisms with distinct organs that carry out different functions. Sediments deposited between 600 Ma and 540 Ma contain fossils of several types of complex organisms that, together, constitute the **Ediacaran fauna** (named for the region in southern Australia where these fossils were first found). Some members of the Ediacaran fauna were soft-bodied invertebrate animals that resembled jellyfish or worms (Fig. 9.7a). Others may have been microbial or fungal communities. How Ediacaran organisms are related to organisms that exist today remains unclear.

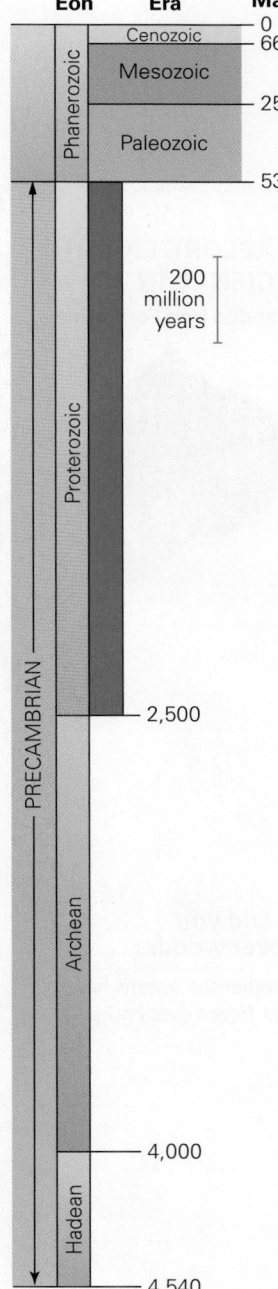

FIGURE 9.7 Late Proterozoic organisms and sediments.

(a) An artist's reconstruction of Ediacaran fauna. These organisms were complex invertebrates without shells.

(b) An outcrop of banded iron formation (BIF) from the Upper Peninsula of Michigan. The gray stripes are hematite and magnetite, and the red stripes are iron-bearing red chert (jasper).

EXPLORE EARTH SCIENCE IN 3D

Banded Iron Formation

Did you ever wonder...

whether the oceans have ever frozen over entirely?

The evolution of life played a key role in changing the composition of the Earth's atmosphere. Before life appeared, the atmosphere contained hardly any molecular oxygen (O_2). With the appearance of photosynthetic organisms, trace amounts of oxygen accumulated in the atmosphere, but it was not until about 2.4 Ga that oxygen concentration increased from less than 1% to 3% or more. That increase, called the **great oxygenation event**, happened when surface rocks and ocean water could no longer absorb or dissolve all the oxygen produced by organisms, so that oxygen began to accumulate as a gas in the atmosphere.

Evidence for the great oxygenation event comes from studying the composition of rocks. Sedimentary strata from before about 2.4 Ga do not contain iron oxide. But when ocean water began to contain sufficient dissolved oxygen, large quantities of iron that had been dissolved in seawater bonded to oxygen atoms to form solid iron oxide minerals that precipitated from the water and settled on the seafloor. In fact, between 2.4 Ga and 1.8 Ga, so much iron oxide settled out of the ocean that thick accumulations of sediment rich in iron oxide accumulated, then lithified to become colorful sequences of sedimentary rock known as **banded iron formation (BIF)**. BIF, which consists of layers of gray iron oxide minerals alternating with layers of red chert, provides most of the iron used by society today (Fig. 9.7b).

The existence of a large supercontinent at the end of the Proterozoic probably affected many aspects of the Earth System. For example, the thick continental lithosphere may have acted as a blanket, preventing escape of heat from the asthenosphere. Researchers hypothesize that as a consequence, both the asthenosphere and the lithospheric mantle beneath the continent warmed up

and became weaker and less dense. Consequently, the lithospheric mantle may have undergone *delamination*, meaning that parts of it peeled off the bottom of the plate and sank. Like the deck of a ship unloading heavy cargo, the surface of the lithosphere rose in order to maintain isostasy (see Chapter 6). Rock exposed at the surface then underwent chemical weathering, which involves chemical reactions that remove CO_2 from the atmosphere. As we will see in Chapter 19, CO_2 is a greenhouse gas, meaning that it traps heat in the atmosphere. Removal of CO_2, therefore, may have triggered cooling of the Earth's climate at the end of the Proterozoic Eon. Due to this cooling, large ice sheets grew on continents.

What's strange about this Proterozoic ice age is that glacially deposited sediment from this time can be found even on land that was located at the equator (Fig. 9.8a). Therefore, it appears that at times, the entire planet was cold enough for glaciers to form. Geologists speculate that when continents became entirely glaciated, the entire ocean surface froze as well, and they refer to the resulting ice-encrusted planet as **snowball Earth** (Fig. 9.8b). A frozen sea surface would have prevented volcanic CO_2 from dissolving in the oceans. Consequently, CO_2 concentrations in the atmosphere rose, causing the atmosphere to warm. This warming melted the ice and brought snowball Earth conditions to a close.

The combination of weathering and glacial erosion that took place near the end of the Proterozoic, by some estimates, shaved a few kilometers of rock off the tops of continents. This erosional surface was covered during the Phanerozoic by sedimentary strata. Because this erosional surface can be found worldwide and represents a major hiatus in the stratigraphic record, it has come to be known as the **Great Unconformity** (Fig. 9.8c).

Strata contain large clasts surrounded by mudstone, for glaciers can carry clasts of all sizes.

Bedding

(a) Layers of glacially deposited sediment (till) occur in regions that were at low elevation and near the equator in the late Proterozoic.

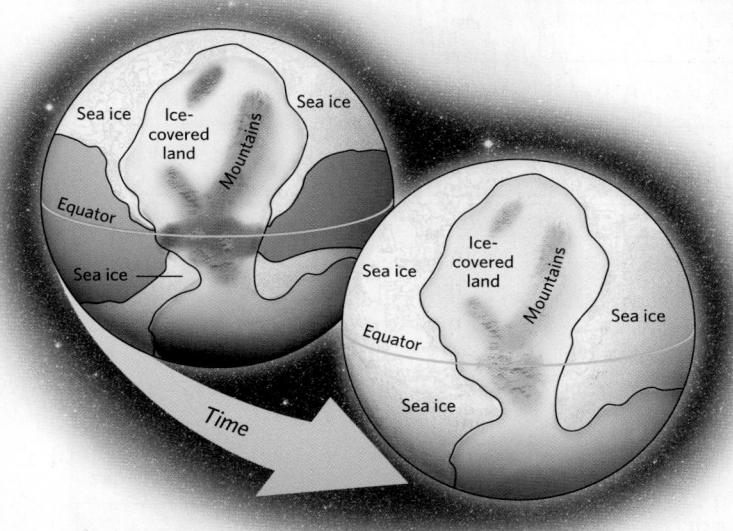

(b) During Proterozoic ice ages, the planet may have frozen over completely to form snowball Earth.

FIGURE 9.8 Snowball Earth.

Paleozoic sandstone (505 Ma)

bedding

Great Unconformity

Hiatus of ~1.25 billion years

Person

Weathered Precambrian gneiss

Unweathered Precambrian gneiss (1.75 Ga)

Gneissic banding

(c) At a cliff exposure of the Great Unconformity in the Grand Canyon, horizontally bedded Cambrian sandstone overlies 1.8 Ga gneiss with vertical gneissic banding. A zone of weathered gneiss lies directly beneath the sandstone. The weathering happened before the sand was deposited.

9.4 The Paleozoic Era (539–252 Ma): Continents Reassemble and Life Diversifies

The end of the Proterozoic Eon defines the end of the Precambrian and the start of the **Phanerozoic Eon**. Geologists studying the fossil record recognized the significance of this boundary long before they could assign it a numerical age (currently, 539 Ma) because it marks the appearance of invertebrates with shells. Shells (external skeletons), which probably evolved as a means of protection against predators, have a much higher preservation potential than do the soft bodies of shell-less invertebrates. The atmosphere also changed as photosynthetic organisms prospered, and O_2 accounted for about 17% of atmospheric gas (compared with 21% today) as the Phanerozoic began. Increasing concentrations of O_2 in the atmosphere had an important effect on the Earth System: chemical reactions in air produced ozone (O_3), a gas that absorbs dangerous solar radiation and therefore shields the Earth's surface from this radiation. Without O_3 in the air, the land was not

Take-home message...

During the Proterozoic (2.5 Ga–539 Ma), large continental crustal areas became cratons. Multicellular organisms appeared, and the atmosphere began to accumulate oxygen. Supercontinents formed and broke apart during the Proterozoic. Global glaciation happened at the end of the eon.

Quick Questions

• Name a Proterozoic supercontinent.

• What was the great oxygenation event?

• What does the term *snowball Earth* refer to?

FIGURE 9.9 Land and sea in the early Paleozoic Era.

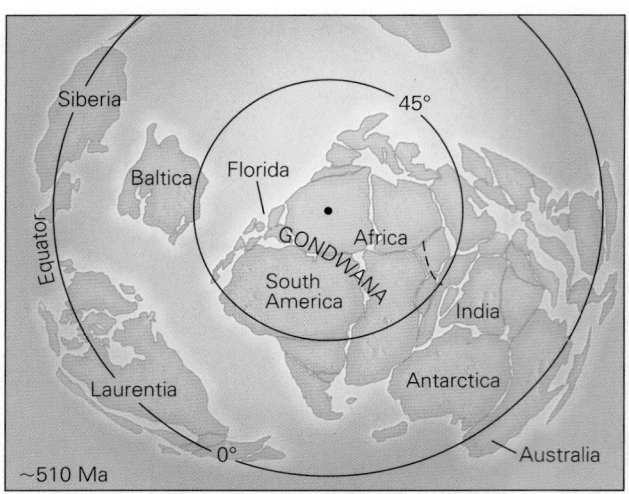

(a) The distribution of continents in the Cambrian Period at about 510 Ma, as viewed looking down on the South Pole.

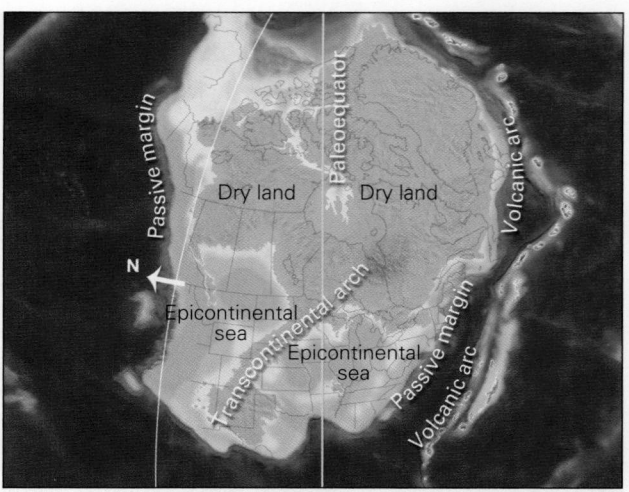

(b) A paleogeographic map of Laurentia at 500 Ma shows that large areas of the continent were covered by epicontinental seas. Present-day state borders provide spatial reference.

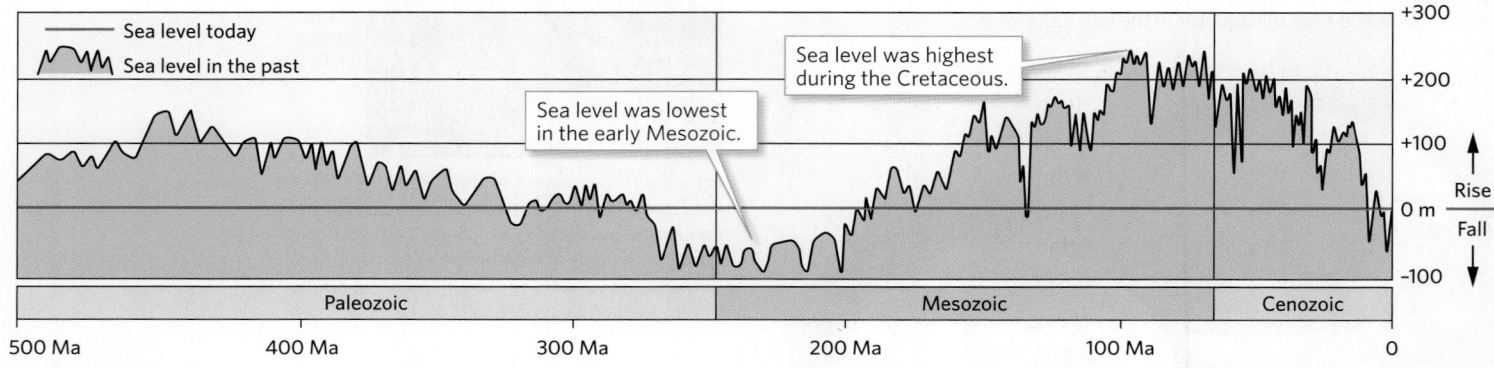

(c) A chart of transgressions and regressions in the stratigraphic record that may indicate relative sea-level change during the past half billion years.

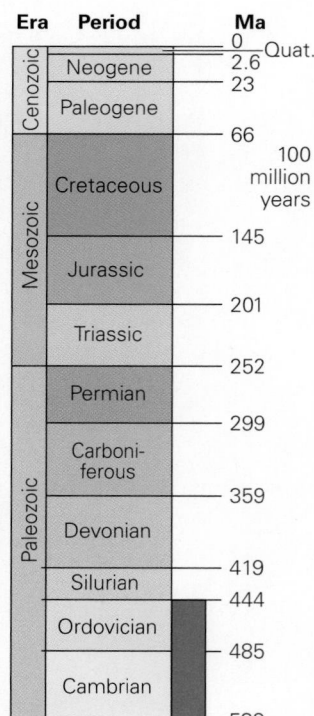

Era	Period	Ma
Cenozoic	Neogene	0 — Quat.
		2.6
	Paleogene	23
		66
Mesozoic	Cretaceous	100 million years
		145
	Jurassic	201
	Triassic	252
Paleozoic	Permian	299
	Carboni-ferous	359
	Devonian	419
	Silurian	444
	Ordovician	485
	Cambrian	539

habitable, but once it had appeared, plants, and then animals, could begin to populate the land.

The Phanerozoic Eon consists of three eras (**Earth Science at a Glance**, pp. 330–331): *Paleozoic* (Greek for ancient life), *Mesozoic* (middle life), and *Cenozoic* (recent life).

The Early Paleozoic Era (Cambrian and Ordovician Periods: 539–444 Ma)

PALEOGEOGRAPHY. Rifting at the end of the Proterozoic had yielded several smaller, but still large, continents. So a map of the Earth in the Cambrian would display the following: Laurentia, **Gondwana** (a combination of South America, Africa, Arabia, Antarctica, India, and Australia), *Baltica* (Europe), and *Siberia* (**Fig. 9.9a**). When rifting ceased, several new passive-margin basins were established. In fact, passive-margin basins existed on both sides of Laurentia during the Cambrian. These basins collected immense amounts of sediment eroded from the

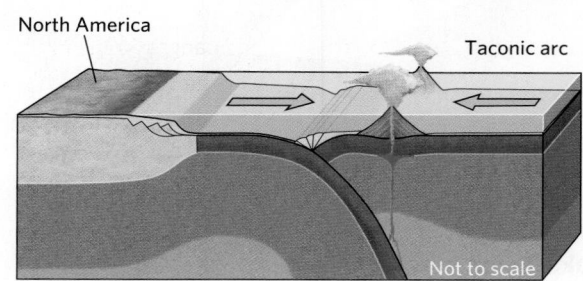

(d) During the Middle Ordovician, an island arc approached the eastern margin of North America as the oceanic plate between the arc and the continent was subducted. Collision of this arc with the east-coast passive margin of North America caused the Taconic orogeny.

continent, and their surfaces became broad continental shelves.

At times during the early Paleozoic, sea level rose as the climate warmed, so large regions of continental interiors became flooded with **epicontinental seas**, shallow seas that submerged continental crust (**Fig. 9.9b**). The existence

BOX 9.1

How can I explain . . .

Sea-level change

What are we learning?

That sea level can rise and fall relative to the land surface over time, and some of the reasons for this change.

What you need:

- A plastic basin about 15 cm (6 in) deep, with an area of about 30 × 40 cm (1 ft²)
- A standard brick
- A small measuring cup and a ladle
- Five stones (each 6–10 cm across)
- Two pencils

Instructions:

- Place the brick in the middle of the basin.
- Fill the basin with water to just slightly below the top surface of the brick.
- To simulate a rise in sea level due to growth of the volume of a mid-ocean ridge, line up the stones in the basin. As the "ridge" grows, the water surface rises. Eventually, the water submerges the brick.
- To simulate a fall in sea level, remove the stones. Water takes the place of the stones, and the top of the brick re-emerges.
- To simulate growth of continental glaciers and the resulting sea-level fall, place the cup on the brick, then ladle some water out of the basin and place it in the cup. Glaciers store water on land, so the amount of water in the sea decreases, and the water surface drops.
- To simulate melting of continental glaciers and the resulting sea-level rise, pour the water from the cup back into the basin.
- To simulate a tectonic rise of the continent relative to sea level, place the pencils underneath the brick so that its surface rises.

What did we see?

Sea-level changes may be due to changes in the capacity of ocean basins to hold water, removal of water from the oceans and its storage on land, or local tectonic events that cause the land to rise or fall.

Brick = continent   Water = oceans

Brick becomes submerged   Rocks = formation of mid-ocean ridge

Cup represents the glacial reservoir

When water is stored in glaciers on land, sea level drops.

of many transgressions and regressions (see Chapter 5) in the stratigraphic record of continental interiors indicates that sea level rose and fell many times, relative to the land surface, over geologic time. Geologists have developed a chart illustrating such changes (Fig. 9.9c). This chart remains controversial, however, because of the difficulty in distinguishing between changes in water level and changes in land elevation. Geologic phenomena that might affect sea level relative to land include the following (Box 9.1):

- *Growth or melting of large glaciers:* Formation of a glacier removes water from the sea and traps it on land, causing sea level to fall. The melting of glaciers causes sea level to rise.

- *A change in the volume of ocean basins:* If the volume of ocean basins decreases, sea level rises and the land floods. Such changes in volume may be due to a change in the number of mid-ocean ridges or in the spreading rate along ridges, because the surface of

The Earth Has a History

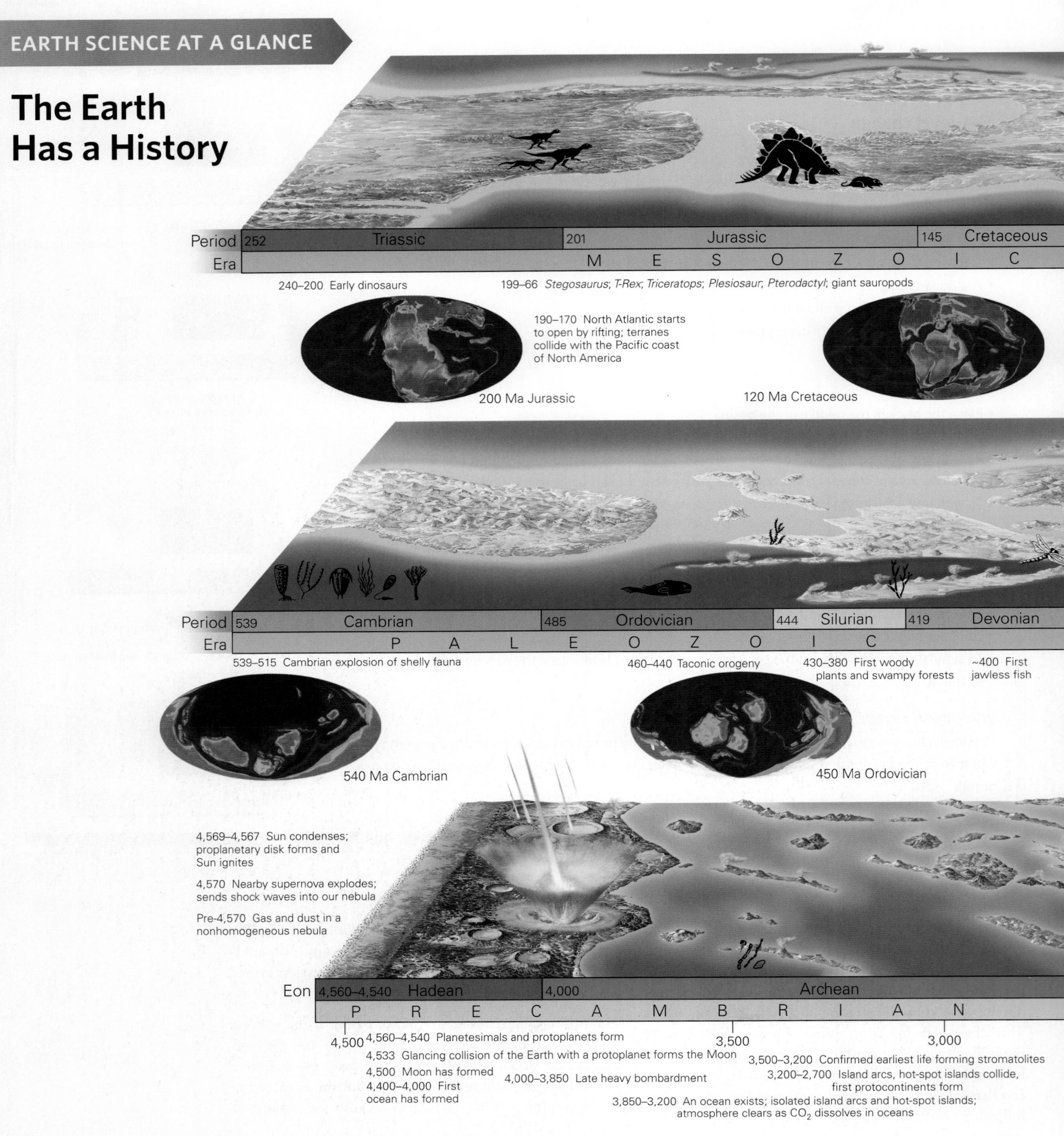

Period | 252 | Triassic | 201 | Jurassic | 145 | Cretaceous
Era | | M E S O Z O I C

240–200 Early dinosaurs

199–66 *Stegosaurus*; *T-Rex*; *Triceratops*; *Plesiosaur*; *Pterodactyl*; giant sauropods

190–170 North Atlantic starts to open by rifting; terranes collide with the Pacific coast of North America

200 Ma Jurassic

120 Ma Cretaceous

Period | 539 | Cambrian | 485 | Ordovician | 444 | Silurian | 419 | Devonian
Era | | P A L E O Z O I C

539–515 Cambrian explosion of shelly fauna

460–440 Taconic orogeny

430–380 First woody plants and swampy forests

~400 First jawless fish

540 Ma Cambrian

450 Ma Ordovician

4,569–4,567 Sun condenses; proplanetary disk forms and Sun ignites

4,570 Nearby supernova explodes; sends shock waves into our nebula

Pre-4,570 Gas and dust in a nonhomogeneous nebula

Eon | 4,560–4,540 | Hadean | 4,000 | Archean
| P R E C A M B R I A N

4,500

4,560–4,540 Planetesimals and protoplanets form

4,533 Glancing collision of the Earth with a protoplanet forms the Moon

4,500 Moon has formed

4,400–4,000 First ocean has formed

4,000–3,850 Late heavy bombardment

3,500

3,000

3,500–3,200 Confirmed earliest life forming stromatolites

3,200–2,700 Island arcs, hot-spot islands collide, first protocontinents form

3,850–3,200 An ocean exists; isolated island arcs and hot-spot islands; atmosphere clears as CO_2 dissolves in oceans

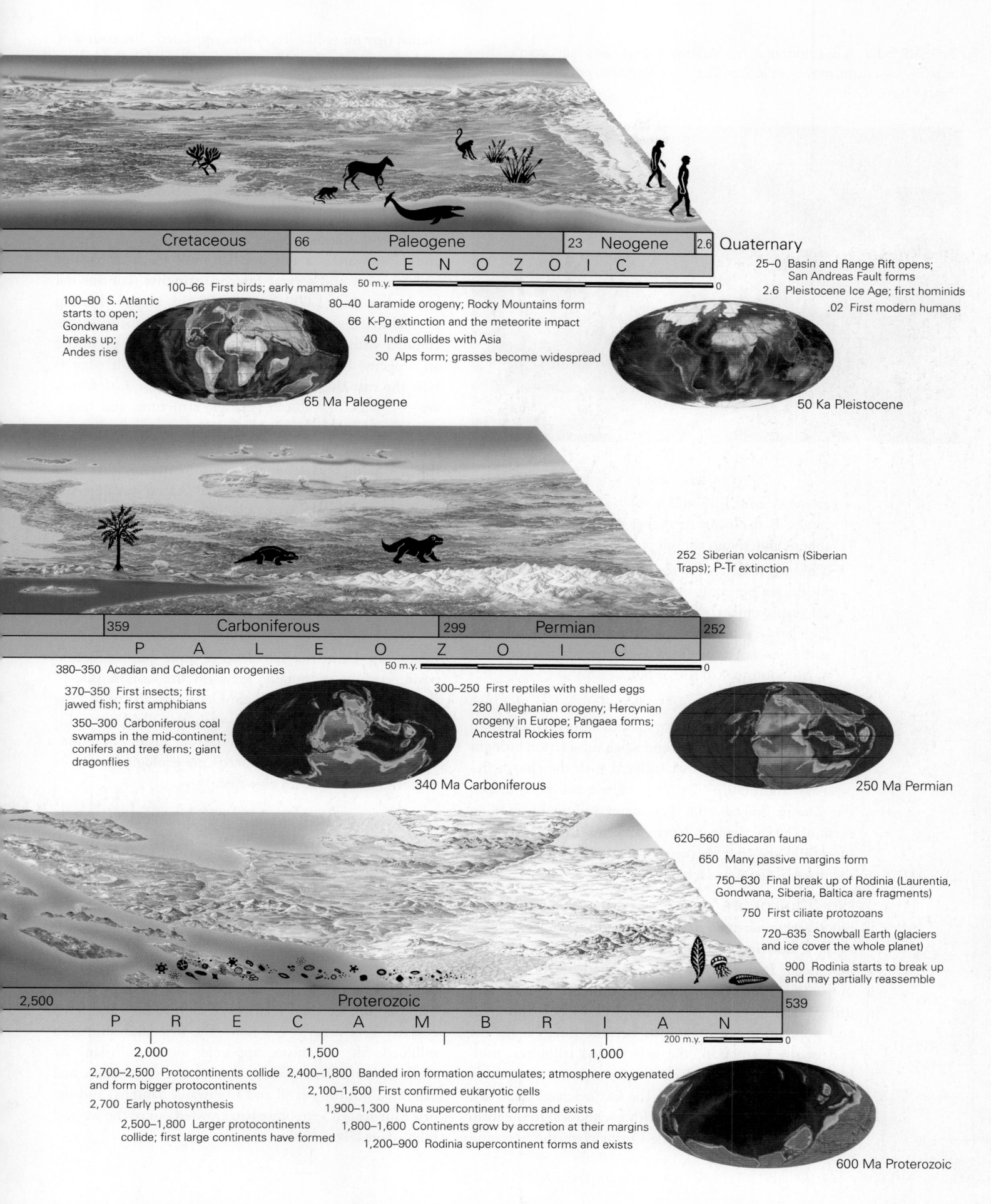

Cretaceous	66	Paleogene	23	Neogene	2.6	Quaternary

C E N O Z O I C

100–66 First birds; early mammals

50 m.y. ⊢━━━━━━┤ 0

100–80 S. Atlantic starts to open; Gondwana breaks up; Andes rise

80–40 Laramide orogeny; Rocky Mountains form
66 K-Pg extinction and the meteorite impact
40 India collides with Asia
30 Alps form; grasses become widespread

25–0 Basin and Range Rift opens; San Andreas Fault forms
2.6 Pleistocene Ice Age; first hominids
.02 First modern humans

65 Ma Paleogene

50 Ka Pleistocene

252 Siberian volcanism (Siberian Traps); P-Tr extinction

359	Carboniferous	299	Permian	252

P A L E O Z O I C

380–350 Acadian and Caledonian orogenies

50 m.y. ⊢━━━━━━┤ 0

370–350 First insects; first jawed fish; first amphibians

350–300 Carboniferous coal swamps in the mid-continent; conifers and tree ferns; giant dragonflies

300–250 First reptiles with shelled eggs

280 Alleghanian orogeny; Hercynian orogeny in Europe; Pangaea forms; Ancestral Rockies form

340 Ma Carboniferous

250 Ma Permian

620–560 Ediacaran fauna

650 Many passive margins form

750–630 Final break up of Rodinia (Laurentia, Gondwana, Siberia, Baltica are fragments)

750 First ciliate protozoans

720–635 Snowball Earth (glaciers and ice cover the whole planet)

900 Rodinia starts to break up and may partially reassemble

2,500	Proterozoic	539

P R E C A M B R I A N

200 m.y. ⊢━━━━━┤ 0

2,000   1,500   1,000

2,700–2,500 Protocontinents collide and form bigger protocontinents

2,700 Early photosynthesis

2,500–1,800 Larger protocontinents collide; first large continents have formed

2,400–1,800 Banded iron formation accumulates; atmosphere oxygenated
2,100–1,500 First confirmed eukaryotic cells
1,900–1,300 Nuna supercontinent forms and exists
1,800–1,600 Continents grow by accretion at their margins
1,200–900 Rodinia supercontinent forms and exists

600 Ma Proterozoic

331

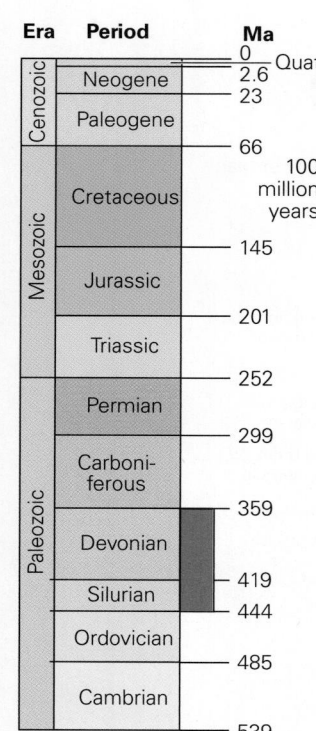

young oceanic lithosphere, such as that of mid-ocean ridges, is shallower than that of older oceanic lithosphere underlying abyssal plains.

- *A change in the area or elevation of continental surfaces:* Mountain building, erosion and deposition, or vertical displacement of a continent caused by processes taking place beneath the lithosphere can cause the surface area of a continent to change and the elevation of the continent to rise or fall.

In the Middle Ordovician Period, however, the geologic tranquility of the passive-margin basin on the eastern side of Laurentia came to an end when subduction brought in volcanic island arcs that collided with the continent's passive-margin basin (Fig. 9.9d). These collisions caused the *Taconic orogeny*, the first major mountain-building event of the Paleozoic in North America. Rocks and structures formed during this orogeny can still be found within the present-day Appalachians.

EVOLUTION OF LIFE. Soon after the appearance of shells, biodiversity increased dramatically. This increase, known as the **Cambrian explosion**, continued over several million years. It may have been triggered by the breakup of Rodinia, for when smaller continents formed and drifted apart, many new ecological niches appeared, and populations of organisms became isolated. By the end of the Cambrian, the seas hosted trilobites, mollusks, brachiopods, nautiloids, gastropods, graptolites, and echinoderms (Fig. 9.10). The Ordovician Period saw the first vertebrates (animals with backbones) in the form of jawless fish. In the Middle Ordovician, the first land

plants, tiny moss-like liverworts, appeared. The course of evolution changed suddenly, however, at the end of the Ordovician because of a mass-extinction event, the first of the *top five mass extinctions* (see Fig. 8.18). Over a short interval of time, 85% of marine species disappeared.

The Middle Paleozoic Era (Silurian and Devonian Periods: 444–359 Ma)

PALEOGEOGRAPHY. When the Silurian Period began, land was divided among two very large continents (Laurentia and Gondwana), a few smaller ones (such as Baltica and Siberia), and several **microcontinents** (continental blocks less than 1,000 km, or 600 mi, across). During the Silurian, smaller continents and microcontinents collided with larger ones, and each collision produced an orogeny. For example, Baltica collided with what is now the east side of Greenland, causing the *Caledonian orogeny*, which produced geologic structures and metamorphic rocks that can be found in western Scandinavia, eastern Greenland, and Scotland (Fig. 9.11a). Soon after this, Avalonia and other microcontinents collided with what is now the eastern United States, causing the *Acadian orogeny* (Fig. 9.11b). Sediment derived by erosion of the resulting mountain belt was carried westward and deposited in huge deltaic fans on continental crust. Red sandstones and conglomerates of one of these sediment fans form the bedrock of the eastern Catskill Mountains of New York State (Fig. 9.11c).

Through most of the middle Paleozoic, the western margin of Laurentia remained a quiet passive-margin basin, even as mountains grew on the eastern side of the continent. But in the Late Devonian, the west-coast passive-margin basin collided with an island arc. This event, the *Antler orogeny*, began the long history of mountain building that has dominated the geologic history of western North America ever since.

EVOLUTION OF LIFE. Middle Paleozoic seas not only welcomed new species of marine invertebrates, which replaced species that disappeared during the mass-extinction event at the end of the Ordovician Period, but also hosted jawed fish such as sharks. Even more dramatic changes took place on land. Vascular plants, which have woody tissues, seeds, and veins for transporting water and food, rooted on land for the first time in the Silurian, and by the Late Devonian, swampy forests made up of tree-sized relatives of club mosses and ferns covered large areas (Fig. 9.12a). The first land animal, a millipede-like organism, appeared in the Silurian. By the Late Devonian, spiders, insects, and crustaceans had exploited both dry-land and freshwater habitats, and the first four-legged vertebrate had crawled out of the sea and inhaled air (Fig. 9.12b).

The geologic time scale table (left column):

Era	Period	Ma
Cenozoic	Neogene	0 — Quat. / 2.6
Cenozoic	Paleogene	23
Cenozoic	Paleogene	66
Mesozoic	Cretaceous	100 million years
Mesozoic	Cretaceous	145
Mesozoic	Jurassic	201
Mesozoic	Triassic	252
Paleozoic	Permian	299
Paleozoic	Carboniferous	359
Paleozoic	Devonian	419
Paleozoic	Silurian	444
Paleozoic	Ordovician	485
Paleozoic	Cambrian	539

FIGURE 9.11 Paleogeography of middle Paleozoic time.

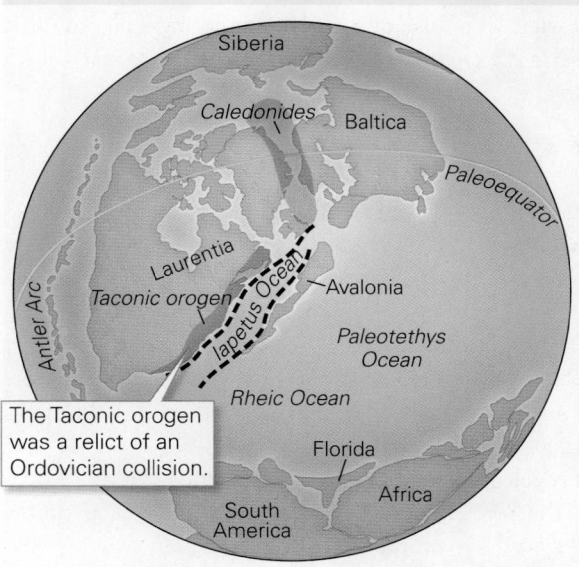

The Taconic orogen was a relict of an Ordovician collision.

(a) During the Caledonian and Acadian orogenies, Laurentia collided with Baltica and Avalonia. Meanwhile, the Antler island arc formed off Laurentia's west coast.

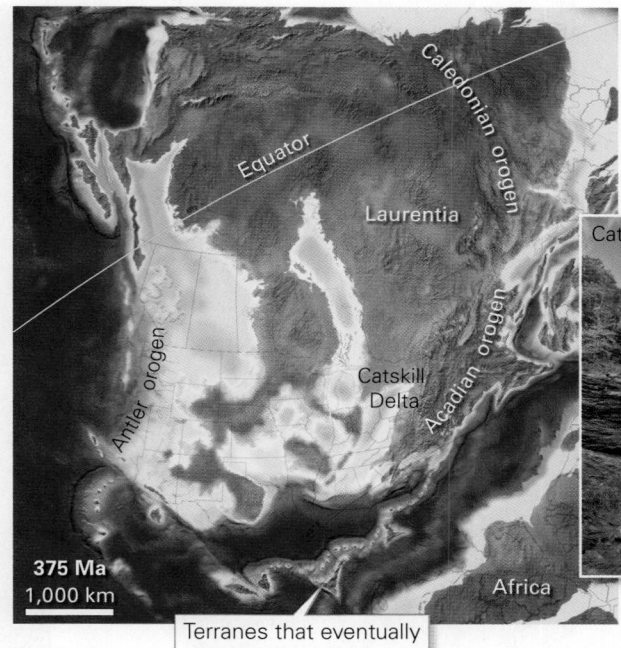

(b) During the Devonian, shallow seas covered parts of North America's interior. The Acadian orogen shed sediment westward to form the Catskill Delta.

Catskill Delta redbeds

375 Ma
1,000 km

Terranes that eventually attach to Laurentia

(c) Strata of the eastern Catskill Mountains consist of redbeds, lithified sediments from river deposits.

FIGURE 9.12 Examples of middle Paleozoic life.

(a) Vascular plants appeared in the Devonian, allowing the first forests to take root.

(b) This Late Devonian lobe-finned fish, *Tiktaalik*, was one of the first vertebrates to walk on land.

FIGURE 9.13 Paleogeography of the late Paleozoic.

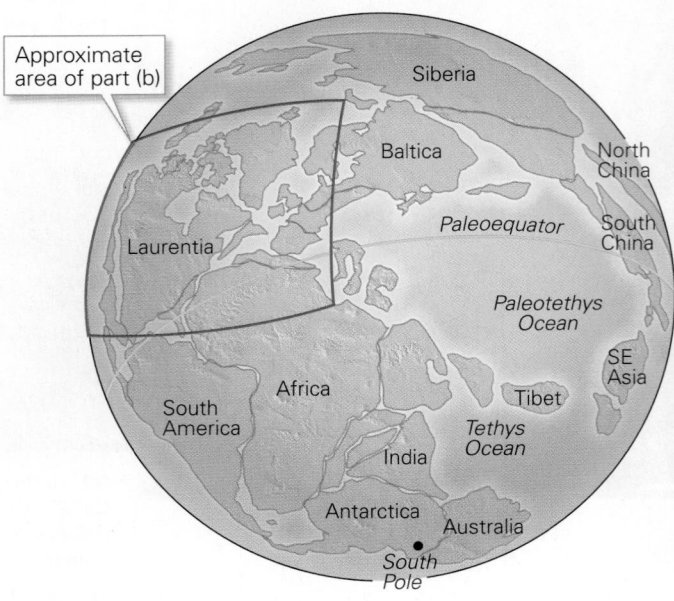

(a) At the end of the Paleozoic, almost all land had combined into a single supercontinent, called Pangaea.

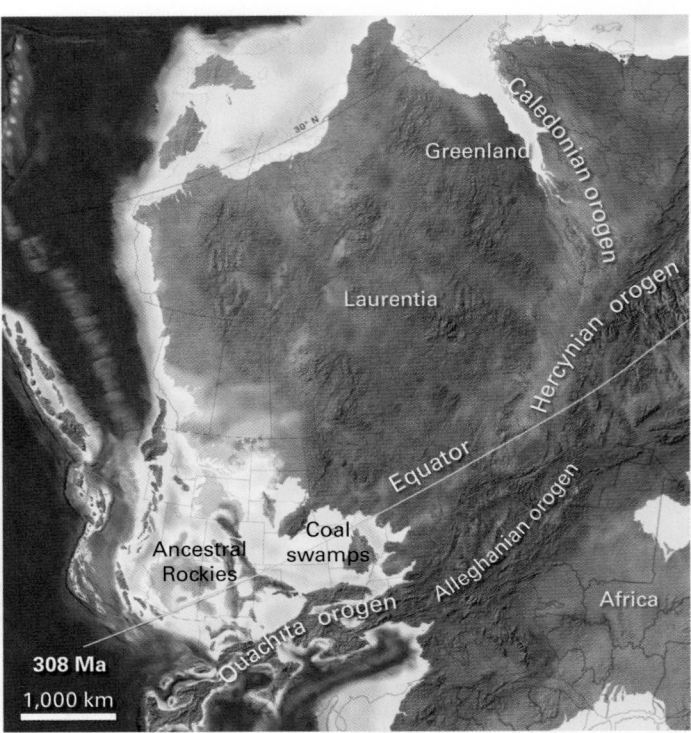

(b) During the Alleghanian and Hercynian orogenies, a huge mountain belt formed. Coal swamps bordered epicontinental seas. The Ancestral Rockies rose in the western part of the continent.

The Late Paleozoic Era (Carboniferous and Permian Periods: 359–252 Ma)

PALEOGEOGRAPHY. The late Paleozoic Era saw another succession of continental collisions, culminating in the formation of Alfred Wegener's supercontinent, **Pangaea** (Fig. 9.13a). The largest and last of these collisions occurred when Gondwana rammed into Laurentia and Baltica. This event caused the *Alleghanian orogeny* and built a vast mountain belt whose eroded remnants crop out in the Appalachian Mountains of the eastern United States (Fig. 9.13b). On the continental side of this mountain belt, compression generated the *Appalachian fold-thrust belt*, in which the sedimentary layer of crust shortened horizontally by as much as 50% (Fig. 9.14). Collision at this time also produced the Ouachita orogen in the south-central United States (the roots of which can be found in the Ouachita Mountains of Arkansas and Oklahoma), and the Hercynian orogen (also known as the Variscan orogen) in Europe. During the Alleghanian orogeny, pre-existing faults in the North American craton became active again. Movement on these faults produced narrow highlands bordered by narrow sediment-filled basins. Locally, this faulting also caused sedimentary layers to fold into monoclines. Several of the highlands and basins developed in the

region presently occupied by today's topographic Rocky Mountains. Because of their location, this set of late-Paleozoic highlands is known as the *Ancestral Rockies* (see Fig. 9.13b). At the conclusion of the Alleghanian and related orogenies, almost all land had coalesced to form Pangaea.

EVOLUTION OF LIFE. The fossil record indicates that during the late Paleozoic Era, plants and animals continued to evolve toward forms that are familiar to us today. The vegetation in huge Carboniferous swamps produced so much O_2 that for a while, this gas accounted for about a third of the gas in the atmosphere. This growth also left thick accumulations of plant debris that were eventually transformed into coal after burial. In these swamps, insects with fixed wings flew through a tangle of ferns, club mosses, and scouring rushes (Fig. 9.15a). The largest insects, giant dragonflies, buzzed through the air with a wingspan of over a meter. Their size was possible because the air was so rich in oxygen. Forests containing *gymnosperms* ("naked-seeded" plants such as conifers) and cycads (trees with palm-like stalks and fern-like fronds) became widespread in the Permian. Amphibians populated the land, and reptiles appeared and rapidly diversified (Fig. 9.15b). The success of reptiles on land relied on a radically new component of

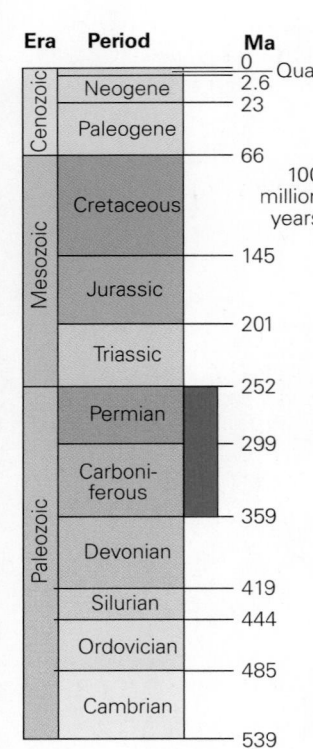

Era	Period	Ma	
Cenozoic		0	Quat.
	Neogene	2.6	
		23	
	Paleogene		
		66	100 million years
Mesozoic	Cretaceous		
		145	
	Jurassic		
		201	
	Triassic		
		252	
Paleozoic	Permian		
		299	
	Carboniferous		
		359	
	Devonian		
		419	
	Silurian		
		444	
	Ordovician		
		485	
	Cambrian		
		539	

FIGURE 9.14 Features of the Appalachian Mountains in the eastern United States.

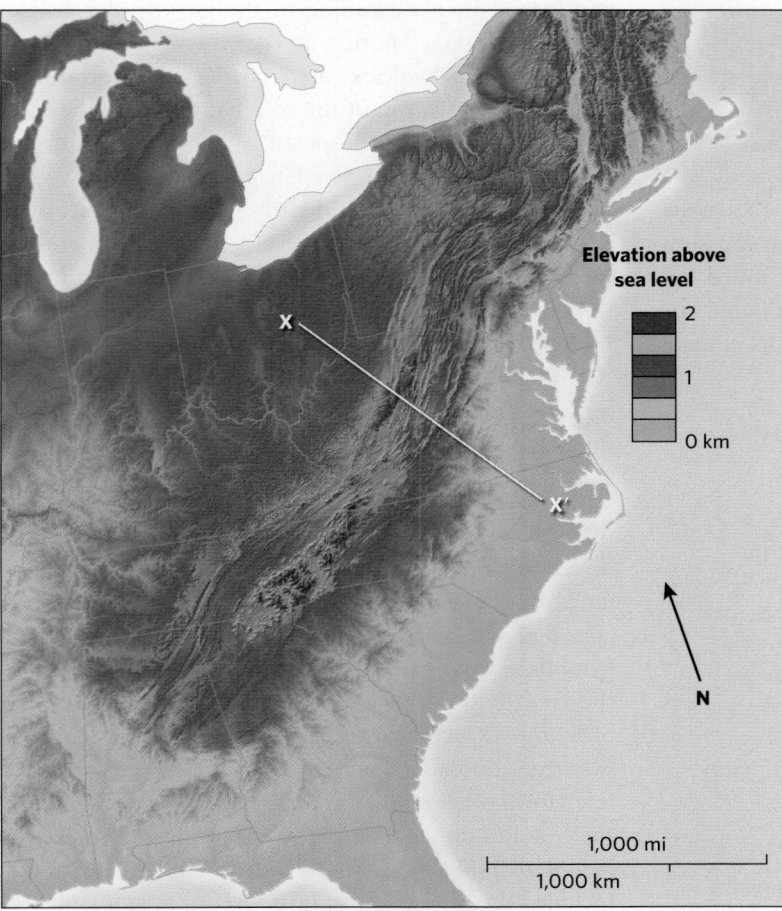

Elevation above sea level

2

1

0 km

1,000 mi

1,000 km

N

(a) The eroded remnants of the Appalachian orogen stand out in this representation of the topography of the eastern United States. Section line X–X' shows the position of the cross section in part (b). Note that the coastal plain, a region of very low relief, overlies part of the orogen's internal zone.

(b) A simplified cross section of the upper crust in the modern Appalachians shows deformation due to Paleozoic orogenies. In the Appalachian fold-thrust belt, strata have been pushed westward in overlapping thrust slices, like layers of sand in front of a bulldozer.

FIGURE 9.15 Life in the late Paleozoic.

(a) A museum diorama of a Carboniferous coal swamp. The inset shows the size of a giant dragonfly from the Paleozoic (wingspan of about 1 m) relative to the size of a human.

(b) By the Permian, reptiles such as *Dimetrodon* inhabited the land.

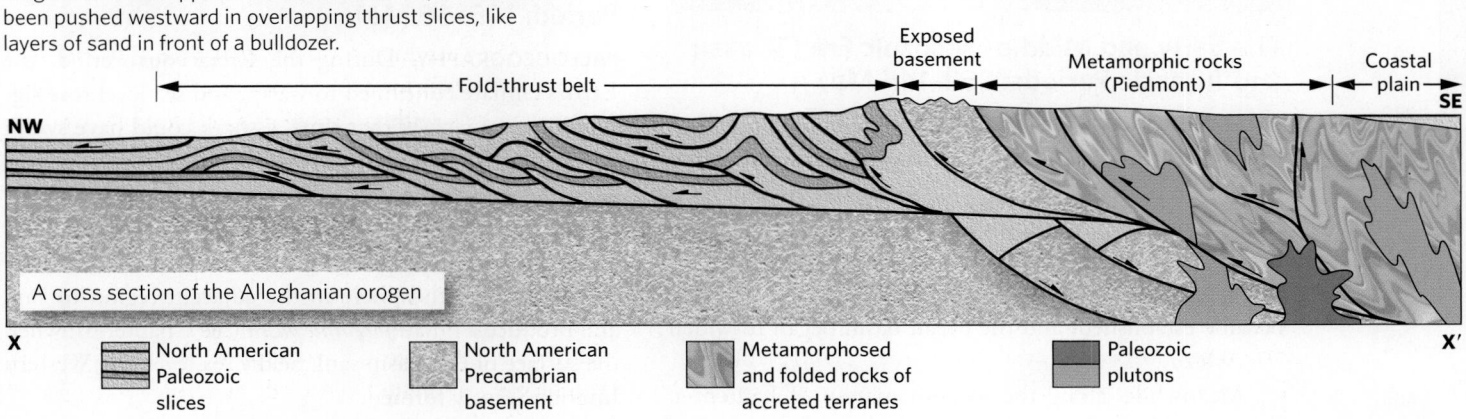

Exposed basement

Metamorphic rocks (Piedmont)

Coastal plain

Fold-thrust belt

NW

SE

A cross section of the Alleghanian orogen

X

X'

North American Paleozoic slices

North American Precambrian basement

Metamorphosed and folded rocks of accreted terranes

Paleozoic plutons

animal reproduction: eggs with a watertight protective covering.

The Paleozoic Era came to a close at about 252 Ma with the devastating *Permian-Triassic mass-extinction event* (known as the *P- Tr extinction*), during which over 96% of marine species and 70% of terrestrial species disappeared. This event has come to be known as the **Great Dying**, for it was the most severe mass extinction of the top five. According to one hypothesis, the event followed the eruption of immense flood basalts in Siberia. This eruption could have led to the emission of gases that clouded the atmosphere and acidified the oceans, making many environments inhospitable for life, and over a longer term, changed the Earth's climate significantly.

Take-home message...

Breakup of a late Precambrian supercontinent ushered in the Paleozoic Era. During the Paleozoic, shallow seas covered continents at times, and life diversified and moved onto land. Several collisional orogenies sutured continents together, and as the era ended, Pangaea had formed. The fossil record shows that the largest mass-extinction event in geologic history marks the end of the Paleozoic.

Quick Questions ─────────────

- Were the boundaries of Laurentia during the Cambrian Period active margins or passive margins?
- What event brought an end to the passive margin of eastern Laurentia?
- During what period did organisms move onto land?

9.5 The Mesozoic Era (252–66 Ma): When Dinosaurs Ruled

The Early and Middle Mesozoic Era (Triassic and Jurassic Periods: 252–145 Ma)

PALEOGEOGRAPHY. As we've seen, supercontinents don't last forever. Pangaea, which had assembled at the end of the Paleozoic, stayed together for about 100 million years until, in the Late Triassic and Early Jurassic, rifting began between North America and Africa. By the end of the Jurassic, the Mid-Atlantic Ridge of the North Atlantic became established, and the ocean basin began to widen (Fig. 9.16a).

Meanwhile, along the western margin of Laurentia, convergent-boundary tectonics became the order of the day. Beginning in the Late Permian and continuing

through the Mesozoic, trenches developed along and off the west coast. As the ocean floor was consumed, offshore island arcs and oceanic plateaus collided with Laurentia's continental margin, leading to the growth of a belt of accreted terranes. Starting at the end of the Jurassic, a large continental volcanic arc, known as the **Sierran Arc**, began to grow on top of the accreted terranes. The roots of this arc, when later exposed by erosion, became the Sierra Nevada batholith (see Chapter 4).

During the Triassic and Early Jurassic, the Earth had a relatively warm climate, so glaciers melted away entirely in polar regions, and sandy deserts covered portions of western North America. The large dunes of these deserts lithified to become the colorful red sandstones exposed in Zion National Park (Fig. 9.16b). In the Middle Jurassic, sea level began to rise, and epicontinental seas once again flooded portions of the continent.

EVOLUTION OF LIFE. During the early Mesozoic Era, many new plant and animal species appeared. Reptiles diversified on land, but some species lived in marine environments and others took to the air. Colonial corals appeared and built extensive reefs in shallow water, and gymnosperms became widespread on land. At the end of the Triassic, small rat-like creatures, the earliest ancestors of mammals, scurried through the underbrush, and the first true dinosaurs appeared (Fig. 9.17). **Dinosaurs** differed from other reptiles in that their legs extended under their bodies rather than off to the sides, and some species were probably warm blooded. This diverse group included the largest land animals ever to exist. By the end of the Jurassic, 100-ton sauropod dinosaurs with long necks and tails, along with other familiar monsters such as *Stegosaurus*, thundered across the landscape, and the first feathered birds, such as *Archaeopteryx*, flew between trees in the forest (see Fig. 8.7d).

The Late Mesozoic Era (Cretaceous Period: 145–66 Ma)

PALEOGEOGRAPHY. During the Cretaceous Period, the Earth's climate continued to warm, and sea level rose significantly. In fact, at that time, a shark could have swum across North America from the Gulf of Mexico to the Arctic Ocean in the *Western Interior Seaway*. The seaway formed, in part, because the loading of the western edge of the continent by growing mountain belts caused the surface of the lithosphere in the adjacent region to sink and produce a broad *foreland basin* (see Chapter 5); when the surface of this basin sank below sea level, the Western Interior Seaway formed.

By the dawn of the Cretaceous, the North Atlantic Ocean had grown to a width of 1,500 km (930 mi),

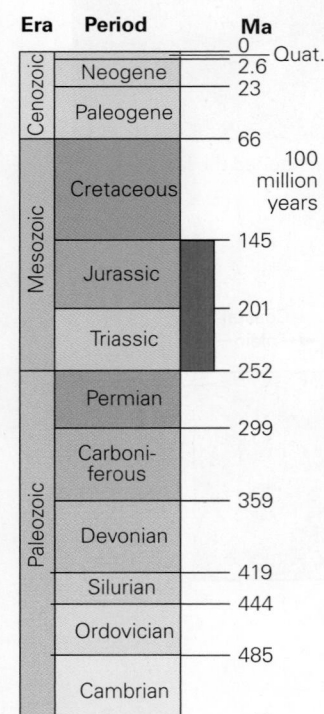

Era	Period	Ma
Cenozoic	Neogene	0 — Quat.
		2.6
		23
	Paleogene	
		66
Mesozoic	Cretaceous	
		145
	Jurassic	
		201
	Triassic	
		252
Paleozoic	Permian	
		299
	Carboniferous	
		359
	Devonian	
		419
	Silurian	444
	Ordovician	485
	Cambrian	
		539

100 million years

FIGURE 9.16 Paleogeography of the early Mesozoic.

(a) Pangaea began to break up in the Triassic. By Jurassic time, a narrow North Atlantic Ocean existed.

(b) During the Jurassic, immense sand dunes blanketed the southwestern United States. These sandstone beds in Zion National Park are the relics of those dunes.

and Pangaea continued its breakup. During the Middle Cretaceous, South America broke away from Africa; a mid-ocean ridge then developed, and the South Atlantic started to grow. And by the Late Cretaceous, Antarctica and Australia had separated, and India began moving rapidly northward, toward Asia, as intervening oceanic lithosphere was subducted (Fig. 9.18a).

The Sierran Arc along the west coast of North America remained active during the Cretaceous. In addition, a long fold-thrust belt, whose remnants crop out in the Canadian Rockies and western Wyoming, developed to the east of the volcanic arc (Fig. 9.18b). Geologists

FIGURE 9.17 During the Jurassic, giant dinosaurs roamed the land. This painting shows several species.

FIGURE 9.18 Paleogeography of the late Mesozoic.

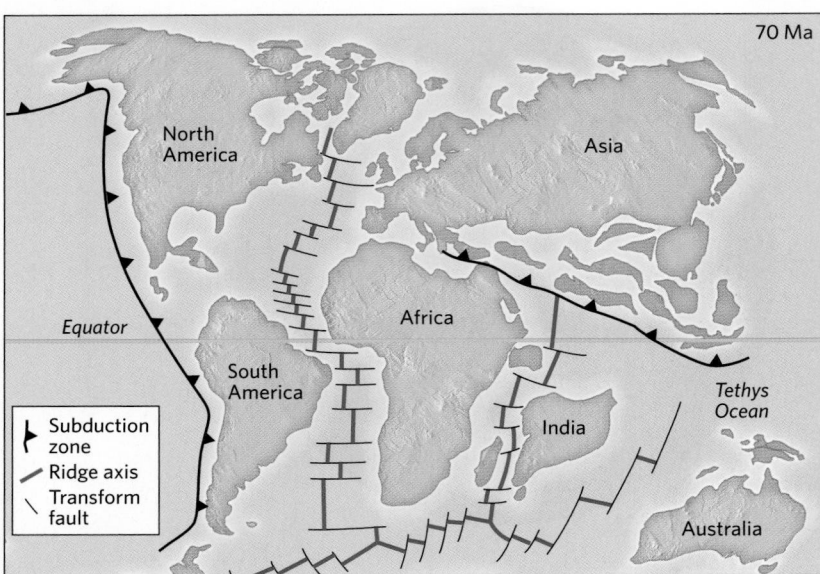

(a) By the Late Cretaceous Period (70 Ma), the Atlantic Ocean had formed, and India was moving rapidly northward toward Asia.

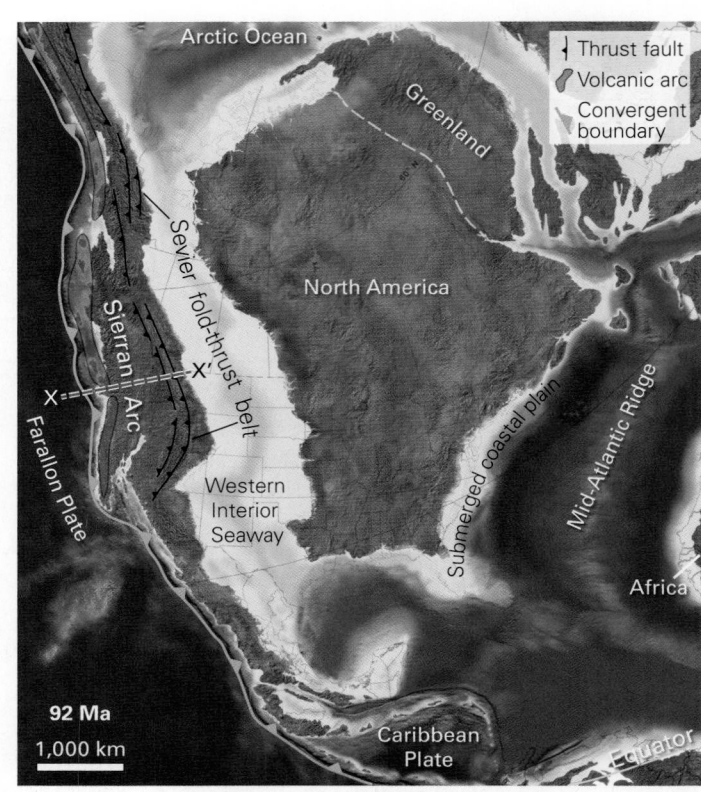

(c) During the Laramide orogeny, at the end of the Cretaceous, deformation shifted eastward in the United States, moving from the Sevier fold-thrust belt into what is now the Rocky Mountains. The red line represents the eastern limit of crust affected by the Laramide orogeny in the Eocene, at about 40 Ma. The inset shows Laramide-style basement-cored uplifts. Movement along faults brings up basement, and a monocline forms in the overlying sedimentary strata.

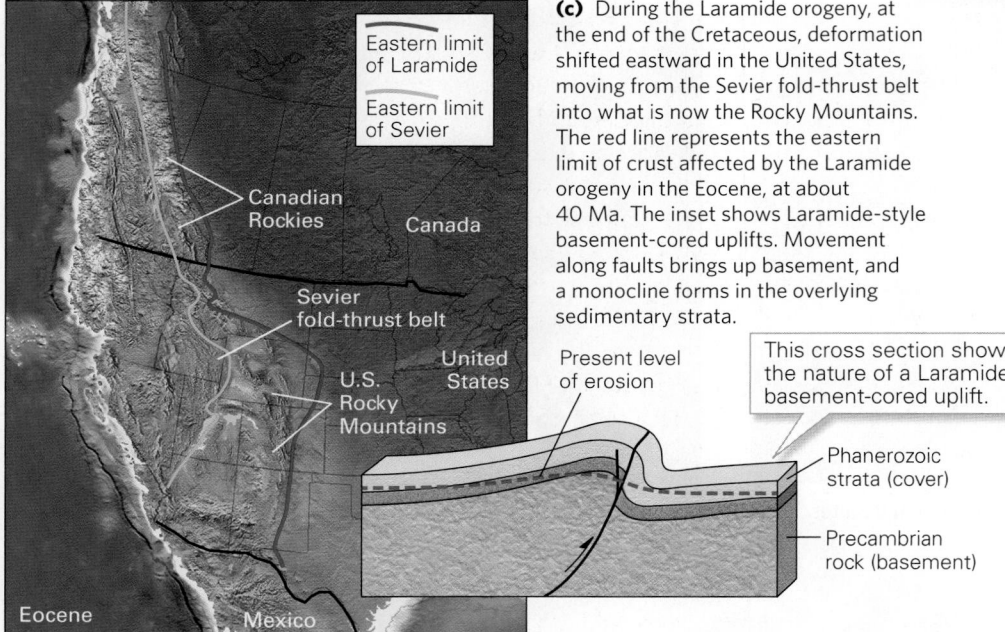

This cross section shows the nature of a Laramide basement-cored uplift.

This diagram (X–X') shows the relation of the Sierran Arc to the Sevier fold-thrust belt.

(b) In the Cretaceous, the Western Interior Seaway flooded the region to the east of the Sierran Arc and the Sevier fold-thrust belt. The inset shows the spatial relationship of the arc and fold-thrust belt.

the base of the continent. Such *shallow subduction* caused orogeny to shift eastward. Compression caused large reverse faults to become active in Wyoming, Colorado, eastern Utah, and northern Arizona. These faults penetrated deep into the Precambrian basement, so movement along the faults pushed basement rocks upward, producing structures named *basement-cored uplifts*. Layers of Paleozoic strata underwent folding, like a carpet pushed up by an opening trap door, resulting in the development of large monoclines (see Chapter 6). This event, which continued into the Cenozoic, is called the **Laramide orogeny**; it marked the growth of the present-day Rocky Mountains in the United States **(Fig. 9.18c)**. Note that the Laramide orogeny did not form basement-cored uplifts in Canada; instead, it caused eastward growth of the Canadian Rockies fold-thrust belt.

refer to the deformation event that produced this fold-thrust belt as the **Sevier orogeny**. The combination of the Sierran Arc and the Sevier fold-thrust belt represented a convergent-boundary orogen, much like the Andean orogen today (see Chapter 6).

At the end of the Cretaceous, the dip of the oceanic plate subducting beneath the western United States decreased, and the downgoing plate started to shear against

FIGURE 9.19 Life in the Cretaceous.

(a) Flowering plants first appeared in the Cretaceous, and were the favored diet of herbivorous dinosaurs, such as *Leptoceratops* (shown here).

(b) This Cretaceous rat-like mammal was carnivorous.

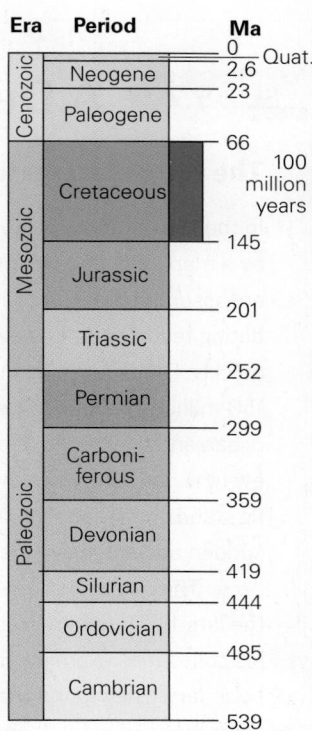

Era	Period	Ma
Cenozoic		0 — Quat.
Cenozoic	Neogene	2.6
Cenozoic	Paleogene	23
Mesozoic	Cretaceous	66 / 100 million years
Mesozoic		145
Mesozoic	Jurassic	201
Mesozoic	Triassic	252
Paleozoic	Permian	299
Paleozoic	Carboni-ferous	359
Paleozoic	Devonian	419
Paleozoic	Silurian	444
Paleozoic	Ordovician	485
Paleozoic	Cambrian	539

EVOLUTION OF LIFE. In the seas of the late Mesozoic world, modern fish appeared and became dominant. They served as prey for huge swimming reptiles and gigantic turtles. On land, angiosperms (flowering plants), including hardwood trees, populated the forests (Fig. 9.19a), and dinosaurs inhabited almost all environments. In fact, the abundance of dinosaurs led to the popular designation of the Mesozoic as the *Age of Dinosaurs*. Mammals also diversified and developed larger brains and more specialized teeth (Fig. 9.19b). Then, at 66 Ma, another mass extinction suddenly changed the course of Earth history. Dinosaurs vanished, along with most other species, during an event now known as the **K-Pg extinction** (for Cretaceous-Paleogene). This event, which resulted from a catastrophic meteorite impact (Box 9.2), defines the end of the Mesozoic Era.

Take-home message...

During the Mesozoic Era, rifting broke Pangaea apart and produced the Atlantic Ocean. In North America, convergent-boundary tectonics along the west coast produced the Sierran Arc, and eventually led to the uplift of the Rocky Mountains. Dinosaurs roamed all continents, but they died off, along with many other species, during the K-Pg mass extinction at the end of the era.

Quick Questions

- Did mammals exist in the Mesozoic?
- Did all of the Atlantic Ocean start to undergo seafloor spreading at the same time?
- Which came first, the Laramide orogeny or the Sevier orogeny?

9.6 The Cenozoic Era (66–0 Ma): The Modern World Comes to Be

PALEOGEOGRAPHY. During the past 66 million years—the Cenozoic Era—the map of the Earth's surface has continued to change, gradually producing the configuration of continents and plate boundaries that we see today. The continents that once constituted Gondwana drifted northward as subduction consumed the Tethys Ocean, which separated Gondwana from Europe and Asia (Fig. 9.20a). Eventually, the southern margins of Europe and Asia collided with various small crustal blocks and then finally with the Indian subcontinent and Africa, resulting in the growth of the largest orogen on the Earth today, the **Alpine-Himalayan chain** (Fig. 9.20b). Continued northward movement of India, as well as possible delamination of lithospheric mantle beneath the Asian side of the chain, led to the uplift of the *Tibetan Plateau* (see Chapter 6). And rifting in the northeastern corner of Laurentia split Greenland away from North America.

As the Atlantic Ocean grew and the Americas moved westward relative to Europe and Africa, active convergent boundaries persisted along their western margins. In South America, convergent-boundary activity produced the Andes, an orogen that remains active today. In North America, the Laramide orogeny continued until about 40 Ma (the Eocene Epoch), causing continued uplift of the Rocky Mountains. At this time, the configuration of plates along the western boundary of North America changed, and by 25 Ma, the convergent boundary was partly replaced by a transform boundary, defined by the trace of the San Andreas Fault. Today, the Pacific Plate

See for yourself

Rocky Mountain Front, Colorado

Latitude: 39°46′2.32″ N
Longitude: 105°13′45.35″ W

Zoom to an altitude of 8 km (5 mi) and look obliquely west.

You can see the steep face of the Rocky Mountains in Colorado. These mountains were uplifted during the Laramide orogeny. During the event, reactivation of large faults thrust Precambrian rocks upward and caused overlying Paleozoic strata to fold.

BOX 9.2 ▶ A Deeper Look

The K-Pg mass-extinction event

In the 18th century, paleontologists defined the end of the Cretaceous by a marked change in fossil species. Until the 1980s, most researchers assumed that the change took place over millions of years. Modern dating techniques, however, indicated that this change happened very quickly. Dinosaurs, which had dominated the planet for more than 150 million years, vanished, along with 90% of plankton species in the ocean and up to 75% of plant species. This abrupt mass-extinction event is now known as the *K-Pg extinction*; K stands for Cretaceous and Pg stands for Paleogene. What kind of catastrophe could cause such a sudden and extensive mass extinction?

The cause of the K-Pg extinction remained a mystery until the late 1970s, when Walter Alvarez, an American geologist, and his colleagues examined a shale layer deposited exactly at the K-Pg boundary. They found that this shale contained relatively high concentrations of iridium, an element that is extremely rare on the Earth, but does exist in meteorites. Further study showed that the clays of this age contained other unusual materials, such as tiny glass spheres that form when a spray of molten rock freezes, grains of coesite (a mineral that forms when intense shock

The Earth at 66 Ma

waves pass through quartz), and even carbon from burned vegetation. All these features pointed to the occurrence of a huge meteorite impact at the time of the K-Pg event (**Fig. Bx9.2a**).

Subsequently, geologists found a meteorite crater, 100 km (62 mi) in diameter and 16 km (10 mi) deep, buried beneath younger strata near the Yucatán Peninsula in Mexico (**Fig. Bx9.2b**). Radioisotopic dating indicated that the crater was formed 66 ± 0.4 Ma, the time of the K-Pg event. Because of its age and size, this crater, known as the Chicxulub crater, appears to be the imprint of the deadly object whose impact with the Earth eliminated so much life.

The K-Pg impact was catastrophic because it not only formed a crater, blasting huge quantities of debris into the sky, but it also generated tsunamis that may have been up to 2 km (1.2 mi) high, and it generated a blast of hot air that set forests on fire worldwide. The blast and the blaze together would have ejected enough debris into the atmosphere to cause months of perpetual night and winter-like cold. In addition, chemicals ejected into the air would have combined with water to produce acid rain. These conditions broke the food chain and triggered a mass extinction.

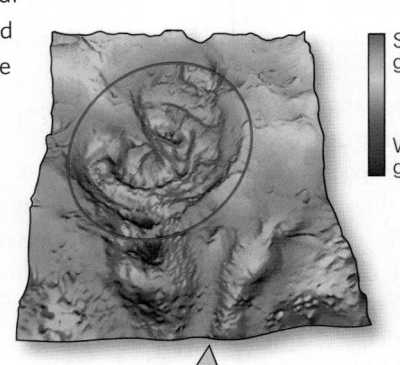

Stronger gravity

Weaker gravity

FIGURE Bx9.2 The K-Pg meteorite impact.

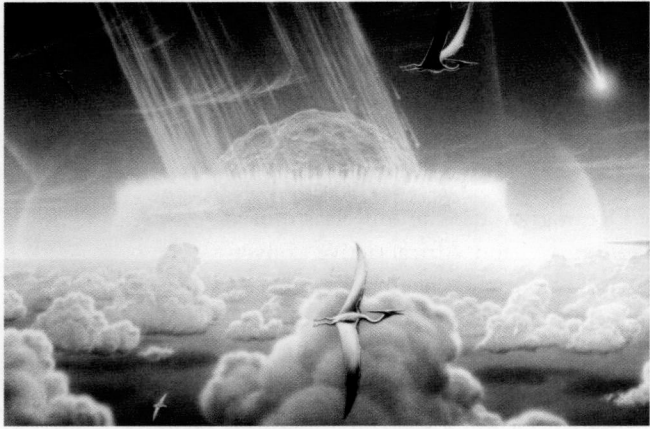

(a) An artist's image of the 13 km wide object as it hit. The inset shows the Earth's paleogeography at the time of impact.

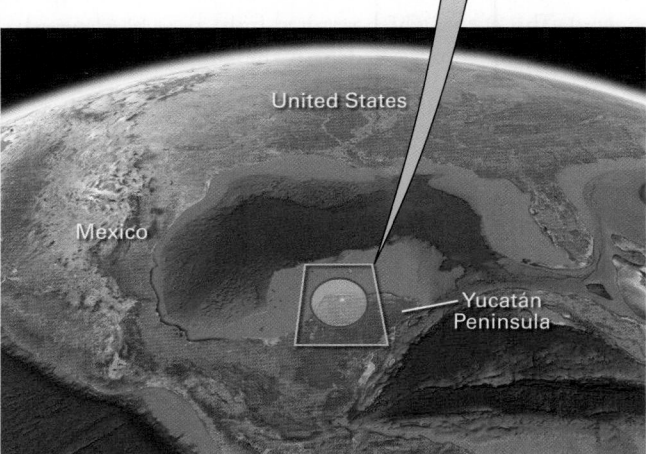

United States

Mexico

Yucatán Peninsula

(b) The location of the Chicxulub crater today. The colors on the inset indicate gravity anomalies, deviations from expected gravitational force. Low-density sediment filled the crater, so the gravitational pull felt at the present Earth's surface over the crater is weaker than in surrounding areas.

FIGURE 9.20 Paleogeography of the Cenozoic.

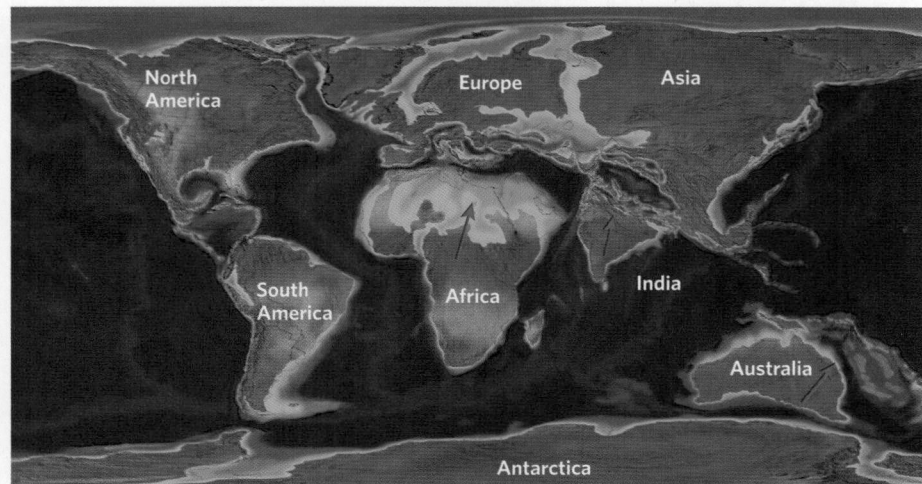

(a) In this Eocene (50 Ma) paleogeographic reconstruction, India is just about to collide with Asia. Africa, along with microplates in the Mediterranean region, is converging with Europe.

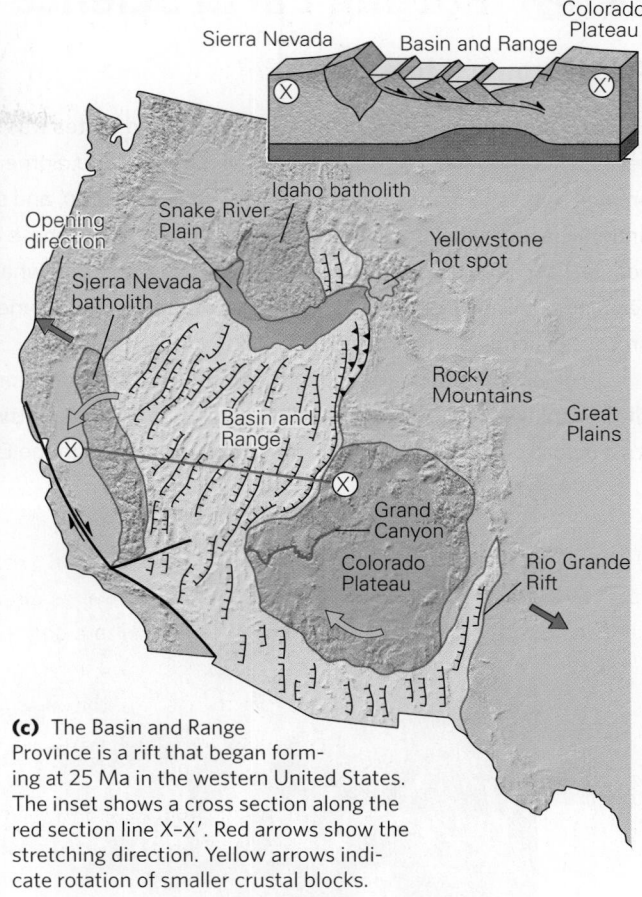

(c) The Basin and Range Province is a rift that began forming at 25 Ma in the western United States. The inset shows a cross section along the red section line X–X'. Red arrows show the stretching direction. Yellow arrows indicate rotation of smaller crustal blocks.

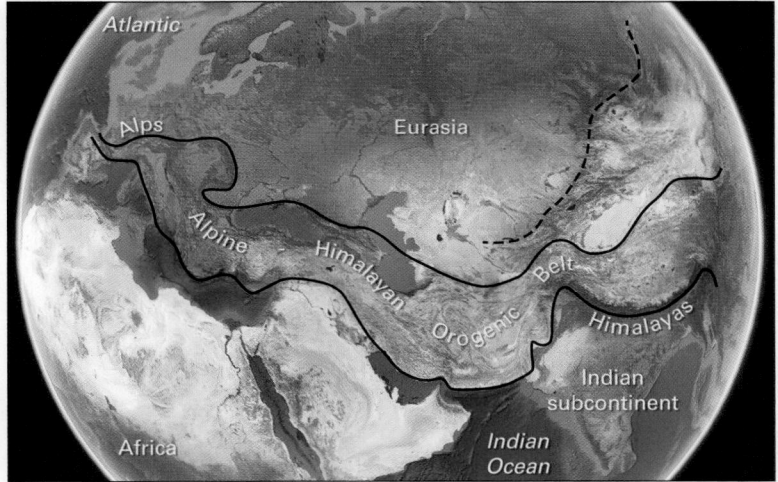

(b) The Alpine-Himalayan orogen, which extends from Spain eastward to eastern China, contains many distinct mountain belts. The solid black lines delineate the northern and southern borders of the orogen. The dashed black line gives the northern limit of the region in which faults outside the orogen have been reactivated significantly by the orogeny.

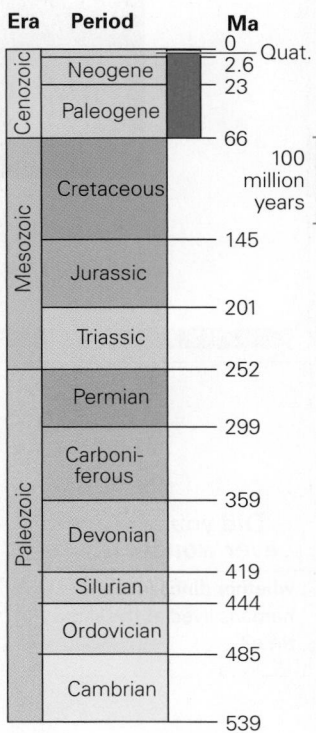

moves northward relative to North America at a rate of about 6 cm/yr (3 in/yr) along this strike-slip fault. This transform boundary traverses only western California; convergent-boundary activity continues in Washington, Oregon, and northern California, where subduction of the Juan de Fuca Plate in the Cascadia subduction zone fuels volcanism in the Cascade Range. The southern end of the San Andreas links to a recently formed divergent boundary that underlies the Gulf of California. The southern end of this divergent boundary, in turn, links to a convergent boundary that runs along the coast of southern Mexico and into South America.

Around the same time that the San Andreas Fault became active, the region that had been affected by the Sevier and Laramide orogenies began to undergo rifting as it was stretched in a roughly northwest-southeast direction. Extensional deformation was probably underway by 25 Ma, but most of the stretching occurred during the Miocene Epoch, and some continues today. The result was the formation of the **Basin and Range Province**, a broad continental rift (Fig. 9.20c) whose name reflects its topography. The region contains long, narrow mountain ranges separated from each other by flat, sediment-filled basins. This topography formed because blocks of crust slipped downward and tilted on the normal faults that broke up the crust (see Chapter 7). The crests of the tilted blocks are the existing ranges, and the depressions between them, which rapidly filled with sediment eroded from the ranges, are the basins. As the Basin and Range Rift opened, the adjacent Colorado Plateau underwent uplift (Box 9.3).

BOX 9.3 **Putting Earth Science to Use**

Adding context to a view

The site of nearly every national park in the United States was selected because it offers spectacular views of the natural world. Some parks provide access to a forest, some to a seashore or swamp, and some to dramatic rocky cliffs. If you have the opportunity to visit these places of beauty, keep in mind that the Earth has a history, and that what you are seeing most likely represents the product of geologic phenomena and processes acting over geologic time.

For example, look at the photo from Bryce Canyon National Park, Utah, displayed in **Figure Bx9.3**. What you see is not just pretty cliffs. It's a record of Earth history from the Cretaceous through the Eocene,

a time when environment and topography changed dramatically in western North America. The oldest rocks in the vicinity, hidden below the base of the cliffs, consist of lithified sand dunes. They are overlain by shallow-marine deposits, and finally by the deposits of lakes and streams. So, the succession of rock layers tells a tale of transgression, regression, and changing climate. The top of Bryce Canyon now lies at an elevation of about 2.7 km (1.7 mi), due to the uplift of the Colorado Plateau. Erosion has attacked the uplifted rocks—frost wedging, flowing water, and gravity have whittled away the bedrock to form the dramatic *hoodoos* (columns of weathered rock) that decorate the park today.

FIGURE Bx9.3 The intricately carved hoodoos of Bryce Canyon. The geologic history of the western United States provides a context for this view.

Did you ever wonder...

whether dinosaurs and humans lived at the same time?

The Basin and Range Province terminates just north of the Snake River Plain, a feature that marks the track of the hot spot that now lies beneath Yellowstone National Park. As the North American Plate drifted westward, volcanic calderas formed along this track. Yellowstone National Park straddles the most recent caldera (see Chapter 4). By about 15 Ma, the mountain belts and plate motions

we see today had been established (**Fig. 9.21**). Around the world, plate interactions continue to drive seismicity, volcanism, and mountain building.

The Cenozoic has also been a time of significant climate change. Global temperatures were quite warm at the end of the Mesozoic, and they continued to warm into the Cenozoic, reaching a peak in the Eocene. The world

FIGURE 9.21 The Alpine-Himalayan orogen formed when Africa, several microcontinents, and India collided with Europe and Asia. The Cordilleran and Andean orogens reflect the consequences of continuing convergent-boundary tectonics along the eastern coast of the Pacific Ocean.

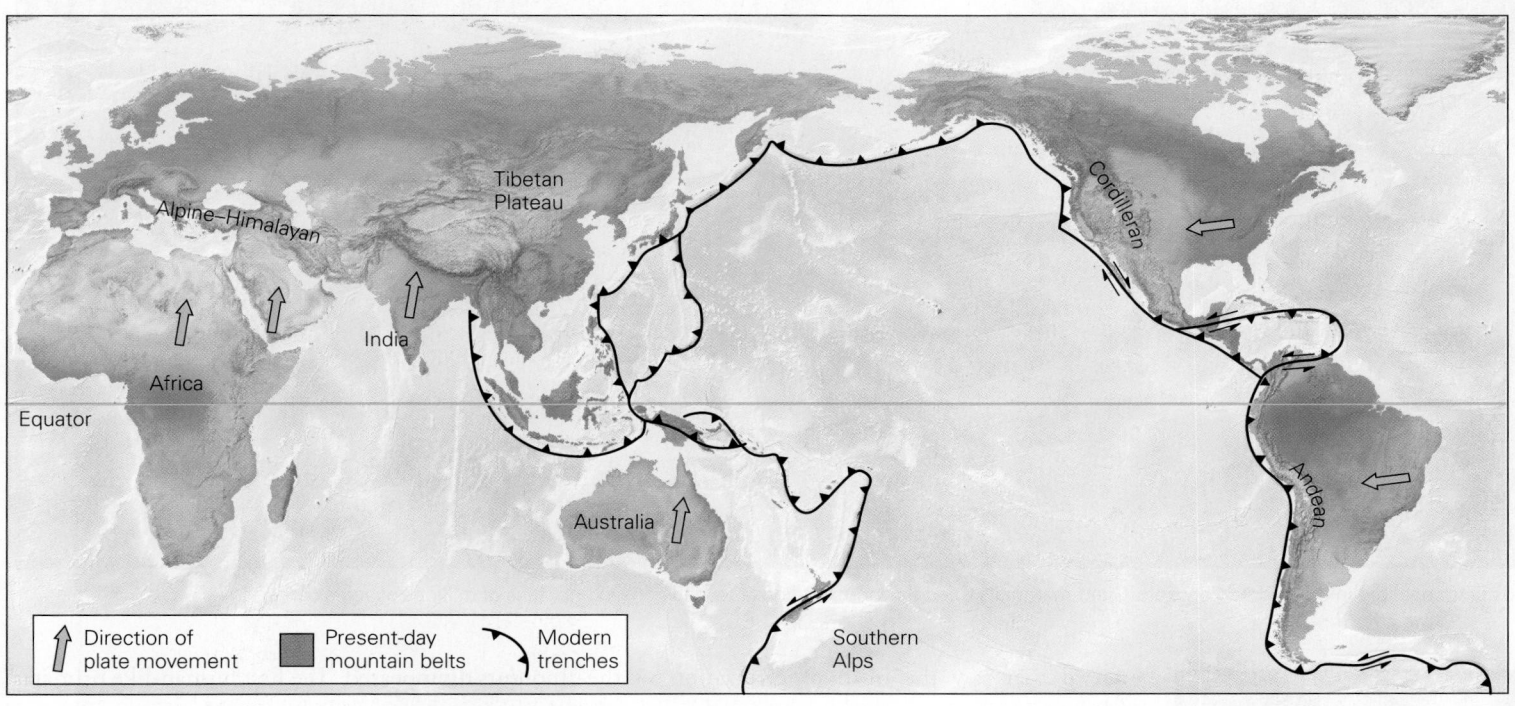

FIGURE 9.22 The maximum advance of continental glaciers during the Pleistocene Ice Age in North America.

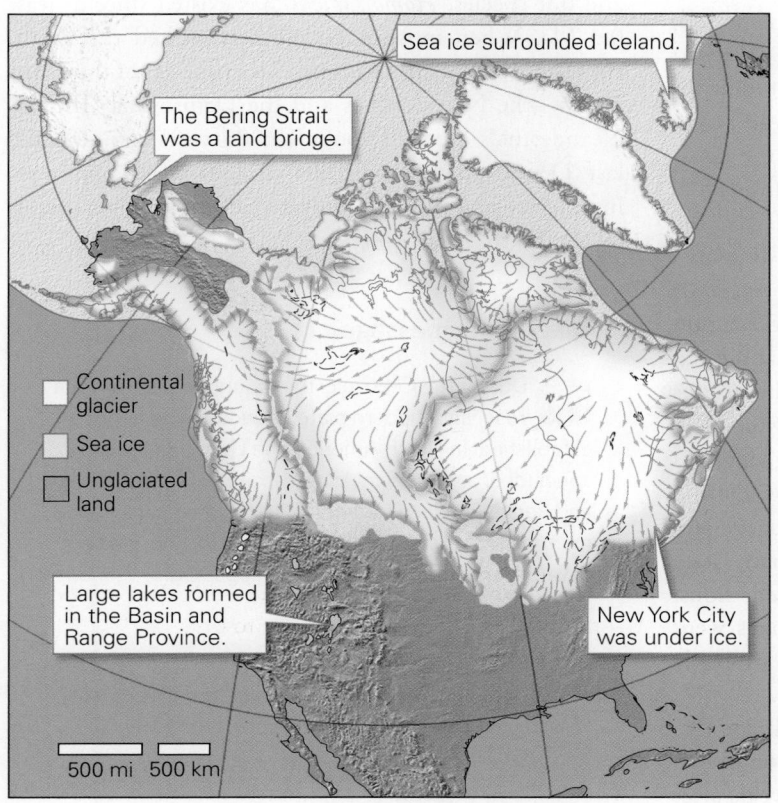

started to cool beginning about 50 Ma, and in the Early Oligocene (about 34 Ma), Antarctic glaciers reappeared for the first time since the Triassic. Cooling continued, and during the past 2.6 million years—the Pleistocene Epoch of the Quaternary Period—huge glaciers expanded and retreated across northern continents at least 20 times (see Chapter 14). Erosion and deposition by glaciers during this **Pleistocene Ice Age** produced much of the landscape we see today in northern temperate regions of North America, Europe, and Asia **(Fig. 9.22)**. The last glacial maximum, when the most recent glaciation reached its southern limit, happened about 20,000 years ago. Subsequently, glaciers retreated. Significant warming beginning about 11,700 years ago led to the disappearance of the last ice sheets in northern continents by about 6,500 years ago. This warming also marks the beginning of the *Holocene Epoch* of the Quaternary Period, the interglacial time

See for yourself

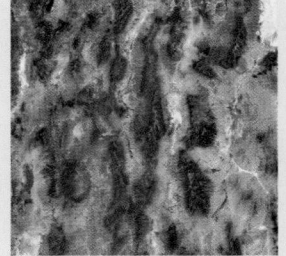

Basin and Range Rift, Utah

Latitude: 39°15'1.83" N
Longitude: 114°38'32.10" W

Zoom to an altitude of 250 km (155 mi) and look down.

In this region of the Basin and Range Rift, the darker bands are fault-block mountains, whereas the lighter areas are sediment-filled basins. The white areas are evaporites, formed where lakes dried up and salt precipitated.

FIGURE 9.23 Life in the Cenozoic.

(a) Grasslands first appeared in the middle Cenozoic. Giant mammals ruled the land, for dinosaurs had gone extinct tens of millions of years earlier.

(b) Our own human lineage appeared in the Cenozoic. Human-like primates such as *Australopithecus*, shown here, first appeared about 4 Ma.

interval that saw the birth of civilization and the transition to a world dominated by *Homo sapiens*.

EVOLUTION OF LIFE. When the skies finally cleared in the wake of the K-Pg meteorite impact, new species of plant life appeared, and forests of both angiosperms and gymnosperms proliferated. By the middle of the Cenozoic Era, grasses had spread across the plains at mid-latitudes. With the dinosaurs gone, birds and mammals diversified. Recent studies suggest that birds are the descendants of small feathered dinosaurs that were related to *Tyrannosaurus rex*. Birds may have survived the K-Pg extinction because of their small size and ability to eat a variety of foods. Most of the groups of mammals that live today originated at the beginning of the Cenozoic, giving this era the nickname *Age of Mammals*. During the latter part of the era (Late Neogene and Early Quaternary), huge mammals, such as mammoths and saber-toothed lions, appeared (Fig. 9.23a). These animals went extinct over the past 10,000 years, probably in part due to hunting by humans.

According to the fossil record, ape-like primates diversified in the Miocene Epoch at about 20 Ma, over 45 million years after

the dinosaurs disappeared. The first human-like primates, *australopithecines*, appeared about 4 Ma (Fig. 9.23b), followed by the first hominids (members of the human genus, *Homo*) at 2.4 Ma. *Homo erectus*, a species capable of making stone axes, appeared in Africa about 1.6 Ma, and our species, *Homo sapiens*, has existed since at least 0.3 Ma. When modern people first walked the Earth, they shared the planet with two other species of the genus *Homo*—the Neanderthals and the Denisovans. The last Neanderthals died off about 40,000 years ago, and the last Denisovans went extinct 25,000 years ago, leaving *Homo sapiens* as the only remaining human species on the Earth.

Take-home message...

During the Cenozoic, the mountain belts of today rose, and modern plate boundaries became established. In North America, the Laramide orogeny concluded, and the San Andreas Fault and the Basin and Range Province formed. With the dinosaurs gone, mammals diversified. During the Pleistocene, glaciers covered large areas, and humans appeared.

Quick Questions

- Did *Homo sapiens* live during the Miocene?
- Where do present-day convergent boundaries exist along the continental margins of North America?
- What event caused the uplift of the Himalayas?

9.7 The Concept of Global Change

Did the Earth's surface look the same in the Jurassic as it does today? Definitely not! As we've seen, in Jurassic time, Africa and South America were part of the same continent, and the call of the wild rumbled from the throats of dinosaurs, whereas today the broad South Atlantic Ocean separates the two landmasses (Fig. 9.24), and the largest animals are mammals (Fig. 9.25). What we see of the Earth today is just a snapshot, an instant in the life story of a constantly changing planet with a long and complex history. This idea arguably stands as geology's greatest philosophical contribution to humanity's understanding of our planet.

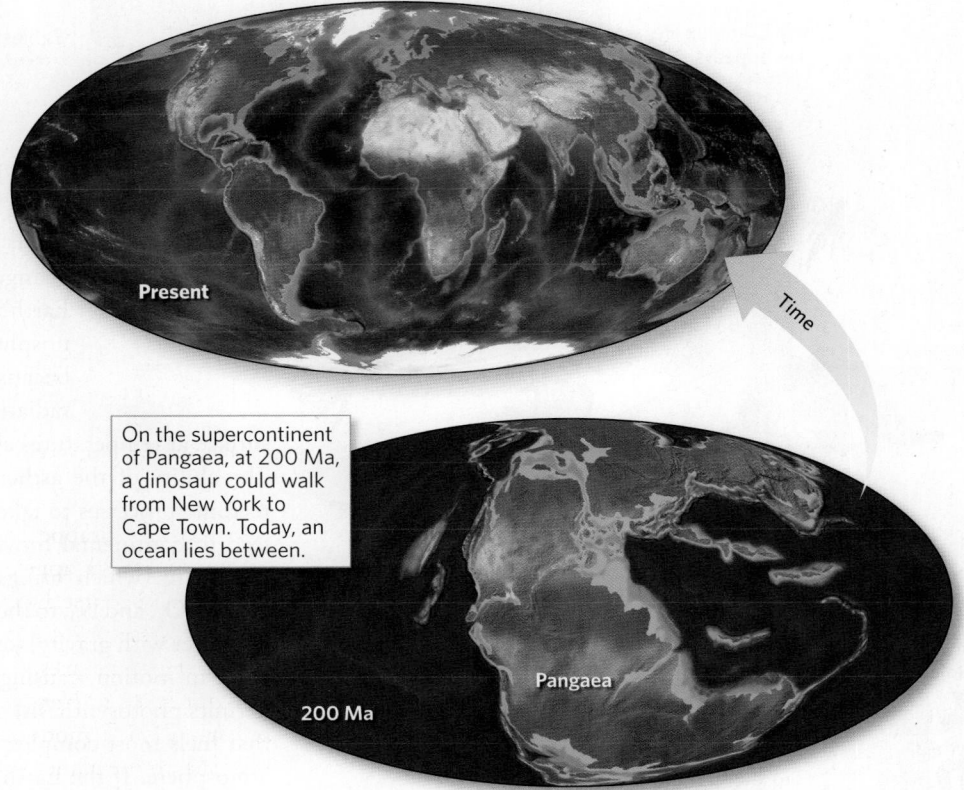

FIGURE 9.24 The map of the Earth's surface changes over time because of plate motions. Here we see the change that has taken place during the past 200 million years.

On the supercontinent of Pangaea, at 200 Ma, a dinosaur could walk from New York to Cape Town. Today, an ocean lies between.

Present

Time

200 Ma

Pangaea

FIGURE 9.25 Evolution is unidirectional change, in that its transformations never repeat. Because of evolution, the species inhabiting the Earth today are not the same as those that inhabited the planet in the past. For example, the largest land animal in the Mesozoic was a reptile, whereas the largest land animal today is a mammal.

The largest Mesozoic land animal was a dinosaur. The inset shows an elephant at the same scale.

The largest land animal today is the elephant.

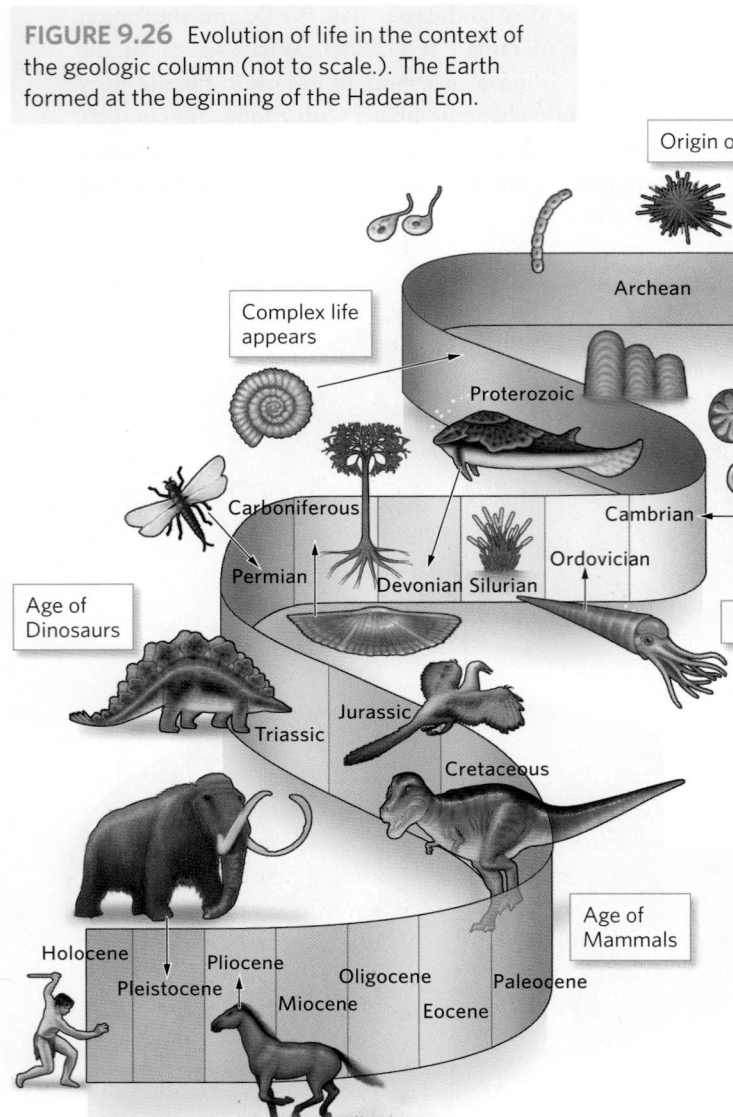

FIGURE 9.26 Evolution of life in the context of the geologic column (not to scale.). The Earth formed at the beginning of the Hadean Eon.

Big Bang

Origin of life

Hadean

Solar System formation

Earliest fossils

Archean

Complex life appears

Proterozoic

Carboniferous

Cambrian

Permian   Devonian Silurian Ordovician

Age of Dinosaurs

Shells appear

Triassic

Jurassic

Cretaceous

Age of Mammals

Holocene   Pliocene   Oligocene

Pleistocene   Miocene   Eocene   Paleocene

To finish our consideration of the Earth's biography, let's focus our attention on the concept of change. In the introduction to this book, we defined the *Earth System* as all the physical and biological realms of the Earth, along with the multitudinous ways in which they interact with one another. In this context, we can define a **global change** as a modification that has an impact on the Earth System worldwide. Researchers distinguish among different types of global change, first, by the rate of change: *gradual change* takes place relatively slowly (over millions to billions of years), whereas *catastrophic change* takes place relatively rapidly (over seconds to millennia). We can also distinguish among different types of global change by the way in which the change progresses: *unidirectional change* involves transformations that never repeat; *cyclic change* repeats the same steps over and over, though not necessarily with the same results or at the same rate; and *periodic change* repeats steps with a definable frequency.

Why has the Earth changed so much over geologic time, and how can it continue to change? Ultimately, change happens both because the Earth's internal heat makes the asthenosphere weak enough to flow and because external energy—the Sun's radiation—keeps most of the Earth's surface at temperatures above the freezing point of water. The ability of the asthenosphere to flow permits plate-tectonic processes to take place. These processes, in turn, lead to continental movement, mountain building, and volcanism (which brings key volatile elements, such as H_2O, CO_2, and N_2, to the Earth's surface). External energy (together with gravity) keeps streams, glaciers, waves, and wind in motion, causing erosion and deposition. It also permits photosynthesis, the foundation of the food chain that fuels most complex life and produces the O_2 in the atmosphere. If the Earth did not have just the right mix of plate tectonics and solar radiation, it would be a frozen dust bowl like Mars, a crater-pocked wasteland like the Moon, or a cloud-choked oven like Venus, and you wouldn't be here to read this book.

Let's now briefly re-examine some aspects of Earth history that we've discussed in this chapter, in the context of global change. We'll see that these aspects illustrate different types of change, and that humans have, in relatively recent time, become an important agent of change.

Unidirectional Changes during Earth History

The Earth System has changed in many irreversible ways over the course of geologic time. For example, differentiation of the Earth's interior into a core and mantle will never happen in the future, because once a core has formed, it can't form again. The filling of the oceans also represents a unidirectional change in that once it took place, liquid water has persisted on our planet's surface ever since. The fossil record, as we have seen, emphasizes that evolution is unidirectional change, punctuated by mass extinctions over the course of geologic time (**Fig. 9.26**).

FIGURE 9.27 The stages of the supercontinent cycle.

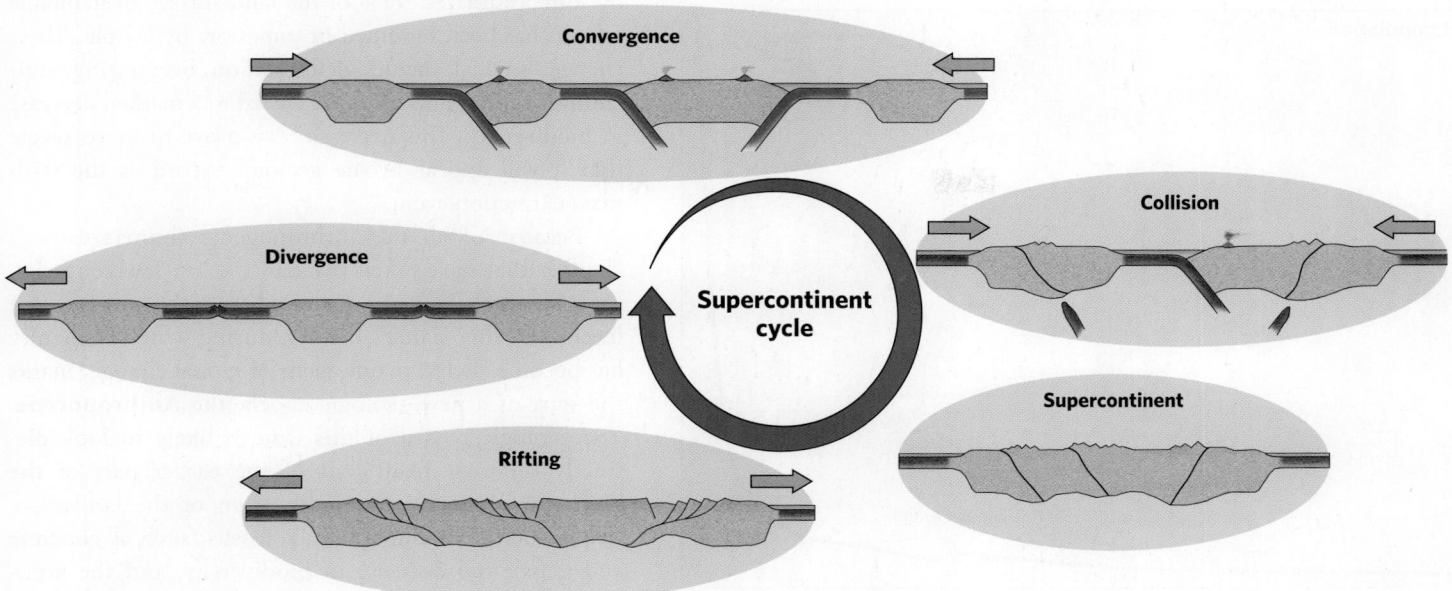

Cyclic Changes in the Earth System

During cyclic change, a sequence of stages repeats over time. Some cyclic changes in the Earth System are periodic, in that cycles happen with a definable frequency, but most are not. Some cycles involve only movements of physical components of the Earth System, while others involve transfer of materials among both living and nonliving components of the Earth System. We consider many examples of cyclic changes in this book:

- *The rock cycle:* In effect, rocks serve as reservoirs of atoms, and movement of atoms from reservoir to reservoir over time constitutes the rock cycle (see Chapter 5).

- *The supercontinent cycle:* On a time scale of hundreds of millions of years, smaller continental blocks collide and collect into large supercontinents. A supercontinent survives for tens to hundreds of millions of years until it undergoes rifting to form smaller continents, which move apart due to seafloor spreading. Eventually, the continents reassemble into a new supercontinent, only to break apart again into different blocks (Fig. 9.27).

- *The sea-level cycle:* Relative sea level has gone up and down over Earth history. When sea level rises, continental interiors may become shallow seas.

- *Biogeochemical cycles:* A **biogeochemical cycle** involves the passage of chemicals among nonliving and living reservoirs within the Earth System. Nonliving reservoirs include the atmosphere, the crust, and the oceans, whereas living reservoirs include plants, animals, and microbes. Examples include the hydrologic cycle (see Chapter 11) and the carbon cycle (see Chapter 19). Some biogeochemical cycles can attain a **steady-state condition**, meaning that the proportions of a chemical in different reservoirs remain fairly constant even though flow among reservoirs continues. A global change in a biogeochemical cycle modifies the proportions of chemicals in different reservoirs, causing a change in the steady-state condition. For example, when vast glaciers grow on land, they store water that has been removed from the ocean, so sea level drops.

The Geologic Record of Our Time: The Anthropocene?

By the dawn of civilization, around 6,000 years ago, the global human population was at most a few tens of millions. But by the beginning of the 19th century, revolutions in industry and hygiene had substantially improved survival rates, so the population grew at accelerating rates and reached 1 billion in 1850. It then took only 80 years for the population to double, reaching 2 billion in 1930. Today, the doubling time is only 44 years, as the population passed the 6 billion mark just before the year 2000 and surpassed 8 billion in November 2022 (Fig. 9.28a).

As the human population grows and the standard of living improves, our use of the Earth's resources increases (Fig. 9.28b). We use soil for agriculture and grazing; wood, rock, and gravel for construction; oil and coal for

Did you ever wonder . . .
whether the Earth will ever look the same in the future as it has in the past?

All we in one long caravan are journeying since the world began, we know not whither, but we know . . . all must go.

—BHARTRIHARI (INDIAN POET, CA. 500 C.E.)

FIGURE 9.28 A sense of the Anthropocene.

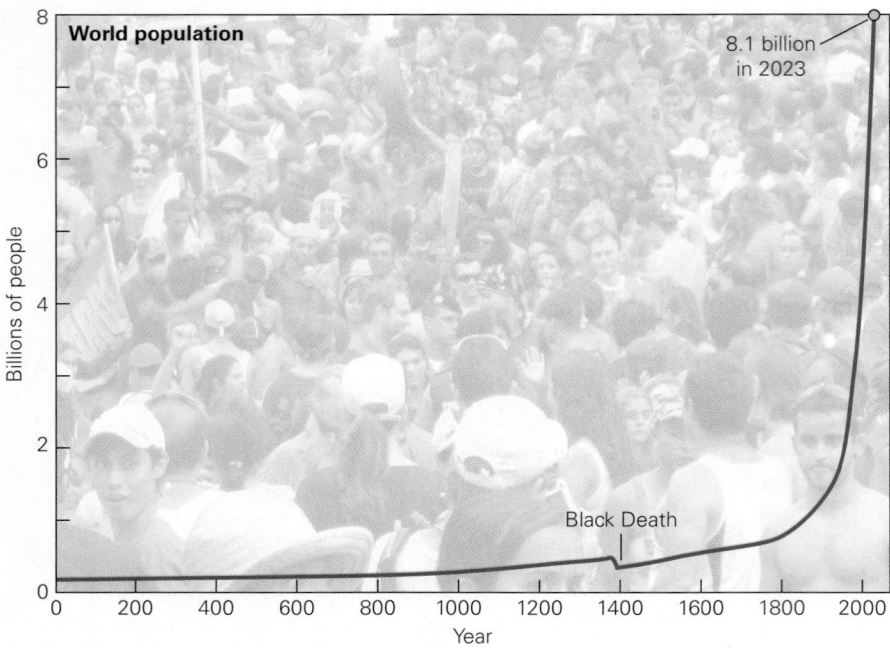

(a) The human population now doubles about every 44 years. The Black Death pandemic in the 14th century caused an abrupt drop that lasted for a few decades.

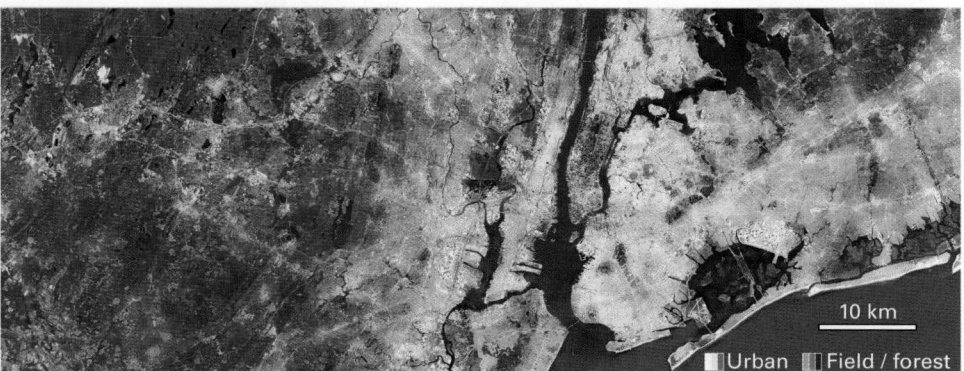

(b) A satellite image shows the radical change in the landscape of New York City and surrounding areas that has happened over the past two centuries. Prior to intense urbanization, the land was covered by forest.

energy and plastics; and ores for metals (see Chapter 10). Every time we move a pile of rock, plow a field, drain a wetland, or pave a road, we change a portion of the Earth's crust. Clearly, our use of resources affects the Earth System profoundly, so humanity has become a significant

agent of global change. Humanity now moves more sediment in a year than gets carried by the largest river, and by some estimates, 95% of the land surface in habitable regions has been modified in some way by people. These changes—which include deforestation, overgrazing, agriculture, and urbanization—have led to a marked decrease in biodiversity. This decrease may prove to be so severe that it will appear in the geologic record as the sixth mass-extinction event.

Because of all the anthropogenic (human-caused) changes that have taken place in the last few centuries, some researchers have suggested that the most recent interval of this planet's history, during which humanity has become such a major agent of global change, marks the start of a new geologic epoch, the **Anthropocene**. The geologic record of this time is likely to look distinctly different from that of the earlier part of the Holocene. Humanity's modification of the landscape, its production of durable trace fossils (such as concrete and glass), the decrease in biodiversity, and the accumulation of long-lived *pollutants* (contaminants that cannot be absorbed or destroyed by natural Earth-System processes) may be preserved as a distinct marker bed, discernible by intelligent life 100 million years in the future. Use of the name *Anthropocene* is gradually becoming accepted, but the question of what year marks its start remains remains under debate. The name does convey the concept that our species will have had a huge impact on the biography of the Earth as read by observers of the future.

Take-home message...

Since our planet first formed, the Earth System has been undergoing major changes. Some of these changes are unidirectional, whereas others are cyclic. Cyclic change can be periodic, but often it is not. In the present day, human activities are a powerful cause of global change.

Quick Questions

- What does global change mean, in a broad sense?
- Do all types of global change happen at the same rate?
- What does the term *Anthropocene* refer to?

9 CHAPTER REVIEW

Objective 9.1

> Characterize how the Earth formed and changed during the Hadean Eon, the time from our planet's birth until the rock record begins.

KEY CONCEPTS

- Radioisotopic dating of meteorites suggests that the Earth formed between 4.56 and 4.54 Ga. For part of its first half billion years, known as the Hadean Eon, the planet was so hot that its surface was molten at times.
- The earliest crust was probably just the solidified surface of the mantle. All crust formed during the Hadean was destroyed during the formation of the Moon, and again during the late heavy bombardment.
- The Earth's first atmosphere consisted of volatile elements and compounds released by volcanoes; it was rich in water and CO_2.
- By the end of the Hadean, the Earth had a solid surface and short-lived oceans.

EARTH-SCIENCE VOCABULARY

Hadean Eon (p. 321)   **late heavy bombardment** (p. 320)

REVIEW QUESTIONS

1. **(a)** Describe the materials and processes involved in the formation of the Earth. **(b)** How did the Earth's interior change as a consequence of differentiation? **(c)** What is the range of dates that geologists consider to be the time when our planet formed? **(d)** Where did the heat that kept the Hadean Earth so hot come from? **(e)** What did the earliest crust of the Earth consist of?

2. **(a)** What is the favored scientific explanation for how the Moon formed? **(b)** Is the Earth likely to have had a solid crust immediately after Moon formation? **(c)** What event does **Figure A** depict?

3. If you were to travel through time and arrive on the Earth at 4 Ga, could you breathe the air? Explain your answer.

A

4. **(a)** Why are there no whole rocks on the Earth that yield radioisotopic dates older than about 4 billion years? **(b)** What date has been assigned to the end of the Hadean, and why? **(c)** What percentage of the Earth's history does the Hadean account for?

Objective 9.2

> Provide a geologic interpretation of how early continents, oceans, and life forms came to be and evolved during the Archean Eon.

KEY CONCEPTS

- The Archean Eon began about 4.0 Ga, a date tied roughly to the time when a permanent rock record begins. The Earth had a solid crust throughout the Archean, and a permanent liquid-water ocean by about 3.85 Ga.
- The early continental crust was assembled out of island arcs and oceanic plateaus, which consisted of mafic rocks that could not be subducted. Collisions brought these blocks together to form protocontinents.
- Metamorphic rocks and sedimentary rocks also existed in early protocontinents. As continents grew wider, their interiors cooled and became stronger.
- The first life forms—bacteria and archaea—appeared in the Archean. The atmosphere of that time contained very little oxygen.

EARTH-SCIENCE VOCABULARY

Archean Eon (p. 321)   **stromatolite** (p. 322)
protocontinent (p. 321)

REVIEW QUESTIONS

5. **(a)** Where did the magma that developed into the earliest igneous rock of continents originate? **(b)** Why do the components of continental crust, once formed, stay at the surface of the Earth instead of being subducted? **(c)** How did protocontinents grow?

6. Label the axes and the color bands on **Figure B**, and interpret the meaning of this graph.

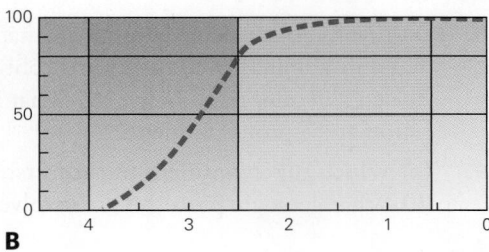

B

7. **(a)** Why did the composition of the atmosphere start to change during the Archean? **(b)** According to fossil evidence, how old was the Earth when the first indisputable cells appeared? **(c)** In what environment might the earliest life have appeared? **(d)** How does a stromatolite form?

Objective 9.3

Explain how life, the atmosphere, and climate changed during the Proterozoic Eon, and how the first supercontinents came into existence.

KEY CONCEPTS

• During the Proterozoic Eon, which began at 2.5 Ga, Archean crustal blocks began to be sutured together to form large continents. The first large supercontinent, Nuna (Columbia), came together by about 1.9 Ga and lasted for a few hundred million years before rifting apart.

• During the great oxygenation event, oxygen began to accumulate in the atmosphere, as the land surface and oceans could no longer absorb it. Addition of oxygen to the ocean caused iron oxide minerals to precipitate out.

• For most of the Proterozoic, life consisted of single-celled organisms. But by the end of the eon, soft-bodied marine invertebrates of the Ediacaran fauna populated the planet.

• By about 900 Ma, continental crust had again accumulated into a supercontinent, known as Rodinia. It broke up as the Proterozoic came to a close. While it was breaking up, a global ice age (known as snowball Earth) covered both continents and oceans with ice.

EARTH-SCIENCE VOCABULARY

banded iron formation (BIF) (p. 326)
craton (p. 323)
Ediacaran fauna (p. 325)
great oxygenation event (p. 326)
Great Unconformity (p. 326)
Grenville orogeny (p. 325)
Laurentia (p. 325)
Nuna (p. 325)
paleogeographic map (p. 325)
Proterozoic Eon (p. 323)
Rodinia (p. 325)
snowball Earth (p. 326)
supercontinent (p. 325)

REVIEW QUESTIONS

8. **(a)** If you look at a current geologic map of the world, you'll find that only about 10% of the Earth's continental crustal surface is labeled "Precambrian." Why? **(b)** What is the difference between a craton and a protocontinent?

9. **(a)** Which supercontinent formed first, Nuna or Rodinia? **(b)** What geologic processes are involved in the formation of

a supercontinent, and what geologic processes happen when a supercontinent breaks apart? **(c)** On **Figure C**, label the cratonic platform, the shield, Phanerozoic orogens, and the coastal plain. Is the portion of the craton at Point X older than or younger than the part at Point Y, and how did the crust at Point X form?

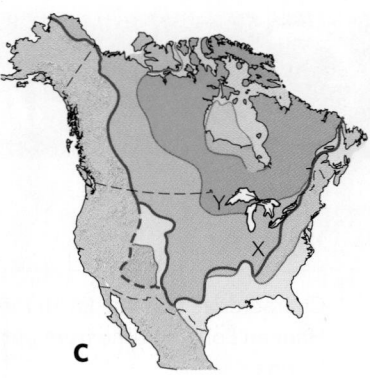

C

10. **(a)** How did the atmosphere change during the Proterozoic, and what role did life play in causing this change? **(b)** As a consequence of this change, what distinctive sedimentary deposits formed?

11. **(a)** What evidence suggests that the Earth nearly froze over during the Proterozoic Eon? **(b)** Did multicellular organisms exist before the Phanerozoic Eon began?

Objective 9.4

Identify major Paleozoic continents and the supercontinent that formed at the end of the era, relate Paleozoic orogenies to plate interactions, and discuss sea-level change and life evolution during this era.

KEY CONCEPTS

• As the Paleozoic Era began, rifting had yielded several separate continents, including Laurentia. The formation of so many new ecological niches may have led to the Cambrian explosion, a major diversification of life. During the early Paleozoic, Laurentia was bordered by passive-margin basins.

• Sea level rose and fell relative to land so that at times, the interiors of continents were submerged by epicontinental seas.

• Several orogenies took place during the Paleozoic, caused by collisions among continents, island arcs, and microcontinents. Strata of the eastern and western passive margins of Laurentia were incorporated in these orogens. By the end of the era, continents had combined to form Pangaea.

• Early Paleozoic life included invertebrates with shells and jawless fish. The first insects and green land plants appeared during the middle Paleozoic.

• In the latter part of the Paleozoic, dense forests appeared. By the end of the era, these forests included gymnosperms, and vertebrate organisms such as reptiles populated the land. The Paleozoic came to an end during the Permian-Triassic mass-extinction event known as the Great Dying.

Cambrian explosion (p. 332)
epicontinental sea (p. 328)
Gondwana (p. 328)
Great Dying (p. 336)

microcontinent (p. 332)
Pangaea (p. 334)
Phanerozoic Eon (p. 327)

REVIEW QUESTIONS

12. (a) How did the Cambrian explosion change the nature of the living world? (b) What key feature did many Paleozoic invertebrates have that Ediacaran ones did not, and what evolutionary purpose might this feature have served? (c) According to the fossil record, during what period did the first vertebrates appear? (d) When did the animal depicted in **Figure D** appear in the fossil record, and what could it do that previous animals could not?

13. (a) Why were new passive-margin basins established at the beginning of the Cambrian? (b) Label the passive margins shown on **Figure E**. (c) What type of map is shown in **Figure E**, and what continent is pictured? (d) Which orogeny 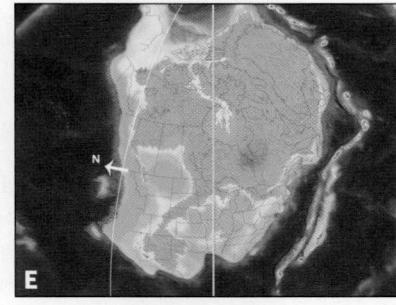 happened during the collision of Africa and North America, and what was a consequence of this event along the western side of this orogen? (e) What supercontinent had formed at the end of this orogeny?

14. How would the land surface at the beginning of the Ordovician have looked different from the land surface at the end of the period?

15. (a) Why did broad areas of continental interiors become covered with layers of marine strata during the Paleozoic? (b) Label an epicontinental sea on **Figure E**. (c) Label the continents and orogens on **Figure F**. What supercontinent resulted from the collisional orogenies depicted?

16. (a) What organism does **Figure G** depict, during what period did it live, and why could such organisms grow so big at that time? (b) What eventually happened to all the vegetation visible in **Figure G**? (c) What is the Great Dying, when did it occur, and what geologic phenomenon might have caused it to happen?

Objective 9.5

Describe how the map of the world changed during the Mesozoic Era, characterize the dominant life forms of the time, and outline a widely accepted hypothesis to explain the sudden end of the era.

KEY CONCEPTS

• In the Mesozoic Era, Pangaea broke apart, and the Atlantic Ocean formed. Convergent-boundary tectonics dominated along the western margin of North America, and the Western Interior Seaway appeared.

• Several pieces of crust accreted to western North America during the Mesozoic, and the Sierran Arc intruded into the accreted terranes. The Sevier fold-thrust belt developed on the eastern side of the arc until the dip of the downgoing plate decreased, and the Laramide orogeny formed the structures of the US Rocky Mountains.

• The first mammals, along with dinosaurs, appeared at the end of the Triassic. Dinosaurs survived throughout the Mesozoic including the largest land animals ever to exist. During the Cretaceous, angiosperms appeared, along with modern fish.

• A huge mass-extinction event, probably due to a meteorite impact, wiped out the dinosaurs at the end of the Cretaceous Period.

EARTH-SCIENCE VOCABULARY

dinosaur (p. 336)
K-Pg extinction (p. 339)
Laramide orogeny (p. 338)

Sevier orogeny (p. 338)
Sierran Arc (p. 336)

REVIEW QUESTIONS

17. (a) How did the position of the western edge of North America change during the Mesozoic, relative to the interior of the continent? (b) What type of plate interaction led to the formation of the Sierran Arc and the Sevier orogeny? What type of orogen did this region represent? (c) Which present-day mountain belt did the Laramide orogeny contribute to forming? (d) Which

Mesozoic orogeny in western North America does **Figure H** depict? Label the key features.

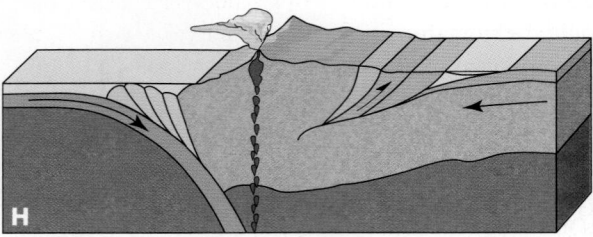

18. **(a)** What major ocean formed as a result of the breakup of Pangaea? **(b)** Did the rifting that led to the formation of that ocean succeed at the same time everywhere?

19. **(a)** What life forms appeared on land during the Triassic? **(b)** Why is the Mesozoic sometimes referred to as the Age of Dinosaurs? **(c)** What event does **Figure I** depict? Explain the consequences of that event.

Objective 9.6

Discuss the origin of present-day mountain belts during the Cenozoic Era, and place the Pleistocene Ice Age and the appearance of *Homo sapiens* in a geologic time context.

KEY CONCEPTS

- The major mountain belts that remain topographically high today were initiated, or continued to develop, during the Cenozoic Era.

- Collision of Africa and India with Europe and Asia produced the Alpine-Himalayan chain, and subduction at a gentle dip during the Laramide orogeny yielded the Rocky Mountains of the United States.

- Convergent-boundary tectonics persisted along the western margin of California until the San Andreas Fault formed; at about the same time, rifting in the western United States produced the Basin and Range Province.

- At times during the Pleistocene, vast ice sheets covered large areas of northern continents. They retreated during interglacial intervals.

- After the extinction of the dinosaurs, birds diversified, and mammals not only diversified, but got bigger. Our genus, *Homo*, appeared at about 2.4 Ma. Modern humans date back to about 0.3 Ma.

EARTH-SCIENCE VOCABULARY

Alpine-Himalayan chain (p. 339)

Basin and Range Province (p. 341)

Pleistocene Ice Age (p. 343)

REVIEW QUESTIONS

20. **(a)** How did the tectonic configuration of the western United States change during the Cenozoic? **(b)** Label the key crustal provinces depicted in **Figure J**, and add arrows to indicate the extension direction related to rifting in this region. **(c)** What plate interaction led to the formation of the Himalayas and the Tibetan Plateau?

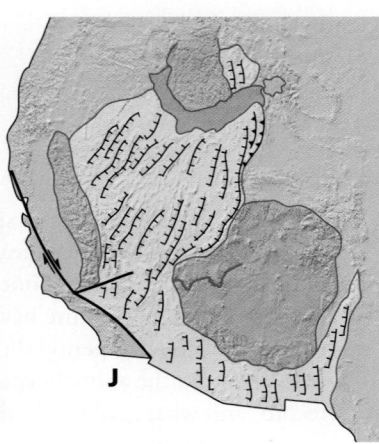

21. **(a)** How did the Earth's climate change through the Cenozoic? **(b)** About when did glaciers reappear in Antarctica? **(c)** When did the Pleistocene Ice Age begin, and about how many times did glaciers advance and retreat during the ice age?

22. **(a)** What animals might birds have descended from? **(b)** According to paleontological data, when did the first human-like primates appear? **(c)** When did modern humans, *Homo sapiens*, appear? Were *Homo sapiens* the only members of the genus *Homo* to have inhabited the Earth during the last 100,000 years?

Objective 9.7

Define global change, provide examples of various types of change that have happened during Earth history, and consider how humanity has become an agent of global change.

KEY CONCEPTS

- Earth history reflects the process of global change, modification of physical and biological components of the Earth System through time.

- Unidirectional change results in transformations that never repeat, whereas cyclic change involves repetition of the same steps over and over. Some changes take place gradually, and some are catastrophic.

- Humans have changed landscapes, caused a decrease in biodiversity, produced durable trace fossils, and added pollutants to the environment. The record of human activity during the present time interval, which has been called the Anthropocene, is significant enough to be preserved in the stratigraphic record.

EARTH-SCIENCE VOCABULARY

Anthropocene (p. 348)
biogeochemical cycle (p. 347)

global change (p. 346)
steady-state condition (p. 347)

23. **(a)** What is global change, and how do unidirectional changes differ from cyclic changes? **(b)** Provide examples of unidirectional changes. **(c)** Do all changes happen at the same rate? **(d)** What is meant by a biogeochemical cycle? **(e)** Provide examples of other cyclic changes besides biogeochemical cycles.

24. **(a)** In what ways have humans become agents of global change? **(b)** How does the Anthropocene differ from the Holocene? **(c)** What might the stratigraphic record of the Anthropocene look like to a geologic observer living 100 million years in the future? **(d)** On **Figure K**, a satellite photo of the New York City area, estimate the percentage of the map area that has been almost completely covered by concrete and asphalt (the light-colored areas). **(e)** Given that the population was only 30,000 in 1790,

this development was the result of construction over a 200-year period. The last ice age glacier covered the area of New York City 20,000 years ago, and was gone by 17,000 years ago. What percentage of the time since 17 Ka did it take for the concrete, brick, and asphalt to cover much of this landscape?

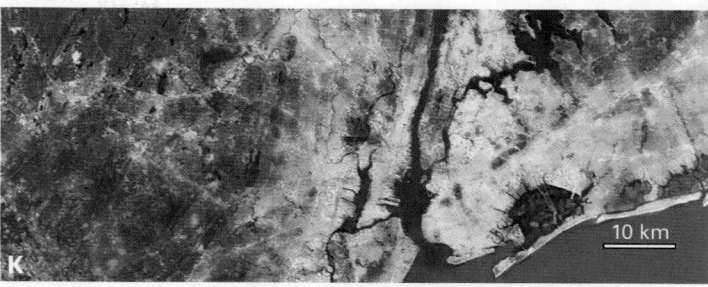

K

10 km

ANOTHER VIEW Layers of grey shale and siltstone form the cliffs of Letchworth State Park, in northwestern New York. Here, the Genesee River has cut a gorge in relatively recent geologic time. The rocks provide a record of marine deltaic deposition about 380 million years ago in an epicontinental sea. The sediments were derived from mountains that grew during the Acadian Orogeny, a Devonian event whose roots are visible in the eastern Appalachians.

10 RICHES IN ROCK
Energy and Mineral Resources

After studying this chapter, you should be able to...

1. explain where the energy stored in biomass and fossil fuels comes from, discuss what the "fossil" in fossil fuels refers to, and distinguish between biological and Earth resources.

2. provide a definition of a hydrocarbon, describe how oil and gas reserves form, distinguish between conventional and unconventional reserves, and explain how each can be found and extracted.

3. describe how coal forms, how it can be classified, and how it can be mined.

4. discuss the environmental and climate-change challenges involved in the use of fossil fuels, and why the Oil Age may end within the next half century.

5. characterize alternative sources of energy that could replace fossil fuels, and explain the pros and cons of developing each source.

6. discuss the nature of metals and their occurrence in ores, how ore deposits form and are processed, and how their mining and processing affect the environment.

7. relate nonmetallic mineral resources to their sources, and discuss how people use these resources.

>> This warehouse in New York City holds thousands of rock slabs that have been cut and polished to be used for kitchen countertops. The textures and colors of real rock are beautiful! In addition, rocks and other Earth materials supply most of the energy and mineral resources that society consumes, after extensive processing.

Picture a wolf living in a forest. To survive, it needs only food and water, which it can find in its immediate surroundings. Humans need more. Even a prehistoric hunter-gatherer used *energy resources* (such as wood or dung) for heat and cooking, and *mineral resources* (such as stone) for temples and tools (Fig. 10.1a). Industrialization in the 19th century greatly increased demands for such *natural resources*, defined broadly as sources of material in the natural world that sustain and enhance life. In fact, a person living in a modern industrialized society uses 100 times more natural resources than did a prehistoric hunter-gatherer, for people in modern industrialized societies live and work in constructed buildings, drive on paved roads, enhance their homes with heat and electric light, and use machines, computers, and phones routinely (Fig. 10.1b). Many of the materials that make all this engineering and technology possible require metals, glasses, and various exotic substances (such as rare-earth elements) that would have been completely unfamiliar to our ancestors (Table 10.1).

Where do natural resources come from? Some, such as wood, are *biological resources* (derived from forests, fields, and animals). All others are *Earth resources*, derived from geologic materials or from geologic, atmospheric, and oceanographic processes. An understanding of where Earth resources come from, and how their consumption impacts our planet, can help citizens of the 21st century make knowledge-based choices when confronting political issues pertaining to resources. This chapter surveys the Earth resources that society now relies on. The first half of the chapter focuses on energy resources. We begin our discussion with *fossil fuels* (oil, gas, and coal), today's most widely used energy resource. We then consider other energy resources of growing importance in modern society. The second half of the chapter focuses on mineral resources. We distinguish between metallic and nonmetallic mineral resources, and for each type, we review both its geologic origin and issues associated with its extraction, production, and use.

FIGURE 10.1 Earth resources, then and now.

(a) These ceremonial standing stones were set in place 5,300 years ago by people living in what is now France. These people also made stone tools, such as the ax head shown in the inset.

(b) A view of New York City. Nearly everything you see—stone, copper sheets, steel, glass, concrete, asphalt—has been constructed from Earth resources.

TABLE 10.1 Yearly Per Capita Use of Earth Resources in the United States

Resource	Amount (kg)	Amount (lb)
Stone	4,100	9,000
Sand and gravel	3,800	8,400
Oil	3,600	7,900
Coal	3,300	7,300
Iron and steel	550	1,200
Cement	360	800
Clay	220	480
Salt	200	440
Phosphate	140	300
Aluminum	25	55
Copper	10	22
Lead	6	3
Zinc	5	11
Rare-earth elements	0.6	1.3

Note: All amounts are approximate.

10.1 Introducing Fossil Fuels

No one knows when a prehistoric genius first figured out how to start a fire and keep it going to cook food and provide heat. Somehow, hunter-gatherers noticed that a stick of wood, a blob of dry dung, or a bundle of straw could be made to emit a bright, hot flame. They were watching the process of **combustion** (burning), during which organic chemicals in the wood, dung, or straw react very rapidly with oxygen. Such reactions break chemical bonds to produce H_2O and CO_2, and to release the energy (heat and light) that had been stored in those bonds.

Initially, people burned **biomass** (living or recently living organisms) to obtain energy. We can consider biomass to be a type of **fuel**, a material that stores energy in a compact and usable form. The energy stored in biomass comes, originally, from the Sun. **Photosynthesis**, a series of chemical reactions that take place in cyanobacteria, green algae, and plants, uses solar radiation to drive the construction of organic chemicals—such as sugar, cellulose, lipids, and carbohydrates—from water, carbon dioxide, and soil nutrients. Chemical bonds in these organic chemicals, therefore, effectively store the Sun's energy as potential energy. During combustion, these bonds are broken to release that energy. Simply put, combustion reverses the work of photosynthesis.

Beginning with the 19th-century *industrial revolution*, society's use of engines and furnaces required more fuel than biomass (at that time, mainly wood) could supply. So people began to use **fossil fuels**, which consist of biomass that has been buried, altered, and preserved ("fossilized") underground. Fossil fuels store solar energy that came to the Earth long ago. For example, when we burn a lump of Carboniferous coal, we are releasing energy originally produced by nuclear reactions in the Sun, say, 320 million years ago, and then stored in plants living at that time. We classify biomass as a **biological resource** (one derived from living or recently living organisms). Fossil fuels, in contrast, are a type of **Earth resource** (one derived from geologic materials or from natural Earth processes such as air movement or water flow). As we noted earlier, both biological and Earth resources are **natural resources**, materials or energy obtained from the natural Earth System that society uses to maintain or improve life.

Globally, fossil fuels remain the world's dominant energy source to this day (Fig. 10.2). Fossil fuels are burned to power engines, operate factories, and produce electricity. Originally, the energy stored in fossil fuels was transformed into kinetic energy that could do work via steam engines. In such engines, the heat released by burning the fuel produced pressurized steam, which drove a piston that, in turn, made wheels turn in locomotives or factory machines. Today, in the engines of cars and trucks, the sudden expansion of gas released by exploding fossil fuels in pistons drives wheels. Fossil fuels are also used

FIGURE 10.2 The proportions of different sources of energy used by human society have changed over time, but the amount of energy used has increased almost continuously. The sudden drops reflect global crises, such as the COVID-19 pandemic. The insets provide a close-up of consumption in 2020.

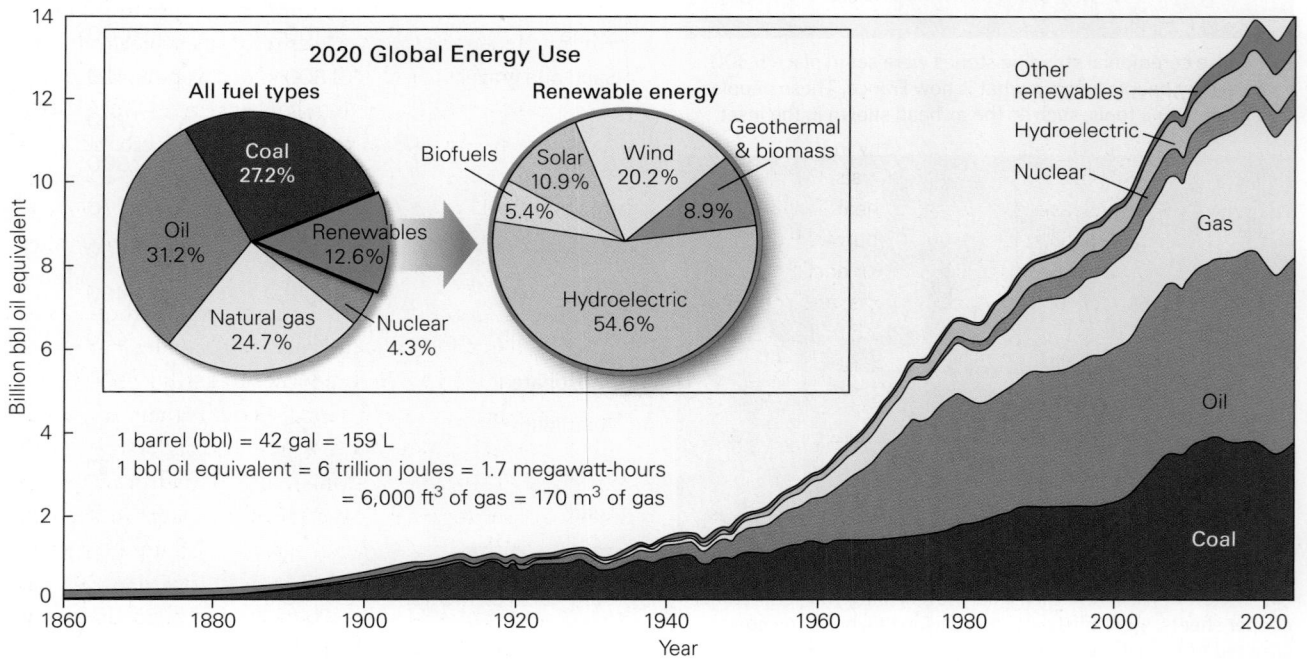

TABLE 10.2 Energy Density of Common Energy Sources

Source	Energy Density (MJ/kg)*
Uranium-235	79,500,000
Gasoline	46
Propane	46
Coal	24
Wood	16
Gunpowder	3
Lithium battery	1.8
Lead-acid battery	0.2

*MJ = megajoules, a unit of energy.

for electricity production; in electrical power plants, the heat released by burning fossil fuels produces pressurized steam, which spins turbines that drive electrical generators (Fig. 10.3).

Why did fossil fuels become so popular, and why have they remained in widespread use? Two reasons stand out. First, they have a high **energy density** (energy stored per kilogram), and second, they can be transported relatively easily. A plane using jet fuel, for example, can carry 500 people halfway around the world, whereas a plane powered by very heavy lead-acid batteries couldn't get off the ground (Table 10.2).

Geologists distinguish between two general categories of fossil fuels: *hydrocarbons* (oil and natural gas) and *coal*. These substances differ significantly from each other, in their composition and origin, as we now see.

Take-home message...

Photosynthesis stores solar energy in the chemical bonds of organic chemicals. Combustion of biomass releases this energy as heat and light. Before the industrial revolution, people burned biomass directly. Because of increased demand for energy, people came to rely on fossil fuels (oil, gas, and coal), which store energy that came to the Earth millions of years ago. Biomass is a type of biological resource, whereas fossil fuel is an Earth resource.

Quick Questions

- What is biomass?
- What is the relationship between photosynthesis and combustion?
- How does the energy density of gasoline compare with that of wood?

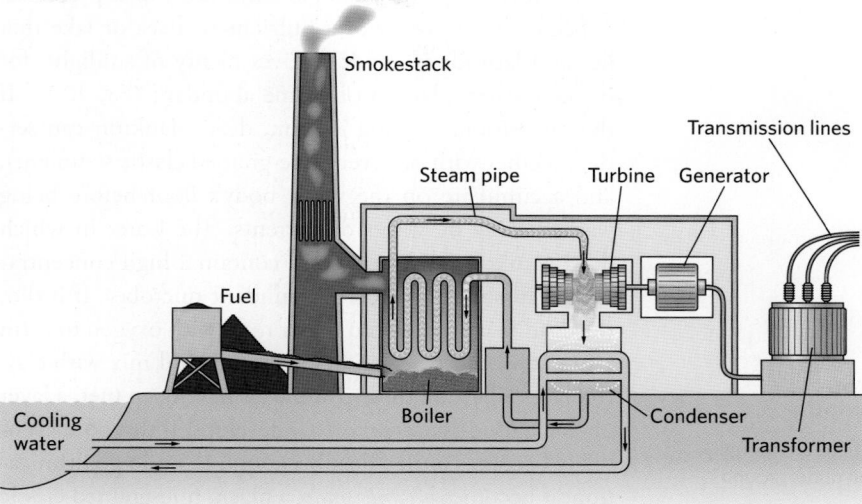

FIGURE 10.3 A simplified schematic diagram showing the configuration of a fossil-fuel power plant that generates electricity. Heat from burning fossil fuel (in this case, coal) converts water to steam, which drives a turbine, which, in turn, spins a generator. Power plants that burn hydrocarbons have the same general design.

10.2 Hydrocarbons: Oil and Natural Gas

What Is a Hydrocarbon?

Gasoline, diesel fuel, cooking gas, lubricating oil, heating oil, tar, kerosene, and jet fuel all consist of **hydrocarbons**, chain-like or ring-like molecules made of carbon and hydrogen atoms. We can distinguish among three general classes of hydrocarbons based on their *viscosity* (ability to flow) and their *volatility* (ability to evaporate). Under the pressure and temperature conditions typically found in your home, **natural gas** exists as a vapor. The natural gas used in homes generally consists of 70%–90% methane, with the remainder consisting of ethane, propane, and butane. Under the same conditions, **oil** exists in liquid form (Fig. 10.4), and **tar** exists as a very viscous liquid or even a solid. The viscosity and volatility of a hydrocarbon product reflect the size of the product's molecules, as determined by the number of carbon atoms in each molecule. Hydrocarbons with smaller molecules have lower viscosity (they flow more easily) and higher volatility (they evaporate more easily) than do those with larger molecules. Natural gas consists of very small molecules (1–4 carbon atoms), liquid hydrocarbons of medium-sized molecules, and tar of very large molecules. Notably, within the general category of liquid hydrocarbons, gasoline, which has relatively low viscosity, has smaller molecules (4–12 carbon atoms) than does, say, lubricating oil (22–70 carbon atoms).

FIGURE 10.4 Motor oil is a liquid composed of hydrocarbons.

Where Do Hydrocarbons Come From?

Most hydrocarbon resources form from the remains of *plankton* (tiny floating aquatic organisms, including algae). They are not, as commonly thought, residues of trees or dinosaur carcasses! Hydrocarbon formation requires a succession of special geologic conditions and processes. It begins in the water of a nutrient-rich sea or lake that lies at a latitude where it receives plenty of sunlight, for in such water, plankton become abundant (Fig. 10.5). If the depositional setting is calm, dead plankton can settle, together with *clay* (very fine-grained clastic sediment), and accumulate on the water body's floor before being washed away by waves or currents. The water in which dead plankton settle must not contain a high concentration of dissolved oxygen or abundant microbes. If it did, the dead plankton would either react with oxygen to form CO_2, or would be eaten before they could mix with clay. In places where all these conditions have been met, a layer of black, mud-like *organic ooze* develops. If this ooze eventually becomes buried deeply enough to undergo lithification, it becomes *organic shale*, a black, fine-grained clastic sedimentary rock containing 25%–75% organic matter. Organic shale (informally known as *black shale*) contains

Did you ever wonder . . .

whether there are actually pools of oil underground?

the raw materials from which hydrocarbons can form, so geologists refer to it as a **source rock**.

Transformation of the organic material of a source rock into hydrocarbons involves chemical reactions that happen only in a relatively narrow temperature range. At a temperature of 50°C–90°C (122°F–194°F), the organic chemicals (fats, sugars, and carbohydrates) in dead plankton transform into waxy molecules of **kerogen**. If the temperature rises to 90°C–160°C (190°F–320°F), kerogen molecules break down into oil and gas molecules, a process known as *hydrocarbon generation*. Oil molecules can survive at temperatures of up to 160°C (320°F). If the temperature of a source rock becomes warmer than that, oil molecules *crack* (split apart) to form smaller molecules of natural gas. And at temperatures over 250°C (480°F), the temperature at which rocks begin to undergo metamorphism, hydrocarbons lose all their hydrogen atoms, leaving behind carbon that crystallizes into graphite. Geologists refer to the limited range of temperatures at which oil can form and survive as the **oil window**, and the broader range of temperatures at which natural gas can form and survive as the *gas window*. At typical geothermal gradients, these windows lie at depths of

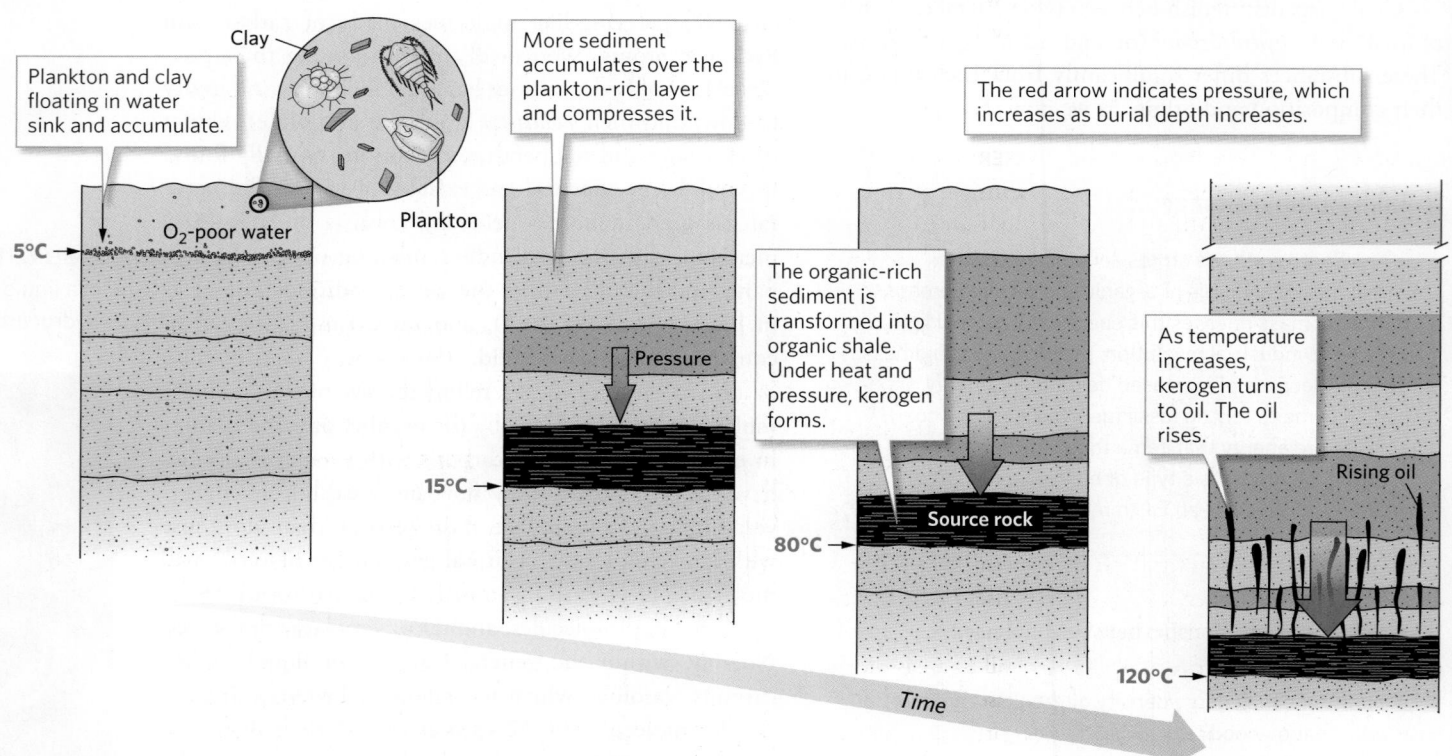

FIGURE 10.5 The process of hydrocarbon formation begins when organic matter and clay settle to form organic-rich ooze. Burial, heat, and pressure transform the sediment into organic shale. At a high enough temperature, the organic matter converts to kerogen, and at still higher temperatures, kerogen transforms into oil and gas, which can seep upward.

less than 10–15 km, so oil and gas can exist only in the Earth's upper crust.

Hydrocarbon Reserves

A significant quantity of potentially extractable oil and gas underground is a **hydrocarbon reserve**. If most of the hydrocarbons are oil molecules, it's an *oil reserve*, whereas if most are gas molecules, it's a *gas reserve*. The world market specifies the volume of oil in a reserve using *barrels* (bbl; 1 bbl = 42 gal = 159 L) and the volume of gas using cubic meters or cubic feet, as measured at 15°C (59°F) and 1.01 bars of pressure.

Hydrocarbons underground occupy cracks and *pores* (tiny open spaces between grains) within solid rock—they do not exist in underground lakes or pools! Geologists distinguish between two categories of reserves based on how easily the hydrocarbons can be pumped from the rock. Hydrocarbons can be pumped relatively easily from a **conventional reserve**, for in such reserves, the hydrocarbons have relatively low viscosity, the rock that holds them contains 10%–30% hydrocarbons, and the hydrocarbons can move relatively easily through the rock to a well. Hydrocarbons cannot be pumped out of the ground easily from an **unconventional reserve**, either because the hydrocarbons are too viscous to flow or because pores in the rock are not connected, so that hydrocarbons have no pathway through which to move. The energy industry uses the adjectives "conventional" and "unconventional" simply to emphasize that conventional reserves have been tapped since the late 19th century, whereas unconventional reserves have been tapped only for about the last 25 years, because their extraction requires advanced technology.

CONVENTIONAL HYDROCARBON RESERVES. Because source rocks are organic shale, which consists of tightly packed clay flakes, hydrocarbons cannot move through them easily (Fig. 10.6a). Therefore, to form a conventional reserve, hydrocarbons must, over geologic time, migrate out of the source rock into a **reservoir rock**, a rock with high **porosity** (percentage of open space between grains) and high **permeability** (the degree to which pores are interconnected). Poorly cemented sandstone exemplifies a good reservoir rock, because it contains lots of interconnected pores (Fig. 10.6b), a pump can suck hydrocarbons out of it relatively easily. Oil and gas, once formed, migrate upward because they are less dense than groundwater, so buoyancy causes them to rise above the water. (To picture the process, imagine oil rising above vinegar in salad dressing.) If both oil and gas occur in reservoir rock, gas rises above oil because it's less dense than oil. This **hydrocarbon migration** process can take thousands to millions of years (Fig. 10.7).

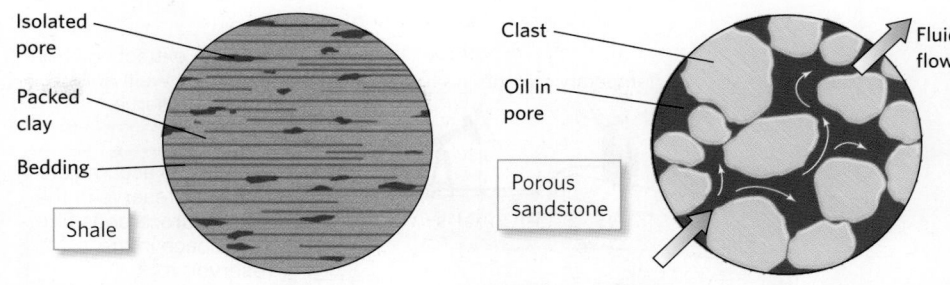

FIGURE 10.6 Porosity and permeability in sedimentary rocks, as seen through a microscope. Hydrocarbons occupy the pore space.

(a) This shale has low porosity and low permeability.

(b) This sandstone has high porosity and high permeability.

If oil and gas move relatively easily through reservoir rock, why don't they reach the surface of the Earth and escape? Sometimes they do, at an outlet known as an *oil seep*. For a conventional underground hydrocarbon reserve to exist, hydrocarbons must migrate into a **trap**. The existence of a trap requires that a *seal rock*, an impermeable rock such as shale or salt, lie above the reservoir rock to prevent hydrocarbons from rising farther, and that the seal rock and reservoir rock be configured in such a way that the hydrocarbons become confined in a relatively restricted area. Geologists recognize several types of trap geometries (Fig. 10.8).

UNCONVENTIONAL HYDROCARBON RESERVES. Formation of an unconventional hydrocarbon reserve does not require migration of hydrocarbons into a reservoir rock, nor does it require confinement within a trap. All it requires is the occurrence of a good source-rock bed, or of a loose sediment containing hydrocarbons. Geologists recognize three types of unconventional reserves: (1) *oil shale* consists of source rock that became warm enough for kerogen, but not oil or gas, to form; (2) *tar sand* consists of weak sandstone whose pores contain hydrocarbons that are too viscous to flow; (3) and **tight oil** and *tight gas* (also known as *shale oil* and *shale gas*) reserves consist of source rock containing oil and gas that has not migrated away.

Finding, Extracting, and Refining Hydrocarbons

THE SEARCH FOR HYDROCARBONS. Prior to the mid-19th century, people used "rock oil," later known as *petroleum* (from the Latin *petra*, meaning rock, and *oleum*, meaning oil), primarily for greasing wagon axles. Such oil was rare, for people could obtain it only at oil seeps. Then, in 1854, a group of investors speculated that kerosene, a hydrocarbon liquid derived from petroleum, could replace increasingly scarce whale oil as a lamp fuel, if petroleum supplies could be found in reasonable quantities. They hired Edwin Drake to drill for oil at a seep near

FIGURE 10.7 Initially, oil resides in source rock. Because it is buoyant relative to groundwater, the oil migrates into overlying reservoir rock. The oil accumulates beneath seal rock in a trap. In this example, the oil migrates along a fault into a fold.

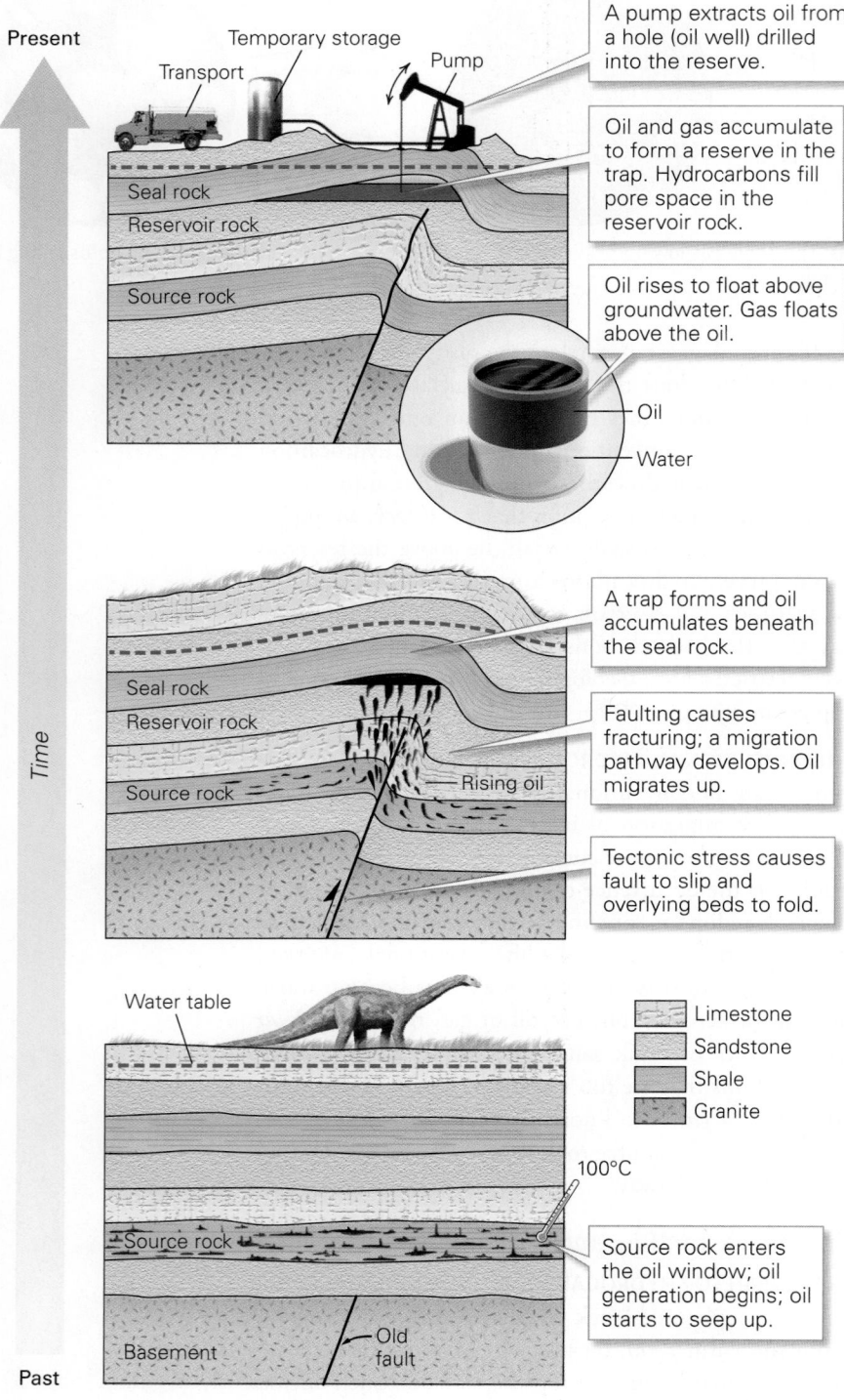

Present

Transport | Temporary storage | Pump

A pump extracts oil from a hole (oil well) drilled into the reserve.

Seal rock
Reservoir rock

Oil and gas accumulate to form a reserve in the trap. Hydrocarbons fill pore space in the reservoir rock.

Source rock

Oil rises to float above groundwater. Gas floats above the oil.

Oil
Water

Time

Seal rock
Reservoir rock

A trap forms and oil accumulates beneath the seal rock.

Source rock
Rising oil

Faulting causes fracturing; a migration pathway develops. Oil migrates up.

Tectonic stress causes fault to slip and overlying beds to fold.

Water table

100°C

Limestone
Sandstone
Shale
Granite

Source rock

Source rock enters the oil window; oil generation begins; oil starts to seep up.

Basement
Old fault

Past

Titusville, Pennsylvania. On August 27, 1859, Drake and his crew succeeded in finding oil underground. Within a few years, thousands of oil wells had been drilled, and by the turn of the 20th century, industrialized societies had become addicted to oil. Initially, most oil was used to produce kerosene, but as electricity took over as the main power source for lighting, people began using oil as a fuel to generate electricity. Demand increased drastically when gasoline became the preferred fuel for powering the cars and trucks that rapidly replaced horses as the main mode of transportation.

The earliest *oil fields* and *gas fields*—areas where numerous wells can profitably extract a hydrocarbon reserve—were found either by searching for seeps or through blind luck. Eventually, energy companies hired geologists to identify source rocks, reservoir rocks, and traps. Most modern exploration relies on **seismic-reflection profiles**, which help to identify hydrocarbon-bearing strata underground (Fig. 10.9). To obtain such a profile (a vertical cross-sectional image) on land, a special truck places a vibrating plate on the ground surface and sends seismic waves into the ground. The waves reflect off contacts between rock layers and return to the ground surface, where they are recorded by seismographs. The time between the generation of each wave and its return indicates the depth of a contact between layers. Computers process the data to produce a profile showing the configuration of subsurface beds. To obtain seismic-reflection profiles offshore, a ship generates seismic signals by releasing high-pressure bursts of air into the water. The energy of the air bursts penetrates the seafloor and passes through sediment and rock beneath the seafloor in the form of seismic waves. This ship also tows an array of seismographs to detect reflected signals. Modern seismic studies can generate three-dimensional models of the subsurface.

DRILLING AND REFINING. To extract oil or gas, workers penetrate underground layers of rock using a *rotary drill*, a hollow metal pipe tipped by a *drill bit*, a metal bulb studded with hard prongs (Fig. 10.10a). As the bit rotates, it grinds the rock into powder and chips, called *cuttings*. Early drilling methods could produce only vertical drill-holes, but today, **directional drilling** allows drillholes to be oriented in any desired direction (Fig. 10.10b). For land-based drilling, engineers use a *drilling derrick*, a tower-like structure used to hoist heavy drill pipe into place, installed on a *drilling pad*, a flat area covered by thick gravel or concrete. Offshore, on the continental shelf, the drilling rig sits on an *offshore-drilling platform* (Fig. 10.11).

During drilling, workers pump *drilling mud*, a slurry of water mixed with clay and various chemicals, down the drill pipe. This mud comes out of holes at the end of the bit, then flows back up to the surface in the space between the pipe and the hole. This mud not only cools the bit and flushes cuttings out of the hole, but its weight produces pressure in the hole, which prevents subsurface hydrocarbons from rushing into the hole and up to the surface. Once drilling has been completed, workers "complete" the hole by removing the drill pipe and replacing it with

FIGURE 10.8 A trap is a configuration of seal rock over reservoir rock in a geometry that keeps the oil underground. There are four basic types.

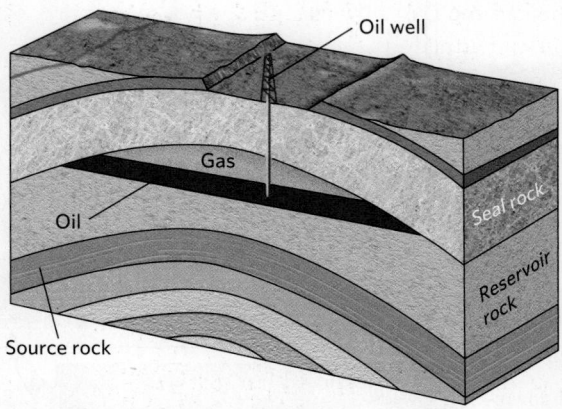

(a) In an anticline trap, hydrocarbons accumulate in the crest of a fold.

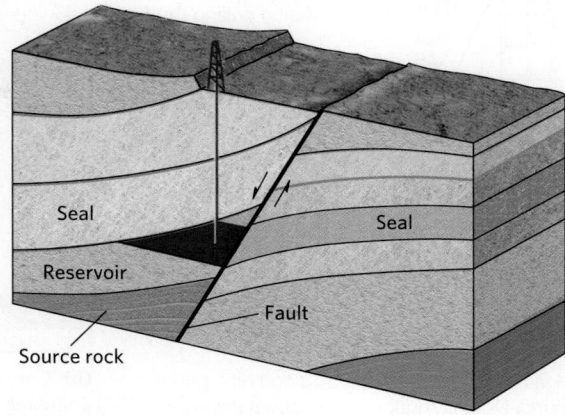

(b) In a fault trap, hydrocarbons collect in tilted strata adjacent to a fault.

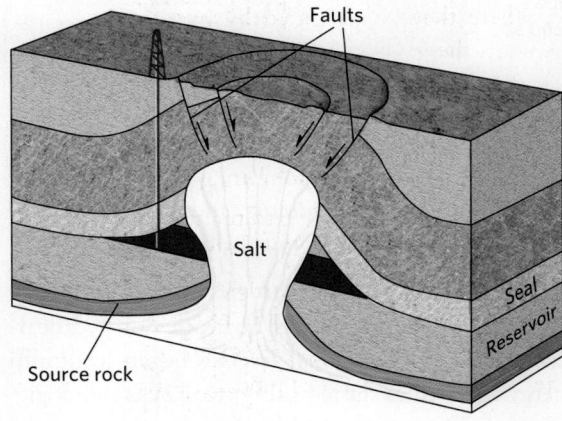

(c) In a salt-dome trap, hydrocarbons collect in strata on the flanks of a salt dome, a bulbous intrusion of salt that has risen into overlying strata. Salt is impermeable.

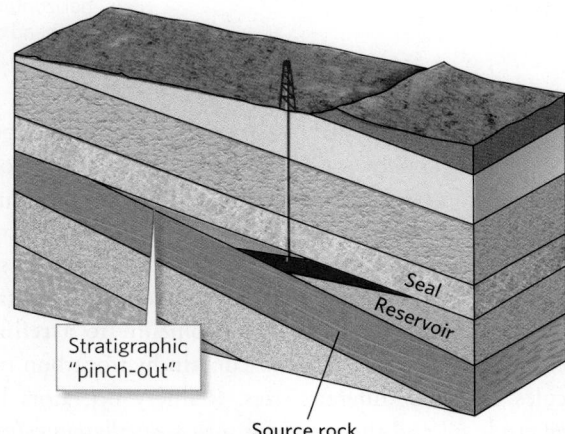

(d) In a stratigraphic trap, hydrocarbons collect where a reservoir-rock layer pinches out.

FIGURE 10.9 Using seismic-reflection profiling to describe underground beds and locate hydrocarbon reserves.

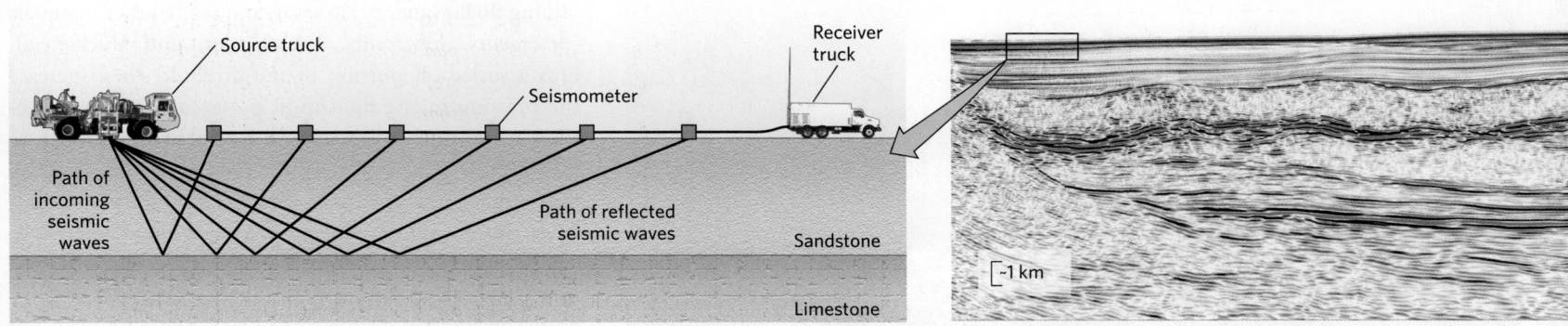

(a) A source truck sends seismic waves into the Earth. The waves are reflected by contacts between rock layers. At the ground surface, they are picked up by seismographs. The waves' travel times indicate the depths of the contacts.

(b) A seismic profile can reveal the presence of geologic structures underground. The colored bands represent layers of strata.

FIGURE 10.10 Drilling for oil.

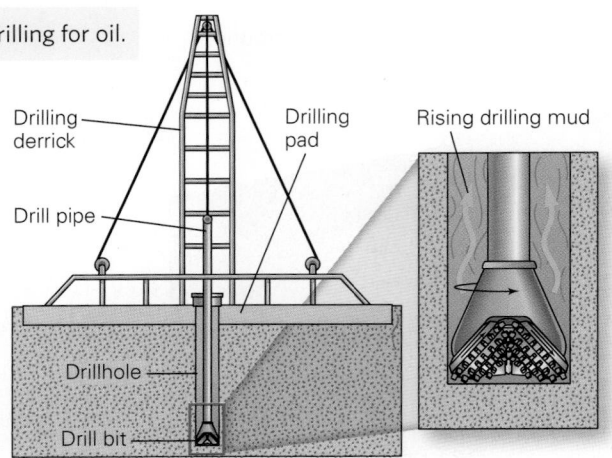

(a) A rotary drill—a rotating pipe tipped by a drill bit—grinds a hole into the ground. Drilling mud, pumped down through the pipe, comes out through holes in the bit.

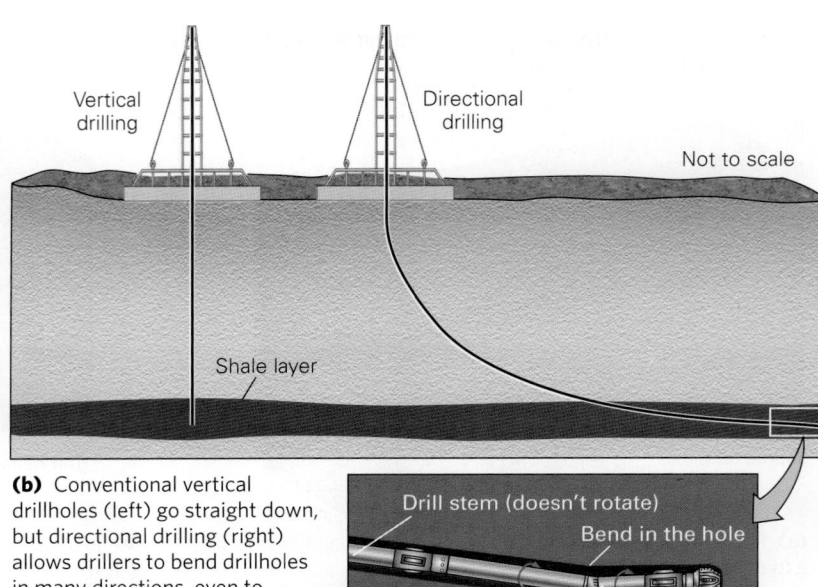

(b) Conventional vertical drillholes (left) go straight down, but directional drilling (right) allows drillers to bend drillholes in many directions, even to horizontal. As the inset shows, the head of the drill can bend at an elbow.

FIGURE 10.11 An offshore-drilling platform. Some platforms are anchored in the seafloor, whereas others float on giant submerged pontoons.

another pipe surrounded by concrete. They then perforate the pipe and concrete at depths where the drillhole crosses hydrocarbon-bearing layers. Finally, they install a pump to pull oil and gas out of the rock (Fig. 10.12a).

Once *crude oil*, the oil coming directly from the subsurface, has been brought to the surface, it can be transported by truck, tanker ship, or pipeline to a **refinery** (Fig. 10.12b, c). Crude oil can contain hydrocarbon molecules of many different sizes. Refinery operators heat the crude oil and then inject it into a *distillation column*. Smaller (lighter) molecules rise in the column while larger (heavier) molecules sink. Consequently, natural gas flows from outlets at the top of the column, gasoline and motor oil flow from outlets in the middle, and tar (used for the production of plastics) comes out the bottom.

HYDROFRACTURING IN UNCONVENTIONAL RESERVES. Starting about 20 years ago, as quantities of hydrocarbons available in conventional reserves began to diminish, energy companies started using two key technologies to extract significant quantities of hydrocarbons from unconventional reserves. First, they started to use directional drilling to bore into a horizontal bed of source rock, and then extend the drillhole for kilometers within the bed. Such a drillhole provides access to a large volume of source rock. Second, they started to use **hydrofracturing** (commonly known as *fracking*) to increase the permeability of the source rock surrounding the drillhole. To "frack" a well, drillers pump *fracking fluid*, a liquid containing 90% water, 9.5% sand, and 0.5% other materials (detergents, lubricants, bactericides, and thickeners), into a sealed-off portion of the drillhole, then increase the pressure in the fluid until pre-existing natural fractures in the rock open up and new fractures propagate (Fig. 10.13). When the drillers pump out the fracking fluid, some sand remains in the cracks and prevents them from closing tightly. The process of fracking, therefore, creates enough new permeability to allow oil or gas to flow from the rock into the drillhole. If workers tried to pump oil directly from a source-rock bed without using fracking to produce and open cracks, they would recover very little oil, because source-rock beds consist of impermeable shale.

FIGURE 10.12 Pumping, transporting, and refining oil.

The head of the pump goes up and down.

Storage tanks

Hole

(a) When drilling is complete, the derrick is replaced by a pump, which pulls oil out of the ground.

(b) The Trans Alaska Pipeline transports oil from fields on the Arctic coast to a tanker port on the southern coast of Alaska.

(c) This tanker can transport a million barrels of oil (enough to supply the entire United States for a few hours).

Directional drilling and hydrofracturing technologies led to a boom in the drilling of unconventional reserves starting around 2007 (Fig. 10.14). In the United States, thousands of new wells were drilled, and the country became a net exporter of oil and gas. But the abundance of drilling has led to significant concerns about contamination of freshwater supplies, leakage of methane (a greenhouse gas) into the atmosphere, and the triggering of small earthquakes (Box 10.1). Because of the law of supply and demand, the price of hydrocarbons remained relatively low, which pleased consumers, but the greater availability of hydrocarbons probably slowed the development of alternative energy sources, which have implications for climate change and, therefore, for meeting future energy needs.

To obtain hydrocarbons from near-surface tar-sand deposits, producers dig open-pit mines (Fig. 10.15), then

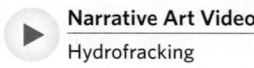

Narrative Art Video
Hydrofracking

FIGURE 10.13 Hydrofracturing increases the permeability of rock by producing new fractures and opening pre-existing fractures. In this example, a horizontal hole is undergoing fracking.

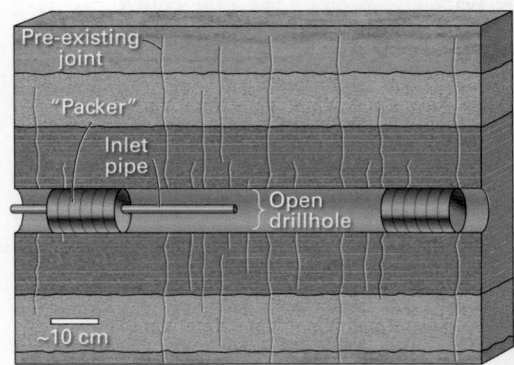

Pre-existing joint

"Packer"

Inlet pipe

Open drillhole

~10 cm

(a) After a hole has been drilled, a portion of the hole is sealed off with expandable cylinders called packers. A pipe passes through one of the packers.

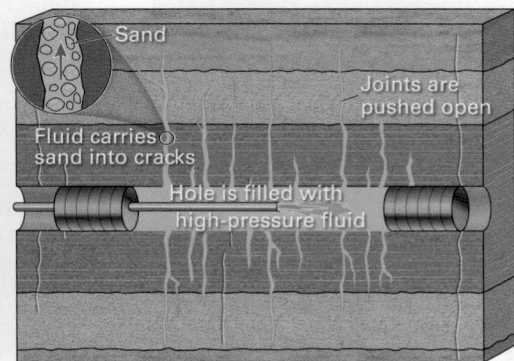

Sand

Fluid carries sand into cracks

Joints are pushed open

Hole is filled with high-pressure fluid

(b) Pumps inject fluid under high pressure into the isolated segment of the hole. The pressure pushes open existing cracks and forms new cracks.

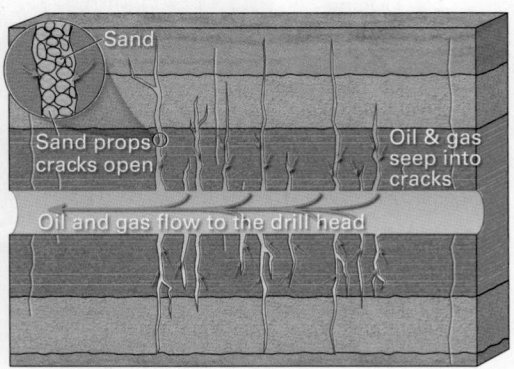

Sand

Sand props cracks open

Oil & gas seep into cracks

Oil and gas flow to the drill head

(c) After the packers and the fluid are removed, sand remains and props open the cracks. Oil or gas can seep into the pipe and flow up to the ground.

BOX 10.1 ▶ Putting Earth Science to Use

Concerns about hydrofracturing

It's likely you've heard of fracking in the news, for it has been the subject of intense political debates. Why is the technology so controversial? Developing an unconventional reserve that couldn't have been developed without fracking has many consequences for a region. For example, developing an oil or gas field provides numerous well-paying jobs, not only in the local energy industry but also in the retail and service industries and in construction. But such a boom may stress the resources of an area. For example, rural roads might not be able to handle the traffic of heavy trucks carrying equipment, water, chemicals, and sand, and small towns might not have sufficient businesses and health-care options to handle a sudden influx of people. Since many of the jobs last only as long as the field remains productive, the population boom may go bust once the reserve has been drained, causing real-estate prices to crash. Residents in an area that's being drilled worry that surface-water supplies may be overused for fracking fluid, or that the chemicals in fracking fluid may spill into surface water or leak into groundwater.

Which aspects of fracking pose the greatest risk for water contamination? Surprisingly, the long horizontal segments of drillholes in source rock are not necessarily the main problem, for many penetrate source-rock beds at depths of over 1.5 km (1 mi), where groundwater tends to be saline and not good for drinking or irrigation; consequently, leakage from the fracked part of the hole into groundwater at depth might not have immediate effects on drinking-water supplies. Fresh, usable groundwater lies at shallower depths, so it's critical that the vertical part of each drillhole be properly sealed to prevent leakage and

that operators take great care not to spill fracking fluid on the ground surface or into surface water **(Fig. Bx10.1)**. If such precautions aren't taken, surface water and well water can be contaminated.

Notably, in some hydrocarbon fields, oil and gas extraction brings up large volumes of saline groundwater along with oil and gas. This water can't be dumped into rivers or used for irrigation, so to dispose of it, drillers pump it back underground into deep wells called *injection wells*. Pumping at very high pressure and at great depth can trigger small earthquakes. Therefore, injection pressures should be closely regulated.

FIGURE Bx10.1 The potential for groundwater contamination by hydrofracturing.

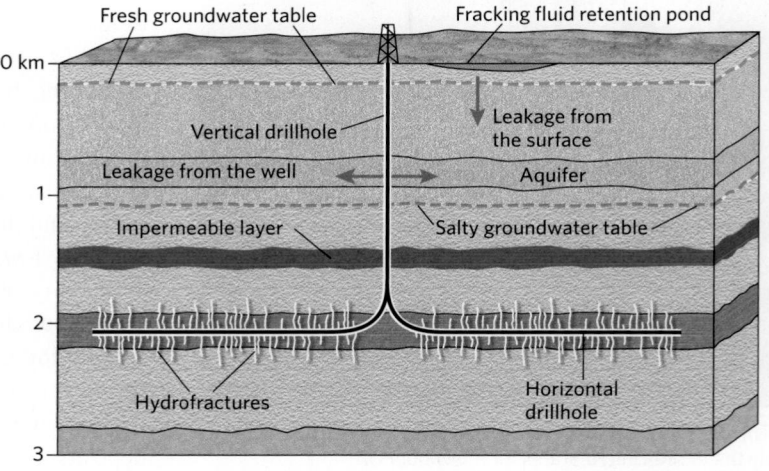

FIGURE 10.14 Drilling and fracking at a well in Pennsylvania. The trucks and the holding pond are used during hydrofracturing.

FIGURE 10.15 An open-pit mine for tar sand in Alberta, Canada.

FIGURE 10.16 The distribution of hydrocarbon reserves.

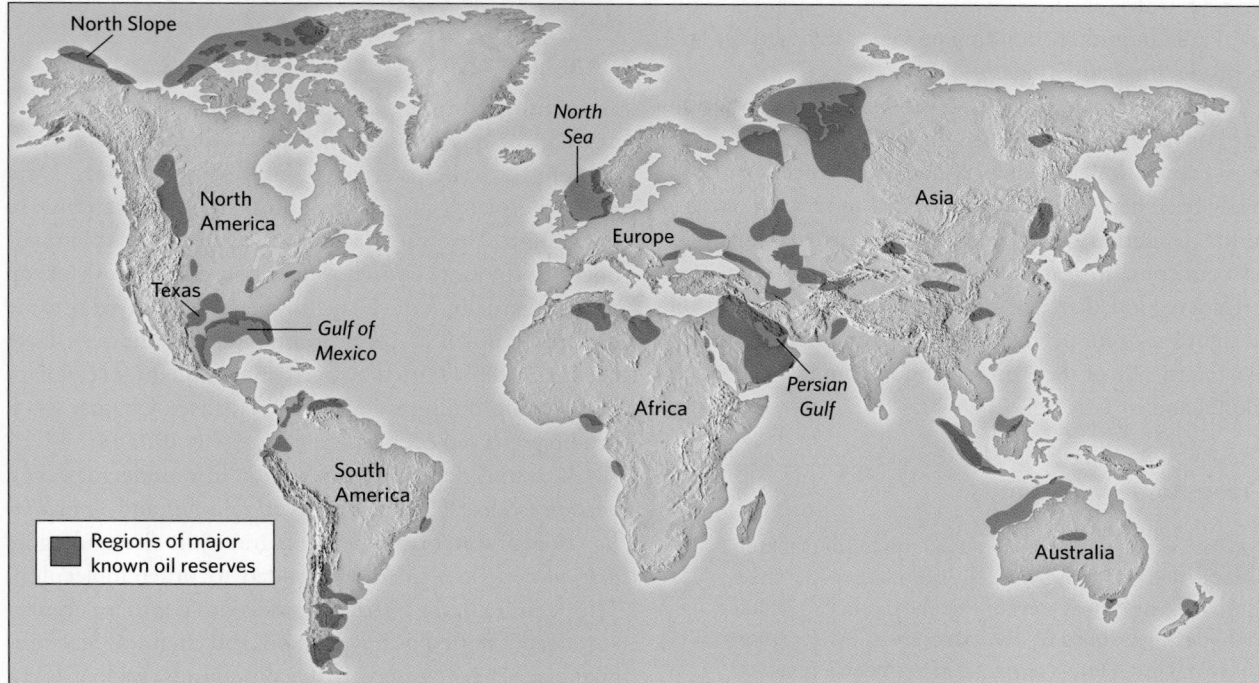

(a) The distribution of conventional oil reserves around the world. Some are onshore and some are offshore.

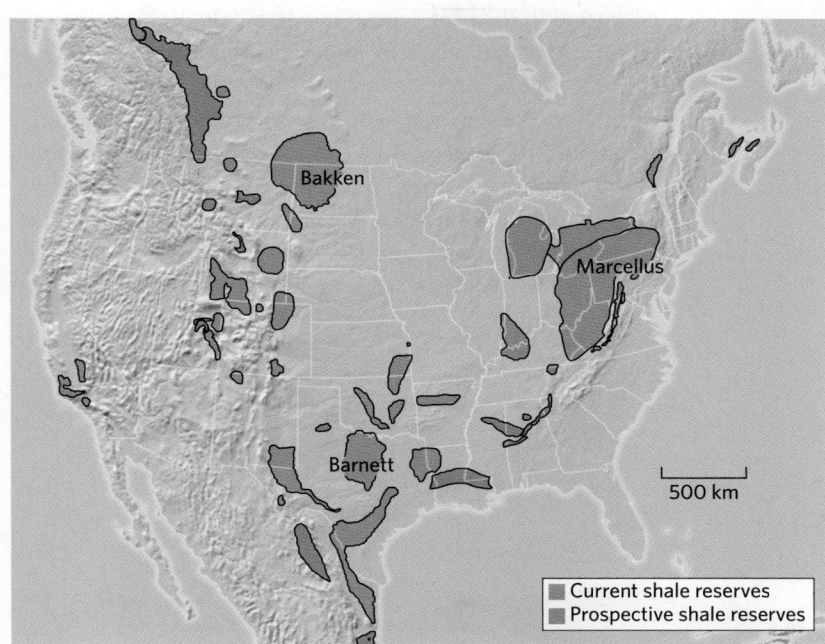

(b) A map of the major unconventional tight oil and gas fields in North America. The three largest are labeled.

heat the extracted sandstone in furnaces until the hydrocarbons become less viscous and can be separated from the sand. These mines damage the landscape, as they can be difficult to reclaim. To extract oil from deep deposits of tar sand, producers drill injection wells into the deposits. They force steam or solvents down the injection wells to liquefy the oil, then pump it up from nearby extraction wells. Extracting kerogen from oil shale requires processes similar to those used to obtain oil from tar sand. The kerogen extracted from oil shale and the tar from tar sand must be processed at a refinery to "crack" (break apart) its molecules and convert it into smaller, usable hydrocarbon molecules. Because of the amount of energy needed to produce usable oil from tar sand or oil shale, the process is quite expensive.

Where Do Oil and Gas Occur?

Oil and gas reserves are not randomly distributed around the planet. Most conventional reserves occur in passive-margin basins, rift basins, intracontinental basins, and fold-thrust belts **(Fig. 10.16a)**. Currently, countries bordering the Persian Gulf contain the world's largest reserves. Why? Much of the continental crust that now lies in the Middle East was situated in tropical areas, and was submerged by shallow seas, during the Cretaceous Period (145–66 million years ago). Biological productivity in the brightly lit waters of these regions was high, so the organic ooze that accumulated

See for yourself

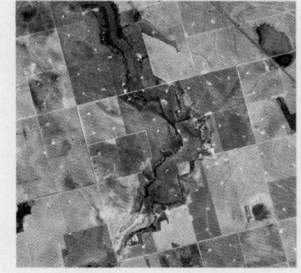

Oil Field near Lamesa, Texas

Latitude: 32°33'18.42" N
Longitude: 101°46'55.39" W

Zoom to an altitude of 8 km (5 mi) and look down.

Within a grid of farm fields, each 1.6 km, or 1 mi, wide, small roads lead to patches of dirt. Each patch hosts or hosted a pump for extracting oil. These wells are tapping an oil reserve in Permian sandstone reservoirs.

there, when lithified, became excellent source rock. Thick layers of sand buried the source rock and eventually became porous sandstone, which serves as ideal reservoir rock. Later, mountain-building processes folded these layers, producing many traps.

Many unconventional reserves occur in association with conventional ones, but they also occur in places where conventional reserves either do not occur or have already been depleted. Notably, several significant unconventional reserves occur near urban areas of industrialized countries, so transport costs to the consumers can be relatively low. Particularly large quantities of tight oil and tight gas occur in shale beds in sedimentary basins on the western side of the Appalachian Mountains, in North Dakota, and in Texas (Fig. 10.16b).

Take-home message...

🏠 Oil and gas form from the remains of plankton that, together with clay, become organic shale, a source rock. Heat can convert the organic molecules into oil or gas. In conventional reserves, hydrocarbons have migrated into a porous and permeable reservoir rock and are held underground by an impermeable trap. In unconventional reserves, the hydrocarbons are too viscous to flow, or the rock containing them is impermeable.

Quick Questions ————————————

- How do natural gas, oil, and tar differ from one another?
- Under what circumstances do drillers use a horizontal drillhole to extract oil?
- Do oil reserves occur everywhere? Explain your answer.

10.3 Coal: Energy from the Swamps of the Past

Coal Formation and Classification

Imagine a swamp in a tropical climate. Trees, shrubs, and other plants grow in abundance, producing an immense amount of cellulose-rich biomass (Fig. 10.17a). Over time, the plants die and fall into the stagnant water of the swamp. This water lacks oxygen, so the dead biomass does not rot away entirely before new plants grow on top of it. Eventually, a thick layer of partially decayed organic material, known as **peat**, accumulates. If the peat layer becomes deeply buried by overlying sediment, as can happen when a transgression takes place (see Chapter 5), it can be preserved over geologic time (Fig. 10.17b).

As burial depth increases, peat first undergoes *compaction* (squeezing into a smaller volume) and *consolidation* (loss of water from between grains). It also undergoes *dehydration* (loss of water bonded to other molecules). Temperature in the Earth increases with depth, so eventually, deeply buried peat warms up, and chemical reactions take place that release volatile compounds, such as CH_4, CO, O_2, and N_2. Peat has a carbon concentration of 30%–45%, but as volatiles escape, the carbon concentration in the peat deposit increases (Fig. 10.18a). In addition, the peat starts to undergo lithification. When the carbon concentration reaches 60%, the deposit can formally be called **coal**. The carbon in coal occurs in huge organic molecules, called *coal macerals*, each of which contains hundreds of carbon atoms.

Coal rank indicates the extent to which coal in macerals has become enriched in carbon by losing volatile organic molecules. Coal containing 60%–70% carbon is a

FIGURE 10.17 Coal forms when plant debris becomes deeply buried.

(a) A museum diorama depicting a Carboniferous coal swamp.

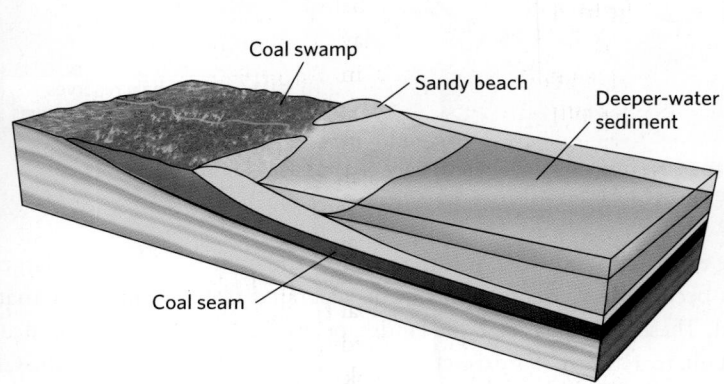

(b) As sea level rises, peat formed in a coal swamp can be buried and preserved.

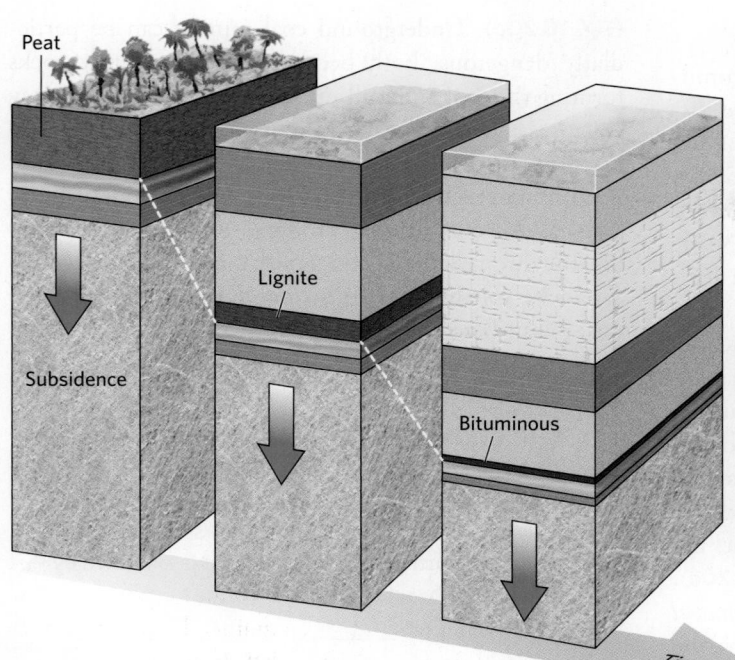

FIGURE 10.18
Formation and appearance of coal seams.

(b) Coal beds interlayered with beds of sandstone and shale.

(a) Burial of peat leads to the formation of coal of progressively higher carbon content, and therefore progressively higher rank. Note that the layer of organic material becomes thinner as it undergoes compaction, consolidation, dehydration, and loss of volatile components.

low-rank coal, known as *lignite* or *brown coal* (Table 10.3). With still deeper burial and associated higher temperatures, additional volatiles escape, yielding even higher concentrations of carbon, so that lignite transforms into dull black intermediate-rank *bituminous coal*, in which the carbon concentration in macerals exceeds 70%. If temperatures become even higher, bituminous coal transforms into shiny black high-rank *anthracite*, in which the carbon content in macerals exceeds 87%. The energy density of coal increases with increasing rank (see Table 10.3). Notably, while lignite and bituminous coal can be formed by burial in a sedimentary basin, the temperatures needed to form anthracite are so high that such coal usually develops only along the margins of mountain belts.

Most of the coal that we use today comes from the deposition of woody plant debris in Paleozoic or Mesozoic **coal swamps**, regions that resembled the woody wetlands found in tropical and subtropical areas today (see Fig. 10.17a). When we burn coal, we are releasing energy first trapped by photosynthesis tens of millions to hundreds of millions of years ago. Because coal forms by deposition in a sedimentary environment, geologists consider it to be sedimentary rock, so coal beds (known as **coal seams**) occur as layers within a sedimentary succession. In an outcrop, a coal seam looks like a black band sandwiched between other kinds of sedimentary

rocks (Fig. 10.18b). Individual seams may be centimeters to meters thick and may underlie a broad region. A coal seam typically does not consist only of coal macerals, however; it also contains clay minerals, as well as quartz silt and sand. Coal deposited in some environments also contains sulfur, commonly in the form of pyrite.

The most extensive deposits of coal in the world occur in strata deposited during the Carboniferous Period—it's the occurrence of so much coal in strata of that period that led to the name Carboniferous. At that time, the supercontinent Pangaea had assembled, and much of it straddled warm tropical regions with high rainfall. In lowlands of equatorial Pangaea (which included much of what is now the United States), land plants flourished in vast coal swamps. The peat of those swamps became the coal found in what is now the midwestern United States and the Appalachian region. The coal of Wyoming formed later, in the Cretaceous Period. The plants that became this coal grew along the margins of the Western Interior Seaway (see Chapter 9).

TABLE 10.3 Coal Rank and Associated Energy Density

Rank	Formation temperature	Carbon concentration	Energy density
Lignite	<100°C	60%–70%	18 MJ/kg
Bituminous	100°C–200°C	70%–87%	27 MJ/kg
Anthracite	200°C–300°C	87%–95%	33 MJ/kg

Mining Coal

Sedimentary strata of continents contain huge accumulations of coal, known as **coal reserves**, which are distributed widely among the sedimentary basins of the world (Fig. 10.19). The methods by which companies extract coal from these reserves depend on the depth of the coal. If coal seams lie within about 100 m (330 ft) of the ground surface, **strip mining** can be the least expensive method. To dig a strip mine, miners use a giant shovel, or *dragline*, to scrape off the soil and the layers of sedimentary rock that lie above the coal seam (Fig. 10.20a). Then they use power shovels to dig out the coal and dump it into trucks, which carry it to a processing facility (Fig. 10.20b). In some strip mines, dragline operators separate out and preserve soil and underlying rock, and when the coal has been scraped out, the operator refills the pit and replants the soil (Fig. 10.20c). In hilly areas, miners may employ *mountaintop removal mining*—during this process, they strip all land above the coal and push it into an adjacent valley.

Deeply buried coal can be obtained economically only by **underground mining**. To develop an underground mine, miners dig a shaft down to the depth of the coal seam and then excavate a maze of tunnels, using huge grinding machines that chew their way into the coal (Fig. 10.20d). Underground coal mining can be particularly dangerous, both because the sedimentary rocks forming the roof and walls of the mine are weak and may collapse, and because methane gas released by chemical reactions in the coal can accumulate in the mine and, if not removed by ventilation fans, will explode if lit by a spark. All coal mining produces dust, so unless they breathe through filters, miners risk contracting *black-lung disease* by inhaling coal dust, or *silicosis* by inhaling the dust of sandstone and shale.

Take-home message...

Coal forms from the remains of plant material. When buried deeply, this organic material undergoes chemical reactions that concentrate carbon. Coal is ranked by its carbon content. It occurs in seams in sedimentary successions, from which it is extracted by strip mining or by underground mining.

Quick Questions

- Why is coal considered to be a type of fossil fuel?
- Are coal and oil formed from the same type of organic material?
- Name the different types of coal, in order from low rank to high rank.

FIGURE 10.19 The global distribution of coal reserves. Most coal has accumulated in intracontinental and foreland basins.

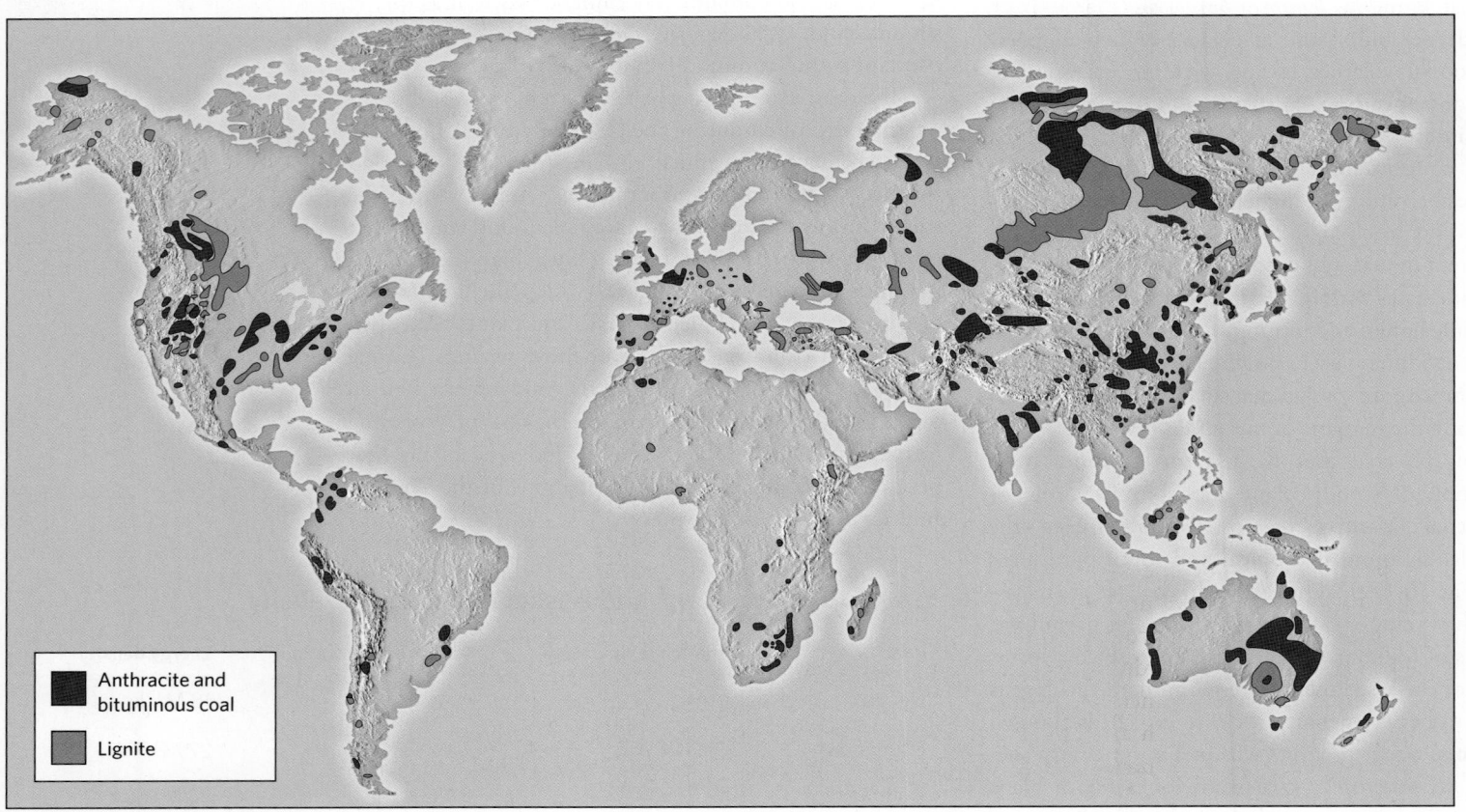

Legend:
- Anthracite and bituminous coal
- Lignite

FIGURE 10.20 Coal mining.

A dragline stripping coal in an Indiana mine.

Large bulldozer   Shovel

(a) A dragline stripping away the soil and sedimentary rock that overlie a coal seam.

(b) A large strip mine exposing horizontal beds of coal in Germany.

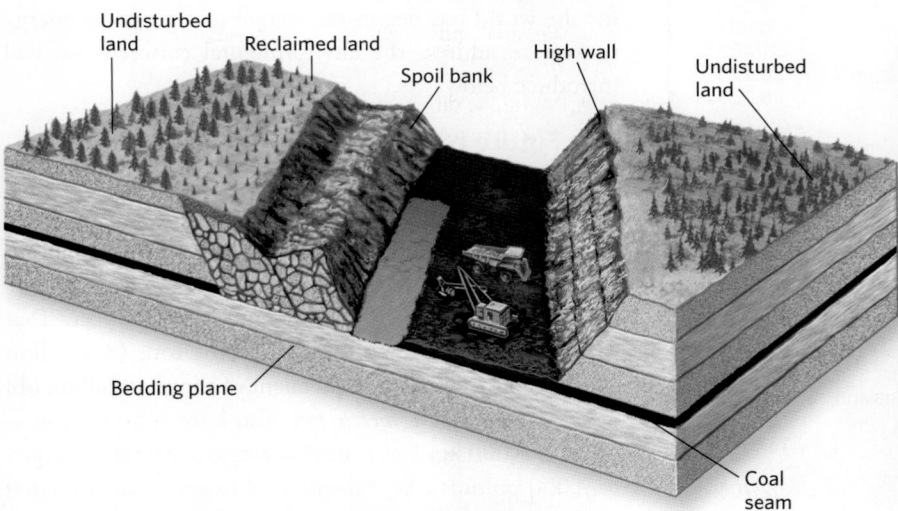

Undisturbed land   Reclaimed land   High wall   Undisturbed land

Spoil bank

Bedding plane

Coal seam

(c) Digging up the coal seam and reclaiming the stripped area.

(d) Underground coal mining.

10.4 The Future of Fossil-Fuel Consumption

The Oil Age

As we've seen, hydrocarbons first became a widely used resource in the latter half of the 19th century and became the major energy source for industrialized countries in the 20th century. Global consumption of hydrocarbons has increased ever since, and continues to increase, in part because densely populated countries such as China and India have undergone rapid industrialization. Consequently, future historians may refer to the 20th and 21st centuries as the **Oil Age** because so much of the world's economy depends on oil.

How long will the Oil Age last? To answer this question, we must keep in mind the distinction between a renewable and a nonrenewable resource: a **renewable resource** can be replaced by nature within months to decades, whereas a **nonrenewable resource** may take centuries to millions of years to replenish. Hydrocarbons are nonrenewable because a reserve takes millions of years to form, so estimates of how long reserves will last assume that the amount of hydrocarbons that now exist represents the total amount available for human consumption.

Did you ever wonder . . .

how much longer the world's fossil-fuel supply could last?

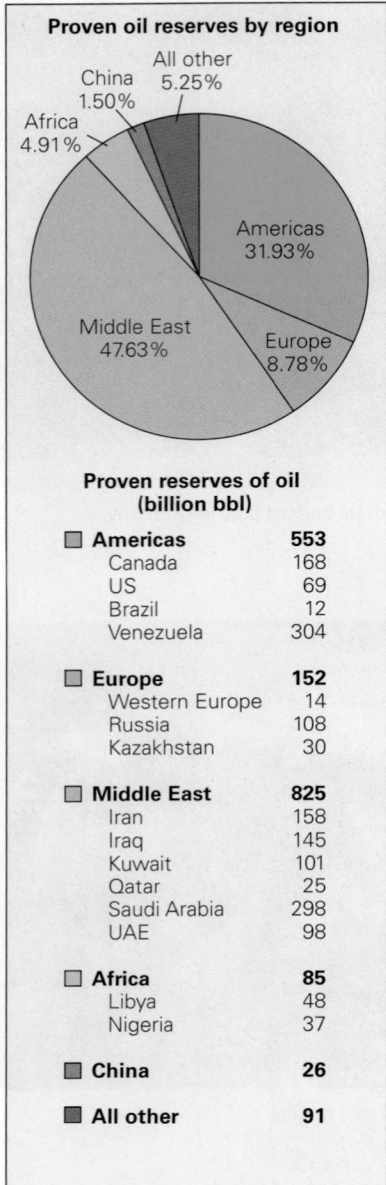

Proven oil reserves by region

Proven reserves of oil (billion bbl)

Americas	553
Canada	168
US	69
Brazil	12
Venezuela	304
Europe	**152**
Western Europe	14
Russia	108
Kazakhstan	30
Middle East	**825**
Iran	158
Iraq	145
Kuwait	101
Qatar	25
Saudi Arabia	298
UAE	98
Africa	**85**
Libya	48
Nigeria	37
China	**26**
All other	**91**

(a) The greatest proven reserves of oil are in the Middle East. According to this estimate, proven reserves of oil total 1,732 billion bbl.

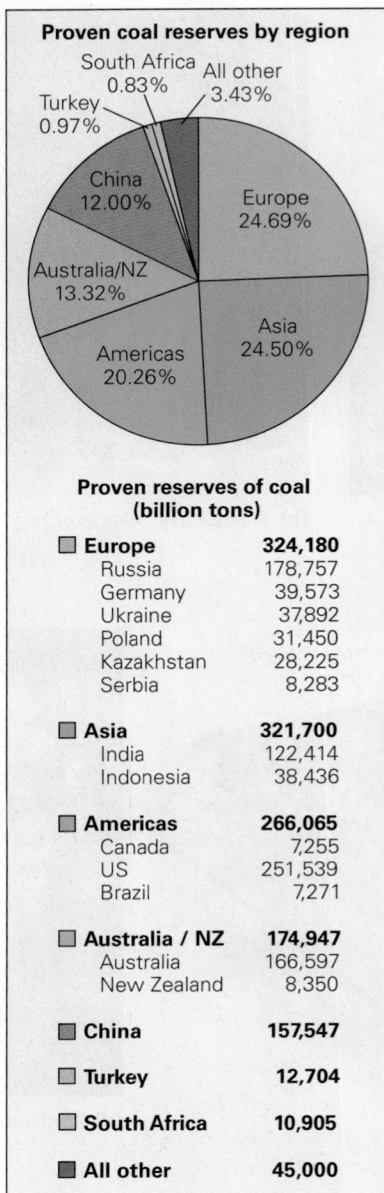

Proven coal reserves by region

Proven reserves of coal (billion tons)

Europe	324,180
Russia	178,757
Germany	39,573
Ukraine	37,892
Poland	31,450
Kazakhstan	28,225
Serbia	8,283
Asia	**321,700**
India	122,414
Indonesia	38,436
Americas	**266,065**
Canada	7,255
US	251,539
Brazil	7,271
Australia / NZ	**174,947**
Australia	166,597
New Zealand	8,350
China	**157,547**
Turkey	**12,704**
South Africa	**10,905**
All other	**45,000**

(b) According to this estimate, proven reserves of coal total about 1.3 trillion tons (= 1.2 trillion metric tons). Probably less than 60% of this amount is minable.

About 1,700 billion bbl of oil resides in *proven reserves*, meaning reserves that have been discovered and measured (Fig. 10.21a). An additional 2,000 billion bbl of oil may exist in conventional reserves, but has not yet been found or confirmed. Unconventional reserves are likely to add significantly to this total. For example, studies suggest that 500 billion bbl of tight oil, 1.5 trillion bbl of oil in tar sand, and 4.5 trillion bbl in oil shale may exist. Of course,

no one knows for sure what proportion of these hydrocarbons can be recovered economically.

Presently, humanity guzzles oil at a rate of about 97 million bbl per day, which works out to about 35 billion bbl/yr (= 5.6 km³/yr). At this rate, proven reserves might run out in about 50 years—within the lifetime of this book's readers. But this interpretation remains highly controversial, because it doesn't consider the changing cost of extraction. Specifically, as extraction of unconventional reserves becomes less expensive, and the price consumers are willing to pay for hydrocarbon products increases, unconventional oil occurrences that are currently considered too expensive to extract profitably may become profitable in the future, so occurrences that are not currently included as reserves could eventually become included. Studies that take price increases and technological improvements into account suggest that perhaps a century or two of accessible oil remains. This means that consumption of hydrocarbons will probably drop off long before the supply has been exhausted, for the world has begun the switch to alternative energy sources to address the environmental concerns we will introduce below.

The Future of Coal

The amount of coal consumed in Europe and North America has dropped in recent years, due in large part to the availability of cheaper and cleaner-burning natural gas. But globally, coal use continues to increase, and more coal was used in 2022 than in any previous year. That year, society consumed 8 billion metric tons (8.8 billion tons) of coal, the energy equivalent of about 40 billion bbl of oil. (Note that 1 *metric ton*, also known as a *tonne*, is equal to 1,000 kg; 1 ton, also known as a *short ton*, is equal to 2,000 pounds.) Worldwide coal reserves are estimated at about 850 trillion tons, which could supply approximately the same amount of energy as 11 trillion barrels of oil (Fig. 10.21b). But such an estimate of coal reserves does not clearly distinguish between accessible (minable) coal and inaccessible coal, which is too deep to mine. Estimating how long coal reserves can last, therefore, remains a challenge. At current rates of use, probably a few centuries' worth of accessible supplies, at a minimum, remain. But as is the case with hydrocarbons, coal use is likely to diminish significantly long before the supply runs out, as society turns to other energy sources.

Environmental Consequences of Fossil-Fuel Use

The extraction of fossil fuels can cause environmental damage. Oil drilling requires construction of drilling pads, pipeline networks, and access roads, all of which disrupt the land surface. *Oil spills* from pipelines or trucks may sink into the subsurface and contaminate soil and groundwater, and leakage at gas wells and refineries releases methane into the air. Also, as we've seen, fracking fluid, if not handled properly, may contaminate water supplies. Offshore, oil spills from ships and tankers, as well as leaks from drilling rigs, can produce oil slicks that spread over the sea surface and foul the shoreline (Fig. 10.22). Marine spills can be devasting to marine life.

If drilling engineers do not adequately counter underground pressure with the weight of drilling mud, or don't properly complete a well, a **blowout** can take place, during which naturally pressurized hydrocarbons from underground rush up the well and spurt out of the ground. In the early days of drilling, oil-rich blowouts—traditionally known as *gushers*—sprayed oil over the countryside. Well drillers might initially have been delighted to strike "black gold," but in reality, a gusher wastes a huge amount of oil and contaminates the landscape. If a blowout catches fire and explodes, it can cost lives. Drilling-rig fires are very difficult to put out; generally, they must be extinguished by a controlled dynamite explosion that momentarily starves the fire of oxygen. And, as demonstrated by the 2010 *Deepwater Horizon* blowout in the Gulf of Mexico, major offshore blowouts can be catastrophic (Box 10.2).

Numerous **air pollution** issues arise from the refining and burning of fossil fuels. Tests reveal that significant quantities of volatile hydrocarbons escape from refineries. Emissions of fossil-fuel combustion products from smokestacks or exhaust pipes introduce soot, carbon monoxide, sulfur dioxide, nitrous oxide, and unburned hydrocarbons into the air, where they can cause health-threatening smog under certain atmospheric conditions. In addition, sulfur-rich emissions can combine with moisture in the air to form dilute sulfuric acid that falls as **acid rain**, causing vegetation and fish to die off in regions downwind of smokestacks.

Acidic water can also be a consequence of coal mining. A dilute solution of sulfuric acid develops when sulfur-bearing minerals (such as pyrite) in coal react with rainwater and oxygen. If the resulting **acid mine runoff** enters streams, it can kill fish and plants, and if it seeps into groundwater, it can contaminate water supplies. Another serious problem arises when coal beds are ignited by arson, accident, or spontaneous processes. Once started, a *coal-bed fire* may burn for years and may be difficult or impossible to extinguish. Such fires produce toxic fumes that rise through joints in rock to make the overlying land uninhabitable. *Coal ash*, the residue from coal combustion, has become an increasing environmental problem. At power plants that burn coal, the ash may

FIGURE 10.22 Marine oil spills can come from drilling rigs or from tankers.

(a) An oil tanker leaking oil on the sea surface.

(b) Oil spills can contaminate the shore and can be very difficult to clean up.

be mixed with water to make a slurry, which is then stored in an ash reservoir. Rupture of the dams or levees holding ash in such reservoirs allows the ash to spread over a landscape or foul a river. Toxic chemicals in the ash may leach into groundwater and end up in streams. Research suggests that some of the pollution problems associated with burning coal could be diminished by transforming solid coal into *syngas*, a mixture of burnable gases, before combustion. This process, called *coal gasification*, allows toxic chemicals to be removed before they go up a smokestack. But coal gasification can be expensive, so it contributes relatively little to global power generation.

Even if new technologies can reduce the sulfur, soot, and poisonous chemicals emitted by burning fossil fuels, extraction and processing of fossil fuels releases methane, and their combustion releases carbon dioxide, into the atmosphere. Measurements confirm that the atmospheric concentrations of these gases have been steadily increasing. By burning fossil fuels at current rates, society has transferred a huge volume of carbon, which took millions of years to accumulate and was then trapped underground for millions of years longer, into the atmosphere (as CO_2 and CH_4) in just a couple of centuries. The hydrosphere, geosphere, and biosphere of the Earth System can't absorb all this excess carbon, so it accumulates in the atmosphere. As we discuss in Chapter 19, carbon dioxide and methane are **greenhouse gases** that trap heat

BOX 10.2 ▶ A Deeper Look

The *Deepwater Horizon* disaster

A substantial proportion of the world's conventional oil reserves resides in passive-margin basins that underlie the continental shelves and are accessible only by offshore-drilling platforms. During both onshore and offshore exploration, drillers worry about the possibility of a blowout, in which hydrocarbons spurt out of a drillhole under their own pressure. A catastrophic blowout occurred on April 20, 2010, when drillers on the *Deepwater Horizon*, a huge offshore-drilling platform, were completing a 5.5 km (18,000 ft) well in water 1.5 km (5,000 ft) deep southeast of the Mississippi Delta. Due to a series of errors, gassy oil under high pressure in reservoir rock punctured by the well rushed up the drillhole. A backup safety device, called a blowout preventer, was supposed to clamp the wellhead (the outlet of the well) shut, but it failed, so the gassy oil reached the platform and sprayed 100 m (330 ft) into the sky. Sparks from electronic gear triggered an explosion, and the platform became a fountain of flame and smoke that tragically killed 11 workers. An armada of fireboats could not douse the conflagration (Fig. Bx10.2), and after 36 hours, the still-burning platform tipped over and sank.

Robot submersibles sent to the seafloor to investigate found oil and gas billowing from the twisted mess of bent and ruptured pipes at the wellhead. On the order of 50,000–62,000 bbl of oil entered the Gulf's water from the well each day. Stopping this underwater gusher proved to be an immense challenge, and initial efforts to block the well or to cover the wellhead with a containment dome failed. Not until July 15 was the flow finally stopped. To prevent future leakage, workers drilled another well from a platform a few kilometers away and, using

FIGURE Bx10.2 Fireboats doused the burning *Deepwater Horizon* platform in vain before it sank.

directional drilling, managed to intersect the 15 cm (6 in) diameter *Deepwater Horizon* drillhole. Finally, on September 19, workers pumped concrete into the failed well and sealed it permanently. All told, about 4.2 million bbl of hydrocarbons from the *Deepwater Horizon* blowout contaminated the Gulf. The spill devastated wetlands, wildlife, and the fishing and tourism industries. Fortunately, in the warm water of the Gulf, bacteria digest the oil, so the region's environment has been recovering very slowly over the course of many years. Catastrophic spills in cold Arctic waters, as happened when the supertanker *Exxon Valdez* ran aground off Alaska in 1989 and spilled 260,000 bbl into Prince William Sound, take much longer to clear up.

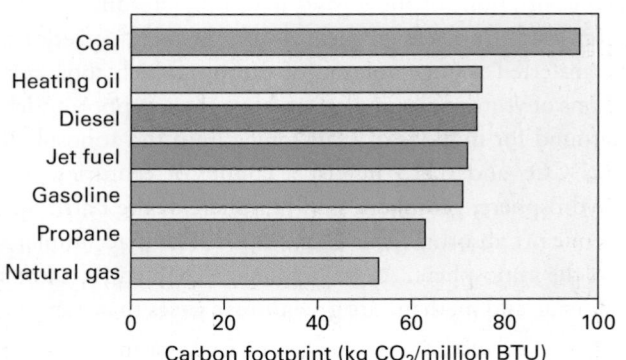

FIGURE 10.23 The carbon footprints of various fossil fuels.

in the atmosphere and thereby cause the Earth's climate, overall, to warm, a change that will have significant consequences for society.

Clearly, society faces difficult choices about where to obtain energy. Because they are increasingly aware of the large **carbon footprint** (amount of greenhouse gases produced) of burning fossil fuels (Fig. 10.23), many governments are encouraging a switch from higher-carbon-footprint fuels (solid coal) to lower-carbon-footprint fuels (natural gas) in the near term, and an eventual switch to carbon-free energy sources. Some researchers estimate that by 2050, most energy will come from alternatives to fossil fuels. In the next section, we'll discuss some of these alternatives.

Fossil fuels are nonrenewable resources. In fact, at current rates of consumption, oil supplies could run out within a century or two. Production and use of fossil fuels lead to environmental consequences, so societies are exploring ways to find alternative energy sources.

Quick Questions

- What does the term "Oil Age" refer to?
- What is a blowout at an oil well?
- Why does acid rain form?

10.5 Alternative Energy Sources

Nuclear Power

When you watch biomass or a fossil fuel burn, you're seeing a *chemical reaction* that releases energy by breaking the chemical bonds that hold atoms together in molecules. Energy produced by a nuclear power plant, in contrast, comes from *nuclear reactions*, which involve breaking the nuclear bonds that hold protons and neutrons together in an atom's nucleus. In most nuclear power plants, uranium-235 (^{235}U), a highly radioactive isotope, serves as the fuel in which the nuclear reactions take place. When struck by a free neutron traveling at very high speed, the nucleus of a ^{235}U atom momentarily absorbs the neutron to become a very unstable nucleus of ^{236}U. This nucleus almost instantly undergoes *fission*, meaning that it splits apart, to produce two smaller daughter nuclei, such as krypton-92 and barium-141, along with three free neutrons. The new free neutrons can then strike other ^{235}U atoms, causing them to undergo fission. This self-perpetuating process, known as a **chain reaction**, produces prodigious amounts of energy, much of it in the form of heat. An extremely rapid, uncontrolled chain reaction in a mass of fissionable material produces an atomic-bomb explosion, like the one that destroyed Hiroshima. In a nuclear power plant, however, the chain reaction releases heat at a controlled rate, so an atomic-bomb-like explosion cannot occur.

HOW DOES A NUCLEAR POWER PLANT WORK? The heart of a nuclear power plant is a **nuclear reactor**, a container holding *fuel rods*, which are metal tubes filled with a mixture of ^{235}U and a minimally radioactive uranium isotope, ^{238}U. Fuel rods alternate with *control rods*, which consist of a material, such as graphite, that absorbs free neutrons. Raising the control rods allows more ^{235}U atoms to undergo fission, and lowering the control rods

prevents ^{235}U atoms from undergoing fission. Therefore, the control rods regulate that amount of heat that the nuclear reactor produces. Heat produced by the reactions in the fuel rods boils water, transforming it into high-pressure steam (Fig. 10.24). This steam, in turn, pushes the blades of a turbine that rotates the shaft of an electricity-producing generator. Note that, except for the source of heat, a nuclear power plant operates in much the same way as a fossil-fuel power plant (see Fig. 10.3). The reactor, and the water that it heats, sit within a containment structure of reinforced concrete, intended to prevent leakage of radioactive material into the environment.

Where does the uranium used in nuclear power plants come from? Rising felsic magma brings uranium atoms into the Earth's upper crust, where various geologic processes, such as groundwater flow, concentrate the uranium in localized deposits. Uranium extracted from these deposits can't be used in a reactor directly, because the fissionable isotope of uranium (^{235}U) that serves as the fuel in reactors accounts for only about 0.7% of naturally occurring uranium; the remainder consists of the isotope ^{238}U, which does not readily undergo fission. To make fuel for use in a reactor, the proportion of ^{235}U relative to ^{238}U must be increased to 3%–5%, an expensive process called **enrichment**. Significantly, the concentration of ^{235}U in nuclear-reactor fuel is much less than that needed for an atomic bomb. To be explosive, uranium must be enriched to over 85% ^{235}U. Therefore, nuclear-reactor fuel cannot explode like an atomic bomb.

CHALLENGES OF USING NUCLEAR POWER. The first nuclear power plants were built in the 1950s, and about 450 nuclear power plants currently operate. To date, two major disasters—meaning events during which containment structures were damaged and significant quantities

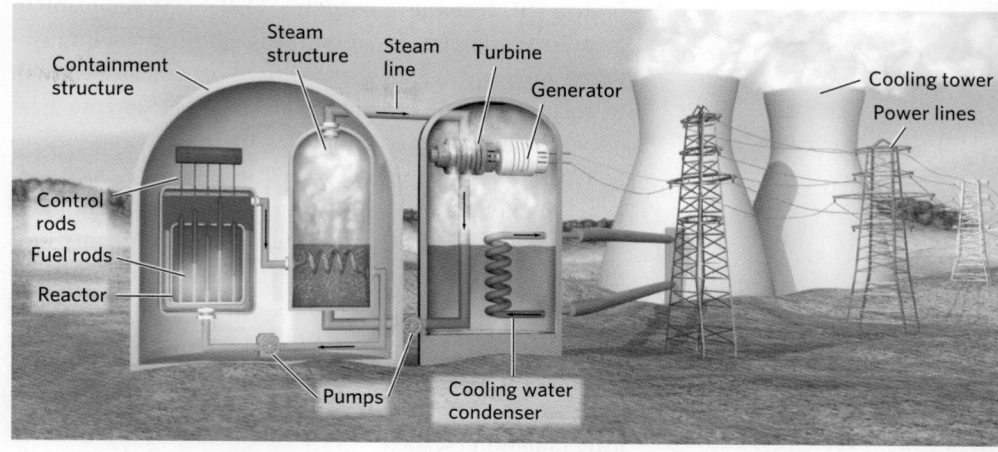

FIGURE 10.24 In a nuclear power plant, a nuclear reactor heats water to produce high-pressure steam, which drives a turbine to generate electricity.

of radiation escaped—have occurred at these plants. The first occurred in 1986 at the Chernobyl power plant in Ukraine (then part of the Soviet Union), and the second in 2011 at the Fukushima Daiichi power plant in Japan (see Chapter 7). In both cases, loss of cooling systems caused the fuel in the reactor to become hot enough to melt, an event called a *meltdown*. The intense heat released caused water molecules in the cooling system to separate into a mixture of hydrogen and oxygen gas, which exploded, causing a rupture of the containment structure and a release of radioactive contaminants into the environment. The Chernobyl disaster, which caused many deaths, was much worse. An accident also happened at the Three Mile Island plant in Pennsylvania in 1979. A partial meltdown took place in the reactor there, but the containment structure was not breached, so very little radiation escaped.

Operation of a reactor produces **nuclear waste**, which includes very radioactive *high-level waste*, such as spent fuel rods, as well as less radioactive *low-level waste*, such as radiation-contaminated water, piping, and concrete. Some radioactive atoms in nuclear waste decay relatively quickly (in decades to centuries), but some will remain dangerous for thousands of years. High-level waste needs to be kept cool by being submerged in water. But even low-level waste cannot simply be buried in a landfill because it might leak into groundwater supplies. Instead, it should be isolated in durable, sealed containers and kept away from corrosive fluids. Appropriate long-term storage sites for nuclear waste have not yet been identified—to date, most such waste remains on the property of the power plants that produced it.

Biofuels

Can we transform modern-day biomass into fuels that have the energy density of fossil fuels? The answer is yes, and the resulting materials are called **biofuels**. For example, *ethanol*, a type of alcohol that can substitute for gasoline in car engines, can be produced commercially either from corn or from sugarcane. More recently, researchers have been developing processes that yield ethanol from cellulose, which will allow perennial grasses to become a source of biofuel. Another promising method uses algae, which naturally synthesize fatty organic chemicals from which hydrocarbons can be produced. Recent technologies also include the commercial production of *biodiesel*, a fuel that comes from chemical modification of used cooking oils. (Note that the term *biofuel* generally refers to liquid products produced by processing biomass, not to the biomass itself.)

Geothermal Energy

The name **geothermal energy** implies the use of energy that comes from inside the Earth. That definition applies to locations where particularly high temperatures occur at relatively shallow depths in the crust due to a steep geothermal gradient. For example, in areas of igneous activity, such as Iceland and New Zealand, accessible groundwater can be so hot that, when pumped from the subsurface and run through pipes, it can heat houses, buildings, and greenhouses directly. And in some locations, commercial operators use very hot groundwater to generate electricity at *geothermal power plants*, for when such groundwater rises to the surface, it undergoes decompression and transforms into steam that can drive turbines (Fig. 10.25a).

FIGURE 10.25 Geothermal energy.

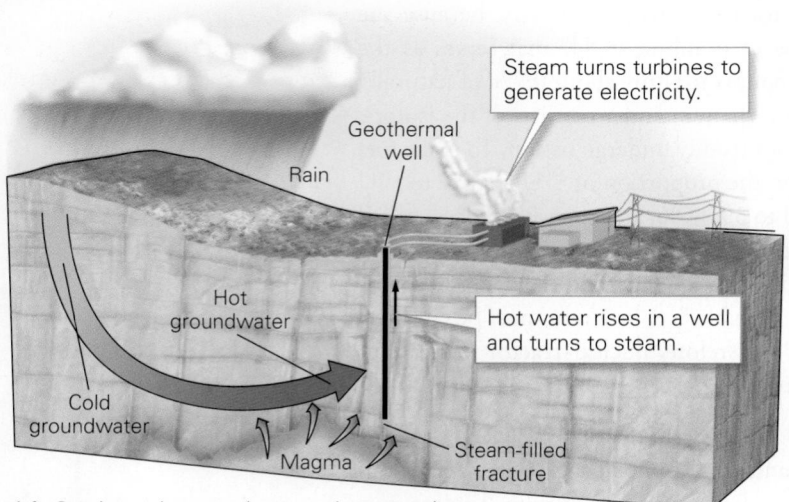

(a) Geothermal power plants use hot groundwater that turns to steam when it reaches the surface and undergoes decompression.

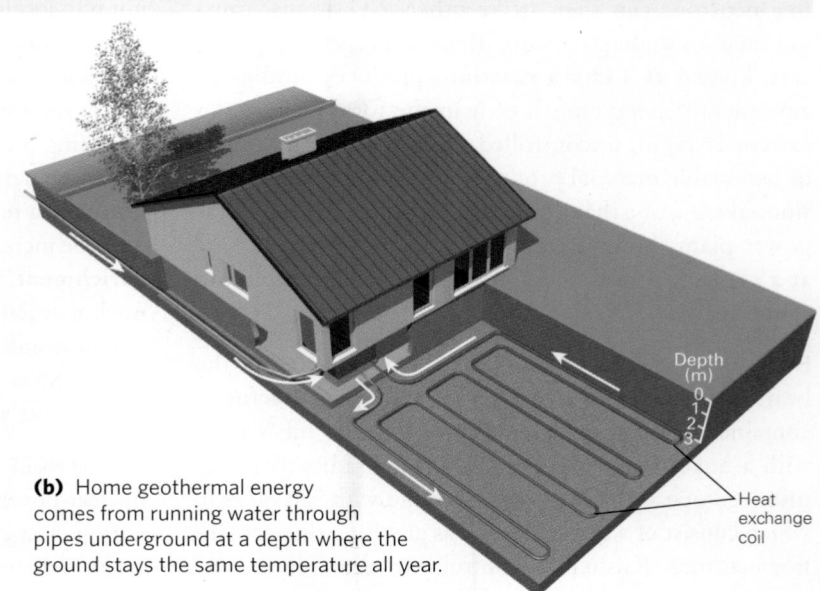

(b) Home geothermal energy comes from running water through pipes underground at a depth where the ground stays the same temperature all year.

Unfortunately, geothermal power plants lose their energy source if the supply of hot groundwater diminishes, as can happen when it is pumped out of the ground faster than new groundwater heats up underground.

Somewhat confusingly, the term *geothermal energy* also applies to heating and air-conditioning systems that use a *ground-source heat pump*. These systems, used for individual buildings or homes, work by circulating a heat-conducting fluid, called a *refrigerant*, through pipes buried in soil or sediment a couple of meters below the ground surface outside the building (Fig. 10.25b). At that depth, soil stays at the *ambient temperature*—the average annual temperature of the region—all year, because the overlying soil serves as an insulator. In the buried pipes, the temperature of the refrigerant approaches the ambient temperature. Because heat always flows from warmer to cooler materials, heat flows out of the refrigerant if the ground's ambient temperature is cooler, and into the refrigerant if the ambient temperature is warmer. The pipes then return the refrigerant to the building, and carry it through a *heat exchanger*. In this device, air from the building's air-circulation system circulates around the refrigerant-carrying pipes. In summer, the air heats the refrigerant, which then carries the heat out of the building to the underground pipes and releases it into the ground. In winter, the exchange works in the opposite direction: refrigerant in the underground pipes absorbs heat from the ground and carries it to the heat exchanger, where it warms the air of the air-circulation system. Because the ground's ambient temperature remains at about 13°C (55°F) in temperate regions, and people prefer buildings to be at about 18°C–22°C (65°F–72°F), a furnace must also operate during the winter to supplement the heat pump. But the furnace burns much less fuel to raise the temperature from, say, 18°C to 22°C than it would take to raise the temperature from, say, –7°C to 22°C, so the building owner saves on energy bills.

Both geothermal power generation and ground-source heat pumps can be considered sources of **green energy** (or *clean energy*). Use of this term implies that the production of energy takes place without releasing chemical or radioactive pollutants or greenhouse gases, and without consuming nonrenewable resources. Of course, because construction and installation of geothermal systems take energy, and because hot groundwater releases some SO_2 and CO_2 when these volatiles come out of solution at the surface, their use is not perfectly "green." But the carbon footprint of geothermal energy production is a fraction of that associated with energy production using fossil-fuel combustion.

Hydroelectric and Wind Power

For centuries, people have used water wheels to power mills and factories directly, for the motion of a rotating wheel driven by water flow, transferred though gears and pulleys, can spin millstones and looms. Modern *hydroelectric power plants* generate electricity by sending flowing water through the blades of a turbine, which in turn drives an electrical generator. Most hydroelectric plants rely on water held in a *reservoir* by a dam (Fig. 10.26a). In effect, **hydroelectric power** comes from the potential energy in the reservoir's elevated water, which converts into kinetic energy when the water is allowed to flow to a lower elevation at the foot of the dam.

Hydroelectric power plants can also be considered a source of green energy. But the construction of dams and reservoirs can cause environmental and societal problems.

(a) The water held back by the Three Gorges Dam, on the Yangtze River in China, flows through turbines to generate electricity.

(b) A wind farm in southwestern England. The towers are about 50 m high.

(c) A giant turbine blade being hauled down a highway, en route to a new windmill installation.

FIGURE 10.26
The kinetic energy of flowing water and air can be transformed into electricity.

For example, damming a river may submerge spectacular scenery, displace towns, and destroy ecosystems. Because the flow velocity of water in a reservoir is so slow, sediment brought into the reservoir by rivers settles out and accumulates on the bed of the reservoir. This accumulation not only decreases the reservoir's volume, but also prevents the sediment from reaching floodplains or deltas downstream. Consequently, the land surface in these downstream regions not only fails to receive new nutrient-bearing sediment, so that farm production diminishes, but may actually sink as organic material in the existing sediment undergoes compaction and dehydration and the sediment itself undergoes consolidation. Finally, the construction of dams requires the production of immense amounts of concrete, a process that, as we'll see, releases significant quantities of CO_2. So, while there are many benefits associated with the construction of hydroelectric dams and associated reservoirs—clean energy, sources of irrigation water, and recreational opportunities, as well as flood control—many downsides need to be considered.

Notably, not all hydroelectric power generation requires the damming of a flowing river. Engineers have been developing new approaches to tap **tidal energy**, the energy associated with the daily rise and fall of tides. One approach involves building a dam, called a *tidal barrage*, across the entrance to a bay. When the tide rises, water flows, via pipes in the barrage that carry it through power-generating turbines, into an estuary or bay. When the tide drops, water trapped behind the barrage flows back to the sea through the same pipes, and again passes through power-generating turbines. So both the rising and falling tides produce power.

Modern efforts to harness **wind power** are developing on a large scale, as technological advances now allow construction of giant wind turbines. To produce wind power, meteorologists identify regions where strong winds are common. In such regions, engineers build *wind farms*, consisting of numerous wind turbines mounted on towers. The blades of each turbine spin when the wind blows, and this rotation, in turn, drives a generator (Fig. 10.26b, c). Friction near the ground slows wind, so the towers need to be tall enough to catch the higher, faster winds. Winds also tend to be stronger over water than over land because there is less friction between water and wind than between land and wind, and because there are no objects such as topography, trees, and buildings in the way. For these reasons, some power companies have developed offshore wind farms, despite the higher cost of their installation.

Wind power, like hydroelectric power, is green, but it does have some drawbacks. Cluttering the horizon with towers may spoil a beautiful view, the loud hum and rotating shadows of the turbines can disturb nearby residents, and the fan blades may be a hazard to migrating birds. Also, because the amount of electricity produced by wind turbines depends on wind speed, the supply is not constant and may not be available when needed. Such a variable energy supply cannot easily be added to a regional or national **electrical grid**, the system that redistributes electrical energy via power lines and converts it to the right voltage via transformers. Currently, natural-gas power, which can be supplied or reduced rapidly, is used in conjunction with wind power to ensure a constant flow of electricity.

Solar Energy

The Sun drenches the Earth with energy in quantities that dwarf the amounts stored in fossil fuels. Such green **solar energy** can be harnessed in two ways. In a *solar collector*, an absorbing material heats up when exposed to solar radiation and uses that heat to warm a fluid. The simplest version of a solar collector consists of a dark surface placed beneath a glass plate. The dark surface absorbs solar radiation and reradiates that energy as infrared radiation, which warms the air trapped between the absorber and the glass. When water passes through pipes placed between the glass and the absorbing surface, it heats up enough to be used as a domestic hot water supply. Such simple solar collectors can be set up for individual households. Similarly, *solar-thermal power plants* use mirrors to reflect solar energy absorbed over a broad area to a receiver. The receiver becomes so hot that it can transform fluid piped through it into high-pressure vapor, which in turn drives electricity-generating turbines.

Alternatively, solar radiation can be used to produce electricity by means of shiny panels containing arrays of **photovoltaic cells** (*solar cells*). Photovoltaic cells convert radiation from the Sun directly into electric current (Box 10.3). They can be installed on individual homes or in vast commercial arrays (Fig. 10.27). But because

FIGURE 10.27 A commercial solar array contains many panels covered with photovoltaic cells.

A commercial solar array

BOX 10.3

A Deeper Look

How photovoltaic cells work

Solar cells are, in many ways, a technical marvel. They could not be produced commercially until many complex problems in physics, chemistry, and materials engineering were solved. Production techniques improve every year, which has led to an impressive drop in price, from more than $5.00 per watt of the solar cell's energy production in 2000 to less than $0.02 per watt in 2023.

The heart of a photovoltaic cell consists of a silicon wafer. But instead of having just silicon atoms, the wafer has been "doped," or contaminated, with tiny quantities of boron ions and phosphorus ions. Boron ions occur throughout the wafer. Each boron ion has lost three electrons, so it has a net positive charge. The upper portion or layer of the wafer has also been contaminated with phosphorus ions. Each phosphorus ion contains five extra electrons, so the upper layer of the wafer, known as the n-type layer, has a net negative charge. The bottom layer, known as the p-type layer, which contains boron ions but no phosphorus ions, has a net positive charge. The n-type and p-type layers merge along a junction. The charge difference between the top and bottom of the silicon wafer, due to the distribution of contaminants, controls the flow direction of any free electrons in the wafer. In the dark, no electrons cross the junction.

When a unit (called a photon) of solar radiation strikes the top surface of the cell during the day, it knocks an electron out of a silicon atom in the wafer. The missing electron, simplistically, can be thought of as a positively charged "hole." This hole effectively migrates from atom to atom until it reaches the back electrode (the conducting bottom surface) of the cell (Fig. Bx10.3). Meanwhile, the electron that was set free migrates toward the top of the photovoltaic cell, where it enters a wire. In the wire, the electron flows out of the cell, through a converter, and then around to the back electrode. At the back electrode, the electron fills a positive hole. Consequently, electrons flow from the top of the cell through the converter and to the bottom of the cell and then back into the cell. This flow establishes an electrical current. (By convention, the direction of electrical current is defined as the direction opposite to the electron flow.) The converter sends this electrical current out of the cell in a usable form. An array of many photovoltaic cells produces enough electricity to power a device directly, or to contribute power to a regional electrical grid.

FIGURE Bx10.3 A cross section of a photovoltaic cell (solar cell).

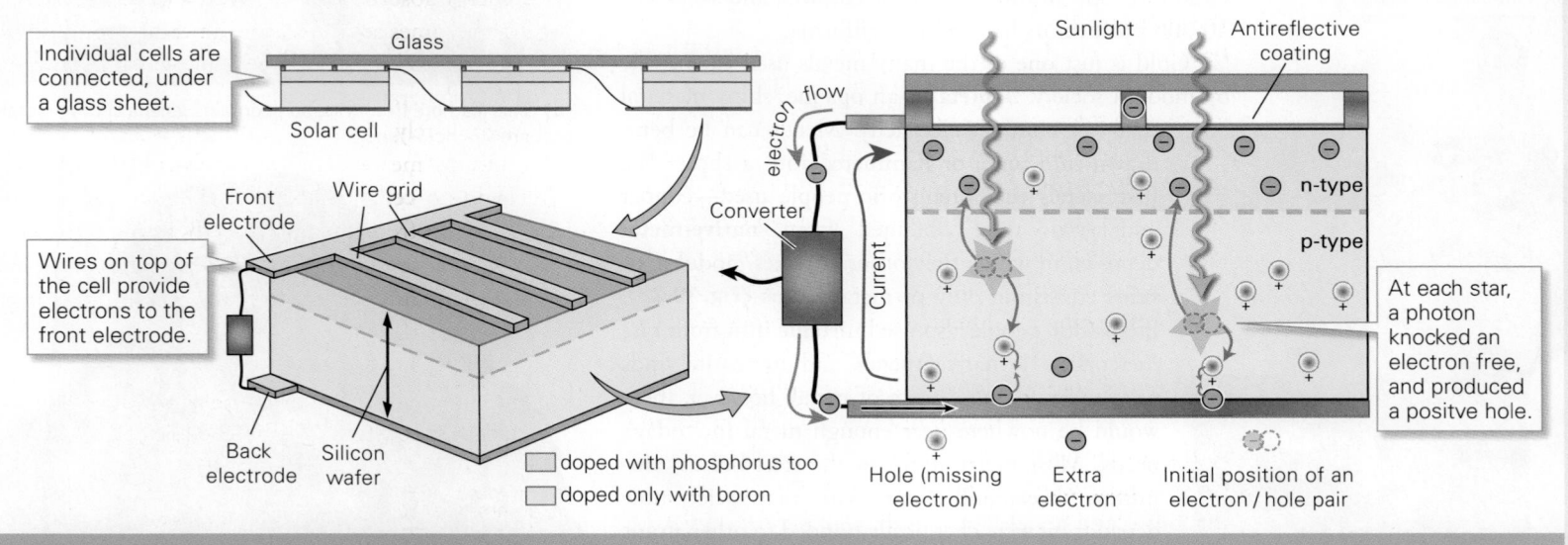

photovoltaic cells can produce power only during the day, their use, like that of wind power, requires a supplemental energy source, such as a fossil-fuel-burning power plant, to maintain a constant energy supply to the electrical grid. Because of its very low carbon footprint, however, net generation of solar energy in the United States has been increasing exponentially since about 2008, and it now yields almost 5% of the electricity used.

10.6 Metallic Mineral Resources

In January 1848, James Marshall and a crew of workers were finishing the construction of a new sawmill in the foothills of the Sierra Nevada. As Marshall stood admiring the building, he noticed a glimmer of metal in the gravel that littered the bed of the adjacent stream. He picked up the metal, banged it between two rocks to test its hardness, and shouted, "Boys, by God, I believe I have found a gold mine!" Word of the discovery soon spread, and within weeks, all the mill's workers had disappeared into the mountains to seek their own fortunes. Gold fever spread throughout the country, and 1849 saw 40,000 prospectors heading to California.

Gold is just one of the many metals used extensively by modern society. A **metal** is an opaque, shiny material that can conduct electricity. Metals can be bent, drawn into wire, or hammered into a sheet. The first metals that prehistoric people used—copper and gold—were obtained from **native-metal** deposits, in which they occur as flakes, nodules, or veins consisting only of metal atoms (Fig. 10.28). Prehistoric people also used metallic iron from rare meteorites. If native metals and meteorite finds were society's only source of metal, however, there would be nowhere near enough metal for today's needs. Most metal atoms in the steel beams, aluminum sheets, and copper wires that society now depends on were chemically bonded to other atoms in minerals. Beginning about 5000 B.C.E., people discovered that if they heated certain rocks to a high enough temperature, some of their minerals would decompose to yield metal plus a nonmetallic residue called *slag*. This extraction process, called **smelting**, along with other metal extraction processes discovered more recently, is what allows us to obtain the variety and quantity of metals in use today.

FIGURE 10.28 Native metals, such as gold, look much the same in their unprocessed form (top) as they do in familiar objects, such as gold bracelets from a Kuwaiti jewelry market (bottom).

Gold embedded in quartz

Ores and Ore Minerals

Geologists use the term **ore** for a rock that contains a high enough concentration of metal atoms to be worth mining. The measured concentration of a useful metal in an ore determines the **grade** of the ore: the higher the concentration, the higher the grade. Whether or not a mining company can profitably mine an ore of a given grade, at a given time, depends on the metal's market price (Box 10.4).

As we've noted, a few metals can occur as native metals, but more typically, the metal atoms of an ore occur in **ore minerals**, so named because they contain a relatively high proportion of extractable metal atoms. For example, galena (PbS) contains 50% lead, so it serves as an ore mineral for lead (Fig. 10.29a). Similarly, hematite (Fe_2O_3) and magnetite (Fe_3O_4) are ore minerals for iron. By comparison, though olivine ($MgFeSiO_4$) contains iron atoms, they make up a relatively small percentage of the mineral.

FIGURE 10.29 Examples of ore minerals.

(a) This lead ore from Missouri contains galena (PbS) crystals that grew in dolostone.

5 cm

(b) A large block of malachite, a green copper carbonate mineral.

BOX 10.4 ▶ How can I explain . . .

The economics of mining

What are we learning?

That a lot of waste rock must be processed to obtain a relatively small amount of ore, and that mining a higher-grade ore deposit can be more profitable than mining a lower-grade deposit.

What you need:

- 4 cups of dry rice
- 1 tablespoon full of m&m's or other small candies
- 2 large bowls and 1 small bowl
- A 1-cup measuring cup

Instructions:

- Pour the rice into one of the large bowls. Add the candies and mix thoroughly.
- Scoop out the rice-plus-candy mixture, one cup at a time.
- Sort through each cup and find the candies. Place the candies in the small bowl and the candy-free rice in the other large bowl. Record the time it takes to recover the candies from each cup.

- Repeat the experiment, but this time place the candies in the center of the rice-filled bowl.
- Then, scoop a cup from the part of the bowl where you know the candies are, and record the time it takes to recover the candies.

What did we see?

When the candy was dispersed throughout the rice, you could recover very little in a given time, for you had to sift through lots of rice to get the candy. If the time you take scooping and sorting represents the time it takes to mine valuable ore minerals, you are spending a lot of time (which equals money) to extract the ore minerals. Of course, it's much easier to extract the candy quickly if it's concentrated and you know where it is. This situation represents the discovery of a high-grade ore deposit. Mining such a deposit is much more efficient.

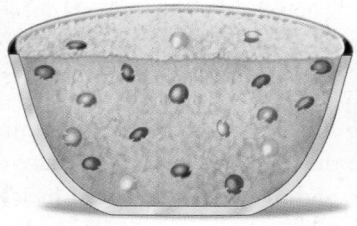 

It's not possible to extract iron from olivine economically, so olivine is not an ore mineral. Some ore minerals, like galena, have a metallic luster, but not all do. Copper, for example, comes from a variety of ore minerals, including chalcopyrite ($CuFeS_2$) and malachite [$Cu_2CO_3(OH)_2$], some of which look nothing like copper (Fig. 10.29b).

Formation of Ore Deposits

Ore minerals do not occur uniformly through the rocks of the Earth's upper crust. Fortunately for society, geologic processes can produce **ore deposits**, concentrations of ore minerals that can be profitably mined. Geologists distinguish among different types of ore deposits based on the process by which the deposits form. Here are a few examples:

- *Magmatic deposits:* In some magmas, ore minerals crystallize and accumulate to form lenses of ore, called *magmatic deposits*, as the magma solidifies into igneous rock (Fig. 10.30a).

- *Hydrothermal deposits:* Hot groundwater circulating through an igneous intrusion and its wall rock can dissolve metal ions. When the resulting high-temperature solution enters cooler rock, or encounters water or rock with certain chemical characteristics, the metals precipitate as ore minerals, either in *veins* (mineral-filled cracks) or in pores as a *disseminated deposit*. An ore that consists of ore minerals precipitated in cracks or pores due to interactions between rock and hot water is a **hydrothermal deposit** (Fig. 10.30b).

- *Seafloor massive sulfide deposits:* Along mid-ocean ridges, black smokers erupt hydrothermal solutions (see Chapter 4). When the solutions mix with cool seawater, the dissolved components form tiny crystals of sulfide minerals (Fig. 10.30c). These minerals settle and accumulate around the hydrothermal vent to form a *seafloor massive sulfide deposit*.

- *Secondary-enrichment deposits:* In the upper crust, groundwater can dissolve ore minerals. When the resulting solution flows into a different chemical environment with appropriate characteristics, it precipitates new ore minerals in a high concentration, generating a *secondary-enrichment deposit*.

- *Sedimentary deposits:* A *sedimentary deposit* accumulates from ore minerals that precipitate and settle out of water at normal temperatures, and become incorporated in layers of sediment.

FIGURE 10.30 Some processes that form ore deposits.

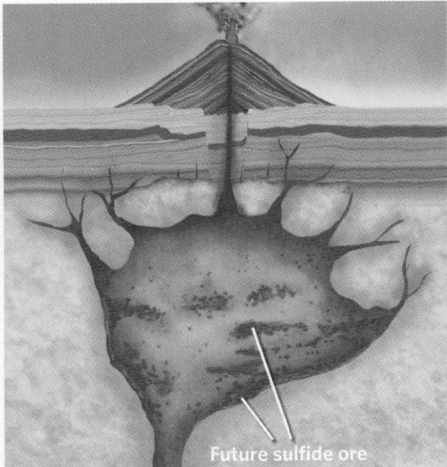

(a) Magmatic ore deposits can form in a magma chamber when ore minerals crystallize.

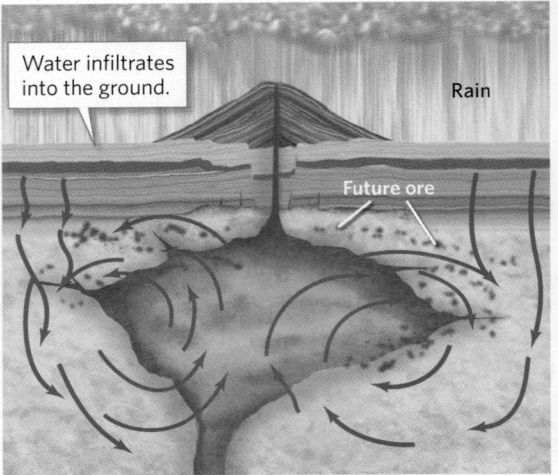

(b) Hydrothermal deposits form when water circulating around and through magma dissolves metals, which then precipitate out of solution elsewhere. (Arrows indicate water flow.)

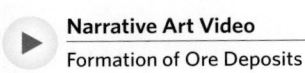
Narrative Art Video
Formation of Ore Deposits

(c) Seafloor massive sulfide deposits form when sulfides precipitate around hydrothermal vents (black smokers) along a mid-ocean ridge.

- *Placer deposits:* When rocks containing native metals erode, the resulting sediment contains rock clasts, metal flakes, and *nuggets* (pebble-sized metal clasts). Moving water preferentially carries away clasts of less dense minerals, leaving behind denser metal flakes and nuggets. A concentration of transported metallic sediment grains constitutes a **placer deposit** (Fig. 10.31). Placer deposits can occur in loose sediment on streambeds or in gravel bars, or they can occur in sediment layers that have been buried and, in some cases, lithified.

- *Residual deposits:* Heavy rains in tropical climates *leach* (dissolve and carry away) soluble minerals from the soil (see Chapter 5), leaving a concentration of insoluble ore minerals called a *residual deposit*. Most of the world's supply of aluminum comes from *bauxite*, a residual deposit containing aluminum hydroxide minerals.

- *Polymetallic nodules:* Photographic surveys reveal that lumps containing high concentrations of metals carpet large areas of the deep seafloor. These *polymetallic nodules* (traditionally known as *manganese nodules*, because they contain particularly high concentrations of manganese) may form partly by chemical precipitation out of seawater and partly by precipitation involving microbial metabolism. They appear to grow very slowly, at 2–15 mm per million years, and they nucleate on shell fragments or sharks' teeth.

Where Are Ore Deposits Found?

The Inca Empire in the Andes built elaborate temples decorated with statues and masks of gold. Then, around 1532, Spanish conquistadors arrived. They cruelly conquered the Incas and transported Inca treasure back to Spain. Why did the Incas possess so much gold? Or, to ask the broader question, what geologic factors control the distribution of ore?

Several of the ore-deposit types mentioned earlier in this chapter occur in association with igneous rocks. As we learned in earlier chapters of this book, igneous activity takes place primarily in the volcanic arcs of convergent boundaries and along mid-ocean ridges. Therefore, magmatic and hydrothermal deposits, as well as

FIGURE 10.31
Placer deposits form where erosion produces clasts of native metals. Sorting by flowing water concentrates the metals.

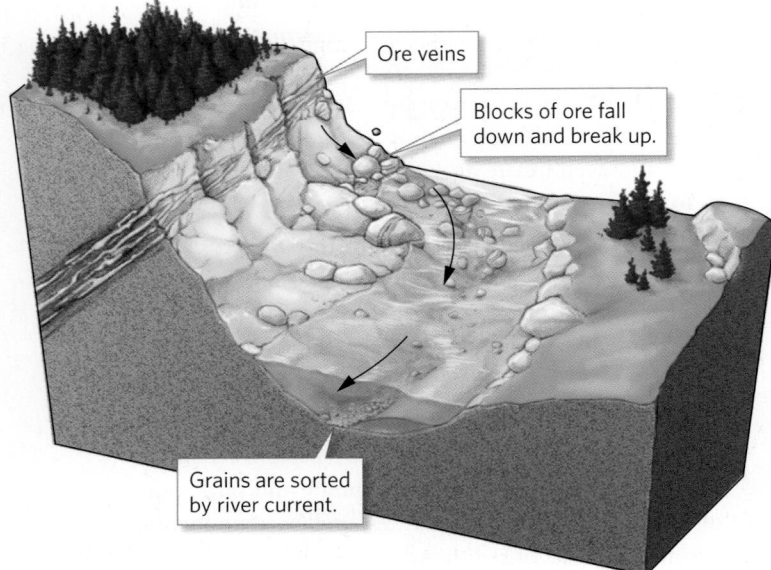

secondary-enrichment deposits and placer deposits derived from them, develop in those geologic settings. Inca gold, for example, came from placer deposits of native gold eroded from hydrothermal deposits formed in the Andean continental arc. Copper ores and many other metal ores also commonly form in association with hydrothermal activity in volcanic arcs.

Major ore deposits also occur in seafloor massive sulfide deposits formed by hydrothermal activity in oceanic crust. At present, these deposits can be mined only in locations where thrust faults that were active during mountain building carried slices of oceanic crust upward and onto continental crust; such sheets of displaced oceanic crust are called *ophiolites*. But mining companies are exploring technologies to mine such deposits where they are currently forming, or recently formed, on the seafloor. Such proposals remain controversial because of their potential environmental impact.

Most iron ore comes from *banded iron formation* (*BIF*), colorful sedimentary deposits of alternating red chert (jasper) and gray iron oxide that precipitated from seawater between 2.4 and 1.8 billion years ago. This time interval represents the portion of Earth history during which the concentration of dissolved oxygen in seawater became sufficient to allow oxygen atoms to bond with dissolved iron atoms to produce insoluble iron oxide minerals (see Chapter 9). Because all BIF is Precambrian, most of the largest mines extracting BIF are located on continental cratons, such as those in southern Canada, Western Australia, and east-central Brazil.

Most aluminum comes from residual ores formed from granite exposed to weathering in tropical climates, where water continuously seeps downward through bedrock and produces a thick zone of leaching. Granite contains feldspars, minerals that contain aluminum. Hydrolysis, a type of chemical weathering, releases aluminum atoms from feldspar, and these atoms bond with oxygen and hydrogen to form insoluble aluminum hydrate minerals. Because these insoluble minerals do not dissolve in downward-percolating water, they accumulate in soil, producing an aluminum-rich residual ore called *bauxite*.

Polymetallic nodules accumulate on the muddy surfaces of abyssal plains (see Fig. 5.25e). In places, these nodules cover up to 70% of the seafloor surface, and could possibly be mined by a vacuum-cleaner-like system that sucks them up to ships at the surface. Such mining, however, would disturb large areas of the seafloor with yet unknown environmental consequences.

Ore Exploration and Production

Imagine prospectors of days past, clanking through the wilderness with worn-out donkeys, searching for ore. What, specifically, were those prospectors hoping to see?

In some cases, they scanned hillsides for outcrops of milky white quartz veins that might contain native metals, or for brightly colored stains caused by oxidation (rusting) of ore minerals. Sometimes they panned streambed gravels, hoping to find gold nuggets. On finding a possible ore deposit, a prospector would take a sample back to town for an *assay*, a test to determine how much extractable metal the deposit contained. If the assay indicated a significant concentration of metal, the prospector might "stake a claim" by literally marking off an area of land with wooden stakes.

These days, commercial mining companies employ geologists to survey potential ore-bearing regions and identify rock units that could host ores. They may also sample rocks, soils, and plants to test for metal concentrations. And they may conduct surveys to find magnetic or gravity anomalies, because ores tend to be especially magnetic and dense. If their observations hint that an *ore body* (a definable volume of ore) lies underground, they drill to sample and assay subsurface rock. If the assays indicate an abundance of ore, economists determine whether mining the deposit, while accommodating environmental concerns and regulations, can yield a profit.

If a company gets the go-ahead to build a mine, miners have to decide on whether to develop an open-pit mine or an underground mine. The decision usually depends on the depth of the ore body. In an open-pit mine (Fig. 10.32a), dug to access a shallow ore body, workers use explosives to shatter rock. Front-end loaders dump the rock fragments into giant trucks that can carry up to 27,000 kg (300 tons) of rock in a single load. The trucks dump waste rock—rock whose grade is so low that it can't be processed economically—on an artificial hill called a *tailings pile*. They carry the sufficiently high-grade ore to a crusher that smashes it into small fragments. Smelting these fragments separates metal atoms from the remaining materials, called *slag*. To develop an underground mine, miners either bore a tunnel into the side of a mountain or sink a vertical shaft (Fig. 10.32b). At the level in the crust where the ore body can be accessed, they excavate a network of tunnels.

Mining and the Environment

Mining activity can leave a big environmental footprint (Earth Science at a Glance, pp. 382–383). Some of the gaping basins that result from open-pit mining are so big

FIGURE 10.32 Mining ore deposits.

(a) An open-pit mine. The steps cut into the wall help ensure its stability. The bluish staining on the rock indicates the presence of copper ores.

(b) An old underground mine tunnel chiseled into a mountainside in Colorado. The rusty staining indicates the presence of ore minerals.

Formation and Processing of the Earth's Mineral Resources

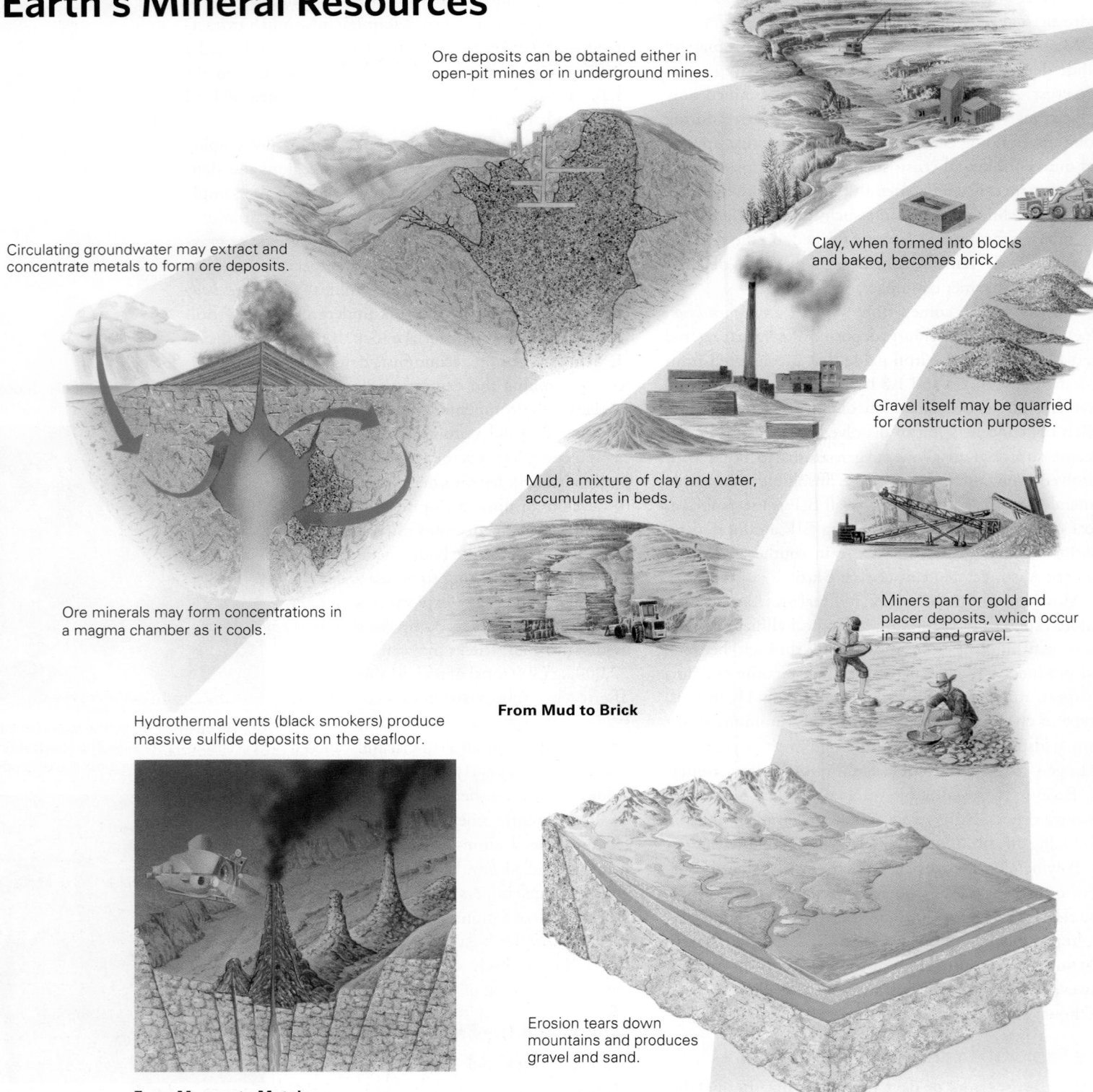

Mining and processing ore has environmental consequences, including acid runoff, acid rain, and groundwater contamination.

Ore deposits can be obtained either in open-pit mines or in underground mines.

Circulating groundwater may extract and concentrate metals to form ore deposits.

Clay, when formed into blocks and baked, becomes brick.

Gravel itself may be quarried for construction purposes.

Mud, a mixture of clay and water, accumulates in beds.

Ore minerals may form concentrations in a magma chamber as it cools.

Miners pan for gold and placer deposits, which occur in sand and gravel.

From Mud to Brick

Hydrothermal vents (black smokers) produce massive sulfide deposits on the seafloor.

Erosion tears down mountains and produces gravel and sand.

From Magma to Metal

From Stream Channel to Roadbed

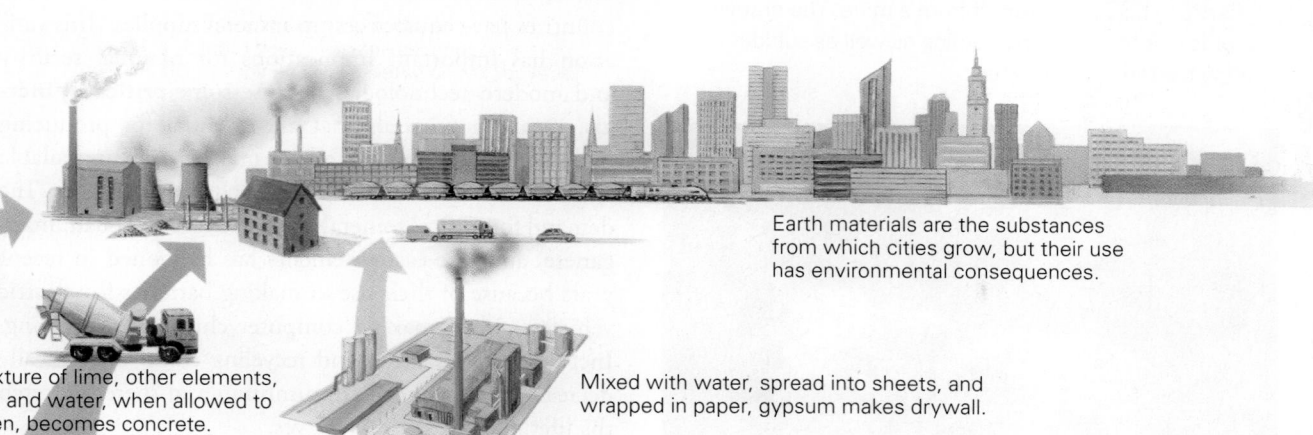

Earth materials are the substances from which cities grow, but their use has environmental consequences.

A mixture of lime, other elements, sand, and water, when allowed to harden, becomes concrete.

Mixed with water, spread into sheets, and wrapped in paper, gypsum makes drywall.

In quarries, operators dig up gypsum, crush it to powder, and ship it to factories.

Quarries extract limestone, some of which becomes building stone and some crushed stone. Some is heated in a kiln to become lime.

Gypsum is an evaporite that precipitates when saline lakes evaporate. It grows as white or clear crystals.

From Lake Bed to Drywall

Over millions of years, shells and shell fragments collect and eventually form beds of limestone.

Organisms extract calcium and carbonate ions from seawater to construct calcite shells.

From Seafloor to Sidewalk

The raw materials from which we manufacture the buildings, roads, wires, and coins of modern society were produced by geologic processes. Concentrations of minerals that provide a source of metals formed during a variety of igneous activities, interactions with groundwater, and sedimentary processes. Limestone, a rock used for buildings and for making concrete, began as an accumulation of seashells. Brick began as clay, a by-product of chemical weathering. And some of the gypsum of drywall began as an accumulation of gypsum-containing evaporites in a desert lake. Metal, lime, clay, and gypsum are all examples of the Earth's mineral resources. We can use some mineral resources right from the Earth, simply by digging them out. But most become usable only after expensive processing.

Mining of metal ores, like the mining of coal, can expose sulfide minerals to air and water. These minerals may dissolve to produce acid mine runoff, which can kill vegetation and pollute water downstream. Similarly, the smoke from ore smelting may contain harmful chemicals, including sulfur, which dissolves in water to produce acid rain. Our need for metals continues to grow, so environmental issues associated with their extraction and use will probably continue to challenge future generations.

FIGURE 10.33 Acid runoff from a mine. The orange color is due to iron oxide staining as well as sulfide-eating bacteria in the water.

See for yourself

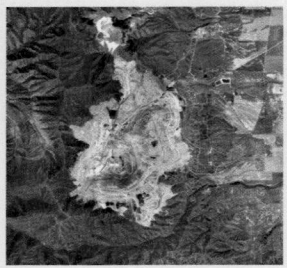

Bingham Copper Mine, Utah

Latitude: 40°31'14.66" N
Longitude: 112°9'1.97" W

Zoom to an altitude of 20 km (12 mi) and look straight down.

The gray patch southwest of Salt Lake City is the largest open-pit mine in the world. Over 17 million tons of copper have been extracted from ore formed when hydrothermal fluids circulated through Cenozoic igneous rock. Zoom in closer to see the pit and tailings pile.

distribution of metallic mineral reserves varies, not all countries have equal access to mineral supplies. This variation has important implications for national security and modern technologies because some **critical minerals**, meaning minerals that are essential for producing high-tech equipment and alloys, may become unavailable to some countries due to supply-chain disruption. The demand for critical minerals such as lithium, cobalt, manganese, and rare-earth elements has ballooned in recent years because of their use in making batteries for electric vehicles and in making computer chips for everything. Increased conservation and recycling could dramatically decrease rates of metal consumption and thereby stretch the lifetimes of existing reserves.

Take-home message...

Ores—rocks that can be processed to produce metals—may contain native metals, but most consist of ore minerals, in which the metal has bonded to another atom. A variety of geologic processes, such as hydrothermal activity, produce ores. Geologists can find ores by studying outcrops, measuring gravitational and magnetic fields, and drilling into bedrock. Mining takes place either in open-pit mines or underground. Most metallic mineral resources are nonrenewable and are not distributed uniformly around the planet.

Quick Questions ——————

- How does a metal differ from other substances?
- Distinguish between a hydrothermal deposit and a placer deposit.
- Why do many ore deposits form along convergent boundaries?

that astronauts can see them from space. And both open-pit and underground mining yield immense tailings piles that, without soil, may remain unvegetated for decades. In some places, mining companies douse tailings with acidic solutions to leach out additional metal ions, and if these acids escape into the environment, they can damage vegetation and contaminate soil and groundwater. Because ores commonly contain sulfide minerals, exposing ore to air and water during mining operations releases acid mine runoff (Fig. 10.33). For a similar reason, smelting of ores can release sulfur-containing smoke, which can lead to acid rain.

How Long Will Metallic Resources Last?

Most metallic mineral resources are nonrenewable, although the metals themselves are recyclable. Geologists have calculated reserves for various minerals, just as they have for fossil fuels, and they are concerned that accessible supplies of some metals may run out in only decades to centuries. Further, because worldwide

10.7 Nonmetallic Mineral Resources

Look around your house or apartment and you'll see a variety of **nonmetallic mineral resources**—concrete, brick, glass, drywall—that come from the Earth and either do not contain metal, or are used for reasons other than the metal they contain. Where do such resources come from?

Dimension Stone

The Parthenon, a colossal stone temple, has stood atop a hill in Athens, Greece, for almost 2,500 years. No wonder: **dimension stone** (or just *stone*, an architect's word for rock) outlasts nearly all other construction materials. The names that contractors give to various types of stone differ from the formal rock names that geologists use. For example, contractors generally refer to any polished

FIGURE 10.34 Stone production in quarries.

(a) An active quarrying operation in Missouri that produces large blocks of cut dimension stone.

(b) Sheets of cut dimension stone being measured for further cutting to become a kitchen countertop.

carbonate rock as "marble," whether or not it has been metamorphosed, and to any crystalline rock containing mostly silicate minerals as "granite," regardless of whether the rock has an igneous or metamorphic texture, or has a felsic or mafic composition. To obtain intact slabs or blocks of dimension stone from a quarry, workers either split the stone from bedrock by hammering a series of wedges into it, causing a crack to propagate, or they slice it off a bedrock wall using various power tools (Fig. 10.34a). Once workers remove blocks of dimension stone from a quarry, the blocks can be cut into smaller pieces to be used in construction, as curbstones, or as gravestones (Fig. 10.34b). Rubbing the blocks with abrasive and water yields a shiny polish. Thin, polished stone slabs are now shipped globally to meet the demand for decorative surfaces such as countertops and building facades (see the chapter opening photo).

Crushed Stone and Concrete

Crushed stone forms the foundations of highways and railroads and serves as the raw material for manufacturing cement, concrete, and asphalt. In crushed-stone quarries, operators use explosives to break up bedrock into chunks, which they then transport by truck to a crusher that breaks it into smaller chunks (Fig. 10.35).

During the past century, concrete has become the dominant material used for the construction of roads and buildings. To make **concrete**, workers mix *aggregate*

(sand, gravel, or both) with cement and water to produce a slurry. When workers pour this slurry into molds, or spread it out in a layer, and let it set, it hardens into a solid rock-like material (Fig. 10.36a). Concrete behaves like a coherent sedimentary rock, for it has the same basic internal structure as a sedimentary rock—both consist of grains bound together by cement. The **cement** used in concrete consists mostly of lime (CaO), with lesser amounts of

Did you ever wonder . . .

how concrete differs from rock?

FIGURE 10.35 A large crushed-stone quarry of Silurian limestone in Illinois. Drillers are working on the shelf in the distance.

FIGURE 10.36 Examples of nonmetallic minerals used in buildings.

(a) The smooth surface coat of this concrete has broken away, revealing rusty "rebar" (reinforcement rods made of steel) and the aggregate fragments that are held to together by cement minerals.

(b) The first step in manufacturing drywall involves spreading a slurry of crushed gypsum and water on a conveyer belt. Next, the slurry will pass through rollers that will flatten it to a uniform thickness.

silica, aluminum oxide, and iron oxide. An assemblage of new, interlocking mineral-like crystals grows from these chemicals in the concrete slurry as it sets.

In the 18th and early 19th centuries, workers produced cement simply by placing chunks of a special type of limestone (that contained quartz and clay in addition to calcite) in a kiln. When heated to a temperature of about 1,450°C (2,640°F), calcite breaks down to form lime and carbon dioxide gas. The breakdown of clay and quartz provided the other oxides for the cement. The specific type of limestone needed to make such *natural cement* is rare, so most concrete structures today use *Portland cement*, a mixture of limestone, sandstone, and shale in just the right proportions to provide the proper mix of chemicals from which the minerals in cement can grow.

Nonmetallic Minerals in Your Home

As we've seen, the concrete in a building's foundation comes from baked limestone mixed with sand or gravel and water. The *bricks* used in the walls consist of clay produced in nature by the chemical weathering of silicate minerals. Workers mold wet clay into blocks and then bake them in order to drive out water and, at higher temperatures, cause metamorphic reactions that lead to

the growth of stronger minerals. Window glass is made by melting and then quickly freezing quartz sand, so that it solidifies without forming crystals. *Drywall* (also known as sheetrock or gypsum board), the solid panels used for interior walls, can be made by mixing crushed gypsum with water to form a slurry which is then spread into a thin layer (Fig. 10.36b). As the slurry dries and sets, new crystals grow and transform the slurry into a solid layer, which is then encased in paper. Gypsum, a sulfate mineral, comes either from natural evaporite deposits or from sulfur scrubbed from the smoke produced by burning sulfur-rich coal. Notably, several other important materials also come from evaporites, such as nitrate for fertilizer, halite for table salt, and lithium for batteries. Lithium can also be obtained from certain types of pegmatite.

Sustainability of Nonmetallic Minerals

Are there enough nonmetallic minerals available to satiate society's need for centuries to come? This isn't an easy question to answer. At first glance, it may seem that our planet contains vast quantities of Earth materials. But in fact, some key nonmetallic resources, such as exotic chemicals in evaporites used for high-technology applications as well as deposits of minerals used for fertilizer, are already in short supply. But even crushed rock, which can be made from vast exposures of bedrock, can be difficult to come by in places where it's needed. The cost of transporting such Earth materials accounts for a major proportion of their total cost, and many cities that consume these materials lie distant from all but a few sources. Even sand, whose quantity seems at first glance to be endless, has become a commodity that people export (for use in concrete), and the amount being mined—in many cases, illegally—from beaches and riverbanks has led to significant loss of land and destruction of the environment. As populations expand and lifestyles improve, society will one day have to face the reality that nonmetallic resources, like most other Earth resources, are limited.

Take-home message...

Society uses a great variety of nonmetallic mineral resources. These materials include dimension stone, crushed stone, cement (made from baked limestone), evaporites (including gypsum), and clay (used to make bricks).

Quick Questions

- Why aren't explosives used to quarry dimension stone?
- What's the difference between natural cement and Portland cement?
- The production of what material involves the melting of pure quartz sand?

10 CHAPTER REVIEW

Objective 10.1

Explain where the energy stored in biomass and fossil fuels comes from, discuss what the "fossil" in fossil fuels refers to, and distinguish between biological and Earth resources.

KEY CONCEPTS

- Photosynthesis stores energy that came from the Sun in the chemical bonds of organic chemicals.
- Combustion of biomass releases water, carbon dioxide, and the energy stored in chemical bonds.
- Society depends on natural resources, which can be classified as biological resources (coming directly from living or recently living organisms) or Earth resources (coming from geologic materials or natural Earth processes). The sources of resources used by society have changed over time.
- Fossil fuels consist of biomass that was buried and altered underground and has been preserved for millions to hundreds of millions of years. There are two types of fossil fuels: hydrocarbons (oil and gas) and coal.

EARTH-SCIENCE VOCABULARY

biological resource (p. 356)
biomass (p. 356)
combustion (p. 356)
Earth resource (p. 356)
energy density (p. 357)

fossil fuel (p. 356)
fuel (p. 356)
natural resource (p. 356)
photosynthesis (p. 356)

REVIEW QUESTIONS

1. **(a)** What component of organic chemicals stores solar energy? **(b)** Explain why combustion can be thought of as the reverse of photosynthesis. **(c)** Give examples to illustrate the difference between a biological resource and an Earth resource.

2. **(a)** Define what a fuel is, and why some types are called fossil fuels. **(b)** Name the two key categories of fossil fuels. **(c)** What does the energy density of a fuel refer to? **(d)** Explain why fossil fuels began to be used, and why they have remained so popular. **(e)** Describe the components of a fossil-fuel-burning power plant like the one shown in **Figure A**.

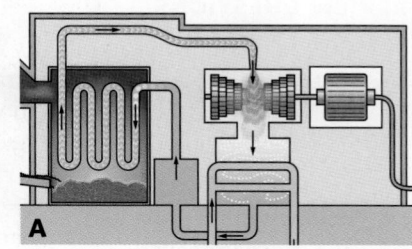

A

Objective 10.2

Provide a definition of a hydrocarbon, describe how oil and gas reserves form, distinguish between conventional and unconventional reserves, and explain how each can be found and extracted.

KEY CONCEPTS

- Hydrocarbons are organic chemicals consisting of hydrogen and carbon atoms. They form from the remains of plankton that accumulate with clay to become organic ooze, which, if buried and lithified, becomes organic shale, a source rock.

- Chemical reactions at a specific range of elevated temperatures underground convert the organic matter in organic shale to kerogen and then to oil or gas.

- To form a conventional oil reserve, oil must migrate from a source rock into a reservoir rock and must then be confined underground by a trap.

- Substantial volumes of hydrocarbons exist in unconventional reserves. These hydrocarbons are difficult to extract, either because they have high viscosity or because the rock holding them has low permeability.

- Obtaining hydrocarbons from tight oil and tight gas reserves requires directional drilling and hydrofracturing.

EARTH-SCIENCE VOCABULARY

conventional reserve (p. 359)	**porosity** (p. 359)
directional drilling (p. 360)	**refinery** (p. 362)
hydrocarbon (p. 357)	**reservoir rock** (p. 359)
hydrocarbon migration (p. 359)	**seismic-reflection profile** (p. 360)
hydrocarbon reserve (p. 359)	**source rock** (p. 358)
hydrofracturing (p. 362)	**tar** (p. 357)
kerogen (p. 358)	**tight oil** (p. 359)
natural gas (p. 357)	**trap** (p. 359)
oil (p. 357)	**unconventional reserve** (p. 359)
oil window (p. 358)	
permeability (p. 359)	

REVIEW QUESTIONS

3. (a) What is a hydrocarbon? (b) What is the source of the organic material in oil, and how did it transform into hydrocarbons? (c) What is the oil window, and what happens to oil at temperatures higher than the oil window?

4. (a) Explain the steps leading to the formation of a conventional oil reserve. (b) Which step does **Figure B** show? (c) Distinguish among a source rock, a reservoir rock, and a trap. (d) Which type of trap does **Figure B** show?

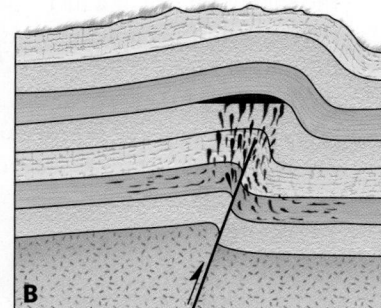
B

5. (a) Explain the difference between a conventional and an unconventional hydrocarbon reserve. (b) Which of the two rock types shown in **Figure C** could contain

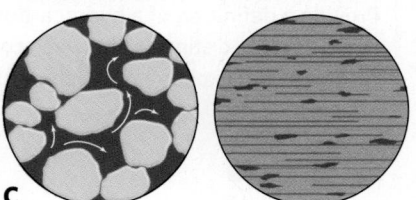
C

a conventional reserve? Explain why. (c) What is the brown fluid rising in **Figure D**, and what is its purpose?

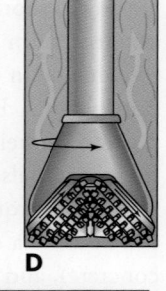
D

6. (a) What are tight oil and tight gas reserves, and what technologies must be used to extract hydrocarbons from them? (b) Which technology does **Figure E** show? Label the features in the image.

7. (a) Where are most of the world's conventional oil reserves found, and what special conditions led to their formation? (b) Are all unconventional oil reserves found in the same places as conventional reserves?

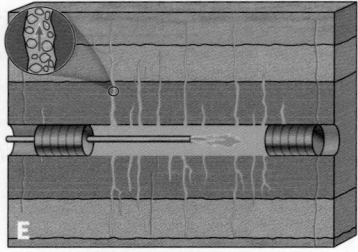
E

Objective 10.3

Describe how coal forms, how it can be classified, and how it can be mined.

KEY CONCEPTS

- For coal to form, abundant plant debris must be deposited in an oxygen-poor environment. Compaction, consolidation, and dehydration of plant debris produces peat, which, when buried deeply and heated, transforms into coal.

- Geologists classify coal into one of three ranks—lignite, bituminous, or anthracite, in order of increasing carbon content. Higher-rank coal forms at higher temperatures and contains more carbon, because volatile chemicals have escaped.

- Coal is a sedimentary rock that occurs in seams that can be mined by either strip mining or underground mining.

EARTH-SCIENCE VOCABULARY

coal (p. 366)	**peat** (p. 366)
coal rank (p. 366)	**strip mining** (p. 368)
coal reserve (p. 368)	**underground mining** (p. 368)
coal seam (p. 367)	
coal swamp (p. 367)	

REVIEW QUESTIONS

8. (a) How does the source material for coal differ from the source material for oil? (b) What changes happen in organic material during the process of forming coal? (c) Taking into account the concept of transgression and regression (see Chapter 5), explain why coal is typically interbedded with sandstone and shale.

9. (a) Which rank of coal forms at the highest temperature? (b) What happens to coal if it gets subjected to the elevated temperature of high-grade metamorphic conditions?

10. (a) Name three different methods used for mining coal, and indicate the settings in which each is used. (b) What is the very large piece of equipment shown in **Figure F**? (c) What phenomena can cause injury to miners in underground mines?

F

Objective 10.4

Discuss the environmental and climate-change challenges involved in the use of fossil fuels, and why the Oil Age may end within the next half century.

KEY CONCEPTS

• We now live in the Oil Age, a time during which hydrocarbon production and consumption dominate many aspects of society. At current rates of consumption, conventional hydrocarbon reserves may last for only another century, but unconventional reserves and coal reserves are sufficient to last longer.

• Production and use of fossil fuels have significant consequences for the environment, including acid runoff and rain as well as air pollution. The release of carbon dioxide from fossil-fuel combustion has significantly increased the concentration of this greenhouse gas in the atmosphere and, thus, is altering the Earth's climate.

EARTH-SCIENCE VOCABULARY

acid mine runoff (p. 371) greenhouse gas (p. 371)
acid rain (p. 371) nonrenewable resource
air pollution (p. 371)   (p. 369)
blowout (p. 371) Oil Age (p. 369)
carbon footprint (p. 372) renewable resource (p. 369)

REVIEW QUESTIONS

11. (a) Distinguish between a renewable and a nonrenewable resource. (b) Why are hydrocarbon and coal reserves considered to be nonrenewable resources?

12. (a) Which of the following represents the largest energy resource: conventional hydrocarbon reserves, unconventional hydrocarbon reserves, or coal reserves? (b) How did the ability to access unconventional hydrocarbon resources profitably impact societies' economic motivation to bring the Oil Age to a close?

13. (a) Where does the acid in acid mine runoff come from? (b) Why has coal ash become an environmental concern? (c) Why does the combustion of fossil fuels produce CO_2, and why is this a problem for society? (d) Label the horizontal axis in **Figure G**. What does the top (longest) bar represent?

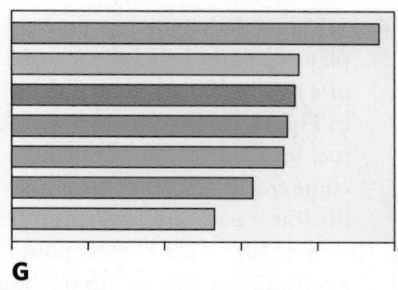
G

Objective 10.5

Characterize alternative sources of energy that could replace fossil fuels, and explain the pros and cons of developing each source.

KEY CONCEPTS

• Nuclear power plants generate electricity using the energy released by nuclear bonds when atoms of radioactive uranium undergo fission. The rate of this reaction must be controlled to avoid meltdown. The safe storage of nuclear waste remains a challenge.

• Commercial geothermal power plants generate electricity using steam produced by pumping hot groundwater to the surface. Home geothermal systems run refrigerant through pipes buried at shallow depths in the ground, where temperature remains constant all year, to remove heat from a building during the summer and add heat to it during the winter.

• Hydroelectric and wind power are produced by using turbines to extract energy from flowing water or air. Each type of energy is green, but there can be environmental problems associated with their use. Photovoltaic cells can capture solar energy to produce electricity.

• It can be challenging to incorporate variable energy sources, such as wind and solar power, into electrical grids. Currently, natural gas is used to maintain a constant supply of electricity.

EARTH-SCIENCE VOCABULARY

biofuel (p. 374) enrichment (p. 373)
chain reaction (p. 373) geothermal energy (p. 374)
electrical grid (p. 376) green energy (p. 375)

hydroelectric power (p. 375)
nuclear reactor (p. 373)
nuclear waste (p. 374)
photovoltaic cell (p. 376)

solar energy (p. 376)
tidal energy (p. 376)
wind power (p. 376)

REVIEW QUESTIONS

14. **(a)** How does a nuclear power plant operate? Label the features of a nuclear power plant shown in **Figure H**. **(b)** Where does the fuel for a nuclear power plant come from? **(c)** What is high-level nuclear waste, and how can it be stored? **(d)** Can a nuclear power plant explode like an atomic bomb? Explain your answer.

H

15. **(a)** How does a biofuel differ from a fossil fuel? **(b)** What can biofuels be made of?

16. **(a)** Explain how a geothermal power plant produces electricity. **(b)** Does installation of a home geothermal energy system require that the home be sited in a region with a very high geothermal gradient? Explain your answer.

17. **(a)** Explain how hydroelectric power plants generate electricity. **(b)** List some of the benefits, and some of the problems, associated with building a hydroelectric dam. **(c)** Why do power companies want to install offshore wind turbines, despite the expense?

18. **(a)** Distinguish between the two ways of harnessing solar energy. **(b)** What is the difference between the n-layer and the p-layer in a photovoltaic cell? **(c)** Why can it be challenging to incorporate wind and solar power into the electrical grid?

REVIEW QUESTIONS

19. **(a)** How does a native metal differ from an ore mineral? **(b)** Why don't we use ordinary granite as a source of metals?

20. **(a)** Describe the different types of ore deposits and how they form. **(b)** Which type of ore deposit is forming in **Figure I**? **(c)** Which type of ore does **Figure J** show? **(d)** Name two types of ores that may occur on the seafloor. **(e)** Where do most hydrothermal ore deposits occur?

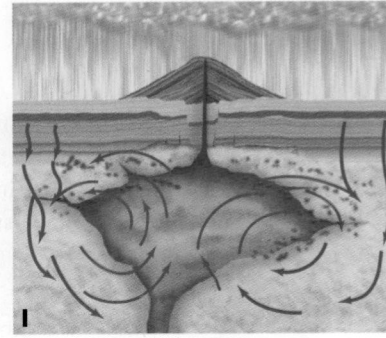

I

J

21. **(a)** What clues can modern exploration geologists use to locate and characterize an ore body today? **(b)** What are the two types of mines used to access ore deposits, and what key factor determines which type is more appropriate for a given location?

22. **(a)** What are some of the environmental consequences of mining? **(b)** What is a critical mineral, and why are governments concerned about their distribution?

Objective 10.6

Discuss the nature of metals and their occurrence in ores, how ore deposits form and are processed, and how their mining and processing affect the environment.

KEY CONCEPTS

• Some metals occur in native form, but most occur in ore minerals. An ore is a rock containing native metals or ore minerals in sufficient quantities to be worth mining.

• Ore deposits form in a variety of ways. Geologists distinguish among types, including magmatic deposits, hydrothermal deposits, secondary-enrichment deposits, sedimentary deposits, and placer deposits.

• Metallic mineral resources are nonrenewable. Many are now or may soon be in short supply.

Objective 10.7

Relate nonmetallic mineral resources to their sources, and discuss how people use these resources.

KEY CONCEPTS

• Many of the materials in your home (such as concrete, brick, glass, and drywall) come from nonmetallic mineral resources, which include dimension stone, crushed stone, clay, sand, and many other materials.

• Bricks are made by molding and heating clay. Cement is made by heating limestone, mixed with lesser amounts of shale and sandstone, to a high temperature. Glass production involves cooling molten SiO_2.

- To produce concrete, workers mix aggregate with cement and allow it to set, so that new mineral crystals grow.
- Supplies of some nonmetallic minerals are becoming scarce.

EARTH-SCIENCE VOCABULARY

cement (p. 385)
concrete (p. 385)
dimension stone (p. 384)

nonmetallic mineral resource (p. 384)

REVIEW QUESTIONS

23. **(a)** How does the production of dimension stone differ from that of crushed rock? **(b)** How does the usage of the terms "marble" and "quartz" by architects and interior designers differ from the usage of those terms by geologists?

24. **(a)** What materials go into the making of concrete? **(b)** In **Figure K**, what are the larger rock fragments called? **(c)** What takes place during the setting of cement that makes concrete hold together like a coherent rock?

25. **(a)** Name some materials in your home that come from nonmetallic mineral resources. **(b)** What does brick consist of, and how does a brick differ from a dried mud block? **(c)** What material can be used to make window glass? **(d)** Name some of the useful materials that come from evaporite deposits.

ANOTHER VIEW This slice of dimension stone contains huge crystals. The polish on the surface makes the rock's texture stand out.

11 SHAPING THE EARTH'S SURFACE
Landscapes, the Hydrologic Cycle, and Mass Wasting

After studying this chapter, you should be able to...

1. explain why the Earth's topography continues to be modified over time, and describe various factors that influence the landforms that develop in a region.

2. sketch a diagram showing processes involved in the hydrologic cycle of the Earth System.

3. describe the characteristics and consequences of different types of mass wasting.

4. explain the factors that cause some regions to be underlain by unstable ground, and discuss the conditions that can trigger a mass-wasting event.

5. use your knowledge of conditions that lead to mass wasting to assess an area's susceptibility to mass-wasting hazards and to evaluate preventive measures.

A dinosaur gazing at the Moon would see nearly the same view that we see today, for most of the Moon's surface has remained unchanged for a few billion years. The Earth's surface, in contrast, has changed radically over time (Fig. 11.1). In addition, the Earth's surface displays vastly more variety in color, shape, and texture than does the Moon's.

Why does the Moon's surface remain static while the Earth's evolves and displays such variation? The Moon's interior has cooled sufficiently so that its asthenosphere cannot flow plastically. Consequently, the Moon does not host plate motion or active volcanism. Without moving plates or mantle plumes, new lunar mountains do not form, and without volcanism, there are no gases to fill a lunar atmosphere. Without an atmosphere, in turn, there can be no lunar hydrosphere and, therefore, no biosphere and no rock cycle. In fact, the Moon's surface changes only in response to meteorite impacts and space weathering (the breakdown of minerals due to the impact of cosmic rays). The character of the Earth's surface is quite different because movement takes place in all realms of the Earth System. Movement of plates in the geosphere, air in the atmosphere, water in the hydrosphere, ice in the cryosphere, and organisms in the biosphere constantly produces, breaks down, displaces, and redistributes our planet's surface materials. This movement can happen, ultimately, because the Earth remains warm enough inside to have a mobile asthenosphere, which not only accommodates plate motion and all its consequences but also leads to magma generation and the associated release of volatile compounds whose presence allows an atmosphere, hydrosphere, and biosphere to exist.

The dynamic character of the Earth System produces an amazing display of *landscapes*, the surface features on land visible within a region. Artists and writers across the centuries have found inspiration in viewing landscapes. Earth scientists, too, feel this inspiration, but they also ask questions: "How did this landscape come to be? How will it change in the future? Do aspects of this landscape pose a threat to life and property?" To set the stage for answering these questions, the first section of this chapter describes land-surface movements and the energy sources that drive them. Then, because water in its various forms plays such a key role in modifying the land surface, we turn our attention to the hydrologic cycle.

FIGURE 11.1 A comparison of the Moon's surface and the Earth's surface, as seen from an altitude of about 50 km.

The final part of this chapter introduces mass wasting, the gravity-driven transport of rock and sediment, and shows how it can alter landscapes over time. Topics in this chapter serve as a foundation for the next two chapters, each of which focuses on characteristics of specific environments in which distinctive landscapes form and evolve.

11.1 The Earth's Ever-Changing Surface

When Earth scientists describe a **landscape**, they are conveying information about the visible form, texture, and composition of the land surface in a region. The character of a given landscape depends on the composition of the **substrate** (the material at and just below the Earth's surface), on the type of vegetation growing on the surface, and on the types of **landforms** (distinct shapes such as beaches, mesas, plains, canyons, valleys, or cliffs) that are visible (Fig. 11.2). We can describe the substrate by characterizing bedrock, soil, and sediment. And we can characterize the vegetation by listing plant species and by noting the percentage of ground covered by plants. Characterizing landforms in a landscape requires documentation of the landscape's shape or topography.

Generation of Topography

If the land surface everywhere were perfectly flat, language might not include a vocabulary for describing landscapes and the landforms they contain. Fortunately, **elevation** (the vertical distance of a point on land above or below mean sea level) varies with location, so our planet hosts mountains, valleys, plateaus, and basins. Elevation

<< In 2017, the red regolith on this partially deforested hillslope in Sierra Leone failed when heavy rain saturated it with water. As it slumped downslope, it disaggregated to form a debris flow that stripped away everything in its path and killed over 1,100 people.

FIGURE 11.2 A great variety of landscapes decorate the Earth's surface.

(a) Rounded mountains border a beach in Rio de Janeiro, Brazil.

(b) Glaciers carved these rugged peaks of the French Alps.

(c) The Amazon River flows through a Peruvian rainforest.

(d) Buttes of sandstone tower above Monument Valley, Arizona.

(e) Cliffs rise from the forest in the Blue Mountains, Australia.

(f) Farm fields checker the plains of the midwestern United States.

> Water flows humbly to the lowest level. In the world, nothing is more submissive or weak than water, yet for attacking what is hard and strong, nothing can surpass it.
>
> —*LAO-TZU*
> (*CHINESE PHILOSOPHER,*
> 604–531 *B.C.E.*)

differences in a region define the region's **topography**, meaning its three-dimensional shape or form. We can portray topography on maps and with digital elevation models **(Box 11.1)**. Such representations indicate the location and steepness of **slopes** (nonhorizontal areas of the land surface), and reveal where hills, valleys, plains, and escarpments lie.

Why do ups and downs develop on the Earth's surface? The first part of the answer comes from the theory of plate tectonics. Tectonic processes, such as subduction, continental collision, rifting, and volcanism, cause the land surface in some regions to undergo **uplift**, upward movement that increases elevation, and in other regions to undergo **subsidence**, downward movement (sinking) that decreases elevation **(Fig. 11.3)**. The second part of the answer comes from considering what happens to land during and after this vertical motion. As soon as tectonic uplift or subsidence produces slopes, other processes of the Earth System kick into action. Specifically, *mass wasting*, the downslope displacement of surface and near-surface material in response to gravity, begins wherever a slope exists, as does **erosion**, the grinding away and removal of material by flowing water, ice, and air. When mass wasting and erosion supply sediment, flowing water, ice, and

air can transport that sediment to another location, where **deposition**—the settling of sediment onto a substrate—takes place (see Chapter 5). Overall, mass wasting, erosion, and deposition act together to redistribute rock and sediment, ultimately stripping it from higher areas and transferring it to fill in lower areas. Thus, tectonic vertical displacements, together with variations in the amount of local erosion or deposition, generate **relief**, differences in elevation between adjacent locations. Mountainous landscapes have *high relief*, meaning a large variation in elevation, whereas plains have *low relief*, meaning slight variations in elevation.

FIGURE 11.3 Uplift raises the land surface, which erodes to form mountains bordered by slopes. Subsidence lowers the land surface to form a basin. Sediment fills the basin as it sinks.

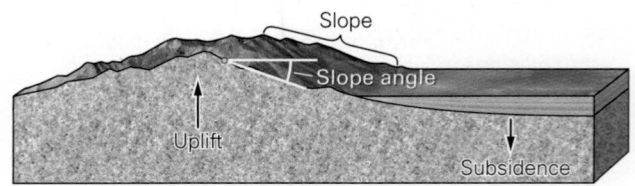

FIGURE 11.4 Uplift, subsidence, erosion, and deposition can be slow or rapid.

(a) Uplifted beach terraces form where the coast rises relative to sea level. The uplifted terrace at this location in California was at sea level about 120,000 years ago. Present-day wave erosion is forming a future terrace surface by eating away at the cliff along the edge of the old one.

(b) So much erosion can take place during a single hurricane that houses built along the beach become undermined. A 2022 hurricane destroyed the fronts of these houses in Florida.

Uplift, subsidence, mass wasting, erosion, and deposition all require lots of energy. To get a sense of what "lots" means in this context, imagine trying to lift a mountain range or carve the Grand Canyon! Where does this energy come from? Three sources of energy drive the evolution of the Earth's surface:

- **Gravitational energy**, the downward pull of gravity, drives mass wasting and participates in driving convection in the mantle, oceans, and atmosphere.

- **Internal energy**, the heat from within the Earth, keeps the mantle soft enough to flow, and therefore allows plate movements to take place. These movements, along with the activity of mantle plumes, drive uplift and subsidence.

- **External energy**, the radiation that comes to the Earth from the Sun, causes air and water near the Earth's surface to become warmer and less dense, and therefore to rise. When air cools at high altitudes, it becomes denser, and it sinks. Differences in heating by the Sun across the world, together with convective movement of air, generate wind, which in turn builds water waves. External energy also causes water on the Earth's surface to evaporate. Water vapor carried over the land by wind provides the rain and snow that fill rivers and build glaciers.

Overall, we can think of landscape evolution as a "battle" between *tectonic processes* (collision, convergence, rifting, and basin formation), driven by internal energy, which build relief by moving the land surface up or down, and *surface processes* (mass wasting, erosion, and deposition), driven by external energy and gravity, which ultimately destroy relief by removing material from high areas

and depositing it in low areas. If, in a particular region, the rate of uplift exceeds the rate of erosion, the land surface rises, but if the rate of erosion exceeds the rate of uplift, the land surface becomes lower over time. Similarly, if the rate of subsidence exceeds the rate of deposition, the land surface becomes lower, but if the rate of deposition exceeds the rate of subsidence, the land surface rises.

How rapidly do vertical movements of the Earth's surface take place? Though the Earth's surface can rise or sink by as much as 3 m (10 ft) during a single major earthquake, rates of vertical surface movement, when averaged over a long time, vary between 0.1 and 10 mm/yr (0.004 and 0.4 in/yr) (**Fig. 11.4a**). Similarly, erosion during a single storm or mass-wasting event can carve tens of meters from the land (**Fig. 11.4b**), and deposition by a single mass-wasting event can produce a layer of debris tens of meters thick in a matter of minutes to days. However, when averaged over a long time, subsidence also takes place at rates between 0.1 and 10 mm/yr. So vertical motions are generally much slower than plate motions. Nevertheless, a change in surface elevation of just 0.5 mm/yr (0.02 in/yr, the thickness of a fingernail) can yield a net change of 5 km (3 mi) in 10 million years. Uplift can build a mountain range, and erosion can whittle one down to near sea level—it just takes time!

Controls on Landscape Evolution

No two landscapes are exactly alike, though there can be similarities that can allow the landscape of one area to resemble that of another. Therefore, Earth scientists distinguish among different types of landscapes by referring to the dominant process or environment that formed the landscape. For example, fluvial landscapes develop due to the flow of rivers, mountainous landscapes host rocky cliffs, glacial landscapes have been modified by ice flow,

Did you ever wonder . . .

how fast the land surface rises or sinks, on average?

BOX 11.1

Science Toolbox

Characterizing topography on a map

A *topographic map* portrays the three-dimensional shape of the land surface on a two-dimensional computer screen or sheet of paper by using contour lines (Fig. Bx11.1a). A *contour line* is an imaginary line on the land surface along which all points have the same elevation. For example, all points along the 100 m contour line lie at an elevation of 100 m above sea level. In effect, a contour line represents the intersection between the land surface and an imaginary horizontal plane (Fig. Bx11.1b).

We refer to the elevation difference between two adjacent contour lines on a topographic map as the *contour interval* for the map. On a given topographic map, the contour interval is constant. So, for example, if the contour interval is 50 m, the next contour line above

the 200 m contour is the 250 m contour, and the one above that is the 300 m contour, and so on. If you walk parallel to the trace of a contour line on the ground surface, you stay at the same elevation. But if you walk at an angle to a contour line, you go upslope or downslope, depending on whether you aim for a higher contour line or a lower one, respectively.

You can picture the angle that a slope makes relative to horizontal just by looking at the spacing between contour lines. Specifically,

FIGURE Bx11.1 Topographic maps and profiles.

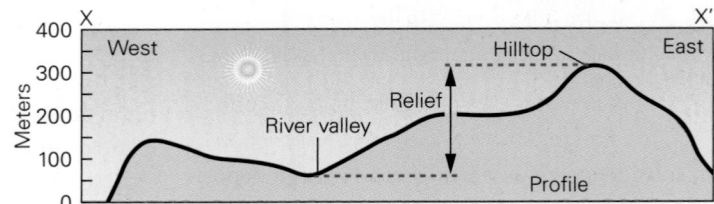

(c) A topographic profile along section line X–X' in part (a) shows the shape of the land surface as seen in a vertical slice.

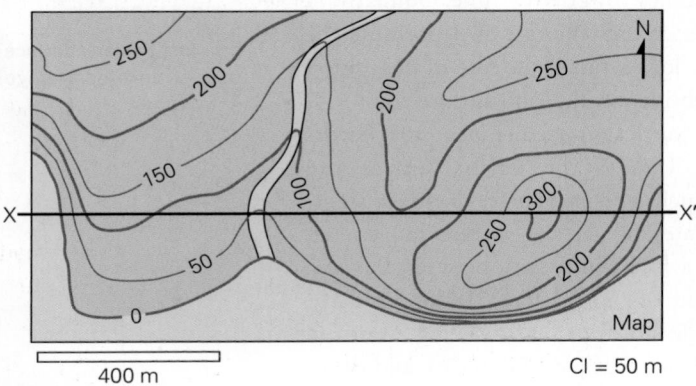

(a) Contour lines, the curving brown lines on this topographic map, depict the shape of the land surface. The elevation of each contour line is provided on this map. The contour interval (CI) indicates the elevation difference between two adjacent lines.

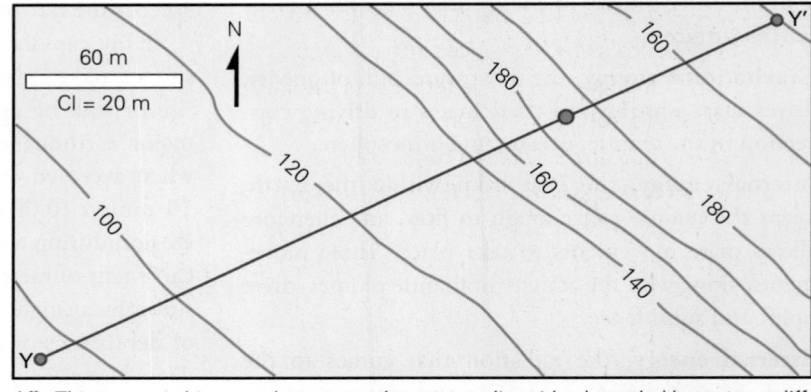

(d) This topographic map shows a northwest-trending ridge bounded by a steep cliff.

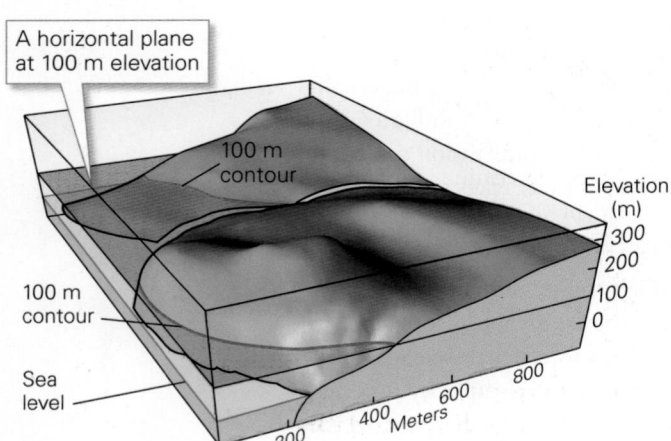

(b) A contour line effectively represents the intersection of an imaginary horizontal plane with the land surface. This block diagram shows the area mapped in part (a).

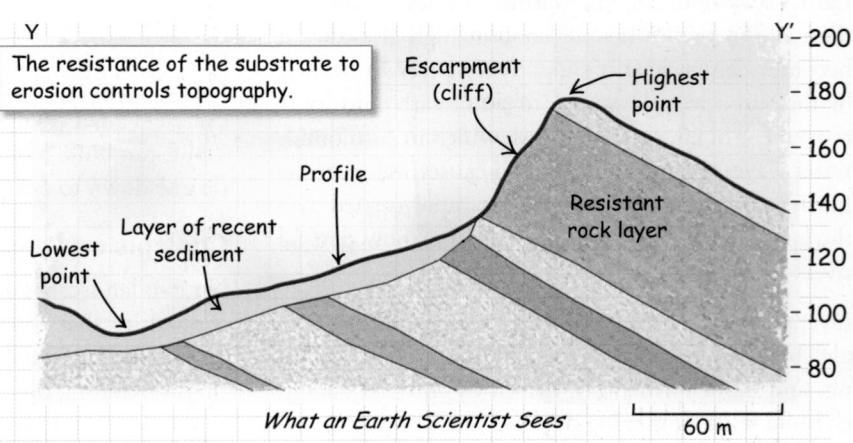

(e) A geologic cross section depicts an Earth scientist's interpretation of the subsurface as represented by a vertical slice along section line Y–Y' in part (d). Note that the cliff follows the edge of a resistant rock layer.

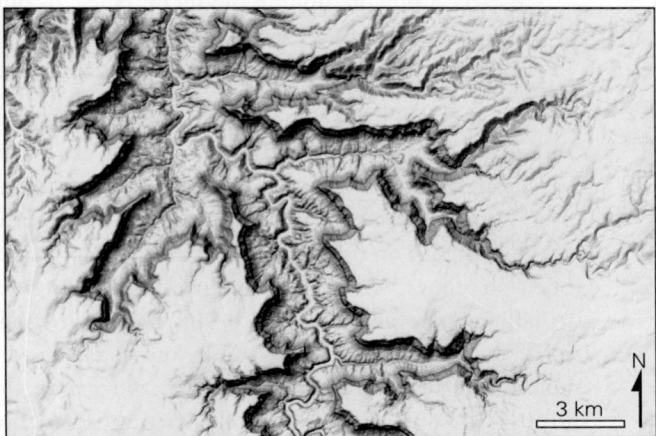

(f) A shaded-relief map of a canyon in Arizona that cut into a flat plateau. Several tributaries connect to the main canyon.

(g) A DEM of the region southwest of Atlantic City, NJ, reveals details of the landscape. All the land depicted on the map lies within 20 m of sea level, so the topography is subtle.

closely spaced contour lines represent a steep slope (where you would cross several contour lines while moving a short horizontal distance on the map), whereas widely spaced contour lines represent a gentle slope. Therefore, on Figure Bx11.1a, the steepest slope shown occurs along the shoreline near the southeast corner.

To represent variations in elevation in a given direction, you can sketch a *topographic profile*, the trace of the ground surface as it would appear on a vertical plane that slices into the ground **(Fig. Bx11.1c)**. Put another way, a topographic profile represents the shape of the ground surface as viewed from the side. On a profile, you can quickly measure the relief and slope angle between two points.

In some cases, the shape of the land surface reflects the distribution of underlying rock units **(Fig. Bx11.1d)**. Geologists can gain insight into subsurface geology by studying a topographic profile. For example,

a steep cliff in a region of sedimentary strata may indicate the presence of beds that are resistant to weathering, whereas low areas or gently sloping areas may be underlain by nonresistant beds. By combining a topographic profile with a representation of geologic features under the ground, geologists produce a **geologic cross section (Fig. Bx11.1e)**.

Shading can help make landscapes stand out. Traditional *shaded-relief maps* add shadows to simulate the appearance of the land if it were lit by the Sun when it's low in the sky **(Fig. Bx11.1f)**. In recent years, Earth scientists have developed methods for representing a landscape with a *digital elevation model* (*DEM*) **(Fig. Bx11.1g)**. Computers construct a DEM from a set of data in which each location on a map has three coordinates: latitude, longitude, and elevation. The resulting map can be presented as a topographic map or, more commonly, as a computer-generated shaded-relief map.

and arid landscapes form in dry regions where plants do not cover the ground. Let's look more closely at the factors that determine whether the region you live in hosts plains, swamps, hills, valleys, mesas, or mountains.

- *Dominant land-surface modification process:* Some landscapes are *erosional,* in that they are primarily a result of the breakdown and removal of rock or sediment, whereas others are *depositional,* in that they are primarily the result of the accumulation of sediment. Not surprisingly, more than one geologic process can happen in a region at the same time, so erosional landscapes and depositional landscapes often occur together.

- *Eroding or transporting agent:* Water, wind, and ice all cause erosion, and all transport sediment. But these three agents do not all produce the same types of

landforms because they differ in their abilities to carve into the land and to carry away debris. Of the three agents, water has the greatest effect on a global basis.

- *Relief:* The elevation difference, or *relief,* between adjacent locations in a landscape determines the height and *steepness* (angle of tilt) of the slope between the two locations. Steepness, in turn, controls the velocity of ice or water flow down the slope, and it determines whether rock or soil will stay in place on the slope or undergo mass wasting.

- *Climate:* The average temperature, the seasonal temperature variation, the overall volume of precipitation in a year, and the seasonal variation in precipitation together define a region's *climate.* Climate not only determines whether running water, flowing ice, or wind serves as the main agent of erosion or deposition, but also controls the lushness of vegetation as well as weathering rates. All of these phenomena can influence landscape evolution.

- *Substrate composition:* The composition and, therefore, the strength of the substrate can control the steepness of slopes and the susceptibility of land to erosion. Strong materials undergo mass wasting and erosion more slowly than do weak materials. Though exposures of strong bedrock occur at many locations, much of the Earth's land surface has a cover of **regolith**, relatively loose material that consists of fragmented rock, chemically weathered rock (including *saprolite*, rock that has been so intensely weathered that grains are barely connected to one another), unconsolidated sediment, and soil. Regolith is weaker than intact bedrock, so it is more susceptible to mass wasting and erosion.

- *Living organisms:* Plants and animals can weaken the substrate (by burrowing, wedging, or digesting) or hold it together (by binding it with roots).

- *Time:* It can take millions to tens of millions of years to carve a canyon, uplift a mountain range, or build a river delta. Therefore, a landscape that we see today represents just an instant in a long evolutionary process.

For almost all of geologic time, water, wind, and ice caused all erosion and deposition, and the land surface was entirely natural. During the past few centuries, however, humans have become a major, if not the dominant, agent of change on this planet's surface. By excavating vast open-pit mines, we produce basins; by dumping debris in tailings piles and landfills, we build hills; by making road cuts, we carve canyons; and by moving soil to build houses, we make steep slopes gentle and gentle slopes steep (Fig. 11.5). Further, by constructing dams, breakwaters, and levees, we modify the shapes of coastlines, change the courses of rivers, and fill new lakes. In cities, buildings and pavement completely seal the ground, forcing water that might have infiltrated the ground to spill instead into a stream, increasing the stream's flow. And in rural areas, agriculture, grazing, water use, and deforestation substantially alter the rates at which natural erosion

FIGURE 11.5 Anthropogenic landforms emphasize the geologic scale of some human activities.

(a) These Egyptian pyramids are human-made hills that have lasted for thousands of years.

(b) Some road cuts, like this one near Denver, are deep valleys cut through ridges.

(c) This stone dam holds back a reservoir in Colorado, modifying the course of a river.

and deposition take place. Because of all these activities, large areas of the Earth's surface have *anthropogenic landscapes*, landscapes whose character, shape, or surface cover developed due to the actions of human society.

Take-home message…

Because of dynamic processes in the Earth System, landscapes and topography on the Earth's surface evolve. Land can undergo uplift or subsidence to yield relief. Landscape evolution is driven by the Earth's internal energy, external energy from the Sun, and gravity. The nature of a landscape depends on eroding or transporting agents, relief, climate, substrate composition, the activities of organisms, and time. Some landscapes are now anthropogenic.

Quick Questions ────────────

• Define relief.

• What happens to land-surface elevation when erosion rates exceed uplift rates?

• Distinguish between tectonic processes and surface processes.

11.2 The Hydrologic Cycle

Because water in its various states—liquid, gas, and solid—plays such an important role in landscape evolution, a study of surface processes and landscapes requires that we first consider how and why water moves around in the Earth System. Our planet's surface and near-surface water occupies several distinct *reservoirs*, a word used here to mean any realm or part of a realm with the capacity to hold water **(Table 11.1)**.

Atmospheric water occurs in vapor (gaseous H_2O), clouds (composed of tiny liquid droplets and/or ice crystals), and precipitation (rain, snow, or hail that falls to the Earth's surface) (see Chapter 16). **Surface water** collects in oceans, lakes, streams, puddles, wetlands, snowfields, and glaciers. Directly below the ground surface, some water adheres to sediment grains and organic material as *soil water*, and some sinks deeper and fills underground pores and cracks as **groundwater**. *Permafrost* exists where soil water and groundwater (down to depths of tens to hundreds of meters underground) remain frozen all year. A significant amount of water also resides in living organisms, for about 70% of a living cell consists of water. Water constantly flows among all these reservoirs, driven by gravity, solar radiation, wind, convection, and biological activity. This never-ending migration is known as the **hydrologic cycle (Fig. 11.6)**.

To get a clearer sense of how the hydrologic cycle operates, let's follow the fate of water molecules that have

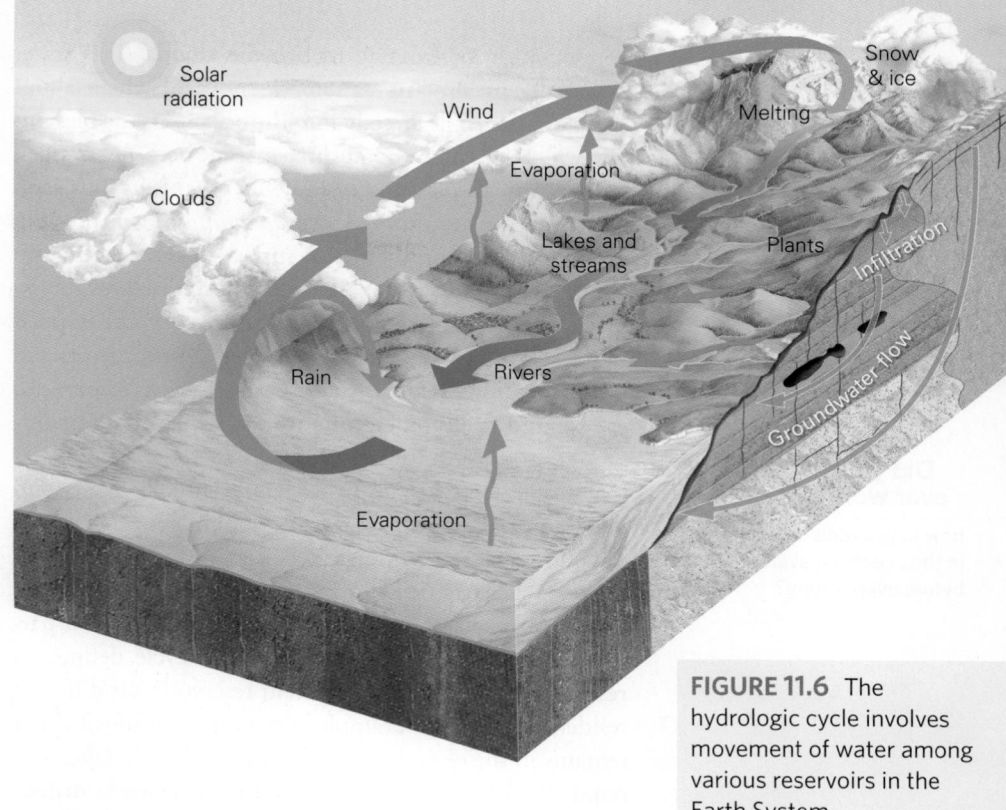

FIGURE 11.6 The hydrologic cycle involves movement of water among various reservoirs in the Earth System.

just reached the surface of the ocean by rising from depth. Solar radiation heats the seawater, and the increased thermal energy of the vibrating molecules causes them to evaporate and drift upward in a gaseous state to become part of the atmosphere. Atmospheric water vapor moves with the wind to higher altitudes, where it cools and undergoes

TABLE 11.1 Major Water Reservoirs of the Earth

Reservoir	Volume (km³)	% of Total Water	% of Freshwater
Oceans and seas	1,338,000,000	96.5	—
Glaciers, ice caps, snowfields	24,064,000	2.05	68.7
Saline groundwater	12,870,000	0.76	—
Fresh groundwater	10,500,000	0.94	30.1
Permafrost	300,000	0.022	0.86
Freshwater lakes	91,000	0.007	0.26
Salt lakes	85,400	0.006	—
Soil moisture	16,500	0.001	0.05
Atmosphere	12,900	0.001	0.04
Wetlands	11,470	0.0008	0.03
Rivers and streams	2,120	0.0002	0.006
Living organisms	1,120	0.0001	0.003

Source: Data from P. H. Gleick, *Encyclopedia of Climate and Weather* (New York: Oxford University Press, 1996).

condensation or freezing to become clouds. This water eventually precipitates as rain or snow, of which about three-quarters falls directly into the ocean. Of the remainder, which falls on the land surface, some becomes trapped temporarily as soil water or in living organisms, but soon returns to the atmosphere by *evapotranspiration* (the sum of evaporation from bodies of surface water, evaporation from the ground surface and soil, and release of water by plants and animals). Water that does not become trapped in the soil or in living organisms enters lakes or rivers and ultimately flows back to the sea. Some sinks deeper into the ground to become groundwater, which flows underground and also ultimately returns to the ocean. Water that becomes frozen into glaciers eventually melts and flows into lakes and streams, or sublimates (transforms directly into gas). Some glaciers survive until they reach the sea, where they finally melt or calve off icebergs.

The average length of time that water stays in a particular reservoir during the hydrologic cycle defines its **residence time**. Water in different reservoirs has different residence times. For example, a typical molecule of water remains in the oceans for 4,000 years or less, in lakes and ponds for 10 years or less, in rivers for two weeks or less, and in the atmosphere for 10 days or less. Groundwater in rock that has at least some permeability (allowing the water to move) stays underground for anywhere from two weeks to tens of thousands of years (see Chapter 12).

Did you ever wonder...

how long a molecule resides in the ocean, on average, before evaporating?

Take-home message...

🏠 Water, which plays a major role in landscape evolution, moves among various reservoirs of the Earth System (the ocean, the atmosphere, the land surface, the subsurface, and living organisms) during the hydrologic cycle.

Quick Questions ————————————

- Distinguish among surface water, soil water, and groundwater.
- Which is the largest reservoir for water on the Earth?
- What does the residence time of water in a reservoir refer to?

11.3 Introducing Mass Wasting

What Is Mass Wasting?

It was Sunday, May 31, 1970, a market day, and thousands of people had crammed into the Andean town of Yungay, Peru, to shop. Suddenly they felt the jolt of an earthquake, strong enough to topple some masonry houses. Unfortunately, worse devastation was yet to come, for the ground shaking had broken a huge ice slab off the end of a glacier.

at the top of Nevado Huascarán, a nearby mountain whose peak reaches an elevation of 6.6 km (4.1 mi). As it tumbled down the mountainside, the glacial ice disintegrated into a turbulent mass of chunks mixed with soil and rock that it had ripped up as it fell (**Fig. 11.7a**). A few

FIGURE 11.7 The landslide disaster in Yungay, Peru.

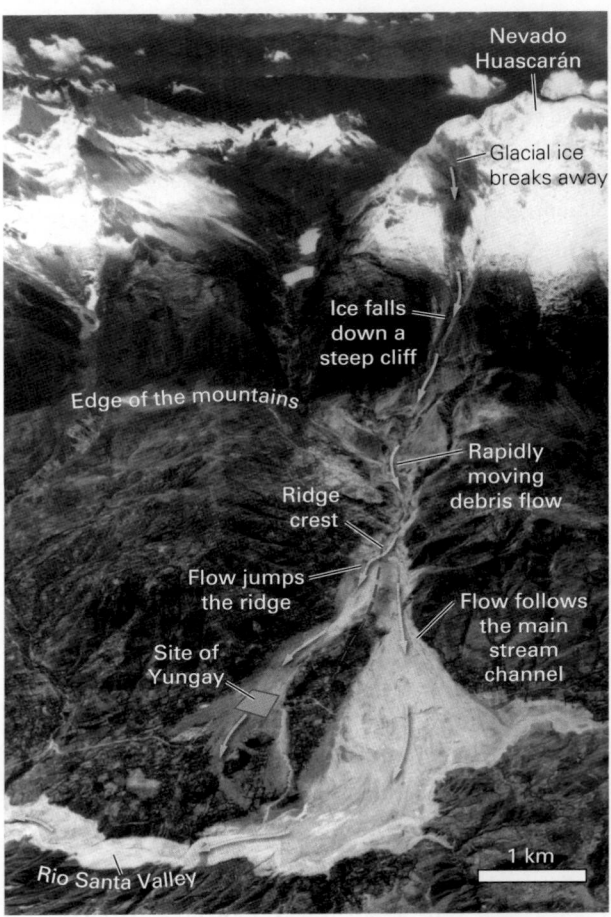

(a) A 1970 air photo showing where the glacier failed, and where the debris ended up.

(b) Before the landslide, the town of Yungay perched on a scenic terrace. After the landslide, no trace of the town remained. The landslide scarred the mountain face in the distance.

Moments later, frictional heating transformed enough ice into liquid water that the mixture transformed into a muddy slurry. After traveling for a few kilometers, the slurry surged into a narrow gorge and became as deep as a 10-story building. When it emerged from the gorge, most of the debris continued down the main valley. But part of it rocketed up the side of the valley and became airborne. This slurry landed on a sloping terrace bordering the main canyon. Yungay stood just downslope. As the town's 18,000 inhabitants stumbled out of earthquake-damaged buildings, they heard a deafening roar, looked upslope, and saw a frothing wall of wet debris rushing toward them. Seconds later, the debris buried the town, along with everyone in it. Today, a grassy meadow topped by a memorial overlies the site where Yungay once bustled (Fig. 11.7b).

The catastrophe at Yungay reminds us that much of the Earth's surface isn't terra firma, but rather hosts **unstable ground**, land whose substrate has the potential to undergo movement, or **failure**, in response to gravity in the not-too-distant future due to a relatively small change in the forces acting on the slope. **Stable ground**, in contrast, has a very low probability of undergoing failure anytime soon. As we've noted, geologists refer to failure involving the tumbling, falling, flowing, or sliding of surface materials in response to gravity as **mass wasting**, or *mass movement*. Mass wasting plays a critical role in the rock cycle (see Chapter 5) as the first step in the transportation of sediment, and it serves as the most rapid process by which the shape of the land surface undergoes modification. Like earthquakes, volcanic eruptions, storms, and floods, mass wasting is a type of *natural hazard*, a feature of the environment that has the potential to cause casualties and property damage. Unfortunately, mass-wasting hazards become more of a threat to communities every year because, as the world's population grows, people have increasingly moved into landscapes underlain by unstable ground. Consequently, several times a year, a mass-wasting event causes substantial casualties and major property damage, and thereby becomes a *natural disaster*.

Most mass-wasting events occur on sloping land due to *slope failure*. In everyday English, these events are known as **landslides**. Geologists and engineers find it useful to distinguish among several specific types of mass-wasting events based on the following factors: (1) the type of material involved (rock or regolith); (2) the velocity of movement (slow or fast); (3) the character of the moving material (coherent or chaotic); (4) the water content of the moving material (wet or dry); (5) the location of the movement relative to sea level (subaerial or submarine); and (6) the slope of the land that moves (vertical, sloping, or horizontal). Let's first examine the different types of mass-wasting events that occur on land (namely, creep,

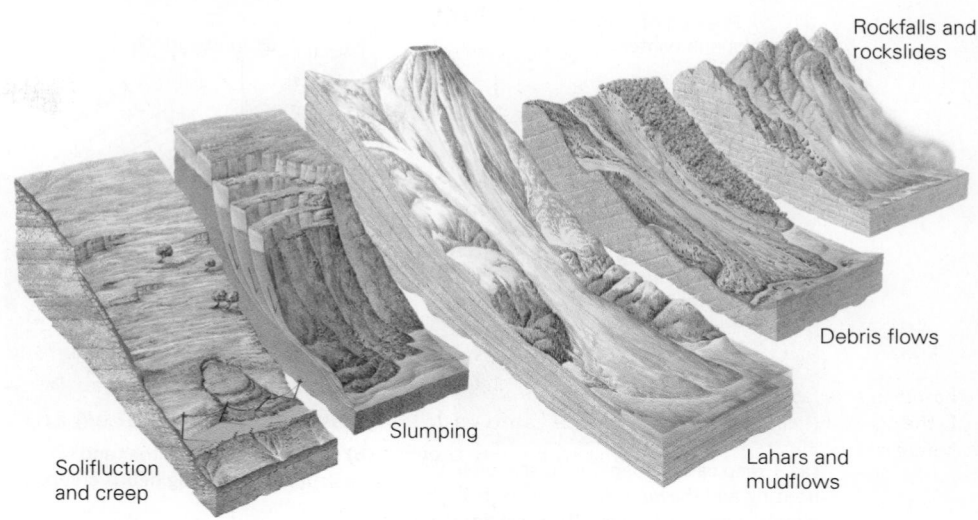

FIGURE 11.8 Different types of mass wasting can modify the landscape. Geologists apply different names to each type.

Rockfalls and rockslides

Debris flows

Slumping

Lahars and mudflows

Solifluction and creep

solifluction, slumping, mudflows, debris flows, lahars, snow avalanches, rockslides, and rockfalls), roughly in order from slow to very fast (Fig. 11.8). Then, we'll briefly digress to describe mass wasting that occurs underwater. Keep in mind that the distinction among types of mass wasting tends to be somewhat fuzzy, because a given event may start out as one type and evolve into another, or it may display characteristics of two or more types.

Creep and Solifluction

Creep, the gradual movement of regolith down a slope, happens when regolith on a slope alternately expands and contracts in response to freezing and thawing, wetting and drying, or warming and cooling. To see how the process of creep works, let's focus on the consequences of seasonal freezing and thawing. In the winter, when water in regolith freezes, the regolith expands, and clasts move outward in a direction perpendicular to the slope. During the spring thaw, when water becomes liquid again, gravity makes the clasts sink vertically, so they effectively migrate downslope slightly (Fig. 11.9a, b). You can't see creep in action by staring at a hillside because it occurs too slowly. But you can see its net consequences over time: creep causes walls, gravestones, and trees to tilt and foundation walls of houses to crack (Fig. 11.9c). Trees that continue to grow while creep takes place develop curved trunks, because as the base of the tree tilts downslope, the top grows straight up (see Fig. 11.9c, inset).

In the Arctic, or in cold high-elevation regions, a warm spell during the summer may cause the uppermost 1–3 m (3–10 ft) of permafrost to thaw. Because meltwater cannot sink into the still-solid permafrost below, the melted layer becomes soggy and weak and flows slowly

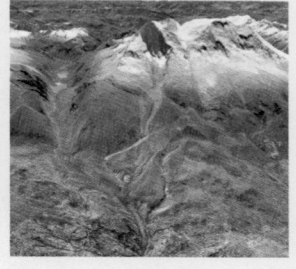

See for yourself

Debris Flow, Yungay, Peru

Latitude: 9°7'21.42" S
Longitude: 77°39'44.86" W

Look NE from an altitude of 6.4 km (4.3 mi)

Here you can see the steep, glaciated face of Nevado Huascarán. In 1970, a debris flow from the mountain rushed down the valley in the foreground and buried the landscape. More recently, smaller debris flows have accumulated on top of the larger one.

FIGURE 11.9 Creep, a type of slow mass wasting.

Position of
slope in winter

Position of
slope in summer

Starting
position

Ending
position

Regolith

Bedrock

Ground surface

Direction
of creep

Regolith

Bedrock

(a) Creep can happen due to seasonal freezing and thawing. A clast rises in a direction perpendicular to the ground during freezing, but sinks vertically during thawing. Here we show the movement after three cycles.

(b) As rock layers weather and break up, the resulting debris creeps downslope.

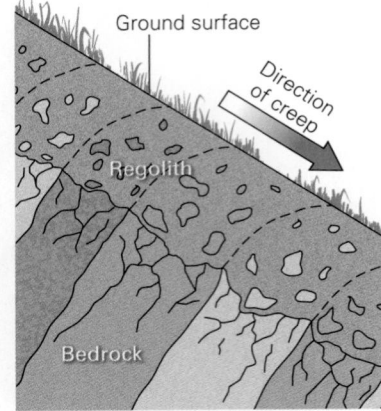

Curving
tree trunk

House with
sagging foundation
and cracked walls

Creep
zone

Intact bedrock

Tilted
telephone
poles

(c) Creep causes walls to bend and crack, building foundations to sink, trees to bend, and utility poles and gravestones to tilt.

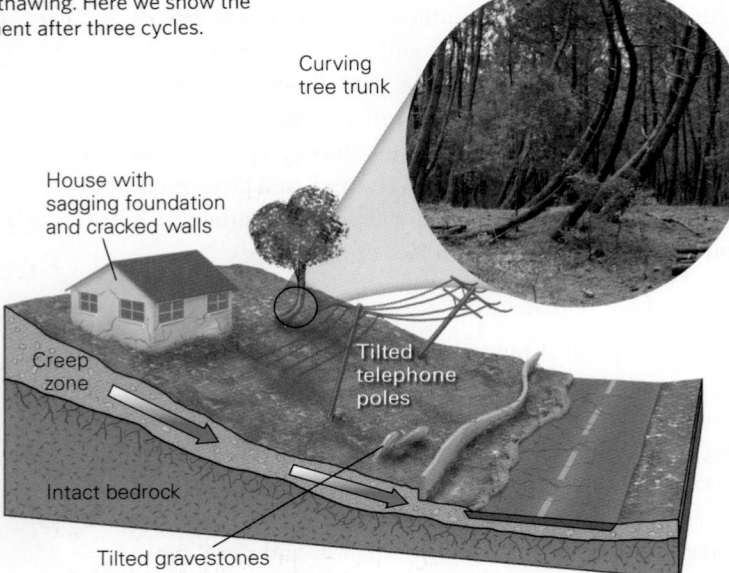

Tilted gravestones

downslope in overlapping sheets. Geologists refer to this kind of creep as **solifluction** (Fig. 11.10).

Slumping and Lateral Spreading

The majestic Holbeck Hall Hotel had perched on a cliff along the eastern coast of England since 1879. Its guests enjoyed spectacular views out across the North Sea—until June 5, 1993. On that day, a block of land bordering the cliff collapsed, taking half the hotel with it (Fig. 11.11a). Fortunately, telltale cracks in the ground weeks before, and a smaller collapse the day before, led officials to evacuate the hotel, so no one was hurt. But the landmark, along with an immense amount of the underlying land, moved out onto a beach far below, where the pounding surf of the North Sea eventually carried it away.

Geologists refer to a relatively slow mass-wasting event during which moving rock or regolith stays somewhat coherent as it slides down a slope as a **slump**. During slumping, the *slump block*, meaning the moving mass, moves down a **failure surface** across which the slump block detached from its substrate. Commonly, the failure surface has a concave-up, spoon-like shape. A cliff-like step, known as a *head scarp*, forms where downslope movement of the slump block exposes the upslope edge of the failure surface (Fig. 11.11b, c). In some cases, the upslope end of a slump block breaks into a series of discrete slices, each separated from its neighbor by a small failure surface resembling a normal fault. Each slump block tends to rotate around a horizontal axis as it moves, forming crescents of ground that tilt toward the head scarp. The downslope end, or *toe*, of the slump may break into discrete overlapping slices or may disaggregate into a chaotic jumble of mud and debris that moves up and over intact ground. If a slump happens along a river, lake, or sea, moving water soon carries away the debris (Fig. 11.11d). Slumps range from a few meters to several kilometers across, and they move at speeds from millimeters per day to meters per minute.

In some cases, slump-like failure takes place even where the land surface over what will become the moving mass

FIGURE 11.10 Solifluction.

Melted layer

Permafrost

Bedrock

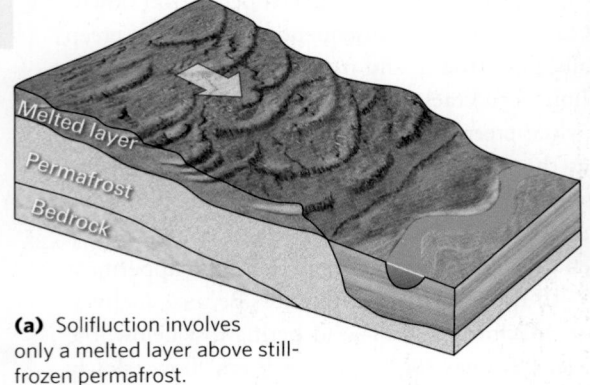

(a) Solifluction involves only a melted layer above still-frozen permafrost.

(b) This entire hillslope is undergoing solifluction.

FIGURE 11.11 Slumping and lateral spreading.

(a) The slump that destroyed the Holbeck Hall Hotel.

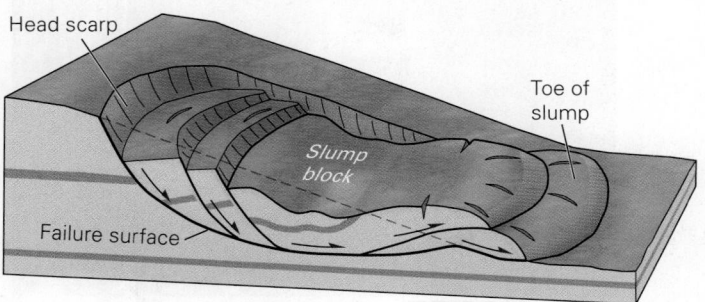

(b) Components of a slump on a spoon-shaped failure surface.

(c) A head scarp on a hillslope.

(d) Sediment slumping into this Costa Rica stream has been washed away.

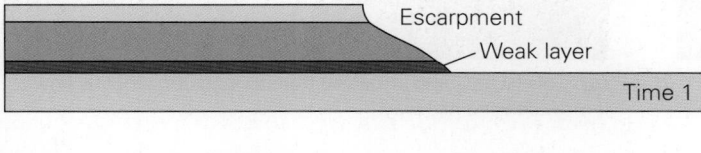

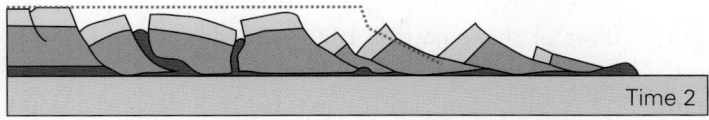

(e) Cross-sectional images before and after lateral spreading. The colored bands represent layers of sediment.

isn't sloping. Such movement happens where a broad area of land underlain by a very weak horizontal failure surface terminates at a steep escarpment. When failure takes place, the horizontal layer of rock and/or regolith above the failure surface slips toward the escarpment. During such **lateral spreading**, the moving layer tends to break up into several separate pieces, with spaces in between, and the land surface, overall, sinks. The displaced rock or regolith spreads out beyond the pre-movement location of the escarpment **(Fig. 11.11e)**.

Mudflows, Debris Flows, and Lahars

Where heavy rains have saturated regolith on a slope, the regolith can transform into a muddy slurry, with the consistency of wet concrete, that flows downslope when failure takes place. The moving slurry is known as a **mudflow**, or *mudslide*, if it consists mostly of clay and water, and is called a **debris flow** if it carries more than 50% sand, cobbles, or boulders mixed in with the mud **(Fig. 11.12a)**. In some cases, mudflows or debris flows start out as slumps, but because of their water content, movement causes them to disaggregate and liquefy as they advance. Alternatively, mudflows or debris flows can form where slumps dump regolith into a flooded river; the debris mixes with the turbulent water to form a slurry that continues to move down the river channel.

Mudflows and debris flows travel at speeds of 20–80 km/h (10–50 mph). The speed at which a given flow moves depends on both the water content and the slope angle: wetter flows (containing >30% water) move faster than drier ones, and flows move faster down steep slopes than they do down gentle slopes. Because mudflows and debris flows have much greater viscosity than clear water, they can carry large rock chunks and huge trees, as well as houses and cars, along with them, and they can apply enough force to crush buildings in their path. Consequently, passage of a mudflow or debris flow may strip away forests, knock down buildings, and bury highways **(Fig. 11.12b-d)**. Mud or debris flows can cause particularly widespread devastation where they flow down a valley that opens onto flatland. Once freed from

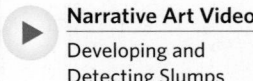

Narrative Art Video
Developing and Detecting Slumps

FIGURE 11.12 Examples of mudflows and debris flows.

(a) A recent debris flow in Utah. Note the chaotic mixture of rock chunks and mud.

(b) These mudflows of 2011 stripped away forests on hillslopes in Brazil.

(c) A 2011 mudflow destroyed houses along this hillside in Brazil.

(d) This large slump in Taiwan transformed into a debris flow as it moved over a highway.

the confines of a valley, the flow spreads out into a fan that can inundate a broad area with mud and boulders. Such an event struck the California city of Montecito in 2018, destroying part of the city and killing 21 people **(Fig. 11.13)**. And in 1999, heavy rains that drenched the coastal mountains of Venezuela produced so many debris

FIGURE 11.13 A debris flow buried this home in California.

flows of this type that 10,000–30,000 people living in towns along the mountain front died.

Mudflows or debris flows that spill down the slopes of volcanoes consist of a mixture of volcanic ash, coarser pyroclastic debris, and water. The water may come from snow or ice melted by the volcano's heat or from heavy rains. As we noted in Chapter 4, such volcanic mudflows are known as *lahars* **(Fig. 11.14a)**. Lahars can travel outward from the base of the volcano along river channels for tens of kilometers. Because lahars can be activated by rainwater, not just meltwater, some happen years after a large eruption of pyroclastic debris. For example, devastating lahars carrying ash from the 1991 eruption of Mt. Pinatubo in the Philippines spilled into the surrounding lowlands in each of the next four rainy seasons after the eruption **(Fig. 11.14b)**.

Rockslides and Rockfalls

In the early 1960s, engineers built a huge new hydroelectric dam across a river on the northern side of Monte Toc, in the Italian Alps. The Vaiont Dam was an engineering marvel, a concrete wall rising as high as an 85-story

(a) A lahar that rushed down the side of Mt. St. Helens, in the state of Washington.

(b) Damage to a town in the Philippines caused by a lahar.

FIGURE 11.14
Examples of lahars.

skyscraper above the valley floor. As soon as the dam was finished, a deep reservoir began to fill behind it. Unfortunately, the dam's builders did not fully appreciate the hazard posed by the unstable face of Monte Toc. The side of the mountain facing the reservoir consists of limestone beds interlayered with weak shale beds, all of which dip parallel to the mountain slope (Fig. 11.15a). As the reservoir

filled, and the *water table* (the top surface of groundwater) rose, the mountain flank began to crack and rumble, and rock chunks tumbled into the reservoir. Local residents began referring to Monte Toc as *la montagna che cammina* (the mountain that walks).

On October 9, 1963, after several days of rain, the reservoir had filled to capacity. Monte Toc started to

FIGURE 11.15 The Vaiont Dam disaster.

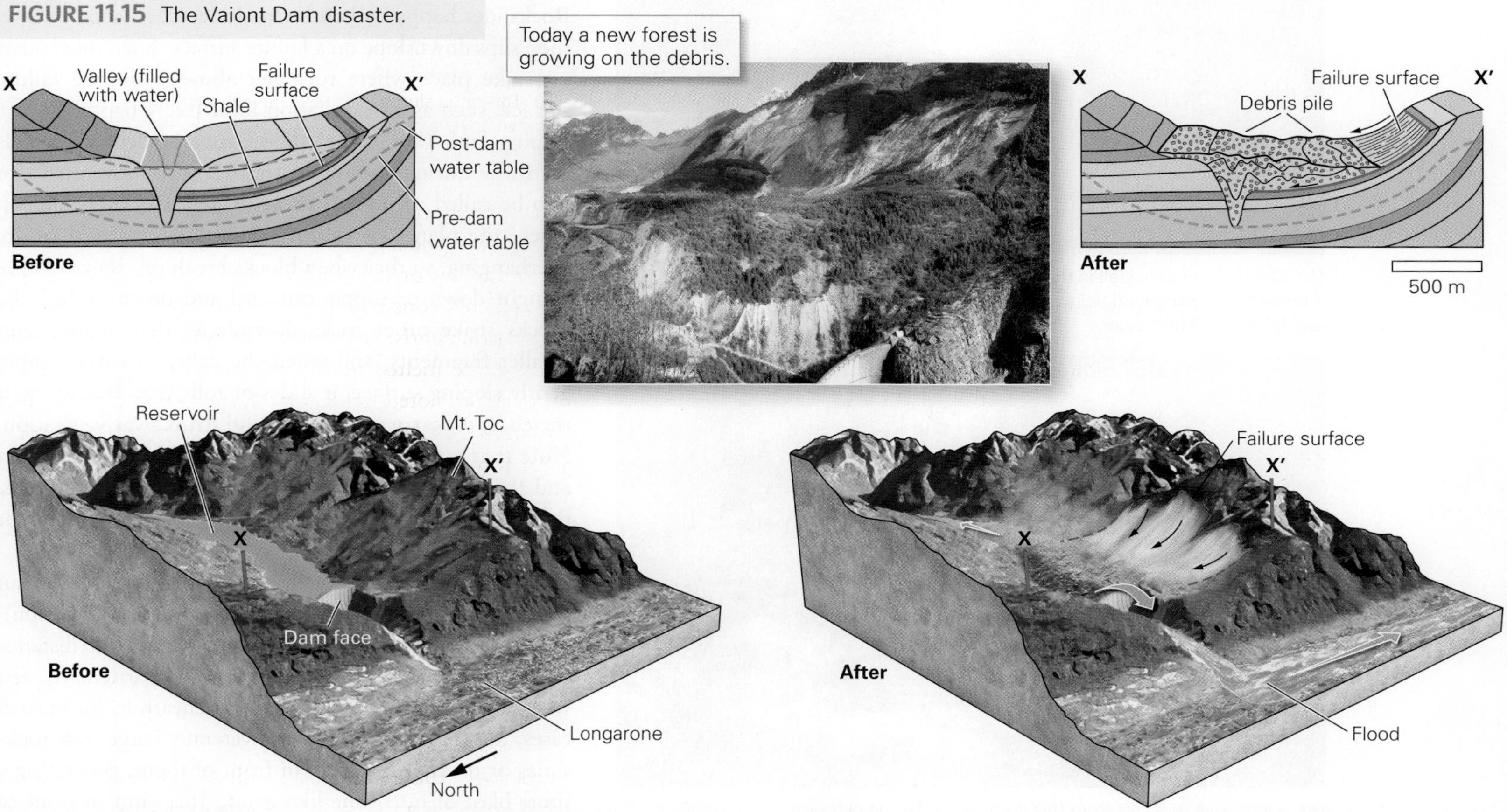

Today a new forest is growing on the debris.

Before
X — Valley (filled with water) — Shale — Failure surface — X′ — Post-dam water table — Pre-dam water table

After
X — Failure surface — X′ — Debris pile — 500 m

Before
Reservoir — Mt. Toc — X′ — X — Dam face — Longarone — North

After
Failure surface — X′ — X — Flood

(a) When the reservoir was filled with water, the water table rose from its pre-dam level (below the valley floor) to the surface level of the newly filled lake. The cross section shows a shale bed that became submerged and turned into a failure surface.

(b) When the slope failed, 260 million m³ (9.2 billion ft³) of rock and debris slid down the mountainside and displaced water in the reservoir. The water surged over the dam and swept away a town in the valley below.

FIGURE 11.16 Examples of rockslides and rockfalls.

(a) This rockslide slid down exfoliation joints.

(b) Debris from a rockfall in the Andes of Peru kept moving after it landed on sloping land below. The fresh rock exposed by the rockfall has a lighter color.

(c) Successive rockfalls from this vertical sandstone cliff sent boulders down a shale slope.

rumble so much that engineers decided to release some water to lower the level of the reservoir. They thought that this lowering might cause some wet ground at the base of Monte Toc to slump into the reservoir, but that such small-scale mass wasting would have only minor consequences. Therefore, authorities did not order an evacuation of the town of Longarone, located a few kilometers down the valley from the dam. This was a grave mistake. At 10:30 that evening, one of the shale layers deep below the mountain face became a failure surface, and an immense chunk of Monte Toc detached from the rest of the mountain and slid into the reservoir. Not only did the debris fill most of the reservoir, but some even shot up the opposite wall of the valley to a height of 260 m (850 ft) above the original reservoir level. As a result, 50 million m³ (1.8 billion ft³) of displaced water splashed up and over the top of the dam, then rushed down the valley below (**Fig. 11.15b**). When the wall of water had passed, nothing remained of Longarone and its 1,500 inhabitants. Though the dam still stands today, it holds back only debris, and has never provided any electricity.

Geologists refer to such a sudden movement of rock down a nonvertical slope as a **rockslide** or, if the mass contains abundant regolith as well as rock, a *debris slide*. Rockslides happen when bedrock detaches from a slope, then slips downslope on a failure surface. Such movement can take place where rock lies above a sloping failure surface, such as an exfoliation joint (see Chapter 5) or a sloping fault (**Fig. 11.16a**). If the rock or debris free-falls vertically during part of its journey, the movement can also be called a **rockfall** or *debris fall*. Rockfalls happen where the planes of weakness are nearly vertical or are overhanging, so that when blocks break off, they collapse straight down or topple outward and down. When the blocks strike other rocks downslope, they shatter into smaller fragments, and when the debris reaches a more gently sloping surface, it slides or rolls (**Fig. 11.16b**). As it travels, it picks up more debris and strips away vegetation. Note that a distinct scar, defined by a lack of vegetation and the presence of fresh rock (typically lighter in color than surrounding rock), remains where a rockslide or rockfall originates.

The moving material in a rockfall or rockslide can accelerate to speeds of up to 300 km/h (185 mph), depending on the steepness of the slope and the distance that it moves. Its movement becomes particularly fast when a cushion of air gets trapped beneath it, for in such cases, the mass moves like a hovercraft. Large, fast rockslides or rockfalls push air in front of them, generating a short blast of hurricane-like wind. The wind in front of a 1996 rockfall in Yosemite National Park, for example, flattened over 2,000 trees. When a fall or slide finally lands on a nonvertical slope, friction slows it down and may bring it to a halt not far from the base of the slope (Fig. 11.16c). In some cases, as we saw at the Vaiont Dam, particularly fast-moving debris may have enough inertia to climb the far wall of a valley.

If fragments from rockfalls accumulate over time at the base of a cliff, they build a sloping apron or pile at the foot of the cliff. Blocks in this pile are called **talus**, and the surface of the pile is called a *talus slope* (Fig. 11.17a). The face of a talus slope—or of any pile of unconsolidated sediment—attains the steepest angle that it can without collapsing. This angle is called the **angle of repose**. While the angle of repose for dry sand typically ranges between

30° and 37°, steeper angles of up to 45° develop where talus consists of irregularly shaped blocks that can interlock (Fig. 11.17b).

Rockslides and rockfalls happen at a variety of scales, depending on the amount of material that detaches and on whether the energy of the tumbling material can set debris in its path in motion. The largest can involve a whole mountainside, as we saw in the example of the Monte Toc catastrophe. Even smaller rockfalls or rockslides, involving just a few cobbles, can be a hazard, particularly along highways, where the debris can tumble onto the road surface or onto cars (see Fig. 11.16a). Hence, highway crews erect "Falling Rock" signs along stretches of highways that border cliffs.

Snow Avalanches

In the winter of 1999, an unusual weather system passed over the Austrian Alps. First it snowed. Then the temperature warmed, and the snow began to melt. But then the weather turned cold again, and the melted snow froze into a hard, icy crust. This cold snap ushered in a blizzard that blanketed the icy crust with tens of centimeters (1–2 ft) of new snow. When the new snow layer became heavy enough, the icy crust acted as a failure surface, and a snow slab high on a mountain slope failed. Downslope movement of snow produces a **snow avalanche**. At the bottom of the slope, the avalanche overran a ski resort, crushing and carrying away buildings, cars, and trees and killing 31 people. It took searchers and their specially trained dogs many days to find buried survivors and victims under the 5–20 m (16–66 ft) thick pile of snow that the avalanche deposited (Fig. 11.18a).

What triggers snow avalanches? As we've seen, some occur when a broad slab of snow on a slope detaches from its substrate along an icy failure surface (Fig. 11.18b). Others take place when a *cornice*, a large drift of snow that builds up on the side of a windy mountain summit, suddenly gives way and falls onto the slopes below, where it knocks additional snow free (see Fig. 11.18b, inset). Avalanches behave differently depending on the temperature of the snow. Dry, cold snow mixes with air and develops into a turbulent cloud of powder called a *dry-snow avalanche*, which surges downslope at hurricane-like speeds of up to 290 km/h (180 mph) (Fig. 11.18c). Wet snow tends to stay closer to the ground, forming a slower-moving slush-like *wet-snow avalanche*, which, because of its viscosity, can knock down structures in its path.

Submarine Mass Wasting

So far, we've focused on mass wasting that occurs on the land surface—we refer to such events as *subaerial mass wasting*, to emphasize that they occur beneath the air. But

Did you ever wonder . . .

why highway engineers erect "Falling Rock" signs?

See for yourself

Rockfalls in Canyonlands, Utah

Latitude: 38°29′50.07″ N
Longitude: 110°0′15.63″ W

Zoom to an altitude of 5 km (3 mi) and look down.

The Green River has carved a deep valley through horizontal strata. Shale forms the slopes, and sandstone forms the top ledge. Erosion of the shale undercuts the sandstone, which breaks away at joints and tumbles downslope to build a talus slope.

FIGURE 11.17 The angle of repose is the steepest slope at which a pile of unconsolidated sediment can remain stable.

(a) This talus slope at the base of a cliff in the Uinta Mountains, Utah, has a relatively steep angle of repose.

30° — Well-rounded sand has a small angle.

45° — Irregularly shaped gravel has a large angle.

(b) The angle of repose depends on the shape and size of the grains in the pile.

FIGURE 11.18 Snow avalanches.

(a) The aftermath of a 1999 avalanche in the Austrian Alps.

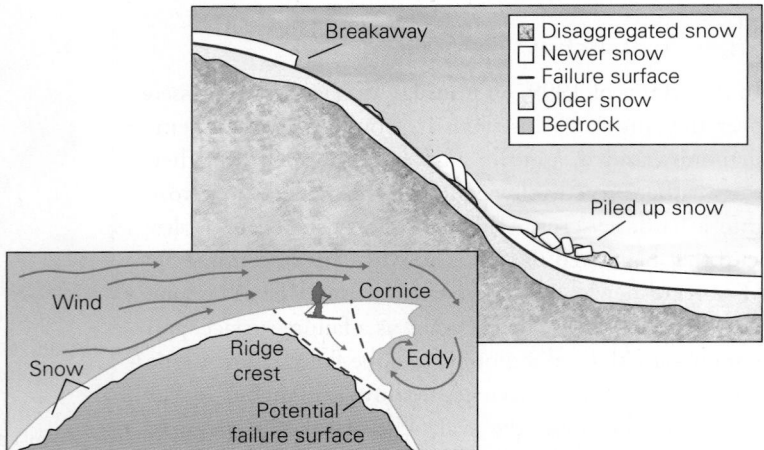

Breakaway

- ☒ Disaggregated snow
- ☐ Newer snow
- — Failure surface
- ☐ Older snow
- ☐ Bedrock

Piled up snow

Wind

Cornice

Snow

Ridge crest

Eddy

Potential failure surface

(b) Some avalanches involve the movement of a snow slab above a failure surface. Others involve the failure of a cornice that was built out by wind blowing over a ridge crest. The inset shows two possible failure surfaces along which a cornice might break away.

Origin of the avalanche

Toe of the avalanche

(c) This dry-snow avalanche in Alaska is a turbulent cloud. The arrow indicates the flow direction.

mass wasting also happens underwater. Geologists distinguish among three types of such *submarine mass wasting* based on whether the mass remains coherent or disintegrates as it moves (Fig. 11.19a). In a **submarine slump**, semicoherent blocks slip downslope along a failure surface. In a **submarine debris flow**, the moving mass breaks apart to form a slurry containing pebbles, cobbles, and blocks suspended in a mud matrix. And in a **turbidity current**, the moving sediment completely disperses in water to yield a turbulent cloud of suspended sand and mud that rushes downslope like an underwater avalanche (Fig. 11.19b). A turbidity current stays close to the seafloor as it moves because the mixture of water and sediment is denser than the surrounding clear water. Recall from Chapter 5 that when a turbidity current finally slows down, its sediment settles to form a graded bed (known as a *turbidite*), and that deposition of successive turbidites produces a *submarine fan*.

Submarine mass wasting can take place anywhere that a submarine slope exists. For example, digital elevation models (see Box 11.1) of the seafloor reveal that submarine slopes bordering active continental margins are scalloped by many immense slumps that have carried displaced material from a steep scarp to a lobe over the seafloor (Fig. 11.19c). The flanks of hotspot islands, such the Hawaiian Islands, have been substantially modified by submarine mass wasting. Large chunks of the islands have slumped and broken up, littering the abyssal plain with an apron of large blocks (Fig. 11.19d). For example, the island of Oʻahu represents a relatively small remnant of what was once a broad shield volcano. Debris from the collapsed volcanic edifice has spread northeastward as far as 180 km (110 mi) across the Pacific Ocean floor. Because a submarine slump can displace a large volume of water, it can trigger large tsunamis (Fig. 11.19e).

Huge slumps have also been mapped along the coasts bordering the Atlantic Ocean, indicating that passive continental margins are not immune to slumping, though it happens much less frequently than it does along active margins. One of the largest slumps known happened about 8,000 years ago, when the 100 km (60 mi) wide Storegga Slide slipped down the face of the passive margin west of Norway. Archaeologists speculate that tsunamis generated by the Storegga Slide washed away Stone Age villages all along the coast of the North Sea.

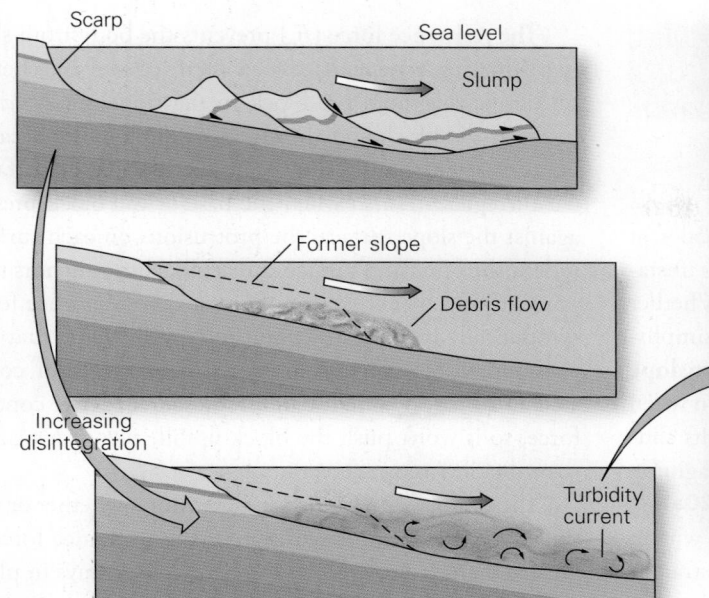

Scarp

Sea level

Slump

Former slope

Debris flow

Increasing
disintegration

Turbidity
current

(a) The name used for submarine mass wasting depends on the degree to which the moving material disintegrates.

FIGURE 11.19
Submarine
mass wasting.

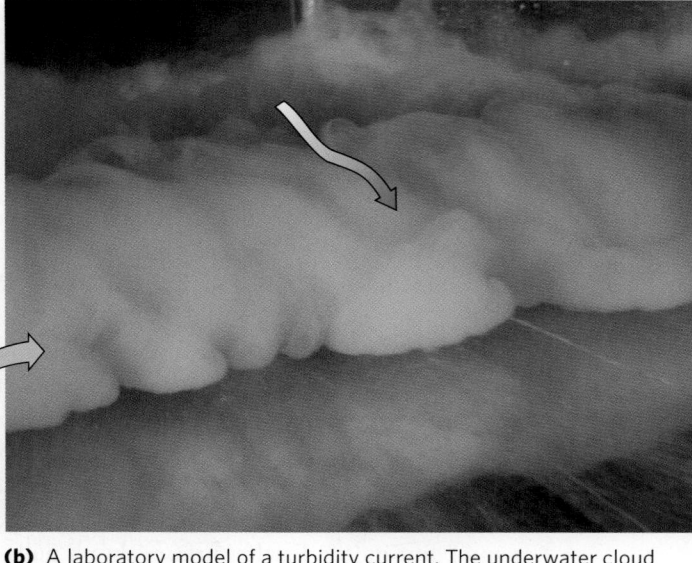

(b) A laboratory model of a turbidity current. The underwater cloud consists of clay.

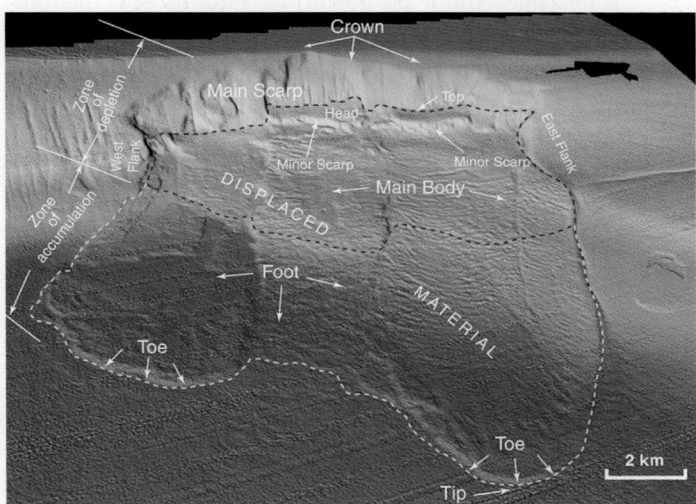

Crown

Main Scarp

Zone of depletion

Top

Head

West Flank

Minor Scarp

Minor Scarp

East Flank

DISPLACED

Main Body

Zone of accumulation

MATERIAL

Foot

Toe

Toe

Tip

2 km

(c) A digital bathymetric model of a submarine slump along the coast of California. The parts of the slump are labeled. Blue is deeper water, orange is shallower.

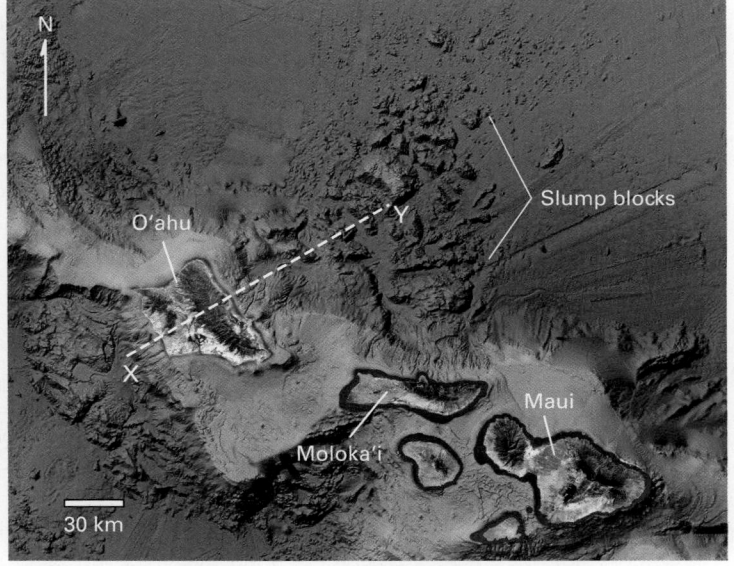

N

O'ahu

Y

Slump blocks

Maui

Moloka'i

X

30 km

(d) Giant underwater slumps border the island of O'ahu, Hawai'i.

Take-home message...

Mass-wasting events differ from one another based on the content, character, speed, and location of the moving material. Creep, solifluction, and slumping are slow. Mudflows and debris flows move faster, and rockslides, rockfalls, and snow avalanches move the fastest. Mass-wasting events, which can occur both on land and underwater, are natural hazards.

Quick Questions

- What is a failure surface?

- How does a slump differ from a debris flow?

- In what way does a turbidity current resemble a snow avalanche?

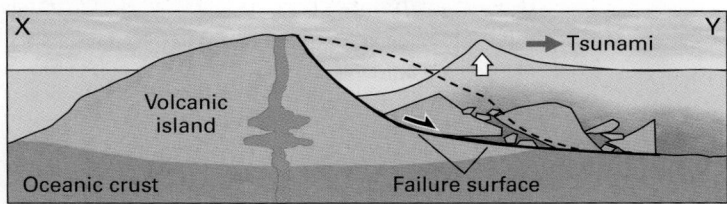

X

Tsunami

Y

Volcanic
island

Oceanic crust

Failure surface

(e) A hypothetical cross section along section line X–Y in part (d), showing a slump that removed what was once part of O'ahu. Movement of the slump probably generated a tsunami.

11.4 Why Does Mass Wasting Happen?

The Battle between Resistance Force and Downslope Force

We've seen that mass wasting can happen at many scales, at a range of velocities, and in many forms. Why does unstable ground undergo failure and start moving? Whether or not an area of unstable ground fails depends, simplistically, on the relative sizes of two forces: the **downslope force**, which will move material to a lower elevation when permitted, and the **resistance force**, which inhibits sliding. Let's look at these two forces in turn by imagining forces acting on a block sitting on a slope (Fig. 11.20a).

The downslope force exists because of gravity, which, at any point on the Earth's surface, always pulls straight down toward the Earth's center. We can represent gravity, therefore, by a vertical vector (F_G). (Recall that a *vector* is an arrow whose length indicates magnitude and whose orientation indicates direction.) The gravity vector, like any vector, can be divided into two mutually perpendicular vectors known as *components*. In this example, the component parallel to the slope represents the downslope force (F_D), meaning the portion of gravity oriented in the downslope direction. The component perpendicular to the slope (here called F_P) represents the push, due to the block's weight, that squeezes the block to the slope.

The resistance force (F_R) prevents the body from sliding. This force depends on *cohesion*, caused by chemical bonds that hold the block to the slope, and *friction* between the block and the slope. Friction exists because no real surface is perfectly smooth. Rather, all surfaces have irregularities, so when the base of the block presses against the slope surface, the protrusions on each surface indent the opposing surface and act like little anchors that prevent movement. We can represent the resistance force symbolically by an arrow pointing uphill. (Note that the resistance force merely represents the strength of cohesion and friction, and does not represent an active contact force, so it won't push the block uphill, even if it's larger than the downslope force.)

The size of F_D relative to F_R determines whether or not the block will move downslope. If the resistance force is greater than the downslope force, the block stays in place (see Fig. 11.20a). If the resistance force and the downslope force are equal, the block may start to move, so the slope is unstable. And if the downslope force exceeds the resistance force, failure takes place, and the block starts to slide downslope (Fig. 11.20b). Therefore, mass wasting may be triggered when the downslope force increases, when the resistance force decreases, or when both forces change by enough that F_D becomes greater than F_R. With these concepts in mind, we see that ground can be considered unstable if a relatively small change would make F_D greater than F_R. Conversely, ground can be considered

---

FIGURE 11.20
The balance between two forces determines whether downslope movement will take place.

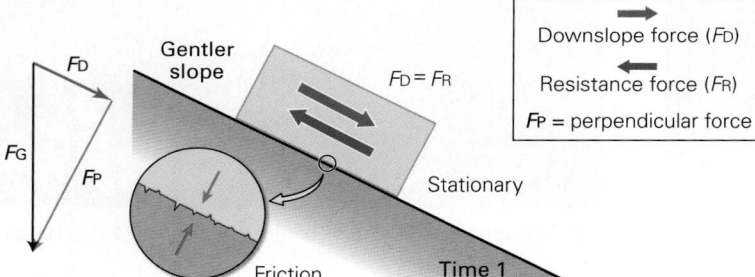

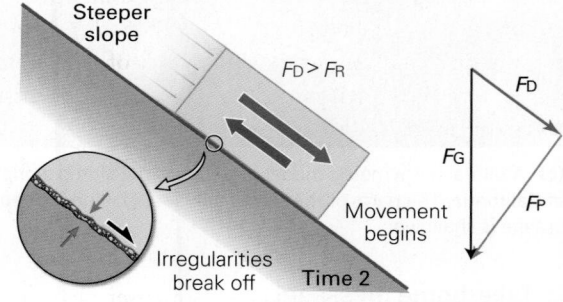

(a) A conceptual model showing the forces acting on a block in contact with a slope. Here, the resistance force, due to friction and cohesion, exceeds the downslope force, so the block remains stationary (Time 1). Friction is due to protrusions across the contact surface.

(b) If the downslope force increases and/or the resistance force decreases, so that the downslope force exceeds the resistance force, protrusions across the contact surface break off, and the block starts to move (Time 2). Here, steepening of the slope has increased downslope force and decreased resistance force.

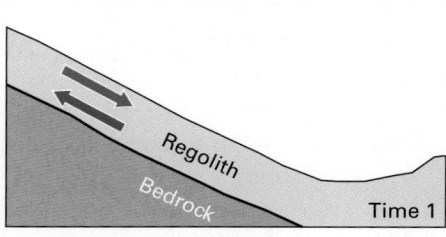
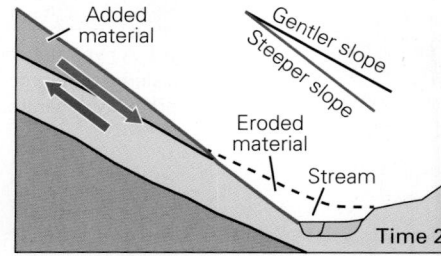
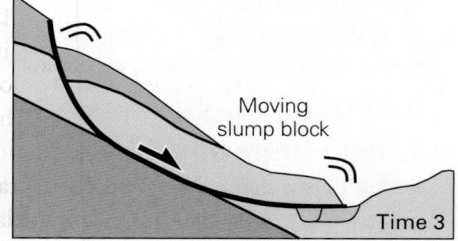

(c) In a more realistic (though still simplified) slope model, regolith rests on bedrock. If the slope is steepened by addition of regolith at the top and/or by erosion of regolith at the bottom, the downslope force increases, the slope becomes unstable (Time 2), and a slump eventually forms (Time 3).

stable if large changes in the relative magnitudes of the forces would be required for failure to occur, and if such changes are unlikely to happen in the near term. Let's examine several natural situations that can lead to failure of unstable ground.

Factors That Trigger Failure

SLOPES BECOME STEEPER OR WETTER. To visualize what happens when a slope steepens, look again at a block sitting on a slope (see Fig. 11.20a). As the slope angle increases, the magnitude of F_D increases. If the steepening continues until F_D exceeds the resistance force (F_R), slip begins (see Fig. 11.20b). Look at the vector diagrams on Figures 11.20a and b, and you'll note that steepening also decreases F_R, because F_p, the force that squeezes the surfaces together and causes irregularities to dig in, decreases, so friction decreases. When F_D becomes greater than F_R, the downslope force has become large enough to break off irregularities along the contact between the block and the slope. Once these tiny anchors have been broken, the block can move.

In the real world, slopes can steepen for several reasons. For example, addition of debris to the top of a slope, or erosion at the base of a slope, causes the slope overall to become steeper and may trigger failure (**Fig. 11.20c**). A slope may steepen for a while, then fail, then steepen again, and fail again. Each time, gravity succeeds in moving mass from a higher to a lower elevation. The model in Figures 11.20a and b, with a block resting on a slope, admittedly shows a somewhat unrealistic situation. More commonly, slope failure happens when the downslope force acting on a slab of rock or a layer of regolith above an underground failure surface exceeds the resistance force (see Fig. 11.20c).

Significantly, the downslope force can be increased, even if the slope angle stays the same, by the addition of water to regolith. If it rains, or if septic systems and pipes leak, water replaces air in the pores of the slope's substrate. Because water is over 800 times denser than air, this replacement causes the regolith to become heavier overall. As the weight of the regolith increases, the downslope force becomes larger and may eventually exceed the resistance force.

WEAKENING OF ROCK, REGOLITH, OR FAILURE SURFACES. If the Earth's surface exposed only fresh, intact, homogeneous bedrock, mass wasting would be of little concern to society, for such rock has great strength and could easily hold up stalwart mountain cliffs. In most places, however, weak regolith, or bedrock that contains weak layers (such as beds of weak rock or foliation planes containing weak minerals), lies beneath the ground surface. In addition, most bedrock contains joints, cracks that separate

rock into discrete pieces between which there is little or no cohesion. Regolith, weak layers in bedded or foliated rock, and joint contacts in fractured rock are all much weaker than intact, homogeneous rock. Therefore, several phenomena that make rock or regolith progressively weaker can lead to failure even if the slope angle does not change:

- *General weathering:* Over time, chemical weathering produces weaker minerals, and physical weathering breaks rocks apart. As weathering happens, a formerly strong rock transforms into regolith, a weaker material. Intact bedrock has strength because the chemical bonds within its interlocking grains, or within the cements between its grains, don't break easily. In regolith, grains are held together by friction, by weak bonds between grains, or by bonds between water molecules, and none of these attachments is as strong as the bonds holding atoms together in the minerals of solid rock.

- *Overall water content:* We've seen that saturating regolith with water makes it heavier. A surplus of water may also cause regolith to disintegrate and transform into a slurry capable of flowing.

- *Weakening of failure surfaces:* Several different kinds of potential failure surfaces, such as wet clay layers, unconsolidated sand layers, exfoliation joints, weak beds of shale, and metamorphic foliation planes, can exist within rock (**Fig. 11.21**). Weak surfaces that are parallel to the land surface of a slope are particularly likely to fail because the downslope force parallels the weak surface. Mass wasting can begin when failure surfaces become even weaker due to weathering. In the case of failure surfaces containing clay, addition of water makes the clay more slippery, or can break bonds holding clay flakes together so that solid clay becomes liquefied. Increasing groundwater pressure in a failure surface may have a similar effect, for such pressure pushes material apart along a failure surface. Such water-related weakening of a potential failure surface triggered the Gros Ventre Slide, the largest observed landslide in American history, in 1925. This debris slide took place on the flank of

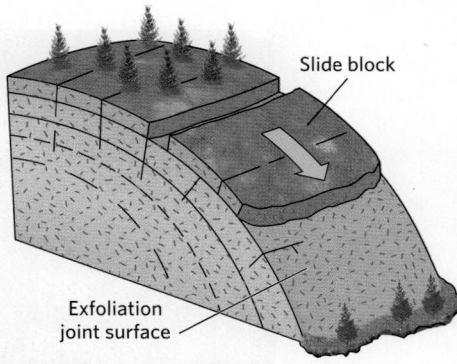

FIGURE 11.21 Several kinds of weak surfaces can become failure surfaces.

(a) Exfoliation joints, which form parallel to slope surfaces in granite, may become failure surfaces.

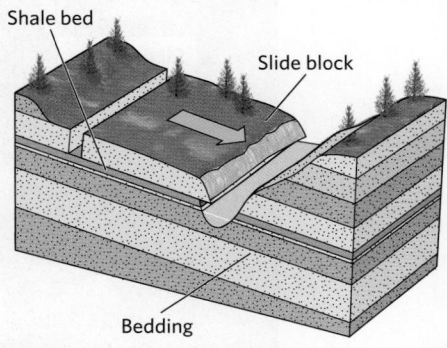

(b) In sedimentary rock, bedding planes (particularly in weak shale) may become failure surfaces.

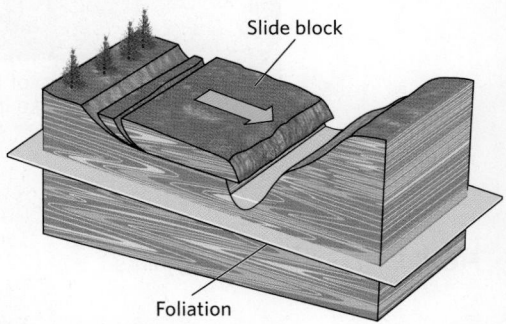

(c) In metamorphic rock, foliation planes (particularly in mica-rich schist) may become failure surfaces.

FIGURE 11.22 Events leading to the 1925 Gros Ventre Slide in Wyoming.

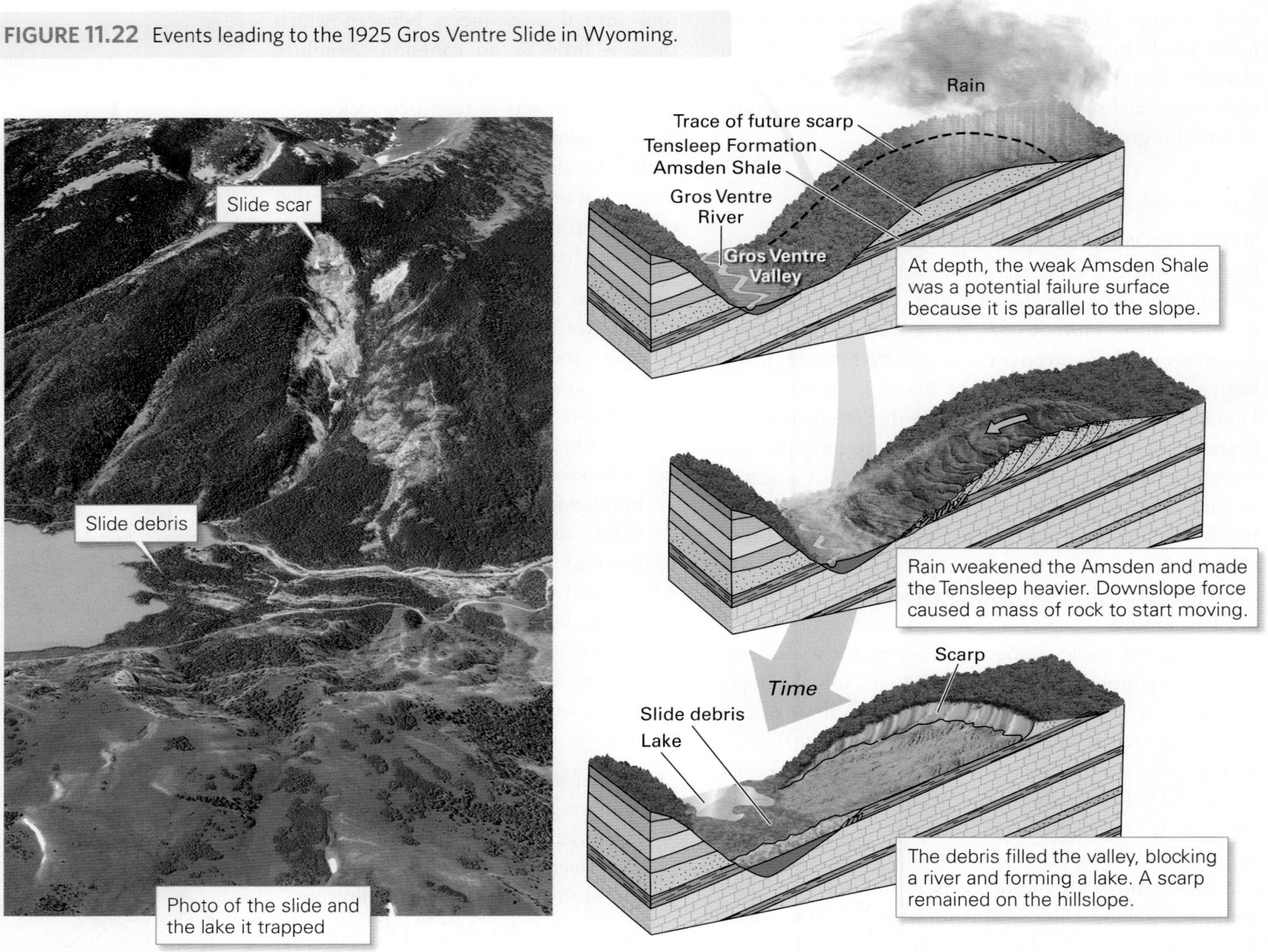

Rain

Trace of future scarp
Tensleep Formation
Amsden Shale
Gros Ventre
River

**Gros Ventre
Valley**

Slide scar

Slide debris

At depth, the weak Amsden Shale
was a potential failure surface
because it is parallel to the slope.

Rain weakened the Amsden and made
the Tensleep heavier. Downslope force
caused a mass of rock to start moving.

Time

Scarp

Slide debris
Lake

Photo of the slide and
the lake it trapped

The debris filled the valley, blocking
a river and forming a lake. A scarp
remained on the hillslope.

FIGURE 11.23 Failure along joints.

(a) Jointing has broken up this thick sandstone bed along a cliff in Utah. Blocks of sandstone break free along joints and tumble downslope.

Bed

Joint

(b) Failure occurred along both bedding and joints on this canyon wall in the Grand Canyon.

Sheep Mountain, in Wyoming, when water seeping into the ground from a heavy rain weakened a shale bed until it failed. The failure allowed almost 40 million m³ (1.4 billion ft³) of rock, as well as the overlying soil and forest, to slide 600 m (2,000 ft) downslope. The debris formed a natural dam across the Gros Ventre River valley, which caused a new lake to fill **(Fig. 11.22)**. The lake has existed ever since.

• *Wedging in joints:* Rockfalls and rockslides typically happen when failure occurs along joints **(Fig. 11.23)**. Frost wedging or root wedging (see Chapter 5) can push existing joints open and cause joints to propagate, allowing rock blocks to detach and topple downslope.

FIGURE 11.24 A slump formed along this deforested slope in Brazil.

Slump scar

• *Removal of vegetation cover:* In the case of slopes underlain by regolith, vegetation tends to strengthen the slope because the roots hold otherwise unconsolidated grains together, and because plants remove water from the ground. Therefore, removal of vegetation has the net result of making slopes more susceptible to downslope movement (Fig. 11.24). In fact, deforestation, due to either logging or fire, can lead to catastrophic mass wasting of the forest's substrate.

GROUND SHAKING. An earthquake, a storm, a large truck driving by, or a blast at a construction site may cause a mass on the verge of failure to begin moving (Box 11.2). Ground shaking can trigger sliding by breaking bonds that hold rock or regolith in place. It can also cause rock above a failure surface to move upward slightly, thereby decreasing friction across the failure surface. If such a decrease in cohesion or friction causes a sufficient decrease in the resistance force, then failure takes place. Unfortunately, after strong earthquakes in mountainous areas, hundreds of mass-wasting events take place, and these events can bury communities at the base of a hill or mountain (Fig. 11.25).

LIQUEFACTION OF SAND AND QUICK CLAY. Ground shaking and water influx can cause wet sand to transform into a slurry of sand suspended in water. This slurry has virtually no strength and flows easily. As a result of this phenomenon, known as **sediment liquefaction**, overlying land breaks up, and buildings constructed on that land crack, crumble, tilt, or tip over (Fig. 11.26a).

BOX 11.2 **How can I explain . . .**

The concept of slope stability

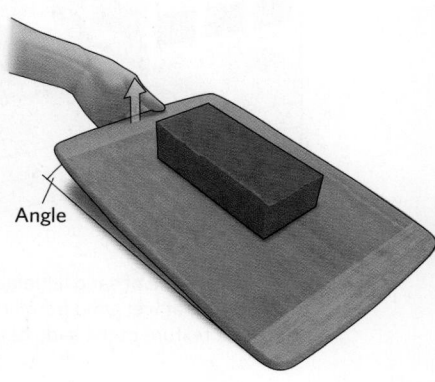

Angle

What are we learning?
How slope angle and ground shaking can affect slope stability.

What you need:
• A brick and a smooth cutting board
• A protractor

Instructions:
• Place the brick on the surface of the cutting board.
• Slowly lift one end of the board until the brick becomes unstable and slips. Measure the angle of the slope at which the brick slips.
• Repeat the tilting experiment, but this time, tilt the board to an angle that is less than the previous angle at which the brick slipped. At this angle, shake the board slightly. Can you get the brick to move?

What did we see?
The surface of the board represents a failure surface. Tilting the board represents steepening of the slope. The brick may be stable on gentle slopes, but it becomes unstable on steeper slopes. Shaking the board (which represents ground shaking) decreases the slope angle at which the brick can remain stable.

FIGURE 11.25 The town of Qushan, China, was destroyed by a combination of seismic shaking and landslides in 2008. Note the landslide scars on the hills and the debris flow that covers the ground in the town.

FIGURE 11.26 When sediment loses its strength.

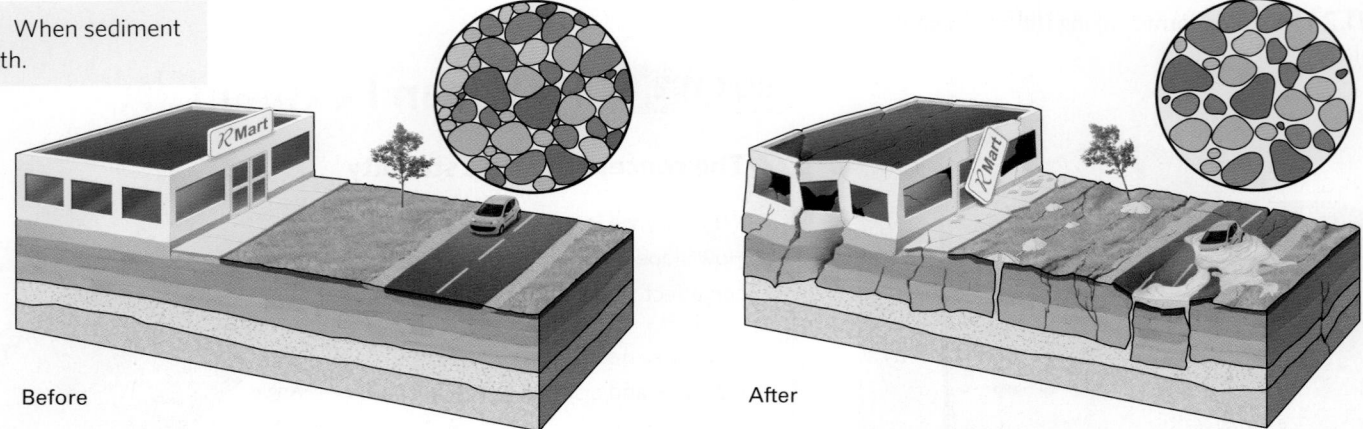

(a) An example of sand liquefaction. The overlying ground and a building standing on it break up, and pieces tilt. Cracks can develop in the ground so that wet sand from underground can squirt upward. In some cases, the sand builds a conical pile called a sand volcano. The insets show the texture of the sand before and after liquefaction.

(b) Before shaking or saturation, clay flakes in quick clay are arranged in a lattice. Shaking or saturation breaks this lattice apart, so the flakes become suspended in water.

(c) After heavy rains, liquefaction of quick clay caused these houses and trees to slide into the water.

Houses broke apart and sank into the mud, which flowed south-west.

The ground and overlying houses have broken up above the liquefied layer.

(d) Widespread sand liquefaction during a 2018 earthquake caused lateral spreading that led to the destruction of whole neighborhoods in Palu, Indonesia. Liquefaction destroyed all homes within the dashed yellow line, an area of about 1 km².

FIGURE 11.27 The Oso, Washington, slump of 2014.

(a) The aftermath of the slump. Note the head scarp.

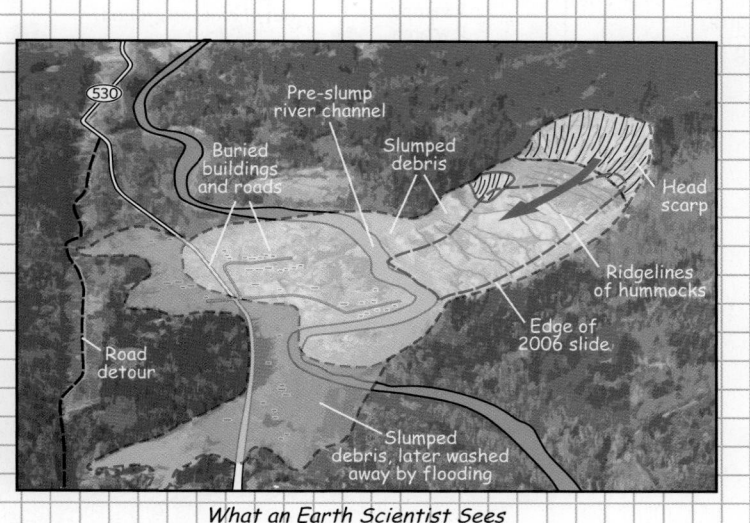

What an Earth Scientist Sees

(b) An Earth scientist's sketch indicates where buildings were overwhelmed.

A similar phenomenon happens in sediment layers formed of **quick clay**, a special type of clay that accumulates in saltwater. When quick clay dries out, it becomes a solid consisting of a lattice of clay flakes held apart by salt crystals. If fresh groundwater later saturates the clay, it dissolves the salt, so that the lattice of clay flakes no longer has internal support. A disturbance—such as a heavy rain or an earthquake—causes the clay flakes to separate and become suspended in water **(Fig. 11.26b)**. Then, the clay behaves like a fluid and moves as a mudflow. Land above the clay layer, consequently, undergoes lateral spreading or downslope movement, sometimes with catastrophic results **(Fig. 11.26c, d)**.

REMOVAL OF LATERAL SUPPORT. Removal of material from the base of a slope, due to erosion or by human excavation of rock or regolith, can play a role in triggering slope failure. The rock or regolith removed in these cases had provided *lateral support* (support from the side); in effect, it served as a natural retaining wall. Removing lateral support allows the existing weight of rock or regolith upslope to trigger failure.

Removal of lateral support played a role in setting the stage for a disastrous slump on a steep slope bordering the Stillaguamish River near Oso, Washington, where, over time, the river had eroded the base of the slope. In 2014, after a heavy rain saturated the regolith of the slope, failure took place, and a large slump slid down toward the river. As it moved, the slump turned into a debris flow that spread across the river and tragically buried a housing development and highway on the other side **(Fig. 11.27)**. In some cases, erosion by a river or by waves carves into the base of a cliff and produces an overhang. When such **undercutting** has occurred, the rock making up the overhang eventually breaks away from the slope and falls **(Fig. 11.28)**.

Take-home message…

🏠 Mass wasting occurs when the downslope force acting on unstable land exceeds the resistance force. Failure can be triggered by changing slope angles; changing water content; weakening of rock, regolith, or failure surfaces; ground shaking; or removal of lateral support.

Quick Questions ────────────

- Distinguish between resistance force and downslope force on an unstable slope.
- How can heavy rainfall cause unstable land to fail?
- Can a slope's strength change over time?

FIGURE 11.28 Undercutting and collapse of a sea cliff.

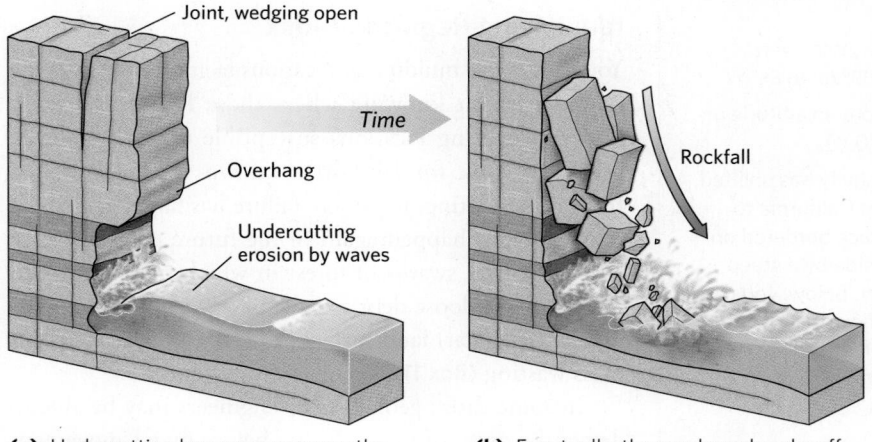

(a) Undercutting by waves removes the support beneath an overhang.

(b) Eventually, the overhang breaks off along joints, and a rockfall takes place.

FIGURE 11.29 Visible clues that a slump has started moving slowly.

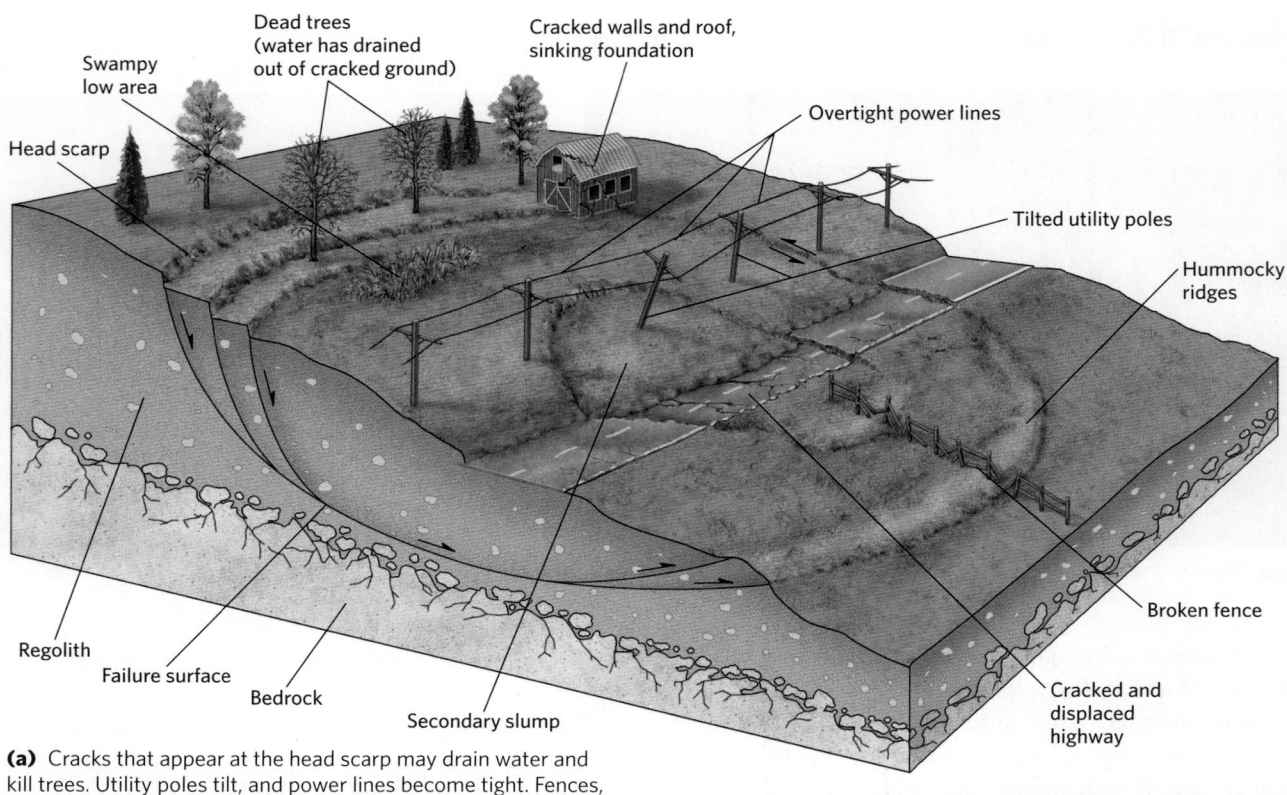

(a) Cracks that appear at the head scarp may drain water and kill trees. Utility poles tilt, and power lines become tight. Fences, roads, and houses on the slump begin to crack.

11.5 Can We Prevent Mass-Wasting Disasters?

Many people live in regions where various types of mass wasting have the potential to cause casualties and destroy property (**Earth Science at Glance**, pp. 418–419). Avoidance provides the surest way not to be a victim of mass wasting—don't build, live, or work on unstable ground. But if avoidance isn't possible, some mass-wasting hazards can be *mitigated* (made less severe) by taking steps, described below, to counter factors that lead to failure.

Identifying Regions at Risk

You can't avoid building in locations facing a mass-wasting hazard without knowing where those locations are. To start pinpointing locations susceptible to mass wasting, geologists look for landforms known to have resulted from mass wasting, for where failure has happened in the past, it might happen again in the future. Features such as head scarps, swaths of forest in which trees have been tilted, piles of loose debris at the bases of hills, and hummocky (irregular) land surfaces all serve as clues to recent mass wasting (**Box 11.3**).

In some cases, geologists or engineers may be able to detect the onset of mass wasting before major movement has taken place. For example, cracks in roads, buildings,

(b) A slump that has started to move on a slope next to a highway. The head scarp has appeared, and the land on the slump block has started to stretch, so more soil is visible between shrubs.

and pipes indicate that the substrate has stretched or shortened, even if movement has been too slight to cause visible *fissures* (open cracks) to develop. Initial mass movement may also cause utility poles to move relative to each other, so that power lines become too tight or too loose (**Fig. 11.29a**). As movement continues, visible fissures eventually develop near the head of a slump, and the

Putting Earth Science to Use

Thinking about slope stability at La Conchita, California

Many factors go into the decision of whether or not to buy or build a home: proximity to family, schools, shopping, jobs, and recreation, as well as cost, often govern such decisions. If the home lies on or near a slope, the stability of the slope should be taken into account as well. After reading this chapter, you might be able to recognize clues that could help you make the right choice about a purchase, but of course you should consult a qualified engineer to fully understand whether slope stability should be a concern.

Evidence of past mass-wasting events provides the most direct evidence that future movements might occur. The history of mass wasting at La Conchita, California, about 100 km (62 mi) northwest of Los Angeles, illustrates this point. La Conchita, a development of about 200 houses, was built on a bench of flat land 500 m (1,600 ft) wide between a beach on the southwest and a steep slope, 180 m (600 ft) high, on the northeast. Above the slope lies a broad terrace covered with irrigated orchards. The terrace was formed long ago, by wave erosion, when its land surface was at sea level. Compression caused by interaction between the Pacific Plate and the North American Plate subsequently uplifted the land to its present elevation. Due to intense weathering, the terrace and slope are underlain by weak, clay-rich regolith.

Look closely at the slope **(Fig. Bx11.3a)**. Notice that its face isn't a smooth, planar surface, but rather hosts several curving, unvegetated cliffs of various sizes. These are the head scarps of slumps that moved recently enough that the exposed failure surfaces have not had time to erode or be revegetated. The shape of the slope suggests that it has undergone mass wasting in the not-too-distant past, implying that it might still be unstable.

A statement that a slope is potentially unstable generally doesn't imply an exact time frame for the next failure. Perhaps the slope will fail within a year, or perhaps it will not fail again for decades. As with any other natural hazard (such as flooding, earthquakes, fire, and severe weather), a statement about risk represents a possibility, not a certainty. At La Conchita, unfortunately, concern about instability was justified. In 1995, following heavy rains that weakened the regolith and made it heavier, part of the slope failed, sending a wet slump block downslope. The slump transformed into a mudflow, which overwhelmed 9 houses at the foot of the slope. Tragically, an even more devastating slump and mudflow happened in 2005, causing 10 fatalities, burying 13 houses, and damaging 23 more **(Fig. Bx11.3b)**.

FIGURE Bx11.3 The 2005 La Conchita mudflow.

(a) An oblique satellite image from Google Earth shows the slope adjacent to La Conchita, and the terrace above. Note the highlighted head scarps.

(b) During heavy rains, the slope gave way, and heavy mud flowed downslope.

Mass Wasting

Avalanche

Slumps and mudflows

Rockslide

Avalanche chute

Rockfall

Solifluction

Tilted graves

Curving tree trunks

Debris flow

Broken power lines

Fissure

Cracked buildings

Communities affected by mass wasting

Nonrotational slump

Mudflow due to quick clay

Sea-cliff landslide

Mass wasting includes many distinct types of downslope movement, such as rotational and nonrotational slumps, soil creep, solifluction, turbidity currents, avalanches, debris slides and flows, lahars, and rockfalls. Many of these events, in everyday English, are known as landslides. Downslope movements can occur where the downslope force, due to gravity, exceeds the resistance forces that hold rock or sediment in place. The type of movement that takes place depends on the steepness of the slope and the character of the movement. Materials may slide down pre-existing weak failure surfaces, such as bedding or foliation planes, or may simply flow. Therefore, during some mass-wasting events, the moving block remains coherent, and during others it disintegrates. Water content plays a key role in mass wasting, as its presence contributes to destabilizing material and to determining if the material flows or stays coherent. All types of mass-wasting events are natural hazards, both for people and for structures on the slope and at the base of the slope.

Submarine slump

Turbidity current

Heavy rain

Source of
the lahar

Head scarp

Talus slope

Failure due
to jointing

Landslide lake

Large debris slide

Toppling
of blocks

Lahar

Head
scarp

Rescue operations

Rotational
slump

Toe of
the slump

Failure surface

419

See for yourself

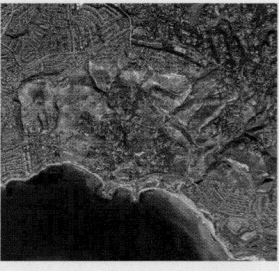

Portuguese Bend, California

Latitude: 33°44′46.94″ N
Longitude: 118°22′7.83″ W

Look down from an altitude of 7 km (4.3 mi).

Housing covers much of the land in southern California, but not on a 3.5 km (2 mi) wide, spoon-shaped depression at Portuguese Bend. In the 1950s, developers built on the hummocky land, but water infiltration weakened a failure surface and reactivated the slump, destroying 150 houses.

ground bulges upward at its toe. Vegetation growing on the incipient slump block may die where subsurface cracks drain water. Eventually, a head scarp becomes clearly defined, and the texture of the land within the slump becomes visibly hummocky **(Fig. 11.29b)**. In recent years, new, extremely precise laser-based surveying methods allow geologists and engineers to detect the beginnings of ground movements that have not yet caused any visible evidence.

Hints to the location of unstable ground on a regional basis come from computer-based mapping programs that plot data concerning factors that influence mass wasting. These factors include slope steepness, substrate strength, degree of water saturation, orientation of potential failure surfaces relative to the slope, nature of vegetation cover, potential for heavy rains, potential for undercutting, and likelihood of earthquakes. The results of such *hazard-assessment studies* permit compilation of **landslide-potential maps**, which rank regions according to the likelihood that mass wasting will occur there **(Fig. 11.30a)**. Detailed mapping and dating of landslides on digital elevation models can also provide insights into mass-wasting hazards by characterizing the dimensions and recurrence interval of landslides in a region **(Fig. 11.30b, c)**.

Mitigating Mass Wasting

In areas where landslide potential exists, individuals or communities can take a number of steps to stabilize unstable land:

- *Revegetation:* Slope stability in deforested areas can be enhanced if landowners replant the region with trees or grasses that send down deep roots to bind regolith together **(Fig. 11.31a)**.

- *Regrading:* A dangerously steep slope can be regraded so that it does not exceed the angle of repose, or it can be terraced to reduce the weight of material acting on failure surfaces at depth. The flat part of a terrace can also serve to catch debris coming down from above **(Fig. 11.31b)**.

FIGURE 11.30 Mapping helps to characterize landslide potential.

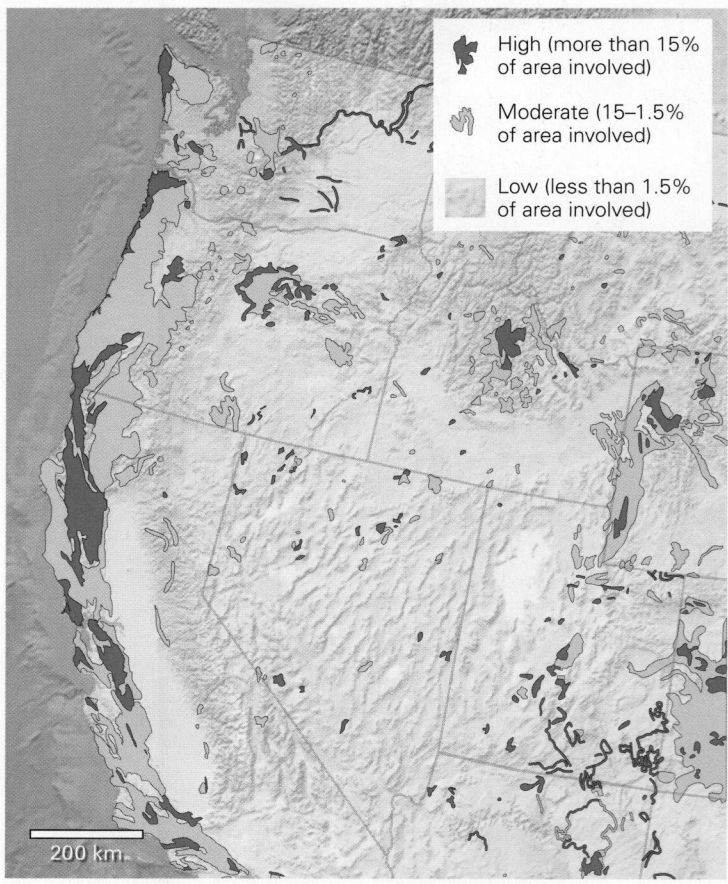

High (more than 15% of area involved)

Moderate (15–1.5% of area involved)

Low (less than 1.5% of area involved)

200 km.

(a) A landslide-potential map of the western United States.

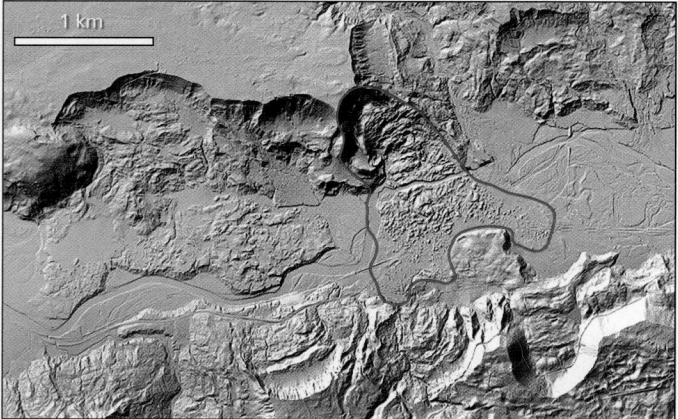

(b) A LIDAR (for Light Detection and Ranging) image, which shows the land surface without vegetation, provides data from which a high-resolution DEM can be made. Landforms in the Oso, Washington, region reveal many slumps along the sides of the Stillaguamish River valley. The 2014 Oso slump is outlined in red.

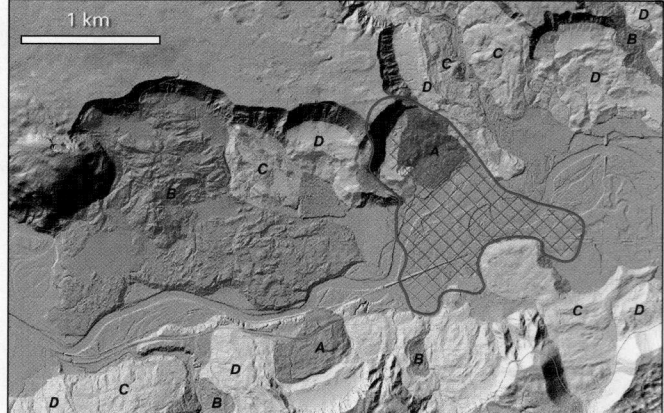

(c) Geologists have determined the ages of slumps that have happened during the past several thousand years along the Stillaguamish River valley. The slumps labeled D are the oldest, and those labeled A are the youngest. The 2014 Oso slump is outlined in red.

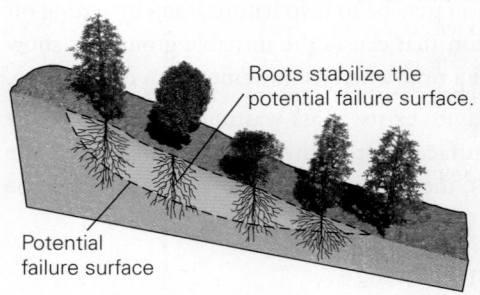

(a) Revegetating a slope results in the growth of roots that can hold the slope together.

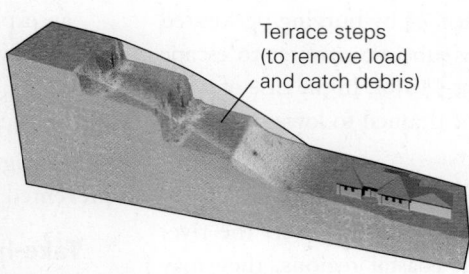

(b) Redistributing the mass on a slope can stabilize it. Terracing can help catch debris.

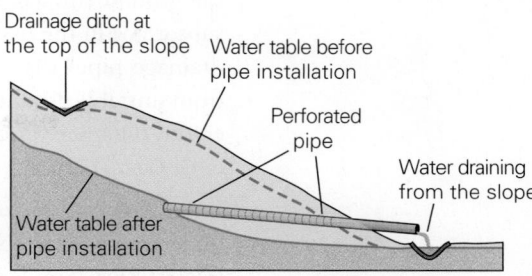

(c) A slope can be drained by digging culverts and installing perforated pipes.

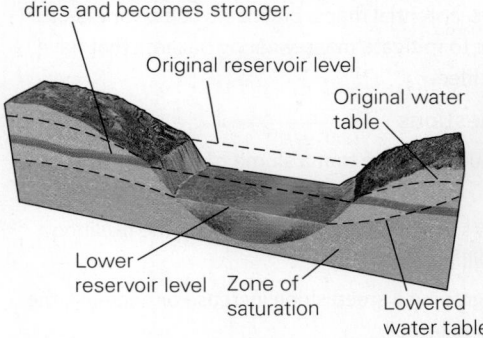

(d) Lowering the water table by draining a reservoir can strengthen a potential failure surface.

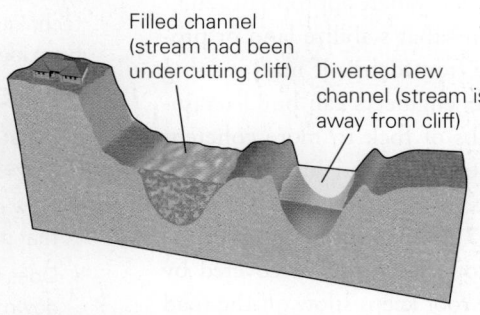

(e) Relocating a river channel can prevent undercutting.

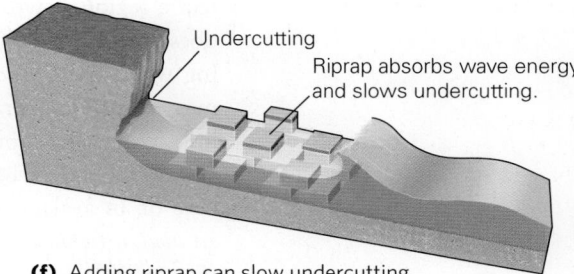

(f) Adding riprap can slow undercutting of coastal cliffs.

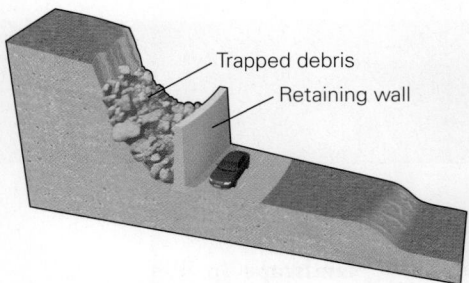

(g) A retaining wall can trap falling rock.

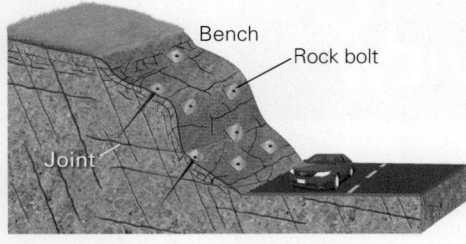

(h) Rock bolts attach fractured material at the surface of a rock face to solid bedrock behind.

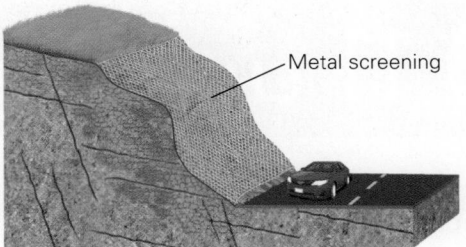

(i) Chain link fencing material can hold loose debris in place.

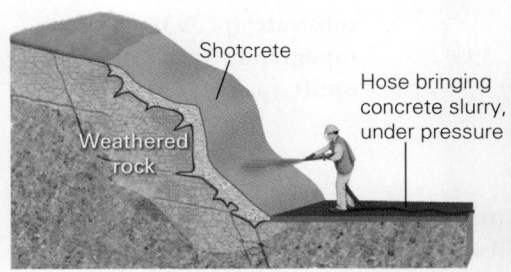

(j) Applying shotcrete (concrete sprayed at high pressure out of a hose) to weathered rock can keep the rock in place.

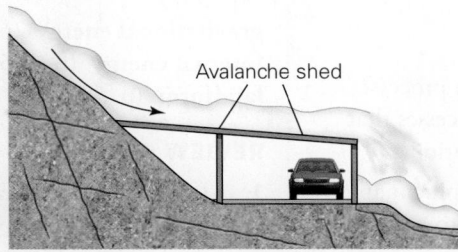

(k) An avalanche shed over a roadway diverts debris or snow.

- *Reducing subsurface water:* Land stability may be improved either by adding a drainage network on the ground surface, so that water does not enter the subsurface in the first place, or by burying perforated drainage pipes that allow subsurface water to escape from unstable ground (Fig. 11.31c). In the case of a reservoir, the reservoir can be drained to lower the water table (Fig. 11.31d).

- *Preventing undercutting:* In places where a river undercuts a cliff face, engineers can divert the river (Fig. 11.31e). Similarly, in coastal regions, they may build an offshore breakwater or pile riprap (loose boulders or concrete) along a beach to absorb wave energy before it can carve into a cliff face (Fig. 11.31f).

- *Constructing safety structures:* Where appropriate, engineers can install structures that stabilize land or protect a region downslope from debris if mass wasting does occur. For example, engineers can build retaining walls, bolt loose slabs of rock to more coherent masses in the substrate, or cover road cuts with chain link fencing or shotcrete in order to prevent failure along highways (Fig. 11.31g–j), and a highway at the base of an avalanche-prone slope can be covered by an *avalanche shed* whose roof keeps snow off the road (Fig. 11.31k).

- *Controlled blasting of unstable slopes:* When unstable rock or snow clearly threatens a particular region, the best solution may be to help nature along by setting off an explosion that causes the unstable ground or snow to move at a time when its movement can do no harm.

As long as gravity exists, mass wasting will take place at the Earth's surface. But with an understanding of the danger signals, the destruction that it causes can often be prevented.

Take-home message...

Various features of the landscape may help geologists to identify unstable ground and estimate the hazard it poses. Systematic study allows production of landslide-potential maps. Engineers use a variety of techniques to mitigate mass-wasting hazards that have been identified.

Quick Questions

- What clues indicate that a slump may be starting to move?
- How does draining a slope help reduce the likelihood that it will fail?
- Does regrading a steep slope increase or decrease the downslope force?

11 CHAPTER REVIEW

Objective 11.1

Explain why the Earth's topography continues to be modified over time, and describe various factors that influence the landforms that develop in a region.

KEY CONCEPTS

- Landscapes represent the outcome of a "battle" between processes that build relief by causing uplift or subsidence and processes that destroy relief through mass wasting, erosion, and deposition.

- The shapes of landscapes can be represented by topographic maps or digital elevation models; such maps can also be used to measure relief and the angles of slopes.

- The types of landforms that develop and evolve at a location depend on many factors, including agents of erosion and deposition, relief, climate, substrate composition, and the nature of living organisms.

EARTH-SCIENCE VOCABULARY

deposition (p. 394)
elevation (p. 393)
erosion (p. 394)
external energy (p. 395)
geologic cross section (p. 397)
gravitational energy (p. 395)
internal energy (p. 395)
landform (p. 393)

landscape (p. 393)
regolith (p. 398)
relief (p. 394)
slope (p. 394)
subsidence (p. 394)
substrate (p. 393)
topography (p. 394)
uplift (p. 394)

REVIEW QUESTIONS

1. **(a)** Why does the Earth's surface constantly undergo change? **(b)** Distinguish between internal and external sources of energy in the Earth System, and characterize the roles that each plays in the development of landscapes.

2. **(a)** On the topographic profile in **Figure A**, determine the slope angle and the relief between Point X and Point Y, and estimate the

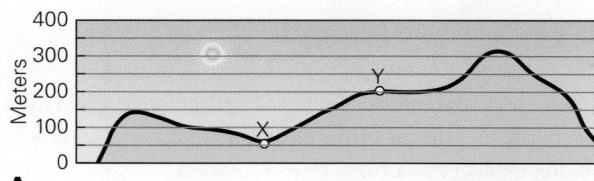

A

highest elevation shown.
(b) As the slope angle increases, do contour lines on a topographic map get closer together or farther apart? **(c)** Explain the role that bedrock resistance to erosion can play in controlling topography. **(d)** **Figure B** shows part of a topographic map (numbers are in meters). What is the contour interval on the map? What is the highest elevation on

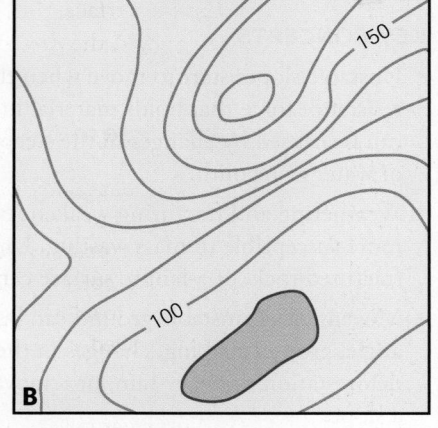

B

the map? Approximately what is the elevation of the lake? Mark the location of the steepest slope with an X.

3. **(a)** Distinguish between a depositional landform and an erosional landform. **(b)** In what way does climate affect landscape development? **(c)** What is an anthropogenic landscape? Provide some examples.

Objective 11.2

Sketch a diagram showing processes involved in the hydrologic cycle of the Earth System.

KEY CONCEPTS

• Water moves among various reservoirs within the Earth System during the hydrologic cycle.

EARTH-SCIENCE VOCABULARY

groundwater (p. 399)   **residence time** (p. 400)
hydrologic cycle (p. 399)   **surface water** (p. 399)

REVIEW QUESTIONS

4. **(a)** What are the various reservoirs for water in the Earth System, as defined in the context of the hydrologic cycle? **(b)** Imagine that it rains at the crest of a hill lying at a distance from the coast. Make a simple cross-sectional sketch showing various pathways that the rain that falls on the ground could take through the hydrologic cycle so as to eventually return to the ocean. **(c)** On your sketch, draw the path that water takes from the ocean to the locality where the rain fell.

5. Is the average residence time of water in a river longer or shorter than that of water in the ocean? Explain your answer.

Objective 11.3

Describe the characteristics and consequences of different types of mass wasting.

KEY CONCEPTS

• Mass wasting refers to the movement of the land surface and its substrate in response to gravity. It most commonly happens on slopes, but it can affect horizontal ground as well.

• A failure surface underground can define the base of the material that moves during mass wasting.

• Failure of an unstable slope causes rock or regolith to move downslope under the influence of gravity. Flat land can also fail due to lateral spreading.

• Mass wasting plays an important role in the evolution of landscapes. It is also a dangerous natural hazard.

• Geologists distinguish among several types of mass wasting based on the type of material involved, the velocity of the movement, and the degree to which the moving material holds together or disintegrates, as well as on the amount of water mixed in with the moving material.

• Mass wasting can take place either on land or underwater. Sudden submarine slumps can generate large tsunamis.

EARTH-SCIENCE VOCABULARY

angle of repose (p. 407)   **slump** (p. 402)
creep (p. 401)   **snow avalanche** (p. 407)
debris flow (p. 403)   **solifluction** (p. 402)
failure (p. 401)   **stable ground** (p. 401)
failure surface (p. 402)   **submarine debris flow**
landslide (p. 401)   (p. 408)
lateral spreading (p. 403)   **submarine slump** (p. 408)
mass wasting (p. 401)   **talus** (p. 407)
mudflow (p. 403)   **turbidity current** (p. 408)
rockfall (p. 406)   **unstable ground** (p. 401)
rockslide (p. 406)

REVIEW QUESTIONS

6. **(a)** Explain why soil creep takes place and indicate what the arrows in **Figure C** represent. Label the winter slope, the summer slope, and the regolith layer. **(b)** What are the manifestations of soil creep? **(c)** How does solifluction differ from soil creep?

7. **(a)** Label the different components of the slump shown in

C

Figure D. (b) What role does a failure surface play during slumping? **(c)** How does lateral spreading differ from a slump? **(d)** What can happen to the debris of a slump when it drops into a flooding river?

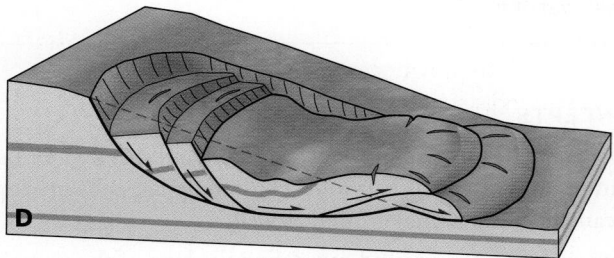

8. **(a)** Distinguish between a rockslide and a rockfall. **(b)** What type of mass wasting does **Figure E** display? What evidence led to your conclusion? **(c)** Which type of mass wasting does **Figure F** display? What type of failure surface was involved?

9. **(a)** What mass-wasting feature does **Figure G** display? **(b)** Where did the material in this feature come from? **(c)** What does the slope angle of the mass-wasting feature represent?

10. **(a)** **Figure H** shows two large slumps: one originated from Oʻahu, and one from Molokaʻi. Draw the outline of each of these slumps, and its head scarp, on the figure, and indicate its direction of movement with arrows. What visual clues led to this interpretation? **(b)** What submarine landforms do turbidity currents contribute to forming?

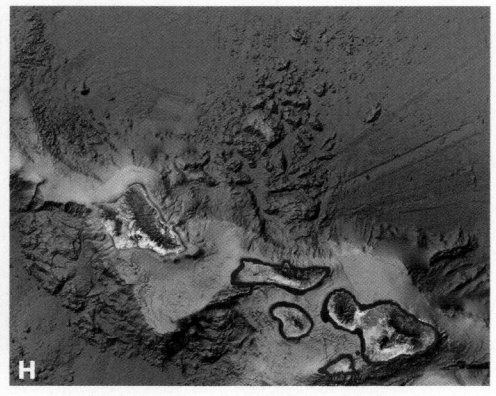

Objective 11.4

Explain the factors that cause some regions to be underlain by unstable ground, and discuss the conditions that can trigger a mass-wasting event.

KEY CONCEPTS

- Unstable slopes start to move when the downslope force exceeds the resistance force that holds material in place. Changes in these forces can be caused by changes in the steepness of a slope or by addition of water to regolith.

- Weathering and fracturing weaken rock and regolith and make it more susceptible to mass wasting. Land underlain by weak regolith, fractured rock, or a failure surface can become unstable ground.

- Movement of unstable ground can be triggered by changes in slope angle, ground shaking, changes in the strength of a slope due to deforestation or heavy rain, or removal of support from the base of a slope.

EARTH-SCIENCE VOCABULARY

downslope force (p. 410)
quick clay (p. 415)
resistance force (p. 410)

sediment liquefaction (p. 413)
undercutting (p. 415)

REVIEW QUESTIONS

11. **(a)** Why is intact bedrock stronger than fractured bedrock? **(b)** Why is intact rock stronger than regolith?

12. **(a)** Explain the difference between a stable and an unstable slope. **(b)** On **Figure I**, what do the two arrows represent? Why has the block shown started to move?

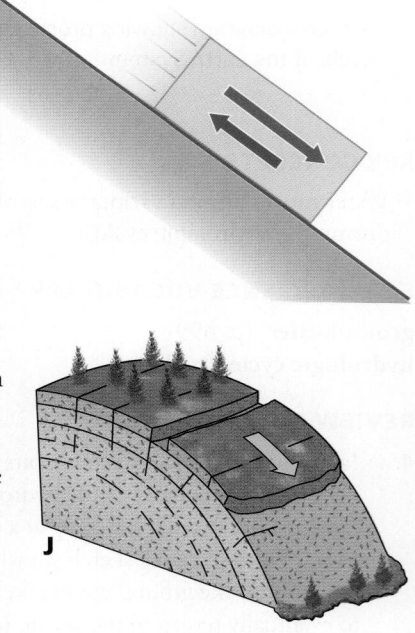

13. **(a)** What features are likely to serve as failure surfaces? **(b)** Which type of failure surface does **Figure J** show? **(c)** How can the slope angle near the top of the photo in **Figure G** be so much steeper than the slope angle near the middle of the photo?

14. **(a)** Discuss the variety of phenomena that can cause a stable slope to become unstable and eventually fail. **(b)** How can ground shaking cause seemingly solid layers of sand or mud to become weak slurries capable of flowing? **(c)** What type of clay can liquefy fairly easily?

15. (a) Discuss the role of vegetation and water in slope stability.
(b) Why can fires and deforestation lead to slope failure?

Objective 11.5

Use your knowledge of conditions that lead to mass wasting to assess an area's susceptibility to mass-wasting hazards and to evaluate preventive measures.

KEY CONCEPTS

- Geologists can sometimes detect unstable ground before it begins to move, and they can produce landslide-potential maps to identify areas susceptible to mass wasting.

- Engineers can help prevent dangerous mass wasting by using a variety of techniques to stabilize slopes.

EARTH-SCIENCE VOCABULARY
landslide-potential map (p. 420)

REVIEW QUESTIONS

16. (a) What clues can indicate that an area of a slope is starting to undergo slumping? **(b)** What factors do geologists consider when producing a landslide-potential map, and how can they detect the beginning of mass wasting in an area?

17. (a) What steps can people take to avoid landslide disasters?
(b) What purpose do the rock bolts shown in **Figure K** serve?
(c) What other protective measure might diminish the hazard of rockfalls at this location?

18. Imagine that you have been hired by a bank to determine whether it makes sense to provide loans to build a dam in the steep-sided, east-west-trending valley shown in the cross section (drawn perpendicular to the valley) in **Figure L**. Initial investigation shows that the rock of the valley wall consists of sedimentary strata in which bedding is roughly parallel to the south wall's slope. A distinctive shale layer crops out on both sides of the valley. Abundant fractures with slip lineations on their surfaces and breccia between surfaces cut across the strata in the floor of the valley. Moderate earthquakes occasionally rattle the region. The contractor proposes to excavate the bases of the hills on both sides of the valley to get rock to make the dam. Explain the potential hazards, what might happen if the reservoir were constructed and filled, and how you would advise the bank.

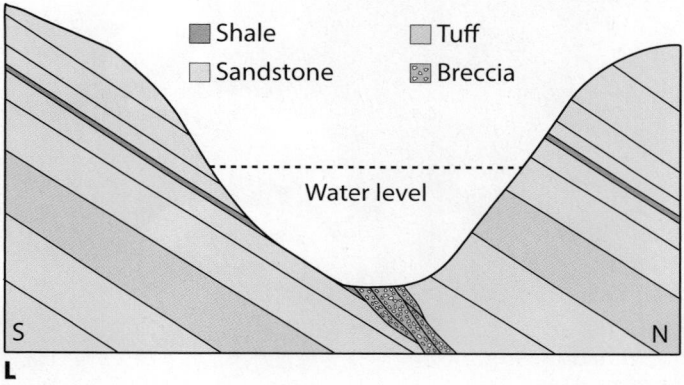

Shale   Tuff
Sandstone   Breccia

Water level

S   N

L

ANOTHER VIEW The Homestake Mine in South Dakota began as an open-pit operation in 1876. Surface excavations mostly ceased in 1945, and mining became almost entirely an underground operation. The mine produced gold out of Precambrian rock. To avoid destructive mass wasting of the mine walls, they are terraced. Terracing makes the walls more stable by decreasing their overall slope and weight. Also, the benches (horizontal ledges) serve as platforms for miners and equipment, and catch debris from small rock falls before it can make it to the mine floor.

12 FRESHWATER
Streams, Lakes, and Groundwater

After studying this chapter, you should be able to...

1. characterize freshwater and its occurrences on land, distinguish between lakes and wetlands, explain how streams and drainage networks develop, and define factors that control stream discharge.

2. explain how streams modify the land surface by erosion and deposition, and describe various landforms that develop in fluvial landscapes.

3. discuss the nature and causes of flooding, and how society can protect against flood damage.

4. characterize groundwater and the water table, describe factors that control groundwater flow, and describe various types of wells and springs at which groundwater can be obtained.

5. identify sustainability and environmental issues that pertain to freshwater resources.

6. discuss the origin of cave networks, and of karst landscapes and the landforms within them.

In the 1880s, developers built a clay-and-gravel dam across Pennsylvania's Conemaugh River, trapping a reservoir of cool water to provide a pleasant setting where wealthy residents of Pittsburgh could spend their summers. Unfortunately, when torrential rains drenched the state on May 31, 1889, the water level in the reservoir rose so much that water flowed over the dam. Eventually, the soggy structure collapsed, and a wall of water roared downstream and slammed into the city of Johnstown. The infamous Johnstown flood killed 2,300 residents of the city and turned its bridges and buildings into chaotic piles of unrecognizable wreckage **(Fig. 12.1)**.

Sadly, as residents of Johnstown learned, the Earth's freshwater occasionally causes catastrophes. But freshwater also plays key roles in the Earth System, both as an essential component of life on land and as a sculptor and transporter of Earth materials. Furthermore, freshwater provides food, transportation, and power to human society. Freshwater exists in several reservoirs in the Earth System: on the land surface as liquid or ice, underground, in living organisms, and in the air. In this chapter, we focus on key occurrences of liquid freshwater: lakes, wetlands, streams, rivers, and groundwater. We consider the roles that these occurrences play in the hydrologic cycle and the ways in which they modify the landscape. We also address the causes and consequences of freshwater floods, when water spreads out over the landscape to submerge areas that are normally dry. The chapter concludes with a look at the sustainability of freshwater resources, and at the hidden wonderland of cave networks.

FIGURE 12.1 A painting depicting the catastrophic flood in Johnstown, Pennsylvania, in 1889.

route died of cold, and some of those taking the desert route died for lack of drinkable water. Humans must have access to **freshwater**, meaning water that contains less than 0.05% dissolved salt, to survive. Desert lakes typically contain only saltwater, which humans cannot drink, because to get rid of so much salt (which our cells can't handle) we would have to expel more water than we drank. Indeed, drinking saltwater will cause a person to die of dehydration! Therefore, seawater can't be used directly for drinking water because it contains 3.5% dissolved salt. Bodies of standing water on the floor of Death Valley are much saltier than that.

Freshwater accounts for just 2.5% of the total volume of water in the Earth's surface and near-surface realms **(Fig. 12.2)**. About 69% of freshwater resides as ice in glaciers and permafrost, and about 30% lies hidden beneath the surface as groundwater. Consequently, fresh liquid water—in lakes, wetlands, streams, soil, air, melted snowfields, and organisms—accounts for only about 1.2% of the total! Geologists and *hydrologists* (scientists who study freshwater in the Earth System) informally refer to fresh liquid water in lakes, wetlands, and streams collectively as terrestrial *surface water*.

12.1 Surface Water: Lakes, Wetlands, and Streams

What Is Freshwater?

In the 19th century, before the completion of the first transcontinental railroad, travelers heading west to California could trek over the high, snowy passes of the Sierra Nevada, or attempt a shortcut and brave the scalding heat of Death Valley. Not all of these travelers, or their pack animals, made it alive to the end of their journeys. Some of those taking the mountain

Lakes and Wetlands

By some estimates, the Earth hosts about 307 million **lakes**, meaning standing bodies of water that occur on land. We add the modifier "on land" to emphasize that lakes do not overlie oceanic crust, and add the modifier "standing" because, to an observer on the shore, water in a lake does not visibly flow over the landscape. Most lakes contain freshwater, because as new water enters a lake via rain, inflowing streams, or *springs* (outlets through which groundwater returns to the surface), some of the

<< A small creek in northern Arizona has carved a canyon in red sandstone layers. When the stream floods, the volume, speed, and turbulence of the water increase, so the stream can carve into bedrock and transport sediment. Streams play a major role in the hydrologic cycle, and in sculpting the Earth's surface.

FIGURE 12.2 The amount and distribution of the Earth's freshwater.

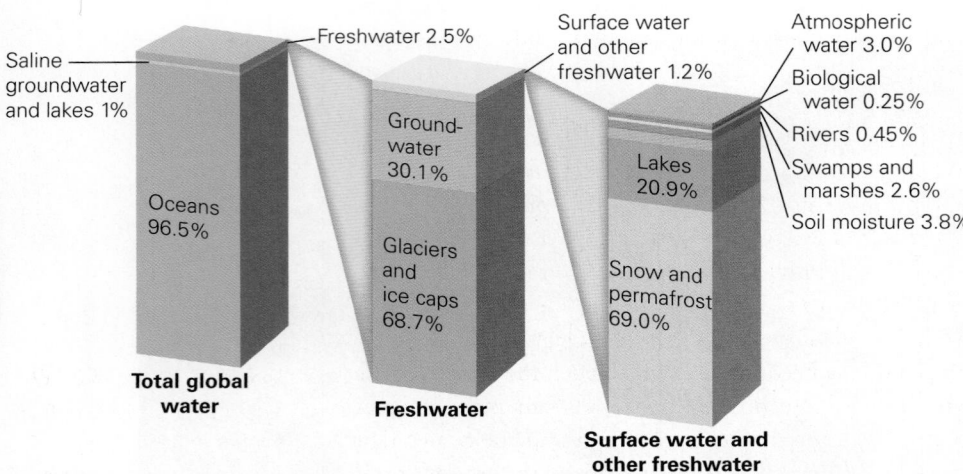

(a) Most water on the Earth is saltwater. Of the freshwater on the Earth, most is frozen, or lies underground.

(b) All the water in the world could be collected in a sphere with a diameter of 1,300 km (800 mi). The freshwater portion could be collected in a sphere with a diameter of only 275 km (170 mi). Surface freshwater would form a sphere with a diameter of less than 70 km (44 mi).

existing water in the lake leaves it via an outlet stream. This exchange of water prevents salt from accumulating. The *residence time* of water in a lake generally ranges from days (for smaller lakes) to years (for larger lakes). A lake's water level can vary over time depending on the balance between inflow and outflow. *Permanent lakes* have enough inflow to keep the bed of the lake filled all year, whereas in *ephemeral lakes*, water dries up during dry seasons.

Some lakes, such as the Great Salt Lake of Utah, contain saltwater. A **salt lake** becomes salty because it has no outlet. Therefore, water leaves the lake only by soaking into the lake bed or by evaporating into the air. Evaporation removes water molecules but leaves behind dissolved ions, so over time, salts become progressively concentrated in the lake. In fact, salt lakes can become saltier than the ocean. Water in portions of the Great Salt Lake, for example, contains 27% salt.

Many lakes exist simply because they are low areas that receive inputs from streams, rain, meltwater, or spring water (**Fig. 12.3a**). Some lakes are a consequence of specific geologic phenomena, such as the collapse of ground over a cave, the isolation of a segment of a river, glacial erosion or deposition, collapse of a volcanic caldera, blockage of a stream by a landslide, or subsidence due to rifting. Small lakes (*ponds*) may be only tens to hundreds of meters across. The largest lakes are hundreds of kilometers across, so wide that you can't see the other side. The Earth's deepest lake, Lake Baikal in Siberia, reaches a depth of 1.6 km (1 mi). And its longest, Lake Tanganyika in Africa, has a length of 660 km (410 mi). Lake Superior, the largest of the Great Lakes along the border between Canada and the United States, has a surface area of 84,000 km² (32,000 mi²), but a maximum depth of only 406 m (1,332 ft).

A **wetland** differs from a lake in that its water is so shallow that vegetation growing in it rises above the

FIGURE 12.3 Surface water in lakes and wetlands.

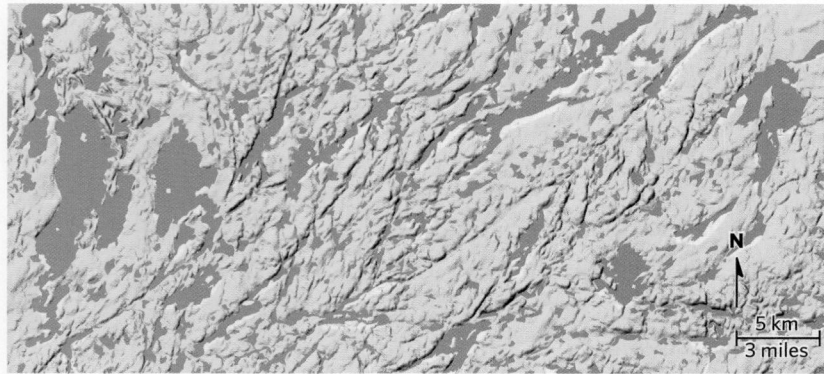

(a) Countless lakes cover the land about 80 km (50 mi) north of the boundary between western Ontario (Canada) and Minnesota.

(b) A swamp in Illinois. The stagnant water is covered with algae.

surface (Fig. 12.3b). Earth scientists distinguish among different types of wetlands by the types of vegetation they host. Specifically, woody plants grow in a *swamp*, grasses grow in a *marsh*, and decaying vegetation and moss fill a *bog*. Wetlands serve as important breeding grounds for birds, fish, insects, and amphibians.

Formation of Streams and Drainage Networks

During the hydrologic cycle, water that entered the atmosphere by evaporation condenses and falls back to the Earth's surface as rain or snow. Some of this *precipitation* accumulates on land as surface water, some sinks into the ground, and some gets absorbed by plants. Gravity causes surface water to flow downslope as **overland flow** (or *surface runoff*). Overland flow commonly occurs in tiny rivulets that follow a tortuous path between ground-surface irregularities, plant stems, and roots. On smooth surfaces, overland flow may occur as a thin film, called

sheetwash (Fig. 12.4a). Where the *substrate* (the material at and just below the land surface) happens to be a little weaker, or water flow happens to be a bit faster, overland flow digs down and eventually produces a trough-like depression, or **channel** (Fig. 12.4b, c). Hydrologists use the word **stream** as a general name for any body of water flowing along a channel, but in everyday English, of course, we refer to large streams as *rivers* and to medium-sized ones as creeks or brooks, and only to small ones as streams. The *streambed* is the floor of the channel, the *channel walls* form the sides of the channel, and the *banks* of a stream are the slopes adjacent to the channel. Stream banks merge with the side of the channel at the water line, and may become submerged when the surface of the stream rises. Streams receive water from many sources (Fig. 12.4d). Inputs come from rain and meltwater that moves as overland flow when it lands on the ground, from water that has passed through soil, and from springs.

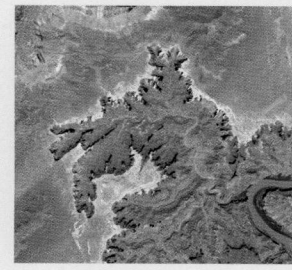
FIGURE 12.4 The formation of stream channels.

Time

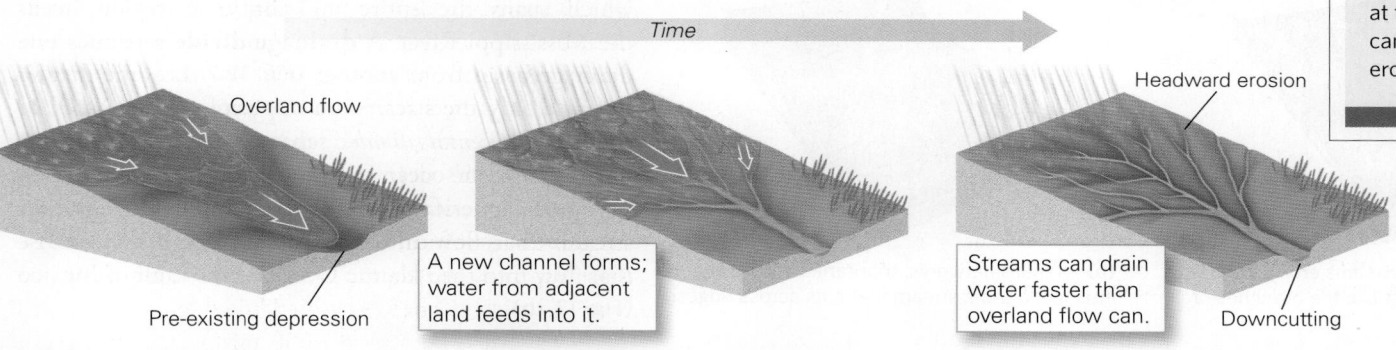

Overland flow

Pre-existing depression

A new channel forms; water from adjacent land feeds into it.

Headward erosion

Streams can drain water faster than overland flow can.

Downcutting

(a) Overland flow is focused in a slight depression.

(b) A new channel starts to form as the flow carves into the substrate.

(c) Tributaries start to develop and feed into the trunk stream.

Puddle

Swamp

①

②

③

④

⑤ Tributary

Trunk stream

Stream bank

⑥

⑦

Water table

2 m

Streambed

(d) The water entering a stream comes from many sources.

①	Melted snow adds water.
②	Swamps and puddles collect water on flat land; water drains into the stream.
③	Some water infiltrates the substrate and becomes groundwater, which flows underground.
④	Sheetwash flows over land into the stream.
⑤	Rain or snow falls directly into the stream.
⑥	Some water entering the stream flows through soil first.
⑦	Groundwater enters the stream via springs.

FIGURE 12.5 An aerial view of headward erosion.

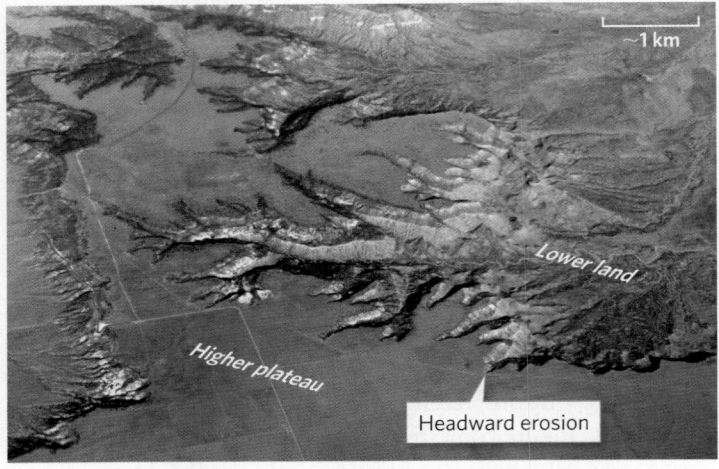

~1 km

Lower land

Higher plateau

Headward erosion

FIGURE 12.6 Five types of drainage networks.

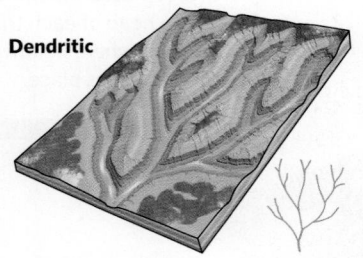

Dendritic

(a) In dendritic networks, streams erode a uniform substrate, and they connect like the branches of a tree.

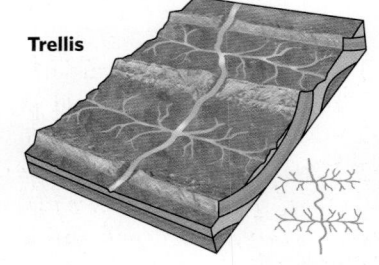

Trellis

(b) In trellis networks, tributaries in valleys intersect a trunk stream that cuts across ridges.

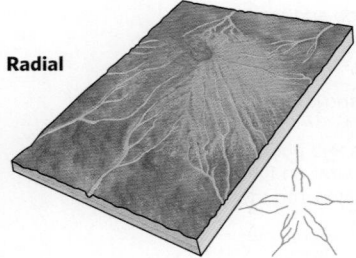

Radial

(c) In radial networks, streams radiate from a central peak like the spokes of a wheel.

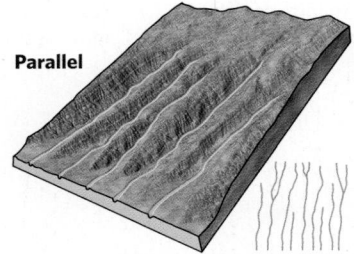

Parallel

(d) In parallel networks, streams form on a uniform slope, and all trend parallel to one another.

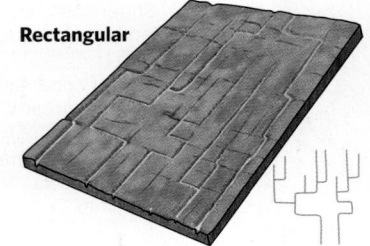

Rectangular

(e) In rectangular networks, streams intersect at right angles because they follow pre-existing cracks.

Over time, continued erosion causes a stream channel to deepen, a process called **downcutting**, as well as to lengthen at its *headwaters* (point of origin), a process known as **headward erosion** (Fig. 12.5). As downcutting progresses along a stream, the surrounding land surface starts to slope toward the stream channel, thereby directing progressively more water into the channel. Eventually, new side channels, or *tributaries*, form on this land surface and flow into the initial, dominant channel, or *trunk stream* (see Fig. 12.4c). This process establishes a **drainage network**, a linked association of channels that remove **runoff** (the combination of stream flow and overland flow) from a *drainage basin* or *watershed*.

Natural drainage networks carry water off the land just as a network of gutters drains a parking lot. Geologists recognize several forms of drainage networks—dendritic, radial, rectangular, trellis, and parallel—by their map patterns (Fig. 12.6a–e). The largest drainage basin in the world covers much of northern South America and provides water to the Amazon River (Fig. 12.6f). North America's largest drainage basin, which spans the entire mid-continent region, feeds the Mississippi River. A **drainage divide** separates one drainage basin from another (Fig. 12.7a). Some divides separate only the streams on opposite sides of a hill or ridge. *Continental divides* separate drainage networks flowing into one ocean from those flowing into another. In North America, the Continental Divide separates streams that flow into the Pacific Ocean from those that flow into the Atlantic Ocean or the Gulf of Mexico (Fig. 12.7b).

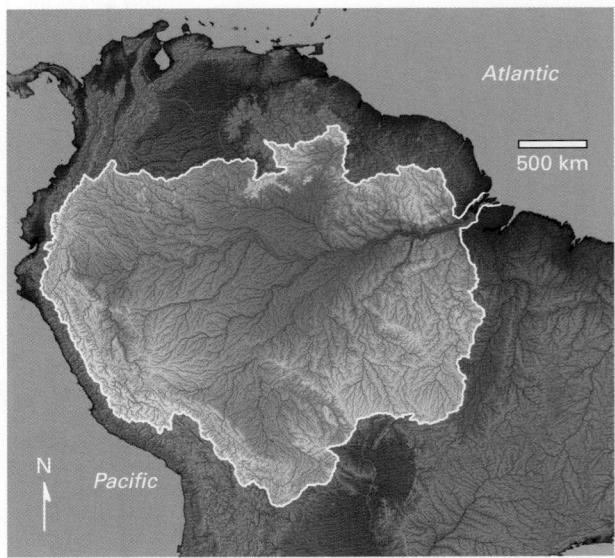

Atlantic

500 km

N

Pacific

(f) The Amazon watershed of South America stretches from the crest of the Andes to the Atlantic Ocean. It contains a dendritic drainage network. The Amazon River is the trunk stream.

FIGURE 12.7 Drainage divides and basins.

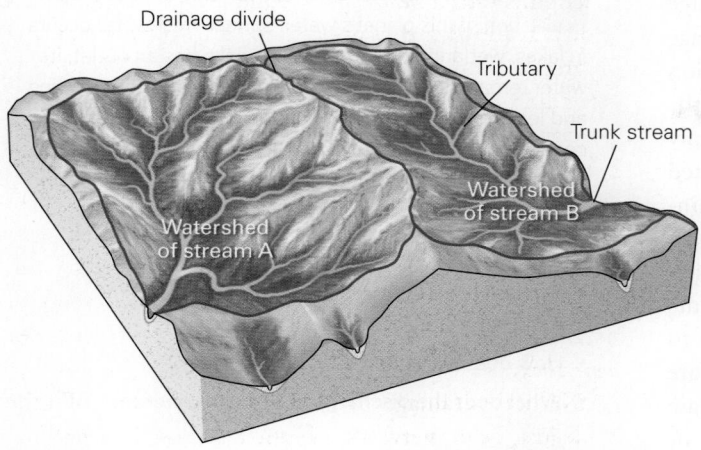

(a) A drainage divide is a ridge that separates two drainage basins.

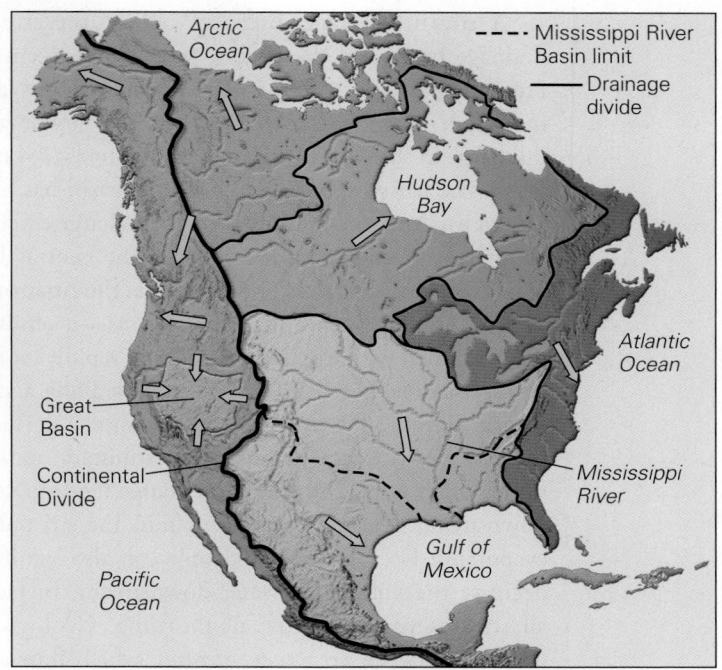

(b) The major drainage basins of North America. The Great Basin is a region of interior drainage with no outlet to the sea.

Stream Discharge

Imagine two streams, a larger one in which water flows slowly, and a smaller one in which water flows rapidly. Which stream carries more water? To answer this question, researchers compare the streams' average **discharge**, meaning the volume of water that passes through a cross section of a stream (a vertical plane perpendicular to the stream channel) in a given time. We calculate discharge by a simple formula: $D = A \times v$. In this formula, A indicates the cross-sectional area of the stream (on a plane perpendicular to the channel), and v gives the average velocity at which water moves in the downstream direction. We specify discharge (D) in units of volume (cubic meters or cubic feet) per unit of time.

The average velocity of stream water (v) can be difficult to calculate because not all water in a stream travels at the same velocity. Why? Friction slows water flow, so water near the banks or near the streambed generally moves more slowly than does water in the middle of the flow (Fig. 12.8a). In detail, **turbulence**—the twisting, swirling motion of a moving fluid—causes water to follow paths that greatly exceed the channel's length. In fact, water may flow upstream as part of an *eddy*, or in a spiral as part of a *whirlpool* (Fig. 12.8b).

FIGURE 12.8 Flow velocity and its measurement in streams.

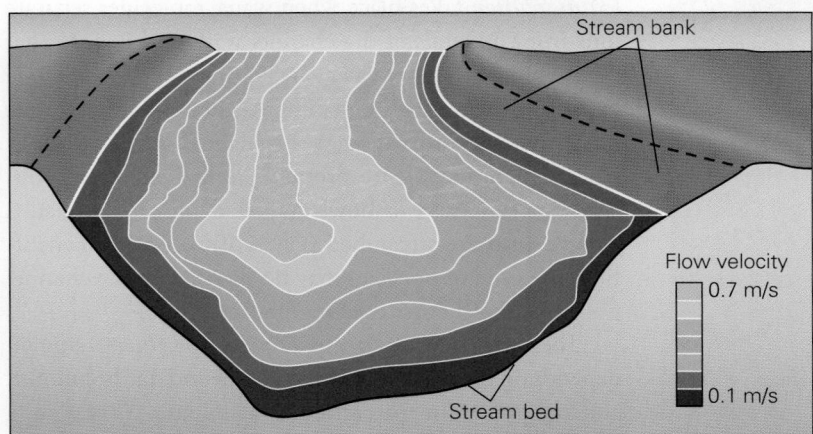

Flow velocity

0.7 m/s

0.1 m/s

(a) Overall, flow velocity varies with location in a stream. Velocity is slower near the banks and the streambed. Along a curve in the stream, the faster flow shifts toward the outer edge of the curve.

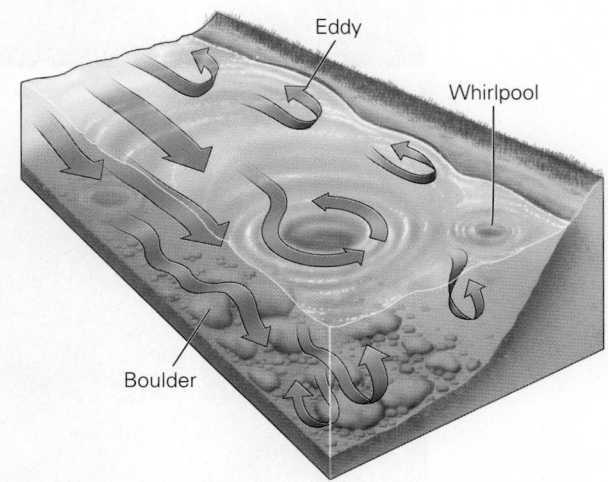

(b) In detail, water in a stream doesn't usually follow a straight path. It swirls and twists, producing turbulence.

A stream's average discharge reflects the size of its drainage basin and the climate where the stream flows. For example, the Amazon River, which drains a huge mountain range and rainforest, has the largest average discharge in the world—about 210,000 m³/s (7.4 million ft³/s). The Mississippi River's drainage basin has half the area of the Amazon's, but its average discharge equals only about 8% of the Amazon's because the central United States receives much less rain than does the Amazon rainforest. Discharge varies with the seasons—it tends to be greatest when winter snows are melting rapidly, or during a rainy season. Discharge may also vary along a stream's length. Specifically, in a wet region, discharge tends to increase downstream because each tributary adds more water, whereas in a dry region, discharge tends to decrease downstream because water seeps into the streambed or evaporates. Use of water by people can also significantly decrease the volume of water downstream. In fact, not all streams contain water all the time. Geologists distinguish between *permanent streams*, which flow all year (Fig. 12.9a), and *ephemeral streams*, which flow only part of the year. When flow in an ephemeral stream vanishes, its channel becomes a *dry wash*, *arroyo*, or *wadi* (Fig. 12.9b).

FIGURE 12.9 The contrast between permanent and ephemeral streams.

Permanent stream in Wyoming

(a) In a permanent stream, typical of temperate or wet climates, streams contain water even between rains.

Dry wash in Arizona

(b) In a dry climate, a stream flows only when water enters the stream faster than it can leave the stream by seeping into the ground or evaporating.

Take-home message…

Liquid freshwater accounts for only a small proportion of this planet's water. Surface freshwater occurs in lakes, wetlands, and streams. If a lake has an outlet, its water remains fresh. Stream channels form by downcutting and lengthen by headward erosion. A drainage network consists of a group of tributaries feeding a trunk stream. Stream discharge depends on the area of a drainage basin and on climate, and can vary along a stream.

Quick Questions ———————————

• By how much does the concentration of salt in freshwater differ from that in the ocean?
• How does a wetland differ from a lake?
• What does the discharge of a stream refer to?

12.2 The Work of Running Water: Fluvial Landscapes

Over time, the flow of water in streams significantly modifies topography to produce *fluvial landscapes* (from the Latin *fluvius*, meaning river). Flowing water accomplishes such changes by eroding the land surface, by transporting sediment elsewhere, and by depositing new accumulations of sediment.

How Do Streams Erode the Earth's Surface?

If you flood the ground with a hose and watch the water dig into the soil and carry it away, you are seeing erosion by running water. In the case of a natural stream, gravity causes water to flow downslope, and this movement erodes the Earth's surface in four ways: (1) *scouring* happens when water picks up and carries away loose sediment; (2) *breaking and lifting* take place when water separates and lifts chunks of rock from the channel; (3) *dissolution* takes place when water molecules separate ions from minerals of the channel walls and bed, so that the stream carries them away in solution; and (4) *abrasion* happens when moving sediment-laden water acts like sandpaper and rasps away at the channel (Fig. 12.10a). In places where turbulence produces long-lived whirlpools, abrasion can carve a bowl-shaped depression, called a *pothole*, into the streambed (Fig. 12.10b). After downcutting, the sides of potholes may be exposed on the channel walls (see the chapter opening photo).

The efficiency of stream erosion (or *fluvial erosion*) depends on the discharge of the stream and on the amount and type of sediment carried in the water. Therefore, a large volume of fast-moving, turbulent, sandy water causes much more erosion than a trickle of quiet, clear water. Notably, the supply of sediment in a stream comes

(a) The walls of this canyon in Arizona have been polished by abrasion.

Pothole

FIGURE 12.10
Erosion by streams.

(b) During floods, turbulent sandy water polished this rock outcrop and carved potholes.

not only from erosion of the stream's channel, but also from landslides that carry debris down slopes bordering the stream and dump it into the stream (see Chapter 11).

How Do Streams Transport Sediment?

The Mississippi River earned its nickname "Big Muddy" for a reason: the water in it tends to be brown because of all the clay and silt it carries. A stream's **sediment load**, meaning the total amount of sediment the stream carries, includes three components (Fig. 12.11a): (1) *suspended load*, consisting of silt- or clay-sized grains that swirl along with the water without settling to the streambed (Fig. 12.11b); (2) *bed load*, consisting of relatively large clasts that bounce or roll along the streambed; and (3) *dissolved load*, consisting of ions in solution.

When describing a stream's ability to carry sediment, geologists distinguish between competence and capacity. Stream **competence** refers to the maximum clast size a stream can carry—a stream with more competence can carry larger clasts, while one with less competence can carry only smaller ones. A stream's competence depends on both the velocity and the viscosity of its water, so a fast-moving, turbulent stream containing suspended sediment has greater competence than does a slow-moving, clear stream. Stream **capacity** refers to the total amount of sediment the stream can carry. A stream's capacity depends on both competence and discharge, so a large, fast-flowing river has more capacity than does a small, slowly flowing creek.

How Do Streams Deposit Sediment?

If the flow velocity of a stream decreases, then the competence of the stream decreases, so the stream's sediment load starts to settle out. The sizes of the clasts that settle at a particular location depend on the flow velocity at that location. So, if the stream slows by a small amount, only large clasts settle, but if the stream slows by a lot,

medium-sized clasts settle, and if the stream slows almost to a standstill, fine-grained sediment settles.

Sediments deposited by a stream are called *fluvial deposits* or **alluvium**. The sizes of clasts in alluvium depend on the velocity of flow. Fast-moving mountain streams, when in flood, can carry coarse gravel or even boulders (Fig. 12.12a). In contrast, alluvium carried by slower streams consists of sand and silt.

Alluvium will be exposed when the water level in the stream drops. Commonly, alluvium collects in elongate

FIGURE 12.11 Streams transport sediment in many forms.

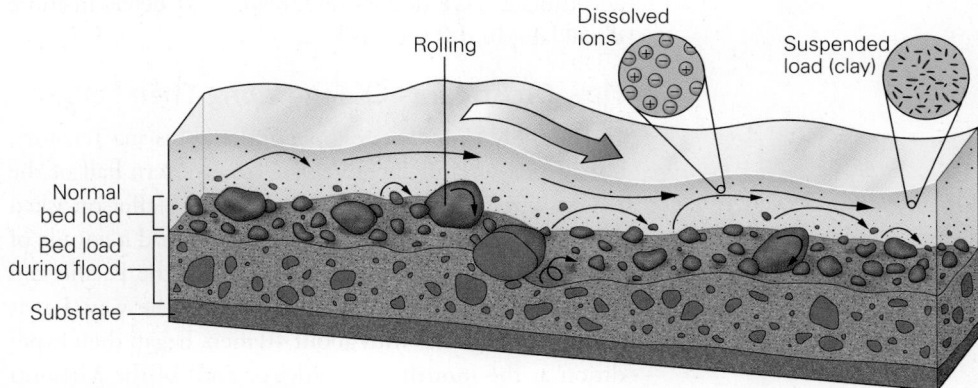

Rolling — Dissolved ions — Suspended load (clay)

Normal bed load
Bed load during flood
Substrate

(a) Dissolved ions are carried in solution; tiny suspended grains are distributed throughout the water; and larger clasts slide or roll along the streambed.

(b) The brown color of this turbulent stream water is due to suspended clay and silt.

FIGURE **12.12** Examples of alluvium deposited by streams.

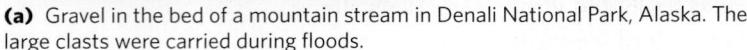

(a) Gravel in the bed of a mountain stream in Denali National Park, Alaska. The large clasts were carried during floods.

(b) Bars of gravel and sand deposited by a stream in the Canadian Rockies.

mounds known as **bars**, or in wedges along the stream's banks **(Fig. 12.12b)**. During floods, a stream may overtop the banks of its channel and spread out over its **floodplain**, a broad, flat area bordering the stream. An escarpment, or *bluff*, at which land elevation abruptly rises, may delineate the outside edges of a floodplain. Friction slows flow on the floodplain, so fine-grained alluvium (silt and mud) settles out to form *floodplain deposits*. Where a stream empties into a standing body of water, the slowdown of flow causes a wedge of sediment called a *delta* to accumulate. (We discuss floodplains and deltas in more detail later in this section.)

How Do Streams Change along Their Length?

In 1803, the United States bought the Louisiana Territory, a vast tract of land encompassing the western half of the Mississippi River drainage basin. President Jefferson asked Meriwether Lewis and William Clark to lead a voyage of exploration from St. Louis, Missouri, to the Pacific and to make a map of what they saw along the way. Lewis and Clark, together with about 40 men, began their expedition at the **mouth** (the outlet or end) of the Missouri River where it flows into the Mississippi. At its mouth, the Missouri is a wide stream of muddy water that can be navigated easily. As the explorers navigated upstream, they found that their journey became more difficult, for the **stream gradient**, meaning the slope of the stream's surface, became progressively steeper, and the stream's discharge diminished. When Lewis and Clark reached North Dakota, they had to abandon their original boats and proceed in smaller vessels. Often, they had to haul their boats up intervals of the stream where fast, turbulent water swirled around protruding rocks. When they reached Montana, they had to abandon boats entirely and

instead had to trudge along the stream's banks on foot, climbing steep gradients up to the Continental Divide. They discovered that not only an individual stream, but the fluvial landscape as a whole, changes from its headwaters, near the drainage divide, to its mouth **(Fig. 12.13a)**. For example, upstream along a tributary, where gradients are steeper, valleys tend to be deeper than they are downstream, where the trunk stream may flow across a nearly horizontal plain.

Geologists can represent the change in stream gradient that Lewis and Clark experienced on a graph that plots elevation on the vertical axis and distance from the mouth on the horizontal axis **(Fig. 12.13b)**; the curving line on such a graph depicts the stream's *longitudinal profile*. An idealized longitudinal profile has an overall concave-up shape, emphasizing that streams have steep gradients near their headwaters and gentle gradients near their mouths. Longitudinal profiles of real streams are not perfectly smooth curves, but rather they display local steps.

The lowest elevation on a longitudinal profile, meaning the lowest possible elevation of a stream's surface, defines the **base level** of the stream. At the mouth of a tributary, the surface of the larger stream acts as the base level for the tributary. A standing body of water (such as a lake or reservoir) or a resistant rock cliff can act as a *local base level* along a stream—local base levels are the local steps, mentioned above, that appear along a longitudinal profile. A standing body of water at the mouth of a trunk stream serves as the *ultimate base level* of a drainage network, the lowest level of its longitudinal profile. For streams that flow into the ocean, sea level represents the ultimate base level. The surface of such a stream can't be lower than sea level, for if it were, the stream would have to flow upslope to reach its mouth.

FIGURE 12.13 The character of a stream, and of fluvial landscapes, changes along its length.

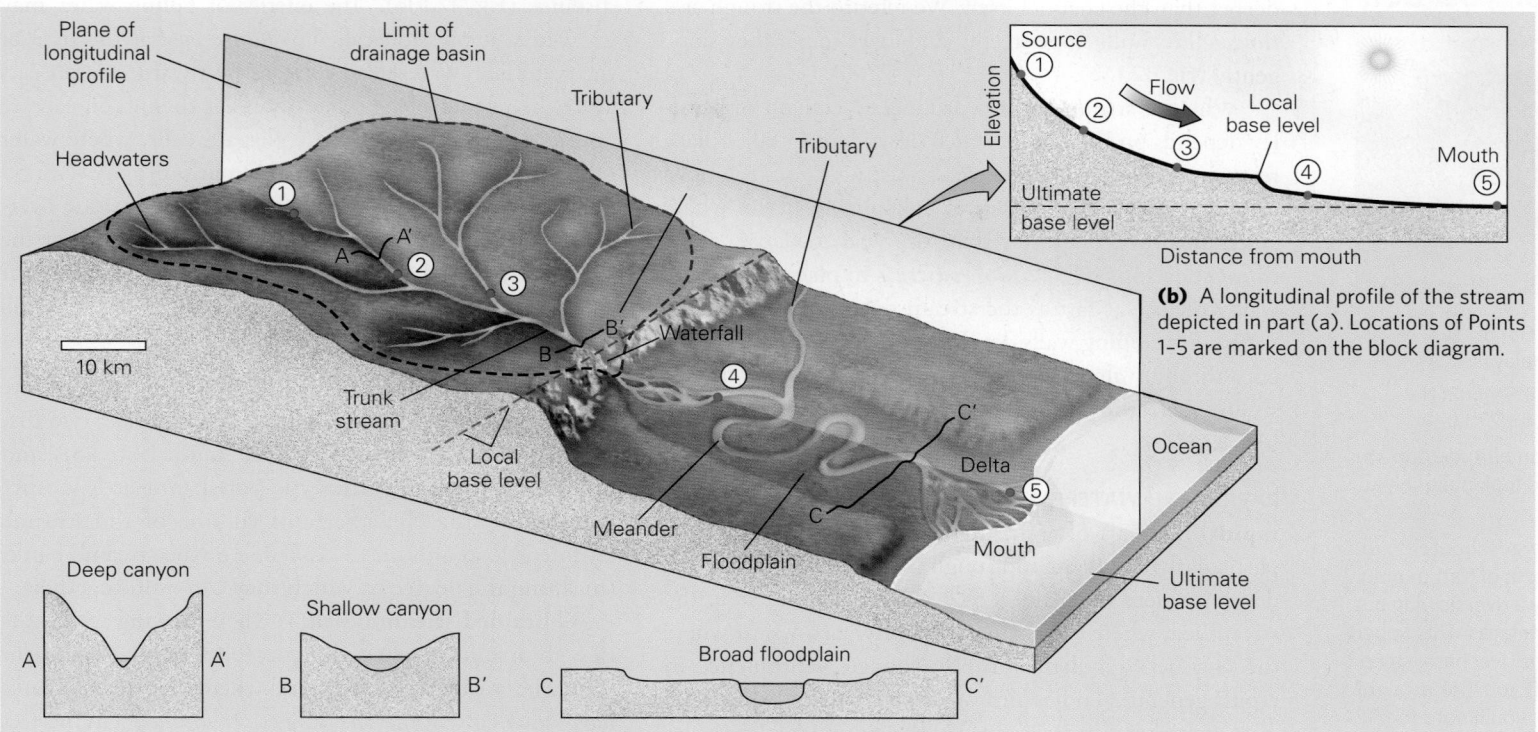

(b) A longitudinal profile of the stream depicted in part (a). Locations of Points 1–5 are marked on the block diagram.

(a) Gradients are steeper and valleys deeper nearer a stream's headwaters. Gradients are gentle and floodplains wide near its mouth. Section lines A–A', B–B', and C–C' show the positions of the profiles below the block diagram.

Fluvial Landforms

As streams flow, they carve into and remove their substrate, and they leave behind sedimentary deposits. These activities yield a variety of distinctive fluvial landscapes. The character of these landscapes depends on location along the stream's longitudinal profile, the nature of the substrate, vegetation, and climate.

VALLEYS AND CANYONS. Millions of years ago, a block of crust in the region of what is now the southwestern United States began to rise; its surface eventually became the Colorado Plateau. As the plateau rose, downcutting by the Colorado River produced the Grand Canyon, whose floor now lies 1.6 km (1 mi) below the plateau's surface (**Fig. 12.14a**). The formation of the Grand Canyon

Did you ever wonder . . .

whether the position of a waterfall can change over time?

FIGURE 12.14 Examples of landscapes shaped by downcutting streams.

(a) The western end of the Grand Canyon, carved by the Colorado River.

(b) A deep valley cut by a stream in the Andes of Peru.

emphasizes that in regions where the land surface lies well above the base level, a stream can carve a trough much deeper than the channel itself. We refer to the trough as a *canyon* if its walls slope steeply, and as a *valley* if they slope gently (Fig. 12.14b).

Whether stream erosion produces a canyon or a valley depends on the rate at which downcutting takes place relative to the rate at which mass wasting causes the walls on either side of the stream to collapse. In places where a stream downcuts faster than the walls collapse, a *slot canyon* develops (Fig. 12.15a), whereas in places where the walls collapse as fast as the stream downcuts, a *V-shaped valley* with sloping walls develops (Fig. 12.15b). If a stream cuts through alternating layers of resistant and nonresistant rock, the resulting canyon has a *stair-step* profile (Fig. 12.15c).

RAPIDS AND WATERFALLS. Rafters and kayakers seek out **rapids**, intervals of a stream where the water surface is particularly turbulent and rough (Fig. 12.16a). Rapids can develop where water flows over ledges or boulders in the streambed, where the channel abruptly narrows, or where the channel's gradient abruptly changes. Turbulence in rapids generates eddies and waves that roil and churn the water surface to yield *whitewater*, a mixture of bubbles and water.

A **waterfall** forms where the stream gradient becomes so steep that the stream's water undergoes free-fall for a distance (Fig. 12.16b). The energy of falling water may excavate a depression called a *plunge pool* at the base of the waterfall. Waterfalls evolve as headward erosion eats away the resistant ledge that underlies them. You can see an example of this process at Niagara Falls, where water flowing from Lake Erie to Lake Ontario drops over a 55 m (180 ft) high escarpment of resistant dolostone overlying weak shale (Fig. 12.16c). Over time, erosion of the shale undercuts the dolostone, leaving an overhang that eventually collapses, causing the position of the waterfall to migrate upstream (Fig. 12.16d).

BRAIDED AND MEANDERING STREAMS. In places where runoff flows over a gently sloping land surface, two distinctive types of streams evolve: braided streams and meandering streams. The type you'll find at a locality depends on the character and volume of the stream's sediment load and on the *cohesion* of the stream's banks (meaning the degree to which they can hold together).

A **braided stream** develops where flowing water carries abundant coarse sediment (because the sediment source lies relatively nearby) and where the stream's banks have low cohesion. The adjective *braided* emphasizes that the stream consists of multiple channels that weave around elongate sediment bars and therefore entwine like strands of hair in a braid (Fig. 12.17). How does a braided stream form? During floods, water in the stream rises and moves faster, so it can pick up and transport large volumes of sediment. When flooding stops and the water slows, sediment settles out in elongate bars, and the remaining water divides to flow through the low areas between the bars.

A **meandering stream** develops where sediment doesn't choke the stream and where the stream's banks have relatively high cohesion, either because their substrate can stick together or because plant roots bind the substrate together. The adjective *meandering* emphasizes that the stream's channel winds back and forth in a succession of snake-like curves called **meanders** (Fig. 12.18a). How do meanders form and evolve? Even if a stream starts out with a straight channel, the location of the strongest current in the stream tends to wander, so that it sometimes lies nearer the center of the channel and sometimes nearer the banks (Fig. 12.18b). Water erodes more rapidly where it flows faster, so when the fastest current runs along the bank, the stream carves into the bank, producing an escarpment called a *cut bank*. (Cut banks have a steep face, so they can form only where the eroding substrate has cohesion.) As the process continues, the deepest part of the channel (known as the *thalweg*) stays close to the cut bank, along the outer arc of the curve. Meanwhile, on the

FIGURE 12.15
The shape of a canyon or valley formed by downcutting depends on the resistance of its walls to erosion.

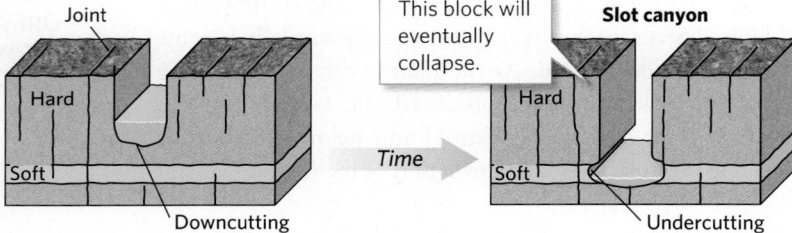

(a) If downcutting by the stream happens faster than mass wasting of the walls, a slot canyon forms. The canyon widens as the stream undercuts the walls.

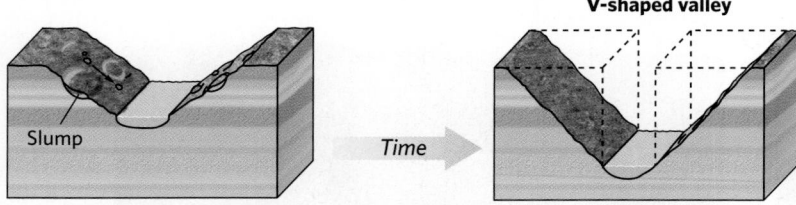

(b) If mass wasting takes place as fast as downcutting occurs, a V-shaped valley develops.

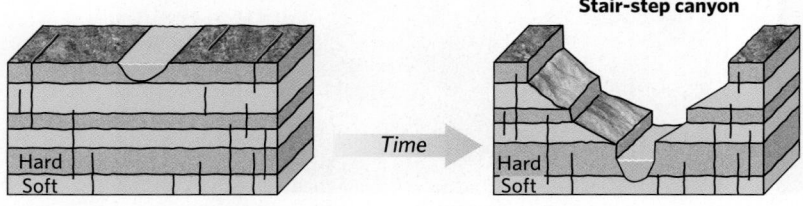

(c) Downcutting through alternating hard and soft layers produces a stair-step canyon.

FIGURE 12.16 Rapids and waterfalls.

(a) These rapids in the Grand Canyon formed when a flood from a side canyon dumped debris into the channel of the Colorado River.

(b) Horseshoe Falls, a part of Niagara Falls.

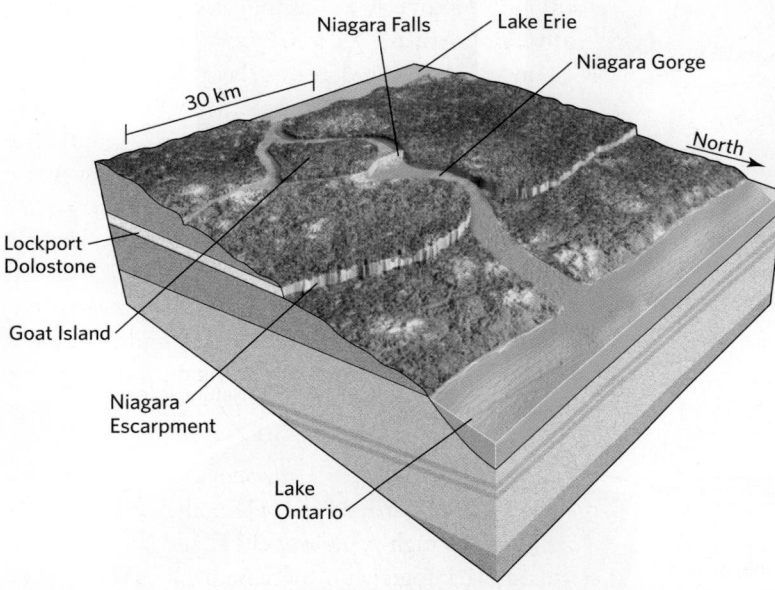

(c) Niagara Falls developed where water spills over the Lockport Dolostone, a resistant rock layer that forms the Niagara Escarpment.

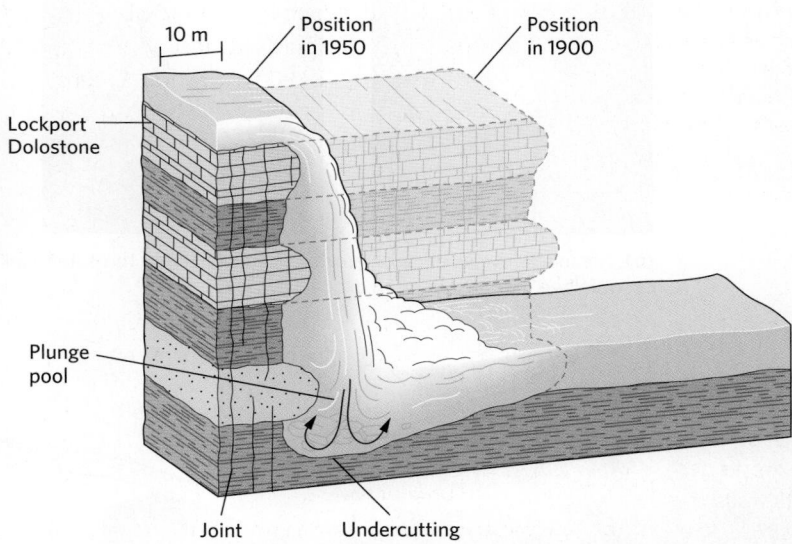

(d) Over time, as the erosion of shale undercuts the dolostone, the position of the falls migrates upstream.

inside edge of the curve, water slows down, its competence decreases, and sediment accumulates in a crescent-shaped deposit called a *point bar* (Fig. 12.18c). As time passes, the cut bank moves outward, the point bar grows wider, and the curvature of the meander becomes more pronounced. Eventually, the outer curves of two neighboring meanders approach each other until only a narrow isthmus, known as a *meander neck*, separates them. When erosion finally eats through a meander neck, a *cutoff* develops. The meander that has been cut off

FIGURE 12.17
Strands of this braided stream carry meltwater from a glacier near Denali, Alaska. The sediment bars were deposited at times when the stream was in flood and had greater competence.

437

FIGURE 12.18 Development of meandering streams.

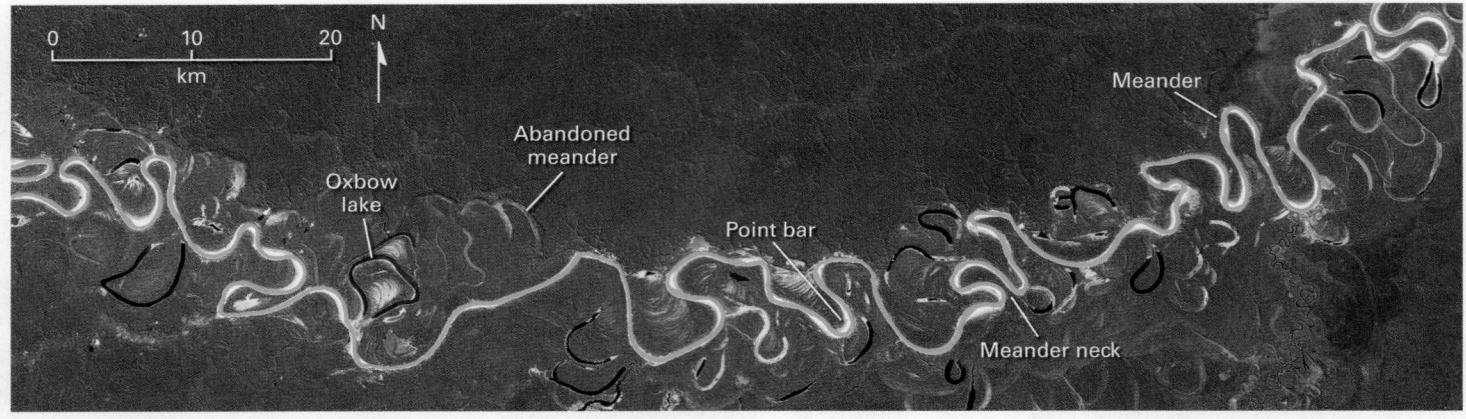

(a) A meandering stream in Brazil, as viewed from space. Note the various landscape features.

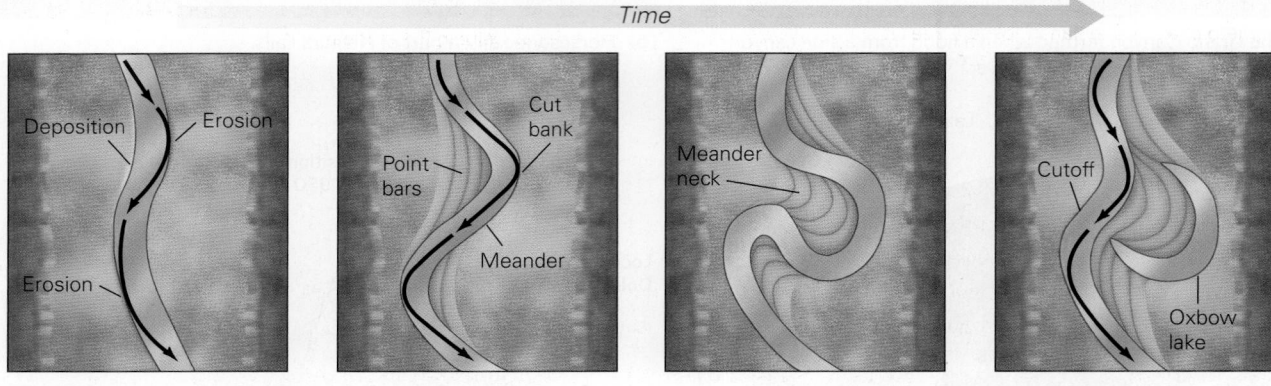

(b) Meanders evolve because erosion occurs faster on the outer bank of a curve, while deposition takes place on the inner bank. Eventually, a cutoff isolates an oxbow lake.

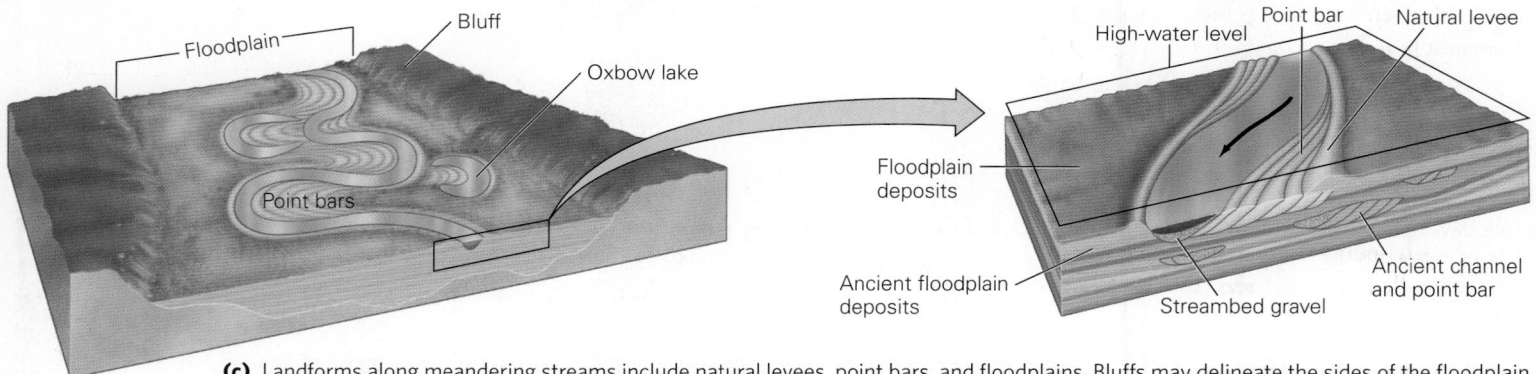

(c) Landforms along meandering streams include natural levees, point bars, and floodplains. Bluffs may delineate the sides of the floodplain.

(d) Two oxbow lakes are visible in this aerial photo of a meandering stream.

(e) The flat land of this floodplain hosts farm fields.

(a) An oblique aerial photo of an alluvial fan in Joshua Tree National Park, California. The fan consists of gravel and debris flows that came out of a canyon. The black line at the lower right is a road.

(b) A small delta in Costa Rica.

then becomes a curving *oxbow lake* **(Fig. 12.18d)**. If sediment fills in the lake, the curving depression remains visible as an *abandoned meander*. Production of new meanders and point bars and abandonment of old ones can cause the course of a meandering stream to change markedly on a time scale of years to centuries.

Most meandering stream channels cover only a relatively small portion of a broad, nearly flat floodplain **(Fig. 12.18e)**. At a given location and time, the stream channel may lie close to one of the bluffs that delimits the flood plain, or it may be out in the middle of the floodplain. As we noted earlier, water overtops the banks of the stream channel and spreads out over the floodplain during a flood, but it rarely, if ever, rises higher than the bluff. Notably, the initial slowdown of water as it overtops the banks causes some sediment to settle out directly along the banks. This sediment builds up to produce a pair of low ridges, called *natural levees*, on either side of the channel.

THE END OF THE LINE FOR SEDIMENT: ALLUVIAL FANS AND DELTAS. We've seen that fluvial sediment gets deposited along streams, in bars between or along channels, or on the floodplain. Some sediment, however, makes it all the way to the mouth of the stream and accumulates there. Where a stream exits a narrow canyon onto a plain in a dry environment, this sediment forms a wedge called an **alluvial fan** **(Fig. 12.19a)**. Where a stream enters a standing body of water, such as a lake or the sea, it forms a **delta** **(Fig. 12.19b)**. On alluvial fans and deltas, streams divide into several branches, known as *distributaries*.

The name *delta* comes from the observation by an ancient Greek historian, Herodotus, that the surface shape of the wedge of sediment deposited by the Nile River where it enters the Mediterranean Sea resembles the shape of the Greek letter delta (Δ) **(Fig. 12.20a)**.

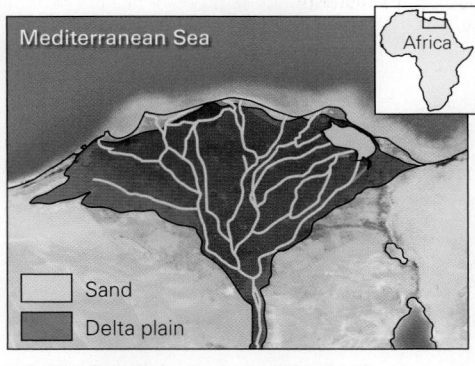

FIGURE 12.20 Not all deltas look the same.

(a) The Nile Delta is shaped like the Greek letter Δ.

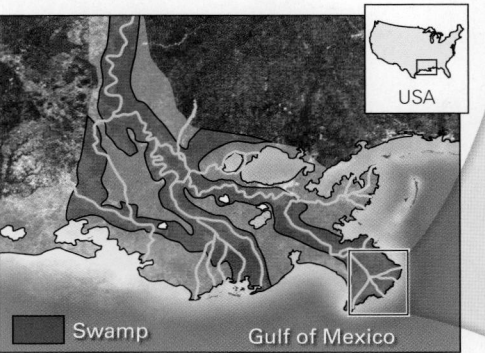

(b) The Mississippi Delta is a bird's-foot delta.

Mouth of the Mississippi as seen from space

Natural levee

FIGURE 12.21 Fluvial landscapes evolve over time.

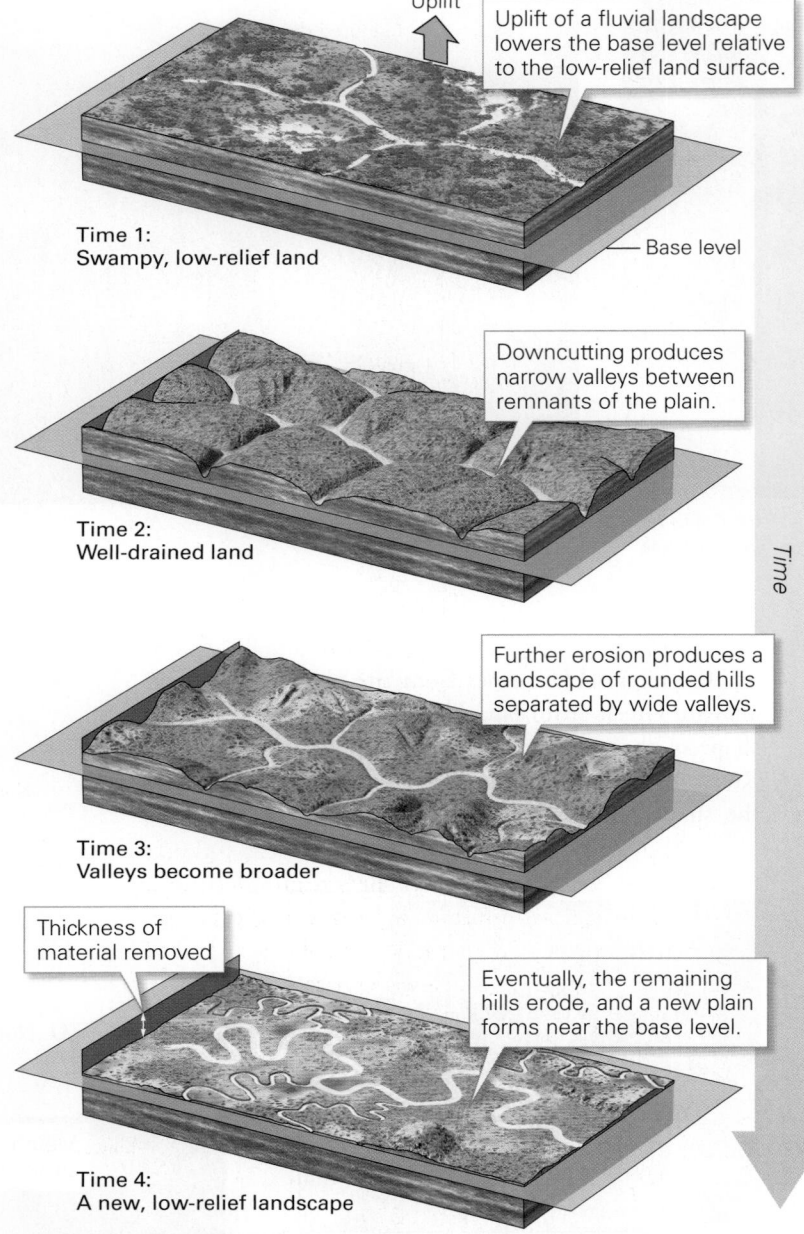

(a) When the base level drops, streams downcut, and a hilly landscape develops. The resulting relief, however, eventually erodes away.

Uplift

Uplift of a fluvial landscape lowers the base level relative to the low-relief land surface.

Time 1: Swampy, low-relief land

Base level

Downcutting produces narrow valleys between remnants of the plain.

Time 2: Well-drained land

Further erosion produces a landscape of rounded hills separated by wide valleys.

Time 3: Valleys become broader

Thickness of material removed

Eventually, the remaining hills erode, and a new plain forms near the base level.

Time 4: A new, low-relief landscape

Time

(b) The "goose-necks" of the San Juan River, Utah, are incised meanders.

Relatively few deltas have the distinct triangular shape of the Nile Delta—some are more rounded (see Fig. 12.19b), and some look like the scrawny feet of birds (Fig. 12.20b). The particular shape of a delta depends on the interplay between the rate at which the river supplies new sediment and the rate at which waves in the standing body of water remove sediment. A delta deposited by a major river along the seacoast may become very thick and wide, so that its surface becomes a broad *delta plain*.

Evolution of Fluvial Landscapes

Imagine a place where the relatively flat land surface of a region, over which a stream flows, begins to rise in response to tectonic processes. When this happens, the landscape evolves (Fig. 12.21a). According to an idealized model, downcutting initially carves deep valleys or canyons into the flat surface, but over time, mass wasting and erosion transform the rugged landscape into one with low, rounded hills and broad valleys with wide floodplains, down which meandering streams flow. Eventually, even the low hills get eroded down, leaving a new, nearly flat land surface at an elevation close to that of the drainage network's base level. On our dynamic planet, however, changes in land elevation or sea level commonly prevent landscapes from ever getting shaved all the way down to the ultimate base level. If the land surface under a drainage network rises or the base level of the network falls, **stream rejuvenation** takes place, meaning that streams begin to downcut again. Rejuvenation of a meandering stream may produce dramatic *incised meanders* (Fig. 12.21b). Notably, if regional uplift causes the tilt direction of the land to change, the flow of streams in a drainage network may eventually change direction as well; this phenomenon is known as a *drainage reversal*.

The details of fluvial-landscape evolution can also be complicated by local uplift or subsidence. For example, if uplift of a mountain range in the path of a stream happens faster than the stream can erode the range, the uplift diverts the path of the stream, but if erosion can keep pace with uplift, the stream may cut across the range. If erosion takes place on one side of a drainage divide faster than it does on the other side, the position of the drainage divide may move over time. And if headward erosion causes the channel of one stream to intersect the channel of another, the water of one stream may start to flow down the channel of the other; this phenomenon is known as *stream piracy* or *stream capture* (Fig. 12.22).

FIGURE 12.22 Stream piracy.

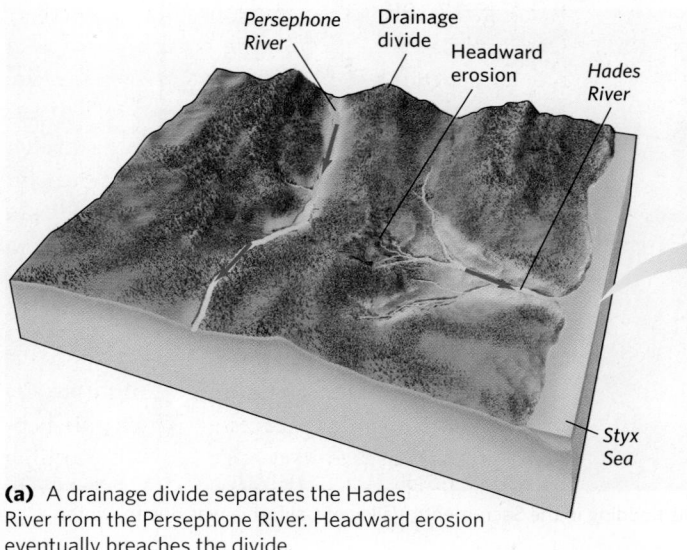

(a) A drainage divide separates the Hades River from the Persephone River. Headward erosion eventually breaches the divide.

(b) The Hades captures the Persephone and carries its water to the Styx Sea. The Hades now flows through a water gap (a channel cut through a ridge), and the former Persephone channel becomes a dry channel.

Take-home message...

Streams erode the Earth's surface and transport sediment. The size of the clasts a stream carries depends on its flow velocity and viscosity. Where velocity decreases, sediment settles out. Typically, the gradient of a stream decreases downstream to a base level; sea level serves as the ultimate base level of drainage networks that flow into the ocean. On gently sloping land surfaces, braided streams or meandering streams develop. Fluvial landscapes evolve over time.

Quick Questions

- Distinguish between abrasion and scouring by a stream.
- Do the meanders of a stream remain in the same location over time?
- Does a stream have the same gradient everywhere along its length?

12.3 Raging Waters: River Flooding and Flood Control

The Danger of Flooding

When large volumes of snow or ice melt, or intense or long-lasting rain falls from the sky, the volume of water entering the streams of a drainage network increases. When the volume of water in a stream exceeds the volume of the stream channel, a **flood** takes place. A flooding stream overtops its banks and covers areas outside its normal channel. On low-relief ground, floodwaters can spread broadly, submerging large areas of a floodplain or delta plain. In a canyon or narrow valley with steep relief, floodwaters can submerge and scour the canyon or valley walls well above the normal stream banks. In some cases, floodwaters inundate towns, turning streets into temporary torrents **(Fig. 12.23)**.

We've seen that hydrologists refer to the amount of water carried by a stream as the stream's discharge. Discharge, in turn, determines the **stage** of the stream, meaning the water level in the stream above a defined reference elevation located below the base of the streambed. Officials report that a stream has risen above **flood stage** when the water level reaches the top of its channel.

Narrative Art Video
River Meanders

FIGURE 12.23 When a thunderstorm caused flooding in Ellicott City, Maryland, its Main Street became a torrent.

FIGURE 12.24 Examples of slow-onset flooding.

(a) This flooding in Dhaka, the capital city of Bangladesh, was caused by monsoon rains. It submerged city streets, making travel hazardous.

(b) This flooding in the Sacramento Valley of California was due to winter rains.

The **crest** of a flood is the highest stage that the stream's level rises to before it starts to fall and floodwaters recede. Hydrologists distinguish between two types of stream-related floods by the rate at which a stream rises and the duration of the flooding: *slow-onset floods* and *flash floods*.

Slow-Onset Floods

A flood that takes days to develop, lasts for days to weeks, and involves the major tributaries or trunk stream of a drainage network where the land has low relief is called a **slow-onset flood** (Fig. 12.24). Such floods take a while to develop because they incorporate water from across a broad drainage area, and it takes time for that water to move overland, collect in tributaries, and reach the trunk stream. Slow-onset floods happen in one or more of the following circumstances: when overland flow increases significantly because the ground has become saturated and cannot absorb any more water; when snowmelt happens quickly during a spring thaw; during sustained rains in a wet season (see Chapter 18); or when a storm system covers a broad region. When slow-onset floods affect large areas, they may also be called *regional floods*, and where they occur only during a particular season, they may be referred to as *seasonal floods*.

Slow-onset floods make headlines every year. For example, during 2022, particularly intense monsoon rains in Pakistan caused the Indus River and its tributaries to rise slowly until water covered about one-fifth of the country's land area. The flooding caused $14.9 billion in damage, left 2.1 million people homeless, and led to about 1,700 fatalities. Similar flooding submerges the floodplain of the Ganges River in India and Bangladesh, and the Yellow and Yangtze Rivers of China. An estimated 3 million people died because of floods along the Yellow and Yangtze Rivers in 1931. This terrible event happened when melting of the winter's heavy snowfall was followed by heavy rains and a succession of typhoons (hurricanes). The ground was saturated, and the melting snow added even more water to the flood than would have come from the rain alone. Water in the Yangtze River crested at 16 m (53 ft) above flood stage. Slow-onset floods affect the Mississippi and Missouri river systems in the United States just about every year. During these floods, large areas of floodplains become submerged in silty water, leaving crops and riverside towns buried in a bed of sediment when the floodwaters recede (Box 12.1).

Flash Floods

An event during which the discharge of a stream increases so fast that people along the stream, or in areas of low-lying land, don't necessarily have time to prepare for the appearance of high water is called a **flash flood** (Fig. 12.25). The US National Weather Service applies the term to floods that happen within six hours of the phenomenon that caused them, and issues a *flash flood warning* when such a flood "has been observed or is imminent" to urge people in the path of the flood to evacuate. In contrast to slow-onset floods, flash floods not only appear suddenly, but may subside in minutes to hours. They also differ from slow-onset floods in that they may affect only a relatively small region and typically involve tributaries instead of a trunk stream.

Particularly intense rainfall causes most flash floods. But flash floods also accompany levee and dam failures (as in the 1889 Johnstown flood) or the sudden breakup of debris or ice jams in stream channels. During a flash

BOX 12.1 ▶ A Deeper Look

The 1993 Mississippi River flood

The Mississippi River drainage network has had its share of slow-onset floods. A terrible flood happened in 1927, and its consequences caused hundreds of thousands of people

FIGURE Bx12.1 The Mississippi River flood of 1993.

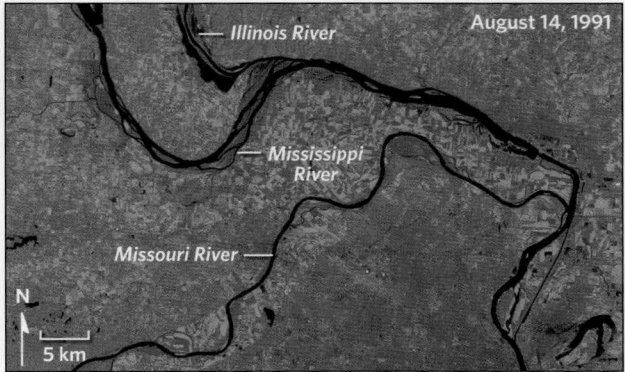

August 14, 1991
— Illinois River
— Mississippi River
Missouri River —
N
5 km

(a) Satellite photo of the area where the Missouri and Illinois Rivers join the Mississippi River at a time of no flooding. The dark color is water.

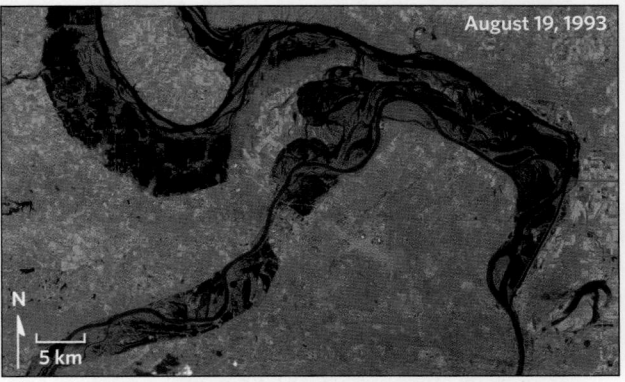

August 19, 1993
N
5 km

(b) Water covers much more of the land in this photo taken during the 1993 flood.

to lose their homes. The worst in recent decades took place in the summer of 1993, when unusual weather patterns caused a whole year's supply of rain to fall across the upper Midwest during a period of a few weeks—some regions received 400% more rain than usual. Immense volumes of water flowed into the Mississippi drainage network. When the water reached the Mississippi and Missouri Rivers, it spread out over these rivers' floodplains, and by July, parts of nine states were underwater **(Fig. Bx12.1)**. Bridges and roads were washed away, and rowboats replaced cars as the favored mode of transportation in towns where only the rooftops remained visible. In St. Louis, Missouri, the river crested at 14 m (47 ft) above flood stage. For 79 days, flooding persisted. When the water finally subsided, it left behind a thick layer of sediment, filling living rooms and kitchens in homes and burying crops in fields. In the end, more than 40,000 km² (15,400 mi²) of the floodplain had been submerged, 50 people had died, and at least 55,000 homes had been destroyed.

Comparable flooding struck the Mississippi River and its tributaries again in 2011 and in 2019. The 2019 floods were caused when heavy winter snow melted suddenly, making the ground soggier than had ever been recorded over the previous 124 years. When spring rains began, and were heavier than usual, water couldn't soak into the saturated ground, so rivers overflowed and floodplains became submerged.

(c) Missouri River floodwaters in 1993.

flood, low-lying land with poor drainage becomes submerged, and streams in narrow canyons or valleys fill to a level well above their normal banks. The turbulent and fast-moving water of a flash flood in a canyon or valley can transport large volumes of sediment, including clasts much larger than those that the stream normally transports. Pressure applied by the fast-moving muddy water

can carry away everything in its path. Flash floods can happen in any climate, but they are especially dramatic in arid regions, where a thunderstorm can send water into dry washes. When such floodwaters flow far downstream from their source, the leading edge of the flood may look like a wall of water as it inundates a previously dry streambed.

FIGURE 12.25 Examples of flash flooding.

(a) Sometimes, people trapped by flash floods can be rescued only by helicopter.

(b) A 2013 flash flood in India washed away bridges and stranded 40,000 people.

(c) This flood ate away at the stream bank along the Big Thompson River, causing a landslide and undercutting a house.

(d) During the 1976 Big Thompson River flash flood, this house was carried off its foundation and dropped on a bridge.

Damage due to Floods

Floods cause damage in many ways. Rising water can threaten the lives of people and animals by trapping them so that they can't reach dry land and perish from drowning. The force of moving water, as well as the buoyancy of objects that become partially or entirely submerged, allows a flooding stream to pick up homes, possessions, buildings, and boulders. Once objects start moving with the water, they become battering rams that can flatten structures in their path. Erosion by floodwaters can undercut land bordering a flooding stream, causing slumps that can drop the land, and any buildings on it, into the water. Such erosion can also undermine and carry away bridges. When floodwaters overwhelm sewage-treatment facilities and industrial sites, they pick up and carry dangerous microbes, viruses, and toxic chemicals that can sicken survivors. And as floodwaters slow and recede, they leave behind layers of sticky—and sometimes toxic—sediment, which buries crops and roads and fills basements and ground floors of buildings that remain standing. The damage to infrastructure (such as water mains, power lines, hospitals, and roads) can take years to repair, so in the aftermath of a major flood, people may suffer from starvation, disease, and loss of income. Major floods are catastrophic natural disasters!

Causes of Floods

Most floods are the result of weather patterns that bring with them particularly moisture-laden air. For example, South Asia and Southeast Asia, regions that have a *monsoonal climate* (characterized by distinct wet and dry seasons), endure floods nearly every year. These floods happen when the *Intertropical Convergence Zone (ITCZ)* moves northward from a location over the Indian Ocean and sweeps over the land (Fig. 12.26a). In the ITCZ, winds from the south collide with winds from the north, causing warm, moist air to rise and produce thunderstorms, as discussed further in Chapter 18. The thunderstorms in the ITCZ drop deluges of rain for days at a time.

In North America and Europe, major flooding commonly occurs in association with huge weather systems called *mid-latitude cyclones.* As we'll see in Chapters 17 and 18, a mid-latitude cyclone, as seen from a satellite, has the shape of a comma, with a broad head and a long tail (Fig. 12.26b). The tail of the comma consists of a chain of thunderstorms, which tend to drift along the tail toward the comma head. Under certain conditions, each thunderstorm pours rain over the same area of land as it drifts along the tail, triggering both flash floods and slow-onset floods.

When mid-latitude cyclones form over the western Pacific Ocean and then move east, winds flowing along the very long tails of such storms transport massive amounts of moisture from the ocean to the west coast of North America. When this moisture reaches the coast, strong storms develop and dump rain on the land, causing widespread flooding. Because of the vast quantities of moisture flowing along the tail of a Pacific mid-latitude cyclone, meteorologists refer to such tails as *atmospheric rivers* (Fig. 12.26c).

Hurricanes, the world's largest and most severe storms, form over tropical oceans (see Chapter 18), where they grow into huge spirals of storm clouds accompanied by intense winds. Hurricanes "feed" on moisture evaporating from warm ocean water. This moisture condenses in the hurricane's clouds and then precipitates back to the Earth's surface. The rain from a hurricane that makes landfall drenches the land and causes flooding. Even the remnants of a hurricane that has moved far inland can retain enough moisture to produce flood-triggering rains.

Living with Floods

Mark Twain once wrote of the Mississippi that we "cannot tame that lawless stream, cannot curb it or confine it, cannot say to it, 'go here or go there,' and make it obey." Was Twain right? Since ancient times, people have attempted to control the courses of rivers to prevent flooding. In the 20th century, such *flood control* intensified as populations living along rivers increased. For example, following the disastrous flood in 1927, the US Army Corps of Engineers began a systematic program to decrease flooding by the Mississippi River. First, they built about 300 dams along the river's tributaries to store excess runoff until it could

FIGURE 12.26 Causes of flooding.

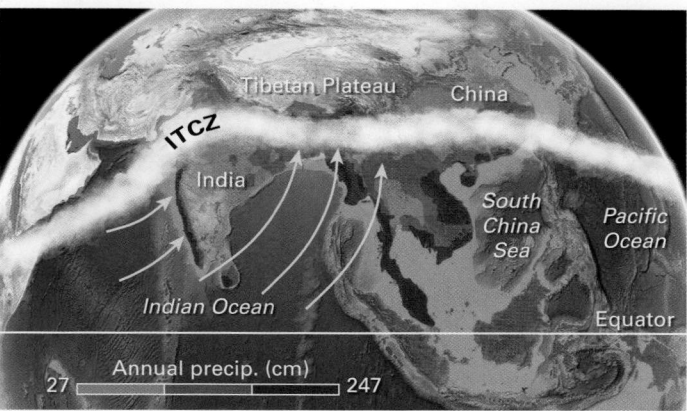

(a) During the wet season in South and Southeast Asia (June and July), the belt of thunderstorms above the ITCZ crosses the region, causing annual flooding.

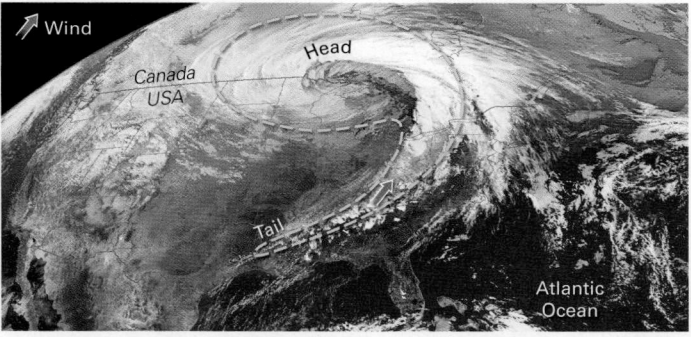

(b) This satellite photo shows a mid-latitude cyclone that covers much of southern Canada and the eastern United States. The yellow dashed line highlights the comma-like shape of the cyclone. Thunderstorms are moving from southwest to northeast along the tail.

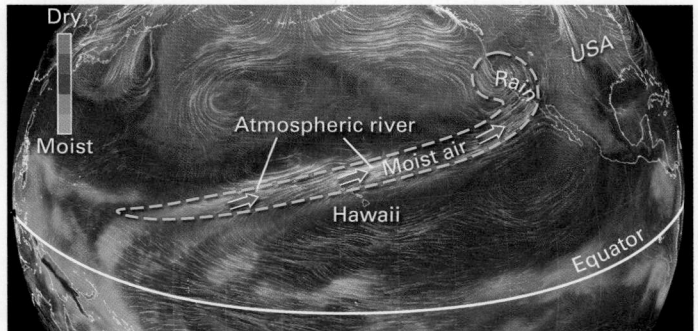

(c) A map showing the direction of winds (the very thin white lines) and the precipitable moisture content of the atmosphere (colors) in mid-March 2023. The band of relatively moist air along the tail of a mid-latitude cyclone (highlighted by yellow dashed lines) is an atmospheric river carrying moisture from the western Pacific to California, where it will drop heavy rains. Blue arrows indicate overall air-flow direction within the atmospheric river.

FIGURE 12.27 Flood-control strategies.

(a) Artificial levees have been built to protect the town of Galena, Illinois.

(b) A concrete floodwall at Cape Girardeau, Missouri. When floods threaten, a crane drops a gate into the slot to keep out the Mississippi River.

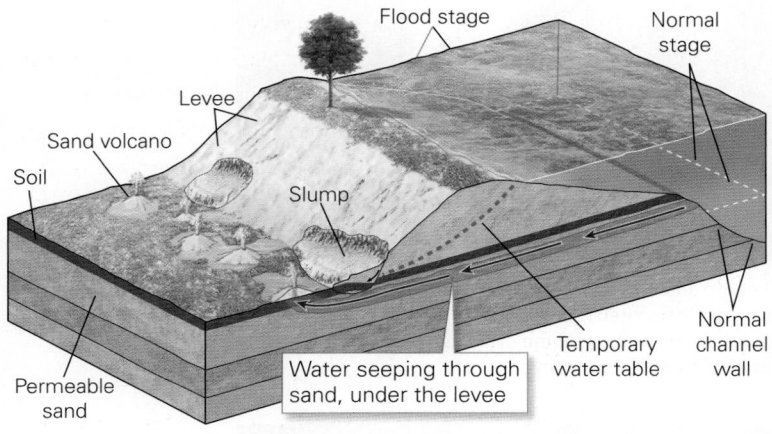

(c) Artificial levees may be undermined when infiltrated by water.

(d) At a levee breach, water flows from the river to the floodplain.

Did you ever wonder . . .

what newscasters mean by a "100-year flood"?

be released safely. Second, they built *artificial levees* of sand and mud, as well as *floodwalls* of concrete (Fig. 12.27a, b). These structures are designed to increase the channel's volume and to isolate portions of the floodplain from flooding.

Although the Corps' strategy worked for floods up to a certain size, it was insufficient to handle the 1993, 2011, and 2019 Mississippi floods, when reservoirs filled to capacity and additional runoff headed downstream. The river rose until it spilled over the tops of some levees and undermined others. *Undermining* occurs when rising water levels increase the water pressure at the base of the river channel, forcing water through sand under the levee (Fig. 12.27c). Water then spurts out of the ground on the floodplain side of the levee, thereby washing away

the levee's support until the levee collapses (Fig. 12.27d). Newer flood-control approaches involve both restoration of wetland areas along rivers—for wetlands can absorb significant quantities of floodwater—and designation of *floodways*, regions in which construction has been banned so that floodwaters have room to spread out without causing expensive damage.

Because flooding is a danger to society, it's important to be able to estimate how susceptible an area might be to flooding. To make this possible, researchers keep records of the variation in a stream's discharge, as well as when and by how much a stream rises above flood stage. Such data help them calculate the average frequency of floods of a given size (Box 12.2).

BOX 12.2 ▶ **Putting Earth Science to Use**

Evaluating flooding hazard

When making decisions about investing in flood-control measures, mortgages, or insurance, you need a basis for defining the hazard or risk posed in an area by flooding. Hydrologists characterize the risk of flooding in two ways: (1) The **annual probability** of, say, a flood that crests 2 m above flood stage at a specified locality indicates the likelihood that discharge sufficient to cause such a flood will occur during the course of any given year. For example, if we say that such a flood has an annual probability of 1%, we mean that there is a 1-in-100 chance that such a flood will happen during the next 12 months. (2) The **recurrence interval** of a flood indicates the average number of years between successive floods of a given size (**Fig. Bx12.2a**). For example, if a flood with a discharge sufficient to cause a stream to rise to a level of 1 m above flood stage happens, on average, once in a century, then it has a recurrence interval of 100 years and can be called a *100-year flood*. Annual probability and recurrence interval are related by the following equation:

$$\text{Annual probability} = 1 \div \text{recurrence interval.}$$

For example, the annual probability of a 50-year flood is $\frac{1}{50}$, or 2%. The recurrence interval for a flood with a larger discharge is longer than that for a smaller flood because large floods happen less frequently than small ones.

Unfortunately, people commonly misunderstand the meaning of a recurrence interval, believing incorrectly that they do not face a flooding hazard for another century if they buy a home within an area where a 100-year flood has just occurred. Their misplaced confidence comes from making the incorrect assumption that because such flooding just happened, it won't happen again until another 100 years have passed. It's important to keep in mind that flooding does not take place periodically. Two 100-year floods can occur in consecutive years, or even in the same year. Alternatively, the interval between such floods could be, say, 210 years, for the recurrence interval only indicates average timing.

By knowing the stage of a flood of a specified annual probability, and by knowing the shape of the stream channel and the elevation of the land bordering the stream, hydrologists can predict the extent of land that will be submerged by a flood that reaches a given stage. Such data, in turn, permit them to produce *flood-hazard maps* (**Fig. Bx12.2b, c**). It's wise to avoid buying a home in a location that has a high probability of flooding!

> **FIGURE Bx12.2** The relationship between flood size and annual probability.

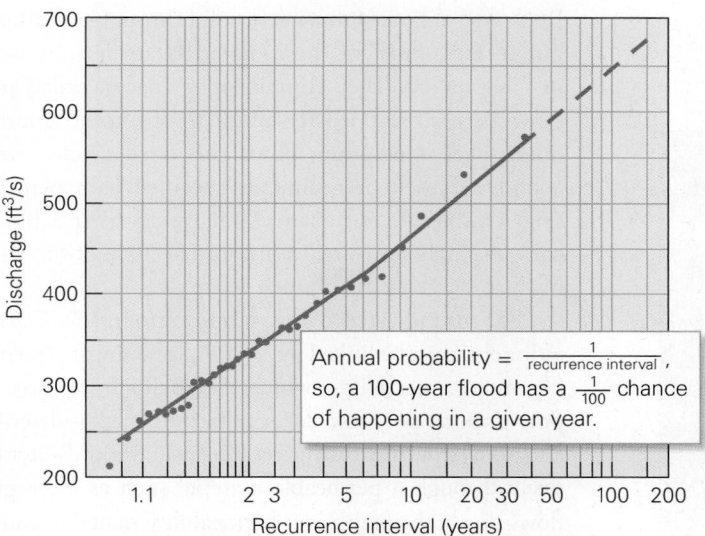

In the figure: Annual probability = $\frac{1}{\text{recurrence interval}}$, so, a 100-year flood has a $\frac{1}{100}$ chance of happening in a given year.

(a) A flood-frequency graph shows the relationship between the recurrence interval and the discharge for an idealized river.

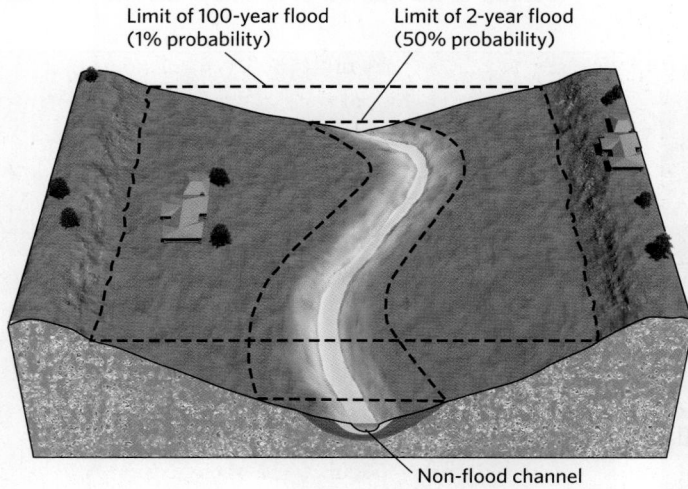

Limit of 100-year flood (1% probability)

Limit of 2-year flood (50% probability)

Non-flood channel

(b) 100-year floods inundate a larger area than 2-year floods, and they occur less frequently.

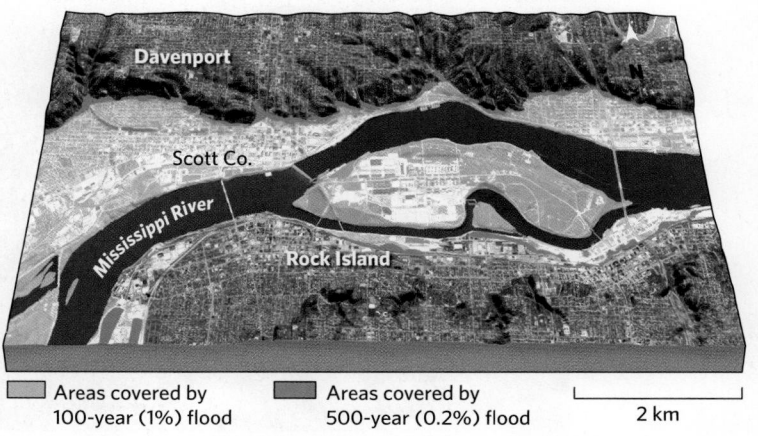

Davenport

Scott Co.

Mississippi River

Rock Island

Areas covered by 100-year (1%) flood

Areas covered by 500-year (0.2%) flood

2 km

(c) A flood-hazard map shows areas likely to be flooded. Here, near Rock Island, Illinois, even large floods are confined to the floodplain.

Take-home message...

🏠 Slow-onset floods submerge broad areas for days or weeks at a time, whereas flash floods are sudden and short lived. Engineers can build reservoirs, levees, and floodwalls to limit casualties and damage due to flooding. The probability of flooding can be predicted.

Quick Questions —

• What does it mean to say, "The river will reach flood stage in an hour."

• Will a slow-onset flood happen at the headwaters of a drainage network?

• Does a 100-year flood happen every 100 years?

12.4 Groundwater: A Hidden Resource

When it rains, some of the water that falls on the land sinks or percolates down into the ground, a process called **infiltration**. Of the water that infiltrates, some remains in the soil. This *soil water* may later evaporate, be absorbed by plant roots, rise back to the ground surface, or seep into streams. The remainder sinks deeper into sediment and rock, where, along with water trapped at the time the sediment accumulated or the rock formed, it makes up **groundwater**. Groundwater exists only in the Earth's upper crust, for below depths of 20–30 km (12–19 mi), metamorphic reactions incorporate water, plastic flow of rock destroys space for water, or water no longer can exist in liquid form.

Porosity and Permeability

Rock and sediment may seem solid through and through, but in fact, most of these materials are not, as we saw in Chapter 10. They contain open spaces, called **pores**, between grains. **Porosity** refers to the total volume of open space—including pores and open cracks—within a body of rock or sediment, specified as a percentage (Fig. 12.28a, b). For example, if we say that a block of rock has 30% porosity, we mean that 30% of the volume of the block consists of open space.

For groundwater to flow, pores and/or cracks in rock or sediment must be linked by *conduits* (openings). The resulting interconnectivity—which allows fluids to pass from pore to pore, pore to crack, or crack to crack—determines a material's **permeability (Fig. 12.28c)**. Groundwater flows easily through a permeable material such as loose gravel, flows slowly through a low-permeability material, and can't flow at all through an impermeable material. The permeability of a material depends on the number of conduits (because as the number of conduits increases, permeability

FIGURE 12.28 Porosity and permeability.

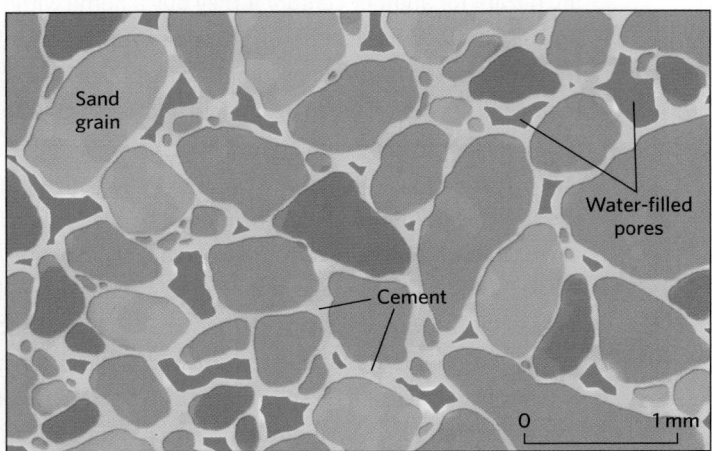

(a) Isolated pores occur in the spaces between grains in a sandstone. Water or air can fill these pores.

These fractures have been enlarged by dissolution.

(b) This limestone outcrop on the coast of Ireland contains abundant joints that provide porosity.

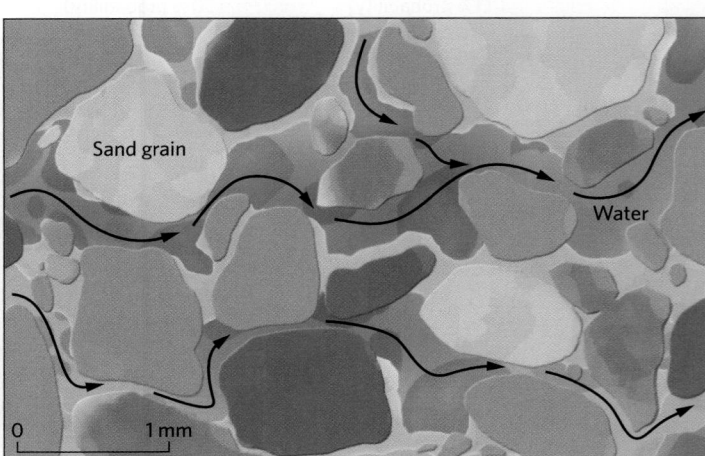

(c) Permeability indicates the degree to which open spaces in a rock or sediment are linked to form conduits through which water can move.

BOX 12.3 ▶ **How can I explain . . .**

Porosity and permeability

What are we learning?
- The difference between porosity and permeability.
- Factors that affect porosity and permeability.

What you need:
- Two large glass beakers, a screen made of soft plastic mesh, and two strong rubber bands
- A supply of pebbles and sand
- A measuring cup
- A stopwatch

Instructions:
- Determine the volume of the beakers (by filling them with water from the measuring cup).
- Drain the beakers, then fill one with pebbles and the other with sand.

- Using the measuring cup, add water to each beaker until the water table in the beaker reaches the top. Compare the volume of water needed to fill the pebble-filled beaker with the volume needed to fill the sand-filled beaker.
- Calculate the porosity of each material. Determine which material has higher porosity, and suggest why.
- Place a screen over the top of each beaker and secure it with a rubber band.
- Tilt the sand-filled beaker so it's horizontal, and record the time it takes for the water to spill out. Repeat for the pebble-filled beaker. Determine which material has higher permeability, and suggest why.

What did we see?
In the ground, water can fill pores between solid grains, but not all materials have the same porosity, nor do all materials have the same permeability. Permeability is determined by the size and continuity of conduits between grains.

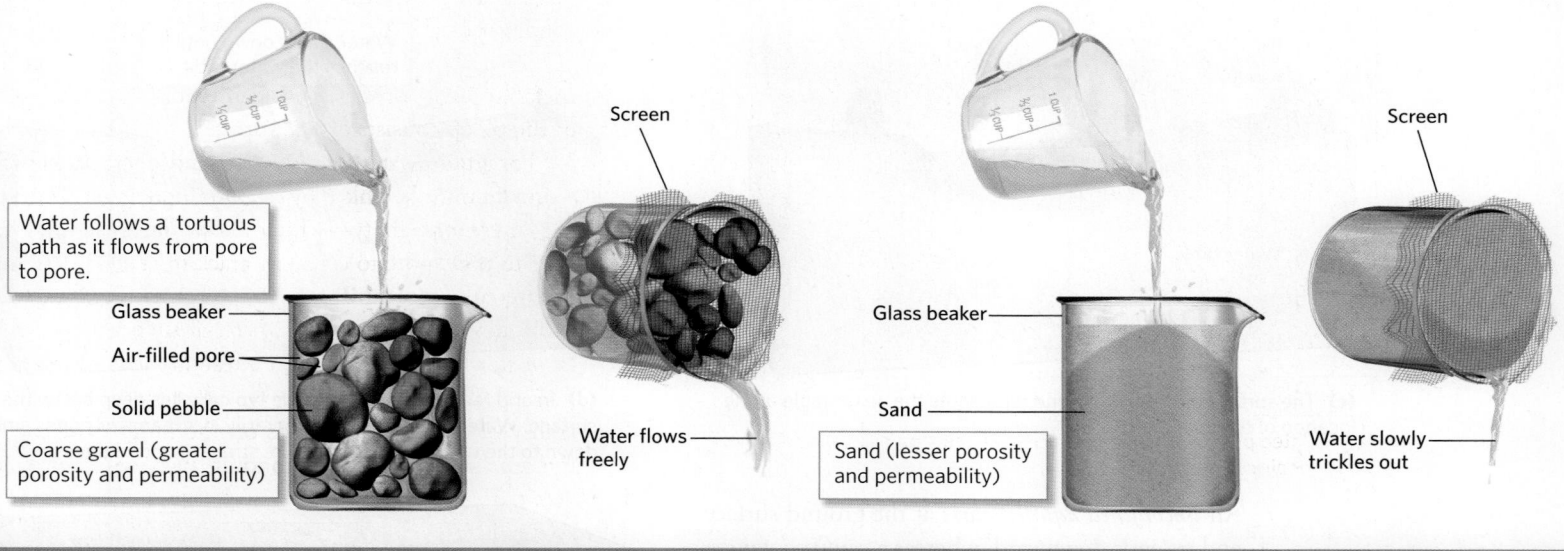

Water follows a tortuous path as it flows from pore to pore.

Screen

Glass beaker

Air-filled pore

Solid pebble

Coarse gravel (greater porosity and permeability)

Water flows freely

Glass beaker

Sand

Sand (lesser porosity and permeability)

Screen

Water slowly trickles out

increases), conduit size (because fluid travels faster through wider conduits than through narrower ones), and conduit straightness (because water flows faster through straight conduits than it does through crooked ones) **(Box 12.3)**. Note that factors controlling permeability resemble those controlling the ease with which traffic moves through a city: traffic can flow quickly through cities with many straight multilane boulevards, whereas it flows slowly through cities with only a few narrow, crooked streets. Significantly, a high-porosity rock or sediment does not necessarily have

high permeability. In fact, a material whose pores do not connect can have high porosity but low permeability.

Aquifers, Aquitards, and the Water Table

With the concepts of porosity and permeability in mind, we can distinguish between an **aquifer**, a body or layer of rock or sediment that has high porosity and permeability, so that it can hold and transmit water easily, and an **aquitard**, a body or layer of rock or sediment that has low permeability, regardless of porosity, so that it cannot.

FIGURE 12.29 Water underground: aquifers, aquitards, and the water table.

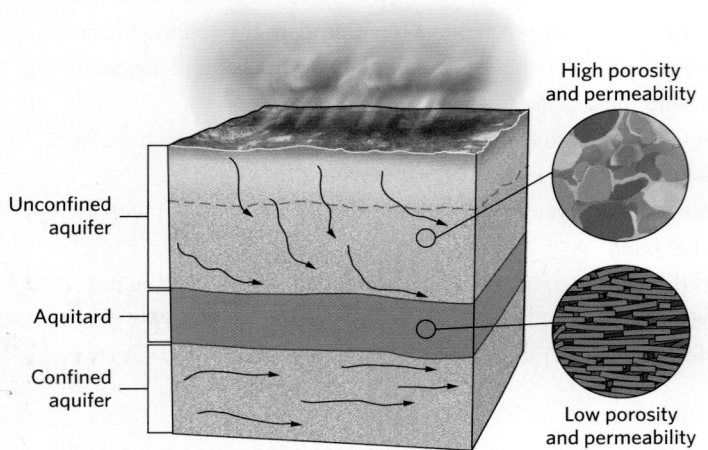

(a) An aquifer consists of high-porosity, high-permeability rock. An aquitard consists of low-permeability rock. Some aquifers are unconfined, and some are confined.

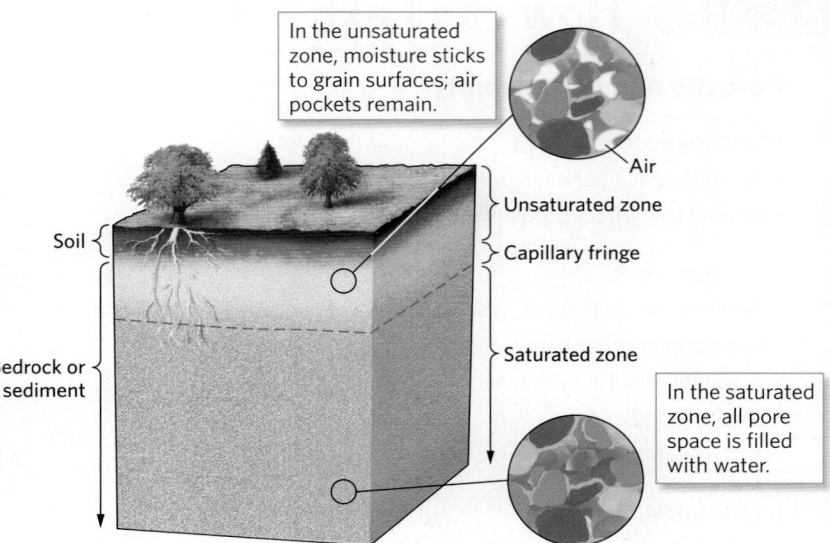

(b) The water table is the top of the groundwater reservoir. It separates the unsaturated zone above from the saturated zone below. The capillary fringe lies along the top surface of the water table.

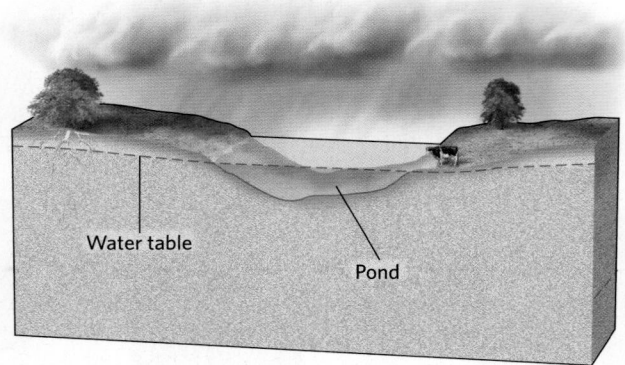

(c) The surface of water in a pond represents the water table at the location of the pond.

(d) In arid regions, the water table typically lies deep below the surface. Water that collects temporarily in streams or ponds sinks down to the water table, so that the stream or pond dries up.

An *unconfined aquifer* starts at the ground surface and extends downward, whereas a *confined aquifer* is separated from the ground surface by an aquitard (Fig. 12.29a).

Significantly, not all pores in an aquifer are filled completely with water. Rock or sediment containing air-filled pores lies above material containing water-filled pores. Hydrologists refer to the portion of the subsurface in which pores are partly or wholly filled with air as the *unsaturated zone*, and to the portion in which pores are completely filled with water as the *saturated zone*. The underground boundary between the unsaturated zone (above) and the saturated zone (below) defines the **water table** (Fig. 12.29b). By convention, the water table forms the top boundary of

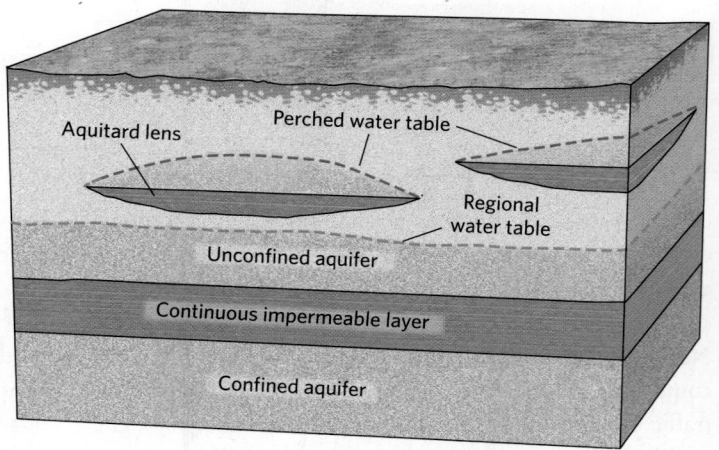

(e) A perched water table occurs where a mound of groundwater becomes trapped above an aquitard lens that lies above the regional water table.

the groundwater reservoir of the Earth System. Typically, some groundwater seeps up for a short distance from the water table due to capillary action (the attraction of water molecules to one another and to grain surfaces), producing a moist *capillary fringe* just above the water table.

The depth of the water table varies with location. In places where it lies at the ground surface, no unsaturated zone exists, and the ground is wet and soggy. Where the water table lies hidden below the ground surface, most of the ground surface remains dry, but depressions or channels where the ground surface lies below the water table fill with water and becomes lakes or streams—in such places, the surface of a lake or stream represents the water table elevation at that location (Fig. 12.29c). In humid regions, the water table generally lies within a few meters or tens of meters of the surface, whereas in arid regions, it may be hundreds of meters below the surface. Rainfall changes the water-table depth at a given locality, so the water table drops during the dry season and rises during the wet season. If the water table drops below the bed of a lake or stream, water infiltrates the bed and sinks down to the water table, so the lake or stream dries up unless replenished with new water (Fig. 12.29d). In some locations, lens-shaped horizons of an aquitard (such as shale) lie within a thick aquifer. If a mound of groundwater accumulates above an aquitard lens so that it lies above the main water table, the top of the groundwater mound is a *perched water table* (Fig. 12.29e).

In hilly regions, the shape of the water table tends to mimic, in a subdued way, the shape of the overlying topography (Fig. 12.30a). This means that the water table lies at a higher elevation beneath hills than it does beneath valleys. This observation may seem surprising, for if you pour a bucket of water into a pond, the surface of the pond immediately adjusts to remain horizontal. But because groundwater moves so slowly through rock and sediment, it cannot quickly assume a horizontal surface. For example, when rain lands on a hill and infiltrates to the water table, the water table rises. Between rainfalls, the water table below the hill sinks, but it drops so slowly that another rainfall will probably make it rise again before it sinks very far.

Groundwater Flow

Above the water table, in the unsaturated zone, water simply percolates downward like the water passing through a drip coffee maker, for this water moves only in response to the downward pull of gravity. Below the water table, however, groundwater moves not only in response to gravity, but also in response to differences in pressure. In fact, pressure can cause groundwater to flow sideways or even upward. Therefore, to understand the nature of groundwater flow, we must first understand the origin of pressure in groundwater. For simplicity, we'll consider only the case of groundwater in an unconfined aquifer.

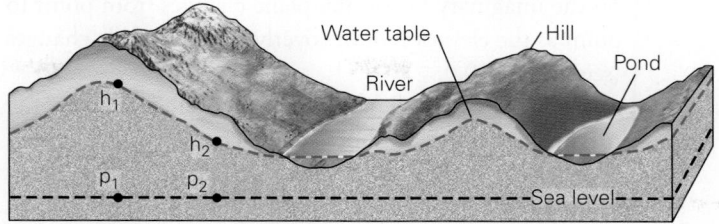

Point h_1 on the water table is higher than Point h_2 relative to a reference elevation (sea level). The pressure at p_1 is, therefore, higher than the pressure at p_2.

FIGURE 12.30 The flow of groundwater.

(a) The water table tends to follow the shape of the overlying topography. Where the water table is higher, the pressure at a reference level below the ground is greater. Since h_1, a point on the water table, is higher than h_2, the pressure at P_1 is greater than the pressure at P_2.

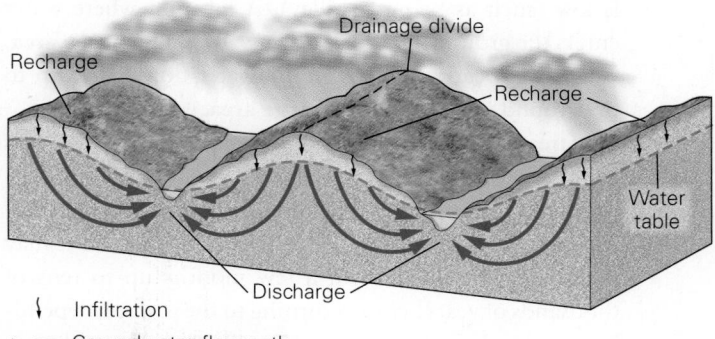

↓ Infiltration

↝ Groundwater flow path

(b) Groundwater flows from recharge areas to discharge areas. Typically, the flow follows curving, concave-up paths.

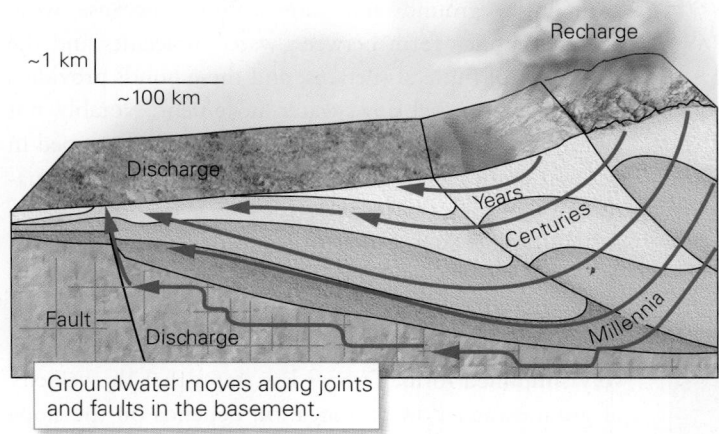

Groundwater moves along joints and faults in the basement.

(c) A large elevation difference in the water table may drive groundwater hundreds of kilometers, across regional sedimentary basins. Groundwater that follows deeper flow paths stays underground longer.

The pressure in groundwater at any specific point underground comes from the weight of all overlying water from that point up to the water table. (The weight of overlying rock does not contribute to the pressure exerted on groundwater because contact points between mineral grains bear the rock's weight.) As a result, a point at a greater depth below the water table experiences more pressure than does a point at a lesser depth. If the water table is horizontal, an imaginary plane at a given depth beneath the water table experiences the same pressure everywhere. But if the water table isn't horizontal, the pressure exerted on the imaginary horizontal plane changes from point to point as the elevation of the overlying water table changes (see Fig. 12.30a).

Both the elevation of a volume of groundwater and the pressure within that water provide energy that, if given a chance, drives groundwater flow. Taking these two energy sources into account, hydrologists calculate that regional groundwater flow typically follows curving, concave-up paths (Fig. 12.30b). These curved paths eventually take groundwater from regions where the water table is high (such as under a hill) to regions where the water table is low (such as below a valley). A location where water enters the ground and flows downward is a **recharge area**, whereas a location where groundwater flows back up to the ground surface is a **discharge area**.

Stream water flowing down a channel can reach speeds of 30 km/h (20 mph). In contrast, mobile groundwater (meaning groundwater residing in interconnected pores) moves at rates of 5–500 m/yr (15–1,500 ft/yr), so it may remain underground from a few months up to tens of thousands of years before returning to the surface, depending on the path that it takes (Fig. 12.30c). Groundwater moves slowly, in part, because it must follow a crooked network of tiny conduits, which make it travel a much greater distance than it would if it followed a straight path. Also, groundwater moves slowly because weak chemical bonds form between water molecules and the molecules of mineral surfaces, and these bonds provide a resistance force that slows water movement. Notably, not all groundwater is mobile; some has remained trapped in isolated pores of low-permeability rock since the formation of that rock.

What key factors determine the rate of groundwater flow at a location? This question intrigued a French researcher named Henry Darcy, and he answered it by developing a formula, now known as *Darcy's law*. In a very simplified form, Darcy's law states that the velocity of groundwater flow at a location depends on the slope of the water table and on the permeability of the material through which the groundwater flows. Consequently, groundwater moves faster through high-permeability rock or sediment than it does through low-permeability rock or sediment, and it moves faster in regions where the water table has a steep slope than it does in regions where the water table has a gentle slope.

Tapping Groundwater Supplies

If you have an opportunity to fly over an arid region in an airplane, look out the window and watch for circular fields of green crops (Fig. 12.31). In the center of each field, a pump sucks up groundwater to supply a sprinkler that pivots around the field to keep crops irrigated. In arid regions where surface water doesn't exist, or in temperate or humid regions where surface water has been polluted, groundwater may be the only source of clean freshwater. People obtain groundwater from *springs* and from *wells*.

Imagine an oasis of green palm trees surrounded by the dry sands of a desert. Where does the water that these plants need to survive come from? It flows from a **spring**, a natural outlet from which groundwater spills or seeps onto the ground surface or into a lake or river. Springs can provide freshwater for drinking or irrigation without the expense of drilling or digging, so many villages and towns worldwide have grown up adjacent to springs. Springs form in a variety of geologic settings (Fig. 12.32).

Springs don't exist everywhere, though. Millennia ago, farmers learned that they could obtain groundwater even where no springs exist by digging a **well**, a hole that reaches down to the water table. Today, people still construct wells by digging holes, but it's usually easier to bore a well by drilling. Hydrologists distinguish between two types of wells.

In an *ordinary well*, the base of the well lies below the water table, so water simply seeps from the aquifer into the well and fills it to the level of the water table (Fig. 12.33a). Drilling into rock or sediment in the unsaturated zone does not provide water, and it yields a *dry well*. To extract water from an ordinary well, you can either pull the water up in a bucket or you can pump the water out of the well. As long as the rate at which groundwater seeps into the well equals or exceeds the rate at which you remove the water, the level of the water table near the well remains about the same. But if you pump water out of the well too fast, the water table sinks around the well—a phenomenon called *drawdown*. When this happens, the water table forms a downward-pointing, cone-shaped surface, called a **cone of depression**, surrounding the well (Fig. 12.33b). Drawdown by a deep well may cause nearby shallower wells to dry up (Fig. 12.33c, d).

An **artesian well**, named for the province of Artois in France, penetrates a confined aquifer in which pressure pushes water up to a level above the top of the aquifer. Artesian wells exist where the recharge area for the confined aquifer lies at a higher elevation than the site of the well, so that if the aquifer were not confined, the water table would be higher (Fig. 12.34a). If the level to which the water table would rise lies below the ground surface, the well is a *nonflowing artesian well*. But if the level lies above the ground surface, the well is a *flowing artesian well*. The water from a flowing artesian well actively gurgles or fountains out of the ground (Fig. 12.34b).

Thousands have lived without love, not one without water.

—*W. H. Auden (English-American Poet, 1907–1973)*

FIGURE 12.31 Irrigated circular fields in eastern Colorado, as seen from a satellite.

1 km

FIGURE 12.32 Springs may form in a variety of geologic settings.

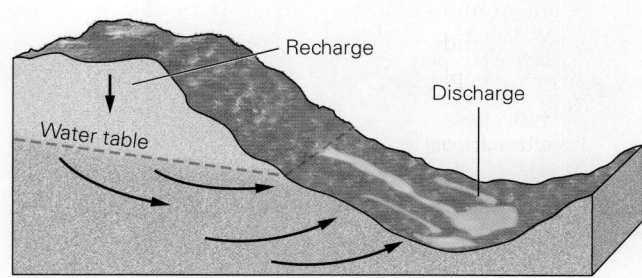

(a) Groundwater reaches the ground surface in a discharge area.

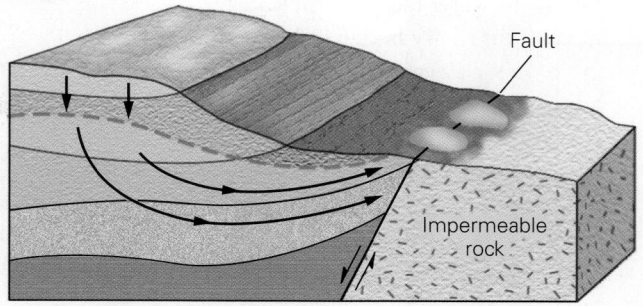

(b) Where groundwater reaches an impermeable barrier, it rises.

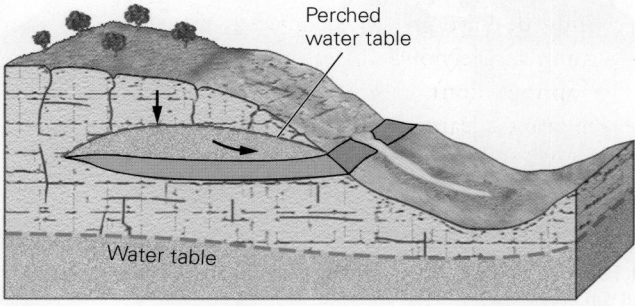

(c) Groundwater seeps from the ground where a perched water table intersects a slope.

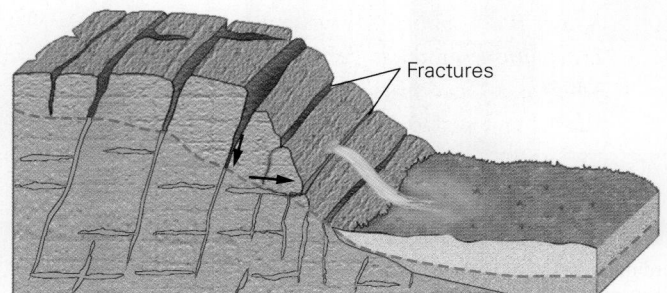

(d) A network of interconnected fractures channels water to the surface of a hill.

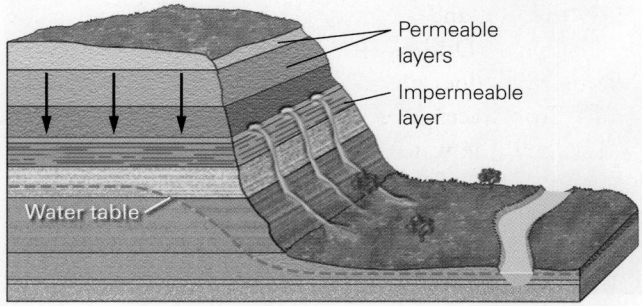

(e) Groundwater seeps out of a cliff face at the top of an impermeable bed.

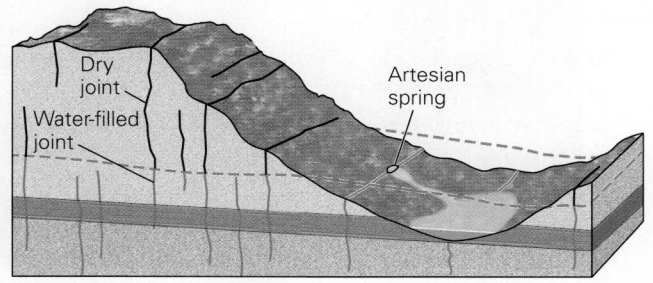

(f) Where water under pressure lies below an aquitard, a crack provides a conduit for an artesian spring.

Hot Springs and Geysers

A natural spring producing water at a temperature above about 30°C (86°F) is called a **hot spring**. Some hot springs occur where groundwater rises relatively rapidly from several kilometers below the surface—depths at which rock is warm because of the normal *geothermal gradient* (the increase in temperature with increasing depth in the Earth). Hot groundwater coming from such springs must have risen rapidly; otherwise, the water would have cooled down before reaching the ground surface. Such flow can take place where fractures provide a high-permeability pathway. Hot springs may also develop in *geothermal areas*, regions of igneous activity where magma-heated rock lies relatively close to the Earth's surface and warms up water that has infiltrated only a short

distance downward **(Fig. 12.35a)**. Many hot springs have become popular spas **(Fig. 12.35b)**.

Numerous distinctive features form in association with hot springs. For example, when hot water that contains a high concentration of dissolved ions spills out of hot springs and then cools, minerals precipitate, forming colorful mounds or terraces of travertine and other chemical sedimentary rocks **(Fig. 12.35c)**. Natural pools fed by mineralized hot springs may harbor brightly colored thermophilic (heat-loving) bacteria and archaea **(Fig. 12.35d)**. In places where the hot water rises into volcanic ash, a viscous slurry forms and fills bubbling *mudpots*.

In some geothermal areas, scalding water spurts out of hot springs to form **geysers**, fountains of steam and

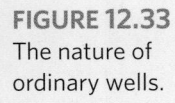

FIGURE 12.33
The nature of
ordinary wells.

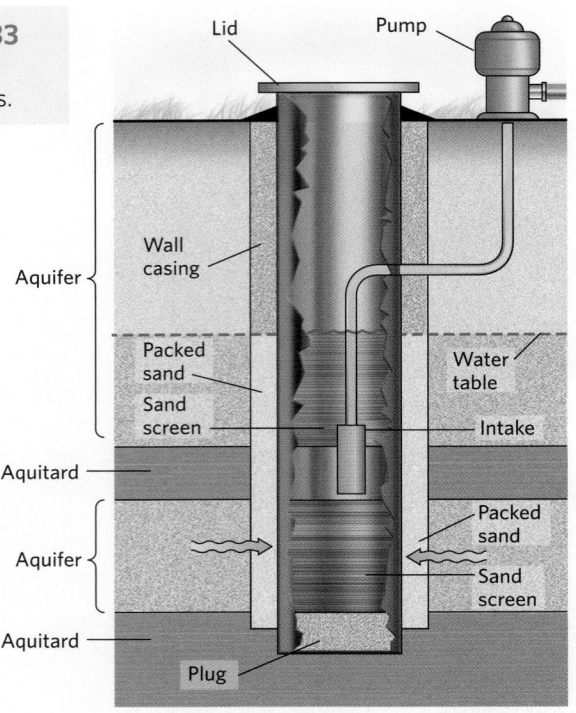

Lid Pump

Wall
casing

Aquifer

Packed
sand

Water
table

Sand
screen

Intake

Aquitard

Aquifer

Packed
sand

Sand
screen

Aquitard

Plug

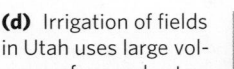

(a) A modern ordinary well sucks up water with an electric pump.
The packed sand surrounding the well filters the water that enters.

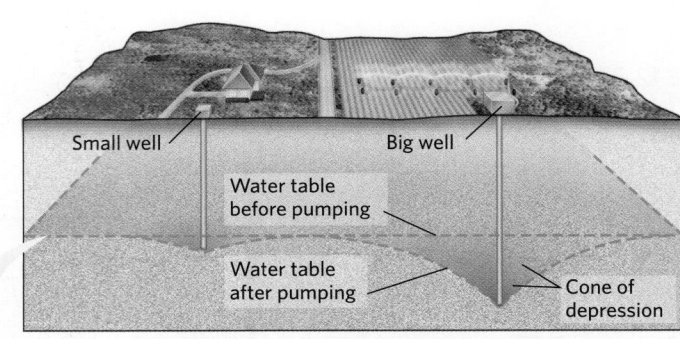

Small well Big well

Water table
before pumping

Water table
after pumping

Cone of
depression

(b) Pumping water from a well faster than groundwater flow can refill
it forms a cone of depression in the water table.

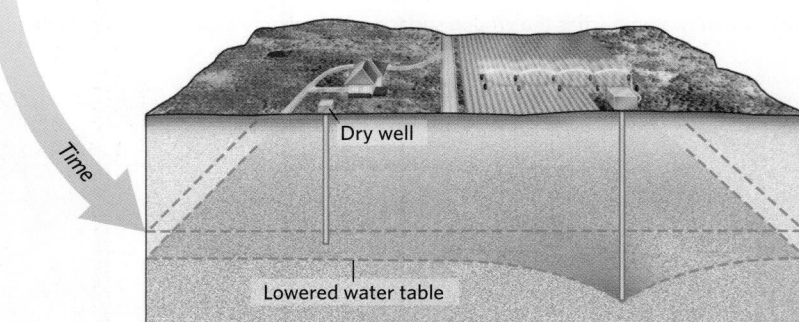

Time

Dry well

Lowered water table

(c) Pumping by the deeper well may be enough to make the shallower
well run dry.

(d) Irrigation of fields
in Utah uses large vol-
umes of groundwater
and causes drawdown.

FIGURE 12.34 The nature of artesian wells.

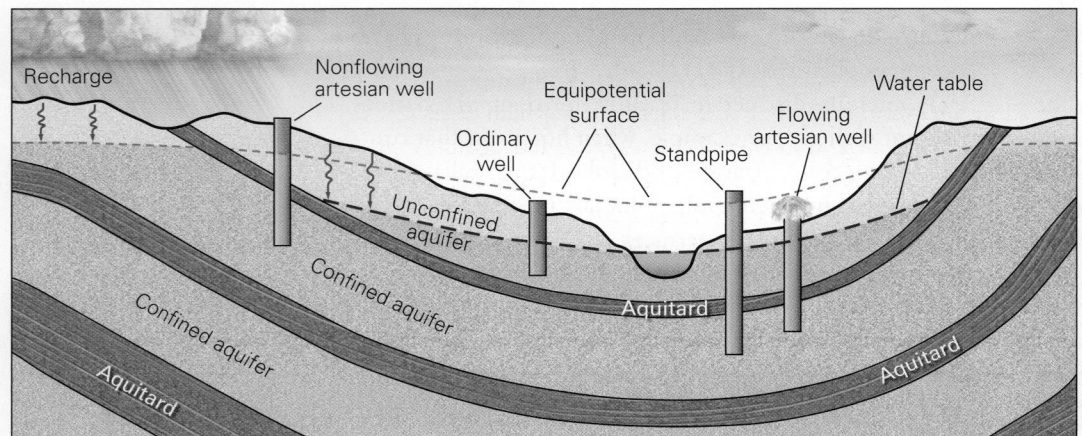

Recharge

Nonflowing
artesian well

Equipotential
surface

Water table

Ordinary
well

Standpipe

Flowing
artesian well

Unconfined
aquifer

Confined aquifer

Aquitard

Confined aquifer

Aquitard

Aquitard

(a) The configuration of a regional artesian system. Water rises above the confined aquifer because the water is
under pressure. The "equipotential surface" represents the level to which groundwater would rise if there were no
aquitards to inhibit flow.

(b) A flowing artesian well in Wisconsin. Groundwater
rises from a confined aquifer into the corrugated pipe
without the need of pumping.

FIGURE 12.35 Hot springs in areas of igneous activity.

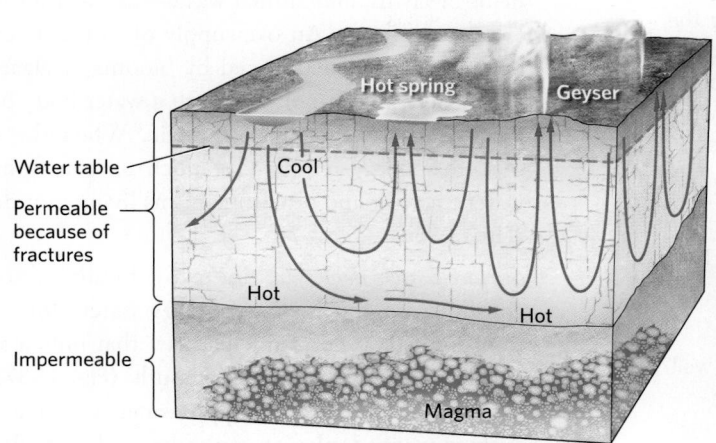

Water table

Permeable because of fractures

Impermeable

(a) Hot springs occur where groundwater, heated at depth by igneous activity, rises to the ground surface.

(b) Hot springs in Iceland, warmed by magma below, attract tourists from around the world.

(c) Terraces of minerals precipitated at a hot spring in Pamukkale, Turkey.

(d) Colorful bacteria- and archaea-laden pools at Yellowstone National Park.

(e) The Old Faithful geyser in Yellowstone National Park erupts predictably.

hot water that erupt episodically **(Fig. 12.35e)**. Geysers form when groundwater seeps into irregular fractures in hot rock. Under the elevated pressure underground, the groundwater can become *superheated*, meaning that its temperature exceeds the temperature at which water boils at the Earth's surface. When superheated groundwater moves upward through a fracture to shallower depths, the pressure acting on it decreases until, eventually, some of it transforms into steam. The resulting expansion pushes some overlying water out of the geyser's vent, causing the pressure in the superheated groundwater to suddenly decrease even more. This new, rapid drop in pressure causes more superheated groundwater to vaporize instantly, forming steam that pushes upward, ejecting all the water above it out of the geyser's vent. Once the vent empties, the eruption ceases. The fracture then fills, and the eruptive cycle begins again.

Take-home message...

🏠 Groundwater resides in pores and cracks in an aquifer, a body of rock or sediment with high porosity and permeability. The boundary between the unsaturated zone and the saturated zone of an aquifer is the water table. Gravity and pressure cause groundwater to flow slowly from recharge areas to discharge areas. Groundwater can be obtained from springs and wells.

Quick Questions

- Does groundwater exist in the Earth's mantle? Explain your answer.
- Is the water table always horizontal?
- What is the difference between a spring and a well?

12.5 Threats to Freshwater Supplies

Surface-Water Challenges

The lands bordering streams and lakes have long been desirable places for people to settle, for these water bodies provide drinking and irrigation water, food, avenues for transportation, sources of power, water for industry, and sites for recreational activities. In recent centuries, unfortunately, the sustainability of clean surface-water supplies has come under threat, for many reasons:

- *Pollution:* Communities dump sewage, garbage, street runoff, spilled hydrocarbons, toxic industrial chemicals, fertilizer, and animal waste into streams (Fig. 12.36a). These pollutants poison aquatic life and make water unsafe for people to drink.

- *Eutrophication:* Clear, well-oxygenated freshwater hosts an amazing variety of life. But an influx of nutrients—from sewage, fertilizer-rich runoff from fields or lawns, and animal waste—can disrupt freshwater ecosystems. An oversupply of nutrients causes **eutrophication**, manifested by blooms of algae and cyanobacteria that not only turn a water body bright green but can also make it toxic. When the algae die, their decay removes so much oxygen from the water that fish and other organisms living in it die off (Fig. 12.36b).

- *Surface-water depletion:* In many localities, society consumes so much of a stream's water (for irrigation, industry, and municipal use) that only a saline trickle reaches the stream's mouth (Fig. 12.37). In locations where streams supply water to a lake, such **surface-water depletion** can cause the lake to dry up.

- *Dam construction:* Dam construction impounds a reservoir. The reservoir may flood upstream towns and forests as well as scenic canyons and rapids. The reservoir also traps sediments and nutrients carried by the stream, preventing them from reaching downstream floodplains and deltas that host farmland and wetlands.

- *Changes in the land surface:* Urbanization covers fields and forests with concrete and asphalt. These impermeable materials decrease rainfall infiltration and therefore increase runoff, a change that can contribute to flooding. Conversion of forests and fields into farmland increases the amount of sediment in streams because farmland hosts relatively little plant cover and may be barren for much of the year, so that overland flow picks up sediment and carries it into streams.

FIGURE 12.36 Sustainability challenges for surface-water supplies.

(a) Large quantities of garbage and various pollutants end up in rivers.

(b) An algal bloom produces green slime on the surface of a lake that has undergone eutrophication.

(a) In the Grand Canyon, the Colorado River is full of water.

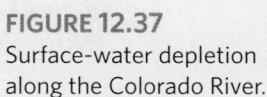

(b) Hoover Dam traps water in Lake Mead. The lake level at the time of this photo was low, so a white stripe of bleached rock is visible along its banks. Much of the water is used by Las Vegas.

FIGURE 12.37
Surface-water depletion along the Colorado River.

(c) A canal brings water from the Colorado River to Phoenix, Arizona, a desert city.

(d) Near its mouth, in Mexico, the river is a saline trickle.

Groundwater Challenges

MINERALIZED AND SALINE GROUNDWATER. Commonly, you can drink groundwater directly from a spring or well, for rocks and sediment serve as natural filters capable of trapping suspended solids, and most well or spring water from relatively shallow aquifers contains low concentrations of dissolved chemicals and salts. In fact, commercial distribution of bottled groundwater (labeled "spring water") has become a major business worldwide. However, dissolved salts and minerals make some natural groundwater, particularly deeper groundwater, unusable. In many locations, fresh groundwater overlies denser saline groundwater. If wells or springs tap into the saline groundwater, they yield water that cannot be used for drinking or irrigation.

Other natural chemicals in groundwater can also make a given supply unusable unless the water is treated. **Hard water**, for example, contains significant concentrations of dissolved calcium, magnesium, and bicarbonate ions, which enter solution when groundwater passes through carbonate rocks. These chemicals can precipitate to form *scale* (a buildup of minerals) that clogs pipes. Similarly, groundwater that has passed through iron-bearing rocks may contain dissolved iron oxide that precipitates to form rust stains. Water softeners and other treatments, fortunately, can remove dissolved chemicals from hard water. Groundwater that passes through sediment containing arsenic-bearing minerals can dissolve the arsenic; such groundwater, in effect, becomes a dilute poison. In Southeast Asia, a significant percentage of wells, drilled so that people could avoid drinking polluted surface water, unfortunately contain dangerous levels of arsenic.

GROUNDWATER CONTAMINATION. In recent decades, increasing amounts of contaminants have been introduced into aquifers by human activities **(Fig. 12.38)**. The underground cloud of contaminated groundwater that flows away from a contaminant source constitutes a **contaminant plume**. Fortunately, in some cases, natural processes can clean up groundwater contamination: clay binds to some contaminants, oxygen in the water reacts with and destroys others, and natural bacteria in the water metabolize still others. But where natural processes can't

Did you ever wonder . . .

where bottled "spring water" comes from?

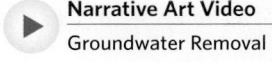

Narrative Art Video
Groundwater Removal

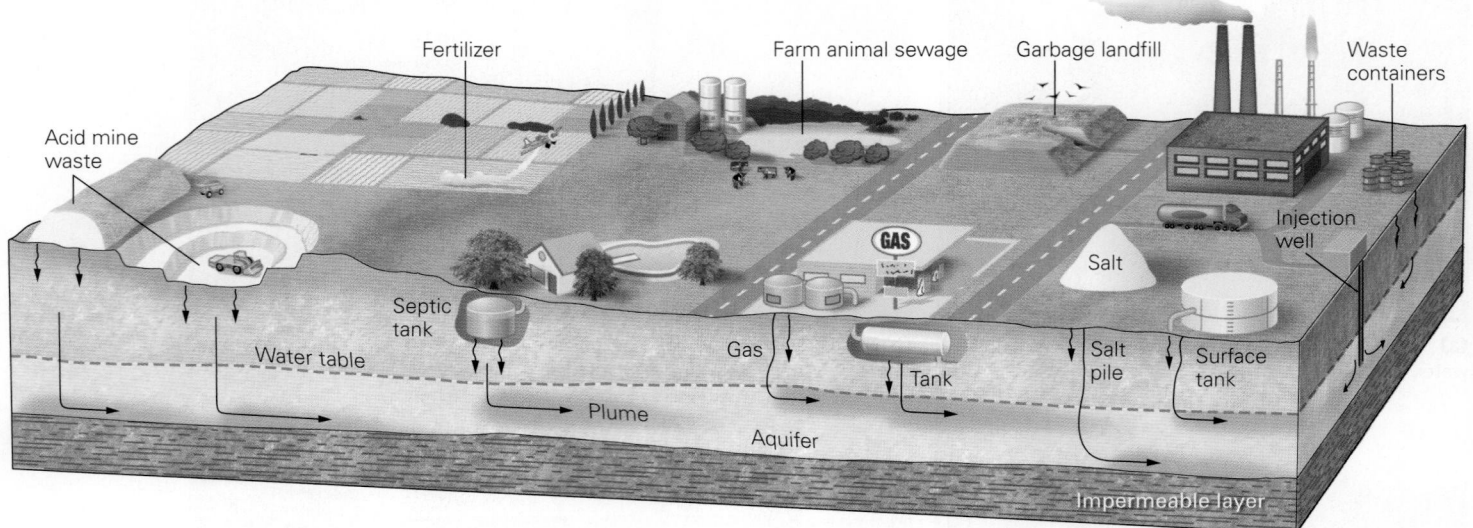

FIGURE 12.38 Contamination of groundwater comes from various sources. The pink areas in the aquifer are contaminant plumes, which move with the overall groundwater flow.

remove all the contamination from groundwater, several approaches can be taken to maintain safe groundwater supplies. First, the direction of plume flow can be tracked so that wells in the path of the contaminant plume can be shut down to prevent consumption of contaminated water. Second, if resources permit, the groundwater can be extracted and purified. If extraction isn't feasible, engineers may install an underground wall of chemicals that will react with the contaminants and cause them to precipitate into solids that won't migrate with groundwater, or they may inject oxygen and nutrients into the plume to foster growth of bacteria that can react with and break down the contaminants. In some cases, however, it's not possible to clean contaminated groundwater. If contaminated groundwater eventually discharges from a spring, it can contaminate surface water.

GROUNDWATER DEPLETION. Groundwater in many heavily populated regions of the world can be viewed as a nonrenewable resource (see Chapter 10). People are pumping water out of the ground so fast that natural recharge will not be able to replace it in a time frame of centuries to millennia. Many problems accompany such **groundwater depletion**:

- *Lowering of the water table:* When we extract groundwater from wells at a rate faster than it can be resupplied by infiltration, the water table drops, and springs, streams, lakes, swamps, and nearby wells dry up (Fig. 12.39a). Deeper wells are needed to access remaining, deeper groundwater, and these wells may, therefore, bring up unusable mineralized or saline groundwater.

- *Reversal of groundwater flow direction:* The cone of depression around a well creates a local slope in the water table large enough to reverse the flow direction of nearby groundwater (Fig. 12.39b). This change may cause contaminant plumes to head toward wells that were once clean.

- *Coastal saltwater intrusion:* In coastal areas, fresh groundwater lies in a layer above denser saltwater that has entered the aquifer from the adjacent ocean (Fig. 12.39c). Pumping water from a well too quickly can suck saltwater up into the well, causing **saltwater intrusion**.

- *Pore collapse and land subsidence:* Water can't be compressed, so groundwater filling pores holds sediment grains apart in the subsurface. Air, however, can be compressed, so removal of groundwater allows grains around pores to pack together more closely. Such **pore collapse** decreases the porosity and permeability of an aquifer, perhaps permanently. It also decreases the volume of the aquifer, causing land above the aquifer to undergo subsidence. Commonly, subsidence results in fissuring of the ground surface (Fig. 12.39d).

Because groundwater, which can be the primary water supply for a community, has become a scarce commodity in many locations, and because groundwater depletion can cause so many problems, rapidly expanding cities in deserts may be facing limits on their growth. For example, in 2023, officials in Phoenix, Arizona, suggested that the planning of new subdivisions be stopped, unless developers can demonstrate that a subdivision will have access to sufficient water.

FIGURE 12.39 Effects of groundwater depletion.

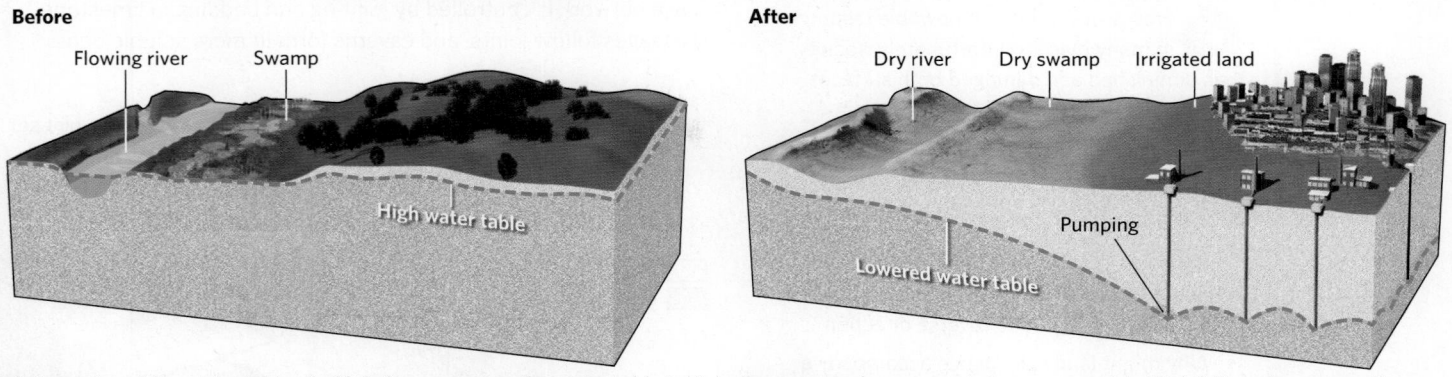

(a) Lowering of the water table. (Left) Before pumping, the water table is high. A swamp and permanent stream can exist. (Right) Pumping causes the water table to sink, so the swamp and stream dry up.

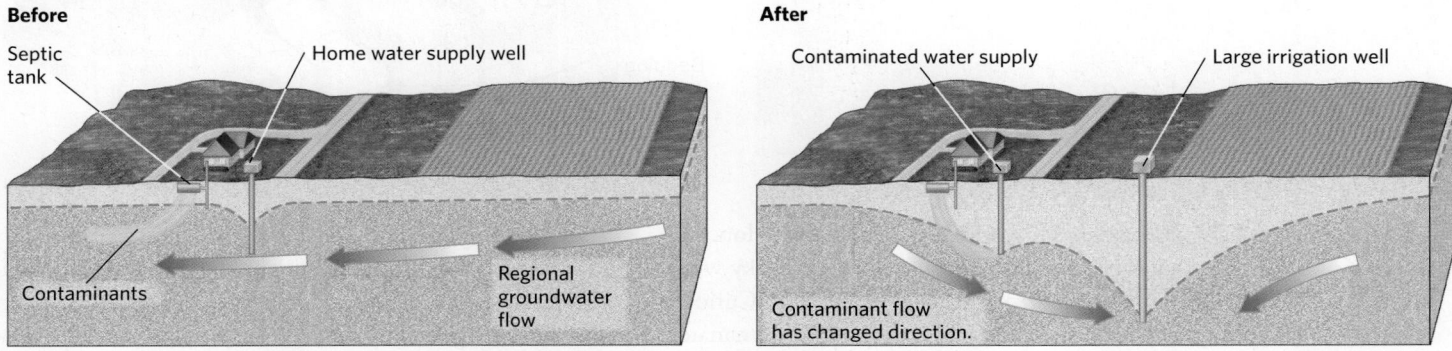

(b) Well contamination. (Left) Before pumping, a plume from a septic tank drifts with the regional groundwater flow. (Right) Pumping creates a cone of depression, causing the plume to flow into the home well.

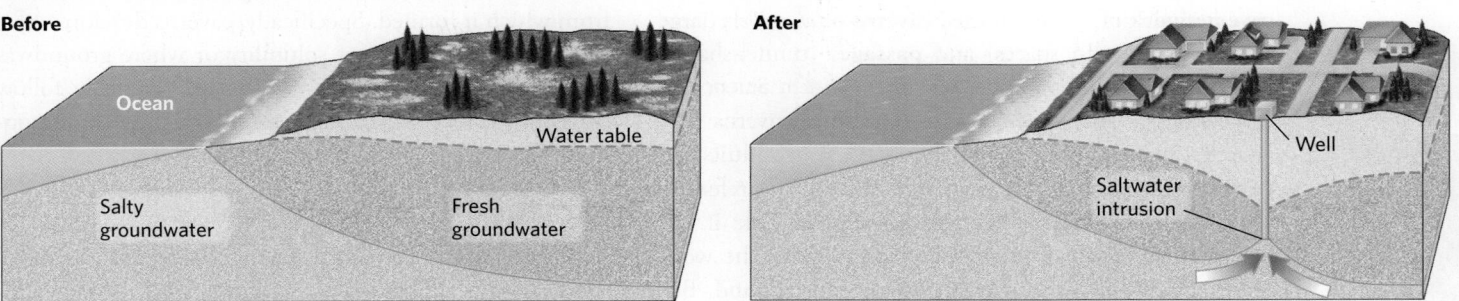

(c) Saltwater intrusion. (Left) Before pumping, fresh groundwater forms a lens above saltwater. (Right) Pumping the fresh groundwater too fast causes saltwater from below to be sucked into the well.

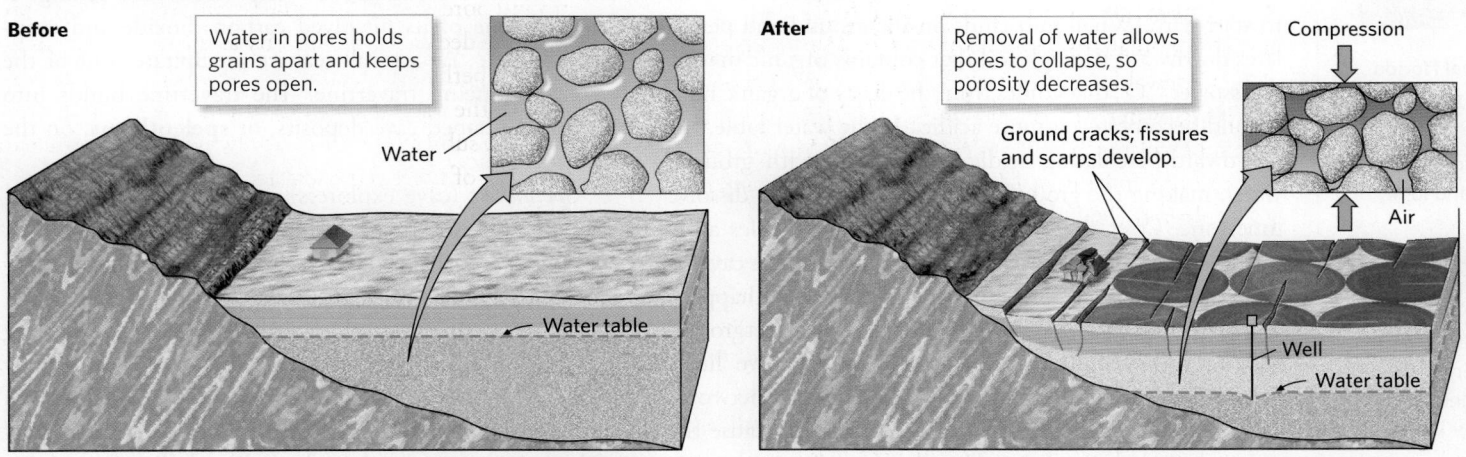

(d) Pore collapse and land subsidence. (Left) Before pumping, groundwater in pores holds sediment grains apart. (Right) Groundwater depletion causes pore collapse. As a result, the land surface sinks, causing ground fissures and foundation cracking.

Take-home message...

🏠 Freshwater is not a renewable resource in many places. Unfortunately, society has diminished and damaged both surface-water and groundwater supplies through overuse and contamination. Such depletion has many serious consequences.

Quick Questions ───────

- What causes eutrophication of a lake?
- How can pumping from a deep well cause a contaminant plume to reverse direction?
- Why might land subsidence accompany a lowering of the water table?

12.6 Caves and Karst Landscapes

Cave Networks

In 1799, as legend has it, a hunter by the name of Houchins was tracking a bear through the woods of Kentucky when he suddenly felt a draft of surprisingly cool air. Curious, he searched for the air's source, and found that it emanated from a dark portal in the hillslope above him. Houchins had discovered an entrance to Mammoth Cave, an immense **cave network** of interconnected **caverns** or *chambers* (large underground open spaces) and **passages** (tunnel-shaped or slot-shaped underground openings) (**Earth Science at a Glance**, pp. 462–463). In some locations, caverns host underground lakes, and passages serve as conduits for underground streams. (Note that the word *cave* refers to any space that naturally remains perpetually dark; it can be either under an overhang or underground. The word *cavern* specifically refers to open space underground. But in everyday English, the words are used interchangeably.)

Why do cave networks form? Rain tends to be acidic to start with. When rain lands on the ground and percolates downward through soil that contains organic matter, it dissolves CO_2 that comes from the decay of organic matter and becomes even more acidic. At the water table, the downward-percolating acidic water mixes with groundwater, making the groundwater acidic enough to dissolve limestone. Over time, flowing groundwater carries away the dissolved ions. Eventually, this process produces caverns and passages. (Acidification of groundwater also happens when groundwater reacts with hydrocarbons underground to yield dilute sulfuric acid, which can dissolve limestone.) Most of the dissolution that forms cave networks takes place at and just below the water table because the water there has not been diluted by mixing with non-acidic groundwater so its acidity remains high, because the downward-percolating rainwater has not become saturated with dissolved ions, and because groundwater flows faster near the water table than it does deeper down.

The configuration of a cave network reflects variations in the composition and the permeability of the rock from which it formed. Specifically, caverns develop where the limestone has greater solubility or where groundwater flow is fastest (**Fig. 12.40**). Passages typically follow pre-existing joints (natural cracks), which provide conduits along which groundwater flows more easily.

Cave Deposits (Speleothems)

When the water table drops below the floor of a cavern, the underground space fills with air. When downward-percolating water—containing dissolved ions of calcium and carbonate from above—drips from the ceiling, it releases some of its dissolved carbon dioxide and evaporates a little. As a result, calcite precipitates out of the water, producing travertine. The travertine builds into intricately shaped cave deposits, or **speleothems**, on the cavern's surface.

Spelunkers (cave explorers) and geologists use different names for different kinds of speleothems (**Fig. 12.41a**). Water dripping from the cavern's ceiling builds an icicle-like cone of calcite called a **stalactite**. Where the drips hit the floor, precipitated calcite builds an upward-pointing cone called a **stalagmite**. Eventually, a stalagmite may merge with an overlying stalactite to form a *column*. In some cases, groundwater flows along the surface of a wall and precipitates to produce drape-like sheets of travertine called *flowstone* (**Fig. 12.41b**).

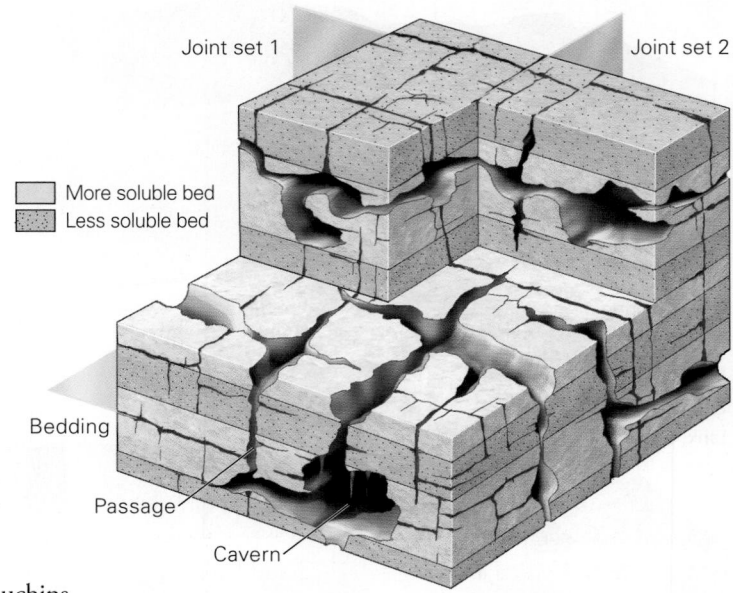

FIGURE 12.40 The configuration of caverns and passages in a cave network is controlled by jointing and bedding in limestone. Passages follow joints, and caverns form in more soluble beds.

Joint set 1　　Joint set 2

☐ More soluble bed
▧ Less soluble bed

Bedding

Passage

Cavern

Did you ever wonder...

why huge underground caves form?

See for yourself

Sinkholes in Central Florida

Latitude: 28°37′50.59″ N
Longitude: 81°23′12.60″ W

Zoom to an altitude of 20 km (12.5 mi) and look down.

You can see several sinkholes, ranging from 100 to 800 m (330–2,600 ft) across, that lie within suburban developments. The sinkholes have filled with water and are now lakes.

FIGURE 12.41 Speleothems in caverns.

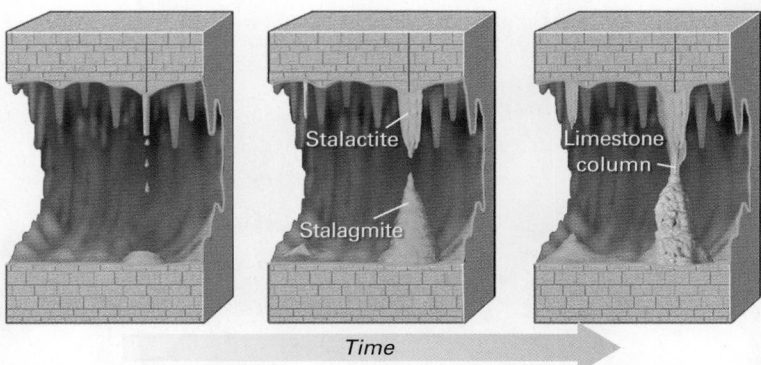

1 m

(a) A drip from the ceiling evolves into a stalactite. Precipitation from water that reaches the floor can form a stalagmite.

(b) Flowstone on the wall of a cave in Vietnam.

Formation of Karst Landscapes

When Mae Owens looked out her window on May 8, 1981, she discovered that the large tree in the backyard of her Winter Park, Florida, home had suddenly disappeared. She went outside to investigate, and found that more than the tree had vanished—her whole backyard had become a deep, gaping pit! The hole continued to grow for a few days until it finally swallowed Owens's house and six other buildings, as well as a municipal swimming pool, part of a road, and several Porsches in a car dealer's lot (Fig. 12.42a).

FIGURE 12.42 Development of sinkholes in central Florida.

Drained pool

(a) The Winter Park sinkhole, as seen from a helicopter.

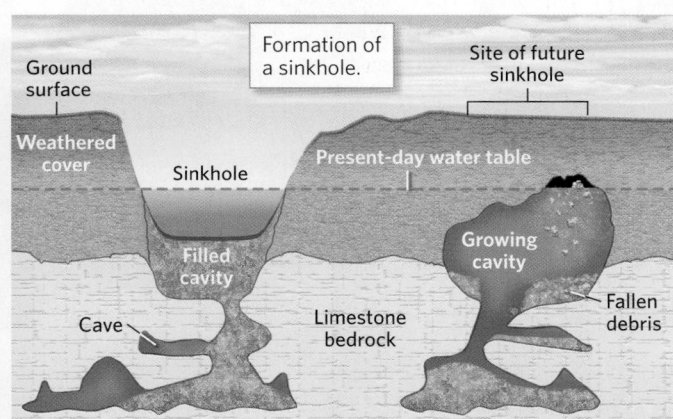

Formation of a sinkhole.

Ground surface

Site of future sinkhole

Weathered cover

Sinkhole

Present-day water table

Filled cavity

Growing cavity

Cave

Limestone bedrock

Fallen debris

(b) The weathered cover of a cavern's roof slowly washes into the cavern, forming a cavity that grows ever closer to the ground surface. When the roof of this cavity collapses, a sinkhole forms. (Not to scale.)

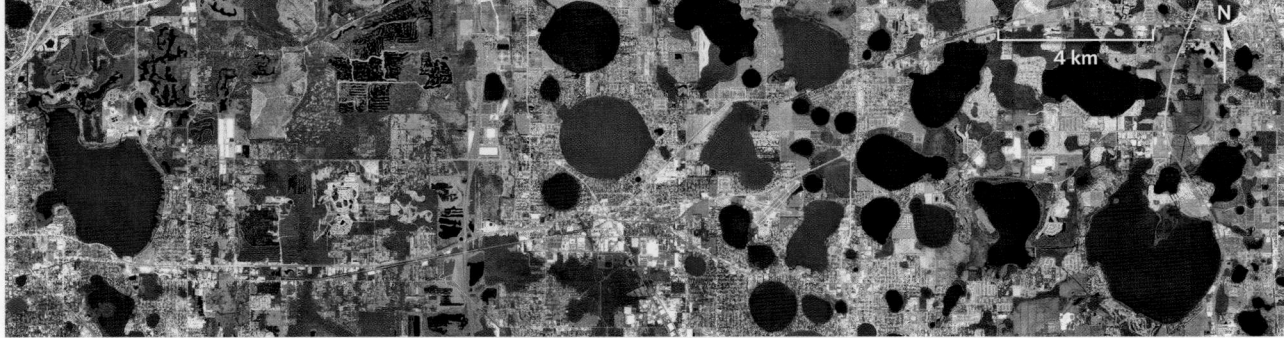

4 km

N

(c) An aerial view of Florida sinkholes that have become lakes.

Cave Networks and Karst Landscapes

Limestone dissolves in acidic water. Much of the water that falls to the ground as rain, or seeps through the ground as groundwater, is acidic, so limestone bedrock in many regions has undergone dissolution. Extensive dissolution produces cave networks, consisting of large, open caverns linked by slot-like passages. Underground lakes and streams occur locally in cave networks.

Disappearing stream

Natural bridge

Sinkhole

Fallen debris

Dissolved joint

Stalactite
Stalagmite

Flowstone   Cavern

Soda-straw stalactite

Underground stream

Limestone column

Underground pool

Passage

Sinkholes

Underground pool, Mexico

Natural Bridge, Virginia

The orientation of bedding and joints in limestone localizes the flow of groundwater and, therefore, controls the configuration of openings in a cave network. Where groundwater drips from the ceiling of a cavern or flows along its walls, calcite precipitates and builds into speleothems, such as stalactites, stalagmites, columns, and flowstone. Distinctive karst landscapes develop at the Earth's surface over limestone bedrock that has undergone extensive dissolution, and where some caverns have collapsed to produce sinkholes.

Spelunker crawling in a cave

Soda-straw stalactites, Utah

5cm

Limestone cliff

Emerging spring

Limestone pavement, Ireland

Limestone pavement

What had happened? The bedrock under Winter Park consists of limestone, which has been dissolved in places to produce caverns. On May 8, the roof of a cavern underneath Owens's backyard began to collapse, forming a circular depression called a **sinkhole** (Fig. 12.42b). Eventually, the depression filled with water, and now it's a circular lake, similar to many others throughout central Florida (Fig. 12.42c). Sinkholes can form without warning, and sadly, their sudden development has caused fatalities.

Geologists refer to regions such as central Florida, where surface landforms develop as cave networks collapse, as **karst landscapes** or *karst terrains*, named for the Kras Plateau east of the Adriatic Sea, a broad region underlain by limestone that has undergone extensive dissolution (Fig. 12.43a). Karst landscapes develop when the rock overlying an air-filled (or partially air-filled) cave network weathers and erodes and the roofs of caverns weaken and start to collapse. Such landscapes display several other distinctive landforms in addition to sinkholes. For example, surface streams may flow into cracks or holes that link to caves below (Fig. 12.43b). Such *disappearing streams* re-emerge from a cave entrance downstream. Karst landscapes evolve over time; in fact, as the water table drops, multiple levels of cave networks develop (Fig. 12.44). In places where most of the ground has

collapsed into sinkholes, the landscape consists of narrow ridges with bowl-like depressions in between. Over time, the lower parts of a ridge may erode, leaving a natural bridge. When these bridges collapse, all that remains are tall spires or pinnacles of limestone (Fig. 12.45a). One of the world's most spectacular examples of such *tower karst*, in the Guilin region of China, has inspired generations of artists (Fig. 12.45b).

Take-home message…

Reaction with acidic groundwater dissolves limestone underground to form caverns and passages, which together form a cave network. Most of this dissolution takes place just below the water table. If the water table sinks, dripping water in caverns can produce speleothems. Collapse of a cave network produces a karst landscape.

Quick Questions

- How does a cavern differ from a passage in a cave network?
- How does a sinkhole form?
- What does the bedrock in karst landscapes consist of?

FIGURE 12.43 Features of karst landscapes.

(a) The rough, rocky surface of the Kras Plateau, for which karst landscapes were named, is formed by numerous sinkholes.

(b) A small disappearing stream in the Hudson Valley of New York; the water is spilling into a subsurface cave.

FIGURE 12.44 The progressive formation of karst landscapes.

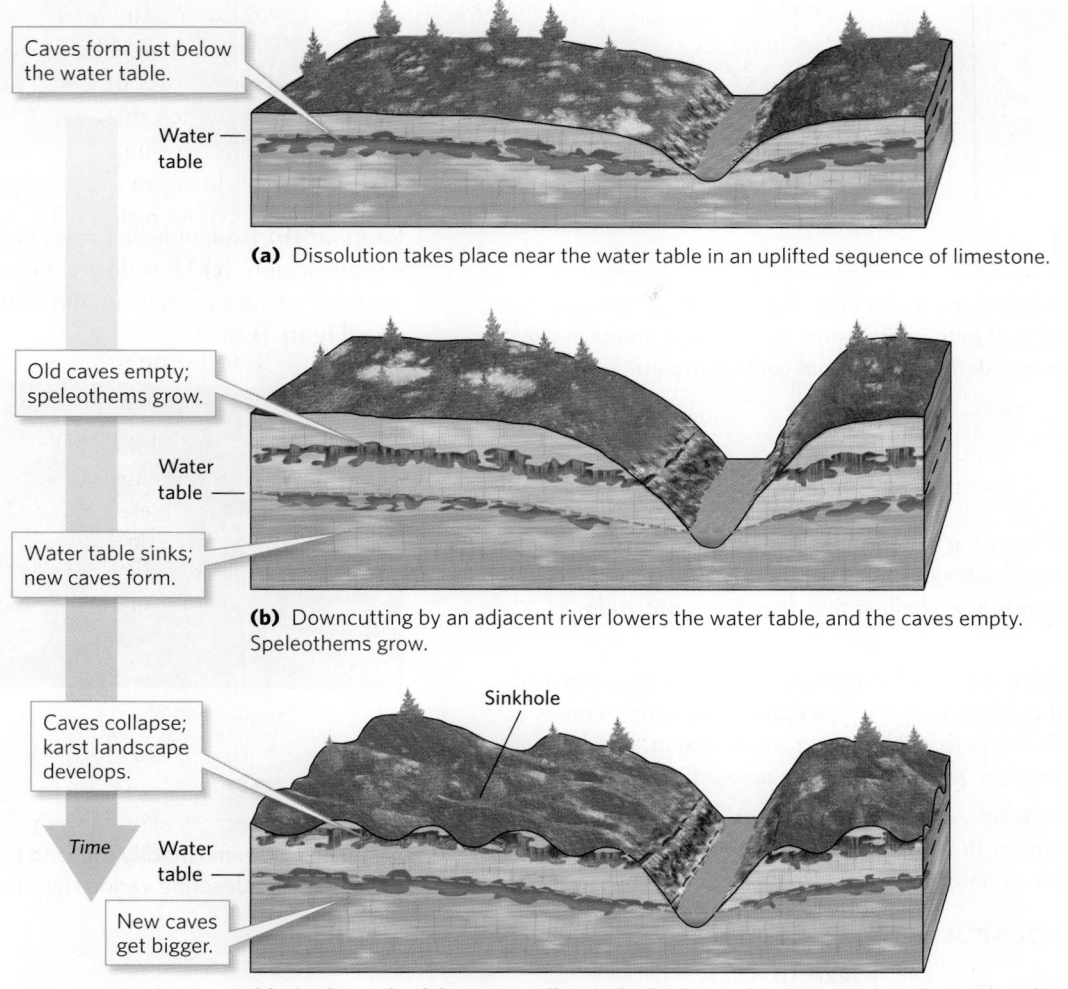

Caves form just below the water table.

Water table

(a) Dissolution takes place near the water table in an uplifted sequence of limestone.

Old caves empty; speleothems grow.

Water table

Water table sinks; new caves form.

(b) Downcutting by an adjacent river lowers the water table, and the caves empty. Speleothems grow.

Sinkhole

Caves collapse; karst landscape develops.

Time Water table

New caves get bigger.

(c) As the roofs of the caves collapse, the landscape becomes pockmarked with sinkholes.

FIGURE 12.45 Tower karst in the Guilin region of China.

(a) The landscape is treeless today, a consequence of industrialization policies in the 1950s.

(b) Chinese artists painted scrolls depicting forested towers of karst.

Objective 12.1

Characterize freshwater and its occurrences on land, distinguish between lakes and wetlands, explain how streams and drainage networks develop, and define factors that control stream discharge.

KEY CONCEPTS

- Lakes are bodies of standing water on land. Lakes that have an inlet and an outlet contain freshwater; if a lake has no outlets, it becomes salty. Water in wetlands is so shallow that vegetation protrudes above water.

- Streams drain runoff from the land surface. They grow by downcutting and headward erosion. Eventually, a drainage network, consisting of many tributaries that flow into a trunk stream, develops. Drainage divides separate adjacent drainage networks.

- The discharge of a stream—the volume of water passing through a cross section of a stream in a given time—varies with location along the stream and with climate and season.

EARTH-SCIENCE VOCABULARY

channel (p. 429)	**lake** (p. 427)
discharge (p. 431)	**overland flow** (p. 429)
downcutting (p. 430)	**runoff** (p. 430)
drainage divide (p. 430)	**salt lake** (p. 428)
drainage network (p. 430)	**stream** (p. 429)
freshwater (p. 427)	**turbulence** (p. 431)
headward erosion (p. 430)	**wetland** (p. 428)

REVIEW QUESTIONS

1. **(a)** What factors determine whether a lake is fresh or salty? **(b)** Distinguish among swamps, marshes, and bogs.

2. **(a)** Define overland flow, and explain how it plays a role in initiating streams. **(b)** How does a stream differ from a lake? **(c)** Explain the difference between headward erosion and downcutting.

3. **(a)** What is a drainage network? **(b)** Which type of drainage network does **Figure A** show? **(c)** Distinguish between a local drainage divide and a continental divide.

4. **(a)** Why does the velocity of water in a stream vary with

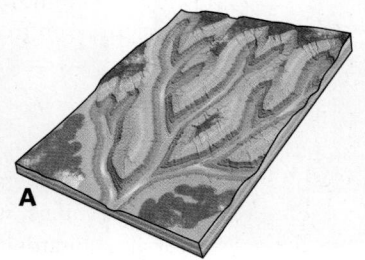

A

location? **(b)** Explain how a stream's discharge can vary along a stream's length. **(c)** How does climate affect discharge? **(d)** Why are some streams permanent and some ephemeral? **(e)** Which type does **Figure B** show?

B

Objective 12.2

Explain how streams modify the land surface by erosion and deposition, and describe various landforms that develop in fluvial landscapes.

KEY CONCEPTS

- A stream carves into the substrate and picks up a sediment load, which it carries as three components: dissolved load, suspended load, and bed load. When a stream's flow slows, its competence decreases, and it deposits alluvium in bars.

- A stream's gradient can be depicted in a longitudinal profile. Typically, a stream has a steeper gradient at its headwaters than near its mouth. A stream's base level limits the depth of downcutting.

- Whether a stream cuts a valley or a canyon depends on the rate of downcutting relative to the rate of mass wasting on bordering slopes. Locally, streams flow over rapids and waterfalls.

- On gentle gradients, streams can be braided or meandering. A meandering stream winds back and forth across a floodplain. Eventually, a meander may be cut off.

- Streams deposit bars within or along channels, alluvial fans at the mouths of canyons, and deltas where the stream enters standing water. Deposition occurs where a stream slows and its competence therefore decreases.

- Over time, fluvial landscapes evolve as the regional landscape gets beveled down. Uplift of land or a drop in the base level can cause stream rejuvenation.

EARTH-SCIENCE VOCABULARY

alluvial fan (p. 439)
alluvium (p. 433)
bar (p. 434)
base level (p. 434)
braided stream (p. 436)
capacity (p. 433)
competence (p. 433)
delta (p. 439)
floodplain (p. 434)

meander (p. 436)
meandering stream (p. 436)
mouth (p. 434)
rapid (p. 436)
sediment load (p. 433)
stream gradient (p. 434)
stream rejuvenation (p. 440)
waterfall (p. 436)

REVIEW QUESTIONS

5. **(a)** Describe how streams erode their substrate. **(b)** What are the three components of sediment loads that a stream can carry? **(c)** Distinguish between competence and capacity. **(d)** Under what conditions will a canyon form instead of a valley? **(e)** What is a floodplain?

6. **(a)** How does a stream's gradient vary along its length? **(b)** What's the difference between a local and an ultimate base level? **(c)** What factors control the locations of rapids and waterfalls?

7. **(a)** How does a braided stream differ from a meandering stream, and what factors determine which forms at a location? **(b)** Describe how meanders form and evolve. **(c)** Label the landforms shown in **Figure C**.

C

8. **(a)** Describe how a drainage network evolves over time. **(b)** What factors cause stream rejuvenation? **(c)** Why does stream piracy take place?

9. **(a)** Why do deltas and alluvial fans form? **(b)** How do deltas differ from alluvial fans?

Objective 12.3

Discuss the nature and causes of flooding and how society can protect against flood damage.

KEY CONCEPTS

• If more water enters a stream than the channel can hold, flooding takes place. Slow-onset floods develop slowly and submerge large regions for days to weeks. Flash floods happen rapidly and affect smaller regions.

• Officials try to prevent floods by building reservoirs, levees, and floodwalls. Nevertheless, it's important to pay attention to flooding hazard (defined by annual probability or recurrence interval) when building in an area susceptible to flooding.

EARTH-SCIENCE VOCABULARY

annual probability (p. 447)
crest (p. 442)
flash flood (p. 442)
flood (p. 441)

flood stage (p. 441)
recurrence interval (p. 447)
slow-onset flood (p. 442)
stage (p. 441)

REVIEW QUESTIONS

10. **(a)** Distinguish between a slow-onset flood and a flash flood. Which type does **Figure D** show? **(b)** Describe phenomena or situations that can cause flooding. **(c)** How does flooding affect communities?

D

11. **(a)** What is the annual probability of a flood, and how is it related to the recurrence interval? **(b)** What information does a flood-hazard map provide? **(c)** How can communities lessen the casualties and damage caused by flooding? **(d)** On **Figure E**, one dashed line represents the area that would be affected by a flood with an annual probability of 1%, and the other represents the area that would be affected by a flood with an annual probability of 30%. Label the two lines, and provide the recurrence interval of each.

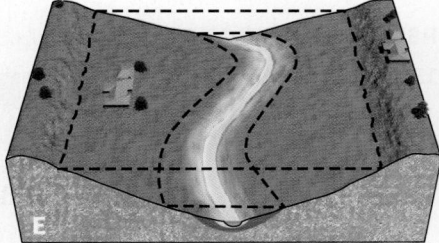

E

12. Records indicate that the height of the Mississippi River surface above flood stage, for a given amount of discharge, has been rising ever since large areas of the floodplain were isolated by the construction of artificial levees. Why?

Objective 12.4

Characterize groundwater and the water table, describe factors that control groundwater flow, and describe various types of wells and springs at which groundwater can be obtained.

KEY CONCEPTS

• Groundwater fills pores and cracks in rock and sediment. It flows more easily through aquifers, which have high porosity and permeability, than through aquitards, which have low permeability.

- The water table separates the unsaturated zone above from the saturated zone below. The elevation of the water table can reflect the shape of overlying topography.

- Groundwater flows from recharge areas to discharge areas. The velocity of flow depends on the permeability of the substrate and on the slope of the water table, as described by Darcy's law.

- Groundwater can be obtained from springs and wells. Springs form in many geologic settings. An ordinary well penetrates below the water table. In an artesian well, pressure causes water to rise on its own.

- If deep groundwater rises quickly to the surface, or groundwater comes close to hot rock in a geothermal area, hot springs can form. These springs can produce distinctive landscapes and geysers.

EARTH-SCIENCE VOCABULARY

aquifer (p. 449)
aquitard (p. 449)
artesian well (p. 452)
cone of depression (p. 452)
discharge area (p. 452)
geyser (p. 453)
groundwater (p. 448)
hot spring (p. 453)

infiltration (p. 448)
permeability (p. 448)
pore (p. 448)
porosity (p. 448)
recharge area (p. 452)
spring (p. 452)
water table (p. 450)
well (p. 452)

REVIEW QUESTIONS

13. **(a)** What is groundwater, where does it come from, and where does it reside? **(b)** How do porosity and permeability differ? **(c)** Using these terms, contrast an aquifer with an aquitard. **(d)** Does

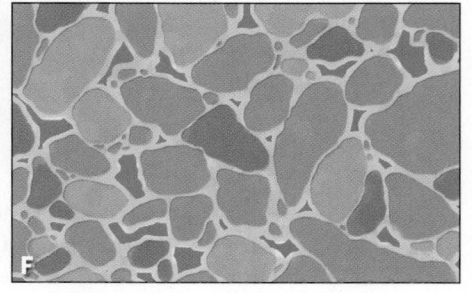

the rock in **Figure F** have high or low permeability? Explain your answer.

14. **(a)** What is the water table, and what factors affect its level? **(b)** Describe the factors that affect the flow rate and flow direction of groundwater. **(c)** On **Figure G**, label the recharge and discharge areas, and indicate what the blue arrows represent.

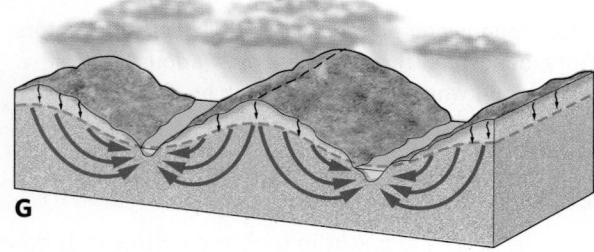

15. **(a)** How does an artesian well differ from an ordinary well? **(b)** What causes a cone of depression? **(c)** Why do natural springs form? **(d)** Describe why hot springs form and why geysers erupt.

Objective 12.5

Identify sustainability and environmental issues that pertain to freshwater resources.

KEY CONCEPTS

- Surface-water supplies have been damaged by pollution, eutrophication, damming, and depletion. Removing vegetation from the land or paving over the ground surface can increase flooding hazards and the sediment load of streams.

- In some localities, groundwater contains dissolved chemicals that make it unusable without treatment. Groundwater deeper underground typically contains salt.

- Groundwater supplies have been severely depleted by overuse and contamination. Depletion leads to a drop in the water table, which can cause pore collapse.

EARTH-SCIENCE VOCABULARY

contaminant plume (p. 457)
eutrophication (p. 456)
groundwater depletion (p. 458)
hard water (p. 457)

pore collapse (p. 458)
saltwater intrusion (p. 458)
surface-water depletion (p. 456)

REVIEW QUESTIONS

16. **(a)** Distinguish between eutrophication and pollution of surface water. **(b)** What clue indicates that the water of a river is being depleted? **(c)** Describe the potential negative consequences of dam construction. **(d)** How does the construction of cities or the development of farms affect streams? **(e)** What does the presence of the white area bordering the lake next to Hoover Dam on **Figure H** indicate?

17. **(a)** Why is groundwater a nonrenewable resource in some locations? **(b)** What do the pink area and the blue arrow in **Figure I** represent? **(c)** What conditions lead to saltwater intrusion in a well? **(d)** Describe potential consequences of groundwater depletion. **(e)** Which of those consequences does **Figure J** represent?

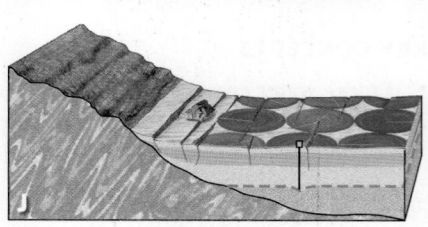

18. The population of a desert town in the southwestern United States has been doubling every 10 years. The town has been growing on a flat, gravel-filled basin between two small mountain ranges. The gravel is an excellent aquifer, and it has served as the main source of water for the city and surrounding farms. What will happen to the water table of the area in coming years, and what will be the consequence of this change?

Objective 12.6

Discuss the origin of cave networks, and of karst landscapes and the landforms within them.

KEY CONCEPTS

- Where limestone dissolves just below the water table, cave networks develop. If the water table drops, the cave networks fill with air, and travertine precipitates out of water dripping from cavern roofs to produce speleothems.

- Regions where cave networks have collapsed to form features such as sinkholes and natural bridges are called karst landscapes.

EARTH-SCIENCE VOCABULARY

cave network (p. 460)	sinkhole (p. 464)
cavern (p. 460)	speleothem (p. 460)
karst landscape (p. 464)	stalactite (p. 460)
passage (p. 460)	stalagmite (p. 460)

REVIEW QUESTIONS

19. (a) Under what conditions can water percolating down to the water table become acidic? **(b)** Why do caverns and passages form at or near the water table? **(c)** What factors control the configuration of caverns and passages in a cave network? **(d)** Would cave networks form in quartzite? Explain your answer.

20. (a) What conditions lead to the formation of speleothems? **(b)** Label a stalactite and a stalagmite on **Figure K. (c)** Name several landforms that occur in karst landscapes.

K

ANOTHER VIEW Water from rain and melting snow rushes down a canyon cut into layers of basalt in Iceland. The torrent spills over a waterfall into a narrower canyon that follows the trace of a large fracture.

⊘13 EXTREME REGIONS
Desert and Glacial Landscapes

After studying this chapter, you should be able to...

1. describe conditions that lead to the classification of a region as a desert, and explain why such conditions develop.

2. explain how weathering, erosion, and deposition lead to the formation of desert landscapes.

3. characterize desert landscapes and landforms, such as rocky cliffs, stony plains, and sand dunes, and explain why some semiarid regions are at risk of becoming deserts.

4. explain how glacial ice forms and flows, distinguish among various kinds of glaciers, explain how glaciers advance and retreat, and describe occurrences of ice in the sea.

5. discuss how glaciers erode the land surface, describe the landforms produced by glacial erosion and deposition, and characterize the various materials deposited by glaciers.

6. describe the effects of continental glaciation on the surface of the lithosphere, water distribution, drainage networks, and sea level.

7. define an ice age, discuss when ice ages took place, and evaluate ideas proposed to explain why ice ages happen and why glacial advances and retreats during an ice age occur periodically.

<< A small, glaciated mountain range rises above parched desert land in north-central China, near the Gobi Desert. Both of these extreme climates produce fascinating landscapes.

By the end of the 18th century, European navigators had compiled maps of coastlines worldwide. But inland regions with particularly inhospitable climates—such as our planet's rainforests, deserts, polar regions, and high mountains—remained largely unknown. Such extreme regions weren't studied in earnest until the 19th century, and it wasn't until the 20th century that photographs obtained from airplanes and satellites allowed researchers to develop a scientific understanding of them. Now, in the 21st century, you can tour even the Earth's most isolated localities on a cell-phone screen.

In this chapter, we introduce two of our planet's extreme regions: desert and glacial landscapes. First, we learn why deserts develop, how erosion and deposition shape their surfaces, and how their bordering regions are, in some places, turning into deserts. Then we turn our attention to landscapes that are, or were, covered by glacial ice. We see how glaciers form and move, and how they sculpt the ground and deposit sediment. This chapter concludes by describing the consequences and causes of ice ages, times when glaciers covered large areas of the continents.

13.1 The Nature and Locations of Deserts

What Is a Desert?

Camels can walk for up to three weeks without drinking or eating because they barely perspire, they can use their own body fat as a water source, and they can withstand severe dehydration. Without such adaptations, camels couldn't survive in a **desert**, a land region that receives average rainfall (or snowfall equivalent) of less than 25 cm/yr (10 in/yr) and, therefore, can sustain vegetation on no more than 15% of its surface. (By comparison, temperate climates such as those of the eastern United States receive an average of 90–180 cm/yr, or 30–70 in/yr.) Because of the *arid* (dry) conditions in deserts, they host no permanent streams or lakes, except for those whose water flows in from a wetter region elsewhere. Note that the definition of a desert depends on a region's *aridity* (dryness), not on its temperature. Therefore, we distinguish between **cold deserts**, where temperatures generally stay below about 20°C (68°F) and, during part of the year, drop below freezing, and **hot deserts**, where daytime temperatures often exceed 35°C (95°F). In fact, daytime highs in some hot deserts can exceed 43°C (110°F) for weeks. Notably, hot deserts don't necessarily stay hot all day, because without moisture, desert air doesn't retain heat. For example, on a day when midday temperatures surpass 38°C (100°F), nighttime temperatures can drop to −3.9°C (25°F).

Where Do Deserts Occur?

Deserts of all types together occupy about 25% of our planet's land surface (Fig. 13.1). And, despite their harsh climates, about a billion people live in or near deserts worldwide. Not all deserts form by the same processes. We distinguish among several different types of deserts by identifying the cause of their aridity:

- *Subtropical deserts:* At equatorial latitudes, intense sunlight strikes the Earth's surface. The surface absorbs this energy, and then reradiates it upward into the base of the atmosphere. Consequently, near-surface air warms and becomes less dense than overlying cooler air. This density difference triggers convective flow: hot, moist air above tropical oceans or rainforests

Did you ever wonder . . .

whether all deserts are very hot?

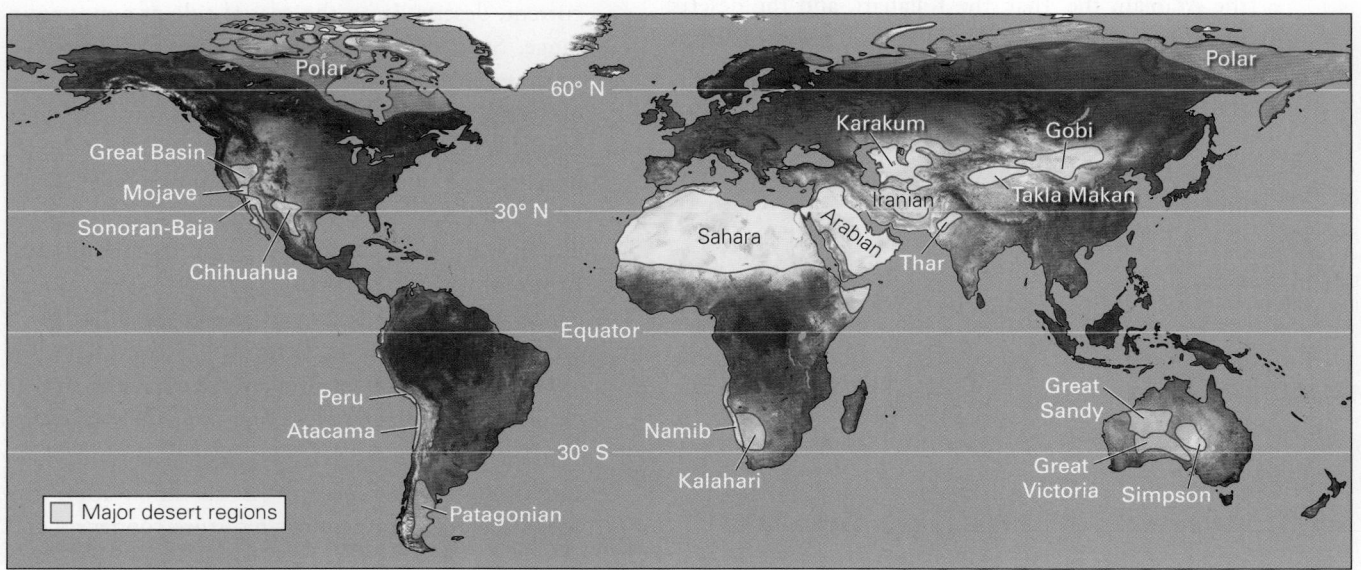

FIGURE 13.1 The global distribution of deserts.

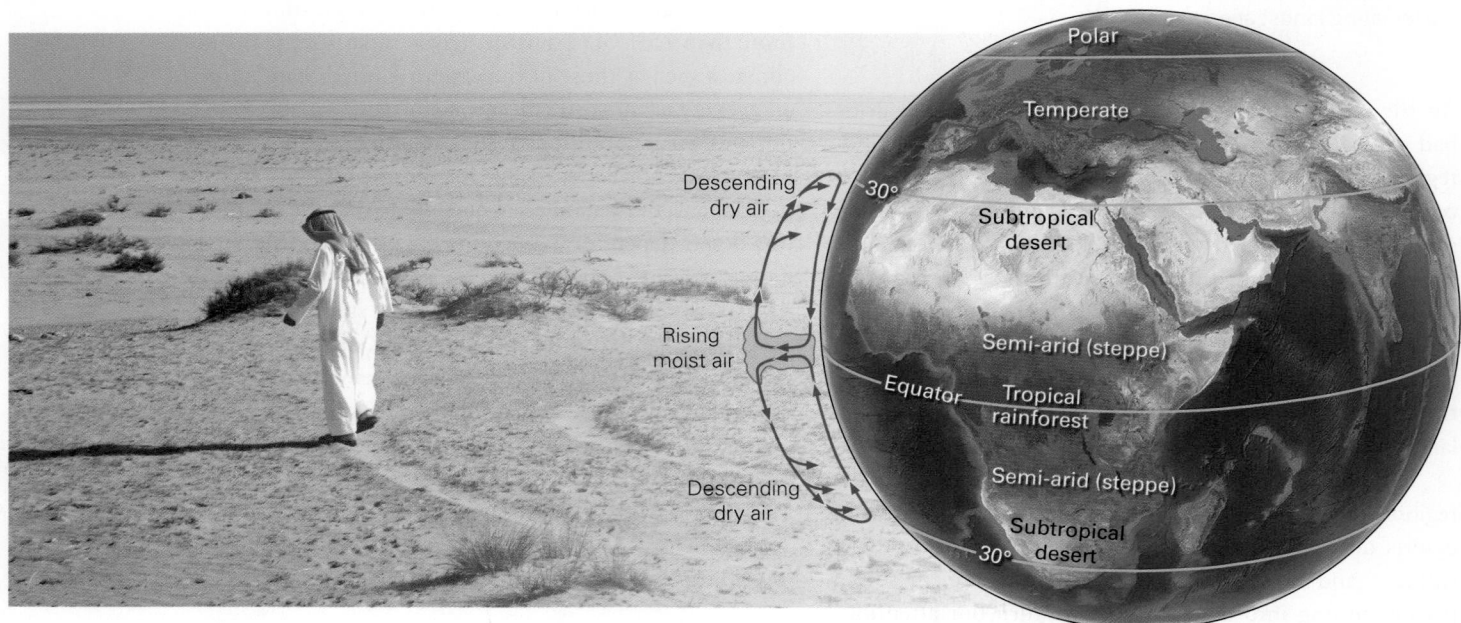

FIGURE 13.2 Formation of subtropical deserts. Subtropical deserts form because they lie beneath the portion of an atmospheric convective cell, the Hadley cell, where dry air descends and warms. As a result, rain clouds rarely form in subtropical latitudes.

rises like a hot-air balloon to high altitudes, where it expands and cools. Cooling of the air as it rises causes the water vapor within it to condense and fall as drenching rain on the tropical rainforests straddling the equator below. At high altitudes, the now drier and cooler air flows away from the equator toward *subtropical latitudes* (geographically defined as lying between 23°–35° N and 23°–35° S), where it sinks. As this air sinks, it undergoes compression, which causes it to heat up, making it even drier. Rain-producing clouds can't form in this very dry descending air, so the land below becomes a **subtropical desert**. Most of the Earth's largest deserts, including the Sahara, the Arabian, the Thar, the Kalahari, and the deserts of Australia, are subtropical deserts. The flow pattern of air that we've just described represents an atmospheric convective cell known as a *Hadley cell* (see Chapter 17). One Hadley cell extends north of the equator, and one south, yielding two subtropical desert belts on the Earth **(Fig. 13.2)**.

- *Rain-shadow deserts:* Air flowing from the ocean toward a coastal mountain range must rise when it encounters the range's slopes. As it does, it expands and cools, so clouds form (see Chapter 16). Rain or snow falls from the clouds on the windward flank of the mountains, nurturing coastal rainforests. When the air moves farther inland and descends on the leeward side of the mountains, it has lost its moisture, warms up, and can no longer provide rain, so land on the leeward side becomes a **rain-shadow desert (Fig. 13.3)**. Examples of such deserts occur to the east of the Cascade and Sierra Nevada Ranges in the United States.

- *Coastal deserts:* Along coasts that border a cold ocean current, deserts lie on the coastal side of a mountain range. Such deserts develop because air above cold ocean water cools and becomes too dense to rise into overlying warmer air. When this cold air blows over coastal lands, it stays low, so the water it contains does not rise and condense to form rain clouds. Therefore, mid-latitude and low-latitude coastal lands bordering cold ocean currents become **coastal deserts**. An example, the Atacama Desert of western South America, holds the record for being the driest place on the Earth **(Fig. 13.4)**. Parts of this desert have not seen a drop of rain for over five centuries! A similar desert, the Namib Desert, occurs along the southwestern coast of Africa.

- *Continental-interior deserts:* As air flows from the ocean across a continent, its moisture condenses and

FIGURE 13.3 Formation of rain-shadow deserts. Moist air rises and drops rain on the windward side of a mountain range. By the time it reaches the leeward side of the mountains, the air has become dry.

FIGURE 13.4 Formation of coastal deserts.

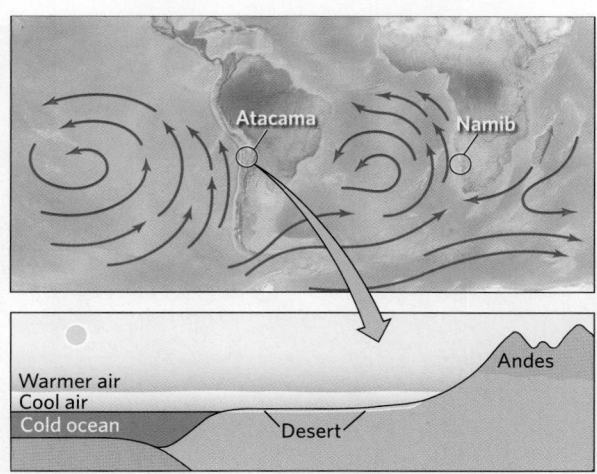

(a) Cold ocean currents cool the air along a coast. This air is too dense to rise and produce rain clouds.

(b) The Atacama Desert is the driest place on the Earth.

falls as rain. If there are no other sources of water to add moisture back to the atmosphere, the air dries out so much by the time it reaches the continent's interior, far from a coast, that it can no longer produce rain, and the land below becomes a **continental-interior desert**. The Gobi Desert of Asia serves as an example.

- *Polar deserts:* The very cold, dense air above a polar region holds very little moisture, so minimal precipitation falls there, making these regions very dry **polar deserts**. What snow does fall, however, can survive, so these deserts have snow and ice cover over large areas. Antarctica, Greenland, and northernmost portions of Canada, Scandinavia, and Russia host polar deserts. In places where land in these regions does not have snow or ice cover, the ground surface either consists of barren bedrock, or has a covering of unvegetated silt and rock fragments.

Take-home message...

Deserts receive an average of less than 25 cm (10 in) of rain per year, which makes them so arid that they host only sparse vegetation. Deserts develop in several settings: subtropical climates, rain shadows, coasts bordered by cold currents, continental interiors, and polar regions.

Quick Questions

- Do all deserts have extremely high daytime temperatures?
- How much of the land surface in a desert has a cover of vegetation?
- Why does the world's largest desert, the Sahara, exist?

13.2 Weathering, Erosion, and Deposition in Deserts

Desert Weathering and Soil Formation

Physical weathering in deserts begins, as it does elsewhere, when joints form and break bedrock into pieces. Though rains are sparse, enough moisture from showers and dew seeps into rock to cause very slow dissolution, hydrolysis, and oxidation. So, over time, desert weathering causes fresh rock to crumble and become sediment (see Fig. 5.3b, c).

Desert soils differ markedly from those formed in other climates because rain occurs so rarely in deserts that vegetation does not grow, and infiltrating water cannot carry dissolved ions completely away. As a result, the light-colored soil, known as an *aridisol*, that develops does not have an organic-rich O-horizon, and calcite and gypsum, even though they are fairly soluble, do not get flushed entirely away. Where significant amounts of calcite and gypsum remain in the soil, they can cement soil clasts together to form a coherent material called **caliche**, which resembles a clastic sedimentary rock.

Notably, the lack of plant cover in deserts allows color variations in bedrock and soil to stand out. In particular, slight variations in the concentration of iron or in the degree of iron oxidation produce spectacular color banding on the land surface. The Painted Desert of northern Arizona earned its name from the brilliant and varied hues of the region's iron-containing shale bedrock **(Fig. 13.5a)**.

EXPLORE EARTH SCIENCE IN 3D
Desert Varnish

BOX 13.1 ▷ Putting Earth Science to Use

Recognizing young landslides and rockfalls in desert regions

In temperate or tropical areas, the character of vegetation on a hillslope can provide a clue to the ages of landslides or rockfalls because mass wasting disrupts vegetation. For example, in an area of recent mass wasting, trees and shrubs tilt chaotically. A somewhat older landslide may host a patch of dead vegetation because mass wasting disrupts plant roots. Can you interpret the relative ages of landslides in desert regions, where vegetation doesn't grow? Yes. The visual cue comes from the color of desert varnish (Fig. Bx13.1). Boulders or cliffs that have existed for a long time have a dark coating of varnish. In contrast, the blocks in a recent landslide, or a new cliff face exposed by a rockfall, don't have a coating of varnish, and therefore they have a lighter color. Intermediate-aged landslides will be darker than new ones, but lighter than very old ones. This visual cue can help you recognize mass-wasting hazards in arid regions.

Landslide scar

FIGURE Bx13.1 On the wall of the Grand Canyon, rocks develop desert varnish over time. Fresh rocks, exposed by a landslide, do not have varnish yet.

Exposed rock surfaces in deserts commonly display **desert varnish**, a dark, shiny, rusty brown coating. It forms when windblown clay settles on the surfaces of rocks in the presence of a small amount of moisture. Over the course of centuries to millennia, chemical reactions and the activity of microbes extract iron and manganese ions from the dust, incorporating them into shiny oxide minerals that bind to the rock surface. Therefore, desert varnish thickens and darkens as time passes (Box 13.1). Traditional cultures created artistic figures, called *petroglyphs*, by chipping away the varnish to reveal the underlying lighter-colored rock (Fig. 13.5b).

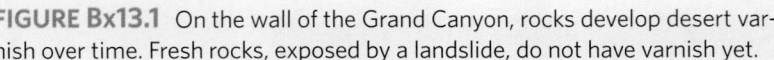

FIGURE 13.5 Desert colors.

(a) The red hues of the Painted Desert in Arizona are due to the oxidation (rusting) of iron in the rock.

(b) By chipping away desert varnish to reveal the lighter rock beneath, Native Americans produced art and symbols.

FIGURE 13.6 Evidence of erosion by water in deserts.

(a) The shapes of these dry hills in the desert near Las Vegas, Nevada, indicate that they were eroded by water. Note the numerous stream channels.

(b) Gravel and sand have been left behind on the floor of a dry wash after a flash flood in Death Valley. Erosion by water is cutting a channel.

(c) Badlands develop where water erodes unvegetated clay-rich sediment or sedimentary strata.

Erosion by Water

Although heavy rains rarely fall in deserts, when they do, they can radically alter a landscape in a matter of minutes, because the lack of vegetation makes surfaces vulnerable to erosion. In fact, in most deserts, water causes more erosion than wind does **(Fig. 13.6a)**. Erosion by water begins with the impacts of raindrops, which knock sediment on the ground into the air. On a hill, this sediment lands downslope. When overland flow across the ground surface begins, it carries away loose sediment. During intense rainstorms, channels called **dry washes** (*arroyos* in Spanish or *wadis* in Arabic), which are usually empty, fill with a turbulent mixture of water and sediment, which rushes downstream as a flash flood, picking up more sediment as it travels. The gooey, dark-colored waters of flash floods in dry washes hint at how much sediment ephemeral streams in deserts can carry away (see Chapter 12). When the water flow ceases, floodwaters seep into the streambed or evaporate, leaving a braided network of gravel-filled channels or, in some cases, steep-walled canyons. Such dry washes provide drainage in most desert landscapes **(Fig. 13.6b)**.

On hillslopes composed of clay-rich rock with no vegetation cover, water erosion may produce **badlands**, a landscape characterized by a parallel drainage network of small, closely spaced gullies and ravines that merge downslope **(Fig. 13.6c)**. If the clay can absorb water, it swells to produce a very weak popcorn-like texture on the land surface. A downpour easily washes away this clay and digs relatively rapidly into the underlying, unprotected substrate.

Erosion by Wind

Winds, even very strong ones, do not change the land surface in temperate or tropical climates because vegetation protects the soil beneath. In deserts, however, winds have direct access to the land surface and can pick up and transport large quantities of sediment. Some of this sediment, the **suspended load**, rises into the air above the ground, while the rest, known as the **surface load**, remains in contact with the ground for part or all of its journey.

Suspended loads consist of fine-grained sediment, such as clay and silt, that can stay aloft for a long time because it is light enough for wind and updrafts to counter gravity. In fact, air currents sometimes lift this sediment—informally known as *dust*—several kilometers above the Earth's surface and carry it completely out of its source region. Satellite photos show that dust blown from the Sahara can cross the Atlantic Ocean **(Fig. 13.7a)**. Particularly strong winds generate dramatic **dust storms**, up to 100 km (60 mi) long and 1.5 km (1 mi) high, that resemble a roiling, opaque wave as they approach. Large dust storms occasionally engulf cities **(Fig. 13.7b)**.

Surface loads, or *bed loads*, are transported when the wind becomes strong enough to move sand grains. This motion occurs by *creep*, which happens when the grains roll along the ground and push one another along, and by **saltation**, which begins when air turbulence lifts sand grains **(Fig. 13.7c)**. Saltating grains follow an asymmetrical, arc-like trajectory as they head downwind. When they return to the ground, they strike other sand grains, causing those grains to bounce up or roll downwind. Saltating grains generally rise no more than 0.5–2 m (1.5–6.5 ft) above the ground **(Fig. 13.7d)**. Only extreme winds can move clasts that are larger than sand.

Just as sandblasting cleans the grime off the surface of a building, windblown sand and dust grind away at surfaces in the desert. Over long periods, such *wind abrasion* can abrade elongate ridges of relatively soft bedrock, producing streamlined landforms called *yardangs* **(Fig. 13.8a)**. At a smaller scale, saltating grains enhance facets on

See for yourself

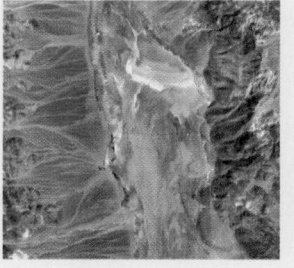

Death Valley, California

Latitude: 36°12′42.34″ N
Longitude: 116°48′25.43″ W

Zoom to an altitude of 35 km (20 mi) and look down.

Death Valley, a narrow basin whose floor lies below sea level, hosts many desert landforms. The white patch is a playa. A bajada forms the slope between the playa and the mountains to the west. Small alluvial fans spill into the basin on the east.

FIGURE 13.7 Wind transport of sediment in deserts.

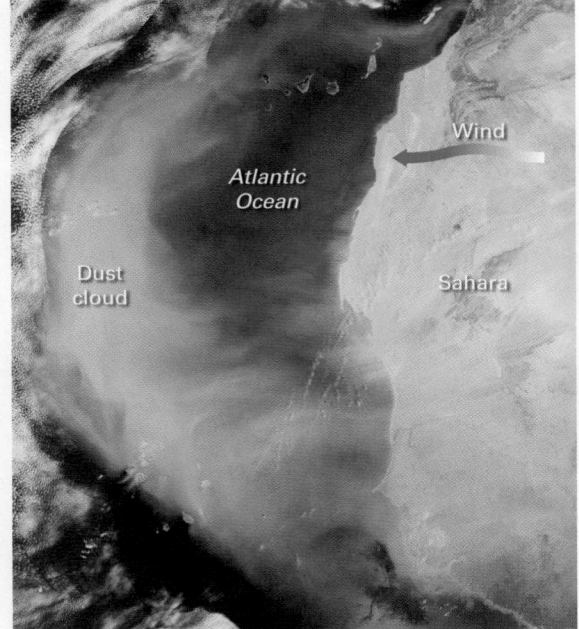

(a) In this satellite image, a huge dust cloud that originated in the Sahara blows across the Atlantic Ocean.

(b) A huge dust storm approaching Phoenix, Arizona.

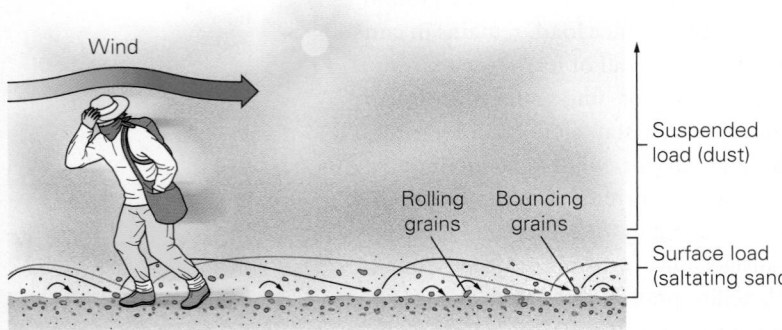

(c) Wind transports desert sediment both in suspension and in a saltating layer.

(d) Sand saltating across a highway in west-central China. Note that it stays close to the ground.

FIGURE 13.8 Examples of wind erosion in deserts.

(a) A yardang in western China, surrounded by coarse sand.

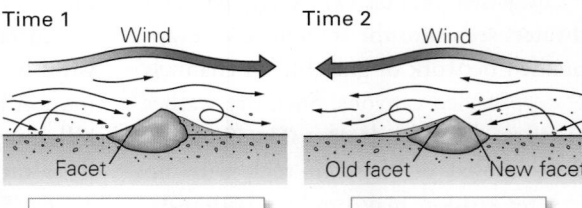

(b) Ventifacts form when windblown sediment erodes the surface of a rock.

Windblown sand abrades the face of a rock, forming a facet.

The wind shifts direction, and a new facet forms.

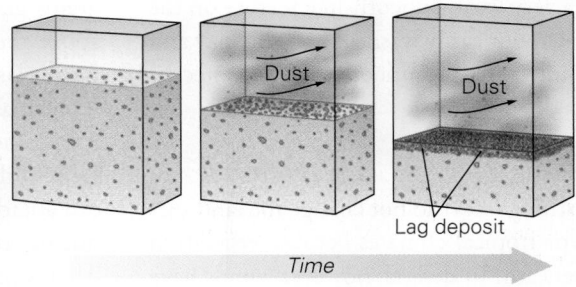

(c) An example of a multifaceted ventifact in the Gobi Desert. The boot provides scale.

(d) A lag deposit develops when wind blows away fine sediment, leaving behind a layer of coarser grains.

pebbles and cobbles, producing *ventifacts* (Fig. 13.8b, c). Locally, wind may carry away so much sand and dust that remaining pebbles and cobbles—clasts that were too big to be moved by wind—become concentrated at the ground surface as a **lag deposit** (Fig. 13.8d).

Deposition in Deserts

Deposition takes place in deserts for the same reasons it takes place in other regions. It happens either when gravity causes rock fragments to tumble from a cliff, and friction keeps them in a pile at the angle of repose at the base of the cliff, or when the fluid carrying sediment slows and loses competence, so that sediment settles out.

Recall from Chapter 11 that the landform that develops where falling rock fragments accumulate at the base of a cliff is called a **talus slope**. The slope represents the angle of repose for the fragments (Fig. 13.9a). In deserts, talus slopes remain unvegetated, and the rock fragments at their surface become coated with desert varnish.

Sediment that tumbles into a dry wash at the floor of a canyon or narrow valley due to a mass-wasting event will eventually be carried downstream by a flash flood. When floodwaters slow and subside, this sediment comes to rest and accumulates to form elongate bars of gravel within the dry wash. If, however, the sediment-laden water of a flash flood emerges from the mouth of a canyon or valley and onto a plain, the water slows, and the sediment settles over a broader area. Eventually, a semicircular (in map view) wedge of sediment, known as an **alluvial fan**, builds out onto the plain (Fig. 13.9b). The channel of the stream, as it flows over an alluvial fan, subdivides into several smaller braided channels, called *distributaries*, that diverge outward

---

FIGURE 13.9 Accumulation of sediment in deserts.

(a) This talus slope along the base of a desert cliff formed from rocks that broke off and tumbled down the cliff. The big boulders came from the thick tan sandstone bed at the top of the cliff. The smaller red clasts came from the underlying, thinner siltstone beds. Joints are more closely spaced in thinner beds, so mechanical weathering produces smaller blocks from the siltstone.

(b) Distributaries carry water and sediment to different parts of an alluvial fan at different times, thereby maintaining the fan shape.

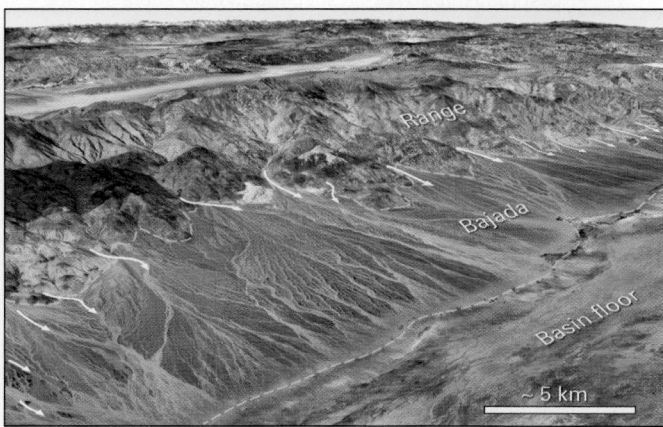

(c) Overlapping fans in Death Valley have produced a bajada. The yellow dashed lines outline the bajada. Arrows indicate where canyons feed sediment onto the bajada.

(d) Farmers have cut terraces into the loess that underlies the Loess Plateau of China. In places, the loess here is 300 m thick. Particles in loess tend to bind together, so escarpments in loess can remain for a long time without collapsing. This loess blew from deserts to the west.

FIGURE 13.10 The formation of playas.

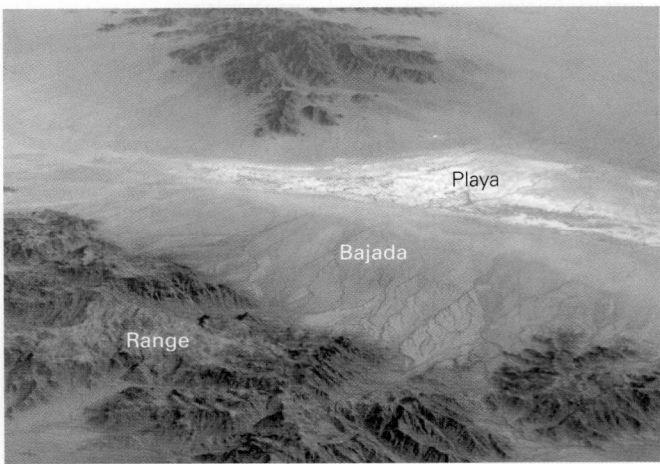

A close-up of salt crystals

(a) This playa in California formed at the base of a bajada, as seen in this aerial, oblique view.

(b) White salt crystals encrust the floor of a playa in Death Valley.

Did you ever wonder . . .

whether all deserts are completely covered by sand?

from the canyon or valley mouth. Not all distributaries on an alluvial fan flow at the same time. Rather, flows carry sediment along one part of the fan for a while until the slope of the distributary becomes too gentle for flow to continue. The next flow finds a steeper path, to the side of the previous path. Repetition of this process over time causes deposition to move from one part of the fan to another, so overall, the fan maintains its symmetrical shape. Alluvial fans emerging from adjacent valleys may eventually merge and overlap along the front of a mountain range, producing an elongate, composite wedge of sediment, known as a **bajada**, bordering the range **(Fig. 13.9c)**.

Windblown sand settles where the wind slows. In some cases, the slowing happens in a wind shadow caused by nearby hills or mountains, so that the area in which sand accumulates is relatively small. But in some deserts, large sand seas cover a broad area. As we will see, wind causes large sand accumulations to develop into dunes. Windblown dust, as we've seen, may travel greater distances, far beyond the limits of the desert. Where deserts border a region with a moister climate, the moisture causes the arriving dust to clump into larger particles, which then settle to form a layer of sediment called **loess**. Dust blown eastward from the Gobi Desert, for example, has accumulated to form a thick layer, the Loess Plateau, in western China **(Fig. 13.9d)**. (As we'll see in Section 13.5, loess also develops in regions bordering glacial environments.)

During particularly wet seasons, water draining from desert mountain ranges during flash floods may cover parts of basins between ranges to form temporary shallow lakes, known as *playa lakes*. Such lakes become salty because they have no outlet. During drier times, the water in the lake evaporates entirely and the salts precipitate, so that the lake bed becomes exposed to form a **playa**,

▶ **Narrative Art Video**
Evolution of Deserts

a flat, dry desert lake bed. Playas typically have a white, flat, crusty surface of precipitated salts and gypsum, mixed with clay **(Fig. 13.10)**. After many cycles of flooding and evaporation, the salt, gypsum, and clay layer forming the floor of a playa can become many meters thick.

Take-home message . . .

Weathering breaks down rocks in deserts, but because of water scarcity, soils are thin, and desert varnish may coat rock surfaces. Wind transports sediment and can erode rock. Heavy rains cause significant erosion and sediment transport in deserts, but most stream channels are dry washes, which rarely fill with water. Some of the clastic sediment carried by streams accumulates in alluvial fans, while salts precipitate on playas.

Quick Questions ———————

- What phenomenon causes most erosion in a desert?
- Distinguish between a suspended load and a surface load of windblown sediment.
- What causes deposition in a desert environment?

13.3 Desert Landscapes and Desert Problems

Movies commonly portray deserts as endless seas of sand. While deserts, in many places, do contain immense piles of sand, they also host other types of landforms, including barren rocky cliffs and stony plains (**Earth Science at a Glance**, pp. 480–481). All these landforms differ from those of temperate and tropical regions, where thick soil and vegetation cover most ground.

Rocky Cliffs, Mesas, and Arches

Because thick soil cover doesn't accumulate on steep slopes in deserts, these slopes become rocky ridges and cliffs of exposed bedrock. At localities underlain by horizontal sedimentary strata, or by horizontal layers of volcanic rock, cliff faces form when rock breaks off along vertical joints and eventually topples and becomes debris at the base of the cliff. By breaking off at successive joints, a cliff face steps back into the land while retaining roughly the same profile, a phenomenon known as **cliff retreat**. Where beds have different resistance to erosion, cliffs develop a step-like shape—the resistant layers become cliffs, whereas the nonresistant layers become debris-covered slopes **(Fig. 13.11a)**. Due to cliff retreat,

FIGURE 13.11 Consequences of cliff retreat.

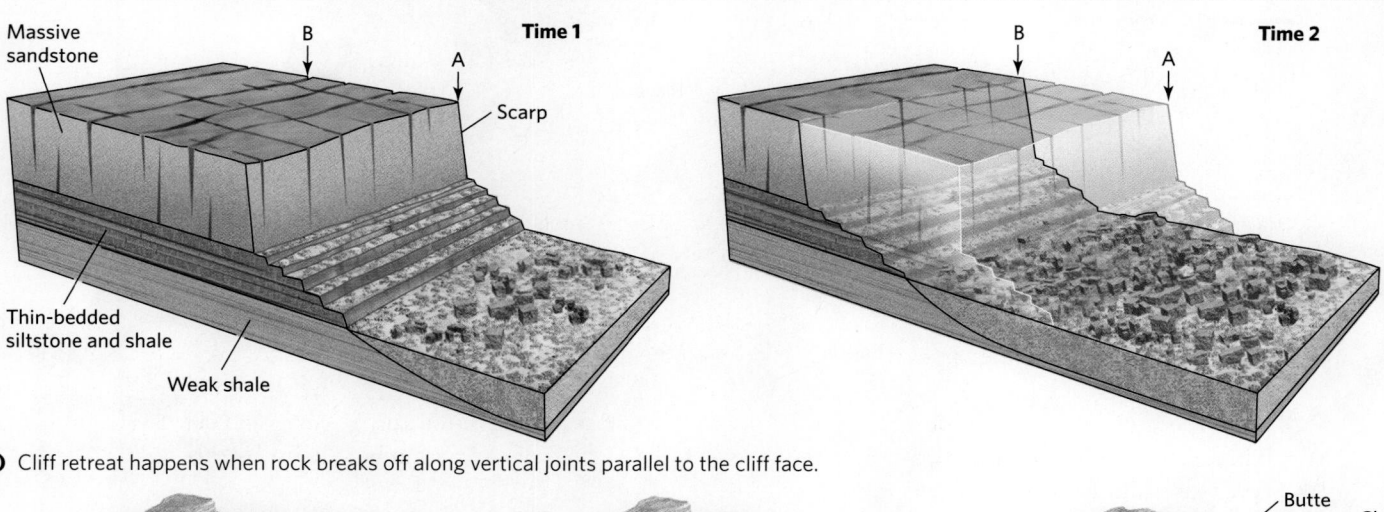

(a) Cliff retreat happens when rock breaks off along vertical joints parallel to the cliff face.

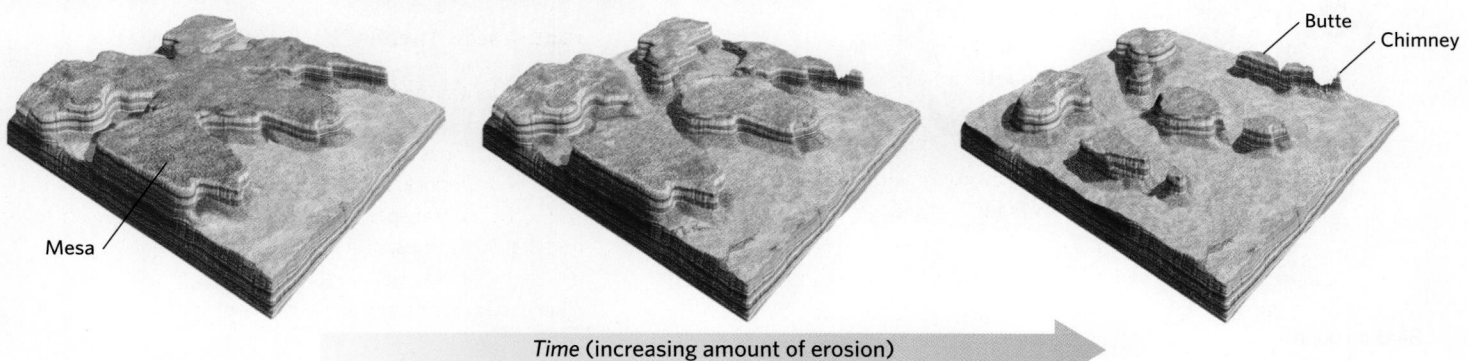

Time (increasing amount of erosion)

(b) A once-continuous layer of rock evolves into a series of isolated mesas, buttes, and chimneys. If the bedding is horizontal, the resulting landforms have flat tops.

(c) Buttes and mesas tower above the floor of Monument Valley, Arizona.

Landforms of Deserts

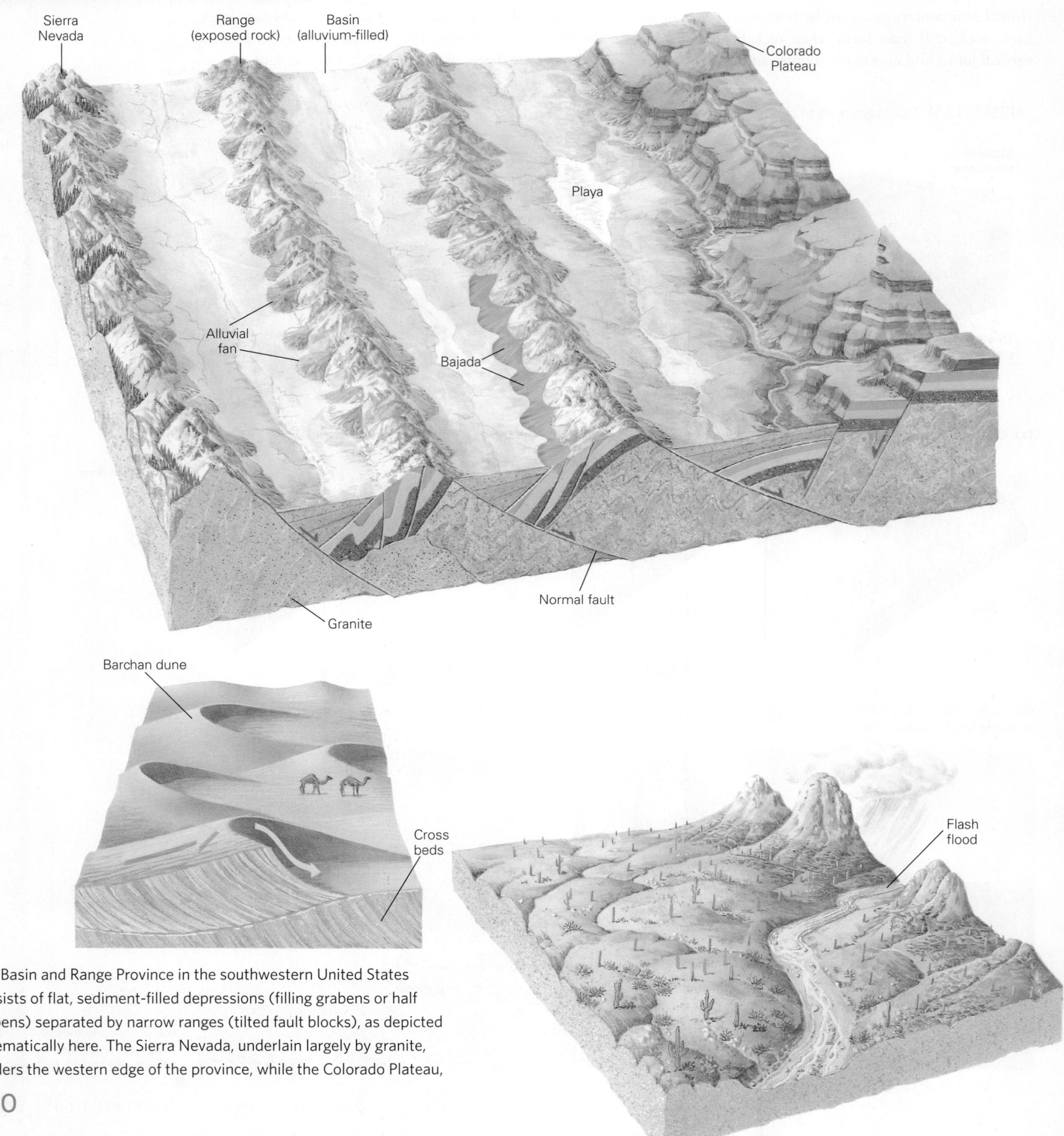

Sierra
Nevada

Range
(exposed rock)

Basin
(alluvium-filled)

Colorado
Plateau

Playa

Alluvial
fan

Bajada

Normal fault

Granite

Barchan dune

Cross
beds

Flash
flood

The Basin and Range Province in the southwestern United States
consists of flat, sediment-filled depressions (filling grabens or half
grabens) separated by narrow ranges (tilted fault blocks), as depicted
schematically here. The Sierra Nevada, underlain largely by granite,
borders the western edge of the province, while the Colorado Plateau,

480

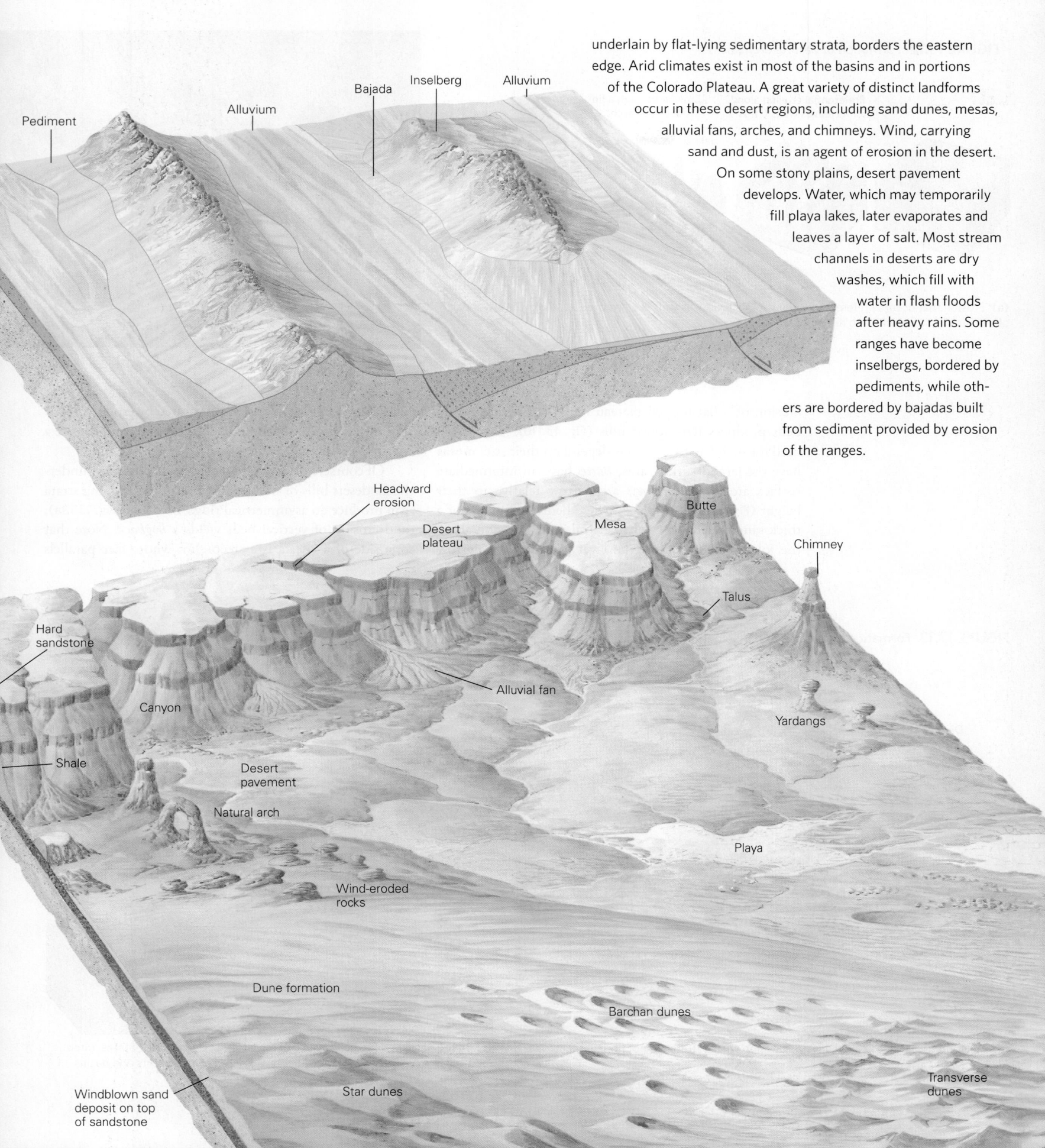

Pediment

Alluvium

Bajada

Inselberg

Alluvium

underlain by flat-lying sedimentary strata, borders the eastern edge. Arid climates exist in most of the basins and in portions of the Colorado Plateau. A great variety of distinct landforms occur in these desert regions, including sand dunes, mesas, alluvial fans, arches, and chimneys. Wind, carrying sand and dust, is an agent of erosion in the desert. On some stony plains, desert pavement develops. Water, which may temporarily fill playa lakes, later evaporates and leaves a layer of salt. Most stream channels in deserts are dry washes, which fill with water in flash floods after heavy rains. Some ranges have become inselbergs, bordered by pediments, while others are bordered by bajadas built from sediment provided by erosion of the ranges.

Headward erosion

Desert plateau

Mesa

Butte

Chimney

Talus

Hard sandstone

Canyon

Alluvial fan

Yardangs

Shale

Desert pavement

Natural arch

Playa

Wind-eroded rocks

Dune formation

Barchan dunes

Windblown sand deposit on top of sandstone

Star dunes

Transverse dunes

FIGURE 13.12 Formation of natural arches.

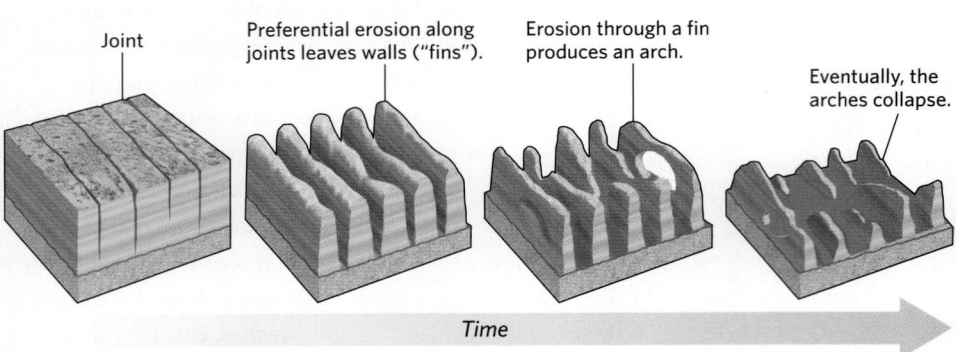

Joint

Preferential erosion along joints leaves walls ("fins").

Erosion through a fin produces an arch.

Eventually, the arches collapse.

Time

(a) Erosion that occurs preferentially along joints produces wall-like fins of rock. When the lower part of a fin erodes, a natural arch remains.

(b) A natural arch in Arches National Park, Utah.

erosion of a flat-topped plateau underlain by horizontal beds produces flat-topped hills **(Fig. 13.11b)**. The names used for such flat-topped hills depend on their size: **mesas** have the largest surface area, *buttes* have an intermediate surface area, and *chimneys* are narrow relative to their height **(Fig. 13.11c)**. In places where bedrock consists of a thick sandstone layer cut by vertical joints, erosion along the joints can transform the layer into a set of narrow sandstone walls known as *fins*, separated by eroded spaces. Localized erosion through the base of a fin can produce a **natural arch (Fig. 13.12)**.

Of course, horizontally layered bedrock doesn't underlie all desert hills or mountains. Erosion of dipping strata can produce an asymmetrical ridge, or *cuesta* **(Fig. 13.13a)**, and erosion of vertical beds yields a *hogback*. Note that one side of a cuesta is a steep cliff, whose face parallels

FIGURE 13.13 Formation of cuestas and inselbergs.

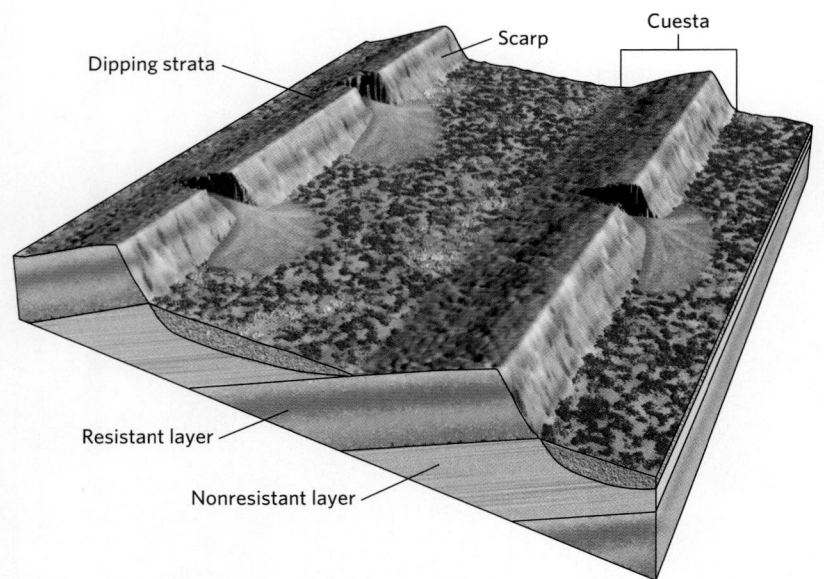

Dipping strata

Scarp

Cuesta

Resistant layer

Nonresistant layer

(a) Asymmetrical ridges called cuestas develop where strata are not horizontal.

(b) Aerial view of an inselberg surrounded by alluvium in China.

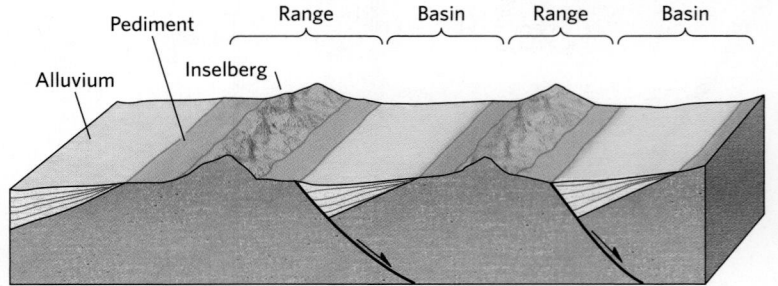

Pediment

Range   Basin   Range   Basin

Alluvium

Inselberg

(c) In the Basin and Range Province of the southwestern United States, tilted fault-block ranges evolve into inselbergs, bordered by pediments and basins.

steep joints. The other side is a sloping surface, called a *dip slope*, that is parallel to bedding. If bedrock consists of rocks that don't contain well-developed sets of parallel joints (such as complexly deformed strata, or intrusive or metamorphic rocks), erosion instead produces jagged rocky ridges. Given enough time, erosion of bedrock in a desert leaves isolated "island mountains," known by their German name, **inselbergs** (Fig. 13.13b). If water and wind strip away the sediment that erosion delivers to the base of an inselberg, then a broad, nearly horizontal bedrock surface, or *pediment*, surrounds the inselberg (Fig. 13.13c). More typically, sediment-filled basins lie between neighboring inselbergs.

Stony Plains and Desert Pavement

The ground surfaces of low-relief, sand-poor desert landscapes typically host a layer of pebble- and cobble-rich sediment, so these landscapes are called **stony plains** (Fig. 13.14a). The clasts may come from

recent rockfalls, or they may be lag deposits left when strong winds or occasional floods carry away fine-grained sediments. Portions of stony plains may develop **desert pavement**, a smooth, flat, desert-varnished surface that consists of countless separate tabular fragments of rock that fit together like pieces of a tile mosaic (Fig. 13.14b, c). Researchers suggest that development of desert pavement involves several processes. The tabular fragments may form when differential heating of rock blocks during the Sun's daily cycle causes the blocks to split along parallel vertical cracks. The resulting pieces eventually fall over and lie flat. Over time, windblown dust settles between them. During rains, clay in the dust absorbs water and swells, and during dry times, the clay shrinks. This repeated swelling and shrinking causes the tabular fragments to shift and settle to produce an overall flat surface. As this process continues, additional clay settles between the fragments, and an aridisol consisting of clay and silt builds up beneath

See for yourself

Uluru, Australia

Latitude: 25°20′47.64″ S
Longitude: 131°2′28.96″ E

Zoom to an altitude of 5 km (3 mi) and look down.

The red sandstone of Uluru (Ayers Rock), an inselberg, rises above the stony plains of the central Australian desert. The NW-trending diagonal lines are traces of vertical bedding in the sandstone.

FIGURE 13.14 Desert pavement and a hypothesis for its formation.

(a) A stony plain near the Gobi Desert. Occasional drainage has carved small channels into the surface.

(b) A well-developed desert pavement in the Sonoran Desert, Arizona. The inset shows a close-up of the pavement.

(c) Students standing at the edge of a trench cut into desert pavement. Note the soil between the pavement and the underlying alluvium.

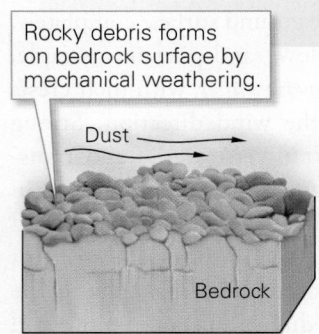
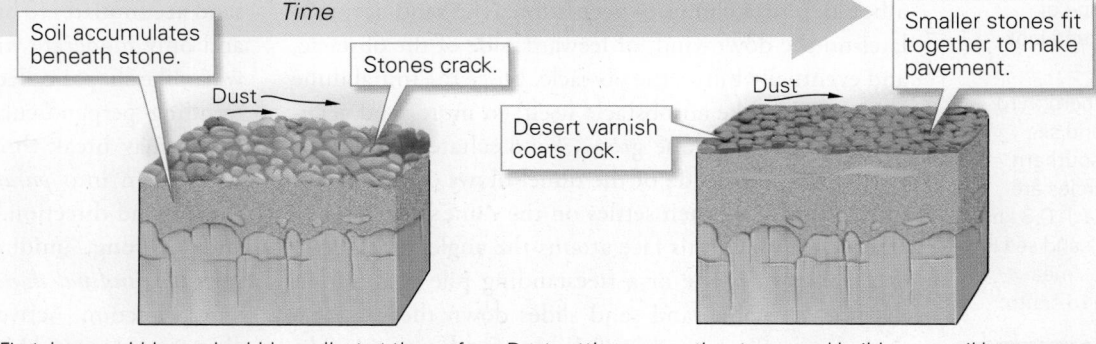

(d) Desert pavement forms in stages. First, loose pebbles and cobbles collect at the surface. Dust settles among the stones and builds up a soil layer below. The stones eventually crack into smaller pieces and settle to form a mosaic-like pavement.

FIGURE 13.15 Sand dunes in deserts.

(a) Relatively small dune fields cover stony plains and mountain flanks in west-central China.

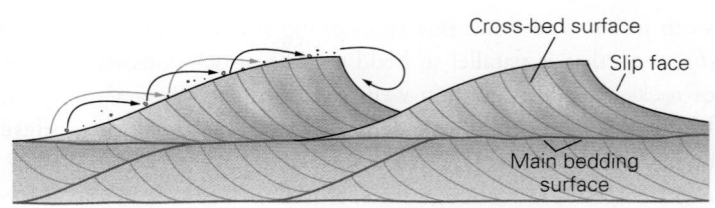

(c) A cross section shows cross bedding inside a dune.

(b) A sand dune with surface ripples.

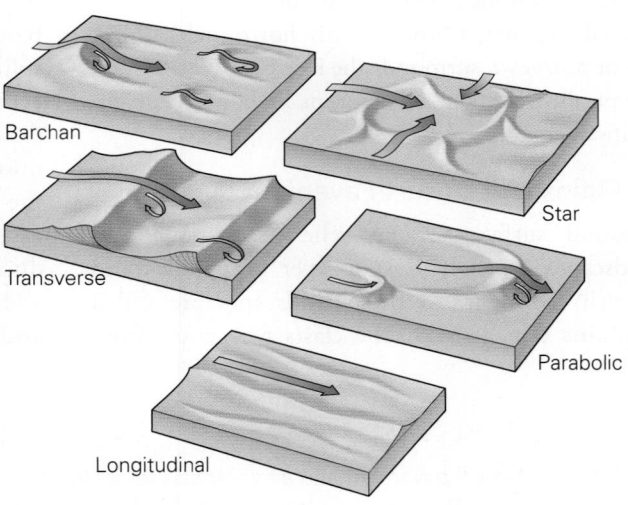

(d) The shapes of sand dunes depend on the sand supply and the character of the wind.

them (Fig. 13.14d). Gradually, desert varnish coats the surface of the pavement.

Seas of Sand: The Nature of Dunes

Places where the wind deposits thick accumulations of sand become sand seas, known as *ergs* in Arabic, which range from just a few kilometers to a few hundred kilometers across (Fig. 13.15a). Within these regions, the wind builds **sand dunes**, localized ridges of sand (Fig. 13.15b). Dune formation begins when sand becomes trapped by an obstacle that causes wind near the ground to slow down. The obstacle may be a pre-existing ridge of rock or sediment on the ground surface, a boulder, or a clump of vegetation. The sand accumulates on the downwind, or leeward, side of the obstacle, and eventually buries the obstacle. Once the initial dune exists, it acts like an obstacle itself, so more sand accumulates, and the dune grows. Sand saltates and creeps up the windward side of the dune, blows over the crest of the dune, and then settles on the dune's steeper, leeward face. When this face attains the angle of repose—the maximum slope of a freestanding pile of sand—it becomes unstable, and sand slides down the slope, so geologists refer to the leeward side of a dune as the *slip face*. Once formed, however, a dune does not stay put. Its crest moves downwind over time, as more and more sand accumulates on the slip face while the windward side erodes. Former slip faces are preserved as cross beds inside the dune (Fig. 13.15c).

Dunes display a variety of shapes and sizes, which depend on the wind direction, wind velocity, and sand supply (Fig. 13.15d). In regions of scarce sand and steady wind in a uniform direction, beautiful crescents called *barchan dunes* develop, with the tips of the crescents pointing downwind. If the wind direction shifts frequently, crescents pointing in different directions overlap one another, producing *star dunes*. Where enough sand accumulates to bury the ground surface completely, and only moderate winds blow, sand piles into simple, wave-like shapes called *transverse dunes*, with their crests trending perpendicular to the wind direction. Strong winds may break through transverse dunes and transform them into *parabolic dunes*, whose ends point in the upwind direction. Finally, if there is abundant sand and a strong, unidirectional wind, the sand streams into *longitudinal dunes*, whose crests lie parallel to the wind direction. Active sand dunes constantly move and change as the wind blows.

FIGURE 13.16 Desertification in the Sahel.

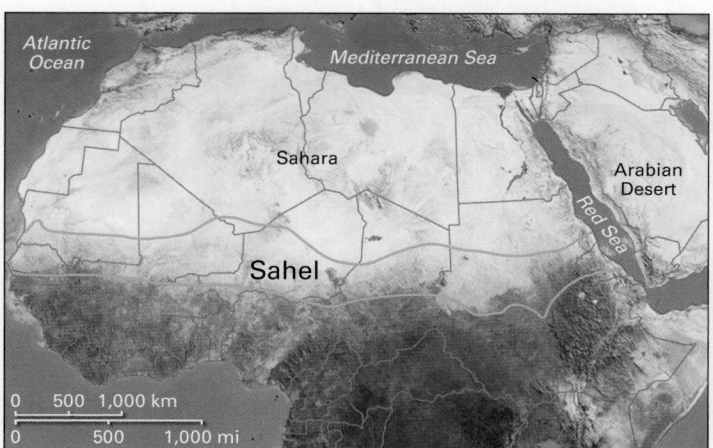

(a) The Sahel is the semiarid land along the southern edge of the Sahara. Large parts have undergone desertification.

(b) In parts of the Sahel, the soil has dried up, and hardly any vegetation survives.

Desert Problems

Many deserts are bordered by *semiarid regions*, dry grasslands or brush-covered areas that receive 25–76 cm (10–30 in) of rain annually. Can these areas bordering deserts transform into true deserts in a human time frame? Unfortunately, yes. Natural *droughts* (periods of unusually low rainfall), overpopulation, overgrazing, intensive farming, and diversion of water supplies can transform a semiarid grassland into a desert in a matter of years to decades. Geologists refer to this transformation as **desertification**. In some cases, a change in rainfall pattern or land use can reverse desertification.

An example of temporary desertification affected the southern Great Plains of the United States during the Great Depression of the 1930s. Banks had failed, workers had lost their jobs, and the stock market had crashed. No one needed yet another disaster—but in 1933, even nature turned hostile, and seasonal rains did not arrive. Farming had broken up soil that had been held together for millennia by the roots of prairie grasses, and without water, the topsoil of the croplands turned to powdery dust. Strong storms blew across the plains and sent the soil skyward to form dust storms that blotted out the Sun. When the dust settled, it buried houses and roads and dirtied every nook and cranny. A region that had once been rich farmland turned into a wasteland that soon acquired a nickname, the Dust Bowl, and it remained that way for almost a decade.

In recent decades, the Sahel, a belt of semiarid grassland that fringes the southern margin of the Sahara (Fig. 13.16a), has endured a fate similar to that of North America's Dust Bowl. The region's growing population has replaced soil-preserving perennial grasses with seasonal crops, and newly introduced herds of grazing animals not only eat grass down to the ground, but also compact the soil as they walk so it can't absorb moisture when it does rain. With the resulting loss of vegetation, the air has grown drier, and portions of the Sahel have turned into desert (Fig. 13.16b). When the changed landscape endures a drought, its inhabitants all too often endure devastating famines.

The recent change in the Aral Sea of central Asia illustrates how desertification can be caused by river diversion. This inland water body was over 250 km (155 mi) wide in 1990. Now, because the water of the rivers that once flowed into it is being diverted to be used for irrigation elsewhere, it has dwindled to a width of only 10 km (6 mi). The fishing boats that once plied its waters lie as rusting hulks amid the sand dunes accumulating on the former lake bed (Fig. 13.17b). The Great Salt Lake in Utah may be facing a similar fate in the next several years (Fig. 13.17c). Water levels in the lake have dropped so low that large areas that were once submerged are becoming playas, because water from the streams that used to flow into the lake has been diverted to other uses, and because climate change has affected precipitation rates. The dust churned up by winds blowing over the now-exposed beds of the Aral Sea and the Great Salt Lake can cause health problems for people living in adjacent towns and cities because it contains salt, clay, and silt as well as toxic chemicals from fertilizers and other sources brought into the lake by streams during the past century.

FIGURE 13.17 Diversion of rivers causes salt lakes to dry up.

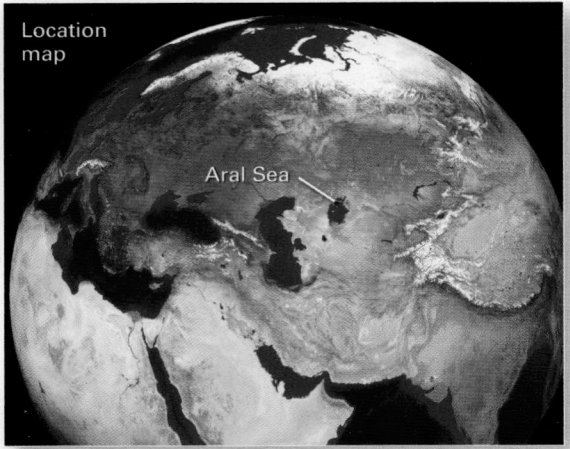
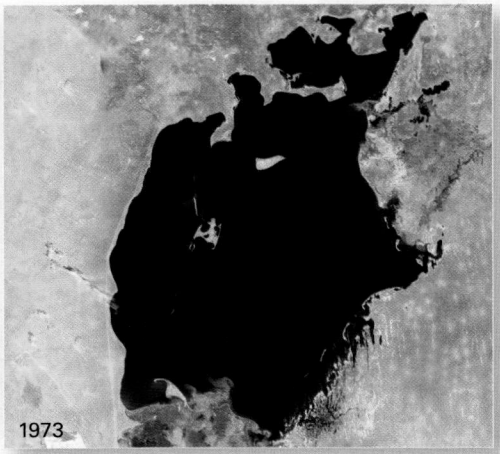
1973
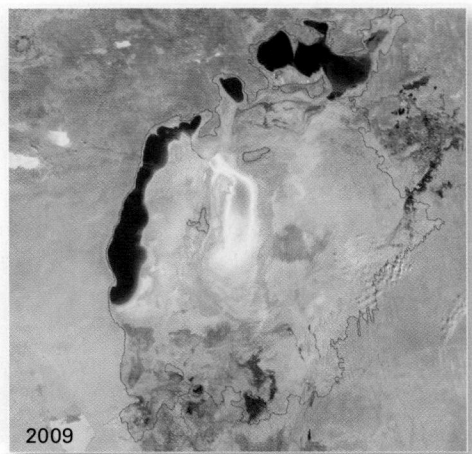
2009

(a) Satellite photos show that between 1973 and 2009, almost the entire Aral Sea dried up.

Boats of the former fishing fleet rust in the sand.

(b) The now-exposed bed of the Aral Sea has become a dusty graveyard for rusty fishing boats.

1989

2023

(c) The Great Salt Lake of Utah has started to dry up, as can be seen by comparing images taken 34 years apart. The portion of the lake bordering Salt Lake City (SLC) has shrunk substantially.

Take-home message...

In deserts, erosion of horizontal beds by cliff retreat yields buttes and mesas, whereas erosion of tilted strata forms cuestas. Stony plains, some of which develop desert pavements, form where deserts lack a substantial sand supply. Where winds build accumulations of sand, deserts contain sand dunes. In lands bordering deserts, droughts, population pressures, and river diversion have led to desertification, the transformation of once-vegetated land into desert.

Quick Questions —————————————

- How does a mesa differ from a chimney?
- Are all deserts covered by sand dunes?
- Why did the Aral Sea dry up?

13.4 Ice and the Nature of Glaciers

We've just discussed some of the hottest places on our planet. Now we shift our attention to the coldest places. In cold regions—the cryosphere of the Earth System—water exists in solid form, as *ice*, for most, if not all, of the year, and can build into **glaciers**, very slowly flowing sheets or streams of ice. Glaciers today cover most of Antarctica and Greenland, and parts of many mountain ranges, a total of about 10% of the Earth's land area. But during the Pleistocene Epoch (2.58 Ma to 11.7 Ka), sheets of ice, at times, covered most of Canada and part of the northern United States, as well as large areas of Europe and Asia, a total of about 30% of the land area. The glaciers of this *Pleistocene Ice Age* sculpted many of the landscapes in regions that are now densely populated, so an understanding of glacial landscapes helps land-use planners, farmers, and others in their work. To begin our discussion of glaciers and how they can modify the landscape, let's explore what they are made of, how they form, and how they move.

Formation of Glaciers

To understand glacial ice in a geologic context, let's apply the concepts introduced in the chapters on minerals and rocks. We can think of a single ice crystal as a mineral specimen—a naturally occurring, inorganic solid with a definable chemical composition (H_2O) and a regular crystal structure (Fig. 13.18a). A layer of snow is comparable to a sediment layer, and a layer of snow that has been compacted so that the grains stick together resembles a sedimentary rock layer (Fig. 13.18b). We can think of the ice on the surface of a pond as an igneous rock, for it forms when liquid water ("molten ice") solidifies by dropping below its freezing temperature (0°C, or 32°F). What about glacial ice? It's comparable to a metamorphic rock, in that it develops when pre-existing ice recrystallizes in the solid state and flows in response to differential stress (Fig. 13.18c).

Glaciers develop in polar regions, even though relatively little snow now falls there, because temperatures remain so cold even in summer that ice survives all year. Glaciers also develop in mountains, even at low latitudes, because temperature decreases with elevation, so that at high elevations, the average annual temperature stays cold enough for ice to survive all year.

How does freshly fallen snow, which consists of about 10% snowflakes and about 90% air, transform into glacial ice? The process begins as snow ages and the points of the snowflakes either *sublimate* (evaporate directly into water vapor) or *melt* (transform into liquid water). As this happens, the flakes pack together more tightly. When packed snow becomes buried deeply enough, the weight of the overlying snow generates sufficient pressure to cause remaining points of contact between snowflakes to melt by pressure solution (see Chapter 5). As a result, the snow transforms into a packed granular material called *firn*, which contains only about 25% air (Fig. 13.18d). Gradually, ice recrystallizes and firn transforms into a solid mass of ice composed of interlocking ice crystals, in which any remaining air exists only in tiny bubbles. Ice that lasts all year, has recrystallized, and has started to flow has become *glacial ice*. Notably, glacial ice may become coarser over time, as some crystals grow at the expense of others. Also, glacial ice preferentially absorbs red light, so light that passes through it has a bluish color. Transformation of fresh snow into glacial ice can take as little as tens of years in regions with abundant snowfall, or as long as thousands of years in regions with little snowfall.

Categories of Glaciers

Geologists distinguish between two general categories of glaciers. **Mountain glaciers**, or *alpine glaciers*, grow in or adjacent to mountainous regions (Fig. 13.19a). This category includes *cirque glaciers*, which fill a spoon-shaped depression, known as a *cirque*, on the flank of a mountain (see Fig. 13.27c); *valley glaciers*, rivers of ice that flow down valleys; mountain *ice caps*, mounds of ice that submerge peaks at the crest of a mountain range; and *piedmont glaciers*, fans or lobes of ice that form where a valley glacier emerges from a valley and spreads out over a plain (Fig. 13.19b-d). Mountain glaciers range in size from a few hundred meters to a few hundred kilometers long, and from tens of meters to 1.5 km (1 mi) thick. **Continental glaciers**, or *continental ice sheets*, in contrast, are vast and thick layers of ice that form at high latitudes and spread over thousands of square kilometers of continental crust. They now exist only on Antarctica and Greenland. Antarctica hosts two main ice sheets, separated by the Transantarctic Mountains (Fig. 13.20). During ice ages, such ice sheets covered large portions of other continents.

The Movement of Glacial Ice

The manner in which ice moves within a glacier typically varies with depth and with temperature. Under the pressures found at depths of greater than about 60 m (200 ft) below the glacier's surface, ice can undergo *ductile deformation*, meaning that the grains within it change shape very slowly without breaking, and new grains grow while old ones disappear (see Chapters 5 and 6). Ductile deformation in ice resembles ductile deformation and recrystallization in metamorphic rocks, except that it takes place at much colder temperatures. In *polar glaciers*,

Did you ever wonder . . .

how a glacier moves?

FIGURE 13.18 The nature of ice and the formation of glaciers.

(a) The shapes of snowflakes reflect the hexagonal crystal structure of ice.

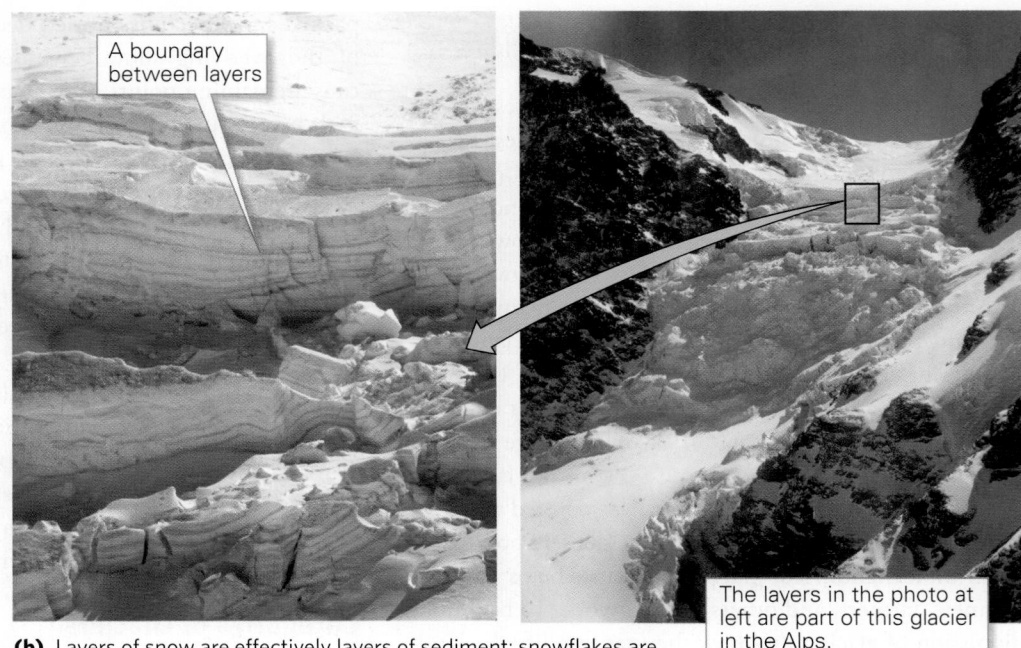

A boundary between layers

The layers in the photo at left are part of this glacier in the Alps.

(b) Layers of snow are effectively layers of sediment; snowflakes are the sediment grains.

The wall of a tunnel bored into a glacier

Loose snow (90% air)

Granular snow (50% air)

Firn (25% air)

Fine-grained ice (< 20% air, in bubbles)

Coarse-grained ice (< 20% air, in bubbles)

10,000 years (250 m)

130,000 years (2,000 m)

Not to scale

(c) Glacial ice is blue. As revealed by a microscope, it has coarse grains. The different colors in this photomicrograph, taken with polarized light, indicate differences in the orientation of crystals.

(d) Snow compacts and melts to form firn, which recrystallizes into glacial ice in a process comparable to metamorphism. Crystal size increases with depth.

FIGURE 13.19 A variety of glacier types form in mountainous areas.

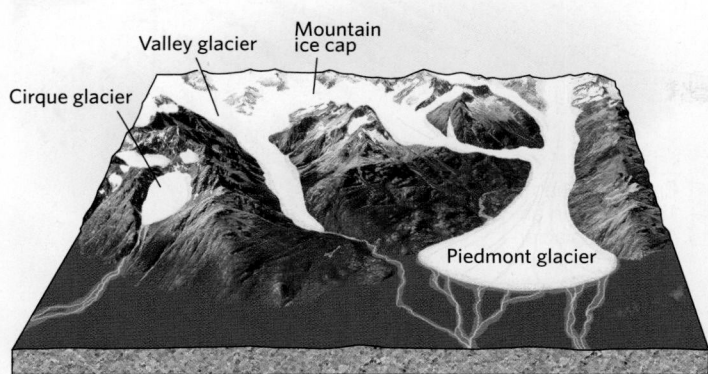

(a) Mountain glaciers are classified by their shape and position.

(b) A valley glacier and cirque glaciers in Switzerland.

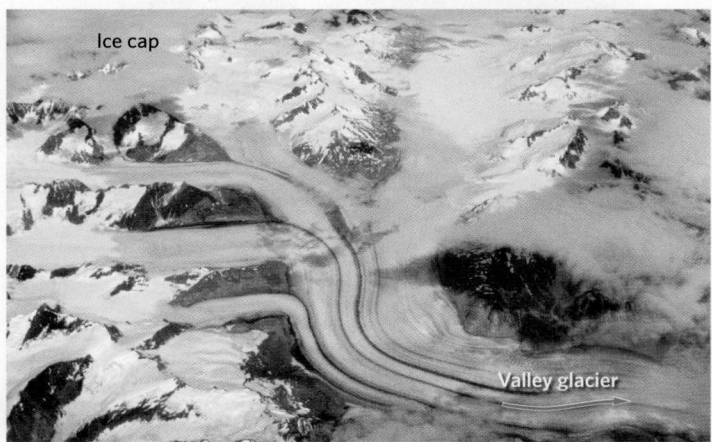

(c) Valley glaciers draining a mountain ice cap in Alaska. A valley glacier that drains an ice cap or ice sheet can also be called an "outlet glacier."

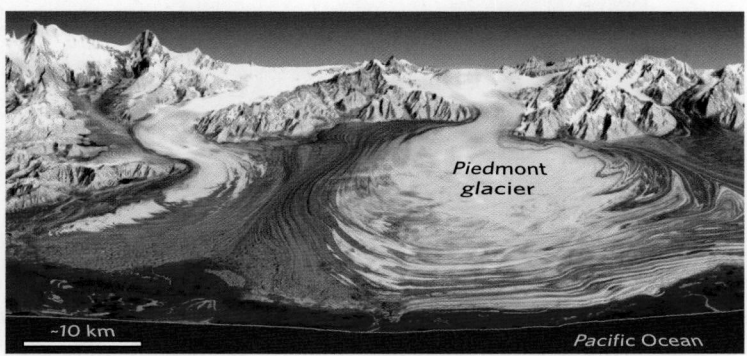

(d) A piedmont glacier near the coast of Alaska.

FIGURE 13.20 Antarctica hosts two ice sheets, separated by the Transantarctic Mountains.

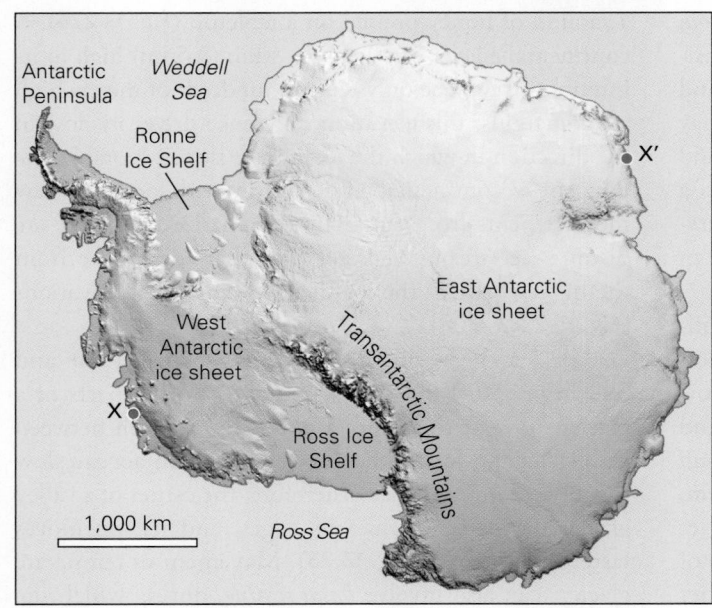

(a) A digital elevation model of Antarctica.

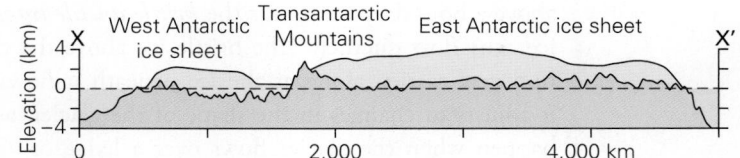

(b) A cross section of Antarctica along a line between X and X' in part (a). The Transantarctic Mountains separate East Antarctica from West Antarctica.

(c) A view across the Antarctic ice sheet looking toward the Transantarctic Mountains. Wind has blown away most loose snow, and the ice surface itself has become rippled.

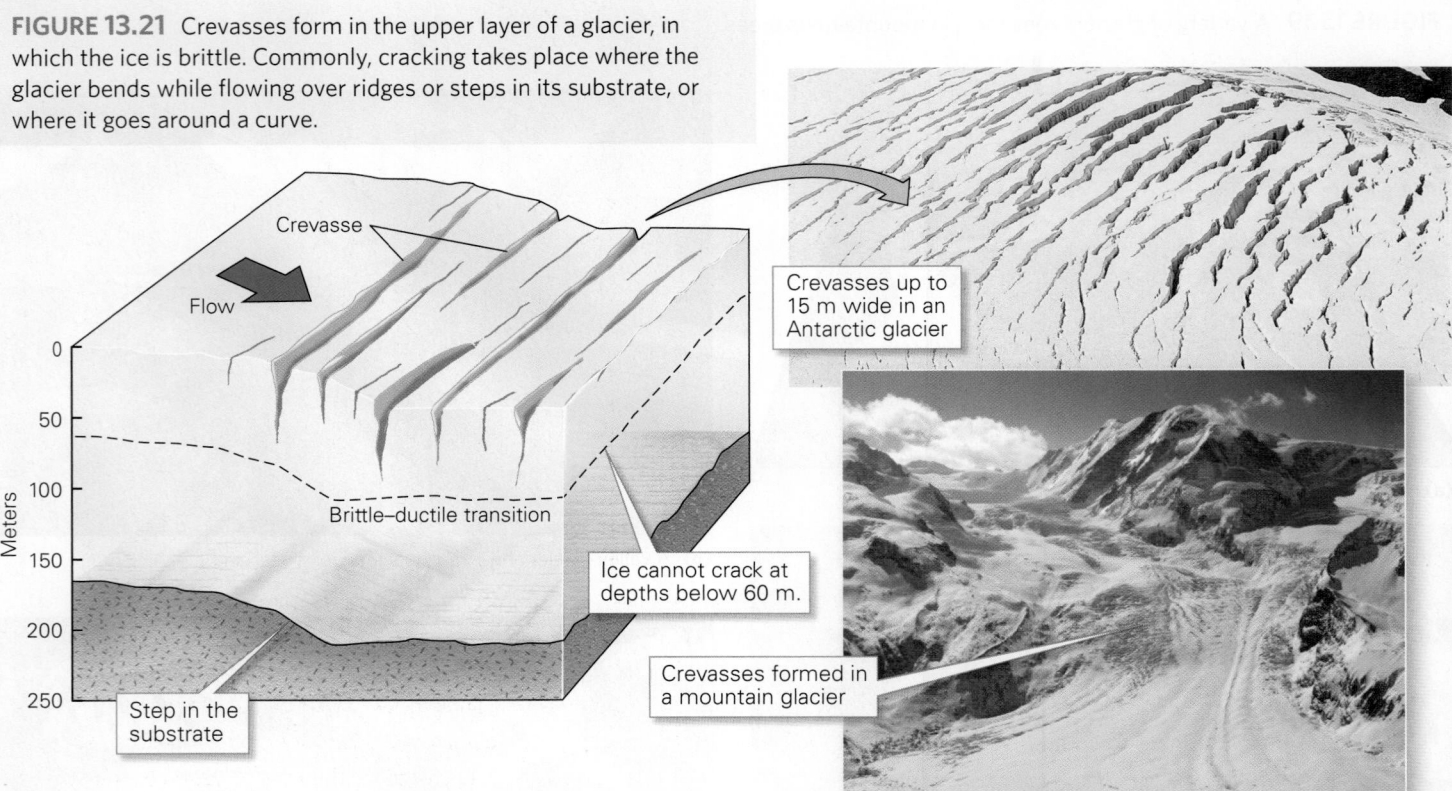

FIGURE 13.21 Crevasses form in the upper layer of a glacier, in which the ice is brittle. Commonly, cracking takes place where the glacier bends while flowing over ridges or steps in its substrate, or where it goes around a curve.

Crevasse

Flow

Brittle–ductile transition

Step in the substrate

Crevasses up to 15 m wide in an Antarctic glacier

Ice cannot crack at depths below 60 m.

Crevasses formed in a mountain glacier

Meters

Animation

Glacier Dynamics

in which ice remains completely frozen all year, ductile deformation involves only plastic processes, the rearrangement of atoms within ice crystals. In *temperate glaciers*, ice reaches its melting temperature for part of the year, so water films develop along grain boundaries. Therefore, ductile deformation may also involve the slip of ice crystals past their neighbors.

At depths of less than about 60 m within a glacier, above a boundary known as the *brittle-ductile transition*, ice can't flow ductilely. The brittle ice above the transition gets carried along as the ice beneath it flows, and it adjusts to changes in the shape of the glacier, as may happen when the glacier flows over a ledge or around a curve, by cracking brittlely. A crack can open into a fissure called a **crevasse** (Fig. 13.21). In large glaciers, crevasses can be hundreds of meters long and up to 15 m (50 ft) across.

A mountain of solid rock can stay standing for millions of years, but a mountain-sized block of ice cannot. Rather, a large mass of ice—a glacier—will slowly flow and change shape. Why? Under the conditions found at the Earth's surface, ice is weak enough that the pull of gravity alone can cause a large block of it to deform. Because gravity provides the driving force, a glacier always flows in the direction in which the top surface of the ice slopes (Box 13.2). Therefore, a mountain glacier flows from its **head** (the origin of the flow) to its **toe** or

terminus (the end of the flow) because ice at the head sits at a higher elevation than ice at the toe. As long as the glacier's surface tips down-slope, the base of the glacier can move up and over underlying small ridges (Fig. 13.22a). The same driving force, gravity, causes a continental glacier to spread outward from the region where its surface has the highest elevation, toward its margins, where the ice lies at lower elevations. In effect, a glacier behaves like a mound of honey poured on a tabletop (Fig. 13.22b). A continental glacier can be up to 4 km (2.5 mi) high in its interior but may be only tens to hundreds of meters high at its margins; this elevation difference drives its flow in the direction in which the ice surface slopes. Notably, the flow rate of continental glaciers varies with location. New measurements from Antarctica emphasize that there are distinct ice streams of faster flow that "drain" ice from the highest part of the ice sheet toward lower elevations (Fig. 13.22c).

Glaciers generally flow at rates between 10 and 300 m/yr (30–1,000 ft/yr). Notably, not all parts of a glacier move at the same rate, because friction between the glacier and rock where they come in contact can slow the movement of the ice. Therefore, the center of a valley glacier moves faster than its edges, and its top moves faster than its base (Fig. 13.23). Movement of temperate glaciers can also involve *basal sliding*, during which the whole mass of the glacier moves relative to its substrate

How can I explain . . .

Flow of a continental ice sheet

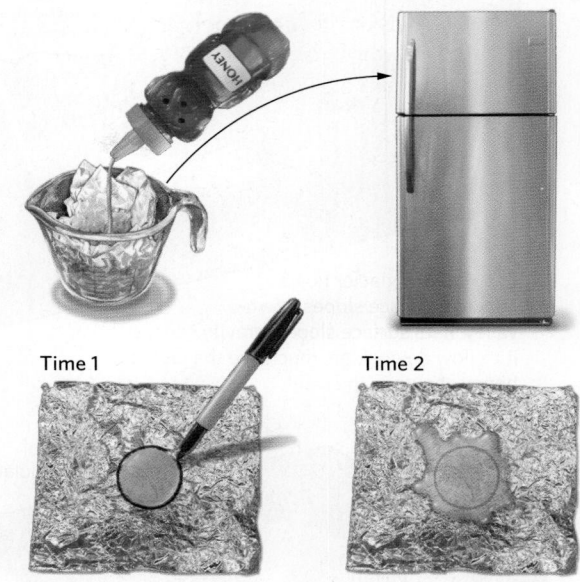

Time 1

Time 2

What are we learning?

- That continental glaciers flow over the landscape because of gravity.
- That an ice sheet spreads out in all directions, but not necessarily uniformly.

What you need:

- A small bowl
- Aluminum foil
- Honey
- A felt-tip marker

Instructions:

- Mold the foil to the inside of the bowl; then pour honey into the foil so that it makes a layer about 2 cm (1 in) thick. Place the foil and honey in the freezer for about a half hour.
- Remove the foil and honey from the freezer and place on a tabletop. Quickly spread out the foil so that it's flat, and draw a line around the edge of the cold honey mound.
- As the honey warms, it gets weaker and flows. Watch as the honey spreads laterally. Look closely at how the honey interacts with irregularities in the foil surface.

What did we see?

- The thick honey mound will flow in response to gravity, even though it is not moving down a surface slope. It moves as long as there is a slope from the center of the honey mound to the edge.
- The honey travels farther in low areas, so distinct lobes develop along its edges.
- The base of the honey can ride up over bumps in the foil.

by sliding on a layer of liquid water or water-saturated sediment. The water or watery sediment holds the glacier above bedrock and can flow faster than can solid ice. Locally, basal sliding may accommodate a *glacial surge*, during which ice accelerates to speeds of 10–110 m (30–360 ft) per day!

Glacial Advance and Retreat

Glaciers resemble bank accounts in that ice can be added or removed, so that the glacier gets larger or smaller. Snowfall adds to the glacier in the *zone of accumulation*, while **ablation**—the removal of ice by sublimation, melting, or *calving* (the breaking off of ice chunks at the toe of the glacier)—subtracts from the glacier in the *zone of ablation*. The **equilibrium line** defines the boundary between these zones of the glacier (**Fig. 13.24**). As the glacier flows, ice follows a curved path, sinking down toward the base of the glacier in the zone of accumulation as new ice accumulates above it, and rising up toward the surface of the glacier in the zone of ablation as overlying ice melts or sublimates.

The balance between accumulation and ablation affects both the position of a glacier's toe and the thickness of the ice. Specifically, if the rate of accumulation equals the rate of ablation, then the position of the toe remains fixed, even though ice within the glacier continues to flow toward the toe, and the glacier's thickness stays the same (**Fig. 13.25a**). If, however, the rate of accumulation exceeds the rate of ablation, then the toe moves forward into previously unglaciated regions—a change known as a **glacial advance**—and the ice thickens (**Fig. 13.25b**). In mountain glaciers, the position of the toe moves downslope during an advance, and in continental glaciers, it moves outward, away from the glacier's origin. This movement allows the glacier's toe to move toward lower latitudes. If the rate of ablation exceeds the rate of accumulation, then the position of the toe moves back toward the glacier's origin—a change known as a **glacial retreat**—and the glacier thins (**Fig. 13.25c**). During a mountain glacier's retreat, the position of the toe moves upslope. During a continental glacier's retreat, the diameter of the glacier (in map view) decreases (**Fig. 13.25d**).

FIGURE 13.22 Forces that drive the movement of glaciers.

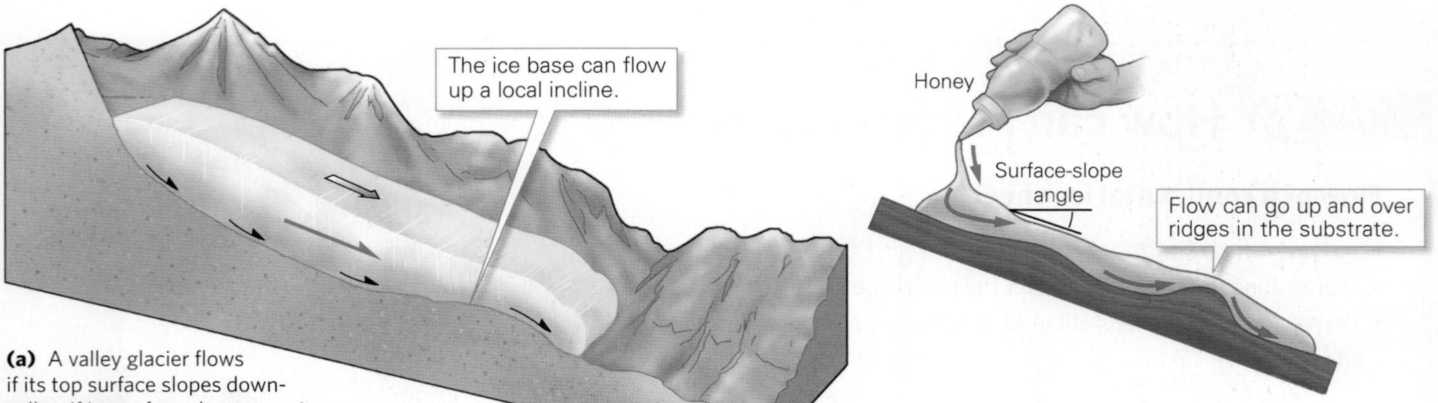

The ice base can flow up a local incline.

Honey

Surface-slope angle

Flow can go up and over ridges in the substrate.

(a) A valley glacier flows if its top surface slopes down-valley. If its surface slopes, gravity causes it to flow downslope, much like the layer of honey shown in the inset.

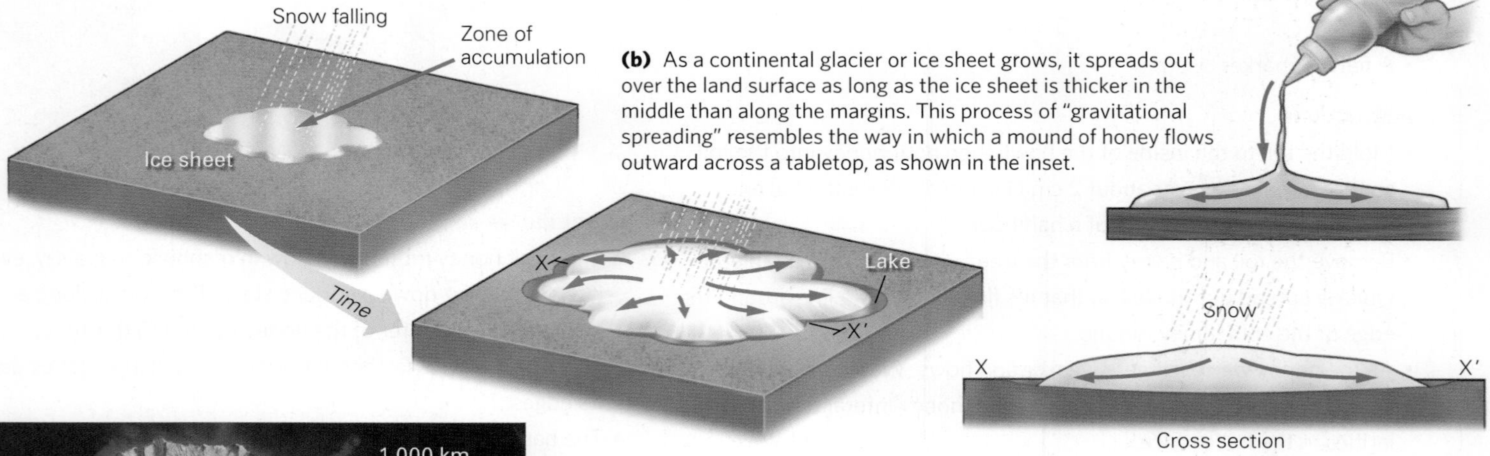

Snow falling

Zone of accumulation

Ice sheet

Time

X

X'

Lake

(b) As a continental glacier or ice sheet grows, it spreads out over the land surface as long as the ice sheet is thicker in the middle than along the margins. This process of "gravitational spreading" resembles the way in which a mound of honey flows outward across a tabletop, as shown in the inset.

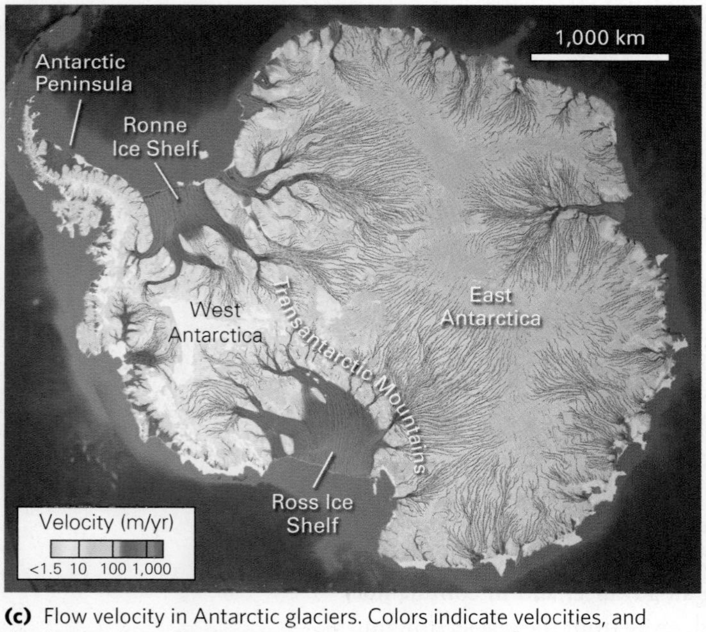

Snow

X   X'

Cross section

Antarctic Peninsula

Ronne Ice Shelf

1,000 km

West Antarctica

Transantarctic Mountains

East Antarctica

Ross Ice Shelf

Velocity (m/yr)

<1.5 10 100 1,000

(c) Flow velocity in Antarctic glaciers. Colors indicate velocities, and thin lines indicate flow direction. Note that valley glaciers carry ice from East Antarctica through the Transantarctic Mountains down to the Ross Ice Shelf and the Ronne Ice Shelf. The valley glaciers passing through the Transantarctic Mountains can be called outlet glaciers of the East Antarctic ice sheet.

FIGURE 13.23 Different parts of a glacier flow at different velocities due to friction with the substrate. The top and center regions flow faster than the bottom and sides.

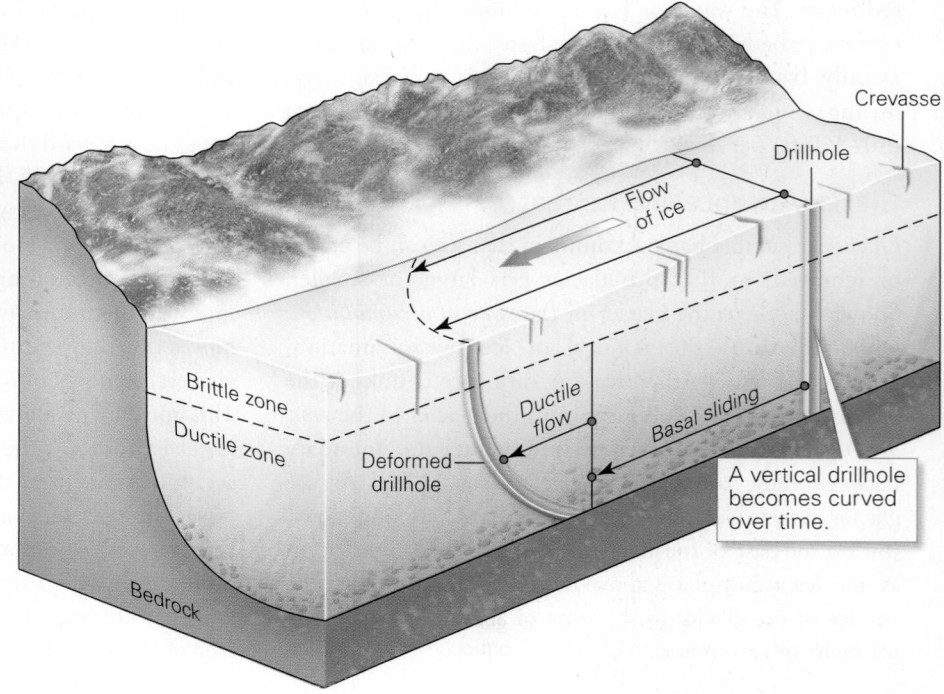

Crevasse

Flow of ice

Drillhole

Brittle zone

Ductile zone

Ductile flow

Deformed drillhole

Basal sliding

A vertical drillhole becomes curved over time.

Bedrock

FIGURE 13.24 The equilibrium line separates the zone of accumulation from the zone of ablation. As indicated by arrows, ice sinks while flowing downhill in the zone of accumulation and rises while flowing downhill in the zone of ablation. Overall, it follows a curving path.

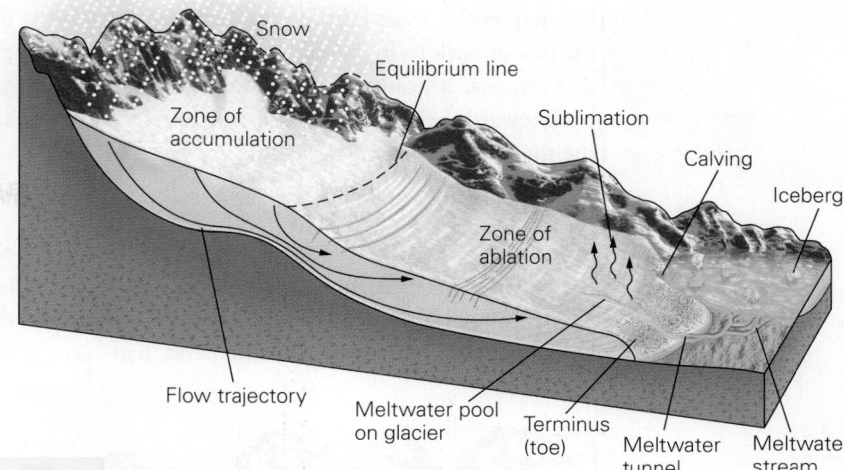

FIGURE 13.25 Glacial advance and retreat.

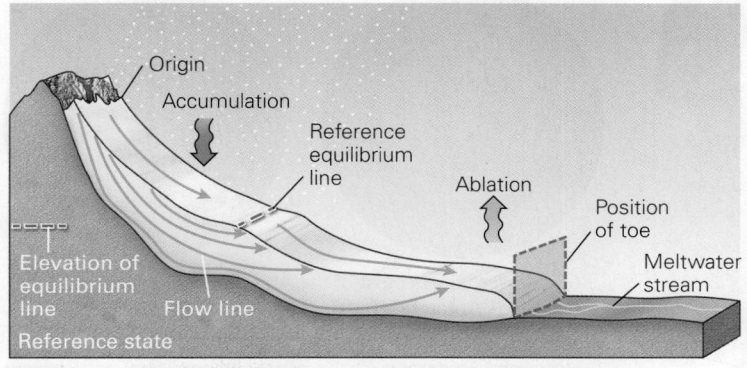

(a) The position of the toe represents a balance between accumulation and ablation.

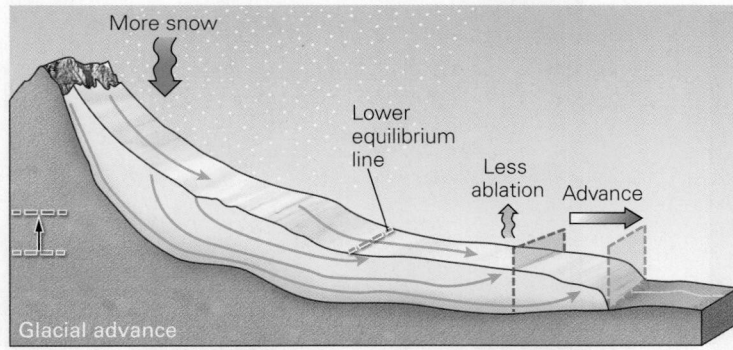

(b) If accumulation exceeds ablation, the glacier advances, the toe moves farther from the origin, and the ice thickens.

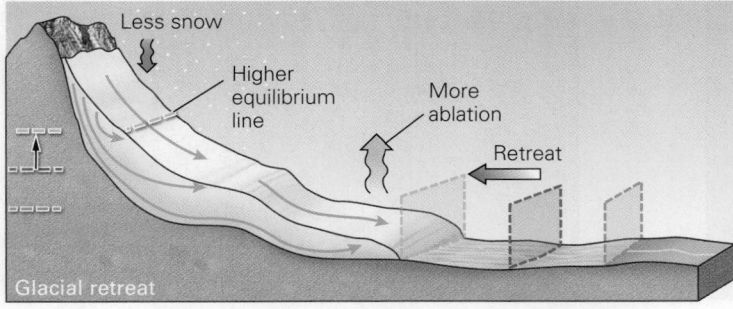

(c) If ablation exceeds accumulation, the glacier retreats and thins. The position of the toe moves back, even though ice continues to flow toward the toe.

(d) A retreating glacier in Switzerland. Note that the areas on the sides of the glacier that were once ice covered are now bare. Lateral and terminal moraines indicate the former extent of the glacier.

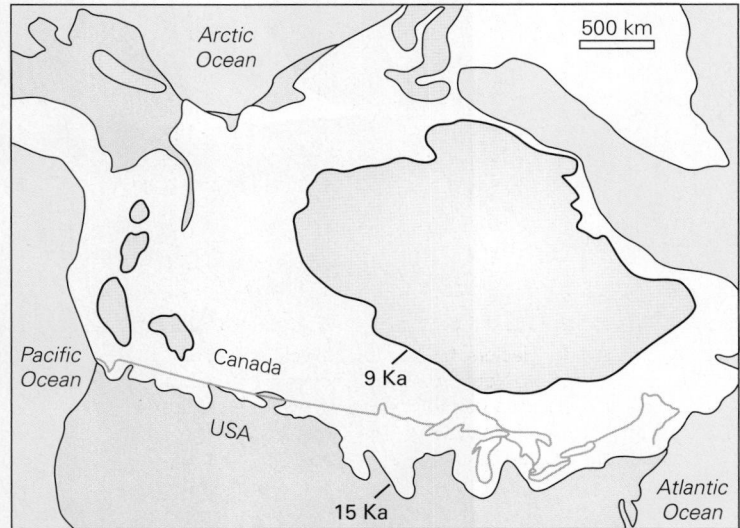

(e) The retreat of the continental glacier that covered parts of North America during the Pleistocene. Between 15 Ka and 9 Ka, it decreased from covering almost all of Canada to covering just a small part of Canada. Note how the diameter of the glacier decreased.

Consequently, lower latitudes undergo *deglaciation* as the toe moves back to higher latitudes. Note that when a glacier retreats, it's only the position of the toe that moves back toward the origin; ice itself continues to flow toward the toe, for it cannot flow against gravity.

Ice in the Sea

On the moonless night of April 14, 1912, the ocean liner RMS *Titanic* struck an *iceberg*, a large block of floating ice, in the frigid North Atlantic. Lookouts had seen the ghostly mass only minutes earlier and had alerted the helm, but the ship couldn't turn fast enough to avoid disaster. The force of the blow split its hull, allowing water to gush in, and less than three hours later, the *Titanic* disappeared beneath the surface.

Where do icebergs like the one responsible for the *Titanic*'s tragic fate originate? At high latitudes, mountain glaciers and continental glaciers flow down to and, in some cases, out into coastal waters to become **tidewater glaciers**. A tidewater glacier that forms where a mountain glacier enters the sea may protrude a few kilometers out as an *ice tongue*, whereas a continental glacier entering the sea becomes a broad, flat **ice shelf**. In shallow water, the glacier remains grounded, but where the water becomes deep enough, the glacier floats (Fig. 13.26a).

(a) Glacial ice is grounded in shallow water, but floats in deep water.

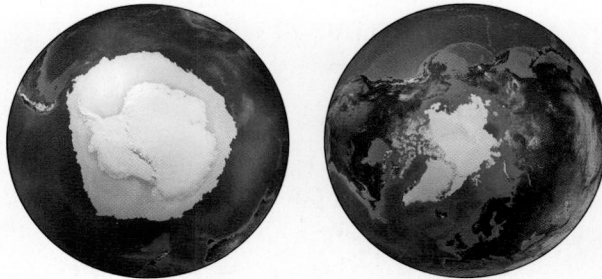

(c) Sea ice covers most of the Arctic Ocean (right) and surrounds Antarctica (left).

(b) This artist's rendition of an iceberg emphasizes that most of the ice is underwater.

(d) In summer, large areas of sea ice around Antarctica break up and melt.

FIGURE 13.26
Ice in the sea.

At the toe of a tidewater glacier, huge chunks of ice calve off, crumble, and tumble into the water with an impressive splash, producing ice fragments in a wide range of sizes. Those fragments that rise at least 5 m (16 ft) above the water and are at least 15 m (50 ft) across are formally called **icebergs**. Since four-fifths of a floating body of ice lies below the surface of the sea, we can estimate that the base of an iceberg that protrudes 40 m (130 ft) into the sky extends 160 m (525 ft) below sea level **(Fig. 13.26b)**. The tallest icebergs protrude skyward by about 100 m (330 ft). Icebergs that break off tidewater glaciers tend to be irregular in shape, with sharp peaks, while those that break off ice shelves can be immense tabular blocks many kilometers across.

Not all ice floating in the sea originates as glaciers on land. In polar climates, the surface of the ocean itself freezes, forming **sea ice** **(Fig. 13.26c)**. (Keep in mind the distinction between sea ice—frozen seawater—and an ice shelf—a floating portion of a continental glacier.) During the southern hemisphere winter, sea ice encircles Antarctica in a belt up to 1,200 km (750 mi) wide, as measured perpendicular to the coast. And during the northern hemisphere winter, nearly the entire Arctic Ocean has a cover of sea ice. Arctic sea ice, which reaches a maximum thickness of 2–6 m (6–20 ft), slowly moves with ocean currents at a rate of about 2 km (1.2 mi) per day. Both Arctic and Antarctic sea ice melt substantially during the summer as the ice breaks up **(Fig. 13.26d)**. Less than 15% of Antarctica's sea ice lasts all year. In the Arctic, about 40% of the sea ice survives the summer, but that amount has been steadily decreasing as the climate warms (see Chapter 19).

Take-home message...

Glaciers form when snow persists all year, gets buried deeply, and turns to ice. Mountain glaciers form at high elevations, whereas continental glaciers form at high latitudes and spread over continents. Ice flows by ductile deformation at depth. The upper, brittle portion of a glacier fractures to form crevasses. The balance between accumulation and ablation controls glacial advance or retreat. In polar regions, sea ice covers large areas.

Quick Questions

- Explain the difference between glacial ice and a layer of snow.
- Can ice flow uphill during retreat of a valley glacier?
- What is an ice shelf?

13.5 Erosion and Deposition by Glaciers

Glaciers serve as very powerful agents of erosion, capable of carving deep valleys and knife-edged ridges in mountains, and of beveling and polishing broad areas of continents. Glaciers also transport vast quantities of sediment, which accumulate to produce distinctive landscapes. Here we look at how glacial erosion and glacial deposition take place, and examine the specific landforms that they produce.

Glacial Erosion and Its Products

Though ice itself is much softer than rock, the movement of glaciers can result in substantial amounts of erosion. Glaciers erode their substrate in three ways:

- *Plucking:* As a glacier moves over its substrate, small quantities of meltwater at the base of the ice can seep into cracks and along bedding planes bounding a block of rock. When this water refreezes, frost wedging pushes up the block. Eventually, the block becomes surrounded by ice, and the glacier lifts it and carries it away. Similar processes allow the glacial ice to incorporate loose sediment from its substrate. Since ice can transport fragments of any size, the entrained sediment is unsorted.

- *Abrasion:* A glacier grinds away at the bedrock that it moves over and turns that bedrock into a clay- or silt-like powder informally called *rock flour*. We've noted that pure ice is too soft to abrade rock, but the hard clasts embedded in the base of a glacier act like the teeth of a giant rasp as they dig into bedrock. Abrasion by incorporated sand- or silt-sized grains polishes bedrock at the base or sides of the glacier **(Fig. 13.27a)**. Larger clasts or trains of clasts embedded in the ice can produce grooves or scratches, called **glacial striations**, that range from 1 mm to 1 m (0.04–40 in) across and may be tens of centimeters to tens of meters long **(Fig. 13.27b)**. Glacial striations trend parallel to the flow direction of the ice.

- *Subglacial water erosion:* Meltwater flows along the base of some glaciers, and like any moving water, this meltwater not only can erode the substrate, but can transport sediment under the ice.

Mountain glaciers contribute to sculpting a variety of distinctive landforms. The process begins high in the mountains, where frost wedging due to freezing and thawing breaks up exposed bedrock bordering the glacier's head. The resulting debris falls on the ice, or gets plucked at the base of the ice, and moves downslope with the

EXPLORE EARTH SCIENCE IN 3D
Glacial Striations

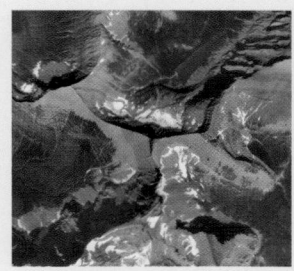

See for yourself

Glaciated Peaks, Montana
Latitude: 48°56′33.66″ N
Longitude: 113°49′54.59″ W

Zoom to an altitude of 8 km (5 mi) and look down.

Three cirques bound a horn in the Rocky Mountains north of Glacier National Park. Note the knife-edged arêtes between the cirques. The glaciers that carved the cirques have melted away. The thin stripes on the mountain face are traces of sedimentary beds that make up the bedrock.

FIGURE 13.27 Glaciers are powerful agents of erosion.

(a) A glacially polished outcrop in Central Park, New York City.

(b) Glacial striations in Victoria, British Columbia.

(c) Examples of a cirque and an arête in the Alps.

(d) The Matterhorn in Switzerland is bordered by cirques.

glacier. Over time, a bowl-shaped depression, or **cirque**, develops on the side of the mountain **(Fig. 13.27c)**. A narrow ridge called an *arête* (French for ridge) separates adjacent cirques. When cirques form on three or more sides of a mountain, the mountain becomes a pointed peak called a **horn (Fig. 13.27d)**.

Glacial erosion dramatically modifies the shape of a valley, as you can see by comparing a river-eroded valley

with a glacially eroded valley **(Fig. 13.28)**. Specifically, if you look along the length of a river in unglaciated mountains, you'll see that it typically flows down a *V-shaped valley*, with the river channel forming the point of the V. The V shape develops because river erosion occurs only in the channel, and mass wasting causes the valley slopes to approach the angle of repose (see Chapter 12). But if you look down the length of a glacially eroded valley, you'll see

FIGURE 13.28 Stages during the development of a glacially carved mountainous landscape.

Time

(a) Before glaciation, valleys are V-shaped, and tributary mouths are at the same elevation as the trunk stream.

(b) During glaciation, ice erodes both the floor and sides of valleys.

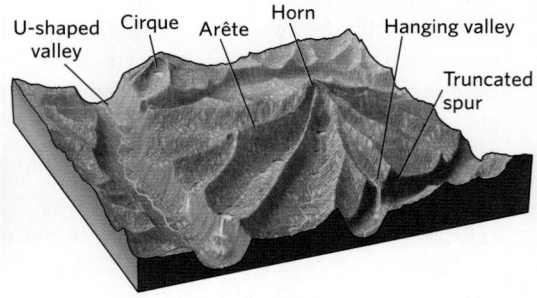

(c) After glaciation, valleys are U-shaped. Hanging valleys and truncated spurs intersect the trunk valley. Some peaks become horns.

FIGURE 13.29 Examples of glacially eroded valleys.

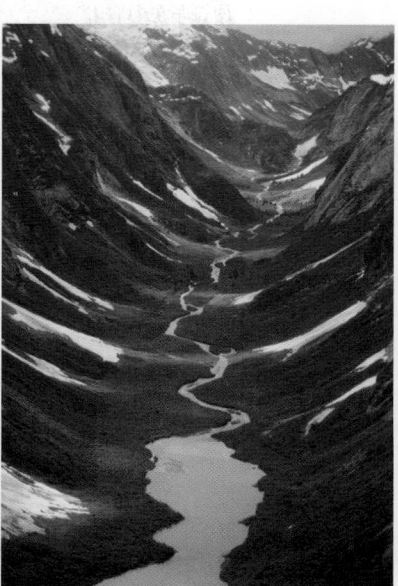

(a) A U-shaped valley bordered by Half Dome, in Yosemite National Park, California.

(b) A U-shaped glacial valley in the Tongass National Forest, Alaska.

(c) A waterfall spilling out of a U-shaped hanging valley in the Sierra Nevada.

that it resembles a U, with steep walls. Such a **U-shaped valley** (Fig. 13.29a, b) forms because glacial erosion not only lowers the valley's floor, but also bevels its sides.

Glacial erosion in mountains also modifies the intersections between tributary valleys and the trunk valley. Recall that the trunk stream of a drainage network serves as the base level for its tributaries, so a tributary stream and a trunk stream intersect at the same elevation. Tributary glaciers, however, do not erode their valleys to the same depth as the valley carved by a trunk glacier at the same elevation. As a consequence, when tributary glaciers melt away, they leave behind **hanging valleys**, whose mouths typically perch well above the floor of the trunk valley. A present-day tributary stream will cascade over a waterfall where it emerges from a hanging valley (Fig. 13.29c). Similarly, glacial erosion carves away the end of an arête, leaving a landform known as a *truncated*

spur (see Fig. 13.28c). (A *spur* is a ridge that intersects the main valley at a high angle; an arête that intersects the main valley is a spur relative to the main valley.)

The erosional features produced by continental glaciers depend, to a large extent, on the nature of the preglacial landscape. Where an ice sheet spreads over a region of low relief, glacial erosion produces a vast area of polished, striated surfaces. Where an ice sheet spreads over a hilly area, it deepens valleys and smooths hills. Glacially eroded hills may be elongate in the direction of ice flow and asymmetrical because glacial abrasion smooths the upstream part of the hill, yielding a gentle slope, whereas glacial plucking eats away at the downstream part, producing a steep slope. Geologists refer to such a hill as a **roche moutonnée**, French for sheep rock, because its profile resembles that of a sheep resting in a meadow (Fig. 13.30).

FIGURE 13.30 The sculpting of a roche moutonnée.

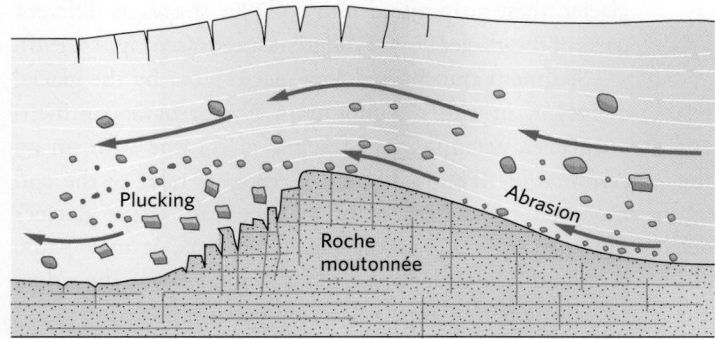

(a) Glacial abrasion rasps the upstream side, and glacial plucking carries away fracture-bounded blocks on the downstream side.

(b) An example of a roche moutonnée in the Sierra Nevada. The glacier that shaped this hill flowed from right to left.

FIGURE 13.31 Glacial fjords.

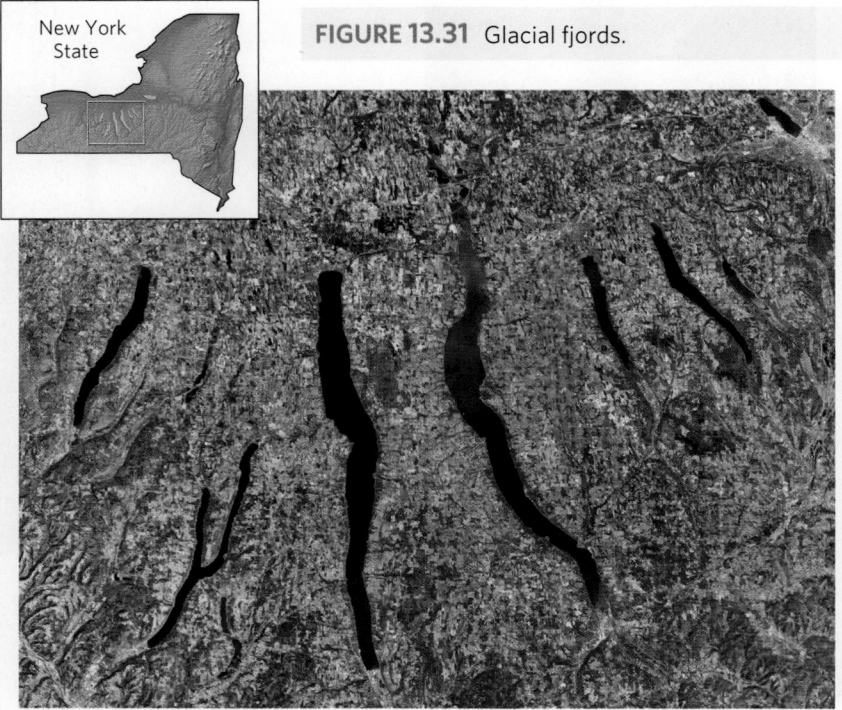

New York State

(a) The Finger Lakes in New York State are freshwater glacial fjords.

(b) One of the many spectacular fjords of Norway. The water is an arm of the sea that fills a U-shaped valley. Tourists are standing on Pulpit Rock (Preikestolen).

See for yourself

Baffin Island, Canada

Latitude: 67°8′19.40″ N
Longitude: 64°59′33.55″ W

Zoom to an altitude of 40 km (25 mi) and look down.

Two valley glaciers draining the Baffin Island ice cap merge into a trunk glacier that flows NE and then into a fjord, partly filling a U-shaped valley. Note the lateral and medial moraines.

Glacially carved valleys that later fill with water produce deep, elongate bodies of water known as **fjords**. Some fjords, such as the Finger Lakes of New York State, are inland freshwater lakes (Fig. 13.31a). Where a glacier carves a valley that extends down to the sea, however, it becomes a *marine fjord*, an elongate bay filled with seawater. The floor of a marine fjord lies below sea level because, at the time the glacier was carving the fjord, its base remained in contact with the substrate until it reached a location at which the sea's depth exceeded four-fifths of the glacier's thickness, so that the glacier started to float and became an ice tongue. As we'll see in Section 13.7, sea level was over 100 m lower during the Pleistocene Ice Age, when coastal fjords were carved, so some of them have depths in excess of 100 m. Spectacular marine fjords with steep walls occur along the coasts of Norway, New Zealand, Chile, Maine, and Alaska (Fig. 13.31b).

The Glacial Conveyor

A glacier, like a conveyor belt, transports the sediment it carries in the direction of ice flow (Fig. 13.32). In the case of mountain glaciers, some of the sediment moving with this *glacial conveyer* comes from the substrate under the ice; rest tumbles onto the glacier during rockfalls or slides down slopes bordering the glacier. All sediment transported by continental glaciers, however, comes from plucking or abrasion of the substrate, because mountains do not rise above the ice, so no debris can tumble onto the ice surface.

Geologists refer to a pile or ridge of sediment carried by a glacier, or left by a melting glacier, as a **moraine**. Sediment that tumbles onto the edge of a mountain glacier forms a debris stripe known as a *lateral moraine* (Fig. 13.33a). When the glacier melts, lateral moraines remain stranded along the sides of the valley it carved, like bathtub rings (Fig. 13.33b). Where two valley glaciers merge, two lateral moraines merge to become a *medial moraine*, a stripe of debris that trends parallel to the ice-flow direction down the interior of the composite glacier (Fig. 13.33c). Trunk glaciers created by the merging of multiple tributary glaciers contain several medial moraines (see Fig. 13.19c). If the ice flows into open land and becomes a piedmont glacier, these stripes fold into complex shapes as different parts of the glacier flow at different speeds (see Fig. 13.19d).

Sediment transported to a glacier's toe by the glacial conveyor, in either a mountain glacier or a continental glacier, accumulates in a pile at the toe to form an *end moraine* (Fig. 13.34a, b). Geologists refer to the end moraine at the farthest limit of ice flow as the glacier's *terminal moraine*. As the glacier retreats, it may pause temporarily and deposit a *recessional moraine*. The ridges of sediment that now form Long Island and Cape Cod in the eastern United States are remnants of the terminal and recessional moraines left by a Pleistocene glacier

FIGURE 13.32 The concept of the glacial conveyor.

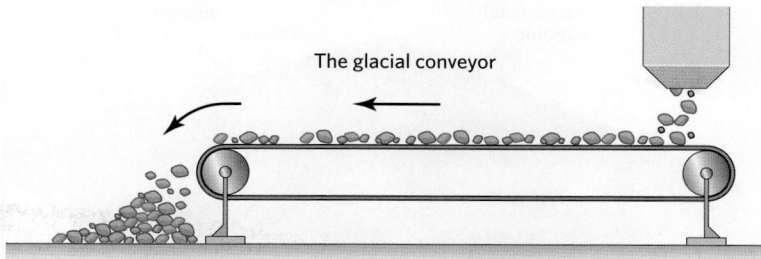

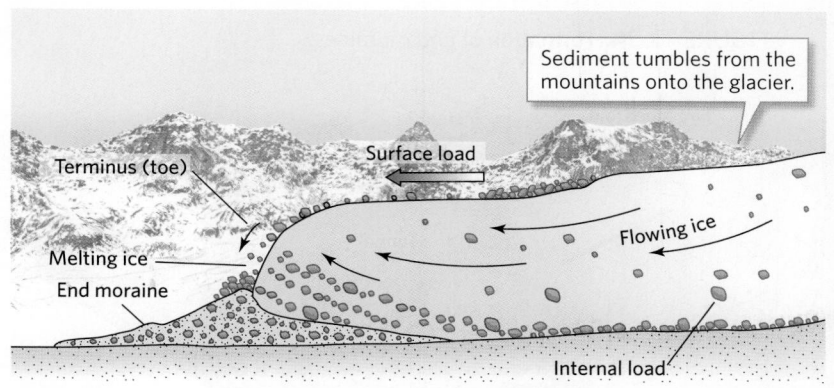

Sediment tumbles from the mountains onto the glacier.

Surface load

Terminus (toe)

Melting ice

End moraine

Flowing ice

Internal load

(a) In effect, a glacier acts like a conveyor belt, moving sediment in the direction of its flow.

(b) Sediment in and on a glacier gets carried to the toe.

FIGURE 13.33 Lateral and medial moraines associated with valley glaciers.

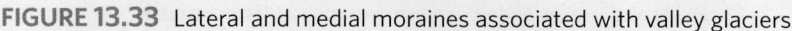

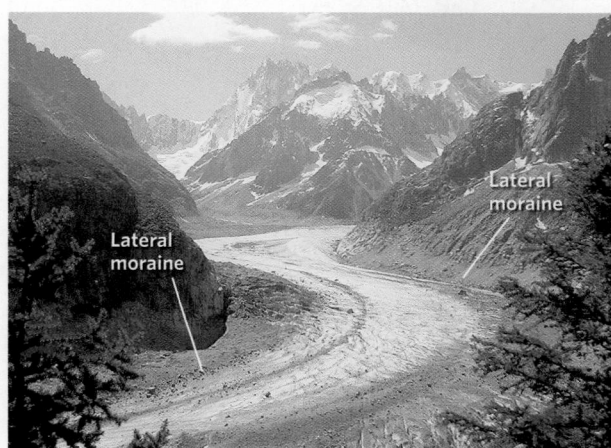

Lateral moraine

Lateral moraine

(a) This glacier in the French Alps has two lateral moraines as well as a medial moraine.

(b) The large ridge of sediment at the base of the mountain slope is a lateral moraine left when the glacier that once filled this valley in the Canadian Rockies melted away.

Rockfall

Medial moraine

Lateral moraine

Origin of medial moraine 1

Lateral moraine

Moraine 2, also a medial moraine, formed to the right of this image.

(c) These medial moraines formed where lateral moraines of two valley glaciers merged in Alaska.

FIGURE 13.34 Formation of end moraines.

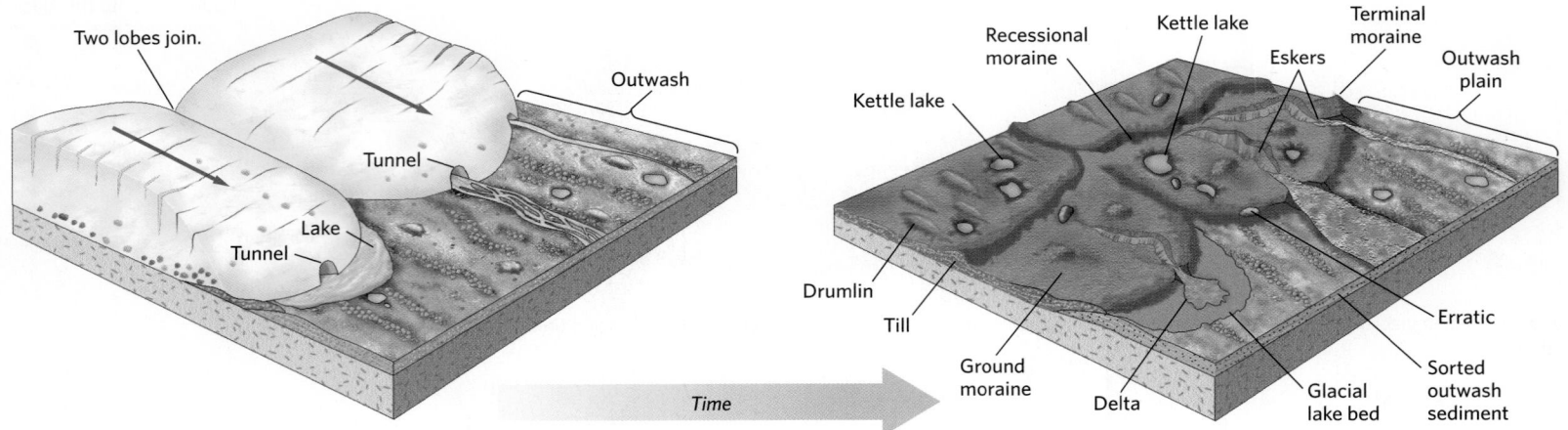

(a) When a glacier begins to retreat, it drops sediment and releases meltwater.

(b) Distinct depositional landforms, including several types of moraines as well as kettle lakes and eskers, remain when a glacier retreats.

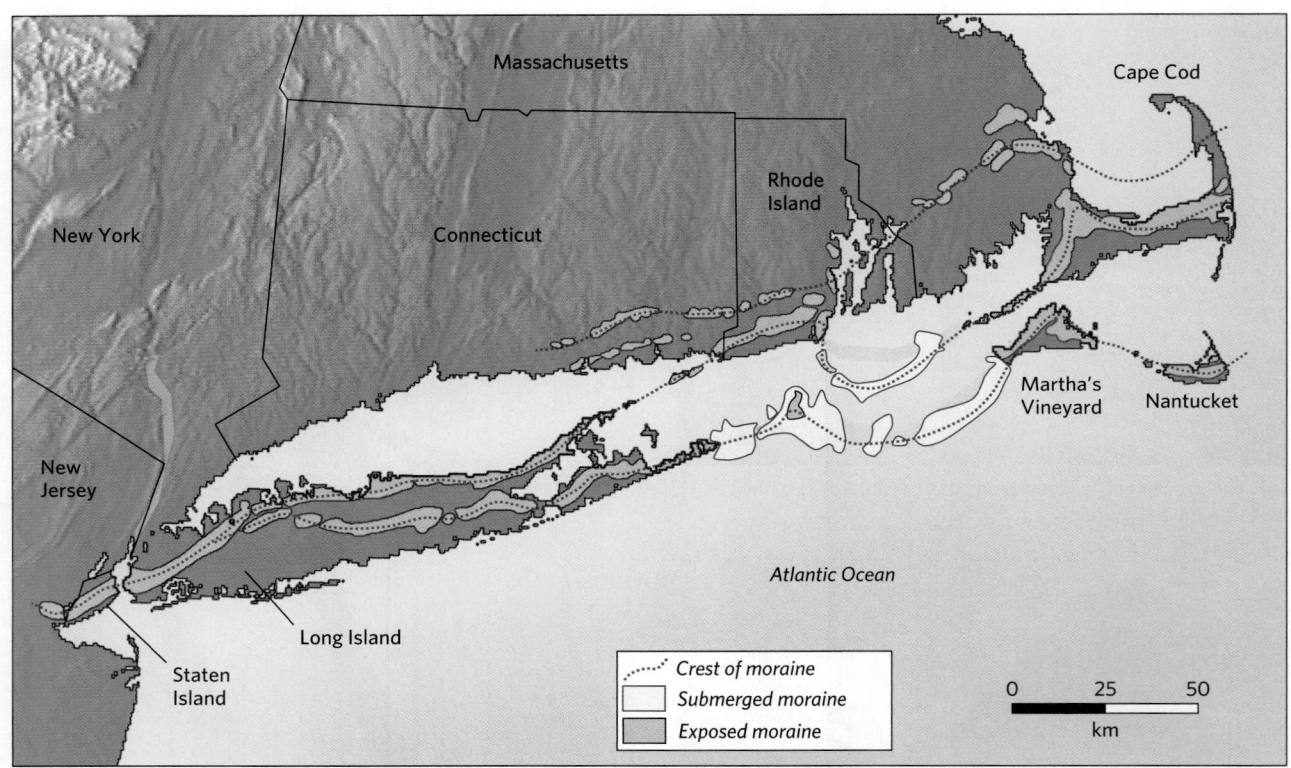

(c) The terminal and recessional moraines of a continental glacier underlie Cape Cod and Long Island in the northeastern United States.

between 20 Ka and 18 Ka **(Fig. 13.34c)**. Sediment left when a glacier retreats without pausing, so that a distinct ridge doesn't form, makes up *ground moraine*.

Types of Glacial Sedimentary Deposits

Geologists distinguish among several different types of sediment that accumulate in glacial environments:

- *Glacial till:* Sediment transported by ice and deposited beneath, at the side of, or at the toe of a glacier is called **glacial till**. Such sediment is unsorted because

the solid ice of glaciers carries clasts of all sizes. Thus, glacial till typically consists of cobbles or boulders suspended in clay and silt **(Fig. 13.35a)**. Moraines are made up of glacial till.

- *Erratics:* A glacial **erratic** is a cobble or boulder that has been transported far from its source and dropped by a glacier. The word comes from the Latin *errare*, which means to wander. Some erratics are embedded in till, while others sit on glacially polished surfaces **(Fig. 13.35b)**.

FIGURE 13.35 Sediment types associated with glaciation.

(a) This glacial till in Ireland is unsorted because ice can carry sediment of all sizes.

(b) Glacial erratics resting on a glacially polished surface in Wyoming.

(c) Braided streams choked with glacial outwash in Alaska. The streams carry away finer sediment and leave the gravel behind.

(d) Thick loess deposits underlie parts of the prairie in Illinois.

(e) (Left) In the quiet water of an Alaskan glacial lake, fine-grained sediments accumulate. (Right) Alternating layers (varves) in glacial-lake sediments, now exposed in an outcrop near Puget Sound, Washington, reflect seasonal changes.

- *Glaciomarine sediment:* Where a sediment-laden glacier flows into the sea, icebergs that calve off the toe float clasts out to sea. As the icebergs melt, the clasts they carry sink to the seafloor. These *dropstones* mix with marine clay and plankton shells to form *glaciomarine sediment* (see Fig. 13.26a).

- *Glacial outwash:* Till deposited by a glacier at its toe may be picked up and transported by braided meltwater streams. Such water-transported sediment, known as **glacial outwash**, tends to be sorted by water to form bars of gravel and sand (Fig. 13.35c).

- *Loess:* Strong winds pick up fine sediment (rock flour) from glacial deposits and transport it away from the glacier. This wind-transported sediment eventually settles out of the air to form thick deposits of loess (Fig. 13.35d). Loess derived from glacial sediments resembles the deposits of fine windblown sediment that accumulate downwind of a desert.

- *Glacial-lake sediment:* Some of the sediment yielded by glaciers accumulates on the floors of **glacial lakes**, bodies of liquid freshwater formed in association with glaciers. These lakes, which contain water from the melting glacier as well as from melting snow, summer rain, or springs fed by groundwater, include *ice-margin lakes* along the toe of a glacier, *kettle lakes* that fill roughly circular depressions on moraines, *moraine-dammed lakes* formed where water collects in depressions between end moraines or in valleys blocked by a moraine, *cirque lakes* that fill the depressions left when cirque glaciers melt away, and *glacier-dammed lakes* filling valleys whose outlets have been blocked by glacial ice. When a glacial lake dries up or drains, it leaves behind thin alternating layers of silt (brought into the lake during spring floods) and clay (which settles from still water when the lake freezes in the winter). A pair of layers, representing the deposits of an entire year, is called a *varve* (Fig. 13.35e).

Depositional Landforms of Glacial Environments

Imagine what you would see if you stood on the toe of a retreating glacier. Looking upstream, at the glacier itself, you would see nothing but ice and the sediment that it carries. Looking in the other direction, you would see a variety of landforms produced by glacial erosion and deposition (Earth Science at a Glance, pp. 504–505). For example, from your perch, you might see a few end moraines. Some moraines are *hummocky*, meaning that they are bumpy topographic surfaces with countless small hills and depressions. This kind of landscape develops both because of variations in the amount of sediment supplied by the ice and because of the development of *kettle holes*, depressions formed when sediment-covered ice blocks, left behind as the glacier retreats, melt away (Fig. 13.36).

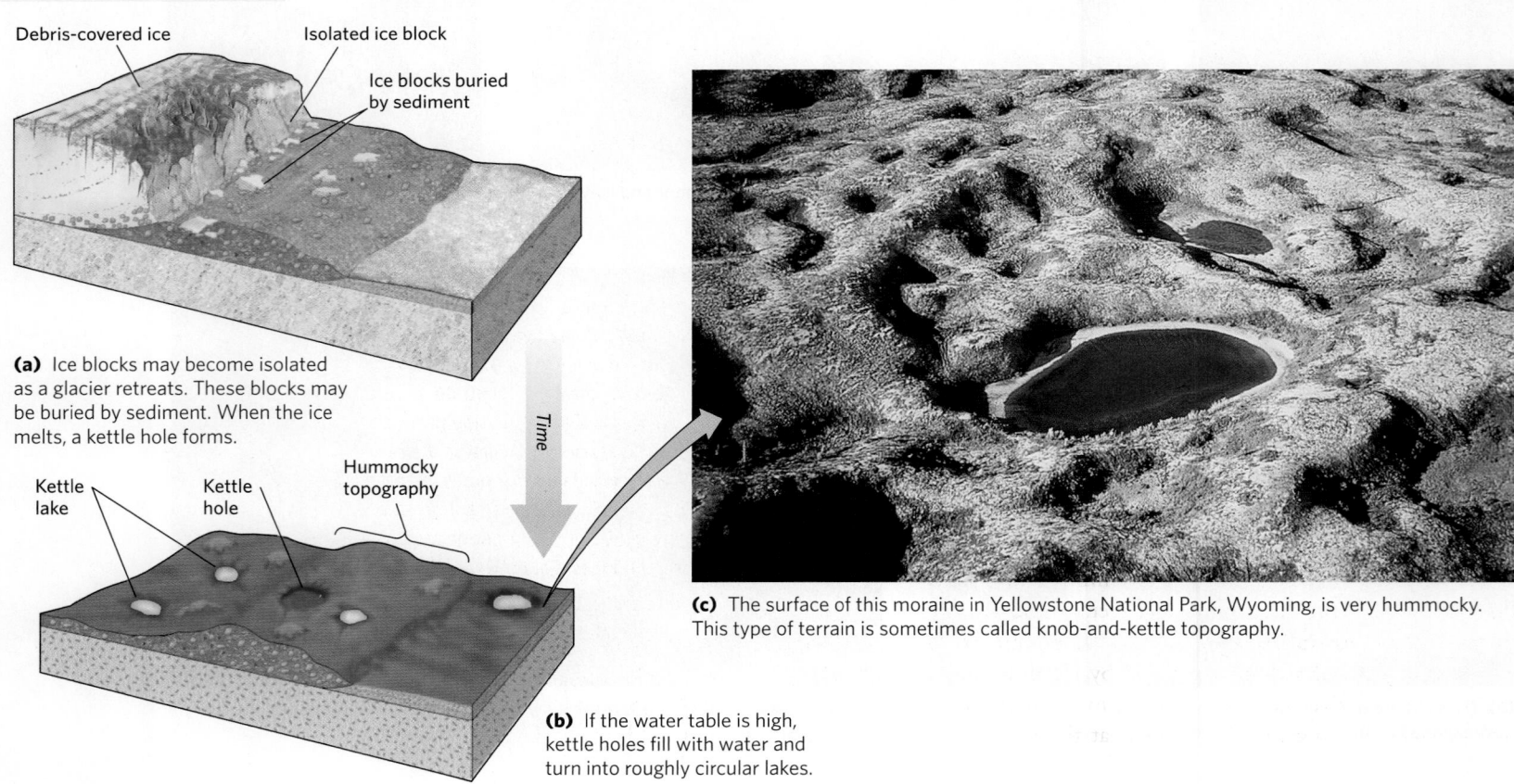

FIGURE 13.36
Hummocky topography characterizes the surfaces of end moraines and ground moraines.

Debris-covered ice   Isolated ice block

Ice blocks buried by sediment

(a) Ice blocks may become isolated as a glacier retreats. These blocks may be buried by sediment. When the ice melts, a kettle hole forms.

Time

Kettle lake   Kettle hole   Hummocky topography

(b) If the water table is high, kettle holes fill with water and turn into roughly circular lakes.

(c) The surface of this moraine in Yellowstone National Park, Wyoming, is very hummocky. This type of terrain is sometimes called knob-and-kettle topography.

FIGURE 13.37 Development of drumlins and eskers.

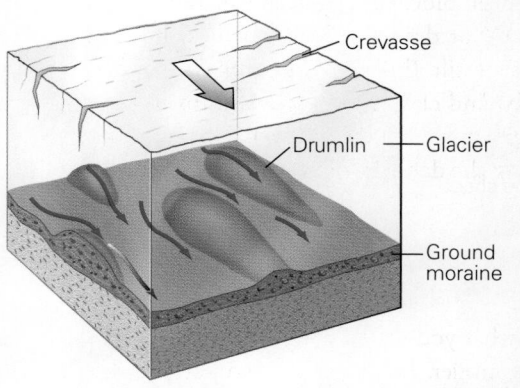

(a) Drumlins are sculpted out of sediment beneath the ice of a glacier and are aligned with its flow direction.

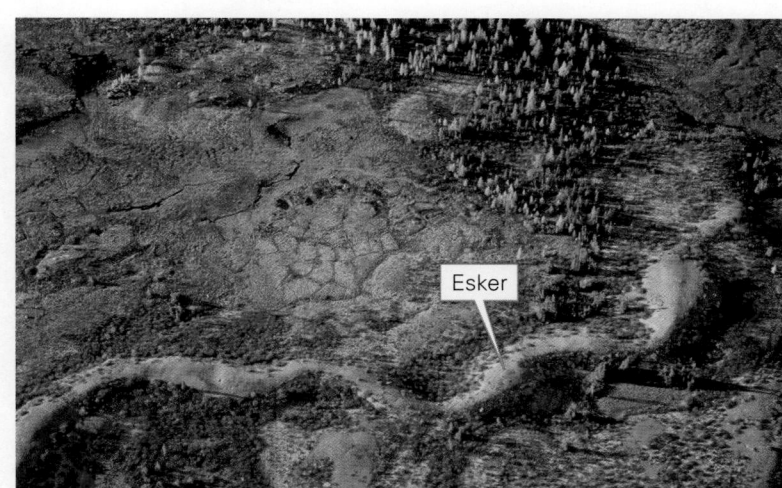

Shaded relief map of the drumlins in central New York. Their SSE angle gives the direction of glacial flow.

Lake Ontario

View looking southeast

(b) Drumlins form a set of hills south of Lake Ontario in central New York State.

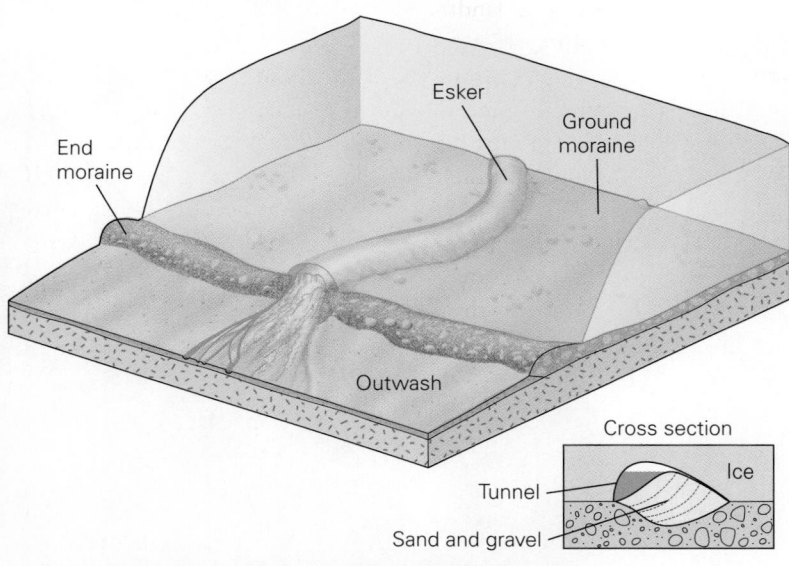

(c) Eskers are snake-like ridges of sand and gravel that form when sediment fills meltwater tunnels at the base of a glacier. The cross section (inset) shows how wedges of sand accumulate in the tunnel.

(d) An example of an esker in an area that was once glaciated.

When kettle holes fill with water, they become the kettle lakes that we mentioned earlier. Beyond an end moraine, the land surface may host braided streams of sediment-choked meltwater. These streams deposit bars of gravel and sand, which build a broad *glacial-outwash plain*. You may also see various types of glacial lakes. Retreating glaciers also leave behind landforms that develop at the base of the glacier. These landforms include **drumlins** (Fig. 13.37a, b), asymmetrical elongate hills of sediment molded by ice flow, and *eskers*, sinuous ridges composed of sediment deposited by meltwater flowing through tunnels at the base of a glacier (Fig. 13.37c, d).

Take-home message...

A glacier plucks and scrapes up rock from its substrate and carries it, along with debris that falls on its surface, in the direction of ice flow. Abrasion by incorporated clasts erodes the glacier's substrate, producing polished rock and striations. Glacial erosion produces distinctive landforms such as U-shaped valleys. The unsorted sediment carried by a glacier becomes glacial till where the glacier melts away. Some of this sediment can be reworked by flowing meltwater. Sediment deposition beneath the glacier also produces distinctive landforms.

Quick Questions

- What is a glacial striation?
- Why do we find hanging valleys in mountains that have been eroded by glaciers?
- What does a moraine consist of?

Glaciers and Glacial Landforms

Continental ice sheet

Crevasses

Ice shelf

Higher sea level

Lower sea level

Dropstones

Iceberg

Horn

Mountain ice cap

Valley glacier

Arête

Cirque

Lateral moraine

Medial moraine

U-shaped valley

Ice-margin lake

Outwash plain

Ground moraine

Recessional moraine

Erratics

Kettle lake

Braided stream

Drumlin

Esker

Roche moutonnée

Note that the terminal moraine here is not visible; it's offshore and is submerged.

Striations

Continental glaciers, vast sheets of ice up to a few kilometers thick, covered extensive areas of land during times when the Earth had a colder climate. These glaciers form from snow that accumulates at high latitudes. When buried deeply enough, the snow becomes tightly packed and recrystallizes into glacial ice. Although it is solid, ice is weak, so ice sheets can spread over the landscape like syrup over a pancake, though much more slowly. When a continental glacier reaches the sea, it becomes an ice shelf. At the edge of the shelf, icebergs calve off and float away, and sediment carried by the ice falls to the seafloor.

A second category of glaciers, called mountain or alpine glaciers, grow in mountainous areas because snow can last all year at high elevations. Valley glaciers are confined to valleys, and ice caps cover the peaks of mountains. These glaciers carve a landscape containing U-shaped valleys, cirques, horns, and hanging valleys. Lateral moraines form along their sides. When the climate is cold, a mountain glacier advances and may eventually flow out onto the land surface beyond the mountain front. When the climate warms, the glacier retreats. The landscape in front of the glacier retains erosional features,

such as striations and roches moutonées, formed when the area was covered by flowing ice. It can also have a cover of till (unsorted glacial sediment) in various types of moraines, such as ground moraines and end moraines. Streams pick up till, sort it, and redistribute it as glacial outwash. Sediment that accumulates in ice tunnels, exposed when the glacier melts, make up sinuous ridges called eskers. Flowing ice can mold underlying sediment into drumlins. Ice blocks buried in till melt away and leave behind kettle holes that fill with water to become kettle lakes. Ice-margin lakes may form along the edge of the ice.

13.6 Additional Consequences of Continental Glaciation

Glacial Subsidence and Glacial Rebound

When a large ice sheet (more than 50 km, or 30 mi, in diameter) grows on a continent, its weight causes the surface of the lithosphere to sink. In other words, a large ice sheet acts as a load that causes **glacial subsidence**. The lithosphere, the Earth's outer shell, can bend or sink in response to such a load because the underlying asthenosphere can flow plastically out of the way **(Fig. 13.38a)**. Large ice-margin lakes develop at the toes of continental glaciers because the land surface tilts toward the glacier as subsidence takes place. The largest of these lakes, Glacial Lake Agassiz, in south-central Canada and the north-central United States, covered an area greater than that of all the present Great Lakes combined. It existed between 11,700 and 9,000 years ago **(Fig. 13.39)**.

When a large continental ice sheet melts away, the surface of the underlying land rises back up, a process called **glacial rebound**, and the asthenosphere flows back underneath to fill the space **(Fig. 13.38b)**. Glacial rebound takes many thousands of years because rock of the asthenosphere flows so slowly. Because of glacial rebound, beaches formed at sea level along the shore of Hudson Bay thousands of years ago now lie well above sea level.

Effects of Continental Glaciers on Drainage

A continental glacier disrupts drainage networks. If, for example, a glacier completely covers a drainage network, a new drainage network, perhaps with a different geometry, may develop. In some cases, a glacier blocks a river, so that the river must find a different path. By the time the glacier melts away, the new path has become so well established that the preglacial path remains abandoned.

When glaciers flowed south and advanced over the Pacific Northwest of North America between 15,000 and 13,000 years ago, glacial ice blocked the outlet of a west-draining valley in what is now western Montana. The valley filled with water to produce a lake 600 m (2,000 ft) deep, known as Glacial Lake Missoula, with a volume equal to the combined volume of Lake Erie and Lake Ontario. When the glacier started to retreat, the ice dam weakened and suddenly broke. The entire contents of Glacial Lake Missoula spilled into what is now the Columbia River drainage network, producing a cataclysmic flood called a **glacial torrent**. Floodwater traveling

FIGURE 13.38 The concept of glacial subsidence and rebound.

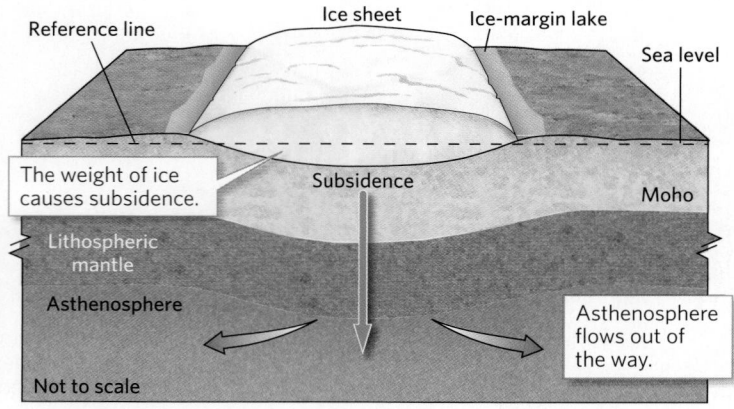

(a) The weight of an ice sheet causes the surface of the lithosphere to sink (subside).

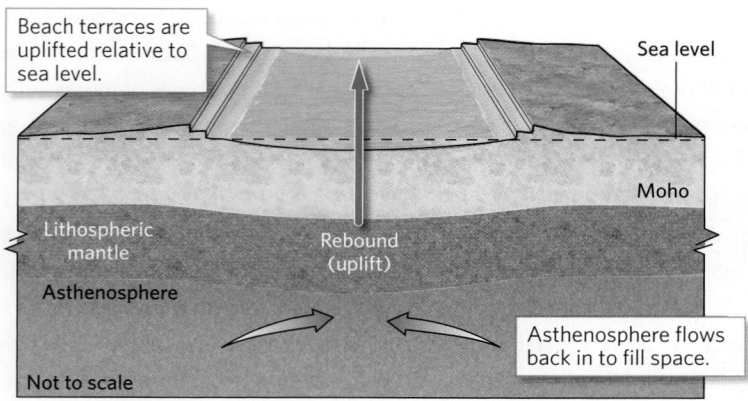

(b) When the glacier melts, the land surface rises (rebounds).

FIGURE 13.39 Glacial Lake Agassiz was a glacial lake that formed near the end of the most recent glaciation of the Pleistocene Ice Age.

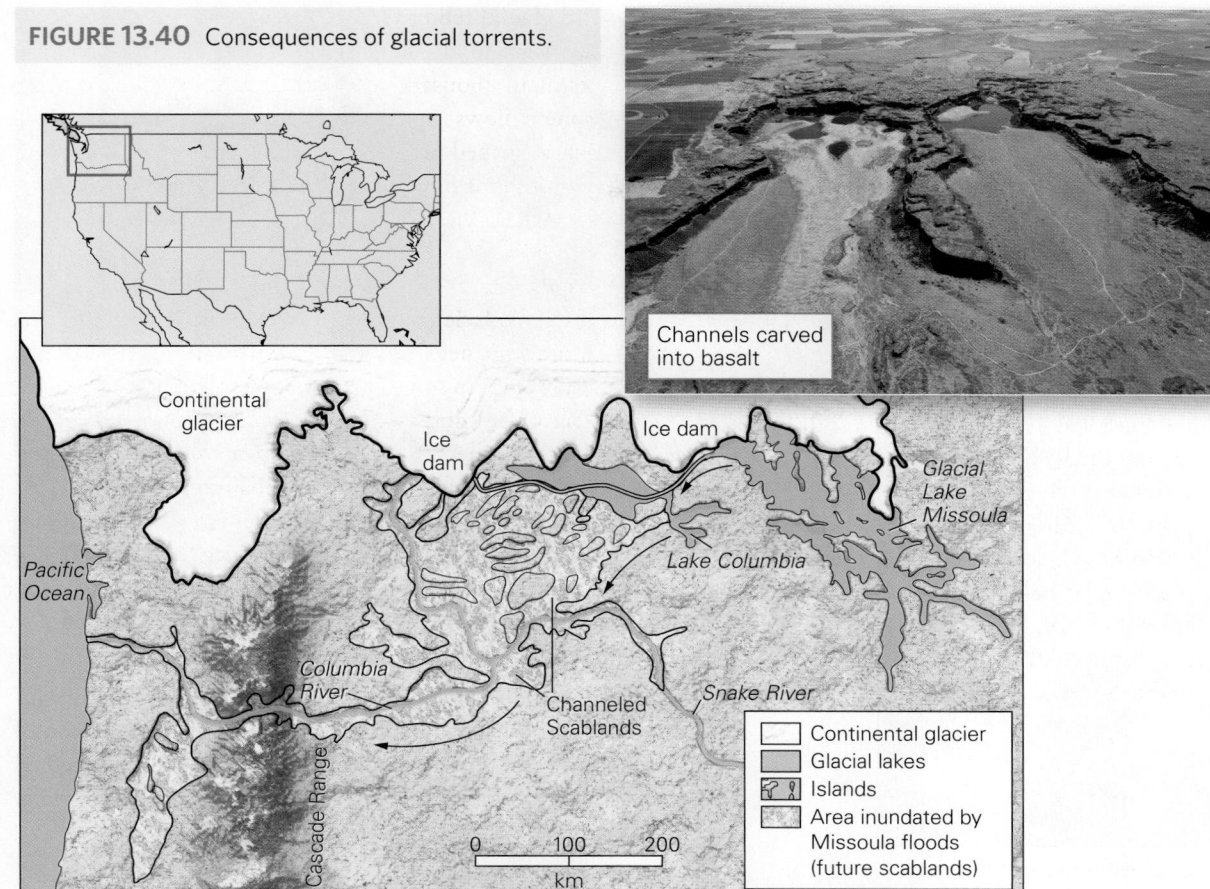

FIGURE 13.40 Consequences of glacial torrents.

Channels carved into basalt

(b) The soil was stripped off a region underlain by basalt bedrock. This area is now called the Channeled Scablands.

Continental glacier

Ice dam

Ice dam

Glacial Lake Missoula

Pacific Ocean

Lake Columbia

Columbia River

Channeled Scablands

Snake River

Cascade Range

	Continental glacier
	Glacial lakes
	Islands
	Area inundated by Missoula floods (future scablands)

0   100   200
km

(a) An ice dam blocked the outlet of Glacial Lake Missoula. When the dam broke, floodwaters swept westward across central Washington State and down the Columbia River valley.

at speeds of up to 100 km/h (62 mph) stripped the soil off the ground, exposing large areas of basalt bedrock; transported house-sized boulders; and left ripple marks 20 m (60 ft) high. Glacial Lake Missoula refilled during glacial advances, and emptied during glacial retreats, up to 25 times during the Pleistocene Ice Age. Erosion by these Great Missoula floods produced a unique landscape known as the Channeled Scablands in eastern Washington, so named because the removal of soil widely exposed crusty-looking basalt bedrock **(Fig. 13.40)**. Glacial torrents happened in many places as continental ice sheets retreated. Some of these floods, like the draining of Glacial Lakes Missoula and Agassiz, happened due to the failure of glacial ice dams. Others, such as the Kankakee torrent in Illinois, were due to the breaching of glacial-moraine dams.

Pluvial Features and Permafrost

When continental glaciers covered most of Canada and part of the northern United States during the Pleistocene Ice Age, the climate in regions to the south of the glaciers was wetter than it is today. Fed by enhanced rainfall, lakes accumulated in low-lying areas at a great distance

from the toe of the glacier. Many such **pluvial lakes** (from the Latin *pluvia*, meaning rain) flooded low areas in the Great Basin, a broad geographic region of Utah and Nevada in the Basin and Range Province, in which lakes do not have outlets (see Fig. 12.7b). The largest of these, Lake Bonneville, covered almost a third of western Utah **(Fig. 13.41a)**. When this lake suddenly drained after a natural dam holding it back broke, it left a bathtub ring of shoreline sediments rimming the mountains near Salt Lake City. Today's Great Salt Lake represents a small remnant (now getting even smaller) of Lake Bonneville **(Fig. 13.41b)**. The now-dry floor of Lake Bonneville has become the Bonneville Salt Flats, famous as a location for high-speed auto racing.

In polar latitudes today, and in regions adjacent to the fronts of continental glaciers during ice ages, the average annual temperature stays low enough (below −5°C, or 23°F) that the ground freezes and becomes **permafrost**. In areas of permafrost, soil water and groundwater remain solid for much of the year. Regions with widespread permafrost that do not have a cover of snow or ice are called *periglacial environments* (the prefix *peri–* implies surrounding). The upper few meters of permafrost may melt

FIGURE 13.41 Pluvial lakes in the western United States.

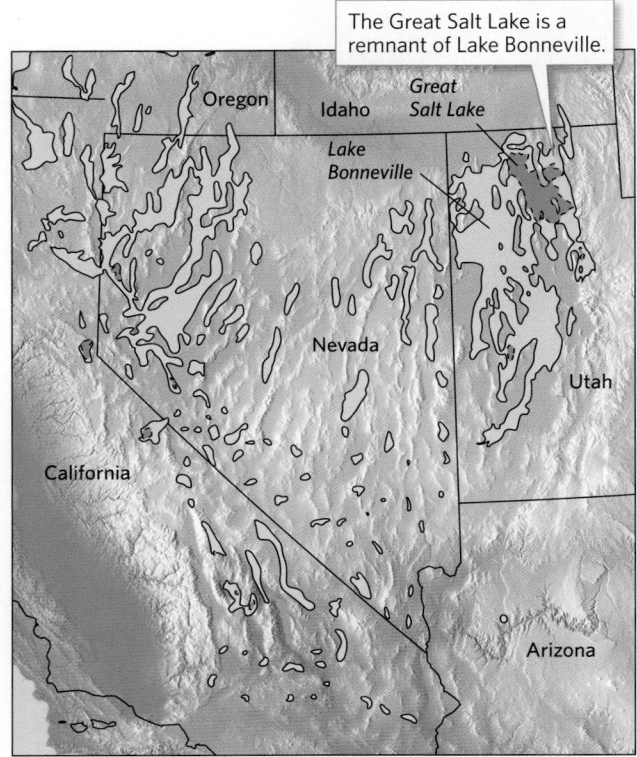

The Great Salt Lake is a remnant of Lake Bonneville.

(a) During the Pleistocene Ice Age, a wetter climate produced pluvial lakes throughout the Basin and Range Province of Nevada and Utah.

Lake Bonneville shoreline

Height above the present lake

Great Salt Lake

(b) The Great Salt Lake of Utah is a small remnant of Lake Bonneville. Subtle horizontal terraces, the remnants of beaches, now lie about 100 m above the present lake level.

during the summer months, only to refreeze when winter comes. Due to the freeze-thaw process, the ground of some permafrost areas splits into pentagonal or hexagonal shapes, producing a landscape known as *patterned ground* (Fig. 13.42).

Sea-Level Changes Related to Glaciation

When glaciers grow, large amounts of water become trapped on land, so sea level drops. During an ice age, when continental glaciers expand greatly, this drop can be as much as 100 m, causing extensive areas of continental shelves to become dry land (Fig. 13.43a, b). During times of low sea level during the Pleistocene Ice Age, people and animals migrated across a **land bridge** that existed at the location of what is now the Bering Strait between North America and northeastern Asia. When continental glaciers melted, sea level rose again and submerged the land bridge (Fig. 13.43c).

Take-home message…

The weight of a continental glacier can cause the lithosphere to subside, and melting of the glacier can allow it to rebound. The land beyond a continental glacier may be covered with permafrost or pluvial lakes. Continental glaciers store large amounts of water, so glacial advance or retreat affects sea level.

Quick Questions

- How can glaciation affect drainage in a region?
- What is a glacial torrent?
- Why did a land bridge develop between Alaska and Siberia at times during the Pleistocene?

FIGURE 13.42 An example of patterned ground in a periglacial environment in Manitoba, Canada. The polygons are tens of meters across.

FIGURE 13.43 The link between sea level and global glaciation.

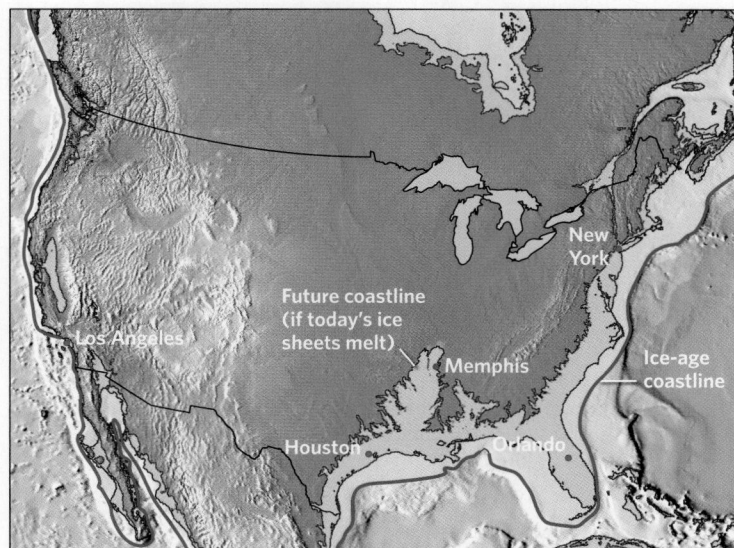

(a) The red line shows the coastline of North America during the most recent continental glaciation, when much of the continental shelf was dry. If the present-day ice sheets melt, areas that are coastal lands today will be flooded.

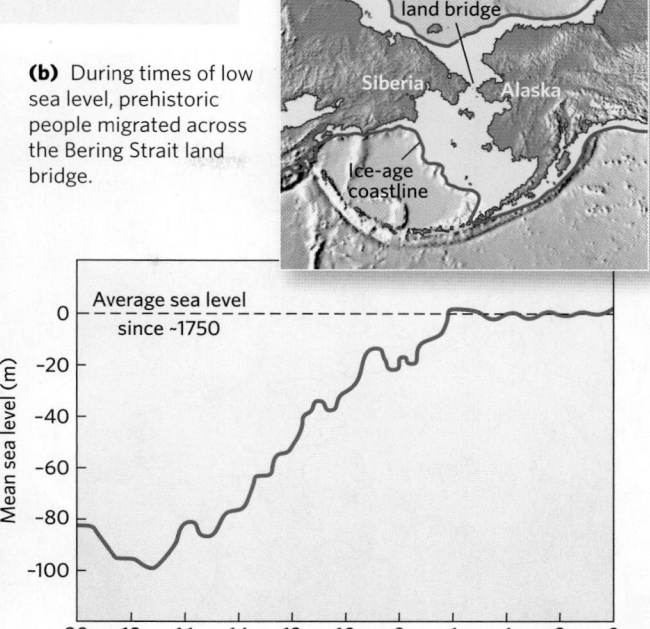

(b) During times of low sea level, prehistoric people migrated across the Bering Strait land bridge.

(c) Sea level rose between 17,000 and 7,000 B.C.E. due to the melting of continental glaciers.

13.7 Ice Ages

Discovery of the Pleistocene Ice Age

When 19th-century farmers in northern Europe prepared their land for spring planting, they occasionally broke their plows on large boulders scattered through the soil of their fields. Some of the boulders consisted of rock that came from bedrock exposures hundreds of kilometers to the north. Such *glacial erratics* fascinated geologists of the day. Most assumed that the erratics had been transported by immense floods. Then, in 1837, a Swiss geologist named Louis Agassiz proposed an alternative solution to the mystery of Europe's wandering boulders. Agassiz lived near the Alps and knew that glaciers could carry boulders as well as sand and mud. He proposed that the erratics had been left by glaciers, implying that Europe had once been in the grip of an **ice age**, a time when climate was significantly colder and vast glaciers covered much of the continent.

Agassiz's ice age had clearly taken place relatively recently in Earth history—otherwise, the landscapes and sediment deposits it produced would have been either eroded away or buried. Twentieth-century studies demonstrated that the deposits were left as glaciers advanced and retreated many times during the Pleistocene, the epoch that, as we noted earlier, spans the time interval between 2.58 Ma and 11.7 Ka. Therefore, geologists refer to this ice age as the **Pleistocene Ice Age**.

The Extent of Pleistocene Glaciers

Today, most of the land surface in New York City lies hidden beneath concrete, but in Central Park, it's still possible to see land in a seminatural state. If you stroll through the park, you'll find that the top surfaces of outcrops are smooth and polished (see Fig. 13.27a) and have been grooved and scratched, and that here and there, glacial erratics rest on the bedrock. You are seeing evidence that an ice sheet once scraped along the bedrock of New York. Geologists estimate that the ice sheet that covered the city area may have been 250 m (820 ft) thick, enough to cover a 75-story building.

Geologists have studied the distribution and orientation of such products of glacial erosion, as well as the distribution of moraines and the sources of erratics, to map out not only the extent of the Pleistocene glaciers, but also their origins and flow directions. In North America, major ice sheets are named for the location at which they originated. On this basis, geologists distinguish among the Labrador, Keewatin, Baffin, and Cordilleran ice sheets (Fig. 13.44a). The first three of these sheets merged to form the giant **Laurentide ice sheet**, which at times covered all of Canada east of the Rocky Mountains and spread southward over the northern United States. The **Cordilleran ice sheet**, which originated in the mountains of western Canada, spread westward to the Pacific coast and eastward until it met the Laurentide ice sheet. Other ice sheets formed in Greenland, Scandinavia, and Russia. During

FIGURE 13.44 Pleistocene Ice Age glaciers.

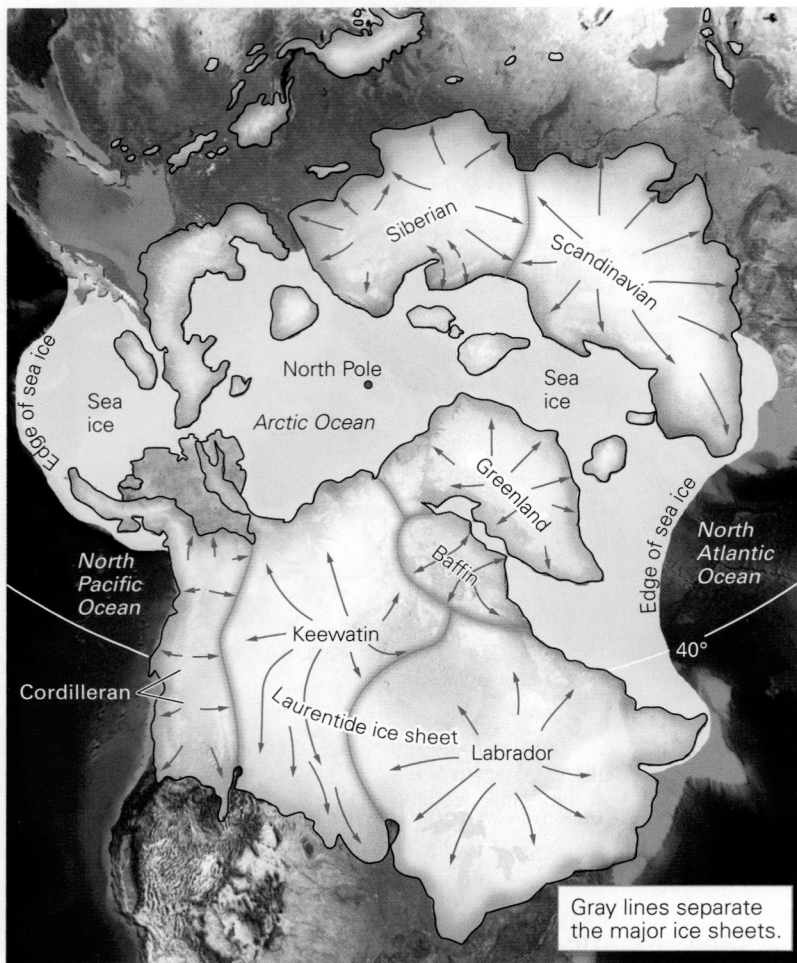

(a) Ice sheets covered large areas of the northern hemisphere continents throughout the Pleistocene Ice Age. The ice sheets are named after the locations where they originated and from which they flowed. Individual ice sheets advanced and retreated several times.

(b) The southward shift of climate belts allowed cold-adapted large mammals, such as woolly mammoths, that are now extinct to roam now-temperate regions of eastern North America.

the Pleistocene Ice Age, polar climates shifted southward, so sea ice not only covered all of the Arctic Ocean, but also part of the North Atlantic (Fig. 13.44b), and *tundra*—land with a very cold climate that may lie over permafrost, but can host low-lying, hardy plants—occurred in regions that now host forests and prairies.

Glacial Advances and Retreats during the Pleistocene

Louis Agassiz assumed that glaciers advanced only once during the Pleistocene. But close examination of glacial deposits on land revealed that **paleosols** (ancient soils preserved in the stratigraphic record), as well as sediment layers containing fossils of warmer-weather animals and plants, occur between layers of glacial till. This observation suggested that several distinct glacial advances happened during the Pleistocene, and that these advances were

separated by times when the ice sheets retreated, warmer climates prevailed, and soil formed where there had been ice. In the second half of the 20th century, researchers used radiocarbon dating methods to determine the ages of wood fragments in glacial deposits, and confirmed that glaciers had advanced and retreated more than once during the Pleistocene. Times during which glaciers covered large areas are called **glaciations**, and times between glaciations are called **interglacials**.

The terrestrial sedimentary record provides evidence for three major glaciations in North America (Fig. 13.45a), named, from youngest to oldest, Wisconsin, Illinoian, and pre-Illinoian. (The pre-Illinoian glaciation appears to be a composite of several separate events that cannot easily be distinguished from one another.) Researchers realized, however, that the terrestrial record is very incomplete, because when a new glaciation takes place, it may erase all or part of the record of previous glaciations. To obtain a more complete record, they began studying changes in the ratio of heavier to lighter oxygen isotopes preserved in plankton shells in marine strata, for this ratio indicates global temperature (see Chapter 19). Glaciations occur during times when the ratio indicates colder temperatures, and interglacials occur at times when the ratio indicates warmer climates. Such analyses indicate that 20–30 distinct glaciations took place during the Pleistocene (Fig. 13.45b). The terrestrial record preserves evidence of only the largest glaciations, or of glaciations whose sediment was not eroded and redeposited by a subsequent glaciation.

Older Ice Ages during Earth History

So far, we've focused on the Pleistocene Ice Age because of its importance in developing our planet's contemporary landscape. Was it the only ice age during Earth history,

FIGURE 13.45 The timing of glaciations.

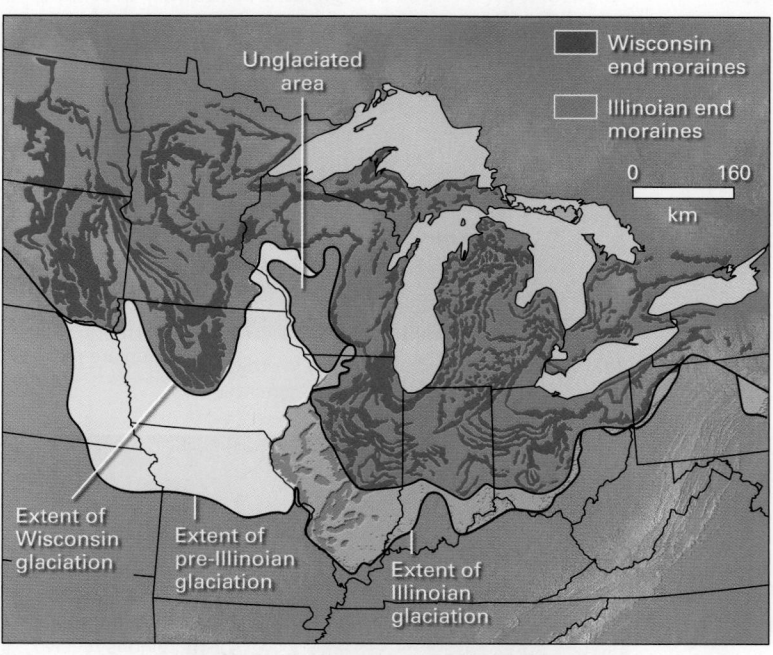

(a) Pleistocene glacial deposits in the north-central United States. The record shows that glaciers advanced more than once. Curving moraines reflect the shape of a glacier's toe.

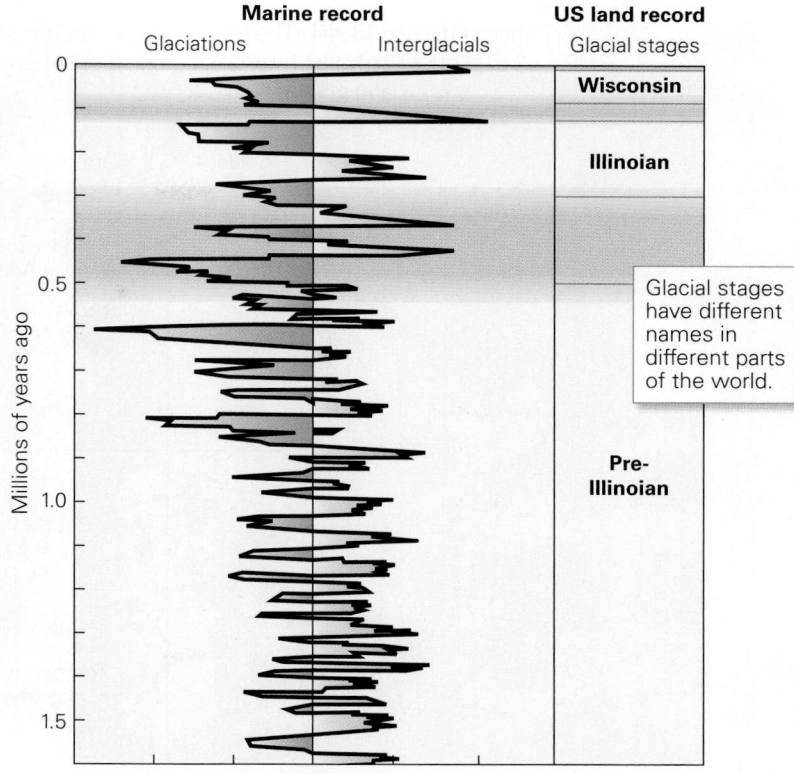

(b) Oxygen-isotope ratios from marine sediment define 20 to 30 glaciations in the Pleistocene. The blue bands represent the traditionally recognized glaciations of the north-central United States.

or did ice ages happen at other times in the geologic past? To answer such questions, geologists studied the stratigraphic record and searched for glacial-till deposits that have lithified to become rock. Such **tillites** consist of cobbles and boulders distributed in a matrix of siltstone and mudstone; in some cases, tillites lie on glacially polished surfaces. Geologists have now found late Paleozoic tillites (the deposits Alfred Wegener studied), multiple late Proterozoic tillites (the ones associated with snowball Earth; see Chapter 9), early Proterozoic tillites, and Archean tillites **(Fig. 13.46)**. Strata deposited at other times in Earth history do not contain tillites, so it appears that major ice ages do not occur frequently or periodically. Rather, only four or five ice ages have occurred, separated by vastly different amounts of time.

Long-Term Causes of Ice Ages

What was special about the time intervals, noted in the previous paragraph, when ice ages occurred? Geologists suggest that several long-term factors contribute to the growth of large ice sheets that characterizes ice ages. For example, since continental ice sheets develop on land, an ice age won't happen if plate motions have transported most continents into warm latitudes. Therefore, only when large areas of land lie at high latitudes, so that they endure long winters, can ice ages begin. Similarly, the movement of continents and the growth of island arcs

can change the configuration of ocean currents; if land distribution guides warm currents to high latitudes, temperatures there may remain too warm for glaciers to grow.

Changes in the concentration of greenhouse gases such as CO_2 probably play the lead role in determining whether our planet will be susceptible to an ice age because these gases regulate atmospheric temperatures (see Chapter 19). For example, when CO_2 concentrations are high, the atmosphere becomes warmer, and ice sheets cannot form, even if continents lie at mid- to high latitudes. What factors cause long-term changes in CO_2 concentrations? Possibilities include changes in the population sizes of organisms that extract dissolved carbonate from water to make shells, or of organisms that extract CO_2 during photosynthesis; changes in the amount of chemical weathering on land (a process that absorbs CO_2); and changes in the amount of volcanic activity (which emits CO_2).

Milankovitch Cycles: Triggers of Short-Term Advances and Retreats

Why do glaciers advance and retreat periodically during an ice age? In 1920, Milutin Milanković, a Serbian geophysicist and astronomer, came up with an explanation.

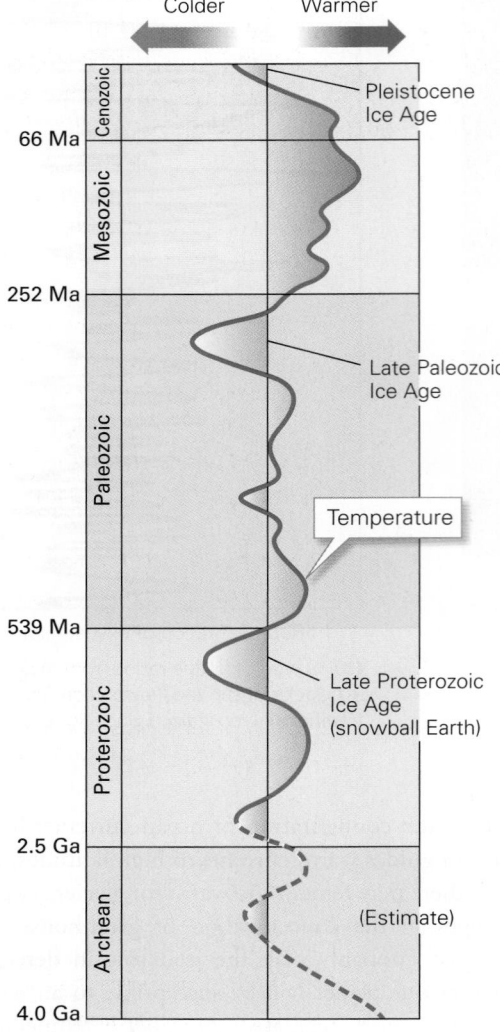

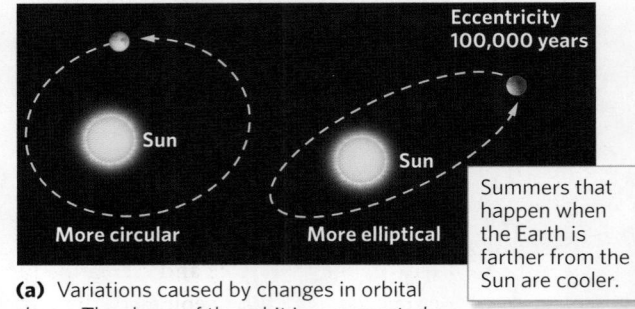

(a) Variations caused by changes in orbital shape. The shape of the orbit is exaggerated.

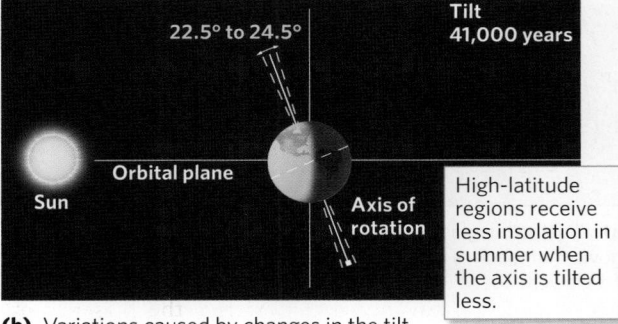

(b) Variations caused by changes in the tilt of the Earth's axis.

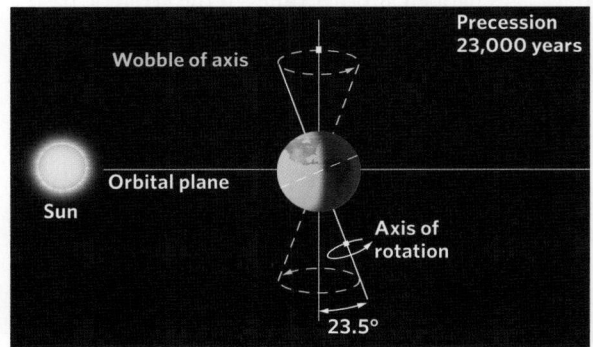

(c) Variations caused by the precession of the Earth's axis.

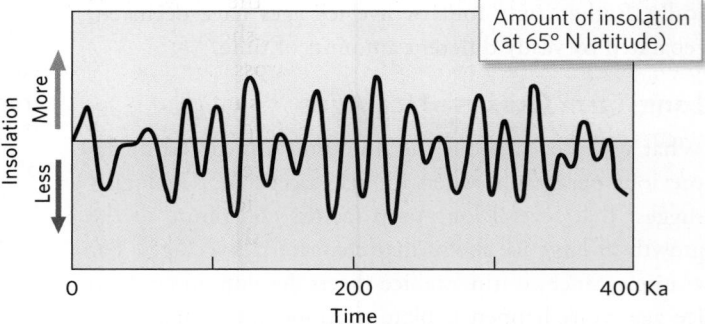

(d) Combining the effects of eccentricity, tilt, and precession produces varying amounts of insolation during the summer at mid- to high latitudes.

He evaluated the Earth's movement through space over time and characterized three types of periodic changes, as follows: (1) The shape of the Earth's orbit (the orbit's *eccentricity*) changes from more circular to more elliptical over a period of around 100,000 years (Fig. 13.47a). (2) The *tilt* of the Earth's axis of rotation varies between 22.5° and 24.5° over a period of 41,000 years (Fig. 13.47b). (3) The Earth's axis wobbles (a movement called *precession*), like a spinning top, over a period of 23,000 years (Fig. 13.47c). Milanković then calculated how the combined effects of eccentricity, tilt, and precession would affect the amount of **insolation** (the solar energy arriving at the top of the Earth's atmosphere) received during the summer at mid- to high latitudes. When these latitudes receive less insolation, cooling takes place, and a glaciation can occur, but when these latitudes receive more insolation, warming takes place, leading to glacial retreat. He found that the three cycles could cause insolation to change by as much as 25%, and therefore could cause significant changes in atmospheric temperature at those latitudes (Fig. 13.47d)!

The three cycles that Milanković described are now called **Milankovitch cycles**, using an English spelling of the researcher's name, and they turned out to be a key to the mystery of why glaciers advance and retreat periodically during an ice age, because geologists discovered that the times of Pleistocene glaciations correlate with the times of minimum insolation at mid- to high latitudes, as predicted by Milanković. This correlation implies that ice-age glaciers advance when mid- to high latitudes have cooler summers, during which ice from the previous winter can't melt completely. Subsequent work, however, suggests that the cooling predicted by Milanković might not be enough to trigger a glaciation, so other factors may come into play during times of minimum insolation to cool the Earth even more and trigger a glacial advance. For example, the presence of snow on the ground, or of high-altitude clouds in the sky, may help cool the Earth by increasing the *albedo* (reflectivity) of the planet, since reflected sunlight doesn't reach the Earth's surface and can't warm the atmosphere.

Taking into account the various inputs that could trigger cooling of the Earth, let's look at a theoretical case history of a single advance and retreat of the Laurentide ice sheet. The process starts when the Earth reaches a point in the Milankovitch cycles when the summer temperature at the latitude of northern Canada becomes cold enough that winter snow does not entirely melt away. Because the snow reflects sunlight, its presence makes the region grow still colder, so more snow accumulates. Eventually, the base of the snow cover turns to ice, and when the newborn ice sheet gets thick enough, it begins flowing laterally across the land. As the ice sheet broadens, it reflects more sunlight, so the climate cools even more, and because of the cold temperatures, broad areas of high-latitude oceans freeze over and develop a cover of very reflective sea ice. Notice that the growth of an ice sheet serves as a **positive feedback**,

a secondary change (in this case, growth of an ice sheet) that amplifies the initial change (cooling related to Milankovitch cycles).

A glaciation, of course, doesn't last forever, because as the Milankovitch cycles bring an increase in insolation, air temperatures rise, and ice sheets start to retreat. Two important *negative feedbacks*, secondary changes that counter the initial change, amplify this warming. First, the growth of the ice sheet causes glacial subsidence, so when an ice sheet has gotten huge, its surface sinks to lower, warmer elevations. And second, the formation of sea ice over large areas of the ocean cuts off the supply of moisture to the air, so snowfall decreases. Taken together, these changes cause the rate of ablation to exceed the rate of accumulation. As retreat starts, more negative feedbacks develop. Specifically, the shrinking of the glacier decreases the area of high albedo, so the climate warms even more. In the warmer air, snowfall turns to rain, which increases the rate of melting because rainwater falling on the ice helps to melt the ice and transport ice chunks away from the toe of the glacier.

Will another glacial advance occur? Considering the periodicity of glaciations predicted by the Milankovitch cycles, we may well be set for a new glaciation to begin. But, as we will discuss in Chapter 19, global temperatures have been warming in the past two centuries, and glaciers worldwide have been retreating. So, whether ice can return in the next few thousand years remains unknown.

Take-home message...

During the Pleistocene Ice Age, continental glaciers advanced and retreated many times, causing many glaciations and interglacials. Ice ages also happened at discrete times earlier in the Earth's history. Ice ages happen when the distribution of continents, the configuration of ocean currents, and the atmospheric concentration of CO_2 are conducive to ice-sheet formation. Within an ice age, advances and retreats of glaciers are triggered by Milankovitch cycles.

Quick Questions ⎯⎯⎯⎯⎯⎯⎯⎯

- When did the Pleistocene Ice Age take place?
- Do ice ages happen periodically over the course of the Earth's history?
- What are the three Milankovitch cycles?

Objective 13.1

Describe conditions that lead to the classification of a region as a desert, and explain why such conditions develop.

KEY CONCEPTS

- Deserts receive less than 25 cm of rain per year, on average, so vegetation covers less than 15% of their surfaces. Use of the word *desert* refers to the aridity of a region's climate, not to its temperature.

- The aridity of deserts can develop for a variety of reasons. For example, deserts form in subtropical latitudes, where dry air descends; in regions where topography or distance from moisture sources prevents rain from arriving; or in regions where the air is cold and holds little moisture.

EARTH-SCIENCE VOCABULARY

coastal desert (p. 472)
cold desert (p. 471)
continental-interior desert (p. 473)
desert (p. 471)

hot desert (p. 471)
polar desert (p. 473)
rain-shadow desert (p. 472)
subtropical desert (p. 472)

REVIEW QUESTIONS

1. **(a)** What factors determine whether a region can be classified as a desert? **(b)** Distinguish between hot and cold deserts. **(c)** What does the adjective *subtropical* mean?

2. **(a)** Explain why subtropical deserts form. **(b)** Distinguish between a rain-shadow desert and a coastal desert. **(c)** Show the air flow pattern in **Figure A**. **(d)** Why can a polar region be classified as a desert even though it has snow and ice cover? **(e)** Why has the Gobi Desert, in the center of Asia, formed?

A

Objective 13.2

Explain how weathering, erosion, and deposition lead to the formation of desert landscapes.

KEY CONCEPTS

- Physical weathering in deserts produces rocky debris. Due to the scarcity of water, chemical weathering happens slowly, but it does

happen, so exposed rocks do eventually disintegrate. Thin soils develop locally in deserts and are different from those formed in temperate or tropical climates.

- Wind picks up dust as suspended load and causes sand at the ground surface to saltate and creep. Windblown sediment abrades desert surfaces.

- Water causes erosion in deserts during downpours. Flash floods carry large quantities of sediment down ephemeral streams.

- Talus slopes form where rock fragments accumulate at the base of a cliff. Alluvial fans accumulate at the mouth of a canyon. When temporary desert lakes dry up, a salty playa remains.

EARTH-SCIENCE VOCABULARY

alluvial fan (p. 477)
badlands (p. 475)
bajada (p. 478)
caliche (p. 473)
desert varnish (p. 474)
dry wash (p. 475)
dust storm (p. 475)

lag deposit (p. 477)
loess (p. 478)
playa (p. 478)
saltation (p. 475)
surface load (p. 475)
suspended load (p. 475)
talus slope (p. 477)

REVIEW QUESTIONS

3. **(a)** How do weathering and soil-forming processes in deserts differ from those in temperate or tropical climates? **(b)** Explain how desert varnish forms, and why its presence allowed people to create petroglyphs. **(c)** What is caliche, and why does it form?

4. **(a)** Describe when and where water erodes the land surface of a desert. **(b)** What are badlands?

5. **(a)** Describe the ways in which desert winds transport sediment. **(b)** Describe landforms that form in response to wind erosion in a desert. **(c)** What is a ventifact?

6. **(a)** Distinguish between a talus slope and an alluvial fan. **(b)** Which landform does **Figure B** show? **(c)** How does a bajada form? **(d)** Why do loess deposits sometimes accumulate adjacent to a desert, and what do they consist of?

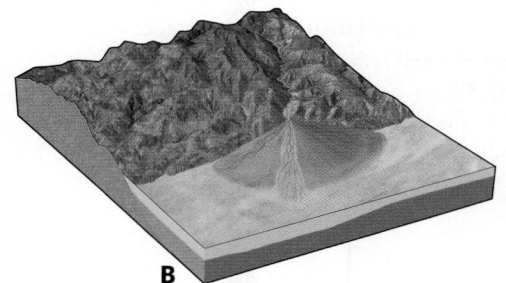

B

Objective 13.3

> Characterize desert landscapes and landforms, such as rocky cliffs, stony plains, and sand dunes, and explain why some semiarid regions are at risk of becoming deserts.

KEY CONCEPTS

- In some desert landscapes, erosion causes cliff retreat, eventually resulting in the formation of landforms such as mesas, buttes, and inselbergs. Large areas of deserts are stony plains, some of which become desert pavements.

- Where sand supplies are sufficient, desert winds can form sand dunes of various shapes, including barchan, star, transverse, parabolic, and longitudinal dunes.

- Drought, overuse or reallocation of water, and destruction of natural vegetation may cause desertification, and can cause lakes to dry up.

EARTH-SCIENCE VOCABULARY

cliff retreat (p. 479)   **mesa** (p. 482)
desertification (p. 485)   **natural arch** (p. 482)
desert pavement (p. 483) **sand dune** (p. 484)
inselberg (p. 483)   **stony plain** (p. 483)

REVIEW QUESTIONS

7. **(a)** Describe the process of cliff retreat and explain how it happens. **(b)** Distinguish between a mesa, a cuesta, and a hogback. Be sure to specify the characteristic of bedding that determines which type of landform develops. **(c)** Label the landforms that appear on **Figure C**.

C

8. **(a)** Why do concentrations of loose pebbles and cobbles form the ground surface in some deserts? **(b)** How does desert pavement develop?

9. **(a)** What key factors determine the shape of sand dunes? **(b)** Distinguish between a barchan dune and a parabolic dune. **(c)** What type of sand dune is pictured **Figure D**? **(d)** What are the dipping surfaces within the sand dunes of **Figure E**? Label the slip face and indicate the wind direction.

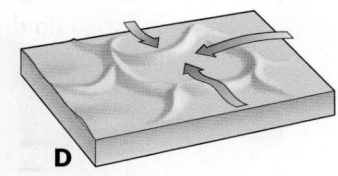

D

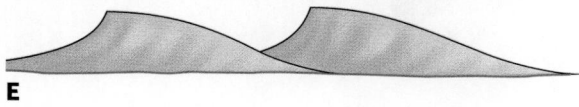

E

10. **(a)** What is the process of desertification, and what causes it? **(b)** What factors transformed the southern Great Plains into the Dust Bowl during the 1930s? **(c)** How has the drying of the Aral Sea affected the local economy and the health of people living nearby?

Objective 13.4

> Explain how glacial ice forms and flows, distinguish among various kinds of glaciers, explain how glaciers advance and retreat, and describe occurrences of ice in the sea.

KEY CONCEPTS

- Glaciers are streams or sheets of ice that flow slowly in response to gravity. They form where snow persists year-round so that it accumulates, compacts, and eventually becomes a layer of recrystallized ice.

- Mountain glaciers exist at high elevations and fill cirques and valleys. Continental glaciers form at high latitudes and spread over substantial areas of continents.

- Gravity drives glacial flow. Glaciers move by ductile deformation or by basal sliding at rates of tens of meters per year. The upper part of a glacier is brittle and fractures to form crevasses.

- Glacial advance or retreat depends on the balance between the rate of accumulation and the rate of ablation. Even a retreating glacier continues to flow toward its toe.

- Icebergs break off glaciers and ice sheets that reach the sea and became ice tongues or ice shelves. Most of an iceberg lies under water. Sea ice forms where the ocean surface freezes.

EARTH-SCIENCE VOCABULARY

ablation (p. 491)   **head** (p. 490)
continental glacier (p. 487) **iceberg** (p. 495)
crevasse (p. 490)   **ice shelf** (p. 494)
equilibrium line (p. 491)   **mountain glacier** (p. 487)
glacial advance (p. 491)   **sea ice** (p. 495)
glacial retreat (p. 491)   **tidewater glacier** (p. 494)
glacier (p. 487)   **toe** (p. 490)

REVIEW QUESTIONS

11. **(a)** Describe the steps leading to the transformation of snow into glacial ice. **(b)** Why is glacial ice blue? **(c)** Why do glaciers form only at high elevations in mountains that lie at low latitudes?

12. **(a)** How does a mountain glacier differ from a continental glacier? **(b)** Distinguish between a mountain ice cap and a piedmont glacier. **(c)** Label the three types of glaciers shown in **Figure F**. **(d)** How does a temperate glacier differ from a polar glacier? **F**

13. (a) Describe the mechanisms that enable glaciers to move. (b) What force drives glacial movement? (c) Why do crevasses form? (d) Where does the fastest motion of a glacier generally occur?

14. (a) Identify the zones of ablation and accumulation, and the equilibrium line, in **Figure G.** Why are the arrows in the cross section curved? (b) What processes take place in the zone of ablation? (c) Explain the difference between glacial advance and glacial retreat. (d) On **Figure H,** there are no striations to the left of the red line, and the toe of the glacier has moved from the red line to the yellow line. Does the figure show advance or retreat? Draw the present flow direction of the ice in the glacier on the surface of the glacier.

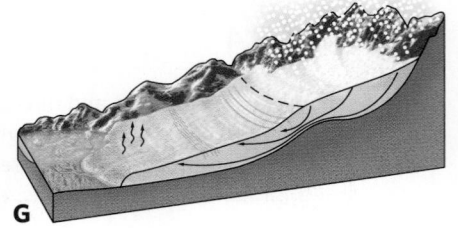

G

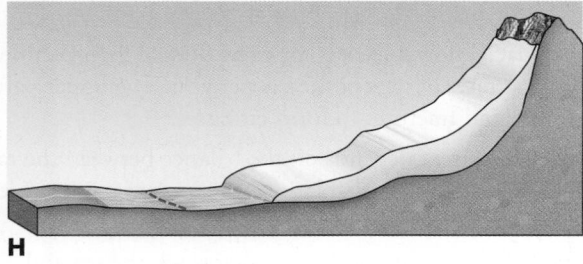

H

15. (a) What is the difference between sea ice and an ice shelf? (b) If an iceberg protrudes above water by 10 m, at what depth in the water will its base be?

Objective 13.5

Discuss how glaciers erode the land surface, describe the landforms produced by glacial erosion and deposition, and characterize the various materials deposited by glaciers.

KEY CONCEPTS

- Glaciers pick up debris by plucking and abrasion. Ice can transport sediment of all sizes. Sediment in moving ice can grind into bedrock and can polish and scratch bedrock surfaces.

- Mountain glaciers carve characteristic landforms, including cirques and U-shaped valleys. Fjords are water-filled U-shaped valleys carved by glaciers.

- Lateral moraines accumulate along the sides of valley glaciers from debris dropped on the glacier, and they merge where glaciers meet to form medial moraines. End moraines form at a glacier's toe.

- Glacial depositional landforms include moraines, kettle lakes, drumlins, eskers, ice-margin lakes, and glacial-outwash plains.

REVIEW QUESTIONS

16. (a) Explain why glacial erosion transforms V-shaped valleys into U-shaped valleys. (b) Name and describe the various landforms formed where mountain glaciers cause erosion, and label them on **Figure I.** (c) What erosional features do continental glaciers produce? (d) Why did coastal fjords fill with water?

I

17. (a) Describe the various kinds of sedimentary deposits associated with glaciation, and note the setting in which each forms. (b) Why were glacial erratics so named when first described? (c) How does glacial till differ from glacial outwash? (d) Why was loess deposited in regions south of the Pleistocene glaciers? (e) Label two types of moraines in **Figure J.** (f) What is an esker and how does it form?

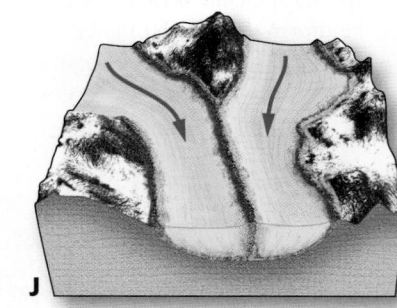

J

18. (a) Label the landforms illustrated in **Figure K.** (b) Why do kettle lakes form? (c) Describe the shape of a drumlin and of a roche moutonnée, and relate their shapes to ice-flow direction. (d) Why do varves form?

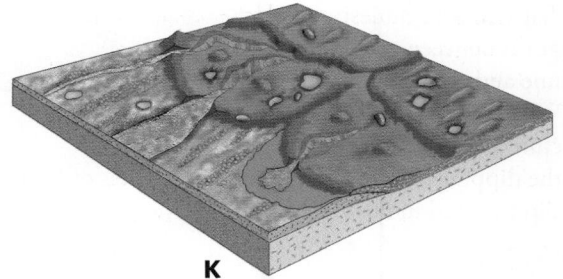

K

Objective 13.6

> Describe the effects of continental glaciation on the surface of the lithosphere, water distribution, drainage networks, and sea level.

KEY CONCEPTS

- The weight of a large ice sheet causes the continental lithosphere to subside. When the glacier melts away, the land surface rebounds.
- Glacial subsidence can lead to the formation of large ice-margin lakes. Ice dams or moraine dams holding back glacial lakes may fail, triggering glacial torrents.
- Permafrost develops where subsurface water remains frozen all year. In regions of the United States that are now deserts, substantially more rain fell during glacial advances.
- Storage of water in continental glaciers causes sea level to drop; when the glaciers melt, sea level rises. During ice ages, part of the continental shelf becomes dry land.

EARTH-SCIENCE VOCABULARY

glacial rebound (p. 506)
glacial subsidence (p. 506)
glacial torrent (p. 506)
land bridge (p. 508)
permafrost (p. 507)
pluvial lake (p. 507)

REVIEW QUESTIONS

19. **(a)** Explain the difference between glacial subsidence and glacial rebound, why each takes place, and why these movements are so slow. Which is happening in **Figure L**? **(b)** Where and why do glacial lakes form? **(c)** What is the evidence that glacial torrents washed over a large area of eastern Washington? **(d)** Can glaciation change the location of a drainage network? Explain your answer.

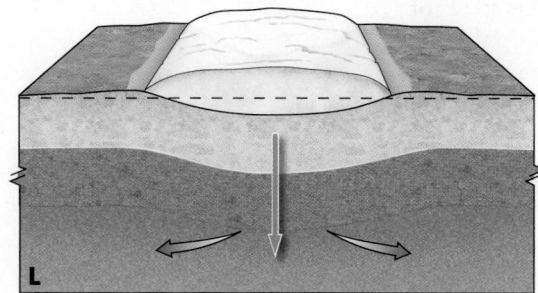

L

20. **(a)** Why did many lakes fill low areas in the Great Basin of the southwestern United States during glacial advances? **(b)** Explain why sea level changed so much when continental glaciers advanced and retreated. **(c)** How did these sea-level changes affect the land area of North America?

Objective 13.7

> Define an ice age, discuss when ice ages took place, and evaluate ideas proposed to explain why ice ages happen and why glacial advances and retreats during an ice age occur periodically.

KEY CONCEPTS

- During the Pleistocene Ice Age, large continental glaciers periodically covered much of North America, Europe, and Asia.
- The stratigraphy of glacial deposits indicates that glaciers advanced and retreated many times during the Pleistocene Ice Age. Other ice ages have happened earlier during Earth history, such as during the latter part of the Proterozoic, when our planet's surface may have frozen entirely to become snowball Earth.
- Long-term causes of ice ages include continental drift and changes in the concentration of CO_2 in the atmosphere. Short-term causes of glacial advances and retreats are linked to the Milankovitch cycles.

EARTH-SCIENCE VOCABULARY

Cordilleran ice sheet (p. 509)
glaciation (p. 510)
ice age (p. 509)
insolation (p. 512)
interglacial (p. 510)
Laurentide ice sheet (p. 509)
Milankovitch cycle (p. 513)
paleosol (p. 510)
Pleistocene Ice Age (p. 509)
positive feedback (p. 513)
tillite (p. 511)

REVIEW QUESTIONS

21. **(a)** Distinguish between a glaciation and an interglacial. **(b)** What evidence indicates that multiple glaciations happened during the Pleistocene Ice Age? **(c)** About how many Pleistocene glaciations were identified using the terrestrial record of North America, and how does this number differ from the number indicated by the study of marine sedimentary deposits? Explain why the numbers are different. **(d)** When did ice ages occur before the Pleistocene, and what evidence points to them in the stratigraphic record?

22. **(a)** What are some of the long-term causes that could set the stage for an ice age to happen? **(b)** How do the Milankovitch cycles affect insolation at mid- to high latitudes? **(c)** What evidence do geologists use to suggest that the Milankovitch cycles control glaciations and interglacials during an ice age?

ANOTHER VIEW Giant sand dunes tower over a high-speed train line and a highway near the Gobi Desert.

14 OCEAN WATERS
The Blue of the Blue Planet

After studying this chapter, you should be able to...

1. place oceanographic discovery in the context of human history, explain why the Earth effectively has one global ocean, and characterize sea-level change over time.

2. explain where the ocean's water and salts come from, describe how the salinity, temperature, and density of ocean water change with location and depth, and characterize pollutants in the ocean.

3. explain what surface currents are and why they develop the pattern of gyres and eddies that we see today, describe how upwelling and downwelling take place, and relate variations in ocean-water density to the formation of deep currents.

4. discuss why waves form and how they behave, explain how to describe waves, and characterize variations in the size and shape of wind-driven waves.

5. distinguish among zones of the ocean by light penetration, water depth, and proximity to land, describe life in the sea in relationship to these zones, and explain the relationship between marine life and dissolved oxygen.

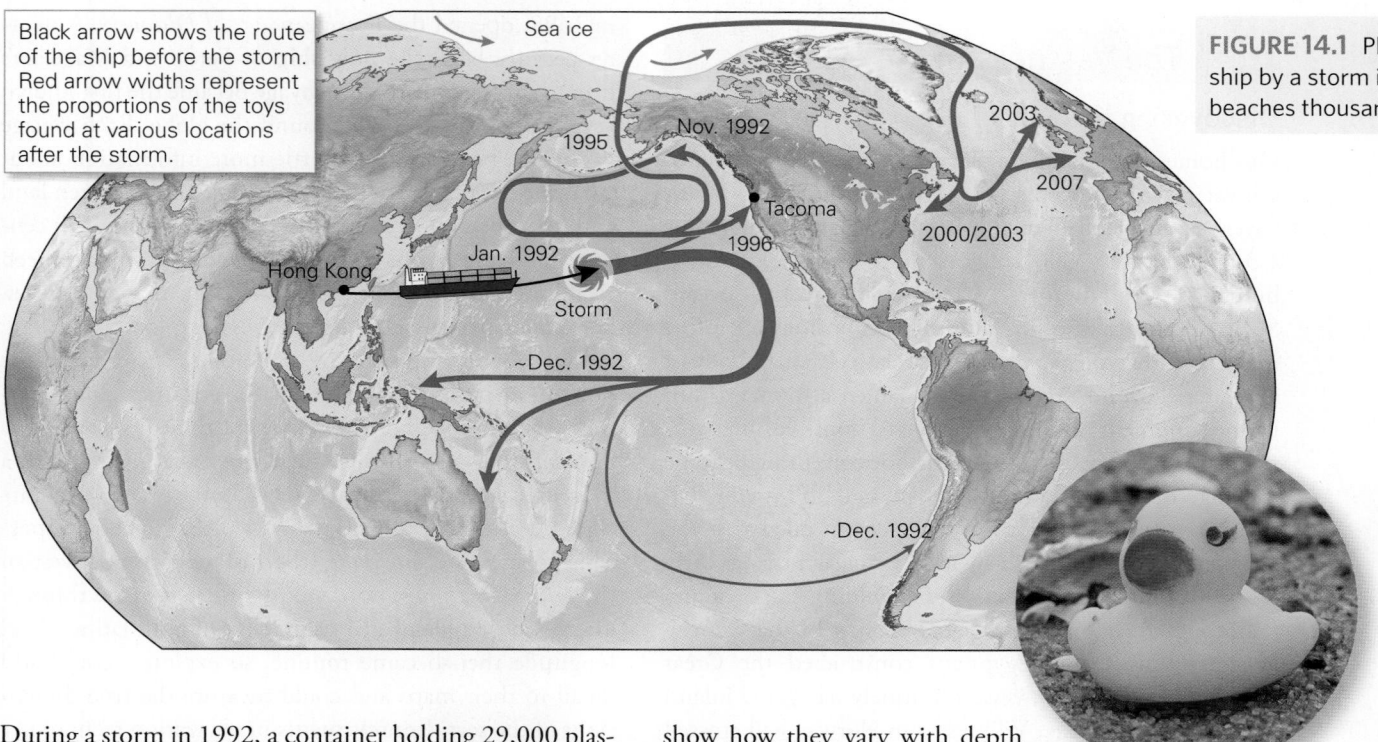

Black arrow shows the route of the ship before the storm. Red arrow widths represent the proportions of the toys found at various locations after the storm.

Sea ice

1995

Nov. 1992

2003

2007

Tacoma

2000/2003

Hong Kong

Jan. 1992

1996

Storm

~Dec. 1992

~Dec. 1992

FIGURE 14.1 Plastic toys, swept off a ship by a storm in 1992, ended up on beaches thousands of kilometers away.

During a storm in 1992, a container holding 29,000 plastic bath toys washed off a cargo ship and into the middle of the Pacific Ocean. Ten months later, little yellow duckies started appearing on beaches all around the Pacific. A small number floated through the Bering Strait, moved with sea ice around the Arctic Ocean to the Atlantic, and years later, ended up on the shores of Great Britain (Fig. 14.1). The amazing voyage of these toys emphasizes that, though we divide it into several geographic regions with different names, there is really only one global *ocean*, a layer of saltwater covering so much of our planet that, from space, the Earth looks like a shiny blue marble (Fig. 14.2).

Water in the Earth's oceans remains in constant motion. It circulates over vast distances in currents (stream-like flows), and its surface elevation changes constantly due to waves (local undulations) and tides (the generally twice-daily rise and fall of sea level). Despite all this motion, ocean characteristics such as salt content, temperature, and light vary regionally and with depth. These contrasts make the ocean a varied environment, hosting an incredible diversity of living species.

In this chapter, we explore the world's oceans. We start with a quick survey of ocean exploration to see how our knowledge of the ocean came to be. Next, we describe the physical and chemical characteristics of seawater and show how they vary with depth and location. We then turn our attention to the movement of seawater by describing currents, upwelling, downwelling, and waves. (Our discussion of tides appears in Chapter 15, because tidal movements primarily affect coasts.) We conclude this chapter by distinguishing the various zones of the ocean, which differ from one another depending on their depth and proximity to shore, and the assemblages of living organisms that inhabit these zones.

The Sea, once it casts its spell, holds one in its net of wonder forever.

— JACQUES-YVES COUSTEAU
(FRENCH EXPLORER, 1910–1997)

FIGURE 14.2 A view of the Earth from space emphasizes that blue ocean water covers most of its surface.

<< Gazing out to sea from the southwest coast of England, we can see how the texture and color of the water surface change constantly as it interacts with wind and light.

14.1 The Vastness of the Ocean

Discovering the Sea

Our home planet is a blue planet, for an **ocean**, a layer of saltwater with an average depth of 4–5 km (2.5–3 mi), covers about 70.8% of its surface. This layer contains 1.3 billion km³ (300 million mi³) of water. Because humans are land dwellers, we can't swim across the ocean. So the extent of the *sea* (a term used both as an informal alternative to *ocean*, and for a relatively small body of saltwater) beyond the horizon remained a mystery until the development of oceangoing boats. Some cultures surmised that the ocean ended at the horizon, thought to be the edge of the world, and then plunged off a cliff. No one knows exactly when the first person sailed past the horizon, but archaeological evidence suggests that as early as 3000 B.C.E., Polynesians were building large canoes that could travel among isolated islands in the western Pacific. By the time Egyptians constructed the Great Pyramid (2570 B.C.E.), vessels routinely navigated inland seas, and by 1000 C.E., Vikings from Norway had crossed the Atlantic. The end of the Middle Ages saw the beginnings of modern ocean exploration. Admiral Cheng Ho of China explored the Pacific and Indian Oceans between 1405 and 1433 C.E., commanding a fleet that included ships up to 120 m (400 ft) long.

Beginning in the late 15th century, European nations became the drivers of global ocean exploration. Christopher Columbus's journey across the Atlantic in 1492, followed by the discovery of sea routes to China in 1498, opened the European *Age of Discovery*. A quarter century later, the only ship of Ferdinand Magellan's fleet to survive a three-year voyage became the first known vessel to journey entirely around the globe. Each voyage of this era brought back a little more information about what the ocean's **shorelines**, the boundaries between land and sea, looked like, so by the beginning of the 17th century, the basic layout of continents and oceans had been established **(Fig. 14.3a)**. Early European maps, however, had major inaccuracies, for although explorers had long known how to measure the angle of the Sun above the horizon at its highest point in the sky to calculate *latitude* (the distance north or south of the equator), it was not until the mid-18th century, when the invention of a mechanical clock permitted navigators to keep accurate time on a moving ship **(Fig. 14.3b)**, that it became possible to measure *longitude* (the distance east or west of the *prime meridian*, a north-south line that passes through Greenwich, England). Measurement of both latitude and longitude then became routine, so explorers could add detail to their maps and could measure the true dimensions of the seas. Improvements in navigation technology, along with aerial photos, satellite images, and GPS, now allow us to characterize every detail of every shoreline.

The early explorers who ventured across the sea undertook their voyages primarily in search of treasure, conquest, and trade routes. Starting in the latter half of the 18th century, however, some European explorers also began to collect data about the sea. Between 1768 and 1780, for example, Captain James Cook of England documented currents, collected biological specimens,

FIGURE 14.3 Early mapping of the ocean.

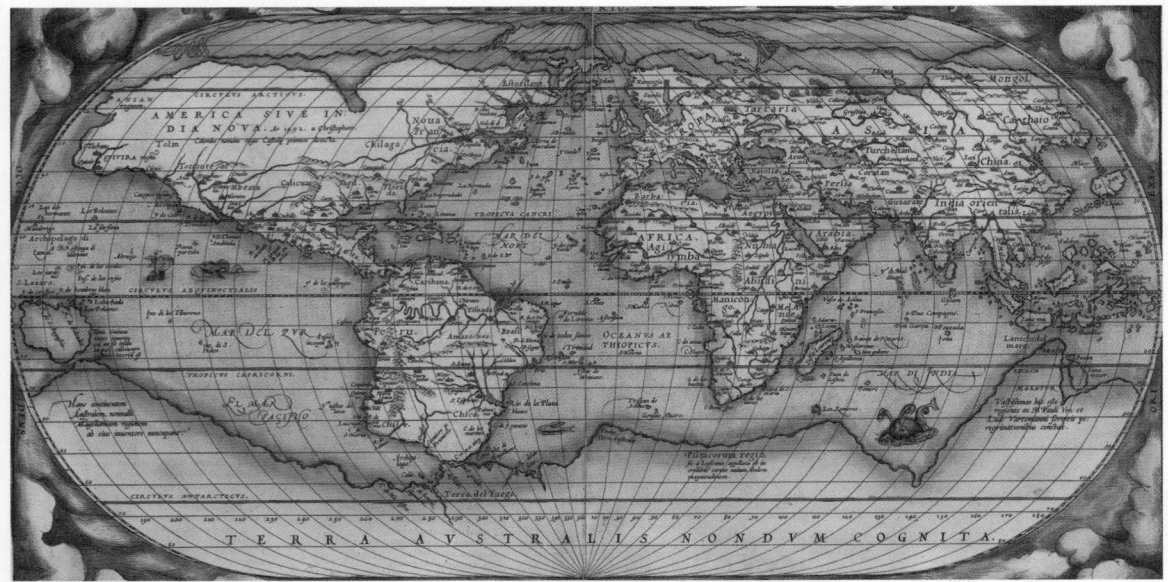

(a) A 1598 map by a Dutch cartographer showing the coastlines as they were known at the time.

(b) A 19th-century chronometer. This mechanical clock was not affected by motion, so it could keep accurate time, to within one second per month, even in stormy seas.

FIGURE 14.4 Many types of research vessels study the oceans.

(a) HMS *Challenger*, the first dedicated oceanographic research vessel, set sail in 1872.

(b) A modern oceanographic research vessel.

(c) Submersibles use spotlights to see at depth.

(d) Robotic submersibles provide information without risking lives.

(e) A single-pilot submersible can navigate along the seafloor.

(f) Seasat was a NASA satellite designed to monitor the world's oceans.

and measured water temperatures in addition to mapping coastlines. The modern science of **oceanography**, the study of the oceans, did not begin until 1872, when Britain's HMS *Challenger* began a four-year cruise during which researchers collected new information about ocean water and the seafloor—enough to fill 50 books (Fig. 14.4a). Now, by using research ships, submersibles, and satellites (Fig. 14.4b–f), modern *oceanographers* (who study the physical and chemical characteristics and movements of ocean water), along with *marine geologists* (who study the seafloor and coasts) and *marine biologists* (who study marine life), continue to enhance our knowledge of our planet's amazing seas.

The Map of the Modern Ocean

Geographers divide the global ocean into five separate bodies—the Atlantic, Pacific, Indian, Arctic, and Southern Oceans—bordered by continents. The equator divides the Atlantic and Pacific into northern and southern portions. In addition, geographers recognize several seas (such as the Mediterranean, Black, Red, and Caribbean Seas),

bays (such as Hudson Bay), and *gulfs* (such as the Gulf of Mexico), smaller regions of seawater partially surrounded by land (Fig. 14.5a). Each ocean fills an **ocean basin**. As we discussed in Chapter 2, most of an ocean basin's floor overlies oceanic lithosphere, which lies at a lower elevation than does the surface of continental lithosphere because of isostasy (see Chapter 6).

Some ocean-basin edges are *passive margins*, formed when rifting succeeded in producing a new mid-ocean ridge (Fig. 14.5b). Passive margins are underlain by the inactive remnants of the rift that accommodated stretching of a continent prior to breakup and ridge formation. Sediment washed off the continent, together with carbonate shells, buries the inactive rift remnant as it cools and slowly sinks, producing a *passive-margin basin*. The seafloor forming the top surface of the thick layer of sediment that accumulates in a passive-margin basin becomes a broad *continental shelf*, so called because the overlying water is relatively shallow. In contrast with passive margins, *active margins* are ocean-basin edges that are plate boundaries. Most active margins coincide with

FIGURE 14.5 Ocean basins and their margins.

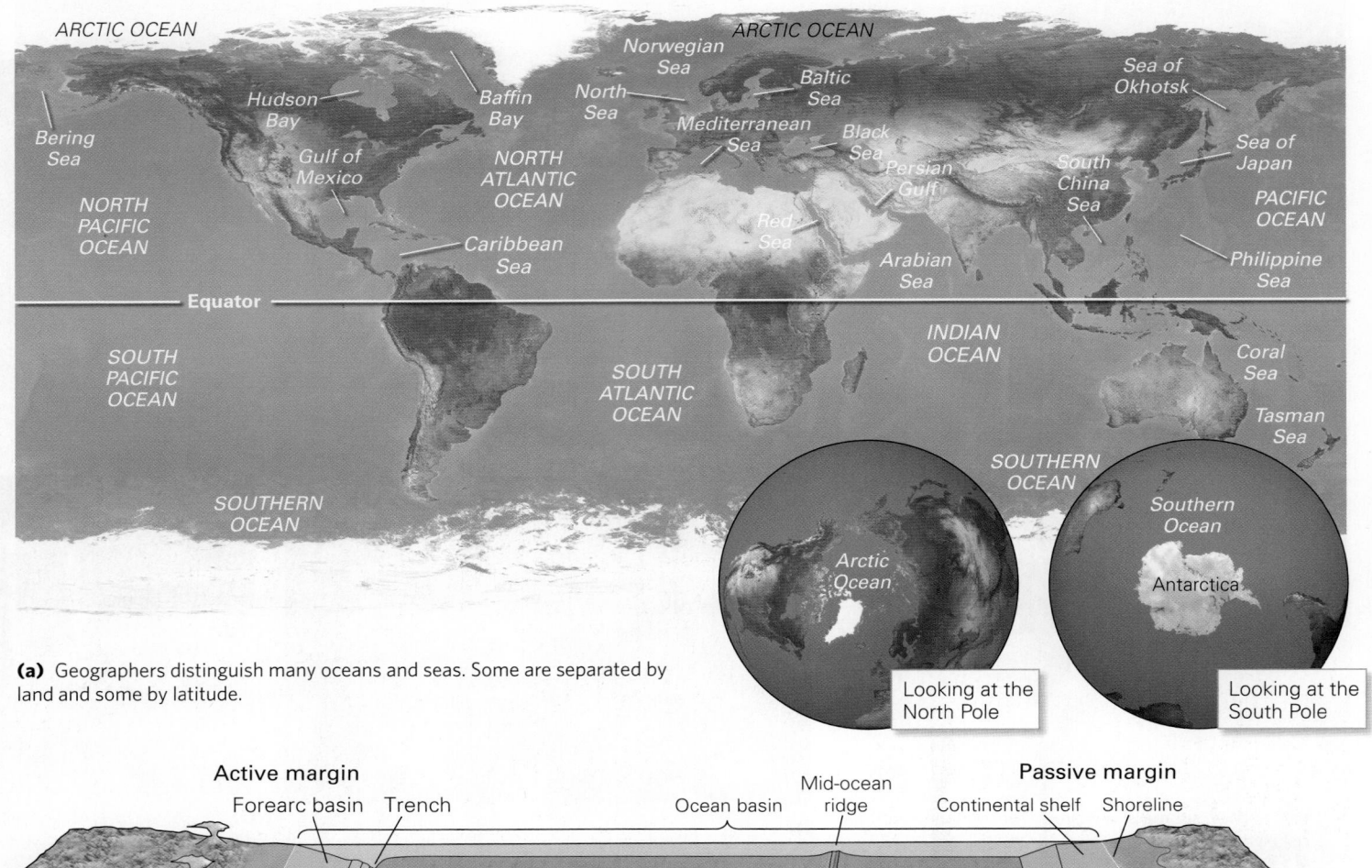

(a) Geographers distinguish many oceans and seas. Some are separated by land and some by latitude.

Looking at the North Pole

Looking at the South Pole

(b) This block diagram distinguishes between an active margin (in this case, a convergent boundary) and a passive margin. Note that oceanic lithosphere underlies most of the ocean floor.

convergent boundaries, delineated by a deep-sea trench and an accretionary prism bordering a volcanic arc. A few active margins follow the trace of a transform fault.

Each ocean basin has had a different geologic history. For example, the largest ocean basin, the Pacific, consists mostly of the Pacific Plate, which started forming by seafloor spreading at about 200 Ma. It also includes the young ocean floor of the Nazca and Cocos Plates on the east, and the Philippine Plate and several smaller plates on the west. The North Atlantic Ocean started opening around 180 Ma, but in the South Atlantic, seafloor spreading didn't begin until after 90 Ma. Of course, the shapes and dimensions of today's oceans and seas are markedly different from the way they were in the past,

and the way they will be in the future, given that subduction consumes some oceanic plates while seafloor spreading produces new ones.

Sea Level and Sea-Level Change

Sea level refers to the elevation of the boundary surface between ocean water and the air above. When we use the term *sea level* in scientific discussion, we imply *mean sea level*, the average height of this boundary over the course of a year, which lies halfway between the average elevation of high tides and that of low tides **(Fig. 14.6a)**. At any given moment, the sea surface at a given locality may be higher or lower than mean sea level because of waves, currents, or tides.

FIGURE 14.6 The concept of sea level and its variation over time.

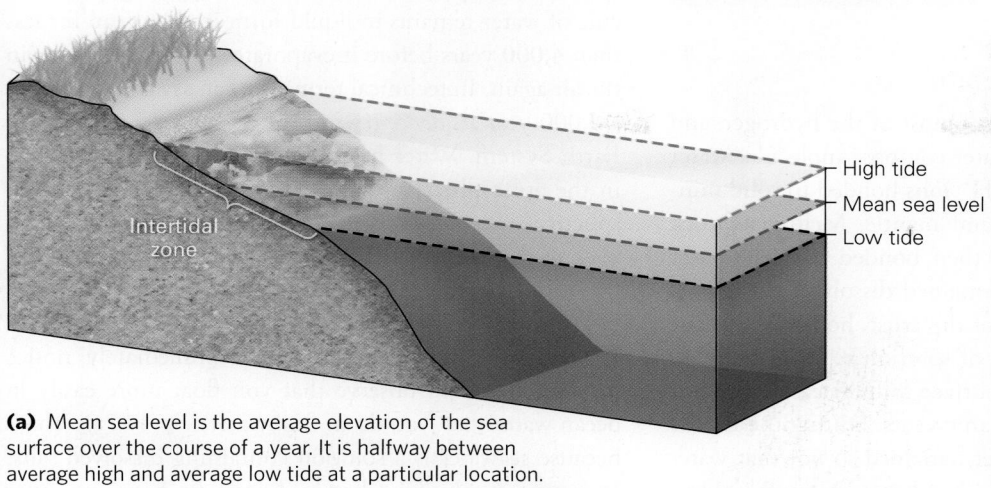

(a) Mean sea level is the average elevation of the sea surface over the course of a year. It is halfway between average high and average low tide at a particular location.

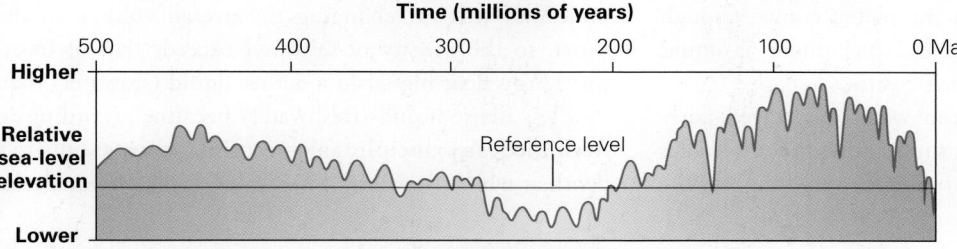

(b) A simplified estimate of sea-level change over the last half billion years. Sea level relative to land has gone up and down by over 100 m during that time.

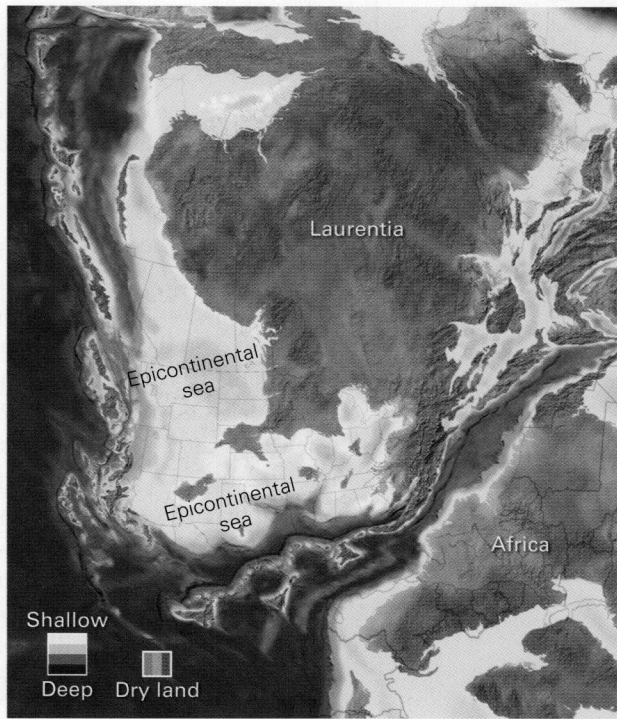

(c) At times, the ocean flooded large areas of continental interiors. During the Mississippian, for example, much of the interior of the United States lay beneath shallow epicontinental seas.

The geologic record indicates that liquid water has filled the oceans at least since the beginning of the Archean—except, perhaps, for "snowball Earth" episodes in the Proterozoic, when the sea surface was frozen—and that the overall volume of surface water on the Earth has stayed roughly constant over time (see Chapter 9). However, *relative sea level*, the position of mean sea level with respect to the land surface, has varied significantly over geologic time, due to a variety of geologic phenomena **(Fig. 14.6b)**. **Sea-level change** occurs due to the following circumstances: (1) When water gets trapped in large glaciers on land during an ice age, sea level falls relative to land everywhere, whereas when large glaciers melt at the end of an ice age, sea level rises. (2) When a region of land undergoes uplift, sea level falls relative to the region, whereas if the region subsides, relative sea level rises. (3) When the volume of ocean basins decreases (due to increases in the volume of mid-ocean ridges or the building of oceanic plateaus), sea level rises globally, whereas when the volume of ocean basins increases, sea level falls. At times when relative sea level rises significantly, ocean water submerges large areas of continents to form *epicontinental seas* **(Fig. 14.6c)**. When sea level falls, portions of continental shelves may be exposed.

Sea level changed significantly during the Pleistocene in association with glaciations and interglacials (see Chapter 13). During the Holocene, as the Laurentide ice sheet melted away, sea level rose by about 120 m (400 ft). In the past 150 years, it has risen by another 23 cm (9 in). Researchers suggest that this recent rise primarily reflects an increase in seawater temperature, for liquid water expands when heated, but it also reflects the continued melting of glaciers (see Chapter 19).

Take-home message…

A layer of saltwater forms a single global ocean. The continents and the equator divide the ocean into distinct geographic regions, and partial enclosure of oceanic regions by land defines seas and bays. Most of the ocean overlies oceanic lithosphere, because its surface sits lower due to isostasy. Mean sea level, the average elevation of the sea surface, has varied significantly over geologic time.

Quick Questions —————————————

- What defines the geographic boundaries of the North Atlantic Ocean?
- Is each ocean basin underlain by a single oceanic plate?
- Has relative sea level remained constant throughout Earth history?

14.2 Characteristics of Ocean Water

Where Does the Water in the Oceans Come From?

When the Earth first formed, most of the hydrogen and oxygen atoms that would later compose molecular water (H_2O) existed instead as OH^- ions bonded to solid minerals in rock of the crust and mantle. Melting of rock released these ions, which then bonded to yield water molecules that, at depth, remained dissolved in magma. As magma rose to the top of the crust, however, the dissolved molecules came out of solution to form bubbles, some of which rose to the surface to be released as vapor into the atmosphere at volcanic vents. During the Earth's very early history, the planet remained so hot that water remained a vapor, mixed with other gases in the atmosphere. Oceans formed when the planet cooled enough for water vapor to condense and precipitate as liquid water that could collect in ocean basins.

While the volume of ocean water has stayed fairly constant for most of geologic time, individual molecules of water do not remain in the ocean for very long, for the ocean serves as one of several reservoirs in the hydrologic cycle (see Chapter 11). In fact, on average, a given molecule of water remains in liquid form in the ocean for less than 4,000 years before it evaporates to become vapor in the air again. In technical terms, we can say that water has a 4,000-year *residence time* in the oceanic reservoir of the Earth System. Water has a much shorter residence time in the atmosphere; usually, within hours to days, atmospheric water vapor condenses or crystallizes and falls back to the Earth's surface as rain, snow, or hail.

The Salt of the Sea

If you swim in the ocean, you'll immediately notice its salty taste and observe that you float more easily in ocean water than you do in freshwater (Fig. 14.7a). That's because seawater is a solution containing dissolved salts. In a water solution, dissolved ions fit between water molecules without changing the overall volume of the water, so the density of saltwater exceeds that of freshwater. You float higher in a denser liquid (seawater) than in a less dense liquid (freshwater) because, according to **Archimedes's principle**, an object sinks in water only to a depth at which the mass of the water displaced equals the

FIGURE 14.7 Salinity varies within the ocean.

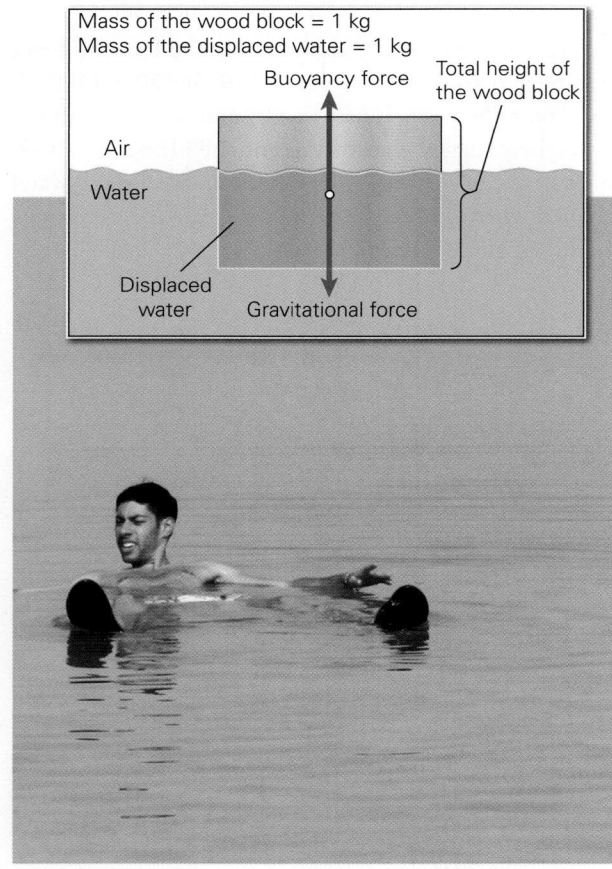

(a) The Dead Sea (which is actually a salt lake, completely surrounded by land) is so salty that people float in it like corks. Its salinity is almost 10 times that of ocean water. The inset illustrates the concept of Archimedes's principle. When the mass of the water volume displaced by the block equals the mass of the whole block, the buoyancy force (upward) balances the gravitational force (downward).

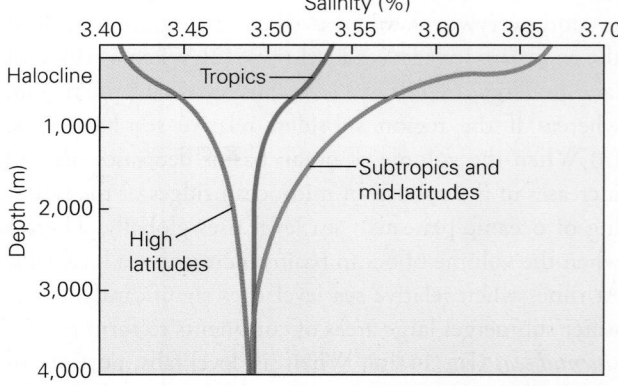

(b) The salinity of seawater varies with location. These regional variations reflect differences in rates of freshwater addition and evaporation.

(c) Salinity varies with depth as well as with climate. Variation in salinity is greatest in the uppermost 1 km of ocean water.

mass of the object, and a given volume of saltwater has a greater mass than the same volume of freshwater.

Typical seawater contains about 96.5% water (H_2O) and 3.5% salt by weight. In contrast, typical freshwater contains less than 0.02% salt. Put another way, seawater contains about 175 times the amount of salt that occurs in freshwater. In fact, there's so much salt in the sea that if all of the sea's water molecules magically evaporated, the layer of salt left behind would be about 67 m (220 ft) thick, the height of a 25-story building.

Oceanographers refer to the concentration of salt in water as **salinity**. Although sea-surface salinity averages 3.5%, it varies with location, ranging from 1.0% to 4.1% (Fig. 14.7b). Salinity depends, in part, on water temperature, because warm water can hold more salt in solution than cold water can. It also depends on the balance between inputs of freshwater from rain and rivers and removal of freshwater by evaporation. For example, in **estuaries**, places where seawater floods the mouth of a river, salty seawater mixes with fresh river water to produce *brackish water*, which has less salinity than seawater. Brackish water also develops where glacial meltwater enters the sea or where heavy rain falls on the sea surface. While in contrast, along the margins of confined seas in hot, arid climates, so much water evaporates that the remaining water becomes *brine*, meaning that its salinity exceeds that of seawater. Note that equatorial waters are less salty than the waters of the subtropics and mid-latitudes. That's because the heavy rains in the tropics dilute surface water. In the subtropics and mid-latitudes, more evaporation than precipitation takes place.

Ocean salinity also varies with depth (Fig. 14.7c). The largest variations in salinity occur in the upper 1 km of the sea, the portion most affected by evaporation or by freshwater inputs. Salinity in deeper water tends to be more homogeneous. Oceanographers refer to the boundary between surface-water salinities and deep-water salinities as the *halocline*.

What does the salt in the ocean consist of? Chemical analyses indicate that the positive ions in seawater include sodium (Na^+), potassium (K^+), calcium (Ca^{2+}), and magnesium (Mg^{2+}), and that the negative ions include chloride (Cl^-) and sulfate (SO_4^{2-}). Of these, sodium and chloride are the most abundant by far (Fig. 14.8a). In fact, if seawater evaporates entirely, about 85% of the salt that precipitates will consist of the mineral halite (NaCl). The remainder consists of other mineral salts, including gypsum ($CaSO_4 \cdot 2H_2O$), anhydrite ($CaSO_4$), magnesium sulfate ($MgSO_4$), magnesium chloride ($MgCl_2$), and potassium chloride (KCl).

Seawater doesn't need to evaporate entirely to precipitate salt (Fig. 14.8b). Different minerals precipitate from brine depending on how concentrated the brine has become. As seawater starts to evaporate, NaCl precipitates first. Most common table salt, the kind in your saltshaker, consists almost entirely of purified salt deposits formed by evaporation of seawater in the geologic past. These deposits occur within successions of sedimentary strata. The "sea salt" that has become popular in gourmet stores is made by evaporating seawater to produce a brine from which salts precipitate. Such salt contains 95%–98% NaCl; the remainder consists of traces of other salts. Evaporation can also be used to extract freshwater from saltwater (Box 14.1).

Where does the salt in the sea come from? In 1715, Sir Edmond Halley (best known for his discovery of Halley's comet) suggested that the rivers flowing on the Earth's surface transported salt to the sea, for he realized

---

FIGURE 14.8 Seawater contains a variety of salts.

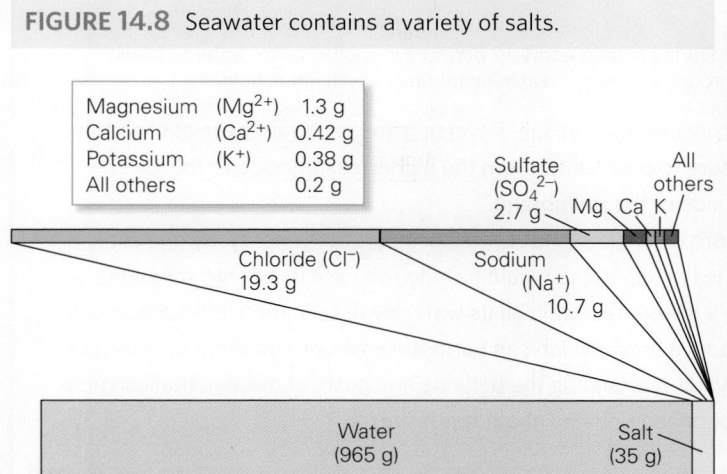

Magnesium	(Mg^{2+})	1.3 g
Calcium	(Ca^{2+})	0.42 g
Potassium	(K^+)	0.38 g
All others		0.2 g

Sulfate (SO_4^{2-}) 2.7 g

All others

Mg Ca K

Chloride (Cl^-) 19.3 g

Sodium (Na^+) 10.7 g

Water (965 g)

Salt (35 g)

(a) The average chemical composition of seawater, in grams per liter.

Close-up of salt crystals

Salt accumulation

(b) Sea salt precipitates in puddles that fill with spray from ocean waves along an arid shore.

BOX 14.1 **Putting Earth Science to Use**

How can seawater be made drinkable?

Human blood has a salinity of 0.9% by weight, much less than that of seawater. Our kidneys regulate the salinity of our blood, so if we drink seawater, our bodies excrete the excess salt in urine. To keep your blood salinity normal, you would have to excrete more water than you drank, so drinking seawater would eventually lead to death by dehydration. That is, of course, if you didn't die first from seizures or heart failure, as excess salt in your blood can disrupt the conduction of electrical signals by nerves. Clearly, we can't drink seawater ... we need freshwater to survive!

Water, water, everywhere, And all the boards did shrink;
Water, water, everywhere, Nor any drop to drink.

—SAMUEL TAYLOR COLERIDGE (ENGLISH POET, 1772–1834)

As use of freshwater—for irrigation, industry, and drinking (see Chapter 12)—has increased, society has been consuming supplies of natural fresh surface water and groundwater faster than it can be replenished. To keep up with freshwater demand, especially in arid parts of the world, can we extract salt from seawater to produce freshwater? The answer is yes. Salt can be removed from water by **desalination**. How? Water can be desalinated either by distillation or by reverse osmosis.

During *distillation*, workers pump seawater into a tank and heat the water to accelerate evaporation. When seawater evaporates, it leaves its salt behind, so the resulting vapor consists of freshwater. Condensation of the water vapor produces liquid freshwater. Most commercial distillation operations use *multistage flash distillation* **(Fig. Bx14.1a)**, a process that takes advantage of the fact that water's boiling temperature depends on atmospheric pressure. Specifically, under lower pressures, water boils at lower temperatures. During multistage flash distillation, seawater that has been warmed in a furnace flows into a chamber from which air has been extracted. At the lower pressure in this chamber, some of the water boils, producing vapor that can be collected and removed. Evaporation also causes the remaining water to cool (see Chapter 18). This slightly cooler water then flows into another chamber with still lower pressure, where it boils and evaporates. By repeating the process in a series of chambers, each with a pressure lower than the previous one, workers can remove about 85% of the water molecules from seawater.

Reverse osmosis purifies seawater without heating it. During this process, workers apply pressure to push seawater through a semipermeable membrane that allows water molecules, but not salt molecules, to pass through it **(Fig. Bx14.1b)**. Why is this process called "reverse" osmosis? During normal osmosis, water from a solution with a lower concentration of salt flows through a membrane into a region with a

FIGURE Bx14.1 Desalination of seawater.

(a) This multistage flash distillation plant on the coast of a desert produces 470,000,000 L (125,000,000 gal) of freshwater a day, enough to supply a small city.

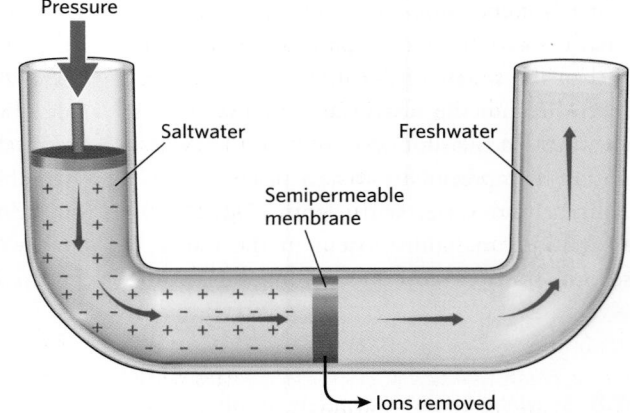

(b) During reverse osmosis, pressure is applied to saltwater to push it through a semipermeable membrane, which filters out salt.

higher concentration of salt. Reverse osmosis requires the application of pressure to push water from the higher-salinity side to the lower-salinity side of the membrane.

Both distillation and reverse osmosis take energy. By one estimate, the United States would have to increase its energy consumption by 10% if it were to obtain all its water by desalination. Both processes also produce brine, which can harm the environment if not disposed of properly. Understanding the benefits and costs of desalination can help you understand debates about this issue.

that even though river water is fresh enough to drink, it contains tiny amounts of dissolved salts. Halley proved to be correct: most ions in the oceans come from the chemical weathering of rock and enter the oceans in flowing groundwater and river water. In fact, rivers deliver over 2.5 billion tons of salt to the sea every year. The rest comes from submarine hydrothermal vents, or from submarine volcanic gases.

Notably, the proportions of ions that enter the ocean from rivers differ from the proportions of ions dissolved in the ocean. Specifically, more than half of the negative ions in river water consist of HCO_3^-, the bicarbonate ion. Ocean water loses its HCO_3^- because organisms extract the ion to build calcite shells, and when the organisms die, the calcite sinks to the seafloor, where it eventually becomes buried. In contrast, Na^+ and Cl^- do not get extracted to make shells, so these ions remain in seawater and become relatively concentrated.

Temperature Variations in the Ocean

When RMS *Titanic* sank after striking an iceberg in the North Atlantic, most of the passengers and crew who jumped or fell into the sea died of cold in less than a half hour. The sea-surface temperature at the site of the tragedy was numbingly frigid, and very cold water removes heat from a human body much faster than the human metabolism can replace it. Yet swimmers can play for hours in the Caribbean, where the sea-surface temperature routinely reaches 28°C (83°F), and their metabolism can keep them warm enough to survive comfortably. Though the global annual sea-surface temperature averages around 17°C (63°F), it ranges from 38°C (100°F) in some tropical seas to −2.2°C (28°F), the freezing temperature of saltwater, in polar seas (Fig. 14.9a).

The correlation of average annual sea-surface temperature with latitude exists because the total amount of solar radiation reaching the Earth's surface varies with latitude. The amount of solar radiation also varies with the seasons, so sea-surface temperature varies with the seasons as well, but not by as much as air temperature does, because water has a high *heat capacity*, meaning that it can absorb or release large amounts of heat without changing its temperature very much. For example, in mid-latitudes, the difference between winter and summer sea-surface temperatures averages about 8°C (14°F). In contrast, average air temperature in the interior of North America changes by 43°C (75°F) over the course of a year.

Water temperature in the ocean varies markedly with depth in tropical and temperate latitudes, for water heated by the Sun expands and becomes less dense than colder water. Therefore, warmer water floats above denser, colder water. The boundary between warmer water

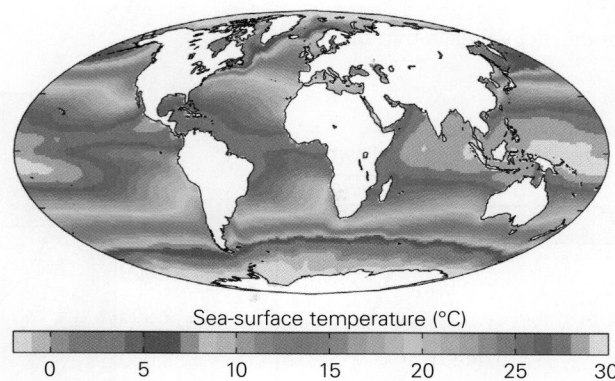

FIGURE 14.9 Temperature varies within the ocean.

Sea-surface temperature (°C)

0   5   10   15   20   25   30

(a) Regional variations in average annual sea-surface temperature correlate roughly with latitude.

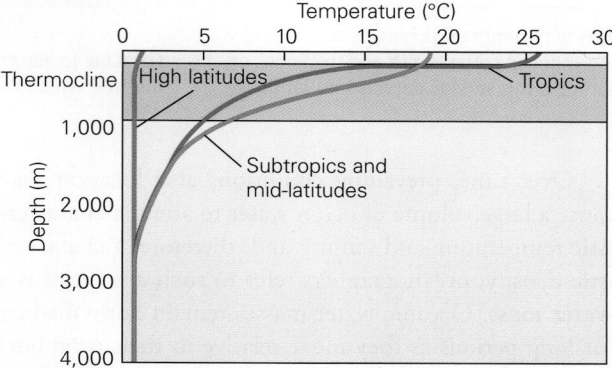

(b) The temperature of seawater changes with depth. Oceanic bottom waters are close to the freezing point of freshwater.

above and colder water below is called the **thermocline** (Fig. 14.9b). In the tropics, the top of the thermocline lies at about 200 m (660 ft), and the bottom at about 900 m (3,000 ft), below the ocean surface. Beneath the thermocline, seawater temperatures remain fairly constant. A pronounced thermocline doesn't develop in polar seas because the surface waters are already so cold that there's hardly any temperature contrast between sea surface and seafloor.

Ocean-Water Density and Water Masses

Both temperature and salinity control the density of water, so seawater density varies laterally and vertically. The density of water at the ocean surface ranges between 1.02 and 1.03 g/cm³, so it's 2%–3% denser than freshwater. Water at the seafloor, due to a combination of the pressure caused by the weight of overlying water, low temperature, and high salinity, can reach a density of 1.05 g/cm³. In some parts of the world, ocean-water density changes relatively rapidly across a boundary, called the *pycnocline*, at a depth of 0.5–1.0 km (Fig. 14.10a). A less dense *shallow layer* extends down from the sea surface to the pycnocline, and a denser *deep layer* continues from the pycnocline down to the seafloor.

FIGURE 14.10 Water density varies within the ocean.

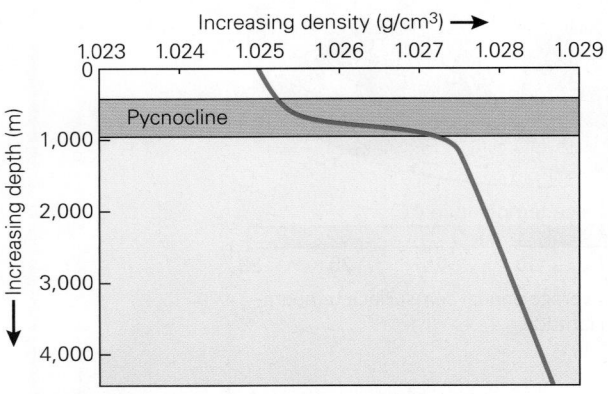

(a) Seawater density varies with depth and changes most rapidly over an interval called the pycnocline. The depth and thickness of the pycnocline vary with latitude; this version depicts mid-latitudes.

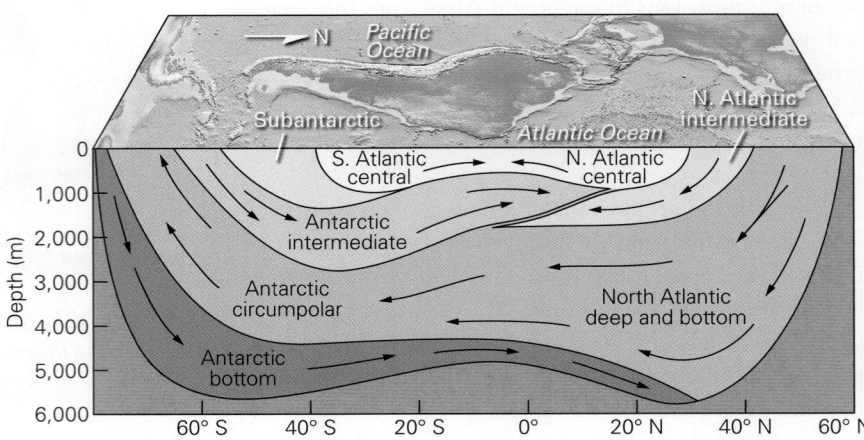

(b) Due to variations in seawater density, the oceans are stratified into distinct water masses, which flow in alternating directions.

Over time, prevailing conditions at a location may cause a large volume of ocean water to attain a characteristic temperature and salinity and, therefore, a characteristic density; oceanographers refer to such a volume as a **water mass**. Oceanic water masses remain fairly distinct for long periods as they move relative to their neighbors because they do not mix easily at their margins. For example, cold water from the coast of Antarctica sinks to the floor of the ocean, forming a layer on the Atlantic Ocean seafloor known as *Antarctic bottom water*. Similarly, cold water from the Arctic region sinks to form the *North Atlantic bottom water*. The latter overlaps the former, and in equatorial regions, other water masses form above these bottom waters. Interleaving of distinct water masses causes an overall layering, or stratification, of the ocean **(Fig. 14.10b)**.

Marine Pollution

So far, we've been discussing clean ocean water. Unfortunately, during the past several centuries, ocean water has undergone significant changes in response to human activity. Agriculture, cities, ships, and industries have all added substances to the sea that would not otherwise be there; of these sources, agricultural runoff is the largest. Some pollutants are harmless, and some act as nutrients (stimulating growth of marine biomass, as we'll discuss in Section 14.5), but many are toxic or hazardous and damage the health of marine life. When people eat contaminated marine organisms, they can absorb some of these toxins, which result in health problems. For example, mercury—a toxic metal that can cause severe health problems in humans—enters the atmosphere due to the burning of coal and during cement production.

When mercury vapor comes into contact with seawater, it dissolves. Measurements indicate that the concentration of mercury in shallow seawater has increased by a factor of 3.5 during the past several decades. Plankton ingest this mercury, so plankton-eating fish ingest it, too. When larger fish eat smaller ones, the larger fish ingest mercury in turn. The US Environmental Protection Agency (EPA) recommends that people eat no more than a few servings of mercury-concentrating fish per month.

The list of chemicals introduced by society into seawater unfortunately gets longer as time passes. Seawater everywhere contains measurable amounts of *persistent toxins*, meaning dangerous "forever chemicals" (such as pesticides, chlorinated hydrocarbons, and heavy metals) that do not degrade quickly even far from shore. Also, over 17 million bbl of oil spill into the ocean, on average, every year. While the volatile components of oil soon evaporate, tar-like residue may survive for years and drift with ocean currents.

In recent years, the dangers of dumping plastic garbage into the ocean have made headlines. Over 8 million tons of plastic debris enters the ocean yearly, and this debris may take centuries to decompose. Much of this debris, such as plastic beverage bottles and plastic bags, can be seen with the naked eye; it accumulates on beaches (see Fig. 15.46) or floats on water. But much of the plastic in the ocean—by some estimates, over 90%—occurs in the form of *microplastics*, particles ranging from about 0.002 to 1 mm in diameter. Plastic debris of all sizes poses a danger for marine life. Turtles, fish, and even whales can become entangled in plastic debris, and even tiny organisms can ingest microplastics.

The ocean's water originated as vapor released during volcanic activity. Salts are carried into the sea by rivers. Ocean salinity, temperature, and density vary with depth and location. Differences in these characteristics cause ocean water to form water masses that do not mix readily with others, but instead form distinct layers. Humanity has added a variety of pollutants to seawater, some of which may persist for centuries.

Quick Questions

• What is the concentration of salt in seawater?

• What factors affect the salinity of seawater?

• How does the density of seawater relate to its temperature?

14.3 Currents: Rivers in the Sea

If you were to tow a raft from a beach in Miami, Florida, eastward for about 20 km, then cut it loose and let it drift, it would end up off the coast of Delaware—1,600 km (1,000 mi) away—in about 250 days, assuming (unrealistically) that winds were calm the whole way. That's because portions of the oceans constantly flow or circulate at speeds of 1–10 km/h (0.6–6 mph). Oceanographers refer to a flowing band of water as a **current**. Unlike a stream on land, an ocean current does not move within a distinct physical channel. But because currents transport water from one environment to another, the temperature and salinity of water in a current generally differs from that of the water outside the current, so currents stand out in certain types of satellite images (Fig. 14.11). Currents flow within two layers in the ocean: **surface currents** affect only the upper 100–400 m (330–1,300 ft) of water, whereas **deep-sea currents** affect water at great depth and may even move water along the seafloor. Let's now consider the factors that drive currents, starting with surface currents.

Surface Currents

THE PATTERN OF SURFACE CURRENTS. When ship captains plan routes between Europe and North America, they pay attention to the directions of surface currents, because sailing with a current speeds up a voyage significantly. For example, when they want to head from North America toward Europe, they steer north so as to stay in the Gulf Stream, a northeastward-flowing surface current. Overall, major surface currents in each ocean display circle-like paths, called **gyres**, each of which follows the margins of the ocean in which it has formed (Fig. 14.12a). Different parts of each gyre have their own unique names;

FIGURE 14.11 The Gulf Stream is a current that carries warm water northward along the southeastern coast of North America and then across the North Atlantic. On this map, the colors represent surface-water temperatures.

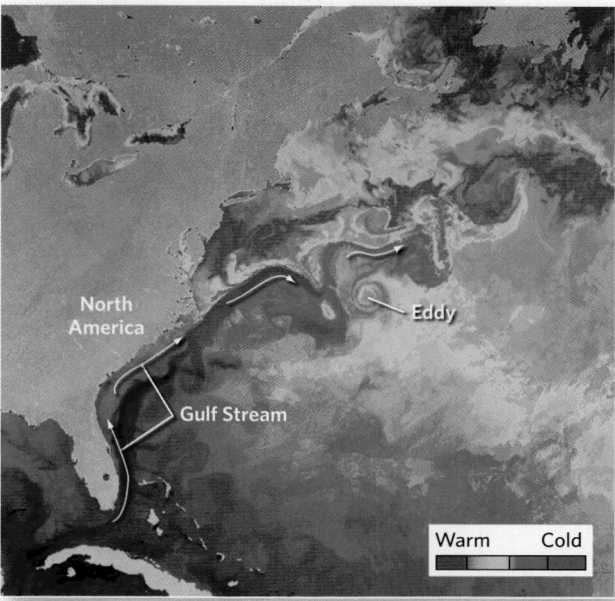

for example, the Gulf Stream is the northwestern part of the North Atlantic Gyre. Where the northern portion of the North Atlantic Gyre reaches Europe, it splits into one branch that heads north into the Arctic Ocean and another branch that remains part of the gyre and heads south toward the equator as the Canary Current. The equator separates the clockwise-flowing gyres of the northern hemisphere from the counterclockwise-flowing gyres of the southern hemisphere. (Later in this section, we will see how this change relates to the Coriolis force.) The currents of a gyre range from 100 to 300 km wide.

Locally, flow within or along the margins of currents defines smaller swirls, or **eddies**, whose circumference can range from a few hundred to over a thousand kilometers. Some eddies form because of shear between the moving water inside the current and the relatively still water outside the current; the eddies on the southeastern side of the Gulf Stream serve as examples (see Fig. 14.11). Other eddies form where currents interact with coastal areas or pass between two landmasses, such as Africa and Madagascar (Fig. 14.12b).

Floating debris gets trapped in the relatively still water that lies in the interior of a gyre. A type of floating seaweed called sargassum covers part of the North Atlantic Gyre's interior, for example, so sailors refer to the region as the Sargasso Sea (see Fig. 14.12a). The interior of the

FIGURE 14.12 Surface currents affect only the upper 100–400 m of seawater.

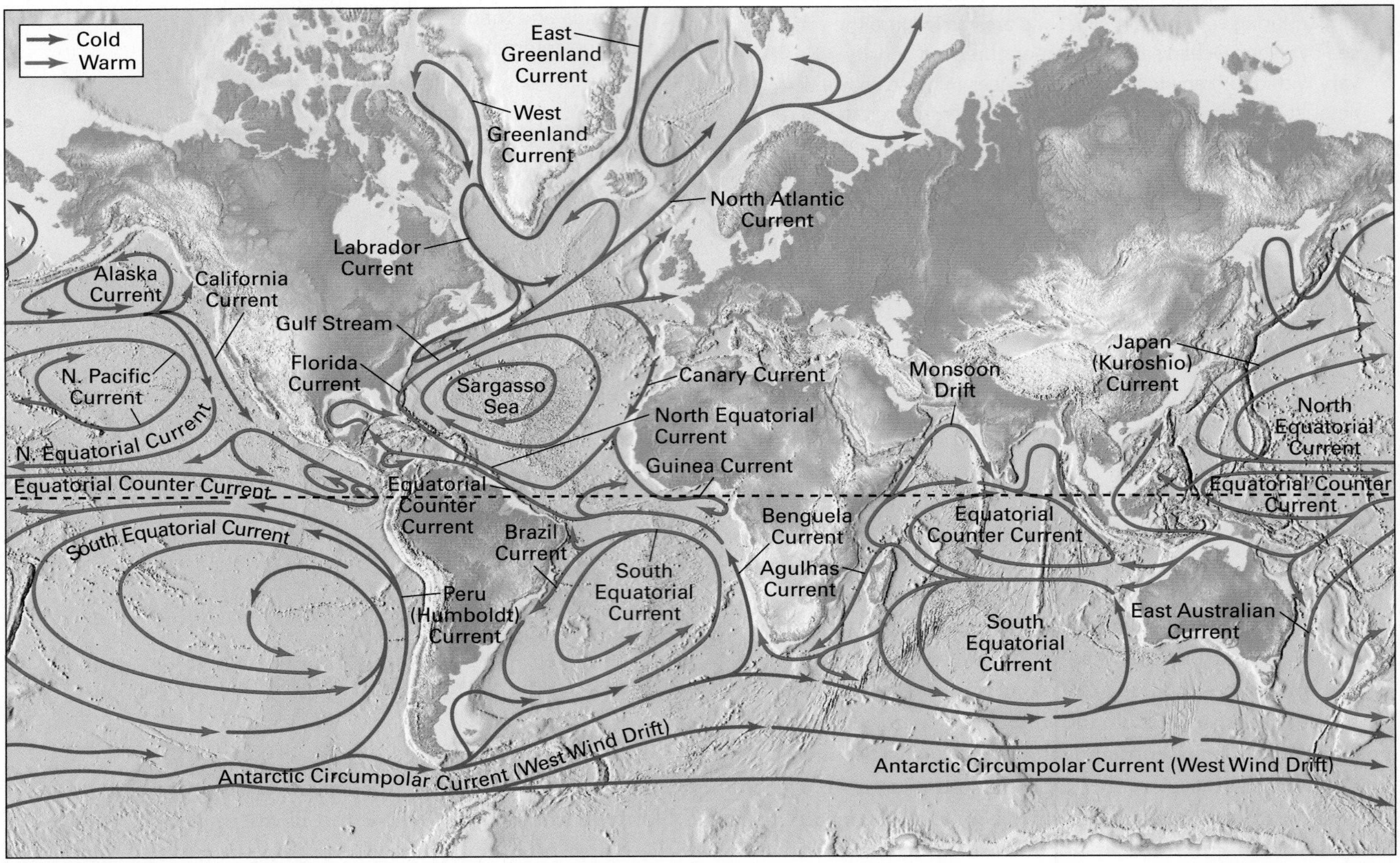

Cold →
Warm →

East Greenland Current

West Greenland Current

North Atlantic Current

Labrador Current

Alaska Current

California Current

Gulf Stream

Florida Current

N. Pacific Current

Sargasso Sea

Canary Current

Monsoon Drift

Japan (Kuroshio) Current

N. Equatorial Current

North Equatorial Current

North Equatorial Current

Guinea Current

Equatorial Counter Current

Equatorial Counter Current

Equatorial Counter Current

South Equatorial Current

Brazil Current

Benguela Current

Equatorial Counter Current

Peru (Humboldt) Current

South Equatorial Current

Agulhas Current

South Equatorial Current

East Australian Current

Antarctic Circumpolar Current (West Wind Drift)

Antarctic Circumpolar Current (West Wind Drift)

(a) In each major ocean basin, surface currents carry water around large circular paths known as gyres.

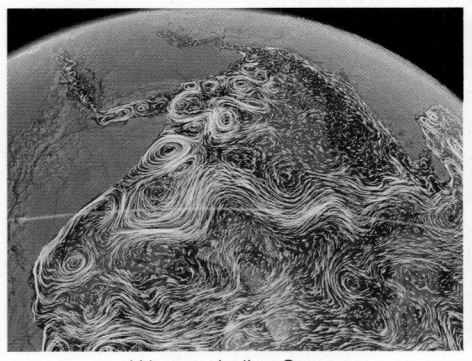
Western Indian Ocean

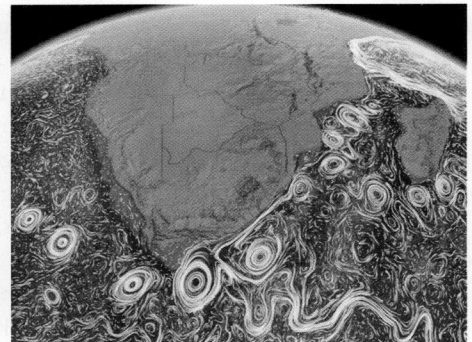
Southern Ocean, south of Africa

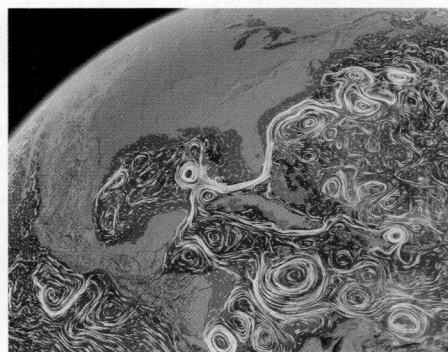
Western North Atlantic and Caribbean

(b) These images from NASA show the presence of many eddies, swirling currents that are smaller than gyres. The white lines represent flow.

North Pacific Gyre has collected vast quantities of floating plastic garbage (possibly over 100,000 tons) and chemical sludge, so the region has come to be known informally as the Great Pacific Garbage Patch **(Fig. 14.13)**.

Why do surface currents form? Why do they flow in the directions that they do? Surface currents begin because of the interaction between moving air and the surface of the sea. But the specific path that a current follows also depends on the influence of two other forces—the Coriolis force and the pressure-gradient force—and on the shape of the ocean basin. Let's examine each of these influences.

WIND SHEARING WATER: FRICTIONAL DRAG. You're already familiar with *friction* as resistance to sliding between two solids; it's the force that slows down a book

FIGURE 14.13 The Great Pacific Garbage Patch.

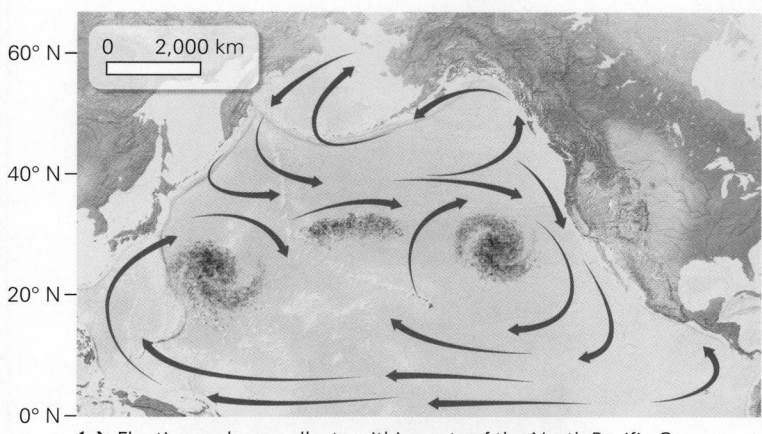

(a) Floating garbage collects within parts of the North Pacific Gyre.

(b) An example of the floating garbage, mostly plastic, as seen from below. (The average concentration of garbage in the Great Pacific Garbage Patch is much less than what appears in this photo.)

when you push it across a table (Box 14.2). A similar force, called **frictional drag**, occurs at the interface between fluids that are moving at different velocities. Simplistically, molecules of the faster-moving fluid shear against and pull along molecules of the slower fluid, causing molecules in the slower fluid to speed up and those in the faster fluid to slow down. Wind, the horizontal flow of air, applies frictional drag to water at the ocean's surface, and the resulting shear causes that water to start moving. The amount of movement depends on the *viscosity* (resistance to flow) of the water, the velocity of the wind, the duration of the wind, and the dimensions of the area affected by the wind. Moving surface water, in turn, shears the water just below it and causes that water to start moving, but not as fast. In a given region, winds tend to blow in the same direction, called the **prevailing wind direction**,

for most of the year, so a large volume of water can be set in motion (Fig. 14.14) (see also Chapter 17). If prevailing winds didn't exist, well-defined surface currents would not develop.

MOVEMENT ON A SPINNING EARTH: THE CORIOLIS EFFECT. If the Earth were stationary, a moving object that started off heading due north would continue heading due north because of its inertia. (Sir Isaac Newton defined *inertia* as the tendency of a moving object to follow a straight path at a constant velocity unless acted on by an outside force.) Our planet, however, isn't stationary, for it rotates on its axis once a day. Because of this rotation, a point on the equator moves in the direction of rotation at a speed of 1,670 km/h (1,037 mph) relative to a stationary observer outside of the Earth. This *rotational velocity*

FIGURE 14.14 Global prevailing winds control the overall pattern of ocean currents.

NAG = N. Atlantic gyre; SAG = S. Atlantic gyre; NPG = N. Pacific gyre; SPG = S. Pacific gyre; IOG = Indian Ocean gyre.

BOX 14.2

How can I explain . . .

Shear caused by frictional drag

What are we learning?

Friction caused by a moving object or fluid can carry underlying material along with it.

What you need:

- A slippery tabletop, masking tape, a sheet of paper, and a textbook
- A shallow pan filled with soapy water with a few small scraps of paper floating in it

Instructions:

- Mark out a reference line on the table with masking tape.
- Align the sheet of paper so that its edge is flush with the line.
- Place the book on top of the paper, then push it forward by about 15 cm (6 in). Measure the distance that the paper moves.
- Place the pan of soapy water on the tabletop.
- Blow hard at a shallow angle to the water surface.

What did we see?

- Friction slows the movement of the book. But frictional drag at the base of the book can move a layer of material beneath it. Materials moved by frictional drag do not move as fast as the materials that drive the motion.
- The same process happens when wind blows on water. You can see this by watching the paper scraps and bubbles move as you blow on the water in the pan. If you add a drop of food coloring, you'll see that the velocity of the water movement decreases with depth.

decreases toward the poles because a point at a higher latitude doesn't have as far to go in a day. In fact, the rotational velocity of a point on the Earth's surface reaches zero at the poles (Fig. 14.15a).

The latitudinal change in rotational velocity of the Earth's surface causes an object moving across the Earth's surface to deflect from a straight-line path and progressively change direction. Such deflection is called the **Coriolis effect**, named for the French physicist who explained the phenomenon in 1835.

Because of the Coriolis effect, any moving object undergoes a deflection regardless of the initial direction in which it was moving—unless the object happens to lie exactly on the equator and happens to start moving exactly parallel to the equator. The sense of the deflection depends on the hemisphere in which the object moves (Fig. 14.15b). Objects that start moving in the northern hemisphere deflect to their right and follow a clockwise path, whereas objects in the southern hemisphere deflect to their left and follow a counterclockwise path. If the object were to keep moving long enough, without friction or external pressure acting on it, it would follow a circular path, called an *inertial circle*, and end up back where it started (Fig. 14.15c). The diameter of this inertial circle depends on the initial speed of the object (faster objects follow larger inertial circles at a given latitude) and on

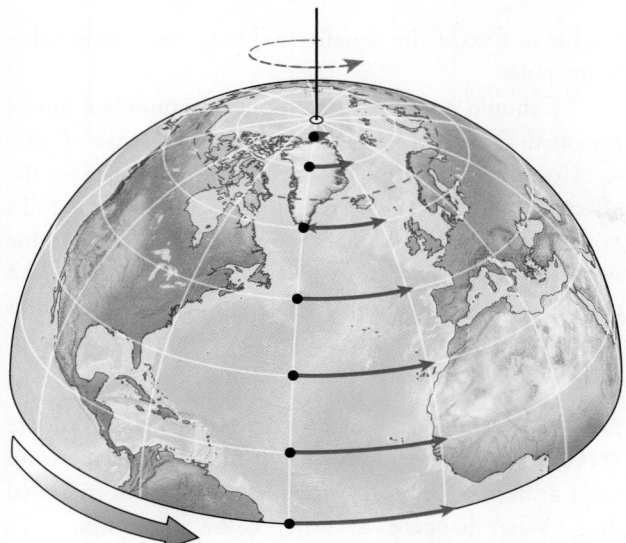

(a) The rotational velocity of a point on the Earth's surface varies with latitude, because all points must make one revolution per day, but points nearer the equator have farther to travel than do points near the poles.

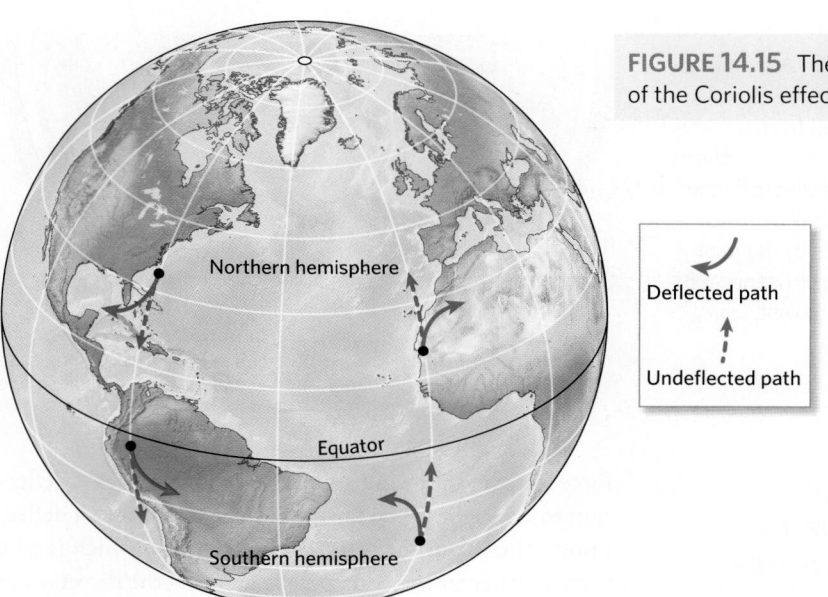

FIGURE 14.15 The concept of the Coriolis effect.

Deflected path

Undeflected path

Northern hemisphere

Equator

Southern hemisphere

(b) Moving objects in the northern hemisphere deflect to the right, whereas moving objects in the southern hemisphere deflect to the left.

its latitude (inertial circles get smaller closer to the poles for objects moving at the same speed). (Note that we use *speed* in this context to refer to the rate of movement of the object relative to the surface below, regardless of the direction in which it moves.)

Why does the Coriolis effect exist? To develop a simplistic model of its origin, imagine that you place a cannon at a latitude of 45° N, aim due north, and fire. Assuming (unrealistically) that no friction acts on the cannonball, the instant that the ball comes out of the cannon's muzzle, it's moving north. But because the cannon itself was moving with the Earth's surface at the time of firing, the ball's movement also has an eastward component. Since the flying ball is not in contact with the Earth's surface, it maintains this initial eastward component of movement as it moves northward. The farther north the ball goes, however, the more slowly the surface of the Earth beneath it is moving in the direction of rotation. Therefore, the ball deflects to the east relative to the Earth's surface beneath it **(Fig. 14.16)**. If you were to aim it south, the opposite situation would occur, in that the eastward component of the ball's movement would be slower than the movement of the Earth's surface beneath it as the ball traveled south, so the ball would deflect to the west relative to the Earth's surface. To apply this concept to ocean currents or air currents, imagine that instead of following the motion of a moving cannonball, you're following the motion of a small volume of water or a small volume of air.

Note that so far, we've considered the Coriolis effect from the perspective of an observer looking down from space. Now let's take a different point of view. Suppose you were sitting on an object (such as the cannonball we described above) as it moved across the Earth in the

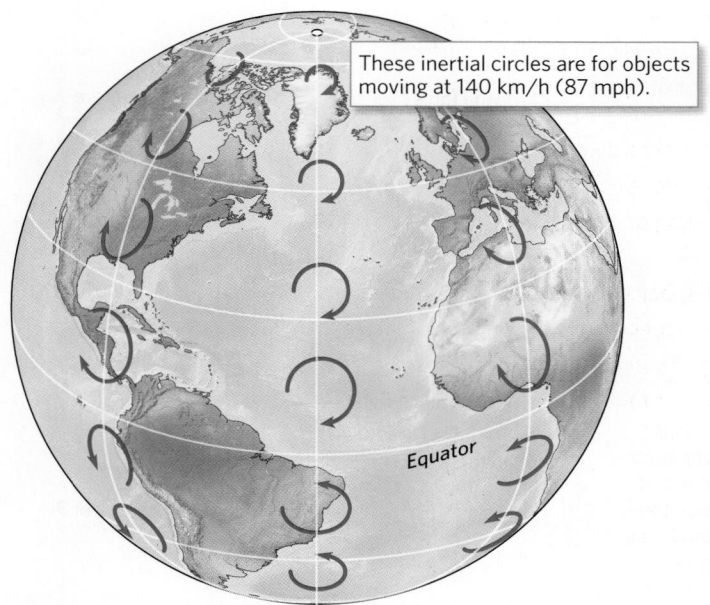

These inertial circles are for objects moving at 140 km/h (87 mph).

Equator

(c) Due to the Coriolis effect, an object that keeps moving on the Earth's surface with no other force (such as friction) acting on it follows a circular path called an inertial circle. The diameter of the circle depends on the object's latitude and speed.

northern hemisphere. It would seem as if an invisible force was pushing you to turn to the right. Physicists refer to the apparent force causing the deflection as the **Coriolis force**. We refer to it as an "apparent force" to emphasize that the change in direction isn't due to the application of a real push or pull on the moving object—the deflection of the object is just a consequence of the Earth's rotation. But by viewing the cause of the Coriolis effect as a "force," it becomes easier to understand how the spin of the Earth affects ocean currents. The Coriolis

FIGURE 14.16 A simplistic analogy for the Coriolis effect. The ball from a cannon fired in the northern hemisphere deflects to the right. So, the ball from cannon A deflects to the east, and the one from cannon B deflects to the west.

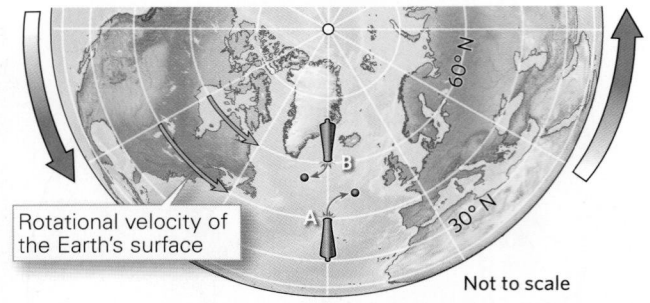

Rotational velocity of the Earth's surface

Not to scale

If you look vertically down at the North Pole, lines of longitude look like spokes of a wheel, and lines of latitude look like circles.

Did you ever wonder...

why currents circle the oceans?

FIGURE 14.17 The concept of Ekman transport.

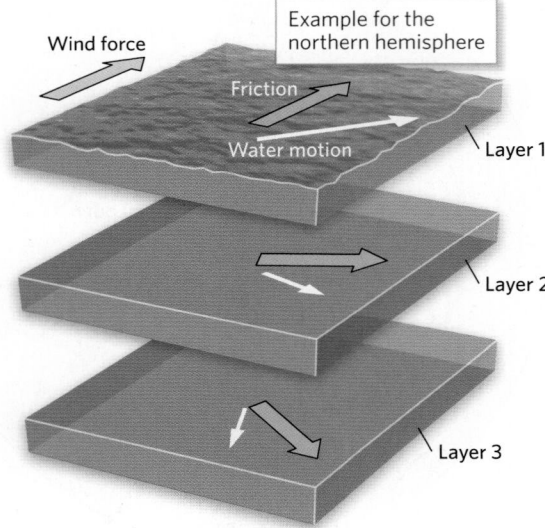

(a) The near-surface realm of the ocean can be pictured as a series of layers. When wind applies a shear force to the top layer, that layer applies a shear force to the layer below, and so on. Due to the Coriolis effect, the motion of each layer is deflected relative to that of the layer above.

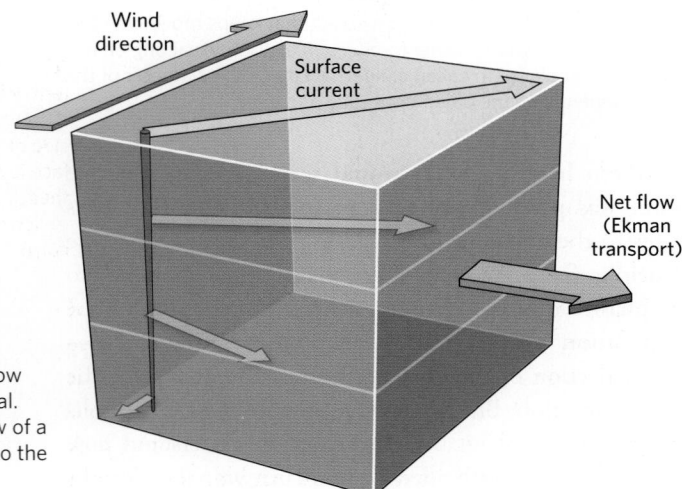

(b) The resulting pattern of flow arrows is called an Ekman spiral. As a consequence, the net flow of a surface current trends at 90° to the wind direction.

force has the following properties: (1) it causes a deflection to the right in the northern hemisphere and a deflection to the left in the southern hemisphere, which is why gyres don't cross the equator; (2) it affects the direction in which an object moves across the Earth's surface, but not the object's overall speed; (3) the magnitude of the force increases as the speed of the object increases; and (4) it has

a value of zero at the equator and has a maximum value at the poles.

We should note that the simplistic cannonball model provided above, while giving an intuitive sense of why the Coriolis effect happens, does not actually explain the effect, or characterize the origin of the Coriolis force, fully. For example, it can't explain why a cannonball fired due west from a point at 30° N deflects to the north. **Box 14.3** provides a more complete explanation of the Coriolis effect that can justify the existence of inertial circles, for those who are interested in the details.

EKMAN TRANSPORT: A CONSEQUENCE OF THE CORIOLIS EFFECT. As we've seen, moving air molecules in the wind exert a shear force on the ocean surface due to frictional drag. What happens to water below the surface? To answer this question, let's picture a volume of ocean water at a location in the northern hemisphere as a stack of thin water layers. We label these as Layers 1, 2, and 3, from top to bottom **(Fig. 14.17a)**. When water in Layer 1 starts to move in response to the wind, the Coriolis effect deflects the flow to the right. The water of Layer 1, in turn, applies shear force to Layer 2 beneath it. Because of the Coriolis effect, the water in Layer 2 deflects to the right of the water in Layer 1. The same effect occurs in Layer 3, and in successive layers below. In other words, the flow in each layer deflects farther to the right than the flow in the layer above it. Furthermore, in each layer, the flow moves more slowly than in the layer above it, because the effect of frictional drag dissipates. Ultimately, at some depth, the wind no longer influences water movement—this depth determines the base of the surface current.

The progressive deflection of water flow with depth in a surface current is called an *Ekman spiral* **(Fig. 14.17b)**, named for its discoverer. The existence of an Ekman spiral in a current causes the current's overall water movement, or *net flow*, to deflect at an angle of about 90° to the direction of the wind. Oceanographers refer to this deflected net flow as **Ekman transport**, and as we will now see, it plays a major role in governing the geometry of regional currents.

A BALANCING ACT: GEOSTROPHIC FLOW AND THE DEVELOPMENT OF GYRES. By this point, you might still be puzzled by the overall flow pattern of surface currents in the ocean. It might seem at first that ocean currents, due to the Coriolis effect, should follow relatively small inertial circles. A map of surface currents, however, shows that they follow gyres that are vastly bigger than inertial circles, and that current directions are not actually at 90° to prevailing wind directions (see Fig. 14.14). What's going on?

The answer comes from considering a third force acting on oceanic surface-water flow; namely, the **pressure-gradient force**. This force develops because Ekman transport causes a slow drift of water toward the center

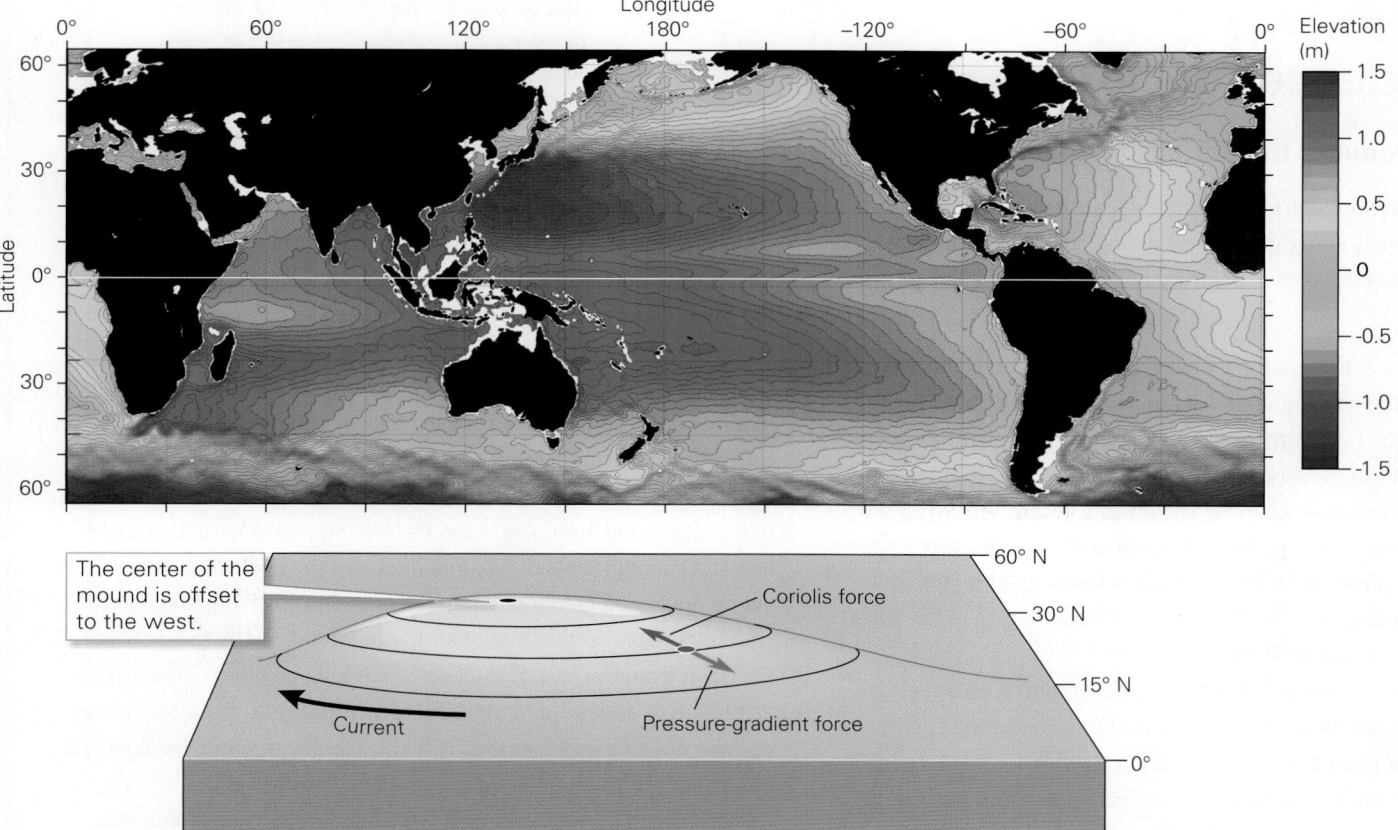

Longitude

Elevation (m)

FIGURE 14.18
Understanding geostrophic currents.

(a) A map showing the subtle surface topography of the oceans. The elevated areas of seawater rise because of Ekman transport. The elevations shown do not include the effects of tides and waves. The highest elevation of a seawater mound lies in the western part of an ocean.

(b) The elevated ocean interior produces an outward-directed pressure-gradient force. (To picture this, imagine a mound of honey sitting on a plate. If the mound is thicker in the interior than at the edges, the pressure acting on the plate due to the weight of the honey is greater beneath the thicker part of the mound and less near the edges. So, a pressure gradient develops between the middle and the edges of the mound. The force due to this gradient tries to push honey outward.) When the pressure-gradient force balances the Coriolis force, a geostrophic current exists.

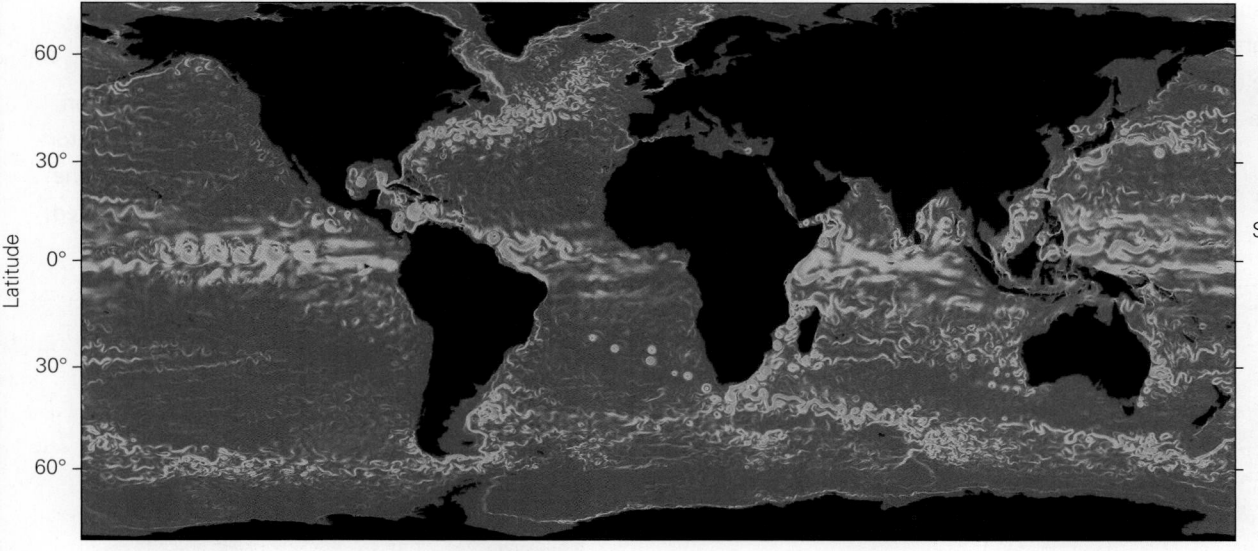

Faster

Slower

(c) A map showing variations in current velocity reveals that geostrophic currents are faster on the western side of an ocean, where the sea surface is highest. Note the many eddies that develop in association with regional currents. The colors indicate current velocity.

of an ocean. As a result of this converging flow, sea level in the interior of an ocean becomes slightly higher (about 1 m, or 3 ft) than at the ocean's margins, independently of waves. It's as if a very broad, but subtle, mound of water rises out in the ocean far from shore. (Of note, the peak of the mound lies to the west of the ocean's center, for reasons beyond the scope of our discussion.) The mound is so broad, and its height so slight, that you don't notice it when you sail across the ocean **(Fig. 14.18a)**, but it is real. Consequently, the pressure acting on an imaginary horizontal surface below sea level is greater under the peak of the mound than at its edges. In other words, the slight

Science Toolbox

What really produces the Coriolis force?

In this discussion, we provide insight into the causes of the Coriolis effect by geometrically illustrating forces acting on a moving object. To streamline our discussion, we begin by reviewing several key terms from physics:

1. *Forces and vectors:* A force, simplistically, is a push or pull in a given direction. Since the consequences of applying a force depend on both its magnitude (strength) and the direction in which it has been applied, we cannot represent a force by a number alone, but have to represent it with a special arrow known as a **vector**. As we have discussed, the length of the vector represents the magnitude of the force, and the orientation of the vector represents the direction of the force. A *component* of a force vector gives the portion of the force that applies in a specified direction.

2. *The Earth's axis of rotation:* The Earth's axis of rotation is an imaginary line, passing through the Earth's poles and the center of the Earth, around which our planet spins. Because the Earth is a sphere, not all points on its surface are equidistant from its axis of rotation. Specifically, points on the equator lie farthest from the axis of rotation, while points at the poles lie on the axis of rotation. This change in distance to the axis of rotation with latitude affects how the centrifugal force, discussed below, acts on objects moving along the Earth's surface.

3. *Rotational vs. angular velocity:* The Earth rotates once a day. This means that all points on the surface of the Earth follow a complete loop around a circle that is parallel to lines of longitude (that is, parallel to the equator). Since there are 360° in a circle, we can say that all points on a circle have an *angular velocity* (the angle that they subtend as they move around a circle in a specified time interval) of 360°/day. A point on the equator must travel along a line that is 40,075 km (24,901 mi) long—the Earth's circumference—in a day, while points at the poles spin in place and don't move at all. Therefore, as we noted earlier, the rotational velocity (velocity along a circular path) of a point fixed to the Earth's surface is greatest along the equator and progressively less toward the poles, reaching zero at the poles. So rotational velocity due to the Earth's spin differs with latitude, but angular velocity has the same value at all latitudes.

4. *Centrifugal force:* The apparent outward-directed force that an object resting on the surface of a spinning disk or sphere experiences is called the **centrifugal force**. We refer to it as an "apparent force" because it exists only from the perspective of the object—it is just a manifestation of inertia, an object's tendency to move in a straight line at a constant speed. The magnitude of the centrifugal force acting on an object at a point on the spinning Earth depends on the square of the object's velocity at that point divided by the distance of the point from the axis of rotation. Therefore, an object moving east at 100 km/h at a latitude of 60° N feels more centrifugal force than does an object moving east at 100 km/h at the equator, because the object at 60° N lies closer to the axis of rotation.

5. *Components of gravity:* Gravity is the attractive force between two objects. Because of gravity, centrifugal force doesn't cause objects to fly off the spinning surface of the Earth. The force of gravity acting on an object at the surface of our planet pulls toward the center of the Earth. The red arrows in **Figure Bx14.3a** represent the direction and magnitude of gravitational force at different points on the Earth's surface. Note that at the equator, the direction of this force is perpendicular to the Earth's axis of rotation, whereas at the poles, it is parallel to the axis, but the magnitude of the force stays the same at all points on the surface. As an object on the Earth's surface moves from the equator toward the poles, the magnitude of the gravitational force acting on the object in the direction pointing toward the axis of rotation (represented by the blue arrows in Fig. Bx14.3a) progressively decreases.

6. *Conservation of angular momentum:* Imagine swinging a weight attached to a string in a circle around your head (**Fig. Bx14.3b**).

FIGURE Bx14.3 The physical origin of the Coriolis effect.

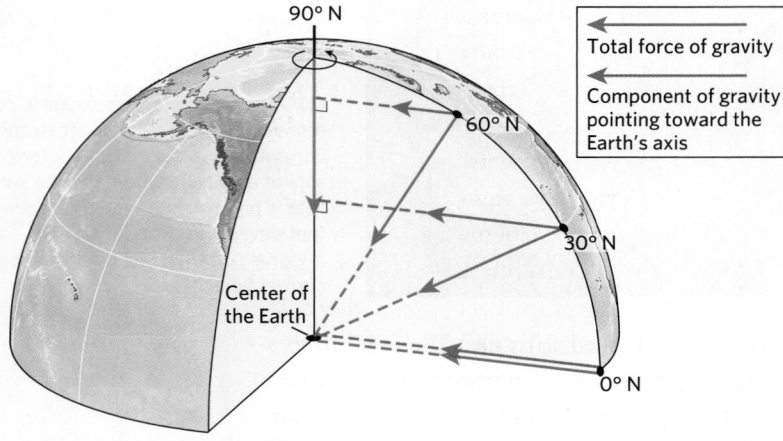

90° N

←	Total force of gravity
←	Component of gravity pointing toward the Earth's axis

60° N

30° N

Center of the Earth

0° N

(a) Gravity (red arrows) pulls toward the center of the Earth at every point on the Earth. The proportion of the gravitational force pointing toward the Earth's rotational axis varies with latitude (as shown by the lengths of the blue arrows). It is strongest at the equator and weakest at the poles.

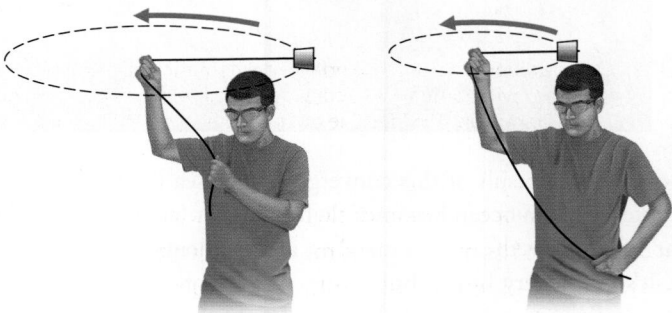

(b) When swinging a weight overhead, shortening the string increases the weight's rotational velocity to conserve its angular momentum.

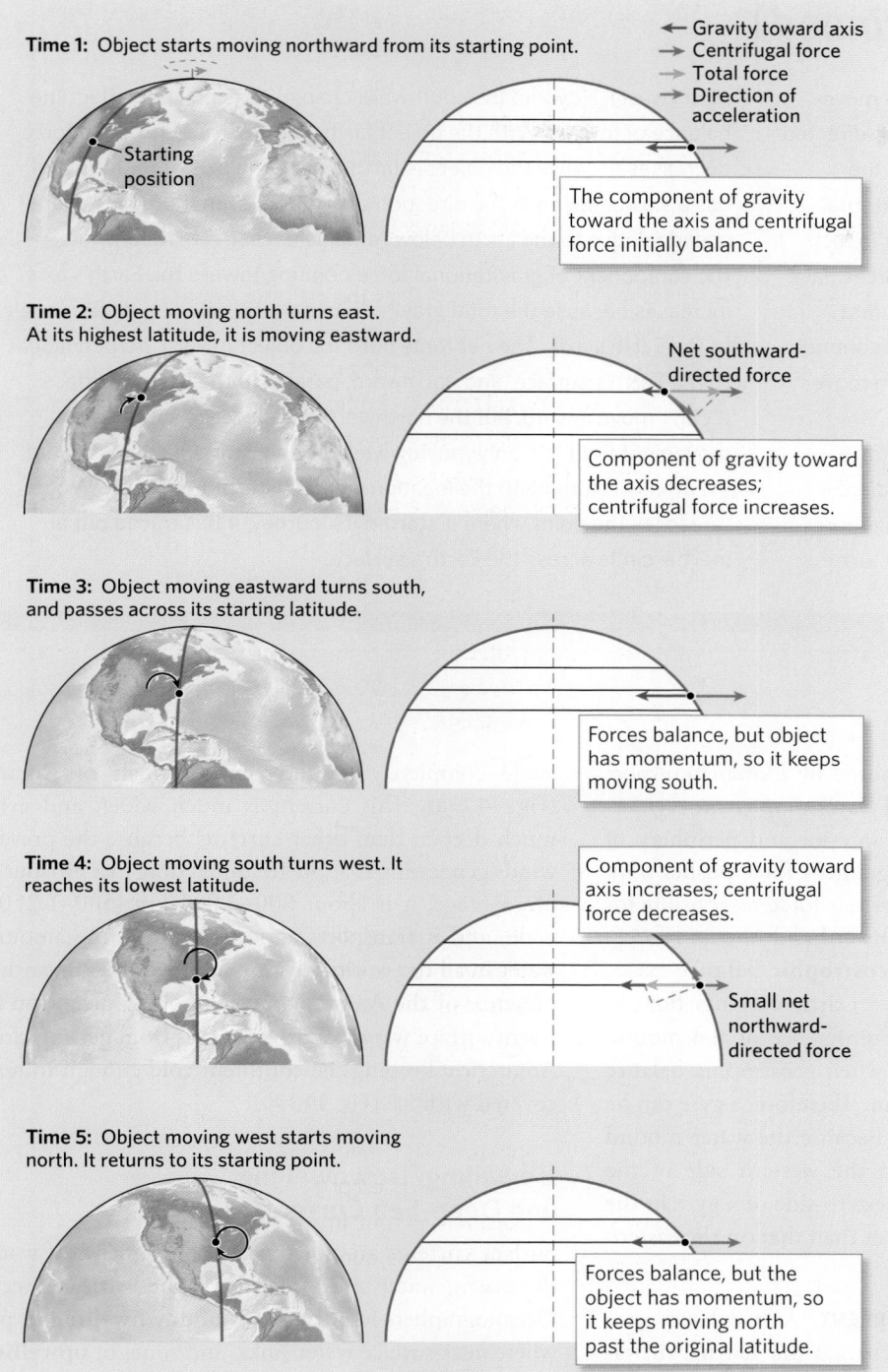

Time 1: Object starts moving northward from its starting point.

← Gravity toward axis
→ Centrifugal force
⟶ Total force
⟶ Direction of acceleration

Starting position

The component of gravity toward the axis and centrifugal force initially balance.

Time 2: Object moving north turns east. At its highest latitude, it is moving eastward.

Net southward-directed force

Component of gravity toward the axis decreases; centrifugal force increases.

Time 3: Object moving eastward turns south, and passes across its starting latitude.

Forces balance, but object has momentum, so it keeps moving south.

Time 4: Object moving south turns west. It reaches its lowest latitude.

Component of gravity toward axis increases; centrifugal force decreases.

Small net northward-directed force

Time 5: Object moving west starts moving north. It returns to its starting point.

Forces balance, but the object has momentum, so it keeps moving north past the original latitude.

(c) As an object moves across the surface of the rotating Earth, the relative magnitudes of centrifugal and gravitational forces acting on it change, so to conserve angular momentum, the object changes direction. The blue line on the globe represents the longitude of the object at the time indicated.

An object moving around a circle has **angular momentum**, defined as the product of its mass (m), its rotational velocity (w), and its distance from the axis of rotation (r). Written as an equation, angular momentum $= m \times w \times r$. Newton showed that if no other force acts on a rotating object, the angular momentum of the object remains the same regardless of its distance from the rotational axis. This law, known as *conservation of angular momentum*, means that as the distance (r) between an object and the axis around which it is rotating decreases, its rotational velocity (w) must increase in order for its angular momentum to stay the same.

Using these terms and concepts, we can develop a clearer sense of the cause of the Coriolis force, the apparent force that causes the Coriolis effect (**Fig. Bx14.3c**). Let's start by considering the forces acting on a stationary object sitting on the Earth's surface at a latitude in the northern hemisphere (Time 1). The object has a rotational velocity in the direction of rotation appropriate for its latitude. Gravity applies a downward force, pulling the object toward the center of the Earth, and the Earth's surface provides an upward force to prevent the object from sinking into the Earth's interior. Meanwhile, centrifugal force also acts on the object due to the Earth's rotation.

Now, imagine that the object starts moving to the north (Time 2). (For simplicity, we'll ignore friction.) In profile, we see that as the object moves north, it gets closer to the Earth's axis of rotation, so the object's speed in the direction of rotation must increase to conserve angular momentum. As it moves north, the rotational velocity of the object stays the same, but the rotational velocity of the Earth's surface beneath the object becomes progressively slower. Since the object now has a greater rotational velocity than does the Earth beneath it, it deflects to the right (east) of its initial trajectory. Note that as the object turns eastward, and therefore changes its movement direction, it still moves at the same overall speed. Therefore, its northward speed decreases as its eastward speed increases.

We can see why the object's northward speed decreases by examining the changing magnitude of the forces acting on it. As the object moves north, the outward centrifugal force increases to conserve angular momentum, because its rotational velocity in the direction of the Earth's rotation is faster, but its distance from the rotational axis is less. At the same time, the component of the gravitational force pointing toward the Earth's axis of rotation decreases, because the total pull of gravity acts at a smaller angle to the Earth's axis. The changes in these two forces result in a net force outward, away from the Earth's axis. The net force pulls the object outward, perpendicular to the Earth's surface, and southward, parallel to the Earth's surface. The outward component is not sufficient to lift the object, but the southward

BOX 14.3 **Science Toolbox** *(continued)*

component of the net force decreases the object's northward movement until the object is moving due east. Then, the southward-directed force continues to deflect the object southward, back toward the equator. (To a person riding on the object, it seems as if an external force—the Coriolis force—has been applied to the object, but in fact, all that's happened is that the relative magnitudes of other forces have changed.) Eventually, the object crosses the latitude where it first started its journey north, this time heading south (Time 3). It continues moving southward because of its inertia, much as a pendulum moves past its equilibrium point as it swings.

What happens next? As the object continues southward, it moves farther away from the Earth's axis, so its speed in the direction of rotation decreases (Time 4). Now, the object is moving more slowly relative to the Earth's surface below, so it continues to turn

right (west). How did the southward change in its position affect the balance of forces? With the object farther south, the centrifugal force decreases because the object is farther from the axis of rotation and has a slower speed in the direction of rotation, even though its overall speed relative to the Earth below remains unchanged. Meanwhile, the component of gravitational force pointing toward the Earth's axis increases because the total gravitational force is now at a greater angle to the Earth's axis. The net force pulls the object inward, perpendicular to the Earth's surface, and northward, parallel to the Earth's surface. It can't move inward, but the northward pull slows its southward movement until it is only moving west. Then it starts back northward, eventually returning to the location where it started (Time 5). When it reaches the point where it started its journey, it has traced out an inertial circle across the Earth's surface.

---

variation in sea-level height caused by Ekman transport produces an outward-directed *horizontal pressure gradient* in the ocean between the interior and periphery of an ocean basin. This pressure-gradient force pushes back against the inward-directed Coriolis force responsible for Ekman transport. When the inward and outward forces become equal and opposite, **geostrophic balance** exists. And when geostrophic balance exists, currents flow in a direction parallel to the circumference of the mound (Fig. 14.18b). A gyre develops when geostrophic balance has been established in an ocean. Therefore, a gyre can be called a **geostrophic current**. Because the water mound in each ocean rises higher on the western side of the ocean, the current along the western side of a gyre in the northern hemisphere flows faster than that on the eastern side (Fig. 14.18c).

ANTARCTIC CIRCUMPOLAR CURRENT. As we have seen, continents and the equator separate the Earth's saltwater layer into separate oceans, and a distinct gyre forms in each of the oceans. Recall that gyres don't cross the equator because the direction of the Coriolis force changes there. The southern edges of gyres in the northern hemisphere and the northern edges of gyres in the southern hemisphere produce westward-flowing equatorial currents, but these, too, are interrupted by continents. Between 55° S to about 65° S, however, no land blocks ocean flow, so a particularly strong current, called the Antarctic Circumpolar Current, carries water in a

circle completely around the continent of Antarctica (Fig. 14.19a). This current is much wider, and extends much deeper, than other currents because the prevailing winds generating it apply frictional drag to a broader area. On average, it is about 800–2,000 km (500–1,250 mi) wide, and it transports about 140 times the amount of water in all the world's rivers combined. Significantly, the presence of the Antarctic Circumpolar Current prevents warm surface waters of other oceans from getting close to Antarctica, keeping the continent cold enough to remain covered with ice (Fig. 14.19b).

Upwelling, Downwelling, and Deep-Sea Currents

Surface currents are not the only movements of water in the ocean; water also circulates in the vertical direction. Oceanographers identify zones of **downwelling** as places where near-surface water sinks, and zones of **upwelling** as places where subsurface water rises. What causes upwelling and downwelling? Several factors come into play, some of which operate along coastlines and some in the interior of the oceans.

Upwelling commonly occurs along coastlines where surface currents carry water away from the shore. As the surface water moves oceanward, a *water deficit* develops, so deep water rises to compensate for the deficit. Such seaward surface currents can be due to Ekman transport associated with a shore-parallel wind in an appropriate

FIGURE 14.19 The Antarctic Circumpolar Current.

(a) The thin white lines are current streamlines (representing surface-water flow direction). The large arrows indicate the overall flow direction of the current. Note the many eddies along the current.

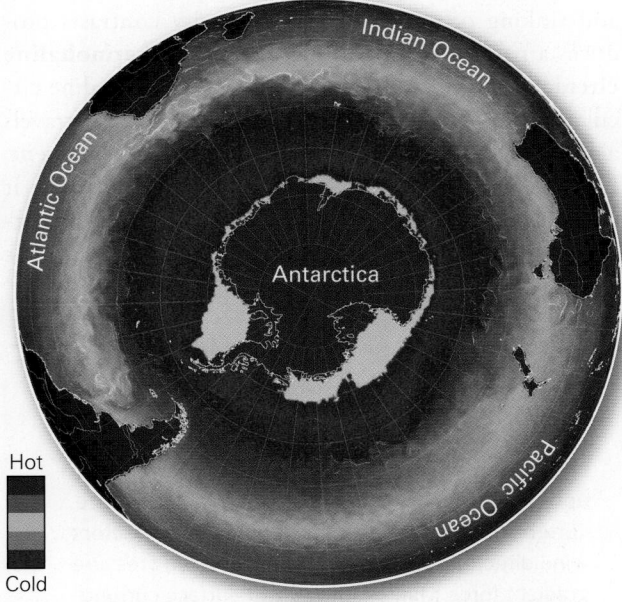

Hot

Cold

(b) A map of surface-water temperature in the southern hemisphere as viewed looking down on the South Pole. Because of the circumpolar current, a band of cold water isolates Antarctica from warmer surface water.

orientation. For example, because the Coriolis force deflects to the right in the northern hemisphere, Ekman transport due to a shore-parallel wind that blows toward the south along a north-south-trending coast will drive a surface current that flows away from the shore (Fig. 14.20a). Because the Coriolis force deflects to the left in the southern hemisphere, Ekman transport

FIGURE 14.20 Coastal upwelling and downwelling in the northern hemisphere.

Seaward Ekman transport due to shore-parallel wind

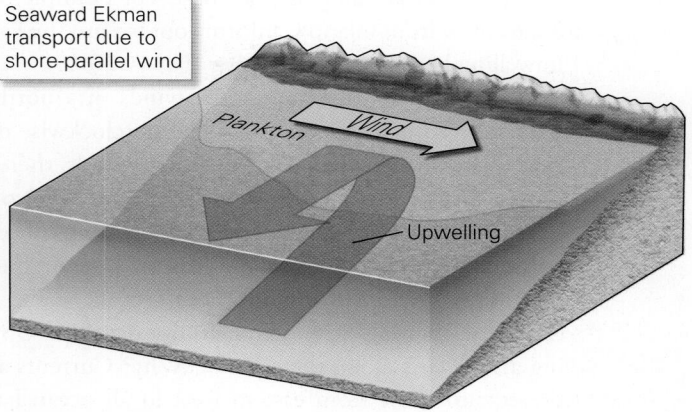

(a) A surface flow that moves water away from the shore causes upwelling. Upwelling brings up nutrients and fosters the growth of algae near the coast.

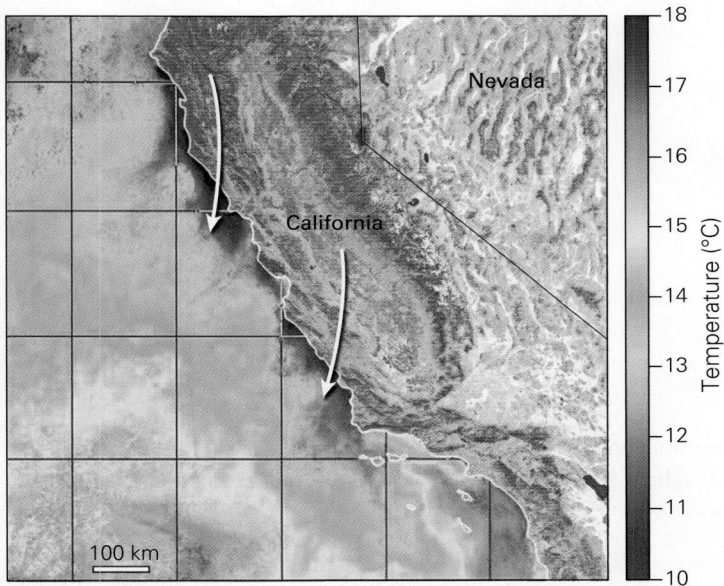

(b) Variations in the surface temperature of the eastern Pacific Ocean show upwelling of cold water along the west coast of the United States. Note the wind direction.

Shoreward Ekman transport due to shore-parallel wind

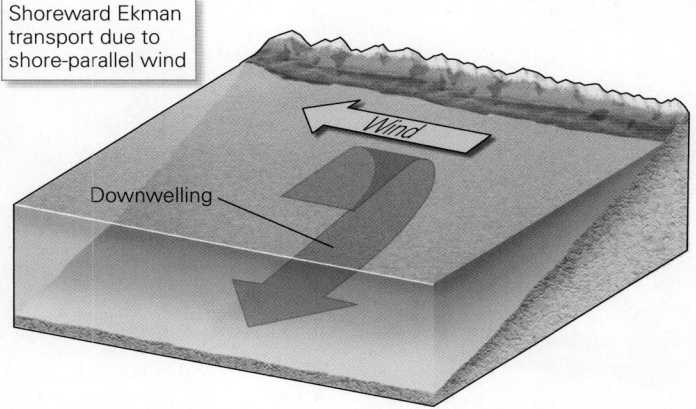

(c) A surface flow that moves water toward the shore causes downwelling.

associated with a shore-parallel wind blowing north produces upwelling along the coast. Ekman-transport-related upwelling will be significantly enhanced if the wind direction trends at an angle to the coast. For example, when winds blow from inland California out toward the Pacific, upwelling becomes even stronger (Fig. 14.20b).

As we discuss in Chapter 18, winds in a northern-hemisphere hurricane flow in a counterclockwise direction. Eckman transport related to these winds, therefore, moves surface water away from the center (eye) of the hurricane, a process called *divergence*, causing upwelling beneath the eye. For the same reason, several zones of upwelling develop in the Southern Ocean, where large clockwise-rotating wind circulations develop. Upwelling also occurs along the equator. Why? Currents along the equator flow from east to west in all ocean basins, for no Coriolis force acts at the equator. However, the Coriolis force causes water just to the north of the equator to deflect northward, and water just to the south of the equator to deflect southward. These two deflections cause divergence of surface water at the equator. Upwelling of deeper water must take place along the equator to compensate for the resulting water deficit. Zones of upwelling are very important for the fishing industry, because the nutrients brought up with water from the deep lead to the growth of algae, which in turn feed fish populations.

Ekman transport that drives water toward the shore can produce a *water excess* that must be compensated for by downwelling. Such a situation can happen, for example, where winds flowing to the north occur offshore of a north-south-trending coast in the northern hemisphere (Fig. 14.20c). For a similar reason, downwelling occurs in regions of the ocean toward the center of the major gyres of the oceans. Ekman transport causes *convergence* of surface water in the center of the gyre, meaning that water moves toward the center, and where this happens, downwelling must happen to compensate for the excess water. Oceanic downwelling regions distribute oxygen absorbed at the Earth's surface or produced by photosynthesis in shallow water throughout the deeper ocean; this oxygen allows organisms to survive in deep ocean water. Downwelling regions do not host large fish populations, because nutrients in the water column are not replenished by rising deep water, so the plankton that serve as food for fish do not thrive.

So far, we've discussed upwelling and downwelling driven by the wind. Such vertical redistribution of water also happens in response to contrasts in seawater density that exist because of variations in temperature and/or salinity. Where density variations exist, denser water sinks relative to less dense water, and less dense water rises buoyantly above denser water. This rising and sinking of water driven by density contrasts produces a pattern of global flow known as **thermohaline circulation** (Fig. 14.21). As a result of thermohaline circulation, cold water in polar regions sinks and travels along the bottom of the ocean as a deep-sea current. Eventually, this water reaches lower latitudes, where it warms and rises. Like a conveyor belt, thermohaline circulation moves water and heat throughout the Earth's ocean basins.

FIGURE 14.21 Thermohaline circulation results in a global-scale conveyor belt that circulates water throughout the ocean system. Because of this circulation, the ocean mixes entirely in a 1,500-year period.

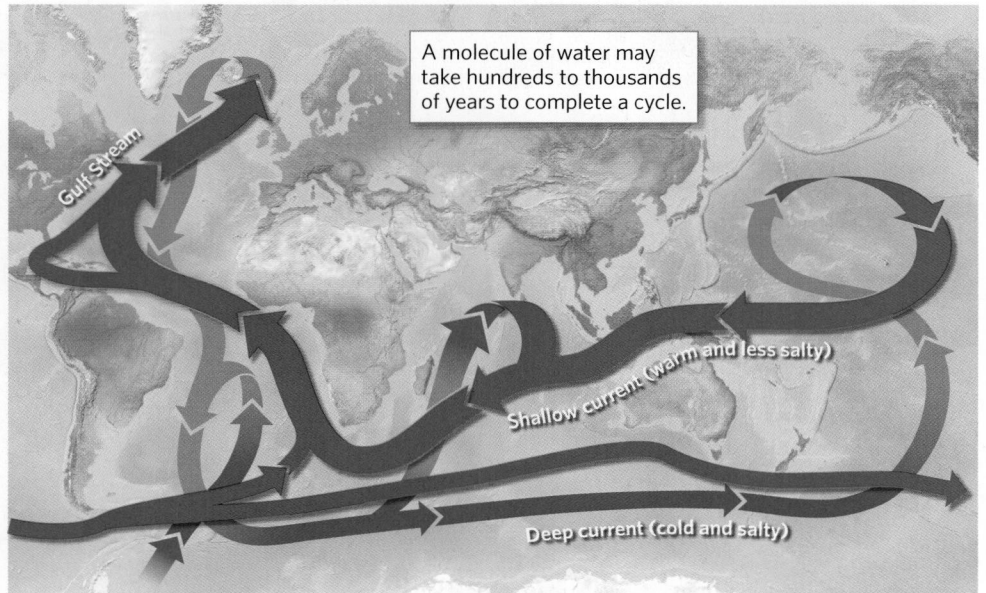

A molecule of water may take hundreds to thousands of years to complete a cycle.

Gulf Stream

Shallow current (warm and less salty)

Deep current (cold and salty)

Take-home message...

Ocean water is in constant motion due to wind-driven surface currents as well as upwelling and downwelling. The paths that surface currents follow depend on a combination of factors: wind direction, the Coriolis effect, and the pressure-gradient force. Overall, the largest surface currents are the gyres, one of which occurs in the northern and southern hemisphere in each ocean. Upwelling and downwelling can be driven either by Ekman transport or by variations in water density.

Quick Questions ————————

• Name the two general categories of currents.

• In which direction will Ekman transport cause water within a current flowing due south in the northern hemisphere to deflect?

• Define thermohaline circulation.

14.4 Wave Action

Why Do Wind-Driven Waves Form?

Water waves—the periodic up-and-down motions of the ocean surface at a scale that generates distinct troughs and ridges visible to your eye—make the ocean surface a restless, ever-changing vista (Fig. 14.22). The waves you see when you look out at the sea from the shore, a plane window, or a ship's deck are **wind-driven waves**. How do wind-driven waves form? To picture the process, imagine still water in a pond on a windless day. The horizontal surface of the water represents an *equilibrium level*—if you pushed the surface up or down with a paddle, it would return to the equilibrium level because of gravity. At the surface, water molecules attract one another more strongly than they attract air molecules above. This attraction produces *surface tension*, which makes the water surface behave somewhat like an elastic sheet. When a breeze starts to blow, frictional drag shears this elastic surface and causes it to stretch. As soon as it stretches, elastic rebound causes it to twang back like a rubber band, so it wrinkles into small waves called **ripples** (Fig. 14.23).

Once ripples exist, wind can push against their sides, causing them to evolve into larger waves. If the wind is strong and steady enough, water in a wave rises above the equilibrium level (Fig. 14.24a). Then, gravity acts on the elevated water in the wave, causing it to sink (Time 1). Inertia causes the sinking water surface to descend below the equilibrium level. As it does so, water gets pushed sideways, so the area that rises above the equilibrium level moves to a new location. Then, the water surface in the initial location rebounds upward and rises above the equilibrium level (Time 2). As the process repeats, the water surface effectively bounces up and down, like a weight suspended from a vertical spring, to yield a **wave train**, which consists of a succession of waves with fairly uniform spacing and dimensions. You can produce a small wave train when you drop a pebble into a pond (Fig. 14.24b). A wave train propagates outward, and it can transmit energy a long distance from the source of the disturbance that produced the waves.

Describing Waves and Wave Motion

We can use the same basic terminology for describing a water-wave train that we use for any type of wave train. The high ridges of a wave train are *crests*, while the low valleys

FIGURE 14.22 Wind-driven waves provide a playground for surfers off Hawai'i.

FIGURE 14.23 Ripples are small waves that develop due to the elastic behavior of the ocean surface.

FIGURE 14.24 Development of wave trains.

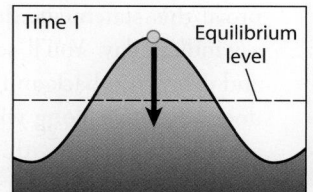 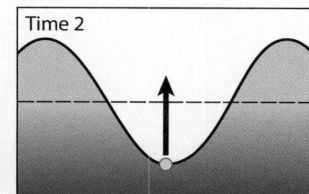

(a) Once a wave builds to a sufficient size, gravity causes its elevated surface to sink below the equilibrium level, as shown here in cross section. Then the water surface rebounds like a weight hanging from a spring.

(b) An example of a wave train formed when a pebble displaces the surface of a pond.

FIGURE 14.25
Terminology
for describing
ocean waves.

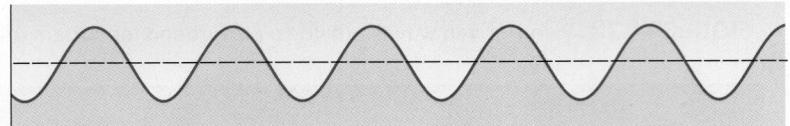

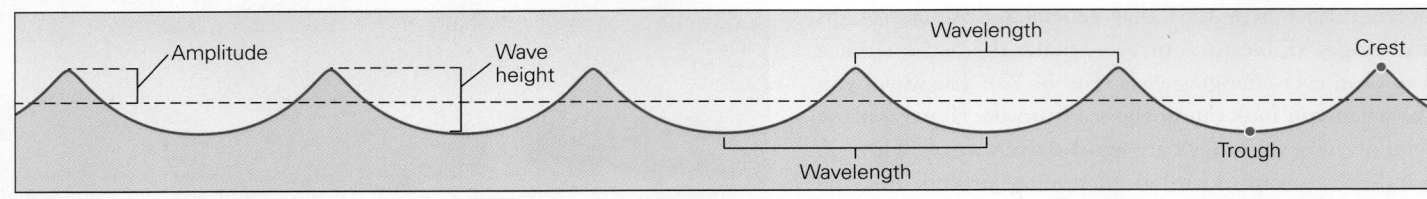

(a) A depiction of ocean waves in which the curvature of a trough and that of a crest are the same. Though this depiction of a wave train is commonly used for simplicity, it is not accurate.

(b) Real water waves have somewhat pointed crests and broad, rounded troughs. Nevertheless, the standard terminology used to describe any wave can be used to describe water waves.

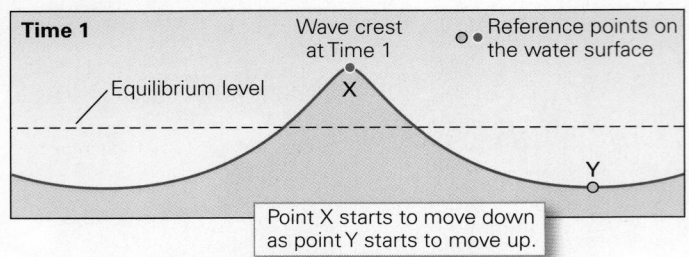

Point X starts to move down as point Y starts to move up.

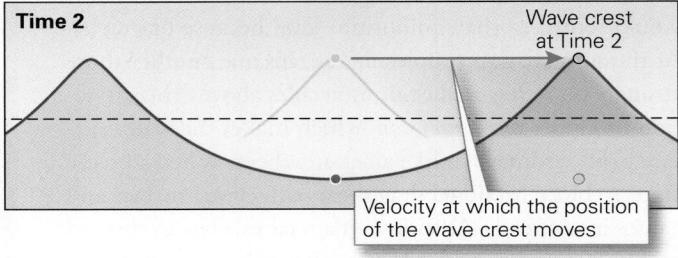

Velocity at which the position of the wave crest moves

(c) The wave velocity indicates that rate at which a given crest (or trough) moves in a given direction.

Animation

Wave Concepts

between crests are *troughs*. The vertical distance between an adjacent crest and trough defines the *wave height*, while half the wave height represents the wave's *amplitude*. The horizontal distance between two successive troughs or two successive crests in a wave train defines the *wavelength*. The number of wave crests (or troughs) that pass a point in a given time gives the *wave frequency*, and the time between the passage of two successive crests (or troughs) is the *wave period*. Water waves, like seismic waves or

electromagnetic waves, are commonly depicted as having rounded crests and troughs that are mirror images of each other **(Fig. 14.25a)**. Real water waves, in contrast, have somewhat pointed crests and broad, rounded troughs **(Fig. 14.25b)**. Crests and troughs of real water waves are not mirror images of each other.

The position of a wave crest (or trough) moves horizontally as time passes **(Fig. 14.25c)**. The speed and direction at which the crest of a wave moves, relative to a fixed location, define the *wave velocity*. Waves always move more slowly than the wind that generates them, but they can still move faster than most people can run. Wave velocity in the open ocean, for example, ranges between 8 and 50 km/h (5–30 mph). Wave velocity depends on wind strength and wavelength: stronger wind produces faster waves, and waves with longer wavelengths travel faster than waves with shorter wavelengths.

When you watch a wave crest travel, you may get the impression that the whole mass of water within the wave moves with the crest at the wave velocity. It does not! To prove this statement, drop a cork into a wave train on a windless day. You'll see that the cork simply bobs up and down and back and forth as individual waves pass; it does not move along with the wave crest. That's because a particle of water within an idealized water wave moves in a circle **(Fig. 14.26a)**. At the ocean's surface, the circle's diameter equals the wave height. With increasing depth, the diameter of the circle decreases until, at a depth equal to about half the wavelength—a depth called the

FIGURE 14.26 The motion in ocean waves.

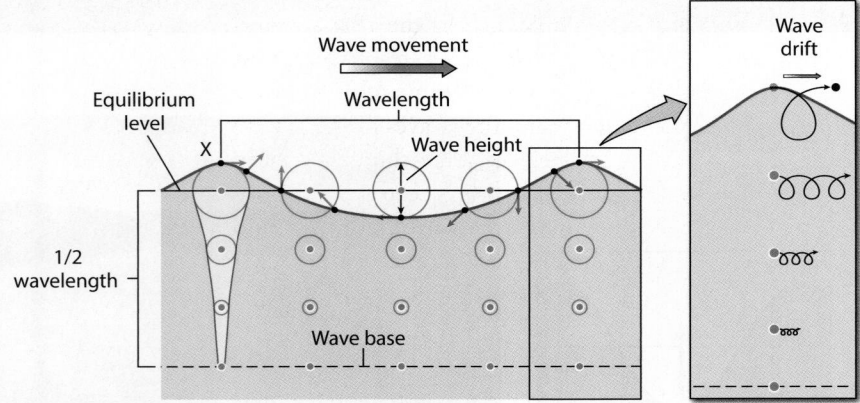

(a) Within a passing wave, a particle of water follows a circular path. The diameter of these circles decreases with depth. The wave base occurs at a depth of one-half the wavelength.

(b) Wind blowing on the surface of the ocean can cause wave drift. When this happens, particles of water follow a looping path instead of a circle. The loops get smaller with increasing depth.

wave base—no wave motion takes place at all. Submarines traveling below the wave base enjoy smooth water while ships toss about on the sea surface above.

While the motion of water molecules in an idealized wave follows a circle, water molecules in real ocean waves tend to display *wave drift*, an overall lateral motion accompanying the passage of a wave (Fig. 14.26b). Due to wave drift, objects on the surface of the water move forward by about one wavelength after four waves have passed. Therefore, wave motion more accurately resembles the shape traced out by a point on a rolling wheel. It's because of wave drift, which is caused in large part by wind, that the surface currents that we discussed in Section 14.3 develop in the ocean. As we'll see in Chapter 15, the path of a water molecule in a wave train also changes as the wave approaches the shoreline and friction with the seafloor slows the base of the wave.

The Size of Waves

The size of waves generated by wind in the open ocean depends on the wind's *strength* (how fast the air moves) and **fetch** (the distance over which it blows), as well as on the wind's *duration*, the length of time during which the wind blows (Fig. 14.27a). During strong winds, not all the energy of the wind goes into building waves; some blows the tops of the waves over, causing water to mix with air and resulting in *whitecaps*. How large can wind-driven waves get in the open ocean? When a strong wind blows over a long fetch for a long time, waves with amplitudes of 2–10 m (6–30 ft) and wavelengths of 40–500 m (130–1,600 ft) can build. There's a limit to how high wind-driven waves can become. In fact, oceanographers calculate that if hurricane-strength winds were to blow across the width of the Pacific for 24 hours, waves reaching 15–20 m (50–70 ft) high would develop.

Notably, when a storm blows over a region, it generates many different wave trains, each characterized by a different wavelength and period (Fig. 14.27b). As we noted above, wave trains with different wavelengths have different periods, so as the wave trains travel, they separate from one another, a phenomenon known as *wave dispersion*. As a result, a ship in a storm gets tossed about on a chaotic sea surface as it rides over wave trains with different wavelengths, whereas a ship sailing far from the source may intersect only regularly spaced waves. Waves that have traveled away from their source, and are not being driven by the local wind, are called **swells**. Such waves may transmit energy across thousands of kilometers of ocean (Fig. 14.27c).

Real wind-driven waves over a given time interval are not all the same height. Studies of wave heights show that they fall along an asymmetrical bell curve (Fig. 14.28a). The *significant wave height* is the mean (average) height of

FIGURE 14.27 Waves build progressively as the wind continues to blow.

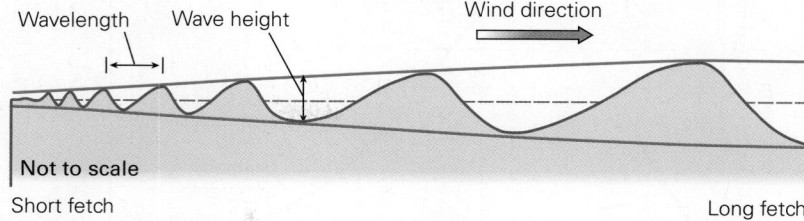

(a) The size of a wave depends in part on the fetch of the wind: the longer the fetch, the larger the wave.

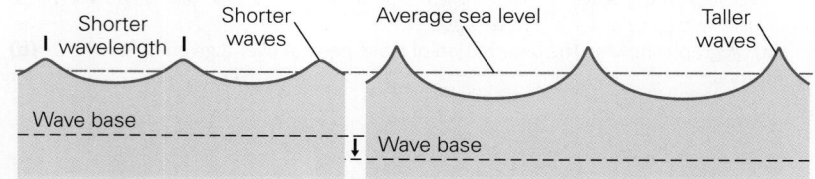

(b) Wave height, the pointiness of the wave crests, and wavelength increase as waves grow.

(c) An example of swells in the ocean. Long-wavelength waves can travel over long distances as swells.

the highest one-third of the waves arriving at a location. Occasionally, a **rogue wave**, defined as an isolated wave that rises to more than twice the significant wave height, develops. Long thought to exist only in the imaginations of sailors, rogue waves have now been documented numerous times. For example, during the 1990s, workers on oil platforms in the North Sea recorded almost 500 encounters with rogue waves—some of which were three to five times higher than other waves present at the time (Fig. 14.28b). The decks of large ships have been swamped by rogue waves in the open ocean (Fig. 14.28c). The largest wave known to have encountered a ship reached a height of 34 m (112 ft). If a rogue wave reaches the shore, it can wash unsuspecting bystanders off piers or beaches.

Particularly large waves, such as those above the significant wave height (see Fig. 14.28a), may form by *constructive interference*, a process that takes place when two wind-driven waves moving from different directions come together in such a way that their crests overlap to form a single crest that is higher than that of either wave

> **Did you ever wonder...**
>
> how big an ocean wave can get?

FIGURE 14.28
Wave height,
rogue waves,
and wave
interference.

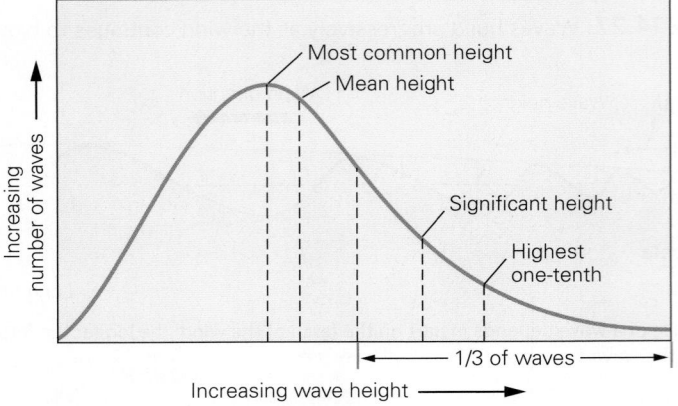

(a) A graph showing the distribution of wave heights over a given time interval.

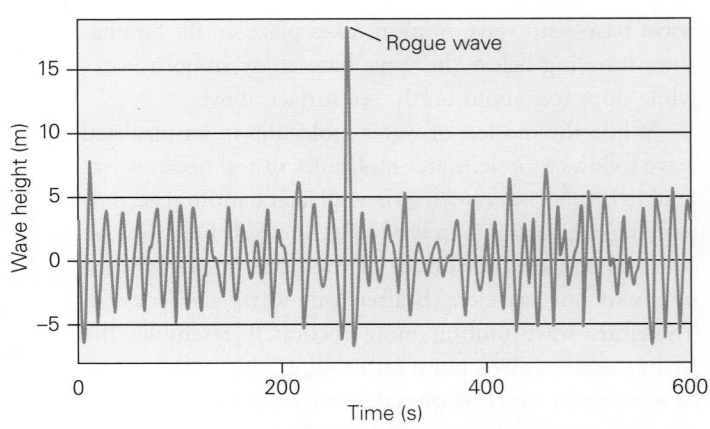

(b) A recording of waves in the North Sea over 10 minutes. Note that most waves are about 5 m (16 ft) high. The rogue wave is 18 m (60 ft) high.

(c) A rogue wave washing over the deck of a large ship.

(d) Interference between waves generated by different storms caused 23 large sailboats in the 1979 Fastnet Race off the coast of Ireland to founder, necessitating dramatic rescues.

(Fig. 14.28d). Interactions of wind-driven waves with strong currents, or the focusing of waves into a small area by the shape of the coastline or seafloor, can also form very large waves. Interference among several waves, along with other conditions producing high waves, may produce rogue waves.

Take-home message...

Frictional drag between wind and the sea surface produces waves. Within a wave, water moves approximately in a circle; the amount of motion decreases with depth. Constructive interference and other factors can generate huge waves.

Quick Questions
- What is a wave train?
- Could the wind produce a wave with a height of 500 m?
- Will all the waves generated by the wind at a location have the same height and wavelength?

14.5 Light and Life in the Sea

Zones of the Ocean Based on Light Penetration

Light, the visible portion of the electromagnetic spectrum that comes from the Sun, provides the external energy that drives surface and near-surface interactions in the Earth System (see Chapter 3). In particular, it provides most of the energy that sustains the biosphere, including the portion of the biosphere in the ocean. The amount of sunlight that ocean water and/or the seafloor receives at a location, therefore, plays a key role in controlling the distribution and character of marine organisms.

When visible light from the Sun enters seawater, it interacts with the atoms in water molecules. Some of the light undergoes **absorption**, meaning that it adds energy to the atoms so that they move and vibrate faster, causing the seawater to heat up. The rest undergoes **light scattering**, meaning that the instant that atoms absorb that light, they immediately re-emit it, but in a different direction. Because

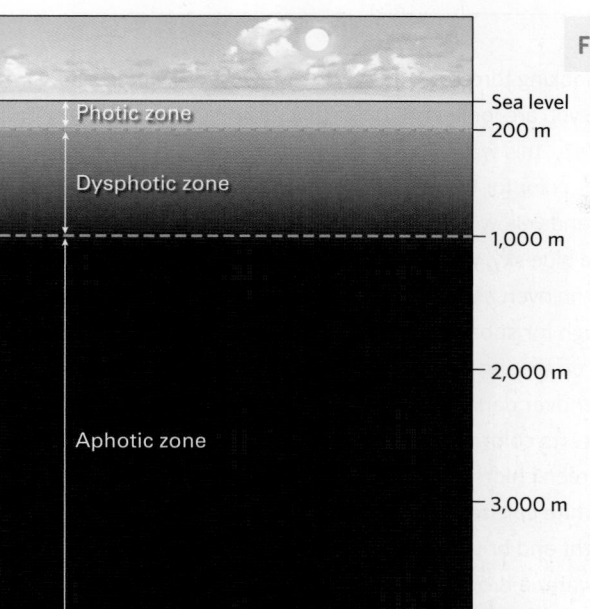

Sea level
200 m

1,000 m

2,000 m

3,000 m

4,000 m

(a) The degree of light penetration serves as one basis for dividing the ocean into zones. In the photic zone, there's enough light for photosynthesis. In the dysphotic zone, some light penetrates, but not enough for photosynthesis. No light penetrates the aphotic zone.

FIGURE 14.29 Zones of the ocean.

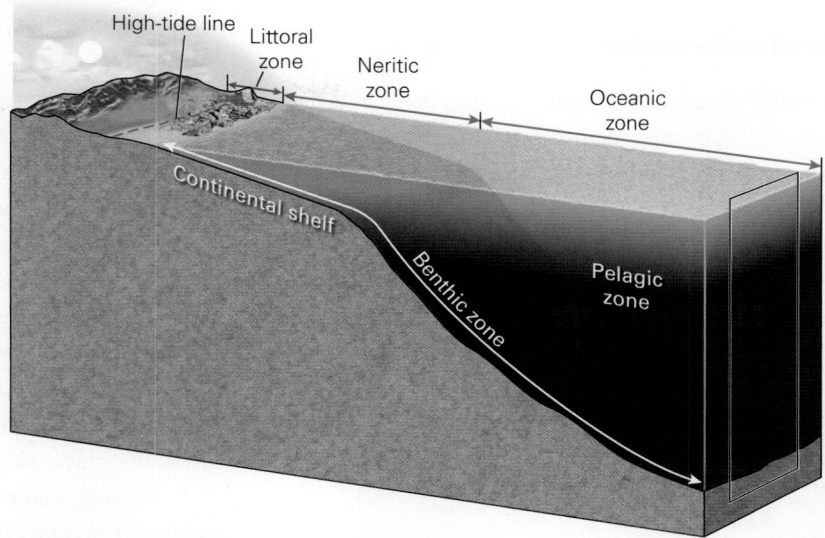

(b) Oceanographers distinguish among littoral, neritic, and oceanic zones based on distance from shore, and between pelagic and benthic zones based on proximity to the seafloor.

of light scattering, some of the incoming radiation arriving from above does not continue deeper into the water column, but rather heads off sideways, obliquely, and even back upward. Much of the scattered light is also absorbed eventually. Because of absorption and scattering, the intensity (brightness) of light decreases progressively with depth. In very clear seawater, about 50% of the incoming light remains at a depth of 30 m (100 ft), about 10% at a depth of 100 m (330 ft), and about 1% at a depth of 200 m. Significantly, different wavelengths of light undergo absorption and scattering by different amounts, and these differences contribute to determining the color of water (Box 14.4).

Oceanographers distinguish between layers of the ocean based on the amount of light penetration (Fig. 14.29a). The **photic zone** (from the Greek *phot*, meaning light) is the upper layer of the ocean in which enough sunlight penetrates the water that photosynthetic organisms can thrive. By convention, the lower boundary of the photic zone is the depth at which light has less than 1% of the intensity that it had before entering the water; this boundary, as we've just noted, lies at a depth of about 200 m in very clear water. The tiny amount of light that penetrates depths between 200 and 1,000 m does not permit photosynthesis, so this depth interval is known as the **dysphotic zone** (from the Greek words meaning bad light), or *twilight zone*. Below 1,000 m, in the **aphotic zone**, it's pitch black. In water that contains suspended

sediment (which blocks light), the base of the photic zone lies at much shallower depths. In fact, in very muddy water, it may lie at depths of less than 20 m.

Zones of the Ocean Based on Proximity to Shore or to the Seafloor

As we've just seen, the degree of light penetration serves as one basis for defining zones of the ocean. Oceanographers also find it useful to distinguish among zones based on proximity to land (Fig. 14.29b): (1) The **littoral zone** includes the *intertidal zone*, in which seafloor emerges above water during low tide and becomes submerged during high tide, and the *shore zone*, coastal land that usually remains above high tide but can become submerged or drenched in saltwater spray during unusually high tides or fierce storms. Notably, both parts of the littoral zone feel the consequences of wave action and receive intense sunlight. (2) The **neritic zone** begins where the shallow seafloor remains submerged during low tide, so tides don't affect it. But water in the neritic zone remains shallow enough (<200 m) that it lies entirely within the photic zone. Because the neritic zone lies relatively close to the shore, sediments and chemicals derived from the land wash into its water. Most continental shelves lie within the neritic zone. (3) The **oceanic zone** consists of the open-water portion of the ocean that lies far enough from shore that it does not receive chemicals or sediment from the land. The upper 200 m of the oceanic zone lies within the photic zone, and the 800 m below that lies in the dysphotic zone. No light at all penetrates below a depth of about 1,000 m,

BOX 14.4 ▶ A Deeper Look

The color of ocean water

If you've looked at photos of the ocean, had the opportunity to walk along its shores, or sailed across it, you may have noticed that the ocean's color can vary greatly. It can be deep blue, turquoise, dark green, or even gray **(Fig. Bx14.4)**. And when churned to a froth by wind or waves, it becomes white. What gives water its color, and why does its color vary so much?

Clean water in a clear glass looks colorless—you can see through it almost as easily as you can see through the glass

itself. That's because you're looking through only about 8 cm (3 in) of water. When you look at the ocean, you are looking at water that can be meters to kilometers deep. Under a sunny sky, this water looks blue, and the deeper the water, the bluer it looks. You see this color for two reasons. First, water molecules preferentially absorb red, green, and yellow light but scatter blue light. Second, when you look at the sea beneath a blue sky, you are seeing the reflection of the blue sky by the water surface. On an overcast day, the water may look greenish-gray.

In water shallow enough for substantial light to penetrate to the seafloor, water color depends in part on the seafloor's color. So, water over white sand looks lighter than does water over dark seaweed. The presence of suspended particles in water also affects its color. For example, the presence of phytoplankton, which contain green chlorophyll, can make water greenish; the presence of plankton with white calcite shells tends to make water turquoise, because the shells reflect light and brighten the water; and the presence of suspended clay makes the water a brownish color.

Because water scatters different wavelengths of light by different amounts, the colors that you see reflected from objects in the water change with depth. For example, the color of a fish as seen by a scuba diver at a depth of 20 m (60 ft) isn't the same as the color of the same fish at the surface.

FIGURE Bx14.4 Colors of the ocean.

(a) On a sunny day, the water off a Caribbean island looks blue. The darker-colored water is deeper.

(b) On an overcast day, the water off Cuba looks gray.

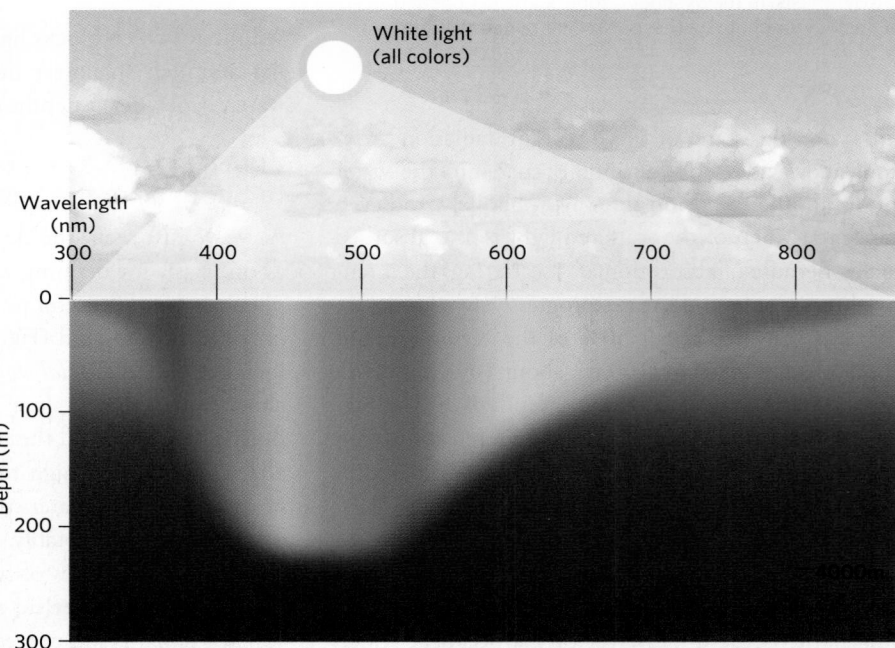

(c) Water absorbs most other colors of light more efficiently than it absorbs blue and bluish-green light. Blue and bluish-green light, therefore, penetrate deeper.

so almost all of the seafloor underlying the oceanic zone remains in the perpetual darkness of the aphotic zone.

In a general sense, oceanographers refer to the open-water portion of the neritic and oceanic zones, regardless of light penetration, as the **pelagic zone**. In contrast,

the seafloor, the sediment just below the seafloor, and the water just above the seafloor are known collectively as the **benthic zone** (see Figure 14.29b). Note that the benthic zone lies at the base of both the neritic and oceanic zones.

The Great Variety of Marine Organisms

The assemblage of living species in the ocean varies from place to place because different organisms have adapted to different conditions of temperature, salinity, light penetration, and nutrient content (**Earth Science at a Glance**, pp. 548–549). Light penetration plays a particularly important role in determining what types of organisms thrive at a location because the availability of light determines whether or not photosynthesis can take place. Therefore, most life of the sea lives in the photic zone. In fact, marine biologists calculate that about 90% of marine biomass occupies the brightly lit uppermost 70 m (230 ft) of the photic zone, partly because of the intensity of light, but also because mixing of water keeps the water well oxygenated. Organisms living in the aphotic zone survive in perpetual darkness and at lower levels of oxygenation. Some creatures that generally live in the photic zone can descend into the aphotic zone, for limited amounts of time, in search of food.

On land, some organisms live on the ground surface, some root in the ground, some fly through the air, and some burrow or subsist underground. The sea hosts such variation as well. In the pelagic zone, some organisms float on the surface of the sea, some remain suspended in the water at various depths, and some swim actively through the water. In the benthic zone, some root in, crawl about on, or swim just above the seafloor, and some burrow or squirm through sediments beneath the seafloor. Marine biologists distinguish among three distinct categories of marine organisms to emphasize these differences.

PLANKTON. All organisms that drift or are suspended in the water column, and cannot move against a current, can be called **plankton** (from the Greek *planktos*, meaning drifting) (**Fig. 14.30a–c**). Plankton include archaea, bacteria, single-celled and multicellular protozoans, and most algae. They also include some larger multicellular organisms, such as krill (tiny shrimp that serve as an important food for many whale species), some species of **seaweed**

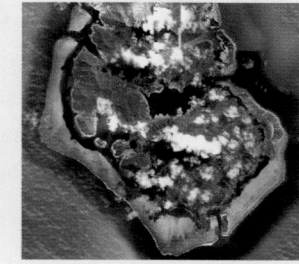
FIGURE 14.30 Examples of marine plankton.

(a) Phytoplankton, which contain chlorophyll, include diatoms.

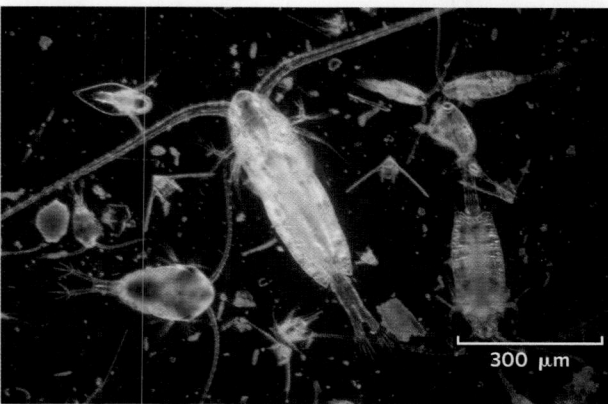

(b) Zooplankton include tiny shrimp, forams, and radiolarians.

(c) The intricate calcareous plates of coccolithophore shells look like wheels.

(d) Jellyfish drift in the water column. Some can move, but not quickly enough to swim against a current.

The Water of the Sea

Our planet's oceans host a variety of realms
that differ in temperature, salinity, and light
penetration. Their water moves constantly,
at a range of rates.

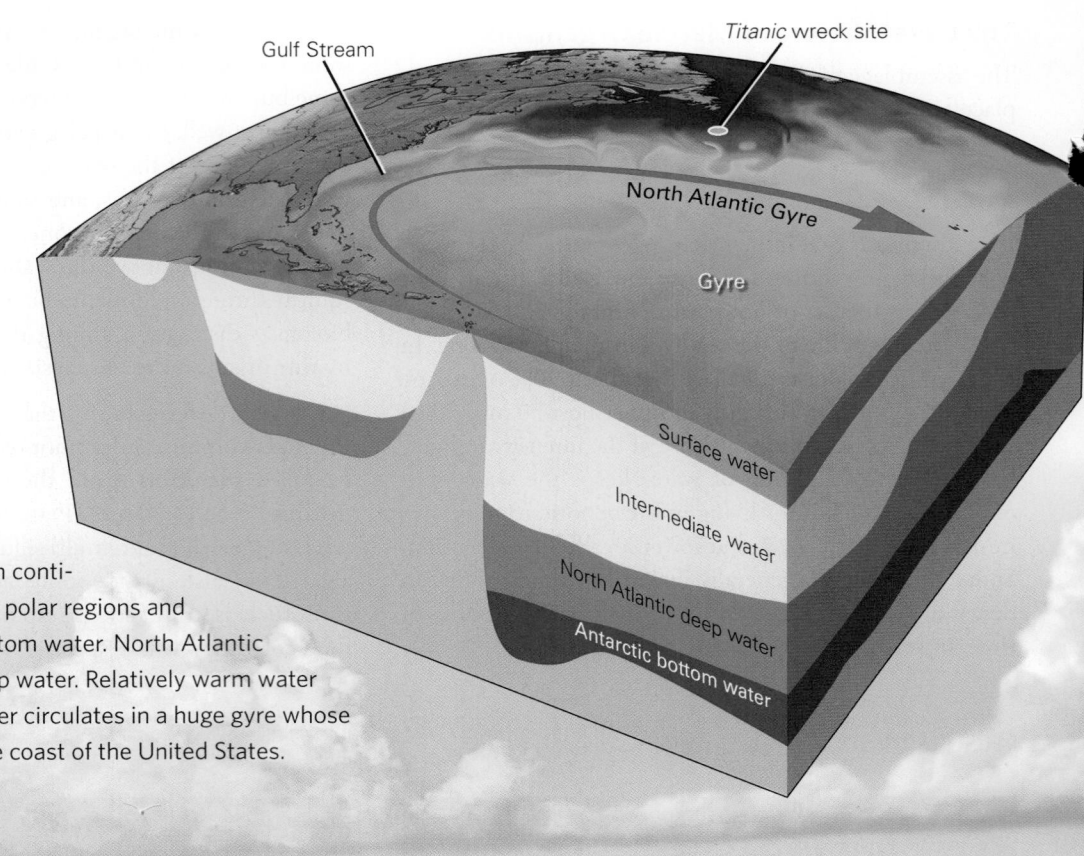

Gulf Stream

Titanic wreck site

North Atlantic Gyre

Gyre

Surface water

Intermediate water

North Atlantic deep water

Antarctic bottom water

Water masses and currents

Oceanographers divide the oceans, at the scale of a whole
ocean basin, into different layers, or water masses. In the
Atlantic Ocean, for example, the Antarctic bottom water, a
cold and saline mass, spreads northward from the southern continent. North Atlantic deep water sinks to depth in the north polar regions and
spreads southward to form a layer above the Antarctic bottom water. North Atlantic
intermediate water, in turn, overlies the North Atlantic deep water. Relatively warm water
forms the top layer. The near-surface portion of the top layer circulates in a huge gyre whose
western margin, the Gulf Stream, flows northeast along the coast of the United States.

Seaweed

Coral reef

The floor of the neritic zone

Zones of the sea

The ocean can be divided into the
littoral, neritic, and oceanic zones. Sunlight
reaches seafloor of the littoral and neritic zones,
but seafloor at the base of the oceanic zone lies in perpetual
darkness. Chemicals and sediments derived from land wash into the
littoral and neritic zones. A great variety of organisms inhabit the sea. The
assemblage found at a location depends on light penetration, temperature,
and nutrient concentration, so different assemblages inhabit different zones.

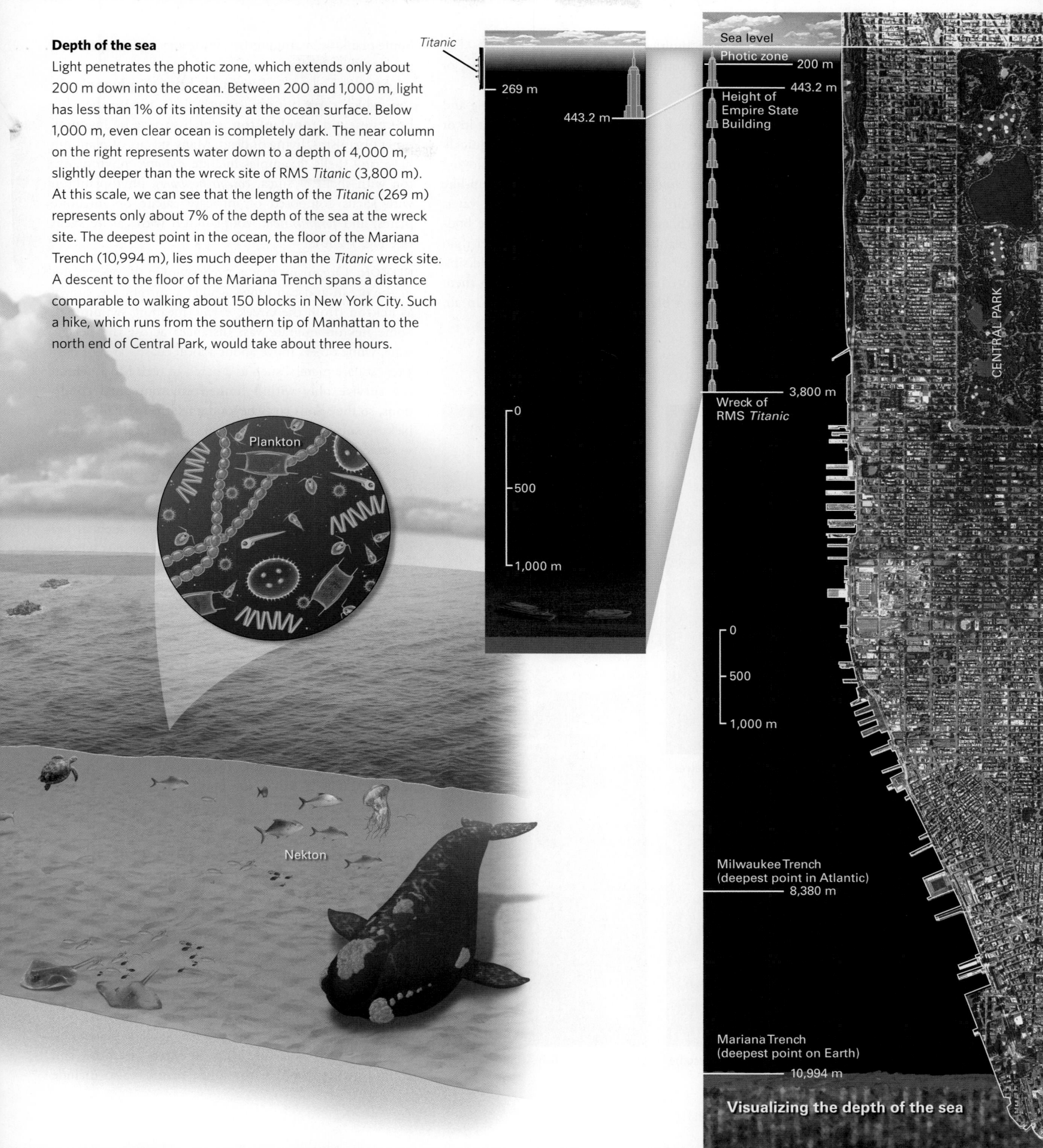

Depth of the sea

Light penetrates the photic zone, which extends only about 200 m down into the ocean. Between 200 and 1,000 m, light has less than 1% of its intensity at the ocean surface. Below 1,000 m, even clear ocean is completely dark. The near column on the right represents water down to a depth of 4,000 m, slightly deeper than the wreck site of RMS *Titanic* (3,800 m). At this scale, we can see that the length of the *Titanic* (269 m) represents only about 7% of the depth of the sea at the wreck site. The deepest point in the ocean, the floor of the Mariana Trench (10,994 m), lies much deeper than the *Titanic* wreck site. A descent to the floor of the Mariana Trench spans a distance comparable to walking about 150 blocks in New York City. Such a hike, which runs from the southern tip of Manhattan to the north end of Central Park, would take about three hours.

Plankton

Nekton

Titanic

269 m

443.2 m

0

500

1,000 m

Sea level

Photic zone — 200 m

443.2 m

Height of Empire State Building

Wreck of RMS *Titanic* — 3,800 m

0

500

1,000 m

CENTRAL PARK

Milwaukee Trench (deepest point in Atlantic) — 8,380 m

Mariana Trench (deepest point on Earth) — 10,994 m

Visualizing the depth of the sea

(nonmicroscopic multicellular algae), and jellyfish (Fig. 14.30d). Taken together, plankton account for most of the ocean's biomass.

Most plankton species have no means of moving and consist of immobile living masses that simply float in or on the water. But some can move, though not quickly enough to overcome currents. Examples of moving plankton include *dinoflagellates*, which have whip-like tails; krill, whose tiny legs propel them through water; and jellyfish, which move by making their whole bodies pulsate. Many planktonic organisms are heavier than water, so if the water were perfectly still, they would sink to the bottom. Currents or local turbulence keep them suspended, just as a breeze keeps dust suspended in air.

Some plankton maintain their buoyancy by incorporating gas into their bodies.

Marine biologists distinguish between **phytoplankton**, which carry out photosynthesis to sustain their life processes, and **zooplankton**, which survive by ingesting other organisms, living or dead. Many species of plankton have tiny shells. Examples include *diatoms*, phytoplankton with shells of silica, *coccolithophores*, phytoplankton with shells composed of intricate calcite or aragonite plates, and *foraminifera*, zooplankton with calcite shells (see Fig. 14.30).

NEKTON. Organisms that actively swim in the open water of the ocean and can move against a current are known as **nekton** (from the Greek *nektos*, meaning swimming). Some nekton migrate vast distances across the pelagic zone, while others move about a local coastal area of only a few square meters, such as a patch of reef. Most nekton can survive only within limited environmental conditions, so they don't stray beyond a relatively small region. Nekton survival also depends on the availability of food, so some species migrate with their food supplies.

Nekton include a great variety of organisms whose sizes range from nearly microscopic to gigantic. Examples of nekton include invertebrates such as squid and octopuses (Fig. 14.31a); swimming reptiles such as sea turtles and sea snakes; marine mammals such as whales, porpoises, and seals (Fig. 14.31b); and all varieties of fish (Fig. 14.31c).

The most populous category of nekton, fish, includes two principal groups. *Bony fish* have a skeleton with a skull, jaws, and spine. Examples include familiar species such as tuna, swordfish, cod, and salmon (Fig. 14.32a). Most bony fish have gas-filled swim bladders that allow them to regulate their buoyancy and remain at a given depth without swimming. The *non-bony fish*, such as sharks and rays, have cartilaginous skeletons, meaning that their skeletons consist of strong but flexible connective tissue (Fig. 14.32b). Non-bony fish do not have swim bladders, so they must actively swim in order to maintain their depth.

BENTHOS. Organisms that live on, just above, or just below the seafloor are known as **benthos** (from the Greek *benthos*, meaning bottom). The seafloor, as we have seen, ranges from fairly shallow depth in the littoral and neritic zones, to moderate and great depths in the oceanic zone. The nature of benthos varies radically among these depths. Neritic benthic communities include a variety of animals such as corals, sponges, and bryozoans, which build spectacular reefs (Fig. 14.33a). In the aphotic depths of the oceanic benthic zone, where no light penetrates, organisms such as snails, crabs, sea stars (starfish), and mollusks creep through the mud. Benthos also include

FIGURE 14.31 Examples of nekton.

(a) Squid, which are marine invertebrates, propel themselves by ejecting jets of water.

(b) Whales, which are marine mammals, can grow to become larger than the largest dinosaur.

(c) A school of fish swimming above a reef.

FIGURE 14.32 Two groups of fish.

(a) The bluefin tuna is a type of bony fish.

(b) A shark is an example of a non-bony fish.

FIGURE 14.33 Examples of benthos.

(a) In the neritic zone, benthos include many varieties of coral and other reef-building organisms.

(b) Kelp, a type of seaweed, can grow to be hundreds of meters long.

(c) Intertidal seaweed can survive exposure to air.

a great variety of seaweed types that anchor to the seafloor. The largest, giant kelp, grows stalks as long as 200 m (Fig. 14.33b). Some smaller species of seaweed can survive in the intertidal zone (Fig. 14.33c).

Food Webs in the Ocean

Who eats whom in the ocean? The answer to this question defines the ocean's **food web**, the group of organisms associated by feeding relationships. In the food web, organisms that survive by extracting nutrients from nonliving energy sources, and thus bring energy into the web, are known as **primary producers**, or *autotrophs*. In the photic zone, phytoplankton serve as the primary producers; these organisms obtain their energy from sunlight via photosynthesis. Undersea exploration in the past few decades has revealed complex benthic food webs surrounding deep-sea hydrothermal vents (black smokers). In these aphotic food webs, the primary producers are microbes that use the chemical bonds in sulfide minerals as their energy source. **Consumers**, or *heterotrophs*, in the food web include a vast variety of organisms, ranging from tiny zooplankton to fearsome killer whales, that survive by eating other organisms.

Today, *Homo sapiens* serves as the ultimate marine consumer. As the global human population exploded, the demand for fish ballooned. Since the 1950s, the fishing industries in many countries have developed mechanized approaches for harvesting fish that can yield huge

Did you ever wonder...

how fish can breathe?

catches. Unfortunately, the catches of many species have been too huge. In fact, many important fish species have been severely depleted by **overfishing**, the extraction of fish from a region at a rate faster than they can be replenished. In some locations, overfishing has driven some species nearly to extinction, causing the collapse of the local fishing industry. For example, in 1992, the cod harvest from the continental shelf east of Canada dropped to just 1% of the size it had been during the previous few years. Due to regulations that now limit catches, it is slowly recovering.

Life's Necessities: Nutrients and Dissolved Oxygen

WHERE DO NUTRIENTS COME FROM? In order to survive, marine primary producers must ingest **nutrients**, elements that organisms use to construct chemicals and carry out metabolic reactions. Nutrients of particular importance to marine organisms include phosphate (PO_4^{3-}), nitrate (NO_3^-), and iron. The presence or absence of nutrients determines the biomass that a volume of water can support at a given location.

Organisms accumulate nutrients during their lifetimes, and when they die and sink, they carry those nutrients with them to the seafloor, below the photic zone in most of the ocean. Over time, decomposition releases the nutrients back into the water. But unless new nutrients flow into the photic zone from rivers or coastal runoff, or rise from the depths back to the ocean surface by upwelling, the concentration of nutrients in the photic zone may become too low for life to thrive. Therefore, if we map the distribution of photosynthetic organisms by means of satellite detection of chlorophyll concentrations at the ocean surface (**Fig. 14.34**), we see that photosynthetic organisms are most abundant near the shore, where nutrient sources enter the sea from land and where upwelling commonly occurs, and in offshore localities where upwelling brings nutrients from the aphotic zone back to the photic zone.

DISSOLVED OXYGEN. Consumers in the food web require molecular oxygen (O_2) for *respiration*, the production of energy by metabolic reactions within a living body. These reactions use O_2 and release CO_2. On land, organisms inhale (breathe in) O_2 in gaseous form from the air, which contains 21% O_2, and use their lungs to extract it. Marine reptiles, birds, and mammals must surface to breathe air, but have adaptations that allow them to hold their breath and stay underwater for relatively long periods—in fact, a sperm whale can stay submerged for 90 minutes,

FIGURE 14.34 The colors on this map represent concentrations of chlorophyll in near-surface seawater. Chlorophyll serves as an indicator of primary production in the ocean's food web. Green represents the highest concentration of chlorophyll.

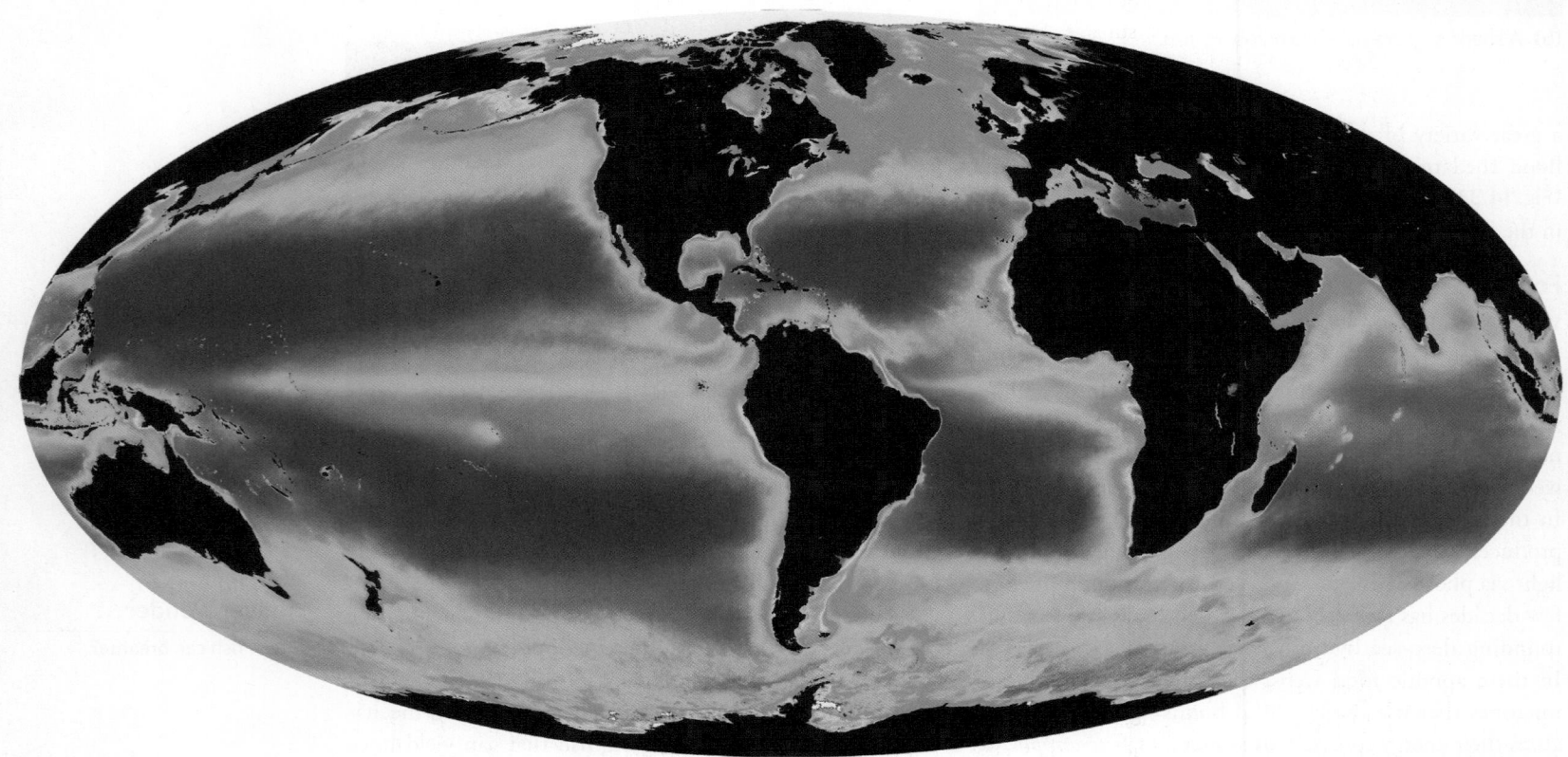

FIGURE 14.35 Consequences of harmful algal blooms.

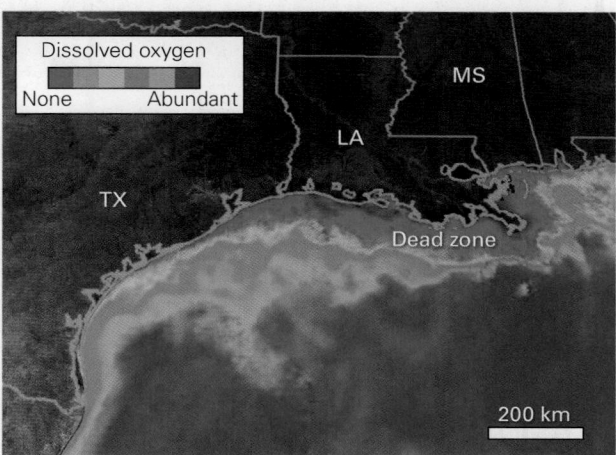

(a) This satellite image shows surface concentrations of algae along the coast of Louisiana and Texas, near the mouth of the Mississippi River. The water underlying the red area will become a dead zone.

(b) A red tide washing onto the shore along the west coast of Mexico.

and during this time can descend to depths up to 3 km (9,800 ft) in search of prey. Fish and marine invertebrates never need to surface, although they also require O_2. They survive by extracting *dissolved oxygen* directly from water by passing the water through their gills.

Ocean water contains about eight molecules of dissolved O_2 for every million molecules of water. (Note that 8 ppm = 0.0008% = 8 mg/L.) Where does the dissolved oxygen in ocean water come from? Some enters the ocean by diffusing into water at the sea surface. Storms can enhance this process by churning surface water into a froth, which has the effect of increasing the surface area of the contact between air and water. The rest enters water as a by-product of photosynthesis carried out by algae and seaweed. Downwelling transports some of the dissolved oxygen into deeper water.

ALGAL BLOOMS. As we have just seen, respiration by consumers removes some dissolved oxygen from seawater. Oxygen also gets removed from seawater during decomposition of dead organisms, for the chemical reactions involved in decomposition incorporate oxygen. In the photic zone, photosynthesis usually produces more oxygen than respiration or decomposition removes. Unfortunately, the addition of excess nutrients can trigger an **algal bloom**, meaning a rapid growth of lots of algae. Ultimately, algal blooms lead to the removal of so much dissolved oxygen from water that the water becomes *anoxic*, meaning that the concentration of O_2 drops below 2 ppm. This drop may seem counterintuitive at first, but decomposition of the algae, along with respiration by zooplankton feeding on the algae, removes more oxygen than the algae can produce by photosynthesis or

can diffuse into water at the sea surface. Most fish die when O_2 concentrations drop below 2 ppm, and almost all oxygen-utilizing organisms die when they drop below 0.5 ppm. Severely anoxic conditions yield a **dead zone** in the water, where almost all fish and shellfish die from lack of oxygen (**Fig. 14.35a**). A large dead zone has developed at the mouth of the Mississippi because the river dumps vast amounts of nutrients into the Gulf of Mexico. These nutrients come from runoff from midwestern farm fields. Combinations of natural and anthropogenic factors also yield **red tides**, algal blooms containing species of algae that not only turn the water red, but also produce dangerous toxins (**Fig. 15.35b**).

Take-home message...

The ocean can be divided into zones based on light penetration, water depth, and proximity to the seafloor. Most marine life lives in the uppermost 70 m (230 ft) of the ocean, the photic zone. Marine organisms can be categorized as plankton, nekton, or benthos, depending on where in the ocean they live and what type of movement they are capable of. Marine life plays an active role in cycling nutrients through the Earth System.

Quick Questions

• If you were in a submersible at a depth of 1,000 m (3,280 ft) with the lights off, and looked out your window, what color would you see?

• Does classification of an organism as plankton depend on its size?

• Are you more likely to find a dead zone in the ocean near the mouth of a major river or along a shoreline far from a river's mouth?

Objective 14.1

Place oceanographic discovery in the context of human history, explain why the Earth effectively has one global ocean, and characterize sea-level change over time.

KEY CONCEPTS

- Mapping of the oceans began by the 16th century, but maps could not be accurate until longitude could be measured. Mappers labeled several geographically distinct oceans, but all oceans are connected, and water circulates among them.

- Oceans fill ocean basins. Most of the area in an ocean basin is underlain by oceanic lithosphere, which sits at a lower elevation than does continental lithosphere due to isostasy. Either active or passive margins can form the border of a basin.

- Mean sea level is the elevation of the air-ocean boundary as averaged over a year; it is halfway between average low and high tide. Sea level has gone up and down over the past half million years by over a hundred meters. At times, epicontinental seas flooded large areas of continents.

EARTH-SCIENCE VOCABULARY

ocean (p. 520)
ocean basin (p. 521)
oceanography (p. 521)

sea level (p. 522)
sea-level change (p. 523)
shoreline (p. 520)

REVIEW QUESTIONS

1. (a) How much of the Earth's surface do the oceans cover? (b) When did the first maps showing the Earth's oceans come into existence?

2. (a) How has mean sea level changed over Earth history, and what is a key consequence of that change? (b) What are some of the causes of sea-level change over geologic time?

3. (a) What is an ocean basin? (b) On **Figure A**, label the ocean basin, the active margin, and the passive margin. How many plates underlie the ocean in this figure?

A

Objective 14.2

Explain where the ocean's water and salts come from, describe how the salinity, temperature, and density of ocean water change with location and depth, and characterize pollutants in the ocean.

KEY CONCEPTS

- Ocean water originally came from water vapor that bubbled out of volcanoes. Rock melting in the mantle released atoms that combined to form molecular water, which dissolved in magma. Rising magma carried this water with it to volcanic vents, where it was released as vapor and later condensed and fell as rain.

- Most salt in the sea consists of NaCl, but seawater contains other salts as well. The salinity of seawater averages about 3.5%, but it varies with location and depth. Most of the ions that make up the salt in seawater get carried to the sea by rivers, although some are released at submarine volcanoes.

- The ocean surface is warmer at the equator than at the poles. A thermocline separates warm surface water from cool deep water.

- Variations in salinity and temperature cause variations in seawater density. These variations lead to the development of distinct water masses the oceans.

EARTH-SCIENCE VOCABULARY

Archimedes's principle (p. 524)
desalination (p. 526)
estuary (p. 525)

salinity (p. 525)
thermocline (p. 527)
water mass (p. 528)

REVIEW QUESTIONS

4. (a) In what form do the atoms that water forms from exist in solid rock? (b) In what form does water exist in magma at depth? (c) In what form does water exist in magma as it approaches the Earth's surface at a volcano?

5. (a) Where does the salt in seawater come from? (b) Which type of salt forms the greatest proportion of the salt in seawater? (c) How does ocean salinity vary with location, and why? (d) Which areas in the bar graph of **Figure B** represent the proportions of chloride and sodium ions in seawater?

B

6. **(a)** How does the temperature of ocean water vary with latitude and depth, and why? **(b)** Why does a distinct thermocline develop at the equator, but not at the poles? **(c)** Which of the lines on **Figure C** represents temperature variation with depth at the equator?

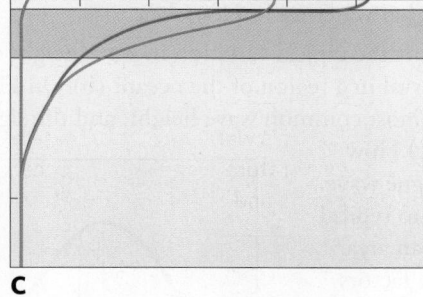

C

7. **(a)** What factors determine seawater density? **(b)** Is ocean water homogeneous or stratified? **(c)** What characteristics distinguish one water mass from another?

8. **(a)** Compare the water of the Persian Gulf, a confined sea at low latitude, with the water of the northern Pacific Ocean, an open ocean at high latitude. **(b)** How can the salt be removed from seawater?

Objective 14.3

> Explain what surface currents are and why they develop the pattern of gyres and eddies that we see today, describe how upwelling and downwelling take place, and relate variations in ocean-water density to the formation of deep currents.

KEY CONCEPTS

- Water in the oceans circulates in currents. Surface currents flow in response to frictional drag applied by prevailing winds. The paths of surface currents are affected by the Coriolis force and the pressure-gradient force.

- Because of Ekman transport, which builds a mound of slightly elevated water in the middle of an ocean, geostrophic balance develops on the perimeter of the mound, which leads to the production of a large gyre in each ocean.

- Eddies develop where a larger current shears against nonflowing water or against land.

- Upwelling and downwelling, caused by surface-water flow divergence or convergence, respectively, as well as thermohaline circulation, allow surface water to exchange with deep water over time.

EARTH-SCIENCE VOCABULARY

angular momentum (p. 537)	**current** (p. 529)
centrifugal force (p. 536)	**deep-sea current** (p. 529)
Coriolis effect (p. 532)	**downwelling** (p. 538)
Coriolis force (p. 533)	**eddy** (p. 529)

Ekman transport (p. 534)	**prevailing wind direction**
frictional drag (p. 531)	(p. 531)
geostrophic balance (p. 538)	**surface current** (p. 529)
geostrophic current (p. 538)	**thermohaline circulation**
gyre (p. 529)	(p. 540)
pressure-gradient force	**upwelling** (p. 538)
(p. 534)	**vector** (p. 536)

REVIEW QUESTIONS

9. **(a)** What forces play a role in establishing surface currents in the ocean? **(b)** On **Figure D**, indicate the direction of circulation in each of the gyres shown by adding appropriate arrowheads, and label each gyre. **(c)** On **Figure E**, indicate the deflection direction of the cannonballs, represented by black dots, if initially sent on the paths indicated by the dashed arrows. **(d)** Why do gyres and eddies form? **(e)** Why does garbage accumulate near the center of the Pacific Ocean?

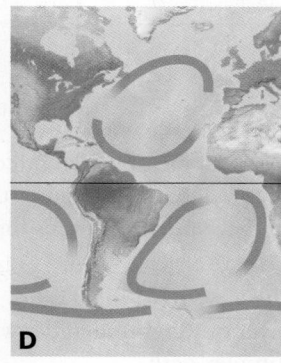
D

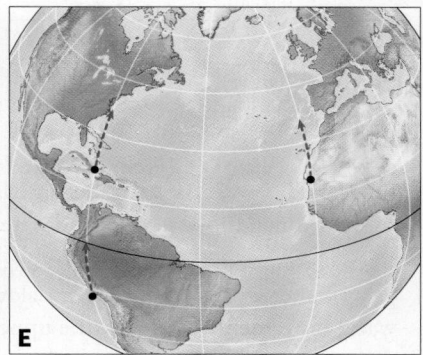
E

10. **(a)** On **Figure F**, the blue arrow shows the prevailing wind direction. Draw an arrow to indicate the direction in which Ekman transport will move water. **(b)** Why does Ekman transport take place?

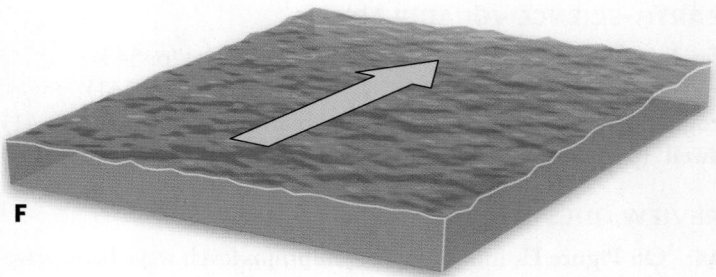

F

11. **(a)** Why are there two gyres in the Atlantic Ocean? **(b)** How does the Antarctic Circumpolar Current flow, relative to Antarctica, and what role does it play in keeping Antarctic glaciated?

12. **(a)** Explain why thermohaline circulation takes place, and describe its role in producing deep-sea currents. **(b)** Why does downwelling of ocean water occur at polar latitudes and upwelling at the equator?

13. (a) Explain why upwelling and downwelling take place along some coasts. **(b)** Which phenomenon (upwelling or downwelling) does **Figure G** show? Indicate the wind direction on the figure. **(c)** If the prevailing wind were to blow north along the west coast of North America, would waters near the west coast become good fishing grounds? Explain your answer.

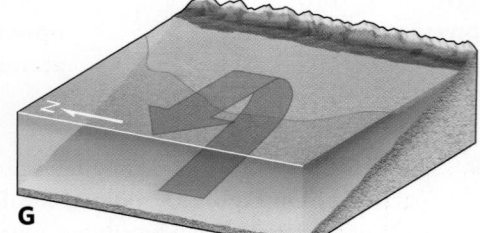

G

Objective 14.4

Discuss why waves form and how they behave, explain how to describe waves, and characterize variations in the size and shape of wind-driven waves.

KEY CONCEPTS

- Waves form in response to frictional drag where the wind shears across the surface of the ocean. They start as ripples, but can build into larger waves. As waves become large enough, gravity causes the up-and-down movement of the water surface, and wave trains form.

- Ideally, water particles move in a circle as a wave passes. The circle gets smaller with depth, so that below the wave base, there is no water movement. In reality, the upper layer of water undergoes wave drift, so water particles follow a more complex path.

- Waves vary in size, with the largest, which form during strong winds that blow over a long fetch, on the order of tens of meters high. Occasionally, much higher rogue waves develop, and these waves can be a hazard for ships.

EARTH-SCIENCE VOCABULARY

fetch (p. 543)   **wave base** (p. 543)
ripple (p. 541)   **wave train** (p. 541)
rogue wave (p. 543)   **wind-driven wave** (p. 541)
swell (p. 543)

REVIEW QUESTIONS

14. On **Figure H**, indicate the equilibrium level, wave base, wavelength, wave height, and amplitude.

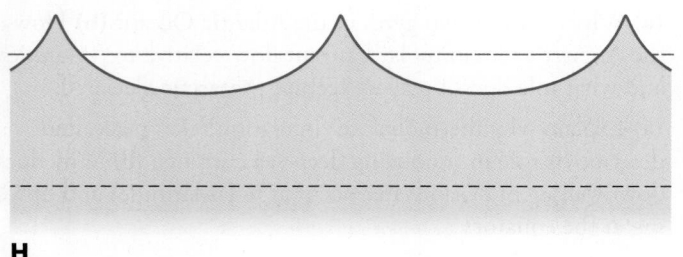

H

15. (a) What forces contribute to the formation of ocean waves? **(b)** Describe the development of waves starting with ripples and ending with large waves. **(c)** What is the shape of a realistic ocean wave?

16. (a) Describe the motion of water molecules as an idealized wave passes. **(b)** If the water in an area is affected by wave drift, what is the path that water molecules take?

17. (a) What is the typical variation in wave height over a given time interval in a region of the ocean? **(b)** On **Figure I**, label the axes, the most common wave height, and the significant wave height. **(c)** How does a rogue wave compare to typical waves in an area? **(d)** What factors can cause particularly large or even rogue waves to develop?

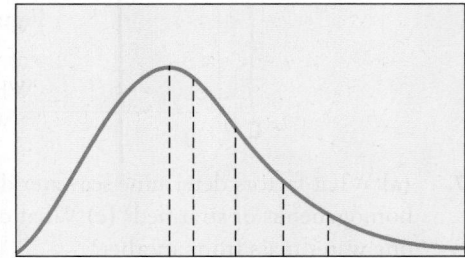

I

Objective 14.5

Distinguish among zones of the ocean by light penetration, water depth, and proximity to land, describe life in the sea in relationship to these zones, and explain the relationship between marine life and dissolved oxygen.

KEY CONCEPTS

- Oceanographers divide the ocean into zones based on light penetration (photic, dysphotic, aphotic), water depth ((littoral, neritic, oceanic), and proximity to the seafloor (pelagic and benthic).

- Life in the sea includes floating organisms (plankton), swimming organisms (nekton), and seafloor-dwelling organisms (benthos). Light penetration controls the distribution of life. Phytoplankton serve as the primary producers in the oceanic food web and account for most biomass. The distribution of nutrients affects the abundance of phytoplankton.

- Consumers, organisms that survive by eating other organisms, need dissolved oxygen to survive. Algal blooms can remove dissolved oxygen and produce a dead zone in the ocean.

EARTH-SCIENCE VOCABULARY

absorption (p. 544)   **light scattering** (p. 544)
algal bloom (p. 553)   **littoral zone** (p. 545)
aphotic zone (p. 545)   **nekton** (p. 550)
benthic zone (p. 546)   **neritic zone** (p. 545)
benthos (p. 550)   **nutrient** (p. 552)
consumer (p. 551)   **oceanic zone** (p. 545)
dead zone (p. 553)   **overfishing** (p. 552)
dysphotic zone (p. 545)   **pelagic zone** (p. 546)
food web (p. 551)   **photic zone** (p. 545)

phytoplankton (p. 550)
plankton (p. 547)
primary producer (p. 551)

red tide (p. 553)
seaweed (p. 547)
zooplankton (p. 550)

REVIEW QUESTIONS

18. **(a)** Explain the differences between the littoral, neritic, and oceanic zones. **(b)** What is the relationship between the neritic and the photic zones? **(c)** On **Figure J**, label the photic zone and the aphotic zone.

19. **(a)** Explain the differences among plankton, nekton, and benthos. **(b)** Which type of organism does **Figure K** show?

20. **(a)** Describe the basic components of the oceanic food web and how light penetration affects it. **(b)** What are dead zones in an ocean, and why do they develop? **(c)** What is a red tide, and why is it dangerous?

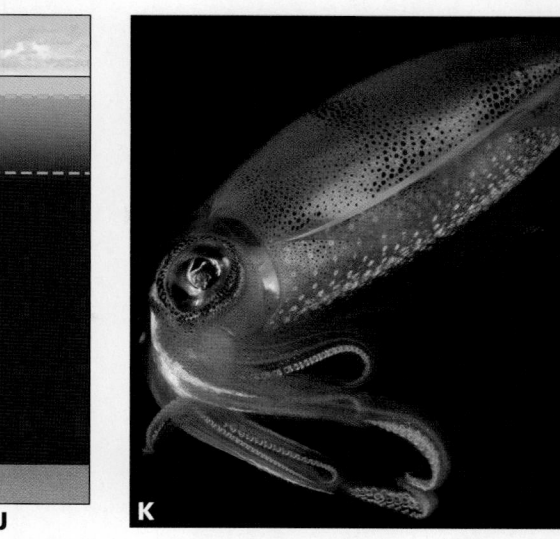

J   K

ANOTHER VIEW The oceans serve as the main reservoir of liquid water in the Earth System. Here, some of that water evaporated, and then condensed to form clouds, which may rain the water back into the ocean.

15 MARINE GEOLOGY
The Study of Ocean Basins and Coasts

After studying this chapter, you should be able to...

1. distinguish among the various bathymetric provinces of ocean basins, and explain how they formed, in the context of plate tectonics.

2. describe ocean tides and explain their causes, define tidal range and characterize how it varies, and illustrate how waves change as they interact with the shore.

3. discuss the factors that control the character of a coastal landscape, describe how beaches and rocky shores evolve, describe estuaries and fjords, and characterize wetlands and reefs.

4. explain why coasts evolve over time, describe efforts undertaken to maintain beaches and protect coastal property, and discuss phenomena leading to the destruction of coral reefs and wetlands.

(a) A crane lowers *Alvin* into the sea from the stern of its mother ship, RV *Atlantis.*

(b) *Alvin* illuminates the seafloor with bright lights.

A thousand kilometers from the nearest land, two scientists and a pilot wriggle through the entry hatch of the research submersible *Alvin,* ready for a cruise to the floor of the ocean and, hopefully, back. *Alvin* consists of a super-strong metal sphere embedded at the front of a cigar-shaped tube **(Fig. 15.1).** The sphere protects its crew from the immense water pressures of the deep ocean, and the tube holds motors and oxygen tanks. A cruise begins when, after sealing the hatch, the pilot releases cables tethering *Alvin* to its mother ship, and the submersible sinks like a stone. Most of this journey takes place in utter darkness, for the photic zone extends down only a couple of hundred meters (see Section 14.5). On reaching the seafloor, the cramped explorers turn on spotlights and gaze at a surreal vista of loose sediment, black rock, and an occasional sea creature. For the next five hours, they take photographs and use a robotic arm to collect samples. When finished, they release some ballast, and *Alvin* rises like a bubble, reaching the surface about two hours later.

In the 21st century, human-piloted submersibles like *Alvin* are but one of many tools employed to collect data about the 70.8% of the Earth's solid surface that lies hidden beneath seawater. Orbiting satellites can characterize regional-scale *bathymetry* (the shape of the seafloor), while shipborne sonar can define local-scale bathymetry in detail **(Fig. 15.2).** Instruments towed by ships, and robotic

<< From the air, a coastal landscape can resemble an abstract painting. Here we see a Caribbean barrier island in water so clear that the texture of the seafloor stands out. Such islands suffer during storms, for storm surge and high waves can wash right over them.

FIGURE 15.2 Bathymetry of the seafloor.

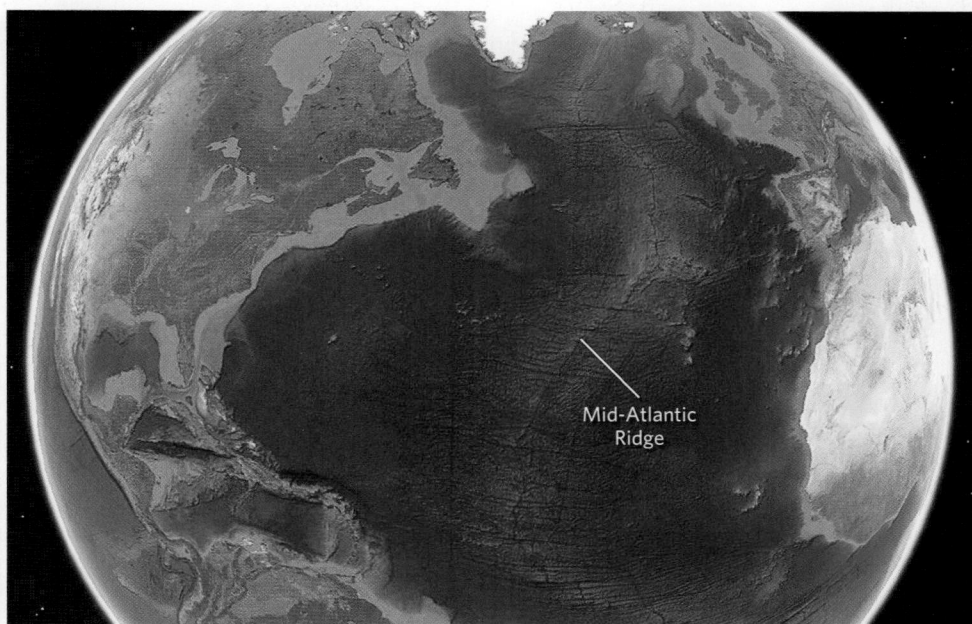

Mid-Atlantic Ridge

(a) Regional bathymetry of the North Atlantic.

~5 km

(b) A three-dimensional oblique digital elevation model of the Turnif Seamount, northwest of Hawai'i. The colors indicate depth; red areas are shallower and blue areas are deeper.

FIGURE 15.3 Seismic-reflection profiling from ships characterizes sedimentary strata beneath the seafloor. The *two-way travel time* gives the time it takes for a seismic pulse to travel from an air gun towed by the ship down to a layer and back up to a hydrophone (a seismometer that detects vibrations passing through water).

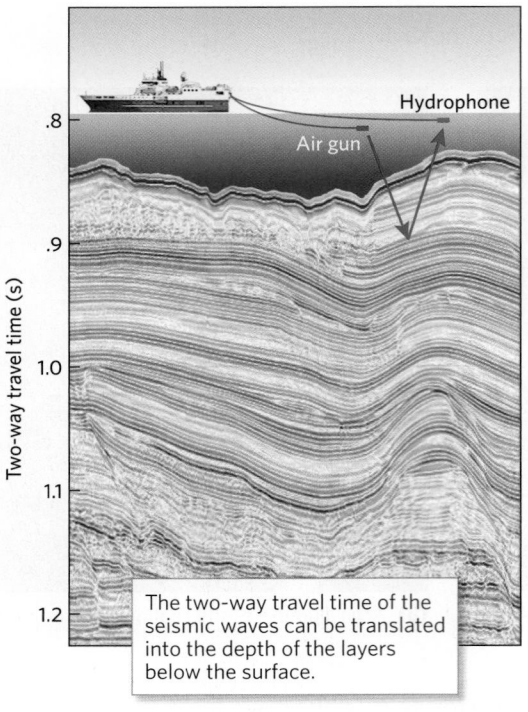

The two-way travel time of the seismic waves can be translated into the depth of the layers below the surface.

FIGURE 15.4 Deep-sea drilling and coring bring up sediment and rock from below the seafloor.

(a) The *JOIDES Resolution* is a research vessel that drills holes into the seafloor.

(b) Sediment cores can be preserved in refrigerated rooms for future analysis.

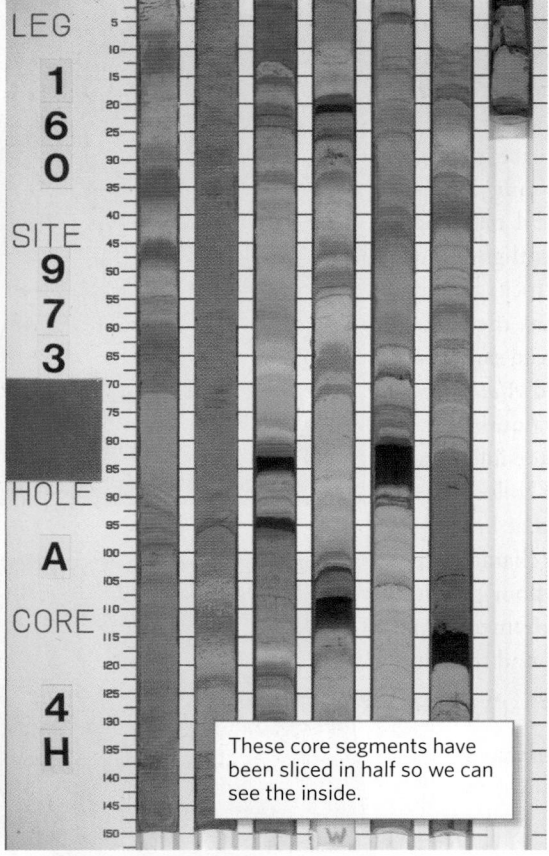

These core segments have been sliced in half so we can see the inside.

(c) The flat surface of a split core reveals distinct sedimentary layers. Numbers on the scale indicate length (in cm).

submersibles that can cruise at great depths without risking lives, can study physical characteristics of the seafloor. Shipborne tools can also produce *seismic-reflection profiles*, cross-sectional images that reveal the configuration of layering under the seafloor (Fig. 15.3). By *dredging* (pulling a chain net along the seafloor) and by *coring* (plunging hollow tubes into the seafloor), researchers can collect samples of seafloor rocks and sediment. And by using *drilling ships* that are capable of boring holes, researchers have recovered oceanic crustal rocks from depths of up to 4 km (2.5 mi) below the seafloor (Fig. 15.4). Geologists use many of the same tools to study the *coast* (the shore and the adjacent areas of land and sea). But on coasts, they can also make direct field observations. The results of all this research—from land, sea, and sky—provide the knowledge base of *marine geology*, the study of the ocean's floor and margins.

In this chapter, we explore the themes of modern marine-geology research. We first review the fundamental bathymetric features of ocean basins; this part of the chapter restates some of the concepts we've introduced earlier in the book, but it also provides additional detail. Then we focus on the landforms and habitats that develop along the coast, where over 60% of the global human population lives today. We'll see how coasts interact with tides and waves, and how they evolve over time in response to geologic phenomena and human activities.

15.1 Character of the Seafloor

If the Earth's solid surface were at the same elevation everywhere, the water currently held in the oceans would cover the surface of the Earth uniformly to a depth of about 2.5 km (1.5 mi). But elevations on the real Earth vary significantly with location. In fact, higher areas (dry land) and lower areas (the seafloor) differ in elevation by an average of 4.5 km (2.8 mi) (Fig. 15.5a). Distinct **ocean basins**, the large areas of the Earth's surface covered with seawater, exist because of this contrast in elevation; gravity causes water to drain from continents and fill the oceans.

Why is average seafloor so much lower than average land? Recall from Chapter 2 that oceanic lithosphere and

FIGURE 15.5 Contrasts between continental lithosphere and oceanic lithosphere.

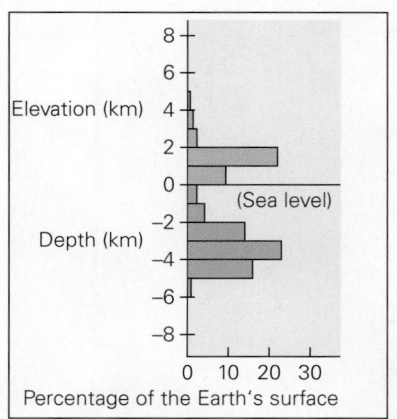

(a) A graph showing the percentage of the Earth's surface at different elevations above sea level or depths below the sea. Note the bimodal distribution of elevation.

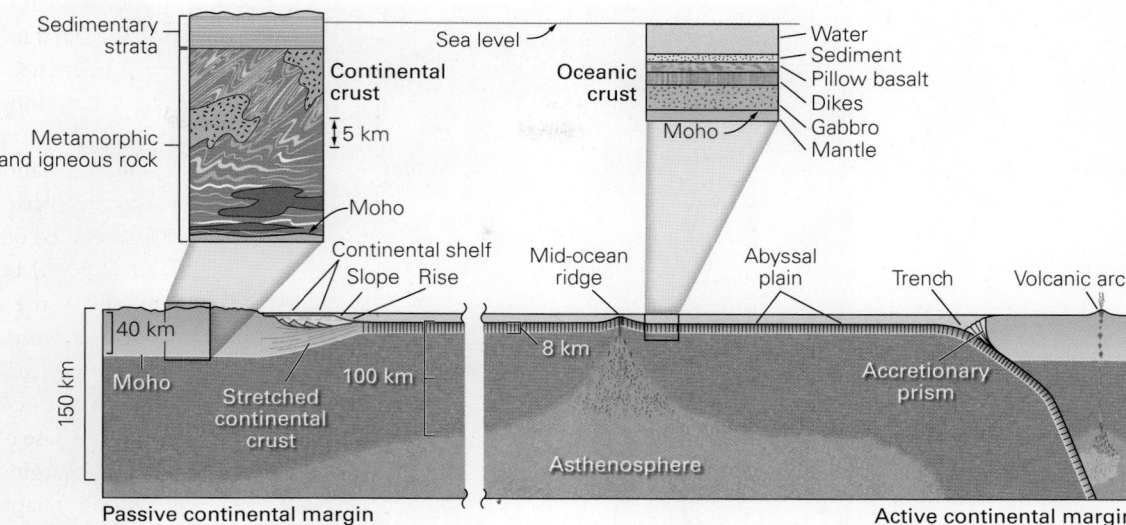

(b) The crustal portion of continental lithosphere differs markedly from that of oceanic lithosphere.

(c) A simple model showing that the surface of a thick, low-density pine block sits higher than the surface of a thin, high-density oak block when both are placed in water.

continental lithosphere differ markedly in terms of both composition and thickness (Fig. 15.5b). Oceanic lithosphere has a maximum thickness of 100 km (60 mi) and includes a 7–10 km (4–6 mi) thick crust of relatively dense basalt and gabbro. In contrast, continental lithosphere reaches a thickness of about 150 km (90 mi) and includes a crust 25–70 km (15–45 mi) thick, composed of relatively less dense felsic and intermediate rock. Because of these differences, seafloor underlain by oceanic lithosphere sits significantly deeper than land (the surface of continental lithosphere), much as the surface of a thick block of low-density wood like pine sits higher in water than the surface of a thin block of dense wood like oak (Fig. 15.5c). In the wood-block model, water flows out of the way so that the blocks attain their equilibrium positions. In the case of the Earth, plastic asthenosphere flows out of the way to let lithosphere achieve *isostatic equilibrium* (meaning that upward-directed buoyancy force and downward-directed gravitational force are in balance; see Box 6.2).

Geologists divide the ocean floor into *bathymetric provinces* based on regional water depth (Fig. 15.6). We'll now revisit these provinces, first mentioned in Chapter 2, to characterize the landscape of the seafloor in a bit more detail.

Continental Shelves and Their Borders

Imagine you're on a ship, heading east from the coast of North America into the North Atlantic. As you sail outward from the shore for a distance of 150–500 km (100–300 mi), you will be above the **continental shelf**, the relatively shallow portion of the ocean that fringes most continents. Over the continental shelf, water depth does not exceed about 200 m (650 ft), and across the width of the shelf, the ocean floor slopes seaward at an angle of only 0.3°. At its seaward edge, the continental shelf terminates at the *continental slope*, a surface that descends at an angle of about 2° from a depth of 200 m down to nearly 4 km (2.5 mi). From about 4 km down to about 4.5–5.0 km (2.8–3.1 mi), the slope angle decreases, defining a region called the *continental rise*. Finally, at a depth of 4.5–5.5 km, you will be above a vast, nearly horizontal surface known as the *abyssal plain*.

Broad continental shelves, like that of eastern North America, form along *passive continental margins* (Fig. 15.7a). Recall that these margins are not plate boundaries, so they lack seismicity. Passive margins originated after horizontal stretching within a rift succeeded in splitting a continent in two, with a new mid-ocean ridge in between. When seafloor spreading began along the new mid-ocean ridge, rifting in continental lithosphere ceased, and the now-inactive remnants of stretched and faulted crust became the continental margins that border the nascent ocean. Stretching of continental lithosphere during rifting not only made the lithosphere thinner, but also made it

FIGURE 15.6 Bathymetric provinces of the South Atlantic Ocean.

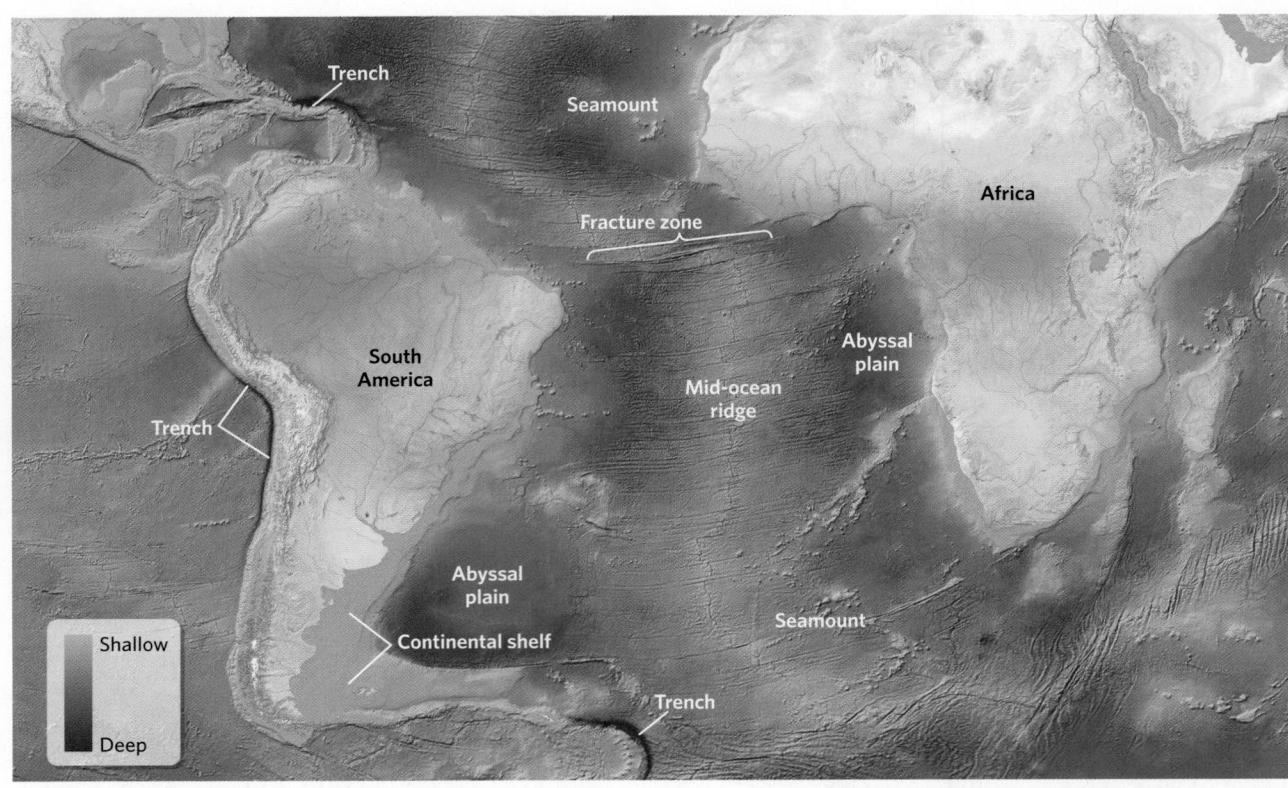

(a) A bathymetric map of the South Atlantic and the eastern South Pacific shows the complexity of the seafloor landscape.

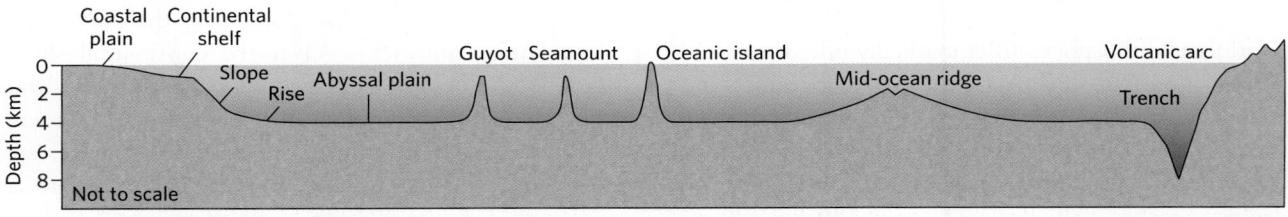

(b) A generalized schematic profile of an ocean floor, showing key bathymetric features.

hotter. When continental rifting ceased, this stretched lithosphere—which now lay between newly formed oceanic lithosphere on one side and unstretched continental lithosphere on the other—gradually cooled, and the underlying lithospheric mantle thickened. To maintain isostatic equilibrium, the surface of the stretched continental lithosphere slowly subsided (sank), just as the deck of a cargo ship moves downward when heavy cargo fills the ship's hold (Fig. 15.7b, c). As it subsided, the stretched lithosphere became a low area that collected sediment washed off the continent, as well as shells of plankton and other marine creatures that lived in the overlying water. Deposition of sediment more or less kept pace with the rate of subsidence, so after tens of millions of years, a very thick layer of sediment covered the region of stretched continental crust. This sediment-filled depression is a **passive-margin basin**,

and the flat surface of this sediment layer constitutes the continental shelf. In some cases, the sediment fill of a passive-margin basin reaches a thickness of 15–20 km (9–12 mi). Note that a passive margin effectively represents a transitional belt between continental lithosphere and oceanic lithosphere.

The bathymetric characteristics of *active continental margins*, those that coincide with a plate boundary (either a convergent boundary or a transform fault) and are seismically active, differ from those of passive margins. For example, if you were to sail into the Pacific Ocean from the west coast of South America, you would cross a convergent boundary where the Pacific Ocean floor subducts beneath the continent (Fig. 15.8a). This active margin has a relatively narrow continental shelf, underlain by a sediment apron that spreads out over the top of an *accretionary prism*, the wedge of deformed sediment scraped off the

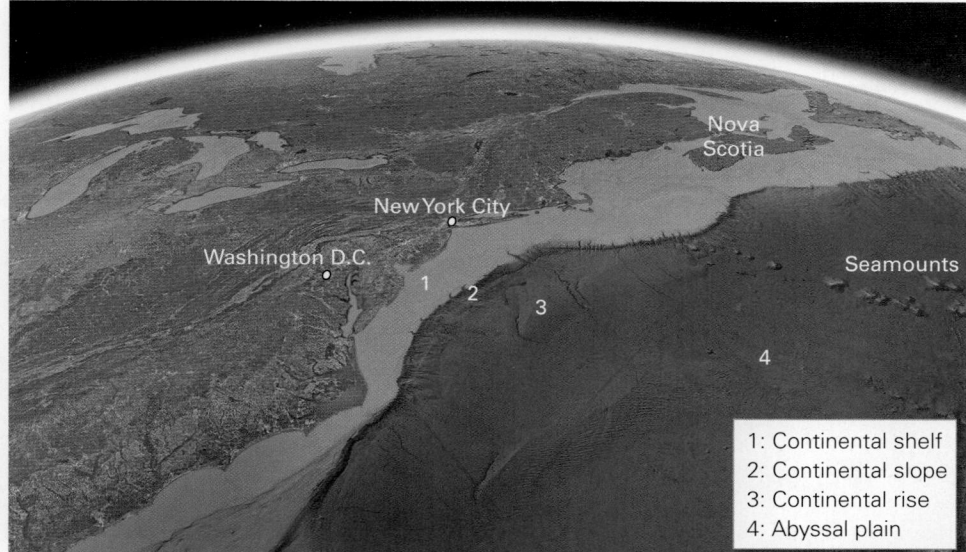

FIGURE 15.7 The continental shelf of a passive margin.

1: Continental shelf
2: Continental slope
3: Continental rise
4: Abyssal plain

(a) An oblique view, looking northwest, of the continental shelf along part of North America's east coast. It is much shallower than the adjacent abyssal plain.

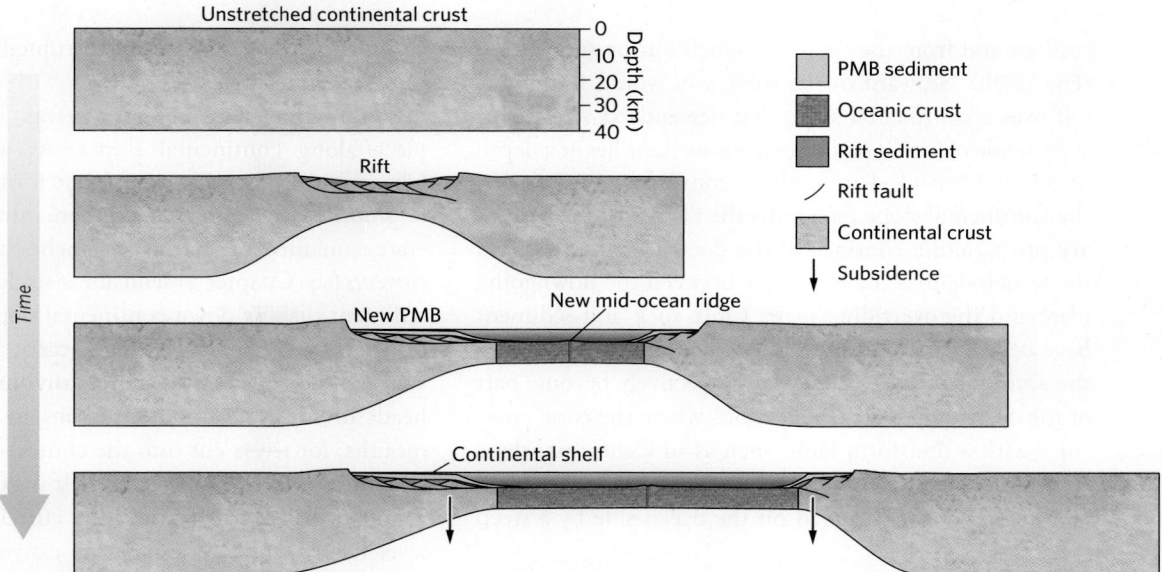

(b) The formation of a passive-margin basin (PMB). The top surface of the PMB is the continental shelf.

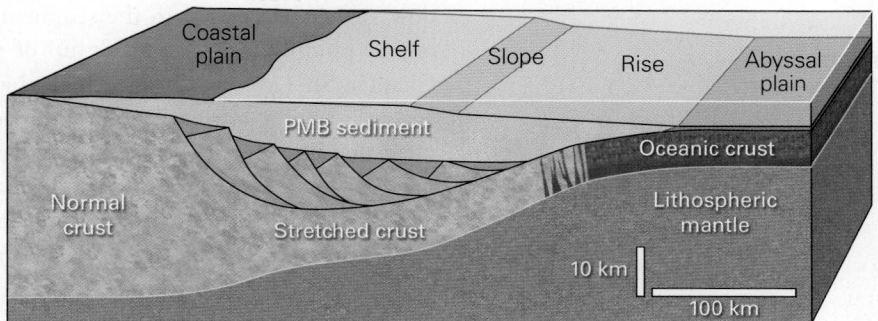

(c) A passive-margin basin forms over lithosphere that underwent stretching prior to the breakup of a continent to form an ocean basin. The remains of the rift underlie the thick sediment of the basin. (The orange areas represent rift-fill sediment.)

FIGURE 15.8 The continental shelf of an active margin.

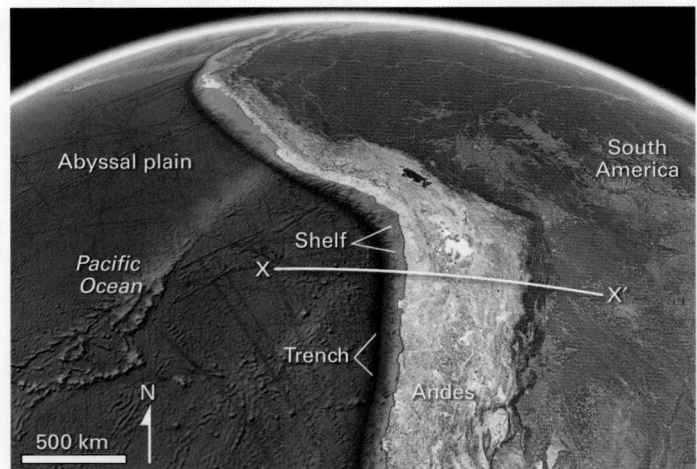

(a) Looking northwards at the west coast of South America, we see a narrow continental shelf, built on top of an accretionary prism. The tan area represents the Andes.

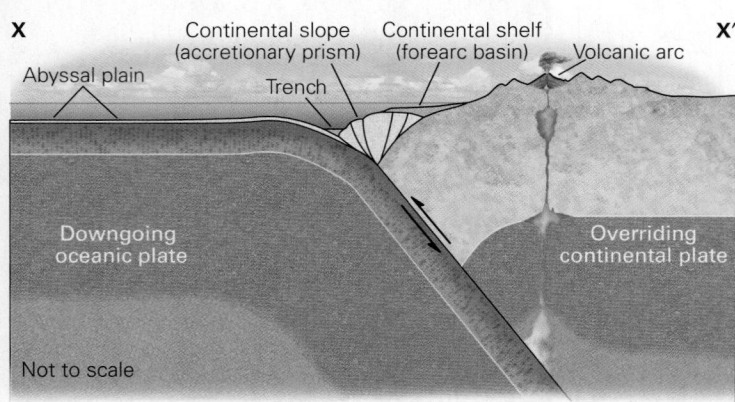

(b) A schematic cross section along section line X–X' in part (a) shows the development of an accretionary prism next to a trench at a convergent boundary.

seafloor and from the deep-sea trench during subduction (Fig. 15.8b). Seaward of the shelf, you would find yourself over a continental slope that descends at a relatively steep angle of 3.5° into a trench whose floor lies at a depth of 5.0–6.5 km (3–4 mi). Along convergent boundaries, the continental slope represents the face of the accretionary prism, and a continental rise doesn't exist. Note that the trench defines the boundary between the downgoing plate and the overriding plate. Once rock and sediment have been scraped off the seafloor and incorporated into the accretionary prism, they have effectively become part of the overriding plate. In locations where the coast coincides with a transform fault, such as in California, there is a narrow continental shelf underlain by submerged continental crust, bordered on the ocean side by a steep

continental slope that may be disrupted by faults that can cause earthquakes.

Submarine mass-wasting events occasionally take place along continental slopes. As we mentioned in Chapter 11, large *submarine slumps* form where continental slopes locally collapse, and these movements may generate tsunamis. Submarine avalanches known as *turbidity currents* (see Chapter 5) send abrasive clouds of suspended sediment rushing down continental slopes. In many locations, these turbidity currents carve deep underwater valleys, known as **submarine canyons** (Fig. 15.9). The heads of some submarine canyons lie offshore of river mouths, for rivers cut into the continental shelf at times when sea level is low and the shelf is exposed, producing a trough that can focus turbidity currents even when sea level rises. Generally, submarine canyons trend roughly perpendicular to the continental margin.

When a turbidity current reaches the mouth of a submarine canyon, at the base of the continental slope, its flow velocity decreases, so the sediments it carries settle out to produce *turbidites* made up of graded beds (see Chapter 5). Layer upon layer of turbidites accumulate and build out a **submarine fan**, a broad wedge of sediment that spreads out over the abyssal plain (along passive margins) or the trench floor (along marine convergent boundaries). Overlapping submarine fans underlie much of the continental rise along passive margins.

Abyssal Plains and Seamounts

Recall from Chapter 2 that, as oceanic lithosphere slowly moves away from the axis of a mid-ocean ridge, it gets progressively older. As it ages, it cools, so the very dense

FIGURE 15.9 A submarine canyon off the coast of New Jersey, New York, and Connecticut. Turbidity currents flow down the canyon and carry sediment to a submarine fan at the base of the continental rise.

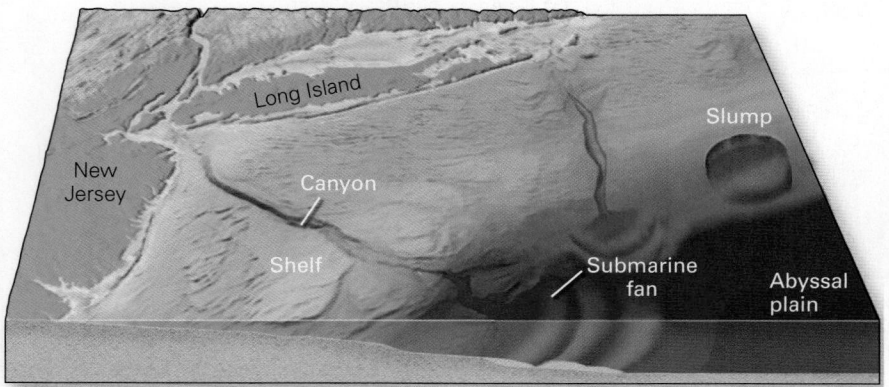

lithospheric mantle gets progressively thicker. Like adding ballast to a ship, thickening of the lithospheric mantle causes the seafloor to become deeper. The rate of cooling, and therefore the rate of subsidence, decreases with increasing distance from the ridge axis, so the average slope of the seafloor decreases away from the ridge. Once oceanic lithosphere has existed for about 80 million years, its cooling rate diminishes to nearly zero. As a result, seafloor older than about 80 million years has a nearly horizontal surface, known as an **abyssal plain**.

As soon as oceanic crust has formed at a ridge, a blanket of **pelagic sediment** settling down from the ocean water above gradually accumulates and covers the basalt of the crust (Fig. 15.10a). This blanket consists mostly of microscopic plankton shells and fine flakes of clay (from volcanic ash or windblown dust), which slowly fall like snow from the ocean water and settle on the seafloor. The older the seafloor, the more time pelagic sediment has had to accumulate. Thus, under an abyssal plain, the pelagic-sediment layer attains a thickness of up to 0.6 km (0.4 mi), so that the flat-lying sediment buries irregularities in the basaltic crust of the seafloor. That's why the surface of an abyssal plain is so smooth (Fig. 15.10b). At the gradual boundary between an abyssal plain and a continental rise, the composition of seafloor sediment begins to change, for it includes not only pelagic deposits, but also deposits of turbidity currents.

Oceanic islands (whose peaks protrude above sea level) and **seamounts** (whose peaks are submerged) rise above abyssal plains or mid-ocean ridges (Fig. 15.11a). These features result from hot-spot volcanic activity. Oceanic islands or seamounts that currently lie over a hot spot are active volcanoes, whereas those that have moved off the hot spot are extinct (see Chapter 2). Some seamounts are hot-spot volcanoes that did not become tall enough to protrude above sea level (Fig. 15.11b), but many started out as oceanic islands that later sank below sea level. Why do they sink? During and after the end of volcanic activity on an oceanic island, the island erodes and undergoes slumping, and over time, the seafloor beneath it ages and subsides. In some cases, erosion bevels the top of a seamount when its summit reaches sea level, or the island may be overgrown by a coral reef as it subsides. When such islands sink below sea level, they become flat-topped seamounts, known as **guyots**. Oceanic islands and seamounts align in a chain parallel to the motion of the oceanic plate over the hot spot—as we saw in Chapter 4, these chains, called *hot-spot tracks*, define the direction of plate motion. Where hot-spot igneous activity becomes particularly voluminous, a broad **oceanic plateau**, composed of a layer of basalt up to 3 km (2 mi) thick, develops (Fig. 15.11c). Oceanic plateaus represent submarine large igneous provinces (LIPs; see Chapter 4).

FIGURE 15.10 The origin and nature of abyssal plains.

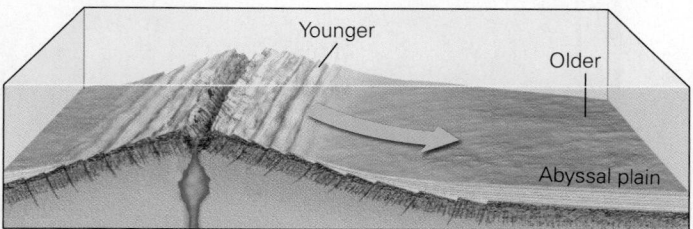

(a) The seafloor sinks progressively as it moves away from a ridge axis. It gets buried by sediment, which thickens away from the ridge axis.

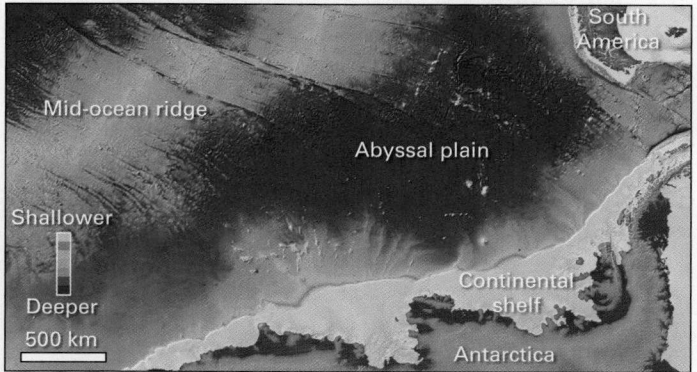

(b) The surface of the abyssal plain has a covering of fine, loose pelagic sediment.

Plate Boundaries on the Seafloor

You can see all three types of plate boundaries by studying the bathymetry of the ocean floor (see Fig. 15.6). Seafloor spreading at a divergent boundary yields a **mid-ocean ridge**, a belt that rises as much as 2 km (1.2 mi) above the depth of the abyssal plain (see Fig. 15.2a). Notably, the bathymetry of *slow-spreading ridges* (spreading rates <4 cm/yr) differs in detail from that of *fast-spreading ridges* (spreading rates >9 cm/yr). Crust at slow ridges stretches and breaks along normal faults, so seafloor near the ridge axis includes numerous step-like fault scarps parallel to the ridge axis (Fig. 15.12a). In contrast, fast ridges, such as the East Pacific Rise, are smoother because they are hotter, so some of the crustal stretching happens plastically, without brittle fault formation. Notably, fast ridges are wider than slow ridges, as measured perpendicular to the ridge axis, because young lithosphere extends farther out from the ridge axis than it does at slow ridges. Put another way, at a given distance outward from the ridge axis, the seafloor of a fast ridge is younger, and therefore has subsided less, than that of a slow ridge.

Oceanic *transform faults* link segments of mid-ocean ridges (see Chapter 2). Seismically active portions of **fracture zones**, narrow belts of broken-up rock, delineate transform faults. Typically, fracture zones are marked by

FIGURE 15.11 Oceanic
islands and seamounts.

(a) An oceanic island protruding from the Pacific Ocean.

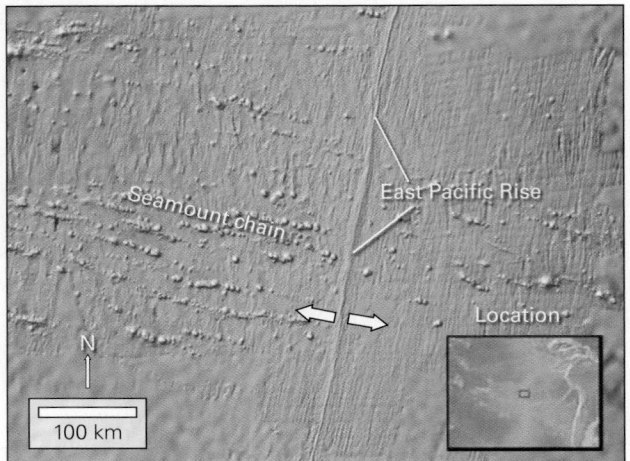

(b) Seamounts adjacent to the East Pacific Rise (112°W, 18°S), a mid-ocean ridge. Note that the seamount chains are perpendicular to the ridge axis and, therefore, are parallel to the direction of plate motion.

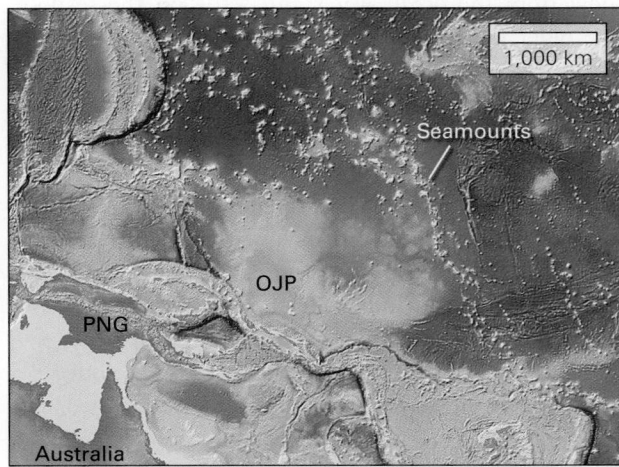

(c) The Ontong-Java Plateau (OJP), a large oceanic plateau, lies about 1,000 km to the east-northeast of Papua New Guinea (PNG). Orange represents shallower water; blue represents deeper water.

FIGURE 15.12 Plate
boundaries on the seafloor.

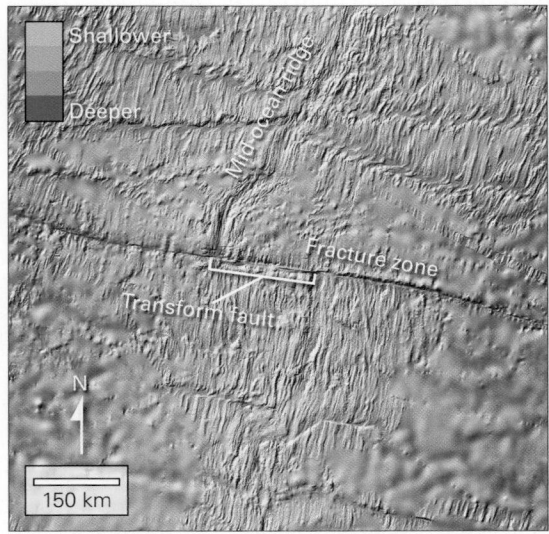

(a) Close-up showing the bathymetry of the Mid-Atlantic Ridge at 46°18′W, 23°15′N.

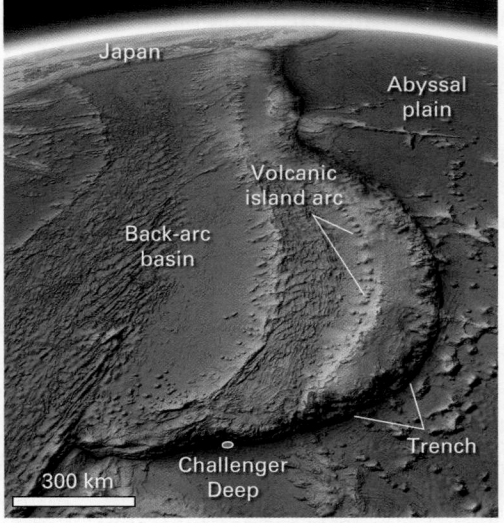

(b) Oblique view of the Mariana Trench, in the western Pacific.

steep escarpments, separated by low areas. Keep in mind that, except between the ends of the two ridge segments that a transform fault links, fracture zones are no longer seismically active, and are not plate boundaries. However, these inactive parts of fracture zones do delineate the boundary between oceanic lithosphere of different ages. Since these different-aged pieces do not move independently, they are part of the same plate (see Fig. 2.22b).

When an oceanic plate bends downward as it enters a convergent boundary, it produces an elongate wedge-shaped trough called a **deep-sea trench** (see Fig. 15.8). The face of the accretionary prism forms one side of the trench, and the surface of the seafloor forms the other side. Some trenches border continents, lying seaward of an active continental volcanic arc, as is the case along the western coast of South America. Others border *island arcs*, curving chains of active volcanic islands such as Alaska's Aleutian Islands. Because they are low, trenches collect sediment carried in by turbidity currents down the face of the accretionary prism.

Deep-sea trenches typically reach depths of over 8 km (5 mi). But their depth varies. The deepest, the Mariana Trench in the western Pacific, attains a depth of 11 km (6.8 mi) at a point called the Challenger Deep (Fig. 15.12b), and the shallowest, the Cascadia Trench off the coast of Washington and Oregon, attains a depth of only 3 km (1.9 mi). The depth of a trench depends on two factors: (1) the age of the seafloor being subducted, for older seafloor is deeper, as we have seen; and (2) the amount of sediment being deposited in the trench by turbidity currents. The Mariana Trench subducts very old oceanic lithosphere, and it does not border a continent, so it receives relatively little sediment input. The Cascadia Trench barely exists as a bathymetric feature because it subducts very young oceanic lithosphere, and has filled with abundant sediment carried into it by rivers draining North America.

Take-home message...

🏠 Ocean basins exist because oceanic and continental lithosphere differ in thickness and density. Broad continental shelves overlie passive-margin basins, and abyssal plains overlie old oceanic lithosphere. Seafloor bathymetric features (such as ridges, trenches, and fracture zones) delineate plate boundaries.

Quick Questions ───────────

- How does the lithosphere beneath a passive margin differ from the lithosphere beneath the ocean floor?
- Which is deepest: the continental slope, the abyssal plain, or the continental rise?
- Name the bathymetric features in ocean basins that delineate plate boundaries.

15.2 Up-and-Down Water Movement along Coasts

For the remainder of this chapter, we will be focusing our attention on coasts, the boundary regions between land and sea. We begin by discussing tides and waves in coastal areas, as these phenomena play a major role in determining the character of a coast. With this background, we will be able to discuss the landscapes that develop along coasts.

The Tides Come In...the Tides Go Out...

Fishermen hoping to sail from a shallow port must pay attention to the **tide**, the generally twice-daily rise and fall of sea level. At *low tide*, when sea level sinks to its lowest elevation, their boats might run aground, whereas at *high tide*, when sea level rises to its highest elevation, they will have no trouble making it out to open water (Fig. 15.13a, b). We mentioned tides in Chapter 14, and their existence is common knowledge. Here, we look more closely at how they are manifested and why they form. The transition from high tide to low tide, or vice versa, displaces the surface of the entire ocean, so this change involves the movement of a huge volume of water on a global scale. We discuss tides in this chapter because tidal movements primarily affect coastal areas.

Tidal range, the difference between sea level at high tide and at low tide (Fig. 15.13c), varies greatly with location. Some regions experience a small tidal range, under 0.5 m (20 in), whereas others experience a large one. The largest tidal range on our planet occurs in the Bay of Fundy (Fig. 15.13d), on the eastern coast of Canada, where the tidal range can be as much as 16.3 m (53.5 ft). Large tidal ranges also occur all along the coast of western Europe. In the open ocean, tidal range averages about 0.6 m (2.0 ft). Notably, at several locations on our planet, the tidal range is zero, so that sea level stays the same all day long.

During a rising tide, or *flood tide*, the **shoreline** (the boundary between water and land) moves inland, and during a falling tide, or *ebb tide*, the shoreline moves seaward. The surface of the seafloor that lies between the shoreline at high tide and the shoreline at low tide constitutes the **intertidal zone** (Fig. 15.14a). The horizontal distance over which the shoreline migrates between high and low tides—and, therefore, the width of the intertidal zone—depends on both the tidal range and the slope of the seafloor surface (Box 15.1).

When a flood tide enters the mouth of a river, the advancing tide moves against the river current and can produce a **tidal bore**, a visible wall of water ranging from a few centimeters to a few meters high. Tidal bores

FIGURE 15.13 Evidence of ocean tides.

(a) Low tide at Perranporth, in southwest England.

(b) High tide from the same viewpoint as in the previous photo.

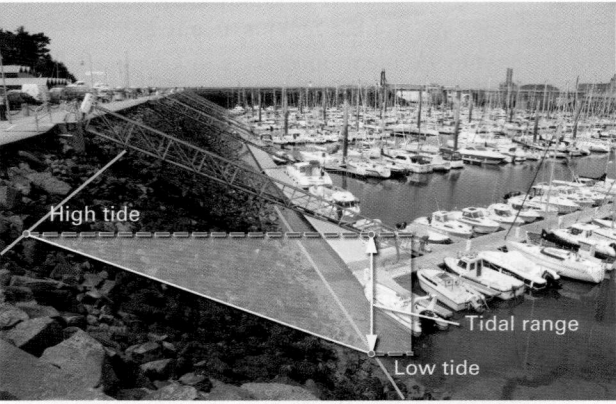

(c) The tidal range is the vertical distance between low tide and high tide. This photo of a French harbor was taken at low tide. Note that the boats are tied to floating piers, which go up and down with the tide.

(d) At low tide, the muddy floor of this small cove along the coast of the Bay of Fundy is exposed.

FIGURE 15.14 The intertidal zone and tidal bores.

(a) The intertidal zone on a beach on Cape Cod. Loose seaweed was deposited at high tide; growing weed is exposed at low tide.

(b) A tidal bore moves up a river that flows into the Bay of Fundy.

can move at speeds of up to 35 km/h (22 mph), faster than a person can run (Fig. 15.14b). In a few places, tidal bores become large enough for surfers to ride. A skilled surfer once rode a tidal bore continuously for 1 hour and 10 minutes, during which he traveled 17 km (11 mi) upstream!

What Causes the Tides?

Nontechnical discussions simply attribute tides to the gravitational pull of the Moon and Sun. Indeed, gravitational pull from these two objects does play a key role in driving tides. In fact, the Moon contributes most of the force that causes tides. (The Sun, even though vastly larger, lies so far away that its contribution to tides is only 46% that of the Moon.) But the gravitational pull of these objects doesn't completely explain why tides happen. In detail, tides take place because the water of the ocean moves in response to two forces: (1) the gravitational attraction of the Moon and the Sun, and (2) centrifugal force caused by the revolution of the Earth-Moon system. Oceanographers refer to the combination of these forces as the **tide-generating force (Box 15.2)**.

The tide-generating force generates two tidal bulges in the global ocean, forming an envelope of water that is slightly more oval shaped than the nearly spherical solid Earth (Fig. 15.15a). One bulge, the *sublunar bulge*, lies on the side of the Earth closer to the Moon, because the gravitational force due to the Moon is greatest at this point. The other, the *secondary bulge*, lies on the opposite side of the Earth, where the outward push of centrifugal force exceeds that due to the Moon's gravity (see Box 15.2). A relatively low area of the global ocean surface separates the two bulges. When a coastal location passes under a tidal bulge as the Earth spins, it experiences a high tide, and when it passes under the low area between bulges, it experiences a low tide. Higher tides develop beneath the sublunar bulge than beneath the secondary bulge. Beneath a given bulge, higher tides occur under the point where the bulge protrudes the most.

If the Earth's solid surface were smooth and completely submerged beneath the ocean, so that there were no continents or islands, the timing and height of tides would be simple to understand—each location would experience two tides a day, and the tidal range would depend on the position of the location relative to the highest part of the bulge. But the story isn't quite that simple, for many other factors affect the timing and magnitude of tides:

• *Tilt of the Earth's axis:* Because the Earth's axis of rotation is not perpendicular to the orbital plane of the Earth-Moon system, any given point on the Earth passes between a higher part of one bulge during one

---

BOX 15.1 ▶ **How can I explain . . .**

Variations in the width of the intertidal zone

What are we learning?
That the slope of the shore determines the width of the intertidal zone at a location on the coast.

What you need:
• A small plastic cutting board
• A rectangular glass pan, large enough to hold the cutting board, deep enough to hold several centimeters (a couple of inches) of water (An aluminum pan can work, too)
• Water that has been dyed blue with food coloring
• A protractor, ruler, and graph paper
• A ladle, a pitcher, and a measuring cup

Instructions:
• Fill the pan with blue water. Place the cutting board and ruler as shown. Record the angle of the cutting board (measured with the protractor) and the height of the water (measured with the ruler).
• Ladle out enough water to lower the water level by half; reserve the water in a measuring cup. Record the height of the water again.
• Repeat the experiment with the cutting board at different slope angles.
• Plot a graph of slope angle against width of the intertidal zone. How are the quantities related?

What did we see?
• Simple geometry requires that the gentler the slope, the wider the intertidal zone. The smallest intertidal zones occur where the shore is a vertical cliff.
• The intertidal zone can be very wide along very gently sloping coasts.

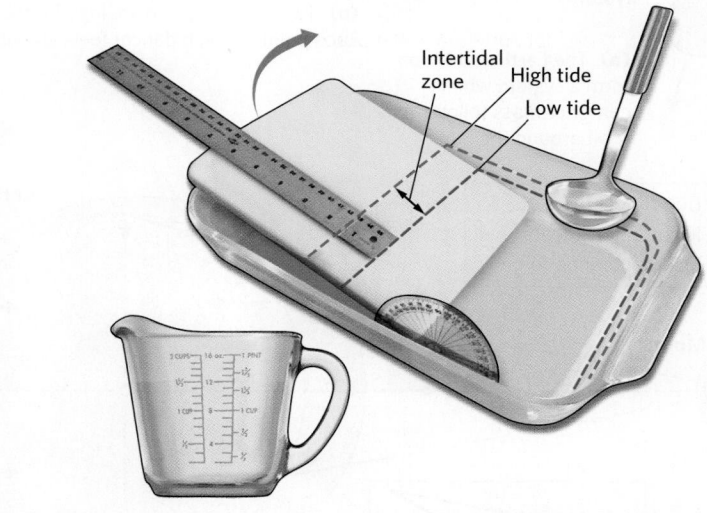

BOX 15.2 # Science Toolbox

The tide-generating force

Tides form in response to two forces acting at the same time: gravitational pull exerted by the Moon and Sun on the Earth, and centrifugal force caused by the revolution of the Earth around the center of mass of the Earth-Moon system. To explain this statement, we must first review a few physics terms:

- *Gravitational pull* is the attractive force that one object exerts on another. Its magnitude depends on the amount of mass in each object and on the distance between the two objects. (The pull increases as mass increases, and the pull decreases as distance between objects increases.)

- *Centrifugal force*, a manifestation of inertia, is the apparent outward-directed (center-fleeing) force that a body on or in an object feels when the object spins or moves in orbit around a point (see Box 14.3). It is an "apparent force" in that it exists only from the perspective of the moving object.

- The **Earth-Moon system** refers to our planet and the Moon, viewed as a gravitationally linked pair as they move together through space.

- The *center of mass* is the point within an object, between objects, or within a group of objects about which mass is evenly distributed. Because the Earth is 81 times more massive than the Moon, the center of mass of the Earth-Moon system lies 1,700 km (1,060 mi) below the Earth's surface.

With these terms on hand, let's begin by considering the origin and consequence of centrifugal force in the Earth-Moon system. To do this, we must observe the way in which the system moves. Although we usually picture the center of the Earth as following a simple orbit around the Sun, it's actually the center of mass of the Earth-Moon system that follows this orbit **(Fig. Bx15.2a)**. To see why, picture the

FIGURE Bx15.2 The concepts of the center of mass of the Earth-Moon system and the tide-generating force.

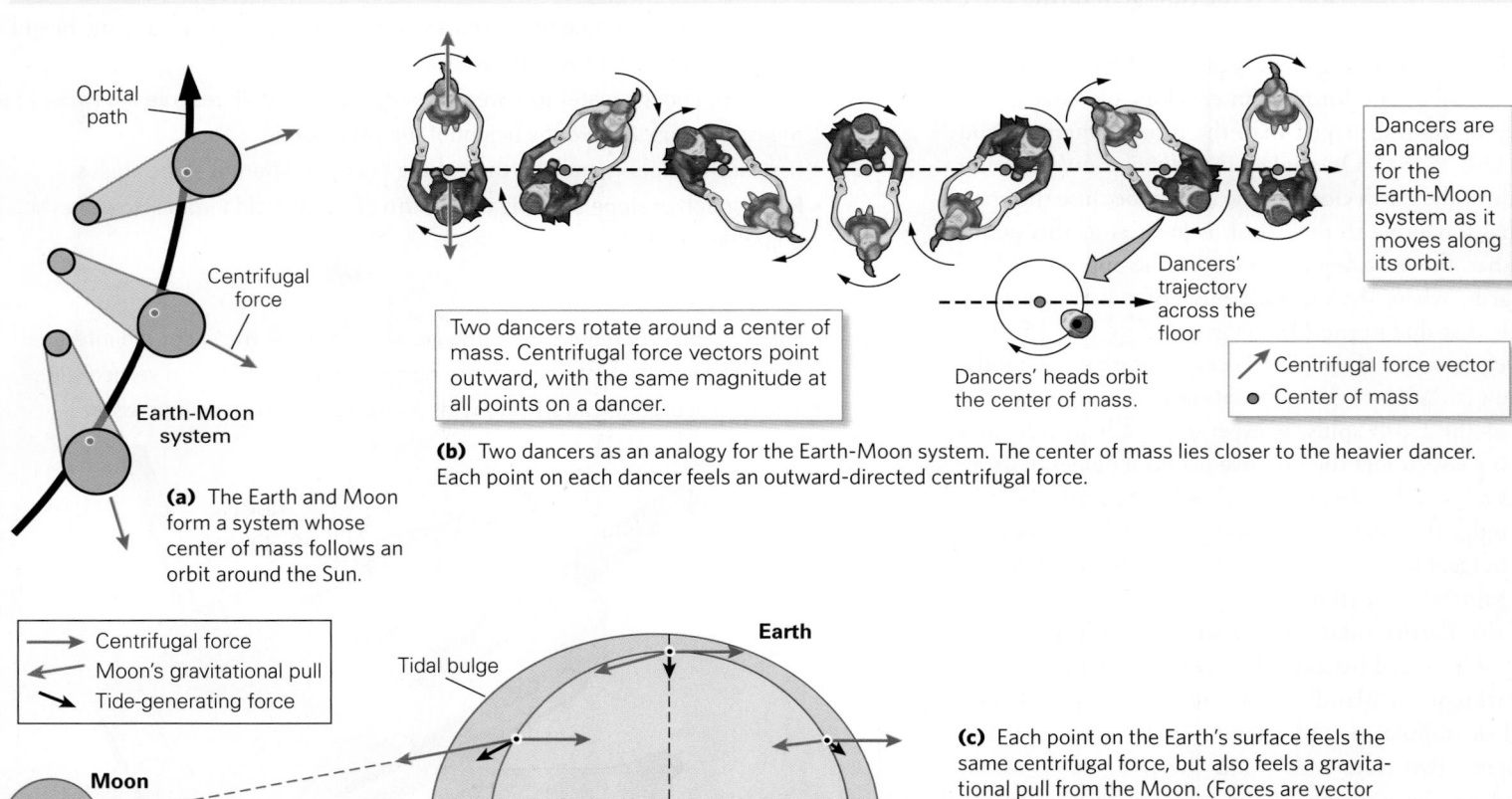

Two dancers rotate around a center of mass. Centrifugal force vectors point outward, with the same magnitude at all points on a dancer.

Dancers' trajectory across the floor

Dancers' heads orbit the center of mass.

Dancers are an analog for the Earth-Moon system as it moves along its orbit.

↗ Centrifugal force vector
● Center of mass

(a) The Earth and Moon form a system whose center of mass follows an orbit around the Sun.

(b) Two dancers as an analogy for the Earth-Moon system. The center of mass lies closer to the heavier dancer. Each point on each dancer feels an outward-directed centrifugal force.

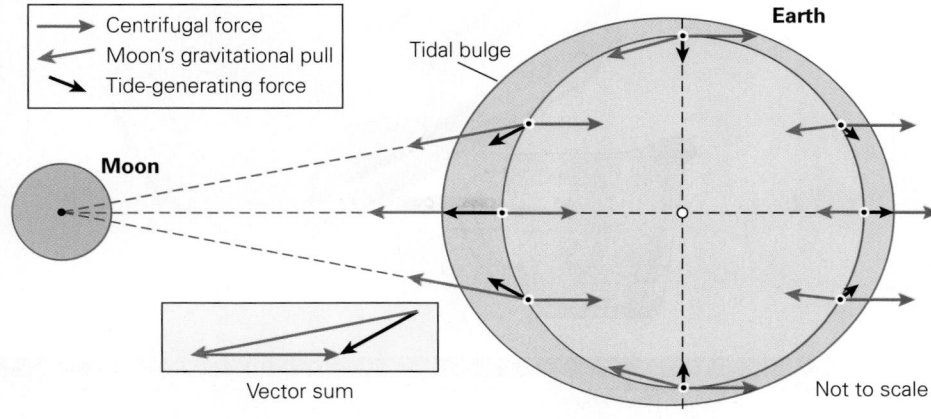

→ Centrifugal force
← Moon's gravitational pull
↘ Tide-generating force

(c) Each point on the Earth's surface feels the same centrifugal force, but also feels a gravitational pull from the Moon. (Forces are vector quantities, for they have both magnitude and direction; see Box 14.3. The inset shows how vectors can be added.) The tide-generating force is the sum of the centrifugal force and the gravitational force vectors. The bulge of the sea surface is greatly exaggerated.

Earth-Moon system as a pair of dancers, one much heavier than the other. The dancers face each other, hold hands, and whirl as they cross the dance floor (Fig. Bx15.2b). The center of mass of the pair defines their overall path—each dancer crosses back and forth across the overall path as the dance progresses.

With this image in mind, let's turn our attention back to the Earth-Moon system. As the Earth revolves around the system's center of mass, centrifugal forces develop on both the Earth and the Moon that would cause the two bodies to fly away from each other, were it not for the gravitational attraction holding them together. We can understand this statement by thinking again of our dancer analogy. We can represent the direction and magnitude of centrifugal force with an arrow, or *vector*, whose length represents the size of the force and whose orientation indicates the direction of the force. The centrifugal force acting on each dancer points outward, away from his or her partner, and is the same for all points on each dancer. Similarly, centrifugal force vectors at all points all around the surface of the Earth point away from the Moon.

Now, let's consider how the force of gravity comes into play in causing tides. To simplify this discussion, we examine only the effect of the Moon's gravity on the Earth. Vectors representing the magnitude and direction of the Moon's gravitational pull, at any point on the surface of the Earth, point toward the center of the Moon. Because the magnitude of gravity depends on distance, the Moon exerts more attraction on the side of the Earth closer to the Moon than on the side of the Earth farther from the Moon. If we draw vectors representing both centrifugal force and gravitational force at various points on or in the Earth, we see that the arrows representing centrifugal force do not have the same length as those representing gravitational attraction, except at the Earth's center (Fig. Bx15.2c). Moreover, the vectors representing centrifugal force do not point in the same direction as the vectors representing gravitational attraction. The force that ocean water feels is the sum of the two forces acting on the water. You can determine the sum of two vectors by drawing the vectors to touch head to tail; the sum is the vector that completes the triangle (see Fig. Bx15.2c, inset). This sum of the gravitational vector and the centrifugal force vector at a point on the Earth's surface is the *tide-generating force*.

The magnitude and direction of the tide-generating force vary with location on the Earth. For example, on the side of the Earth closer to the Moon, gravitational force is greater than centrifugal force, so adding the two gives a net tide-generating force that causes the sea surface to bulge toward the Moon. On the side of the Earth farther from the Moon, centrifugal force is the larger vector and causes the surface of the sea to bulge away from the Moon. Therefore, the ocean has two tidal bulges—one on the side of the Earth close to the Moon, and one on the opposite side (see Fig. Bx15.2c). The bulge closer to the Moon is always larger.

FIGURE 15.15 Tides.

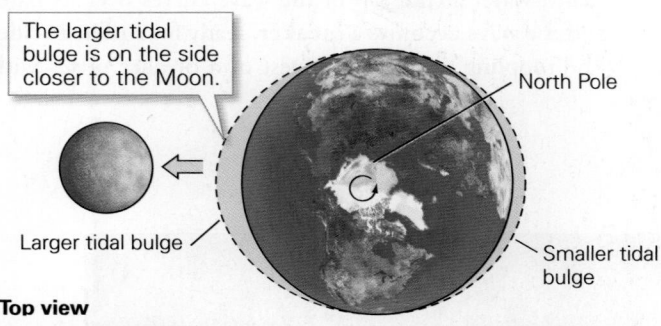

The larger tidal bulge is on the side closer to the Moon.

North Pole

Larger tidal bulge

Smaller tidal bulge

Top view

(a) Tidal bulges as viewed looking down on the North Pole. The larger (sublunar) tidal bulge always faces the Moon, and the smaller (secondary) tidal bulge is always on the side of the Earth opposite the Moon. Bulges are greatly exaggerated.

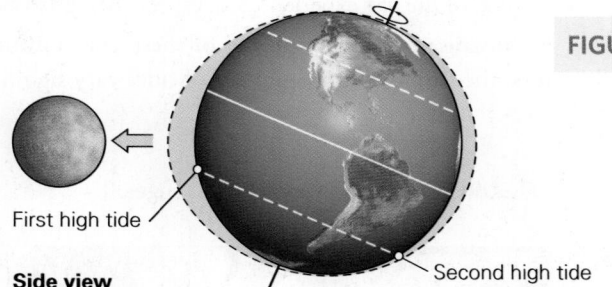

First high tide

Second high tide

Side view

(b) Viewed from the side, the sublunar bulge does not align with the equator.

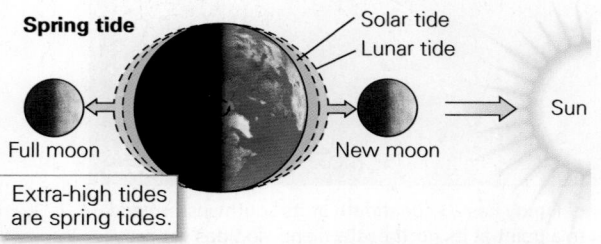

Spring tide

Solar tide
Lunar tide

Full moon

New moon

Sun

Extra-high tides are spring tides.

(c) When the Sun is aligned with the Moon, stronger, higher tides, called spring tides, result.

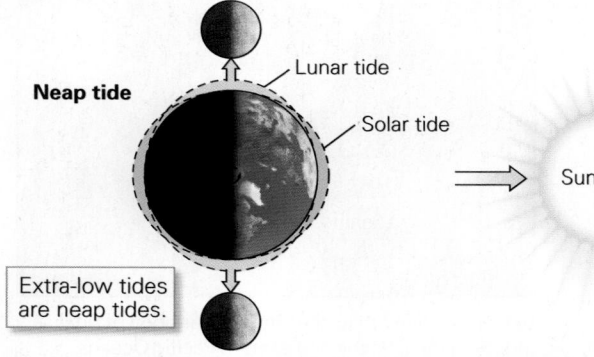

Neap tide

Lunar tide

Solar tide

Sun

Extra-low tides are neap tides.

(d) When the Sun is at right angles to the Moon, weaker, lower tides, called neap tides, result.

part of the day and a lower part of the other bulge during another part of the day (Fig. 15.15b).

- *The Moon's orbit:* The Moon progresses in its 28-day orbit around the Earth in the same direction as the Earth rotates. High tides arrive 50 minutes later each day because of the difference between the time it takes for the Earth to spin on its axis and the time it takes for the Moon to orbit the Earth.

- *The Sun's gravity:* When the Sun lies on the same side of the Earth as the Moon (a new Moon) or on the side opposite the Moon (a full Moon), we experience particularly high flood tides, called *spring tides*, because the Sun's gravitational attraction adds to that of the Moon (Fig. 15.15c). When the Moon and Sun are 90° apart relative to the Earth (quarter Moons, when we on the Earth see half of the Moon lit), we experience lower flood tides, called *neap tides*, because the Sun's gravitational attraction counteracts that of the Moon (Fig. 15.15d).

- *Ocean-basin shape:* The shape of an ocean basin influences the sloshing of water back and forth as tides rise and fall. Depending on its timing and magnitude, this sloshing can add to or subtract from the tidal bulge on a local scale. In some locations, sloshing entirely cancels out one of the daily tides.

- *Focusing effect of bays:* On a still more local scale, the shape of the shoreline influences the tidal range. In a bay that narrows to a point, the flood tide brings a large volume of water into a small area, so the point at the end of the bay experiences an especially high flood tide.

Because of the complexity of these contributing factors, the timing and magnitude of tides vary significantly around the ocean and along the coast (Fig. 15.16). At certain times during the year, these factors add to one another, producing especially high tides, known as *king tides*. At any given location, high and low tides are periodic and can be predicted (Box 15.3). For early civilizations, tides provided a rudimentary way to tell time. In fact, in some languages, the word for tide is the same as the word for time.

Waves Approaching the Shore

In Chapter 14, we discussed the motion of waves in the open ocean and saw that, unless wind drift takes place, water molecules follow circular paths when a wave passes. (When wind drift happens, surface water moves in the direction of the wind by about one wavelength for every four waves that pass; for simplicity, we'll ignore wind drift in our discussion here.) The diameter of these circles decreases with depth until, at a distance below the surface equal to half a wavelength, water remains stationary (see Fig. 14.26). The *wave base*, meaning the lowest level at which water moves in a wave, lies far above the seafloor of the deep ocean, so wave motion has no effect on the seafloor far from shore. In the *shoaling zone*, nearer shore, however, the wave base touches the ocean floor. At the seaward edge of the shoaling zone, a passing wave causes a slight back-and-forth motion of sediment. Still closer to shore, as the water gets shallower, shear between the wave and the seafloor stirs sediment significantly and slows the deeper part of the wave, so the circular motion in the wave becomes more elliptical (Fig. 15.17a). Eventually, water at the top of the wave curves over its base, and the wave becomes a **breaker**, ready for surfers to ride. The toppling water at the crest of a breaker mixes with

Did you ever wonder...

why waves become breakers near the shore?

FIGURE 15.16 Variation in tidal ranges.

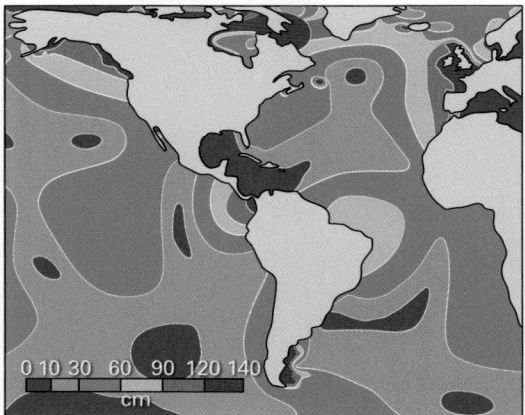

(a) A simplified map showing the variation of tidal ranges in the Atlantic and eastern Pacific Oceans. Note the particularly high tides along the west coast of Europe.

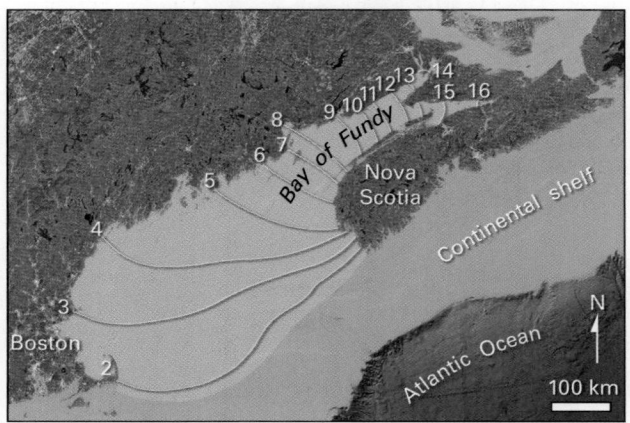

(b) The Bay of Fundy has a wide mouth at its southwestern end, and narrows to a point at its northeastern end, so tides get progressively higher from the southwest to the northeast. The contour lines indicate height (in meters) above mean sea level at high tide.

BOX 15.3 ▶ **Putting Earth Science to Use**

Planning for clamming: Reading tidal charts

Imagine that you are an independent shell fisherman, or a tourist taking a seaside holiday, and you want to harvest clams from tidal mudflats (Fig. Bx15.3a). If the mudflats occur in an area with a significant tidal range, you'll need to pay attention to when highs and lows occur. Clams lie buried in the mud, so it makes sense to dig for them at low tide, when you can walk farther offshore to find clams without having to wade through deep water. But pay close attention to when the tide starts coming in! If you get caught by a rising (flood) tide, you may find yourself having to swim in cold water to get back to shore, and risk exhaustion or hypothermia, even death. Similarly, if you anchored a boat close to shore at high tide, you may find your boat stuck in the mud during a falling (ebb) tide, and you'll have to wait many hours before getting underway again.

In the United States, NOAA uses computers that consider the many variables affecting tides in order to predict when highs and lows will take place. You can find NOAA's data, in the form of tables and charts, online at https://tidesandcurrents.noaa.gov. The charts depict water level on the vertical axis and time on the horizontal axis, allowing you to read off the water level at a specific time and to see when tides will rise or fall (Fig. Bx15.3b). A tidal chart representing tides over the course of a month emphasizes how tidal range varies, primarily due to the relative positions of the Sun and the Moon (Fig. Bx15.3c).

FIGURE Bx15.3 Reading tidal charts.

(a) Digging for clams at low tide.

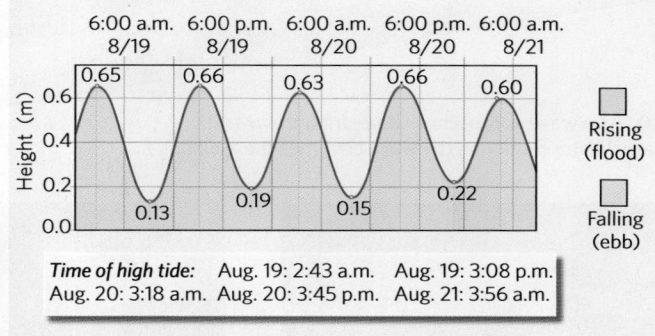

Time of high tide: Aug. 19: 2:43 a.m.   Aug. 19: 3:08 p.m.
Aug. 20: 3:18 a.m.   Aug. 20: 3:45 p.m.   Aug. 21: 3:56 a.m.

(b) The NOAA tidal chart for Schooner Bay, Virginia, for August 19–21, 2019. Note that high tides arrive a little later each day.

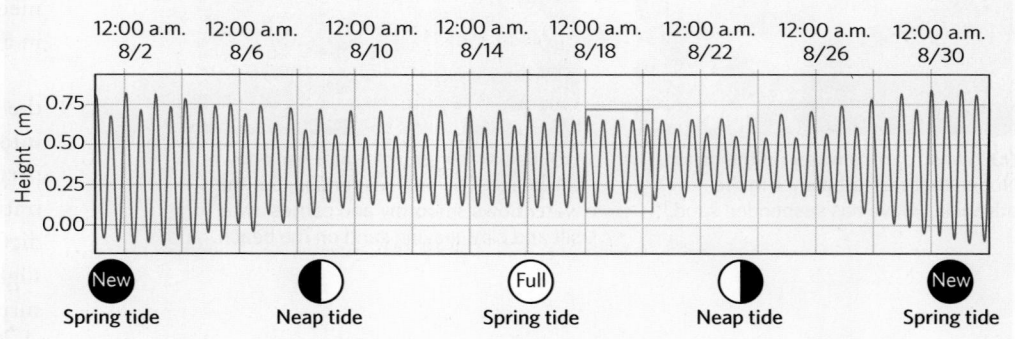

(c) The tidal range varies with the phases of the Moon. Note that the tidal range during neap tide is less than that at spring tide, and that tides are particularly high when the Moon is new. The red rectangle shows the days depicted in part (b).

---

air to form a white froth. In the **surf zone**, breakers roll ashore, and ultimately send a surge of water up the beach (Fig. 15.17b). This upward surge, or **swash** (Fig. 15.17c), continues until friction and gravity bring water motion to a halt. Then gravity draws the water back down the beach as **backwash** (Fig. 15.17d).

If the crests of waves move toward the shore at an angle, they bend as they approach the shoreline, causing

a phenomenon known as **wave refraction**. Typically, by the time the wave reaches the shore, the angle between the crest and the shoreline has decreased to less than 5° (Fig. 15.18a). To understand why wave refraction happens, imagine a wave approaching the shore so that its crest initially makes an angle of 45° with the shoreline. The end of the wave closer to the shore touches bottom first and slows down as it shears against the seafloor, whereas

The three great elemental sounds in nature are the sound of rain, the sound of wind in a primeval wood, and the sound of the outer ocean on a beach.

—*HENRY BESTON (AMERICAN NATURALIST, 1888–1968)*

FIGURE 15.17 The interaction of waves with the shore.

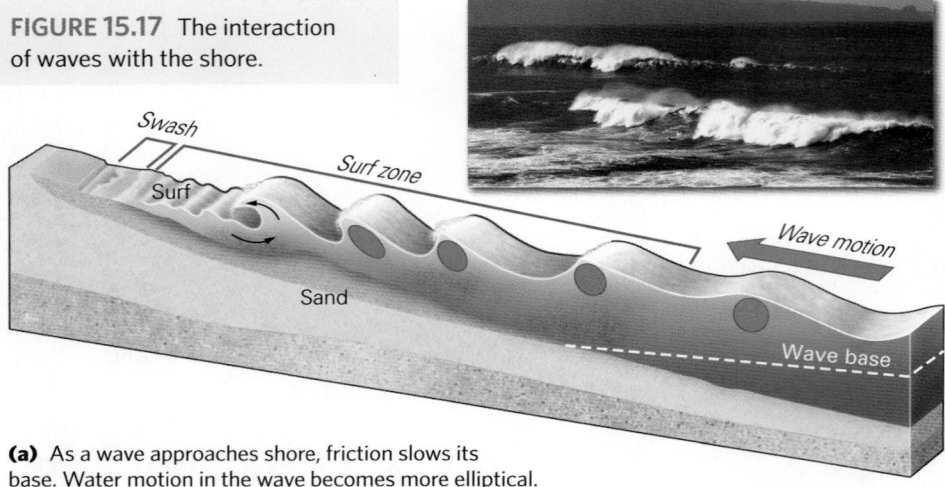

(a) As a wave approaches shore, friction slows its base. Water motion in the wave becomes more elliptical.

(b) Waves become breakers as they approach a beach, forming a surf zone.

(c) Swash carries water up the beach. Notice that the turbulence in the advancing wave has suspended sand.

(d) Backwash returns water back down the beach. The bubbles remain, but the water flows smoothly and carries away silt and clay, leaving sand on the beach.

the end farther offshore continues to move at its original velocity. This difference swings the whole wave around so that it becomes more parallel with the shoreline.

Where waves arrive at the shore obliquely, water in the nearshore region has a component of motion that trends parallel to the shore. This **longshore current** causes swimmers floating in the water just offshore to drift gradually in a direction parallel to the shoreline (Fig. 15.18b).

Waves push water toward the shore incessantly. As the backwash moves back to the sea, it may concentrate into a strong seaward flow, called a *rip current*, that moves perpendicular to the shoreline (Fig. 15.19). This concentration occurs over an elongate depression, also perpendicular to the shoreline, that develops due to erosion. The slightly deeper water over the depression flows faster than surrounding water. Rip currents cause many drownings along beaches every year because they can carry unsuspecting swimmers out into deeper water.

FIGURE 15.18 Wave refraction and its consequences along the shore.

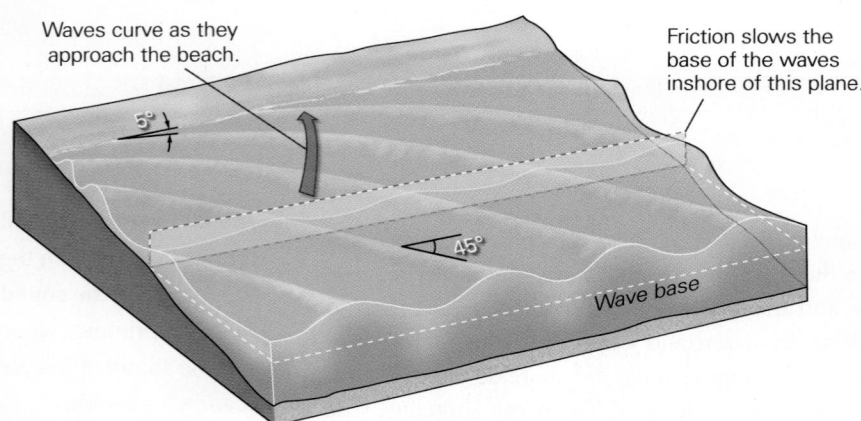

(a) Wave refraction occurs when waves approach the shore at an angle. Waves arriving at the shore obliquely can cause a longshore current.

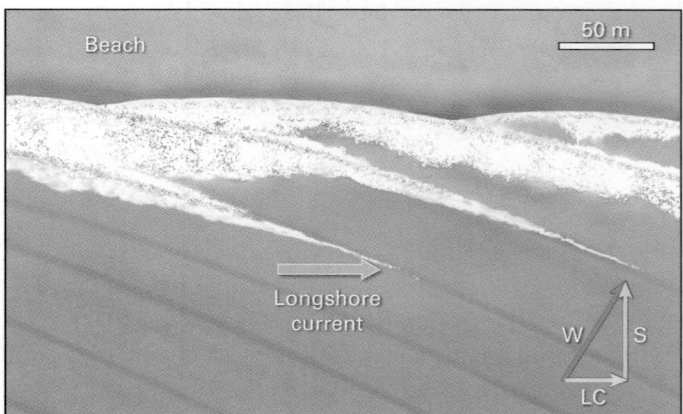

W = velocity of incoming waves; S = velocity perpendicular to shore; LC = resulting longshore-current velocity.

(b) An example of waves refracting along a beach, as viewed from above.

Take-home message...

Tides are the generally twice-daily rise and fall of the sea surface. They are caused by the tide-generating force, a consequence of gravitational pull by the Moon and Sun and of centrifugal force due to the revolution of the Earth-Moon system. Many factors affect the specific tidal range—the difference between the elevation of high tide and that of low tide—at a given location. Waves are affected by their interaction with the seafloor as they enter shallow water and evolve into breakers. If waves approach the shore at an angle, they undergo refraction.

Quick Questions

- What factors determine the width of the intertidal zone in a region?
- What is the strongest component of the tide-generating force?
- Why does a longshore current form?

15.3 Where Land Meets Sea: Coastal Landforms

Tourists along the Amalfi Coast of Italy thrill to the sound of waves crashing on rocky shores, but in the Virgin Islands, sunbathers can find seemingly endless white sand beaches. A fisherman plying the channels of the

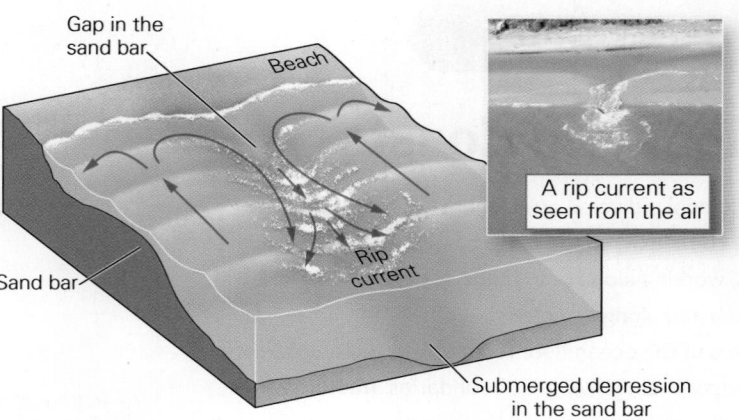

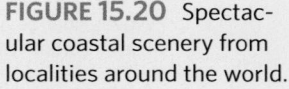
A rip current as seen from the air

Gap in the sand bar

Beach

Sand bar

Rip current

Submerged depression in the sand bar

FIGURE 15.19 A rip current forms when backwash concentrates into a narrow local seaward flow that carries water farther offshore.

Mississippi Delta will find vast swamps along the seacoast, but a ship approaching Rio de Janeiro, Brazil, will come face-to-face with granite domes rising directly from the sea (see Figure 11.2a). While a sailor landing on the coast of Florida could walk ashore without getting winded, one arriving at the southern shore of central Australia would have to climb a 100 m (330 ft) vertical cliff to get up to the land surface of the Nullarbor Plain. As these examples illustrate, coasts vary dramatically in terms of their topography and associated landforms (**Fig. 15.20**). Some are dominated by depositional features, some by rocky cliffs, and some by expanses of vegetation. All are influenced by the climate in which they form, as we now explore (**Earth Science at a Glance**, pp. 576–577).

FIGURE 15.20 Spectacular coastal scenery from localities around the world.

(a) Rocky cliffs form the Amalfi Coast of western Italy.

(b) Steep green slopes form the Nā Pali coast of Kaua'i, Hawai'i.

(c) A sandy beach along the coast of St. John, US Virgin Islands.

(d) The abrupt edge of the Nullarbor Plain forms the southern coast of Australia.

The Seafloor and Coasts

The oceans of the world host a diverse array of environments and land-scapes, illustrating the complexity of the Earth System. Tectonic processes and surface processes, working alone or in tandem, generate unique features beneath the sea and along its coasts.

The major structures of the ocean floor reflect plate tectonics. For example, mid-ocean ridges define divergent boundaries, fracture zones delineate transform faults, and trenches mark subduction zones. Oceanic islands, seamounts, and plateaus build above hot spots. Along passive continental margins, broad continental shelves develop, locally incised by submarine canyons.

Coastal landscapes reflect variations in sediment supply, tides, and climate. Tides cause sea level to rise and fall, generally twice daily, and wind builds waves that churn the sea surface, erode shorelines, and transport sediment. In locations not receiving abundant sediment, rocky shores with dramatic cliffs and sea stacks may evolve. Abundant sediment can yield sandy beaches and offshore bars. Regions where glaciers carved deep valleys now feature spectacular fjords. Coastal areas, especially those in warm climates, host unique coastal ecosystems, such as coral reefs.

Temperate and Tropical Coastal Landforms

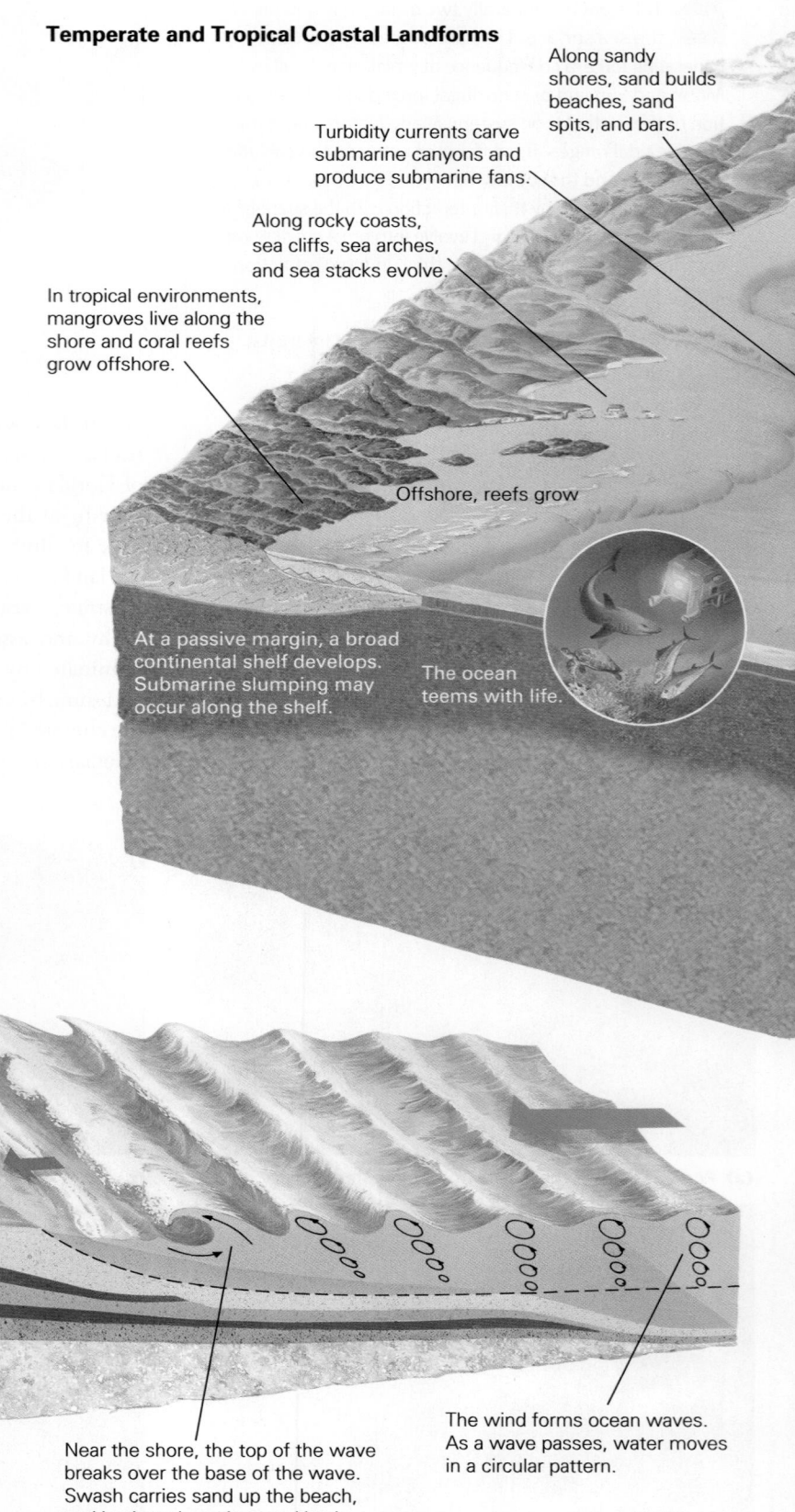

Along sandy shores, sand builds beaches, sand spits, and bars.

Turbidity currents carve submarine canyons and produce submarine fans.

Along rocky coasts, sea cliffs, sea arches, and sea stacks evolve.

In tropical environments, mangroves live along the shore and coral reefs grow offshore.

Offshore, reefs grow

At a passive margin, a broad continental shelf develops. Submarine slumping may occur along the shelf.

The ocean teems with life.

Waves and Beaches

Waves change as they approach the shore. They interact with sediment at a beach.

A lagoon, in which mud accumulates, may be trapped behind the dunes.

Sand may pile into dunes.

Beach

Near the shore, the top of the wave breaks over the base of the wave. Swash carries sand up the beach, and backwash carries sand back.

The wind forms ocean waves. As a wave passes, water moves in a circular pattern.

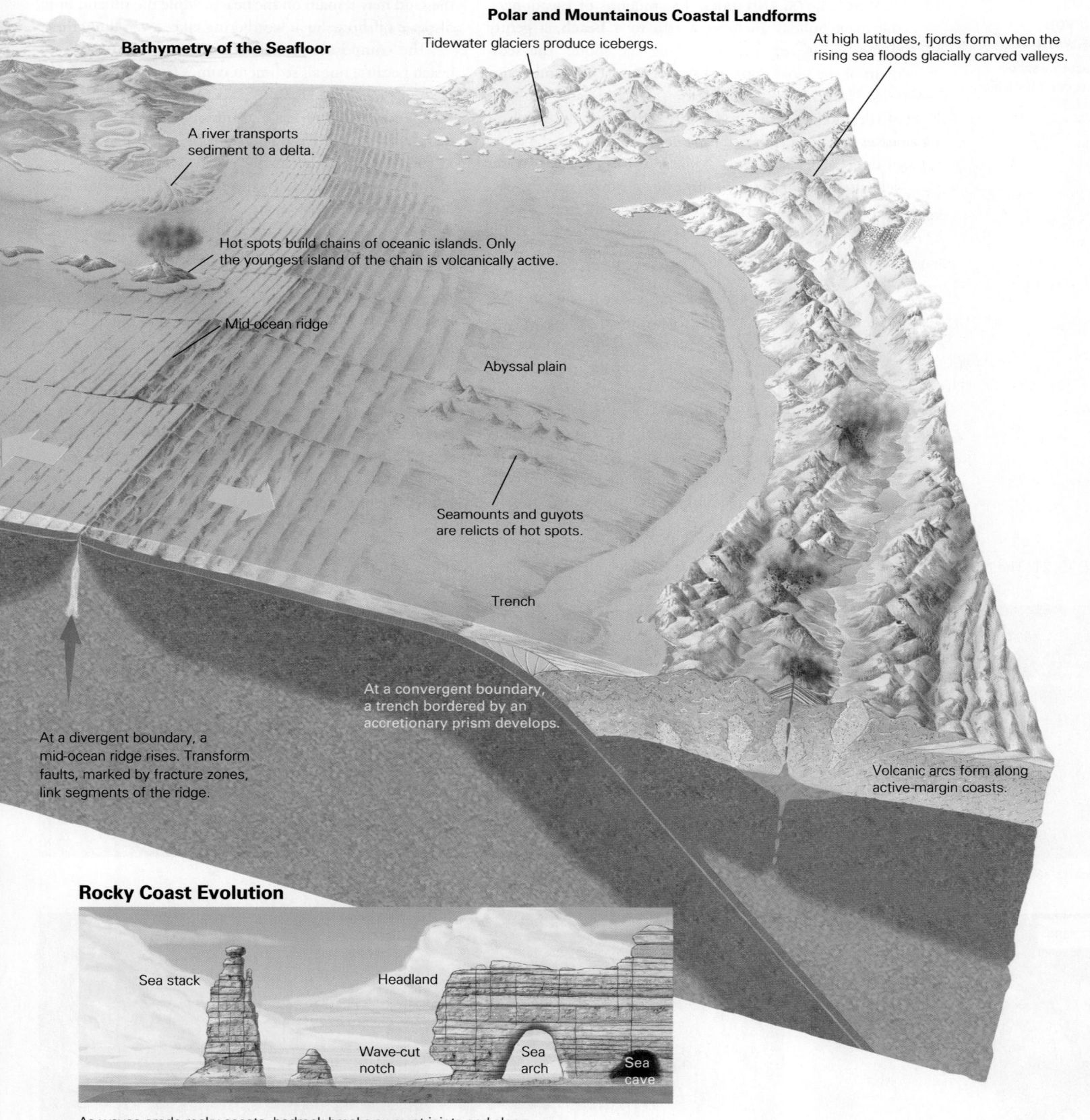

Bathymetry of the Seafloor

Polar and Mountainous Coastal Landforms

Tidewater glaciers produce icebergs.

At high latitudes, fjords form when the rising sea floods glacially carved valleys.

A river transports sediment to a delta.

Hot spots build chains of oceanic islands. Only the youngest island of the chain is volcanically active.

Mid-ocean ridge

Abyssal plain

Seamounts and guyots are relicts of hot spots.

Trench

At a convergent boundary, a trench bordered by an accretionary prism develops.

At a divergent boundary, a mid-ocean ridge rises. Transform faults, marked by fracture zones, link segments of the ridge.

Volcanic arcs form along active-margin coasts.

Rocky Coast Evolution

Sea stack

Headland

Wave-cut notch

Sea arch

Sea cave

As waves erode rocky coasts, bedrock breaks away at joints and along bedding planes. As a result, several distinct landforms evolve over time.

577

Sedimentary Coasts

BEACHES, SPITS, AND BARS. For millions of vacationers, the ideal holiday includes a trip to a **beach**, a gently sloping fringe of sediment along the shore. The sediment of the most popular beaches consists of sand grains (Fig. 15.21a). The accumulation of sand, and the lack of silt and mud, occurs for a reason. After the swash of waves on a beach has transported sediment up the slope of the beach, the backwash starts to carry it back. Water velocity in the backwash, however, is slower, so sand settles out while still on the beach, while silt and clay get carried offshore, where they settle in quieter water (see Fig. 15.17c, d). This winnowing leaves mostly sand on sandy beaches.

Not all beaches have a comfortable cover of sand, however. *Pebble* or *cobble beaches* (known also as *shingle beaches*) consist of coarse clasts (Fig. 15.21b). Such beaches develop where nearby rocky cliffs, deposits of glacial till, or the outlet of a debris-choked stream supply rock fragments. Resistant rocks roll around in the surf and bash into one another, so that eventually they end up being rounded and of roughly uniform size. Nonresistant rocks break into finer sediment that eventually gets washed offshore. If they are not replenished, the cobbles and pebbles will eventually weather to produce sand, silt, and mud; the sand may remain on the beach, while the silt and mud disperse offshore. Such weathering takes a very long time.

The composition of sediment varies from beach to beach because not all sediment comes from the same type of source rock (Fig. 15.21c). Sand derived from the weathering and erosion of felsic to intermediate rocks consists mainly of quartz, a durable mineral; the other minerals in such rocks weather to form clay, which washes away in the waves. Beach sediment produced by the erosion of limestone, coral reefs, or shells consists of carbonate sand, whereas beaches derived from the erosion of basalt boast black sand that consists of tiny basalt grains.

A beach can be divided into several distinct zones, as illustrated by a **beach profile** drawn perpendicular to the shore (Fig. 15.22). Starting from the sea and moving landward, the *foreshore zone* includes the intertidal zone, across which the tide rises and falls, as well as the *beach face*, a steeper, concave-up area that forms where the swash of the waves frequently scours the sand. The *backshore zone* extends from a small step, cut by high-tide swash, to the far edge of the beach farther inshore. The backshore zone includes one or more *berms*, horizontal to slightly landward-sloping terraces made up of sediment deposited

---

FIGURE 15.21 The type of sediment that makes up a beach varies with location and source-rock type.

(a) A quartz-sand beach in Puerto Rico.

(b) Cobble-sized clasts make up a beach in Oregon.

Quartz sand

Carbonate sand

Basalt sand

(c) Different kinds of sand make up different sandy beaches. Quartz sand comes from erosion of felsic rock, carbonate sand from fragmentation of shells, and black sand from basalt.

FIGURE 15.22 Interaction with waves, over time, produces a distinct beach profile, with several different zones.

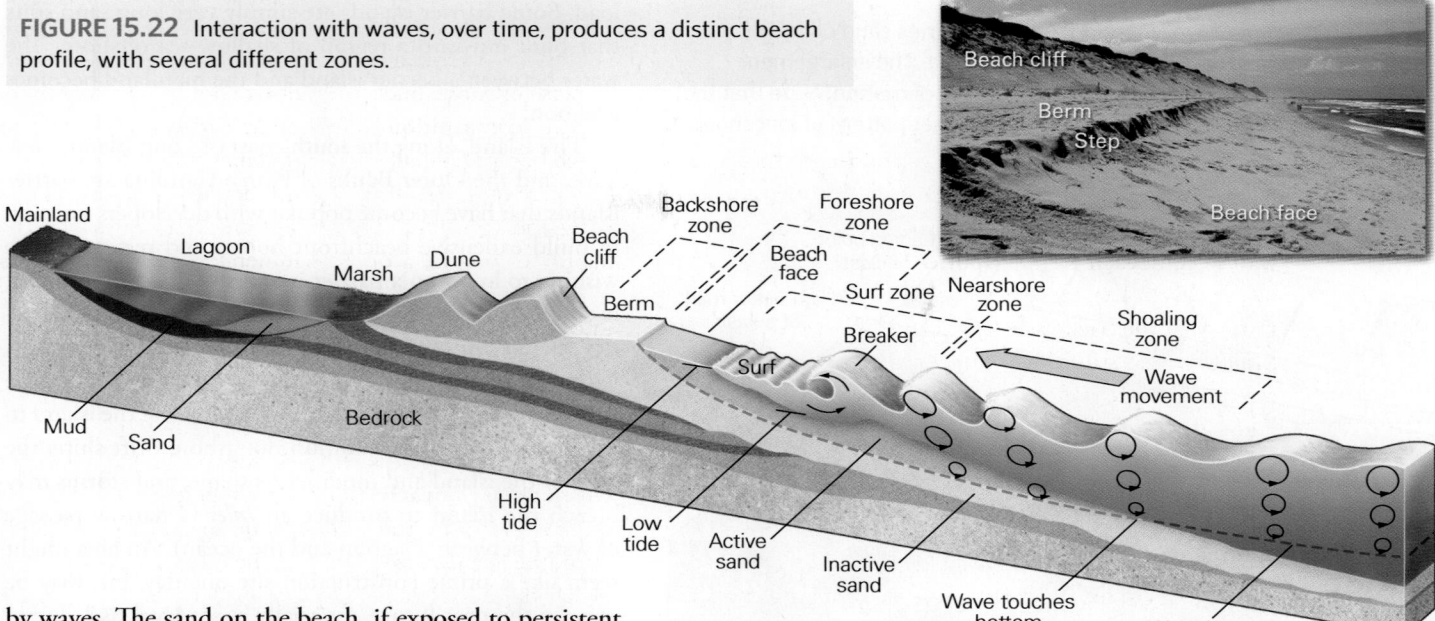

by waves. The sand on the beach, if exposed to persistent winds, may build into sand dunes, similar to those of deserts, along the landward edge of the beach. The inshore edges of some beaches, however, border cliffs or vegetated land not affected by waves. Behind the dunes, in some localities, a **lagoon**, a relatively quiet body of shallow seawater separated from the open ocean, develops. In temperate to tropical environments, the lagoon may host abundant vegetation.

Typical wave action moves an *active sand layer* of a sandy beach back and forth. (The underlying *inactive sand layer* moves only during severe storms or not at all.) Therefore, the width of a beach, as measured perpendicular to the shore, represents the area subjected to frequent wave action; put another way, beaches remain beaches only because wave action displaces their sediment frequently enough that vegetation cannot take root in it.

Beyond the limit of frequent wave action, in appropriate climates, beach grasses and shrubs cover much of the sediment. If it's not confined by sea cliffs, the width of a beach depends on the tidal range and on the *significant wave height* (the mean height of the highest third of waves; see Chapter 14) of the waves arriving at the beach.

Beach profiles may vary seasonally in response to variation in the strength of storms affecting a beach during different seasons. For example, in mid-latitudes, winter storms tend to be stronger and more frequent than summer ones. The larger, shorter-wavelength waves of winter storms wash beach sand into deeper water, making the beach narrower. But smaller summer waves with longer wavelengths bring sand in from offshore and deposit it on the beach **(Fig. 15.23)**.

FIGURE 15.23 The width of a beach can change seasonally.

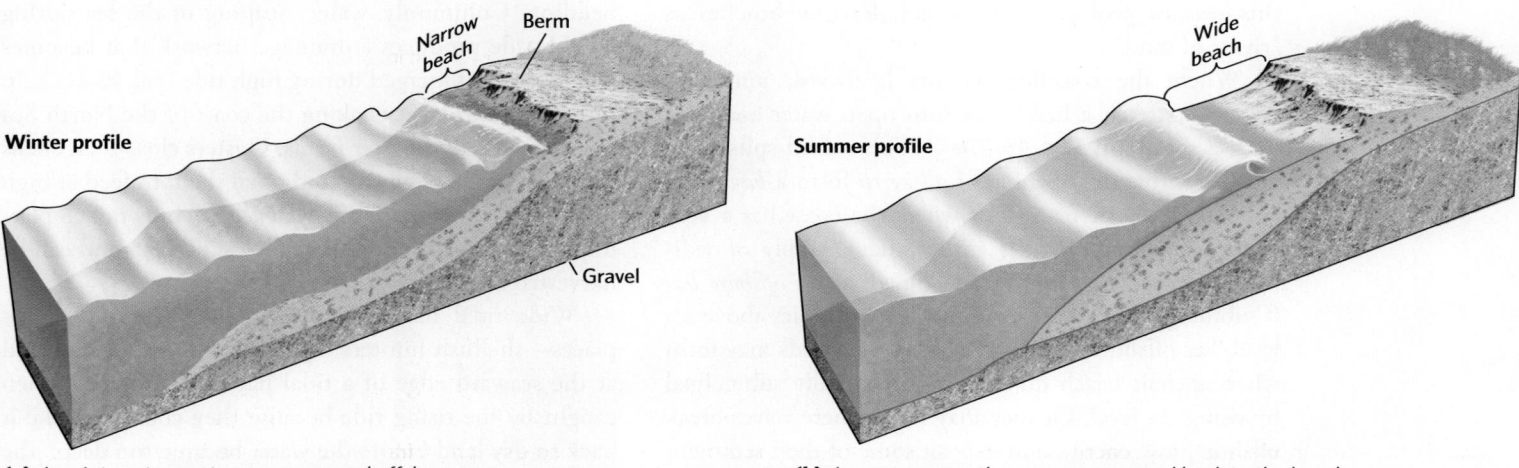

(a) In winter, strong storms move sand offshore.

(b) In summer, gentler waves carry sand back to the beach.

FIGURE 15.24 The concept of longshore drift. Swash carries sand obliquely up the beach, whereas backwash carries it straight downslope. The enlargement shows the motion of a representative clast after six cycles of motion. Note that the sand grains follow a blunted sawtooth pattern, yielding a net pattern of longshore drift (brown arrow).

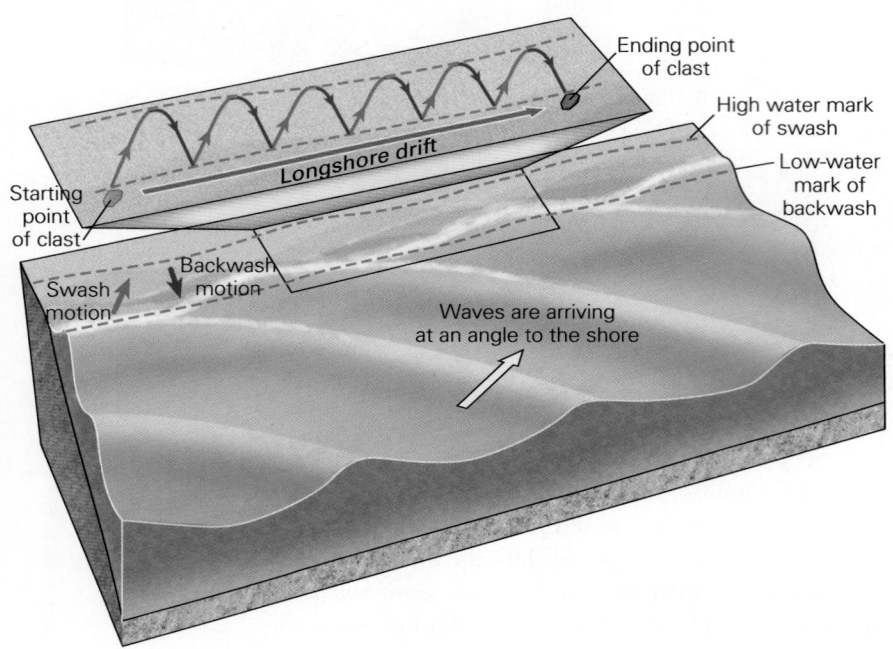

load. Some barrier islands are simply very long sand spits that built out into a region of shallow sea offshore. The water between a barrier island and the mainland becomes a lagoon.

Fire Island, along the south coast of Long Island, New York, and the Outer Banks of North Carolina are barrier islands that have become popular with developers wanting to build expensive beachfront homes and resorts. Those wishing to build on a barrier island should keep in mind, however, that on a time scale of decades to millennia, barrier islands are temporary geologic features. Commonly, wind and waves pick up sand from the ocean side of the island and drop it on the lagoon side, causing the island to migrate landward. In addition, longshore drift shifts the sand of the island and modifies its shape, and storms may breach the island to produce an *inlet* (a narrow passage of water between a lagoon and the ocean). An area might seem like a prime construction site one day, but may be gone in the near future, especially as sea level rises due to climate change.

TIDAL FLATS. The land of the Netherlands consists mostly of silt, sand, and clay deposited by four rivers to form a broad, nearly flat complex of deltas along the coast of the North Sea. Tidal range, in places, exceeds 6 m (20 ft), so at low tide the sea retreats by a few kilometers and the intertidal zone becomes a broad **tidal flat** exposed to the air. At high tide, the sea returns and submerges the tidal flat. Farms and towns have been built on portions of the delta complex, and remain dry only because dikes, built over the centuries, keep high tides out (Fig. 15.26a).

Tidal flats protected from strong wave action by barrier islands or sand spits typically consist of sticky mud and silt (Fig. 15.26b), while those washed over by waves may be fairly sandy. Tidal flats provide a home for burrowing organisms such as clams and worms, so *bioturbation*, the churning and burrowing of sediment by organisms, mixes tidal-flat sediments and disrupts bedding. Commonly, water escaping to the sea during an ebb tide produces a drainage network that becomes completely submerged during high tide (Fig. 15.26c). In recent years, tidal flats along the coast of the North Sea have been used as oyster farms. Oysters close their shells when exposed at low tide and open them to feed at high tide. In an oyster farm, the oysters grow on wooden platforms above the mud, so that they stay clean and can be harvested easily.

Wide tidal flats can, unfortunately, be dangerous places—shellfish hunters digging for clams in the mud at the seaward edge of a tidal flat have drowned when caught by the rising tide because they couldn't make it back to dry land before the water became too deep. The

Where waves roll onto the shore obliquely, they not only produce a longshore current offshore, but also cause **longshore drift** (or *beach drift*), which means that sediment in the active sand layer follows a sawtooth pattern of movement that yields gradual net transport of the sediment parallel to the beach. This sawtooth pattern happens because the swash of a wave moves perpendicular to the wave crest, so an oblique wave carries sediment diagonally up the beach; backwash, however, flows nearly straight down the slope of the beach face due to gravity. Over time, longshore drift may move sand tens to hundreds of kilometers along a coast (Fig. 15.24). For this reason, geologists sometimes describe beaches as "rivers of sand."

Where the coastline indents landward, longshore drift can stretch a beach out into open water and produce a **sand spit** (Fig. 15.25a–c). Some sand spits grow entirely across the opening of a bay to form a *baymouth bar* (Fig. 15.25d). In regions where the coast has a very gentle slope and there is an abundant supply of sediment, a narrow ridge of sand, known as an *offshore bar* if submerged, or a **barrier island** if its crest lies above sea level, lies offshore (Fig. 15.25e). Barrier islands may form when ancient beach dunes become partially submerged by rising sea level. Or they may form where waves break offshore, lose energy, and deposit some of their sediment

FIGURE 15.25 Sand spits, bars, and barrier islands.

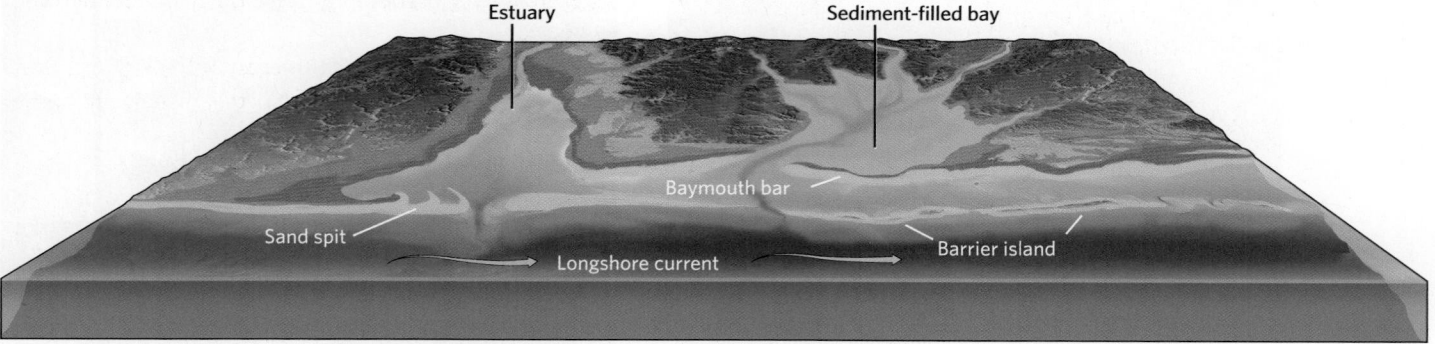

(a) Longshore drift can generate sand spits and baymouth bars. Sedimentation may fill in the region behind the bar.

(b) An example of a sand spit.

(c) A sand spit growing along the eastern Canadian coast.

(d) An example of a baymouth bar.

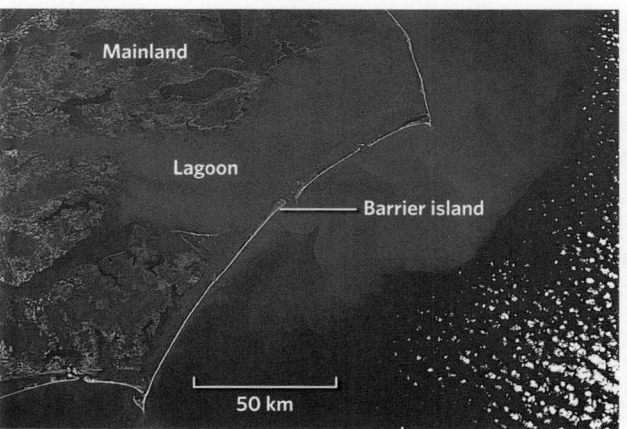

(e) Barrier islands may be separated from the mainland by a lagoon.

risk of being caught by high tide provided a way for monks living in the monastery of Mont-Saint-Michel, which was built on a small granite island protruding from the tidal flats of western France, to defend themselves from invaders, who risked being stuck in the mud or drowned by high tide if they tried to invade by foot or horse (Fig. 15.26d).

Rocky Coasts

More than one ship has met its end, smashed and splintered in the spray and thunderous surf, along a **rocky coast**, where bedrock cliffs rise directly from the sea (Fig. 15.27). Lacking the protection of a beach, rocky coasts feel the full impact of ocean breakers. The water pressure generated during the impact of a breaker can

FIGURE 15.26 Tidal flats.

(a) Broad tidal flats along the coast of the Netherlands.

(b) A small, very muddy tidal flat on the coast of France.

(c) Aerial view of a drainage network on a tidal flat near Boston, seen at low tide. Seaweed delineates some of the channel boundaries.

(d) Mont-Saint-Michel, a monastery on the coast of France, was built on an island surrounded by a tidal flat.

FIGURE 15.27 Rocky coasts, such as this one on the south coast of England, develop unique landforms as a result of wave erosion.

pick up boulders and smash them together until they shatter, and it can squeeze air into cracks, creating enough pressure to pluck chunks free from bedrock. Further, because of its turbulence, the water hitting a cliff face carries suspended sand that can abrade the cliff. The combined effects of shattering, plucking, salt and ice wedging, boring by animals, and abrasion by wave-carried sand together produce **wave erosion**.

If wave erosion attacks the base of a cliff face, it can produce a *wave-cut notch* (Fig. 15.28a, b). Undercutting continues until the overhanging rock becomes unstable and breaks away at a joint, toppling down to form a pile of rubble at the base of the cliff (Fig. 15.28c). Over time, wave action breaks up the rock until fragments become small enough to be transported offshore by moving water. In this way, wave erosion continually eats away at a rocky coast, so that the cliff gradually migrates inland. Such *cliff retreat* leaves behind a *wave-cut bench*, or *wave-cut platform*, that becomes visible at low tide (Fig. 15.28d).

Typically, rocky coasts have *headlands* that protrude into the sea and *embayments* that are set back from the sea (Fig. 15.29). Such irregular coastlines evolve over time

FIGURE 15.28 Wave erosion of a rocky coast.

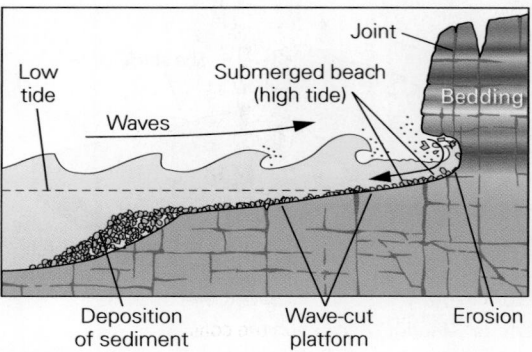

(a) Formation of a wave-cut notch.

(b) This wave-cut notch has developed in sedimentary strata.

(c) At this locality in Norfolk, England, the red siltstone is weaker than the white limestone. When wave-cut notches get deep enough, the limestone beds collapse and form rubble piles.

(d) A wave-cut bench exposed at low tide at Étretat, on the coast of France.

FIGURE 15.29 Wave refraction at headlands.

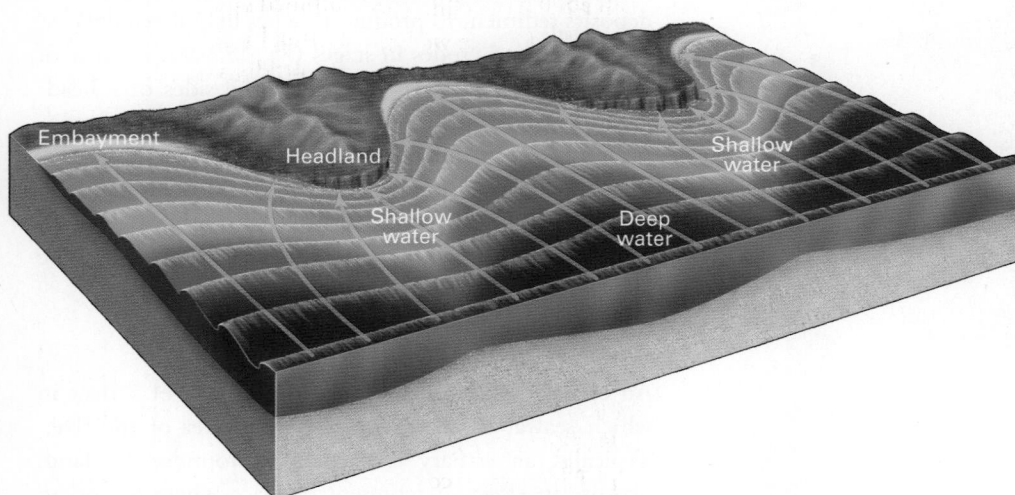

(a) Like a lens, wave refraction focuses wave energy on a headland, so erosion occurs there. Wave energy weakens in embayments, so deposition occurs there.

(b) An example of waves eroding a headland on the coast of Hawai'i.

FIGURE 15.30 Erosion of a headland.

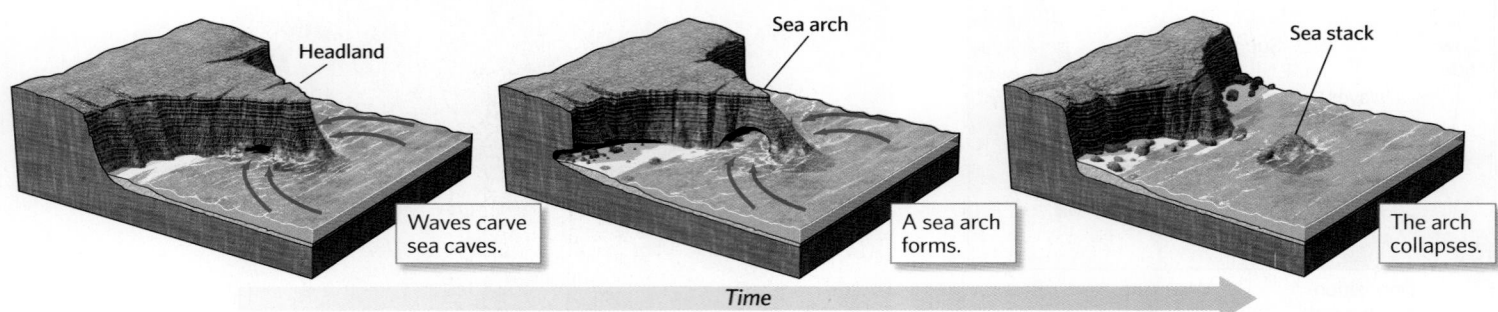

Headland · Sea arch · Sea stack

Waves carve sea caves. · A sea arch forms. · The arch collapses.

Time

(a) Stages in the erosion of a headland.

(b) Sea arches (top) eventually collapse, leaving sea stacks (bottom) on the southern coast of Australia.

(c) A sea arch, next to an embayment, on the south coast of England.

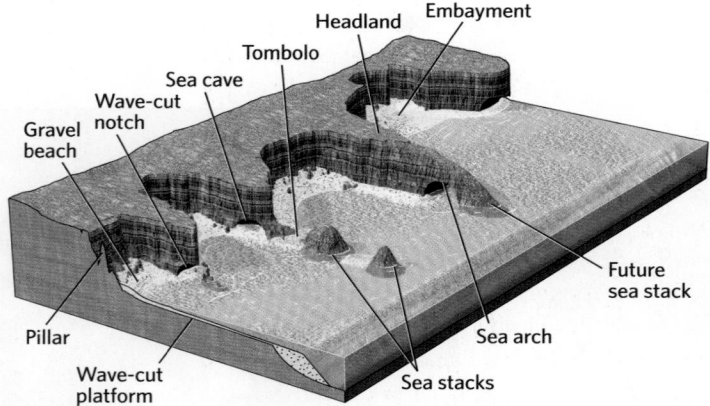

Headland · Embayment · Tombolo · Sea cave · Wave-cut notch · Gravel beach · Future sea stack · Pillar · Sea arch · Wave-cut platform · Sea stacks

(d) A summary of landforms along a rocky shore.

because, due to wave refraction, wave energy focuses on headlands and disperses in embayments. The resulting pattern of wave erosion removes debris at headlands and deposits sediment to produce beaches in embayments.

A headland erodes in stages (**Fig. 15.30a**). Because of refraction, waves curve and attack the sides of a headland, slowly eating through it to create a *sea arch* connected to the mainland by a narrow bridge. Eventually, the arch collapses, leaving isolated *sea stacks* just offshore (**Fig. 15.30b, c**). Sea stacks eventually crumble and disappear. At any given time, all of these landforms may be visible along a rocky coast (**Fig. 5.30d**).

Estuaries and Fjords

An **estuary** is a coastal area at the mouth of a river in which seawater mixes with the freshwater of the river. Typically, an estuary is partially surrounded by land, because in effect, it represents a place where the ocean has partially drowned a river and its valley. An estuary

FIGURE 15.31 River valleys that are flooded by sea-level rise are called estuaries.

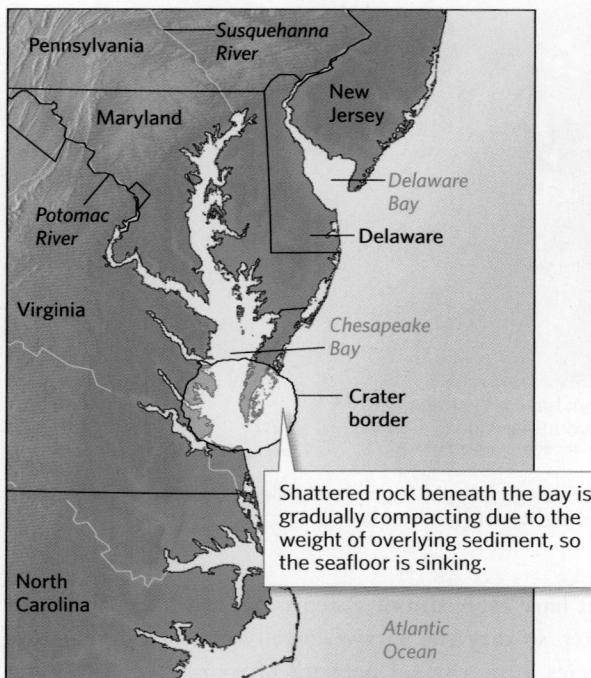

Shattered rock beneath the bay is gradually compacting due to the weight of overlying sediment, so the seafloor is sinking.

(a) Chesapeake Bay is an estuary at the mouth of the Susquehanna and Potomac Rivers. Recent research suggests that the region became an estuary because the land beneath has been undergoing extra subsidence. This subsidence is occurring because a large meteorite crater, filled with breccia, underlies the floor of the estuary. The breccia within the crater has compacted more than has the sediment elsewhere along the Atlantic coast.

(b) The land surrounding the main channel of an estuary floods during high tide along the coast of the Isle of Wight, England.

develops where a river cut a valley and floodplain at a time when sea level was lower, or where the region containing a river and its valley is slowly subsiding. You may be able to recognize an estuary on a map by the branching pattern of its river-carved coastline, inherited from the shape of a dendritic drainage network (**Fig. 15.31a**). The mouths of estuaries are commonly bordered by marshes that flood at high tide (**Fig. 15.31b**).

Seawater and river water can interact in two different ways within an estuary. In quiet estuaries that are protected from wave action or river turbulence, the water becomes stratified, so during a flood tide, relatively dense seawater flows upstream as a *saltwater wedge* beneath the less dense freshwater. As the tide rises, the water surface of the estuary rises, so what may be a flowing river at low tide becomes submerged at high tide. In places where the river flows at or near its base level for a significant distance upstream, a saltwater wedge can migrate upstream for long distances. For example, a saltwater wedge can be detected about 100 km (62 mi) up the Hudson River from New York City, and another can be detected about 40 km

(25 mi) up the Columbia River in Oregon. In turbulent estuaries such as Chesapeake Bay, which is wide enough to be stirred by storms, seawater and river water mix to produce brackish water. Estuaries typically host complex ecosystems inhabited by unique species of shrimp, clams, oysters, worms, and fish that can tolerate large changes in salinity.

During the Pleistocene Ice Age, glaciers carved deep valleys in coastal mountain ranges. When the last glaciation came to an end, the glaciers melted away, leaving deep, U-shaped valleys (see Section 13.5). As the water stored in the glaciers returned to the sea and sea level rose, it flooded glacial valleys to produce marine **fjords**, narrow bays of the sea bordered by steeply sloping walls. Because of their deep-blue water and steep walls of polished rock, fjords are distinctively beautiful (**Fig. 15.32**). Their steep walls are relics of glacial erosion. Ribbon-like waterfalls locally spill into fjords from hanging valleys along the side of the main valley.

Organic Coasts

So far, we've focused on coasts where the distribution of sediments and rocks primarily controls the landforms at or near the shore. Along many coasts, however, living organisms significantly modify nearshore regions. Such **organic coasts** include coastal wetlands and coral reefs. Since the character of an organic coast depends on the types of organisms that live there, their growth and survival depend on climate.

COASTAL WETLANDS. It would be hard to find a more dramatic contrast to the crashing waves of rocky coasts than the gentlest type of nearshore environment, a **coastal wetland**, which consists of a vegetated, flat-lying

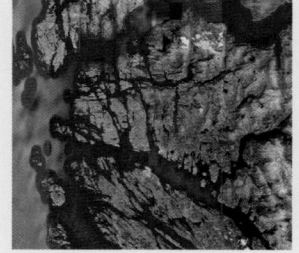

FIGURE 15.32 Marine fjords decorate formerly glaciated coasts.

A fjord in Norway

(a) Fjords are so beautiful that they are popular destinations for tourist cruises.

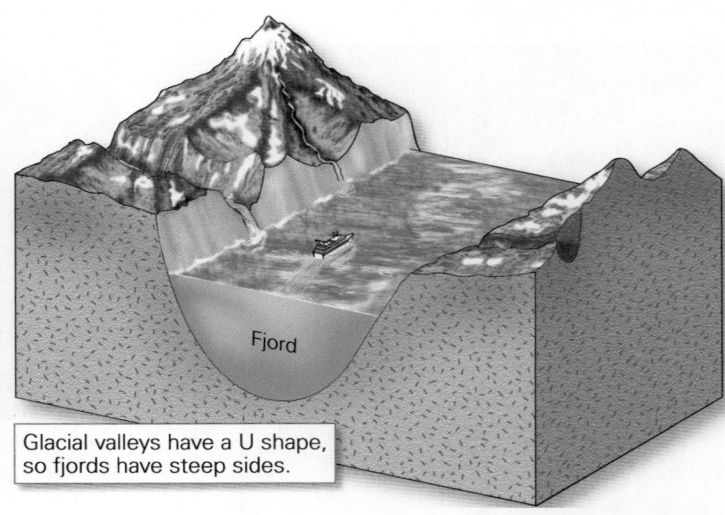

Glacial valleys have a U shape, so fjords have steep sides.

Fjord

(b) Fjords are submerged, glacially carved valleys. Their steep walls and great depth are relics of glacial erosion.

region of coast that floods at high tide, becomes partially exposed at low tide, and does not feel the impact of strong waves. Wetlands typically grow in tidal flats of lagoons or estuaries that are well protected from wave action. Vast numbers of marine species spawn in wetlands. In fact, wetlands account for 10%–30% of all marine organic productivity, even though they constitute just a tiny fraction of the oceans.

In mid-latitude climates, coastal wetlands include *swamps* (wetlands dominated by trees), *marshes* (wetlands dominated by grasses) (Fig. 15.33a), and *bogs* (wetlands dominated by mosses and shrubs). In tropical or subtropical climates (between 30° N and 30° S), *mangrove swamps* thrive along the shore (Fig. 15.33b). Mangroves are trees that have evolved root systems that can filter salt out of water, so they can survive in saltwater. Some mangrove species form a broad network of roots above the water surface, making the tree look like an octopus standing on its tentacles, and some send up small protrusions from the roots that rise above the water and allow the plant to breathe. Dense stands of mangroves absorb the impact of waves and, by doing so, prevent coastal erosion.

CORAL REEFS. Along the azure coasts of Hawai'i, mounds of coral form shallow reefs just offshore (Fig. 15.34a). Snorkelers swimming over such a reef may see a great variety of different living coral species. Some species build mounds that look like brains, others like elk antlers, and

FIGURE 15.33 Coastal wetlands.

(a) A marsh along the coast of Cape Cod. The tide is low, exposing the mud in which the grass grows.

(b) A mangrove growing in a swamp along the shore of southern Florida.

FIGURE 15.34 The character of coral reefs.

(a) A reef off the shore of Honolulu, Hawai'i.

(b) A healthy reef as seen underwater; corals have a great variety of colors and shapes.

(c) A close-up of living coral polyps.

FIGURE 15.35 What's inside a reef?

(a) A limestone quarry in Florida cuts into fossil coral that grew thousands of years ago. (The vertical ridges on the outcrop face are drillholes made during the quarrying operation.)

(b) A close-up shows the internal structure of a long-dead coral mound, now part of a limestone bed.

still others like delicate fans (Fig. 15.34b). Sea anemones, sponges, clams, and many other organisms grow on and around the coral. Though at first glance, living coral looks like a plant, it is actually a colony of tiny invertebrates related to jellyfish. An individual coral animal, or *polyp*, has a tube-like body with a head of tentacles (Fig. 15.34c). Corals obtain some of their nutrition by filtering nutrients from seawater; the remainder comes from algae that live within the corals' tissues. Corals have a *symbiotic*

(mutually beneficial) relationship with these algae, in that the photosynthetic algae provide nutrients and oxygen to the corals while the corals provide carbon dioxide and other nutrients to the algae.

Coral polyps secrete calcite shells, which gradually build into a mound of solid limestone. At any given time, only the surface of the mound consists of living corals; the mound's interior consists of shells from previous generations of corals, as we can see in quarries that have been cut into fossil reefs (Fig. 15.35). The realm of shallow water underlain by coral mounds, associated organisms, and debris constitutes a **coral reef**. Living corals must remain submerged, so the tops of the shallowest reefs lie just below the level of low tide, and they must receive abundant sunlight, so most growing reefs lie at depths of less than 60 m (200 ft), the upper part of the photic zone. Coral reefs absorb wave energy and, like mangrove swamps, serve as a living buffer that protects coasts from erosion.

Corals need clear, warm (18°C–30°C, or 64°F–86°F) water with normal ocean salinity, so most coral reefs grow along unpolluted coasts at latitudes below about 30° (Fig. 15.36a). The largest reef in the world, the Great Barrier Reef, extends for close to 2,000 km (1,200 mi) along the northeastern coast of Australia and reaches a width of 120 km (75 mi) (Fig. 15.36b). Most reefs are much smaller.

Marine geologists distinguish three different shapes of coral reefs (Fig. 15.37). A *fringing reef* forms directly along the coast, a *barrier reef* lies offshore (separated from the coast by a lagoon), and an *atoll* forms a circular ring surrounding a lagoon. As Charles Darwin first recognized

See for yourself

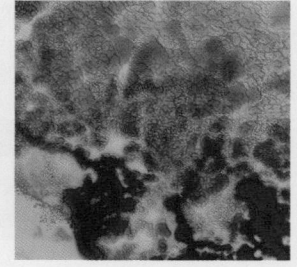

Coral Reefs, New Caledonia

Latitude: 21°43'51.51" S
Longitude: 165°33'14.17" E

Zoom to an altitude of 3 km (2 mi) and look down.

Coral buildups lie beneath the surface of this lagoon, giving the shallow seafloor a complex texture. A barrier island lies about 7 km (4 mi) offshore.

FIGURE 15.36 Reefs of the world.

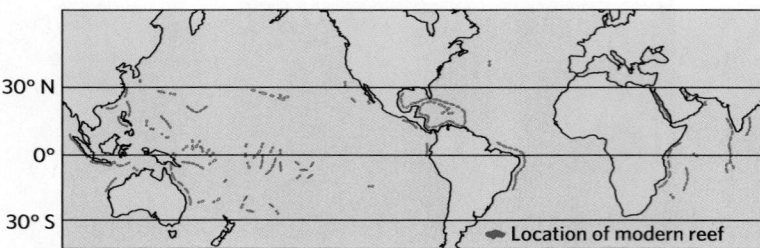

(a) Most coral reefs lie between 30° N and 30° S latitude.

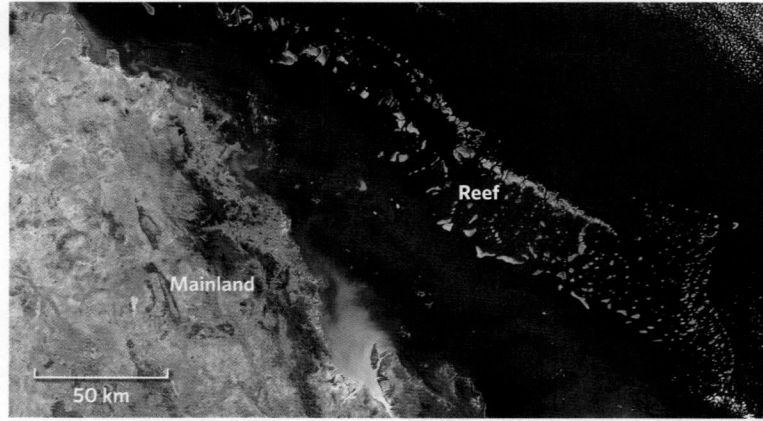

(b) A satellite image of the Great Barrier Reef, Australia. Extensive reefs such as this one are hazards to ships.

back in 1859, coral reefs associated with oceanic islands in the Pacific start out as fringing reefs, then later become barrier reefs and, finally, atolls. Darwin suggested, correctly, that this progression reflects the continued growth of the reef as the island around which it formed gradually subsides. Eventually, the reef itself sinks too far below sea level to remain alive, and the limestone that remains becomes the flat cap of a guyot.

Take-home message...

In regions with ample sediment, beaches develop, each with a distinctive profile. Longshore drift can produce sand spits. In areas protected from wave action, tidal flats develop. Rocky coasts evolve over time due to wave erosion. Where the sea floods the mouth of a river valley, an estuary develops, and where the sea floods a glacially carved valley, a fjord results. Living organisms modify landforms along organic coasts; some form coastal wetlands. Coral reefs, which grow primarily in warm, clear water, consist of generations of coral shells.

Quick Questions

- What is the difference between a berm and a beach face?
- How does a wave-cut bench develop in bedrock?
- Is coral more closely related to a tree or to an octopus?

FIGURE 15.37 Evolution of reefs around oceanic islands.

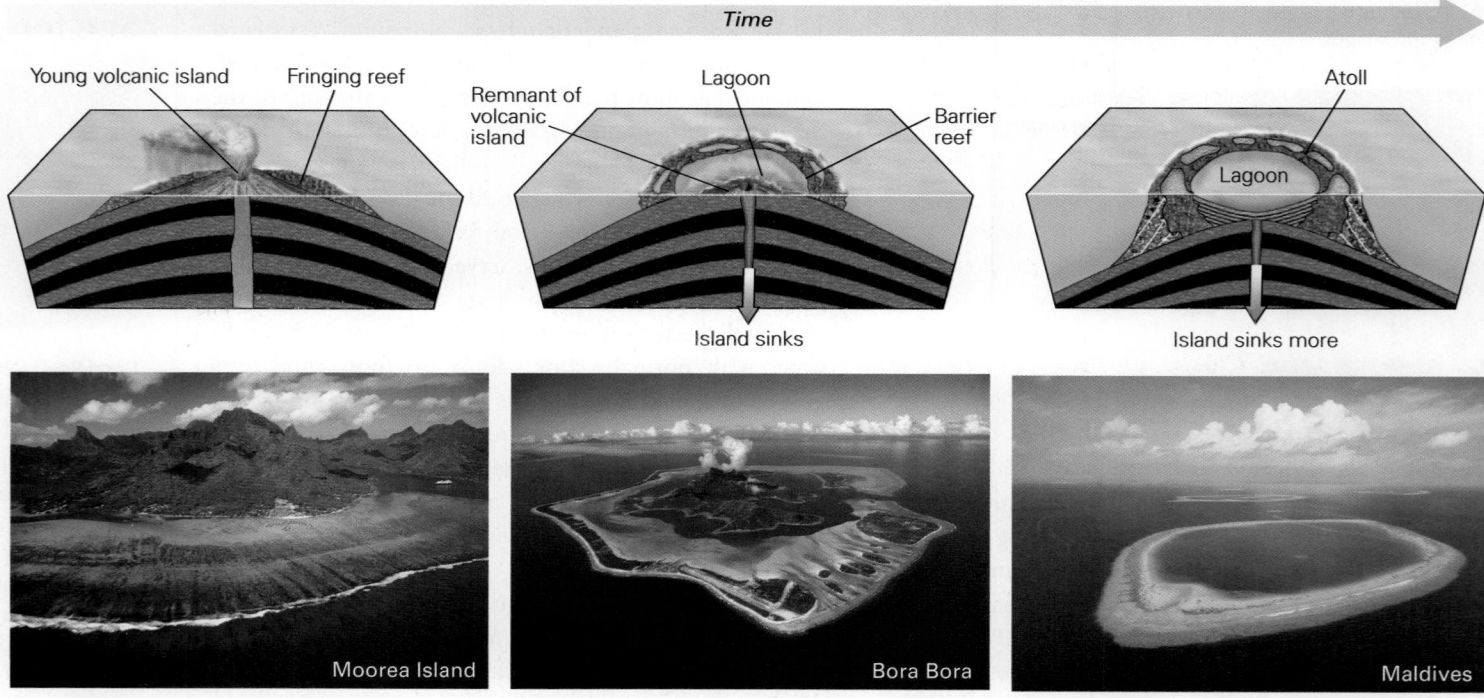

(a) A fringing reef forms around an island.   **(b)** As the island subsides, a barrier reef remains.   **(c)** This island has sunk below sea level. An atoll remains.

FIGURE 15.38 Features of emergent and submergent coasts.

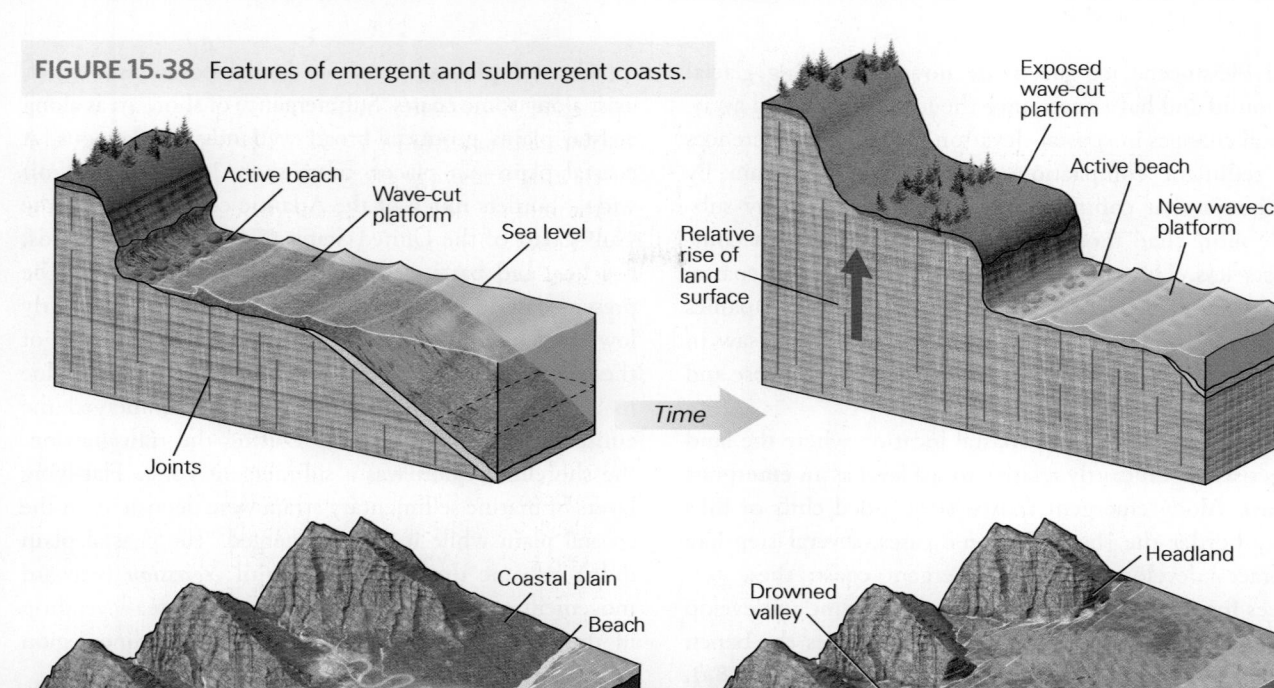

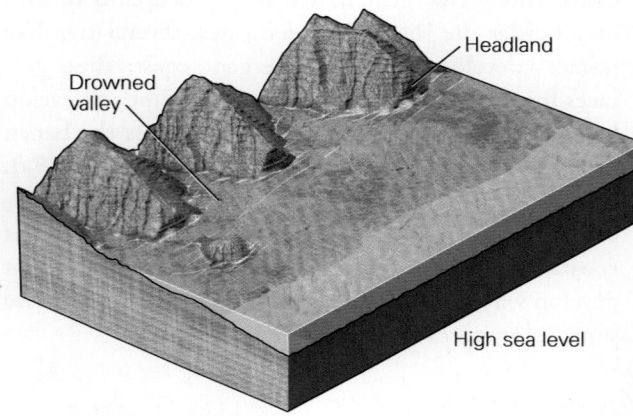

Time

(a) Wave erosion produces a wave-cut bench along an emergent coast. As the land rises, the bench becomes a terrace, and a new wave-cut bench develops.

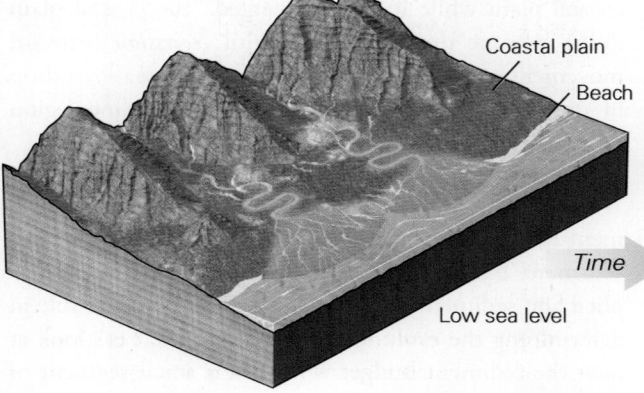

Time

(b) Rivers drain valleys and deposit sediment on a coastal plain along a submergent coast. As the land sinks, sea level rises and floods the valleys, and waves erode the headlands.

15.4 Changes and Challenges along Coasts

People tend to view the location and character of a shoreline as permanent geologic features. But, as we have already seen in this chapter, they are not. Shorelines, like many geologic features, are ephemeral—they can change on a time scale of hours to millennia. Let's first review the geologic causes of these changes, and then we'll examine how people try to mitigate ongoing changes.

Long-Term Causes of Coastal Variation

SEA-LEVEL CHANGE. Changes in **relative sea level**, meaning the position of sea level relative to the land surface at a given location, can result either from *global sea-level change* (the rise or fall of sea level worldwide; see Chapter 9) or from vertical movement (uplift or subsidence) of land areas. Vertical movement of the land can happen for many reasons. In some cases, it reflects tectonic activity due to plate interactions. For example, in locations where the San Andreas transform boundary in California is not oriented exactly parallel to the movement of the Pacific Plate relative to the North American Plate, some crustal shortening takes place

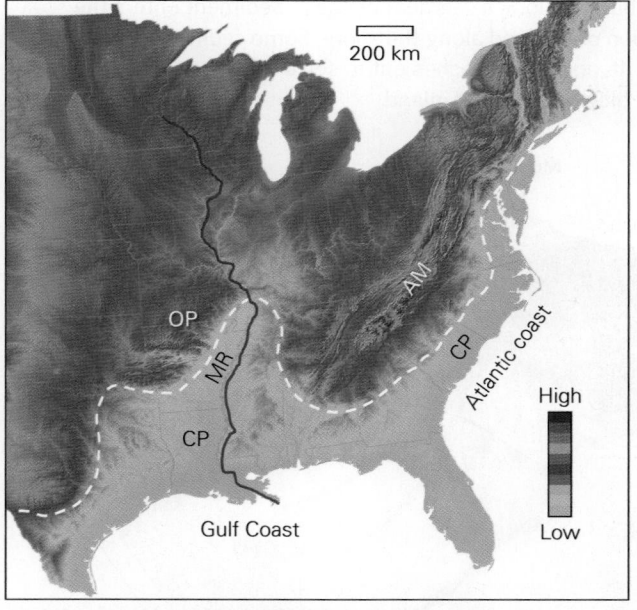

(c) The coastal plain of the Atlantic and Gulf coast region of the United States. (CP = coastal plain; MR = Mississippi River; OP = Ozark Plateau; AM = Appalachian Mountains.)

across the boundary, and these strains result in uplift of the California coastline. Vertical movement may also reflect the addition or removal of a load placed on the surface of the lithosphere. For example, regions of eastern North America that subsided due to the growth of the Pleistocene ice sheets are now undergoing glacial rebound and have risen since the ice sheets melted away. Local changes in coastal elevation may reflect differences in sediment compaction: coastal areas underlain by sediment that compacts more during burial may subside more than those underlain by sediment that compacts less. Human activity also affects vertical coastal movement in places where communities or companies are pumping out groundwater or oil, for as we saw in Chapter 12, such pumping can lead to pore collapse and the subsidence of overlying land.

Geologists refer to a coastal location where the land has risen significantly relative to sea level as an **emergent coast**. Along emergent coasts, steep-sided cliffs or hills may border the shore. In some cases, several step-like terraces develop along an emergent coast; these terraces form where a wave-cut bench has time to develop before uplift or a change in sea level causes the bench to rise higher than the high-tide level (Fig. 15.38a). Coasts where the land undergoes significant subsidence relative to sea level are known as **submergent coasts** (Fig. 15.38b). Estuaries and fjords, landforms that develop when the sea floods coastal valleys, characterize some submergent coasts.

Coastal plains, nearshore landscapes of low relief, exist along some coasts. Submergence of shore areas along coastal plains produces broad wetlands and lagoons. A coastal plain—in places, more than 200 km (125 mi) wide—borders much of the Atlantic coast and all of the Gulf Coast of the United States (Fig. 15.38c). Wetlands, beaches, and barrier islands occur along its shore. The present-day landscape of this coastal plain has particularly low relief because a *transgression* (landward migration of the shoreline), which began in the Mid-Cretaceous due to a global rise in sea level, eventually submerged the entire plain (see Chapter 5). During the transgression, the shoreline region was a submergent coast. Flat-lying layers of marine sedimentary strata were deposited on the coastal plain while it was submerged. The coastal plain didn't become dry land again until *regression* (seaward movement of the shoreline) accompanied a sea-level drop in the Eocene. During regression, the shoreline region was an emergent coast.

THE COASTAL SEDIMENT BUDGET. Because of the movement of sediment on beaches and in adjacent areas, the **sediment budget**—the balance between sediment supplied and sediment removed—plays an important role in determining the evolution of a coastal area. Let's look at how the sediment budget works for a small segment of beach (Fig. 15.39). Sand may be added to the segment by local rivers. It may also be brought from just offshore by waves and wind, or from far away by longshore drift. (In fact, the large quantity of sand along the beaches of the southeastern United States may have been carried there by longshore drift from the outwash plains and moraines of Pleistocene glaciers in coastal New England.) Conversely, sand may be removed from the segment by longshore drift, be carried offshore by backwash or rip currents, or be blown away by wind. Sediment carried offshore settles in deeper water below the wave base; some of this sediment avalanches down the continental slope in turbidity currents. If new sediment does not replace lost sediment, the beach segment grows narrower. If, however, the supply of sediment exceeds the amount that washes away, the beach becomes wider. **Erosional coasts** form where wave erosion washes sediment away faster than it can be supplied; such coasts recede landward and may become rocky if there's insufficient sand to supply a beach. In contrast, **accretionary coasts**, those that receive more sediment than they lose by erosion, grow seaward and develop broad beaches and associated landforms.

Climate plays a key role in controlling the sediment budget of a coast. Coasts that occur in climates characterized by generally calm weather erode less rapidly than those frequently subjected to ravaging storms. A sediment supply that could generate an accretionary coast in a calm

FIGURE 15.39 The sediment budget along a segment of coast. Sediment enters the system from rivers and by erosion of the land along the shore. Some sediment is moved along the coast by longshore drift, and some washes out to sea or avalanches down submarine canyons. Wind may blow beach sand inland.

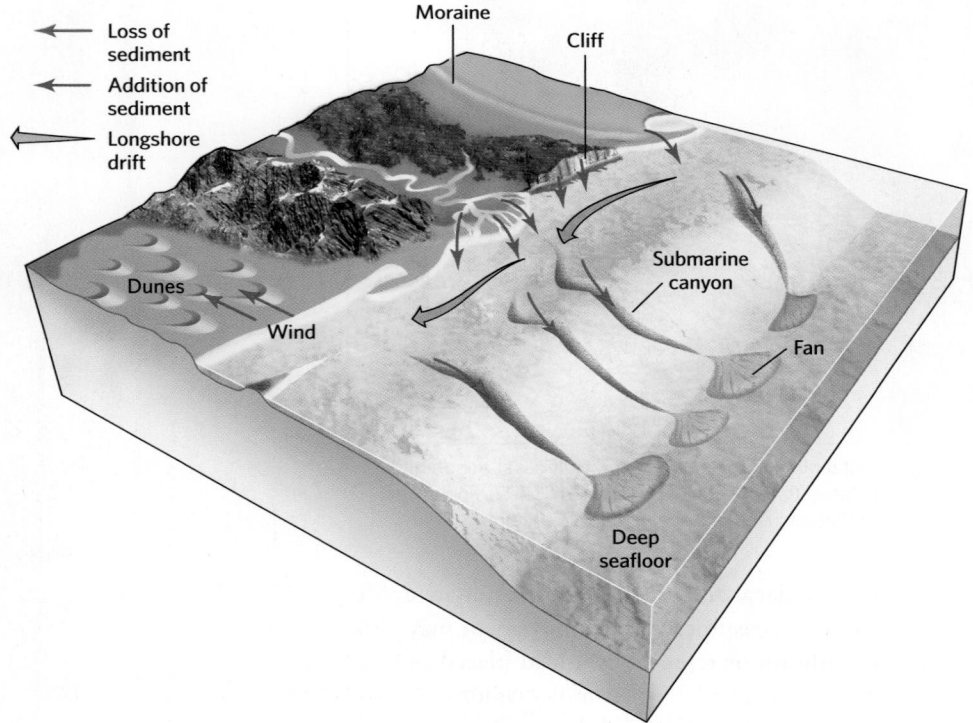

environment might be insufficient to prevent the development of an erosional coast in a stormy environment. Similarly, if climate change increases the discharge of rivers bringing sediment to the coast, an erosional coast could transform into an accretionary coast.

Coastal Changes in a Human Time Frame

THE CONSEQUENCE OF COASTAL STORMS. When the winds of large storms, such as hurricanes, blow across the surface of the sea, they generate immense waves. When these waves reach the shore, they become giant breakers that cause intense erosion and can destroy buildings, ships, and ports. **Storm surge**, the bulge of water that builds in front of a strong storm (see Chapter 18), increases the width of the area affected by storm waves, as measured perpendicular to the shoreline (Fig. 15.40a). The high water of a surge can overwhelm barrier islands and can inundate land far inland of a beach, allowing waves to

batter areas normally far from the shoreline. In a matter of hours, storm surge and waves together can radically transform a beach that took centuries or millennia to build into a narrow skeleton of its former self. Storm waves and wind also rip out mangrove swamps, scour wetlands or bury them in sand, and break up coral reefs, thereby destroying organic buffers that protects a coastal region. The affected region, consequently, becomes vulnerable to accelerated erosion for years to come.

Major storms also destroy human constructions: erosion undermines seaside buildings, causing them to collapse into the sea; wave impacts smash them to bits; and storm surge floats them off foundations (Fig. 15.40b, c). Storm waves also undermine sea cliffs and trigger landslides (Fig. 15.40d). Because storms can be immensely destructive, coastal population centers need safety and evacuation plans, which should ensure that residents know what to do when storms approach.

FIGURE 15.40 Coastal destruction due to storms.

(a) Waves and storm surge submerge and batter a coast.

(b) A community in Galveston, Texas, before Hurricane Ike.

(c) The same location, after destruction by storm surge and storm waves.

(d) A sea cliff composed of weak sediment has slumped into the sea, taking houses with it.

FIGURE 15.41 Present and future submergence of coasts.

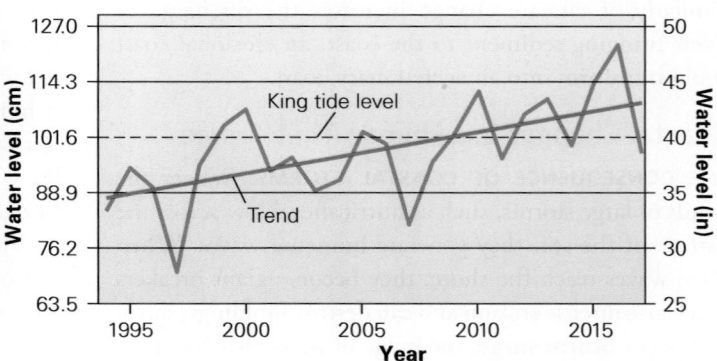

(a) Nuisance tides in San Juan, Puerto Rico. The water submerging the street is saltwater, which advanced overland from a nearby lagoon.

(b) Due to sea-level rise, nuisance tides are becoming more common.

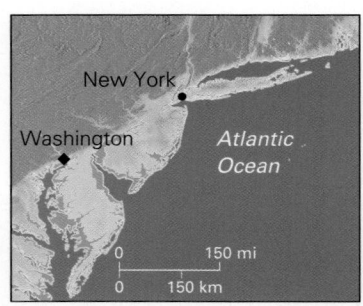

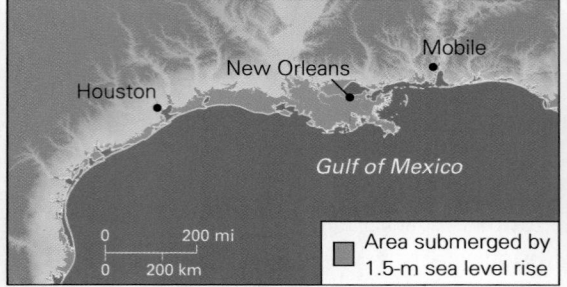

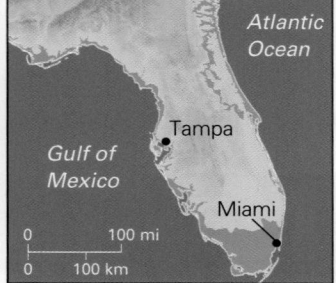

Area submerged by 1.5-m sea level rise

(c) Areas of the United States coast that may be submerged if sea level rises by just 1.0–1.5 m.

ONGOING SEA-LEVEL RISE. Global sea level is rising today at an estimated rate of about 3.2 mm/yr (0.13 in/yr), as we discuss further in Chapter 19. Rising seas, coupled with land subsidence due to pumping of groundwater and oil, have already led to the submergence of large areas of coastal Louisiana. An estimated 5,200 km² (2,000 mi²) of Louisiana land has vanished since 1930. Already, sea-level rise has led some coastal cities to flood not only during storm surges, but also during times of particularly high king tides. Such king tides have become known as **nuisance tides** (Fig. 15.41a). During a nuisance tide, saltwater fills basements, covers roads, and damages plants (such as grass) that do not have salt tolerance. And over the years, saltwater corrodes construction materials (such as reinforced concrete), making seaside buildings potentially unsafe. Due to sea-level rise, the frequency of nuisance tides has been increasing (Fig. 15.41b). For example, in Norfolk, Virginia, there were generally fewer than 5 nuisance tides per year before 1980, but since 2015 there have been 10–15 per year. Extreme nuisance tides happen so frequently on some oceanic islands in the Pacific that the islands have become uninhabitable. Many of the world's large cities have been built on coastal land less than a few meters above sea level, and are therefore very vulnerable to the consequences of sea-level rise. On a time scale of centuries, the shoreline of the United States will change significantly (Fig. 15.41c).

How will society deal with coastal flooding due to this ongoing sea-level rise? Solutions remain a work in progress. Some options, all of which would come at great cost, include moving towns inland, adding fill to make the land surface higher, constructing barriers (long dikes or seawalls) against the sea, installing gates that close off harbors during high tide, and installing pumps and drainage canals to return rising waters to the sea. New York City, for example, is exploring the cost of building 4 m (12 ft) high concrete floodwalls along the shoreline to prevent storm surge from flooding streets, buildings, and subways, as happened during Hurricane Sandy in 2012. A rough estimate of the cost of the walls is about $53 billion. (By comparison, Hurricane Sandy cost the city about $23 billion, in 2023 dollars.)

Beach Destruction and Beach Protection

As we've seen, storms can cause major changes to beaches overnight. But even less dramatic events, such as the loss of river sediment, a gradual rise in sea level, or the

(a) Along the shore at Palm Beach, Florida, all the sand has been removed, and a ledge of rock has been exposed. Waves are undercutting the rock layer, which is now collapsing.

(b) On the coast of Cape Cod, wave erosion has removed the entire beach right up to the coastal dunes.

FIGURE 15.42 Examples of beach erosion.

destruction of coastal vegetation, can alter the sediment budget of a beach. If the sediment supply decreases, erosion eventually removes so much sediment that the beach becomes much narrower—a process known as **beach retreat**. Sometimes beach erosion exposes underlying bedrock (Fig. 15.42a). If erosion attacks cliffs at the landward edge of a beach, cliff retreat takes place (Fig. 15.42b). Surveys indicate that beach retreat and cliff retreat at some locations take place at rates of 1–2 m/yr (3–6 ft/yr). Some building owners have responded by literally picking up and towing their buildings inland.

Because of the value of beachfront property, many communities try to prevent beach and cliff retreat. To do this, they construct artificial barriers to protect their stretch of coast or to shelter the mouth of a harbor from waves. These barriers alter the natural movement of sand and can change the shape of a beach, not always with desirable results.

- **Groins** are concrete or stone walls built perpendicular to the shore. They impede longshore drift, so sand accumulates on the updrift side of the groin, forming a triangular wedge, but erodes away on the downdrift side (Fig. 15.43a). Property owners on the downdrift side don't appreciate the loss.

- **Jetties** are concrete or stone walls located at the entrance to a harbor or at the mouth of a river (Fig. 15.43b). They can prevent baymouth bars from closing off a harbor. In the case of rivers, they extend the river channel into initially deeper water farther offshore. But by doing so, they may cause an offshore sandbar to grow.

- **Breakwaters** are offshore walls that trend parallel to the beach. They prevent the full force of waves from

reaching the beach (Fig. 15.43c). Over time, sand may build up on the landward side of the breakwater.

- **Seawalls** are walls built of steel, concrete, or *riprap* (a chaotic jumble of stone or concrete blocks) on the landward side of the backshore zone (Fig. 15.43d–f). They prevent, or at least slow, wave erosion. But they also reflect wave energy, which increases the rate of erosion on the neighboring beach. During a large storm, erosion may undermine a seawall so that it collapses (Fig. 15.44).

In some places, people have sought alternatives to barrier construction. For example, they may cover the land with fabric, or reinforce escarpments with sandbags. Planting beach grass on areas rarely affected by surf can help stabilize the sand (Fig. 15.45a). Other communities have given up trying to decrease the rate of beach erosion and instead undertake efforts to increase the sediment supply. To do this, they pump sand from farther offshore, or bring in sand from elsewhere by truck or barge, to replenish a beach (Fig. 15.45b, c). This procedure, called **beach nourishment** or *beach replenishment*, can be hugely expensive and at best may provide a temporary fix, for the backwash and longshore drift that formerly removed the sand continue unabated as long as the wind blows and the waves break. But for heavily used tourist beaches, even a temporary fix may be worth it.

Pollution Problems

Bad cases of beach pollution create headlines. Because of longshore drift, garbage dumped in the sea in an urban area may float along the shore and be deposited on a beach far from its point of introduction. For example,

See for yourself

Chicago Lakefront Preservation

Latitude: 41°54′59.76″ N
Longitude: 87°37′36.41″ W

Zoom to an altitude of 3 km (9,800 ft) and look down.

Sandy beaches fringe the shores of Lake Michigan at Chicago. Waves strike the shore obliquely, and longshore drift carries sand southward. The city has built a series of groins to prevent beach erosion.

FIGURE 15.43 Techniques used to preserve beaches.

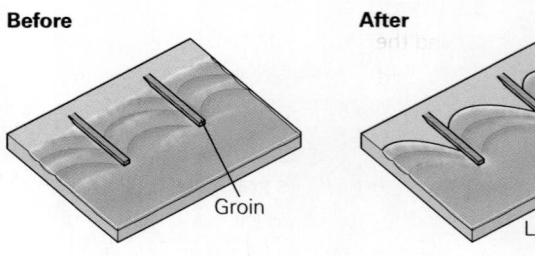

Before | After

Groin | Longshore drift

(a) The construction of groins redistributes sand.

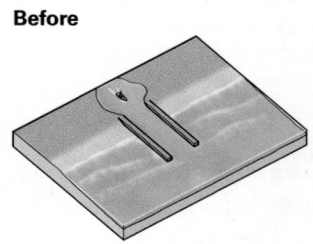

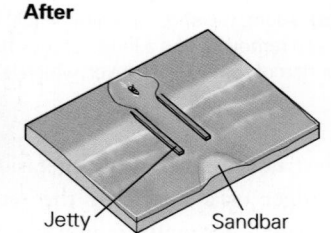

Before | After

Jetty | Sandbar

(b) Jetties extend a river channel farther into the sea.

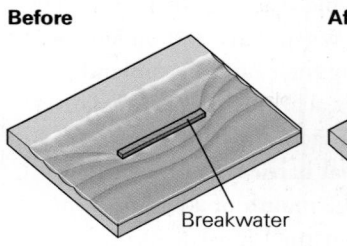

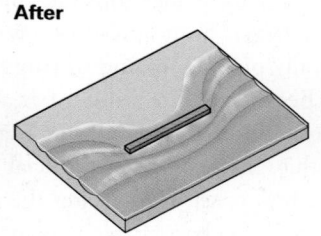

Before | After

Breakwater

(c) A beach may grow seaward behind a breakwater.

(d) A steel seawall.

(e) A carefully constructed stone seawall.

(f) Riprap protecting the edge of a parking lot.

FIGURE 15.44 A seawall can protect a sea cliff under most conditions, but during a severe storm, wave energy reflected by the seawall helps scour the beach. As a result, the wall may be undermined and may collapse.

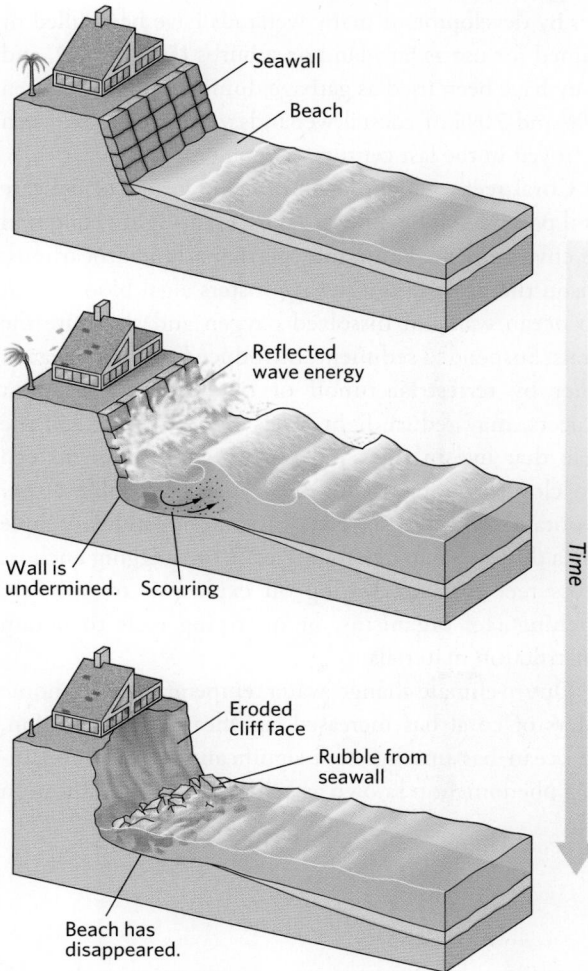

Seawall

Beach

Reflected wave energy

Wall is undermined. Scouring

Time

Eroded cliff face

Rubble from seawall

Beach has disappeared.

FIGURE 15.45 Beach protection and nourishment.

(a) Beach grass has been planted to stabilize the sand.

(b) The ship offshore sucks up sand, which is sent, via a pipe, to a beach, where it gushes out. Excavators and bulldozers redistribute the sand.

Before restoration

Narrow beach

After restoration

Wide beach

(c) Before-and-after views of a stretch of beach near Miami that underwent restoration.

hospital waste from New York City has washed up on beaches tens of kilometers to the south. Plastics, in particular, do not decay rapidly and can wash up on beaches on the other side of an ocean **(Fig. 15.46)**. Recent studies emphasize that *microplastics*, particles that are less than a couple of millimeters across, now occur virtually everywhere in the oceans and on beaches. Oil spills, which may come from ships that flush their bilges with seawater, from tankers that have run aground or foundered in stormy seas, or from blowouts at offshore wells (see Box 10.2), have also contaminated shorelines around the world. In some cases, oil mixes with sand on a beach, and when the more volatile components of the oil evaporate, the remaining mixture of tar and sand resembles asphalt.

FIGURE 15.46 Garbage washing up on a tropical beach. It typically includes objects that float, including plastic bottles and tarps, as well as tires. The sediment consists of coral fragments.

▶ **Animation**
Living with the Coast

Threats to Organic Coasts

In wetlands and tidal flats, sewage, chemical pollutants, and agricultural runoff cause havoc. Toxins, along with clay, settle and concentrate in sediment, where they contaminate or kill burrowing marine life as well as birds and the eggs of marine organisms. Wetlands also face destruction by development: many wetlands have been filled or drained for use as farmland or suburbs (Fig. 15.47a), and many have been used as garbage dumps. In fact, between 20% and 70% of coastal wetlands worldwide have been destroyed in the last century.

Coral reefs, which depend on the health of delicate coral polyps, can be devastated by even slight changes in the environment. Pollutants, particularly hydrocarbons, poison the polyps, and sewage fosters algal blooms that rob ocean water of dissolved oxygen and suffocate the corals. Suspended sediment, introduced to coastal waters either by terrestrial runoff or by beach-nourishment projects, may reduce light levels sufficiently to kill the algae that live in the coral polyps. Fine sediment can also clog the pores that the polyps use to filter water to obtain food and oxygen. In some locations, people play a direct role in destroying reefs by dragging anchors across reef surfaces, setting off explosives to kill fish, touching reef organisms, or quarrying reefs to obtain construction materials.

Due to climate change, water temperature in the home waters of coral has increased significantly. In addition, the ocean has undergone a significant increase in acidity, a phenomenon known as **acidification**. Acidification

FIGURE 15.47 Destruction of landforms on organic coasts.

(a) Urban encroachment in a wetland near Boston.

(b) Coral bleaching killed this coral reef. Compare this image to that of Figure 15.34b to see the difference in color between a healthy reef and a dying one.

happens because some of the CO_2 that has entered the atmosphere due to fossil-fuel combustion and other industrial processes has dissolved in seawater; the seawater turns it into weak carbonic acid, which inhibits shell formation. By some estimates, climate change has led to the death of over 50% of reefs worldwide during the past 20 years. Reef death usually begins with **reef bleaching**, the loss of coral color that happens when corals expel their symbiotic algae **(Fig. 15.47b)**. Bleaching appears to be triggered when corals undergo environmental stress, which has been attributed both to rising temperature and acidification, as well as to the settling of toxic dust carried by winds from deserts or from agricultural areas that are drying out. Exceedingly high temperatures (up to 38°C, or 100°F) in coastal waters off Florida during the early summer of 2023 caused widespread bleaching of Florida's reefs much earlier in the season than usual. Weakened corals also become more susceptible to diseases that can quickly kill off a whole colony. Reef death has become a global crisis. Clearly, preserving the fragile realm of the coast will be a challenge for society in the decades to come.

Take-home message...

Coastal areas evolve over time for many reasons. Along emergent coasts, land rises relative to sea level, while along submergent coasts, land sinks. Depending on its sediment budget, a coast may be accretionary and accumulate sediment, or it may be erosional and lose sediment. The character of a coast can change rapidly in response to storms, sea-level change, and human activity. Communities may try to control this change by constructing barriers or replenishing sand. Pollution and warming of seawater, which have worsened over recent decades, are destroying wetlands and coral reefs.

Quick Questions

- Describe landscape features that may develop along a submergent coast.
- Which coast is likely to have a wider beach: an accretionary coast or an erosional coast?
- How can a beach become polluted even if it is far from a source of pollution?

15 CHAPTER REVIEW

Objective 15.1

Distinguish among the various bathymetric provinces of ocean basins, and explain how they formed, in the context of plate tectonics.

KEY CONCEPTS

- Ocean basins are distinct from continents because they are mostly underlain by oceanic lithosphere, which sits lower to maintain isostatic equilibrium.
- Passive-margin basins, filled with up to 15–20 km of sediment, overlie stretched continental lithosphere that subsides after the rifting that forms a new ocean ceases. The surface of a passive-margin basin is a continental shelf.
- The continental slope descends from the seaward edge of the continental shelf to the landward edge of the continental rise. The continental rise merges with an abyssal plain. Turbidity currents flow down submarine canyons and deposit submarine fans at the base of the continental rise.

- An abyssal plain is underlain by old oceanic lithosphere. Its surface layer consists of pelagic sediment. Seamounts and oceanic islands rising from abyssal plains form due to hot-spot volcanism.
- Plate boundaries are defined by distinctive bathymetric features; namely, deep-sea trenches, the active parts of fracture zones, and mid-ocean ridges.

EARTH-SCIENCE VOCABULARY

abyssal plain (p. 565)
continental shelf (p. 561)
deep-sea trench (p. 567)
fracture zone (p. 565)
guyot (p. 565)
mid-ocean ridge (p. 565)
ocean basin (p. 560)
oceanic island (p. 565)
oceanic plateau (p. 565)
passive-margin basin (p. 562)
pelagic sediment (p. 565)
seamount (p. 565)
submarine canyon (p. 564)
submarine fan (p. 564)

REVIEW QUESTIONS

1. **(a)** How does the lithosphere beneath the interior of a continent (far from the ocean) differ from that beneath a continental shelf along a passive margin? **(b)** Explain how and why a passive-margin basin forms. **(c)** Label the geologic and bathymetric

features displayed in **Figure A**. **(d)** Where does the sediment filling a passive-margin basin come from?

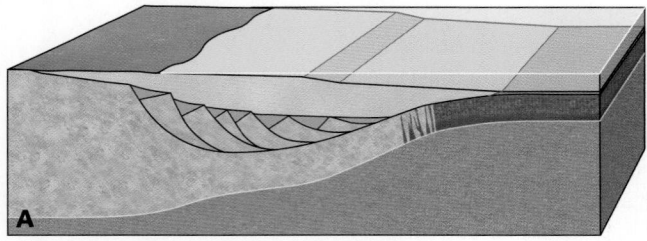

2. **(a)** How do the continental shelf and slope of a convergent boundary (a type of active continental margin) differ from those of a passive continental margin? **(b)** Would a 10 km long vertical drillhole passing down through the continental shelf of a convergent boundary intersect the same material as one passing down through the continental shelf of a passive continental margin? Explain your answer. **(c)** Where and why do submarine canyons form?

3. **(a)** Why is the seafloor over old oceanic lithosphere deeper than that over young oceanic lithosphere close to a mid-ocean ridge? **(b)** Why is the sediment on the right side of **Figure B** thicker than that near the ridge axis? **(c)** How do seamounts and oceanic islands form? **(d)** What does pelagic sediment consist of, and how does it differ from sediment of the continental rise? **(e)** How does a fast-spreading ridge differ, bathymetrically, from a slow-spreading ridge?

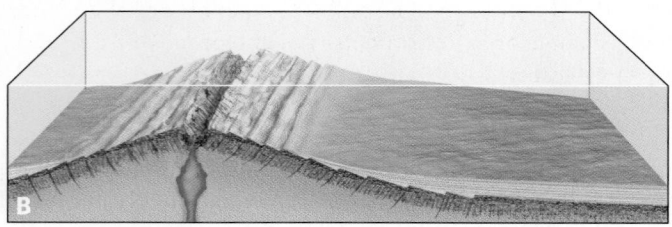

4. **(a)** What factors play a role in determining the depth of a deep-sea trench? **(b)** What does a fracture zone look like on a bathymetric map? **(c)** Which part of a fracture zone hosts earthquakes?

Objective 15.2

Describe ocean tides and explain their causes, define tidal range and characterize how it varies, and illustrate how waves change as they interact with the shore.

KEY CONCEPTS

• Tides are the generally twice-daily rise and fall of sea level. Tidal range, the difference between low and high tide, varies significantly depending on location.

• Where coastlines have a gentle slope, the position of the shoreline can change by kilometers between high and low tide. The intertidal zone becomes submerged at high tide, and is exposed at low tide.

• Tides are caused by the tide-generating force, a combination of gravitational force due to the Moon and Sun and centrifugal force due to the rotation of the Earth-Moon system around its center of mass. High tides happen when a location passes beneath a tidal bulge.

• The motion of waves changes as they approach the coast, so that they transform into breakers. Where waves intersect the shore at an angle, wave refraction and longshore drift of beach sediment take place.

EARTH-SCIENCE VOCABULARY

backwash (p. 573)   **swash** (p. 573)
breaker (p. 572)   **tidal bore** (p. 567)
Earth-Moon system (p. 570)   **tidal range** (p. 567)
intertidal zone (p. 567)   **tide** (p. 567)
longshore current (p. 574)   **tide-generating force** (p. 569)
shoreline (p. 567)   **wave refraction** (p. 573)
surf zone (p. 573)

REVIEW QUESTIONS

5. **(a)** What is the greatest tidal range, and what is the smallest tidal range, on the Earth? **(b)** Why are there generally two high tides a day at a given location along a coast? **(c)** Which tidal bulge on **Figure C** lies closer to the Moon? **(d)** Indicate the position of the Sun on **Figure C** at the time of a neap tide.

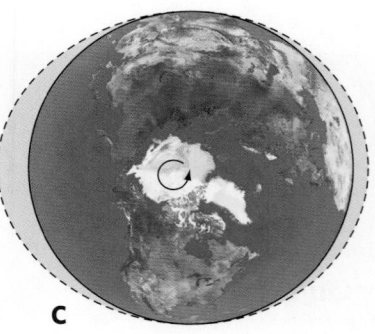

6. **(a)** What factors control the width of the intertidal zone? **(b)** Label the intertidal zone and the tidal range on **Figure D**. **(c)** What factors play a role in producing the tide-generating force?

7. **(a)** Describe how waves change as they approach the shore. **(b)** Explain why wave refraction occurs along some beaches. **(c)** What is the difference between swash and backwash?

8. **(a)** Explain the difference between a longshore current and a rip current. **(b)** Draw the direction of the longshore current on **Figure E**.

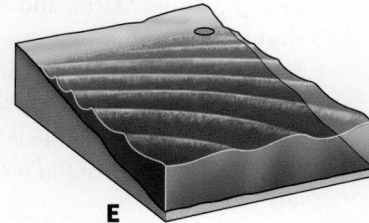

E

Objective 15.3

Discuss the factors that control the character of a coastal landscape, describe how beaches and rocky shores evolve, describe estuaries and fjords, and characterize wetlands and reefs.

KEY CONCEPTS

- The swash and backwash of waves move sand on a beach and yield a distinctive beach profile. In some locations, barrier islands develop offshore.

- On rocky coasts, waves grind away at rocks, yielding such features as wave-cut notches and benches, sea arches, and sea stacks. Because of wave refraction, wave energy focuses on rocky headlands.

- Coastal wetlands, including marshes and mangrove swamps, host salt-resistant plants. Coral reefs grow offshore in warm, clear water. Reefs around oceanic islands evolve as the islands slowly sink.

EARTH-SCIENCE VOCABULARY

barrier island (p. 580)
beach (p. 578)
beach profile (p. 578)
coastal wetland (p. 585)
coral reef (p. 587)
estuary (p. 584)
fjord (p. 585)

lagoon (p. 579)
longshore drift (p. 580)
organic coast (p. 585)
rocky coast (p. 581)
sand spit (p. 580)
tidal flat (p. 580)
wave erosion (p. 582)

REVIEW QUESTIONS

9. **(a)** Why don't sandy beaches also include a significant proportion of clay and silt? **(b)** On **Figure F**, label the landforms visible. **(c)** Under what circumstances will beach sediment consist mostly of pebbles or cobbles?

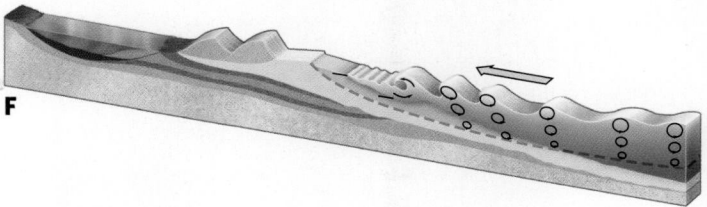

F

10. **(a)** Describe sand motion during longshore drift. **(b)** What landforms can be produced by longshore drift? **(c)** What is a barrier island, and how might one form? **(d)** Label landforms on **Figure G** and indicate longshore current direction.

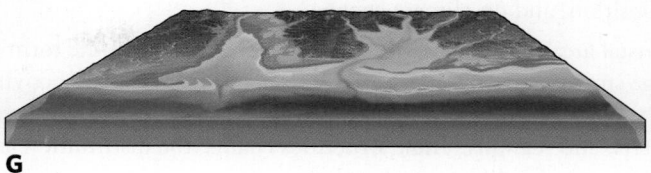

G

11. **(a)** Describe how the headlands of a rocky coast evolve over time. **(b)** Why do headlands erode more rapidly than embayments? **(c)** How does a wave-cut bench develop? **(d)** Label the landforms displayed on **Figure H**.

H

12. **(a)** What is the difference between an estuary and a fjord? **(b)** Why do fjords have such steep walls? **(c)** How do seawater and river water interact in an estuary?

13. **(a)** Describe and name the different types of coastal wetlands. **(b)** What factor(s) determine(s) which type of wetland develops at a given locality?

14. **(a)** How does a coral reef form, and what material composes the interior of a reef? **(b)** Why do reefs primarily form at low latitudes? **(c)** How does a coral reef surrounding an oceanic island change over time? **(d)** Which type of reef does **Figure I** show?

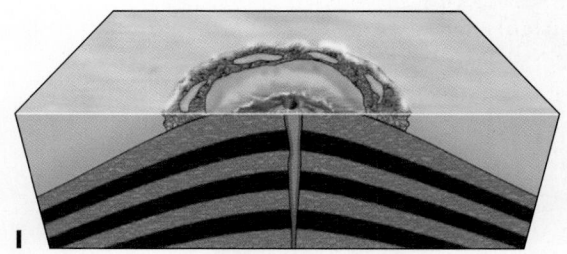

I

Objective 15.4

Explain why coasts evolve over time, describe efforts undertaken to maintain beaches and protect coastal property, and discuss phenomena leading to the destruction of coral reefs and wetlands.

KEY CONCEPTS

- Differences in coastal landscapes depend on whether relative sea level is rising or falling (which may reflect global sea-level change or local uplift or subsidence), on whether the coast is undergoing erosion or deposition, and on climate change.

- Coastal areas face many threats, both from nature (in the form of large storms and ongoing sea-level rise) and from human activities. To protect beachfront property, people build groins, jetties, breakwaters, and seawalls. These structures change the distribution and movement of sediment on beaches.

- Human activities have led to the pollution of coastal regions. Changes in oceanic environments have killed about half of the world's coral reefs.

EARTH-SCIENCE VOCABULARY

accretionary coast (p. 590)
acidification (p. 596)
beach nourishment (p. 593)
beach retreat (p. 593)
breakwater (p. 593)
coastal plain (p. 590)
emergent coast (p. 590)
erosional coast (p. 590)
groin (p. 593)

jetty (p. 593)
nuisance tide (p. 592)
reef bleaching (p. 597)
relative sea level (p. 589)
seawall (p. 593)
sediment budget (p. 590)
storm surge (p. 591)
submergent coast (p. 590)

REVIEW QUESTIONS

15. **(a)** Explain how emergent and submergent coasts differ from each other. **(b)** What landscape features might occur along an emergent coast? **(c)** Label the landscape features displayed on **Figure J**. Is this coast undergoing uplift or subsidence?

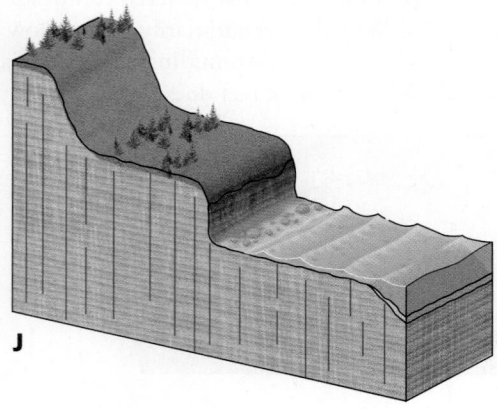

J

16. **(a)** What is a coastal plain? **(b)** What is the approximate age of the sedimentary strata underlying the coastal plain of the eastern United States, and what does the presence of these strata tell us about how sea-level elevation relative to North America at the time of their deposition differed from sea-level elevation today? **(c)** Steep cliffs border much of the coast of California, while Florida is on the coastal plain. Will sea-level rise affect more land in Florida or in California? Explain your answer.

17. **(a)** What factors influence the sediment budget of a coastal area? **(b)** How does the sediment budget of an erosional coast differ from that of an accretionary coast? **(c)** What phenomena can cause the sediment budget of a given coastline to change over time?

18. **(a)** What phenomena can cause relative sea level to change along a coastline? **(b)** Why are nuisance tides considered to be a nuisance? **(c)** What is the relationship between rising sea level and the frequency of nuisance tides? **(d)** What is beach retreat, and why does it happen?

19. **(a)** In what ways do people try to stabilize coasts? **(b)** Are such coastal protection efforts always successful? Explain your answer. **(c)** Distinguish between a groin and a breakwater. Which does **Figure K** show? Label the direction of longshore drift and the direction of the longshore current on **Figure K**. **(d)** What does beach nourishment accomplish?

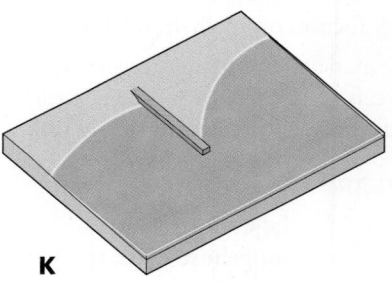

K

20. **(a)** How can plastic garbage that enters the ocean from a city end up on an isolated beach hundreds of kilometers away? **(b)** How do mangrove swamps affect the ability of waves to erode the shore? **(c)** What phenomena cause reef death?

ANOTHER VIEW Since 1791, this lighthouse at Portland Point, Maine, has helped ships avoid running aground on jagged shores.

◉16 THE AIR WE BREATHE
Introducing the Earth's Atmosphere

After studying this chapter, you should be able to...

1. explain how the atmosphere evolved since the Earth first formed, describe how photosynthesis contributed to the Earth's modern atmosphere, and list the gases and other materials found in today's atmosphere.

2. define properties that meteorologists use to describe the atmosphere, list instruments they use to monitor atmospheric conditions, and explain how the data are used in weather forecasting.

3. draw a diagram that displays atmospheric layers based on temperature, composition, and electric charge.

4. identify clouds you see in the sky, describe how they form, and explain how they produce precipitation.

5. classify various types of electromagnetic radiation, describe why optical effects such as rainbows occur, and depict how interactions between radiation and the atmosphere control the Earth's surface temperature.

FIGURE 16.1 Atmospheric phenomena can be both beautiful and dangerous.

(a) A storm in the midwestern United States pours rain from the base of a swirling cloud.

(b) Arcs and halos at sunrise form by the interaction of sunlight with ice crystals in the air.

(c) An ice storm can cause power lines to collapse.

(d) Fog at the entrance to San Francisco Bay, California, hides part of the Golden Gate Bridge.

Breathe deeply. You've just inhaled air, the unique mixture of nitrogen, oxygen, water vapor, carbon dioxide, and other gas molecules that makes up the Earth's atmosphere. Amazingly, some of the nitrogen molecules that entered your lungs were exhaled by Madame Curie, George Washington, Julius Caesar, Confucius, and Cleopatra; some of the water molecules once flowed down the Congo River, rained from Hurricane Katrina, and watered the gardens of Versailles; and some of the oxygen molecules came from plants on your windowsill, algae in the ocean, trees of the rainforest, and grasses of the prairie. That's because the atmosphere flows and stirs constantly on local to global scales and interacts with many other components of the Earth System.

Without the atmosphere, the Earth would be as barren as the Moon. The atmosphere provides plants with carbon dioxide for photosynthesis and animals with oxygen for metabolism, shields our planet's surface from dangerous ultraviolet radiation in sunlight, and carries water from where it evaporates to where it falls as rain, snow, or hail. The atmosphere also serves as the medium in which birds, insects, and airliners fly and through which our voices carry. In human culture, the atmosphere's visual splendor—its aurorae, rainbows, sunsets, and clouds—has inspired artists and sparked romances. The atmosphere can also unleash violent *storms*, intense episodes of locally strong winds and precipitation (Fig. 16.1).

Though we are merely bottom dwellers in this shared sea of air, we have become its caretakers, for our

<< The beauty of an Arizona sunset results from the interaction between sunlight and the atmosphere, the blanket of gases and aerosols that surround our planet and allow life to inhabit the land.

603

activities have modified, and will continue to modify, the composition of the atmosphere. Furthermore, our vulnerability to storms and other atmospheric events increases as cities and farms spread into lands susceptible to flooding and drought. To prevent calamities due to atmospheric phenomena, and to ensure that the atmosphere can continue hosting our planet's rich biosphere into the future, it is essential that we strive to understand our atmosphere's character and behavior and learn how to predict its short-term and long-term future. This chapter provides a foundation for addressing these challenges by discussing several basic questions: What is the atmosphere? How did it form and attain its present composition? What instruments can we use to measure atmospheric properties and describe atmospheric conditions? How does the atmosphere change as we travel upward from the Earth's surface to space? What are clouds, why are there so many different types, and how do they produce rain and snow? How does the Sun's energy warm the Earth, and in what ways does the atmosphere interact with light?

16.1 Atmospheric Evolution during Earth History

The Earth's **atmosphere**, the envelope or blanket of gas that surrounds our planet, has not always hosted the rich amount of oxygen that we inhale with each breath today. In fact, if you stepped out of a time machine into the air of the Precambrian, you would suffocate instantly. Let's look briefly at how our atmosphere has changed over time, from the Earth's birth to the present.

The First Atmosphere

When the Earth first formed, between 4.56 and 4.54 Ga, molecules of hydrogen (H_2) and atoms of helium (He), the most common gases of interstellar nebulae, along with molecules of ammonia (NH_3) and methane (CH_4), surrounded our planet (**Fig. 16.2**). This *primordial atmosphere* didn't last long. The gases were blown away by particles streaming from the newly formed Sun, and any that remained were blasted away during the Moon-forming collision. After the collision, volcanic eruptions, as well as the arrival of comets, added molecules of many other gases (such as H_2O, CO_2, N_2, and SO_2), so that by about 4.0 Ga, the end of the Hadean, the Earth's air contained an assemblage of gases markedly different from that of the nebula. Specifically, this *first atmosphere* consisted of about 55% H_2O, 15% CO_2, 15% N_2, 15% NH_3, and traces of CH_4, gases that came mostly from *volcanic outgassing*—the emission from volcanoes of gases once

bonded to minerals inside the Earth—and some from impacting comets.

The Second Atmosphere

Between about 4.0 and 3.9 Ga, the beginning of the Archean, so many meteorites and comets pummeled the Earth that its surface may have temporarily become molten, and liquid water could not survive. Any ocean that had existed before this *late heavy bombardment* would have evaporated. The geologic record shows that by 3.85 Ga, after major impacting had ceased, our planet's surface once again cooled below the temperature at which water vapor condenses into liquid. When condensation became possible, H_2O precipitated to fill rivers, oceans, and lakes. H_2O also seeped into the crust to become groundwater, and in regions with cold temperatures, it froze into ice. Therefore, at the beginning of the early Archean, the concentration of H_2O in the atmosphere started to drop substantially, and the hydrologic cycle began.

As liquid water accumulated, the concentration of atmospheric CO_2 began to decrease, for CO_2 dissolves in water. Dissolved CO_2 can, in turn, be removed from water by bonding with other ions to produce solid carbonate minerals, so once it enters the ocean, much of it does not return to the atmosphere. As CO_2 concentration decreased, the proportion of another gas, N_2, increased. Some N_2 came directly out of volcanoes, but a significant proportion occurred in NH_3, once a major component of the atmosphere. NH_3 eventually decomposes to form N_2 and H_2. Lightweight hydrogen soon escaped into space, leaving behind N_2. Once it enters the atmosphere, N_2 remains there, because it is an *inert gas*, meaning that it does not react with other chemicals. Due to all these chemical changes, the first atmosphere of the Earth was replaced, during the Early Archean, by the second atmosphere, in which N_2 and CO_2 were the major gases. In fact, by the middle Archean, water accounted for only a small percent of the atmosphere.

The Third Atmosphere

If the Earth were devoid of life, the atmosphere would contain hardly any *free oxygen* (O_2 molecules) because volcanoes do not emit O_2 gas. Fortunately for us, the Earth abounds with life. The first organisms appeared at about 3.8 Ga, and by 2.7 Ga, cyanobacteria, tiny blue-green single-celled organisms that carry out photosynthesis, were thriving in the ocean. *Photosynthesis*, a chemical process that takes place within certain types of living cells, extracts CO_2 from the atmosphere or ocean, converts it into organic chemicals, and releases O_2 as a by-product. At first, nearly all of the released O_2 was absorbed by rocks or dissolved in water, so the atmosphere retained only a trace

Did you ever wonder...

whether you could have breathed the atmosphere of the early Earth?

Did you ever wonder...

where the oxygen in air comes from?

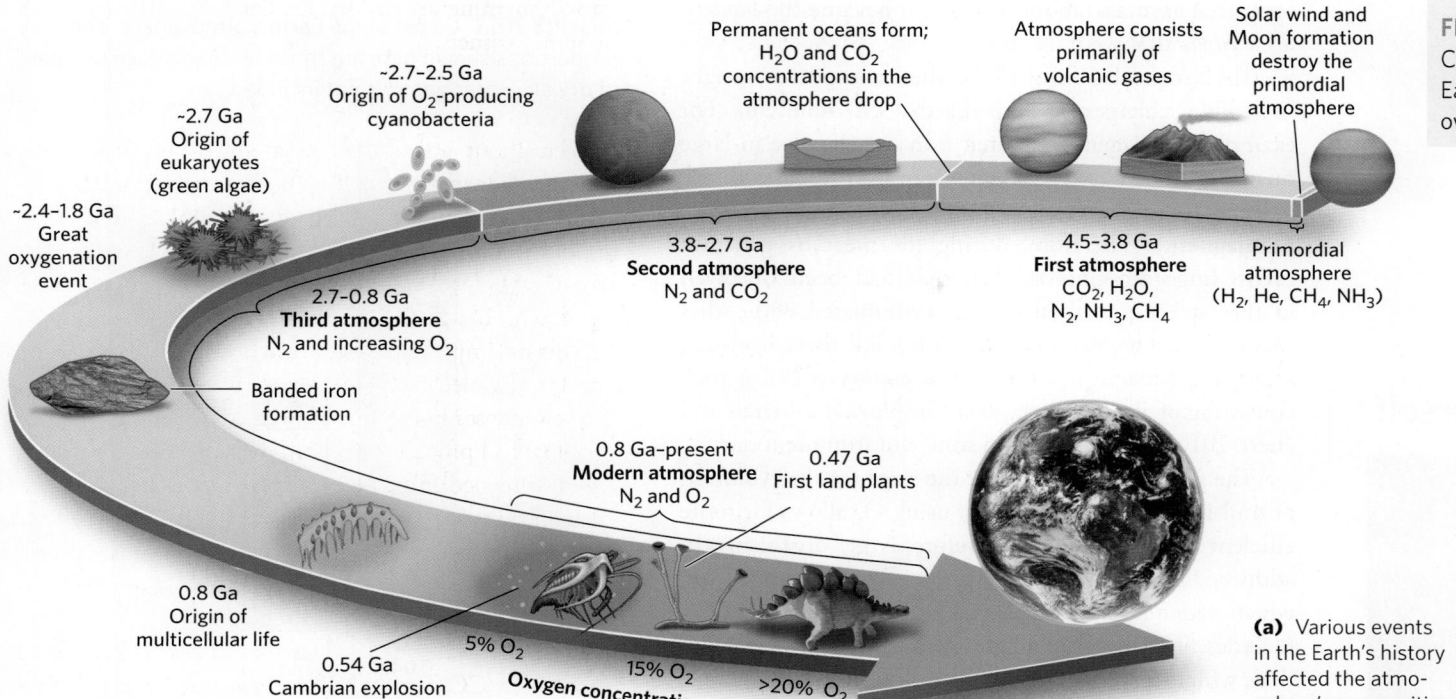

FIGURE 16.2
Changes in the Earth's atmosphere over geologic time.

(a) Various events in the Earth's history affected the atmosphere's composition.

of O_2. Then, at 2.7 Ga, a dramatic transition took place on the Earth: organisms with eukaryotic cells (cells with a nucleus) appeared. Eukaryotic cells are more efficient than bacterial cells at carrying out photosynthesis, so the global rate of O_2 production started to increase. Between 2.4 Ga and 1.8 Ga, a time known as the **great oxygenation event**, O_2 began to accumulate to yield more than trace quantities in the atmosphere. This change marks the appearance of the Earth's *third atmosphere*, an atmosphere influenced by life.

Photosynthetic organisms not only release oxygen—they also absorb CO_2 that had dissolved in the ocean. Some of the dissolved CO_2 became incorporated in shells, which sank to the seafloor when organisms died, and became sediment that later turned into limestone. The rest went into the organisms' *biomass* (the material within organisms). If all of this biomass decayed when the organisms died, the atmosphere's CO_2 concentration would have remained at about 20%, because decay returns CO_2 to the atmosphere. In the Earth System, however, not all biomass decays. Instead, some gets deposited, along with sediment, on the seafloor or on lake beds. If this organic-rich sediment becomes buried deeply enough to turn into rock, the biomass it contains effectively becomes locked underground. The trapping of organic matter in sedimentary rocks led to a further decrease in the atmosphere's CO_2 concentration, from about 20% to a trace amount. As a result,

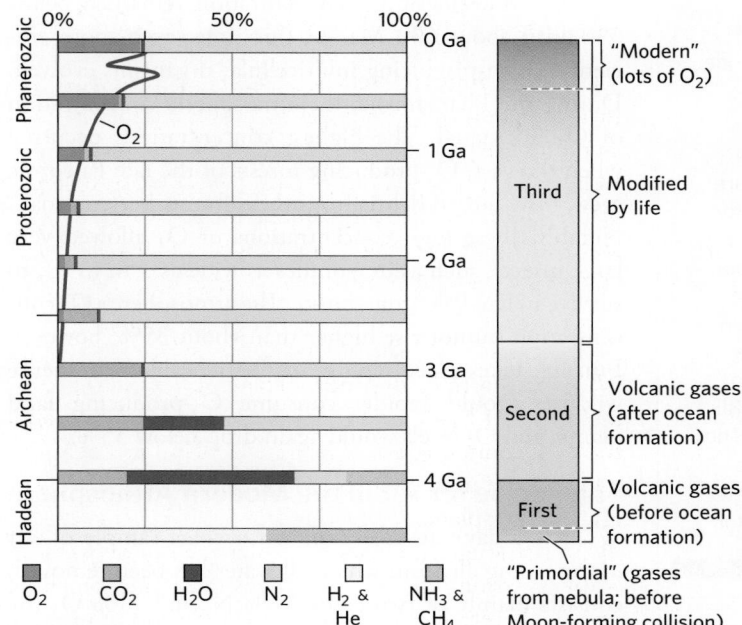

(b) A model for the compositional change of the atmosphere over geologic time. The first atmosphere may have changed many times, due to temporary episodes of ocean formation, and due to collision with comets. These changes aren't shown.

the third atmosphere became composed predominantly of N_2 and O_2. This atmosphere became progressively more oxygen-rich as life evolved and proliferated. When the O_2 concentration rose to a level that could sustain terrestrial animals (about 0.5 Ga), it became the *modern atmosphere* that we have today.

The accumulation of O_2 in the atmosphere brought incredible changes to the Earth's environment. For example, in oxygen-free water, iron can dissolve in large quantities. But in oxygen-rich water, iron reacts with the oxygen to produce iron-oxide minerals such as hematite or magnetite. Therefore, during the great oxygenation event, huge amounts of iron that had been dissolved in the oceans precipitated and accumulated with other sediments on the seafloor. When buried, these iron-rich sediments became *banded iron formation* (BIF), a rock consisting of alternating layers of iron oxide minerals and chert. BIF serves as our main source of iron ore today.

The presence of O_2 also set the stage for the evolution of multicellular life: respiration using O_2 allows for more efficient metabolism and, therefore, larger organisms. In addition, the O_2 in the atmosphere provided atoms from which ozone (O_3) molecules could form. Solar radiation includes not only visible light, but also ultraviolet radiation, which can be harmful to life. The addition of ozone to the atmosphere provided a protective shield by absorbing ultraviolet radiation and made it possible, eventually, for plants and animals to leave the ocean and live on land.

The atmospheric O_2 concentration remained below 5% until about 600 Ma. At this time, even more efficient oxygen-producing multicellular organisms evolved. During the Phanerozoic, the atmospheric concentration of O_2 fluctuated. The highest concentrations occurred when the vast, O_2-producing forests of the late Paleozoic grew. (The buried debris from these forests became coal.) Notably, these high concentrations of O_2 allowed very large insects, such as dragonflies as wide as 1 m (3 ft), to survive in late Paleozoic forests. The atmosphere's O_2 concentration cannot rise higher than about 35%, however. Burning takes place rapidly in O_2-rich air, so immense wildfires would rapidly consume O_2-producing land plants, and O_2 levels would again drop below 35%.

The Recipe for Air in the Modern Atmosphere

What is today's air made of? An average sample of *dry air*, meaning air from which all water has been removed, consists mainly of two gases: 78% N_2 and 21% O_2 (by volume) (Fig. 16.3). The remaining 1% consists of **trace gases**, including argon (Ar), CO_2, CH_4, O_3, and SO_2. Although these other gases occur in such minuscule quantities, some, as we'll see, play key roles in controlling atmospheric temperature.

Note that when we describe the composition of the Earth's atmosphere, we consider only dry air. That's because in nature, the amount of water in air varies from place to place at the same time, and from time to time in the same place. In general, water concentration ranges

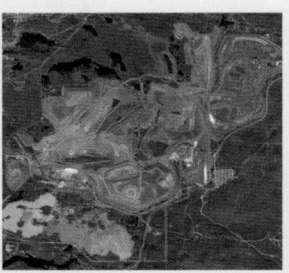
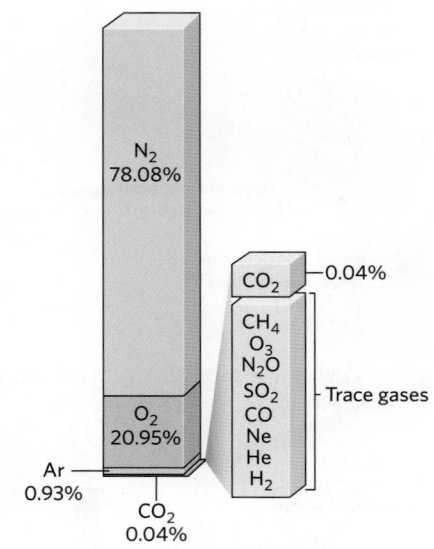

FIGURE 16.3 Gases of the Earth's atmosphere. The proportions shown here are those for an average sample of dry air; water vapor is not included.

between 0.1% in very dry air to 5.0% in humid tropical air. Water in the atmosphere occurs in three forms: as an invisible gas called **water vapor** (Fig. 16.4a), as a liquid (in tiny *droplets* or larger *drops*) (Fig. 16.4b), and as a solid (in the form of ice, snow, or hail) (Fig. 16.4c). Looking up, we often see **clouds**, collections of countless water droplets or tiny ice crystals suspended in air. Clouds generally float above the Earth's surface; when they come in contact with the ground, they are called *fog*.

Atmospheric Aerosols

Press the button on a can of air freshener, and countless droplets spray into the air, droplets so small that they can be wafted away by even gentle air currents. As the droplets evaporate, what is left behind are microscopic solid particles called aerosols. Researchers refer generally to any tiny solid particle suspended in air, regardless of its source, as an **aerosol** particle. Air always contains many types of aerosols (Fig. 16.5a). Inorganic aerosols include specks of mineral dust lofted by winds blowing over soil, salts and sulfates carried into the air by sea spray, fine ash injected into the atmosphere by volcanic eruptions, and soot (carbon) particles billowing from forest fires (Fig. 16.5b). Organic aerosols include pollen, bacteria, molds, and viruses, as well as detritus from decaying organisms. Many aerosols are produced by human activities such as burning fossil fuels, spraying pesticides on crops, or using household products such as oven cleaner or deodorant.

Atmospheric aerosol concentrations vary with location. They tend to be greater over land surfaces than over the sea because there are more aerosol sources on land.

FIGURE 16.4   Water in the atmosphere occurs in three forms.

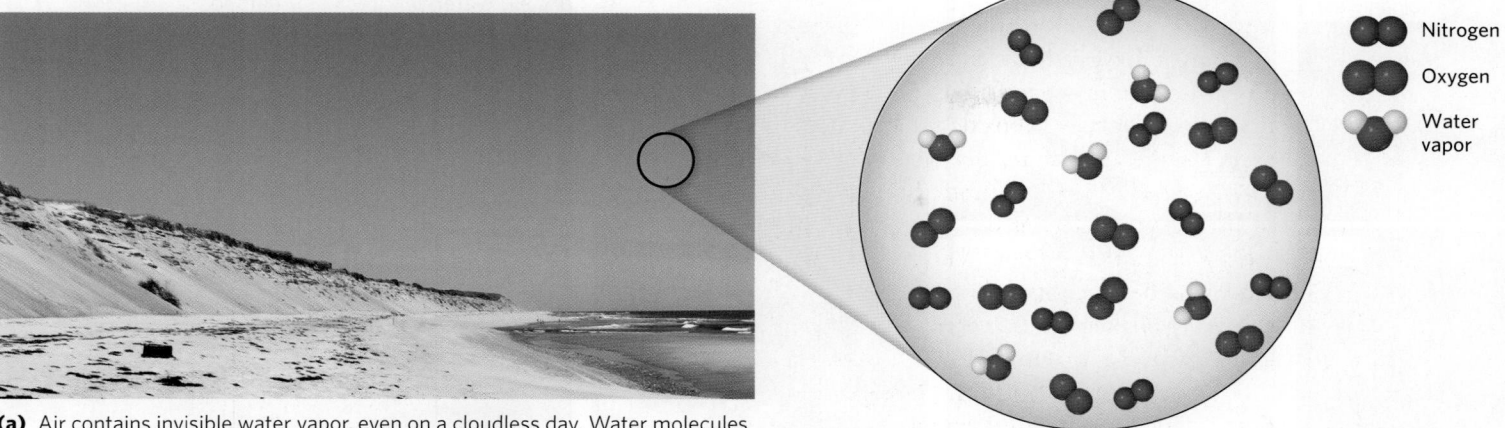

(a) Air contains invisible water vapor, even on a cloudless day. Water molecules in vapor are mixed in with other gas molecules.

Nitrogen
Oxygen
Water vapor

(b) Clouds in the warm sky over Brazil consist of liquid water droplets.

(c) High clouds over the University of Illinois consist of tiny ice crystals.

Aerosols from land sources differ in composition from their oceanic counterparts. Most aerosols found over the oceans are salts and sulfates, whereas over land, most arise from windblown dust, fires, and human activities. Aerosols can travel long distances. Saharan dust blown from Africa (Fig. 16.5c), for example, can reach the Caribbean. Aerosol particles absorb light, so an abundance of aerosols in the air limits our ability to see into the distance. Under high-humidity conditions, some aerosols capture water molecules in the air and dissolve in the water to form microscopic liquid solution droplets known as haze droplets. The presence of many haze droplets causes **haze**, which decreases visibility on sultry days (Fig. 16.5d).

Aerosols, along with certain gases such as ozone, contribute to air pollution. In discussions of pollution, aerosols may be referred to as *fine particulate matter*. The dark *smog* that engulfed industrial cities of the 19th century formed when factories and heating stoves produced coal smoke that mixed with fog. In modern cities, *photochemical smog* develops when exhaust from cars and trucks reacts with air in the presence of sunlight to produce an ozone-rich brown haze (Fig. 16.5e). Government agencies monitor the aerosol concentrations in air to characterize air quality—high concentrations lead agencies to post health-hazard warnings. These warnings typically distinguish among aerosols of different size. Particles smaller than PM_{10} (meaning particulate matter with a diameter less than 10 μm, or 0.01 mm) can be inhaled into the lungs (Fig. 16.5f). Fine particulate matter ($PM_{2.5}$) can be inhaled more deeply into the lungs. Both PM_{10} and $PM_{2.5}$ can remain in the lungs and cause lung disease. During the summer of 2023, aerosols composing the smoke from forest fires in Canada blew south over

FIGURE 16.5 Atmospheric aerosols.

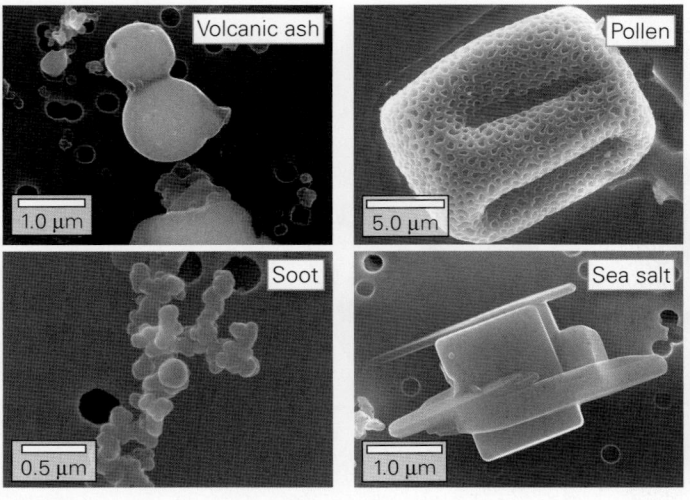

(a) Aerosol particles viewed through a scanning electron microscope (1 μm = 0.001 mm).

(b) Smoke rising from a forest fire.

(c) Dust blowing from the Sahara out over the Atlantic, as viewed from space.

(d) Haze over the Blue Ridge Mountains, part of the Appalachians.

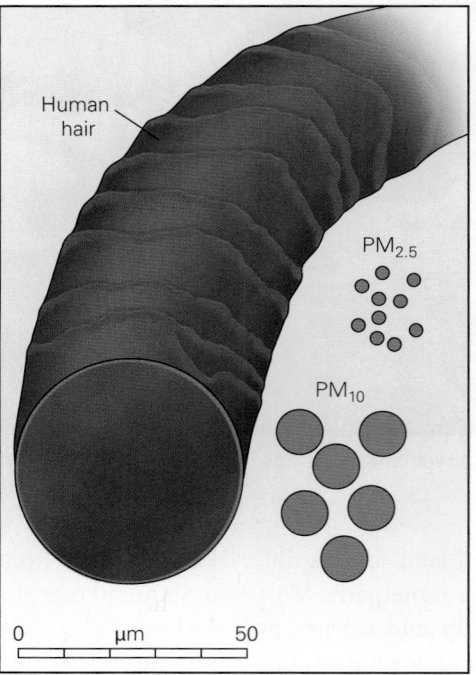

(f) Relative size of aerosol particles compared with a human hair. PM_{10}, particulate matter 10 μm or less in diameter; $PM_{2.5}$, fine particulate matter 2.5 μm or less in diameter.

(e) Photochemical smog over Los Angeles, California, includes particles from car exhaust and other sources.

midwestern and eastern states of the United States and turned the air golden brown. The $PM_{2.5}$ concentration in the air of New York City reached historic highs; the daily average for June 7 was 204 μg/m³. By comparison, the Environmental Protection Agency (EPA) considers a daily average value less than 35 μg/m³ to be a safe level, and the previous record for New York City, set in 2003, was 86 μg/m³.

Take-home message...

🏠 The composition of the Earth's atmosphere has changed radically since the Earth first formed. A primordial atmosphere of nebular gases was quickly replaced by one dominated by volcanic gases including N_2, H_2O, and CO_2. The concentration of H_2O decreased when oceans formed, and CO_2 decreased gradually, as it dissolved in the ocean and was trapped in sediment, leaving N_2 as the dominant gas. Photosynthesis added O_2 to the atmosphere, and today, air consists of nitrogen (78%), oxygen (21%), and trace gases (1%). The amount of water in the atmosphere, which can occur as a gas, liquid, or solid, varies considerably over time and space. The atmosphere also contains aerosols, tiny particles that may be solid or dissolved in tiny droplets. A high concentration of aerosols can cause haze or smog.

Quick Questions

- Could you have survived in the atmosphere on Earth that existed 3 billion years ago? Explain.

- Is the concentration of oxygen in today's atmosphere the highest that it has ever been?

- What does smog consist of?

16.2 Describing and Measuring Atmospheric Properties

Everywhere in the world, people complain about, marvel at, or just simply button up and prepare for the atmospheric conditions they find outdoors. We refer to the specific atmospheric conditions at a given time and place as **weather**. Some of the common properties that **meteorologists**—atmospheric scientists who focus on describing and predicting the weather—use to describe atmospheric conditions include the following:

- *Temperature:* a measure of hotness or coldness
- *Atmospheric pressure:* a measure of the weight of a column of air above a location
- *Relative humidity:* a measure of the amount of water vapor in the air
- *Wind speed:* a measure of how fast the air moves horizontally
- *Wind direction:* a measure of the compass direction from which the wind blows
- *Visibility:* a measure of how far one can see through the air
- *Cloud cover:* a measure of the portion of the sky covered by clouds
- *Precipitation:* a measure of the amount of water falling from the air and reaching the ground during a time interval

Let's look at each of these properties in more detail and learn how they can be measured.

Air Temperature

We all pay attention to the air temperature—if it's hot, you might stroll about in shorts, but if it's cold, you need to put on a coat. But what exactly does *temperature* measure? In the air, gas molecules move rapidly in random directions, vibrate, and occasionally collide with one another. **Air temperature** represents the average speed at which these molecules move. The faster the average speed of the molecules, the higher the temperature. Note that temperature is not synonymous with heat (see Box 1.3). *Heat* refers to the total thermal energy contained in a material, meaning all the energy due to vibrations and movements of all atoms or molecules in the material; this thermal energy can be transferred to other materials. A volume of air near the top of the atmosphere contains few, but very fast-moving, gas molecules. It has a very high temperature, but it does not contain as much heat as the same volume of air at the same temperature at sea level, for the sea-level air contains a great many more molecules.

We measure temperature with a **thermometer**, calibrated in degrees according to standard scales (Fig. 16.6). Originally, thermometers consisted of a thin column of mercury in a glass tube. Today, most thermometers use alcohol because mercury is toxic. When the alcohol warms, it expands, so it rises in the tube. Meteorologists can also measure temperature with a *thermistor*, a sensor containing a conductor in which the resistance to the flow of electric current depends on temperature. Most thermometers in the United States still use the *Fahrenheit scale*, whereas those in the rest of the world use the *Celsius scale*, also known as the *centigrade scale*. At sea level, water boils at 212°F and freezes at 32°F on the Fahrenheit scale,

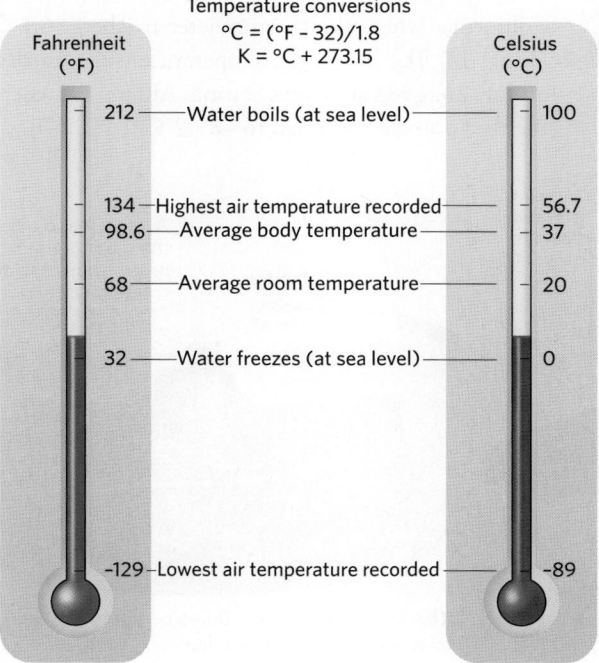

Temperature conversions
$$°C = (°F - 32)/1.8$$
$$K = °C + 273.15$$

Fahrenheit (°F)		Celsius (°C)
212	Water boils (at sea level)	100
134	Highest air temperature recorded	56.7
98.6	Average body temperature	37
68	Average room temperature	20
32	Water freezes (at sea level)	0
–129	Lowest air temperature recorded	–89

FIGURE 16.6 Temperature scales and the relationship between Celsius (centigrade) and Fahrenheit values for common temperatures.

TABLE 16.1 Common Units Used for Measuring Atmospheric Pressure

Unit	System	Average atmospheric pressure at sea level
Pounds per square inch	English	14.7 lb/in²
Pascals (Pa)	Metric	101,325 (= 1,013.25 hPa)
Atmospheres (atm)	By convention*	1.000 (= 1.013 bars)
Bars (= 1,000 Pa)	By convention*	1.01325 (= 1 atm)
Millibars (mb) (= 0.001 bar)	By convention*	1,013.25 (= 1,013.25 hPa)
Inches of mercury	Traditional	29.92 in

*"By convention" means that the number was defined by an international agency.

and it boils at 100°C and freezes at 0°C on the Celsius scale. An increment of temperature, a *degree*, represents a greater temperature change on the Celsius scale than on the Fahrenheit scale. Scientists commonly use another scale, the Kelvin scale. *Absolute zero*, the lowest temperature possible, is 0 K (–273.15°C)—at 0 K, molecules stop moving or vibrating. (Note that when we write a temperature using the Kelvin scale, we do not use the word *degree* or the degree symbol.)

At any particular location, air temperature fluctuates daily as the Sun rises and sets. It also changes seasonally because the number of daylight hours and the Sun's position in the sky vary with the time of year. Temperature at a location can also change when volumes of cooler or warmer air move in from elsewhere. During the 20th century, the average annual temperature on the Earth, worldwide, was 13.9°C (57.0°F). In 2022, it was 0.86°C (1.55°F) higher. The highest air temperature ever recorded on land was measured on July 10, 1913 in Death Valley, California, where the thermometer reached 56.7°C (134.1°F). The lowest air temperature was recorded on July 21, 1983 at Vostok Station, Antarctica, where the thermometer dropped to –89.2°C (–128.6°F).

Atmospheric Pressure and the Height of the Atmosphere

Atmospheric pressure refers to the force applied by the overlying air on a surface of a specified area, such as a square meter, a square inch, or a square centimeter. You can picture atmospheric pressure as the weight of a column of air over a unit area at the base of the column. For example, using English units, imagine a column of air that is 1 in by 1 in at its base and extends from the ground at sea level all the way up, through the atmosphere, to the edge of space. On average, this column weighs 14.7 lb, so we can say that the average atmospheric pressure at sea level is 14.7 lb/in². Atmospheric pressure exists because air, like any substance, has mass, so in the Earth's gravitational field it has weight.

How is pressure different from the *differential stress* (see Chapter 6) that you feel when your friend pushes you? Pressure, by definition, acts in all directions equally. You don't get knocked over by air pressure in still air because the air pressure pushing on your front is the same as that pushing on your back. Similarly, an empty can doesn't collapse because air pressure pushing on its inside walls equals that pushing on its outside walls. If a fluid (liquid or gas) flows, however, dynamic pressure develops and pushes in the direction that the fluid moves. When you face into the wind, it applies dynamic pressure, in the sense that you feel more pressure on your front than on your back.

Note that pressure affects the density of a gas. As pressure increases, air molecules get pushed closer together, and as pressure decreases, air molecules move farther apart. The density of a gas also depends on temperature. If you heat a given volume of air, it expands, because the molecules start moving faster. So, if the air is confined in a container, the pressure in the container increases. Similarly, if you cool air, it contracts, and if the air is confined in a container, its pressure decreases (Box 16.1).

As Table 16.1 shows, we can use several different units to specify atmospheric pressure. These days, meteorologists prefer either *millibars* (mb) or the numerically equivalent *hectopascals* (hPa). In this book, we'll use the more common unit, millibars. The numbers given for "average atmospheric pressure at sea level" in Table 16.1 are just that—averages. The specific value of atmospheric pressure varies with time and location at a given elevation, but it always decreases as elevation increases because the higher you go, the less gas lies above you.

Meteorologists measure atmospheric pressure with a **barometer**. A traditional mercury barometer consists of a dish of mercury into which a vertical glass tube has been inserted (Fig. 16.7a).

FIGURE 16.7 How a barometer works.

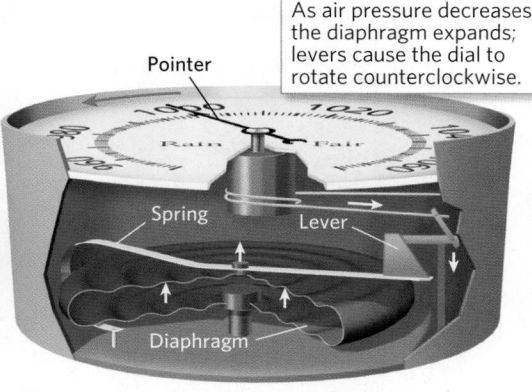

(a) In a traditional mercury barometer, air pressure pushes mercury up a vacuum tube. The greater the pressure, the higher the mercury rises.

(b) In an aneroid barometer, compression or expansion of a diaphragm moves a dial.

BOX 16.1 ▶ **How can I explain . . .**

Atmospheric pressure

What are we learning?

- That the force exerted by atmospheric pressure is real.
- That atmospheric pressure is related to the density of air.

What you need:

- An open, empty aluminum soda can; a bowl filled with water and ice; a small hot plate; and a pair of tongs

Instructions:

- Put a small amount of water in the can. Set the hot plate on low heat and place the can, open top up, on the heating element. Wait for the air in the can to heat and the water to boil.
- Using the tongs so you don't burn yourself, grasp the can, invert it, and place it open-side-down into the ice water.
- The can will immediately crumple.

What did we see?

- Before the can is heated, the atmospheric pressure exerted on the inside of the can wall and the outside of the can wall is the same. When you put the can on the hot plate, two things happen: the gas in the can heats up and begins to expand, and the water at the bottom begins to boil, releasing water vapor.

- During boiling, most of the air molecules are pushed out the top opening, and they are replaced with water vapor molecules.
- The can doesn't crumple yet, because as air escapes, water vapor takes its place, so the total pressure inside the can is equal to the atmospheric pressure outside the can.
- When you invert the can in ice water, the small amount of air inside the can immediately cools. At the same time, the water vapor inside the can condenses, further reducing the interior air pressure. When the pressure inside the can no longer equals the air pressure outside the can, the outside pressure crushes the can until the pressures inside and outside the can are the same. The fact that the can is crushed emphasizes that air, although invisible, exerts pressure.

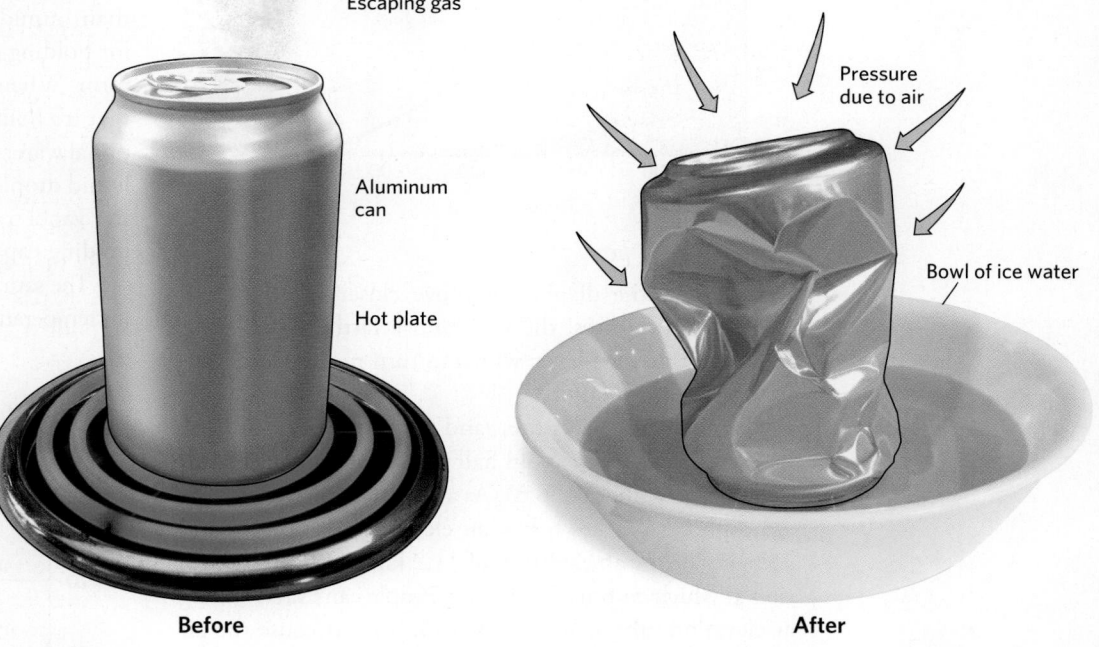

Escaping gas

Aluminum can

Hot plate

Pressure due to air

Bowl of ice water

Before

After

The tube is open on the bottom and closed at the top, and the air has been pumped out of the tube, so it contains a vacuum. As the atmosphere pushes down on the surface of the mercury, the mercury rises into the tube. The higher the air pressure, the greater the weight of air pushing on the mercury, so the higher the mercury rises in the tube. When the pressure equals the average (mean) atmospheric pressure at sea level (1,013.25 mb), the column attains a height of 29.92 in (see Table 16.1). Most television meteorologists and newspaper weather pages in the United States still report pressure in "inches of mercury." When atmospheric pressure decreases, meteorologists say that "the barometer is falling" because the height of the mercury column drops. Conversely, if air pressure increases, they say that "the barometer is rising." Mercury barometers were actually replaced long ago by *aneroid barometers*, which consist of a flexible metal diaphragm containing a vacuum **(Fig. 16.7b)**. As pressure increases,

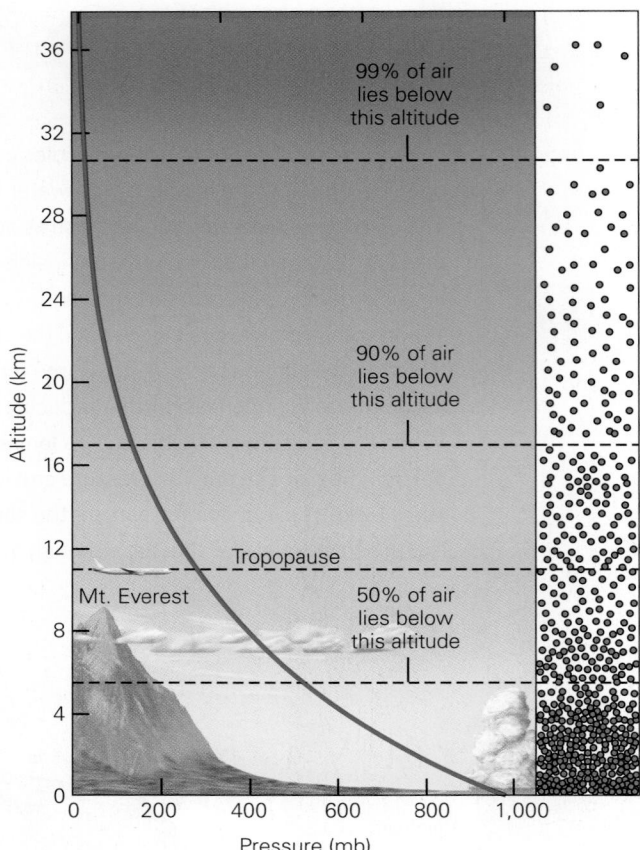

FIGURE 16.8 Atmospheric pressure and air density decrease with increasing altitude in the air.

99% of air lies below this altitude

90% of air lies below this altitude

Tropopause

Mt. Everest

50% of air lies below this altitude

Altitude (km)

Pressure (mb)

therefore denser, air, whereas regions of lower pressure occur beneath warmer, less dense air. Corrected for elevation, the highest sea-level pressure ever recorded was 1,083.8 mb in Siberia on December 31, 1968, whereas the lowest ever recorded occurred in the center of a typhoon (a west Pacific hurricane) on October 12, 1979, where the pressure dropped to 870 mb. As we'll see in Chapter 17, horizontal variations in atmospheric pressure control the speed and direction of the wind.

Moisture and Relative Humidity

Water vapor, an invisible gas, consists of freely moving water molecules. These molecules enter the atmosphere by evaporation from the Earth's oceans, lakes, and rivers and from plants by transpiration. Meteorologists measure the water vapor content of the atmosphere in several ways.

The part of total atmospheric pressure exerted by water vapor molecules alone is called the **vapor pressure**. At a given temperature, dry air has a lower vapor pressure than humid air. The atmosphere has a limited capacity for holding water vapor—otherwise, clouds would never form. When the atmosphere becomes *saturated*, meaning that it's holding as much water vapor as it can, any additional water vapor added to the air will condense and form liquid droplets or solid ice crystals. Meteorologists refer to the vapor pressure at which the atmosphere reaches its holding capacity as the **saturation vapor pressure**.

The saturation vapor pressure of air depends strongly on its temperature **(Fig. 16.9)**. Warm air, in which molecules

the walls of the diaphragm move closer together, and if pressure decreases, the walls move farther apart. This movement drives a lever, which in turn moves a dial calibrated in pressure units.

Atmospheric pressure, and therefore atmospheric density, decreases by about half for every 5.6 km (3.5 mi) of elevation gain **(Fig. 16.8)**. As a result, about 50% of the atmosphere's gas lies below an elevation of 5.6 km, and 75% lies below an elevation of 11.2 km (7 mi), the elevation at which commercial jets fly. People can't survive long at elevations above about 6 km (3.7 mi) because the air doesn't contain enough oxygen.

Continuing upward, we find that 99% of gas in the atmosphere lies below 31 km (19 mi). Nevertheless, there's enough gas above that altitude that meteors begin to heat up and vaporize when they pass below 70–120 km (43–75 mi). The "top" of the atmosphere, where the density of the atmosphere equals the density of interplanetary space, actually varies between 350 and 800 km (217–497mi) in altitude, depending on variations in solar activity and its interaction with the Earth's magnetic field.

Sea-level air pressure varies across the globe at any given time, so rarely will a direct measurement of atmospheric pressure at sea level at a particular location yield a number exactly equal to average sea-level pressure. Regions of higher pressure develop beneath colder, and

FIGURE 16.9 Saturation vapor pressure (SVP) decreases rapidly as temperature (T) decreases.

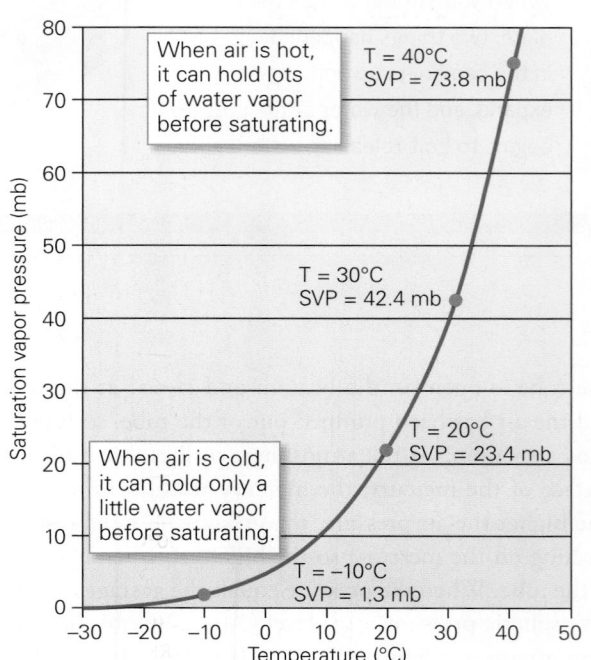

When air is hot, it can hold lots of water vapor before saturating.

T = 40°C
SVP = 73.8 mb

T = 30°C
SVP = 42.4 mb

T = 20°C
SVP = 23.4 mb

When air is cold, it can hold only a little water vapor before saturating.

T = −10°C
SVP = 1.3 mb

Saturation vapor pressure (mb)

Temperature (°C)

jostle rapidly, can hold more water vapor than cold air can hold. In fact, the atmosphere can hold 84 times more water vapor molecules at 30°C (86°F) than it can at −30°C (−22°F). At 30°C, a cubic meter of dry air weighs 1,160 g (2.6 lb). If the air were saturated, 28 g (0.06 lb) of water molecules would replace some of the nitrogen and oxygen molecules, because at a given temperature and pressure, two equal volumes of gas must contain the same number of molecules, regardless of their composition. Surprisingly, the saturated air would weigh a bit less than the dry air, because water molecules (H_2O) weigh less than the replaced nitrogen (N_2) and oxygen (O_2) molecules.

Humans can sense the air's moisture content because our bodies cool themselves by perspiring. Perspiration, which consists of water and a trace of salt, must absorb heat from your body (and from the air) to evaporate. The heat absorbed from your body cools your body temperature. When hot air is humid, our perspiration can't evaporate quickly, so we feel hot and sticky. In dry air at the same temperature, our perspiration evaporates quickly, so we feel cool and comfortable. Significantly, it's not the absolute amount of water in the air (as indicated by the vapor pressure) that makes it feel humid, but rather how closely the air approaches saturation. To convey a sense of how damp air feels, meteorologists characterize the air's moisture content by specifying its **relative humidity** (RH), defined as the amount of water vapor in the air (the measured vapor pressure) divided by the air's capacity for holding water vapor (the saturation vapor pressure). We can express this quotient as a percentage:

$$RH = \frac{\text{vapor pressure}}{\text{saturation vapor pressure}} \times 100\%$$

Clearly, relative humidity depends on both the amount of water vapor in the air and the air's holding capacity. For example, if two volumes of air contain the same number of water vapor molecules, the warmer one will have a lower relative humidity (Fig. 16.10a). Consequently, a given location may feel cool and be foggy because the air there has a high relative humidity, while another location may feel hot and dry because the air has a low relative humidity, even if it contains the same number of water vapor molecules as the cool, foggy air does (Fig. 16.10b). When the relative humidity is 90%, an actual temperature of 29°C (84°F) might feel like 37°C (98°F) to you because your perspiration won't evaporate.

FIGURE 16.10 Relative humidity.

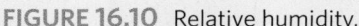

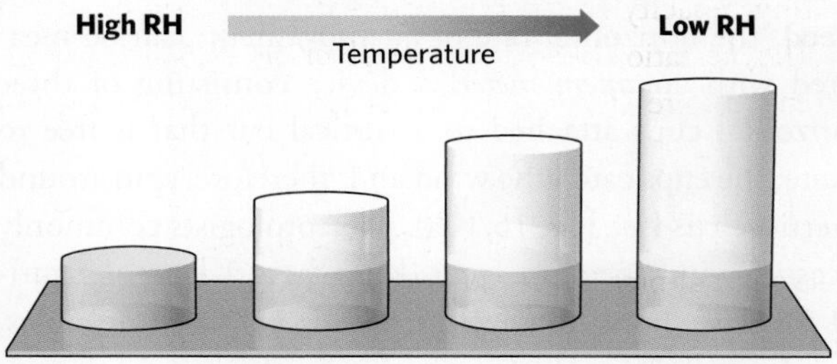

High RH → **Low RH**
Temperature

(a) Because cold air can hold less water vapor than warm air, the same volume of water vapor in the air makes cold air more humid.

Hot, dry day

Warm, humid day

Cool, dry day

Foggy, chilly day

| 10°C | 20°C | 30°C | 40°C |
| RH = 100% | RH = 20% | RH = 80% | RH = 5% |

(b) In the real world, both the amount of water vapor in the air and the temperature vary from place to place. The size of each beaker represents the saturation vapor pressure. The ratio of the amount of water in the beaker to the beaker's capacity represents the relative humidity (RH).

FIGURE 16.11 The dew-
point is the temperature
at which air, when cooled
at constant pressure,
becomes saturated with
water vapor.

(a) The air where this spider built its web has cooled to the dewpoint, and water has condensed on the web.

(b) On a cold day, frost forms on leaves.

**Did you
ever wonder...**

why you feel hotter on a
steamy, humid day in Florida
than on a hot, bone-dry day
in the Arizona desert?

The relationship between human comfort, temperature, and relative humidity can be expressed by the *heat index*, a calibration of how hot air feels, given its humidity.

While vapor pressure provides a good way to describe the actual amount of moisture in the air, it's very difficult to measure. Instead, meteorologists rely on another variable, the dewpoint, to characterize the absolute, as opposed to the relative, amount of water vapor in the air. The **dewpoint** is the lowest temperature to which air can be cooled, at constant pressure, before becoming saturated. To measure the dewpoint, we simply cool air at constant pressure and measure the temperature at which liquid droplets first appear. When temperatures cool overnight, air may cool to the dewpoint and become saturated, so that tiny liquid drops, called *dew*, form on exposed surfaces **(Fig. 16.11a)**. If the process happens when the temperature is below freezing, *frost*, a thin ice coating on solid surfaces, will form. The temperature at which frost appears is called the *frost point* **(Fig. 16.11b)**.

Note that the dewpoint is always lower than, or equal to, the air temperature. We can estimate relative humidity by comparing the dewpoint with the air temperature. When the dewpoint approaches the air temperature, the relative humidity is high, as happens on muggy days. In contrast, when the two temperatures are far apart, the relative humidity is low, as happens on crisp, dry days.

Wind Speed and Direction

Everyone knows the feel of flowing air. It can cool your skin, rustle the leaves, or—if stronger—rip off your roof. Atmospheric scientists define **wind** as the horizontal movement of air. As we noted earlier, when the wind blows, it exerts *dynamic pressure* that applies a push to an object. Of course, air in nature doesn't just move horizontally—it can also flow up as an **updraft** or down as a **downdraft**. While up-and-down air movements have real consequences, such as triggering storms or causing discomfort for airline passengers, vertical air movements are not represented by a standard "wind" measurement.

To describe wind, we use two familiar quantities: wind direction and wind speed. By convention, the **wind direction** refers to the direction from which the wind blows. For example, in the northern hemisphere, a north wind brings colder air from the north toward the south. We can measure wind direction with a *wind vane*, mounted so that it swings about a vertical axis; the push of the wind reorients an arrow so that the arrowhead points in the direction the wind comes from **(Fig. 16.12a)**. **Wind speed**, the horizontal rate of air movement, can be measured with an *anemometer*, a device consisting of three horizontal cups attached to a vertical bar that is free to rotate; the cups catch the wind and, therefore, spin around a vertical axis (see Fig. 16.12a). Meteorologists commonly measure wind speed in *knots* (kt), where 1 kt = 1 nautical mile per hour. A nautical mile is 1.15 statute miles, so 1 kt = 1.15 mph. Converting to the metric system, 1 kt = 1.85 km/h. Meteorologists display winds by placing *wind barbs* on a weather map **(Fig. 16.12b)**. They may also display winds on a map by drawing lines parallel to the wind direction **(Fig. 16.12c)**, with the line segments brightening progressively in the direction that the air flows.

Visibility and Cloud Cover

The old phrase "On a clear day, you can see forever" may be an exaggeration, but it does draw attention to the fact that air doesn't always have the same transparency. Meteorologists describe air transparency by specifying **visibility**, the distance from which a person with normal vision can identify objects while looking through the air. In the past, visibility measurements were made by a human observer looking at objects at known distances. Today, meteorologists measure visibility automatically with a sensor that detects air transparency. What factors affect visibility? Pure air is virtually transparent, so visibility decreases

FIGURE 16.12 Measuring wind speed and direction.

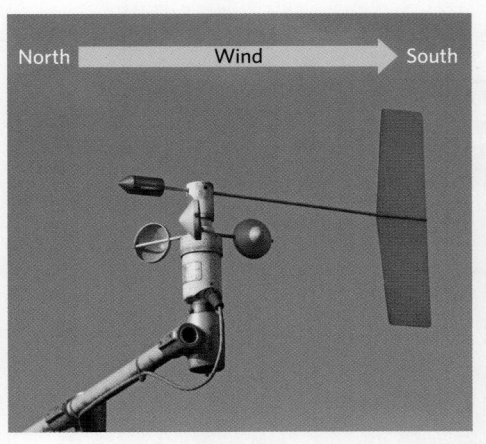

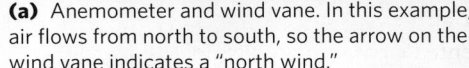

North   Wind   South

⊙   Calm

From the W at 10 kt

From the SW at 5 kt

From the E at 25 kt
(5 + 10 + 10 = 25)

From the NW at 50 kt

From the SE at 110 kt

(a) Anemometer and wind vane. In this example, air flows from north to south, so the arrow on the wind vane indicates a "north wind."

(b) Key to wind barbs used on a weather map.

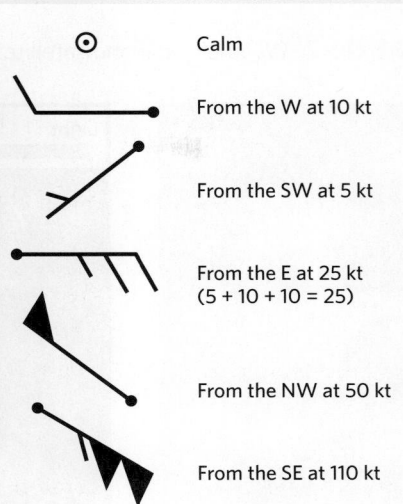

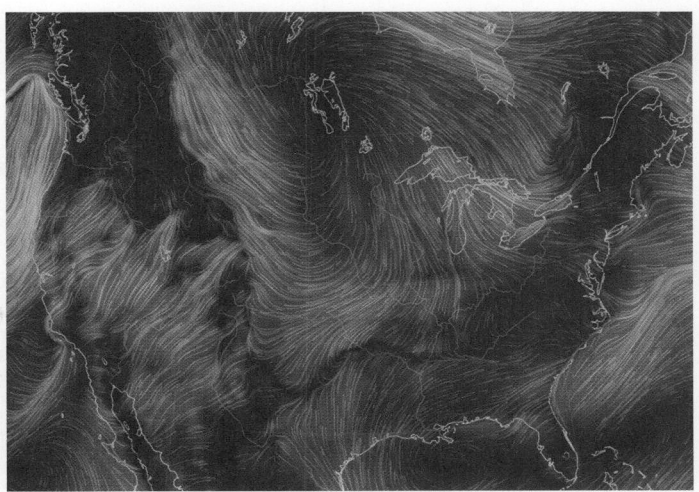

(c) On this wind map, winds are represented by swirling lines; overall, the lines are brighter and more closely spaced where the wind is faster. The orientation and the direction in which the brightness of the colored lines tapers represent wind direction.

when air contains aerosols, haze droplets, water droplets, or ice particles. A fog can reduce visibility to near zero, making driving hazardous, whereas on a clear day, visibility may be "unlimited" in that you can see to the horizon and beyond. Visibility measurements are provided in units of distance. For example: a dense fog means visibility is less than 0.40 km (0.25 mi), while very clear air means visibility is unlimited to the horizon and beyond.

Meteorologists define **cloud cover** as an estimate of how much of the sky is obscured by clouds at a given time. In the past, human observers would determine cloud cover by visually estimating what fraction of the sky contained clouds. Today, laser technology allows us to determine what fraction of the time clouds lie directly overhead. When describing cloud cover, we typically use familiar descriptive terms (clear skies, few clouds, scattered clouds, broken clouds, or overcast skies) to indicate progressively more cloud cover. Cloud cover varies radically with location and time of year. At any given time, clouds hide nearly 70% of the Earth's surface (Fig. 16.13).

FIGURE 16.13 Composite satellite image showing the Earth's cloud cover.

BOX 16.2 **A Deeper Look**

Interpreting meteorological radar and satellite information

Almost every television newscast or website has a segment on the weather. During those segments, on-air meteorologists almost always show animations of weather radar and satellite data. Free apps are also available that can display the latest weather radar animations from the United States, Europe, and other regions, as well as worldwide satellite images of clouds, to anyone with a smartphone.

Every year, floods, tornadoes, and other types of hazardous weather cause loss of life and property damage. Meteorologists can warn the public of these impending hazards by using **weather radar**. What is weather radar, and how do we interpret the images it provides? Radar systems transmit energy in the form of *electromagnetic radiation*. As we discuss later in this chapter, electromagnetic radiation travels in the form of invisible waves. Different types of radiation are distinguished from one another by their wavelength. Radar waves have a wavelength of 1–10 cm (0.4–4 in), similar to that of the microwaves in a microwave oven **(Fig. Bx16.2a)**. But unlike your oven, a radar transmitter sends out microwaves in pulses that last only about one millionth of a second. When the microwaves encounter raindrops and hailstones, some of the microwaves scatter back to the radar system's antenna. The antenna collects this energy, called the *radar echo*, and measures the time it took the microwave pulse to travel out to the precipitation and back. Because microwaves travel at the speed of light, we can calculate the distance to the precipitation from the following equation: distance = velocity × time. By knowing the angle at which the antenna points to the sky, we can precisely determine the location and altitude of the precipitation within the atmosphere.

The character of precipitation affects the strength of the returned signal, or the *radar reflectivity*. For example, the greater the size and number of particles the pulse intercepts, the greater the radar reflectivity will be. High radar reflectivity values, depicted with red and orange colors on a display, indicate heavy rain or hail. In contrast, lower values, depicted with green, indicate moderate rain, while blues indicate light drizzle or nonprecipitating clouds **(Fig. Bx16.2b)**. By examining radar reflectivity measurements over time, meteorologists can estimate the total amount of rain that fell during the period of observation. These data help predict flash floods. On television, meteorologists normally show the radar reflectivity in a loop to give viewers an idea of where rain is falling, the intensity of the rainfall, and where the rain is heading next.

FIGURE Bx16.2 Weather radar and satellites.

(a) A radar antenna used in research.

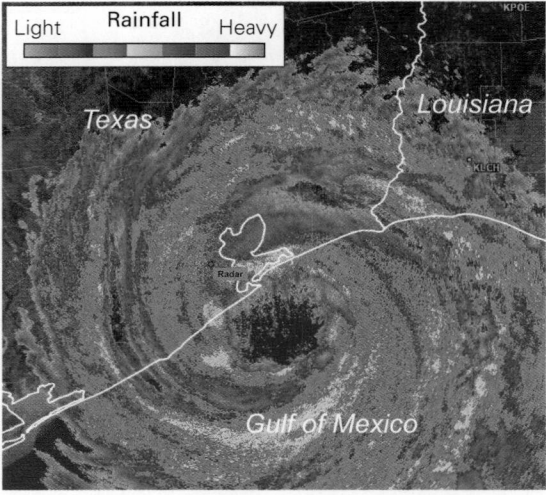

(b) Radar reflectivity from Hurricane Ike as it made landfall near Galveston, Texas, in 2008.

(c) This image shows cloud cover as revealed in the wavelengths of visible light.

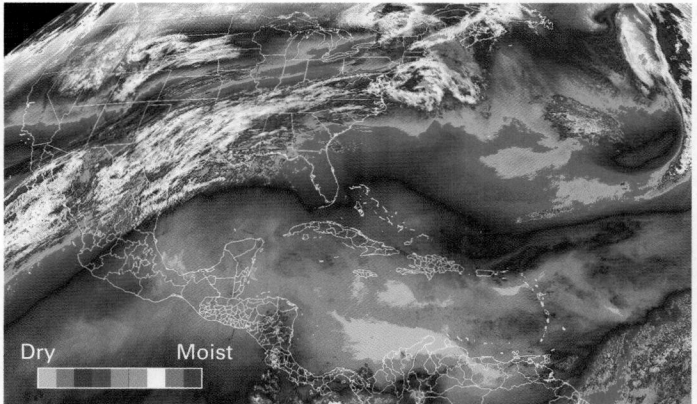

(d) By sensing certain wavelengths of infrared light, satellite images reveal variations in atmospheric moisture content. Reds and browns in the image correspond to dry air, while whites and blues correspond to moist air.

Radar systems transmit microwave energy at specific frequencies (the number of waves that pass a point in one second). When the radar echo returns to the antenna, the frequency of the returned energy typically shifts to a slightly different value as a result of the movement of the raindrops or ice particles along the direction of the radar beam. **Doppler radar** systems measure this shift and use it to determine the speed of the wind in the direction parallel to the beam. Unfortunately, Doppler radar cannot measure the motion of air moving perpendicular to the beam. Nevertheless, by mapping the wind in the direction of the beam, Doppler radar can identify strong straight-line winds as well as rotation in the air flow, which may be evidence of a tornado. Therefore, meteorologists who monitor severe weather commonly use Doppler radar to detect these hazards and to provide early warning to the public. The development of polarization Doppler radar, which employs beams in which the microwaves move only in a given plane, can now even help meteorologists identify tornadoes that have lofted airborne debris.

Meteorologists supplement radar measurements from ground-based locations with data collected by orbiting *weather satellites*, which can acquire images of the atmosphere over a wide region, or even the entire world. Animated satellite images allow meteorologists to watch storms develop and monitor the progress of a storm as it moves across the Earth. Weather satellites use different wavelengths of radiation to gather different kinds of information. The *visible channel* uses reflected sunlight to show clouds as you would see them from space, in black and white or even in color **(Fig. Bx16.2c)**. The *infrared channel* shows the temperature of the cloud tops or the ground. The *water vapor channel* uses reds and blues to indicate dry and moist regions of the atmosphere **(Fig. Bx16.2d)**. Meteorologists on television often show satellite data animations to inform viewers about clouds moving into or out of the regions they serve.

> **▶ Animation**
> Radar Monitors a
> Landfalling Hurricane

> **▶ Animation**
> Weather Monitoring
> with Satellites

Precipitation

In the geology chapters of this book, we generally use the word *precipitation* to refer to the formation of a solid when a liquid solution becomes supersaturated. In atmospheric science, **precipitation** refers to water vapor that has condensed to a liquid or solid state and falls from the sky. Precipitation includes sprinkles to heavy downpours of rain, flurries to heavy snow, and small to giant hailstones.

Meteorologists traditionally measure precipitation with a *precipitation gauge*, a container that collects either rain or snow and measures its weight or volume **(Fig. 16.14)**. The number recorded by the gauge can be converted to inches or millimeters, to represent the thickness of the water layer or snow layer that landed on the ground. Today, we can also use information from *weather radar* to estimate precipitation over large regions **(Box 16.2)**.

To characterize precipitation, meteorologists specify both the *precipitation rate*, given in inches per hour or millimeters per hour, and the *total precipitation*, the cumulative amount that falls during a specified time period, such as a single storm, a single month, a season, or a whole year. We can report total rainfall in inches or millimeters. For total snowfall, the measurement indicates either an *accumulated depth*, the thickness of the snow layer, or its *water equivalent*, the depth of the water that would be formed if the snow melted completely. Water-equivalent measurements are helpful in comparing snowfalls that accumulated under different temperature conditions,

FIGURE 16.14 Measuring rainfall with a tipping-bucket rain gauge.

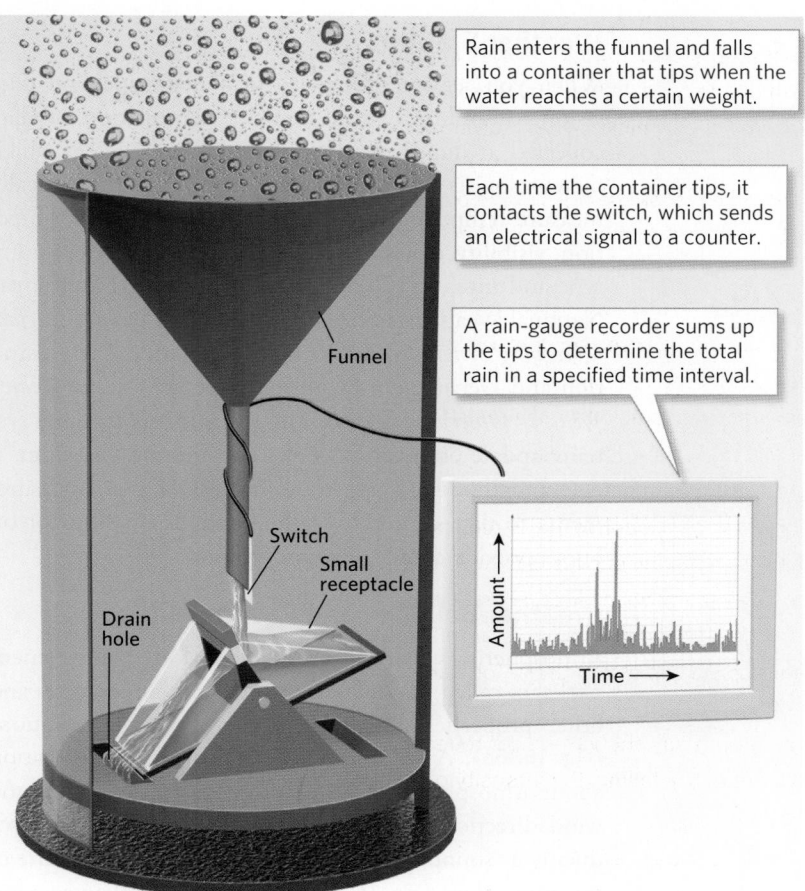

Rain enters the funnel and falls into a container that tips when the water reaches a certain weight.

Each time the container tips, it contacts the switch, which sends an electrical signal to a counter.

A rain-gauge recorder sums up the tips to determine the total rain in a specified time interval.

Funnel

Switch

Small receptacle

Drain hole

Amount

Time

FIGURE 16.15 Surface weather stations in North America.

(a) Each white dot on this map represents a weather station.

(b) Equipment set up at a weather station.

because snow density can vary significantly depending on temperature: warmer snow tends to be dense, whereas colder snow tends to be light and fluffy.

Weather Stations

When you see weather maps in a newspaper, on television, or online, you're seeing a compilation of data collected at hundreds of *weather stations*, sites at which meteorologists have set up instruments that automatically measure temperature, dewpoint, wind speed and direction, visibility, cloud type and coverage, and precipitation type and intensity (Fig. 16.15). In the United States, the National Weather Service operates the *Automated Surface Observing System*, while the Federal Aviation Administration and Department of Defense operate the *Automated Weather Observing System*. The instruments in these systems update observations every five minutes, 24 hours a day, and they send electronic records of their measurements to data centers. Weather stations on ships or on buoys provide data for oceanic regions.

Weather Balloons and Rawinsondes

Atmospheric scientists use balloon-lofted instrument packages called *rawinsondes* to measure temperatures and other properties of the atmosphere at high elevations (Fig. 16.16a). A rawinsonde transmits back information about atmospheric pressure, temperature, dewpoint, and wind direction and speed continuously as it rises, producing a "sounding" that depicts the vertical structure of the atmosphere (Fig. 16.16b). By international agreement,

rawinsondes are launched simultaneously twice a day at 00:00 (midnight) and 12:00 (noon) Universal Coordinated Time (the time in London, England) at hundreds of localities worldwide. On average, rawinsonde launch sites on land are located about 500 km (~300 mi) apart, so their spacing is much greater than that of ground-based weather stations. As a result, high-elevation data are not as detailed as ground-based data. Because atmospheric pressure decreases with elevation, weather balloons expand as they rise. When a balloon reaches about 20 km (12.5 mi), it breaks, and the rawinsonde it carried parachutes back to the ground.

Take-home message...

To characterize weather, meteorologists measure temperature, atmospheric pressure, dewpoint, wind speed and direction, cloud cover, visibility, and precipitation. These properties vary with location at a given time and with time and elevation at a given location. Meteorologists also use Doppler radar systems and satellites to track storms and identify hazardous weather conditions.

Quick Questions ——————————

- Is there a temperature that is numerically the same on both the Fahrenheit and Celsius temperature scales?
- Would an unopened bag of potato chips expand or contract if you carried it up a mountain?
- What does wind speed refer to?

(a) Students launching a weather balloon with a rawinsonde attached. The inset shows the rawinsonde before it was launched.

Symbols indicate direction (relative to north) and wind speed. At 12 km elevation, wind comes from the west, and wind speed is 50 knots.

Tropopause

Cloud layer

Temperature

Dewpoint

Wind

FIGURE 16.16 A rawinsonde unit can be used to obtain information about the atmosphere at high elevations.

(b) A vertical profile of atmospheric temperature, dewpoint, and wind speed and direction from a rawinsonde. See Figure 16.12 for an explanation of the wind symbols at the right. Note that the orientation of the wind symbol gives the compass orientation of the wind, at each indicated elevation, relative to the north arrow.

16.3 Vertical Structure of the Atmosphere

Imagine that you're sitting outside reading this book on a warm, calm summer day. Right now, only 10 km (6 mi) directly above you, the wind may be blowing at over 160 km/h (100 mph), the temperature may be hovering at a frigid –50°C (–58°F), and the gas in the air has such low density that you couldn't survive by breathing it. If you're a passenger in an airliner that has climbed to cruising altitude, these conditions exist right outside your window. Clearly, atmospheric conditions change rapidly with elevation. Based on these changes, atmospheric scientists have divided the atmosphere into layers.

Thermally Defined Layers

Temperature changes rapidly with altitude. Research shows that between certain altitudes, temperature increases as altitude increases, whereas between others, temperature decreases as altitude increases. Altitudes at which the direction of change reverses (from decreasing to increasing upward, or vice versa) delineate boundaries that separate four distinct atmospheric layers **(Fig. 16.17)**. Let's examine these layers in sequence, beginning at the Earth's surface.

Energy from the Sun reaches the Earth in the form of visible light and ultraviolet (UV) light, both of which are

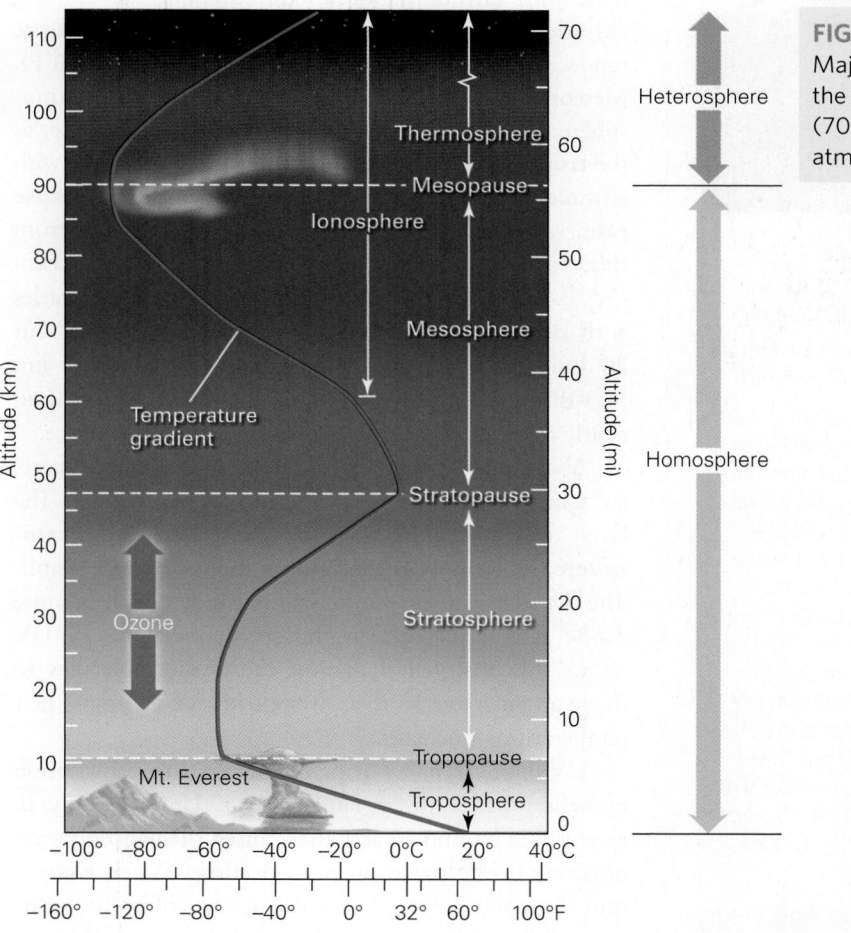

FIGURE 16.17 Major layers in the lower 110 km (70 mi) of the atmosphere.

types of electromagnetic radiation. (As we discuss later in this chapter, visible light and UV light have wavelengths that are shorter than the microwaves of the radar beams discussed in Box 16.2.) Since solar energy reaches the atmosphere from above, it's a common misperception that the air around you becomes warm because it's "broiled from above." In fact, air is transparent to most incoming solar energy that arrives as visible light—that's why you can see through air! The Earth's surface absorbs the visible light that reaches it, and then radiates the energy back into the air in a different form, known as *infrared radiation*, that you can't see but can feel as heat. (The wavelength of infrared radiation is between that of visible light and microwaves.) Some of this infrared radiation gets trapped by specific gases in the air and causes the air to warm, a phenomenon known as the *greenhouse effect*. Put another way, the Earth's surface acts as a heater that warms the air at the base of the atmosphere—in effect, air is "baked from below," not "broiled from above."

As we move upward, away from the Earth's surface, the distance from the Earth-surface heat source increases, so the air gets cooler. So, if you climb a tall mountain, you'll find that as your elevation increases, air temperature decreases. This decrease in temperature with increasing altitude continues up to about the cruising altitude of commercial airliners, where air temperature ranges between –40°C and –60°C (–40°F and –76°F). Meteorologists refer to this lowest layer of the atmosphere, in which temperature decreases with altitude, as the **troposphere**. The rate of change in temperature with altitude, a quantity known as the **environmental lapse rate**, varies substantially from place to place and from time to time, but on average has a value of 6.5°C/km (3.5°F/1,000 ft). The thickness of the troposphere varies with latitude and time of year: it decreases from about 20 km (12.5 mi) above the equator to about 7 km (4.3 mi) above the frigid poles. The Earth's storms and nearly all clouds occur entirely within the troposphere.

Above the *tropopause*, the top of the troposphere, air temperature increases with increasing altitude. The layer in which this increase takes place, the **stratosphere**, reaches up to a height of about 50 km (31 mi). The increase in temperature with altitude occurs because ozone (O_3) molecules in the stratosphere absorb UV light. This absorbed radiation causes air molecules to move about faster, so their average kinetic energy—their temperature—increases.

UV light can harm living organisms, so its absorption by ozone makes life on land possible. Unfortunately, in recent decades human activities caused the concentration of ozone in the stratosphere to decrease. When emitted into the atmosphere, chemicals such as chlorofluorocarbons (CFCs) react with and break up ozone molecules.

This reaction happens most rapidly on the surfaces of tiny ice crystals in polar stratospheric clouds, so an **ozone hole**, a region of diminished ozone concentration, forms over the South Pole during the southern hemisphere spring (Fig. 16.18a). Because of the importance of stratospheric ozone, a 1987 international summit in Montreal led to an agreement to reduce CFC emissions globally.

Above about 50 km, an altitude called the *stratopause*, there's not enough ozone absorption of solar UV light to heat the air, so air temperature again begins to decrease with increasing altitude. This altitude marks the bottom of the **mesosphere**, which continues upward to the *mesopause*, at an altitude of about 90 km (56 mi). Above the mesopause lies the outermost layer of the atmosphere. In this layer, the **thermosphere**, temperature increases with increasing elevation, reaching a maximum of 1,500°C (2,700°F), as the few remaining air molecules absorb high-energy radiation from the Sun. Note that the entire thermosphere lies above the *Kármán line*, a conventional boundary between the Earth's atmosphere and outer space set at an altitude of 100 km by an international record-keeping body. To an astronaut, therefore, the thermosphere is effectively part of space. These high temperatures in the thermosphere are deceptive, for an astronaut entering the thermosphere without first donning a space suit would freeze instantly. That's because the air density in the thermosphere is so low that a given volume of air holds hardly any heat. All the air molecules colliding with the astronaut, though they are traveling very fast, could not provide enough heat to counteract the loss of heat by infrared radiation coming from the astronaut's body. The density of air in the thermosphere decreases upward until, at an altitude that ranges from 350 to 800 km (217–500 mi), it equals that of interplanetary space. This altitude, the highest definition of the top of the atmosphere, varies depending on the quantity of incoming atoms brought to the Earth in the solar wind.

Layers Defined by Composition and Electric Charge

Below an altitude of about 90 km, air circulation mixes the atmosphere so thoroughly that its composition remains mostly the same from place to place. Researchers refer to this well-mixed layer as the **homosphere** (see Fig. 16.17). We use the qualifier "mostly" because the concentration of certain gases varies. For example, ozone forms and occurs primarily in the stratosphere, and gases that enter the atmosphere from localized sources at the Earth's surface tend to decrease in concentration with distance from the source. Also, water vapor, which has a short *residence time* in the atmosphere (meaning that it enters and leaves the atmosphere relatively quickly), varies substantially with location and time. Note that the homosphere includes all of the troposphere, stratosphere, and

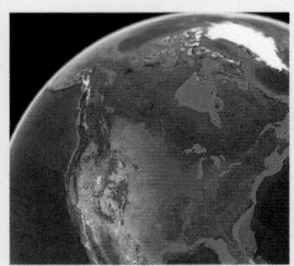

mesophere. Above the homosphere, gases sort gravitationally according to their molecular weight, with lighter gases rising and heavier ones sinking. This stratified layer, the **heterosphere**, coincides with the thermosphere.

Meteorologists also recognize another layer, not based on temperature or composition, but rather on electrical characteristics. In this layer, known as the **ionosphere**, the atmosphere contains a relatively high concentration of *ions*, molecules that have gained or lost electrons and thus have a negative or positive electric charge, respectively. The ionosphere, which occurs above elevations of 60 km (37 mi) during the day and 100 km (62 mi) at night (see Fig. 16.17), serves an important role in human communications because it reflects certain wavelengths of radio transmissions. These transmissions, notably AM radio and shortwave radio, bounce back and forth between the ionosphere and the Earth's surface, which allows them to travel over long distances (Fig. 16.18b).

Incoming high-energy ions from the Sun channel along the Earth's magnetic field lines, producing electrical currents that cause high-energy ions trapped within the Earth's magnetic field to flow into the atmosphere toward the poles of the Earth. When they interact with certain molecules in the ionosphere, these molecules emit light. We see this light as a shimmering **aurora**, a gauzy curtain of undulating green and red color in the sky (Fig. 16.18c). We refer to the aurora above northern polar regions as the *aurora borealis*, or northern lights, and that over southern polar regions as the *aurora australis*, or southern lights.

Take-home message…

Atmospheric scientists divide the atmosphere into layers based on changes in its temperature, composition, and electric charge. The layers defined by changes in temperature are the troposphere, stratosphere, mesosphere, and thermosphere. The layers defined by consistency of composition are the homosphere and heterosphere, and the layer defined by electric charge is the ionosphere.

Quick Questions

- If you viewed a thunderstorm towering above you while standing on the top of a tall mountain, would you be in the troposphere or the stratosphere? How about if you flew an SR-71 spy plane, to its maximum cruising altitude of 90,000 ft?

- When and where does the ozone hole appear in Earth's stratosphere?

- Why would an (accidentally) exposed astronaut freeze in the thermosphere, despite the high temperature of the air there?

- What role does the ionosphere play in radio transmission?

FIGURE 16.18 High-altitude phenomena of the atmosphere.

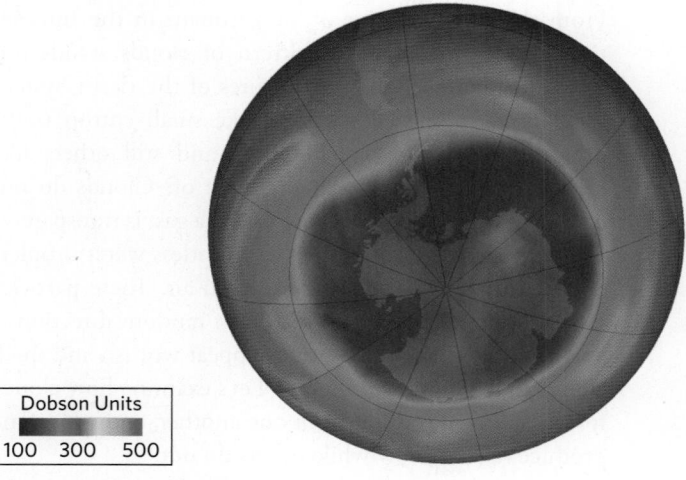

Dobson Units
100   300   500

(a) The ozone hole over Antarctica in August 2018. A Dobson unit measures the concentration of ozone in the atmosphere; larger numbers mean more ozone.

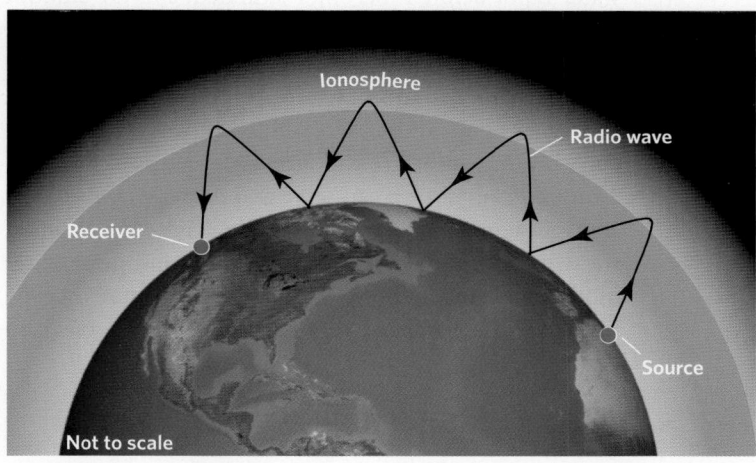

(b) The ionosphere, a layer containing charged ions, causes certain radio waves to reflect back to the Earth, from which they bounce up again. These waves can be detected beyond the horizon.

(c) The aurora borealis, as seen over Alaska.

16.4 Clouds and Precipitation

From the vantage point of an astronaut in the International Space Station, the pattern of clouds stands out as one of the most striking features of the Earth System (Fig. 16.19). Some clouds look like small cotton puffs, others like broad white blankets, and still others like elegant swirls. What do they consist of? Clouds do not consist of water vapor, because vapor, a gas, is transparent. Rather, clouds are collections of countless water droplets and/or tiny ice crystals suspended in air. These particles reflect, scatter (absorb and reemit in random directions), and absorb light, so thin clouds appear whitish and thick clouds look gray and ominous. Let's examine how clouds form, how clouds differ from one another, and why some produce rain or snow while others do not.

Cloud Formation

As we have seen, air can hold only a certain amount of water vapor before it becomes saturated, and the amount it can hold depends on its temperature. If saturated air cools, it becomes *supersaturated*, meaning that it contains a greater concentration of water vapor than it can hold at a given temperature. When this happens, water molecules begin to attach to the surfaces of aerosols. In effect, a tiny aerosol particle can serve as a *condensation nucleus*. If the air temperature is above freezing, a droplet of liquid water grows around the nucleus as more and more water molecules attach. If, however, the air temperature is below freezing, an aerosol can serve as a *deposition nucleus*, to

▶ **Animation**
Dewpoint and
Cloud Formation

FIGURE 16.20 When air containing water vapor rises, it expands, cools, and if it rises far enough, forms a cloud.

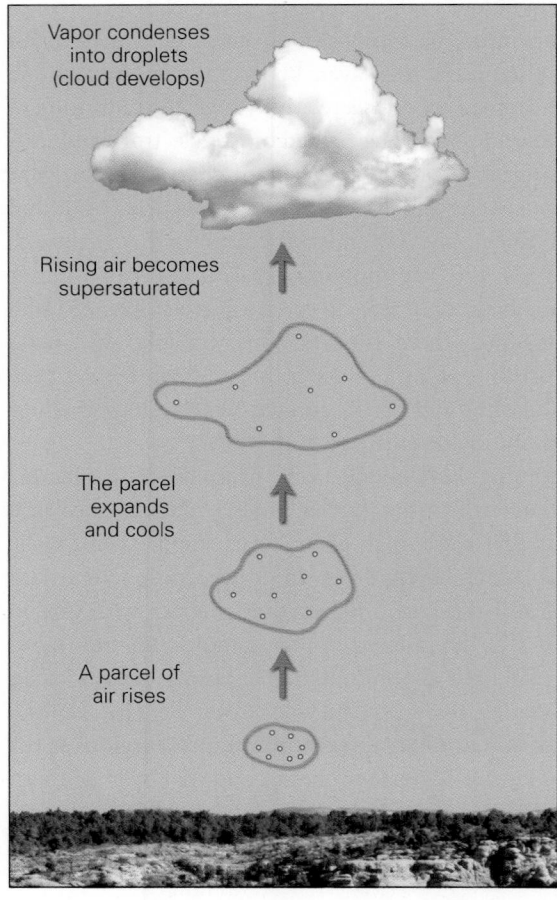

Vapor condenses into droplets (cloud develops)

Rising air becomes supersaturated

The parcel expands and cools

A parcel of air rises

which water molecules attach to produce a crystal of solid ice. Water droplets and ice particles are visible, so when condensation or deposition takes place, a cloud appears. Note that clouds form only when water molecules in supersaturated air collect on aerosols to form liquid droplets and ice particles. Therefore, every raindrop and snowflake you see initially formed on an aerosol particle, so if there were no aerosols, there would be no rain, and life on land would never have evolved.

How does air become locally supersaturated? In general, supersaturation happens where air rises (Fig. 16.20). Since atmospheric pressure decreases with altitude, rising air expands as it moves upward. In order to expand, the air must push against surrounding air molecules. Such pushing takes energy, so the air loses heat energy and becomes cooler (as we'll see in Section 18.3). Cooling, in turn, reduces the air's capacity to hold water vapor, so if its temperature decreases enough, the air cools to its dewpoint. At this point, any further rise leads to further cooling, causes supersaturation and, therefore, cloud production.

FIGURE 16.19 A view of the Earth and its clouds from the International Space Station.

FIGURE 16.21 Some of the mechanisms that cause air to rise.

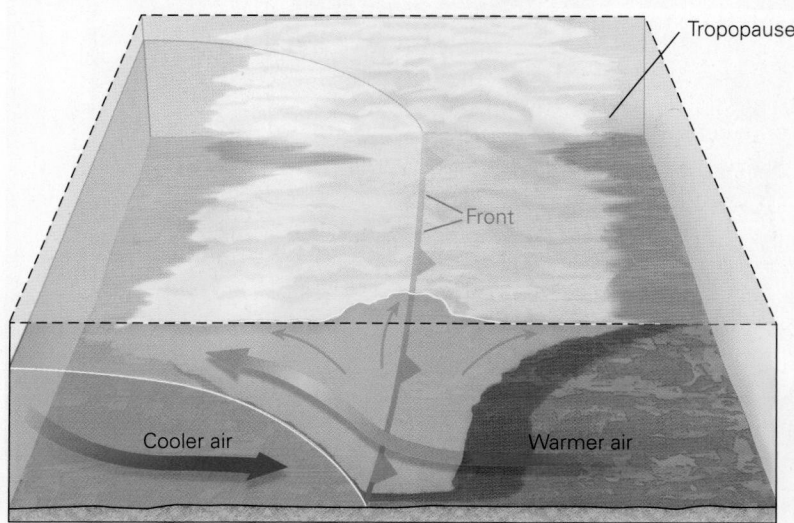

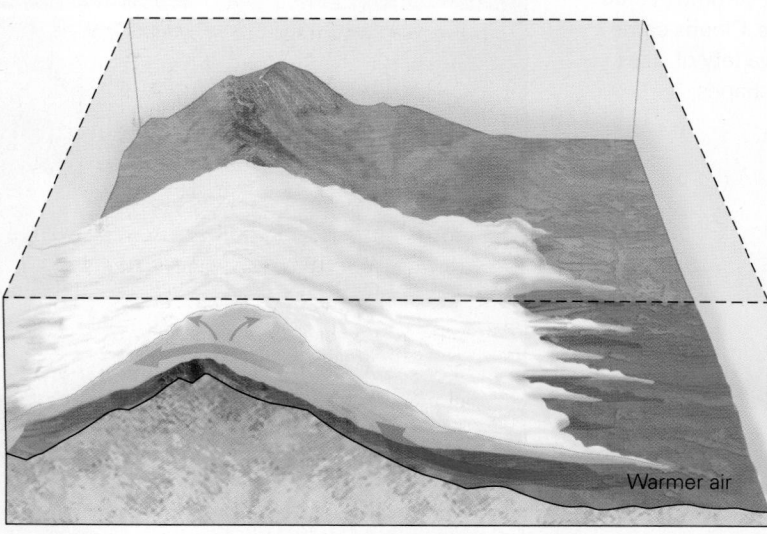

(a) When a mass of warm air meets an advancing mass of cold air, the warm air is forced to rise. The leading edge of the cold air mass defines the location of the front.

(b) When flowing air reaches a mountain range, it is forced to rise.

What phenomena cause air to rise? Air rises when it undergoes **lifting** by any of several processes that force air upward. Lifting can happen, for example, when a mass of less dense, warmer air encounters and flows up and over a mass of denser, cooler air (Fig. 16.21a). The boundary where two such contrasting air masses meet is called a **front**. A similar phenomenon happens along a shoreline when cool air moves onshore and displaces warm air, so the warm air has nowhere to go but up. Air can also rise as it encounters and flows over a mountain range (Fig. 16.21b). And air can rise when it becomes *buoyant*, meaning that it becomes less dense than its surroundings (Fig. 16.21c). Buoyancy can develop, for example, when heat rising from the ground below warms air near the ground surface. You've seen this process happening if you've ever seen a hot-air balloon head skyward. We'll discuss additional ways in which air can be lifted in Chapter 18.

Types of Clouds

Clouds fascinate people because of their beauty and the way they constantly change. They come in many shapes and sizes and form at many altitudes within the troposphere. Meteorologists classify clouds by their altitude and shape and by whether or not they produce precipitation, using terminology first proposed in 1802 by a British pharmacist named Luke Howard (Fig. 16.22). Howard, adapting Latin vocabulary, coined the terms *cirrus* for high, wispy clouds; *stratus* for sheets of clouds that cover vast areas; and *cumulus* for tall, puffy clouds with a cauliflower-like appearance. Combining these terms with others provides additional detail about clouds. For example, adding the prefix or suffix "*nimbus*" to a cloud

(c) Air becomes buoyant and rises over warmer ground.

name indicates that the cloud produces rain, so a *cumulonimbus cloud* is a towering puffy cloud that produces rain. Today, we recognize many more types of clouds, but still use many of the names Howard first proposed in 1802.

Why are there so many types of clouds? The character of a cloud depends on the amount of water vapor available, the nature of the lifting mechanisms that drive cloud formation, the temperature where the cloud forms in the troposphere, and the velocity of winds at various altitudes.

HIGH-ALTITUDE CLOUDS. As we've noted, wispy, elongated clouds that develop very high in the troposphere (above 6 km, or 20,000 ft) are known as **cirrus clouds** (Fig. 16.23a). The temperature is so cold where cirrus clouds form that these clouds consist entirely of tiny ice crystals. As the ice crystals fall through the atmosphere,

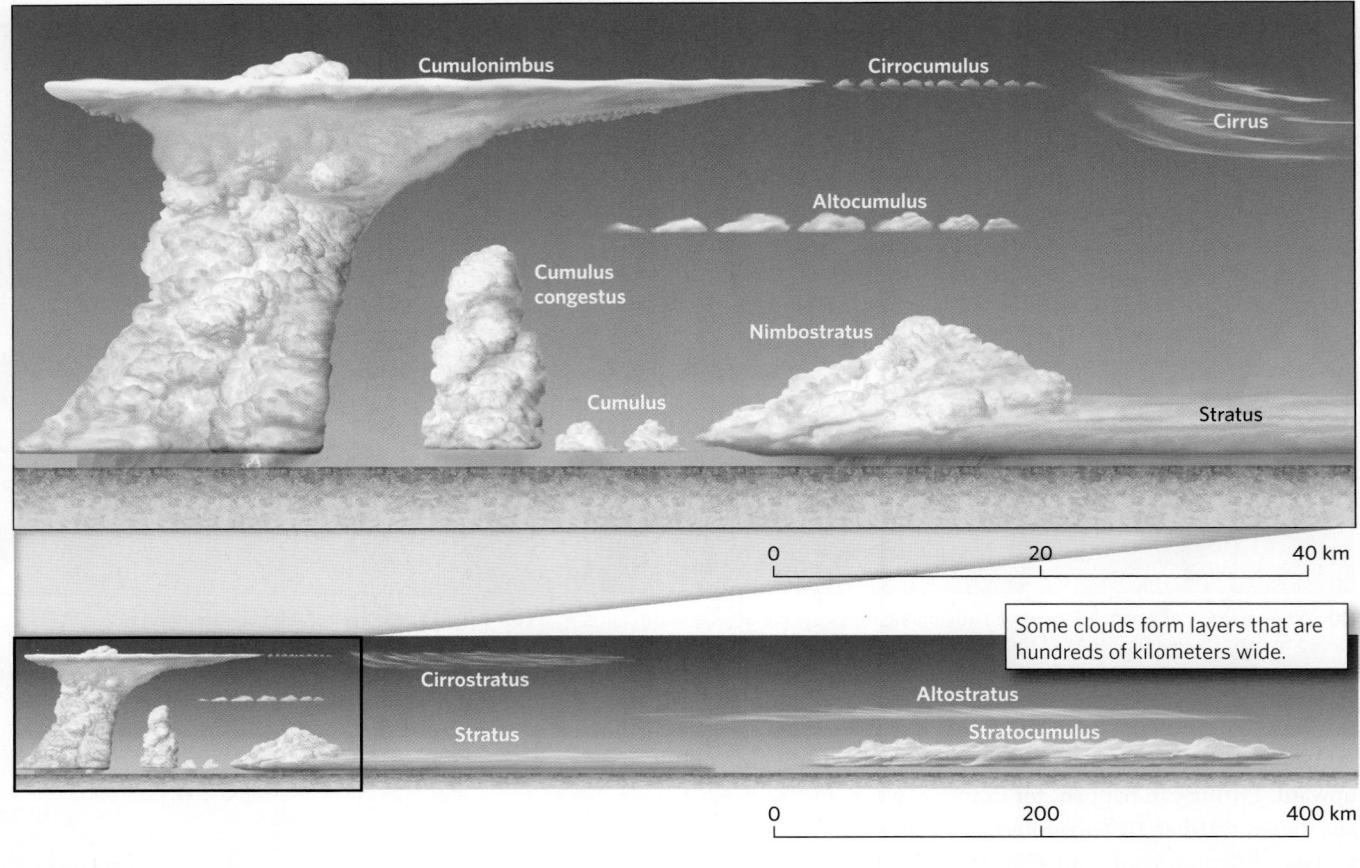

Some clouds form layers that are hundreds of kilometers wide.

Did you ever wonder...

why clouds have different shapes?

winds at different altitudes cause them to line up in wisp-like strands. Cirrus clouds are too high and thin to produce precipitation. If they do accumulate into layers thick enough to dim the Sun, they become *cirrostratus clouds*, and if they grow into puffy billows, they become *cirrocumulus clouds*.

MID-ALTITUDE CLOUDS. Mid-altitude clouds form between 3 and 6 km (10,000–20,000 ft) above the Earth's surface (Fig. 16.23b). Meteorologists refer to them as *altostratus clouds* if the clouds are layered, uniformly gray, and thick enough to hide the Sun. If the clouds resemble roll-like patches or puffs, they are *altocumulus clouds*. Altostratus and altocumulus clouds, like cirrus clouds, are too high and too thin to produce precipitation.

LOW-ALTITUDE CLOUDS. Widespread, layered clouds that spread out like great blankets at relatively low altitudes (below 3 km, or 10,000 ft) are called **stratus clouds** (Fig. 16.23c). The width of stratus clouds greatly exceeds their thickness. If portions of stratus clouds billow into puffy shapes, they become *stratocumulus clouds*, and if stratus clouds thicken sufficiently that precipitation falls from them, they become *nimbostratus clouds*.

TOWERING CLOUDS. Unlike the altitude-defined clouds described above, *towering clouds* rise across a range of altitudes (Fig. 16.23d). The largest of these extend from

low altitudes to the tropopause. Small towering clouds are called **cumulus clouds** if they are growing vertically and have cauliflower-like lobes. Cumulus clouds form in strong updrafts and, in general, they have comparable horizontal and vertical dimensions. If cumulus clouds grow very large, they become *cumulus congestus clouds*, and if they produce precipitation, they become *cumulonimbus clouds*. Some cumulonimbus clouds rise into the base of the stratosphere, where strong winds blow their tops into a broad sheet; these cloud sheets are known as *anvil clouds*. Notably, if you film a cumulus cloud and watch it in fast motion, you will see it change shape and size, rapidly rising upward as it develops cauliflower-like lobes. You may also see rain falling from the cloud, and the cloud later evaporating. The rising cloud consists of complex updrafts, downdrafts, and winds that push out new cauliflower-like lobes as earlier ones dissipate, all while raindrops continually form and fall.

UNUSUAL CLOUDS. Some clouds take on strange shapes. For example, when air flows over mountains, lens-shaped clouds, known as *altocumulus lenticularis clouds*, may develop; these clouds have been mistaken for flying saucers (Fig. 16.24a). Thunderstorms can produce ominous cloud formations with names such as wall clouds, shelf clouds, and mammatus clouds (from the Latin word *mamma*,

FIGURE 16.23 Photos of some common cloud types.

Cirrus

Cirrostratus

Cirrocumulus

(a) High-altitude clouds.

Altostratus

Altocumulus

(b) Mid-altitude clouds.

Stratocumulus

Stratus

(c) Low-altitude clouds.

Cumulus

Cumulus congestus

Cumulonimbus

(d) Towering clouds.

FIGURE 16.24 Examples of strangely shaped clouds.

(a) Lenticular clouds (altocumulus lenticularis) may look like flying saucers.

(b) Mammatus clouds over Ohio at sunset.

FIGURE 16.25 Three types of fog.

meaning udder) (Fig. 16.24b), which we'll examine in our discussion of thunderstorms in Chapter 18.

Fog

On a foggy day, visibility can become so poor that you can't see more than a few meters in front of you. Sadly, many deadly vehicle pile-ups happen along foggy stretches of superhighways when drivers can't see stopped cars in front of them. For example, over 300 vehicles collided in a chain reaction that took place in a fog along a Brazilian highway in 2011; from front to back, the pile-up stretched for 1.6 km (1 mi). **Fog** exists when the base of a cloud lies at the Earth's surface and cloaks the ground. Three types of fog develop under different conditions. The first type, known as *radiation fog*, tends to form at night when, in the absence of sunlight, the ground cools by radiating infrared energy upward (Fig. 16.25a). The cooled ground absorbs heat

(a) Radiation fog.

(b) Advection fog.

(c) Evaporation fog.

from the air above it, causing the air to cool and reach saturation so that water condenses. The second type, *advection fog*, forms when warm, humid air flows over a cooler surface, as can happen when warm air moves over cold ocean water or a field of snow (Fig. 16.25b). The third type, *evaporation fog*, forms where enough surface water evaporates into air that the air becomes saturated (Fig. 16.25c). Such fog might develop over a warm lake or marsh.

Precipitation and Its Causes

You've probably experienced *precipitation* (rain, hail, or snow) many times, and you are likely familiar with the common terms (drizzle, downpour, flurries, heavy snow) used to describe precipitation. What conditions lead to precipitation, and why do some clouds pass overhead without dropping any?

WARM-CLOUD PRECIPITATION. Meteorologists define a *warm cloud* as one in which the temperature stays above 0°C throughout, so that the cloud consists only of liquid water droplets. In such a cloud, the individual *cloud droplets* can remain so small that they remain suspended in air until they evaporate. If, however, the droplets grow sufficiently large, they become **raindrops**, spheres of water heavy enough to overcome air resistance and fall toward the ground. *Rain* consists of a vast collection of raindrops—during a downpour, trillions of raindrops fall on a square kilometer of the Earth's surface.

To understand how raindrops form in warm clouds, we must first see how tiny cloud droplets, which range from 0.01 to 0.1 mm (0.0004–0.004 in) in diameter, grow big enough to fall as raindrops (Fig. 16.26a). The formation of raindrops begins in a cloud with the formation of tiny cloud droplets while the air remains supersaturated. In such a cloud, random motions of water molecules cause additional water molecules to collide with and bond to those at the surfaces of the existing droplets. Such migration of atoms or molecules due to their random motions is called *vapor diffusion*. Vapor diffusion operates very slowly, however, so that process alone can't produce raindrops in the lifetime of most clouds. For raindrops to form, the cloud droplets themselves must come in contact with one another and merge, a process called **collision-coalescence**. Once formed, a somewhat larger droplet starts to fall, and as it does so, it collides with neighboring droplets, incorporates their water, and continues to grow, falling even faster. Small raindrops range from 0.5 to 1.0 mm (0.02–0.04 in) in diameter, large ones range from 1.0 to 4.0 mm (0.04–0.16 in) across, and very large ones are 4.0–6.0 mm (0.16–0.24 in) across. For a single large (4-mm) raindrop to form, it must collect 8 million 0.01-mm cloud droplets during its fall!

FIGURE 16.26 Raindrop growth and breakup in a warm cloud.

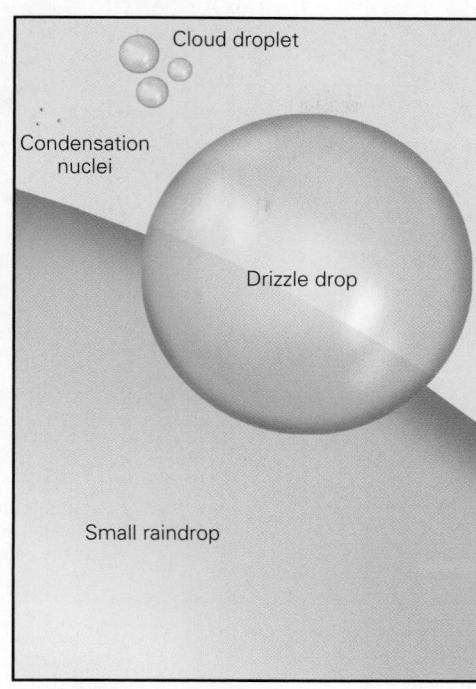

(a) Cloud droplets form when water vapor condenses around a condensation nucleus. Rain forms as droplets collide with one another and grow.

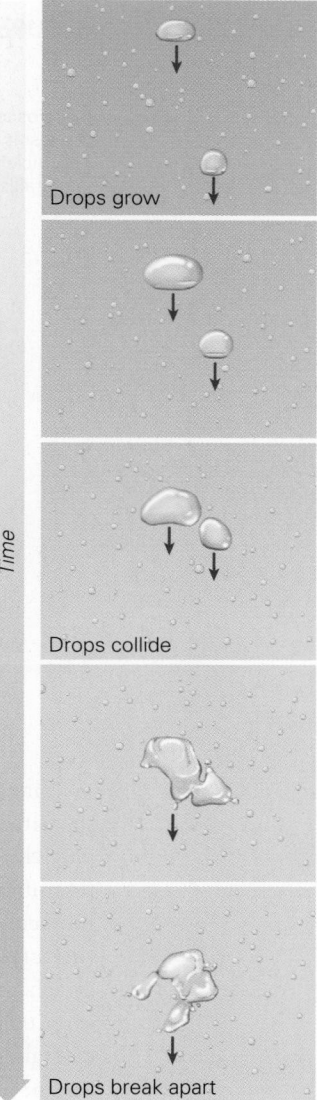

(b) When large raindrops collide, they can break up into many more, smaller raindrops.

Popular media typically depict raindrops as having a teardrop shape—round on the bottom, tapering to a point at the top. Real raindrops don't look like teardrops at all. In fact, cloud droplets and small raindrops have a spherical shape because of *surface tension*. Surface tension exists because the water molecules within a drop bond to other water molecules in all directions around them, but can't bond to air molecules around the drop. Because the bonds on the surface of the droplet are all oriented inward, the water drop assumes the shape of a sphere. Due to air resistance, large raindrops flatten into a hamburger shape as they fall. Raindrops larger than about 6.0 mm rarely develop because when such large drops collide with their neighbors, they break apart (Fig. 16.26b).

COLD-CLOUD PRECIPITATION. Many clouds extend to altitudes where atmospheric temperatures fall below 0°C. In these *cold clouds*, ice particles grow by one of two paths, depending on whether or not supercooled droplets are present (Fig. 16.27). *Supercooled droplets*, which are droplets that remain in a liquid state even though their temperature is well below 0°C (32°F), can exist to temperatures as cold as about –30°C (–22°F).

FIGURE 16.27 Ice particle growth in a cold cloud.

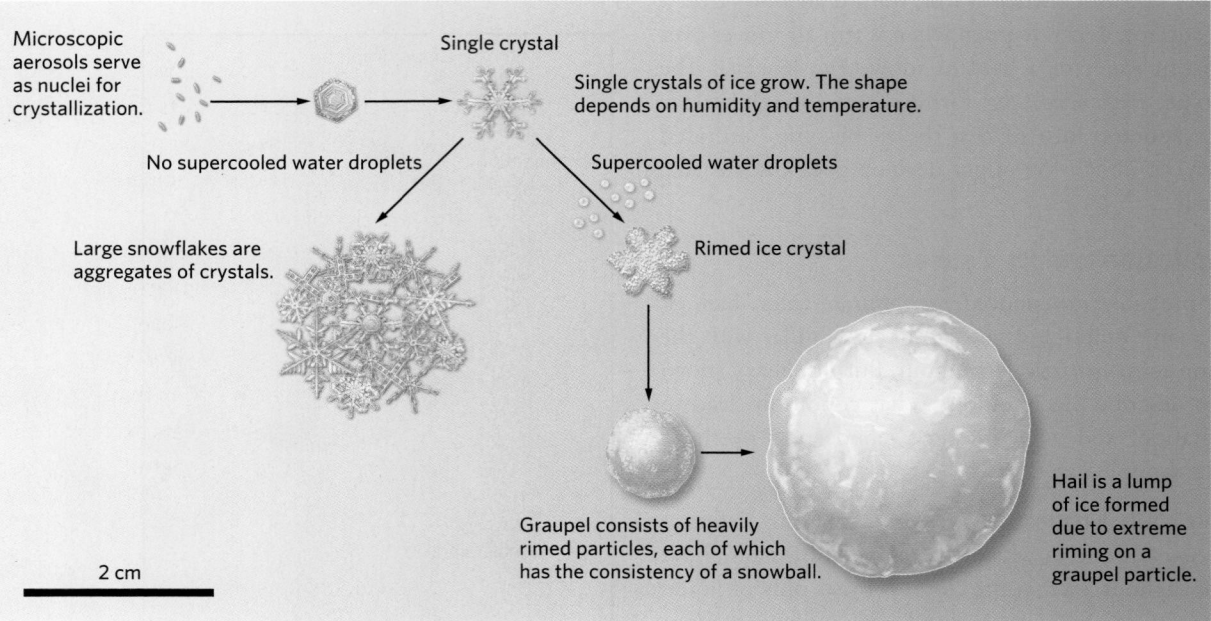

First, tiny ice crystals grow on special types of aerosol particles called *ice-nucleating particles*. Ice-nucleating particles are mineral grains whose crystal structure resembles that of ice. Because of this similarity, water molecules can find favorable attachment sites. Once these tiny crystals have formed, more water vapor molecules attach to their solid surfaces; this process of deposition eventually builds beautiful hexagonal crystals. These crystals grow into a wide variety of shapes, depending on the temperature and humidity. Some crystals have branched arms, so that when they collide with neighboring ice crystals, they latch onto one another and clump together. The resulting composite of numerous ice crystals becomes a **snowflake** (see Fig. 16.27). Some snowflakes fall all the way to the ground in cold weather, but others melt to form rain, if the air at lower altitudes is above freezing. Notably, in an effort to increase snowfall to provide more available water and energy for consumers, companies or organizations sometimes undertake cloud seeding, the process of adding ice-nucleating particles to a cloud by either introducing them from the ground or from an airplane (Box 16.3).

Not all ice particles grow entirely by deposition of water vapor directly on ice-nucleating particles, however. If sufficient ice-nucleating particles do not exist in a cloud, water droplets can remain in a liquid state, but they become supercooled. When an ice crystal falling through the cloud does encounter these supercooled droplets, the droplets instantly freeze onto the ice and form *rime* (a coating of ice). As more supercooled droplets eventually strike the rime-coated crystal, the rime layer grows until the crystal has become a soft sphere of ice called **graupel**.

PRECIPITATION INVOLVING BOTH WARM- AND COLD-CLOUD PROCESSES. As we learned earlier in this chapter, temperature decreases with altitude in the troposphere, so even if warm temperatures exist at ground level, the upper parts of towering clouds have temperatures well below freezing. Therefore, water vapor may attach to ice-nucleating particles in the upper part of a tall cumulus congestus cloud to form ice crystals. These ice crystals collect supercooled water droplets as they fall, and eventually become graupel. When the graupel particles fall to lower altitudes, where the temperatures rise above freezing, they melt to form raindrops. However, if strong updrafts exist in the cloud, graupel may remain aloft long enough to collect more water droplets and grow into *hailstones*, which are harder and larger than graupel. Graupel are about 5 mm (0.2 in) or less in diameter, while hailstones range from the size of a pea to the size of a grapefruit. If hailstones become large enough and, therefore, heavy enough to fall through the updraft, or if the updraft weakens, they can reach the ground before completely melting, producing **hail**, a downpour of hailstones.

In winter, a warm, above-freezing layer of air may exist aloft while the air beneath it is below freezing. If this happens, raindrops falling from the warm layer aloft sometimes freeze into tiny ice pellets called **sleet** when they enter the cold air near the ground. At other times, they arrive as **freezing rain**, supercooled water droplets

BOX 16.3 ▶ Putting Earth Science to Use

Cloud seeding

Water is a precious commodity across the western United States, with most of the supply coming from melting snow that fell in the mountains during the winter season. In years when winter storms are infrequent, and not much snow has accumulated, water for home use and irrigation, as well as hydroelectric energy, can be in short supply during the summer months. To increase water supplies for municipal, agricultural, and energy production, companies or government agencies sometimes carry out cloud seeding, spreading tiny aerosol particles into an existing cloud to produce ice crystals that fall as additional snowfall. Can such efforts work?

First, let's consider the conditions in clouds that might be amenable to seeding. Clouds that form over mountains in winter have temperatures below freezing, and often contain both ice crystals and supercooled water droplets. Under similar temperature and humidity conditions, ice crystals grow rapidly while supercooled water droplets grow very slowly, if at all. These supercooled water droplets never grow large enough to precipitate and are simply swept along with air currents as air passes over the mountains. Ice crystals, on the other hand, grow quickly, so they have a good chance of getting large enough to fall to the ground at the mountain crests.

In the 1940s, scientists discovered that ice crystals could be induced to form from the supercooled water droplets by introducing microscopic aerosol particles of silver iodide into a cloud. Silver iodide aerosol works to seed crystal growth because the crystalline structure of the aerosol particles is similar to ice crystals. Once formed, the ice crystals then can grow large enough to fall as snow. Because of the need for water and hydroelectric energy in the western United States, mountain cloud systems serve as a natural laboratory to test the technology. The government as well as water and power industries wanted to determine if enough supercooled water passing over the mountains could be converted to snow to make cloud seeding cost-effective. That extra snowpack represents irrigation water, water for home use, and hydroelectric power later in the year when the snow melts.

During a cloud-seeding operation, silver iodide can be introduced into clouds in one of two ways: (1) ground-based generators: these are placed at high-elevation locations upstream of the target mountains, where they burn an acetone solution containing silver iodide; or (2) silver-iodide flares: these are mounted on aircraft, and are burned during flight, releasing silver-iodide crystals (Fig. Bx16.3). Scientific studies have clearly shown that seeding can generate ice crystals from supercooled water droplets in clouds under the right conditions. But whether the process is cost-effective remains uncertain.

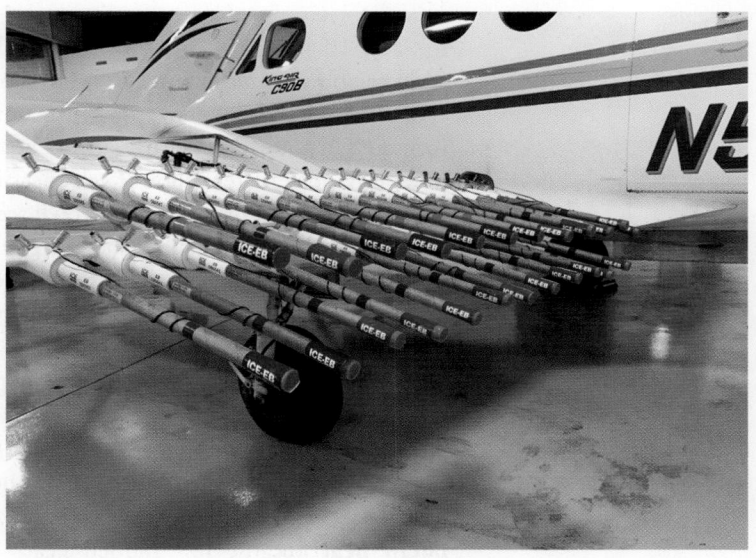

FIGURE Bx16.3 Flares mounted on an aircraft used to seed clouds over the Idaho Mountains in wintertime.

that freeze on contact with the ground, trees, and wires, to produce an ice-covered landscape. We discuss conditions leading to these types of precipitation in Chapter 18.

Clouds and Energy

Recall that atmospheric water can exist in three states: gas (vapor), liquid, and solid (ice). A **phase change** happens when water in one state converts to another, such as when ice *melts* (solid to liquid) or *sublimates* (solid to gas), or a water droplet *freezes* (liquid to solid) or *evaporates* (liquid to gas). Phase changes can happen when air containing water cools or warms (Fig. 16.28). Because phase changes don't take place instantly, and because temperature varies with altitude, a cloud can contain all three phases at once.

Phase changes absorb energy from, or release energy to, the atmosphere. To understand the relationship between a phase change and energy, imagine a simple experiment. Take a large, full pot of water out of the refrigerator and place it on the red-hot burner of a stove. Heat the water

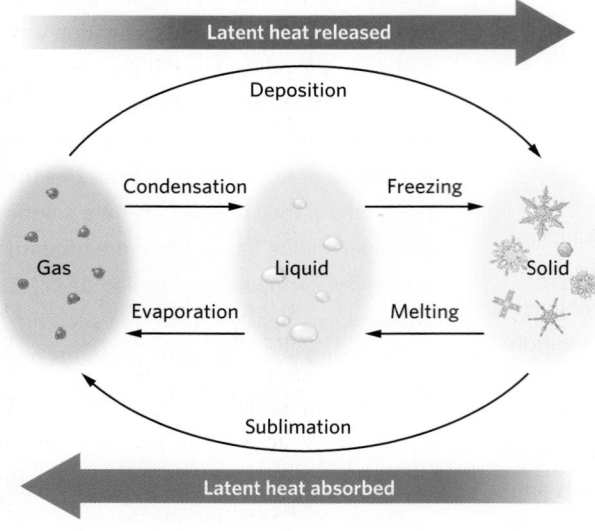

This vapor stores the solar energy as latent heat. When clouds form, water vapor condenses to liquid, thereby releasing its latent heat. This release, in turn, warms the air. Similarly, since it's necessary to add heat to ice to get it to melt, liquid water stores latent heat that will be released when the liquid water freezes. We'll see in Chapter 18 that the energy from phase changes is necessary for the development of storms.

Take-home message...

Clouds form when water molecules attach to aerosols and form either liquid droplets or ice crystals. Droplets and crystals grow both by absorbing water molecules from the air and by collecting and aggregating. When they become large enough, they fall as precipitation. Processes involved in cloud formation and precipitation absorb or release large amounts of energy.

Quick Questions —————————

- On what basis do meteorologists classify clouds?
- A raindrop plops on your head. Describe two growth paths that the raindrop may have taken.
- When you step out of a swimming pool, you often get the shivers as water evaporates from your skin. Why?

for, say, 10 minutes—time for enough heat to transfer from the burner to the water to bring the water to a boil. Once the water has come to a boil, continue heating it for another hour, until nearly all the liquid water in the pot has become water vapor. If you measure the temperature of the water during boiling, you'll see that it remains unchanged, meaning that all the energy from the burner goes into driving the phase change. This experiment tells us that it took an hour's worth of energy from the red-hot burner to convert the pot's water into vapor—that's a lot of energy! Where did all that energy go? It went into accelerating the movement of the water molecules to a high-enough speed that they could break the bonds holding them in the liquid and escape into the air. Therefore, water vapor contains a lot of energy. We call this energy **latent heat** because it's "hidden heat" that a material inherits or incorporates from a phase change.

We've just seen how water absorbs latent heat when it changes from liquid to gas. What happens when water condenses again? To answer this, let's continue our experiment. Imagine that you could magically force all the water vapor molecules that had boiled out of the hot pot back into the pot instantly, so that they all became part of a liquid again. The latent heat in the water vapor (all the heat added to the water by the burner during an hour's time) would be suddenly released all at once, and the pot (and probably the kitchen) would explode!

The thought experiment we've just described may seem odd, but the processes it describes occur in the atmosphere all the time. Water in the oceans absorbs solar energy, evaporates, and enters the atmosphere as vapor.

16.5 The Atmosphere's Power Source: Radiation

The Earth's envelope of gas does not sit still—winds blow, updrafts and downdrafts develop, and precipitation forms and falls. Furthermore, phase changes of water in the air absorb or release vast amounts of latent heat. Where does the energy driving all these processes come from? In a word, the Sun. The Sun emits electromagnetic radiation, a small percentage of which reaches the Earth. But that's enough to keep not only the atmosphere, but also the hydrosphere and biosphere, in motion. We've already mentioned, briefly, aspects of how this radiation interacts with the atmosphere. In this section, we complete the story.

The Nature of Electromagnetic Radiation

All solar energy arrives at the top of the Earth's atmosphere as radiation. In a general sense, **radiation** is energy that travels away from its source either through a material or through a vacuum. Radiation comes in two general forms: *electromagnetic radiation* (such as visible light) and *particulate radiation* (such as particles emitted from radioactive elements). Here, we focus our attention on electromagnetic radiation—specifically, the solar energy you feel warming (or burning) you on a bright sunny afternoon.

All objects—the Sun, you, the ground beneath you—emit electromagnetic radiation as long as they have a temperature above absolute zero (0 K). Physically, **electromagnetic radiation** is a form of energy that travels invisibly through the vacuum of space. In the early 20th century, Albert Einstein proposed that the smallest unit of energy in radiation is a massless particle, which later came to be known as a *photon*. Groups of photons moving through space constitute *electromagnetic waves*, and groups of electromagnetic waves constitute *beams*. Electromagnetic waves can be depicted as a set of periodic up-and-down curves that, on a drawing, resemble the shape of ocean waves. But physically, electromagnetic waves differ greatly from ocean waves. They are vibrations of electric and magnetic fields that travel through a vacuum at the speed of light. Though invisible, they can be detected by instruments and by our senses.

We can characterize electromagnetic waves by specifying their **wavelength**, the distance between wave crests, and their **frequency**, the number of wave crests passing a point in space in one second (**Fig. 16.29**). Electromagnetic waves come in a huge range of wavelengths and, by implication, frequencies. Because all electromagnetic waves travel at the same velocity in a vacuum—the speed of light (300,000 km/s, or 186,000 mi/s)—frequency increases as wavelength decreases. The full range of possible wavelengths constitutes the **electromagnetic spectrum**. Physicists use different names for different parts of the spectrum. *Shortwave radiation* includes gamma rays,

X-rays, and ultraviolet (UV) radiation, as well as visible light and some infrared radiation. Infrared is divided into shorter-wavelength near-infrared and longer-wavelength far-infrared. *Longwave radiation* includes far-infrared radiation as well as microwaves (used in radar) and radio waves. The energy carried by electromagnetic radiation depends on its wavelength and frequency. Shortwave (high-frequency) radiation contains more energy than longwave (low-frequency) radiation.

Energy Balance in the Earth-Sun System

THE CONCEPT OF BLACKBODY RADIATION. To understand why the Earth's atmosphere has the temperature that it does, we have to dig into a few background concepts from physics. When discussing the behavior of electromagnetic radiation, physicists picture an object, called a **blackbody**, that absorbs all radiation that shines on it, and then reradiates energy across the electromagnetic spectrum. The energy emitted at each wavelength of the resulting *blackbody radiation* depends only on the object's temperature. The Earth, the Sun, and most other objects in space radiate energy very nearly like blackbodies. Three laws govern the behavior of blackbodies:

- A warmer object emits more radiation, at all wavelengths, than a colder object.
- The total amount of radiation an object emits increases as its temperature increases.
- As the temperature of an object increases, the wavelength of maximum radiation emission shifts toward shorter wavelengths.

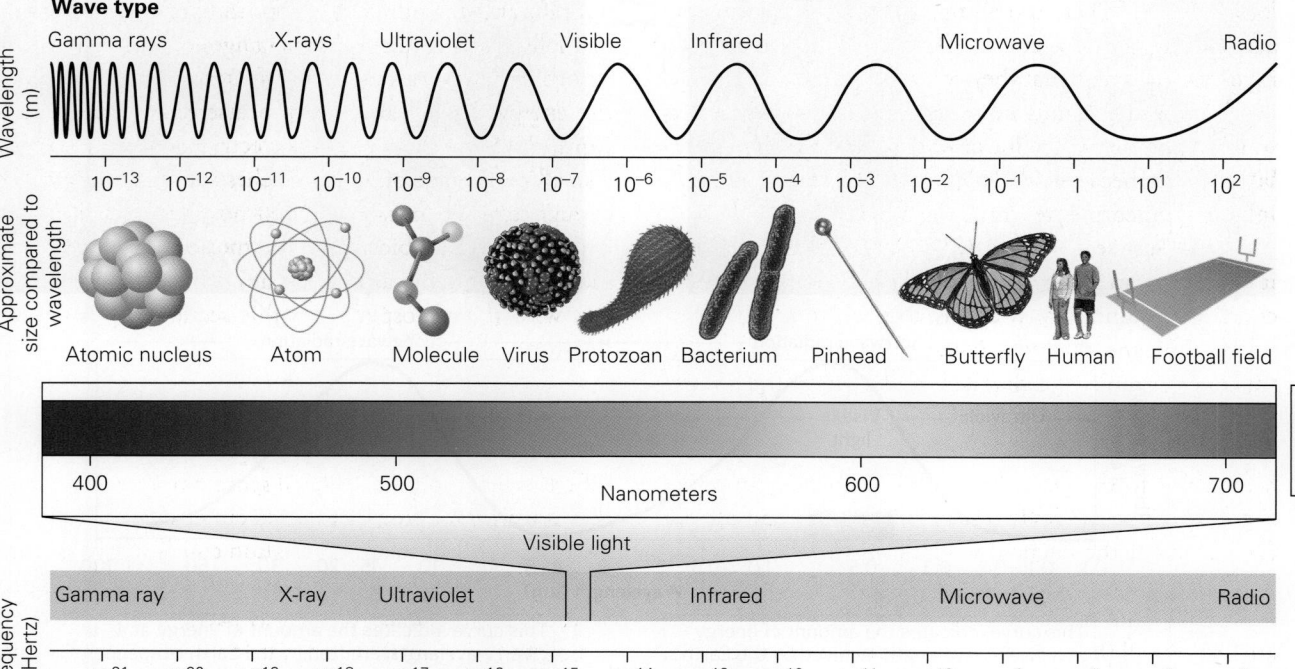

FIGURE 16.29 The electromagnetic spectrum.

A more advanced book can explain why these laws operate as they do, but for our purposes, we'll simply keep them in mind as we explore how sunlight adds energy to the Earth's atmosphere.

RADIATION IN EQUALS RADIATION OUT. The Sun's surface reaches a temperature of about 5,600°C (10,000°F), hot enough to vaporize the Earth (see Chapter 22). But because the Sun sends energy in all directions and lies so far from the Earth, our planet intercepts only a tiny fraction (0.00000015%) of the Sun's total radiation. Nevertheless, if the Earth were to absorb all the energy that it receives from the Sun, without releasing any back into space, it would eventually vaporize. Of course, this doesn't happen, because the Earth, like any blackbody, constantly reradiates energy into space. Over time, the total energy arriving at the Earth from the Sun exactly balances the amount of energy reradiated from the Earth into space. On the time scale of a decade, the rate at which solar energy reaches the Earth, a quantity called the *solar energy flux*, remains nearly constant. If the Earth had no atmosphere, the solar energy flux would cause the Earth to stay at a constant, but very chilly −18°C (0°F), a temperature known as the Earth's *radiative equilibrium temperature*.

Because the Sun is so hot and the Earth is so cool, the spectrum of the solar radiation arriving at the Earth differs greatly from the spectrum of the radiation that the Earth reradiates into space, as the laws of blackbody radiation predict (Fig. 16.30). Specifically, solar energy reaching the Earth arrives predominantly as shortwave radiation—UV light (the radiation that can give us sunburns), visible light (the radiation our eyes detect), and near-infrared radiation (the radiation we can feel as heat but cannot see). In contrast, the radiation emitted by the Earth consists of longwave radiation (far-infrared radiation).

THE ATMOSPHERIC GREENHOUSE EFFECT. If the Earth's surface were at its radiative equilibrium temperature, you wouldn't be reading this book, for life as we know it today would not exist. Fortunately, the actual average surface temperature of our planet, 15°C (59°F), is much warmer, so liquid water can exist and life can flourish. This difference is due entirely to the *atmospheric greenhouse effect*—also called, simply, the **greenhouse effect**. We commonly hear about the greenhouse effect in news concerning the Earth's changing climate (see Chapter 19). Here, let's explore how it works and why it is important to our survival.

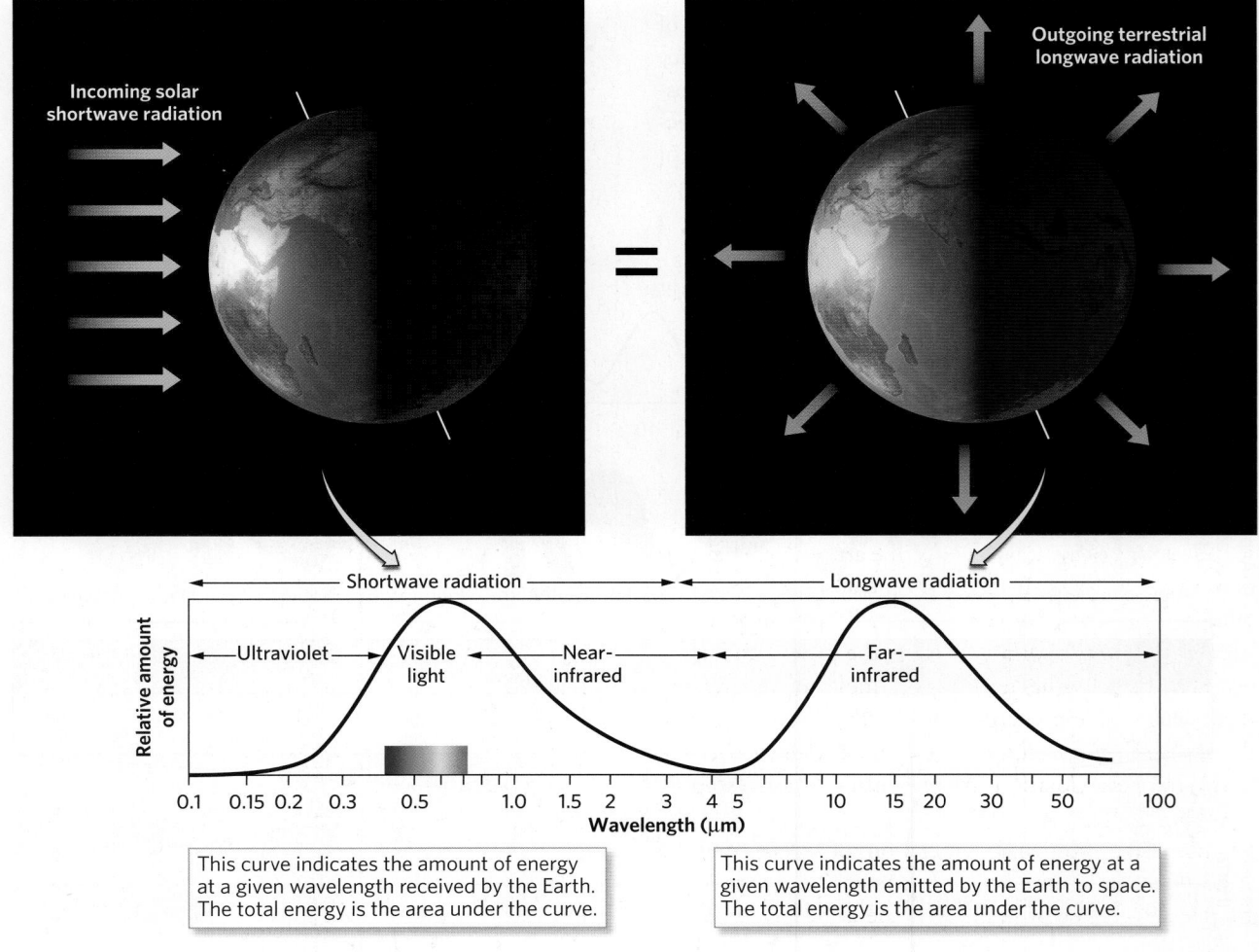

FIGURE 16.30 In order for the Earth to be in equilibrium, the shortwave radiation that the Earth receives from the Sun must equal the longwave radiation that the Earth emits to space.

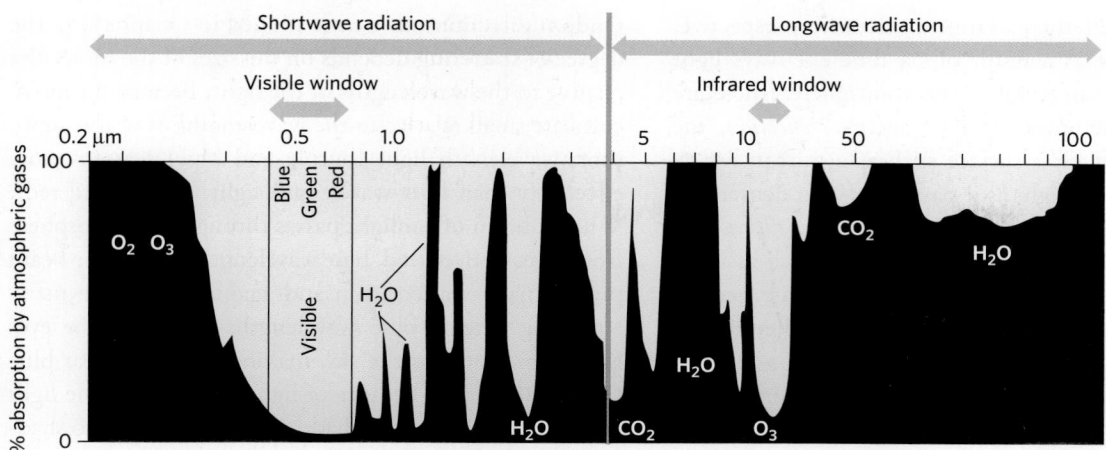

FIGURE 16.31 Different atmospheric gases absorb radiation of different wavelengths. Low-absorption regions of the spectrum are called atmospheric windows.

While the Earth itself acts like a blackbody, its atmosphere does not. Rather, each atmospheric gas absorbs and emits radiation selectively, meaning that it absorbs only certain wavelengths and emits only certain wavelengths. Why do certain gases in the atmosphere exhibit this behavior? It's because of the nature of the molecules that these gases contain. Molecules such as O_2 and N_2, which consist of only two atoms apiece, do not absorb or emit very much solar radiation or terrestrial radiation. In contrast, molecules with more than two atoms (such as H_2O, CO_2, O_3, CH_4, and N_2O) are excellent absorbers and emitters of radiation at certain wavelengths. This difference in behavior can be traced to how the molecules vibrate when exposed to radiation.

Because of the different absorption and emission characteristics of the various gases in the atmosphere, the atmosphere is transparent to two ranges of solar and terrestrial wavelengths (Fig. 16.31). Put another way, the atmosphere has two **radiation windows** through which certain wavelengths of solar and terrestrial radiation can pass between the Earth's surface and space without interference. The first window is transparent to the visible wavelengths (allowing incoming sunlight to reach the Earth's surface and allowing us to see), and the second window is transparent to certain infrared wavelengths (allowing radiation from the Earth to escape back into space). For all other wavelengths, the atmosphere behaves as a strong absorber. It reradiates the absorbed energy in random directions. Some of the reradiated energy heads back to the Earth's surface, heating the planet, and some gets absorbed by other atmospheric molecules, increasing the average kinetic energy (that is, the temperature) of the air. Thus, the atmosphere acts like a blanket over the Earth, trapping radiation and preventing it from escaping to space (Fig. 16.32).

The absorption and reradiation of energy by the atmosphere has come to be known as the *atmospheric greenhouse effect* because the way the atmosphere retains heat makes people think of the way in which a glass greenhouse retains heat. The analogy is not perfect, however, because the heat in a real greenhouse gets trapped simply because the warm air inside cannot escape through glass windows to the outside. In contrast, the atmospheric greenhouse effect works because, although solar energy can pass through the atmosphere to the ground, certain atmospheric gases prevent the Earth's infrared radiation from passing through the atmosphere directly into space. The atmospheric gases that absorb and re-emit infrared radiation are known as **greenhouse gases**.

Optical Effects in the Atmosphere

Light, as it passes through the atmosphere, interacts with air molecules, aerosols, water droplets, and ice crystals to produce a bewildering array of optical phenomena that delight the eyes, warm the heart, and astonish those lucky

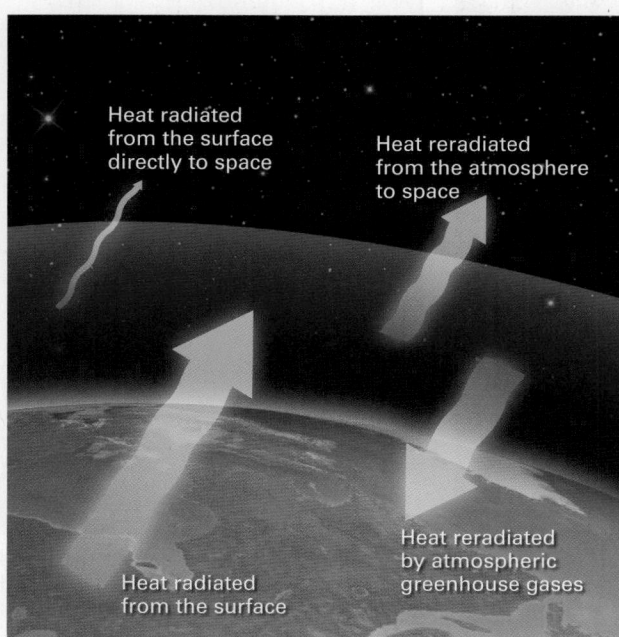

FIGURE 16.32 A simple depiction of the atmospheric greenhouse effect.

enough to witness them. From a scientific perspective, these effects arise as a result of six different ways light interacts with the material in the atmosphere. These are *absorption*, the removal of light energy; *reflection*, the bouncing of light off a surface such as a mirror; *refraction*, the bending of light as it passes between denser and less dense substances, such as water and air; *diffraction*, the bending of light waves around objects such as small raindrops; *scattering*, the absorption and immediate re-emission of light in a different direction; and **emission**, the production of light by atoms as their electrons give up energy. With these phenomena in mind, we can interpret a variety of optical features.

BLUE SKY. When astronauts walked on the Moon and looked upward, they saw a black sky, even though they were bathed in sunlight. Similarly, when they looked into the shadows behind boulders, they saw only black. In contrast, when you look up from the surface of the Earth on a sunny day, you see a blue sky, and if you look into the shadow behind a boulder, you can still see the ground. Why?

Blue skies exist because air molecules cause **light scattering**, a phenomenon that takes place when molecules absorb visible light and then immediately re-emit it in

random directions, as we mentioned in Chapter 14. The degree of scattering depends on the sizes of the molecules relative to the wavelengths of the light. Because air molecules are small relative to the wavelengths of visible light, short-wavelength light (purple and blue) scatters more effectively than long-wavelength light (orange and red). When a beam of sunlight passes through the atmosphere from directly overhead, blue wavelengths within the beam preferentially scatter again and again. After these many scattering events, blue wavelengths arrive at our eyes from somewhere in the sky, making the sky appear blue (Fig. 16.33a). Also because of light scattering, some light energy makes it into the shadows behind objects, so shadows on the Earth aren't completely black.

When the Sun lies low in the sky, the distance through the atmosphere that a beam of sunlight must travel to reach our eyes increases. As a result, nearly all blue, green, and yellow wavelengths scatter back into space, leaving only reds and oranges, the color of the rising or setting Sun (Fig. 16.33b, c).

MIRAGES. Desert travelers sometimes see what looks like a lake of shimmering water on parched land in the distance. Sailors might see ships floating upside down in the sky, while polar explorers look twice when they see

FIGURE 16.33 The many colors of the sky.

More blue light than red light reaches the observer's eyes.

More red light than blue light reaches the observer's eyes.

(a) At midday, the sky appears blue.

(b) When sunlight passes through the atmosphere, short wavelengths (primarily blue) are scattered out of the main solar beam. At noon, the observer sees blue light coming from many directions. At sunrise or sunset, only the red light reaches the observer.

(c) At sunrise or sunset, the sky appears orange or red.

FIGURE 16.34 Mirages.

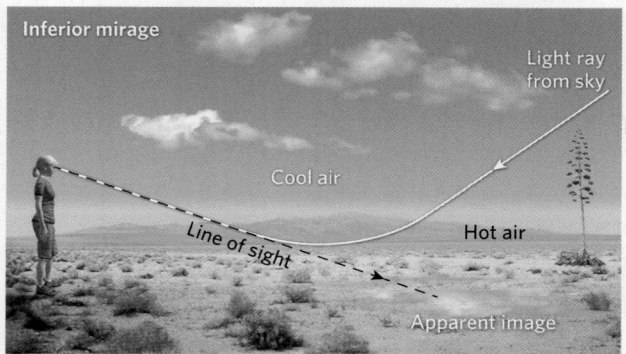

(a) An inferior mirage forms when hot air is present near the ground and bends light upward. The mirage looks like shimmering water because the hot air at ground level is not still. The shimmering water is actually refracted blue light from the sky.

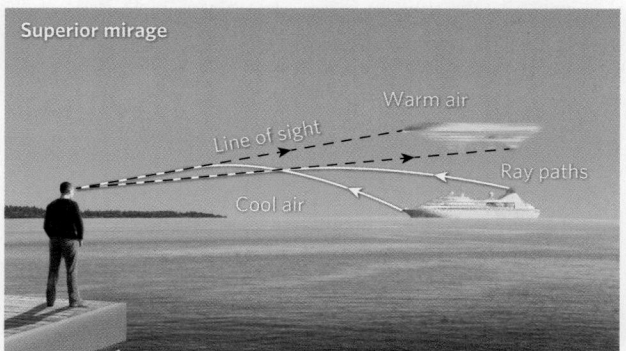

(b) A superior mirage forms when warm air overlies cooler air and bends light downward. Generally, but not always, the mirage is upside down.

ice castles floating above the horizon. Such images are mirages, and they disappear when you approach them. Simply put, a **mirage** is an image formed when the light reflected from an object bends, or *refracts*, through the atmosphere so that the object seems to appear where it doesn't actually exist.

Mirages develop when the air near the ground surface is either exceptionally warm or exceptionally cold relative to the air above. As light passes from less dense warm air into denser cold air (or vice versa), it undergoes refraction. The presence of a hot layer of air beneath a cool layer produces an *inferior mirage* (meaning that the image you see lies below the real object) **(Fig. 16.34a)**, and the presence of a cold layer produces a *superior mirage* (meaning that the image you see lies above the real object) **(Fig. 16.34b)**. In deserts, where a layer of very hot air lies at ground level, you'll see inferior mirages of the sky on the ground surface. Because the hot air moves, the light of the mirage shimmers, so it looks like water. Superior mirages develop mostly in the Arctic and Antarctic, where a layer of cold air develops at the Earth's surface, so that light bends downward as it travels from colder air upward into warmer air. There, distant objects appear to be floating above the ground.

RAINBOWS. When sunlight passes through air containing water droplets or ice particles, a variety of colorful optical phenomena may develop. For example, when the Sun lies close to the horizon and its rays pass through clear air behind you into rain in front of you, you'll see a **rainbow**, an arc displaying all the colors of visible light in sequence. If you look closely, sometimes you'll see two rainbows, a lower one called the *primary bow* and an upper one called the *secondary bow* **(Fig. 16.35a)**. Note that the order of colors in the secondary bow is the opposite of that in the primary bow.

Rainbows develop when light reflects within and refracts through raindrops. Specifically, when an incoming beam of sunlight strikes the spherical surface of a raindrop, the drop acts as a prism and splits the light into colors. The light then reflects off the interior of the drop's back surface and refracts again as it exits the drop **(Fig. 16.35b)**. Light in a secondary bow undergoes an additional reflection.

OTHER OPTICAL EFFECTS. Many other, less frequently observed optical phenomena develop locally in the atmosphere (**Earth Science at a Glance**, pp. 636–637). Examples include *cloud iridescence*, bands of shimmering colors

Atmospheric Optics

Light, as it passes through the atmosphere, interacts with aerosols, water droplets, and ice crystals to produce a bewildering array of optical phenomena that delight the eyes, warm the heart, and astonish those lucky enough to witness them. From a scientific perspective, these effects arise as a result of six different ways light interacts with the material in the atmosphere. These are absorption, where light is removed from a beam; reflection, where light bounces off a surface such as a mirror; refraction, where light bends as it passes between denser and less dense substances such as water and air; diffraction, where light waves bend around objects such as small raindrops; scattering, where light is absorbed and then immediately emitted in a different direction; and emission, where light is produced as high energy particles from the Sun crash into oxygen and nitrogen molecules high in the atmosphere.

Iridescence and glory are diffraction effects caused by light waves bending around tiny water droplets.

Crepuscular rays form when clouds absorb sunlight, shadowing some, but not all solar beams.

Glory appears when looking down at the shadow of an aircraft passing over water clouds.

Water

Shadow

Sun pillar
(caused by reflection off
bottom of plate-like crystals)

Arcs, halos, and sundogs
form as light refracts through
columnar and plate-like ice
crystals falling with different
orientations.

Arcs

Halos

Sundogs

Ice

Auroral colors are caused
when light is emitted during
collisions between solar
particles and molecules in the
Earth's upper atmosphere at
different altitudes.

Oxygen;
higher altitudes

Oxygen;
lower altitudes

Nitrogen;
much lower altitudes

Upper atmosphere

FIGURE 16.35 Rainbows delight the eye.

(a) If you're at the right position, with the Sun behind you and rain in front of you, a rainbow will appear. Sometimes you'll see a double rainbow.

A secondary rainbow forms when light reflects off back of drops twice.

Water drops

Incoming light

Secondary rainbow

A primary rainbow forms when light reflects off back of drops once.

Light to observer

Primary rainbow

(b) A rainbow forms because light refracts and reflects within raindrops as it interacts with the drops.

seen when light passes through thin clouds; *glory*, a series of colored rings that surround the shadows of aircraft on the tops of clouds when viewed from above; and *halos* and *arcs*, circles and arches of light and color that develop when light passes through a thin cloud or fog of ice crystals (see Fig. 16.1b). All of these phenomena are associated with the manner in which light passes around and through water droplets and ice crystals in the air.

Take-home message...

The Sun provides energy to the Earth and its atmosphere in the form of electromagnetic radiation. Visible light and some ultraviolet radiation that pass through the atmosphere are absorbed by the Earth's surface, which in turn reradiates this energy as infrared radiation. Greenhouse gases, which trap some of this reradiated energy, keep the atmosphere warm enough for liquid water—and life—to exist on the Earth. Radiation interacting with the atmosphere produces an array of optical phenomena, such as rainbows, mirages, arcs, and halos.

Quick Questions

- If there were no atmosphere, what would the temperature of the Earth's surface be?
- What atmospheric gases are most important to the Earth's atmospheric greenhouse effect?
- What is the difference between longwave and shortwave radiation? Which reaches the Earth from the Sun?
- Where is the Sun, relative to you, when you see a rainbow?

Objective 16.1

Explain how the atmosphere evolved since the Earth first formed, describe how photosynthesis contributed to the Earth's modern atmosphere, and list the gases and other materials found in today's atmosphere.

KEY CONCEPTS

- The Earth's atmosphere evolved over time. Before the appearance of oceans, it consisted of volcanic gases. Formation of the oceans removed much of the atmosphere's water vapor and some CO_2, leaving N_2 behind. When photosynthetic organisms evolved, they released O_2 and absorbed most of the remaining CO_2, which became buried either as shells or organic matter.

- Addition of O_2 to the ocean and atmosphere through photosynthesis provided breathable air. It also led to the formation of ozone, which absorbs harmful components of sunlight, allowing life to flourish on land.

- Dry air consists of nitrogen (78%), oxygen (21%), and other gases such as argon and carbon dioxide (1%). The amount of water vapor in the atmosphere varies over time and space.

- In addition to gases, the atmosphere contains tiny particles called aerosols.

EARTH-SCIENCE VOCABULARY

aerosol (p. 606)
atmosphere (p. 604)
cloud (p. 606)
great oxygenation event
  (p. 605)

haze (p. 607)
trace gas (p. 606)
water vapor (p. 606)

REVIEW QUESTIONS

1. **(a)** What major changes occurred in the Earth's atmosphere over the Earth's lifetime as a planet? **(b)** If you used a time machine to go back 3 billion years, what gases would you find in the Earth's atmosphere?

2. **(a)** At what point in Earth history did the atmosphere become hospitable to life on land? **(b)** What role did ozone play in allowing plants and animals to move on to land? **(c)** What would happen to life on the Earth if ozone were to disappear from the atmosphere?

3. **(a)** What are the main two components of today's atmosphere? **(b)** What are common trace gases? **(c)** When did the concentration of oxygen in the atmosphere first exceed about 19.5%, the minimum concentration needed for people to survive? **(d)** When, during Earth history, did oxygen reach its peak concentration, and why?

4. **(a)** Which gas in the modern atmosphere varies rapidly over time and from location to location? **(b)** What role did plants play in forming the atmosphere we have today? **(c)** Explain why the atmosphere could never contain 100% O_2.

5. **(a)** What phenomena caused most of the carbon dioxide to be removed from the early atmosphere, and when did this change happen? **(b)** Where does the carbon from this carbon dioxide currently reside?

6. **(a)** What is the term used for the murkiness of the atmosphere visible in **Figure A**? **(b)** What component of the atmosphere causes this phenomenon? **(c)** What does this component consist of?

A

Objective 16.2

Define properties that meteorologists use to describe the atmosphere, list instruments they use to monitor atmospheric conditions, and explain how the data are used in weather forecasting.

KEY CONCEPTS

- Air temperature represents the average speed at which gas molecules move in the atmosphere.

- Atmospheric pressure can be considered the weight of a column of air over a unit area at the base of the column.

- Relative humidity represents the amount of water vapor in air, relative to its holding capacity. Another quantity, the dewpoint, represents the actual amount of water vapor in air. The atmosphere's capacity to hold water vapor increases as its temperature increases.

• Atmospheric pressure decreases rapidly with altitude. Half of the atmosphere's mass lies below an elevation of 5.6 km (3.5 mi).

• Properties of the atmosphere, such as temperature, pressure, humidity, and wind direction and speed, are measured just above ground level using a variety of instruments at a weather station. To measure conditions high in the atmosphere, meteorologists launch rawinsondes attached to weather balloons.

• Radar systems and satellites are used by forecasters to monitor weather patterns and anticipate hazardous weather. Modern weather forecasts result from worldwide data collection and predictions of computer models that use the data as input.

EARTH-SCIENCE VOCABULARY

air temperature (p. 609)
atmospheric pressure
  (p. 610)
barometer (p. 610)
cloud cover (p. 615)
dewpoint (p. 614)
Doppler radar (p. 617)
downdraft (p. 614)
meteorologist (p. 609)
precipitation (p. 617)
relative humidity (p. 613)

saturation vapor pressure
  (p. 612)
thermometer (p. 609)
updraft (p. 614)
vapor pressure (p. 612)
visibility (p. 614)
weather (p. 609)
weather radar (p. 616)
wind (p. 614)
wind direction (p. 614)
wind speed (p. 614)

REVIEW QUESTIONS

7.   (a) At what Fahrenheit and Celsius temperatures does ice start to melt? (b) Which represents a greater change in temperature, 1°C or 1°F? (c) What does the term "absolute zero" refer to, and on what temperature scale is it measured? (d) At noon during the day on the Moon, the temperature reaches 120°C, and at midnight, it reaches –130°C. Is that range greater than or less than the range between the highest and lowest temperatures ever measured on the Earth?

8.   (a) What is the device shown in **Figure B** called, and what does it measure? (b) How will the measurements indicated by this device change if you transport it from sea level to Denver, Colorado at

B

an elevation of 1,609 m (5,280 ft)? (c) If you placed the device at a location in Denver, would the measurement it shows remain exactly the same all year?

9.   (a) Explain why the absolute amount of moisture in the atmosphere over a hot desert can be larger than that over the Arctic tundra, even though the relative humidity over the tundra may be 90%, while that over the desert may be 10%. (b) What is the relative humidity of saturated air? (c) Say that Tampa and Miami each have a temperature of 90°F, but Tampa's dewpoint is 75°F and Miami's is 73°F. Which location has a higher relative humidity?

10.   (a) The weather report says that there is a south-southwest wind. The wind blows from where to where? (b) What is the wind speed and direction indicated by the wind barb in **Figure C**?

C

11.   (a) How many times a day are rawinsondes launched worldwide? (b) What local time of the day are they launched where you live? (c) What information does a rawinsonde tell us that a ground-based weather station does not?

12.   (a) What does radar reflectivity measure, and what can it tell us? (b) What can Doppler radar systems detect that makes them so important for monitoring weather hazards?

13.   (a) What does the visible channel of a weather satellite detect? (b) How does that differ from the infrared channel?

Objective 16.3

Draw a diagram that displays atmospheric layers based on temperature, composition, and electric charge.

KEY CONCEPTS

• Four atmospheric layers are distinguished by patterns of temperature change with elevation: the troposphere, stratosphere, mesosphere, and thermosphere.

• Temperature decreases with altitude in the troposphere and mesosphere, and it increases with altitude in the stratosphere and thermosphere.

• The distribution of temperature with altitude results from the Earth's surface heating the base of the troposphere, and absorption of ultraviolet solar radiation by ozone heating the stratosphere.

• The atmosphere can also be divided into layers that are well mixed (homosphere) and unmixed (heterosphere).

• The ionosphere occurs high in the atmosphere and is recognized by its high concentration of ions.

• In recent times, the ozone concentration in the stratosphere has decreased due to manmade chlorofluorocarbons introduced into the atmosphere by human activities.

REVIEW QUESTIONS

14. **(a)** Label the four major thermal layers of the atmosphere in **Figure D**. **(b)** What is the name of the boundary between the troposphere and the stratosphere? **(c)** What does the green wispy color near an altitude of 90 km represent? **(d)** What do the thick purple arrows indicate?

15. **(a)** What is the importance of the tropopause in relation to the Earth's storms? **(b)** Is the tropopause higher or lower in altitude over the tropics compared with the polar regions? Why?

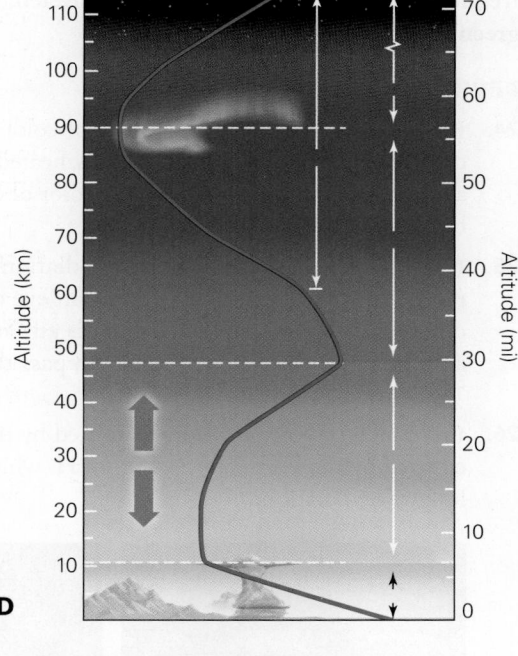

D

16. **(a)** In what layer is a high concentration of ozone found? **(b)** Why is this ozone layer important to life on the Earth?

17. The thermosphere hosts the hottest temperatures in the Earth's atmosphere. **(a)** Why would an astronaut freeze if accidentally exposed to the thermosphere without a space suit? **(b)** Would the astronaut be able to breathe if her helmet came off?

18. **(a)** What is the phenomenon shown in **Figure E**, and why does it occur? **(b)** When, where, and at what altitude on Earth does this phenomenon occur?

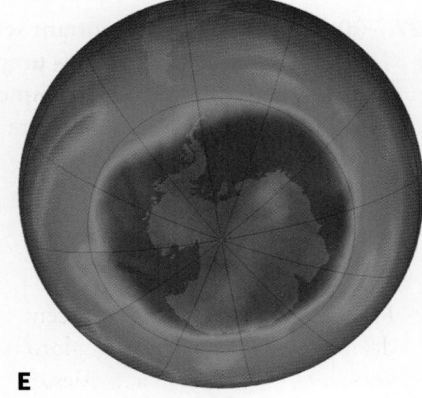

E

Objective 16.4

Identify clouds you see in the sky, describe how they form, and explain how they produce precipitation.

KEY CONCEPTS

- Clouds form when air rises, expands, and cools, as happens when air is lifted or when it becomes buoyant. Clouds are classified based on their altitude, shape, and whether or not they produce precipitation.

- Cirrus are high clouds, stratus are widespread low clouds, and cumulus are towering clouds. Fog is a cloud in contact with the ground. Nimbostratus and cumulonimbus are clouds that produce precipitation.

- Cloud droplets nucleate around aerosol particles. They grow into raindrops by first absorbing water molecules from water vapor, and then by coalescing with neighboring droplets. Ice particles first form when water molecules deposit on specific aerosol particles called ice-nucleating particles. Ice grows to precipitation size, first by diffusion and later when several crystals link together to form snowflakes or when supercooled water droplets attach to form rimed ice particles, graupel, or hail.

- Water in the atmosphere constantly undergoes phase changes, releasing or absorbing latent heat in the process.

REVIEW QUESTIONS

19. **(a)** List two ways in which air can be forced to rise and produce clouds. **(b)** Are you more likely to find clouds on the upwind or downwind side of a mountain range? Why?

20. **(a)** Name and describe the primary cloud types. **(b)** What type of cloud appears in **Figure F**? **(c)** Is a cirrus cloud typically higher or lower than a stratus cloud? **(d)** You wake up in the morning to see the sky covered by a uniform sheet of cloud stretching from horizon to horizon. What type of cloud is overhead?

F

21. **(a)** What type of fog was present when the photograph shown in **Figure G** was taken in the early morning, just as the Sun was rising? **(b)** San Francisco is famous for the fog that frequently envelops the Golden Gate Bridge. The ocean west of the city is quite cold. What type of fog typically occurs in this region?

G

22. **(a)** What types of precipitation can form when supercooled water is present in clouds? **(b)** What weather hazards are associated with these precipitation types? **(c)** What does supercooled water have to do with cloud seeding?

23. **(a)** Is latent heat released to the atmosphere or absorbed from the atmosphere when raindrops evaporate? **(b)** How about when snowflakes melt? **(c)** How about when a cloud forms?

Objective 16.5

Classify various types of electromagnetic radiation, describe why optical effects such as rainbows occur, and depict how interactions between radiation and the atmosphere control the Earth's surface temperature.

KEY CONCEPTS

- The Sun provides energy to the Earth in the form of electromagnetic radiation, which includes many different wavelengths. An object that absorbs all the radiation it receives is called a blackbody. A blackbody re-emits radiation at wavelengths that depend on its temperature.

- Solar energy arrives at the Earth's surface as shortwave radiation, but the much cooler Earth reradiates this energy as longwave radiation. Certain gases in the atmosphere trap some of this energy, causing the greenhouse effect, without which the Earth's surface would be below freezing.

- Visible light interacts with air molecules and scatters, making the sky appear blue. Mirages form when sunlight bends due to variations in air temperature with elevation. Rainbows, halos, and other brilliant phenomena form when light interacts with water droplets and ice crystals in the atmosphere.

EARTH-SCIENCE VOCABULARY

blackbody (p. 631)
electromagnetic radiation (p. 631)
electromagnetic spectrum (p. 631)
emission (p. 634)
frequency (p. 631)
greenhouse effect (p. 632)

greenhouse gas (p. 633)
light scattering (p. 634)
mirage (p. 635)
radiation (p. 630)
radiation window (p. 633)
rainbow (p. 635)
wavelength (p. 631)

REVIEW QUESTIONS

24. **(a)** Is the surface of the Sun hotter or colder than the electric coils on a kitchen stove when they're fully heated? **(b)** Explain your answer, considering the dominant color of each and considering blackbody radiation.

25. **(a)** What is meant by shortwave radiation and longwave radiation? **(b)** Which can the human eye detect? **(c)** Which does your body radiate? **(d)** What is an "atmospheric window," and which wavelengths of light can pass through a such a "window"?

26. **(a)** Which type of radiation is emitted by the Sun? **(b)** Which is emitted by the Earth? **(c)** In **Figure H**, which panel represents longwave radiation?

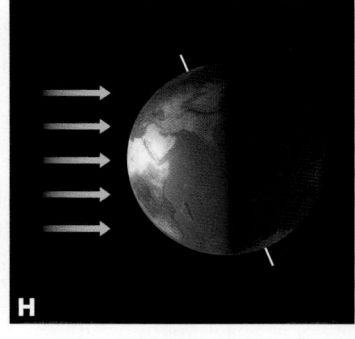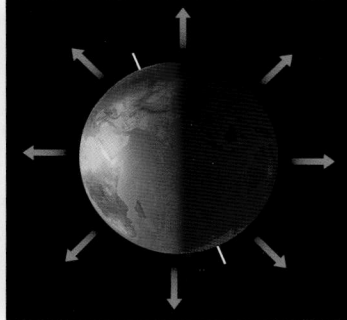

H

27. **(a)** Which gases are important selective absorbers of radiation? **(b)** What relevance does this property of gases have to the atmosphere? **(c)** What is the atmospheric greenhouse effect? **(d)** Which atmospheric gases are most responsible for the greenhouse effect?

ANOTHER VIEW The interaction of clouds and light can add drama to a scene. Here, cumulus clouds cast shadows on the land, as a large cumulonimbus cloud appears on the horizon of Arizona.

17 WINDS OF THE WORLD
Global and Local Wind Systems

After studying this chapter, you should be able to...

1. understand the forces that control air movement, how they combine to create the world's winds, and how winds flow around high-pressure and low-pressure centers in both hemispheres.

2. describe the major circulations of the tropical atmosphere, including the Hadley cells, El Niño, the Southern Oscillation, and monsoons; explain why they exist, and why they are important to society.

3. explain why the polar front exists in the lower troposphere, how it is related to the polar-front jet stream in the upper troposphere, and how the jet stream flows around the globe in a wave-like pattern.

4. discuss how local weather systems develop near the boundaries between land and water, and in the vicinity of mountain ranges.

When Sir Francis Drake set sail with the *Golden Hind* and its sister ships in 1577 (Fig. 17.1) to begin what would become the second circumnavigation of the globe, he and his crews were literally at the mercy of the wind, and during their three-year voyage they experienced its wild variations. Sailing south, they blew right into the clutches of a swirling storm that battered the ships so badly that Drake had to return to England for repairs.

Drake's second effort proved more successful, and his ships reached the belt of trade winds—westward-blowing winds so named because of their importance to transatlantic commerce. These winds filled the explorers' sails and blew them steadily southwestward. When the ships reached the equator, fortunately missing a fierce hurricane over the Caribbean to their west, they entered the sultry doldrums, a belt near the equator where the winds are weak and the weather is rainy. After bobbing helplessly with limp sails for days, they managed to cross the equator and enter another belt of trade winds. These winds blew toward the northwest, so it took all the crew's skill to maintain an overall southward route. When the ships reached the southern tip of South America, they encountered perpetually strong winds, known as the prevailing westerlies because they usually blow from west to east. Sea spray doused the ships and, in the freezing temperatures of low latitudes, the ships' decks and sails were soon coated with ice. The ice disabled part of the fleet, but the luckier ships made it into the Pacific and headed north along the west coast of the Americas. Drake and his remaining crew then turned west. As they crossed the Pacific and Indian Oceans, they encountered many challenging weather conditions, particularly winds associated with typhoons (Pacific hurricanes) and monsoons (seasonally changing tropical winds driven by temperature differences between land and sea). Finally, after rounding the southern tip of Africa, the *Golden Hind*, now the only remaining ship, headed home, its hold filled with plundered gold and its sailors full of tales about the winds of the world.

Drake had only a vague sense of the flow pattern of air around the globe. Today, you can see what Drake could not, by looking at satellite images, and animations made from them, on a computer screen. Atmospheric scientists can now generate maps that display all sorts of global characteristics of the atmosphere—such as wind velocity, temperature, moisture content—at any desired altitude (Fig. 17.2). These are used to understand the complexity of atmospheric circulations, from the largest global circulations to the smallest swirls of leaves.

In this chapter, we focus on *wind*, our atmosphere's horizontal currents of air, both large and small. We begin by investigating the nature of wind itself. We consider the conditions that cause the wind to blow, and we examine factors that control its direction and intensity. With this background, we then show how the characteristics of the winds vary with latitude. We start with the tropics, the region that straddles the equator, where intense solar heating drives the planet's largest distinct air circulations. We then shift to the Earth's mid-latitudes. These regions, where strong winds occur at the altitude where commercial jets fly, play a major role in controlling the weather. Finally, we wrap up by examining some notable local wind systems around the globe. As we will see, the strengths of the world's winds, and the directions in which they blow, vary substantially worldwide. Today, knowledge of these wind patterns is helping power companies tap the wind's energy, by using wind turbines, to provide alternative energy (**Earth Science at a Glance**, pp. 646–647).

FIGURE 17.1 The *Golden Hind*, the second ship to circumnavigate the world, was driven by wind power.

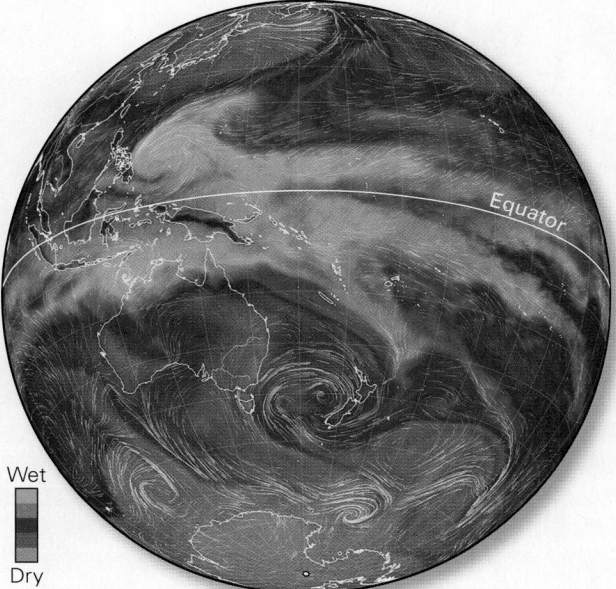

FIGURE 17.2 A map showing the distribution of total precipitable water vapor in the atmosphere (the amount of water vapor that is available to condense and fall out as rain). The thin white, curving lines are *wind streamlines*, which indicate the velocity (direction and speed) of wind flow.

Equator

Wet

Dry

<< Turbines in Germany capture the energy of the Earth's winds, converting wind power into electricity.

645

Winds of the World: A Resource for Power

Wind maps

The four global wind maps illustrate the direction and speed of surface winds at different locations all at the same time, emphasizing that winds vary dramatically around the world. The line segments are *wind streamlines*, which give a visual image of horizontal air flow at the time the image was taken; the line orientation indicates the compass orientation of the wind, the taper in brightness of the lines indicates which way along the line the air moves, and the overall brightness, color, and spacing represent speed, with brighter and more closely spaced lines indicating faster speeds. Areas where the lines have a reddish tint are regions where winds are particularly fast (>70 km/h, >43 mph). On computer animations, the line segments appear to move to give a sense of the flow. But keep in mind that they only show flow at a given time, so they do not indicate the source of air. Note that surface air flows in a wavy loop around Antarctica, and at several locations, it curves into eddies. In the northern hemisphere, several distinct "cyclones" (counterclockwise flows) stand out, as do the trade winds and the doldrums. Overall, surface winds over land are much slower than those over the ocean, due to frictional interaction between air and land.

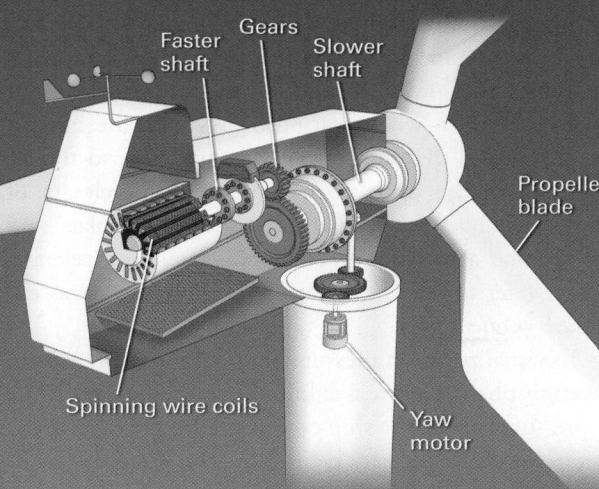

Faster shaft
Gears
Slower shaft

Propeller blade

Spinning wire coils

Yaw motor

Map of potential wind power

Because of the variation in the strength and steadiness of winds, some regions have the potential to generate more wind power than do others. On the map below, darker colors represent more potential power. The strongest steady winds are over the oceans, but the technology for building wind farms beyond the continental shelf does not yet exist.

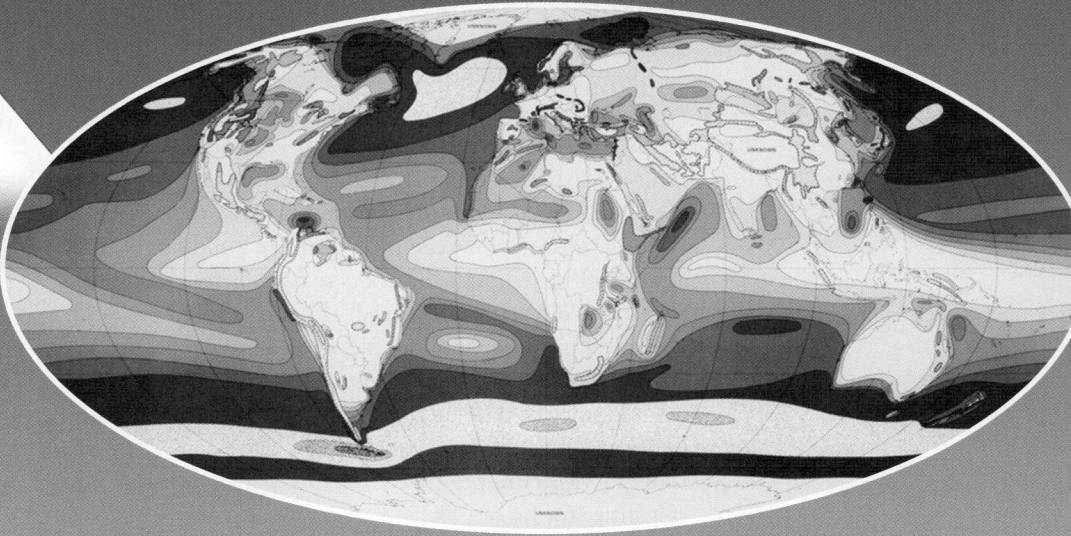

The inside of a wind turbine

The blades of modern wind turbines translate the flow of air in the wind into the rotation of a shaft. Gears shift the relatively slow rotation of the shaft connected to the spinning blades into the relatively fast rotation of a shaft that enters a generator. In the generator, magnets attached to the spinning shaft rotate within a coil of conducting wire. The movement of the magnets induces an electrical current in the wire.

647

17.1 Why Do the Winds Blow?

Our world would be a very different place without **wind**, the horizontal component of air flow. Air also moves upward, in *updrafts*, and downward, in *downdrafts*, but these nonhorizontal movements are not normally considered wind. For example, when a TV meteorologist says that winds will gust to 50 mph, they are referring to horizontal air motion only. You can feel the wind, for when it blows, air molecules bombard your skin. Even though you can't sense an individual molecule, your nerves can sense the combined impact of the septillion air molecules that batter you when you step into the wind. The faster the molecules move, the more force they exert, so during storms, winds can rip roofs from buildings or blow trains off their tracks. In *steady winds*, air flow maintains the same direction and speed; in *gusts*, a wind suddenly increases in speed, and then slows; and in *turbulent winds*, the flow of air twists and turns, so the speed and direction at a locality change constantly. Winds form at all scales, from small breezes that rustle a few leaves on a summer day, to vast currents that circle the planet or swirl into gigantic storms.

Have you wondered why the wind blows? Three forces play a key role in determining the direction and speed of the wind: (1) the pressure-gradient force, (2) the Coriolis force, and (3) the frictional force. Let's look at these forces and their consequences in turn. You'll recall that we introduced some of these forces in Chapter 15, in our discussion of ocean currents.

The Pressure-Gradient Force

In the atmosphere, the air pressure at one location may differ from that at another. We define a **pressure gradient** as a change in pressure divided by the distance over which

that change occurs. Where a pressure gradient exists, air tries to move from the location of higher pressure toward the location of lower pressure. To picture why, think of a bicycle pump—when you push down on the handle, the pressure in the pump's cylinder increases, so the air flows out of the open nozzle. Because the velocity of air flow changes when this movement takes place, we can say that a pressure gradient causes air to accelerate, and we refer to the force causing this acceleration as the *pressure-gradient force*.

To visualize how a pressure-gradient force can exist in the atmosphere, picture a sailing ship on the ocean. If the air pressure behind the ship exceeds the air pressure ahead of it, then the air density behind the ship exceeds the air density in front. Air molecules are in constant motion, so more air molecules strike the back side of the sails than the front side, and the ship moves forward (Fig. 17.3). The same concept applies even if the "ship" is the size of a single air molecule. More collisions happen on the side of the air molecule where the pressure is greater, so the molecule, on average, moves in the direction of lower pressure. When this phenomenon affects a vast number of air molecules, the whole air volume moves in the direction of lower pressure, and the wind blows.

The wind's speed—measured in meters per second, kilometers per hour, miles per hour, or *knots* (nautical miles per hour)—depends on the magnitude of the pressure gradient. The wind blows faster where the air pressure changes more quickly over a given horizontal distance than where it changes more slowly. To picture why, imagine the difference between stomping on an air-filled plastic bag and squeezing the bag gently with your hands. If the bag has an opening, the air escapes faster when you stomp on the bag than when you squeeze it gently.

We can represent pressure gradients on a map of constant elevation by plotting **isobars**, contour lines along which all points have the same air pressure. Where isobars lie closer together, the pressure gradient is larger (or steeper), so stronger winds blow (Fig. 17.4a). Where isobars lie farther apart, the pressure gradient is smaller (or gentler), so weaker winds blow (Fig. 17.4b). Note in Figure 17.4 that the wind is not moving in the direction of the pressure-gradient force, from high pressure to low pressure. Why? In nature there are three forces governing wind flow. The other two, the Coriolis force and the frictional force, work together with the pressure-gradient force to determine the wind direction, as we will see.

Why does air pressure differ across locations in the atmosphere? One primary reason involves heating and cooling of air. As we saw in Chapter 16, the Earth's surface absorbs incoming solar energy and then reradiates it as infrared energy, which heats air at the base of the atmosphere. Consequently, the near-surface air expands and

FIGURE 17.3 The relationship between wind and a gradient in air pressure. There are more air molecules to the left of the ship, so the air pressure is greater on the left, and the wind blows from left to right.

FIGURE 17.4 The relationship between wind and the spacing of isobars. The red arrows indicate the pressure-gradient force, with longer arrows indicating a stronger force. The blue arrows indicate the actual wind velocity. (Note that winds are not oriented down the pressure gradient. We will examine, later in the chapter, why winds blow in the direction shown when we consider the other two forces that affect the wind.)

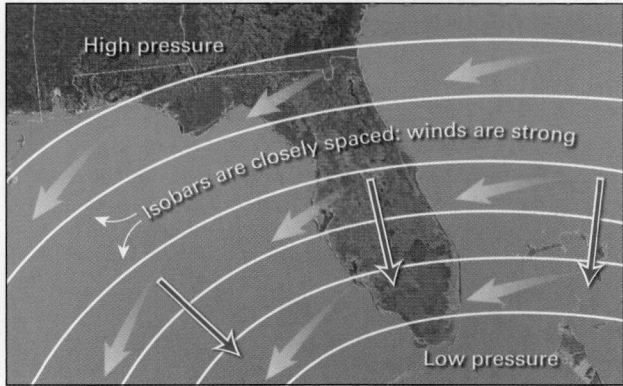

(a) Closely spaced isobars depict a strong pressure gradient and fast winds.

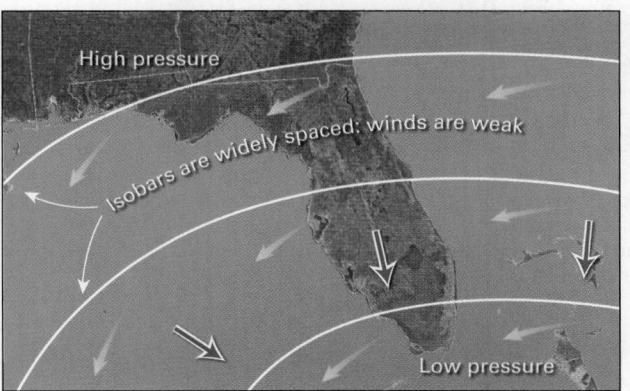

(b) Widely spaced isobars depict a weak pressure gradient and slow winds.

decreases in density. This change can create low pressure at the Earth's surface if it occurs over a localized region. To understand why, we need to introduce the concept of an *isobaric surface*. Imagine that we ascend in the atmosphere to an altitude where the pressure has a certain value, say 500 mb. (Recall that at sea level, the average air pressure is about 1,000 mb, or 1 bar. So, 500 mb occurs at an altitude of about 5.5 km or 3.4 mi, on average.) Now imagine that we place a floating marker there. Then, we move to different locations at 500 mb and do the same, again and again. Imagine now that we connect all the markers to form an **isobaric surface**, a level above the Earth where the pressure is the same everywhere.

Now imagine that at Time 1, temperature at any altitude is uniform, so isobaric surfaces are horizontal (Fig. 17.5a). In this situation, pressure decreases uniformly with increasing altitude. At Time 2, imagine that a region of the atmosphere near the ground warms up, as might happen over a hot desert. Air in the heated region expands and causes the pressure in a column of air *aloft* (meaning

above it) to become greater than in surrounding regions at the same altitude, so isobaric surfaces bulge upward (Fig. 17.5b). In the upper atmosphere, this situation produces an outward-directed pressure-gradient force, and air starts to flow out of the column. As a result, at Time 3, sometime later, the total mass of air within the column has decreased. Because air pressure at the Earth's surface represents the weight of the overlying column of air, the expansion associated with heating ends up decreasing the air pressure at the surface, so isobaric surfaces near the ground bow downward. This results in a near-ground low-pressure center and an inward-directed pressure gradient just above the ground.

The opposite effect happens when air cools (Fig. 17.5c). Once again, imagine air at Time 1 with uniform temperature at a given elevation. Large-scale cooling—for example, over Canada or Siberia in winter—causes near-surface air to contract and increase in density, which makes the pressure in a column of air above it lower than in surrounding regions, so isobaric surfaces bow down. In

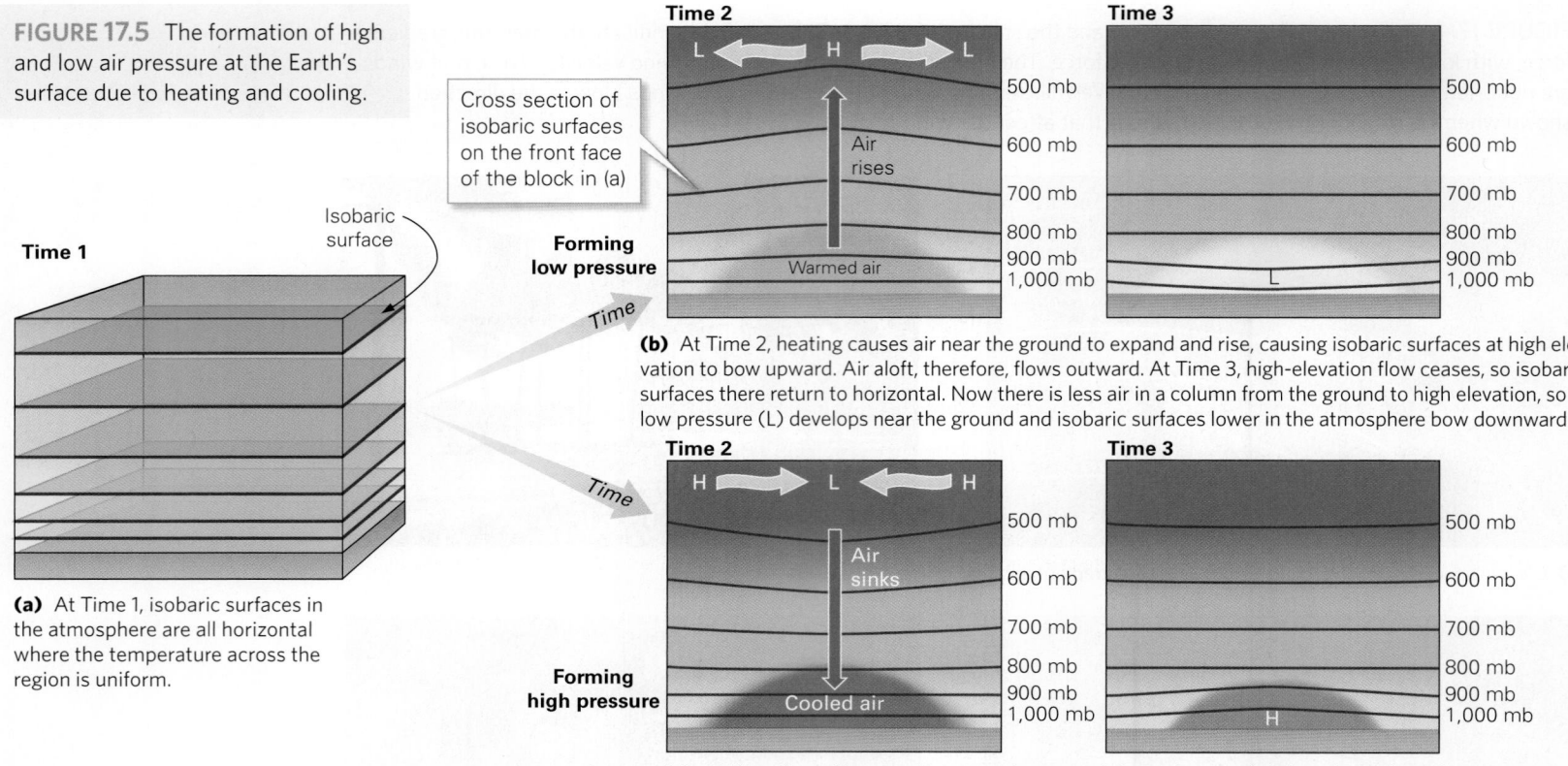

FIGURE 17.5 The formation of high and low air pressure at the Earth's surface due to heating and cooling.

Cross section of isobaric surfaces on the front face of the block in (a)

Isobaric surface

Time 1

Forming low pressure

Time 2

L ← H → L

Air rises

Warmed air

500 mb
600 mb
700 mb
800 mb
900 mb
1,000 mb

Time 3

500 mb
600 mb
700 mb
800 mb
900 mb
1,000 mb

L

(a) At Time 1, isobaric surfaces in the atmosphere are all horizontal where the temperature across the region is uniform.

(b) At Time 2, heating causes air near the ground to expand and rise, causing isobaric surfaces at high elevation to bow upward. Air aloft, therefore, flows outward. At Time 3, high-elevation flow ceases, so isobaric surfaces there return to horizontal. Now there is less air in a column from the ground to high elevation, so low pressure (L) develops near the ground and isobaric surfaces lower in the atmosphere bow downward.

Forming high pressure

Time 2

H → L ← H

Air sinks

Cooled air

500 mb
600 mb
700 mb
800 mb
900 mb
1,000 mb

Time 3

500 mb
600 mb
700 mb
800 mb
900 mb
1,000 mb

H

(c) At Time 2, cooling near the ground causes air to contract and sink, so isobaric surfaces at high elevation bow downward, producing a pressure gradient aloft that causes air to flow inward. At Time 3, high-elevation flow ceases, and isobaric surfaces aloft return to horizontal. Since there is more air in a column from the ground to the top of the troposphere, high pressure (H) develops at low elevation, and isobaric surfaces bow upward near the ground.

the upper atmosphere, this results in an inward-directed pressure-gradient force so that air flows into the column aloft (Time 2). As a result, the total mass of the air column increases, high pressure develops at the Earth's surface (Time 3), and isobaric surfaces near the ground bow upward. The development of pressure gradients at the Earth's surface due to heating or cooling is one process that causes winds to blow. We will explore others when we study storms in Chapter 18.

The Coriolis Force in the Atmosphere

In Chapter 14, while discussing ocean currents, we introduced the Coriolis force, which develops because the Earth rotates on its axis (see Box 14.3). The Coriolis force also influences wind. Because of the Coriolis force, an *air parcel* (a small volume of air) moving at a constant speed across the Earth (in the absence of the frictional force and pressure-gradient force) will follow a curved path. Significantly, the Coriolis force has four important properties that affect air-parcel movement:

1. it causes moving parcels to veer to the right of their direction of motion in the northern hemisphere, and to the left in the southern hemisphere;

2. it affects the direction in which a parcel moves across the Earth's surface, but has no effect on its speed;

3. it is stronger for parcels that move fast relative to a point on the surface of the Earth, and has a value of zero for stationary parcels; and

4. it has a value of zero on the equator and a maximum value at the poles.

The Frictional Force

When you picture a *frictional force*, you might think of the force that slows a book that you've slid across a table. Friction occurs in the atmosphere, too, either when a volume of faster-moving air molecules shears against a volume of slower-moving air molecules, or when a volume of air molecules shears against the Earth's surface. The frictional force always acts in the direction opposite to that of the moving object, so friction always reduces the speed of air flow. Near the Earth's surface, the frictional force increases, both because moving air interacts with the surface and because air density increases toward the base of the atmosphere. The rougher the surface, the greater the frictional force. Therefore, more friction exists between moving air and a city of tall buildings or a forest of trees

(a) A wind-flow map of Lake Superior and its surroundings. The brightness and density of the wind-flow lines indicate speed. Note how air accelerates as it passes over the lake, because there is less friction between air and water than between air and land.

(b) The boundary layer's thickness depends on the roughness and temperature of the surface below; it can vary from a few hundred meters to a few kilometers. Here, warmer air over cities and hillslopes produces updrafts that increase the thickness of the boundary layer and can cause clouds to form.

(c) Pollution or low-level clouds can sometimes show where the top of the boundary layer lies.

than between moving air and a lake or the ocean. For this reason, winds tend to be stronger (faster) over water bodies than they are over land (Fig. 17.6a).

Atmospheric scientists refer to the layer of air adjacent to the Earth's surface, where friction significantly affects air movement and generates turbulence, as the **boundary layer**. The frictional force can generate turbulence in the boundary layer. *Eddies* (whirlpool-like swirls) tend to develop, for example, when wind speed changes rapidly with distance, or wind shears against the ground, or air passes between buildings or over mountain peaks. The thickness of the boundary layer depends on several

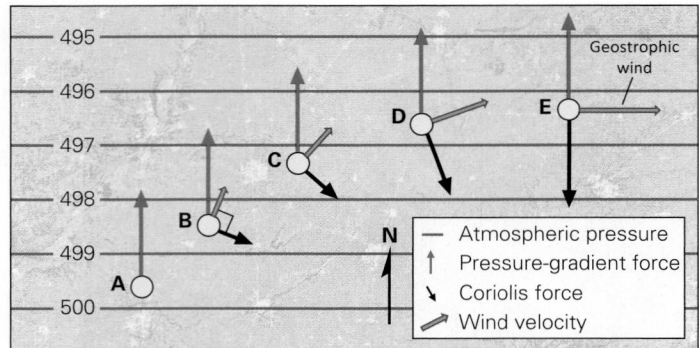

(a) This map shows isobars ranging from a high of 500 mb at the south to a low of 495 mb at the north. At Point A, the air parcel is at rest. As the parcel accelerates from Points A to E, the Coriolis force turns the air's path to the right until it flows parallel to the isobars. Note that the pressure-gradient force is constant over the map. As air accelerates the Coriolis force increases, and because the wind direction changes, the orientation of the Coriolis force changes so that the Coriolis force remains perpendicular to the wind direction. Eventually, air flow becomes parallel to the isobars.

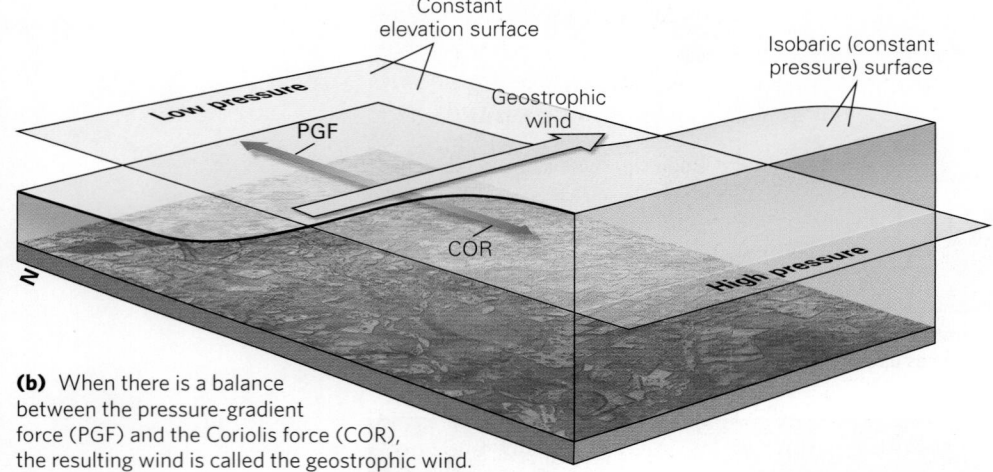

(b) When there is a balance between the pressure-gradient force (PGF) and the Coriolis force (COR), the resulting wind is called the geostrophic wind.

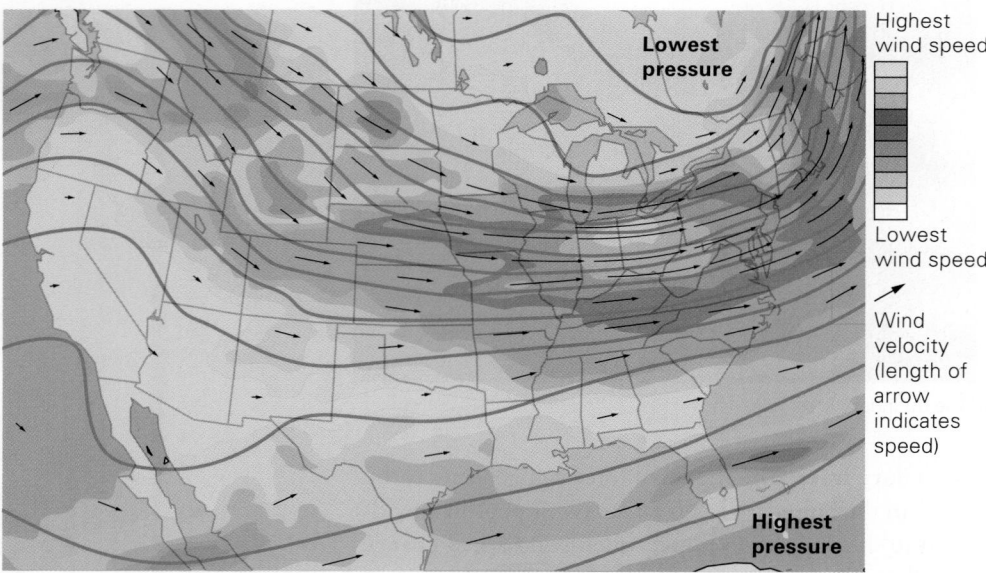

(c) A map of the wind speed at an altitude of about 10 km (6 mi). At this altitude, wind blows nearly parallel to the isobars, so the wind is nearly, but not exactly, in geostrophic balance. Note that air flow is faster where the pressure gradient is greater.

factors: the roughness of the underlying surface, surface heating (which causes the turbulent layer to loft to higher elevations), the presence of updrafts and downdrafts, and the wind speed. On a cold, calm winter morning at sunrise, the boundary layer over a large ice-covered lake extends upward only a few hundred meters. In contrast, on a hot, windy summer afternoon over a city or over hills, the boundary layer extends upward for a few kilometers (Fig. 17.6b). The boundary layer is sometimes visible as a layer of pollution or shallow clouds capped by clear air when air in the boundary layer does not mix immediately with the air above (Fig. 17.6c). Clouds form at the top of the boundary layer in buoyant air, when rising air contains moisture, as we will see in Chapter 18.

What Controls Wind Direction?

If the temperature of the atmosphere at each altitude across the Earth were the same everywhere, and if the Earth's surface were uniform, atmospheric pressure at any given altitude wouldn't vary from location to location, and the wind wouldn't blow. In reality, this doesn't happen for several reasons: solar radiation provides more heat in the tropics compared with polar regions; land and sea absorb and release heat at different rates; and the density of air changes with elevation at different rates in different locations. Given all these variables, the atmosphere never achieves a balanced state, and therefore it remains in constant motion.

At first glance, one might think that winds should flow from high pressure to low pressure, in a direction perpendicular to isobars, in locations where pressure gradients exist in the atmosphere. But because the Earth's rotation results in the Coriolis force, and interaction with the Earth's surface leads to a frictional force, the simple concept of air flowing directly from high pressure to low pressure in the Earth's atmosphere does not occur, as we will now discover. We first consider air above the boundary layer, where friction can be ignored, and then we focus on air nearer the surface, where the frictional force influences air motion significantly.

ABOVE THE BOUNDARY LAYER. Atmospheric conditions above the boundary layer approach a near-balanced state where the horizontal pressure-gradient force and the Coriolis force are equal and opposite. When they are exactly equal and opposite, a condition called **geostrophic balance** exists. To understand how conditions approaching geostrophic balance develop in the atmosphere, picture an initially stationary air parcel floating high above the ground at a location in the northern hemisphere where a pressure gradient exists (Fig. 17.7a). The pressure-gradient force accelerates the air and, initially, causes it to move from a region of higher pressure toward one of lower pressure. As soon as the parcel of air starts to accelerate, however, the Coriolis force deflects it to the right, causing it to start moving at a small angle relative to the isobars. As the air parcel continues to accelerate, the Coriolis force increases (because the magnitude of the Coriolis force depends on the speed of a moving object), causing the amount of deflection to increase. Acceleration and deflection continue until the Coriolis force and the pressure-gradient force become equal and opposite. When this happens, the air has attained geostrophic balance (Fig. 17.7b), and the wind that results, the **geostrophic wind**, flows parallel to the isobars. When wind flows parallel to the isobars, its velocity (speed and direction) stays constant, provided the isobars are straight. If the isobars curve, the wind has to accelerate (in this case, change direction) to follow the curve so the air will no longer be in geostrophic balance. In nature, air is nearly but not exactly in geostrophic balance at elevations above the boundary layer over much of the Earth, because isobars are curved and the spacing between them changes with distance (Fig. 17.7c). We'll see in the next chapter how deviations from geostrophic balance contribute to the formation of large storms.

WITHIN THE BOUNDARY LAYER. The force of friction significantly affects air flow within the boundary layer. As noted earlier, the strength of the Coriolis force depends on wind speed, and friction always reduces the speed of the wind. By slowing the wind, friction weakens the Coriolis force. How do all these changes affect wind direction? When friction slows the wind, the pressure-gradient force, which hasn't changed, exceeds the Coriolis force. The difference causes the wind to turn and, therefore, slant across the pressure gradient in a direction toward lower pressure (Fig. 17.8). This turning is strong where the frictional force is strong, particularly over land with obstructions like forests or buildings, and less so over water, where there is less friction. The wind direction is established when the combined Coriolis and frictional forces together balance the pressure-gradient force.

High-Pressure and Low-Pressure Systems

CONVERGENCE VERSUS DIVERGENCE. The development of a low-pressure center near the Earth's surface leads to the formation of a pressure gradient. As we noted above, if the Earth were not rotating, this pressure gradient would cause air from surrounding regions to flow inward toward the low-pressure center. On a rotating Earth, with the associated Coriolis force but no friction, airflow would adjust to be parallel to the isobars, circling the low-pressure center in a counterclockwise direction (in the northern hemisphere); this motion is called *cyclonic flow*. (In the southern hemisphere, cyclonic flow is clockwise.) Due to friction, air spirals inward toward the low-pressure center. Inward-spiraling flow causes **convergence**, meaning that air within the boundary layer flows into the low-pressure center (Fig. 17.9a). As air molecules converge toward the low-pressure center, air also rises (Box 17.1). A low-pressure center, together with the associated three-dimensional pattern of air circulation, is called a **low-pressure system** (Fig. 17.9b). We'll see in the next chapter that rising air in a low-pressure system can produce clouds and storms—in other words, stormy weather. A falling barometer, therefore, portends worsening weather.

When a high-pressure center develops, it generates a pressure gradient that causes air to flow outward from the center into surrounding regions. Again, if the Earth were not rotating and there were no friction, this gradient would cause air to flow outward from the middle of the high-pressure center into surrounding regions. If the pressure-gradient force and the Coriolis force were the only forces affecting airflow, air would circle the high-pressure

FIGURE 17.8 Effect of friction on airflow.

Above the boundary layer

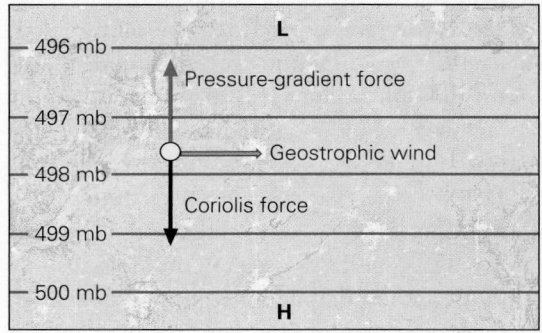

Within the boundary layer

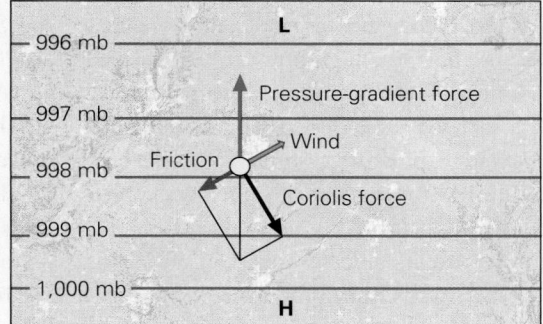

(a) At high altitude (above the boundary layer) geostrophic balance can exist, so air flows parallel to the isobars.

(b) At low altitude (in the boundary layer), friction slows air flow, effectively reducing the Coriolis force. Air flow, therefore, slants across isobars. Note that the friction force direction is opposite to the wind direction.

FIGURE 17.9 The effect of friction on air flow in the boundary layer.

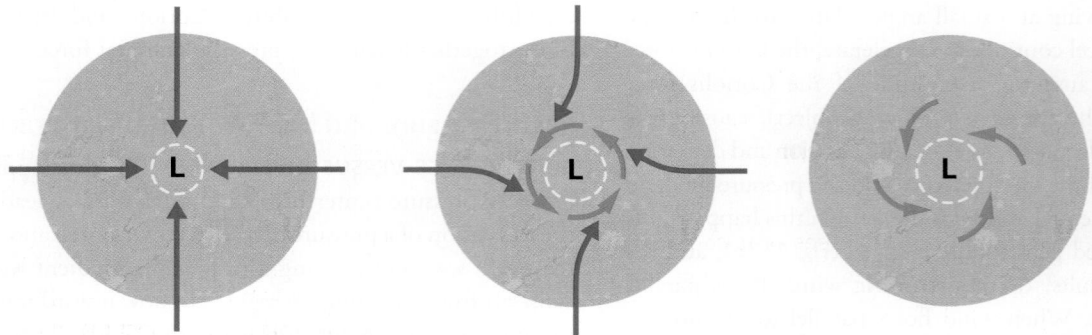

(a) If the Earth didn't rotate and there was no friction, air would flow directly toward the low-pressure center, and it would converge and rise (left). Because the Earth does rotate, the Coriolis force deflects air to the right, producing an overall counterclockwise rotation of the wind (middle). At low elevation, friction slows the flow and reduces the Coriolis force, so air spirals inward toward a low-pressure system (right).

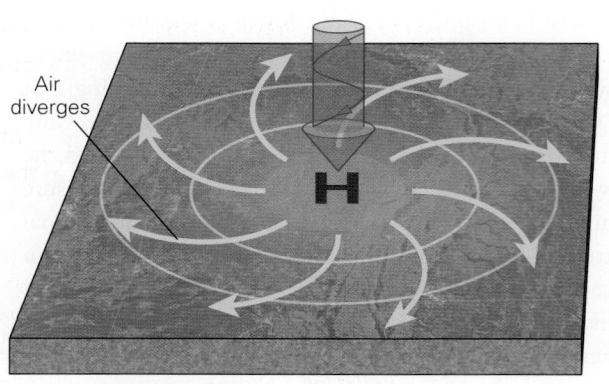

Low-altitude divergence

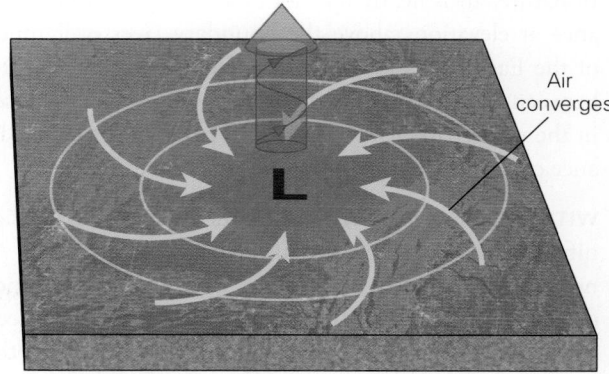

Low-altitude convergence

(b) Air descends as it spirals outward from a high-pressure system (left). Air ascends as it spirals inward toward a low-pressure system (right).

center in a clockwise direction (in the northern hemisphere); such movement is called *anticyclonic flow*. Note that flow around a high-pressure center is opposite to the motion that occurs around a low-pressure center. Because friction slows air motion, and the air's speed determines the magnitude of the Coriolis force, the pressure-gradient force exceeds the Coriolis force, and air spirals outward from the high-pressure center, in a clockwise sense in the northern hemisphere. Such outward-spiraling clockwise flow generates **divergence** in the boundary layer. Therefore, a deficit of air molecules develops near the Earth's surface, so air descends from higher elevations above the high-pressure center (see Fig. 17.9b). Meteorologists refer to a high-pressure center, together with the associated three-dimensional pattern of wind circulation, as a **high-pressure system**. When the air within a high-pressure system descends, it undergoes compression and warms, so its relative humidity decreases. As a result, high-pressure systems tend to host relatively dry, cloud-free air—in other words, they produce *fair weather*. A rising barometer, therefore, portends improving weather.

SEMIPERMANENT PRESSURE SYSTEMS. The Earth's atmosphere hosts two types of low-pressure systems: migrating

and semipermanent. The great storms of the Earth's atmosphere, tropical and mid-latitude cyclones, are examples of migrating low-pressure systems, as we will discuss in Chapter 18. The remaining low-pressure systems, and most high-pressure systems, are semipermanent, meaning that they form and re-form over the same general geographic area in a given season. Here we consider these *semipermanent pressure systems* and how they affect the Earth's winds and weather.

What happens to the temperature outdoors after the Sun goes down? We know from common experience that air starts cooling and we feel a chill in the air. At night, the Sun no longer heats the land, so the Earth's surface and the air above radiate infrared energy upward into the higher atmosphere and, therefore, the Earth's surface and the near-surface air become cooler. Now imagine what happens during the winter in large regions of Canada, Siberia, and Antarctica, regions that are in nearly perpetual darkness because of their high latitude. These vast landmasses continually cool for months at a time during winter. The resulting broad, long-lived pool of cool air, in turn, generates a winter **semipermanent high** near the Earth's surface

BOX 17.1

How can I explain . . .

Flow in a low-pressure system

What are we learning?

- How friction near the ground causes air to converge in the center of a low-pressure system

What you need:

- A transparent cup filled about three-fourths of the way with tea
- Loose tea leaves at the bottom of the cup
- A spoon to stir the tea

Instructions:

- Let the liquid come to rest so that the tea leaves are spread out and resting at the bottom of the cup.
- Stir the tea vigorously in one direction so that the fluid rotates in the cup.
- Observe the tea leaves.

What did we see?

- This experiment illustrates fluid flow in a low-pressure system. As the tea in the cup rotates, the top surface of the tea becomes funnel-shaped. As a consequence, less liquid is in the middle of the cup than along the sides, so there's less pressure on the bottom of the cup in the center than along the sides. This distribution of mass must mean that low pressure exists at the bottom of the cup at its center. One might expect the tea leaves to follow a circular path around the edge of the cup due to the centrifugal force.
- Meanwhile, however, the fluid experiences a pressure gradient that should drive it toward the low pressure at the center of the cup. If a balance were to be established between centrifugal force and pressure-gradient force, the tea leaves would follow a circular path as you stirred. But toward the bottom of the cup, friction between the tea and the cup slows the flow, so the inward-directed pressure-gradient force (toward the center) exceeds the centrifugal acceleration of the spinning fluid, and the leaves spiral toward the center of the cup. The converging liquid then rises in the middle of the cup.
- A similar process occurs in low-pressure systems in the atmosphere. Air near the Earth's surface converges toward the low-pressure center, and as the air approaches the center, it rises, causing clouds and precipitation.

FIGURE Bx17.1 Stationary tea leaves sit at the bottom of the cup (left). Stirring the cup disperses the tea leaves and produces a low-pressure zone (middle). The tea leaves start to spiral inward and upward around the resulting whirlpool (right).

Upward spiraling tea

(Fig. 17.10), for the reasons we described in Figure 17.5. Even if air flows out of these regions to lower latitudes, a high-pressure system quickly re-forms, because new air moving into the regions continues to cool.

During the summer, nonequatorial oceans are cool relative to land. Therefore, air over the oceans cools relative to air over the continents. This cooling causes summer semipermanent highs that form and re-form throughout the season over temperate-latitude oceans

(see Fig. 17.10). (Similar semipermanent highs form in subtropical latitudes during the winter.) In the North Atlantic and North Pacific, the summertime semipermanent high-pressure systems are called the *Bermuda High* and the *Pacific High*, respectively.

As you might anticipate, **semipermanent lows** develop in the lower atmosphere where the Earth's surface warms and continually heats the air above it, relative to surrounding air. Notably, such lows develop over high-latitude oceans

FIGURE 17.10 Semipermanent high- and low-pressure systems.

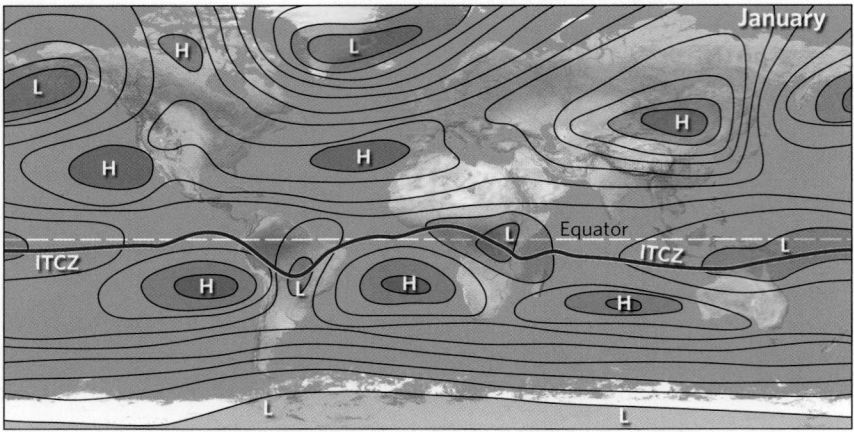

(a) In January (the northern hemisphere winter), semipermanent highs develop over Siberia and Canada, caused by preferential cooling over continents. Semipermanent lows occur over the northern Pacific and Atlantic, because the oceans are warmer than adjacent lands. In the southern hemisphere it is summer. Nevertheless, cold ocean currents keep the ocean waters sufficiently cool, compared with land, that high pressure develops over the oceans in subtropical latitudes. Central Antarctica is always cold, and waters off its coast are warmer, so relative low pressure develops along the Antarctic coast.

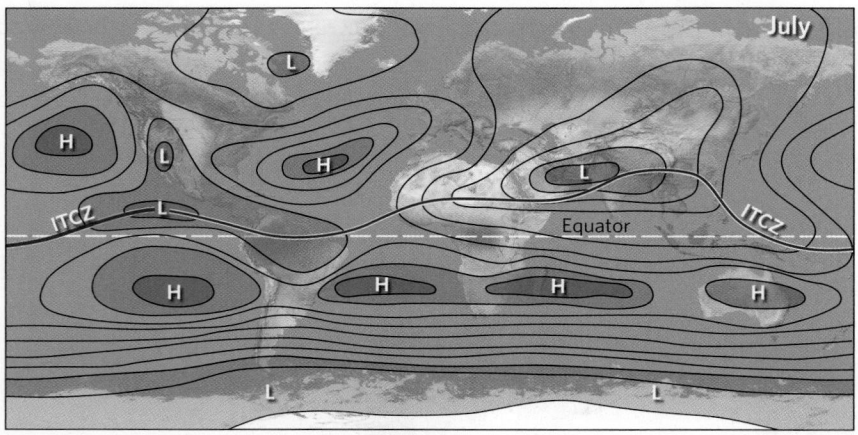

(b) Semipermanent highs strengthen over the central north Pacific and Atlantic in the northern hemisphere summer while low pressure develops over the tropical and subtropical landmasses, which are heated.

during the winter, because ocean water retains more heat than the land can retain. Heat rising from oceans warms overlying air, so this air becomes warmer than the air over high-latitude continents. Such a situation develops over the North Atlantic south of Iceland, and over the North Pacific south of the Aleutian Islands, in November through March, producing the *Icelandic Low* and the *Aleutian Low*, respectively. Low pressure also develops over desert regions, such as the southwestern United States, in summer because the strong solar heating that occurs during the daytime makes the air above the ground particularly hot. Low pressure also develops along the equator in a belt around the world. This belt is called the *intertropical convergence zone* (ITCZ), for reasons we discuss next.

Take-home message...

The pressure-gradient, Coriolis, and frictional forces determine the speed and direction of the wind. Air accelerates toward regions of lower pressure. As it does so, the Coriolis force increases and eventually balances the pressure-gradient force, yielding geostrophic balance, so wind flows parallel to isobars. In the boundary layer, friction causes air to spiral in toward low-pressure centers, and out from high-pressure centers. Heating or cooling at the base of the atmosphere generates semipermanent high- and low-pressure systems.

Quick Questions ———————————

- If the Earth were not rotating, which force would no longer exist?
- Why does the thickness of the boundary layer vary with location?
- Would strong high-pressure centers be more common in summer or winter over Canada?

17.2 Air Circulation in the Tropics and Subtropics

What image comes to mind when you hear the word *tropics*—warm sunny breezes and swaying palm trees? In reality, the **tropics**—which lie between the Tropic of Cancer and the Tropic of Capricorn (about latitude 23° N and 23° S, respectively)—cover two-thirds of the Earth's surface and are regions that have summer-like weather all year long. North of the Tropic of Cancer and south of the Tropic of Capricorn lie the **subtropics**, latitudes (from 23° N to 30° N in the northern hemisphere, and from 23° S to 30° S in the southern hemisphere) that typically have dry, hot, desert climates that transition to temperate mid-latitude climates in each hemisphere. Overall, air circulation patterns in the tropics and subtropics are driven by differential solar heating, and by related variations in sea-surface temperatures. Let's explore how these circulations operate.

Hadley Cells and the Intertropical Convergence Zone

George Hadley, a British lawyer with a strong interest in meteorology, wondered why winds blow. In 1735, he published an article in which he suggested that warming the air at the equator—where the Sun heats the Earth most intensely—would cause air to rise from low elevations to the top of the atmosphere, where, unable to rise higher, it would start to flow poleward. As the pole-ward-moving air aloft cooled and became denser, Hadley proposed that it would sink back toward the surface and, nearer the ground, would flow back toward the equator. Such a circulation caused by convective flow produces a **convective cell** (see Chapter 2).

Hadley's model was partially correct, but he did not take the Earth's rotation, or the decrease in atmospheric temperature with altitude, into account. If the Earth did not rotate, the poleward-flowing component of Hadley's proposed convective cell might theoretically extend farther toward the polar regions. But the Earth does rotate, so the Coriolis force comes into play and deflects the high-altitude poleward flow. The air also stays confined to the troposphere. Therefore, high-altitude air rising in the convective cell to the tropopause initially flows northeastward, and then eastward in the northern hemisphere, and southeastward, and then eastward in the southern hemisphere. As a consequence, air flowing poleward at the top of the troposphere reaches a latitude of only about 30° before it begins to follow a path roughly parallel to that line of latitude **(Fig. 17.11)**. Furthermore, by 30° latitude,

much of the air that rose to the top of the convective cell near the equator has sunk back toward the Earth's surface. When the sinking air reaches a low altitude, it flows back toward the equator. The Coriolis force deflects this near-surface air flow westward, so that in the northern hemisphere it heads southwest, and in the southern hemisphere it heads northwest, at a decreasing angle to lines of latitude as it approaches the equator. By the time the air reaches the equator, it follows a path almost parallel to the equator. This tropical surface flow constitutes the **trade winds**, the same winds that sent Sir Francis Drake westward across the oceans. Atmospheric scientists refer to these convective cells in each hemisphere, which extend from the equator to a latitude of about 30°, as **Hadley cells**, in honor of George Hadley.

Near the Earth's surface, the trade winds of the northern hemisphere and those of the southern hemisphere converge (flow toward each other), producing the **intertropical convergence zone**, or **ITCZ**. The ITCZ

FIGURE 17.11 The Hadley cells control the circulation that leads to the distribution of rainfall in low latitudes. They determine where the world's tropical rainforests, grasslands, and deserts occur. The inset shows a cross section of the two Hadley cells, with the ITCZ at the equator.

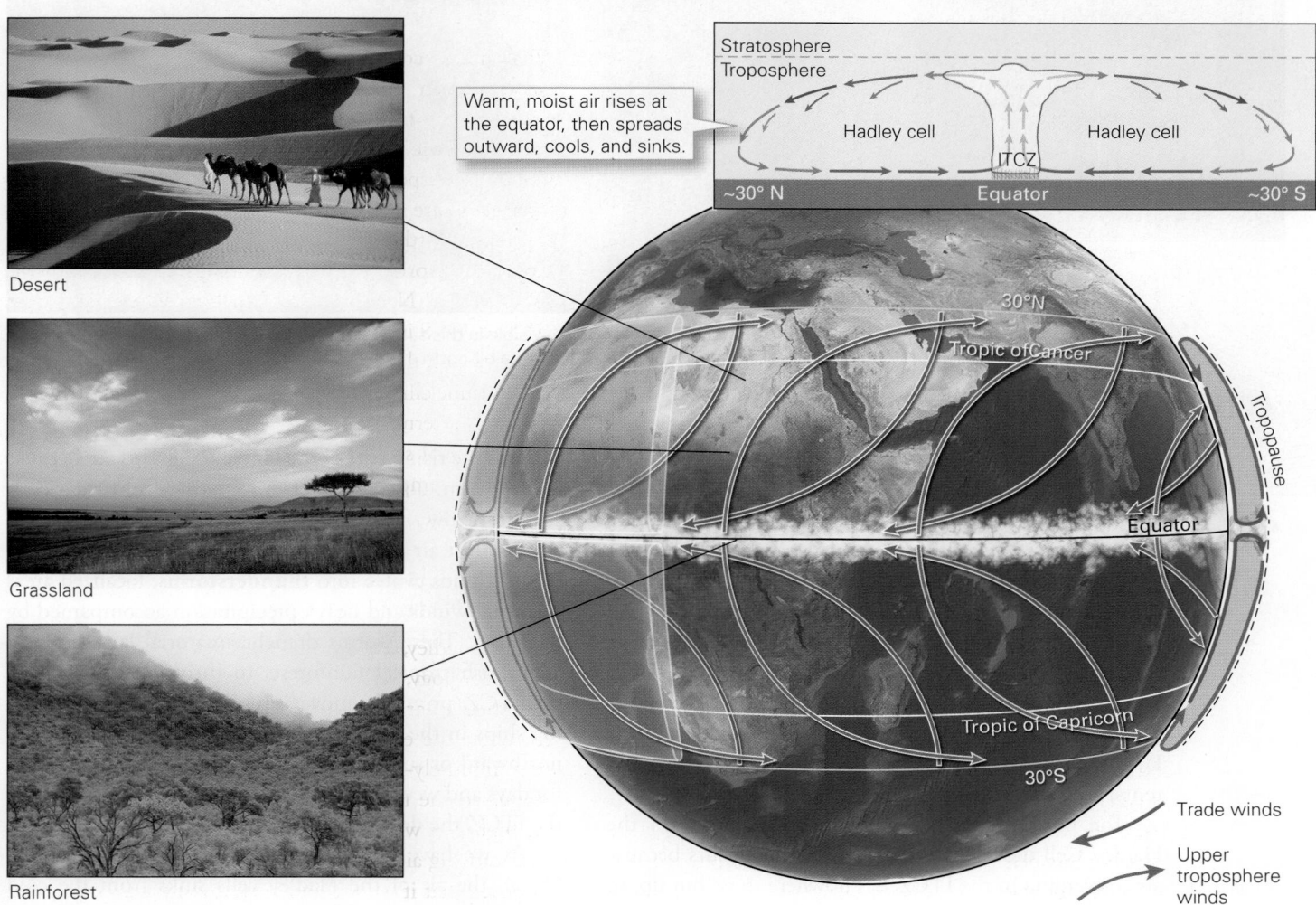

Desert

Grassland

Rainforest

Stratosphere
Troposphere

Warm, moist air rises at the equator, then spreads outward, cools, and sinks.

Hadley cell   Hadley cell

ITCZ

~30° N   Equator   ~30° S

30°N

Tropic of Cancer

Tropopause

Equator

Tropic of Capricorn

30°S

Trade winds

Upper troposphere winds

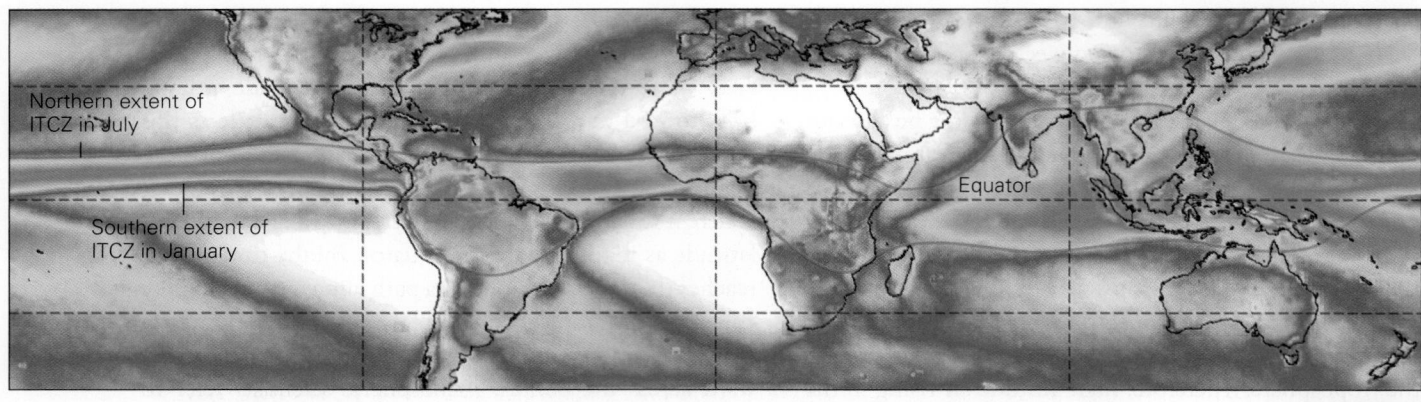

(a) The position of the ITCZ changes over the course of a year. Heavy rains fall in regions over which the ITCZ passes.

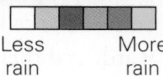

Less rain   More rain

(b) Clouds forming along the ITCZ are visible in this satellite image. The inset shows wind streamlines converging at the ITCZ, when the ITCZ lies a bit north of the equator.

approximately follows the line along which the Sun's heat is most intense. Because of the Earth's tilt, the position of this line changes over the course of a year. In general, the ITCZ lies north of the equator during the northern hemisphere summer and south of the equator during the northern hemisphere winter. (We will see later that this movement of the ITCZ plays a major role in monsoonal weather.) The exact position of the ITCZ also depends on the temperature of the Earth's surface below, so it can be affected by ocean currents, land cover, and topography (Fig. 17.12a).

The position of the zone of rising air within the Hadley Cell moves with the ITCZ. This occurs because air converging in the ITCZ has nowhere to go but up, so it forms the rising part of the Hadley cell. This air, because it's so warm, absorbs water evaporating from the ocean surface below. For reasons we'll discuss in Chapter 18, rising moist air produces clouds (Fig. 17.12b). Some clusters of clouds evolve into **thunderstorms**, localized areas of strong winds and heavy precipitation accompanied by lightning. These storms drench equatorial latitudes and allow lush tropical rainforests to thrive. Because air in the ITCZ primarily moves upward and westward, sailing ships in the past had a hard time crossing the ITCZ northward or southward. Stuck under the rainy weather for days and weeks, sailors named the belt of storms along the ITCZ the **doldrums**.

As we have noted, to the north and south of the ITCZ, the air of the Hadley cells sinks from the cold upper troposphere to the middle and lower troposphere. This air contains very little water, for most of its moisture rained out in the ITCZ. As this dry air descends, it compresses and warms, so its relative humidity decreases even more. Skies of the northern and southern portions of the tropics, together with much of the subtropics, therefore, tend to host the clear, hot, dry weather leading to the formation of the world's largest deserts. The *steppes*, the grasslands between deserts and rainforests, have a long dry season and a short rainy season. The rainy season occurs during the summer, when the ITCZ lies overhead.

The Subtropical Jet Stream

A strong temperature contrast exists in the upper troposphere above subtropical latitudes, where the poleward-flowing part of the Hadley cell meets the cooler air of the mid-latitudes. As we will see in Section 17.4, this temperature change leads to a pressure gradient at the top of the troposphere that accelerates high-elevation air northward. The Coriolis force, in turn, deflects the moving air eastward as it approaches its geostrophic value. As a result, a very fast (>160 km/h, or 100 mph), high-altitude river of air, called the **subtropical jet stream**, encircles the globe at a latitude, on average, of about 30°, at the northern and southern end of the Hadley cells in each respective hemisphere. It tends to be relatively narrow, only between 300 and 500 km (180–300 mi) wide, measured perpendicular to the air-flow direction, and its strength is tied to the amount of air flowing poleward locally from the Hadley cells. This jet stream can affect weather in the mid-latitudes as well as in the tropics. At any given time, two jet streams exist in each hemisphere; we'll discuss the second jet stream, the polar-front jet stream, later in this chapter.

The Southern Oscillation and El Niño

Often, when a major weather event happens, such as the intensely hot weather in the southern United States during the summer of 2023, someone in the media ties the event to El Niño or La Niña. What are these phenomena, and how do they affect the world's weather? To address this question, we must first introduce the *Walker circulation*, a cell of flowing air that spans the Pacific Ocean along the equator.

The **Walker circulation**, named after British physicist Gilbert Walker (1868–1958), extends vertically between the Earth's surface and the tropopause and horizontally from South America's western coast to Australia and Indonesia, a distance of more than 16,000 km (9,940 mi). This circulation, which appears as a distinctive feature only within about 5° of the equator, can be thought of as a secondary, east-west-flowing convective cell within the Hadley cells. Within the Walker circulation, air arriving from north and south in the trade winds flows westward in the lower troposphere, rises over the western equatorial Pacific Ocean, returns eastward in the upper troposphere, and sinks over the eastern equatorial Pacific (Fig. 17.13a).

Why does the Walker circulation exist? The western Pacific tends to be warmer than the eastern Pacific because of the configuration of surface currents in the ocean. As a consequence, air over the western Pacific basin becomes warmer than that over the other parts of the Pacific, generating a region of low pressure at the Earth's surface. Meanwhile, in the eastern Pacific, where ocean currents bring cold water up from the depths in a process called *upwelling* (see Chapter 14), a region of high pressure develops at the base of the atmosphere as air temperature decreases above the cooler water (see Fig. 17.5). Because the Coriolis force near the equator is very weak—it's zero at the equator—the pressure-gradient force drives air flow from the high-pressure zone of the eastern Pacific toward the low-pressure zone of the western Pacific.

When the temperature contrast between the eastern and western Pacific becomes large, strong westward-blowing trade winds push surface water westward (see Chapter 15). These winds cause sea level to rise by 10 to 20 cm (4–8 in) in the western Pacific relative to the eastern Pacific. Because the ocean currents carry surface water away from the eastern Pacific, cool water upwells along the coast of South America and westward to replace the surface water (Fig. 17.13b). This water brings up nutrients that nourish plankton. Fish eat plankton, so fish populations thrive and fishing crews have successful hauls.

Every few years, for reasons that atmospheric scientists are still studying, the intensity of the Walker circulation weakens. This weakening reduces the speed of the winds blowing west across the equatorial Pacific. When this happens, the tilted sea surface flattens, and warm ocean water from the western Pacific spreads eastward along the equator all the way to the South American coast (Fig. 17.13c, d). As a result, the temperature of the ocean water becomes more uniform across the Pacific, so the pressure-gradient force decreases. In fact, surface water in the eastern Pacific may become warmer than that in the western Pacific, leading to a reversal in the pressure gradient that, in turn, causes a brief reversal in the direction of the equatorial winds. Upwelling along the coast of South America then shuts off. Over the course of a year or so, the Walker circulation slowly strengthens, low pressure again develops in the western Pacific, high pressure develops in the eastern Pacific, and the trade winds again begin to blow strongly westward.

Atmospheric scientists refer to the east-west seesaw in surface air pressure across the equatorial Pacific that drives these changes in the Walker circulation as the

Did you ever wonder . . .

what El Niño is, and why it may affect your local weather?

FIGURE 17.13 The Walker circulation and the Southern Oscillation.

-5　-4　-3　-2　-1　0　1　2　3　4　5

Sea-surface temperature anomalies (°C)

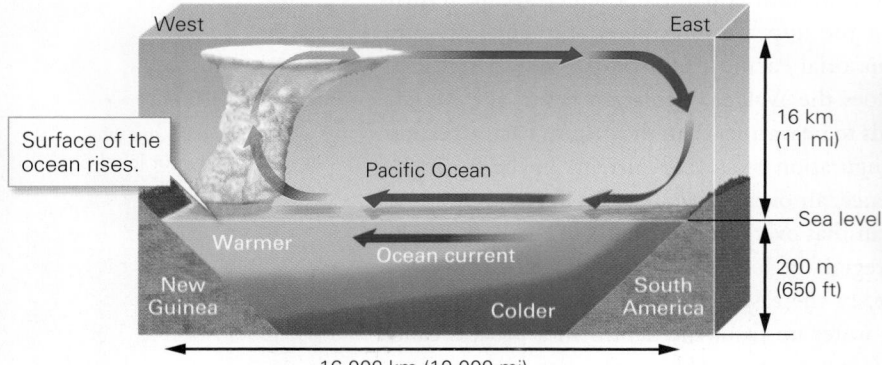

(a) Normal Walker circulation. During La Niña, this circulation strengthens.

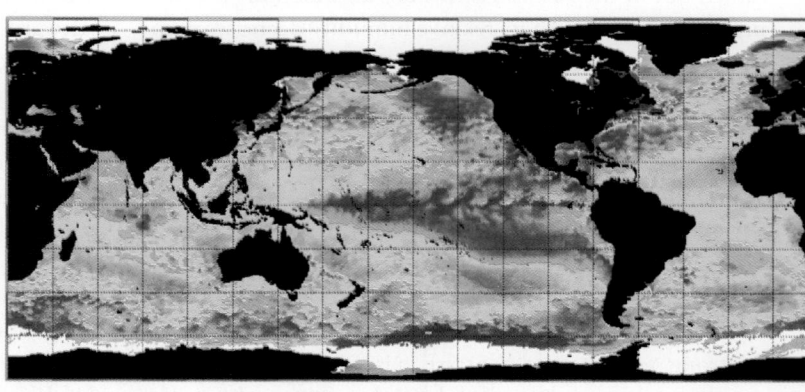

(b) Sea-surface temperature anomalies during La Niña.

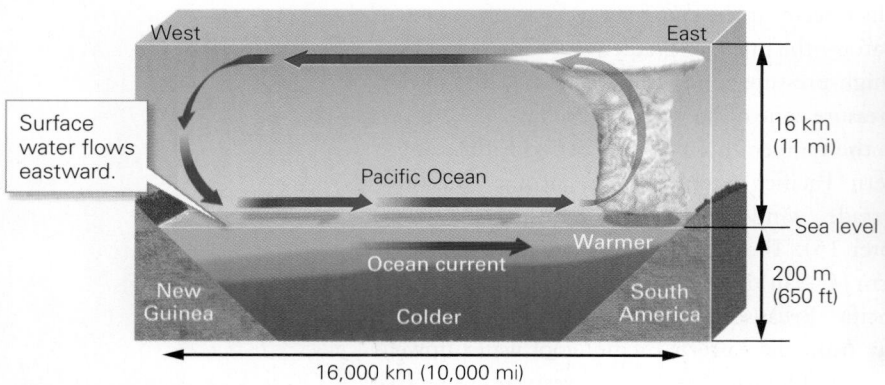

(c) During El Niño, the Walker circulation weakens or reverses.

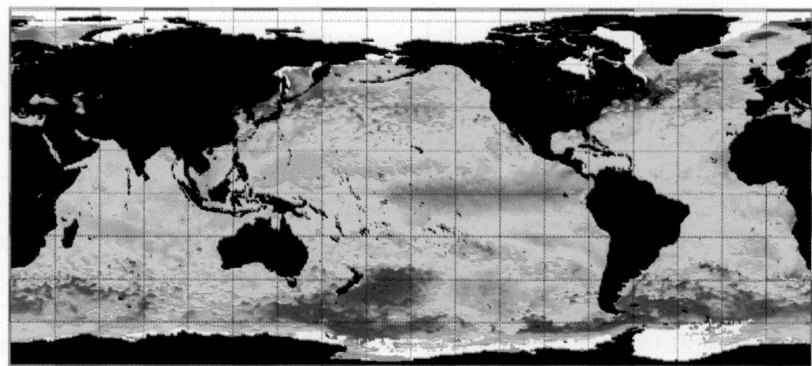

(d) Sea-surface temperature anomalies during El Niño.

Southern Oscillation. The warming of ocean temperatures in the eastern equatorial Pacific has come to be known as **El Niño**, a term originally coined by Peruvian fishermen because the onset of the warming often coincides with the Christmas season (*El Niño* is Spanish for little boy, meaning the Christ Child). El Niño threatens the livelihoods of fishermen because the shutoff of upwelling nutrients decreases fish populations. Atmospheric scientists coined the term **La Niña** (Spanish for little girl) to describe times when the Walker circulation becomes very strong and the upwelling of cold water along coastal South America increases. The acronym **ENSO** (pronounced enn-soh), which stands for El Niño/Southern Oscillation, refers to the overall seesaw pattern and its consequences. During the past century, more than two dozen El Niño events have taken place; one of the strongest happened in 2015, while another began in 2023.

If it were just a local phenomenon, El Niño would not have become so famous. But the rise of sea-surface temperatures in the eastern Pacific modifies atmospheric circulations around the globe, even in the mid-latitudes, influencing weather worldwide. For example, El Niño triggers droughts in Australia and Indonesia, particularly rainy weather in the eastern Pacific, and flooding along South America's northwestern coast. In North America, El Niño can cause storms that normally move from the Pacific into Washington and Oregon to move northward instead, toward the Gulf of Alaska, or can cause storms that normally migrate to the south of the United States to move north instead, triggering floods in California and winter storms farther east. El Niño years are typically associated with warmer temperatures globally because of the broad area of warm water across the equatorial Pacific. For example, the 2023 El Niño, combined with overall global warming due to increased CO_2, produced record heat waves across the northern hemisphere. Strong La Niña conditions also affect weather around the globe, but in different ways.

Monsoons

The word *monsoon*, in everyday English, carries with it an image of drenching rains for weeks on end, leading to catastrophic flooding, such as happened during July of 2023 in India. In atmospheric science, the word has a more general meaning. Technically, a **monsoon** is a twice-a-year reversal of wind direction, in tropical climates, that causes a transition between a distinct dry season and a distinct rainy season. The reversal in wind direction reflects the annual migration of the ITCZ over a region, as the ITCZ migrates across the equator during the seasons. This migration happens because, due to the tilt of the Earth's axis, latitudes south of the equator receive more solar energy during northern hemisphere winter months (December–February), whereas latitudes north of the equator receive more solar energy during northern hemisphere summer months (June–August).

Let's examine how the transition takes place over southern Asia, a highly populated region that lies in the northern hemisphere (Fig. 17.14). During winter, land cools faster than the ocean because water does not lose heat as quickly as rock or soil. As a result, air over the land cools, forms a region of high pressure, and descends toward the Earth's surface. The resulting pressure gradient generates a wind that blows seaward. This is the near-surface flow of the northern Hadley cell. Because air sinks in the high-pressure system over the land, it becomes clear and dry (see Fig. 17.14a). This is the Southeast Asian dry season. In the ITCZ, the air warms and rises, forming a belt of storms that pour rain into the sea, and on to islands in the equatorial southern hemisphere (Fig. 17.15a). As spring progresses, the ITCZ drifts north, across the equator and into the northern hemisphere. Solar energy heats the land and mountain slopes more rapidly than it heats water in the ocean in the summer months. As a result, air over land becomes warmer than air over the adjacent ocean; it therefore expands and rises, and the air pressure over land at the Earth's surface becomes lower than that over the ocean. The resulting pressure gradient causes moist air to flow from the ocean over the land, resulting in the ITCZ migrating northward over southern Asia (Fig. 17.15b). This onshore flow of wind represents the near-surface flow of the southern Hadley cell. Once over the hot surface of the land, the air warms, becomes buoyant, and rises in the ITCZ. The rising moist air produces thunderstorms that drench the land beneath the ITCZ, causing the infamous summer flooding of southern Asia (Fig. 17.15c) (see also Fig. 17.14b).

The most intense monsoonal rainy seasons in Southeast Asia begin when the ITCZ drifts over the land. Over a period of two months, thunderstorms spread northward across the continent (Fig. 17.16a). The water they supply causes rivers to rise and flood large areas (Fig. 17.16b). Monsoonal rains of this region are the most intense in the world because the moist air blowing from the Indian Ocean northward undergoes orographic lifting along the front of the Himalayan Range. Monsoons determine the success or failure of agriculture in southern Asia—abnormally wet or dry monsoon years can destroy a region's crops and livestock and lead to economic disaster (Fig. 17.16c).

FIGURE 17.14 Seasonal transitions in the southern Asian monsoon.

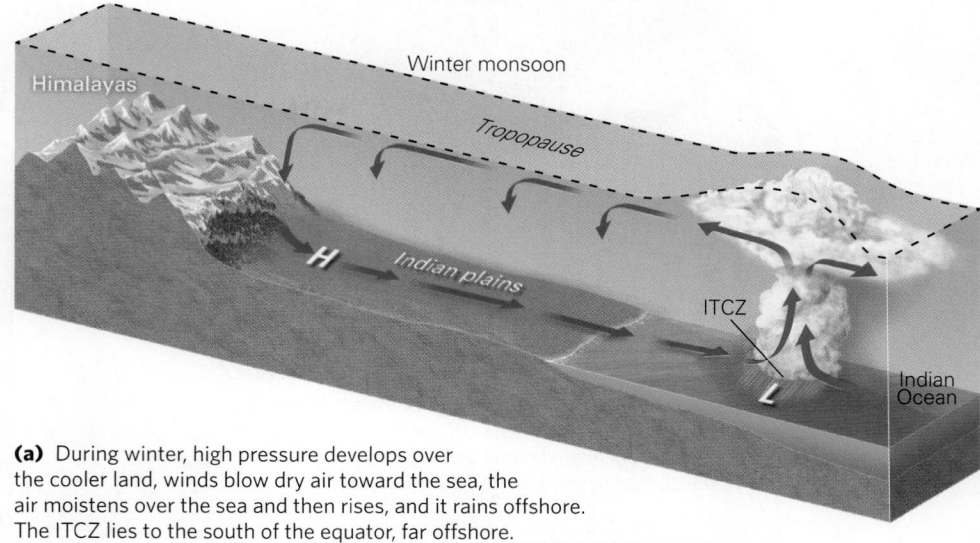

(a) During winter, high pressure develops over the cooler land, winds blow dry air toward the sea, the air moistens over the sea and then rises, and it rains offshore. The ITCZ lies to the south of the equator, far offshore.

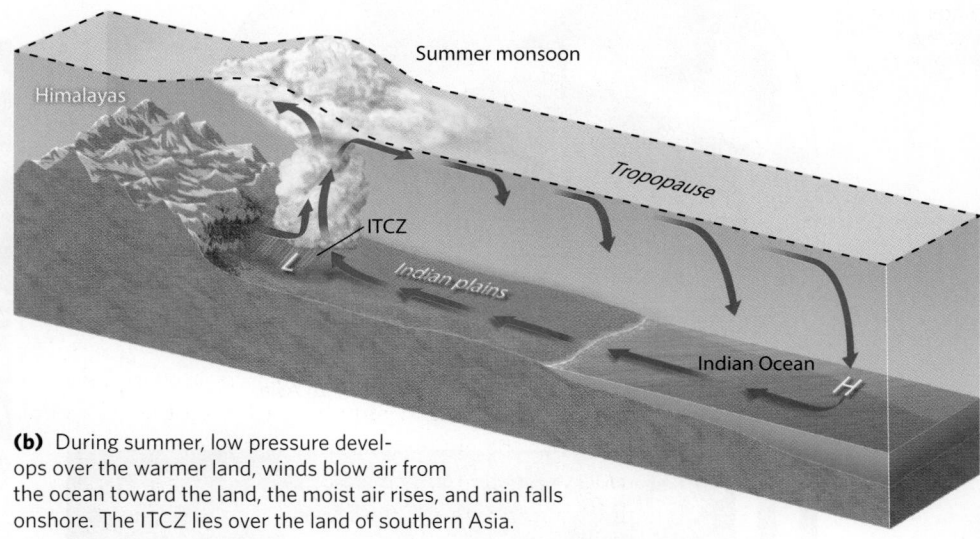

(b) During summer, low pressure develops over the warmer land, winds blow air from the ocean toward the land, the moist air rises, and rain falls onshore. The ITCZ lies over the land of southern Asia.

Monsoons also occur in other tropical and subtropical regions. For example, the rainy season in South America develops in the southern hemisphere summer, when the Sun heats the Amazon basin and eastern Andes highlands and large thunderstorms are triggered along the ITCZ in those regions. In the southern hemisphere winter, the ITCZ moves into the northern hemisphere, the Sun is not as strong over the Amazon basin, and that area experiences its annual dry season. The summer wet season in the southwestern United States and northwestern Mexico develops when summer sunlight heats the deserts of this region while the waters of the eastern Pacific remain relatively cool. The rains that fall when moist winds blow eastward over the deserts provide nearly 70% of the region's summer rainfall.

FIGURE 17.15 The southern Asian monsoon.

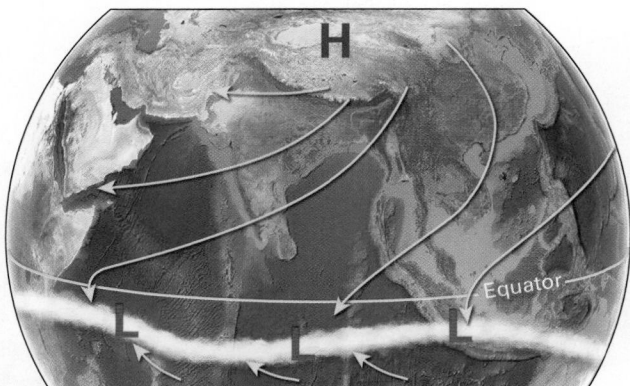

(a) During the winter, the ITCZ lies south of the equator and northeasterly winds blow from Asia out over the Indian Ocean. Note that the winds change direction where they cross the equator, due to the change in the direction of the Coriolis force.

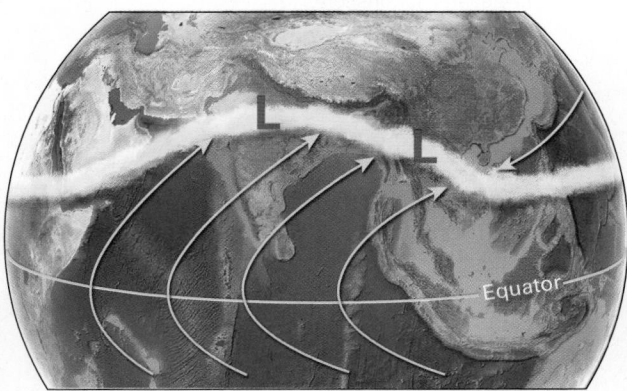

(b) During the summer, the ITCZ lies over India and Southeast Asia. Winds blow from the Indian Ocean onto southern Asia.

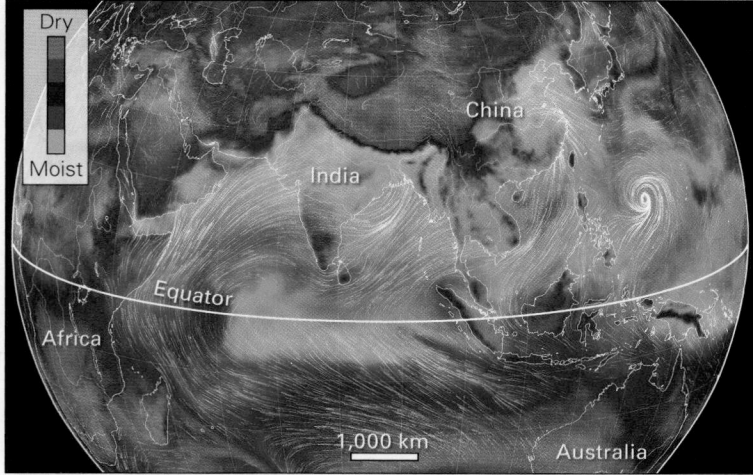

(c) A map of total water in an atmospheric column (called the *precipitable water*) from mid-July, 2023. Note that very moist air within the ITCZ covers much of southern Asia (south of the Himalayas) and Southeast Asia. Monsoonal rains represent the precipitation of this moisture. Also note that the trade winds of the southern hemisphere shift direction when they cross the equator, because the deflection due to the Coriolis force changes at the equator.

FIGURE 17.16 The position of the heaviest rain, which occurs along the ITCZ, migrates over the course of a year. When the ITCZ lies over the land during the summer, moist air blows in from the Indian Ocean and nourishes storms that drop heavy rains. Regions of greatest total rainfall are highlighted.

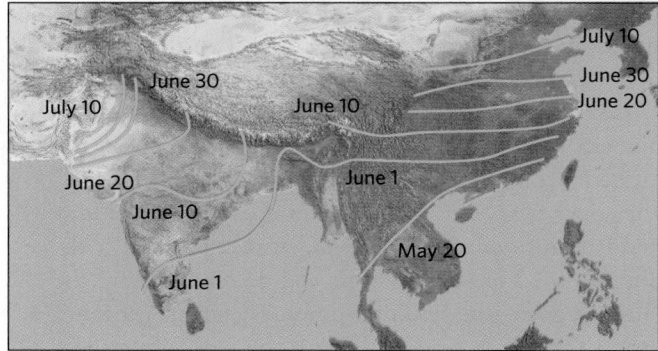

(a) The typical advance of monsoons in the summer season. The yellow lines represent the position of the ITCZ and, therefore, the heaviest rainfall.

(b) Monsoon thunderstorms bring heavy floods.

(c) When the monsoon fails to bring rain, major droughts occur.

Take-home message...

Atmospheric circulations in the tropics are associated with the formation of large convective cells in the atmosphere. Air warmed by strong solar heating near the equator rises to the tropopause along the ITCZ and flows poleward, forming two Hadley cells, one north and one south. The surface flows of these cells are the trade winds. The annual migration of the ITCZ from a latitude south of the equator to a latitude north of the equator produces monsoons. An equator-parallel circulation over the Pacific, the Walker cell, causes the Southern Oscillation between El Niño and La Niña conditions.

Quick Questions

- Is it easier to sail eastward or westward in tropical waters? Why?
- What feature of the Hadley cell is associated with heavy rainfall?
- Which of the following is due to a shift in the latitude of the ITCZ: El Niño or a monsoon?

17.3 Air Circulation at Mid-Latitudes

People living at tropical and subtropical latitudes typically experience weeks or months of the same weather—dry during the dry season and wet during the wet season—and do not endure any days of cold and ice. This is not so in **mid-latitude regions**, the latitudes between 30° N and 60° N and between 30° S and 60° S. Large portions of these regions have *temperate climates* (see Chapter 19), in which weather changes from very cold to very hot as seasons progress, but can change even during the course of a single day. A typical late summer afternoon in the central United States, for example, may feel hot and humid. In a matter of minutes, skies grow overcast and towering thunderstorms roll by. An hour later, the Sun shines again, and the air feels cool and dry. In winter, a warm southerly breeze bringing mild air can quickly change to a northern blast of frigid air, followed by heavy snow.

Why do such radical changes in weather happen in mid-latitude regions? To develop an answer, we need to step back and think again about how the Sun heats the Earth's surface. During the course of a year, much less solar energy reaches the surface in the polar regions than in the tropics because of the tilt of the Earth's axis and Earth's curvature. Because of this imbalance, polar regions remain cool, and in winter, when they experience nearly perpetual night, they become extremely cold. In the tropics, the Sun rises high in the sky all year, so incoming solar energy always exceeds the energy lost as outgoing infrared radiation. If there were no atmosphere or oceans, the temperature differences between the poles and the equator would be much greater than they are. But the atmosphere and oceans reduce these differences by transporting warm air and water from the tropics toward the poles, and cold air and water from the poles toward the tropics. In other words, circulation patterns in atmospheric winds and ocean currents act like giant conveyor belts that redistribute heat between the hot tropics and the cold polar realms in a never-ending attempt to balance the temperature differences across latitudes.

The Earth's mid-latitudes, which lie between the tropics and the poles, serve as the battleground in the war to balance temperature extremes. When cold polar air wins a battle, it pushes back warm tropical air and envelops a region of the mid-latitudes. When tropical air wins, the reverse happens. Interactions between bodies of cold and warm air, called *air masses*, generate the large storms of the mid-latitudes (which we examine in Chapter 18). Here, we explore air masses and their relationship to the wind systems in the complex zone of air circulation that characterizes the mid-latitudes.

What Is an Air Mass?

An **air mass** is a broad body of air within which temperature and humidity are relatively uniform. Major air masses are typically several kilometers thick and a thousand or more kilometers wide, so a single major air mass can cover a quarter to a half of a continent or ocean. Consequently, interactions between air masses can affect weather over a broad area. Numerous distinct air masses exist worldwide at any given time. Meteorologists characterize air masses by their source region (continental or maritime) and their latitude of origin (tropical, polar, or arctic).

Seven major air masses typically lie over and around North America and influence its weather (Fig. 17.17). A *continental polar air mass* develops over southern and central Canada and the northern United States during the

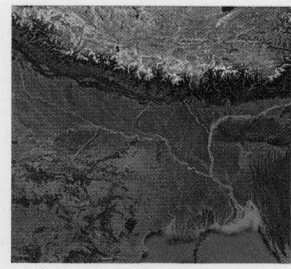

See for yourself

The Southern Asian Monsoon and Flooding in Bangladesh

Latitude: 23°57′37.21″ N
Longitude: 89°03′26.15″ E

Look down from an altitude of 1,609 km (1,000 mi).

You are looking at the Himalayas to the north, the drainage basin of the Ganges River to the west, the Brahmaputra River to the east. These rivers join and flow into the Indian Ocean through the Ganges Delta, which is located at the north end of the Bay of Bengal. During the summer, torrential monsoon rains fall on the plains south of the Himalayas, triggering huge floods along these rivers.

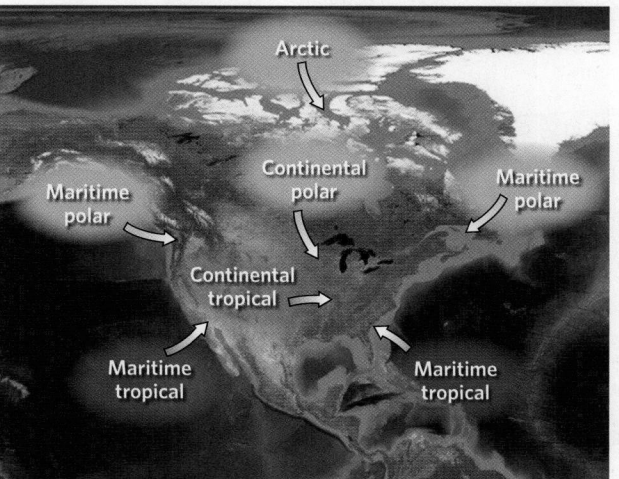

FIGURE 17.17 Major air masses affecting North America. Maritime air masses are moist, while continental air masses are dry. Tropical air masses are warm, while arctic and polar air masses are cool.

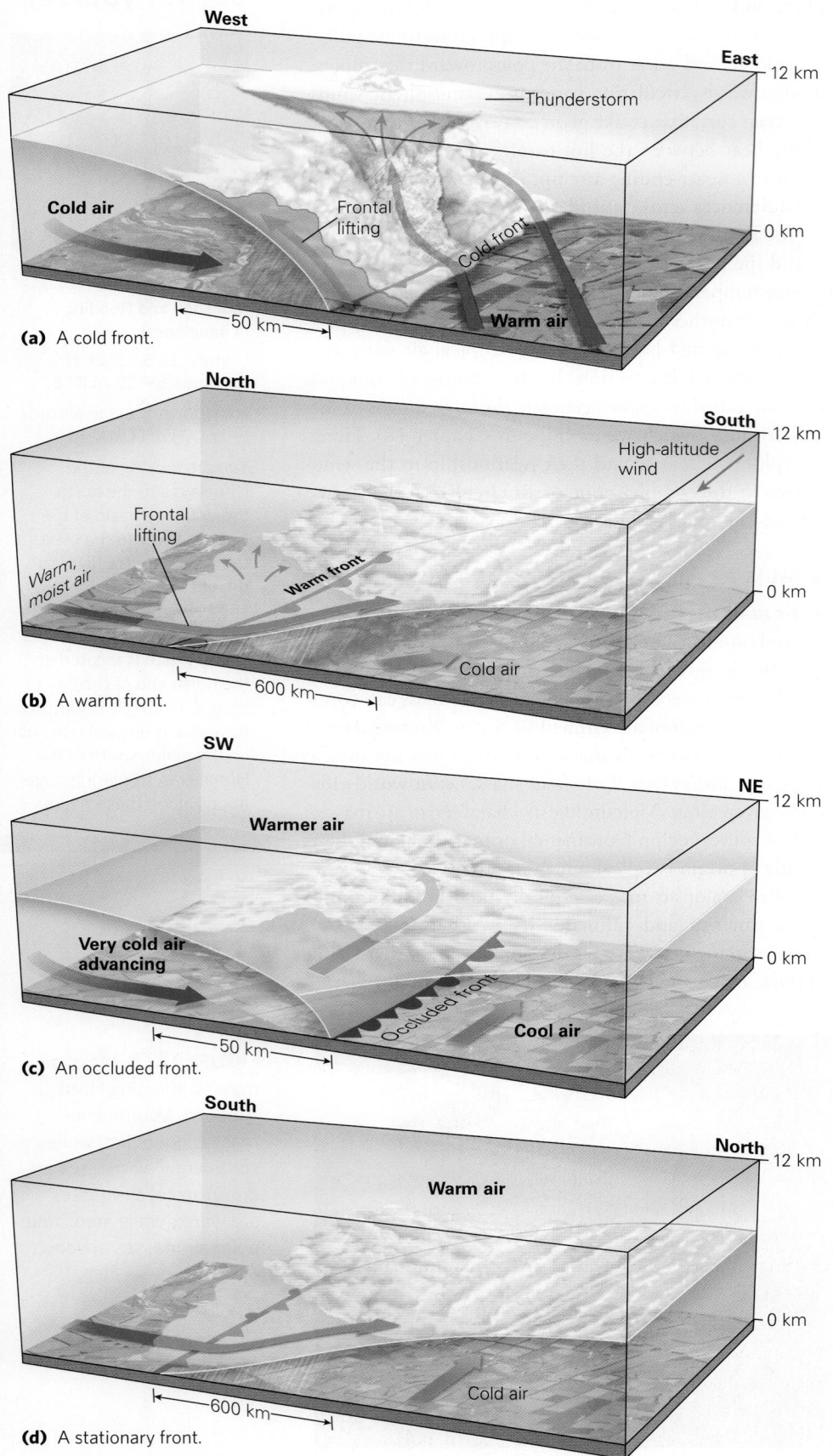

(a) A cold front.

(b) A warm front.

(c) An occluded front.

(d) A stationary front.

winter, while an *arctic air mass* lies over the Arctic Ocean and northern Canada. Cool and humid *maritime polar air masses* develop over the North Atlantic and North Pacific Oceans, while warm and humid *maritime tropical air masses* develop over the tropical North Atlantic, the Gulf of Mexico, and the tropical North Pacific. A *continental tropical air mass* forms in summer over the western United States and Mexico. Because of the global atmospheric circulation, air masses may remain nearly stationary over a period of days, move slowly, or sweep rapidly (at speeds up to 60 km/h, or 40 mph) across a continent or ocean. When an air mass moves out of its source region, a new one soon regenerates there.

Fronts

The boundary between two air masses typically occurs along a narrow zone called a **front**. Because the air masses on either side of a front have different characteristics, fronts delineate an abrupt contrast in temperature and, commonly, relative humidity as well. Most stormy weather in the mid-latitudes, particularly in cooler seasons, develops along fronts, because air on the warm side of a front flows up and over air on the cold side, and this rising air produces clouds and precipitation. Depending on how the air masses in contact along a front are moving, the front may remain stationary, or it may move. When a front moves, meteorologists refer to its direction of movement as the direction in which the colder air mass travels, relative to a specified location on the ground on the warm side of the front. An *advancing* cold air mass moves toward the location, whereas a *retreating* cold air mass moves away from the location. When a front passes over a location, weather conditions can change very rapidly. For example, temperatures can change by over 8°C (15°F) when a front passes.

Meteorologists classify fronts based on the direction in which the cold air mass moves. At a **cold front**, a cold air mass advances under a warm air mass, forcing the warm air to rise (Fig. 17.18a). Typically, the top surface of the cold air mass has a dome-like shape, with the boundary between warm and cold air sloping upward over the cold air. Meteorologists represent the trace of the front where it intersects the ground as a blue line, along which blue triangles, placed on the warm side of the blue line, point in the direction in which the front is advancing. A **warm front** exists where a cold air mass retreats and a warm air mass advances (Fig. 17.18b). In general, the slope of a warm front is gentler than that of a cold front. On a map, meteorologists represent the trace of a warm front on the ground by drawing a red line decorated by red half circles; the half circles lie on the cold side of the front and indicate the direction in which the front moves and, therefore, the cold air mass is retreating. At a warm front, the warm air flows toward and then over the cold air mass.

In some cases, an advancing cold front overtakes a warm front. When this happens, the cold front lifts the cool air behind the warm front, so the warm front no longer intersects the ground surface, and two cold air masses come into contact at the ground surface. The warmer air, therefore, forms a layer over the top of both fronts (Fig. 17.18c). Meteorologists indicate the presence of such an **occluded front** by a purple line, with alternating purple triangles and half circles, pointing away from the coldest air and indicating the direction in which the front moves.

At a **stationary front**, the position of a front does not move, even though air on both sides of the front keeps flowing. At such a front, air on the cold side of the front flows nearly parallel to the front, but air on the warm side rises over the cold air (Fig. 17.18d). Along stationary fronts, the same weather may remain over a location for a long time. Meteorologists represent a stationary front by a line of alternating blue and red segments. Blue triangles on the blue line segments point toward the warm air, and red half circles on the red line segments point toward the cold air.

If we take a global view, we can think of the warm air over the tropics and subtropics and the cold air over the polar regions as two large masses of air in each hemisphere. In each hemisphere, the boundary between these two air masses, which meteorologists call the **polar front**, is often quite distinct and can be traced around the globe in the mid-latitudes (Fig. 17.19). We can think of the polar front as linking the many cold and warm fronts associated

with weather systems that are moving across the mid-latitudes. The air typically circulates in complex paths along the polar front.

The Polar-Front Jet Stream

The contrast in temperature across the polar front exerts a major influence on wind velocities at the top of the troposphere. Why? Warm air occupies more volume than cold air, so the vertical distance between isobaric surfaces is greater on the warm side of the front than on the cold side. Isobaric surfaces, therefore, are lower on the cold side of the front than on the warm side. At the front, isobaric surfaces slope downward from the warm side toward the cold side. Their slopes become progressively steeper at higher altitudes, and they are steepest near the tropopause (Fig. 17.20a). As a result, the horizontal pressure-gradient force near the tropopause over the polar front is very strong, much stronger than it is at lower altitudes or far away from the front. Since pressure gradients, as we have seen, drive the wind, this steep pressure gradient generates very fast winds. The river of fast winds that is present over the polar front is called the **polar-front jet stream** (Fig. 17.20b). Note in Figure 17.20b that the winds of this jet stream generally flow from west to east, so the mid- to upper-tropospheric winds are known as the *mid-latitude westerlies*. The strength of the polar-front jet stream globally at any one time depends on the strength of the *temperature gradient*, the change in temperature with distance, across the polar front.

Note that we have introduced two jet streams in this chapter: the subtropical jet stream and the polar-front jet stream. Both result from strong pressure gradients that develop across a boundary between different air masses. The subtropical jet stream forms at the boundary between the high-latitude end of a Hadley cell and the cooler air of the mid-latitudes. This boundary, called the *subtropical front*, exists only in the upper troposphere. The polar-front jet stream, in contrast, lies over the polar front, an air mass boundary that extends from the Earth's surface upward to the tropopause. Temperature differences across the polar and subtropical fronts increase in winter, so both jet streams strengthen during winter (Fig. 17.20c). Because of the near balance between the pressure-gradient force and the Coriolis force, each jet stream becomes a nearly geostrophic wind that flows more or less parallel to isobars. The polar-front jet stream tends to occur at an altitude of about 10 km (6 mi), lower than the 10–13 km (6–8 mi) altitude of the subtropical jet stream. The difference results from the decrease in the altitude of the tropopause with increasing latitude that exists because air overall gets cooler toward high latitudes.

As the atmosphere attempts to balance the results of differential heating in the tropics and in polar regions,

FIGURE 17.19 The polar front separates colder air over the polar regions from warmer air over the subtropics and tropics. Colors on the map indicate surface temperature.

FIGURE 17.20 The relationship between the polar front and the polar-front jet stream.

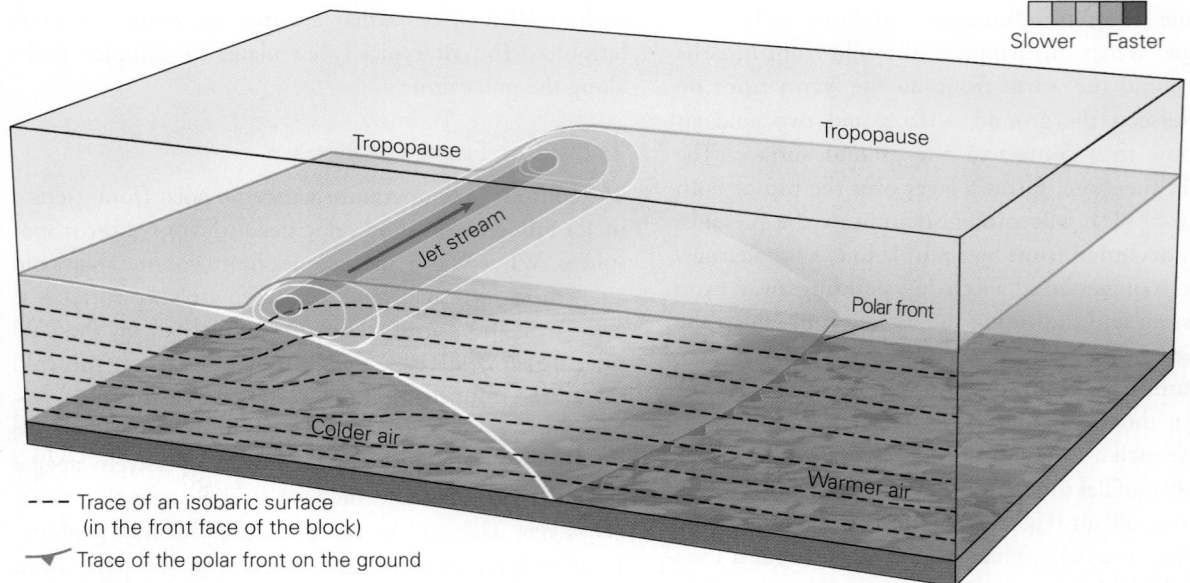

Slower — Faster

--- Trace of an isobaric surface
(in the front face of the block)

∕ Trace of the polar front on the ground

(a) The polar-front jet stream is found near the tropopause over the polar front. The strongest winds are at the center of the jet stream, located at the top of the troposphere where the pressure gradient is steepest.

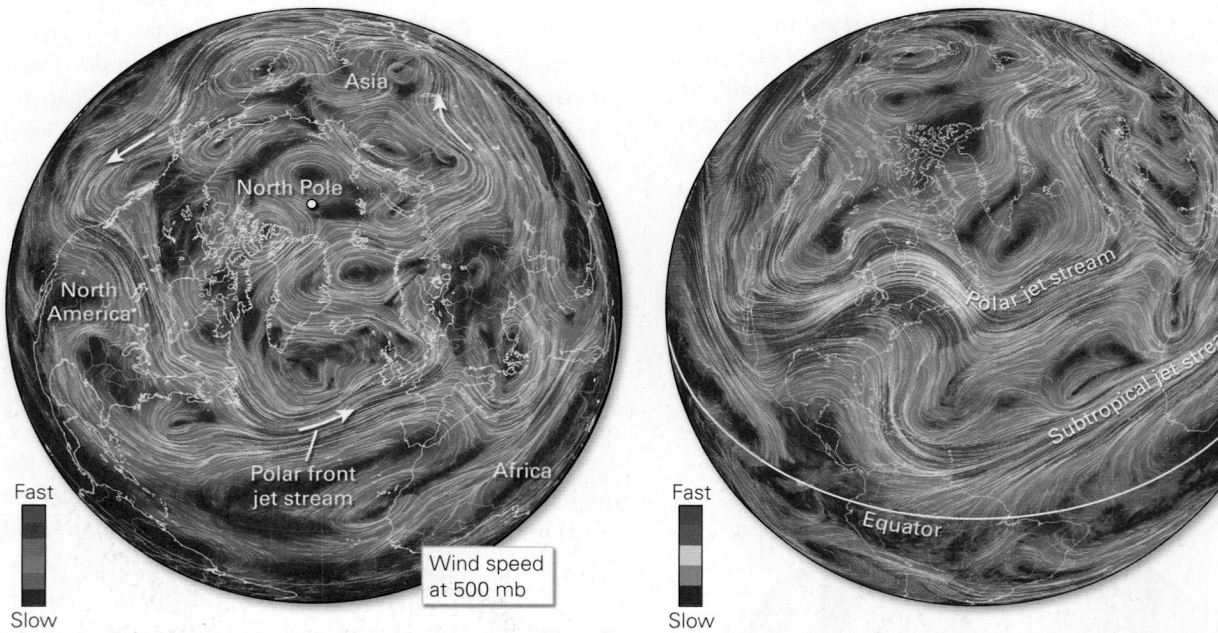

(b) As viewed looking obliquely down toward the North Pole, a wind streamline map shows that the polar-front jet stream flows around the pole in a wave-like pattern, with some regions moving faster and some slower.

(c) This view shows both the polar front and subtropical jet streams. These jet streams sometimes merge as air moves about the globe.

cold air masses north of the polar front frequently advance southward, spread outward, with the air higher in the airmass descending. Warm air masses south of the polar front move northward over the cold air masses and ascend in altitude. Viewed at a global scale, the polar front jet stream typically follows an undulating, wave-like, north-south pattern that marks the warm air–cold air boundary (Fig. 17.21a). For example, over North America, the polar-front jet stream may flow northward over the Pacific Northwest, continue southward along the Rockies into

the southeastern United States, then turn northeastward, and finally, exit the United States flowing northward toward the North Atlantic (Fig. 17.21b). Meteorologists describe a region where the jet stream bows toward the poles as a **ridge**, and one where it bows toward the equator as a **trough**. Warm air lies beneath a ridge, and cold air lies beneath a trough. Why are they called ridges and troughs? Picture an imaginary isobaric surface near the top of the troposphere. Because of variations in air temperature below this isobaric surface, the surface has ups

FIGURE 17.21 Ridges and troughs along the polar-front jet stream.

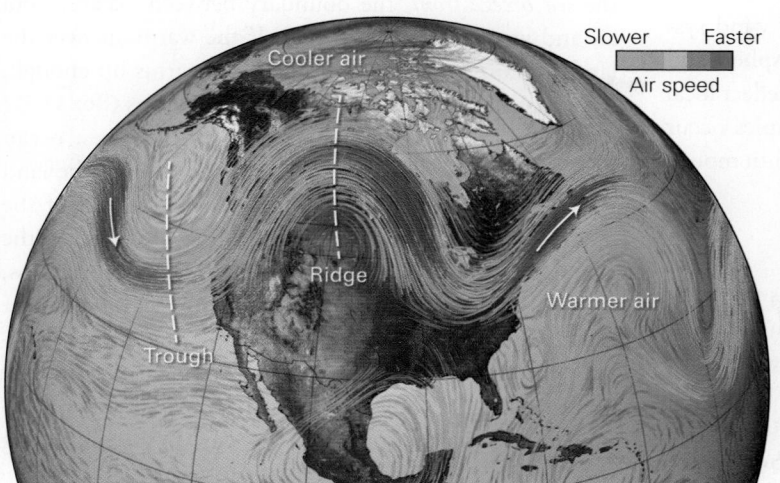

(a) The polar-front jet stream takes on a wavy pattern as it circles the globe. The waves change position over time, as they gradually move eastward and as their amplitude changes.

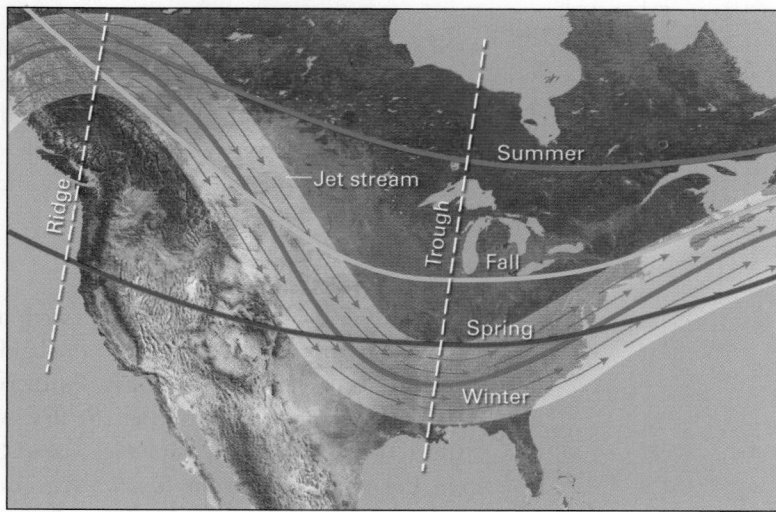

(b) The polar-front jet stream's latitude changes with time. Over North America, it tends to shift north and become weaker in summer; in winter, it shifts south and becomes stronger.

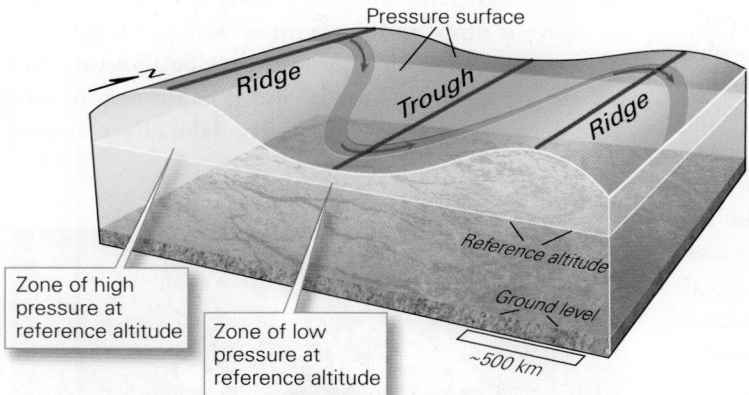

(c) The shape of a surface of constant pressure in the upper troposphere. A ridge is a region where this isobaric surface is at a high altitude, and a trough is a region where it forms a valley.

and downs, and in three dimensions, it has a wave-like shape. A line following this isobaric surface's high points corresponds to a ridge, and a line following the low points corresponds to a trough (Fig. 17.21c).

The location and intensity of the polar-front jet stream change over time in response to the movements of air masses and the positions of fronts. As we've noted, the jet stream's path, in map view, displays wave-like undulations as it curves around ridges and troughs, and these features migrate around the planet. As air masses move, the positions of troughs and ridges, and therefore the positions and shapes of the waves in the jet stream, change. Sometimes the jet stream shows many low-amplitude ridges and troughs, whereas at other times it shows relatively few large-amplitude ridges and troughs. Therefore, the latitude at which the polar-front jet stream flows varies across

the mid-latitudes around the planet, and at any given time it may carry air from polar latitudes into the mid-latitudes, or vice versa.

Air traffic must account for the position and direction of the polar-front jet stream. Specifically, eastbound aircraft flying with the high-altitude westerlies can take advantage of the significant tailwind that the jet stream provides to save on time and fuel costs. Planes flying westward, however, face into these westerlies and experience a strong headwind. As a consequence, travel times for westbound and eastbound flights crossing a continent or ocean on the same route can differ by more than an hour!

Take-home message...

Air masses are broad bodies of air within which temperature and humidity are relatively uniform. The boundaries between air masses, called fronts, are the locations of most stormy weather. This is because at fronts, warm, moist air flows upward over colder air, and this movement can produce clouds and precipitation. The polar-front jet stream flows in a wave-like pattern, generally from west to east, in the upper troposphere.

Quick Questions

- Why does the weather in the mid-latitudes change so frequently?
- What is the difference between a warm front and a cold front?
- Why does the polar-front jet stream form?

17.4 Local Wind Systems

Many regions of the world have unique local wind systems that reflect interactions between the atmosphere and specific features on the Earth's surface, and/or reflect local seasonal conditions. Particularly notable examples occur along land-water boundaries and along and within mountain belts.

Sea Breezes

Oceans have a huge *heat capacity*, which means that they can hold a lot of heat relative to the land and that their temperature changes very slowly. Furthermore, land surfaces warm up rapidly during the day, while the ocean does not—you'll burn your feet walking on the beach in the tropics, but your feet will be quite comfortable immersed in the adjacent sea. Why? When solar radiation strikes the opaque surface of immobile rock and soil, only an extremely thin layer absorbs the energy. In contrast, when solar radiation strikes the transparent sea surface, it can penetrate into deeper layers, and because water circulates, the energy can be transported vertically or horizontally. The contrast between warmer land surfaces and cooler water surface can drive local air circulation along coastlines, and can generate local wind systems. Specifically, during the afternoon, air rises over the hot land, flows seaward aloft, descends over the cooler ocean, and flows landward just above the ocean surface to form a **sea breeze** (Fig. 17.22a, b). At night, some coastlines experience a *land breeze*, a mirror image of a sea breeze, because the land cools off faster than the sea does, so the direction of the breeze reverses.

Weather associated with a sea breeze depends on the characteristics of the warm and cool air on either side of the *sea breeze front*, the boundary between oceanic cool air and warm air over the land. If the warm air over the land contains sufficient moisture and warms up enough, it rises and triggers clouds and thunderstorms (Box 17.2).

Along the Pacific coast of the United States, ocean water remains cold in summer, and the air over the land is dry. When the land heats up in the early part of the day, the sea breeze causes cool air to blow onshore. In the Los Angeles basin, which is surrounded by mountains, the cool, dense marine air collects at low elevations over the city and remains there for most of the day, particularly during peak traffic hours. This cool marine air is not buoyant and does not rise, so urban air pollutants accumulate within it throughout the day, producing smog (Fig. 17.23).

Winter Winds over the Great Lakes

Because water has a high heat capacity, the water of a large lake may remain liquid in early winter even if the air above it drops to temperatures well below freezing. As a consequence, the lake can provide moisture and heat to air flowing across it, and it can therefore produce local weather systems called **lake-effect storms**

FIGURE 17.22 The sea breeze.

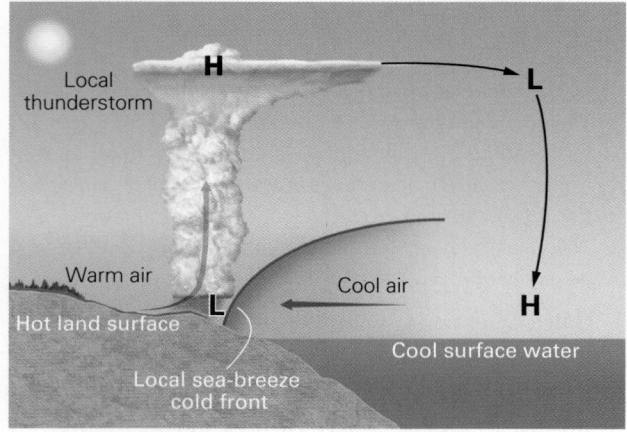

(a) Sea breeze circulation during the daytime along a coastline can trigger thunderstorms.

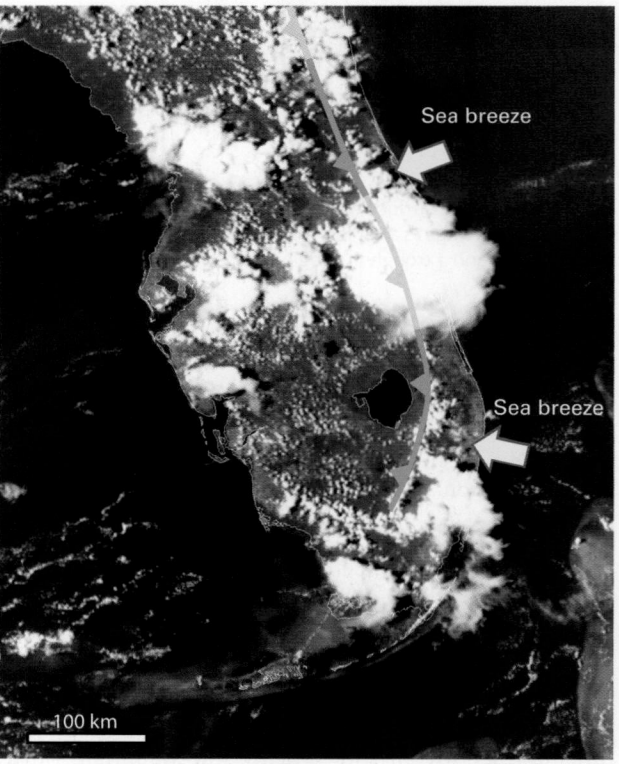

(b) Storms form along the sea breeze fronts on both Florida coasts.

BOX 17.2 ▶ Putting Earth Science to Use

Beach vacations and the sea breeze

Along the east and Gulf coasts of the United States, vacationers flock to the shore to enjoy summer sunshine and warm ocean water. Some choose to reside along the coast during their vacations, while others pitch tents or pull up trailers at campgrounds farther inland to have outdoor fun and reduce expenses. These coastlines host a strong sea breeze on many summer days, particularly on clear mornings when the Sun heats the land quickly, while ocean water temperatures barely change. Clouds and thunderstorms can develop along the sea breeze front as cool oceanic air flows onshore and the morning sea breeze intensifies. The storms typically develop several miles inland, and they often drift farther inland as the sea breeze strengthens during the afternoon (Fig. Bx17.2). From a vantage point on the coast, vacationers can gaze inland and watch these storms clatter and bang while enjoying an otherwise sunny day at the beach. Unfortunately, for folks inland, where storms develop overhead, the day appears washed out by the rain, thunder, and lightning. Discouraged, many never head to the beach. Those few vacationers who understand the sea

FIGURE Bx17.2 Sea breeze clouds forming over central Florida, as seen looking inland from the shore.

breeze know that more pleasant weather is only a short distance away along the shore.

(Fig. 17.24a). Lake-effect storms commonly occur over and adjacent to the Great Lakes of North America when cold air from northern Canada flows southward over the lakes. As this air passes, heat and moisture move from the warm lake surface into the boundary layer, while the air above remains cold. The warmed air starts to rise to form cumulus clouds. These clouds become taller as the layer of warm air reaches higher elevations as it moves across the lake. Eventually, the clouds start to precipitate snow.

When the relatively warm, moist air arrives at the downwind shore, friction with trees and buildings reduces the wind speed near the ground. Effectively, the near-surface wind piles up as it reaches the shore and is forced to rise. This additional lifting triggers production of even more clouds and yields heavy snowfalls downwind for several kilometers inland from the shore. As a consequence, regions downwind of the Great Lakes sometimes receive very heavy snow—up to 150 cm (60 in) may fall during a single lake-effect storm (Fig. 17.24b). Such lake-effect snowfall can continue for days, sometimes at a rate of more than 2.5 cm/h (1 in/h), resulting in enormous accumulations in areas within a few tens of kilometers of

FIGURE 17.23 Smog trapped in the marine air over the Los Angeles basin.

FIGURE 17.24 Lake-effect storms.

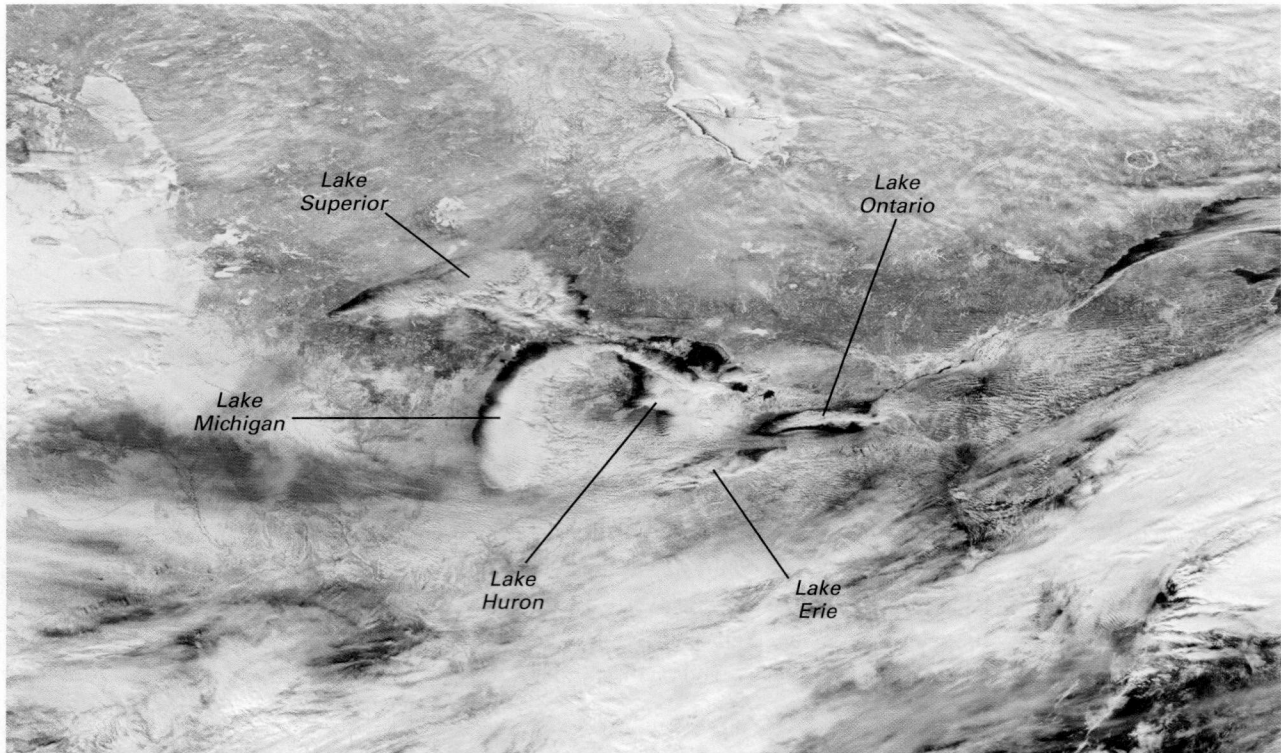

(a) Lake-effect clouds develop when cold air flows over the Great Lakes.

See for yourself

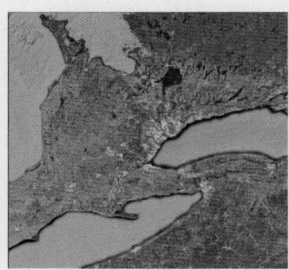

Lake-Effect Regions

Latitude: 43°05′27.42″ N
Longitude: 79°05′02.27″ W

Look down on this scene from an altitude of 650 km (400 mi).

Below you are Lakes Erie and Ontario. During cold air outbreaks in winter, frigid air travels from west to east along the lakes. On its way, the air picks up moisture and heat from the lakes' surfaces, so the air rises and clouds form. As these clouds reach the downwind shores around Buffalo, New York, and the Tug Hill Plateau east of Lake Ontario, they dump their loads of snow. Like a conveyor belt, cold air keeps coming, and snow keeps falling. These areas receive enormous amounts of snow during winter. The average snowfall for the Tug Hill region is about 5 m (200 in) each year, enough to bury a small house!

(b) Lake-effect storms produce heavy snowfall.

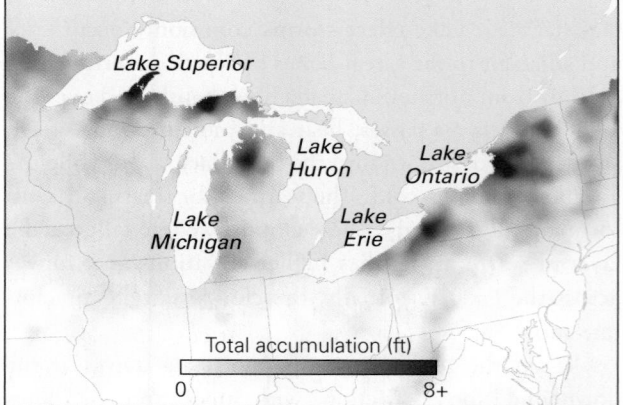

(c) Snow accumulation around the Great Lakes over a winter season.

the shore **(Fig. 17.24c)**. During the early winter of 2022, Buffalo, New York, was buried by as much as 2 m (6 ft) of snow, while surrounding regions had hardly any.

Mountain-Valley Winds

Winds in mountain valleys during cloud-free days and nights flow in response to the heating of the valley and slopes in the daytime and the cooling that occurs at

night **(Fig. 17.25)**. Specifically, when the Sun heats the slopes just after sunrise, air above the slopes warms, becoming warmer than air at the same elevation over the central part of the valley. The warm, buoyant air starts to rise along the surface of the slopes, producing low pressure at the Earth's surface near the top of the valley, relative to the downstream end of the valley. The resulting pressure gradient causes wind to blow up the valley.

FIGURE 17.25 Mountain-valley wind systems.

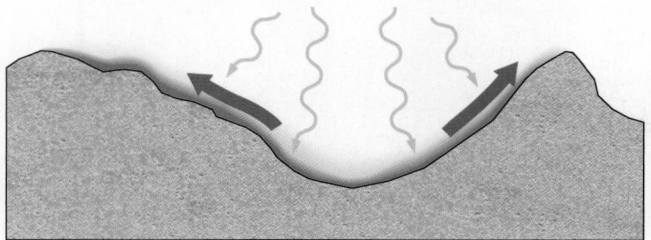

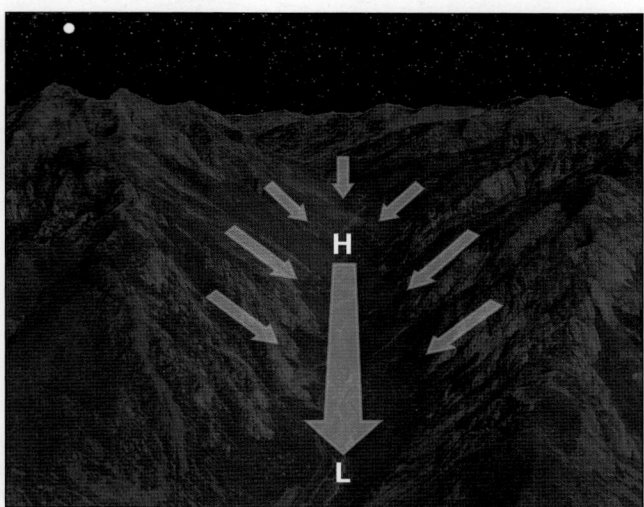

(a) Solar heating of the ground causes the air just above the ground to warm and flow upslope. As the photo shows, the upslope movement also causes wind to blow up-valley.

(b) At night, the slopes cool, and cooler, denser air flows downslope. This, in turn, causes wind to blow down the valley axis.

As the day continues to warm, wind continues to flow up the valley and becomes stronger. However, when the Sun starts to set, heating stops, and the slopes begin to radiate heat back to space. Air near the slopes now cools. The resulting cool, dense air along the slopes descends into the valley, producing high pressure at the top of the valley. The resulting pressure gradient causes wind to blow down the valley, a flow that continues through the night. The up- and down-valley seesaw of the winds is a common feature of mountain valleys that experience clear skies.

Downslope Windstorms in Mountains

Because air moving over a mountain range loses most of its moisture as rain or snow on the upwind, or *windward*, side of the range, it is relatively dry as it descends on the downwind, or *leeward*, side of the range. In addition, as it descends, the air compresses, becomes denser, and warms. Consequently, the leeward side of the range may be a rain-shadow desert, even where the windward side is wet and vegetated (see Chapter 13). In addition

to dry weather, the leeward side of a mountain range occasionally experiences a **downslope windstorm**. This type of windstorm has different names in different locations: along the Rocky Mountains, it's a *chinook wind*; in the Alps, it's a *foehn wind*; and in California, it's a *Santa Ana wind*.

Why do downslope windstorms develop? Wind will blow across a mountain range if higher pressure is present on the windward side than on the leeward side. This situation might occur along the Rockies, for example, if a low-pressure system lies east of the Rockies at the same time a high-pressure system develops west of the Rockies. As we saw earlier, stronger horizontal pressure gradients cause faster winds. When a particularly strong pressure gradient exists across a mountain range, the stage is set for a downslope windstorm.

In the case of chinook and foehn winds, air flow becomes more rapid as air rises over the crest of the mountains in the shape of a wave, much like the wave that forms when a stream of water flows up and over a large boulder. Clouds form on the upwind side of

FIGURE 17.26
Development of a
downslope windstorm.

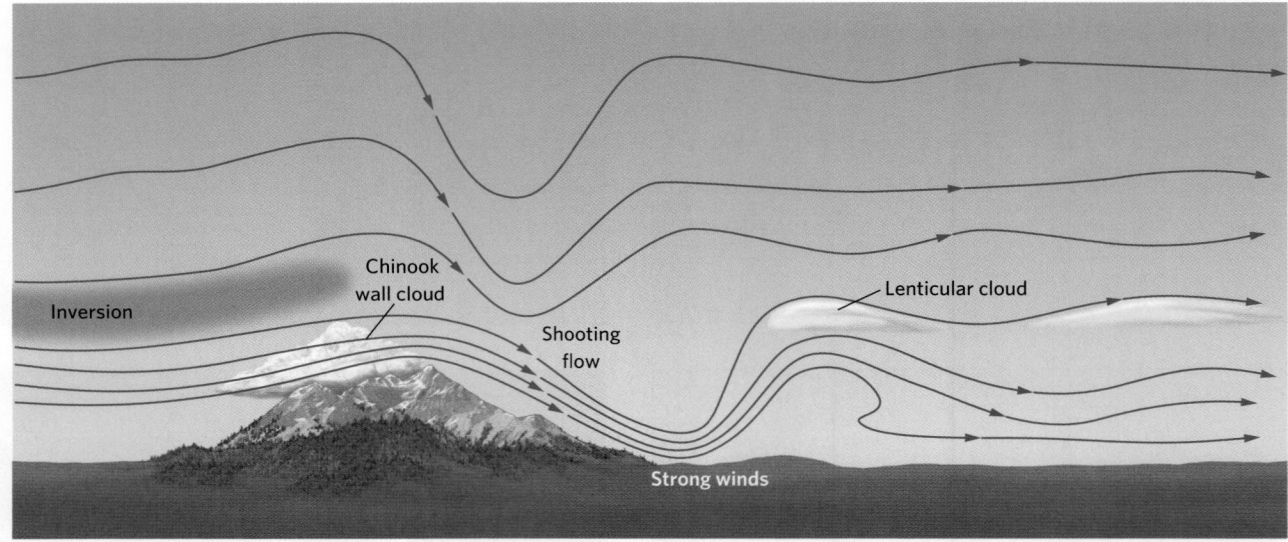

(a) Winds blow faster where the streamlines (blue lines) are closer together. Interaction of the flowing air with the mountains generates a wave in the flow and produces locally faster winds.

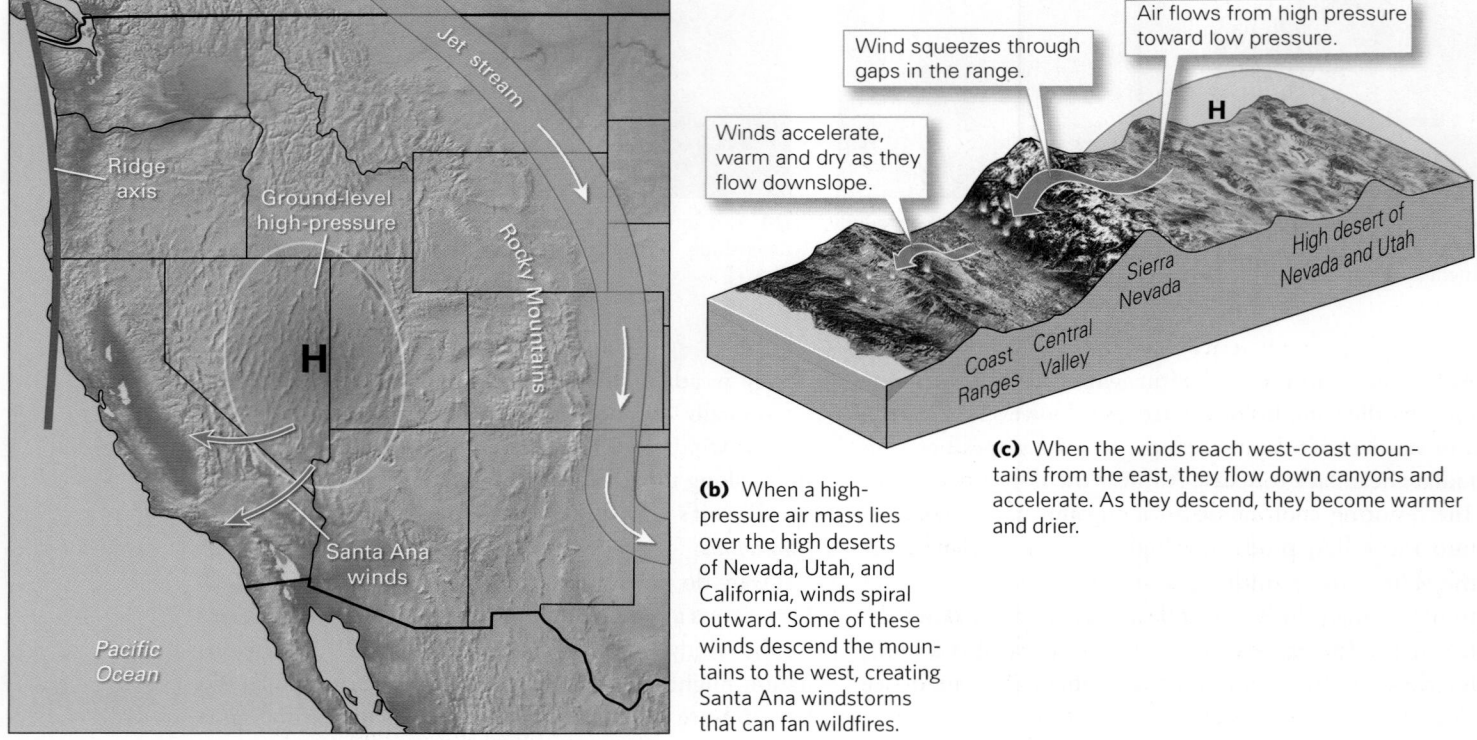

(b) When a high-pressure air mass lies over the high deserts of Nevada, Utah, and California, winds spiral outward. Some of these winds descend the mountains to the west, creating Santa Ana windstorms that can fan wildfires.

(c) When the winds reach west-coast mountains from the east, they flow down canyons and accelerate. As they descend, they become warmer and drier.

the mountains as the air rises. Viewed from the downwind side, the clouds appear like a wall, called a *chinook wall cloud*. The background pressure gradient above the mountains accelerates the air passing across the range. This wave of air takes on a special shape when an **inversion**—a layer in the atmosphere in which the temperature increases with altitude—forms upwind of the mountain crest at an elevation just above it (Fig. 17.26a). The inversion acts like a lid, confining the wind to a narrow layer above the mountaintop. As a consequence, the wind accelerates even more. You may have seen a similar phenomenon if you've blocked the end of a water hose with your thumb so that water spurts out of the hose more quickly. This strong wind comes down the leeward side

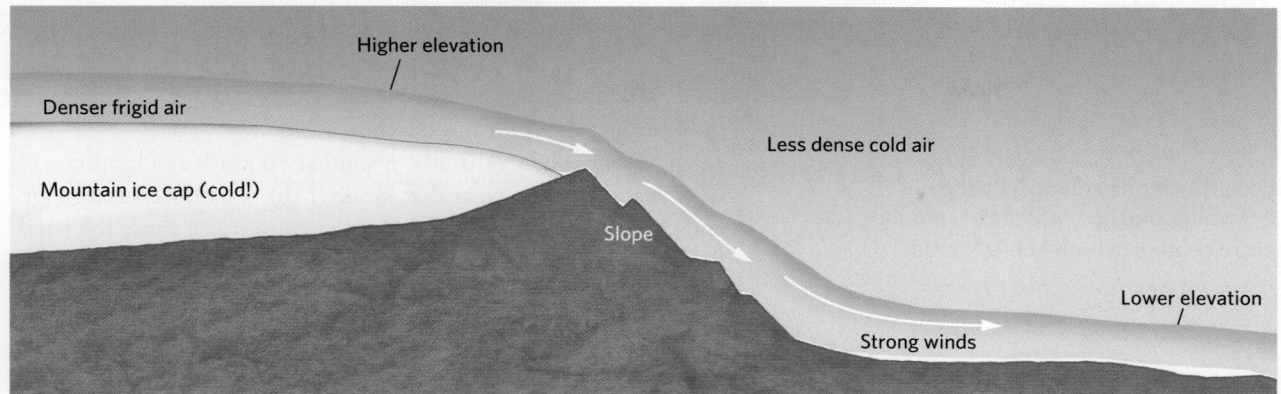

Katabatic winds in Antarctica

katabatic winds develop because air above the mountain range or ice sheet becomes very cold and therefore relatively dense. As a result, it flows downslope, sometimes reaching speeds up to 160 km/h (100 mph). Katabatic winds can begin abruptly when a dome of extremely cold, dense air has built up over an ice sheet. The winds begin when gravity finally causes the cold air to spill off the ice sheet and rush outward onto the land or sea below (Fig. 17.27). Katabatic winds are common in Antarctica, along the coast of Greenland, and in mountainous regions of Canada and Alaska that have permanent glaciers.

of the mountain range in a *shooting flow* and descends to lower elevations, where it accelerates further—the strongest downslope winds can reach 160 km/h (100 mph). Because the air warms as it descends, it can melt snow rapidly. It can also fan wildfires once a fire starts, such as the Marshall fire in Superior, Colorado, on December 30, 2021, that destroyed 1,084 homes and damaged another 149.

Santa Ana winds blow when high pressure develops over the desert regions to the east of California's mountain ranges (Fig. 17.26b, c). If a low-pressure zone exists to the west, on the coastal side of California, the pressure gradient causes winds to blow from the desert westward over the mountains. These winds rush down the valleys of southern California, and as with chinook winds, the air warms as it descends. During a dry season, these winds often fan disastrous wildfires.

The downslope windstorms we have just described are driven by strong pressure gradients. A second type of strong windstorm occurs in cold mountainous regions and polar regions where the top surface of a large ice sheet has a higher elevation than its surroundings. These

Take-home message...

The contrast in heat capacity between land and water can drive sea breezes. Lake-effect snowstorms form in winter when large lakes are warm and the air crossing them is cold. Mountain topography can lead to valley flows, as well as to the development of strong downslope winds, such as chinook and Santa Ana winds. Cold air pools over ice sheets and flows to lower elevations to produce katabatic winds.

Quick Questions ————————————

- Why does a sea breeze in Florida flow from the ocean toward land?

- Why do sea breezes contribute to causing air pollution in the Los Angeles basin?

- If you planned to tour a mountain valley on a bicycle, would the wind make it easier to ride up or down the valley in the morning?

Objective 17.1

Understand the forces that control air movement, how they combine to create the world's winds, and how winds flow around high-pressure and low-pressure centers in both hemispheres.

KEY CONCEPTS

- Three forces—the pressure-gradient force, the Coriolis force, and the frictional force—together generate and control winds in the atmosphere.

- The pressure-gradient force acts from high pressure toward low pressure, its strength proportional to the magnitude of the pressure gradient.

- The Coriolis force causes air to veer to the right of its direction of motion in the northern hemisphere and to the left in the southern hemisphere; but it has no effect on the air's speed. The force is zero at the equator, increases with latitude, and increases with wind speed at a given latitude.

- The frictional force always acts to slow the winds, and it is important in the lower atmosphere, in a region called the boundary layer.

- Above the boundary layer, wind speed and direction are largely controlled by the balance between the pressure-gradient and Coriolis forces. Where this balance has been attained, winds are geostrophic and blow parallel to isobars.

- In the northern hemisphere, air flows counterclockwise around low-pressure centers, while spiraling inward. Air flows clockwise around a high-pressure center, while spiraling outward. The opposite occurs in the southern hemisphere.

- Semipermanent high-pressure systems develop over large regions where air is cooled, such as Canada in winter. Semipermanent low-pressure systems develop over regions that are warmed, such as the oceans south of Iceland and the Aleutian Islands in winter.

EARTH-SCIENCE VOCABULARY

boundary layer (p. 651)
convergence (p. 653)
divergence (p. 654)
geostrophic balance (p. 652)
geostrophic wind (p. 653)
high-pressure system (p. 654)
isobar (p. 648)

isobaric surface (p. 649)
low-pressure system (p. 653)
pressure gradient (p. 648)
semipermanent high (p. 654)
semipermanent low (p. 655)
wind (p. 648)

REVIEW QUESTIONS

1. **(a)** If air only flows vertically upward at 10 km/h at a locality, what is the wind speed at that locality? **(b)** What forces act on air and cause it to flow? **(c)** Which of these forces can cause the wind to increase in speed? **(d)** Which forces control wind speed in the boundary layer along the equator?

2. **(a)** What is the relationship between wind direction and isobars when geostrophic balance exists? **(b)** Can air be in geostrophic balance in the boundary layer, if there is friction between the ground and the air? **(c)** In **Figure A**, which diagram most closely shows the force balance above the boundary layer?

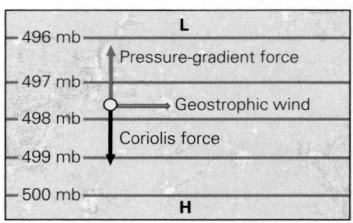 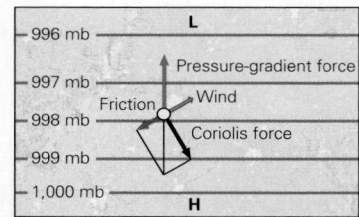

A

3. **(a)** What is the general direction of the pressure-gradient force, toward the north or south, in the upper troposphere of the northern hemisphere? **(b)** If you are standing with the wind at your back in the northern hemisphere and hold out your arms, does your left hand point toward low or high pressure?

4. **(a)** In the southern hemisphere winter, would you expect a semipermanent high-pressure system or low-pressure system to develop over central Antarctica? **(b)** In summer, the east coast of the United States is often warm, while the west coast is cool. How might this difference be related to air flow around the Pacific High and the Bermuda High shown in **Figure B**?

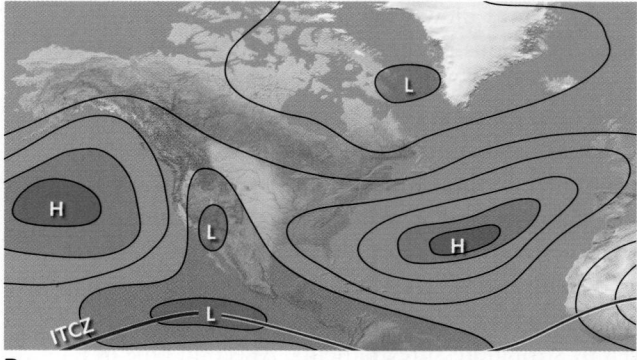

B

Objective 17.2

Describe the major circulations of the tropical atmosphere, including the Hadley cells, El Niño, the Southern Oscillation, and monsoons; explain why they exist, and why they are important to society.

KEY CONCEPTS

- Hadley cells consist of air that rises at the ITCZ near the equator, then flows poleward while descending from the upper troposphere of each hemisphere, and finally returns toward the equator in winds near the Earth's surface. The Hadley cells strongly influence the distribution of tropical rainforests, steppes, and deserts, and their air flow near the surface causes the trade winds.

- The Walker circulation is an east-west-flowing convective cell covering the equatorial Pacific. An El Niño event occurs when the Walker circulation weakens, and a La Niña event occurs when it strengthens. The resulting pattern of changes in surface air pressure, called the Southern Oscillation, can affect weather around the world.

- A monsoon is a seasonal change in wind direction. The largest monsoon is located over southern Asia, where it brings heavy rains in summer and dry weather in winter. Monsoons also occur in other regions of the world, such as the southwest United States and northwest Mexico.

EARTH-SCIENCE VOCABULARY

convective cell (p. 657)	**Southern Oscillation** (p. 660)
doldrums (p. 658)	**subtropical jet stream**
El Niño (p. 660)	(p. 659)
ENSO (p. 660)	**subtropics** (p. 656)
Hadley cell (p. 657)	**thunderstorm** (p. 658)
intertropical convergence zone	**trade wind** (p. 657)
(**ITCZ**) (p. 657)	**tropics** (p. 656)
La Niña (p. 660)	**Walker circulation** (p. 659)
monsoon (p. 660)	

REVIEW QUESTIONS

5. **(a)** Identify the Hadley cells in **Figure C**. What determines their northern and southern limits? **(b)** Explain the relationship between the locations of tropical rainforests and subtropical deserts. **(c)** Where are you more likely to need an umbrella?

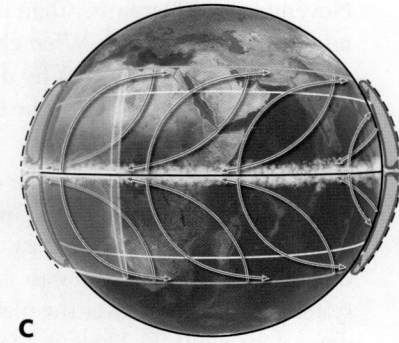

C

Under the south end or the north end of the Hadley cell in the northern hemisphere? Explain. **(d)** What is the relationship between the ITCZ and the Earth's Hadley cells?

6. **(a)** What is the relationship between the trade winds and Hadley cells? **(b)** Do winds at high elevation above the trade winds blow in the same direction as the trade winds? **(c)** When the ITCZ lies at the equator, why do trade winds curve and become parallel to the equator? **(d)** What are the doldrums, and why do they occur?

7. **(a)** Is the sea surface off the shores of Peru warmer or colder than normal during an El Niño event? Why? **(b)** How do sea-surface temperatures change during the Southern Oscillation, and why?

8. **(a)** Do you always get wet during a monsoon? **(b)** How is the Southeast Asian monsoon related to the ITCZ?

9. **(a)** The summer-winter transition in weather in the southwestern United States and northwestern Mexico has been described as the "North American monsoon." Based on your understanding of monsoons, why might meteorologists use this terminology? **(b)** Do monsoons occur in the mid-latitudes?

Objective 17.3

Explain why the polar front exists in the lower troposphere, how it is related to the polar-front jet stream in the upper troposphere, and how the jet stream flows around the globe in a wave-like pattern.

KEY CONCEPTS

- The Earth's middle latitudes, between latitudes of 30° and 60° in both hemispheres, are affected by air from both the subtropics and polar regions, and are zones of rapidly changing weather, with hot summers and cold winters.

- The boundary between air from polar regions and air from the subtropics is called the polar front. The polar front can be traced around the globe in the mid-latitudes.

- The polar-front jet stream is found in the upper troposphere over the cold air and along the polar front. This jet stream exists as a result of the strong pressure gradient generated in the upper troposphere by the strong temperature contrast across the polar front.

- The polar-front jet stream flows around the world in a wave-like pattern. A region where the jet stream bows toward the poles is called a ridge, and where it bows toward the equator is called a trough. Warm air lies beneath a ridge, and cold air lies beneath a trough.

air mass (p. 663)
cold front (p. 664)
front (p. 664)
mid-latitude region (p. 663)
occluded front (p. 665)
polar front (p. 665)

**polar-front jet
  stream** (p. 665)
ridge (p. 666)
stationary front (p. 665)
trough (p. 666)
warm front (p. 664)

REVIEW QUESTIONS

10. **(a)** Why are the mid-latitudes sometimes described as the location of a battle between air masses? **(b)** The polar front is the global boundary between which air masses?

11. **(a)** What causes a jet stream to form? **(b)** Which jet stream occurs at a higher altitude, the polar-front jet stream or the subtropical jet stream? **(c)** Airplanes flying from Los Angeles, on the west coast of the United States, to Boston, on the east coast, take about five hours to make the trip in winter. When they return to Los Angeles, they take about seven hours. Explain why these times are so different. Does the difference change with the season?

12. **(a)** Label the ridges and troughs in the jet stream in **Figure D**. **(b)** In the northern hemisphere, will air temperature at the ground likely be warmer or colder beneath a ridge or beneath a trough, at the same latitude? **(c)** Label the legend on **Figure D**, and indicate the flow directions in the jet stream by adding arrows.

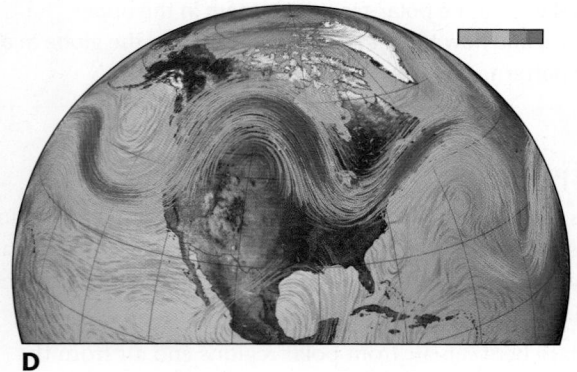

D

Objective 17.4

Discuss how local weather systems develop near the boundaries between land and water, and in the vicinity of mountain ranges.

KEY CONCEPTS

- A sea breeze is a circulation driven by the difference between day-time heating of the shore and of the ocean. It consists of air flowing onshore at low altitudes, rising over land, flowing oceanward aloft, and sinking over the ocean. It may trigger thunderstorms in some environments, and trap pollution in others.

- Lake-effect storms develop over large lakes during winter as moisture and heat rise from the lake and moisture condenses to form clouds above the lake and its downwind shores.

- Winds in mountain valleys on clear-sky days undergo a distinct daily cycle, with upslope and up-valley winds during the day, and downslope and down-valley winds at night.

- Chinook, foehn, and Santa Ana winds are strong downslope winds that develop occasionally on the leeward side of a mountain range when a strong pressure gradient exists across the range.

- Katabatic winds develop when exceptionally cold air cascades off a high-elevation ice sheet.

EARTH-SCIENCE VOCABULARY

downslope windstorm
  (p. 671)
inversion (p. 672)

katabatic wind (p. 673)
lake-effect storm (p. 668)
sea breeze (p. 668)

REVIEW QUESTIONS

13. **(a)** Identify the flow of warm and cold air in **Figure E**, a diagram of a sea breeze. **(b)** Are sea breezes more common in low latitudes or high latitudes? Explain your reasoning.

E

14. **(a)** Lake-effect storms typically produce greater snowfall in November and December than in January and February. Why might this be so? (Hint: What changes are likely to occur in a lake as winter progresses?) **(b)** Why do areas most affected by lake-effect snow on the shores of the Great Lakes generally lie on the east side of the lakes?

15. **(a)** What conditions lead to the development of strong downslope winds on the leeward side of a mountain range? **(b)** Denver is located on the plains just east of the Rocky Mountains in Colorado. Tomorrow's forecast for Denver calls for a high-pressure system to be located over the plains just east of Denver. Will downslope winds be likely in Denver while these conditions exist? **(c)** Are katabatic winds more common at the top or the base of a mountain located in polar regions?

Mt. Everest in Nepal rises to 8,850 m (29,032 ft) above sea level. At that altitude, the top of the world's tallest mountain is located within the jetstream, appearing visually as a cloud streaming over the mountain peak.

18 DANGER IN THE AIR
Stormy Weather

After studying this chapter, you should be able to...

1. understand the basic conditions that lead to the formation of thunderstorms, and describe the difference between ordinary, squall-line, and supercell thunderstorms.

2. explain how thunderstorm hazards develop, and how to protect yourself against them.

3. discuss how tornadoes form, how they are classified, and how they cause destruction.

4. explain how mid-latitude cyclones form and evolve, and the relationships among weather, fronts, and jet streams within these cyclones.

5. illustrate how hurricanes form in the tropics, move across the oceans, and produce massive destruction as they make landfall.

Many develop over the tropics, where they provide the downpours that sustain rainforests. During the warm season, they occur fairly frequently in mid-latitudes, but they rarely take place in high latitudes, in subtropical deserts, or over large parts of the world's oceans. In the United States, thunderstorms primarily affect regions east of the Rocky Mountains, and they are particularly frequent over the southeastern states. In fact, in the eastern two-thirds of the United States, a given location will experience an average of 30 to 50 *thunderstorm days* (days on which at least one thunderstorm occurs) per year (Fig. 18.4). Note that during a thunderstorm day, a particular location may underlie a thunderstorm for as little as a half hour. Florida holds the record for thunderstorm days in the United States, because some locations in Florida endure 65 to 80 thunderstorm days per year—or more!

Why don't thunderstorms occur everywhere all the time? They form only when three special conditions exist: (1) the lower atmosphere must hold *moist air*; (2) a *lifting mechanism* must be present to initiate an **updraft**, an upward-moving air flow that can transport the moist air to a higher elevation; and (3) *atmospheric instability* must exist so that the air keeps rising buoyantly to the top of the troposphere once it has started to rise. Let's consider each of these conditions in turn.

The Source of Moist Air

Moisture fuels a thunderstorm, and without it, a storm can't grow. Air in the lower atmosphere must have a high relative humidity in order for a thunderstorm to develop. Once formed, a typical thunderstorm contains over 3.7 million tons of water—a weight equal to the weight of oil carried by 10 supertankers—at any given time. Over its life, many times that amount pass through the storm.

Where does all the moisture in a thunderstorm come from? Most evaporates from the world's oceans, the largest reservoir of water in the Earth System. For example, moisture fueling thunderstorms in the central and eastern United States typically comes from water that evaporated from the Gulf of Mexico and the Atlantic Ocean and was then blown over the land by wind. Water that enters thunderstorms can also come from lakes and wetlands, from soils that have absorbed water from earlier rainfalls, and from transpiration by plants.

Lifting Mechanisms

Development of a thunderstorm begins when moist air starts to rise as an updraft from the lower troposphere. Any process that initiates this upward motion can be called a **lifting mechanism**. Atmospheric scientists distinguish

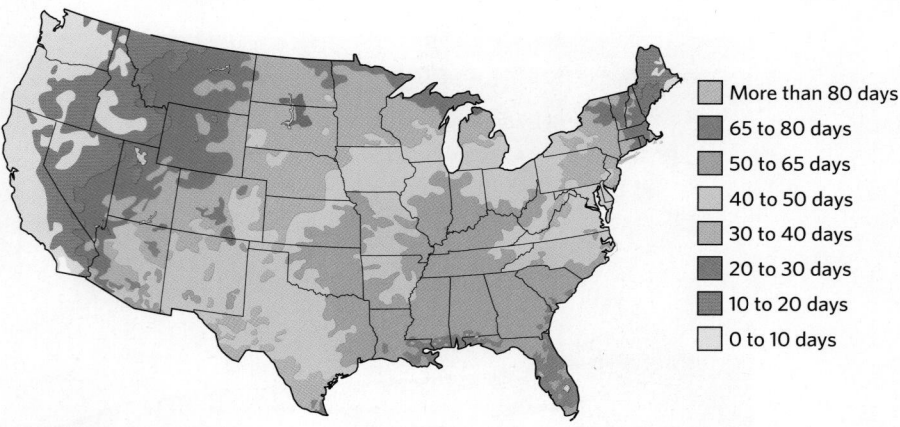

FIGURE 18.4 Numbers of thunderstorm days per year in the United States.

	More than 80 days
	65 to 80 days
	50 to 65 days
	40 to 50 days
	30 to 40 days
	20 to 30 days
	10 to 20 days
	0 to 10 days

among several types of lifting mechanisms, including the following:

1. *Lifting due to movement of a front:* Many thunderstorms develop when an advancing cool air mass moves toward a warmer air mass, so that warm, moist air flows up the sloping face of the dome-like cold air mass. As we discussed in Chapter 17, the boundary between two air masses where cold air is advancing toward warm air is a *cold front*. The process by which warm, moist air physically moves up, due to the advancing front, is called **frontal lifting** (Fig. 18.5). Such upward movement of air represents an updraft. Lifting by an advancing cold front often produces a line of several thunderstorms along the front. Although less common, the lifting of moist air that triggers thunderstorms can also take place along a warm front.

2. *Lifting due to gust fronts:* As we'll see later in this chapter, when rain starts to fall in a thunderstorm, it can produce an intense **downdraft**, a downward flow of air, that rushes from the storm to the ground below. The downdraft in a thunderstorm contains relatively cool air, so when it reaches the ground and spreads outward to the sides of the storm, it produces a small dome of cold dense air called a *cold pool*. As the cold

Animation
Fronts: Frontal Lifting

FIGURE 18.5 A thunderstorm forming due to frontal lifting.

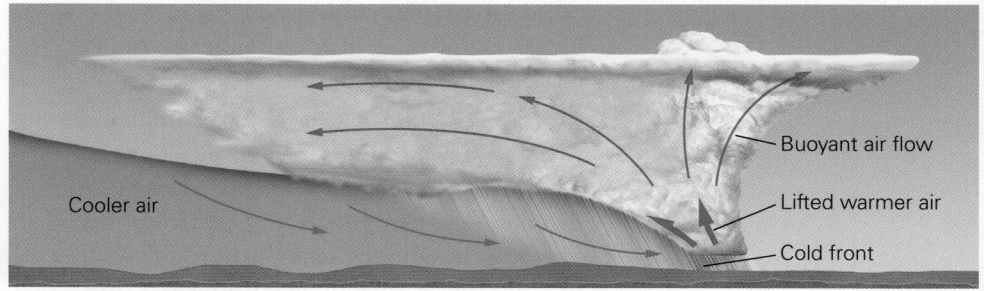

Buoyant air flow

Cooler air

Lifted warmer air

Cold front

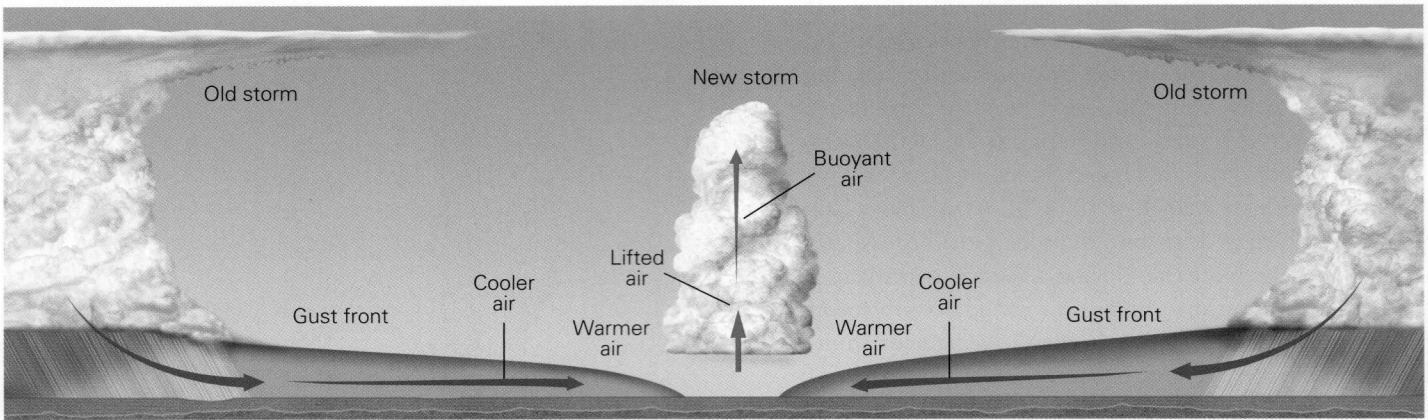

FIGURE 18.6 Development of thunderstorms along colliding gust fronts. Each storm sends out a gust front from its base. Where the gust fronts collide, air rises, and a new storm forms.

pool spreads outward into areas beyond the storm's base, its leading edge produces a **gust front**. When a gust front passes, winds suddenly strengthen and blow outward from the base of the storm, bringing cooler air with it. The gust front, in effect, acts like a local cold front, lifting warm air ahead of it, and this lifting can set the stage for the development of new thunderstorms. At places where gust fronts from different earlier storms collide, or where a gust front and a cold front collide, particularly intense lifting can lead to the evolution of strong thunderstorms **(Fig. 18.6)**.

3. *Lifting along less distinct boundaries:* Lifting can take place not only along fronts, but also along less distinct boundaries in the lower troposphere, where two local air masses of different temperature come in contact. Such a situation might arise, for example, where the character of the Earth's surface within a region varies, so that the overlying air warms by different amounts in different locations. For example, air above a fallow field of dark soil may be warmer than air over a forest, and air above an island may be warmer than air above the surrounding ocean. At such boundaries, the cooler, denser air can flow under and lift the warmer air.

4. *Lifting due to high relief:* When wind reaches hills or mountains, the moving air can no longer travel horizontally and instead, must flow upslope. This process, called **orographic lifting**, can carry moist air to higher altitudes, leading to thunderstorm formation above the mountains.

5. *Convergence lifting:* Convergence can occur when air flowing from different directions collides, as happens in summer when the sea breezes that flow from the western and eastern coastlines of Florida meet in the central part of the peninsula and trigger large thunderstorms **(Fig. 18.7)**.

When the two currents of air collide, the excess air has nowhere to go but up.

Air does not always have to be physically lifted in order to rise. In some cases, moist air near the ground starts to rise simply because the hot ground beneath it warms the air enough to make it buoyant relative to surrounding air, like the air in a hot-air balloon. This process typically happens during the summer when intense sunlight strikes the ground. Such warming can take place above plains as well as along mountain ranges. For example, on a hot afternoon, the slopes of steep mountains warm air adjacent to the slopes. This causes updrafts that eventually trigger late afternoon thunderstorms **(Fig. 18.8a)**. People living near mountain ranges witness this process many days during the summer **(Fig. 18.8b)**.

FIGURE 18.7 Convergence of sea breeze flows in mid-afternoon often triggers thunderstorms over central Florida.

▶ **Animation**
Fronts

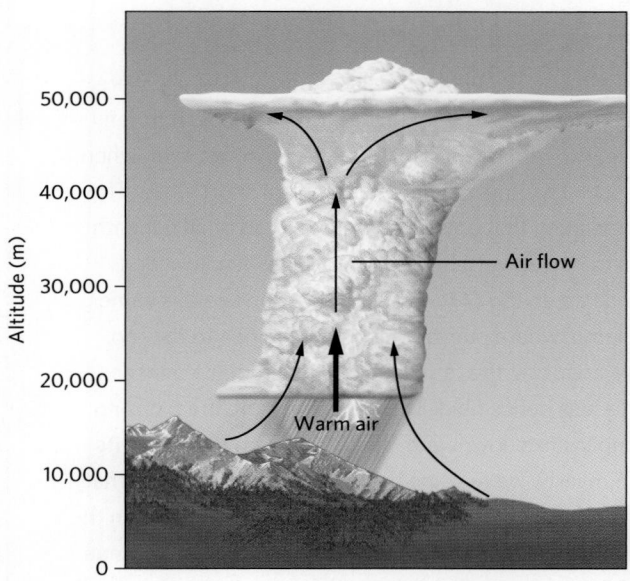

(a) Updrafts can begin over mountain ranges when, during a hot afternoon, the slopes of steep mountains warm the adjacent air.

(b) An example of thunderstorms over mountains in Arizona.

Development of Atmospheric Instability

Lifting of air alone, by any of the mechanisms described above, will not produce a thunderstorm. The location where the lifting takes place must also have the potential to develop **atmospheric instability**. This means that it must be possible for the lifted air to become less dense than the air that surrounds it so that it can continue to rise buoyantly as an updraft to even higher altitudes. For a volume of lifted air to become *unstable*, meaning that it's less dense than the air surrounding it, it must become warmer than the surrounding air. If a volume of air does not become warmer than its surroundings after lifting, we say that the air is *stable*, and the lifted air will stop rising or sink back down (Box 18.1). Thunderstorms develop only where air becomes unstable. To understand how atmospheric instability develops, let's first follow the changes that take place in an *air parcel* (a small volume of air) that has just undergone lifting.

As an air parcel rises to progressively higher altitudes, the air pressure and, in most situations, the temperature in the air parcel's "environment" (the surrounding air) decreases. As a result of the decrease in pressure, the rising air parcel expands, just as a balloon would expand as it rises to higher altitudes (Fig. 18.9). If no heat or mass exchanges between the air parcel and its environment during this expansion—that is, if the air parcel undergoes what physicists call **adiabatic expansion**—the temperature of the parcel decreases as it rises. Why? In order for the parcel to expand, air molecules within it must push aside the surrounding air, and this pushing requires energy. Because

the energy that does this work comes from heat energy inside the parcel, the parcel must give up some heat energy as it expands, causing the air in the parcel to cool.

After undergoing adiabatic expansion the rising air parcel is now less dense than it was, but it is also cooler. Whether the air parcel stops rising, or remains buoyant and continues to rise, depends on two factors. First, stability depends on the **environmental lapse rate**, the rate at which the temperature of the air in the environment surrounding the rising parcel decreases with altitude. Typically, environmental lapse rates range between 4°C/km and 9°C/km (12–26°F/mi), depending on location. Second, it depends on the moisture content of the rising air parcel, as we now see.

Unsaturated air contains some water vapor, but not enough to form water droplets as the air rises. During such *dry adiabatic expansion*, clouds do not form. Since no condensation takes place, no release of latent heat happens in this air. However, cooling due to expansion does take place. In fact, dry adiabatic expansion causes the temperature of an air parcel to decrease by about 10°C for each kilometer (29°F/mi) that the parcel rises. This cooling rate, known as the **dry adiabatic lapse rate**, does not depend on altitude: an air parcel undergoes the same temperature change rising between altitudes of 1 and 2 km (0.6–1.2 mi) as it does rising between 9 and 10 km (5.6–6.2 mi). Now here's a key point: if the dry adiabatic lapse rate exceeds the environmental lapse rate, meaning that a lifted unsaturated air parcel cools faster than the surrounding air as the parcel rises, it will eventually become colder, and

FIGURE 18.9 A parcel of air expands as it rises because the pressure in the environment around the parcel decreases with increasing altitude. Note that the number of air molecules (symbolized by dots) in the parcel doesn't change as the parcel rises.

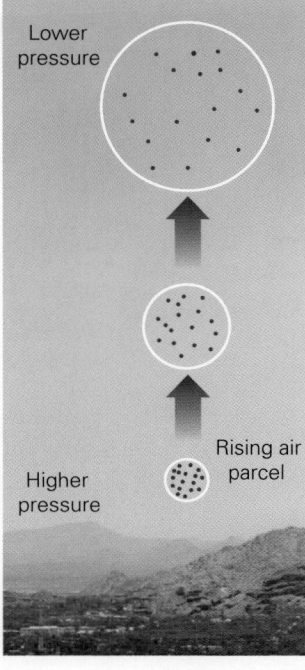

Lower pressure

Rising air parcel

Higher pressure

Atmospheric instability

What are we learning?
- The concept of atmospheric instability.

What you need:
- Vinegar; salad oil; a jar or other glass container; a spoon

Instructions:

We can simulate the concept of atmospheric instability by doing two experiments, using familiar kitchen materials.

- Start by filling a jar a third of the way with vinegar, which has a density of 1.05 g/cm³. Then add the oil, which has a density of 0.85 g/cm³. Because it's less dense, the oil forms a horizontal layer floating above the vinegar. Now, plunge a spoon down below the boundary and lift a blob of vinegar up. What happens to the blob of vinegar?

- Starting again with an empty jar, invert the layering by pouring the oil in before the vinegar. Carefully add vinegar on top of the layer of oil. What happens to the oil?

What did we see?
- In the first experiment, if you remove the spoon, the vinegar sinks back down, so the boundary between oil and vinegar returns to horizontal. We say that the fluid in the jar is stable because if we displace some of the fluid upward, it returns to its original state, even though it underwent lifting.

- In the second experiment, the fluid in the jar is unstable. If you were careful when adding the vinegar, you might have been able to briefly get a layer of vinegar to remain on top of the oil. However, even when left undisturbed, such layering can persist for only a short while. The fluid isn't stable, meaning that any place where a blob of oil beneath the vinegar starts to rise upward the rest of the oil will follow, and soon all the oil ends up on top of the vinegar. You can see this same process in a lava lamp, where buoyant blobs of fluid rise to the top.

- In this context, we are using the term *stable* to mean that a denser fluid lies beneath a less dense fluid. The denser fluid, if lifted, has no buoyancy and simply sinks downward, and the system returns to its initial condition. Similarly, we're using the term *unstable* to mean that a less dense fluid underlies a denser fluid. If the less dense fluid on the bottom get an initial upward push, its buoyancy causes it to continue to rise upward while the denser fluid above sinks to take its place.

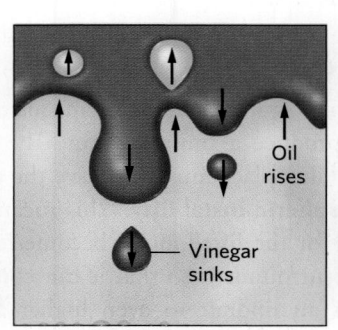

| Salad dressing | Stable condition | Unstable condition | Lava lamp |

therefore denser, than the surrounding air. Therefore, the lifted parcel cannot rise farther on its own, so atmospheric conditions at the site of lifting can be considered stable.

Saturated air—meaning air that holds as much water as possible—forms clouds as it rises and expands. During such *moist adiabatic expansion*, water vapor condenses into water droplets, and condensation releases latent heat (see Chapter 16). At the same time, however, expansion of the air causes it to cool. In contrast to the dry adiabatic lapse rate, rising saturated air cools at a rate of only 6°C/km (17°F/mi) of altitude, because the heating due to condensation partially counteracts the cooling due to expansion. Note that this rate, known as the **moist adiabatic lapse rate**, is less than the dry adiabatic lapse rate. Because the moist adiabatic lapse rate can be less than the environmental lapse rate, a saturated, cloudy air parcel has the potential to become buoyant, and it can remain buoyant as it continues to rise. Meteorologists refer to such air as *conditionally unstable*, the condition being that the air is saturated.

Pulling all these ideas together, we see that the buoyancy of lifted air, which determines whether or not it can rise to the tropopause, thereby leading to thunderstorm formation, depends on the environment surrounding the lifted air, and on the moisture content of the lifted air. If lifted air is unsaturated, it virtually always cools more quickly with increasing elevation than does the surrounding air of the environment. Consequently, the lifted air becomes denser than the surrounding air, remains stable, and cannot rise further, so thunderstorms will not develop (Figure 18.10a). In contrast, if lifted air is saturated, the water vapor within begins to condense as the air rises, and condensation releases heat (Figure 18.10b). Whether the lifted air becomes unstable at this point depends on the lapse rate of the air in the environment surrounding it. If air in the environment remains warmer and less dense than the lifted air, the lifted air will be denser than the surrounding air. So while the lifted air may become cloudy, it remains stable and won't rise further, and thunderstorms won't develop. But if the lifted air becomes warmer than the surrounding air in its environment, it becomes unstable and will rise buoyantly as an updraft. If the updraft is strong enough, and carries enough moisture to high elevations, towers of billowing cumulonimbus clouds form, and a thunderstorm develops.

Note that moisture drawn into the base of the storm effectively provides the "fuel" to drive the storm. Conditions leading to atmospheric instability and thunderstorm formation, therefore, exist only when air in the lower troposphere is warm and moist while air in the upper troposphere is cool. These conditions can happen throughout the year in the tropics, but in mid-latitudes they occur primarily during late spring through early fall.

Ordinary Thunderstorms

Towering dark clouds, flashes of lightning, claps and rumbles of thunder, torrential rains, and sometimes the clatter of hail—these are the signatures of thunderstorms. But even though all thunderstorms share these basic characteristics, not all are alike. The most common type of thunderstorm, known as an **ordinary thunderstorm**, is an isolated thunderstorm that does not rotate—meaning that the air in the storm does not spin around a vertical axis. Such storms are also known as *single-cell thunderstorms*, because each has one updraft and one downdraft, or as *air-mass thunderstorms*, because they tend to form in

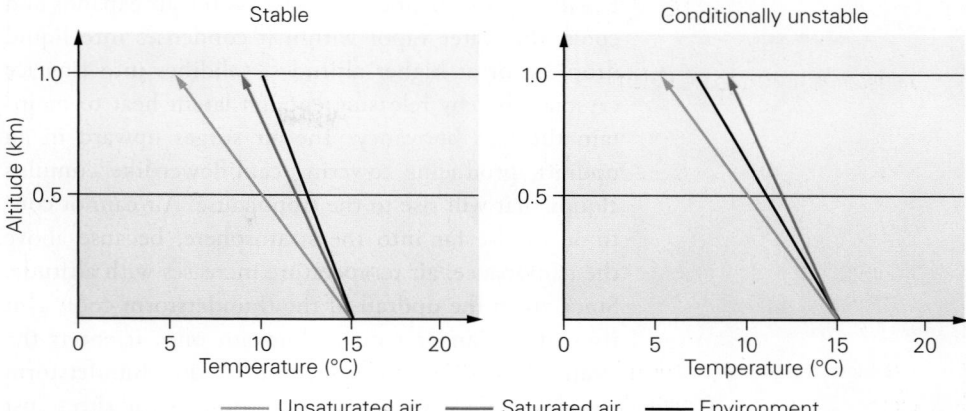

FIGURE 18.10 Stability under different environmental conditions.

(a) In this environment, both moist and dry air parcels, when lifted, would return to their original altitude, because both remain cooler and denser than the environmental air around them.

(b) In this environment, the environmental lapse rate is greater than in part (a). An unsaturated air parcel, when lifted, remains cooler and denser than the environment, so it would remain stable. A lifted saturated air parcel will become warmer and less dense than its environment, so it becomes unstable and rises buoyantly without further lifting.

the middle of an air mass rather than along a front. Note that a *cell* in the context of a thunderstorm refers to a convection cell composed of an updraft and a downdraft.

Ordinary thunderstorms form where, at the time of storm formation, regional winds do not change substantially in direction or speed with altitude. For example, ordinary thunderstorms may happen where air undergoes orographic lifting or gets warmed by the ground. In general, ordinary thunderstorms develop during hot afternoons and last for an hour or so before they dissipate. Only rarely do they produce severe weather such as hail, strong winds >50 knots (92 km/h, or 57 mph), or tornadoes. Worldwide, as many as 2,000 ordinary thunderstorms are active at any given time.

The life cycle of an ordinary thunderstorm can be divided into three stages (Fig. 18.11). During the

FIGURE 18.11 Stages in the evolution of an ordinary thunderstorm.

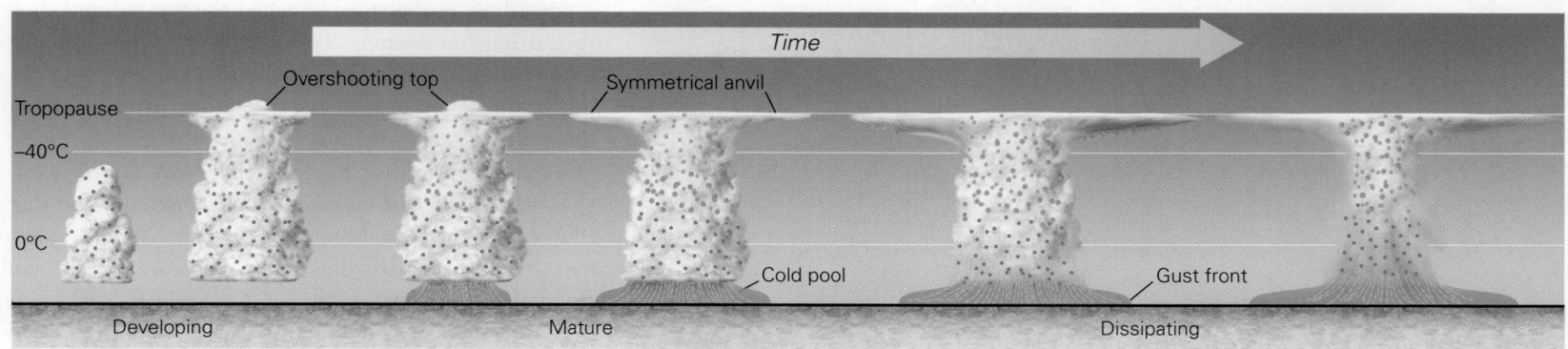

developing stage, warm, moist air undergoes lifting to an altitude at which it becomes buoyant and unstable. Recall that this happens because, as the air expands and cools, the water vapor within it condenses into liquid droplets, or at higher altitudes, solidifies into tiny ice crystals, thereby releasing enough latent heat to maintain the air's buoyancy. The air surges upward in an updraft, producing towering cauliflower-like cumulus clouds that will rise to the tropopause. Air cannot continue to rise far into the stratosphere, because above the tropopause, air temperature increases with altitude. Since air in the updraft of the thunderstorm cools as it rises, it no longer remains buoyant once it enters the stratosphere. Therefore, the top of the thunderstorm cloud spreads laterally into a symmetrical sheet just below the tropopause. The resulting flat-topped cloud is called an **anvil cloud**, because of its resemblance to an anvil used by a blacksmith (Fig. 18.12). Notably, some updrafts are so strong that they surge briefly above the tropopause, producing a bulge known as an **overshooting top** that protrudes upward from the anvil cloud. As the updraft weakens, the overshooting top sinks back into the anvil.

With the formation of the anvil cloud, the thunderstorm enters the *mature stage* of its life cycle. During this stage, ice crystals aloft grow into snowflakes and graupel (small, rounded, soft ice particles) as they cascade downward through the cloud, then fall into the warmer lower atmosphere, where they melt into raindrops (see Fig. 18.11). As this precipitation falls, it causes a downdraft to develop, for two reasons. First, as rain falls, it encounters drier air mixing in from the sides of the cloud, so some of the raindrops evaporate. Evaporation of rain cools the air, making it denser so that it begins to descend, sinking faster as rain continues to evaporate. Second, falling raindrops or ice particles push underlying air downward. When the downdraft reaches the ground, the air within it is cooler than its surroundings, so it produces a *cold pool*. As the cold pool spreads out laterally, its leading edge becomes a gust front.

The downdraft that forms during the mature stage of an ordinary thunderstorm opposes the updraft that drove development of the storm: because both the updraft and downdraft are vertical, they occur sequentially in the same volume. As the downdraft develops, the updraft dissipates. Consequently, the overshooting top sinks and disappears. As the downdraft intensifies, it eventually suppresses the lower part of the updraft entirely and shuts off the source of the moisture that feeds the storm. Without this fuel, the thunderstorm weakens and enters its *dissipating stage* (see Fig. 18.11). The storm's clouds begin to evaporate and disperse. Even as the storm is dissipating, however, the cold pool, bordered by a gust front, spreads out and lifts warm air ahead of it, sometimes triggering the formation of new thunderstorms.

Squall-Line Thunderstorms

Commonly, thunderstorms don't occur in isolation but rather in a distinct line that may be tens to hundreds of kilometers long. If you live in a region affected by such thunderstorms, you may have noticed that just before a line of thunderstorms arrives, a *shelf cloud*, a dark and ominous cloud that has a ragged bottom and a flat, somewhat smooth top, may approach (Fig. 18.13). The shelf cloud lies over a long gust front that marks the edge of a broad cold pool, in which air flow that started as a downdraft in the storm reached the ground and then spread out horizontally. Typically, only a small amount of precipitation falls from the shelf cloud as it passes, but immediately behind the gust front come **straight-line winds**, strong winds sometimes exceeding severe criteria of 92 km/h or 57 mph, that blow in one direction. These straight-line winds can be hazardous in that they can damage buildings, topple trees, flatten crops, and pose a hazard for low-flying airplanes. When you feel the straight-line winds, you have been enveloped by the storm's cold pool, and within a few minutes you will be drenched by rain. Some of the rain within the

FIGURE 18.12 When an ordinary thunderstorm rises to the tropopause, the top of the cloud spreads out into a flat layer, making the cloud overall resemble a classic blacksmith's anvil.

FIGURE 18.13 A shelf cloud develops when warm air is lifted over an approaching gust front that is rushing out from beneath a thunderstorm. The inset shows a closeup of a shelf cloud. The heavy rain is to the rear.

Shelf cloud

Heavy rain

Research radar truck

straight-line winds appears to move nearly horizontally. Meteorologists define a sudden increase in wind speed, to a sustained value above 22 knots (41 km/h, or 24 mph) for more than one minute, generally accompanied by driving rain, as a **squall**. Therefore, the line of thunderstorms accompanying the broad straight-line winds are called **squall-line thunderstorms**. Meteorologists also refer to squall-line thunderstorms as *multicell thunderstorms*, to emphasize that many updrafts and downdrafts occur along the line. As these thunderstorms arrive and pass overhead, lightning streaks the sky, heavy rains fall, and the wind blows in strong individual **gusts**, short-lived wind bursts that are variable in speed and/or direction, because directly beneath the storms you may encounter winds generated by several different downdrafts. After the heavy rain passes, lighter rain may continue for hours, with an occasional flash of lightning and crack of thunder. As a squall line advances, it produces rain over a broad area.

Squall-line thunderstorms can develop in either tropical or temperate climates. In tropical climates, squall lines can happen any time of the year. In temperate climates they develop during both the warm season (late spring through early fall) and cool season (late fall through early spring), but conditions leading to their formation differ in the different seasons.

In the temperate warm season and in the tropics, the process of forming a squall line begins when a disorganized cluster of several ordinary thunderstorms grows (Fig. 18.14a). As time progresses, downdrafts from these storms coalesce to form a single large cold pool at the ground. This cold pool spreads outward, and its leading edge forms a strong gust front. As the initial thunderstorms go through their life cycle, the cold pool continues to expand outward. Lifting along the gust front defining the edge of the cold pool triggers the formation of new thunderstorms, and since these new storms align with a very long gust front, they define an organized squall line (Fig. 18.14b). Eventually, as the cold pool advances forward into warmer air, the squall line becomes wider, moist air flows over the top of the cold pool, and rainfall becomes more widespread. Eventually, as the cold pool spreads out so much that the squall line weakens, but lighter rain may continue for several more hours as the storms eventually dissipate. Squall-line thunderstorms are responsible for much of the summer rainfall over the interior plains of the United States.

During the cooler season in temperate climates, lifting leading to the formation of squall-line thunderstorms becomes focused along strong cold fronts. In this case, very long lines of thunderstorms, known as **frontal squall lines**, develop (Fig. 18.15). (As we'll discuss later in this chapter, frontal squall lines lie along the "tail" of a comma-shaped mid-latitude cyclone.) Typically, frontal squall lines move with the front as it crosses a continent. Because they have relatively long lives (many hours to more than a day), frontal squall lines can yield significant wind damage and flooding.

FIGURE 18.14 A squall line develops when ordinary thunderstorms coalesce and a large cold pool forms.

Time 1
Individual storm
Small cold pool

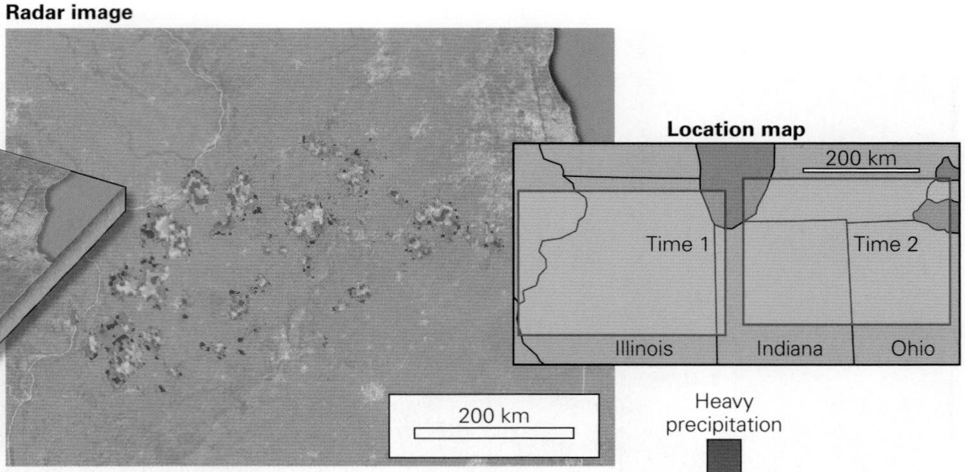

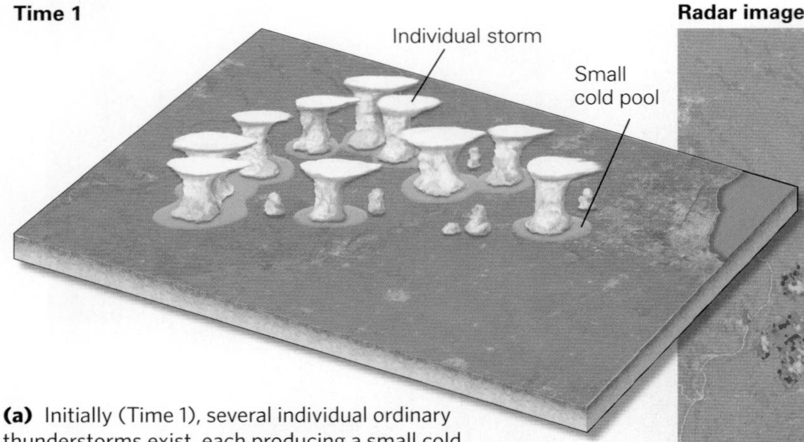

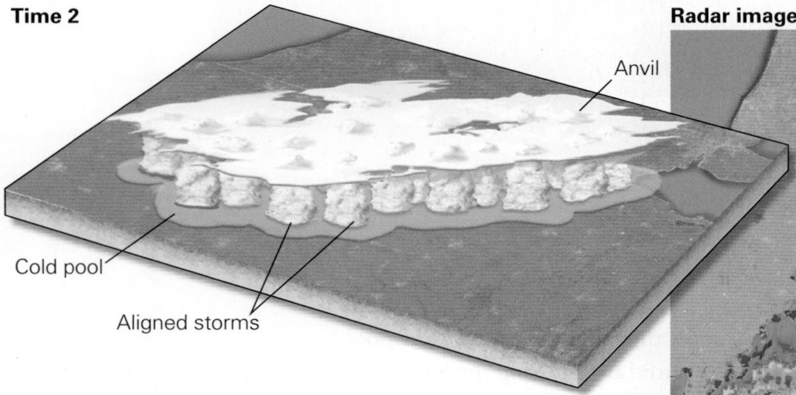

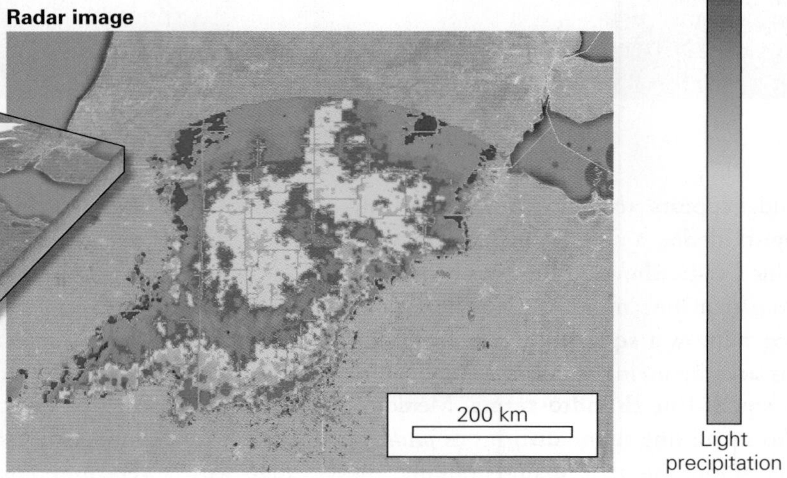

Radar image
200 km
Location map
200 km
Time 1
Time 2
Illinois
Indiana
Ohio

(a) Initially (Time 1), several individual ordinary thunderstorms exist, each producing a small cold pool. The radar shows that the storms are isolated from each other.

Time 2
Anvil
Cold pool
Aligned storms

Radar image
200 km
Heavy precipitation
Light precipitation

(b) Four hours later (Time 2), new storms have organized in a line formed by merging cold pools. The resulting squall line stands out on radar.

FIGURE 18.15 A frontal squall line.

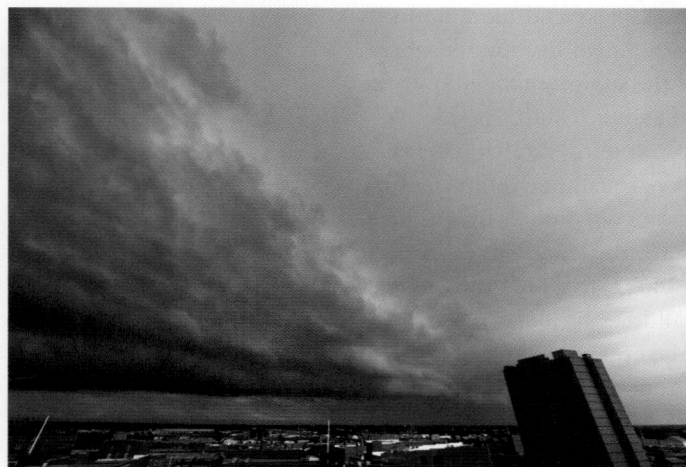

(a) A shelf cloud ahead of a frontal squall line.

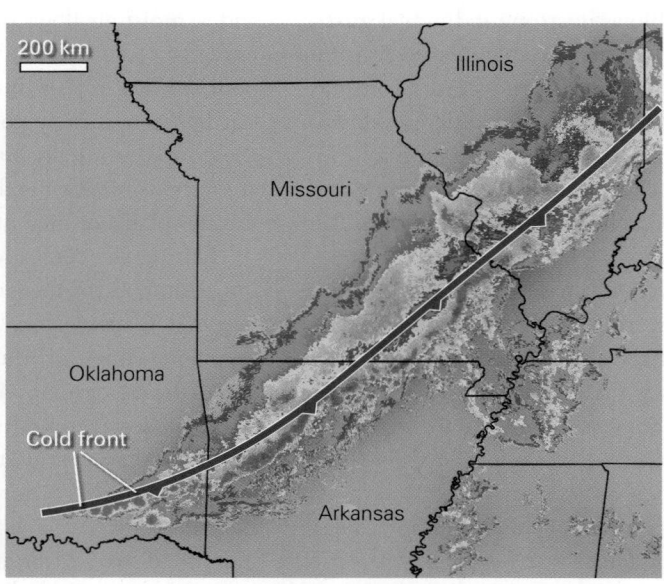

200 km
Illinois
Missouri
Oklahoma
Cold front
Arkansas
Light precipitation
Heavy precipitation
Radar reflectivity

(b) A radar reflectivity image showing thunderstorms along an advancing cold front in the central United States. The colors, indicating rainfall intensity, highlight the position of the storms.

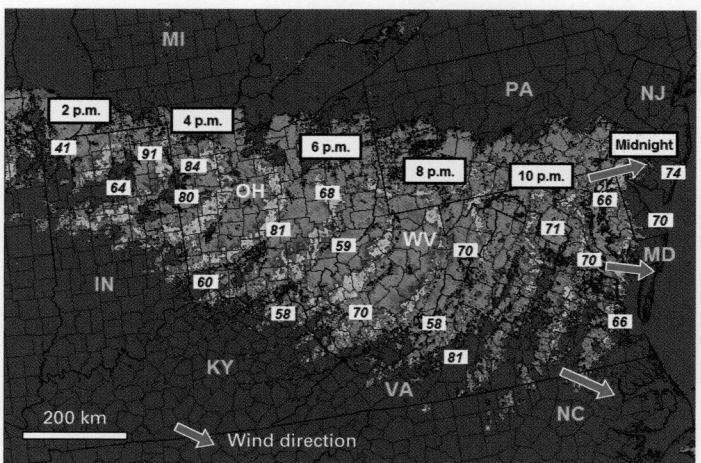

(a) A composite of successive superimposed radar images as a squall line moved from western Indiana to the US east coast in 10 hours. Colors indicate the intensity of rain (red is strongest), and numbers in the boxes indicate wind speeds (in mph).

(b) Tree damage like that seen here occurred all along the path of the derecho.

During some squall-line events, straight-line winds become severe and long-lasting, and therefore affect a large geographic region. Such a widespread thunderstorm-generated windstorm is called a **derecho** (from the Spanish, meaning direct, or straight). One of the worst derechos on record occurred June 29–30, 2012 (Fig. 18.16a). As the squall line moved, the derecho it produced raced eastward from western Indiana along the Ohio River Valley, over the central Appalachians, and into the mid-Atlantic states. It caused 22 fatalities, most due to falling trees, and damage totaling $2.9 billion (Fig. 18.16b).

Supercell Thunderstorms

Between April 25 and 28, 2011, violent thunderstorms developed and continually moved across the southeastern United States, producing the country's worst outbreak of tornadoes. An estimated 360 tornadoes emerged from the storms, producing $10.2 billion in damage and breaking the record both for the number of tornadoes produced in a four-day period and the number produced in the month of April. Despite advance warnings, 346 people were killed.

Meteorologists describe the violent thunderstorms that can produce deadly tornadoes as supercell thunderstorms. Formally defined, a **supercell thunderstorm** (or *supercell*) is a thunderstorm containing a particularly strong, long-lasting, updraft that rotates around a somewhat tilted axis. They differ from ordinary thunderstorms in that updrafts in ordinary thunderstorms do not rotate. Supercells, which occur much less frequently than ordinary thunderstorms and are typically isolated from neighboring thunderstorms, usually have an overall base

diameter of about 50 km (30 mi), much larger than that of ordinary thunderstorms. The most dangerous portion of the storm, which lies beneath the storm's updraft, has a diameter of 5–10 km (3–6 mi). What conditions lead to supercell formation? Supercells develop only when strong **vertical wind shear**—a condition in which wind near the ground differs in speed and direction from wind at a higher altitude—exists in the lower part of the troposphere, and strong winds are present in the upper troposphere.

Supercell thunderstorms develop multiple times a year in the central United States. Why? In this region, vertical wind shear typically develops because slow winds near the ground come out of the south-southeast, bringing in warm, moist air from the Gulf of Mexico, while faster winds 1–2 km (1 mi) above the ground come out of the west-southwest. The resulting wind shear causes air in the lower troposphere to start swirling around a horizontal axis. The shape of this movement can be visualized by considering a rolling horizontal flexible tube (Fig. 18.17a). When a strong, thunderstorm-producing updraft develops, it draws up the horizontally rotating air. To get a sense of this phenomenon, imagine that the tube shown in Figure 18.17a gets pulled upward in the middle and becomes an arch. Because the updraft also triggers cloud formation, most of the rotating air cylinder lies hidden within a growing cumulus cloud (Fig. 18.17b). Air in the now steeply tilted sides of the arched tube continues to rotate and, therefore, spirals upward. As soon as precipitation begins and the cloud becomes a thunderstorm, a downdraft forms, driving the center of the arch downward. This process separates the initial arch into two separate rotating updrafts. Very quickly, turbulence causes

FIGURE 18.17 Stages in the development of a supercell thunderstorm.

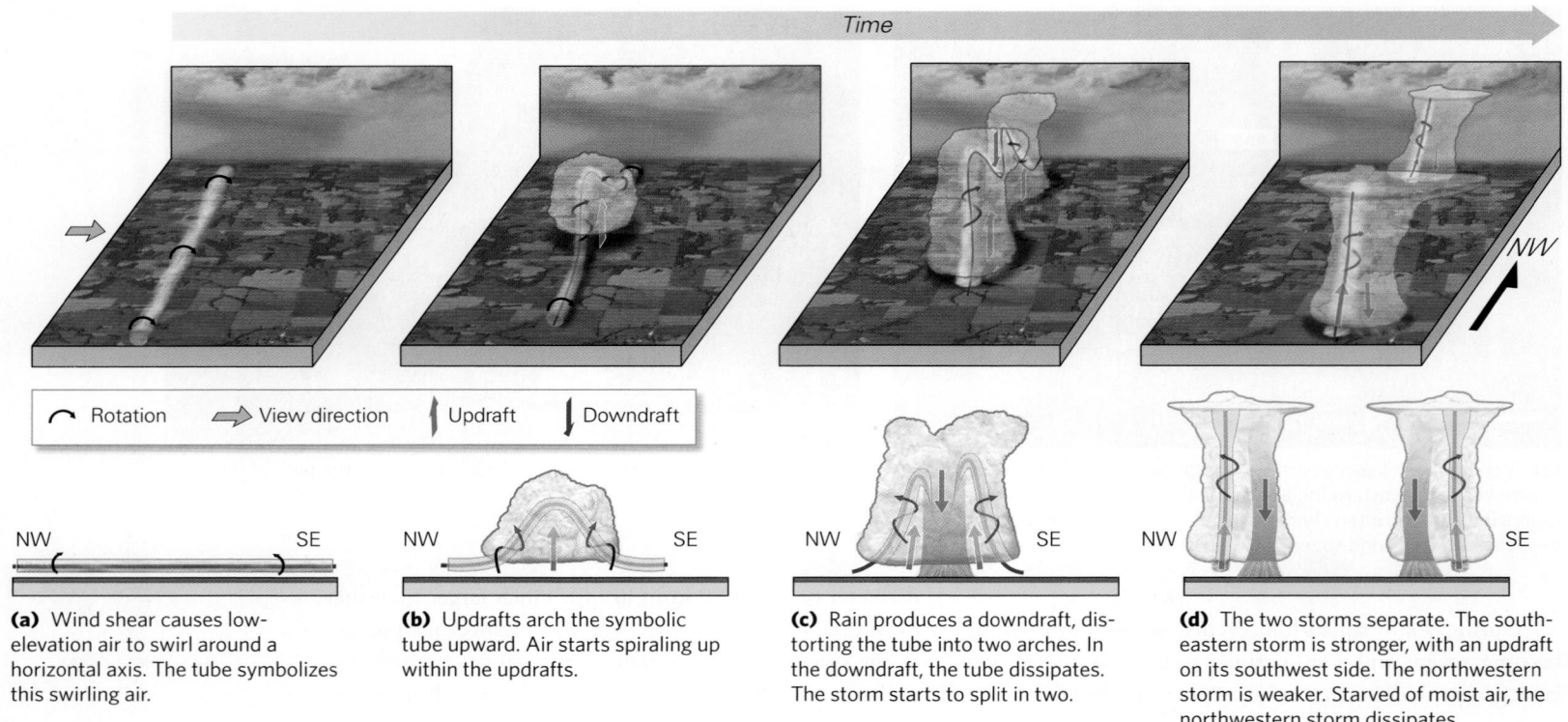

Time

Rotation | View direction | Updraft | Downdraft

(a) Wind shear causes low-elevation air to swirl around a horizontal axis. The tube symbolizes this swirling air.

(b) Updrafts arch the symbolic tube upward. Air starts spiraling up within the updrafts.

(c) Rain produces a downdraft, distorting the tube into two arches. In the downdraft, the tube dissipates. The storm starts to split in two.

(d) The two storms separate. The southeastern storm is stronger, with an updraft on its southwest side. The northwestern storm is weaker. Starved of moist air, the northwestern storm dissipates.

rotation within the central downdraft to weaken, so the original storm splits in two (Fig. 18.17c).

Once the original storm splits, two distinct thunderstorms exist. In the northwestern storm, air rotates clockwise as it spirals up the updraft, whereas in the southeastern storm, air rotates counterclockwise. Because the warm, moist air that feeds the storms comes from the south-southeast, the southeastern storm draws in moisture and grows, while the northwestern storm, starved of warm, moist air, normally weakens and dies out (Fig. 18.17d). Therefore, almost all supercells that pose a hazard in the central United States rotate counterclockwise.

Meteorologists refer to the rotating updraft at the heart of a supercell as a **mesocyclone**; the term means medium-sized (much smaller than a hurricane or a mid-latitude cyclone, but bigger than a tornado), counterclockwise-rotating winds. Within the rotating updrafts of supercells, vertical wind velocity can reach 160 km/h (100 mph). The strong winds of the middle and upper troposphere push the axis of the updraft slightly sideways, so it tilts somewhat away from vertical.

When the updraft of a supercell thunderstorm nears the tropopause, the high-altitude winds of the polar-front jet stream blow the upper part of the storm downwind. Therefore, supercell thunderstorms have a distinctive structure characterized by a large, asymmetrical anvil at the top, and a strong mesocyclone within (Fig. 18.18a, b). The updraft in the mesocyclone can be so strong that it forms a particularly large overshooting top (Fig. 18.18c). The lower end of the rotating updraft extends beneath the base of a supercell's cumulonimbus cloud, and appears as a cylindrical rotating cloud known as a **wall cloud** (Fig. 18.18d).

A radar image, in which colors indicate the intensity of precipitation, highlights the key features of a supercell (Fig. 18.19a). Meteorologists refer to the northeastern side of a northeast-moving supercell, such as those of the central United States, as the *forward flank,* and to the southwestern side as the *rear flank.* Heavy rain on the forward flank produces an intense downward-directed air flow, called the *forward-flank downdraft (FFD),* which descends to the ground and spreads out as a cold pool (Fig. 18.19b). Meanwhile, high-altitude winds bring dry air from the southwest into the high clouds on the rear flank of the storm. Consequently, droplets and ice particles in these high clouds evaporate or sublimate, a process that cools the air. The resulting cooled air sinks rapidly, producing a strong *rear-flank downdraft (RFD).* Because of the tilt of the mesocyclone within a supercell, the FFD lies to the northeast of the storm's updraft. Therefore,

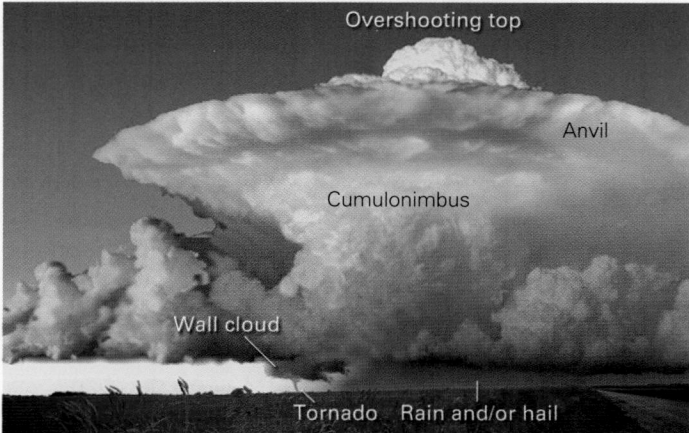

Overshooting top

Anvil

Cumulonimbus

Wall cloud

Tornado   Rain and/or hail

(a) A composite image, showing a supercell thunderstorm as viewed from the southwest.

(b) A huge anvil cloud over a supercell.

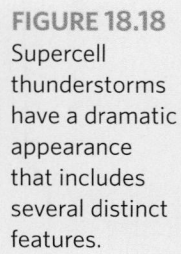

(c) This view of a supercell thunderstorm from a high altitude shows the anvil and a large overshooting top.

Wall cloud

(d) A close-up view of the base of a supercell thunderstorm, showing the wall cloud and a tornado beneath it.

FIGURE 18.19 Structure of a supercell thunderstorm.

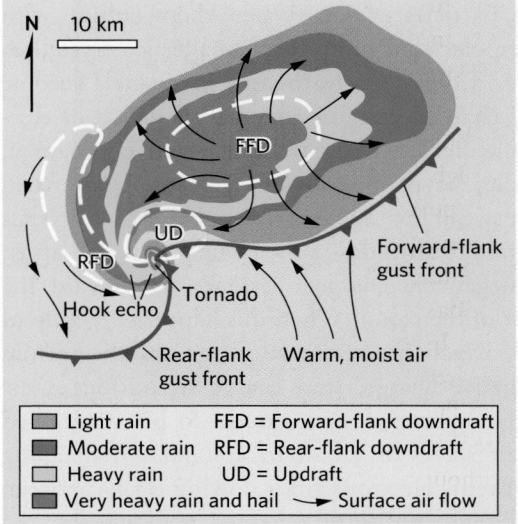

N   10 km

FFD

UD

RFD

Hook echo

Tornado

Rear-flank gust front

Warm, moist air

Forward-flank gust front

☐ Light rain   FFD = Forward-flank downdraft
☐ Moderate rain   RFD = Rear-flank downdraft
☐ Heavy rain   UD = Updraft
☐ Very heavy rain and hail ➝ Surface air flow

(a) A map-view diagram of a radar reflectivity image of a supercell. White dashed lines encircle areas affected by updrafts and downdrafts. The red spot near the tip of the hook echo is caused by debris created by a tornado.

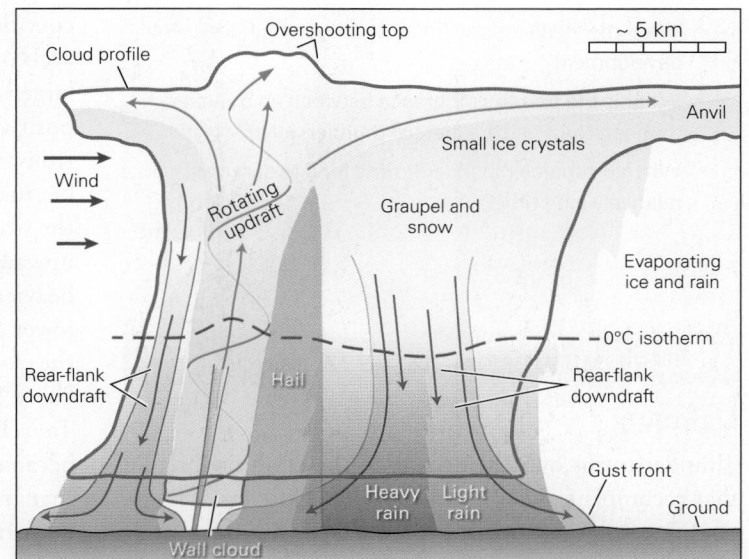

Cloud profile

Overshooting top

~ 5 km

Wind

Rotating updraft

Anvil

Small ice crystals

Graupel and snow

Evaporating ice and rain

0°C isotherm

Rear-flank downdraft

Hail

Rear-flank downdraft

Heavy rain   Light rain

Gust front

Ground

Wall cloud

(b) The internal structure of a supercell storm is rather complex. In this cross section, we see warm, moist air spiraling up the rotating updraft, and two downdrafts, one on the forward flank and one on the rear flank. Different types of precipitation occur in different parts of the storm. Ice (graupel, snow, tiny crystals) occurs above the 0°C isotherm. When it falls below that isotherm, it melts to become rain. Hail falls in the area adjacent to the updraft, and if large enough, can reach the ground. In this example, a tornado extends to the ground beneath the wall cloud.

unlike the updraft of an ordinary thunderstorm, which tends to be destroyed by the downdraft, the FFD of a supercell thunderstorm does not suppress the updraft. As a result, the supercell's updraft can survive for a relatively long time (a few hours), and the storm can maintain its identity as it moves hundreds of kilometers across the land. A distinctly curving *hook echo* outlines the southwestern side of the updraft. The updraft itself appears as an echo-free region within the hook because no rain falls in the updraft. The large echo to the northeast of the updraft indicates the presence of heavy rain in the FFD. Hail, if it forms, falls primarily on the northeastern side of the updraft, and a tornado, if it develops, occurs at the tip of the hook echo.

Take-home message...

Three conditions are necessary for a thunderstorm to form: moist air near the Earth's surface, a lifting mechanism, and local atmospheric instability. Once lifted, a moist air parcel can become buoyant when its moisture condenses into cloud droplets because the release of latent heat makes the parcel warmer and less dense than its surroundings. Such instability can produce an updraft that carries rising air up to the tropopause, producing large billowing clouds that evolve into a thunderstorm. Ordinary thunderstorms have a vertical, nonrotating updraft and tend to dissipate fairly quickly. Squall-line thunderstorms form in a row along storm-generated cold pools of air or along strong cold fronts. The most violent thunderstorms, supercells, form where vertical wind shear exists. A supercell has a rotating updraft, a particularly broad, asymmetrical anvil cloud, and an overshooting top.

Quick Questions —————————————

- What mechanisms can lift air to trigger thunderstorm development?

- What is the primary difference between an ordinary thunderstorm and a supercell thunderstorm?

- Why do supercell thunderstorms tend to survive for a relatively long time?

18.2 Thunderstorm Hazards

Lightning

Thunderstorms owe their name to the crash of thunder that accompanies them. But thunder, while frightening, doesn't result in casualties or damage. Thunder rumbles as a consequence of lightning. It's the lightning in a storm that can cause damage and death. What causes lightning? If you rub your shoes on a carpet and then touch a metal doorknob, you'll see and feel an electrical spark (Fig. 18.20a). This spark occurs because electrons (negatively charged particles) from the carpet's atoms flow into your body, so that your skin becomes negatively charged relative to the doorknob. Put another way, a **charge separation** develops between the doorknob and your finger. When your finger gets close enough to the doorknob, electrons suddenly flow from your skin to the positively charged doorknob, and this short-lived current generates a spark that arcs across the charge separation and tingles your nerves. Physicists refer to such a spark as an *electrostatic discharge*. **Lightning** represents a huge electrostatic discharge caused when a large charge separation develops within a thunderstorm's clouds (Fig. 18.20b).

The difference between the spark that jumps from your finger to a doorknob and a typical **lightning stroke**, or *lightning bolt*, the jagged spark produced by a thunderstorm, is the amount of electrical current that flows during a discharge, and the distance between the ends of the discharge. The spark at a doorknob jumps about 1 mm (0.04 in), has a diameter less than 0.1 mm (0.004 in), and transmits a current of about 0.0001 amp. (An *amp*—short for *ampere*—is a measure of the quantity of electrons flowing in a current during a one-second interval.) In contrast, a lightning stroke can be more than 5 km (3 mi) long, has a diameter of about 2–3 cm (about an inch), and it can carry a current of 15,000 to 30,000 amps. To put this energy release in perspective, a single lightning stroke produces enough energy to power a typical house for 2 months.

Why does the charge separation that triggers lightning develop? A tall thunderstorm cloud can contain countless particles of ice. When these collide with each other, electrons jump from smaller ice crystals to larger ones. Consequently, the tiny ice crystals become positively charged, because they have a deficit of electrons, while the larger ones become negatively charged, because they have an excess of electrons. Updrafts in the storm sweep the tiny, positively charged ice crystals upward to high altitudes in the cloud, while the larger, heavier, negatively charged particles fall toward the lower part of the cloud. When this happens, the top of the cloud, overall, has a positive charge, and the bottom has a negative charge, so a charge separation exists. The charge separation in a cloud can be maintained because air serves as an *electrical insulator*, meaning that it prevents electrons from flowing easily between negatively and positively charged regions. As the charge separation develops, the *electrostatic potential*,

FIGURE 18.20 Lightning is an electrostatic discharge.

(a) An electrostatic discharge from a finger to a door handle resembles a tiny lightning stroke.

(b) Some lightning strokes go from cloud to cloud, or stay within a cloud. Some go from cloud to ground.

meaning the potential energy that could be available to drive electrons from the region of negative charge to the region of positive charge, builds. A lightning bolt forms when the electrostatic potential becomes great enough to drive electrons across a large expanse of insulating air.

Most lightning occurs when electrons from a negatively charged part of a storm cloud jump to a positively charged part of the same cloud or of a nearby cloud. Such **cloud-to-cloud lightning** may hit an airplane flying through a storm. In fact, lightning strikes a commercial airliner, on average, once or twice a year. Fortunately, electricity conducts along the metal skin of an airplane without penetrating its interior, so the strike typically causes minimal damage, if any, to the airplane. **Cloud-to-ground lightning**, which strikes buildings, trees, people, or other objects, poses the greatest hazard to communities. Cloud-to-ground lightning requires charge separation to build between the cloud and the ground. Such separation develops when negative charges accumulate toward the base of a cloud. Then, just as the negative end of a magnet repels the negative ends of other magnets, the negatively charged cloud base drives away negative charges on the ground, leaving a region of the ground with a net positive charge.

A cloud-to-ground lightning stroke begins when electrons surge first toward the cloud base and then toward the ground in a series of steps, producing a *stepped leader* (Fig. 18.21a). Each step is about 50–100 m (160–330 ft) long and takes only a few millionths of a second to form. Typically, different streams of electrons

take different paths downward, so the stepped leader has several branches. The branch that happens to approach the ground surface first attracts positively charged particles. These particles jump upward from an object on the ground, producing a feature called a *positive streamer*. Such streamers typically rise from a narrow high point, such as a tower, a tree, a chimney, or an unfortunate person. When the positive streamer connects with the stepped leader, a channel for conducting electron flow suddenly exists. This channel behaves like an electrical conductor because, unlike the air around it, the channel contains charged particles. Therefore, the instant that the positive streamer connects with the stepped leader, a powerful *return stroke*—the main electrostatic discharge—flashes (Fig. 18.21b). The return stroke, which runs from the cloud to the point where the positive streamer started, transmits vast numbers of electrons streaming from the cloud to the ground. Since nothing actually returns to the cloud when lightning flashes, the name "*return stroke*" is a misnomer; it remains from a time before researchers understood why lightning initiates.

The process of forming a leader, positive streamer, and return stroke typically repeats multiple times, generally within a second or two (Fig. 18.21c, d). Subsequent leaders, after the first stepped leader, are called *dart leaders*. They follow the previously established channel through the air, since it has become an electrical conductor. The whole process happens so fast that you may see it as a single bolt that flashes like a strobe light.

Meteorologists refer to the type of lightning stroke that we've just described as a *negative polarity*

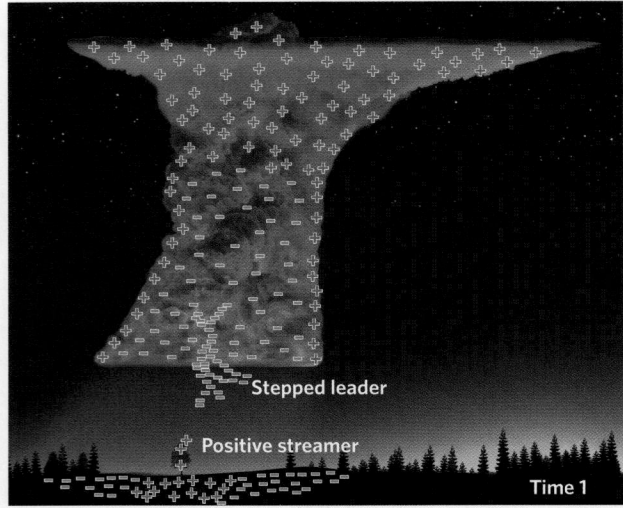

(a) A stepped leader descends from the cloud, while a positive streamer rises from the ground below.

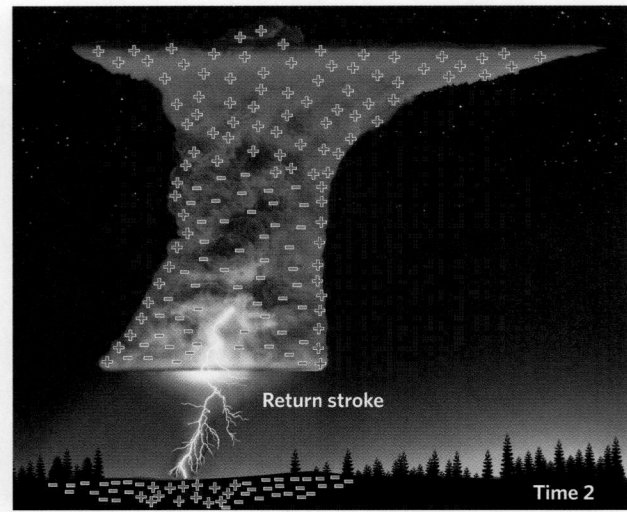

(b) A return stroke makes a brilliant flash.

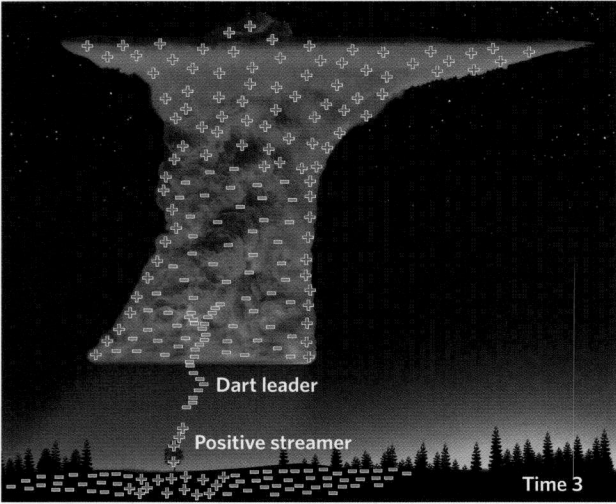

(c) A dart leader descends from the cloud, while a positive streamer flows upward.

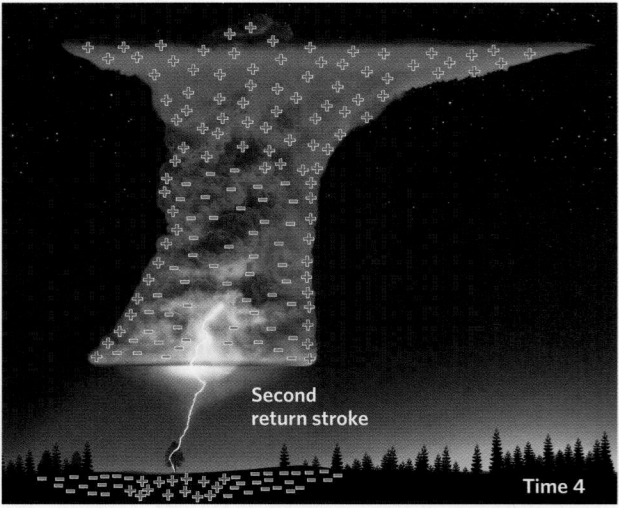

(d) The second return stroke flashes.

cloud-to-ground stroke (Fig. 18.22). As we've seen, it occurs when the negative charge in the base of a storm jumps to the positive charge on the ground. A rarer *positive polarity cloud-to-ground stroke* occurs when electrons in the negatively charged ground jump all the way to the positively charged top region of the storm cloud. Because the distance between the higher regions of the storm and the ground is greater than that between the base of the storm and the ground, a very large electrostatic potential must develop for such a stroke to take place. As a result, these strokes can release an immense amount of energy (300,000 amps), so they are particularly dangerous. Positive polarity cloud-to-ground strokes sometimes strike 10–25 km away from the storm, in advance of, or after the departure of the storm's rain, even when some blue sky may be visible. Such strokes where blue sky is

visible are informally known as "bolts from the blue." Because of the lack of rain when they strike, bolts from the blue are notorious for triggering forest fires.

Needless to say, lightning is dangerous, both because of the electrical shock it represents, and also because of the thermal energy that it produces. If lightning strikes power lines, it produces an electrical surge that blows out transformers and causes a blackout. If lightning strikes a tree, it can cause the sap in the tree to vaporize instantly, so that the tree explodes. And if it strikes a house, it can pass through the frame of the house and ignite flammable materials inside. While lightning strikes cars, it rarely does them harm because their metal skins, like those of airplanes, conduct the electricity and keep it from entering the vehicle. Unfortunately, when lightning strikes a person, it can cause permanent harm or even instant death. Although people struck by lightning can survive, many suffer burns, neurological damage, and other lasting effects. Injury can happen even if a lightning stroke hits the ground or a tree near a person, for the electricity can flow through the ground to the person. In the United States, lightning strokes kill, on average, about 50 people per year. To avoid being a victim, go inside (and stay away from electrical conductors) or hop in a car when a thunderstorm approaches. If you can't take cover, don't stand in the middle of an open area or near isolated tall objects, or be on open water where you would be the highest object. If there's no place to go and you feel your hair standing on end (because it's becoming positively charged), crouch down as close to the ground as possible with your feet together and your heels up to minimize ground contact.

All lightning produces **thunder**, a loud crack or boom. Thunder develops because the intense energy in a lightning stroke instantly heats adjacent air to a temperature five times hotter than the surface of the Sun. This heating causes the air to expand explosively. When the stroke ceases, the air contracts suddenly and generates sound waves, much like the sound produced when you clap your hands. Since sound travels so much more slowly than light, you can estimate the distance to a lightning stroke by counting the seconds that pass between the flash and the thunder—if you divide by three, you get the distance in kilometers, and if you divide by five, you get the distance in miles.

Downbursts

As we noted earlier, very strong downdrafts develop within thunderstorms when it rains, for two reasons—cooling of air due to evaporation and the downward push of the falling rain—and when a downdraft reaches the ground, the air must turn and flow outward. The outrushing wind emerging from a short-lived intense downdraft is called a **downburst**. A downburst flows radially outward from the base of the downdraft, but at any given location, the wind of a downburst flows straight outward. The resulting straight-line winds may travel a few kilometers outward away from the center of the downdraft, generally at speeds of 30–100 km/h (19–62 mph). In severe downbursts, they may be as fast as 100–130 km/h (62–81 mph). Note that the short-lived straight-line winds of a downburst differ from the straight-line winds associated with a squall line or derecho only in terms of duration and area affected.

A small, intense downburst, known as a **microburst**, can develop in association with rain falling from a tall cumulonimbus cloud, even if the cloud hasn't become a thunderstorm. Microbursts affect an area only about 5 km² (2 mi²), but even so, they can be dangerous to aircraft

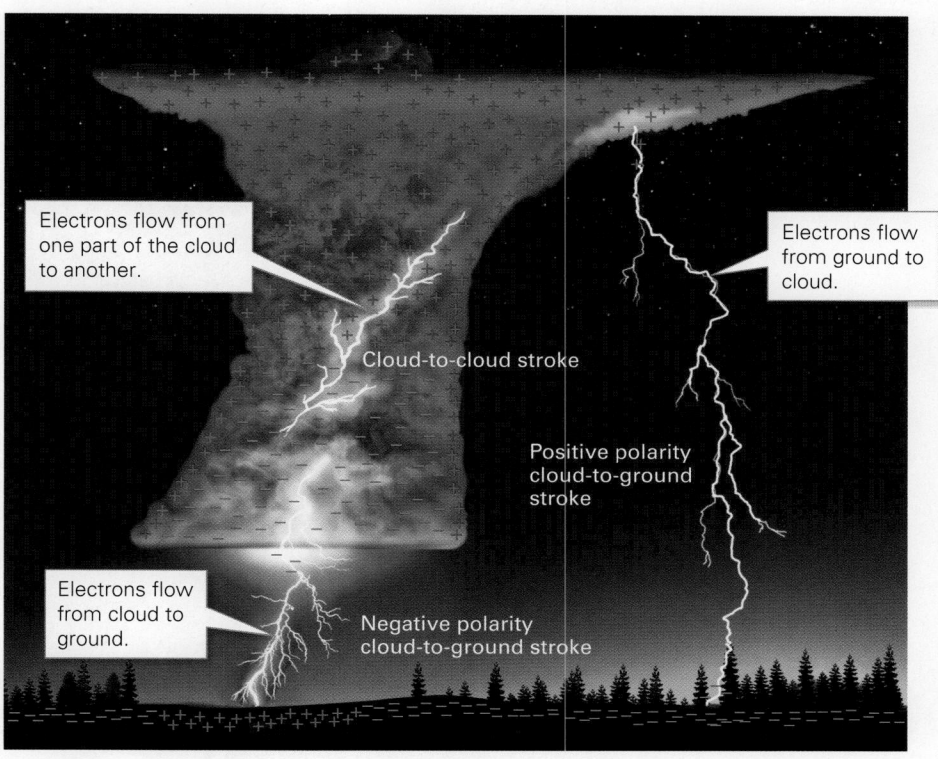

FIGURE 18.22 Lightning discharges can occur in three ways: as negative polarity cloud-to-ground strokes, positive polarity cloud-to-ground strokes, and cloud-to-cloud strokes.

approaching or departing airports (Fig. 18.23a). In news stories describing air traffic hazards, microbursts are commonly referred to as *wind shear*, because wind speed and direction change rapidly across a microburst. If an airplane takes off into approaching winds from a microburst, it lifts into the sky easily. But when the plane crosses into the downdraft itself, it gets pushed down, and when it enters the portion of the microburst blowing in the same direction that it's flying, it loses lift (Fig. 18.23b), which can cause the aircraft to go below its stall speed and potentially lose control.

Hail and Flooding

In nearly all thunderstorms, **hail**, which consists of spherical or irregularly shaped lumps of solid ice (each of which is a *hailstone*), forms at high altitudes. Most hail is small (pea-sized to marble-sized) and melts before reaching the ground, but some hailstones can grow large enough to survive all the way down to the ground. The heaviest hailstone ever measured in the United States, weighing 0.8 kg (1.94 lb), fell during a 2010 storm in South Dakota (Fig. 18.24a). Hail, which can cause major damage by flattening crops, denting cars, and breaking glass (Fig. 18.24b), generally falls intensely only during a limited period of time as the thunderstorm travels; the

FIGURE 18.23 Microbursts can be a serious hazard in the vicinity of an airport.

(a) A microburst over Bangkok, Thailand.

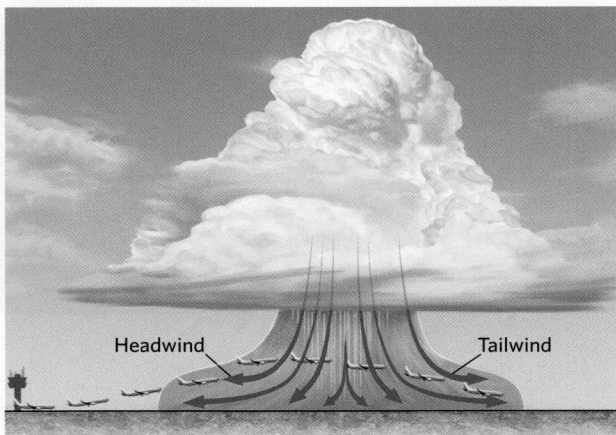

(b) As an airplane approaches a microburst, it encounters strong headwinds, then strong tailwinds after it passes the microburst center.

FIGURE 18.24 Hail occurs over a relatively small area, but can cause significant damage.

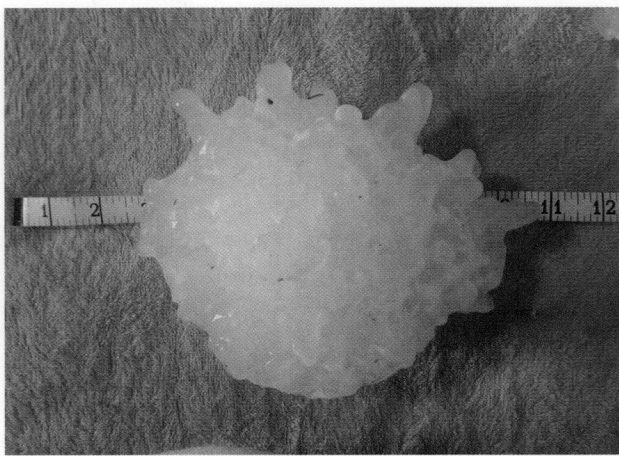

(a) A hailstone that fell in Vivian, South Dakota, on July 23, 2010, holds the US record for diameter (20.3 cm, or 8.0 in).

(b) Hail damaged this car's back window.

Did you ever wonder...

how big a hailstone can grow?

area affected by hail, therefore, occurs in a band called a *hail streak*. Hail streaks can be up to 10 km wide and 60 km long.

The process of hail formation begins when updrafts carry liquid cloud droplets upward to altitudes where temperatures are well below freezing. These droplets become *supercooled*, meaning that they remain liquid despite the subzero conditions. When ice particles falling from higher altitudes collide with these supercooled droplets, the droplets bond to the ice particles and immediately freeze, so an ice particle quickly grows into a hailstone (see Chapter 16). In an updraft, upward air flow keeps hailstones floating in the cloud, where they can keep growing. Eventually, however, a hailstone drifts out of the updraft, or its weight overcomes the force of the updraft, and it starts to fall. It collects more supercooled droplets on its way down, so it continues to grow until it encounters above-freezing temperatures in the lower part of the cloud and starts to melt. Hailstones reach the ground only if they grow large enough to avoid melting

away as they fall. This happens when a thunderstorm contains particularly strong updrafts, so the most damaging hail tends to develop in supercell thunderstorms.

Thunderstorms can produce a lot of rain in a short time, which can trigger **flash floods**, localized short-duration flooding that happens rapidly with little warning (see Chapter 12). Flooding can be a particular problem when thunderstorms move slowly, as storms formed by orographic lifting along mountain slopes sometimes do. Extreme rainfall can also develop along fronts, particularly when the front becomes stationary and winds in the middle troposphere flow nearly parallel to the front. With these conditions, thunderstorms that develop along the front move along the front and reach maturity over the same region, so that rain falls over the region for an extended period as a succession of thunderstorms pass (Fig. 18.25a). Meteorologists refer to this phenomenon as *thunderstorm training* because the storms are visually analogous to train cars passing over the same spot on a track (Fig. 18.25b).

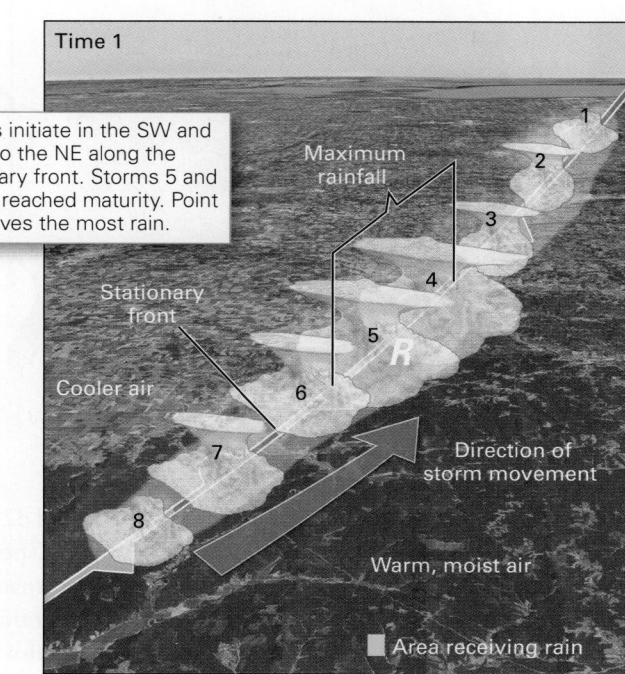

Time 1

Storms initiate in the SW and move to the NE along the stationary front. Storms 5 and 4 have reached maturity. Point R receives the most rain.

Maximum rainfall

Stationary front

Cooler air

1
2
3
4
5
R
6
7
8

Direction of storm movement

Warm, moist air

■ Area receiving rain

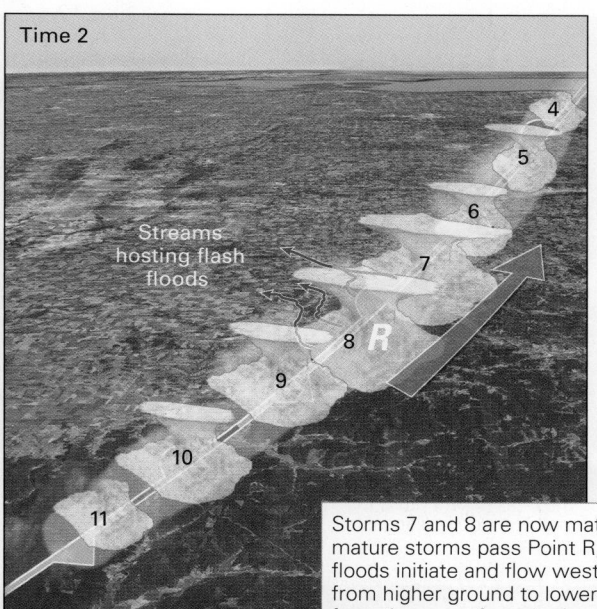

Time 2

FIGURE 18.25 Training of thunderstorms along a stationary front.

Streams hosting flash floods

4
5
6
7
8
R
9
10
11

Storms 7 and 8 are now mature. As mature storms pass Point R, flash floods initiate and flow westward from higher ground to lower ground from the area of most rain.

(a) Storms form on the south end of a stationary front, move northward, mature by Point R, where they dump the most rain, and dissipate farther north.

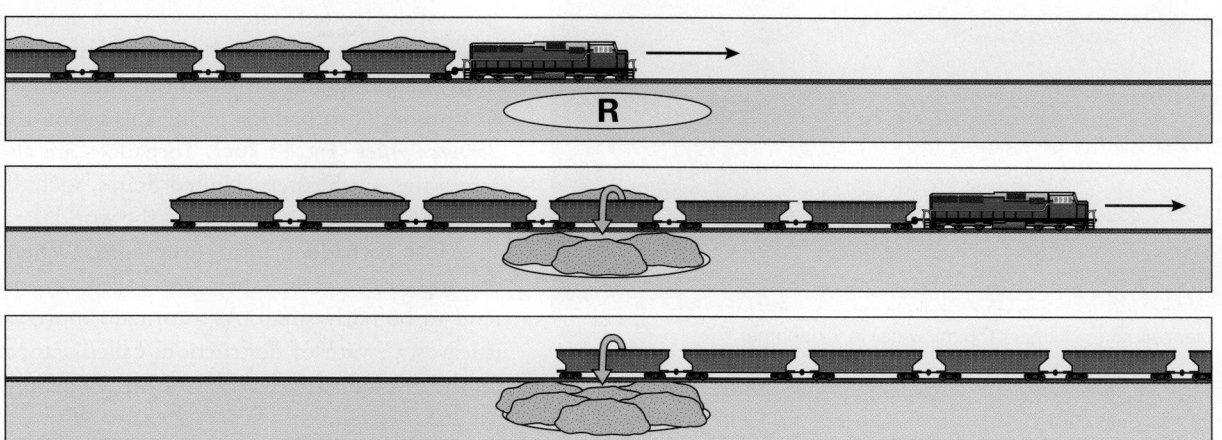

R

(b) The term "training" is an analogy to a freight train with full boxcars (storms) dumping a load (rain) at a location (Point R), while moving along a track (stationary front).

Take-home message...

Thunderstorms produce several dangerous phenomena. When the electric charge separation in a thunderstorm becomes great enough, a lightning stroke flashes from one part of a cloud to another or from the cloud to the ground. Strong downdrafts in thunderstorms can cause damaging straight-line winds. Thunderstorms with strong updrafts can produce large hail. Slow-moving thunderstorms can produce heavy rains that may trigger flash floods.

Quick Questions

- Is it safe to be in a parked car during a lightning storm?

- Why is it unsafe for a plane to pass through a microburst on final approach to an airport?

- Relative to common fruits, how big are the largest hailstones?

18.3 Funnels of Destruction: Tornadoes

What Is a Tornado?

Perhaps no other weather phenomenon is as terrifying as a **tornado**, a violently rotating funnel, or *vortex*, of air that extends from the ground into the lower part of a severe thunderstorm (Fig. 18.26a, b). The strongest tornadoes can damage or destroy steel-reinforced concrete structures, can throw trucks long distances, and can sweep houses completely away. At its base—the end touching the ground—a tornado typically ranges in width from 50 m to 1 km (150 ft–0.6 mi), but some are larger. For example, the largest tornado ever recorded reached a width of 4.2 km (2.6 mi) at its base when it struck El Reno, Oklahoma, in 2013.

(a) Small tornado in Kansas.

(b) Giant tornado (1 km, or 0.6 mi, wide) in Oklahoma.

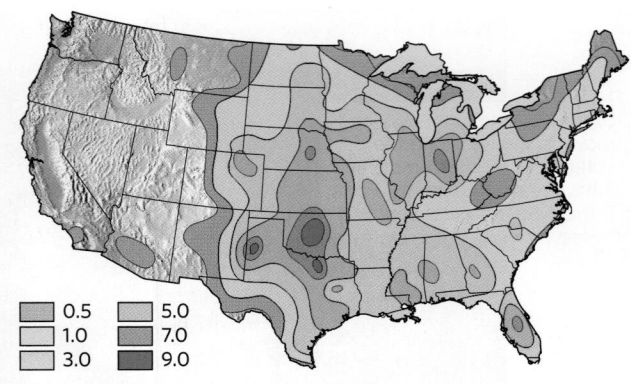

0.5	5.0
1.0	7.0
3.0	9.0

(c) Number of tornadoes per year (per 26,000 km², or 10,000 mi²) in the United States over a 27-year period.

Tornado winds generally blow between 100 and 320 km/h (65–200 mph), but the fastest tornado wind speed ever recorded with radar, also in the El Reno tornado, was 484 km/h (301 mph). A tornado moves across the countryside along with its host storm, so the speed at which its base moves across the ground can vary from near zero to 110 km/h (70 mph). The midwestern and Great Plains states of the United States, where warm, moist air from the Gulf of Mexico meets cold air from Canada along strong fronts in spring and early summer, host the world's largest number of tornadoes, giving this region the nickname *Tornado Alley* (Fig. 18.26c). Tornadoes are also common across the southeastern United States, including Florida, and they can occur elsewhere in the world.

Most tornadoes form over land. When tornadoes develop over water, they are called *waterspouts*, and tend to be fairly small. As a tornado moves across land, it leaves a swath of destruction, called a **tornado track**, on the ground surface (Fig. 18.27a). Generally, tornadoes

FIGURE 18.27 Tornado tracks.

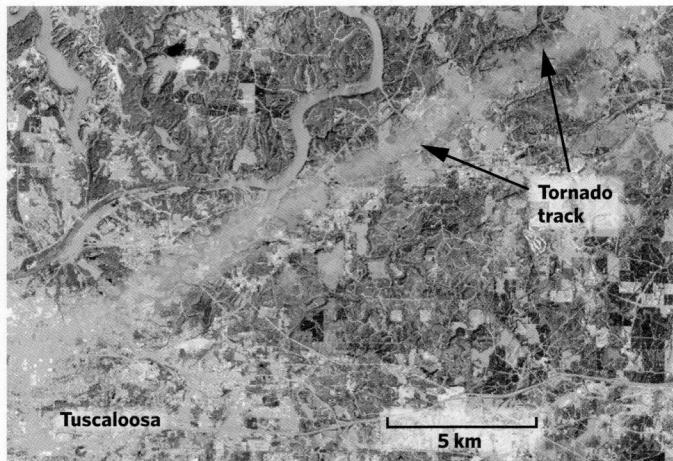

(a) This false-color satellite image shows the track of a tornado near Tuscaloosa, Alabama. Red areas are forested—the winds ripped out the trees along the tornado track.

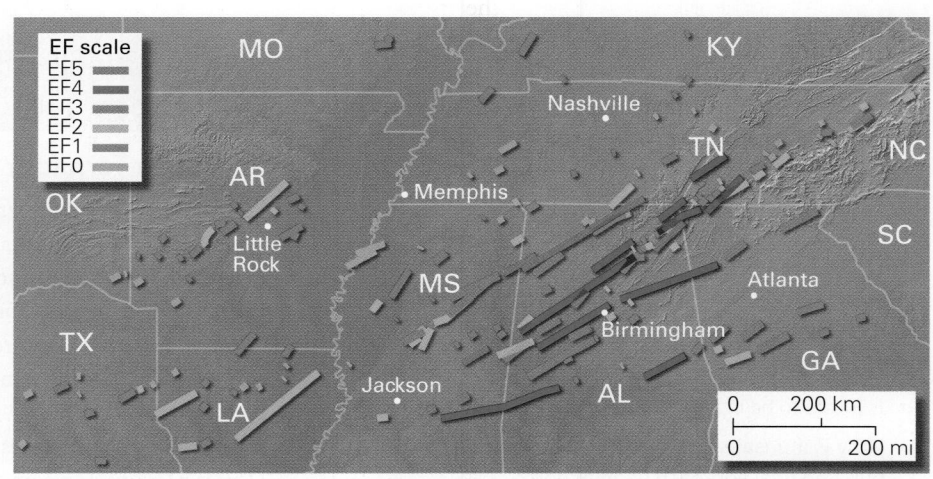

(b) A series of tornado tracks from an outbreak of tornadoes in April 2011. EF scale = Enhanced Fujita scale; see **Defining the Intensity of a Tornado**, below.

are short-lived and leave a track only a few kilometers long, measured from where they touch down to where they lift and dissipate. But occasionally, a tornado can survive for over an hour and produce a very long track. The great Tri-State Tornado of 1925, for example, left a track 352 km (219 mi) long across Missouri, Illinois, and Indiana. On occasion, a *tornado outbreak* develops, during which numerous tornadoes develop in a set of related storms within a relatively short time. The world's largest tornado outbreak, which took place in late April 2011, produced 360 tornadoes over the course of three days (Fig. 18.27b). Tornadoes can form for different reasons, as we will now see.

Supercell Tornadoes

To understand tornado formation in supercell thunderstorms, let's focus on how a tornado develops in a single supercell as it moves from southwest to northeast across the midwestern United States (Fig. 18.28a). Recall that a supercell contains a strong rotating updraft, called a mesocyclone, and that strong downdrafts form on both the forward (northeastern) flank and the rear (southwestern) flank of the mesocyclone (Fig. 18.28b). Tornado development appears to occur most often in association with the rear-flank downdraft. As the rapidly downward-flowing air of this downdraft approaches the ground, it spreads outward and produces a roll of rotating air near the ground along the downdraft's spreading outer boundary (Fig. 18.29a). The air in this roll spins around a horizontal axis. A **supercell tornado** forms when part of the roll passes under the base of the mesocyclone and is pulled upward into the mesocyclone (Fig. 18.29b). Initially, the roll warps into a tight arch, but the clockwise-rotating arm, rotating in the opposite direction of the larger mesocyclone, quickly weakens and dissipates, while the counterclockwise-rotating arm, the one rotating in the same direction as the mesocyclone, gets carried into the updraft, where it stretches and narrows to become a spiral or vortex of spinning air. If the spinning vortex is in contact with the ground, it forms a tornado (Fig. 18.29c). The immense wind speed of the tornado develops because, as the vortex is stretched vertically by the accelerating updraft, it narrows, and the air within it rotates faster and faster, just like a figure skater begins to spin faster when she pulls in her arms. Note that although the process by which a tornado forms resembles the way a mesocyclone forms, a tornado and a mesocyclone are different phenomena, and they have different dimensions. A tornado extends downward from within a wall cloud, and it is much narrower than the wall cloud.

If a tornado were composed only of air, it would be invisible. Some tornadoes appear only as a dust swirl below the wall cloud of a supercell. Most tornadoes become

(a) A supercell storm migrates across the plains of the central United States.

(b) Strong winds aloft cause the updraft (orange arrows) to tilt. Both forward-flank and rear-flank downdrafts develop. When the rear-flank downdraft reaches the ground, it spreads outward, with some of the air flows beneath the mesocyclone.

visible because the rapidly rotating air within them creates very low air pressure. The low pressure at the center of the tornado forms as air is centrifuged outward by the rotation, particularly closer to the ground. As moist air gets drawn into the tornado, the pressure decrease causes the

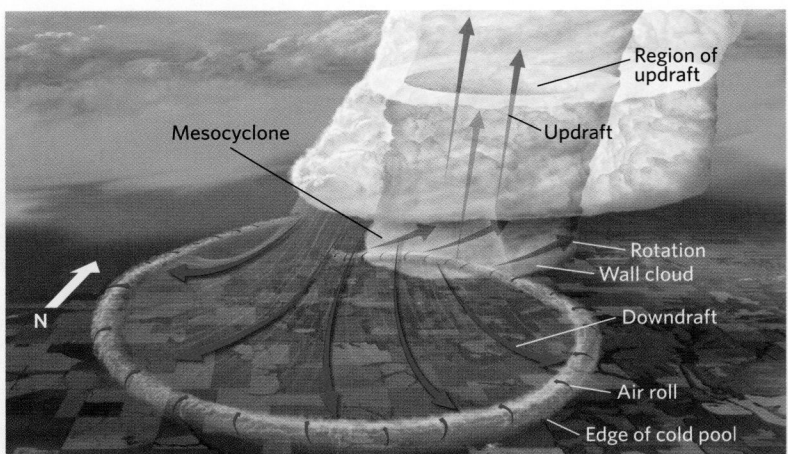

(a) The boundary of the rear-flank downdraft forms a horizontal roll of rotating air close to the ground.

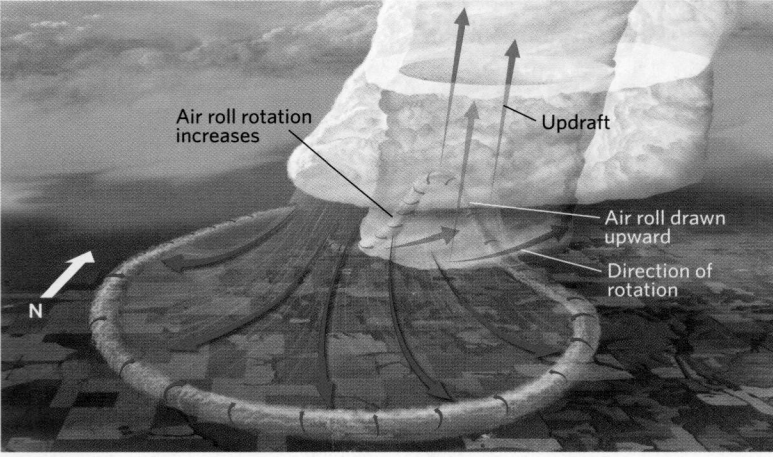

(b) The part of the roll under the updraft gets drawn upward and tilted into the updraft. As it stretches and thins, it spins faster.

air to expand and cool so that moisture condenses and forms the visible **funnel cloud** that we see. When a tornado moves across the ground, dirt and debris it rips from the ground become incorporated into the funnel, making it even more visible and causing its dark appearance.

As we've noted, most tornadoes are small, with short lives, but some grow to be over 1 km wide. Very large tornadoes may include several small vortices (known as **suction vortices**) spinning around the main vortex **(Fig. 18.30)**. How can a tornado become particularly wide, and how do wide ones evolve into *multiple-vortex tornadoes*? Tornadoes, as we've seen, consist of rapidly rotating, rising air. In some tornadoes, the extremely low pressure that develops at the base of the vortex can draw air downward within the tornado core. In other words, a downdraft forms aloft within the tornado circulation and descends, inside the tornado, toward the ground. This process pushes the sides of the tornado outward, so that the tornado becomes wider. When the downdraft reaches the ground, the difference in wind direction between the tornado's downdraft outflow and the rotating winds causes the main tornado to break down into several smaller vortices moving around the main tornado. The most violent winds in the tornado occur in these suction vortices, because the speed of the wind in the main tornado is enhanced by the rotational speed of the suction vortices. The most destructive and intense tornadoes are nearly always multiple-vortex tornadoes, which have long lifetimes and produce particularly wide swaths of devastation.

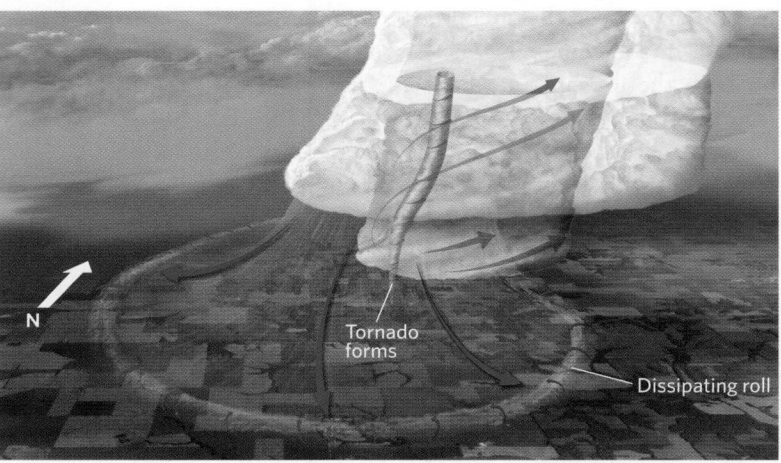

(c) The air roll tightens into a tornado and protrudes from the base of the wall cloud.

Non-supercell Tornadoes

The largest tornadoes occur in association with supercells, but not all tornadoes form in such storms. Tornadoes can also develop along squall lines and in hurricanes. Such **non-supercell tornadoes** develop where significant *horizontal wind shear* has developed. This means that air on one side of a front is moving horizontally in a direction opposite to the air on the other side.

In the case of a squall line formed along a northeast-southwest-trending cold front in the northern hemisphere, for example, the wind on the warm side of the front blows toward the northwest, whereas the wind on the cold side may be blowing toward the east **(Fig. 18.31a)**. If it is strong enough, shear between these two wind flows causes air to start rotating around a vertical axis, producing small vertical vortices **(Fig. 18.31b)**. The updraft of an overlying thunderstorm can pull a vortex upward,

FIGURE 18.30 Formation of a multiple-vortex tornado.

(a) A small tornado, consisting of a single vortex, forms beneath the wall cloud at the base of a supercell.

Updrafts form along the side.

A downdraft forms in the center, with eddies along its edge.

(b) A downdraft forms in the middle of the tornado and pushes the sides of the tornado outward.

(c) The central downdraft reaches the ground, so the whole tornado is now almost as wide as the wall cloud.

Map view at ground looking down

1 km

(d) Within the overall huge tornado, a number of short-lived but very intense suction vortices form. These vortices contain the most intense winds.

FIGURE 18.31 Formation of non-supercell tornadoes along a squall line.

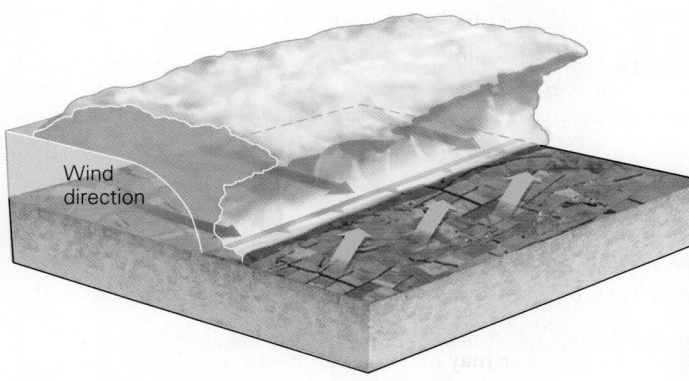

Wind direction

(a) Horizontal wind shear along a front.

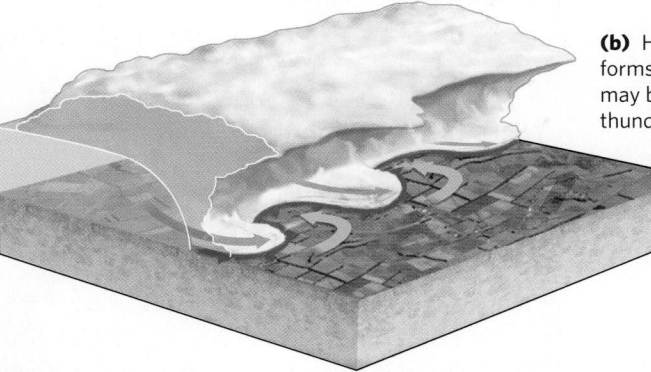

(b) Horizontal wind shear forms small vortices that may be pulled upward by thunderstorms.

(c) Vortices can form at several locations, as in this row of tornadoes.

causing it to stretch vertically and spin faster to become a short-lived tornado. This process can take place simultaneously at many locations along a front—in rare cases, producing a line of tornadoes (Fig. 18.31c). The winds within these tornadoes are generally weaker than those of supercell tornadoes, but they can cause local damage.

Defining the Intensity of a Tornado

The damage caused by a tornado depends on wind velocity, so it would be convenient if we could classify tornadoes based on their wind speed. However, since tornadoes are relatively small, instruments that can measure wind speed rarely lie in the path of a tornado, and they would likely be destroyed anyway if a tornado struck them. To address this classification challenge, Ted Fujita, a professor at the University of Chicago, proposed classifying the intensities of tornadoes on the basis of the damage they caused. His system, published in 1971, came to be known as the *Fujita scale*. In 2007, a more accurate version of the Fujita scale, called the **Enhanced Fujita (EF) scale**, became the standard in use today (Table 18.1). The EF scale takes into account laboratory studies that better constrain the relationship between wind speed and observed damage.

The EF scale divides tornadoes into six categories, labeled EF0 through EF5. All tornadoes can be dangerous to people who are not in adequate shelters (Box 18.2). But the damage produced by an EF0 tornado is relatively minor when compared with that produced by an EF5 tornado whose wind speeds exceed 322 km/h (200 mph). An EF5 tornado causes catastrophic damage; it can sweep even the sturdiest house away and can rip up the ground like a monstrous bulldozer.

Use of the EF scale still presents some challenges. First, the EF rating does not characterize the size of a tornado, but in general, more intense tornadoes are larger. Second, as a tornado evolves, its EF rating may change. For example, it may start out by causing EF1 damage, accelerate to causing EF5 devastation, and then weaken to an EF1 before dissipating entirely. Its EF rating in the record books is based on the worst damage the tornado causes along its track (Earth Science at a Glance, pp. 704–705). Finally, not all tornadoes cross through areas containing buildings that may be damaged. For example, even large tornadoes with powerful winds have been classified as EF0 or EF1 because they remained over open fields and, therefore, did little damage.

Take-home message...

A tornado is a rapidly rotating vortex of air that extends from the ground into the base of a severe thunderstorm. Most tornadoes form when a rotating cylinder of air along the boundary of the rear-flank downdraft of a supercell gets drawn into the updraft of the supercell's mesocyclone. As it's pulled up, the cylinder stretches, narrows, and spins faster. The Enhanced Fujita scale classifies tornado intensity based on the damage caused. Doppler radar can detect tornadoes.

Quick Questions

- What produces the visible funnel of a tornado?
- On what basis are tornadoes classified in the EF scale?
- How does a tornado warning differ from a tornado watch?

TABLE 18.1 The Enhanced Fujita Scale

EF rating	Wind (km/h)	Wind (mph)	Typical damage
EF0	105–137	65–85	*Minor damage:* Peels off some shingles; breaks branches; topples weak trees
EF1	138–177	86–110	*Moderate damage:* Severely strips roofs; overturns mobile homes; breaks windows
EF2	178–217	111–135	*Considerable damage:* Tears roofs off; shifts houses off foundations; destroys mobile homes; uproots large trees; flings debris and lifts cars
EF3	218–266	136–165	*Severe damage:* Destroys upper stories of well-built houses; severely damages large buildings; overturns trains; throws cars; debarks trees
EF4	267–322	166–200	*Extreme damage:* Completely levels well-built houses; cars and trucks thrown about
EF5	>322	>200	*Catastrophic damage:* Tall buildings collapse; structures made of reinforced concrete severely damaged; cars and trucks carried for more than a kilometer

BOX 18.2 ► A Deeper Look

Modern tornado detection and your safety

Tornadoes destroy property and landscapes, and they often injure or kill (Fig. Bx18.2a). On average, about 60 people per year die in tornadoes in the United States, but sadly, in some years, death tolls are much higher. For example, in 2011, the year of the Joplin tornado, 553 people lost their lives due to tornadoes. Most injuries and fatalities happen when people are struck by flying debris, but casualties also happen when vehicles carrying people are picked up and tossed in the air, or when roofs, walls, and trees collapse onto people.

Given the devastation wrought by tornadoes, accurate prediction of their arrival can be a key to survival. Fortunately, with Doppler radar, tornado signatures commonly appear clearly on a radar screen. Specifically, since flying debris scatters radar energy, a *tornado debris signature* appears as a bright spot on the screen at the tip of a hook echo (Fig. Bx18.2b). Doppler radar can also detect rotating winds because the side of a vortex moving toward the observer looks different from the part moving away. The rotating winds appear on the radar screen as bright green pixels, representing winds moving toward the radar installation, next to bright red pixels, representing winds moving away. Therefore, when a meteorologist sees a supercell thunderstorm with a hook echo on the southwestern side, rotation within the hook, and a debris signature at the tip of the hook, it's a sure sign that a tornado is on the ground and that a warning should go out.

The National Weather Service (NWS) Storm Prediction Center issues a tornado watch for specific areas when weather conditions favor tornado formation. A local NWS office issues a tornado warning either when observers have spotted a tornado visually or when one has been detected by Doppler radar. NWS warnings trigger weather alerts on radio, TV, and cell phones. In some communities, emergency managers activate warning sirens.

The NWS recommends that families, schools, and businesses prepare safety plans and hold regular tornado drills. In general, when a tornado warning goes out, you should move to a predetermined safe location, preferably a basement. If an underground space is not available, the safest choice is an interior room or a hallway on the building's lowest floor. In an interior bathroom, plumbing provides additional wall support. Take shelter with as many walls as possible between you and the outdoors, and avoid windows. Mobile homes are dangerous, as they may be crushed or rolled. Abandon your car, especially in an urban or congested area, and seek a sturdy shelter. If you are caught outdoors, move away from potential sources of airborne debris, and lie flat in the lowest spot available. In all cases, when severe weather threatens, early action can be the key to safety and survival.

FIGURE Bx18.2 Tornado damage.

(a) Above: A tornado leaves a swath of destruction. Below: During a strong tornado, trees snap off and even brick walls fall.

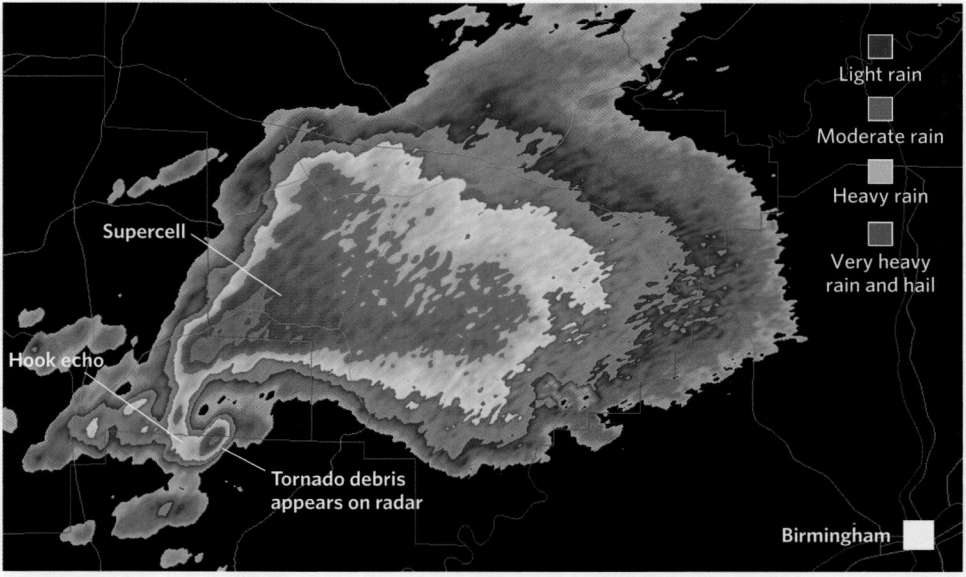

(b) Radar signature of a tornado that struck Alabama in 2013. The bright spot at the tip of the hook echo comes from debris thrown upward by the tornado.

Life Cycle of a Large Tornado

EF2 and EF3

EF0

EF1

EF0

EF1

EF2

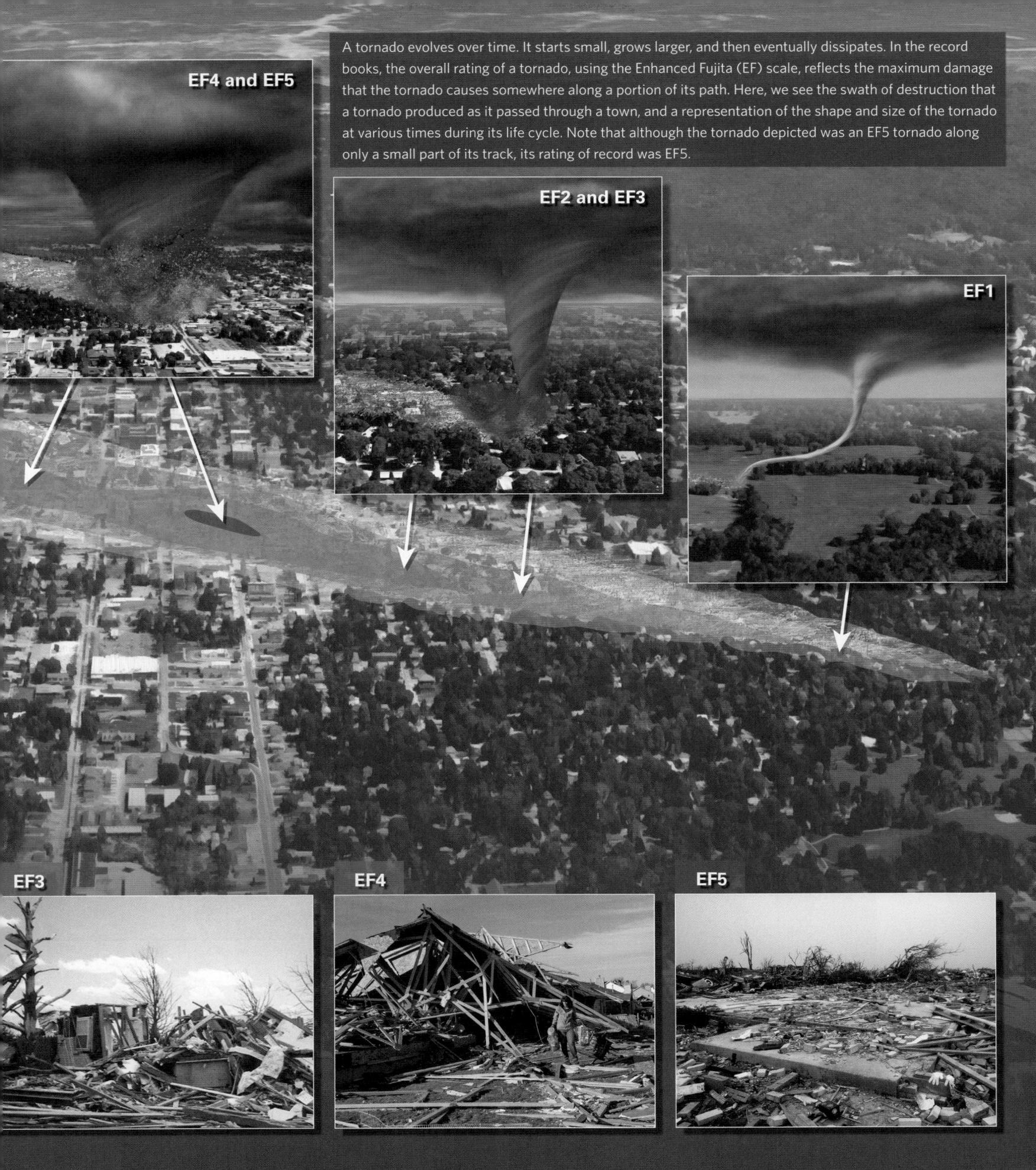

EF4 and EF5

EF2 and EF3

EF1

A tornado evolves over time. It starts small, grows larger, and then eventually dissipates. In the record books, the overall rating of a tornado, using the Enhanced Fujita (EF) scale, reflects the maximum damage that the tornado causes somewhere along a portion of its path. Here, we see the swath of destruction that a tornado produced as it passed through a town, and a representation of the shape and size of the tornado at various times during its life cycle. Note that although the tornado depicted was an EF5 tornado along only a small part of its track, its rating of record was EF5.

EF3

EF4

EF5

18.4 Mid-latitude Cyclones

A Name to Remember

Most people have heard of hurricanes, huge spiral-shaped storms that can span a region hundreds of kilometers to over a thousand kilometers wide. As we will see, hurricanes form only in tropical climates. Much larger swirling currents of air develop in temperate climates, at latitudes between 30° and 60° north or south of the equator, where subtropical air masses and polar air masses interact. In the northern hemisphere, such currents are known as *cyclones* if they rotate counterclockwise, and as *anticyclones* if they rotate clockwise; in the southern hemisphere, cyclones rotate clockwise and anticyclones counterclockwise. As we discussed in Chapter 17, cyclones form in association with low-pressure centers, while anticyclones form in association with high-pressure centers, so the former are associated with stormy weather, while the latter are associated with fair weather.

In some cases, cyclones grow into immense storm systems that can span half of North America. These large cyclones, whose cloud cover when viewed from a satellite has a shape resembling that of a huge comma (Fig. 18.32a), are called **mid-latitude cyclones**, because of the latitudes in which they form. They are also known as *extratropical cyclones*, to emphasize that they develop outside of tropical latitudes.

The name "mid-latitude cyclone" may be unfamiliar to you because TV and radio meteorologists don't often use the term. But if you live in a temperate latitude you have likely endured their consequences many times.

The fronts along which many severe thunderstorms and associated tornadoes develop are commonly associated with mid-latitude cyclones. Such thunderstorms impact large population centers and agricultural areas in North America. Snow and ice storms that interrupt commerce during the winter are also a manifestation of mid-latitude cyclones. Indeed, the infamous "Storm of the Century," three deadly days of thunderstorms, tornadoes, and blizzards that impacted 40% of the American population in March 1993, was due to a mid-latitude cyclone. This weather event killed over 200 people, knocked out the power to 10 million homes, raked the east coast with storm surge and high waves, brought transportation to a standstill, and caused over $3 billion in damage.

Mid-latitude cyclones not only affect continental interiors, but they can also ravage coasts, where they have different local names. For example, a mid-latitude cyclone that develops over the northeastern Pacific Ocean, off the west coast of North America, is called a *Pacific cyclone*. When such storms reach the mountain ranges of the west coast, the moisture-laden air that they carry flows up the slopes of the mountains. This orographic lifting triggers precipitation in the form of torrential rains or heavy snows, depending on the air temperature and altitude. A mid-latitude cyclone formed along the east coast of North America is called a **nor'easter** because its strongest winds come from the northeast, due to the overall counterclockwise flow in the storm (Fig. 18.32b). Because nor'easters blow water and waves toward the shore, they can cause significant coastal damage.

To understand how mid-latitude cyclones can cause so much damage, we must first consider how they form

FIGURE 18.32
Examples of mid-latitude cyclones.

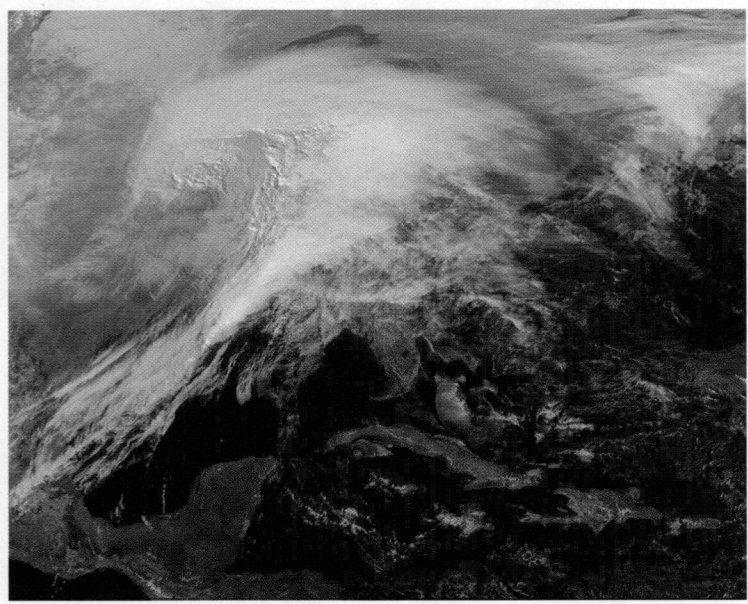

(a) A mid-latitude cyclone covers the upper midwestern and southeastern United States in February.

(b) A nor'easter moves up the east coast of North America.

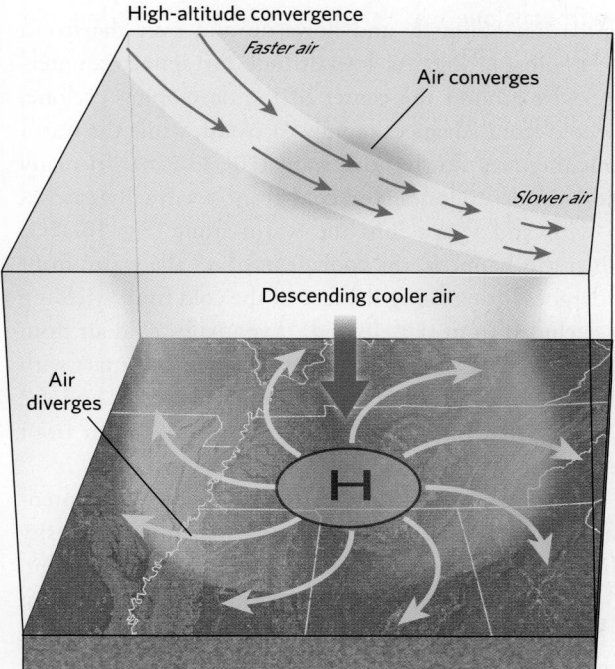

High-altitude convergence

Faster air

Air converges

Slower air

Descending cooler air

Air diverges

H

Low-altitude divergence

(a) Convergence aloft generates a surface high-pressure center.

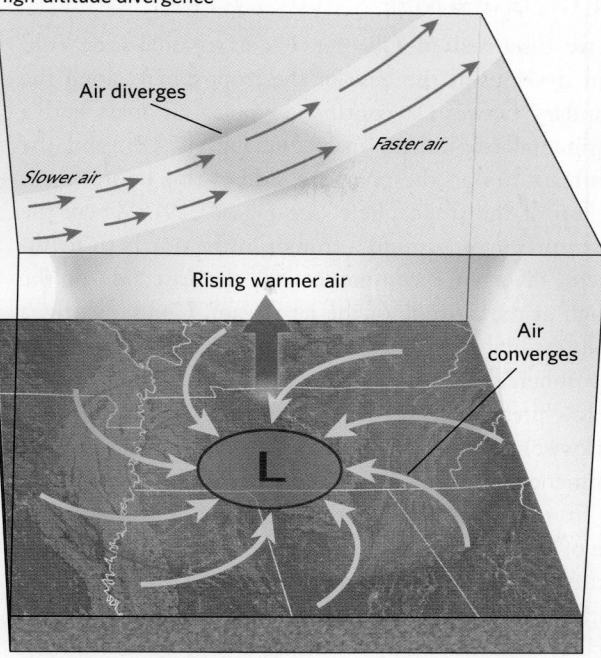

High-altitude divergence

Air diverges

Slower air

Faster air

Rising warmer air

Air converges

L

Low-altitude convergence

(b) Divergence aloft generates a surface low-pressure center.

FIGURE 18.33 The relationship between convergence and divergence in the jet stream and the development of high- and low-pressure centers at the surface.

and evolve. This, in turn, requires us to examine the relationship between air flow at the jet-stream level and the formation of high- and low-pressure centers in the lower atmosphere.

The Relation of High- and Low-Pressure Centers to the Jet Stream

In the previous chapter, we learned that heating and cooling of the Earth's atmosphere in specific locations lead to the formation of high-pressure and low-pressure centers at the Earth's surface. The flow of the polar-front jet stream can also contribute to the development of high- and low-pressure centers. Where horizontal air flow in the jet stream slows down, air converges, or "piles up," meaning that the flow carries more air into the locality (in this case, a segment of the jet stream) than it carries away. Such **convergence** of air at the top of the troposphere increases the mass within the column of air between the tropopause and the Earth's surface, so a high-pressure center develops at the base of the troposphere (**Fig. 18.33a**). Air cannot rise into the stratosphere, so it must descend from the convergence zone aloft down toward the high-pressure center at the Earth's surface below. As the descending air approaches the ground, it starts to flow outward from the high-pressure zone toward lower pressure regions. Due to the Coriolis and frictional forces (see **Fig. 17.8**), the air flow soon deflects and spirals outward, rotating in a clockwise sense, forming an *anticyclone*.

In other segments of the jet stream, the flow carries more air out of a locality than has entered that locality (**Fig. 18.33b**). This **divergence** of air, and the resulting air deficit at the top of the troposphere, causes the mass of the underlying column of air to decrease, so a low-pressure center develops at the base of the troposphere. To replace the deficit of air at jet-stream altitude, air from lower in the troposphere rises in and around the low-pressure center. As air approaches the low-pressure center, the Coriolis and frictional forces cause air to deflect, so that it rotates in a counterclockwise sense while spiraling inward, forming a *cyclone*. The high-altitude convergence and divergence zones arise from changes in jet-stream velocity at places where the jet stream curves around ridges and troughs, and where the jet stream accelerates and decelerates (**Fig. 18.34**).

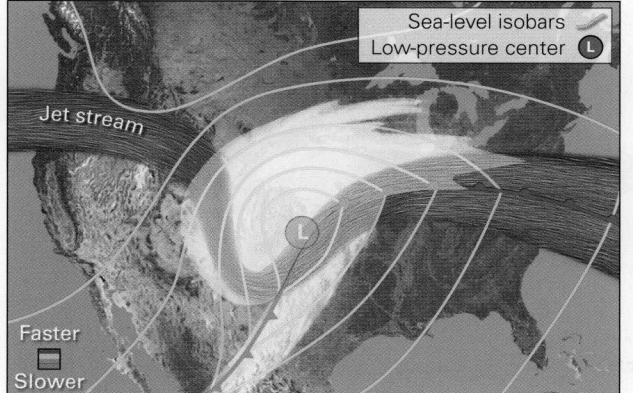

Sea-level isobars
Low-pressure center **L**

Jet stream

L

Faster

Slower

FIGURE 18.34 The relationship between the polar-front jet stream and the formation of a mid-latitude cyclone. Low pressure develops at the Earth's surface below a zone of divergence, which lies to the east of a trough and just north of the strongest winds in the jet stream. The isobars show how sea-level air pressure decreases toward the low-pressure center.

Life Cycle of a Mid-latitude Cyclone

As we discussed in Chapter 17, a regional-scale cold front develops at the base of the troposphere along the boundary between a continental polar air mass and a continental tropical air mass (see Fig. 17.18), and the jet stream lies on the poleward side of this boundary at the top of the troposphere (see Fig. 17.19). Where the jet stream curves around a trough in the northern hemisphere, divergence commences on the east side of the trough at the altitude of the jet stream. Consequently, a low-pressure center begins to develop at the base of the troposphere, beneath the strongest divergence. As soon as the low-pressure center starts to form, cyclonic (counterclockwise) flow of the air about the low-pressure center commences, and this flow immediately begins to distort the front (Fig. 18.35a). Specifically, on the northeastern side of the low-pressure center, the front starts to retreat northward, becoming a warm front. Meanwhile, on the southern side of the low-pressure center the front starts to advance southeastward, becoming a cold front. Initially, warm air to the southeast of the cold front tends to be flowing north to northeast, so it approaches the front at an angle. Thus, as the cold front advances, the warm air approaches obliquely and flows up and over the front, undergoing lifting. As low-altitude air spirals counterclockwise around the center of the developing cyclone, the cold front advances rapidly eastward, while the warm front migrates slowly northward (Fig. 18.35b). In many mid-latitude cyclones, the cold front eventually catches up with and wraps into the warm front (Fig. 18.35c). When this happens, the cool air north of the warm front undergoes lifting along the face of the cold front, yielding an occluded front (Fig. 18.35d). Eventually, cold air from the western side of the cold front comes into contact with cold air from the north side of the warm front, and the warm air that once lay to the south of the warm front becomes a layer that overlies both cold air masses.

A typical mid-latitude cyclone develops and intensifies rapidly, generally reaching its maximum intensity (the lowest air pressure at the low-pressure center) within 36–48 hours of initiation. The storm system can remain at maximum intensity for another day or two, but then it begins to weaken. This weakening, which may take about a week, happens when low-altitude convergence feeds more air into the air column in the lower troposphere than can be removed by jet-stream divergence in the upper

FIGURE 18.35 Evolution of a mid-latitude cyclone.

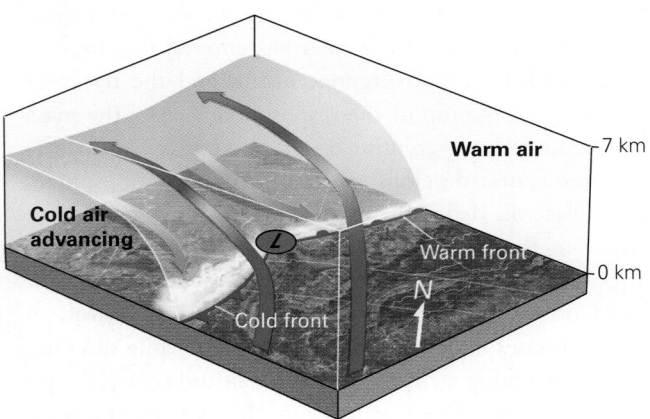

(a) Warm air initially flows upward over the fronts.

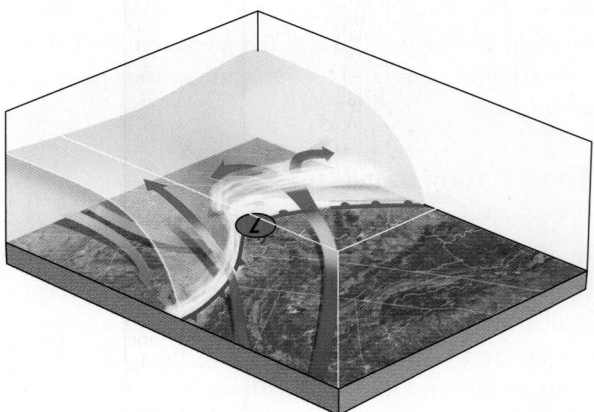

(b) The cold front advances southward and eastward around the center of low pressure.

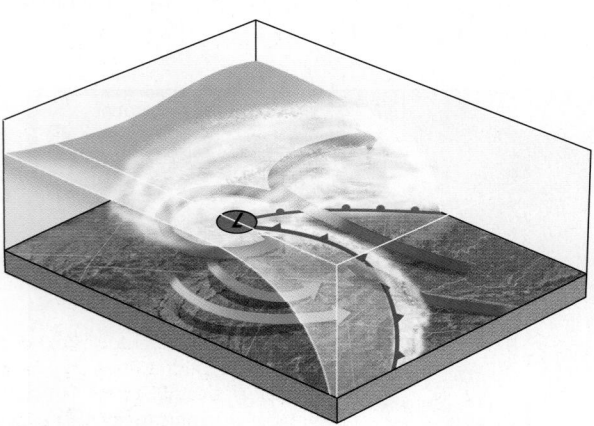

(c) Eventually, the cold front reaches the warm front.

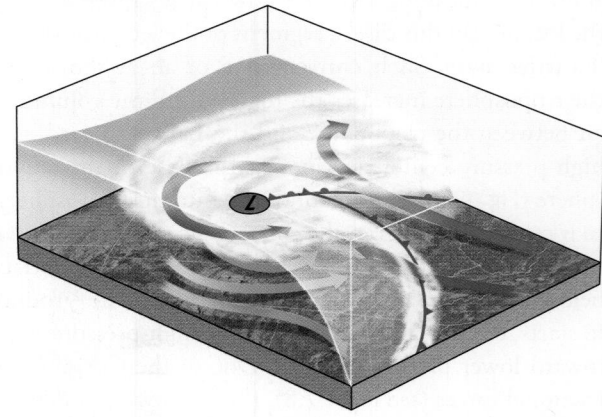

(d) The cold front overtakes the warm front, creating an occluded front.

troposphere. The addition of air to the column causes the air pressure to rise at the center of the cyclone and, therefore, causes the surrounding pressure gradient to diminish. When the low-pressure center weakens sufficiently, the storm dissipates. Overall, a mid-latitude cyclone can survive as a distinct circulation for up to two weeks, during which it drifts eastward along with the trough with which it is associated. As a consequence, a single cyclone can cross a large part of a continent or ocean.

The Characteristic Comma Shape of Mid-latitude Cyclones

The pressure gradients that develop in a mid-latitude cyclone generate strong winds that flow counterclockwise (in the northern hemisphere) at ground level within the cyclone's overall spiral flow (Fig. 18.36a). Convergence within the low-pressure system, and lifting of air along and over the system's fronts, produce broad areas of cloud cover and precipitation in the shape of a comma (Fig. 18.36b). The "tail" of the comma lies over the cold front and typically consists of a line of precipitating clouds or, in some cases, thunderstorms. In many cyclones, a frontal squall line will develop along the cold front, and some of the thunderstorms may become severe. The "head" of the comma surrounds the low-pressure center's west and north sides, lying over the occluded front and the warm front. It, too, produces large amounts of precipitation. In warmer weather, the precipitation falls as rain, but when the weather is cold enough, the precipitation falls as snow or freezing rain. During fall, winter, and spring, rain may be falling along the comma's tail, while snow falls around the comma's head, due to the poleward decrease in temperature characteristic of mid-latitude regions (Fig. 18.36c). Next, we'll turn our attention to the unique weather features associated with mid-latitude cyclones that form in winter.

FIGURE 18.36 Features of a large mid-latitude cyclone.

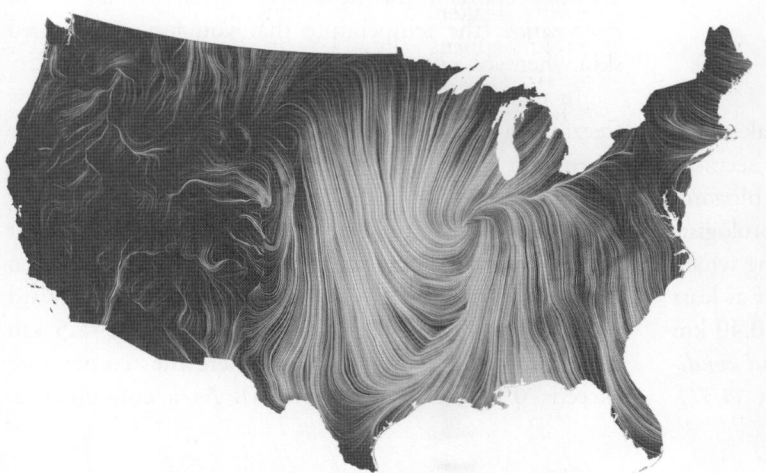

(a) A map showing near-surface wind streamlines (the path of the wind at an instant in time) in and around a mid-latitude cyclone.

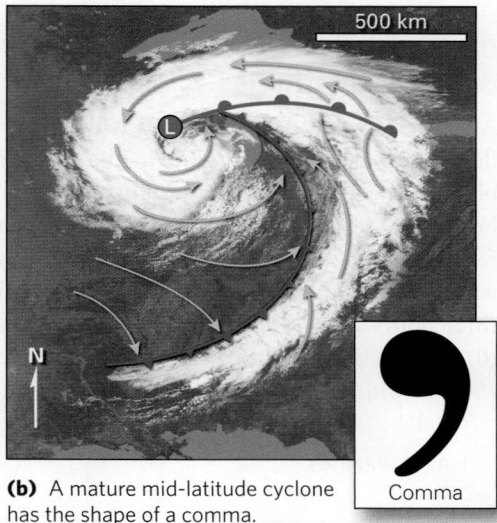

(b) A mature mid-latitude cyclone has the shape of a comma.

Comma

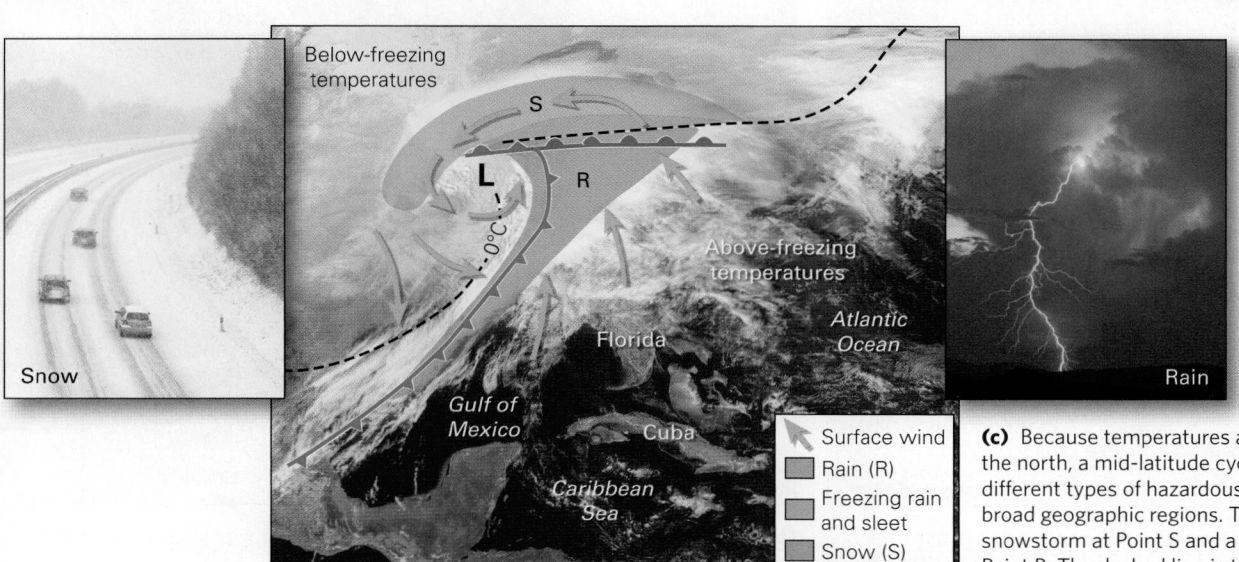

(c) Because temperatures are cooler to the north, a mid-latitude cyclone can bring different types of hazardous weather over broad geographic regions. The insets show a snowstorm at Point S and a thunderstorm at Point R. The dashed line is the 0°C isotherm.

FIGURE 18.37 During a blizzard, visibility falls to nearly zero.

Note that the definition of a blizzard does not specify an amount of snowfall—in fact, a blizzard does not necessarily produce a lot of new snow, although some blizzards can drop 30 cm (1 ft) of snow or more. Heavy snowfall in a blizzard can accumulate quickly, and when blown by the wind—either during or after a snowfall—can build into high *snowdrifts* (dunes made of snow). The combination of snowfall, drifting, and decreased visibility can force airports to close, trap cars on highways, and prevent people from leaving their homes.

The extreme winds in a blizzard can also cause people to suffer from *frostbite* (freezing skin) or *hypothermia* (lowering of body temperature), for the heat loss that people experience when it's windy is greater than that at the same air temperature in still air. For example, an air temperature of –7°C (20°F) feels like –16°C (4°F) when the wind blows at 32 km/h (20 mph). This **wind chill** happens because flowing cold air removes heat more efficiently from the human body than does calm cold air. Weather reports in the media often specify the *wind-chill temperature*, the temperature that you feel on exposed skin when you go outside, so that you can be prepared.

Blizzards and Ice Storms during Mid-latitude Cyclones

A gentle snowfall brings delight, as the shiny flakes that float down from the sky glint in the light and seem to blanket the surroundings with calm and quiet. A blizzard, however, is anything but calm and quiet. Meteorologists formally define a **blizzard** as a snowstorm during which winds exceed 56 km/h (35 mph) for a period of at least three hours and visibility decreases to less than 0.40 km (0.25 mi). Blizzards sometimes lead to *whiteout conditions*, in which visibility drops to nearly zero (Fig. 18.37).

In regions where the temperature hovers just below freezing, a mid-latitude cyclone may produce an **ice storm**, an event in which precipitation lands on the ground, trees, houses, and cars as water but instantly freezes to ice. Such **freezing rain** produces an ice glaze that makes roads extremely slippery. Ice storms happen when snow falling from high in the clouds passes downward into a lower atmospheric layer—typically 0.5–1.5 km (0.3–1 mi) above the ground—where the temperature exceeds 0°C (32°F) (Fig. 18.38a). As a consequence,

FIGURE 18.38 Ice storms are another winter hazard associated with mid-latitude cyclones.

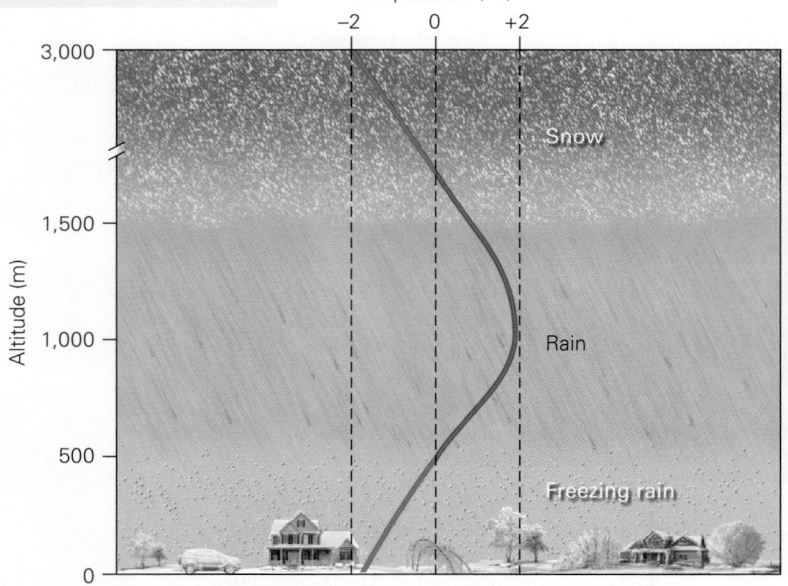

(a) When surface temperatures are below freezing, but a layer of air aloft is above freezing, snow from high in the clouds can melt into rain and then freeze when it hits the ground.

(b) The weight of accumulated ice can cause trees and power lines to collapse.

FIGURE 18.39 A mid-latitude cyclone acts like a complicated series of conveyor belts, bringing warm air northward, cold air southward, high-altitude air downward, and low-altitude air upward as the fronts evolve.

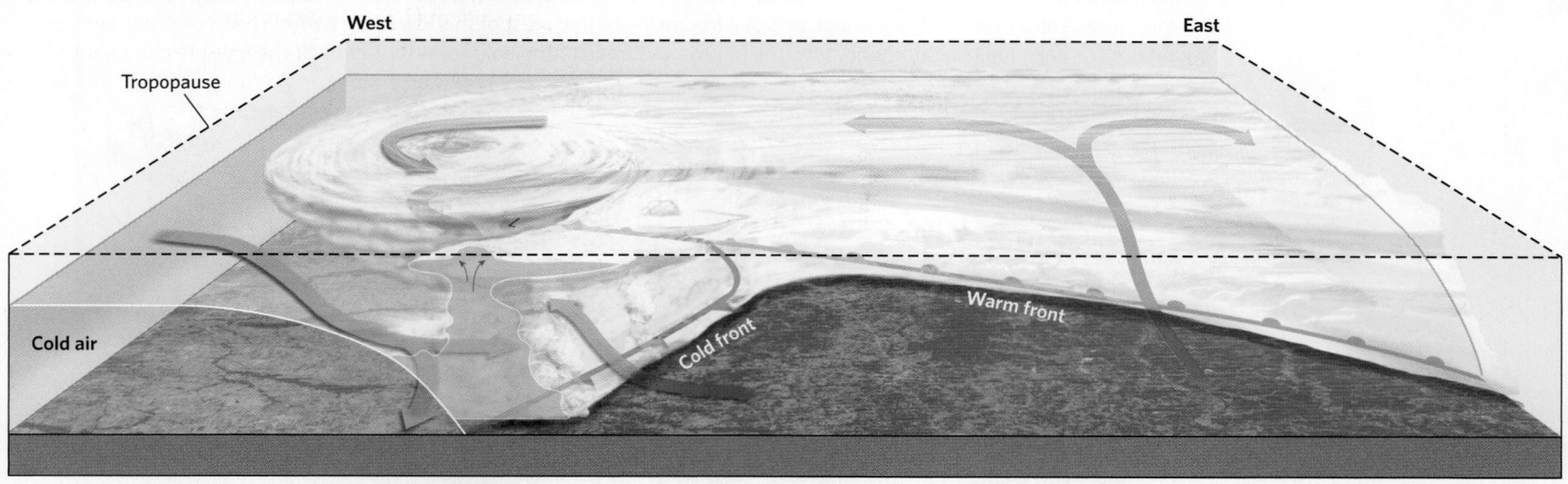

(a) The complex circulation in a mid-latitude cyclone.

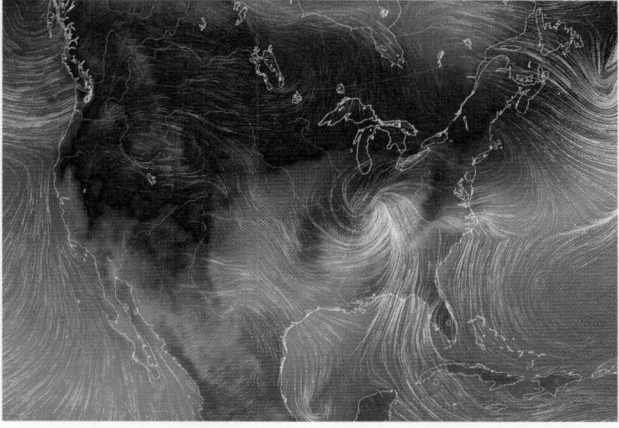

(b) A map of wind streamlines (thin white lines) and temperature at ground level (colors). Note how air of different temperatures is transported north and south by the cyclone's circulation.

the snowflakes melt and become very cold raindrops. If a *temperature inversion* exists in the lowest part of the atmosphere—meaning that the temperature increases with altitude—the temperature in the layer of air below an altitude of about 0.5 km can remain below freezing. If this happens, the raindrops refreeze when they strike the ground. The ice that builds up on trees and power lines can become so heavy that it causes branches to snap off or, in some cases, whole trees to topple, and may cause power lines to collapse, producing blackouts that last for days **(Fig. 18.38b)**.

Redistribution of Heat by Mid-latitude Cyclones

Before finishing our discussion of mid-latitude cyclones, let's consider the role that these storms play in redistributing atmospheric heat across the Earth. Prior to the formation of a mid-latitude cyclone, warm, humid air lies near the Earth's surface at subtropical latitudes, while cold air lies near the surface at high latitudes. In addition, cold air occurs high in the troposphere. When a mid-latitude cyclone forms, warm air flows northward as well as upward, while cold air flows southward as well as downward **(Fig. 18.39)**. This conveyor-belt-like redistribution of heat reduces the temperature difference between polar and subtropical regions, as well as between the lower and upper troposphere. In this manner, each cyclone contributes to a never-ending process of rebalancing the heat distribution in an atmosphere that continually becomes unbalanced due to uneven solar heating between the polar and tropical regions of the Earth.

Forecasting Mid-latitude Cyclones and Their Associated Weather

How do meteorologists forecast the weather? And how do they come up with statements like "Tomorrow there is a 60% chance of rain" or "Tuesday will be partly cloudy"? Weather forecasting is a complicated process that begins with the collection of weather data worldwide from airports and cities, weather balloons, satellites, radar systems, and other instruments. The data flow continuously into national and international data centers, such as the National Centers for Environmental Prediction in College Park, Maryland. These centers process the data using supercomputers, arranging the information on orderly grids that cover all or parts of the planet. Numerical models, sophisticated computer codes consisting of mathematical equations that describe the forces that govern atmospheric motion as well as other factors that control the behavior and evolution of the atmosphere, use these

FIGURE 18.40 A typical weather map, from a newspaper, TV, or website, combines many types of data. This map shows the predicted distribution of temperature and precipitation, where high-pressure and low-pressure centers lie, and the location of fronts. Note the association of key weather features with the blue band, the track of the jet stream. For example, the low-pressure center over Indiana lies just east of a trough. Note that west of this low is the comma head of a mid-latitude cyclone. A frontal squall line of thunderstorms follows the trace of the cold front that ends at the low-pressure center.

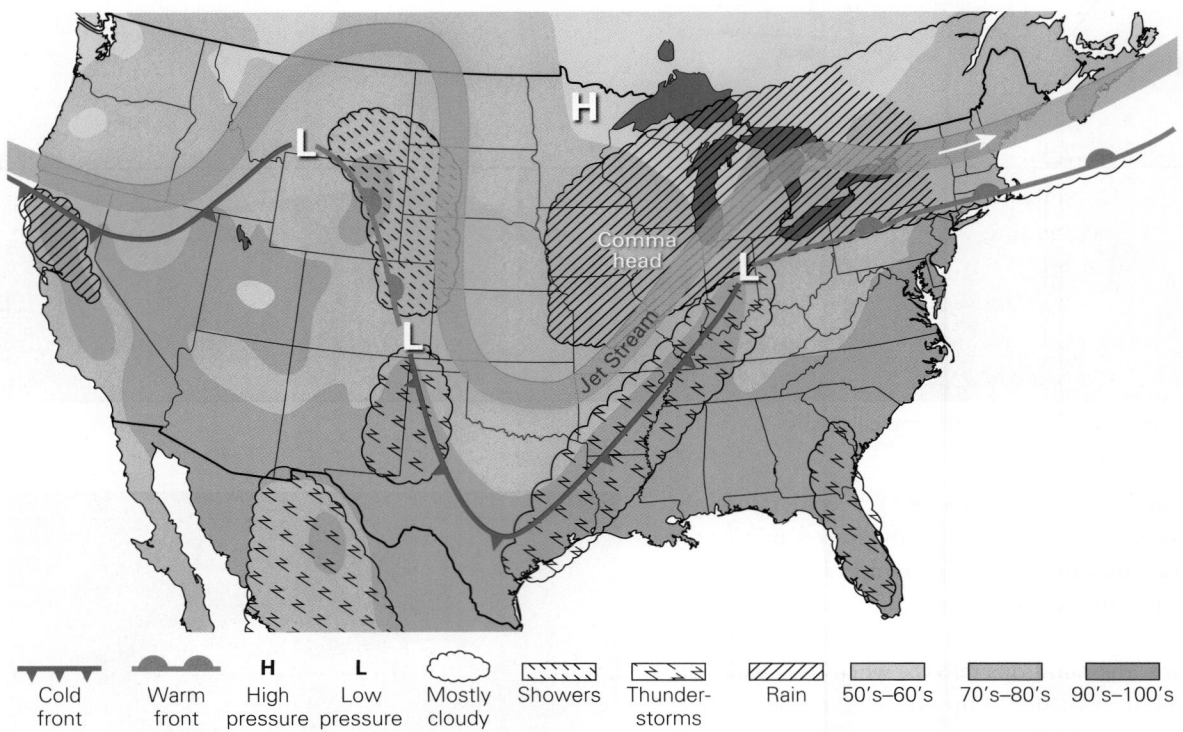

| Cold front | Warm front | H High pressure | L Low pressure | Mostly cloudy | Showers | Thunder-storms | Rain | 50's–60's | 70's–80's | 90's–100's |

data to calculate the future state of the atmosphere as far out as 14 days. Such predictions can be quite accurate within a three-day range, but models can help estimate general weather conditions for even longer intervals.

So how does all that relate to the chance of rain in Cincinnati at noon on Tuesday? To pinpoint that type of information, meteorologists compare, using statistics, data from the current model forecast with thousands of past model forecasts and historical data about rainfall in Cincinnati (and all other weather data relevant to people living at thousands of other locations!). These model output statistics provide the information you actually receive on your television and internet-based forecasts. For example, when a meteorologist states that there is a 60% chance of rain tomorrow afternoon in Cincinnati, the forecaster means that, historically, rain occurred 60% of the time that the weather conditions forecast by the model were present in the afternoon in the Cincinnati area. A meteorologist might forecast clear, partly cloudy, mostly cloudy, or cloudy weather (sometimes meteorologists use sunny and partly sunny for daytime hours—these terms have the same meaning as clear and partly cloudy). These predictions, again, result from model output statistics, which provide the forecaster with the percentage of

the time when clouds are expected at a specific location. Forecasters always have the option of "tweaking" the model predictions based on their local knowledge, but their forecasts always start with the computer-based predictions and model output statistics. The data are also compiled to create the weather maps and forecasts that you see on television, on the internet, or in newspapers (Fig 18.40).

Take-home message...

Mid-latitude cyclones are large, counterclockwise-rotating systems that can result in hazardous weather across the mid-latitudes. They form in association with low-pressure systems and generally move eastward. Their most dangerous weather occurs along fronts, which move as the result of air flow within the cyclone.

Quick Questions

- What is the shape of a mid-latitude cyclone, as seen from a satellite?

- What type of front is typically found south of the low-pressure center of a mid-latitude cyclone?

- How do mid-latitude cyclones play a role in redistributing atmospheric heat?

BOX 18.3 **Putting Earth Science to Use**

Interpreting weather maps

Weather impacts us daily, whether it's a hot day outdoors, a brief rain shower requiring an umbrella, heavy snow that clogs a driveway, or severe weather that threatens a neighborhood. Today, most people get weather information from television or the internet, presented in the form of maps (Fig. Bx18.3a, b). Armed with the information you've learned in this and previous chapters, you should now be able to interpret these maps and understand the weather impacting your life. The maps of Figure Bx18.3, which depict different aspects of the weather at the same time, illustrate a mid-latitude cyclone, whose low-pressure center lies over Lake Michigan. Note that the pattern of isobars (showing air pressure at sea level) forms the basis for identifying the low-pressure center.

Let's examine the maps and see what they tell us about weather at particular locations. Note that a cold front extends southward from the low-pressure center. Across the front, wind shifts direction (indicated by a change in the symbols used for indicating wind direction; see Fig. 16.12), but temperature changes only slightly. As a result, clouds and some precipitation happen right along the frontal boundary, but no severe thunderstorms develop. So, if you lived in Ohio, Indiana, or Tennessee, you would experience a few light showers, followed by a slight temperature drop and a wind shift from southerly to westerly as the front passed by. If you lived farther north, in southern Minnesota or Wisconsin, your weather experience would be very different. There, temperatures are below freezing, snow is falling, and a much stronger pressure gradient exists. This gradient drives strong winds, which have generated blizzard conditions. Farther east, north of the warm front, the pink colors in Figure Bx18.3b indicate mixed precipitation, either rain mixed with snow, or possibly freezing rain. Given your knowledge at this point in the chapter, you should now be able to look at these maps and give a description of the weather at any given location.

FIGURE Bx18.3 Interpreting weather maps.

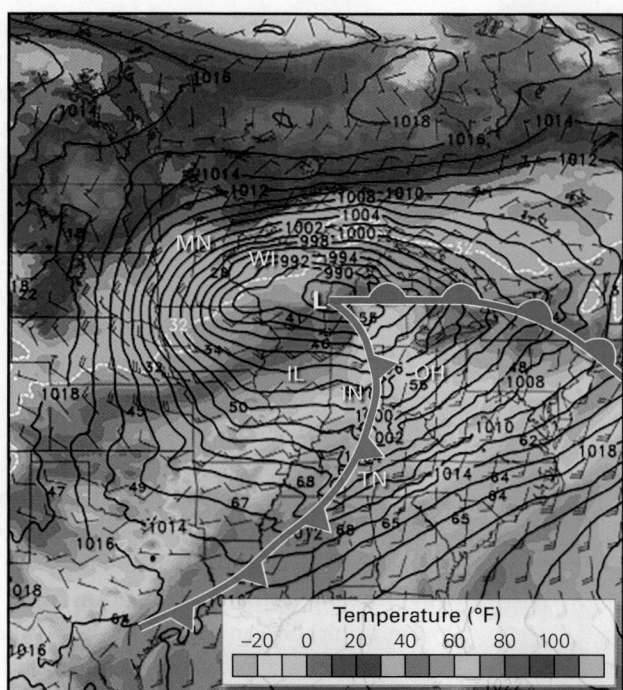

(a) A weather map of the eastern United States and Canada showing isobars of sea-level pressure (black lines), temperatures (colors), and fronts.

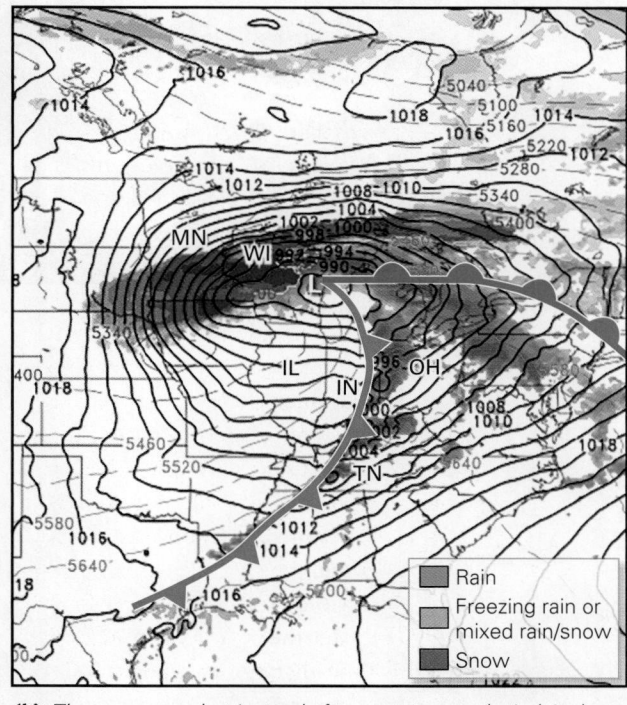

(b) The same area, but instead of temperature and wind, it shows precipitation type.

FIGURE 18.41 Destructive hurricanes in 2017.

(a) Flooding in Port Arthur, Texas, near Houston, during Harvey.

(b) Buildings destroyed by Irma in the US Virgin Islands.

(c) Power lines downed by Maria along a street in Puerto Rico.

18.5 Tropical Cyclones: Hurricanes and Typhoons

On August 26, 2017, Hurricane Harvey slammed into the Texas coastline. The swirling mass of clouds, wind, and rain covered an area of more than 400 km (250 mi) in diameter, and caused wide-spread damage near the shore. Once over land, the storm drifted along the coastline until it stalled over Houston. There it remained for days, and during that time, dumped up to 150 cm (60 in) of rain, setting a US record for rainfall from a single storm. Water filled low areas and submerged over a quarter of the metropolitan area (Fig. 18.41a). Harvey killed 108 people and displaced thousands more. It caused damage to infra-structure and to commercial, public, and private property with economic losses totaling $105 billion.

Unfortunately, the meteorological disasters of 2017 were far from over. As Harvey *dissipated* (weakened and lost its structure) over Louisiana, another storm grew on the eastern side of the Atlantic. As regional winds carried this storm westward, its *intensity* increased (meaning, its winds accelerated and its rainfall rates increased), and it became Hurricane Irma, one of the strongest hurricanes ever observed over the Atlantic. Irma packed **sustained winds**—winds that maintain their velocity for over a minute, as opposed to gusts, which may last for just seconds—as high as 300 km/h (185 mph). These winds, together with pounding waves and torrential rain, left a trail of destruction across several Caribbean islands (Fig. 18.41b). Irma eventually crossed the Florida Keys and damaged Florida's west coast before dissipating. Immedi-ately following Irma, yet another storm formed over the Atlantic and grew to become Hurricane Maria. Maria slammed into Puerto Rico and Dominica, causing these islands' worst-ever natural disasters. Parts of Puerto Rico remained without power more than a year later, and the economic losses to Puerto Rico and the nearby US Virgin Islands exceeded $90 billion (Fig. 18.41c).

As the deadly storms of 2017 made clear, hurricanes are the most destructive and costliest storms on the Earth. What is a hurricane? Meteorologists consider hurricanes to be a type of **tropical cyclone**, a term that applies to any rotating spiral-shaped storm that originates over warm tropical ocean waters. In the northern hemisphere, these storms rotate counterclockwise, and in the southern hemisphere, they rotate clockwise, as a consequence of the Coriolis force (see Chapter 17). Not all storms that develop in the tropics are cyclones, however. Specifically, meteorologists refer to a single thunderstorm that devel-ops over tropical oceans as just "a thunderstorm," and to a cluster of thunderstorms that have begun to interact with one another as a *tropical disturbance*. Only when a

TABLE 18.2 The Saffir-Simpson Scale for Hurricane Intensity

Category	Sustained wind speed	Types of damage resulting from wind
1	74–95 mph 64–82 knots 119–153 km/h	**Dangerous winds produce some damage:** Well-constructed frame homes could have damage to roof shingles, vinyl siding, and gutters. Large branches of trees snap, and shallowly rooted trees may topple. Some power outages occur.
2	96–110 mph 83–95 knots 154–177 km/h	**Extremely dangerous winds cause extensive damage:** Well-constructed frame homes sustain major roof and siding damage. Many shallowly rooted trees are snapped or uprooted, blocking roads. Near-total power outages occur.
3	111–129 mph 96–112 knots 178–208 km/h	**Devastating damage occurs:** Well-built frame homes may incur major damage or removal of roof decking. Large trees snap or are uprooted. Numerous roads become blocked. Electricity and water are unavailable for days to weeks.
4	130–156 mph 113–136 knots 209–251 km/h	**Catastrophic damage occurs:** Well-built homes sustain severe damage. Most trees snap or are uprooted, and most power lines are downed. Debris isolates residential areas, and power outages last weeks to months. The area becomes temporarily uninhabitable.
5	≥157 mph ≥137 knots ≥252 km/h	**Total catastrophic damage occurs:** Most homes are destroyed. Only strongly reinforced buildings remain standing. Debris isolates large areas, and power outages last for weeks to months. Most of the area remains uninhabitable for weeks or months.

low-pressure center develops, at the Earth's surface, within a tropical disturbance does it become a tropical cyclone. Weak tropical cyclones, with sustained winds of less than 63 km/h (39 mph), are *tropical depressions*, and those with sustained winds between 63 and 118 km/h (39–73 mph) are **tropical storms**. Only if a tropical storm strengthens to the point that it has sustained winds of over 119 km/h (74 mph) does it formally become a **hurricane**. Note that meteorologists use the name *hurricane* only for such storms in the North Atlantic or eastern Pacific. If the same type of storm forms in the northwestern Pacific, it is called a **typhoon**, and if it forms over the Indian Ocean or anywhere in the southern hemisphere, it is simply called a **cyclone**. For simplicity, we'll generally refer to strong tropical cyclones as hurricanes in this chapter, unless we are discussing a particular storm that did not form in the North Atlantic or eastern Pacific.

In a hurricane, only the central region, within a circular region 200 km (120 mi) in diameter or less, hosts hurricane-force winds. But the entire storm, in which strong winds spiral around the center and dense clouds hide the sky, may cover a circular area with a diameter between 500 and 1,500 km (300–900 mi). About a hundred tropical cyclones originate every year over the world's tropical oceans and seas. The World Meteorological Organization assigns names to tropical storms worldwide. The storms retain those names, whether or not they grow into hurricanes. The first tropical storm of the season gets a name that starts with the letter A, and each successive storm gets a name that starts with the next letter of the alphabet, skipping some letters, such as Q, that

are uncommon for names. If the number of hurricanes in a season exceeds the letters available in the alphabet, a second set of names starts over with the letter A. And if a particular hurricane causes historic damage, its name is retired. To communicate a hurricane's severity to the public, meteorologists use the **Saffir-Simpson scale**, which rates hurricanes based on their maximum sustained wind speed (Table 18.2). In this scale, the strongest hurricanes are Category 5.

What's Inside a Hurricane?

When viewed from space, hurricanes appear as a broad sheet of cirrus clouds at the level of the tropopause (Fig. 18.42a). A nearly cloud-free circular area, the top of the hurricane's **eye**, lies at the center of the storm. In three dimensions, a hurricane's eye tapers downward, extending from the top of the storm down to the low-level clouds located at an altitude of less than 1 km (0.6 mi) above sea level (Fig. 18.42b). At its base, the eye ranges between 10 and 65 km (6–40 mi) in diameter; the top of the eye is roughly twice the diameter of the base. The **eye wall**, a ring of dense, rapidly swirling clouds, surrounds the eye. The eye wall, which typically ranges between 10 and 20 km (6–10 mi) in thickness as measured horizontally, hosts the fastest winds of a hurricane. Air spirals upward within the eye wall, slowing with altitude, so that at high altitude in the eye wall the winds are not as fierce (Fig. 18.42c, d). As a very intense hurricane evolves, the eye wall contracts in diameter, while a second eye wall forms around the first. The first eye wall then dissipates as the second one contracts, a process called *eye-wall replacement*.

Anyone who says they're not afraid of a hurricane is either a fool or a liar, or a little of both.

—ANDERSON COOPER (AMERICAN NEWSCASTER, 1967–)

FIGURE 18.42
Structure of a hurricane.

(a) Hurricane Irma (2017), as viewed from space, near its time of maximum intensity.

(b) The eye and eye wall of a hurricane, as viewed from an aircraft inside the eye.

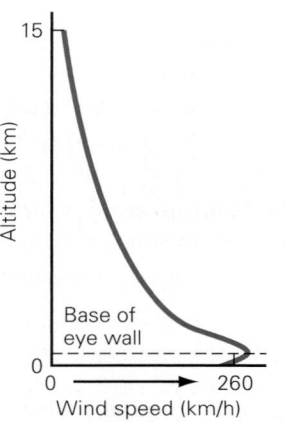

(c) Variation in eye wall wind speed with altitude.

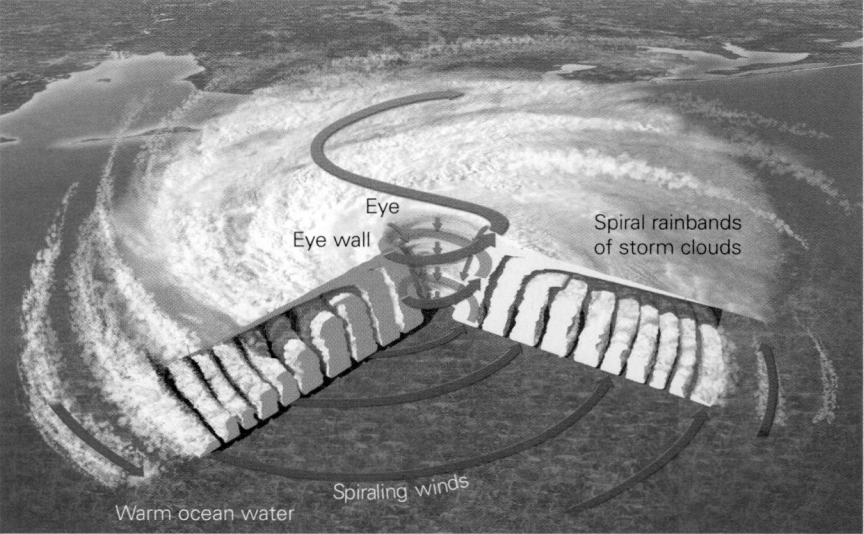

(d) Spiral rainbands surround the eye wall of a hurricane.

Because of the importance of pressure gradients in controlling wind velocity, a reading of sea-level air pressure in the eye of a hurricane reflects the intensity of the storm. The lowest air pressure at sea level ever measured (870 mb) occurred beneath the eye of Typhoon Tip, in 1979. The resulting gradient between pressure at the eye's center and pressure surrounding the storm caused Tip's sustained winds to reach an amazing 305 km/h (190 mph)—equivalent to those of an EF4 tornado, but covering a much broader area.

Spiral rainbands, consisting of tall, aligned thunderstorms, curve outward from the eye to define a spiral shape when the storm is viewed from above (see Fig. 18.42d). Note that a rainband is not a continuous arcing line of clouds, but rather consists of a great many individual thunderstorms. As each of these storms evolves, it may send an overshooting top that rises above the layer of cirrus clouds that lies over the top of a hurricane; when viewed in a time-lapse video taken by a satellite, the overshooting tops look like short-lived bubbles pushing into the stratosphere above the hurricane. Torrential downpours, and rare tornadoes, occur beneath the rainbands. The overall air flow in a rainband carries clouds and thunderstorms along the curve of a spiral arm, toward the eye wall. The heaviest rain falls from the eye wall, and rainfall rates decrease outward in a hurricane, but very heavy rain can still be falling in rainbands as far as a few hundred kilometers from the eye. In the gaps between rainbands, there's less rain or no rain, and no rain at all falls beneath the eye (Fig. 18.43a). Air rises in updrafts within rainbands and slowly sinks in downdrafts in the regions between rainbands (Fig. 18.43b). At any given time, a line

FIGURE 18.43 Images of the eye wall and spiral rainbands in a hurricane.

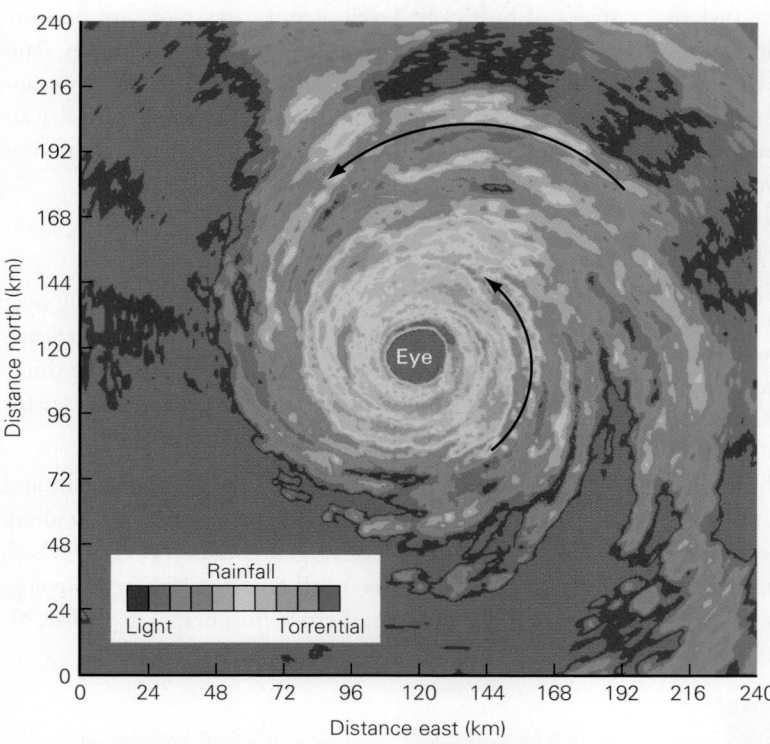

(a) A radar reflectivity map view of a hurricane as viewed from above. Greater reflectivity means more rain, so on this image, the most intense rain is in the red area along the north side of the eye.

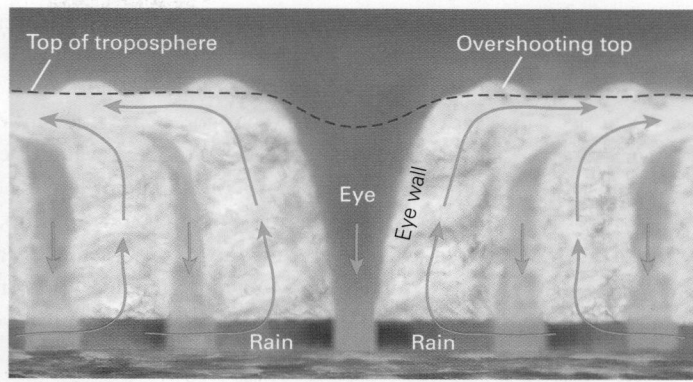

(b) A simplified cross section through the center of a hurricane shows the pattern of air flow within and between the spiral rainbands.

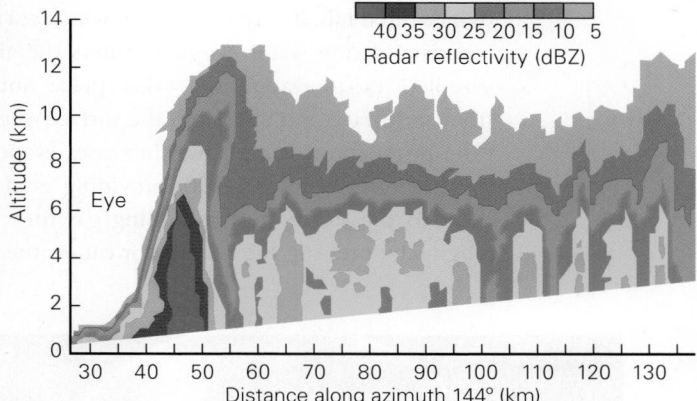

(c) A radar reflectivity cross section through Hurricane Andrew (1992) shows that rain occurs in narrow bands, separated by regions of less or no rain. The heaviest rain falls under the eye wall.

drawn outward from the eye to the margin of a hurricane may cross several rainbands, depending on the size of the storm (Fig. 18.43c).

A hurricane drifts along with the regional winds in its environment. This forward motion typically takes place at a speed that averages 20–35 km/h (12–22 mph), although speeds of forward motion exceeding 100 km/h (62 mph) have been recorded. Some hurricanes stall, causing long-lasting downpours leading to flooding beneath the storm, as was the case with Hurricane Harvey, discussed earlier. In 2019, Hurricane Dorian, one of the strongest hurricanes to develop in the Atlantic, stalled near Grand Bahama, and its winds ravaged the island for 24 hours.

Where Does the Energy in a Hurricane Come From?

Hurricanes develop only over tropical oceans with water temperatures above 26°C (79°F) down to a depth of at least 60 m (200 ft). That's because it takes thermal energy to drive the air flow in a hurricane, and only warm ocean waters can provide enough heat to accelerate winds to hurricane speeds. Ocean waters are at their warmest

during late summer and fall, a time range which meteorologists traditionally refer to as *hurricane season*. In the Atlantic and eastern Pacific, hurricane season traditionally runs from June 1 to November 30, but climate change has lengthened its duration. About half of the tropical storms that form during this time develop into hurricanes. In the western Pacific, most hurricanes occur between July and October, but some may form throughout the year because the ocean waters of that region stay sufficiently warm even in the colder months.

How does warm ocean water transfer heat into the atmosphere to provide energy that drives a hurricane? Only a very small portion conducts directly from the ocean water into the overlying air, because the air and water have a similar temperature. Nearly all the heat comes from the condensation of water that evaporated from the sea surface. Specifically, the water vapor produced by evaporation contains vast amounts of latent heat, and when the vapor rises and condenses into clouds, it releases this heat, which then warms the surrounding air (see Chapter 16). As air warms, its density decreases relative to its surroundings; it therefore becomes unstable and continues to rise, building towering storm clouds. In effect, water vapor that evaporated off the ocean serves as the fuel of a hurricane. Consequently, a hurricane can only form over warm oceans, because only warm water can evaporate fast enough to provide sufficient fuel to drive a hurricane. Since cold ocean water does not evaporate much, it can't provide sufficient latent heat to power a hurricane. Because of this relationship, hurricanes strengthen when they move over warmer ocean water and weaken when they move over colder water or over land.

Growth of a hurricane involves positive feedbacks. This means that once the storm has formed, it produces conditions that provide it with even more energy, so that it grows and intensifies. The first positive feedback comes into play due to the winds that the storm generates. Those winds build tall, frothy waves from which sea spray, which consists of tiny water droplets, enters the air. Once the droplets exist, evaporation takes place not only from the sea surface, but also from the surface of each droplet. The production of sea spray increases evaporation rates by a factor of 100 to 1,000, providing even more latent heat to the storm. Just as adding gas to a running car engine by pressing the accelerator causes the car to speed up, adding latent heat to a storm by increasing evaporation rates causes the storm to intensify. A second positive feedback happens because the updrafts during the initial stages of hurricane development carry moisture into the middle and upper troposphere, areas that normally tend to be fairly dry. In this dry air, clouds and rain can evaporate, extracting heat from the air, but as the once-dry air becomes humid, evaporation is suppressed and more heat can be retained to power the storm.

How Do Hurricanes Form and Evolve?

A cluster of thunderstorms growing and interacting over a warm sea serves as the "seed" from which a hurricane can grow. The convergence of air that initiates the thunderstorms of a tropical depression typically takes place for one of two reasons, depending on location:

- Convergence triggering the development of Pacific and Indian Ocean tropical cyclones occurs mainly along the intertropical convergence zone (ITCZ), where trade winds of the northern hemisphere converge with those of the southern hemisphere (Fig. 18.44a, b).

FIGURE 18.44 Formation of tropical thunderstorm clusters; these may evolve into tropical cyclones.

(a) At the intertropical convergence zone (ITCZ), the southwest-flowing trade winds of the northern hemisphere and the northwest-flowing trade winds of the southern hemisphere converge.

(b) In the western Pacific, convergence lifting happens over the ITCZ.

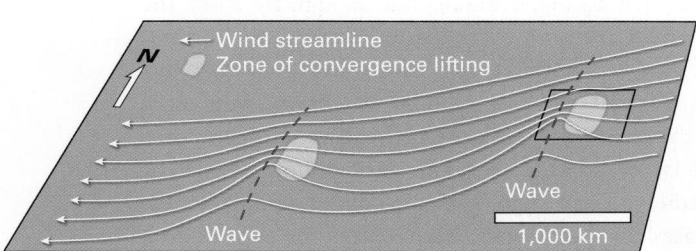

(c) Easterly waves develop in the trade winds, as seen here in map view. Convergence happens where air flow turns north on the eastern side of a wave.

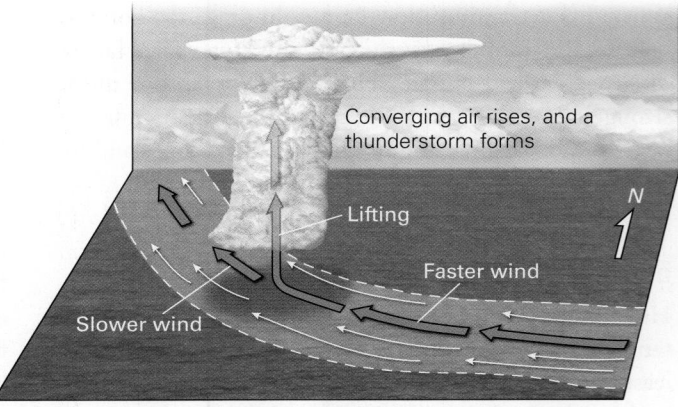

(d) As air flow in the trade winds turns north, it slows. Faster winds approaching from the east converge with these slower winds.

Convergence triggering the development of Atlantic hurricanes occurs mainly where *easterly waves* develop within the trade winds. These air circulations are called "easterly" because their wave-like shape, as seen in map view, moves progressively from east to west, and "waves" because they represent north-south oscillations in the direction of the wind (similar to the ridges and troughs of a jet stream, but occurring instead in the lower atmosphere). Easterly waves originate over Africa and eventually move westward over the Atlantic. Convergence of air within easterly waves takes place because (in the northern hemisphere) air moves somewhat faster through the southern part of the wave than through the northern part, causing air to pile up and, therefore, rise in the branch of the wave flowing northward (Fig. 18.44c, d).

When a tropical disturbance moves over warm ocean water, the cluster of thunderstorms within it strengthens, and the updrafts within the storms carry moist, unstable air to high elevations. The release of latent heat causes air to expand and diverge at high elevation, resulting in the initiation of a low-pressure center at low elevation (see Fig. 17.5). The thunderstorms now begin to rotate around the developing low-pressure center. At this point, the tropical disturbance has evolved into a tropical depression. As the rotation rate increases, air pressure in the low-pressure center decreases further, so winds accelerate (Fig. 18.45). When winds within the circulation exceed 63 km/h (39 mph), the tropical depression has become a tropical storm, and receives a name. The pressure at the center of the developing tropical storm continues to decrease not only because of divergence at high elevation, but also because an outward-directed centrifugal force, due to the rotation of air in the storm, drives air outward, away from the center of the storm, much as a ball on a string moves outward when you start swinging it around your head. This outward air movement further reduces the weight of the air column at the center of the storm, and therefore reduces the pressure at the base of the storm's center. The lower the pressure becomes, the faster the winds circulating the low-pressure center become.

Under normal circumstances, air pressure decreases with elevation so that there is a vertical pressure gradient in the atmosphere. But as the windspeed in a tropical storm increases, and the centrifuge effect increases, air pressure in the low-pressure center decreases sufficiently that the downward pull of gravity slightly overcomes the upward push due to the vertical pressure gradient in the storm's center. As a result, air from high elevations begins to sink slowly downward in the storm's center. As

FIGURE 18.45 Evolution of a thunderstorm cluster into a tropical cyclone.

(a) Release of latent heat causes air in the growing thunderstorm cluster to expand and flow outward at high altitudes.

(b) As the thunderstorms start rotating around the low-pressure center, a spiral-shaped storm forms.

(c) Centrifuging of air in the rotating storm further lowers the central pressure, and an eye wall and spiral rainbands begin to form.

this air sinks, it undergoes compression and therefore warms and dries. The warming causes the air to expand, and the expansion moves more air molecules sideways, out of the storm's center, further reducing the air pressure below the center. Meanwhile, the drying causes clouds to dissipate, so the center becomes a visible eye. When air pressure has become low enough for an eye to form, an extreme pressure gradient exists between the storm's center and the region outside. The developing, extreme pressure gradient across an eye wall accelerates air sufficiently to create hurricane-strength winds, at which point the tropical storm has now become a hurricane.

Once a hurricane exists, its "engine" will run as long as it has fuel (warm ocean water) and there are no strong winds in the upper troposphere to break apart the storm's spiral circulation. The temperature and supply of warm water, the character of the surrounding winds, the thickness of the troposphere at the latitude where the hurricane resides, and friction with the Earth's surface all control the strength of a hurricane. If the hurricane moves over warmer water, it can strengthen to become a higher category storm, whereas if it moves over cooler water, where evaporation rates decrease, it will weaken. Once a hurricane moves over cold water, or over land, it loses its fuel supply and weakens rapidly, downgrading to a tropical storm, or tropical depression.

Hurricane Tracks

Tropical cyclones do not develop directly over the equator, and cannot cross the equator, despite its plentiful supply of warm seawater. In fact, for a strong cluster of thunderstorms to start rotating and become a tropical cyclone, the cluster must lie at least 5° north or south of the equator. Why? The rotation of a cyclone is due to the Coriolis force. The Coriolis force is zero at the equator, and it only becomes strong enough to cause rotation at a latitude of 5° or greater. During and after their formation, tropical cyclones move over the surface of the Earth, tracing out a path known as a **hurricane track** (provided the storm strengthens to become a hurricane for part of its life). On a map of a hurricane track, different colors indicate the strength of the storm at a given location (Fig. 18.46a).

As an example, let's follow a hurricane track in the Atlantic. Many Atlantic hurricanes start out as thunderstorms that move across the Sahel of Africa, and emerge out over the eastern Atlantic near Cape Verde as a tropical disturbance. If conditions are appropriate, the disturbance first evolves into a tropical depression and then into a tropical storm (Fig. 18.46b). Initially, tropical storms move west within the belt of westward-flowing trade winds. During this westward track across the Atlantic, a tropical storm can evolve into a hurricane. In the northern hemisphere, hurricanes also drift northward as they move westward, a consequence of the Coriolis force. When hurricanes reach a latitude of about 30° N, they come under the influence of mid-latitude winds, which generally blow from west to east. As a result, hurricane tracks in the northern hemisphere typically curve eastward after passing a latitude of 30° N. Worldwide, hurricanes generally follow similar tracks, initially moving westward in the trade winds, but taking a poleward turn later in their lifetime as they move toward the middle latitudes. Exact paths vary from storm to storm. For example, some Atlantic hurricanes remain at sea, while others cross into the Caribbean and then move westward into Mexico or northward into the United States (Fig. 18.46c). Some even cross into the Pacific before heading north.

Meteorologists use computer models, supplemented by measurements made by special aircraft that fly into the storms, to predict hurricane tracks. With the models, experts can even estimate the width of the belt in which hurricane-force winds will occur (Fig. 18.46d, e). Many countries have now set up monitoring centers that use data from satellites, radar, weather stations, and aircraft to produce computer models of hurricane tracks. The reliability of these tracks decreases with time from the prediction, but they provide emergency managers with guidance on where and when to begin evacuations and deploy resources as these monstrous storms approach the coast.

How Do Hurricanes Cause Damage?

WIND DAMAGE. We've already seen that the winds in the eye wall of a hurricane have immense power. During Hurricane Patricia in 2015, winds over the eastern Pacific reached a record 345 km/h (215 mph)—virtually no building can remain standing in such winds (see Table 18.2). Even milder winds can cause damage in many ways (Fig. 18.47a). The wind speed at a location within a hurricane depends on two factors: first, the speed due to the spiraling of the air around the hurricane, which decreases with distance from the eye wall; and second, the speed of the overall movement of the hurricane as the entire storm drifts along, propelled by the regional atmospheric circulation. Typical forward speeds of hurricanes range between 15 and 90 km/h (10–60 mph), though some hurricanes stall so that the eye remains stationary for a while. As a consequence of a hurricane's forward movement, winds on the right side of a counterclockwise-rotating hurricane (as viewed looking in the direction the hurricane is moving) are stronger than those on the left side. For example, if a hurricane eye wall rotates at 150 km/h while the storm overall moves forward at 40 km/h, then the wind speed on the right side will be 150 + 40 = 190 km/h, and 150 − 40 = 110 km/h on left side (Fig. 18.47b). For this reason, the side of a hurricane rotating in the direction of its forward motion causes more wind damage than the side rotating in the direction opposite its forward motion.

DAMAGE DUE TO WAVES AND STORM SURGE. Because hurricanes come from the sea, they cause immense damage to the coast, not only because of towering waves, but also because of **storm surge**, a dome of water that rises in front of a hurricane (Fig. 18.48a). Storm surge develops because the wind of the storm drives water in front of it, effectively building a pile of water. To a lesser extent,

FIGURE 18.46 Hurricane tracks.

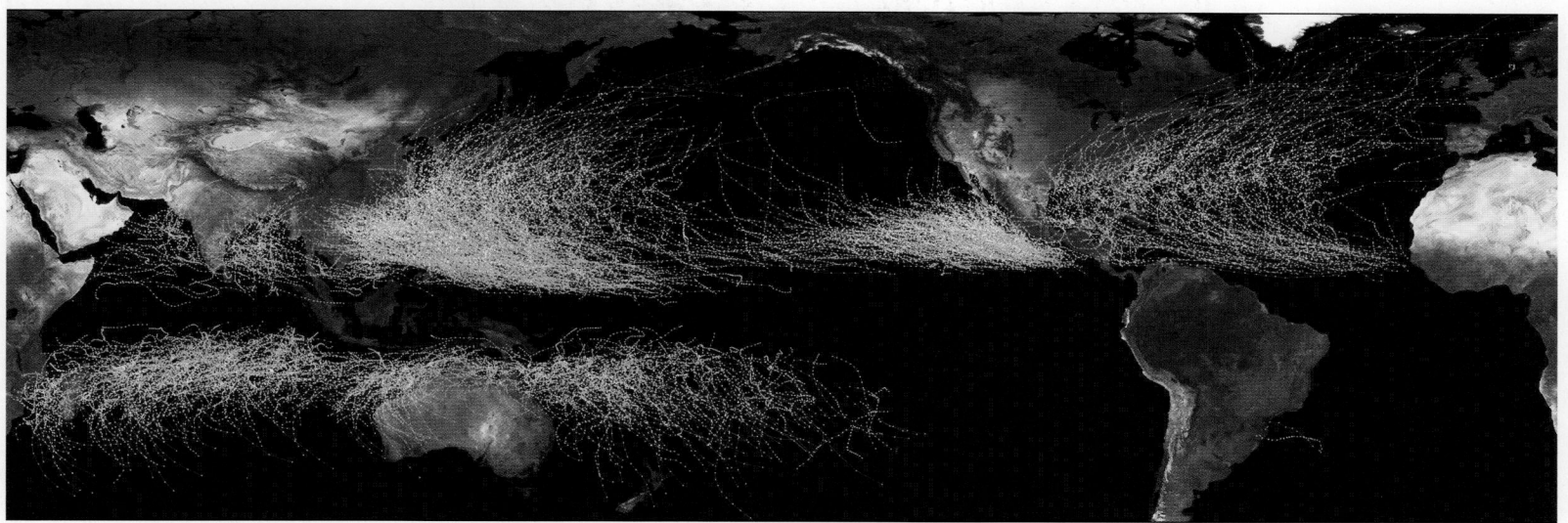

(a) Tracks of all documented hurricanes. Light blue lines are weaker storms, and yellow lines are stronger storms. Note that none of these tracks crosses the equator.

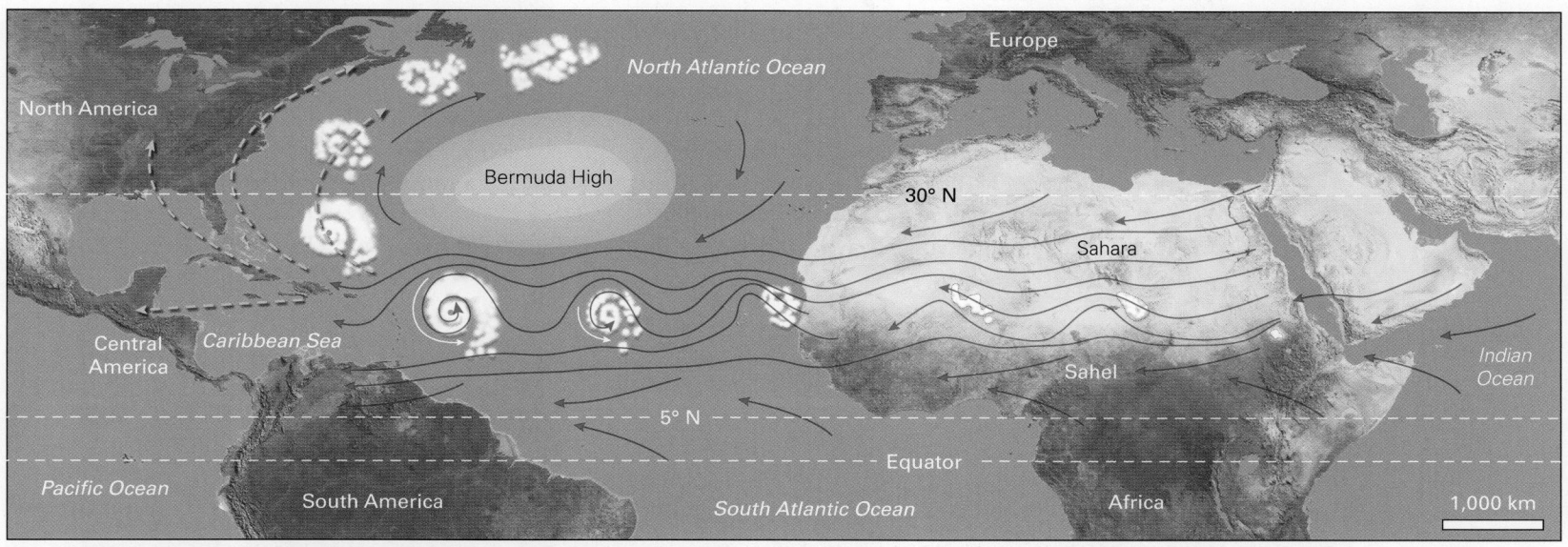

(b) A typical path during the life cycle of an Atlantic hurricane remaining over the ocean. In this example, a tropical depression forms off the west coast of Africa, becomes a tropical storm in the mid-Atlantic, and becomes a hurricane north of South America. It dies out in the North Atlantic.

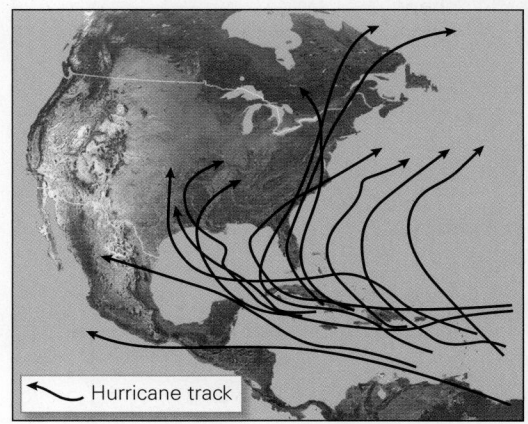

(c) Tracks of representative Atlantic hurricanes. Most head west and then curve north.

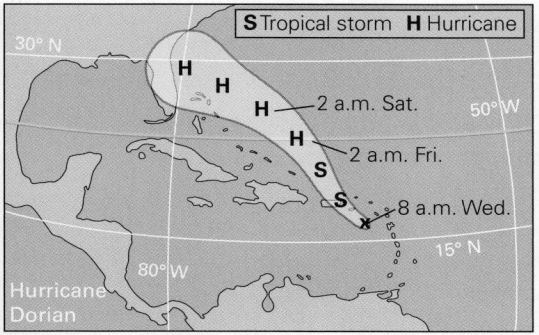

(d) Prediction of the track of Hurricane Dorian in 2019, five days before it reached the US coast. The width of the track indicates uncertainty, not the storm size.

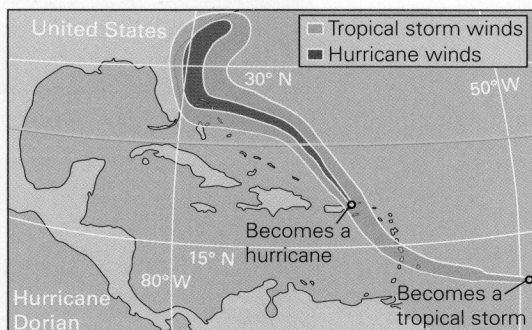

(e) A wind-swath map indicates the actual track of the storm after it had passed. The colors represent the areas affected by winds of a given strength.

FIGURE 18.47 Wind during a hurricane.

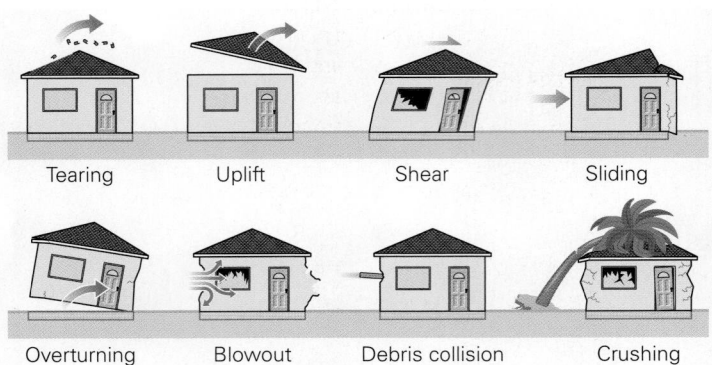

Tearing   Uplift   Shear   Sliding

Overturning   Blowout   Debris collision   Crushing

(a) Hurricane winds, even in weaker storms, can damage buildings in many ways.

Stronger winds

N

Weaker winds

(b) The wind speed in a hurricane at a specific location (red dots) is the sum of the hurricane's rotational wind speed (red arrows) and the overall forward motion of the hurricane (blue arrows). Therefore, as Hurricane Katrina, shown here, moved north, the wind speed on the east side of the storm (green arrow) exceeded that on the west side (pink arrow).

it develops because the low atmospheric pressure beneath the storm reduces the weight of the atmosphere at the sea surface, allowing the sea surface to rise, as water does in a straw when you remove air. The consequences of storm surge can be catastrophic (Fig. 18.48b–d). When a cyclone moved over the Ganges River delta, at the head of the Bay of Bengal, in 1970, storm surge lifted sea level by about 6 m (20 ft), and high waves (10 m, or 33 ft) built on top of the surge. As a result, water inundated the delta, and 300,000 people drowned. If storm surge arrives at high tide, the water level will become even higher. Storm surge was responsible for the failure of New Orleans's levees during Hurricane Katrina, and the resulting catastrophic flooding of New Orleans, as well as for the destruction along Florida's west coast during Hurricane Ian in 2022.

FIGURE 18.48 Storm surge and its consequences.

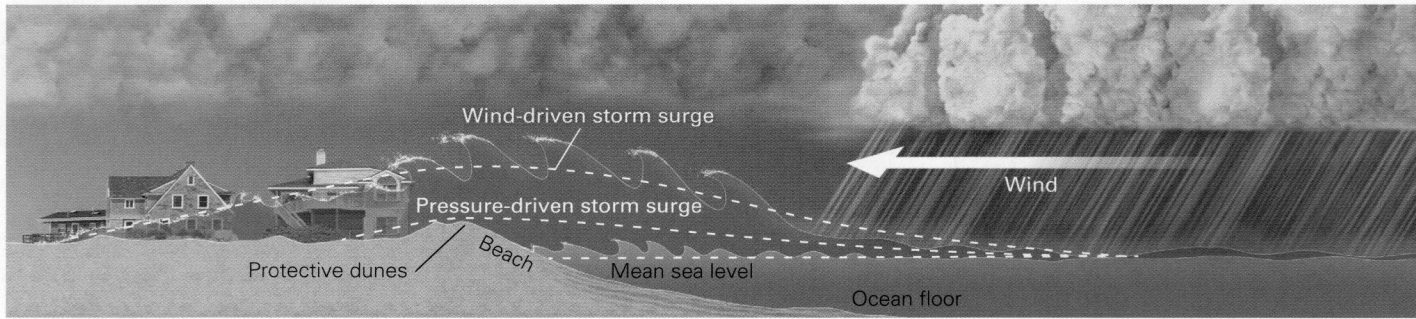

Wind-driven storm surge

Pressure-driven storm surge

Wind

Protective dunes   Beach   Mean sea level   Ocean floor

(a) Wind and low atmospheric pressure combine to produce a bulge of water.

(b) Storm surge arriving onshore, with waves on top.

(c) Due to storm surge, coastal areas can be submerged.

(d) Storm surge and waves together can devastate coastal communities.

DAMAGE DUE TO INLAND FLOODING AND MUDSLIDES.
Although hurricane winds slow when the storms head over land, their clouds can still dump torrential rains. For example, when Hurricane Mitch stalled over the mountains of Central America in 1998, the downpours it released caused numerous flash floods and mudslides. Some of the mudslides buried whole towns. In 2022 Hurricane Ian poured 25–37 cm (10–15 in) of rain over parts of central Florida. Remnants of Hurricane Ike traveled northward from the Gulf of Mexico, and then northeastward across the midwestern United States. It brought with it so much moisture that it drenched the Chicago area—2,250 km (1,400 mi) inland of its landfall—with 13 cm (5 in) of rain. It extensively ruined crops and caused severe flooding along its route.

Take-home message...

Tropical cyclones—hurricanes and typhoons—are the most destructive storms on the planet. They originate from clusters of thunderstorms and grow into giant spiraling storms, nourished by latent heat in water evaporating from warm seas. The strongest winds occur beneath the eye wall. The winds, waves, storm surge, flooding, and mudslides created by these storms produce devastation.

Quick Questions

- What is the difference between a hurricane and a tropical depression?
- Where are the strongest winds found in a hurricane?
- Why don't hurricanes form over the equator or over the Arctic Ocean?

18 CHAPTER REVIEW

Objective 18.1

Understand the basic conditions that lead to the formation of thunderstorms, and describe the difference between ordinary, squall-line, and supercell thunderstorms.

KEY CONCEPTS

- For a thunderstorm to form, three conditions must exist: abundant moisture in the lower atmosphere, a lifting mechanism, and atmospheric instability. Severe thunderstorms also require the existence of vertical wind shear.

- Lifting occurs when air is forced upward by a regional front or a gust front, by the convergence of air flow, or by the presence of high relief.

- Air can also rise buoyantly due to heat rising from the Earth's surface.

- Rising moist air, because of the release of latent heat due to condensation, can become buoyant relative to its surroundings and produce atmospheric instability. Updrafts in unstable air can sometimes extend all the way to the tropopause. In stable air, lifted air sinks back down.

- Ordinary thunderstorms form where winds do not vary substantially with altitude, so these storms do not rotate. They evolve through three stages (development, mature stage, and dissipation).

- In squall-line thunderstorms, several thunderstorms develop and organize into a line. The combined outflow of cool air from such storms can trigger straight-line winds and the formation of new thunderstorms.

- Supercell thunderstorms develop if wind changes rapidly in speed and direction with elevation, causing the updraft of the storm to rotate. Rain falls downwind from the updraft. As a result, the updraft is not destroyed by the downdraft, and the storm can survive for several hours.

- Supercell updrafts are so strong that a large overshooting top rises above the anvil cloud of the storm.

EARTH-SCIENCE VOCABULARY

adiabatic expansion (p. 683)
anvil cloud (p. 686)
atmospheric instability (p. 683)
derecho (p. 689)
downdraft (p. 681)
dry adiabatic lapse rate (p. 683)
environmental lapse rate (p. 683)
frontal lifting (p. 681)
frontal squall line (p. 687)
gust (p. 687)

gust front (p. 682)
lifting mechanism (p. 681)
mesocyclone (p. 690)
moist adiabatic lapse rate (p. 684)
ordinary thunderstorm (p. 685)
orographic lifting (p. 682)
overshooting top (p. 686)
saturated air (p. 684)
severe thunderstorm (p. 680)
squall (p. 687)

squall-line thunderstorm (p. 687)
straight-line wind (p. 686)
supercell thunderstorm (p. 689)

thunderstorm (p. 680)
unsaturated air (p. 683)
updraft (p. 681)
vertical wind shear (p. 689)
wall cloud (p. 690)

REVIEW QUESTIONS

1. **(a)** What distinguishes a thunderstorm from other kinds of rainstorms? **(b)** What characteristics does a thunderstorm need to have for it to be classified as severe? **(c)** On a "thunderstorm day," do the thunderstorms last all day?

2. **(a)** List the three conditions that must exist for a thunderstorm to develop. **(b)** Where does the moisture that fuels thunderstorms come from? **(c)** Describe the evolution of an ordinary thunderstorm.

3. **(a)** Explain the difference between stable and unstable atmospheric conditions. **(b)** How does the rate at which temperature changes with altitude influence whether lifted air will become buoyant and trigger a thunderstorm?

4. **(a)** What are squall-line thunderstorms? **(b)** How can gust fronts from squall-line thunderstorms produce new thunderstorms? **(c)** What is a derecho?

5. **(a)** Explain how a supercell thunderstorm forms. **(b)** Why can a supercell thunderstorm last so much longer than an ordinary thunderstorm? **(c)** Why do supercell thunderstorms typically have large overshooting tops and asymmetric anvil clouds?

6. **(a)** On **Figure A**, label the updraft, wall cloud, downdraft, anvil, and overshooting top. **(b)** Describe the sequence of weather conditions that you would experience as a supercell thunderstorm, moving southwest to northeast, approaches and then passes over your home.

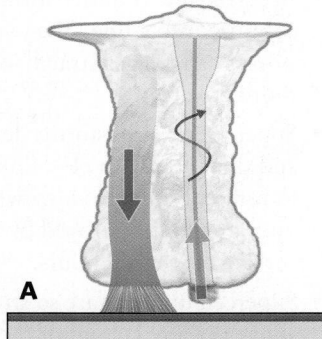

A

Objective 18.2

Explain how thunderstorm hazards develop, and how to protect yourself against them.

KEY CONCEPTS

- Lightning happens because electric charge separation develops in a thunderstorm: the cloud base accumulates negative charge, and the upper parts of the cloud accumulate positive charge.

- Because of charge separation, a spark—lightning—can jump across the cloud, or between the cloud and the ground.

- Downdrafts develop within thunderstorms when it rains, for two reasons: cooling of air due to evaporation, and the downward push of the falling rain. When a downdraft reaches the ground, the air turns and flows outward.

- Downbursts are particularly hazardous for approaching and departing aircraft near airports. Outrushing straight-line winds from downbursts, or squall lines, are strong enough to produce damage.

- Hailstones form when supercooled water droplets freeze onto ice particles in clouds. The largest hailstones, which can grow to the size of grapefruits, form in supercell thunderstorms.

- Flash floods occur when slow-moving thunderstorms, or a succession of thunderstorms, dump a large amount of rain over an area in a short time.

EARTH-SCIENCE VOCABULARY

charge separation (p. 692)
cloud-to-cloud lightning (p. 693)
cloud-to-ground lightning (p. 693)
downburst (p. 695)

flash flood (p. 696)
hail (p. 695)
lightning (p. 692)
lightning stroke (p. 692)
microburst (p. 695)
thunder (p. 695)

REVIEW QUESTIONS

7. **(a)** Identify the positive polarity and negative polarity cloud-to-ground strokes on **Figure B**. **(b)** When charge separation occurs in a thunderstorm, in which parts of the clouds do the positive and negative charges accumulate? **(c)** Describe the steps that lead to a cloud-to-ground lightning stroke.

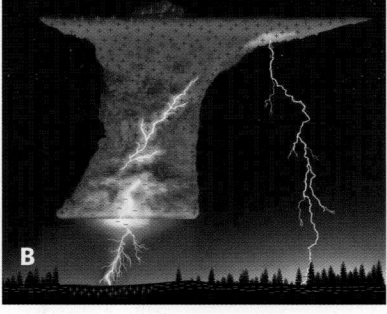

B

8. **(a)** During a lightning storm, are you safer inside or outside a car? **(b)** What is a "bolt from the blue"? **(c)** What is a downburst, and how can it pose a hazard to aircraft?

9. **(a)** What is supercooled water, and what is its role in hail formation? **(b)** Why is hail more common in supercell thunderstorms than in other types of thunderstorms? **(c)** Describe circumstances that can lead to flash flooding related to thunderstorms.

Objective 18.3

Discuss how tornadoes form, how they are classified, and how they cause destruction.

KEY CONCEPTS

- A tornado is a violently rotating funnel, or vortex, of air that extends between the ground and the lower part of a severe thunderstorm.

- Supercell tornadoes typically emerge from the wall cloud, a cloud within the storm's updraft that typically protrudes below the main storm cloud on the southwest side of a supercell thunderstorm.

- Tornadoes typically develop on the rear flank of a supercell thunderstorm. They form when part of the rear-flank downdraft gets drawn into the supercell's updraft.

- Large tornadoes form when a downdraft develops within the vortex of a small tornado, forcing the sides of the tornado to expand outward. Large tornadoes often include suction vortices, which appear like narrow tornadoes within the large tornado. Suction vortices can produce severe damage.

- Tornadoes can cause immense damage along a tornado track. The Enhanced Fujita (EF) scale rates their intensity based on the damage the tornado causes.

- The EF scale classifies tornado intensities from 0 to 5. An EF0 tornado produces relatively minor damage, while an EF5 tornado produces catastrophic damage.

- To avoid injury during a tornado, seek shelter in a basement or strong building. Put as many walls as possible between yourself and the tornado. Abandon cars and mobile homes. If outside, lie in the lowest spot available.

EARTH-SCIENCE VOCABULARY

Enhanced Fujita (EF) scale (p. 702)

funnel cloud (p. 700)

non-supercell tornado (p. 700)

tornado (p. 697)

tornado track (p. 698)

suction vortex (p. 700)

supercell tornado (p. 699)

REVIEW QUESTIONS

10. (a) Explain the basic process of tornado formation. (b) Why do tornadoes typically develop on the rear flank of a supercell thunderstorm? (c) Label the features displayed by the Doppler radar image in **Figure C**.

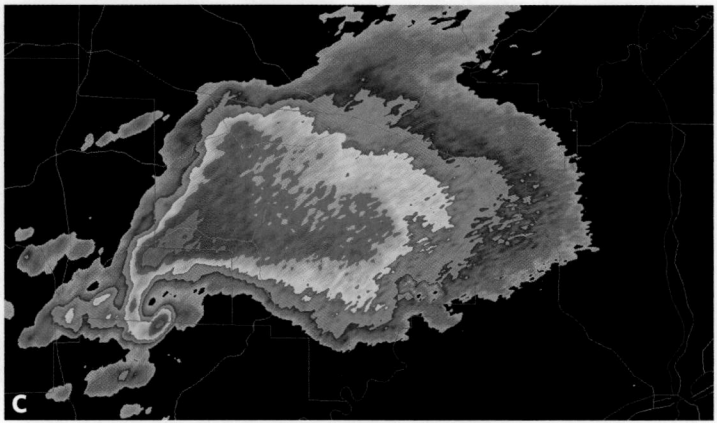

11. (a) On what basis do meteorologists classify tornadoes? (b) Which of the two tornadoes shown in **Figure D** is more likely to contain suction vortices? (c) In what range of widths do tornados occur, and does width have a relationship to strength?

12. (a) Is it possible for a giant tornado with 250 mph winds to be classified as an EF1? Explain your answer. (b) Does a tornado cause the same EF damage for its entire length?

13. (a) What phenomenon causes the most injury to people who encounter a tornado? (b) What should you do if a tornado is coming your way? (c) If a tornado is about to strike your house and it has no basement, where is the safest location to go?

Objective 18.4

Explain how mid-latitude cyclones form and evolve, and the relationships among weather, fronts, and jet streams within these cyclones.

KEY CONCEPTS

- Mid-latitude cyclones develop between latitudes of 30° and 60°, when a large low-pressure system develops at ground level beneath an area of divergence in the polar-front jet stream.

- Air spirals counterclockwise around the center of a developing cyclone in the northern hemisphere. As a cyclone develops, the cold front advances rapidly eastward, while the warm front migrates slowly northward.

- In many mid-latitude cyclones, the cold front eventually catches up with the warm front and wraps into it, yielding an occluded front.

- When viewed from a satellite, the clouds of a mid-latitude cyclone have the shape of a comma. The comma tail follows the cold front, and along it, thunderstorms often form. The comma head lies northwest and northeast of the low-pressure center and the warm front. Rain, snow, or freezing rain may fall within the comma head and over the warm front.

- Blizzards and ice storms can occur under the clouds of the comma head in winter.

EARTH-SCIENCE VOCABULARY

blizzard (p. 710)

convergence (p. 707)

divergence (p. 707)

freezing rain (p. 710)

ice storm (p. 710)

mid-latitude cyclone (p. 706)

nor'easter (p. 706)

wind chill (p. 710)

14. **(a)** What is a mid-latitude cyclone? In your answer, describe the basic shape of the storm system, and provide a sense of its dimensions. **(b)** Describe the stages during the formation and evolution of a mid-latitude cyclone. **(c)** How is the formation of a mid-latitude cyclone related to the polar-front jet stream?

15. **(a)** What kinds of weather occur in association with mid-latitude cyclones, and how can the type of weather vary with location in the storm? **(b)** On **Figure E**, label the low-pressure center, label the cold front and the warm front, and use arrows to indicate the direction that the two fronts are moving.

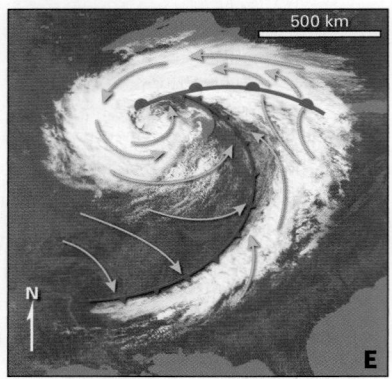

16. **(a)** Define a blizzard and explain how a blizzard differs from an ice storm. **(b)** **Figure F** shows a large mid-latitude cyclone that formed in January. Label the region of the storm where you expect blizzard conditions. **(c)** What is a nor'easter, and why is it called that?

Objective 18.5

Illustrate how hurricanes form in the tropics, move across the oceans, and produce massive destruction as they make landfall.

KEY CONCEPTS

- Tropical cyclones are rotating storms that develop in tropical latitudes at least 5° north or south of the equator. Weak cyclones are called tropical depressions, and stronger ones are called tropical storms. When sustained winds exceed 119 km/h (74 mph), they are called hurricanes, typhoons, or cyclones depending on location.

- A hurricane's maximum sustained wind speed determines its classification on the Saffir-Simpson scale, which rates hurricanes from Category 1 (the least strong) to Category 5 (the strongest).

- Hurricanes have a nearly cloud-free eye surrounded by an eye wall, a ring of towering clouds. Beyond the eye wall, spiral rainbands extend outward for hundreds of kilometers. Torrential rain and wind occur beneath the eye wall and rain bands. Winds are strongest in the eyewall.

- Energy powering a hurricane comes from the condensation of water vapor following evaporation of water at the sea surface. Latent heat released during condensation produces instability that leads to storm development.

- Development of frothy waves increases evaporation rates, providing more latent heat to a hurricane. Evaporation of early-formed clouds moistens the troposphere, so that as the storm builds, rainfall increases.

- Clusters of thunderstorms that can evolve into tropical cyclones commonly form in easterly waves over the Atlantic, and within the ITCZ over the western Pacific.

- Tropical cyclones only form over warm water (which evaporates more easily). The Atlantic and eastern Pacific hurricane season occurs when tropical ocean water temperatures are at their warmest. The season is longer in the western Pacific, where ocean waters retain heat throughout the year.

- In the tropics, hurricanes typically move westward with the trade winds while drifting poleward due to the Coriolis effect. As hurricanes cross into the mid-latitudes, they turn eastward.

- Winds, waves, storm surge, inland flooding, mudslides, and occasional tornadoes triggered by hurricanes cause the devastation associated with these storms. Forecasters use computer models to estimate the most likely hurricane path.

EARTH-SCIENCE VOCABULARY

cyclone (p. 715)
eye (p. 715)
eye wall (p. 715)
hurricane (p. 715)
hurricane track (p. 720)
Saffir-Simpson scale (p. 715)

spiral rainbands (p. 716)
storm surge (p. 720)
sustained wind (p. 714)
tropical cyclone (p. 714)
tropical storm (p. 715)
typhoon (p. 715)

REVIEW QUESTIONS

17. **(a)** Label the key structural features of the storm shown on **Figure G**. **(b)** Name the three types of tropical cyclones, and the basis for distinguishing among them. **(c)** Where does the heaviest rain and strongest wind occur in a hurricane? **(d)** How does the Saffir-Simpson scale differ from the Enhanced Fujita scale?

18. **(a)** What does **Figure H** show? Label the various components of a hurricane visible on this figure, and indicate where winds are strongest. About how many rainbands are visible in **Figure H**? **(b)** Describe the stages that take place during the

formation and evolution of a hurricane. **(c)** What is the energy source for a hurricane? **(d)** Describe at least one positive feedback that occurs as a hurricane intensifies.

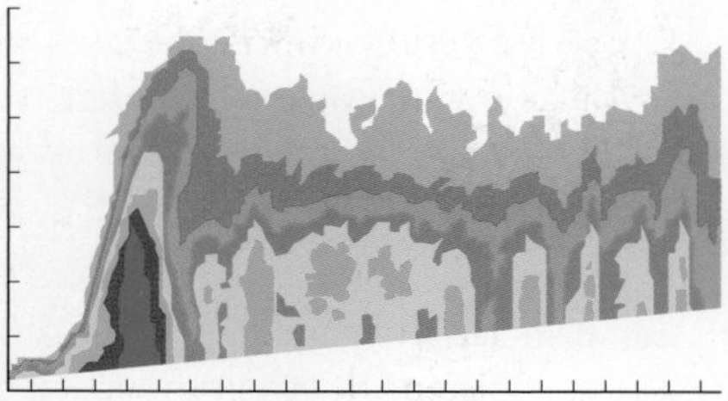

H

19. **(a)** How does the eye of a hurricane develop? **(b)** Describe how the wind direction would change if you were located on the Gulf of Mexico coastline and a hurricane traveling due north approached, with the eye passing directly over you. What if the eye passed 50 km west of your location? **(c)** What is the difference between a hurricane and a tropical storm? What is the difference between a hurricane and a typhoon?

20. **(a)** In what atmospheric feature does convergence leading to hurricane formation most commonly occur over the Atlantic Ocean?

(b) In what atmospheric feature does such convergence occur over the western Pacific Ocean?

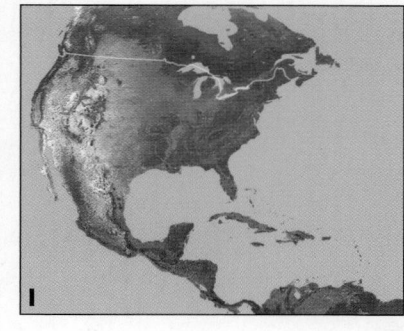

I

21. **(a)** On **Figure I**, draw the typical track of a hurricane that forms in the Atlantic and reaches the southeast United States. Why do hurricanes follow such tracks? **(b)** Does a hurricane have the same strength along its entire track?

22. **(a)** What is storm surge? **(b)** What else causes destruction when a hurricane moves across a coastline? **(c)** If a hurricane in the Gulf of Mexico tracked due west and made landfall in Mexico, which side of the storm (north or south) would produce the strongest winds and the highest storm surge?

ANOTHER VIEW A cloud-to-cloud lightning bolt in a thunderstorm brightens the night sky, as rain drenches the town below in northern Italy.

◉19 CLIMATE AND CLIMATE CHANGE

After studying this chapter, you should be able to...

1. identify the Earth's major climate zones, and relate them to latitude, altitude, and proximity to the ocean.

2. describe the factors that control climate conditions prevailing in a region.

3. understand how scientists determine past changes in the Earth's climate.

4. explain possible causes of both long-term and short-term climate change.

5. summarize the evidence for recent climate change, and discuss causes of this change.

6. assess the impacts of future global climate change on the Earth System and society.

(a) The Mauna Loa Observatory operated from 1959 until a volcanic eruption in 2022.

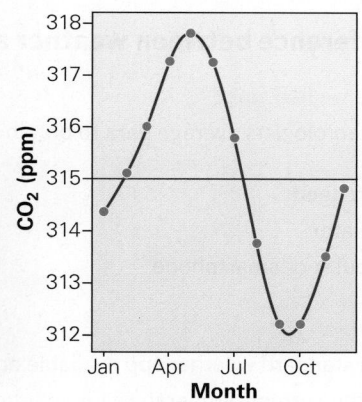

(c) The variation in CO_2 concentration during 1959, the first full year in which Keeling measured it.

(b) Because of a recent eruption, the observatory was moved to a site at the Mauna Kea Observatories.

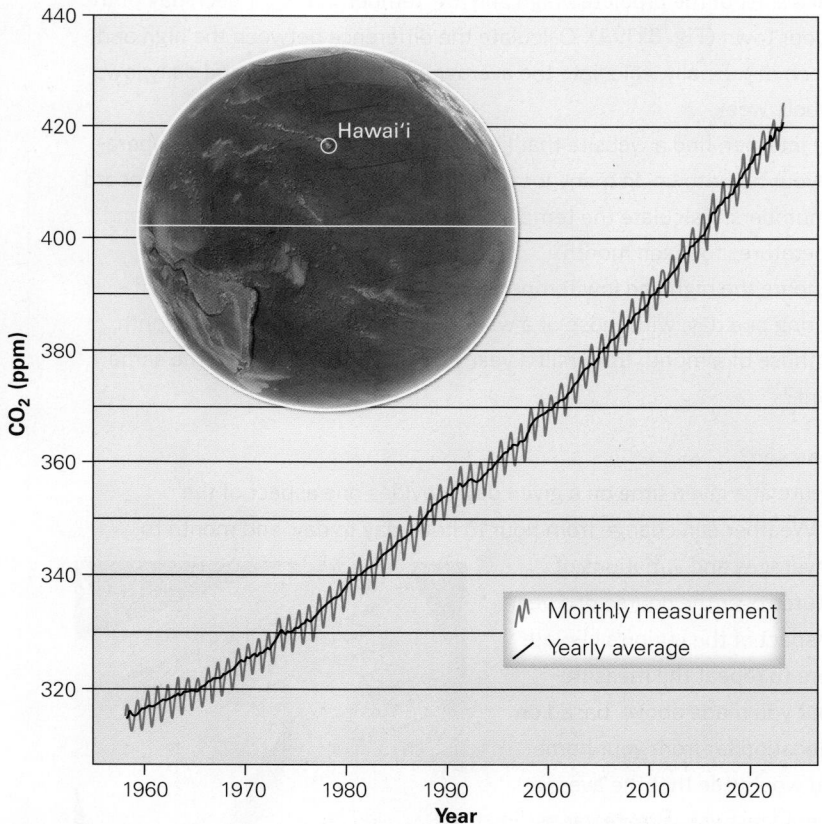

(d) The variation in CO_2 concentration up through 2023.

In 1955, Charles Keeling constructed an instrument that could measure the concentration of carbon dioxide (CO_2) in air very accurately. This instrument had to be very sensitive because CO_2 is a *trace gas*, one that accounts for far less than 1% of the gas in air. Keeling made a set of measurements in California and learned that the average of his measurements was higher than the average of measurements made in the mid-19th century. Three years later, to better understand global variations in atmospheric CO_2 concentration, Keeling installed a monitoring station at an elevation of 3.4 km (11,141 ft) on the flank of Mauna Loa, a volcanic peak on Hawai'i **(Fig. 19.1a)**. He

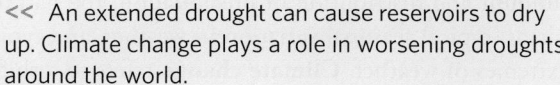

<< An extended drought can cause reservoirs to dry up. Climate change plays a role in worsening droughts around the world.

chose this location—far from cities, factories, forest fires, and other sources of the gas—so that his measurements would represent only CO_2 that had mixed thoroughly into the atmosphere. He found that the CO_2 concentration decreased in the summer, when plants of the northern hemisphere absorbed the gas during photosynthesis,

How can I explain . . .

The difference between weather and climate

What are we learning?
How meteorologists average data to obtain climate statistics

What you need:
- A calculator
- A computer or smartphone

Instructions:
- Using a standard weather app available on a smartphone, or a local newspaper, record the current temperature for your hometown, and the predicted hourly temperatures for that whole day.
- Next, make a list of the predicted high and low temperatures for each day of the week in your town (Fig. Bx19.1). Calculate the difference between the high and low for each day. Finally, calculate the average of the daily highs and daily lows for the whole week.
- Using the internet, find a website that lists the monthly high and low temperatures for your hometown. In many locations, a climatology office keeps a record of these numbers. Calculate the temperature difference between the high and low temperatures for each month.
- Now, compare the high and low temperatures (and the difference between them) during one day, with those of a week, with those of the current month, and with those of a month from half a year earlier. Are the numbers the same or different?

What did we see?
- Temperature at a given time on a given day provides one aspect of the weather. Weather can change from hour to hour, day to day, and month to month. Averages and variations of temperatures over the course of a year are one aspect of the region's climate. If you were to repeat the measurements that you made above, based on another location far from your hometown, you would see that the averages are different, because climate varies with location.

FIGURE Bx19.1 A table of temperature data for the month of January.

and it rose in the winter, when those plants shed their leaves, which then decayed and released CO_2 (Fig. 19.1c). In effect, Keeling had observed the Earth System inhaling and exhaling CO_2!

Keeling continued making measurements on Mauna Loa during the remainder of his career. Since then, other researchers maintained the CO_2-monitoring station at the Mauna Loa Observatory without interruption until 2022, when a volcanic eruption of Mauna Loa shut down the station so that it had to be moved to Mauna Kea, a less active volcano to the northwest (Fig. 19.1b). The multidecade record of CO_2 concentration, now called the *Keeling curve*, reveals that the concentration of atmospheric CO_2 steadily goes up, year after year (Fig. 19.1d). In fact, the concentration of CO_2 has increased by about 35% since 1960. Why should society care about the atmospheric concentration of CO_2? Earlier in this book, we pointed out that CO_2 is a *greenhouse gas* that traps infrared radiation in the atmosphere, keeping the Earth's surface realm warm enough for liquid water, and life, to exist. But can there be too much of a good thing? When CO_2 concentrations increase, the air and oceans get warmer, and this change has had a noticeable impact on global climate and, therefore, society. Addressing climate change has become a challenge to society, worldwide.

To help you understand climate and climate change, this chapter begins by describing the Earth's present climates and the factors that control them. Then we look back in time to examine changes in climate over the Earth's history and the phenomena that may have caused these changes. We conclude by examining how human activities modify the Earth's climate today, and how climate change might affect the Earth System and society in the future.

19.1 The Earth's Major Climate Zones

How does climate differ from weather? **Weather** refers to the conditions of air temperature, relative humidity, wind speed, cloudiness, and precipitation at a given location at a given time. For example, if a radio announcer says, "Right now, it's overcast with a temperature of 30°F at the airport," they are reporting the current weather. A description of **climate**, in contrast, characterizes the average weather conditions and the range of weather conditions for a region over the course of decades to millennia (Box 19.1). More specifically, a region's climate reflects its average temperature and typical temperature range, the amount and distribution of precipitation, the nature of storms, typical seasonal variation in weather, and typical extremes of weather. **Climate change** refers to a shift in the average of one or more of the characteristics that define climate.

By comparing your perceptions of a rainforest, a hot desert, and a tundra, you'll realize that climate affects many of our senses. In a rainforest, the air feels sultry, nearly everything in sight has a lush covering of green shrubs, trees, and vines. Downpours drench the land, and vibrant sounds rise from countless species of insects, monkeys, and birds. In a hot desert, the Sun beats down relentlessly, the air is so dry that your lips are chapped, and if you look to the horizon you see mostly rocks and sand in hues of brown and tan. Spiny plants may poke up here and there, but you hear hardly any animals. And, in the northern Canadian tundra, the ground remains frozen solid almost all year, it's so cold that your breath freezes, and while the ground has plant cover, most of it is scruffy and low. Note that the major factors that contribute to our experience of climate reflect surface temperature, humidity, and precipitation. To a large extent, these variables correlate with latitude, but many other factors, such as elevation and proximity to oceans, influence them as well. The nature of plant cover serves as the most obvious visual consequence of the climate.

Global Temperature and Rainfall Patterns

When you check the weather report in the morning to decide how you should dress, you typically see or hear the day's expected high and low temperatures. If you check the weather report for another city, far away, you'll likely see a different high and low. Clearly, temperatures vary from place to place. To characterize such variations, a key indicator of climate, meteorologists produce maps showing global temperature variation. The data used to make these maps come from hourly temperature measurements, made using a *thermistor* (an electronic device containing a wire whose electrical resistance depends on temperature) mounted at the top of a 2 m (6 ft) high pole at a weather station. Such data can be averaged for a designated time period (such as one month).

To make these temperature maps easier to interpret, they are contoured and colored such that different colors represent different temperatures. A map of average air temperature for the month of January shows a distinct pattern (Fig. 19.2a); bands of warm temperature occur at low latitudes, bands of cold temperatures occur at high latitudes, and bands of intermediate temperatures lie in between. If we now look at a global temperature map for July (Fig. 19.2b), we see the same overall relationship between average temperature and latitude, but the color bands have shifted to different latitudes. Clearly, low-latitude regions are always warmer than high-latitude regions, and temperature at a given latitude varies during the course of a year. Closer examination of the maps in Figure 19.2 reveals that boundaries between color bands

FIGURE 19.2 Global map of average surface temperatures.

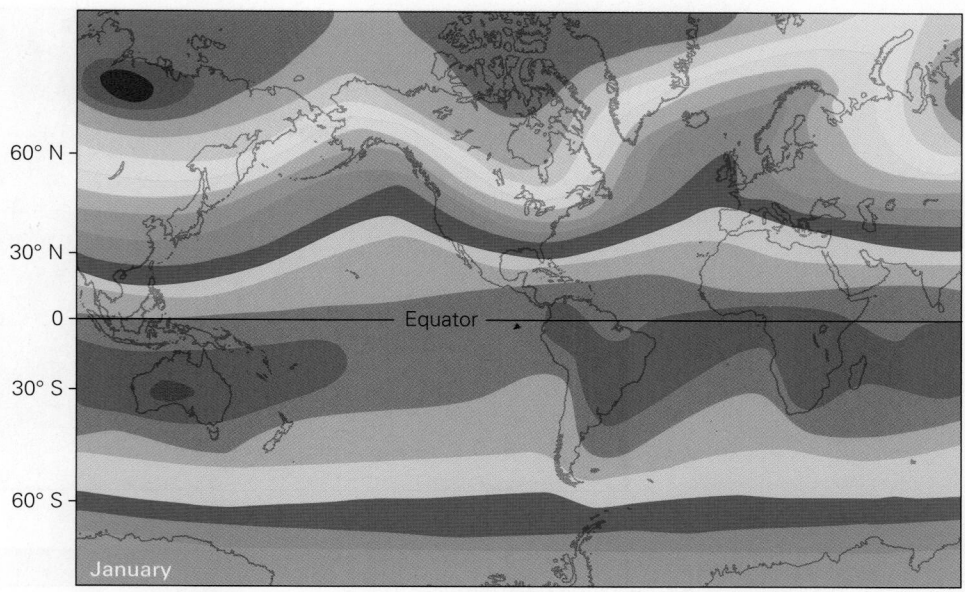

(a) In January, northern mid-latitudes are cool and southern mid-latitudes are warm.

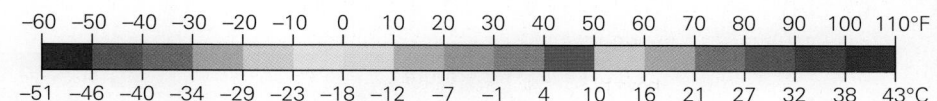

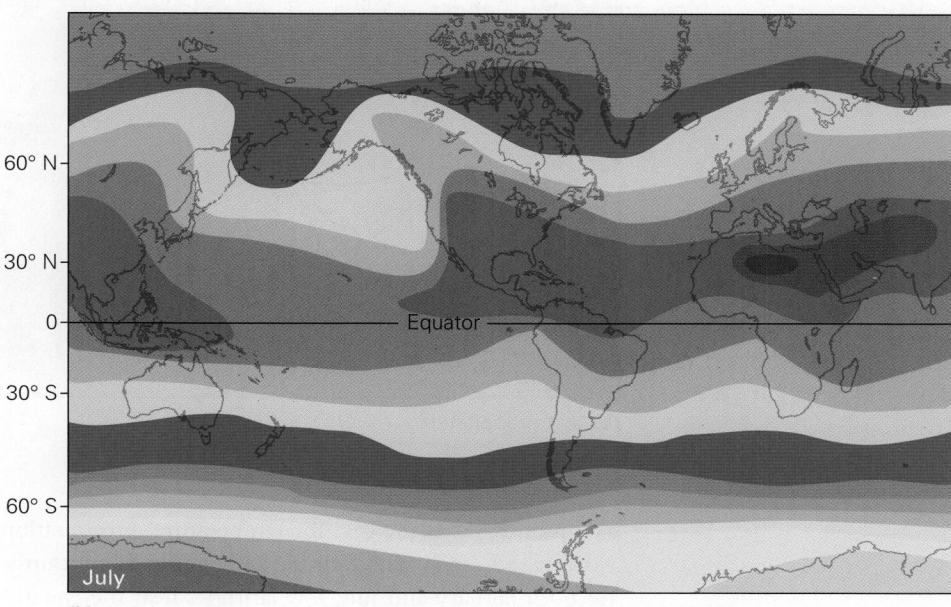

(b) In July, northern mid-latitudes are warm and southern mid-latitudes are cool.

do not exactly parallel latitude lines, and they are not all continuous around the planet because, as we'll see, several other factors besides latitude and season affect temperature.

Using the methods for measuring precipitation described in Chapter 16, scientists can also calculate **average precipitation** for a defined time period, such

FIGURE 19.3 Global map of average precipitation.

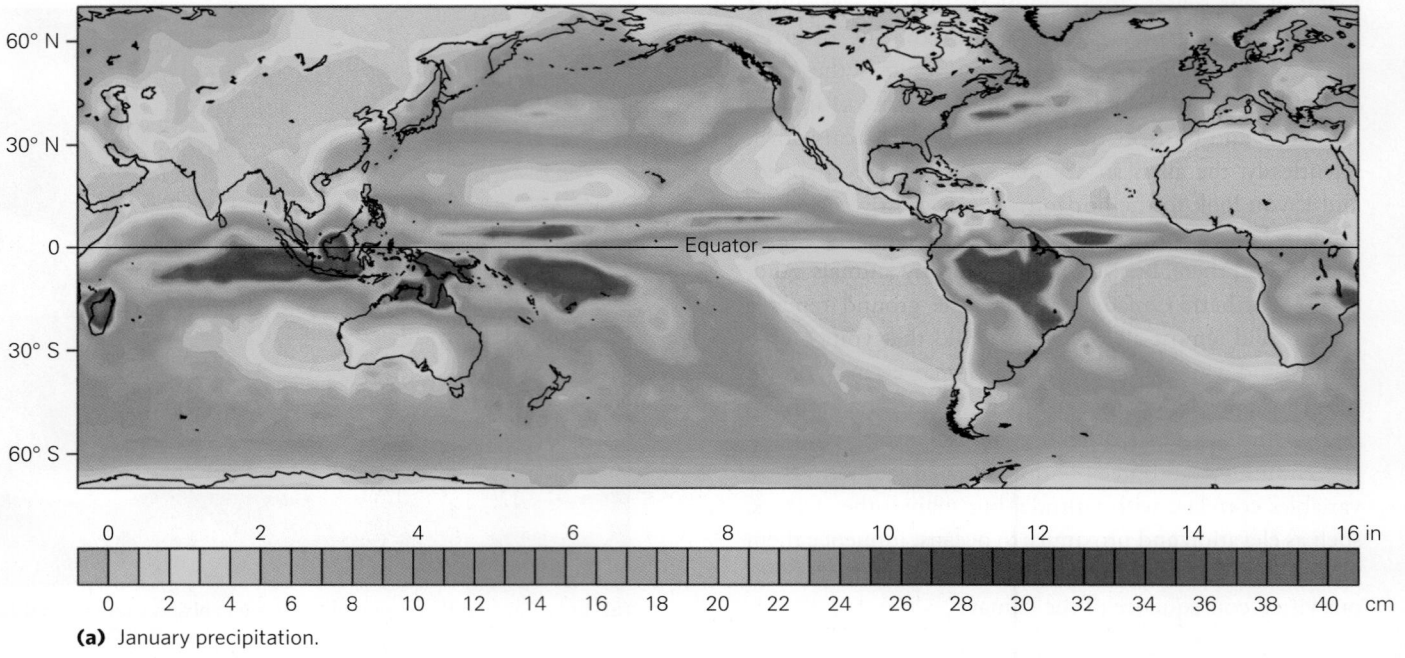

(a) January precipitation.

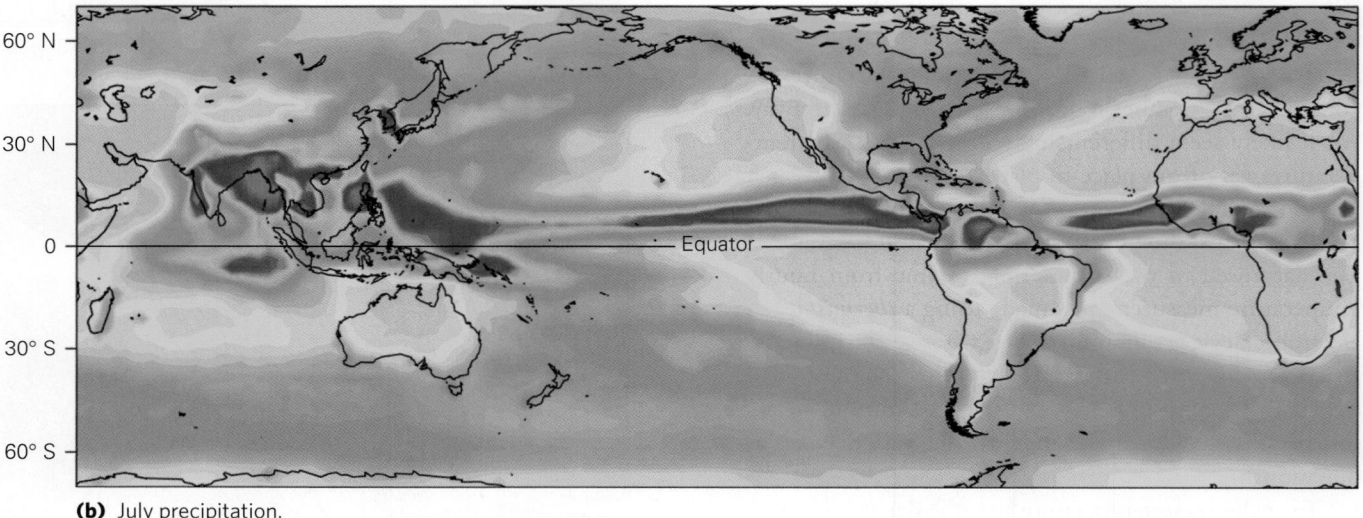

(b) July precipitation.

as a month. As is the case for temperature, precipitation varies significantly with latitude and season. Significantly, for both January and July, low latitudes near the equator host more precipitation (Fig. 19.3). But the belt of highest precipitation does shift over the course of a year.

The Köppen-Geiger Climate Classification

Wladimir Köppen spent most of his adult life in Germany, where he became one of the first *climate scientists* (researchers who focus on understanding climate). In 1884, Köppen published a classification scheme of the Earth's climate zones. Later in life, Köppen worked with

Rudolf Geiger to refine the original scheme. The resulting **Köppen-Geiger climate classification (KGCC)** remains in use today (Table 19.1). Köppen and Geiger used natural vegetation as the primary criterion to characterize the climate of a given area, because, as we noted earlier, the assemblages of plant species that thrive in an area reflect the average, range, and seasonal variation of weather conditions over a period of many years.

The KGCC recognizes five main **climate groups**— tropical, arid, temperate, cold, and polar—which differ from one another in their average temperatures, in the temperature range between their hottest and coldest

TABLE 19.1 The Köppen-Geiger Climate Classification (KGCC)

Group	Type	Subtype	Description
A			**TROPICAL:** Temperature of the coldest month is 18°C or higher.
	f		Rainforest: Precipitation in driest month is at least 6 cm.
	m		Monsoonal: A short dry season, with precipitation in the driest month of <60 mm, but ≥[100 − (R/25)] mm.*
	w		Savannah: Well-defined winter dry season, with precipitation in the driest month of <60 mm or <[100 − (R/25)] mm.
B†			**ARID:** Either ≥ 70% of the annual precipitation falls in the summer half of the year and R < [20T + 280] mm, or ≥ 70% of the annual precipitation falls in the winter half of the year and R < 20T mm. *Alternatively*, neither half of the year has ≥ 70% of annual precipitation and R < [20T + 140] mm.
	W		Desert: R < one half the upper limit for classification as a B type.
	S		Steppe: R < the upper limit for classification as a B type, but is > 50% of that amount.
		h	*Hot:* T ≥ 18°C
		k	*Cold:* T < 18°C
C			**TEMPERATE:** Temperature of the warmest month is ≥10°C, and temperature of the coldest month is <18°C but >−3°C.
	s		Dry summer (Mediterranean): Precipitation in the driest month of the summer half of the year is <30 mm and is <33% of the amount in the wettest month of the winter half.
	w		Dry winter: Precipitation in the driest month of the winter half of the year is <10% of the amount in the wettest month of the summer half.
	f		No dry season: Precipitation is more evenly distributed throughout the year; criteria for neither s nor w satisfied.
		a	*Hot summer:* Temperature of the warmest month is ≥22°C.
		b	*Warm summer:* Temperature of each of the four warmest months is ≥10°C, but the warmest month is <22°C.
		c	*Cold summer:* Temperature of one to three months is ≥10°C, but the warmest month is <22°C.
D			**COLD:** Temperature of the warmest month is ≥10°C, and temperature of the coldest month is not more than −3°C.
	s		Same as for Group C
	w		Same as for Group C
	f		Same as for Group C
		a	Same as for Group C
		b	Same as for Group C
		c	Same as for Group C
		d	*Very cold winter:* Temperature of the coldest month is ≤−38°C (d designation is then used instead of a, b, or c).
E			**POLAR:** Temperature of the warmest month is <10°C.
	T		Tundra: Temperature of the warmest month is >0°C but <10°C.
	F		Frost (ice cap): Temperature of the warmest month is ≤0°C, and ice covers most land.

*In the formulas, R refers to annual rainfall in millimeters, and T refers to average annual temperature in degrees centigrade. The summer half of the year is April through September for the northern hemisphere and October through March for the southern hemisphere.
†Any climate that satisfies criteria for designation as a B type is classified as such, regardless of its other characteristics.

FIGURE 19.4 The Köppen-Geiger climate classification (KGCC).

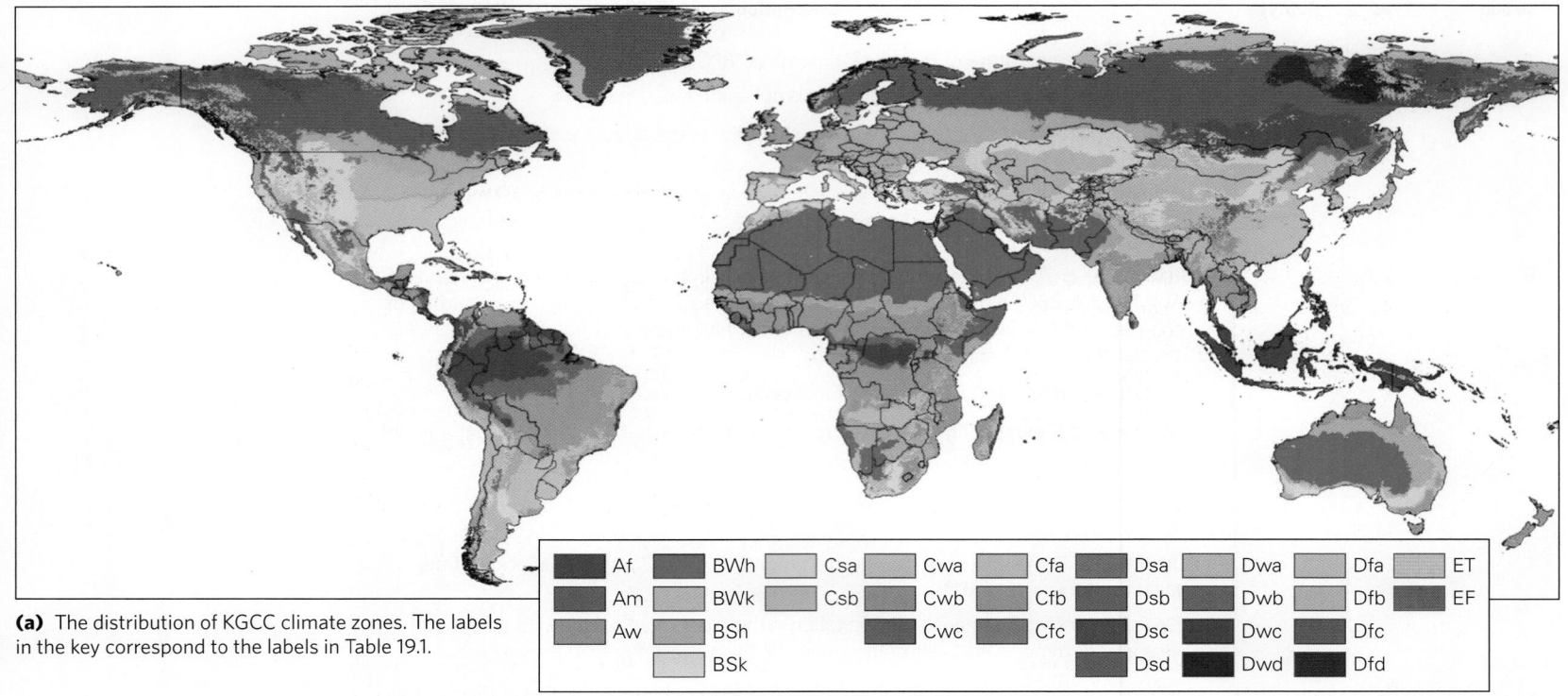

(a) The distribution of KGCC climate zones. The labels in the key correspond to the labels in Table 19.1.

Af	BWh	Csa
Am	BWk	Csb
Aw	BSh	
	BSk	

Cwa · Cfa · Dsa · Dwa · Dfa · ET
Cwb · Cfb · Dsb · Dwb · Dfb · EF
Cwc · Cfc · Dsc · Dwc · Dfc
Dsd · Dwd · Dfd

Tropics · **Mid-latitude** · **Pole**

BWh · Dfb · EF

Af · Cfa · ET

(b) Examples of climate zones. The inset in each photo indicates the KGCC climate zone it represents.

months, and in their average annual precipitation. Each climate group, in turn, includes a few *climate types* based on additional characteristics. For example, the three types of tropical climates (rainforest, monsoonal, and savannah) can be distinguished from one another based on whether or not rain falls throughout the year. A third level of subdivision (*climate subtypes*) distinguishes among regions, such as deserts, that can be either hot or cold. Climate groups and their subdivisions, as defined by the KGCC, can be portrayed on a map (Fig. 19.4).

Take-home message...

🏠 The Earth hosts many different climates. A region's climate primarily reflects its temperature and precipitation and their variation during the year. Local vegetation serves as the primary basis for defining climate groups and types in the Köppen-Geiger climate classification.

Quick Questions

- How does climate vary with latitude and with proximity to the ocean?
- Do you think the climate might be different in a city located upwind of a very large lake, compared with a city located downwind of the lake? Why?
- What is the Köppen-Geiger climate classification for the location where you live?

19.2 What Factors Control the Climate?

Like Baby Bear's porridge in the tale of *Goldilocks and the Three Bears*, the Earth is not too hot and not too cold . . . it's just right for liquid water and, therefore, for life to survive. This **Goldilocks effect** exists on the Earth for four reasons. First, the Earth lies within the Sun's **habitable zone**, a distance range that receives neither too much nor too little **insolation** (the amount of solar energy per unit area) at the planet's surface. Second, the concentration of greenhouse gases in the Earth's atmosphere is just right to trap enough heat so that air becomes warm, but not so that the air becomes scalding. Third, since the oceans can store heat, the circulation of water and air can redistribute heat to regulate temperature extremes. Fourth, the Earth

rotates fast enough that the Sun's energy is distributed around the Earth rather than on one side. To fully understand these statements, let's look in more detail at how the Sun's energy interacts with the Earth's surface and atmosphere. We can then explore how climates naturally vary around the world today, and why this variation exists.

The Earth's Energy Balance

The Sun floods the Earth with an immense amount of energy—mostly as visible light and UV radiation—every second. What happens to this energy? If the Earth had no atmosphere, all the energy would reach its surface. Some of this energy would instantly be reflected back into space from vegetation, rock, or water, but most would be absorbed by the Earth's surface. However, as we discussed in Chapter 16, the existence of our atmosphere significantly changes the fate of both solar energy reaching the Earth and reradiated energy leaving the Earth (Fig. 19.5). Specifically, some incoming radiation reflects off clouds directly back into space, and some gets absorbed by the air itself as well as by the water droplets and ice particles the air contains. Consequently, only about half of the Sun's energy that arrives at the top of the atmosphere reaches the Earth's surface to be absorbed. A small amount of the energy absorbed by the Earth's surface rises into the atmosphere by conduction and convection, or by evaporation of water and its recondensation into clouds. However, most of the absorbed energy reradiates as infrared radiation. Of this reradiated energy, a small amount travels directly back into space, but most gets absorbed by clouds or by gas molecules—H_2O, CO_2, and methane (CH_4)—in air, causing the atmosphere to warm further. Reradiation from the surface provides most of the heat energy stored in the atmosphere, so, as we noted in Chapter 16, the lower atmosphere is baked from

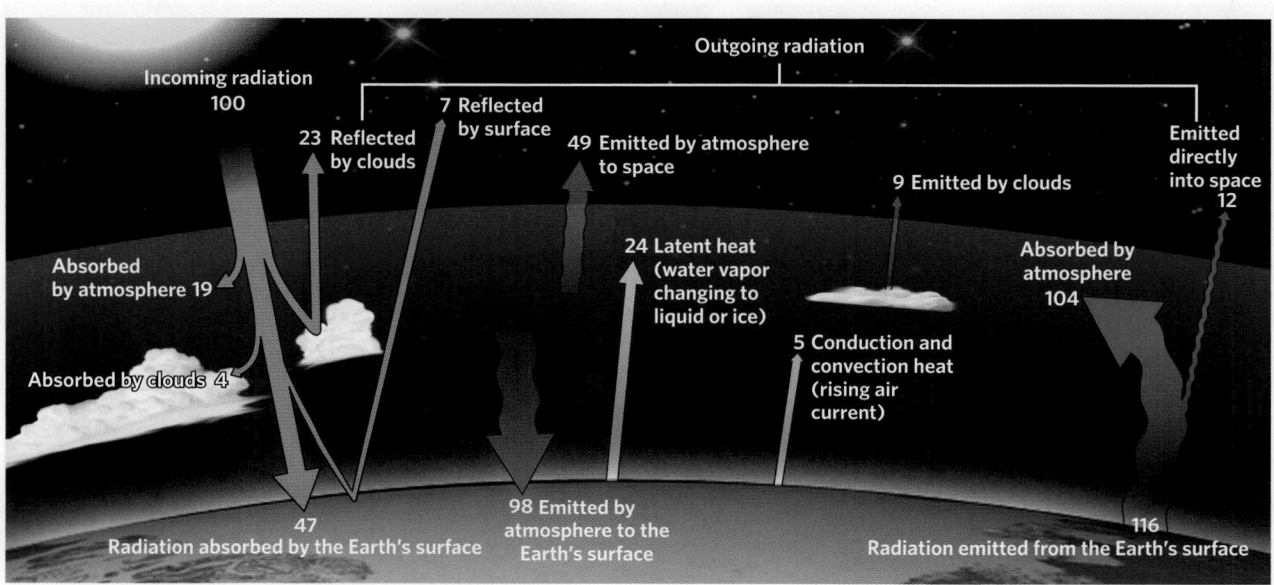

FIGURE 19.5 The energy balance in the Earth-atmosphere system. Each arrow represents the transfer of energy. The numbers represent arbitrary units of energy and show the relative amounts of energy in each transfer.

Incoming radiation 100

Outgoing radiation

23 Reflected by clouds

7 Reflected by surface

49 Emitted by atmosphere to space

9 Emitted by clouds

Emitted directly into space 12

Absorbed by atmosphere 19

24 Latent heat (water vapor changing to liquid or ice)

Absorbed by atmosphere 104

Absorbed by clouds 4

5 Conduction and convection heat (rising air current)

47 Radiation absorbed by the Earth's surface

98 Emitted by atmosphere to the Earth's surface

116 Radiation emitted from the Earth's surface

below, not broiled from above. The atmosphere doesn't hold onto reradiated infrared energy permanently. Some returns to the Earth's surface and contributes again to surface warming, while some escapes into space. When averaged over time, the total amount of energy arriving at the Earth from the Sun equals the total amount emitted back into space by the Earth; this relationship is called the Earth's **energy balance**.

Atmospheric scientists refer to the trapping of reradiated energy by the atmosphere as the **atmospheric greenhouse effect** (see Chapter 16), and the specific gases in the atmosphere that absorb reradiated energy from the Earth's surface as **greenhouse gases**. Because of the greenhouse effect, the Earth's atmosphere acts like a blanket, keeping the Earth's surface warmer than it would be if the Earth had no atmosphere. In fact, researchers estimate that without the greenhouse effect, the Earth's average global surface air temperature would be about 33°C (59°F) lower than it is today, and our planet's surface would be a frozen wasteland. Greenhouse gases make the Earth's climate habitable!

Latitudinal Control of Climate Belts

We noted earlier that the basic pattern of global temperature variation on the Earth depends on the latitude. This relation is a consequence of how the **angle of incidence**, meaning the angle between light beams from the Sun and the surface of the Earth, varies on a sphere and therefore affects the amount of insolation that an area of the ground receives. To see why, let's do a simple experiment (Fig. 19.6a). Aim a flashlight straight down on a tabletop from a distance of about 10 cm (4 in). The beam intersects

the tabletop at an angle of 90° and forms a small circle of bright light. Now, tilt the flashlight and aim it so that it intersects the tabletop at a shallower angle; the beam spreads out over an ellipse. Even though the amount of light energy emitted by the flashlight remains the same, the light in the ellipse looks dimmer because the light has spread out over a larger area. With this concept in mind, look at a cross section of the spherical Earth and note that sunlight strikes different latitudes at different angles (Fig. 19.6b). Locations in low latitudes, where sunbeams reach the Earth at a high angle, receive more insolation than do locations at higher latitudes, where sunbeams reach the Earth's surface at a low angle.

Notably, the energy that reaches the Earth's surface at a given latitude also depends on the amount of atmosphere through which the Sun's rays must pass, for air molecules scatter some radiation back into space before it can reach the Earth's surface. When the Sun's rays intersect the atmosphere at a steep angle, they encounter fewer air molecules before reaching the ground than when the rays arrive at a shallow angle, because the thickness of air as measured along the light beam depends on the angle between the atmosphere and the light beam. Because of the combined effect of insolation intensity and atmospheric thickness, a point where the angle of incidence is 90° receives much more insolation than a point where the angle of incidence is at a low angle.

Seasons and Their Consequences

If the Earth's axis of rotation were exactly perpendicular to the plane of the Earth's orbit around the Sun (its *orbital plane*), the climate at a given latitude would stay

Did you ever wonder...

why tropical rainforests lie near the equator?

FIGURE 19.6 The amount of insolation depends on the angle at which sunlight strikes the Earth.

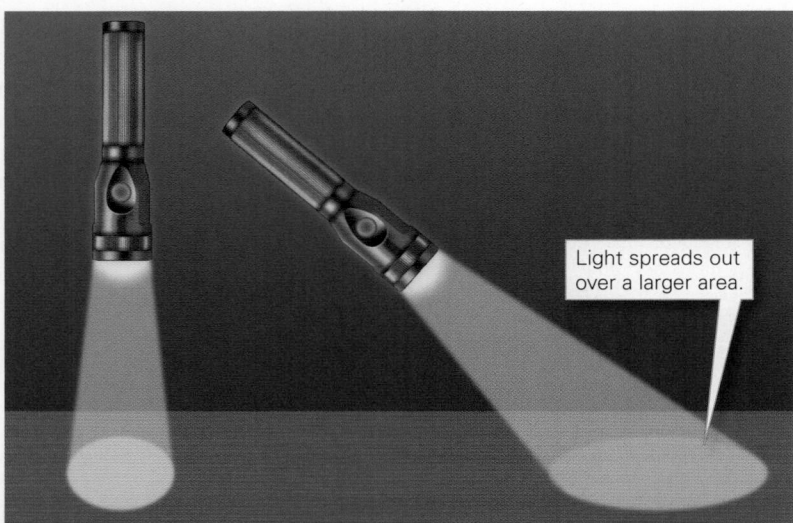

(a) The difference between the intensity of a vertical flashlight beam and a tilted one is an analogy for why insolation varies with latitude.

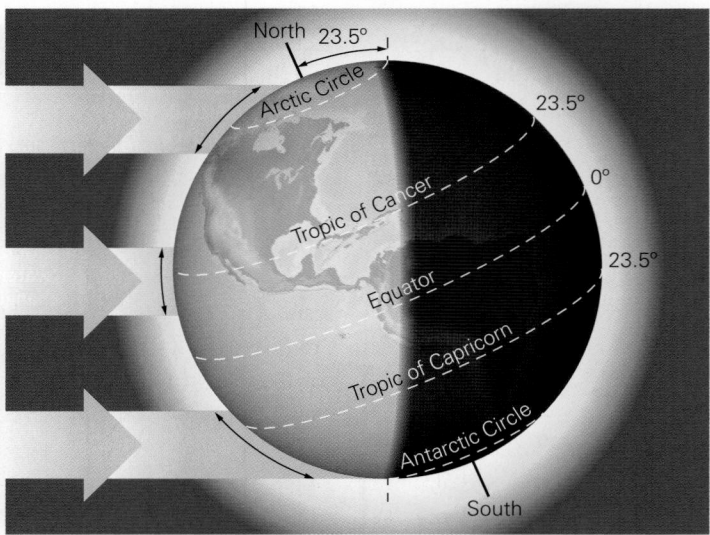

(b) High latitudes receive less insolation per square meter than do low latitudes.

FIGURE 19.7 The reason for the seasons. The Earth's axis of rotation tilts at an angle of 23.5° to a line perpendicular to the Earth's orbital plane and stays in that orientation as the Earth orbits the Sun. The seasons are labeled for the northern hemisphere.

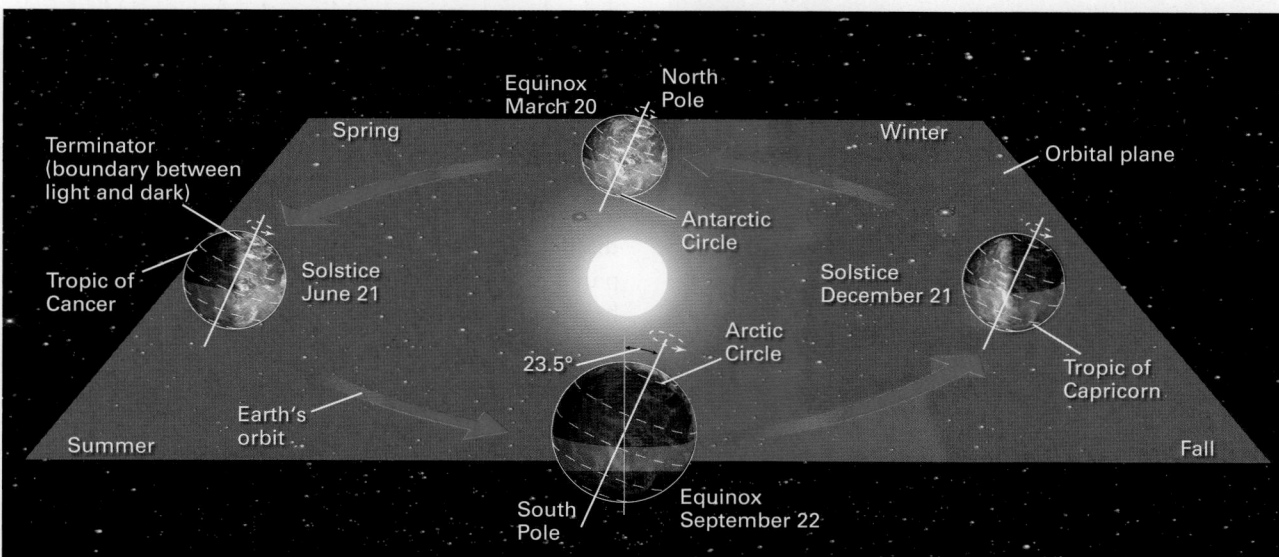

roughly the same all year. Instead, much of the Earth outside the tropics experiences distinct **seasons** during the course of the year, each characterized by different overall weather conditions: days are longer and warmer during the summer and shorter and colder during the winter. Fall and spring represent transitions between summer and winter.

Some people mistakenly believe that seasons reflect changes in the distance between the Earth and the Sun over the course of a year. That is not the case. Though the Earth's orbit is elliptical, it is only very slightly so (see Chapter 20). As a result, the distance between the Earth and the Sun varies by only 3% over the course of a year, and this variation has almost no effect on climate. In fact, the Earth lies closest to the Sun during the northern hemisphere's winter. Rather, the Earth has seasons because its axis of rotation tilts at an angle of 23.5° from a line drawn perpendicular to its orbital plane.

Why does the tilt of the Earth's axis control seasons? The Earth's axis stays in the same orientation, relative to an observer outside of the Solar System, as the Earth goes around the Sun (Fig. 19.7). As a result, during one half of the year, the southern hemisphere faces the Sun more directly, and during the other half of the year, the northern hemisphere faces the Sun more directly. For example, when the northern hemisphere tilts toward the Sun, the angle of incidence of incoming light is steeper, so the hemisphere receives more insolation, becomes warmer, and experiences summer. When the northern hemisphere tilts away from the Sun, the angle of incidence is

less steep, so the hemisphere receives less insolation and becomes colder and experiences winter.

The amount of insolation over the course of a season, at a location, primarily depends on the number of daylight hours that the region receives during the season. If you live in the temperate climates of the mid-latitudes, you'll notice that at the *summer solstice* (the day when the Sun rises highest in the sky at noon), the days are long and the nights are short, while at the *winter solstice* (the day when the Sun is lowest in the sky at noon), days are short and nights are long, a direct consequence of the tilt of the Earth's axis (Fig. 19.8a, b). North of the Arctic Circle, no solar radiation arrives at all during mid-winter in the northern hemisphere, because the North Pole tilts away from the Sun; but during mid-summer in the northern hemisphere, the Sun shines there 24 hours a day. Only on the spring and fall equinoxes does the Sun illuminate both hemispheres equally.

The Effect of Oceanic and Atmospheric Circulation on Climate

THE ROLE OF OCEANS AND THEIR CURRENTS. As Figure 19.2 shows, average air temperature depends not only on latitude and season, but also on proximity to the ocean, because oceans serve as vast heat reservoirs that slowly absorb a large amount of heat during the summer and slowly release it during the winter. Consequently, ocean water moderates the temperature of the air directly above. When the overlying air is colder than the ocean, the water transfers heat into the air by both conduction

Did you ever wonder...

why the Earth has seasons?

FIGURE 19.8 Solar heating of the Earth's surface varies with the seasons because of the angle of the Earth's axis relative to incoming beams of sunlight.

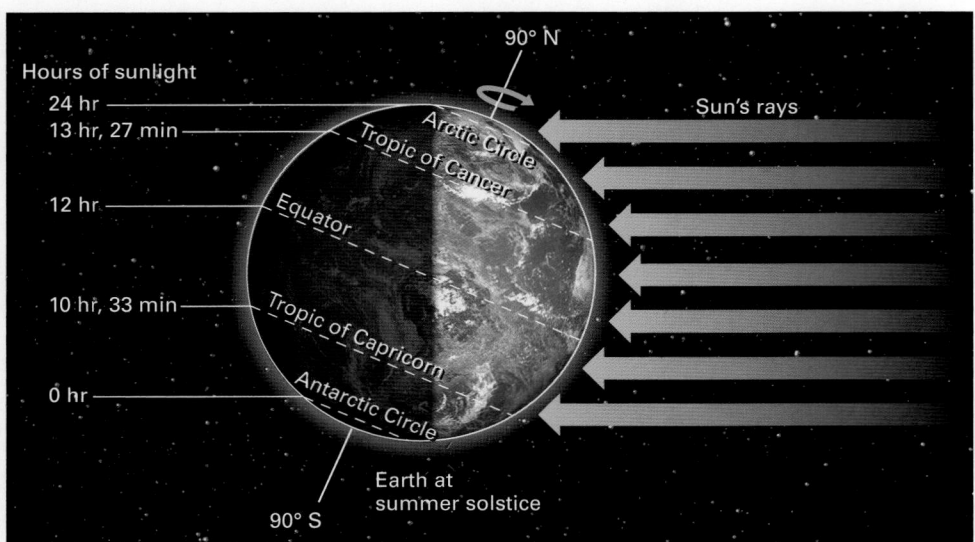

(a) At the northern hemisphere summer solstice, the North Polar region above the Arctic circle receives sunlight for a full 24 hours a day, and northern mid-latitudes have long, warm days.

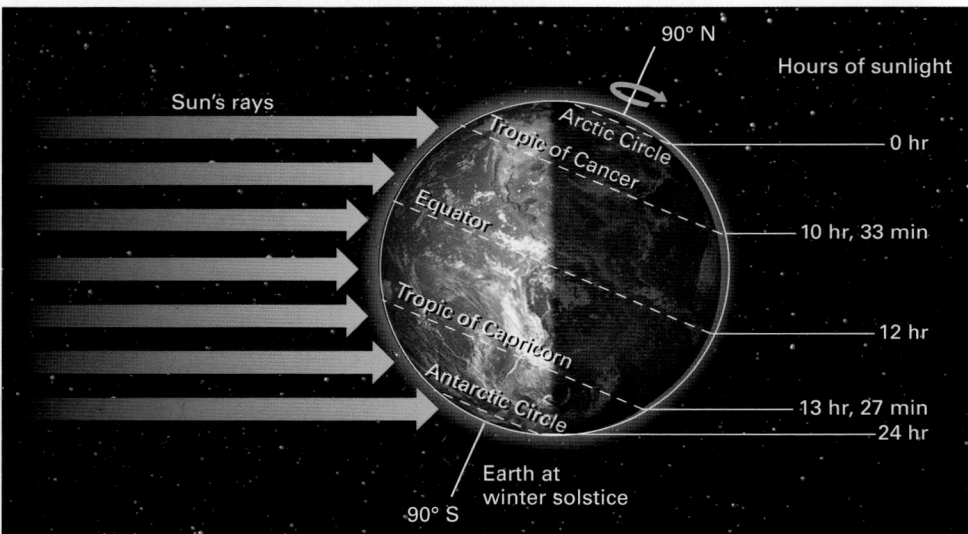

(b) At the northern hemisphere winter solstice, the North Polar region above the Arctic circle is dark all day, and northern mid-latitudes have short, cool days.

and evaporation, followed by the release of latent heat in clouds. In contrast, when the air is warmer than the ocean, air loses heat to the water. Oceans have such an important influence on overlying air temperature because water has a high **heat capacity**, meaning that it can absorb a large amount of heat without changing temperature very much. Also, since water is transparent and circulates, ocean water becomes warm down to the depth of the thermocline, 100–200 m (330–660 ft) below the surface. This means that a great volume of water can serve as a heat reservoir.

Because of the ocean's ability to moderate temperature, annual temperature variation tends to be greater inland, far from the coast, than in coastal areas where winds carry air from over the ocean onto the land. For example, a location in Missouri, in the middle of the United States, experiences seasonal variation of 28°C (50°F), whereas temperatures at the same latitude along the Pacific coast of California display seasonal variation of only 7°C (13°F). On all continents, the moderation effect of oceans decreases progressively inland from the coast.

Because the warm water serves as a thermal reservoir that loses or gains heat slowly, ocean currents can transport heat from lower latitudes to higher latitudes. This heat produces warmer climates in nearby coastal lands. For example, Ireland and the United Kingdom have relatively mild winters, despite their high-latitude locations, because the Gulf Stream carries warm water to high latitudes in the Atlantic (**Fig. 19.9**; compare with Fig. 19.2a). The average January temperature along the coast of Labrador, Canada, which lies at the same latitude as Ireland but borders cold-water surface currents, is about 15°C (27°F) cooler than that of Ireland.

THE ROLE OF AIR CIRCULATION. Global air circulation patterns strongly influence climate (see Chapter 17). For example, in the tropics, large quantities of rain fall along the *intertropical convergence zone (ITCZ)*, where trade winds at the base of the southern Hadley cell converge with trade winds at the base of the northern Hadley cell. This convergence produces updrafts that carry moisture to high elevation, where it fuels thunderstorms along the ITCZ. The migration of the ITCZ from just south of the equator to, in the case of Southeast Asia, a latitude of about 25° N (compare the two parts of Figure 19.3) causes monsoonal climates with distinct summer wet seasons and winter dry seasons. In contrast, in the subtropical latitudes, air in the descending parts of the Hadley cells dries out and warms as it descends, causing subtropical desert climates to develop; grasslands known as *steppes* delineate the transition zone between rainforests and deserts. At equatorial latitudes (up to 5° N and 5° S of the equator), the Walker circulation in an east-west direction drives ENSO, the alternation between El Niño and La Niña years, which affects climate on a multiyear time frame. Effectively, the Hadley cells and the Walker circulation redistribute heat and moisture in the atmosphere, so they control the amount of rainfall and the timing of wet and dry seasons at tropical and subtropical latitudes.

Atmospheric circulations at mid-latitudes, a region that includes temperate climate belts, are more complicated than those of lower latitudes, because *air masses*—bodies of air within which temperature and humidity

FIGURE 19.9 Atlantic Ocean temperatures and associated currents. The white arrows indicate the direction of surface-current flow.

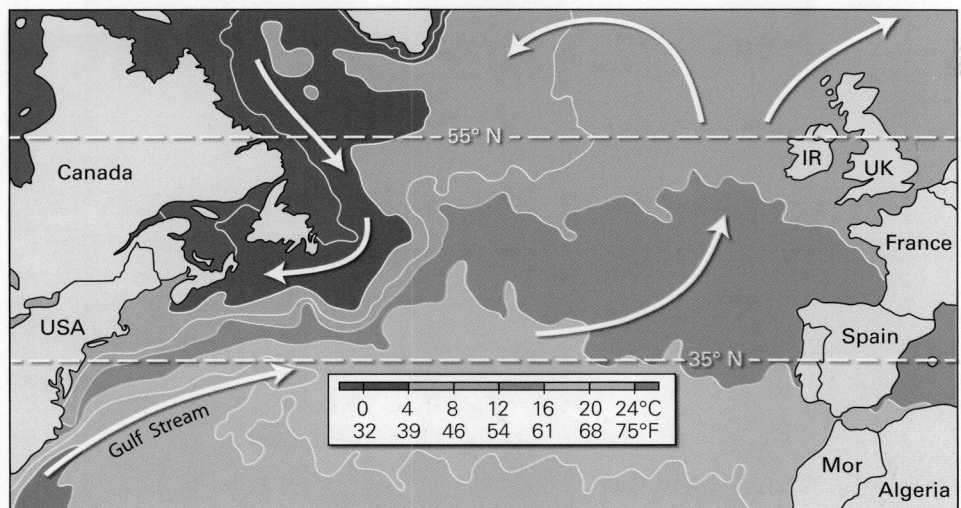

are relatively uniform—form in certain locations (see Chapter 18). Notably, semipermanent (long-lasting) regions of high atmospheric pressure develop at the Earth's surface beneath large cold air masses, whereas semipermanent regions of low atmospheric pressure develop beneath large warm air masses (**Fig. 19.10**). These *pressure systems*

have a dramatic effect on local climates at mid-latitudes because the pressure gradients between air masses drive the winds. At the Earth's surface, winds blow counterclockwise around low-pressure systems and clockwise around high-pressure systems in the northern hemisphere (the opposite is true in the southern hemisphere). In addition,

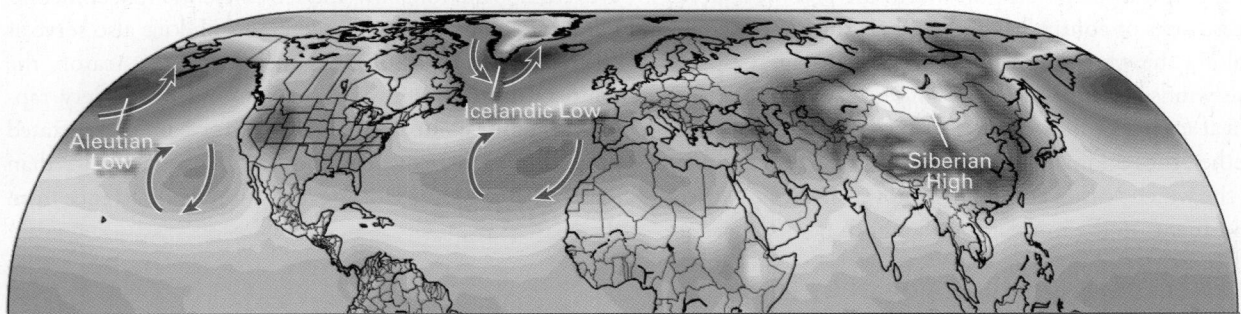

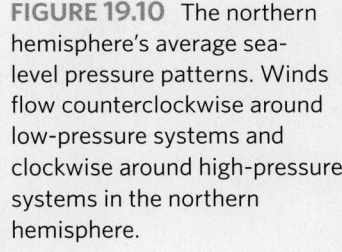

FIGURE 19.10 The northern hemisphere's average sea-level pressure patterns. Winds flow counterclockwise around low-pressure systems and clockwise around high-pressure systems in the northern hemisphere.

(a) Northern hemisphere winter (December through February). This map indicates names of semipermanent highs and lows.

996 | 1,000 | 1,004 | 1,008 | 1,012 | 1,016 | 1,020 | 1,024 | 1,028 | 1,032
Mean sea-level pressure (mb)

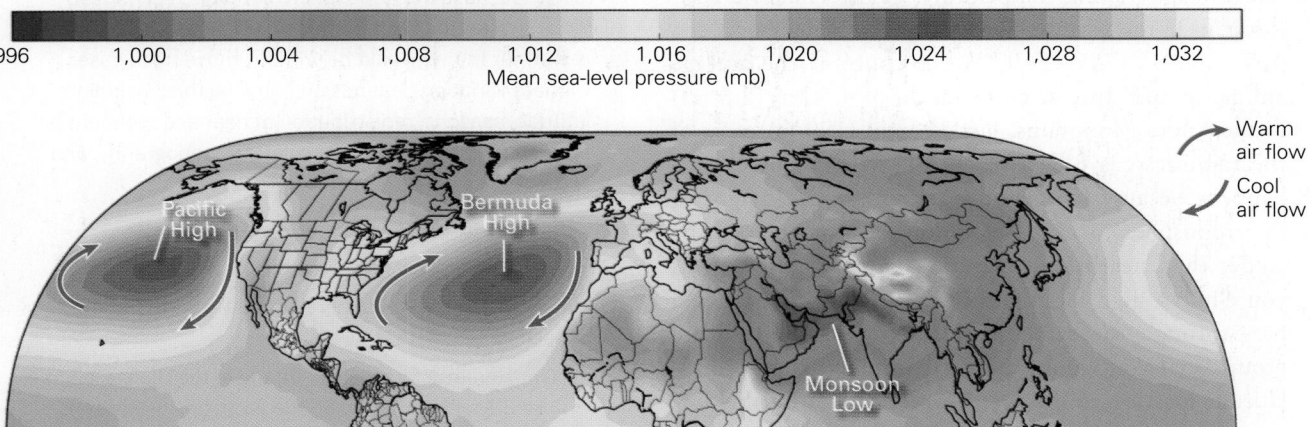

(b) Northern hemisphere summer (June through August). The Bermuda High and the Pacific High become pronounced during the summer.

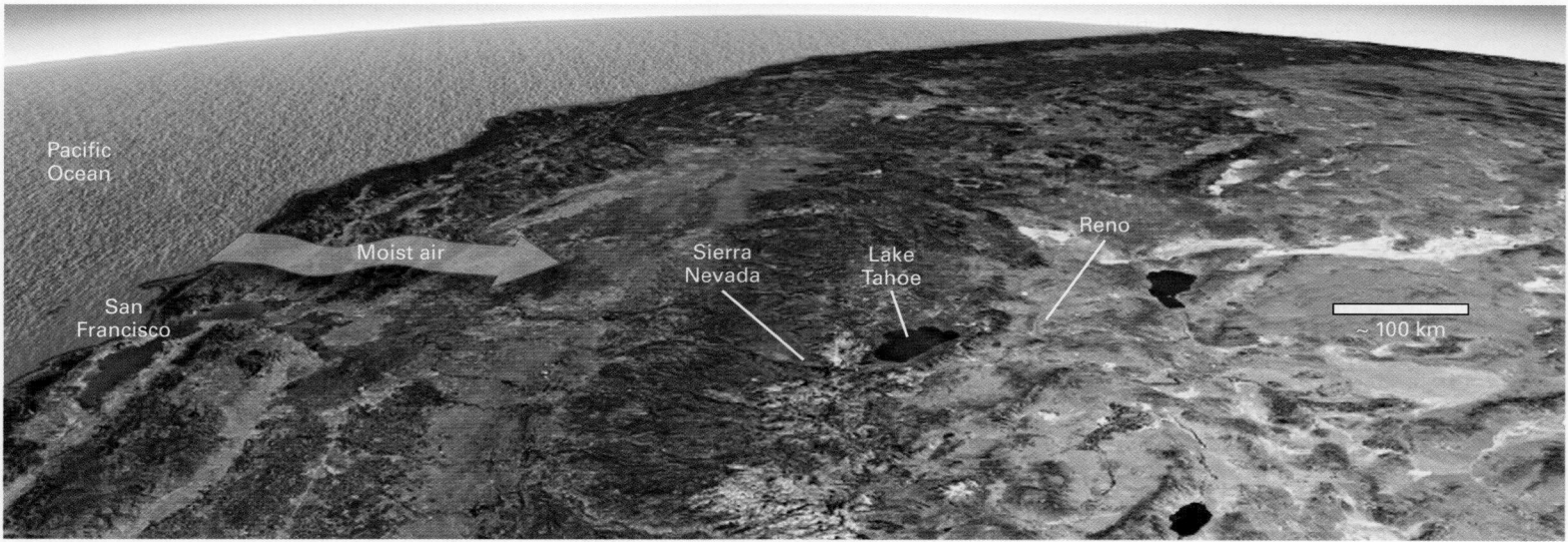

strong migrating low-pressure zones occur, producing huge mid-latitude cyclones, which dramatically transport cold air from higher to lower latitudes and warmer air from lower to high latitudes. Notably, regional winds associated with the semipermanent high-pressure systems over the North Atlantic and North Pacific in the northern hemisphere summer transport warm air poleward along the east sides of continents and cold air toward the equator along the west sides of continents (see Fig. 19.10b). These winds, in turn, drive ocean currents that accelerate heat exchange between the tropics and polar regions. Together, the regional winds and associated ocean currents cool the west coasts of continents and warm the east coasts of continents in the mid-latitudes of both hemispheres.

The Effect of Mountain Topography and Snow Cover on Climate

The average annual temperature in the Amazon rainforest near the eastern border of Peru ranges between 32°C and 36°C (90°F–97°F). In contrast, the average annual temperature at an elevation of 4.5 km (2.8 mi) in the Andes Mountains, just 300 km (190 mi) west of the rainforest, is 8°C–12°C (46°F–54°F). This change happens because temperature decreases with height in the troposphere, so high-elevation locations tend to be colder than nearby low-elevation locations. In fact, as you climb a high mountain in the mid-latitudes, you'll pass through the same succession of KGCC climate groups that you would cross if you hiked from a mid-latitude location to the Arctic Ocean at a uniformly low elevation. Recall that the presence of a mountain range

also controls distribution of rainfall (see Chapter 13), so the climate on the windward side of the range may be very different from that in the rain shadow on the leeward side (Fig. 19.11).

If you try to traverse a snowfield or ice field on a sunny day, you have to wear very dark glasses to protect your eyes from glare, for snow and ice efficiently reflect incoming solar energy back to space. Snow and ice also serve as efficient radiators of infrared energy. For this reason, the temperature over snow-covered ground drops very rapidly at night, compared with that over bare or vegetated ground, so snow-covered land has a cooler climate than dry land at the same latitude, elevation, and distance from the coast.

Take-home message...

Because the atmosphere contains greenhouse gases, the Earth's surface temperature overall is warmer than it would be without those trace gases. Climate variation depends on many factors, including latitude, seasons, redistribution of heat and moisture by atmospheric and oceanic circulations, topography, and snow and ice cover.

Quick Questions ———

- Which gases are important greenhouse gases?
- How does the atmosphere affect the Earth's energy balance?
- Name three factors that contribute to the change in temperature with latitude and the Earth having seasons?

19.3 Climates of the Past

How often have you seen a news report proclaiming, "Record High Temperature Expected Tomorrow"? Does such a headline mean that the climate is changing? Not necessarily—a single hot spell or cold snap could simply represent a *natural variation* in the weather, meaning a change that happens simply because weather responds to input from several causes that sometimes reinforce each other. But when a new set of conditions—such as an overall increase in average temperature or an overall change in precipitation—becomes the norm for a region for a relatively long time, then climate change has occurred. An increase in average global atmospheric and sea-surface temperatures represents **global warming**, and a decrease represents **global cooling**.

Strong evidence indicates that contemporary climate change, manifested by global warming during the recent past, has been taking place, will continue to occur into the near future, and has consequences for the Earth System, economies, communities, and lifestyle. Consequently, climate change has become a high-profile political issue in recent decades. In the remainder of this chapter, we first describe studies that characterize climates of the past, and then we examine the record of climate change through Earth history. Finally, we focus on contemporary climate change (change that has happened during the past century and is ongoing).

Characterizing Paleoclimate

What we know about the Earth's past climate, or **paleoclimate**, comes from studying a variety of *paleoclimate indicators*, meaning clues to paleoclimate preserved in geological materials or in fossils. By determining the numerical ages of these indicators, researchers can compile a record of how climate has changed over time. Such information provides a sense of how much climate could change in the future. Climate change takes place at two different scales: *long-term climate change*, which happens over millions to hundreds of millions of years, and *short-term climate change*, which happens over decades to centuries or millennia. The *resolution* (detail) of the paleoclimatological record allows long-term variations to be recognized back through at least the Phanerozoic. Short-term variations, however, can be recognized only for relatively recent geologic time.

Long-Term Climate Change

The depositional settings in which sediment accumulates, and the assemblages of organisms that live in or on sediment, depend on the climate. Therefore, the stratigraphy of sedimentary successions, the fossils preserved in the strata, and the ratio of isotopes in certain minerals in sediment, provide a record of long-term paleoclimate.

Taken as a whole, the stratigraphic record shows that the Earth's surface temperature has stayed between the freezing point of water and the boiling point of water for most of Earth history since the beginning of the Archean. The Earth's temperature has not remained uniform, however. Geologists refer to a time when it has been warmer as a **hothouse period** (or *greenhouse period*) and a time when it has been cooler as an **icehouse period** (Fig. 19.12a). During hothouse periods, lands at polar latitudes were largely ice free, whereas during icehouse periods, polar regions were ice covered. The more familiar term, *ice age*, refers to a time interval during an icehouse period when ice sheets advanced and covered substantial areas of continents at mid-latitudes (see Chapter 13). Not every icehouse period includes an ice age. Geologists recognize about a half-dozen ice ages in the record of Earth history. Notably, during the late Proterozoic ice age, glaciers covered all land—even at the equator—and the oceans may have frozen over. Researchers refer to the resulting frozen globe as *snowball Earth* (see Chapter 9).

If we focus on the climate record of the past 100 million years, we can get a sense of the range of typical long-term climate changes that have happened in the Earth System. The climate of the late Mesozoic was much warmer than that of today. In fact, temperatures may have been 2°C–6°C (4°F–11°F) warmer than our current climate at the equator, and 20°C (36°F) warmer at the poles. Consequently, dinosaurs could live at latitudes equivalent to what is ice-covered central Greenland today. In fact, it was so warm that no polar ice caps existed, and average sea-surface temperatures in some parts of the ocean may have been as much as 17°C (31°F) warmer than today. Starting about 80 Ma, however, the Earth's surface and atmosphere began to cool. The cooling trend continued through the Cenozoic, except for about a 7-million-year-long interval (56–49 Ma), known as the *Eocene climatic optimum*, when hothouse conditions briefly returned and, by some estimates, the CO_2 concentration reached 1,000 ppm (Fig. 19.12b). Following the Eocene climatic optimum, cooling began again, and our planet entered an icehouse period at about 34 Ma, the time when the Antarctic ice sheet started to form. Beginning at about 2.6 Ma, the Pleistocene Ice Age began, and glaciers advanced repeatedly over large areas of the continents of the northern hemisphere.

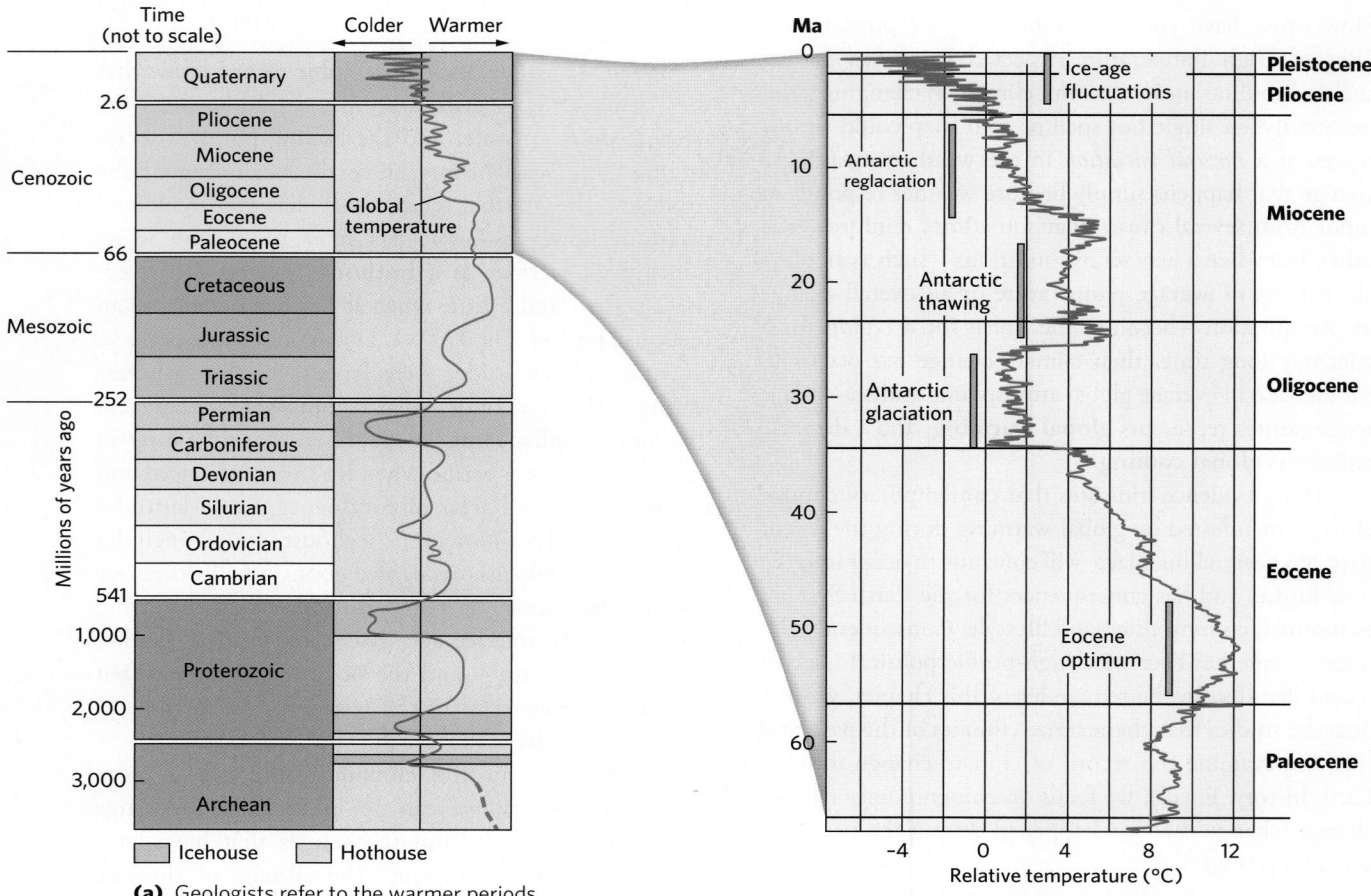

(a) Geologists refer to the warmer periods as hothouse periods and to the colder ones as icehouse periods.

(b) The peak temperature of the Cenozoic occurred during the Eocene, between 55 and 48 Ma, a time called the Eocene optimum. The horizontal axis indicates temperature relative to average global temperature of the past several decades.

Short-Term Climate Change

When farmers cultivated wheat 4,500 years ago in what is now Iraq, the region enjoyed frequent rains and hosted lush vegetation, but today, it has an arid climate and desert landscapes. When Queen Elizabeth I reigned over England 450 years ago, glaciers were advancing down valleys in the Alps, and the canals of the Netherlands froze over every winter (so ice skating became a tradition). Today, the glaciers have retreated and the canals rarely freeze. Clearly, climate can change noticeably on a scale of centuries to millennia. To characterize short-term climate change over the past 800,000 years, paleoclimatologists examine several types of paleoclimate indicators:

- *Historical records:* Because human life depends directly on climate conditions, people tend to record unusual conditions such as droughts, floods, long cold snaps, or severe storms. Prior to the development of writing, these memories were held within oral traditions,

but since the advent of writing, they have appeared in diaries and histories.

- *Microfossils:* Fossilized marine plankton in Pleistocene and Holocene deposits provide a record of climate change because the assemblage of species living when the deposits formed reflects water temperature at that time, and because these plankton have distinctive shells that allow the identification of species. Similarly, different plant species produce pollen grains with distinctive shapes, so the study of pollen preserved in lake deposits allows researchers to determine how the ecosystem on land and, therefore, the climate at a given latitude, has changed (Fig. 19.13).

- *Oxygen-isotope ratios in ice:* The ratio of ^{18}O to ^{16}O isotopes in glacial ice provides a clue to the atmospheric temperature in which the snow that makes

(a) Example of a lake sediment core.

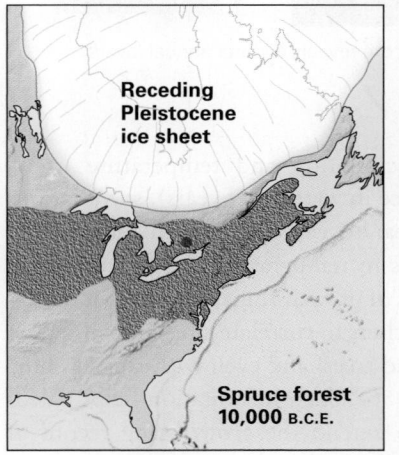

Receding Pleistocene ice sheet

Spruce forest 10,000 B.C.E.

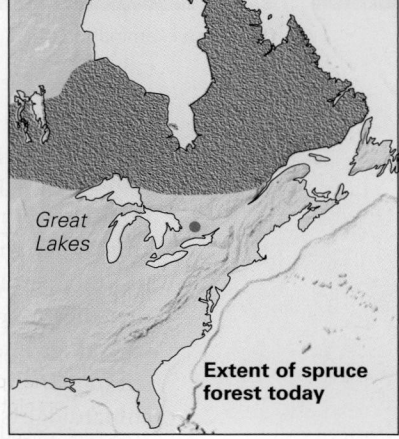

Great Lakes

Extent of spruce forest today

Grass dominates when it's warmer.

Spruce dominates when it's cooler.

Tree pollen
Grass pollen

(b) The pollen preserved in a lake sediment core can be used to determine the atmospheric temperatures at the time the sediment was deposited.

(c) As we see from the pollen analysis, spruce forests (green) grew farther south 12,000 years ago (10,000 B.C.E.) than they do today. The red dots point out the location where the core sample was taken.

up the ice formed. Therefore, the ratios of these isotopes measured in a succession of ice layers in a glacier can indicate temperature change over time. Paleoclimatologists have obtained *ice cores* in Antarctica that provide a temperature record spanning over 800,000 years (Fig. 19.14a–c). Ice records preserved in mountain glaciers can characterize the climate at nonpolar latitudes, but unfortunately, these glaciers are rapidly disappearing, so researchers are rushing to collect cores from them. Not surprisingly, the logistics of transporting ice from a remote mountaintop to a laboratory halfway around the world can be extremely complicated.

- *Air bubbles in glacial ice:* When glacial ice forms, the ice crystals within it surround tiny pockets of air that become preserved as bubbles in the ice. By sampling these bubbles, researchers can measure

the composition of air at the time the ice formed. Bubble studies document changes in CO_2 concentrations over time.

- *Oxygen-isotope ratios in plankton shells:* The $^{18}O/^{16}O$ ratio in the calcite ($CaCO_3$) that makes up plankton shells in marine sediments provides an indication of past ocean-surface temperatures. Therefore, study of marine sediment cores can extend the record of short-term temperature change back over millions of years (Fig. 19.14d).

- *Growth rings:* Trees, corals, clams, and many other organisms develop distinct rings as they grow because their rate of growth varies with the season (Fig. 19.15a). These **growth rings** can be used as markers for a given year and, in some cases, to characterize climate conditions during that year. The thickness of growth rings in trees, for example,

FIGURE 19.14 Paleoclimate data can be obtained by studying oxygen-isotope ratios in ice and sediment cores.

(a) Researchers drilling an ice core.

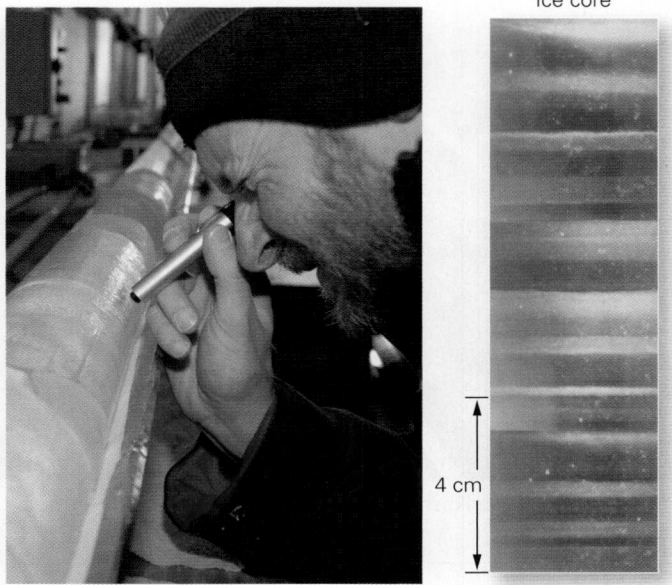

Ice core

4 cm

(b) Close examination of an ice core reveals distinct annual layers.

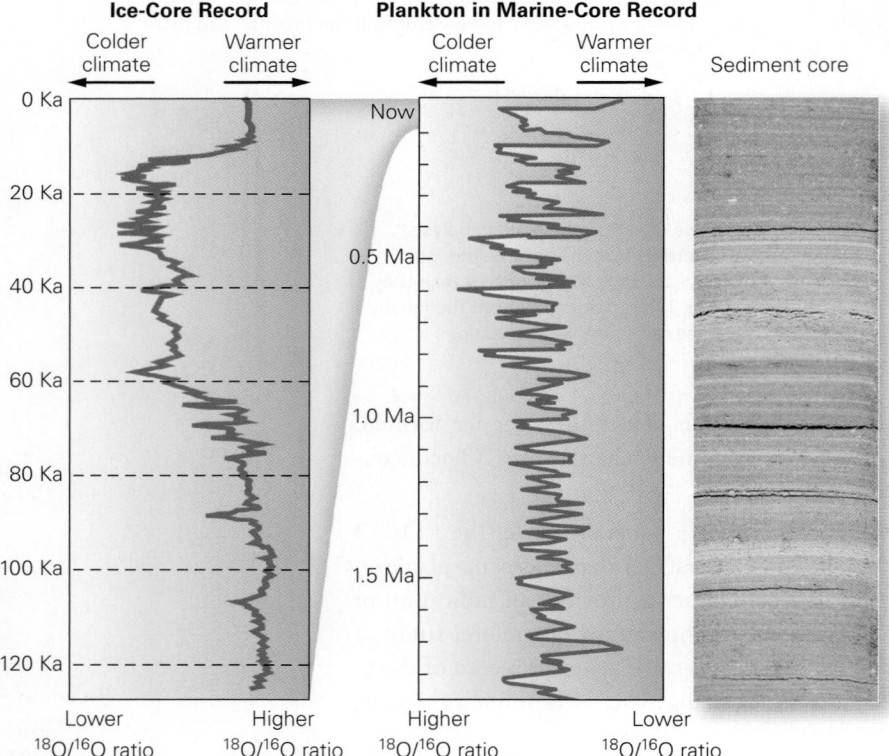

Ice-Core Record

Colder climate → Warmer climate

0 Ka
20 Ka
40 Ka
60 Ka
80 Ka
100 Ka
120 Ka

Lower
$^{18}O/^{16}O$ ratio

Higher
$^{18}O/^{16}O$ ratio

Plankton in Marine-Core Record

Colder climate → Warmer climate

Now
0.5 Ma
1.0 Ma
1.5 Ma

Higher
$^{18}O/^{16}O$ ratio

Lower
$^{18}O/^{16}O$ ratio

Sediment core

(c) $^{18}O/^{16}O$ ratios in the ice indicate temperatures over the past 120,000 years. The isotope ratio depends on many factors, but overall, lower ratios in ice correlate with colder temperatures.

(d) $^{18}O/^{16}O$ ratios in plankton shells in layers in a core of marine sediment record past temperatures over longer time scales. Note that lower ratios in shells correlate with warmer temperatures.

depends on precipitation and temperature (more growth takes place in warm, wet years), so variations in rings from year to year serve as an indicator of annual variations in temperature and precipitation. Distinct patterns of growth rings serve as "bar codes" that allow researchers to correlate rings in living trees with rings in dead trees, and even with rings in buried logs (Fig. 19.15b). Such information helps paleoclimatologists to extend the growth-ring record of climate variation far back in time.

By compiling data from the study of the paleoclimate indicators described above, researchers have identified several alternating warming and cooling intervals during the past 800,000 years (Fig. 19.16). These temperature changes controlled the advances and retreats of glaciers during the Pleistocene Ice Age (see Chapter 13). The resolution of data allows temperature changes lasting only for a few centuries to be recognized during the past 15,000 years. For example, in the northern hemisphere, rapid warming began about 15,000 years ago, during which the last continental glaciers of the ice age substantially melted (Fig. 19.17a). But this warming trend was short-lived; between about 13,000 and 11,700 years ago, cold temperatures returned and much of Europe became a treeless tundra. Researchers have named this cold interval the *Younger Dryas*, after an Arctic flower that became widespread during that time. Temperatures warmed again starting about

FIGURE 19.15 Tree rings provide a record of past climate. More growth, leading to wider rings, happens in warm, wet years than in cold, dry years.

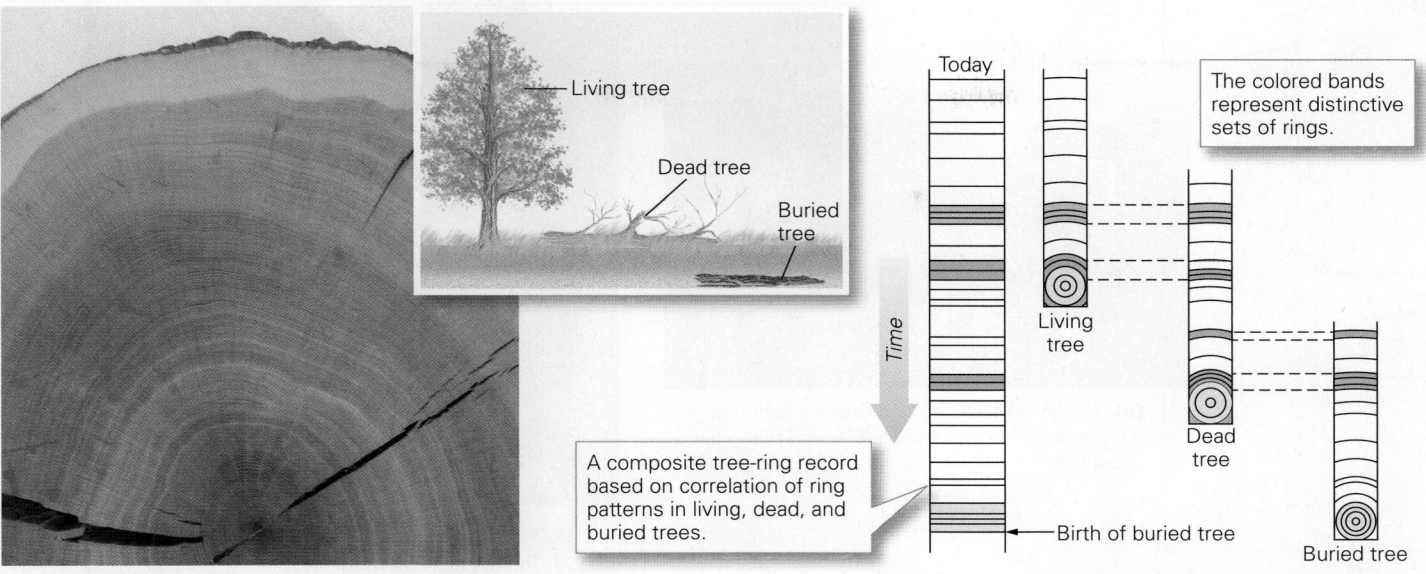

(a) Each ring represents growth in one year. Ring width reflects average precipitation and length of the growing season.

The colored bands represent distinctive sets of rings.

A composite tree-ring record based on correlation of ring patterns in living, dead, and buried trees.

(b) By correlating distinctive ring patterns, researchers can obtain a composite tree-ring record going back thousands of years.

11,700 years ago, and this change marked the beginning of the Holocene. Between 9,500 and 5,500 years before the present—an interval called the *Holocene maximum*—average temperatures were a little higher than today and the Middle East was unusually wet and fertile; these conditions may have contributed to the rise of agriculture in what historians call the "fertile crescent." Another temperature peak happened during the *Medieval Warm Period*, between 950 and 1250 C.E., allowing Vikings to settle along the coast of Greenland (Fig. 19.17b). Temperatures dropped again between 1500 C.E. and 1800 C.E., a period known as the *Little Ice Age*. This was the time interval during which alpine glaciers advanced (Fig. 19.17c) and the canals of the Netherlands froze over in winter. Overall, climate has warmed since the end of the Little Ice Age.

Take-home message...

Climate change over geologic time includes episodes of both global warming and global cooling. Researchers use the stratigraphic record to characterize long-term changes (taking place over millions to hundreds of millions of years). They use other indicators, such as plankton species, tree rings, pollen, or variations in isotope ratios, to document short-term changes that have happened over the past 800,000 years.

Quick Questions

- How has climate changed in the northern hemisphere since the end of the last glaciation?
- Are we in an icehouse or hothouse period today?
- What types of paleoclimate indicators might be used to determine climate 100,000 years ago?

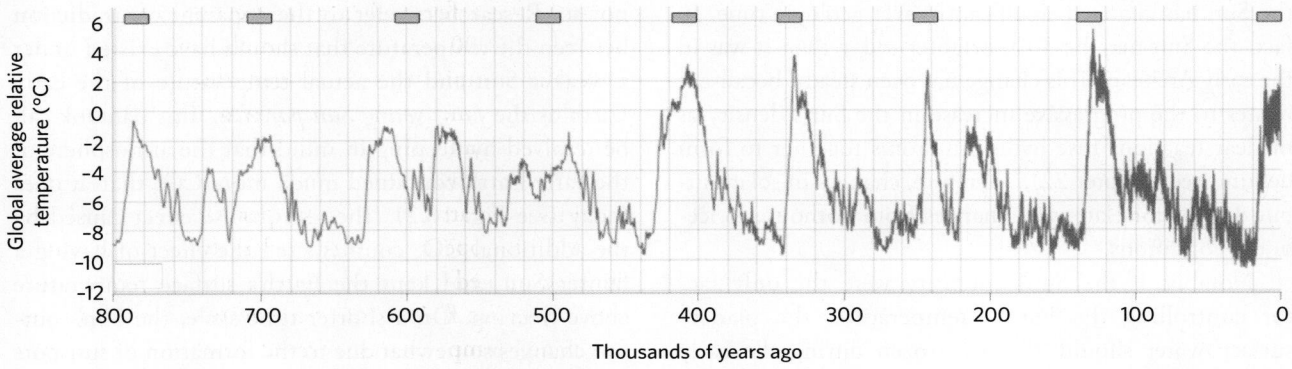

FIGURE 19.16 Temperature record of the past 800,000 years obtained by analyzing an ice core extracted from the Antarctic ice sheet, relative to the average temperature of the past several decades. Horizontal bars at the top of the graph indicate the duration of interglacials.

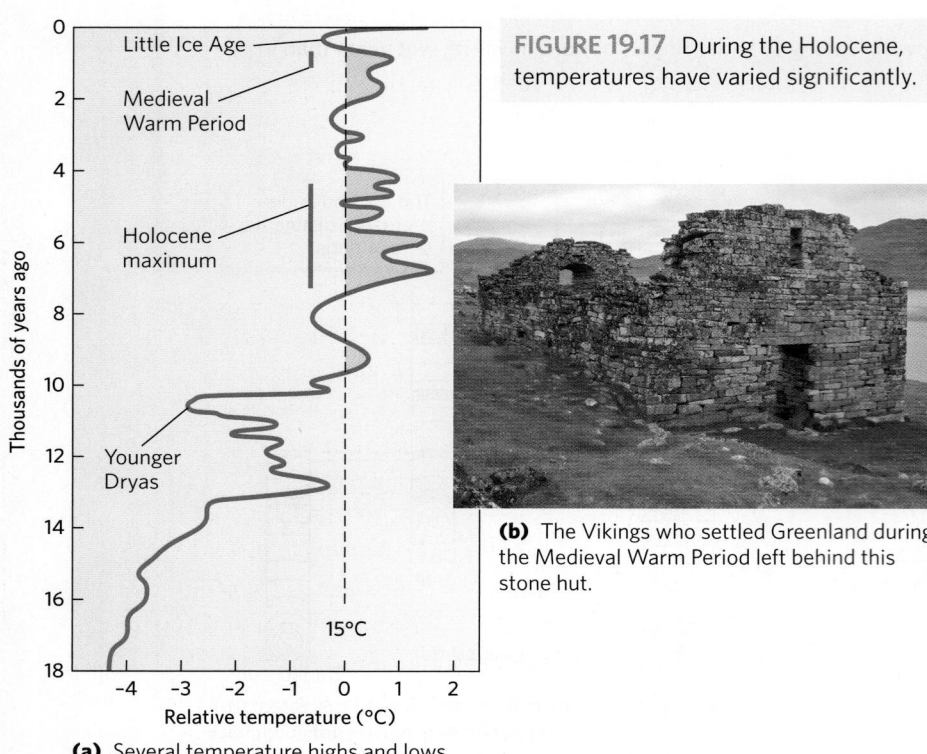

FIGURE 19.17 During the Holocene, temperatures have varied significantly.

(a) Several temperature highs and lows have taken place since the most recent glaciation.

(b) The Vikings who settled Greenland during the Medieval Warm Period left behind this stone hut.

At the end of the Little Ice Age, this tributary glacier in France reached the main valley floor.

Today, glaciers extend only partway down the side valleys.

(c) During the Little Ice Age, glaciers in the Alps advanced. They have since melted away.

19.4 Causes of Climate Change

Though the Earth has remained habitable for life forms for over 3 billion years, atmospheric temperature has changed significantly over that time. Why do these changes take place? The answer depends, in part, on the time frame under consideration. We will see, however, that the *carbon cycle* plays an important role, regardless of the time frame (Box 19.2).

Long-Term Climate Controls

No single factor determines when the Earth's climate, overall, shifts from hothouse to icehouse conditions. We discuss several possible factors below.

SOLAR ENERGY OUTPUT. Astronomers have determined that the intensity of radiation reaching the Earth from the Sun has changed significantly over geologic time. In fact, the Sun may be 30% brighter today than it was in the early Archean. This change has been steady, because it relates to the progressive increase in the Sun's density as nuclear reactions fuse hydrogen atoms together to form helium (see Chapter 22). Therefore, changes in solar output do not correlate with changes from hothouse to icehouse conditions.

Notably, if the Sun's intensity were the only factor controlling the Earth's temperature, the planet's surface water should all have frozen during the early Archean, but stratigraphic records prove that this was

not so. Researchers refer to the apparent contradiction between the temperature that should have existed under a weaker Sun and the actual temperature of the early Earth as the *faint young Sun paradox*. This paradox can be resolved by keeping in mind that the atmosphere of the early Earth contained much more CO_2 than it does today (see Fig. 16.2). The greenhouse effect caused by the additional CO_2 counteracted the effect of having a fainter Sun, and kept the Earth's surface temperature above freezing. On a shorter time scale, the Sun's output changes somewhat due to the formation of sunspots (see Chapter 22). This event occurs on an 11-year cycle,

BOX 19.2 **A Deeper Look**

The carbon cycle

In Chapter 9, we pointed out that global change includes *unidirectional changes*, the type of changes that don't repeat during Earth history, and *cyclic changes*, those that repeat, though not always with the same outcome or with a definable periodicity. Certain types of cyclic changes, known as *biogeochemical cycles*, involve exchanges of chemicals among a variety of living and nonliving reservoirs in the Earth System. In Chapter 11, we discussed an important example of a biogeochemical cycle, the *hydrologic cycle* (cycling of H_2O). Here, we introduce another such example, the **carbon cycle**, which involves the transfer of carbon in its various forms—such as coal, CO_2, CH_4, calcite ($CaCO_3$), dolomite [$CaMg(CO_3)_2$], dissolved HCO_3^- ions, oil, gas, graphite, and biomass—among various reservoirs **(Fig. Bx19.2)**. This cycle plays a key role in climate change because carbon in the atmosphere occurs as CO_2 and CH_4, both of which are greenhouse gases.

To follow the stages of the carbon cycle, let's begin at the beginning. Carbon, which was incorporated in rock inside the Earth during the planet's formation, diffuses into magma when melting takes place. It initially enters the carbon cycle in the surface and near-surface realms of the Earth System when it bubbles out of volcanoes and into the atmosphere as CO_2. Some of the CO_2 in air diffuses into seawater at the ocean surface, where it reacts to form bicarbonate (HCO_3^-) ions in dissolved form. This ion later bonds with calcium ions and precipitates as solid calcite ($CaCO_3$), or as calcite and aragonite in the shells of organisms such as plankton, clams, and coral. If buried, shells get cemented together and recrystallize to form limestone, which, if buried deeply and heated, metamorphoses

into marble. So, we can see that the atmosphere, seawater, shells, and carbonate rocks are all reservoirs of carbon. Note that carbonate rocks trap carbon as solid rock in the Earth's crust for millions to even billions of years. Transforming carbonate rocks into the cement used to make concrete for constructing buildings and roads transfers carbon from a long-term geologic reservoir back to the surface where it can enter into the atmospheric reservoir.

CO_2 can also be extracted from the air by plants, which incorporate it into biomass through the chemical reactions of photosynthesis. Biomass may be eaten, so effectively it can be incorporated in a succession of animals up the food chain. Therefore, at any given time, biomass (both plant and animal) serves as a reservoir of carbon. Burning biomass returns carbon to the atmospheric reservoir in the form of CO_2 gas. And the decay of biomass releases both CO_2 and CH_4 back into the atmospheric reservoir, as does the respiration and flatulence of animals. Organic matter, however, may also be stored in sediment, soil, peat, permafrost, and *methane hydrates* (a frozen material in which methane molecules occupy positions in the crystal lattice of water ice). If organic-rich sediment becomes deeply buried, it can transform into fossil fuels (coal and hydrocarbons), which may remain underground in rocks for geologic time. When permafrost melts and rots, peat decays or burns, or methane hydrates melt, carbon returns to the atmospheric reservoir as CH_4 and CO_2. Burning fossil fuels, of course, transfers carbon from long-term storage in geologic reservoirs back into the atmosphere as CO_2.

In sum, we see that various reservoirs—underground, on land, in the ocean, and in the air—can store carbon. During the carbon cycle, carbon moves among these reservoirs. At times during Earth history, the rate of transfer of carbon into the atmosphere can be balanced by the amount removed, so the concentration of carbon (as CO_2 and CH_4) in the air remains constant. If something changes to upset this balance, the concentration of CO_2 and CH_4 in the atmosphere can change, and that change can impact climate because both of these trace gases are greenhouse gases.

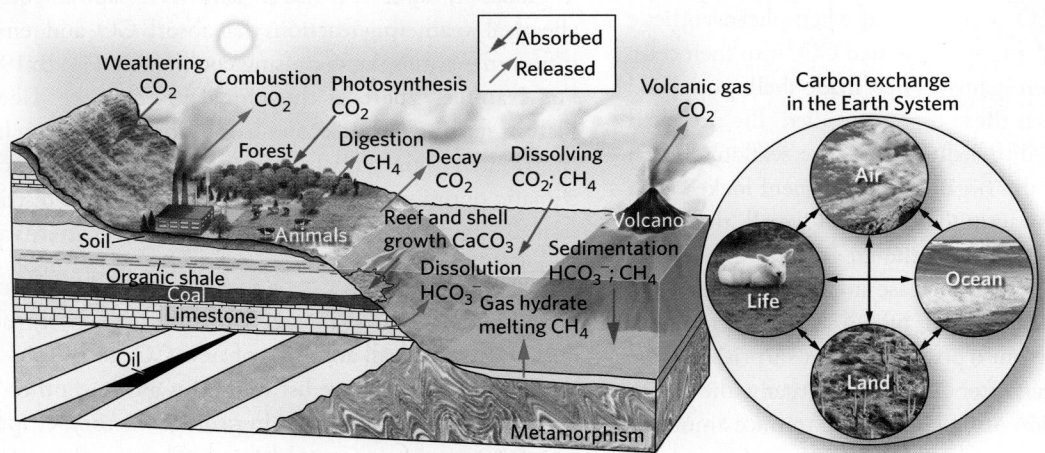

FIGURE Bx19.2 The Earth System contains many reservoirs among which carbon is exchanged during the carbon cycle.

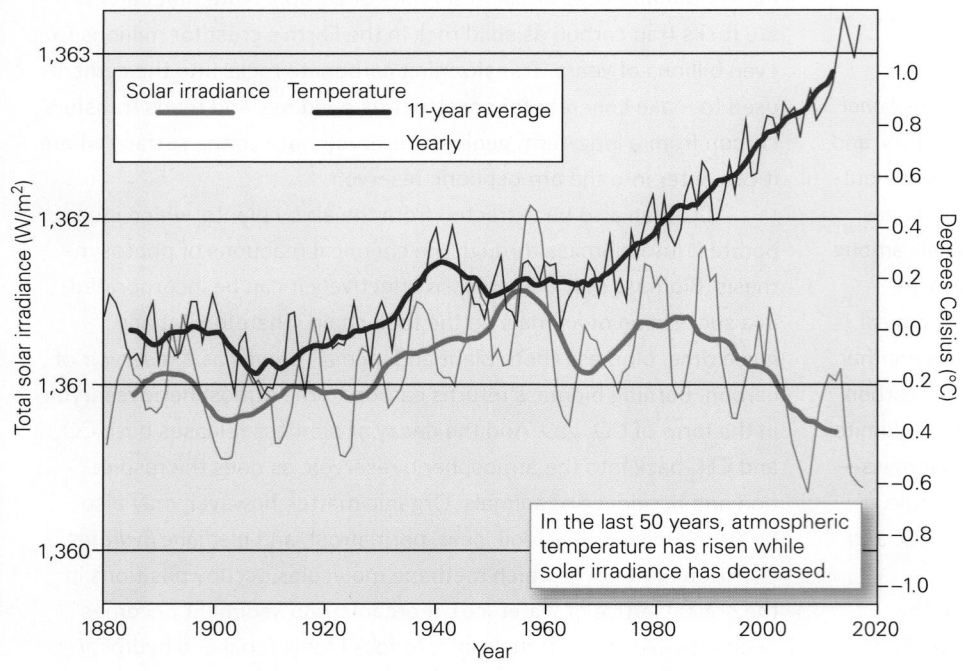

record as well. For example, some researchers speculate that the growth of immense coal swamps during the Carboniferous decreased atmospheric CO_2 concentrations enough (by transferring carbon from the atmospheric reservoir into the biomass reservoir, and then into the rock reservoir; see Box 19.2) to cause global cooling and the initiation of the late Carboniferous Ice Age. Overall, therefore, changes in greenhouse gas concentration can contribute to long-term climate change.

DISTRIBUTION AND AMOUNT OF LAND AND SEA. The late Proterozoic, late Paleozoic, and Pleistocene Ice Ages all occurred when substantial areas of land lay at high latitudes. Why? When large continents drift to high latitudes, the area of oceans at high latitudes decreases. As we have discussed, water has a much higher heat capacity than land, so without oceans at high latitudes, the land at high latitudes can become colder, setting the stage for icehouse conditions. In addition, the distribution of land affects ocean currents, which control heat distribution around the planet. In some cases, even a small amount of land, when strategically located, can block major currents. For example, when the Central American volcanic arc grew in the Cenozoic, it blocked currents from transporting warm Pacific water into the Atlantic, and this change affected temperatures around the Atlantic.

At times during Earth history, the interiors of continents have been flooded by shallow seas (see Chapter 9). When water covers significant portions of the land, the distribution of heat around the globe changes, and therefore the climate changes. Similarly, tectonic events that lead to the long-term uplift of large areas of land to form mountains or plateaus may affect land temperatures, by lifting land to higher elevations. Both uplift and sea-level change may affect atmospheric CO_2 concentration, because a drop in global sea level or a rise in land level exposes rock to chemical weathering reactions that absorb CO_2 and remove it from the atmosphere, causing cooling (see Fig. Bx19.2). For example, uplift of the Himalayas and the Tibetan Plateau starting in the second half of the Eocene may have contributed to the global cooling that followed the Eocene climatic optimum. Mountain building and land distribution also affect global atmospheric circulation patterns and, therefore, the distribution of temperature and rainfall.

VOLCANIC ACTIVITY. Generally, volcanic activity adds only relatively small amounts of new CO_2 to the air, and the Earth System can redistribute this gas into other reservoirs. But at certain times in geologic history, eruption of *large igneous provinces* (LIPs; see Chapter 4) increased the amount of volcanic-gas emission dramatically. This change may have been sufficient to cause a substantial increase in the atmospheric concentration of CO_2, which in turn would have caused global warming. For example,

but no evidence points to a significant impact on global temperature (Fig. 19.18).

CHANGES IN ATMOSPHERIC COMPOSITION. The Earth's early atmosphere consisted mostly of H_2O and CO_2 gas as well as NH_4 and N_2 (see Chapter 16). An atmosphere so rich in greenhouse gases made the Earth's surface realm the equivalent of an oven, like that of Venus today. Fortunately, the Earth cooled sufficiently for water to precipitate, and once the oceans had formed, some CO_2 dissolved in the water and then became incorporated into rock. Even more CO_2 was removed when photosynthetic organisms appeared and incorporated CO_2 into their cells, and more still when organisms that made shells populated the oceans, for when those organisms died, the extracted carbon was mixed into sediment on the seafloor. Burial and lithification of this organic-rich sediment locked even more carbon into strata underground. Overall, the transfer of greenhouse gases from the air into the liquid and then solid Earth led to a decrease in global temperature. Only after this had happened, by 2.4 Ga, could icehouse conditions even become a possibility. Not surprisingly, the earliest ice age began after 2.4 Ga. Eventually, the atmospheric concentration of CO_2 decreased to trace amounts. Enough CO_2 has remained in the atmosphere, fortunately, to keep the oceans above the freezing temperature of water.

Later changes in atmospheric greenhouse gas concentrations during Earth history have caused subsequent climate changes that were recorded in the stratigraphic

the eruption of the Siberian Traps, a LIP that produced immense amounts of flood basalt and pyroclastics, may have emitted enough CO_2 to cause an intense greenhouse effect that led to the Permo-Triassic mass extinction event. Similarly, widespread rift-related volcanism associated with the breakup of Pangaea, along with the eruption of large igneous provinces, may have contributed to Cretaceous warming.

Natural Short-Term Climate Controls

During the Pleistocene Ice Age, continental glaciers advanced and retreated perhaps 20 to 30 times. The most recent advance ended only about 20,000 years ago, and since then, the continental glaciers retreated so that by 11,700 years ago, the planet entered an *interglacial*. What factors can cause such changes, or comparable ones, that have relatively short durations? Some changes of very short duration (a year or decade) may simply reflect natural variation. But to explain events that last centuries to millennia, researchers point to the following mechanisms:

MILANKOVITCH CYCLES. As Milutin Milanković recognized in 1920, the shape of the Earth's orbit (eccentricity) changes over a period of 100,000 years, the tilt of its axis changes over a period of 41,000 years, and the axis undergoes precession (wobble) over a period of 19,000 to 23,000 years. Such periodic changes are called *Milankovitch cycles* (see Chapter 13), and each affects the amount of insolation received at high latitudes during the summer (Fig. 19.19). The combined effect of the cycles yields times of warmer temperatures, which lead to interglacials, and times of cooler temperatures, which may trigger glacial advances.

CHANGES IN OCEAN CURRENTS. Ocean currents serve a major role in redistributing heat in the Earth System. Recent studies suggest that the configuration of ocean currents can change quickly, and that these changes can in turn affect heat redistribution and, therefore, climate. The Younger Dryas, for example, may have resulted when a sudden release of water from melting glaciers (a glacial torrent; see Chapter 13) spread a layer of freshwater over the North Atlantic and temporarily slowed or stopped the thermohaline circulation throughout the oceans (see Chapter 14). Rapid growth or loss of sea ice may also disrupt currents. For example, when sea ice spreads to lower latitudes, warm currents flowing north cannot reach high latitudes, so the moisture source for snowfall may be cut off.

CHANGES IN THE EARTH'S ALBEDO. The degree of reflectivity, or **albedo**, of the atmosphere increases if the concentration of volcanic aerosols in the atmosphere increases. Specifically, large eruptions can emit large amounts of sulfur-bearing gases, such as sulfur dioxide (SO_2), that rise into the base of the stratosphere. The gases react with atmospheric moisture to produce aerosols consisting of

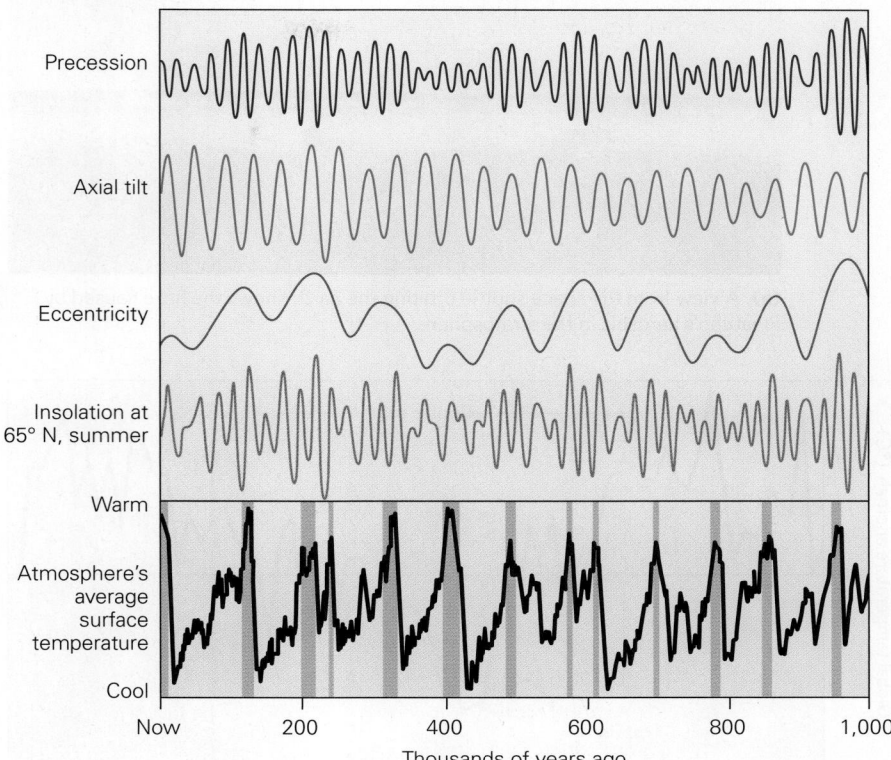

FIGURE 19.19 When the effects of the Milankovitch cycles—precession, axial tilt, and eccentricity—are combined, they cause changes in the daily insolation (the amount of heat added to the atmosphere), here shown at 65° N. The brown stripes indicate the warmest times of interglacials.

frozen sulfuric-acid droplets, which eventually sublimate to produce aerosols containing sulfate. Sulfur-bearing aerosols circulate worldwide in the lower stratosphere and can remain there for years, where they scatter incoming solar radiation, much of which returns to space. Consequently, these aerosols decrease the amount of insolation at the Earth's surface, and cause temporary global cooling.

Several examples of cooling due to explosive eruptions have happened in historic time. In 1815, Mt. Tambora in Indonesia exploded violently. This eruption followed other large volcanic eruptions in 1812 and 1814. Together, these eruptions ejected huge quantities of aerosols into the stratosphere and caused several degrees of global cooling. It remained so cold that 1816 came to be known as the *year without a summer*. A hard freeze that June killed animals and destroyed crops in the United States. In fact, temperatures in July didn't rise above 7°C (45°F) on many days. Northern New England even experienced crop-destroying frosts in mid-July and mid-August, when the region should have had its hottest days. Northern Europe suffered similar cold during the summer of 1816, and monsoons in India and China were severely disrupted, events that set the stage for famine and outbreaks of disease that killed millions. Recent

FIGURE 19.20   Effects of the eruption of Mt. Pinatubo in 1991.

(a) A view from the space shuttle orbiting the Earth shows the haze caused by Pinatubo's aerosols in the stratosphere.

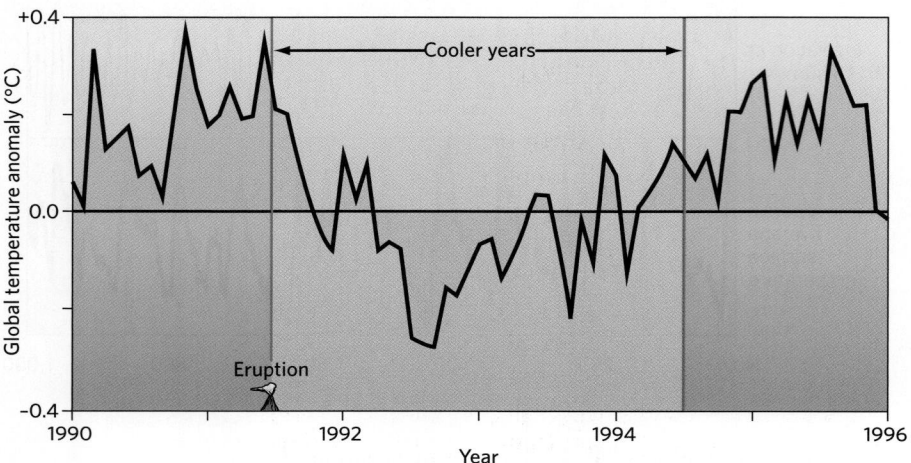

(c) The large injection of dust and aerosols into the atmosphere caused the average global temperature to drop for a few years after the eruption.

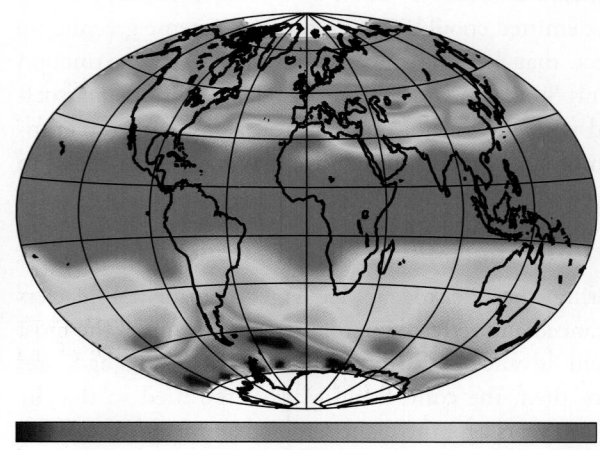

Low   Aerosol concentration   High

(b) Two months after the eruption, volcanic aerosols surrounded the Earth.

lesser volcanic eruptions have caused similar responses in the temperature record. For example, when Mt. Pinatubo in the Philippines erupted in 1991 (see Fig. Bx4.3c), researchers documented a three-year-long global decrease in temperature and precipitation (Fig. 19.20).

Reflection of solar energy into space doesn't happen just at the top of the atmosphere. Regional-scale changes in vegetation cover, in the proportion of snow and ice cover, or in the distribution of light-colored volcanic ash can also affect the albedo of the Earth's surface. Increased albedo causes cooling, whereas decreased albedo causes warming. The eruption of Toba, a volcano in Indonesia, about 74,000 years ago may have produced a widespread ash layer that amplified cooling due to aerosols and led to centuries of cold weather, conditions that may have killed off all but about 10,000 of the few million humans alive before the eruption.

ABRUPT CHANGES IN CONCENTRATIONS OF GREENHOUSE GASES. A relatively sudden change in greenhouse gas concentrations in the atmosphere can affect climate. Such a

change might happen in several ways. For example, huge quantities of **methane hydrates** (in which methane molecules are surrounded by water molecules in the crystal lattice of ice) formed in seafloor sediment when decaying organic matter combined with water under temperature and pressure conditions that exist at depths of 300–500 m. Even a slight rise in water temperature, or a decrease in pressure, could cause methane hydrates to melt, releasing bubbles of methane gas that would escape and enter the atmosphere. A similar release of CH_4 might happen if permafrost started to melt over broad areas, and the organic matter within it began to rot. Finally, because growth of photosynthetic organisms removes CO_2 from the atmosphere, global algal blooms and changes in forest cover could also conceivably change atmospheric CO_2 concentrations.

Take-home message…

Factors causing long-term climate change include changes in atmospheric CO_2 concentrations, changes in the distribution of continents and oceans, and uplift of land surfaces. Those causing short-term climate change include the Milankovitch cycles and large volcanic eruptions.

Quick Question

- What factors lead to changes in atmospheric CO_2 concentrations over geologic time?

- Why were icehouse conditions unlikely early in Earth's history?

- If a large volcanic eruption occurred tomorrow, would the next few years be warmer or colder globally?

19.5 Evidence for and Causes of Recent Climate Change

How has the Earth's climate changed over the past few centuries? Based on measuring the duration of interglacials over the past 400,000 years, the beginning of the current interglacial that began 11,700 years ago could have been drawing to a close, and we might expect to see global cooling in this century. That, however, isn't what measurements show. Rather, the Earth appears to be undergoing global warming at a pace that is unprecedented in the climate record. Let's look at some of the evidence that has led to this conclusion.

Discovering Recent Climate Change

In 1896, scientists published the first articles describing the atmospheric greenhouse effect and explaining how it works. This understanding eventually led others to explore the possibility that humans could cause global warming by adding CO_2 to the atmosphere. The **Keeling curve**, the detailed plot of CO_2 concentration over time that we mentioned at the beginning of this chapter (see Fig. 19.1d), together with other data, sparked growing interest in studying *recent climate change*. The time period under discussion begins with the industrial revolution of the 18th and 19th centuries. Beginning in the 1970s, growing numbers of researchers directed their attention to testing the relationship between greenhouse gases and climate change, and to exploring potential consequences of climate change. By the 1980s, evidence indicated that climate change could lead to profound consequences affecting both humans and the environment, and the

topic entered the public arena. Scientific journals began to publish so many climate-change studies that individual researchers struggled to keep track of them all. Consequently, in 1988, the World Meteorological Organization in collaboration with the United Nations Environment Program founded the **Intergovernmental Panel on Climate Change (IPCC)**. The IPCC evaluates published climate-related studies and summarizes its conclusions in a report intended for a broad audience, publishing a new edition every five years, the latest in 2023. In each report, statements have become increasingly certain that global warming is underway, that human activities contribute to the problem, and that the consequences are dire. Below, we examine evidence for recent global warming, as summarized by the IPCC and others.

Observed Recent Temperature Change

We noted earlier that the average global temperature during the Pleistocene (2.6 Ma-15 Ka) bounced up and down significantly, causing advances and retreats of continental glaciers. Zooming in on the past thousand years, researchers found that the average global temperature rose and fell within a range of about 1°C (1.7°F) until about 1880, when it began to rise relatively rapidly and steadily. The original graph showing this change became known informally as the **hockey-stick diagram** because of its shape (Fig. 19.21a). Graphs extending the record back for another thousand years—through the *Medieval Warm Period* and the *Little Ice Age*—emphasize that at no time during the past 2,000 years was the global average temperature as high as it is now, nor has it increased at such a rapid rate (Fig. 19.21b).

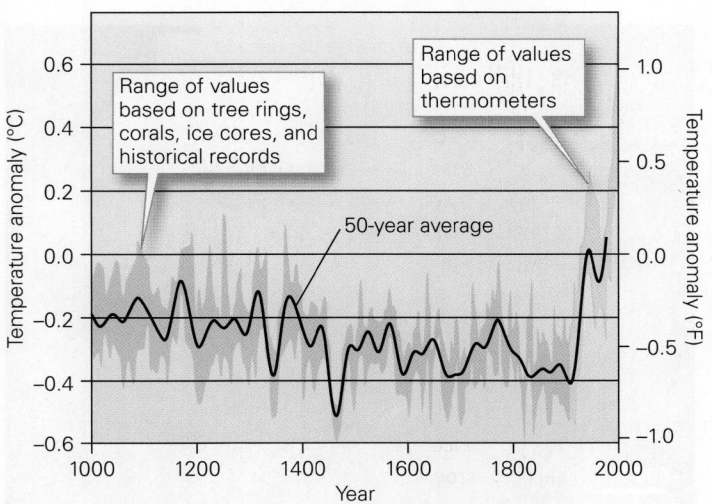

(a) The original hockey-stick diagram, published in 1998. The data are for the northern hemisphere. The temperature anomaly is indicated relative to the average 1961–1990 average temperature.

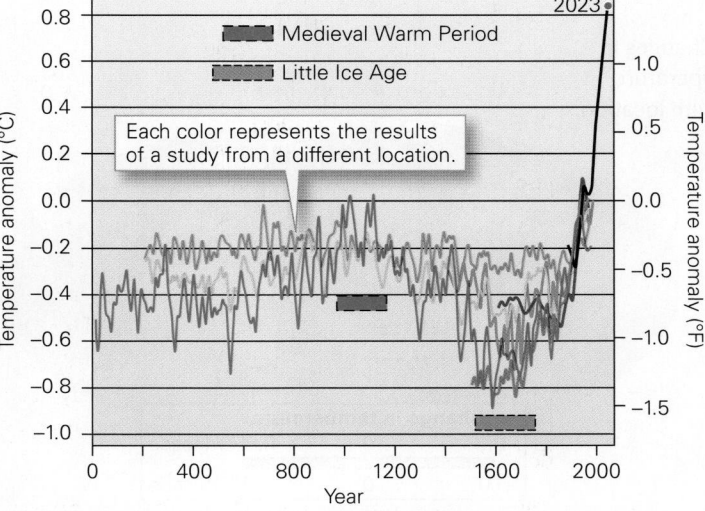

(b) Expanding the graph back to 2,000 years ago and forward to today reveals the Medieval Warm Period, the Little Ice Age, and the current warming trend.

FIGURE 19.21 Change in average global temperature over the past 1,000 years.

FIGURE 19.22
Changes in average global temperature between 1880 and 2022.

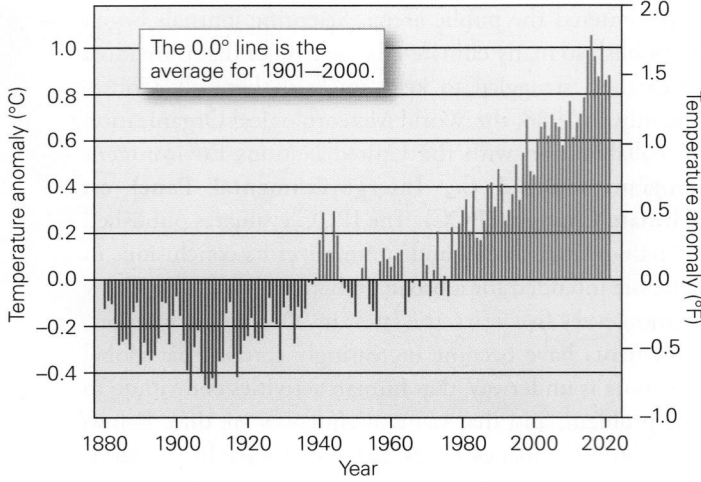

If we zoom in closer to the present, focusing on 1880–2022, scientists have determined that the global mean temperature in the period 2017–2022 has warmed over 1.5°C compared with the average of the period from 1901 to 2000, based on multiple independently produced analyses (Fig. 19.22). The warming hasn't been continuous; occasional multiyear intervals show no significant change, while others show large changes. Concern about global warming is based on examining trends lasting 20 years or more. Notably, in every year since 1980, temperatures have been higher than the long-term average. The five warmest years since 1880 are the most recent. Recent statistics show that what had been considered "normal" in past decades is no longer true. We are living in a "new normal," when record-breaking heat waves, such as occurred during the summer of 2023, begin to happen all too frequently.

Maps of temperature change emphasize that the amount of warming varies with location. The northern hemisphere has warmed more than the southern hemisphere and air over continents has warmed more than air over oceans because, as we've seen, large bodies of water moderate temperature change (Fig. 19.23). The greatest warming during the past half century has taken place in northern North America, northern Asia, and the Arctic. In fact, in 2018, temperatures rose above freezing at the North Pole, and in 2021 rain fell at the highest elevation of the Greenland ice sheet.

Observed Recent Sea-Level Change

The geologic record indicates that over the course of geologic time, global sea level has gone up and down significantly relative to the land surface. At times, shallow seas covered large areas of continents (see Chapter 9) (Fig. 19.24a). Significant changes in sea level have also happened during the past few million years. As an example, during Pleistocene glacial advances, a huge volume of water evaporated from the ocean and became trapped in continental glaciers. Sea level went down so far that large areas of continental shelves were exposed (Fig. 19.24b). During the most recent glacial retreat, which started between 15,000 and 20,000 years ago, the return of water to the ocean caused sea level to rise by about 120 m (394 ft) (Fig. 19.24c). This rise tapered off about 8,000 years ago, when the last North American and Asian ice sheets had completely melted.

The changes that we just reviewed happened over geologic time and reflect a response to natural processes. In the past 140 years, global sea level has started to rise again. Sea level rose at a rate of about 1.4 mm (0.06 in) per year between 1880 and 2006. Since then, the rate of rise increased to 3.6 mm (0.14 in) per year. Consequently, sea level today measures about 23 cm (9 in) higher than it did

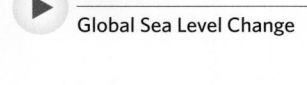

Animation

Global Sea Level Change

FIGURE 19.23 Changes in temperature (temperature anomalies) vary with location.

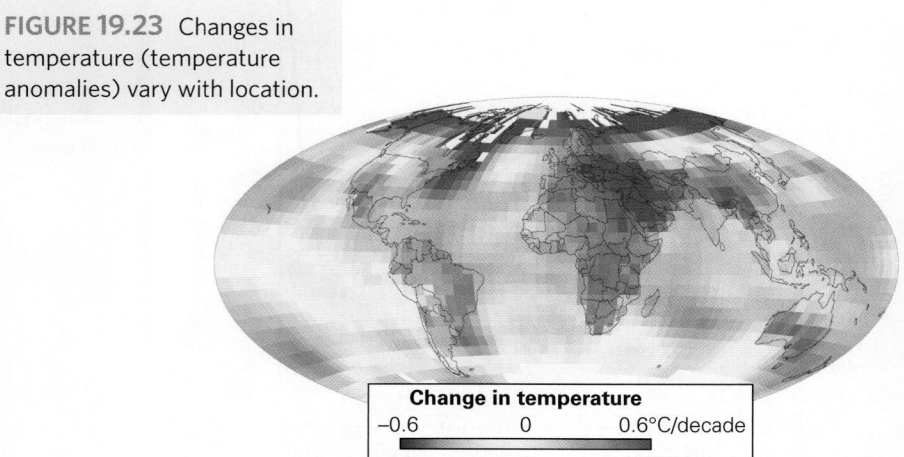

(a) Global temperature change between 1990 and 2021. Note that continents and Arctic regions are warming the most.

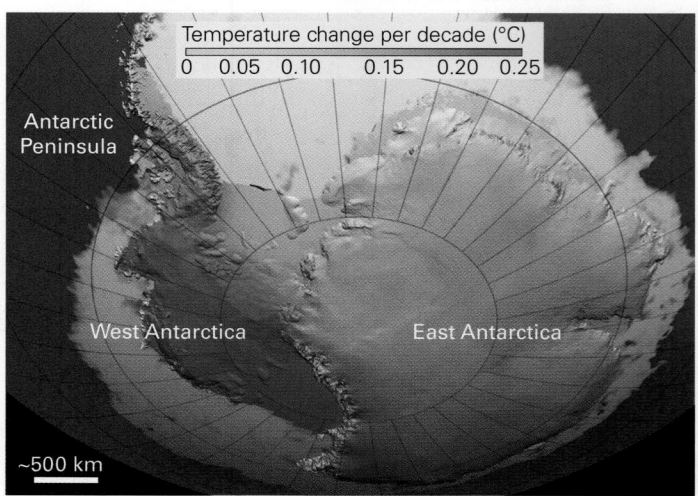

(b) Warming of Antarctica between 1957 and 2006. West Antarctica is at a lower elevation, overall, than East Antarctica.

FIGURE 19.24 Changes in relative sea level over geologic time.

FIGURE 19.25 Changes in sea level in the past 140 years.

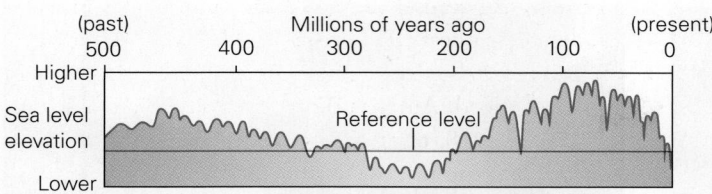

(a) Sea level has changed over the past 500 million years. The reference level is the current sea level value.

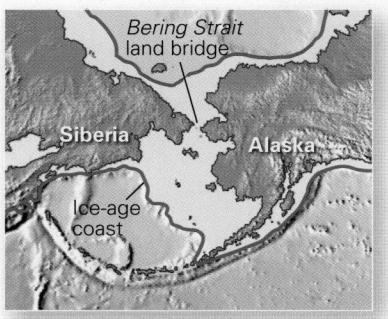

(b) At the peak of the most recent glaciation, sea level was so low that people could walk from Siberia to Alaska.

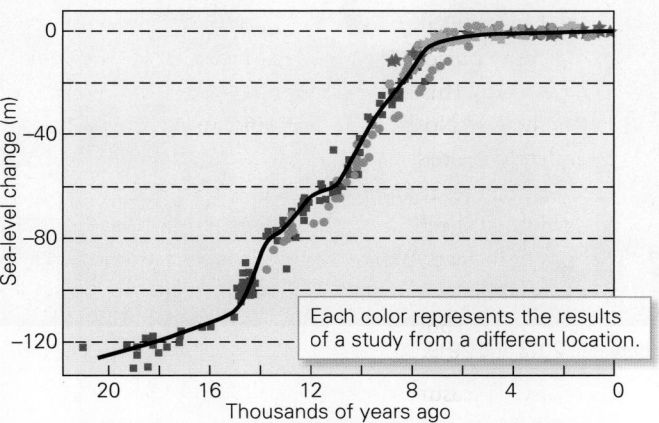

(c) During the most recent glacial retreat, sea level rose rapidly as continental glaciers melted.

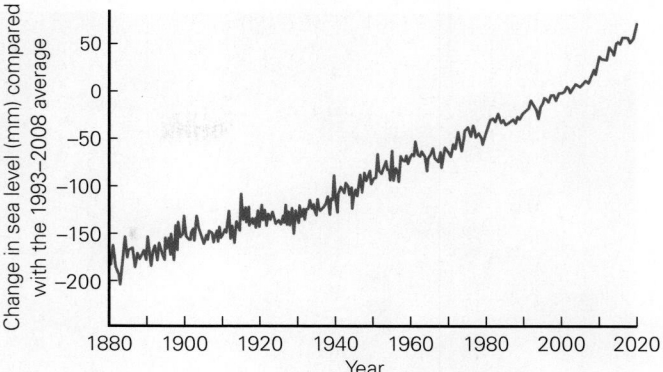

(a) Records from tidal gauges show that sea level has risen fairly steadily during the past 140 years. The rate has increased since 1970. This graph shows sea level relative to the average for 1993–2008.

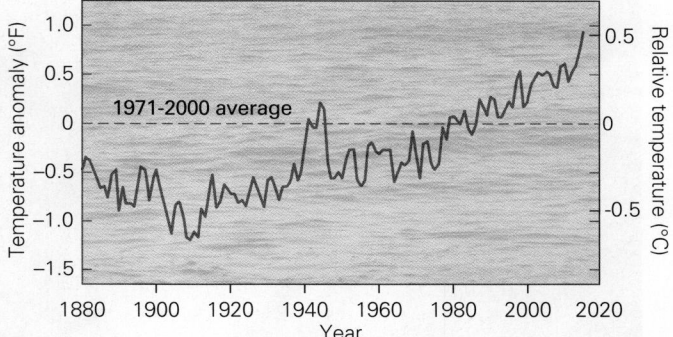

(b) Average global sea-surface temperatures have increased since 1880. The temperature anomaly is indicated relative to the average 1971–2000 temperature.

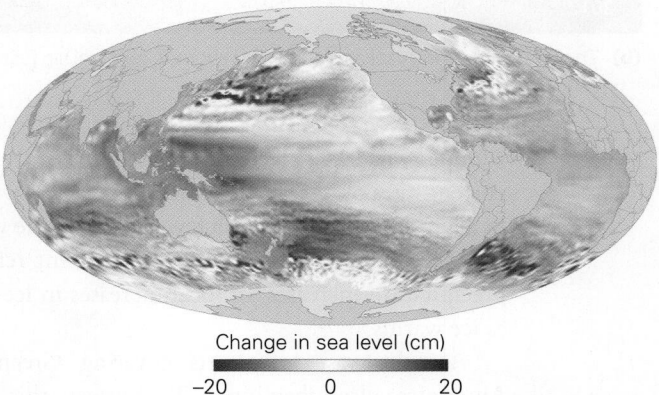

(c) Sea-level changes between 1993 and 2019. Regional variation occurs because winds and currents affect locations of sea level rise.

in 1880 **(Fig. 19.25a)**. Researchers link this rise to global warming because rising temperature affects sea level in two ways. First, liquid water expands (by a few percent) as it warms, and measurements of sea-surface temperature show that the oceans have indeed warmed **(Fig. 19.25b)**. Given that oceans are confined on their bottoms and sides by the seafloor, this *thermal expansion* forces the sea surface upward. The warming hasn't been uniform, however, and currents and winds transport water around the ocean, so sea-level change varies across ocean basins **(Fig. 19.25c)**. Second, when temperatures rise, glaciers melt. Water that had been locked within ice on land then returns to the oceans, causing the amount of water in the oceans to increase.

Melting Glacial and Sea Ice

Most mountain glaciers—such as those of the Alps, Andes, Himalayas, and Rockies—have retreated and thinned dramatically since the Little Ice Age **(Fig. 19.26)**. In many locations, glaciers that once filled valleys have disappeared entirely. For example, in 1850, Glacier National Park,

(a) The Muir Glacier in Alaska retreated 12 km (7 mi) between 1941 and 2004.

(b) The Bear Glacier in Alaska retreated by about 2 km (1 mi) in just seven years, between 2012 and 2019.

Montana, had 150 glaciers, but now it has only 25, and at current rates of retreat the remaining glaciers will be gone in a few decades. Notably, a few glaciers in the world have advanced in recent years. Their movement reflects local precipitation increases, or local increases in ice-flow rates as ice warms.

Are the vast ice sheets covering Greenland and Antarctica also shrinking? To answer this question, researchers use data from the GRACE and GRACE-FO satellites, whose instruments can measure the Earth's gravitational pull very precisely. The resulting measurements indicate that the pull exerted by the masses of ice in the Antarctic and Greenland ice sheets has decreased (Fig. 19.27a). By converting these changes in gravitational pull into changes in ice mass, and then into ice volume, researchers estimate that the Greenland ice sheet lost about 90 km³ (22 mi³) of ice in 2003 and 535 km³

(128 mi³) in 2019, and that it's now thinning by about 1 m (3 ft) per year (Fig. 19.27b). This result confirms conclusions from studies of satellite images, which show that the summer melt zone of Greenland's ice sheet has widened and that its duration is now longer. Meltwater first pools in lakes on the ice surface, and then drains through cracks to the base of the ice sheet, where it flows through tunnels to the sea (Fig. 19.27c). Antarctica's ice sheets have also been diminishing—the continent now loses about 270 km³ (65 mi³) of ice per year.

In the 19th century, almost all of the Arctic Ocean was covered by *sea ice* (a layer of ice formed by freezing of the ocean surface) all year, and explorers' goal of finding a "Northwest Passage" that would allow a ship to sail north of North America from the Atlantic to the Pacific was unattainable. So much ice now melts away in summer that cruise ships can now make the

FIGURE 19.27 Changes in the Antarctic and Greenland ice sheets.

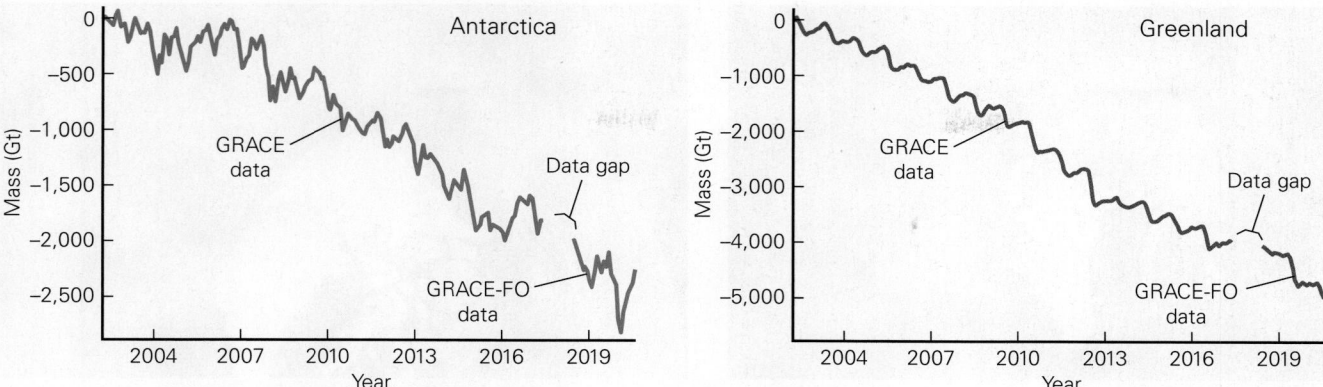

(a) Ice-mass loss in Antarctica and Greenland, measured in gigatons (billions of tons, abbreviated Gt), as measured by the GRACE and GRACE-FO satellites for 2002–2020.

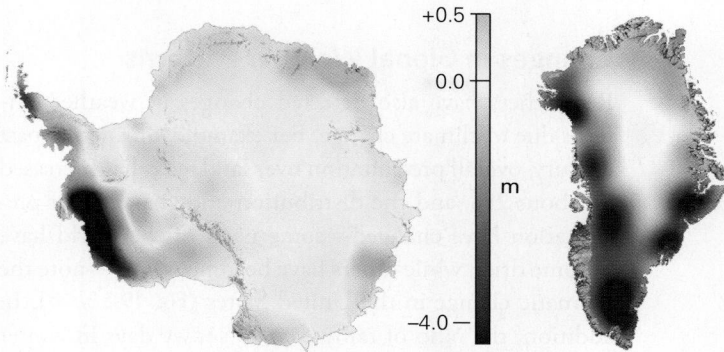

(b) Ice thickness has decreased significantly along ice-sheet margins in Antarctica and Greenland over the course of a decade. The scale shows the change in ice thickness.

(c) Meltwater lakes and streams on the Greenland ice sheet often disappear suddenly as they drain through cracks to the base of the ice sheet.

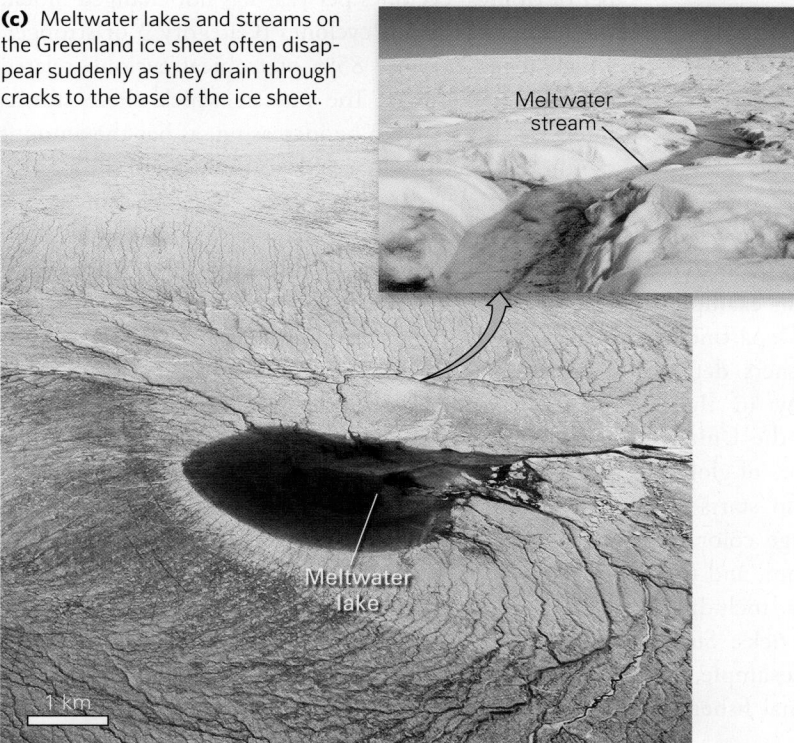

journey (Fig. 19.28a), and long-term trends in midsummer sea-ice cover show a definite decrease (Fig. 19.28b). Satellite measurements indicated that the average annual coverage of sea ice in the Arctic Ocean has decreased by about 20% between 1980 and 2022. Sea ice has also been disappearing in Antarctica, and *ice shelves* (regions where ice sheets have flowed over the sea adjacent to the coast) have started breaking up. In 2002, the Larsen B Ice Shelf, with an area of 3,250 km² (1,255 mi²), broke up (Fig. 19.29a). In 2017, a 5,800 km² (2,239 mi²) section of the Larsen C Ice Shelf—85% larger than Rhode Island—separated from the ice sheet and drifted out to sea. In recent years, 150 m (500 ft)-thick icebergs the area of large cities have been breaking off of Antarctic glaciers that flow into the sea (Fig. 19.29b).

Combining data from observations of glacial and sea-ice loss shows that the Earth's total ice volume has been diminishing in the past half century (Fig. 19.30). Since it takes time for glaciers to grow or melt, changes in the overall volume of glacial ice on land and of sea ice in the ocean effectively represent temperature trends averaged over several decades. The loss of ice, therefore, confirms that warming represents a trend, not just an annual anomaly.

See for yourself

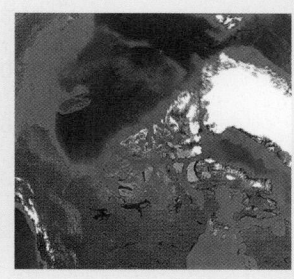

Northwest Passage, Arctic Ocean

Latitude: 74°22′33.45″ N
Longitude: 101°51′31.42″ W

Look down on this scene from an altitude of 5,246 km (3,260 mi).

You will see the Northwest Passage, the famed path between the Atlantic and the Pacific sought by ships over the centuries to avoid sailing all the way around the tip of South America. The passage has been blocked by sea ice for centuries. That is changing as the Arctic has warmed and Arctic sea-ice coverage has diminished. In some recent years, open water has been found in late summer, stretching along the coasts from Greenland to Alaska's northern coastline.

FIGURE 19.28 Sea-ice cover in the Arctic Ocean has decreased.

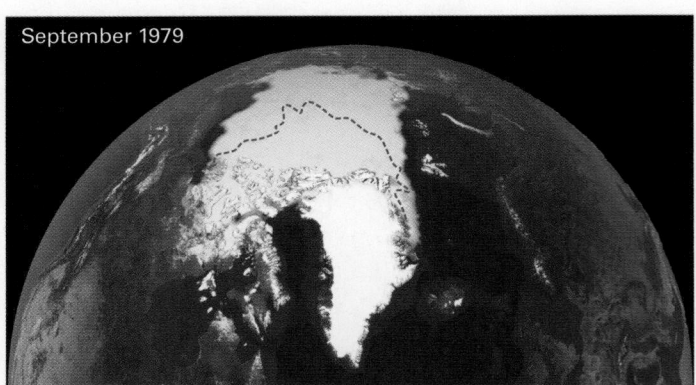
September 1979

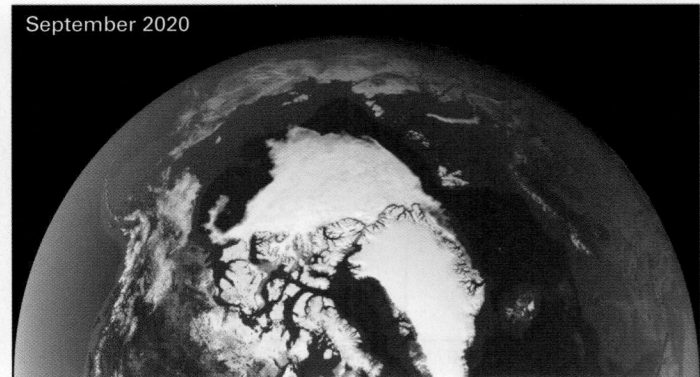
September 2020

(a) A comparison of satellite images of Arctic Ocean sea ice in September 1979 and in September 2020. The blue dashed line on the 1979 image is the approximate trace of the 2020 ice extent.

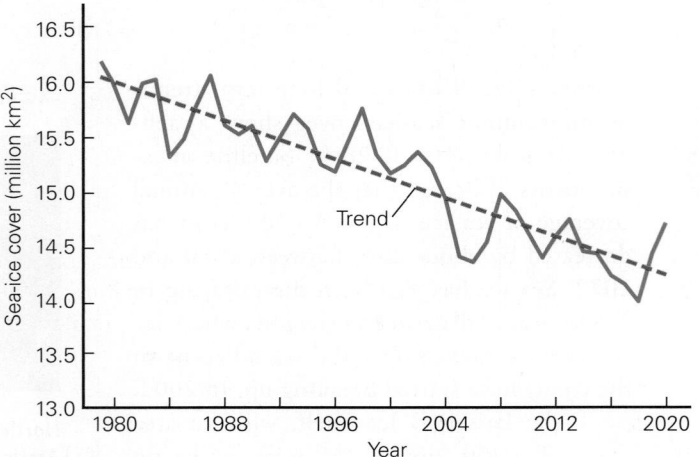

(b) A graph of sea-ice cover relative to the average for 1979–2020 shows fluctuations, but also shows an overall downward trend (blue line).

Changes in Global Weather Patterns

Researchers have also detected changes in weather patterns due to climate change. For example, during the past century, overall precipitation over land areas has increased by about 2%, and the distribution and character of precipitation have changed—some parts of the world have become drier, while others have become wetter—note the dramatic change in the United States (Fig. 19.32a, b). In addition, the ratio of rainy days to snowy days in winter in the northern hemisphere has increased.

Recent studies now show that the severity of storms has also increased. For example, while the total number of tropical cyclones per year has not changed much, the number of major cyclones (Category 3 or stronger, which together cause 85% of tropical cyclone–related damage) has grown. The rate at which these storms intensify also seems to be increasing, as has the amount of rain that they produce and the duration of the storms. By some estimates, each 0.55°C (1°F) increase in average global ocean temperature may cause a 7% increase in the potential maximum wind speed of tropical cyclones. Overall, the land area experiencing extreme one-day heavy rainfall events in the United States appears to be increasing (Fig. 19.32c). Researchers relate such changes to global warming, because warmer sea-surface temperatures increase evaporation rates, which puts more water vapor (the fuel of storms) into the atmosphere.

Causes of Recent Climate Change

Measurements of the composition of air bubbles in Antarctic ice cores provide a record of atmospheric CO_2 concentrations back through almost 800,000 years. This record indicates that during the alternating glaciations and interglacials of the late Pleistocene, CO_2 concentrations

Biological Indicators of Climate Change

Some organisms that are sensitive to temperature and the length of seasons serve as *biological indicators* of climate change. Biological indicators also provide evidence that global warming is happening. For example, *plant hardiness zones*, defined by the US Department of Agriculture to help farmers and gardeners determine which plants will successfully grow in their region, have been migrating northward in the United States (Fig. 19.31). Other biological indicators of global warming include the date when maple sap starts to flow in spring, the date when leaves change color in fall, the date at which cherry blossoms bloom, and the latitudinal ranges of birds, fish, and insects, including species of disease-causing mosquitoes and ticks. Such changes have economic consequences. For example, as fish migrate to cooler water, some traditional fisheries have been ruined.

Earth's climate system is an ornery beast which overreacts even to small nudges.

—*WALLACE BROECKER (AMERICAN GEOCHEMIST, 1931–2019)*

FIGURE 19.29 Ice-shelf collapse and iceberg formation in Antarctica.

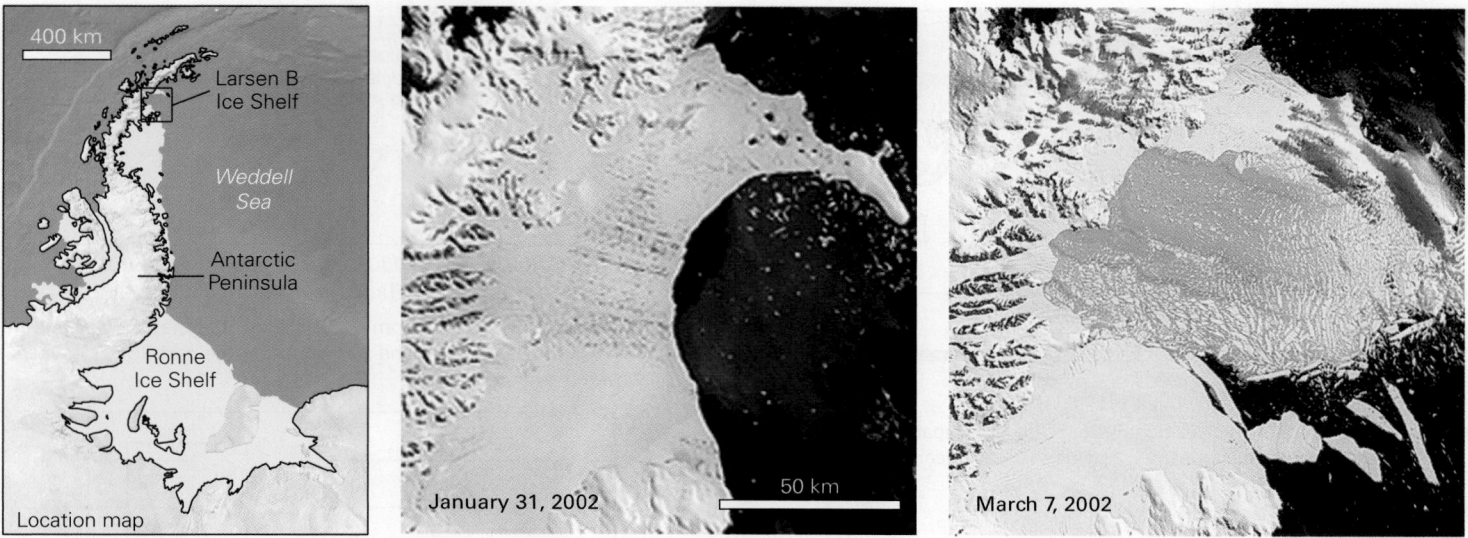

(a) A large portion of the Larsen B Ice Shelf on the coast of the Antarctic Peninsula disintegrated in 2002.

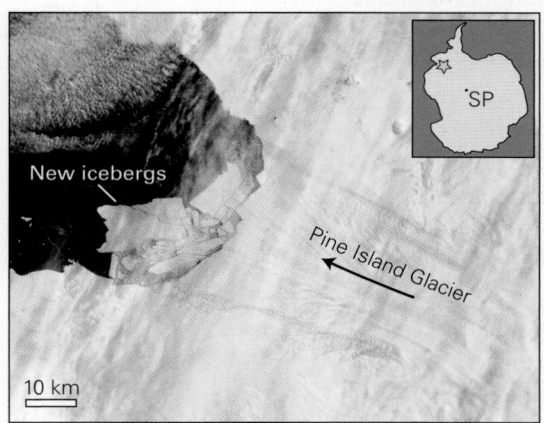

(b) Large icebergs calving off the end of the Pine Island Glacier in Antarctica. SP = South Pole.

FIGURE 19.30 Global ice loss (from mountain glaciers and continental glaciers combined), given in water-equivalent depth. This measure is obtained by converting ice mass to water volume, and then dividing by the total area of ice.

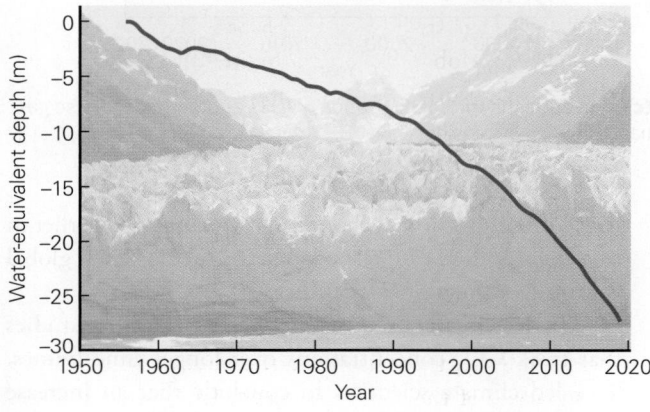

FIGURE 19.31 The boundaries between plant hardiness zones, as defined by the US Department of Agriculture, shifted northward between 1990 and 2012. Blues denote shorter growing seasons, and reds indicate longer growing seasons.

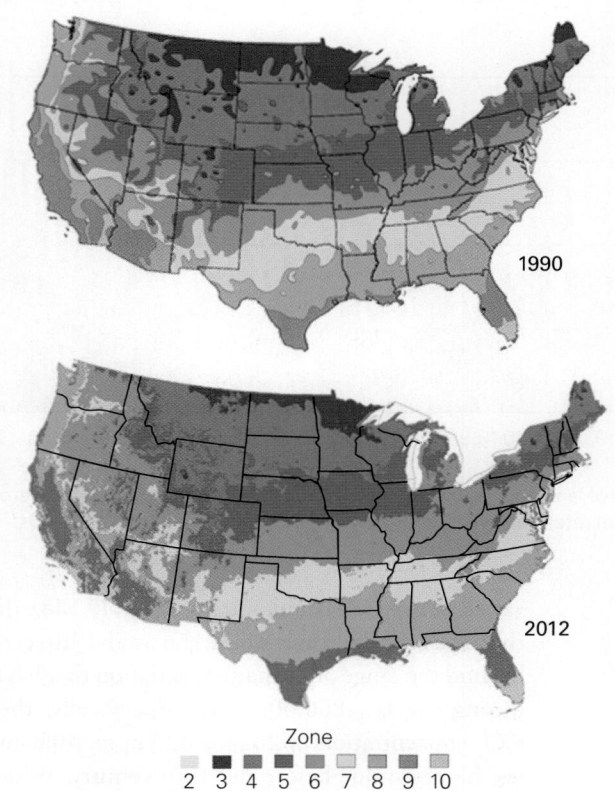

FIGURE 19.32 Changes in precipitation patterns.

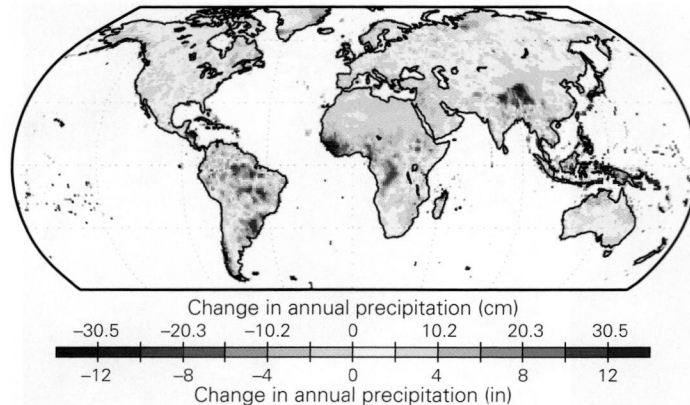

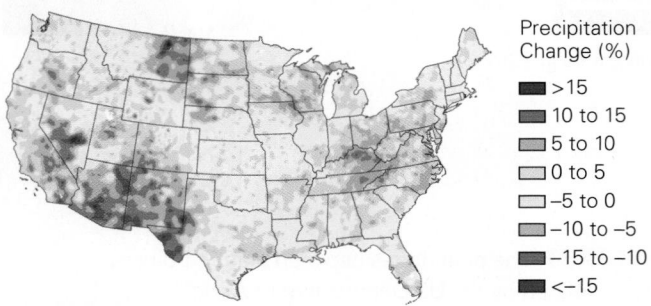

(a) Changes in average annual precipitation, 1986–2015, as compared with the mid-20th century. Brown areas became drier, and green areas became wetter.

Precipitation Change (%)
- >15
- 10 to 15
- 5 to 10
- 0 to 5
- -5 to 0
- -10 to -5
- -15 to -10
- <-15

(b) Precipitation changes between 1991 and 2020, as compared with the average precipitation of 1981–2010. Brown areas are getting drier, and green areas are getting wetter.

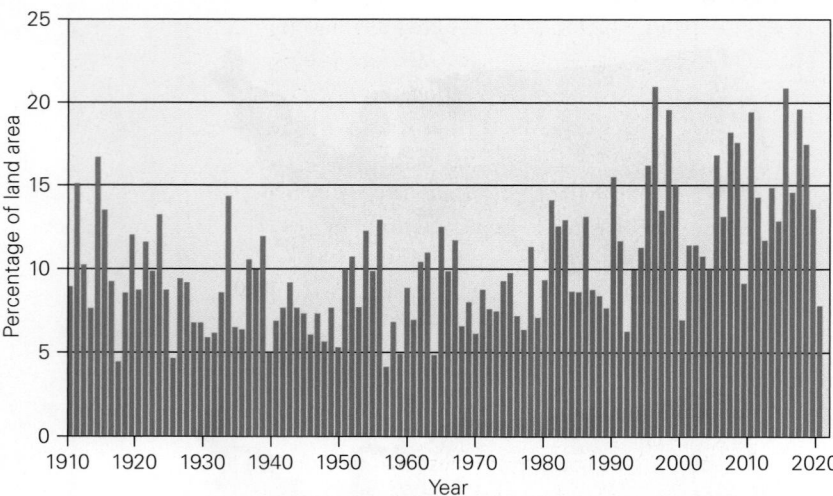

(c) The percentage of land area in the continental United States experiencing extreme one-day precipitation rates each year. An increase is evident beginning in the 1970s.

FIGURE 19.33 Changes in atmospheric carbon dioxide and methane concentrations over time.

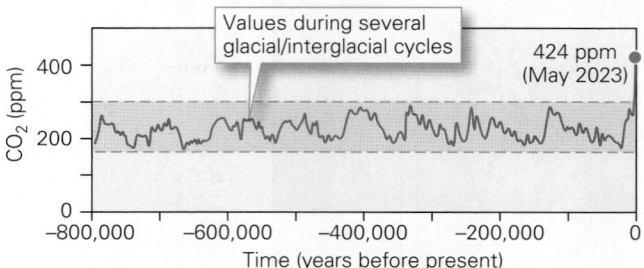

(a) Studies of ice cores from glaciers show that the CO_2 concentration varied between 180 and 300 ppm over the past 800,000 years. Now it is over 420 ppm.

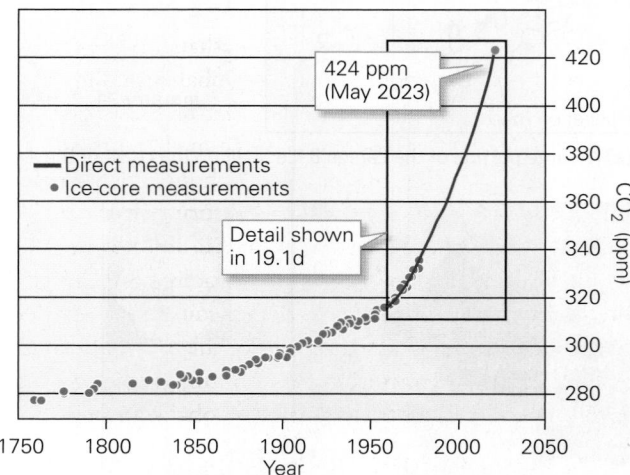

(b) Since the industrial revolution, the atmospheric CO_2 concentration has steadily increased.

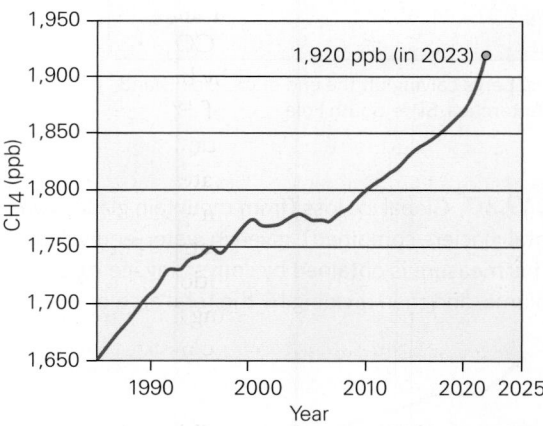

(c) The atmospheric concentration of CH_4, another greenhouse gas, has also been increasing.

varied between 180 and 300 ppm **(Fig. 19.33a)**. Therefore, the observed increases since the mid-19th century are beyond the range of the natural variation that has occurred during the last 800,000 years. Specifically, the average CO_2 concentration in 2023 is 420 ppm, 40% more than the highest value before the 19th century. Which of the

many causes of climate change that we discussed earlier in this chapter might be responsible for the observed global warming of the past few centuries?

The Keeling curve in Figure 19.1d, as well as studies that track CO_2 concentrations over longer time frames, have led climate scientists to conclude that an increase

TABLE 19.2 Sources and Sinks of Carbon-Containing Greenhouse Gases

Sources	Causes
Natural CO_2	Volcanoes; decay of organic matter; animal respiration
Anthropogenic CO_2	Fossil fuel burning; cement production; industrial output
Natural CH_4	Decay of organic matter; microbial respiration
Anthropogenic CH_4	Fossil fuel burning; venting or leakage of natural gas from oil and gas fields; rice paddy decay; landfill decay; livestock flatulence
Sinks	
Natural CO_2	Dissolution in the ocean; weathering reactions with rocks and soils; incorporation by photosynthetic organisms
Natural CH_4	Chemical reactions in soil and the atmosphere

in the concentrations of CO_2 and methane in the atmosphere serves as the main driver of global warming (and associated sea-level rise) since the start of the industrial age (Fig. 19.33b, c). This conclusion immediately leads to a question: Does the change in greenhouse gas concentrations observed during the past century reflect natural causes, or does it reflect **anthropogenic sources** (inputs into the Earth System from human activities)? Insight into this question comes from two key sources: (1) studies of carbon isotopes, and (2) studies of the carbon budget. Both lead to the conclusion that anthropogenic sources are the primary cause of present-day global warming.

INSIGHT FROM ISOTOPIC ANALYSES. The element carbon exists as three different isotopes (^{12}C, ^{13}C, and ^{14}C). The ratio of these isotopes in a volume of CO_2 gas defines the *isotopic composition* of the gas. Careful analyses demonstrate that the isotopic composition of CO_2 produced by volcanoes, and that of CO_2 produced by burning organic matter, are different: the proportion of ^{13}C tends to be higher in organic matter than in volcanic gas because photosynthesis preferentially incorporates ^{13}C. Studies of the isotopic composition of CO_2 in glacial air bubbles show that the proportion of ^{13}C in atmospheric CO_2 has increased since the industrial revolution. This change shows that much of the new CO_2 entering the atmosphere must come from the burning of organic matter (biomass and fossil fuels), not from volcanoes.

INSIGHT FROM CARBON BUDGET CALCULATIONS. As we discussed in Box 19.2, carbon passes through many components of the Earth System during the carbon cycle. If carbon transfers into the atmosphere in the form of CO_2 and CH_4 at a rate faster than it transfers out of the atmosphere into the hydrosphere, biosphere, or geosphere, then the concentration of these greenhouse gases in the atmosphere will rise. Understanding whether greenhouse gases are being added at a rate faster than they are being removed challenged researchers to characterize the Earth's

carbon budget by quantifying *carbon sources* that add carbon to the atmosphere and *carbon sinks* that remove carbon from the atmosphere (Table 19.2).

Calculations summarized by the IPCC indicate that most of the anthropogenic CO_2 entering the atmosphere today comes from two sources, namely: burning fossil fuels, because combustion reactions release CO_2, and cement production, because obtaining cement (a key component of concrete) involves heating limestone to such a high temperature that calcite ($CaCO_3$) breaks down into solid CaO (lime) and gaseous CO_2 (Fig. 19.34a). Sinks include the following: the ocean, for about 30%–40% of the CO_2 added to the atmosphere dissolves in ocean water; incorporation in plants and algae, which use CO_2 during photosynthesis; and incorporation into minerals formed by the chemical weathering of rocks, for certain weathering reactions absorb CO_2. Notably, anthropogenic destruction of sinks, mainly by deforestation, has decreased the amount of atmospheric carbon that can be absorbed by plants. Careful comparison of the weight (in gigatons) of greenhouse gas released by sources with the amount absorbed by sinks, for a given year, indicates the amount of carbon that remains in the atmosphere in the form of CO_2 and CH_4 at the end of that year (Fig. 19.34b). Such a determination is called a **mass-balance calculation**. A review of mass-balance calculations published by many different research groups has led the IPCC to attribute the rise in CO_2 concentrations during the past two centuries to industrialization, urbanization, and deforestation.

It may seem strange that anthropogenic inputs of greenhouse gases are so significant relative to natural sources. But a comparison of human production of CO_2 relative to volcanic production shows that indeed they are. In an average year, all volcanic eruptions together, including both submarine and subaerial eruptions, emit 0.15–0.26 gigatons of CO_2. By comparison, human society emits about 35 gigatons of CO_2 each year (about 135 times as much). Simply put, human society is transferring carbon

FIGURE 19.34 Annual anthropogenic carbon emissions since 1870.

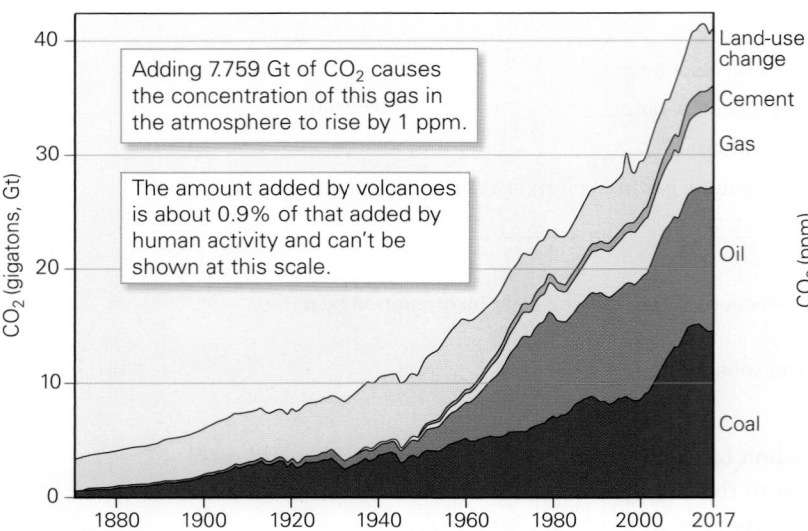

(a) The quantity of carbon emissions, overall, has increased dramatically since the start of the industrial age. Beginning in the mid-20th century, hydrocarbons dominated energy consumption.

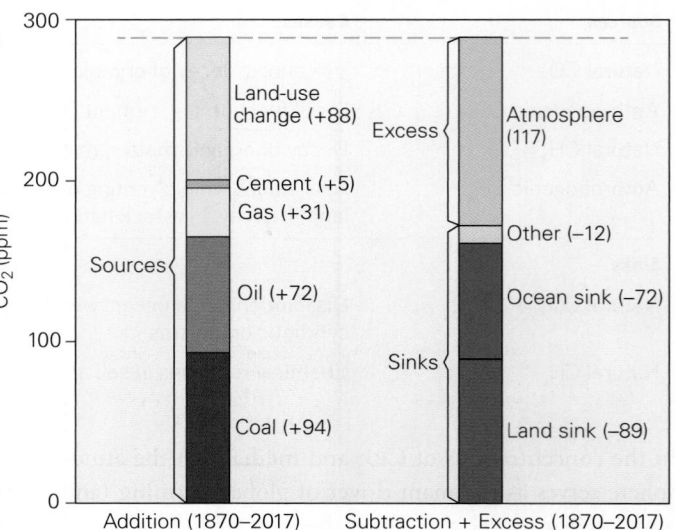

(b) Sources and sinks of atmospheric carbon since the start of the industrial age. The left column shows the relative proportions of different sources (in ppm). The right column shows where the added carbon goes. The "excess" is the amount remaining in the atmosphere, where it can contribute to global warming.

that took millions to tens of millions of years to accumulate, and that has been stored in geologic reservoirs underground for tens to hundreds of millions of years, back into the atmosphere in a matter of decades to centuries. The rate of transfer far exceeds the capacity of sinks to remove it.

Take-home message...

Global temperature and sea level have been rising since the mid-19th century. Many observed changes in the Earth System, such as changes in the volume of glacial ice and in the distribution of vegetation belts, support the conclusion that global warming is taking place. This warming corresponds to the largest increase in atmospheric CO_2 that has happened during the last 800,000 years. Researchers attribute this temperature increase to anthropogenic CO_2 and CH_4 production beginning with the industrial revolution.

Quick Questions

- What two factors cause sea level to rise?

- What common change has been observed in the world's glaciers, ice sheets, and sea ice over the last 100 years?

- Name one biological indicator that suggests the global climate has been warming.

- What are the primary sources of anthropogenic greenhouse gases?

19.6 Climate in the Future, and Its Consequences

Global Climate Models and Their Complexities

DEVELOPMENT OF GCMs. The development of high-speed supercomputers allowed climate scientists to create **global climate models (GCMs)**, computer programs that simulate the behavior of climate over time scales of decades to centuries. These models, which use equations to represent physical and chemical interactions among the atmosphere, ocean, and land, are similar to models used for weather forecasting, but they cover much longer time spans. Using GCMs, researchers can compare climate behavior in an imagined situation in which greenhouse gas concentrations remain at preindustrial levels with the way the climate actually behaved for the past two centuries. These simulations show that, if greenhouse gas concentrations had remained constant, the Earth would have experienced a slight cooling trend over the last century, caused by a decreasing insolation at high latitudes associated with the Milankovitch cycles. When observed greenhouse gas concentrations are added, however, the models predict the warming trend that we have observed (Fig. 19.35a). Such results support the interpretation that addition of anthropogenic CO_2 has significantly influenced climate.

FIGURE 19.35 Using GCMs to make predictions of future global warming.

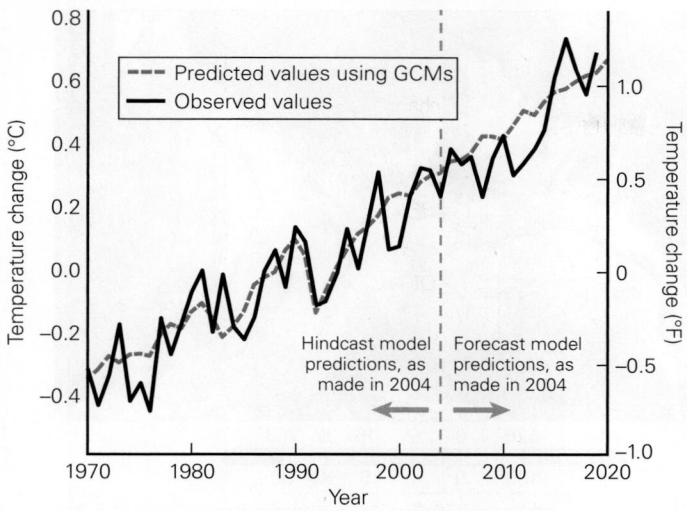

(a) In 2004, researchers used GCMs to predict temperature changes from 1970 to 2020. Comparison of the predictions with the observed values shows a good match. Temperature changes are indicated relative to the average 1980–1999 temperature.

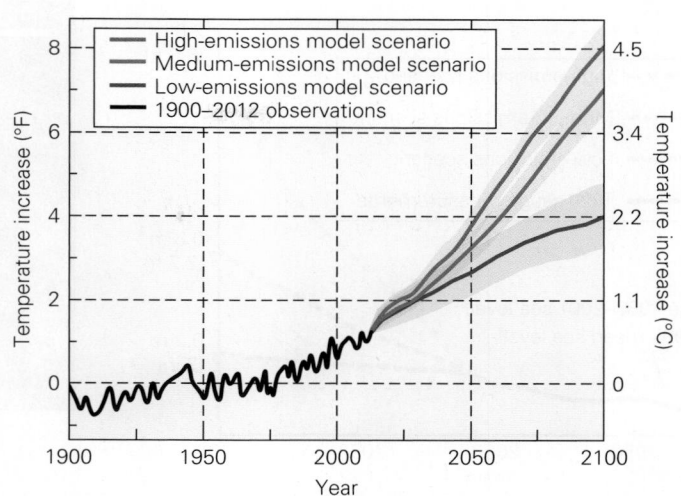

(b) GCM projections of temperature change, made in 2012 assuming different rates of anthropogenic CO_2 emission. All three simulations suggest a significant global temperature increase by 2100.

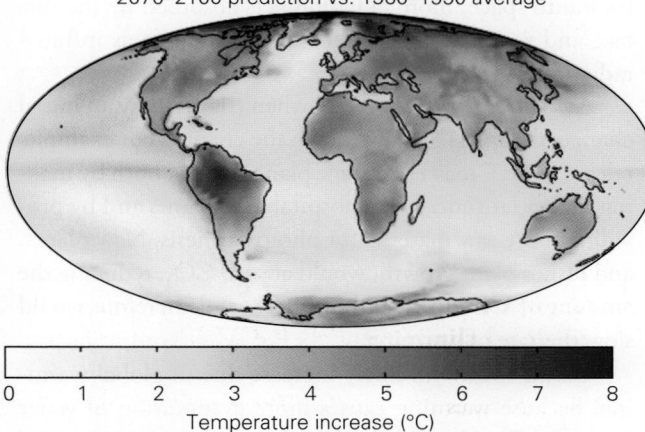

2070–2100 prediction vs. 1960–1990 average

Temperature increase (°C)

(c) Global warming varies with location. This map shows a GCM prediction of possible regional variation in temperature.

By adjusting the variables programmed into a GCM, researchers can also use the models to provide insight into what future climate conditions may be. Since no one knows what the rate of emissions will be in the future, scientists provide a range of possible answers by producing three models: a *high-emissions model*, in which greenhouse gas emissions increase above current rates; a *medium-emissions model*, in which the rate of greenhouse gas emissions stays about the same as today; and a *low-emissions model*, in which the rate of greenhouse gas emissions decreases below current rates (Fig. 19.35b). The results are concerning. For example, the medium-emissions model predicts that atmospheric temperatures will increase by 2°C–4°C (3°F–7°F) by the end of the 21st

century, with the greatest warming taking place at high latitudes (Fig. 19.35c).

FEEDBACKS IN CLIMATE MODELS. Work on developing GCMs continues because they remain far from complete. There are many aspects of the atmosphere, and its interactions with other components of the Earth System, that remain poorly understood. Also, several important **feedbacks**—phenomena that are themselves triggered by a change, and then modify the change—that influence the climate can't easily be incorporated into models. Because feedback processes are so important, let's look at them more closely.

A *positive feedback process* amplifies the consequence of the initial change and may cause the rate of change to accelerate. The most important example of a positive feedback process during global warming involves the effect of warming on the concentration of water vapor in the atmosphere. Even though individual molecules of CO_2 or CH_4 are efficient in trapping heat, water vapor overall causes most of the Earth's atmospheric greenhouse effect because its concentration in the atmosphere far exceeds that of CO_2 or CH_4. On a hot, humid day, water vapor accounts for about 2.3% of air, whereas CO_2 accounts for 0.04% and CH_4 accounts for 0.0002%. Warming of the atmosphere leads to an increase in its water vapor concentration, which leads to more warming and therefore to more evaporation, which in turn leads to a further increase in atmospheric water vapor—and this process goes on and on. This repeating cycle is a positive feedback process that can accelerate global warming.

FIGURE 19.36 Predictions of future sea-level change.

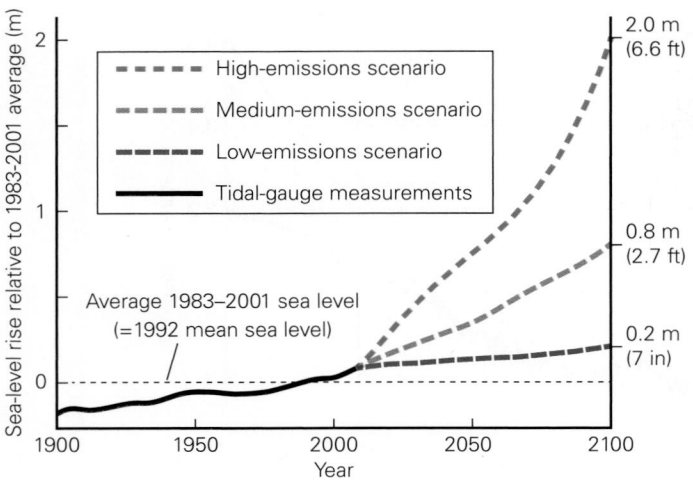

(a) GCMs suggest that sea level will continue to rise, increasing as much as 2.0 m by 2100.

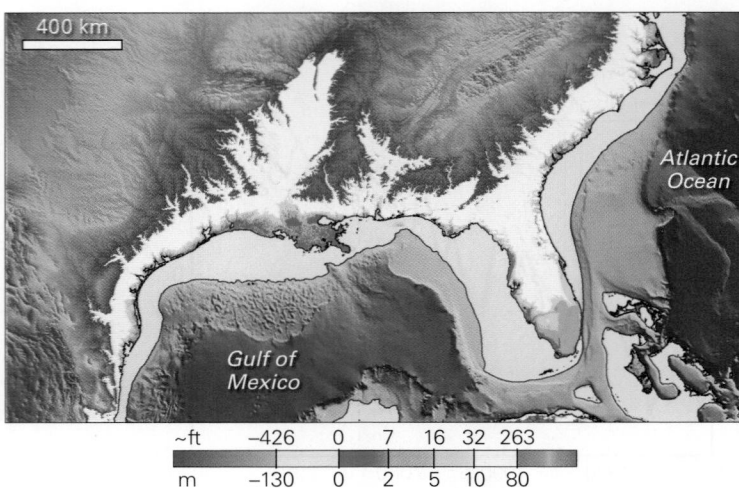

(b) Land elevations are low along the US Gulf and Atlantic coasts. The lowest of these areas will be flooded by rising seas.

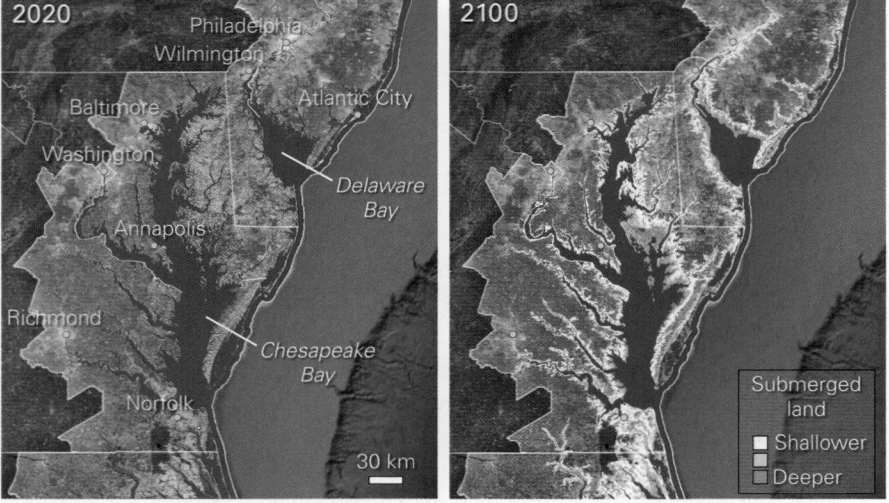

(c) Sea-level rise predicted for a high-emissions scenario will affect the densely populated Chesapeake and Delaware Bay areas by 2100. Dark blue shows present seas; light blue shows the area that will be covered by higher seas.

Note that H_2O concentration, while amplifying global warming, does not drive global warming. That's because a given molecule of H_2O has a short residence time, meaning that it remains in the atmosphere only for a matter of days, before it becomes part of a raindrop or snowflake that falls to the Earth's surface. CO_2 and CH_4, in contrast, have a very long residence time, so once in the air, these greenhouse gases influence climate for a long time.

The melting of permafrost and of methane hydrates also serves as a positive feedback, for melting these substances adds even more CO_2 and CH_4 to the atmosphere. The decrease in ice and snow cover, which decreases the albedo of the Earth's surface, represents another positive feedback, because when less solar radiation reflects back into space, more radiation gets absorbed by the surface and reradiates back into the atmosphere as infrared radiation.

A *negative feedback process*, when triggered by an initial change, dampens or slows down the change. For example, release of CO_2 may encourage plant growth, both by making higher latitudes more hospitable to plants and by providing more raw material for photosynthesis. New plants, and lusher plant growth, would absorb CO_2, reducing the amount of CO_2 in the atmosphere, and therefore would slow the rate of warming.

Cloud cover will likely increase due to global warming, because warming causes more evaporation of water from the Earth's surface. Whether an increase in cloud cover serves as an overall negative or positive feedback remains uncertain. Clouds scatter sunlight back to space, a cooling effect, but also absorb infrared radiation from the surface, a warming effect. Globally, the answer, cooling or warming, depends on which effect dominates across the planet. Determining which of these feedbacks—positive or negative—related to cloud cover will dominate has not been determined yet because so many different types of clouds occur, and each type may behave differently.

Predicted Consequences of Global Warming

Recent news reports describing natural disasters—floods, severe storms, crop failures, coastal flooding, heat waves, water shortages, wildfires, droughts, and more (see **Earth Science at a Glance**, pp. 764–765)—commonly attribute the severity of the event to climate change. While it's probably not realistic to associate every disaster with climate change, the underlying basis for such attributions makes sense in many cases, as we will see. Let's look more closely at some of the many impacts that climate change may bring in the not-so-distant future.

FUTURE SEA-LEVEL RISE. Sea-level rise has already impacted low-lying coastal communities, as manifested by the increasing frequency of nuisance tides and the progressive inland migration of storm surge. GCMs predict an additional rise in sea level of 0.2–2.0 m (0.6–7 ft) by 2100, depending on which emissions scenario comes to pass (Fig. 19.36a). The high-emissions scenario suggests that storm surge during hurricanes will flood areas as much as 16 m (52 ft) above present sea level. Because 10% of the world's population lives near a coast on land less than 10 m (33 ft) above sea level, large, populated areas could be inundated. Even a sea-level rise of 1.5 m (5 ft) would cause storm surge to flood many coastal cities and destroy beaches, damage coastal wetlands, and ruin groundwater supplies (Fig. 19.36b, c). Projecting further into the future, sea-level rise could inundate much of the world's low-lying coastal area. Overall, the economic impacts of sea-level rise will be staggering.

REDISTRIBUTION OF WATER RESOURCES. People need clean freshwater to live. Our major sources of freshwater are rivers, lakes, underground aquifers, and the snow and ice of mountains. In highly populated regions, some of these sources have already shrunk due to overuse and pollution. GCMs indicate that, in the future, global warming will cause the amount and timing of precipitation to decrease in some locations and to increase in other locations, and will cause the average volume of snow that collects in the winter to decrease nearly everywhere. The resulting redistribution of water resources will impact large populations.

How will climate change affect precipitation? GCMs predict that precipitation in certain subtropical and temperate climate types will decrease by about 20%–40%, while precipitation in other temperate climate types,

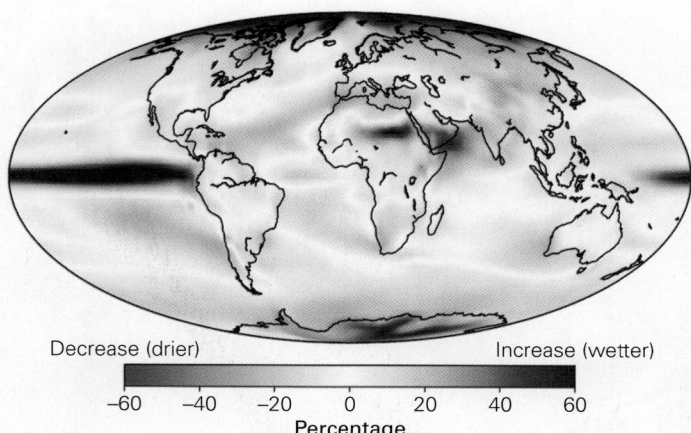

FIGURE 19.37 GCM prediction of regional variation in precipitation change by 2081–2100, relative to average global precipitation for 1981–2000, under a high-emissions scenario.

Decrease (drier)   Increase (wetter)

−60   −40   −20   0   20   40   60
Percentage

in polar climates, and along the ITCZ may increase by 10%–60% due to changes in atmospheric flow patterns and the overall increase in atmospheric water vapor (Fig. 19.37). In the United States, precipitation is predicted to decrease in the southwestern and southeastern United States, and to increase in the Midwest. Clearly, the water supplies of many large desert cities, such as Los Angeles, Las Vegas, and Phoenix, will be stressed even more than they are already.

Significantly, in regions near mountain ranges, or at mid- to high- latitudes, changes in the **snowpack** (the net accumulation of snow over the course of the winter) will play a key role in limiting surface water supplies. Snow accumulates above an elevation called the *snow line*, due to the decrease in temperature with elevation. In cold climates, the snow line can drop down to sea level during the winter, so broad areas receive a temporary covering of snow. In warmer climates, the snow line occurs on mountain slopes, well above sea level. As climate warms, the snow line retreats to higher elevations (Fig. 19.38). A rise

(a) If the elevation of the snow line rises by a relatively small amount, the volume of the snowpack can decrease substantially.

FIGURE 19.38 Mountain snowpacks store water, so a reduction in snowpacks diminishes water supplies.

(b) Diminished snowpack on a mountain in Switzerland.

Ongoing Climate Change

Forest-fire smoke

Desertification

Irrigation-dependent agriculture

Drying-up reservoirs

Receding glaciers

Dying cattle

Dried-up river

Abandoned town (buried by dunes)

Diversion canal

Desert city with inadequate water

Empty reservoir

Need for trucked-in water

Irrigation in deserts

Lowered water table due to groundwater depletion

Intensification of storms

Dust storms

Withering crops

Climate refugees

Forest fires

Eutrophication

Groundwater contamination

Sinkhole collapse due to lowered water table

Wind erosion of soil

Soil erosion

Sea-level rise (nuisance tides)

Dead zones and red tides

Broken levees and flooding

Saltwater intrusion

Transfer of atmospheric heat into ocean water, along with the melting of glaciers, has caused a rise in sea level sufficient to cause nuisance tides along the coast. In tropical regions, warming has caused coral reef destruction. Trouble in the sea has been made worse where rivers add contaminants and nutrients to the water, causing dead zones in the sea and eutrophication of lakes. On land, heat and drought make forests more susceptible to wildfires, some of which fill normally clear air with smoke. In addition, dust storms, soil erosion, and severe storms have worsened. Dangerously hot weather has led to heat illness in people and livestock; some people have had to become climate refugees. Changes in the global water distribution have caused rivers and reservoirs to dry up and have lowered the water table.

765

FIGURE 19.39 Warming of the Arctic, and its consequences.

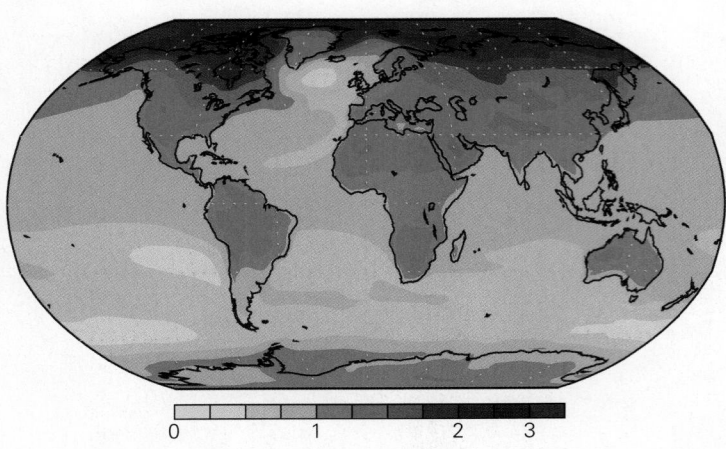

Predicted temperature change by 2100,
relative to the average of 1981–2000 (°C)

(a) GCMs predict that the Arctic will warm more than lower latitudes, particularly under a high-emissions scenario, as shown here.

(b) Permafrost thawing makes the ground uneven and has caused the foundation of this house to sink.

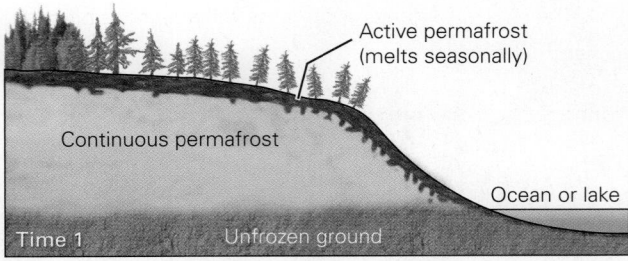

Before warming, permafrost forms a continuous layer down to a depth where the Earth's internal heat keeps ground temperatures above freezing.

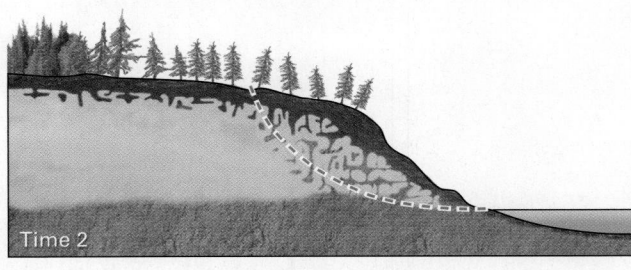

When the climate warms, the permafrost layer thins, and the upper part starts to break up. Slopes underlain by permafrost become unstable.

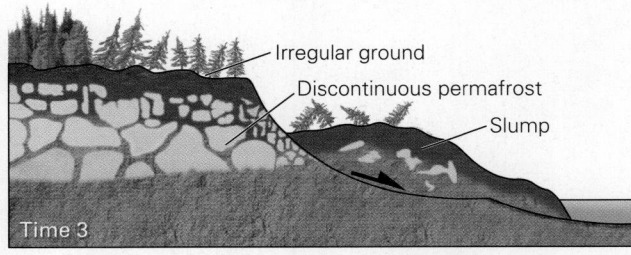

Eventually, permafrost breaks up entirely and becomes discontinuous. The land surface becomes irregular, and slumps carry melted permafrost downslope.

(c) When permafrost thaws, the land sinks irregularly, so trees tilt; slumps develop because melting weakens sediment.

in the snow line will cause the volume of the snowpack to shrink. Snowpack volume will also shrink because the cold season during which snow accumulates will be shorter. Snow won't accumulate until later in the season, and it will start melting earlier in the spring. Such changes mean that the total volume of water in the snowpack decreases. Also, more meltwater will enter streams in winter and early spring, and less will be available in summer and fall, resulting in more frequent spring floods and increased summer and fall water shortages. Loss of mountain snowpack is especially serious for California, where virtually all precipitation arrives in the winter months. Snowpack reduction has led to water shortages, as demonstrated by the 2021 water crisis in California's agricultural districts. When heavy snows do occur, melting of an early,

heavy snowpack can contribute to disastrous flooding, as occurred in California in 2023.

FUTURE CHANGES IN THE ARCTIC. GCMs indicate that the planet's most dramatic warming will happen in the Arctic and subarctic regions during winter (Fig. 19.39a). This phenomenon, known as **polar amplification**, occurs because warming air temperatures are melting ice cover on the Arctic Ocean, and snow cover on the land. This melting causes a significant drop in albedo and, therefore, a significant increase in the amount of heat that the land and sea can absorb to be reradiated back into the atmosphere. By some estimates, the Arctic Ocean could be ice free in mid-summer by 2050, and it may then be possible to sail to the North Pole in summer.

FIGURE 19.40 Changing climates will affect the biosphere.

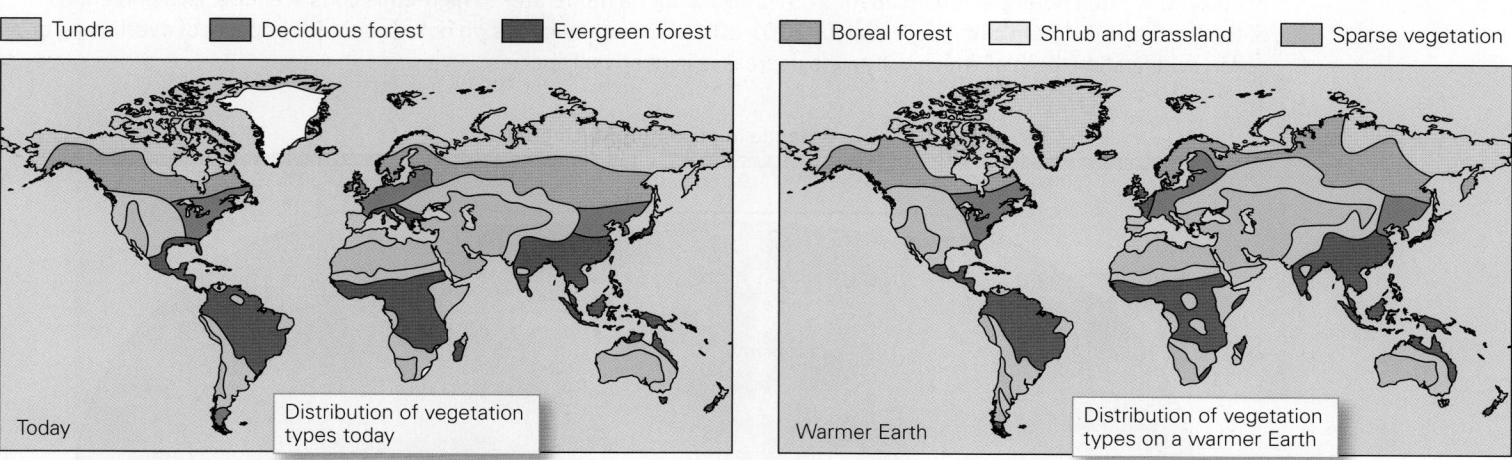

Tundra   **Deciduous forest**   **Evergreen forest**   **Boreal forest**   **Shrub and grassland**   **Sparse vegetation**

Today — Distribution of vegetation types today

Warmer Earth — Distribution of vegetation types on a warmer Earth

(a) Global warming will cause the latitude of climate zones to move poleward, and arid areas to expand.

(b) Polar bears rely on sea ice to gain access to food sources. Sea-ice cover has diminished greatly in the past few decades, so polar bear populations have decreased.

Arctic warming will also lead to a longer period of **permafrost thawing** during the year, allowing the organic material within the permafrost to decay, a process that releases both CO_2 and CH_4 and causes a positive feedback. Warming also increases the susceptibility of tundra and bordering forests to pests and wildfires. In places where people live on permafrost, thawing damages foundations for homes, roadbeds, and infrastructure, because it results in uneven sinking of ground surfaces and can cause landslides (Fig. 19.39b, c). Notably, responses to climate change for the Antarctic and the Arctic regions will differ, because the former is a continent, isolated by the circum-Antarctic current, while the latter is an ocean covered by relatively thin sea ice.

CHANGES IN THE BIOSPHERE. As global warming progresses, GCMs indicate that isotherms will slowly shift poleward. Several lines of evidence, such as the plant hardiness zones we discussed earlier, indicate that this shift has been taking place. Eventually, forests currently growing at temperate latitudes will be growing in areas that are now tundra (Fig. 19.40a). In fact, by 2100, the north-central United States may have a climate resembling that of the Gulf Coast. The implications of such **climate zone migration** for agriculture are profound. For example, crops such as corn and soybeans that now grow in the US Midwest will be better suited, climatically, for southern Canada. But because southern Canada has hosted cool climates for thousands of years, that region

FIGURE 19.41 The predicted redistribution of fish populations due to the warming of seawater as indicated by change in maximum catch potential predicted for 2051–2060, under a moderate- to high-emissions scenario, as a percentage of the average maximum catch potential for 2001–2010. These predictions do not account for impacts of overfishing or ocean acidification, both of which will cause decreases in nearly all species.

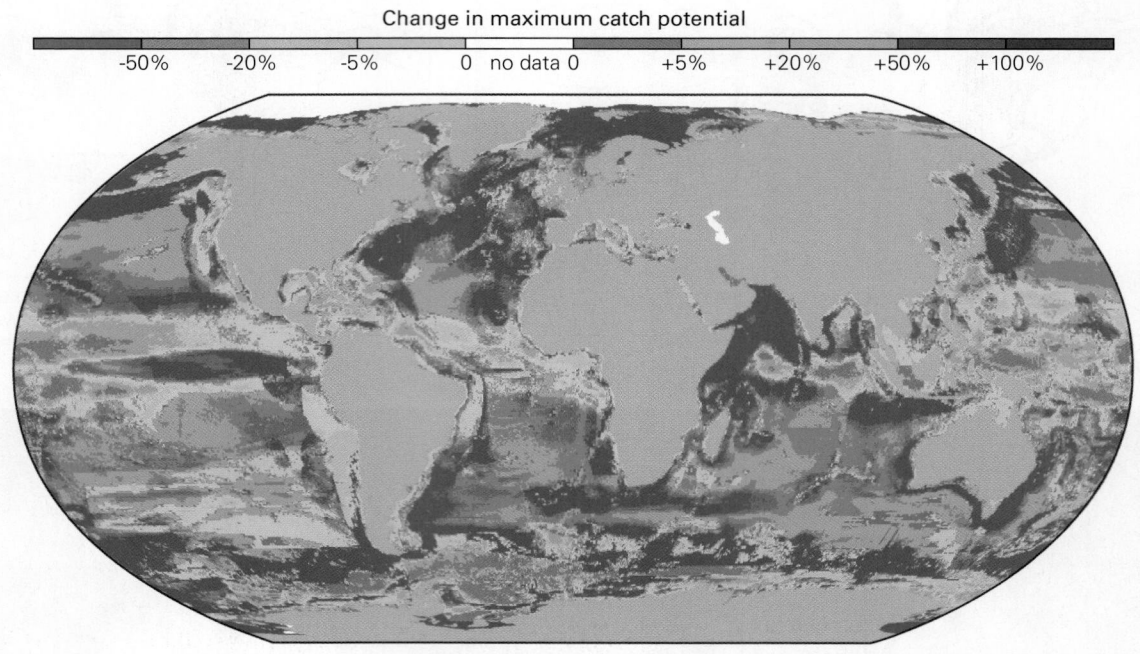

Change in maximum catch potential

–50% –20% –5% 0 no data 0 +5% +20% +50% +100%

has not developed thick soils like those of the midwestern United States. Therefore, the productivity of areas with appropriate climates for economically viable crops in the future might be substantially less than the productivity of areas that now produce those crops.

Animal life in the biosphere will also feel the threat of climate change. Populations of climate-sensitive wildlife and bird species have already been shrinking, and in some cases going extinct, as their habitable range diminishes. For example, declining sea ice in the Arctic has already resulted in a reduced survival rate for polar bears and walruses, because these predators need full sea ice cover to hunt seals, their primary food source (Fig. 19.40b). Extinction rates have increased to such an extent that some scientists suggest that we are living during a time that will be recorded in the geologic column as the *sixth extinction*, a mass extinction comparable to that which occurred at the end of the Permian and at the end of the Cretaceous. But while some species go extinct, others have been expanding their territories. For example, insects that once lived exclusively in tropical climates are now appearing in warming temperate lands. These include species, such as mosquitoes, that carry disease. Consequently, the global range in which tropical diseases are common has been expanding.

Zooplankton form the base of the marine food chain, and their numbers and locations are changing with the climate. As oceans warm, cool-water zooplankton species migrate to higher latitudes. Fish and other marine animals that feed on zooplankton must move with the organisms. These changes have led to widespread changes in the productivity and distribution of fish populations. Continued warming will amplify these changes, so fishing grounds will have to shift to new locations. The shift can be visualized by considering the change in *maximum catch potential* (the maximum exploitable catch of a species) (Fig. 19.41). Many areas, like the Southern Ocean, show steep declines, while others, like the North Atlantic, show increases due to migration of warm-water fish to higher latitudes.

Temperature rise in shallow oceanic ecosystems has already had a major impact on the life of coral reefs. As we noted earlier, many coral reefs are now undergoing **coral bleaching**, a process that occurs when corals under environmental stress expel the algae that live in their tissues, and turn white (Fig. 19.42). Although corals can survive initial bleaching, they will die if the conditions that cause the bleaching persist. Scientists suggest that rising water temperatures are probably the main cause of coral bleaching, but changes in water chemistry may also play a role.

FIGURE 19.42 Coral bleaching.

(a) A living reef with various colored corals.

(b) A dead reef with bleached corals.

The rates of bleaching and reef death are astounding. A 2019 survey of Australia's Great Barrier Reef showed that close to half the reef died in just a few years, with nearly a 90% drop in the number of new corals growing. The future of coral reef systems, and of the marine organisms that depend on them, will be further threatened as ocean temperatures rise.

FIGURE 19.43 Ocean acidification is taking place worldwide.

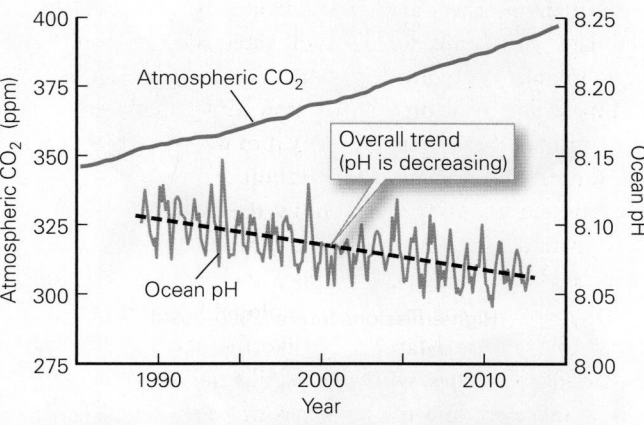

(a) As the CO_2 concentration in the atmosphere rises, more CO_2 dissolves in the ocean, where it produces carbonic acid.

CHANGES TO OCEAN WATERS. Chemists express acidity using a measure called *pH*. A material with a pH less than 7 is an *acid*, and one with a pH greater than 7 is a *base*. Pure water has a pH of 7, so it's *neutral*. Seawater, prior to the industrial revolution, had an average pH of about 8.25, meaning that it was slightly basic. In the past few decades, however, the pH of seawater has already dropped to about 8.1, and by 2100 it may have dropped as low as 7.8 (Fig. 19.43). This phenomenon, known as **ocean acidification**, happens because some of the CO_2 that enters the atmosphere diffuses into seawater and dissolves to produce carbonic acid, as we noted in our discussion of the carbon budget. While this process, fortunately, decreases the concentration of greenhouse gas in the atmosphere, it unfortunately harms certain marine organisms by depressing their metabolic rates, their ability to grow shells, and their immune responses. Ocean acidification, for example, makes it more difficult for zooplankton to form calcite shells, because the solubility of calcite in seawater increases as pH decreases.

Global warming has an important indirect impact on oceanic currents. As we've seen, the thermohaline circulation transfers heat across latitudes. According to some models, if global warming melts enough polar ice, or if precipitation increases significantly, freshwater will dilute

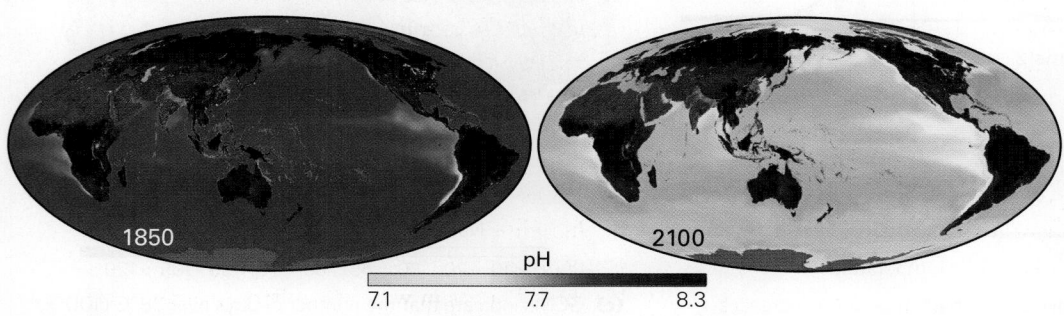

(b) Changes in ocean pH will vary with location. But by 2100, ocean water is predicted to be below pH 7.8 worldwide.

surface seawater at high latitudes so that the water will not be dense enough to sink, and the thermohaline circulation will be shut off. By some estimates, the thermohaline circulation in the Atlantic has already slowed by about 15% since the mid-20th century, perhaps due to melting of the Greenland ice sheet. If the thermohaline circulation did stop, a result might be a substantial cooling of high-latitude regions, a negative feedback on global warming.

FUTURE CHANGES IN HAZARDOUS WEATHER. As we noted earlier, climate change already seems to be increasing the severity and duration of hurricanes, so the destructive capability of these storms will be amplified. GCMs suggest that climate change will also impact major climate patterns in temperate latitudes of the northern hemisphere, because as global warming progresses, Arctic and subarctic regions will warm more than the tropics (due to the loss of sea ice and land albedo). Therefore, the average temperature gradient across the polar front (see Chapter 17) will decrease during winter, and the polar front will shift northward. Researchers expect that, as a result, mid-latitude cyclones tracking across continents of the northern hemisphere will decrease in number, and those that do form will shift farther north. By some estimates, 5%–10% fewer winter mid-latitude cyclones will occur per year during the late 21st century, compared with the 20th century. The storms may also decrease in intensity over land. However, coastal mid-latitude cyclones, such as the nor'easters discussed in Chapter 18, may tap energy from warmer seas and increase in strength.

Continued global warming will overall increase the intensity of heavy rain events worldwide, since warmer air can hold more water vapor. Consequently, rainfall in many flood-producing weather systems will be more intense. For example, flooding associated with monsoonal weather in southern Asia could get worse. Although researchers cannot attribute any single weather event to climate change, society has endured several record rainfalls in recent years, an expected consequence of a warming climate.

FUTURE INCREASES IN HEAT WAVES AND DROUGHTS. During a **heat wave**, the temperature remains significantly above normal for days or weeks. Heat waves become news when the temperature rises high enough for many people to suffer from *heat illness* (such as heat stroke and heat exhaustion). The number of heat waves has increased in recent years, which isn't surprising, given that recent years, especially 2023, have seen many records for average global temperature (Fig. 19.44a). More global warming will further increase the number of heat waves. To see why, consider a graph showing the range of temperatures for a given location, recorded over many years, with each temperature plotted against the number of days during which that temperature occurred (Fig. 19.44b). The result is a bell curve, with the right-hand limb representing extreme heat. If global warming takes place, the whole curve shifts to the right, so the number of days with extreme heat increases.

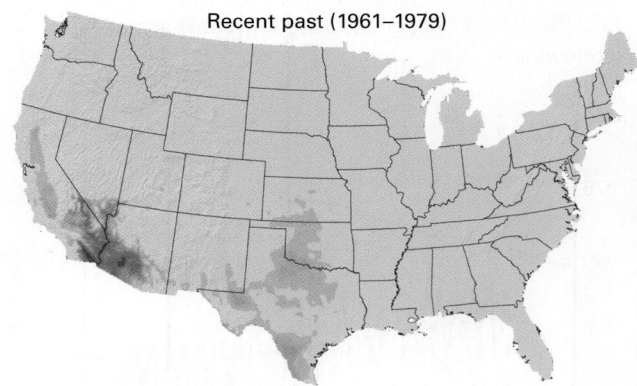

Recent past (1961–1979)

FIGURE 19.44 Heat waves will be more common in the future, due to global warming.

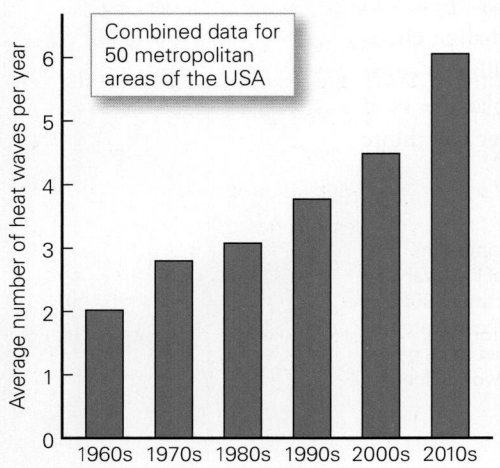

(a) The average number of heat waves per year in the United States has nearly tripled since the 1960s.

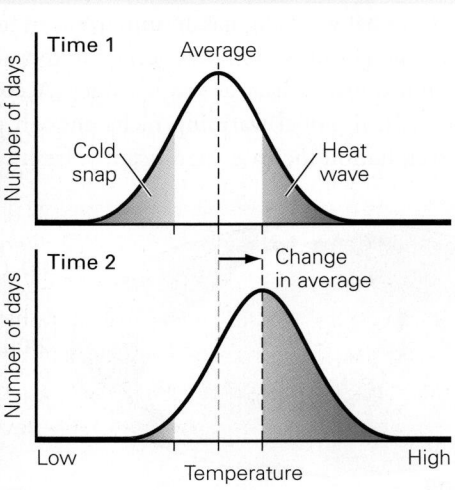

(b) If average global temperature increases slightly, the number of heat waves increases a lot.

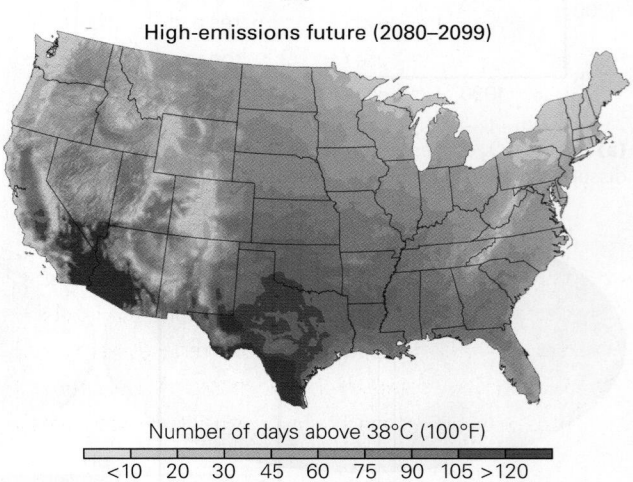

High-emissions future (2080–2099)

Number of days above 38°C (100°F)

<10   20   30   45   60   75   90   105   >120

(c) GCMs indicate that the number of days over 38°C (100°F) will increase significantly by 2100.

GCMs predict that under a high-emissions scenario, large areas of the United States will suffer from frequent heat waves by 2100, so that the number of days with a **heat index** (a value that takes into account humidity, and therefore represents human perception of temperature and associated discomfort) of over 41°C (105°F) will increase threefold by 2050 in the north-central United States (Fig. 19.44c). In fact, some locations that now have only a few such days of extreme heat per year may have more than 40 such days per year by the end of the century. Places that are normally hot, such as Phoenix, Arizona, are already setting records, not just for high temperatures, but also for the number of consecutive days with extreme heat.

Increases in extreme heat will have many effects on society. Severe levels of heat over extended time periods decrease worker productivity, increase wildfire hazard, decrease water supplies, and cause more frequent crop failures. Such heat also causes a rise in heat illness (heat stroke and heat exhaustion). For example, a heat wave that struck Europe in 2003 caused 70,000 deaths.

A continued rise in global temperature will also worsen droughts because heat increases soil-moisture evaporation rates; without soil water, crops and natural plants will wither. Loss of plant cover will allow more radiation to reach and heat the ground. This is a positive feedback, as the extra heat will cause still more soil moisture to evaporate. On a broad scale, global warming increases the land areas affected by drought for two reasons. First, the north-south width of the Hadley cells (see Chapter 17) will increase because, as oceans warm, more air will rise along the ITCZ, so air will spread outward to higher latitudes at the tropopause. Such **Hadley cell expansion** means that hot, dry air will descend at latitudes that are now semiarid, causing those regions to undergo desertification. Second, the northward shift of Pacific storm tracks will decrease the supply of rain and snow to the western United States, reducing water supply to the region.

CLIMATE POLICY. The GCM predictions described above, along with estimates of the large economic cost of global warming, explain why the issue of climate change has become a subject of intense international discussion. But what can be done? As a first step, nations need to build a consensus about the nature of the problem and about tolerable target values for the temperature rise. An international agreement, the *Kyoto Protocol*, signed at a 1997 summit meeting held in Japan, proposed that the first step would be to slow the input of greenhouse gases into the atmosphere by decreasing the burning of fossil fuels. This approach was reinforced by the 2015 *Paris Agreement*, and most recently by the *Sharm El-Sheikh Climate Change Conference* in Egypt in 2022. All such agreements committed signatories to efforts that will keep the global average temperature from exceeding 2°C above the preindustrial temperatures. Achieving the goals of these agreements, unfortunately, has been elusive, and as of 2023, the concentration of greenhouse gases continues to increase at or above high-emission scenarios of GCMs.

BOX 19.3 ▸ **Putting Earth Science to Use**

Reducing your personal carbon footprint

Combating climate change requires a global effort. What can we do individually to contribute to a reduction in greenhouse gas emissions? More than you might think! In our homes, everyone can implement easy changes that decrease energy demand. For example, replacing incandescent light bulbs with LED bulbs, regularly replacing filters in furnaces, using programmable thermostats that reduce heating and cooling in our houses when nobody is home, unplugging electronic devices when not in use, adding insulation, purchasing energy-efficient appliances and vehicles, using properly inflated tires, and, when possible, walking, biking, or using public transportation instead of driving all reduce energy usage. Recycling also saves energy, because producing new materials from recycled materials uses less energy than it takes to extract new resources from the ground.

Communities, as a whole, can also save energy by improving public transportation, improving traffic flow, replacing unneeded parking lots and abandoned buildings with green spaces, lining streets with trees, developing building codes and reasonable regulations that lead to energy-efficient construction, advocating for increases in non-fossil-fuel energy production networks, and improving energy grids (for distributing energy). Planting trees can be particularly helpful. Not only do they shade buildings and, therefore, reduce air-conditioning costs, but they incorporate CO_2 to make wood. By one estimate, the growth of 1 trillion more trees (130 trees per person) could remove 25% of atmospheric CO_2, thereby returning atmospheric concentration to 1960s levels.

But some progress has been made. Worldwide, individuals, private companies, nongovernmental organizations, and some governments have found ways to decrease greenhouse gas emissions. Approaches to this problem focus on switching to alternative energy sources with a smaller **carbon footprint** (the total amount of greenhouse gas produced by a process, phenomenon, or the production of a material), and by increasing the efficiency of vehicles, factories, homes, and offices (Box 19.3). Another, more controversial, approach involves collecting CO_2 produced at large sources, such as power plants, so that it can be condensed and then injected into deep wells to be stored in rock underground; this overall process is called *carbon capture and sequestration*. Regardless of these efforts, with warming trends already happening, society must adapt. Hopefully, research in the coming decades will lead to the discovery of innovative solutions to the challenges of energy production, water and food security, and coastal preservation.

Take-home message...

The impacts of climate change are far-reaching. The potential economic and social costs of sea-level rise and changes in the environment pose threats to future generations. Predictions of future climate change based on global climate models remain complicated because of the complexity of possible feedback processes. Nevertheless, significant changes are already underway in the distribution of climate belts and in sea level. Such changes will need to continue in the coming decades, particularly if society fails to decrease its carbon footprint.

Quick Questions

- How might global warming affect global agricultural productivity?
- Why does global warming cause a rise in sea level?
- Name one positive and one negative feedback process.

19 CHAPTER REVIEW

Objective 19.1

Identify the Earth's major climate zones, and relate them to latitude, altitude, and proximity to the ocean.

KEY CONCEPTS

- Weather refers to atmospheric conditions at a given location and time. Climate refers to the average and range of variation of weather conditions in a region over the course of decades or more. Climate change is a shift in average climate conditions.

- Two variables, temperature and precipitation, largely determine the climate a region will experience. Daily and seasonal changes in temperature and precipitation depend on latitude, proximity to the ocean, and elevation.

- In general, it is warmer and wetter at low latitudes. Researchers classify the Earth's climates, using the Köppen-Geiger climate classification, into five climate groups—tropical, arid, temperate, cold, and polar. These groups are divided into several subcategories. Vegetation provides a key clue to the climate of a region.

EARTH-SCIENCE VOCABULARY

average precipitation (p. 731)
climate (p. 730)
climate change (p. 730)
climate group (p. 732)

Köppen-Geiger climate
    classification (KGCC) (p. 732)
weather (p. 730)

REVIEW QUESTIONS

1. **(a)** A TV weather reporter states "the average seasonal snowfall in our city is 20 inches, but today's storm dumped 25 inches!" Which part of this statement refers to the city's weather and which to the city's climate? **(b)** What factors control the climate where you live? **(c)** How would you describe the climate of your region?

2. **(a)** What characteristic (evident in **Figure A**) did Köppen and Geiger use to map out climate zones? **(b)** What are the five major climate categories specified in the Köppen-Geiger climate classification? **(c)** Using Table 19.1, name the climate zones shown in **Figure A**.

3. **(a)** What factors control a region's rainfall? **(b)** What factors control a region's temperature and temperature variation?

Af

Cfa

ET

A

Objective 19.2

> Describe the factors that control climate conditions prevailing in a region.

KEY CONCEPTS

- The Earth lies in the habitable zone of the Solar System, due to the amount of insolation. Over time, the amount of solar energy arriving at the Earth from the Sun must balance the amount of infrared energy radiated away from the Earth.

- The atmosphere traps some of the radiated energy leaving the Earth's surface and reradiates it back to the Earth, keeping the Earth's surface warmer than it might otherwise be. This process is called the atmospheric greenhouse effect.

- Latitude plays a key role in controlling climate. Because of the tilt of the Earth's axis of rotation, the amount of solar radiation arriving at a particular latitude changes over the course of the year; this change causes the Earth's seasons.

- Atmospheric circulations, ocean currents, distribution of land and oceans, elevation, and position relative to a mountain range also affect climate in a region.

EARTH-SCIENCE VOCABULARY

angle of incidence (p. 736)	**Goldilocks effect** (p. 735)
atmospheric greenhouse effect (p. 736)	**habitable zone** (p. 735)
energy balance (p. 736)	**heat capacity** (p. 738)
greenhouse gases (p. 736)	**insolation** (p. 735)
	season (p. 737)

REVIEW QUESTIONS

4. **(a)** What does the Goldilocks effect refer to? **(b)** Explain the cause of the atmospheric greenhouse effect and how it affects the Earth's energy balance. **(c)** Why do we say that the atmosphere is baked from below rather than broiled from above? **(d)** What processes not related to radiation are required to achieve an energy balance at the Earth's surface?

5. **(a)** What three factors determine how much solar energy will reach a given latitude? **(b)** Why does the Earth have seasons?

6. **(a)** Explain why the southern hemisphere experiences summer when the northern hemisphere experiences winter. **(b)** How does **Figure B** illustrate why the Earth has seasons? **(c)** How many hours of sunlight per day does the South Pole receive on December 21? **(d)** Where would you be if you received 12-hour-long days and nights every day of the year?

7. **(a)** Why does the distribution of landmasses and mountain ranges affect climate in a region? **(b)** Why is the average temperature of Ireland milder than that of eastern Canada at the same latitude? **(c)** Would you expect more rain to fall on the windward or leeward side of a mountain range? **(d)** Why does climate vary from the bottom to the top of a large mountain?

8. **(a)** How are the Hadley cells related to the climate zones of the tropics? **(b)** Can both rainforests and deserts exist in a tropical climate? **(c)** Would you expect more extreme summer-to-winter climate variations in the middle latitudes along a coastline or in the mid-continent? **(d)** What phenomena affect air circulation in the mid-latitudes?

Objective 19.3

> Understand how scientists determine past changes in the Earth's climate.

KEY CONCEPTS

- Various paleoclimate indicators allow researchers to characterize the climates of the past. Paleoclimate studies indicate that there has been both long-term and short-term climate change over the course of Earth history.

- The stratigraphic record, which permits reconstruction of long-term climate change, indicates that icehouse and hothouse periods alternate over geologic time, but that the durations of these periods vary.

- The record of short-term climate change comes from studies of fossils, isotope ratios, and pollen in sediment cores. Evidence also comes from studies of isotope ratios and air bubbles in ice cores extracted from glaciers, as well as from the study of growth rings in trees and other organisms.

- Since the most recent glaciation, there have been significant episodes of global warming and global cooling, including the Medieval Warm Period and the Little Ice Age.

EARTH-SCIENCE VOCABULARY

global cooling (p. 741)	**hothouse period** (p. 741)
global warming (p. 741)	**icehouse period** (p. 741)
growth ring (p. 743)	**paleoclimate** (p. 741)

REVIEW QUESTIONS

9. **(a)** What is meant by "paleoclimate"? **(b)** Why are depositional settings of the past important for understanding past climates? **(c)** What does the stratigraphic record imply about the presence of liquid water on the Earth over its history?

10. **(a)** What is meant by a hothouse and icehouse period in Earth's history? **(b)** What must occur for a period of the Earth's history to be characterized as an icehouse period? **(c)** What paleoclimate indicators would be best for studying climate change over the last 100,000 years?

11. **(a)** Why are bubbles in ice cores studied to understand climate change? **(b)** What information do oxygen isotopes give about past climates? **(c)** What do the colors, and the horizontal axis, on **Figure C** represent? Indicate the times on the graph when temperatures were cooler, and explain the basis for your conclusion.

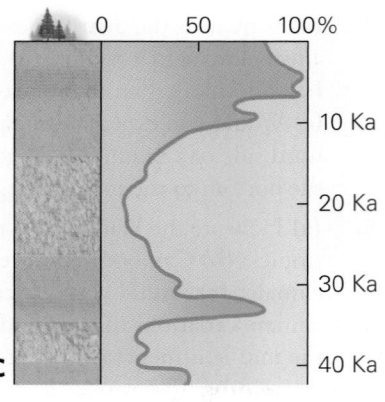

C

14. **(a)** What are the Milankovitch cycles, and what role do they play in controlling climate? **(b)** Name three other factors that impact short-term climate change. **(c)** Contrast the key factors that play a role in causing long-term climate change with those involved in short-term climate change.

15. **(a)** Based on paleoclimate records, how has climate changed over the course of the last few millennia? **(b)** Is the Earth currently in a glacial or interglacial period?

Objective 19.5

Summarize the evidence for recent climate change, and discuss causes of this change.

KEY CONCEPTS

- Since the industrial revolution, average global temperature has increased at rates much greater than in the previous millennia.

- Observations of glacial and sea-ice retreat, sea-level rise, and shifting plant hardiness zones support the conclusion that global temperatures have risen. The atmospheric CO_2 concentrations observed during the past two centuries are much greater than at any other time during the past 800,000 years.

- The increase in global temperature is directly related to the increase in greenhouse gases in the atmosphere, a result of human activities.

EARTH-SCIENCE VOCABULARY

anthropogenic sources (p. 759)
carbon budget (p. 759)
hockey stick diagram (p. 751)

Intergovernmental Panel on Climate Change (IPCC) (p. 751)
Keeling curve (p. 751)
mass-balance calculation (p. 759)

REVIEW QUESTIONS

16. **(a)** What is the mission of the IPCC? **(b)** How often have they published reports concerning climate change?

17. **(a)** What has been the global temperature trend over the last 100 years? **(b)** What has been the trend in the last 20 years? **(c)** What years have been the warmest since records have been kept?

18. **(a)** What changes have occurred to most glaciers around the world? **(b)** How are the ice sheets of Greenland and Antarctica changing? **(c)** What other evidence led researchers to the conclusion that the Earth's climate has warmed during the 20th and 21st centuries?

19. **(a)** Which coasts of the United States will suffer the worst impacts of sea level rise and why? **(b)** Will the melting of Arctic sea ice cause sea level to rise? How about the melting of continental glaciers in Greenland?

Objective 19.4

Explain possible causes of both long-term and short-term climate change.

KEY CONCEPTS

- Long-term climate change is a consequence of changes in atmospheric concentrations of greenhouse gases, the positions of continents and oceans, the flow of ocean currents, and the uplift of mountain ranges.

- The primary controls on short-term climate change appear to be the Milankovitch cycles, volcanic aerosols in the atmosphere, modifications of surface albedo, and abrupt changes in greenhouse gas emission or absorption.

EARTH-SCIENCE VOCABULARY

albedo (p. 749)
carbon cycle (p. 747)

methane hydrates (p. 750)

REVIEW QUESTIONS

12. **(a)** Explain the difference between long-term and short-term climate change. **(b)** What factors control long-term climate change? **(c)** Do you think an icehouse period is more likely when continents are closer to the poles or closer to the equator?

13. Label the axes on **Figure D**, and describe what the black line shows. Explain the cause of the phenomenon depicted.

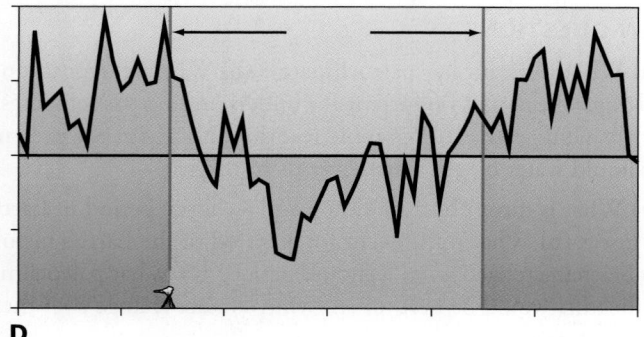

D

20. (a) What is meant by a "biological indicator" of global warming? (b) Provide two examples of biological indicators. (c) What areas in **Figure E** have warmed the most over the past 50 years? (d) Why have these regions warmed more dramatically than others?

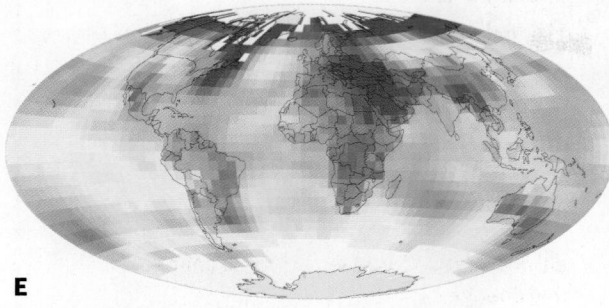

E

Objective 19.6

Assess the impacts of future global climate change on the Earth System and society.

KEY CONCEPTS

• Global climate models (GCMs) suggest that, in the absence of anthropogenic CO_2 emissions, the climate might be experiencing a weak cooling trend rather than the observed warming. Global warming, if continued at the current rate, will significantly affect the intensity of storms, precipitation patterns, and intensity of drought and heat waves.

• Sea level rise will impact the habitability of coastal areas. Shifts in climate belts will impact agricultural productivity, while melting glaciers and higher snow levels in mountains will reduce freshwater supplies.

• Positive feedback processes (increasing evaporation, releases of CH_4, decreasing albedo) may be accelerating warming. Negative feedbacks, such as expansion of regions where plants can grow, may be

slowing warming. The role of clouds in future climate change is still being assessed because clouds can produce both positive and negative feedbacks.

• Because of the role that anthropogenic CO_2 plays in driving global warming, nearly all governments are negotiating polices to decrease emissions of greenhouse gases worldwide.

EARTH-SCIENCE VOCABULARY

carbon footprint (p. 772)
climate zone migration (p. 767)
coral bleaching (p. 768)
feedback (p. 761)
global climate model (GCM) (p. 760)

Hadley cell expansion (p. 771)
heat index (p. 771)
heat wave (p. 770)
ocean acidification (p. 769)
permafrost thawing (p. 767)
polar amplification (p. 766)
snowpack (p. 763)

REVIEW QUESTIONS

21. (a) Give reasons why freshwater supplies may be reduced in some areas of the world if the Earth's atmosphere continues to warm. (b) Give reasons why flooding might become more common in other regions. (c) Aside from the immediate effects of coastal flooding, what long-term effects might sea-level rise have on coastal populations? (d) If current observed trends in global temperature change continue, will the number of people at risk of contracting mosquito-borne diseases increase or decrease during the next century? Explain your answer.

22. (a) What kinds of feedback may amplify the process of global warming, and why? (b) Are the feedbacks you identified positive or negative feedbacks? (c) List as many impacts to society that you can think of that will result from climate change. (d) Which of these will likely cause human migration? (e) Which will impact global food supplies?

23. (a) What efforts may help to diminish human society's carbon footprint? (b) Which of these can be done in your local community?

ANOTHER VIEW The sprawl of Phoenix, Arizona, covers land that was once desert. Covering land with asphalt causes local warming, because asphalt absorbs heat. Water supplies are under so much stress that officials are trying to slow construction of new housing developments.

◉20 INTRODUCING ASTRONOMY
Looking Beyond the Earth

After studying this chapter, you should be able to...

1. describe how our ideas about the organization and movement of celestial objects evolved from ancient to modern times.

2. relate the observed movements of celestial objects to the Earth's rotation on its axis and to its orbit around the Sun.

3. explain the path of the Moon across the sky, why the Moon has phases, and why eclipses take place.

4. discuss how astronomers use electromagnetic radiation to determine a star's temperature, composition, velocity, and distance.

5. list the units that astronomers use to describe vast distances between celestial objects, and describe the tools that they use to determine these distances.

6. distinguish how various kinds of telescopes function, and explain why we use space telescopes.

Centuries before Columbus, Polynesian navigators guided sailing canoes across vast reaches of the Pacific Ocean and colonized over one hundred island groups **(Fig. 20.1)**. Their accuracy was amazing! For example, when a canoe journeyed 4,200 km (2,600 mi) from Tahiti to Hawai'i, the navigators had to aim for a target only 120 km (75 mi) across—a feat equivalent to hitting a small coin from a distance of 35 km (22 mi)—and they succeeded without using GPS, compasses, clocks, or charts. How did they do it? They merged their knowledge of the night sky and how it changed during the year with their knowledge of how ocean swells move, birds migrate, and clouds drift. Successful navigators described routes in songs and myths so that the information could be remembered by the next generation.

If you gaze upward on a dark, clear night, you'll see many of the objects that guided Polynesian navigators across the ocean. The display includes both *stars*, twinkling points that stay fixed in position relative to their neighbors even as they make their nightly journey across the sky **(Fig. 20.2a)**, and *planets*, which move relative to the backdrop of stars and relative to one another **(Fig. 20.2b)**. If it's a particularly dark night, you'll also see the *Milky Way*, a wispy, glowing haze that spans the sky **(Fig. 20.2c)**. On many nights, however, you'll be able to see only the brighter stars and planets, for the glow of the Moon masks fainter lights, and during the day, of course,

<< This image from NASA's James Webb Space Telescope reveals emerging stellar nurseries and individual stars in the Carina Nebula.

FIGURE 20.1 Polynesian sailing canoes similar to this modern replica were used for trans-Pacific voyages.

the Sun's intense light completely obscures all planets and stars.

Over many millennia, people have struggled to describe the objects of the night sky and to characterize the paths those objects followed. They have also speculated about the nature and origin of the objects. Eventually, knowledge about the night sky coalesced into the discipline of *astronomy*, the study of what lies outside the Earth System. In the past few centuries, various tools have

FIGURE 20.2 Stars, planets, and galaxies.

(a) Look into the sky on a dark night, and you'll see a myriad of stars.

(b) On a June day in 2022, planets appear near the horizon just before dawn.

(c) On a particularly clear, moonless night, in a dark location, you can see the Milky Way.

allowed astronomers to refine their descriptions, and in the past century, modern instruments, satellites, and space probes have answered many fundamental questions about the workings of space and the objects within it.

In this final part of this book, we offer a brief survey of astronomy to provide a context within which you can understand the origin and evolution of our home planet. This introduction will help you complete your personal image of your natural surroundings by explaining what you see when you look up and beyond the Earth. This chapter starts by tracing the evolution of ideas that philosophers and, later, scientists considered as they tried to fathom humanity's place in the cosmos. We then introduce the modern concept of how celestial objects move in the sky. This concept relies on the study of electromagnetic radiation that comes from space, so we next explore the basic properties of this radiation and describe the tools astronomers use to measure it in ways that allow us to peer at unfathomably distant places and see events that happened in the distant past.

20.1 Building a Foundation for the Study of Space

Humans have wondered about the nature of **celestial objects**—the Sun, Moon, stars, planets, and other objects in the night sky—since conscious thought began. The oldest hint of humanity's interest in the heavens dates to 35,000 B.C.E., when cave dwellers chiseled representations of celestial objects into rock surfaces. By 14,000 B.C.E., cave paintings found at many localities portrayed distinctive patterns of dots that resemble star clusters **(Fig. 20.3)**. As civilization dawned, observers knew that the Sun, Moon, and stars rose at specific points on the

horizon at times of seasonal change, so the study of the sky became a basis for calendars that people used to plan the sowing or reaping of crops. By 1,000 B.C.E., scholars in Mesopotamia were carefully recording the movements of stars and planets. Such work evolved into the science of **astronomy**, the systematic study of the **Universe** (or *cosmos*), meaning all of space and all the celestial objects within it. In some cultures, people thought these movements could somehow serve to predict or influence events that would happen to people on the Earth; such magical thinking became the basis of *astrology*, a nonscientific assemblage of speculations with no provable basis in fact.

By watching the motion of the Sun, Moon, and stars across the sky, many cultures came to believe that the Earth lay at the center of the Universe, so that all celestial objects followed orbits around the Earth, and all stars lay on the surface of a distant sphere outside it. (Note that an **orbit**, in a general sense, is a closed path in space that one object takes around another; the complete orbit of one object around another is called a *revolution*.) This idea, a **geocentric model** of the Universe, was promoted by Ptolemy (90–186 C.E.), a mathematician living in Egypt, and came to dominate European thinking about the heavens through the Middle Ages.

Astronomical thinking in Western society largely stagnated between the time of Ptolemy and the European Renaissance, when new ideas began to take hold. Renaissance thinking led Nicolaus Copernicus (1473–1543) to re-examine planetary orbits. He concluded that the geocentric model could not be correct, and in 1632, he published a book—*De Revolutionibus Orbium Coelestium* (*On the Revolutions of the Heavenly Spheres*)—in which he demonstrated that a **heliocentric model**, in which objects orbited the Sun, provided a better explanation of celestial motions. This idea, which also came to be known as the *Copernican model*, did not gain widespread acceptance for many decades. It took the work of Galileo Galilei (1564–1642), the great Italian scientist, to provide the observations that ultimately led to acceptance of the Copernican model. Galileo, the first astronomer to examine the skies with a telescope, saw that Jupiter has moons, and that the moons cross the face of Jupiter and disappear behind the planet. He reasoned, correctly, that if moons orbit Jupiter, then not all celestial objects orbit the Earth. It took the work of Johannes Kepler to understand planetary orbits, and the brilliance of Isaac Newton to provide the fundamental laws of physics, to explain planetary motions and why those motions are what they are.

Kepler and the Laws of Planetary Motion

Johannes Kepler (1571–1630), a German astronomer, made the next great astronomical discovery. Kepler developed a set of simple mathematical relationships that

FIGURE 20.3 Early observations of celestial objects include cave paintings thought to depict groups of stars.

Star cluster in the night sky

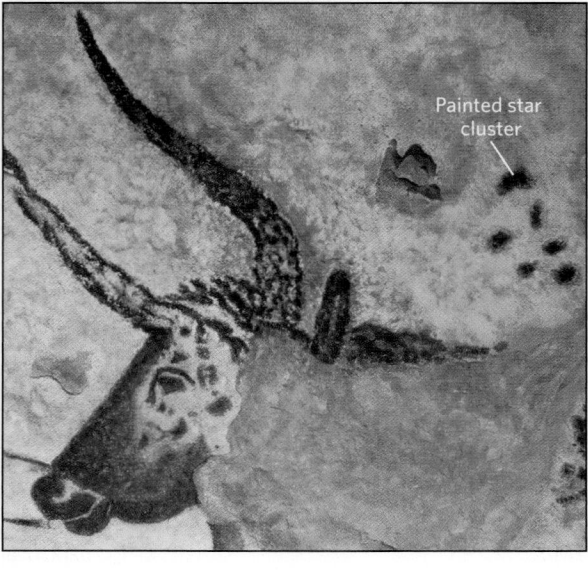

Painted star cluster

BOX 20.1   A Deeper Look

The geometry of an ellipse

An **ellipse** resembles a flattened circle or the outline of an egg **(Fig. Bx20.1)**. The *major axis* stretches across the widest part of the ellipse, and the *minor axis* stretches across the narrowest part of the ellipse. A line running from the center of the ellipse to the rim along the major axis is called the *semi-major axis*, and one running from the center to the rim along the minor axis is the *semi-minor axis*. Note that the major axis and minor axis are perpendicular to each other. An ellipse has two *foci*, positioned along the major axis. You can draw an ellipse by first sticking two pushpins along the major axis, spaced several centimeters apart, into a sheet of cardboard, then attaching one end of a piece of string to one pin and the other end to the other pin, allowing the string between the pins to remain loose. Hold a pencil perpendicular to the cardboard, with the point touching the cardboard, and press against the string to make the string taut. If you now move the pencil in an orbit around the pins, you'll trace out an ellipse. Each pin is a focus of the ellipse. Note that the center of the ellipse lies on the major axis halfway between the two foci, and that the center of the ellipse bisects the minor axis. The distance between the foci determines the ellipse's *eccentricity*. An ellipse with high eccentricity looks flattened and elongate, and one with low eccentricity looks more circular. A circle, in effect, is a special ellipse—it's one whose major and minor axes have the same length, and whose foci both lie at the circle's center.

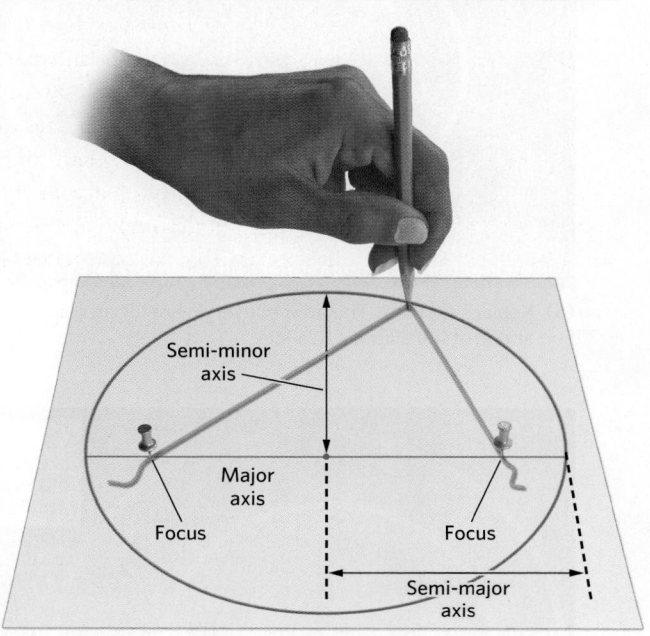

FIGURE Bx20.1   The geometric features of an ellipse. You can draw an ellipse by tying a loose piece of string to two pushpins.

completely explained the motions of planets in the context of the Copernican model. These relationships, now known as **Kepler's laws** of planetary motion, portrayed planetary orbits as ellipses instead of as circles **(Box 20.1)**. Kepler's laws can be stated as follows:

- *First law:* The orbit of every planet is an ellipse, not a circle, with the Sun at one of the ellipse's two foci **(Fig. 20.4a)**.

- *Second law:* An imaginary line extending from the Sun to a planet sweeps out equal areas during equal intervals of time as a planet moves along its orbit. Therefore, a planet's speed along its orbit changes over the course of a revolution **(Fig. 20.4b)**. It moves fastest at *perihelion*, the point at which the planet lies closest to the Sun, and slowest at *aphelion*, the point at which the planet lies farthest from the Sun. (Note that the points representing perihelion and aphelion lie at the end points of the major axis.)

- *Third law:* The farther a planet lies from the Sun, the longer its *orbital period* (the time it takes for the planet to go around the Sun). Put simply, this law states that we can calculate a planet's orbital period if we know the dimensions of its orbit **(Fig. 20.4c)**. Mathematically, the square of a planet's orbital period is directly proportional to the cube of the semi-major axis of the orbit.

Newton's Laws of Physics

Isaac Newton (1642–1726) was born in England in the year of Galileo's death. Newton contributed more to the establishment of modern science than any other person. He is one of the inventors of calculus, he discovered that white light can be broken into a spectrum of colors, and he invented the type of telescope favored by research astronomers today. Newton's most important contributions appeared in *Philosophiae Naturalis Principia Mathematica*. In this book, generally known simply as *Principia*, Newton proposed the three *laws of motion* as well as the *law of gravity*. Using these laws, listed below, he mathematically derived Kepler's laws, and by doing

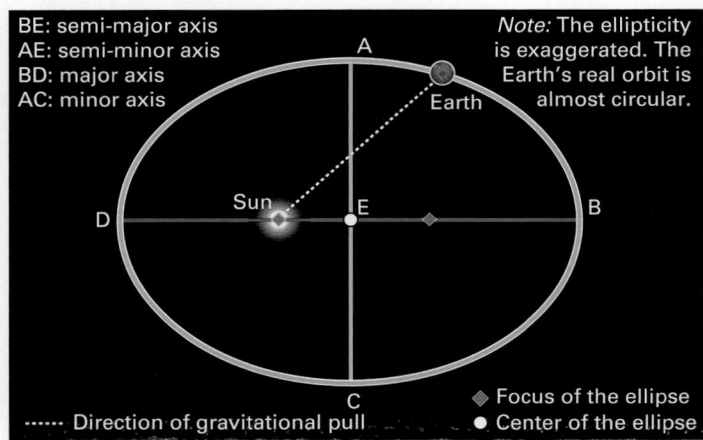

(a) Kepler's first law: The orbit of every planet is an ellipse, with the Sun at one of the ellipse's two foci.

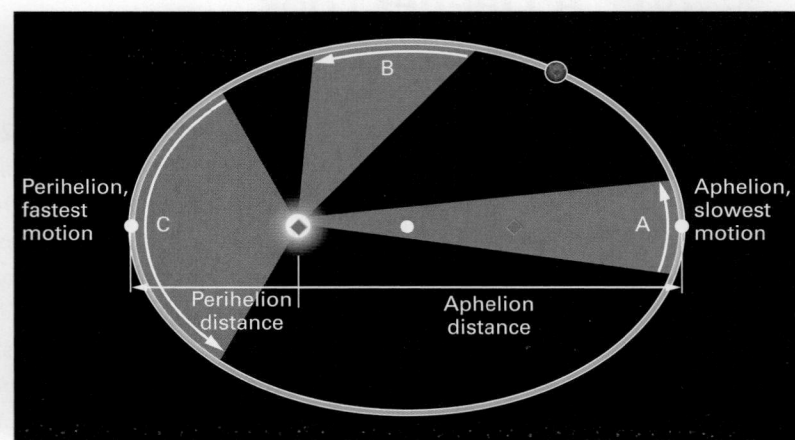

(b) Kepler's second law: As a planet orbits the Sun, an imaginary line joining that planet and the Sun sweeps out equal areas (labeled A, B, and C) during equal intervals of time. As a result, planets move fastest at perihelion and slowest at aphelion.

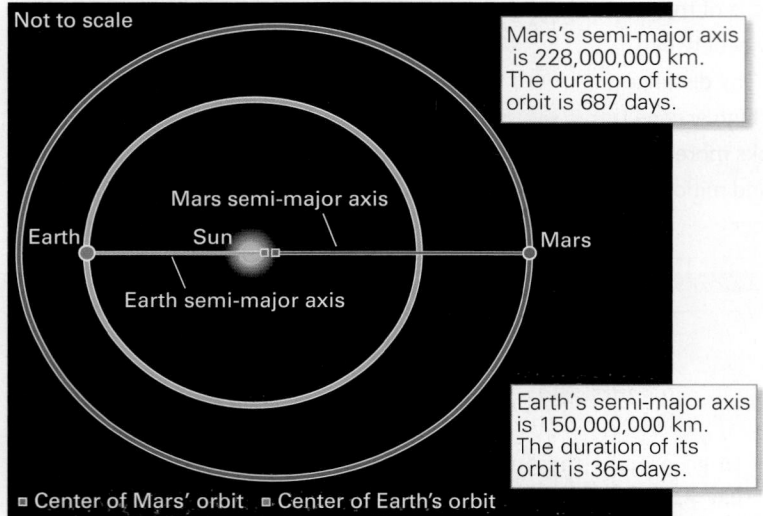

(c) Kepler's third law: The longer the semi-major axis of a planet's orbit (a measure of the planet's distance from the Sun), the longer the planet's orbital period (the time it takes the planet to orbit the Sun).

so, he showed why planets have the motions and orbits that they do **(Box 20.2)**.

- *First law of motion:* Every object has **inertia**, meaning that an object will remain in a state of rest, or in a state of uniform motion in a straight line, unless the application of a force causes it to change.
- *Second law of motion:* When a force (F) acts on a mass (m), it causes the mass to change its *velocity* (that is, its speed and/or direction), meaning that it causes an *acceleration* (a) (that is, a change in velocity). This law can be written as $F = ma$.
- *Third law of motion:* For every action, there is an equal and opposite reaction. So, for example, when we apply a force to a ball by hitting it with a bat, the ball applies the same force to the bat, but in the opposite direction.
- *Law of gravity:* Every object in the Universe attracts every other object. Objects with larger masses produce a greater gravitational force than objects with smaller masses, and objects closer to each other attract each other more strongly than do those that are farther apart.

Astronomy Enters the Modern Era

In the 18th and 19th centuries, astronomers developed new, more powerful telescopes that allowed them to see celestial objects more clearly and to see deeper into space. Such work provided details of the structure of our **Solar System** (the Sun and all the objects that orbit around it) and provided new information about the physical nature of celestial objects. As the 20th century dawned, astronomers learned how to measure distances to celestial objects. With this knowledge, they came to realize that our Sun and almost all the stars we can see in the sky without a telescope lie in the Milky Way, and that the Milky Way is a **galaxy**, a collection of hundreds of billions of stars distributed in a spiral or disk that slowly spins around a *galactic center* **(Fig. 20.5)**. Clearly, as astronomers came to realize, Copernicus's heliocentric model applies only to objects within the Solar System. Astronomers also discovered that some of the objects that earlier observers knew only as puzzling hazy patches of light were actually **nebulae**, vast clouds of dust and gas, and that others were independent galaxies located far beyond the edge of the Milky Way. Eventually, astronomers learned that the

A Deeper Look

Launching a satellite into orbit

What does it take to put an object into orbit around the Earth **(Fig. Bx20.2)**? Imagine that we attach a gun to a post that extends high above the Earth, and fire projectiles out of it horizontally at different speeds. Gravity pulls all the projectiles back toward the Earth. A slow projectile follows a curved path and lands not far from the post. The faster the projectile, the farther it lands from the post. If a projectile moves fast enough, it keeps falling, but it never gets back to the Earth's surface, because the planet has a spherical shape and curves away from the falling projectile.

So how do rocket scientists place a satellite into orbit? Speed! If a rocket propels a satellite into space fast enough, the arc of the satellite's fall due to the pull of gravity no longer brings it back to the Earth's surface, but rather takes it entirely around the planet. At this speed, called *orbital velocity*, gravity still pulls the satellite downward, but the Earth curves away from the satellite just as quickly, so the satellite stays at a constant altitude above the Earth's surface. For example, to stay in orbit around the Earth, satellites travel about 8 km/s (18,000 mph) and orbit the Earth every 90 minutes! Astronauts in a satellite like the International Space Station are continuously falling with the satellite, so they are weightless, and they can float in space as if gravity did not exist.

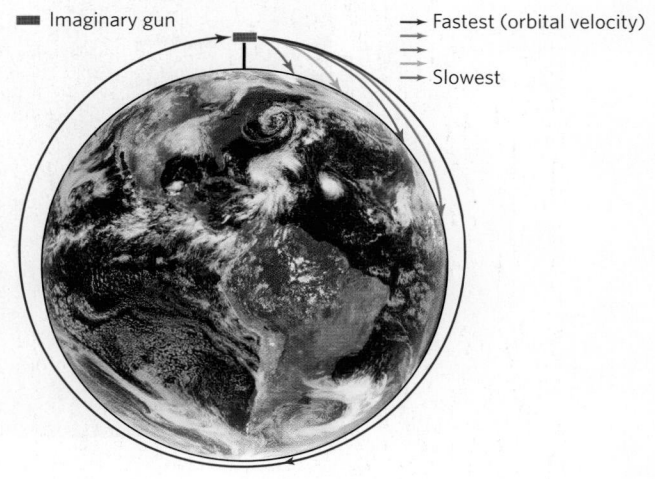

FIGURE Bx20.2 Paths of projectiles launched at different speeds.

■ Imaginary gun
→ Fastest (orbital velocity)
→ Slowest

Universe contains over a trillion galaxies, and that the distant galaxies are all moving away from one another, meaning that the Universe is expanding. This realization set the stage for the development of the *Big Bang theory*, a scientific model of the origin of the Universe (see Chapter 1). Meanwhile, physicists developed new theories to explain the behavior of matter and energy, which provided astronomers with ways to use measurements of electromagnetic radiation emitted by celestial objects to characterize those objects.

We now know that our Solar System lies on a relatively small arm of the Milky Way. Together with neighboring stars and their planets, the Sun orbits the galactic center about once every 230 million years. This means that the Sun zips through space at about 250 km/s (155 mi/s) relative to an observer at the center of our galaxy, and that it has orbited the center of the Milky Way about 24 times since it first formed. The Milky Way is one of many galaxies, all moving relative to one another. Our image of the Earth's place in Universe has evolved radically since the days of Ptolemy. Rather than being the motionless center

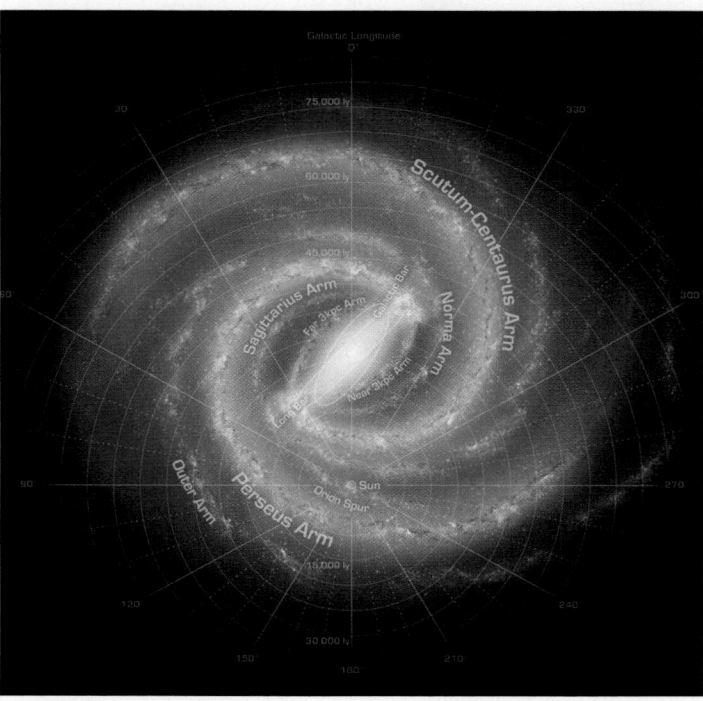

FIGURE 20.5
An artist's reconstruction of the Milky Way Galaxy based on observations from the Spitzer Space Telescope. The middle bars and spiral arms are dense collections of stars. Our Sun resides on a minor arm called the Orion Spur.

FIGURE 20.6 The two tiny spots of light close to each other in this image are the Earth and the Moon, as viewed by the *Messenger* space probe orbiting Mercury.

of everything, we spin around the axis of a planet that orbits the Sun, which in turn orbits the Milky Way, which itself traverses space. A photo of the Earth and Moon as viewed from Mercury emphasizes that we are all riding on a small speck on a journey through the vast reaches of space **(Fig. 20.6)**. But because we live here, it is a very important speck indeed!

Take-home message…

Early astronomers thought the Earth lay at the center of the Solar System, with planets and the Sun following circular orbits around it. This idea had to be discarded in the wake of work by Copernicus, Galileo, Kepler, and Newton. Astronomers now know that planets follow elliptical orbits around the Sun, and that those orbits can be explained by Newton's laws of physics. Modern discoveries reveal that the Sun is one of the hundreds of billions of stars within one of more than a trillion galaxies.

Quick Questions

- The planet Mars is farther from the Sun than the Earth. According to Kepler's laws, is a Mars year (one trip around the Sun) longer or shorter than an Earth year?
- Does the velocity of a planet along its orbit remain the same all year?
- Is the Sun at the center of the Milky Way Galaxy?

20.2 Motions of Stars and Planets

The Celestial Sphere

To portray the locations of celestial objects relative to one another, as we see them in the night sky, astronomers mark points representing these objects on the **celestial sphere**, an imaginary spherical surface in space that surrounds the Earth. We can picture the celestial sphere as a giant glass shell **(Fig. 20.7)**. To determine the position of an object on the celestial sphere, astronomers draw a line from the center of the Earth to the object. The point where the line pierces the celestial sphere represents the object's position. Note that specifying this position does not provide any information about the distance of the object from the Earth. Astronomers orient the celestial sphere so that its axis aligns with the Earth's **axis of rotation**, the imaginary line passing through the center of the Earth around which our planet spins. The axis of rotation intersects the surface of the Earth at our planet's **geographic poles**. On the celestial sphere, we position the *north celestial pole* as the point directly above the north geographic pole of the Earth, and the *south celestial pole* as the point directly above the Earth's south geographic pole.

The Journey of Helios: A Consequence of the Earth's Rotation

In ancient times, people thought that the Sun moved along the celestial sphere while the Earth stood still, and they concocted various myths to explain this movement. The Greeks, for example, attributed it to the god Helios, who drove the chariot of the Sun across the sky once a day. Now, of course, we understand that the Sun's apparent movement occurs because our planet rotates on its axis **(Fig. 20.8a)**. Because the Earth rotates counterclockwise, as viewed looking down on the North Pole, the **terminator**, meaning the boundary between night and day, moves from east to west **(Fig. 20.8b)**. When the terminator moves past a point on the Earth's surface, the Sun rises or sets at that point. Because the atmosphere scatters sunlight, sunset and sunrise do not take place instantaneously. Rather, we experience a period of time, called *twilight*, before sunrise and after sunset during which the sky glows near the horizon. The duration of twilight at a location on the Earth depends on the angle at which the Sun approaches the horizon as seen from that location.

How long does the Earth take to make one **rotation**, one complete turn on its axis? If we use the Sun as a reference and measure the amount of time it takes for the Sun to appear again at the same point in the sky, the Earth rotates, on average, once every 24 hours. Astronomers call this measure of time a **solar day**, the basis for a day on a calendar. A **sidereal day** (pronounced side-e-re-al), in contrast, represents the time it takes for a distant star to appear at the same point in the sky on successive days. A sidereal day has four fewer minutes than a solar day because the Earth changes its location relative to the Sun over the course of a day as it orbits the Sun. A sidereal day provides the true measure of one full rotation of the Earth.

The Earth's Orbit and the Ecliptic

Our planet orbits the Sun once each year, traveling a distance of 940 million km (584 million mi) over a period of 365.256 solar days. So, when you think you're sitting still, you—and the ground beneath you—are actually zipping along an orbit at an average speed of about 107,000 km/h (66,000 mph) relative to the Sun. To accommodate the extra quarter day, the *Gregorian calendar*—the standard calendar of the Western world since its introduction by Pope Gregory in 1582—includes a *leap year* every four years, when we add an extra day to February. At the turn of each century, except those centuries divisible by 400, the calendar skips adding the leap day to take into account the extra 0.006 days.

As Kepler recognized, the Earth follows an elliptical orbit, and the Sun lies at one focus—not at the center—of the ellipse (see Fig. 20.4a). The distance between the Earth and the Sun, therefore, changes over the course of a year, ranging from 147.1 million km (91.4 million mi) at perihelion to 152.1 million km (94.5 million mi) at aphelion (see Fig. 20.4b). Note that these distances differ by only about 4%, so the Earth's orbit is actually fairly close to being circular **(Fig. 20.9)**.

As viewed from space, the Earth's orbit around the Sun defines the Earth's **ecliptic plane (Fig. 20.10a)**. Put another way, we can picture the ecliptic plane as an imaginary surface that contains the ellipse of the Earth's orbit. The intersection between the ecliptic plane and the celestial sphere is a circle called the *ecliptic*. Note that, because the Earth's axis of rotation tilts at 23.5° relative to the ecliptic plane, the trace of the ecliptic tilts at an angle of 23.5° relative to the celestial equator (see Fig. 20.7). Note, too, that during half of the year, the northern hemisphere inclines toward the Sun and the southern hemisphere inclines away from the Sun; during the other half of the year, the northern hemisphere inclines away from the Sun and the southern hemisphere inclines toward it **(Fig. 20.10b)**. As we discussed in Chapter 19, the Earth's tilt gives our planet its seasons (see Fig. 19.7). In the northern hemisphere, the time when the Earth's axis transitions from tilting toward the Sun to tilting away defines the **autumnal equinox** (September 22), and the time when it transitions from tilting away from the Sun to tilting

FIGURE 20.7 The concept of the celestial sphere.

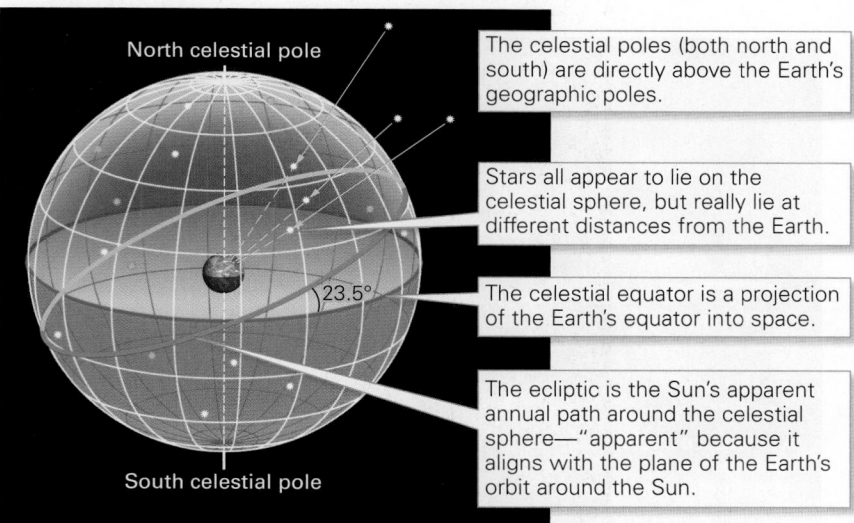

The celestial poles (both north and south) are directly above the Earth's geographic poles.

Stars all appear to lie on the celestial sphere, but really lie at different distances from the Earth.

The celestial equator is a projection of the Earth's equator into space.

The ecliptic is the Sun's apparent annual path around the celestial sphere—"apparent" because it aligns with the plane of the Earth's orbit around the Sun.

FIGURE 20.8 The path of the Sun across the sky.

The path of the Sun, as seen from a high northern latitude during the winter solstice

(a) If you watch the Sun during the course of the day, you see that it rises at dawn, arcs across the sky, reaching its highest point near noon, and sets at dusk.

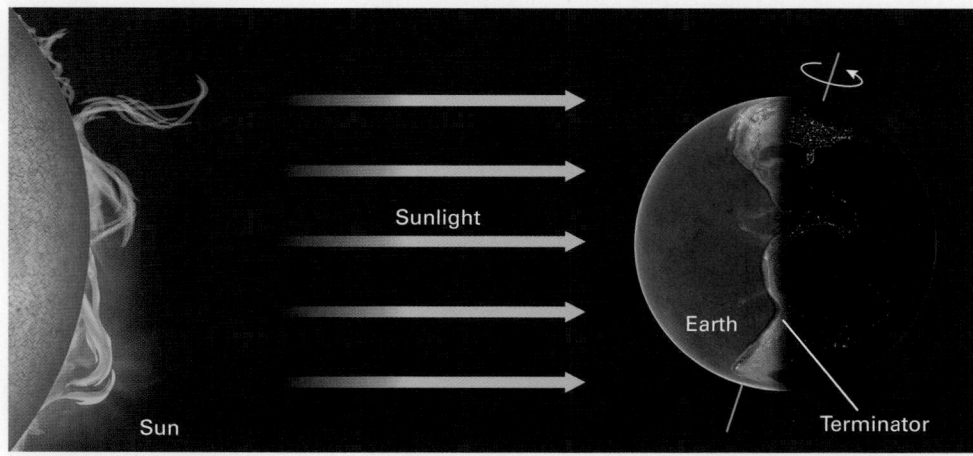

(b) The terminator is the boundary between the sunlit and dark sides of the Earth. As the Earth rotates, the terminator moves from east to west.

FIGURE 20.9 The eccentricity of the Earth's real orbit is so small that the orbit is almost circular.

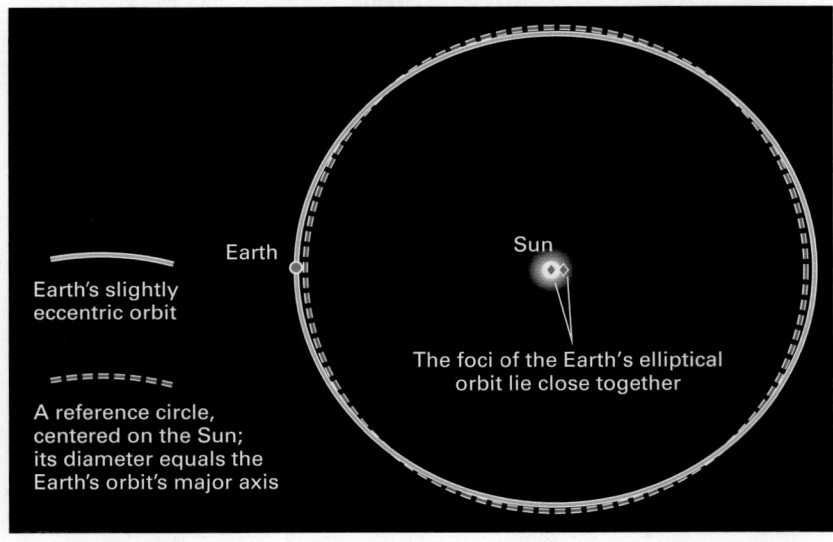

Earth's slightly eccentric orbit

- - - - - - - -
A reference circle, centered on the Sun; its diameter equals the Earth's orbit's major axis

Earth

Sun

The foci of the Earth's elliptical orbit lie close together

FIGURE 20.10 The Earth's ecliptic plane and orbital tilt.

Ecliptic plane

N

N

(a) The ecliptic plane contains the Earth's elliptical orbit.

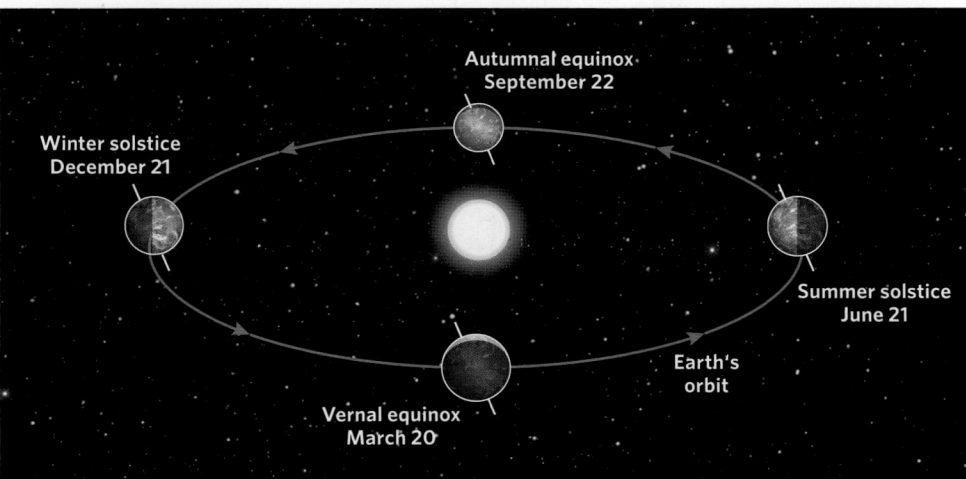

Autumnal equinox
September 22

Winter solstice
December 21

Summer solstice
June 21

Earth's
orbit

Vernal equinox
March 20

(b) The Earth revolves around the Sun once a year. The Earth's axis tilts 23.5° with respect to the Earth's ecliptic plane, so that the Earth's polar regions are in perpetual darkness at the winter solstice and in perpetual light at the summer solstice.

toward it defines the **vernal equinox** (March 20). At an equinox, the Sun is directly overhead to an observer at the equator. The northern hemisphere's **summer solstice**, when that hemisphere reaches its maximum tilt toward the Sun, occurs on June 21. The **winter solstice**, when the northern hemisphere reaches its maximum tilt away from the Sun, occurs on December 21. At the northern hemisphere summer solstice, the Sun is directly above the Tropic of Cancer, and at the northern hemisphere winter solstice, the Sun is directly above the Tropic of Capricorn.

Because the tilt of the Earth's axis does not change as the Earth orbits the Sun, the arc that the Sun takes across the sky, from the perspective of an observer on the Earth, varies with both latitude and time of year. Therefore, the duration of daylight also varies with latitude and time of year. To understand this concept, consider how the Sun progresses across the sky from the perspectives of three observers at different latitudes—one standing on the equator, one at mid-latitude in the northern hemisphere, and one at the Arctic Circle—at three different times of year (**Fig. 20.11**).

Navigating Darkness: The Night Sky and Its Constellations

Imagine yourself on a boat with no lights, floating in the middle of the ocean on a moonless, cloudless night. How many stars can you see? Astronomers estimate that 6,000 stars can be seen from the Earth in the entire night sky. At any given moment on any given night, you can see, at most, about half of these, because you see only about half of the celestial sphere from wherever you are on the Earth's surface.

Now imagine that you're standing at the North Pole, so that the top of your head aligns with the north celestial pole. If you were to aim a camera straight up and set it to record starlight on a single image all night long as the Earth spins, the image would show circular *star streaks*. In the northern hemisphere, the star Polaris, positioned directly above the North Pole, would lie closest to the center of the star-streak circles, so we refer to Polaris as the *North Star* (**Fig. 20.12a**). If you were to move to a lower latitude, you would find that the North Star does not lie directly overhead, but rather sits lower in the northern sky. Furthermore, you would record star streaks that are partial circles that run into the horizon. If you were to go farther south, you would see the position of the North Star progressively lower in the sky, and when you reached the equator, you'd see the North Star right at the northern horizon (**Fig. 20.12b**). You wouldn't see the North Star at all from the southern hemisphere.

The identity of the star that serves as the North Star changes over millennia because the Earth's axis of rotation undergoes **precession**. This means that the axis wobbles

FIGURE 20.11 The paths the Sun takes across the daytime sky at three times of year, each as viewed from three different points: on the equator, at about 35° N, and at the Arctic Circle. At the Arctic Circle, the Sun is up for 24 hours at the summer solstice, and it only briefly peeks above the southern horizon at noon at the winter solstice before disappearing for the remainder of the day.

Viewed from Arctic Circle, 66.5° north

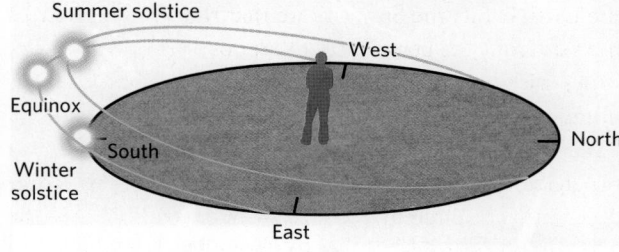

Viewed from 35° north

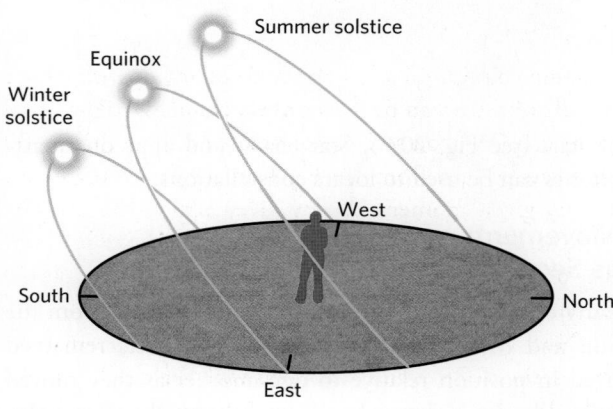

Viewed from equator

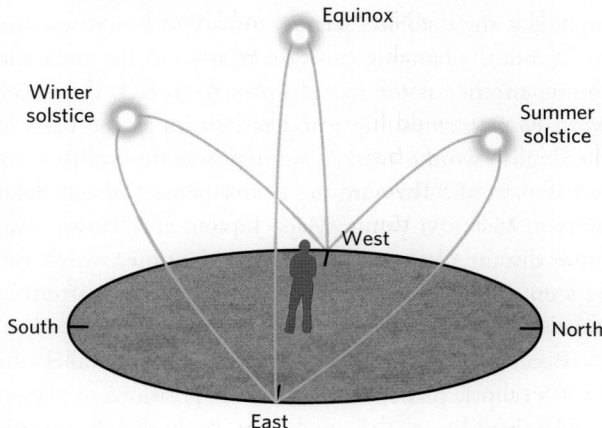

FIGURE 20.12 An artist's rendering of star streaks around Polaris.

(a) In the northern hemisphere, star streaks form circles around the North Star, Polaris. At latitudes south of the North Pole, some star streaks intersect the horizon.

(b) The distance of Polaris above the horizon depends on latitude, so in the southern United States, Polaris lies closer to the horizon than it does in Canada.

like a top, so that it points to different stars on the celestial sphere **(Fig. 20.13)**. It takes about 23,000 years for the Earth's axis to make a complete circle defining its precession. Because of precession, about 12,000 years from now, the North Star will be Vega, not Polaris. Precession contributes to the Milankovitch cycles discussed in Chapters 13 and 19.

Did you ever wonder . . .

why the North Star is called the North Star?

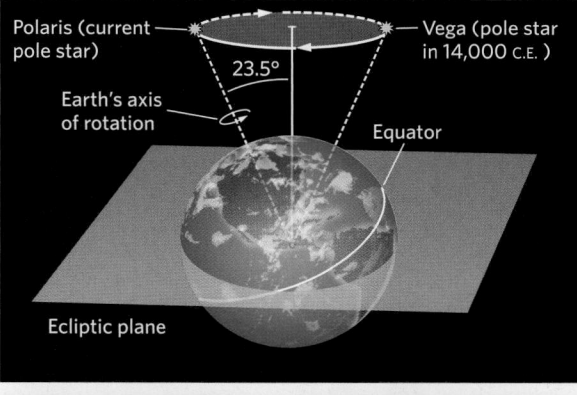
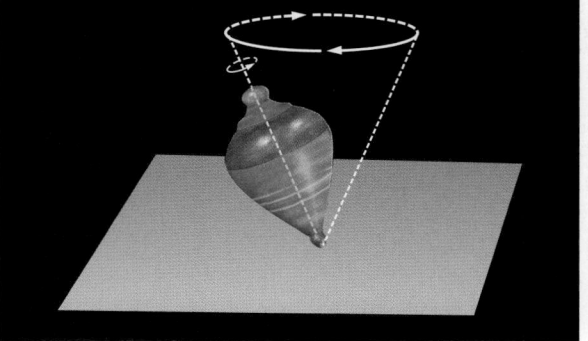

FIGURE 20.13 The Earth's axis of rotation wobbles, much like a top. This movement is called precession.

FIGURE 20.14 Constellations.

(a) The constellation Orion, as seen in a starry sky above a birch forest in winter. The lines connect the stars that make up the constellation. Orion, in Greek myth, was a giant hunter. The three stars in the middle of the constellation represent Orion's belt.

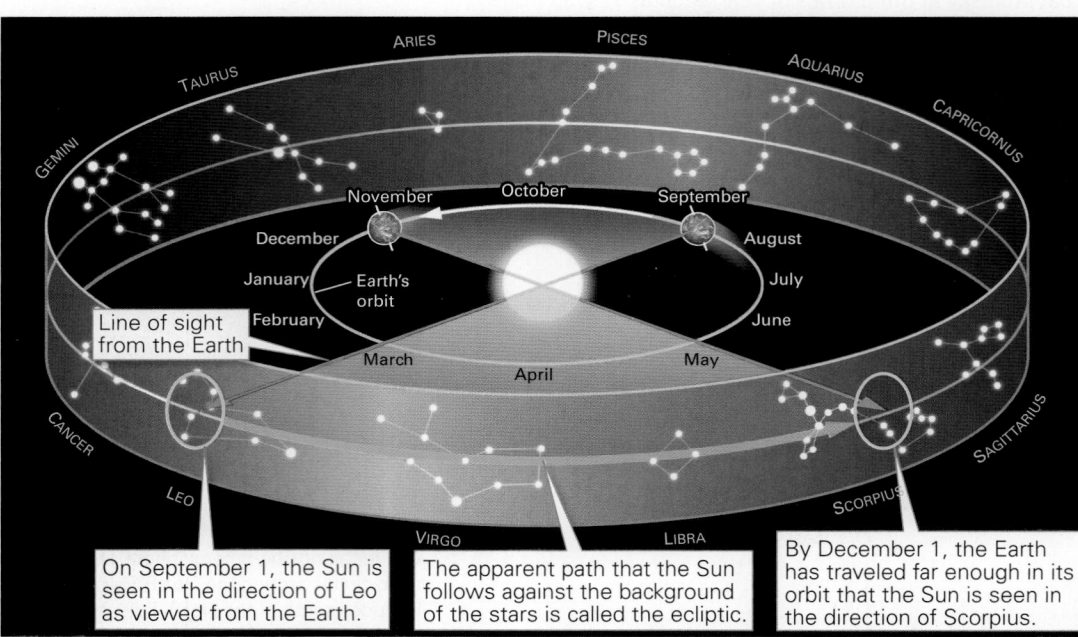

(b) The zodiac is the set of 12 constellations that lie along the ecliptic.

Ancient observers in many cultures found that if they wished to identify specific locations in the night sky, it helped to group stars into recognizable patterns or arrangements, which came to be known as **constellations (Fig. 20.14a)**. These patterns were typically named after animals, warriors, or mythical beings. Each culture came up with its own set of constellations—Greek constellations are not the same as Chinese constellations—emphasizing that constellations are entirely a human invention.

The times and locations at which specific constellations appear at the horizon after sunset provide a useful basis for navigation and calendars. The International Astronomical Union (a nongovernmental organization based in Paris) recognizes 88 constellations, which stargazers use like ZIP codes to direct attention not just to the stars in the constellation, but to a given sector of the celestial sphere. On successive months throughout the year, different constellations make their appearance in the night sky because our planet has moved to a different position along its orbit. The 12 constellations that lie along the ecliptic constitute the **zodiac (Fig. 20.14b)**. Each of these constellations lies within a 30° sector of the circle corresponding to the ecliptic. (Note that 30° × 12 = 360°, the number of degrees in a complete circle.) The ecliptic, therefore, can also be thought of as the path that the Sun appears to take against the backdrop

of stars (see Fig. 20.7). Star charts and apps on smartphones can be used to locate constellations.

Movements of the Planets as Seen from the Earth

Early observers of the heavens knew that, aside from the Sun and Moon, all but five celestial objects remained fixed in position relative to one another as they moved smoothly across the night sky, and that night after night, each of these objects retained the same brightness. The five exceptions, however, were not so well behaved. Although their overall paths remained close to the ecliptic in the night sky, these objects seemed to vary in brightness and to "wander," changing position relative to the stars and to one another as the months passed. In fact, the Greek word *planetes*, meaning wanderer, serves as the basis of the English word *planet*. As we've noted, these objects are not stars at all. They are the nearer planets of our Solar System: Mercury, Venus, Mars, Jupiter, and Saturn. Two more distant planets, Uranus and Neptune, which can be seen only with telescopes, follow the same pattern of motion as the visible planets.

Because their orbital planes roughly parallel the Earth's ecliptic plane, you can use the positions of planets to find the ecliptic in the night sky. To do this, go outside right after sunset or right before sunrise and find one or more planets. Today, you can easily download a smartphone app to help you find them. Because planets are brighter than most stars, they are among the first celestial objects (not counting the Moon) to be visible at night and the last to be visible in the morning, so you can identify them more easily at dusk or dawn. Once you've picked out a planet, draw an imaginary line connecting it with the position of the setting or rising Sun, or other planets. This line shows the trace of the ecliptic on the celestial sphere **(Fig. 20.15)**.

All of the planets orbit the Sun in a counterclockwise direction, as viewed looking down on the Sun's north pole. From an earthbound observer's viewpoint, therefore, the planets usually display **prograde motion**, meaning that they traverse the celestial sphere from west to east. Occasionally, however, a planet displays **retrograde motion** (east to west) on the celestial sphere **(Fig. 20.16a)**. Why do we see retrograde motion? You can visualize the

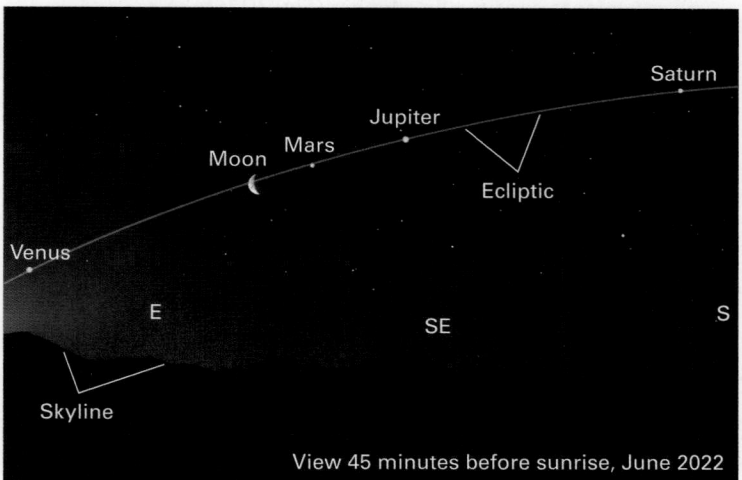

FIGURE 20.15 You can find the ecliptic in the night sky by drawing an imaginary line connecting visible planets with the setting or rising Sun.

FIGURE 20.16 Why do planets display retrograde motion?

(a) Observed apparent retrograde motion of Mars. The numbers indicate the month and day in 2010. Sometimes, but not always, the path of Mars looks like it makes a complete loop.

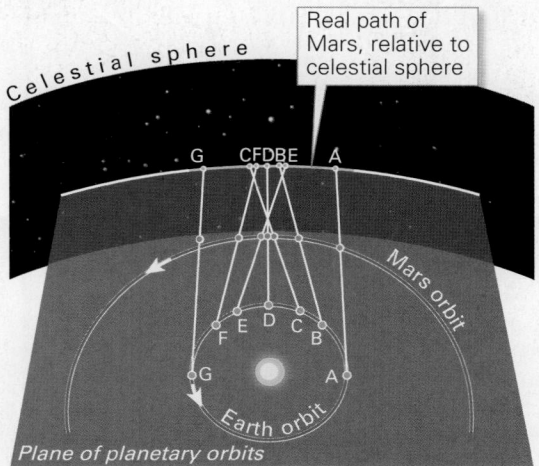

(b) The retrograde motion of Mars is displayed in the night sky because the Earth and Mars orbit the Sun at different speeds, so the view of Mars from the Earth changes. Its retrograde motion can be observed from the Earth about every two years. The retrograde path followed does not look exactly the same every time.

(c) Apparent path of Mars across the sky, relative to the Earth's horizon. Its path appears to go up and down because, due to the Earth's axial tilt, the trace of Mars's orbital plane above the horizon moves up and down over the course of a year.

answer by keeping in mind that planets orbit the Sun at different distances, and therefore, at different velocities. As a result, the line of sight from the Earth to another planet continually changes **(Fig. 20.16b, c)**.

Why does the brightness of planets, as seen from the Earth, vary over time? Unlike stars, planets do not produce their own light. The light that we see when looking at them comes from reflected sunlight. The brightness of a planet as viewed from the Earth, therefore, depends on several factors: its distance from the Sun, which determines how much sunlight it receives; its size, which determines the dimensions of its reflecting surface; its *albedo*, meaning the reflectivity of its surface; and its position relative to the Earth at a particular time, which determines its distance from the Earth and how much of its lit side faces the Earth. As the distance to the planet from the Earth increases, or as the planet reaches a position that allows us to see light reflected from less of its Sun-facing surface, the planet's light appears fainter to an Earth-based observer.

Take-home message...

🏠 To represent the positions of celestial objects as seen from the Earth, astronomers plot them on the celestial sphere. We see the movement of these objects in the night sky because the Earth rotates on its axis, which is tilted with respect to its orbital plane. The Earth revolves around the Sun once a year, causing an observer on the Earth to see the constellations move across the sky in predictable paths. The planets move independently of the stars in the night sky. All lie in roughly the same ecliptic plane.

Quick Questions

- Does the Earth, as viewed from above the North Pole, rotate clockwise or counterclockwise?
- If you lived on the equator, where would the Sun appear in the sky at noon on June 21, the summer solstice?
- Why do planets sometimes display retrograde motion?

20.3 The Moon: The Earth's Traveling Companion

The Moon, our planet's closest neighbor, has a fascinating surface that displays both smooth, dark areas called *maria* (singular: *mare*, Latin for sea) and rugged, light-colored highlands **(Fig. 20.17a)**. If you watch the Moon night after night, you'll notice that from the Earth, we always see the same face of the Moon. That's because, as the Moon moves along its orbit around the Earth, the Moon spins on its axis at just the right rate to keep the same side continually visible **(Fig. 20.17b)**. This synchrony developed because

FIGURE 20.17 The Moon and its orbit.

(a) Distinct features of the Moon's surface are visible from the Earth with a telescope.

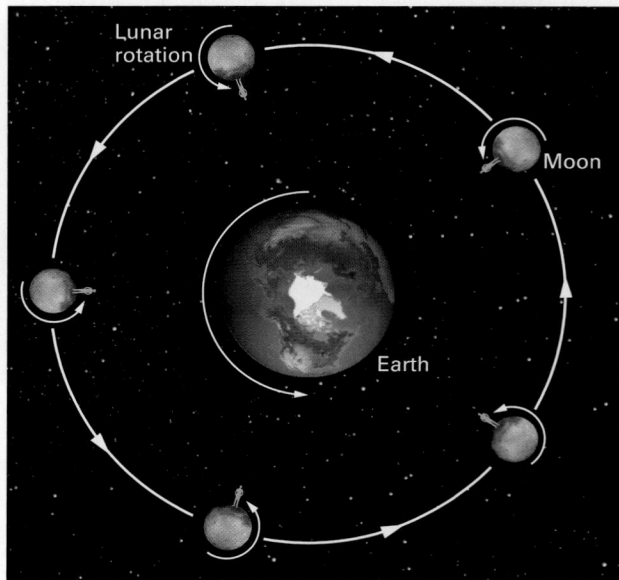

(b) The synchrony between the Moon's rotation and its orbit around the Earth causes the same side of the Moon to face the Earth continuously.

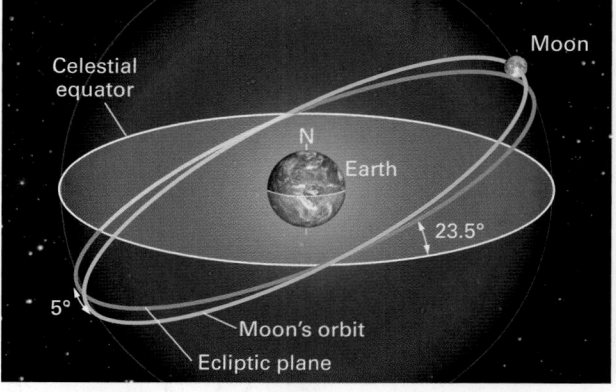

(c) The Moon's orbit is 5.2° off the ecliptic plane.

the tug of the Earth's gravity generates a small tidal bulge on the Moon's surface. Over geologic time, friction didn't allow this bulge to move as fast as the Moon spun, so the Earth pulled on one side slightly more than on the other. The pull slowed the Moon's spin until it locked into a rate that keeps one side facing the Earth. The synchrony of this *tidal locking* isn't perfect, though, so we actually see 59% of the Moon's surface over the course of a month. Because of tidal locking, humans didn't see the *far side* of the Moon until 1959, when the Soviet *Luna 3* space probe sent back pictures.

Ancient Greeks used geometric methods to determine that the Moon lies at a distance of about 30 times the Earth's diameter away from the Earth. Modern measurements indicate that, on average, it's 384,400 km (238,900 mi) away, although its distance varies somewhat during its orbit. Because of its orbit around the Earth, the Moon's position in the sky arcs from east to west along a path defining an orbital plane tilted at about 5.2° relative to the ecliptic plane **(Fig. 20.17c)**. Relative to a fixed point on the Earth, the Moon speeds along its orbit at 1 km/s (3,600 km/h, or 2,200 mph).

Phases of the Moon

Even casual watchers of the sky know that if you look at the Moon night after night, you'll see the shape of the Moon's illuminated face change in a predictable cycle. Each successive stage of this cycle is known as a **phase** of the Moon. A single cycle between identical phases takes 29.5 days, a period called a **lunar month**. Lunar months were once used as the basis for calendars, but they don't work well over multiple years because they get out of synchrony with the seasons.

Why do phases of the Moon happen? Except during times when the Moon passes into the Earth's shadow—an event called a *lunar eclipse*, which we will discuss shortly—sunlight always shines on the Moon, lighting up half of its surface. This light, which reflects from the Moon's surface, is what we see when looking at the Moon **(Fig. 20.18a)**.

When a *new Moon* occurs, the Moon rises at approximately the same time as the Sun, so the Sun is behind the Moon, and the side of the Moon facing the Earth is shaded. Every night after the new Moon, the visible part of the Moon *waxes*, or grows, as sunlight illuminates a progressively larger part of the Moon's surface visible from the Earth **(Fig. 20.18b)**. The Moon passes sequentially through phases known as the *crescent Moon*, *quarter Moon*, and *gibbous Moon*. Halfway through a lunar month, the Moon rises as the Sun sets, and the entire face of the Moon that we see from the Earth becomes lit, producing a *full Moon*. On subsequent days, the Moon rises progressively later than the Sun sets, so the lit portion

FIGURE 20.18 The phases of the Moon.

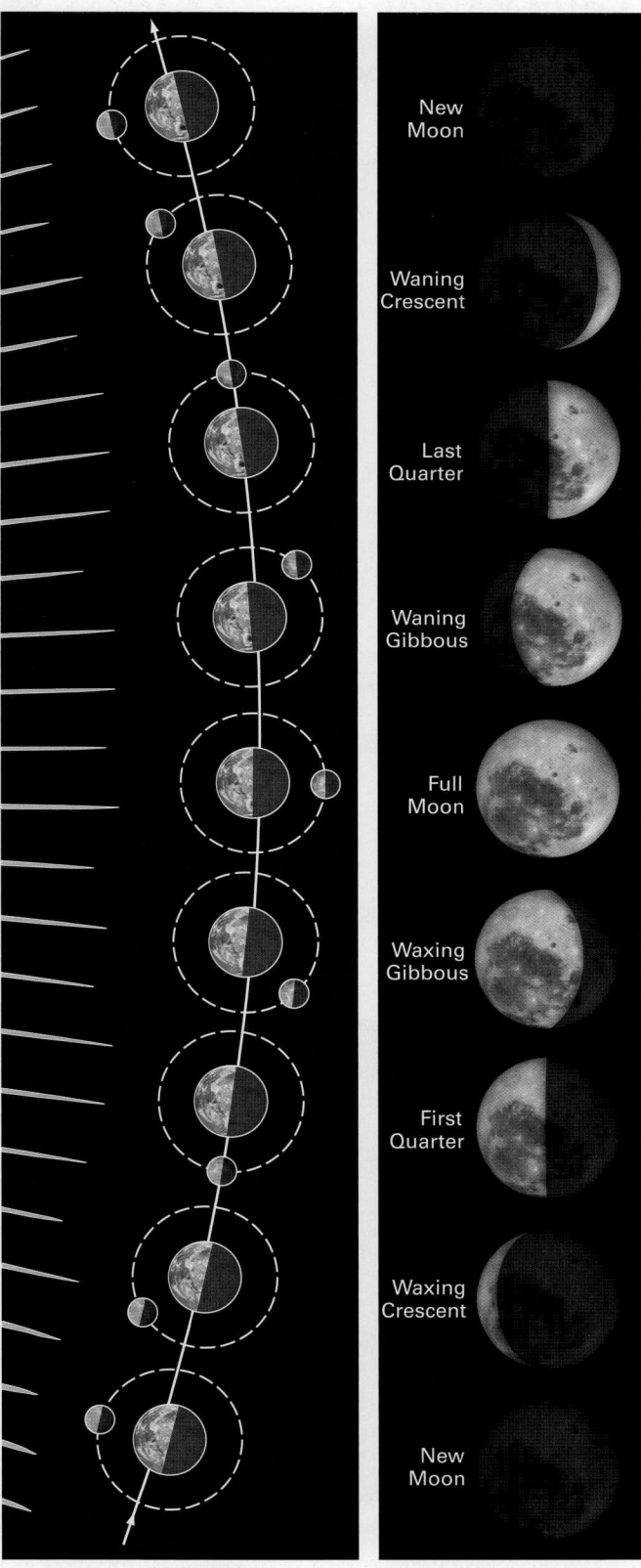

New Moon

Waning Crescent

Last Quarter

Waning Gibbous

Full Moon

Waxing Gibbous

First Quarter

Waxing Crescent

New Moon

(a) The Moon's position relative to the Sun and the Earth determines how much of the illuminated part of the Moon can be seen from the Earth.

(b) The Moon completes its phases in 29.5 days.

The relative sizes of the Sun and the Moon

What are we learning?

How the apparent sizes of celestial objects in the sky depend on their distance from the Earth.

What you need:

- A basketball
- A measuring device that uses millimeters
- A piece of paper containing a typewritten period
- A calculator

Instructions:

- Use the basketball as your scale model for the Sun. Measure the diameter of the basketball and convert it to millimeters. (If you don't have a basketball, look up its diameter on the internet.)
- To calculate the scale-model size of the Moon, we need to calculate proportions:

$$\frac{\text{true diameter of Sun (km)}}{\text{true diameter of Moon (km)}} = \frac{\text{measured diameter of basketball (mm)}}{\text{scaled diameter of Moon (mm)}}$$

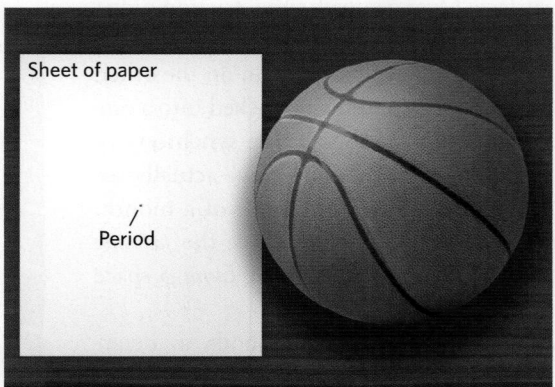

The unknown in this equation is the scaled diameter of the Moon. Now calculate what the diameter of the Moon would be if the Sun were the size of a basketball.

What did we see?

If you did the calculation correctly, you arrived at a very small number (0.5 mm). That is the diameter of a period at the end of a sentence, so if the Sun were the size of a basketball, the Moon would be the size of a period. Yet they appear the same size in the sky because the Moon lies much closer to the Earth.

FIGURE 20.19 The causes of eclipses. Because the Moon's orbital plane is oriented at an angle to the ecliptic plane, the Moon's orbit and the ecliptic plane intersect at only two times each year. Only at these brief times is an eclipse possible—the Moon must be new (for a solar eclipse) or full (for a lunar eclipse).

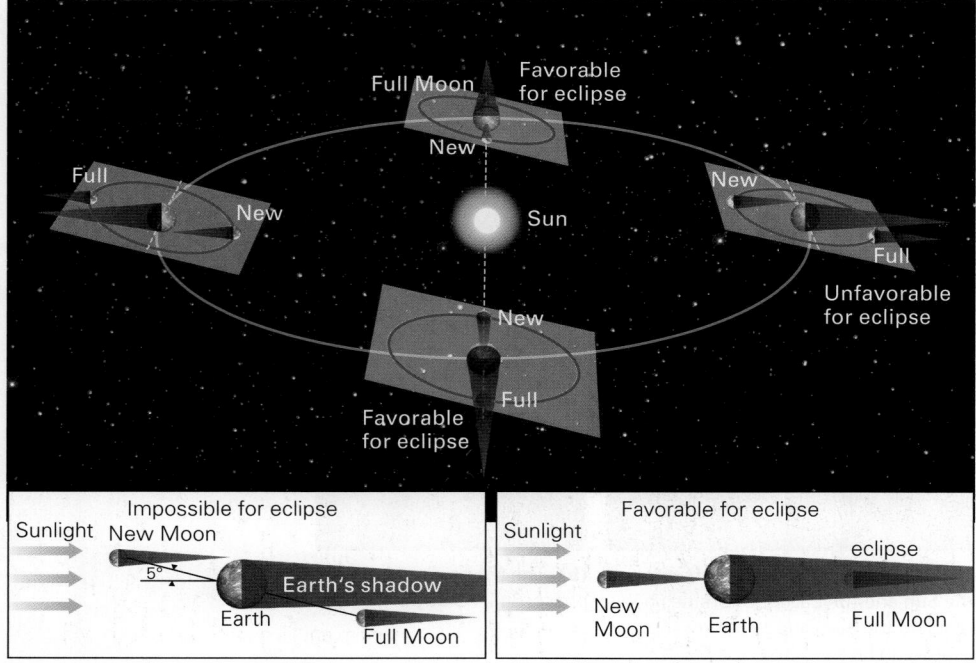

seen from the Earth *wanes*, or shrinks, through gibbous, quarter, and crescent phases. The cycle comes to completion when the Moon once again enters its new phase.

The Nature of Eclipses

The Moon and the Earth occasionally pass through each other's shadows. When they do, an *eclipse* takes place. If the Moon blocks the light of the Sun from reaching the Earth, we see a **solar eclipse**, whereas if the Earth blocks the light of the Sun from reaching the Moon, we see a **lunar eclipse**. Because such shadows can be cast only when the Earth, Sun, and Moon line up, and because the Moon's orbit inclines at 5.2° to the ecliptic plane, eclipses do not happen frequently. They take place only when a line drawn from the Sun to the Earth crosses the orbital plane of the Moon and when the Moon is either full or new **(Fig. 20.19)**.

Our Moon, with a radius of 1,737 km (1,079 mi), is clearly much, much smaller than the Sun, which has a radius of 695,700 km (432,300 mi). But because the distance to the Moon from the Earth is much less than the distance to the Sun, the two objects, when viewed from the Earth, seem almost identical in size **(Box 20.3)**. As a result, the Moon can occasionally block all the light of the Sun. In a general sense, the shadow cast when an object obscures light coming from a nonpoint source (meaning a source that has a significant width) has two parts: an *umbra* of complete darkness and a *penumbra* of partial darkness. We see a *total solar eclipse* when the Earth passes through the Moon's umbra, where the Sun is completely obscured **(Fig. 20.20a)**. The path of the umbra on the Earth's surface is quite small, so relatively few locations on the Earth experience a particular total solar eclipse; furthermore, the locations of umbra paths on the Earth are different for different eclipses. The movement of the Moon and the Earth along their orbits prevents a total solar eclipse from lasting more than 7.5 minutes. For observers in the penumbra, the Moon only partially obscures the Sun, producing a *partial solar eclipse*. When the Moon lies in a position far from the Earth, it appears smaller in the sky, so that if the Moon blocks the Sun, an outer ring of the Sun is still visible around the Moon's shadow, which is called an *antumbra*. This phenomenon is called an *annular eclipse*. Except during *complete totality*, when the viewer is in the umbra, enough radiation reaches the Earth during a solar eclipse that viewing it directly can severely damage your eyes. Never look at a solar eclipse without taking precautions **(Fig. 20.20b)**!

A lunar eclipse takes place when the Moon passes through the Earth's shadow. The Earth has a large shadow relative to the Moon, so total lunar eclipses are seen more

FIGURE 20.20 The nature of eclipses.

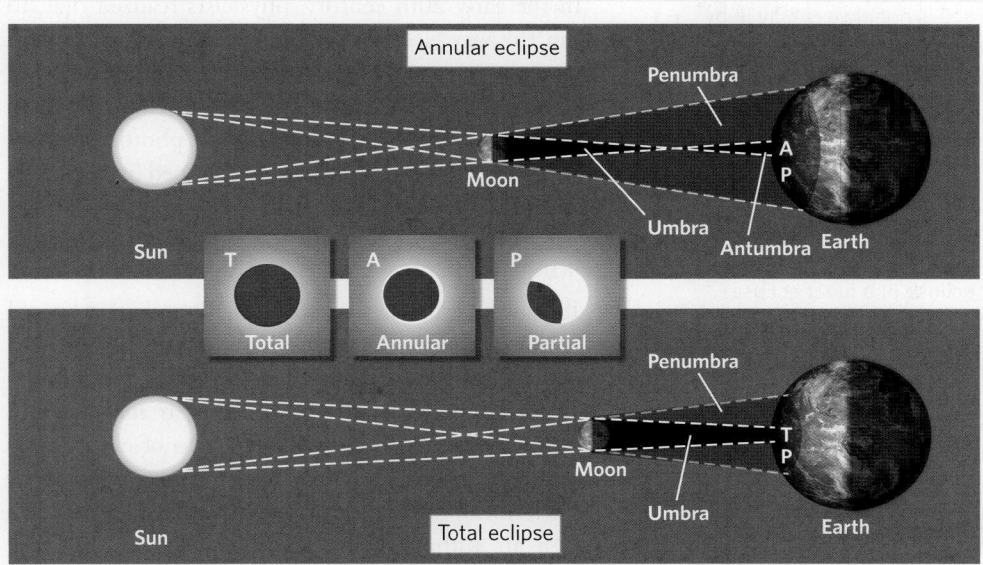

(a) We see a solar eclipse when the Moon blocks the Sun. The distance between the Earth and the Moon determines the sizes of the umbra and penumbra (or antumbra, in the case of an annular eclipse), and thus whether a solar eclipse will be seen from the Earth as partial, total, or annular.

(b) During the total eclipse of the Sun in 2017, visible across the United States, observers could see gases streaming from the Sun into space. Millions looked up at the amazing sight when the sky went dark.

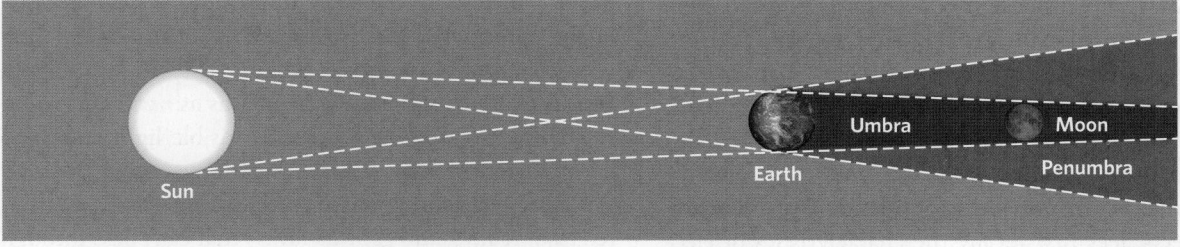

(c) We see a total lunar eclipse when the Moon passes through the Earth's umbra and a partial lunar eclipse when it passes through the penumbra. The top diagram is not drawn to scale, so that the components of the shadow are clearer, whereas the bottom diagram is drawn to scale.

(d) During a total lunar eclipse, the Moon appears red because light from the Sun passes through the Earth's atmosphere and is bent toward the Moon.

frequently than total solar eclipses, since they can usually be seen from anywhere during nighttime on the Earth **(Fig. 20.20c)**. During a lunar eclipse, the Moon doesn't go black, but rather turns a ruddy red **(Fig. 20.20d)**. Why? Sunlight passing through the Earth's atmosphere undergoes scattering and bending, and some of this light, after leaving the atmosphere, follows a path to the Moon. Because the atmosphere preferentially scatters blue light (see Chapter 16), the light that remains after passing through the atmosphere and reaching the Moon is red.

Take-home message...

The Moon rotates around its axis at the same rate that it orbits the Earth, so we always see the same side of the Moon. We see phases of the Moon because the side that we do see is not always fully lit by the Sun. Occasionally, the Moon casts a shadow on the Earth, so we see a solar eclipse, and occasionally the Earth casts a shadow on the Moon, so we see a lunar eclipse.

Quick Questions

- How long does it take for the Moon to go through all its phases?

- Would you expect a solar eclipse or a lunar eclipse when the Moon is full?

- What is the difference between a total and a partial eclipse?

20.4 Light from the Cosmos

Our ancestors knew that celestial objects existed because they could see the visible light that travels from these objects to the Earth. **Visible light** refers to the wavelengths of electromagnetic radiation that can be detected by the human eye. In Chapter 16, we introduced electromagnetic radiation to explain how energy from the Sun heats the Earth's atmosphere. Modern astronomers study electromagnetic radiation—not just visible light, but other wavelengths as well—in order to understand the composition, size, temperature, and movement of celestial objects. To provide a foundation for discussing these objects, let's first review the characteristics of electromagnetic radiation.

The Electromagnetic Spectrum, Revisited

Recall that electromagnetic radiation travels across space in the form of **electromagnetic waves**, which, unlike sound or water waves, can move through a vacuum. We distinguish among different types of electromagnetic waves by specifying their **wavelength** (the distance

between wave crests) or their **frequency** (the number of wave crests passing a point in space in 1 second). In a vacuum, all wavelengths of electromagnetic radiation travel at the same speed, the speed of light, so we can relate frequency (f) to wavelength (w) by the equation $f = c \div w$. (In this equation, c stands for the speed of light.) Electromagnetic radiation occurs in a vast range of wavelengths, which together constitute the **electromagnetic spectrum** **(Fig. 20.21a)**. Visible light represents only a small part of this spectrum. Radio waves, microwaves, and infrared radiation have longer wavelengths than visible light, whereas ultraviolet (UV) light, X-rays, and gamma rays have shorter wavelengths. The amount of energy carried by electromagnetic waves depends on their frequency: higher-frequency (shorter) waves carry more energy than do lower-frequency (longer) waves.

In the early 20th century, physicists realized that, to explain the way electromagnetic radiation interacts with atoms, radiation must be pictured not only as a succession of waves, but also as a stream of particles. These particles, known as **photons**, have no mass. Each photon contains a very specific amount of energy determined by the wavelength of the radiation.

Looking through Atmospheric Windows

If all wavelengths of electromagnetic radiation could pass through the Earth's atmosphere, we would not need to send up satellites to study celestial objects. Unfortunately for ground-based astronomers, gas molecules in the atmosphere absorb most wavelengths of electromagnetic radiation. In fact, there are only two *atmospheric windows*, meaning ranges of wavelengths that can pass through air without being absorbed **(Fig. 20.21b)**. The first, the **optical window**, is *transparent* to visible light and *translucent* to some infrared energy. This means that nearly all visible-light waves penetrate the atmosphere, but only some infrared radiation does. The second, the **radio window**, is transparent to radio waves. Because of the names used for these windows, astronomers refer to the study of visible wavelengths as *optical astronomy*, and to the study of radio wavelengths as *radio astronomy*. Similarly, *optical telescopes* capture visible light and *radio telescopes* record radio signals. As we'll see, *space telescopes* can detect other wavelengths of radiation—such as infrared rays, X-rays, and gamma rays—that can't pass through the atmospheric windows, so space telescopes provide data that cannot be obtained from Earth-based telescopes.

Using Spectroscopy to Answer Questions about Celestial Objects

WHAT'S IT MADE OF? If you've ever played with a prism, you've seen how it spreads a beam of white light into a spectrum of different colors **(Fig. 20.22a)**. That's

FIGURE 20.21 The electromagnetic spectrum.

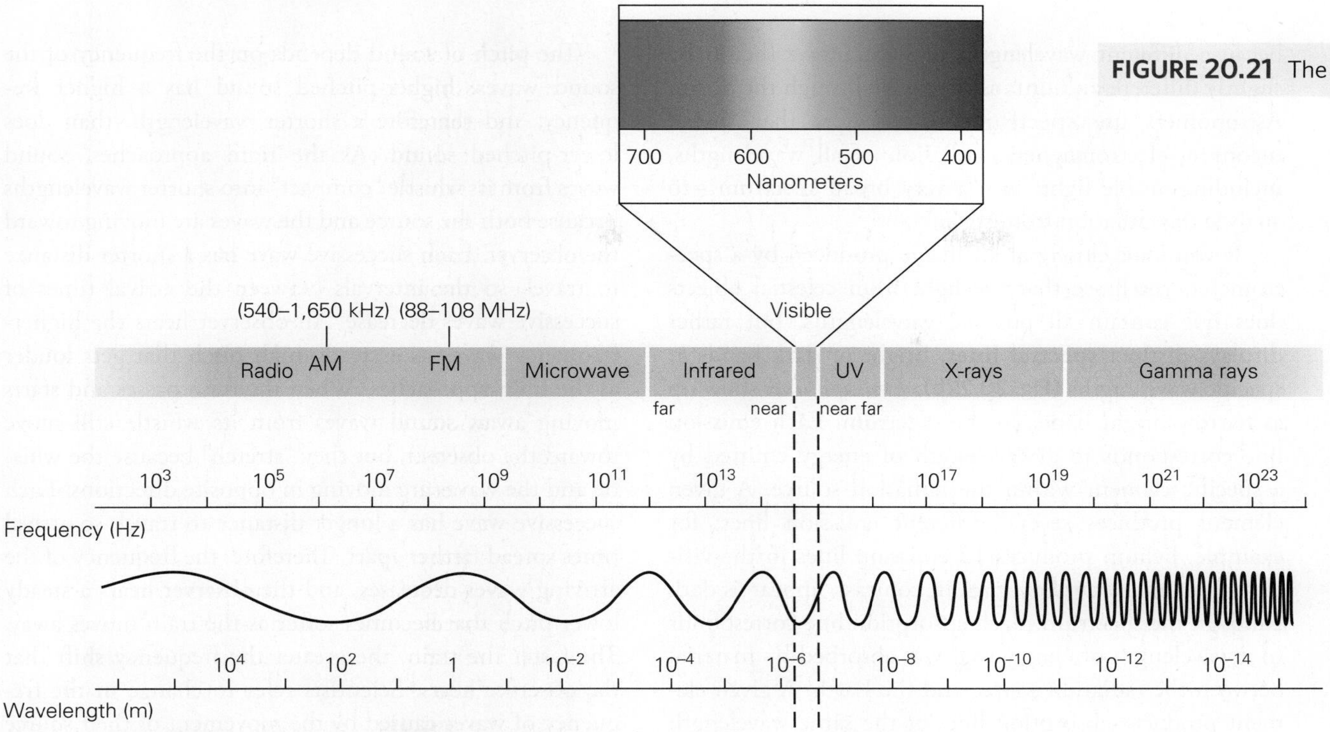

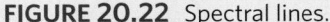

700
600
500
400
Nanometers
Visible
(540–1,650 kHz)
(88–108 MHz)
Radio
AM
FM
Microwave
Infrared
UV
X-rays
Gamma rays
far
near
near
far
Frequency (Hz)
10^3 10^5 10^7 10^9 10^{11} 10^{13} 10^{17} 10^{19} 10^{21} 10^{23}
Wavelength (m)
10^4 10^2 1 10^{-2} 10^{-4} 10^{-6} 10^{-8} 10^{-10} 10^{-12} 10^{-14}

(a) Common names for different parts of the spectrum.

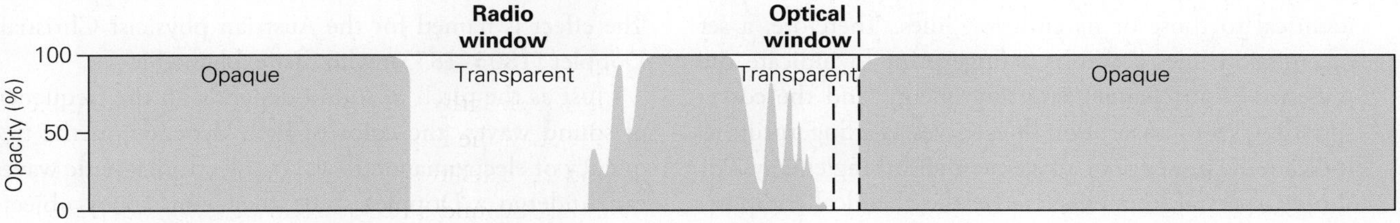

Radio window
Optical window
100
50
0
Opacity (%)
Opaque
Transparent
Transparent
Opaque

(b) Absorption of specific wavelengths by the atmosphere. Electromagnetic radiation from space can reach the Earth only through the radio and optical windows.

FIGURE 20.22 Spectral lines.

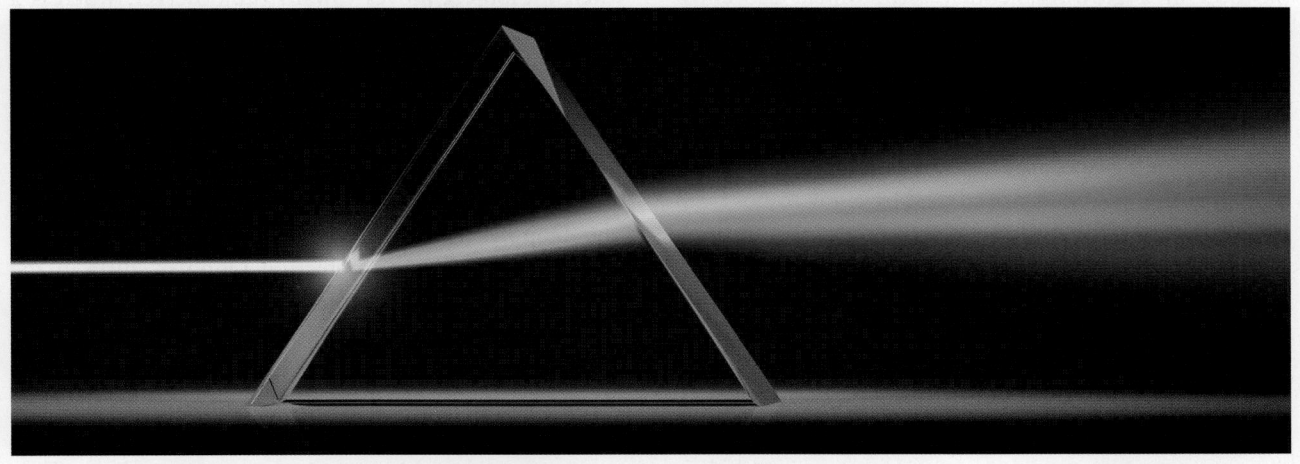

(a) A prism can be used to break white light into a spectrum of colors.

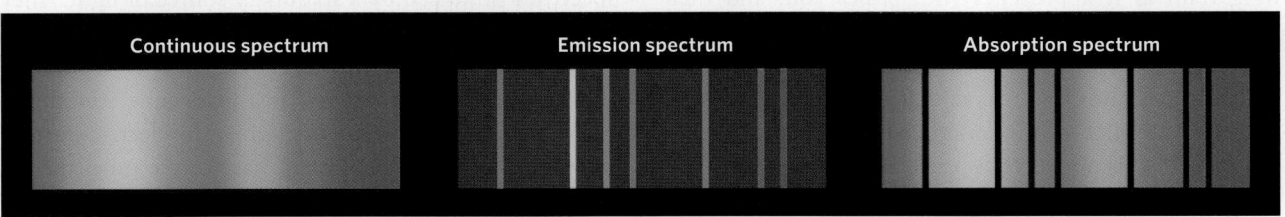

Continuous spectrum
Emission spectrum
Absorption spectrum

(b) The spectral lines of a specific element compared with the continuous visible spectrum. Bright lines are emission lines; dark lines are absorption lines. Note that emission and absorption lines occur at the same wavelengths for a given element.

because different wavelengths of light *refract* (bend) by slightly different amounts as they pass through the prism. Astronomers use **spectrometers**—devices that spread incoming electromagnetic radiation of all wavelengths, including visible light, into a very broad spectrum—to analyze the radiation from stars.

If you look closely at an image produced by a spectrometer, you'll see that the light from celestial objects does not contain all possible wavelengths, but rather displays distinct **spectral lines**, bright or dark bands at specific wavelengths **(Fig. 20.22b)**. *Emission lines* show up as narrow bright bands in the spectrum. Each emission line corresponds to a wavelength of energy emitted by a specific element within the radiation source. A given element produces several different emission lines; for example, helium produces 12 emission lines in the visible spectrum. *Absorption lines*, in contrast, appear as dark bands in the spectrum. Each absorption line corresponds to a wavelength of energy that was absorbed by material between the radiation source and the Earth. A given element produces absorption lines at the same wavelength as its emission lines; helium, for example produces 12 absorption lines in the visible spectrum at wavelengths identical to those of its emission lines. Therefore, a set of emission lines serves as a fingerprint to indicate the presence of an element emitting energy, and the corresponding set of absorption lines serves as a fingerprint to indicate the presence of an element absorbing energy. Put simply, spectral lines characterize the chemical composition of celestial objects.

Astronomers use spectral lines to study celestial objects. For example, to learn about the chemical composition of a glowing nebula, astronomers use emission lines. The specific wavelengths the nebula radiates into space depend on the identity of the gas atoms or molecules producing energy in the nebula. In contrast, clues to the chemical composition of stars come mostly from examining the absorption lines in their spectra. That's because the energy produced by a star comes from its very hot surface, and this energy must pass through the star's cooler outer layers before heading into space (see Chapter 22). The specific wavelengths of radiation absorbed by the outer layers identify the gases in these layers.

HOW FAST IS IT MOVING? Listen to the sound of a whistling train as it approaches you, passes you, and then moves away **(Fig. 20.23a)**. As it approaches, the sound gets louder, but its *pitch*, the note on the musical scale, remains constant. The instant that the train passes you, however, the pitch of the sound changes to a lower note and stays at this note even as the train rumbles away and the sound gets softer. What's happening?

The pitch of sound depends on the frequency of the sound waves: higher-pitched sound has a higher frequency, and therefore a shorter wavelength, than does lower-pitched sound. As the train approaches, sound waves from its whistle "compact" into shorter wavelengths because both the source and the waves are moving toward the observer. Each successive wave has a shorter distance to travel, so the intervals between the arrival times of successive waves decrease. An observer hears the higher-frequency waves as a steady high pitch that gets louder as the train approaches. When the train passes and starts moving away, sound waves from its whistle still move toward the observer, but they "stretch" because the whistle and the waves are moving in opposite directions. Each successive wave has a longer distance to travel, so arrival times spread farther apart. Therefore, the frequency of the arriving waves decreases, and the observer hears a steady lower pitch that becomes softer as the train moves away. The faster the train, the greater the frequency shift that the observer hears. Scientists refer to change in the frequency of waves caused by the movement of their source as the **Doppler effect** and a specific change of frequency, as in the example of a train whistle, as a *Doppler shift*. The effect is named for the Austrian physicist Christian Doppler (1803–1853), who first explained it.

Just as the pitch of sound depends on the frequency of sound waves, the color of light depends on the frequency of electromagnetic waves. Electromagnetic waves can undergo a Doppler shift when emitted by objects moving toward or away from an observer. The faster the object travels, the greater the Doppler shift. A shift toward a higher frequency (shorter wavelength) happens when the object moves toward the observer, whereas a shift toward a lower frequency (longer wavelength) happens when the object moves away from the observer **(Fig. 20.23b)**. Because the wavelength of blue light is shorter than that of red light, a light source moving toward an observer exhibits a **blue shift**, whereas a light source moving away from an observer exhibits a **red shift**. The Doppler radar used by meteorologists (described in Box 16.2) relies on detecting Doppler shifts in the wavelengths of reflected radar beams. Similarly, astronomers use these Doppler shifts in wavelengths of visible light to determine the movements of celestial objects. To detect blue or red shifts, astronomers need to compare the spectrum received from a celestial object with the spectrum produced by a stationary source of light. They can use the spectrum of light from the Sun **(Fig. 20.23c)**, or more sophisticated instrumentation, such as a spectrum tube. (A *spectrum tube* is a glass tube containing gas of a known composition; when an electric current flows through a filament at the base of the tube, the gas glows.) For example, when we say that a distant galaxy

FIGURE 20.23 The Doppler effect.

(a) The wavelength of the sound waves emitted by a stationary train whistle is the same in all directions. The sound waves behind a moving train whistle have longer wavelengths than those in front of it.

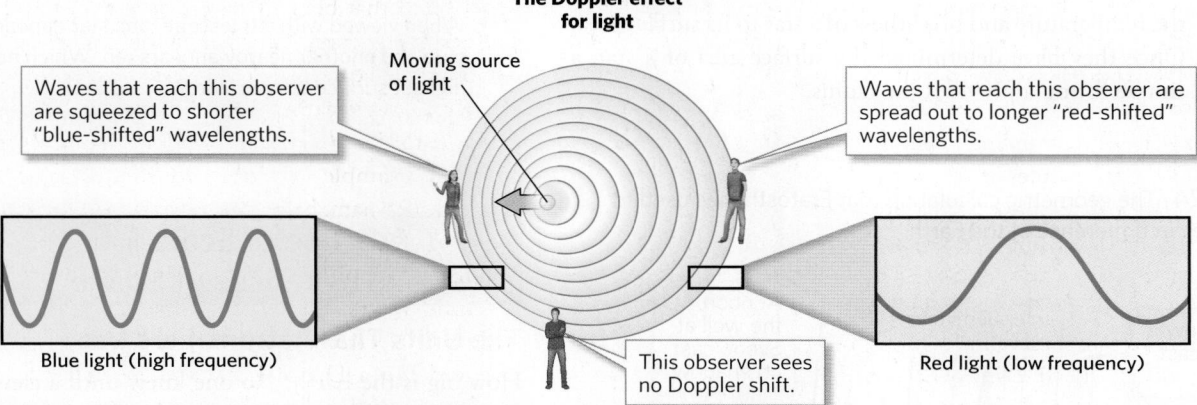

(b) If a light source moves toward an observer, the radiation detected by the observer will be shifted toward shorter wavelengths (blue-shifted). If the light source moves away, the radiation will be shifted toward longer wavelengths (red-shifted).

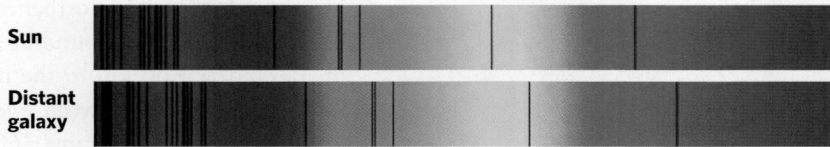

(c) The difference in wavelengths between radiation from a distant source, such as a galaxy or a star, and from the Sun can be used to determine the velocity of the source's movement toward or away from the Earth.

exhibits a red shift, we mean that its spectral lines have shifted toward the red end of the spectrum, relative to equivalent spectral lines displayed on the spectrum of light from the Sun.

How do astronomers use Doppler shifts to determine the velocities of celestial objects? Analysis of the Doppler shift for an individual star provides a basis for characterizing the star's rotation speed because, as the star spins on its axis, the side of the star moving toward us exhibits a blue shift, while the side moving away from us exhibits a red shift. Doppler shifts can also indicate how fast a star is moving toward or away from the Earth. The same type of analysis can be applied to galaxies. When astronomers

studied spectra from distant galaxies, they discovered that all display a red shift, indicating that all distant galaxies are moving away from the Earth. They also learned that galaxies farther from the Earth are moving away faster. This realization led Edwin Hubble to propose the theory of the expanding Universe (see Chapter 1).

WHAT'S ITS TEMPERATURE? According to the laws of *blackbody radiation* that we discussed in Chapter 16, the wavelength at which a glowing object emits the greatest amount of energy depends on the surface temperature of the object. Hotter objects emit the most energy at shorter wavelengths of light (toward the blue end of the spectrum), while cooler objects emit the most energy at longer wavelengths of light (toward the red end of the spectrum). By using this concept, astronomers have determined that blue stars have a surface temperature of 20,000°C (36,000°F), yellow stars have a surface temperature of 6,000°C (10,800°F), and red stars have a surface temperature of 3,000°C (5,400°F). (We'll take a closer look at these spectral classes of stars in Chapter 22.)

WHAT'S ITS SIZE? Modern telescopes are sensitive enough to allow direct measurement of the sizes of a few nearby stars. Astronomers studying nearby stars using telescopes discovered that stars come in a huge range of sizes. Telescopes cannot, however, be used to measure the radius of a distant star directly, because all distant stars, no matter how big, look like a point of light to even the largest telescope. Instead, astronomers calculate the dimensions of distant stars indirectly, using equations that relate the temperature and brightness of a star to its surface area. Once they have determined the surface area of a star, a simple formula provides the radius.

Take-home message...

Energy from stars passes through space in the form of electromagnetic radiation, but only part of this radiation makes it through the Earth's atmosphere. The electromagnetic energy emitted by a given element occurs in distinct spectral lines. By studying spectral lines, astronomers can determine the composition of celestial objects, how fast they are rotating, and how fast they are moving toward or away from the Earth. Using the laws of blackbody radiation, astronomers can determine the surface temperatures of stars.

Quick Questions

- Is an astronomer on the Earth's surface likely to use a telescope to study X-rays coming from the Sun? Why or why not?

- What does the Doppler shift tell us about motion of a star or galaxy relative to the Earth?

- When viewed with a telescope, one star appears blue and another nearby appears red. Which has the highest surface temperature?

FIGURE 20.24 The geometric calculation that Eratosthenes used to determine the circumference of the Earth.

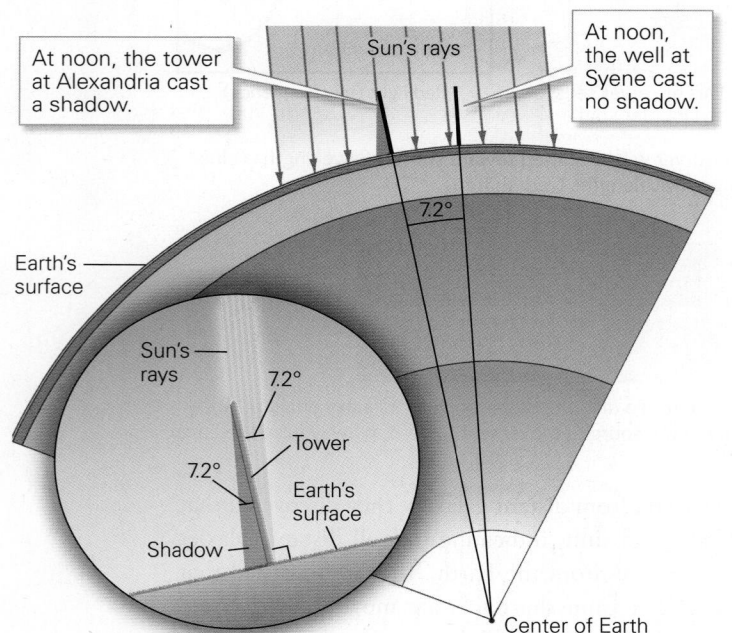

At noon, the tower at Alexandria cast a shadow.

Sun's rays

At noon, the well at Syene cast no shadow.

7.2°

Earth's surface

Sun's rays

7.2°

Tower

7.2°

Earth's surface

Shadow

Center of Earth

Eratosthenes's calculation:

$$\frac{360°}{x} = \frac{7.2°}{5,000 \text{ stadia}}$$

$$x = \frac{360° \times 5,000 \text{ stadia}}{7.2°}$$

$$x = 250,000 \text{ stadia}$$

250,000 stadia × 0.1572 km/stadium = 39,300 km (or, 24,421 miles)

20.5 A Sense of Scale: The Vast Distances of Space

The Units That Astronomers Use

How big is the Earth? No one knew until a clever Greek astronomer, Eratosthenes (ca. 276–194 B.C.E.), figured it out. He read that at noon on the first day of summer, sunlight lit the bottom of a deep well in the town of Syene, in southern Egypt. Eratosthenes found that at noon, the Sun's rays strike the ground at an angle 7.2° off vertical in Alexandria, 800 km to the north. Since Eratosthenes knew the distance between Syene and Alexandria, and knew that a circle contains 360°, he was able to calculate the Earth's circumference using a simple geometric equation and came up with a number within 2% of the modern accepted value of 40,008 km (24,865 mi) **(Fig. 20.24)**.

To describe the mind-boggling distances between celestial objects, familiar units such as kilometers and miles aren't big enough to be practical. So instead, astronomers use three extremely large units of distance. The first, called an **astronomical unit** (**AU**), which is the average distance from the center of the Earth to the center of the Sun, equals about 150 million km (93 million mi). The second, called a **light-year**, became available when researchers determined that light travels at a constant speed of about 300,000 km/s (186,000 mi/s) in a vacuum. Because light moves at a constant speed, it travels a specific distance in a given time. A light-year, the distance

FIGURE 20.25 Looking into deep space and back in time.

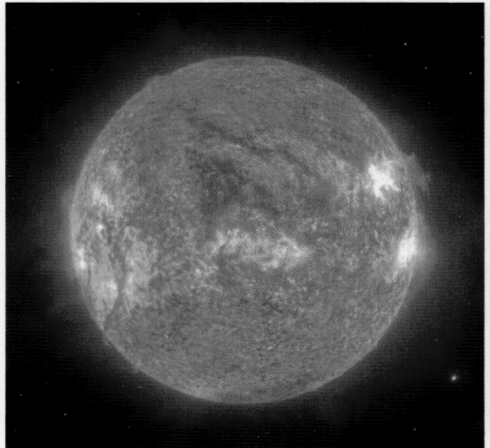

(a) The Sun, our nearest star, is about 150,000,000 km (93,000,000 miles) away. Light from the Sun takes about 8 minutes to arrive at the Earth.

(b) Jupiter, the largest planet in the Solar System, is about 778,000,000 km (483,000,000 miles) from the Sun. Light from Jupiter takes between 35 and 52 minutes to reach the Earth, depending on where the planets are in their orbits.

(c) By contrast, Alpha and Beta Centauri, the nearest stars to our Sun, are 39,740,000,000,000 km (4.2 light-years) from the Earth. Light from Alpha and Beta Centauri takes 4.2 years to reach the Earth.

(d) Our home galaxy, the Milky Way, is about 100,000 light-years wide. Light departing from one side of the galaxy would take 100,000 years to arrive at the other side.

(e) Magnification of a tiny area of space in a Hubble Space Telescope image of distant galaxies reveals the farthest object yet found in the known Universe. The light from that object took 13.5 billion years to reach the Earth.

light travels through the vacuum of space in 1 year, equals about 9.5 trillion km (5.9 trillion mi). Nontechnical discussions of astronomy generally describe distances using light-years. Professional astronomers, however, use an even larger unit, the *parsec*—roughly 3.3 light-years—for measuring distances. We'll discuss the basis for defining a parsec later in this section.

Because light takes time to travel, we're looking into the past when we look into space **(Fig. 20.25)**. The light we see when looking at the Moon left the Moon 1.3 seconds ago, light arriving from the Sun left the Sun 8 minutes ago, and light arriving from Alpha and Beta

Centauri left those stars 4.2 years ago. The farther we peer into deep space, the farther we peer back into deep time. Light arriving from the most distant visible object in the Universe left that object billions of years before the Earth even existed.

How Can We Measure Distances in Space?

Modern astronomers can measure distances from the Earth to the planets, to stars, and even to galaxies that are millions of light-years from the Earth. How do they do it? Astronomers today rely on five basic tools for measuring the great distances of space.

Did you ever wonder . . .

how astronomers can tell how far a star is from the Earth?

RADAR. Astronomers can use radar to determine the distances to objects in the Solar System. To make a radar measurement, they use a transmitter to send a pulse of radio waves out from an antenna. When the pulse bounces off an object, it returns to the antenna, where it is detected by a receiver. Because a radar pulse travels at the speed of light, the distance to the object can be calculated by precisely measuring the time between the transmission and return of the pulse.

GEOMETRIC PARALLAX. If you hold your thumb out at arm's length, and first close one eye and then the other, you'll see an example of **geometric parallax**, the apparent displacement of a foreground object relative to background objects when an observer's position, and therefore their line of sight, changes **(Fig. 20.26a)**. The movement of the Earth from one side of its orbit around the Sun to the other side provides enough distance so that astronomers can detect the slight geometric parallax that closer stars display relative to more distant stars **(Fig. 20.26b)**. By precisely recording the position of a given nearby star (meaning one less than 500 light-years away) relative to the backdrop of very distant stars, at two times separated by six months, astronomers can use the principle of geometric parallax to calculate the distance to the nearby star.

The concept of geometric parallax provides the basis for defining a parsec. A **parsec** (short for parallax second) represents the distance that an imaginary star would have to be from the Earth for the angle between two measurements of its position relative to the background of distant stars to be exactly 1 arc-second ($\frac{1}{3,600}$ of a degree) if an observer were to move from a point at the Sun's center to a point on the Earth's surface, 1 AU away.

STELLAR BRIGHTNESS. The geometric-parallax method can't be used to determine how far away a distant star lies,

FIGURE 20.26 Geometric parallax.

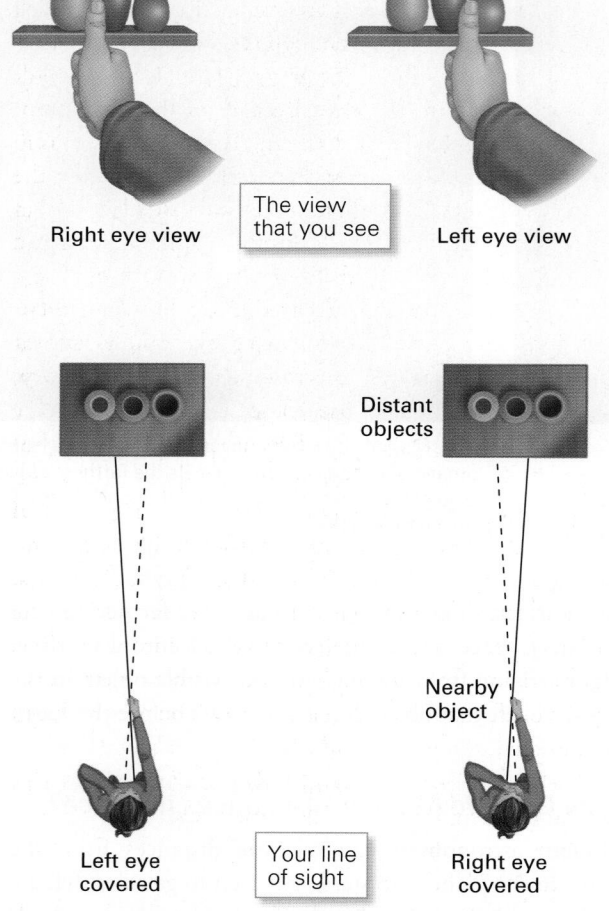

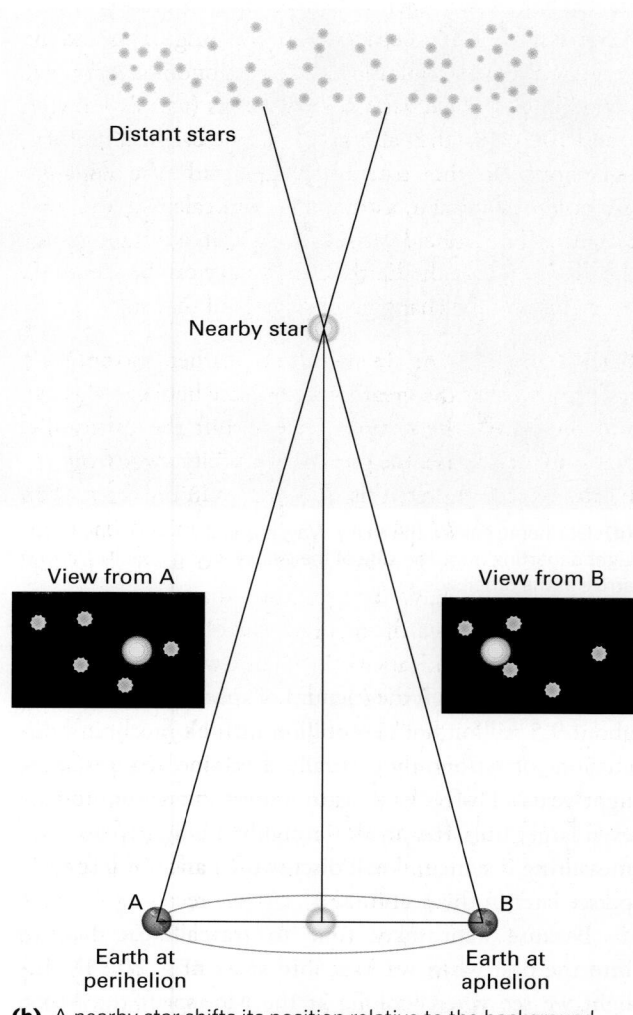

(a) You can observe geometric parallax by holding your thumb out at arm's length. When you close one eye and then the other to shift your line of sight, your thumb covers different distant objects.

(b) A nearby star shifts its position relative to the background stars when viewed from opposite sides of the Earth's orbit (in this example, at the perihelion and aphelion).

because the displacement of the star against its background is too small to detect. For distant stars, astronomers rely on the simple relationship between apparent brightness and distance. The **apparent brightness** of a light source (the brightness you see) decreases by the square of its distance from an observer because its light energy spreads over an increasing area, even if its *true brightness* (the amount of energy emitted) remains unchanged. We'll discuss these measurements of brightness in more detail in Chapter 22. If astronomers can estimate the true brightness of a star and measure its apparent brightness in the Earth's sky, they can calculate the star's distance from the Earth.

CEPHEID VARIABLES. Measurements of brightness can determine stellar distances up to about 10,000 parsecs. Measurements of even greater distances come from the study of special kinds of stars that pulsate, meaning that they expand and contract periodically. When a pulsating star expands, it has a larger surface area, but its temperature doesn't change, so its apparent brightness increases. Henrietta Leavitt (1868–1921) discovered that pulsating stars of one class, known as *Cepheid variables*, display a predictable relationship between their true brightness and the periodicity of their pulsations. So, if astronomers know the period of a Cepheid variable's pulsations (usually between 1 and 100 days), they also know its true brightness. Then, by comparing this true brightness with the apparent brightness of the star, astronomers can calculate the star's distance. If a Cepheid variable lies within a distant galaxy, the distance from the Earth to the galaxy can be measured by examining the changing brightness of that star.

DOPPLER SHIFTS. As we mentioned earlier, astronomers determined that the greater the distance between a galaxy and the Earth, the greater the red shift the galaxy displays, and therefore, the greater its velocity away from the Earth. We can express this relationship in the form of an equation, $v = Hd$, where v is velocity, H is a constant, and d is distance. Edwin Hubble, who used the study of red shifts as a basis for proposing that the Universe is expanding, provided the first estimate of a value for H. To recognize the importance of Hubble's work, astronomers now refer to H as *Hubble's constant*, and to the equation relating velocity to distance as **Hubble's law**. Astronomers realized that they could determine the distances to relatively nearby galaxies by measuring the pulsations of Cepheid variables in those galaxies. Using these measurements, they could associate the distance to a specific galaxy with the red shift, and thus the receding velocity, displayed by that galaxy. This discovery allowed them to determine the value of Hubble's constant. Then, by calculating the velocity associated with the specific red shift of a very distant galaxy, astronomers could determine the distance to that galaxy using Hubble's law.

Take-home message...

Astronomers use three units of measure to describe the vast distances in the Universe: astronomical units, light-years, and parsecs. Astronomers measure the distance from the Earth to closer objects using radar or geometric parallax and the distance to more remote objects using stellar brightness. For very distant objects, they use Hubble's law, which states that the velocity of a distant galaxy, as indicated by its red shift, depends on its distance from the Earth.

Quick Questions

- What is an astronomical unit?
- If a galaxy's true distance from the Earth was 11 billion light-years, would you use geometric parallax or Hubble's law to determine its distance from the Earth?
- Why are Cepheid variables important for determining stellar distances?

20.6 Our Window to the Universe: The Telescope

The human retina—the light-sensing tissue in the back of our eyes—doesn't have enough size or sensory capability to detect very faint light or to *resolve* (see the details of) objects that are far away. Because of this limitation, earthbound human observers can see only about 3,000 stars, even though countless stars exist, and can't see the volcanoes of Mars, even though they are vastly larger than those on the Earth. To detect distant celestial objects and to resolve the surface features of planets, we must use a **telescope**, an instrument designed to collect and focus electromagnetic radiation so as to make objects appear brighter and larger. At a ground-based **observatory**, astronomers set up and use telescopes anchored to the ground to observe the sky. Because the Earth's atmosphere permits transmission of visible light and radio waves, ground-based astronomers can use **optical telescopes**, which collect and magnify visible light, and **radio telescopes**, which collect and amplify radio waves. The telescopes can be rotated to observe any part of the sky. **Space telescopes**, which observe space from satellites orbiting above the atmosphere, can detect other wavelengths of the electromagnetic spectrum, such as infrared rays, X-rays, and gamma rays (**Earth Science at a Glance**, pp. 800–801). To understand how telescopes work, let's look at the designs of a few different types.

Optical Telescopes

TYPES OF OPTICAL TELESCOPES. Optical telescopes work by gathering all the light striking a broad area and concentrating it into a small area. Telescopes can be configured

Observatories of the World.

Hubble Space Telescope:
April 24, 1990

James Webb Space Telescope:
December 25, 2021

Visible

Infrared

Radio

Spitzer Space Telescope:
August 25, 2003

Astronomical observatories on the Earth and in space observe the Universe using a wide array of telescopes that detect different wavelengths of electromagnetic radiation.

Ground-based *optical telescopes* can see only energy arriving at wavelengths in the optical window. Such telescopes can make images of the wide array of celestial objects that emit light in wavelengths that human eyes can detect. Optical telescopes have a long history of opening windows to the Universe, from the first observations by Galileo to the sophisticated mountaintop observatories of today. Ground-based *radio telescopes* employ large antennas to detect strong radio-wave sources. They allow us to study radio-wave emitters, such as stars, nebulae, and galaxies, in the distant reaches of the Universe. Notably, radio telescopes can be used in the daytime as well as at night.

Chandra X-ray Observatory:
July 23, 1999

Gamma

X-ray

UV

Fermi Gamma-ray Space Telescope:
June 11, 2008

Space telescopes, on the other hand, orbit above the Earth's atmosphere, so they can be designed to detect wavelengths of electromagnetic radiation that can't pass through the optical or radio windows. Each space telescope specializes in detecting a different region of the electromagnetic spectrum. The Spitzer Space Telescope has used the infrared spectrum to reveal the existence of exoplanets and has allowed scientists to study forming stars. The Hubble Space Telescope, which uses the visible spectrum, has astounded the world with fantastic views of space, from stellar nurseries shrouded with dust to clusters of galaxies at the far reaches of the Universe. The Chandra X-ray Observatory allows scientists to study very hot regions of the Universe, such as exploding stars, clusters of galaxies, and matter around black holes. The Fermi Gamma-ray Space Telescope targets strange phenomena of the Universe, such as supermassive black holes, neutron stars, and streams of hot gas moving close to the speed of light. The new James Webb Space Telescope, which detects infrared energy, can see through nebulae to detect forming stars, and can detect galaxies at the far reaches of the observable Universe, reaching back in time nearly to its birth.

Space telescopes have provided insight into the architecture of galaxies and other celestial objects, and they have yielded thousands of mind-boggling images of distant galaxies and nebulae. With every new image released, people the world over marvel at the astounding beauty and complexity of the Universe.

Ground-based optical telescopes

FIGURE 20.27 How a refracting telescope works.

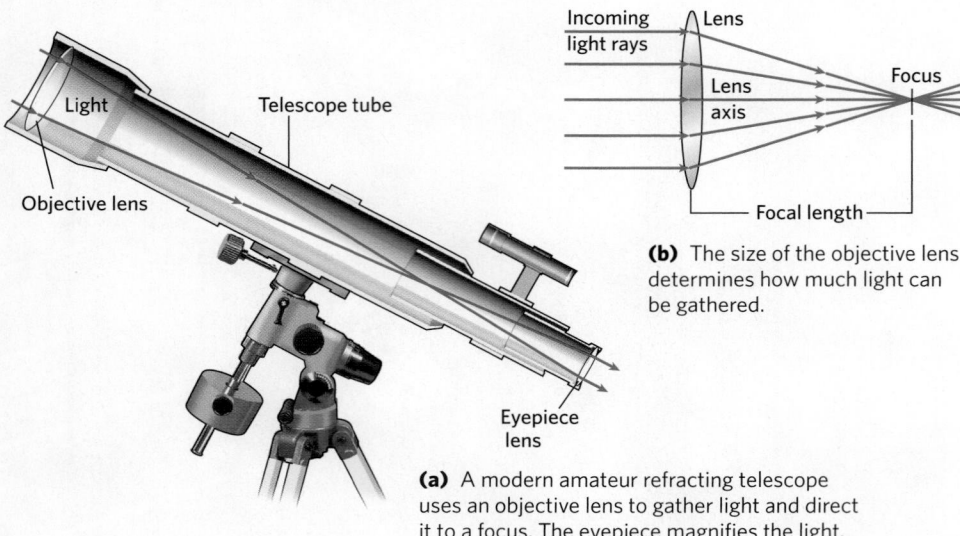

Incoming light rays

Lens

Lens axis

Focus

Focal length

(b) The size of the objective lens determines how much light can be gathered.

Light

Telescope tube

Objective lens

Eyepiece lens

(a) A modern amateur refracting telescope uses an objective lens to gather light and direct it to a focus. The eyepiece magnifies the light.

to collect light in two different ways: *refracting* telescopes were developed first (in 1608), followed by *reflecting* telescopes (in 1668).

When Galileo spotted the moons of Jupiter for the first time, he used a **refracting telescope**. In this type of telescope, an *objective lens*—a disk of glass that has been ground and polished so that it has curved surfaces on its top and/or bottom—refracts incoming light and

concentrates it at a *focus* behind the lens **(Fig. 20.27)**. A second lens, the *eyepiece*, magnifies the concentrated light so that objects appear larger. Most handheld telescopes are refracting telescopes. The quality of a refracting telescope depends on how clear and how perfectly curved its lenses are, and its light-gathering ability depends on the size of the objective lens.

When Newton began studying optics, he developed the **reflecting telescope**, in which a concave mirror reflects light to a focus in front of it **(Fig. 20.28)**. The larger the area of the mirror, the more light the telescope can gather. A smaller mirror at the focus sends the concentrated light to an eyepiece at the back or at the side of the telescope.

Astronomers using a modern research telescope don't stare through the eyepiece directly, but rather examine photographs or computer images of the light coming through the eyepiece. Large research telescopes **(Fig. 20.29)** can be moved by motors so that they stay aimed at exactly the same spot in space for a long time, even as the Earth spins. Therefore, the camera can accumulate all the light collected for many minutes or even hours. Prior to the electronic age, cameras used film to record light coming through the eyepiece. Today, film has been replaced with an electronic screen containing millions of light-sensitive pixels. Such a device can be configured to produce a digital visual image or to serve as a photon counter that can accurately determine the brightness of an object.

CONSIDERATIONS IN TELESCOPE DESIGN AND LOCATION.
The ability of a ground-based optical telescope to collect light depends on the size of the objective lens (in a refracting telescope) or mirror (in a reflecting telescope). Because it's difficult and expensive to grind lenses, and because lenses are very heavy and can warp due to their own weight, the largest objective lens ever made for a refracting telescope has a diameter of only 100 cm (40 in). It's much easier and cheaper to produce a curved mirror than it is to produce an objective lens. In addition, a large mirror can be supported from behind so that it won't warp. As a result, reflecting telescopes can be much bigger than refracting telescopes, and today's major observatories rely on reflecting telescopes for their research. The largest reflecting-telescope mirror has a diameter of 10.4 m (409 in).

Even if a telescope has great light-gathering power, the telescope's **resolving power**—its capacity to distinguish between two faraway objects that lie close together—limits an observer's ability to discern details. The resolving power of a telescope can be affected by several phenomena, including *light diffraction*, which happens when straight waves bend into curves due to interaction with an obstacle. The best resolution possible in a modern

FIGURE 20.28 How a reflecting telescope works.

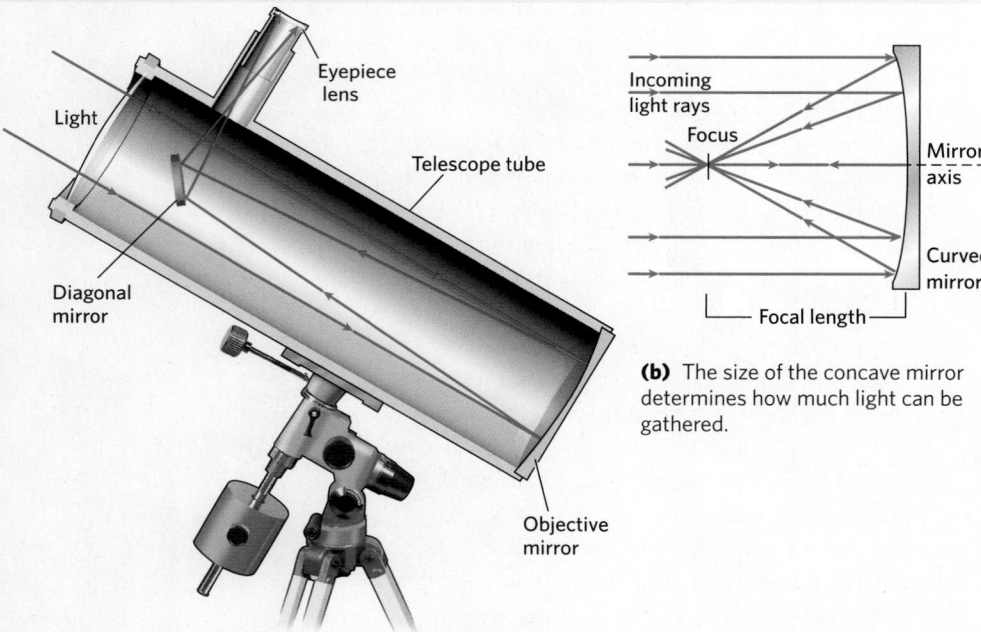

Eyepiece lens

Light

Telescope tube

Diagonal mirror

Incoming light rays

Focus

Mirror axis

Curved mirror

Focal length

(b) The size of the concave mirror determines how much light can be gathered.

Objective mirror

(a) A reflecting telescope uses a concave mirror to gather light and direct it to a focus.

telescope permits an observer to distinguish two coins placed side by side at a distance of 100 km (62 mi).

Even with the best telescopes, ground-based astronomers must contend with **atmospheric seeing**, the blurring or fuzzing of an object's image that happens because light from the object had to pass through the atmosphere **(Fig. 20.30)**. Random variations in air density associated with turbulence cause incoming light to refract in an irregular way. Atmospheric seeing can be reduced by using *adaptive optics*, a technology that uses computers to warp mirrors slightly as needed to compensate for distorted light. This technology has vastly improved the quality of images obtained by ground-based telescopes.

Any atmospheric phenomenon that decreases atmospheric clarity, telescope stability, or the darkness of the sky limits the image quality provided by ground-based telescopes. Snow, rain, clouds, and wind, for example, obscure views and blur images. In recent decades, **light pollution**—the glow of the sky due to reflection and scattering of light from anthropogenic sources such as streetlights, windows, and billboards—has become a significant problem for observatories, for this glow can wash out faint starlight coming from space **(Fig. 20.31)**. For all these reasons, when locating optical observatories, astronomers take into account elevation, climate, and proximity to urban areas. The largest observatories built in the last century perch on isolated mountaintops in arid climates. Such locations offer less atmospheric seeing, less bad weather, and less light pollution **(Fig. 20.32)**. Examples include the Paranal Observatory of Chile, the Mauna Kea

FIGURE 20.29 The Hale Telescope at the Palomar Observatory has a mirror that is 5.1 m across.

Observatories of Hawai'i, the Kitt Peak National Observatory of Arizona, and the Roque de los Muchachos Observatory on the Canary Islands. All of these observatories lie at elevations between 2 and 5 km (1.2–3.1 mi) above sea level.

Radio Telescopes

In the late 1920s, Karl Jansky, a scientist at Bell Labs, began to investigate sources of *static* (unwanted electronic noise) that interfered with ground-based radio transmissions. Jansky built a dish-shaped antenna capable of collecting very weak radio signals and pointed it skyward. The dish, like the mirror of a reflecting telescope, collected and concentrated signals. He discovered that some of the static came from thunderstorms. But he also recorded a faint, mysterious background hiss. After measuring this signal over a period of several days, he realized that a particularly strong hiss repeated every 24 hours, and that this hiss appeared when the telescope faced a particular location in space. Jansky concluded that the hiss must be coming from space, and he eventually determined that it originated in the galactic center of the Milky Way. In effect, Jansky's radio antenna was detecting electromagnetic radiation that had come from space through the radio window of the atmosphere. Science historians now consider Jansky's antenna to be the first radio telescope.

Modern radio telescopes work much like reflecting telescopes. They consist of a dish that gathers radio-wave energy and reflects it to a focus, where instruments record and analyze that energy. Because radio-wave energy coming from space is weak and its wavelengths are long, radio

Did you ever wonder . . .

why astronomers like to locate observatories on isolated mountaintops?

See for yourself

Mauna Kea Observatories, Hawai'i

Latitude: 19°49'23.89" N
Longitude: 155°28'20.54" W

Look obliquely down from an altitude of 5 km (16,400 ft).

The peak of Mauna Kea is home to a number of astronomical research facilities and large telescope observatories that are used to study the stars and galaxies.

FIGURE 20.30 Atmospheric seeing. The path of light passing between space and an observer at the ground is altered by atmospheric turbulence, causing images to go out of focus.

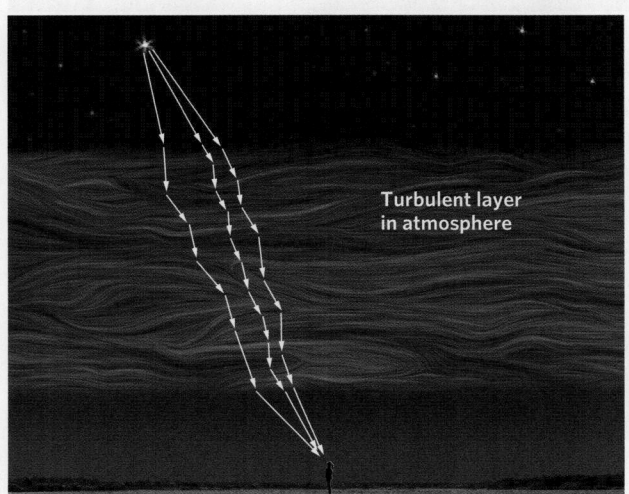

Turbulent layer in atmosphere

FIGURE 20.31 Light pollution.

(a) A photo of the United States from space at night shows the extent of urban areas with bright lights.

(b) From the ground, in a brightly lit city, only the brightest celestial objects are visible, even on a clear night.

See for yourself

Paranal Observatory, Chile

Latitude: 24°37'38" S
Longitude: 70°24'15" W

Look down from an altitude of 3 km (10,000 ft).

Paranal Observatory is located in the Atacama Desert of northern Chile, on Cerro Paranal, at 2,635 m (8,645 ft) elevation. The telescopes are operated by the European Southern Observatory (ESO). This observatory studies stars and galaxies that are visible from the southern hemisphere.

FIGURE 20.32 Examples of large observatories that employ optical telescopes. All of these observatories are located at high elevations in arid climates.

Paranal Observatory, Chile

Mauna Kea Observatories, Hawai'i

Roque de los Muchachos Observatory, Canary Islands, Spain

FIGURE 20.33 Radio telescopes and radio interferometers.

(a) The radio telescope at the National Radio Astronomy Laboratory in West Virginia has a dish 105 m (344 ft) in diameter.

(b) The Very Large Array, a radio interferometer in New Mexico.

telescopes need to be very large **(Fig. 20.33a)**. In recent decades, astronomers have built **radio interferometers**, which superimpose the waves collected by numerous radio telescopes arranged in an array **(Fig. 20.33b)**. Radio interferometers effectively act like a single huge radio telescope whose diameter equals the distance from one side of the array to the other.

Even with their huge antennas, the resolution of radio telescopes can't come close to that of optical telescopes. Nevertheless, radio telescopes do have advantages. The Sun does not emit significant radio-wave energy, so radio telescopes can be used both day and night. Also, since radio waves pass through clouds and precipitation, radio telescopes work in all types of weather. Finally, radio telescopes can detect some objects that cannot be seen with optical telescopes. For example, nebular dust absorbs visible light coming from the center of the Milky Way, so we can't see the galactic center optically. Radio waves pass through this dust, so radio telescopes can detect energy sources at the galactic center.

Space Telescopes

A *space telescope*, one placed on a satellite or spacecraft that orbits outside the Earth's energy-absorbing atmosphere, has the distinct advantage of being able to detect any part of the electromagnetic spectrum, including wavelengths that can't pass through the optical or radio windows to the ground **(Box 20.4)**. Each space telescope has been designed to detect a specific region of the spectrum. Some focus on the shortest wavelengths (X-rays and gamma rays), some on familiar visible wavelengths, and some on infrared radiation. Because radio waves have particularly long wavelengths, space telescopes designed

to detect these wavelengths would have to be enormous, and none have yet been built.

Because of their capabilities, astronomers use space telescopes to detect particularly hot objects (which emit short-wavelength energy), particularly cold objects (which emit long-wavelength energy), and objects at the farthest reaches of the Universe that are moving away so fast that their radiation is red-shifted into the infrared. Furthermore, because space telescopes do not have to contend with atmospheric seeing, they can produce amazingly sharp images. Space telescopes are, however, expensive and difficult to operate.

Take-home message...

A telescope can gather much more electromagnetic radiation than a human eye can see, allowing us to detect and resolve objects deep in space. Optical telescopes collect visible light and provide visual images. Radio telescopes collect radio waves and can detect some objects that cannot be seen with optical telescopes. Space-based telescopes have allowed astronomers to view all parts of the electromagnetic spectrum and have led to revolutionary discoveries about the nature of the Universe.

Quick Questions

- Why are refracting telescopes seldom used in scientific investigations?

- Why can we detect energy from the center of the Milky Way with a radio telescope, but not with an optical telescope?

- Why aren't radio telescopes practical for use in space?

Putting Earth Science to Use

FIGURE Bx20.4 Images from the Hubble and James Webb Space Telescopes.

Awesome views of our Universe from NASA's space telescopes

The US National Aeronautics and Space Administration (NASA) has placed several scientific telescopes in space. The most famous space-based observatory, the Hubble Space Telescope (HST), was launched in 1990. It's the only space telescope to be serviced in space by astronauts. The HST has given us amazingly sharp visual images of celestial objects **(Fig. Bx20.4a)**, as well as improved measurements of the distances to stars and a better estimate of the rate at which the Universe is expanding. When the HST imaged patches of the night sky once thought to be empty, it found countless new galaxies, billions of light-years away.

The James Webb Space Telescope (JWST), launched on December 25, 2021, is the largest telescope ever flown in space. JWST's primary mirror consists of 18 hexagonal segments made of gold-plated beryllium, which together create a mirror 6.5 m (21 ft) in diameter, much larger than the HST's 2.4 m (7 ft 10 in). JWST primarily studies long-wavelength infrared radiation. (The HST, in contrast, operates in wavelengths spanning the near-ultraviolet through the visible and near-infrared.) To properly detect long infrared wavelengths coming from distant objects in space, the JWST must be maintained at the frigid temperature of −223°C (−370°F). This prevents infrared radiation emitted by the telescope itself from contributing to the radiation it collects. To maintain this extreme temperature and still use solar power, the side of the spacecraft facing the Sun must be shielded from the side hosting the telescope. The JWST can view faint and faraway objects that formed during the early days of the Universe **(Fig. Bx20.4b)**. During its multiyear mission, it will observe the earliest stars and the formation of the earliest galaxies, and may even be able to characterize the atmospheres of rocky planets orbiting other stars.

Engineers have designed each space telescope for a distinct mission. For example, the Compton Gamma Ray Observatory, in operation from 1991 to 2000, detected gamma rays, the most powerful form of radiation, generated by gases with temperatures exceeding 10 billion °C (18 billion °F). It was followed by the Fermi Gamma-Ray Space Telescope, launched in 2008 and still operating. These two space observatories opened a window of understanding to the most energetic objects in the Universe, such as black holes (see Chapter 22). The Chandra X-ray Observatory, placed in orbit by the space shuttle *Columbia* in 1999, also detects radiation at the high-energy end of the spectrum and has provided new insight into the nature of exploding stars and their products. The Spitzer Space Telescope, launched in 2003, detects infrared radiation, at the cooler end of the spectrum, so it has the capability of studying the nebulae in which new stars form. The Kepler Space Telescope has made headlines since its 2009 launch, for its instruments have detected thousands of *exoplanets*, planets orbiting other stars. On the basis of observations made by the Kepler Space Telescope, astronomers now estimate that as many as 40 billion Earth-sized planets exist in the Milky Way.

(a) A spectacular Hubble image of the Pillars of Creation, a stellar nursery (a portion of a nebula where stars are forming).

(b) The first image from the JWST encompasses a section of the sky approximately the size of a grain of sand held at arm's length. The tiny piece of sky is full of galaxies, each containing billions of stars. Some of the light from galaxies has been distorted when passing through the gravitational fields of nearby stars or other galaxies.

Objective 20.1

Describe how our ideas about the organization and movement of celestial objects evolved from ancient to modern times.

KEY CONCEPTS

- Prior to the Renaissance, most people thought that the Earth sat at the center of the Universe. Scientists eventually showed that the Sun lay at the center of the Solar System.

- Kepler showed that planets follow elliptical orbits, that a planet's orbital period depends on its distance from the Sun, and that an imaginary line extending from the Sun to a planet sweeps out equal areas during equal intervals of time as a planet moves along its orbit.

- Newton developed laws governing forces, including gravity, and their relationship to the motion of objects such as planets and moons in space.

- Discoveries in the modern era have shown that the Sun is a star in the Milky Way Galaxy, one of more than a trillion galaxies, and that the Universe is expanding over time.

EARTH-SCIENCE VOCABULARY

astronomy (p. 778)
celestial object (p. 778)
ellipse (p. 779)
galaxy (p. 780)
geocentric model (p. 778)
heliocentric model (p. 778)

inertia (p. 780)
Kepler's laws (p. 779)
nebula (p. 780)
orbit (p. 778)
Solar System (p. 780)
Universe (p. 778)

REVIEW QUESTIONS

1. **(a)** Describe the contributions of Copernicus and Kepler to modern astronomy. **(b)** Contrast the geocentric model of the Solar System with Copernicus's heliocentric model. **(c)** What did Galileo discover that provided evidence that the geocentric model was not correct?

2. **(a)** What do Kepler's laws tell us about the shape of planetary orbits? **(b)** Which law does **Figure A** illustrate? According to the figure, does the Earth move at a constant speed as it orbits the Sun? **(c)** Is the Earth's velocity along its orbit faster or slower than that of Mars? How about that of Venus?

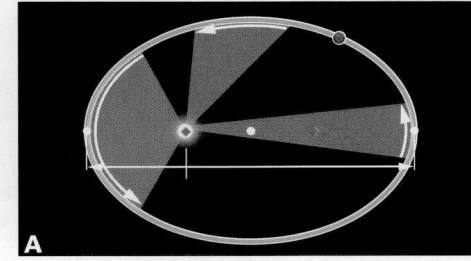

A

3. **(a)** A pool ball moves in a straight line across a pool table, unless it collides with the edge of a table or another ball. Which of Newton's laws does this interaction illustrate? **(b)** Imagine that you apply the same force to a baseball and a bowling ball. Which ball will undergo a greater acceleration? Write an equation that illustrates this concept. **(c)** Which exerts a larger gravitational force, the Sun or the Earth? **(d)** Does the International Space Station experience the force of gravity when it orbits the Earth? Why do astronauts float around within the ISS?

4. **(a)** What is a galaxy? **(b)** What is the name of our home galaxy? **(c)** About how many galaxies populate the Universe, according to contemporary astronomers?

Objective 20.2

Relate the observed movements of celestial objects to the Earth's rotation on its axis and to its orbit around the Sun.

KEY CONCEPTS

- The celestial sphere is an imaginary surface in space that surrounds the Earth and has the same center as the Earth. Astronomers plot the relative positions of celestial objects on the celestial sphere to map their locations.

- The Sun appears to move across the sky daily because the Earth rotates around its axis, which intersects the Earth's surface at its geographic poles. The location of the Sun's path across the sky depends on the observer's latitude and the time of year.

- The plane containing the orbit of the Earth is known as the ecliptic plane. It lies at an angle of 23.5° relative to the plane containing the celestial equator because the Earth's axis of rotation is tilted. The intersection of the ecliptic plane with the celestial sphere is a circle called the ecliptic.

- We can divide the night sky into recognizable groups of stars called constellations. The zodiac consists of 12 constellations located at about 30° intervals along the ecliptic.

- All of the planets orbit the Sun in a counterclockwise direction as viewed from the celestial north pole, but because of the movement of the Earth relative to the other planets, those planets can occasionally appear to an observer on the Earth to move in the opposite direction in the night sky.

EARTH-SCIENCE VOCABULARY

autumnal equinox (p. 783)
axis of rotation (p. 782)

celestial sphere (p. 782)
constellation (p. 786)

ecliptic plane (p. 783)
geographic pole (p. 782)
precession (p. 784)
prograde motion (p. 787)
retrograde motion (p. 787)
rotation (p. 782)
sidereal day (p. 783)

solar day (p. 783)
summer solstice (p. 784)
terminator (p. 782)
vernal equinox (p. 784)
winter solstice (p. 784)
zodiac (p. 786)

REVIEW QUESTIONS

5. **(a)** Does the celestial sphere lie at a specific distance from the Earth? Explain your answer. **(b)** What is the ecliptic plane, and how can you determine the location of the ecliptic on the celestial sphere? **(c)** Are the orbital planes of the other seven planets fairly similar to or radically different from the Earth's ecliptic plane?

6. **(a)** What is the difference between perihelion and aphelion? **(b)** Identify the perihelion and aphelion locations on **Figure B**. **(c)** Does the Earth move faster in its orbit when it is at perihelion or aphelion? Which of Kepler's laws predicts this?

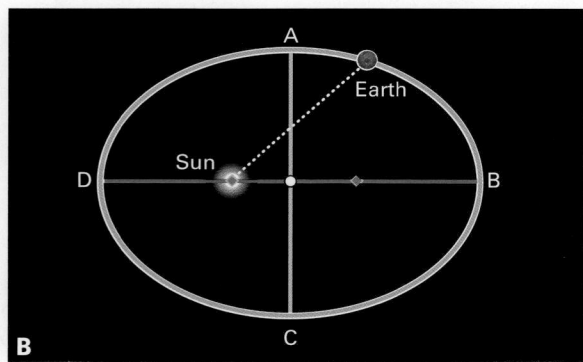

B

7. **(a)** What is the difference between a sidereal day and a solar day? **(b)** Why do we have a leap year every four years? **(c)** What are the exceptions to adding a leap year every four years?

8. **(a)** What is the difference between a solstice and an equinox? **(b)** At the southern hemisphere's winter solstice, is the Sun high or low in the sky at noon as viewed from a city in northern Canada? **(c)** If an observer saw the Sun's path across the sky on the solstices and the equinox as shown in **Figure C**, where on the Earth would that observer be located?

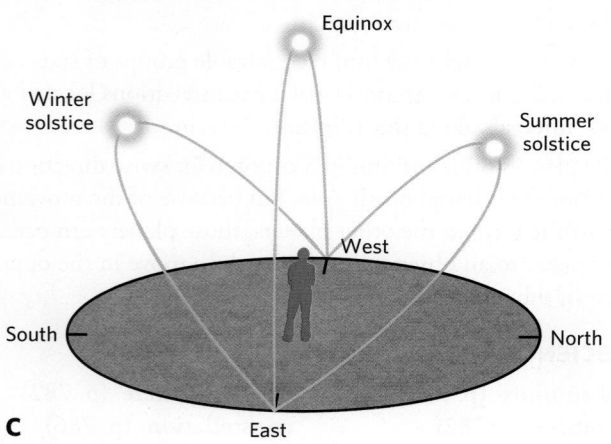

C

9. **(a)** What are constellations? **(b)** What determines which constellations constitute the zodiac?

10. **(a)** How can an observer identify a planet among the stars in the night sky? **(b)** Why do planets sometimes exhibit retrograde motion?

Objective 20.3

Explain the path of the Moon across the sky, why the Moon has phases, and why eclipses take place.

KEY CONCEPTS

- The Moon is tidally locked to the Earth, meaning that one side always faces the Earth and one side always faces away.

- The phases of the Moon are referred to as new, crescent, quarter, gibbous, and full Moon; when the sunlit side of the Moon faces the Earth, the Moon is full, but when it faces away from the Earth, the Moon is new. The Moon completes a cycle of phases from new to full and back again about every 29 days.

- Eclipses can occur only when the Moon's orbit and the ecliptic plane intersect. Solar eclipses occur when the Moon lies between the Earth and the Sun. Lunar eclipses occur when the Earth lies between the Sun and the Moon. During a total solar eclipse, the Moon completely blocks the Sun, while during an annular eclipse, the outer edge of the Sun remains visible.

- Only a small strip of the Earth's surface experiences a total solar eclipse when one occurs; lunar eclipses, in contrast, can be seen over much of the Earth because the Earth's shadow is large.

EARTH-SCIENCE VOCABULARY

lunar eclipse (p. 790)
lunar month (p. 789)

phase (p. 789)
solar eclipse (p. 790)

REVIEW QUESTIONS

11. **(a)** Why can we see only one side of the Moon (59%, to be exact) from the Earth? **(b)** What phase of the Moon does **Figure D** show? **(c)** Why don't modern societies use a lunar month as the basis for the standard calendar?

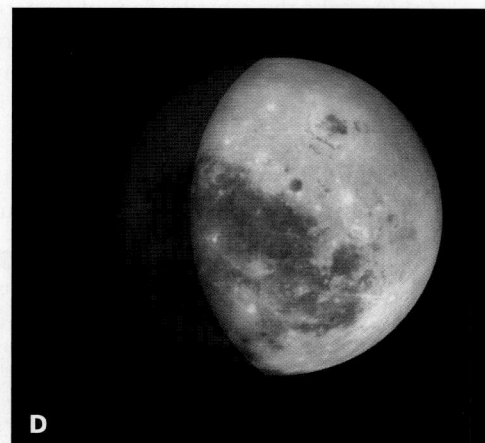

D

12. **(a)** Is the Moon closer to or farther from the Earth during an annular eclipse than during a total solar eclipse? **(b)** Identify the areas where a total and a partial eclipse would be seen in **Figure E**. **(c)** What is the difference between an umbra and a penumbra?

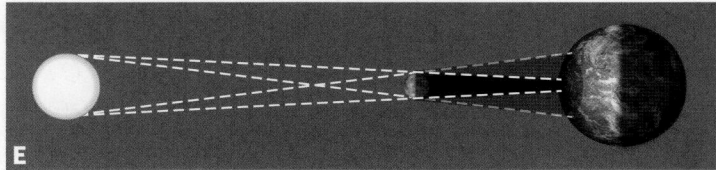
E

13. **(a)** What is a lunar eclipse? **(b)** Why can observers in most locations on the Earth see a total lunar eclipse? **(c)** Why does the Moon appear ruddy red, rather than black, during a lunar eclipse?

Objective 20.4

Discuss how astronomers use electromagnetic radiation to determine a star's temperature, composition, velocity, and distance.

KEY CONCEPTS

- The electromagnetic spectrum encompasses all wavelengths of radiation. Only two portions of the electromagnetic spectrum can pass through the Earth's atmosphere. These ranges of wavelengths define the optical window and the radio window.

- Scientists study the spectra of light coming from celestial objects by using a spectrometer. This instrument displays spectral lines (either absorption or emission lines), which provide information about the chemical composition of the radiation source and/ or the material that the radiation passes through before reaching the Earth.

- Spectral lines from moving sources of radiation shift due to the Doppler effect. Scientists measure the shift to determine how fast stars or galaxies are moving toward or away from the Earth. All distant galaxies are moving away from the Earth, at a speed that depends on their distance from the Earth.

- The brightness of stars, together with their spectra and distance from the Earth, can be used to estimate their temperature and size.

EARTH-SCIENCE VOCABULARY

blue shift (p. 794)
Doppler effect (p. 794)
electromagnetic spectrum (p. 792)
electromagnetic wave (p. 792)
frequency (p. 792)
optical window (p. 792)

photon (p. 792)
radio window (p. 792)
red shift (p. 794)
spectral line (p. 794)
spectrometer (p. 794)
visible light (p. 792)
wavelength (p. 792)

REVIEW QUESTIONS

14. **(a)** What is meant by an atmospheric window in the electromagnetic spectrum? **(b)** What atmospheric windows do astronomers on the Earth use to peer into space? **(c)** Why do you think cell phones transmit signals in the radio region of the electromagnetic spectrum, rather than the infrared or ultraviolet regions?

15. **(a)** What is a spectrometer? **(b)** What are the spectral lines visible in the spectrum produced by a spectrometer related to, and what do those spectral lines tell us about the composition of stars?

16. **(a)** **Figure F** shows a shift in the spectral lines between the Sun and a distant galaxy. What caused the shift? **(b)** Imagine you are an astronomer who has just used a spectrometer to measure the emission lines of radiation from a distant galaxy. When you look at the emission lines of hydrogen, you note that they are shifted from their expected positions toward the red end of the spectrum. Is the galaxy moving toward you or away from you?

Sun

Distant galaxy
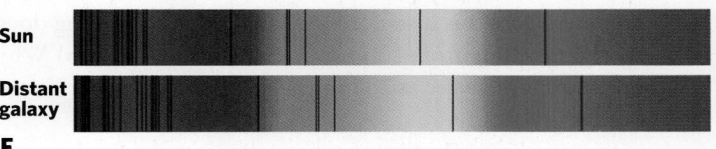
F

17. **(a)** How do astronomers determine the surface temperature of a star? **(b)** Is a blue star hotter or colder than a red star? **(c)** Can astronomers directly measure a star's size? If not, how do they determine its size?

Objective 20.5

List the units that astronomers use to describe vast distances between celestial objects, and describe the tools that they use to determine these distances.

KEY CONCEPTS

- To describe vast distances in space, we can use one of three units of measure: the astronomical unit or AU, the light-year, or the parsec.

- An astronomical unit is equal to the distance from the Earth to the Sun. A light-year is the distance light travels through a vacuum in a year. It is the common unit used in nontechnical discussions of astronomical distance.

- A parsec is the distance that an imaginary star would have to be from the Earth for the angle between two measurements of its position relative to the background of distant stars to be exactly 1 arc-second (1/3,600 of a degree) when viewed from two points exactly one AU apart. It is equivalent to 3.3 light-years.

- Astronomers use five methods to determine distances to celestial objects: radar for objects in the Solar System, geometric parallax for nearby stars, stellar brightness for stars up to 10,000 parsecs distant, Cepheid variables for those stars and galaxies farther still, and Hubble's law for the most distant galaxies.

EARTH-SCIENCE VOCABULARY

apparent brightness (p. 799)   **Hubble's law** (p. 799)
astronomical unit (AU)   **light-year** (p. 796)
   (p. 796)   **parsec** (p. 798)
geometric parallax (p. 798)

REVIEW QUESTIONS

18. **(a)** What is the numerical value of the speed of light in a vacuum? **(b)** How far does light travel in a year? **(c)** Light arrived on the Earth from a star called Gliese 514 that emitted that light about 28 years ago. How far is that star from the Earth?

19. **(a)** Neptune orbits the Sun at a distance of 30 AU. How many kilometers does this distance represent? About how long does it take light coming from Neptune to reach the Earth? **(b)** Why don't astronomers use astronomical units to specify the distances to other galaxies?

20. **(a)** Which technique for measuring the distances to stars does **Figure G** illustrate? How long was it between the times when the two photographs were taken? Why? **(b)** Is the method you just identified used to measure distances to nearby stars or to distant ones? **(c)** What is a Cepheid variable, and how is it used to measure distances? **(d)** You are an astronomer

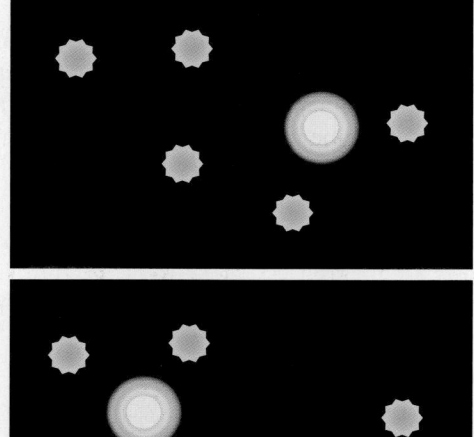

G

measuring the distance to a galaxy that you suspect is near the far edge of the observable Universe. Which method would you use to determine its distance from the Earth?

Objective 20.6

Distinguish how various kinds of telescopes function, and explain why we use space telescopes.

KEY CONCEPTS

- An optical telescope gathers and magnifies visible light. Refracting telescopes use an objective lens to focus light, whereas reflecting telescopes use a mirror.

- Reflecting telescopes with large mirrors are the best light gatherers, but diffraction and atmospheric seeing limit their resolution. Modern astronomers collect light coming through an optical telescope over time by using digital technology.

- Most observatories using optical telescopes are built on isolated mountaintops in arid climates to minimize the effects of weather, light pollution, and atmospheric seeing.

- Radio telescopes use dish-like antennae to gather weak radio-wave energy from sources throughout and beyond our galaxy.

- Telescopes on satellites, such as the James Webb Space Telescope, allow astronomers to use wavelengths of the electromagnetic spectrum that cannot be detected by ground-based telescopes. They can provide spectacular images of deep space.

EARTH-SCIENCE VOCABULARY

atmospheric seeing (p. 803)   **reflecting telescope** (p. 802)
light pollution (p. 803)   **refracting telescope** (p. 802)
observatory (p. 799)   **resolving power** (p. 802)
optical telescope (p. 799)   **space telescope** (p. 799)
radio interferometer (p. 805)   **telescope** (p. 799)
radio telescope (p. 799)

REVIEW QUESTIONS

21. **(a)** Describe the difference between the two basic types of optical telescopes. **(b)** Which type can be larger? Explain why. **(c)** Give two reasons why the world's major astronomical observatories are all located on isolated mountaintops.

22. **(a)** How does a radio telescope work? **(b)** Why must radio telescopes be so large? **(c)** Would it make sense to build an infrared telescope on the Earth? Explain your answer.

23. **(a)** What advantages do space telescopes have over Earth-based telescopes? **(b)** List some differences between the Hubble and James Webb space telescopes. **(c)** Why is it that space telescopes can detect very hot celestial objects, whereas Earth-based telescopes cannot?

ANOTHER VIEW The Southern Ring Nebula imaged in mid-infrared light (*top*) and near-infrared light (*bottom*) by NASA's James Webb Telescope. At the center is a white dwarf star, the remains of a star like our Sun, after it stopped burning fuel through nuclear fusion and ejected its outer layers to create the Nebula.

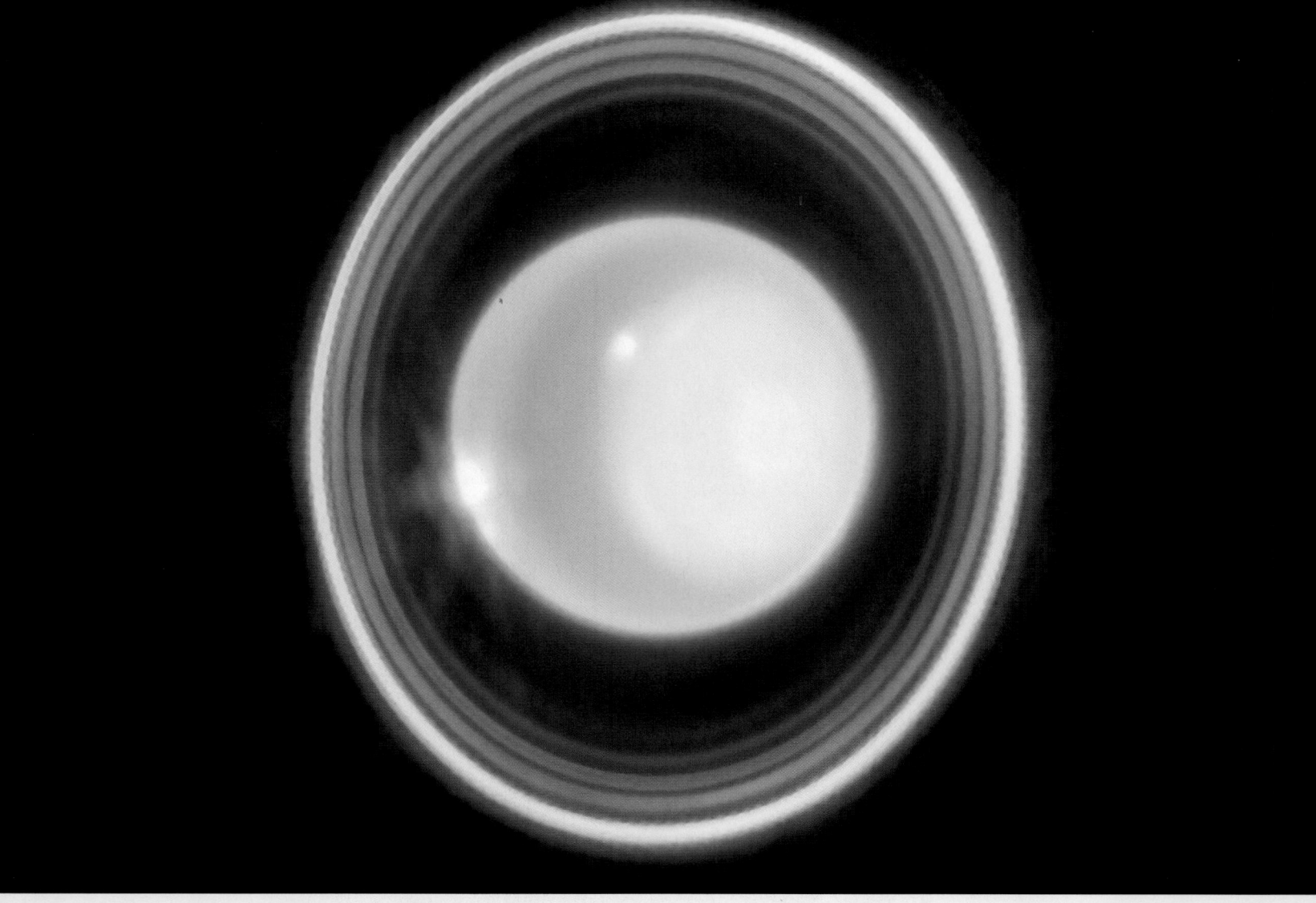

21 OUR NEIGHBORHOOD IN SPACE
The Solar System

After studying this chapter, you should be able to...

1. explain the overall structure of the Solar System, and describe the major features of the planets.

2. discuss how the Moon formed, and interpret its surface features.

3. contrast the physical characteristics of the terrestrial planets, and explain why they are so different from one another.

4. identify the key features of the gas-giant and ice-giant planets and their moons.

5. outline what astronomers have learned about dwarf planets, and evaluate the threats to the Earth posed by asteroids, comets, and meteoroids.

On August 24, 2006, the International Astronomical Union (IAU), the world's principal organization of professional astronomers, held a fateful vote. When the ballots were counted, Pluto had been demoted. Henceforth, it would no longer be a planet. For three generations, students had memorized the names of the "nine planets" of the Solar System in order of their distance from the Sun: Mercury, Venus, Earth, Mars, Jupiter, Saturn, Uranus, Neptune . . . and Pluto. The next generation of students would stop at eight **(Fig. 21.1)**.

Clyde Tombaugh, a 23-year-old astronomer, discovered Pluto in 1930. Headlines blared that "Planet X" had been found, and a schoolgirl, Venetia Burney Phair, thinking about how lonely it must be so far from the Sun, suggested naming it after the Greek god of the underworld. Pluto might still be the ninth planet were it not for the discovery in the 1990s that Pluto is, in fact, only one of many icy objects that lie beyond the orbit of Neptune. If Pluto is a planet, then should all these other objects be called planets, too? This was the dilemma facing the IAU, so their vote on Pluto was really a decision on the fundamental question, "What is a planet?"

After much deliberation, the IAU decided that to be a planet, an object must pass the following tests: (1) it must orbit a star directly, so moons, natural objects that orbit a planet, are not planets; (2) it must be nearly spherical, which implies that it's large enough for internal gravitation to smooth out bumps and dimples on its surface; and (3) it must have cleared its orbit of other objects, either by incorporating them through collision or by trapping them as moons. Pluto passed the first two tests, but not the third—it has not cleared its neighborhood of other objects. The demotion of Pluto didn't dim its mystery, however, and in 2006, NASA launched a space probe, *New Horizons*, to study it. In July 2015, humanity received its first close-up photos of Pluto, which revealed beautiful but baffling surface textures **(Fig. 21.2)**.

In this chapter, we explore all components of the Solar System except the Sun. (We'll describe the Sun and other stars in Chapter 22.) We begin by characterizing the overall structure of the Solar System. With this background, we analyze distinctive features of each planet, starting with those closest to the Sun. We finish the chapter by discussing the smaller objects of the Solar System, such as asteroids and comets. As you wander through our Solar System, keep in mind the following themes: (1) There's no place like home! Each object in our Solar System is unique, and none shares the features of the Earth that make our planet livable. (2) Without data returned to us from space probes, we would know very little about our neighbors. (3) While many discoveries have been surprising, established scientific

FIGURE 21.1 The eight planets of the Solar System.

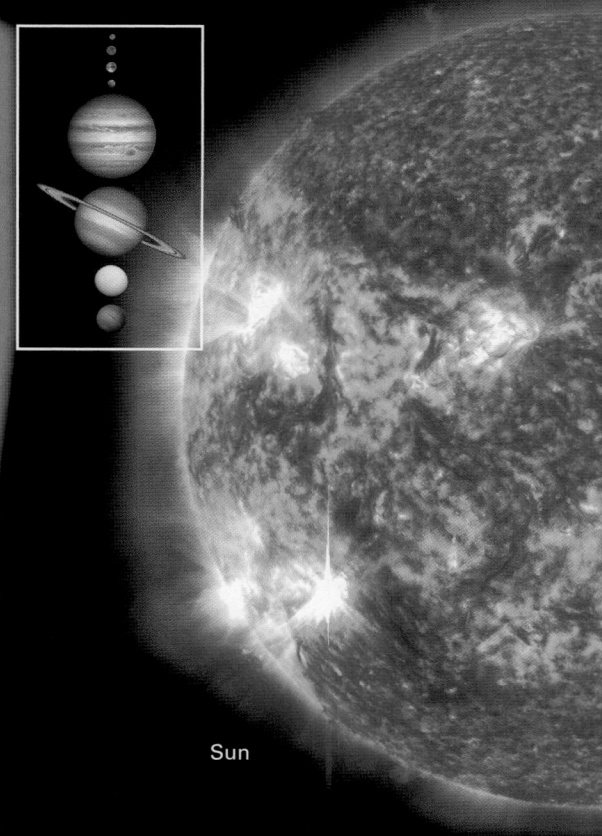

(b) All eight planets are much smaller than the Sun.

(a) Images of the eight planets, to scale, emphasize that each is unique.

principles provide a basis for understanding newly discovered features of the Solar System.

21.1 Structure of the Solar System

The inventory of objects that make up the Solar System continues to grow as astronomers make new discoveries. The larger objects include the Sun, eight **planets** (objects that orbit the Sun, grew large enough for gravity to make them nearly spherical, and have cleared their orbits of other objects), five **dwarf planets** (smaller spherical objects that orbit the Sun, but have not cleared their orbits), and many **moons** (objects that orbit planets). The number of dwarf planets and moons increases with new discoveries. In fact, in 2016, scientists who had performed computer simulations of the orbits of small objects beyond the orbit of Neptune announced that a Neptune-sized, yet-unseen planet may be lurking in the far reaches of the Solar System. As of 2023, 290 moons have been identified. Astronomers recognize many other types of objects, distinguished from one another by their size, composition, and orbital characteristics, as summarized in **Table 21.1**.

Table 21.1 Objects in the Solar System

Objects	Description
1 star	Our Sun, a nuclear inferno emitting intense energy.
8 planets	Relatively large spherical objects that orbit the Sun and have cleared their orbits of other objects. The orbital planes of all planets are within 7° of each other.
5 dwarf planets	Spherical objects that orbit the Sun, but have not cleared their orbits of other matter; as many as 100 more may yet be discovered.
290 moons	Objects orbiting planets; at least 20 moons are large enough to be spherical. The rest are too small. Many additional moons orbit dwarf planets or large asteroids. And many very small moons that have not yet been detected may orbit large planets.
Asteroids	Rocky and/or metallic objects orbiting the Sun out to the orbit of Neptune. Most lie in the asteroid belt between Mars and Jupiter. Three have diameters over 500 km (300 mi), and one (Ceres) is also a dwarf planet; millions more are small and irregularly shaped.
Kuiper Belt	A disk-shaped band of icy objects orbiting the Sun beyond the orbit of Neptune; the densest part of the belt lies 50 AU from the Sun. Thousands of objects in the belt have been discovered, over 100 of which have diameters exceeding 300 km (180 mi); perhaps 100,000 more, with diameters over 100 km (60 mi), may exist, but most are much smaller.
Oort Cloud	A spherical cloud of icy objects, held by the Sun's gravity but extending from a distance of 2,000 AU out to 100,000 AU from the Sun; objects in the Oort Cloud have not been directly observed.
Comets	Icy and dusty objects, with diameters from tens of meters to a few kilometers, that revolve around the Sun in highly elliptical orbits; comets emit a tail of glowing gases and a second tail of dust particles when they approach the Sun. Nearly 4,000 have known orbits.
Meteoroids	Solid objects that orbit the Sun and are less than 1 m (3 ft) in diameter.

FIGURE 21.3 The arrangement of the planets and the orientations of their orbital planes.

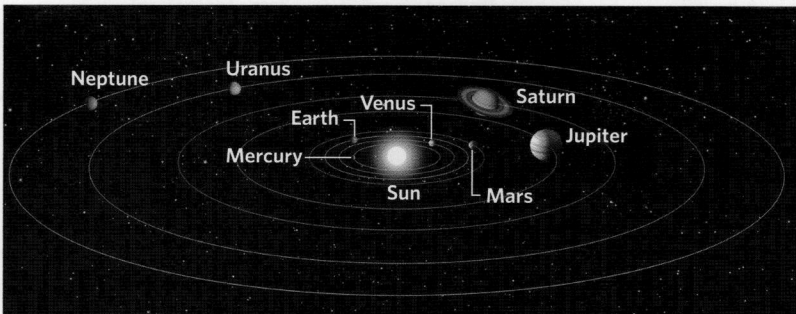

(a) Oblique view of all eight planetary orbits (not to scale).

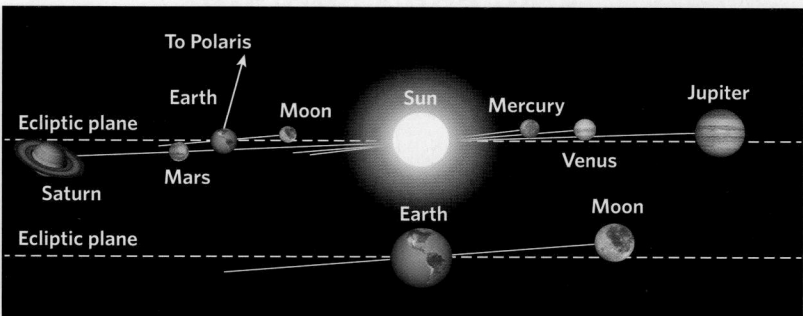

(b) A cross section showing the orientation of the planetary orbits (top) and of the Moon's orbit (bottom) relative to the ecliptic plane.

How are the objects of the Solar System arranged? The planets follow slightly elliptical orbits, and as we noted in Chapter 20, they display *prograde motion*, which in this context means that they move counterclockwise around the Sun as viewed from above the Sun's north pole. All planets except Mercury have orbital planes that lie within 3° of the ecliptic plane **(Fig. 21.3)**; Mercury's orbit deviates 7° from the ecliptic plane.

When astronomers defined the orbits of asteroids and Kuiper Belt objects, they realized that the majority of these objects also display prograde motion, but that some follow highly elliptical paths, called **eccentric orbits**. Also, while some of these objects have orbital planes close to the ecliptic plane, others follow paths that incline significantly to the ecliptic plane.

Table 21.2 reveals that the planets differ from one another in many ways. As they orbit the Sun, all but one of the planets rotate counterclockwise on their axes—in the prograde direction. The exception, Venus, rotates in a *retrograde*, or clockwise, direction. Astronomers suggest that Venus displays such a spin because early in Solar-System history, it collided with a protoplanet, and the collision caused Venus's axis of rotation to reverse. Astronomers describe the orientation of a planet's axis of rotation by specifying its *axial tilt*: an axis perpendicular to the planet's orbital plane has a 0° tilt, and an axis parallel to the orbital plane has a 90° tilt. If we compare the planets, we see that they display a range of tilts **(Fig. 21.4)**. The tilt determines whether or not a planet has seasons, because the axis stays in the same orientation as the planet orbits the Sun. When a planet's axis has a tilt, the *insolation* (the amount of incoming solar radiation) striking a location on the planet varies over one revolution of the planet around the Sun. Note that Venus, because its axis is reversed, has an axial tilt of 177°.

Table 21.2 Comparing the Planets

Name	Duration of day*	Axial tilt	Rotation direction[†]	Duration of year[‡]	Known moons	Mass (× Earth)	Radius (km)	Radius (× Earth)	Distance to Sun (AU)	Density (g/cm³)
Mercury	59 d	0.1°	CCW	88 d	0	0.06	2,440	0.38	0.39	5.4
Venus	243 d	177°	CW	224 d	0	0.8	6,052	0.95	0.72	5.2
Earth	1 d	23°	CCW	365 d	1	1.0	6,371	1.00	1.00	5.5
Mars	1 d	25°	CCW	687 d	2	0.1	3,390	0.53	1.52	3.9
Jupiter	10 h	3°	CCW	12 yr	95	317.8	69,911	10.97	5.20	1.3
Saturn	10 h	27°	CCW	29 yr	146	95.2	58,232	9.14	9.54	0.7
Uranus	17 h	98°	CCW	84 yr	27	14.5	25,362	3.98	19.18	1.2
Neptune	16 h	28°	CCW	165 yr	14	17.2	24,622	3.86	30.06	1.7

*A day is the time it takes for a planet to rotate once on its axis, specified in Earth days or hours (h = hours, d = days).
[†]CW = clockwise; CCW = counterclockwise.
[‡]A year is the time it takes a planet to complete one orbit around the Sun, specified in Earth days or years (d = days; yr = years).

BOX 21.1 How can I explain . . .

Distances in the solar system

What are we learning?

How to visualize the distances between planets and their distances from the Sun.

What you need:

- A 100-yard or 100-meter line on a sidewalk, a parking lot, or, ideally, a football field
- One soccer ball
- Eight golf balls

Instructions:

- Place the soccer ball at one end of the marked line.
- Then, use the measurements in the accompanying table to place a golf ball at the location of each planet (numbers have been rounded). The scale has been provided by setting 100 yards, or meters, equal to 30.1 AU, the distance from the Sun to Neptune.
- To highlight the locations, have a student stand next to each ball.

What did we see?

This exercise provides a simple but powerful visual illustration that contrasts the proximity of the terrestrial planets to the Sun and to one another with the vast distances of the Jovian planets from the Sun.

Planet	Distance from Sun (AU)	Yards (or meters) from soccer ball
Mercury	0.39	1
Venus	0.72	2
Earth	1.00	3
Mars	1.52	5
Jupiter	5.20	17
Saturn	9.54	32
Uranus	19.18	63
Neptune	30.06	100

Different planets take different amounts of time to make a complete rotation. The Earth, as we know, takes one day (24 hours) to spin on its axis. In comparison, Venus is a slowpoke, taking 243 Earth days to rotate. Jupiter, Saturn, Uranus, and Neptune whirl completely around in less than 17 hours. Different planets also take different amounts of time to orbit the Sun because their distances from the Sun range so widely **(Box 21.1)**. As Kepler's third law states, the farther a planet lies from the Sun, the longer the planet takes to orbit the Sun (see Chapter 20). For example, Mercury makes its journey in just 88 Earth days, but Neptune takes 165 Earth years.

FIGURE 21.4 Each planet's axis of rotation tilts differently with respect to the ecliptic plane. Uranus's axis is almost parallel to the ecliptic plane, and Venus's axis is reversed. Although all the outer planets have rings, only Saturn's are visible without image enhancement (see the chapter opening photo).

Red arrow = counterclockwise rotation
Green arrow = clockwise rotation

When we consider planetary size and density, we see that the planets fall into two categories, as Table 21.2 makes clear. The *inner planets*, those closer to the Sun (Mercury, Venus, Earth, and Mars), are relatively small (Earth-sized or smaller) and have relatively high average densities (greater than the density of basalt). A high density indicates that the mass of a planet consists mostly of rock and metal. Because of the similarities of the other three inner planets to the Earth, the four inner planets are known collectively as the **terrestrial planets**. The *outer planets*, those farther from the Sun (Jupiter, Saturn, Uranus, and Neptune) are all much larger than the Earth. In addition, they all have much lower average densities, indicating that they are composed primarily of gases or ices. In this context, *ice* refers not only to frozen water, but also to the solid versions of other volatile compounds (such as carbon dioxide, methane, and ammonia). Because of the similarity of the other outer planets to Jupiter, the four outer planets have been traditionally known as the **Jovian planets**. Notably, all planets except Mercury and Venus have moons, but no two planets have the same number of moons. Pluto, although no longer classified as a planet, has five moons. In addition, all the Jovian planets have **rings**, thin bands of tiny objects in orbit along each planet's equator.

The Solar System contains many other objects in addition to the planets (see Table 21.1). All display prograde motion, but some have very eccentric orbits, some of which are inclined to the ecliptic plane. Despite their great numbers, all the objects of the Solar System, other than the Sun, together account for only a tiny fraction (0.15%) of the Solar System's mass; 99.85% of the mass lies within the Sun itself. We'll now take a tour of these objects, starting with our own Moon, then moving out to the planets and other objects.

Take-home message...

🏠 The Solar System consists of the Sun, 8 planets, 5 dwarf planets, 290 moons, and numerous asteroids, Kuiper Belt and Oort Cloud objects, meteoroids, and comets. The terrestrial planets, including the Earth, are small and rocky, while the Jovian planets are composed of gases and ices. Planets' rotational axes display a range of tilts, and they rotate about their axes at different speeds. Mercury, the closest planet to the Sun, orbits the Sun every 88 days, while the outer planets take many years to orbit the Sun once.

Quick Questions

- What is a dwarf planet?
- Which planets have rings orbiting their equators?
- Which planet rotates in the opposite direction from the others?

21.2 Our Nearest Neighbor: The Moon

History of Lunar Exploration

Long before the invention of the telescope, people realized that the Moon orbits the Earth, that we see it because it reflects sunlight, and that its surface has dark and light patches. Early astronomers were able to measure its dimensions, so they knew that the Moon has a radius of 1,737 km (1,079 mi), about one-fourth that of the Earth. But beyond that, they knew very little about our nearest neighbor—it was beyond humanity's reach.

Scientific understanding of the Moon took a great leap forward in the 1960s, when unmanned space probes and, later, Apollo spacecraft orbited the Moon **(Fig. 21.5a)**.

(a) The 35-story-high *Saturn V* rocket that launched the Apollo missions.

(b) The lunar module and the lunar rover on the Moon in 1972.

(c) The Artemis rocket, before a test launch in 2022. It is smaller than the *Saturn V*, but packs more thrust.

FIGURE 21.5 The Apollo and Artemis Moon missions.

FIGURE 21.6 The near and far sides of the Moon.

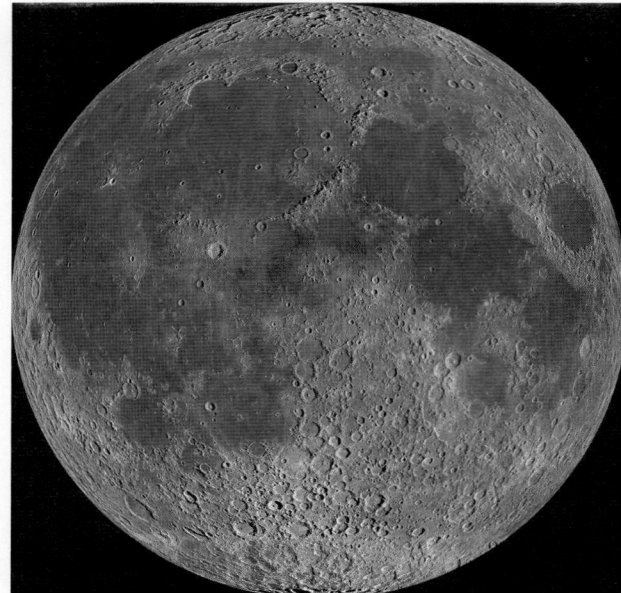

(a) The near side of the Moon hosts the maria, which lie at relatively low elevations.

(b) The far side of the Moon consists entirely of heavily cratered highlands.

That's one small step for a man, one giant leap for mankind.

—NEIL ARMSTRONG (AMERICAN ASTRONAUT, ON TAKING HIS FIRST LUNAR STEP, 1969)

In 1966, an unmanned Soviet space probe made the first successful landing on the Moon. Then, between 1969 and 1972, the United States landed six manned spacecraft on the lunar surface **(Fig. 21.5b)**. Altogether, the astronauts brought back 381 kg (840 lb) of Moon rocks. More recently, in 2020, the Chinese National Space Administration landed an unmanned probe on the Moon, collected 1.7 kg (3.7 lb) of lunar samples, and returned it to the Earth. By analyzing lunar rocks, researchers were able to gain insight into the chemical composition and the numerical age of the Moon. Modern orbiting space probes have produced high-resolution digital topographic maps of the entire lunar surface, and seismometers placed on the Moon have even detected moonquakes. Now, NASA and several collaborators are preparing a new manned mission to the Moon, using the new Artemis *Orion* spacecraft **(Fig. 21.5c)**. Let's look at the lessons learned from lunar mapping and sampling so far.

General Characteristics of the Moon

Unlike the Earth, the Moon has no active volcanoes or plate-tectonic processes, because most of its mantle has become too cool for the conditions leading to melting to occur. Without active volcanism to supply gases, the Moon has no atmosphere, and therefore, no oceans, rivers, glaciers, or life. And without an atmosphere or ocean to hold and redistribute heat, ground temperatures on the Moon undergo violent swings, ranging from over 120°C (248°F) on the sunlit lunar equator to −175°C (−283°F) on the Moon's dark side.

The Moon's lack of an atmosphere, and of the light scattering that it causes, also means that we can see the

lunar surface clearly. This surface displays varied landscapes with elevations ranging from a high of 10.8 km (6.7 mi) above mean (average) elevation to 9.1 km (5.7 mi) below. Lower-elevation regions of the Moon, called **maria** (pronounced MAR-ee-ah; singular: **mare**, pronounced MAR-ay), have fairly smooth, dark surfaces **(Fig. 21.6a)**. Their name comes from the Latin word for sea because early astronomers thought that maria were filled with water. Through the study of Moon rocks, we now know that the maria consist of plains underlain by basalt lava flows, which have a low *albedo* (reflectivity). Notably, maria lie only on the side of the Moon that faces the Earth, so the near and far sides of the Moon look very different from each other **(Fig. 21.6b)**. The regions at higher elevations, the **lunar highlands**, are relatively rugged and have a high albedo, so they display a whitish glow when viewed from the Earth. They are underlain by a light-colored rock, called *anorthosite*, that consists largely of the mineral plagioclase.

Because so much of the Moon consists of anorthosite, a relatively low-density rock, the Moon has a lower average density than the Earth (3.4 g/cm³ for the Moon vs. 5.5 g/cm³ for the Earth). This characteristic, along with its smaller size, means that the Moon has a mass only about $\frac{1}{80}$ of the Earth's mass. Less mass means weaker gravity, so a 68 kg (150 lb) astronaut would weigh only 11 kg (25 lb) on the Moon.

Lunar Craters and Regolith

When Galileo turned his telescope toward the Moon in 1609, he discovered that bowl-shaped indentations, or *craters*, pockmark the surface. For centuries after Galileo, researchers assumed that these craters formed during volcanic eruptions. In 1892, however, an American geologist, G. K. Gilbert, argued that they were **impact craters**, excavated by meteorite impacts. To prove his point, he made laboratory models of impact craters by dropping clay balls onto wet sand. Lunar exploration has confirmed that the lunar craters formed as a result of meteorite impacts. There are over 100,000 impact craters on the Moon with a diameter of 1 km (0.6 mi) or more, and probably millions of smaller ones.

Why do we see so many more impact craters on the Moon than on the Earth? First, fewer objects strike the Earth than the Moon because many of those that reach the Earth burn up or explode in its atmosphere. Second, plate-tectonic processes and erosion destroy craters on the Earth over time. Such processes don't happen on the Moon, so a lunar crater, once formed, can survive for billions of years, and we see craters that have accumulated over long periods. Though cratering has happened throughout lunar history, and continues today, the most intense cratering occurred during the **late heavy bombardment**, a time of abundant impacts beginning about 4.1 Ga and tailing off by about 3.8 Ga. During this interval, the gravitational pull of the outer planets sent huge swarms of objects careering across the orbits of other planets. Almost all of the Earth has been resurfaced since that time, for seafloor spreading replaces 70% of the Earth's surface at least once every 200 million years, and continental surfaces undergo constant change due to plate-tectonic processes (such as continental collision) and the many steps of the rock cycle.

Not all lunar craters look the same. Their character depends on the size and speed of the impacting object. Smaller *simple craters* tend to be smooth-floored bowls, whereas larger *complex craters* have a central uplift, formed when an impact disrupts the crust deeply enough that it rebounds upward, and may be surrounded by concentric ridges, which are traces of normal faults **(Fig. 21.7a)** (we'll describe the processes responsible for these features in Section 21.5). Impacts also send out debris, which forms an *ejecta blanket* around the crater. In some cases, the debris forms a pattern of spoke-like streaks.

As the first astronauts to walk on the Moon discovered, a layer of **lunar regolith**, consisting of dust and debris, covers the Moon's surface **(Fig. 21.7b)**. This regolith consists of fragmented rock blasted out of impact craters, glass droplets frozen from rock melted by impacts and sent upward as a molten spray, and accumulations of *micrometeorites* (meteorites less than 2 mm, or 0.08 in, in diameter). The top part of the regolith tends to be very fine, probably due to the pulverization of surface material by micrometeorite impacts and by its absorption of *cosmic rays* (high-energy atomic nuclei from space that can disrupt crystal structures). Researchers refer to such breakdown of material as **space weathering**. Because of space weathering, older craters tend to be blunter and somewhat smoothed, while newer ones have sharper features.

FIGURE 21.7 The lunar surface.

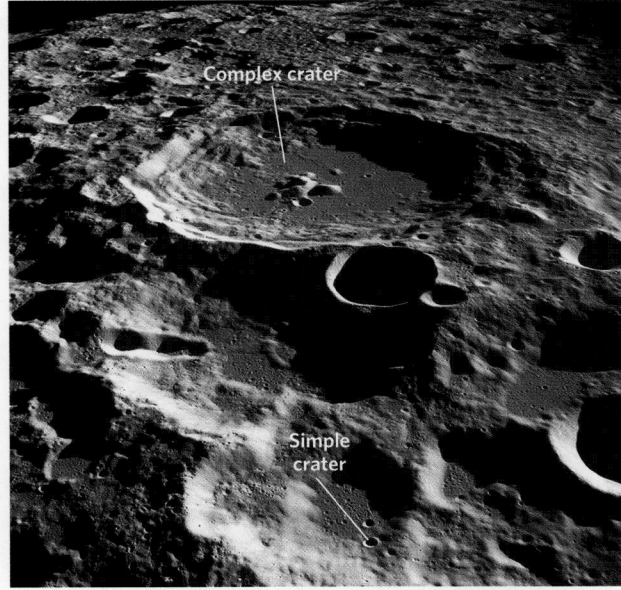

(a) Lunar impact craters, as seen from an orbiting spacecraft. Larger complex craters have a central uplift and several ridges. Smaller simple craters are smooth bowls.

(b) An astronaut's footprint in lunar regolith.

FIGURE 21.8 The formation of the Moon may have begun with an immense impact that blasted debris into orbit.

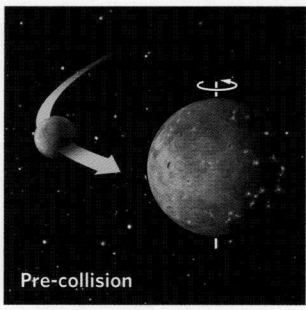

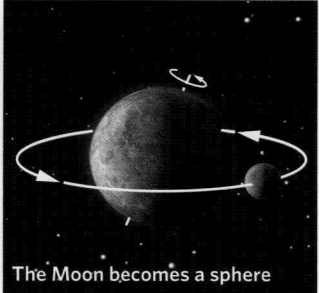

Pre-collision | Collision | Debris sprays into orbit | Orbiting debris coalesces | The Moon becomes a sphere

Time

Age and Origin of the Moon

Most researchers favor a model, based on analyses of Moon rocks together with other data, in which the Moon formed from debris ejected into orbit around the Earth following a catastrophic collision of the newborn Earth with a protoplanet named Theia. This event happened at about 4.53 Ga, shortly after the interior of the Earth had undergone differentiation. The debris, mostly from the mantles of the colliding object and the Earth, first formed a ring of particles around the Earth. Gravity soon caused them to coalesce into a sphere—the Moon **(Fig. 21.8)**.

When the Moon first formed, its surface was probably a magma sea. As the Moon cooled, denser mafic minerals formed and sank, while less dense felsic minerals, such as plagioclase, rose. The outer layer of less dense rock solidified into an anorthosite crust, exposed in the lunar highlands today. Geologists use standard radioisotopic dating methods to determine the ages of Moon rocks. The dates obtained so far fall between 4.36 Ga and 3.16 Ga. How the maria formed remains a subject of research. According to one model, they began as immense basins carved by meteorite impacts between 4.3 and 3.8 Ga. Melt from the Moon's mantle eventually broke through the thin crust on the floors of these basins in a succession of volcanic eruptions. These eruptions began about 100 million years after the basins had formed and continued on and off for the next several hundred million years. The crust of the Moon's near side is thinner and weaker than that of its far side, so while the crust could crack and permit volcanic eruptions on the near side, it could not on the far side, which explains the difference between the two sides. No one knows for sure why the melt formed. Some melt may have remained from the Moon's formation, but some may have formed due to decompression melting or frictional heating associated with tidal forces.

Take-home message...

The Moon has no atmosphere, ocean, or active plate-tectonic processes. Its surface displays dark, low-lying maria and light-colored, elevated lunar highlands. Impact craters pockmark the Moon's surface, and a layer of regolith covers most regions. The Moon formed around 4.53 Ga, most likely from debris blasted into space when a protoplanet struck the Earth.

Quick Questions

- Which side of the Moon has the most impact craters?
- Why are there so many more impact craters on the Moon than on the Earth?
- When astronauts walked on the Moon, did they find its surface hard or powdery?

21.3 The Terrestrial Planets

The inner Solar System is the realm of four planets, including the Earth. Aside from having similar overall compositions and internal structures, the other three terrestrial planets, Mercury, Venus, and Mars, barely resemble our home planet. In this section, we explore these three planets. We'll see that our neighbors have surfaces that range from Moon-like to volcanic, and temperatures that range from cold enough to freeze CO_2 to hot enough to melt lead **(Fig. 21.9)**.

Mercury: A Moon-Like Landscape

THE FAST PLANET. The English name for Mercury came from the Roman messenger god known for his speed, because ancient astronomers saw that the planet moves across the sky faster than any other planet. Because it's closest to the Sun, it takes Mercury only 88 days to complete an orbit. Mercury's orbit has greater eccentricity than

Mercury | Venus (without clouds) | Earth | Moon | Mars

(a) The surfaces and diameters of the terrestrial planets differ markedly from one another.

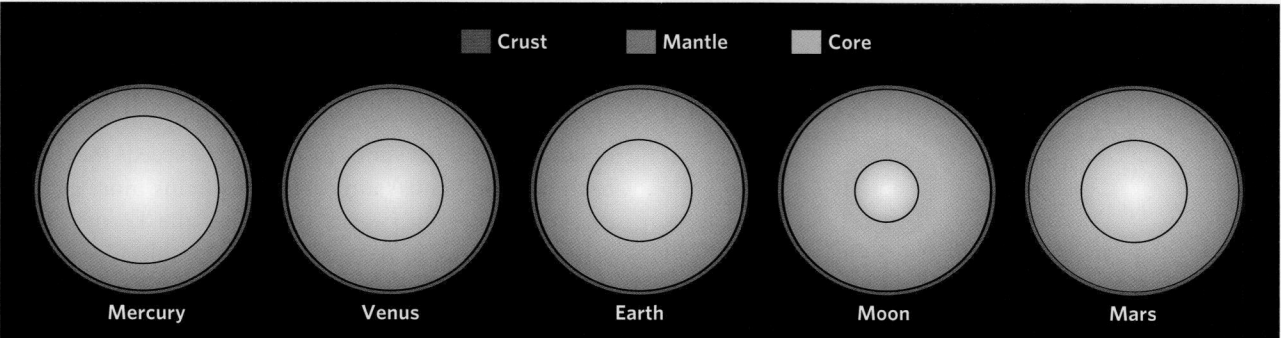

Crust | Mantle | Core

Mercury | Venus | Earth | Moon | Mars

(b) The interiors of the terrestrial planets differ from one another, too. In this figure, all the planets have been resized to the same diameter so as to emphasize differences in their proportions of crust, mantle, and core.

FIGURE 21.9 Mercury, Venus, Earth, Earth's Moon, and Mars, with representations of their internal structures.

that of any other planet. At perihelion, it lies 70 million km (43.5 million mi) from the Sun, and at aphelion, it lies 74 million km (46 million mi) from the Sun. Though it revolves around the Sun quickly, it rotates on its axis relatively slowly; one day on Mercury equals 59 days on the Earth.

Because Mercury orbits so close to the Sun, and because its surface consists of dark rock that absorbs radiation, surface temperatures on the planet can rise far above the boiling points of volatile materials. These characteristics, along with the lack of volcanic activity to produce new gas, mean that the planet has no ocean and virtually no atmosphere. The trace quantities of gas surrounding Mercury consist of atoms released by space weathering, and the strong solar wind that engulfs the planet eventually blows these atoms into space. Without an ocean or atmosphere to transport heat around the planet, the temperature at a location depends only on the angle of insolation. Mercury's axis lies almost exactly perpendicular to its orbital plane, so at its poles, the temperature remains a constant −93°C (−136°F) all year, whereas at its equator, the temperature ranges from 427°C (800°F) at noon to −173° (−280°F) at midnight; this difference represents the largest temperature range of any planet in the Solar System.

Mercury's mass is only about 5.5% that of the Earth, so a 68 kg (150 lb) astronaut would weigh about 26 kg (57 lb) on Mercury. Researchers suggest that Mercury's metallic core accounts for a large portion of its interior (see Fig. 21.9b), perhaps because the accretion disk from which the Solar System formed contained a higher proportion of refractory material closer to its center than existed farther away. Mercury is the only terrestrial planet besides the Earth to possess a magnetic field. Recent studies have shown that Mercury's inner core is solid and has a diameter very nearly the same as that of the Earth's inner core. Like the Earth, Mercury has an outer core composed of liquid metal, which generates its magnetic field.

SURFICIAL SIMILARITIES TO THE MOON. Mercury resembles the Earth's Moon in many ways. The two bodies are similar in size (Mercury's diameter is 1.4 times that of the Moon), both have heavily cratered surfaces, both lack active volcanoes and plate-tectonic processes, both have a surface layer of regolith, and neither has an atmosphere or ocean. The *Messenger* space probe, which began to orbit Mercury in 2011, has mapped the planet's topography in exquisite detail **(Fig. 21.10)**, revealing that Mercury hosts wide plains that look like lunar maria. Mercury's plains,

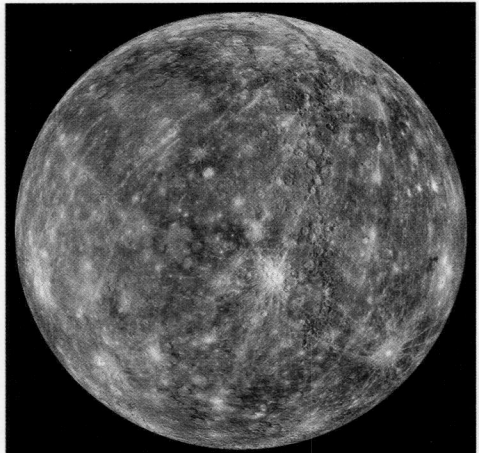

(a) A photograph, in false color, of the cratered surface of Mercury.

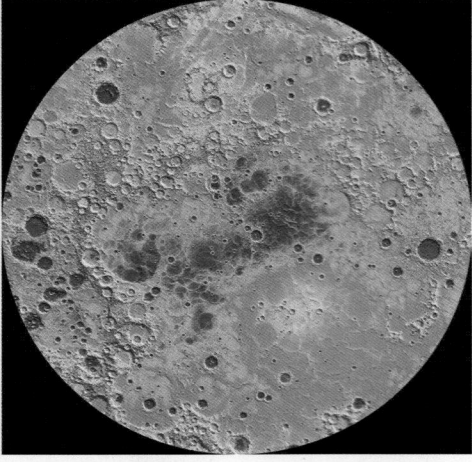

(b) A digital topographic map of Mercury's surface. Blue is lowest; red is highest.

however, may be covered with pyroclastic debris instead of lava flows. Close-up photos indicate that large faults have offset craters on Mercury **(Fig. 21.11)**.

Venus: Veiled by Clouds

THE RETROGRADE PLANET. Venus, the second planet from the Sun **(Fig. 21.12a)**, shines brighter than any star in the Earth's night sky. Named for the Roman goddess of beauty, Venus glows so brilliantly because it has a high albedo and lies fairly close to the Earth. And because it orbits relatively close to the Sun, it receives lots of solar radiation. The planet resembles the Earth in size, but it rotates much more slowly on its axis, taking 243 Earth days to spin once. Because Venus takes 224 days to orbit

**Did you
ever wonder...**

if you could breathe the air on Venus?

FIGURE 21.11 Part of Mercury's surface dropped to a lower elevation along a large fault.

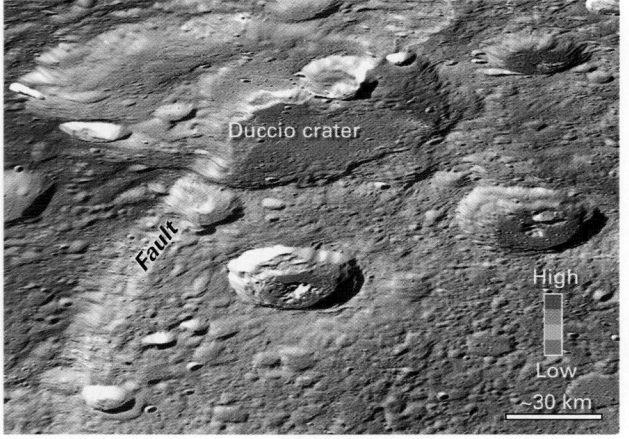

the Sun, its rotation rate exceeds its orbital period, so a day is longer than a year on Venus. Venus and the Earth have about the same average density, which suggests that Venus, like the Earth, has an iron core. In addition, as we noted earlier, Venus differs from all the other planets in that it spins clockwise around its axis (see Fig. 21.4).

A CAUSTIC ATMOSPHERE. Venus hosts an atmosphere consisting almost entirely of carbon dioxide (96.5% by volume), with the remainder made up of nitrogen molecules (3.5%) and trace amounts of water vapor, sulfur dioxide, and carbon monoxide **(Fig. 21.12b)**. This atmosphere has a much greater density than that of the Earth. In fact, it's so dense that if you were to stand on Venus's surface, the air pressure (93 atm) would equal the pressure you would feel at a depth of 800 m (0.5 mi) beneath the Earth's ocean surface. Such pressure would crush most submarines, for they descend only to depths of less than 650 m. Venus's air pressure exceeds the Earth's both because Venus's atmosphere contains more gas and extends farther into space than does the Earth's atmosphere, and because the weight of CO_2 molecules exceeds that of N_2 or O_2 molecules. The presence of sulfuric acid clouds makes Venus's atmosphere so caustic that it would burn through metal. These clouds also give Venus its high albedo.

The intense greenhouse effect caused by the abundance of CO_2 in Venus's atmosphere keeps the planet's surface temperature at about 450°C (840°F), hot enough to melt lead. Very early in its history, Venus, like the Earth, had water in the form of vapor in its atmosphere. Unlike that on the Earth, Venus's water never condensed into oceans because of the higher temperatures, and the initial CO_2 in its atmosphere never had a chance to dissolve in water and become locked in carbonate rocks, or become incorporated in organisms. Rather, on Venus, the additional water vapor in the atmosphere amplified the greenhouse effect. The result—a **runaway greenhouse effect**—made the planet's atmosphere so hot that water molecules, under constant bombardment by the solar wind, broke apart. The hydrogen atoms then drifted into space, while the oxygen atoms were absorbed by weathering reactions with surface rocks. As a result, Venus's water disappeared for good. Winds blow so strongly on Venus that its atmosphere circulates very rapidly. Due to this circulation, which transports heat across latitudes, hardly any temperature difference exists between Venus's poles and equator or between its night and day sides.

A WORLD OF VOLCANOES. We can't see the surface of Venus from the Earth, or even from orbit around Venus, because of the planet's dense, cloudy atmosphere. In fact, it has been photographed only twice, by Soviet space probes that landed on the Venusian surface **(Fig. 21.13a)**.

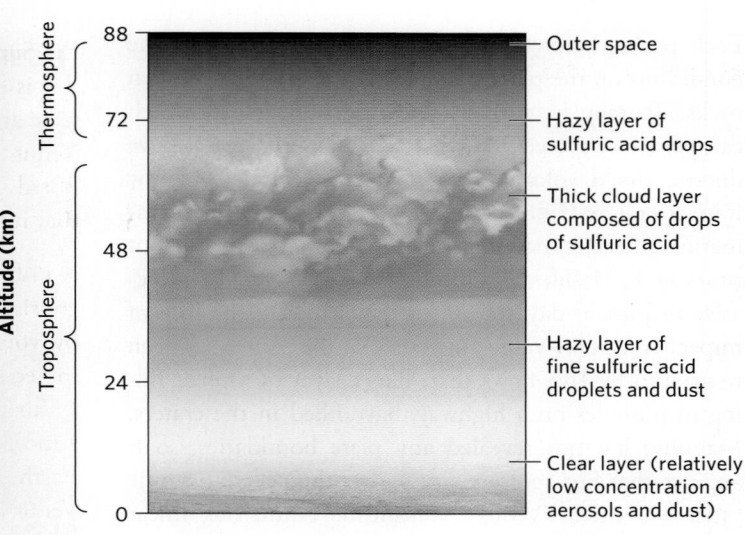

(a) A *Mariner 10* image of Venus shows a thick, cloudy atmosphere.

Outer space

Hazy layer of sulfuric acid drops

Thick cloud layer composed of drops of sulfuric acid

Hazy layer of fine sulfuric acid droplets and dust

Clear layer (relatively low concentration of aerosols and dust)

Thermosphere

Troposphere

Altitude (km)

88

72

48

24

0

FIGURE 21.12 Venus has an atmosphere 93 times as dense as the Earth's.

(b) A cross section of Venus's atmosphere. The atmosphere consists mostly of carbon dioxide. Its clarity depends on the concentration of sulfuric acid aerosols and dust.

(a) An artist's depiction of the ground surface on Venus, based on images from USSR's *Venera 13* space probe.

(b) An oblique radar image of Maat Mons, a volcano on Venus.

FIGURE 21.13 Images of the surface of Venus.

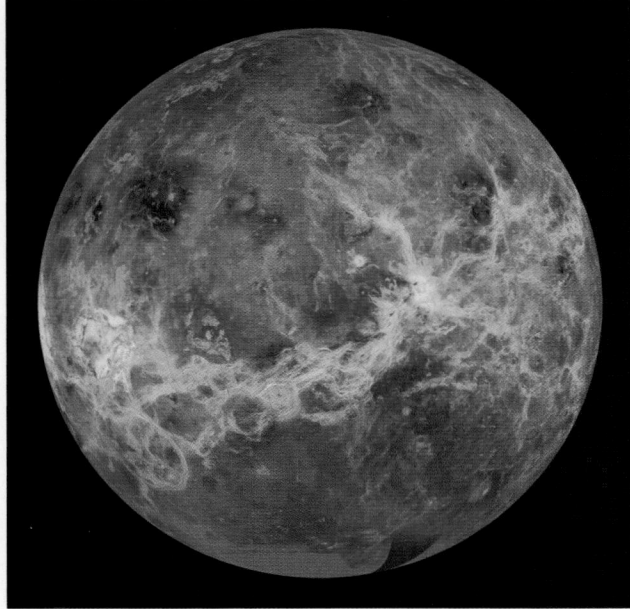

(c) A radar image of the planet from the *Magellan* space probe.

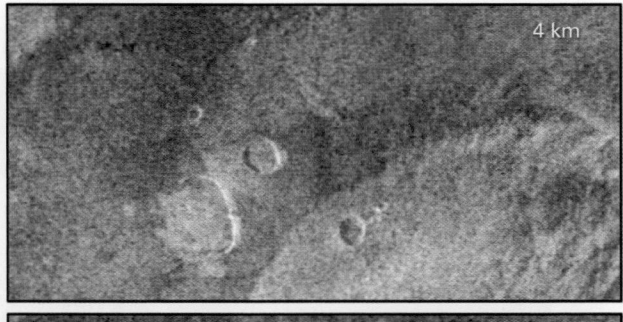

4 km

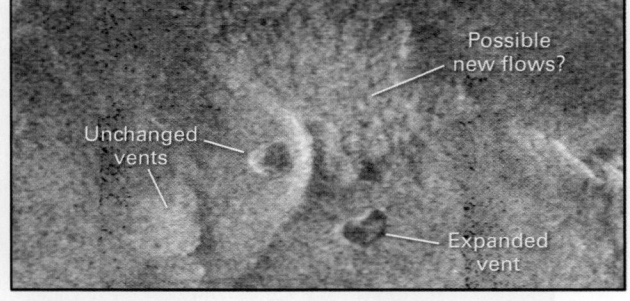

Possible new flows?

Unchanged vents

Expanded vent

(d) Close-ups showing subtle changes in the Venusian surface that may be due to volcanic activity.

Each probe, before rapidly succumbing to the extreme conditions on the planet, returned images of flat, broken rocks. To map its surface, researchers use radar, which can penetrate the veil of gas. Radar mapping reveals lava domes, shield volcanoes, and volcanic craters, but few impact craters **(Fig. 21.13b, c)**. In 2023, a careful examination of radar mapping images taken at two different times in 1991 showed a local change that may be indicative of present-day volcanism **(Fig. 21.13d)**. The lack of impact craters on Venus suggests that the planet has been resurfaced by lava flows that, like a layer of asphalt filling in potholes on a highway, have filled in the craters. Mapping has not revealed any plate boundaries, so it appears that hot-spot activity, rather than plate-tectonic processes, drives Venus's volcanism. Numerous linear structures, considered to be faults, cut across the planet's surface, indicating that the crust of Venus locally stretches or shortens.

Mars: Could Life Exist on the Red Planet?

Mars has never hosted life in the form of animals or plants. However, researchers wonder whether microbial life may have existed there in the past, or could still exist underground. Because life as we know it requires water, much of the effort to search for life on Mars has focused on finding evidence that liquid water existed there. Interest in the possibility of life on Mars, on moons of Jovian planets, and elsewhere in the galaxy has led to the development of a field of study called **exobiology**.

Very strong evidence suggests that long ago, Mars indeed had liquid water, as well as a thicker atmosphere. That environment has long since vanished, however, and

FIGURE 21.15 The Mars rover *Curiosity*.

the Mars of today is a dry, cold, desolate place, covered with rock and dust. This material has a red tint because it contains oxidized iron, presumably due to interaction with oxidizing air or water in the past. The rust color of the planet led to the name Mars, after the Roman god of bloodshed, and to the planet's nickname, the Red Planet.

EXPLORING MARS. Because of its proximity to the Earth and its mostly transparent atmosphere, we can see fascinating rocks and structures on Mars **(Fig. 21.14)**. Several space probes have landed successfully on the planet, and several have surveyed Mars from orbit. The United States has so far deployed five Mars rovers (*Sojourner* landed in 1997, *Spirit* and *Opportunity* in 2004, *Curiosity* in 2012, and *Perseverance* in 2021) that could travel across the Martian surface **(Fig. 21.15)**. *Curiosity* and *Perseverance* were still operating in 2023. *Ingenuity*, the first helicopter on Mars, accompanied *Perseverance* on some of its forays. The Mars rovers have discovered minerals and sedimentary structures that may have formed underwater in the distant past. Since 2006, the *Mars Reconnaissance Orbiter* has sent back high-resolution images of landscape features on Mars, and in 2008, *Phoenix*, a stationary lander,

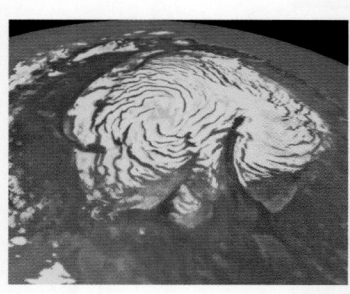

FIGURE 21.14 Mars, the best-studied planet beyond the Earth, displays many notable topographic features. The inset shows the planet's north-polar ice cap.

touched down in Mars's north-polar region and drilled into the surface, confirming the presence of subsurface water ice. The *Orbiter* has now identified surface water ice in five locations. The largest accumulation of surface water ice forms the north-polar ice cap, which contains about a third as much ice as the Earth's Greenland ice sheet (see Fig. 21.14, inset).

A THIN ATMOSPHERE. Air on Mars is much less dense than the Earth's air, so the air pressure on Mars's surface reaches only about 0.6% of that on the Earth; that means that the air pressure at the surface of Mars resembles that of the Earth's atmosphere at an altitude of about 19.5 km (64,000 ft), about twice the altitude of cruising jet aircraft. Compositionally, the Martian atmosphere resembles that of Venus: it contains 95.3% carbon dioxide, 2.5% nitrogen, 1.6% argon, and just traces of oxygen, carbon monoxide, and water vapor. Even though CO_2 makes up most of the Martian air, there's not enough to cause a significant greenhouse effect, so the surface of Mars remains frigid, with an average surface temperature of −55°C (−67°F). The air might have been denser earlier in the history of the planet when volcanoes were active, but most of the gas that formed then froze, reacted with the surface, or escaped into space. The thin air does move, and researchers have clocked winds in the range of 30–100 km/h (20–60 mph). These winds can churn up fine dust, producing haze that can last for months, but because the air has such low density, the winds cannot move larger objects.

GENERAL CHARACTERISTICS OF MARS. A day on Mars (24.6 hours) approximately equals a day on the Earth, and because Mars has an axial tilt of 25°, it has seasons. But seasons on Mars are more extreme than those of the Earth, for the orbit of Mars is more eccentric. (Recall that the eccentricity of the Earth's orbit is so slight that it has no effect on the Earth's seasons.) Summer in the Martian southern hemisphere occurs when the planet lies closest to the Sun, so the southern hemisphere summer is warmer than the northern hemisphere summer.

The radius of Mars is only about half that of the Earth. Mars has an iron core whose radius accounts for about half that of the entire planet. But because of its smaller size, Mars has only about 10% of the Earth's mass, so a 68 kg (150 lb) astronaut would weigh just 26 kg (57 lb) on Mars. Why is this weight the same as the astronaut's weight on Mercury, which has about half the mass of Mars? The gravitational pull at a planet's or moon's surface depends not only on its mass, but also on its radius. Because Mars has a radius 39% greater than Mercury's (see Table 21.2), Mars's surface is farther from its center, so the pull of gravity at Mars's surface is less than it would be if both planets were the size of Mercury.

Mars has two moons, Phobos (Fear) and Deimos (Panic), only 22 km (13.7 mi) and 13 km (8 mi) in diameter, respectively **(Fig. 21.16)**. Because of their small size, they have large bumps and dimples that gravity didn't smooth out. Numerous craters speckle their surfaces. The moons, which are probably asteroids that were captured by Mars's gravity, now orbit Mars at such low altitudes (9,400 km and 23,500 km, or 5,841 and 14,600 mi, respectively) that a human standing on the Martian surface could easily see them with the naked eye, even though they're so small.

THE SURFACE OF MARS. The surface of Mars displays many distinctive topographic features **(Fig. 21.17)**. Researchers divide the planet into two topographic

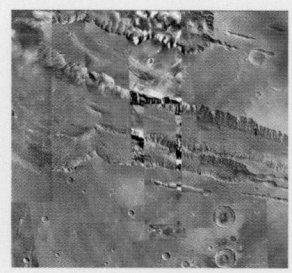
FIGURE 21.16 The moons of Mars.

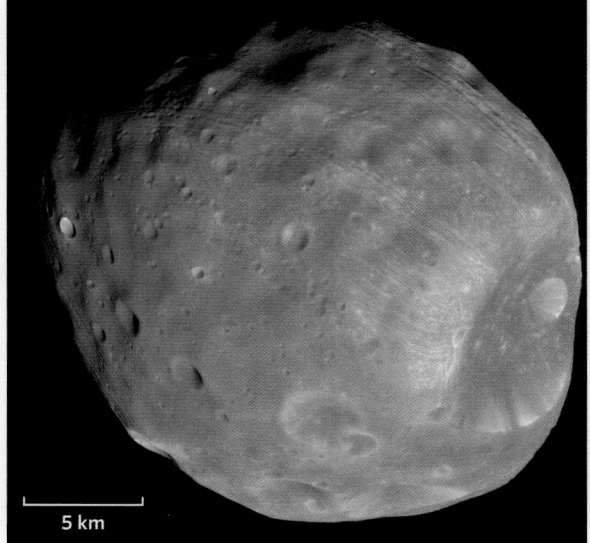

(a) Phobos.

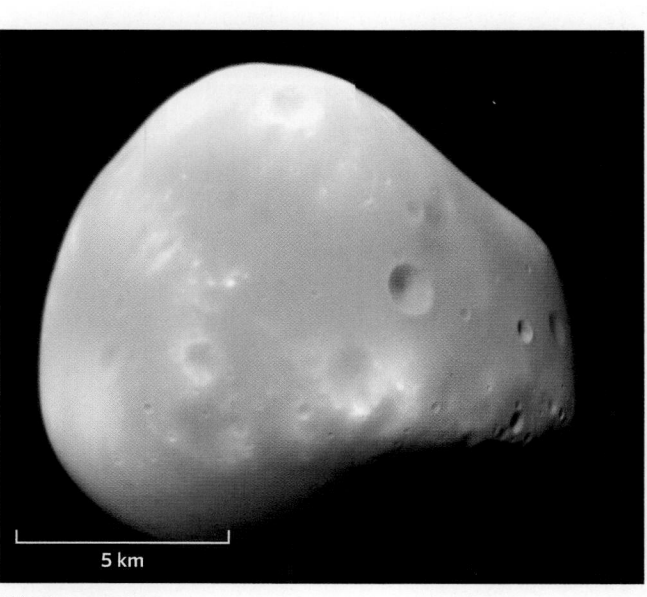

(b) Deimos.

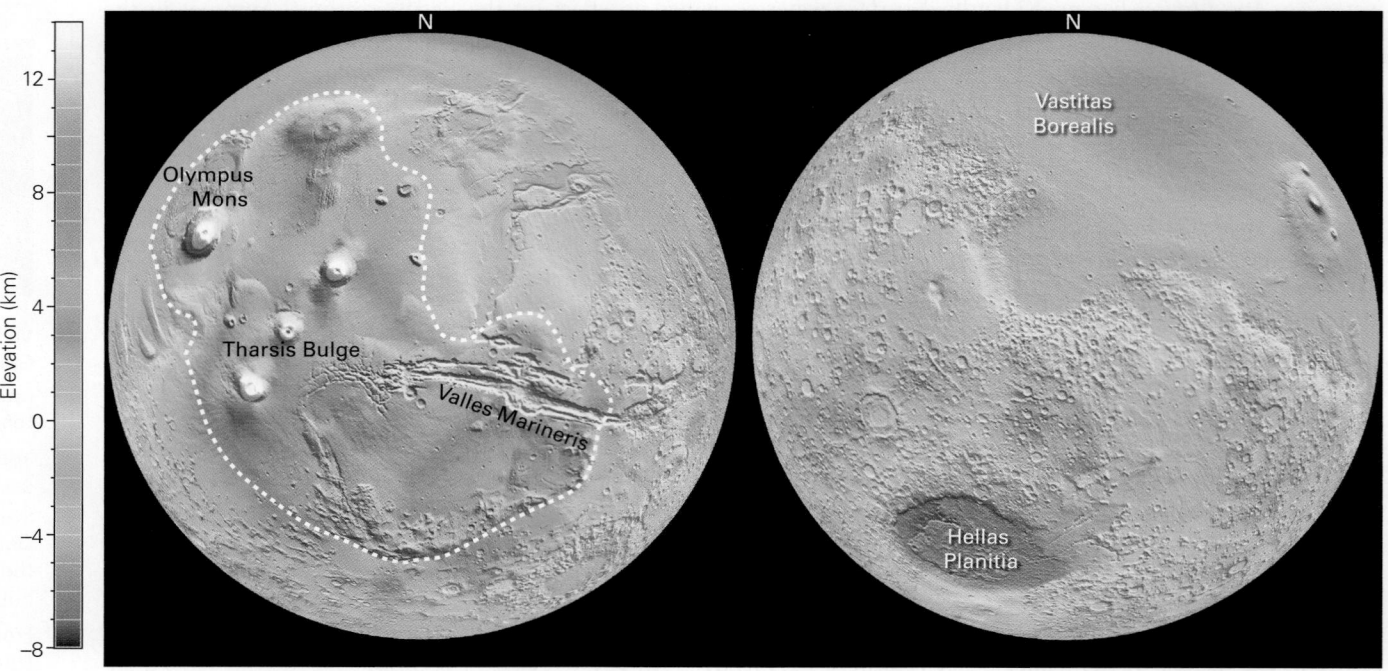

provinces. The northern lowlands province, Vastitas Borealis, a gigantic lava-covered plain encompassing 40% of the planet's surface, lies at a low elevation. It has fewer impact craters than the rest of the planet, suggesting that basaltic lava covered it after the late heavy bombardment. The rest of the Martian surface consists of cratered highlands, with the exception of two huge impact basins, one of which—Hellas Planitia—includes the lowest point on Mars. Near the equator, the Tharsis Bulge, a plateau as big as North America, rises above the other highlands. Relatively few impact craters exist on the plateau's surface, suggesting that it's younger than the rest of the highlands. An immense canyon, Valles Marineris, defines part of the northern edge of the Tharsis Bulge **(Fig. 21.18a)**. This canyon, which probably formed as a tectonic rift, is 4,000 km (2,500 mi) long, 600 km (373 mi) across, and up to 8 km (5 mi) deep, so it dwarfs the Earth's Grand Canyon, which is only 250 km (155 mi) long, 10–15 km (6.2–9.3 mi) wide, and up to 1.6 km (1 mi) deep.

FIGURE 21.18 Dramatic landscape features on Mars.

(a) Valles Marineris, the largest canyon in the Solar System.

(b) Oblique view of Olympus Mons, with a caldera visible at its summit.

FIGURE 21.19 Martian landscapes and features that could have been formed by flowing water in the past.

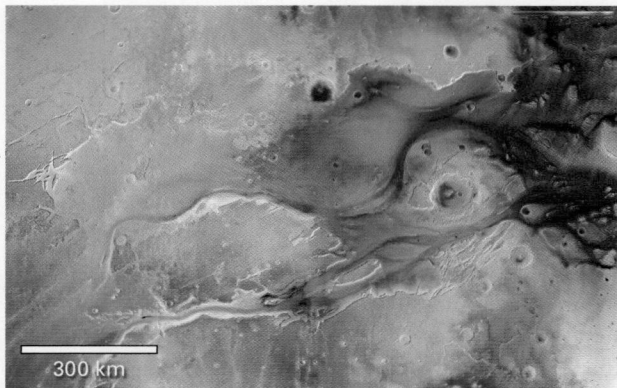

(a) Kasei Valles, showing what appears to be an outflow channel that, at times, contained flowing water.

(c) The thin laminations in this rock look like cross beds deposited by currents.

(b) Close-up of stream-like channels, as viewed obliquely from a space probe.

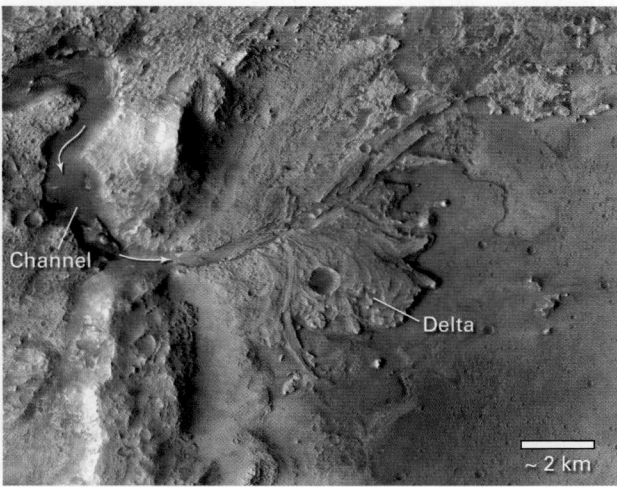

(d) A channel carried water into Jezero Crater, which at the time may have been a lake filled with water. The sediment carried by the water was deposited in a delta.

Olympus Mons, the largest volcano in the Solar System, rises from the plains to the west of the Tharsis Bulge **(Fig. 21.18b)**. With a base diameter of 700 km (435 mi) and an elevation of 25 km (15.5 mi), this volcano is 2.5 times the size of the Earth's largest volcano, Hawai'i's Mauna Loa. Like the volcanoes of Hawai'i, Olympus Mons formed as a shield volcano that erupted relatively low-viscosity basalt. Notably, some of the flows on the flanks of Olympus Mons have so few impact craters that they may have spilled out of the volcano within the last 100 million years. A caldera 80 km (50 mi) wide formed by collapse at the crest of the volcano.

Mars has an ice cap at each pole. *Martian ice caps* consist of seasonal ice—frozen CO_2 (dry ice) that accumulates when temperatures at the poles drop in the winter—on top of permanent ice, which lasts all year. The permanent ice of the north pole consists of a sheet of water ice 1–3 km (0.6–1.9 mi) thick, while that of the south pole consists of frozen CO_2.

Data collected by probes suggest that large amounts of liquid water existed on Mars in the past. For example, surface soils contain clay, calcite, and gypsum, minerals that form only in the presence of water. Images reveal landscape features that look like river-carved channels, drainage networks, flow-eroded islands, and even deltas **(Fig. 21.19)**. The presence of deltas and of terraces that look like remnants of beaches along the edge of Vastitas Borealis suggests that the plain was once an ocean. Does any liquid water exist on Mars today? Maybe. In

2015, the *Mars Reconnaissance Orbiter* obtained images of streaks on hillslopes that become darker, as if damper, during part of the year. In 2018, scientists used radar to detect a 20 km wide lake of liquid water underneath solid ice in the Planum Australe region, a plain that starts at a Martian latitude of 75° S and extends under the planet's south pole. The water is probably kept from freezing by dissolved salts and the pressure of the ice above.

Take-home message...

The terrestrial planets, although similar in their overall composition, differ in many ways. Mercury resembles the Earth's Moon, but is hotter on its sunlit side. A very dense, CO_2-rich atmosphere cloaks dry, hot Venus. Mars is now a dry, cold planet with a very thin, CO_2-rich atmosphere. All but Venus contain abundant impact craters. Venus and Mars also host large volcanoes.

Quick Questions

- Do plate-tectonic processes happen on all terrestrial planets?
- Why is the surface of Venus so much hotter than the surface of the Earth?
- Which planets in the inner Solar System host volcanoes?

21.4 The Jovian Planets and Their Moons

Each of the Jovian planets that lie beyond Mars is a giant, compared with the terrestrial planets, and each has many moons. Because of the state of the material within the Jovian planets, astronomers refer to the two nearer to the Sun (Jupiter and Saturn) as the **gas giants** and the farther two (Uranus and Neptune) as the **ice giants (Fig. 21.20)**. Although Jupiter and Saturn can be seen in the night sky, we knew little about these distant orbs until space probes sent back stunning images and other data. Some observations raise the possibility that liquid water may lie beneath the icy surfaces of some of the moons circling the Jovian planets . . . and if there's water, could there be life? This mystery remains to be solved.

Jupiter: The Giant among Giants

Jupiter, the Roman king of the gods, serves as an apt namesake for the fifth planet from the Sun, for Jupiter contains more matter than all the other planets combined. In fact, 70% of our Solar System's total mass, not including the Sun, lies within Jupiter. This huge planet rotates much faster than any other planet and completes a day in only 10 hours. The centrifugal force associated with this

FIGURE 21.20 Characteristics of the Jovian planets.

(a) The diameters and colors of the giant planets differ markedly from one another. All of the giant planets are much larger than the Earth.

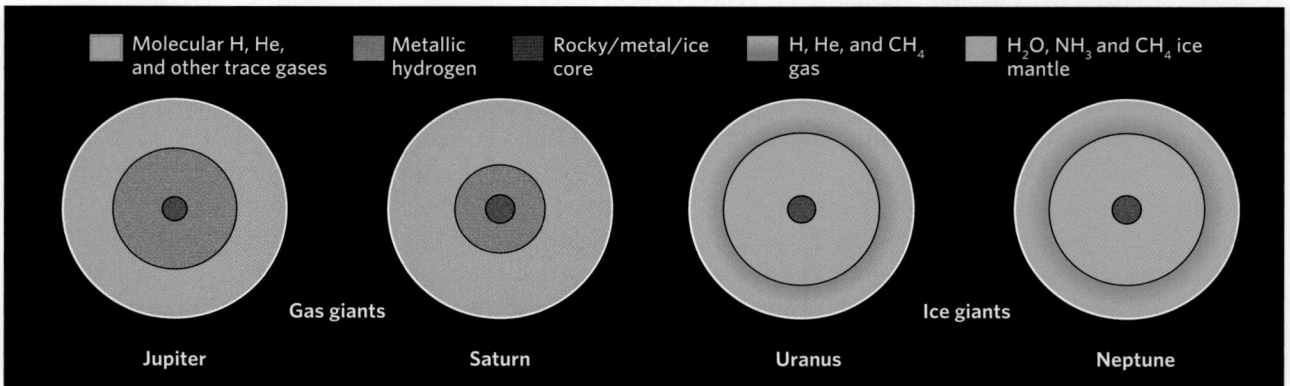

(b) The interiors of the Jovian planets differ from one another, too. In this figure, all of the planets have been resized to the same diameter so as to emphasize the differences in their interior components.

fast rotation, together with the weakness of the materials within the planet, makes the planet's equator bulge, so its equatorial radius (11 times the Earth's) exceeds its polar radius by 7%. Jupiter produces its own internal heat, radiating twice as much energy as it receives from the Sun. This heat does not come from nuclear fusion inside Jupiter, for the planet would have to be a hundred times more massive for fusion to occur. Rather, it's left over from the compression of gas during Jupiter's formation.

THE MOST COLORFUL ATMOSPHERE OF ALL. Even an amateur astronomer looking through a telescope at Jupiter from the Earth can see that the giant planet's atmosphere differs markedly from the Earth's. Orange and tan bands, oriented parallel to the equator, surround the planet **(Fig. 21.21a)**. The composition of these bands remained uncertain until 1994, when *Galileo* entered orbit around Jupiter and dropped a probe into the atmosphere. The probe sent back data for 78 minutes as it descended for a few hundred kilometers, until its instruments succumbed to the intense temperature and pressure of its surroundings.

The *Galileo* orbiter and its probe revealed that Jupiter's outer atmosphere consists primarily of hydrogen (86.1% by mass) and helium (13.8% by mass). The trace gases that constitute the remainder include methane (CH_4), ammonia (NH_3), ammonium hydrosulfide (NH_4SH), and water vapor (H_2O). Different components concentrate in distinct layers. Jupiter has no solid surface, but the gases become denser and denser toward the center of the planet. Researchers arbitrarily define the base of the atmosphere—and, therefore, the "surface" of the planet—as the elevation at which the atmosphere has a pressure of 10 atm (10 times that of the Earth's atmosphere). Temperatures at this elevation reach about 300°C (570°F). The orbiter also discovered that Jupiter has storms that produce lightning discharges. These storms are many times larger than those on the Earth.

The colors of the visible atmospheric bands come from trace elements, such as sulfur and phosphorus, in the cloud layers. Researchers suggest that the bands indicate the presence of alternating upwelling (brighter colors) and downwelling (darker colors) regions of the atmosphere. Due to the Coriolis force, these bands align parallel to the equator. The winds within these bands flow at different speeds and even in different directions, so swirling eddies form at the boundaries between bands **(Fig. 21.21b)**.

A giant ellipse of reddish gas, the **Great Red Spot**, dominates images of Jupiter's southern hemisphere **(Fig. 21.21c)**. Detailed cloud patterns in and around the spot indicate that it's a huge, counterclockwise-rotating storm, big enough to contain two to three Earths. The Great Red Spot has existed for centuries and may be permanent. Its origin and longevity remain a mystery. Other

FIGURE 21.21 Color bands, winds, and storms on Jupiter.

Great Red Spot

(a) Jupiter hosts distinct color bands.

Wind velocity

−250 km/h   0   +250 km/h

West   East

Equator

(b) Wind velocity and direction on Jupiter vary with latitude.

(c) A close-up of the Great Red Spot, a giant storm that rotates counterclockwise. The spot is bigger than the Earth.

huge storms form on Jupiter, but they exist for a shorter time before dissipating or combining with other storms.

WHAT'S INSIDE JUPITER? While the volume of 1,320 Earths could fit inside the volume of Jupiter, the giant planet's mass is only 318 times that of the Earth (see Table 21.2). Jupiter contains mostly hydrogen and helium, whereas the Earth consists mostly of rock and metal, so Jupiter has an average density that's only about one-fourth that of the Earth.

What would we see if we could slice through Jupiter? At the base of the atmosphere, the planet consists of compressed hydrogen and helium gas. Perhaps 100 km (60 mi) farther down, the pressure becomes great enough for hydrogen to become liquid. This liquid hydrogen still consists of H_2 molecules, but the molecules can get close enough to one another to interact and bond. The liquid hydrogen layer extends downward for perhaps 20,000 km (12,400 mi). Liquid helium, which is denser than liquid hydrogen, fell like rain out of the liquid hydrogen layer and accumulated as a relatively thin layer of liquid helium beneath the liquid hydrogen layer. Below the liquid helium layer, pressures reach 3 million atmospheres and temperatures reach 11,000°C (20,000°F). Under these conditions, hydrogen undergoes a transition into a strange material, called **metallic hydrogen**, that doesn't exist naturally on the Earth. (Metallic hydrogen was created in a laboratory for the first time in 2017.) It flows like a liquid, but the atoms have been squeezed together so tightly that they can't hold their electrons in orbit, so the electrons flow freely, as in a metal. The flow of electrons in Jupiter's metallic hydrogen layer generates a very strong magnetic field around the planet and causes brilliant aurorae near its poles. The magnetic field, in turn, traps charged particles to produce intense radiation belts thousands of times stronger than the Earth's Van Allen belts. Because of these radiation belts, space probes exploring Jupiter must have strong shielding to prevent their electronics from frying. The metallic hydrogen layer continues down to a depth of 60,000 km (37,300 mi). Below that depth, researchers speculate that Jupiter has a core consisting of refractory material (rock and metal), which may contain 12–45 times the mass of the Earth, though it is only a small part of Jupiter's mass.

JUPITER'S AMAZING MOONS. Jupiter has 95 moons (**Earth Science at a Glance**, pp. 832–833). The count is complicated because thousands of small objects orbit this planet, and astronomers have to decide which are large enough to be called moons. The four largest, Io, Europa, Ganymede, and Callisto—known as the *Galilean moons* because Galileo observed them in 1610—reach planetary dimensions (**Fig. 21.22**). Ganymede, Jupiter's largest moon, has a diameter 8% larger than that of Mercury and has twice the mass of the Earth's Moon, which makes it the largest moon in the Solar System. The other Galilean moons are comparable to the Earth's Moon in size, but all four differ markedly from our Moon and from one another in several ways.

The surface of Io, the closest Galilean moon to Jupiter, looks like a weirdly colored pizza, with splotches and spots of rich reds, golds, yellows, browns, and blacks (see Fig. 21.22a). The colors come from sulfur and other elements in the ash and lava erupted by hundreds of volcanoes. The *Voyager* space probe caught eight of these volcanoes in the act of erupting. In fact, Io's volcanic activity, in terms of output, exceeds that of the present-day Earth by about 100 times. Beneath the colorful surface, Io probably resembles the terrestrial planets in having a rocky mantle and an iron core. What provides the heat that drives Io's volcanism? The gravitational pull of two other Galilean moons, Ganymede and Europa, forces Io into a highly elliptical orbit around Jupiter. When Io reaches its closest point to Jupiter, the planet's immense gravitational pull produces a 100 m (330 ft) tidal bulge on Io's surface. The same side of Io always faces Jupiter, so when Io reaches its farthest point from the planet, the tidal bulge sinks away entirely. Because Io orbits Jupiter in only 1.8 Earth days, the moon's solid-rock crust tidally bulges up and down by 100 m every 1.8 days. Such movement generates a lot of friction within the moon, which in turn generates internal heat. Volcanoes on Io erupt explosively, constantly resurfacing the moon, so it retains no impact craters.

Unlike Io, the remaining three Galilean moons, Europa, Ganymede, and Callisto, contain significant amounts of water. Ganymede and Europa each appear to have an icy crust up to 25 km thick. These moons' icy crusts surround underground oceans of liquid saltwater, about 100 km (62 mi) thick on Europa and even thicker on Ganymede (**Fig. 21.23**). The crust of Ganymede displays dark regions covered with impact craters and light regions with ridges and grooves (see Fig. 21.22b). Between 50% and 90% of Ganymede's crust is water ice. Recent measurements suggest that Ganymede's underground ocean contains more water than all the oceans on the Earth. Scientists speculate that primitive life might exist in these hidden oceans because the environment may resemble deep-ocean environments on the primitive Earth. A rocky mantle underlies the hidden oceans of both Ganymede and Europa, and this mantle surrounds a metallic core. The core of Ganymede produces a strong magnetic field, so the outer part of its core is molten; consequently, the moon displays bright aurorae. Callisto differs from the other large moons, and from any terrestrial planet, in an important way: its interior did not differentiate into a metallic core and rocky mantle. Rather, the interior, beneath water and ice layers, is a mixture of rock and metal. The lack of differentiation suggests that Callisto accumulated from debris orbiting Jupiter later in Solar System history, when the accreting material was too cool to melt even when squeezed together tightly. Callisto's surface is more heavily cratered than any other object in the Solar System. Higher elevations within complex craters appear as bright peaks on the moon's surface

FIGURE 21.22 Jupiter's large (Galilean) moons.

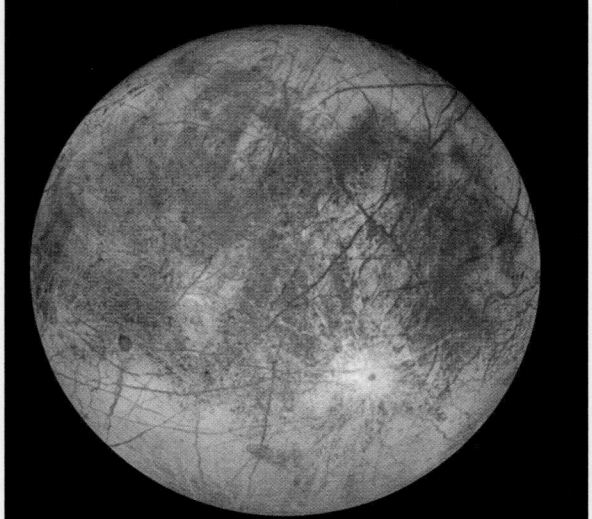

(a) Io has a diameter of 3,636 km (2,259 mi).

(b) Ganymede has a diameter of 5,268 km (3,273 mi).

(c) Europa has a diameter of 3,100 km (1,930 mi).

(d) Callisto has a diameter of 4,821 km (2,995 mi).

FIGURE 21.23 Speculative models of the possible watery interiors of Ganymede and Europa.

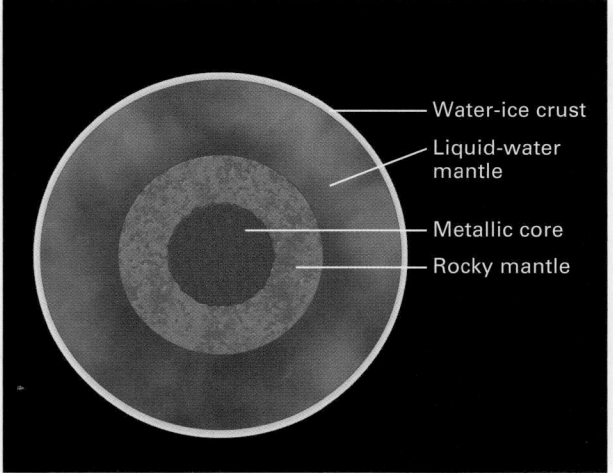

Water-ice crust
Liquid-water mantle
Metallic core
Rocky mantle

(a) Ganymede may have a liquid-water mantle.

Metallic core
Water layer
Rocky mantle

(b) An ocean may exist beneath an ice crust on Europa.

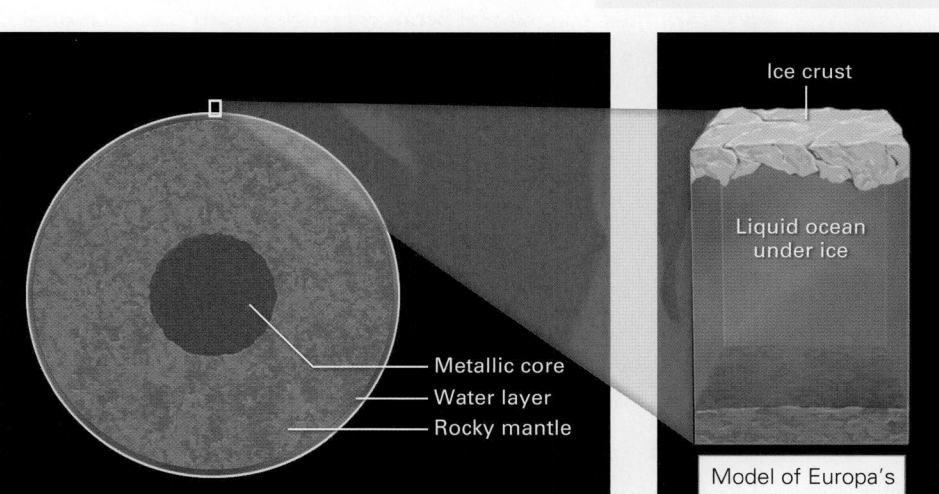

Ice crust
Liquid ocean under ice
Model of Europa's outer layers

A Menagerie of Moons

The moons of the Solar System vary in size, shape, composition, and surface characteristics. Some, like Io, have active volcanoes. Others, like Enceladus, are covered with water ice, and may harbor an ocean beneath the icy surface. Some, like Mimas and Iapetus, are cratered, while others have smooth surfaces. One, Titan, has a dense atmosphere. Many are spherical, but the smallest, such as Deimos, are not large enough for gravity to compress them into spherical shapes. The objects within the curving band are shown to scale.

Deimos,
Mars's small moon

Mimas,
Saturn's cratered moon

Titan

Mercury

Callisto

Io

Moon

Europa

Iapetus,
Saturn's icy and
cratered moon

Triton

Pluto

Titania

Oberon

Tethys

Enceladus

Enceladus,
Saturn's icy
moon

Mimas

Ganymede

Mars

Venus

Earth

Io,
Jupiter's erupting
moon

Titan,
Saturn's moon
with an atmosphere

Europa
Jupiter's
ice-covered moon

FIGURE 21.24 *Cassini* views of the rings of Saturn.

(a) Picture of the rings of Saturn during an eclipse (the Sun is behind Saturn). Note the fainter, more distant rings.

(b) False colors represent different particle sizes. Purple means that particles larger than 5 cm (2 in) are common; green means that particles smaller than 5 cm are common. White indicates dense rings.

Saturn's rings, in contrast to those of Jupiter, are bright enough to be seen from the Earth.

SATURN'S ATMOSPHERE AND INTERIOR. Like Jupiter, Saturn consists mostly of hydrogen (75% by mass) and helium (25% by mass), along with several trace gases. Saturn has the lowest average density of all the planets (0.7 g/cm³, less than that of water). Also like Jupiter, Saturn lacks a solid surface, so researchers arbitrarily define the base of its atmosphere as the elevation where pressure is 10 atm. And, like Jupiter, Saturn rotates so fast—a day on Saturn lasts 10.75 hours—that the planet's equatorial radius exceeds its polar radius (by 11%). Saturn's thick atmosphere of hydrogen and helium gas contains cloud layers that resemble Jupiter's in composition, but Saturn's atmosphere doesn't have convection as vigorous as Jupiter's, so Saturn has duller atmospheric bands and an overall tannish hue **(Fig. 21.24)**. Nevertheless, Saturn has intense winds that race along at up to 1,500 km/h (930 mph).

The core of Saturn, which probably consists of rock and metal, has a mass 10–20 times that of the Earth and a temperature of about 11,700°C (21,000°F). Outside this refractory core, Saturn consists of a layer of dense ice (composed mostly of ammonia and water), a layer of metallic hydrogen, a layer of helium (which probably developed when drops of liquid helium formed and fell like rain toward the interior), and a layer of liquid molecular hydrogen. Because Saturn has less internal pressure than Jupiter does, Saturn's metallic hydrogen layer is thinner than Jupiter's, and Saturn has a weaker magnetic field. Saturn generates about 2.5 times as much energy as it receives from the Sun. While some of this energy comes from compression of the planet's gases, astronomers speculate that some comes from the friction generated by the downward rain of helium.

(see Fig. 21.22d), perhaps due to ice cover. Like Io and Europa, Callisto may have a subsurface liquid ocean.

Jupiter is encircled by four rings of debris, but they can't be seen from the Earth. They were detected for the first time by the *Voyager* space probe in 1979. These rings, which lie in the plane of Jupiter's equator, range from 6,500 to 97,000 km (4,000–62,000 mi) wide and from 30 to 12,500 km (19–7,770 mi) thick. Jupiter's ring system consists of dust and larger fragments that are believed to have been kicked into orbit when interplanetary meteoroids smashed into the planet's four small inner moons.

Saturn: The Ringed Planet

Saturn, named for the Roman god of agriculture, has such a high albedo and is so big that you can see it without the aid of a telescope, even though it's so far away. Saturn resembles its bigger brother Jupiter in many ways, but

Did you ever wonder...

why Saturn has rings?

AMAZING RINGS. When Galileo first looked at Saturn, he saw ear-like protrusions to either side of the planet, but he had no idea what they could be. A half century later, an astronomer using a more powerful telescope recognized the protrusions as rings of matter that surround the planet and have a high albedo. In 1675, Giovanni Cassini showed that Saturn has several distinct rings extending over 100,000 km (62,000 mi) out into space. No other planet in our Solar System displays rings as bold and beautiful as Saturn's.

Our understanding of the rings improved greatly when the *Cassini* space probe sent back high-resolution photographs revealing fainter rings, which extend beyond the main rings, and showing that each main ring consists of hundreds of ringlets 10–15 m (32–50 ft) wide, with tiny gaps in between. The rings, which are exceptionally thin (only 5–30 m, or 2–10 ft, in thickness), consist of countless particles of water ice and rocky dust, ranging in size from less than a millimeter to a few meters wide. Why does Saturn have such pronounced rings? Astronomers suggest that the rings were formed by tidal forces acting on a moon or moons that once orbited fairly close to the planet's surface. These close-in object(s) were subjected to such strong tidal forces that eventually, they shattered into fragments, which dispersed into rings. Some gaps between rings exist because they have been cleared by small moonlets, bodies 10–20 km (6–12 mi) in diameter that gravitationally vacuum in debris. Other gaps have formed because Saturn's moons act as shepherds that gravitationally "herd" particles into rings and out of gaps.

THE MANY MOONS OF SATURN. Though moons orbiting close to Saturn's surface were shattered by tidal forces, Saturn still hosts 146 known moons. Most of these moons are quite small; only 13 have diameters larger than 50 km (30 mi). Most of Saturn's smaller moons don't spin on an axis, but rather tumble through space as they follow their orbit. Astronomers suggest that these objects represent either captured planetesimals or fragments produced by collisions of planetesimals.

Each of Saturn's seven largest moons—those that grew big enough to become roughly spherical—has unique characteristics **(Fig. 21.25a)**. Titan, the largest by far, has a diameter larger than Mercury's and an atmosphere 50%–60% denser than the Earth's at ground level. Unlike the atmosphere of Saturn itself, Titan's atmosphere consists of nitrogen (98%), methane (2%), and other trace gases, including a variety of hydrocarbons. It appears as an orange haze that prevents space probes from taking photographs of Titan's surface. Images made by the *Huygens* space probe, which parachuted to the surface of Titan in 2005, suggest that methane clouds hover in Titan's lower atmosphere, and that liquid methane rains from these clouds, collecting to form hydrocarbon lakes and rivers

FIGURE 21.25 The moons of Saturn.

Mimas (363 km)

Titan (5,152 km)

Enceladus (505 km)

Dione (1,122 km)

(a) Four of Saturn's 62 known moons.

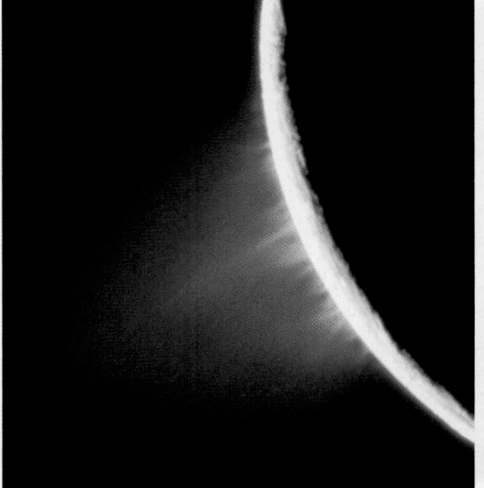

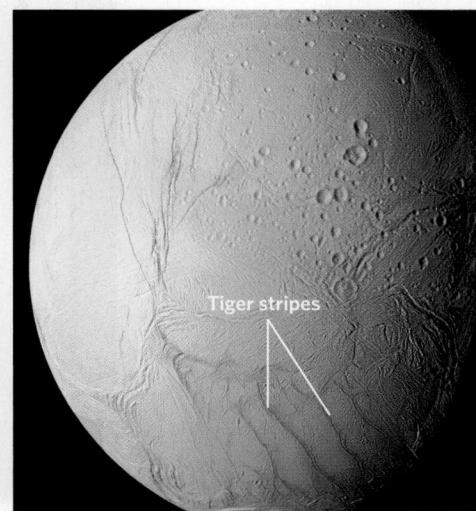
Tiger stripes

(b) Jets of volatiles (left) erupt from the "tiger stripes" of Enceladus's south pole (right).

on the surface. Radar mapping reveals volcanic vents that emit gases, as well as huge dunes, presumably composed of ice and frozen methane grains. Titan's surface is mostly made of water ice, and scientists believe that under its surface, Titan has a liquid-water ocean.

The next six largest moons (Rhea, Iapetus, Dione, Tethys, Enceladus, and Mimas) are much smaller than Titan. None have atmospheres, and most appear to be similar to one another in composition, with rocky interiors beneath a water-ice shell. Regions on the surfaces of most moons have high crater densities, suggesting that these areas have been tectonically inactive since the early days of the Solar System. But some of the moons display cracks or ridges indicative of tectonic activity in the past, and one, Enceladus, has been substantially resurfaced by younger material. Photographs of Enceladus reveal geysers of water vapor and ice crystals erupting from a set of beautiful blue fissures, informally called tiger stripes, near the moon's south pole **(Fig. 21.25b)**. The *Cassini* space probe discovered that some of the icy particles jetting out of Enceladus leave the moon fast enough to populate one of Saturn's rings that coincides with Enceladus's orbit. *Cassini* also revealed that Enceladus might have a habitable internal ocean, which led to speculation that it might host primitive life forms, but this proposal remains unsubstantiated.

Uranus and Neptune: The Ice Giants

DISCOVERY AND BASIC CHARACTERISTICS. Uranus and Neptune lie so far from the Earth that they cannot be seen with the naked eye, even on the darkest night. Astronomers using telescopes first detected Uranus in 1781 and named it after the Greek god of the sky. Irregularities in Uranus's orbit, which could only have been caused by the pull of yet another planet farther out, led astronomers to find Neptune in 1846. Uranus and Neptune are both much larger than the Earth—their radii are 4.0 times and 3.9 times the Earth's radius, respectively—but they are only about one-third the size of Jupiter. The first detailed information about Uranus and Neptune became available when the *Voyager 2* space probe passed them in 1986 and 1989.

Uranus's axis of rotation has a tilt of 98° (see Fig. 21.4), so it lies almost parallel to the planet's orbital plane! In effect, Uranus lies on its side. As a result, Uranus has the most extreme seasons of any planet in the Solar System. During its orbit, which takes 84 Earth years, its polar regions remain in complete darkness for 42 years and then in constant sunlight for 42 years. Neptune's axis also tilts, but at an angle of 28°, just slightly greater than that of the Earth's axis. Both planets rotate faster than the Earth: a day lasts 17.2 hours on Uranus and 16.1 hours on Neptune.

INTERIORS AND ATMOSPHERES OF THE ICE GIANTS. On Uranus and Neptune, interior pressures don't become great enough for metallic hydrogen to form, as on Jupiter and Saturn. The relatively small rocky and metallic interiors of Uranus and Neptune appear, instead, to be surrounded by a thick layer of slush composed of water, ammonia, and methane ice. Most of the planet's mass (80% or more) is made up of this hot dense slush. A layer of molecular hydrogen, helium, and methane surrounds the slush (see Fig. 21.20b). The presence of icy slush inside Uranus and Neptune led to their designation as ice giants.

From space, all we can see when looking at Uranus is the thick, frigid (−215°C, or −355°F) haze that forms the top of its atmosphere, so Uranus appears as a teal-colored globe without banding or storm spots **(Fig. 21.26a)**.

FIGURE 21.26 The ice giants.

(a) The atmosphere of Uranus appears featureless.

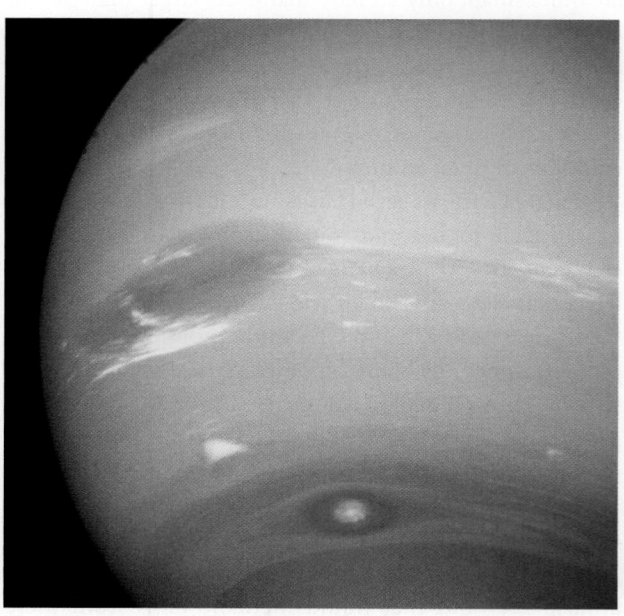

(b) Neptune's atmosphere displays bands and storms.

FIGURE 21.27 Rings of the ice giants.

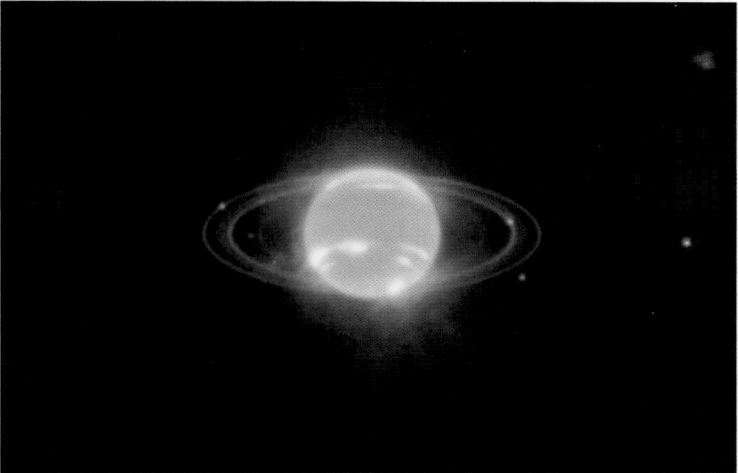

(a) The rings of Uranus. These rings could not be seen using visible light sensors without image enhancement. However, using the new infrared sensors on NASA's James Webb Space Telescope, the rings stand out much like Saturn's rings (see the chapter opening photo).

(b) The rings of Neptune. This image, too, is from the James Webb Space Telescope.

Neptune, in contrast, has a deep-blue atmosphere with identifiable features, including bands and white streaks caused by clouds (Fig. 21.26b). Uranus's and Neptune's atmospheres consist primarily of hydrogen (84% by volume) and helium (14% by volume). The concentration of methane determines the richness of the blue color of these planets, so Neptune (with 3% methane) appears deep blue, while Uranus (with 2% methane) appears paler.

STILL MORE MOONS AND RINGS. At present, astronomers have identified 27 moons orbiting Uranus and 14 orbiting Neptune, as well as thin rings around each planet (Fig. 21.27). The five largest moons of Uranus are all smaller than the Earth's Moon. They appear to be composed of ice and rock and are heavily cratered (Fig. 21.28a). Neptune has one large moon, Triton, which has a diameter of about 2,700 km (1,680 mi), and is the only large moon in the Solar System to orbit in a retrograde direction (Fig. 21.28b). The frigid −236°C (−393°F) surface of Triton consists of water ice with polar caps of nitrogen ice.

FIGURE 21.28 Moons of the ice giants.

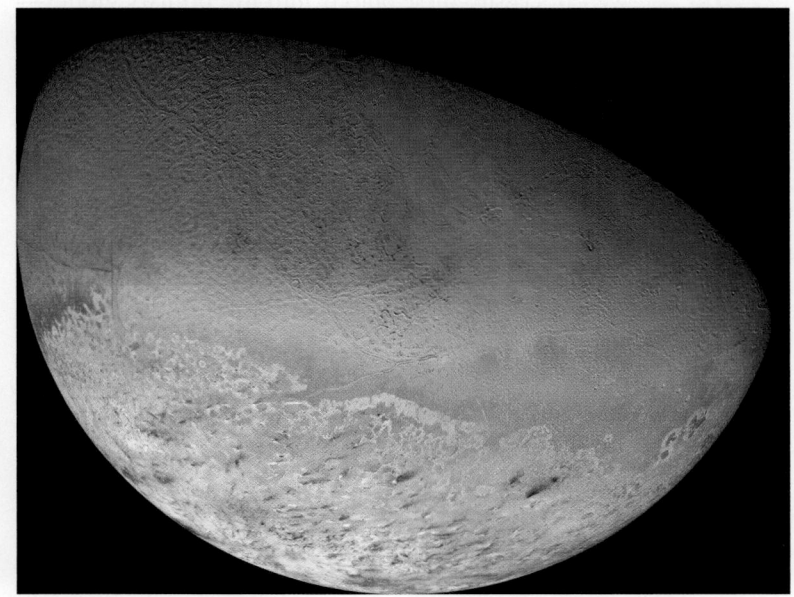

(a) Titania, a moon of Uranus.

(b) Triton, a moon of Neptune.

21.5 Other Objects within the Solar System

The eight planets and their moons represent the vast majority of the nonsolar mass in the Solar System, but only a tiny fraction of the total number of objects. Researchers estimate that billions of other objects—nearly all too tiny or too far away to observe—move with the Sun as it orbits the Milky Way's galactic center. We can classify these objects into five primary groups—asteroids, Kuiper Belt objects, Oort Cloud objects, comets, and meteoroids—that differ from one another in terms of size, orbital characteristics, and location relative to the planets. (As we've seen, astronomers now use the term *dwarf planet* for larger, spherical asteroids or Kuiper Belt objects.) An understanding of these objects not only enhances our understanding of how the Solar System developed, but also allows astronomers to evaluate the small, but real, risk that a large object may someday collide with our planet.

Asteroids

By convention, astronomers consider a rocky or metallic object larger than about 100 m (328 ft) across to be an **asteroid**. There are many millions of small asteroids—about 1.5 million have diameters greater than 1 km (0.6 mi), about 200 have diameters greater than 100 km (62 mi), and only 4 have diameters greater than 400 km (249 mi). Despite their great numbers, the combined mass of all the asteroids equals only about 3% of the Moon's mass. Most asteroids lie within the **asteroid belt**, a band between the orbits of Jupiter and Mars **(Fig. 21.29a)**. Those asteroids not within the asteroid belt include the *Trojans* and *Greeks*. These asteroids lie in the orbit of Jupiter, but because of a balance between the gravity of the planet and the gravity of the Sun, they are locked in a position that prevents them from colliding with the planet. Asteroids outside the asteroid belt also include four small groups closer to the Earth. Two of these groups, known as the *Atens* and *Apollos*, have been nudged by Jupiter's gravity into paths that cross the Earth's orbit. Astronomers have so far identified approximately 1,500 Atens and 10,500 Apollos, of which about 2,000 may represent a threat to our planet. An individual Aten or Apollo asteroid survives for only a few million years before it collides with a planet or is gravitationally thrown out of its orbit, but the tug of Jupiter's gravity on other objects constantly replenishes the supply.

Researchers who have studied meteorites derived from asteroids recognize three distinct compositional classes of asteroids. *Chondritic asteroids*, the most abundant type, are planetesimals or fragments of planetesimals that never attached to others during the formation of the Solar System. Therefore, they represent the primitive solid material from which the terrestrial planets and the solid deep interiors of the giant planets formed. *Stony asteroids* are composed primarily of silicate minerals, and *metallic asteroids* are composed of iron alloy. Stony and metallic asteroids probably represent fragments of planetesimals that were large enough (>400 km, or >250 mi, in diameter) to warm up inside, soften, and differentiate into a metallic core and a rocky mantle before breaking apart during collisions with other bodies.

Three asteroids are large enough to have become spherical. Of these, Ceres, the largest asteroid, has a diameter of 945 km (587 mi) and qualifies as a dwarf planet **(Fig. 21.29b)**. Ceres probably has a rocky interior, surrounded by a shell of water ice, which in turn is coated with a surface layer of micrometeorite dust. The *Dawn* space probe went into orbit around Ceres in 2015 and sent back images of its intensely cratered surface, on which a highly reflective mountain stands out. This mountain may represent an eruption of ice. Almost all other asteroids are irregular in shape because they are too small to have been remolded into spheres by gravity. Some irregularly shaped asteroids are roughly equant, but some are distinctively elongate **(Fig. 21.29c, d)**. Space probes have taken close-up images of a number of asteroids. In some images, asteroids appear to be conglomerations of many smaller pieces **(Fig. 21.29e**; also see Figure 1.9d). In 2001, a space probe landed on Eros, and one landed on Itokawa, grabbed a sample, and in 2010, returned it to the Earth.

Pluto, Other Kuiper Belt Objects, and the Oort Cloud

Even before the dramatic vote demoting Pluto from planet to dwarf planet, astronomers realized that Pluto was an oddball among the planets. Most notably, Pluto moves differently. Its orbit not only lies at an angle of 17° relative to the ecliptic plane, but is also highly eccentric: at its aphelion, Pluto lies 1.7 times farther from the Sun than it does at its perihelion. In fact, for part of a Pluto year, Pluto lies closer to the Sun than Neptune does, although their orbits never cross.

Pluto's surface characteristics remained unknown until 2015, when the *New Horizons* probe sent back amazingly detailed photos **(Fig. 21.30a, b)**. The photos reveal that this small world, with a diameter about a fifth of the Earth's, has a surface with plains that display complex textures resembling snakeskin and mountains that rise to elevations of 3.5 km (2 mi) (see Fig. 21.2). Pluto has a low density (1.9 g/cm³), which results in a mass that's about 0.2% of the Earth's; it probably consists mostly of water ice and frozen nitrogen. Five moons orbit Pluto, of which the largest, Charon, has a diameter of 1,300 km (800 mi) **(Fig. 21.30c)**. All the moons appear to consist of water ice.

Pluto is one of several objects with diameters in the range of 1,000–2,500 km (600–1,500 mi) that lie outside the orbit of Neptune. One of these, Eris, may contain more mass than Pluto. Pluto and Eris, both dwarf planets, are the largest known objects of the **Kuiper Belt**, named for the astronomer Gerard Kuiper (1905–1973). These objects are icy bodies that circle the Sun in a disk-shaped region that extends from the orbit of Neptune (30 AU) out to a distance of about 50 AU from the Sun. Millions of these bodies are tiny, but perhaps over 100,000 have a diameter larger than 100 km (60 mi). Overall, many objects in the Kuiper Belt orbit near the ecliptic plane, but many other individual objects, such as Pluto, have orbits that are inclined to the ecliptic plane.

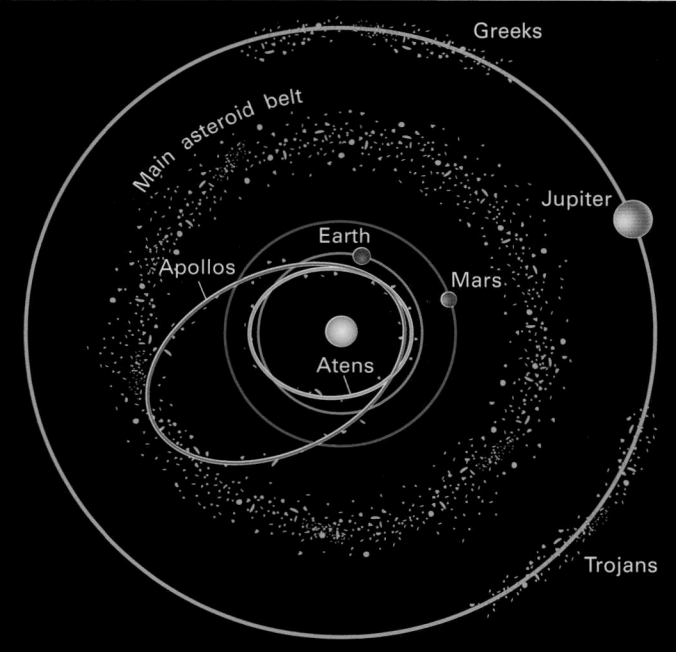

FIGURE 21.29 Asteroids and their distribution.

(a) The distribution of asteroids. The Apollo and Aten asteroid groups follow orbits that cross that of the Earth.

(b) The largest asteroid, Ceres, which is spherical and considered a dwarf planet, has a diameter of 945 km (587 mi).

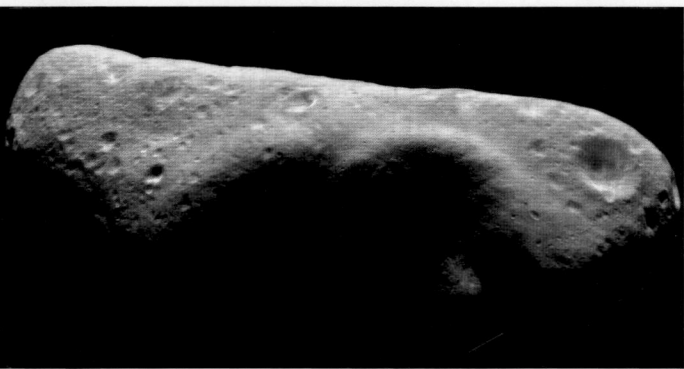

(c) Some asteroids, such as Eros, are oblong. Eros is 21 km (13 mi) long.

(d) Lutetia, a nonspherical asteroid, has a diameter of 121 km (75 mi).

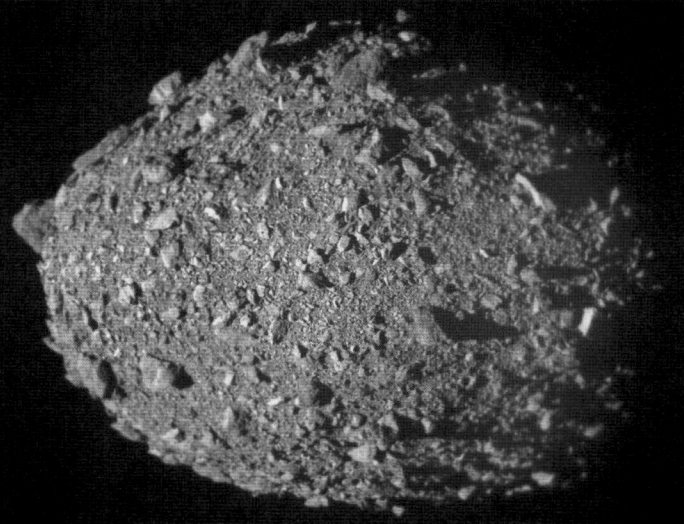

(e) Some asteroids, such as Dimorphos, are conglomerations of small pieces. Dimorphos, which orbits another, larger asteroid, is about 177 m (580 ft) long. Seconds after taking this photo, NASA's DART space probe slammed into Dimorphos. The planned collision tested the idea that such an impact could change the orbit of an asteroid.

FIGURE 21.30 *New Horizons* images of Pluto and Charon.

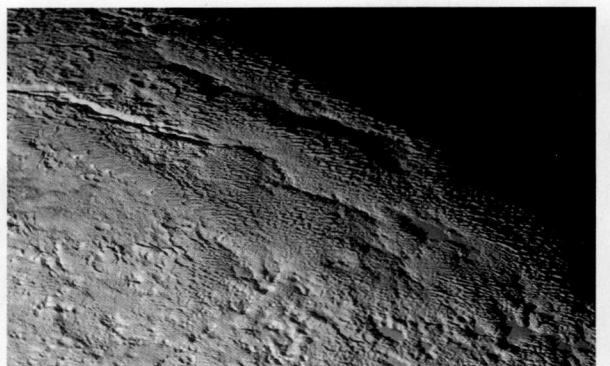

(a) Pluto's surface displays a variety of textures and colors. The field of view in this image is about 330 km (200 mi) across.

(b) A close-up of the surface shows blocky mountains and hexagonally fractured plains. The field of view in this image is about 80 km (50 mi) across. Note that the plains are covered by dunes of ice particles.

(c) Charon has few impact craters, suggesting that the moon has been resurfaced. The canyon visible on the lower half of this photo is over 1,000 km long and 3 km deep.

In 1950, a Dutch astronomer, Jan Oort (1900–1992), proposed that even more icy objects lie beyond the Kuiper Belt but remain gravitationally attached to the Sun. Researchers envision these objects as occupying a somewhat spherical shell, now called the **Oort Cloud**. Its inner edge lies at a distance of about 2,000 AU from the Sun, and its outer edge extends to a distance of 100,000 AU (1.5 light-years), more than a quarter of the distance to the nearest star. The Oort Cloud probably consists of debris that condensed during Solar System formation but never fell into the accretion disk.

Comets

Comets mystified ancient observers because, unlike other celestial objects, they produce a glowing tail as they travel across the sky, relative to the backdrop of stars, over the course of weeks to months **(Fig. 21.31a)**. A **comet** differs from an asteroid because it follows a highly elliptical orbit with the Sun at one focus **(Fig. 21.31b, c)**. Modern space probes have provided spectacular images of comets **(Fig. 21.31d)**. In 2005, debris blasted off a comet by the *Deep Impact* probe was analyzed remotely, and in 2014, the *Rosetta* probe went into orbit around a comet and successfully placed a lander on its surface.

A comet consists of a relatively loosely bound aggregate of rock, dust, and various ices (water ice, frozen methane, dry ice, and frozen ammonia). It emits gas and dust, which form tails when the comet approaches the Sun inside a radius of about four times the Earth's orbit. At that distance, solar radiation causes the comet's interior to undergo **ablation**, the erosion of material from an object by any externally driven process **(Fig. 21.31e)**. In the case of comets, solar radiation causes ablation because refractory materials (dust) that were cemented onto the comet by ices become loose and get carried away with gases as they seep out of cracks in the comet (a process called *outgassing*). Various ices also sublimate (transform directly into gas), which can loosen the dust particles and contribute to overall erosion. Together, the gases and dust form a haze called a *coma*, which surrounds the comet's solid *nucleus*. Gases derived from the coma form an *ion tail* whose orientation lies parallel to the direction of the solar wind, so that it always points away from the Sun. The dust forms a *dust tail* that tends to curve, because the dust particles in it partly follow the comet's orbit but are also affected progressively by the solar wind. Some tails can be longer than the distance from the Earth to the Sun. The closer a comet comes to the Sun, the longer and brighter its tails become. The formation of tails causes the comet to ablate each time it approaches the Sun. Small comets probably have little chance of survival beyond a few orbits, and even a large comet probably disintegrates after a few thousand orbits.

Because comets have such short lifetimes, relative to the age of the Solar System, a source must exist that, over time, sends new icy bodies into the inner Solar System. Researchers hypothesize that gravitational interactions can eject objects from both the Kuiper Belt and the Oort Cloud into orbits that take them close enough to the Sun so that they become comets. Astronomers classify comets based on the time they take to orbit the Sun. *Short-period comets*, which complete an orbit in less than 200 years, probably come from the Kuiper Belt, while *long-period comets*, which complete an orbit in more than 200 years (in some cases, thousands of years), probably come from the Oort Cloud. Notably, 106 existing comets have orbits that bring them relatively close to the Earth's orbit.

Did you ever wonder...

why a comet has a tail?

FIGURE 21.31 Comets consist of ice, rock, and dust.

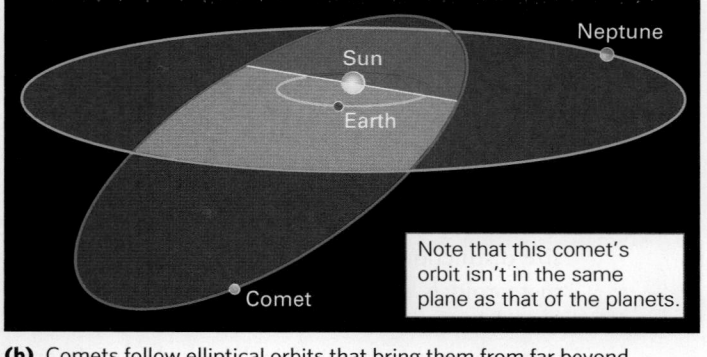

(a) The comet Hale-Bopp, a particularly dramatic comet, was visible for 18 months beginning in May 1996.

(b) Comets follow elliptical orbits that bring them from far beyond Neptune into the inner Solar System.

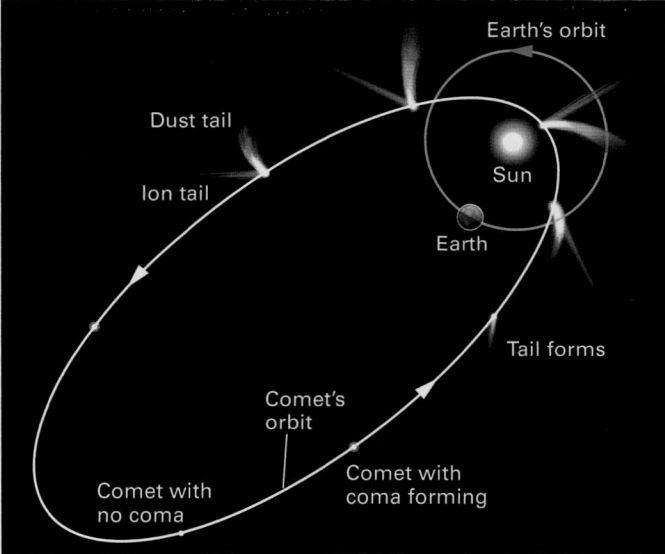

(d) The irregular solid nucleus of a comet.

(c) When a comet approaches the Sun, it develops two tails. The ion tail points straight away from the Sun, whereas the dust tail curves along the comet's orbit.

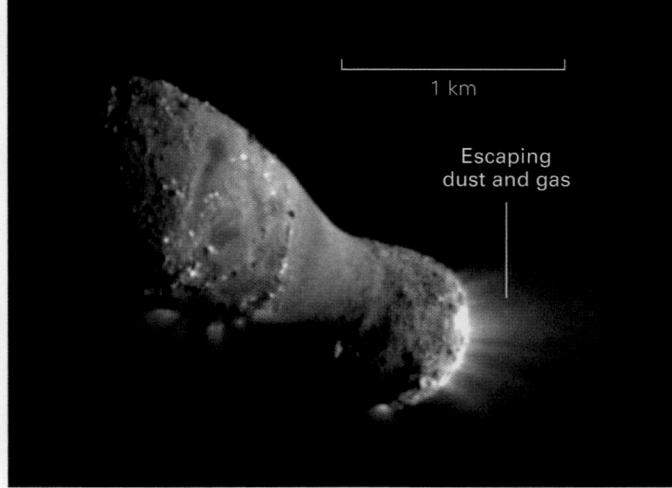

(e) Outgassing releases gas from a comet, which also sheds dust.

When Celestial Objects Collide with the Earth

THE SIZE AND SPEED OF CELESTIAL OBJECTS. Science-fiction TV shows and movies sometimes depict spacecraft of the future passing through asteroid belts. As red lights flash and alarm bells clang, spinning boulders slowly bounce and roll off the hull. Such images stray very far from reality. The chunks of rock and/or metal that fly through space do so at speeds of 40,000–250,000 km/h (25,000–160,000 mph). At such speeds, a space chunk travels 11–72 km (7–45 mi) in just 1 second—you wouldn't even see it go by. While the hulls of real spacecraft, such as the International Space Station (ISS), can handle impacts of objects that are less than 1 cm (0.4 in) in diameter, anything larger will make a hole, and a boulder-sized chunk would probably cause complete destruction.

Clearly, size and composition, as well as speed, matter when we consider the energy represented by a moving celestial object. Astronomers make the distinction between asteroids (large solid objects composed of rock and/or metal), comets (large solid objects composed of ice and rock), and **meteoroids** (small solid objects, regardless of composition, that are less than about 1 m, or 3 ft, in diameter), as noted in Table 21.1. Notably, the size distinction between a small comet or asteroid and a meteoroid is somewhat arbitrary and depends on context. Meteoroids include small asteroids or comets that were never incorporated in larger objects, fragments of larger objects produced during collisions, and fragments ejected into space from planets or moons by the impact of another object. Recent evidence suggests that rare, very high-speed meteoroids may come from outside the Solar System.

METEORS AND METEORITES. A meteoroid whose path crosses the Earth's orbit may collide with the Earth. The speed of collision depends, in part, on the velocity the meteoroid attained when entering an orbit that crosses the Earth's, and in part on whether it has a component of motion in the same or in the opposite direction as the Earth's orbital motion. On average, meteoroids reach the Earth at a speed of around 20 km/s (50,000 mph)—100 times the speed of a jet plane. By comparison, the ISS orbits the Earth at a speed of about 8 km/s (18,000 mph). When a meteoroid enters the Earth's atmosphere, it compresses the air in its path very quickly and intensely **(Fig. 21.32a)**. Compression of a gas heats it up, so the meteoroid generates immense heat, enough to cause some or all of the meteoroid to vaporize. As this happens, some of the meteoroid undergoes ablation, both because sublimation loosens particles, and because solid particles become red-hot and spall off (meaning that they split or peel off the meteoroid in small fragments). Glowing gases from the meteoroid vapor, along with glowing superheated air molecules and glowing solid particles, produce a blazing light streak behind the object. The light streak is called a **meteor** (or, misleadingly, a "shooting star"); by this definition, a meteor is an atmospheric phenomenon. To produce a meteor, an incoming meteoroid must be larger than a sand grain. (Note that frictional heating does

FIGURE 21.32 Meteors.

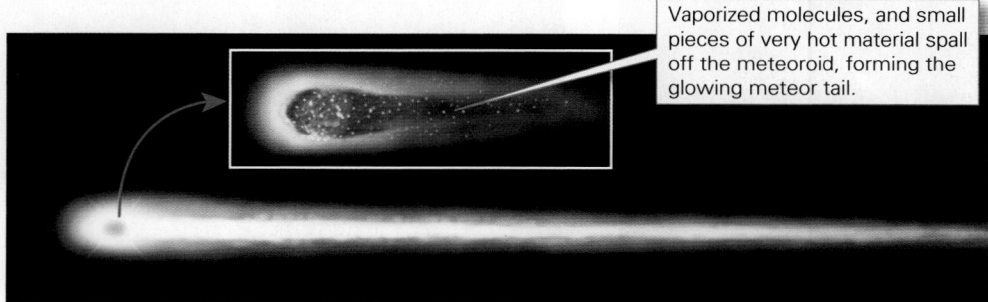

Vaporized molecules, and small pieces of very hot material spall off the meteoroid, forming the glowing meteor tail.

(a) At the head of a meteor is a solid meteoroid undergoing sublimation and spalling. The temperature is so high that the resulting vapor and dust glow.

(b) Several meteors of the Perseids meteor shower. Note that all the meteors appear to be originating from the same point.

(c) The Chelyabinsk bolide explosion was as bright as the Sun.

not produce most of a meteor's light, for air in the upper atmosphere isn't dense enough for friction to have much of an effect.)

During a **meteor shower**, an observer can see from 10 to over 100 meteors per hour, all of which appear to emerge from a common point in the sky **(Fig. 21.32b)**. These showers, which may last as long as 1–3 days, happen when the Earth passes through the orbit of a comet and intersects its dust tail. The Earth passes specific cometary orbits at the same time every year. Occasionally, meteor showers become *meteor storms*, during which over 1,000 meteors streak across the sky every hour. Not all meteoroids arrive in showers or storms, however; astronomers refer to those that arrive independently as *lone meteoroids*.

When a large meteoroid enters the atmosphere, it becomes a very bright *fireball* that leaves a visible smoke-like trail across the sky. If it explodes, a fireball becomes a *bolide*. During the last century and a half, two bolides have exploded over Siberia. The first, known as the Tunguska bolide, released about 10 megatons of energy—over 650 times more than that released by the Hiroshima atomic bomb—and flattened about 2,100 km^2 (830 mi^2) of Russian forest—an area almost twice the size of New York City. The bolide was probably about 120 m (400 ft) across and blew up at an altitude of 5–10 km (3–6 mi). A similar but smaller (0.5-megaton) bolide explosion took place over Chelyabinsk, Russia, in 2013 **(Fig. 21.32c)**. Dashboard video cameras recorded the streak of the fireball's tail, the blinding light of its explosion, and the chaos caused when its shock wave blew out windows and knocked down walls.

If a meteoroid makes it through the atmosphere without completely burning up, it strikes the Earth. The meteoroid then becomes a **meteorite (Fig. 21.33)**.

FIGURE 21.33 Examples of meteorites.

100 mm

(a) A chondrite, a type of stony meteorite, contains chondrules.

2 cm

(b) Metal crystals inside an iron meteorite.

20 cm

(c) An iron meteorite with an ablated surface.

10 cm

(d) A stony-iron meteorite contains dark metal and lighter-colored silicate minerals.

Researchers recognize three classes of meteorites, distinguished from one another by composition: 93% are *stony meteorites*, which consist of silicate rock; 6% are *iron meteorites*, which consist of metallic iron-nickel alloy; and 1% are *stony-iron meteorites*, which contain both metal and rock. Most stony meteorites (87%) are *chondrites*, so named because they contain colorful, rounded fragments, known as *chondrules*, that represent some of the original material from which the Solar System formed. The remaining stony meteorites are formed from the mantles of differentiated asteroids, and a small minority are fragments of the Moon or Mars. Iron meteorites and stony-iron meteorites are also fragments, probably from the cores of differentiated bodies. Meteorites typically have lumpy shapes and surfaces smoothed by the process of ablation.

CONSEQUENCES OF COLLISION. Meteoroids actually collide with the Earth quite frequently. The damage caused by an impact depends on the size, speed, composition, and impact angle of the incoming object; for example, larger, denser, and faster objects do more damage. By one estimate, about 100 million to 1 billion kg of space material fall to the Earth, on average, in a year. (That's roughly the weight of two very large cruise ships.) Most of this material consists of *micrometeorites* no bigger than a sand grain or a dust flake so these collisions don't make headlines. Some micrometeorites are tiny when they arrive at the top of the Earth's atmosphere, whereas others are fragments that have spalled off larger meteoroids that blazed through the atmosphere. Marble-sized to softball-sized meteoroids slow down due to air resistance, so they strike the Earth at a speed of only 320–1,300 km/h (200–800 mph). Such objects can excavate small indentations when striking the ground, or, in the rare cases when they land in populated areas, punch holes through roofs or crush the trunk of a car.

How big does a meteoroid need to be to excavate a recognizable impact crater? A meteoroid weighing about 1,000 kg (1.1 tons), equivalent in volume to a sphere with a diameter of about 0.3 m (1 ft), does not slow down significantly as it passes through the atmosphere. If it does not break apart before impact, it can produce an impact crater on the order of 20–30 m across and 2–3 m deep (that is, about 10 times the diameter of the meteoroid). The 1.2 km (0.75 mi) diameter Meteor Crater in Arizona was formed about 50,000 years ago by the impact of a meteorite that was 50 m (160 ft) in diameter, weighed about 11,000 kg (12,000 tons), and was traveling at a speed of about 48,000 km/h (30,000 mph) **(Fig. 21.34a)**. The largest impact crater on the Earth,

the Vredefort crater, was formed in South Africa about 2 billion years ago by the impact of an asteroid 15 km (9 mi) in diameter. Before erosion, the crater was about 300 km (200 mi) across (about 20 times the diameter of the asteroid). Immediately after impact, the crater was about 40 km deep, meaning that the impact excavated crust down to the Moho. But moments later, the floor of the crater rebounded and rose so much that it formed a bulge, known as a *central uplift*, in the center of the crater. The impact also caused layers of rock beneath and around the crater to tilt steeply, and produced normal faults with circular traces that surround the crater like rings; displacement on these faults dropped crust down toward the interior of the crater.

Recall that relatively small impact craters such as Meteor Crater, are known as *simple craters*, in that they are simply bowl-shaped depressions underlain by fractured rock and surrounded by an ejecta blanket **(Fig. 21.34b)**. Large impact craters that contain a central uplift and are surrounded by ring-shaped faults are called *complex craters* **(Fig. 21.34c)**. The rock beneath large simple craters and all complex craters contains *shatter cones*, nests of cone-shaped fractures (see Box I.2), as well as glass formed when melts produced by the impact fill cracks and quickly freeze. Very large impacts, such as the one that formed the Vredefort crater, also trigger melting in the underlying mantle. These melts rise and solidify to produce igneous rocks, some of which contain rich metal ores.

Very large impacts have left their mark in the geologic record. For example, the impact of an asteroid 10 km (6 mi) in diameter at the site of what is now the Yucatán Peninsula produced an impact crater 180 km wide and 20 km deep. The impact sent immense amounts of debris into the atmosphere, caused forest fires worldwide, and swamped coastlines with tsunamis. As a consequence, it caused the mass-extinction event (including the loss of all dinosaur species except those that evolved into birds) that defines the Cretaceous-Paleogene boundary (see Box 9.2). Such events can be called *extinction-level catastrophes*.

Even smaller impacts could be devastating. For example, if an asteroid 500 m in diameter, meaning one about four times as large as the one that blew up as the Tunguska bolide, were to strike a major city today, much of the city would be replaced by an impact crater, and the blast produced by the impact would flatten buildings in the surrounding area out to a distance of 100 km (60 mi) **(Fig. 21.34d)**. For this reason, governments are beginning to invest in efforts to detect **near-Earth objects**—comets, asteroids, and large meteoroids—that cross the path of the Earth and, therefore, have the potential to collide with our home planet, and to test

FIGURE 21.34 Meteoroid impacts.

(a) Aerial view of Meteor Crater, Arizona, which is 1.2 km (0.75 mi) across.

Simple crater

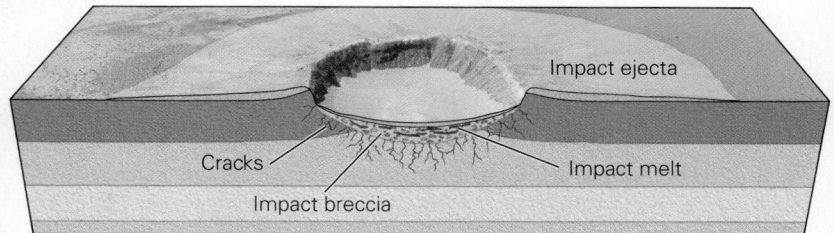

(b) Formation of a simple crater during a small impact.

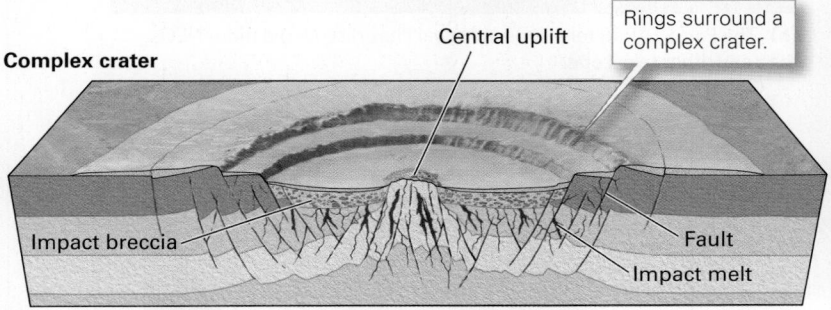

Rings surround a complex crater.

(c) Formation of a complex crater during a large impact.

Small asteroid striking a city

(d) If an asteroid 500 m wide were to land on the Empire State Building, it would produce an impact crater 10 km (6 mi) wide. At a distance of 100 km (62 mi), wind from the blast would be blowing at 400 km/h (250 mph), and the heat would be so intense that everything would instantly ignite.

approaches that might deflect them before they reach the planet (**Box 21.2**).

The Heliosphere, and Beyond

The *solar wind*, a continuous stream of plasma consisting of charged particles (primarily protons, helium nuclei, and free electrons), flows outward from the Sun at about 1.6 million km/h (1 million mph). Because its charged particles are flowing, the solar wind produces a magnetic field. As the solar wind blows through the Solar System, the particles slow due to the Sun's gravitational pull. In addition, the particles spread out into a progressively larger volume, so the number of particles per unit volume becomes less. Consequently, the pressure produced by the solar wind, as well as the strength of the magnetic field it produces, decreases with increasing distance from the Sun. Like the Sun, other stars spew stellar winds,

and some of this material moves off into space toward our Solar System. Astronomers refer to particles zooming through space from sources outside the Solar System as the **interstellar medium**. The interstellar medium has a density of only 1,000 atoms per liter; in contrast, interplanetary space contains 5,000 to 100,000 atoms per liter, and the Earth's atmosphere at sea level contains 2.5×10^{22} molecules per liter. Nevertheless, the interstellar medium exerts pressure and produces a magnetic field, both of which appear to be relatively constant everywhere.

Our Sun's solar wind blows toward the interstellar medium. But because the solar wind not only slows, but also spreads out over a greater volume, as it moves away from the Sun, its density decreases with distance from the Sun. Therefore, the pressure produced by the solar wind at a great distance from the Sun becomes small enough that it equals the pressure produced by the interstellar

BOX 21.2 Putting Earth Science to Use

Early detection of near-Earth objects

Could a near-Earth object (NEO)—defined as an asteroid, comet, or large meteoroid that comes within 50 million km (31 million mi) of the Earth's orbit—collide with the Earth sometime during the next century? To answer this question, the US Congress commissioned a study in 1992 to clarify the risks of such an impact. This study, known as the *Spaceguard Survey Report*, required NASA to locate 90% of NEOs with diameters greater than 1 km (0.6 mi). When the fragments of comet Shoemaker-Levy 9 collided with Jupiter in 1994, producing huge explosions visible to observatories on the Earth, public awareness of the potential for a catastrophic impact increased, and the International Astronomical Union encouraged Spaceguard-like efforts worldwide. In 2005, Congress expanded the Spaceguard Survey goals and asked NASA to identify 90% of potentially hazardous objects with a diameter greater than 140 m (459 ft). Efforts continue under the auspices of the Center for Near-Earth Object Studies, overseen by NASA's Planetary Defense Coordination Office, which is also responsible for issuing warnings about possible impacts. Several large telescopes have been employed in the effort **(Fig. Bx21.2a)**. In the years since the systematic search for NEOs began, astronomers believe they have documented over 97% of objects larger than 1 km, and almost 90% of those larger than 140 m **(Fig. Bx21.2b, c)**. Over 20,000 NEOs have been identified, and new ones are discovered every year **(Fig. Bx21.2d)**. About 10% of NEOs might represent a threat to our planet, and of these, approximately 5,000 have diameters greater than 300 m (984 ft). The impact of one of these objects would produce a crater about 8 km (5 mi) across.

To track NEOs, astronomers (including amateurs) compare photographs of the night sky taken at different times to see if any of the points of light have moved. Computers can aid in this search. Since stars are so far away, they do not appear to move with respect to one another, so any object that does move, and is not a known planet, moon, asteroid, or satellite, becomes a focus of attention **(Fig. Bx21.2e)**. By comparing the positions of the object at different times, astronomers can calculate its orbit, and they can decide whether the object qualifies as an NEO. The accuracy of orbital calculations depends on how far away the object is from the Earth and on the time between observations, for these factors determine the change in position of the object between sequential images. In most cases, initial calculations are not accurate, as the observation window for seeing small objects from the Earth is minimal. With additional data from subsequent near approaches of the object,

FIGURE Bx21.2 The search for near-Earth objects.

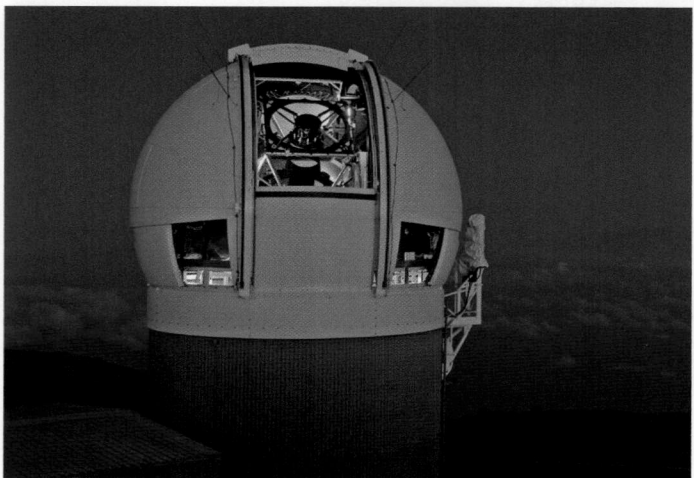

(a) The Pan-STARRS telescope in Hawai'i has discovered more NEOs than any other telescope.

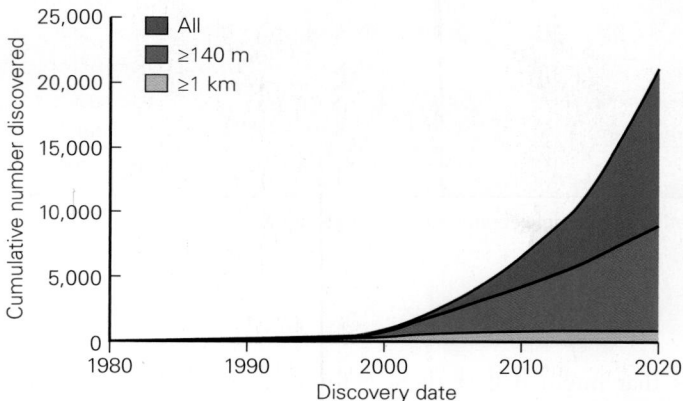

(b) The number of NEOs discovered has increased dramatically since the search began.

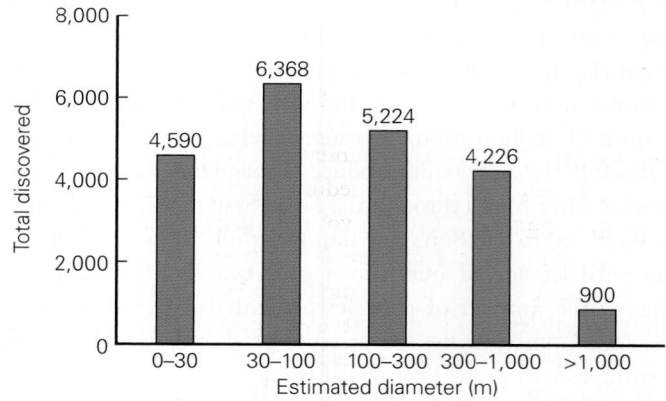

(c) Most of the NEOs discovered range from 30 m to 100 m (98–328 ft) in diameter. Additional small ones probably exist, but they are hard to detect.

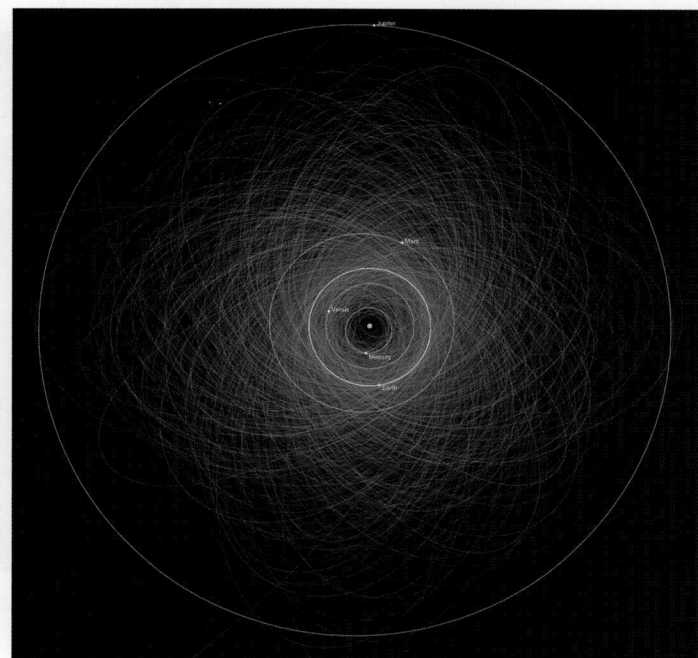

orbital calculations become more accurate. For example, in 1998, when astronomers discovered an asteroid 1 km in diameter, initial calculations of its orbit suggested that it might collide with the Earth in 2028. Needless to say, a media circus followed, raising public awareness of destructive impacts. However, refined calculations showed that the object, fortunately, will not collide with the Earth.

As we discussed earlier in this book, collisions with NEOs such as the one that took place at 66 Ma have had globally catastrophic consequences (see Box 9.2). So, in addition to efforts to detect NEOs, efforts to develop strategies to prevent collisions with them have begun. For example, in 2021, NASA conducted the first Double Asteroid Redirection Test (DART) on the asteroid Dimorphos (see Fig. 21.29e), which orbits another asteroid, Didymos. They deliberately smashed a space probe into Dimorphos, and its impact successfully altered the asteroid's orbit around Didymos. This test was humanity's first effort to purposely change the motion of a celestial object and to demonstrate asteroid-deflection technology. The targeted asteroid posed no threat to the Earth, but the test was meant to determine whether humanity could deflect an asteroid in the future that was hurtling toward a collision with our planet.

(d) The orbits of thousands of NEOs have been calculated.

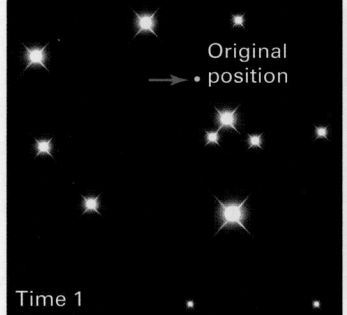
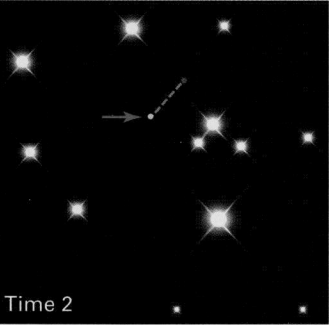
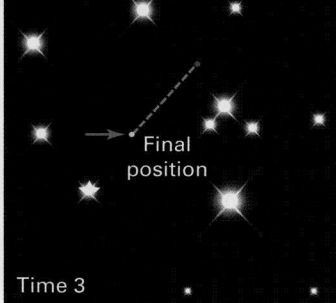

(e) By comparing the positions of an object at different times, astronomers can calculate its orbit.

medium. Astronomers refer to the boundary at which solar-wind pressure and interstellar-medium pressure are equal as the **heliopause**, and to the volume of space within the heliopause as the **heliosphere**. Notably, since both the solar wind and the interstellar medium contain flowing charged particles, they produce electrical currents which, in turn, produce magnetic fields. Within the heliosphere, the magnetic field far from the Sun is due to flow of ions in the solar wind. Outside of the heliosphere, the magnetic field is due to the flow of ions in the interstellar medium. The solar wind's magnetic field plays a role in keeping ions of the interstellar medium out of the

heliosphere, but they can't prevent some neutral particles of the interstellar medium from crossing the heliopause and passing through the heliosphere. The solar wind's magnetic field also deflects some cosmic rays.

Despite its name, the heliosphere is not spherical. Why? As the Sun and its solar wind move through the interstellar medium as they revolve around the center of the Milky Way, the force applied by the interstellar medium on the solar wind compresses the heliosphere on the side toward which the Sun moves through the galaxy. The interstellar medium then flows around the solar wind, somewhat like water flowing around the prow of a ship,

FIGURE 21.35 The heliosphere and its relationship to the Kuiper Belt and Oort Cloud.

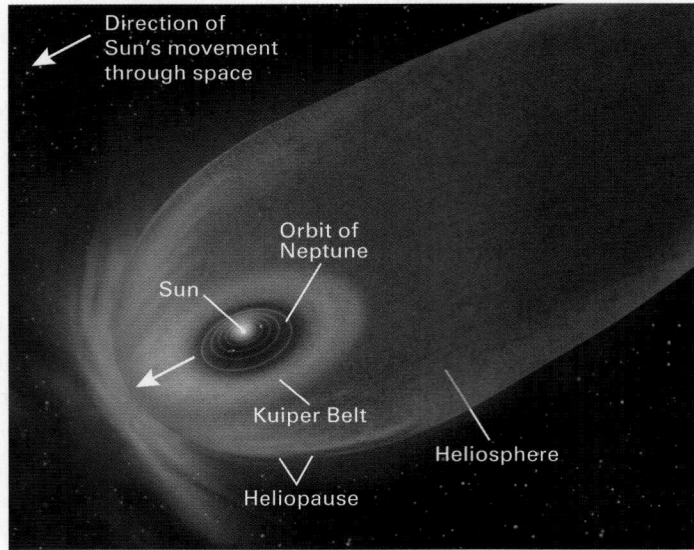

(a) The heliosphere encompasses all the planets of the Solar System and the Kuiper Belt.

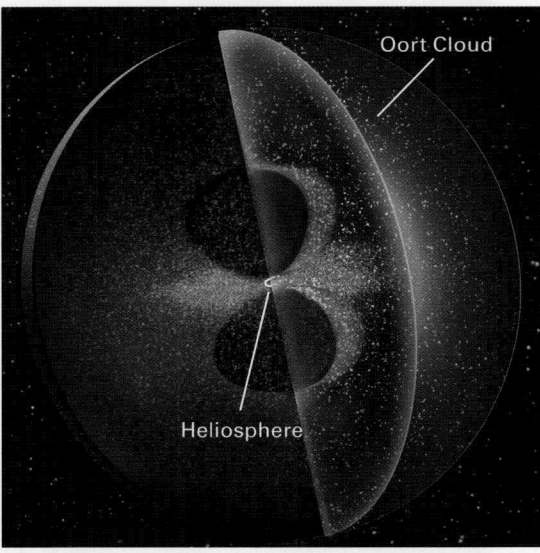

(b) The heliosphere, in turn, lies at the center of the much larger Oort Cloud.

so that the heliosphere stretches downwind far beyond the Sun in the direction away from which the Sun moves. Solar-wind particles are also deflected by this interaction, and some start to move along the heliopause in a direction headed downwind of the Sun as well. Therefore, the heliopause has an elongate, sock-like shape, somewhat like that of the Earth's magnetosphere; the distance from the Sun to the "bow" (leading edge) of the heliopause is much less than that from the Sun to the trailing edge of the heliopause **(Fig. 21.35a)**.

At its closest, the heliopause lies about 123 AU from the Sun. Therefore, the Kuiper Belt lies entirely within the heliosphere, but the Oort Cloud, which occupies a volume between 2,000 AU and 100,000 AU from the Sun, lies well outside of it **(Fig. 21.35b)**. Therefore, defining the "edge" of the Solar System isn't straightforward. If it's defined as the boundary between the region of space in which the concentration of the interstellar medium exceeds that of the solar wind, and the region in which the magnetic field of the Sun is stronger than that of the interstellar medium, then the heliopause can be considered to be the edge. But if we define the edge of the Solar System as the boundary within which the Sun's gravity has a sufficient grip on solid particles that they remain in orbit around the Sun, then the outer limit of the Oort Cloud represents that boundary.

Only two human-made objects, the *Voyager 1* and *Voyager 2* space probes, both launched in 1977, have crossed the heliopause. As of 2024, *Voyager 1* will have traveled more than 160 AU from the Earth, and *Voyager 2*

will have gone 135 AU. At their current speed of 17 km/s (38,000 mph), they will reach the inner edge of the Oort Cloud (at a minimum distance of about 2,000 AU) in about 575 years, and unless they collide with an icy chunk in the Oort Cloud, they will pass its outer edge, at a maximum distance of about 100,000 AU, in about 20,000 years. They will traverse a distance equal to the distance between the Earth and the nearest star in about 40,000 years, although, given their trajectory, neither will be near any stars at that time. In the next chapter, we'll focus our attention on stars, galaxies, and other celestial objects, from beyond the Oort Cloud out to the edge of the Universe.

Take-home message...

The Solar System encompasses an enormous number of small objects in orbit around the Sun. They occur primarily in three locations: the asteroid belt, between Jupiter and Mars; the Kuiper Belt, beyond Neptune; and the Oort Cloud, which lies beyond the outer fringes of the Solar System but remains tied to the Sun by gravity.

Quick Questions

- Why is Pluto classified as a Kuiper Belt object rather than a planet?

- How is a comet different from an asteroid?

- What is the difference among a meteoroid, a meteor, and a meteorite?

Objective 21.1

Explain the overall structure of the Solar System, and describe the major features of the planets.

KEY CONCEPTS

- Astronomers classify objects in the Solar System based on their size, composition, and orbital characteristics.

- A planet orbits the Sun, has a spherical shape, and has cleared its orbit of other objects. A dwarf planet is a body large enough to be spherical but has not cleared its orbit. A moon orbits a planet or dwarf planet.

- A comet is an icy and dusty object that has a highly elliptical orbit around the Sun and emits glowing gas and dust particles that appear as tails behind the comet when it gets close to the Sun. An asteroid is a rocky and/or metallic object orbiting the Sun; most asteroids occur in a belt between Mars and Jupiter.

- Kuiper Belt objects orbit the Sun beyond the orbit of Neptune and have diameters ranging from less than a millimeter up to 2,370 km (1,473 mi) in the case of the dwarf planet Pluto. Oort Cloud objects lie even farther out.

- The Solar System's eight planets follow slightly elliptical orbits that lie within 7° of the Earth's ecliptic plane. All planets revolve counterclockwise around the Sun when viewed from above the Sun's north pole.

- The four smaller inner (terrestrial) planets are composed of rock and/or metal. Mercury and Venus have no moons, the Earth has one, and Mars has two small ones. The four larger outer (Jovian) planets are composed of gases and ices, have rings, and have many moons.

EARTH-SCIENCE VOCABULARY

dwarf planet (p. 814) **planet** (p. 814)
eccentric orbit (p. 815) **ring** (p. 817)
Jovian planet (p. 817) **terrestrial planet** (p. 817)
moon (p. 814)

REVIEW QUESTIONS

1. **(a)** List the succession of planets in the Solar System, starting from the Sun. **(b)** Which of these planets do not have moons? **(c)** Besides planets, what other types of objects exist in the Solar System?

2. **(a)** Are the axes of all the planets parallel to one another? Explain your answer. **(b)** Which planet takes the longest time to orbit the Sun? **(c)** Which planet orbits the Sun in the least time?

3. **(a)** How do the orientations of the orbital planes of other planets compare with the Earth's ecliptic plane? **(b)** Identify each of the planets in **Figure A**. **(c)** Which planets have rocky, solid surfaces, and which planets have surfaces of gas and/or ice?

4. **(a)** What characteristics must an object have to be formally considered a planet? **(b)** How does a dwarf planet differ from a planet? **(c)** How does an asteroid differ from other solid objects of the Solar System?

A

Objective 21.2

Discuss how the Moon formed, and interpret its surface features.

KEY CONCEPTS

- Between 1969 and 1972, the United States landed six manned spacecraft on the lunar surface. Samples that astronauts brought back have allowed scientists to determine the Moon's chemical composition and age.

- Unlike the Earth, the Moon hosts no active volcanoes or plate-tectonic processes, no atmosphere or ocean, and large swings in temperature from its sunlit to its dark side.

- The surface of the Moon consists of lower-elevation, relatively flat regions called maria and higher-elevation regions called lunar highlands. Its mass is only about 1/80 of the Earth's.

- Impact craters of all sizes, formed by meteorite impacts, pockmark the Moon's surface. Smaller craters are bowl-like depressions, while larger ones have central uplift and are surrounded by concentric ridges. A layer of regolith coats most of the Moon's surface.

- Scientific evidence suggests that the Moon formed from debris ejected into orbit around the Earth following a catastrophic collision of the early Earth with a protoplanet named Theia.

- Maria occur only on the side of the Moon facing the Earth. They may have formed from basaltic lava flows that came from volcanic eruptions early in the Moon's history.

EARTH-SCIENCE VOCABULARY

impact crater (p. 819)
late heavy bombardment (p. 819)
lunar highlands (p. 818)

lunar regolith (p. 819)
mare (plural: maria) (p. 818)
space weathering (p. 819)

REVIEW QUESTIONS

5. **(a)** What event does **Figure B** depict? **(b)** What sequence of events happened after the event shown in the figure that led to the formation of the Moon? **(c)** According to the model in the figure, why is the average density of the Moon so much less than that of the Earth?

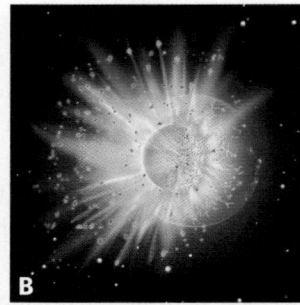
B

6. **(a)** What do the different surface landscapes of the Moon consist of, and how did they form? **(b)** How can we determine which surfaces are older than others? **(c)** How does the surface of the Moon's far side differ from that of its near side?

7. **(a)** Why do astronauts on the Moon weigh less than they do on the Earth? **(b)** When astronauts first stepped on the Moon, they left footprints. What was the material they stepped in to form the footprints, and how did it form?

8. **(a)** Why is the Moon's surface covered with impact craters, while hardly any can be found on the Earth's surface? **(b)** When do astronomers suggest that most of these craters were formed? **(c)** Are impact craters still forming on the Moon today?

9. **(a)** What two rock types are common on the Moon? **(b)** Which is common in maria, and which in the lunar highlands?

Objective 21.3

Contrast the physical characteristics of the terrestrial planets, and explain why they are so different from one another.

KEY CONCEPTS

- Mercury resembles the Earth's Moon in that it hosts no volcanic activity, no ocean, and no atmosphere, and has a heavily cratered surface. It differs from the Earth's Moon in that it is slightly larger and hosts a magnetic field. The day side of the planet (facing the Sun) is extremely hot, while its night side (facing away from the Sun) is frigid.

- Venus is almost the same size as the Earth, but it rotates so slowly on its axis that it takes 243 Earth days to spin once. Venus needs 224 days to orbit the Sun, so a day on Venus takes longer than the planet's year. Venus is the only planet to rotate clockwise around its axis, probably due to a collision that altered its rotation direction.

- The atmosphere of Venus consists almost entirely of CO_2 and is 93 times denser than the Earth's atmosphere. A runaway greenhouse effect keeps surface temperatures hot enough to melt lead and prevents water from existing on the planet.

- The dense atmosphere of Venus hides the planet's surface; however, radar mapping reveals lava domes, volcanoes, and volcanic craters, but few impact craters. Recent evidence suggests that the planet remains volcanically active.

- Mars has been explored using robotic probes, rovers, and a helicopter. Mars once hosted liquid water and a thicker atmosphere, but today the planet is dry, cold, and covered with red rock and sediment. Some researchers speculate that it may have hosted primitive life.

- Mars has a thin, largely CO_2 atmosphere, with a surface pressure that is only 0.6% of the Earth's, and has an average surface temperature of $-55°C$. Winds on Mars can loft dust, producing haze that can last months.

- Like the Earth, Mars has seasons. The planet's radius is half of the Earth's. Mars hosts two small, nonspherical, cratered moons. The planet's surface includes a gigantic lava-covered plain, cratered highlands, huge impact basins, a huge plateau, an immense canyon, huge volcanoes, ancient stream channels, and polar ice caps consisting of frozen CO_2 and some water ice.

EARTH-SCIENCE VOCABULARY

exobiology (p. 824)

runaway greenhouse effect (p. 822)

REVIEW QUESTIONS

10. **(a)** What characteristics do Mercury and the Earth's Moon share, and in what ways are they different? **(b)** Mercury is much closer to the Sun than the Earth is, yet parts of Mercury are colder than any location on the Earth. Why?

11. **(a)** Why do the Earth and Venus have vastly different atmospheres and surface temperatures? **(b)** Why do astronomers have to use radar to determine characteristics of Venus's surface? **(c)** Why can't surface rovers survive on Venus?

12. **(a)** Which of the terrestrial planets have atmospheres? **(b)** Which has the thickest atmosphere? **(c)** Which has the thinnest? **(d)** Which have a greenhouse gas as the atmosphere's primary component?

13. **(a)** Why is a windstorm on Mars unlikely to move a heavy object like a rover? **(b) Figure C** is a photo taken by a Mars rover. Why did Mars investigators consider this geologic feature significant? **(c)** Describe each of these features: Olympus Mons, the Tharsis Bulge, and Valles Marineris.

20 cm

14. **(a)** What is the evidence for water on Mars in the present or at least in the planet's past? **(b)** Was it possible that Venus had water early in its history? If so, where did the water go? **(c)** What material makes the surface of Mars red?

Objective 21.4

Identify the key features of the gas-giant and ice-giant planets and their moons.

KEY CONCEPTS

- Jupiter, the largest gas planet, rotates faster than any other planet, and produces its own internal heat. Its outer atmosphere consists of hydrogen and helium, with atmospheric bands made up of trace elements such as sulfur and phosphorus. Winds within these bands flow at different speeds and in different directions, with swirling eddies at their boundaries.

- Jupiter's interior consists of metallic hydrogen surrounded by liquid hydrogen; the liquid hydrogen generates a very strong magnetic field around the planet. A rock and metal mass probably lies at the planet's center.

- Jupiter hosts four large moons—Io, Europa, Ganymede, and Callisto—as well as many small moons, and faint rings along its equator. Io has active volcanoes, while the other three moons have water-ice surfaces surrounding liquid water layers, which in turn surround rocky interiors.

- Saturn, the second largest planet, consists mostly of hydrogen with some helium. Its interior probably consists of a rock and metal center surrounded by layers of dense ice, metallic hydrogen, and liquid

helium and hydrogen. Like Jupiter, Saturn has a magnetic field and generates more energy than it receives from the Sun.

- Saturn's brilliant rings consists of hundreds of thin ringlets made up of countless particles of water ice and rocky dust. The rings circle Saturn's equator and are extremely thin. Saturn's largest moon, Titan, hosts an atmosphere containing methane clouds that produce liquid methane rain, and its surface has active volcanoes. Both Titan and Enceladus have a water-ice crust encasing an internal ocean.

- Uranus and Neptune, called the ice giants, have rocky and metallic centers surrounded by a thick layer of slush composed of water, ammonia, and methane ice. Uranus's axis of rotation is sideways, so that planet has extreme seasons, each lasting 42 Earth years. Uranus has a featureless teal-colored atmosphere, while Neptune has a deep-blue atmosphere with bands and white streaks caused by clouds. Both Uranus and Neptune have atmospheres consisting primarily of hydrogen, helium, and methane, and both planets host moons and rings.

EARTH-SCIENCE VOCABULARY

gas giant (p. 828) **ice giant** (p. 828)
Great Red Spot (p. 829) **metallic hydrogen** (p. 830)

REVIEW QUESTIONS

15. **(a)** Which of the Jovian planets have moons? **(b)** Why do Jupiter and Saturn have pole-to-pole diameters that are different from their equatorial diameters? **(c)** Are Jupiter and Saturn composed entirely of gas? Explain your answer.

16. **(a)** Explain the features on the moon shown in **Figure D**. **(b)** Which moon is it, and which planet does it orbit?

D

17. **(a)** Which has a greater thickness, Saturn's rings or a football stadium? **(b)** How might Saturn's rings have formed? **(c)** What causes the dark bands in Saturn's rings?

18. **(a)** Why do Jupiter and Saturn have distinctly different colors in their atmospheres? **(b)** Why do Uranus and Neptune appear bluish, while Saturn appears tan and Jupiter is multicolored?

19. **(a)** Summarize the distinctive characteristics of the largest moons of Jupiter and Saturn. **(b)** Are all of the moons composed of rock and metal?

Objective 21.5

Outline what astronomers have learned about dwarf planets, and evaluate the threats to the Earth posed by asteroids, comets, and meteoroids.

KEY CONCEPTS

- Asteroids are rocky or metallic objects with a size larger than about 100 m across. Some are remnants of planetesimals that never differentiated into a metallic core and a rocky mantle, while others are fragments of bodies that had differentiated before breaking up due to collisions. There are millions of asteroids, most occupying a zone between the orbits of Jupiter and Mars known as the asteroid belt, but some lie in planetary orbits and some cross the Earth's orbit. Only three asteroids are large enough to be spherical, and only one, Ceres, qualifies as a dwarf planet.

- The dwarf planets Pluto and Eris are two of hundreds of thousands of objects lying beyond the orbit of Neptune in a zone called the Kuiper Belt. The Oort Cloud, which hosts many more icy objects, lies even farther from the Sun.

- Comets consist of rock, dust, and ice. They follow highly eccentric orbits that take them close enough to the Sun to start vaporizing and develop a coma and tails. Comets probably come from the Kuiper Belt or Oort Cloud.

- A meteoroid is a solid object that is smaller than a comet or an asteroid and orbits the Sun. When a meteoroid enters the Earth's atmosphere, it generates a streak of light known as a meteor. A meteorite is a meteoroid that strikes the planet's surface; NASA monitors near-Earth objects (asteroids, comets, or meteoroids) that have the potential to cross the Earth's orbit.

- The heliosphere is the region in which the pressure exerted by solar-wind particles exceeds the pressure of the interstellar medium. It surrounds the Earth like a giant bubble in the shape of a sock, and deflects the interstellar medium along its boundary, the heliopause.

EARTH-SCIENCE VOCABULARY

ablation (p. 840)
asteroid (p. 838)
asteroid belt (p. 838)
comet (p. 840)
heliopause (p. 847)
heliosphere (p. 847)
interstellar medium (p. 845)
Kuiper Belt (p. 839)

meteor (p. 842)
meteorite (p. 843)
meteoroid (p. 842)
meteor shower (p. 843)
near-Earth object (NEO) (p. 844)
Oort Cloud (p. 840)

REVIEW QUESTIONS

20. **(a)** Between which two planets are most asteroids in the Solar System found? **(b)** Are any of them spherical? **(c)** Do all asteroids lie in the asteroid belt? **(d)** Why are Aten and Apollo asteroids important to people on the Earth?

21. **(a)** Distinguish between the Kuiper Belt and the Oort Cloud. **(b)** In which of these entities do Pluto and Eris lie? **(c)** What does the surface of Pluto look like?

22. **(a)** What does a comet consist of? **(b)** How do the orbits of comets compare to those of planets? **(c)** Where do comets come from? **(d)** Why do comets have two tails, as shown in **Figure E**? **(e)** Label the tails in **Figure E**. Which most aligns with the direction of the solar wind?

E

23. **(a)** What is the distinction between a meteoroid and an asteroid, and among a meteoroid, a meteor, a meteorite, a fireball, and a bolide? **(b)** What causes the high temperatures of a meteor? **(c)** Why do meteor showers occur at predictable times? **(d)** Describe the differences between a simple crater and a complex crater. Which does **Figure F** show? **(e)** About how big is the largest impact crater on the Earth, and how big was the object that caused it?

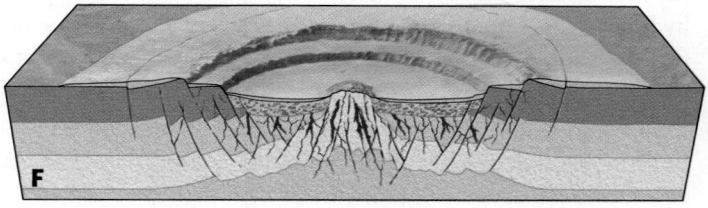
F

24. **(a)** What does NASA consider a near-Earth object to be, and why are officials concerned about these objects? **(b)** How do astronomers detect near-Earth objects?

25. **(a)** What characteristic defines the heliopause? **(b)** Do Oort Cloud objects lie within or outside the heliosphere? **(c)** Have any human-made objects passed through the heliopause?

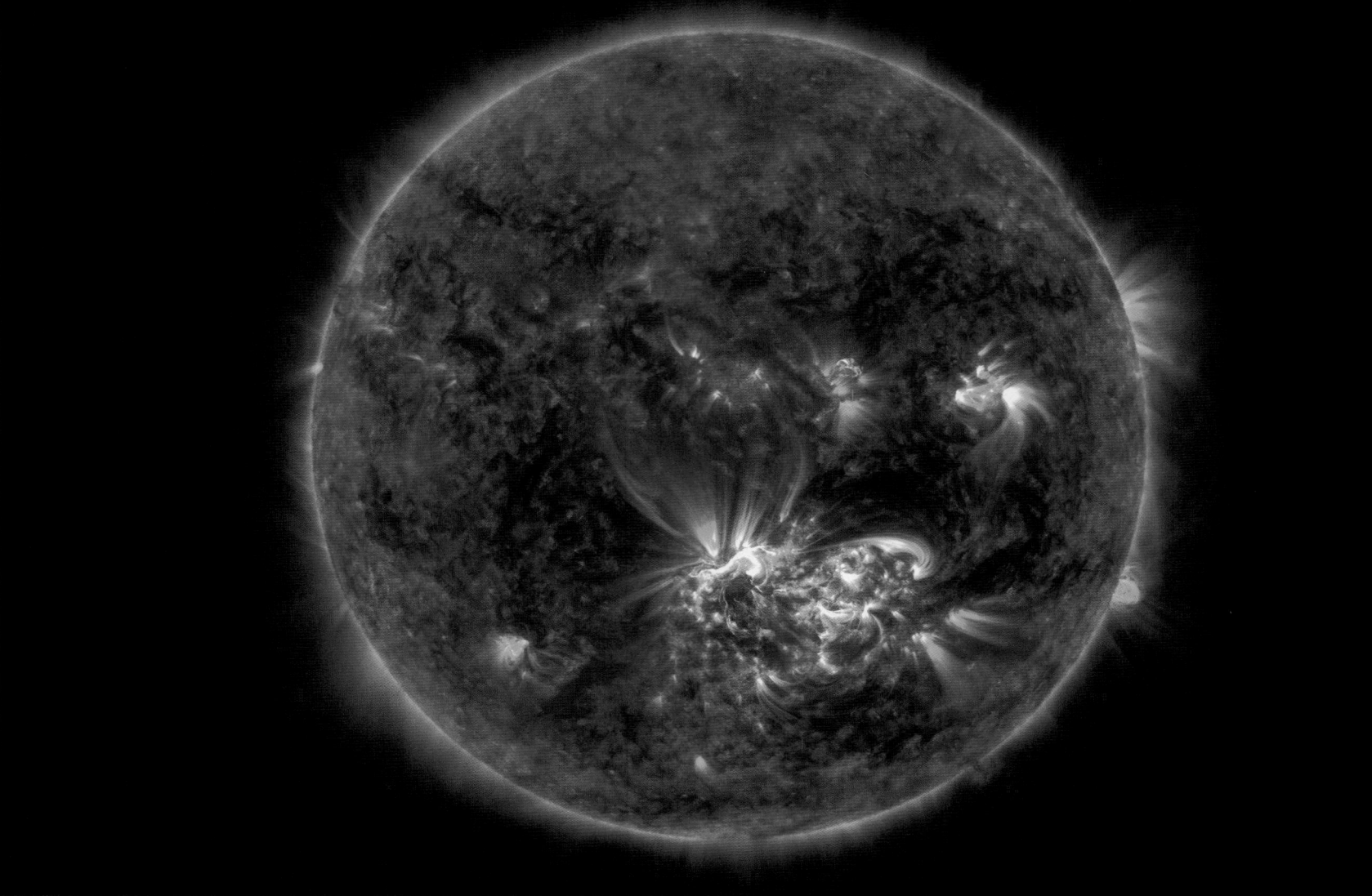

⊙ 22 THE SUN, THE STARS, AND DEEP SPACE

After studying this chapter, you should be able to...

1. describe the structure and composition of the Sun and its atmosphere, and explain how the Sun produces energy.

2. describe how the Sun's magnetic field evolves, and how changes in its magnetic field cause solar storms.

3. characterize how stars can differ from one another, and show how stars can be classified on an H-R diagram.

4. discuss how stars form, and how the evolution of a star depends on its mass.

5. explain what happens when stars run out of fuel, and distinguish among the various objects that represent remnants of dead stars.

6. describe the overall structure of a galaxy, the differences among galaxies, and the relationship of a supermassive black hole to a galaxy.

7. summarize current ideas concerning the overall structure and fate of the Universe.

<< A close-up image of the Sun, taken by the Solar Orbiter spacecraft, emphasizes the turbulence of a star's surface as the immense energy rising from a nuclear furnace at depth finally reaches the surface and escapes into space.

Gazing at a moonless black sky awash with bright stars makes for a truly memorable night. Such a view may be hard to come by for an urban dweller, given the light pollution and haze above cities, but look up from a desert mountain peak on a cloudless summer night and prepare to be amazed. In addition to stars and planets, you'll see the central part of the Milky Way Galaxy slicing across the sky from horizon to horizon. The Sun, and every individual star that you can see from the Earth, lies within the Milky Way. However, not all the points of light in the night sky are single stars, nor do they all lie within the Milky Way. Some are nebulae, some are clusters of stars, some are galaxies, and some are unusual objects that astronomers still don't completely understand. Observations first made with the Hubble Space Telescope, and more recently with the James Webb Space Telescope, emphasize that even a tiny part of the sky that looks dark to the naked eye contains billions upon billions of celestial objects **(Fig. 22.1)**. Most are so far away that the light they emit traveled for millions or even billions of years since it began its journey across deep space to the mirror of a telescope.

In the previous chapter, we focused on the planets and smaller objects that circle our Sun in the Solar System. In this chapter, we complete our introduction to astronomy by exploring the vast variety of other objects in the known Universe. We begin with the Sun, because studies of our own star provide the basis for understanding all other stars. Next, we turn our attention to those other stars, to galaxies, and finally to a host of objects, some of which have strange-sounding names—nebulae, dwarf stars, giant stars, neutron stars, quasars, novae, supernovae, and black holes—that no one had even dreamed of before the 20th century. This chapter, and this book, ends by pondering the possible fate of the Universe in a time long after the Earth will have ceased to exist.

22.1 Our Home Star: The Sun

Our Sun, the daily source of light that makes the Earth habitable, was worshipped as a god by many ancient cultures. It's actually just a fairly ordinary star—not too big or too small, and not too hot or too cool. Because of its proximity to our planet **(Fig. 22.2)**, astronomers can study our Sun in detail. Their observations have led to an understanding of its energy source, its composition, and its internal structure. This knowledge, in turn, provides a basis for interpreting distant stars.

FIGURE 22.1 What you can see when you magnify a dark spot in the night sky.

(a) A tiny part of the sky, 5% of the area covered by the Moon, looks black when viewed from the Earth.

(b) When the Hubble Space Telescope magnifies this tiny area, it reveals thousands of galaxies, each containing hundreds of billions of stars. The point of light with spikes is a relatively nearby star in our own galaxy.

FIGURE 22.2 The Sun, as seen by astronauts orbiting the Earth.

Composition of the Sun

In 1925, a British-American graduate student, Cecilia Payne-Gaposchkin, who was studying the spectra of sunlight and starlight (see Chapter 20), demonstrated that the Sun consists primarily of hydrogen (71.0% by mass) and helium (27.1% by mass). Prior to her work, astronomers had assumed that the Sun had the same chemical composition as the Earth but, for unknown reasons, was much hotter. Because hydrogen atoms weigh less than helium atoms, 91.2% of the individual atoms in the Sun are hydrogen, and only 8.7% are helium. The next eight most common elements in the Sun, in order of their contribution to its mass, are oxygen, carbon, nitrogen, silicon, magnesium, neon, iron, and sulfur. The Sun contains traces of all the other naturally occurring elements as well.

Elements in the Sun do not exist in the same form in which we find them on the Earth, for at the temperatures reached within the Sun, atoms lose some or all of their electrons and become free, positively charged nuclei. A material consisting almost entirely of such *ionized atoms*, circulating along with *free electrons* (which move independently of nuclei), constitutes **plasma**, the fourth state of matter.

The Source of the Sun's Energy

Discoveries made during the first half of the 20th century first provided a path to explaining the Sun's energy generation. The key to solving this mystery came from the work of Albert Einstein (1879–1955). According to his iconic equation, $E = mc^2$, annihilation of a tiny amount of mass produces an enormous amount of energy. In the 1920s, physicists discovered that when temperatures are high enough that nuclei move about at immense speed, two atomic nuclei can collide with enough force to fuse together to form a single nucleus. During this process, now called **nuclear fusion**, a tiny amount of matter converts into a large amount of energy, as Einstein's equation predicted. During the 1930s, physicists realized that nuclear fusion takes place in the Sun, and by 1939, researchers had worked out the specific series of fusion reactions that occur inside the Sun and explain its energy production.

Almost all nuclear fusion that happens in the Sun involves the bonding of protons (the nuclei of hydrogen atoms) to form helium nuclei. This reaction, known as the *proton-proton chain reaction*, involves several steps **(Fig. 22.3)**. Simplistically, the reaction begins when two protons fuse. During this reaction, one of the colliding protons transforms into a neutron. Therefore, the product of the reaction is *deuterium* (^{2}H), an isotope of hydrogen whose nucleus contains one proton and one neutron. Next, a deuterium nucleus collides with another proton to form ^{3}He, an isotope of helium containing two protons and only one neutron. In the final step, two ^{3}He nuclei collide to form a normal helium nucleus (^{4}He), containing two protons and two neutrons, plus two excess protons (which can eventually be involved in other fusion reactions). Each step in the chain releases energy.

Because of nuclear fusion, the Sun produces an inconceivable 4×10^{26} (four followed by 26 zeros) watts of power. (A *watt* [W] is a unit of power; it represents the expenditure of 1 joule of energy in 1 second.) By comparison, burning all the hydrocarbon reserves on the Earth in 1 second would yield 10^{21} W, and a typical nuclear power plant produces 5×10^8 W. If all of the Sun's energy were focused on the Earth, the planet would vaporize. Fortunately for us, the energy spreads out over a progressively larger area of space as it travels away from the Sun, so the Earth receives only a tiny fraction of it.

Internal Structure of the Sun

The Sun's radius, defined as the distance from its center to its visible surface, is 696,000 km (432,000 mi), 110 times that of the Earth. Researchers distinguish three concentric layers within the Sun: the solar core at the center, the radiative zone, and the convective zone **(Fig. 22.4)**. Let's examine the characteristics of these layers, starting from the center.

THE SOLAR CORE. The **solar core** extends from the center outward for 160,000 km (100,000 mi), and therefore it accounts for about 23% of the Sun's radius. Temperatures reach 15 million °C (27 million °F) at the center of the

FIGURE 22.3 A series of fusion reactions in the Sun produce He nuclei from H nuclei. The reactions release energy (gamma rays) and extremely tiny bits of matter similar to electrons, but with opposite charge (called positrons) and even smaller neutral particles (called neutrinos).

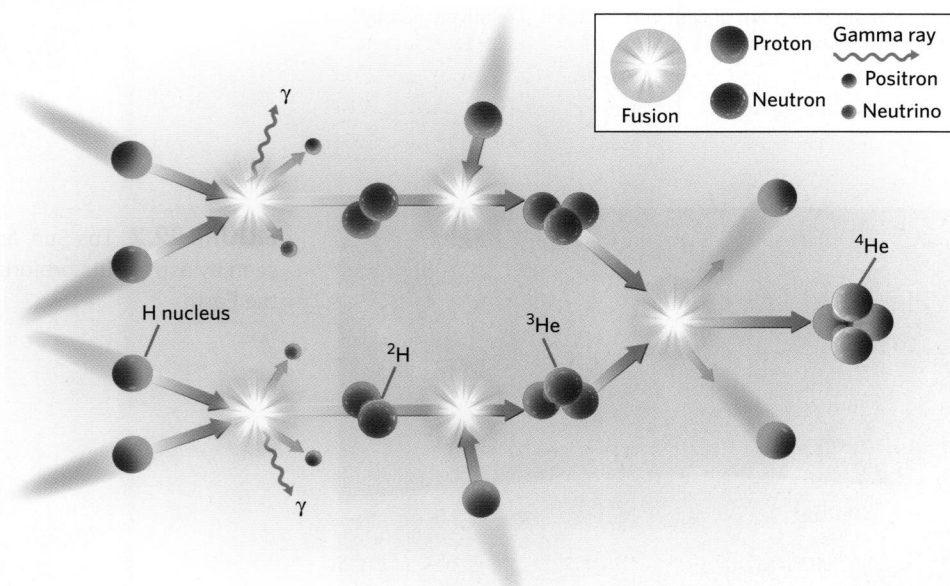

FIGURE 22.4 The internal structure of the Sun. The axes show how physical characteristics of the Sun change from the core to the top of the convective zone.

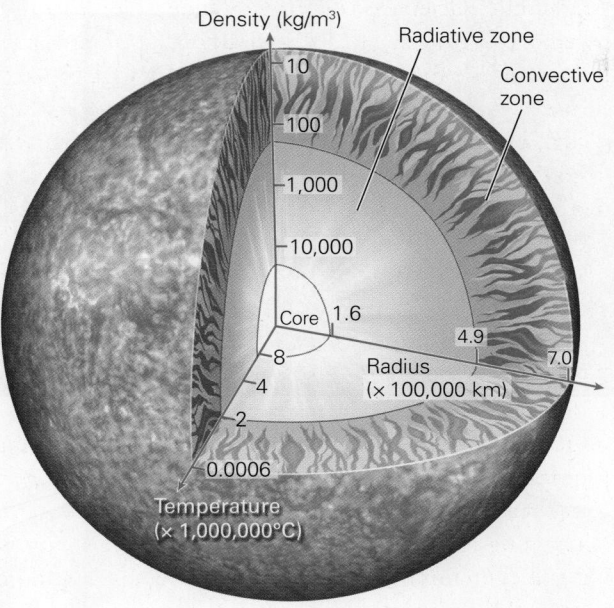

THE CONVECTIVE ZONE. When energy reaches the outer edge of the radiative zone, it heats the base of the overlying layer of plasma to about 2 million °C (3.6 million °F). This heat decreases the density of the plasma, making it relatively buoyant. The buoyant plasma rises upward and carries thermal energy with it. As it rises toward the Sun's surface, the plasma cools, and it eventually sinks to replace hotter, buoyant plasma. Astronomers refer to the region of the Sun in which this convective transport of energy takes place as the **convective zone**. This zone, which surrounds the radiative zone, has a thickness of about 200,000 km (125,000 mi).

The upwelling and downwelling currents of plasma in the convective zone organize into very tall convective cells called *convective columns*. Plasma rises at the center of each convective column and sinks back down along the sides. Each column ranges from 500 to 2,000 km (300–1,200 mi) across, a tenth to a third the width of North America. The column's center, where hot plasma rises, is hotter, and therefore brighter, than the sides, where cooler plasma sinks. This pattern of light and dark gives the Sun's surface a grainy appearance, known as **solar granulation (Fig. 22.5)**. At any given time, about 4 million granules (the visible tops of convective columns) exist, but the pattern constantly evolves, since each granule lasts for only a few minutes before dissipating as new granules form.

The Solar Atmosphere

The Sun doesn't have a solid crust like that of the Earth. Astronomers, therefore, place its surface boundary at the depth where the density of material in the Sun becomes great enough to be opaque to visible light. The material above this boundary, known as the **solar atmosphere**, includes the distinct layers we describe next.

THE PHOTOSPHERE. All the light that we see from the Sun comes from glowing gases within the **photosphere**, a 50–500 km (30–310 mi) thick layer between the top of the convective zone and the rest of the solar atmosphere. Transparency decreases gradually from the top to the base of the photosphere. At the base, the plasma within the convective zone is too dense for light to escape, so the base of the photosphere serves as the visually sharp boundary of the Sun that you see at sunset or sunrise **(Fig. 22.6a)**. The temperature range between the top of the photosphere, at 4,300°C (7,800°F), and the base of the photosphere, at 5,700°C (10,300°F), resembles the temperature range within the Earth's core. Electromagnetic energy emitted by the photosphere radiates into space at the speed of light, reaching the Earth about 8 minutes later.

THE CHROMOSPHERE AND CORONA. If you observe the Sun during a total eclipse (with appropriate eye protection,

core and diminish to about 8 million °C (14.4 million °F) at its outer edge. Because of the inward gravitational pull of the Sun's immense mass, the pressure at the center may be over 10,000 times the pressure at the Earth's center. About 99% of the Sun's energy production comes from nuclear fusion reactions in the solar core. These convert 600 million tons of hydrogen into helium every second. (This weight is comparable to the combined weight of all the buildings in New York City.) The remaining 1% of the Sun's energy comes from the base of the radiative zone. Temperatures farther out in the Sun are too cool for fusion to be possible.

THE RADIATIVE ZONE. Researchers refer to the thick layer surrounding the solar core as the **radiative zone** because energy passes through this zone in the form of electromagnetic radiation. The radiative zone extends from the core out to about 490,000 km (300,000 mi) from the center of the Sun, so it has a thickness of about 330,000 km (205,000 mi). It accounts for about 43% of the Sun's radius and about 48% of the Sun's mass. In the radiative zone, photons of energy emitted by fusion reactions in the core travel only a short distance before they interact with other particles of matter and are redirected in a new, random direction. These photons then travel a short distance before they, too, encounter another particle and are redirected in a new direction. Energy, in effect, wanders nearly randomly as it travels through the radiative zone, so a given unit of energy may take about 170,000 years to reach the zone's outer surface.

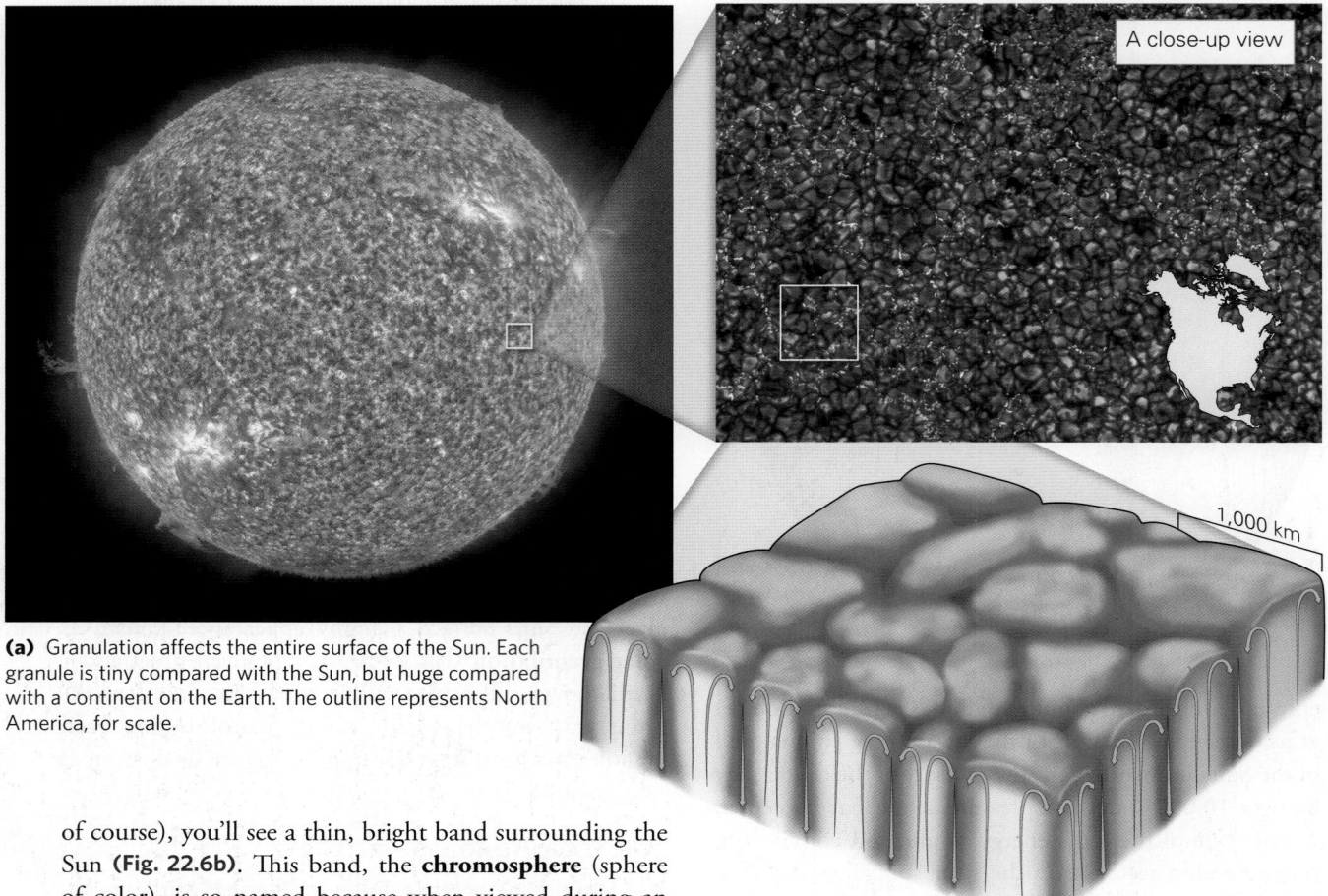

(a) Granulation affects the entire surface of the Sun. Each granule is tiny compared with the Sun, but huge compared with a continent on the Earth. The outline represents North America, for scale.

(b) In three dimensions, we see that the lighter areas are rising hot plasma. The plasma cools and sinks in the darker areas.

of course), you'll see a thin, bright band surrounding the Sun **(Fig. 22.6b)**. This band, the **chromosphere** (sphere of color), is so named because when viewed during an eclipse, it displays flashing tints of brilliant red. The chromosphere, with a thickness of 3,000–5,000 km (1,800–3,000 mi), consists of glowing plasma and lies just above the photosphere. Its density is only about 0.00000001 that of the Earth's atmosphere at sea level.

At the top of the chromosphere, across a thin (100 km, or 60 mi) *transition zone*, the temperature increases dramatically. In the overlying **corona**, the outer layer of the solar atmosphere, temperatures rise to 1 million °C (1.8 million °F) at an elevation of about 10,000 km (6,200 mi) above the photosphere. (Note that the corona has such low density that, though much hotter, it contains much less heat than the photosphere.) We can see the inner part of the corona during an eclipse, when it appears as a wispy, glowing cloud **(Fig. 22.6c)**. When viewed in ultraviolet light, the outer part of the corona appears as bright streaks radiating far out from the Sun's surface **(Fig. 22.6d)**.

The Solar Wind

At high temperatures, particles move very fast, so it's no surprise that some of the particles in the extremely hot corona achieve escape velocity and head off into space. The high-speed stream of particles—mostly protons and electrons, along with some helium nuclei—that flows outward from the Sun and into space makes up the **solar wind**. Note that the solar wind consists of moving matter, so it really is a "wind," like the winds of the Earth's atmosphere. But solar wind and winds on the Earth differ dramatically, in that solar-wind particles move at enormous speeds and are ionized, whereas molecules in the Earth's wind have a neutral charge and travel relatively slowly. Because of the solar wind, the Sun loses about 1.8 billion kg (2 million tons) of matter every second. But even at this seemingly huge rate, only about 0.1% of the Sun has been lost since its nuclear furnace first ignited. Solar-wind particles travel at 250–750 km/s (560,000–1,700,000 mph), much slower than the speed of light, so they take days to reach the Earth. Because the particles have an electric charge, the Earth's magnetosphere deflects most of them. A strong solar wind will trigger aurorae as it passes along, and interacts with, the Earth's magnetic field (see Chapter 16).

FIGURE 22.6 The solar atmosphere.

(a) At sunset, the edge of the Sun's surface that you see is the base of the photosphere.

(b) The chromosphere, the thin red and white ring close to the Sun's surface, as seen in visible light during an eclipse. To make this layer visible, the light of the corona has been blocked.

(c) This photo shows the gases of the inner part of the corona, which are visible during a solar eclipse.

(d) In this photo, the light of the inner corona has been blocked. The outer corona, as viewed in ultraviolet light, streams far into space.

Take-home message...

The Sun is a star, and like other stars, it produces huge amounts of energy by nuclear fusion reactions. These reactions can take place because it's so hot inside the Sun that nuclei can collide with enough force to fuse together. The interior of the Sun can be divided into layers. Most nuclear fusion takes place in the solar core. The resulting energy passes through the radiative zone as electromagnetic energy and then heats the base of the convective zone, producing convective columns that carry energy to the Sun's surface. In the photosphere, energy is emitted into space as electromagnetic radiation. Beyond the photosphere, the solar atmosphere consists of the chromosphere and the corona. Particles escaping from the corona head into space as solar wind.

Quick Questions

- What does the Sun's atmosphere consist of?
- What is the source of the Sun's energy?
- How is the Sun's convective zone different from its radiative zone?

22.2 The Energy That the Sun Sends into Space

Solar-energy production, measured as insolation arriving at the top of the Earth's atmosphere, varies between 1,360.3 W/m² and 1,362.1 W/m², a difference of only 0.13%. This variation takes place in 11-year cycles. In addition to this periodic variation, the Sun can emit short-lived pulses of energy that can approach 15% of the total energy production of the Sun over the same 11-year cycle. The periodicity of energy production, as well as the occurrence of short-lived high-energy pulses, relates to the behavior of the Sun's magnetic field.

Why Does the Sun Have a Magnetic Field?

The flow of plasma, which consists of charged (ionized) particles, produces electric currents in the Sun. (This phenomenon resembles the flow of electrons through a wire to provide electric current to gadgets in your home.) All electric currents, in turn, generate magnetic fields. Not surprisingly, therefore, the immense amount of plasma moving within the Sun yields an intense magnetic field. This magnetic field is more complex and variable than the Earth's because at least two types of plasma flow take place in the Sun, each yielding a different component of the field.

The first type of flow develops in response to the Sun's overall rotation around its axis. This flow causes magnetic-field lines to extend outward at the Sun's north magnetic pole (also known as its *positive pole*), and inward at the Sun's south magnetic pole (also known as its *negative pole*). Consequently, scientists refer to the magnetic field generated by the Sun's rotation as the Sun's *polar field* **(Fig. 22.7a)**.

Recall that magnetic-field lines between the Earth's poles are roughly parallel to lines of longitude (see Chapter 2). Magnetic-field lines between the poles of

FIGURE 22.7 The Sun's magnetic field.

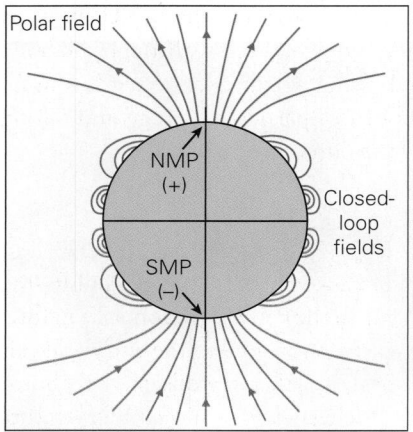

(a) A simplified sketch illustrating the difference between the polar field and the closed-loop fields of the Sun. (NMP = north magnetic pole; SMP = south magnetic pole.)

(d) The purple lines (positive polarity) and the green lines (negative polarity) represent the polar field. During a reversal, the polarity of the polar field reverses.

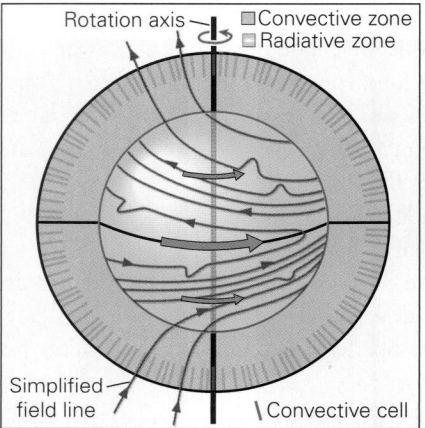

(b) A cross section through the convective zone and the spherical radiative zone beneath it. The large green arrows indicate the variation in plasma flow velocity with latitude. Because of this variation, magnetic-field lines near the base of the convective zone bend and become nearly parallel to the Sun's equator. Note that the polarity of field lines in the southern hemisphere is opposite to that of field lines in the northern hemisphere.

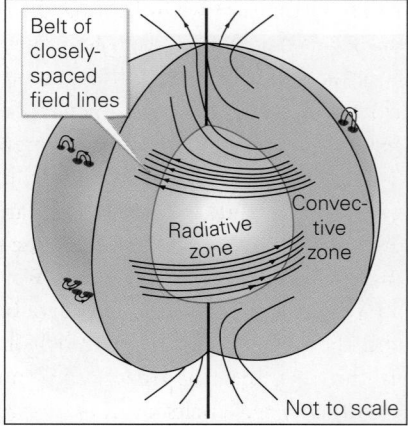

(c) Eventually, magnetic-field lines get wrapped around the Sun, forming two belts of closely spaced field lines.

(e) A computer model of the Sun's magnetic field shows the polar field (indicated by colored lines shooting into space from the Sun's poles), and closed-loop fields (indicated by white lines that form arcs that intersect the Sun's surface at both ends).

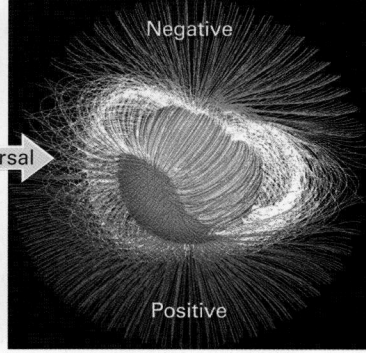

the Sun, in contrast, are not parallel to lines of longitude because the speed of plasma rotation at one latitude differs from its speed at another. Specifically, plasma at low latitudes completes one rotation around the Sun's axis in 25 days, while plasma at high latitudes completes one rotation in 38 days. At depth in the Sun, near the boundary between the convective zone and the radiative zone, shear between bands of plasma moving at different speeds causes magnetic-field lines to bend, twist, and become contorted **(Fig. 22.7b)**. (To picture this phenomenon, imagine a line of corks floating on a river in which flow velocity varies with location. If the line starts out perpendicular to the river, variations in flow velocity will cause the line to curve, and if there is turbulent flow, the line will become contorted.) In fact, some of these deep magnetic-field lines wrap completely around the Sun to form belts of closely spaced field lines, which trend nearly parallel to lines of latitude: one such belt forms in the northern hemisphere, and the other in the southern hemisphere **(Fig. 22.7c)**. These belts have opposite polarity, meaning that arrows indicating the direction of the field point in a clockwise direction in one hemisphere, and in a counterclockwise direction in the other hemisphere. The close spacing of the field lines indicates that these belts represent regions of particularly intense magnetic fields.

The belts of closely spaced field lines first form in the Sun's mid-latitudes. When at mid-latitudes, the field lines are particularly close together, so the polar field attains its maximum strength there. Then, over the course of about 11 years, the northern hemisphere belt drifts southward, and the southern hemisphere belt drifts northward, and as the two belts approach the equator, they weaken. Eventually, new belts of closely spaced field lines form in the mid-latitudes of the two hemispheres, causing the polar field to strengthen again.

Soon after the polar field attains its maximum strength, the polarity of the field starts to flip. Such reversals may take less than one year or as long as five years. The drift of magnetic-field-line belts toward the equator and weakening of the field take place during the reversal. As the field weakens, the configuration of field lines reorganizes. During this reorganization, field lines in space around the Sun become somewhat jumbled. When a reversal is complete, the polarity of the polar field is the opposite of what it had been previously **(Fig. 22.7d)**. This means that if the north magnetic pole (positive pole) was at the north geographic pole for most of one 11-year cycle, it will be at the south geographic pole for most of the next 11-year cycle. Note that the magnetic-field lines at depth in the Sun also switch polarity during a reversal, so if the arrows indicating the orientation of magnetism in a belt of field lines were pointing clockwise before the reversal, they point counterclockwise after the reversal. The Sun's polar field

maintains its polarity through the next time it reaches its maximum strength. Then another reversal begins.

The second type of flow that produces the Sun's magnetic field involves the relatively narrow vertical convective cells of plasma at shallower levels of the convective zone (represented by the short red lines in Fig. 22.7b). This flow generates complicated, rapidly changing magnetic-field lines near the Sun's surface in the region between the poles. In this region, magnetic-field lines arc upward to form arc-like curves within the corona. Notably, arcing field lines point outward at one end of the arc and inward at the other end. The length of an arc, from exit point to entrance point, is much less than the Sun's circumference, so scientists refer to the arcs as closed loops. (In this regard, they differ from the field lines that enter or exit from the poles, shown in Fig. 22.7a.) The resulting maze of field lines resembles a jumble of spider legs **(Fig. 22.7e)**. Some of the loops are probably generated by the plasma flow in the vertical convective cells. Others come from deeper down, where contortions of deep field lines produce powerful loops, which pass up through the convective zone and break through the Sun's surface when the polar field becomes particularly strong. These deep field lines form particularly large and strong magnetic arcs above the Sun's surface.

Sunspots

At the entry and exit points of strong magnetic arcs, the magnetic field can be so intense that it inhibits the upwelling of hot plasma in the convective zone. Therefore, patches of the photosphere surrounding the entry and exit points become cooler than other regions of the photosphere. These cooler patches—which have a temperature of about 4,200°C (7,600°F), as compared with about 6,000°C (10,800°F) for normal photosphere—are known as **sunspots** because their lower temperatures make them look darker than the surrounding regions **(Fig. 22.8a)**. Individual sunspots move with the Sun's rotation and can survive for days. They average 10,000 km (6,000 mi) across, but some grow to be up to 50,000 km (30,000 mi) in diameter **(Fig. 22.8b)**. During their lifetimes, they change in shape and dimensions, but they always occur in pairs, with one member of the pair located at the point where magnetic-field lines point upward, away from the Sun's surface, and the other at the point where they point downward toward the surface **(Fig. 22.8c)**. Bursts of radiation and plasma from sunspots briefly increase the luminosity of the Sun. When many sunspots exist and frequent energy bursts take place, the Sun's average luminosity increases by about 0.1%. As we'll see shortly, these energy and plasma bursts are known as *solar storms*.

The number of sunspots varies over time **(Fig. 22.8d)**. On average, this variation takes place over an 11-year

FIGURE 22.8 Sunspots and the solar cycle.

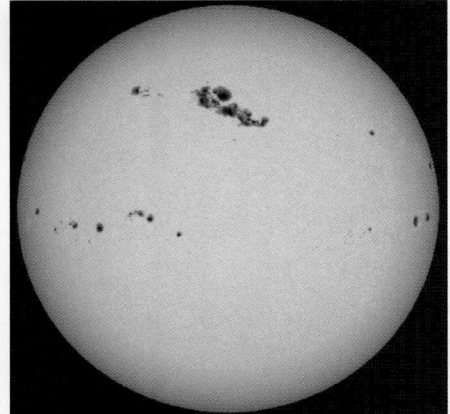

(a) Sunspots are dark patches on the photosphere.

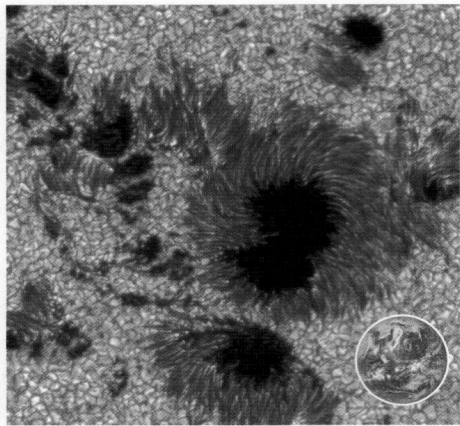

(b) A close-up shows that a sunspot has a cooler, darker central area and a warmer outer area. The white circle represents the Earth's diameter, for scale.

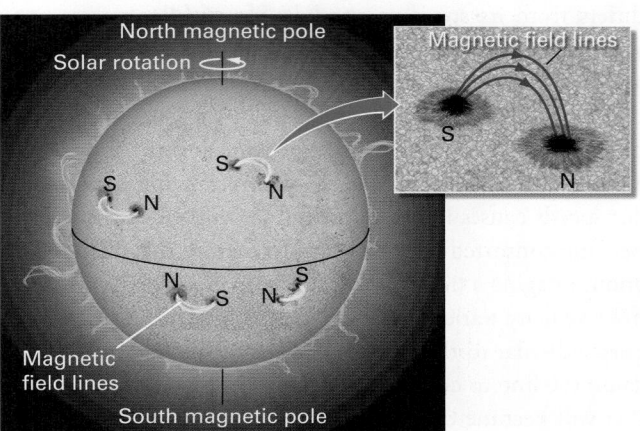

(c) Sunspots form where strong magnetic-field lines arc out from the solar surface. The inset emphasizes that the field lines point outward from one spot and inward at the other.

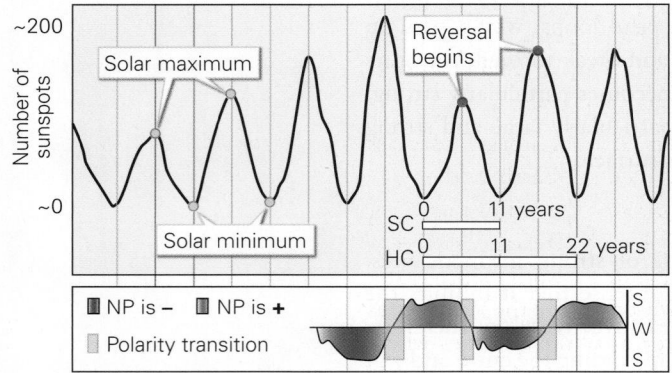

(d) The upper graph shows that the number of sunspots varies over time: they are most numerous at the solar maximum and least numerous at the solar minimum. A solar cycle runs from minimum to minimum and lasts, on average, 11 years. Reversals begin just after the solar maximum. (NP = north magnetic pole; SC = solar cycle; HC = Hale cycle.) The lower graph shows polarity reversals, and variations in field strength between reversals. (S = strong; W = weak)

(e) Solar luminosity varies with the number of sunspots. When there is "low activity" (few sunspots), the Sun is slightly dimmer than when there is "high activity" (many sunspots).

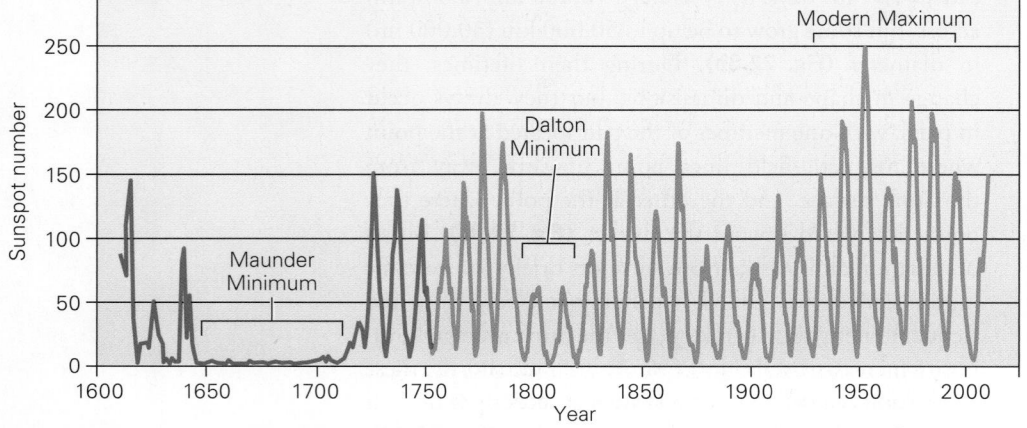

(f) Over the course of centuries, the number of sunspots at the solar maximum has varied. During the Maunder Minimum, solar maxima were nearly nonexistent.

cycle, known as the **solar cycle**, or *sunspot cycle*. The solar cycle starts at the **solar minimum**. At this time, when very few sunspots exist, those that occur are mostly at low latitudes, because the belts of deep magnetic-field lines lie near the equator. The number of sunspots increases until the **solar maximum**, when the most sunspots exist. At the solar maximum, sunspots lie mostly at mid-latitudes, as does the belt of deep magnetic-field lines. Following the solar maximum, the band of sunspots, along with the deep belt of magnetic-field lines, migrates back toward the equator. A plot of sunspot number versus time over a solar cycle yields an approximately bell-shaped curve. Note that the Sun's average luminosity reaches its greatest value at the solar maximum and its smallest value at the solar minimum **(Fig. 22.8e)**, and that the Sun's polarity reverses at the beginning of the interval between a solar maximum and a solar minimum (see the lower graph of Fig. 22.8d). The Sun's complete magnetic-field cycle, during which polarity reverses and then reverses again, is called the **Hale cycle**, or *magnetic cycle*. It includes two solar cycles and therefore lasts, on average, about 22 years (see Fig. 22.8d).

Sunspots have been recorded over the course of several centuries, and scientists have kept detailed records of solar cycles since 1755. Solar cycle 25 began in December 2019 and reaches solar maximum in about mid-2024. While solar cycles have an average duration of 11 years, they are not strictly periodic; the shortest on record lasted 9.0 years and the longest lasted 13.7 years. Over the centuries, the number of sunspots at the solar maximum has sometimes attained large values for several successive cycles, and at other times it has remained low for several successive cycles **(Fig. 22.8f)**. In fact, during an interval called the *Maunder Minimum* in the second half of the 17th century, solar maxima were hardly detectable.

Solar Storms

Occasionally, local intensification or disruption of the Sun's magnetic field triggers the ejection of particularly large amounts of high-energy particles into space. These events, called **solar storms**, usually take place in association with sunspot maxima in the solar cycle. Astronomers distinguish three types of events associated with solar storms. During a **solar prominence**, outbursts of glowing gas and plasma emerge from the Sun's surface at one sunspot in a pair and follow arcing magnetic-field lines to the other member of the pair, where they return to the Sun's surface **(Fig. 22.9a)**. Solar prominences can last for hours, days, or weeks. Occasionally, a **solar flare**—an intense flash of electromagnetic radiation—bursts from the Sun's surface, generally from an area of sunspots. A solar flare, which can last minutes to hours, releases tremendous amounts of electromagnetic energy. Traveling at the speed of light, that energy takes 8 minutes to reach the Earth **(Fig. 22.9b)**. Energy from a solar flare can disrupt the part of the Earth's atmosphere through which radio waves travel, degrading or, at worst, temporarily blacking out navigation and communications signals. A third type of event, called a **coronal mass ejection (CME)**, hurls a huge volume of solar plasma directly into space **(Fig. 22.9c)**. The plasma of a CME moves much more slowly than the speed of light, but it travels significantly faster than the typical solar wind—some particles reach speeds of over 3,000 km/s (6.7 million mph). The high-energy plasma streams into space in a specific direction **(Fig. 22.9d)**, so it affects the Earth only when our planet enters the path of the ejected particles **(Fig. 22.9e)**. When particles from CMEs reach the Earth, they produce strong disturbances of the Earth's magnetic field that can pose a significant hazard to human activities.

Space weather refers to the character of electromagnetic radiation and the abundance and speed of charged particles in the region of space around the Earth. Milder space weather happens when only weak or moderate solar wind passes through the region, whereas strong or intense space weather occurs when the output of solar flares or CMEs reaches the Earth. The Earth's magnetic field and atmosphere fortunately shield the Earth's surface from most of the particles and electromagnetic energy of space weather, but spacecraft in orbit can be directly impacted. In fact, during a 1989 episode of intense space weather, astronauts in the space shuttle reported a burning sensation in their eyes, so Mission Control ordered them to retreat to the most shielded part of the spacecraft. While the particles from the solar wind and solar storms themselves are not a direct health risk to people on the Earth, they do produce several notable effects:

- *Low-latitude aurorae:* During intense space weather, incoming solar particles have enough energy to produce aurorae much farther from the poles than usual. For example, during the 1989 event, aurorae were visible in Florida and Texas, and in 2023, aurorae could be seen as far south as the Carolinas.

- *Heating of the thermosphere:* A solar storm's particles and radiation, when they reach the Earth, add energy to the thermosphere, the layer of the atmosphere that lies above about 90 km (56 mi). This energy causes the thermosphere to expand outward, since gases expand when heated. Consequently, the concentration of air molecules increases at the altitudes where over 1,000 satellites, and more than 10,000 fragments of *space junk* (defunct satellites and loose pieces of equipment), currently orbit. As satellites and space junk interact with air molecules, they slow down and lose altitude. When this happens, their orbits become less predictable, and collisions between objects may be more frequent. In fact, some satellites or pieces of space junk may slow so much that they fall toward the Earth, where they may become hazards to surface inhabitants.

- *Disruption of radio transmissions:* CMEs generate radio waves that interfere with radio transmissions from the Earth. Radio noise generated by CMEs can disrupt the communications of satellites, cell phones, and the global positioning system (GPS). GPS not only serves as the basis for navigation, but also provides essential time signals to numerous computer applications, so such disruptions can have severe consequences.

- *Electrical discharges on satellites:* The influx of charged particles causes a buildup of static electricity in and on satellites, eventually leading to sparks that may damage electronics in the satellites.

FIGURE 22.9 Examples of events caused by solar storms.

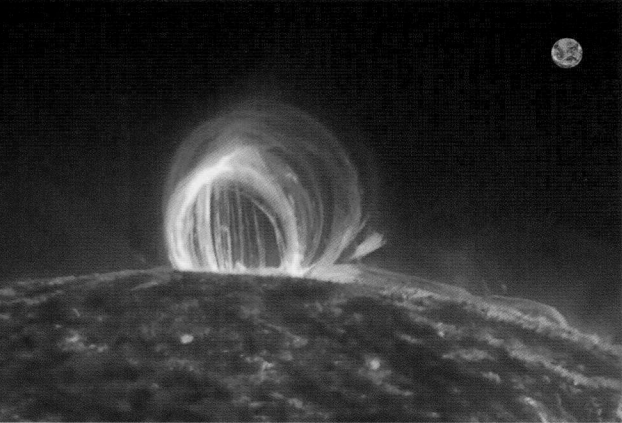

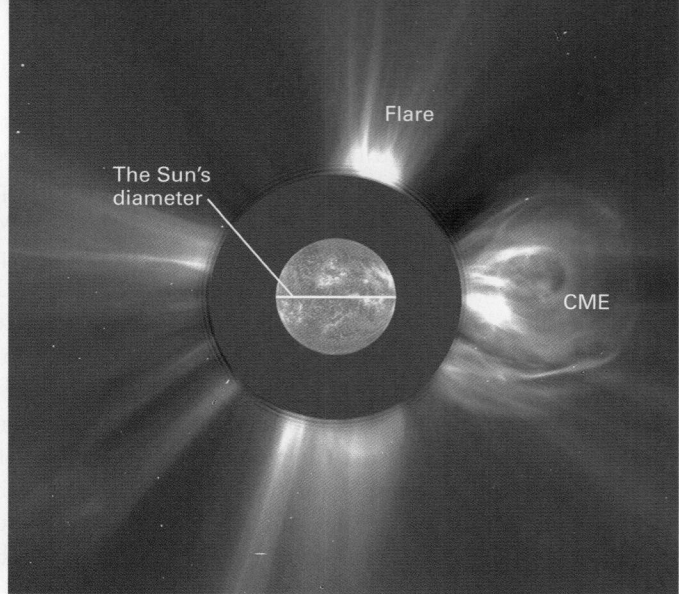

(a) Solar prominences follow arcing magnetic-field lines between sunspots and therefore form a loop. The Earth is shown for scale.

(b) Solar flares are giant flashes of electromagnetic energy.

(c) Coronal mass ejections (CMEs) release a huge number of charged particles into space.

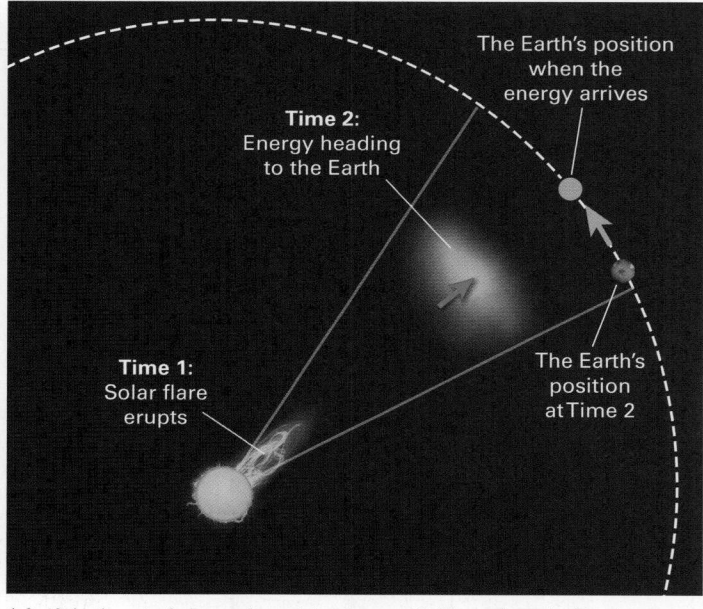

(d) CMEs eject plasma in specific directions.

(e) If the burst of charged particles from a CME reaches the Earth, severe space weather develops.

- *Disruption of electrical grids:* Solar particles moving in the Earth's magnetic field can induce rogue electric currents in electrical transmission lines, producing a power surge that triggers a shutdown of an electrical grid. Because the electrical grids of different regions are interconnected, one failure can lead to a cascade of failures affecting a very broad region. In fact, the 1989 failure of the grid in Quebec almost knocked out other grids across North America. Loss of power shuts off lights, refrigerators, computers, and anything else that runs on electricity. The resulting blackout can trigger chaos at airports, crises in financial markets, and looting in cities.

- *Damage to electronic equipment:* Electric currents generated by intense space weather can cause local power surges, putting electronic equipment out of commission. The consequences of such widespread failure are almost inconceivable, for nearly every aspect of our modern society relies on computers and the internet.

The most intense episode of space weather on record happened in 1859. This episode, known as the *Carrington event,* caused amazing aurorae and triggered electrical surges in telegraph lines that started fires. But fortunately, in the pre-electronic world of 1859, most people didn't notice it. The risk that space weather poses to society increases every year as we become more and more dependent on electric power, electronic equipment, satellite transmissions, and the internet. What can be done to prevent a space-weather catastrophe? Satellite-based electronic equipment can be *radiation hardened*, meaning that it can be protected from an intense influx of charged particles and the currents they generate. Methods of protection include mounting computer chips on insulated surfaces, covering chips with a resistant coating, or encasing equipment in resistant shells. On the Earth's surface, electrical grids can be designed to be more resistant to surges. For example, *smart electrical grids* can redirect excess electricity into other grids or into capacitors (electronic devices that store energy). Power companies can prepare replacement *transformers* (devices that transfer electrical energy from one electrical circuit to another) so that they can be activated the instant an active transformer gets knocked out. To meet this challenge in the United States, the US National Weather Service (working with the military, NASA, and other agencies) has set up a *Space Weather Prediction Center* to watch for solar storms and send out warnings to utilities and government agencies so that they can take action to prevent damage to electrical grids, electronic equipment, and satellites.

Take-home message...

🏠 The Sun generates an intense magnetic field. Local intensification of this field causes magnetic-field lines to arc into space. Solar prominences are arcs of plasma that flow between sunspots. Particularly large releases of energy, generally emerging from clusters of sunspots, are called solar flares. During a coronal mass ejection, huge bursts of highly energetic plasma erupt and head into space. Space weather refers to the character of radiation and concentration of plasma in the vicinity of the Earth. Intense solar storms have the potential to disrupt satellite operations and electrical grids.

Quick Questions

- What causes sunspots?
- What is the difference between a solar flare and a coronal mass ejection?
- How can space weather affect human society?

22.3 A Diversity of Stars

From the Earth, all stars look like points of light, but some appear brighter than others, and some emit what appears to be red light, others blue light, and others at all other wavelengths in between. What factors cause these differences? Research reveals that differences in brightness are due to differences in the amount of energy that a star emits, the distance of the star from the Earth, and the size of the star, whereas differences in color depend on the temperature of the star's photosphere.

Variations in Stellar Brightness

Astronomers refer to the amount of electromagnetic energy emitted by a star in a specified amount of time as the star's **luminosity**, or *true brightness*. You can think of luminosity as a measure of *power* (the amount of energy transmitted per unit of time): a more luminous star produces more power than a less luminous star does, just as a more powerful light bulb releases more light than a less powerful light bulb. Luminosity does not depend on the distance between an observer and the star.

Stellar brightness as viewed by an earthbound observer depends not only on a star's luminosity, but also on the star's distance from the Earth. That's because, as energy radiates from a source, it spreads out. You see the same phenomenon when you look toward a light bulb at night. At a distance, the light looks small and dim, but close up, it appears large and bright **(Fig. 22.10a)**. The size and power of the light haven't changed, but your eye detects a smaller amount of the energy from a distant light than from a nearby light.

The amount of energy from a star that passes through a square meter at the surface of the Earth in 1 second defines the star's **apparent brightness**. Put another way, apparent brightness refers to the brightness of a star, relative to others, when you look at the night sky. When astronomers describe the apparent brightness of stars, they use a scheme that dates back to the ancient Greek astronomer Hipparchus (ca. 190–120 B.C.E.). Hipparchus classified stars visible to the naked eye by using a scale of **apparent magnitudes**, with 1 being the brightest and 6 being the faintest **(Fig. 22.10b)**. Modern astronomers have extended the scale in both directions to include fainter objects that Hipparchus couldn't see, as well as much brighter objects such as the Sun and the Moon. The lower the number, the brighter the object is, and the higher the number, the dimmer it is. On this logarithmic scale, every five steps in magnitude equals a hundredfold decrease in brightness. For example, magnitude 10 is 100 times dimmer than magnitude 5, which is 100 times dimmer than magnitude 0. Catalogs of stars used by amateur and professional astronomers today provide the apparent magnitudes of stars and other objects in the sky. Objects such as the Sun that are brighter than the brightest stars have a negative apparent magnitude. Specifically, the Sun's apparent magnitude is −26.7, the full Moon's is −12.5, and Venus's is −4.4. Stars that you can't see with the naked eye, but that can be detected with telescopes, have apparent magnitudes

Did you
ever wonder . . .

whether all stars are the same?

FIGURE 22.10 The concept of stellar magnitude.

(a) The same light bulb looks dimmer to you as its distance from you increases.

(b) Sirius looks particularly bright relative to the other stars around it in the night sky, so its apparent magnitude is greater than that of its neighbors.

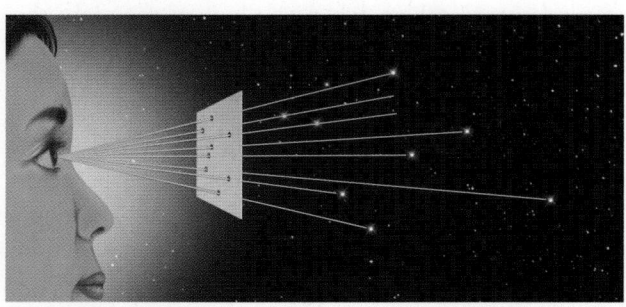

Observer on Earth   Projections of stars   Stars at different distances

(c) The absolute magnitude is the brightness a star would have if positioned at a specific distance (32.6 light-years) from the Earth.

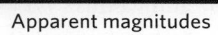
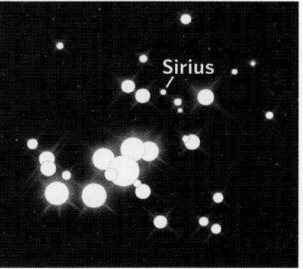

Apparent magnitudes   Absolute magnitudes

(d) A comparison of apparent magnitude and absolute magnitude for the stars around Sirius.

greater than 6. The Hubble Space Telescope can detect objects with an apparent magnitude of 30.

To distinguish among stars by luminosity (true brightness), astronomers use two different scales. The **absolute magnitude** scale tells us what a star's apparent magnitude would be if the star were magically placed at exactly 32.6 light-years (10 parsecs) from the Earth **(Fig. 22.10c, d)**. This scale accommodates the effect of distance on stellar brightness. For example, the Sun has an apparent magnitude of −26.7 (as the brightest object in the sky by far), but it's actually a fairly ordinary star with an absolute magnitude of only +4.9. In contrast, Rigel, a star that sits 870 light-years away from the Earth in the constellation Orion, emits 47,000 times as much light as the Sun does, but because of its distance, its apparent magnitude is +0.1, much less than that of the Sun. Because of its luminosity, though, Rigel has an absolute magnitude of −7.84. Astronomers also express a star's luminosity using the *solar unit scale*, in which 1 **solar unit** equals the average luminosity of our Sun. A star's luminosity, expressed in solar units, equals that star's true luminosity divided by the Sun's luminosity. Rigel, for example, has a luminosity of 120,000 solar units. Our nearest neighbor, Proxima Centauri, has an absolute magnitude of 15.6, and a luminosity of less than 0.0015 solar units, because it generates only one-thousandth as much energy per second as the Sun.

Variations in Stellar Surface Temperature and Color

Different stars emit visible light of different colors, meaning that they emit different wavelengths of radiation. As we discussed in Chapter 16, the wavelengths emitted by an object depend on the temperature of the object, according to the laws of *blackbody radiation*. The color of starlight, therefore, depends on the temperature of the star's photosphere. The coolest stars emit maximum radiation in the red part of the spectrum, while the hottest emit maximum radiation at blue wavelengths. Note that our Sun emits radiation at all wavelengths, with its maximum radiation in green wavelengths. Yet, to astronauts in space, the Sun appears white, while to us at the ground, the noontime Sun appears yellow, and the setting Sun appears red. From the information provided by the cells in the eyes, the human brain interprets the full spectrum of colors emitted by the Sun as white. The atmosphere at noon scatters the blues and violets out of direct sunlight, and our brains interpret the remaining spectral colors as yellow. At sunset, the green and yellow wavelengths are also scattered out of the sunlight, and the Sun appears red.

Astronomers divide stars into **spectral types** on the basis their surface temperatures. These types are named, from hottest to coldest, O, B, A, F, G, K, and M **(Table 22.1)**. There are gradations between these spectral color categories, so each category includes stars in a range of temperatures. To add greater specificity, each letter category can be divided into 10 subcategories, numbered 0 through 9 (with 0 hotter than 9). According to this classification scheme, our Sun, a G2 star, is cooler than an A0 star and hotter than an M2 star.

Variations in Stellar Radius and Mass

Stars not only come in a wide range of colors, but also in a tremendous range of sizes **(Fig. 22.11)**. Astronomers calculate a star's radius from its temperature and luminosity using the laws of blackbody radiation. Analysis of thousands of stars reveals that the smallest stars are Earth-sized (but are much denser than the Earth), whereas one of the largest stars, Betelgeuse, a bright star visible to the naked eye, has a radius 1,000 times that of our Sun. In fact, recent studies with the James Webb Space Telescope show that the largest known star, UY Scuti, which lies about 10,000 light-years from the Earth, has a radius that is 1,700 times that of the Sun. **Dwarf stars** that have fusion reactions occurring in their cores range in size from 10 times the radius of the Sun down to 30% of the solar radius. So, while the Sun seems huge, it's actually considered to be a dwarf star.

Notably, dwarf stars occur in a range of colors and sizes, and not all are true stars (in the sense that not all produce energy via nuclear fusion). A *white dwarf*, for example, is the super-dense remnant of a certain type of dead star in which nuclear fusion reactions no longer occur. It glows because of the stored thermal energy left over from an earlier stage of its life. White dwarfs can have a radius as small as 1% that of the Sun. Another type of dwarf, called a *brown dwarf*, is a huge gas-giant planet that emits detectable light, but never became large enough for nuclear fusion to begin in its core; in effect, it is a small protostar.

Giant stars, most of which are *red giants*, typically have radii 10–50 times that of the Sun, and **supergiant stars** have radii 50–500 times that of the Sun. Smaller stars are much more abundant than large stars. In fact, about 99% of stars are dwarfs.

Stars also come in a huge range of masses. The first calculations of stellar mass came from studying **binary stars**, pairs of stars that orbit around a common center of mass, for the distance between the stars in a binary pair, and the velocity of their orbits, depend on the masses of the two companions within the binary star. Later, astronomers found that a star's mass can be related to the star's luminosity, so an observation of luminosity can provide an estimate of mass. Astronomers describe the masses of stars by comparing them with our Sun: one **solar mass** (M_S) represents the mass of our Sun. *Low-mass stars* contain less than 0.5 M_S, *intermediate-mass stars* contain

0.5–8 M_S, and *high-mass stars* contain 8–40 M_S. The boundaries between these groups are based on both the lifetime of the stars and their fate once nuclear fusion ceases in their cores. There are relatively few *supermassive* stars, with masses greater than 40 M_S; the largest supermassive star has a mass of 150 M_S. Such stars are rare because their lifetimes are exceptionally short relative to other stars. Because the density of stars can vary so much, the mass of a star does not correlate with its radius. For example, a planet-sized white dwarf can contain as much as 1.2 times the mass of our Sun, and a red giant with the same mass as our Sun can have a radius over 200 times greater. Notably, calculations suggest that the minimum mass required for a body to become hot enough inside to initiate fusion reactions, and therefore transform from a brown dwarf into a star, is about 85 times that of Jupiter.

The H-R Diagram: A Way to Classify Stars

In the early 20th century, two astronomers, Ejnar Hertzsprung of Denmark and Henry Norris Russell of the United States, independently examined the relationship between stellar luminosity (expressed either as solar units or as absolute magnitude) and surface temperature. The diagram they developed, now known as the

TABLE 22.1 Classification of Stars into Spectral Types

Spectral type	Approximate surface temperature	Color	Example
O	30,000°C (54,000°F)	Electric blue	Mintaka (O9)
B	20,000°C (36,000°F)	Blue	Rigel (B8)
A	10,000°C (18,000°F)	White	Vega (A0)
F	7,000°C (12,600°F)	Yellow-white	Canopus (F0)
G	6,000°C (10,800°F)	Yellow	Sun (G2)
K	4,000°C (7,200°F)	Orange	Arcturus (K2)
M	3,000°C (5,400°F)	Red	Betelgeuse (M2)

FIGURE 22.11 The radii and spectral colors of stars vary significantly, as we can see by comparing several stars that astronomers have named. Betelgeuse, a red supergiant, has a radius 1,000 times that of our Sun.

FIGURE 22.12 Hertzsprung-Russell diagrams.

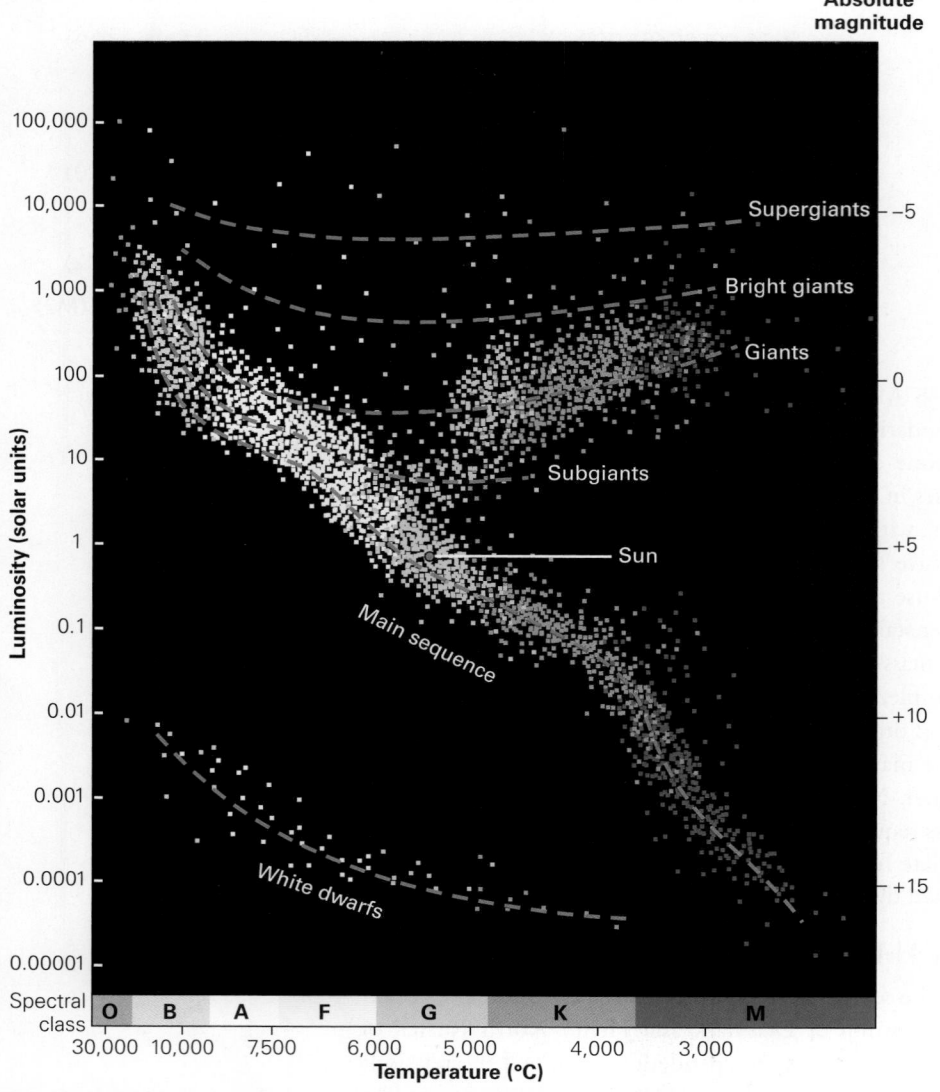

(a) An H-R diagram shows the relationship between luminosity and temperature (represented by spectral type). Most stars, including the Sun, fall on the main sequence.

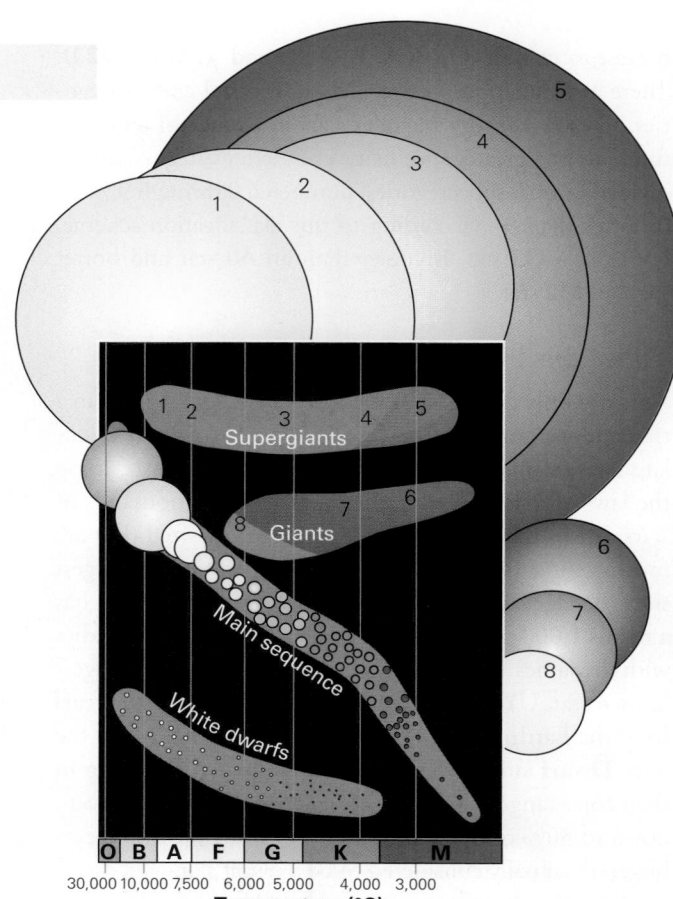

(b) The different groups of stars on the diagram correspond to different star sizes. The spheres show the range of sizes of stars, to scale. Numbers indicate where those stars appear on the H-R diagram.

Hertzsprung-Russell diagram (commonly abbreviated as **H-R diagram**), has become one of the most important tools for classifying stars and, as we will see, for understanding stellar life cycles **(Fig. 22.12a)**. An H-R diagram is a graph with surface temperature (corresponding to spectral type) on the horizontal axis and luminosity on the vertical axis. Data obtained for thousands of stars, when plotted on the diagram, show a relationship between luminosity and spectral type. Stars fall into distinct groups, each of which appears as a band spanning part of the diagram. Because of the interrelationships among luminosity, size, and temperature, each group on an H-R diagram consists of stars that lie within a particular range of radii **(Fig. 22.12b)**.

Most stars, including our Sun, lie in a narrow band that stretches diagonally across the diagram from the upper left to the lower right. This band is called the **main sequence**, and a star that falls within it is called a *main-sequence star*. Most main-sequence stars have a mass less than 8 M_S and are classified as dwarf stars. Our Sun, a yellow dwarf of spectral type G, falls at roughly the midpoint of the main sequence. The hottest, brightest stars (O and B) on the main sequence are also the most massive stars. The coolest, faintest stars (K and M) are the least massive. Astronomers can make a reasonable estimate of a star's mass by locating its position on the main sequence in the H-R diagram.

While the main sequence of an H-R diagram includes a great variety of stars that differ from one another in terms of mass, radius, temperature, and luminosity, not all stars fall on the main sequence. Smaller numbers of stars fall into three other areas (see Fig. 22.12b). *Giants* populate an area above and to the right of the main sequence. These stars have radii 5–100 times that of the Sun, but have surface temperatures of only 3,000°C–5,500°C (5,400°F–9,900°F). The red giants are the stars with the lowest temperatures in this group. Between the giants and main-sequence stars are *subgiants*, most of which are

In a diffuse nebula, stars cannot start to form.	A shock wave from a supernova compresses the nebula.	Some regions of the nebula become dense enough for stars to form.

FIGURE 22.13 The evolution of a nebula into a star nursery may begin when shock waves from a supernova pass through the nebula.

Time →

(a) Compression due to a shock wave causes pockets of matter to become dense enough for gravitational collapse to begin.

spectral type G. Above the red giants lie the even rarer *bright giants* and *supergiants*. White dwarfs, which are small, very dense stars, the remnants of low and intermediate mass stars after nuclear fusion ceases, fall below the main sequence, in the lower left part of the H-R diagram.

Take-home message...

The apparent brightness (apparent magnitude) of a star that we see from the Earth depends not only on the star's luminosity (the amount of energy it emits), but also on its distance from the Earth. Absolute magnitude—the brightness we would see if the star were exactly 32.6 light-years away—provides a better characterization of a star's luminosity. Astronomers commonly indicate luminosity using solar units, a comparison with the luminosity of the Sun. Stars vary greatly in their luminosity, temperature, radius, and mass. We can represent variations in luminosity and temperature on an H-R diagram; this diagram serves as a key to classifying stars. Most stars fall on the diagram's main sequence.

Quick Questions

- Why is the Sun considered a dwarf star?
- What characteristic defines a star's spectral type?
- Is the Sun a main-sequence star?

22.4 The Life of a Star

A single glance at the sky provides only an instantaneous snapshot of the Sun and stars. If you could observe stars over the course of billions of years, and in some cases, just millions of years, you would see significant changes in their character, manifested primarily by changes in their luminosity, radius, and mass. In other words, stars evolve over time—but the rate of change isn't uniform. Stellar evolution happens rapidly during the birth and infancy

(b) The Henize 206 Nebula, 163,000 light-years from the Earth, contains stars that are less than 10 million years old.

of a star, slowly during most of the star's life, and then very rapidly at the end of the star's life. In this section, we consider stellar evolution from a star's birth through the beginning of its end. Astronomers have deciphered this process, even though the steps take vastly more time than all of human history, by examining different stars, each at a different stage of life.

The Evolution of a Star from Birth through Middle Age

The process of star formation begins when a disturbance causes a nebula (composed of gases, ices, and dust) to become relatively denser in a localized region (see Chapter 1). Such a disturbance might be caused, for example, by the passage of a **shock wave**—a pulse of pressure—sent outward by an immense supernova explosion, which locally compresses mass **(Fig. 22.13a)**. Typically, many distinct pockets of higher density develop within a given nebula, and each pocket may become a separate star **(Fig. 22.13b)**.

Animation
Stellar Evolution

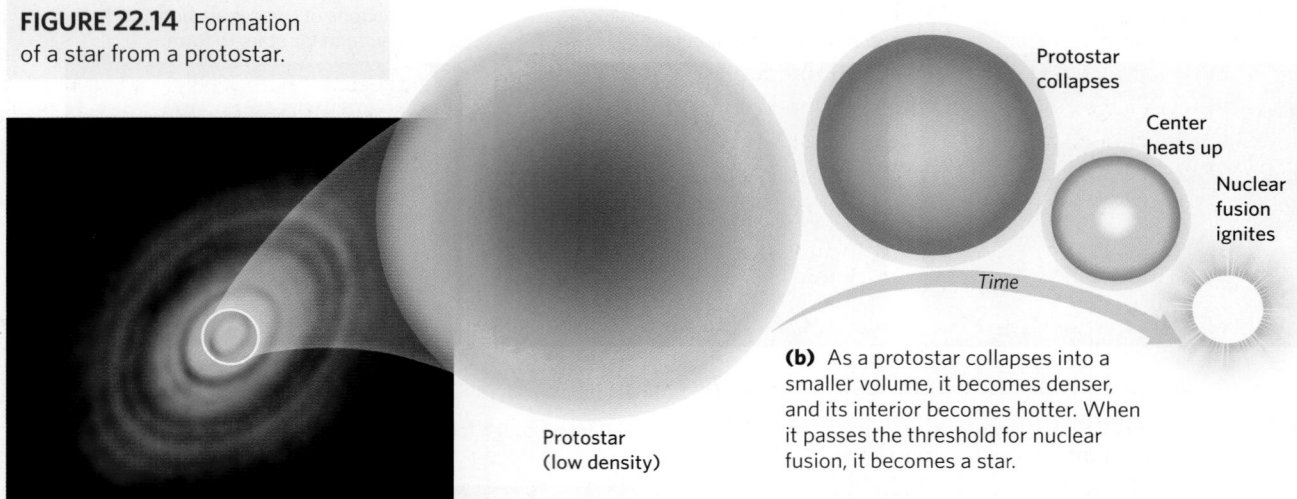

FIGURE 22.14 Formation of a star from a protostar.

Protostar collapses

Center heats up

Nuclear fusion ignites

Time

Protostar (low density)

(b) As a protostar collapses into a smaller volume, it becomes denser, and its interior becomes hotter. When it passes the threshold for nuclear fusion, it becomes a star.

(a) A Hubble Space Telescope photo of HL Tauri, an accretion disk about 450 light-years from the Earth. Astronomers estimate that this disk is only 100,000 years old.

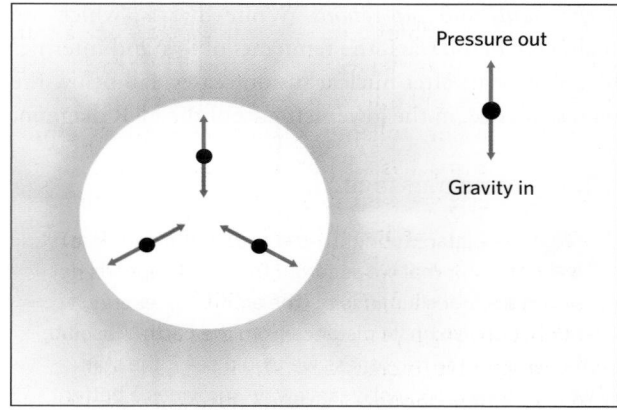

Pressure out

Gravity in

(c) The concept of equilibrium pressure in a star.

When a shock wave causes local compression in a nebula, the compressed region won't undergo gravitational collapse to form stars if the nebula consists only of gas. Recall that, according to **condensation theory** (see Chapter 1), gravitational collapse to form stars like our Sun also requires the presence of dust and ice, for these materials radiate heat out of the nebula more efficiently than does gas alone, allowing the nebula to remain cool enough for long enough to begin inward collapse. These particles also provide condensation nuclei on which other particles collect, building into masses that become large enough for their gravity to pull in other matter. As a patch of a nebula collapses inward, the rotation rate of the original nebula increases to conserve angular momentum. Gravity, together with the centrifugal force caused by rotation, flattens the collapsing material into an *accretion disk* **(Fig. 22.14a)**. At the center of this disk, a nearly spherical mass—a *protostar*—grows.

Compression of a gas increases its temperature, but in the opaque interior of the protostar, radiation can't escape, so as gravitational collapse progresses and the radius of the protostar shrinks, the temperature in its core progressively rises. A protostar that contains an amount of mass comparable to that of our Sun passes a milestone about 20 million years after gravitational collapse begins: its core temperature reaches the threshold—10 million °C, or 18 million °F—for nuclear fusion reactions. When this happens, hydrogen atoms undergo fusion to form helium, releasing vast amounts of energy in the process, and the protostar becomes a true star that emits bright light into space **(Fig. 22.14b)**. Over the next

30 million years or so, an intermediate-mass star like our Sun contracts further, its core temperature increases to about 15 million °C (27 million °F), and its photosphere temperature reaches about 6,000°C (10,800°F). At this point, the star reaches an *equilibrium condition*, in that its temperature and radius stabilize and remain much the same for a long time (billions of years, for a star such as our Sun). The radius of the star when it achieves equilibrium is called the *equilibrium radius*, and its luminosity, represented by its surface temperature, is the *equilibrium luminosity*. The equilibrium radius of a star depends on the balance between two forces: the inward pull of gravity, and an outward-directed *thermal pressure* caused by the tendency of hot material—the plasma in a star's interior—to expand **(Fig. 22.14c)**. Notice that, at equilibrium, the star's radius, mass, temperature, and luminosity are related. The temperature of a star in equilibrium depends on the star's mass. A more massive star has a denser core and, therefore, a hotter core temperature, so fusion reactions happen more quickly. Mass controls a star's equilibrium luminosity as well: a star with greater mass produces more energy.

Notably, not all accretion disks produce a single star. Quite commonly, a gas-giant planet forming from rings of matter in the accretion disk around one forming star becomes large enough that nuclear fusion begins within it. In that case, it, too, will become a star. This situation yields a binary star **(Fig. 22.15)**. More than half of all known stars are members of a binary-star pair.

Depicting the Evolution of a Star on an H-R Diagram

Changes that take place as a star forms and evolves can be depicted on an H-R diagram. Let's summarize these changes for a main-sequence star of intermediate mass, like our Sun. The numbered points on the H-R diagram in **Figure 22.16** refer to the numbers indicated in the following steps.

Initially, a patch of a nebula that will become a star, but has not yet undergone gravitational collapse, has low luminosity and a low surface temperature. As time passes, this volume collapses inward and becomes a protostar. A protostar has a large radius, and because it has so much surface area, it can have a relatively high luminosity (Point 1). As it continues to contract, but does not yet contain a nuclear furnace, it has a low surface temperature (about 2,700°C, or 4,900°F), so it appears red (Point 2). As a protostar contracts further, its surface temperature increases while its radius decreases. Therefore, even though its temperature is greater, its luminosity decreases (Point 3). Once fusion reactions begin, the star's surface temperature and luminosity both increase, and its position on the H-R diagram begins to move toward the main sequence (Point 4). Eventually, the star achieves an equilibrium size and luminosity. At this time, the star arrives at a position on the main sequence (Point 5). This position won't change much until the star begins to die. In other words, the main sequence represents stars within a certain range of mass that have achieved equilibrium. Stars stay on the main sequence for most of their lifetimes, as long as they remain in equilibrium.

The specific spot where a given star falls on the main sequence of an H-R diagram depends on the mass of the star (which also determines the spectral type). Intermediate-mass stars comparable to the Sun follow tracks similar to that shown in Figure 22.16 and end up as G or F stars. Those with a mass greater than the Sun's have a greater equilibrium size and luminosity, and they end up at the upper left of the main sequence as B or O stars. (Notably, B and O stars are much larger than G and F stars.) Stars with a mass less than the Sun's have a smaller equilibrium size and luminosity and end up at the lower right of the main sequence, as M or K stars.

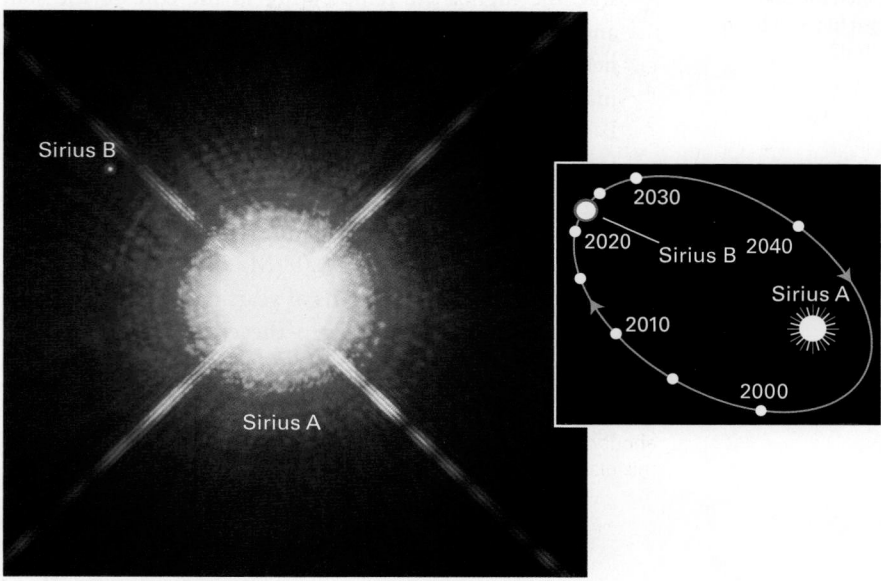

FIGURE 22.15 An example of a binary star. To the naked eye, Sirius A looks like a single star. A Hubble Space Telescope image reveals two separate stars that are orbiting each other. The second, Sirius B, although it is much smaller than Sirius A, is a white dwarf with a mass that is half that of Sirius A. The inset shows their orbit.

FIGURE 22.16 The beginning of the life of a star with the mass of the Sun can be depicted by a succession of points on an H-R diagram. The star moves from the upper right of the diagram down to the main sequence.

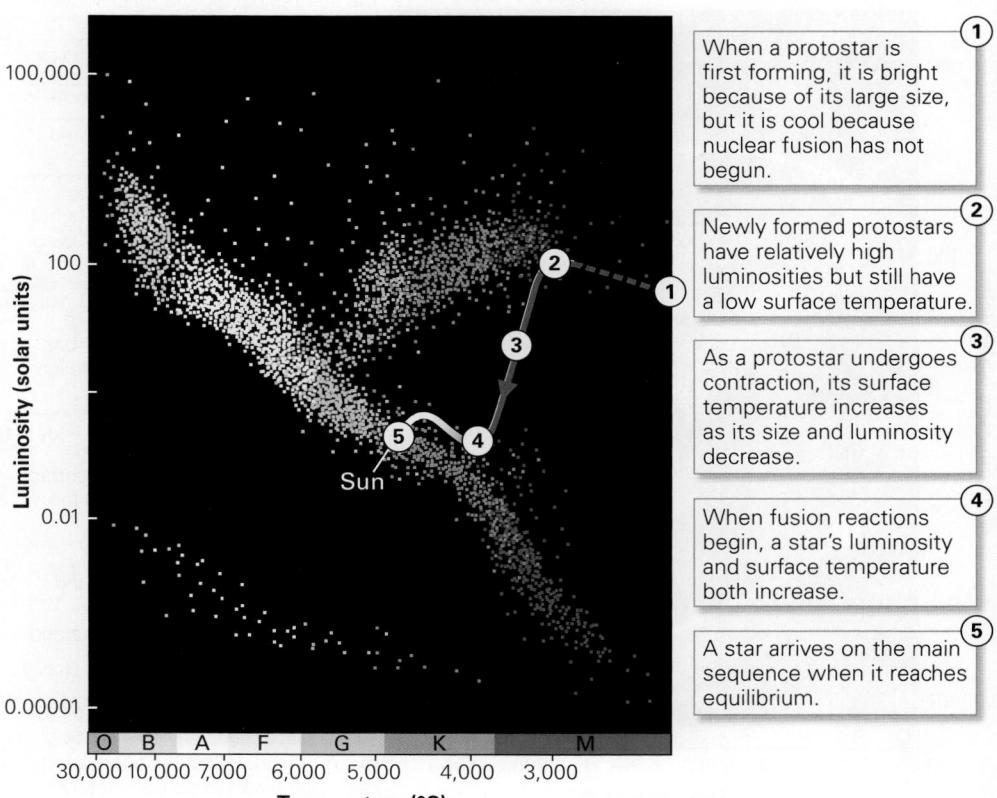

① When a protostar is first forming, it is bright because of its large size, but it is cool because nuclear fusion has not begun.

② Newly formed protostars have relatively high luminosities but still have a low surface temperature.

③ As a protostar undergoes contraction, its surface temperature increases as its size and luminosity decrease.

④ When fusion reactions begin, a star's luminosity and surface temperature both increase.

⑤ A star arrives on the main sequence when it reaches equilibrium.

How Long Does a Star Survive?

Did you ever wonder...

what will happen to the Earth when the Sun's hydrogen fuel has been exhausted?

A star remains in equilibrium until the hydrogen fuel in its core has been used up. How long this equilibrium lasts depends on the star's total mass. The more massive the star, the higher the temperature in the core of the star, and the faster the rate at which hydrogen fuses to form helium. A G star with a mass of 1 M_S can remain on the main sequence for about 10 billion years. In contrast, a B star with a mass of 5 M_S resides on the main sequence for only about 100 million years, and an O star with a mass of 10 M_S exhausts its hydrogen fuel in just 20 million years. Clearly, lower-mass stars have longer lives. In fact, stars with masses between 0.08 M_S and 0.25 M_S can survive for hundreds of billions of years. Such low-mass stars are called **red dwarfs**. Given the current estimated age of the Universe (13.8 billion years), no star with a mass this low has yet consumed its hydrogen and left the main sequence.

Take-home message...

A star forms from the protostar at the center of an accretion disk when the core of the protostar becomes hot enough for fusion reactions to begin. During its infancy, a star evolves rapidly, but eventually its luminosity and surface temperature reach an equilibrium that depends on its mass. Such stars lie on the main sequence of an H-R diagram. High-mass stars have shorter lives than low-mass stars because they consume their hydrogen fuel at a faster rate.

Quick Questions

- What forces balance each other inside a star when the star reaches an equilibrium condition?
- What variables are on the horizontal and vertical axes of the H-R diagram?
- Why do red dwarfs reside on the main sequence longer than stars with a mass similar to that of the Sun?

FIGURE 22.17 The death of a star with the mass of the Sun can be depicted by a succession of points on an H-R diagram. The star leaves the main sequence and follows a large loop around the diagram.

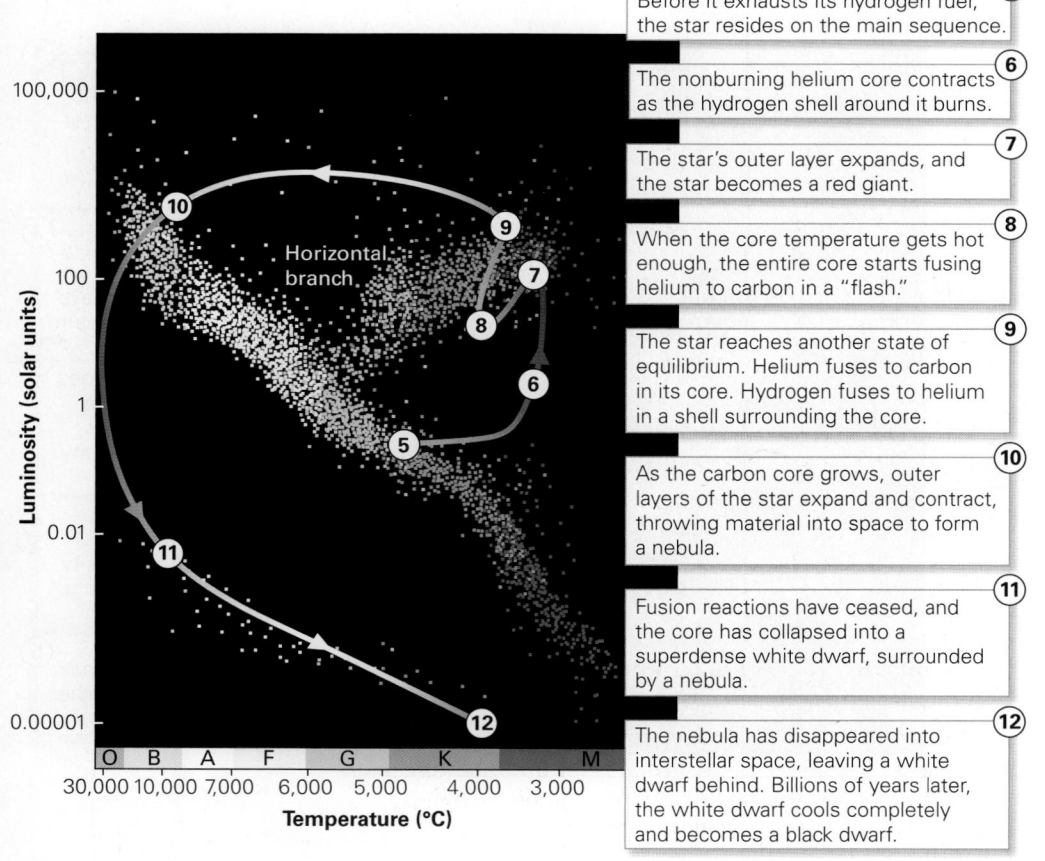

⑤ Before it exhausts its hydrogen fuel, the star resides on the main sequence.

⑥ The nonburning helium core contracts as the hydrogen shell around it burns.

⑦ The star's outer layer expands, and the star becomes a red giant.

⑧ When the core temperature gets hot enough, the entire core starts fusing helium to carbon in a "flash."

⑨ The star reaches another state of equilibrium. Helium fuses to carbon in its core. Hydrogen fuses to helium in a shell surrounding the core.

⑩ As the carbon core grows, outer layers of the star expand and contract, throwing material into space to form a nebula.

⑪ Fusion reactions have ceased, and the core has collapsed into a superdense white dwarf, surrounded by a nebula.

⑫ The nebula has disappeared into interstellar space, leaving a white dwarf behind. Billions of years later, the white dwarf cools completely and becomes a black dwarf.

22.5 The Deaths of Stars and the Unusual Objects Left Behind

In this section, we explore the strange and cataclysmic events that take place when a star ages and finally dies. We'll see that the products of stellar death, which can be quite spectacular, depend on the mass of the star.

Death of an Intermediate-Mass Star and White-Dwarf Formation

As we have seen, low-mass stars burn so slowly that none have yet exhausted their hydrogen fuel, and they remain red dwarfs. An intermediate-mass star, such as our Sun, in contrast, eventually exhausts its hydrogen fuel so that only helium remains in its core. It then begins to undergo a succession of major changes in luminosity and temperature, as indicated by the numbered points on the H-R diagram in **Figure 22.17**.

Just before it begins to die, a Sun-like intermediate-mass star lies on the main sequence of an H-R diagram (Point 5 in Fig. 22.17). Once the hydrogen fuel in the star's core has been exhausted, in that all hydrogen atoms in the core have undergone fusion reactions to form helium, the outward-directed thermal pressure decreases, so the inward-directed gravitational force causes the star to contract. The resulting compression raises temperatures high enough for fusion to take place in the layer of hydrogen surrounding the helium core (Point 6).

Once ignited, fusion reactions in this hydrogen shell continue at a furious rate. The thermal pressure generated by this burning shell causes the outer layers of the star to expand, even though the helium core continues to

FIGURE 22.18 Development of a nova from a binary star.

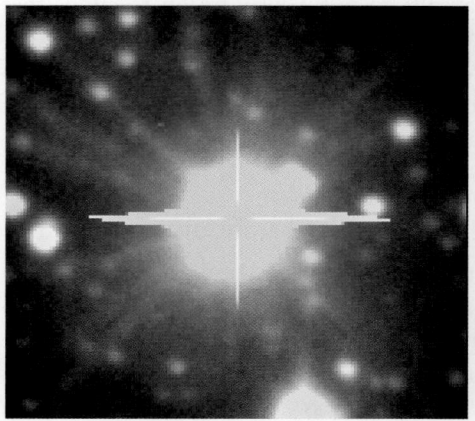 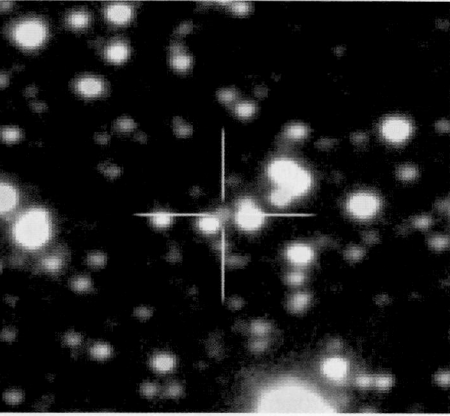

(a) This nova, seen here through a telescope, suddenly appeared in May 2009 and maintained intense brightness for about a week. It continued to fade over the next seven years. The left image shows a sector of space during the nova, whereas the right image shows the same sector after the nova.

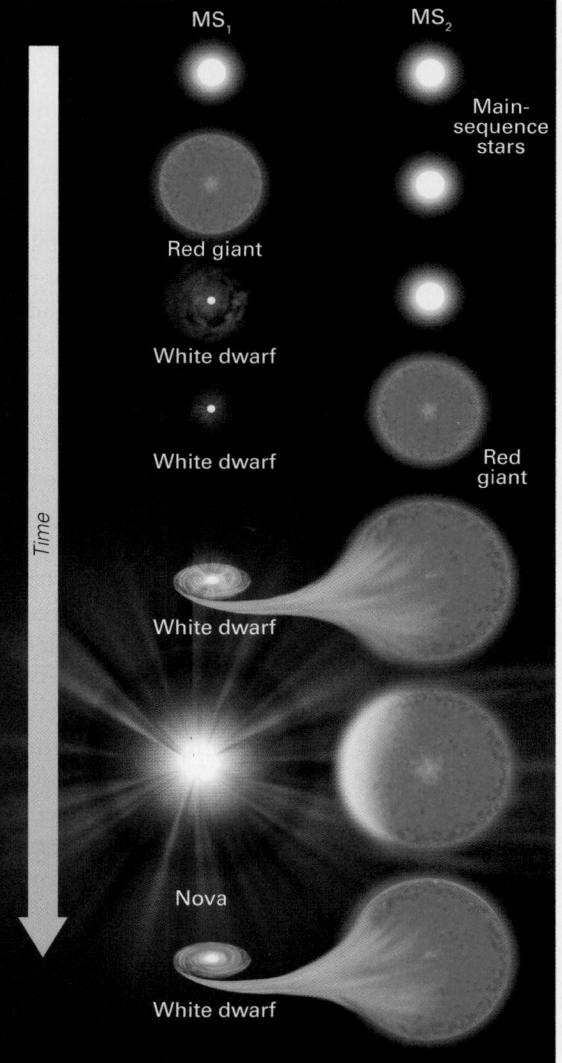

(b) Stages during the development of a nova, which form when one star in a binary pair gravitationally draws off the mass of its partner. If runaway nuclear fusion occurs on the surface of the white dwarf, but the star is not destroyed, a nova occurs, as shown here. If the mass is sufficient to cause gravitational collapse of the white dwarf, it explodes and is destroyed, resulting in a Type I supernova. After a nova explosion, the stars return to their original states for a while.

contract. As the diameter of the expanding star increases, its luminosity increases with its surface area, even though its surface temperature decreases. Over the next 100 million years, the hydrogen shell burns outward, like a spreading forest fire, and the outer layers of the star continue to expand even more. As this happens, the helium core continues to collapse inward, so its temperature rises. When the core's temperature reaches 100 million °C (180 million °F), fusion reactions begin within it, and these reactions bind helium nuclei together in a series of steps to produce carbon. As these reactions proceed, the helium core transforms into a carbon core. Meanwhile, both the outer surface of the helium core and that of the surrounding hydrogen shell move outward (Point 6), and the star continues to expand. Eventually, the star grows to 100 times its former size, and its photosphere cools to about 3,000°C, so it glows red. Such a dying star has become a **red giant** (Point 7). When our Sun passes through the red-giant stage, some 5 billion years from now, its photosphere will extend beyond the orbit of Mars, and the Earth will have been vaporized.

As the red-giant stage progresses, the star's luminosity decreases as its temperature decreases (Point 8). At the end of the red-giant stage, the burning helium and hydrogen shells become unstable and start to alternately expand and contract (Point 9). This process spews matter from the star's outer layers into space, where it becomes a nebula. Finally, with its nuclear fuel exhausted, the carbon core cools and thermal pressure decreases, so the star contracts. Over tens of thousands of years, the nuclear reactions of this residual carbon core die, but the outer large, highly luminous nebula of expelled gas keeps the white-hot core from becoming visible (Point 10). Eventually, the outer layers of gases dissipate into space and become sufficiently

transparent that the still white-hot core of the star becomes visible as an Earth-sized remnant known as a **white dwarf** (Point 11). A white dwarf contains a mass equal to about half that of the Sun packed into a volume only about 1% of the Sun's, which gives it an immense density: 1 cm³ (0.06 in³, or $\frac{1}{5}$ of a teaspoon) of a white dwarf contains about 1,000 kg (1.1 tons) of matter. Ultraviolet radiation emitted by the white dwarf causes the gas of the nebula surrounding it to continue to glow.

As the energy of a white dwarf radiates into space, the star cools and dims, for nuclear reactions are no longer taking place within it. Over the course of about 10 trillion years, a white dwarf will darken until it disappears from the night sky, becoming a cold cinder known as a *black dwarf* (Point 12). No black dwarfs currently exist because the Universe, with an age of 13.8 billion years, simply isn't old enough for any to have evolved.

Death of a Binary Star: Novae and Type I Supernovae

The term *nova*, from the Latin word for new, refers to a bright star that suddenly appears where there had been no bright star before. Specifically, a **nova** (plural: **novae**) is a star that was once too dim to see, but for up to 80 days, it may attain a luminosity of 50,000–100,000 solar units **(Fig. 22.18a)**.

Modern research indicates that novae develop when mass passes from one star to another in a dying binary star in which one of the companions has already become a white dwarf and the other has started to become a red giant. The exchange of mass typically begins when the red giant swells in size **(Fig. 22.18b)**. When the surface of the red giant gets close enough to the white dwarf, the white dwarf's gravity starts pulling matter (most of which is hydrogen) away from the red giant. This matter accretes on the surface of the white dwarf, which gains mass and becomes denser. Suddenly, the high temperature of the white dwarf causes the dense, accreted material to ignite, so that fusion reactions take place throughout the accreted material all at once. The sudden and pervasive onset of fusion in the captured hydrogen results in a violent hydrogen-bomb-like explosion on a stellar scale, causing the star's luminosity to increase quickly and dramatically, and blasting much of the remaining accreted matter back into space. It's this explosion that we see from the Earth as a nova.

The nuclear fuel that brightens a nova burns up quickly. Once the hydrogen that was captured by the white dwarf has been converted to helium, both the white dwarf and its red-giant companion return to their original states. Later on, the white dwarf again draws in mass from its companion red giant until another nova occurs. White dwarfs can undergo several novae during the lifetime of a binary star.

If, in a binary star, the white dwarf and its companion star are particularly close together, the amount of matter transferred to the white dwarf by its gravitational pull becomes exceptionally large. When this occurs, gravity causes the white dwarf to collapse inward rapidly. The rapid increase in temperature and pressure associated with the collapse triggers runaway fusion reactions that release so much energy that the star explodes and undergoes complete destruction, leaving only a remnant nebula. This type of event, which involves the explosion of an intermediate-mass star, is called a **Type I supernova**.

Death of a High-Mass Star and Type II Supernova Formation

The evolution of a high-mass star differs dramatically from that of a low- or intermediate-mass star because so much gravitational compression takes place within the high-mass star that its core becomes much hotter than that of a star with less mass. As a high-mass star's core temperature rises, the rate at which hydrogen fusion takes place in its core increases. In fact, as we've seen, fusion takes place so rapidly in a high-mass star that the star can exhaust its hydrogen in just a few million years.

The early stages in the death of a high-mass star resemble those for an intermediate-mass star: when the core's hydrogen fuel has been used up, gravity compresses the helium that remains until helium fusion ignites and produces carbon. Meanwhile, fusion reactions begin in a shell of hydrogen surrounding the core, and rising temperatures cause the outer layers of the star to expand. Because the star contains so much mass, however, the process yields a **red supergiant**, vastly larger than a red giant (see Fig. 22.12b).

Because of its immense mass, the late stages of a supergiant's evolution differ from those of a giant. Gravity causes much more inward compression within the core of a supergiant than within the core of a giant. This intense compression, in turn, drives the temperature of the supergiant's core so high that carbon nuclei can collide with enough energy to fuse; such reactions can't take place in the somewhat cooler core of a giant. Carbon fusion reactions produce nuclei of even heavier elements, such as neon, silicon, magnesium, and oxygen. When carbon fuel runs out, gravity again overcomes outward pressure, and the core of the red supergiant collapses even more and becomes even hotter. As a result, the products of carbon fusion themselves begin fusing to produce still heavier elements. In fact, a succession of fusion reactions, each producing progressively heavier nuclei, continues until iron nuclei have been produced. At each stage, fusion reactions involving successively lighter nuclei take place in concentric shells around the star's core. So, for example, a

FIGURE 22.19 Production of elements in shells within the core of a high-mass red supergiant just before it becomes a supernova.

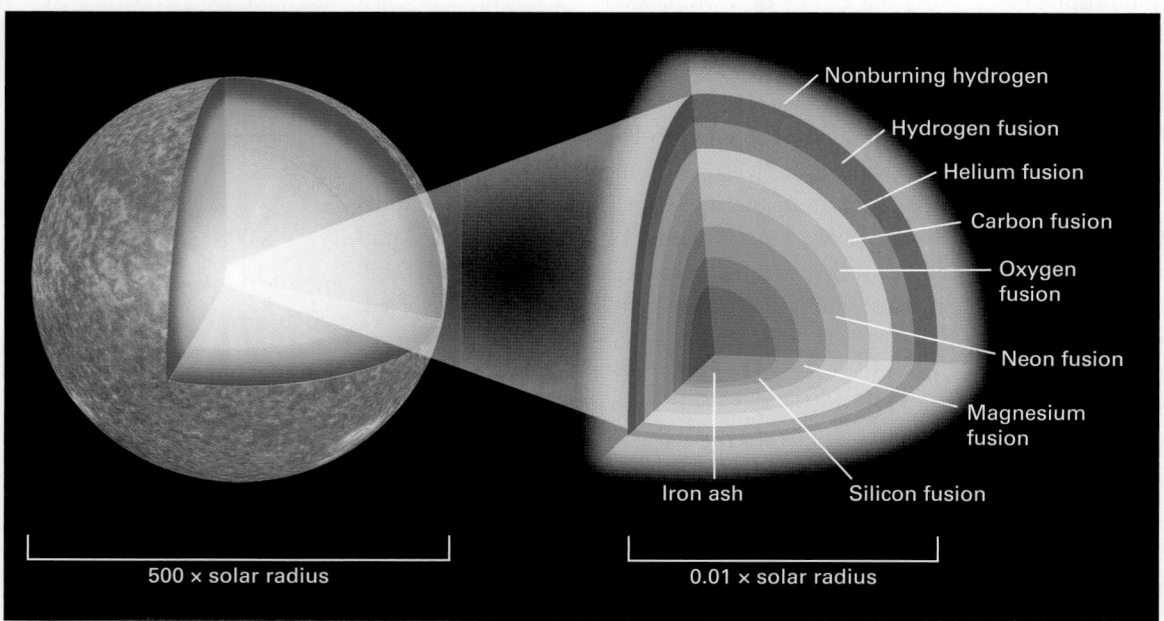

Nonburning hydrogen
Hydrogen fusion
Helium fusion
Carbon fusion
Oxygen fusion
Neon fusion
Magnesium fusion
Silicon fusion
Iron ash

500 × solar radius

0.01 × solar radius

shell of oxygen fusion surrounds a shell of neon fusion, and a shell of carbon fusion surrounds a shell of oxygen fusion. In fact, while heavy nuclei are fusing in the core of a red supergiant, hydrogen fusion (to form helium) still takes place in outer shells of the star **(Fig. 22.19)**.

Once the core of a high-mass star contains only iron, its fusion reactions end. Without the outward thermal pressure produced by the energy of fusion reactions, nothing can counter the relentless inward pull of gravity, and the star suddenly collapses. During this *implosion*, matter in the core undergoes such intense compression that its temperature rises to 10 billion °C (18 billion °F). At these incredible temperatures, it takes only about 1 second for all the core's atomic nuclei to split apart into free protons and neutrons. Further compression of the core causes protons to merge with electrons to form neutrons until the core consists entirely of neutrons packed tightly together in an inconceivably dense mass.

But gravity hasn't finished with the star yet. The neutrons rush inward into a progressively smaller space until they compact into a tiny volume with a density approaching 1 trillion kg/cm^3 (67 million $tons/in^3$). This mass suddenly becomes catastrophically unstable and rebounds outward, generating a violent shock wave that destroys the star and blows all the mass lying outside the core into space. This explosion generates a brilliant but short-lived light in the night sky called a **Type II supernova (Fig. 22.20a, b)**. Recorded Type II supernovae have had apparent magnitudes of up to -7.5. Some have been so bright that they can be observed in the daytime sky, even though they have all been very far away.

The matter blasted into space by a supernova forms an expanding nebula, traditionally known as a *planetary nebula* because it surrounds the location of the exploded star; despite its name, it actually contains no planets **(Fig. 22.20c)**. For example, the spectacular Crab Nebula,

FIGURE 22.20
A supernova produces an expanding nebula.

(a) The M-82 Galaxy (11 million light-years from the Earth) before and after a supernova. The supernova was visible in 2014.

(b) An artist's rendering of an immense supernova explosion.

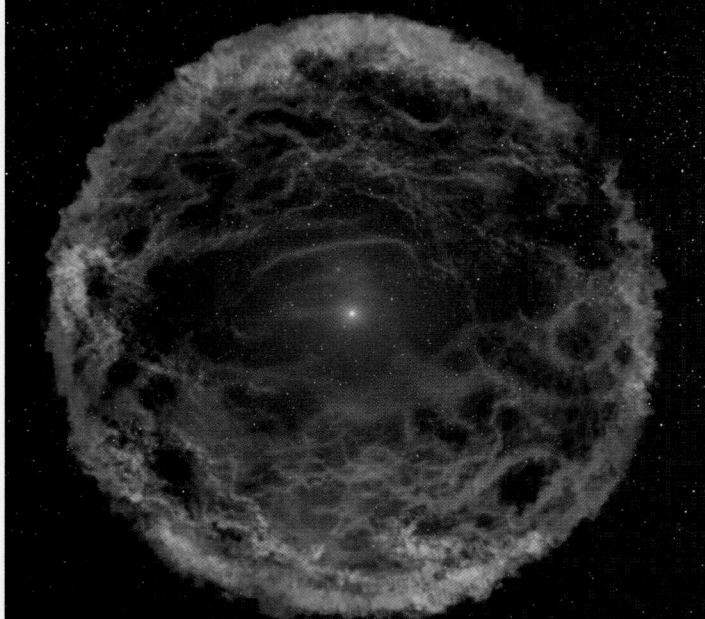

(c) Gas and dust from the explosion expand outward to form a nebula. A tiny, but supermassive, remnant of the star remains.

(d) The Crab Nebula, shown here, was formed by a supernova explosion in 1054 C.E.

so named because the radiating streaks of glowing gas within it look like the legs of a crab (Fig. 22.20d), consists of gases blown into space by a supernova observed in 1054 C.E. During supernova explosions, subatomic particles such as protons and neutrons move so fast that when they collide, the resulting fusion reactions can produce all the natural elements, including heavy elements such as uranium. These newly formed heavy elements become part of the expanding nebula and become available to be incorporated into the next generation of stars—stars like our Sun. Indeed, most of the elements that compose the Earth today originated in supernovae that occurred earlier in our Universe's evolution (see Chapter 1). Astronomers estimate that supernovae in the Milky Way have a recurrence interval of about 100 years. Hundreds of supernovae have been observed over the centuries, because even in distant galaxies, supernovae become bright enough to be seen from the Earth with telescopes. In fact, some supernovae in other galaxies outshine their parent galaxies.

Corpses of High-Mass Stars: Neutron Stars and Black Holes

What's left after a supernova blast? We've seen that much of the star's matter spreads out into space as a nebula. However, the star's innermost core, composed of highly compressed neutrons, survives. This remnant object, known as a **neutron star** if its mass is less than about 2.5 solar masses, is tiny by stellar standards—only 10–20 km (6–12 mi) in diameter, about the width of Manhattan Island in New York City—but its mass can be 1.1–3.0 M_S. A cubic centimeter of neutron star weighs about 100 billion kg (100 million tons)!

Astronomers can detect neutron stars, despite their small size, because neutron stars generate very strong magnetic fields and rotate rapidly. As a result, they emit radio waves at regular intervals. The pulses of energy from rotating neutron stars arrive on the Earth with clockwork accuracy, so researchers also refer to these objects as **pulsars**. There are well over 1,000 pulsars known in the Milky Way, including one at the center of the Crab Nebula. A pulsar's rotation rate slows, and its magnetic field weakens, over millions of years.

What happens if the mass of the remnant object exceeds 2.5 solar masses? In this case, the overwhelming inward pull of gravity catastrophically crushes the neutrons into an even smaller volume, so that the gravitational force becomes so great that, within a region whose boundary is called the **event horizon**, nothing can escape, not even electromagnetic radiation or high-speed atomic particles. Because no light comes through the event horizon, the region looks completely dark to an observer. Astronomers refer to the region within the event horizon as a **black hole (Fig. 22.21)**. Note that the body of matter

Did you ever wonder...

what's black about a black hole?

FIGURE 22.21 An artist's rendition of a black hole. No radiation can escape past the event horizon, the boundary of the dark region on the image. The inset shows the first photo of a real black hole, captured by the Event Horizon Telescope in 2019.

Event horizon

within a black hole is much smaller than the black hole itself, but the actual size of such a body remains unknown.

Black holes do not all contain the same amount of matter. *Stellar black holes*, formed during the deaths of individual stars, typically contain 5–20 M_S. Currently, the smallest known black hole, which lies 1,500 light-years from the Earth, contains 3 M_S; astronomers estimate that the diameter of its event horizon is about 9 km (5.6 mi). The largest black holes, which we examine in Section 22.6, are called *supermassive black holes* and are a product of galaxy formation. They can contain the mass of millions of stars within the volume of a single star or less.

Because black holes have mass, they are real physical objects. But the formation of a black hole destroys matter as we know it; in fact, scientists currently cannot describe the state of matter in a black hole, because it is even denser than that within a neutron star. According to one hypothesis, within a black hole, subatomic particles such as neutrons and protons separate into even smaller particles, known as *elementary particles*, which can then pack together even more tightly than they do in normal matter. A black hole can also be thought of as a hole in space, because no communication can take place between it and the space outside.

Because black holes emit no light or particles, the only way to detect them is by identifying their gravitational influence on other bodies in their vicinity. A lone black hole can be completely undetectable, but if a black hole has a binary companion, the companion's behavior can be used to determine the black hole's mass and location. Science-fiction stories sometimes depict black holes as deadly galactic scavengers, lurking in space and gobbling hapless planets that pass near their location. In reality,

FIGURE 22.22 Life cycles of stars with different masses. The top row shows how an average star becomes a red giant, and eventually, a white dwarf. The bottom row shows how a massive star becomes a supergiant, then explodes as a supernova, leaving behind a neutron star or a black hole.

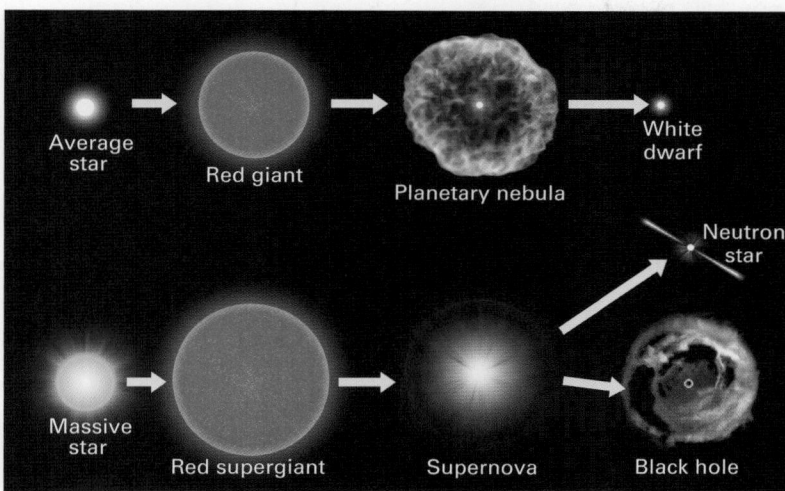

black holes act like all other objects in space: beyond the event horizon, objects orbit black holes just as they do other celestial bodies.

Clearly, the way in which a star's demise takes place depends on the star's mass. Less massive stars end up as white dwarfs, ones with greater mass as neutron stars, and the most massive as black holes **(Fig. 22.22)**.

Take-home message...

🏠 No low-mass stars (red dwarfs) have yet exhausted their nuclear fuel and died. Intermediate-mass stars die by expanding to become red giants and, eventually, collapsing to become dense white dwarfs surrounded by a nebula consisting of expelled gas. White dwarfs that are part of a binary star may produce novae or Type I supernovae through interaction with their companion stars. A high-mass star becomes a red supergiant before eventually dying in a violent Type II supernova, which creates a nebula rich in heavy elements. After a Type II supernova, an extremely dense object, known as a neutron star or pulsar, remains if the remnant object is smaller than 2.5 solar masses. A more massive remnant object will continue to collapse into a tiny object of incredible density called a black hole, from which not even light can escape.

Quick Questions

- What's the difference between a nova and a Type I supernova?
- How were the heaviest elements formed?
- Does a black hole consist of empty space?

22.6 Galaxies

Stars, along with their planets and moons, don't travel through space alone. Rather, nearly all reside within galaxies—collections of hundreds of billions of solar systems that, together with nebulae of interstellar gas, dust, and ice, travel through the Universe together. Over a trillion galaxies speckle the vastness of space **(Box 22.1)**. In this section, we look at the character of galaxies, starting with our own, the Milky Way.

The Milky Way Galaxy: A Closer Look

When we look at our home galaxy, the Milky Way, from the Earth, we can't see its overall form **(Fig. 22.23a)**. But, by analyzing images and comparing them with images of other galaxies, astronomers conclude that the Milky Way is a *barred spiral galaxy* **(Fig. 22.23b)**. The term *barred* refers to a concentration of stars that form a straight bar-like shape along the **galactic center**, the midpoint of the galaxy. The term *spiral* refers to the existence of concentrations of stars in bands, or arms, that curve into the center of the galaxy. The *spiral arms* of the Milky Way lie within a plate-like **galactic disk (Fig. 22.23c)**. The Sun resides along the outer edge of one of those spiral arms, about 27,000 light-years from the galactic center. The Milky Way slowly rotates, so our Sun makes one orbit around the galactic center about once every 230-million-year-long *galactic year*. Overall, the Milky Way, which has a diameter of 130,000 light-years and a thickness, on average, of 1,200 light-years, contains between 200 and 400 billion stars, most of which host several planets and their moons. Notably, recent research suggests that galaxies also include *rogue planets*, free-floating planets that do not orbit a star but rather cruise through space on their own. By one estimate, over a trillion rogue planets occur within the Milky Way. These planets probably went rogue when they were ejected from orbit due to collisions with other objects or due to the gravitational tugs from other planets or stars that accelerated the soon-to-be rogue planet to escape velocity.

Toward its center, the galactic disk of the Milky Way thickens into a **galactic bulge**, with a diameter of about 6,500 light-years and a thickness of about 4,000 light-years. The concentration of stars in the bulge greatly exceeds that in the rest of the disk; in fact, so many stars would fill the night sky of a planet orbiting a star within

The vastness of space

What are we learning?

How to think about the vast distances of interstellar space.

What you need:

- A small ball to represent the Sun
- A tiny ball to represent the Earth
- A small ball to represent the Sun's nearest neighbor, Proxima Centauri
- A map of your local area that extends at least 97 km (60 mi) beyond your location

Instructions:

- Place the ball representing the Sun on the floor.

- Now carefully place the ball representing the Earth 30 cm (1 ft) away from the first ball. Note that that this 30 cm distance represents 148 million km (92 million mi).
- So where would the nearest star be located? That star is Proxima Centauri, and it is 4.3 light-years, or 40 trillion km (25 trillion mi), away from the Earth. Scaled to the two balls, Proxima Centauri would be 82 km (51 mi) away from the ball representing the Sun. Find a familiar spot on the map that is 82 km (51 mi) away from where you are. That is where the next ball should be placed!
- Now, using the same scale (30 cm = 148 million km), determine where the nearest galaxy, Andromeda, would be located. (Andromeda is 2.5 million light-years from the Sun.)

What did we see?

- The goal of this exercise is to get some sense of the vastness of space. Picturing the distance you would have to travel from your scale-model Earth to your scale-model Proxima Centauri provides a dramatic example of the vastness of our Universe.
- Amazingly, if 30 cm represents 148 million km, the nearest large galaxy, Andromeda, would still be about 254 billion km (158 billion mi) away from the ball representing the Sun!

the galactic bulge that night on that planet would be as bright as day on the Earth. The galactic bulge also contains an abundance of interstellar gas and dust and serves as a nursery for new stars, and it's a violent place, where stars collide and supernovae explode.

Do stars occupy the entire galactic bulge? No. The *galactic center*, the region at the center of the bulge, consists of a **supermassive black hole** that emits no light at all **(Fig. 22.23d)**. This black hole, known as Sagittarius A, is called "supermassive" because it contains the mass of 4 million Suns in a volume that has a diameter of only about 52 million km (32 million mi)—the diameter of Mercury's orbit. Its incredibly strong gravitational pull effectively holds the Milky Way together. Recent research indicates that a supermassive black hole like the one at the center of the Milky Way lies at the center of every large galaxy in the Universe.

A **galactic halo**, a sphere with a diameter twice that of the galactic disk, surrounds the Milky Way Galaxy. This halo, whose total mass equals about 1% that of the

galactic disk, contains hardly any dust or gas. Rather, it hosts widely separated, very old stars as well as over 150 **globular clusters**, spherical accumulations of hundreds of thousands of stars that may be as close as 1 light-year apart **(Fig. 22.24)**. Most stars in globular clusters are red dwarfs that have probably been burning since the formation of the galaxy. New stars do not form in these clusters, and all large stars that once existed there have long since died.

Other Galaxies, and Quasars

With the spectacular images of galaxies that we now have, it's hard to imagine that about a century ago, astronomers didn't even realize that galaxies outside the Milky Way existed. They had seen objects they called "spiral nebulae," but thought that these objects lay within the Milky Way. It was only in the 1920s, when astronomers determined that spiral nebulae were actually galaxies, that astronomers began to catalog distant galaxies. Edwin Hubble classified galaxies into four basic types

(a) The central arc of the Milky Way Galaxy graces a sky filled with stars.

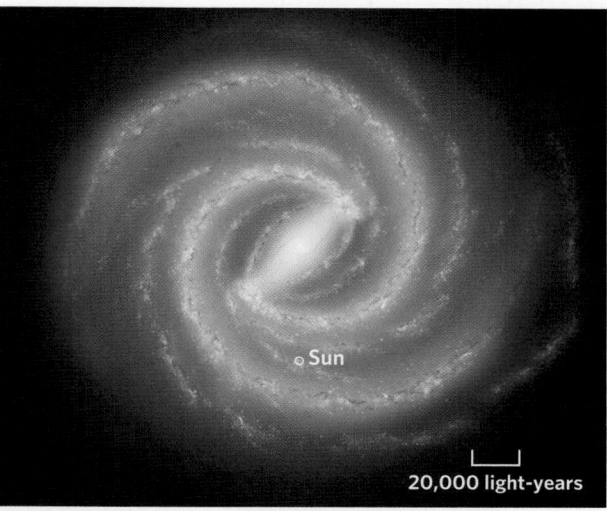

FIGURE 22.23
The Milky Way
Galaxy.

20,000 light-years

(b) Computer graphic of the Milky Way, as viewed from the top. Note the position of the Sun.

Globular clusters

Galactic
halo

Galactic disk

Galactic bulge

O, B stars

Galactic
center

13,000
light-years

Sun

Gas and dust

26,000
light-years

Nebula

100,000 light-years

(c) Artist's rendering of the Milky Way as viewed from the side, with components labeled. The galactic bulge lies at the center.

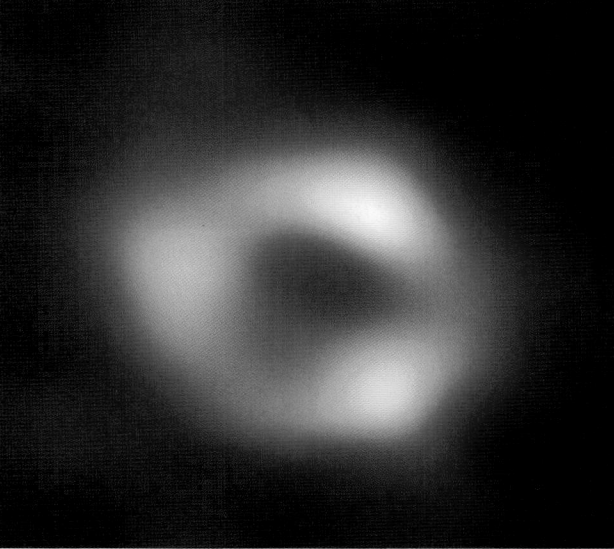

(d) The supermassive black hole at the center of the Milky Way, surrounded by glowing gases, as imaged by the Event Horizon Telescope in 2019.

FIGURE 22.24 This globular cluster, known as M-13, is 25,000 light-years from the Earth and lies within the galactic halo of the Milky Way. It has a diameter of about 145 light-years and contains about 300,000 stars.

(a) Spiral galaxy NGC 5457, known as the Pinwheel Galaxy.

(b) Barred spiral galaxy NGC 1300.

(c) Irregular galaxy NGC 1569.

(d) Elliptical galaxy ESO 325-G004.

based on their shapes: *spiral*, *barred spiral*, *elliptical*, and *irregular* **(Fig. 22.25)**.

In addition to distinguishing among galaxies by shape, astronomers now categorize them by the nature of the energy they emit. About 90% of known galaxies, our own Milky Way included, are **normal galaxies** in that they emit their greatest amounts of radiation at or near visible wavelengths. Normal galaxies range in size from *dwarf galaxies*, with luminosities starting at about 1 million solar units, to *giant galaxies*, with luminosities of up to 1 trillion solar units.

The remaining 10% of known galaxies are called **active galaxies** because, unlike normal galaxies, they emit not only visible light, but also an enormous amount of energy in the X-ray and radio regions of the spectrum. In fact, the centers of these galaxies radiate more than 100 times the electromagnetic radiation (including visible light) of all other stars of the galaxy combined. An active galaxy's X-ray and radio energy does not come from stars in the central bulge, but rather from a relatively small volume surrounding the supermassive black hole at the galactic center.

Why do normal and active galaxies differ from each other? In a normal galaxy, a void (a vacuum or gap) surrounds the supermassive black hole at its center. Any matter that once existed in this void has already been pulled inward, across the event horizon and into the supermassive black hole, from which neither matter nor energy can escape. Most stars outside of the void are in equilibrium and simply orbit the void. Because no energy escapes from the supermassive black hole itself, the energy of a normal galaxy comes only from stars, mostly in the visible and ultraviolet wavelengths.

In the case of an active galaxy, immense amounts of matter—derived from stars that have been violently torn apart by tidal forces as they approach the supermassive black hole—form an accretion disk surrounding the event horizon. As matter in this accretion disk spirals inward toward the black hole, it undergoes intense compression and releases immense amounts of radiation (including in the radio and X-ray wavelengths) until it crosses the event horizon and disappears forever **(Fig. 22.26)**. Not all matter of the accretion disk reaches the event horizon, however. Some atomic particles, such as protons and helium nuclei, get caught in intense magnetic fields and accelerate to speeds up to 80% that of light, thereby becoming **cosmic rays** (extremely high-energy ions). Cosmic rays, as well as some more slowly moving particles, shoot out into space in *astrophysical jets* that are parallel to the axis of rotation of the supermassive black hole. The jets, which extend into space for distances greater than the diameter of the whole active galaxy, also act as focused sources of extreme electromagnetic radiation in the X-ray and radio-wave portion of the spectrum. But active galaxies don't stay active forever. Once all the stellar material of the accretion disk in the vicinity of the supermassive black hole has been drawn into it, energy production decreases, and the active galaxy becomes a normal galaxy. This was undoubtably what happened in our own Milky Way in the distant past, long before our Sun ever formed.

Astronomers categorize active galaxies based on the intensity of their radio-wave emissions. (These galaxies also emit X-rays, but only radio waves can be detected by ground-based radio telescopes.) Some active galaxies emit particularly intense radio waves, while others—known as "radio-quiet" active galaxies—emit relatively weak radio waves. It appears that the intensity of the radio signal we detect depends simply on the orientation of a galaxy's astrophysical jets relative to the Earth. Active galaxies whose jets point straight at the Earth are strong radio emitters, while those whose jets are directed away from the Earth are weak radio emitters.

In the 1950s, long before the relationship between galaxies, supermassive black holes, and accretion disks was understood, radio astronomers detected intense radio-wave emissions coming from distant points in space. They named these sources *quasi-stellar radio sources*, or **quasars**. Astronomers now consider quasars to be active galaxies whose jets happen to be pointing toward the Earth. The intensity of the electromagnetic radiation emitted by some quasars approaches that of 1,000 Milky Ways, making them among the brightest objects in the known Universe. Many quasars display an extreme red shift, indicating that they are the most distant objects detectable within the Universe **(Fig. 22.27)**. The energy we receive from these very distant quasars today started its journey at a very early stage in the Universe's history, long before our Sun formed. Their existence records the early stage of a galaxy's life cycle, the time when an accretion disk is being pulled into the galaxy's newly formed supermassive black hole.

Do all active galaxies represent an early stage in galaxy formation? No, because they can develop for other reasons. For example, some active galaxies form in the wake of a collision between two normal galaxies **(Fig. 22.28)**. As the two galaxies pass through each other's space, gravitational forces may bring stars close enough to a supermassive black hole that they get torn apart and become incorporated in an accretion disk, which provides material to spiral into the black hole. (Stars themselves rarely collide directly during galaxy collisions because they are so far apart, but they do exert gravitational pulls on one another.) In some cases, the supermassive black

FIGURE 22.26 An artist's depiction of the accretion disk around a supermassive black hole in an active galaxy. Note the jets of energy streaming outward, focused along the axis of rotation.

Jet

Event horizon

Supermassive black hole

FIGURE 22.27 Quasars.

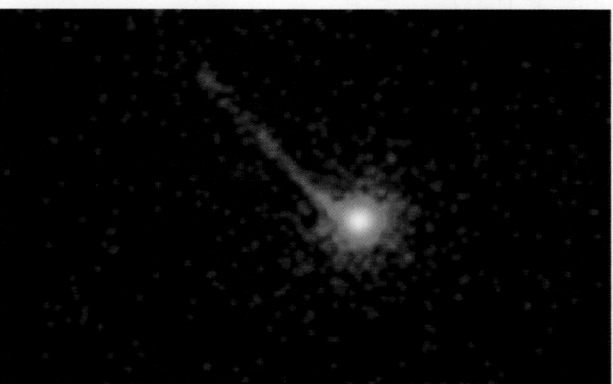

(a) A quasar located 10 billion light-years from the Earth, as seen in an X-ray photograph by the Chandra X-ray Observatory.

(b) An artist's conception of a quasar.

holes of both galaxies exert so much gravitational attraction on each other that they eventually merge. Most active galaxies that lie relatively close to the Milky Way formed as a consequence of galactic collisions. To date, more than a million quasars have been found; the nearest one lies only about 600 million light-years from the Earth.

FIGURE 22.28 Examples of colliding galaxies.

(a) These colliding galaxies (NGC 4676B and NGC 4676A, 300 light-years from the Earth) are called The Mice because of their long tails of stars and gas.

(b) Two spiral galaxies (NGC 2207 and IC 2163, 80 million light-years away) interacting.

Take-home message...

🏠 A galaxy is an assemblage of stars, their planets and moons, and interstellar gas, dust, and ice, all bound together by gravity. Most, if not all, galaxies have a supermassive black hole at the center. Whether or not this supermassive black hole is able to pull in matter from a surrounding accretion disk determines whether a galaxy is normal or active.

Quick Questions

- What type of galaxy is the Milky Way?

- About how many solar masses are contained in the Milky Way's supermassive black hole?

- What is a quasar?

FIGURE 22.29
Andromeda, the nearest large galaxy to the Milky Way, lies 2.5 million light-years from the Earth.

22.7 The Structure and Evolution of the Universe

When astronomers mapped the distribution of galaxies, they discovered that galaxies are not randomly scattered across the Universe and that they do not all move independently from one another. For example, several relatively small *satellite galaxies*, including the Large and Small Magellanic Clouds, orbit the Milky Way Galaxy. The Milky Way and its satellites, together with the Andromeda Galaxy **(Fig. 22.29)** and its satellites, as well as a few smaller galaxies, constitute the **Local Group**, a 10-million-light-year-wide, gravitationally linked association of galaxies **(Fig. 22.30a)**. Of the 54 galaxies in the Local Group, Andromeda is the largest; the Milky Way, the second largest, has a diameter that is about 57% that of Andromeda. Our Local Group serves as an example of a **galactic group**, an association of generally fewer than 100 galaxies that gravitationally interact with one another.

Within a galactic group, galaxies may be moving toward or away from one another. Andromeda and the Milky Way, for example, zip toward each other at about 400,000 km/h (250,000 mph). At this rate, the two galaxies will collide in about 5 billion years, about the time our Sun becomes a red giant. As we have seen, stars in a galaxy are so far apart that during a collision, individual stars generally don't bash into one another. Rather, gravitational forces merge the two galaxies into a single elliptical galaxy. Notably, the supermassive black holes of the two parent galaxies may combine, or pull stars into their accretion disks. As a result, the colliding system becomes active.

Larger associations of galaxies, which may contain hundreds to thousands of galaxies, are known as *clusters*, and even larger associations, containing on the order of 100,000 galaxies, are called *superclusters*. The Milky Way is part of the Virgo Supercluster, which in turn is part of the even larger Laniakea Supercluster, a region of galaxies that is over 500 million light-years across and contains about 100,000 times the mass of the Milky Way (**Earth Science at a Glance**, pp. 884–885).

At a still larger scale, astronomers visualize the distribution of galaxies in the Universe as a three-dimensional mesh composed of string-like *filaments* and screen-like *walls*, whose long dimensions are 30 to 300 million light-years (**Fig. 22.30b**). Where walls intersect, lumpy concentrations of galaxies occur. Walls and filaments surround *voids*, bubble-like regions of space on the order of 300 million light-years across in which galaxies do not occur.

Many puzzles remain about the structures of both galaxies and their clusters and superclusters. One of the most vexing of these puzzles comes from comparing the amount of gravitational force needed to hold together galaxies and, on a larger scale, clusters and superclusters, with the amount of gravitational force produced by all detectable matter. Astronomers conclude that the gravitational attraction of all detectable matter simply isn't enough to hold galaxies and clusters together. This realization has led astronomers to speculate that some mysterious **dark matter**, which does not appear to absorb, reflect, or emit electromagnetic radiation, must be present throughout the Universe. In fact, calculations suggest that as much as 85% of all matter in the Universe may be dark matter. If dark matter exists, then each galaxy, and the Universe as a whole, contains vastly more mass than telescopes can detect. If it does not, our current understanding of gravity must be quite incomplete.

As we noted earlier, astronomers have observed that all distant galaxies display a red shift and are therefore moving away from the Earth at a high rate of speed. (Note that galaxies in a galactic group can be moving toward each other.) If all distant galaxies are moving away from one another, then earlier in the history of the Universe, galaxies must have been closer together. Mathematical simulations led astronomers to conclude that about 13.8 billion years ago, all the matter and energy in the Universe existed within a **singularity**, a presumably very small region in space whose temperature and density must have been so great that they can't be described by known theories of physics. For reasons not yet understood, the Universe came into being when this singularity exploded and began to expand. As we discussed in Chapter 1, this

FIGURE 22.30 Galaxies are organized in space at different scales.

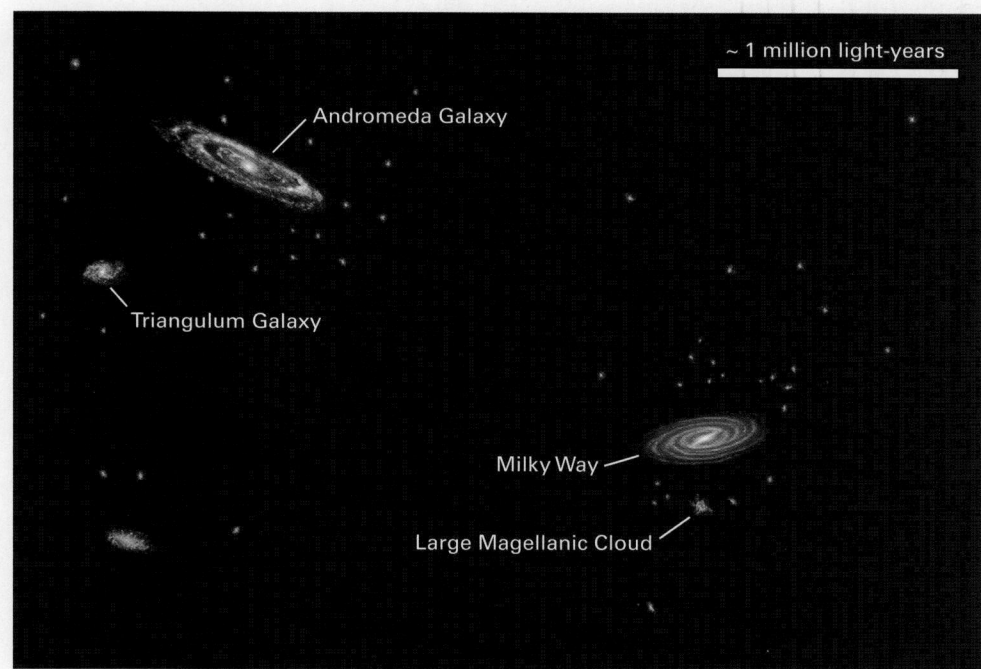

(a) The Local Group of galaxies in the vicinity of the Milky Way.

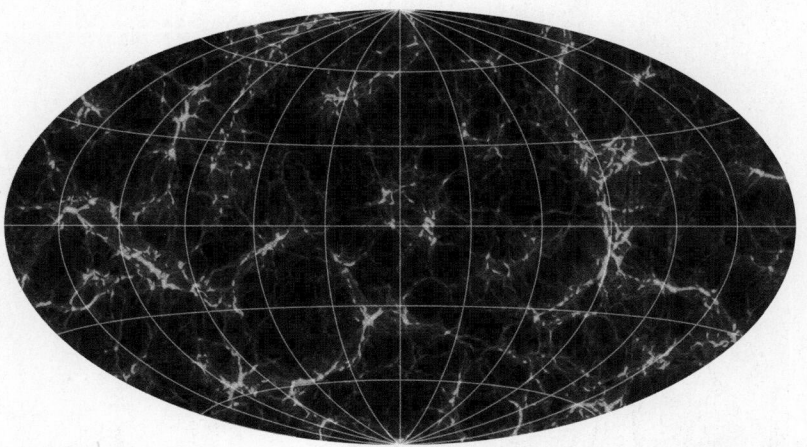

(b) At the scale of the entire Universe, the distribution of superclusters displays a web-like configuration, in which filaments and walls surround voids.

Big Bang marks the beginning of the Universe as we know it. Since that time, space itself has enlarged as the Universe has expanded, and galactic groups and clusters of galaxies have moved farther apart (**Fig. 22.31a**).

What does the future of the Universe hold? Will the Universe continue to expand forever, and if so, will the expansion rate stay the same, slow down, or speed up? Or, at some point in the future, billions and billions of years from now, will the Universe start to contract, eventually collapsing back into a singularity (**Fig. 22.31b**)? This question lies at the heart of cosmologic research today. The answer depends on the answer to another question:

The Fabric of Space

Our Sun resides in a tiny part of a vast Universe whose farthest detectable reaches lie some 13 billion light-years from the Earth. Within this Universe, the distribution of objects and energy isn't random. At each scale, astronomers observe organization. Where does our home planet lie within this amazing structure? At the largest scale (the background of this image) the known matter of the Universe glows within vaguely defined filaments and walls that surround relatively matter-free voids. If we zoom in on a small area of one wall, we find the Virgo Supercluster, composed of more than 100,000 galaxies whose gravity influences one another. A supercluster includes dozens to hundreds of clusters, each of which contains on the order of 1,000 galaxies.

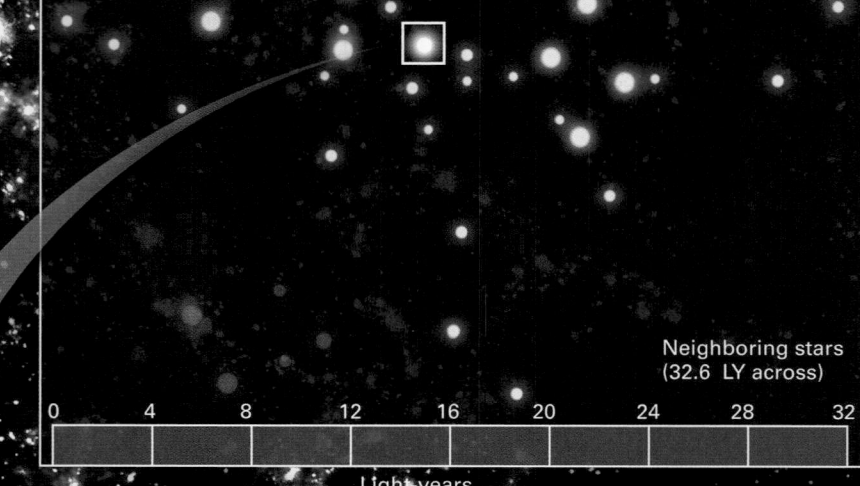

Neighboring stars
(32.6 LY across)

0   4   8   12   16   20   24   28   32

Light-years

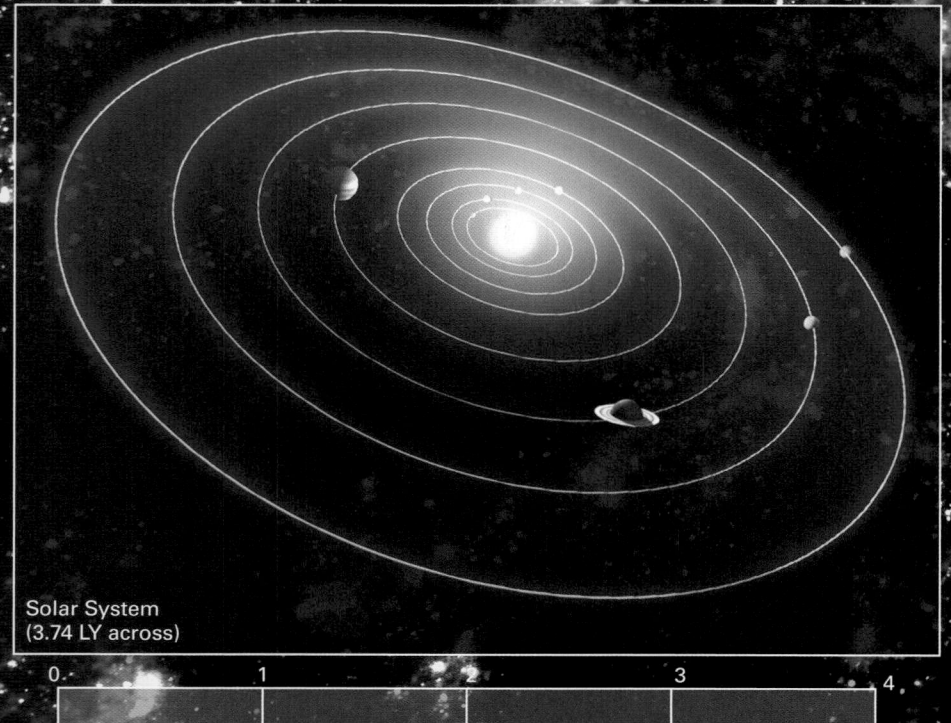

Solar System
(3.74 LY across)

0   1   2   3   4

Light-years

If we zoom in more closely, we find our local galactic group, dominated by two large galaxies, the Milky Way and Andromeda, which, together with their orbiting globular clusters, are moving toward one another. Zooming in on the Milky Way, we see its barred spiral structure, slowly swirling around a supermassive black hole. Our Sun and its neighboring stars occupy a volume slightly more than 30 light-years across. Zoom in on the Sun, and you see its eight planets, all orbiting in roughly the same plane. We live on the blue planet, the third out from the Sun.

Observable Universe (91 billion LY)

10   20   30   40

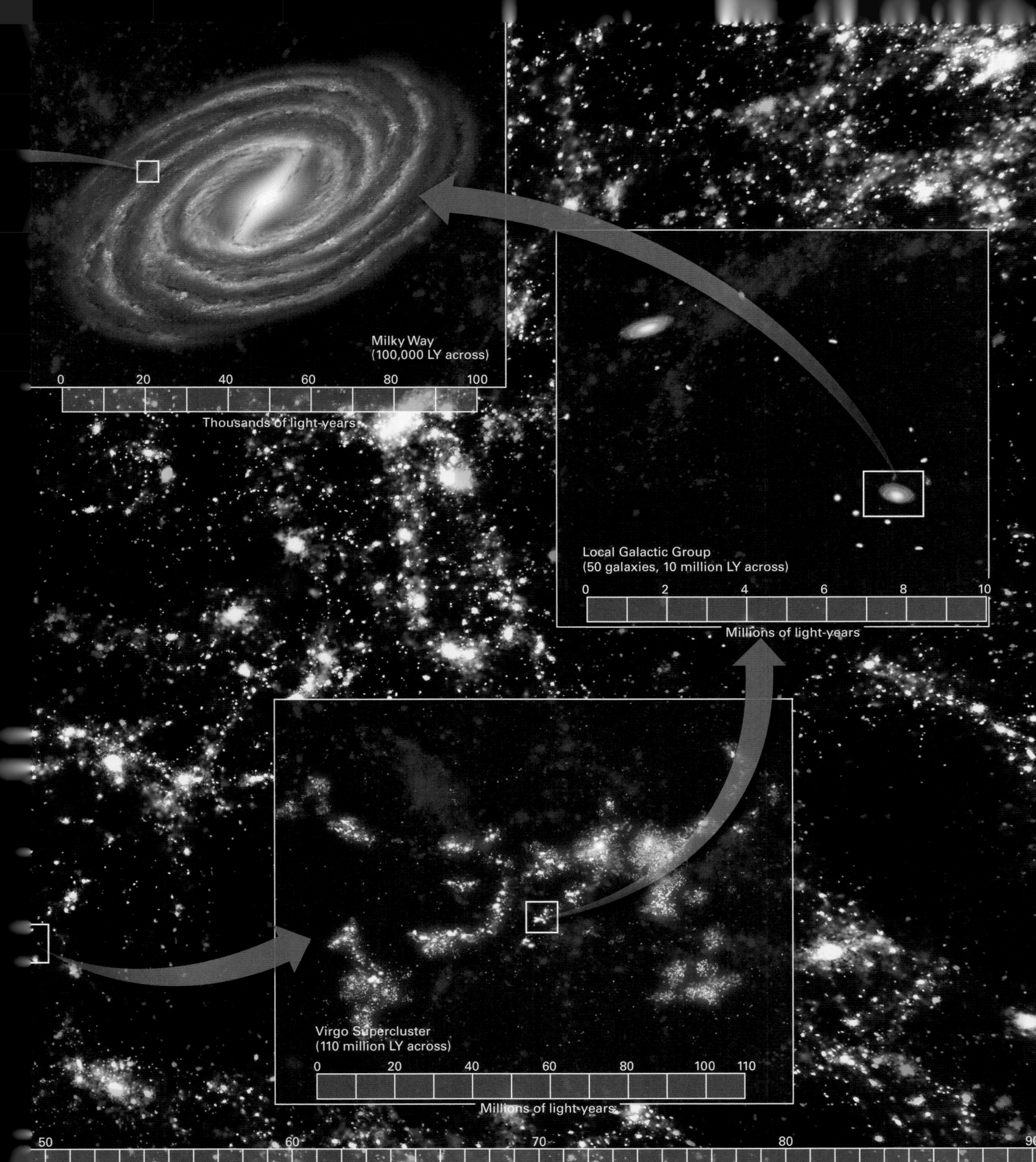

Milky Way
(100,000 LY across)

0   20   40   60   80   100
Thousands of light-years

Local Galactic Group
(50 galaxies, 10 million LY across)

0   2   4   6   8   10
Millions of light-years

Virgo Supercluster
(110 million LY across)

0   20   40   60   80   100   110
Millions of light-years

50   60   70   80   90

FIGURE 22.31 Evolution of the Universe.

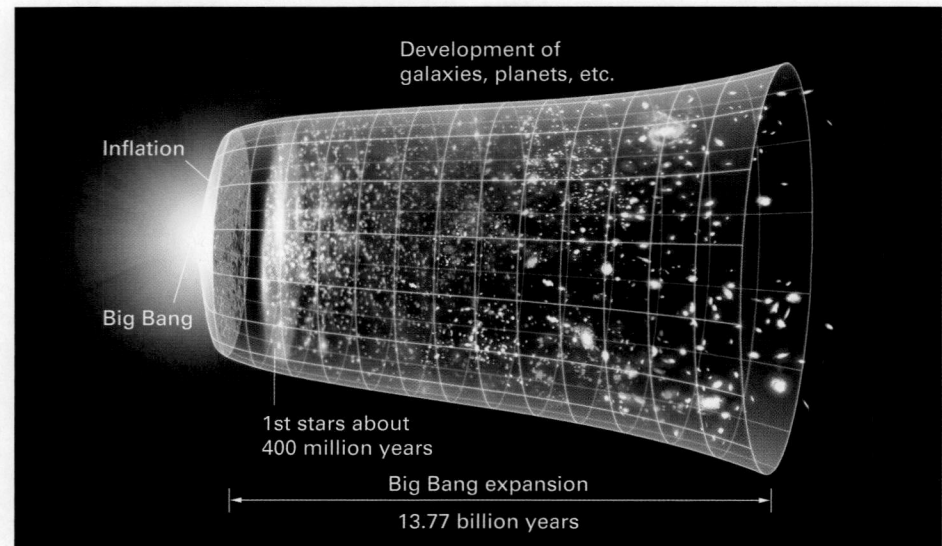

(a) The Universe expanded rapidly after the Big Bang and has continued to expand ever since.

Development of galaxies, planets, etc.

Inflation

Big Bang

1st stars about 400 million years

Big Bang expansion

13.77 billion years

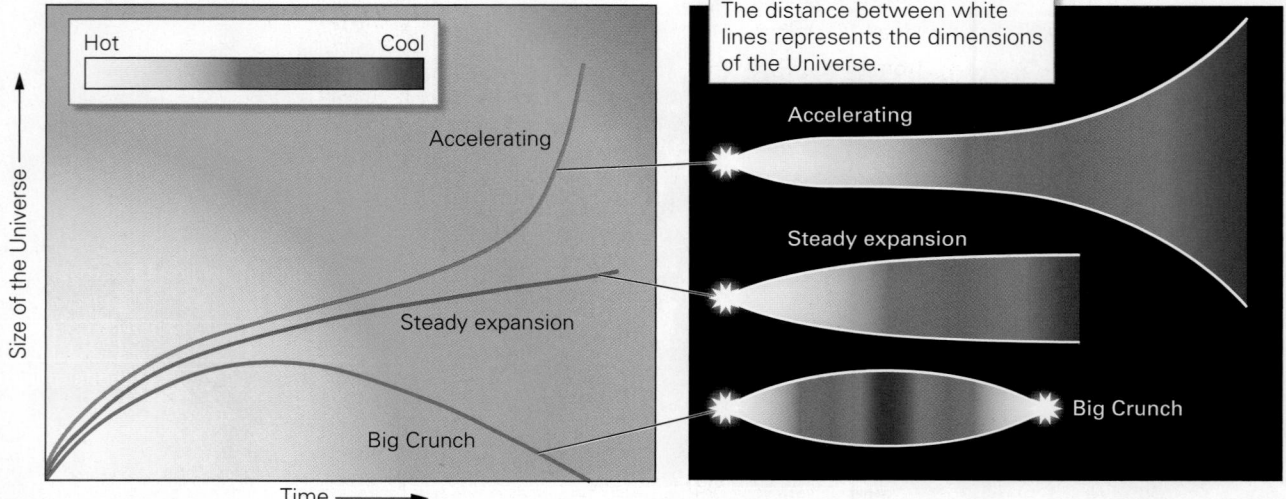

The distance between white lines represents the dimensions of the Universe.

Hot   Cool

Size of the Universe

Accelerating

Steady expansion

Big Crunch

Time

Accelerating

Steady expansion

Big Crunch

(b) Alternative scenarios for the future of the Universe. Its expansion may accelerate, it may expand steadily, or if a Big Crunch happens, it may collapse back into a singularity. Currently, most astronomers believe that the universe will continue to expand forever.

Is gravity strong enough to slow and eventually reverse the expansion of the Universe, or not? This question remains unanswered, so the fate of our Universe cannot yet be definitively predicted. Whatever the long-term fate of the Universe turns out to be, it will happen billions of years after our Sun dies and our galaxy collides with Andromeda.

We hope that the explorations of this chapter, this book, and your Earth-Science course have left you with a new, if partial, understanding of our beautiful planet and its place in the Universe **(Box 22.2)**. Perhaps this knowledge will help you to address the challenges of living in an ever-evolving Earth System. The process of research and discovery will continue—perhaps with your participation!

Take-home message...

Galaxies are arranged in clusters and superclusters, which in turn form vaguely defined filaments and walls. Distant galaxies are moving away from one another due to the expansion of the Universe following the Big Bang. Researchers continue to gather data to determine whether this expansion will continue forever or reverse in the future.

Quick Questions

- What is the largest galaxy other than the Milky Way in our Local Group?
- Are galaxies randomly arranged throughout the Universe?
- What is the evidence for the Big Bang?

Putting Earth Science to Use

Exploring the Universe

For nearly all of history, the mysteries of space beyond the Earth lay far beyond our reach, except within the imagination. Human exploration of the Universe was limited to going outdoors on dark evenings and staring at the heavens. Today, in the information age, a few clicks on your smartphone or taps on your computer keyboard bring you a far-field snapshot showing thousands of galaxies (see Box 20.4), or an edge-on image of the Sombrero Galaxy, as seen from a space telescope **(Fig. Bx22.2)**. You can watch the clouds of Jupiter spin about the Great Red Spot, marvel at Saturn's rings, or explore the icy landscape of Pluto. Planetariums around the world project these glorious images on their domes. Large-screen theaters can put you on the launch pad in a rocket to the Moon or Mars. Hobbyists can now purchase amateur telescopes with sufficient power and stabilization to provide clear images of galaxies 100 light-years from the Earth. People today have an open door to our vast, incredible Universe. Walk through and explore it, and learn about the strange and fascinating objects—black holes, quasars, and pulsars—that speckle the real cosmos. They can make the realm of science fiction seem tame by comparison!

FIGURE Bx22.2 The Sombrero Galaxy.

22 CHAPTER REVIEW

Objective 22.1

Describe the structure and composition of the Sun and its atmosphere, and explain how the Sun produces energy.

KEY CONCEPTS

- The Sun consists primarily of hydrogen and helium. Other elements make up only about 1% of its mass.
- The Sun has a core in which fusion of hydrogen nuclei to form helium nuclei generates immense amounts of energy. The radiative zone, through which radiation from the core passes, surrounds the core. In the outer layer, the convective zone, convection transports energy outward.

- The Sun has an atmosphere consisting of the photosphere, the chromosphere, and the corona. The Sun's light radiates from the photosphere. The corona is an extremely hot, low-density gas layer that can be seen only during an eclipse. High-energy particles flowing outward from the Sun's atmosphere at high speeds constitute the solar wind.

EARTH-SCIENCE VOCABULARY

chromosphere (p. 858)
convective zone (p. 857)
corona (p. 858)
nuclear fusion (p. 856)
photosphere (p. 857)
plasma (p. 856)

radiative zone (p. 857)
solar atmosphere (p. 857)
solar core (p. 856)
solar granulation (p. 857)
solar wind (p. 858)

REVIEW QUESTIONS

1. **(a)** What element makes up most of the Sun? **(b)** What element does nuclear fusion of hydrogen produce in the Sun? **(c)** Describe the reactions leading to the production of this element. **(d)** Explain the difference between plasma and gas, and why the Sun consists mostly of plasma.

2. **(a)** Name the Sun's three internal layers. **(b)** Which layer produces energy? **(c)** How does energy move across the layers to the surface? **(d)** What causes granulation of the Sun's surface?

3. **(a)** Which layer of the Sun produces the radiation that we see as visible light, and how does the temperature in this layer compare with that of the Sun's core? **(b)** What defines the edge of the Sun's disk, as viewed from the Earth? **(c)** Which is thinner, the Sun's chromosphere or its corona? **(d)** Which layer of the solar atmosphere appears in **Figure A**?

A

4. **(a)** What does the solar wind consist of? **(b)** How fast does the solar wind typically move through space?

space weather, which happens when plasma from coronal mass ejections arrives at the Earth, can cause disruptions of satellites and electrical grids.

EARTH-SCIENCE VOCABULARY

coronal mass ejection (CME) (p. 863)
Hale cycle (p. 863)
solar cycle (p. 862)
solar flare (p. 863)
solar maximum (p. 862)
solar minimum (p. 862)
solar prominence (p. 863)
solar storm (p. 863)
space weather (p. 863)
sunspot (p. 861)

REVIEW QUESTIONS

5. **(a)** Explain why the Sun has a magnetic field, and distinguish between the field's two components. **(b)** Explain how the overall pattern of magnetic-field lines on the Sun changes, and on what time scale. **(c)** What is a sunspot, and why is it darker than its surroundings?

6. **(a)** What is the solar cycle? **(b)** What are solar prominences? **(c)** Where on **Figure B** would you expect to see solar prominences?

7. **(a)** At what part of a solar cycle are solar flares most likely to be seen? **(b)** How do coronal mass ejections affect space weather, and what threats does space weather pose to people and infrastructure on the Earth?

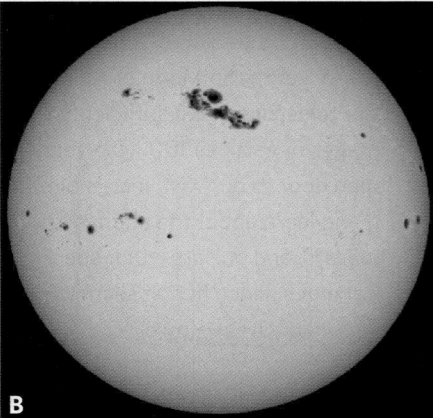

B

Objective 22.2

> Describe how the Sun's magnetic field evolves, and how changes in its magnetic field cause solar storms.

KEY CONCEPTS

- The Sun has a very strong polar magnetic field that reverses polarity every 11 years. Because the Sun's rotation rate is faster at the equator than at the poles, its magnetic-field lines are contorted and constantly changing. Convection within the Sun produces complicated arcing field lines at the Sun's surface.

- When the Sun's magnetic field becomes particularly intense, magnetic-field lines arc upward from the solar surface; they rise from one sunspot and re-enter at another. Sunspots are visible because the patches of the Sun's surface beneath the exit and entry points of the magnetic-field lines become cooler.

- A solar prominence is a stream of plasma that follows the arc of magnetic-field lines between sunspots. Occasionally, a solar flare sends a burst of electromagnetic energy into space. Coronal mass ejections send high-energy plasma into space at speeds exceeding those of normal solar wind.

- Space weather refers to the presence of electromagnetic radiation and charged particles from the Sun in the vicinity of the Earth. Severe

Objective 22.3

> Characterize how stars can differ from one another, and show how stars can be classified on an H-R diagram.

KEY CONCEPTS

- Stars vary in luminosity, a measure of the energy they emit, and in surface temperature. These two quantities are related to a star's size and mass, which also vary greatly.

- Apparent magnitude indicates how bright a star appears to be from the Earth; absolute magnitude indicates how bright a star would appear to be if it were located 32.6 light-years from the Earth. Astronomers often compare the luminosity of a star with that of the Sun using the solar unit scale.

- An H-R diagram, which plots the luminosities of stars against their spectral types, provides a tool for distinguishing among different groups of stars.

absolute magnitude (p. 866)   **luminosity** (p. 865)
apparent brightness (p. 865)   **main sequence** (p. 868)
apparent magnitude (p. 865)   **solar mass** (M_s) (p. 867)
binary star (p. 867)   **solar unit** (p. 866)
dwarf star (p. 867)   **spectral type** (p. 866)
giant star (p. 867)   **supergiant star** (p. 867)
H-R diagram (p. 868)

REVIEW QUESTIONS

8. **(a)** Distinguish among luminosity, apparent magnitude, and absolute magnitude. **(b)** When you classify a star's magnitude by viewing it with your eyes, are you classifying it based on its apparent magnitude or its absolute magnitude? **(c)** Which looks brighter to the naked eye, a star with an apparent magnitude of 1, or a star with an apparent magnitude of 3?

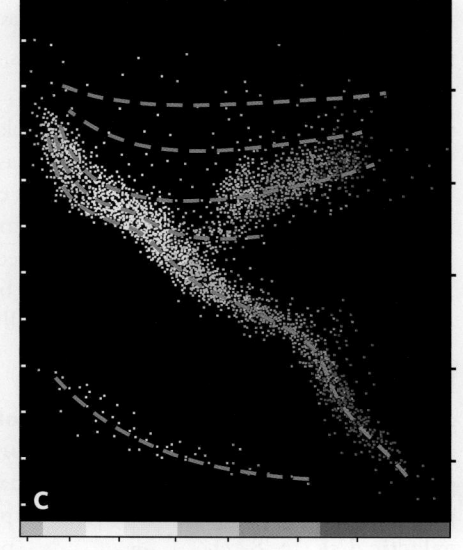

9. **(a)** What are the two axes of an H-R diagram? **(b)** Label the principal groups of stars and the spectral classes on **Figure C**. **(c)** Where in **Figure C** do the largest stars occur? Where do the smallest occur?

10. **(a)** Which star is a dwarf star, the Sun or Betelgeuse? **(b)** What is the luminosity of the brightest star, in solar units? **(c)** How large is the radius of the largest star, compared with that of the Sun? **(d)** About how large is a white dwarf?

Objective 22.4

Discuss how stars form, and how the evolution of a star depends on its mass.

KEY CONCEPTS

- Stars form from the gravitational collapse of gas, dust, and/or ice within nebulae, which may be triggered by disturbances, such as shock waves, that pile up mass locally. According to the condensation theory, particles of ice and dust must be present for stars like the Sun to form.

- As gravity compresses a protostar's mass into a smaller and smaller volume, its core temperature increases. A protostar becomes a star when its core reaches a temperature of 10 million °C, the temperature at which nuclear fusion can begin. When the star reaches its equilibrium radius, it lies on the main sequence of the H-R diagram.

- The life history of a star depends on its mass. Less massive stars reside much longer on the main sequence than do more massive stars, because fusion reactions happen faster in more massive and, therefore, hotter stars.

condensation theory (p. 870)   **shock wave** (p. 869)
red dwarf (p. 872)

REVIEW QUESTIONS

11. **(a)** What is a nebula? **(b)** What roles do nebulae play in the life cycle of stars? **(c)** In the context of condensation theory, why must dust and gas exist within a nebula for stars like our Sun to be able to form? **(d)** What might the role of a shock wave be in triggering new star formation?

12. **(a)** When astronomers plot the positions of thousands of stars on an H-R diagram, most of those stars fall along the main sequence. Why are so many stars found along this region rather than in other parts of the diagram? **(b)** When, in its lifetime, does a star arrive on the main sequence? **(c)** Which kind of star spends the most time on the main sequence, a low-mass star or a high-mass star?

13. **(a)** Describe the path that a star with the mass of our Sun takes on an H-R diagram during its lifetime, before it runs out of fuel. **(b)** Does the luminosity of a star change during its lifetime? Explain your answer.

Objective 22.5

Explain what happens when stars run out of fuel, and distinguish among the various objects that represent remnants of dead stars.

KEY CONCEPTS

- After a star consumes the hydrogen in its core, the core collapses and heats up enough to trigger nuclear reactions involving larger atoms. Meanwhile, a layer of hydrogen surrounding the core begins to burn; the added heat causes the outer layers of the star to swell, so that an intermediate-mass star becomes a red giant, and a high-mass star becomes a red supergiant.

- A star's life ends when it burns through its nuclear fuel. In smaller stars, the chain of fusion reactions stops with carbon, whereas in larger stars, it stops with iron. An intermediate-mass star's core collapses into an extremely dense white dwarf, while its outer layers disperse into space.

- When one member of a binary star is a white dwarf, it may capture matter from its companion; when this matter ignites, a nuclear explosion called a nova takes place. If the white dwarf captures a large amount of material from its companion, the white dwarf will explode and be obliterated in a Type I supernova.

- When a high-mass star runs out of nuclear fuel, the star's core collapses to become a neutron star or black hole, depending on the star's mass, while most of the star's mass blasts into space in a titanic explosion called a Type II supernova.

- White dwarfs, neutron stars, and black holes contain inconceivably dense matter. A black hole's gravity is so strong that no matter or energy can escape from it. The boundary surrounding a black hole is called the event horizon.

EARTH-SCIENCE VOCABULARY

black hole (p. 876)
event horizon (p. 876)
neutron star (p. 876)
nova (plural: novae) (p. 873)
pulsar (p. 876)

red giant (p. 873)
red supergiant (p. 874)
Type I supernova (p. 874)
Type II supernova (p. 875)
white dwarf (p. 873)

REVIEW QUESTIONS

14. **(a)** What property of a star is most important in determining whether it will end its life as a white dwarf, a neutron star, or a black hole? **(b)** Describe the succession of nuclear reactions that takes place after the core of a star runs out of hydrogen fuel. **(c)** When our Sun runs out of fuel, will it produce a black hole, a neutron star, a pulsar, or a white dwarf?

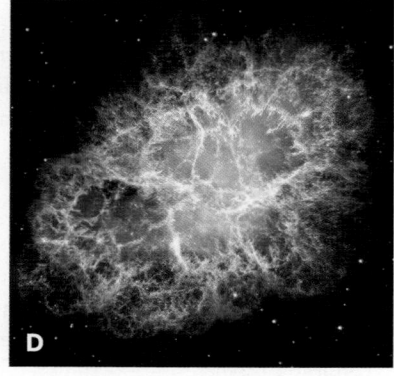

15. **(a)** Which is most likely to produce a black hole: a nova, a Type I supernova, or a Type II supernova? **(b)** How did the nebula in **Figure D** form?

16. **(a)** Describe the characteristics of a black hole. **(b)** Identify the event horizon in **Figure E**. **(c)** Can a planet orbit a black hole within the event horizon?

Objective 22.6

Describe the overall structure of a galaxy, the differences among galaxies, and the relationship of a supermassive black hole to a galaxy.

KEY CONCEPTS

- A galaxy is a collection of stars, planets, and moons and interstellar gas, dust, and ice bound together by gravity. Over a trillion galaxies exist.

- The Milky Way, our home galaxy, consists of a galactic disk with spiral arms surrounding a central star-filled galactic bulge, anchored by a supermassive black hole at the galactic center.

- Galaxies are categorized by their shapes, which include spiral, barred spiral, elliptical, and irregular.

- All large galaxies have a supermassive black hole at their center. In normal galaxies, stars orbit the supermassive black hole. In active galaxies, the supermassive black hole is consuming matter, and emitting X-ray and radio energy. Active galaxies that generate particularly intense radio-wave emissions are referred to as quasars. Most quasars formed near the time of the Universe's birth, but some form later in the Universe's history when galaxies collide.

EARTH-SCIENCE VOCABULARY

active galaxy (p. 880)
cosmic rays (p. 880)
galactic bulge (p. 877)
galactic center (p. 877)
galactic disk (p. 877)
galactic halo (p. 878)

globular cluster (p. 878)
normal galaxy (p. 880)
quasar (p. 881)
supermassive black hole (p. 878)

REVIEW QUESTIONS

17. **(a)** What are the key structural features of the Milky Way Galaxy? **(b)** Where is the Sun located in the Milky Way Galaxy? **(c)** What is a galactic year? **(d)** How do galaxies differ from one another in shape? **(e)** According to recent research, do all planets in the Milky Way orbit stars?

18. **(a)** Distinguish between a normal and an active galaxy. **(b)** What is the difference between a quasar and a pulsar? **(c)** Is it possible that a quasar will develop in the Milky Way Galaxy in the very distant future? If so, when? Justify your answer.

Objective 22.7

Summarize current ideas concerning the overall structure and fate of the Universe.

KEY CONCEPTS

- Gravity binds galaxies together. At a small scale, large galaxies are orbited by satellite galaxies. Several galaxies and their satellites may form a galactic group; the Milky Way and Andromeda are two of about 54 galaxies in our own Local Group. Galaxies in a galactic group can move toward one another, and may collide and combine.

- At a larger scale, galactic groups occur in clusters, which in turn occur in superclusters. At the largest scale, galaxies appear to be arranged as a three-dimensional mesh with string-like filaments and screen-like walls. Large volumes of the Universe are voids in which few, if any, galaxies occur.

- Distant galaxies all move away from one another, implying that the Universe is expanding. The expansion of the Universe implies that about 13.8 billion years ago, all of the Universe's mass and energy emerged from a singularity during the Big Bang.

- The quantity of observable matter in the Universe cannot explain the strength of the gravitational force binding galaxies together and to one other, suggesting that the Universe also contains undetectable dark matter. The future of the Universe—whether it will expand forever or eventually collapse—remains uncertain.

EARTH-SCIENCE VOCABULARY

Big Bang (p. 883)   **Local Group** (p. 882)
dark matter (p. 883)   **singularity** (p. 883)
galactic group (p. 882)

REVIEW QUESTIONS

19. **(a)** Why are the Magellanic Clouds considered to be satellite galaxies to the Milky Way? **(b)** What is the other major galaxy in the Local Group, and in which direction is it moving relative to the Milky Way? **(c)** Roughly how many galaxies occur in a supercluster? **(d)** In what direction do distant galaxies move relative to the Milky Way? Why?

20. **(a)** Describe the current model of the overall structure of the Universe in terms of how galaxies are arranged over vast distances. **(b)** What are the alternative distant futures proposed for the Universe, and what will determine which fate actually happens?

ANOTHER VIEW The bright remnant of a supernova in the Milky Way's satellite galaxy, the Large Magellanic Cloud. This image combines data from the Chandra X-ray Telescope, displayed here in blue, and NASA Spitzer Space Telescope, displayed in red.

ADDITIONAL CHARTS

Metric Conversion Chart

Length

1 kilometer (km) = 0.6214 miles (mi)

1 meter (m) = 1.094 yards = 3.281 feet

1 centimeter (cm) = 0.3937 inches

1 millimeter (mm) = 0.0394 inches

1 mile (mi) = 1.609 kilometers (km)

1 yard = 0.9144 meters (m)

1 foot = 0.3048 meters (m)

1 inch = 2.54 centimeters (cm)

Area

1 square kilometer (km^2) = 0.386 square miles (mi^2)

1 square meter (m^2) = 1.196 square yards (yd^2)

= 10.764 square feet (ft^2)

1 square centimeter (cm^2) = 0.155 square inches (in^2)

1 square mile (mi^2) = 2.59 square kilometers (km^2)

1 square yard (yd^2) = 0.836 square meters (m^2)

1 square foot (ft^2) = 0.0929 square meters (m^2)

1 square inch (in^2) = 6.4516 square centimeters (cm^2)

Volume

1 cubic kilometer (km^3) = 0.24 cubic miles (mi^3)

1 cubic meter (m^3) = 264.2 gallons

= 35.314 cubic feet (ft^3)

1 liter (L) = 1.057 quarts

= 33.814 fluid ounces

1 cubic centimeter (cm^3) = 0.0610 cubic inches (in^3)

1 cubic mile (mi^3) = 4.168 cubic kilometers (km^3)

1 cubic yard (yd^3) = 0.7646 cubic meters (m^3)

1 cubic foot (ft^3) = 0.0283 cubic meters (m^3)

1 cubic inch (in^3) = 16.39 cubic centimeters (cm^3)

Mass

1 metric ton = 2,205 pounds

1 kilogram (kg) = 2.205 pounds

1 gram (g) = 0.03527 ounces

1 pound (lb) = 0.4536 kilograms (kg)

1 ounce (oz) = 28.35 grams (g)

Pressure

1 kilogram per

square centimeter (kg/cm^2)* = 0.96784 atmospheres (atm)

= 0.98066 bars

= 9.8067×10^4 pascals (Pa)

1 bar = 0.1 megapascals (Mpa)

= 1.0×10^5 pascals (Pa)

= 29.53 inches of mercury (in a barometer)

= 0.98692 atmospheres (atm)

= 1.02 kilograms per square centimeter (kg/cm^2)

1 pascal (Pa) = 1 kg/m $\cdot$ s^2

1 pound per square inch = 0.06895 bars

= 6.895×10^3 pascals (Pa)

= 0.0703 kilograms per square centimeter

Temperature

To change from Fahrenheit (F) to Celsius (C):

$$°C = \frac{(°F - 32°)}{1.8}$$

To change from Celsius (°C) to Fahrenheit (°F):

$$°F = (°C \times 1.8) + 32°$$

To change from Celsius (°C) to Kelvin (K):

$$K = °C + 273.15$$

To change from Fahrenheit (°F) to Kelvin (K):

$$K = \frac{(°F - 32°)}{1.8} + 273.15$$

*Note: Because kilograms are a measure of mass whereas pounds are a unit of weight, pressure units incorporating kilograms assume a constant value for the pull of gravity on the Earth's surface. In reality, gravitational pull on the Earth varies slightly with location.

Periodic Table of the Elements

Legend:
Symbol | He | 2 — Atomic number
Helium — Name
4.002 — Atomic weight

Alkali metals · Inert gases · Nonmetals · Transition elements (metals)

Group	1	2	3	4	5	6	7	8	9	10	11	12	13	14	15	16	17	18
	H 1 Hydrogen 1.007																	He 2 Helium 4.002
	Li 3 Lithium 6.941	Be 4 Beryllium 9.0121											B 5 Boron 10.811	C 6 Carbon 12.011	N 7 Nitrogen 14.006	O 8 Oxygen 15.999	F 9 Fluorine 18.998	Ne 10 Neon 20.179
	Na 11 Sodium 22.989	Mg 12 Magnesium 24.305											Al 13 Aluminum 26.981	Si 14 Silicon 28.085	P 15 Phosphorus 30.973	S 16 Sulfur 32.066	Cl 17 Chlorine 35.452	Ar 18 Argon 39.948
	K 19 Potassium 39.098	Ca 20 Calcium 40.078	Sc 21 Scandium 44.955	Ti 22 Titanium 47.88	V 23 Vanadium 50.941	Cr 24 Chromium 51.996	Mn 25 Manganese 54.938	Fe 26 Iron 55.847	Co 27 Cobalt 58.933	Ni 28 Nickel 58.693	Cu 29 Copper 63.546	Zn 30 Zinc 65.39	Ga 31 Gallium 69.723	Ge 32 Germanium 72.64	As 33 Arsenic 74.921	Se 34 Selenium 78.96	Br 35 Bromine 79.904	Kr 36 Krypton 83.80
	Rb 37 Rubidium 85.467	Sr 38 Strontium 87.62	Y 39 Yttrium 88.905	Zr 40 Zirconium 91.224	Nb 41 Niobium 92.906	Mo 42 Molybdenum 95.94	Tc 43 Technetium 98.907	Ru 44 Ruthenium 101.07	Rh 45 Rhodium 102.905	Pd 46 Palladium 106.42	Ag 47 Silver 107.868	Cd 48 Cadmium 112.411	In 49 Indium 114.82	Sn 50 Tin 118.710	Sb 51 Antimony 121.757	Te 52 Tellurium 127.60	I 53 Iodine 126.904	Xe 54 Xenon 131.29
	Cs 55 Cesium 132.905	Ba 56 Barium 137.327	La 57 Lanthanum 138.905	Hf 72 Hafnium 178.49	Ta 73 Tantalum 180.947	W 74 Tungsten 183.85	Re 75 Rhenium 186.207	Os 76 Osmium 190.2	Ir 77 Iridium 192.22	Pt 78 Platinum 195.08	Au 79 Gold 196.966	Hg 80 Mercury 200.59	Tl 81 Thallium 204.383	Pb 82 Lead 207.2	Bi 83 Bismuth 208.980	Po 84 Polonium 208.982	At 85 Astatine 209.987	Rn 86 Radon 222.017
	Fr 87 Francium 223.019	Ra 88 Radium 226.025	Ac 89 Actinium 227.027															

Lanthanides:

Ce 58 Cerium 140.15	Pr 59 Praseodymium 140.907	Nd 60 Neodymium 144.24	Pm 61 Promethium 144.912	Sm 62 Samarium 150.36	Eu 63 Europium 151.965	Gd 64 Gadolinium 157.25	Tb 65 Terbium 158.925	Dy 66 Dysprosium 162.50	Ho 67 Holmium 164.930	Er 68 Erbium 167.26	Tm 69 Thulium 168.934	Yb 70 Ytterbium 173.04	Lu 71 Lutetium 174.967

Actinides:

Th 90 Thorium 232.038	Pa 91 Protactinium 231.035	U 92 Uranium 238.028	Np 93 Neptunium 237.048	Pu 94 Plutonium 244.064	Am 95 Americium 243.061	Cm 96 Curium 247.070	Bk 97 Berkelium 247.070	Cf 98 Californium 251.079	Es 99 Einsteinium 252.083	Fm 100 Fermium 257.095	Md 101 Mendelevium 258.10	No 102 Nobelium 259.100	Lr 103 Lawrencium 262.11

The modern periodic table of the elements. Each column groups elements with related properties. For example, inert gases are listed in the column on the right. Metals are found in the central and left parts of the chart. Elements heavier than uranium do not occur naturally.

GLOSSARY

ablation (1) The erosion of material from an object by any externally driven process. (2) The removal of ice from a glacier by melting, sublimation, or calving.

absolute magnitude A measure of the luminosity of a star; specifically, the apparent magnitude a star would have if it were located at a standard distance of exactly 32.6 light-years (10 parsecs) from the Earth.

absolute plate velocity The speed and movement direction of a plate relative to a fixed reference point.

absorption The addition of energy to molecules by light so that their atoms move and vibrate faster.

abyssal plain A broad, sediment-covered, relatively flat region of the seafloor that lies 4.5–5.5 km (2.8–3.4 mi) below sea level.

accretion The addition of material; can pertain to the suturing of a crustal block to the margin of a continent, incorporation of sediment to the accretionary prism at a subduction zone, deposition of sediment along a coast, or collision of matter with an accretion disk around a protostar.

accretionary coast A coast that is widening because it receives more sediment than erodes away.

accretionary prism A wedge-shaped mass of sediment and rock formed along a convergent boundary; it consists of material scraped off the top of the downgoing plate, as well as trench sediment; this material accretes to the overriding plate.

accretion disk A flat, rotating disk of gas, dust, and ice surrounding a celestial object, such as a protostar, a white dwarf in a binary star, or a black hole.

accuracy The closeness of a measured value to its true value.

acidification See *ocean acidification*.

acid mine runoff A dilute solution of sulfuric acid, produced when sulfur-bearing minerals in a mine react with rainwater or groundwater and flow out of a mine.

acid rain Rain formed from atmospheric water that has reacted with pollutant aerosols to become a weak solution of sulfuric acid.

active galaxy A galaxy that emits a large amount of its energy in the X-ray and radio regions of the electromagnetic spectrum, with most of that energy coming from the galactic center.

active margin A continental margin that is also a plate boundary, and therefore hosts earthquakes.

active volcano A volcano that has erupted within the past few centuries or millennia and has the potential to erupt again.

adiabatic expansion Expansion of a volume of air during which no mass or energy is exchanged with the air in the environment surrounding the volume.

aerosols Microscopic particles, either solid or dissolved in liquid droplets, that are small enough to remain suspended in the atmosphere.

air The mixture of gases that make up the Earth's atmosphere.

air mass A large body of air (typically several kilometers thick and hundreds to thousands of kilometers wide) within which temperature (at a given elevation) and humidity are relatively uniform.

air pollution Atmospheric aerosols and toxic gases that have short- or long-term harmful effects on humans or other organisms.

air temperature A measure of the average kinetic energy of air molecules, as measured by a thermometer or thermistor.

albedo The reflectivity of a surface.

algal bloom A rapid growth of algae due to abundant nutrients in water; subsequent death and decomposition of the algae lowers the concentration of dissolved oxygen in the water in which the bloom occurred.

alloy A metal containing more than one metallic element.

alluvial fan A gently sloping wedge of sediment dropped by an ephemeral stream at the base of a mountain in arid or semiarid regions.

alluvium Sediment deposited by a stream (also called a *fluvial deposit*).

Alpine-Himalayan chain The collisional orogen that formed when the Indian subcontinent, Africa, and intervening crustal blocks collided with the southern margin of Eurasia during the Cenozoic; includes the Alps and the Himalayas.

amplitude One-half the vertical distance from crest to trough of a wave; that is, one-half the wave height.

angle of incidence The angle between incoming light from the Sun and another surface, such as the Earth's surface.

angle of repose The angle of the steepest slope that a pile of unconsolidated material can attain without collapsing.

angular momentum A property of an object moving around a circle, defined as the product of its mass (m), its rotational velocity (w), and its distance from the axis of rotation (r).

angular unconformity An unconformity in which the strata below have an orientation different from that of the strata above, because the strata below were tilted or folded before the strata above were deposited.

annual probability The likelihood of an event, such as a flood or an earthquake, happening in a given year, specified by 1 divided by the recurrence interval, expressed as a percentage.

Anthropocene A name for the time during which human activity has had a major influence on climate, the environment, landscapes, and biogeochemical cycles in the Earth System.

anthropogenic change Change in the Earth System caused by human activity.

anthropogenic source An input into the Earth System from human activities.

anticline A fold with an arch-like shape, such that the limbs dip away from the hinge.

anvil cloud The flattened top of a large cumulonimbus cloud that has spread laterally at the tropopause.

aphotic zone The layer of the ocean below the level to which sunlight penetrates.

apparent brightness The amount of energy from a star that passes through 1 m^2 at the surface of the Earth in 1 second.

apparent magnitude A measure of the apparent brightness of a celestial object as viewed from the Earth. On the scale for defining the apparent magnitude of a star, a magnitude 6 star is dimmer than a magnitude 1 star, and very bright objects can have negative magnitudes.

apparent polar-wander path A path on the globe along which a magnetic pole appears to have moved over time; the word "apparent" is used because, in fact, it's the continents that move, while the magnetic pole stays fairly fixed.

aquifer A body or layer of sediment or rock that transmits water easily.

aquitard A body or layer of sediment or rock that does not transmit water easily and therefore slows or prevents the motion of groundwater.

Archean Eon The middle eon of the Precambrian; it spans the time from 4.0 Ga to 2.5 Ga.

Archimedes's principle The principle that an object sinks in water to the depth at which the mass of the water it displaces equals the mass of the object.

artesian well A well, tapping a confined aquifer, in which water rises without pumping due to pressure within the aquifer.

ash Very fine particles of glass or pulverized rock erupted by a volcano.

asteroid A rocky or metallic celestial object larger than 100 m (330 ft) across.

asteroid belt The zone, between Mars and Jupiter, in which most asteroids orbit the Sun.

asthenosphere The layer of the mantle that lies directly beneath the lithosphere, in which the mantle can flow plastically and, therefore, can undergo convection.

astronomical unit (AU) The average distance from the Sun to the Earth; approximately 150 million km (93 million mi).

astronomy The scientific study of celestial objects (such as planets, stars, galaxies, and nebulae) and of the Universe as a whole (also called *space science*).

atmosphere A layer of gases that surrounds a planet.

atmospheric greenhouse effect See *greenhouse effect*.

atmospheric instability Atmospheric conditions in which a parcel of air, after undergoing initial lifting, can remain buoyant and continue to rise.

atmospheric pressure The force applied by air on a unit area of a surface; equivalent to the weight of a column of air above the unit area.

atmospheric science The study of the layer of gases that surrounds the Earth.

atmospheric seeing The blurring or fuzzing of a celestial object's image in a telescope caused by turbulence in the atmosphere.

atom The smallest piece of an element that has the properties of the element; consists of a nucleus surrounded by an electron cloud.

atomic mass The amount of matter in an atom; approximately equal to the total number of protons plus neutrons in the atom's nucleus. Different isotopes of an element have different atomic masses.

atomic number The number of protons in an element's nucleus. All isotopes of an element have the same atomic number.

aurora Undulating translucent curtains or streaks of varicolored light that appear across the night sky, generally at high latitudes, when charged particles from the Sun interact with ions in the ionosphere.

autumnal equinox The day of the year when the Earth's axis of rotation transitions from tilting toward the Sun to tilting away from it. In the northern hemisphere, it is September 22.

average precipitation The amount of precipitation falling in a location averaged over a specific time period, such as a month, year, decade, or century.

axial plane An imaginary surface that contains the hinges of successive layers of rock affected by a fold; it effectively defines the boundary between the two limbs of the fold.

axis of rotation An imaginary line around which an object rotates.

backwash The gravity-driven flow of water back down the slope of a beach after swash from a wave has surged up the beach.

badlands A landscape, formed in vegetation-free, very weak sediment or rock, characterized by a parallel drainage network of small, closely spaced gullies and ravines that merge downslope.

bajada A composite wedge of sediment formed when alluvial fans emerging from adjacent valleys merge and overlap along the front of a mountain range.

baked contact A contact between wall rock and an igneous intrusion at which the wall rock has been altered by heat emitted from the igneous intrusion.

banded iron formation (BIF) Iron-rich sedimentary layers consisting of alternating beds of gray iron oxide minerals and reddish iron-rich chert.

bar An elongate mound of sediment deposited by flowing water.

barometer An instrument that measures atmospheric pressure.

barrier island A partially submerged offshore sandbar or sand dune that forms an elongate offshore island parallel to the shore.

basalt A fine-grained, dark-colored igneous rock with a mafic composition.

base level The lowest elevation to which a stream's water can flow at a given locality.

basin A fold or depression in strata whose shape resembles a right-side-up bowl. See also *sedimentary basin*.

Basin and Range Province A broad Cenozoic continental rift located in the western United States, characterized by tilted fault blocks that form ranges separated by alluvium-filled basins.

batholith A composite body of igneous rock, generally felsic to intermediate, up to several hundred kilometers long and tens of kilometers wide, formed by the intrusion of numerous plutons.

bathymetry Variation in depth of a water body.

beach A belt of sediment, bordering the shoreline, that undergoes sorting and shifting by the swash and backwash of waves.

beach nourishment The practice of adding sediment, generally pumped or dredged from offshore or trucked in from elsewhere, to replace beach sediment that has been eroded away.

beach profile The variation in elevation of a beach, as measured along a line perpendicular to the shoreline.

beach retreat The inland movement, and sometimes narrowing, of a beach due to erosion; due to beach retreat, the position of the shoreline moves inland.

bearing The compass orientation of a linear structure's projection on a horizontal plane.

bed An individual layer of sedimentary rock.

bedrock Rock still attached to the Earth's crust.

benthic zone The region encompassing the seafloor and the region just above and below the seafloor.

benthos Organisms that live on, just beneath, or just above the seafloor (that is, in the benthic zone).

Big Bang The cataclysmic explosion of a singularity that, according to theory, represents the formation of the Universe.

Big Bang nucleosynthesis The formation of low-mass nuclei (mainly H and He) during the first few minutes after the Big Bang.

binary star A pair of stars that are close enough to each other to be bound to each other gravitationally so that they orbit around a common center of mass.

biochemical sedimentary rock Sedimentary rock formed from solid material, such as shells, produced by living organisms.

biodiversity The number of different types of organisms that exist on the Earth at a given time.

biofuel Gas or liquid fuel made from plant material (biomass); examples include ethanol (made from fermented sugar) and biodiesel (made from vegetable oil).

biogeochemical cycle A succession of chemical exchanges among living and nonliving reservoirs in the Earth System.

biological resource A resource derived from living or recently living organisms.

biomass The total mass of biological material (consisting of living or recently living organisms) in a given volume.

biosphere All living organisms on the Earth, together with the portions of the Earth in which they live; includes the portion of the Earth System from a few kilometers below the Earth's surface to a few kilometers above it.

blackbody An object that absorbs all radiation incident upon it and reradiates energy according to the three laws of blackbody radiation.

black hole An extremely massive object so dense that its gravitational force does not allow even light to escape.

black smoker A vent along a mid-ocean ridge that spews hot, mineral-rich water; its dissolved sulfide components instantly precipitate when the water mixes with seawater and cools.

blizzard A snowstorm in which winds exceed 56 km/h (35 mph) for at least 3 hours, causing reduced visibility due to falling or blowing snow.

blowout An uncontrolled event during which naturally pressurized hydrocarbons spew from a well; the hydrocarbons spill onto the land or into the sea, or vent into the air, and may explode.

blue shift A phenomenon in which light emitted by an electromagnetic source moving toward an observer shifts to a higher frequency.

body wave A seismic wave that passes through the interior of the Earth.

boundary layer The layer of the atmosphere adjacent to the Earth's surface, in which friction with the surface significantly affects air movement.

braided stream A stream in which the water divides into strands that weave back and forth among bars of alluvium.

breaker A water wave, close to shore, in which water at the top of the wave curves over its base and falls to form spray and foam; it forms because friction with very shallow seafloor slows the base of the wave.

breakwater An offshore wall of rock or concrete, oriented parallel to a beach, that is constructed to prevent the full force of waves from eroding the beach.

brittle deformation A process by which a material subjected to stress permanently separates into pieces along fracture planes.

buoyancy force A force that drives a less dense material upward if located within a denser material.

caldera A large circular or elliptical depression with steep walls and a fairly flat floor, formed after a volcanic eruption causes the central area of the volcano to collapse into the drained magma chamber below.

caliche A rock-like material formed in soils of arid climates due to cementation of clasts by calcite formed from residual ions.

Cambrian explosion The rapid diversification of life, as indicated by the fossil record, that occurred at the beginning of the Cambrian Period.

capacity The total quantity of sediment a stream can carry; the capacity of a stream depends on both its competence and its discharge.

carbon budget The balance between additions and subtractions of carbon within the different reservoirs of the Earth System.

carbon cycle The succession of exchanges among the Earth System's various living and nonliving reservoirs of carbon; the carbon may occur in atmospheric gases, biomass, soil, fossil fuels, and limestone.

carbon footprint The total amount of greenhouse gases produced by an anthropogenic process or activity, such as burning fossil fuels, or by the production of anthropogenic materials, such as cement.

cave network A network of interconnected caverns and passages underground, formed by the dissolution of limestone bedrock by acidic water.

cavern A large underground open space in a cave network (also called a *chamber*).

celestial object An object or feature visible in the night sky.

celestial sphere An imaginary sphere surrounding the Earth, oriented so that its axis aligns with the Earth's axis of rotation; it has no physical existence, but is a convenient reference frame for specifying the positions of celestial objects.

cement (1) Mineral material that precipitates from water and fills the spaces between sediment grains, holding the grains together. (2) A powder, produced by roasting limestone and other rocks, that when mixed with aggregate and water, yields solid concrete.

cementation A process of lithification in which cement precipitates from groundwater and binds sediment grains together.

centrifugal force The apparent outward-directed force experienced by an object in rotation, as when resting on the surface of a spinning disk or sphere.

chain reaction A succession of fission reactions that takes place when particles released from one reaction trigger further reactions.

channel A trough in the ground surface carrying flowing water.

charge separation A process by which positive ions and electrons are physically separated in a cumulonimbus cloud during the collision of large and small ice particles; it can result in the production of lightning.

chemical bond A bond that attaches atoms to one another through interactions between the electron clouds of the atoms, such as the sharing or exchange of electrons.

chemical formula A recipe that specifies the composition of a chemical compound by using abbreviations to represent the elements in the compound as well as their relative proportions.

chemical sedimentary rock Sedimentary rock made up of minerals that precipitate directly from water solutions.

chemical weathering A process in which chemical reactions alter or dissolve minerals when the minerals come in contact with water solutions and air.

chert A type of sedimentary rock composed of cryptocrystalline quartz (quartz grains that are too small to be seen even with a petrographic microscope).

chromosphere The region of the Sun's atmosphere located between the photosphere and the corona; it displays color during an eclipse.

cinder cone A subaerial volcano consisting of a cone-shaped pile of lapilli whose slope approaches the angle of repose.

cirque A bowl- or spoon-shaped depression carved by a glacier on the side of a mountain.

cirrus cloud A wispy, elongate cloud that develops very high in the troposphere (above 6 km, or 20,000 ft).

clast A single fragment or grain of rock.

clastic rock Sedimentary rock formed when a layer of clasts (fragments), derived from the weathering of pre-existing rock, undergoes compaction, consolidation, and cementation.

clay Extremely fine-grained sediment; its grains, most of which consist of sheet-like clay minerals, are too small to see with an optical microscope.

cleavage (1) The tendency of a mineral to break along preferred planes whose orientation is defined by the internal crystal structure of the mineral. (2) A type of tectonic foliation in rock formed by pressure solution under low-grade metamorphism.

cliff retreat Change in the position of a cliff face over time due to erosion.

climate The average, range, character, and seasonality of weather conditions in a given region, as recorded over the course of decades to millennia.

climate change A long-term shift in the average of one or more climate conditions.

climate group One of five major climate types in the Köppen-Geiger classification, based on precipitation, temperature, and seasonal variation of weather; each group contains several subgroups.

climate zone migration The slow poleward or equatorward shift of isotherms.

closure temperature The temperature below which atoms are locked into a mineral grain.

cloud A visible mass of condensed water vapor, consisting of tiny water droplets, ice crystals, or both, floating in the atmosphere, typically high above the ground.

cloud cover The percentage of the sky covered by clouds.

cloud-to-cloud lightning An electric discharge that jumps between clouds and does not come into contact with the ground.

cloud-to-ground lightning An electric discharge that extends from a cloud to the ground.

coal A black organic sedimentary rock consisting of more than 50% carbon, formed from the lithified remains of plant material.

coal rank A measure of the carbon content of coal; higher-rank coal contains a higher concentration of carbon and forms at higher temperatures.

coal reserve An economically valuable accumulation of coal in sedimentary strata.

coal seam A layer or bed of coal in strata.

coal swamp A tropical or subtropical region that contains woody wetlands in which substantial quantities of plant debris can accumulate in oxygen-poor water.

coastal desert A desert along a coast that borders a cold ocean current, where air cooled by the current does not rise and, therefore, does not release precipitation.

coastal plain A low-relief region of land adjacent to a coast.

coastal wetland A vegetated, low-lying region of coast that becomes submerged by shallow water for all or part of the day; such wetlands host salt-resistant vegetation and many other organisms.

cold desert A desert where temperatures generally stay below about 20°C (68°F) and, during part of the year, can drop below freezing.

cold front A boundary between a cold air mass and a warm air mass, at which the cold air advances forward and lifts the warmer air mass.

collision In the context of plate tectonics, the process during which two buoyant pieces of crust merge after subduction of the intervening ocean lithosphere.

collisional orogen A mountain belt formed by collision.

collision-coalescence The process by which different-sized cloud droplets come into contact and merge with one another to form raindrops.

color The portion of the visible spectrum that your eyes see when looking at something; for example, the color of a mineral depends on wavelengths of visible light that the mineral does not absorb, and the color of a star depends on its surface temperature.

combustion A rapid reaction of a material with oxygen, which releases the potential energy in chemical bonds in the form of heat and light; combustion of organic material breaks apart its chemicals and produces H_2O and CO_2 as well as energy.

comet An object composed of ice, dust, and rock that revolves around the Sun in a highly elliptical orbit and produces a dust tail and a gas tail as it nears the Sun.

competence The maximum clast size that a stream can carry.

compression Stress that pushes on or squeezes a material.

concrete A construction material made by mixing sand, gravel, or both with cement and water to produce a slurry that hardens into a solid rock-like material.

condensation theory The theory describing how a patch of a nebula containing dust and ice contracts to form a star and planetary system.

cone of depression The downward-pointing, cone-shaped surface of the water table in a location where the water table has experienced drawdown because of pumping at a well.

conglomerate A clastic sedimentary rock formed from rounded pebbles, cobbles, and/or boulders.

constellation A recognizable pattern of stars in the night sky that suggests an image; Western cultures recognize 88 constellations.

consumer In the context of the biosphere, an organism in the food chain or food web that eats other organisms.

contact In a geologic context, the boundary surface between two rock bodies; such as that between two stratigraphic formations, between an igneous intrusion and wall rock, between two igneous rock bodies, or between rocks juxtaposed by a fault.

contaminant plume Pollution that has mixed with or dissolved in groundwater, and has moved with groundwater flow so that it has traveled away from the pollution source.

continent A very large land area underlain by continental lithosphere.

continental arc A volcanic arc formed at a continental margin along which an oceanic plate subducts beneath the continent.

continental drift The idea that continents have moved, and are still moving, slowly relative to one another and relative to a fixed reference frame in the mantle. The "drift" of a continent happens because it is a part of a moving lithosphere plate.

continental glacier A vast sheet of ice that spreads over thousands of square kilometers of continental crust (also called a *continental ice sheet*).

continental-interior desert A land area that receives hardly any precipitation because it is located so far from an ocean that the air it receives has dried out.

continental margin The boundary between an ocean basin and a continent.

continental shelf A broad region of shallow sea along a continental margin. The widest continental shelves occur over passive margins.

convection A process of circulation and heat transfer that happens when a deeper layer of material warms, expands, and becomes less dense than the overlying layer of cooler material, and therefore rises, and cooler material sinks to take its place.

convective cell A circulation pattern caused by convection.

convective plume In the context of volcanic eruptions, a buoyant, cloud-like plume of hot ash and air that can rise as high as the stratosphere; it forms when the eruptive jet emitted by an explosive volcanic eruption heats the surrounding air.

convective zone In the context of astronomy, the layer of the Sun in which convection transports energy between the radiative zone and the photosphere.

conventional reserve A hydrocarbon reserve that can be accessed simply by drilling into it and pumping the hydrocarbons from the reservoir rock.

convergence (1) In the context of meteorology, the net inflow of air molecules into a region of the atmosphere, caused by slowing wind speed or changes in wind direction, which increases atmospheric pressure at the Earth's surface. (2) In the context of plate tectonics, the relative movement of two plates toward each other.

convergent boundary A plate boundary, associated with a deep-sea trench, where one plate (the downgoing plate) sinks and slips into the mantle beneath the other (the overriding plate).

coral bleaching A process that occurs when corals under environmental stress expel the symbiotic algae that live in their tissues, and turn white (also called *reef bleaching*).

coral reef A marine realm of shallow water underlain by coral mounds, many other types of shell-secreting organisms, and sediment formed from the shells of the organisms.

Cordilleran ice sheet A North American ice sheet of the Pleistocene Ice Age that originated in the mountains of western Canada, then spread westward to the Pacific coast and eastward until it met the Laurentide ice sheet.

core-mantle boundary The interface at a depth of 2,900 km (1,802 mi) below the Earth's surface that separates the core from the mantle.

Coriolis effect The deflection of a moving object or material relative to the surface of a rotating object beneath it.

Coriolis force The apparent force that causes the Coriolis effect.

corona The hottest, outermost layer of the Sun's atmosphere.

coronal mass ejection (CME) An event that ejects a huge stream of high-energy solar plasma directly into space in a specific direction, at speeds of over 3,000 km/s (6.7 million mph).

correlation The process of defining the age relationships between strata at one locality and strata at another.

cosmic rays High-energy ions moving at speeds up to 80% that of light; they originate from the Sun, from other sources in the Milky Way, and from other galaxies.

cosmologic time The span of time since the origin of the Universe.

cosmology The study of the overall structure, formation, and evolution of the Universe.

crater A bowl-shaped depression at the summit of a volcano. See also *impact crater*.

craton A region of Precambrian continental crust consisting of rock that has not been affected by orogenies for at least the last 1 billion years, and behaves as a relatively rigid, stable block.

cratonic platform A portion of a craton in which Phanerozoic strata bury most of the underlying Precambrian basement.

creep In the context of mass wasting, the gradual downslope movement of regolith.

crest In the context of floods, the highest level that a stream rises to.

crevasse A large crack that develops by brittle deformation in the top 60 m (197 ft) of a glacier.

critical mineral A mineral that is essential for producing high-tech equipment and alloys; supply-chain disruption may limit the availability of these minerals to countries without reserves.

cross beds A sedimentary structure within a bed represented by thin layers inclined at an angle to the top and bottom of the bed; it represents the preserved slip faces of dunes or ripple marks.

cross-cutting relations The geologic principle stating that if one geologic feature cuts across another, the feature that has been cut must be older than the one that cuts it.

crust In the context of the Earth's structure, the solid outermost layer of the Earth.

crustal root A downward bulge of crust that results from horizontal shortening and vertical thickening of the crust in a collision zone.

cryosphere The frozen component of the Earth's hydrosphere.

crystal A single, continuous piece of a mineral that grew to achieve its present shape; inside a crystal, atoms, molecules, or ions are fixed in an orderly arrangement.

crystal face A flat surface on a crystal, which intersects with other such surfaces at sharp edges, that developed naturally as the crystal grew.

crystal habit The general shape of a crystal or cluster of crystals that grew unimpeded.

crystalline (1) Having atoms that are arranged in a specific, orderly pattern; (2) a rock texture characterized by the presence of interlocking crystals.

crystalline rock A rock with a crystalline texture, meaning a texture characterized by interlocking crystals.

crystal structure The orderly arrangement of atoms inside a crystal.

cumulus cloud A vertically growing cloud with cauliflower-like lobes.

current A band of flowing water that moves distinctly faster than adjacent water.

cyclone The name used specifically for a hurricane that forms over the Indian Ocean or affects Australia. See also *mid-latitude cyclone*, *tropical cyclone*.

dark matter Hypothetical matter in the Universe that does not emit or absorb electromagnetic radiation, so it cannot be observed directly. Dark matter is thought to constitute most of the mass of the Universe, but its character remains unknown.

daughter atom An atom that is the product of radioactive decay.

dead zone A region of the sea near the mouth of a river where nutrients brought in by the river cause algal blooms, and the subsequent death and decomposition of the algae deplete the water of oxygen so that it cannot sustain marine organisms.

debris flow A downslope movement of mud mixed with larger rock fragments.

decompression melting Melting that occurs in hot mantle rock that rises to shallower depths in the Earth, so that pressure acting on it decreases but its temperature remains unchanged.

deep-sea current A current that flows far below the ocean's surface; such currents play an important role in thermohaline circulation.

deep-sea trench An elongate trough where the ocean floor may reach depths of 7–11 km (4–7 mi); deep-sea trenches form at a convergent boundary.

deformation A change in the shape, position, or orientation of a material by bending, breaking, or flowing.

delamination Removal of a layer of lithospheric mantle from the base of a lithosphere plate.

delta A wedge of sediment formed at the mouth of a stream where the stream enters standing water and its current slows, so that the stream loses competence, and sediment settles out.

density Mass per unit volume.

deposition The process by which sediment settles out of a transporting medium.

depositional setting Environmental conditions at the time and place at which sediment accumulated.

derecho Severe and long-lasting straight-line winds formed in association with a squall line of thunderstorms; these winds can affect a large geographic region and cause significant damage.

desalination The process of removing salt from seawater to make freshwater.

desert A region that receives average rainfall (or snowfall equivalent) of less than 25 cm/yr (10 in/yr), and is therefore so arid that it contains no permanent streams and supports vegetation on no more than 15% of its surface.

desertification The conversion of a semiarid region into a desert by natural drought or human activities.

desert pavement A mosaic-like surface of rock fragments that covers the ground in a desert.

desert varnish A dark, rusty brown coating of iron oxide and magnesium oxide that forms on the surface of rock in a desert.

detachment A gently dipping to subhorizontal fault that forms the base of a system of more steeply dipping faults. Deformation due to the faulting above the detachment does not affect the rock below the detachment.

dewpoint The temperature at which air will become saturated if it is cooled at constant pressure.

differential stress A condition in which the push or pull in one direction differs from that in another direction, and/or a condition that produces shearing.

differentiation In the context of planetary interiors, the process by which denser materials sink toward the center of a molten or fluid planetary body, resulting in compositional layering.

diffusion The migration of atoms or molecules through a material.

dike A vertical, wall-shaped tabular intrusion that cuts across a layer of wall rock.

dimension stone Rock used for architectural purposes.

dinosaur One of a group of reptiles of the Mesozoic Era whose legs extended under the body rather than off to the sides.

dip The angle of a planar structure with respect to the horizontal, as measured in a vertical plane; that is, the angle at which a planar structure tilts. Strike and dip together uniquely define the orientation of a planar structure.

dipole A magnetic field with two poles, which can be represented by an imaginary arrow that points from one pole to the other.

dip-slip fault A fault in which sliding occurs up or down the slope (dip) of the fault.

directional drilling A drilling technology that allows the drill bit to be tilted so that the drill hole can curve and be angled or even horizontal.

discharge The volume of water that passes through a cross section of a stream in a given time (usually 1 second).

discharge area A location where groundwater flows back up to the land surface, where it may emerge from springs.

disconformity An unconformity separating two sedimentary units, in which the strata above and below have the same orientation.

displacement The amount of movement or slip across a fault (also called *offset*).

divergence In the context of atmospheric science, the net outflow of air molecules from a region of the atmosphere, caused by increasing wind speed or changes in wind direction, which decreases atmospheric pressure at the Earth's surface.

divergent boundary A plate boundary where two oceanic plates move apart by the process of sea-floor spreading (also called a *spreading boundary*).

doldrums A belt along the equator where rain is frequent and where little horizontal air movement occurs.

dolostone A carbonate rock that contains a substantial proportion of the mineral dolomite.

dome A fold or warping of strata whose shape resembles an overturned bowl.

Doppler effect The change in the frequency of observed wave energy that takes place when the source of the energy moves toward or away from an observer.

Doppler radar A radar system that detects precipitation and wind by transmitting microwave energy and measuring both the intensity and the shifts in the frequency of the returned energy.

dormant volcano An active volcano that does not exhibit current activity and has not erupted for a long time, but does have the potential to erupt again in the future.

downburst A straight-line wind that emerges when air in a short-lived, intense downdraft produced by a thunderstorm reaches the ground and flows radially outward.

downcutting The process by which water flowing along a stream channel cuts into the substrate and deepens the channel relative to its surroundings.

downdraft A downward-moving air flow.

downslope force The portion of gravity oriented in the downslope direction, which will move material to a lower elevation when permitted.

downslope windstorm A strong flow of wind down the leeward side of a mountain range; North American examples include the chinook and Santa Ana winds.

downwelling The sinking or downward flow of near-surface water to greater depth in the ocean.

drainage divide A boundary that separates one drainage basin from another.

drainage network An array of interconnecting tributary streams that all flow into the same trunk stream.

drumlin A streamlined, elongate hill formed when a glacier overrides and molds glacial till; its long axis is parallel to the flow direction of the glacier.

dry adiabatic lapse rate The rate (about 10°C/km) at which an unsaturated parcel of air will cool as a result of expansion during its ascent if there is no exchange of heat or mass with the surrounding environment.

dry wash A stream channel that fills with water only rarely.

ductile deformation A process during which a material subjected to stress changes shape permanently without forming visible cracks or breaking apart.

dune A large pile of sediment, deposited by a current of wind or water, whose shape depends on current direction and velocity. See also *sand dune*.

dust Tiny solid particles made up of refractory materials; on the Earth, dust generally refers to tiny mineral specks, whereas in space, the term applies to a broader range of materials.

dust storm An event in which strong winds strip fine-grained sediment from unvegetated soil and send it skyward to form rolling dark clouds that block out the Sun.

dwarf planet A spherical object that orbits a star directly, but has not cleared its orbit of smaller bodies.

dwarf star A star of relatively small size and low luminosity. The majority of stars on the main sequence of an H-R diagram, including the Sun, are dwarf stars.

dynamic pressure The pressure exerted on an object by a flowing fluid such as air or water.

dynamothermal metamorphism Metamorphism that occurs over a large region as a consequence of heating, pressure, and differential stress; generally associated with mountain building (also called *regional metamorphism*).

dysphotic zone The "twilight" layer of the ocean that is penetrated by some sunlight, but not enough to permit photosynthesis.

Earth-Moon system The Earth and its Moon, viewed as a gravitationally linked pair rotating around a common center of gravity.

earthquake The production of seismic waves, or an episode of ground shaking, caused by a sudden movement underground; most earthquakes are caused by slip on faults.

earthquake engineering The design of buildings that can withstand ground shaking.

earthquake preparedness Precautions designed to mitigate the dire consequences of earthquakes; examples include seismic retrofitting.

earthquake zoning Regulations limiting the construction of buildings in seismically active areas.

Earth resource A resource derived from geologic materials or from natural Earth processes such as air movement or water flow.

Earth System The geosphere, hydrosphere, atmosphere, and biosphere, along with the intricate ways in which those realms interact with one another over time.

eccentric orbit A highly elliptical path taken by one object revolving around another.

ecliptic plane The plane defined by the Earth's orbit around the Sun.

eddy A somewhat swirling portion of a flowing fluid, in which movement is opposite to overall flow.

Ediacaran fauna An assemblage of shell-less invertebrates that lived during the last 100 million years of the Proterozoic.

effusive eruption A volcanic eruption dominated by the release of lava; most effusive eruptions release low-viscosity mafic lava.

Ekman transport The overall movement of a mass of water at a 90° angle to the wind direction, due to the progressive change in motion of water with depth in response to the Coriolis effect.

elastic deformation A change in the shape of a solid due to bending and stretching, but not to breaking of the chemical bonds in the solid; the change disappears when the stress causing it ceases to exist.

elastic-rebound theory A theory proposing that elastic deformation occurs in rock adjacent to a fault prior to an earthquake, and that the rock then rebounds when the fault slips, generating seismic waves.

electrical grid A regional or national system that redistributes electrical energy via power lines and converts it to a usable voltage via transformers.

electromagnetic radiation Energy that travels in the form of oscillating electric and magnetic fields and can pass through the vacuum of space.

electromagnetic spectrum The full range of possible frequencies or wavelengths of electromagnetic radiation, from gamma rays through radio waves, including the visible radiation our eyes can detect.

electromagnetic wave A wave consisting of oscillations in electric- and magnetic-field strength.

electromagnetism A field force produced by magnetic objects or by electric currents.

element Matter consisting entirely of one kind of atom; elements cannot be subdivided or changed into other elements by chemical reactions.

elevation The vertical distance of a point on land above or below mean sea level.

ellipse A shape, resembling a flattened circle, produced by the intersection of a plane with a cone when the plane is passed through the cone at an angle to the axis other than 0° or 90°.

El Niño The phase of ENSO that occurs when, at low latitudes, the wind direction changes so that warm water that had formed a mound in the western Pacific flows from west to east; this reversal shuts off the upwelling of cold water along the western coast of South America and causes significant global changes in weather patterns.

emergent coast A coast where the land is rising relative to sea level, or sea level is falling relative to the land.

emission In the context of electromagnetic radiation, the production of light by atoms as their electrons give up energy.

energy The capacity to do work. Energy can exist in a variety of forms.

energy balance In the context of atmospheric temperature, the relationship between the amount of solar energy arriving at the Earth and the amount of infrared energy radiated by the Earth back to space, averaged over time.

energy density The amount of energy stored per kilogram of a fuel.

Enhanced Fujita (EF) scale A scale for classifying the intensity of a tornado; it is based on assessing the damage caused by the tornado, which in turn has been correlated with wind velocities.

enrichment The process of concentrating radioactive isotopes of uranium.

ENSO Short for *El Niño/Southern Oscillation.* The overall seesaw pattern of shifts in sea-level atmospheric pressure across the Pacific Ocean, in association with the Walker circulation, that causes alternations between El Niño and La Niña conditions.

environmental lapse rate The rate at which the temperature of the atmosphere changes with altitude, outside of an updraft.

eon The largest subdivision of geologic time.

epicenter The point on the surface of the Earth directly above the focus of an earthquake.

epicontinental sea A shallow sea overlying a continent.

epoch An interval of geologic time representing the largest subdivision of a period.

equant Having similar dimensions in all directions.

equilibrium line On a glacier, the boundary between the zone of accumulation and the zone of ablation.

era An interval of geologic time representing the largest subdivision of the Phanerozoic Eon.

erosion The grinding or beveling away of material at the Earth's surface by moving water, air, or ice.

erosional coast A coast that is receding landward because wave erosion washes sediment away faster than it can be supplied, or causes cliff retreat.

erratic A boulder or cobble picked up by a glacier and deposited far from the outcrop from which it detached.

eruptive jet A blast of gas, ash, and fragmented rock emitted by an explosive volcanic eruption.

estuary An inlet in which seawater and river water mix; an estuary can form when a coastal valley floods because of either rising sea level or land subsidence.

eutrophication The transformation of a well-oxygenated body of water into a poorly oxygenated one by an influx of nutrients that causes an algal bloom.

evaporite A sedimentary deposit formed by accumulation of salts, such as halite and gypsum, that precipitate when salty water evaporates.

event horizon The effective "surface" or boundary of a black hole; once inside that boundary, nothing—not even light—can escape.

exhumation The process, involving uplift and erosion, that brings deeply buried rocks to the Earth's surface.

exobiology The study of life beyond the Earth.

expanding Universe theory A theory, based on the observation that galaxies are moving away from one another, that the dimensions of the Universe increase over time.

explosive eruption A violent volcanic eruption that produces a towering plume of ash and avalanches of pyroclastic debris.

external energy Energy that enters the Earth System in the form of electromagnetic radiation from the Sun.

extinction The dying off of the last representative of a species.

extinct volcano A volcano that was active in the past, but will not erupt in the future because the geologic conditions that could cause an eruption no longer exist.

extrusive igneous rock Rock that forms by the freezing of lava above ground, after it flows or explodes (extrudes) onto the Earth's surface and comes into contact with the atmosphere or ocean.

eye The relatively calm, clear region in the center of a hurricane, surrounded by the intense winds of the eye wall.

eye wall The slightly downward-tapering cylinder of clouds that surrounds the eye of a hurricane.

fabric In the context of rocks, the degree to which grains or clasts are equant or inequant, and the degree to which inequant grains align parallel to one another.

facet A smooth, shiny face on a gem that forms sharp angles with its neighbors, created by a gem cutter by grinding and polishing; it is not a natural cleavage plane or crystal face.

failure In the context of mass wasting, the movement of material in response to gravity due to a relatively small change in the forces acting on a slope.

failure surface The plane on which a mass of material moves downslope during mass wasting.

fault A fracture on which one body of rock slides, or slips, past another.

fault scarp A step-like feature that offsets the ground surface; it develops when slip on a fault causes the ground on one side of the fault to move vertically with respect to the ground on the other.

feedback A phenomenon, produced by a process, that then modifies the rate at which the original process takes place.

fetch The distance across a body of water along which wind blows.

fissure eruption The ejection or flow of lava from an elongate crack rather than from a circular vent.

fjord A deep, glacially carved, U-shaped valley flooded by the sea to form a narrow, elongate bay, or filled with freshwater to form a lake.

flash flood A flood during which the discharge of a stream increases rapidly, as may occur during unusually intense rainfall or as the result of a dam collapse.

flood An event during which the volume of water in a stream becomes so great that water covers areas outside the stream's normal channel.

flood basalt Rock formed when hot, low-viscosity mafic lava erupts and spreads over a particularly broad area; it may form where a hot spot underlies a continental rift or a mid-ocean ridge.

floodplain The broad, flat area on either side of a stream that becomes covered with water during a flood.

flood stage The level at which a rising stream reaches the top of its channel.

flux melting Melting that occurs when volatile material is added to a solid, effectively lowering its melting temperature.

focus In the context of seismology, the location on a fault where slip that causes an earthquake begins (also called a *hypocenter*).

fog A cloud that forms at ground level.

fold A bend or wrinkle in a rock layer that forms as a consequence of deformation.

fold-thrust belt A region containing an assemblage of thrust faults and related folds; such belts typically form during mountain building.

foliated metamorphic rock Metamorphic rock that contains a planar fabric.

foliation Any type of planar fabric that forms during metamorphism.

food web A network of organisms defined by their feeding relationships.

force A push, pull, or shear that can cause an object to move, change shape, or rotate.

fossil A remnant or trace of a living organism that has been preserved in rock or sediment.

fossil assemblage A group of fossil species that occur together.

fossil fuel An energy resource, such as oil or coal, that comes from the altered remains of organisms that lived long ago; it stores solar energy that reached the Earth at the time the organisms lived and was turned into organic material via photosynthesis.

fossilization The process of fossil formation.

fossil succession The geologic principle stating that distinct assemblages of fossils are found in the same order at different locations because the assemblages are of different ages (younger over older, as required by the principle of superposition), and because as time passes, some species go extinct while others appear.

fracture zone A narrow band of vertical fractures in the ocean floor, lying roughly at right angles to a mid-ocean ridge. The actively slipping part of a fracture zone is a *transform fault*.

fragmental igneous rock An igneous rock formed from pyroclastic debris that is either cemented or welded together.

freezing rain Supercooled raindrops that freeze on contact with the ground to produce an ice-covered landscape.

frequency The number of waves that pass a point in a given time interval.

freshwater Water that contains very little, if any, salt.

friction Resistance to sliding caused by irregularities on a surface that act as tiny anchors and resist shear.

frictional drag The force applied by a moving material to another material across an interface.

front A boundary between air masses that differ in temperature and humidity.

frontal lifting Lifting of warm, moist air as the air flows up the face of a front.

frontal squall line A line of thunderstorms that forms along a frontal boundary such as the cold front of a mid-latitude cyclone.

fuel A substance that stores energy in a concentrated form and can release that energy; fossil fuels, for example, store energy that is released when they burn.

funnel cloud A visible rotating cloud extending down from the cloud base of a thunderstorm.

gabbro A coarse-grained mafic igneous rock.

galactic bulge The central region of a spiral galaxy where stars occur in a protrusion that extends above and below the galactic disk.

galactic center The midpoint of a galaxy, where its supermassive black hole is located, and around which the rest of the galaxy revolves.

galactic disk The plane in which most stars and spiral arms of a galaxy reside.

galactic group An association of generally fewer than 100 galaxies that gravitationally interact with one another.

galactic halo An extended, roughly spherical component of a galaxy that extends beyond, and contains much less mass than, the galactic disk.

galaxy An immense accumulation of hundreds of billions of stars that revolves around a supermassive black hole.

gas A state of matter in which atoms and molecules move about freely. A gas can flow, and can expand or contract when the size of its container changes.

gas giant A giant planet formed mostly of hydrogen and helium. In our Solar System, Jupiter and Saturn are gas giants.

gem A mineral specimen that has been cut and polished and is particularly beautiful or valuable.

geocentric model An old, incorrect idea suggesting that the Earth sits motionless at the center of the Universe, while the stars, other planets, and the Sun orbit around it.

geochronology The science of assigning numerical ages to geologic events.

geode A cavity in rock lined with euhedral crystals that precipitate out of water solutions passing through the rock.

geographic poles The locations (north and south) where the Earth's axis of rotation intersects the planet's surface.

geologic column A composite stratigraphic chart that represents the entirety of the Earth's history.

geologic cross section A graphic representation of a vertical slice through the Earth that displays the configuration of rock units and geologic structures underground.

geologic history A description of the sequence of geologic events that have taken place in a region.

geologic map A two-dimensional representation showing the distribution of rock units and geologic structures across a region as they would appear if projected on a horizontal plane.

geologic time The span of time since the formation of the Earth.

geologic time scale A geologic column on which the intervals of geologic time have been assigned numerical ages.

geology The study of the Earth, including our planet's composition, behavior, and history.

geometric parallax The apparent displacement of a foreground object relative to a background object when an observer's line of sight changes.

geosphere The component of the Earth System from the Earth's solid surface to its center.

geostrophic balance The balance that exists when the pressure-gradient force and the Coriolis force are equal and opposite.

geostrophic current A current that results from geostrophic balance in an ocean basin.

geostrophic wind A wind that flows parallel to isobars when air in the atmosphere is in geostrophic balance.

geotherm A line on a graph that shows the change in temperature with depth in the Earth.

geothermal energy Heat and electricity produced by using the internal heat of the Earth.

geothermal gradient The rate of change in temperature with depth in the Earth.

geyser A fountain of steam and hot water that erupts periodically from a vent in the ground in a region of hot springs.

giant star A star with a substantially larger radius and luminosity than a star with the same surface temperature on the main sequence of an H-R diagram.

glacial advance The forward movement of the position of a glacier's toe when the rate of accumulation exceeds the rate of ablation.

glacial lake A body of liquid freshwater formed in association with a glacier, which contains meltwater from the glacier as well as water from other sources.

glacial outwash Sand and gravel deposited on a low-relief landscape by meltwater streams flowing from a glacier.

glacial rebound The rising of the land surface after a large overlying ice sheet melts away and the weight of the ice is removed.

glacial retreat The movement of the position of a glacier's toe back toward the glacier's origin when the rate of ablation exceeds the rate of accumulation.

glacial striation A scratch or groove on rock produced by clasts embedded in flowing glacial ice.

glacial subsidence The sinking of the land surface due to the weight of a large ice sheet.

glacial till Unsorted sediment transported by flowing ice and deposited beneath a glacier, along its margins, or at its toe.

glacial torrent A cataclysmic flood that results when the outlet of a large glacial lake is suddenly unblocked.

glaciation An interval of time during which glaciers grow and cover substantial areas of continents.

glacier A sheet or stream of ice that slowly flows across the land surface and lasts year-round.

glass An inorganic solid in which atoms are not arranged in an orderly pattern.

glassy rock Igneous rock consisting entirely of glass, or of tiny crystals surrounded by a glass matrix.

global change Transformations or modifications of the Earth System over time that have an impact on the Earth System worldwide.

global climate model (GCM) A computer model that uses mathematical equations to simulate changes in global climate conditions over time.

global cooling A fall in the average global atmospheric temperature over time.

global positioning system (GPS) A satellite network that allows determination of the location of a receiver with great accuracy; it can be used by geologists to measure rates of movement of portions of the Earth's crust relative to one another.

global warming A rise in the average global atmospheric temperature over time.

globular cluster A spherical accumulation of hundreds of thousands of stars.

gneiss A high-grade metamorphic rock composed of alternating bands of dark- and light-colored minerals.

Goldilocks effect The influence of factors that make the Earth "not too hot, not too cold, but just right" for liquid water and, therefore, life to exist; named for the tale of *Goldilocks and the Three Bears*.

Gondwana A supercontinent of the Paleozoic Era that consisted of today's South America, Africa, Antarctica, India, and Australia (also called *Gondwanaland*).

grade In the context of mining, the measured concentration of a useful metal in an ore. See also *metamorphic grade*.

graded bedding A pattern of deposition in which grain size changes progressively within a bed from coarser at the base to finer at the top; it characterizes beds deposited by turbidity currents.

grain Any small, natural solid particle, such as a crystal, a part of a crystal, or a fragment of glass or rock.

granite A coarse-grained, felsic intrusive igneous rock.

graupel Small (<3–4 mm), soft ice balls that form when supercooled water droplets attach to the surfaces of ice crystals and freeze.

gravitational energy The potential energy stored in a mass if the mass lies within a gravitational field.

gravity The attractive force that one mass exerts on another; its magnitude depends on the amount of matter in the objects and the square of the distance between them.

Great Dying The Permian-Triassic mass-extinction event that ended the Paleozoic Era; during the event, over 96% of marine species and 70% of terrestrial species became extinct.

great oxygenation event The increase in atmospheric oxygen between 2.4 and 1.8 Ga, when the Earth's surface rocks and water could no longer absorb or dissolve all the oxygen produced by photosynthetic organisms.

Great Red Spot The reddish, oval, counterclockwise-rotating storm, big enough to contain two or three Earths, that is visible in the atmosphere of Jupiter's southern hemisphere.

Great Unconformity An erosional surface, formed by weathering and glacial erosion near the end of the Proterozoic and found worldwide, that represents a hiatus in the stratigraphic record; in many places, it is the boundary between crystalline basement and sedimentary cover.

green energy Energy produced without the release of chemical or radioactive pollutants or greenhouse gases, and without consuming nonrenewable resources (also called *clean energy*).

greenhouse effect The trapping of heat in the Earth's atmosphere by carbon dioxide, methane, water, and other greenhouse gases, which absorb infrared radiation.

greenhouse gas A gas that retains heat in the atmosphere by absorbing and re-emitting infrared radiation.

Grenville orogeny The last major collision during the formation of the supercontinent Rodinia, which produced the rocks of the Grenville orogen that now underlie parts of eastern and southern North America.

groin A concrete or stone wall built perpendicular to the shore to impede longshore drift.

ground motion Shaking of the land surface during an earthquake.

groundwater Water that resides under the surface of the Earth in pores or cracks in rock or sediment.

groundwater depletion The pumping of water from aquifers so rapidly that natural recharge will not be able to replace it in a time frame of centuries to millennia.

growth ring One of a series of distinct rings that develop in trees and shelly organisms whose growth rates vary with the seasons.

gust A short-lived burst of wind.

gust front The leading edge of the outflow of a thunderstorm's rain-cooled downdraft, marked by a sudden increase of wind speed and the arrival of cooler air.

guyot A seamount with a flat peak, formed by erosion or by the growth of a coral reef over it before the peak subsided below sea level.

gyre An ocean-surface geostrophic current with a somewhat circular path that follows the margins of the ocean basin in which it has formed.

habitable zone The range of distances from the Sun at which a planet's surface receives neither too much nor too little insolation for life to exist.

Hadean Eon The earliest eon of the Precambrian; roughly, the time between the Earth's origin and the formation of the first rocks that have been preserved.

Hadley cell A low-latitude convection cell in the troposphere that extends from the equator to a latitude of about 30° in each hemisphere; one lies to the north of the ITCZ, and one to the south.

Hadley cell expansion An increase in the north-south width of Hadley cells as oceans warm, which may cause hot, dry air to descend at higher latitudes, causing desertification in regions that are now semiarid.

hail Balls or clumps of ice, with diameters ranging from 3 mm to 20 cm, resulting from the freezing of supercooled liquid droplets when they attach to graupel in the strong updrafts of thunderstorms.

Hale cycle The Sun's complete 22-year magnetic-field cycle, during which the polarity of the field reverses and then reverses again (also called the *magnetic cycle*).

half-life The time it takes for half of the atoms of a radioactive isotope to decay.

hand specimen A roughly fist-sized piece of rock that a geologist collects to examine with a hand lens or to take to a lab for thin sectioning or further examination.

hanging valley A glacially carved tributary valley whose floor lies at a higher elevation than the floor of the trunk valley with which it intersects.

hardness A measure of the relative ability of a mineral to resist scratching; it represents the resistance of bonds in the mineral to being broken.

hard water Groundwater that contains significant concentrations of dissolved calcium, magnesium, and bicarbonate ions, which enter the water when it passes through carbonate rock; these ions can precipitate out to clog pipes.

haze An atmospheric condition that exists when the concentration of aerosols, or of tiny droplets formed when aerosols dissolve in water, is sufficient to cause a decrease in visibility.

head In the context of glaciers, the location at which a mountain glacier originates.

headward erosion The lengthening of a stream channel by erosion at the stream's origin.

heat The total thermal energy contained in a material or transferred from one material to another.

heat capacity The amount of heat a material can absorb to produce a specific temperature change.

heat flow The rate at which heat transfers from a warmer to a cooler material.

heat index A measure that takes humidity as well as temperature into account, and therefore represents human perception of temperature and associated discomfort.

heat-transfer melting Melting that results when heat from a very hot magma that has intruded a rock with a lower melting temperature causes melting of that rock.

heat wave An interval of time during which the temperature remains significantly above normal for days or weeks.

heliocentric model A model of the Solar System in which the Sun lies at the center, and all other objects, including the Earth, orbit the Sun (also called the *Copernican model*).

heliopause The boundary of the heliosphere, where pressure produced by the solar wind equals the pressure produced by the interstellar medium; sometimes thought of as the edge of the Solar System.

heliosphere An elongate, sock-shaped volume in space within which most particles (not counting Oort Cloud objects) come from the solar wind; outside it, most particles come from the stellar winds of other stars or from stellar explosions and active galaxies.

heterosphere The upper portion of the Earth's atmosphere, in which gases separate into distinct layers on the basis of composition and, therefore, density.

high-pressure system The anticyclonic circulation of air surrounding a center of high atmospheric pressure; generally associated with sinking air and clear skies.

hinge In the context of geologic structure, the portion of a fold where its curvature is greatest.

hockey-stick diagram A graph of global temperature change over the past thousand years, so called because it shows that temperature was fairly constant until its relatively recent rapid rise, so that the line representing temperature change over time resembles a hockey stick lying on its side.

homosphere The lower part of the atmosphere, in which the gases have been stirred into a homogeneous mixture by air circulation.

horn In the context of topography, a pointed mountain peak where glaciers have carved cirques on three or more sides.

hornfels A fine- to medium-grained, nonfoliated metamorphic rock formed by thermal metamorphism, in which inequant grains are randomly oriented.

hot desert A desert where daytime temperatures often exceed 35°C (95°F).

hothouse period An interval of geologic time during which the Earth's atmosphere is warmer than average and ice sheets do not cover polar regions.

hot spot A location where igneous activity occurs independently of plate interactions; a hot spot may develop where a mantle plume causes melting.

hot-spot track A chain of extinct volcanoes, each of which formed over a hot spot, then moved away from it due to the movement of the lithosphere plate on which the volcano formed.

hot spring A source at which very warm to boiling groundwater spills out of the ground.

H-R diagram A Hertzsprung-Russell diagram, which shows the relationship of a star's luminosity to its surface temperature. Stars fall into distinct groups on the diagram.

Hubble's law A simple equation ($v = Hd$) relating the speed (v) at which a galaxy is moving away from the Earth to the distance (d) of that galaxy from the Earth. H is called Hubble's constant.

hurricane A tropical cyclone in which sustained wind speeds reach 119 km/h (74 mph); also used more specifically to refer to such a storm over the North Atlantic or eastern Pacific Ocean.

hurricane track The path a hurricane follows.

hydrocarbon A substance made of chain-like or ring-like molecules containing hydrogen and carbon atoms; petroleum and natural gas are hydrocarbons.

hydrocarbon migration The upward movement of oil and gas, driven by their buoyancy, from source rock to reservoir rock.

hydrocarbon reserve A significant quantity of potentially extractable oil or gas underground.

hydroelectric power Electricity produced by converting the potential energy of water held in a reservoir to kinetic energy by allowing the water to flow to a lower elevation through the blades of a turbine.

hydrofracturing The process of injecting water and other chemicals into a drillhole under high pressure to generate cracks in the surrounding rock that provide pathways for the release of hydrocarbons; informally known as *fracking*.

hydrologic cycle The flow of water from reservoir to reservoir in the Earth System.

hydrosphere The Earth's solid and liquid water, including surface water (lakes, rivers, and oceans), groundwater, glaciers, and permafrost.

hydrothermal deposit An ore that consists of ore minerals precipitated in cracks or pores in rock due to interactions between rock and a hot-water solution.

hydrothermal fluid A hot-water or steam solution that circulates through rock underground.

hypothesis A possible natural explanation for a set of data.

ice The solid state of a volatile material; familiar examples include water ice (H_2O) and dry ice (CO_2), but many other compounds can form ice in the cold of space.

ice age An interval of geologic time in which the climate was cold, so that ice sheets grew to cover large areas of the continents and mountain glaciers advanced; an ice age can include many glaciations and interglacials.

iceberg A large block of ice floating in the sea, typically formed by calving from a tidewater glacier or an ice shelf.

ice giant A giant planet formed mostly of solid volatile substances (ices); in our Solar System, Uranus and Neptune are ice giants.

icehouse period An interval of geologic time when the Earth's atmosphere is cooler than average and ice ages are possible.

ice shelf A broad, flat region of ice along the edge of a continent, formed where a continental glacier flows into the sea.

ice storm A winter storm that produces an accumulation of ice at the ground surface, as well as on trees, power lines, and buildings.

igneous activity A set of phenomena including melting deep underground, the rise of magma and volcanic gas, formation of intrusions, volcanic activity, and the production of igneous rock.

igneous intrusion A body of igneous rock that formed when magma solidified underground after the magma intruded cooler rock.

igneous rock Rock that forms when magma or lava freezes solid, or when pyroclastic debris lithifies.

impact crater A depression excavated by the impact of a meteorite.

inclusion A fragment of one material contained within another; according to geologic principles, such a fragment must be older than the material that contains it.

inequant Having different dimensions in different directions; for example, having a length, width, and/or thickness that are not the same.

inertia The tendency of an object to remain in a state of rest, or in a state of uniform motion in a straight line, unless the application of a force causes its state to change.

infiltration In the context of groundwater, the migration of a liquid into or through a porous material.

inner core The central portion of the Earth, extending from a depth of 5,155 km (3,203 mi) to the Earth's center at 6,371 km (3,159 mi); it consists of solid iron alloy.

inorganic chemical A chemical that does not contain carbon, or if it does contain carbon (such as $CaCO_3$), does not fit other aspects of the definition of an organic chemical.

inselberg An isolated "island mountain" in a desert, formed when the erosional remnant of a larger mountain has become surrounded by alluvium.

insolation The amount of solar energy per unit area that arrives at the top of the Earth's atmosphere.

intensity In the context of seismology, the degree of ground shaking caused by an earthquake at a locality, as indicated by damage done and by human perception.

interglacial An interval of time between two glaciations.

Intergovernmental Panel on Climate Change (IPCC) An international group of scientists charged with evaluating and summarizing research pertaining to how the Earth's climate is changing and how humans are influencing those changes.

internal energy Thermal energy that rises from the interior of the Earth.

interstellar medium The gas, dust, and/or ice that occurs in the space between stars.

intertidal zone The area of coastal land across which the tide rises and falls.

intertropical convergence zone (ITCZ) The zone where the trade winds (the surface flow of the

Hadley cells) of the northern and southern hemispheres converge in the vicinity of the equator.

intraplate earthquake An earthquake that occurs away from plate boundaries.

intrusive igneous rock Rock formed by the freezing of magma underground.

inversion In the context of meteorology, a layer in the atmosphere in which the temperature increases with altitude.

ion An atom that has lost or gained one or more electrons, and therefore has a positive or negative electrical charge.

ionosphere The region of the upper atmosphere, above 60–100 km (37–62 mi), that has a relatively high concentration of ions.

island arc A chain of volcanic islands that forms on the overriding plate along a convergent boundary where one oceanic plate subducts beneath another.

isobar A contour line on a map of air-pressure variations at a given altitude; that is, a line along which all points have the same atmospheric pressure.

isobaric surface An invisible surface above the Earth at which atmospheric pressure has the same given value.

isostasy The condition that exists when the buoyancy force pushing lithosphere up equals the gravitational force pulling lithosphere down.

isotope A version of a given element that has the same atomic number as other versions, but a different atomic mass. Some elements come in only one atomic mass, and thus have no isotopes, but most have two or more isotopes.

jetty A concrete or stone wall placed at the entrance to a harbor or at the mouth of a river to prevent a baymouth bar from closing off the harbor; a jetty may extend the river channel into deeper water farther offshore.

joint A naturally formed crack in rock; there is no shear displacement on a joint.

Jovian planet One of the four outer planets of the Solar System—Jupiter, Saturn, Uranus, and Neptune—all of which lack a solid surface and are much larger and less dense than any of the terrestrial planets; the term is used for both gas giants and ice giants.

karst landscape A region underlain by caverns formed in limestone bedrock, some of which have collapsed, so that the land surface has sinkholes separated by limestone ridges or spires.

katabatic wind Downslope near-surface wind caused by gravitational drainage of very cold, dense air from a high elevation to a lower one; such winds are common along the edges of ice sheets and large glaciers.

Keeling curve A graph, begun by Charles Keeling in 1958, indicating the change in carbon dioxide concentration in the atmosphere over time; it reveals both seasonal changes and the overall increase that takes place from year to year.

Kepler's laws The three rules of planetary motion inferred by Johannes Kepler.

kerogen The waxy molecules into which the organic material in shale transforms on reaching 50°C–90°C. At higher temperatures, kerogen transforms into oil.

Köppen-Geiger climate classification (KGCC) A characterization of the Earth's climate zones based on temperature and precipitation, as manifested by the character of vegetation, developed by Wladimir Köppen and Rudolf Geiger.

K-Pg extinction A sudden mass die-off of all species of dinosaurs, as well as many other species, at the end of the Cretaceous Period (66 Ma), which has been attributed to an immense meteorite impact (formerly called the *K-T extinction*).

Kuiper Belt A diffuse ring of icy objects that orbit the Sun outside the orbit of Neptune; it includes Pluto and some other dwarf planets, as well as countless small objects.

lag deposit The coarse sediment left behind in a desert after wind erosion has removed finer sediment.

lagoon A body of shallow seawater separated from the open ocean by a barrier island.

lahar A flowing slurry composed of volcanic ash and debris mixed with water, formed when rain or melting snow and ice provide abundant water on the flank of a volcano during or after an ash-rich eruption.

lake A standing body of water located on land.

lake-effect storm A snowstorm over and immediately downwind of a large lake, triggered by the flow of very cold air over relatively warm lake water.

land bridge An area of continental shelf exposed as dry land during an ice age, when continental glaciers expand so that sea level drops.

landform A distinct, visible natural feature of a landscape.

landscape The character and shape of the land surface in a region.

landslide A general term for a mass-wasting event during which rock and/or regolith moves downslope.

landslide-potential map A map, based on hazard-assessment studies, on which regions are ranked according to the likelihood that mass wasting will occur there.

La Niña A strengthening of the Walker circulation and the trade winds that causes a flow of water from east to west in the equatorial Pacific Ocean and, therefore, increases the upwelling of cold water along the western coast of South America; the phase of the Southern Oscillation opposite to an El Niño event.

lapilli Pyroclastic particles 2–64 mm in diameter (marble- to golf-ball-sized), consisting of frozen lava clots, fragments of pre-existing rock, or ash clumps.

Laramide orogeny A mountain-building event, which began in the Cretaceous Period and continued into the Cenozoic Era, that resulted in the formation of the present-day Rocky Mountains in the United States.

large igneous province (LIP) A region in which huge volumes of extrusive igneous rocks formed over a relatively short interval of geologic time; examples include regions where low-viscosity mafic lava spread over broad areas and built up a plateau.

late heavy bombardment The battering of the inner planets of the Solar System by meteorites between 4.1 Ga and 3.8 Ga; it may have pulverized and remelted the Earth's crust, thereby destroying the rock record of the Earth's history during that time.

latent heat Energy required for, or released by, a transition between two states of matter (solid, liquid, gaseous); condensation of gas to form liquid releases latent heat.

lateral spreading Mass wasting that occurs where a broad area of relatively horizontal land that terminates at a steep escarpment, and is underlain by a weak failure surface, slips toward the escarpment, spreads out, and breaks into pieces.

Laurentia An early Paleozoic continent that consisted of today's North America and Greenland.

Laurentide ice sheet A North American ice sheet of the Pleistocene Ice Age that at times covered all of Canada east of the Rocky Mountains and spread southward over the northern United States.

lava Molten rock that has flowed or fountained out onto the Earth's surface.

lava flow A sheet or mound of lava that flows onto the ground surface or seafloor in molten form; also, a layer of igneous rock formed from such a flow.

lava fountain An eruption of lava that is under sufficient pressure to spurt into the sky.

lifting In the context of atmospheric science, the raising of air to a higher altitude by any of several mechanisms.

lifting mechanism A process that causes air near the ground surface to rise to a higher altitude.

lightning Electrostatic discharge in the atmosphere that occurs when a charge separation develops within a cloud and/or between a cloud and the ground.

lightning stroke A giant spark and pulse of current produced by a thunderstorm (also called a *lightning bolt*).

light pollution Light produced by human activities that reduces the visibility of stars in the night sky.

light scattering The absorption of light, and its immediate re-emission in random directions, by molecules and particles in the atmosphere or ocean.

light-year The distance that light travels in one Earth year (about 9.5 trillion km, or 6 trillion mi).

limb The part of a fold outside of the hinge zone; typically, a limb has less curvature than the hinge zone.

limestone Sedimentary rock composed mainly of calcium carbonate; it may form by biochemical or chemical processes, or by the accumulation of shell fragments.

liquid A state of matter in which atoms or molecules can move relative to one another. A liquid can flow and conform to the shape of its container.

lithification The transformation of loose sediment into solid rock through compaction, consolidation, and cementation.

lithosphere The relatively rigid outer layer of the Earth, 100–150 km (62–93 mi) thick, that consists of the crust and the lithospheric mantle.

lithospheric mantle The uppermost layer of the Earth's mantle that is cool enough so that it cannot flow easily, and therefore forms part of the lithosphere.

littoral zone The nearshore area of a body of water. In the ocean, it includes the intertidal zone and the shore zone just above it, which may be submerged during storms and receives salt spray.

Local Group The cluster of galaxies relatively close to the Milky Way; it includes Andromeda.

loess A deposit of fine-grained sediment (clay and silt) picked up by wind in a desert or glacial environment and transported elsewhere, where it accumulates to form cohesive layers.

longshore current A nearshore flow of water parallel to the shore that develops when waves move toward the shore obliquely.

longshore drift The net transport of sediment laterally along a beach that occurs when waves wash up a beach obliquely.

lower mantle The deepest section of the Earth's mantle, extending from a depth of 670 km (416 mi) down to the core-mantle boundary.

low-pressure system The cyclonic circulation of air surrounding a center of low atmospheric pressure; rising air in such a system commonly produces clouds and precipitation.

low-velocity zone (LVZ) A region of the Earth's mantle, just below the oceanic lithosphere, in which seismic waves travel more slowly than they do in the lithospheric mantle above or in mantle deeper down; the cause may be slight partial melting of mantle rock in that zone.

luminosity The amount of electromagnetic energy emitted by a star in a specified amount of time (usually 1 second) (also called *true brightness*).

lunar eclipse A darkening of the Moon's surface, as seen from the Earth, that occurs when the Moon is partially or entirely in the Earth's shadow.

lunar highlands High-elevation regions on the Moon that appear light-colored from the Earth relative to the darker maria.

lunar month The interval of time between successive new Moons (roughly 29.5 days).

lunar regolith The loose surface material found on the Moon, including micrometeorites, pulverized rock scattered by impacts, and dust produced by space weathering.

luster The way a mineral's surface interacts with light; the two main categories are metallic luster and nonmetallic luster.

magma Molten rock beneath the Earth's surface.

magma chamber A volume below ground, typically under a volcanically active region, consisting mostly of magma or a mush of magma and crystals.

magma migration The upward movement of magma away from its source toward a cooler location due to buoyancy force and pressure at depth.

magnetic anomaly The difference between the expected strength of the Earth's magnetic field at a certain location and the actual measured strength of the field at that location.

magnetic pole The end of a magnetic dipole. All magnetic dipoles, including the Earth, have a north magnetic pole and a south magnetic pole; on the Earth, the magnetic poles lie near the geographic poles, but do not exactly coincide with them.

magnetic reversal A change in the Earth's magnetic polarity in which the magnetic field flips from normal to reversed polarity, or vice versa.

magnetosphere A teardrop- or sock-shaped region in which the strength of the Earth's magnetic field exceeds the strength of other magnetic fields in space.

magnitude (1) A number, on a logarithmic scale, representing the amount of energy released by an earthquake. (2) The luminosity of a star; see *absolute magnitude, apparent magnitude*.

main sequence The diagonal band on an H-R diagram that contains stars that are fusing hydrogen in their cores to form helium; stars on the main sequence are in equilibrium and can remain on the main sequence for a long time.

mantle plume A relatively narrow column of very hot rock rising up through the asthenosphere to the base of the lithosphere.

marble A metamorphic rock composed of calcite, transformed from a protolith of limestone.

mare (plural: maria) A broad, dark area on the Moon's surface, consisting of flood basalts that erupted over 3 billion years ago and spread out across a lunar lowland.

marine deposition Sediment deposition that occurs in oceans or along the coasts of oceans.

marine magnetic anomaly One of a series of alternating positive and negative magnetic anomalies on the seafloor, which together define a pattern of alternating stripe-like bands parallel to mid-ocean ridges.

mass-balance calculation An arithmetic calculation that assumes that the volume of matter in a defined system remains constant, so that matter removed from one part of the system must add to another part of the system; in other words, the amount of mass removed from a source should balance the amount added to a sink.

mass-extinction event A relatively brief interval of geologic time when vast numbers of species abruptly vanish.

mass wasting The downslope transport of rock, regolith, snow, or ice by gravity.

matter The material substance of the Universe, which consists of atoms and has mass.

meander A snake-like curve in a stream channel.

meandering stream A stream that contains meanders.

megathrust earthquake A huge ($\geq M_W$ 8.0) convergent-boundary earthquake that occurs when a large area of the relatively shallow thrust fault that delineates the boundary between the base of an overriding plate (and its accretionary prism) and the top of a downgoing plate suddenly slips.

melt The process of transforming solid into liquid (e.g., by heating or decompression); a magma formed when a solid rock transforms into liquid.

mesa A large, flat-topped hill (with a surface area of several square kilometers) in an arid region; the surface may be a nearly horizontal erosion-resistant bed or a lava flow.

mesocyclone A strong, long-lasting, rotating updraft within a supercell thunderstorm, typically 5–10 km (3–6 mi) in diameter; normally coincides with the region of the storm in which tornadoes form.

mesosphere The cooler layer of atmosphere overlying the stratosphere.

metal A solid composed of atoms of metallic elements bonded to one another by metallic bonds; a metal is opaque, shiny, smooth, malleable, and can conduct electricity.

metallic hydrogen A form of hydrogen that occurs only under the tremendous pressures found near the core of a giant planet such as Jupiter; it has properties similar to those of a metal.

metamorphic aureole The region around an igneous intrusion in which heat transferred into the wall rock has metamorphosed the wall rock.

metamorphic fabric A directionality in a rock with metamorphic texture, that exists when inequant minerals are aligned; it can form due to pressure solution, plastic deformation, or growth of new minerals.

metamorphic facies A set of metamorphic mineral assemblages indicative of metamorphism under a specific range of metamorphic conditions; the specific minerals in the assemblage depend on the chemical composition of the rock.

metamorphic grade An informal indication of the intensity of metamorphism, characterized primarily by metamorphic temperature, to which a rock has been subjected.

metamorphic mineral A mineral that was not present in the protolith but grows during metamorphism.

metamorphic rock Rock that forms when pre-existing rock undergoes transformation into new rock, without first becoming melt or sediment, in response to changes in temperature, pressure, and/or interaction with chemically active fluids.

metamorphic texture A crystalline texture, in a metamorphic rock, that was not found in the protolith.

metamorphism The process by which one kind of rock is transformed into a different kind of rock by an increase in pressure and/or temperature, by shearing under elevated temperatures, or by interactions with chemically active fluids.

metasomatism A change in a rock's overall chemical composition during metamorphism caused by interactions with hydrothermal fluids that speed diffusion and bring in or remove elements.

meteor The incandescent trail produced by a small piece of interplanetary debris as it travels through the Earth's atmosphere at very high speeds and compresses air just in front of it.

meteorite A piece of rock or metal alloy that fell from space and landed on the Earth.

meteoroid A small fragment of planetary debris (less than 1 m, or 3 ft, across) that enters the Earth's atmosphere.

meteorologist A person who studies and forecasts weather and its consequences.

meteor shower A larger-than-normal display of meteors, which occurs when the Earth passes through the orbit of a comet and intersects its dusty tail.

methane hydrate A compound consisting of methane molecules surrounded by water molecules in the crystal lattice of water ice; it can form in seafloor sediment at depths of 300–500 m, and its melting releases bubbles of methane gas.

microburst A strong, intense downdraft produced by evaporating rain falling from a tall cumulonimbus cloud.

microcontinent A block of continental crust too small to be considered a continent; geologists use the term in reference to relatively small blocks of continental crust that collide with a larger continent.

mid-latitude cyclone A large, comma-shaped (as viewed from space), low-pressure system that forms along the polar-front jet stream between 30° and 60° latitude and produces many types of weather (also called an *extratropical cyclone*).

mid-latitude region The latitudes between 30° N and 60° N and between 30° S and 60° S; large portions of this region have temperate climates.

mid-ocean ridge A submarine belt of elevated seafloor that forms along a divergent boundary; seafloor spreading occurs along the axis of a mid-ocean ridge.

Milankovitch cycle Any of three cycles of change—in the shape of the Earth's orbit, in the tilt of its axis of rotation, and in the wobble of its axis—that occur over tens to hundreds of thousands of years and influence climates on the Earth.

Milky Way Galaxy The galaxy in which our Sun and Solar System reside.

mineral A naturally occurring, homogeneous, crystalline solid that has a definable chemical composition and, in most cases, is inorganic.

mineralogist A researcher who studies minerals.

mirage An optical illusion that an observer at the Earth's surface sees, caused by specific atmospheric conditions that cause light rays to bend; mirages can be reflections of the sky or of features on the surface, depending on the way that light bends.

model A simulation developed by laboratory and computational scientists to visualize processes that take place too slowly or too quickly to see in real time, or to characterize objects that are too small or too large to examine directly.

Modified Mercalli Intensity (MMI) scale A scale for assessing an earthquake's intensity by the damage it causes and by people's perception of the shaking.

Moho The seismic-velocity discontinuity that defines the boundary between the Earth's crust and mantle.

Mohs hardness scale A list of 10 minerals, in a sequence of relative hardness, with which other minerals can be compared.

moist adiabatic lapse rate The rate (in °C/km) at which a water-saturated parcel of air will cool as a result of expansion during its ascent if there is

no exchange of heat or mass with the surrounding environment.

molecule A particle that consists of two or more atoms attached by chemical bonds.

moment magnitude scale A logarithmic scale for assessing the energy released by an earthquake by measuring the amplitudes of seismic waves, as well as taking into account the dimensions of the slipped area and displacement on the fault; it has replaced the Richter scale in modern descriptions of earthquake size because it more accurately characterizes very large earthquakes.

monocline A fold whose shape resembles that of a carpet draped over a stair step.

monsoon A seasonally changing air circulation in tropical climates in which summer winds blow from the ocean toward the land, bringing heavy rain over land, and winter winds blow from the land toward the ocean and yield drier weather on land.

moon A sizable solid object that orbits a planet; the term is capitalized when referring to the Earth's Moon.

moraine A pile or ridge of glacial till deposited by a moving or melting glacier.

mountain glacier A glacier that grows in or adjacent to a mountainous region (also called an *alpine glacier*).

mouth The outlet or end of a stream.

mudcrack An open crack formed when mud dries, shrinks, and breaks into polygonal plates.

mudflow A downslope movement of a slurry at slow to moderate speed; it consists mostly of clay and water (also called a *mudslide*).

native metal A solid mass of metal atoms in elemental form, meaning that the atoms are not bonded to other elements.

natural arch A bridge-like span of rock that forms when erosion along joints leaves a narrow wall of rock, and the lower-middle part of the wall then erodes while the upper part remains intact; natural arches form in desert and karst landscapes.

natural change Change in the Earth System due to geologic, meteorological, or astronomical events that would happen whether or not people inhabited the Earth.

natural gas A compound, such as methane or propane, composed of hydrocarbons that can exist in gaseous form at the Earth's surface.

natural hazard A natural phenomenon that has the potential to cause human casualties and property damage.

natural resource A material or energy source from the natural Earth System that human society uses to maintain or improve life, built environments, and transportation.

near-Earth object (NEO) An asteroid, comet, or large meteoroid whose orbit intersects the Earth's orbit.

nebula (plural: nebulae) A cloud of gas, ice, and dust in space; often a location of star formation.

nebular theory The theory that the first stars formed within patches of nebulae that collapsed in response to gravity, resulting in protostars, some of which were massive enough to continue to collapse to form true stars.

nekton Organisms, such as fish, that actively swim in open water and can move against a current.

neritic zone The zone of the ocean that extends from the low-tide line to the edge of the continental shelf, in which the water is less than 200 m (660 ft) deep and sunlight can penetrate to the seafloor.

neutron star The superdense core of a high-mass star left behind after a supernova explosion.

nonconformity An unconformity in which sedimentary strata are in contact with underlying igneous or metamorphic rock; that is, the contact between basement and cover.

nonfoliated metamorphic rock A metamorphic rock in which metamorphic minerals are none equant or, if inequant, have a random orientation.

nonmarine deposition Sediment deposition that occurs on land or in or along lakes and streams.

nonmetallic mineral resource A mineral resource that does not contain metals; examples include stone, gravel, sand, gypsum, phosphate, and salt.

nonrenewable resource A resource that nature will take a long time (hundreds to millions of years) to replenish, or may never replenish.

non-supercell tornado A tornado that develops along a squall line or in a hurricane, rather than in a supercell thunderstorm; it forms where significant horizontal wind shear is occurring.

nor'easter A strong mid-latitude cyclone along the east coast of North America, so called because its strongest winds come from the northeast due to the overall counterclockwise flow in the storm.

normal fault A dip-slip fault on which the hanging-wall block moves down the slope (dip) of the fault.

normal galaxy A galaxy that emits most of its radiation at or near visible wavelengths of light.

nova (plural: novae) A stellar explosion that results from runaway nuclear fusion in a layer of accreted material on the surface of a white dwarf in a binary star.

nuclear bond An attractive force that keeps protons and neutrons together in the nucleus of an atom despite electrical repulsion between protons; it acts only over subatomic distances.

nuclear fusion A reaction in which the nuclei of atoms bond together, yielding new, larger atoms.

nuclear reaction A reaction in which nuclear bonds form or break.

nuclear reactor The part of a nuclear power plant where nuclear fission takes place in nuclear fuel.

nuclear waste Radioactive by-products, or radiation-contaminated materials, that could pose a danger to people; used nuclear fuel as well as water and other materials within a reactor become nuclear waste.

nuisance tide An especially high tide that floods coastal communities and damages property.

numerical age The age of a geologic material or feature specified in years (also called *absolute age*).

Nuna The earliest known supercontinent, which existed between 1.9 Ga and 1.3 Ga (also called *Columbia*).

nutrient A chemical essential for the survival of living organisms.

observatory In the context of space science, a place where astronomers use telescopes to observe objects and features in space.

occluded front A boundary between air masses that forms when a cold front overtakes a warm front, so that the cool air behind the warm front is lifted over the cold air behind the cold front; consequently, the base of the warm front no longer intersects the ground surface.

ocean A large body of saltwater between continents.

ocean acidification A significant increase in the acidity of the oceans, caused by the dissolution of atmospheric CO_2 in seawater and the resulting production of carbonic acid.

ocean basin A broad, low area of the Earth's surface mostly floored by oceanic crust and filled with saltwater.

oceanic island A volcanic peak that has built up above the surrounding seafloor and rises above sea level.

oceanic plateau A submarine large igneous province (LIP), where a hot spot erupted voluminous hot mafic lava, which formed a thick layer of basalt on the seafloor.

oceanic zone The deep, open-water portion of the ocean that does not interact with the coast.

oceanography The study of water and life in the oceans as well as the way in which ocean water moves and interacts with land and air.

oil A viscous liquid composed of hydrocarbons (also called *petroleum*).

Oil Age A period of human history, starting in the mid-19th century and likely to continue decades into the future, during which the economy has depended on oil and gas.

oil window The relatively narrow range of temperatures under which oil can form in source rock.

Oort Cloud A cloud of icy objects that orbit the Sun in a region mostly outside of the heliosphere.

optical telescope A type of telescope that allows an observer to see objects in space that emit visible light.

optical window The range of visible and near-infrared wavelengths to which the Earth's atmosphere is transparent.

orbit The path taken by one object revolving around another object under the influence of their mutual gravitational or electrical attraction.

ordinary thunderstorm An isolated thunderstorm that does not rotate (also called a *single-cell thunderstorm* or an *air-mass thunderstorm*).

ore Rock containing metal atoms in sufficient concentration to make it worth mining.

ore deposit A concentration of ore minerals that can be profitably mined.

ore mineral A mineral that contains metal atoms in relatively high concentrations and in a form that can be extracted.

organic chemical A carbon-containing compound that occurs in living organisms or resembles compounds produced in living organisms, in which the carbon atoms are typically arranged in chains or rings and are bonded to hydrogen, nitrogen, or oxygen atoms.

organic coast A coast where living organisms, such as mangroves or corals, control landforms along the shore.

organic sedimentary rock Sedimentary rock (such as coal or oil shale) that contains a high concentration of organic material.

original horizontality The geologic principle stating that sedimentary beds, when originally deposited, are nearly horizontal.

orogen A linear belt in which mountain building has produced deformation, metamorphism, igneous rock, and/or elevated land; even after the elevated topography has eroded away, the remnants of an orogen can be identified.

orogeny Mountain building; that is, the process of producing an orogen.

orographic lifting The forced ascent of air when wind reaches the face of a mountain range and must go up because it can't pass through the mountains.

outcrop An exposure of bedrock at the Earth's surface.

outer core The portion of the Earth's core that lies between the mantle and the inner core, extending from 2,900 to 5,150 km (1,802–3,200 mi) in depth; it consists of liquid iron alloy.

overfishing The extraction of fish from a region at a rate faster than the rate at which the fish population can replenish itself.

overland flow The non-channelized downslope flow of water over the land surface due to gravity (also called *surface runoff*).

overshooting top A bulge at the top of an anvil cloud produced by an updraft strong enough to surge through the tropopause.

ozone hole An area of diminished ozone concentration in the stratosphere; it occurs over the Antarctic polar region.

paleoclimate The past climate of the Earth.

paleogeographic map A map that portrays the distribution of land and sea at times in the past.

paleomagnetism A record of the orientation of the Earth's magnetic field preserved in rock.

paleontology The study of fossils for the purpose of understanding Earth history and the evolution of life.

paleopole The supposed position of the Earth's magnetic poles in the past with respect to a particular continent; it is now understood that apparent movement of paleopoles is actually due to movement of continents relative to the magnetic poles.

paleoseismology The study of earthquakes in the past by examining evidence preserved in the geologic record.

paleosol Ancient soil preserved in the stratigraphic record.

Pangaea A supercontinent that assembled at the end of the Paleozoic Era and included nearly all land; it broke up in the Mesozoic when the Atlantic Ocean formed.

parent atom A radioactive atom that undergoes decay.

parsec Short for *parallax second*. The distance to a star with a geometric parallax of 1 arc-second ($\frac{1}{3,600}$ of a degree) using a base of 1 astronomical unit (AU). One parsec is approximately 3.3 light-years.

partial melting Melting in rock during which only a small percentage of the rock actually becomes liquid; magmas formed by partial melting are more felsic than the rock from which they were derived.

passage In the context of cave networks, a tunnel-shaped or slot-shaped underground opening connecting caverns.

passive margin A continental margin that is not a plate boundary.

passive-margin basin A region along a passive margin that was stretched during successful continental rifting; the stretched lithosphere subsided due to cooling after rifting ceased and filled with a very thick accumulation of sediment.

peak ground acceleration (PGA) The maximum acceleration of the land surface during an earthquake; the greater the acceleration, the greater the intensity of the ground motion.

peat Compacted and partially decayed vegetation that accumulates beneath a swamp or bog.

pelagic sediment Microscopic plankton shells and fine flakes of clay that settle out of ocean water like snow and accumulate on the seafloor.

pelagic zone The zone of the ocean, above the benthic zone of the open ocean, that is far enough from the shore that it is not influenced by coastal processes.

peridotite A type of dense, coarse-grained ultramafic rock; rocks with a composition like that of peridotite make up most of the Earth's mantle.

period In the context of geologic time, the largest subdivision of an era.

permafrost Ground in which soil water and groundwater remain frozen for much or all of the year.

permafrost thawing The melting of water within permafrost; due to thawing, permafrost can become soft, and organic material in it may start to decay.

permeability The degree to which a material allows fluids to pass through it via an interconnected network of pores and/or cracks.

Phanerozoic Eon The most recent eon; an interval of time from 539 Ma to the present.

phase In the context of the Earth-Moon system, one of the various shapes of the sunlit surface of the Moon; these shapes change over a 29.5-day cycle due to change in the location of the Earth relative to both the Sun and the Moon.

phase change (1) A change in the state of a substance, such as from water to ice, or water to vapor. (2) The change that takes place during metamorphism when the internal arrangement of atoms in a mineral changes to form another mineral, in the absence of a compositional change.

photic zone The upper layer of the ocean that is penetrated by enough sunlight to permit photosynthesis.

photomicrograph A photograph taken through the lens of a microscope.

photon A discrete unit or particle of electromagnetic radiation that contains energy, but no mass.

photosphere The visible layer of the Sun's atmosphere, located between the top of the convective zone and the rest of the solar atmosphere; it radiates the sunlight that we see, and is the surface of the Sun when we look at the Sun near sunset.

photosynthesis The process by which plants, algae, and cyanobacteria convert energy from sunlight into food, which the organisms use to sustain life.

photovoltaic cell A silicon wafer modified to transform solar radiation directly into electricity (also called a *solar cell*).

phyllite A fine-grained metamorphic rock with foliation defined by the preferred orientation of very fine-grained light-colored mica.

physical weathering Processes at or near the Earth's surface that break intact rock into smaller grains or chunks.

phytoplankton Plankton that can carry out photosynthesis.

pillow basalt A submarine lava flow consisting of glass-encrusted basalt blobs that form when mafic magma extrudes under water and cools very quickly.

placer deposit A concentrated deposit of clasts, flakes, and nuggets of a native metal that have been eroded from rock, transported and sorted, and deposited on a streambed or gravel bar.

planet A large spherical mass that directly orbits a star and has incorporated all the matter that lies in or near its orbit.

planetesimal A solid body with a diameter greater than 1 km that forms in a protoplanetary disk.

plankton Organisms that float in or on a water body, but are not capable of moving against a current.

plasma A material consisting almost entirely of ionized atoms and free electrons; considered by physicists to be the fourth state of matter.

plastic flow A process by which a solid material subjected to stress can move and change shape slowly, without cracking or breaking, due to internal movement of atoms.

plate In the context of plate tectonics, one of about 20 distinct pieces of the Earth's lithosphere (also called a *lithosphere plate*).

plate boundary The border between two adjacent lithosphere plates, defined by seismicity and other geologic features.

plate tectonics The theory that the outer layer of the Earth (the lithosphere) consists of separate plates that move with respect to one another (also called the *theory of plate tectonics*).

playa A flat, dry lake bed that remains when all the water evaporates from a shallow, typically salty desert lake, and the salts precipitate.

Pleistocene Ice Age The interval of time, from 2.6 Ma to 11.7 Ka, during which the Earth experienced several glaciations.

plunge The angle that a linear structure makes with respect to the horizontal, as measured in a vertical plane; that is, the angle at which a linear structure tilts.

pluton An irregular or blob-shaped igneous intrusion; plutons can range in size from tens of meters to tens of kilometers across.

pluvial lake A lake formed at a distance from a continental glacier as a result of enhanced rainfall during an ice age.

polar amplification The increase in average temperature in the north polar region, greater than the change at other latitudes, due to the reduction in albedo associated with the loss of ice cover there.

polar desert A desert in a high-latitude region that forms because cold air holds very little moisture, and therefore does not release much precipitation.

polar front The boundary between the cold air mass located over the polar and subpolar regions and the warm air mass located over the tropics and subtropics; it encircles the globe in each hemisphere.

polar-front jet stream A band of strong winds at the top of the troposphere, above the polar front, that encircles each hemisphere at mid- to high latitudes. Its trace defines a wave-like pattern in map view; it is most prominent during the winter months, although it exists in all seasons.

pore A small, open space within sediment or rock.

pore collapse A decrease in the porosity and permeability of an aquifer caused when groundwater depletion allows grains around pores to pack together more closely.

porosity The total volume of empty space (pore space) between grains in rock or sediment, usually expressed as a percentage.

positive feedback A change that increases the rate at which the original process that caused that change occurs.

Precambrian The interval of geologic time between the Earth's formation, at 4.56–4.54 Ga, and the beginning of the Phanerozoic Eon, at 539 Ma.

precession Movement of the Earth's axis of rotation that traces out a conical path; simply put, it is the "wobble" of the axis.

precipitate (1) A solid formed when ions in a supersaturated solution bond together and separate from the solution (for example, salt is a precipitate obtained by evaporating seawater). (2) To form a solid from ions in solution (for example, salt precipitates from saturated seawater). (3) To fall as rain or snow from a cloud.

precipitation (1) The process by which a solid forms from ions in solution, or by which liquid drops or ice particles form from moisture in a cloud. (2) The liquid drops or ice particles that fall from the sky as a result of condensation of water vapor in clouds.

precision How close a succession of measurements of the same feature are to one another.

preferred orientation A metamorphic fabric in which platy grains lie parallel to one another and/or elongate grains align in the same direction.

preservation potential The likelihood that an organism will be buried and transformed into a fossil.

pressure The push (force per unit area) acting on a material when the push is the same in all directions.

pressure gradient The rate of change in pressure over a given distance.

pressure-gradient force The force applied to a small region of air or water due to a difference in pressure over a small distance around it.

prevailing wind direction The direction in which winds in a given region tend to blow during most of the year.

primary producer An organism in the food web that produces food and energy by photosynthesis or by extracting chemical energy from minerals.

prograde motion (1) Rotational or orbital motion of a moon that is in the same direction as the planet it orbits. (2) The counterclockwise orbital motion of Solar System objects as seen from above the Earth's orbital plane.

Proterozoic Eon The latest eon of the Precambrian.

protocontinent A composite crustal block, formed early in the Earth's history, that was large enough to remain at the Earth's surface and contained rocks of all three groups.

protolith The original rock from which a metamorphic rock formed.

protoplanet A body that is growing in a protoplanetary disk by the accumulation of planetesimals, but has not yet become a planet.

protoplanetary disk The portion of the accretion disk around a young star from which a planetary system may form.

protostar A dense ball of gas at the center of an accretion disk that is collapsing inward because of gravitational forces and has become hot enough to emit radiant energy, but has not yet started nuclear reactions in its core.

pulsar A rapidly rotating neutron star that emits radio waves at regular intervals.

P-wave A seismic compressional body wave, which causes rock to vibrate back and forth in a direction parallel to the direction in which the wave itself moves.

pyroclastic debris Material, ejected by an erupting volcano, that lands on the ground or seafloor in solid form. Pyroclastic debris includes ash, lapilli, blocks, and bombs; it also includes fragments of materials that had already been part of the volcanic edifice.

pyroclastic flow A fast-moving, extremely hot avalanche that occurs when erupting volcanic ash mixes with air and flow down the side of an erupting volcano.

quartzite A metamorphic rock composed of interlocking quartz grains, formed by recrystallization of a quartz-sandstone protolith.

quasar Short for *quasi-stellar radio source*. A source of intense radio emissions; now believed to be an active galaxy whose supermassive black hole is emitting astrophysical jets pointed toward the Earth.

quick clay A type of clay that solidifies into a lattice of clay flakes held apart by salt crystals; it can liquefy if fresh groundwater dissolves the salt.

radiation Energy that can travel away from its source through a medium or a vacuum.

radiation window A range of radiation wavelengths that can pass between the Earth and space without being absorbed by the atmosphere (also called an *atmospheric window*).

radiative zone A thick layer of the Sun surrounding the solar core, through which energy passes in the form of electromagnetic radiation.

radioactive decay The release of subatomic particles by an atom of a radioactive element, which changes the atom's atomic number and thus changes it into an atom of a different element.

radioactive element An element whose atoms decay by releasing subatomic particles from the nucleus and, in a few elements, by fission.

radio interferometer An instrument that superimposes the waves collected by numerous radio telescopes arranged in an array, effectively acting like a single huge radio telescope whose diameter equals the distance from one side of the array to the other.

radioisotopic dating A method that uses the known rate of decay of radioactive elements contained in minerals to provide a rock's numerical age (also called *radiometric dating*).

radio telescope An instrument for detecting and measuring radio-frequency emissions from celestial sources.

radio window The range of radio wavelengths to which the Earth's atmosphere is transparent.

rainbow An arc in the sky displaying all colors of visible light, formed by the refraction and reflection of light by water droplets in the atmosphere.

raindrop A small volume of liquid water that is heavy enough to fall through the atmosphere; initially, it has a spherical shape due to surface tension, but as it falls and grows, it becomes somewhat hamburger-bun-shaped, then warps into an upside-down bowl shape due to air resistance.

rain-shadow desert A desert located on the leeward side of a coastal mountain range, formed because air moving toward it from an ocean has released most of its precipitation on the windward side of the range.

rapid In the context of streams, a portion of a stream in which the water surface is particularly turbulent and the water mixes with air to form whitewater.

realm A region or volume of the Earth System with a distinct character, composition, and behavior (also called a *domain*).

rebound The sudden movement of an elastic material back to its original shape in response to the relaxation of stress.

recharge area A location where water enters the ground, infiltrates downward to the water table, and adds to the groundwater supply.

recurrence interval (RI) The average time between successive geologic events of a certain size; often used to characterize the frequency of events such as floods and earthquakes.

red dwarf A small, relatively cool star that resides on the main sequence of an H-R diagram.

red giant A large, relatively cool star that forms when an intermediate-mass star starts to die and expands; it resides in an area above and to the right of the main sequence of an H-R diagram.

red shift A phenomenon in which light emitted by a source moving away from an observer appears to shift to a lower frequency.

red supergiant A huge star that forms when a high-mass star starts to die and expands.

red tide An algal bloom containing species of algae that turn the water red; such algae can produce dangerous toxins.

reef bleaching See *coral bleaching*.

refinery A facility where crude oil is separated into various petroleum products by heating it, then injecting it into a distillation column, where lighter molecules rise and heavier molecules sink.

reflecting telescope An optical telescope that uses mirrors for collecting and focusing incoming light to form an image.

refracting telescope An optical telescope that uses objective lenses for collecting and focusing incoming light to form an image.

refractory material A substance that has a relatively high melting point and tends to exist in solid form at the Earth's surface and in space.

regolith A general term for unconsolidated material at the Earth's surface, including soil, uncemented sediment, and weathered and broken-up rock.

regression The seaward migration of a shoreline caused by a lowering of sea level relative to land.

relative age The age of one feature or event with respect to another.

relative humidity (RH) The ratio of the amount of water vapor in the atmosphere to the atmosphere's capacity for holding water vapor at a given temperature; it is the vapor pressure divided by the saturation vapor pressure, expressed as a percentage.

relative plate velocity The rate and direction of movement of one lithosphere plate with respect to another lithosphere plate.

relative sea level Sea-surface elevation specified with respect to that of the land surface.

relief The difference in elevation between adjacent high and low locations on the land surface.

renewable resource A resource that can be replenished by nature in a time span of years to decades.

reservoir rock Rock with high porosity and permeability that can contain a large amount of easily accessible oil and gas.

residence time The average length of time that a material stays in a particular Earth System reservoir.

resistance force In the context of mass wasting, the strength of cohesion and friction represented as a force that inhibits downward movement.

resolving power A telescope's ability to see separate objects that appear close together or are very small.

retrograde motion (1) Revolution of a moon in a direction opposite to the rotation of the planet it orbits. (2) The clockwise rotation of a Solar System object as seen from above the Earth's orbital plane; Venus exhibits such motion. (3) The apparent motion of a planet in a direction opposite to its actual direction of revolution around the Sun, as seen by an observer on the Earth, due to changing lines of sight from the Earth.

reverse fault A dip-slip fault on which the hanging-wall block moves up the slope of the fault.

Richter scale A logarithmic scale for assessing the size of an earthquake using the amplitude of the largest ground motion that would be recorded by a seismograph at a specified distance from the earthquake.

ridge In the context of atmospheric science, an elongate area of high atmospheric pressure; the polar-front jet stream bows toward the poles as it crosses a ridge.

ridge-push force The outward-directed push on a lithosphere plate caused by the gravitational potential energy of the elevated lithosphere at a mid-ocean ridge.

rift On the Earth, a distinct linear belt in which rifting (stretching perpendicular to the belt) has occurred or is taking place.

rifting A process that stretches a continent and may eventually break it apart.

rigid Of or relating to a material that can bend and break, but cannot flow.

ring In the context of space science, a thin band of tiny objects in orbit around a planet's equator; all of the Jovian planets have rings.

ripple A very small wave formed by the frictional drag of the wind on the surface of a water body.

ripple marks Small wave-like ridges and troughs on the surface of a sedimentary layer, formed by flowing air or water; ripple marks can be preserved during lithification.

roche moutonnée A glacially eroded hill that is elongate in the direction of glacial flow and asymmetrical in a cross section drawn parallel to its length; the upstream end has been smoothed by glacial rasping into a gentle slope, while glacial plucking has steepened the downstream end.

rock A coherent, naturally occurring solid, consisting of an aggregate of minerals or a mass of glass.

rock cycle The progressive transformations that result in the passage of atoms through different rock types over geologic time.

rockfall Mass wasting that involves the sudden drop of a mass of rock from a vertical cliff or overhang, so that for part of its downward journey, the rock free-falls through air.

rockslide Mass wasting that occurs when a mass of rock detaches from its substrate on a failure surface and moves down a nonvertical slope.

rocky coast An area of coast where bedrock rises directly from the sea and beaches are absent.

Rodinia A supercontinent that had assembled by about 1 Ga, and had entirely broken apart by the end of the Precambrian.

rogue wave An isolated water wave that rises more than twice as high as the mean height of the highest one-third of the waves passing a location during a specified time interval.

rotation The process of turning around an axis.

runaway greenhouse effect A situation that develops when warming due to the greenhouse effect causes the quantity of water in the atmosphere to keep rising until the atmosphere becomes warm enough that water vapor can no longer precipitate.

runoff Water that flows downslope over the land surface as overland flow or in stream channels.

Saffir-Simpson scale A scale used by meteorologists to communicate the strength of a hurricane; it rates hurricanes based on their maximum sustained wind speed.

salinity The concentration of salt in water.

saltation The downwind movement of sediment grains that are pushed by air so that they bounce and roll along their substrate, and knock other grains into the air in the process.

salt lake A lake that contains a high concentration of salt, typically because the lake has no outlet, so that evaporation of its water concentrates salt in the water that remains.

saltwater intrusion The movement of salty groundwater into an aquifer that had contained fresh groundwater; it occurs when the fresh groundwater overlying the saltwater has been depleted.

sand dune In the context of deserts, a relatively large ridge of sand formed by wind deposition, within which cross bedding occurs. See also *dune*.

sand spit A ridge of sand, parallel to the shore, that stretches out into open water, typically across part of the mouth of a lagoon or bay.

sandstone A clastic sedimentary rock formed by lithification of a sand layer.

saturated air Air that is holding as much water vapor as it can, so that the vapor will condense if the air rises, cools, and expands.

saturation vapor pressure The vapor pressure at which air is saturated at a given temperature.

schist A medium- to coarse-grained metamorphic rock that possesses schistosity, a type of foliation defined by the preferred orientation of visible crystals of mica (generally muscovite or biotite).

science The systematic analysis of natural phenomena based on observation, experiment, and calculation.

scientific method A sequence of steps for systematically analyzing scientific problems in a way that leads to verifiable results.

scientist A person who searches for ideas that can explain the way natural phenomena on the Earth, or elsewhere in the Universe, operate.

sea breeze A circulation that develops along shorelines during the daytime as warm air over land rises and moves seaward aloft, while cool air offshore descends and moves onshore to replace the warm air.

seafloor spreading The gradual widening of an ocean basin as new oceanic lithosphere forms at a mid-ocean ridge axis and then moves away from the axis.

sea ice Ice formed by the freezing of the sea surface.

sea level The elevation of the sea surface; the term usually refers to mean sea level, which is the average height of the sea surface over a year, lying roughly halfway between the levels at high tide and at low tide.

sea-level change Changes in relative sea level over geologic time.

seamount A volcanic edifice whose peak lies below sea level.

season One of the four divisions of the year (spring, summer, autumn, and winter), each characterized by different weather.

seawall A wall built of steel, concrete, stone blocks, or riprap on the landward side of the backshore zone to slow the erosion of a beach by waves.

seaweed Nonmicroscopic multicellular algae that grow in the sea.

sediment An accumulation of unconsolidated clasts of any size that are not cemented together.

sedimentary basin A depression in the lithosphere that fills with sediment.

sedimentary rock Rock that forms at or near the Earth's surface by the compaction, consolidation, and cementation of rock fragments or of shells and shell fragments; by the accumulation and alteration of organic matter; or by the precipitation of mineral crystals directly from water solutions.

sedimentary structure A shape, fabric, or layering that develops in sedimentary rock as a consequence of depositional processes; examples include bedding, ripple marks, cross beds, mudcracks, and graded beds.

sediment budget The difference between the input of sediment into a region and the removal of sediment from a region.

sediment liquefaction A phenomenon that happens when seismic shaking causes wet sediment that has been consolidated, but not cemented, to disaggregate and form a slurry of sediment and water that cannot support weight.

sediment load The total volume of sediment carried by a stream.

seismic belt An elongate region, in map view, in which earthquakes occur relatively frequently; if the region is more equant in shape, it can be called a *seismic zone*.

seismic-intensity map A contour map portraying the variation in an earthquake's intensity over a region.

seismicity The occurrence of earthquakes.

seismic-reflection profile A cross-sectional image of subsurface layering made by measuring the time it takes for artificial seismic waves to be reflected by boundaries between different layers of rock.

seismic retrofitting The practice of strengthening existing structures, that were not originally designed to be earthquake resistant, by methods such as adding internal braces to a building's structure.

seismic risk The probability that an earthquake of a given size (defined as producing a specified peak ground acceleration or a specified intensity) will happen during a specified time period in a given region.

seismic tomography A method of producing three-dimensional images of the Earth's interior by measuring variations in seismic-wave velocity.

seismic waves Vibrations generated by an earthquake that pass through the Earth or along its surface.

seismogram The record of an earthquake produced by a seismograph.

seismograph A device, consisting of a seismometer and a recording device, that documents ground movements.

seismologist A researcher who studies earthquakes.

seismometer An instrument that detects ground motion caused by seismic waves.

semipermanent high An atmospheric high-pressure system that develops as a result of air cooling over a broad region, lasts for a relatively long time, and is regenerated in roughly the same place when the air mass containing it moves away.

semipermanent low An atmospheric low-pressure system that develops as a result of air warming over a broad region, lasts for a relatively long time, and is regenerated in roughly the same place when the air mass containing it moves away.

severe thunderstorm A thunderstorm with the potential to produce hail larger than 2.5 cm (1 in) in diameter, winds that exceed 50 knots (92 km/h, or 56 mph), or a tornado.

Sevier orogeny A mountain-building event that produced a fold-thrust belt to the east of the Sierran Arc in western North America during the Cretaceous Period.

shale A very fine-grained clastic sedimentary rock that splits into thin sheets; it is produced by the lithification of mud.

shear The movement of one material, or part of a material, sideways past another.

shear stress Differential stress that causes shear.

shield A portion of a craton in which Precambrian basement is exposed over a broad area of subdued topography.

shield volcano A subaerial volcano shaped like a broad, gentle dome, constructed mainly from multiple mafic lava flows; thin layers of pyroclastic debris occur only locally, mainly near the summit.

shock wave An intense change in pressure caused by a very rapid (greater than the speed of sound) push against a material; for example, a meteorite impact produces shock waves.

shoreline The boundary between land and sea.

sidereal day The Earth's period of rotation, as measured by the time it takes for a distant star to return to the same point in the sky (about 23 h 56 min), and thus the true measure of one full rotation of the Earth on its axis; it differs slightly from a *solar day* because of the Earth's motion around the Sun.

Sierran Arc A large continental volcanic arc that formed on the western margin of Laurentia during the Jurassic Period due to subduction of the Farallon Plate; its exposed roots make up the Sierra Nevada batholith.

silicate mineral A mineral composed principally of silicon-oxygen tetrahedra linked in various arrangements; examples include quartz, feldspar, pyroxene, mica, and amphibole. Groups of silicate minerals differ from one another based on how the tetrahedra are linked.

silicon-oxygen tetrahedron The fundamental building block of silicate minerals, which consists of one silicon atom surrounded by four oxygen atoms, which together produce a pyramid-like shape with four triangular faces (also called a *silica tetrahedron*).

sill A tabletop-shaped tabular intrusion that forms where magma intrudes between layers of wall rock; if the wall rock is not layered, the term generally implies that the intrusion is roughly horizontal.

singularity In the context of space science, the inconceivably dense and hot point that exploded during the Big Bang.

sinkhole A circular depression in the land surface that forms when an underground cavern collapses.

slab-pull force A plate-driving force, caused by the sinking of a downgoing plate at a subduction zone, which pulls the rest of the plate behind it; it exists because the downgoing slab is denser than the surrounding asthenosphere.

slate Very fine-grained, low-grade metamorphic rock that splits into thin sheets, formed by the metamorphism of shale or mudstone.

sleet Precipitation consisting of frozen raindrops.

slickenside A polished fault surface produced by slip on the fault.

slip lineation A linear groove or mineral fiber on a fault surface produced by slip on the fault.

slope A nonhorizontal area of the land surface or seafloor.

slow-onset flood Flooding that develops over several days and take days to weeks to subside, as can happen when snow melts or rainfall occurs over a long time.

slump Mass wasting that involves the displacement of rock or regolith, typically above a spoon-shaped failure surface, at slow to moderate rates.

smelting The heating of an ore to a temperature at which the ore decomposes to yield metal plus a nonmetallic residue called slag.

snow avalanche The downslope movement of a mass of snow, either as a turbulent cloud of powder or as a slurry of snow and water.

snowball Earth A condition thought to have happened in the late Proterozoic when nearly all of the Earth's land surface was covered by glaciers and nearly all of the sea surface was covered by sea ice.

snowflake A composite of hexagonal ice crystals that forms in the atmosphere.

snowpack The net accumulation of snow in a region over the course of a winter.

soil Sediment that has undergone changes at the Earth's surface due to interactions with air, infiltrating water, and life and, in some climates has mixed with organic material.

soil erosion The removal of soil by water or wind.

soil horizon One of several distinct zones within a soil, distinguished from one another by factors such as chemical composition and organic content.

solar atmosphere The photosphere, chromosphere, and corona of the Sun.

solar core The central part of the Sun where nuclear fusion occurs.

solar cycle The 11-year cycle of variation in sunspot frequency on the Sun's surface (also called the *sunspot cycle*).

solar day The Earth's period of rotation, as measured by the time it takes for the Sun to return to the same point in the sky (24 h); the concept of a *day* in common English.

solar eclipse A darkening that occurs when the Sun, as seen from the Earth, is partially or entirely blocked by the Moon.

solar energy The energy of the Sun; the term is commonly used in reference to solar radiation harnessed by human society to produce heat or electricity.

solar flare An explosion of plasma on the Sun's surface that sends an intense burst of radiation and high-energy particles into space; it is a type of solar storm.

solar granulation The grainy appearance of the Sun's surface caused by temperature variations within convective cells in its convective zone.

solar mass (M_s) A unit of mass equal to the mass of the Sun.

solar maximum The portion of the solar cycle in which the most sunspots exist.

solar minimum The portion of the solar cycle in which the fewest sunspots exist.

solar prominence A large, bright arc of plasma that extends outward from the Sun's surface between a pair of sunspots; it is a type of solar storm.

solar storm An event in which the Sun ejects particularly large amounts of radiation and high-energy particles into space, which can result in a disturbance of the Earth's magnetic field.

Solar System The Sun and all of the bodies that orbit it, including planets, moons, asteroids, meteors, Kuiper Belt objects, and Oort Cloud objects.

solar unit A unit of luminosity equal to the luminosity of the Sun.

solar wind A stream of charged particles with enough energy to escape from the Sun's gravity and flow outward into space.

solid A state of matter in which atoms or molecules remain fixed in position. A solid retains its shape regardless of changes in the size of its container.

solidification The process by which a melt (a liquid) transforms into a solid, in which atoms or ions moving about in the melt progressively bond together to form solid crystals or glass (also called *freezing*).

solid-state diffusion The slow movement of atoms or ions through a solid as bonds break and reattach.

solifluction A type of creep characteristic of tundra regions that occurs during summer when the uppermost layer of permafrost melts and the resulting soggy, weak layer of ground flows slowly downslope in overlapping sheets.

sorting The degree to which rock or sediment contains clasts of a single grain size or a range of grain sizes.

source rock Shale containing the organic raw materials from which hydrocarbons form.

Southern Oscillation The shift in sea-level atmospheric pressure back and forth across the Pacific Ocean in association with the Walker circulation.

space telescope A telescope mounted in or on a satellite in orbit above the Earth's atmosphere.

space weather Conditions in the region of space surrounding the Earth that result from interactions with the solar wind and solar storms.

space weathering The breakdown of a planet's or moon's surface material by micrometeorite impacts and by absorption of *cosmic rays*.

specific gravity A number representing the density of a mineral, as specified by the ratio between the weight of a volume of the mineral and the weight of an equal volume of water.

spectral line A dark or bright line representing a specific wavelength of light in an image produced by a spectrometer.

spectral type Any of the categories (O, B, A, F, G, K, M) in a classification system for stars based on a star's surface temperatures, as determined by the character of its spectrum; different spectral types have different colors.

spectrometer A device that spreads incoming electromagnetic radiation into a broad spectrum to characterize the chemical composition of the radiation source, or of nebula-absorbing radiation.

speleothem A body or sheet of travertine that grows in a limestone cavern by the precipitation of calcite from a water solution that drips from the ceiling or flows down the wall of the cavern.

spiral rainband A spiral arm of a hurricane, consisting of tall, aligned thunderstorms, that curves inward to the eye of the storm.

spring A natural outlet from which groundwater flows to the ground surface.

squall A sudden increase in wind speed to a sustained value above 22 knots (41 km/h, or 24 mph) for more than 1 minute, generally accompanied by driving rain.

squall-line thunderstorm A thunderstorm in a distinct line of several thunderstorms that marks the edge of a front or gust front.

stable ground Land whose substrate has a low probability of undergoing failure in the near future.

stage In the context of streams, the elevation of the water level above a reference elevation; the reference elevation lies below the streambed.

stalactite A downward-pointing, cone-shaped speleothem attached to the ceiling of a cavern that forms when a water solution drips from the ceiling.

stalagmite An upward-pointing speleothem that forms when a water solution drips from the ceiling of a cavern and lands on the cavern floor.

star A celestial object in which fusion reactions produce vast amounts of energy; our Sun is a star.

stationary front A boundary between air masses where the colder air mass is neither advancing nor retreating at the Earth's surface.

steady-state condition The condition of a cycle when the proportions of a material in different reservoirs remain fairly constant even though material continues to move among reservoirs; for example, if the concentration of CO_2 in the atmosphere remained constant, the carbon cycle would be in steady state.

stellar nucleosynthesis The fusion of smaller nuclei to form larger ones within a star, which yields atoms up to the size of iron (atomic number 26).

stick-slip behavior Stop-start movement along a fault due to friction; it prevents slip until stress builds up sufficiently to overcome that friction.

stony plain A landscape found in low-relief, sand-poor deserts where a layer of pebble- and cobble-rich sediment forms on the ground surface.

storm surge A rise in local sea level caused by a storm, primarily because the storm's strong winds push the sea into a mound, and secondarily because the low atmospheric pressure beneath the storm allows the sea surface to rise above its normal level.

straight-line wind A strong wind produced by a downdraft that blows in a single direction at the ground surface, rather than in a rotational pattern.

strain A measure of the change in the shape of a material in response to the application of a stress.

strata A succession of sedimentary beds.

stratigraphic column A chart representing the succession of stratigraphic formations in a region and their relative thicknesses, which displays the formations in order from oldest at the bottom to youngest at the top.

stratigraphic formation A recognizable layer consisting of a specific sedimentary rock type, or set of types, deposited during a certain time interval; it can be traced over a broad region.

stratosphere The layer of the Earth's atmosphere directly above the troposphere, in which the temperature increases with altitude.

stratovolcano A large, cone-shaped, subaerial volcano consisting of alternating layers of lava and pyroclastic debris.

stratus cloud A sheet-like, widespread cloud below 3 km (10,000 ft) in altitude.

streak The color of the powder produced by pulverizing a mineral on an unglazed ceramic plate.

stream A body of flowing water confined to a channel, except during floods; a large stream is commonly known as a river.

stream gradient The slope of a stream's channel in the downstream direction.

stream rejuvenation The renewed downcutting of a stream into the land surface caused by uplift of the land surface relative to the base level, or a drop in the base level relative to the land.

stress The push, pull, or shear that a material experiences when subjected to a force; formally, the force applied per unit area.

strike The compass orientation of a horizontal line on a plane; strike and dip together uniquely define the orientation of a planar structure.

strike-slip fault A fault in which one block slides horizontally past another (that is, in a direction parallel to the strike line of the fault surface) so that there is no relative vertical displacement.

strip mining A method of mining a resource, such as coal, when layers of the resource lie within about 100 m (330 ft) of the ground surface, in which the surface rock that lies above the resource is stripped away and the resource is dug out.

stromatolite One of the earliest fossil types, consisting of a layered mound made up of alternating mats of cyanobacteria and layers of sediment.

subduction The process by which a downgoing lithosphere plate slides beneath the edge of an overriding plate and sinks into the asthenosphere beneath.

subduction-related orogen A mountain belt formed on the overriding plate at a convergent boundary by horizontal compressive stress in the continental crust (also called a *convergent-boundary orogen*).

subduction zone The region along a convergent boundary where one lithosphere plate sinks beneath another.

submarine canyon A canyon cut into a continental shelf and slope, entirely beneath sea level.

submarine debris flow Mass wasting that occurs when a slurry of pebbles, cobbles, and boulders suspended in a mud matrix flows down a submarine slope.

submarine fan A wedge-shaped accumulation of sediment at the base of a submarine slope, usually at the mouth of a submarine canyon.

submarine slump Mass wasting that occurs when a semicoherent block of sediment slips down a submarine slope.

submergent coast A coast where the land is subsiding relative to sea level, or sea level is rising relative to the land, so that coastal features become submerged.

subsidence The slow vertical sinking of the Earth's surface in a region.

substrate A general term for material at and just below the ground surface.

subtropical desert A desert that lies in the subtropics, beneath the portion of the Hadley cells where dry air descends, warms, and therefore releases little precipitation.

subtropical jet stream A band of strong (>160 km/h, or 100 mph) winds at the top of the troposphere typically found between 25° and 35° latitude.

subtropics Latitudes between 23° and 35° in each hemisphere, where climates transition between the uniformly hot tropics and the temperate mid-latitudes.

suction vortex One of several small vortices spinning around the main vortex of a very large tornado, in which particularly violent winds can occur.

summer solstice The day of the year when the Sun rises highest in the sky at noon.

sunspot A relatively cool region on the solar surface produced when arcs of the Sun's magnetic field break through the Sun's surface.

supercell thunderstorm A thunderstorm that rotates, meaning that it contains a mesocyclone; such storms often become severe and can produce hail, strong straight-line winds, and tornadoes.

supercell tornado A tornado that develops in a supercell thunderstorm, most often in association with its rear-flank downdraft, when the downdraft forms a roll of rotating air near the ground that is then pulled upward into the mesocyclone.

supercontinent A huge landmass that forms when several large continents, as well as numerous smaller crustal blocks, collide and are sutured together.

supergiant star A star with a radius 50–500 times that of the Sun and thousands of times brighter than the Sun.

supermassive black hole A huge black hole, with a mass of 200,000 solar units or more, that resides in the center of a galaxy.

supernova A short-lived, very bright object in space that results from the cataclysmic explosion of a star; the explosion ejects large quantities of matter into space to form new nebulae.

supernova nucleosynthesis The fusion of smaller nuclei to form larger ones during a supernova explosion, which can yield atoms heavier than iron (with atomic numbers of 27 and above).

superposition The geologic principle stating that, in a succession of sedimentary rock, younger rocks are deposited over older rocks.

supersaturated (1) Of or referring to a solution that contains more ions than can remain dissolved in it, in which case precipitation takes place. (2) Of or referring to air whose relative humidity exceeds 100%, in which case water droplets form.

surface current A current flowing in the top 100–400 m (330–1,300 ft) of the ocean.

surface load Sediment transported by rolling or saltation along a surface; its movement can be caused by wind on land or by a current at the base of a water body (also called *bed load*).

surface water The portion of the hydrosphere that occurs in oceans, lakes, streams, rivers, swamps, snow, and glaciers on the Earth's surface; also used informally by geologists and hydrologists to refer to fresh liquid water in lakes, wetlands, and streams, as distinct from groundwater.

surface-water depletion Consumption of so much freshwater from streams, lakes, and/or wetlands that those water bodies shrink or even dry up.

surface wave A seismic wave that travels along the Earth's surface.

surf zone The portion of the shore above the shoaling zone, where waves break and roll onto the shore.

suspended load Fine-grained sediment carried by water or wind without touching the ground.

sustainability The capacity to maintain or improve society's standard of living without running out of resources or ruining the environment.

sustained wind A wind that maintains a velocity for at least one minute; or, the velocity of wind averaged over a time interval (typically two minutes).

suture The boundary formed during a collision between the two formerly separate, relatively buoyant crustal blocks.

swash The upward surge of water on a beach produced by breaking waves.

S-wave A seismic shear body wave, which causes rock to vibrate back and forth in a direction perpendicular to the direction in which the wave itself moves.

swell Long-wavelength, periodic ocean waves that have traveled a long distance from their source and are not being driven by the local wind.

syncline A fold with a trough-like shape, such that the limbs dip toward the hinge.

talus Rock fragments that have fallen from a cliff or steep slope.

talus slope A landform that develops where falling rock fragments accumulate at the base of a cliff to form a wedge-shaped apron whose surface lies at the angle of repose.

tar A substance composed of hydrocarbons that exists in solid form at room temperature.

taxonomy The study and classification of the evolutionary relationships among organisms.

tectonic geologic structure A feature that has formed due to deformation caused by crustal stresses; examples include folds, cleavage, faults, and joints.

telescope The basic tool of astronomers, designed to collect and focus electromagnetic radiation in order to make celestial objects appear brighter and larger.

temperature A measure of the average kinetic energy of atoms in a material; it represents the relative hotness or coldness of a material as measured with a thermometer.

tension Differential stress that pulls on a material and may lead to stretching.

terminator The boundary between night and day, which moves from east to west as the Earth rotates counterclockwise.

terrestrial planet One of the four inner planets of the Solar System—Mercury, Venus, Earth, and Mars—all of which are made of rock and metal and have a solid surface.

texture In the context of rocks, the manner in which the material in a rock is connected to provide coherence.

theory A scientific idea, supported by an abundance of evidence, that has passed many tests and has failed none.

theory of evolution The idea that new species of organisms appear as others go extinct, so that the assemblage of species on the Earth changes over time by the process known as "natural selection by the survival of the fittest."

theory of plate tectonics See *plate tectonics*.

thermal energy The total kinetic energy in a material due to the vibration and movement of atoms in that material.

thermal metamorphism Metamorphism caused by the heat of an igneous intrusion unaccompanied by change in pressure or differential stress (also called *contact metamorphism*).

thermocline A boundary between layers of water with differing temperatures.

thermohaline circulation A global oceanic circulation that involves the upwelling and sinking of ocean water and the flow of surface and deep-sea currents, driven by contrasts in the temperature and salinity of ocean water, which determine its density.

thermometer An instrument used to measure temperature.

thermosphere The hot outermost layer of the Earth's atmosphere, containing very few gas molecules.

thin section A slice of a rock sample thin enough for light to pass through (about 0.03 mm), which can be examined with a petrographic microscope.

thrust fault A reverse fault with a gentle dip (less than 30°) on which the hanging-wall block moves up the slope of the fault.

thunder The booming or rumbling sound created by the rapid expansion and abrupt contraction of air that has been instantly heated by a lightning stroke.

thunderstorm A storm consisting of tall cumulonimbus clouds that produces localized areas of strong winds and heavy precipitation, accompanied by lightning and thunder.

tidal bore A visible wave of water produced when a flood (rising) tide arrives at the mouth of an estuary, then moves upstream against a river current.

tidal energy Electricity produced by using tidal flow to drive generators.

tidal flat A broad, nearly horizontal plain of mud and silt that is exposed, or nearly exposed, at low tide but completely submerged at high tide.

tidal range The difference in sea level between high tide and low tide at a given location.

tide The rising or falling of sea level that generally occurs twice daily.

tide-generating force The force that generates tides, caused in part by the gravitational attraction between the Moon and the Earth and between the Sun and the Earth, and in part by the centrifugal force created by the revolution of the Earth-Moon system.

tidewater glacier A mountain glacier that has entered the sea along a coast.

tight oil Oil that has not migrated away from its source rock; considered an unconventional hydrocarbon that can be extracted only by hydrofracturing (also called *shale oil*).

tillite A rock formed from lithified deposits of glacial till, consisting of cobbles and boulders distributed through a matrix of sandstone and mudstone.

tilted fault block A hanging-wall block that rotates and becomes tilted as it slips down the dip of a normal fault; the ranges in the Basin and Range Province are tilted fault blocks.

toe The downstream end of a glacier (also called the *terminus*).

topography Variations in the elevation of the land surface.

tornado A vertical to tilted, funnel-shaped cloud in which air rotates violently around the axis of the funnel and can cause catastrophic damage; most tornadoes form at the base of a supercell thunderstorm.

tornado track The path a tornado takes across the ground.

trace gas A gas found in a planet's atmosphere in very small quantities.

trade winds Prevailing surface winds that blow from northeast to southwest between 30° N and the equator in the northern hemisphere and southeast to northwest between 30° S and the equator in the southern hemisphere; in both hemispheres, the winds curve to blow nearly from east to west near the equator.

transform boundary A nearly vertical plate boundary where one plate slips sideways relative to the other, so no new lithosphere forms and no old lithosphere subducts (also called simply a *transform*).

transform fault A strike-slip fault marking a transform boundary. Most transform faults are the actively slipping segment of a fracture zone between two segments of a mid-ocean ridge. Some cut through continental crust, and can be dangerous natural hazards.

transgression The inland migration of a shoreline caused by a rise in sea level relative to land.

transition zone The lower portion of the upper mantle, extending from 410 to 660 km (255–410 mi) in depth, in which several jumps in the velocity of seismic waves occur due to mineral phase changes.

transportation The carrying away of sediment from its site of origin by a medium such as moving air, water, or ice.

trap A configuration of impermeable seal rock lying over reservoir rock that confines hydrocarbons in a restricted area underground.

travel time The interval of time that it takes for a seismic wave to travel from the focus of an earthquake to a given seismograph.

triple junction A point at which three plate boundaries intersect.

tropical cyclone Any large, spiral-shaped low-pressure system that originates over warm tropical ocean waters and has an organized rotation around a surface low-pressure center.

tropical storm A tropical cyclone with sustained winds between 63 and 118 km/h (39–73 mph).

tropics Latitudes of the Earth between 23.5° N and 23.5° S, where cold weather is rare.

troposphere The layer of the Earth's atmosphere that extends from the Earth's surface to the tropopause and contains all the Earth's weather.

trough An elongate area of low atmospheric pressure that causes a downward bulge of isobaric surfaces at high altitude; the polar-front jet stream bows toward the equator as it crosses a trough.

tsunami A water wave that is produced by displacement of seawater by a large mass; a tsunami has a much greater wavelength, and thus contains a much greater volume of water, than does a wind-driven wave.

turbidity current A submarine avalanche of sediment mixed with water that flows down a submarine slope.

turbulence The chaotic twisting, swirling motion of a flowing fluid.

Type I supernova A runaway nuclear fusion event on the surface of a white dwarf in a binary star, resulting in the explosion of the white dwarf.

Type II supernova The explosion of a high-mass star after its nuclear fuel runs out, in which the star's core collapses into a neutron star or black hole, and the outer layers of the star are blown into space to form an expanding nebula.

typhoon A tropical cyclone over the western Pacific Ocean, north of the equator, in which sustained wind speeds reach 119 km/h (74 mph) or higher; that is, it is a hurricane in the western Pacific north of the equator.

unconformity A depositional contact between sedimentary strata that represents an interval of time during which no new strata were deposited and/or older strata were eroded away.

unconventional reserve A hydrocarbon reserve that can be accessed only by using expensive technology such as directional drilling and hydrofracturing.

undercutting Erosion by river flow or ocean waves that carves into the base of a cliff and produces an overhang, which may eventually break away from the cliff and fall.

underground mining A method of mining energy or mineral resources lying underground (generally 100 m or more below the Earth's surface) using a system of tunnels and shafts.

uniformitarianism The geologic principle stating that the same physical processes that we observe today happened in the past at roughly the same rates; or, simply put, that the present is the key to the past.

Universe All of space, and everything within it.

unsaturated air Air that contains some water vapor, but not as much as it can hold; therefore, water droplets do not form as unsaturated air rises.

unstable ground Land whose substrate has the potential to undergo failure in response to gravity in the not-too-distant future if there is a relatively small change in the forces acting on the slope.

updraft An upward-moving air flow.

uplift The upward vertical movement of the Earth's surface in a region.

upper mantle The upper section of the Earth's mantle, extending from the Moho down to a depth of 660 km (410 mi).

upwelling The rising of deep, cold ocean water to the ocean surface.

U-shaped valley A valley carved by glacial erosion so that its profile has the shape of a U, with very steep sides and a curved floor.

vacuum A volume that contains very little matter.

vapor pressure That part of the total atmospheric pressure exerted by water vapor molecules.

vector An arrow whose length represents magnitude, and whose orientation represents direction; force can be represented by a vector.

vein A mineral deposit formed when a water solution flows through a crack in rock and minerals precipitate from the solution to fill the crack.

vernal equinox The day of the year when the Earth's axis of rotation transitions from tilting away from

the Sun to tilting toward it; in the northern hemisphere it occurs on March 30.

vertical wind shear A condition in which wind at a lower altitude differs in speed and direction from wind at a higher altitude.

vesicle An open hole in igneous rock formed by the preservation of a gas bubble in lava as the lava cools into solid rock.

viscosity The resistance of material to flow.

visibility In the context of meteorology, the maximum distance at which a person with normal vision can distinguish objects when looking through the atmosphere.

visible light Electromagnetic radiation that can be detected by the human eye.

volatile material An element or compound, such as H_2O or CO_2, that can melt at relatively low temperatures and can exist in gaseous form at the Earth's surface.

volcanic arc A chain of active volcanoes formed on the overriding plate adjacent to a convergent boundary.

volcanic edifice A hill or mountain formed from pyroclastic debris and lava flows that build up around a volcanic vent.

volcanic eruption An event during which lava and/or pyroclastic debris is expelled from a volcanic vent.

volcanic explosivity index (VEI) A logarithmic scale that classifies volcanic explosions based on the volume of material expelled by an eruption.

volcanic gas Volatile materials released from minerals during melting that rise with, and in some cases bubble up through, magma.

volcanic-hazard map A map that delineates those areas near a volcano that lie in the path of potential lava flows, lahars, or pyroclastic flows and should thus be evacuated if an eruption seems imminent.

volcano (1) A vent from which melt from inside the Earth spews out onto the planet's surface. (2) A volcanic edifice.

Walker circulation An atmospheric circulation in the equatorial regions of the Pacific Ocean characterized by rising air over the western Pacific, eastward air flow near the tropopause, sinking air over the eastern Pacific, and westward air flow near the Earth's surface.

wall cloud A region of rotating clouds that extends below the rain-free base of a supercell thunderstorm; it delineates the base of the storm's updraft. The formation of a wall cloud often precedes tornado formation.

warm front A boundary between air masses where the colder air mass is retreating and the warmer air mass is advancing at the Earth's surface.

waterfall A feature that forms where the gradient of a stream becomes so steep that some or all of the water free-falls above the streambed.

water mass In the context of oceanography, a large volume of water in the ocean with relatively

uniform density characteristics due to its relatively uniform temperature and/or salinity.

water table The underground boundary, approximately parallel to the Earth's surface, that separates the unsaturated zone (nearer the ground surface), in which air partially fills pores, from the underlying saturated zone, in which groundwater completely fills pores.

water vapor Water in its gaseous form.

wave base The depth in a body of water, approximately equal to half a wavelength, below which there is no wave motion.

wave erosion The grinding or breaking away of a coast by the action of water waves.

wavelength The horizontal difference between two adjacent wave troughs or two adjacent wave crests in a wave train.

wave refraction The bending of water waves as they approach a shore so that their crests make no more than a 5° angle with the shoreline.

wave train A succession of waves with fairly uniform spacing and dimensions.

weather Local-scale atmospheric conditions as defined by temperature, atmospheric pressure, relative humidity, wind speed, and precipitation.

weathering The processes that gradually modify, weaken, and break up rock, eventually transforming it into sediment.

weather radar A radar system used to detect the location and intensity of precipitation.

well A hole in the ground dug or drilled in order to obtain water or hydrocarbons.

wetland A body of water so shallow that vegetation growing in it rises above the surface.

white dwarf A stellar object formed when an intermediate-mass star has exhausted its hydrogen fuel, has undergone gravitational collapse to develop a dense core, and has lost its outer layers.

wind The horizontal movement of air.

wind chill The additional heat loss that people experience in cold, windy weather compared with what they would experience in still air at the same temperature.

wind direction The direction from which the wind blows; for example, a westerly wind blows from the west to the east.

wind-driven wave A water wave formed by the interaction of moving air with the surface of a water body.

wind power Electricity produced by using the kinetic energy of wind to drive generators.

wind speed The speed at which air moves horizontally across the Earth's surface, as measured by an anemometer.

winter solstice The day of the year when the Sun is lowest in the sky at noon.

zodiac The array of 12 constellations, recognized in Western culture, that lie along the ecliptic.

zooplankton Plankton that survive by ingesting other organisms, living or dead.

CREDITS

p.148: Stephen Marshak; **p. 149:** Stephen Marshak; **p. 152 (a):** USGS; **(b):** Vittoriano Rastelli/Corbis via Getty Images; **(c):** Mike Goldwater/Alamy Stock Photo; **(d):** Stephen Marshak; **(e):** Stocktrek Images, Inc./Alamy Stock Photo; **(f):** Rudianto/ZUMA Press/Newscom; **(g):** Magnus T. Gudmundson, University of Iceland; **(h):** USGS; **p. 153:** Google Earth; **p. 154 (b-c):** Stephen Marshak; **p. 156 (b):** Google Earth; **(c-d):** Stephen Marshak; **p. 158:** NASA Image Collection; Alamy Stock Photo; **p. 159:** Vittoriano Rastelli/Corbis/Getty Images RF; **p. 161 (e):** Dirk Wiersma/Science Source; **(f):** Stephen Marshak; **(h):** Stephen Marshak; **p. 162 (j):** USGS; **p. 163:** Stephen Marshak.

Chapter 5

Page 164: Stephen Marshak; **p. 165 (a-c):** Stephen Marshak; **p. 166 (a):** Cordelia Molloy/Science Source; **(b-c):** Stephen Marshak; **p. 167 (all):** Stephen Marshak; **p. 168:** Stephen Marshak; **p. 169 (all):** Stephen Marshak; **p. 171 (c):** Stephen Marshak; **p. 172 (both):** Stephen Marshak; **p. 173:** Martin Shields/Alamy; **p. 174:** Google Earth; **p. 176 (all except arkrose):** Stephen Marshak; **(arkrose):** Scottsdale Community College; **p. 177 (all):** Stephen Marshak; **p. 178 (b-c):** Stephen Marshak; **p. 179 (all):** Stephen Marshak; **p. 180 (top):** Google Earth; **(bottom both):** Scott Wilkerson; **p. 181 (a, c):** Stephen Marshak; **p. 182 (a-d):** Stephen Marshak; **p. 183 (a, d):** Stephen Marshak; **(b):** Tom Grubbe/Getty Images; **p. 184 (c-d):** Stephen Marshak; **p. 185 (a):** Emma Marshak; **(b):** Stephen Marshak; **(c):** Marli Miller/Geology Pics; **(bottom):** Google Earth; **p. 186 (b):** Google Earth; **(d):** Manfred Gottschalk/Alamy Stock Photo; **(e):** Diva Amon and Craig Smith, University of Hawai'i at Manoa; **(f):** The Natural History Museum/Alamy Stock Photo; **p. 187 (a):** Stephen Marshak; **(right):** Google Earth; **p. 188:** Stephen Marshak; **p. 191 (left a):** Scott Camazine/Science Source; **(right a):** John Serrao/Science Source; **(left b):** Science Stock Photography/Science Source; **(right b):** Science Stock Photography/Science Source; **(left c):** Zoonar GmbH/Alamy Stock Photo; **(right c):** agefotostock/Alamy Stock Photo; **(left d):** Scenics & Science/Alamy Stock Photo; **(right d):** Roger Allen Photography/Alamy Stock Photo; **p. 194 (all):** Scott Wilkerson; **p. 195 (top b):** Stephen Marshak; **(right d):** Emma Marshak; **(left d):** Stefano Clemente/Alamy Stock Photo; **(bottom a-b):** Stephen Marshak; **p. 196 (a-b):** Stephen Marshak; **p. 197 (top a):** Fossil & Rock Stock Photos/Alamy Stock Photo; **(top right):** Scott Wilkerson **(left b inset):** Steve Lowry/Science Source; **(left b):** Stephen Marshak; **(right b inset):** Dirk Wiersma/Science Source; **(right b:)** Stephen Marshak; **(left c inset):** Zoonar GmbH/Alamy Stock Photo; **(left c):** Stephen Marshak; **(right c inset):** Stephen Marshak; **(right c):** Stephen Marshak; **(bottom):** Stephen Marshak; **p. 198 (a):** lillisphotography/Getty Images; **(b):** Google Earth; **(c):** Stephen Marshak; **p. 201:** Stephen Marshak; **p. 203 (both):** Google Earth; **p. 204 (a):** Stephen Marshak; **(b):** Danita Delimont/Alamy Stock Photo; **(d):** All Canada Photos/Alamy Stock Photo; **(e):** Michael Stewart, University of Illinois; **p. 205 (a):** Stephen Marshak; **p. 206 (c-f):** Stephen Marshak; **p. 207 (h):** Google Earth; **(i):** Stephen Marshak; **p. 208 (k-l):** Stephen Marshak; **p. 209:** Stephen Marshak.

Chapter 6

Page 210: Stephen Marshak; **pp. 212-213:** NOAA; **p. 215 (all):** Stephen Marshak; **p. 216 (a-b):** Stephen Marshak; **p. 217 (a):** Aurora Photos/Alamy Stock Photo; **(b, d):** Stephen Marshak; **(c):** James Brunker UK/Alamy Stock Photo; **p. 218:** Google Earth; **p. 219:** Stephen Marshak; **p. 220:** Scott Wilkerson; **p. 221 (a):** Lloyd Cluff/Corbis Getty Images; **(b-c):** Stephen Marshak; **(d):** USGS; **p. 222 (a):** Chris Jackson; **(b-c):** Stephen Marshak; **(d):** Marli Miller/geologypics; **(bottom):** Scott Wilkerson; **p. 223 (top):** Scott Wilkerson **(bottom):** Google Earth; **p. 224 (a-d):** Stephen Marshak; **(e):** Landsat/USGS; **p. 225:** Scott Wilkerson; **p. 226 (a, d):** Stephen Marshak; **(left):** Google Earth; **p. 227 (c):** Google Earth; **p. 230:** Google Earth; **p. 231 (both b):** Stephen Marshak; **p. 235 (a-b):** Stephen Marshak; **p. 236 (left and right):** Stephen Marshak; **(center):** Google Earth; **p. 239 (c):** Stephen Marshak; **p. 241:** Stephen Marshak.

Chapter 7

Page 242: Xinhua/Alamy Stock Photo; **p. 244 (a):** AFP/Getty Images; **(b):** JIJI Press/AFP/Getty Images; **(c):** AP Photo/Kyodo News; **(bottom):** Stephen Marshak; **p. 245:** NOAA/NGDC, USGS; **p. 253 (b):** USGS; **p. 257 (b):** National Archives; **(c):** AP Photo/Paul Sakuma, file; **p. 258:** Google Earth; **p. 260 (b):** Guillaume/ZUMA Wire/ZUMAPRESS.com/Alamy Live News; **(c):** Tim Dirven/Panos Pictures; **Fig. 7.18a (left),**

p. 260: Figure made with GeoMapApp (www.geomapapp.org)/CC BY/CC BY (Ryan et al., 2009). http://creativecommons.org/licenses/by/4.0/; **p. 261 (a):** Google Earth; **(b):** Dr. Ross W. Boulanger; **p. 265 (a):** © Tolga Ildun/ZUMA Press Wire/Alamy; **(b):** Pacific Press Service/Alamy; **(c):** M. Celebi/USGS; **(d):** Javier Casella/AFP/Getty Images; **p. 266 (a):** AP Photo/Vincent Yu; **(b):** Joseph Sohm/Shutterstock; **(c):** NOAA/National Geophysical Data Center (NGDC); **(d):** AP Photo/New Zealand Herald, Geoff Sloan; **(e):** Adek Berry/AFP via Getty Images; **p. 267 (a):** National Geophysical Data Center (NGDC); **p. 268 (c):** Brendan Hoffman/Alamy Stock Photo; **(d):** Stocktrek Images, Inc./Alamy Stock Photo; **p. 271 (a):** AFP/Getty Images; **(b):** Photo by David Rydevik. 2004. Wikimedia; **(c, both):** Jessica Wilson/NASA/Science Source; **(bottom):** Google Earth; **p. 272:** AP Photo/Kyodo News; **p. 275 (b):** NOAA/NOAA Center for Tsunami Research; **(insert b):** Stephen Marshak; **p. 276:** Visions of America, LLC/Alamy Stock Photo; **p. 277 (a):** Nik Wheeler/Getty Images; **p. 284 (e):** USGS; **p. 286 (m):** © Tolga Ildun/ZUMA Press Wire/Alamy; **p. 287:** Stephen Marshak.

Chapter 8

Page 288: Stephen Marshak; **p. 289:** Stephen Marshak; **p. 290 (both):** Stephen Marshak; **p. 291 (a-b):** Stephen Marshak; **p. 293 (a-b):** Stephen Marshak; **(c):** Reynolds Sumayku/Alamy Stock Photo; **p. 295 (a):** VPC Travel Photo/Alamy Stock Photo; **(b):** Dirk Wiersma/Science Source; **(c,d,g,h):** Stephen Marshak; **(e):** Naturfoto Honal/Getty Images; **(f):** Kevin Schafer/Corbis/Getty Image; **(i):** Biophoto Associates/Science Source; **p. 297:** John Sibbick/Science Source; **p. 299:** Google Earth; **p. 300:** Stephen Marshak; **p. 301 (a-b):** Stephen Marshak; **p. 303:** 1980 Grand Canyon Natural History Association; **p. 308:** Google Earth; **p. 312:** Stephen Marshak; **p. 314:** Stephen Marshak.

Chapter 9

Page 318: Stephen Marshak; **p. 320 (a):** Art Collection/Alamy; **(d):** artwork © Don Dixon; **(e):** Richard Bizley/Science Source; **p. 323 (a):** Courtesy of Dr. J. William Schopf/UCLA; **(c):** Stephen Marshak; **(d):** Bill Bachman/Alamy Stock Photo; **p. 324:** USGS; **p. 326:** John Sibbick/Science Source; **(b):** Stephen Marshak; **(left):** Scott Wilkerson; **p. 327 (a):** Courtesy of Dr. Paul Hoffman, Harvard University. Photo credit: Gabrielle Walker; **(c):** Stephen Marshak; **p. 328 (b):** Ronald Blakey, Colorado Plateau Geosystems, Deep Time Maps; **p. 330 (all):** Ronald Blakey, Colorado Plateau Geosystems, Deep Time Maps; **p. 331 (all):** Ronald Blakey, Colorado Plateau Geosystems, Deep Time Maps; **p. 332:** Tom McHugh/Science Source; **p. 333 (b):** Ronald Blakey, Colorado Plateau Geosystems, Deep Time Maps; **(insert b):** Alamy; **(bottom a):** Chase Studio/Science Source; **(bottom b):** Science Photo Library/Alamy Stock Photo; **p. 334 (b):** Ronald Blakey, Colorado Plateau Geosystems, Deep Time Maps; **p. 335 (left a):** Mackenzie, J. 2012. Hillshaded Digital Elevation Model of the Continental US; **(right a):** Science History Images/Alamy Stock Photo; **(b):** Stephen J. Krasemann/Science Source; **p. 337 (left b):** Stephen Marshak; **(right a-b):** Ronald Blakey, Colorado Plateau Geosystems, Deep Time Maps; **(bottom):** Richard Bizley/Science Source; **p. 338 (top right):** Ronald Blakey, Colorado Plateau Geosystems, Deep Time Maps; **p. 339 (a):** Mohamad Haghani/Stocktrek Images/Science Source; **(b):** De Agostini Picture Library/Getty Images; **(bottom):** Google Earth; **p. 340 (left):** Ronald Blakey, Colorado Plateau Geosystems, Deep Time Maps; **(bottom a):** NASA, JPL; **(bottom b):** Google Earth; **p. 341 (a):** Ronald Blakey, Colorado Plateau Geosystems, Deep Time Maps; **(b):** Google Earth; **p. 342:** Stephen Marshak; **p. 343 (bottom right):** Google Earth; **p. 344 (a):** Mauricio Anton/Science Source; **(b):** P. Plailly E. Daynes/Science Source; **p. 345 (top both):** Ronald Blakey, Colorado Plateau Geosystems, Deep Time Maps; **(bottom left):** Daniel Eskridge/Alamy Stock Photo; **(bottom right):** Janvdwest/Dreamstime; **p. 348 (a):** Stephen Marshak; **(b):** Google Earth; **p. 349 (a):** Richard Bizley/Science Source; **p. 351 (d):** Science Photo Library/Alamy Stock Photo; **(e-f):** Ronald Blakey, Colorado Plateau Geosystems, Deep Time Maps; **(g):** Science History Images/Alamy Stock Photo; **p. 352 (i):** NASA, JPL; **p. 353 (top):** Google Earth; **(bottom):** Stephen Marshak.

Chapter 10

Page 354: Stephen Marshak; **p. 355 (all):** Stephen Marshak; **Fig. 10.2, p. 356:** Reused with permission from Sato, M. H. (2021, July 20). *Energy Consumption.* Updating the Climate Science. http://www.columbia.edu/~mhs119/EnergyConsump/;

p. 357: choja/Getty Images; **p. 361:** Virtual Seismic Atlas, http://www.seismicatlas.org; **p. 362:** Kanok Sulaiman/Getty Images; **p. 363 (a-b):** Stephen Marshak; **(c):** Calvin Larsen/Science Source; **p. 364 (left):** Andrew Harrer/Bloomberg via Getty Images; **(right):** dan prat/Getty Images; **p. 365:** Google Earth; **p. 366:** Field Museum Library/Getty Images; **p. 367:** Francois Gohier/Science Source; **p. 369 (a):** Stephen Marshak; **(b):** Alamy; **(d):** Cultura Creative/Alamy; **p. 371 (a):** RGB Ventures/SuperStock/Alamy Stock Photo; **(b):** Jonathan Plant/Alamy Stock Photo; **p. 372:** Science Source; **p. 375 (all):** Stephen Marshak; **p. 376:** Heino Kalis/REUTERS/Alamy; **p. 378 (left):** Stephen Marshak; **(left inset):** Layne Kennedy/Getty Images; **(right a):** Richard P. Jacobs/JLM Visuals; **(right b):** Stephen Marshak; **p. 381 (a-b):** Stephen Marshak; **p. 384 (left):** Google Earth; **(right):** Boyd Norton/Alamy Stock Photo; **p. 385 (a):** Richard P. Jacobs/JLM Visuals; **(b and bottom):** Stephen Marshak; **p. 386 (a):** Stephen Marshak; **(b):** Dzmitry Rishchuk/Dreamstime; **p. 389 (f):** Stephen Marshak; **p. 390 (j):** Layne Kennedy/Getty Images; **p. 391 (both):** Stephen Marshak.

Chapter 11

Page 392: Xinhua/Alamy Stock Photo; **p. 393 (top):** NASA; **(bottom):** Google Earth; **p. 394 (all):** Stephen Marshak; **p. 395 (a):** G. R. Roberts/Science Source; **(b):** Joe Tabb/Alamy Stock Photo; **p. 397 (f):** Google Maps; **(g):** NOAA; **p. 398 (all):** Stephen Marshak; **p. 400 (a):** Science History Images/Alamy Stock Photo; **(b):** Lloyd Cluff/Getty Images; **p. 401:** Google Earth; **p. 402 (inset):** Stephen Marshak; **(b):** Marli Miller/Geology Pics; **p. 403 (a):** Mick Jack/Alamy Stock Photo; **(c):** Tom Myers/Science Source; **(d):** Stephen Marshak; **p. 404 (a):** Stephen Marshak; **(b):** Cascades Volcano Observatory/USGS; **(c):** BrazilPhotos.com/Alamy Stock Photo; **(d):** Ho New/Reuters/Redux; **(bottom):** Mike Eliason/Santa Barbara County Fire Department via AP, File; **p. 405 (top a and center):** Stephen Marshak; **(top b):** Hemis/Alamy Stock Photo; **p. 406 (a):** AP Photo; **(b-c):** Stephen Marshak; **p. 407 (a):** Stephen Marshak; **(right):** Google Earth; **p. 408 (a):** Jacques Lange/Paris Match via Getty Images; **(c):** Design Pics Inc/Alamy Stock Photo; **p. 409 (b):** Jerome Neufeld and Stephen Morris (c) 2002 Nonlinear Physics, University of Toronto; **(c):** USGS/Barry W. Eakins; **(d):** Google Earth; **p. 412 (top):** Google Earth; **(a-b):** Stephen Marshak; **p. 413 (top):** Stephen Marshak; **(bottom):** David Wald, USGS; **p. 414 (c):** AP Photo; **(bottom both):** Google Earth; **p. 415 (both):** AP Photo/Ted S. Warren; **p. 416 (left):** Google Earth; **(b):** Stephen Marshak; **p. 417 (a):** Google Earth; **(b):** Bob Schuster/USGS; **p. 420 (left):** Google Earth; **(b):** Photo courtesy Harvey Greenberg; **(c):** Ralph A. Haugerud/USGS; **p. 424 (e, g):** Stephen Marshak; **(h):** Google Earth; **p. 425:** Stephen Marshak.

Chapter 12

Page 426: Stephen Marshak; **p. 427:** Wikimedia; **p. 428 (a):** Google Earth; **(b):** Stephen Marshak; **p. 429:** Google Earth; **p. 430 (top):** Stephen Marshak; **(f):** NASA; **p. 432 (a-b):** Stephen Marshak; **p. 433 (all):** Stephen Marshak; **p. 434 (a-b):** Stephen Marshak; **p. 435 (a-b):** Stephen Marshak; **p. 436:** Google Earth; **p. 437 (a-b):** Stephen Marshak; **(bottom):** imageBROKER/Alamy Stock Photo; **p. 438 (a, e):** Stephen Marshak; **(d):** P.A. Lawrence, LLC./Alamy Stock Photo; **p. 439 (a):** Google Earth; **(b):** Christina Neal/Alaska Volcano Observatory/USGS; **(bottom):** NASA/GSFC/meti/ersdac/jaros, and US/Japan ASTER Science Team; **p. 440:** Google Earth; **p. 441 (bottom):** Libby Solomon/The Baltimore Sun via AP; **p. 442 (a):** NurPhoto SRL/Alamy Stock Photo; **(b):** Randy Pench/The Sacramento Bee via AP; **p. 443 (a-b):** NASA; **(c):** The Missouri Department of Transportation; **p. 444 (a):** The Asahi Shimbun via Getty Images; **(b):** Reuters/Alamy; **(c):** NOAA; **(d):** USGS; **p. 445 (a):** NASA; **(b):** NASA Earth Observatory imagery created by Jesse Allen, using imagery provided courtesy of the NASA GOES Project Science Office; **(c):** Cameron Beccario; **p. 446 (a-b):** Stephen Marshak; **(d):** AP Photo/The News-Star, Margaret Croft; **p. 448:** Stephen Marshak; **p. 452:** Google Earth; **p. 454 (d and bottom b):** Stephen Marshak; **p. 455 (b):** Allan Tuchman; **(c):** Thom Foley; **(d-e):** Stephen Marshak; **p. 456 (a):** Les Gibbon/Alamy Stock Photo; **(b):** Stephen Marshak; **p. 457 (a-b):** Stephen Marshak; **(c):** Photo courtesy of the Bureau of Reclamation; **(d):** Google Earth; **p. 460:** Google Earth; **p. 461 (b):** Stephen Marshak; **(center a):** USGS; **(c):** Google Earth; **(right):** Scott Wilkerson; **p. 462 (sinkholes):** David Wall/Alamy Stock Photo; **(pool):** SCPhotos/Alamy Stock Photo; **p. 463 (Ireland):** Stephen Marshak; **(Virginia):** Design Pics Inc/Alamy Stock Photo; **(spelunker):** Ashley Cooper/Corbis/Getty Images; **(Utah):** Stephen Marshak; **p. 464 (top):** Google Earth; **(a-b):** Stephen Marshak; **p. 465 (a):** INTERFOTO/Alamy Stock Photo; **(b):** Christie's Images/Bridgeman Images; **p. 466:** Stephen Marshak; **p. 467:** The Missouri Department of Transportation; **p. 468:** Stephen Marshak; **p. 469 (k):** Francesco Tomasinelli/Science Source; **(bottom):** Stephen Marshak.

Chapter 13

Page 470: Stephen Marshak; **p. 472:** Stephen Marshak; **(b):** Adrian Wojcik/Alamy Stock; **(bottom):** Scott Wilkinson; **p. 474 (all):** Stephen Marshak; **p. 475 (a-c):** Stephen Marshak; **(bottom):** Google Earth; **p. 476 (a):** Image courtesy Jacques Descloitres MODIS Rapid Response Team/NASA; **(b):** AP Photo/Mark J. Terrill; **(d, bottom a, c):** Stephen Marshak; **p. 477 (a):** Stephen Marshak; **(c):** Google Earth; **(d):** John Holmes/Alamy Stock Photo; **p. 478 (all):** Stephen Marshak; **p. 479:** Stephen Marshak; **p. 482 (top b):** Whit Richardson/Alamy Stock Photo; **(bottom b):** Stephen Marshak; **p. 483 (top):** Google Earth; **(a-c):** Stephen Marshak; **p. 484 (a-b):** Stephen Marshak; **(bottom):** Google Earth; **p. 485 (b):** Eye Ubiquitous/Newscom; **(bottom):** Google Earth; **p. 486 (top left a):** Stocktrek Images, Inc./Alamy; **(top a center and right):** USGS; **(b):** BDR/Alamy; **(c both):** Google Earth; **p. 488 (a):** Shutterstock; **(b both, c top):** Stephen Marshak; **(c bottom):** USGS; **p. 489 (top b-c, bottom c):** Stephen Marshak; **(d):** SRTM Team NASA/JPL/NIMA; **p. 490 (top):** Stephen Marshak; **(bottom):** National Geophysical Data Center/NOAA; **p. 492:** NASA; **p. 493 (d):** Alberto Perer/Alamy Stock Photo; **p. 494 (b):** Tetra Images/Alamy Stock Photo; **(c both):** NASA; **(d):** Stephen Marshak; **p. 495 (top):** Scott Wilkerson; **(bottom):** Google Earth; **p. 496 (a-d):** Stephen Marshak; **p. 497 (a):** Shutterstock; **(b):** Stephen Marshak; **(c):** 1986 Keith S. Walklet/Quietworks; **(bottom b):** Marli Miller/Geology Pics; **p. 498 (a):** Google Earth; **(b):** Hemis/Alamy Stock Photo; **(bottom):** Google Earth; **p. 499 (a-c):** Stephen Marshak; **p. 501 (a-e left):** Stephen Marshak **(e right):** Kevin Schafer/Alamy; **p. 502:** Stephen Marshak; **p. 503 (b):** Stephen Marshak; **(d):** All Canada Photos/Alamy Stock Photo; **p. 507 (b):** National Geographic Image Collection/Alamy Stock Photo; **p. 508 (b):** Google Earth; **(bottom):** Photo by Lynda Dredge © His Majesty the King in Right of Canada, as represented by the Minister of Natural Resources; **p. 510:** PA Images/Alamy Stock Photo; **p. 515 (c):** Stephen Marshak; **p. 517:** Stephen Marshak.

Chapter 14

Page 518: Stephen Marshak; **p. 519 (inset):** Ni_racha/Shutterstock; **(bottom):** NASA; **p. 520 (a):** Library of Congress; **(b):** Stephen Marshak; **p. 521 (background):** Stephen Marshak; **(a):** Actep Burstov/Alamy Stock Photo; **(b):** Woods Hole Oceanographic Institution; **(c):** National Geographic Image Collection/Alamy Stock Photo; **(d):** Ingo Wagner/picture-alliance/dpa/AP Images; **(e):** Adam Soule, WHOI, WHOI-MISO Facility, Nadir submersible Pilots, Dalio Explore Fund, Woods Hole Oceanographic Institution; **(f):** NASA; **p. 523 (c):** Ronald Blakey, Colorado Plateau Geosystems, Deep Time Maps; **p. 524 (a):** Stephen Marshak; **p. 525 (both b):** Stephen Marshak; **p. 526:** Juan José Pascual/agefotostock; **p. 529:** NOAA; **p. 530 (b all):** NASA/SVS; **p. 531 (b):** NOAA; **p. 535 (a):** The 1992-2002 mean ocean dynamic topography data has been obtained from Nikolai Maximenko (IPRC), who developed it in collaboration with Peter Niiler (SIO); **(b):** Los Alamos National Laboratory; **p. 539 (left a):** Los Alamos National Laboratory; **(left b):** Cameron Beccario; **p. 541 (top and center):** Stephen Marshak; **(bottom b):** Blend Images/Alamy Stock Photo; **p. 543:** Stephen Marshak; **p. 544 (c):** NOAA; **(d):** IWM/Royal Navy Archive, courtesy of the UK Ministry of Defence; **p. 546 (a):** Emma Marshak; **(b):** Stephen Marshak; **p. 547 (top right):** Google Earth; **(a):** Andrew Syred/Science Source; **(b):** D.P. Wilson/FLPA/Science Source; **(c):** Steve Gschmeissner/Science Source; **(d):** Seaphotoart/Alamy Stock Photo; **p. 549 (inset):** Manhattan map: Google Earth; **p. 550 (a-b):** WaterFrame/Alamy Stock Photo; **(c):** Martin Habluetzel/Alamy Stock Photo; **p. 551 (left a):** Lorna Roberts/Alamy Stock Photo; **(left b):** Andrew J. Martinez/Science Source; **(right a):** Reinhard Dirscherl/ullstein bild via Getty Images; **(right b):** Mark Conlin/Alamy Stock Photo; **(right c):** Stephen Marshak; **p. 552:** SeaWiFS Project and NASA/Goddard Space Flight Center; **p. 553 (a):** Goddard SVS/NASA; **(b):** Cathyrose Melloan/Alamy Stock Photo; **p. 557 (k):** WaterFrame/Alamy Stock Photo; **(bottom):** Stephen Marshak.

Chapter 15

Page 558: Stephen Marshak; **p. 559 (top a):** Mountains in the Sea Research Team; the IFE Crew; and NOAA/OAR/OER; **(top b):** Ralph White/Corbis/Getty Images; **(bottom a):** Google Earth; **(bottom b):** Chris Kelley/HURL, data from Schimdt

Ocean Institute's ship Falkor; **p. 560 (a):** International Ocean Discovery Program (IODP), JOIDES Resolution Science Operator (JRSO), Leg 160 science party; **(b):** International Ocean Discovery Program (IODP); **(c):** International Ocean Discovery Program (IODP), JOIDES Resolution Science Operator (JRSO), William Crawford; **p. 562:** NOAA; **p. 563:** Google Earth; **p. 564 (a):** Google Earth; **(bottom):** NOAA; **p. 565 (b):** GeoMapApp; **p. 566 (a):** © Jose Fuste Raga/Age Fotostock; **p. 566 (b, c, bottom a):** Lamont-Doherty Earth Observatory; **(bottom b):** Google Earth; **p. 568 (top a-b):** © Michael Marten; **(c-d, bottom a-b):** Stephen Marshak; **p. 572 (b):** NOAA/Google Earth; **p. 573:** Mark Benham/Alamy Stock Photo; **p. 574 (a-d):** Stephen Marshak; **p. 575 (top inset):** NOAA; **(a-c):** Stephen Marshak; **(d):** Manfred Gottschalk/Alamy; **p. 578 (all):** Stephen Marshak; **p. 579:** Stephen Marshak; **p. 581 (b):** Panther Media GmbH/Alamy Stock Photo; **(c):** Stephen Marshak; **(d):** David Wall/Alamy Stock Photo; **(e):** NASA; **p. 582 (a):** frans lemmens/Alamy Stock Photo; **(b-c, bottom):** Stephen Marshak; **(d):** Wikimedia; **p. 583 (top b):** Wirestock, Inc./Alamy; **(c):** Daryl Bowers/Alamy Stock Photo; **(d, bottom b):** Stephen Marshak; **p. 584 (top b, c):** Stephen Marshak; **(bottom b):** Emma Marshak; **p. 585 (b):** The National Trust Photolibrary/Alamy Stock Photo; **(bottom):** Google Earth; **p. 586 (top a):** Cody Duncan/Alamy; **(bottom a-b):** Stephen Marshak; **p. 587 (left a, right a-b):** Stephen Marshak; **(left b):** Reinhard Dirscherl/ullstein bild via Getty Images; **(c):** Steve Bloom Images/Alamy; **(bottom):** Google Earth; **p. 588 (b):** NASA/GSFC/LaRC/JPL, MISR Team; **(bottom a):** DeAgostini/Getty Images; **(bottom b):** Marcello Bertinetti/Science Source; **(bottom c):** Shutterstock; **p. 589:** USGS; **p. 591 (a):** Avpics/Alamy Stock Photo; **(b-c):** USGS; **(d):** Stephen Marshak; **p. 592:** Stephen Marshak; **p. 593 (a-b):** Stephen Marshak; **(bottom):** Google Earth; **p. 594 (a-f):** Stephen Marshak; **p. 595 (a):** Stephen Marshak; **(b):** imageBROKER/Alamy Stock Photo; **(c both):** Google Earth; **p. 596 (all):** Stephen Marshak; **p. 598 (d):** Stephen Marshak; **p. 601:** Stephen Marshak.

Chapter 16

Page 602: Stephen Marshak; **p. 603 (a):** Sean Heavey; **(b):** Desintegrator/Alamy Stock Photo; **(c):** Gary Walts/ZUMA Press/Newscom; **(d):** Ai Shieu/Snapwire/Alamy Stock Photo; **p. 605:** NASA; **p. 606:** Google Earth; **p. 607 (a-b):** Stephen Marshak; **(b inset):** Nigel Cattlin/Alamy Stock Photo; **(c):** Jeffrey Frame; **(c inset):** Ted Kinsman/Science Source; **p. 608 (a 4):** James Anderson, Arizona State University; **(b):** Luis Sinco/Los Angeles Times via Getty Images; **(c):** NASA image courtesy Jeff Schmaltz, LANCE/EOSDIS MODIS Rapid Response Team at NASA GSFC; **(d):** Travelart/Alamy Stock Photo; **(e):** blickwinkel/Alamy Stock Photo; **p. 613 (all but noted):** Stephen Marshak; **(foggy):** Dan Lloyd/Alamy Stock Photo; **p. 614 (a):** Roger Dauriac/Science Source; **(b):** Catherine Hoggins/Alamy Stock Photo; **p. 615 (a):** Mark Boulton/Alamy Stock Photo; **(c):** Cameron Beccario; **(bottom):** NASA; **p. 616 (a):** © 2016 UCAR, photo by Scott Ellis; **(b):** Robert Rauber; **(c-d):** NOAA/CIRA, courtesy Prof. Chris Kummerow; **p. 618 (a):** Larry Oolman; **(b):** Stephen Marshak; **p. 619 (a both):** Troy Zaremba; **p. 620:** Google Earth; **p. 621 (a):** NASA; **(c):** Suranga Weeratuna/Alamy Stock Photo; **p. 622 (top):** Stephen Marshak; **(bottom):** NASA; **p. 625 (Cirrus):** John Spragens, Jr./Science Source; **(Cirrostratus):** Craig Joiner Photography/Alamy Stock Photo; **(Cirrocumulus):** Pekka Parviainen/Science Source; **(Altostratus):** Lesley Pardoe/Alamy Stock Photo; **(Altocumulus):** Doug Allan/Science Source; **(Stratocumulus):** Bildarchiv Okapia/Science Source; **(Stratus):** Robert Rauber; **(Cumulus):** Jim Corwin/Science Source; **(Cumulus Congestus):** Dr. Jeffrey Frame; **(Cumulonimbus):** Howard Bluestein/Science Source; **p.626 (top a):** Ron Sanford/Science Source; **(top b):** Stephen Marshak; **(center a):** William Brooks/Alamy Stock Photo **(bottom b):** Sylvia Schug/Getty Images; **(bottom c):** Rob Crandall/Alamy Stock Photo; **p. 629:** Robert Rauber; **p. 634 (a):** Mike Hill/Alamy Stock Photo; **(c):** Joanne Weston/Alamy Stock Photo; **p. 635 (a):** Kent Wood/Science Source; **(b):** Jack Stephens/Alamy Stock Photo; **p. 636 (water background):** Jaap Kranenburg/Alamy Stock Photo; **(glory):** Stephen Marshak; **(shadow background):** Elizabeth Debenham/Alamy Stock Photo; **p. 637 (sun pillar):** Mats Lindberg/Alamy Stock Photo; **(ice backgroud):** Desintegrator/Alamy Stock Photo; **(upper atmosphere background):** Rana Umair/Getty Images; **p. 638:** Science Source; **p. 639 (a):** blickwinkel/Alamy; **p. 641 (e):** NASA; **(f):** Howard Bluestein/Science Source; **p. 642 (g):** William Brooks/Alamy Stock Photo; **p. 643:** Stephen Marshak.

Chapter 17

Page 644: imageBROKER/Alamy Stock Photo; **p. 645 (top):** Pictorial Press Ltd/Alamy Stock Photo; **(bottom):** Cameron Beccario; **pp. 646-647 (4 wind maps):** Cameron Beccario; **p. 649 (a):** Greg Vaughn/Alamy Stock Photo; **(b):** JeffG/Alamy Stock Photo; **p. 651 (a):** Cameron Beccario; **(left c):** Sergio Mendoza Hochmann/Getty Images; **(right c):** blickwinkel/Alamy Stock Photo; **p. 657 (desert):** Frans Lemmens/Alamy Stock Photo; **(grasslands):** FLPA/Alamy Stock Photo; **(forest):** Hemis/Alamy Stock Photo; **p. 658:** NASA; **p. 660 (b, d):** NOAA; **p. 662 (left a-b):** Google Earth; **(left c):** Cameron Beccario; **(right b):** STR/AFP/Getty Images; **(right c):** Sam Panthaky/AFP/Getty Images; **p. 663:** Google Earth; **p. 665:** Cameron Beccario; **p. 666 (b-c):** Cameron Beccario; **p. 667 (a):** NASA; **p. 668 (b):** NOAA; **p. 669 (top):** Tiffany Marie Green/Dreamstime; **(bottom):** Peter Carey/Alamy Stock Photo; **p. 670 (a):** NASA/NOAA image; **(b):** John G. Walter/Alamy Stock Photo; **(c):** NOAA; **(left):** Google Earth; **p. 673:** Greg Dimijian/Science Source; **p. 676 (d):** NASA; **p. 677:** Anton Sokolov/Alamy Stock Photo.

Chapter 18

Page 678: NASA; **p. 679 (a-b):** NOAA; **p. 680 (top):** John Sirlin/Alamy Stock Photo; **(bottom):** NASA image by Marit Jentoft-Nilsen, based on data provided by the Global Hydrology and Climate CenterLightning Team; **p. 683 (b):** Wild Horizons/UIG via Getty Images; **(bottom):** Stephen Marshak; **p. 686:** Andreas Thalier/Alamy Stock Photo; **(inset):** Judith Collins/Alamy Stock Photo; **p. 687:** Ryan McGinnis/Alamy Stock Photo; **(inset):** Domenic Di Girolamo; **p. 688 (radar images):** NOAA; **(bottom a):** LorenRyePhoto/Alamy Stock Photo; **(bottom b):** Iowa Environmental Mesonet, Iowa State University; **p. 689 (a):** Modified from base image by Gregory Carbin, NOAA Storm Prediction Center; **(b):** AP Photo/Evan Vucci; **p. 691 (a):** NOAA; **(b):** Tsado/Alamy Stock Photo; **(c):** NASA; **(d):** Howard Bluestein/Science Source; **p. 693 (a):** Robert Rauber; **(b):** Mike Hollingshead/Science Source; **p. 696 (left a):** Natapat Ariyamongkol/iStockphoto/Getty Images; **(right a):** NOAA; **(b):** Tribune Content Agency LLC/Alamy Stock Photo; **p. 698 (top a):** Roger Hill/Science Source; **(b):** Dan Ross/Shutterstock; **(bottom a):** NASA; **p. 701:** NOAA/Science Source; **p. 703 (top):** Christine Prichard/ZUMA Press, Inc./Alamy Stock Photo; **(center):** AP Photo/Gerald Herbert; **(bottom):** From "Understanding basic tornadic radar signatures" by Kathryn Prociv, 2/14/2013; **pp. 704-705 (background spread):** Google Earth; **p. 704 (EF-0):** Brendan Fitterer/Zuma Press/Newscom; **(EF-1):** Gary Coronado/ZUMA Press/Newscom; **(EF-2):** Benjamin Simeneta/Dreamstime; **p. 705 (EF-3):** NOAA; **(EF-4):** US Department of Labor/ZUMA Press/Newscom; **(EF-5):** RGB Ventures/SuperStock/Alamy Stock Photo; **p. 706 (a):** NASA; **(b):** NOAA; **p. 709 (a):** Cameron Beccario; **(left c):** philipus/Alamy Stock Photo; **(right c):** Charles Floyd/Alamy Stock Photo; **p. 710 (top):** Mandel Ngan/AFP/Getty Images; **(bottom b):** AP Photo/Joel Page; **p. 711:** Cameron Beccario; **p. 713 (a-b):** College of DuPage Meteorology – NEXLAB; **p. 714 (a):** U.S. Air National Guard photo by Staff Sgt. Daniel J. Martinez; **(b):** Erika P. Rodriguez/The New York Times/Redux; **(c):** AP Photo/Carlos Giusti; **p. 716 (a-b):** NOAA; **p. 717 (a):** NOAA; **Fig. 18.43c, p. 717:** Adapted from Gall, R., Tuttle, J., & Hildebrand, P. (1998). Small-scale spiral bands observed in Hurricanes Andrew, Hugo, and Erin. *Monthly Weather Review*, 126, 7. © American Meteorological Society. Used with permission.; **p. 721 (a):** Nilfanion & NASA; **p. 722 (top b):** NASA/Goddard Space Flight Center Scientific Visualization Studio; **(bottom b):** Guy Reynolds/Dallas Morning News/MCT/Newscom; **(c):** U.S. Air Force photo/Staff Sgt. James L. Harper Jr; **(d):** Smiley N. Pool/Rapport Press/Newscom; **p. 725 (c):** From "Understanding basic tornadic radar signatures" by Kathryn Prociv, 2/14/2013; **(left d):** Roger Hill/Science Source; **(right d):** Dan Ross/Shutterstock; **p. 726 (f):** NOAA; **(g):** Image courtesy Jacques Descloitres, MODIS Rapid Response Team at NASA GSFC; **p. 727:** Massimo De Candido/Alamy Stock Photo.

Chapter 19

Page 728: Justin Sullivan/Getty Images; **p. 729 (a-b):** NOAA; **p. 730:** Stephen Marshak; **p. 732 (a-b):** Cooperative Institute for Climate, Ocean, and Ecosystem Studies at the University of Washington; **p. 734 (b all):** Stephen Marshak; **Fig. 19.4a, p. 734:** From Beck, H. E., Zimmermann, N. E,, McVicar, T. R. et al. (2018). Present and future Köppen-Geiger climate classification maps at 1-km resolution. *Scientific Data*, 5, 180214. https://doi.org/10.1038/sdata.2018.214. Copyright © 2018 The Authors.

INDEX

Note: *Italicized* page numbers refer to illustrations, tables, and figures. **Bold** page numbers refer to Key Terms.

plastic behavior, 60, *62*
plastic deformation, 191–192, *192*, 216, *216*, 490
plastic flow, 47
plastic material, 60
plate boundaries
　about, 65–80
　continental collision, *74*, 74–75
　convergent, 61, *63*, *64*, 65, 68–71, *69*, *70*, *71*, 143–146, *144*, *145*
　defined, **60**
　divergent, 61, *63*, *64*, 65–68
　identifying, 61, *64*, 65
　map of, *15*, *61*
　rifting, *75*, **75**–77, *77*
　on the seafloor, 565–567, *566*
　spreading boundary, 65–66
　transform, 61, *63*, *64*, 65, 71–73, *73*
　triple junctions, *63*, **77**
　types of, *64*
plate interior, 61
plate motion
　convective flow, 80, *81*
　GPS measurement of, 82, *82*
　maps of, *15*, *82*
　paleogeography, 87–88, *88*
　reasons for, 14, 80
　relative *vs.* absolute plate velocity, 80–82, *81*
　ridge-push force, **80,** *81*
　slab-pull force, **80,** *81*
　velocity of, 80–82, *81*
plate tectonics. *See also* plate boundaries; plates
　about, 14, 52–88
　asthenosphere, **60,** *61*, *62*, 65
　continental drift, 53, 54–56, *54–56*, 65, 87–88, *88*
　defined, **14, 60**
　Earth history, Archean Eon, 321–322, *322*
　lithosphere, 53, **60,** 60–61, *61*, *62*
　modern theory of, 60–65
　paleomagnetism, 83–88, *88*
　plate motion, 14, *15*, 80–82, *81*, *82*, 87
　rifting, *75*, **75**–77, *77*, 111
　rock cycle and, *112–113*
　seafloor spreading, 53, 57–**59**, *57–59*, 72
　seismic belts, **58,** *64*, 65, **255**–256
　subduction, 53, *64*, **69**–70, *71*, 227, 230, *231*
　synopsis of, *62–63*, 65
　volcanic activity and, 148, *149*
plateaus, 42, *43*
plate-boundary earthquakes, 255–256
plate-boundary volcanoes, 77
plates. *See also* plate boundaries
　defined, 14, 53, **60,** 62
　downgoing plates, *71*
　hot spots, **77**–79, *78*, *79*
　lithosphere plates, 60, *61*, 67, *69*
　map of major plates, *64*
　microplates, 61
　motion of, 14, *15*, 80–82, *81*, *82*, 87
　subduction, 53, *64*, **69**–70, *71*, 227, 230, *231*
Plato, 381
platy grains, 193, *194*
playa, **478,** *478*, *480–481*, 485
playa lakes, 478
Pleistocene Epoch, *308*, *313*, *346*
Pleistocene Ice Age, **343**–344, *344*, **509**–511, *510*, 741, 748, 749
Pliocene Epoch, *308*, *313*, *346*, *742*

plunge, **219,** *219*
plunge pool, 436
plunging, folds, 222, *223*
Pluto, 813, *814*, 839, *840*
plutons, **126,** 127, *128*, 146, *292*, 335
pluvial lakes, **50,** *508*
point bar, 437, *438*
polar amplification, **766**
polar climate, *733*, *734*
polar deserts, **473**
polar field, Sun, 860, *860*
polar front, **665**
polar glaciers, 487, 490
Polar Plateau (Antarctica), 3–4
polar-front jet stream, **665**–667, *666*, *667*
Polaris (star), 784–785
polarity chrons, 85, *86*, 87
polarity intervals, 85
polarity subchron, 85, *86*
polarization Doppler radar, 617
polarized light, studying rocks with, 109, *110*
pole climate, *733*
pollen, in lake deposits as paleoclimate indicators, 742, *743*
pollutants, defined, 348
pollution. *See also* air pollution
　of beaches, 593–597
　of groundwater, 456–458, *458*
　marine, 528
　in rivers, 456, *456*
　threat to reefs, 596–597
　in wetlands, 596
polymetallic nodules, 380
polymorphs, 95
Pompeii (Italy), *119*, 119–120
ponds, 428
pore collapse, **458,** *459*
pores, 359, **448**
porosity, **359, 448,** *448*, 449, 452
porous rocks, 359
porphyritic rocks, 133
Port Arthur (Texas), Hurricane Harvey, *714*
Portland cement, 386
Portuguese Bend (California) slump, *420*
positive charge, 26
positive feedback, 761–762
positive magnetic anomalies, **85**–87, *86*, *87*
positive polarity cloud-to-ground stroke, 693
positive pole, Sun, 860, *860*
positive streamer, 693
potential energy, 6, 11
pothole, 432, *433*
Powell, John Wesley, 289
power plant diagram, *357*
Precambrian, *305*, **308,** *313*, *330–331*
precession, 512, *512*
　of Earth's axis, **784**–785, *785*
precipitable water, *662*
precipitate (noun), **94**
precipitate (verb)
　about, 94, 399
　in caves, 460
　from a gas, 96, *97*
　from liquid solution, 96, *97*, 98
precipitation (rain/snow), **609.** *See also* rain; snow
　causes of, *603*, 627–629
　cold-cloud precipitation, 627–629

　defined, 617–618
　global rainfall patterns, 731–732, *732*
　global warming and, 763, 766, 769–770
　graupel, **628**
　hail, **628**
　satellite and radar images of, *616*, 616–617, 618
　sleet, **628**
　trends in measured annual precipitation, 756, *758*
　warm-cloud precipitation, 627
precipitation gauge, 617, *617*
precipitation rate, 617
precision, **10**
preferred orientation, **193,** *194*
preservation potential, **294**
pressure
　atmospheric pressure, 38, *40*, *610*, **610**–612, *619*, 619–620, 622
　defined, 38, **213**–214
　deformation and, 214, 216
　in Earth's crust, 192, *193*
　inside Earth, 45
　metamorphism caused by changes in, 192, *193*, *194*
pressure gradient, **648**–650
pressure solution, metamorphism and, 191, *192*
pressure-gradient force
　surface currents and, 538, *539*
　wind and, **648,** *648*–650, *652*, 652–654
prevailing westerlies, 645
prevailing winds, surface currents and, **531,** *531*
primary producers, **551**
primary rainbow, 635, *638*
prime meridian, 520
Principia (Newton), 779
principle of baked contacts, 290, *291*
principle of cross-cutting relations, 290, *291*, 312, *312*
principle of inclusions, 290, *291*
principle of original horizontality, 290, *291*
principle of superposition, 290, *291*
prograde motion, **787,** 815
projection, 219, *219*
prokaryotic life, 325
propagate, 167
Proterozoic Eon
　about, *14*, 308, *308*, *313*
　climate in, 325–328, *326*, 741
　defined, **323**
　Earth history in, *322*, 323–328, *324–329*, *331*, 741
　evolution of life, 325–326, *346*
　global climate during, 741, *742*
protocontinents, **321**
protoliths, **190**–191, *191*, *192*, 194, 203
proton, 26, 94
proton-proton chain reaction, 856, *856*
protoplanetary disk, 32, *32*, *33*
protoplanets, *32*, *33*, 34–35, 36
protostar, **30,** *870*, *870*, 871, *871*
Proxima Centauri, 878
Ptolemy, 778
Pueblo dwellings, *172*
Puerto Rico, *464*
　Hurricane Maria, 714, *714*
Puget Sound (Washington), *501*
pulsars, **876**
pumice, 134, *135*
pumice lapilli, *135*
pump for oil, *363*
P-wave shadow zone, 280, *281*

spiral rainbands, **716,** *719,* 720
Spirit (rover), 824
Spitzer Space Telescope, *800, 801,* 806
spreading boundary, 65–66
spring tides, *571, 572*
spring water, 457
springs, 427, **452–455,** *453*
spur, 497
squall lines, 687, *688,* 695
squall-line thunderstorms, 686–689, **687**
squid, *550*
St. Elmo's fire, *151*
St. Helens, Mt. (Washington), 138, *139,* 149, *153, 405*
St. John (US Virgin Islands), 575, *575*
stable air, 683, 685, *685*
stable isotopes, 310
stair-step profile, 436, *436*
stalactites, *460, 461, 463*
stalagmites, **460,** *461*
star dunes, *481,* 484
star streak, 784, *785*
stars, *884–885*
    about, 777, 865–869
    absolute magnitude, **866**
    apparent brightness, **799, 865,** *866*
    apparent magnitude, **865,** *866*
    binary stars, 871, *871, 873,* 873–874
    brightness of, *865–866, 866*
    constellations, **786,** *786*
    death of, *872,* 872–873, *877*
        neutron stars, **876**
        novas formed from, *873,* 873–874, *874, 875*
        supernovas formed from, *873,* 873–874, *875*
    defined, **23**
    dwarf stars, 867
        black dwarfs, 873
        brown dwarfs, 867
        red dwarfs, 867, **872**
        white dwarfs, 867, *868,* **873,** *873,* 874, *877*
        yellow dwarfs, 867
    equilibrium luminosity, 870
    equilibrium radius, 870
    first-generation stars, 31
    giant stars, *867,* 873
        red giant, 31
        red supergiants, **874–876,** *877*
        supergiant stars, **867,** *868,* 874–876
    globular clusters, **878**
    H-R diagram, 867–**868,** *868,* 869, 871, *871*
    life cycle of, 869–872, 876–877, *877*
        death of, *872,* 872–873, 876–877, *877*
        evolution of, 871
        formation of, 29–30, *30,* *869,* 869–871, *870*
        nebulae, formation from, 30, *30,* 31, *31, 869,* 869–870
        survival of, 872
    luminosity, **865,** *866,* 867, 869
    main sequence, **868**
    mass of, 867, 870
    neutron stars, **876,** *877*
    novas formed from, *873,* **873–874**
    nuclear fusion, 31
    protostars, **30,** *870, 870,* 871, *871*
    red supergiants, **874–876,** *877*
    second-generation stars, 31
    size of, 795–796, 867
    spectral classification of, *867*

spectral types, **866–867,** *867*
stellar nucleosynthesis, **31**
supergiant stars, **867,** *868,* 869, 874–876
supermassive stars, 867
supernovas, **31,** *874,* **874–876,** *875, 877*
    temperature of, 795–796, 868, *868,* 870
    zodiac, **786,** *786*
states of matter, 6, *6*
static electricity, lightning and, 692
stationary front, *664,* **665**
steady winds, 648
steady-state condition, **347**
Stegosaurus, 336
stellar black holes, 876, *876*
stellar brightness, 798–799, 865–866, *866*
stellar nucleosynthesis, **31**
stellar wind, 31
stepped leader, 693, *694*
steppes, 659
stick-slip behavior, **246**
stone, quarrying, 385, *385*
stone production, *385*
stone tools, 355, *355*
stony meteorite, *843,* 844
stony plains, *483,* **483–484,** *484*
stony-iron meteorite, *843,* 844
Storegga Slide, 408
storm surge, 591, *591,* **720,** 722
storm waves, *vs.* tsunamis, 269, *270*
storms. *See also* cyclones; hurricanes; typhoons
    coastal, 591, *591*
    global warming and, *762, 763,* 770
    image of, *603*
straight-line winds, **686,** 695
strain, 214, **214,** *215*
strata, **180,** *327, 333*
stratification, 180, *181*
stratification formation, 180, *181*
stratigraphic column
    about, 299–300
    defined, **302**
    unconformities, **300,** *301,* 302, 304, *306–307,* 308, *308*
stratigraphic formations
    about, 299–300, 308
    correlation of, 304, *304, 305,* 308
    defined, **299**
    geologic maps, 303, *303*
    Great Unconformity, **326,** *327*
    unconformities, **300,** *301,* 302, 304, 307, 308
stratigraphic group, 299
stratigraphic trap, *361*
stratocumulus clouds, 624, *624, 625*
stratopause, 620
stratosphere, **620**
stratovolcanoes, **141,** *141, 142,* 146
stratus clouds, 623, *624, 624, 625*
streak, of minerals, **99,** *99*
stream gradient, **434**
stream piracy, 440, *441*
stream rejuvenation, **440**
streambed, *431,* 432
stream-eroded landscapes, 440, *441*
streamers, 693, *694*
streams
    base level, **434**
    capacity of, **433**

competence of, **433**
defined, **429**
depositional processes, *189,* 433–434, *434*
discharge, 431–432
drainage divide, **430,** *431*
drainage networks, 429–430, *430, 431*
dry washes, 432, *432*
erosion, 432–433, *435,* 435–436
flooding, **441–443,** 445–446
flow velocity, *431,* 431–432
formation of, 428–430, 432
landscapes
    alluvial fans, 185, *186, 439,* **439–440,** *477, 481*
    braided streams, **436,** *437, 501*
    deltas, *186,* 434, *439,* **439**–440
    meanders, **436,** *438*
    rapids and waterfalls, **436,** *437*
    stream-eroded landscapes, 440, *440, 441*
    valleys and canyons, *435,* 435–436, *436*
rejuvenation of, **440**
sediment load, **433,** *433*
sediment transport in, 433, *433*
stream gradient, **434**
stream piracy, 440, *441*
sustainability challenges to, 456–458
turbulence, **431,** *431,* 432–433, *433*
types of
    braided streams, **436,** *437, 501, 505*
    disappearing streams, 464, *464*
    ephemeral streams, 432, *432*
    meandering streams, *437, 438,* 440
    permanent streams, 432, *432*
    trunk stream, *429,* 430, 497
underground, *463*
stress
    defined, **213**
    deformation and, 213–216, *214*
    differential, 194, **213**
    metamorphism caused by changes in, **193,** *194,* 214, 216, *216*
stretching, 214, *215*
strike, **219,** *219*
strike line, 219, *219,* 220
strike-and-dip symbol, 219, *219*
strike-slip faults, **220,** *220,* 256–257
strip mining, **368,** *369*
stromatolites, *322, 323*
Styx Sea, *441*
subaerial mass wasting, 407–408
subaerial volcanoes, 144–146
subatomic particles, 26
subduction, *128*
    about, *64,* 69–70, *71*
    of convergent boundaries, 567
    defined, 53, **69**
    mountain belts related to, 157, 227, 230, *231*
subduction plates, metamorphism in, 203
subduction zones
    about, 69–70, *71*
    defined, **69**
    oceanic trenches and, *576*
subduction-related orogen, 227, **230,** *231,* 232
sublimation, 491, 629
sublunar bulge, 569, *571*
submarine canyons, **564,** *564*
submarine debris flow, 407–408, *409*
submarine fans, 183, *188,* **564**